International Scientific Conference "High-Rise Construction 2017" (HRC 2017)

E3S Web of Conferences Volume 33 (2018)

Samara, Russia
4 – 8 September 2017

Part 1 of 2

Editors:

D. Safarik
Y. Tabunschikov
V. Murgul

ISBN: 978-1-5108-5945-6

Printed from e-media with permission by:

Curran Associates, Inc.
57 Morehouse Lane
Red Hook, NY 12571

Some format issues inherent in the e-media version may also appear in this print version.

Printed by Curran Associates, Inc. (2018)

For additional information, please contact EDP Sciences – Web of Conferences
at the address below.

EDP Sciences – Web of Conferences
17, Avenue du Hoggar
Parc d'Activité de Courtabœuf
BP 112
F-91944 Les Ulis Cedex A
France

Phone: +33 (0) 1 69 18 75 75
Fax: +33 (0) 1 69 28 84 91

contact-edps@webofconferences.org

Additional copies of this publication are available from:

Curran Associates, Inc.
57 Morehouse Lane
Red Hook, NY 12571 USA
Phone: 845-758-0400
Fax: 845-758-2633
Email: curran@proceedings.com
Web: www.proceedings.com

TABLE OF CONTENTS

PART 1

PART 2

https://doi.org/10.1051/e3sconf/20183300001

PREFACE
High-Rise Construction-2017 (HRC-2017)

SAMARA | RUSSIA
International Scientific Conference
HIGH-RISE CONSTRUCTION-2017

In September Samara State Technical University (SamGTU) hosted the international scientific conference "High-Rise Construction-2017" (HRC-2017). The event lasted for five days (from Septermber 4 to September 8) and was a great success. The conference was initiated by the Department of the Architecture of Residential and Public Buildings, headed by professor Victor P. Generalov.

In the modern world, high-rise buildings are indicators of economic progress, they form an idea about the prestige of leading and successful firms, about the status and the scale of cities.

The international Conference "High-Rise Construction-2017" has brought together theoretical and practical experience on designing high-rise buildings and sustainable urban development of the scientists from the United States, European countries and a number of Russian cities.

The delegation from Chicago, USA, was most representative and included Jason Gabel, Communications Manager, the Council on Tall Buildings and Urban Habitat (CTBUH); Daniel Safarik, Editor, CTBUH journal; Sara Beardsley, Senior Architect, and Adrian Smith from Gordon Gill Architecture. Anastasia Kucherova came as a representative of Stefano Boeri Architetti, an architectural firm form Italy, Milan. There also were many presenters from other countries, associations and universities.

The participants discussed such issues as the integration of high-rise buildings into urban environment, typological innovations in high-rise buildings construction, the design of structures for high-rise construction, engineering systems and technology. They also shared their best practices and experiences in the field of high-rise construction, discussed a wide range of questions on the formation of dense comfortable vertical urban environment and on innovative models of high-rise buildings adapted to the unique conditions of a particular construction site. The conference became a networking platform for scientists, businessmen and representatives of governmental and non-governmental organizations, making it possible to carry out critical comprehensive analysis of the state of modern science and industry in the field of high-rise construction.

We would like to express our gratitude to Editorial Board of E3S Web of Conferences and personally Isaline Augusto for her attention to all our questions and great support.

Organizing Committee

SAMARA | RUSSIA

International Scientific Conference

HIGH-RISE CONSTRUCTION-2017

Samara State Technical University

ORGANIZER:

SAMARA POLYTECH
Flagship University

PROMOTIONAL PARTNER:

Council on Tall Buildings and Urban Habitat

HRC 2017 SSTU | SAMARA | RUSSIA HRC 2017 SSTU | SAMARA | RUSSIA

edp sciences

Statement of Peer review

In submitting conference proceedings to E3S Web of Conferences, I certify to the Publisher that I adhere to the **Policy on Publishing Integrity** of the journal in order to safeguard good scientific practice in publishing.

1. All articles have been subjected to peer review administered by the proceedings editors.
2. Reviews have been conducted by expert referees, who have been requested to provide unbiased and constructive comments aimed, whenever possible, at improving the work.
3. Proceedings editors have taken all reasonable steps to ensure the quality of the materials they publish and their decision to accept or reject a paper for publication has been based only on the merits of the work and the relevance to the journal.

Title, date and place of the conference

HIGH-RISE CONSTRUCTION 2017 (HRC 2017)
4-8 September 2017
Samara State Technical University, Russia

Proceedings editor(s):

Daniel Safarik
Yury Tabunschikov
Vera Murgul

Date and editor's signature

18.12.2017

Megacities and tall buildings: symbiosis

Daniel Safarik[1*], *Shawn* Ursini[1] *and Antony* Wood[1]

[1]Council on Tall Buildings and Urban Habitat, 104 South Michigan Ave., Suite 620, Chicago, IL, 60603, United States

Abstract. Anyone concerned with the development of human civilization in the 21st Century will likely have heard the term «megacity». It is – as it should be – increasingly prevalent in both mainstream and academic discussions of the great trends of our time: urbanization, rising technological and physical connectivity, increasingly polarized extremes of wealth and poverty, environmental degradation, and climate change. It is a subject as large and far-reaching as its name implies. This paper sets the scene on how megacities and the built environment are growing together, and examines the implications for those who plan, design, develop and operate tall buildings and urban infrastructure.

1 What is a Megacity?

In order to rationalize the data CTBUH collects – predominantly on skyscrapers and large urban developments – with that collected by other organizations, first a definition that reflects a distillation of the prominent literature on the subject should be set forth:

A megacity is an urban agglomeration with a total population of 10 million people or greater, consisting of a continuous built-up area that encompasses one or more city centers and suburban areas, economically and functionally linked to those centers.

A megacity is typically, though not always, polycentric, with multiple nodes of concentrated urban activity and high-density development, rather than being centered around one large primary central business district (CBD). Indeed, a telltale sign of a megacity, and an indicator of its polycentric nature, is the tendency of residents and urban planners alike to refer to more than one «CBD». Even if there is a consensus about the location of the «center of town», development and transportation patterns strongly suggest otherwise; it should be thought of as an interwoven web, rather than a series of concentric zones.

The polycentric pattern is often the result of established urban centers traditionally separated by distance and their own identities eventually merging together through a continuous spread of urban and suburban development. A key aspect of the megacity is that these linkages of urbanity fuse the agglomeration together, not only physically, but also economically, functionally, and often, culturally.

In a megacity, the extent of urban development spread will not be described by a single radius or a compact, circular shape; in other words, it is asymmetrical and polymorphic.

* Corresponding author: dsafarik@ctbuh.org

This is due to a variety of factors, including but not limited to; uneven development patterns, geographic obstacles, transport corridors and political boundaries. While green spaces and «undeveloped» land may separate urban centers, this does not necessarily indicate that there is a definitive economic, cultural or political division between cities and their relationship within a megacity.

In other words, there may be considerable amounts of open space contained within a megacity (see Fig. 1).

Open spaces could be the result of geological features such as mountains and bodies of water, military installations or protected greenbelts. Meanwhile, «leapfrog» development has a tendency to create long, narrow strands of development along transportation routes, which then fill in perpendicularly to those corridors over time.

Fig. 1. Example of an undeveloped space comprised of mountains and farms within an urbanized area, in this case, Los Angeles. (Source: CC BY-SA Doc Searls).

For the purposes of the 45 megacities noted in this study (see Table 1), it should be clearly noted that the cited population, area and density figures are the result of existing political boundaries which can dilute density numbers, because they may encompass open spaces and adjacent hinterlands potentially available for future development lying beyond highly built-up areas.

For example, if a district, county, prefecture or other political jurisdiction adjoins a heavily built-up area, and a distinctly dense tendril of urban land penetrates into what is otherwise a rural political unit, along a watercourse, highway or railway, the entire surrounding political unit is typically counted in area and population figures.

Thus, the «Los Angeles» megacity in this study extends all the way through open desert to the Colorado River and the border with Arizona, because the political entity of Riverside County, California – heavily urbanized in the west and sparsely populated in the east – is included.

Table 1. List of 45 megacities ranked by population. (Source: See «Population and Area Sources» in References)

Rank, (by Pop.)	Megacity	Country	Combined Pop	Area (Sq Km)	Density (Pop/SqKm)	# of 200m+ Buildings	Cities & Administrative areas within
1	Pearl River Delta	China	64,899,778	56,217	1,154	220	Guangzhou, Shenzhen, Hong Kong, Foshan, Dongguan, Huizhou, Zhaoqing, Zhongshan, Zuhai, Jiangmen, Macau
2	Shanghai-Changzhou	China	50,302,212	28,010	1,796	90	Shanghai, Suzhou, Wuxi, Changzhou, Jiaxing
3	Tokyo (Kanto Region)	Japan	42,797,000	32,424	1,320	29	Prefectures of Tokyo, Kanagawa, Saitama, Gunma, Tochigi, Ibaraki & Chiba
4	Beijing-Tianjin	China	40,594,839	34,588	1,174	50	Beijing, Tianjin, Langfang
5	Delhi	India	34,397,873	15,562	2,210	3	Delhi, Nodia, Gurgaon, Ghaziabad, Rohtak, Meerut
6	New York-Philadelphia	USA	30,907,175	54,880	563	96	New York, New Haven, Jersey City, Philadelphia, Wilmington, Trenton, Atlantic City
7	Chongqing	China	30,165,500	82,403	366	46	Chongqing Province
8	Sao Paulo	Brazil	29,740,692	23,556	1,263	0	Sao Paulo, Campinas, Baixada Santista, Santos, Sorocaba, Sao Jose dos Campos
9	Jakarta	Indonesia	28,424,717	6,438	4,415	46	Jakarta, Depak, Bogor, Tangerang, Bekasi
10	Mumbai	India	26,136,721	17,313	1,510	38	Districts of Mumbai, Mumbai Suburban, Thane, Pulghar & Raigad
11	Seoul-Incheon	South Korea	25,524,572	11,807	2,162	39	Seoul, Incheon, Gyeonggi Province
12	Manila	Philippines	25,169,197	8,113	3,102	30	National Capitol Region and Provinces of Rizal, Leguna, Cavite, Bulacan
13	Dhaka	Banglades h	24,952,038	9,353	2,668	0	Districts of Dhaka, Narayanganj, Mymensingh, Munshiganj & Gazipur within Dhaka Division.
14	Karachi	Pakistan	23,500,000	3,527	6,663	1	Karachi Administrative District
15	Mexico City	Mexico	23,492,352	11,317	2,076	6	Metropolitain areas of Mexico City, Toluca, Tianguistenco, Tula and the municipality of Tepeji del Río de Ocampo
16	Cairo	Egypt	21,455,656	6,649	3,227	0	Cairo, Al Qalyubia & Giza Governorate
17	Hangzhou-Ningbo	China	21,218,301	34,936	607	24	Hangzhou, Shaoxing, Ningbo
18	Osaka	Japan	20,750,000	27,351	759	6	Prefectures of Osaka, Shiga, Nara, Kyoto, Wakayama & Hyogo; including the cities of Kobe, Hemeji and Izumisano
19	Kolkata	India	20,608,327	18,885	1,091	1	Districts of Kolkata, Howrah, Hooghly, North 24 Parganas, South 24 Parganas
20	Lahore	Pakistan	20,530,000	12,631	1,625	0	Districts of Lahore, Sheikhupura, Gujranwala, Kasur

Rank, (by Pop.)	Megacity	Country	Combined Pop	Area (Sq Km)	Density (Pop/SqKm)	# of 200m+ Buildings	Cities & Administrative areas within
21	Moscow	Russia	19,002,220	33,262	571	19	Moscow City and the more urbanized portions of the Moscow Oblast
22	Los Angeles	USA	18,679,763	87,944	212	13	Los Angeles, Long Beach, Riverside, Oxnard
23	Ho Chi Minh	Vietnam	18,051,200	23,724	761	7	Ho Chi Minh City & Provinces of Ba Ria-Vung Tau, Dong Nai, Tien Giang, Long An, Binh Duong, Tay Ninh
24	Bangkok	Thailand	17,718,258	21,028	843	20	Provinces of Bangkok, Chon Buri, Nakhon Patham, Pathum Thani, Samut Sakhon, Samout Prakan, Nonthaburi, Chachoengsao, Rayong
25	Chengdu	China	17,663,383	18,115	975	24	Chengdu, Deyang
26	Xiamen	China	16,469,863	25,792	639	20	Xiamen, Zhangzhou, Quanzhou
27	Istanbul	Turkey	16,437,489	8,808	1,866	7	Istanbul and Kocaeli provinces, including the districts of Izmit and Gebze
28	Tehran	Iran	15,450,000	18,814	821	0	Provinces of Tehran and Alborz, including the cities of Karaj, Varamin and Eslamshahr
29	Buenos Aires	Argentina	15,333,035	11,134	1,377	1	Greater Buenos Aires and La Plata Metropolitan Areas
30	London	United Kingdom	14,031,830	12,091	1,161	8	London and the districts of Surrey, Kent, Essex, Hertfordshire
31	Shantou	China	13,943,141	10,660	1,308	0	Shantou, Jieyang, Chaozhou
32	Johannesburg-Pretoria	South Africa	13,937,500	22,017	633	1	Gautang Province (including Johannesburg, Pretoria, Midrand) and the municipality of Madibeng
33	Bangalore	India	13,093,168	13,139	997	0	Bangalore, Ramanagara, Krishnagiri Districts
34	Kinshasa	Democratic Republic of Congo, Republic of the Congo	13,271,392	10,229	1,297	0	Kinshasa, Brazzaville
35	Rhine-Ruhr	Germany	12,695,656	14,160	1,154	0	Bonn, Colonge, Dusseldorf, Duisburg, Essen, Wuppertal, Mönchengladbach
36	Chicago-Milwaukee	United States of America	11,970,050	37,324	321	31	Chicago, Milwuakee, Naperville, Michigan City, Shaumburg, Kankakee
37	Lagos	Nigeria	12,864,745	20,107	640	0	Lagos State, Ogun State
38	Rio de Janeiro	Brazil	12,678,779	7,249	1,749	0	Rio de Janeiro, San Goncalo, Dudue de Caxias, Nova Iguacu, Belford Roxo
39	Chennai	India	12,373,088	8,052	1,537	0	Chennai, Thiruvallur, Kancheepuram Districts
40	Hyderabad	India	12,273,352	17409	705	0	Districts of Hyderabad, Rangareddy, Medak
41	Paris	France	12,073,914	12,011	1,005	2	Departments of Paris, Seine-et-, Essonne, Seine-Saint-Denis Marne,

Rank, (by Pop.)	Megacity	Country	Combined Pop	Area (Sq Km)	Density (Pop/SqKm)	# of 200m+ Buildings	Cities & Administrative areas within
							Yvelines, Val-de-Marne, Val-d'Oise
42	Nagoya	Japan	11,321,000	21,567	525	4	Prefectures of Aichi, Gifu, Mie; including the cities of Nagoya, Tsu and Toyohashi
43	Wuhan	China	10,834,056	10,088	1,074	29	Wuhan, Ezhou
44	Taipei	Republic of China (Taiwan)	10,280,569	5,209	1,974	6	Taipei, New Taipei City, Keelung, Taoyuan, Hsinchu
45	Shenyang	China	10,244,261	24,132	425	41	Shenyang, Fushun
	Totals:		**958,258,662**	**990,025**	968	**958**	***958 = 55.34% of all 200 meter + buildings**
							***958,258,662 = 13% of global population**
							***990,025 sq. km. = 0.66% of global land surface area**

In step with the theme and site of the CTBUH 2016 Conference, the primary benchmark for a megacity in this study is the Pearl River Delta region of southern China, the world's largest megacity (see Fig. 2).

Fig. 2. The Pearl River Delta megacity boundaries (top) and the skylines of the three largest cities in it (left: Hong Kong and Guangzhou; right: Shenzhen). (Source: (top) Google Maps, citypopulation.de; (photos) James Antrobus, Tansri Muliani, and Popolon).

Drawing a line around the boundaries of the Pearl River Delta's urban centers would encompass a span of up to 367 km from southwest to northeast (that is, from the southwestern-most corner of Jiangmen to the northeastern-most corner of Huizhou) and 331 km from northwest to southeast (that is, from the northwestern-most corner of Zhaoqing to the southeastern-most corner of Hong Kong). This boundary would give an area of 56,217 km2, which would actually rank it 127th on the list of country areas around the world, just below Croatia, and above Costa Rica, Denmark, and Israel, for sheer size. It also would be the 12th largest country in terms of gross domestic product (GDP), lying between South Korea and Australia on the GDP per capita scale. Thus, as we can see, the Pearl River Delta megacity is comparable to numerous countries in terms of physical size, and far greater than many in terms of attributable economic output [10, 11, 12]. While many studies consider the Pearl River Delta and Hong Kong to be separate urban entities, due to Hong Kong's special administrative status within the People's Republic of China, this study includes Hong Kong, as all indicators point to the former British colony becoming more integrated with mainland China, and more to the point, with its immediate neighbors. The fact that it is currently a Special Administrative Region with a quasi-national boundary, a different political system, different currency – and, not insignificantly, left-hand driving – acutely underscores one central quandary of the megacity: the economic and functional realities of these cities are often several steps ahead of their political realities [8, 9].

Estimates of the Pearl River Delta's population vary. The estimated current population for the Pearl River Delta, including Hong Kong, is 64.9 million. This would make it the 22nd-largest country in the world by population, just below the United Kingdom and just above France [1].

The Pearl River Delta encompasses an urbanized area spreading into a total of 11 different municipalities whose administrative divisions define the overall boundary of the megacity (see Fig. 2). Using this criterion, the second largest megacity would be Shanghai-Changzhou, which also includes the adjacent cities of Suzhou, Wuxi and Jiaxing for a total population of approximately 50.3 million. The third largest megacity is Tokyo, but under this study's parameters, «Tokyo» should be considered as the larger Kanto Region, which encompasses neighboring prefecture-level political divisions and balloons the population to about 42.8 million.

The methodology of this study attempts to normalize data collection across cities that vary greatly in terms of topography, organization and available information. As much previous study and literature has noted, it is unwise to assume that metropolitan planning models that work in one country, region, or even in a single city, will necessarily work in the next. The patterns of urbanization and densification are substantially different in different parts of the world. But as a means of identifying significant trends and issues, all megacities have something to teach each other, and our urbanizing world at large.

2 Where are Megacities?

Of the 45 megacities identified in this research, 29 (64%) are in Asia, and 10 are in China alone (22%). Another six are in India. Europe has five (11%), North America and Africa have four (9% each), and South America has three (7%) (see Fig. 3, 4).

There is no evidence that a megacity, purely based on size, is inherently ungovernable; for example, Tokyo is one of the best-run and most thoroughly-integrated urban areas anywhere, of any size [2, 7] – but it's also an outlier in almost every category. The much more typical scenario – found in a dozen or so megacities near the Equator- sees some of the bleakest poverty, deepest corruption, most daunting environmental circumstances, and most chaotic daily life on the planet.

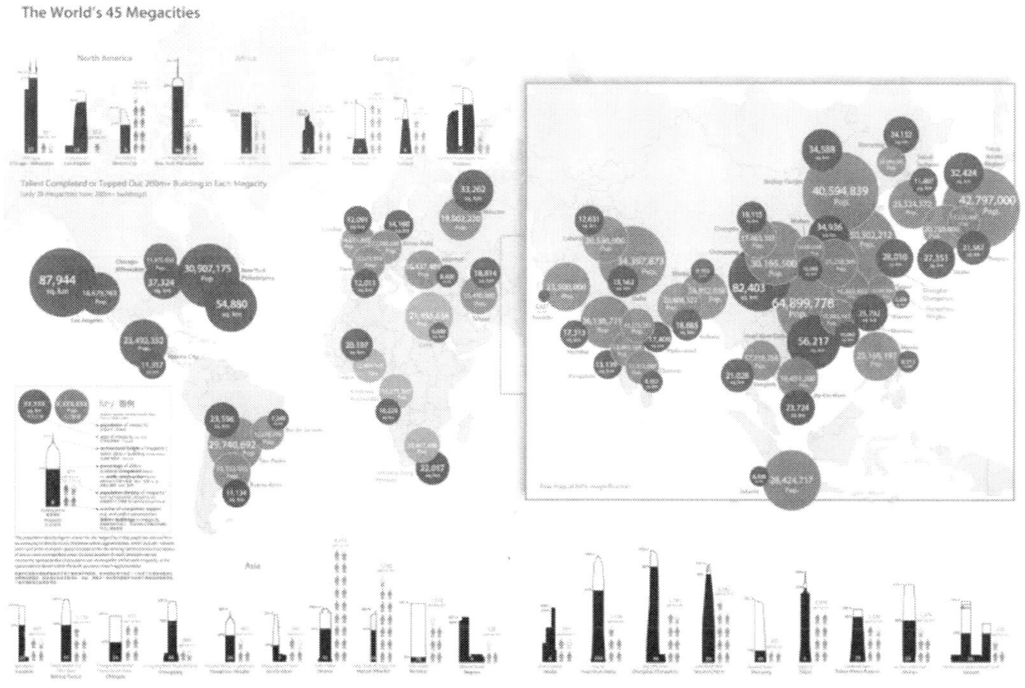

Fig. 3. Location, population, area, and population density of the world's 45 megacities. (Source: CTBUH Skyscraper Center and see «Population and Area Sources» in References).

Based on the United Nations Human Development Index (HDI) figures for 2015, 11 megacities are in the top quartile (or «Very High» HDI), including those in Germany, the United States and Japan; 18 are in second quartile or «High» category, including those in Russia, Brazil and China; 12 are in «developing» countries with «medium» HDI scores, including Indonesia, the Philippines, and India; and four are in «undeveloped» countries, including Pakistan, Nigeria, and the Democratic Republic of the Congo [3] (see Table 2).

Bearing in mind that 16 of the 45 megacities are in underdeveloped or developing countries, consider this: «Of the global urban population of 730 million in 1950, 300 million (42%) lived in developing countries. By 2010, with a global urban population of 3.5 billion, 2.6 billion (73%) lived in developing countries. In 2050, the figures will be 7.3 billion, 5.2 billion and 83%, respectively. In other words, the urban population of developing countries will grow 15 times as much as in developed countries» [2].

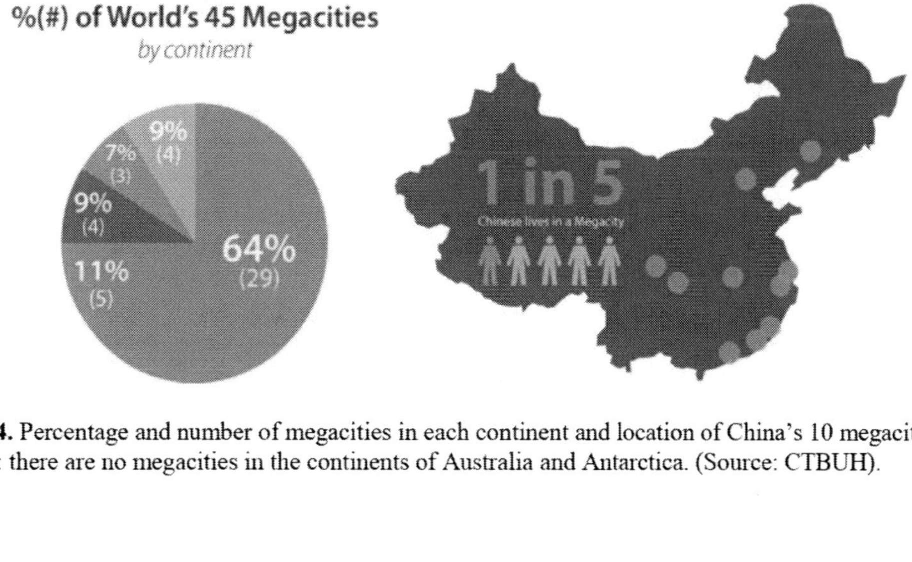

Fig. 4. Percentage and number of megacities in each continent and location of China's 10 megacities. Note: there are no megacities in the continents of Australia and Antarctica. (Source: CTBUH).

In addition to the predominance of urban population growth in developing countries, a substantial portion of the largest megacities are doubly affected by their location. Put another way, the fastest-growing places are those in the most vulnerable positions geographically, with respect to natural disasters: «A ranking of world cities by size of population vulnerable to a range of natural disasters found that the six most vulnerable urban areas in the world were in East Asia» [4]. «Today, more than 1.5 billion of Asia's 4 billion people live within 100 km of the Indian or Pacific oceans, where rising sea levels could overwhelm existing coastal barriers. Mankind's voluntary concentration into a dense, coastal civilization is certainly efficient, but it may not be very wise» [5].

It is already well-acknowledged that coastal defenses must be a part of any long-range plan for most of the megacities of the world; the experience of New York City during Hurricane Sandy in 2012 – with billions of dollars of damage and disruption, and from which recovery continues to this day – underscored this. Practitioners in the tall-building community have already begun to respond to this issue.

Table 2. List of 45 megacities ranked by Country's Human Development Index (HDI). HDI is a composite statistic of life expectancy, education, and income per capita indicators, which are used to rank countries into four tiers of human development. A country scores higher HDI when the lifespan is higher, the education level is higher, the GDP per capita is higher, the fertility rate is lower, and the inflation rate is lower. (Source: World Bank 2, Wikipedia)

HDI Rank	HDI Value	Quartile	Country	Megacities
5	0.916	1-Very High	Germany	Rhine-Ruhr
8	0.915	1-Very High	USA	New York-Philadelphia
8	0.915	1-Very High	USA	Los Angeles
8	0.915	1-Very High	USA	Chicago-Milwaukee
14	0.907	1-Very High	UK	London
17	0.898	1-Very High	South Korea	Seoul-Incheon
20	0.891	1-Very High	Japan	Tokyo (Kanto)
20	0.891	1-Very High	Japan	Osaka
20	0.891	1-Very High	Japan	Nagoya
22	0.888	1-Very High	France	Paris
40	0.836	1-Very High	Argentina	Buenos Aires
50	0.798	2-High	Russia	Moscow
69	0.766	2-High	Iran	Tehran
72	0.761	2-High	Turkey	Istanbul
74	0.756	2-High	Mexico	Mexico City
75	0.755	2-High	Brazil	Rio de Janeiro
75	0.755	2-High	Brazil	Sao Paulo
90	0.727	2-High	China	Pearl River Delta
90	0.727	2-High	China	Shanghai-Changzhou
90	0.727	2-High	China	Beijing-Tianjin
90	0.727	2-High	China	Chongqing
90	0.727	2-High	China	Hangzhou-Ningbo
90	0.727	2-High	China	Chengdu
90	0.727	2-High	China	Xiamen
90	0.727	2-High	China	Shantou
90	0.727	2-High	China	Wuhan

HDI Rank	HDI Value	Quartile	Country	Megacities
90	0.727	2-High	China	Shenyang
90	0.727	2-High	Taiwan	Taipei
93	0.726	2-High	Thailand	Bangkok
108	0.69	3-Medium	Egypt	Cairo
110	0.684	3-Medium	Indonesia	Jakarta
115	0.675	3-Medium	Philippines	Manila
116	0.666	3-Medium	South Africa	Johannesburg-Pretoria
116	0.666	3-Medium	Vietnam	Ho Chi Minh City
130	0.609	3-Medium	India	Delhi
130	0.609	3-Medium	India	Mumbai
130	0.609	3-Medium	India	Kolkata
130	0.609	3-Medium	India	Bangalore
130	0.609	3-Medium	India	Chennai
130	0.609	3-Medium	India	Hyderabad
142	0.57	3-Medium	Bangladesh	Dhaka
147	0.538	4-Low	Pakistan	Karachi
147	0.538	4-Low	Pakistan	Lahore
152	0.514	4-Low	Nigeria	Lagos
176	0.433	4-Low	DR Congo	Kinshasha

3 How are Megacities Growing?

In general, urban land cover – the amount of built-up area – has been increasing faster than urban population. The variance in settlement patterns contributes greatly to this. The global average density across all 45 megacities in this study is 963 people per square kilometer, a number that is highly skewed by the patterns of urbanization in the two nations with the greatest influence over urban planning worldwide, which are also the world's two largest economies – China and the United States.

For example, the urbanized area within the Shanghai municipality (not the megacity as this study has defined it) grew from 1,600 square kilometers to nearly 3,500 square kilometers from 2000 to 2010, but its population density decreased from 8,700 to 6,900 people per square kilometer during the same period. This is despite the fact that Shanghai added 29 buildings of 200 meters or greater, and 115 tall buildings of all heights during the same period [6]. This pattern has been repeated across China – despite its urban population having increased by more than 130 million people between 2000 and 2010, average density has remained almost constant, at 5,300 people per square kilometer, with much construction occurring in places with declining populations [4].

The knowledge that urban land coverage significantly outpaces density in the US comes as less of a surprise. Its overall average is 110 people per square kilometer. America's densest metropolitan area, the New York-Philadelphia megacity as defined in this study, had an average of only 563 people per square kilometer. Having said this, as a new generation of Americans rejects suburban living for urban lifestyles (and workplaces follow suit), and as the aging baby-boomer population downsizes its dwellings and moves into downtowns and closer to amenities, the US is the only «highly developed» country that can expect a significant new wave of urbanization. It will add more than 100 million people to cities from 2010 to 2050, increasing its urban population by 40% [4].

It is significant to note the development patterns that persist in the US and China, because these are the countries where most of the urban planning «prescriptions» for the rest of the world have thus far been devised. Also, much of the design, construction and economic power of the world is housed in, and deployed by, these countries. But both countries are outliers in terms of their spatial development, when compared to the rest of the world.

The US is well known for having devised a particularly expansive form of auto-centric urban sprawl. Long-distance commuting patterns have created far-flung suburbs in semi-rural settings, and as a result, have created expansive metropolitan areas recognized by the US Census Bureau. This is one reason why the study includes the urbanized area of southeastern Pennsylvania, all of New Jersey, and significant portions of southwest Connecticut in the «New York – Philadelphia» megacity, an area of 54,880 square kilometers. Even more dramatically, the «Los Angeles megacity» includes substantial areas of open desert between settlements within the Los Angeles – Long Beach, CA Combined Statistical Area (CSA), which encompasses 87,945 square kilometers – bigger than Austria and about the same size as Serbia.

China, «may also be an outlier, because its urban expansion has been accompanied by rapid economic growth, strong intervention by the central government in the urban expansion process, and massive investment in public infrastructure» [4]. In other words, most countries do not have characteristics that would easily yield a similar or appropriate response to their localized forms of urban growth.

4 How Dense and Tall are Megacities?

Of the 1,731 buildings measuring 200 meters or higher, that are currently completed or under construction worldwide, 958, or 55% of these, are in the 45 megacities. Asia also claims the lion's share of 200-meter-plus buildings in megacities, with 774 (81%) of the world's 958 (see Fig. 5). In addition to being the world's largest megacity by population, with nearly 65 million people, the Pearl River Delta also has the greatest number of buildings 200-meters and taller – 220 – and the highest number of 200-meter-plus tall buildings per capita – one for every 295,000 people. Nearly 30% of the 200-meter-plus buildings in the world's megacities are here. But is it the densest?

Measured as a single entity, the answer is surprising. The Pearl River Delta contains some of the densest places in the world, in particular the Kowloon area of Hong Kong, which holds up to 32,100 people per square kilometer; but, overall, its vast 56,217-square-kilometer area has an average density of 1,154 km/m2, ranking the megacity as the 23rd-densest in this study (see Table 1). The twenty densest megacities are shown in Figure 6, with Karachi, Pakistan being the densest on average. Karachi has an average of 6,663 people for each of its 3,527 square kilometers – but only one building of 200 meters or taller (see Fig. 7).

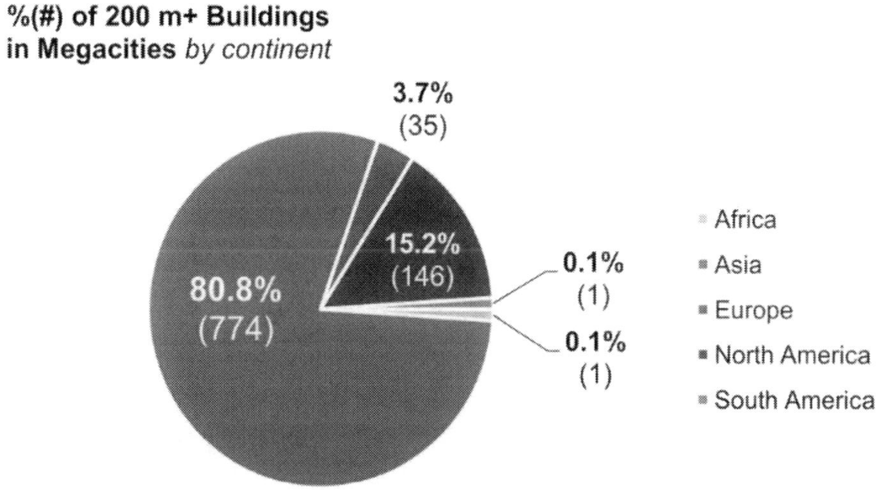

Fig. 5. Percentage and number of 200m+ buildings in megacities for each continent. (Source: CTBUH Skyscraper Center).

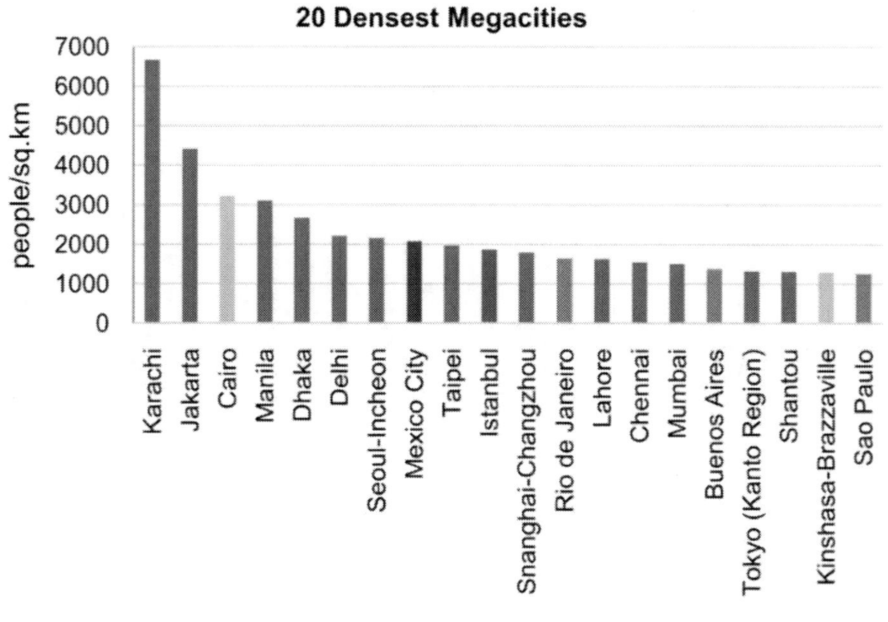

Fig. 6. Population density figures shown for the 20 densest megacities are derived from an averaging of density across the entire urban agglomeration. (Source: CTBUH and see «Population and Area Sources» in References).

Meanwhile, the vast informal settlements and transient populations of cities such as Mumbai and Dhaka make precise counts almost impossible, but it is generally accepted that these cities contain within them some of the most densely populated urban land on earth, in excess of 40,000 people per square kilometer in some places – and mostly in desperate urban conditions, in terms of space and infrastructure provisions. Even within this cohort – low-lying cities on the Indian subcontinent – generalization is difficult. Mumbai is the world's 11th-biggest megacity in terms of population, and has 38 buildings of 200 meters or higher, while Dhaka is number 13, and has none (see Fig. 8, 9).

Fig. 7. Karachi's cityscape. (Source: CC BY-SA Flickr account with username, «No Real Name»).

Fig. 8. (Left) New tall buildings rising above a laundry district in Mumbai and (Right) Mumbai cityscape. (Source: CC BY-SA Aleksandr Zykov and Vidur Malhotra).

In many ways, the Jakarta megacity is representative of the megacities still to come. By drawing a line around its urbanized area, from the Jakarta coastline to southern edge of Bogor, and from the Tangerang Regency boundary on the west to the Bekasi Regency boundary on the east, 28 million people fall within its orbit (see Fig. 10).

This makes Jakarta the ninth-largest megacity in this study in terms of population. Significantly, Jakarta has recently seen some of the most rapid tall-building construction outside of China. In 2015, the city saw seven buildings over 200 meters completed – the largest number of any city worldwide [6]. It also has a high population density, at 4,415 people per square kilometer (rank no. 2), and has forty-six 200-meter-plus buildings, ranking it no. 6 in this megacity survey. It's in a populous, economically-growing but still largely rural and poor country, with many low-lying areas susceptible to flooding. It is highly fragmented across multiple jurisdictions and travel corridors, and its infrastructure has not generally kept up with the speed of urbanization.

Fig. 9. 2004 flooding in the streets of Dhaka. (Source: CC BY-SA dougsyme).

Fig. 10. Newer high-rise buildings juxtaposed with the traditional, low-rise urban context of Jakarta. (Source: CC BY-SA Magdalinski).

Cities like Jakarta, and those in less-developed conditions across South Asia and Africa, demonstrate both the enormous potential of, and the dire need for, heavy investments in transportation, infrastructure, and comprehensive regional planning. Without such improvements, the traditional, low-slung kampungs and the gleaming new towers alike will both be inundated by one or more vectors of a catastrophic combination: traffic, flooding, pollution, sewage, and social unrest. If they do not face total destruction, at the very least, large swaths of some megacities are at risk of becoming disconnected from the rest of the city and the world. Perhaps counterintuitively, some of the densest cities have the fewest tall buildings, and they tend to be comparatively under-developed economically and in terms of infrastructure. The majority of the world's poor are not living in tall buildings, but in low-rise, tightly packed dwellings with poor sanitation and building services. It is thus unsurprising that the default built «solution» to poverty for many cities, when the means become available, is to construct as many instances of a typology as close to the opposite of the sprawling shantytown as possible.

The result, to a large degree, has been the seemingly endless repetition of tall (though mostly sub-200-meter) modern apartment blocks, set in a «tower-in-the-park» model amidst broad roadways straight from Le Corbusier's visions. This model, devised for Europe in the 1920s but not implemented until after World War II, then repeated in America with public housing and Interstate highways in the 1950s and 1960s, and now in China, has gone virtually unchallenged for almost 100 years. Given the choice, of course, many of the urban poor would choose to live in dignity. But many believe the tower-in-the-park model isolates people and destroys the cultural integrity of their close-knit neighborhoods. The world clearly has much more densifying, vertical construction and poverty alleviation to do – the big question is, what shape will it take?

5 Conclusions

Megacities comprise 13.1% of global population and occupy 0.66% of its land surface area. From this, it would be logical to infer that megacities demonstrate a high degree of sustainability with respect to land use. But their energy use, their disproportionate location along coastlines and deltas, and their population growth rates will have much wider implications than their built-up surface areas would suggest.

Even if megacities do not yet contain anything near a majority of the world's population, their strategic importance to national and global economies is difficult to overstate. The Shanghai and Pearl River Delta clusters account for less than 25% of China's population, but account for 80% of its exports [5]. The implications of having to «move» even one of these megacities inland, or undertake massive coastal defense works, or elevate the ground plane to safety beyond new flood zones, are mind-boggling – but that is not an excuse for not thinking about the possibility. What happens in megacities matters disproportionately to the rest of the world, which is well on its way to becoming 70% urban. Any interventions of design, planning, construction or governance in these cities will not only affect a great number of people in the first instance, but their effects will be amplified across the world due to the disproportionate significance of these cities as economic engines. The world is becoming an increasingly interdependent place. As we know only too well from revelations about greenhouse gases, climate change and the rapid spread of viruses – both biological and technological – the choices we make in one part of the world are likely to affect many others. Nowhere is this more the case than when discussing pieces of critical infrastructure set in the world's megacities. There will not be one «template» that can be copied around the world to solve all of its problems, but that does not mean we should not think «big». This is a community that has no difficulty thinking big. The «big thinking», however, needs to extend beyond the height and style of individual buildings, and to instead think in terms of systems – tall buildings need to become part of a three-dimensional, globally-connected, and locally-vital, infrastructure. Tall buildings are critical infrastructure, not just users of it. Those who finance, design and operate them, and decide where they should be built, need to think of them this way, if they are truly to be part of the solution.

References

1. http://www.nationsonline.org/oneworld/countries_by_area.htm, last accessed 2017/08/25
2. S. J. Angel, *Planet of Cities* Cambridge (MA: Lincoln Institute of Land Policy 2012)
3. http://hdr.undp.org/sites/default/files/hdr_2015_statistical_annex.pdf, last accessed 2017/08/25
4. www.worldbank.org/eap/MeasuringUrbanExpansion, last accessed 2017/08/25
5. P. Khanna, *Connectography* (New York: Random House, 2016)
6. www.theskyscrapercenter.com, last accessed 2017/08/25.

7. http://www.lincolninst.edu/subcenters/atlas-urban-expansion/, last accessed 2017/08/25.
8. http://www.mckinsey.com/industries/infrastructure/our-insights/the-power-of-collective-action-forging-a-global-role-for-mayors?cid=other-eml-alt-mip-mck-oth-1606, last accessed 2017/08/25.
9. R. Florida, T. Gulden, C. Mellander, *The Rise of the Mega-Region* (Cambridge Journal of Regions, Economy and Society 2008) **1** Issue 3, 459-476 pp
10. www.hsbc.com/~/media/hsbc-com/investorrelationsassets/.../prd-factbook, last accessed 2017/08/25.
11. http://china-trade-research.hktdc.com/business-news/article/Fast-Facts/PRD-Economic-Profile/ff/en/1/1X000000/1X06BW84.htm, last accessed 2017/08/25.
12. http://data.worldbank.org, last accessed 2017/08/25.

Spatially organized «vertical city» as a synthesis of tall buildings and airships

Olga Gagulina[1], *Sergei* Matovnikov[1]

[1]Volgograd State Technical University, Volgograd, Akademicheskaya St., 1, 400047, Russia

Abstract. The paper explores the compact city concept based on the «spatial» urban development principles and describes the prerequisites and possible methods to move from «horizontal» planning to «vertical» urban environments. It highlights the close connection between urban space, high-rise city landscape and conveyance options and sets out the ideas for upgrading the existing architectural and urban planning principles. It also conceptualizes the use of airships to create additional spatial connections between urban structure elements and high-rise buildings. Functional changes are considered in creating both urban environment and internal space of tall buildings, and the environmental aspects of the new spatial model are brought to light. The paper delineates the prospects for making a truly «spatial» multidimensional city space.

1 Introduction

In the 20th century, ultra-large cities became a unique phenomenon rather than simply vast living areas shared by huge numbers of people. The society living in large urban agglomerations proved most sensitive to social changes, while large cities themselves continued to evolve in line with the outdated architectural principles. The key problems, among which there were transport crisis and growing detachment between people or even hostility of cities to their inhabitants, were increasingly aggravating, and the attempts to resolve them using the then available planning techniques failed to improve the quality of life, causing further crisis situations. The compact city concept that surfaced yet in the 60s of the previous century was among the first insights in the true needs of consumers. It was obvious that the city is to remain human-scaled and comfortable regardless of its size. Without slowing down their horizontal expansion, the cities started crawling upward. But a simple change in the plane and vector of growth was insufficient to reconcile citizens and cities. High-rise buildings offered by large cities and, in a sense, becoming their identity elements greatly simplified the settlement of growing population, but skyscrapers as they now exist turned into «ivory towers» ineffective in helping people interact with the city environment. In order to overcome this «isolation» and allow high-rise structures to be active participants of the city life, it is necessary not only to revise the architectural and urban planning principles and techniques, but also to consider the means for connection

[1] Corresponding author: o.gagulina@yandex.ru

with the city plane and between parts of standalone buildings. The development and introduction of air transport would have become a step change in improving the functionality and appearance of large cities and resolved numerous problems faced by urban planners today and in the future.

The issues of organizing a new type compaction-focused urban environment and the aspects of high-rise construction discussed by such authors as J. Dancig and T. Saati [1], J. Jacobs [2], M. Jenkins, E. Burton, K. Wiliams, M. Brehani, N. Dempsey [3], P. Neuman [4], E. Mellows, J. Gehl, I. Lezhava [5], A. Golovin, A. Krasheninnikov [6], A. Magai [7, 8], as well as the emerging conceptual developments in the field of architecture and advanced transport service, formed the basis for this study aimed at rethinking the contemporary urban planning reality and fleshing out the prospects for spatial development of large cities aided by air transport.

Although aviation has long been in good and extensive use for long-distance communication, the use of air transport in the «body» of an urban settlement would seem exotic. Since the beginning of the 2000s, there has been a growing interest toward lighter-than-air vehicles. There are even a few concepts for the use of airships relevant in terms of municipal connections and architectural opportunities. Still, such projects and suggestions seem fantastic to most specialists. Transport is inseparable from the city life, just as city architecture, which is formed within the context of the city while shaping and determining its appearance. These three notions are so closely interconnected that they cannot be modified separately and require a holistic view. Each component of the «urban environment — architecture — transport» triad has an identical value and needs commensurate attention.

2 Materials and Methods

In a number of countries, the «new urbanism» concept has already been assigned a government policy status and become the basis for the modern urban development mentality. The core rule of compact cities is the intensive use of the existing city spaces, which explains their main «compact» characteristics.

First, this is the optimal density of development achieved through the use of various techniques and means, where the major role is that of tall buildings subject to their competent and multi-optional deployment.

Second, this is about building close connections between various parts of urban structure and improving the accessibility aspect.

Third, there is multi-functionality of each particular city district as well as the functional saturation of both the urban environment and buildings, while preserving the integrity of both a confined territorial unit and the city as a whole.

Fourth, there is a clear focus on environmental conservation and accessibility of natural resources for citizens.

The optimal density, according to most urban planners, can be achieved through the even spread of population over the city territory excluding overpopulated or underpopulated areas, coupled with a gradual transition from the highest-density city center to the more discrete periphery.

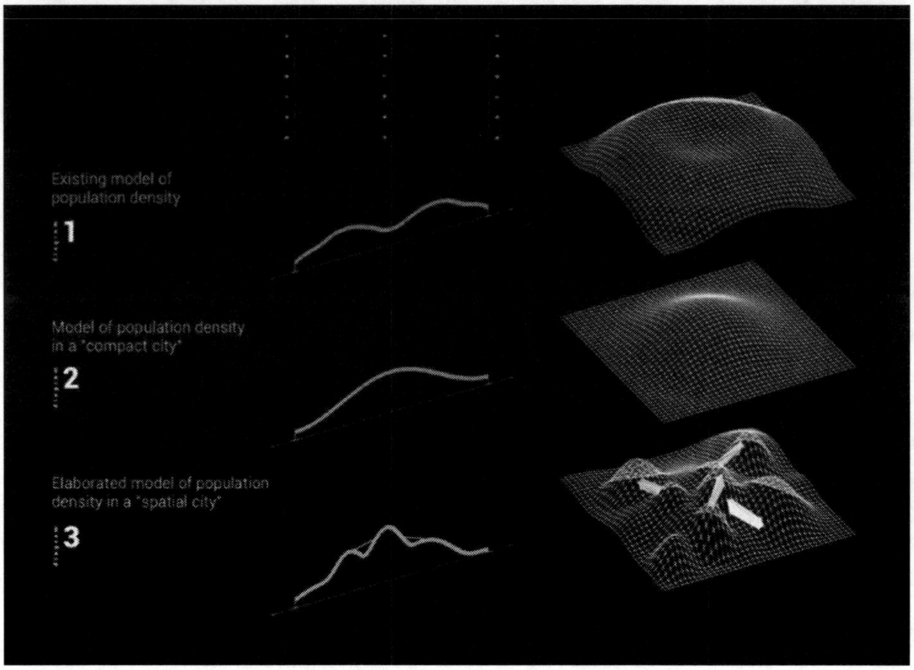

Fig. 1. Density of settlement in various urban organization models

The first diagram shows a «conventional model» with the core slightly «subsiding» in terms of the population density. The core part accommodates various «centers»: office spaces, shopping malls, cultural and administrative facilities surrounded by a vast residential area with so called «dormitory» communities. This structure, based on the functional zoning principle, presents the «variable concentration» scenario. Because the bulk of population live outside the city center and use it as a platform for work, study and leisure, the periphery remains «empty» for the greater part of the day. This model is characterized by the presence of areas with a nearly zero density – extensive park areas or abandoned industrial facilities if located within the city limits. The second diagram demonstrates the principle for building a «compact» model, which can be defined as a principle of uniform concentration. The concentration refers to all aspects — residential space, social functions and spaces, as well as resources. The degree of concentration is high, and its percentage varies negligibly in all areas. The system of strict functional zoning associated with «daily commute» of citiziens is turned away. The third diagram depicts the suggested concept featuring a sophisticated structure with a transition to the «spatial» model based on the «compact city» principles, but adopted for ultra-large municipalities (megapolises) subject to the use of public air transport (airships). This model, as well as the «compact» one, eliminates «complete sparsity» and «empty zones», but allows for certain gradation with reasonable «surges» leading to higher concentrations, and «admissible sparsity» zones. In turn, «higher concentration» zones may have additional spatial connections and allow for certain «variability» of the daily concentration. The last model suggests the blending and smooth coexisting of the two previous models to form a new flexible and utterly variable urban organization pattern. It avoids negating the predecessors, and is rather based on the principle of adapting all the preceding experience, taking into account the specifics and differences of organizing cities with various population sizes.

While criticizing the rigid planning structure and framework in which «conventional» urban planners were trying to arrange the life of citizens, the second model also tries to offer its own scenario, undoubtedly more democratic and humane, stemming from a deep analysis and relying on a solid array of consumer data, but still «common for all» even if fairly flexible. The «compact city» concept does not employ a rigid framework and facilitates diversity in creating urban environments, but sets the clear «preferences». For example, a priority in urban development is placed on low and medium-rise buildings of various types. Gigantism in all its forms is alien to both «new urbanism» and «compact city» concepts which became synonyms. The compact deployment is seen not only in the insightful urban planning solutions, but also in the architectural style. A human being, who according to Le Corbusier's modular has become the measure of all things in urban planning, sets the scales on all levels of environmental organization. This is why sky scraping appears almost barbarian when applied to people and their habitat. Today, tall buildings, which are not a guarantor of «correct» compactness, often become an object of architectural and social criticism [8, 9].

Many urban planners have come to see tall buildings as one of the major architectural and social threats rather than an efficient settlement arrangement. Thus, the Dutch architect Jan Gehl underlines the extremely negative, in the social respect, architectural elements of skyscrapers [10]. This is first of all isolation, both with regard to other people and the city as a whole. Isolating people from the neighborhood and the city life, skyscrapers provoke detachment and cause social differentiation. The «closed» system of high-rise buildings does not help to extend social contacts and deforms the basic urban life ideas for their inhabitants. The ultra-large scale basically falls outside the pedestrian's perception. A significant concentration of skyscrapers in one area disorients and creates a feeling of a human-alien space. By changing the prevailing approach to urban planning and deploying tall buildings within city space it is possible to at least neutralize the negative aspects as described above.

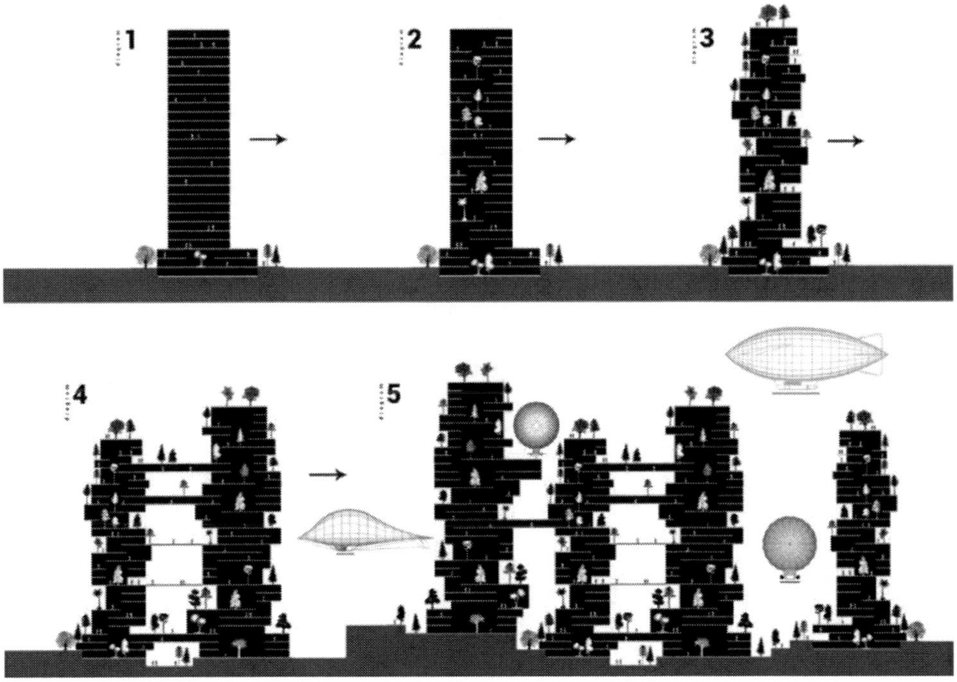

Fig. 2. Evolution of the structure of high-rise buildings

Figure 2 shows the transition from strictly floor-by-floor planning to the «vertical spatial city» system. Even now many projects forbear strict level-based division when a person living in the first floor can use the inner vertical transport to reach the required floor and avoid interacting with the «bulk» of the building (first diagram). The elaborated structure of the internal space allows organizing platforms for social interaction, use of new functions and getting more diverse visual impressions (second diagram).

A more sophisticated internal organization, penetrating through the structural boundaries of a tall and mainly «smooth» building, imparts the necessary amount of «porosity» by providing open zones inside the building normally used for green spaces. This allows creating additional social platforms and differentiating the configuration of tall buildings, but more important, combining the outer (urban) environment with the inner building space. As a result, there is a certain effect of blurring boundaries between architecture and the surrounding space. Such belts and areas inside buildings stimulate the search for new engineering, structural and architectural solutions changing the appearance and image of tall buildings (third diagram).

The comprehensive approach to urban development envisages «ensembles» of high-rise structures in contrast to stand-alone skyscrapers [11]. Despite having functional and visual cohesion, buildings may still need «hardwired» structural ties (fourth diagram). Various functional bridges and passages, as well as certain protruding parts of buildings, when mutually connected, encourage people traffic and provide additional access to nearby structures.

The use of airships to provide «movable» non-structural links between buildings and the ground can help integrate separate structures into a common spatial environment. Air transportation between various points of this environment located at various heights will form a sort of «spatial» streets. «Infusion» of urban environment in buildings will make the city truly spatial (fifth diagram).

In compact cities, all principal objects being the gravity points of everyday life, are to be within easy reach. Reaching all key targets should not take more than half-an-hour. The intrinsic urban planning criteria are quick relocation and enhanced mobility. Cities employing the new urbanism ideology set their sights on the new means of transport [12]. The introduction and development of a municipal network of airships could allow connecting any points of the city landscape. Modern airships materially differ from their predecessors thank to suspension in the air and precision navigation. Airships do not need clumsy and complex mooring structures any longer. With their new capabilities, they could be «moored» to the existing structures, while mooring systems and light mooring structures could be integrated with tall buildings.

Such architectural projects appeared already in the beginning of the previous century. This brings to mind the images of one of the first landmark skyscrapers — Empire State Building with an airship moored to its spire. Similar ideas were promoted by the Russian architect Ivan Leonidov. The blueprint prepared by the People's Heavy Industry Commissariat in 1934-1936 featured an airship site on the roof of one of the towers. The prototypes of the modern concepts of «soaring» skyscrapers can be seen in the works of G. Krutikov and I. Iozefovich [13]. These projects remained utopia, but the technologies for engineering both airships and tall buildings have taken a huge step forward, and the 21st century already saw an array of new diverse concepts. The most renowned system of airships is Airbia of Aleksandros Tsolakis and Iren Shamma designed for carrying passengers from suburban homes to city centers [14]. This concept is set to support air transportation between various parts of the city environment helping to extend communication to its remote areas.

Due to their ability to lift and accommodate a large number of passengers, airships, along with their immediate transportation function, can play a role of mobile architecture. A whole segment of soaring buildings and structures relies on the use of various type airships. For example, the «Hydrogenase» project of the architect Vinsent Callbaut has a clear environmental focus. He offers to create a soaring vertical space accommodating residential quarters, offices, scientific laboratories, manufacturing centers, as well as gardens and plantations of green macro-algae producing bio-hydrogen as the main fuel for airships. The examples here illustrate a sustained interest in the air as a potential space for future urban development.

The integration of transport and city structures will transform the conveyance function from a regional «dictator» into an «option» adjusted and catered for the needs of citizens.

In order to avoid making each standalone high-rise building an isolated «city in the city», it is important to balance the functional distribution between the open «street» environments and «inner streets» built in large architectural objects. «Outer streets» penetrating and transfusing buildings should remain functionally and visually attractive to citizens.

The correct scale and structure of tall buildings, their specifically optimal height and competent functional filling, as well as the justified layout of skyscrapers in the city landscape will help to eliminate idle areas both inside buildings and within adjacent urban environments [15].

For many decades, the planning practice was building on the clear time and space differentiation of the key routine functions of people: work, dwelling, leisure [16]. These totally separated functional pieces were «attached» to the common social center. The elimination of this marked structuring and the blurring of functional boundaries have also led to «diffuse» the center and create a polycentral planning system [17].

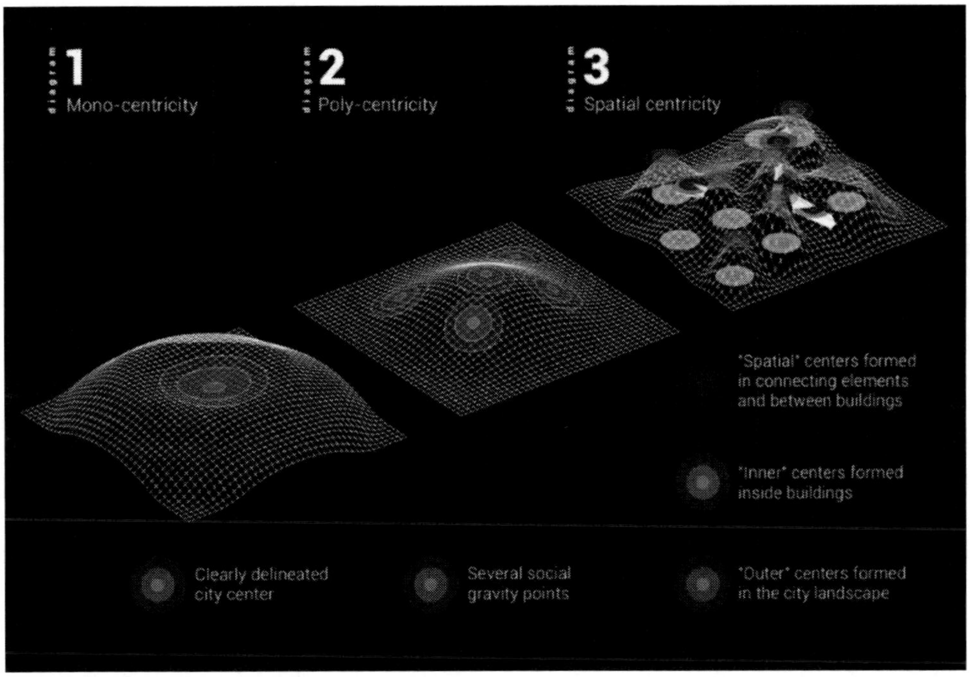

Fig. 3. Distribution of social centers in the city structure

Figure 3 illustrates the transition from a single-core structure to a system of a few city gravity centers evenly distributed within the city limits. The next stage can include, along with elevating the «street» and its partial lifting from the ground, the transfer of «social magnets» associated solely with the «street» to the inner space of buildings built in accordance with the urban development system. With the appearance of social centers developing both horizontally and vertically a spatial poly-centricity model can be developed. The special «porous» structure of buildings will create platforms existing at the junction of architecture and city environment (third diagram).

Similarly to functional diversity inside confined city areas, the internal structure of new tall buildings should combine functions without clearly dividing them by levels or zones. High-rise buildings and skyscrapers should not become living areas for a specific class of citizens only, as they can encourage good communication between neighbors and social interaction of people with different status and level of income.

New type high-rise buildings, developing the new urbanism theory, can encompass, in addition to the traditional forms, essentially different functional zones: residential, cultural and educational, «agricultural», industrial etc.

Caring for the environment has long stopped being just a popular trend and turned into an understanding that sustainable development is an integral part of urban planning. As such, solutions concerning the distribution of «green spaces» in large and ultra-large cities are of great importance no less than the transport regulation or settlement system. Great successes achieved in developing the vertical gardening technology will allow replacing idle park areas to follow the compact trend. The transfer of parks inside buildings, the system of diffused green spaces available not only in the city, but as far as possible everywhere inside high-rise buildings and accessible by everybody will make «nature» closer to people. Being so handy, elements of natural environment can serve as objects of closer and more frequent contact for megapolis inhabitants. A bright example of such continuous interaction is the Bosco Verticale skyscraper in Milan offering yet another example of vertically organized environment — «Vertical Forest» [18]. Such projects can be efficient environmental conservation tool on densely populated grounds.

One should remember that high-rise construction has a significant environmental impact [8]. The negative aspects associated with tall buildings, in particular, the behavior of air masses near buildings and insolation of the adjacent territory, make this type of construction an object of thorough analysis and balanced deployment [19].

Avoiding extensively spread urban districts with discrete development will also help to draw urban territory closer to the natural environment. The use of airships for transportation not only between city parts, but also beyond their limits will improve communication between urbanized and natural territories.

3 Results

The research conducted by the authors has revealed that the compact city concept should be adjusted in line with the previous urban development experience. To ensure efficient functioning of a city, today and in the future, it is necessary to switch from a flat planning model to the spatial system.

The spatial organization of a city structure requires the consideration of a vertical layout of city environment, including vertical streets, gardens and squares.

In this vertical context, a special part will be played by high-rise construction. Buildings of a «spatial city», instead of being closed systems, will become a part of the city landscape and allow for «penetration» of some functions, spaces and elements inside them. High-rise structures should be built following the same principles as the «street» city environment. The city and its buildings should become a single architectural and structural organism.

Spatial transport connections can be built using the public air transport system. Mooring structures for airship network, being integrated with high-rise structures, will allow moving «from building to building» and provide immediate access to the ground level and to social spaces placed at a certain height.

Additional transport connection will help to avoid «conservation» of people inside autonomous buildings, given a competent and versatile functional arrangement of territories and spaces «inside» and «outside» them.

In order to improve the city compactness and natural resources accessibility for residents of high-rise structures, it is necessary to develop vertical gardening technologies for buildings and structures. Also, an efficient tool for providing closer ties between urbanized territories and the natural environment as one of the components of a sustainable city development is the use of environmentally friendly air transport (airships).

4 Discussion

The earlier studies of various authors in the field of high-rise architecture considered not only technical aspects but also problems of integration of high-rise buildings into the existing urban environment. In turn, the theoretical foundations of modern urban planning concepts try to avoid super-large vertical settlement, treating it as an inhumane way of organizing urban and architectural environments. The presented work makes an attempt to remove this contradiction and identify the directions of changes in urban planning systems, means of transport communication and high-rise construction.

The researched showed that in the issue of the formation of large-scale urban education, one should not only unite and reconcile the past practice of territorial planning but also come up with truly innovative solutions in the environmental organization. Such a decision can be a transition to a system of spatial movement and formation of territories and spaces of social interaction, and not only on the plane but also on various elevations and levels. The life and principles of the organization of large cities can (and perhaps should) be different from their less densely populated counterparts. Spatial development will contribute to the formation of the specifics of the environment conditioned by the necessity of living together and the constant social interaction of a large number of people.

The verticality of certain areas and some urban processes multiplied by the system of spatial transport links will

- solve many urban problems;
- more compactly accommodate the numerous residents of large cities'
- provide increased mobility;
- help to change the idea of a super-large scale of architecture considered now as something "frightening," "alien" and "enslaving."

The introduction of air transport will not only promote spatial communication but also solve the problem of organizing roads, partially unload urban traffic and, most importantly, help in mastering the air environment and adapting it to the needs of residents.

Polyfunctionality of both external environment and internal space of buildings, as well as the functional mobility, will make the city interesting and comfortable for each citizen. In turn, the development of the vertical distribution system of green plantations can change the idea of megacities as stone jungles with scanty vegetation and large park oases, the scale and value of which is sometimes difficult to assess because it is quite hard to visually cover the horizontal green areas. Also, the air transport network, which helps residents of metropolitan areas move relatively quickly and comfortably from a densely populated large city to areas of natural wild or less dense urban formations, can eliminate the feeling of detachment from natural resources.

To develop and implement the concept of a modern spatial city, it is necessary not only to change the skeptical attitude towards the issue of the ubiquitous use of air transport but also conduct a number of additional studies on the effective interaction of the principles of the new urbanism and modern methods of organizing of the high-altitude construction and vertical settlement.

5 Conclusions

One can certainly state that, by changing the existing image, high-rise construction can be an effective way of forming a stable, humane and diverse in its structure living environment. Together with the air transport system, high-rise buildings transform the flat structure of city into a volumetric one. However, the nature of the human and society is such that both the individual and the society are afraid and resist to global changes. In this respect, the concepts of the "new urbanism" and the "compact city" in their simplified form create around the person the most comfortable world, a kind of modernized semblance of ancient European cities, but with the support of the latest technologies. However, the metropolis is not a place of silence and sleepy comfort but rather a territory of challenges. The big city has always become a platform for the birth of revolutions, namely architectural, transport, technological and social. This is its main feature, which should be maintained and stimulated at all levels of urban planning.

Soaring skyscrapers and flying vehicles, which now exist only in literary fiction and cinema, still look like rather naive dreams than practical solutions to problems of settlement and urban organization, although some of the technologies exist today. However, the limitedness, depletion and high value of land plots push city planners and architects towards fresh, creative, not mundane solutions.

Spatial urban development, which is based on high-altitude architecture and air transport, will make urbanized areas truly compact. A smart growth of urban education in modern realities is not about the universal verticalization of construction but rather the appearance of a multitude of vectors for its spatial development.

References

1. J. Jacobs, *The Death and Life of Great American Cities* (Moscow, New Publishing house 2011)
2. M. Jenks, N. Dempsey, *Future Forms and Design for Sustainable Cities Routledge* 444 (2005)
3. M. Neuman, The Compact City Fallacy Journal of Planning Education and Research **25**, 11-26 (2005)
4. G. Lezhava, Modern architecture and city. Academia. Architecture and construction **4**, 5-14 (2014)
5. A.V. Krasheninnikov *Town-planning development of the housing estate Research of experience of the western countries* (Moscow, Architecture-S 2005)
6. A. Magaj High-rise buildings as the vertical integrated inhabited environment, Bulletin of the construction equipment, **6**, 64-65 (2016)
7. A. Magaj, Modeling of functional structures of high-rise buildings. Housing construction **12**, 1-5 (2016)
8. URL: www.asrmag.ru/article/vysotnye-zdaniya/ (last accessed 2017/08/05)
9. V. Korotich Image of modern high-rise architecture, Academic messenger UralNIIproekt RAASN, **3**, 36-39 (2011)
10. J. Gehl, *Cities for People*, (Island Press, 2010)
11. P. P. Zueva, The high-rise building in the urban environment. Proc. of the Conf. Topical problems of construction, ecology and energy saving in the conditions of Western Siberia 98-104 (Russia, Tyumen, 2014)

12. D. Ju. Lomakina, Spatial concepts and images of the future in degree projects on town planning. All-Union research institute of problems of scientific and technical progress of information in construction **4**, 42-55 (2015)
13. N. A. Saprykina, Development of the concept of mobile architecture: the flying objects. Architecture and modern information technology, **34** 1, 1 (2016)
14. URL: www.airbia.wordpress.com/2010/01/22/airbia-a-suburban-airship/ (last accessed 2017/08/15)
15. O. M. Shencova, High-rise buildings in the volume and spatial urban environment. Architecture. Construction. Education, **7, 1**, 102-110 (2016)
16. D. S. Parygin, New urbanism in the large industrial city, Regional economy: theory and practice, **38**, 35-40 (2013)
17. F. Dieleman, M. Wegener Compact City and Urban Sprawl, Built Environment, **30(4)**, 308-323 (2004)
18. URL: www.archdaily.com/777498/bosco-verticale-stefano-boeri-architetti (last accessed 2017/08/17)
19. S. V. Nikolaev, *High--Rise Manual: Typology and Design, Construction and Technology* (Moscow, Atlant-Stroy 2006)

Three centuries of multi-storied St. Petersburg

Leonid Lavrov[1], *Fedor* Perov[1,*], *Aleksandra* Eremeeva[1] and *Vladimir* Temnov[1]

[1]Saint Petersburg State University of Architecture and Civil Engineering (SPSUACE), 2-nd Krasnoarmeiskaya St. 4, 190005, St. Petersburg, Russia

Abstract. The article is devoted to assessment of the role of high-rise buildings in the St. Petersburg historic city's ensemble. Features of formation of city architectural look, the conditions of city typical silhouette's appearance which is characterized by the contrast of a small number of high-rise structures with a low horizontal mass building are observed. The consequences of the emergence of a significant number of great height buildings, the silhouette of which conflicts with the traditional St. Petersburg landscape's compositional principles, are analyzed. The economic reasons of high-rise construction of residential and office buildings are given. The conclusions about the prospects of St. Petersburg high-rise construction in the light of city-building and economic factors are made.

1 Introduction

The problem of transformation of the historic St. Petersburg's specific silhouette, which evolved over three centuries, appeared at the beginning of the new century. The weakening of height regulations in the mid of 1990-ies allowed to place buildings up to a height of 28-40 meters in the city center and in the depth of the districts which led to the fact that there are more and more objects in the classic urban panoramas, which destroy the "skyline" beauty.

The research urges to turn to the assessment of the role of high-rise structures in the ensemble of the historic St. Petersburg, the search for causes of spontaneous expansion of high-rise constructions at the turn of XX-XXI centuries, finding the ways for further effective management of Saint-Petersburg building activity on the basis of the lessons of the past and present. The scientific base of the research consists of the works of V. G. Lisovsky, S. V. Sementsov, P. Nikonov, dedicated to the history of St. Petersburg's architecture, stages of formation of urban spatial environment.

The history of regulation of St. Petersburg's building, including high-rise characteristic, has been examined in these works. A special place in the article is the analysis of the economic features of high-rise constructions in St. Petersburg. City-planning errors can't be corrected, but it is possible to adjust the further course of the St. Petersburg high-rise policy, to prevent the disappearance of the typical city silhouette.

* Corresponding author: dekanaf@gmail.com

2 Methods

Methodological basis of the research is a comprehensive approach that includes study and generalization of literature and Internet sources on the history of high-rise construction in St. Petersburg; analysis of the legal framework regulating the height of the building for three centuries in the history of the city. The survey and photofixation of St. Petersburg's urban landscapes characterized by a pronounced silhouette were conducted; comparison of historical and modern city views, identifying the consequences of the current high-rise construction. The classification of buildings of St. Petersburg, exceeding the height of 100 m and functional affiliation of the building was made; the role of the buildings as urban dominants was defined.

3 Results

3.1 1712. The first high-rise building in Russia is an instrument of state policy

The first high-rise building in St. Petersburg was built in 1712, when the city became the Russian capital. The city was only nine years old, and its buildings were mainly one-story wooden or wattle and daub constructions. They were lost among the stunted forest and did not violate the horizontality of the Neva Delta's landscape. Such image of the Northern capital did not meet the changed political situation. Peter I ordered to build a major stone Cathedral in the heart of the city, in the Peter and Paul's fortress, to replace the existing wooden one. The priority object for the Emperor was the high bell tower. Its architectural image had to reflect the idea of a new time, the appearance of Russia on the European arena. The works were headed by Domenico Trezzini. Lack of materials (bricks were brought from abroad partly), workers were escaping, but building continued. The stonework was 50 meters up, more lightweight wooden structures were used higher: the frame was mounted above them to accommodate the large clock and bell tower, and then - the belfry. The installation of the spire's wooden framework was started at the height of 72 meters, which was prepared for the 1720-th year. The Cathedral's construction was completed in 1733. The cross on the top of the spire reached the height of 112 meters [1].

Even according to modern classification the bell tower could be considered as high-rise building. Its vertical was in contrast with the predominant horizontality of the city landscape and became a visual dominant on the banks of the Neva River. It surpassed the height of the Moscow Ivan the Great bell tower by one third and showed the potential of the "new" Russia. Cathedral's vertical exceeded the height of surrounding buildings in ten-twenty times. The size of the Cathedral was paralleled only by the Twelve collegiums' building: the length of this government complex, was about 400 meters. Peter's enormous construction projects were implemented on "empty space". Two huge buildings (one was the Church building, another was the government) contrasted sharply with the environment and looked like structures transported from another world. There was nothing like this for hundreds of miles around. However, Peter I didn't have a chance to see the results of planned projects' implementation: he died in 1725.

3.2 From the XVIII to the XXI century. Regulation of horizontal position of Saint-Petersburg

V. G. Belinsky wrote that "Petersburg, left by Peter the Great, was too poor and insignificant town to speak about it as something important, but its plans, its destiny was to last everlastingly" [2]. The presence of the huge constructions in the center had an enormous influence on the city urban development. Their appearance defined the image of the development for centuries to come. Contrast of the few high-rise buildings with calm horizontal lines and contour lines of low buildings had become the hallmark of St. Petersburg's silhouette. The useful legacy after Peter the Great was the special technique of strict control of the urban development process. Regulation of the height of buildings was one of the main directions, taking into account the level of technical building capacities, urban planning, compositional and social aspects.

The system of "exemplary" projects was the method of regulation during the reign of Peter the Great. It was supposed to build the houses for "eminent" people with a height of two storeys on sites in the town center (particularly on the embankments of the Neva River). In the peripheral areas projects of one or one and a half storeys houses applied.

With the city evolution, the degree of beautification was increasing, the development nature was changing. It became clear that the front parts of two-storey houses needed desired representation, therefore, it was decided to raise the height of buildings in the city center. In 1765 the Building Committee requested Catherine II to establish the height of 10 fathoms (21,336 m) for embankments' development that "buildings on the Neva bank at least a little match stony shore constructed on the River". In the same year, the rule to build houses of the same height along the street ("under the common cornice") was confirmed [3]. The first wave of existing buildings' completing was in 1820-1830-ies. Three or four-storey houses dominated on Nevsky prospect in 1830; the height of buildings increased also in other critical areas. However, when five-storey residential buildings began to appear, City authorities decided to limit the private developers' ambitions. Nikolay I, who partly liberalized some aspects of St. Petersburg construction, found it necessary to stick to a strict line in relation to the height of the buildings. In 1844 the decree "On the restriction of the high buildings' construction in St. Petersburg, and limitation of the storey add-ins on existing buildings" was issued, according to which the height of private buildings was not to exceed the cornice of the Winter Palace, that is 23,47 m [4].

It should be noted that authoritative specialists considered it a more moderate parameter reasonable. They reminded that K. Rossi had taken the mark of 22 meters to the buildings on Teatralnaya Street, and then recommended to stay within 21 meters for St. Petersburg (Fig. 1) [5].

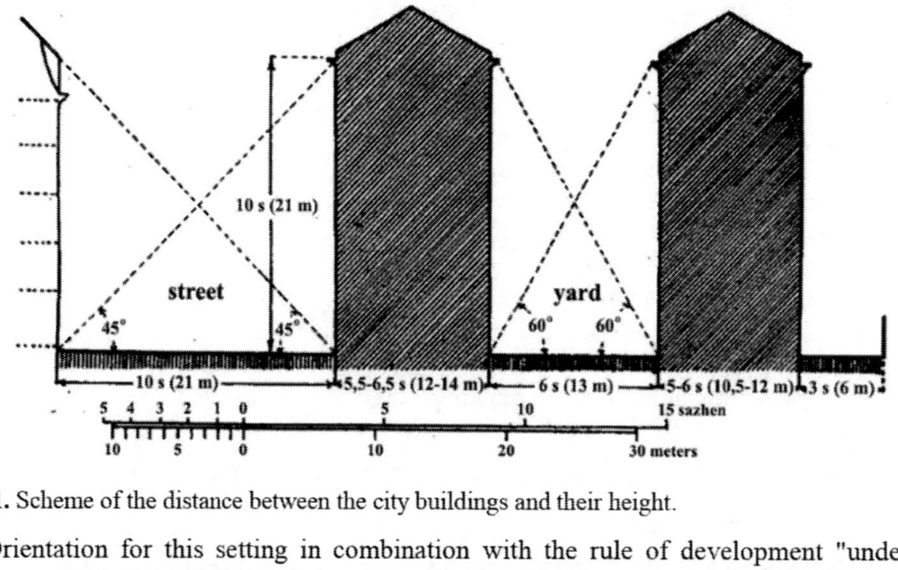

Fig. 1. Scheme of the distance between the city buildings and their height.

Orientation for this setting in combination with the rule of development "under the common cornice" were defining in urban planning in subsequent years as well. This rule was adhered to during the construction of new buildings both in the historic center and on the state highway of "socialistic" Leningrad – Moscovsky Avenue. When restoring destroyed buildings and repairing damaged ones, the desire to link their height with the eaves' level of adjacent buildings was decisive (house №16 on the Palace embankment, houses №19, №107, №184 on Nevsky prospect). The figures, set in 1844, remain a determining factor for the Central part of the city at the present time as well. Even with the reduction of protection zone, stipulated by the law of St. Petersburg №728-99 from 22.12.2005 about approving of the new General plan, they are obligatory on the territory of 1267 hectares [6].

3.3 XIX century. Above 23.47 meters

In Nikolay's decree deviations from the established level allowed, in particular during the construction of religious buildings. Church bell towers, steeples and domes, rising above the horizon, created the system of high-rise dominants in Saint-Petersburg. In 1857-1858 increasing the height of the bell tower of Peter and Paul's Cathedral at 10,5 meters was made (architect K. Ton, engineers - D. I. Zhuravsky, P. P. Melnikov, A. S. Rekhnevsky). When the spire's construction was changed and metal profiles were used instead of the wooden rafters, the height of the building became 122 meters. In 1809 it was decided to build a Church on St Isaac's square. Construction began in 1818 and lasted for 40 years. Foundation preparation was unique, it included more than 12000 piles with a length of 6,5 meters (however, later St. Isaac's Cathedral leaned over and settled on a few tens of centimeters).

Fig. 2. St. Isaac's Cathedral in the panoramic view of Neva River.

New technologies applied in the mining and processing of various types of stones, which were used both for wall covering and for the columns' production. The main dome of the Cathedral was built with the use of metallic structures and shells. The gilding of the domes was performed by the method of "fire gilding", associated with the emergence of dangerous mercury vapor, which killed 60 artisans. It is considered that the total number of dead and injured on the construction site was a few tens of thousands of people. Construction costs exceeded 23 million of silver rubles, huge amount at that time [7]. Suffice it to say that it was spent over 400 kilograms of gold on the site [8]. With the advent of St. Isaac's Cathedral, St. Petersburg had acquired the second building exceeding a height of 100 meters. Its developed form was fundamentally different from the compact vertical of Peter and Paul's Cathedral. Important compositional center appeared in the silhouette of the city, which recognized the connection of two banks of Neva River (Fig. 2).

3.4 XX century. Deformation of "horizontal landscapes"

In the late XIX century, increasing the height of buildings in the historic core of the city was allowed. The house of six stories high appeared downtown. In Soviet times the silhouette of the historical centre had lost a lot of the local verticals - Church domes and bell towers. Instead, in the 1930-ies only functionally necessary technical facilities were successfully built the way they were conceived, such as the water towers of the factory "Red nailer" (under the sketch of Y. G. Chernikhov). In the postwar years it was planned to build several residential buildings crowned with spires (in imitation of Moscow «Stalin skyscrapers»). The most significant realization was on the Moscow Prospect, near the Victory Park: by 1953 the height of the angular scope was increased and have crowned it with colonnade and the spire [9].

Since 1970-ies, 9-, 12 -, and partly 16-storey buildings, which supplanted typical five-storeys in a short time, were being spread in the peripheral areas. Several 22-storey residential buildings were constructed in those places, where important urban compositions were being created – on Victory square, on Smolenka River's embankment. Over time, the inclusion of high-rise buildings in the city's skyline began to cause a public's negative reaction. The building of the 18-story hotel in the Western part of the Fontanka River evoked sharp criticism. Academician D. S. Likhachev found 52-meter high building "violating the typical Leningrad "skyline" [10].

The appearance of new high-rise peripheral areas and emergence of new lifestyle in 1970-80-ies were considered as inappropriate St. Petersburg traditions. Sociologists and hygienists drew their attention to the lack of proper hygienic conditions of the upper floor apartments. Difficulties with foundation construction began to appear. Ancient, strong enough soils lie at a depth of 20-30 m, and closer to the surface there are weak and heterogeneous water-saturated silty-clay soils. Such soils cause significant and differential subsidence of the buildings. It was not always possible to use piles up to 20-30 meters: in severe cases they were "sucked". In the 1970-ies the construction of the 18-storey tower on the Constitution square had to be stopped, as its foundation began to sink.

Significant impact on the silhouette of the city center was provided by new buildings of industrial and technical purposes. In 1962 the tallest building in the city appeared - the television tower. Its height was set by political reasons (Leningrad tower had to be higher than Eiffel tower in Paris). The disadvantage of the architectural appearance was some similarity of its proportions with the bell tower of Peter and Paul's Cathedral. It continued to remain the subject of pride in 2002 as well: in connection with the approaching 300th anniversary of St. Petersburg it was additionally highlighted in the panorama city with light illumination [11].

In 1973-1986 years another high-rise technical structure was built in the peripheral area (the architects Artushin B. I., Savin S.) – tower, which serves as a testing laboratory. Built of modern materials, it has a two-tier design concept, like the bell tower of Peter and Paul's Cathedral. Ferroconcrete wall, having the shape of a ring, raises from the ground up to 77 m and above is the antenna, reaching a height of 104,6 m.

3.5 After 1990 year: prestige, comfort, reliability and economy

Since the mid 1990-ies the horizontal silhouette of the city historic core started to deform as well. Within the law "About the borders of areas of cultural heritage objects' protection on the territory of St. Petersburg and the usage mode of lands within the borders of the specified zones" the opportunity to place building up to a height of 28-40 meters in the depth of the blocks appeared [12]. Above the formed front cornice line, which still kept the traditional mark "not above 24 metres", mansard and intra corps were being appeared here and there. It was the market economy, where the key role was kept by construction firms and wealthy investors, backed by leading banks (Fig. 3).

Fig. 3. Panoramas of the Neva, 2000 years: a - panorama of the North Bank of Neva River, b - panoramic view on the Strelka of Vasilyevsky island.

Modern technologies were quickly mastered, and residential buildings of 22-25 and more floors were used in new housing. However, by the early 2000-s it became clear that doubts about the feasibility of high-rise building, that occurred in the 1970-s, had a number of reasons. It was understood that it is very difficult to construct a building of a great height: the technology of high-rise construction is very different from the standard one - different foundation and different frame are required. Increasing wind loads and large amounts of energy should be considered. It's necessary to book additional electricity and heat supply from the city network, to install emergency power sources. Cost increase of high-rise building is also caused by expensive engineering and fire protection systems. The city emergency services should be also re-equipped (there was the experience in 1991 in Saint-Petersburg, when the fire in the hotel "Leningrad" killed 16 people. The victims could have been avoided if rescuers had mobile emergency stairs of necessary length).

There are more elevators and stairs in the high-rise building. As a result, the part of unsaleable areas is growing. According to "L1 Building company №1", in the construction of buildings higher than 75 m, the cost of the project increases by 40-50% [12].

The cost of an apartment in multi-storey building in St. Petersburg was higher than in most European countries. It was recognized that "high-rise and low-rise buildings belong to different price ranges". Experts report that it is quite difficult to find a buyer for expensive apartments above the 17th floor, and the high-rise residential construction of more than 30 floors (i.e. above 100 m) is economically inefficient [13].

In these conditions, leading developers and the administration began to show cautious about high-rise urban construction. In 2006 in St. Petersburg territorial building standards "Residential and public high-rise buildings. Saint-Petersburg" (TSN 31-332-2006) came into force [14], developed with the participation of specialists from SPSUACE. The standards apply to the construction of residential and public buildings up to 150 m (residential buildings with the height of more than 75 m, public - more than 50 m) and raise the questions of control, automation and mechanism of elevators, communication systems, design and architectural-planning decisions, fire prevention, systems of water supply and sewerage, bases and foundations. The implementation of some projects of high-rise buildings has been rejected, but some of them have been redesigned and realized in a shortened version [12].

3.6 2000-s. Apartments with panoramic view - commercial realization of the "horizontal landscape" potential

Discussion about the reasonability of high-rise construction in St. Petersburg continues. Meanwhile now there are about 30 projects of residential buildings above 25 floors in the city in the implementation stage [12]. Investors take into account the exclusive attractiveness of luxury apartments with panoramic views of the city historic center. Unique horizontal position of Petersburg landscape reveals many architectural dominants from the height of «bird flight». Wealthy clients would pay well for the apartment, if it overlooks the dome of St. Isaac's Cathedral or Peter and Paul's fortress.

Story of the multifunctional complex "Mont Blanc" can be demonstrative. It began in 2003, when the corpus of the new building began to grow at the intersection of Neva and Small Nevka Rivers, just outside the two-storey building of the military hospital, built in the early XIX century, and the advertisement of luxury apartments appeared in the city, the important advantage of which was wonderful view of the wide panorama of the center. During the construction it became clear, what damage the invasion of the enormous building to the historic landscape made (Fig. 4).

In May of 2006, the City Planning Committee expressed their disagreement with the project of building "Mont Blanc" and announced that the new building irreversibly distorted the historical and architectural panorama of Neva banks. By that time, the height of the building was 74,5 meters, which exceeded the height of the bell tower of Peter and Paul's Cathedral. The attempt to make some adjustments to the architectural image of the facility (without reducing the height) was made, but it failed. The facility's construction was recognized as one of the worst urban planning mistakes of St. Petersburg. The financial results of sale of these apartments with panoramic view were never reported. Likely, opportunities to achieve similar economic effect are not exist currently - legal framework have changed, and places with such potential are not seen.

However, it is known, that apartments with the view in the residential complex "Premier Palace", where the surrounding landscapes are not comparable to the unique situation around "Mont Blanc", are 20-30% more expensive than ordinary ones. "High-rise housing is oriented primarily to people with income far above the average. The apartments on the upper floors generally are more expensive than the same ones on lower floors" - the expert says [12, 15].

Fig. 4. View on Peter and Paul's fortress and the complex "Mont Blanc".

The problem of distortion of Saint-Petersburg's traditional silhouette has worsened since the beginning of the conversion of "industrial zone". "Industrial zone" began to emerge in the XIX century. At that time the territory of St. Petersburg was located only on the islands in the Neva Delta and was limited by Obvodny canal on the South. Subsequently, the city grew. Areas, which were considered peripheral previously, have appeared on the border with the center. Today they possess exceptional investment attractiveness. These areas lie outside the zone of UNESCO protection, the height regulations here are less strict what allows to build luxury houses with magnificent panoramic views of St. Petersburg's downtown. The lack of a unified conversion concept of the "industrial zone" led to the spontaneous development of individual sites. The "industrial zone" has grown up; the tall ring of parallelepipeds and towers began to form around the center of Saint-Petersburg.

3.7 XXI century. High-altitude records of Saint-Petersburg

The intention to build skyscraper with a height of 396 meters called "Okhta-center", close to baroque Smolny Cathedral, have caused an international protest. It was decided to hold an international design competition. The fierce public debate began. In July of 2006 the appeal of St. Petersburg Union of architects, who claimed that the skyscraper would destroy the image of the city, was published. This position was supported by the Russian Union of architects, noted that this contest violated all international and national rules. In protest, the Union of architects boycotted the competition. Three of the four invited foreign architects (N. Foster, K. Kurokava and R. Vinoli) got out of jury, expressed their negative attitude to a competitive task and paid attention on the danger of "aggressive effect of tall building on the city skyline". In June of 2007, the UNESCO session also negatively assessed the possible consequences of the emergence of high-rise housing on this place. In the end, the project was not implemented [16].

In 2011 «Gazprom» acquired a piece of land in the new district on the Northern shore of the Gulf of Finland. The skyscraper "Lahta-center" – the company's headquarters - is created at the distance of 9 km by air-line from the city historic center (Fig. 5). The authors of the project tried to shape the tower, resembling a silhouette of the bell tower of Peter and

Paul's Cathedral: spire vertical monolith. Obviously, that the 462-metre tower with distinctive vertical silhouette will be perceived as dominant even against the backdrop of the prevailing 25-storey buildings. The large size of the structure is in tune with multikilometer aquatory of the Gulf of Finland - just as the vertical of Peter and Paul's Cathedral contrasted with 500-meter Neva River.

Fig. 5. The construction of a skyscraper "Lahta-center".

"Lahta-center" will be the unique building: other high-rise buildings in Saint-Petersburg are significantly conceded to it in many ways. The creators of this complex faced the same problems as the builders of St. Isaac's Cathedral, but more complicated. In accordance with this, equivalent architectural and technical decisions, regarding supporting structure, glass facades and engineering equipment, were developed and implemented. For example, the core of "Lakhta-center", in which emergency exits are, is able to withstand flame exposure during 4 hours without changing the properties of concrete and steel. The building is installed on 264 piles, each with a depth of 82 meters and with a diameter of 2 meters. In 2015, the project "Lahta-center" received "gold" certificate "LEED for Core & Shell" on the complex's environmental efficiency [17].

By 2018 "Lahta-center" should reach 462 meters in height, and it will exceed by 1/3 the highest European skyscraper «The Shard» in London. It is obvious, that the construction of such buildings is determined by prestige, among other things, and it will remain unique, like "Stalin's skyscrapers" in Moscow. It is possible to use buildings with a height of 100-150 meters in St. Petersburg's further construction. The examples of such high-rise buildings are:

- "Leader tower" – the tallest office building in St. Petersburg. The skyscraper is located at a distance from the city center, but its silhouette can be seen from one of the three Central radial avenues. The building, compact in plan, was placed on the territory, where the 18-storey one was tried to build in Soviet times. The foundation of the Soviet high-rise construction was dismantled. Now, the observation platform, overlooking the historical center, is equipped on the last (42-nd) floor of the skyscraper; it is also possible to land helicopters there [18].

- The highest city residential complex "Prince Alexander Nevsky" is located far from the center, upstream of Neva Rover, right on the riverside. From the top floors apartments in a good weather you can see the Peter and Paul's fortress and St. Isaac's Cathedral. These are luxury apartments which cost more $1 million [19].

The classification of buildings of St. Petersburg, exceeding a height of 100 meters, are presented in the Table 1.

Table 1. The classification of buildings of St. Petersburg, exceeding a height of 100 meters.

№ п/п	Year of Building	Height, number of floors	Name	Address	Function	Notes
1703-1990-s years						
1	1733	112 m	Bell tower of St. Peter and Paul's Cathedral	St. Peter and Paul's Fortress	Religious building	after replacing the construction of the spire (1858) the height reached 122,5 m
2	1958	101,5 m	St. Isaac's Cathedral	St. Isaac's Square	Religious building	
3	1962	301, 5 m (without antenna)	TV tower	Academic Pavlov st., 3	Technical construction	after the installation of new antenna (2011) the height reached 326 m
4	1985	77 m, with antenna 105 m	Tower of CSII RTC	Tikhorecky pr., 21	Technical construction	
after 1990 year						
5	2018	462 m, 87 floors	Lakhta-center	near Lakhtinsky spill	Multifunctional Complex	
6	2013	145,5 m, 42 floors	Leader Tower	Constitucii Square, 7	Office Building	
7	2012	126 m, 37 floors	«Prince Alexander Nevsky»	Obukhovskoy Oboroni pr., 138	Residential Building	
8	2009	110 m, 27 floors	«Atlantic-City»	Savushkina st., 126 a	Office Building	
9	2003	108 m, 25 floors	«Bogatirsky, 2»	Bogatirsky pr., 2	Residential Building	
10	2009	105 m, 25 floors	«House near Kommendant Square»	Komendantsky pr., 13	Residential Building	
11	2012	104,5 m	«Dominanta»	Kosmonavtov pr., 37a	Residential Building	
12	2012	101 m, 27 floors	«Poem near Three Lakes»	Lunacharskogo pr., 11-15	Residential Building	
13	2008	100 m, 28 floors	«Twin Peaks»	Bukharestskaja st., 110	Residential Building	
14	2010	100 m, 28 floors	«Rainbow»	Zagrebsky boulevard	Residential Building	

St. Petersburg high-rise buildings can be divided into two groups:

1. The unique objects. There are three of them: the bell tower of Peter and Paul's Cathedral, St. Isaac's Cathedral and «Lahta-center».

Common features:

- their construction had not pragmatic incentive (it was not caused neither a necessity nor an expectation of economic benefit). These projects were justified by political or corporate challenges,

- these objects' construction was not based on existing construction basis, it required study of advanced experience and innovative ideas (which required time and money),
- the silhouette of St. Isaac's Cathedral enriches the landscape of the city center, organically interacting with the surrounding buildings; dominant value of the bell tower of Peter and Paul's Cathedral and «Lahta-center» is determined by the contrast of their strongly pronounced verticality with the horizontal silhouette of the city,
- currently, the probability of emergence of equivalent structures is not visible.

2. "Ordinary" high-rise buildings (from 75 to 100 meters), which were total 353 on the 15th January of 2015:

- projects of residential buildings are calculated for commercial success, while for office buildings prestige is important,
- the decisive role in making profit is played by the specific view potential of the apartments, manifested at the level of the upper floors,
- the tendency of formation of characteristic volume-spatial solution of such buildings causes alertness. The investment calculations stimulate the release of maximum number of "view apartments". Projects of 25-30-storey buildings with a silhouette in the shape of a rectangle are born (like residential complex "Prince Alexander Nevsky"), or the groups of towers that conflict with the compositional basics of traditional St. Petersburg landscape (a small amount of vertical dominants over the horizontal low mass construction).

4 Conclusion

1. Formed over the centuries, the silhouette of the historic center is of exceptional value. It is characterized by the predominance of calm horizontals in combination with a small number of high-rise dominants. These traditional features have the strong impact on the perception of architectural solutions in Saint-Petersburg.
2. The experience in design, construction and operation of the residential high-rise buildings over the past decade has shown that the problems, emerging with increasing of the height of the buildings, had been underestimated both in St. Petersburg and in Russia.
3. Providing necessary sanitary-hygienic norms, required home comfort, the security of living in multi-storey apartment houses is possible only with significant expenses, which do not allow to attribute these buildings to mass housing type. The number of high-rise residential buildings will depend on the indicators of effective demand.
4. The greatest damage to St. Petersburg's silhouette is caused by "ordinary" high-rise buildings - residential and office complexes. Economic reasons (the desire to get more expensive view apartments in the crowning part of a tall building and, as a consequence, facades of rectangular shape) are in contradiction with the traditional St. Petersburg's formation (tower spire above the horizontal mass of the buildings).
5. At a moderate altitude, the interaction of old and new buildings through the silhouette's elaboration should be searched. However, we must remember that there have traditionally been not many dominants in Saint-Petersburg.

References

1. V. G. Lisovsky, Architecture of St. Petersburg, *Three centuries of the history. St. Petersburg: AO Slaviya*, 416 (2004)
2. URL: http://az.lib.ru/b/belinskij_w_g/text_1250.shtml (last accessed 28.02.2017)

3. S. V. Sementsov, Stages of formation of the spatial environment of St. Petersburg. *Bulletin of Civil Engineers*, **2(7)** (2006)
4. URL: http://www.nlr.ru/e-res/law_r/search.php?part=483&regim=3 (last accessed 05.03.2017)
5. E. Y. Kupffer, Residential house: a guide for design and construction of modern dwellings, 273 (Moscow: M.O. Wolf, 1914)
6. URL: http://nikonovpn.spb.ru/?p=3298 (last accessed 06.03.2017)
7. URL: https://ru.wikipedia.org/wiki/%D0%98%D1%81%D0%B0%D0%B0%D0%BA%D0%B8%D0%B5%D0%B2%D1%81%D0%BA%D0%B8%D0%B9_%D1%81%D0%BE%D0%B1%D0%BE%D1%8 0 (last accessed 05.03.2017)
8. URL: http://www.spb.aif.ru/culture/event/1187704 (last accessed 05.03.2017)
9. L. P. Lavrov, 1000 addresses in St. Petersburg: a short architectural guide, 416 (St. Petersburg: Eklektika, 2008)
10. D. S. Likhachev, "Skyline" of the city on the Neva River. Our Heritage, **1** 8-13 (1989)
11. URL: https://en.wikipedia.org/wiki/Saint_Petersburg_TV_Tower (last accessed 05.03.2017)
12. URL: http://kommersant.ru/doc/2223512 (last accessed 27.02.2017)
13. URL: http://ppt.ru/news/106118 (last accessed 14.04.2016)
14. URL: http://ohranatruda.ru/ot_biblio//normativ/data_normativ/47/47488/index.php (last accessed 08.02.2017)
15. URL: http://rway.ru/publication/publication71-2180.aspx (last accessed 07.05.2016).
16. URL: https://en.wikipedia.org/wiki/ Okhta_Center (last accessed 06.05.2016)
17. https://ru.wikipedia.org/wiki/Lakhta-tsentr (last accessed 06 03 2017)
18. https://ru.wikipedia.org/wiki/lider-Tauer (last accessed 06 03 2017)
19. https://ru.wikipedia.org/wiki/Kniaz_Aleksandr_Nevskii_zhiloi_kompleks (last accessed 06.03.2017)
20. https://ru.wikipedia.org/wiki/Spisok_samykh_vysokikh_zdanii_Sankt-Peterburga (last accessed 28.02.2017)

Green infrastructure in high-rise residential development on steep slopes in city of Vladivostok

Alla Kopeva[1*], *Olga* Ivanova[2], and *Olga* Khrapko[3]

[1]Far Eastern Federal University, 690900, Sukhanova st. 8, Vladivostok, Russia
[2]Vladivostok State University of Economics and Service, 690014, Gogolya St. 41, Vladivostok, Russia
[3]Botanical Garden-Institute of Far Eastern Branch of Russian Academy of Sciences, 690024, Makovskogo st. 142, Vladivostok, Russia

Abstract. The purpose of this study is to identify the facilities of green infrastructure that are able to improve living conditions in an urban environment in high-rise residential apartments buildings on steep slopes in the city of Vladivostok. Based on the analysis of theoretical sources and practices that can be observed in the world, green infrastructure facilities have been identified. These facilities meet the criteria of the sustainable development concept, and can be used in the city of Vladivostok. They include green roofs, green walls, and greening of disturbed slopes. All the existing high-rise apartments buildings situated on steep slopes in the city of Vladivostok, have been studied. It is concluded that green infrastructure is necessary to be used in new projects connected with designing and constructing of residential apartments buildings on steep slopes, as well as when upgrading the projects that have already been implemented. That will help to regulate the ecological characteristics of the sites. The results of the research can become a basis for increasing the sustainability of the habitat, and will facilitate the adoption of decisions in the field of urban design and planning.

1 Introduction

The authors introduce the hypothesis that innovative facilities of green infrastructure (GI) can be used in high-rise residential apartments buildings on steep slopes in the city of Vladivostok, enhancing the ecological sustainability and aesthetic expressiveness of the urban environment. These facilities include green roofs, green walls, as well as the methods of greening of disturbed hills.

This study introduces a brief conceptual view of the aspects of landscape design in the modern urban environment, including high-rise buildings. A number of theoretical studies are devoted to the integration of nature into urban design and planning. They include the publications of H.L. Venhaus [1], T. Beatley [2-5], the research study of J. Krcmarova that is devoted to E.O. Wilson's concept of biofilia [6]. J. Pomeroy considers skycourts and

* Corresponding author: 457594@mail.ru

skygardens in terms of social, cultural, economic, environmental, technological and indirect benefits that they provide to the urban habitat [7].

There are studies devoted to the benefits of nature-based solutions for health and well-being. They include a scientific report of Shanahan, Bush & 5 more authors [8]. Fuller, Irvine & 3 more have identified the psychological benefits of greenspace [9]. Carrus, Scopellity & 8 more established the positive effects of biodiversity on the human well-being [10]. As Jayasooriya, Ng & 2 more have found out, Green Infrastructure (GI) practices have shown to be promising in mitigating the air pollution in urban areas of several cities across the world [11]. Shahli, Hussain & 2 more have figured out that landscape planning and projects that base on the concept of sustainable development assist in creating of a favorable and responsive environment for housing construction [12]. Dobbs, Nitschke and Kendal [13] have studied the landscape structure of 100 cities all over the world, and came to the conclusion that a social system and climate influence the urban landscape and human well-being. Vavilova notes that the solution of the problem of a habitat sustainable development can be achieved with the help of new innovative technological facilities [14].

The experience of using such innovative technological GI facilities as green roofs and green walls in high-rise apartments buildings in many cities all over the world demonstrates that these facilities are widely used. These promising tools are able to significantly improve the urban landscaping. Generalov and Generalova highlight that the process of designing and constructing of high-rise residential apartments buildings in Singapore is based on the principles of sustainable urban development, minimizing the environmental impact [15]. The authors have also identified that a clear standard of comfort for public housing has been developed in the city, including a set of required improvements of the site [16].

Singapore is one of the world's leaders in urban greenery on buildings. In 2009, the National Parks Board of Singapore developed the "Skyrise Greenery Incentive Scheme" (SGIS) [17], and in 2014 the Ministry of the Environment and Water Resources and the Ministry of National Development designed the "Sustainable Singapore Blueprint 2015" [18], which promotes greenery on high-rise buildings, and, thus, contributes towards Singapore's vision of a City in a Garden. Additionally, Skyrise Greenery under the SGIS offers environmental, economic, social and aesthetic benefits. Because of limited land space in Singapore, the SGIS became an important element in sustainable urban development. There are two categories of urban greenery actions funded by the SGIS: the rooftop greenery and vertical greenery on facades and walls of existing and new buildings.

In Singapore, there is a shortage of land to respond to population and economic growth, and a number of authors (Yuen, Hien [19]; Wai [20]) have substantiated the possibility of the existence of gardens on the roofs of public buildings, private houses and other sites not only as the part of urban design, but also as vertical farms that supply residents with food. Despommier [21] and a number of other researchers (Russo, Escobedo & 2 more [22]) also consider that the vertical farm will have an incredible impact on changing the face of this planet for future generations.

During the last decades several researches were conducted proving that green walls can contribute to enhancing and restoring the urban environment and improving construction building performance. First of all, this is the work of Manso and Castro-Gomes, who presented the review of all types of green wall systems in order to identify and systematize their main characteristics and technologies involved [23].

However, Russia is still behind in adapting the GI facilities mentioned above. Nevertheless, there are works by Russian authors (Khrapko, Kopeva and Ivanova) investigating the application of vertical gardening in the cities of the Russian Far East [24]. Maslovskaya and Chernyavina [25, 26] refer to the statistics on the use of flat roofs inThe city of Vladivostok, and consider them as places for social interactions that can increase the number of public activity sites. Khusnutdinova, Alexandrova and Novik [27] have determined the most economical vertical green system (VGS), the most user-friendly VGS and the easiest to implement VGS in the conditions of St. Petersburg.

The situation with the adaptation of GI facilities is complicated by the location of high-rise residential apartments buildings on steep slopes. Improper planning and rapid development in the management of high-rise construction on steep hills often leads to the destruction of natural elements and the loss of green space. The assortment of plants for greening of disturbed steep slopes has not been sufficiently studied. In this regard, the work of Chinese researchers Chau and Chu [28] is very interesting. They have identified that the fern cover can reduce soil erosion on steep slopes.

2 Materials and Methods

The city of Vladivostok is the administrative, economic and cultural center of the Primorsky Krai, the core of the largest agglomeration in the settlement system of the Far East, the most important center of the Pacific regional settlement system. The authors examined all sites of high-rise residential apartments buildings that have been constructed in the city of Vladivostok after 2000. The city is located in the southern part of Primorsky Krai. The Amur and Ussuri Bays that are part of the Peter the Great Bay of the Sea of Japan wash it. The study examined the territory of the city within the boundaries of the Muravyov-Amursky Peninsula. The following materials have become the theoretical basis for the study:

- Data on the analysis of hill slope and orientation maps in the city of Vladivostok (2004) [29];
- Data on the current state of the territory, found in the *Master Plan of the city of Vladivostok Municipal District*, designed by the Primorgrajdanproekt project organization until 2025 (2012);
- Data on the conditions of nature on the territory of the city of Vladivostok urban district, provided in the report of the Pacific Institute of Geography of the Far Eastern Branch of the Russian Academy of Sciences titled as *Integrated assessment of the territory of the city of Vladivostok and agglomeration* (2012).

In accordance with SR 131.13330.2012, the climate of the city of Vladivostok urban district is a moderate monsoon. The relief is low-mountainous, strongly dissected with absolute elevations of 250-350 m, with separately rising domed mountain peaks 430-475 m high. The relief of the city has a shallow structural module, with small areas of homogeneous slope and orientation characteristics. The city of Vladivostok area is seismically active with the possible force of earthquakes reaching 6-7 points (according to SR 14.13330.2014). The city's territory can be exposed to dangerous natural and man-made impacts, causing emergency situations. The dangerous natural phenomena occurring in the city include periodic tropical cyclones (typhoons). Tsunamis and earthquakes are not excluded.

In Russia, high-rise buildings are denoted as those that are taller than 75 m (starting from 25 floors) [30]. High-rise buildings on the steep of the city of Vladivostok slopes are located on high stylobates, therefore, this study has considered sites with high-rise residential apartment buildings of 18 to 52 floors. The sites for the research were symbolically divided into two groups: set in operation and under construction. In the first group, 27 sites were examined. The second group has 36 sites.

In the process of research, the photos of high-rise residential buildings were made, the slope of the territory and orientation of the hills were determined, microclimatic conditions were examined, aesthetic and ecological characteristics were described, and federal and local regulations concerning urban landscaping were analyzed. In order to determine the

 https://doi.org/10.1051/e3sconf/20183301004

range of plants for the greening of high-rise apartments buildings on the slopes, the authors conducted an introductory experiment at the Laboratory of Far Eastern Flora of the Botanical Garden Institute of the Far Eastern Branch of the Russian Academy of Sciences.

Based on the data obtained, it was suggested that the percentage of green areas of the high-rise residential construction sites in the city of Vladivostok does not meet regulatory requirements, the original natural elements such as relief and landscapes are not preserved, they are disturbed in the process of construction. The first bases concerning the necessity to solve the problem of high-rise building greening on the steep slopes of the city of Vladivostok were presented by the authors in previous works [31, 32, 33].

3 Results

The authors investigated the conditions of all high-rise residential sites in the city of Vladivostok It has been identified that the majority of high-rise residential sites are located in the central part of the city of Vladivostok on slopes up to 30% at the top elevations of the relief (figure 1).

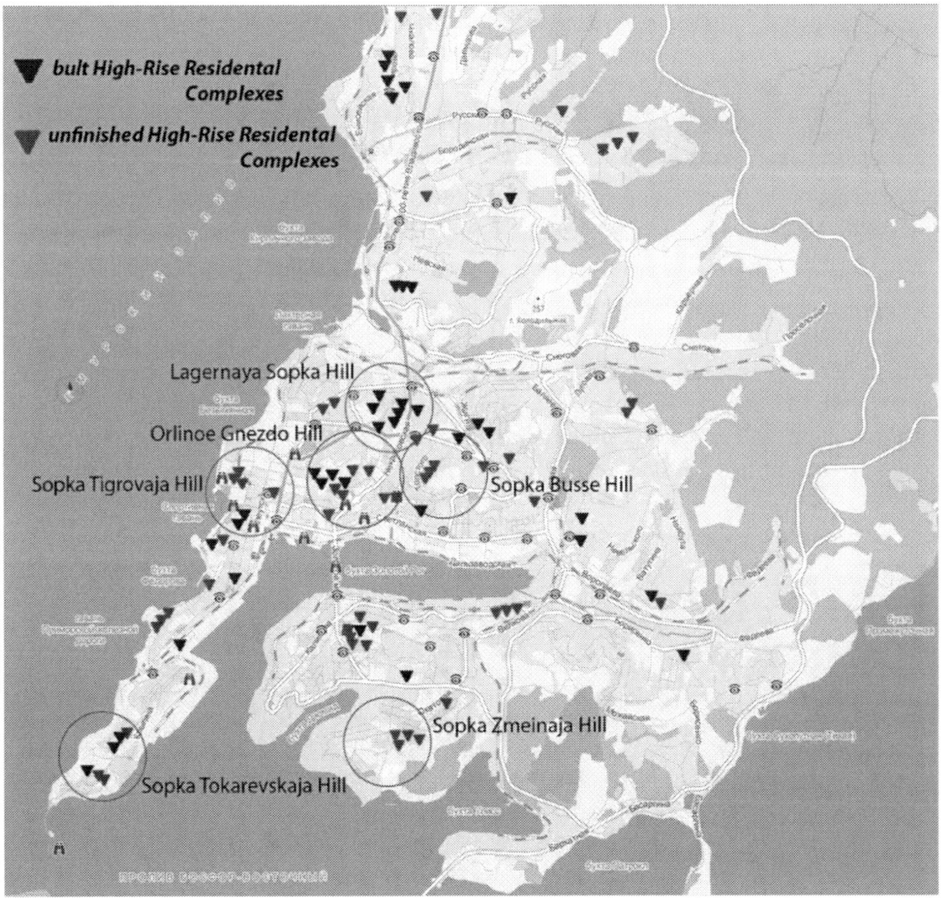

Fig.1. Layout of high-rise residential complexes in the city of Vladivostok

The reasons why the initial natural relief is damaged on these territories have been indicated as the following: a large structural module of construction does not correspond to a small structural module of a relief; the application of construction methods that are used in building on flat relief (figure 2).

Fig.2. Examples of high rise residential complexes construction on the slopes in the city of Vladivostok: a – *Kaskad*, b – *Orlinoe Gnezdo-I*, c – *Orlinoe Gnezdo-II*, d – *Alye Parusa*

The reasons why the regulatory requirements for greening of high-rise residential sites on steep hills are not observed have been identified as the following: the failure to comply federal and local construction rules with the features of steep slopes greening - the current standards do not consider green roofs and green walls as a landscaping facility of compensatory nature; the disturbance of the relief leads to the disturbance of soil cover on dumps and slopes, the destruction of primary plant groupings, the changes in the dynamics of plant communities. Single examples of the use of green roofs have been identified (figure 3).

Fig.3. Examples of area landscaping of high-rise residential apartment buildings in the city of Vladivostok: a – *Orlinoe Gnezdo-I*, b – *Armada*

Examples of the use of green walls and slope greening are absent. Proposals on the use of GI facilities in high-rise residential sites on steep slopes in the city of Vladivostok have been formulated and the first proposals on the assortment of plants have been given.

4 Discussion

4.1 Theoretical studies of GI facilities implementation

A review of previous studies, journals and documents provides a theoretical justification for the importance of applying innovative GI facilities as elements of landscape design to replace the original natural environment in dense high-rise residential sites on steep slopes. This study is based on research materials and documents on the integration of nature into urban design and planning [1-7]; the benefits of nature-based solutions for health and well-being [8-11]; the importance of landscape design in the formation of a stable habitat [12-13]; the solution of the problem of ecologization and humanization of the environment [14]; the results of using GI facilities in dense high-rise sites [17, 18], including the foreign experience of using green roofs, vertical trusses [19-22] and vertical green systems (VGS) [15, 16, 23], as well as Russian experience of using GI facilities [24-27] and the experience of restoration of disturbed landscapes [28].

4.2 The practice of high-rise residential construction on the city of Vladivostok slopes

There is a unique practice of high-rise residential construction on the city of Vladivostok slopes. It supports the image of the Primorsky Krai capital as the Russian Pacific Gate, and preserves the peculiarity of its seaside landscape.

For a long time, high-rise construction was banned in the city of Vladivostok due to seismic hazard. In recent years, high-rise residential apartments buildings are being built, and this tendency will continue due to such reasons as the high price of land and its deficit, the economic attractiveness for investors and developers, the continuing demand for housing, and the willingness of the authorities to accept and approve such projects. There are other reasons why high-rise construction will continue. Firstly, after the construction of the Zolotoy Bridge (length - 1400 meters, width - 30 meters, desk level above the water level - 65 meters, height of the pylons - 230 meters), the view of the central part of the city has significantly changed. Giant bridge requires noticeable buildings on nearby hills. Secondly, according to the current land use and construction plans, almost all the city of Vladivostok refers to the zone of high-rise buildings. This means that the whole city can be built up with high-rise apartments buildings.

In 1997, the first 25-storey high-rise residential apartment building of a monolithic structure, 80.5 meters high, was built at Fastovskaya Street. The next high-rise residential complexes in the city of Vladivostok appeared in the first decade of the XXI century: 2006 - Atlantis-I, 2007 - Arcade House. In 2010-2013, there was a significant increase in the construction of high-rise apartments buildings. During this period, most of the high-rise residential complexes were built. The largest of them are: Metallist 5 (2010), six residential complexes in the residential district Snegovaya Pad (2011), Admiral (2012), Pallada (2012). The city administration plans to continue the construction of high-rise residential apartments buildings. Currently, dozens of projects have been approved, and the construction of new buildings has already begun with the completion dates planned in 2017-2020: Alye Parusa, Atlantix, Zolotoy Rog, Magnum, etc.

Until the 2000s, the upper parts of the city of Vladivostok slopes were not built up. Multi-apartment houses up to 16 floors were located along the relief, along the slopes, encircling them and leaving its upper part with a natural landscape. At the moment, the slopes of all the hills located in the central part of the city of Vladivostok are built up: Sopka Orlinoe Gnezdo (143 m), Sopka Tigrovaja (79 m), Sopka Busse (206 m), Sopka Burachek (138 m) and others. All the hills have a rocky base, and were originally covered by local plants. The relief in the central part of the city has slopes of up to 30%, in some areas it is up to 60% [29].

The authors examined all high-rise residential sites that have been constructed in the city of Vladivostok after 2000. The authors identified that high-rise buildings destroy the unique natural landscape of the city of Vladivostok. There is a negative impact of new high-rise apartment buildings on the local areas. As the scale of modern construction in the city of Vladivostok is comparable to the scale of the hills, the new buildings dramatically change the wind and insolation regime. Those locations that used to be sunny and warm are now covered with shadow and blown with the wind due to high-rise apartment buildings. According to the laws of aerodynamics, the elite high-rise residential buildings in the area of Sopka Orlinoe Gnezdo hill redirect the north and south winds from the upper floors down to pedestrians.

The authors figured out that due to large deviations of the relief with some slopes reaching up to 30% or more, the problem of disturbing the natural relief and its further transformation into anthropogenic one is acute. In the central part of the city, the retaining walls transform high-rise building adjacent territories into non-aesthetic and inhuman stone landscape, free from plants.

4.3 Greening of high-rise residential sites on steep slopes in the city of Vladivostok

The reason why the regulatory requirements for greening of high-rise residential sites on steep slopes in the city of Vladivostok are not universally observed, have been identified: the failure to comply federal and local construction rules with the features of steep slope greening. The current standards do not consider green roofs and green walls as a landscaping facility of compensatory nature. According to SR 42.13330.2011, where federal norms are specified, the area of green roofs should not be included in the indicator of the territory green space when calculating the balance of the area of the projected site, while the area of the exterior surfaces of buildings and constructions for vertical gardening should be indicated in the "Accomplishment" section. This requirement corresponds to construction works on flat relief, but not on steep slopes. It is impossible to plant trees and shrubs in a traditional manner on steep slopes due to the lack of free horizontal areas in the ground level. Therefore, the use of green roofs and green walls is a necessity.

According to the current land use and construction plans for the city of Vladivostok, the area of the building site is determined in accordance with SR 30-11-98: the minimum area of the multi-apartment residential building is calculated at the rate of 17.3 sq.m. per person in case of the compaction of the existing building and 11.7 sq.m. per person in the projected building. The area of green territory should be at least 5 sq.m. per person. It is impossible to comply with these standards on dense high-rise sites on steep slopes in the central part of the city.

The favorable location of a building in the city center forces developers to apply to the city authorities for declining the percentage of green areas, or include play and sports grounds, parking lots, green roofs and vertical gardening as their part. In existing conditions, when creating green roofs and vertical landscaping, mobile greening is used, namely growing plants in special mobile containers.

Modern methods of construction and technology do not allow preserving the original natural relief and greening on steep slopes. A building can be inscribed in the existing relief only when erecting a stylobate with a height of 5-6 floors. This entails extensive work on excavating the soil, erecting retaining walls and strengthening slopes. The broken relief is strengthened with the help of gabions, geo-grids, non-woven geo-textiles, prefabricated concrete-lawn elements in combination with greening. It has been identified that while constructing high-rise apartments buildings in the city of Vladivostok, trees are cut down and the fertile soil is replaced with a clay substrate. Steep hills are either filled up with soil unsuitable for growing plant, or are filled up with fertile soil without preliminary preparation. The root system of bushes and lawn grass does not retain soil, therefore slopes are eroded. As a rule, construction companies leave slopes without greening in the hope that nature will take care of them, and they will cover themselves with green plants.

The authors of the study conducted an introductory experiment in the Laboratory of Far Eastern Flora of the Botanical Garden Institute of the Far Eastern Branch of the Russian Academy of Sciences. The experiment demonstrated the first results in determining the range of plants for the greening of high-rise areas on the slopes. Local perennials and shrubs are at the top of the list for planting roofs, while the key part of the list for vertical landscaping and slope greening are vines as Far Eastern and introduced species and forms.

5 Conclusion

Currently, there is an increase in the rates of environmental degradation and a reduction in the area of gardening in areas of high-rise residential sites on steep slopes in the city of Vladivostok. It is caused by the application of construction methods that are suitable for building on flat relief, but when applied on steep relief they lead to its disturbance, the disappearance of the original natural elements of gardening, and ultimately to the deterioration of the ecological situation and the dehumanization of the habitat. Taking this into account, the use of GI facilities such as green roofs, green walls and slope greening is necessary. Further research is recommended to develop plant assortments for green roofs, green walls and slope greening, taking natural and climatic conditions of the city of Vladivostok into account. The results of this study can support local policies aimed at providing green space for all residents of the city.

References

1. H. L. Venhaus. *Designing the Sustainable Site: Integrated Design Strategies for Small Scale Sites and Residential Landscapes* 256 (Publisher Wiley, 2012)
2. T. Beatley. Biophilic Cities: Integrating Nature into Urban Design and Planning 208 (Publisher Island Press, 2010)
3. T. Beatley, L. Laurian, D. Medearis, W. Daseking, M. Bruel, M. Jaakkola, M. Moretti, L. A. Orive, R. D. Lema, C. Ween. Green Cities of Europe: Global Lessons on Green Urbanism 248 (Publisher Island Press, 2012)
4. http:// www.mdpi.com/journal/sustainability (last accessed 21.03.2017)
5. T. Beatley. Handbook of Biophilic City Planning & Design 312 (Publisher Island Press, 2017)
6. J. Krcmarova, E.O. Wilson's concept of biofilia and the environmental movement. Klaudyan: Int. J. of Historical Geography and Environmental History **6/1-2**, 4-17 (2009)
7. J. Pomeroy. The Skycourt and Skygarden: Greening the urban habitat. Publisher: Routledge, 304 p. (2014)
8. https://www.nature.com/articles/srep28551 (last accessed 20.03.2017)
9. http://rsbl.royalsocietypublishing.org/content/3/4/390.figures-only (last accessed 21.03.2017)

10. http://www.sciencedirect.com/science/article/pii/S0169204614002552 (last accessed 21.03.2017)
11. V. M. Jayasooriya, A.W.M Ng, S. Muthukumaran, B. J. C. Perera, Green infrastructure practices for improvement of urban air quality Urban Forestry & Urban Greening **21**, 34-47 (2017)
12. F. M. Shahli, M. R. M. Hussain, I. Tukiman, N. Zaidin, The Importance Aspects of Landscape Design on Housing Development in Urban Areas APCBEE Procedia 10, 311-315 (2014)
13. C. Dobbs, C. Nitschke, D. Kendal, Assessing the drivers shaping global patterns of urban vegetation landscape structure Science of The Total Environment 592 pp 171-177 (2017)
14. T Ya Vavilova, Retrospective review of UN documents on sustainable development of the living environment Vestnik SGASU. Town Planning and Architecture **1**, 24-28 (2011)
15. E. M. Generalova, V. P. Generalov, Designing high-rise housing: The Singapore experience CTBUH journal 4, 40-45 (2014)
16. E. M. Generalova, V. P. Generalov, N. D. Potienko, Affordable housing under shaping dense vertical urbanism, Cities to Megacities. Shaping Dense Vertical Urbanism Proceedings of the CTBUH 2016. Council on Tall Buildings and Urban Habitat, Chicago, 650-659 (2016)
17. https://www.skyrisegreenery.com (last accessed 12.03.2017)
18. http://www.mewr.gov.sg/ssb/files/ssb2015.pdf (last accessed 12.03.2017)
19. B. Yuen, W.N. Hien. Resident perceptions and expectations of rooftop gardens in Singapore Landscape and Urban Planning, **73/4**, 263-276 (2005)
20. W. T. D. Wai. Beyond Skyrise Gardens: The Potential of Urban Roof Top Farming in Singapore, CTBUH research Paper, 412-427 (2011)
21. D. Despommier. The Vertical Farm: Feeding the World in the 21st Century, 336 (Publisher Picador, 2011).
22. A. Russo, F. J. Escobedo, G. T. Cirella, S. Zerbe. Edible green infrastructures: An approach and review of provisioning ecosystems services and disservices in urban environment. Agriculture, Ecosystems and Environment, **242**, 53-66 (2017)
23. M. Manso, J. Castro-Gomes, Green wall systems: A review of their characteristic, Renewable and Sustainable Energy Reviews, **41**, 863-871 (2015)
24. O. V. Khrapko, A. V. Kopeva, O. G. Ivanova, Vertical green walls as an element of landscape design The New Ideas of New Century-2016 - The Sixteenth International Scientific Conference Proceedings **3**, 159-163 (Khabarovsk, 2016)
25. O. V. Maslovskaya, L.A. Chernyavina. Special features of functioning of flat roofs as the places for social communication (on the example of VSUES campus in the city of Vladivostok), Sociology of City **4**, 53-67 (2014)
26. http://elibrary.ru/item.asp?id=24921741, pdf. (last accessed 28.03.2017)
27. A. I. Khusnutdinova, O. P. Aleksandrova, A. N. Novik, Vertical greenery systems, Construction of unique buildings and structures, **12(51)**, 20-32 (St. Petersburg, 2016)
28. N. L. Chau, L. M Chu. Fern cover and the importance of plant traits in reducing erosion on steep soil slopes, CATENA, **151**, 98-106 (2017)
29. http://elibrary.ru/item.asp?id=24507861, pdf. (last accessed 25.03.2017)

30. V. P. Generalov, E. M. Generalova. Problems in the classification of high-rise residential buildings, Urban planning and architecture, **2**, 12-14 (2011)
31. O. V. Khrapko, A. V. Kopeva. Prerequisites for the formation of the concept of landscaping of the city of Vladivostok, Problems of landscaping of settlements - Materials of the scientific-practical conf. 149-156 (2013)
32. O. V. Khrapko, A. V. Kopeva, M. N. Koldaeva, E. N. Golovan Informal style in landscape architecture of Far-Eastern cities, The New Ideas of New Century-2013 – The Thirteenth Int. Scientific Conf. Proc. 3, 395-399 (2013).
33. http://elibrary.ru/item.asp?id=24848604, pdf. (last accessed 28.03.2017)

Conceptions of Height and Verticality in the History of Skyscrapers and Skylines

Oksana Maslovskaya[1*], and *Grigoriy* Ignatov[1]

[1]Far Eastern Federal University, 690900, 8 Sukhanova St., Vladivostok, Russia

Abstract. The main goal of this article is to reveal the significance of height and verticality history of skyscrapers and skylines. The objectives are as follows: 1. trace the origin of design concepts related to skyscraper; 2. discuss the perceived experience of the cultural aspects of skyscrapers and skylines; 3. describe the differences and similarities of the profiles of with comparable skylines. The methodology of study is designed to explore the perceived theory and principals of skyscraper and skyline development phenomenon and its key features. The skyscraper reveals an assertive creative form of vertical design. Skyscraper construction also relates to the origin of ancient cultural symbolism as the dominant vertical element as the main features of an ordered space. The historical idea of height reaches back to the earliest civilization such as the Tower of Babel. Philosophical approaches of elements of such post-structuralism have been included in studying of skyscraper phenomenon. The analysis of skyscraper and their resulting skyline are examined to show the connection to their origins with their concepts of height and verticality. From the historical perspective, cities with skyscrapers and a skyline turn out to be an assertive manifestation of common ideas of height and verticality.

1 Introduction

Creating a city image has always been one of the primary concerns of architectural thinking, because cities of the world, like people, want to be significant and recognizable. Figuratively speaking, sometimes they want to be ascended to the sky. If historical towns are satisfactorily served by the clock and bell towers, fire watchtowers and church steeples, the key to the image of the megacities is the system of very tall buildings such as skyscrapers. The skyscraper cities are easily distinguishable, because they, like Gothic cathedrals, rush into the sky.

The main goal of this article is to reveal the significance of height and verticality history of skyscrapers and skylines. The objectives are as follows: 1. trace the origin of design concepts related to skyscraper; 2. discuss the perceived experience of the cultural aspects of skyscrapers and skylines; 3. describe the differences and similarities of the profiles of with comparable skylines.

This study is based on works of theorists devoted to problems of contemporary high-rise architecture and urban-planning. The problems of the form and stylistics, symbolic values,

* Corresponding author: oxym69@gmail.com

and ideology of skyscraper are discussed in critical and theoretical researches of such authors as Louis Sullivan [1], Hugh Ferriss [2], Alfred Bossom [3], Charles Jencks [4], Ada Louise Huxtable [5] and others. History of skyscraper and skyline phenomena is considered in works of Francisco Mujica [6], Christopher Tunnard [7], Judith Dupre [8], Paul Goldberger [9], James Giblin [10], John Zukowsky [11], Joseph Korom [12], etc.

Groups of skyscrapers as a whole, as well as 'skyline' as the concept for its aesthetic and socio-cultural interpretations, are analyzed in works of Wayne Attoe [13], L.R. Ford [14], Thomas A. P.van Leeuwen [15] and others. A number of 21st-century researchers deal with skylines of certain cities: London [16, 17], Kuala Lumpur [18], Saint Petersburg [19, 20], Moscow and other Russian cities [21], Istambul [22], Dubai [23], etc. The problems of height are highlighted in works of Louis Sullivan [1], Thomas A. P.van Leeuwen [15], Joseph Korom [12] and others. The problems of verticality are emphasized in works of various authors, among them Thomas A. P.van Leeuwen [15], Ada Louise Huxtable [5], Philip Nobel [24] etc.

2 Methods

The methodology of the study is based on principles of the semiotic approach developed by Roman Jakobson, J.M. Lotman, O.M. Freidenberg [25] etc., opening semantic features and the importance of the studied phenomenon in the system of culture. The wider phenomenological approach of Christian Norberg-Schulz [26] and Juhani Pallasmaa [27], which is taking into account not only realized architectural and town-planning objects but also non-realized or not intended for realization architectural projects, images, and ideas expressed in literary works, graphic and painting, movies and a theatrical art. Philosophical approaches of elements of Roland Barthes [28] post-structuralism have been included in studying of skyscraper phenomenon as well.

3 Results

The skyscraper reveals an assertive creative form of vertical design. Skyscraper construction also relates to the origin of ancient cultural symbolism as the dominant vertical element as the main features of an ordered space. The historical idea of height reaches back to the earliest civilization such as the Tower of Babel. Philosophical approaches of elements of such post-structuralism have been included in studying of skyscraper phenomenon. The analysis of skyscraper and their resulting skyline are examined to show the connection to their origins with their concepts of height and verticality.

Table 1. Comparative analysis of *skyline*, *silhouette*, *horizon*, and *panorama* notions.

Image character	Skyline	Silhouette	Horizon	Panorama
High (H) vs. Wideness (W)	H	H	W	W
Dynamism (D) vs. Statics (S)	D	D	S	S
Verticality (V) vs. Horizontality (Z)	V	V	Z	Z
Multi-Layered (L) vs. Flat (F)	L	F	F	L
Includes City Scenes	-	-	-	+
Involves Architectural Matter	+	-	+	+

4 Discussion.

4.1 Semantization of skyscraper and skyline notions

According to Bible and Russian researcher Olga Freidenberg in her book *Image and Concept: Mythopoetic Roots of Literature* [25], in beginning was the Word, a sign-symbol without pragmatic content. Then it becomes a conventional notion or an operational concept, and only after this transformation is possible to reach a prosaic non-symbolic reality. Let us consider this rule concerning several notions linked with the conception of verticality.

Terms most commonly associated with the idea of height and conception of verticality in architecture are skyline and skyscraper. "Skylines are largely a 20th-century concern" [13]. However, this word was used earlier to describe not architectural, but the natural phenomenon. Originally, in the English language *skyline* meant "horizon" or "outline on the background of the sky." The first general use of the word *skyline* is recorded in 1824 when it appeared as a synonym for the *horizon*, the line where earth and sky meet [29]. "In travel literature – journals, diaries, and guidebooks – and in novels about travel and exotic places, 'skyline' was employed in reference to the meeting of sky and land" [13].

In American culture, the word *skyline* has gained a new and more specific architectural meaning: "outline of skyscrapers on the background of the sky." The use of the word *skyline* as "the outline or silhouette of a building or number of buildings or other subject seen against the sky" for the first time appeared in writings of George Bernard Shaw at the end of the 19th century [30]. *Skyline* has gotten a new underlying sense, a secondary architecture-poetical meaning.

Undoubtedly, this derivative developed only after the emergence of such specific architectural term as the *skyscraper*. The word *skyscraper* was used in 18-19th centuries to describe notions linked with human activity, but not yet in the architectural sense. *Skyscraper* was originally used in various senses: high-standing horse (1788); a triangular sky sail (1794); a tall hat or bonnet (1800); an exaggerated or "tall" story (1841); a very tall man (1857); in baseball a ball propelled high into the air, a towering hit, a skier (1866); a tall ornament on top of a building (1883); a rider on one of the high cycles formerly in use around 1892 [30]. Old and contemporary senses of the word *skyscraper* have in common the idea of height and verticality.

The term *cloud scraper* as an alternative notion for tall building description was introduced by Minneapolis architect L. S. Buffington in 1880-s. He studied the idea of steel skeleton application for construction of high-altitude buildings at a level of sketches and engineering calculations of 30, 50 and even 100-storeyed towers [10]. In 1887, the inventor has patented his *cloud scrapers*. Buffington "brought the potential of the iron skeletal frame to the attention of the national architectural and building communities" [31], but the neologism *cloud-scraper* has not got accustomed.

Charles Jencks in his book *Skyscrapers-Skyprickers-Skycities* [4] introduced some more notions for different types of high-rise buildings and complexes. *Skypricker* stands for a "skyscraper with a pointed spire" [32]. However, this term has not got accustomed. *Skycity* is determined by Jencks as a "the compound tall building" [4]. Contemporary using the word *skycity* includes names of games (one of the science fiction tabletop games has name "War of the Sky Cities"), films, firms, restaurants, casinos, big skyscrapers, airports etc.

We consider *skycity* as an opposition to solitaire, stand alone building (even if it is big and composite *vertical city* as Abeno Harukas in Japan, discussed by K. Mizutani, K. Hirakawa, M. Nakashima [33]); *skycity* is rather a city with skyscraper's clusters. Some contemporary researchers use the term *Skyscraper City* in similar meaning. For example, Prof. Terri Boake writes, "The increase in density of our urban environment as we create "skyscraper cities" is inevitable and desired as a means to limit urban sprawl and preserve valuable agricultural land" [34].

4.2 Idea of height

Modern skyscrapers and their clusters had emerged at the turn of the 19th and the 20th century in America. At that date architects and artists of America conducted active searches of the national identity and "new" city's architectonic. The idea of height was principally actual. In the search of skyscraper's style, the central attention was given to the height of its structure. In 1896 Louis Sullivan wrote, "What is the chief characteristic of the tall office building? And at once we answer, it is lofty. This loftiness is to the artist-nature its thrilling aspect. It is the very open organ-tone in its appeal. It must be in turn the dominant chord in his expression of it, the true excitant of his imagination. It must be tall, every inch of it tall. The force and power of altitude must be in it the glory and pride of exaltation must be in it. It must be every inch a proud and soaring thing, rising in sheer exultation that from bottom to top it is a unit without a single dissenting line, - that it is the new, the unexpected, the eloquent peroration of most bald, most sinister, most forbidding conditions" [1].

Dutch researcher Thomas A. P. van Leeuwen in his book *The Skyward Trend of Thought: Metaphysics of the American Skyscraper* considers skyscraper not only as innovative structure, but as well "as a late, but rightly timed, example of paradigmatic architecture; indeed as the tower of cosmogony" [15]. He believes that skyscraper is a modern successor of Tower of Babel: "Skyscrapers were the fulfillment of the Babylonian promise; the realization of both its technical enigma and its utopian-cosmopolitan objective" [15].

One of the first examples of such stepped form and skyrocket striving we can see in the fantastic view of Erastus Salisbury Field's painting of 1876 named "The Historical Monument of the American Republic." One of the latest examples is the Le Royal [35], 7th tallest building in Amman, Jordan, completed in 2003.

As Thomas A. P. van Leeuwen notes, "The non-conscious knowledge or intuition is probably what the old tower builders and originators of the first American skyscrapers had in common: an enthusiastic, non-rational, desire to build high in celebration of something that probably they were not aware of" [15].

The concept of height becomes the basic sense of skyscraper. As J.J. Korom shows, "We celebrate the push upward to be nearer the clouds and we celebrate our tussle with nature, gratified it allows us the notion of supremacy. We can now build higher than ever before and realize the celebration of the height is primal and constant" [12]. The concept of height summarizes the essence of all possible kinds of height. More precisely, it embodies the idea of height. The form of skyscraper itself does not make any deep spirit, it is included in one or other pragmatic context directed on the realization of certain practical function. It is possible to say that *height* is concept in the absolute. The interconnected idea of being "No. 1" is embodied in architecture in the constantly updated lists of the word tallest buildings – symbols of maximum achievements.

The idea of height is excluded from any context; it is organic, immanent to the human, deep architectonic sense as such. The basic concept of *height* is used not only in architecture; it is performed in other arts. As Roland Barthes wrote, "This common domain of the signifieds of connotation is that of *ideology*, which cannot but be single for a given society and history, no matter what signifiers of connotation it may use" [28]. Thus, the concept of height belongs to the ideology extending on all manifestations of culture because the area of ideology is common for all signified senses.

Moreover, as Thomas A. P. van Leeuwen writes, “19th-century Americans, as well as Babylonians, had shared the common realm of platonic ideas, from which they had derived their ambitions as well as their concept of verticality” [15].

4.3 Verticality as a creative assertive gesture

Many theorists connect vertical character of skyscrapers with the deep initial, primitive senses of architecture. As Philip Nobel notes, “there’s something driving the height race. Something primal, competitive, romantic, eternal: some consequence of human minds, hard-wired to love a prospect, a mountain, needing to be high to feel safe from threat, congenitally driven to manufacture a marker on the land – the higher the better, and better still if it can be the highest of them all” [24].

Ada Louise Huxtable asserts, “From the Tower of Babel onward, the fantasies of builders have been vertical rather than horizontal” [5]. By themselves, the tower and skyscraper due to the rising form, reveal the phenomenological meaning of verticality since it represents a creative assertive gesture. It fulfills the requirement of certain efforts to maintain the stability of a vertical position, which means activity and strength in comparison with the less “energy-consuming” horizontal position symbolizing calm and relaxation.

This phenomenological meaning of skyscraper’s and medieval tower’s cities is expressed also, for example, in such tectonic games as the erection of a tower from playing cards, tower piles of children's cubes or sand castles. As Thomas A. P. van Leeuwen asserts, “From that days to this, whenever men have become skillful architect at all, there has been a tendency in them to built high; not it any religious feeling, but in mere exuberance of spirit and power – as they dance or sing – with a certain mingling of vanity – like a child builds a tower of cards” [15].

Hence, skyscraper construction also relates to the origin of ancient cultural symbolism as the dominant vertical element as the main feature of an ordered space. The vertical development of the city space gives it a “new” architectonics – the symbolic dimension, outwardly invisible to the eye, which can be comprehended by the set of additional analytical interpretations.

4.4 Comparative analysis of skyline, silhouette, horizon, and panorama notions

The wide range of terms related to observing cities from different points of view exists in the English language. *View of city*, *bird’s eye view*, and wor*m’s eye view*, *silhouette of city*, *panorama*, *vista*, *cityscape*, *skyline* and *horizon* are among them. The objective of this paragraph is to state differences and similarities of some this notions.

Beside *panorama, skyline* is characterized by larger height, compactness, and dynamism of depictured urban architectonic. At the same time *panorama* has large spaciousness, the wide-format scope of territories and static character of depictured urban form.

Skyline in its contemporary meaning and *horizon* are the characteristic projections of the vertical and horizontal, volumetric and linear, dynamic and static city’s views.

Skyline is closest to the *silhouette* because of its dynamism and other qualities of the described forms. Nevertheless, the *silhouette* is the undifferentiated mass, filled outline, while *skyline* is multi-layered, the mosaic collage of forms, ornamentation, textures, styles. *Silhouette* can be applied to any, even not architectural object, for example, *silhouette of vase* or the *silhouette of man*, while *skyline* already implies architectural matter, since it means the “outline of the groups of skyscrapers against the background of sky.”

Panorama can be interpreted as *panorama of urban life*, *city scene* or *urban picture*, where the everyday or holiday life of city may be depictured, including the typical urban elements, the motion of people and automobiles. *Skyline* and *silhouette of city* are more outlines of cityscape than a city scene in this matter.

Skyline comprises the wide range of the city's views: with upper, frontal, internal, even worm's eye points of observation. *City's silhouette* can be described only as a special mode of skyline in this matter, because it cannot be, for example, the upper view.

Most significantly, *skyline* belongs to the non-classical system of esthetic values. New symbolism and the artistic methods of *collage* (as discussed by Colin Rowe in *Collage city* [36]) *accumulation* and *assemblage* lie at the basis of its polycentric organization.

5 Conclusion

For a long time, skyscraper cities and skyline was characterized as an object of the American national and territorial identity. Since the 1950s, this city planning phenomenon as the unconscious realization of the archetype of 'tower city' was extended throughout the world. From the historical perspective, cities with skyscrapers and a skyline turn out to be an assertive manifestation of common ideas of height and verticality.

References

1. L. Sullivan, *The tall office building artistically considered* The public papers, R, Jwombly (ed.). (University of Chicago Press, Chicago, London, 1988)
2. H. Ferriss, *The metropolis of tomorrow* (Princeton Architectural Press, Princeton, NJ, 1986)
3. A. Bossom, *Building to the Skies, the Romance of the Skyscraper* (The Studio, New York, 1934)
4. Ch. Jencks, *Skyscrapers-Skyprickers-Skycities* (New York, Rizzoli, 1980)
5. A. L. Huxtable, *The tall building artistically reconsidered: the search for a skyscraper style* (Pantheon Books, New York, 1984)
6. F. Mujica, *History of the Skyscraper* (Archeology and Architecture Press, New York, 1930)
7. Chr. Tunnard, H. H. Reed, *American skyline; the growth and form of our cities and towns* (Houghton Mifflin, Boston, 1955)
8. J. Dupre, *Skyscrapers,* (Black Dog & Leventhal, New York, 1996)
9. P. Goldberger, *The skyscraper* (Knopf, New York, 1981)
10. J. C. Giblin, *The skyscraper book* (Crowell, New York, 1981)
11. J. Zukowsky, *The Sky's the limit: a century of Chicago skyscrapers* (Rizzoli, New York, 1990)
12. J. J. Korom, *The American Skyscraper, 1850-1940: A Celebration of Height* (Branden Books, 2008)
13. W. Attoe, *Skylines: understanding and molding urban silhouettes* (Wiley, Chichester, New York, 1981)
14. L. R. Ford, *The Urban Skyline As A City Classification System,* Journal of Geography **75.3**, 154-164 (1976)
15. T. A. P. van Leeuwen, *The skyward trend of thought: the metaphysics of the American skyscraper* (MIT Press, Cambridge, Mass., 1988)

16. R. Tavernor, *Visual and cultural sustainability: The impact of tall buildings on London,* Landscape and Urban Planning **83.1**, 2-12 (2007)
17. R. Tavernor, G. Gassner, *Visual consequences of the plan: Managing London's changing skyline*, City, Culture and Society **1.2**, 99-108 (2010)
18. N. A. H. Yusoff, A. M. Noor, R. Ghazali, City Skyline Conservation: Sustaining the Premier Image of Kuala Lumpur, *Procedia Environmental Sciences*, **20**, 583-592, (2014)
19. N.S. Trumbull, *St. Petersburg, Russian Federation,* Cities, **31**, 469-490, (2013)
20. M. Dixon, Gazprom versus the Skyline: Spatial Displacement and Social Contention in St.Petersburg, *International Journal of Urban and Regional Research* **34.1**, 35-54, (2010)
21. G.A. Ptichnikova, New Century High Risers in the Core Areas of Historic Cities in Russia, *Procedia Engineering* **165**, 1903-1910, (2016)
22. C. Guney, S.A. Girginkaya, G. Cagdas, S. Yavuz, Tailoring a geomodel for analyzing an urban skyline, *Landscape and Urban Planning*, **105**, 160-173, (2012)
23. M. Acuto, High-rise Dubai urban entrepreneurialism and the technology of symbolic power, *Cities* **27.4**, 272-284, (2010)
24. P. Nobel [Editor], *The Future of the Skyscraper* (Metropolis Books, 2015).
25. O. Freidenberg, *Image and Concept: Mythopoetic Roots of Literature*, (Taylor & Francis, 1997)
26. Chr. Norberg-Schulz, *Genius Loci: Towards a Phenomenology of Architecture* (Rizzoli, New York, 1979).
27. J. Pallasmaa, The geometry of feeling: a look at the phenomenology of architecture *Theorizing a new agenda for architecture: an anthology of architectural theory* 1965-1995, Nesbitt, Kate (ed.). (Princeton Architectural Press, New York, 1996)
28. R. Barthes, Rhetorique de l'image, *Communications* **4**, 49, (1964)
29. R.K. Barnhart [editor], and S. Steinmetz [managing editor]; *The Barnhart dictionary of etymology* (Wilson Co., Bronx, N.Y., H.W., 1988)
30. J.A. Simpson, and E.S.C. Weiner [prepared by]; *The Oxford English dictionary; 2nd ed.* (Oxford University Press, Oxford, Clarendon Press; New York, 1989)
31. http://www.minnesotainventors.org/inductees/leroy-buffington.html (last accessed 2017.07.16)
32. http://www.skyscraperdictionary.com/?project=skypricker (last accessed 2017.07.16)
33. K. Mizutani, K. Hirakawa, and M. Nakashima, Construction of a 300-Meter Vertical City: Abeno Harukas, *International Journal of High-Rise Buildings* **4.3**, 199-207, (2015)
34. T.M. Boake, It's Not About the Skyline, It's About the Base Condition, *Global Interchanges: Resurgence of the Skyscraper City, CTBUH 2015 New York Conference*, 494-501, (2015)
35. S. Abu-Ghazalah, Le Royal in Amman: A new architectural symbol for the 21st century, *Cities*, **23**.2, 149-159 (2006)
36. C. Rowe, F. Koetter, *Collage city* (MIT Press, Cambridge, Mass., 1995)

The problems of designing a multifunctional courtyard space of high-rise buildings by the example of residential development in Volgograd

Sergei Matovnikov[1*], *Natalia* Matovnikova[1], *Polina* Samoylenko[1]

[1]Volgograd State Technical University, 400047, Volgograd, Akademicheskaya St., 1, Russia

Abstract. The paper considers the issues of designing a modern courtyard space for high-rise buildings in Volgograd to obtain a multifunctional environment through the arrangement of new recreational territories and the search of innovative planning methods for urban landscape design. In professionals' opinion, the problem concerning the design and construction of recreational zones and greenery planting is very acute for Volgograd, such territories are often absent in many districts of the city. Generally, the decrease in the natural component and a low level of recreational territories improvement are typical for Volgograd. In addition, the problem of designing a modern urban courtyard space for high-rise buildings to obtain a multi-functional environment exists and requires a thorough investigation. The question is if there is a possibility to solve these difficult tasks by means of local design methods only or whether there should be a complex approach at the stage of the formation of master plans for modern residential areas and which modern design methods can ensure the creation of a courtyard space as a multi-functional environment. These questions as well as some other ones will be the topic of our paper.

1 Introduction

The relevance of the chosen topic of designing a modern courtyard space for high-rise buildings as a multifunctional environment is determined by the urgent problem of shortage or deficiency of well-organized courtyard spaces around high-rise buildings in Volgograd. Moreover, the city is particularly concerned about the lack of places for the recreation of citizens in the residential zone [1, 2]. The needs of city dwellers for recreation are set by the urban planning standards. However, the decrease in the natural component, failure of greenery plantations, a low level of recreational territories improvement are characteristic of the Volgograd Agglomeration [1]. In professionals' opinion, the problem concerning the recreational zones and greenery planting is very acute, they are absent in many districts of Volgograd [2, 3]. The living conditions get worse due to urban development densification, various types of pollution, ultrahigh concentration of cancer-inducing compounds of sulfur, nitrogen, dust, heavy metals in the air, due to electromagnetic, radioactive, visual and

* Corresponding author: matovnikov4@yandex.ru

thermal pollution, vibration, noise, spontaneous origination of dumps etc. [4,5]. All these make the issue of increasing the level of urban development comfort in the Volgograd agglomeration the one of the most immediate interest.

Today the recreational territories are more complicated and dynamic in structure, more interactive and mobile from the point of view of the contents. The creation of provocative architectural landscape objects, thematic installations, the application of "performance" design methods in the form of an architectural landscape show are welcome. The new recreation space plays the role of an instrument which inculcates new behavior norms to a consumer, with the help of which new cultural attitudes are announced, old life style stereotypes are ruined [17, 18]. The modern urban planning principles of urban landscapes design have changed. If earlier recreational territories acted as "concentrated nature" in a city, today the world practice indicates changes in the very philosophy of the creation of urban recreation [17, 18]. The "clear" boundaries between residential and recreational zones are blurred. The analysis of innovative urban planning methods of design, of modern domestic and foreign projects allowed us concluding that a modern recreational territory "captures" urban space and transforms from a local environment object into an integral urban environment [19]. Designers suggest mobile green lawn parkings for dwellers' recreation, as well as greened walls, roofs, sheds, bridges and modern building envelopes of glass, water and greenery. "Green mazes" made of trees, shrubs, sculptures as well as installations, unusual water reservoirs, fountains, green sculptures of ornamental shrubs and trees are arranged in recreational territories. When an urban landscape is designed, the whole complex of components is taken into account: real and virtual space, its external and internal life processes, the items used, the whole socially oriented infrastructure of a recreational territory. The ways are determined how to create an interactive environment, a visitor is involved into a gaming communication, cooperation, training. It is necessary to note that the arrangement of urban recreational territories as multifunctional urban complexes increases their "recreational capacity". However, it should be recognized that construction in the recreational territories of residential districts increases the load on the territory and requires additional efforts aimed at the natural landscape preservation [19, 20].

In the new Master plan of Volgograd, the "green frame" concept of the Volgograd agglomeration has been developed, since the system of green wedges which plays the role of protective zones preventing the negative influence of urbanization drawbacks on the urban landscapes is the optimal structure of massive green plantations for most large cities [21, 22, 23, 24]. Public green plantations are designed at the area of 2570 ha, which ensures the urban greening level of 23 sq.m. per citizen. The increased norm of greening in the complicated climatic conditions of Volgograd will contribute to the creation of favourable sanitary-and-hygienic situation, to the improvement of the urban microclimate. However, the territory of the Volgograd agglomeration is ill-adjusted to the fulfillment of the stated task, which is determined by a number of reasons. According to the climatic conditions, it is situated in the dry steppe zone and is characterized by extremely arid climate with explicit continentality. The mean annual duration of sunshine is 2221 hours. [7]. The mean value of total solar radiation equals to 113,4 kcal/cm^2 per year. The annual radiation balance is positive and averagely makes up to 50,1 kcal/cm^2. The openness of the territory contributes to the barrier-free penetration of cold air masses both from the north and the east in winter. In summer dry air comes from the Middle East and Central Asia when very hot and dry weather is set: the air temperature reaches 34-38°, the relative humidity decreases to 10-12%. Increased wind velocities are conductive to the origination of dust storms which happen 5-14 days per year [8]. According to the hydrometeostation data of Volgograd State Agrarian University, the absolute minimum for the Volgograd region is 41°, while the maximum is +45°.

The soils in Volgograd and its surroundings formed mainly on loess loams and clays, and on loamy sands and sands at some plots. The soils show low-level fertility, the humus content is usually not high (up to 2 %). Concerning the fitness for forests, meadow-chestnut soils are the best ones in the chestnut zone. They stand out for darker color, higher humus content, larger number of nutrition elements and have a better structure. In the right-bank part of the Volgograd territory, the main relief forms determining the soils formation are gullies, cloughs, ravines, while in the left-bank part, in the territory of the Volga-Akhtuba flood plain, they are saucer-shaped depressions. The average depth of humus horizon is 40-45 cm, reaching 70-75 cm at particular plots. The humus amount in upper horizons varies from 2,5% to 4,5%. These soils can be used for recreational and protective forestation when appropriate irrigation regime is followed. A significant part of the Volgograd agglomeration soils are polluted to a high degree [9,10]. The local vegetation is rather poor as according to the forest zoning the Volgograd agglomeration belongs to the territory of the Northern semi-desert of Yergeny and Eastern Ciscaucasia in the province of the southern part of the Russian plain of the Eurasian steppe zone [11]. The natural tree vegetation of the zone is represented by ravine forests of pedunculate oaks (Quercus robur) with aspens (Populus tremula), smooth-leaved elms (Ulmus carpinifolia), common elms (Ulmus laevis) and wild apple-trees (Malus sylvestris). Typical steppe shrubs: crenulated spirea (Spiraea spiraca) and hypericifolia one (Spiraea hypericifolia), ground cherry (Cerasus fruticosa), dwarf almond (Amyqolalus nana) form tangled vegetation at forest verges. Oak ravine forests grow in cloughs with a lower location and the most favourable soil and hydrological conditions [12]. It is necessary to note that the forests of the Volga-Akhtuba flood plain situated on the Volga bank opposite to Volgograd are of extraordinary value. They are of great recreational, climate-regulating, sanitary-and-hygiene, water-protective and erosion-resisting significance. Here, oak forests are the most valuable – pure or with an admixture of common elms [4,12]. However, this territory of the Volgograd agglomeration is the National Park and therefore belongs to the recreational territories of restricted usage. The artificially planted urban greenery in Volgograd is represented to a great degree by the plantations of alien forms of Bolle's poplar and Black poplar, which are characterized by a relatively short life time [13]. In addition, the geomorphologic structure, the surface waters and hydrogeology of the Volgograd territory also cause problems when urban parks are designed. According to the geomorphologic zoning it belongs to the accumulation-denudation southern part of the Volga Upland [14]. The characteristic feature of the northern part of the city is a high ledge of the Yergeny surface comprised of a layer of white inequigranular sands up to 8 m deep. To the south of it at the Volga bank, there is an abrasion-and-accumulation marine terrace (20 – 40 m.) and two terraces above the flood plain with mean relative heights of 10 – 15m and 50 – 60m. In the urban territory, in the interstreams of the Sukhaya Mechetka and Mokraya Mechetka rivers as well as in the mouth of the Tsaritsa river, low-rise terrace-like steps are composed of chocolate clays. In the rest of the territories, the chocolate clays surface merges invisibly with the more abrasive surface of the terrace of 20 – 40 m [8, 15]. The surface waters are represented by the rivers of Volga and Akhtuba as well as by small rivers (Tsaritsa, Mokraya Mechetka, Sukhaya Mechetka), the Volgo-Don Shipping Canal, a series of Sarpinskie lakes, and numerous lakes of the Volga-Akhtuba flood plain [16]. The depth of location and the chemical properties of the subterranean waters depend on the degree of the relief roughness, at the watershed and near-watershed areas the subterranean waters are located at the depth lower than 10 metres. At the bottoms of cloughs, gullies and ravines, in the river flood plains and the lower terraces of lakes they can occur at the depth (up to 1-2 metres) available for roots or even come to the surface during spring time. In the north of the Yergeny Upland, subterranean waters are located in sands and loams of the Yergeny suite at the depth of 30 - 40 metres which can be reduced to 1 meter in gullies. In the surroundings of Volgograd and Volzhsky, near detention-evaporation basins, the subterranean waters are highly polluted. The area of the polluted subterranean waters spreading is more than 970 sq. km. All these stipulate definite difficulties for the creation of comfortable urban landscapes and make us search for alternative ways of urban recreational territories arrangement, use innovative urban planning methods of urban landscape design [20].

2 Methods

In this regard, a range of scientific and applied issues concerning the design of courtyard space for high-rise buildings aimed at obtaining a multifunctional environment was identified. The study of the relevant normative documents, the analysis of the modern urban planning situation in Volgograd and experts' advice allowed us determining the content of the first stage of the research: to carry out a preliminary analysis of the existing courtyard spaces of high-rise buildings in Volgograd in terms of their functioning as a multifunctional environment. We conducted observations of certain situations typical for Volgograd in the existing courtyard spaces of various types of high-rise buildings. In addition to the examination and observation, the initial stage of the research included the aerospace photography and field photo survey of the courtyard spaces. The residents of the given buildings were also offered a questionnaire. The data collected helped us conduct an objective and adequate assessment of the existing territory improvement and of the functioning of modern courtyard spaces serving high-rise residential buildings and draw some preliminary conclusions that will be covered in this article.

According to the type, courtyard spaces in modern construction are divided into monofunctional and polyfunctional (multifunctional) ones. The first subgroup includes, for instance, the courtyards where only a children's playground is located, i.e. the territories that perform a strictly defined function and are designed for a narrow audience. Such spaces, in our opinion, are intended only for a certain group of the inhabitants of a particular residential development. The creation of a monofunctional space is, indeed, a justified, logical step dictated by objective reasons of the lack of space that could be used for a courtyard. Oftentimes, however, it becomes an example of explicit discrimination of groups of people who live in these buildings (actually, children's interests are often neglected). The second subgroup of polyfunctional courtyard spaces refers to the spaces that include various functional zones, it means the territory is designed and zoned in accordance with the interests of the major part of the residents. In our opinion, nowadays, it is the multifunctional courtyard space that most accurately meets various needs of a modern city resident. Only the construction and arrangement of a multifunctional courtyard territory can create a unique, exceptional and comfortable urban courtyard environment as a part of sound urban recreation.

The urban environment of a modern courtyard should be built on the principles of accessibility, equality of interests of all population groups, their social, ecological and ethical safety. Residents of a city should feel that "their" territory is not only the one that is located next to their house but the whole city as well. Furthermore, the main visitors and users of courtyard spaces are children, parents, their supervisors, and elderly people, that is, groups of residents who cannot stay at a large distance from their homes for a long time. Besides, the process of disunity for the majority of urban residents, which is currently taking on alarming proportions, is largely rooted in the unsettledness and inconvenience of courtyard areas, and, oftentimes, in their complete absence. Today urban courtyards are often deserted or are instead filled with personal vehicles of the residents. Therefore, we believe that it is solely the design of a courtyard as a multifunctional environment that allows not only taking into account the interests of the most, if not all of the resident groups, but also returning the interest of the residents to their courtyards, and inspiring a new life in those spaces.

For this study, we selected courtyards typical of the Soviet period of territory development, and several modern variants of courtyards of high-rise buildings. Historical background: the construction of high-rise buildings (12, 14, 16 floors) in Volgograd began in the late 60s of the XXth century. At that time, many courtyards for Volgograd were designed by the Leningrad Institute of General Planning. The attachment to the terrain was carried out by the workshop No. 1 of Volgograd Grazhdan Proekt Institute (the head of the workshop V.P. Statun). According to A. M. Vyazmin, the former chief architect of Volgograd, the old courtyards were designed in compliance with the SNiP (Construction Rules and Regulations); that is, the organization carried out territory zoning, designed a path-and-road network, provided options for improvement, etc. The committee did not allow for a constructed building acceptance without these requirements being fulfilled. The approval of the courtyard draft at the Sanitary and Epidemiological Station was an obligatory condition. Some buildings in the Central district of Volgograd were even designed with underground parking spaces (for up to 54 cars per building). Typical equipment for children's playgrounds (slides, swings, carousels, benches, etc.) was also designed, and was consequently installed in each courtyard. However, the construction of courtyard territories was not always completed. There were "inconsistencies" in the plans concerning the schedule of commissioning, or financing. Design and construction of courtyards were often funded by a residual principle. Nowadays, in the opinion of the expert community, courtyards of high-rise buildings are not designed; they "remain at the mercy of the developer" and the expert commission does not pay any attention to the condition of the courtyard territory.

3 Results

Two "historical" examples of the organization of courtyard space located near sixteen-storey buildings in the Central District of Volgograd were considered: 1) Courtyard No.1 in Dnestrovskaya Street (Fig. 1,a) - a courtyard of a section-type multi-storey panel building (the general layout of the courtyard executed by Volgograd Grazhdan Proekt Institute); 2) Courtyard No.2 in Parkhomenko Street (Fig. 1,b) -a courtyard of a section-type multi-storey brick building (the general layout of the yard executed by Leningrad Institute of General Planning).

Courtyard No. 1: a lived-in one, but the existing situation and improvement efforts make a rather modest impression (if not less). The courtyard area is quite large (more than 0.5 hectares), it is open in the direction of the roadway (intra-quarter passage). There is no either sufficient isolation from the noisy transit zone, or clear and definite zoning of the courtyard space, the vertical layout of the territory is missing or minimized, the equipment is obsolete, the options for improvement are minimized, etc. According to the survey of residents, the courtyard territory does not meet modern requirements, it looks deserted and abandoned. Another drawback of this yard is its extremely low isolation and minimal privacy. For family recreation, citizens prefer to go out into the city, where places for recreation by interests are available (e.g. cinemas, cafes, sports grounds, etc.).

Courtyard No. 2: smaller in size (about 0.25 hectares), open from both sides, both towards the city roadway and the passage. However, the design tasks are solved more successfully here. The courtyard is perceived as private and is isolated from the noisy transit zone due to the construction of a decorative 1.5-meter high wall delimiting the courtyard. This method made it possible to clearly divide the area into two partially isolated territories (a children's zone and the recreation zone for the other residents of the building). Besides, "pockets" for car parking and some household needs of residents, such as garbage collection, carpet cleaning, etc., are organized in the local area. There are various forms of landscaping present in the courtyard. Thus, such a drawback of the territory as its small area did not prevent from designing a cozy and quite a private courtyard. The shortage of parking places for tenants' vehicles, the outdated equipment of the children's playground and utility zones can be considered disadvantages of this yard (during the Soviet time an underground parking was planned to be built in this yard).

Fig. 1. "Historical" examples of the organization of courtyard space: a - courtyard No.1 in Dnestrovskaya Street,
b - courtyard No.2 in Parkhomenko Street

In current development, the following urban planning situations in the Central district of Volgograd were taken to investigate courtyard spaces of high-rise houses: 1. Kalinin street - courtyard of a two-section high-rise building (an individual design project of "Volzhskie Parusa" (Fig.2, a); 2. Pugachevskaya Street - courtyard of a corridor-type multi-storey panel building (Fig.2, b); 3. Nevskaya Street - courtyard of a two-section high-rise building (Fig.2,c); 4. Donetskaya Street - courtyard of a multi-sectional high-rise residential buiding (Fig.2, d);

4 Discussion

The feasibility study of the chosen topic was shifted from the plane of scientific debate to the plane of the study of the actual needs of residents of high-rise buildings. The survey conducted among the inhabitants of these buildings showed that there are various age groups, various social and national groups of residents living in the aforementioned residential buildings, in the other words, a regular segment of population of a modern city. Here are the basic wishes of the residents that they would like to see realized in the territory of their courtyard spaces. In a modern courtyard, according to the tenants who took part in the survey, there should be: 1. Sidewalks for mothers walking with children in baby carriages; 2. Children's playgrounds for children with parents accompanying them; 3. Bike paths and tracks for skateboards; 4. Playgrounds for team games; 5. Sportsgrounds with sports equipment; 6. Isolated zones for quiet secluded rest; 7. Pavilions for communication by interests; 8. A place for general meetings; 9. Territory for dog walking; 10. Parking spaces located away from children's playgrounds. This list of wishes could be expanded. In our opinion, it is convenient to have a summer stage or an open-air cinema in the courtyard area, places for private entrepreneurship or home-based work (e.g. mini bakery, tailor's shop, hairdresser's, shoe repair shop, etc.), green park islands and small decorative ponds, etc. [11, 19].

Fig.2. Examples of the existing courtyard spaces of high-rise buildings, typical for Volgograd: a - Kalinin street - courtyard of a two-section high-rise building (an individual design project of "Volzhskie Parusa"; b - Pugachevskaya Street - courtyard of a corridor-type multi-storey panel building; c - Nevskaya Street - courtyard of a two-section high-rise building; d - Donetskaya Street - courtyard of a multi-sectional high-rise residential buiding.

5 Conclusions

The carried out initial observation of the considered courtyard territories allowed for a preliminary conclusion that, at the present time, no modern courtyard space under investigation has a design which meets the needs of all the groups of tenant living in the buildings nearby. Thus, we can see that the problem of designing a modern urban courtyard space for a high-rise building as a multifunctional environment exists and requires a thorough investigation. The question is if there is a possibility to solve these difficult tasks by means of local design methods only or whether there should be a complex approach at the stage of the formation of master plans for modern residential areas. Which modern design methods can ensure the creation of a courtyard space as a multifunctional environment. These and other questions will be a topic for our further research.

References

1. L. Kremer, G. Vinter, Environmental Law of the European Union, 144 (Moscow, Gorodets, 2007)
2. G. N. Kovalyova, State of recreational spaces in Volgograd we present stage and their impact on the socio-cultural life of the city **3(13)** (Polythematic, 2010) URL: http; // www.vestnik.vgasu.ru/
3. E. V. Shagieva, E. A. Baranska, O. G. Chesnokova, Environmental problems of the city of Volgograd, *Construction and architecture*, **39(58)**, 110-117 (2015)
4. The report "On the state of the environment of the Volgograd region in 2011" 325 (Volgograd "SEE", 2012)

5. The report "On the state of the environment of the Volgograd region in 2010" 352 (Moscow "Globus", 2011)
6. V. N. Anopin, Geographical bases of forest reclamation of degraded urbolandscapes Lower Volga region, 169, (Volga-grad, 2005)
7. A. N. Sazhin, Modern climate change and zonal agri - ecology problems (for example, the Lower Volga region), *Problems of regional ecology*, **1**, 14-20 2001
8. The report "On the state of the environment of the Volgograd region in 2009", 304 (Moscow "Globus", 2010)
9. V. A. Brylyov, The natural conditions of the Volgograd region (Ed. Brylëv V.A.) **122**, 264 (Volgograd, "Change", 1996)
10. State report "On the state and Environmental Protection of the Volgograd region in 2002", 300, (Moscow, NIA Nature REFIA, 2003)
11. N. D. Sashnin, Research of ways of the basic processing of light-chestnut soils in combination with saline soils for growing protective forests on the right bank of the Volgograd Volga Abstract of thesis ... cc-Agricultural Sciences, 25 (Volgograd, 1981)
12. S. Shmelyov, I. Shmelyova, Stable city: problems of interdisciplinary research. Int. *Journal of Stable Development*, **12**, 23 (2009)
13. E. E. Krasilnikov, Landscape Urbanism, Theory and practice. Part 1: the scientific and practical bases of landscape Urbanism: monograph, 156 (Volgograd: Lld "IAA" Regional news ", 2015)
14. E. S. Migunova, Plantations on saline soils, **99**, 140 (Moscow, Forestry, 1978)
15. V. N. Anopin, N. G. Matovnikova, Condition and methods to improve the attractiveness urbolandscapes South Russian Gazette geology, *geography and global ecology* **15(18)**, 99-102 (Astrakhan, 2006)
16. N. G. Matovnikova, Biogeographical bases landscaping of recreational areas of the Volgograd agglomeration, *Abstract PhD of Sciences in geography*, 25 (Volgograd, 2007)
17. V. N. Anopin, S. A. Matovnikov, N. G. Matovnikova The Features of the Establishment and Maintenance of Parkland in the Solution of the Problems of Urban Agglomerations in Volgograd, *Procedia Engineering*, **150**, 1972-1977 (2016)
18. S. A. Matovnikov, N. G.Matovnikova. Innovative Urban Planning Methods for the Urban Landscape Design in the Volgograd Agglomeration, *Procedia Engineering, "2nd International Conference on Industrial Engineering"* ICIE2016, **2**, 1966-1971
19. E. E. Krasilnikov, Landscape Urbanism. Theory and practice.Part 1: the scientific and practical bases of landscape Urbanism: monograph, 156 (Volgograd: OOO "IAA" Regional news ", 2015)
20. S. A. Matovnikov, N. G. Matovnikova, Some current trends in the theory and practice of designing urban parks Science and education: architecture, urban planning, construction of: *Proceedings of the International meeting*., 6-10 September 2010, Volgograd, 386-391 (Volgograd: VolgGASU, 2010)
21. S. A. Matovnikov, N. G. Matovnikova, Creating a comfortable multisensory environment as the actual problem of the modern park construction, **22(41)**, 155-164 (2011)
22. V. N. Anopin, N. G. Matovnikova, S. A. Matovnikov, Architectural planirovochnve and adaptive landscape basics of green building in the Volgograd agglomeration, 159 (VolgGASU, Volgograd, 2012)

23. The decision of the Volgograd City Duma from 29.06.2007 № 47/1112 «On the approval of the Master plan of Volgograd".
24. A. V. Antyufeev, "Large Volgograd" in the program of development of Russian megalopolises century, the annual scientific materials, Architecture, urban planning and construction, 9-11 (Volgograd: VolgGASU, 2008)
25. L. I. Abakumov, Influence of industrial emissions on the stability of trees and shrubs, In sb. gardening problems, *Gradostroitelnye, environmental, sanitary and hygienic aspects*, 36-37 (Volgograd, 1995)
26. V. A. Gorohov, The green nature of the city: a textbook for high schools. Ed. 2nd, 528 (Moscow,.Architecture, 2005)

High-rise architecture in Ufa, Russia, based on crystallography canons

Ildar Narimanovich Sabitov[1*], *Dilara Radikovna* Kudasheva[1], *Denis Yaroslavovich* Vdovin[1]

[1]Ufa State Petroleum Technological University, Ufa, 450062, Kosmonavtov st., 1, Russia

Abstract. The article considers fundamental steps of high-rise architecture forming stylistic tendencies, based on C. Willis and M. A. Korotich's studies. Crystallographic shaping as a direction is assigned on basis of classification by M. A. Korotich's. This direction is particularly examined and the main high-rise architecture forming aspects on basis of natural polycrystals forming principles are assigned. The article describes crystal forms transformation into an architectural composition, analyzes constructive systems within the framework of CTBUH (Council on Tall Buildings and Urban Habitat) classification, and picks out one of its types as the most optimal for using in buildings-crystals. The last stage of our research is the theoretical principles approbation into an experimental project of high-rise building in Ufa with the description of its contextual dislocation aspects.

1 Introduction

Today, equally with such unquestionable factors as: economical susceptibility, high cost of lots within the city boundaries, approach to raise image constituent and attraction for tourists, growth of the city status and economic efficiency of projects, one of the most basic factor of the high-rise buildings forming is an object's spatial composition, its appearance and design. According to that, not for a long time there has been created a big quantity of stylistic methods and tendencies of the high-rise architecture cover design, using mathematical functions and formulas, contextual specifics, method of the form following the function, metaphorical transfer of the natural specifics to the architectural form, etc [1, 2, 3, 4]. However, what "high-rise building" is? According to CTBUH classification (Council on Tall Buildings and Urban Habitat) [5] there are three wraparound groups of high-rise buildings: Tall to 300 m, Supertall from 300 to 600 m, Megatall (from 600 m and higher). The high-rise or Tall building – is rather subjective conception, regarding to the context, proportions and technological fitting out. In New York the 20-storey building would be difficult to name as a high-rise, but, for example, in Prague it would be obviously higher than the main mass of buildings. Also, there exist buildings with an expressed proportional ratio of the base sides to the vertical axis. A building can reach 100 m along

* Corresponding author: ufasabitov@gmail.com

the vertical, and have ribs of the base for 50 m. In such buildings there also can be used technological systems and elements, inherent in high-rise. If an object gets into one of these categories, it could be high-rise. Objectively, any object could be the high-rise, if its top level gets 50 m from the ground. There is a classification of constructive systems:

- A steel framework – vertical and horizontal constructive elements composed of steel elements. If a ceiling construction is a concrete slab, laid on steel beams, then a system would relate to the steel framework.
- A concrete framework – vertical and horizontal elements consist of concrete.
- A mixed framework – using two full systems in one buildings – steel and concrete frameworks with an interchange.
- A composite framework – combination of steel and concrete elements in one constructive scheme, for example, composite columns, steel columns with reinforced concrete and a core, etc.

Today there exist also «wooden skyscrapers» – high-rise buildings, in which the main constructive element is a wooden beam and steel fastening elements.

Carol Willis describes minimalistic and the most rational periodization of the high-rise architecture development: the first period (before 1940), including «historicism» architecture, the Chicago school of high-rise architecture and modern [6]. By the second period, she ascribes modernism and post-modern, that included most of all modern developing form-building tendencies. Nowadays the last landmark of architectural styles development is post-metabolism or, in other words, «authoring architecture». This tendency is hard to ascribe to the one period or style, because it includes complex relationships of a multitude of stylistic tendencies, national cultural specifics, authoring solutions, etc. M. A. Korotich described compositional specifics of high-rise architecture development [7, 8]. He detected two basical periods: the initial one, during which have formed the main prototypes of the skyscrapers forms, and the modern one, within the framework of which Korotich ascribes three basic compositional form-shaping tendencies of high-rise architecture. The first tendency includes a period from the end of 20 century to 1980-1990-s. Its characteristics are:

- Decorative elaboration of elemental prismatic generalized forms of high-rise buildings.
- Division or shaping using vertical bay windows, pylons.
- «Telescopic» buildings.
- A wide use of historical architectural styles elements.

The second tendency originates in the second half of the 20 century and is actively used by architects nowadays. It originates from Soviet avant-garde, conceptual projects by I. Leonidov, V. Krinsky and N. Ladovsky (fig.1).

Its characteristics are:

- Abundance of plane-face truncation of the initial prismatic forms with different transversal sections.
- Combination of separate objects into complex compound compositions.
- Full or primary use of glass as a material of the skyscrapers cover.

The third tendency begins its development from the second half of 20 century (1970-s) and is actual till nowadays. Its characteristics are:

- Development of the list of high-rise architecture initial forms.
- Use of different forms of form-development (twist, bend, etc.).
- A wide use of the high-rise architecture complexes.
- Use of decoratively-plastic solutions of national architectural motives.
- Combination of structurally heterogeneous, but compositionally interchangeable elements of composition.

Fig.1 Project of the new building of National commissariat of heavy industry,1934

The architectural filling of Ufa is a complex stratified structure that has passed a style mixing, such as: classicism, Stalinist architecture, Soviet modernism, brick eclecticism, the post-modernism. Nowadays in the city there prevail a quoting of the West European architecture – modernistic methods and rethinking of Soviet modernism, less commonly – using neoclassical methods. There are also used: constructivism, adapted to modern materials, «facadizm», characterized by the illusion of integrity, sharp collision of styles – classical facades combined with glass up-and-over walls, steel and concrete constructions [9]. Just now, during the rising of architectural aesthetics level, when the building culture of the region endures its growth and new technologies implementation, and the normative legal base of the Russian Federation at last has filled with the document of high-rise architecture erection rules (Joint venture 267.1325800.2016 «High-rise buildings and complexes. Rules of planning»), it is necessary to form a flexible language for all future buildings, giving a development vector in this direction.

2 Materials and methods

Crystallography is nonrandom chosen as a base for an architectural image creating – under the city there is an advanced system of gypsum and other mineral-based materials generation. The main type of the city industry – oil production and refining, and geological explorations are a compound part of this process [10]. Crystallography is known as a part and parcel of geology. Also, Ufa is associated with water, two rivers that go about the city in the east and west and one of the water conditions – hard – is a crystal. Crystal bodies, consisted of some grown together single crystals, «polycrystals» are the architectural composition prototype. Single crystals, growing in the structure of these formations begin to compete with each other, trying to fix their developing vector, thus, mass of this shoot grows, that influences on its dimensions. Other single crystals that didn't make progress have to change their vector of development for a growth continuation. At the expense of this anisochronous development, every polycrystal formation would get a unique appearance. Despite the fact that most of these formations have an evident vector of growth, or «development axis», not all polycrystals develop the same way, there also exist formations, where a quantity of single crystals are developed rather equally, that makes the formation silhouette less contrast from the spatial composition point of view [11]. That property of crystals, in view of their geometrical foundation and naturalness, can be adapted to architectural forming principles, taking account of the spatial composition of architectural objects. Then there would be given the main points of natural crystal transforming into an architectural spatial composition algorithm:

1. Correction of the crystal's «growth axis» deviation to acceptable altitudes. It is important to take into consideration that for hardness of the expressed growth axis would be given by the constructive stiffening core, it is necessary to hold perpendicularly to the basement (fig. 2).
2. Forming of the object's stylobate fragment. Optimization and generalization of the small single crystals form and their merging into the unified mass. Small single crystals develop and grow chaotically in a polycrystal structure, based on the principle that is true for the all polycrystals. It is necessary to run processes of generalization and optimization to get a tectonically correct spatial composition. (fig. 3).
3. Forming a spatial composition from the skyscraper's dominant part. A natural single crystal, more developed in a polycrystal structure, is equal to a high dominant of the complex – a skyscraper. A natural form of an ideal single crystal is a prismatic polyhedral quantum with different degree and angle of the top borders truncation that could be conditionally compared to a pencil. This form is often complicated by the small single crystals that have ingrown. For transforming into the tectonic spatial composition it is necessary to get that form through the generalization, primitivization, decompozition processes (fig. 4).
4. The parts conjugation into a united structure. For the united complex construction it is necessary to form a rhythmical structure, that, on the one hand, should quote some natural crystal formations properties, and on the other hand – be an architectural building with rational tectonic structure and spatial composition, that is peculiar to it. It is possible by finding a rhythmical language of separate blocks linking into a united spatial composition (fig. 5).

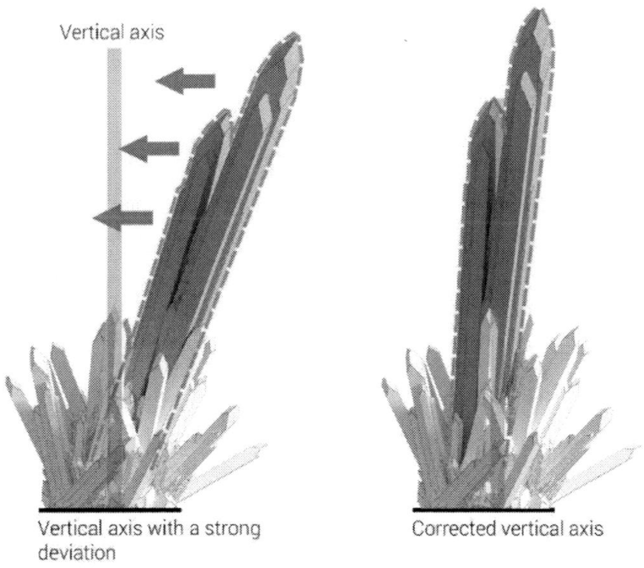

Fig. 2 Scheme of the crystal's «vertical axis» deviation.

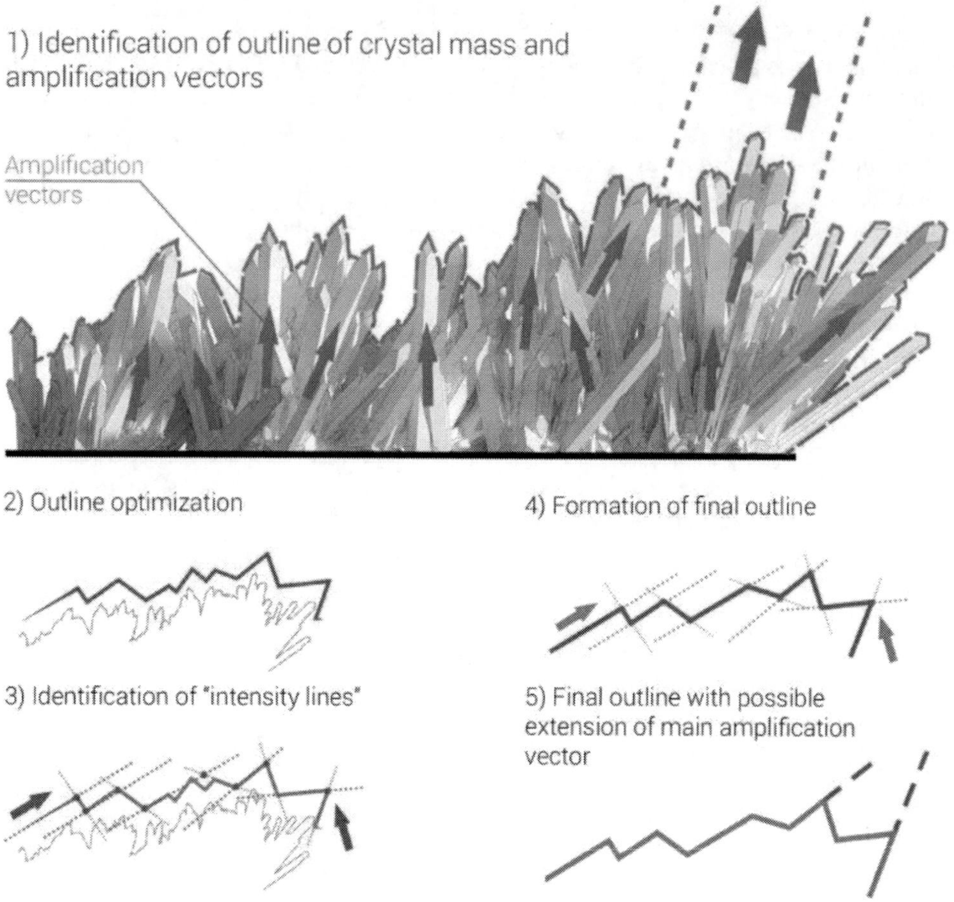

Fig. 3 The polycrystal mass generelization scheme.

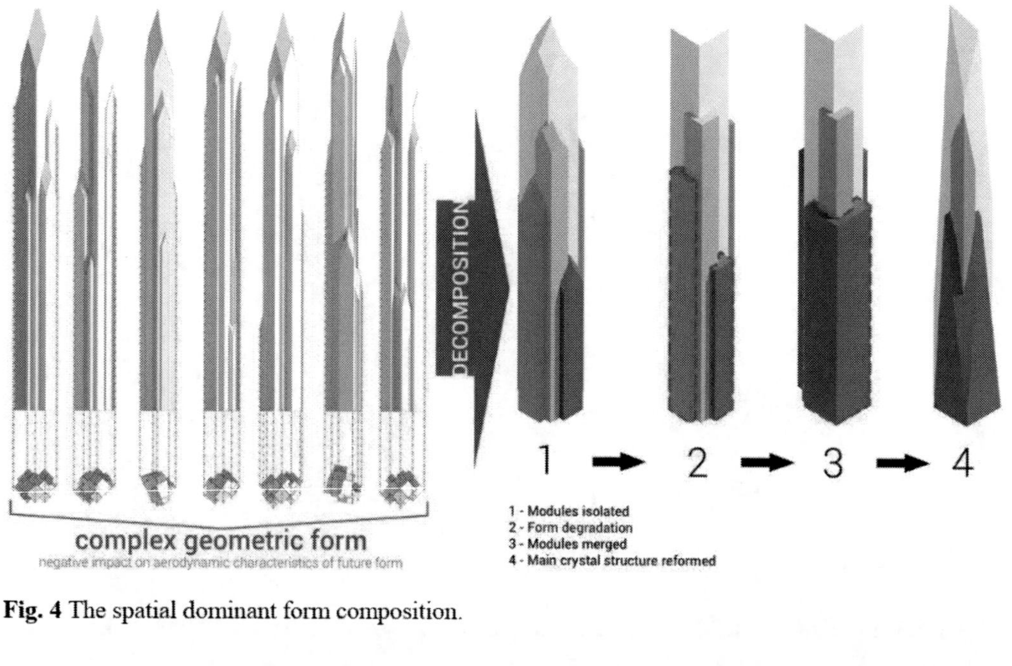

Fig. 4 The spatial dominant form composition.

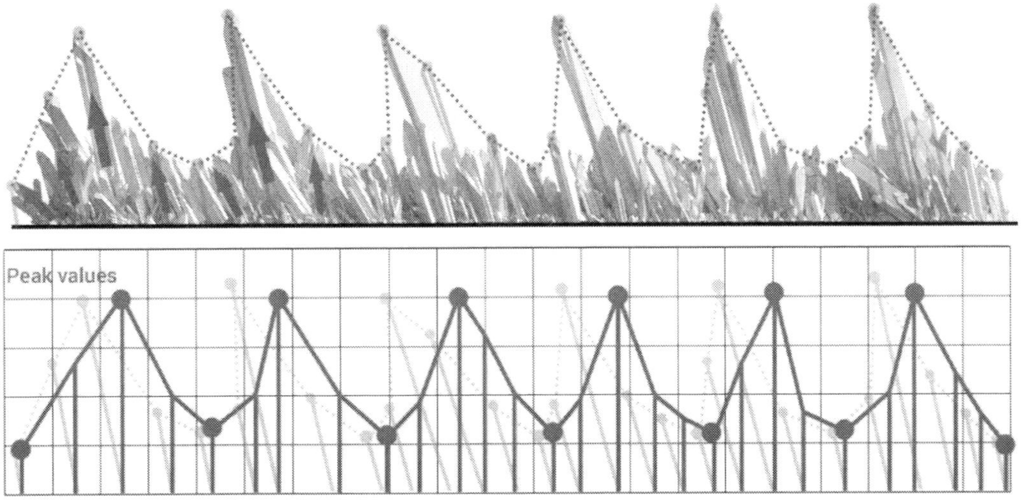

Fig. 5 The rhytmical structure analysis.

At this stage of development, this technology of erection is not a lack of shortages – buildings are limited enough in proportions and height [12]. During the long time, there has been used a standard inner framework with a core – the ladder and elevator block in which also were all engineer systems. Today is widely used a type of external framework – in this system the core is also situated inside the building (rarely displaced from the center, more rarely – gets outside), and the main supporting member are taken out to the facade, underlining structural properties and tectonics of the building, releasing big squares inside [43, 14, 15]. Such solution is recommended to use in crystal form transmission, to get the most expressed and tectonic appearance of a spatial composition.

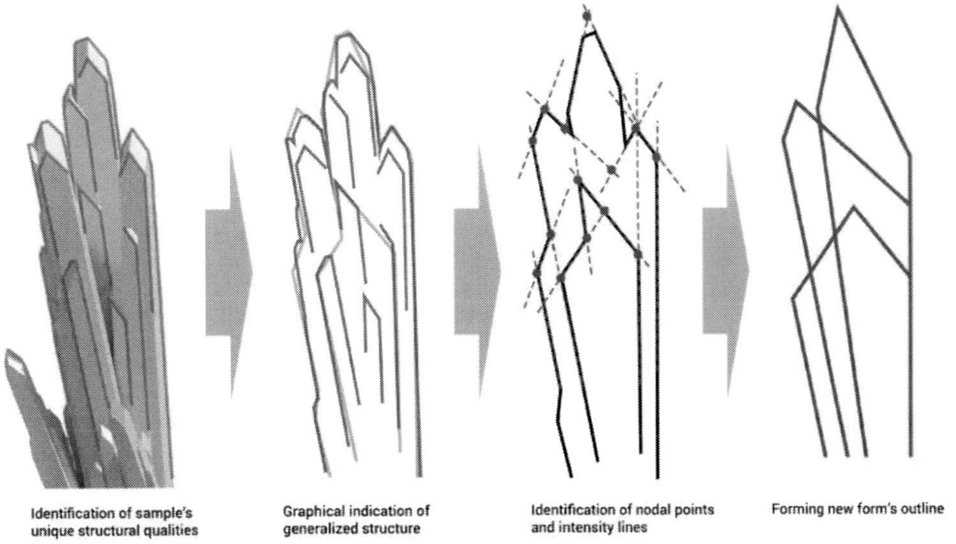

Fig. 6 The compositional structure of the crystal form exposure.

Crystal forms have such a high visual expression degree due to its hardly ribbed structure – arranging borders at different angles to the light, every crystal has its own individual appearance and constructive system, based on hard conjugation of its form ribs. Adapting a natural crystal form to the architectural spatial composition it is necessary to understand that the form needs alignment and rethinking at the stage of early sketches for the most optimal inner spaces distribution and adequate constructive scheme behavior. At the same time it is necessary to represent an individual image and a visual effect from an initial crystal prototype and study axonometric projection of the form to reveal force points of ribs coupling and the main compositional efforts in a crystal silhouette (fig. 6). For the research approbation there has been taken a regional city of Ufa – the capital of the Bashkortostan Republic.

3 Results

The Ufa city planning structure is a linear city, spreading from the south to the north and bordered by the Belaya River from the south to the west and the Ufa River from the east. A rapidly growing level of the city urbanization, actively projects under construction, business-centers and public buildings burden the transport infrastructure. As a result, there is a multitude of 10-point traffics at the main city roads. All this is aggravated by the main infrastructure problems – weak transport links between the city districts, distant from each other [16, 17, 18, 19, 20]. There is a way to enhance this links by bringing an overhead monorail, new in principle transport system.

The overhead monorail advantages are: relatively rapid velocity of travel – up to 50 km/h, a car capacity – up to 65 people (20 sitting places), 250 people is a maximum for 4 cars - rolling stock capacity. Therefore, it is possible at rush-hours to transport up to 5000 people per hour in one direction with intervals for 3 minutes. This system doesn't have a bad effect on the environment, is virtually noiseless, the permissible distance from dwellings windows is 15 meters.

The first step of this transport system introduction into the Ufa city is to strengthen the west and east city parts transport links by laying a number of new streets, which are perpendicular to the main transport arteries (Oktyabrya Prospect, Salavata Yulaeva Prospect). Strengthening of these links would serve as a basis of the next step – the Ufa overhead monorail's main branch (fig. 7). The central branch would stretch out along one of the main city roads – Salavata Yulaeva Prospect. The overhead monorail stops must be located in points of intersections with streets, linking east and west parts of the city, such as: 50 let SSSR street, Sagit Agisha – Revolutsionnaya street, Ayskaya – Bakalinskaya, Zaki Validi – Sochinskaya. The overhead monorail first stage development will stretch along the largest city districts and link the most remote from each other territories – Ufa international airport in the south and the industrial estate «Ufaneftehim» in the north. The second step permits to link a city base with new perspective districts, developing from the former suburban settlements. Creating intercepting parkings at the terminus would shorten the transport stream that enters the city.

The overhead monorail stations should be linked with bus-stops by pedestrian crossings (the fixed-run bus taxi, the state transport system), that will directly transport to the destination. These links are necessary for indissoluble and non-suture transporting city system creation. Arrangement of the stations a little way from the carriage-way, at the favourable places for little public spaces organization and parting passenger traffics would be the most optimal method.

The overhead monorail is suggested to make by means of the H-Bahn technology («hanging railway»), which is the double-track overhead system on rod posts. The advantages of this solution are: an economical profit (this type is easier, faster, cheaper in building); an aesthetics (convinient rod posts construction); safety; simplicity in use (the construction is protected from weather conditions influence, sabotage and vandalism).

Fig. 7 The overhead monorail arrangement scheme.

With the correct interaction of different types of public transport, such as a monorail, a bus, a trolleybus, a tram, would be created an advanced transport network that links all city districts. The suggested transport structure would minimize the passengers provisional cost for transportation within the frameworks of the city, shorten the number of traffics; optimize the routes for dwellers and city guests. The main transport road territory – Salavata Yulaeva Prospect is not ennobled because of it's strictly transit function. In the perspective, this city artery would become the main city axis, its gate, because the main entry in the city is the beginning of that road. Erection of the suggested experimental project would develop territories, contiguous to the prospect that would be a positive point of a visual perception of the city. A beautifying level growth would uplift the district environment quality; reduce the exhaust gases influence upon the citizens and the administrative buildings employees. At a later date, perspective plans of this aspect development are linked with the creation of the green belt along Salavata Yulaeva Prospect that would help to link this district with the main recreational framework of Ufa (fig. 9)

4 Conclusion

On the basis of the aforesaid, it is possible to ascertain that the form-building aspect of the high-rise building is one of the most important factors of its future development, because, taking into consideration a level of the building science and technologies all in all, today every architectural idea can practically realize. That is why it is important to pay attention to the form-building and spatial modelling, as well as to the economical or technological factors. Crystallography is an inexhaustible source of impression for architectural forming, because it contains a great resource of canons and rules, that, with due notice and examination would accommodate to the spatial composition, at that, there is created an individual object appearance that possesses high aesthetic properties, because a crystal is a natural formation and the nature is the best artist and engineer. That compositional development direction in high-rise architecture and skyscrapers is one of those vectors that would be actively develop in future and change the city image, we are living today.

https://doi.org/10.1051/e3sconf/20183301007

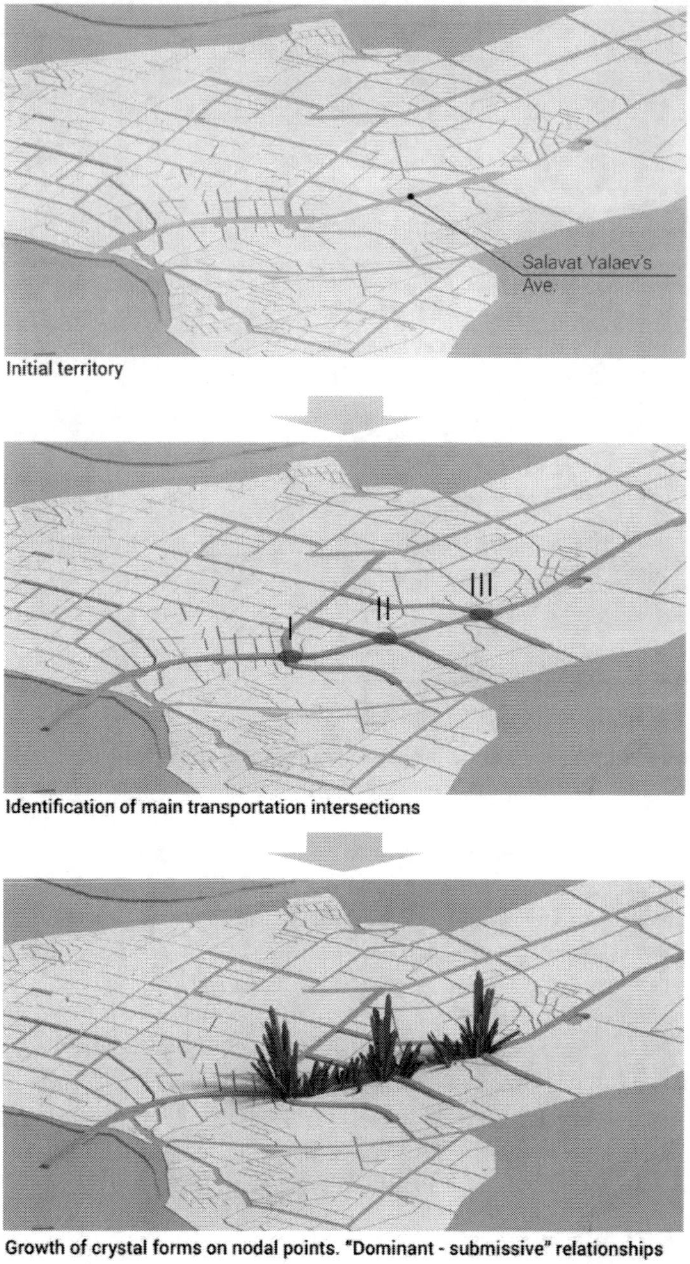

Fig. 8 The image forming in the context.

Fig. 9 The recreational framework development scheme.

References

1. Ch. Jenks, Yazyk arkhitektury postmodernisma (The language of Post-Modern architecture) (Moscow, Stroyizdat, 1985)
2. O.V. Maslovskaya, Formoobrazovanie i arkhitekturno-hudozhestvennaya interpretaciya vysotnoj zastrojki na primere gorodov SSHA, PhD Dissertation, 221 (Russia, Vladivostok, 2002)
3. O.V. Orelskaya, *Sovremennaya zarubezhnaya arkhitektura: uchebnoe posobie dlya vuzov*, 272 (Moscow, Akademiya, 2006)
4. G.I. Revzin, *Ocherki po filosofii arkhitekturnoj formy* (Moskva, OGI, 2002)-4
5. http://www.ctbuh.org/. (last accessed: 31st October 2016)
6. C. Willis, Form follows Finance: Skyscrapers and skylines in New York and Chicago, 221 (NY, Princeton Architectural Press, 1995)
7. M.A. Korotich, Kompozicionnoe razvitie vysotnoj arkhitektury. Akademicheskij vestnik URALNIIPROEKT, (Ekaterinburg, 2010)
8. M.A. Korotich, Kompozicionnye osobennosti strukturnogo formoobrazovaniya obolochek vysotnyh zdanij, *Akademicheskij vestnik UralNIIproekt RAASN*, **2**, (2009)
9. V.P. Generalov, *Osobennosti proektirovaniya vysotnyh zdanij* (Samara, Samarskij gosudarstvennij arkhitekturno-stroitelnij universitet 2009)
10. Yu.K. Egorov-Tismenko, Kristallographiya I kristallohimiya, 592 (Moskva, KDU 2005)
11. Sila diagonalej, *Zhurnal «ARX.»* **1**, 34-40 (2006)
12. http://naukovedenie.ru/PDF/32TVN515.pdf. (last accessed 21st April 2016)
13. Dmitrieva, Sovremennoye vysotnoe stroitel'stvo: ehffektivnye tekhnologii I materialy, *Tekhnologii stroitelstva*, **7**, 4-10 (2005)
14. Yu.G. Granik, A. A. Magay Arkhitekturno-konstruktivnye osobennosti vysotnyh zdanij za rubezhom. Nnform, *Unikalnye I specialnye tekhnologii v stroitelstve* **1** (Moscow, Arkhitekturno-stroitelnyj centr «Dom na Brestskoj» 2004)
15. Rukovodstvo po vysotnym zdaniyam tipologiya I dizajn, stroitelstvo I tekhnologiya Edited by S. V. Nikolaev, 228 (Moscow, OOO «Atlant-Stroy», 2006)
16. Yhttp R. F. Bajmuratov, S. Kh. Bajmuratova, Resursy krupnogo goroda. Problemy stroitelnogo kompleksa Rossii: materialy Mezhdunarodnoj nauchno-prakticheskoj konferencii pri XIII specializirovannoj vystavke «Arkhitektura. Stroitelstvo. Kommunalnoe hozyajstvo-2009», 162-165, Ufa
17. http://archyort.ru/article/view/1869. (last accessed 21st April 2016)
18. U.V. Rasuleva, R. F. Bajmuratov, I. N. Sabitov, K. A. Donguzov, Imploziya Schemaforma, 100 (Ufa, Neftegazovoe delo, 2012)
19. I.N. Sabitov, R. F. Bajmuratov Vliyanie transporta na sovremennoe zhilishchnoe stroitelstvo v g. Ufe. Problemy stroitelnogo kompleksa Rossii, *materialy Mezhdunarodnoj nauchno-prakticheskoj konferencii pri XI specializirovannoj vystavke «Stroitelstvo. Kommunalnoe hozyajstvo-2007»*, 232 (Ufa, 2007)
20. N. Sabitov, Yu. V. Rasuleva, E. G. Mazina, Implozivnyi metod I pyatoe izmerenie v arkhitekture. Arkhitekturnoe intersprostranstvo XXI veka: opyt, problemy, perspektivy: materialy Mezhdunarodnoj nauchno-prakticheskoj konferencii, 404 (Sankt-Peterburg, SPGASU, 2013)

High-rise construction in the Saint Petersburg agglomeration in 1703–1950s

Sergey Sementsov[1*], *Nadezhda* Akulova[1], and *Severina* Kurakina[1]

[1]Saint Petersburg State University of Architecture and Civil Engineering, Department of Architectural and Urban-Planning Heritage, 4 Vtoraya Krasnoarmeyskaya str., Saint Petersburg, Russian Federation

Abstract. Regularities of high-rise construction (implemented projects and developments) in Saint Petersburg and the Saint Petersburg agglomeration since the foundation of the city in 1703 till the 1950s are considered. Based on these regularities, a single spatially developed system of vertical dominants is formed. High-rise construction in the city and its suburbs started in the 1710s and continues up to the present time. In the considered decades (1703–1950s), high-rise construction mostly performed urban-planning functions (with vertical and symbolic dominants), relying on patterns of the visual perception of man-made landscapes under development. Since the 1710s, the construction of vertical dominants (mainly temples, spires of towers, lighthouses, etc.) of five ranks (depending on the altitude range and in relation to the background development) was conducted in territories of the entire agglomeration. These dominants were arranged in landscapes of the city and suburbs with almost mathematically precise accuracy and according to special regulations. Such dominants obtained particular descriptive and silhouette characteristics in accordance with the conditions of spatial perception. In some periods of city development, attempts were made to create monuments (symbolic dominants) of specific height and include those in the spatial system of high-rise dominants as significant elements of the city silhouette.

1 Introduction

Several stages in the creation of the system of high-rise facilities can be distinguished with regard to Saint Petersburg and its agglomeration. Even during the development of the flat, swampy landscape, where the Saint Petersburg agglomeration consistently formed since 1703, numerous waves in the creation of the system of high-rise facilities can be traced. Revealing regularities in the spatial construction of such high-rise facilities is one of the tasks of historical and urban-planning studies regarding Saint Petersburg.

2 Methods

[1] Corresponding author: s.sementsov@mail.ru

Concurrent comprehensive studies of vast arrays of urban-planning and architectural archive materials, comparative full-scale studies of survived and/or lost facilities.

3 Results

High-rise construction in territories of the Neva plain, whereon the capital city of Saint Petersburg and surrounding suburban flat lands formed since 1703, started since the 1710s, according to personal orders of Peter the Great [Figure 1].

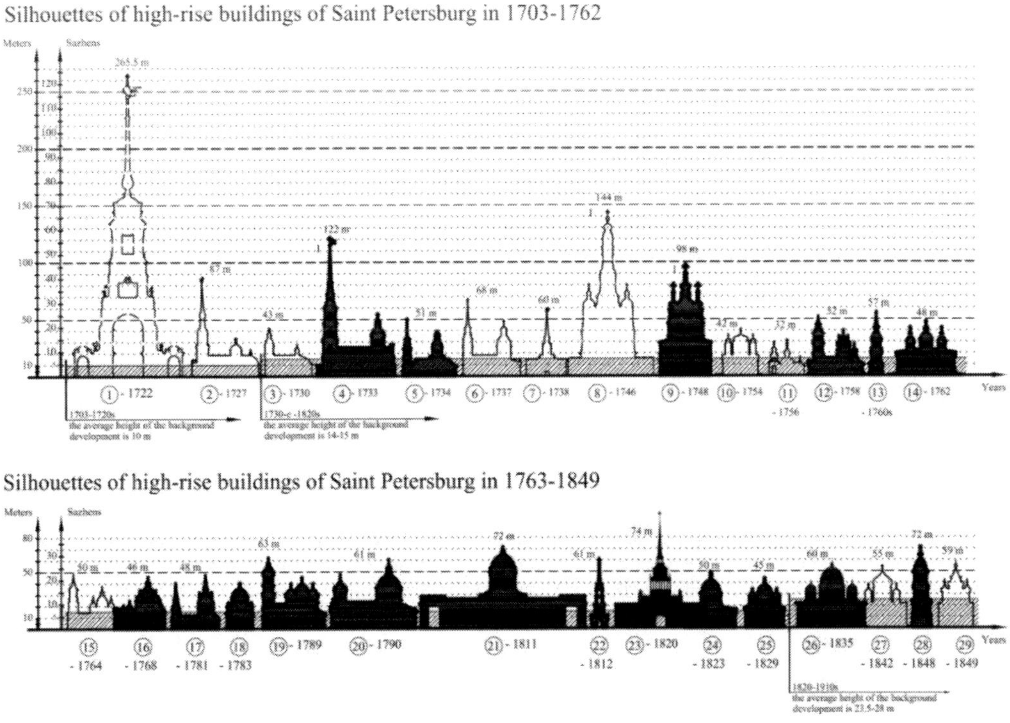

Fig. 1. Construction of vertical dominants in Saint Petersburg in 1703–1850s

Up until the middle of the 1720s, highest buildings were erected not only in the city, but also in its suburbs according to his initiatives: the Peter and Paul Cathedral (with the bell tower, the spire of which reached the height of 122 m, No. 4 on schemes), the original building of the Admiralty Shipyard (with the spire of 40 m in height), as well as few buildings with spires of lower height. As is obvious, these were not only temple buildings, but also public and industrial buildings (Foundry Yard (Liteiny Dvor), Particular Shipyard, lighthouses at the western end of Vasilyevsky Island, etc.). In the years of 1721–1723, the construction of a high-rise lighthouse (with the design spire height of 266 m, No. 1 on schemes), which contemporaries compared with the ancient Colossus of Rhodes (which was "only" 37–40 m in height), was started by the main ship channel to the south of Kronstadt (however, it was not completed). This height was predetermined in decrees as a necessary condition (and a requirement) for creating high-rise compositional dominants on axes of main streets (roads, main lines) of urban and suburban territories and main ship channels in the waters of the Gulf of Finland, as main dominants for the visual arrangement of formed urban and suburban spaces [1].

Silhouettes of high-rise buildings of Saint Petersburg in 1850-1920

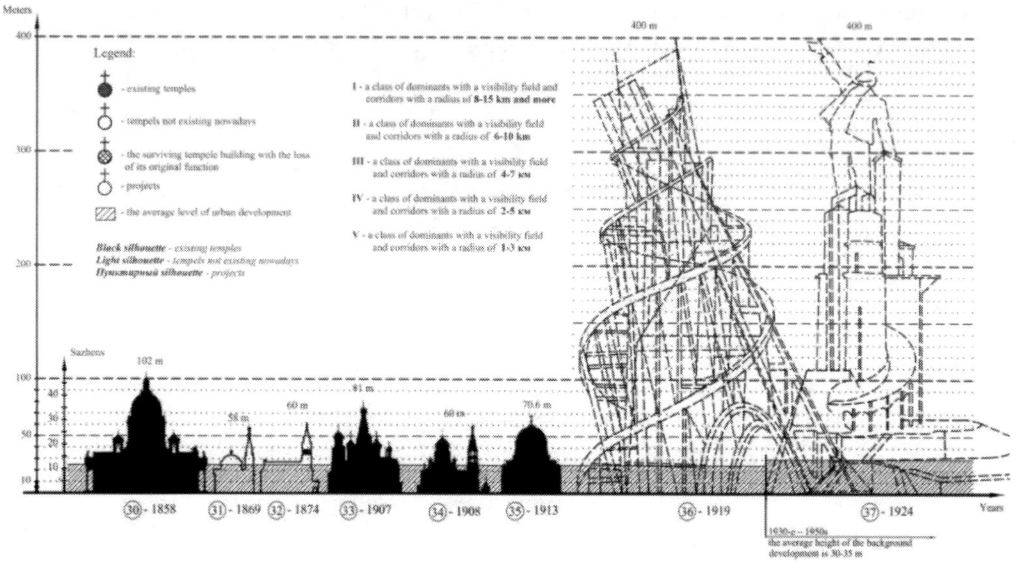

Fig. 2. Construction of vertical dominants in Saint Petersburg – Petrograd in 1851–1924

At the same time, the formation of the system of symbolic dominants was conducted in the territories of the city and suburbs. Their creation was carried out according to another principles and did not always take into account patterns of vertical perception and requirements for the altitude of facilities, remaining often within the framework of the symbolic field "only". For example, among main symbolic dominants of Petrine times not only main vertical dominants (e.g. the Peter and Paul Cathedral), but also less higher objects (but with equally powerful symbolic aura) (e.g. the original Trinity Cathedral on Troitskaya Square, the height of which did not exceed 15 m) were distinguished. In terms of the structure, such constructions were erected with brick walls, with the use of a log-frame system of high spires and domes, with pile wooden foundations. And as for the first erected buildings, the use of log-crib constructions in foundations can be noted [2].

Silhouettes of high-rise buildings of Saint Petersburg in 1921-1950s

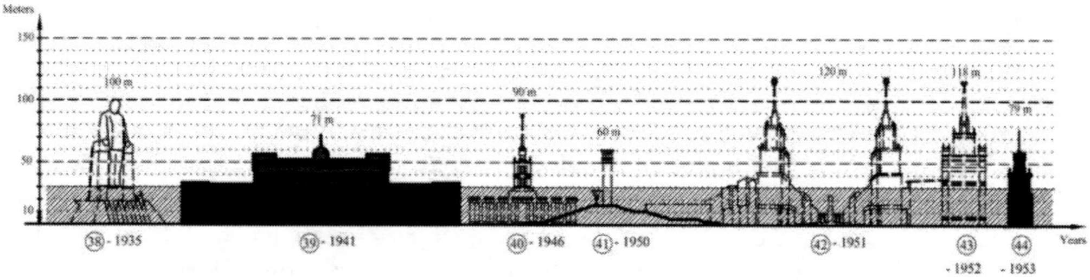

Fig. 3. Construction of vertical dominants in Leningrad in 1925–1950s.

Thus, as early as in those decades, high-rise construction started ("high-rise" in terms of Saint Petersburg conditions) as the system of relatively small vertical and symbolic dominants rising above the standard "background" development to various extents.

Later, under the reign of Anna Ioannovna, in the 1730s, temples and objects of civil and industrial architecture of lower height (with spires of only up to 30–40 m in height) were mainly erected as dominants in the city and suburbs. However, the Admiralty Shipyard was reconstructed and the height of its timber spire increased up to 60 m (No. 7), becoming one of the largest verticals. Under the reign of Elizaveta Petrovna during 1740–1750s, the group of the highest objects of the city and its suburbs (in the framework of the single Saint Petersburg agglomeration) was supplemented by the Smolny Cathedral of the Resurrection (height of the cross over the central dome — 98 m, No. 9) and the bell tower, the construction of which was started but not completed (height of the hipped spire — 144 m, No. 8) [3].

Until the end of the 18th century, the large-scale vertical construction in the city and suburbs was performed with objects of relatively low height. The majority of temples and industrial objects built in those decades, with overall dimensions of vertical elements not exceeding 30–40–50 m, had fairly average height by standards of the time.

By the beginning of the 19th century, the mathematically precise and spatially accurate system of vertical dominants formed in the city, which is clearly seen on the famous Vitztum plan (1821) [Figure 4]. Not only temples themselves (with domes and bell towers), but domes of civil buildings, the highest tops (attics), other elements of development, vertical in terms of semantics and morphology, served as visual vertical dominants [4]. This direction of mathematically precise arrangement of high-rise dominants in the spatial system of the entire city not only continued up to 1917, but also preserved traditions till the 1950s.

Throughout the whole 19th century, objects of vertical construction included numerous temples, keeping the overall height within 50–60 m. They also included several largest and, at the same time, highest buildings, i.e. the Kazan Cathedral (built in 1801–1811, 72 m high, No. 21) and Saint Isaac's Cathedral (erected in the 1830's–1850's, with the height up to the cross top of 102 m, No. 30). These objects also became new main symbolic dominants of the Russian capital. At the beginning of the 20th century, the Naval Cathedral (with the height up to the cross top of 71 m, No. 35) was erected on Kotlin island in the center of Kronstadt, 35 km from the city, which officially became the main naval cathedral in Russia [5, 6].

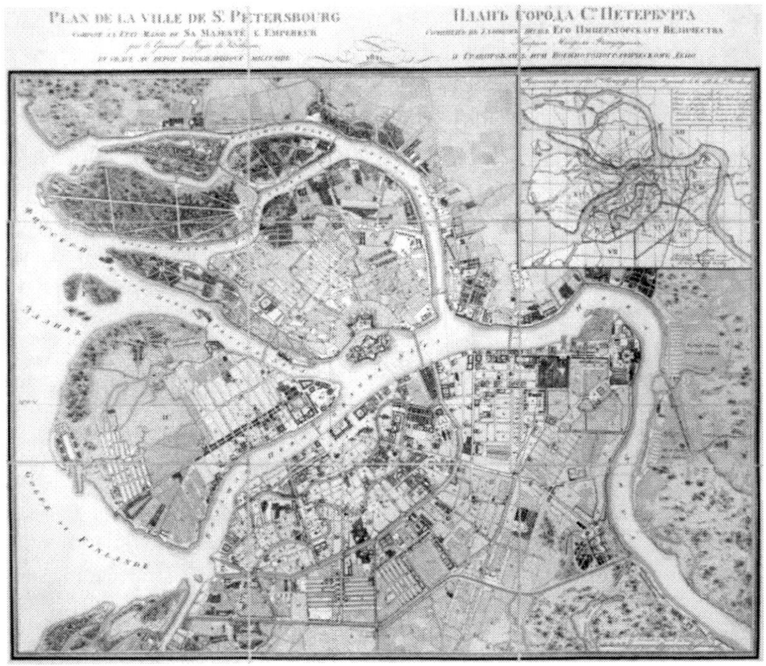

Fig. 4. PLAN OF THE CITY OF SAINT PETERSBURG *compiled in the General Headquarters of* HIS IMPERIAL MAJESTY *by Major General I. Vitztum AND ENGRAVED IN THE MILITARY TOPOGRAPHICAL DEPOT.* 1821 / PLAN DE LA VILLE DE St. PETERSBOURG *Composé à l'Etat-Maior de* SA MAJESTÈ L' EMPEREUR *par le Général-Major de Vitzthum, ET GRAVÈ AU DEPOT TOPOGRAPHIQUE MILITAIRE.* 1821. Saint Petersburg. RNb K1-Pb 6/72.

By the beginning of the 20th century, construction on Saint Petersburg soils and under Saint Petersburg natural and climatic conditions led to the use of not only brick walls and vault structures, but also monolithic concrete and reinforced concrete frames and load-bearing brick walls. Foundations were based on solid wooden pile fields (of larch and oak).

Such large-scale development of territories led to the fact that even suburban territories were densely developed with vertical dominants which were erected in centers of settlements, at crossroads and in road bends, in elevated areas, in visual centers between settlements, i.e. in centers of spatial perception of man-made and natural landscapes. Plain natural and man-made landscapes of suburban districts (uyezds) (e.g. Tsarskoye Selo district (uyezd)) allowed erecting temples (as vertical and symbolic dominants) of 40–50–60 m in height. The location of these dominants in the district territories was quite precise in terms of spatial arrangement. It is clearly seen in the map of the Tsarskoye Selo district (uyezd) (as of 1916). It shall also be note that the spatial arrangement of temples in Saint Petersburg, in suburban districts (uyezds) (through the example of the Tsarskoye Selo district (uyezd)) and in the territories of the Saint Petersburg governorate (guberniya) (as a whole) was carried out according to unified principles and patterns [7].

At the beginning of the 20th century (after 1917), during the first stage of socialist state formation and the spread of "avant-garde" ideas, a wave of proposals to create the largest world-class revolutionary monuments emerged. The powerful burst of symbolic energy led to the fact that, during the first years under the Soviet power, unusual grand projects (in terms of spatial design) with eclectic symbols were developed and proposed for construction [8]. Those projects included proposals for erecting the Tower to the Third International (Tatlin's Tower with the height of 400 m, No. 36) on Troitskaya Square in Petrograd in 1919 (author — architect and designer V. Ye. Tatlin) and the monument to V. I. Lenin, proposed by architect V. A. Shchuko (with the height of 400 m, No. 37), in the same place in the beginning of 1924 [9, 10, 11] [Figure 2].

Even projects of applied technical facilities of those times were distinguished by gigantic sizes. For example, during the contest for the construction of a lighthouse in the Sea Port of Leningrad, a lighthouse in the form of a sculpture of V. I. Lenin (100 m high, No. 38) was proposed [12]. It was proposed to build those objects in bearing structures of the metal and reinforced concrete frame, but with foundations based on solid wooden pile fields (of larch and oak).

In the 1930s–1950s, within the framework of developed master plans for the formation of the new, socialist Leningrad, proposals for erecting tens of high-rise buildings, which were to form the system of high-rise dominants in the single silhouette field of the entire city, primarily within the borders of the new, socialist Leningrad, were made. Projects for the arrangement of dominants (of various functions: public, residential, and industrial) were developed, their urban-planning significance and altitude range (in relation to the significance and altitude range of other dominants) were determined. It could be clearly seen in design documents for the urban planning of Leningrad main lines (Moskovsky prospect, Stachek prospect, Zanevsky prospect, Kirovsky prospect, etc.), main entrances to Leningrad. Among those, projects of the House of Soviets (spire height — 71 m, No. 39), Finland Railway Station (spire height — 90 m, No. 40), residential buildings on the routes along Neva embankments, public buildings can be noted [13].

Not many of numerous projects for the creation of vertical dominants were implemented. Among outstanding examples of this project-based approach, numerous options for the reconstruction of Moskovsky prospect (Stalin prospect at the time) with proposals for the polyrhythmic arrangement of stand-alone dominants (with height of up to 110–120 m, Nos. 42, 43) and dominants, integrated in the system of residential buildings (with height of up to 70–80 m, No. 44), can be noted [14, 15, 16] [Figure 3].

The altitude range of all these buildings was not random. It corresponded to the requirements for the implementation of spatial visibility and symbolic perception tasks, as well as possibilities of building construction on Saint Petersburg – Leningrad swampy soils, ensuring the erection of vertical and symbolic dominants. It was possible to perform such tasks by creating buildings of special type rising above the standard "background" development to various extents.

"Standards" of the altitude range of the Saint Petersburg background development were changed several times according to imperial decrees and officially approved urban-planning requirements. The average height of the background development depended on the formation period and created a plain silhouette for the main part of urban and suburban development. For example, in the 1703–1720s, the average height of the background development (taking into account the height of roofing to the ridge top) was about 10.0 m (one floor, high basement, high roofing). In the 1730s–1820s, the average height of the background development was approximately 14–15 m (2 floors, high basement, high roofing). In the 1820s–1910s, it was 23.5–28.0 m (5–6-story buildings with the roofing at an angle of 18–30°, or with attics, which provided the height of the ridge top at the same level), in the 1930s–1950s, it was 30–35 m (5–7-story buildings, high basement, with the roofing at an angle not exceeding 30°).

With the beginning of the era of industrial housing construction and wide use of industrial methods of construction, the number of floors (altitude range) of the large-scale standard background development began to increase rapidly in the city. In the 1950s–1960s, the altitude range of the background development was 16–17 m (5 floors of industrial buildings of the first large-scale prefabricated housing, low basements, flat roofing), in the 1970s–1980s, the altitude range of the standard background development was already 37–38 m (9–12 floors with flat roofing), and in the 1990s it rapidly and significantly increased and reached 72–75 m (24–25 floors with flat roofing).

The absolute size and height of vertical dominants in comparison with the background development played a big role, predetermining their altitude and visual significance, as well as the spatial and compositional role. As early as in Petrine times, in territories of Saint Petersburg formation and development of suburban areas, i.e. in the zone of the development of flat landscapes, deliberate formation of a system of vertical dominants, designed for various zones of perception and spatial arrangement, started. Different absolute heights, different "excess" over the background development, and, at the same time, different characteristics of "visual work" in the area being formed were stipulated (by design solutions and construction capabilities) for various elements of the system of high-rise dominants. Absolute dimensions (absolute height) made it possible to provide the visibility of objects from different distances. The relative altitude range (relative to the background development) also helped to ensure contrast (nuances) in the perception of dominants, their visual orientation features.

Class (rank) 1. Dominants of regional significance. The designed and implemented height of objects reached 100 m and more. Facilities of such height were constructed throughout the entire development of the city. This group includes: the lighthouse by the Kronstadt canal (1721–1723, the construction was not completed, height — 265 m, No. 1 on schemes); the Peter and Paul Cathedral (the construction was completed in 1733, spire height of the bell tower — 122 m, No. 4); the Smolny Cathedral of the Resurrection (built in 1748, height of the main dome — 98 m, No. 9) and the unfinished bell tower of the Smolny Monastery (design height — 144 m, No. 8); the Saint Isaac's Cathedral (built in 1858, height of the cross over the dome — 102 m, No. 30).

A few more objects were slightly lower, but due to special architectural solutions (sharp silhouette, gilding, etc.) they can also be referred to the group of major dominants of the historic city. Among such objects: the Kazan Cathedral (with the dome top at the height of

72 m, No. 21); the spire of the Main Building of the Admiralty (with the spire top at the height of 74 m, No. 23); the Bell Tower of the Vladimirskiy Cathedral on Vladimirskaya square (72 m high, No. 28); the Church of the Savior on Blood (81 m high, No. 33); the Naval Cathedral of St. Nicholas the Wonderworker in Kronstadt (71 m high, No. 35).

In terms of height, the first group of objects is in large excess over the background development: during the construction of high-rise objects, their height was 7–10 times more than the height of the background development. And in the next decades (up to the end of the 20th century), upon the stable preservation of the altitude range with regard to the background development of the historical center of the city at the level of 24–30 m, it was 4–6 fold higher. The second group was initially 2–4-fold higher than the background development, and at the beginning of the 20th century it was 2–3-fold higher.

The main high-rise dominants (dominants of the first group) provided spatial perception (and visual arrangement) at distances of up to 30 km and more, and dominants of the second group — at distances of up to 8–15 km.

Class (rank) 2. Dominants of city significance. With the height of domes and spires in the range of 50–60 m. They were erected both in the city and in extensive suburban territories, especially in the zones of main lines and routes, within the borders of the entire historical Saint Petersburg agglomeration. Among them: the Church of the Holy Righteous Simeon and Anna the Prophetess (height of the bell tower spire — 51 m, No. 5); the St. Nicholas Naval Cathedral (height of the central dome — 48 m, No. 14) and the stand-alone bell tower (height — 57 m, No. 13); the Church of the Vladimir Icon of the Mother of God (height of the central dome — 47 m, No. 16); the Holy Trinity Cathedral in the Alexander Nevsky Lavra (height of the cross above the main dome — 62 m, No. 20), the bell tower of the Cathedral of the Exaltation of the Holy Cross (spire height — 61 m, No. 22). During construction, they were 3–4-fold higher than the background development, and when the height of the background development increased up to 24–30 m, they were 2–2.5-fold higher. Such altitude characteristics of buildings and their prominence relative to the background development provided visual arrangement of spaces with a radius of 6–10 km.

Class (rank) 3. Dominants of district significance. With the height of domes and spires in the range of 40–50 m. Such buildings and groups of buildings were widely spread. The designed and implemented height of objects 2.5–3-fold exceeded the average height of the standard (background) surrounding development (in decades before the 1820s) and 2-fold exceeded the average height up to 1917, providing visual arrangement of spaces with a radius of 4–7 km.

Class (rank) 4. Dominants of local significance. Such dominants also represent one of the most common options of historical "high-rise development"; at the stage of the development of the historical Saint Petersburg with the height of the background development of 15 m (in 1703–1820s), such "dominants" had a height of 25–35 m, and when the background development exceeded 24 m (in the 1820–1910s), such dominants had a height of up to 40–45 m. Thus, even such height 1.5–2-fold exceeded the height of the background development. With regard to plain areas of the suburban zone with local rural settlements (and the average height of buildings of 10–15 m), they certainly represented dominants, providing spatial and orientational distinctions of the perception environment. Thus, such dominants were also well perceived in the visual field, providing visual arrangement of spaces with a radius of 2–5 km.

Dominants of classes (ranks) 3 and 4 were of the largest scale in the city's development and in the development of the suburban zone landscapes (agglomeration territories). In the city, they prevailed mainly in the first half of the 18th century, and in vast territories of the Saint Petersburg governorate (guberniya) they were the most widespread until the beginning of the 20th century. This can be clearly seen through the example of the entire Saint Petersburg governorate (guberniya) (on a scheme demonstrating its state as of 1850) [Figure 5] and, in particular, the "exemplary" (as it was indicated in documents of those years) suburban Tsarskoye Selo district (uyezd) (on a scheme demonstrating its state as of 1916) [Figure 6].

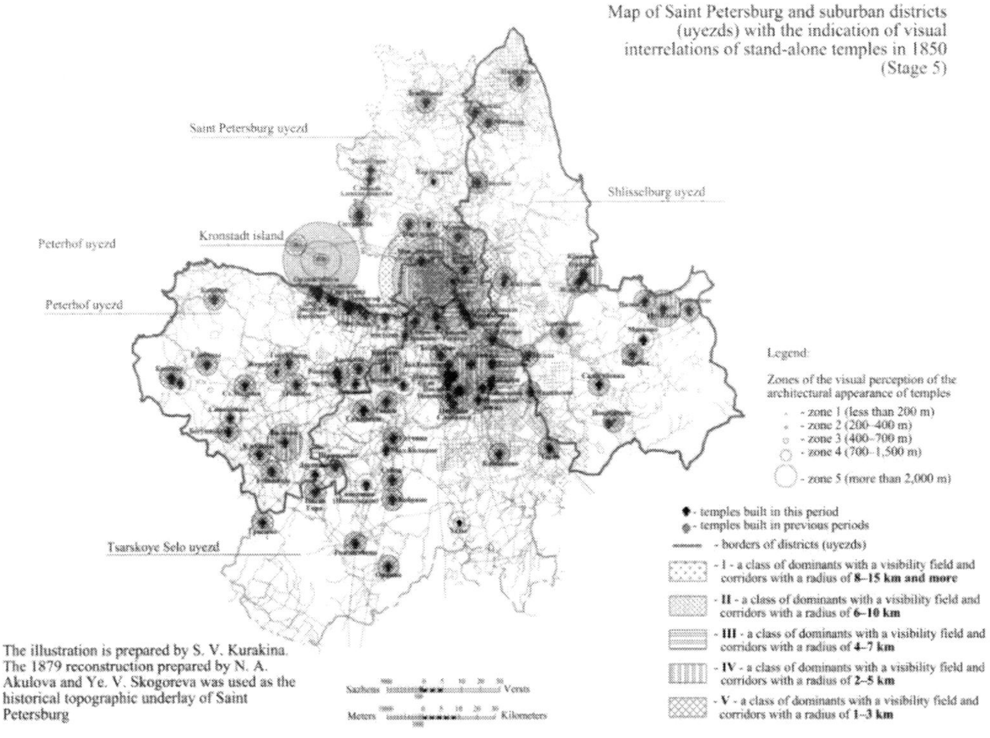

Fig. 5. Map of Saint Petersburg and suburban districts with the indication of the arrangement of temples and their visibility zones (as of 1850).

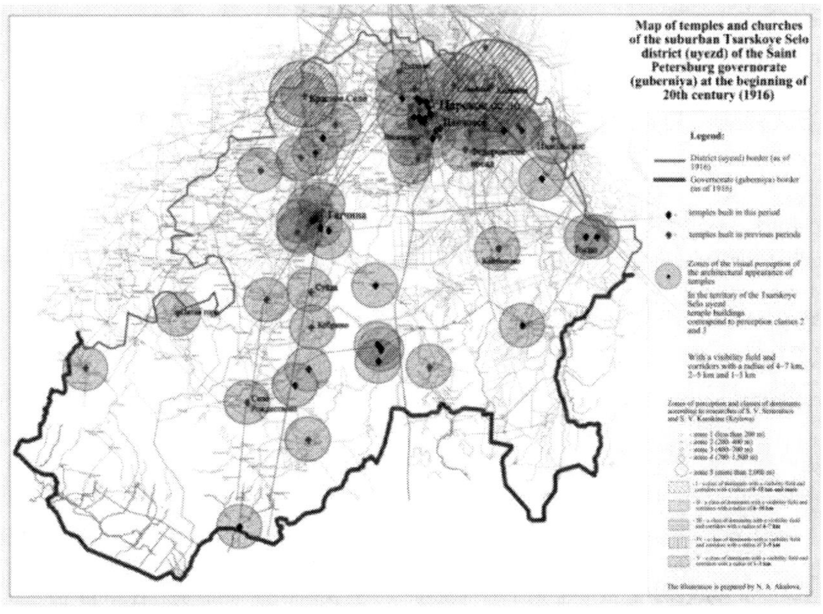

Fig. 6. Map of suburban Tsarskoye Selo district (uyezd) with the indication of the arrangement of temples and their visibility zones (as of 1916)

Such high-altitude arrangement of suburban temples provided the spatial development of visual fields of plain spaces in the governorate [17]. Text: "On the 2nd day of this September, the Emperor, according to the regulation of the Minister of Internal Affairs, commands to: for the lifetime of the Governorate Committee, established by His Majesty on May 15, 1845 to handle common affairs concerning the exemplary arrangement of the Saint Petersburg and Tsarskoye Selo districts (uyezds), appoint a special official for documentation management in such committee, with position of grade 9, and with grade 9 as for the uniform and pension, with a scribe to assist; the remuneration shall be allocated from the local tax of the Saint Petersburg governorate in the amount of 200 rubles for salary and 143 rubles for accommodation (for the special official), in total: 343 rubles, and 173 rubles in silver per year for the scribe" [17]. "On the appointment of a special official for documentation management in the Governorate Committee established to handle affairs concerning the arrangement of the Saint Petersburg and Tsarskoye Selo districts (uyezds)".

Class (rank) 5. Local dominants. Dominants in the form of towers, spires, domes, etc., which had a height of 2–5–8 m above the level of eaves in historical buildings. They were widely spread in the urban development and in suburban towns in the middle and end of the 19th century. The designed and implemented height of objects only 1.1–1.2-fold exceeded the height of the background development, reaching 20–22.0 m in the 18th century, and 30–35 m — by the end of the 19th century, providing visual arrangement of spaces with a radius of 1–3 km [18, 19, 20].

In general, this multi-level system of vertical dominants had significant features of vertical construction. However, only designed (unrealized) and erected objects of class (rank) 1 with a height of 70–100 m and more can be classified as high-rise buildings. What features of the system were the most "in-demand" during the creation and functioning of the city and its suburbs? Probably, both features in their unity and inseparability.

This spatial high-altitude system formed in the course of the entire development of Saint Petersburg and the Saint Petersburg agglomeration, emerging gradually, step by step.

Chronologically, such five-level system of vertical visual and symbolic dominants, including real high-rise objects, began to form as early as in the first decades of the foundation of Saint Petersburg, under the reign of Peter the Great. As far back as at that time, objects of classes (ranks) 1–5 were created. Initially, under the reign of Peter the Great, the system of vertical dominants of all five ranks almost immediately and systematically started to form, ensuring the network multi-layer development of formed spaces. However, at that time the large-scale development had a height of only 1–2 floors (4.5–10.0 m taking into account the height up to the roof ridge). Even palaces of the nobility were not higher than 10.0–12.0 m (according to the height of the roof ridge) in Petrine times. Therefore, all dominants of various ranks rising above them represented very important visual references under conditions of flat terrains of the Prinevskaya Plain. Efforts to erect objects with a height of 122–266 m were, obviously, predetermined by the need to solve more complicated tasks than those related to the provision of the visibility of developed spaces.

4 Conclusions

Until the 1950s, in territories of Saint Petersburg – Leningrad and its extensive agglomeration, a single tradition of creating the system of vertical dominants formed (mainly by the historical system of temples and vertical spires, etc.), which, in its turn, was based on common principles of the visibility and formation of visual zones (visibility zones) and visibility axes, and resulted from the system of five classes (ranks). Among those, objects of class (rank) 1 were 70–120 m high or more and actually represented historical high-rise objects, i.e. dominants having an orientational and/or symbolic component. They formed in a single spatial field of the city–governorate. In the history of the urban-planning development of Saint Petersburg, periods when large-scale proposals were formulated to erect super-high buildings 260–400 m high, can be noted as well. And that was against the background of the large-scale standard development, altitude range of which increased from the beginning of the 18th century to the beginning of the 20th century from 7–10 m to 24–30 m. Since the 1930s, a new tradition for the arrangement of high-rise dominants (with a height of up to 12–130 m) formed in the urban planning of Leningrad. Thus, high-rise construction, occurred in the Saint Petersburg region as far back as in the 1710s, solved tasks of forming the single visual urban-planning space.

References

1. Edict "On the Construction of Houses in Saint Petersburg" signed by the Tsar on November 4 (15), 1714. *Complete collection of laws of the Russian Empire*, first collection, **2855**, Saint Petersburg, 128 (1830).
2. S. V. Sementsov, Traditions of Peter the Great in the formation of the permanent Saint Petersburg agglomeration in the 18th century. *Faces of Petrine times — 2011. To the commemoration of the 30th anniversary of the Menshikov Palace department of the State Hermitage (1981–2011). Proc. of the scientific conf.*, 329–341 (2011)
3. S. V. Sementsov, Krasnikova O. A., Mazur T. P., Shrader T. A. *Saint Petersburg on maps and plans of the first half of the 18th century*, 436 (2004).
4. PLAN OF THE CITY OF SAINT PETERSBURG compiled in the General Headquarters of HIS IMPERIAL MAJESTY by Major General I. Vitztum AND ENGRAVED IN THE MILITARY TOPOGRAPHICAL DEPOT. 1821 / PLAN DE LA VILLE DE St. PETERSBOURG Composé à l'Etat-Maior de SA MAJESTÈ L' EMPEREUR par le Général-Major de Vitzthum, ET GRAVÈ AU DEPOT TOPOGRAPHIQUE MILITAIRE. Saint Petersburg. RNb K1-Pb 6/72 (1821).
5. B. A. Rozadeyev, R. A. Somina, L. S. Klescheva, *Kronstadt,* 144 (1977).
6. N. A. Akulova, Stages of urban-planning development in the city of Kronstadt in the 19th century. Bulletin of Civil Engineers, **4 (33)**, 5–8 (2012).
7. S. V. Krylova, Architectural-spatial and compositional features of the arrangement of temples in Saint Petersburg and surrounding districts (uyezds) in the 18th – first half of the 19th centuries. Moscow Bulletin of Engineers, **3 (28)**, 15–22 (2013).
8. S. V. Sementsov, Development of the revolutionary architectural avant-garde in Petrograd–Leningrad according to urban-planning standards of the Imperial Saint Petersburg. Part 1: from Peter the Great to V. I. Lenin. Vestnik of Saint Petersburg University. Arts, **7 (2)**, 249–272. (2017).
9. N. N. Punin *The Monument to the Third International. Project by artist V. Ye. Tatlin*, 1–8 (1920).
10. N. A. Semashko (ed.). *The Great Leader: art and literary collection*, (1924).
11. *About the monument to Lenin*, 112 (1924)
12. Competitive design of the lighthouse-monument to V. I. Lenin in the Leningrad port, executed on behalf of the Society of Architectural Artists by Langbard I. G. Annual publication of the Society of Architectural Artists, **14**, 281 (1935).
13. N. N. Baranov, V. G. Isachenko *Chief architect of Leningrad Nikolay Baranov: career and destiny*, 127 (2001).
14. Architecture and construction of Leningrad. *Collection 13*, 11 (1950).
15. Architecture and construction of Leningrad. *Collection 15*, 11 (1951).

16. Architecture and construction of Leningrad. *Collection 17*, 10 (1952).
17. Imperial regulation of the Committee of Ministers, issued by the Minister of Internal Affairs, September 2, 1847 Complete collection of laws of the Russian Empire, second collection, **22 (1)**, No. 21.515, 699–700 (1847).
18. S. V. Sementsov, Ranks of detalization and zones of medieval Novgorod architecture perception. Aesthetic problems of Soviet architecture. Inter-university collected works, 117–124 (1979).
19. S. V. Krylova, Silhouette and visual interrelations of temples in Saint Petersburg and its suburbs in 1800–1835. *Challenging Issues of Modern Construction. Proc. of the Int. Scientific and Practical Conf. of Students, PhD Students, Young Scientists and Second Doctorate Students*. **1**, 156–159 (2012).
20. N. Akulova, Y. Voznyak, Y. Ardasheva, History of Holy Trinity Church in Pyataya Gora homestead of Tsarskoselsky County and choice of method of its restoration. Architecture and Engineering, **1 (3)**, 3–13 (2016).

Architecture of high-rise buildings as a brand of the modern Kazakhstan

Gulnara Abdrassilova[1,*], *Nina* Kozbagarova[1], *Ainagul* Tuyakayeva[1]

[1]Kazakh Leading Academy of Architecture and Civil Engineering AO (JSC), Ryskulbekov street, 28, Almaty, Republic of Kazakhstan, 050043

Abstract. Using practical examples article reviews urban-planning and space-planning features of design and construction of high-rise buildings in Kazakhstan conditions; methods are identified that provide for structural stability against wind and seismic loads based on innovative technical and technological solutions. Article authors stress out the fashion function of high-rise buildings in the new capital of Kazakhstan, the Astana city.

1 Introduction

High-rise buildings in modern cities are active elements of the space environment, supporting the formation of the city's unique silhouette. At the same time they are the objects representing the technological progress, social-economic achievements of individual construction companies, as well as the country's achievements as a whole.
Each high-rise building solves unique spectrum of tasks based on the urban conditions and individual features of the construction site [1,2]. Urban tasks are connected with formation of the cities' environment, functional organization of public spaces, tasks of creating the compositional dominants and landmarks, important issues of increasing the attractiveness of the construction, etc. Apart from those named are engineering tasks related to composition of the soil, temperature fluctuations, wind power, seismic conditions [3,4]. High-rise building also requires significant resources during its operation in order to maintain communication systems, firefighting systems, people evacuation systems in case of emergency, optimization of resources' energy consumption, etc.

Article authors aim to identify features of designing and constructing high-rise buildings in the climatic conditions of Kazakhstan.

Sharply continental climate over the whole territory of Kazakhstan and high seismic activity in the South and East parts of the country are serious challenges for high-rise construction. However, prestige issues and the need to increase financial-economical, investment and tourist potential of the young country dictate the breakthrough development in this branch of architecture.

Up until now architecture theory of Kazakhstan has not comprehensibly studied high-rise architecture. Materials primarily focus on individual aspects, descriptions of

* Corresponding author: g.abdrassilova@kazgasa.kz

architectural features of particular high-rise buildings [5-7]. The issue of high-rise construction is reflected in depth in the works of engineers specializing in the issues of seismic stability of structures [8-13].

Modern Kazakhstan has become a platform for implementing fresh new engineering and architectural solutions in high-rise buildings. High-rise buildings are mostly concentrated in Astana, the new capital of the country, which through the use of world-renowned technologies in design and construction positions itself as a modern, innovative city [14, 15].

2 Methods

During the study of the high-rise buildings architecture in Kazakhstan the authors of the article had used literary sources, project materials, regulatory documents. Research methods are: comparative analysis of the sources, on-site assessment of the objects.

3 Analysis

Almaty is located at the foothill of Zailiyski Alatau mountains. Mountain gorges supply the air mass, ventilating the city and releasing it from smog. The location of Almaty in a semi-closed intermountain valley in the wind-shadow region is the reason for city's windless weather almost all year round, which contributes for build-up of the exhaust gasses in the form of smog in the city's air basin. In recent years the issue is getting more complicated by the move of construction zone up the mountain slopes obstructing the natural movement paths of mountain air into the valley towards the city.

Peculiarities of the urban aerodynamics related to katabatic drain of mountain air masses; limitedness of the city territories amplified by the high seismic activity have long determined the number of storeys in urban construction – not higher than 12 storeys.

First high-rise building in the city, the 25-storey hotel 'Kazakhstan', was erected in 1977 defying the opinion of the impossibility of city's vertical growth. Hotel's ellipsoid building had become the impressive dominant against the backdrop and had aptly complimented the city center's silhouette (Fig. 1). Vertical array of bow windows, giving the building its simplicity and delicacy, ends with high attic in the form of aureate crown made of anodized aluminium [16, 17]. When designing the hotel, factors were taken into account determining its shape and comfort of residency (direction of the winds, angle of sunlight incidence, etc.) Hotel was planned to accommodate 1,000 guests, included 497 one-, two- and three-room suites. Certain tightness of the rooms (according to the contemporary standards one-room suite area was 12 square meters) was due to predefined dimensions of the hotel's rigid structure. For de luxe suites cavities in the rigid structure were used [18].

Construction of the hotel in the 9-point seismic zone mandated use of then contemporary achievements of construction science and technology. To provide for hotel's earthquake-resistance the then innovative construction method was used - reinforced concrete monolith, made by climbing formwork technology which was actively used in Japan at the time. Main structural element is the oval-shaped rigidity core. Inside the core are the elevator shafts and engineering communications. From the oval structure the diaphragms are propagated, which are the suite walls [18].

The high-rise part of the hotel is installed onto a two-storey socle building embracing the lobby, savings bank branch, post office, café, restaurant and a conference-hall. The hotel was equipped with modern engineering facilities and high-speed passenger elevators. The lobby is made in the form of open space. Aluminium structures resembling a crown at the roof were fixed at the ceiling. By the registration desk the space theme panel was hung. The floor was laid out with black gabbro, labradorite.

Expensive construction materials were used in interior: marble, aluminium, synthetic materials. Walls finished with 'Sochi' plasterwork (two layers of marble chips of the same colour but different shades) created an elegant pattern. Walls on each floor had their own distinctive colouring. Space theme became primary in the interior design of the high-rise building's hotel beginning from entry lobby on the first floor to the 'Space' restaurant on the hotel's top floor at the mark of 86.1 meters, offering magnificent panorama of the Zailiyski Alatau mountain range.

The then modern construction materials were used in exterior finishing: aluminium panels, limestone. On the main and rear facades light bars were installed onto which aluminium sheets were hung. A layer of insulation material was laid between the bars. Bars shaped zigzag in terms of plasticity gave building the slenderness and lightness. Side facades were coated with light pink Mangyshlak limestone.

The surmounting 'crown' which became the distinctive feature of the hotel have raised a lot of debate [9], but time have confirmed the relevance and artistic value of that component. 'The crown' has appeared as part of the camouflage: the elevator shafts' structure was supposed to be two storey higher than the main building, so that two-storey structure appeared in the very centre of the rooftop. In order to disguise it a metal structure was introduced, 'the crown', which encircled the roof and repeated its facade shape. Each individual toothed element of the metal 'crown' was 10 meters high. At night the hotel's 'crown' lit up with ruby lights which could be seen from every observation point in the city [19]. Not long before the construction was over in 1977 seismic stability tests were held: up on the rooftop the machines were mounted that imitated vibrations, special sensors on the building floors detected the tremors. The building did not get any damage as a result of tests.

The hotel is situated in one of the important urban hubs, at the intersection of Dostyk and Abay avenues. Architectors regard the complex of buildings comprising this urban hub as an example of good solution of urban and composite tasks. The building is orientated along the meridian which provides for good aerodynamic conditions of the area. Searching for idea architectors pursued visual connection of the 'Kazakhstan' hotel's vertical with spread out volume of Palace named after V.I. Lenin (now named the Palace of the Republic) revealing the uniformity of the ensemble of Abay square based on the contrast of the forms. In order to stress out the method of visual interconnection facades' finishing included alike construction materials: light limestone along with finishing by anodized aluminium, that allowed uniformity and lightness to the massive structures.

Thanks to its distinctive look and unorthodox solution in facade finishing the 102-meter high building to these days remains one of the most well-known objects in Almaty and Kazakhstan. The authors (architectors Y.Ratushny, L.Ukhobotov, A.Anchugov, V.Kashtanov, engineers A.Deyev, N.Matviets, T.Zhunussov, A.Tatygulov) created an object that became not only a sight and a symbol of the city but also an example of a unique engineering solution. In 1980 authors were honoured with the State award of Kazakh SSR named after C.Valikhanov. 'Kazakhstan' hotel is an example of engineering art which was not overcome until 2008 when another skyscraper appeared in Almaty, the 'Esentai Tower'.

In recent decade economical conditions and engineering technologies made it possible to construct new high-rise buildings in Almaty, namely the 25-storey multifunctional business complex 'Almaty Towers' (100 meters high) and the 168 meter high 'Esentai Tower'.

Esentai' complex's architecture designed by world-renowned company Skidmore, Owings and Merrill (SOM) and implemented in Almaty in 2008 serves as a vivid example of the use of innovative technologies and accounting for the 'spirit of the place'. Project aimed at solving main issues of the construction in Almaty conditions - high seismic activity and low ventilation of city's air basin.

The location of Almaty in a semi-closed intermountain valley in the wind shadow region is the reason for having windless weather in the city almost all year round, which contributes to accumulation of the exhaust gasses in the form of smog in the city's air basin. In recent years the issue is getting more complicated by the move of construction zone up the mountain slopes obstructing the natural movement paths of mountain air into the valley towards the city.

Fig. 1. Hotel 'Kazakhstan', Almaty [22]

'Esentai' complex situated in the foothill region of Almaty solves problem of 'air shadow' of the city with use of architectural tools. SOM proposed use of spacial-dimensional solutions forming movement of the air masses. Complex comprises high-rise tower and several residence blocks. Primary element of the complex is the 'Esentai tower'. Lower 14 storeys are occupied by commercial offices, above them is the hotel Ritz-Carlton, upper eight storeys host serviced apartments [20].

Spacial-dimensional solution of the complex is associated with the gorge: narrow gaps between buildings create unique tunnels shaded or lit depending on the time of the day. Interaction of cooled and heated parts of the complex territory creates breezes which move in the space between blocks and form air flows. At the same time between the block gaps the mountain fragments appear and become the part of artistic perception of the complex.

In its project SOM combined progressive western engineering solutions and local natural conditions of Almaty. Composite solution of the complex parts is based on revealing the look of natural surroundings: building with its intriguing shape resembling rocks spreads out under the pressure of mountain breeze. Game of light on the refractive surfaces brings out strict plasticity of the facades. Lapidary shapes of the glass tower, fullness of the territory with sunlight, panoramic view at mountains create special perception of space environment of the complex (Fig. 2).

As in any high-rise building, here in the basis of the composite effect of the form lie innovative engineering-design solutions. Constructively 'Esentai tower' is made in the shape of dimensional system of the core, developed in terms of multisectional rigidity, and rigid frames located along the building's perimeter, that have high resistance against seismic loads. Specially for 'Esentai Park' complex Leslie E. Robertson Associates RLLP designed concrete structure with the consideration of probability of highly destructive earthquakes. Dampers were installed in the building allowing it to sway and quench the horizontal loads in case of seismic vibrations. Strict norms and requirements used in the project are significantly tighter than those used when constructing such high-rise buildings in seismically dangerous cities of the world like San Francisco, Los Angeles and Tokyo [21].

In 'Esentai Tower's' architecture the innovative glass was used which does not overheat the building and does not let harmful ultraviolet light through, is of high degree of transparency, made by means of modulating natural crystallized quartz from which the iron was extracted. Glass in the exterior of 'Esentai Tower' accomplished artistic tasks as well: for example, in the lower part of the western facade the glass is as if giving the glare of shimmering water.

Fig. 2. 'Esentai Tower', Almaty: a – South-West view from Al-Farabi avenue, b – North-West view, c – South-West view, the photos was taken in 2017 [author's source]

Today the highest buildings in Kazakhstan of 100+ meters in height are mostly concentrated in new capital, Astana (Fig.3). In harsh economic conditions in December, 1997 the move of capital from Almaty to Astana have set an enormous task before architects and constructors - to create a new city meeting the requirements of world standards, that shall become a well-recognizable brand on international arena. It was a method to make way to the world's top rapidly developing cities, win political and economical dividends for the country, a fragment of the mighty soviet empire.

The climate of Astana is sharply-continental, with cold winter (lower than -40ºC) and hot summer (up to +40ºC). Besides intense gusty winds and dramatic yearly temperature fluctuations constructing high-rise buildings in Astana faced yet another huge problem - elevated standing groundwater and aggressive soil. Theoretically, harsh climatic conditions of Astana did not encourage the birth of a new city. But past 20 years have proved rightness of the President of Kazakhstan N.A. Nazarbayev who had initiated the move of the capital.

In 1998 after the contest for the master-plan of Astana in which Japanese architect Kisho Kurokawa have won, the new capital began its development based on the metabolic model aimed at forming a symbiosis of natural and artificial environments, respecting history of the place, creating green zones inside the city and forest plantations around it [22].

In January, 2002 in order to create a good investment climate and attract money the free economic zone 'Astana - new city' was created, which led to increase in number of local and international companies that invested in building the new capital of Kazakhstan, modern residences, service, trade and amusement sites. Private investors have played a significant role in forming the distinctive architectural look of the new capital of Kazakhstan.

Designers and constructors of Astana had been solving a super-task of building a 21st century city based on achievements of a modern architectural and urban-planning ideas. To have the capital's architectural look match a model city of the third millennium the adopted decisions must have been made in the fairway of world's innovative trends. With this aim prominent modern architects have been invited to design the most significant objects to be erected in Astana. Construction of significant objects forming the look of main avenues and squares was carried out only according to projects that went through the competitive selection process. After the official approval of the new master plan of the capital in 2001 as a 'project representing the philosophy and theoretical postulates of urban planning of 21st century vigorous works have started to actively implement them, form new planning structure of the city and its architectural look' [23]. Main scope of work was taken to the left bank of Ishim river to form the New center, erect complex of Water-green boulevard buildings, Round plaza, government buildings of the Main square. In 2003 first 36-storey 155-meter high kazakhstani skyscraper, the building of Ministry of transport and communications, was constructed.

In 2006 main architectural ensembles of Astana were finalized (Water-green boulevard, Main square, etc.), construction have started of multifunctional high-rise residences 'Emeral block', 'Northern lights', 'Triumph of Astana' [23].

Height of some of the 'Emerald block' buildings is 200+ meters. Its concept was designed by the architect Roy Varacalli and the Zeidler Partnership Architects (Canada) together with Bazis development company (Kazakhstan). Complex consists of three buildings, 37, 43 and 54-storey high. Starting from 32nd floor area of each consecutive floor increases, building becomes asymmetrical with the rooftop deviating for 15 meters. Given strong wind loads of Astana, Canadian engineers conducted a significant amount of project testing in aerodynamic tube. The highest building of the complex weighing 186 thousand tons is sitting on bored piles at the depth of up to 20 meters and a foundation plate around 3 meters thick.

'Northern Lights' residence situated on Astana's Water-green boulevard includes three units (34, 39 and 44 storeys) and is up to 180 meters high.

Besides residences list of high-rise buildings of Astana includes commercial buildings, trade-amusement centers. Office of the National company 'Kazakhstanskiye zhelezniye dorogi' (Kazakhstan railways), the 'Temir Zholy' tower, a 44-storey 175-meter high skyscraper.

One of the iconic objects of Astana is the 150-meter high trade-amusement complex 'Khan-Shatyr' (architector Norman Foster) which was opened on July 6, 2010. The core of the building's framework is a unique system constituting a tripod - the support for the steel guards system's upper ring with ETFE fixed transparent coating, and lower ring fixating guards at the base of the building. Thanks to special chemical composition coating protects the inner space of the complex from dramatic temperature fluctuations and creates a comfortable micro-climate inside the complex.

Total area of the complex is 127,000 square meters. Here numerous trade and amusement complexes are located: supermarket, family part, café and restaurants, cinemas, gyms, aqua park with artificial beach and water pools with wave effect, administrative and commercial offices, 700 car parking lot and many more. Main landmark of 'Khan-Shatyr' is the beach with sea sand, tropical plants and +35 temperature all year round.

'Grand Alatau' complex built in 2009 in historical center of Astana consists of four high-rise residence buildings (20, 28, 38 and 43 storeys) connected by a common stylobate. The 43-storey tower is 144 meters high.

Fig. 3. Examples of high-rise buildings of Astana: a – 'Emerald block' residence complex, 2013, Arch. R.Varacali / Zeidler Partnership Architects; b – 'Temir Zholy' tower, 2009, Arch. T.Abilda [7]; c –'Transport Tower', Ministry of transport and communications of the Republic of Kazakhstan, 2013, Architicture – Bazis company; d – 'Northern Lights' residence complex, 2009, Arch. A.Saumenov, Y.Ezzauu, T.Abilda; e – 'Khan-Shatyr' trade-amusement center, 2010, Arch. Norman Foster; f – 'Grand Alatau' residence, 2008, Arch. S.Mataibekov; the photos was taken in 2017 [author's source]

First decade of the new capital witnessed an unprecedented amplitude and pace of construction works. To allow forming holistic space environment, interconnection of municipal improvement and construction, architects shifted from the 'spot' construction of local parcels to the complex construction of the whole block by single investor. Examples of such complex construction are the residences 'Highvill' and 'Grand Astana' constructed with social and cultural-resort related objects that are necessary for servicing the local residents.

During the years of Astana construction numerous foreign architects participated in designing unique sites of the capital of Kazakhstan: Palace of Peace and Reconciliation (arch. N.Foster), 'Kazakhstan' Central Concert Hall (arch. M.Nicoletti), trade-amusement complex 'Khan-Shatyr' (arch. N.Foster), multifunctional hotel-residence skyscraper complex 'Abu-Dhabi Plaza Astana' (arch. N.Foster), a 216-meter high hotel-commercial

building (arch. K.Kikutake); 280-meter high hotel-commercial complex (arch. R.Stern), etc. These super-modern buildings and structures of unique architecture implemented at world standard give unmistakable look to the new capital of Kazakhstan [23].
A lot of unique projects were carried out in Astana by the projects of kazakhstani architects: House of the Ministries, construction of the Main square and Round plaza, Water-green boulevard, multistorey residences, museums, commercial buildings, hotels and trade-amusement centers, universities, medical centers and many more [23]. Among them are: 142-meter high residence complex 'Triumph of Astana' (arch. A.Zuyev, N.Boriskin), 180-meter high complex 'Northern Lights' (arch. A.Saumenov, Y.Ezzauu, S.Mataibekov), 175-meter high tower 'Temir Zholy' (arch. T.Abilda), 144-meter high residence complex 'Grand Alatau' (arch. S.Mataibekov) and others [22].

4 Results

Analysis of modern high-rise architecture in Kazakhstan has shown their disposition only in two cities, Almaty and Astana.

High-rise construction in Almaty is complicated due to 9-point seismic activity of the foothill zone. Despite the experience of constructing and operating the hotel 'Kazakhstan' for 40 years has shown that special engineering solutions allow erecting skyscrapers in this city, huge economic expenditures are necessary to support sustainable operations.

An example of innovative approach is the 'Esentai' complex with 'Esentai Tower'. Taking into account the probability of a highly destructive earthquake company Leslie E. Robertson Associates RLLP during construction of the 168-meter high tower used new engineering solutions - multilayer rigidity core and stiff frameworks, placed along the building perimeter and possessing increased endurance against seismic loads; dampers allowing the building to sway in case of seismic vibrations and quench horizontal loads. Authors of 'Esentai Tower' project inserted the site into the local context having reflected in the tower's look mountain ranges, deep gorges, raging rivers [21].

In Astana more than 20 buildings of 100+ meters high are constructed. Astana skyscrapers not only fulfil their primary function (residence, offices, trade centers) but also demonstrate achievements of the country: in 20 years of building its new capital Kazakhstan has obtained significant political, economical and technological dividends. During that time Astana has become a world-wide recognizable brand symbolizing the dynamic development of Kazakhstan. Country's economy has attracted colossal foreign investments, the infrastructure development has been boosted country-wide, new jobs and production companies have been created. High level of technologies implemented by foreign companies in Astana provides for increase in quality of architectural and engineering solutions.

5 Discussion

In today's scientific literature in Kazakhstan the issues of high-rise buildings are considered mainly from the point of view of their engineering stability. Nevertheless skyscrapers are of interest not only in the aspect of innovative construction solutions but also as distinctive architectural objects that serve as prominent dominants in the space environment of the cities, optimize functional organization of valuable city areas.

There are a lot of opponents to building high-rises in Kazakhstan who support their opinion with complex seismic, hydrogeological and other conditions. Contemporary engineering technologies allow implementing the most audacious architectural ideas even in complex seismic conditions. During the construction of high-rise buildings advanced geotechnologies were implemented for arranging driven and bored piles in special ground conditions of Astana (Palace of Peace and Reconciliation, Khan-Shatyr, Ministry of transport and communication, etc.) [11]. All of the aforementioned allows forecasting expansion of the high-rises 'park' in the big cities of Kazakhstan.

6 Conclusion

The conducted analysis of the experience in design and construction of high-rise buildings in Kazakhstan revealed main requirements to formation of architecture of high-rise buildings:

- the architecture of high-rise buildings is impacted in a complex way made by methods that allow resistance against wind and seismic loads;
- in seismic conditions most widespread structural solution is monolithic framework consisting of the rigidity core, fortified foundation and rigidity diaphragms, as well as tools 'quenching' seismic tremors - moving, swivel or kinematic systems, etc.;
- formation of the architecture of high-rise buildings is based on creating spacial-dimensional solutions and use of ecological methods of construction to provide comfortable microclimate in their closest vicinity and city-wide: accounting climatic conditions of the landscape by orientating in the building structure, selecting forms and other plastic tools;
- functional content of the high-rise buildings includes following types: hotel, commercial building, trade-amusement complex, residence complex, etc.;
- overall trend is towards sophistication of services from multifunctional building to multifunctional object, that allows considering high-rises as a promising type of buildings introducing solutions to many urban planning, architectural-constructional and social-economical tasks in the given construction, in the areas of reconstruction with a deficit of unoccupied city space and on the territories of new construction;
- in the conditions of global competition the construction of high-rise buildings is a way of promoting the brand of construction companies, investors, implementing innovative technologies. Astana's case once again proves a well-known truth - high-rise construction is a reputational project that provides multiplicative effect at the city level and even country level.

Some of the cons to constructing high-rises in Kazakhstan are: absence of the comprehensible energy saving architectural-constructional and energy-effective engineering-technological solutions that provide notable reduce in their operational costs; low use of regional imagery in the architecture of high-rise buildings.

Considering all of the aforementioned, the article uncovers and formulates urban planning and space-planning issues of forming the architecture of high-rise buildings in Kazakhstan: construction taking into account climatic, ecological requirements; need for compliance with urban planning requirements, providing architectural-spacial uniformity of the building, developing functional zoning and conditions of visual perception, etc.

References

1. V. A. Kharitonov, *Design, construction and operation of high-rise buildings: Monograph*, (Moscow, ASV publishing house, 2014)
2. Specifics of design and construction. High-rise buildings and other distinctive structures of China. Academic edition by: P.A. Akimov, V.N. Sidorov, A.R. Tusnin Translated from Chinese, 808 (Moscow, ASV publishing house, 2013)

3. A. V. Dukart, A. I. Oleynik, Dynamic dampers of structure vibrations Edited by A.V. Dukart, 248 (Moscow, ASV publishing house, 2015)
4. G. E. Shablinski, Monitoring distinctive high-rise buildings and structures for dynamic and seismic effect *Scientific edition,* 328 (Moscow, ASV publishing house, 2013)
5. B. A. Glaudinov, M. G. Seydalin, A. S. Karpykov, Architecture of the Soviet Kazakhstan, 319 (Moscow Stroyizdat, 1987)
6. G. M. Kisamedin, Architecture of distinctive buildings, 212 (Almaty, Stroitelstvo i arkhitectura, 2012)
7. Meuser Ph. Kasachstan, 539 (Berlin: DOM publishers, 2015)
8. V. A. Khomyakov, *Research of strength and stability of soil arrays autoref. diss... doctor of engineering*, 37 (Almaty, 2010)
9. V. A. Khomyakov, *Researches of Properties of Soil at Underground Construction in the City of Almaty Korea-Kazakhstan Joint Geotechnical Seminar Aug 21 – 22*, 52-57 (Incheon, Republic of Korea, 2012)
10. V. A. Khomyakov, *Researches of deformational characteristics of loess soil during three-dimensional compression for use in calculations of buildings and structures on seismic loads Theses of 7th Kazakhstani-Chinese int. symposium 'Forecast of earthquakes, assessment of seismic danger and seismic risk of Central Asia',* 237-239 (Almaty, June 2-4, 2010)
11. Zhusupbekov A.Z. Research of bearing property of driven piles using dynamic probing // Bases, foundations and mechanics of soil. 2014. No. 2. - pp. 2932.
12. M. S. Abakanov, Low-cycle strength of reinforced concrete structures of framework buildings under loads like seismic, 132 (Almaty, KazNIISSA, 2016)
13. I. I. Vedyakov, D. V. Konin, P. D. Odesskiy, Steel frameworks of high-rise buildings: Scientific edition, 272 (Moscow, ASV publishing house, 2014)
14. Ph. Meuser, A. Dalbai, G. Herz Astana, 223 (Berlin, DOM publishers, 2015)
15. S. Kuc (CUT, Krakow, Poland), S. S. Sadykova (ENU Astana, Kazachstan), W.Kuc (AFM Krakow, Poland), *The Architecture of Astana as the Example of Contemporary City Creation, Proc. of 3nd Annual International Conference of Architecture and Civil Engineering (ACE-2015),* ISSN: 2301-394X, doi: 10.5176/2301-394X_ACE15.154, **2**, 683-690 (Singapore, 2015)
16. A. Galimzhanova, M. Glaudinova, The history of the arts of Kazakhstan, **2**, 192 (Almaty, Oner, 2011)
17. http://kazakhstanhotel.kz/ru/551-istorija.html http://www.voxpopuli.kz/history/2249-pervyy-neboskreb-alma-aty.html (last accessed 2017.04.25)
18. http://www.almatymadeniet.kz/ru/home/pamyatniki/mestnogo-znacheniya/arkhitekturnye/zdanie-vysotnoj-gostinitsy-kazakhstan (last accessed 2017.04.20)
19. The Registry of historical and cultural monuments of Almaty, Ed.in chief B.G. Ayagan, (Almaty, 'Kazak Entsiklopediyasi' LLP, 2006)
20. http://capitalpartners.ws/ru/index.php/snapshot/news/capital-partners-opened-esentai-tower (last accessed 2017.04.29)
21. http://files.gisap.eu/sites/files/digest/95.indd%20Internet.pdf (last accessed 2017.04.27)

22. Meuser Ph. Aesthetik der Leere: modern Architektur in Zentralasien, 151 (Berlin, Braun, 2002)
23. A. S. Tchikanayev, Astana: architectural symphony of the Great Steppe, 270 (Astana, Delovoi Mir, 2008)
24. The master plan of Astana. 10 years, Ed.in chief S.Zhunysov, B.Dosmagambetov, 272 (Astana, Delovoy Mir Astana, 2011)

Gardening as vector of a humanization of high-rise building

Nina Lekareva[1]* and *Anna* Zaslavskaya[1]

[1]Samara State Technical University, 443100, Molodogvardeyskaya St., 244, Samara, Russia

Abstract. Article is devoted to issue of integration of vertical gardening into structure of high-rise building in the conditions of the constrained town-planning situation. On the basis of the analysis of the existing experience of design and building of "biopositive" high-rise building ecological, town-planning, social and constructive advantages of the organization of gardens on roofs and vertical gardens are considered [1]. As the main mechanism of increase in investment appeal of high-rise building the principle of a humanization due to gardening of high-rise building taking into account requirements of ecology, energy efficiency of buildings and improvement of quality of construction with minimization of expenses and maximizing comfort moves forward. The National Standards of Green construction designed to adapt the international requirements of architecture and construction of the energy efficient, eco-friendly and comfortable building or a complex to local conditions are considered [2,3].

1 Introduction

The issue of the most effective introduction of gardening to our cities doesn't lose the relevance for more than 100 years. Projects and Howard and Le Corbusier's theories considering a city conglomerate as the developing organism which is separately taken complex, unfortunately, are almost not applicable to already developed infrastructures of the modern cities. How to be with dense multystoried building of such megalopolises as Shanghai, Tokyo, New York, Kuala Lumpur, Moscow, London? It is almost impossible to find free areas for high-quality and effective gardening in such cities. Here theoretical researches of Le Corbusier find new value: "gardens on roofs" extend to facades, "green fingers" enter interiors of skyscrapers, turning them into dynamically changing architectural organisms with the energy saving and the developing resources characteristics.

2 Materials and methods

The world community recognized efficiency of roof gardens for a long time. So for example, in Germany one of indispensable conditions at design of buildings – gardening of the roof including having a considerable slope. In Switzerland allocated under lawns 25%

* Corresponding author: nalec1950@bk.ru

of flat roofs while in Japan owners of the real estate are obliged to lay out gardens on flat roofs more than 100 sq.m .

Today technologies have allowed to go further to the fields of gardening of architectural objects and to transfer a landscape from horizontal to the vertical plane [4,5]. Peter Blanca who has presented to the world Vertical Gardens technology in the project of gardening of walls of the Museum of Modern Art in Paris on a wall with a total area of 800 sq.m has placed more than 170 types and 15000 plants. Ecological efficiency of similar projects strikes. For example, application as dressing of a facade, vertically focused "green" modules of Emilio Llobato has allowed to disinfect 2,6 million air cubic meters a year thanks to what 400 thousand people in 200 buildings could breathe fresh air.

Example of use of natural gardening in high-rise building are so called "green" high-rise buildings of Ken Yeng, the creator of bioclimatic approach in design of high-rise buildings (figure 1). According to Yeng: "The building must be integrated in environment in physical, system and time aspect. I mean integration into physical characteristics of the place: topography, underground waters, hydrology and vegetation. System integration is rational use of water systems, energy, waste, the sewerage and so on. Time integration is a ratio of use of resources and construction materials with their ability to be restored or be replenished" [6].

a b

Fig.1: Bioclimatic approach in creation of contemporary buildings: a - EDITT office complex in Singapore, arh. Ken Young [2], b - Multi-storey Garden in Osaka. Japan [4].

3 Results

The organization of gardens on roofs and vertical gardens in high-rise buildings most often is connected with rise in price of construction which is perceived by the customer as excessive expenses. However these expenses are often justified not only for esthetic or ecological reasons. There are several exclusively practical aspects which can often convince the customer. And it is first of all the design features keeping designs of a roof and walls from temperature changes [7,8].

First, "green roofs" with respect to technology are often much more durable than usual due to multilayered modern materials "pie" with a covering from plants serves as the best waterproofing and thermal insulation for the rooms located under him. Usual roofs are getting dry, become wet, roof slabs experience contrast change of heating and cooling and in such conditions are exposed to far bigger corrosion stress, than the roof which is constantly covered with a lawn blanket. In gardens on roofs the amount of soil is minimized, because soil – the heaviest part of a garden giving big overloads on roof slabs design. Therefore the first step when developing a roof garden is determination of the maximum loads which the roof design can sustain. According to minimum settlement load of 500 kg of sq.m (the soil, a drainage, snow) the scheme of the device of "a green roof" can be the following:

- a waterproofing with root protection layer.
- the drainage plates representing plates of the perforated polystyrene passing moisture.
- a filter-bed for prevention of a contamination of a drainage particles of the vegetable soil. Geotextiles.
- a soil layer with vegetation: qualitative and light soil.

Ecological advantages of gardening of roofs and the inclusion of gardens in structure of skyscrapers whith concern to increase in microclimatic and sanitary and hygienic indicators of building aren't less important [9,10].

Plants on roofs promote dust content reduction. In hot days air streams with dust are caught in crowns and the bushes having a rough surface of leaves. Even in a dry state small strips of a lawn with a bush catch up to 50% of dust from the air streams passing over them, and when watering and moistening plants the effect is even higher. The vegetation layer, besides, reduces also harmful electromagnetic radiations, protecting roof designs from ultraviolet rays. Researches proved also considerable noise-protective effect of the planted trees and shrubs roofs which is a little decreasing, but not eliminated in the winter. Sound waves are absorbed by soft and rough plant material, and it is especially essential to the top floors of buildings

The advantages of the social plan, including esthetic comfort, come down to obtaining the additional area of a recreation and communication. The flat planted trees and shrubs roof or a green terrace besides that is pleasing to the eye of those who look at them from higher floors, become useful area and can be used depending on gardening type for the most different purposes. Besides it is a possibility of disclosure of panoramas on the environment, increase in esthetic indicators of the interior spaces opened on the planted trees and green spaces [11].

Savings of building area due to the lack of territorial reserves comes to the forefront now. In this regard economic and town-planning advantages of the roof gardens and vertical gardens which aren't demanding additional urban areas are available. Being the powerful compensator in the field of an environment humanization in the conditions of close building, vertical gardens and gardens on roofs make the building more attractive to investors.

4 Discussion

The humanization of high-rise building due to ecocentric approach to formation of its structure regarding her gardening and the organization of additional recreational spaces increases its investment appeal. Considering the large volume of input of the new areas and reduction of free platforms under construction, the investor needs to take all factors influencing final project cost into consideration: requirements of ecology, energy efficiency of buildings and improvement of quality of construction at minimization of expenses and maximizing comfort.

The huge areas of roofs industrial, inhabited and public the building, underground constructions represent an irreplaceable reserve of the urban area. They can become some kind of artificial basis for gardens, boulevards, squares and other objects of a landscape city architecture. Protecting designs of a roof of buildings from damages, and having raised as they say, their wear resistance, green plantings absorb moisture, reduce loading to the stormwater drainage system and as a result it will be able to prevent floods.

Absolutely not incidentally program point of new architecture" the world famous architect Le Corbusier who, like many outstanding architects of the present, has repeatedly confirmed this principle in the creativity considered use of roofs., having declared that: "Really it contradicts any logic when the area equal to the whole city isn't used, and slate needs to admire stars!"

For formation of a complex of requirements and an assessment of design of the steady, ecologically healthy and comfortable environment for accommodation of the person introduction of voluntary certification of real estate objects - Green Standards are offered. Green Standards are designed to establish a level of "complex efficiency of all life cycle of the artificial habitat of the person integrated in natural". The objects certified according to "The green standard" provide the minimum environmental pollution and high level of ecological safety for people. At the same time they increase investment appeal of the territory of their dislocation[12]. The term "green building" was strongly included into professional terminology.

Use of "green" technologies in construction and operation of high-rise buildings bears unconditional economic benefits as energy consumption level decreases. reduction of costs of the electric power is reached, and also water consumption decreases. Use of environmentally friendly materials and high level of comfort of green buildings promotes preservation of health to their inhabitants. Besides tax benefits and grants are presented to the buildings which have undergone certification on Green standards that promotes increase in their investment appeal and forms public opinion.

As a result of modeling with use of the innovative methods directed to decrease in dependence of consumers on energy resources the architectural concept of an object with system integration – rational and autonomous use of water systems, energy, heat, waste, the sewerage and so on with necessary reduction of scales of environmental pollution can be created. Similar experience of design with use of methods of replacement of traditional sources of heat has been applied in the project of the Energy efficient building "Ekody Solar-5".

Increase in attention and interest in creation of artificial ecosystems which could imitate properties, processes and the device of ecological systems in the nature, including creation of autonomous energy efficient buildings with inclusion of gardens in its structure is steady tendencies in the field of green construction [13].

Gardening of "the fifth facade" or construction of gardens on roofs belongs to the hi-tech decisions increasing the level of comfort of high-rise buildings, durability of their designs and answering to a number of ecological requirements of Green Standards. Construction of gardens on roofs is for the present poorly widespread in our country though in the world there is a vast experience of such construction. Our severe climate brings the restrictions and narrows possibilities of gardening of a roof. Besides because of a set of agrotechnical and engineering devices it is quite expensive. However, these expenses are often justified not only for esthetic or ecological reasons, but also from the point of view of a practicality.

Green Standards assume rather high level of efficiency of design and construction which achievement is very complicated. However the possibility of creation of energy saving, ecologically effective buildings in the cities is absolutely real irrespective of climatic conditions [14,15].

Thus, investment appeal of urban areas and the architectural objects located on them can be considered as a complex multicriteria and multiple-factor challenge which solution will demand step by step repetition of calculations of urban planning and economical value of the city earth on the basis of the updated data of town-planning monitoring. It is proved by the fact that eventually requirements to the factors entering a method of calculation of economical value of urban areas to ponderability of each of them in the general assessment change, and also that standards of norms, and first of all the ecological, focused on improvement of qualityof lives of citizens change [16,17].

5 Conclusions

Creation of the native habitat in the conditions of high-rise construction requires complex gardening in and outside of the building which becomes possible due to use of the planted trees and shrubs atriums and the yards, green floors, green walls, green roofs, artificially created natural landscapes and other landscape elements uniting the main directions and ways of gardening of high-rise buildings [18].

The terrace solution of the gardens included in skyscrapers is the most widespread and functionally justified way of gardening of high-rise buildings of terrace type. Terraces carry out a role of open recreational space.

Gardens - "whatnots" are a way of deeper integration of gardening into structure of the high-rise building. Development of this direction gives the chance of the construction of "through gardens" on certain floors of the high-rise building and additional spaces for communication and a recreation of its inhabitants.

a

b

Fig. 2. Types of city gardening: a - Roof Garden - MOMA museum in New York [10], b - Vertical Garden - L'Oasis D'Aboukir. Paris [12].

Gardening of roofs – perspective and now actively mastered way of gardening of high-rise buildings. Among a set of examples of the planted trees and shrubs roofs, it is possible to note the modern project on a roof of the Museum of Modern Art in New York created by the landscape architect Ken Smith. The new roof garden of the building belonging to the Museum of Modern Art is an innovative construction which successfully supplements green space of the territory of the Museum with a garden of sculptures. It is available to visitors and is well looked through from surrounding high-rise buildings of Manhattan (figure 2, a).

Vertical gardening. The live green carpet promotes increase in microclimatic indicators of building and according to increase in comfort of the building, reducing thermal radiation, increasing relative humidity of air that improves climate feeling of the person. Vertical gardening plays also decorative and esthetic role as allows to overcome monotony of urban development substantially. Application of the way "green wall" can turn the building into the present work of art (figure 2, b).

The principle **of modular gardening** of high-rise buildings is the most economic and universal way of gardening as it can be applied in various climatic conditions and various constructive systems of the building. Use of ready modules allows to save means and time when gardening buildings [19].

In conclusion, there is a wish to note the general world tendency to a humanization of the architectural environment of the large megalopolises striving for eco-friendly technologies, "clean" materials and correct design methods, delicate in relation to nature. For already developed multy-storied architectural complexes the technology of vertical gardening of facades as it is almost autonomous seems to be the most perspective, and the constructive decision can be realized with the minimum transformation of the existing planning of buildings [20]. At the same time the effect of similar decisions is maximum when it is used for a "green" surface, turned vertically, it is perceived by pedestrians and inhabitants of the next high-rise buildings entirely and practically without distortions, unlike horizontal surfaces of the operated roofs. however, it wouldn't be desirable to exclude the principles of creation of many-tier gardens and gardening of roofs from the general scenario of return of "human face" to the urbanistic centers. Only an integrated approach in a humanization of the urban environment, taking into account ecological, esthetic and technological aspects in the choice of constructive and composite decisions is capable to give "the second life" to densely built up multilevel megalopolises [21, 22].

References

1. V. A. Nefedov, City landscaping, 318 (Saint- Petersburg, 2012)
2. http://greenroofing.ru/novosti/2014 (last accessed 2017.06.26)
3. N. A. Lekareva, Territorial and spatial resource of the city, *Volga scientific magazine*, **3(31)**, 107-110 (2014)
4. N. A. Lekareva, Landscape architecture and design, *Unity and variety*, 247 (Samara, 2011)
5. N. A. Lekareva, Green standards in training architectural town-planning to design. In the collection: Questions of science and education: theoretical and methodical aspects the collection of scientific works on Proc. of Int. Sci. Conf.: in 11 parts, 94-95 (2014)
6. Richards Ivor. Groundscrapers + subscrapers of Hamzah & Yeang. - John Willey & Sons Ltd., 84 (Great Britain, 2001)
7. V. P. Generalov, E. M. Generalova, Prospects of development of typology of high-rise buildings, *Future of the cities*, **4** (SGASU Vestnik, 2014)
8. E. M. Generalova, V. P. Generalov, A. A. Kuznetsova, Modular Buildings in Modern Construction, *Procedia Engineering*
9. A. Y. Zaslavskaya, E. M. Zaslavskiy Natural analogies in a design of contemporary architectural objects, *Town planning and architecture*, **2(19)**, 19-23 (2015)
10. A. Y. Zaslavskaya, Evolution of organic architecture, *Architecton: news of higher education institutions*, **19**, 3 (2007)

11. A. A. Romanov, S. V. Evdokimov , A. A. Orlova, Results of Survey of Reinforced Concrete Structures of Drainage Gallery in Earh-fill Dam, *Procedia engeneering*
12. www.verticalgardenpatrickblanc.com (last accessed 2017.06.23)
13. Berrizbeitia Anita and Pollak Linda, Inside Outside, 189 (Rockport Publishers, Inc. USA, 1999)
14. D. V. Zelentsov, K. L. Chertes, O. V. Tupicina, Theoretical Basis and Experimental Study of the Aeration Characteristics of the Composting Mixtures during the Design and Construction of the Aeration System of the Oily Waste Biodegradation Complex. *Procedia Engineering* **153**, 903-908
15. Advanced architectural studio 6. Merging Landscape and Building Abstract, *Columbia Architecture Planning Preservation*, 62 – 65 (New York, 2001)
16. Architecture and Landscape *the Design Experiment of the Great European Gardens and Landscapes,* 410 (THOTH publishers, Bussum, 1996)
17. M. Simonova, N. Shekhova, S. Kolesnikov, Strategic directions of personnel potential forming of a building complex, *MATEC Web of Conferences* 86, 05008
18. Betsky Aaron, Landscrapers, *Building with the land*, 387 (Thames & Hudson, London, 2002)
19. Jodidio Philip, *Building a new millennium*, 559 (TASCHEN, Rotterdam, 1999)
20. Landscape (9+1 young dutch landscape architects) *essays by Henk van Blerck and Jorg Dettmar,* 137 (NAI Publishers, Rotterdam, 1999)
21. A. O. Mateyko, History of development and current trends in high-rise construction, **3** (Samara, SGASU Vestnik, 2016)
22. The organic Approach to Architecture *edited by Deborah Gans, Zehra Kuz,* 190 (willey Academy, 2003)

High-rise buildings in the structure of an urbanized landscape and their influence on the spatial composition and image of the city

Valery K. Moor[1,*], Elena A. Erysheva[1]

[1]Far Eastern Federal University, 690950, Vladivostok, Russia

Abstract. High-rise buildings play an important role in the spatial structure, composition and image of the city, as they are its spatial dominants and landmarks. The conditions of high-rise buildings perception in an urbanized landscape are changing in depending on the landscape-morphological conditions (the relief, the existence of the water area and the character of the coastline). Regularities of high-rise buildings' spatial organization in various natural-landscape situations are considered in the research; the basic principles of high-rise elements inclusion in structure of the urbanized landscape providing their harmonious interaction are formulated.

1 Introduction

High-rise buildings have significant role in the spatial structure, composition and image of the city because of their expression in the surrounding buildings. They represent spatial landmarks and often are original symbols of the urbanized landscape in the process of perception and comprehension of the city. In case if the natural landscape does not have a clearly defined structure, it does not affect the high-rise structure of the city, which is formed on the basis of other considerations (functional, constructive, economic, etc.). However, if the original natural landscape has a well-defined geomorphologic structure, it inevitably affects the morphology of the building, and the problem of their mutual coordination arises. In this case the natural landscape has a certain effect on the structure and image of the city, sets some "rules of behavior" that must be taken into account by the urban architect if he strives for harmony of the natural landscape and buildings. The aim of the research is to formulate the principles of including high-altitude elements in the structure of an urbanized landscape with a clearly defined relief, ensuring their harmonious interaction.

The problem of the relationship between the natural landscape and buildings is considered in a series of works [1-4]. Basic concepts are formulated and ways of urban development of the natural landscape are considered in these works. The influence of the natural landscape on the structure and image of the city, including its vertical composition is considered in other works [5-8]. Recently high-rise construction is increasingly being

* Corresponding author: moorv@rambler.ru

considered in the context of sustainable development of the city's cultural landscape. In particular this problem is reflected in the works of a number of Russian and foreign authors [9-13]. The ongoing researches provide a rather wide field of modern ideas about the strategy, methodology and principles of sustainable development of the modern city's urban landscape. Despite the large amount of works in this direction, the problems of the relationship between high-rise construction and a sustainable, harmonious development have not been adequately studied. Base on this, it seems important to systematize this problems and formulate the basic principles of including high-rise elements in the structure of the urbanized landscape, ensuring their harmonious interaction, and hence, sustainable development.

2 Methods

The structural-morphological and architectural-composite analysis of urbanized landscapes of cities located in the water area and the clearly defined relief are used as the main research methods. The methods of urban planning analysis of the urbanized landscape were considered in the works of a number of authors [14-16]. The analysis and generalizing of these works, as well as carried out theoretical studies and experimental work, allowed to formulate the author's methods of analysis of the urbanized landscape, to reveal the basic principles of high-rise elements including in the structure of the urbanized landscape with clearly defined relief, ensuring their harmonious interaction [17, 18].

Let's consider in more detail the relationship of the natural landscape and buildings, primarily its vertically expressed elements (buildings and structures). High-rise buildings can play a different role in the spatial structure and architectural image of the city depending on the initial natural landscape situation. The main natural factors determining the role of high-rise buildings in the spatial composition of the city are as follows:

- the nature of the geomorphologic structure of the natural landscape;
- the existence of the water area and the nature of the coastline.

As for the nature of the geomorphologic structure of the natural landscape, it is largely determined by the relief. There are a large number of natural landscapes classifications. However it is sufficient to distinguish three different relief situations for purposes of subsequent analysis:

- the city is located on a flat or slightly defined relief;
- the city is located on medium and strongly defined, but undivided relief;
- the city is located on medium and strongly defined relief, which has a large degree of horizontal and vertical dissection (hilly relief).

These factors determine the conditions for the perception of high-rise buildings in the structure of the urban landscape in addition to directly influencing the shape and location of buildings. In some cases they make possible only a near and angular perception of high-rise buildings; in other cases they make conditions for the long-term perception of high-rise buildings in the structure of panoramic views or urban facades. Consider these three relief situations, from the point of view of the role and importance of high-rise buildings in the composition structure of the urbanized landscape, as well as the impact on the formation of the architectural and artistic image of the city.

The coastline is the most important composite element for the first relief situation. The nature of its irregularity determines the spatial structure and composition of the city, especially the part of the building that directly is located near the sea. The sea panorama of the city in this case is one-tier, the silhouette of the city is determined by the silhouette of the building. The influence of the coastline is particularly strong in cities located on the banks of deep-seated coves and bays, as well as in their location on peninsulas and islands.

As the distance from the sea is increasing, the influence of the coastline on the structure and composition of the city is weakening (figure 1).

Fig. 1. High-rise buildings in conditions of unexpressed relief by the example of Singapore (photo by V.K. Moor).

The character of the water area, the outline of the coastline in this relief situation determine the shape, size and location of high altitude dominants in relation to the shore: they are usually located at key points, on the characteristic protrusions or bends of the shore, which allows them to occupy the most expressive positions in the spatial composition of the panorama. For example, in Venice, due to the lagoon's peculiarity, the dominant accents of the city - the tower of San Marco, the churches of San Giorgio Maggiore and Santa Maria Della Salute form a spatial triangle, the unchanging "frame" of the panorama as the points of perception of the city change.

Actively defined relief (the second relief situation) is an essential architectural component of the volumetric-spatial structure of the city. Along with the coastline it serves as the most important compositional precondition for the formation of the specificity of the spatial structure and artistic image of the seaside city. The spatial structure of a complex, active relief can be expressed by a system of axes, boundaries, centers, etc. They create a kind of "force fields" that penetrate the territory and determine the composition and spatial structure of the city. In turn, the nature of the coastline has a similar effect on the spatial organization of the city. In the context of complex relief, the value of the coastal strip increases. It captures not only the front of the building closest to the water area, but spreads through a visual interrelation much further (figure 2).

Fig. 2. High-rise buildings in a complex terrain based on the example of Hong Kong (photo by V.K. Moor).

Clearly defined relief creates conditions for a wide survey of urban formations, both from the sea side and (in case of a sloping and rugged coastline) from within the urban space. In this situation the city is perceived multilevel, and its silhouette is determined by imposing a silhouette of the building on the silhouette of the natural landscape. Architectural dominants play the role of landmarks of a considerably larger range of action than in plain conditions. They often represent a coherent system of spatial orientation of the city.

The panorama of such city is multilevel and multifaceted and quite dynamic when perceived in movement. The coastline forms a clearly defined border of the city, forming the first plan of the panorama, predetermines the direction of development of the urban composition. We can identify certain spatial "layers" of the city, differing in the level of communication with the sea (direct communication, visual communication), which play a different role in the panorama's formation. Especially acute and emotional it happens when approaching the city from the sea, when the foreground, "floating into the frame," constantly changes panoramic pictures.

The third relief situation - an actively expressed and complexly dissected relief - largely retains the space-organizing qualities described above. At the same time, a number of new qualities and compositional features appear here. A complexly dissected relief, such as a hilly terrain with significant elevations, is characterized by the polycentricity of the "force field", the complex imposition of composite form-building forces. In addition, in this situation, we are faced with slopes of different orientations to the sides of the world, as well as in relation to the sea. In each such case, there are regularities and methods for the formation of urban development. Complexly dissected relief is characterized by a large number of visual perspectives and interrelations between the structural elements of the city. The variety of species, sometimes unexpected, makes the perception dynamic and emotionally saturated (figure 3).

Fig. 3. High-rise buildings in difficultly dissected hilly terrain on the example of San Francisco, USA (photo by Vishnu Vardhan)

Of course, the formation of the city planning structure takes place under the influence of a number of factors, and in reality there are situations when the possibility and necessity of a visual relationship with the sea is "crossed out" by other circumstances (for example, the orientation of the slope in the sides of the world, the unfavorable effects of prevailing winds, Etc.). However, practice shows that the visual relationship with the sea is one of the most important conditions for the formation of the seaside city, its artistic image. Therefore, there are resourses, methods and forms of resolving this contradiction in the direction of preserving such a relationship even in the presence of unfavorable factors.

The examples considered above characterize the simplest forms of the relationship between the natural landscape and the planning structure of the seaside city. In reality, more complex situations arise, as, for example, in the Far Eastern seaside cities, Vladivostok is a representative of which. First, here the natural landscape is a difficult dissected, hilly terrain; different parts of this territory have different slopes, different orientations in the sides of the world and differently correlate with the sea. Secondly, the natural landscape is characterized by a complex coastline, deep bays and protruding peninsulas, as well as the island part. Consequently, the spatial and visual connections of the city and the sea are more complex here (figure 4). It is necessary to keep certain principles in order to ensure harmony of the natural landscape and high-rise buildings. Those principles are formulated below according to the results of the research.

Fig. 4. The urban landscape of Vladivostok (photo by V.K. Moor).

As the analysis shows, the natural landscape of Vladivostok has a complicated polycentric character. In this regard, the planning structure of the buildings and its vertical composition should be coordinated with the natural landscape in order to be in harmony with it. High-rise buildings will form complex harmonious compositions organizing and identifying numerous hills and creating the basis for the sustainable development of the cultural landscape.

3 Results

Researches and experimental projects were conducted at the Department of Architecture and Urban Design of the Far Eastern Federal University (FEFU) for a long time, the purpose of which was the development of methodology, strategy and principles for improving and humanizing the existing environment of the Far Eastern seaside cities. The researches and experimental projects allowed formulating some general principles of renovation of the urban environment, including those relating to high-rise buildings [19-21].

The basic principles of including high-rise elements in the structure of an urbanized landscape with a clearly defined relief, ensuring their harmonious interaction, can be characterized as follows:

1. Principle of structural coordination of the natural landscape morphology and vertical buildings elements (high-rise buildings and structures). In accordance with this principle, high-rise buildings, due to their compositional significance, should be located on characteristic areas of the terrain, emphasizing and revealing the structure of the landscape. In reality, this requirement is often not observed, hence one of the reasons for the disharmony of buildings and landscape.

2. The principle of modular coordination of the natural landscape morphology and the geometric shape of high-rise building elements. The lack of modular coordination of the landscape and high-rise buildings leads to their disharmony, and we also often observe this shortcoming in practice.

3. The principle of differentiation of town-planning methods of construction and the plastic structure of high-rise buildings, depending on the development of certain tiers of the natural landscape. This means that in order to preserve the aesthetically valuable elements

of the natural landscape, it is necessary to apply building methods specific for each tier of the relief, including high-rise buildings.

4. The principle of grouping high-rise elements into an integral composite system that provides artistic unity and expressiveness of the high-rise structure and silhouette of the urbanized landscape. As practice shows, solitary high-rise buildings often do not create expressed rhythmic structures in space and, thus, do not form an artistically expressive composition. It is no accident therefore high-rise buildings are very often grouped into clusters.

5. The principle of active semantic interpretation of high-rise elements as the most significant elements of the urbanized landscape. High-rise buildings must concentrate in themselves figuratively-symbolic content, be original "tower-symbols of the place" to become dominant and spatial reference points in the cultural landscape of the city.

4 Discussions

The above principles of including high-rise elements in the structure of the urbanized landscape have been tested in a series of experimental architectural and town-planning projects carried out at the Department of Architecture and Urban Planning of the FEFU in the framework of the graduation project. Project proposals have been elaborated for the development and improvement of the spatial structure and architectural image of Vladivostok In these works. One of the main tasks in these projects was the improvement of the high-rise structure and silhouette of the city on the basis of the principles formulated above, ensuring a harmonious interaction of high-rise buildings and natural context.

The problem is that the natural landscape of Vladivostok has a complex polycentric character. In this regard, the planning structure of the buildings and its vertical composition should be coordinated with the natural landscape in order to be in harmony with it on the base of certain principles and methodology. The hypothesis was that the formulated principles would solve this design problem. And, thus, the applying of these principles will improve the high-rise composition of the city, creating a basis for the sustainable development of the cultural landscape.

Let us consider several examples of these projects. A new proposal for the formation of spatial structure of the southern coast of the Golden Horn Bay in Vladivostok, which has been intensively developing recently, offered in the project by O.V. Zvereva. To this end, the author conducts a complex analysis of the natural landscape, reveals its compositional structure, planning and volume module, the main compositional axes and characteristic points of perception. The functional-planning structure of this part of Vladivostok was also analyzed. A detailed study of the initial environment is a prerequisite for the sustainable development of the cultural landscape. Based on this analysis, a more effective solution to the spatial structure and architectural image of the Golden Horn Bay's southern slope is proposed. The project also developed a specific architectural and spatial solution of one of the city modules, including a principal solution to its vertical structure (figure 5a).

A new public and business center on Russkiy Island in Vladivostok was developed in the project by S.F. Valeev. The territory of the social and business center has a unique natural landscape; this place plays an active role in the formation of the maritime facade of Vladivostok. The project was preceded by a multifactor analysis of the natural landscape, as well as a comparative analysis of analogues. The proposed planning solution emphasizes the natural peculiarity of the site; it fits organically into the natural landscape. At the heart of the spatial solution is the idea of a developed town-planning structure, solved in the form of several terraces placed on the natural tiers of the landscape. The structure of the town-planning complex includes three spatial dominants, each of which is a group of high-rise buildings, original "crystals" of development. The complex consists of the following functional zones: business, residential, entertainment, park, cultural. Detailed pre-project analysis, rather organic interaction with the natural landscape, an interesting interpretation of the spatial composition, silhouette and panorama of the public-business complex are the advantages of the project (figure 5b).

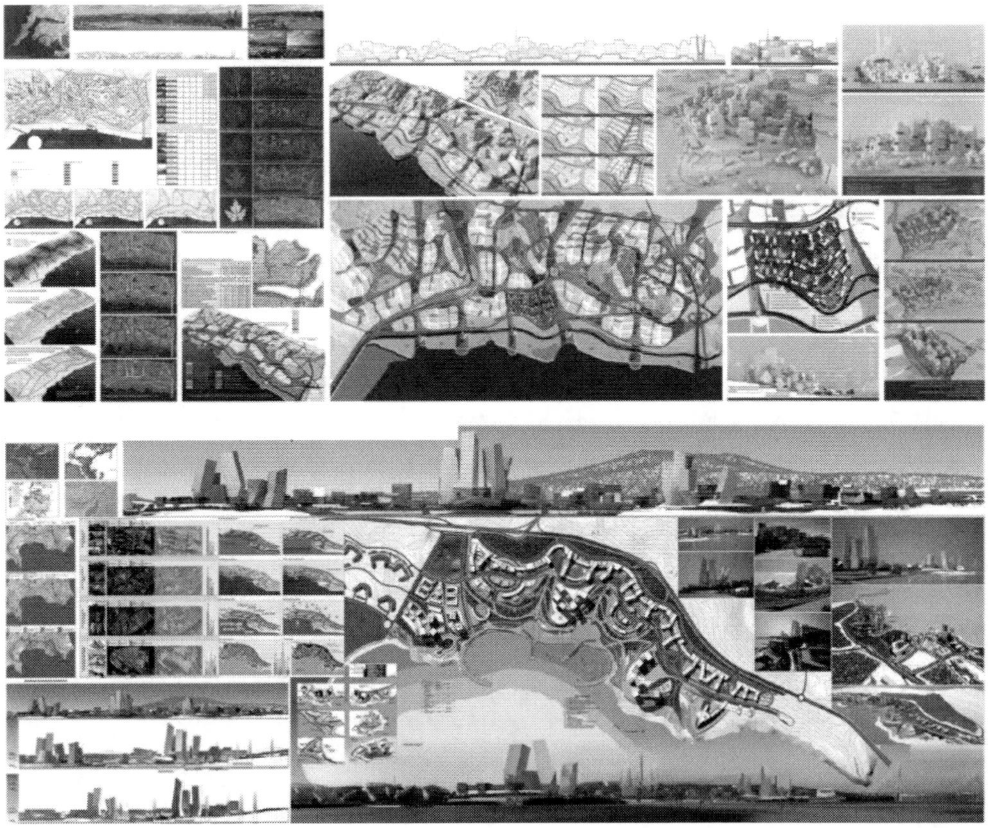

Fig. 5. Examples of Graduation architectural projects:
a - "Formation of the spatial composition of Golden Horn Bay's southern coast, Vladivostok" (Author O.V. Zvereva, tutors Professor V.K. Moor, Professor V.I. Smotrikovskiy, Associate Professor A.G. Gavrilov); b - "New public and business center on Russkiy Island in Vladivostok (Author S.F. Valeev, tutors Professor V.K. Moor, Associate Professor A.G. Gavrilov)

In addition to the above works on the FEFU Department of Architecture and Urban Design, a large number of similar architectural projects were carried out, in which the problem of high-rise buildings in a complex relief was solved. The approbation showed that taking into account the basic principles of including high-rise elements in the structure of the urbanized landscape in a series of experimental projects allowed ensuring a harmonious interaction of high-rise buildings and the natural context. The projects were highly appreciated at the state attestation; they were repeatedly exhibited at different architectural competitions and were awarded professional awards and prizes.

5 Conclusions

High-rise buildings play an important role in the spatial structure and image of the city, as they are significant elements of the cultural landscape. Sustainable development of the city's cultural landscape implies adherence to certain principles for including high-rise buildings in urban context. These principles depend on the structure of the natural landscape (the relief, the existence of water and the character of the shoreline), as they determine the conditions for the perception of high-rise buildings in an urbanized landscape. Analysis of regularities in the spatial organization of high-rise buildings in three different natural landscape situations has revealed specific techniques and principles for including high-rise elements in the structure of the urbanized landscape that ensure their harmonious interaction. The basic principles of including high-rise elements in the structure of an urbanized landscape with an actively expressed hilly terrain: structural and modular coordination of the morphology of the natural landscape and high-rise buildings; tier differentiation of town planning techniques for high-rise buildings; grouping of high-rise elements in an integrated composite system; active semantic interpretation of high-rise elements as the most significant elements of an urbanized landscape.

References

1. T. G. Maklakova, Visotnyie zdaniya. *Gradostroitelnyie i arhitekturno-konstruktivnyie problemyi proektirovaniya* (Moskva, ASV, 2008)
2. Yu. V. Razumovskiy, L. M. Fursova, V. S. Teodoronskiy, *Landshaftnoe proektirovanie*, (Moskva, Forum, 2016)
3. S.I. Palienko, Arhitekturnyie zadachi vertikalnoy planirovki territoriy so slozhnyim relefom, (Vladivostok, DVGTU, 2007)
4. A. V. Gorodkov, *Arhitektura, proektirovanie i organizatsiya kulturnyih landshaftov* (Sankt-Peterburg, Prospekt Nauki, 2013)
5. V.P. Generalov, *Osobennosti proektirovaniya vyisotnyih zdaniy* (Samara, SamGASU, 2009)
6. Proektirovanie sovremennyih vyisotnyih zdaniy, Pod red. Syuy Peyfu (Moskva, ASV,2008)
7. A. A. Magay, Arhitekturnoe proektirovanie vyisotnyih zdaniy i kompleksov (Moskva, ASV, 2015)
8. Osobennosti proektirovaniya i vozvedeniya, Vyisotnyie zdaniya i drugie unikalnyie sooruzheniya Kitaya (Moskva, ASV, 2013)
9. Y. Jabareen, Sustainable urban forms: Their typologies, models, and concepts, 26, 38–52 (J. Plan. Educ. Res. 2006)
10. O. A. Lavrenova, Prostranstva i smyislyi: Semantika kulturnogo landshafta (Moskva, Institut Naslediya, 2010)
11. UNESCO. Recommendation on the Historic Urban Landscape (UNESCO World Heritage Centre, Paris, France, 2011)
12. T. A. Smolitskaya, T. O. Korol, E. I. Golubeva, Gorodskoy kulturnyiy landshaft: traditsii i sovremennyie tendentsii razvitiya (Moskva, Liberkom, 2014)
13. V. K. Moor, E. A. Erysheva, Ustoychivoe razvitie kulturnogo landshafta goroda. Ustoychivaya arhitektura: nastoyaschee i buduschee, Proc. of Int. Sci. Conf., 17-18 noyabrya 2011 (Moskva, OOO «Adelant», 2012)
14. Lorraine Farrelly, *Drawing for Urban Design* (London, Laurence King Publishing, 2011)
15. V. A. Sosnovskiy, N. S. Rusakova, *Prikladnyie metodyi gradostroitelnyih issledovaniy* (Moskva, «Arhitektura S», 2006)
16. M. V. Shubenkov, *Strukturnyie zakonomernosti arhitekturnogo formoobrazovaniya* (Moskva, «Arhitektura S», 2006)

17. V. K. Moor, A. G. Gavrilov, E. A. Erysheva, *Arhitekturnaya masterskaya V. Moora: shkola, laboratoriya, studiya* (Vladivostok, DVFU, 2013)
18. A. G. Babenko, A. G. Gavrilov, E. A. Erysheva, G. E. Ignatov, A. V. Kopeva, O. V. Maslovskaya, V. K. Moor, S. I. Palienko, Rekonstruktsiya slozhivsheysya zhiloy sredyi primorskogo goroda (Vladivostok, DVGTU, 2004)
19. V. K. Moor, A. G. Gavrilov, A. V. Kopeva, Formirovanie vladivostokskoy arhitekturnoy shkolyi, Spetsialnost «Arhitektura» v DVGTU (Vladivostok, DVGTU, 2009)
20. V. K. Moor, A. G. Gavrilov, Katalog luchshih diplomnyih proektov, Spetsialnost «Arhitektura», DVPI-DVGTU, 1980-2009 (Vladivostok, DVGTU, 2010)
21. E. A. Erysheva, V. K. Moor, T. A. Demidova, *The best Projects of Far Eastern Architectural School* (Russia, Vladivostok. Harbin, Hitpress, 2007)

Tall trends: quantifying the skyscraper phenomenon

Jason Gabel[1*]

[1]Council on Tall Buildings and Urban Habitat, 104 South Michigan Ave., Suite 620, Chicago, IL, 60603, United States

Abstract. The world continues to witness an explosion of growth in the number of 200-plus-meter skyscrapers, with three straight years of record-breaking completions (from 2014 to 2016) and a 441 percent increase on the total number of such towers in the 21st century, from 265 in 2000 to 1,168 at the end of 2016. Fueled largely by strong economic performance, much of this activity is centered in Asia and the Middle East, upending longstanding geopolitical trends. China in particular has dominated worldwide skyscraper construction, accounting for two-thirds of all completions in the last calendar year (2016). Further, the traditional role of the skyscraper has diversified, with residential and mixed-use buildings accounting for a greater share of 200-plus-meter buildings. This paper explores these interconnected trends in detail and analyzes both the causes and impacts of an evolving skyscraper industry.

1 Overview – Trends in 2016 and 2017

The Council on Tall Buildings and Urban Habitat (CTBUH) releases an annual study of yearly 200-plus-meter skyscraper completions, investigating year-over-year changes in skyscraper locations, functions, and construction material. These reports create a snapshot of an evolving skyscraper industry, one that has seen exponential growth in recent years, driven primarily by a massive shift in skyscraper construction from the United States to Asia and the Middle East, particularly China [1].

Most recently, CTBUH determined that 128 buildings of 200 meters' height or greater were completed around the world in 2016 – setting a new record for annual tall building completions and marking the third consecutive record-breaking year (see Fig. 1). Of those 128 buildings, a total of 10 supertalls (buildings of 300 meters or higher) were completed in 2016, the third-largest number of supertall completions of any year [2].

The tallest building to complete in 2016 was Guangzhou CTF Finance Centre, which stood at the time as the tallest building in Guangzhou, the second-tallest building in China, and the fifth-tallest building in the world at 530 meters. However, the first few months of 2017 witnessed the completion of two even taller skyscrapers; namely Ping An Finance Center (599 meters) and Lotte World Tower (555 meters), which are currently the fourth- and fifth-tallest towers worldwide. The 2015 CTBUH report had originally predicted that

* Corresponding author: jgabel@ctbuh.org

these supertalls would complete in 2016, illustrating the difficulty of anticipating skyscraper completions and, of course, the technical and logistical challenges inherent in completing these behemoth structures [3].

Many of the major trends in 2016 remain consistent through the first part of 2017, confirming long-term tendencies that have shaped the industry over the last few decades. Although it is very difficult to predict completion dates of buildings at this scale, the CTBUH predicts that 2017 will see the completion of anywhere between 126 and 150 buildings of 200 meters or greater and 12 to 20 supertalls, meaning that the records set in 2016 are likely to be broken for a fourth straight year.

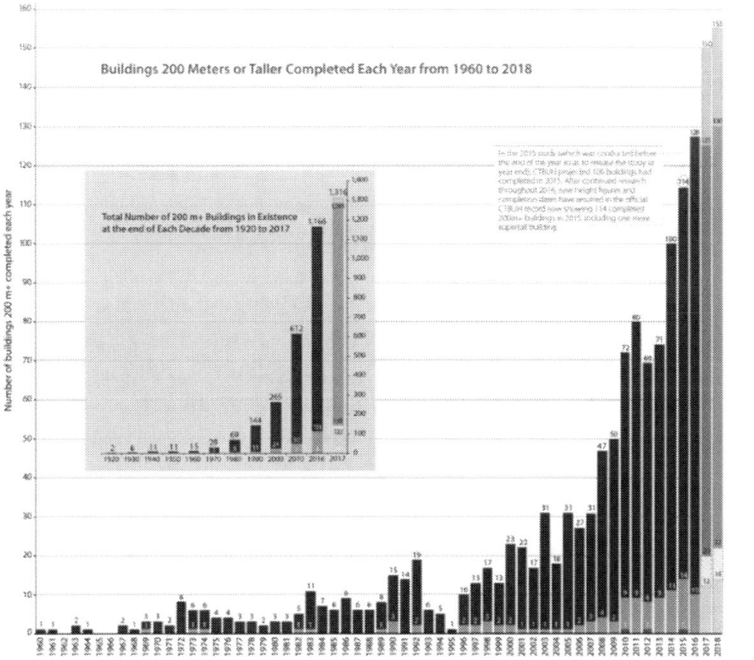

Fig. 1. Buildings 200 meters or taller completed each year from 1960 to 2018.

2 Key Worldwide Market Snapshots

Although Asia continues to dominate the skyscraper industry, other regional markets are experiencing growth in the typology, some for the first time. The Middle East continues to draw attention to itself with some of the tallest buildings completed, while Europe, Oceania, and South America are just now waking up to the benefits of building tall. Meanwhile, North America, once the unrivaled leader in skyscraper construction, is slowly reversing a long-term decline in overall completions.

2.1 Asia (Excluding Middle East)

The momentum of Asia has been unyielding for many years, and 2016 only served to further reinforce this trend. The region recorded 107 of the 128 completions for that year, or 84 percent of the total (see Fig. 2). A majority of those buildings are located in China, which completed the most 200-meter-plus buildings (84) of any country in the world for the ninth straight year (see Fig. 2). Thirty-one cities in China had at least one 200-meter-plus building completion, with Shenzhen outperforming any other city in the world, with 11 (to put that in perspective, that is more than any other country completed in 2016. The US completed seven 200-plus-meter buildings, and was second only to China).

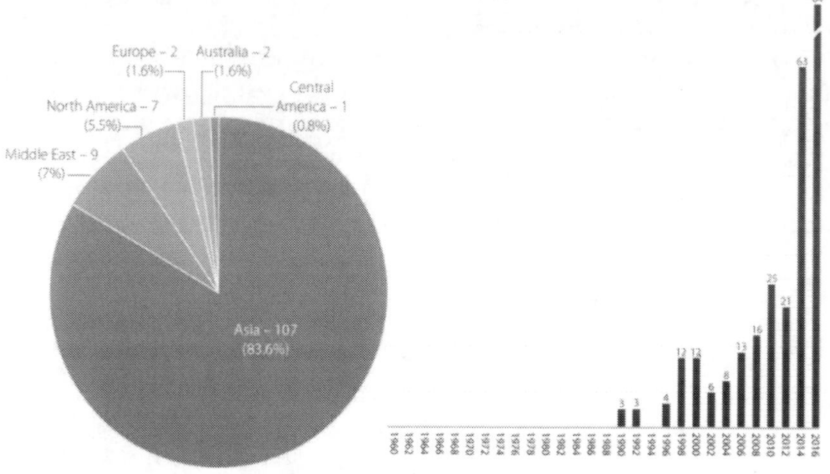

Fig. 2. (Left) Buildings 200 meters or taller completed in 2016 by Region and (Right) Annual 200-meter-plus building completions in China since 1960.

To date, 2017 has seen the continuation of Asia's and China's dominance. Although Asia accounts for just over half of all confirmed completions through May 2017, the continent accounts for 163 out of the 230 buildings that CTBUH has flagged for possible completion in the calendar year. Of those 163 potential buildings in Asia, 115 are located in China. Of course, it is highly unlikely that all of those buildings will complete before the end of the year, but ongoing construction is generally a good indicator of future trends.

Beyond China – South Korea, Indonesia, and The Philippines led the way in Asia with six, five, and four completions, respectively, in 2016 (see Fig. 3).

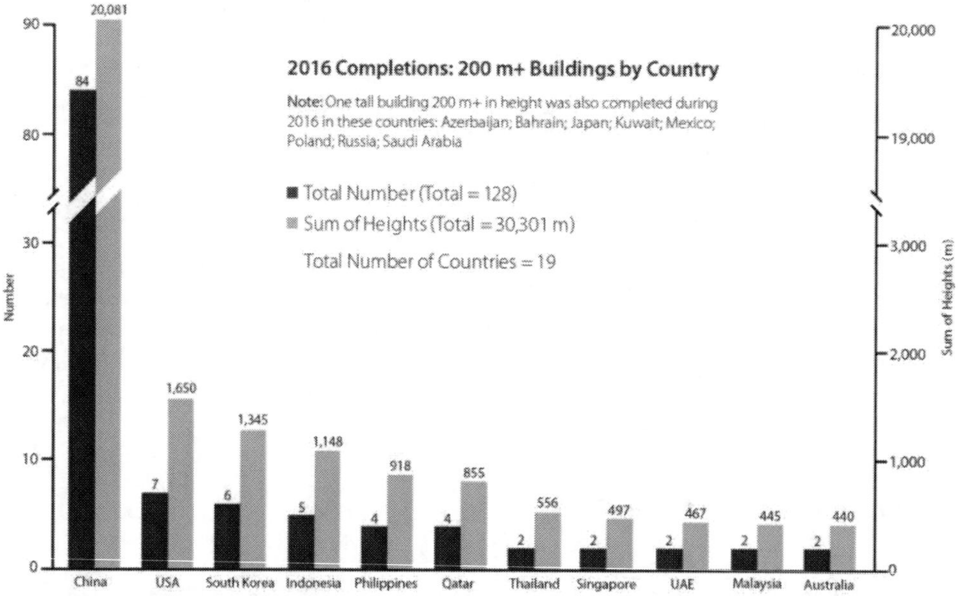

Fig. 3. 2016 completions by country.

South Korea's six completions doubles its count of three from 2015, but all six can be attributed to a single development, the Ilsan Yojin Y-City Complex in Goyang, suggesting that the country's overall completion numbers are highly dependent on a handful of major skyscraper complexes. For example, only one 200-plus-meter building was completed in 2014, but nine were completed in 2013, with all but one of those linked to a single development.

So far, Lotte Tower is the only 200-plus-meter building that has completed in South Korea in 2017; however, it is a major accomplishment as the fifth-tallest building worldwide. A further seven buildings are targeted for potential 2017 completions, meaning that South Korea will likely find itself near the top of the country list at the end of the year. Indonesia and The Philippines are similarly poised to retain their status as secondary drivers to Asia's high-rise boom behind China, with eight and six expected 2017 completions, respectively.

2.2 Australia

The land down under boasted two completions in 2016, Vision Apartments in Melbourne at 223 meters and International Towers – Tower 1 in Sydney at 217 meters. The annual trickle of tall buildings coming online in the country is expected to continue in the years ahead, with eleven 200-meter-plus buildings currently under construction or topped out, with completion years ranging from 2017 to 2020 [2]. So far in 2017, the country has already recorded one such completion with news that the 203-meter EQ Tower in Melbourne finished.

2.3 Central and South America

The Central America region had only one 200-meter-plus completion in 2016, in the form of Mexico City's Torre Reforma. The 246-meter tower was recognized as a CTBUH 2016 Awards finalist for the Americas region for its innovative structural configuration, significant opacity along one of its façades, and the preservation of a historic building at its base [4]. The tower is now the tallest building in Mexico. Two towers of similar height could complete in Mexico in 2017, Torre Paradox in Mexico City and Torre KOI in San Pedro Garza García. At 279 meters, the latter is poised to become the tallest building in the country. Although South America saw no 200-plus-meter completions in 2016, 2017 could reverse that trend, with a number of currently rising towers that could finish within the year [2]. Colombia leads the way with three under-construction skyscrapers, Hotel Estelar Bocagrande in Cartagena and BD Bacata Torre 1 and 2 in Bogota. BD Bacata Torre 1 (260 meters) will be the tallest building in the country once complete. It is notable for its unique crowd-sourced development strategy. Meanwhile, Venezuela, Argentina, and Brazil are also expected to contribute to South America's resurgence in 2017.

2.4 Europe

Despite having a significant influx of tall building completions in 2015, 2016 saw the lowest number of 200-meter-plus completions in Europe (two) since 2009, in which it had zero (see Fig. 4). Notably, the year saw the completion of the 374-meter Vostok Tower, Moscow, the tallest component of the Federation Towers complex. After having begun construction in 2005, the tower had been 11 years in the making. It is now the tallest building in Europe, a title that previously belonged to OKO Residential Tower, a development that completed in 2015 just blocks away from Vostok Tower in the Moscow City district.

Although there are no 200-plus-meter skyscrapers expected to complete in Russia in 2017, several buildings are in the pipeline for the remainder of the decade, including the country's future tallest, Lakhta Center in St. Petersburg. At 462 meters, it will be almost 100 meters taller than Vostok Tower. Also on the rise is Akhmat Tower in Grozny, which will also top the 400-meter mark.

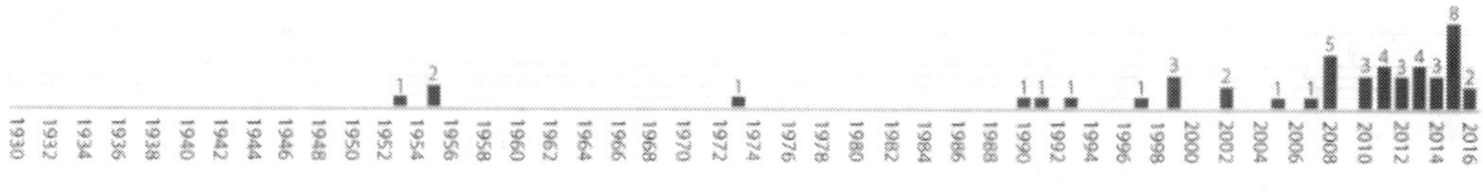

Fig. 4. Annual 200-meter-plus building completions in Europe since 1930.

The other 2016 European completion was Warsaw Spire, a 220-meter tower that represents the tallest of several new projects recently rising in the Polish capital. In terms of its height on the Warsaw skyline, the tower ranks second only to the Soviet-era Palace of Culture and Science, built in 1955. The majority of potential 2017 completions are found in Turkey, where six 200-plus-meter buildings are topped out or nearing that point, most notable the Skyland Istanbul complex, which will deliver two 284-meter towers. Additionally, Italy buoys the list with Turin's Regione Piemonte Headquarters, a 209-meter government high-rise.

2.5 Middle East and Africa

For the second year running, the Middle East ended 2016 with nine 200-meter-plus completions (see Fig. 5). This continues a steady trend of completions in the region, but pales in comparison to its all-time high of 23 in 2013, a spike that was attributed to a global post-recession recovery in tall building construction [5]. This is the first year since 2006 that the Middle East has not seen the completion of a supertall (300-plus-meter) building, but one should be wary of assuming that this is indicative of a regional swing away from the supertall height threshold. Projections show as many as five supertall buildings completing in the Middle East in 2017 [2].

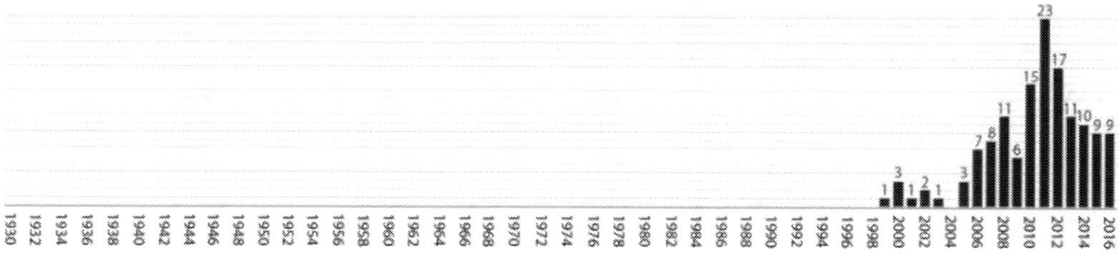

Fig. 5. Annual 200-meter-plus building completions in the Middle East since 1930.

In an unusual turn for the region, the United Arab Emirates did not have the greatest number of completions in the region for 2016. That accomplishment belonged to Qatar, which saw four towers completed in the calendar year. The UAE followed with just two completions, and Saudi Arabia, Kuwait, and Bahrain tied with one completion each. The tallest building to complete in the Middle East in 2016 was Regent Emirates Pearl, a 255-meter tower in Abu Dhabi that twists along its height at a rate of approximately 0.481 degrees per floor.

Africa is expected to get points on the board in 2017 with the completion of Britam Tower in Nairobi. Notably, the topped-out tower is just the second 200-plus-meter building ever built on the continent, following only the Carlton Centre in Johannesburg, which remains Africa's tallest building at 223 meters.

2.6 North America

Bouncing back from just four completions in 2015, North America saw the finalization of seven 200-meter-plus buildings in 2016, the greatest number since 2010, in which it also had seven (see Fig. 6). All of these buildings are located within the United States, and four are in New York City alone. Most of the buildings completed in the US were residential in function, with six of the seven incorporating living components, and just one incorporating office uses (10 Hudson Yards, 268 meters).

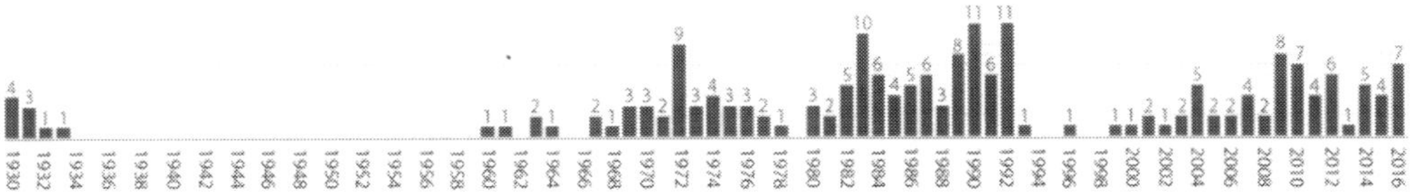

Fig. 6. Annual 200-meter-plus building completions in North America since 1930.

The tallest building to complete in the US was 30 Park Place, a 282-meter residential / hotel tower in Downtown Manhattan, accommodating a Four Season Hotel in the lower portion and luxury apartments above the 40th floor. Notably, the year also saw the opening of 56 Leonard Street, a tower that has been tracked closely throughout its development by the media given its unique «Jenga-like» design that extrudes upper-level penthouses on cantilevering floor slabs. 2017 seems likely to eclipse the success of the previous year with 14 potential completions in the United States in cities across the country, including New York, Chicago, Philadelphia, Miami, Seattle, San Francisco, Houston, and Los Angeles. An additional five buildings could complete in Canada as well.

3 Completions by Function

Interestingly, the functional split for tall building completions in 2016 remained almost perfectly consistent with that of previous years (see Fig. 7). Office functions have by far the highest share, representing 52 percent of completions with 67, achieving a consecutive all-time record over 2015, in which there were 53. Meanwhile, 37 mixed-use buildings were completed, representing 29 percent of the total; while 20 residential buildings came online with a share of 16 percent. Only three all-hotel towers were completed in 2016, representing just three percent of the total. Given that hotel functions are critical components of many mixed-use developments, the low number of completions for hotel-only towers does not indicate a downward trend for the function altogether. While that completion split is expected to remain relatively consistent in 2017, the share of all office buildings could fall as residential and mixed-use buildings become more popular. In particular, the share of residential buildings completing in 2017 could rise as high as 26 percent.

4 Completions by Structural Material

The number of buildings completed in 2016 with composite structural systems hit an all-time high, with 68, or 53 percent of the total (see Fig. 7). Composite construction represents an effort to use the optimal properties of steel and concrete and counteract their respective weaknesses. Thus, this material trend is expected to continue to dominate, particularly as tall buildings continue to proliferate in seismically active regions like China, which mandate robust structural responses [6].

Concrete construction took the remaining share of completions with 58, or 45 percent of the total. The significant use of concrete can be attributed to a combination of concrete's relative ubiquity and lower cost in many regions, as well as its comparative simplicity in construction, which would increase its appeal in regions with lower-skilled labor pools. In 2016, not a single building of 200 meters or higher was completed with an all-steel structural system. The use of steel in tall buildings has almost completely transitioned into composite construction. As of this writing, there are only twenty 200-meter-plus buildings currently under construction that employ all-steel structural systems, with the tallest being Akhmat Tower [2].

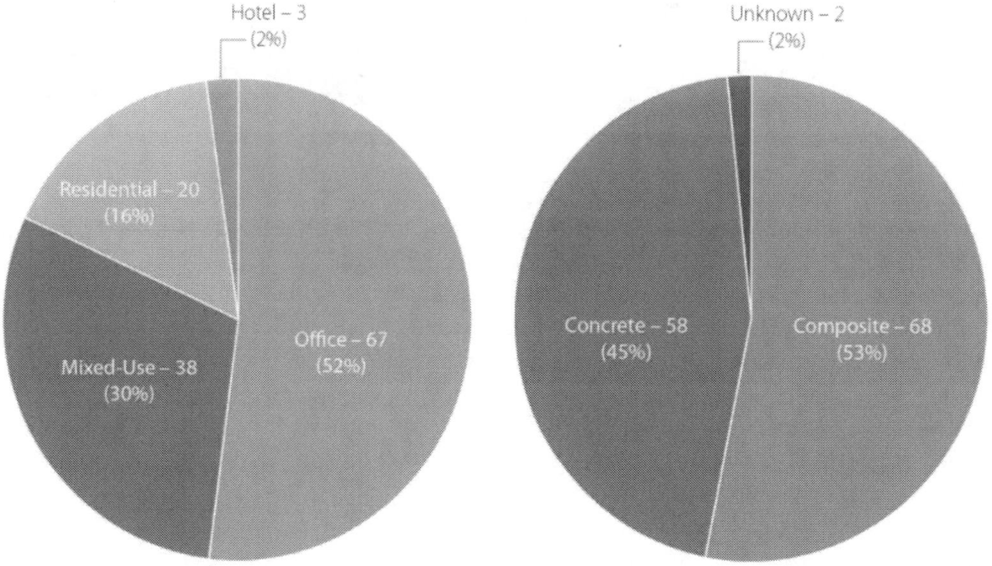

Fig. 7. (Left) Buildings 200 meters or taller completed in 2016 by Function and (Right) Buildings 200 meters or taller completed in 2016 by Structural Material.

5 The World's 100 Tallest Buildings: 2016–2017 Impact

In 2016, 10 buildings entered the list of the World's 100 Tallest Buildings (see Fig. 8). 2016 also marked the first year that every building on the 100 Tallest list was a supertall (300-plus meters tall). It follows that all 10 buildings entering the list also lie above this threshold, increasing the average height of the 100 tallest buildings from 357 meters in 2015 to 362 meters in 2016.

Through the first part of 2017, four more buildings have entered the list, including the fourth and fifth tallest (Ping An Finance Center and Lotte World Tower). The shortest building on the list is now Diwang International Fortune Center, Liuzhou, at 303 meters.

Mixed-use functions continue to enjoy a plurality in the 100 Tallest list, with 44 buildings. Meanwhile, office functions account for 37 buildings, down slightly from 40 at the end of 2016 (see Fig. 9). Residential and hotel functions have smaller shares, with 12 and seven buildings respectively.

The 100 Tallest list continues to be dominated by composite construction, with 52 buildings now employing composite systems (52 in 2016), up from 46 in 2015 (see Fig. 10). All-concrete and steel primary structures have experienced slight losses, down to 35 buildings for concrete and nine for steel.

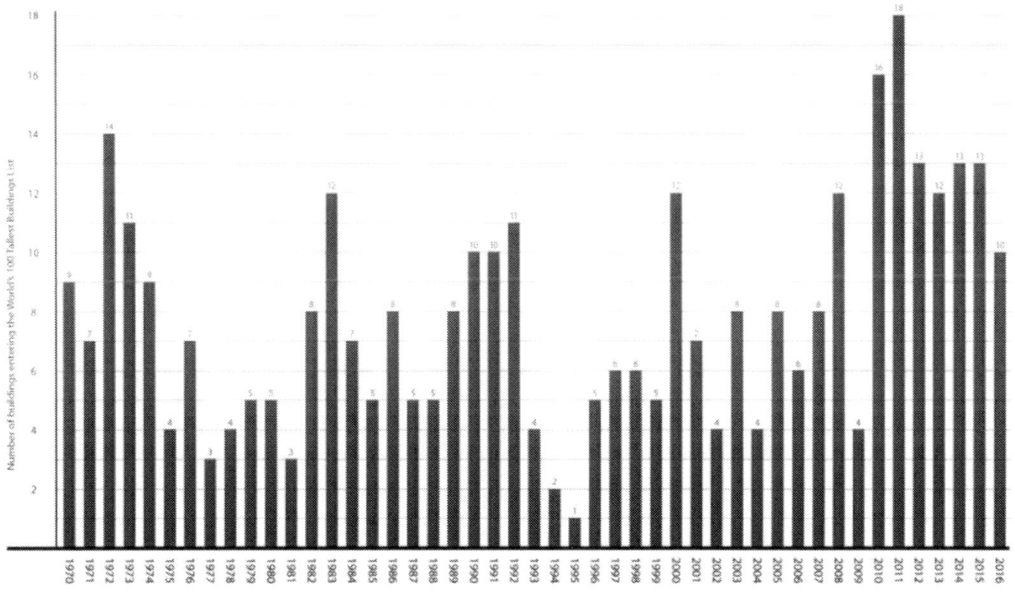

Fig. 8. Number of buildings entering the world's 100 tallest by Year.

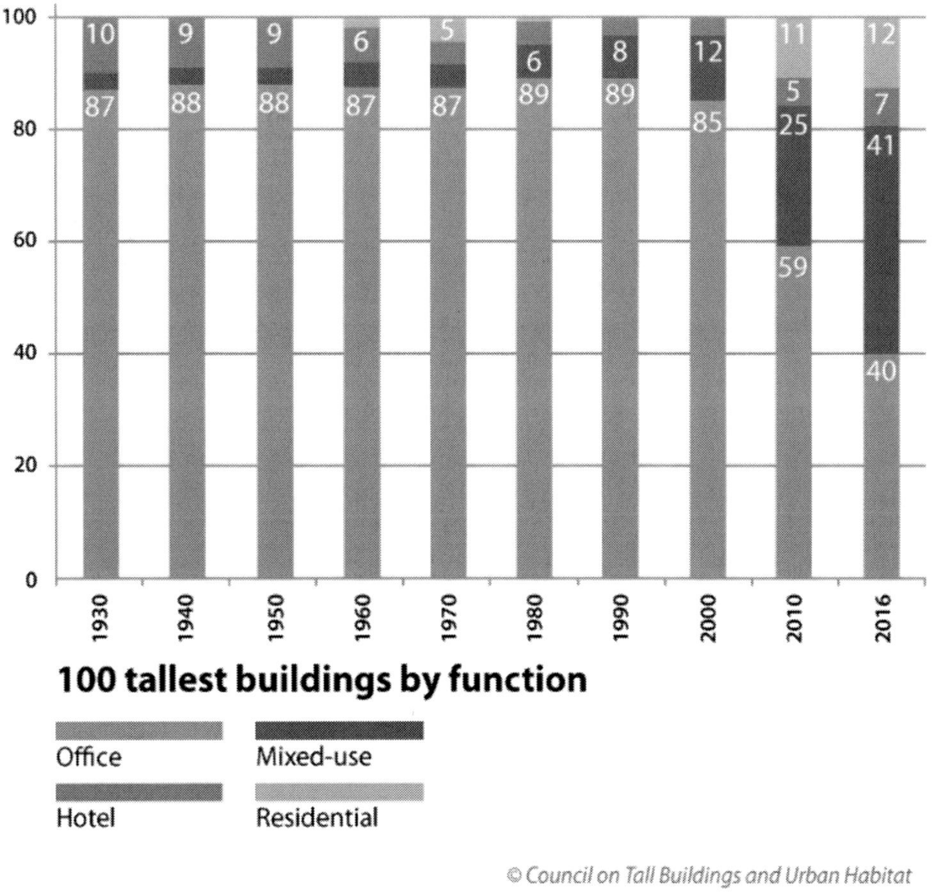

Fig. 9. The world's 100 tallest buildings over time by Function.

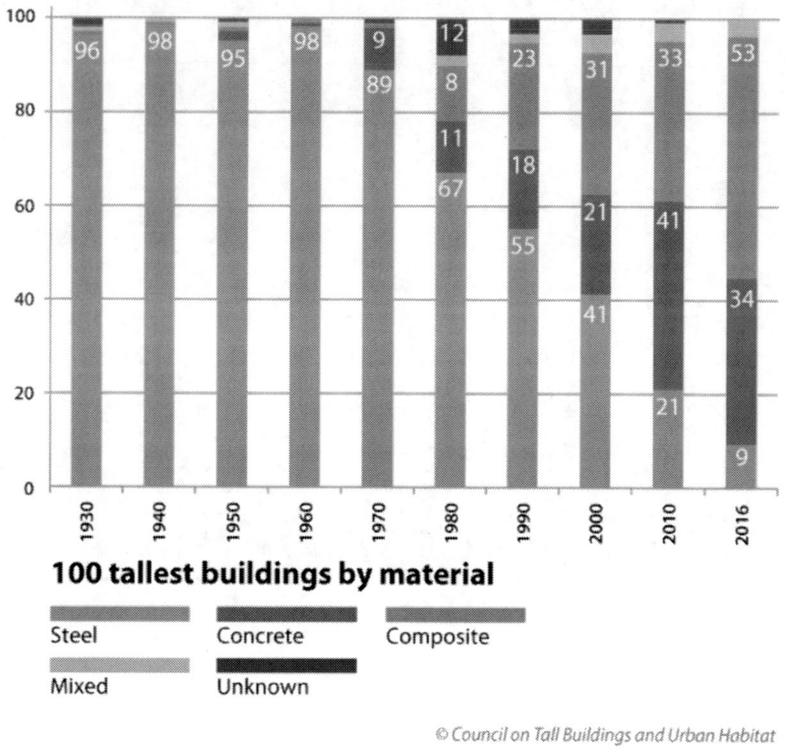

Fig. 10. The world's 100 tallest buildings over time by Material.

6 Analysis

For the third year running, the world has built more 200-meter-plus buildings in 2016 than any other year (128), beating the previous record of 114 in 2015. In 2015, we predicted that between 110 and 135 buildings would complete within this range in 2016, so it comes as no surprise that it was yet another record-breaking year in the skyscraper industry.

It's impossible to decouple the enormous boom in global skyscraper construction from the rapid urban development of China. Additionally, over the past several years, there has been a major uptick in the amount of real estate investment coming out of China, rather than remaining within it. This, in combination with a still-cooling national economy has caused some to assume that the era of unprecedented construction is on a downward decline. While that may be the case in the long run, short-term prospects remain high, with 328 buildings in the 200-plus-meter range currently under construction in the country.

It's safe to say that the role of the tall building in China will never truly fade, the only question that remains is how long it will continue to keep a majority share of tall building completions annually. With the closest contender in 2016 having only seven completions and China having 84, it is clear that the gap will take a number of years to close. Perhaps the most indicative trend from this report highlighting China's status is the fact that the city of Shenzhen completed 11 buildings that were 200 meters or higher. This is more than any country other than China managed to complete, and represents a record number of 200-meter-plus buildings completed by a single city in any calendar year. With nine of these buildings accommodating pure office functions, and the remaining two combining office and either hotel or residential, the priorities for Shenzhen are clear: These buildings are strategic endeavors to stimulate the local and national economy with commerce, some of it originating internationally as a result of Shenzhen's status as a Special Economic Zone. The ethos is still very much in the realm of «if you build it, they will come»; we'll just have to wait and see if the wager is fruitful.

«Tallest» titles also reigned supreme in 2016, with 17 buildings completing that became the tallest in a city, country, or region. Just to name some of the most significant superlatives: Vostok Tower (374 meters) completed to become the tallest building in Europe, Russia, and Moscow; MahaNakhon (314 meters) completed as the tallest building in Thailand and Bangkok; Guangzhou CTF Finance Centre (530 meters) completed as the tallest building in Guangzhou, the second-tallest (now third) in China, and the fifth-tallest (now seventh) in the world; Torre Reforma (246 meters) became the tallest in Mexico and Mexico City; Tajong Pagar Centre (290 meters) is now the tallest in Singapore; Gama Tower became the tallest in Indonesia and Jakarta…the list continues.

2017 has already continued this trend with the completion of Lotte World Tower (555 meters) as the tallest in Seoul and Ping an Finance Center (599 meters) as the tallest in Shenzhen and second tallest in China. In each instance, the drivers behind these buildings remain constant: the desire to establish identity (whether for a commercial business or an entire locale), to enhance the skyline, to respond to urban population growth, and to maximize functional utility upon minimal land area.

References

1. J. Gabel, et al: CTBUH Year in Review: Tall Trends of 2016, CTBUH Journal 2017 Issue I, 38–47 (2017)
2. http://www.skyscrapercenter.com/, last accessed 2017/01/10
3. J. Gabel, et al: CTBUH Year in Review: Tall Trends of 2015, CTBUH Journal 2016 Issue I, 40–49 (2017)
4. A. Wood, H. Steven, Best Tall Buildings: A Global Overview of 2016 Skyscrapers, Council on Tall Buildings and Urban Habitat (2016)
5. D. Safarik, et al: CTBUH Year in Review: Tall Trends of 2014, CTBUH Journal 2014 Issue I, 38–47 (2017)
6. S. Zhang, et al: Tall Building Seismic Design Case Study per China and U.S. Codes. In: Safarik, D, CTBUH 2014 Shanghai Conference Proceedings, 617–624 Council on Tall Buildings and Urban Habitat (2014)

Urban Forest Revolution

Anastasia Kucherova[1] , *Hana* Narvaez[1,*]

[1]Stefano Boeri Architetti, Via G. Donizetti 4 20122, Milan, Italy

Abstract. The tendency of the last 100 years shows the radical change of the proportion between urbanized areas and those dedicated to wild nature and agriculture. Whether we agree with it or not, cities are our future. Also it is confirmed by the annual reports of UNEP that cities are responsible for 75% of CO_2 emissions. Among negative consequences of the uncontrolled urban sprawl on the ecology there are: greenhouse effect leading to global warming; increasingly growing number of people affected by respiratory diseases due to the smog; drastic reduction of the biodiversity. The environmental impact and the look of the city of the future is in the hands of the architects of today. At Stefano Boeri Architetti we are exploring one of the possible way of reducing the buildings' ecological footprint and improving the quality of air in cities. It is called "Vertical Foresting", an evolution of the revolutionary idea of bringing the trees to the sky, far beyond its first realization – it became a philosophy, a new lifestyle. But first of all, Vertical Forest is a big question mark, addressed to the global community, to the politicians and common people, asking each and every one living in our planet – what can you do for our common home?

1 Introduction

Stefano Boeri Architetti based in Milan, with offices in Shanghai and Tirana, (called Boeri Studio until 2008) is dedicated since 1993 to the research and practice of architecture and urbanism. Among the most known projects there are: the Vertical Forest in Milan, the General Local Plan of Tirana 2030 in Albania, the Villa Méditerranée in Marseille and the House of the Sea at La Maddalena.

Stefano Boeri Architetti has provided services for the architectural and urban design for over 20 years, especially on large scale projects and public space renewal. It develops projects and regeneration strategies in complex environments, outlining and supporting synergies between the various stakeholders, public and private entities. Counting on a staff of over 40 co-workers, in the design process SBA collaborates with a wide network of professionals, from engineering consultants and landscape architects, to social scientist specialist, to provide ad hoc solutions over a wide range of territorial and socio-economic contexts. This approach towards collaboration has allowed the studio to spread its practice and to pair with professionals such as *Jeremy Rifkin*, to develop a concept of urban planning for the third industrial revolution.

*Corresponding author: a.kucherova@stefanoboeriarchitetti.net

Over the last years, the studio has developed a know-how in the new typology of the high rise building, integrating the trees and other plants with traditional construction materials. The first propotype of Vertical Forest, completed in 2014 in Milan, is a new model of sustainable residence in height, with two towers completely covered with more than 700 trees and 20.000 plants, The Vertical Forest has won numerous international awards, including the High Rise Building Award sponsored by the Museum of Architecture in Frankfurt in 2014 and in 2015 the CTBUH Award, as Best Tall Building Worldwide sponsored by the Council for Tall Building and Urban Habitat and the Illinois Institute of Technology both based in Chicago. After this first sustainable model of housing, the studio recently won a competition, in Lausanne, Switzerland, to further develop this model and build a new 117 meters tall **residential tower**, which will host more than 100 cedar trees and will be covered by shrubs and plants over an area of 3.000 sqm; this construction is scheduled to start in 2017. In the same time the third Vertical Forest is rising up in Nanjing, China, its completion is scheduled for 2018.

But the idea of urban forestation is now being developed to another level. The studio is working on several Forest cities around the world, starting from China, one of the most polluted countries. This practical experience, together with research activities both of our studio and other architects and environmentalists led us to formulate the basic principles of Urban Forest revolution.

1.1 Why should we act immediately? Some considerations on the climate change

"The Metropolis strives to reach a mythical point where the world is completely fabricated by man, so that it absolutely coincides with his desires" [14]. Nowadays the professional competences of an architect go far beyond just designing and building. It becomes more important than ever the ethics, the ability of team work in multidisciplinary groups, the ability of observing and analyzing the ever-changing conditions in which your building is going to exist, because the lifespan of a building is now sufficient to witness the drastically changing background. But, above all, not only as architects but as inhabitants of the Earth we should be sensitive to the problems of ecology and climate change which our planet is facing.

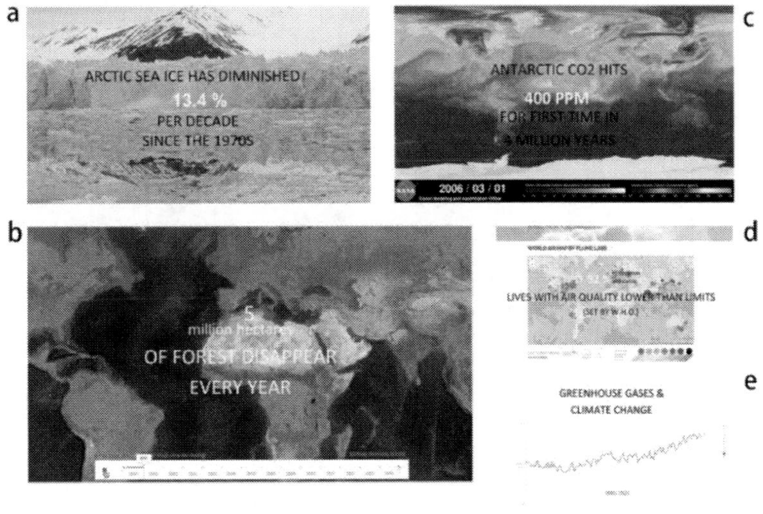

Fig. 1. Climate change and negative environmental effects of the cities: a – Arctic sea ice reduction; b - Air quality graphics in the cities of the world; c – Antarctic CO_2 level change; d – Disappearance of the cities; e – Greenhouse gases and climate change.

According to the estimations of UNEP cities are responsible for 75 % of global CO_2 emissions, vehicles and buildings are among the major contributors. This means that the architects and urban planners are more responsible than ever for the future of our cities. The environmental impact, the quality of life and the look of the city of the future is in the hands of the architects of today.

2 Materials and Methods

2.1 Vertical Forest

"A house for trees and birds, inhabited also by humans" Stefano Boeri [4]. The Bosco Verticale in Milan, Italy holds one of the most extensive living green facades ever realized. The combination of its sophisticated plant selection, the deployment of greenery in all orientations, the structural design to accommodate the plants, and the maintenance, safety and irrigation systems, represent one of the most innovative tall building projects in recent memory [6].

Vertical Forest is a model for a sustainable residential building, a project for metropolitan reforestation contributing to the regeneration of the environment and **urban biodiversity** without the implication of expanding the city upon the territory. It is a model of **vertical densification of nature** within the city that operates in relation to policies for reforestation and naturalization of large urban and metropolitan borders. The first example of the Vertical Forest consisting of two residential towers of 110 and 76 m height, was realized in the centre of Milan, on the edge of the Isola neighborhood, hosting **900 trees** (each measuring 3, 6 or 9 meters) and over 20,000 plants from a wide range of shrubs and floral plants distributed according to the sun exposure of the facade. On flat land, each Vertical Forest equals, in amount of trees, an area of **20,000** square meters of forest. In terms of urban densification, it is the equivalent of an area of a single family dwelling of nearly 75,000 sq.m. The vegetal system of the Vertical Forest contributes to the construction of a microclimate, produces humidity, absorbs CO_2 and dust particles and produces oxygen.

Fig. 2. Overview of the integration of the trees into built environment. Inspirations of Vertical Forest: a – Study of the mineralized urban surfaces (glass facades of the skyscrapers); b – Trees positive effects on heat reduction in urban environment; c – Vertical Forest introduction, Milan, Italy; d – Friedensreich Hundertwasser in Milan in 1972, planting the trees on the balconies; e - Joseph Beuys artwork in Berlin, 1982.

2.1.1 Inspirations of Vertical Forest

As any other architecture, Vertical Forest took shape under various influences, it grows on the soil of Italian and, in particular, Milanese architectural tradition, and sometimes, the inspiration comes from the most unexpected directions. "In 1972 I was 16 years old and I had no idea that travelling around the streets of my city, Milan, there was a bizarre character – the Austrian artist Friedensreich Hundertwasser – who while holding a tree was preaching the idea of a new style of architecture, built around the presence of trees in houses, courtyards and rooms.

In the middle of the streets with a small oak tree in his hand and just a few metres from the La Scala Opera House, Hundertwasser was making the case for organic architecture, based on a standard that governed the relationship between the number of humans and the number of trees in any space where people lived" [4]. Another strong impression that pushed Stefano Boeri to a revolutionary idea of bringing the trees to the sky, was the view of Dubai he was exploring in 2007 and its skyline composed entirely of glass and concrete skyscrapers. A triumph of artificial and mineral amidst the desert.

In 1982 in front of the Fridericianum Museum in Kassel the German artist Joseph Beuys built a triangle of 7,000 basalt stones, each of which was intended to be used for the planting of a tree. Anyone paying a sum of money, could "adopt" one of the seven thousand stones; the money raised would be used to plant an oak tree. So, day after day, the pile of stones diminished until it finally disappeared, and seven thousand new oak trees, each with one of the basalt stones at its base, appeared along the streets and avenues and in the squares of the city of Kassel [4].

2.1.2 Living facades

The real revolutionary element of the project – its trees– has influenced every part of the design. The vegetation is the outermost exterior element of the envelope and represents a filter between the interiors of the towers and the urban environment.

Fig. 3. Vertical Forest in Milan - first prototype of the new sustainable building typology: a – Irrigation system of the building and planting scheme; b – Planting of the trees on balconies; c - Trees in the nursery before being planted; d – Wind-resistance test of the tree species in Florida, USA.

The water distribution network for irrigation is composed of three main elements: the principal network, which brings water from underground to the terraces; the control group in the plant containers, which regulates the water supply; and the drip line, which provides widespread distribution throughout the plant containers.

Pre-Cultivation: the trees were selected based on a consideration of the most advanced principles of pre-cultivation in nurseries. The trees, which are at least 15 years old, have precisely calibrated sizes (obtained thanks to a professional pruning) and are free of any "defects of shape," which means that they have a single trunk, the lateral branches do not compete with each other and are radial, well-spaced, symmetrical and lightly webbed.

2.1.3 Manifesto

- The Vertical Forest is an anti-sprawl project and a densification project to reduce urban expansion while increasing the quantity of green within the built city. A model that connects to the policies for reforestation and naturalization of the edges of large urban and metropolitan areas. Each tower of the Vertical Forest is the equivalent in height of an outdoor area of single-family houses and tree-lined gardens of approximately 50,000 square meters.
- The Vertical Forest is a forestation project which helps to demineralize urban surfaces. It is a project that helps to regenerate the environment by increasing urban biodiversity without implying an expansion of the city territory.
- The Vertical Forest sets a new urban standard in terms of balance between the human and the natural sphere. In a Vertical Forest, per each human there are 2 trees, 40 bushes, 8 shrubs, 30 birds and butterflies.
- The Vertical Forest increases biodiversity. The Vertical Forest favours the formation of an urban ecosystem – rocky and mixed - in which various types of green (tree, shrub, grass roof, green vertical connections between the arboreal nucleuses) create a sort of vertical environmental network that could also be colonized by wildlife rocky environment dwelling species (such as redstart or kestrel) and/or other species (such as chickadees, goldfinch, doves), but also by insects (such as diptera and some beetles, spiders, butterflies if there are corners of hedges and flowering lawns).
- The Vertical Forest is a project that reduces pollution and the environmental footprint. The vegetation on the Vertical Forest is designed in order to form a continuous vegetative filter between the interior and the exterior. Thanks to this filter of leaves on the facades, the Vertical Forest absorbs the urban pollutants produced by private vehicles in the urban environment.
- The Vertical Forest is a project of environmental sustainability. The irrigation of the plants is carried out to a large extent through the filtering and reuse of grey water produced from the building. The shielding produced by the arrangement of plants along the perimeter of each floor, promotes proper air-conditioning of apartments and reduces the consumption of electricity for air conditioning.
- The Vertical Forest is a living architecture. The green filter of the Vertical Forest is to be considered as a living façade of the building, therefore its maintenance needs to be controlled in a systematic way.
- The Vertical Forest increases the urban ecological culture. The service "cell" that will ensure the functionality and long-term maintenance of the vertical green can become a landmark for the city and the schools in the district as a parameter for urban ecosystems. Each "cell for maintenance of the vertical green" can in fact be used for the collection and dissemination of data needed to evaluate its ecological function over time (consensus of spaces and fluctuations in the colonization, etc.).
- The Vertical Forest is a new landmark in the urban landscape. The towers of the

Vertical Forest do not offer only to those who live within an unusual perspective on the urban landscape from inside the apartment. By cyclically changing the skin according to the different type of plants and their different arrangement with respect to the solar thermal axis, the Vertical Forest also offers a changing panorama to the metropolitan crowd looking up.

- How to export Vertical Forest - adapt to local biodiversity. The Vertical Forest provides a model of combination between the urban and the natural world that can be adapted to different environmental conditions according to the local climatic zones. The vegetative and animal species populating the Vertical Forest in a cold climate will be different from the ones of a temperate or dry or tropical climate. Each climatic zone provides a set of local criteria to adapt the prototype to the local conditions.

2.2 Evolution Of The High-Rise Forest Building

2.2.1 Tour Des Cedres

"Les Terraces des Cedres" is not just a tower, nor simply a tall residential building. It is the prototype of a new dwelling model, a building that combines the comfort of the apartments, the wonderful view of the lake and nearby areas, with the vegetal world; especially with the various species of cedars that already inhabit the territory. A landmark that can be spotted from Lausanne northern border.

Fig. 4. Other high-rise buildings by Stefano Boeri Architetti in the world: a – Tower of cedars, Lausanne, Switzerland; b – Tower of cedars planting scheme and tree species used; c – Nanjing Towers, China; d – Urban tower, Antwerp, Belgium; e – Wonderwood, Utrecht, Netherlands.

The construction is a 117-meter-tall tower, with a rectangular plan of 52 x 15.5 meters, oriented north-south, composed of 36 floors that include a gym, offices, housing, and a restaurant. A structure that will host more than 100 trees (4 typologies of cedars) covered by shrubs and other plants over an area of 3,000 square meters, the volume of the tower is characterized by the presence of cantilever elements, of variable length that animate the facades and break the succession of levels. These prominent elements are cantilevered over 3 m mainly on the south facade and punctually on the east and west facades. Since the apartments loggias offer an extension of the outward space to fully enjoy the scenery. The residential tower is accompanied by a 5,000 square meters shopping mall.

The project establishes a clear separation between the residential, and the commercial spaces. The distinction of these main functions resulted in the drawing of three different buildings in close relationship with each other, connected by an open public space. The design of the plan integrating the three main buildings establishes a mass / void relationship where architecture and public open space are alternated. These spaces between buildings are living environments, full of activity and sociability. The choice to separate the tower building from the base, designed to accommodate public activities, favors the creation of urban relationships between the elements that shape the neighborhood, both in public and residential focal point.

The implantation of the tower site is the most developed of any public intervention, where new buildings are implanted along the main north-south directions. They redefine the view and the course of the whole intervention and the relationship between the site, the city, the highway, the ground.

2.2.2 Nanjing Towers

Nanjing Green Towers, promoted by Nanjing Yang Zi State-owned National Investment Group Co.ltd, will be the first Vertical Forest built in Asia. Along the facades, 600 tall trees, 500 medium-sized trees (for a total amount of 1,100 trees from 23 local species) and 2,500 cascading plants and shrubs will cover a 6,000 sqm area. A real vertical forest, contributing to regenerate local biodiversity, that will provide a 25 tons of CO_2 absorption each year and will produce about 60 kg of Oxygen per day.

The taller tower, 200 metres high, crowned on the top by a green lantern, will host offices – from the 8th floor to the 35th – and it will include a museum, a green architecture school and a private club on the rooftop.

The second tower, 108 metres high, will provide a Hyatt hotel with 247 room of different sizes (from 35 sqm to 150 sqm) and a swimming pool on the rooftop. The 20 metres high podium, will host commercial, recreational and educative functions, including multi-brands shops, a food market, restaurants, conference hall and exhibition spaces.

2.2.3. Other Typologies of Sustainable Urban High-Rise Buildings

The latest research of the studio is studying the possibilities of exporting the concept of green buildings in different contexts, in terms of climate and socio economical background. The towers developed for Antwerp (Belgium), Utrecht (Netherlands), Treviso (Northern Italy) represent different branches of evolution of urban high-rise building. In case of northern countries, the importance of sunlight entering the apartments is crucial. This requirement was both challenging and fortunate for the final design solution, since it encouraged us to push the limits of internal and external spaces, exploring the theme of terrace, balcony, under a completely different light.

The balconies in their northern version, from Switzerland to Netherlands are turned into extensions of the interior, bringing the residents closer than ever to the surrounding environment. In one way or another the high rise residential buildings explore and reinvent the border between private and public, between natural and artificial, between inside and outside.

3 Results

3.1 Towards the Forest City

"When we dream alone it is only a dream, but when many dream together it is the beginning of a new reality".
Friedensreich Hundertwasser

In 2014, the first two towers completely covered by trees and shrubs, equivalent to 20.000 sq m of real forest, have been completed in Milan. Since then this idea of densification of greenery in urban environment received scientific and political approval of its efficiency. Just two towers are able to absorb 19 tons of CO_2 a year and to produce the comparable amount of oxygen. The presence of trees improves considerably the quality of living environment, both for the residents and other citizens, by humidifying the air, filtering the sun rays in summer, reducing the temperature inside the apartments by 2-3 grades. But how can this concept help to fight the air pollution on urban scale?

Fig.5. Forest cities of Stefano Boeri Architetti: a – Shijazhuang Forest City conceptual vision, China; b - Shijazhuang Forest City bird view; c – Liuzhou Forest City; d – Liuzhou green buildings and transport of the future

3.1.1 Shijazhuang Forest city

The answer, or rather the first hypothesis of expansion of the concept has been first developed for Shijazhuang. Shenzhen and other cities, are reaching a population of 60 million people. Shijiazhuang for its part, has a sad record: it is the city with the highest rate of air pollution in China. A pall of smog and soot rising only a few days a year. The Forest City in Shijiazhuang will be a new city for 100,000 inhabitants. A city of a new generation, capable of becoming a model of sustainable growth in a large country seeing, each year, 14 million farmers migrating to the cities.

ForestCityShijiazhuang© is:

- **The prototype of a city composed by Vertical Forests**. ForestCityShijiazhuang© (FCS) is the prototype of a new generation of small, compact and green cities, composed by dozens of tall and middle size buildings – the so called "VerticalForests"– all surrounded by the leaves of trees (ranging from 3-9 meters in hight), shrubs and flowering plants. Every VF grafts the equivalent of thousands of 20.000 square meters (2 ha) of a real forest.

- **An urban ecosystem**. ForestCityShijiazhuang© is an urban ecosystems which hosts 100,000 inhabitants and occupies a land of 225 hectares. ForestCityShijiazhuang© is composed by 5 districts and one central Park (carpel). Every district (petal) is hosting ca. 20,000 inhabitants. Every Petali is a mix-use social environment, with residential housings, offices, retails, malls, public spaces and gardens. The central Park (carpel) is the place for the main Public Facilites of FCS: the Hospital, the School, the Cultural Activities.
- **An anti-sprawl device**. Every Forest City will concentrate in a vertical dimension – and within a perimeter of 1,5 for 1,5 square kilometers (225 hectares) – the urban volumes that normally are hosted in 25 hectares of land.
- **A sustainable city, with low energy consumption**. The vegetative filter on the buildings balconies creates a reduction – in the difference between the outside and inside temperature – of about 3 degrees.
- **An absorber of CO_2 and the dust of urban pollution**. FCS cleans the air. The vegetation within FCS is designed in such a way as to form a continuous green filter between inside and outside of inhabited areas, able to absorb the fine particles produced by urban traffic, to produce oxygen, to absorb CO_2, and to shield the balconies and interiors from very high pollution of Chinese cities.
- **A multiplier of biodiversity of the living species**. FCS will be the home of hundreds of different spices of plant life, including trees, shrubs and perennials. FCS will host many spices of birds and domestic animals.
- **An ever-changing urban landmark**. Thanks to the variety of plant spices housed along the balconies – and in the presence of deciduous trees –, FCS will change its skin and the color composition of its leaving facades, according to changing seasons and weather conditions.
- **The basic element of a new model of urbanization in China**. FCS is a model of urbanization and the basic element for a large number of settlement combinations. Different FC could be assembled in a cluster or along a line, creating a major conurbation, but always respecting the maintenance of the standard established in FCS: for 225 hectares of a city in 25,000 hectares of a green (agricultural, natural, sport) permeable land.

Urban Forest Revolution is just one of the measures that must be applied within a complex strategy of environmental protection. Globally speaking, it integrates the 5 pillars of the Third Industrial Revolution formulated by Jeremy Rifkin. “The five pillars of the Third Industrial Revolution are (1) shifting to renewable energy; (2) transforming the building stock of every continent into micro–power plants to collect renewable energies on site; (3) deploying hydrogen and other storage technologies in every building and throughout the infrastructure to store intermittent energies; (4) using Internet technology to transform the power grid of every continent into an energy-sharing intergrid that acts just like the Internet (when millions of buildings are generating a small amount of energy locally, on site, they can sell surplus back to the grid and share electricity with their continental neighbors); and (5) transitioning the transport fleet to electric plug-in and fuel cell vehicles that can buy and sell electricity on a smart, continental, interactive power grid” [1].

3.1.2 Liuzhou Forest City

Liuzhou has a unique landscape of karst, the design inspiration for the design blueprint and for building integration into the local landscape comes from the characteristics of this mountain soil. *Vertical forest* technology will help us to achieve this vision. *Liuzhou City*, located in the northeastern part of the Guangxi Zhuang Autonomous Region, is a unique landscape of the historical and cultural city. In recent years, the urban construction of Liuzhou has been accelerated, and the urban spatial pattern of "one city, two cities" has been gradually forged with the Lianhua Mountain Scenic Area as the "green heart" and the main city and the Liudong New Area.

This living settlement follows the Vertical Forest and Forest City ideas introduced both in horizontal and vertical distribution of green facade surfaces. The living environment creates an ideal space for symbiosis between Man and Nature. Function: purifying city air, absorbing Carbon dioxide and dust particles, and producing oxygen; blocking radiation and noise; creating more living space also for birds, insects, and small animals.

Structural design: planting suitable local tree species along the outer wall, building a watering system, using recycled water to maintain plant growth, and producing wind energy and solar energy to increase energy self-sufficiency.

4 Conclusions

"Generations of human beings were transformed into machines in the relentless pursuit of material wealth: We lived to work" [1].

Building always more and more, conquering the soil from nature has been the objective of architecture in the same way it worked for the world in general. But nature is warning us that we have to reconsider the long-term strategy. It is no longer a question of respect of our habitat, it's a matter of survival of mankind and the planet. That's why sustainable strategies are nowadays acquiring the same importance for architecture as structural design and engineering. As J. Rifkin says, the buildings must be converted from the major polluters into natural filters, from principal energy consumers into energy factories. Realization of the first Vertical Forest established a new era in the relationship between mankind and other species.

Vertical ForestING – is a worldwide trend, generated by the first Vertical Forest built by Boeri Studio in Milan, in 2014. When Stefano Boeri came up with the idea of giving back to nature the space we are taking from it with continuous urban sprawl, he conceived Bosco Verticale as a prototype of the skyscrapers of the future. Today we see that the Future is now**.** Every day architects all around the world rethink and reinterpret the idea of the Trees in the sky. Each of these attempts is extremely important for the future of architecture and the Future of our planet, because the single thought can have certain impact, but thousands of thoughts can change the world. Slowing down the climate changes, reducing CO_2 emissions, making our living sustainable and in harmony with nature.

References

1. J. Rifkin, The Third Industrial Revolution: How Lateral Power Is Transforming Energy, *the Economy, and the World* (2011)
2. S. Boeri, *Biomilano. Glossary of ideas for a metropolis based around bio-diversity* (Corraini edizioni, 2011)
3. S. Boeri, *La città scritta* (Quodlibet, 2016)
4. S. Boeri, *A Vertical Forest. Instructions booklet for the prototype of a forest city* (Corraini edizioni, 2015)
5. Harry Rand, *Hundertwasser* (Taschen, 2007)
6. Elena Giacomello, Massimo Valagussa, *Vertical Greenery. Evaluating the High-Rise Vegetation of the Bosco Verticale* (CTBUH Research Report, 2015)
7. I. Calvino, *Il Barone Rampante*, Einaudi (1957)

8. E. Thomas, R. Jalonen, J. Loo, D. Boshier, L. Gallo, St. Cavers, S. Bordács, P. Smith, M. Bozzano, Genetic Considerations in ecosystem restoration using native tree species, Forest Ecology and Management, **333**, 66-75 (2014)
9. Peter Droege, *La città rinnovabile* (Edizioni Ambiente, 2008)
10. Richard Register, *Ecocities. Rebuilding cities in Balance with Nature* (2006)
11. E. Howard, *A Peaceful path to Real Reform* (1898)
12. Maurizio Corrado, *Vegetecture* (2011)
13. http://greenevolution.ru (last accessed 2017.04.27)
14. R. Koolhaas, *Delirious New York* (2001)

The problems and perspectives for the introduction of high-rise construction in Russian cities

Anna Pershina[1,*], *Mehman* Radzhabov[1], and *Tatyana* Dormidontova[1]

[1]Samara State Technical University, 194, Molodogvardeyskaya St., 443001, Samara, Russia

Abstract. The propose of academic affairs is discovery the principal areas of concern high-rise construction in Russia. Examples of modern Russian and foreign high-rise construction are considered in the work. The most important problems and their solutions for Russia are identified on their basis. The everyone area of concern is considered separately. Ecology problems and influence of high-rise construction for the healthy and psychological effect of people are considered special. High-rise constructions influence negative and positive for urban environment in Moscow and Samara cities. The experience lack, defects in requirements document, which don't include all high-rise constructions specific, system problem of construction and often non-availability of proper control at the existing requirements document result for complexity of designing, construction and operation. At this moment, high-rise constructions temp is increasing in Moscow. Feasibility of high-rise buildings come up in regions of Russia. The reasons include high material inputs, irregularities of requirements network and utility lines and maintenance problems. The researching follow up of conclusions and recommendations for high-rise constructions development in Russia. The reasons of high-rise buildings are urbanization of people and necessary of concentration labor supply. The important tasks for organization are creating compact urban environment, decrease urban area for development, using an innovative technology for construction and properly maintenance. The balance between the preference of high-rise construction, inputs for construction and influence for ecology are resolve for this task.

1 Introduction

In the modern world, urban process has brought to appearance a lot of big city. An enormous number of people were reasons for appearance of high-rise buildings. Today skyscrapers are inherent part of megapolis. The progress of the construction and research and technology base, the deficit of developed infrastructure and communication systems and the increase of the value of land plots, as well as an increase of the prestige of skyscrapers have been the reason for this. The skyscraper is height more than 75 meters or more than 25 floors at the Russian classification, In the world classification height is more than 100 m meters.

* Corresponding author: mironova163@mail.ru

The first high-rise buildings in the world appeared at the end of the XIX century in Chicago. The huge theoretical and practical experience have found, the main problems have been discovered for all ways of high-rise buildings design and construction for the more than one hundred years. There are variety skyscrapers stagger everyone's imagination. The enormous height, plasticity form, flexible planning, construction complexity brings them to the innovation area in the realm of construction. The high-rise buildings have distinct functions such as hotels, offices, apartment houses.

In Russia, the first high-rise buildings were constructed in thirty years of twenty ages. From eight scheduled high-rise buildings, just seven were constructed in Moscow. The project "Counsel palace" is height 430 meters wasn't constructed (picture 1). The details of Stalin-style skyscraper have been Moscow symbols, namely individual architectural style, arrangement and sharp distinction from early constructed buildings. The main building is Moscow Government University in Sparrows mountains. (picture 2). The next years there were general buildings and feasibility of skyscrapers were postponed on indefinite period. This decision brought to intense lag in the realm of construction from lead country of the world.

Fig. 1. The project "Counsel palace".

Fig. 2. Moscow Government University.

The Moscow City is the first skyscraper in the modern Russia. The first planned heights of new modern blocks of buildings were from 240 to 612 meters, but built height are from 85 to 317 meters. The foreign specialists were involved in design and construction. The many problems were caused during the construction process. The deadlines weren't observed. The original project was rejected. The highest building "Russia" wasn't constructed as a result. This ensured that the problems of high-rise buildings were discovered and gotten valuable researching. The experience lack, defects in actual requirements document for high-rise buildings and the system problems of construction followed to the problems on all stages of the project. This ensured that the first requirements document for high-rise buildings were approved and published in Moscow, such as Moscow City Construction Regulations 4.19-2005 "Temporary requirements document for multifunction high-rise buildings and complexes in Moscow city" and official regulations 267.1325800.2016 "Buildings and high-rise complex" was published in 2016 year. The new high-rise buildings and complexes often appear. The skyscrapers are quite appropriate, but have something issues. [4, 5].

The high-rise buildings have slow temps in the others Russian cities. The buildings mainly are 100-150 meters' height in the big cities. The highest appreciable building "Lahta Center" was have started to construct in 2012 year in Sankt-Petersburg. It will be the highest building in Europe. The building will 462 meters and will be able supposed in 2018 year.

2 Materials and Methods

The high-rise construction methods were analyzed for them estimate realm. The most problem aspects were discovered, estimated their actual and difficulty of architecture and engineering solution. The big worldwide experience is good for prosses solution problems of high-rise construction. High-rise construction problems shall be resolve by complex methods of attack. The non-typical and unique methods are used. The worldwide experience is transformed to surrounding realty. The Europe experience is more useful and applicable. Each problem aspect and examples will be able examined.

The many materials and technical resources needs to high-rise construction, such as a lot of metals, concreter, expensive engineering equipment, manpower, transports and others. The less than 70% of inside volume of skyscraper might be used if design has been performed good, other part is used for support elements, stairwells and wells. The 80 % of volume usually is used in the common building. Until 30 engineering systems are used in skyscrapers. The strongest concreter sort was used for construction of tower "Federation" (Moscow-City) for example. The more improved concreter already was used for tower "Eye" and unique concreter line was used for blend delivery at a great height. The use of secondary materials, construction automation, design is done with accordance requirements document, recycled materials resolve these problems [6, 7].

Profitability of the high-rise construction hold a special main place in analyses. The accommodation in skyscrapers are elite realty for many reasons and it limit them expansion. The big company offices more often locate in skyscrapers. One of more problem is location as usual in the center of city. This is main reason for expensive accommodations price. Despite the fact, the experience was and the social accommodations located in the skyscrapers. The burning issue was due to the slums were pulled down, including the famous Kowloon. That problem was resolved by a lot of built skyscrapers. These were very compacted building up. The accommodation standard was 2,2 square meters for one adult. A lot of quantity of accommodation were built in a brief time due to a plot of land were saved a lot. The living space was increased in percent to 10 m2. An analogous situation was observed in Singapore. Here were built HDB residential complexes, which are social housing. From the Hong Kong variant they were distinguished by more spacious apartments, better landscaping and complex building up [8, 9].

Ensuring the competent development of transport infrastructure shall be commensurate with the built upness and the flexible distribution of automobile traffic. It is known that building high-rise buildings are reason of a significant increase in population density, which requires the presence on the inner yard of enough parking spaces. There is a need for more parking, the expansion of road networks is required due to the increased number of cars. Huge areas of fertile land roll under the asphalt. There is less space for landscaping and site improvements. A big traffic leads to a high concentration of exhausts in the air. Skyscrapers can accommodate an enormous number of people, and their movement requires a well-developed network of public transport. The construction of an underground metro or overhead monorail allows to resolve most of the transport problems. But the developed metro network requires a lot of material resources. There are several ways to solve these problems in the worldwide experience. Laws are introduced restricting the use of a private car, as well as their importation in Singapore. A well-developed network of public transport, together with a decrease in the number of cars, helped to avoid a transport collapse. Dubai has another situation; the high-rise buildings were located at a considerable distance from each other. Multi-level road junctions were built at the appeared large spaces. They were designed for the growth of the city in the future.

In many Russian cities, the problem of parking spaces shortage is becoming pressing problem with annual increase of cars on the roads. While dense urban building up does not allow to locate enough of parking on the street especially in the city center. One way of the solution is creating various systems of multistage parking lots that occupy the minimum area. The simplest solution is a multistage covered parking lot, which is in inside of new residential areas. The covered parking lots usually don't enough in case of a dense building up construction. Another wildly solution is underground parking. The most part of constructed high-rise office centers have underground parking. The modern automated parking lots are even more rational from the point of view of minimizing the economy occupied area (Picture 3), because such systems do not need the area under the ramp and reversal platforms, and the erection is much cheaper than capital construction. These parking lots are mobile systems that can be dismantled and moved to another location [10, 11].

The design of parking is carried out in accordance with the requirements of 113.13330.2012 "Parking", as well as considering the specifics and style solutions inherent in each object in Russia. Aboveground parking shall be provided with a height of no more than 9 floors, underground parking shall be no more than 5 underground floors.

The most compact and at the same time capacious parking "Car Tower" (Picture 4) is in Wolfsburg city in the German near the factory of the company "Volkswagen". It is made in the form of a giant glass pillar with cells, which one car can accommodate each of. Cars are delivered to a free cell with a special lift. As a result of to this solve, up to 800 cars can fit in a small area.

Recently, the problems of ecology are becoming increasingly important, there is a tendency to develop economic and environmental technologies. The special attention is devoted to energy efficiency and environmental protection. This is reflected in all stages of construction and maintenance of the high-rise building. There are very different approaches to this issue. This can be the introduction of such technologies in existing buildings, as well as completely new projects, originally created under the influence of "green" standards and technologies.

Fig. 3. Automated parking

Fig. 4. Parking "Car Tower", German. Wolfsburg city.

Almost all modern high-rise buildings pass environmental certification according to international standards (LEED, BREEAM and others). Concentration of people in a relatively small area leads to the accumulation of a lot of daily trash. A garbage is used technically very difficult in skyscrapers. The most common collection of trash in skyscrapers' apartments occurs floor by floor. Services need for the timely collection of trash and further utilization. Sorting of trash is especially actual for skyscrapers. The question of shading is also very important, there are standards of insolation that shall be met. Large shaded areas appear in conditions of high-rise buildings up. Sunlight may not be enough due to it will negatively affect to the human health and some plant species [12].

Important is the impact on the physiological and psychological health of a person, statistics are known from the results of research by scientists from the University of Canada, Victoria, that people living above the 9th floor are three times more likely to turn to neuropathologists and psychiatrists than those who live in low-rise buildings[13]. Many doctors have opinion that rarefied air and a sharp change in atmospheric pressure at high altitude are not useful for the heart and blood vessels. There are high-speed elevators, which accelerating up to 16 meters per second in most skyscrapers. Some people do not have time to adapt when lifting a sudden change in pressure therefore they have darkening in the eyes and a ringing in the ears. Video ecologists argue that the psychosomatic man is negatively affected by the very faceless appearance of high-rise buildings with a lot of identical windows, monotonous inaccessible walls, depressing the gray veil in the windows (the lower boundary of the rain clouds is just at an altitude of about 100 m). An uncomfortable visual environment is created provoking the "big city syndrome", which is often manifested in the aggressiveness of a person. Psychologists have proved that the percentage of suicides is always greater and 80% of those who commit suicide live above the 9th floor in places of mass building up skyscrapers.

Security issues are importance in high-rise buildings, because organization of evacuation routes requires compliance with the mass of regulatory sources, due to the complexity and length of the paths providing access to the exits. Experts argue that, the use of noncombustible materials is impossible, despite the high development of technologies in construction. Therefore, even in the newest houses a fire is a real threat, for example, the burning of the CCTV tower in Beijing in 2009 and the fire in the tower "Olympus" in "Grozny-City" in Chechnya in 2013. Fire danger for people in high-rise buildings is exacerbated by the fact that unlike low-rise buildings, evacuation is greatly hampered, and the complexity of fighting fires increases. The solution is considered by the example of the highest building of the Burj Khalifa. Burj Khalifa has nine special shelter areas (safe areas), which are protected by layers of reinforced concrete and sheet refractory coatings. Their walls can withstand by the fire for 2 hours. Each safe area is equipped with a special ventilation system. Sealed fireproof doors prevent the penetration of smoke. People can hide in safe areas until the emergency services take control of the situation or their turn for evacuation approaches. The fire lift is designed to move 5,500 kg (50-60 people). The early warning system works continuously for 24 hours a day. As soon as a signal is received to the smoke detector and the temperature sensor, an automatic water extinguisher is triggered and a system of powerful fans. They pour clean cool air into the building through refractory air ducts. Fresh air displaces smoke from the stairwells, ensuring the safety of evacuation routes [14].

For high-rise construction in modern Russia has not yet fully approved the requirements document. It is impossible to build obviously quality facilities that meet modern standards of safety and comfort, while using outdated technologies. For sustainable development high-rise construction requires new regulation, simplification of the approval procedure for the joint requirements document, GOST and technical regulations. Today every development project becomes a testing ground for designers and builders in fact. Almost all high-rise under construction buildings are innovative and each of them strives to become better and more difficult than others in Russia.

The main problematic aspects and some examples of their solutions were considered, we will analyze a concrete example of the introduction of a high-rise building in Russia. An erected three-section high-rise building will be considered in the city of Samara at the corner of the crossing of Vilonovskaya and Sadovaya streets (Picture 5). Its height is 140 m and this is the tallest building in Samara now. It is noteworthy against the background of the Samara building, but among modern high-rise buildings it does not stand out. The surrounding buildings are much smaller by number of floors s, this makes it the dominant one in the area. Many residents of Samara are negative about the appearance of this building. What are the reasons? They are many, both objective and not very much. The building contrasts strongly with the surrounding buildings, creating the impression of randomness. The building is in a cramped environment, the area of the adjacent territory is too small for such a large object. The lack of parking spaces, the total area of which is 3200 m2 according to the developer, and in fact a two-storey underground parking with a total capacity of 66 cars and a closed courtyard area, which is prohibited from parking, which clearly does not meet the needs of the tenants of the house. Residents use nearby roads to park their cars due to lack of parking spaces, thereby the carrying capacity dramatically reduces of already big traffic. It should be noted that such buildings create some additional problems on the transport network, because there are trade and office premises on the first floors for which parking spaces have not been designed, which additionally creates congestion of the transport network and creates an unfavorable road situation. A lot of residents in this building also needs social facilities in the area (kindergartens, schools, hospitals). The improvement of the territory has only a children's playground above the underground parking (Picture 5) and a place for placing trash containers. The cost of apartments is one of the highest in this house in the city despite the above data. This building is an unfortunate example of high-rise construction, because aerodynamic vortex flows are created due to improper location on the ground constant that create an uncomfortable environment for staying at the playground, lack of landscaping, parking spaces, recreational areas and sports. These problems are typical not only for Samara and for many Russian cities, where high-rise buildings are erected [15-20].

Fig. 5. The facade of a three-section building at the crossing of Vilonovskaya and Sadovaya streets in the city of Samara. Children's playground in the courtyard of a three-section building at the crossing of Vilonovskaya and Sadovaya streets in Samara.

3 Results

The Russian construction industry has a lack of experience in high-rise construction and generally has more problems in their construction than in developed countries in this area. The problems appear because of the climate, the proper location in the existing urban development, the lack of materials, technologies and construction equipment suitable for building high-rise buildings. As well as there are lack of right requirements documentation and proper compliance control, the existing requirements for territorial planning and zoning in the city, construction and maintenance of the building.

There are various approaches to solving the problems discussed in world practice. Any approaches however aren't applied for Russian reality. The infrastructure of a high-rise building solution is more urgent for Russia. Among them are modernization of engineering systems, transport accessibility, improvement of the adjacent territory, application of modern systems for building maintenance. We conclude that the correct location of a high-rise building in the urban area is important on the example of a building in Samara.

4 Conclusion

The importance of this topic is that any building in the city directly or indirectly affects its inhabitants. The growth of the largest Russian cities leads to a shortage of territory, an increase in the population, the need for its placement. There are all prerequisites for the development of high-rise construction. An analysis of this area problems allows to find optimal solutions for a certain territory, design features for each building separately.

It is impossible to make a conclusive conclusion about the necessity or inadmissibility of high-rise construction in the Russian realities. The introduction of high-rise complexes into the largest Russian cities seems justified. High-rise buildings allow to avoid the widening of the city territory, creating a "compact" urban environment in the conditions of a lack of territories. The concentration of the working population in the business center is also justified for economic reasons. A compromise is needed between the advantages of high-rise construction and the costs of building them. The construction of high-rise buildings leads to even greater problems and only strengthens the negative attitude of the population towards high-rise construction in an incorrect place for their construction for example, among low-rise buildings surrounded by narrow streets. The issue of optimizing the relationship between the density of building up, the width of road gaps, quantity of parking spaces and make these adjustments to the requirements documents ensuring proper control over the compliance of high-rise buildings with all regulatory requirements, especially at the stage of erection and putting into the operation of buildings it is necessary to regulate now in Samara, however, as in other Russian cities with a million population.

References

1. V. P. Generalov, The special aspects of designing high-rise buildings: a study guide, 296 (2009)
2. V. P. Generalov, E. M. Generalova, The development perspective of high-rise buildings topology, *The cities future*, **1(18)**, 13-18 (2015)
3. V. P. Generalov, E. M. Generalova, The skyscrapers era megatall, 334-335 (2013)
4. V. P. Generalov, E. M. Generalova, Prerequisites for the formation of high-rise social housing complexes in Hong Kong, 379-380 (2014)
5. E. M. Generalova, K. E. Galstyan, Analysis of the existing regulatory structure for the construction of high-rise buildings in Russia, 52-55 (2015)
6. A. S. Mironova, Aspects of utilization of nanotechnological wastes in the construction industry, **1(120)**, 58-59 (2009)
7. A. S. Pershina. Aspects of the Formation of Modern Architectural and Expressive Facade Designs of Civil Buildings Taking into Account Application of Nano-Materials, 132-135 (2016)
8. E. M. Generalova, V. P. Generalov, Specificity of modern architectural and spatial environment formation of cities (illustrated by South Korea), **11**, 46-51 (2015)
9. V. P. Generalov, E. M. Generalova, Energy efficiency and environmental friendliness of mass affordable housing in the case of Hong Kong, Sustainable architecture. **4(21)**, 32-38 (2015)
10. V. R. Vuchic, Transport in cities that are convenient for living, 413 (2011)
11. S. F. Zhilkin, E. A. Akhmedova, T. B. Karakova, Urban Strategic planning (2001)
12. T. Y. Vavilova, The principle of environmental compliance as a condition for the high technology industries development, **1 (2)**, 110-116 (2010)
13. D. Cappon Mental Health in the High-Rise CJPH, **62(5)**, 426-431 (1971)
14. V. P. Generalov, E. M. Generalova, Identifying the distinctive features of the concepts of "comfort of living" and "comfortable living environment", **2(23)**, 85-90 (2016)
15. T. V. Filanova, V. M. Melnikova, Functions Evolution and structures in the largest city (For example by Samara), **4**, 118-123 (2010)
16. N. A. Lekareva, Planting standards and planting construction development, 6-9 (2011)
17. Henley Simon. *The Architecture of Parking*, (New York, Thames and Hudson, 2007)
18. A. A. Kayasov, Perspectives for the business centers development in Russian regions, 16-20 (2011)
19. V. Ya. Lyubovny, Cities of Russia. Problems, possible ways of development and improvement of management, 14-16 (2012)
20. E. A. Akhmedova, Planning aspects of the perspective agglomerations transformation. 40-45 (2009)

The Notion of "High" and commitment to excellence in contemporary Russian architecture. History and project: looking into future

Yuri Volchok[1,2*]

[1]Scientific Research Institute of Theory and History of Architecture and Urban Planning, 111024, Dushinskaya St., 9, Moscow, Russia
[2] Moscow Institute of Architecture (State Academy), 107031, Rozhdestvenka St., 11, Moscow, Russia

Abstract. The article covers the issue of high-rise buildings (skyscrapers) construction in Russia as the dialogue of artistic image and intellectual idea. The study shows that the professional commitment to skyscrapers erection brings to the foreground the comprehension of magnitude of the notion of "contemporary" in terms of time. It is important from methodical point of view to return to the initial meaning of the notions that provide for adding authentic meaning to the words "suprematism", "commitment", "excellence", "new", "high", and other determinants of creativity capable of going beyond "flying geese" development pattern in architectural shaping. It is well known that V.G. Shukhov's patents of 1896 were widely used in contemporary morphology of shaping. The heritage of Russian Avant Garde of 1910-20ies serves as an inspiration from methodological point of view (it is more and more evident from foreign master's creative experience). This is why it is important to return, first of all, to comprehension of the author's version of the notion "suprematism" ascending to Malevich - meaning commitment to excellence and not the "emblem" of preferences in style. The article includes the arguments providing for the capture of the 2010ies and, especially, 2015-17ies as the years of critical changes in history. Russian masters of architecture started as equals the stage of cooperative creative work with foreign architects erecting skyscrapers.

1 Introduction

"The City is a vault full of architectural jewelry
- it is higher than the highest mountains in the World in its grandeur
and loftier than all blossoming landscapes of the nature in its beauty".
K.S. Melnikov

* Corresponding author: yvolchok@gmail.com

"The nature is not at all obliged to start
with things simplest for our understanding"
E. Mach

This article is in memory of Alvin Toffler who died in June of the last year (1928-2016), and was, perhaps, the most recognized futurologist in today's world. He is the author of the key phrase that serves as a camertone for the text: "The future calls for attentive and careful attitude" [1, 2].

As far as the issue of the future is discussed in connection with the architecture of high-rise buildings, the focus is made, of all possible levels of engagement with futurology issues, on the notion of "project". The project of "the future" is shaped as a dialogue of an art image and an intellectual idea. The emphasis should be made in this case on "maturation" of this dialogue into a latent (invisible from outside) stage of appearance of the new in shaping.

High-rise building construction has become, perhaps, the major (priority) feature of today's world predetermining the condition and the scale of global architectural trends. This is why the erection of literally every skyscraper in any point of the world cannot but be noticed. It becomes, as a rule, a global-scale event, the subject of public debate, academic discussions and a wide variety of interdisciplinary studies. Russian professional community does not neglect construction of skyscrapers. Thus, we can refer to the annual large international "Forum 100+" in Yekaterinburg, covering not only the current trends and new examples of high-rise constructions all over the world, but also serving as the platform for specific creative proposals for high-rise buildings and large complexes construction in various Russian cities and former USSR countries. The feasibility of skyscrapers erection not only in Moscow and St.-Petersburg but also in other large cities is substantiated as a resolute step towards their development.

At this point, the key issue of today comes to the foreground: How can we overcome the limitations imposed on creative efforts aimed at high-rise construction in this country by the "flying geese" development pattern?

The existing circumstances result in the need to stratify the issue comprehension into the two relatively independent sections. Beside the evident revelation and methodologically focused systematization of the trends changing one by one in the contemporary world, the attention fixes more and more on accumulated shaping regularities providing for the discussion of the mere possibility of (readiness for) independent solution, comprehensive from the creative point of view, of the issues and professional challenges resulting from the search for original concepts of high-rise construction in Russia. The readiness of the professional community for true skyscrapers construction shifts the focus to comprehension of magnitude of the notion of "contemporary". The regularities emerging in the process provide for revelation of the trends predetermining the future direction of an architect's professional creativity.

Methodologically, the implementation of these objectives in this work is performed through a gradual immersion into time depositions of architectural expertise in Russia. However, it is very important from methodological point of view to try and return to the initial meanings of the notions which have become overgrown, in impatient sequence of every-day life, with new historical details, particularities, limitations, often standing far away from "the original". As far as the focus in this article is made on high-rise construction issues: The way high-rise buildings arrange urban space and implement the notion of "the new" in shaping, it is necessary from methodological point of view to fix the attention on the fact that the altitude of multi-storey buildings is the last thing associated with the perception of "high".

It would be reasonable to understand that the term "high" means, first of all, a "high level" associated with quality... "high" meaning "elevated"; however, there are also high pitch (of voice), high pain threshold and many other derivatives from the notion, including: higher education, elevated temper, and, surely, high-tech, etc.

The notion of "the New" as a strive for perfection in this atmosphere of diversified understanding what is "high", as applicable to high-rise buildings, provides for considerable extension of the meaningful creative attitude to fully perceived 3-dimensional urban structure.

It is worth noting at this point that this issue is interesting to the author of the article as a process, in dynamics (evolution) of the notions which "do not stand still" but improve their performance. Thus, it is difficult today to speak on a 3-dimensional arrangement of urban space without taking into account the notion of "3D language space" [3]. The understanding of the subjects embraced by grammatology, syntax and other liberal science disciplines gives the opportunity for better correlation of general and specific in the course of Russian and foreign experience comparison than the "direct" comparative analysis of high-rise construction. In this context, the widely discussed modern architecture of skyscrapers in different countries based on the morphology lying on (ascending to) V.G. Shukhov's patents of 1896, acquires, inter alia, a deeper meaning. The patents provided for one of the main trends lasting over a century [4]. (It would be proper to remember that Gothic cathedrals erection based on concepts shaped as projects lasted even longer - at least for 150 years).

2 Methods

The magnitude of contemporary is, perhaps the most sensitively and speculatively perceived phrase in the every-day life of all decades that have passed since the beginning of the 20th century. And this is not a surprise: The last century faced so many "tectonic shifts" and revolutionary transformations in Russian history (and architecture was not an exception) that no immediate evaluations (or priorities, to be more exact) can obscure real values which make it into history. It is especially painful for architecture and urban planning, and cannot but be harmful for the comprehension thereof. Permanent changes to creative direction which often do not result from the logics shaping professional trends but are rather susceptible to various passing mindsets, however, break "the life and fate" not only of masters in architecture but also of cities and towns. As the result, they do not allow to bring together the fully functional ideas of coherence, magnitude, regularities of shaping transformations, while the contemporary architecture as a consistent stage of the architectural profession's merging into the future, as a meaningful "morphological carcass" for futurologist concepts, is impossible without them.

In view of such attitude to the discussion of the issue of place and role of Russian high-rise construction experience as a fully implemented dialogue between artistic image and intellectual idea as perceived from architectonic point of view of architectural shaping, the previous years, or, to be more exact, 2014-2017, will make it into the history of contemporary architecture as "a break-through". The rules of the "flying geese" game have been overcome. Russian architecture came up to the level of common creative work with its foreign counterparts.

As it should be, the process started somewhat earlier: In 2013, the house in the Mosfilmovskaya Street by S.A. Skuratov was marked with the prestigious international prize Emporis Skyscraper Award. Skuratov was awarded the fifth prize. The third prize was awarded to Jean Nouvele for the skyscraper in Doha (Qatar) [5]. In autumn 2016, Skuratov's Architectural Bureau promulgated the concept of a new quarter of skyscrapers in the Kransnopresnenskaya Embankment, in the immediate vicinity of the Moscow City [6]. The expert community appreciated the deep, comprehensive proposal, and its perhaps main advantage - its prospects and promises. The logic of Moscow City creation is developed reasonably and consistently.

But the main subject of our discussion will be the other key urban-planning space of the modern Moscow. I mean the territory of the former automotive plant ZIL (Likhachev Plant). The urban and architectural concept of its renovation was created by Yu.E. Grigoryan's Architectural Bureau Meganom Project. The concept was announced in December 2014. The active design works started in the municipal territory divided into construction lots. Two of them, the key to the perception of the "ZILART Peninsula", are designed by S.A. Skuratov's bureau. The housing complexes within the project do not rush upward. However, there would be a skyscraper in this, practically, new city center. It is being designed by the bureau Asymptote of the architect Hani Rashid [7]. We do not need a detailed analysis of the skyscraper's structure here. It is only notable that the author, in the course of its architectural solution presentation, emphasized that such masters of the 1920ies culture as Dziga Vertov, El Lisitskiy, Gustavs Klucis, would like his concept. This remark by Hani Rashid is the result of the naturally perceived desire to root his work as much as possible in the culture of the place where it is erected. It is to the point to remind that P. Eisenman invited in 2010 to ARCH Moscow Festival, when asked from whom he studied architecture, named J. Derrida and T. Adorno. Derrida came to Moscow in 1990. In numerous lectures and discussions he demonstrated deep knowledge of the shaping experience of OPOYAZ (Society for Poetic Language Studies) and emphasized its integral and significant role for contemporary philosophy of culture and morphological creativity [8].

It is not necessary to speak in detail on the close connection between Zaha Hadid (and other masters of deconstructivism) and Jaques Derrida, on one part, and her infatuation with Malevich's suprematism, works by Lisitskiy. Zaha Hadid turned to their heritage for the creative driver when devising the concept of her (master) diploma's concept (1977) [9].

Many of similar "strokes" taken from every-day architectural practice form the "architectural fabric" of high professional level, the only possible basis for cooperative creative work of most different architectural schools and experts. The New York architect Hani Rashid erects in Moscow the skyscraper which will be the dominant point of ZILART district, based on the urban-planning idea by Yu. Grigorian. The Moscow architect Yu. Grigoryan and his bureau "Meganom Project" approved the concept of the 305-metre-high skyscraper in Manhattan (262 Fifth Avenue Tower). The skyscraper construction with preservation of the existing 12-storey building as its foundation is one of the numerous creative insights drawing attention to the project. (We may even speak about the attitude to heritage preservation in urban environment trained by the Moscow practice of the authors of the skyscraper implemented in the concept devised for New York).

3 Results

This is why I find it important to fix (not only for future history but also for our every-day life) an entirely new phase of professional relations between Russian experts in architecture and their star foreign colleagues. The extreme purpose orientation of objectives emerging in the course of modern skyscrapers development provides for deeper understanding of the fact that the creative heritage from Russian Avant-Garde "holds hard" the most bright and deep masters of architecture and thinkers today, after many decades.

Suprematic Architecture. Commitment to Excellence.

World perception today is more and more often based on non-linear regular trends (which are not evident, i.e. are not on a surface) in world outlook. However, the dialogue of an art image and an intellectual idea focused on acquisition of authentic new things in the course of creative activity.

This is why the interest to suprematic approach to creative process is growing. K.S. Malevich, being both an artist and an art academician, created suprematism as an intellectual art [10]. Basing on cubism, he set out an author's triplicity of conscious attitude to creative work: to see, to know and to build. However, it is notable that in the beginning of the 1910ies construction was perceived as synonymous to creative work (V.Ya. Brusov) [11]. It made creative process definite and shifted the focus from the visionary perception and superficial style formation to the need to perceive creatively and to go deep into the pattern, the structure of spacial integrity. At the same time, tectology as a general science of organization, or, up to its author, "the humanitarian science of construction" (A.A. Bogdanov) appears [12].

The establishment of suprematic architecture took place in the environment of perception characteristic of that time, as an intentional dialogue of propaedeutics and prolegomena to the notions of "space" and "time". Up to Malevich, suprematism emerges in "the spatial time of the new architectural structuring". In 1928 he wrote in his letter to the editorial board of the journal "Modern Architecture": 'Suprematism appeared in 1913... Form 1918 spatial suprematism, the elements of which appeared in 1915, started its development' ... Therefore, all the way of the new art in all fields of culture came to the modern art, which is architecture" [13].

Unique character of suprematic art is rooted in the perception of its author (founder) — K.S. Malevich which developed to become the belief that it is necessary to create a suprematic method — the way of targeted movement towards perfection, constructed as a mental projection from the surface "painted on a canvas" onto the space of the world order.

K. Malevich fixed attention on "all types of culture" jointly making their way to new architecture. Beside his interest to the opportunities of painting without a subject, he went deeply into contemporary scientific advances and engineering developments. However, he was interested not in the specifics of each scientific or engineering disciplines but in the universal character of the world order, its most general regularities, It was natural for Malevich to understand the arrangement of a city, the Universe and a man as the wholesome integral system. Search for interpretation of deep layers on humanitarian and natural-science knowledge with help of images gradually led Malevich to understanding of methodical opportunities of architectonic thinking and activities, as a "pure artistic form". The art, embracing new perception of the world order following the logic of suprematic thinking, advances to understanding of spacial element's role in the structure of contemporary life. Geometric surfaces as suprematic elements interact with the space, organize and customize it. Stylistic regularities go "to the shadow" letting to the foreground the problems of architectural shape creation.

At the same time, the artist El Lisitskiy, in Vitebsk as well, taken away by the charm of Malevich's works, starts to create axonometric suprematic compositions thus translating suprematisn from surface to a space and making a decisive step on the further way to architectons and suprematic architecture itself. Malevich does not simply work on sculptural compositions named "Architecton", he shows at the same time how it can exist in a city [14].

Suprematism managed to lead the architectural Avant Garde, fertilising one of the most significant architectural trends forming the life style of that time - constructivism. Suprematism gave it an additional momentum and professional tools necessary on the way to perfection, search for one's creative maximum (first of all, in the works by L.M. Khidekel from the workshop of A.S. Nikolskiy and I.I. Leonidov, an apprentice of A.A. Vesnin).

Later on, the suprematic architecture was not in demand in Russia for some decades. Scientific interest to it as well as to other Avant Garde culture manifestations showed itself in the 1960ies, and in the 1980ies some major exhibitions and publications provided for new discussions of suprematic architecture values. The first full-fledged experience of implementation of suprematic architecture ideas ascending to Malevich in our time was Skolkovo campus in Moscow Region (2006-2010). The architect David Adjaye working in England managed to win the international competition for this project as he combines several meaningful planes in his concept, both properly and successfully [15].

Malevich distinguished two types of architectural activities: "architecture as a problem" and "architecture in life". It will be proper to compare the philosophic foundation of Malevich's artistic and theoretical endeavors with noospheric thinking of V.I. Vernadskiy [16]. However, the artist and the scientist moved not towards each other but along parallel ways in the same direction: towards perfect knowledge, and believed wholeheartedly that the parallel ways of the art and the science would merge.

In 1927 in Berlin, L. Lisitskiy put some efforts in the arrangement of the large personal exhibition of the works by Malevich with a great number of publications about the artist, his lectures and other events accompanying any considerable exhibition. Europe saw the art by Malevich first-hand, and he was introduced everywhere as the architect Kazimir Malevich. The suprematic architecture was successfully actualized.

"Figurines" by El Lisitskiy and the idea of the opera

"Victory over the Sun" in 1923: projection from the plane to the space. 2007.

The concept of the opera "Victory over the Sun" conceived as a futuristic work, turned for the future in 1913, as long as I understand, was devised by M. Matushin and A. Kruchenikh not "from scratch". The mute film shot in 1913 and based on A.S. Pushkin's poem "The Small House in Kolomna" written as long ago as in 1830, might serve as a momentum.

It is well known that the authors team of the opera consisted in 1913 of three participants of the First (National) Congress of Russian Futurists. It is of no importance that all three (K. Malevich worked together with the abovementioned authors) were the only delegates to the Congress and "the founding fathers" of both the movement and the futuristic theatre [17].

The problem of the new future construction in the concept of the opera as a futuristic phenomenon made the emphasis on the notion of "the new", and it was the new of the same scale as the maximum manifestation of creative grandeur. A.S. Pushkin was perceived in the context as the only "measure" of perfection - the benchmark. Or rather the personification of the absolute.

The fact that the idea of the opera direction was revisited in 1923 confirms such interpretation of the role "Victory over the Sun" played in the creative life of its authors. I do not think it was only by coincidence that in 1922 Igor Stravinskiy wrote the comic opera "Mavra" based on the same poem by A.S. Pushkin - "The Small House in Kolomna" [18].

The figurines (stage outfits) for that performance were made by El Lisitskiy. However, the outfits by Lisitskiy were not for actors but for puppets which substituted the actors as per the author's concept. It was a great conception. But it was not destined to be implemented. However, inn 1923 in Berlin Lisitskiy published the book of sketches of the outfits. Who could imagine that this story would so symbolically continued in 80 years...

But first let me say some words about the poem "The Small House in Kolomna". Pushkin wrote it not in his traditional iambic rhythm. It was the first time he used **octaves**!

It was the presentation of innovation in Pushkin's poem in the logic of his time with active focus on the new in art that served as a powerful provocation for the authors of "Victory over the Sun".

From this point of view, "the sophistry" proposed by Kruchenikh is similar to Eisenstein's methods. Numerous reviews of Dziga Vertov's news reels ("cinema eye") also shot in 1920ies, reported by Hani Rashid, is also rooted there.

In the modern art museum Danubiana Meulensteen near Bratislava one of the figurines from the sketchbook published in Berlin with the self-explanatory name "Coffinmaker" was translated into space. In the park in front of the museum there is the sculpture authored (in accordance with the museum's curators' concept) by El Lisitskiy: "Victory over the Sun", 1923.

It is not by chance that Hani Rashid has permanent reminiscences on behalf of El Lisitskiy and G.G. Klucis. The artist's creative interests merging into the architectural problems was, in this case as well, associated with his strive for the high art. The contemporary master of architecture working in Moscow took interest in the establishment of the creative method born by its time and aimed at excellence in the city.

4 Conclusions

Yu. Grigoryan, along with the urban concept, designs large spaces for ZILART. Residential houses along the designed central prospect are located in place of the blacksmith workshop erected against the project by the Vesnin brothers. The spatial solution of the new complex also refers to the works by the Vesnins: Their project for Narcomtyazhprom (People's Commissariat of Heavy Industry) tender in 1934.

It would be up to the point also to remember the conceptual theoretic heritage of the constructivist leaders inseparable from their design studies. In he same issue of the magazine "Architecture in the USSR" (USSR - the Union of Soviet Socialist Republics) (No. 2 of 1934) that published the Decree of the Council for the Palace of Councils under the Presidium of CEC of the Union of SSR (CEC - the Central Executive Committee) on approval of the architectural project and the authors team for the Palace design including: B.I. Iofan, V.G. Gelfreich and V.A. Shuko, published the large (in volume) and undoubtedly conceptual (as for its content) article by A.A. Vesnin, V.A. Vesnin and M.Ya. Ginzburg "Problems of Soviet Architecture" (under the heading "Creative Platform") [19]. The authors closed the discussion on certain then-relevant issues of architect's creative work with the question about "the adequacy of an architect's creative range". "We (Russia - Yu.V.) have become wealthier, we have more opportunities, we can now afford to renunciate ascetism (the reference to artistic opportunities of constructivism in that time's discussions - Yu.V.) and much wider scale. It is quite natural that our palette should become an adequate creative palette. But we should find a real delicacy here. Both economic and artistic delicacy".

The issues raised in that article has were developed in the text by I.I. Leonidov, who took part in the later discussion in the same magazine, also under the heading "Creative Platform", on the subject "Technology - Material - Style". His article's title was "Architect's Palette" ... Leonidov and his senior tutors still heard and supported each other. The article by Leonidov looked like a continuation of the conceptual article by the Vesnins and Ginsburg in the framework of the new discussion.

"Has the contemporary architecture mastered new materials and construction not only in engineering but also in artistic sense? What should be the way of further creative work of the Soviet architecture in mastering new construction techniques?" Those we the questions put by the magazine's editorial board before the prospective participants of the discussion [20].

Let us see how I.I. Leonidov exposed the metaphor "architect's palette" in a rather compact (less than one magazine spread) text.

We cannot but note the form of the text presentation. The author laid it out as arguments - strokes on a palette" of an artist. It (palette) unites and arranges the meanings, while the combination of strokes give the opportunity to "mix" them discovering new shades in the essence of the same problem discussed together with other participants of the discussion by I.I. Leonidov.

I.I. Leonidov does not speak about the ways an artist-architect can use the palette but about the ways to create it. And this is. I believe, the key to the never dying interest to Avant Garde heritage after a century.

The first argument presented by the author reads as follows: "one on the essential factors predetermining the style and the general composition of an architectural work is, along with social and household conditions, the technology of construction materials and structures. It is well known that this provision was worded for the first time in all its relevance by our functionalists".

It does not matter that after this I.I. Leonidov had to remark on the "mechanistic" approach to this problem in our country and in the West during the previous decade. The mere fact of the reference to the heritage of functionalism (constructivism) and inevitable continuity in the problem discussion speaks for itself. It sets not only "the starting point" but also a precisely aligned ethical level off common discussion, conscious professional responsibility for one's (recent) past. For the author, it was the conversation about the continuity of the world perception worded by constructivism, the return to recently neglected values in its essence. It sounded especially bright and decisive when voiced out by the architect resuming his active professional life.

The first argument set the tone to the rest of comprehension of the subject under discussion by Leonidov. "An architect should not approach construction technologies only from a merely constructive point of view. He should... Master the opportunities given by construction technologies philosophically. He should create new forms and construction of the given material. It is a basic creative need calling for some courage in the search for new forms".

This argument by I.I. Leonidov is supported by profound experience in both philosophical comprehension and shapes creation in Russian architecture and engineering. Today, 80 years later, the results of those speculations have served as the basis of architectural shapes creation adequate to our time and allowed to say that nowadays architectural shapes creation becomes the result of full-fledged cooperation on a parity basis of Russian and foreign experts in architecture.

References

1. A. Toffler, The Third Wave: The Glassic Study of Tomorrow, 544 (New York. Bantam Books 1980)
2. A. Toffler, The Third Wave, 784 (Moscow, AST, 2010)
3. Yu. S. Stepanov, In 3D Language Environment *Semiotic Problems of Linguistics, Philosophy and Art*, 335 (Moscow, Nauka Publ. 1985)
4. Genius of V.G. Shukhov and Contremporary Epoch: International Consgress Materials, 319 (Moscow, Bauman MSTU, 2015)
5. URL:https://www.emporis.com/awards/2012 (last accessed on 17.06.2014)
6. URL:https://naked-science.ru/article/concept/plyus-tri-neboskreba-dlya-moskvy (last accessed on 16.04.2017)
7. URL:http://archi.ru/projects/world/9194/bashnya-na-territorii-byvshego-zavoda-zil-zil-gateway-tower (last accessed on 25.04.2016)
8. J. Derrida, Jacques Derrida in Moscow: Deconstruction of the Journey. Translated by M. Ryklin, 198, (Moscow, Ad Marginem, 1993)
9. Zaha Fadid at the Hermitage, 208 (St. Petersburg, Arca, 2015)

10. Yu. P. Volchok, Suprematic Architecture: Style. Method. Commitment to Excellence *Televant Problems of Theory and History of Art. Issue VII,* 332-333 (St.Petersburg, Publishing House of St.Petersburg University, 2016)
11. V. Ya. Brusov, Keys to Mysteries, *Magazine Vesy* **1** (Moscow, Scorpio Publishing House 1904)
12. A. A. Bogdanov, Tectology. *Comprehensive Science of Organization*, 300 (Leningrad-Moscow, Kniga Pubishing House 1925)
13. K. S. Malevich, Letter to Editorial Board of Modern Architecture Magazine, *Modern Architecture*, **5**, 156 (1928)
14. S. O. Chan-Magomedov, Idols of Avant, Kazimir Malevich, 217 (Moscow. R.E. Gordeev Publishing House, 2010)
15. URL:http://archi.ru/projects/world/4835/moskovskaya-shkola-upravleniya-skolkovo (last accessed on 25.04.2016)
16. V. I. Vernadskiy, The Biosphere and the Noosphere, (Ayris Press, Moscow, 2013)
17. K. S. Malevich, Articles, Manifests and Other Works, 1913-1929, 5-volume set of works, 395 (Hylaea Publishing House, 1995)
18. A. S. Pushkin, Small House in Kolomna, 10-volume set of works, **3**, 539, Poems, Fairy Tales (Moscow, Fine Literature State Publishing House, GIHL 1959-1962)
19. A. A. Vesnin, V. A. Vesnin, M. Ya. Ginzburg, Problems of Soviet Architecture, Magazine Architecture in the USSR, **2**, 64-69 (1934)
20. I. I. Leonidov, Architect's Palette, Magazine Architecture in the USSR, **4**, 33-34 (1934)

Organizing vertical layout environments: a forward-looking development strategy for high-rise building projects

A.A. Magay[1], *E.A.* Bulgakova[2,3*], *S.A.* Zabelina[2,3]

[1]Institute for complex design of residential and public buildings
Dmitrov highway,house 9, building 3, Moscow,127434

[2]Moscow Information and Technological University
32 Volgogradsky Prospekt, building 11 , Moscow, Russian Federation, 109316

[3] Moscow Architecture and Construction Institute
32 Volgogradsky Prospekt, building 11 , Moscow, Russian Federation, 109316

Abstract. The article highlights issues surrounding development of high rise buildings. With the rapid increase of the global population there has been a trend for people to migrate into megacities and has caused the expansion of big city territories. This trend, coupled with the desire for a comfortable living environment, has resulted in numerous problems plaguing the megacity. This article proposes that a viable solution to the problems facing megacities is to create vertical layout environments. Potential options for creating vertical layout environments are set out below including the construction of buildings with atriums. Further, the article puts forth suggested spatial organization of the environment as well as optimal landscaping of high-rise buildings and constructions for the creation of vertical layout environments. Finally, the persuasive reasons for the adoption of vertical layout environments is that it will decrease the amount of developed urban areas, decrease traffic and increase environmental sustainability.

1 Introduction

Despite the variety of works devoted to residential high-rise buildings, there are no complex researches on the specifics of the formation of a vertical habitat with the development of terminology on this topic. For centuries, construction of buildings had a horizontal layout. First, there were caves altered for habitation. Then simple structures, that were constructed using the landscape features, such as leaning tree. Following this, building

* Corresponding author: traumeels@mail.ru

materials came into place and the first single-story buildings started to appear. With the invention of bricks (an artificial material), it became possible to construct building of three to five levels. If single-story buildings called for only a simple roof covering, multi-story buildings required floor slabs. To reach the second floor and higher, the stairway was invented - and that was the first step to introducing a vertical layout environment.

Increasing the height of buildings came with new issues, mainly involving construction. The issues included; defining the bearing, constructing protective structures, pumping water to the higher floors and developing water management, heating and lighting systems. These same problems were encountered with the appearance of taller buildings with between nine and sixteen floors. With buildings that had many levels, new challenges were encountered such as; ensuring water supply and transportation to the top floors, that required more advanced technology. Despite the increase in multi-level buildings, the general direction of urban and rural development was to remain following the horizontal layout. Following this trajectory, there would be an expansion in the occupation of more and more of the planet's territory [1].

2 Materials and Methods

Further expansion of vertical layout environments required special conditions, which arose at the end of 19th century in Chicago, USA, when massive fire destroyed most of the important strategic infrastructure of the city. After the fire, land speculators and various business owners started to rebuild the city, the price of land increased, which in turn caused the height of the buildings to increase too. Building developers began to favor multi-story houses that simultaneously used less land and achieved the required number of houses for residents.

In 1891, the tallest load-bearing brick building called "The Monadnock" was constructed, but it remains a sole example of this record-breaking height [2]. Without steel reinforcement, the walls of the first floor were 1.8 m thick, which naturally caused problems with lighting and inhibited the use of wide glazing. The problem was solved by specialists from the famous "Chicago School" of architecture, who implemented steel frames for high-rise buildings. This marked a revolutionary transition from load-bearing walls to steal frame construction of skyscrapers.

One of the pioneering ideologists of high-rise buildings was Louis Sullivan, who set out the architectural principle that "form follows function". Later, he divided the vertical space of skyscrapers into functional zones, stating that: "At first, a skyscraper needs an underground floor, where boiler rooms, power engines and other systems that supply resources for the building will be placed. Second, banks, shops and other institutions that require wide space, a lot of lighting, bright display cases and easy access from the street should be placed on the first floor. Third, the second floor should not have any less lighting or space than first floor, because it is easily accessible by stairway. Fourth, in between the second and the last floor, there should be numerous offices that may not differ in layout. Lastly, the top floor, just as the underground, should be technical and contain ventilation systems" [3]. Until recently, his division was applied by all architects. Modern architects implemented functions of the urban habitat such as: living place, work, trade, study, vacation and amusement. Yet with the change in the desired functions of the urban habitat, it became necessary to modify the concept of vertical zoning layouts with a new wider and forward-looking approach.

According to projections made by the United Nations Department of Social and Economic Affairs (UN DESA) and UN-HABITAT, 80% of people will be living in cities and the world's total population will reach nine billion by 2050, up from the current estimated population figure of 7.1 billion. This will result in the expansion of urban territories and an increase in population density, especially for existing large cities. Therefore, issues of vertical environment development for existing megacities becomes especially topical [4].

One of the major techniques used in creating vertically-oriented environmental layouts was the implementation of atriums that ensured the organization of multi-storey internal space with natural lighting and natural ventilation. A primary example of this technique is clearly seen in the construction of the Guggenheim Museum in New York City, USA. Implementation of an atrium into construction, besides creating a distinctive layout and composition (see Fig. 1a), works to organize traffic of visitors inside the space on the ramp, from top to bottom (see Fig. 1b).

a) b)

Fig. 1. The organization of vertical environment space in the Guggenheim Museum, New York City, USA.

Another noteworthy example is the experimental 16-story residential building with vertically composed public spaces was built in Tashkent, Uzbekistan. In this monolith construction, two buildings are connected at certain floors (2, 5, 8, 11 and 14) by traditional Uzbek open galleries [5]. Multi-story spaces were aimed to establish public zones for rest and residential interaction (see Fig. 2).

Fig. 2. Organization of public vertical space in the residential house. Tashkent, Uzbekistan.

The next step in the evolution of vertical environment layouts was the construction of Commerzbank in Frankfurt, Germany. Architect N. Foster designed an atrium that spread through the entire 260-meter building, and created additional open space with a 4-story winter garden, where light shines through internal windows into the building. Moreover, all of these spaces have natural ventilation because of the atrium construction.

At present, various designs and layouts are being developed and built, in which all kinds of human living activities could be efficiently carried out without leaving the building or complex. One common thread of all these current projects is that they implement vertical spaces. An example of this would be two residential towers "Bosco Verticale" (Vertical forest) in Milan, Italy, that represent possibilities of vertical environment layout for everyday living [6]. Specialists from "Boeri Studio" created a concept of space where high rise building are integrated with the greenery and landscape. This takes the shape of "Bosco Verticale" not only becoming a concrete building with potted plants on its sides, but a real forest with different species of bushes and trees intertwined. There are more than 800 trees, up to 5000 bushes and more than 10000 perennial plants and flowers in "Bosco Verticale" (see Fig. 3).

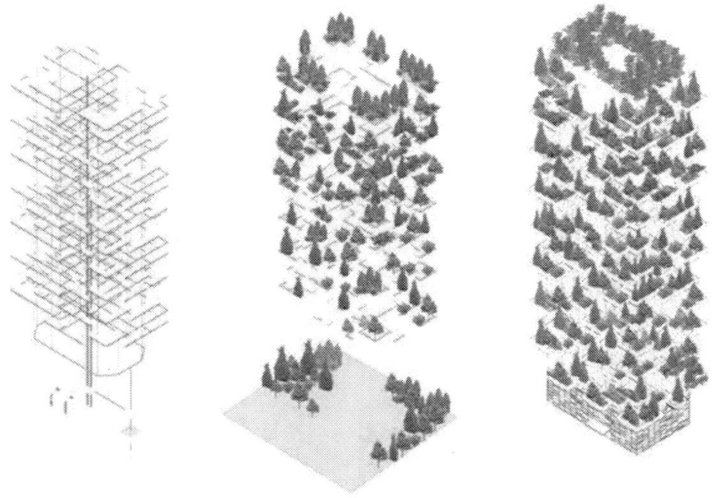

Fig. 3. Sustainable spatial landscaping of high rise residential building "Bosco Verticale", Milan, Italy.

There are also examples of modern high-rise office buildings with vertical environments. An example is the 330-meter tall "Chinese World Trade Tower III" skyscraper in Beijing, China [7]. In this multifunctional building there are hotels, offices, apartments, restaurants and different commercial spaces. The lower four floors are connected to the underground trade center, which expands the vertical environment for living activities. On the roof of the stylobate, there are pines growing with water cascades fitted into the green landscape. This building is a clear example of uniting the internal vertical space with the external urban environment.

A future project to construct a sustainable skyscraper with greenery and 26-stories of vertical space was recently proposed by Malaysian architect K. Young. The "EDITT Tower" skyscraper aims to bring living nature to the area of development that is currently lacking plants and vegetation. Architects are attracted to the visual effect of transitioning from a horizontally-oriented surrounding landscape to a vertical one, thus creating an organized vertical environment. A diverse vertical living environment can be attained by using wide open ramps that will connect higher floors with the first level.

A concept for a residential pyramid was designed for the Parisian exhibition complex of Porte de Versailles. It is suggested that a pyramid skyscraper will incorporate whole town districts with a full range of services for its inhabitants. There will be residential blocks, offices, restaurants and recreational areas [8].

A different take on the vertical layout is in the 196-floor skyscraper-city «Tokyo's Sky City 1000» project, developed by Japanese company "Mori". The skyscraper is divided into 14 platform sectors, each the size of a stadium, which can accommodate 35,000 people for living and further 100,000 for working. This city-skyscraper is designed to incorporate full-circle urban infrastructure and to satisfy all human needs including; accommodation, work and recreation. Being a full-fledged city, this skyscraper will help its inhabitants save a great amount of time. This is an especially pertinent factor considering that nowadays, most of Tokyo's citizens spend anywhere from two to four hours commuting [9]. Combining the hours spend commuting each day, in an average lifetime, commuting takes up six years.

Fig. 4. Project for organizing an autonomous vertical city space in skyscraper. Tokyo, Japan.

Creators of the futuristic project "Endless City", went even further and developed an idea of turning a skyscraper into a true vertical sustainable system (see Fig. 4) [10]. It is designed as a multifunctional ramp tower that is surrounded by a gigantic atrium. Ramps represent streets and connect various horizontal spaces into one vertical body of green landscapes, shops, restaurants and promenades - all together forming a comprehensive vertical-horizontal system. The design of the building responds to the needs for natural ventilation and natural lighting, that are aimed at decreasing the tower's energy consumption. It is important to highlight that numerous floors of this construction will have the ability to transform according to needs. Further details stipulate ramp platforms serving as open parking lots and public spaces. Projects like this futuristic skyscraper actually shifts urban life from horizontal living space that is close to the ground into a vertical habitat that provide all necessary functions for human living [11,12]

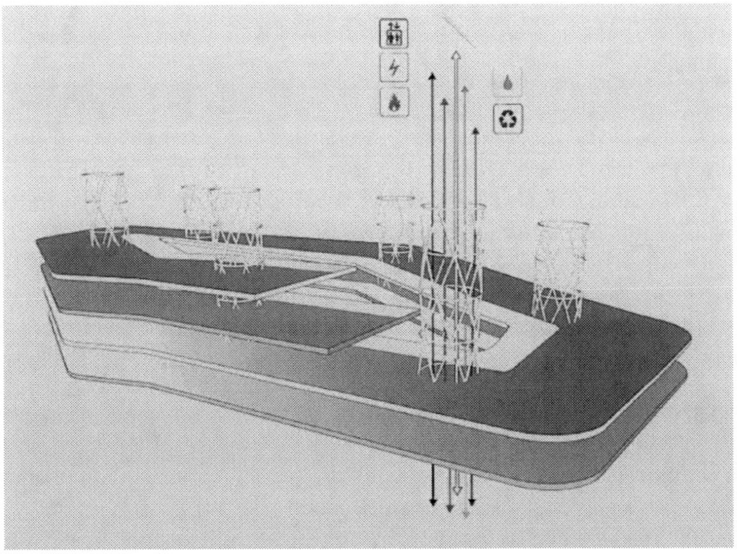

Fig. 5. Organization of vertical-horizontal living environment. Futuristic project.

World design experience shows that the main focus of high-rise building in the future is the creation of vertical cities [13, 15]. The projects are still in the nature of offers, but the main trend is aimed at providing everything necessary for a large number of people, which may live and work there; and at the same time envisage the use of solar and wind energy.

The mandatory consideration of environmental standards has revealed an actual perspective for creating intellectual self-sufficient and self-regulating houses [14]. The projects envisage the maximum use of natural daylight, reducing of energy consumption and reducing of solar overheating [16].

The best environmental direction is manifested in the (so-called) "green skyscrapers". The main feature of such constructions is that one-third of the total area of the high-rise building is allocated to gardens with live plants and a natural irrigation system, which forms a general view of the building [17,18]. The main idea is the effect, when the surrounding landscape turns from the horizontal plane into the vertical one.

Analysis of the submitted material makes possible to single out several trends of long-term development of the vertical habitat`s creation, including: the possibility of transformation in function and structure, the formation of multifunctional complexes, essentially "vertical cities" and an environmental component, that combines several functions in a vertical structure [19].

3 Results

As aforementioned, it could be argued that one of the most progressive directions for architecture and high-rise construction development is the organization of vertical layout environments. Different projects reviewed in this article let us define trends of development.

From buildings already existing, such as Guggenheim Museum and the 16-floor residential house, "Bosco Verticale", to projects that are ready to be implemented, such as the "Chinese world trade tower III", and the "EDITT Tower". Additionally, conceptual projects such as a pyramid construction in Paris, and futuristic projects such "Tokyo's Sky City 1000" and "Endless City" let us envisage what is to come [20].

4 Discussion

The theoretical basis for the study were the work of Y. Bocharov, A. Bokov and others. In the work on this topic were used the design materials of the Central Research and Design Institute for Residential Construction, where the questions of typology and social aspects of construction of high-rise buildings were analyzed in the works of A. Magay [5,6], T. Maklakova, A. Rubanenko, H. Saprykina, and others. Foreign researches of high-rise buildings, with the linkage of bioclimatic design, are associated with the works: K. Yang, Alexander, Masetti and others.

5 Conclusions

Design and engineering of buildings and constructions with vertical environment layouts will help to accomplish several tasks at once such as: improvement of ecology through introduction of greenery and vegetation in buildings, reduce the spread of megacities into wider territories, diminish traffic and commuting time and conserve energy consumed by shortening time spent travelling to services. Beyond the positives of implementing vertical layout environments, distinctive architecture of high-rise buildings and constructions may make the urban landscape more diverse and create new urban visual accents.

References

1. A. A. Magay, Architectural design of high rise building and complexes: Learning material, 248 (Moscow, ASV Publishing, 2015)
2. A. A. Magay, Modelling of functional structures in high rise buildings, *Housing construction*, **12**, 17-21 (2016)
3. A. A. Magay, Normative basis for high rise building in Russia, ed. Nikolaev S.V. *Housing construction*, **1(2)**, 3-6 (2016)
4. Skyscraper, Wikipedia — the free encyclopedia-2007. [Electronic resource]
5. A. A. Magay, High rise buildings as vertical integrated residential environment, *BST: bulletin for building technic,* **6**, 64-65 (2016)
6. A. A. Magay, Specificities of architectural typology of high rise buildings, *Architecture and construction in Russia*, **4**, 22-29 (2009)
7. B. Gaiboullaev, I. I. Tojiyev, Principles for projecting energy-active buildings, *Young Scientist,* **7**, 55-58 (2016)
8. http://www.membrana.ru/ particles/tag/151 (last accessed 2017.08.22)
9. E. A. Boulgakova, A. A. Savicheva, Modern trends in engineering of transport hubs in megapolis infrastructure, *Eurasian scientists union*, **4**, 155 (2015)

10. Y. G. Repin, Spatial city: theory and practise, Yuri Gavrilovich Repin, 267 (2009)
11. http://skyscrapernews.com (last accessed 2017.08.22)
12. S. V. Nikolaev, V. I. Travush, Y. A. Tabunshikov, Regulatory Framework of High-Rise Construction in Russia, *Housing construction*, **1-2**, 3-6 (2016)
13. A. A. Magay, V. S. Zyrianov, E. Y. Shalygina, Importance of Special Technical Conditions for Design of High-Rise Buildings *Housing construction,* **11**, 17-20 (2015)
14. A. A. Magay, Architectural and compositional features high-rise buildings, *Akademicheskij vestnik uralniiproekt RAASN*, **4**, 25-30 (2015)
15. G. P. Vasilyev, V. A. Leskov, N. V. Mitrofanova, V. F. Gornov, Technical solution for protection of heat pump evaporators against freezing the moisture condensedmatec web of conferences Cep, *"2015 International Conference on Mechanical Engineering and Electrical Systems, ICMES 2015"* (2016)
16. K. Yeang, Eco Skycrapers, 160 (Hong Kong, Images Pudlishing, 2007)
17. A. A. Magay, Prospects of development of the regulatory framework of high-rise construction in Russia, *Housing construction*, **12**, 3-6 (2016)
18. A. A. Magay, The peculiarities of architectural typology of high-rise buildings, *Architecture and construction of Russia*, **4**, 22-29 (2009)
19. N. A. Saprykina, Architecture on the edge integrated technologies, *Housing construction*, **7**, 14-16 (2003)
20. http://www.jbdesign.it/ (last accessed 2017.08.22)

The Global Experience of Deployment of Energy-Efficient Technologies in High-Rise Construction

Natalia D. Potienko[1,*], *Anna A.* Kuznetsova[1], *Darya N.* Solyakova[1], and *Yulia E.* Klyueva[1]

[1]Samara State Technical University, 443001, Samara, Molodogvardeyskaya st, 194, Russia

Abstract. The objective of this research is to examine issues related to the increasing importance of energy-efficient technologies in high-rise construction. The aim of the paper is to investigate modern approaches to building design that involve implementation of various energy-saving technologies in diverse climates and at different structural levels, including the levels of urban development, functionality, planning, construction and engineering. The research methodology is based on the comprehensive analysis of the advanced global expertise in the design and construction of energy-efficient high-rise buildings, with the examination of their positive and negative features. The research also defines the basic principles of energy-efficient architecture. Besides, it draws parallels between the climate characteristics of countries that lead in the field of energy-efficient high-rise construction, on the one hand, and the climate in Russia, on the other, which makes it possible to use the vast experience of many countries, wholly or partially. The paper also gives an analytical review of the results arrived at by implementing energy efficiency principles into high-rise architecture. The study findings determine the impact of energy-efficient technologies on high-rise architecture and planning solutions. In conclusion, the research states that, apart from aesthetic and compositional interpretation of architectural forms, an architect nowadays has to address the task of finding a synthesis between technological and architectural solutions, which requires knowledge of advanced technologies. The study findings reveal that the implementation of modern energy-efficient technologies into high-rise construction is of immediate interest and is sure to bring long-term benefits.

1 Introduction

Rapid industrial development alongside with urban sprawl that led to a high urban population density has prompted the grand-scale construction first of multistory, and then of high-rise buildings. In the early stages of its development high-rise construction was primarily focused on the problems of structural design, space planning, massing and engineering, but later on a great number of prominent architects turned to extensive research on structural morphology [1, 2]. This led to the establishment of the International Style, as well as to the formation of various conceptions concerning the location of high-rises in the layout of American, European and Asian cities. There emerged a necessity to introduce new massing solutions in order to make monotonous residential areas more varied and resident

* Corresponding author: natalia.potienko@mail.ru

friendly, which in its turn gave rise to the new types of spectacular and eye-catching high-rise buildings (Petronas Towers, Burj al-Arab, Burj Khalifa, Taipei 101, etc.).

The 1974 and 1992 energy crises caused the shifting of emphasis in the sphere of building design, with high-rise design being no exception. The first steps were to make the best use of such architectural techniques as the orientation of a building along the cardinal directions and to the prevailing winds, and the application of different types of façade glazing: maximum on the south side and minimal on the north. These passive methods depend upon the climate of a certain region and are characterized by low energy consumption. However, the very attitude towards the usage of energy was reconsidered later on. This problem is covered in a considerable number of research studies [3-22]. Today the energy policies of all countries in the world are aimed at designing energy-efficient and comfortable buildings through the implementation of efficient architectural and technical solutions. The resiliency of a building is improved by harnessing alternative energy sources – solar and wind energy, geothermal energy and hydropower, as well as by the conservation of natural resources and the application of modern technologies (intelligent systems). All this results in the transformation of exterior design and makes it necessary to reconsider the traditional functional, planning, design and engineering structures of a building.

2 Materials and Methods

It should be noted that the grand-scale construction of energy-efficient high-rise buildings is carried out worldwide. The USA along with a number of Asian countries and the UAE are considered to be the world`s leading nations in the sphere of high-rise construction. The climates of the abovementioned countries comprise a wide range of weather conditions, which influence the implementation of certain energy-efficient technologies in different ways. Many experts claim that the methodology of energy-efficient high-rise design should be based on a systematic and comprehensive analysis of a building as a unified energy system. Nowadays the reduction of energy consumption and a negative environmental impact has become one of the ways to improve the comfort level of a building. From this perspective, various energy-saving methods and techniques are employed around the world to improve the energy performance of high-rise buildings and complexes. These methods rely not only on the use of modern engineering systems, but also on the implementation of efficient space planning, architectural and design solutions.

Due to the vastness of its territory, Russia is located in four climatic zones: arctic, subarctic, temperate and subtropical. The temperate zone occupies the largest part of the country, while only a small portion of the Black Sea coast, namely the Caucasus Riviera and Crimea, possesses a subtropical climate. As a consequence, significant weather contrasts can be observed in the country. For instance, the average temperature of the warmest month ranges from +1 °C in the Far North to +25 °C in the Caspian Depression and +40 °C in the south-west of Siberia. The same pattern is witnessed in the temperature of the coldest month – from +6 °C on the Black Sea coast to -50 °C in north-eastern Siberia. Many factors influence the shape of a building, hence its natural lighting, ventilation and thermal efficiency. Solar activity, for one. Also, the prevailing winds and wind force, the amount and pattern of precipitation, as well as mistiness and air humidity.

The concept of energy-efficiency has finally started to gain recognition in Russia. Many Russian companies, both those who construct and those who use buildings, have come to understand that there exists an acute problem of heat loss and that it is necessary to implement the most advanced energy-saving solutions and modern thermal protection materials, sandwich wall constructions, energy-saving plumbing systems and engineering equipment into high-rise construction. Moreover, the country is witnessing significant changes in the legislative and regulatory frameworks. A bill has been passed recently to increase the share

 https://doi.org/10.1051/e3sconf/20183301017

of nonconventional and renewable energy sources up to 10 per cent. Besides, documents were developed providing for regulations concerning the implementation of energy-efficient technologies in Russia. In this regard, it is of vital importance to examine and analyze the global expertise in the use of energy-efficient technologies in order to apply it in high-rise construction and adjust it to the Russian climate, as well as to develop our own unique solutions.

3 Results

The research offers an analytical review of the results arrived at by implementing the principles of energy-efficient architecture.

The **Pearl River Tower** (310 m, 71 stories, 2013) is considered to be the first genuinely green skyscraper in China. The tower is also known as the most sustainable building in the Middle Kingdom (figure 1a). The building was deliberately oriented along the cardinal directions and to the prevailing winds [23].

Fig. 1. Examples of buildings: a - Pearl River Tower, an administrative office centre, Guangzhou, China. Authors: Adrian Smith + SOM [23], b - Edith Green-Wendell Wyatt Federal Building, an office building, Portland, USA. Authors: Skidmore, Owings and Merrill [24].

The tower`s triple-glazed façade blocks external heat, thus eliminating the need for air cooling. Lighting control sensors and mechanized blinds are programmed to maximize natural lighting. The south façade has three main structural parts formed by glass panels curved both vertically and horizontally. The façade is shaped so that air masses can accumulate in the cavities between the parts. Convex glass panels acting as deflectors are located in the middle of the cavities. Air masses hit the surface of the deflector and move into air inlets where wind turbines are located. The Pearl River Tower design uses some elements of the exterior enclosure as a surface for accumulation and deflection. Besides, the enclosure also functions as a part of the building`s wind turbine system, which characterizes it as a polyfunctional element. The tower owes its unique shape to the synthesis of design, material and technological solutions. Another innovative aspect of the building`s design was the use of wind mills: four wind turbines with 6m diameter impellers are incorporated into the openings of the mechanical floor enclosures – two mills in each mechanical floor. Furthermore, the building`s power supply relies on the use of photovoltaic panels integrated into the east and west façades. The most notable energy savings are achieved via cooling ceiling systems, which are remarkable for using water instead of air. In order to prevent condensation the incoming air is dehumidified in special heat transfer apparatuses located in the mechanical floors. Fresh air is provided via the system of displacement ventilation integrated into the floor construction. Another notable feature is that air ventilation between the glass panels of the façade constitutes a part of the overall aerodynamic system of the building.

The **Edith Green-Wendell Wyatt Federal Building** (110 m, 18 stories, (1974) – 2013) is a high-rise structure completed in the 1970s and transformed into a sustainable tower in the course of a three-year renovation project (figure 1b). A new glass and metal enclosure completely changed the appearance of this modernist looking concrete parallelepiped [24].

The building`s façades were tuned to respond to the unique solar exposure of each side. The reed-like shades affixed to the west façade of the building provide 50 percent shading on the west side, where the sun is low and light comes in at a small angle. The 'reeds' have a trapezoidal cross section, the narrow part of which faces the interior. This was done in order to optimize shading and to reduce both the size of 'reeds' and the total weight of the construction. The round edges of the 'reeds' do not produce sharp shadows. The south and east façades are equipped with both horizontal and vertical shading systems with a depth of 60 cm. While the bottom part of the horizontal shelves casts a slight downward shadow, the upper part bounces daylight 9-10.5 m toward the interior spaces, thus making the most of natural light. Apart from optimizing the building`s solar exposure, these structures also help to reduce cooling costs.

Two layers of insulation, each 10 cm wide, were incorporated into window sill panels to protect the building`s envelope from heat loss. Air circulation is provided via the dual system of radiant heating and cooling reinstalled above the drop ceiling. Greater than 65 percent water savings were made possible due to implementing water conserving plumbing fixtures together with a rainwater system. A 770-liter tank allows rainwater to be stored and used for technical purposes such as toilet flushing, irrigation and cooling. The single-pitched roof of the building supports a 180 photovoltaic panel that ensures additional energy savings (4-15 percent). Another aspect of the 'green modernization' is the use of energy-efficient elevators with regenerative drives that restore potential energy when the carriage travels down.

The **Al Bahar Towers** (145 m, 29 stories, 2012) is a high-rise structure consisting of two towers that use the motifs of traditional architecture on their façades (figure 2a). In order to protect the interior spaces from 50 degrees Celsius heat, the design team installed a shading system above the building`s façade on the south, east and west sides. This screening system comprises 2000 umbrella-like components that open and close in response to the sun`s path and external lighting conditions [25, 26]. The shading units are controlled via the building management system so that the devices are fully open in the morning and fully closed at noon. The system runs on renewable energy generated by photovoltaic panels that are located on the southward sloping roofs of the towers. The protective screen helps to maintain a comfortable room temperature and eliminates the building's need for air-conditioning, which significantly reduces energy consumption. It also ensures indoor air ventilation and reduces the need for artificial lighting. The glass used for the façade glazing was slightly tinted in order to improve the comfort level for office workers and enhance the protective effect. The building design is a perfect example of how modern technologies can be integrated with ancient Arabic traditions that let local people survive extreme weather conditions.

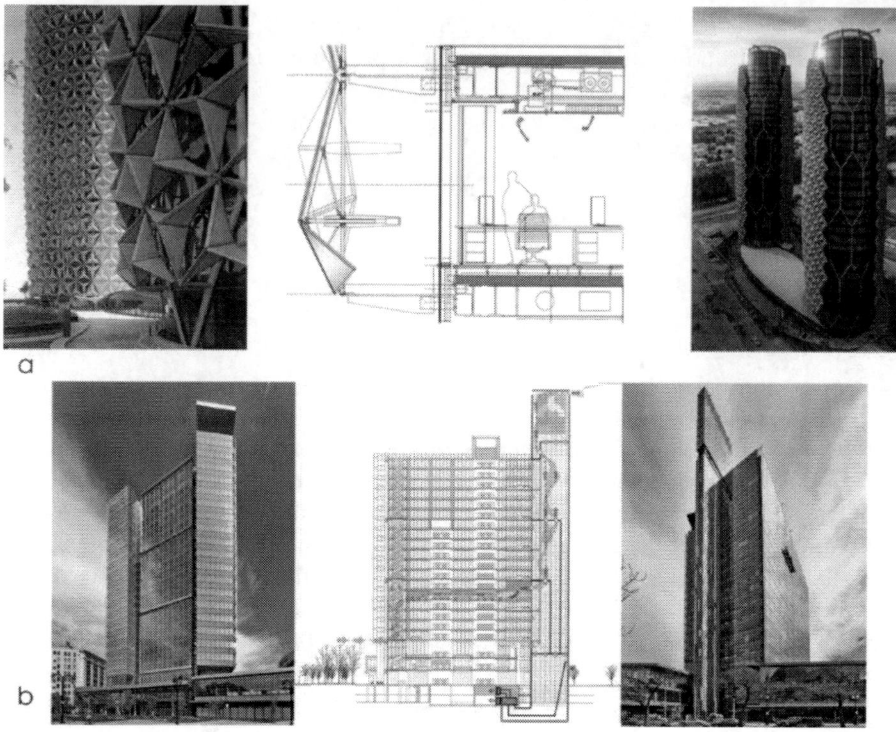

Fig. 2. Examples of buildings: a - Al Bahar Towers, an administrative office centre, Abu-Dhabi, UAE. Authors: Aedas Architects Ltd [25, 26], b - Manitoba Hydro Place, an office centre, Winnipeg, Canada. Authors: KPMB Architects with Smith Carter Architects [27].

Manitoba Hydro Place (115 m, 22 stories, 2008) is a building, the architectural form, orientation and size of which take into account Winnipeg`s harsh continental climate (figure 2b). A 115-meter tall solar chimney makes the main entrance and is rotated to position the multilevel atria southward, thus allowing solar radiation and wind energy to be used during winter [27]. The tower itself functions as a passive solar collector which relies on the natural stack effect. In winter, exhaust air is drawn to the bottom of the solar chimney by fans. The recovered heat is then used to warm the parkade and to preheat the incoming cold air in the south arcade. During summer months used air is drawn out of the building. East and west oriented office spaces are splayed southward, which significantly reduces heat loss. Floor-to-ceiling windows and narrow floor plates maximize natural lighting. Among other functions, winter gardens located on the south side pre-condition the incoming air, while an artificial waterfall provides humidification. A green roof supports plants and harnesses rainwater. The active systems of the building include light-regulating, computer-controlled and energy-saving fluorescent lamps. The reduction in heating demand for the building is largely due to a double façade on the west, east and south sides. In summer, operable windows on the outer (double-glazed) façade open and allow wind and convective air flows to ventilate the double façade. Mechanized window blinds control heat gain, while exterior blinds are manually operable and can be open for the purpose of air ventilation. Thus, the façade contributes to reducing thermal load. Heating and cooling is achieved primarily through the use of active radiant slabs (more than 290 000 m of plastic tubing is embedded in the concrete). Besides, the building also harnesses the internal heat of its geothermal system. The massive all-concrete building reduces temperature fluctuations through the occupied hours and allows heating and cooling systems to be turned off in the evenings and at weekends. Two on-site weather stations collect climate data (temperature, solar radiation, wind velocity and direction, rainfall intensity) and use it to optimize the performance of engineering equipment (operable windows, the angle of the shading devices, slab temperatures, etc.).

The **Bosco Verticale** (115.9 m, 30 stories, 2008-2014) consists of two residential towers, 80 and 112 meters high respectively (figure 3a). The façades are covered with plants which vary in form and size and provide dust absorption, solar protection and a favourable microclimate [28, 29]. This building can be considered a kind of biological architecture which refuses to take a strictly technological and mechanical approach to environmental sustainability. Both towers are oriented exactly along the cardinal directions. Due to their compact layout, the towers minimize heat loss during the cold period. A special kind of soil was selected in order to reduce the structural load of the cast-in-place reinforced concrete building. On the one hand, the soil is not very heavy. On the other, it meets the nutritional requirements of the plants. Experimental tests performed in a wind tunnel showed that the plants did not add to the overall wind load on the building. Centralized automatic drip irrigation is provided via a system of water storage and treatment which collects and filters grey water and rainwater. Photovoltaic panels and solar collectors provide the towers with power and hot water supply, while geothermal energy is exploited for heating.

The building of the **Strata Tower** (147.9 m, 44 stories, 2005-2010) is oriented along the cardinal directions and to the prevailing winds so as to maximize the use of daylight and south wind (figure 3b).

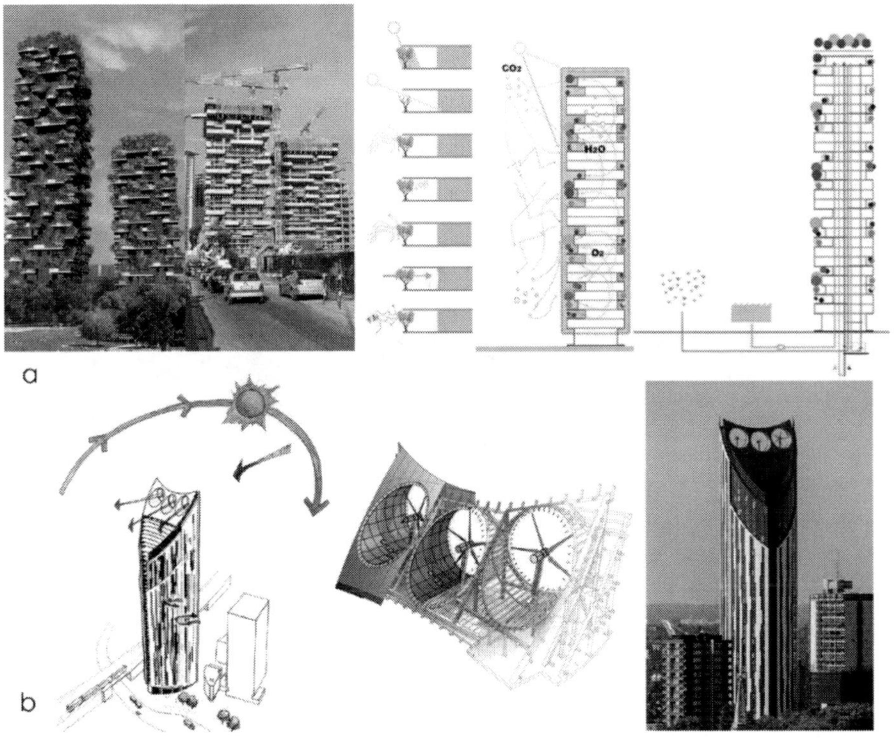

Fig. 3. Examples of buildings: a - Bosco Verticale, a residential complex, Milan, Italy. Authors: Stefano Boeri Architetti [28, 29], b - Strata Tower, a residential complex, London, Great Britain. Authors: BFLS [30].

The form of the building helps to cut lighting costs and reduces heat loss due to a smaller cladding area. Basement and roof spaces also contribute to the reduction of heat loss [30]. The roof space is occupied by three turbines, since the Strata Tower is the world`s first building to integrate wind turbines in its design and architecture. The aerodynamics of the construction was designed to maximize wind power throughout the year. The turbines produce only 8 percent of the building`s total energy consumption; for this reason, natural ventilation, heat recovery ventilation systems and high-performance insulating glass units are used to reduce heating and cooling costs. Heating is provided via boilers with low gas flow rate, while grey water is employed for toilet water supply. Thus, the building primarily aims at conserving its internal energy.

The **Wuhan Greenland Center** building (636 m, 131 stories, 2010-2018) has the shape of a triangle, the softly rounded corners of which are oriented towards the prevailing north-east, south-east and south-west winds (figure 4a). The building`s uniquely streamlined form makes a great contribution to wind load reduction [31]. In addition, apertures have been placed in the curtain wall to assist in venting wind pressure against the tower. Six basement floors and a mechanical floor in the upper level constitute 'buffer' zones and reduce the total heat loss. The building uses heat recovery ventilation, pretreated grey water for its evaporative cooling system, an automatic lighting system with low-energy-consuming lamps, and water-conserving low-flow plumbing fixtures (air balloons). Every detail of the building`s design is aimed at preserving internal energy and reducing carbon footprint.

Fig. 4. Examples of buildings: a - Wuhan Greenland Center, a mixed-use complex, Wuhan, China. Authors: Adrian Smith + Gordon Gill Architecture [31], b - Citadel Skyscaper, a mixed-use complex, Japan. Authors: Victor Kopieikin, Pavlo Zabotin [32].

The **Citadel Skyscaper** (500 m, 2012) project was developed for the coastline of Japan, a country affected the most by natural disasters (figure 4b). The design is actually a symbiosis of a skyscraper and a bunker able to protect the inhabitants of the island against natural and anthropogenic influences [32]. Within the framework of the project, all functions of the urban environment are performed by autonomous towers installed in the sea. The citadels themselves are connected by the coastline and create a defensive shield functioning both above and below ground. It is proposed to locate the skyscrapers at a distance of 2-3 km from the shore. The towers will be able to withstand tsunami waves up to 50 meters. The foundation is planned to be laid at a depth of 1200 m, while the height of the buildings will constitute 500 m, with a guarantee of seismic stability. It is worth mentioning that the citadels are intended to be completely self-sustaining in terms of energy: tidal energy is planned to be used as the principal source of power, drinking water supply will be provided via boreholes, and food is planned to be stored in the basement floors.

4 Discussion

The examples considered above make it possible to speak about the successful implementation of modern energy-efficient technologies into high-rise construction. In this respect, alongside with traditional technologies used in low- and mid-rise buildings, new specific technologies are created that are characteristic of high-rise buildings and result from their design, massing and planning solutions. The range of technologies applied is wide and continues to develop.

Recent decades have witnessed the rapid development of energy efficient and alternative energy sources technologies, which continue to evolve and generate more and more concepts of energy efficient buildings with respect to the number of stories. Some ideas have already become a reality, while others remain on paper in the form of futuristic projects which will be realized in the near future. One can say with certainty that energy-efficiency issues are of vital importance today in relation to any construction. However, in the case of high-rises it commonly happens that not the technologies form a building, but rather a conceptual idea gives rise to technology. The key aspects of the projects include lower energy consumption and lesser environmental impact of high-rise construction and operation.

5 Conclusions

Energy efficient technologies introduce new forms and technological elements into architecture. Furthermore, new building materials continue to appear, and new compositional and artistic solutions have been found for façade silhouette and articulation. Apart from aesthetic and compositional interpretation of architectural forms, an architect nowadays has to address the task of finding a synthesis between technological and architectural solutions, which requires knowledge of advanced technologies. In this regard, an architect should collaborate with other specialists, conduct experiments and research, analyze the findings and choose such methods to improve energy efficiency (or the combination of such methods) that will meet the given parameters (aesthetic, ecological, energy-efficiency, etc.).

References

1. V. P. Generalov, The History of High-Rise Construction (Samara, 2011)
2. V. P. Generalov, The Characteristic Features of High-Rise Building Design (Samara, 2009)
3. V. Ivanov, I. Bakhtina, T. Ivanova, S. Ilinykh, Urban Construction and Architecture, **2**, 88-93 (2015)
4. O. S. Popel, Energosvet, **5**, 22-26 (2011)
5. T. Ya. Vavilova, N. D. Potienko, I. V. Zhdanova, *Procedia engineering*, **153**, 938-943 (2016)
6. G. Sarkisian, Ya. Gorbatenko, Energy efficiency in Russia: hidden reserve (Moscow, Center for energy efficiency, 2009)
7. V. Generalov, E. Generalova, Urban Construction and Architecture, **4**, 23-29 (2015)
8. G. V. Esaulov, Proceedings of TSUAB, *Architecture and urban planning*, **6**, 9-24 (2014)

9. A. N. Remizov, Sustainable Architecture: Present and Future, *Proceedings International Symposium* 17-18 November 2011, 40-50 (2012)
10. E.Yu. Vityuk, Architecton, **13**, 45-50 (2010)
11. J. W. Lund, Design of closed-loop geothermal heat exchangers in the U.S. Course on geothermal heat pumps (2002)
12. H. Ehhorn, J. Reiss, H. Kluttig, R. Hellwig, AVOK, **2**, 36 (2006)
13. B. Sanner, Ground Heat Sources for Heat Pumps (classification, characteristics, advantages) (2002)
14. N. V. Shilkin, Methods to improve the thermal efficiency of buildings and their economic evaluation: dissertation in support of candidature for a technical degree (Moscow, 2007)
15. G. P. Vasiliev, Heating and cooling of buildings using low-grade thermal energy of the surface layers of the Earth: dissertation in support of candidature for a technical degree (Moscow, 2006)
16. B. Kuwabara, T. Auer, T. Akerstream, G. Klym, M. Pauls, K. Opie, J. Peterson, Harnessing climate (High Performing Buildings, 2011)
17. V. Generalov, E. Generalova, Urban Construction and Architecture, **4**, 32-38 (2015)
18. A. Yu. Zhigulina , Urban Construction and Architecture, **1**, 29-30 (2011)
19. B. Holdsworth, Refocus, **6**, 58–60 (2005)
20. P. Lotfabadi, Renewable and Sustainable Energy Reviews, **38**, 285–295 (2014)
21. A. Wood, Rethinking the Skyscraper in the Ecological Age: Design Principles for a New High-Rise Vernacular, *Proc. of the CTBUH 2014 Shanghai Conf.* (2014)
22. P. Lotfabadi, H. Z. Alibaba, A. Arfaei, Renewable and Sustainable Energy Reviews, **57**, 1337–1346 (2016)
23. URL: http://capitagreensingapore.com/green-features/ (last accessed 25.03. 2017)
24. URL: http://www.architectmagazine.com/design/buildings/edith-greenwendell-wyatt-federal-building-designed-by-cutler-anderson-architects_o [electronic resource] (last accessed 12.04. 2017)
25. URL: http://www.theplan.it/eng/webzine/international-architecture/al-bahr-towers#sthash.RcKFbuA9.dpbs (last accessed 02.04. 2017)
26. URL:http://www.ctbuh.org/TallBuildings/FeaturedTallBuildings/AlBaharTowersAbuDhabi/tabid/3845/language/en-US/Default.aspx (last accessed 02.04. 2017)
27. URL: http://www.archdaily.com/44596/manitoba-hydro-kpmb-architects (last accessed 02.04. 2017)
28. URL: http://www.archdaily.com/777498/bosco-verticale-stefano-boeri-architetti (last accessed 02.04. 2017)
29. URL: http://archi.ru/projects/world/8818/zhiloi-kompleks-bosco-verticale (last accessed 02.04. 2017)
30. URL: https://www10.aeccafe.com/blogs/arch-showcase/2011/04/15/strata-se1-in-elephant-castle-uk-by-bfls-architects/ (last accessed 02.04. 2017)
31. URL: https://ecotechnica.com.ua/arkhitektura/992-wuhan-greenland-center-samyj-vysokij-neboskreb-kitaya-budet-samym-energoeffektivnym.html (last accessed 02.04. 2017)
32. URL: http://www.archdaily.com/221560/citadel-skyscraper-victor-kopeikin-pavlo-zabotin (last accessed 02.04. 2017)

The development of the architectural form of a tower derived from a traditional and philosophical symbol, realized by solutions of high-class technologies. The case of the Bitexco Financial Tower

Tran Van Khai[1,*]

[1]Van lang University, 45 Nguyen Khac Nhu, District No1, Ho Chi Minh City 700000, Vietnam

Abstract. The Bitexco Financial Tower, majestically standing tall in the heart of Ho Chi Minh City, Vietnam, rejects the box-shaped, abstract forms of modernism, incorporating an innovative idea of contemporary architecture. Based on the inspiration from the Bitexco Group, a renowned architect designedthe tower that became an iconic landmark of the city in the form of a lotus bud, one of the most iconic symbols of Vietnamese culture since ancient times. High class structural system solution designed by top international professional teams enable the building to rise high with its graceful, statuesque design of the lotus flower shape. CNNGo recently ranked the Bitexco Financial Tower fifth in their listing of the world's 20 most-iconic skyscrapers.

1 Introduction

When many trends of contemporary architecture turn away from the pure geometrical forms of box-shaped abstraction developed by the masters of modernism, the architects of the 21st century are increasingly challenged to find symbolic ideas to effectively address the changing needs of a dynamic society. In this context, the developer Bitexco Group with Mr. Vu Quang Hoi as Chief of Board, laid down the policy for the formation of the Bitexco Financial Tower to become an iconic embodiment which represents the energy and aspirations of the Vietnamese people.

The selection of the form of a tower from traditional symbols was carried out in a broad philosophical context. Reviews show the meaning of the lotus flower, a symbol of Buddhism and enlightenment, was that the flower grows in the mud, but when it reaches the light, becomes the most beautiful flower ever, while purifying the water wherever it grows. In the imagination of the people in Vietnam, the lotus flower symbolizes beauty, and the spiritual and pure nature of ethnicity.

With AREP world-renowned architect Carlos Zapata behind the delicate lotus flower form design, Bitexco Financial Tower became memorable and iconic by encapsulating an image that symbolises the national flower of the Vietnamese people.

* Corresponding author: angela-1309.m@yandex.ru

Wind-induced building sway was a major concern, as the tower is tall with a rounded lotus form while the perimeter tapers inward at the lower floors. Top international structural and construction teams partnered to meet the challenge of technological complexity of the lotus tower form. A structural system was designed using concrete cast-in-place outrigger and belt trusses which stiffen the tower against the swaying motion while cantilever beams beyond the primary girders effectively support the circular edge of the helipad which was suspended high above the ground. This was analized as the most constructible and economical structural solution in the context of the central business district of the city here. This research is intended to describe very briefly how architecture and traditional symbols and high-class structural solutions have functioned in relation to each other in a project through the dynamic expressive features of its form. In a broader philosophical context, the finding hopes to consider that the constant development of design through architectural representation with the help of high-class technologies is critical to its success, the Bitexco Financial Tower project become a true example of forward thinking and design, an unmistakably world-class presence.

2 Literature review

2.1 Review of the theories of contemporary architecture on the creation of new effective design forms to address the changing needs of society

An article with the title of: 7 things to discover about contemporary architecture, November 21, 2014/in Articles, Architecture/by BONE Structure [1] argued that: Contemporary architecture aims to break away from the processes and ways of thinking that have become standard. It is innovative. All we need to do is to look around to see that the dominant line in architecture is the straight line. Contemporary architecture tends to distance itself from this habit by opting more often for curved lines, instead. In some cases, a building is entirely designed around curved lines. This Article cited that: The main thing to remember about contemporary architecture is that it tries to distinguish itself from conventional architecture through new architectural ideas. For example, the goal of giving a building an unusual form is obviously to surprise, to astonish, and to alter perspectives. The use of volumes other than cubes makes it possible to create different interior spaces.

The Article, Architecture/by BONE Structure [1] illustrates the tendencies of contemporary architecture by the example of the Capital Gate tower in Abu Dhabi, United Arab Emirates (fig. 1) This circular, 35-story tower has an 18-degree tilt in the middle, making it the most inclined building in the world, a design by the Scottish architecture firm RMJM. Serena Love [2] cited: Furthermore, the choice of building materials and their use in architecture can be considered codes of social practice and even ideology. As material culture, architecture becomes a metaphor for human engagement and symbolic communication. Selay Yurtkuran, Gözde Kırlı and Yavuz Taneli Uludağ [3] argue that: Designers draw their inspiration from multiple sources to address challenging design problems. One method is to study nature, and attempt to comprehend the ways in which it has evolved to address environmental challenges. This practice has been coined biomimetic, a term introduced in the 1950s by Otto Schmitt.

Fig. 1. Capital Gate tower in Abu Dhabi

Fig. 2. The BITEXCO Finacial Tower in the midst of box shape buildings of Modern architecture in dist.1, Hochiminh City

[Source: BONE StructureNov.21.2014).

Biomimetic has increasingly been employed in architecture and engineering. Nikos A. Salingaros & Kenneth G. Masden [4] argued that: A creative revolution is now possible, one that will transform the built environment in wonderfully human ways. A new intelligent architecture can impart a greater sense of humanity to the city and to the world.

The above research shows that the main tendencies of contemporary architecture turn away from the pure geometrical forms of box-shaped abstraction developed by modernism, and aims through studying nature to discover ways to effectively address the changing needs of the developing society.

2.1 Review of the approaches to the symbolic form of the lotus flower

Pedram Rezania [5] examined historyand concludesthat the lotus in the ancient East is as important as the Rose in the West, and in the 8th BC the image of lotus was transferred from Egypt into Phoenicians and then to Mesopotamia and Iran. The Phoenician goddess has lotus in her hands as the symbol of their power of creation. Rezania described that this plant was worshipped in ancient Egypt and in many parts of Asia. The holiness of the lotus goes back to its aquatic environment because the water was the ancient symbol of an old ocean, which the world was created from and the floating lotus on that was the symbol of the uterus. Since the lotus opens at the rising of the sun and closes at sunset, it seems to be like the sun.

M.A. Carrillo de Albornoz & M.A. Fernández [6] suggested that in the Orient, this flower has various symbolic aspects. But its main characteristics stem from the fact that it blossoms on stagnant waters. This is why the lotus is seen as a symbol of purity; although blossoming on murky waters it remains immaculate. M.A. Carrillo de Albornoz & M.A. Fernández [6] further described the characteristic of the lotus flower as: "Rising out of the darkness to blossom in full sunlight, this flower is the symbol of spiritual fulfilment. It represents manifestation that emerges on the surface of the primeval waters, like the Cosmic Egg. The closed bud contains the realization of the potential hidden in the initial seed".

Rev. Katherine Aryshta Dean [7] stated that for over 3,000 years, the Sacred Blue Lotus (actually a water lily) was used medicinally and spiritually by the priesthoods of Egypt, Greece and Rome. So important was this flower that the Egyptians dedicated a god, Nefertem, the son of Sekhmet and Ptah, to protect and care for it. The Blue Lotus, or Nymphaea Caerulea by its scientific name, was revered as a symbol of something sacred and holy to the ancients – and the fact that this plant was seen as a god itself is an indicator of its spiritual significance to the Egyptians.

The above discussions show that national symbols can fulfill certain unifying and divisive functions, which affect feelings of common identity. The image of the symbols could be derived from nature such as the national animal, bird, flower, fruit tree and particularly in this case the lotus flower.

2.2 Review of the concept of the lotus as a symbol of divine beauty and the national flower of Vietnam

Cutemintran [8] describes the role of the lotus flower in Vietnam: 'Lotus flowers play an important role and special position in both Vietnamese spirit and culture. From a long time ago up to now, lotus flowers have been in Vietnamese people's hearts, life and culture. Contemplating lotus flowers, we can see and realize images of Vietnamese people'.

The article from Administrator [9] on 28 August, 2008 describes the One Pillar Pagoda as a historic Buddhist temple in Hanoi, the capital of Vietnam. This article presented the legendary story: According to legend, ageing emperor Ly Thai To of the Ly dynasty, who had no children, used to go to pagodas to pray to Buddha for a son. One night, he dreamt that he was granted a private audience to the Bodhisattva Avalokiteshvara, who was seated on a great lotus flower in a square-shaped lotus pond on the western side of thang long citadel, and who gave the king a baby boy. Months later, when the queen gave birth to a male child, the emperor ordered the construction of a Pagoda supported by only one pillar to resemble the lotus seat of his dream in honour of the Bodhisattva Avalokiteshvara. According to one theory, the pagoda was built in the a style of a lotus emerging out of the water.

In an article concerning the lotus flower (scientific name N. nucifera) Share Siwek Share [10] said that the N. nucifera is widely naturalized. It is the national flower of Vietnam and India. Many times, the term "lotus" is applied to water lilies as well as the true lotuses, although their growing habits and uses are very different.

Pham Ha [11] in A Brief History of the Luxurious Pink Lotus, Vietnam's National Flower, Jun. 22nd, 2017 shared on Facebook and Twitter, cited that professor To Ngoc Thanh, a member of the Advisory Council on Vietnam's National Flower has answered in an interview with Labour newspaper: "As I understand, India and Sri Lanka also have their lotus chosen as national flowers, but theirs is the white lotus, while we choose our pink lotus flower. Our lotus represents the noble mind and pure spirit of Vietnamese and is the best symbol to characterize Vietnam." According to a poll organized by the Ministry of Culture, Sports and Tourism (MoCST), the lotus ranks first on the list with 71.1%, out of a total 60,000 votes. Le Tien Tho [12], Deputy Minister of MoCST, clarified that the lotus was chosen because it originates from Vietnam and has been grown in most parts of the country for a long time to embody both cultural identity and national spirit. The lotus also has a pleasant fragrance and appeared in many traditional literatures, cultural and architectural works of Vietnam. Furthermore, after receiving the majority of votes for the lotus as national flower, the pink lotus was featured at another exhibition in the city of Danang).

It is clear that all the above research assumes that the Vietnamese love the lotus because it symbolizes the gracefulness, commitment and spiritual knowledge that it was regarded as Vietnam's national flower.

2.3. Review of the of structure solutions appropriate for the tall buildings with an unconventional tube shape of the lotus flower.

Er. Nishant Rana, Siddhant Rana [13] examines the main structure systems of the high-rise tower building and summerized that there are 4 main structural systems that have recently been applied for high-rise tower buildings: (1) Braced-Frame Structural System; (2) Framed Tube Structural System; (3) Outrigger-Braced Structural System, and; (4) Shear Wall Structural System. Nishant Rana and Siddhant Rana [13] also described that the 'Braced-Frame Structural System' is both an efficient and economical way for improving the lateral stiffness and resistance of rigid frame system. The 'Framed Tube Structural System' evolved from a quest to develop a bracing configuration that would place as much gravity load on the exterior columns as possible to help counter the overturning effects of lateral loads while taking advantage of exterior.

Bryan Stafford Smith and Alex Coull [14] examined and described the characteristics of the 'Shear Wall Structural System' as follows: Concrete or masonry continuous vertical walls may serve both architecturally as partitions and structurally to carry gravity and lateral loading. Bryan Stafford Smith and Alex Coull found that because of their stiffness, the shear wall structural system can be economical up to 35 stories. It is especially important in the shear wall system to try to plan the wall layout so that the lateral load tensile stresses are suppressed by the gravity load stresses.

Hi Sun Choi, Leonard Joesph, and Neville Mathias [15] described the 'Outrigger-Braced Structural System' as: Outrigger are rigid horizontal structures designed to improve building overturning stiffness and strength by connecting the core or spine to distant columns. Outrigger systems function by tying together two structural systems- typically a core system and a perimeter system to yield whole structural behaviors that are much better than those of component system. Outrigger system performance is affected by outrigger locations through the height of the building, the number of levels of outrigger provided, their plan locations, outrigger truss depths and the primary structural materials used.

Catrina Grünbaum [16] stated that: The building should not get too heavy for economical and practical reasons, so it is important to find an optimal equilibrium. The heaviest structure is the one with two cores and the least heaviest is the one with trusses on the outside of the building and with the core placed in the middle, which is also the stiffest structure. This conclusion is taken from the case study.

Solutions of structural technologies for high-rise buildings are not new, but the above professionals conceptualize the advantage of the Outrigger-Braced Structural System with trusses on the outside of the building which we can conclude is more appropriate for unconventional tube shaped towers.

3 Materials and methods

3.1 Materials from the project location context

Located in the Central Business District of a City of one of the world's fastest-growing economies and strategically positioned to attract leading companies, Bitexco Financial Tower stands at the heart of Ho Chi Minh's financial centre.

In the CBD in District 1 of the city, it is carefully positioned on many urban composition axises. Situated adjacent to Nguyen Hue boulevard, it is well connected to the Thu Thiem Tunnel and two main bridges, in close proximity to the Tan Son Nhat International Airport. With this advantage in location, the tower is often chosen as the place to hold fireworks and laser display shows, the tower shines resplendent in this event, a worthy symbol of the development of the city.

Fig. 3. Bitexco Financial Tower

[Source: Talk Vietnam]

Fig. 4. Bitexco Financial Tower in the CBD

[Source: Bitexco Group]

3.2 Materials from the required functions of the tower:

According to the information provided from the project developer, the Bitexco Financial Tower is a mixed-use project which includes office and retail space. The tower has more than 39,000 square meters of office space, from 7th to 64 floors, and the six-story retail podium Icon68, with around 10,000 square meters. Around 464 square meters of space is available for a casual dining restaurant on level 50 and around 568 square meters of space is used for a fine dining restaurant on level 51, as well as 300 square meters on level 52 for a VIP & Bar Lounge. A heli-pad located on the southern side of the Bitexco Financial Tower extends 22 meters out from the main structure. The heli-pad cantilevers from the 52nd floor are strong enough to carry a helicopter under 3 tons of weight and resembles a blossoming lotus bud. It required a structural and economical structural solution.

3.3 Methodology

The methodology here describes the selection of the form of a tower from traditional symbols which was carried out in a broad philosophical context using both qualitative and quantitative methods, including a mixture of both as the topics discussed here comes from an architectural object that is both technology and social and the social aspect focusing on art.

Documentary analysis carried out mainly from reliable research sources is the main basis of the methods.

The qualitative methods of the article were based on the analysis of the nature of the lotus flower symbol and its possibility of its application in the design of the Bitexco Financial Tower. The quantitative methods of the article were based on an analysis of dimensions of the building, number of stories to have a broad but clear academic justification of the technological solutions.

4 Discussion and analysis of the design philosophy

To critically examine how an architectural project represents its form, the significance and objective of representation must be questioned. The developer and professionals who created the Bitexco Financial tower intended that the architecture of this lotus building will be both a symbol, which represents Vietnam's vitality and its rapid development towards full global integration. Architecture must be considered for both its physical value – form, measure, scale, materiality and its poetic nature – symbolism, sign and meaning [17].

Considered as the national flower of Vietnam, the lotus represents the open, generous spirit and the beauty of the Vietnamese people. The meaning of the poetic form of the lotus in Vietnamese culture has been providing inspiration for art, literature and architecture ever-present in ceremonies. The botanical name of the lotus is Nelumbo Nucifera of the Nelumbonaceae flower family. Originating in Vietnam a long time ago, the lotus is highly-adaptive in almost all parts of the country. The lotus grows in water and has primarily pink or white flowers. It has a special position in life and culture representing the beauty of the Vietnamese soul and the Vietnamese identity. The optimistic outlook of the flower can adapt easily to any situation whilst simultaneously purifying the water wherever it grows. It is usually placed in the most spiritual area of the Vietnamese house as decoration.

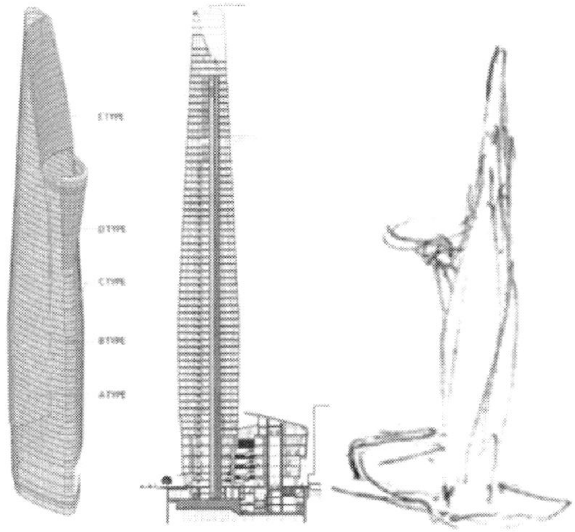

Fig. 5. Tower form and section. **Fig. 6.** The inspiration from the lotus bud.

[Source: AEWORLDMAP.COM][Source: Bitexco]

Effective symbolism in architecture is rooted in the strongest and most universal experiences. Amongst these are religious beliefs and philosophical ideas, but also the perceptual dynamics of spontaneous symbolism deriving from daily experiences [18]. In Buddhism: The lotus represents refinement, mystery, contemplation and the cleanliness of the soul despite its earthly surroundings. [18]. The lotus plant's light, sweet fragrance evokes a noble spirit [18],with its white or pink petals and yellow stamen, the lovely flower discreetly grows up from the ground to proudly display its full beauty under the sun [18]. The lotus is considered to be the national flower, not just because of its indomitable spirit, purity and beauty, but also because of the admiration the Vietnamese people have for the flower.

The contradiction between form and function also has to be solved in this case by the developer as follows: Architecture today is necessarily "complex and contradictory to address the changing needs of a dynamic society" [19]. The city authority has allowed the tower to have a hotel and apartments to recover its capital in less time. But Mr. Vu Quang Hội had the foresight to see that the design will be more complex if the building has many different functions, so it was determined that the tower should only be financial, office and commercial to maintain the specific architectural curved lotus form as every floor must be different from the next. The world renowned architect Carlos Zapata from AREP (France) designed the delicate curved lotus flower form in a poetic discipline to enable Bitexco Financial Tower to become a memorable icon. As a retrogressive idea, representation can be flawed if it does not come from the architectural discipline itself [20].

The organization of the functional spaces that is to be in accordance with the lotus form of the tower. The Bitexco Financial Tower stands 262 meters high at the centre of the city's business district. It is a mixed-use building with more than 37,000 square meters of office space from the 7th to 65th floors highly occupied by finance organizations, banking, and multi-national corporations, and the six-story retail podium with around 10,000 square meters. Bitexco Financial Tower with its 68-story tower houses offices, shops, restaurants, and a helipad offering visitors an unparalleled city view from its Sky Deck. Around 464 square meters of space is available for a casual dining restaurant on level 50 and around 568 square meters of space is used for a fine dining restaurant on level 51, and approximately 300 square meters on level 52 for a VIP & Bar Lounge. The building is highly occupied by finance organizations, banking and multi-national corporations.

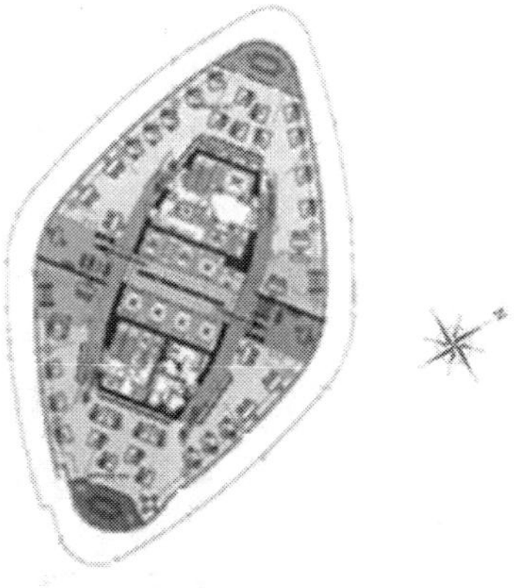

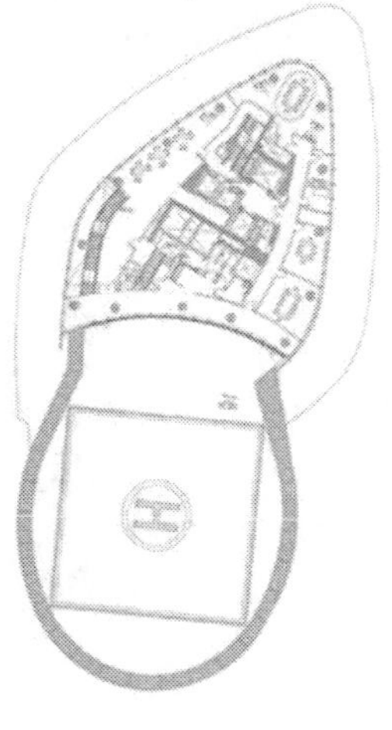

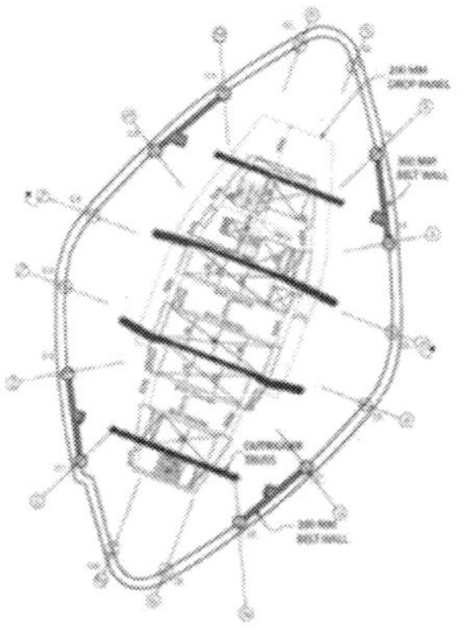

Fig 7: Typical floor plan belt walls. **Fig 8**: Helipad floor plan **Fig 9**: Outrigger trusses .
[Source: AEWORLDMAP.COM] [Source: STRUCTURE magazine]

The Structural system and technological issues. To meet the challenge of technological complexity of the tower, top quality professional consultants from DSA Engineering (M&E), LERA structural engineers and VNCC local partner, Hyundai Engineering and Construction Main contractor and other experts, partnered in the design and construction process. The office tower is essentially an all-concrete structure. Initially, a mixed system comprised of structural steel floor framing, composite perimeter columns and concrete core walls was developed [21]. Office floors are framed with reinforced concrete flat plate system.

The tower is tall and slender, core aspect ratio is 20. An innovative structural design using concrete outriggers and belt trusses will stiffen tower against swaying motion. [21] The structure received an 'Excellence in Structural Engineering' award from NCSEA in 2011 and is currently the 124th tallest building in the world.

The weight of this tall building required very deep foundations. Poor soil conditions due to proximity to the Saigon River and the Mekong Delta require very long foundation piers. Comprehensive initial studies and subsurface explorations were performed to form a basis of the foundation design and to estimate foundation settlement. The tower foundation is a bored pile supported mat which effectively distributes the tower loads from the walls of the narrow services core over the extent of the mat [22].

Fig. 10. The helipad of the tower. *[Source: DESIGN.VN]*

Fig. 11. View of the Tower. *[Source: DESIGN.VN]*

The helipad was located on the southern side of the tower and suspended 191 meters above the ground, echoing the main form of the building with the shape representing a blossoming lotus bud. The total length of the axis of the helipad is 40 meters, including 18meters extends into and 22 meters extend out from the structural system of the tower. The cantilever of the helipad was a challenge for the installation in the implementation process. The helipad central structures had to be special designed to accommodate and balance loads and forces, was constructed from more than 250 tons of structural steel and 4,000 ultra bolts. Before it could be hoisted up in parts to the 52th floor, the helipad structures had to be preassembled on the ground at a local factory.

Conclusion

Built at a time of unprecedented growth in the Vietnamese economy, the Bitexco Financial Tower is designed using both sophisticated form and high technology, which enables the tower to represent the energy and aspirations of the people and its form is an iconic embodiment of the new and dynamic Vietnam. Rising amidst several other buildings adapting the abstract geometrical glazed box form, the Bitexco Financial tower is distinguished notably because of its specific curved lines.

The case of the Bitexco Financial Tower revealed some of the considerable principles behind the process of the evolution of a shape generation strategy with respect to original traditional forms that gives a specific possibility of representation. That is only with the careful blending of both iconic form and objective can architecture succeed.

The result is that the tower was officially inaugurated on October 31, 2010. Bitexco Financial Tower has been listed as one of the 25 great skyscrapers - Icons of construction (by CNN - 2013); among the world's Top 20 iconic skyscrapers (by CNNGo - 2011); and has been awarded by NCSEA for "The excellence in structure engineering 2011" CNNGo recently ranked the Bitexco Financial Tower fifth in their listing of the world's 20 most iconic skyscrapers and is currently the 124th tallest building in the world.

References

1. 7 things to discover about contemporary architecture, BONE Structure (2014)

2. S. Love, Elsevier, Journal of Anthropological Archaeology **32**, 746-758 (2013)
3. S. Yurtkuran, G. Kırlı, Y. Taneli, Uludağ University, Faculty of Engineering and Architecture, Görükle, Bursa 16059, Turkey . *Learning from Nature: Biomimetic Design in Architectural Education* . 2nd Cyprus International Conference on Educational Research, (CY-ICER **2013**).
4. N.A. Salingaros, K.G. Masden, International Journal of Architectural Research, **1** (2007)
5. P. Rezania, *The Symbol of lotus in ancient world* . Department of Cultures and languages of ancient Iran, Science and Research branch, Islamic Azad University, Tehran, Iran, **2011**
6. M.A. Carrillo de Albornoz & M.A. Fernández .*The Symbolism of the Lotus* .Article posted by India North, Aug. 26, 2014
7. Rev. Katherine Aryshta Dean. *The Sacred Blue Lotus – Flower of the Gods.*Excerpts from "Priestess of the Blue Lotus", 2017.
8. Cutemintran .*Lotus – Vietnam's national flower, a symbol of divine beauty.* Culturehttps:// cutemintran.wordpress.com Source from Http://vietnam-online
9. ADMINISTRATOR .*One-Pillar Pagoda (Chùa Một Cột)* posted on May 11, 2013.
10. Le Tien Tho , Deputy Minister of Ministry of Culture, Sports and Tourism. *Lotus voted as Vietnam National Flower* Vietnam online,2011.
11. Share Siwek *The Sacred Lotus, Nelumbo nucifera*. A Plant Study attended the 2007 Flower Essence Society (FES) Professional Course,2007.
12. Pham Ha . *A Brief History of the Luxurious Pink Lotus, Vietnam's National Flower.* Share on Twitter Jun. 22nd, 2017.
13. Er. Nishant Rana, Siddhant Rana . Structural Forms Systems for Tall Building Structures SSRG International Journal of Civil Engineering (SSRG-IJCE) – volume1 issue 4 September 2014.
14. B. Stafford Smith, Alex Coull , Tall Building Structures-Analysis and Design, John Wiley & Sons, 2006.
15. Hi Sun Choi, Leonard Joesph, Neville Mathias. Outrigger Design for High-Rise Buildings, An output of CTBUH 2017Images Publishing, Victoria 3170, Australia,
16. Catrina Grünbaum. *Structures of tall buildings* Avdelningen för Konstruktionsteknik. Lunds Tekniska Högskola .Lund Universitet, 2008Rapport TVBK - 5156 (Department of Structural Engineering Lund Institute of Technology, Box 118, S-221 00 LUND, Sweden).
17. A. Hiller, B. Walters, (mentor) Honoris Thesis. Architecture Representation: Abstraction and Symbol within Design. Aprill 2012 .
18. C. Carter, Department of Philosophy, Marquette University. Symbol and Function in Contemporary Architecture for Museums,2010.
19. Buddhist studies. *Buddha*. BDEA/ Dharma Education Association & BuddhaNet (Copyright 1999 – 2017).
20. R. Venturi, Iconography and Electronics: Upon a generic Architecturep 88. Cambridge, Massachusetts/London, England: MIT press, 1966.
21. P. Eisenman The End of the Classical: End of the Beginning, the End of the End". In Architecture Theory since
1968, Edited by Michael Hays: Cambridge Mass: The MIT Press, 1968.
22. W. Faschan, PE and.Trivedi, Nayan B, PE*Bitexco Financial Tower* . STRUCTURE magazine, June 2009.

Collective space of high-rise housing complex

Tatyana Bakaeva[1,*]

[1]Samara State Technical University, Molodogvardeyskaya St, 194, Samara, Russia

Abstract. The article considers the problems of support of citizens a comfortable living environment in the conditions of the limited territory of the megalopolis, the typological principles of formation of space-planning structure high-rise residence complexes with public space. The collective space for residents of high-rise housing estates on the example of international experience of design and construction is in detail considered. The collective space and the area of the standard apartment are analysed on comfort classes: a social - complex Pinnacle @ Duxton, a business - Monde Condos and an elite - Hamilton Scotts. Interdependence the area of the standard flat and the total area of housing collective space, in addiction on the comfort level, is revealed. In the conditions of high-density urban development, the collective space allows to form the comfortable environment for accommodation. The recommendations for achievement of integrity and improvement of quality of the city environment are made. The convenient collective space makes a contribution to civil policy, it creates the socializing sense of interaction of residents, coagulates social effect.

1 Introduction

Relevance of a research is caused by need of improvement of consumer qualities of the inhabited environment. Carrying out in-depth studies in the field of typology of the dwelling and on formation of structural elements of the inhabited environment depending on economic characteristics and requirements of various social groups becomes the most important. Now there was a certain stereotype of perception of comfortableness of housing depending on its area. This article demonstrates other spatial and environmental ideas of the qualitative inhabited environment in the conditions of megalopolis hyper density. Problem of this research is to analyse international experience of design and construction. The purpose of the current research to prove that comfort of the apartment isn't in direct dependence on its area.

The research is based on theoretical works of such authoritative researchers of the dwelling as Generalov V. P., Generalova E. M., Potiyenko N. D., Grigoriev I. V., Durmanov V. Yu., Kartashova K. K., Kaydalova E. V., Kiyanenko K. V., Krasheninnikov A. V., Molchanov V. M., Ovsyannikov V. A., Yargina Z. N. and others.

* Corresponding author: 881019@mail.ru

According to the aim it is necessary to solve research problems. Not always it is possible to call housing, big on the area, comfortable and, on the contrary, the apartment very modest can meet all requirements imposed to the modern dwelling by means of set of services and the services, provided by a housing estate, which part it is for metric area [1-4].
For detailed consideration of international experience of construction of high-rise housing we will concentrate on such indicators as the area of the standard flat and the total area of housing collective space.

We will consider the dwelling as set of manifestations of activity of the person, various on degree of a privacy, in the course of accommodation. As a part of an inhabited cell it is possible to allocate, directly, residential zone where private individual needs are carried out, there are also communicative needs of the person to self-realization through socialization and interaction with society [5-10].

2 Methods

The research leans on the following methods: 1. Studying and the analysis of references, results of scientific developments and design decisions on the chosen subject; 2. Consideration of international experience of design and construction of inhabited houses complexes of the increased number of stories with multilevel public space.

2.1. Collective space in a social high-rise housing complex

Striking example of comfortable housing with the developed collective space as a part of a housing estate in the conditions of hyper dense building of the megalopolis is **Pinnacle @ Duxton** (fig. 1a). This complex is an award-winning 50-storey residential development in Singapore's city center, next to the business district. All seven connected towers are collectively the world's tallest public residential buildings. Unique amongst Housing and Development Board (HDB) projects, it is the design winner of a worldwide competition [11]. The Pinnacle @ strikes Duxton designed by ARC Studio Architecture + Urbanism with altitude (163 m for Asian skyscrapers – very average height) and scale (1848 apartments on the area of 2,5 hectares, the total area of a complex about 280 000 m^2. The characteristic of this HDB precinct is that as of now, it is one of the highest public housing precinct in Singapore. The building offers the unique concept of mass housing for the densely populated cities. It is the courageous experiment in the field of typology of high-rise housing resolving functional, economic and social issues, confirming in practice that high-rise building can form the comfortable inhabited environment [1]. The Pinnacle@Duxton is sated with all necessary public functions: shopping centers, enterprises of food, educational center, kindergarten, sports both playgrounds and etc. In addition are in the radius of pedestrian availability: police, hospital, restaurants, bars, cafe, shops, etc. It is competently organized traffic and pedestrians, level for the movement of fire trucks, and also special transport for garbage removal is provided. The bend of towers forms extensive domestic space, keeping communication with the existing public center of regional value "Tanjong Pagar". In the territory of the yard are provided: children's playgrounds, a set of zones of quiet rest with gardening, the covered arbors and benches, the universal sports ground, and also historical park. The yard represents the raznourovnevy space united by system of ramps that makes him easily available for handicapped groups of the population (fig. 1b).

The project features the world's two longest sky gardens of 500 meters (1,600 ft) each, on both the 26th and 50th floors. Uniting seven high-rise blocks, these platforms form a laconic sculptural silhouette of a complex, being his business card. They not only expand public space, but also perform function of a comfortable recreation area for residents and visitors of the building. Heavenly galleries give in addition to the land level 1 more hectare of the well-planned territory. It offers three levels of public spaces whereby of 50% of the site on the ground is public while the 26th floor and 50th floor sky bridges account each for 21 of % of the total site. On them the survey platforms, jogging track and zones created by various thematic landscape compositions are organized. In total There are 12 heavenly gardens of All exists 12 sky gardens. There are sky gum, meadows, space net, crater, padi field, observation deck on the 26th floor. There are cottage, rocky, beach, lounge, hillock, fiesta on the 50th floor. At the same time these collective spaces of sky bridge perform function of zones for evacuation during the fire, and also facilitate engineering service of a facade. A variety of apartments is presented by 35 different version of planning of inhabited cells. Also a possibility of re-planning exists, because internal partitions are made of lightweight concrete. Smaller flats measuring 93 to 97 m^2 and bigger ones measuring 105 to 108 m^2 [12].

All public space of the precincts is located at ground level, raised platform level and skybridges. Total multilevel provision of public space: Ground open area incl. pavillions 9600 m^2, Elevated open area 24000 m^2, Ground level void deck 4000 m^2, Upper level void deck 4000 m^2 [13].

To understand the opportunities of access to public space as a part of a housing complex, necessary enter the coefficient of saturation collective space.

$$K_{SCS} = S_{TF} \,/\, S_{CS} \tag{1}$$

The coefficient of saturation collective space (K_{SCS}) is equal to the relation of the area of the typical flat (S_{TF}) to the total area of housing collective space (S_{CS}).

Calculation for the complex Pinnacle@Duxton:

$$K_{scs} = 100.5\, m^2 / \,41600\, m^2 = 0.0024$$

a b

Fig. 1. Pinnacle@Duxton: a - look outside, b - the collective space of platform

2.2. Collective space in a high-rise housing complex of a business class

Monde Condos (fig. 2a) is a dramatic 40-storey building of a business-class at Queen's Quay and Bonnycastle, Toronto. The main entrances to both the tower and the podium are off of Bonnycastle Street. The architecture of Monde, a groundbreaking building targeting

LEEd Gold certification comes from internationally recognized architect Moshe Safdie in collaboration with Quadrangle Architects. These masters of design worked with Cecconi Simone on the interiors and Janet Rosenberg & Studio on the landscape design to create a building with premium quality standards a design aesthetic inspired by nature, the lake and the city. Monde features a breathtaking double-height lobby with a living green wall, along with extensive amenities designed by acclaimed interior design firm Cecconi Simone. The collective space (fig. 2b) will be located on the entire tenth floor and will include yoga/pilates/cardio/kinesis training and weights studios, hot plunge pool and his and hers sauna, steam and change-rooms with experience showers. Relax and entertain in the kitchen/dining/bar/lounge and play in the billiards room. Outside enjoy the outdoor lounge with indoor/outdoor fireplace or relax in a cabana deck side next to the outdoor pool with infinity views to Lake Ontario. Extensive landscaping by celebrated Janet Rosenberg & Studio complements the pool deck.

There are many amenities: 24 hour and 7 day a week concierge, 2 guest suites provided for residents' use, resident superintendent for on-site maintenance, underground parking equipped with security monitoring system, underground garage painted white and well-lit for added safety and security, 5 custom-designed elevators - 3 designated for the tower and 2 for the podium, vistor parking on the uppermost parking level, separate lobby access for podium and tower with mailrooms conveniently located beside each.

Monde is a member of the "Intelligent Community" at Waterfront Toronto, an open-access network that provides internet speeds of up to 100 times faster than anywhere in the city. Total number of suites is 516, suite size range 45-92 m^2. The total are of architects is 3200 m^2 [14-19]

Calculation for the complex Monde Condos by means formula (1):

$$\boldsymbol{K}_{scs} = 68\ m^2 /\ 3200\ m^2\ = 0.021$$

a b

Fig. 2. Monde Condos: a - look outside, b - the collective space of Monde Condos

2.3 Collective space in an elite high-rise housing estate

Hamilton Scotts (fig. 3a), a high-class apartment condominium in Singapore, one of the most innovative residences in the country and perhaps the whole of Asia, is offering its residents the ultimate luxury in terms of parking – an individual two-car garage located inside each apartment. A glass wall separates the garage from the living room, offering a lovely view of their expensive cars from the couch.

The system is quite simple – when residents enter the building, they need to park their cars inside a glass elevator shaft. The elevator then raises the vehicle up to the corresponding condo and parks it within a glass enclosure that is visible from the living room. Although the Hamilton Scotts condo numbers 30 floors and tens of thousands of square meters of living space, there just wasn't enough to build a multi-level parking for the residents. Faced with this challenge, the team of architects and engineers in charge of development came up with something so much better. The unique design enables supercar owners to showcase their prized possessions, instead of hiding them away in basement garages. Hamilton Scotts is the world's first building to offer such an advanced system of parking [20]. This single block residential property offers a full range of facilities to future residents along with an impressive array of outside amenities for people who want to live it up and enjoy the fruits of their labors. There are many facilities such as a gym, conference rooms, jacuzzis, 24 hours security, a barbeque corner and a residents' lounge, Meeting Rooms, Swimming Pool (fig. 3b) [21, 22].

On floors from the 2nd for the 27th 54 apartments with three bedrooms (on two apartments on the floor) of 256 m^2 are located. Two penthouses finish a tower, with a one area of 300 m^2 occupies the 28th floor, other more magnificent of 661 m^2 - two-level, it was located on the 29th and 30th floors. In penthouses at the disposal of owners on four parking spaces. All bedrooms have personal bathing and wardrobe [1, 3]. Apartments are characterized spacious layout with luxury finishing, private lift and modern interiors with 3 en-suites designed with meticulous attention. There are fully equipped kitchen and premium fittings, large spacious balcony with unobstructed greenery and city views. Each unit will have 3 parking lots. (2 lots at Sky Garage + 1 Basement lot) [23-27]. The additional entrance to the apartment is carried out from the public ladder and lift knot, performing function of an emergency and economic exit. The total area of housing collective space is 1540 m^2.

Calculation for the complex Hamilton Scotts by means of formula (1):

$$\boldsymbol{K}_{scs} = 256\ m^2/\ 1540\ m^2\ = 0.16$$

a

b

Fig. 3. Hamilton Scotts: a - look outside, b - the collective space of Hamilton Scotts

3 Results

The coefficient of saturation collective space ($\boldsymbol{K}_{SCS}$) is equal respectively: in the social class 0.0024, in the business-class 0.021 and in an elite-class 0.16.

3.1 The saturation collective space in the social high-rise housing complex

Take into consideration the calculations for the complex Pinnacle@Duxton by means of formula (1). Thus, a ratio of the considered categories of the areas:

$$S_{TF}: S_{CS} = 41600\, m^2 \,:\, 100.5\, m^2 \; = 1{:}\,411$$

Calculations show that, having in property, or, being a tenant of one conventional unit of the private area, the resident can take the benefits and advantages of the collective area equal to 411 conventional units.

3.2 The saturation collective space in the high-rise housing complex of a business -class

In view of the calculations for the complex Monde Condos by means of formula (1). Thus, a ratio of the considered categories of the areas:

$$S_{TF}: S_{CS} = 3200\, m^2 \,:\, 68\, m^2 \; = 1{:}\,47$$

Calculations show that, having in property, or, being a tenant of one conventional unit of the private area, the resident can take the benefits and advantages of the collective area equal to 47 conventional units.

3.3 The saturation collective space in an elite high-rise housing estate

Take into consideration the calculations for the complex Hamilton Scotts by means of formula (1). Thus, a ratio of the considered categories of the areas:

$$S_{TF}: S_{CS} = 1540\, m^2 \,:\, 256\, m^2 \; = 1{:}\,6$$

Calculations show that, having in property, or, being a tenant of one conventional unit of the private area, the resident can take the benefits and advantages of the collective area equal to 6 conventional units.

4 Discussion

In the social high-rise housing complex the public places, that used complex residents are opposite to inhabited cells concernly the square at the person. The most part of all public area of the complex using how a zone of storage of bicycles and reception of guests, gym, a laundry, green terrace on a roof, trade, offices. It is supposed, for example, that residents will receive guests in special guest game rooms, access to which is provided according to the preliminary application.

In the high-rise housing complex of a business class the collective space, available only to residents of a complex, is less contrast on the metric area of standard apartment, however, rather brightly expressed in composition of inhabited structure.

Elite housing assumes, as a rule, the communicative spaces for the inhabitants developed on the areas. However distinctive feature is comparative small functional variability of such zones. In view of extremely wide quantity of the opportunities which are maked directly inside private apartments. The most various functions in this segment of the real estate are drawn, generally, towards individual manifestation, emphasizing, thereby, exclusiveness, uniqueness and privileges of the owner.

Future researches have to develop the idea of interdependence the area of the standard flat and the total area of housing collective space. Study in the long term must show, in addiction on the comfort level, the position of development high-rise living collective space changes.

5 Conclusions

Depending on the level of comfort of the dwelling, the ratio of living space to the total area of the apartment, changes. So in the social class an inhabited part occupies 80% of the general space, and in elite – only 30% of total area, i.e. with increase in the area appears many additional rooms allowing to carry out various functions. But apartments the social - and comfort - classes can be not less representative because some functions, except for private, can be carried out by residents directly in the territory of a living complex.

It should be noted that functional zoning and degree of a saturation of services of the serving structure and private space also variously depending on the level of comfort of a housing complex. At the same time the serving infrastructure is subdivided into the public space focused on the city, and on the collective space intended only for building residents. The qualitative collective space makes a contribution to civil policy, it makes the socializing sense of interaction of residents, coagulates social effect the fact of the existence in architectural planning solution of the high-rise housing complex.

References

1. V. P. Generalov, *High-rise residential buildings and complexes. Singapore. Experience of design and construction of high-rise housing* (Samara, Ltd Kniga, 2013)
2. V. P. Generalov, E. M. Generalova, High-rise residential complex houses as the element of creating highly comfortable living environment, *SGASU vestnik*, **2**, 12-16 (2013)
3. V. P. Generalov, Architectural design of an apartment house of section type, V.P. Generalov, E. M. Generalova, **17**, 164 (Samara, Ltd Kniga, 2013)
4. I. A. Bondarenko, Architecture of the changing Russia: state and prospects *Russian academy architectural and construction sciences, scientific research institute of the theory and history architecture and town planning,* 464 (Moscow, Komkniga, 2011)
5. V. A. Nefyodov, *Landscaping and stability of the environment*, (St.-Petersburg, 2002)
6. V. A. Nefyodov, *City Landscaping* (St.-Petersburg, 2012)
7. Design of modern high-rise buildings, ACB, 330 (2012)
8. A. A. Magaj, Features of architectural typology of high-rise buildings, *Architecture and construction of Russia*, **4**, 22-29 (2009)
9. I. Chizhov, Skyscrapers and people, *Architecture and construction*, **1**, 36-42 (2009)
10. M. F. Utkin, *Architectural design of the inhabited environment* (Moscow, Architecture-C, 2010)
11. https://en.wikipedia.org/wiki/The_Pinnacle@Duxton (last accessed 04/15/2017)
12. http://eresources.nlb.gov.sg/infopedia/articles/SIP_1779_2011-02-18.html (last accessed 04/19/2017)

13. http://wohnforum.arch.ethz.ch/sites/default/files/dateien/fcl_booklet_final_0915_light.pdf (last accessed 04/10/2017)
14. http://urbantoronto.ca/database/projects/monde (last accessed 04/10/2017)
15. https://condonow.com/Monde-Condos (last accessed 04/11/2017)
16. http://www.pacecondo.ca/monde (last accessed 04/11/2017)
17. https://www.condosky.com/condo/monde-condos/ (last accessed 04/11/2017)
18. http://www.blogto.com/city/2012/11/new_in_toronto_real_estate_monde_condominiums/ (last accessed 04/12/2017)
19. http://www.listofcondo.com/Monde_Condos/page_2443300.html (last accessed 04/12/2017)
20. http://www.odditycentral.com/architecture/luxury-singapore-skyscraper-allows-residents-to-park-their-cars-in-the-balcony.html (last accessed 04/14/2017)
21. http://www.designboom.com/architecture/hamilton-scotts-singapore-reignwood-tower-sky-garages-12-03-2014/ (last accessed 04/14/2017)
22. https://www.propertyreview.sg/hamilton-scotts/ (last accessed 04/17/2017)
23. https://vincentongvirtualhomes.com/hamilton-scotts-scotts-road/#Project Details (last accessed 04/17/2017)
24. https://en.wikipedia.org/wiki/Tower_block (last accessed 04/20/2017)
25. http://global.ctbuh.org/resources/papers/download/294-humanizing-high-rise-urbanism-design-strategies-and-planning-tools.pdf (last accessed 04/20/2017)
26. http://www.kopproperties.com/developments/hamilton-scotts (last accessed 04/20/2017)
27. http://www.stproperty.sg/condominium-directory/reignwood-hamilton-scotts-condo/11811(last accessed 04/20/2017)

Typological diversity of tall buildings and complexes in relation to their functional structure

Viktor P. Generalov[1*], *Elena M.* Generalova[1], *Nadezhda A.* Kalinkina[1] and *Irina V.* Zhdanova

[1]Samara State Technical University, 443001, Samara, Molodogvardeyskaya St, 194, Russia

Abstract. The paper focuses on peculiarities of tall buildings and complexes, their typology and its formation in relation to their functional structure. The research is based on the analysis of tall buildings and complexes and identifies the following main functional elements of their formation: residential, administrative (office), hotel elements. The paper also considers the following services as «disseminated» in the space-planning structure: shops, medicine, entertainment, kids and sports facilities, etc., their location in the structure of the total bulk of the building and their impact on typological diversity. Research results include suggestions to add such concepts as «single-function tall buildings» and «mixed-use tall buildings and complexes» into the classification of tall buildings. In addition, if a single-function building or complex performs serving functions, it is proposed to add such concepts as «a residential tall building (complex) with provision of services», «an administrative (public) tall building (complex) with provision of services» into the classification of tall buildings. For mixed-use buildings and complexes the following terms are suggested: «a mixed-use tall building with provision of services», «a mixed-use tall complex with provision of services».

1 Introduction

There is a process of intensive construction of tall buildings as well as a tendency to strengthen research, design, construction bases and to increase investment in this area of industry in some countries.

The concept of «sustainable development of vertical urbanism», which is closely connected with the search for new types of tall buildings, is also being developed. Tall buildings, that are rising up, are a particular social and economic indicator of progress, of the prestige of cities, firms, the political power of the state [1]. Besides, implementation of tall buildings developments leads to efficient use of urban land, reduced traffic flow, easy access to points of attraction for residents of cities, to creating so-called «compact cities». World experience of embedding tall buildings and complexes in the urban development shows a positive shift in terms of increasing the level of comfort, creating a modern

* Corresponding author: vp_generalov@mail.ru

comfortable residential environment not only for inhabitants of skyscrapers, but also for the population in general. As new tall buildings appear and this trend in construction goes on and develops, there is also a debate about their typological diversity. Thus, there is a need for further study, research and systematization of these unique typological constructions [2-11]. Functional-typological characteristics of tall buildings and complexes have their own specific features, which are closely connected with the space-planning structure of tall buildings or complexes.

Currently, both domestic and foreign regulatory framework for the design and construction of tall buildings does not give clear recommendations concerning typological diversity of tall buildings. There is no classification of their types, either. The only exception is the international standards for measuring and defining tall buildings, which were developed by the Council on Tall Buildings and Urban Habitat (CTBUH). In accordance with this standard, buildings above 100 m high are recognized as tall buildings. In Russia, tall buildings are those which are at least 75 m high. When buildings with a height of 300 meters and higher appeared, CTBUH suggested additions to the standards for defining tall buildings. Thus, buildings of 300 to 600 meters are recognized as Supertall skyscrapers, and those over 600 meters – as Megatall skyscrapers. According to the latest data, more than 250 tall buildings will have a height of over 300 meters by 2020, and there will be 8 buildings over 600 meters. As of today, there are only 3 megatalls completed globally. They are «Burj Khalifa» in Dubai, 828 meters high, «Shanghai Tower» in Shanghai, which is 632 meters and «Makkah Clock Royal Tower» in Mecca, which is 601 meters.

2 Methods

It should be noted that there is a certain clarity as far as the height of buildings is concerned. As to their functional structure, there is no regulation or complete research yet. At the same time, the nature and character of regulatory requirements for the design, construction and operation of a building are determined by characteristics of functional elements of the building. When analyzing the existing normative foreign and domestic regulations of tall buildings construction, it is difficult to find clear recommendations on typological features of tall buildings in relation to their functional purpose. Now, three main functional types of tall buildings are mentioned: residential, office and mixed-use tall buildings. The analysis of skyscrapers built in different countries reveals that they hold such main functional elements as residential, administrative (office), hotel elements. These features form and define the type of a tall building by its functional use. In domestic typology they can be divided into three main types: residential, public (administrative, office, hotel) and of mixed use.

In addition, in practice there are various «blotches» and «blendings» of main functional elements in a single space-planning decision. E.g. there might be offices or hotel rooms in residential buildings. At the same time, there might be a certain percentage of residential apartments in an office building. Which building can be considered to be a single-function residential or public (office) building, and which can be referred to as a mixed-use building? Apart from that, there are service functions which may be (and may be not) included in their space-planning structure.

To introduce an adequate, user-friendly regulatory framework, it is necessary to create a clear tall buildings subdivision into types depending on their functional properties. Such a subdivision would solve many challenges in the design, construction and operation of tall buildings. Sad to say, there is no clarification of concepts such as «a single-function building» and «a mixed-use building or complex» in existing regulatory documents.

E.g., SP 160.1325800.2014 «Mixed-use buildings and complexes. Design rules» gives definitions to a mixed-use building and a mixed-use complex. But the interpretation which is given in this document is not, in our view, quite accurate and raises several questions. According to his document, a mixed-use building is a building in which two or more functionally-planning components are interconnected by using planning techniques. A definition of a mixed-use complex is not accurate, either. It says that such a complex holds two or more buildings for different functional purposes, which may be of mixed use and interconnected by planning techniques. It is clear, that this definition of multi-functionality of the building lacks clarity. Besides, this set of rules is designed only for the design, construction and operation of buildings and complexes up to a height of 75 meters with the allowed location of public spaces not higher than 55 meters. At the same time, SP 267.1325800.2016 «Tall buildings and complexes. Design rules» does not even mention tall buildings and complexes, as well as their functional-typological structure.

3 Results

Let us study foreign literature to find a solution to this problem. It should be noted that in foreign literature this issue is examined more clearly and thoroughly. As for the definitions of «single» and «mixed use» tall buildings, there are clear recommendations given. To «single-function» buildings they attribute buildings in which 85% or more of the total area is used for a basic function (residential, administrative, or office, hotel, etc.). To «mixed-use» buildings they refer buildings in which two or more functions are combined. The share of these functions here should be more than 15% of the total area of the building.

This classification of single and mixed-use buildings, in our opinion, could easily be fitted into the Russian Classification of functional characteristics of tall buildings. This would provide a basic framework for a general classification of tall buildings and complexes by their functional use. But, as we noted earlier, apart from the main functional constitutive elements, there are also functions for infrastructure maintenance and service provision implemented in these buildings structure. They are shopping, medical, library, sports and physical training facilities, institutions, cultural establishments, etc. It is a modern approach to the formation of living environment to provide services implemented into the building structure. According to their space-planning structure, these elements can be either embedded into a tall building or attached to it. Our analysis of existing buildings and complexes revealed that embedded-or-attached elements (as a base part of the building or complex) are the most common elements at the moment.

To what type of buildings should we refer a tall building, which performs a dominating function of residential or office use but also provides a service function? The same question relates to mixed-use buildings or complexes. The main purpose of including service functions into the structure of a building or complex is to create a comfortable, modern, high-quality living environment for human habitation. The number of serving functional elements and their intensity is not taken into account in this case, but they directly affect qualitative characteristics of a living environment in a tall building or complex. The only criterion to be taken into account when determining such types of buildings is the accessibility to elements of service and infrastructure facilities and convenience of their use by people regardless of time of day, year, weather conditions, etc.

4 Discussion

Summing up the results of the study, the authors propose creating a typological range of this type of buildings and complexes. Thus, it seems appropriate to determine that «a single-function building (complex)» is a building (complex), in which the main function covers 85% or more of the total area of this tall building (complex), and other functions (residential, office, hotel), interconnected by planning techniques, cover 15% or less of this building area. Besides, this «single-function building or building complex» can be **a** building with provision of services. In real life, there exist single-function buildings either with provision of services or without them. Depending on the basic functional element in the structure of a tall building, a single-function tall building with provision of services can be either «a residential tall building with provision of services» or «a public (office) tall building with provision of services».

If an architectural structure consists of two or more tall buildings performing one function it is recognized as «a tall single-function complex». If this structure holds service functions, it is «a single-function complex with provision of services». For example, the main function of the building is residential. Thus, this is a residential tall complex with provision of services.

To «mixed-use buildings» we refer buildings in which additional functions cover more than 15%of the total area of the building, and the dominating function covers less than 85% of the total area of the building. If the structure of a building or complex provide service functions, it is a mixed-use building or complex. It can be either «a mixed-use tall building with provision of services», «a mixed-use tall complex with provision of services».

Fig. 1. «30 St Mary Axe» (179.8 m), London.

«30 St Mary Axe», better known under the name of «The Gherkin» or «Swiss Re» (see Fig. 1), a 40-storey tall building of a unique round shape in London, can serve as an example of «a public single function (office) tall building». This glass skyscraper is 179.8 meters (40 floors) high. It is the project of the architect Norman Foster. This is an entirely office building with a small restaurant and bar occupying the top 39 and 40th floors which is only small part of the total area of the building. It is open only for office staff working in the building. There are no service functions at the ground level [12].

«Greatwall Tower 1», an office complex in Wuhan, China, is can be described as «a single-function office complex with provision of services» (see Fig. 2). This complex consists of two towers with offices and a 4-storey podium which holds a shopping mall with elements of service, terraces for relaxing walks, an amphitheater, open and accessible to the public [13].

 https://doi.org/10.1051/e3sconf/20183301020

Fig. 2. «Greatwall Tower 1» (240.0 m), Wuhan (China) [http://skyscrapercenter.com/building/greatwall-tower-1/26388].

Jiangxi Nanchang Greenland Zifeng Tower in Nanchang, China, a 56-storey (268 meter) building, should be referred to «mixed-use tall buildings with provision of services». Two-thirds of the building is occupied by offices, and the upper, more prestigious floors, are occupied by a luxurious hotel occupies [14]. This building stands on the podium which houses service functions. There are open terraces, rest areas, etc. (see Fig. 3a) on its accessible roof.

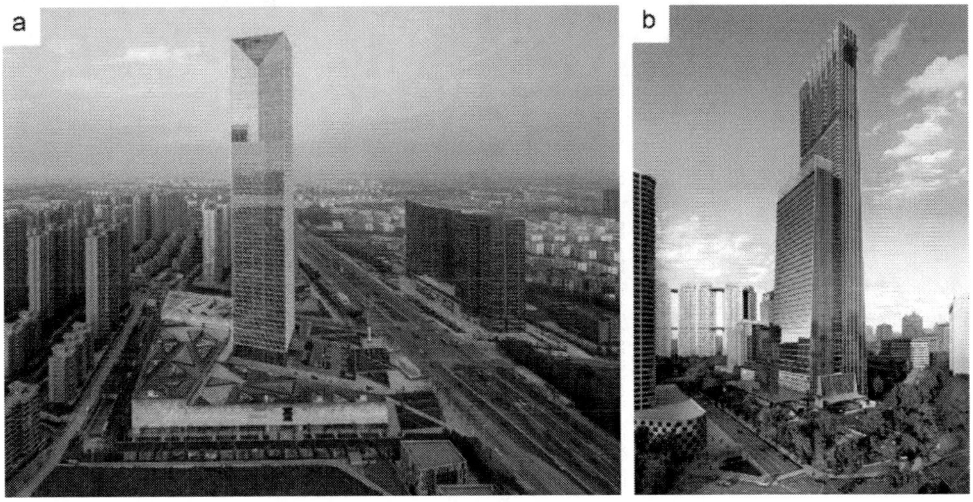

Fig. 3. Mixed-use tall buildings, a - «Jiangxi Nanchang Greenland Zifeng Tower» (268.0 m), Nanchang (China), b - «Tanjong Pagar Centre» (290.0 m), Singapore.

«Tanjong Pagar Centre», an unique skyscraper built in Singapore over the Central MRT station, can serve as a vivid example of «a mixed-use tall building with provision of services». MRT station provides easy access to any part of the city. There are office, residential and hotel functions in the structure of this building providing its multifunctionality [15]. Service functions, which are located in both the 6-storey podium and in a tall building bulk, include: restaurants, a conference center, a gym, a swimming pool, parking lots, and numerous shops (see Fig. 3b). Accommodation service facilities on the upper floors, stretching them vertically in the building is one of new trends in the typological diversity of tall buildings for various purposes. This technique makes it possible to bring service elements closer to working areas in office facilities or to living apartments [16].

«ICE Condominiums at York Centre», consisting of two residential tall buildings located on one podium, can be referred to «a single-function residential tall complex with provision of services». The complex is built in Toronto, Canada (see Fig. 4). The height of residential buildings is 234.1 and 202.3 meters, 67 and 57 floors, respectively. There are medical and sport facilities such as a gym with the latest cardio-training equipment, a swimming pool, leisure and entertainment zones, playgrounds for children, etc. in the structure of the podium [17, 18]. The inclusion of service functions into the structure of the podium creates a comfortable modern living environment for residents of the complex.

Fig. 4. «ICE Condominiums at York Centre 2» (234.1 m), Toronto (Canada) [http://skyscrapercenter.com/building/ice-condominiums-at-york-centre-1/11636].

5 Conclusions

In conclusion, let us emphasize that at the moment there is still a search for more advanced types of tall buildings and complexes. New space-planning decision are introduced. There are also attempts to implement production functions into the structure of buildings. Besides, there are attempts to create a comfortable living environment while developing underground space and including service elements not only into the ground floor level, but also placing these functional zones in the vertical structure of tall buildings.

There is another conclusion based on the materials of CTBUH studies which were presented at international conferences held in New York and Shenzhen. According to these studies, the typology of tall buildings at the present stage reaches a new level of space-planning and functional structure. Currently, development constructors and architects of skyscrapers promote a new typology of a **«3D»**, **«multi-layer tall building»** which will fundamentally change the urban vector frame. According to these developments, urban environment will be repeatedly **«replicated»** vertically in these tall complexes, which will include different functions, e.g. sports, leisure, educational, medical, health-giving, shopping facilities, etc.

The typological structure classifying tall buildings in relation to their functional-organizing elements, introduced in this paper, makes it possible to introduce regulatory requirements both to separate tall buildings and to tall complexes, as well as to solve problems related to the regulatory framework in the design, construction and operation of tall buildings. The classification also provides conditions for type-design practice of tall buildings along with their sophisticated functional planning structures.

References

1. E. M. Generalova, V. P. Generalov, N. D. Potienko, Affordable housing under shaping dense vertical urbanism: Cities to Megacities: Shaping Dense Vertical Urbanism. Proceedings of the CTBUH. *Council on Tall Buildings and Urban Habitat*, 650-659 (Chicago, 2016)
2. V. P. Generalov, E. M. Generalova, Development Prospects for a typology of Hi-rise Buildings. The future of cities, *Vestnik of SGUACE, Town Planning and Architecture*, **1(18)**, 13-18 (2015) DOI: 10.17673/Vestnik.2015.01.2.
3. A. A. Kolosovskaya, N. D. Potienko, Evolution of the Typological Structure of Rental Accommodation Abroad, *Town Planning and Architecture*, **3(24)**, 109-115 (2016) DOI: 10.17673/Vestnik.2016.03.12.
4. V. P. Generalov, E. M. Generalova, Sustainable architecture. Energy Efficiency and Sustainability of Affordable Housing on the Example of Hong Kong, *Vestnik of SGUACE. Town Planning and Architecture*, **4(21)**, 23-29 (2015) DOI 10.17673/Vestnik.2015.04.3.
5. E. M. Generalova, V. P. Generalov, A. A. Kuznetsova, Modular Buildings in Modern Construction, *Procedia Engineering*, **153**, 167-172 (2016)
6. A. Y. Zhigulina, Foreign and home experience in designing energy efficient residential houses, *Vestnik SGASU, Town Planning and Architecture*, **1**, pp. 29-30 (2011)
7. A. Wood, Rethinking the Skyscraper in the Ecological Age: Design Principles for a New High-Rise Vernacular, *Proceedings of the CTBUH Shanghai Conference «Future Cities: Towards Sustainable Vertical Urbanism»*, 26-38 (Shanghai, China, 16th-19th September 2014)
8. M. Balzannikov, V. Alpatov, I. Kholopov, A. Saharov, A. Lukin, Usage of spatial lattice metal structures as roofing for mechanical equipment rooms of hydroelectric power stations, *MATEC Web of Conferences*, **73**, 01012 (2016).
9. A. O. Mateyko, Development History and Current Trends in High-rise Construction, *Town Planning and Architecture*, **3(24)**, 74-78 (2016) DOI: 10.17673/Vestnik.2016.03.12.
10. S. A. Kolesnikov, Town-Planning Principles of High-Urbanized Multi-Functional Units Formation of the largest city urban structure, *Vestnik MGSU*, **3**, 25-29 (2009)

11. T. Y. Vavilova, N. D. Potienko, I. V. Zhdanova, On Modernization of Capital Construction Projects in the Context of Sustainable Development of Social Sphere, *Procedia Engineering,* **153**, 938-943 (2016)
12. http://archi.ru/projects/world/155/bashnya-30-st-mary-axe (last accessed Apr. 18, 2017)
13. https://skyscrapercenter.com/building/greatwall-tower-1/26388 (last accessed Apr. 20, 2017].
14. http://www.archello.com/en/project/jiangxi-nanchang-greenland-zifeng-tower (last accessed Apr. 20, 2017)
15. https://skyscrapercenter.com/singapore/guoco-tower/15186/ (last accessed Apr. 20, 2017)
16. V. P. Generalov, E. M. Generalova, High-Rise Complexes with Serving Zones of Vertical Configuration, *Scientific Survey*, **3**, 163-167 (2015)
17. http://urbantoronto.ca/news/2015/05/%C3%AFce-condominiums-east-tower-nearing-completion (last accessed Apr. 20, 2017)
18. http://www.skyscrapercenter.com/complex/172 (last accessed Apr. 20, 2017)

Mixed-use development in a high-rise context

Elena M. Generalova[1,*], *Viktor P.* Generalov[1], *Anna A.* Kuznetsova[1], and *Oksana N.* Bobkova[1]

[1]Samara State Technical University, Institute of Architecture and Civil Engineering, 443001, Samara, Molodogvardeyskaya St, 194, Russia

Abstract. The article deals with an actual problem of finding techniques and methods to create a comfortable urban environment. The authors emphasize that in the existing conditions of intensive urban development greater attention should be given to spatial concentration based on and more compact distribution of population in urban space. It is stressed that including mixed-use facilities into urban realm results in a significant improvement of living environment qualitative characteristics. The paper also examines modern approaches to constructing a «compact city» for comfortable and convenient living with a mixed-use tall building development. The authors explore the world's experience of designing tall mixed-use buildings and reveal modern trends in their construction. The statistics given is based on the data analysis of a group of tall mixed-use buildings consisting of more than 400 objects, constructed in 2007-2016. The research shows functional and architectural peculiarities of this typology of tall buildings and investigates a mechanism of creating zones of mixed-use tall building development in the urban structure. In conclusion, the authors consider prospects of development and major directions of improvement of mixed-use tall building parameters for a reasonable territorial urban growth and creation of high-density and comfortable building development.

1 Introduction

Massive world urbanization is accompanied by an alarming tendency for urban sprawl of city areas. Leading specialists in architecture, urban planning, economics, sociology, ecology, etc. are worried about it and try to find new, more rational urban planning and space-planning solutions to meet the challenges of modern society. In these circumstances, the concept of **«sustainable vertical urbanism»** based on finding a totally new typology of tall buildings seems appropriate and develops rapidly. This typology is based on **the principle of multi-functionality**. The term «multifunctionality» denotes the ability of a large system to perform many functions. In this paper, «functions» are urban services which are different types of activities in urban area (housing, industry, health care, trade, education, etc.). «Multifunctionality» as a phenomenon of urban environment has been studied by many researchers [1-13].

* Corresponding author: generalova-a@yandex.ru

In the 80-ies of XX century, A.E. Gutnov (a Soviet architect and a theoretician of architecture and urbanism) created a term «urban layout», saying that it was «a relatively separate, functionally connected area of spatial environment, within which a set of basic types of social activity, resulting from the present level of society development, is performed». He stressed that the basis of an urban layout was a «framework», which defined the spatial form of the modern city and had an ability for growth and transformation. Such a framework included «highways and transport nodes, multilevel spatial structures with an extensive network of internal communications» [13]. A.E. Gutnov also emphasized the degree of a city's «communication characteristics» meaning the intensity of its territory development and its connection with the efficiency of location and spacing as a value measured by time spent on traveling around the city because of basic social necessities.

Some earlier statements about the necessity of an urban environment multi-functional development sounded mostly as criticism of cities territorial zoning (based on functional processes typical for a city). Recent studies, however, see the «framework» of the city in a new light as there is an urgent need for spatial concentration of urban realm in vertical direction. There is an active search for a modern innovative concept of a «compact city» with new types of tall buildings, which are developed as integrated mini-cities providing high density residential development. Here, we find it important to emphasize that high density residential development will be able to satisfy requirements for high quality of life only through the integration of various urban services that contribute to the formation and full development of the personality.

2 Methods

In view of the above, this study aims to analyze the multi-functional typology of mixed-use tall buildings and its current state of development as well as to estimate its prospects. The analysis was made on the interactive database «CTBUH Skyscraper Center» containing information about more than 10 000 tall buildings. Therefore, it is appropriate to accept a definition of tall mixed-use buildings given by CTBUH (The Council on Tall Buildings and Urban Habitat): «A mixed-use tall building contains two or more functions (or uses), where each of the functions occupies a significant proportion (15 percent or greater of either) of the tower's total space. Support areas, such as car parks and mechanical plant space, do not constitute mixed-use functions. Functions are denoted on CTBUH «Tallest» lists in descending order (e.g., «hotel/office» indicates hotel function above office function)» [14].

The first building, which satisfies to CTBUH criteria for mixed-use tall buildings, was built in Chicago in 1924. It was «Chicago Temple Building» (office/religious) 173.1 m high. Apart from offices, it fulfills another independent function – a religious one. On the first floor of this building there is the First United Methodist Church of Chicago that can accommodate nearly 1000 people. On the second floor, there is Dixon Chapel and the church administration. There are rooms for a Sunday school, a Conference Hall and a room for the church choir on the third and fourth floors. Subsequent levels (Floors 5 to 23) hold offices which are rented to various commercial organizations. There is also the so-called "Sky chapel" at a height of 120 meters at the base of the Spire. It has room for 30 people.

In the XX century, within the period of 1924-2000 (76 years), 96 mixed-use tall buildings above 150 m high were constructed all over the world. In the XXI century, within the period of 2000-2016 (16 years), 499 mixed-use tall buildings above 150 m high were built. These figures prove that there is definitely a growing interest in mixed-use tall buildings). However, if we relate mixed-use tall buildings with the total number of skyscrapers built during this period, we can conclude that single-function tall buildings prevail. Thus, over the past 10 years (2007-2016) **2071** tall buildings above 150 m high

 https://doi.org/10.1051/e3sconf/20183301021

were built in the world. From that number, only 410 objects can be described as mixed-used buildings. That is only 20% of the total number. The study focuses further on these 410 skyscrapers, which we refer to as called «the analyzed group» (see Fig. 1).

The research identifies a current trend of the intensity of construction of mixed-use tall buildings during this ten-year period in the following years:

- 2007 – 43 buildings, 22 being of mixed-use (15.4%);
- 2008 – 218 buildings, 38 being of mixed-use (17.5%);
- 2009 – 203 buildings, 37 being of mixed-use (18.3%);
- 2010 – 195 buildings, 36 being of mixed-use (18.5%);
- 2011 – 200 buildings, 32 being of mixed-use (16.0%);
- 2012 – 170 buildings, 33 being of mixed-use (19.5%);
- 2013 – 178 buildings, 35 being of mixed-use (19.7%);
- 2014 – 261 buildings, 50 being of mixed-use (19.2%);
- 2015 – 253 buildings, 61 being of mixed-use (24.2%);
- 2016 – 250 buildings, 668 being of mixed-use (26.4%).

It can be seen there is a definite increase both in the intensity of high-rise construction and in the percentage of mixed-use tall buildings. Functions, mixed-use tall buildings can perform, include: office, residential, serviced apartments, hotel, retail, exhibition, casino, etc. According to the degree of their development, these functions can be divided into 3 main types: «dominating», «additional» and «accompanying». According to their «dominating» function, 410 mixed-use tall buildings in the analyzed group are of the following types: the «residential» function (including serviced apartments) prevails in 50% of buildings (202 buildings); «hotel» function goes second with 34% (138 buildings); the «office» function with 17% (70 buildings) is the third.

Buildings combining only two main functions are most common, their proportion is about 85%. Only 60 objects from «the analyzed group» combine 3 or more functions. «three-dimensional» multi-layer urban environment development and serve as basis for creating a new typology of tall buildings.

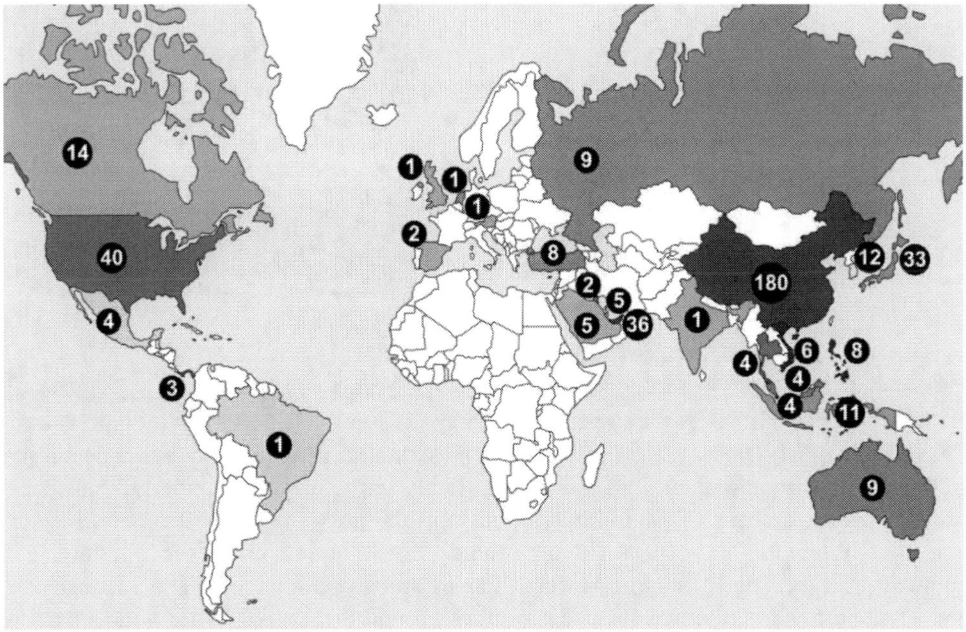

Fig. 1. Mixed-use tall buildings above 150 m high, built in 2007-2016 on the world map.

3 Results

From a bird's perspective or on a cityscape many mixed-use tall building look like conventional high-rise towers, often significantly dominating above the surrounding buildings. Their true **environment-forming** role can be seen only when you thoroughly analyses how these large-scale objects have been integrated into the urban environment, especially at the level of its connection to pedestrian and traffic flows. Talking about the effective integration of mixed-use tall buildings in the current building development, one of the most important criterion – the degree of «openness» of a building into the urban space – should be mentioned. It is very important that the building, which includes many functions and services, does not become an unassailable fortress where housing is available only for the tenants, offices – for the employees and a hotel – for its guests. Thus, a reasonable balance between the **«isolation» and «openness» is required.** «Isolation» is necessary for a few processes, ensuring their safety and optimum performance. «Openness», which means availability in general, availability for transit, etc., is important both for the dissolution of the object in the urban space, and for the improvement of urban infrastructure efficiency.

Fig. 2. «Toranomon Hills», Japan, Tokyo, 2014, 258 m, 52 floors, designed by Nihon Sekkei [12].

«Toranomon Hills» (office/residential/hotel, 258 m, 52 floors), a complex built in Tokyo in the 2014 upon the project of Nihon Sekkei, is a remarkable example of quality architecture. A notable feature of this project is the active use of its underground space for additional compactness of planning solutions and the integration of all functions with the transport system of the city. An underground tunnel, buried under the building, with a highway in it (see fig. 2) is an important part of this design concept. The space above the underground track is released for pedestrians at ground level; a 40-meter wide avenue with wide sidewalks about 13 meters wide is constructed there. To avoid any vibration from traffic on this tall building, the tunnel is designed independent of the building structure and is further isolated by using a special vibration-damping material. There 52 floors above ground and 5 underground levels, serving as a parking lot, in this building. Floors 1-5 hold shops, conference halls, cafes and restaurants. Offices, located on Floors 6-35, are of the «dominant» function. Floor 36 is a technical floor which in itself is a unique structural frame serving to change column spacing from Floor 37 and above. Floors 37-46 serve as housing accommodations, and Floors 47 and above are given to a hotel. «Toranomon Hills» is integrated into the urban environment through a massive green public space used as the center of social activity area. The park is located at the bottom of the tower on the basement roof. There is a rooftop bar, available to everyone, on top of the tower [15].

This is not the only example, where tall mixed-use buildings designs are actively linked with the transport system of the city. «Abeno Harukas», **a complex built in Osaka** in 2014, is considered to be the world's only building of this scale (300 m high), located directly above a train station (the Abenobashi Station). This train station, «Kintetsu», is located on the second floor, its capacity exceeding 70000 passengers per day. The complex is a real high-rise vertical city with a maximum capacity of 110 000 people. It has 60 floors and 5 underground levels with various functions: a terminal station, a department store, an art museum, offices, a hotel, an observatory, a clinic, a school, parking spaces, etc. The department store occupies sixteen floors – two underground levels and 14 floors above ground. The art museum is located on the sixteenth floor. There is a wide variety of works, including the Western and modern art, in it. The observation deck «Harukas 300» (Observatory) is one of the main attractions of this skyscraper, it is located on the 58-60 floors and holds a garden and an open-air café. The internal volume of the building is filled with free public spaces that are also involved in the ventilation system and heat transfer of the whole construction. A stepped shape of the building allowed to create additional public green areas at different levels (see Fig. 3). The building construction is integrated with several technologies aimed at reducing emissions of carbon dioxide into the atmosphere, an innovative technology of urban biogas system for the building being among them. The tower is connected with the existing urban realm through public pedestrian bridges, built on different floors [16, 17].

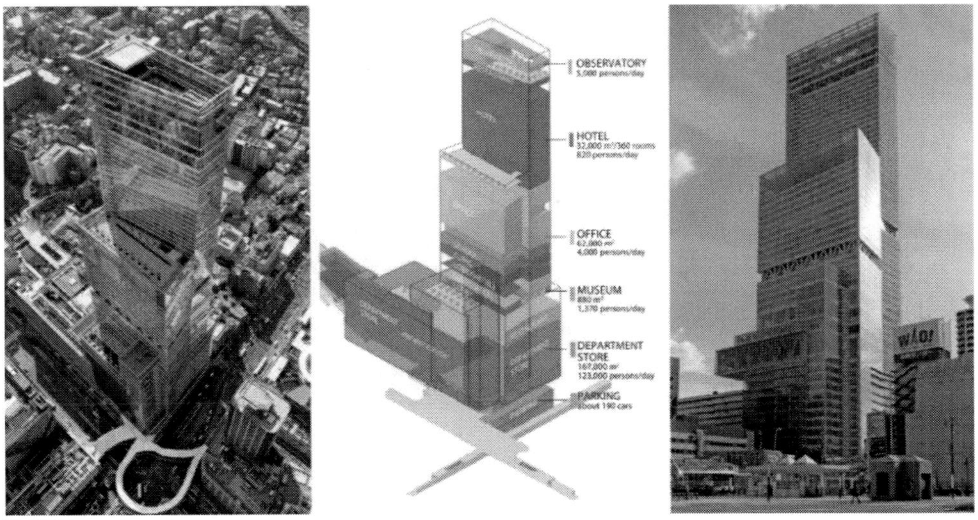

Fig. 3. «Abeno Harukas» (Osaka, Japan, 300 m, 60 floors, 2014, hotel/office/retail. Takenaka Corporation [http://skyscrapercenter.com/building/abeno-harukas/533].

Talking about the methods of tall buildings integration into the urban environment from the point of view of their connection to the transport system of the city, one should mention «The Shard» (306 m, 73 floors, residential/hotel/office), a skyscraper in London. It was built in 2013, on the former territory of Southwark Towers office complex, and is located next to a train station. In this tall building, Floors 3-28 hold offices, Floors 31-33 house restaurants, the 200-room Shangri-La Hotel is situated on Floors 34-52, and the next block is for housing accommodations (Floors 53-65). A public observation deck (Floors 68-72) crowns the building. The construction of this super tall building next to a major transportation hub, in a city that has a rich historical heritage, introduced several innovations. The construction team had to reconsider basic principles of construction and make use of entirely new methods. Thus, while constructing the concrete core, a top-down construction technology was used for the first time. Because of this technology the first 23 floors of the concrete core and much of the surrounding tower had been built before the basement was fully excavated. Besides, a special drilling rig, which allowed to place piles in underground communications of the Victorian era with an extremely high precision, was specially design for the construction. Today, «the Shard" is a new symbol of London, a key link in the newest commercial quarter giving powerful energy to revive the entire southern shore of London [18].

4 Discussion

The result of the work on the analysis of the structure of mixed-use tall buildings and methods of their integration into the urban environment would be incomplete, if we do not consider the future to consider development prospects of this typology of buildings. It is planned that in 2017 new 108 mixed-use tall buildings will be erected. Among these buildings, 13 have already been built. Three buildings here can be referred to the category of super tall buildings:
1 – «Lotte World Tower» (Seoul, 554,5 m, 123 fl., hotel/office/retail);
2 – «Address Boulevard Dubai» (Dubai, 368 m, 72 fl., serviced apartments/hotel);
3 – «Zhuhai IFC», (Zhuhai, China, 330 m, 67fl. hotel/office).

Fig. 4. «Raffles City Hangzhou Tower» (250,0 m, 60 floors, China, City Hangzhou (residential / hotel / office), UN Studio [http://www.unstudio.com/projects/raffles-city-hangzhou].

However, we would like to describe a few sites which are not that tall. The first here is a complex named «Raffles City Tower Hotel Hangzhou» (China, City Hangzhou, 250.0 m, 60 fl.) with its unique space-planning decision. The building is in its final stages of construction and incorporates retail, offices, housing and hotel facilities. This project consists of two tall intertwining twin towers at the podium and a three-storey underground parking. According to the designers, its characteristic feature is a close interconnection of the landscape aspect with the urban context (see Fig. 4). The project involves a huge amount of technical and engineering features that are implemented while using digital engineering through building information modelling, virtual construction planning and simulation [19].

Singapore complex «**Marina One**» (see fig. 5) will also make a significant contribution to the formation of a new typology of tall buildings after its completion. It is a high-density, mixed-use building complex in the heart of a new financial district. The basis of its concept is turning Singapore into a «Garden City». This concept is implemented through the creation of internal public space fringed by office buildings and housing accommodations, as well as a podium with shopping and entertainment function. It is the largest public Plaza in the Central Business District) which is supplemented by green areas at different levels of the complex. The perimeter formed by the buildings is not completely closed. The towers have a porous structure, which provides good air circulation, ensures natural ventilation and sustains a comfortable internal climate. Environmental sustainability and energy efficiency are achieved though introducing a variety of technologies: external sun-protection devices and high performance glazing, centrally provided recycled water for toilet flushing, rainwater harvesting system, photovoltaic cells, and more. The complex is connected to the four lines of the Metro (MRT) and bus terminals [20].

Fig. 2. «Marina One» (Singapore, 225,4 m, 30 floors, residential/office/retail). Ingenhoven Architects & A61. [http://www.ingenhovenarchitects.com/projects/more-projects/marina-one-singapore/]

5 Conclusions

Summarizing the results of the analysis, it should be noted that the typology of mixed-use tall buildings is being actively developing. This development is aimed at searching a modern interpretation of the concept of the vertical city, where people can live, work and relax. Even though today among mixed-use tall buildings three buildings have surpass the mark of 600 m and there are 47 buildings above 300 m high, we can confidently say that «a race in the vertical direction» is not a philosophical basis of design concepts discussed above and implemented in the unique complexes under analysis. The most important criteria for the quality of these buildings architecture are the following approaches, techniques and methods of their integration into the urban environment: town planning; space-planning; engineering; technological; integration through transport infrastructure;

integration through the creation of public spaces and pedestrian links with surrounding buildings; because of a reasonable balance between the «isolation» and «openness» of different functional areas within buildings, etc. In conclusion, we would like to stress that a proper integration of mixed-use tall buildings will allow to create a convenient and comfortable living environment and to improve environmental and economic status of existing cities.

References

1. V. P. Generalov, E. M. Generalova, Sustainable architecture, Energy Efficiency and Sustainability of Affordable Housing on the Example of Hong Kong, *Vestnik of SGUACE. Town Planning and Architecture*, **4(21)**, 23-29 (2015) DOI 10.17673/Vestnik.2015.04.3.
2. E. M. Generalova, V. P. Generalov, A. A. Kuznetsova, Modular Buildings in Modern Construction, *Procedia Engineering*, **153**, 167-172 (2016)
3. T. Y. Vavilova, N. D. Potienko, I. V. Zhdanova, On Modernization of Capital Construction Projects in the Context of Sustainable Development of Social Sphere, *Procedia Engineering,* **153**, 938-943 (2016)
4. V. P. Generalov, E. M. Generalova, Revealing the Special Features of the Concepts "Comfortable living" and "Comfortable living environment", *Vestnik of SSUACE, Town Planning and Architecture,* **2(23)**, 85-90 (2016) DOI 10.17673/Vestnik.2016.02.16.
5. E. M. Generalova, V. P. Generalov, Specific Features of Modern Architectural and Spatial Environment of Cities and Its Formation (with South Korea as an Example), *Scientific Survey*, **11**, 46-51 (2015)
6. E. M. Generalova, V. P. Generalov, Super Thin Residential Skyscrapers in New York as a New Trend in the Typology of Tall Buildings, *Town Planning and Architecture*, **4 (25)**, 85-91 (2016) DOI: 10.17673/Vestnik.2016.04.16.
7. A. Y. Zhigulina, N. G. Chumachenko, Choosing Building Materials to Improve Comfort and Ecological Safety of Residential Buildings, *Vestnik of SGUACE. Town Planning and Architecture*, **4(21)**, 94-99 (2015) DOI 10.17673/Vestnik.2015.04.12.
8. A. A. Kolosovskaya, N. D. Potienko, Evolution of the Typological Structure of Rental Accommodation Abroad, *Town Planning and Architecture*, **3(24)**, 109-115 (2016) DOI: 10.17673/Vestnik.2016.03.12.
9. M. Balzannikov, V. Alpatov, I. Kholopov, A. Saharov, A. Lukin, Usage of Spatial Lattice Metal Structures as Roofing for Mechanical Equipment Rooms of Hydroelectric Power Stations, *MATEC Web of Conferences*, **73**, 01012 (2016)
10. A. O. Mateyko, Development History and Current Trends in High-rise Construction, *Town Planning and Architecture*, **3(24)**, 74-78 (2016) DOI: 10.17673/Vestnik.2016.03.12.
11. S. A. Kolesnikov, Town-Planning Principles of High-Urbanized Multi-Functional Units Formation of the largest city urban structure, *Vestnik MGSU,* **3**, 25-29 (2009)
12. T. Ya. Vavilova, International Experience in Rehabilitation of Depressed Residential Areas for Sustainable Development, *Architeckton: News of Higher Educational Institutions*, **49**, 4 (2015)
13. A. E. Gutnov, Evolution of Urban Planning, 256 (Moscow, Stroiizdat, 1984)

14. http://www.ctbuh.org/TallBuildings/HeightStatistics/Criteria/tabid/446/language/en-US/Default.aspx (last accessed Apr. 3, 2017)
15. Y. Hitomi, H. Takahashi, H. Karasaki, Toranomon Hills – Super High-Rise Building on Urban Highway, *International Journal of High-Rise Buildings*, **3(3)**, 167-171 (2014)
16. K. Hirakawa, K. Saburi, S. Kushima, K. Kojima, Performance-Based Design of 300 m Vertical City «Abeno Harukas», *International Journal of High-Rise Buildings*, **3(1)**, 35-48 (2014)
17. T. Harada, M. Yonezu, Forging a Supertall Compact City, *CTBUH Journal*, **II**, 12-20 (2015)
18. I. Sellar, Developing an Icon – The Story of the Shard. *Proceedings of the CTBUH 2015 International Conference «Global Interchanges: Resurgence of the Skyscraper City» 26th-30th October,* 138-145 (New York, USA, 2015)
19. A. Wang, Raffles City in Hangzhou China – The Engineering of a 'Vertical City' of Vibrant Waves, *International Journal of High-Rise Buildings*, **6(1)**, 33-47 (2017)
20. http://skyscrapercenter.com/building/marina-one/16878 (last accessed Apr. 3, 2017)

The Design of Akhmat Tower

Sara Beardsley[1*], *Alejandro* Stochetti[1], and *Marc* Cerone[1]

[1]Adrian Smith + Gordon Gill Architecture, 111 West Monroe, Suite 2300, Chicago, Illinois 60603, USA

Abstract. Akhmat Tower is a 435m supertall building designed by Adrian Smith + Gordon Gill Architecture. It is currently under construction in the city of Grozny, in the Chechen Republic, in the North Caucasus region of Russia. The design of the tower was done during a collaborative process by a multi-disciplinary architectural and engineering team, based primarily in the United States and Russia. During this process, the designers considered many factors including, most primarily, the cultural and historical context, the structural requirements given the high seismicity of the region, and the client's programmatic needs. The resulting crystalline-shaped tower is both an aesthetic statement and a performative architectural solution which will be a new landmark for Chechnya. "The Design of Akhmat Tower" describes in detail the design process including structural considerations, exterior wall design, building program, interior design, the tuned mass damper, and the use of building information modeling.

1 Introduction

In 2014, Adrian Smith + Gordon Gill Architecture (AS+GG) was commissioned to design a signature tower adjacent to the Suzha river in Grozny, Chechnya - a Republic of Russia located in the north Caucasus region near the Caspian Sea. This tower is to be a new icon specifically for the region – and is named after the first President of Chechen Republic, Akhmat Haji Kadyrov [1]. The design was done by a team of over one hundred people, based primarily in the United States and Russia, and utilized expertise from architects and engineers with vast experience in super-tall building design. The 435m tower, currently under construction, is planned to be completed in 2020 and at that time will become the tallest building in Chechnya and the second tallest building in Russia [2]. The only tower that will exceed its height is the Lakta Tower in St. Petersburg, which also under construction [3]. This case study, «The Design of Akhmat Tower», documents the design process for Akhmat Tower, the resulting design solution and the many factors that contributed to its development.

* Corresponding author: sarabeardsley@smithgill.com

2 Process & Methodology

2.1 Design Team

Adrian Smith + Gordon Gill Architecture (AS+GG), the design architect, is a firm which was established in 2006 and is dedicated to high performance, energy efficient and sustainable architecture on an international scale [7]. The firm's portfolio features several super-tall buildings including as the 1000m Jeddah Tower, which is to be the world's tallest building when complete, and the 600m Wuhan Greenland Center. Adrian Smith, one of the design partners, has over 40 years of experience specifically working on super-tall projects, including the Burj Khalifa. He is known for 'global environmental contextualism' [4] and follows the principle that each design should be a specific response to its site, country of origin, and environment.

Thornton Tomasetti, the structural engineering firm responsible for Akhmat Tower, is also world-renowned in the design of super-tall buildings and for their experience in designing tall buildings high-seismic areas, such as Taipei 101, completed in 2004 [5]. The design team also included the civil engineers Langan, Inc.; wind engineers Rowan Williams Davies and Irvin, Inc (RWDI); vertical transportation consultants Fortune Shepler Saling; and the architect of record, Gorproject, based in Moscow, Russia.

2.2 Design Process

The design team worked closely with the client, Smart Building, Inc. through a series of workshops during all phases of the project. During the early concept design phase, Adrian Smith + Gordon Gill Architecture proposed three independent massing concepts. The client selected a tapering pyramidal shaped tower as the basic architectural massing form, and from that point, the complex geometry of the tower base and tower crown was further studied and developed. The design was optimized with input from both the structural engineers and the wind tunnel test results, and in response to the needs of the interior programmatic elements of the building. It was an iterative process which took place over a period of over 1 year. The development of the architecture was led by the design team in Chicago with input from the architect of record, Gorproject, especially concerning compliance with local codes and Russian standards. The Thornton Tomasetti team included engineers based in Russia. This close collaboration between local and international architects and engineers is very beneficial when performing a project of this scale.

2.3 Design Criteria

The design of Akhmat Tower required a strong historical and cultural reference. The client requested that he tower be influenced by the traditional Nakh watchtowers which are commonly found in traditional Chechen medieval architecture (see Fig. 1). This medieval watchtower is the most recognizable symbol of historical Chechen architecture and represented advanced building techniques for their time. The traditional four-sided watchtowers were typically square or nearly square in plan, always tapering, with a prominent series of window openings on the south façade. The construction of these towers reached its height from the 13th to the 17th centuries [6], as dwellings and for military defense. The defensive towers were often beacons of communication from one village to another, as they were typically found on a hill or mountain top. While the form of the watchtower was square in plan, the pinnacle of the was either flat, or a slightly pyramidal shape with a capstone. The client's desire for a strong historical reference as a design criteria for the project was a challenge and an opportunity for the design team, which had the task of creating a contemporary building that reflected the technology of the future, but also had reverence to the past.

Fig. 1. (Left) Tsoi-Pheda Tower, Chechnya and (Right) Guard Tower in Ingushetiya.

The client's program for the building, which is mixed-use consisting of office, hotel, residential, observation, restaurant and retail uses – required a strict separation between functions in order for the building to operate smoothly and accommodate the arrival sequence for all of the building users and visitors. These programmatic needs, consideration for security and access, and the spatial needs of the hotel amenities heavily influenced the design of the building base, which incorporates the mixed-use functions of the building in multiple levels. The topography of the landscape design was used to reinforce the separation of building uses. The development of the building core and the placement of the technical floors was carefully coordinated to facilitate the most optimal layouts, lease spans, and elevator zoning. Building performance and energy use criteria were also of significant consideration.

Structural requirements strongly influenced the architectural design. The design process required very close coordination between the architectural and structural design needs, due primarily to the high seismic zone where the project is located, and its height of over 400 m. The architects had to create a design that both worked from a planning point of view and was cost effective and efficient from a structural point of view. The design team extensively used Building Information Modeling (BIM) to document the design and share information between members of the design team. The complexity of the exterior wall design pushed the envelope in terms of three-dimensional modeling and required the use of innovative software tools in order to create the production drawings.

3 The Design of Akhmat Tower

3.1 Design Concept

In response to the historical reference of the medieval watchtower, AS+GG's design vision for the Akhmat Tower shares a strong geometric relationship in its architectural proportion, the use of a four-sided pyramidal shape and monumental top. The tower base design incorporates flared geometry and earth berms - a reference to the traditional watchtowers which were always on a hilltop or mountain (see Fig. 2). The division of the tower form into a distinctive top, middle and bottom zone parallels the watchtower design which also had the same three parts. The size of the pinnacle of Akhmat Tower utilizes a similar proportion to the top of the traditional watchtower. During the design, the prominence and size of the pinnacle was carefully considered in order to reinforce the relationship to the traditional tower. The crystalline shape of the pinnacle has a specific indented center to its geometry whose location relates to the location of the window opening that is typically present in Nakh watchtowers. Four sloping walls taper to reach the observation and museum floors at the top of the building, taking advantage of the pinnacle's volume, which features distant views of the presidential palace, the Akhmad Kadryov Mosque, and the distant mountains. Akhmat Tower's pinnacle is used as an observation space where one can overlook the city of Grozny, just as the medieval watchtowers were used to survey distant lands.

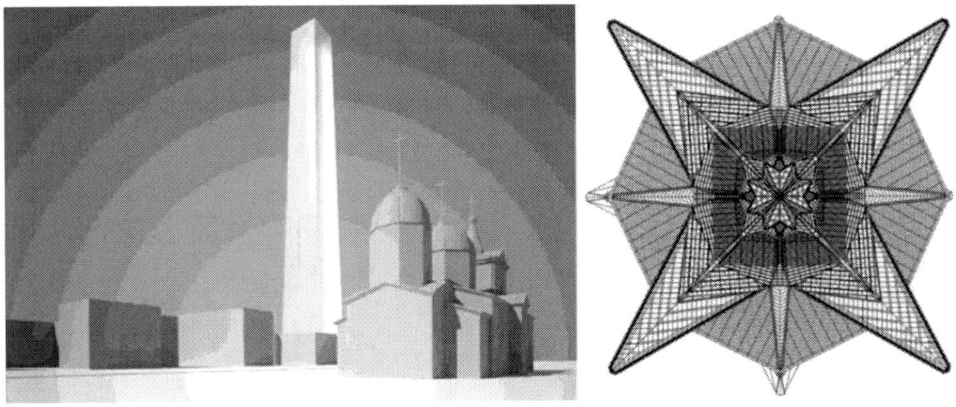

Fig. 2. (Left) Early Conceptual Massing View from Mosque, © AS+GG 2017 and (Right) Geometry Plan of Akhmat Tower, © AS+GG 2017

The use of the square plan, and the symbolism of the geometry formed by two intersecting squares, is very important in terms of the historical reference to Chechen architecture, which used square plans in nearly all of the historical buildings [6]. The shape formed by two intersecting squares, one rotated at 45 degrees to the other, is also seen in traditional Islamic patterns. The idea of the square, diamond, or rotated square geometry is repeated throughout different elements of the tower and in the exterior wall panels.

Similar to the hilltop setting of Nakh Towers, the landscape design of Akhmat Tower situates the building and its multi-level lobbies atop a landscaped hill. Drop-off zones for the different uses of the tower are reached through roads at various levels, accessing the different sides of the tower with elegant earth berms negotiating the elevational differences. The free-flowing berms contrast with the sharp geometry of the tower design itself, in the same way the historical towers were juxtaposed against the green hills of the North Caucasus region. Native plantings and trees are used throughout the design. Water features welcome visitors in the drop-off areas of the tower, recalling the rivers found in the region.

3.2 Tower Program

Akhmat Tower has an approximate gross area of 250,000 square meters, containing office, hotel, residential, observation and retail spaces (see Fig. 3). Because the building tapers, the program is organized in order to utilize ideal lease spans in each zone. The building has one basement level, containing parking and mechanical spaces. The main lobbies exist on two levels with the office lobby located at level 1 and the hotel lobby located at level 3. Both lobbies have visual access to the 8-story atrium above. Each program type has separate drop-offs on various sides of the building. The observation drop-off, at the north side of level 1, is adjacent to the public retail entry and retail mall connection. The hotel ballrooms, wedding suite, grand lobby, and restaurants are all located at the upper level 3 lobby.

Within the tower, the office program occupies the lower 10 levels of the building, followed by 7 hotel guest room floors, and 11 levels of serviced apartments. The upper 49 levels of the tower are residential units culminating in 2 levels of penthouse units at the top. The pinnacle features a signature restaurant, an observatory, and exhibit space.

The elevator system for Akhmat Tower is divided into zones by program. All tower functions are served with a local elevator system which takes passengers directly from the lobby to the floor of occupancy. While express elevator shuttles and transfer floors are often seen in a building of this size, it is most efficient, luxurious, and convenient to use a system that does not require elevator transfers. Therefore, Akhmat Tower was designed with direct elevator service from the lobby to the occupied floors. Dedicated parking, public, hotel amenity, and retail elevators serve the base of the building, ensuring security and privacy for the main tower elevators.

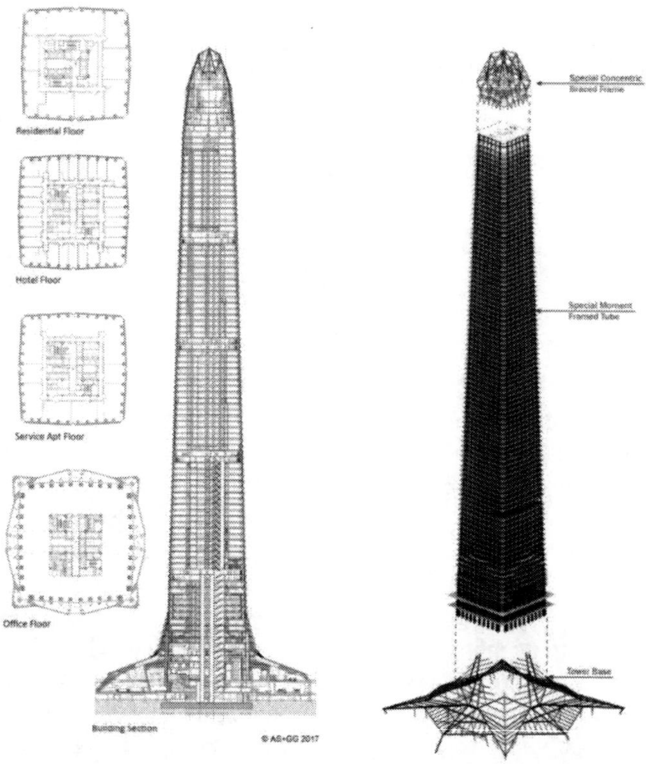

Fig. 3. (Left) Plans and Section of Akhmat Tower, ©AS+GG 2017 and (Right) Structural Design Diagram, © Thornton Tomasettti, 2017

3.3 Integration of Structural Design and Architecture

The seismicity of Grozny is one of the highest in the world, exceeding San Francisco by fifty percent and the seismic criteria used in Taipei 101. Most new supertall buildings utilize either completely concrete structures, with shear-walls on every floor, or composite concrete and steel structures, which allow the buildings to have very few large columns and rely on outrigger trusses located at the technical floors. Neither of these solutions was suitable or efficient given the seismic loads in Grozny.

A building which behaves well in a highly seismic area needs to be sufficiently ductile, and should have a uniformity to the structural design in order to minimize any areas where loads could be concentrated during an earthquake event. Therefore a customized, evolved version of the steel 'perimeter tube' system was chosen for Akhmat Tower. The perimeter tube system, which was utilized in the Willis Tower in Chicago and the original World Trade Towers in New York, requires a closely spaced set of columns at the perimeter of the building. These perimeter columns and coupling beams will create the world's tallest «special moment frame» seismic resisting system once realized [5] (see Fig. 3).

Akhmat Tower's perimeter columns, spaced at 4.5m on center, are sufficiently small as to minimize their impact on the usable interior space. The steel perimeter columns are typically less than 1m in size, which is an advantage over a mega column system, which commonly has columns of two meters or more. The column layout of Akhmat Tower is designed so that there is no corner column, allowing the corners to be chamfered and include windows that offer views out diagonally. The use of a special moment frame structure meant that the technical floors could be designed without large outrigger trusses, making them faster and more simple to construct, and easing the layout in terms of fitting the mechanical equipment on those floors.

3.4 Performative Massing and Exterior Envelope Design

The design of the final shape of Akhmat Tower was influenced by wind performance. From the conceptual phase of the project, a square-shaped building was desired. However, it was also noted early on by the design team that square-shaped plan with sharp square corners was not ideal for wind performance. Thus, as design refinements were done, consideration was given to optimizing the form in a way that improved its behavior. This optimization reduced wind loads on the building, making the structural members more efficient, saving significant tonnage and cost in steel.

The earliest conceptual design included a completely flat top pinnacle and a completely square plan, yielded a wind-load response that was estimated to be up to 15% greater than the more refined final design. The final design incorporated two-meter chamfered corners and a forty-fiver meter tall, sculpted, crystalline top. The shaping of the base of the building also showed a measurable benefit. It was found through wind test data that pedestrian wind comfort in the spring, summer and fall was in an acceptable range with very little mitigation. This performance is attributed to the shape of the flared base and canopies, which shield pedestrians from wind coming down the face of the tower.

In the design of Akhmat's building envelope, the stone wall patterns of traditional architecture of Chechnya are reinterpreted and materialized in a three-dimensional developed glass, aluminum and stainless-steel wall. The exterior wall has a three-dimensional shape that protrudes at each floor and slopes back toward the main volume, thereby reducing glare and solar gain to the façade. The shingled wall has a curved corner, which forms a stainless steel accent piece that catches the sunlight. Operable mullions are fully integrated with the wall design, enabling the façade to have a strong consistency. Specifically designed picture windows at the corners offer enhanced views while also improving the tower's response to wind loads.

The exterior wall of Akhmat Tower is a next-generation development of the architects' Federation of Korean Industries (FKI) Headquarters exterior wall in Seoul, Korea, which was completed in 2014 and was tilted downward 10 degrees (see Fig. 4). During the design of the FKI project, solar analysis showed that the greater the wall's downward tilt, the less glare and solar heat gain would be realized versus a flat curtainwall. The typical Akhmat wall has a 14-degree downward tilt. The incorporation of the 80-millimeter-tall operable mullions was driven by a programmatic need for natural ventilation up to nearly 400 meters in height for the residential units. On windy days, especially on the upper floors of supertall buildings, traditional operable windows tend to be used less frequently by occupants due to powerful wind gusts flowing through the occupied floors, and difficulty opening or closing the windows. The operable mullion provides controlled natural ventilation, utilizing a button to open and close the device. This encourages natural ventilation to be used during more times of the year and alleviates some of the perceived safety concerns associated with open windows at such heights.

Fig. 4. (Left) Typical Exterior Wall, © AS+GG 2017 and (Right) Corner Exterior Wall, ©AS+GG 2017

The texture and tilt of the exterior wall is continuous from the bottom of the tower pinnacle to the canopies at Level 2. As the wall reaches the bottom of the tower, it flares dramatically at the base, forming the eight-story atria that frame each lobby of the building. The atria create a double-skinned wall – a monumental volumetric space that can be viewed from the lobbies below and from the lower office floors. The atrium base extends out to form glass wings that act as sky-lit spaces, which frame the entries to the lobbies. Triangular, shingled canopies extend dramatically over each drop-off. Operable louvers within the atrium exterior wall system are linked to the smoke evacuation system, using the tapering shape of the atrium to help draw air upward, and out.

The crystalline geometry of the tower pinnacle is a special feature in the exterior wall design. The many facets catch the light as it hits the building at different angles. The volumetric spaces inside the pinnacle include the two-story tuned mass damper, which was designed as part of the public observation area, as well as a multi-story volume at the top that forms the museum and restaurant. The steel tube structure is minimally clad, in order to expose the complex geometry; fritted glass is used to control glare. The building's maintenance system for cleaning the peak is strategically integrated into the top of the tower, above the core, using operable swing panels that open when the unit is deployed.

4 Interior Design and Planning

The form of the tower base informed the unique spatial opportunities of the interior design of the office, apartment, and hotel lobbies and became a primary influencer to the design. The office and apartment drop-off elevation is situated below the hotel lobby on level 1, allowing for a continuous space for the hotel program above. Taking advantage of the soaring and sculptural spaces created by the tower base form, the office and apartment lobbies are outside of the tower columns, further separating the hotel by walls of stone at the office and wood at the apartment lobby. The spaces are naturally lit, the angular feature walls as a back-drop. The structure creates a sense of a screen to the office and apartment's inner lobbies, respectively, which naturally form more intimate spaces. Materials become warmer and more tactile as building users progress to the lift lobbies in the core.

The hotel lobby program benefits from the ability to have an almost totally circular flow around the core of the building on level 3 (see Fig. 5). The reception lobby is an impressive 50 meter tall winter garden with light dappling in from the tower base's roof skylights. Exaggerated column enclosures allow passive air to circulate by a series of integrated grilles and ducts that move the air though natural convection. A series of interior trees create a canopy over the seating and social areas of the lobby, bringing these areas down to a more human scale, enhancing guest comfort and functionality of the grand space.

Fig. 5. (Left) Office Lobby Interior, © AS+GG 2017 and (Right) Hotel Lobby Interior, ©AS+GG 2017

Restaurants are situated in two of the four arms of the public floor of the hotel, while the ballroom pre-function and bridal support spaces are housed in the other two arms. The tower base form and structure create a virtually column-free space for a 700 square meter ballroom that has a column free and sun lit pre-function area.

Additional personal and attentive functions of the guest experience are tucked away on the 21st floor of the 165-key hotel. Spaces such as guest check in, guest lobby and lounge, cocktail lounge and all-day dining are arranged in a club-like atmosphere, providing a sense of exclusivity and privacy for guests. Spa functions on level 4 are accessible to the hotel as well as to the apartments and serviced apartments.

The interior design of the building takes much inspiration from a local traditional dance called The Lezginka, which means "the dance of the eagle and the swan." This national dance, shared by people in the Caucasus region, is performed during special occasions and weddings. In it, a male dancer becomes an eagle with sharp, angular motions and a woman

 https://doi.org/10.1051/e3sconf/20183301022

dancer becomes a swan with soft, fluid circular movements. This memorizing juxtaposition of totems motivates the dialogue of the tower's interior design forms and furnishings. The frenzied dance with its sharp, strong, and angular form, further enhances the narrative of the architectural design.

5 Tuned Mass Damper

Akhmat Tower requires a tuned mass damper as part of its design. The damper is required not for reasons of structural strength, but instead to retain occupant comfort. A steel building is designed to have more movement than a concrete structure, which is one of the reasons it performs so well in an earthquake. The damping is required in order to make sure the upper floors of the building are very comfortable for occupants as the building moves. This is especially true given that the upper floors are residential, where comfort perception limits are the most strict.

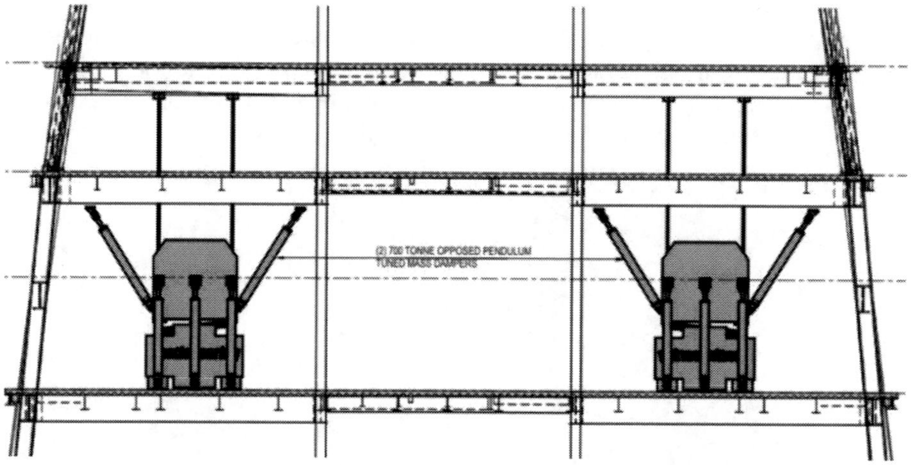

Fig. 6. Tuned Mass Damper Section Diagram, © RWDI 2017

The two tuned-mass dampers (TMDs) are located on Level 96 - above the residential floors and directly below the observation floor (see Fig. 6). Because of the tapering shape of the tower, floors become quite small near the top of the building and finding enough space was a significant challenge. The layout of the core required that the express elevators and fire-fighting elevators be located at the center of the building, necessitating a non-centralized location for the damper in plan. It was decided early in design that the damper would be split up into two locations on each side of the core. Although two dampers are not as space-efficient as a single damper overall, it was found that the two dampers did satisfy the required performance and that it was a feasible approach. The layout of the two dampers also leaves ample space for the building maintenance machines to co-exist on the double height technical floor.

In addition to the challenges of fitting the dampers in plan, the height requirement of the damper in order to achieve the necessary swing of the cables presented another challenge. The solution was to allow the cables to extend up through floor 98 – into the observation deck. Glass handrails will be built around the cables and a glass floor will be installed so that visitors to the observation deck can view tuned mass damper. In this way, the tuned mass damper will be used as an architectural feature, and educational exhibit, to show building visitors how dampers are used in super-tall buildings.

6 Building Information Modeling (BIM)

Adrian Smith + Gordon Gill Architecture uses building information modeling extensively in its production of drawings. While the early design process is done with Rhino, a three-dimensional modeling design program, the working drawings are done in Autodesk Revit. The issuance of drawings for Akhmat Tower was done entirely from Revit. The model took the team over a year to develop from start to finish. The use of Revit was multi-disciplinary, including the structural and mechanical engineers, interiors team, lighting design and landscape. The various models were separated based on sub-teams within the project; the largest models being the «core» and «exterior wall». As the design evolved, translation between Rhino and Revit was required on a regular basis.

The creation of the accordion-shaped exterior wall in Revit was a complex process, which was done using Rhino, various plug-ins, and translated to Revit using adaptive components. The typical exterior wall family was a panel that contained twelve main geometry points. The most complex part of the exterior wall modeling process was to build the tapering corners and the trapezoidal panels that flared and tapered as they connected the main exterior wall to the atrium. By the end of the project, the entire exterior wall was usable directly from Revit without use of masking or placeholder geometry, so that any part of the model could be cut accurately in plan and in section. One advantage of the design was the symmetry – as it was only necessary to build one eighth of the main exterior wall, which could then be mirrored 8 times.

7 Discussion: Why Supertall?

Akhmat Tower is an example of a supertall building that integrates the latest technology, and will become a new landmark for the region of Chechnya. It will attract visitors to its museum and restaurants, and luxuriously accommodate its hotel guests, office tenants and residents. It is important for the designers of super-tall buildings to consider the overall impact of tall buildings on the environment, and examine the question: Why build a super-tall building? Adrian Smith + Gordon Gill Architecture has done extensive research into the energy impact of tall buildings and analysis of how these buildings fit into the urban fabric of our cities. While super-tall buildings are energy intensive, they can be designed with excellent energy performance, and energy standards are becoming more strict as technology advances.

AS+GG is currently performing research called «The Density Study» whereby different buildings were analyzed in terms of their impact to carbon emissions. Nine typologies of residential buildings, from a single family home, to low-rise, to mid-rise, to supertall, were modeled to find out which is the most ideal density in terms of creating the lowest environmental impact (see Fig. 7). The findings showed that the density with the greatest negative effect on the environment is a single-family home, but the second most impactful was the supertall, which for the purposes of the study, is categorized as a building from 300m to 600m in height. The extra energy used in a supertall building is primarily due to the space and energy requirements for the elevator system, and to move mechanical services, electricity and water throughout the building. The initial study found that the most efficient or 'ideal' urban density [in terms of carbon emissions] would be a 3-4 story walk up apartment building, which has a very low energy load from its building systems and low use of land area. Why then, if super-tall buildings are an energy intensive typology, should they continue to be built? The answer to this question is complex and has to do with deeper principles of urban planning and the way that we live in the 21st century.

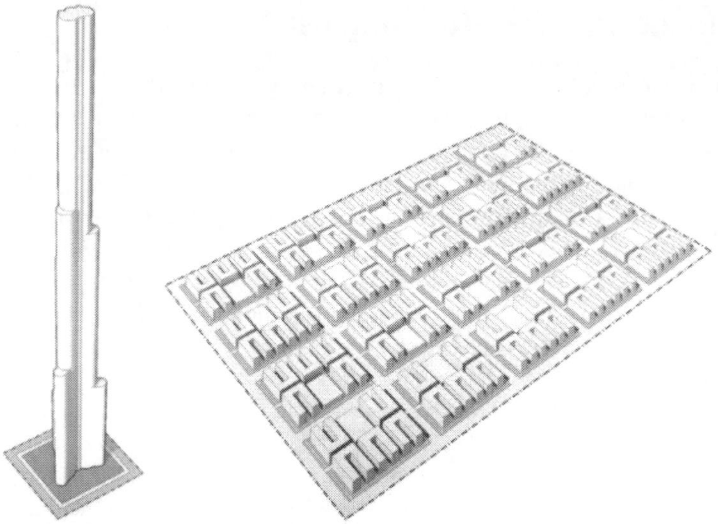

Fig. 7. (Left) Supertall Prototype Diagram, © AS+GG 2017 and (Right) Courtyard. Prototype Diagram (illustrating the number of courtyard buildings equivalent to a supertall prototype), ©AS+GG 2017.

Supertall buildings were first invented in the 20th century as the technologies that were necessary to build them were developed. These structures became important to cities as our modern landmarks, and symbols of prosperity and identity. They are also important as catalysts for development of urban centers. A supertall building is most often located at the center of a successful business district of a city, where land value is the greatest. Typically the buildings directly around the super-tall become commercial centers for those cities. Supertall buildings also attract culture, tourism and public amenities at their base and the areas direction around them, often incorporating major transit hubs. Therefore the supertall building cannot be looked at in isolation – it must be considered for the important part it plays in the urban environment and for its role as an icon or symbol for the city where it is located – and as a catalyst for the city around it.

Even though a low-rise building may be the most energy efficient density, it is important for major cities to have a mix of densities and a very dense urban core, in order for those cities to prosper. In the 21st century, more people are living in cities than ever before. Increasing density is advantageous in the future, because it will minimize sprawl as the world population grows. Supertall buildings have become an important part of our society and are part of the growth of urban density that will continue over the next century and beyond.

8 Conclusion

Super-tall buildings are important landmarks that have special meaning to the cities where they are built. Akhmat Tower is a building that will honor the past and represent the culture of Chechnya, while also becoming a symbol of hope and future prosperity (see Fig. 8). Akhmat Tower is a unique building which responds to the environment where it is located, the requirements of the client's program and structural challenges. The crystalline shape of the pinnacle, the monumental tower, and the soft curves of the site design give the tower its unique identity and create architecturally significant public spaces which will be enjoyed by the occupants as a new center for the city of Grozny.

Fig. 8. Overall View, Akhmat Tower, © AS+GG 2017.

Acknowledgements

We would like to acknowledge the entire Akhmat Tower AS+GG Design team: in particular, Adrian Smith; Gordon Gill; Robert Forest; Sae Hwang Oh; Alejandro Stochetti (Design Director); Myongki Seong; Nathan Diestelkamp; Sung Hak Ko; James Struthers; Yuchen Liu; Marc Cerone.

Additionally, the AS+GG Density Study research team, including Chris Drew, Director of Sustainability, and Natalia Quintanilla, energy modeler.

The Akhmat Tower consulting team: Thornton Tomasetti Engineers; Langan Engineering & Environmental Services; Positivenergy Practice MEP engineers; FS2 Vertical Transportation Consultants; RWDI Wind Engineers; Lerch Bates Façade Access; Garth A Wemmer Landscape Architecture, LLC; Forcade Associates, Graphics; Gorproject, Architect of Record.

References

1. www.smbuild.ru (last accessed 2017/08/26)
2. www.skyscrapercenter.com (last accessed 2017/08/26)
3. www.lakhta.center/en (last accessed 2017/08/26)
4. Smith, Adrian. The Architecture of Adrian Smith, SOM: Towards a Sustainable Future, (Images Publishing Group, 2007)
5. www.thorntontomasetti.com (last accessed 2017/08/26)
6. http://lechailyasov-eng.com «Medieval Chechen Architecture» (last accessed 2017/08/26)
7. http://smithgill.com/news/csr_2013/ (last accessed 2017/08/26)

Regional approaches in high-rise construction

O.G. Iconopisceva[1], *G.A.* Proskurin[1*]

Orenburg State University, 460018, Orenburg, prospect Pobedy, 13, Russia

Abstracts. The evolutionary process of high-rise construction is in the article focus. The aim of the study was to create a retrospective matrix reflecting the tasks of the study such as: structuring the most iconic high-rise objects within historic boundaries. The study is based on contemporary experience of high-rise construction in different countries. The main directions and regional specifics in the field of high-rise construction as well as factors influencing the further evolution process are analyzed. The main changes in architectural stylistics, form-building, constructive solutions that focus on the principles of energy efficiency and bio positivity of "sustainable buildings", as well as the search for a new typology are noted. The most universal constructive methods and solutions that turned out to be particularly popular are generalized. The new typology of high-rises and individual approach to urban context are noted. The results of the study as a graphical scheme made it possible to represent the whole high-rise evolution. The new spatial forms of high-rises lead them to new role within the urban environments. Futuristic hyperscalable concepts take the autonomous urban space functions itself and demonstrate us how high-rises can replace multifunctional urban fabric, developing it inside their shells.

1 Introduction

It's impossible nowadays to imagine the world without high-rises, which is the mainstream in the forming of high dense urban landscapes. Since 2000 the number of buildings above 200m has grown in the world by 325%. According to annual research of the CTBUH 2011 and 2014 were the record for the number of completed skyscrapers. In 2016 there were 3720 high-rise buildings in the world (height above 150m), 110 of them – "supertall" and 1153 – above 200m. Today's high-rise leader is China with 66% of all buildings 200+ over the world [1]. In this study the generalized analysis of high-rise construction is represented according to timeline divided by well-known key evolutionary stages. It allows us to note the features and typology diversity.

High-rises of the 21st century are high-tech intellectual objects. Energy efficient technologies developed in high-rise construction explain its extravagant sculptural plastics, complex surface design and individual space-planning solutions. Archphytomelioration transforms high-rises into oases, which different them on a qualitative new typological and structural level from the 20th century high-rise architecture.

* Corresponding author: georgepro@yandex.ru

The presented material is based on the data of the CTBUH [1-10] and the work of the world-famous bureaus like Skidmore, Owings & Merill (SOM), Foster and Partners (F&P), Office for Metropolitain Architecture (OMA), Zaha Hadid (ZH), Llewelyn Davies (LD), T. R. Hamzah & Yeang Sdn Bhd, Stefano Boeri Architetti (SBA), Studio Liebeskind (SL), Adrian Smith+Gordon Gill Architecture, Herzog & de Mueron, MVRDV, SLCE Architects, Eric Parry Architects, Gensler, UNStudio , P&T Group, NBWW, Perkins+Will, Emaar Properties, Kohn Pedersen Fox Associates (KPF), Hariri Pontarini Architects, IAD, BSB, Wilkinson Eyre Architects, Archgroup Consultants, BFLS, Oiio, Hufton+Crow , NBBJ, MAD.

Materials of scientists and researchers in the field of high-rise and energy-efficient construction K. Yeang, S. Boeri, [2], K. Al-Kodmany are discussed in the article as well as N. Foster, S. Luff, D. Visco, M. Keeler, B. Burke, J. Wines.

Russian scientists who study the high-rise urban, architectural and structural features in Russia and abroad - T.G. Maklakova, Y.G. Granik, V.P. Generalov, E.M. Generalova, Y.A. Tabunschikov, M.M. Brodach, N.V. Shilkin, A.N. Tetior. Materials of 'Tall Buildings' magazine are in use within the article as well as data of the Museum of High-Rises and web resources dedicated the high-rise features.

2 Methods

The article contains complex analysis of the high-rise evolution. Graph-analytical method allows to differentiate the study tasks clearly. And the material divided by three key periods and each structured by the global region.

1st Period. The Origin. 1880s-1930s.

The urban conditions of Manhattan determined the high-rise origin – the brightest achievement of construction of the 20th century. In this period the classical structure and image of skyscraper were formed within rationalism and Art Deco. The iconic objects are Empire State Building, 1931, arch. Shreve, Lamb and Harmon; Chrysler Building, 1930, arch. W. Van Alen; Rockefeller Center, 1933, arch. R. Hood. etc.

2nd Period. 1950s – mid. 1990s. High-Rise Development as Multifunctional Centers. Since the middle of the 20th century from Canada to South Africa the multifunctional high-rise buildings being appeared such as Price Tower, 1956, arch. F.L. Wright; MetLife (Pan Am) Building, 1963, arch. W. Gropius, E. Roth, P. Belluschi; John Hancock Center, 1969, arch. SOM [2]. The rethinking of high-rise accents comes, as major, thanks to using curtain glass walls. In the 1960s it was determined as "International style". The iconic objects were Lever House, 1952, arch. SOM [2]; UN Secretariat in New York, 1952, arch. O. Niemeyer, Le Corbusier, W. Harrison. In 1958 after the Seagram Building was built it became the golden standard in office real estate of the US. Since 1960s the structural solutions were radically changed because the invention and implementation of tubular and shell constructions [3]. The landmark objects were Renaissance Center in Detroit, 1977, arch. SOM; AT&T Building in New York, 1984, arch. P. Johnson; Bank of America Plaza in Dallas, 1985, arch. JPJ Architects, HLM Design [4]; Lloyd of London, 1986, arch. R. Rogers. In the mid-1980s postmodernism gradually changed to high-tech. Skyscrapers of the mid-1980s are monumental and heavyweight. In the 1960s, the growth in the number of skyscrapers stopped. Up to the end of 20th century the height of high-rises declined. At the end of the 1980s the industrial and business transfer to Southeast Asia helped the beginning of high-rise development within these countries. Tubular and shell constructions enriched the architecture with tectonics.

Europe. In 1960s-1970s the urban concept of "integrated urbanism" and clear high-rise approach was formed. The principal feature of high-rise construction became the complexity and miltifunctionality as well as for business zones the exterritorial location out of historical centers. In 1950s the first business district was born of 20-30-story offices of "International Style" in Milan. In 1958 in Paris was adopted the plan of 'La Defence' business district [3]. In 1970s Defence - the geometrical architecture of glass and metal. Tour Elf, Tour Areva, Tour Gan, Tour Defence 2000, CNIT Building were the classical example of American business style. The construction of multifunctional Tour Montparnasse damaged the panoramic view of Paris. In 1980s in London the reconstruction of Docklands showed the creation of multifunctional urban environment with concentrated location on the north bank of Thames. In 1991 Canary Wharf (245m) was built first, arch. C. Pelli being the tallest high-rise building of tower-type in Europe [2]. The example of the complex approach to reconstruction by multifunctional high-rises in Germany is the development of financial district in Frankfurt-am-Main. The erection of high-rises above 60 floors in Europe continued till 1970s.

Russia. High-rise construction in USSR had the symbolic and ideological mission. In Soviet Russia the first buildings breaked mark of 100+ were the Stalin's Skyscrapers called "Seven Sisters". Among them was Kotelnicheskaya Embankment Building, 1952, arch. D. Chechulin, A. Rostkovsky. One year later the highest building in Moscow – the Main Building of Moscow State University (236m), arch. L. Rudnev. In 1967 the construction of New Arbat's Skyscrapers (below 100) was completed. In 1968 the 100m Hydroproject Building was built, then in 1970 – SEV Building (Comecon) – now the Moscow City Hall. Just another example of international style is Intourist Hotel, end 1960s, arch. M. Vinogradsky. In 1970s-1980s in Moscow 8 buildings 100+ in addition were built more. The House of Russian Government (102m), 1980, is noted among them.

Middle East. The beginning of high-rise development is the end of 1970s. In 1980-1990s the Dubai structure as a settlement included in fact two districts – "historical" center Deira and buildings along Sheikh Zayed Road, where were concentrated all the high-rises of office and multifunctional complexes. The first splash of high-rise construction in this region was at the end of 1970s – beginning of 1980s. The first skyscraper in Dubai the 39-story Dubai World Trade Center, 1979, arch. J. Harris, was adopted at once as the tallest in Middle East (149m), set the further trend. Architecture of 1980s was stand out with its faithfulness to postmodernism according to world trends.

3rd Period. 1990s-2017. New High-Rise Paradigms.

The changed architectural fashion flew the new aesthetic orients. By thy beginning of the 21st century postmodern aesthetics transformed to neo-modern laconic and tech oriented trends. After the Energy crisis of 1974 the basic principle was to use energy-efficient technologies. The result of such development was the practice of "sustainable buildings" where the merge of architecture and energy-efficient technologies was reached [5-7]. The high-rises to be considered as energy-efficient and bio positive objects [8].

USA. 9 years passed after the Commerzbank in Germany the first "green" skyscraper was built in Manhattan. It was Hearst Tower, 2006, arch. F&P. Frontage design used in large amount of skyscrapers by F&P represents the diagonal mesh using modular energy-efficient glazing [9]. Bank of America Tower is considered the most green in the US. The Building works as giant air filter for Midtown Manhattan with its double-sided filtration system [10]. Urban conditions in Manhattan and the demand for luxury real estate near the Central Park led the way to new skyscraper types being introduced into super-dense development [11]. Like a point injection, ascetic "super-slenders" "punctured" urban fabric. Some of them have ratio of the base width to height 1/23 [12]. The other one in a row of skyscraper typology is 'courtscraper' VIA 57 West, 2016, arch. BIG – a hybrid of traditional courtyard and high-rise tower.

Europe. The principles of bioclimatic energy-efficient architecture were presented in 1997 by Commerzbank in Frankfurt-am-Main, arch. F&P [13]. The building was the first 'green' skyscraper in the world and still the tallest in Europe till 2005. During the design process computer modeling and aerodynamic methods were used [7]. The principles of energy efficiency continue to realize – Main Tower, City Hall, Post Tower. N. Foster in the project of Swiss Re, 2004 in London moved to the system of round-hydrodynamic outlines. The building became the new city symbol and recognized as the best one of the decade. Foster's "Gherkin" idea was picked up by J. Nouvel in the bright "Torre Agbar", 2005, in Barcelona. But the Foster's ideas introduce slowly and office buildings with landscape bureau plan continue to erect all over the world. The landmark new generation bioclimatic object with the high-rise apartments is Bosco Verticale in Milan, 2014, arch. SBA [14]. It recognized the best skyscraper 100+ taking the International Highrise Award. This project opened the new perspectives for greening the skyscrapers and set an example for the development future cities.

China. The nowadays leader in mass, developing the whole cities. The individual iconic objects stand out of contemporary Chinese architecture in Beijing, Shanghai and Guangzhou. One of the first iconic was high-tech Chinese Bank in Hong Kong, 1985, arch. F&P. Also the landmark is Jin Mao Tower [2] by SOM in Pudong, Shanghai. Its outline was inspired by the opening petals of lotus and imitates the length of the pagodas.

In 2008 the best building under the CTBUH rating was Shanghai World Financial Center by KPF, called 'the bottle opener' [15]. In 2010 the leader under energy-efficiency was supertall building with open-type engineering Pearl River Tower (309m) in Guangzhou by SOM [16]. Ping An Financial Center (599m) by KPF is the tallest in Shenzhen, 2nd tallest in China and the 4th tallest in the world. Today the Shanghai Tower demonstrates the most faithfulness to common space in high-rise, which was built in 1997 for the first time. Shanghai Tower is the new prototype of high-rise building, the tallest in China, the 2nd tallest in the world and the 3rd 600+ skyscraper within the "Megatall" category [17]. As early as 2018 the first vertical forest to be in China. It will be the Green Towers in Nanjing by SBA also designed on principles of Bosco Verticale. By 2021 S. Boeri to design in Liuzhou the whole series of high-rises vertical gardens within. It will be the new 'clever environment' and the answer to 'crazy city widening' in advanced but wasted China [18].

Australasia. By the mid. 1990s the high-rise buildings had the stepped structure and intricate top as usual – Republic Plaza (280m), 1995 in Singapore or Millenia Tower (223m), 1996, Capital Tower (254m), 2000, – composition of buildings eased as in United Overseas Bank Plaza One (280), 1992.

The major Taiwan's skyscraper Taipei 101, 2004, arch. C.Y. Lee & Partners, with the national romantic forms is figuring in ultramodern materials. One of the international landmarks of the 20th century expressed in general postmodern design were the Twin Towers of Petronas, 1996, arch. C. Pelli.

In Seoul, South Korea, Lotte World Tower is the 5th tallest building in the world now.

Malaysian architect dr. Ken Yeang is the inventor of bioclimatic approach within the design of high-rises [19]. Over 10 buildings realized by himself with such features as vertical gardens – the buildings of new eco generation – Menara Mesiniaga, EDITT Tower, Putrajaya 2C5 in Singapore [20]. After this approach Marina Bay Sands in Singapore was erected as green landmark led to new typology of multidimensional high-rise structure (200m), arch M. Safdie [21]. The giant object has three 55-story hotel towers connected by a roof terrace. The observation deck provides panoramic views.

The tallest Australian skyscraper is Q1 Tower (323m) in Gold Coast, 2005, arch. Atelier SDG. To date, in Brisbane is located the number of tallest buildings of this continent. Nowadays the primacy belongs to Infinity Tower (249m), 2013, arch. SOM. In 2014 CTBUH noted the One Central Park residential complex as the best high-rise of 2014. Apartments in Sydney by J. Nouvel is the major trend of future agriculture – the vertical farms within the city. Such an experimental farms in the courtyards and on the roofs of some residential and office buildings being exist [22, 23].

Middle East. Since 2005 several city-forming district have emerged. Dubai Marina is 20 km from Deira, here there was a number of high-rises being erected, including Burj Khalifa. High-rises are the main attractors of the whole state. With such settings all the new skyscrapers must be the most of the most, the unique of the unique and irresistible in any parameters. Here different high-rise typology being erected claiming the primacy in many parameters or unique tech specs. The tallest residential building in the world is Infinity Tower. The tallest hotel Marriott Marquis (355m) or the largest shopping center Dubai Mall. Unique buildings that made revolution in the field of high-rise, engineerings and urban planning as the Burj Khalifa (828m), arch. SOM, are marked by the CTBUH with the title of Global Icon. The competition to have the title of the tallest high-rise is being developing exactly in this region [24]. Spectacular in terms of attraction are kinetic elements on the facades of buildings and modular mesh coverings, as in Al-Bahar and Burj Qatar by J. Nouvel.

Russia. The first building 100+m in Russia was completed in 1995 – the Gazprom Tower in Moscow (150m). Outside Moscow, the first building of 100+ was Uralsib Bank's office, Ufa, 1999. By the 21st century, there were 25 100+ buildings in Russia. Modern high-rise construction began after the 1998 crisis. The first new ones were three 115-m towers of residential complexes at Leninsky Prospect in Moscow. Triumph Palace residential complex (264m), 2005, Naberezhnaya Tower (268m), 2007, City of Capitals complex (302m), 2009, Mercury City Tower (339 m), 2013, OKO complex (about 354m)... Now the highest in Russia and Europe is the Federation Tower (364 m), 2017, arch. nps+partner, Schweger Associated Architects. As a result, an area with ultra-high buildings appeared in Moscow, which is not found elsewhere in the world, except for the Dubai-Marina area. Nowhere else man build a 'wall to wall; higher than 300m. This fact reduces the quality of environment making impossible to fully inspect the interesting ones [25]. From Russian implemented buildings the CTBUH Awards received a residential building on Mosfilmovskaya Street, 2010, arch. S. Skuratov. In Yekaterinburg, the tallest symbols of the Ural region are the Iset Tower (209m), 2016, arch. W. Zobek, and the Vysotsky Tower (196m), arch. S. Bashkareva.

3 Results

The complexity of high-rise construction, increasing proportionally to the height, doesn't stop the geography of its spread. Each country is trying to form the image of the main regional as well as world center. Today high-rises are the new paradigms of urban life as an effort to solve the environmental problems. In countries with a high population density, high-rises are the solution to housing. In different countries the sum factors for the high-rise appearance is almost the same. The differencies are exactly in national self-identity, expressed in architectural image and relation to the context.

USA. Homeland of skyscrapers where the main stages of high-rise evolution were passed. The leadership with massiveness transferred to China and with uniqueness and advanced technologies to the Middle East. The existing urban super density and planning regulations helped to arise the new typologies of high-rises.

China. The high-rise leader with pace and massiveness. The search for a new typology is aimed at positioning the region as a global financial center, to find the solution of a complex ecological situation that doesn't exist anywhere else in the world and to provide super-dense population with living space. Australasia: The region has a wide range of typological diversity in high-rise construction, due to climatic conditions helping to upgrade the typology and to introduce the techs for eco buildings. The program of mass high-rise housing is being successfully implemented. Middle East. The region's economic strength and ambitious positioning in the world economic space is fixed in maximum and unique technological and architectural achievements in the field of high-rise construction. Climatic conditions determine the search for energy-efficient solutions. The region provided the world with innovations and paradigms of eco buildings.

Europe. The approach of 'integrated urbanism' dictates a complex and exterritorial development in high-rise construction. No race to build higher compensated by the development of advanced energy-efficient solutions and its rapid implementation to reduce dependence on external energy sources.

Russia. Up to the end of the 20th century the high-rise objects were rarely, due to political reasons and no fixed place of the state within the global economy. High-rise objects appeared after Millennium are the effort to fix the country within the full-fledged participants within the world economy. The European selective approach and lack of regionalism are traced.

4 Discussion

The American urban planning practice, created the image of a high-rise business center, adjusts throughout the world according to urban conditions, positioning and strategy of the regions, national mentality. Urban landscapes in New York, similar to fields of stalagmites constantly growing up, continue to arise new typological and structural units. Super-slenders realize the ambitions, connected with the best 'window view' buying, and fix New York on the 2nd position after Dubai as the tallest building owner. Attempts to enter the super dense fabric of Manhattan sometimes reach absurd concepts of hanging from asteroids high-rises, in which elementary human comforts are perceived as blasphemy.

In European cities, skyscrapers are an exception and there are some confrontations with a historical context that hardly admits large-scale neighbors. But it doesn't block the extravagant high-rise silhouettes appearance with its advanced tech solutions. In China, Middle East and Asian region, skyscrapers are actually the context, and confrontation with the environment is impossible. Race for height, especially in the Middle East, led to the term 'useless height'. So, the Burj Khalifa has 244m of useless height, what might be an independent skyscraper. Nevertheless, the designers don't stop. Another problem is the balance of investments and the resulting dividends from ultra-high buildings. Sometimes streaming to uniqueness, economic expediency and common sense are lost. So Burj Khalifa was realized, both through the personal investments of the owner, and through the sale of real estates overlooking the tower in the surroundings [4].

Eco problems in high-rise megacities are trying to solve with the help of high-rise construction. In one case, compensating the lack of natural environment by vertical gardens, as S. Boeri suggests, calling modern megacities a 'true nightmare'. According to him architects nowadays thinking of what might be in cities without the greens hundred years after [14]. Such position may leads in near future to reduction of stylistic diversity in favor of architectural biopositivity in regions with strong eco troubles. S. Ban to erect high-rises using wood structures. James Law Cybertecture International creates the concept of a "cybertectural world" based on the symbolic balance between space and techs [26].

In other cases, the solution to over-urbanization is a concept of 'multidimensional city' [27]. Using the example of contemporary conceptual projects of Cloud Citizens, The Bride, Sky Village, Endless City in Height, one can see a different typology of high-rises representing mega structures of an autonomous urban ecosystem that can return more to the environment than they consume. [28]

5 Conclusion

The results of the study have made it possible to differentiate the existing regional approaches, connected with environment context, positioning and image of states, national self-identity, which reflected in architecture and spatial planning. The design of advanced high-rise objects is impossible without the eco strategies which solve regional problems. In large variety of completed objects, it is possible to single out the set of stylistic and tech methods in high-rise as well as basic missions preserved in the near future.

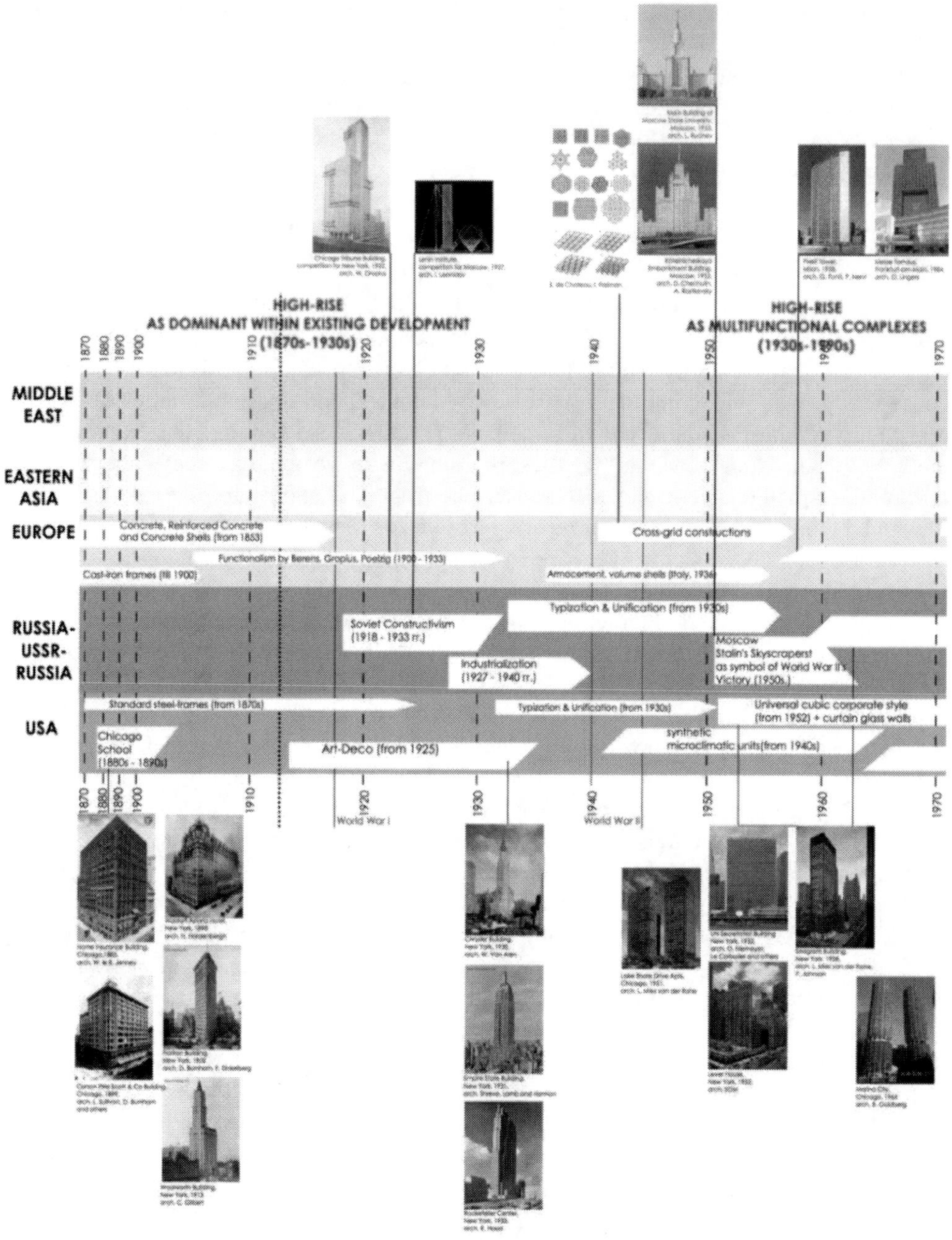

Fig. 1. Evolutionary Matrix of High-Rises by the Global Regions. Part1

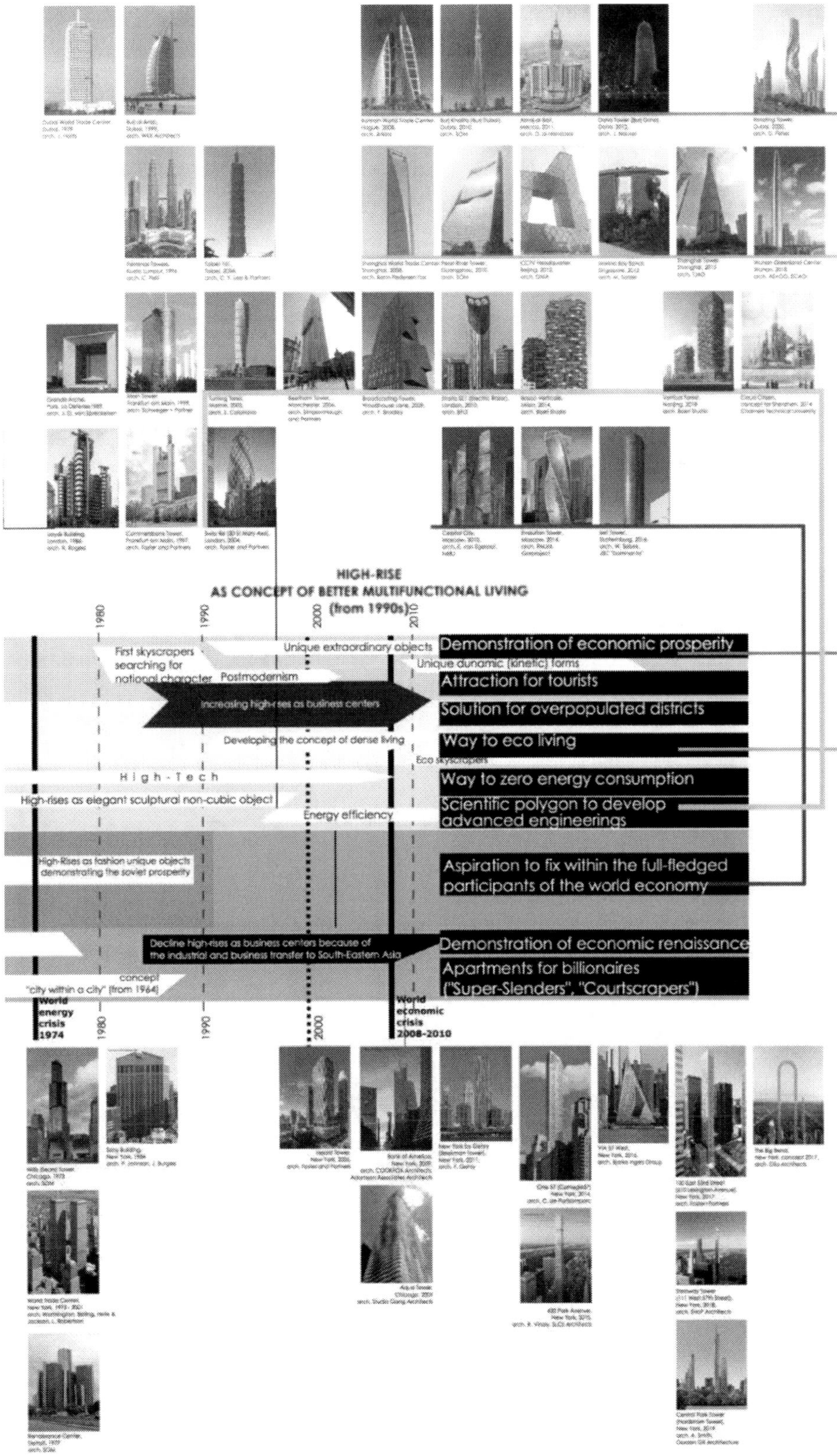

Fig. 2. Evolutionary Matrix of High-Rises by the Global Regions. Part2

High-rises as symbols. The birth of each new skyscraper is a unique event. Unique high-rise buildings have become not only the symbols of their countries, but revolutionize in the field of architecture and engineerings, as Burj Khalifa, for example. Here they are: Swiss Re, London; Torre Agbar and Porta Fira in Barcelona; Burj Qatar in Doha; Shanghai Tower; Absolute World Towers in Toronto; Capital Gate in Abu-Dabi, Makkah Clock Royal Tower in Mecca and hundreds of other in row.

High-rises as advanced tech solutions. The priority of energy efficiency in modern high-rise is provided by optimal space-planning solution and energy-efficient engineering and materials. The smallest perimeter, for example, with the common useful area, leads to a reduction in heat and streamlining forms. The appearance of "super-slenders" with a base of up to 12m and a high flexibility of the structures shows how compact a high-rise building plan can be. Atrium spaces allow us to save on heating and create a natural environment within. Amazing twisting forms with different floor shifts is justified by the reduction of aerodynamic loads and makes them spectacular objects. Designing streamlines, it is possible to minimize the wind effect. Objects of high-rise construction are a field for the modern facade systems. One of the most popular techniques has been the use of membranes, which give sculptural freedom. There is a revolution from traditional methods to information techs, among them is BIM (Building Information Modeling) technology.

Acknowledgments

Special thanks to professor V.P. Generalov - Head of the department ARPB in Samara State Tech University and assistant professor E.M. Generalova for the promotion of advanced high-rise construction in the field of architectural education in Russia.

References

1. http://www.ctbuh.org/ (last accessed 02.05.2017)
2. http://www.som.com/ (last accessed 06.06.2017)
3. T. G. Maklakova, High-Rise Buildings, Urban and architectural-structural problems in design, Monography 2nd edition, supplemented, 160 (Moscow, ASV, 2008)
4. http://tallbuildings.ru (last accessed 12.05.2017)
5. Y. A. Tabunshikov, From Energy-Efficient to Life-Retaining Buildings, **3**, 8 (AVOK, 2003)
6. Y. A. Tabunshikov, Energy-Efficient Buildings, 8-76 (Moscow, AVOK-Press, 2003)
7. M. M. Brodach, HIgh-Rise Engineerings, 320 (Moscow, AVOK-Press, 2007)
8. Kheir Al-Kodmany, Green towers and iconic design: Cases from Three Continents, *International Journal of Architectural*
9. Ahmad Rahimian, Yoram Eilon, New York's Hearst Tower. A Restoration, an Adaptive Reuse and a Modern Steel Tower Rolled Into One, *Structure magazine*, **2**, 25-29 (2006)
10. https://courses.cit.cornell.edu/crp384/2008reports/18Green_Skyscrapers (last accessed 12.05.2017)
11. E. M. Generalova, V. P. Generalov, Super-slender residential skyscrapers in New York as a new direction in high-rise buildings typology, *Urban construction and architecture*, **4(25)**, 85-91 (2016) DOI: 10.17673 Vestnik.2016.04.16.
12. https://static.dezeen.com/uploads/2016/06/new-york-super-slender-skyscrapers_dezeen_large.jpg (last accessed 11.05.2017)

13. http://www.fosterandpartners.com/Projects/1405/Default.aspx 23 November 2017
14. htt://web.utk.edu/~archinfo/a489_f02/PDF/commerzbank.pdf (last accessed 30.05.2017)
15. https://www.stefanoboeriarchitetti.net/en/portfolios/taranto-calling-2/ (last accessed 30.05.2017)
16. https://www.kpf.com/ (last accessed 26.05.2017)
17. http://www.som.com/content.cfm/pearl_river_tower (last accessed 26.05.2017)
18. http://members.ctbuh.org/shanghai-tower-construction-development (last accessed 26.05.2017)
19. www.stefanoboeriarchitetti.net/en/portfolios/sustainable-dystopias/ (last accessed 25.04.2017)
20. I Richards, T. R. Hamzah & Yeang: ecology of the sky (The Millennium Series) Images, 247 (Australia, 2001) ISBN 1864700955, 9781864700954
21. https://books.google.ru/books?id=QitFFq7Ybg0C&printsec=frontcover&hl=ru#v=one page&q&f=false (last accessed 28.05.2017)
22. Generalova E., Generalov V. Designing High-Rise Housing: The Singapore Experience, *CTBUH Journal*, **4**, 40-45 (Chicago, Illinois Institute of Technology, 2014)
23. http://grandengineer.ru/skyscrapers/one-central-park-luchshij-neboskreb-2014-goda/ (last accessed 06.05.2017)
24. http://www.jeannouvel.com/ (last accessed 06.05.2017)
25. http://skyrisecities.com/database/cities/dubai (last accessed 07.05.2017)
26. http://www.forum-100.ru/pressroom/vysotnoe-stroitelstvo-rossii-na-rubezhe-vekov/ (last accessed 12.05.2017)
27. http://www.jameslawcybertecture.com/ (last accessed 11.05.2017)
28. V. P. Generalov, E. M. Generalova The Perspectives to Develop New Types of High-Rises. Urban Future, *Vestnik SGASU, Urban Planning and Architecture*, **1(18)**, 13-18 (2015)
29. http://www.archdaily.com/549665/cloud-citizenawarded-joint-top-honors-in-shenzhen-bay-super-citycompetition/ (last accessed 12.05.2017)
30. http://oma.eu/ (last accessed 16.05.2017)

Use of high-rise structures for sustainable tourism

Tatiana Ya. Vavilova[1*], and *Efim G.* Vyshkin[1]

[1]Samara State Technical University, 194, Molodogvardeyskaya St., 443001, Samara, Russia

Abstract. The paper deals with such issues as formation and development of the infrastructure of objects for serving tourists in urban environment and specially protected natural areas with particular focus on open tower structures - a type of object which is so popular in Russia. The authors systematize international experience of integrating watchtowers in natural and anthropogenic environment as well as specific features of their modern architectural solutions. A number of examples are given. Summing up the results of the analysis we have come to conclusion that in the field of tourism the most promising tendency in functional use of vertical structures is the demonstration of cultural and natural attractions. It is also noted that in national and natural parks objects of the tower type can be built for other purposes, e.g. for conducting research, monitoring weather conditions and emergency situations. It is shown that the development of infrastructure of high-rise buildings for educational tourism contributes to sustainable development of territories and settlements.

1 Introduction

The Russian Federation has high potential for developing domestic and international tourism [1]. Throughout the country there are more than 40 historical settlements of federal importance, objects of the world cultural and natural heritage, more than 13070 specially protected natural areas (SPNAs) of federal, regional and local importance. The increase in the number of foreign tourists visiting our country is confirmed by selective official data of the Federal State Statistics Service. The overall number of inbound tourists per year reached about 26.852 million in 2015. This exceeds the ratio of 2014 for more than 1.4 million. It allows making a conclusion that there is some growth of attractiveness of tourism as a sphere of entrepreneurial activity.

In many countries educational tourism is considered to be one of the priority sectors of the economy. In particular, all business activities in SPNAs can be developed on its basis [2, 3]. The major obstacle to the improvement of tourism service in Russia is the poor quality of infrastructure providing services for tourists [4]. According to the Department of State Policy and Environmental Regulation of Ministry of Natural Resources (2013) up to 75% of the NP potential is used inefficiently. The types and number of buildings and small architectural forms do not currently meet modern requirements neither in organization of

* Corresponding author: vatatyan63@yandex.ru

scientific research, nor in the variety of services and level of comfort. Moreover, the condition of engineering structures, branching and methods of improving routes and places of visitors' concentration don't allow minimizing environmental impact under increasing recreational pressure [5-7].

In the era of sustainable development of natural and urbanized areas the erection of observation towers (watchtowers) and other high-rise structures has become the main trend in improving the quality of educational touristic infrastructure. Their design and construction, which can be called one of the drivers to increase popularity of guided tours, are most widely-spread in Europe, China and the United States. Let us consider the possibility of using tower structures for developing tourism in the Russian Federation.

2 Materials and Methods

2.1 Scientific basis for developing educational touristic infrastructure and constructing open high-rise structures

In the contemporary history of Russia at the time when market relations have started to form, the most attention to science-based development of tourism objects infrastructure has been paid by economists and architects.

The economic value of tourism in Russia and the strategy of its development was considered in the thesis works by D. A. Kovalev (2006), Yu. V. Chernyavsky (2011) and M. V. Vinogradova (2013), as well as in the thesis work by the specialist in economic geography M. S. Bezuglova (2007). The contribution of the tourism sector to sustainable development of regions was studied in the papers by R. V. Hachmamuk (2004), L. V. Vasilyeva (2006), S. A. Sevastyanova (2006), K. V. Maslennikova (2007), E. M. Maksarova (2009), O. A. Bunakov (2011), E. G. Kiyakbaeva (2015), D. S. Khasova (2015) and others. The issue of objects' infrastructure development was touched upon by A. S. Levizova (2008), A. V. Kuchumov (2011), D. F. Vasilikha (2012), O. B. Evreinov (2012), V. E. Kotelnikova (2014), etc.

Urban and architectural-typological features of tourism objects became the subject of thesis studies by Yu. S. Fedorova (1998), V. A. Antyufeev (2007), N. F. Vdovin (2009), N. V. Morozova (2012), L. A. Fedotova (2013), O. A. Antyufeeva (2014), E. K. Bulatova (2014), A. N. Azizova-Poluektova (2015), A. F. Perova (2015), M. E. Pechenik (2016). Scientific background for the principles of spatial organization of the tourism objects network has been developed in a number of theses by foreign authors, for example, by A. Sh. Sultanalieva (1997) for Kyrgyzstan, by Zhu Binsy (2009) and Ven I (2009) for China.

It should be noted that developing the infrastructure of educational tourism in Russia is still associated with improving objects related to the tourists' accommodation (hotels, motels, campsites, etc.). However, buildings and structures on the routes that provide contemplation, identification and memorization of cultural or natural attractions have remained beyond designers and investors' vision for years [8, 9]. Among them there are information and visitor centers, observation towers and sightseeing platforms, small architectural forms and elements of branding. Meanwhile, these objects are designed to increase the effect of perception of the unique urban or natural environment, to enhance understanding of the aesthetic uniqueness of a particular area [10-12].

Modern tourism is an interdisciplinary phenomenon. International and Russian practice of its development is based on the integration of various production and service industries, aimed at meeting the leisure demands of people. It employs the theory of cluster development and the idea of forming an integrated system of objects and services. The key element of this system is destinations – territories possessing a certain potential for tourism development [13, 14]. Cluster-destination approach allows taking into account the uneven distribution of tourism resources, specific character of the area climatic conditions, their cultural-historical and socio-economic characteristics. In this regard, improving the competitiveness of the tourism sector of the economy is associated with the creation of favorable investment climate for infrastructure formation. In the international theory and practice at the turn of the 20th and 21st centuries, the issues of its development have become in the focus of interest [15, 16].

Probably one of the first attempts to achieve comprehensive understanding of architectural and design approaches to the development of touristic infrastructure has become a planning and design guidebook edited by M. Baud-Bovy, F. Lawson (the United States) and first presented in 1977. In its updated version (1998) the authors took into account environmental, social and economic aspects, actualizing sustainable development issues [17]. Then in 1999 the book about tower structures by J. Kleinmanns was published in Germany. The author noted that observation towers are mostly used for tourism. In addition, they can still be used to prevent fires, and to accommodate transmitting and receiving TV and radio systems. The author gave examples of multifunctional operation of observation tower structures, co-located with a restaurant and a mini-hotel. Besides, the types of structural schemes and building materials were systematized and the principles of ensuring personal safety of tourists were identified [18]. At present this book remains the only publication in which the typological features of open towers and tower structures with internal space were classified and analyzed.

Undoubtedly more recent investigations related to regulation of the architectural environment of touristic infrastructure objects are of great interest, for example, the U. S. Army Corps of Engineers (USA) and the Department of National Parks, Sport and Racing (Austria). They contain standards of spaces development for different types of recreational activities [19, 20].

2.2 Review of traditions and modern experience in the design of open high-rise structures

2.2.1 Historical prototypes of modern observation towers

Numerous observation structures and bell towers in fortresses and monasteries, dominants at city halls and towers for engineering purposes built in the 19th and 20th centuries can rightfully be considered as prototypes of modern observation towers. Many of these objects are actively used for sightseeing in many countries. The most vivid example is the Eiffel tower, its observation platforms are annually visited by millions of tourists.

In the 20th century in connection with the development of recreation and entertainment industry high-rise objects began to be built purposely. One of their functions was to regulate the recreational load on the system of nature conservation areas (figure 1).

The steel observation tower on the Hot Springs mountain (figure 1a) also used for wireless telegraph communication was built in the eponymous national park (Arkansas, the USA) in 1906. Its height is 50.3 m. It is interesting to mention that since 1877 in the same place there had been a wooden dominant about 23 m high. In 1975, the metal tower in poor condition was demounted, but in 1983 and in its place a new building 65.8 m high was built. It was designed exclusively for tourism purposes [25].

Fig. 1. Examples of tower structures of the 20th century:
a - national park Hot Springs, USA, the photo was taken in1932 [21], b - Central Park of Culture and Leisure named after Gorky, Moscow, USSR, the photo was taken in 1937 [22]; c - Shark Valley, Miami, United States [23]; d - Kobe, Japan [24]

Wooden tower structures approximately 30 m high, built in the parks of the Soviet Union in 1930s, are of particular interest (figure 1b). They were used as entertainment extreme rides for parachute jumping. Their spectacular appearance, that vaguely resembles the symbol of the revolutionary era – the Tatlin Tower, was conditioned by spirals of stair railings. In the upper part there was an observation platform.

The principle of the spiral ascent was used while erecting the tower in Shark Valley of Everglades National Park in Miami, USA (Figure 1c). Its observation platform, built in 1984 above the habitat of reptiles – crocodiles and turtles, is on the height of twenty meters. This also allows creating a panoramic view of the area. It should be particularly noted that there are no steps on the flat gradient, which is essential for disabled visitors [26].

In Japan the 90-meter hyperboloid tower in Kobe city built in 1963 has become such a unique object (Figure 1d). The designers surrounded the inner cylindrical core with the constructive system, which was invented by the great Russian engineer V. G. Shukhov and applied by him in the construction of the famous building in Moscow. The strength properties of the shell allow this Japanese construction to withstand seismic vibrations of great power. On the upper levels there are shops, cafes, and an observatory. One of the tiers of the service zone rotates and floor slabs have holes filled with break-resistant glass. All these details attract tourists from all over the world, and it is not surprising that every day the object is visited by 3 thousand tourists [24].

The second half of the 20th century was a period of conducting experiments. The profitability of constructing observation towers was proven. Here is just one example. Built in 1959, a small elevated observation platform in the Great Smoky Mountains of the National Park (the USA), which vaguely reminds the tower in Everglades park is annually visited by about 600 thousand tourists from April to November [27].

2.2.2 Modern experience of designing and constructing touristic high-rise buildings

In the 21st century professional architects and designers have been actively involved in designing open high-rise buildings for tourism purposes. They have created objects that, from the very beginning, have played the role of "magnets" that attract tourists. This is of particular importance for developing tourism in the province. For example, the construction of an observation tower in the Latvian village of Shashkes located on the border with Lithuania and Belarus was planned in order to attract tourists [28].

Modern observation towers ceased to be the ordinary engineering structures long time ago. Unique technical and architectural solutions allow them to become city or area symbols, identifiers. This is most evidently seen in the tower construction in the urban environment (figure 2).

Fig. 2. Examples of observation towers located in the urban environment:
a - Beijing, China [29]; b - Madrid, Spain [30]; c - Glasgow, Scotland [31]; d - Brighton, England [32]

In most cases observation towers constructed in cities and suburbs, play the role of amusements. Going upstairs, staying on the sky deck and going downstairs are often enhanced by rotation of the platform (Düsseldorf, Singapore, Brighton, etc.). The height of such structures is taken considering the conditions of panoramic perception, and is limited mainly to the load-bearing capacity of structures and the necessity of taking into account wind pressure.

The highest contemporary observation tower with a rotating platform is British Airways i360 located above the beaches of Brighton (England). It was commissioned in 2016. The core reaches the height of nearly 162 m (Figure 2d). In the glass capsule up to 200 passengers can rise to the elevation of 138 m simultaneously. Technically and technologically advanced solution had been developed and implemented for 11 years. Project authors – architects of Marks Barfield Architects Bureau are famous for construction of the Millenium Observation Wheel (London Eye) [32].

In contrast to solutions suitable for the urban environment, objects for national and natural parks are maximally adapted to the landscape features of the area. Traditional materials aremainly used: concrete, metal, wood, glass, and their combinations. The degree of openness of structures may vary depending on the functional purpose. In composite solutions three main approaches can be traced – contextual (figure 3), accentual (figure 4) and contrasting (figure 5).

Fig. 3. Examples of the contextual approach to integration of observation towers in the natural environment: a - Lithuania [33]; b - Germany [34]; c - Latvia [35]

Fig. 4. Examples of the accentual approach to integration of observation towers in the natural / natural-anthropogenic environment : a - USA [36]; b - Austria [37]; c - United Arab Emirates [38]

Fig. 5. Examples of the contrasting approach to integration of observation towers in the natural / natural-anthropogenic environment: a - Netherlands [39]; b - Belgium [40]

3 Results and discussion

Analysis of the theoretical base, historical and contemporary experience of design and construction of open structures of the tower type allows making a number of conclusions related to their use. Due to intensification of social processes, raising importance of ecological security on the international and national levels and increasing opportunities for introduction of scientific and technological innovations, the functions of observation towers can vary a lot, including multi-purpose towers. For example:

- in the natural and urban environment the most promising tendency in functional use of open high-rise objects is the demonstration of cultural and natural attractions; for the erection of these buildings in cities the areas of parks and garden squares adjacent to zones of cultural heritage objects' concentration can be used as well as special sites located within easy reach. In particular, it refers to the acquaintance with the monuments of the archaeological heritage;

- high-rise structures can serve as a kind of landmarks in space, which help tourists to orient on the routes;
- in national and natural parks the construction of tower-type objects on the sites of panoramic perception provides maximum aesthetic effect – the attractiveness of the landscape;
- towers tend to be like attractive "magnets", in the vicinity of which the tourist accommodation objects and rest stops, visitor centers, trade and service companies, in particular, objects of public catering, health care, equipment repair or transport can be build; this approach allows localizing the zone of increased recreational load in close proximity of towers;
- the use of high-rise dominants as structures to place engineering systems of alternative energy – solar and wind can reduce the consumption of non-renewable environmental resources and the ecological balance disruption;
- open high-rise structures can serve to shelter people during bad weather or when they observe objects of the natural world;
- observation towers, that rise predominant above trees, are located in the areas of improved radar systems; they are used for communication with groups or individual travelers and their movement control;
- high-rise dominants designed for conducting research (aero-labs), monitoring weather conditions and emergency situations can be erected in nature conservation areas;
- towers can also perform the traditional role of the water retaining construction, and become part of the reserve system for collecting rainwater for later use during dry periods.

Many of functions performed by towers contribute to adherence to the principle of comprehensiveness, which is considered a prerequisite for sustainable tourism development.

Perhaps, the most complicated problems include regulation of height, choice of composition and structural layout of these structures. Focusing on the principle of diversification, however, a number of corrective restrictions should be mentioned. They include, above all, the following:

- necessity of taking into account the statutory and regulatory requirements for organizing protected zones of cultural heritage objects with regard to their valuable status;
- compliance of the altitude dominate role with the category and zone mode of the nature conservation area;
- specificity of natural and climatic conditions;
- landscape and geological features of the area;
- technological base for construction, allowing to reduce environmental impact during the construction and operation of tower structures;
- tourists safety during the tour;
- possibility of meeting the needs of disabled categories of population.

4 Conclusion

The use of high-rise structures of open type for educational tourism purposes is one of the most controversial. On the one hand, the construction of high-rise dominants is constrained by the conservatism of tradition, and existing regulations concerning the development of the urban environment and nature conservation areas. On the other hand, observation towers give more opportunities to educate, to familiarize tourists with unique cultural objects and natural phenomena, they enhance the aesthetic impact of their perception, promote the development of entrepreneurial initiatives and creation of additional jobs. At this stage we can conclude that only experimental design and construction will determine the possible implementation of open high-rise structures in the environment and allow making final conclusions.

References

1. Yu. A. Barzykin, Current problems and main developmental perspectives of the Russian tour industry, *Vestnik of National Tourism Academy*, 2, 7-8 (2010)
2. L. D. Zhigula, Nature protected areas as a resource of ecological tourism, *Vologdin recitations*, 6-8 (2008)
3. K. A. Makarova, Assessment of the environmental and recreational potential of the national parks of Russia, *Newsletter of North-Caucasus Federal University*, **2(41)**, 19-23 (2014)
4. A. Yu. Aleksandrova, The newest comprehension of tourism as a system, Modern challenges of service and tourism, **1**, 24-38 (2014)
5. T. Ya. Vavilova, I. S. Chakina, Prospects of development of ecological tourism infrastructures in protected natural areas, *Urban construction and architecture*, **3(24)**, 97-102 (2016)
6. Samarskaya Luka: current state and sustainable development of landscape-urban complex, 376 (Samara, 1997)
7. A. V. Shabanova, Restrictions on environmental requirements arising from the development of tourism in the village of Shiryaevo (Zhiguli Pearl), *Ecological challenges of the post-Soviet region: international collection of scientific articles*, 117-128 (Lipetsk, 2014)
8. M. I. Balzannikov, E. G. Vyshkin, Yu. M. Galitskova, Use of hydraulic engineering structures as recreational and tourist objects, *Bulletin of the Volga regional branch of the Russian Academy of Architecture and Construction Sciences*, **19**, 189-195 (2016)
9. T. Ya. Vavilova, I. S. Chakina, Functional design features of infrastructure objects of educational tourism for national parks. Visit centers. Collection of articles: Traditions and innovations in construction and architecture, *Architecture and design*, 44-48 (Samara, 2016)
10. E. Ju. Kolbovsky, Aesthetic Estimation of Landscapes: Problems of Methodology, *Yaroslavl Pedagogical Bulletin*, **4**, 161-166 (2011)
11. T. M. Krasovskaya, Aesthetic Landscapes Functions: Methodology of assessment and preservation, *Geopolitics and Ecogeodynamics of regions*, **2(13)**, 51-55 (2014)
12. V. A. Nikolaev. Problems of landscape aesthetics, *Geography, society, environment*, 276-283 (2004)
13. S.Y. Grishin. Methodological foundations for sustainable development of the tourism industry, *Izvestiâ Sankt-Peterburgskogo gosudarstvennogo èkonomičeskogo universiteta*, **4**, 7-10 (2014)
14. T. M. Krivosheeva. Tourist destination: the questions of the marketing strategy forming, *Services in Russia and abroad*, **6(53)**, 217-229 (2014)
15. D. Yu. Desyatnichenko, O. Yu. Desyatnichenko, A. D. Shmatko. *Journal of Economy and entrepreneurship*, **3-2(68-2)**, 328-332 (2016)
16. M. S. Oborin. Tourist infrastructure: social and economic analysis of the basic concepts and definitions, *Scientific notes of Orel state University. Series: Humanities and social Sciences*, **5**, 87-93 (2014)

17. M. Baud-Bovy, F. Lawson. Miscellaneous: Tourism and recreation: handbook of planning and design, 304 (1998)
18. J. Kleinmanns. Schau ins Land. Aussichtstürme, 152 (Marburg, Jonas-Verlag, 1999)
19. URL: http://www.npsr.qld.gov.au/recreation/pdf/development-of-outdoor-rec-facility.pdf. (last accessed 21.06.2017)
20. Planning and Design of Outdoor Recreation Facilities, *U. S. Army Corps of Engineers*, 192 (2001)
21. URL: http://thepostcardemporium.ecrater.com/p/3336704/steel-observation-tower-in-hot-springs. (last accessed 21.06.2017)
22. URL: http://art-oboz.ru/uncategorized/fotografii-sdelannye-vo-vremia-pyteshestviia-geologa-arnolda-haima-po-sovetskomy-souzy.html. (last accessed 20.06.2017)
23. URL: http://americanbutler.ru/majami/sady-parki-i-zapovedniki-v-majami/park-dolina-shark-valley-v-majami. (last accessed 26.06.2017)
24. URL: https://tury.club/sight/1606 (last accessed 29.06.2017)
25. URL: http://www.hotspringsar.com/info/tower/tower.htm. (last accessed 29.06.2017)
26. URL: http://www.fodors.com/world/north-america/usa/florida/the-everglades/things-to-do/sights/reviews/observation-tower-428531 (last accessed 01.07.2017)
27. URL: http://archive.knoxnews.com/news/local/park-competes-for-grant-to-restore-clingmans-dome-tower-33ad8a45-54e2-3d18-e053-0100007ffd4f-380991721.html (last accessed 11.07.2017)
28. URL: http://old-news.tts.lt/site/ru/V_Visaginase/19786-Gorjachaja_linija_TTS_Obzornojj_bashne_da_da_da_.htm (last accessed 09.07.2017)
29. URL: https://c2.staticflickr.com/6/5474/11368480306_2e4343e288_b.jpg (last accessed 10.07.2017)
30. URL: http://cortijo-los-almendros.co.uk/things-to-do/granada/trip-parque-de-las-ciencias-granada/ (last accessed 23.06.2017)
31. URL: https://gorsociety.wordpress.com/travel-numc-2012/about-glasgow/ (last accessed 24.06.2017)
32. URL: http://www.marksbarfield.com/#/projects/brighton-i360/?null (last accessed 21.06.2017)
33. URL: http://www.balticlakes.lt/en/entertainment/panoramic-tower-rumsiskes-1783/ (last accessed 21.06.2017)
34. URL: http://ecology.md/page/nacionalnyj-park-bavarskij-les-foto (last accessed 18.06.2017)
35. http://zvejai.ltic.lt/ru/tourism-list_poi/obzornaya-bashnya-labanorskogo-regionalynogo-parka/ (last accessed 03.07.2017)
36. URL: http://niagarawatch.com/portfolio/niagara-falls/ (last accessed 04.07.2017)
37. URL: http://venagid.ru/4867-pyramidenkogel/ (last accessed 21.06.2017)
38. URL:http://www.archdaily.com/24089/zaabeel-park-observation-tower-xten-architecture (last accessed 20.06.2017)
39. URL:http://www.amusingplanet.com/2015/12/20-most-beautiful-observation-towers.html (last accessed 19.06.2017)
40. URL: http://www.archdaily.com/787423/vlooyberg-tower-close-to-bone (last accessed 15.06.2017)

High-rise housing in the city of Samara: the first steps on the path to sustainable development

Tatiana Ya. Vavilova[1*], *Elena Yu.* Makeeva[2]

[1]Samara State Technical University, 443100, Molodogvardeyskaya St., 244, Samara, Russia
[2]Samara State University of Education and Social Sciences, 65 / 67, M. Gorky St., 443099, Samara, Russia

Abstract. This paper outlines theoretical background of high-rise housing and discusses its design experience. It particularly focuses on environmental, social and economic aspects which are among crucial sustainable development issues. The authors dwell upon the implementation of innovative solutions that meet principles and goals of sustainable development and take construction objects built in Samara (which is one of the largest metropolises in Russia) as an example. The research also investigates the quality of project designs and reveals techniques corresponding to the "green standards". It considers the issues of practicing high-rise building construction in specific urban conditions and identifies unresolved architectural problems.

1 Introduction

The largest cities of Russia hold considerable potential for urban construction growth. These cities historical development came together with changes in their ecological state, with transformation of social processes and with adjustment of principles of territorial development control to new economic conditions. The period of industrialization turned out to be the most difficult. This period was associated not only with the emergence of various manufactures and plants, but also with the introduction of typical projects into house construction practice. By 1980s new residential areas (the so-called micro-districts) were built in most Russian cities. These micro-districts designers did not aim at increasing living standards but just at meeting the standards of living space. It should be noted that, to a certain extent, such an approach corresponded with capabilities of the Soviet economy, in which the concepts of "quality of life" or "living standards" did not exist.

The stage market relations formation in Russia coincided with the development of the concept of sustainable development in western countries. The concept of sustainable development inextricably links the issue of improving the quality of life with the improvement of ecological condition, social services infrastructure, and with small business growth. The revival in the Russian Federation such concepts as "ground rent" and "property value" led to the creation of new mechanisms for controlling the growth of cities and to the introduction of urban land use effective management techniques. At present, raising the

* Corresponding author: vatatyan63@yandex.ru

number of storeys and socio-infrastructural saturation of newly constructed residential buildings and complexes are considered to be promising trends in urban development [1, 2].

2 Methods

High-rise residential buildings design used to be an important issue in Russia even in Soviet times. The so-called "Stalin's skyscrapers", built in Moscow in the late 40s – early 50s of the XX century, are known as the most ambitious project of that time. Two of these skyscrapers (on the Kudrinskaya Square and on the Kotelnicheskaya Embankment) are residential buildings. The building on the square near Red Gate is an administrative-and-residential one, and the other two buildings are the hotels "Ukraine" and "Leningradskaya". There are also apartments located in the lateral wings of the main building of Moscow State University and the hotel "Ukraine".

In the late 50s political and economic priorities in the USSR changed. It led to the development of low-rise mass house construction. Only in the late1970s – early 1980s building technologies were and designers returned to the idea of erecting residential high-rise buildings. A new scientific basis of high-rise buildings architectural design appeared in the end of this period.

2.1 Retrospective review of the scientific basis for the design and construction of high-rise residential buildings in Russia

In the Soviet times, one of the first dissertations in architecture considering the evolution of the number of storeys in residential buildings was the work of D. Fisenko (1990). In 1992 I. Melnikova introduced principles of forming compositions of multi-storey residential houses facades, and in 1994 L. Petrova offered a range of techniques to improve living environment in such houses. B. Gandelsman (1997) studied the problem of center-pieces (dominants) in the historical part of Moscow. N. Demidov in 2005 raised issues of the formation of multi-storey centers in historical parts of Ekaterinburg. S. Lyzhin's dissertation (2006) became one of the most comprehensive analytical studies on the architecture and structure of the Urals cities housing stock. D. Sin considered the aspects of evolution of Novosobirsk multi-storey buildings functional features starting his research in the beginning of the XX century up to the early XXI century (2007). In 2009 P. Zueva made an attempt to analyse genesis and evolution of high-rise buildings construction in the United States. I. Zhdanova examined industrial residential high-rise buildings in her dissertation in 2013. She introduced architectural methods of evaluation and improving consumer properties of living environment (2013). Yu. Skoblitskaya (2013) analyzed architectural and space-planning design of service infrastructure in multi-storey residential complexes of the largest cities.

There were also several theses defended by foreign scientists in Russia which were devoted to the architecture of apartment buildings. These dissertations were written by Sarath Boram (1997) and Lee Rasmey (2009) from Cambodia, Nguyen Khuyen (2004), Nguyen Van Tien (2013) and Nguyen Hong Loan (2013) from Vietnam.

In recent years, special attention has been given to the introduction of technologies for sustainable development. The first steps were made in dissertations of S. Molodkin ("Architectural formation principles of energy-efficient high-rise residential buildings", 2007), A. Voronin ("Formation principles of green spaces in residential multi-storey buildings", 2012), P. Semikin ("Architectural formation principles of high-rise buildings with renewable energy sources", 2014) and M. Blagova ("Architectural formation of commercial housing on social-functional basis with Rostov-on-Don as an example", 2015). Technical and technological aspects of high-rise housing construction were considered in dissertations of engineers, for example, in the works of O. Goryachev (2004), E. Sultanov (2006), O.

Djachkova (2009), D. Strigin (2009), S. Dubinskiy (2010), V. Shcherba (2010), I. Sychev (2014), as well as in the thesis of Li Ruixin (2013) from China. Engineering geological problems in construction of high-rise buildings were examined by A. Zhukova (2011) and O. Ovechkina (2013). Some Russian researchers and economists focused on multi-storey real estate in new economic and environmental conditions (A. Melentyev, 2009, A. Oleinik, 2010, A. Balberov, 2011, E. Trushkovskaya 2011, E. Matveeva, 2012, and S. Shikhaliev, 2012).

Traditional and contemporary approaches to design and construction of multi-storey tall buildings are described in detail in works of Russian scientists V. Generalov along with E. Generalova, and O. Maslovskaya along with G. Ignatov. Their extensive researches generalized international experience [3-6].

2.2 International practices and Russian prospects for the design and construction of multistory housing in the context of sustainable development

Formation of the concept of sustainable development has stimulated new design and construction technologies and their development. Now systems of environmental certification based on the standards of BREEAM (Building Research Establishment Environmental Assessment Method, United Kingdom, 1990), LEED (Leadership in Energy and Environmental Design, United States, 1998) and DGNB (Deutsche Gesellschaft für Nachhaltiges Bauen, Germany, 2007) are frequently used for assessing the quality of high-rise residential buildings. They have become a basis for the formation of national "green" standards in many countries, such as China and Russia. Besides, the World Green Building Council (WorldGBC) pays much attention to this trend. WorldGBC representative offices are currently located in 74 countries.

There are a few examples below illustrating successful implementation of some projects. For example, LEED Gold certificates were awarded to such American residential complexes as Macallen in Boston (architectural firm office dA) and Via Verde in New York (Dattner Architects, Grimshaw). A residential complex Triumph Park, Russia (Yashar Architects) and wooden residential buildings of a multipurpose complex in Bordeaux, France (Sou Fujimoto Architects) were designed in accordance with the requirements of BREEAM standard. DGNB silver certificate was awarded to the project of Gezhouba Hongqiao Purple County, a residential complex in Shanghai (China), and DGNB gold certified was earned by a project of a high-rise apartment house Skyline in Stuttgart (Germany).

There is scientific research underway in many countries of the world which is aimed at finding new technologies of sustainable architecture. According to the principles of sustainable development the main lines of research are environmental, social and economic. However, the most promising is a synergistic approach which combines these three lines. Indicators of sustainable development and sustainability indicators used in various fields and at different levels of detail are becoming now a kind of guidance system of urban environment systemic changes [7-9]. In 2010 Binh K. Nguyen and Hasim Altan, scientists from the University of Sheffield (United Kingdom), introduced a ranking system for assessing sustainability of existing high-rise construction projects [10].

According to it, principles of "green" architecture should take into account the entire life cycle of a building: construction, maintenance and demolition. At the same time, a complex of ecological problems has the most significant influence on the development of housing design principles in the early XXI century. Global climate change makes many countries to pay more attention to reducing energy consumption in the housing sector. External wall insulation systems which can be adapted to regional specific features are generally considered to be the most effective means here [11]. There are also experiments aimed at finding solutions to prevent formation of heat bridges, in particular, in window openings [12].

In connection with this, BIM-technologies of energy modeling are being introduced. For example, in Hong Kong, they are being spread into mass residential high-rise projects [13].

Attention to innovative solutions for high-rise buildings with the use of wood as renewable raw low-emission material grows every year [14, 15]. Recently, an international team of researchers from Germany, the Netherlands, Italy and China has proved that a combination of CLT-panels with a concrete core and concrete pillars possess advanced capabilities [16].

In many countries introduction of typological innovations and quality of affordable housing improvement have become a priority approach to reduce social tension. More than thirty public energy efficient housing projects for low-income people are being implemented in several cities of Thailand. Application of even minimal social standards there makes it possible to improve the quality of life and cultural context [17]. Results of some other studies show economic efficiency of industrial methods used in high-rise housing construction [18, 19]. In Malaysia and Indonesia, for example, "green" technologies of resource conservation are applied not only to commercial housing, but also to rental housing that is built for the needs of young and elderly people [20-22].

However, it is known that one of the related problems that prevents an increase in construction of high-rise housing is its fire hazard. The solution for this problem is traditionally associated with special engineering facilities [23, 24], which make construction more expensive and, therefore, restricts the amount of potential buyers. Similar contradictions arise when designing ventilation and air conditioning systems in high-rise buildings [25, 26]. In Russia, the introduction of sustainable technologies is in a formative stage. Despite the fact that in 1999 Russian representatives together with those from Australia, Great Britain, Spain, Canada, UAE, the United States and Japan were one of the founders of the World Green Building Council (WorldGBC), the first "green" standards were implemented in practice only in 2011, with a huge delay if compared with other countries [27]. Currently, rating systems for evaluating sustainability of residential and public buildings, designed by the State Corporation "Olimpstroy", by the National Association of builders (NOSTROY standards) and by an autonomous non-profit organization "Research Institute for sustainable development in construction" (GREEN ZOOM standards), are used on a voluntary basis. In actual practice, challenges of sustainable development, assessed by these systems, confront mainly energy efficiency activities, which have become one of the priorities of the national economy from 2008. In this respect, it is interesting to explore the real experience of designing and building high-rise housing in provincial cities of Russia. In these cities, there are no buildings, which would be nominated for compliance with "green" standards. At the same time, designers and builders there consistently introduce technologies that integrate environmental, social and economic aspects.

3 Results

The city of Samara, founded in 1586, is a typical Russian metropolis, emerging from the industrial era. There are neighbourhoods, zones of mixed construction, abandoned factory sites there alongside with the historic core area. There is a number of multistory buildings and residential complexes being erected in different parts of the city. In some cases, designers and developers introduce innovations corresponding to the goals of sustainable development of life activities. Let's look at some examples.

3.1. Environmental aspects

Over a period of several decades, the approach to designing the outer shell of the building has radically changed. Apartment blocks, built of reinforced concrete with the use of sliding formwork can serve as one of the earliest examples of the construction of buildings without thermal bridges in joints of slabs and enclosing walls. The first residential building was built in Samara in 1987-1988 as an experiment for realization of a typical project developed in Central Research Institute of Housing Engineering Design (Moscow). The authors of this project are such architects as A. Belokon, O. Tretyakov and V. Brandenburg, as well as such engineers as A. Lurie and E. Zaferidi. Similar buildings were erected in other cities of the USSR: in Alma-Ata, Minsk and Ufa. The first floor is designed to accommodate trade and service functions. There are eight apartments located on each residential floor. These apartments are two studios, four double and two triple flats. There are 152 apartments in this 20-storey building. At its gable facades there are smoke controlled evacuation stairs. Brutal faces of the building are non-decorated and filled with balconies in staggered arrangement. Trying to reduce heat losses, many residents decided to glaze these summer spaces after several years of living in apartments (fig.1A).

At the turn of the XX century compact monolithic 18-storey towers were built in different part of Samara by means of the same sliding formwork technology. They were the last examples of the use of solid wall construction with no insulation.

"Ladya" (2004-2008), a residential complex designed by the architects Yu. Koryakin, B. Zhukov, A. Koryakina, T. Vasilyeva, T. Khimchak, N. Ilyasova, D. Lychev, I. Vinshu with Volgatransstroy as a developer, became a breakthrough project. It was the first high-rise residential complex in Samara in which monolithic framework and energy efficient ventilated facades were purposefully introduced. The complex consists of two 25-storey towers with a height of 95 meters and one 18-storey tower with a height of 70 m, which are connected with the help of stylobate with social infrastructure facilities. In its underground level, there is a parking for 289 vehicles and an autonomous roof boiler room that allows precise control of the flow of heat-transfer medium (fig.1B).

Fig. 1. The first high-rise residential buildings in the city of Samara.

This project made it possible to start massive erection of high-rise residential buildings with fundamentally new approaches to resources saving. Currently, most residential buildings of 12 floors and above are erected on the basis of monolithic framework with the use of two main types of facade systems –"wet" and "ventilated". The choice is mostly dictated by the target consumer group. The first technology is frequently used in the construction of affordable housing, the second is used in the construction of premium-class objects.

Such geometrical features of the residential complex "Ladya" as the streamlined shape of the towers and the absence of large consoles also help reduce energy consumption. A similar formation is used, for example, in the residential complex "Argo", designed by A. Gerasimov's project team (with A. Gerasimov, A. Nikolaeva, V. Charkov as architects and

A. Gribanov, D. Gerasimov as engineers and "Stroyresurs" as a construction company). Its three 24-story towers house a total of 552 apartments.

3.2. Social aspects

Changes in national housing policy are clearly reflected in the development of a typology of innovations which take into account diversity of social needs of the population. These innovations occur primarily in housing construction of different price categories, in diversity of growing social infrastructure, as well as in compliance with the principles of forming an inclusive living environment.

Real estate agencies in Samara define the following categories: "economy", "comfort", "business" and "elite". Now "economy" and "comfort" categories of real estate are most popular on the market. Such housing is most commonly built in the central or fringe parts of the city. Let us go through Samara Regional Fund of housing and real estate mortgages experience. Local architects were invited to develop space-planning decisions of high-rise housing. Thus, "Oberon", a local architectural firm, designed a project of a 21-storey residential building with 256 apartments within the boundaries of Antonova-Ovseenko, Blucher and Soviet Army streets (with E. Krupnova, E. Kazachkova as architects and S. Dudarenko as an engineer). It is important that apartments of "economy" class are often commissioned in final finishing. Buyers are offered a choice of several design options.

The economy-class housing group also include a residential complex «Dimitrova 74A" (by an architectural firm "Sreda", its chief architect A. Suvidova, and its chief engineer K. Zhuchkov), a residential complex "Raduzhniy Elite" (2014, chief architect T. Khimchak, chief engineer V. Fedoseev and an association of construction companies "Bereg" as a developer), a residential complex "Favorit" (designed by an architecture-construction company "Arkada" with S. Balabanov being an architect and "Premier-Investstroy" being a developer), as well as a residential complex "Olimpia Park", located in the immediate vicinity of the World Cup 2018 football stadium construction site (by A. Gerasimov's project team with V. Lavrentiev and A. Nikolaeva as architects, D. Gerasimov as an engineer and a construction company "Russian Bazaar" as a developer). Number of stories in these sites ranges from 16 to 25, the height of buildings is from 2.24 to 2.7 m. There are public accommodations and facilities in the lower floors of some complexes. Indoor or underground parking is generally not provided. In the outdoor parking, parking spaces number does not exceed 20% of the number of apartments (fig.2).

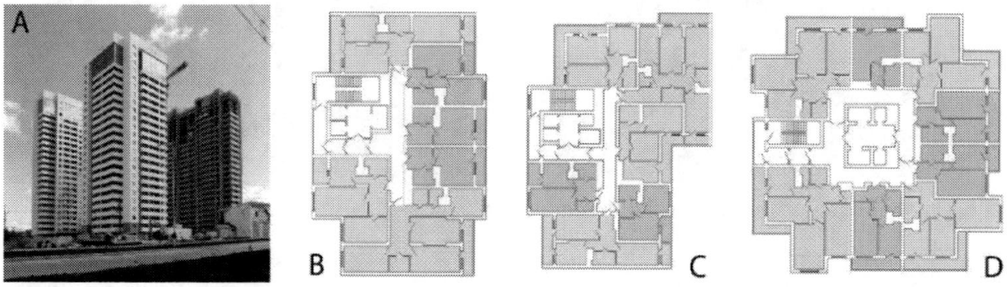

Fig. 2. Residential complex "Olimpia Park". A - overall view; B, C, D – variants of the residential units (materials provided by the authors of the design).

A business-class group includes such residential complexes as "Step Green House" (by an architectural firm "Arthouse" with T. Rychkova, A. Barannikov, A. Varenov as architects, E. Bogdanov as an engineer, and "Samarazhilinvest" as a developer, fig. 3), "Panorama" (Samaragorproekt project team with D. Khramov as an architect and "Stroykom" as a developer), "Imperial" (with I. Khakhalin as an architect and "Stroyekonom" group as a developer), a residential complex "Fregat" (designed by an architectural firm "Oberon" with E. Krupnova, A. Semenov as architects, E. Nikitin as an engineer and "Alliance-Management" as a developer), a residential complex "Parus" (designed by an architectural firm «Oberon» with A. Semyonov as a chief architect, K. Tracevskiy as a chief engineer and "Alliance-Management" as a developer).

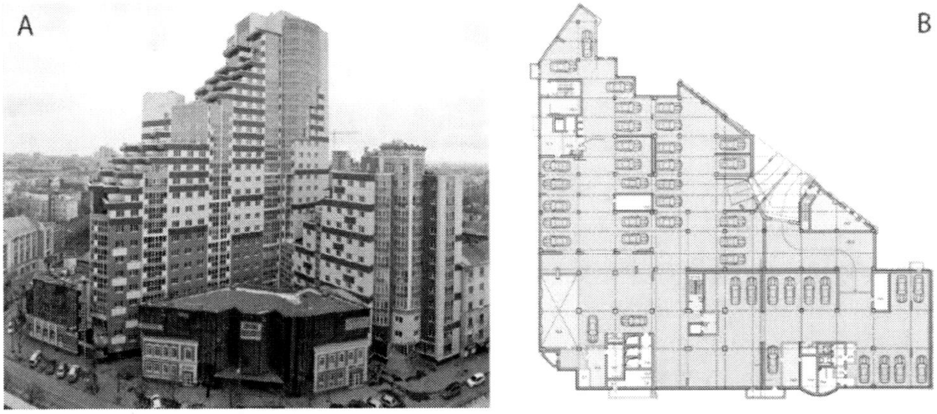

Fig. 3. Residential complex " Step Green House". A - overall view; B - plan of the underground floor (materials provided by the authors of the design).

These residential complexes are located either in the historical part of Samara, or in areas with picturesque landscapes, not far from the banks of the Volga River. The eight of residential premises there are ranges from 2.85 m to 4.0 m. In most cases, lower floors are for rent, and parking lots are located in the underground levels. Final finishing in the apartments is not provided. There is an option of re-planning and exclusive design solution given to owners.

It is interesting to compare sizes of apartments in new residential high-rise buildings in Samara with federal rules of designing housing facilities intended for social needs (Table 1).

Table 1. Comparative analysis of apartments space in residential complexes of Samara

Address	Types of apartments			
	Studio	1-room apartment	2-room apartment	3-room apartment
	Space, m^2			
A set of rules (SP 54.13330.2011, Table 5.1)	-	28,0 – 38,0	44,0 – 53,0	56,0 – 65,0
"ECONOMY»"				
Dimitrova 74A		≥ 39.8	≥ 58.2	≥ 107.1
Boundaries of Antonova-Ovseenko, Blucher and Soviet Army streets	≥ 51.1	≥ 36.3	≥ 58.1	
"Olympia Park"		≥ 30.0	≥ 58.2	≥ 96.0
"Raduzhniy Luxe"		≥ 39.0	≥ 55.9	≥ 82.1
"Favorite"		≥ 39.3	≥ 60.4	≥ 75.7
"COMFORT"				
"Argo"		≥ 31.0	≥ 70.0	≥ 99.0
"Volga"		≥ 49.0	≥ 69.9	
"Raketa"		≥ 45.7	≥ 71.5	≥ 108.0
"BUSINESS"				
"Brig"		≥ 57.0		≥ 102.7
"Ilyin Plaza"		≥ 45.5	≥ 57.8	≥ 98.2
"Imperial"		≥ 52.0	≥ 76.4	
"Liverpool"		≥ 33.0	≥ 65.7	≥ 70.0

"Panorama"		≥ 56.5	≥ 92.4	≥ 102.3
"Fregat"		≥ 30.0	≥ 60.0	
"ELITE"				
"Parus"		≥ 44.1	≥ 62.8	≥ 93.5

Now, two residential complexes of elite class in Samara (that is "Ladya" and "Step Green House") have adopted an innovative solution: an operated ballasted roof increasing the comfort of living (fig.1). There is a general recreational space constructed on the stylobate of the first complexes and individual terraces are built on the ledges of the second.

After regulations on needs of people with limited mobility have been introduced in Russia, all residential buildings regardless of their category of comfort have to be equipped with entrance ramps. Vertical inter-floor transfer is made with the help of elevators.

3.3. Economic aspects

From the viewpoint of "green" economy, the growth in efficiency of urban land use, cost reduction during construction and operation of buildings [28, 29] as well as infrastructure development for small businesses are the most appropriate means to solve economic problems in Russian architectural sphere.

High-rise housing construction allows to increase population density, to reduce the length of transport communications and mainstream engineering systems. It also raises the possibility of violations of human-induced load values allowable for urban environment. In the Russian Federation this problem is still in the discussion stage [30-32], and regulations for territory and premises insolation serve as the main constraining factor at the moment.

The real way to reduce the cost of construction, for both "business" and "elite" segments in Samara is to launch on and the same project several times. One of the most replicated projects here is the project of a dwelling-house of tower-type, included into the residential complex "Argo". It has been lately replicated in residential complexes "Volga", "Pushkinskiy" and "Raketa" (by A. Gerasimov's project team with construction companies "Stroyproektservis", "Samara project" and "Grad" as developers, respectively). In different conditions, these frame-monolithic buildings with protecting ceramic brick walls are of 17 or 25 floors.

Many designers achieve savings in operating costs by increasing the width of buildings, by optimizing the degree of compactness of the planning decision, by rejecting the use of open summer spaces that are not protected from the effects of precipitation [33].

Over the past twenty years in Russia, as well as in Samara, there is a change in forms of organization of social services. For example, built-in premises for rent are included into the structure of new residential complexes already at the design stage [34]. Shops, pharmacies, hairdressers, kindergartens, banks, travel companies, medical clinics, etc. come closer to the consumer and increase the urban environment sustainability.

4 Discussion

A study of modern experience of designing and building high-rise residential complexes in Samara makes it possible to recognize a whole set of unsolved problems that prevent ecological and social comfort of accommodation from progressive improvement, as well as decrease economic efficiency of using territories and objects.

First of all, these problems are connected with the development of high-rise building in the historical part of the city. On the one hand, it obvious consumer appeal come into collision with established morphological randomness, the presence of old housing stock and a significant number of buildings which are of arguable historical and cultural value [35]. All this does not correspond to the concept of sustainable urban development. On the other hand, attempts to include multistory (above 20 floors) housing into the structure of the historical urban fabric while maintaining the traditional quarterly schematics and observing "the red lines" of construction lead to the transformation of aerodynamic conditions, increase transport loads, contribute to spontaneous parking formation, to the reduction and extinction of green spaces in courtyards.

The main problems of center and fringe parts of the city are those of insufficient saturation of residential spaces with small businesses, as well as the use of obsolete composite and technological solutions typical of architecture of the late socialism era. A functional reserve here lies in active introduction of free planning into the practice of economy-class housing, which allows to take into account the needs of people in a wide range of ages, from youth to seniors.

5 Conclusions

The research carried out in Samara demonstrates that various architectural and spatial, functional planning and construction-technological solutions are being introduced in the architecture of modern high-rise residential buildings and housing complexes. This fact underlines positive dynamics of the principles of sustainable development implementation. Resource-saving technologies here are used most widely. In particular, the process of increasing thermal comfort and energy-efficiency of buildings affected housing in all price categories.

However, the analysis of the current situation makes it possible to conclude that, despite significant changes in the formation of service infrastructure of residential areas, there is an open question of real estate proposals and their correspondence with socio-demographic needs and economic opportunities of society.

References

1. L.A. Viktorova, Tall Buildings – the Pros and Cons Of Building, *Architecture and Construction of Russia*, **10**, 2-11 (2012)
2. P. P. Semikin, P. V. Semikin, Perspective Types of High-Rise Buildings, *Regional Architecture and Art Schools*, **1**, 132-138 (2015)
3. V. P. Generalov, Features of Designing High-Rises Houses, 296 (Samara, 2009)
4. V. P. Generalov, E. M. Generalova, High-Rise Residential Buildings and Complexes. Singapore, *Experience in High-Rise Housing Design and Construction*, 400 (Samara, 2013)
5. V. P. Generalov, E. M. Generalova, Perspectives of High-Rise Housing Typology Development. Future of Cities. Vestnik SGASU, *Town Planning and Architecture*, **1(18)**, 10-18 (2014)
6. O. V. Maslovskaya, G. E. Ignatov, Conception of Height in the Design of Urban Environment, *VSUES,* 212 (Vladivostok, 2012)
7. D. Kyrkou, R. Karthaus. *Procedia Engineering*, **21**, 204 – 211 (2011) doi:10.1016/j.proeng.2011.11.2005.
8. G. Barbano, A. Egusquiza. *Energy Procedia*, **78**, 1853 – 1858 (2015) doi: 10.1016/j.egypro.2015.11.332

9. T. Ya. Vavilova, Foreign Experience in Using the Sustainable Development Indicators to Assess the Quality of Urban Environment, *Sustainable Development of Urban Environment*, 6-9 (SGASU, Samara, 2016)
10. B. K. Nguyen, H. Altan, *Procedia Engineering*, **21**, 387 – 394 (2011) doi:10.1016/j.proeng.2011.11.2030.
11. S. K. Saha, Cost Effective Thermal Wall System for Residential Housing, *Procedia Engineering*, **14**, 1913–1919 (2011) doi:10.1016/j.proeng.2011.07.240.
12. Pham Thi Hai Ha, *Procedia Engineering*, **142**, 54 – 160 (2016) doi: 10.1016/j.proeng.2016.02.026
13. Y. Zhao, W. Pan, Y. Ning, *Procedia Engineering*, **118**, 614 – 621 (2015) doi: 10.1016/j.proeng.2015.08.494
14. Roman Milwicz, Piotr Nowotarski. *Procedia Engineering*, **122**, 158 – 165 (2015) doi: 10.1016/j.proeng.2015.10.020.
15. J. L. Skullestad, R. A. Bohne, J. Lohne. *Energy Procedia*, **96**, 112 – 123 (2016) doi: 10.1016/j.egypro.2016.09.112.
16. J. W. G. Van de Kuilen, A. Ceccotti, Zhouyan Xia, Minjuan He, *Procedia Engineering*, **14**, 1621 – 1628 (2011) doi:10.1016/j.proeng.2011.07.204.
17. W. Tanaphoom, D. Bart, *Procedia Environmental Sciences*, **28**, 689 – 697 (2015) doi: 10.1016/j.proenv.2015.07.081
18. Le Minh Ngo, *Procedia Engineering*, **142**, 42–47 (2016) doi: 10.1016/j.proeng.2016.02.011.
19. Malay Dave, Bruce Watson, Deo Prasad, *Procedia Engineering*, **180**, 676 – 686 (2017) doi: 10.1016/j.proeng.2017.04.227.
20. M. S. Suhaida, N. M. Tawil, N. Hamzah, A. I. Che-Ani, H. Basri, M.Y. Yuzainee. *Procedia Engineering*, **20**, 345 – 353 (2011) doi:10.1016/j.proeng.2011.11.176
21. A. R. Musa, N. M. Tawil, S. M. Sood, A. I. Che-Ani, N. Hamzah, H. Basri. *Procedia Engineering*, **20**, 466-373 (2011) doi:10.1016/j.proeng.2011.11.189
22. Ed. E. Pandelaki, W. S. B. Pribadi, *Procedia Environmental Sciences*, **20**, 146 – 153 (2014) doi: 10.1016/j.proenv.2014.03.020
23. Adam Cowlard, Adam Bittern, Cecilia Abecassis-Empis, Jose Torero, *Procedia Engineering*, **62**, 169 – 181 (2013) doi: 10.1016/j.proeng.2013.08.053
24. T. Yu Eremina, I. A. Egorov, Problems of evacuate people from high-rise buildings during fires, *Technology of technosphere safety*, **3(55)**, 21 (2014)
25. W. Sujatmiko, H. K. Dipojono, F.X. Nugroho S., Soegijanto, *Procedia Environmental Sciences*, **28**, 360 – 369 (2015) doi: 10.1016/j.proenv.2015.07.045
26. Xi Chen, Hongxing Yang. *Energy Procedia*, **88**, 748 – 753 (2016) doi: 10.1016/j.egypro.2016.06.065.
27. T. Ya.Vavilova, I. V. Zhdanova, Consumer Properties of Architectural Solutions for Housing and "Green" Standards, *Tradition and Innovation in Construction and Architecture*, 333 – 334 (SGASU. Samara. 2013)
28. Fam Khak K, Innovative methods of solution designs assessment of efficiency in high-rise construction, *"FES: Finance. Economy. Strategy"*, **9**, 19a – 23 (2011)
29. L. P. Lavrov, F. V. Perov, High-Rise Buildings: the Wrong Vector of Housing Construction Development, *Bulletin of Civil Engineers*, **5(58)**, 16 – 27 (2016)
30. K. A. Karelina, High-rise buildings: the opinion of experts and residents to high-rise construction, *APRIORY: Natural and Technical Sciences*, **1**, 11 (2016)

31. A. I. Androsov, A. V. Kopyeva, O. V. Maslovskayam Analysis of Foreign and Domestic Experience in the Design of Housing for the Elderly, *The New Ideas of New Century,* **2**, 12 – 16 (Pacific National University, 2017)
32. G. A. Ptichnikova, "Unfair City": Urban Space as a Reflection of Public Relations, *Sociology of City*, **3**, 47 – 54 (2012)
33. I. V. Zhdanova, T. Ya. Vavilova, N. A. Kalinkina, Life Cycle Extension of Mass Residential Housing Development, *Architectural Innovations. Sustainable Development of Urban Environment,* 124 – 127 (SGASU, Samara, 2016)
34. A. A. Kolosovskaia, N. D. Potienko, Evolution of Topological Structure of Rental Housing in Foreign Countries, *Urban Construction and Architecture*, **3(24)**, 109 – 115 (2016)
35. G. A. Ptichnikova, Architectural Symbols of Globalization in the Russian Cities: Skyscrapers and City, *Vestnik VolgGASU, Construction and architecture*, **15**, 173 – 178 (2009)

Design of high-rise dwelling houses for Ho Chi Minh City within the framework of the "smart city" concept

Nguyen Hong Loan[1,*], *Nguyen* Van Tin[1]

[1]University of Architecture Ho Chi Minh City, 196 Pasteur, District 3, Ho Chi Minh City, 700000, Vietnam

Abstract. There are differences in the concepts of smart cities, which are reflected in many ideas and solutions. Globally one of the similarities of the goals for achieving smart cities is sustainable developmentwith the provision of best living conditions for people beingthe first priority. Ho Chi Minh City is not out of trend, taking the planning steps for the goal of becoming a smart city. It is necessary that design and construction of high-rise dwelling houses meet the criteria of "smart city" concept. This paper explores the design of high-rise dwelling houses forHo Chi Minh City with regards tothe framework of "smart city" concept. Methods used in the paper includedata collection, analytical – synthetical and modeling method.In order to proposedesign tasks and solutions of high-rise dwelling houses forHo Chi Minh Cityinthe concept "smart city"in the current period and near future, we present new approach, whichcan alsobe applied in practice for different cities in Vietnam.Moveover, it can also establishinformation resources, which areuseful in connecting and promotingfurther development for the success of a "smart city" program.

1 Introduction

Smart city is a new concept of urbanization in the world. Many studies have demonstrated thatdefinitions of "smart city" show differences in approach, scale and performing steps in various territories. Through review of the studies, there were somesimilarities of the criteria of a smart city. A development process of a "smart city" needs to be organized, designed and built in successionto ensure its success [1].

Vietnam has been under a comprehensive change with rapid urbanization. There are many studiesfrom the government and experts to provide solutions for the successful implementation of urbanization. Although the current condition in Ho Chi Minh City has some cons in terms of infrastructure, urban planning, energy, living quality, etc., it also has a lot of potential to become a smart city in the future. This could be explained by the viewpoint "smart city concept can be used for transforming any city into a smart city. Smart city has various overwhelming benefits and it a win – win situation for both, government and the citizens" [2].

* Corrwsponding author: loanru@mail.ru

Those studies showed that the concept and criteria for a smart city are very diverse.However,by reviewing the world experiences and practicalconditions ofHo Chi Minh City, the followingmissions can be proposed to convertHo Chi Minh City into a smart city: (1) to develop and improve infrastructure; (2) to ensure the best living conditions for citizens; and (3) to build basis for implementingprogram of a smart city in the future. Toachieve these goals, there should be a concerted participation of many different areas, subjects and levels, where the role of people will be the key to the success. This can be done in a variety of ways depending on the actual conditions and capabilities of the participants, where good living conditions will always be a top priority.

This paper demonstrates thathigh-rise dwelling houses havemany advantages in addressing problems of housing demand in Ho Chi Minh City with the new missions. The objectives of this paper areto build design tasks and to propose design solutions for high-rise dwelling houses in Ho Chi Minh City within the framework of "smart city" concept. The ability to apply new design solutions of this study plays a vital role in addressing people's living needs. The requirements of comfort, safety, humanity and sustainability of living space in the high-rise condominiums will be met.In addition, its results can also contribute to build a "smart city" program forHo Chi Minh City and other cities of Vietnam.

2 Literature review

There are many different definitions of smart cities. For example, in 2014 International Telecommunication Union found out over 100 definitions of smart cities[3]. The most common definition of "smart city" is "the effective integration of physical, digital and human systems in the built environment to deliver sustainable, prosperous and inclusive future of its citizens" [2]. Through the review of scientific papers and reports six components of a smart city are drawn, including: smart governance, smart economic, smart mobility, smart environment, smart people and smart living [2-9]. It was presented in"Research report on Smart City"by Government of Hong Kong.In content of the smart environment, the requirements for resource management, urban planning and smart buildingswere set [9].

In the paper "Smart city in Europe" of the3rd Central European Conference in Regional Science – CERS 2009, the author focused"on the role of infrastructure, although much research has also been carried out on the role of human capital/education, social and relational capital and environmental interest as important drivers of urban growth"[10].The report of the United Nationsat the Eighteenth Session in May 2016, presented by Commission on Science and Technology for Development, mentioned key urbanization trends and their links to the 2030 Agenda Sustainable Development.Thisreport identified opportunities and challenges for the different cities when implementing the program urbanization. For developing countries, a complete infrastructure system is in immediate need. The role of infrastructure is defined as "smart infrastructure provides the foundation for all of the key themes related to a smart city, including smart people, smart mobility, smart economy, smart living, smart governance and smart environment" [3].

The deployment of the smart city program in the world had many challenges, including the problem of abusing of technology which causes incompatibility in harmony with human life. In June 2015, the Government of India developed the program "100 smart cities" that aimed to improve the quality of life for its citizens [11]. This program was built in 5-year period, which emphasizedthe role of government and civil society. Many challenges were set out in this process, in which the development and improvement of infrastructure played an important role. The paper "Evaluating the need for smart cities in India" have been pointed out the current conditions and causes of incompletion in cities. The simultaneousdevelopment of all areashassynchronously created prosperity for citizens [4]. Taking the urban planning in Singapore as an example, it setted up four basic missions including smart environment, smart life, smart service and smart urban planning to satisfy the requirements in comfort, efficiency, safety and sustainability [12].

In addition, in the report "Smart cities in Italy", the author emphasized: "Smart cities offer revolutionary opportunities for economic, social and technological change". The author also listed several definitions of smart city and pointed out that "a smart city is a sustainable city – for everyone" and "the only unifying factor seems to be the concept of sustainability" [1].

The world's urbanization hasalready been underway with 66% of the population expected to live in urban areas by 2050, in which population will be mainly concentrated in Africa, Asia and Latin America [13, 14]. The development context of cities in Vietnam is reflected by a rapid urbanization (approximately 3% per year) witha low starting point. Vietnam still has many opportunities for new construction because it is only the half way through urbanization [15]. Ho Chi Minh City is the largest city in Vietnam with a population of about 10 million andapproximately 3 million immigrants. GDP per capita in Hochiminh City is the highest in the country, about 5500USD/year, which accounts for 30% of the country[16]. By statistics, by 2035 the city will be expanded to a metropolitan area equivalent to a population of 26 million. This issue possibly poses many challenges to the government and its citizens.

Socio-economic factors have been the barriers to many solutions in the implementation of urban development strategies for Ho Chi Minh City. According to Vietnamese architectural experts and sociologists, there is a need for a new governance approach with the aim of coordination between human capital, social capital and technological strength to create a better and more sustainable life.The Government of Ho Chi Minh City has planned to build thecity to become a smart city by 2025 to promote economic development and improve the quality of living environment and service forcitizens[17].During the project, many seminars and cooperation agreements have been organized. "Smart city must be based on the technology development with the human factor as a focal point and the solution is build a modern city with safety and healthy in sustainable development" a Microsoft Corporation's delegate mentioned in one of the seminars[18].

However, there have not been studies in designing of high-rise dwelling houses in "smart city" concept for Ho Chi Minh City. Current proposalshave focused to mobility, grid power, communication, water supply and drainage system, ecology and governance. The implementing programsneedmore participation from experts and citizens.This also contributes to the evaluation of the program's effectiveness and feasibility, as well as to the determining of future plan.

In terms of the natural and climate conditions of Vietnam, especially in Ho Chi Minh city, there is a high requirement for ventilation and heat insulation for buildings. The architectural and technical solutions for high-rise dwelling houses are based on a combination of traditional architectural values and modern options [19]. Compliance with environmental parameters of microclimate by means of costly structures and materials or the use of energy-intensive equipment is not recommended [20].For example, Niu specifiedmain issues of designing of high-rise condominiums to ensure the best microclimate in living environment for the users[21]. The experience of designing, constructing and operating high-rise dwelling houses in Singapore demonstrates the success of the projects to ensure the living conditions of citizens. Generalove and Generalov identified six typical operational planning diagrams of high-rise dwelling housesin Singapore, which can be considered for Hochiminh City[22].

By statistics of Department of Constructionof Ho Chi Minh City, there are about 500 thousandsof households do not have their own houses, 300 thousandsneed to rent apartments and 143 thousand need to buy condominiums or houses. About the structure of condominiums in the real estate marketin Vietnam, by reports CBRE Co. Ltd., in 2016 the investors focus on providing the middle class wih three types of products: Type A (1 bedroom) 45 - 65m^2 with the rate of 10-20%; Type B (2 bedrooms) 70-90m^2 with the rate of 50-60%; Type C (3 bedrooms) 95 - 120m^2 with the rate of 10-30%. It reflects the market demand and the ability to meet the living conditions of different types of condominiums. Besides, there are special condominiums offered by high-quality projects within the city center: Studio, Loft, Duplex and Penthouse with the rate of 5-10% [23]. The growth of new types specifically has represented the change of the market, the trust and favourite of citizens. In general, it will encourage the propose of smart solutions in the building and completeness the infrastructure in a smart way.

The studyof Le Van Dung researched the living environment in urban residences in Vietnam and the ability to improve the quality of comfort and health in three levels: apartment, building and residential area. The author presented the concept and principle of the intergrated model of comfort in high-rise dwelling housesfor conditions of urban Vietnam. Directionfor developmentand improvement the quality of life was proposed for more comfortable and smarter in the future [24].Another issue of designing of condominiums issolvethe needs for privacy and comfort in living environment, particularly in tropical and sub-tropical climates, where design solutions for microclimate conditions, can strongly affect the organization of living spaces. This issue was considered in the paper by Rosemary Kennedy, Laurie Buys and Evonne Miller as an example of approach and solve the problem in other areas such as Ho Chi Minh City [25].

By experiences, electricity is considered the main source of energy and will continue playing a key role in smart cities: "no electricity, no smart city" [13]. Vietnam has achieved some remarkable results in bringing green technology into construction works. Technological solutions which focus on the green energy (solar and wind energy), unbaked materials, technology saving and filtration water for reuse, etc could participate in the operation of the buildings step by step in order to advance the living condition [26]. Smart management and engineering technologyhavebeen chosen for different projects, for example, BMS/BAS-Building Management System/Building Automation System–has beenused in the integration of building systems to manage the engineering and automatical systems[27].In addition, design requirements for urban planning, fire protection system, standards of microclimate and guidelines for encouraging the use of energy-efficient materials are widely disseminated and implemented effectively through the formalities permission of the project.

Theapplication of green building rating systems has been implemented in many countries with different views and aspects of preference. Vietnam has released own green building rating system - Lotus.It is a set of market-based green building rating tools developed by the Vietnam Green Building Council specifically for construction in Vietnam. Rating systemLotus has been developed through long-term research, with the expert advice of specialists giving particular consideration to Vietnam's economic and natural characteristics and to existing Vietnamese standards and policies [28].The criteria of system Lotus are considered appropriate for the organization and formation of smart cities in the future.

In the program of sustainable development, there are many workshops and studies to provide direction for managers and designers, contributing to encouraging people and investors to build modern, efficient and national architecture. In declaration "Green Architecture of Vietnam", published on 27 April 2011, Vietnam Architects Association issued five criteria [29]: (1) sustainable location; (2) use resources and energy efficiently; (3) the quality of microclimate in the building; (4) advance and identityof architecture; (5) sustainability of society and humanity.

In the aspect of tradition and culture in architecture, especially in architecture of housing in new design and construction, there are many assertions by experts. For example, "for achievement of architectural identity, modern Vietnamese architecture must ensure technical and cultural sustainability. It should research into the root source of traditional architecture, which is appropriate for the policies, so new architectural spaces are formed in a dynamic and attractive environment in terms of tradition and culture"[30].In addition, trends and solutions for the development of new urban centers for Ho Chi Minh City must be promotedspecific cultural values of urban architecture[31].Nguyen Song Hoang Nguyen provided many diagrams and conclusions about the traditional cultural values that can be exploited and promoted for the architecture of urban residence. The author proposed to build models and methods of exploiting the value of culture in the architecture of dwelling houses[32].

3 materials and methods

The paper was organizedin a logical order with references to different scientific papers, reports and opinions of experts. Common assertions and trends were emphasized. The main ideas and scientific bases were guided to increase access to the author's suggestions and conclusions. Methodsused in the paper include as follows:

Data collection method: Reviewscientific papers on the theme "smart city" to give a general viewwhich is the basis for determining the direction of the article.Determinationof general criteria in "smart city" concept is important. The views and experiences of implementing program of a smart city in developing countries and in the territories with natural and historical similarities with Ho Chi Minh City were thoroughly studied.

Urbanization trends in the world and Vietnam were studied to provide appropriate conclusions to Vietnam in the current period and near future. The actual conditions of Ho Chi Minh City were updated and reviewed. The main aspects of interest were (1) urbanization trend of the city, (2) socio-economic conditions, and (3) requirements of the climate in architecture. In addition, information on the apartment market and users' demand was listed to ensure the actualization of the proposed solutions. Condominiums have considered as the new for whole Vietnamese, so their design, construction and operation should be realized with experiences from other countries, especially from territories with similar conditions with Ho Chi Minh City. The important aspects are asfollows: implementation at different stages of each project and housing program, design goals, users' request, design standards, feasibility of solutions. General tendencies and perspectives of sustainable architecture were studied for the designing of high-rise dwelling houses in Ho Chi Minh City. In addition, new solutions in the direction of smart operation were identified.

Analytical – synthetical method: The collected information was analyzed and synthesized to clarify the ideas, perspectives and practical experiences. "Smart city" concept has been new for Vietnam, so the chosen informattion is analysed to identify general criteria of a smart city in the world, that can be basic for Vietnam. Real conditions in Ho Chi Minh City were analysed and synthesized with regards to the current and future demands of citizens. In addition, information on the current state of urban planning and architecture was analyzed to address their strengths and weaknesses, particularly the types of housing such as townhouses and high-rise condominiums. These specific types of housing in urban Vietnamwere compared, in that had been shown many advantages of high-rise dwelling houses in the program ofturnning Ho Chi Minh City into a smart city. Aspects, which effect to the design of high-rise dwelling houses were considered to ensure the success of project and the "smart city" program.

Modeling method: A colorful diagram was used to show causal links between the main contents of this article.

4 Results

The trend of high-rise dwelling design to convert Ho Chi Minh City to a smart city is based on theoretical and practical evidences. Relevant aspects are presented as following:(1) the actual conditions of Ho Chi Minh City; (2) the requirements of a successful real estate project,

(3) the requirements of sustainable development; (4) preparation for smart operation and management; (5) forecast for steps toward smart city improvement.

Aspects of a project of high-rise dwelling housesin Ho Chi Minh City within the framework of the "smart city" conceptare recommended: (1) urban planning; (2) architecture-interior; (3) structure and system mechanical, electrical andplumbing (MEP); (4)sustainable development. These factors play an important role in the success of a project, demonstrating the combinationin large-scale housing projects. It meets the needs of users in the direction of more comfortable and smarter. Moreover, it acts as a stepping stonefor the next steps of the "smart city" program.

Their characteristics can be shownin follows:

(1) urban planning: it is the weakness of the status of urban Vietnam, such as: high construction density, small area of sites, low proportion of green and traffic area, and facilities do not meet the needs of modern life;

(2) architecture- interior: the structure of apartments is an important problem of any project. The structure and layout must be reasonable to meet market demand on the basis of ensuring the architectural and interior design standards. These criteria should be implemented at a higher level and more flexible, meeting the requirements of living environment in the current period and near future;

(3) structure and system MEP:technical issues affect the safety and economic problems of the project. Flexibility in planing of architectural spaces and convenience in the use of interior spaces must be emphasized. New smart technological solutions should be applied in accordance withthe staged division of program "smart city";

(4) sustainable development: in the context of rapid urbanization and economic development of Vietnam sustainable development is necessary. Demand for specific culture and traditional values has been one of interest from citizens and experts. It is important to put them in housing architecture, especially high-rise dwelling houses, which are a modern type of housing for most Vietnamese people. It answers the question of people's attachment to the living environment, enabling smart solutions to be realized and replicated.

Design tasks for high-rise dwelling housesin Ho Chi Minh City within the framework of the "smart city" concept are divided according to the mentioned aspects so as to facilitate the approach of design solutions.

Table 1. Design tasks for high-rise dwelling houses in Ho Chi Minh City within the framework of the "smart city" concept.

Aspects	Design tasks
Urban planning	• Identify convenient areas and locations for high-rise dwelling houses. • Determine suitabilityof parameters of urban planning and architecture for high-rise dwelling houses in the direction of smartcity.
Architecture - Interior	• Ensurethe structure of the apartments inthe efficiency of business management. • Improve the quality of the residential area through the general plan. • Enhance the area, quantity and quality of facilities in the buildings. • Ensure the microclimate conditions in the residential areasand in each interior space.
Structure and system MEP	• Enhance efficiency, economy and safety. • Ensure the convenience in the layout and reconstruction for the future. • Apply smart solutions for operation and maintenance of buildings.
Sustainable development	• Ensure the criteria of sustainable architecture, which has been applied in Vietnam. • Meet the requirements of regional architectural identity and promote specific cultural values.

	• Apply technological solutions for operation and maintenance of buildings in the smart direction.

The proposed design solutions are based on the principles of designing of high-rise dwelling house. In addition, in view of "smart city" conceptthey must focus on improving the quality of living space, utility space, comfort and the smart solution. The design solutions must ensure the ability to perform under the actual conditions in Ho Chi Minh City in the current period and near future. Based on the recommendations of this paper, more sensible and smart proposals for high-rise dwelling house will be realized in the future when practical conditions will be met.

Table 2. Design solutions for high-rise dwelling houses in Ho Chi Minh Citywithin the framework of the "smart city" concept.

Aspects	Design solutions
Urban planning	• The role of location must beshown in the ensuring convenient access and effective infrastructure connection. • Reasonable functional areas of facilities in daily activities of citizens are appropriately determined. • Density and land use of constructionmust beeffectivecoefficient.
Architecture - Interior	• Utilities and living environment are ensured through solutions on functional plan and operation management. • The structure of apartments is divided by market segment and level of investment to ensure the rationalityin the combination of following factors: standard area, comfort, health and cost. • The microclimate conditions in the residential areas and in each interior space are ensuredthrough the solutions of functional areas, form and enclosure.
Structure and system MEP	• Structural solutions, which are appropriate to the architecture of the building must been selectedin the calculation of economic and safety factors. • Solutions of structural model can provide appropriate columnar steps for plan and flexibility in each binding position. • SystemMEP must be built in a large scale for serving community groups with long term effectiveness. • The ability to link and update technology will be ensured in different stages of development. • It is necessary to apply smart operation and maintenance solutions in each stage, especially in the current period couldbe thought of smart solutions as BMS, BAS, etc. • It is recommended toapply solutions for connecting the technical system of operation of the building, connectproject to the general infrastructure, build database sources, create the premise for the future "Internet of Things" system. • Smart operational models for each apartment or building can be proposed, for example: automatic control system, security system, air quality control system, sensor system, etc.
Sustainable development	• Cultural values must be integrated into methods of spaces organization, models of activity, solutions of forms and details, as well as in the organization of small functional spaces in the interior. • The core values of ethnic culture is identified and promoted to create an identity. • It is necessary todetermine solutions of energy saving through layout, forms, structures and using energy-saving materials, etc. • The role of using different types of clean energy sources such as solar and wind energy, etc must be shown.

	• Water resources can be saved and utilized in combine architecture and technology. • It must be minimized impact on the surrounding environment. • And it is necessary toimprove the quality of the environment and ecology.

5 Discussion

Application ofsmart city model to Ho Chi Minh Cityis very complex. Thiscity is the fastest growthof Vietnam's economic, thereforethe forecast, organization and implementation to the development stages of this citywill play an important role. The synchronous implementation of all items of the smart city development model could be considered prioritizing areas based on the following factors: (1) necessity; (2) actual demands; (3) ability to realize in stages; (4) the potential of update and synchronization with the general system of the smart city model in the implementation phases.

In addition, the experience of urban development in developing countries demonstrates many issues that need to be addressed. In building the program of organization and formation of smart city for Ho Chi Minh City, there are many criteria to be implemented and the focus is smart infrastructure. The solutions for renovation and development of urban planning and architecture of Ho Chi Minh City are considered and proposed by many experts and managers, in which the priority are developing infrastructure and construction modern buildings with large scale.In that picture, determining the imperative and long-term factors in organizing implementation of a smart city has a practical basis. It needs to be affirmed that the role of citizens is always concerned and enhanced in all urban development activities, reflecting the humanity and the tendency of sustainable development. The requirement to ensure the best living conditions with a modern, comfortable, safe and ecological living environment is a priority.

One of the most enormous problems of architecture in Vietnam is the poor management – the people build housesthemselves, the government only manages. That leaves the city incomplete and in chaos[30]. In that picture, townhouses with major proportions in area and quantity have shown many inadequacies. By statistics, amajority of residential areas in Vietnam has beenlocatedin townhouses. The characteristics of the urban planning and architecture of townhouses in terms of comfort, efficiency, safety and sustainability spesify many disadvantages such as: (1) each household builds and uses a house on its own site, that splits the land into small lots and put high pressure on the infrastructure system; (2) each house usually has only one facade, the case for having a backyard is not the popular, so hence the requirement of microclimate is not guaranteed; (3) the need to solving the quality of living environment may lead to the growth in the use of space, the costs for facades and roof as well as the equipments for improving microclimate; (4) status quo of urban planning, architecture and landscapes reflect the dislocation, lack of uniformity and aesthetics; (5) there are difficulties in identifying urban architectural identity.

In order to solve the problem of crowded land, the decision to move into high-rise dwelling houses has considered as most feasible, even this trend has gradually become popular in urban areas. Advantages of high-rise dwelling houses are shown in following aspects:

(1) location: citizens have a variety of choices for convenient and optimal locations in terms of infrastructure;

(2) cost: buy a condominium is more reasonable than buy a townhouse;

(3) utilities: many utilities and services for living are organized according to standards and market demands;

(4) security: fire protection system and security are ensured through housing management;

(5) health - comfort: the microclimate is better than most townhouses;

(6) supporting policy: liquidity is supported through sales policies and programs promoting the living conditions of the government.

The inadequacies of living environment, public space and microclimate in each small space of townhouses make it difficult to improve the quality of life and urban embellishment. Moreover, the environmental and energy factors in the type of individual housing can cause many disadvantages to overcome.While, the advantages of high-rise dwelling houses have been shown to represent the alternative and pioneer of revolutionary solutions for the development of housing funds and corresponding quality. Another advantage of high-rise dwelling houses is the ability to manage and operate for community groups when proposing innovative alternatives. Revolutionary solutions of design, construction and operation will be provided a foundation for linking the city's management and operation systems, creating the premise for applying the technological advances of the "smart city" concept. This will also effective in dividing the urban development phases on a predictable basis.

The interest and investment for the development of high-rise dwelling houses is considered suitable for the reality of Ho Chi Minh City.Through the diagram, it can be seen the location and role of high-rise dwelling houses in the scheme of organization and formation of a smart city.

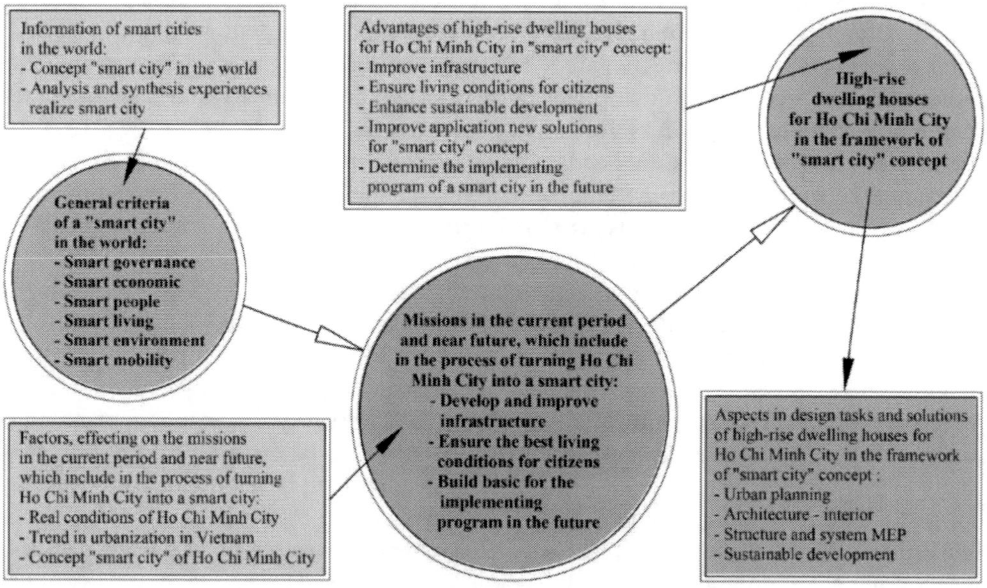

Fig. 1. Interrelated factors to access to the results of the paper.

Determining the trend in the development of high-rise dwelling houses design, serving the organization and formation Ho Chi Minh City into a smart city in the current period and near future it should follow the points of view: (1) high-rise dwelling houses need to meet requirements of smart cities on the condition that ensuring the best living conditions for citizens; (2) the overall strategy for the implementation process and the staged division should be worked out; (3) in the current period and the near future of Ho Chi Minh City the appropriate criteria must be choosen to ensure the success of the project, (4) ensuring the quality of living environment in the direction of sustainable development; (5) creating a database and a basis for connecting to the general management system of the smart infrastructure system; (6) ensuring the adaptability of the buildings in the direction of smart system in the future.

Suggestions of this paper represent the generality of architecture and are appropriate for the type of housing project with many factors to address, especially large-scale buildings, serving for community groups.Requirements in design tasksand solutions are based on the research and experience in designing, constructing and operationof high-rise dwelling houses, especially in Ho Chi Minh City.It demonstrats the practicality of the proposals, which are based on the design experience of real projectsin Ho Chi Minh City.Results of this paperwere applied on different scales: planning, architecture - interior. In the technical aspect, solutions of structure and system MEP were considered in high level. This approach represents the synthesis of aspects of architecture: functional, technical, economical, and aesthetical.

Moreover, the cultural identity and ecological aspects of sustainable development are considered as mandatory factors. It should be noted that in the "smart city" concept the role of science and technology is vital, especially the process of management with information and the performance of various tasks will minimize direct contact. All will be more automated and useful - so that it can create challenges for the future when it comes to cultural and human factors .One of the aspects of sustainable development, which has been emphasized by designers is the need to maintain and promote traditional values in creating a unique identity for the building. The specific cultural values need to be expressed in different solutions to create a connection between the people and the living space, to be promoted the interplay between people and people in the context of modern life.

The development of high-rise dwelling houses for Ho Chi Minh City within the framework of "smart city" concept is more likely to succeed for the following reasons: (1) ensuring the social security issue and promoting the improvement of infrastructure, thus contributing to successful implementation of many criteria for a smart city; (2) the highly revolutionary of proposed solutions, which facilitate the application of many synchronous design solutions; (3) the large scale of applicationof project to the community groups; (4) the potential to bring advances of smart technology into the buildings.

This paper provides the new approach in design of high-rise dwelling houses for Ho Chi Minh City.Requirements of a better life and tendency of perfection living conditions in the direction of smart way will be ensured. This approach has potential inother criteria of "smart city" concept.

One of issues of this paper can be the fact that higher level of finacial investment for high-rise dwelling houses will be enhanced more comfortable and smarter for living conditions.It could influence the realization of project in cause of real conditions in Ho Chi Minh City. However, proposed solutions can be chosen in caculation of large scale and long-term investment for groups communities.And there are various options of smart solutionsfor each holdshousewith regards to demand and ability of users. In addition, smart solutions can be upgrated and extended on the basic of suggestions in current period and near future of Ho Chi Minh City, where has been demonstrated the fast growth of standard of living.

6 Conclusions

Smart cities are the target of many cities in the world, where citizens are ensured the best living with all their daily activitieswill be raised to a higher and smarter level. That will provide the foundation and motivation of the whole society to pursue larger, higher and more meaningful goals. It is a modern city where people live in a comfortable, safe, intelligent and promising environment with better opportunities for future generations.

High-rise dwelling houses are designed, constructed and operated by "smart city" concept are of great significance as they play a pioneering role in organiziation and implementation of a smart city model for Ho Chi Minh City. Results of this paper are based on a strong scientific basis, expressed through specific discussion, in order to show the development trend of high-rise dwelling houses in the current period and near future of Ho Chi Minh City. The proposed design tasks and solutions are highly realistic, ensuring the success of determined objectives.

Results of this paper can also be applied in design of real projects for other cities of Vietnam. Specific and diverse requirements of design will stimulate other sectors to participate in the design, construction and operation of high-rise dwelling houses to improve the effectiveness of projects. Recommendations of this article canalso stimulate the basis for management and operation ofbuildings of different scales. In addition, they will provide the premise for the development and perfection of other criteria of a smart city.

References

1. Research summary. *Smart cities in Italy: an opportunity in the spirit of the Renaissance for a new quality of life.* ABB e The European House-Ambrosetti, 2012.
2. P.R. Patel, H.J. Padhya, *Review paper for smart city.* International Journal of Advance Research In Engineering, Science and Management IJARESM (2011)
3. United Nations. *Smart cities anh infrastructure.* Report of General Secretary. Commission on Science and Technology for Development Nineteenth session. Geneva, (2016)
4. M.F. Jawaid, S.A. Khan, International Journal of Advance Research In Science and Engineering IJARSE, **4**, 990-996 (2015)
5. Center of Regianal Science (SRF). *Smart cities – Ranking of European medium-sized cities.* Vienna University of Technology (2007)
6. O. Parishwad, T. Singh, Journal of Civil and Environmental Technology, **1(6)**, 54-59 (2014)
7. A. Monzon, *Smart cities concept anh challenges. Bases for the assessment of smart city projects.* International Conference on Smart Cities and Green ICT Systems (SMARTGREENS) (2015)
8. K. Singh, N. Neha Sharma, International Journal in Management and Social Science IJMSS, **4** (2016)
9. Central policy unit. *Research report on smart city.* The government of the Hong Kong special administrative region (2015)
10. A. Caragliu, C. Del Bo, P. Nijkamp, 3rd Central European Conference in Regional Science – CERS, **1**, 45-59 (2009)
11. R. Aijaz, *Challenge of making smart cities in India*. Asie. Visions, **87**, Ifri, (2016)
12. M.H. Nguyen, http://ashui.com/mag/congnghe/xuhuong/13473-thanh-pho-thong-minh-khuynh-huong-tat-yeu-hay-su-lua-chon-khon-ngoan.html (2017)
13. International Electrotechnical Commission. *Smart cities.* IEC, Geneva, Switzerland (2016)
14. United Nations, Department of Economic and Social Affairs. *World urbanization prospects - The 2014 Revision – Highlights.* New York (2014)
15. N.H. Nguyen, Vietnamese Journal of Urbanism, **26+27**, 82-87 (2017)
16. *Vietnam population projection 1014-2049* (Vietnam news agency publishing house, Hanoi, 2016)
17. H. Phan, http://tphcm.chinhphu.vn/xay-dung-tphcm-tro-thanh-%E2%80%9Cthanh-pho-thong-minh%E2%80%9D-vao-nam-2025 (2016)

18. http://ictnews.vn/cntt/cuoc-song-thong-minh/tp-hcm-co-the-tro-thanh-thanh-pho-thong-minh-trong-10-nam-toi-144831.ict (2016)
19. V.T. Nguyen, *Development of high-rise dwellinghouses for low- and middle-income people in big cities of Vietnam (for example Ho Chi Minh City)*. Doctoral thesis of architecture. Saint-Petersburg State University of Architecture and Civil Engineering (2013)
20. H.L. Nguyen, *Development of energy-efficient architecture in big cities of Vietnam(for example Ho Chi Minh City)*. Doctoral thesis of architecture. Saint-Petersburg State University of Architecture and Civil Engineering (2013)
21. J. Niu, Energy and Buildings, **36**, 1259–1263 (2004)
22. E. Generalova, V. Generalov, CTBUH Journal, **IV** (2014)
23. H.K. Nguyen, https://www.tapchikientruc.com.vn/chuyen-muc/hinh-thuc-thiet-ke-nha-cao-tang-thuc-trang-dinh-huong.html (2016)
24. V.D. Le, *Architectural formation of urban dwelling houses in Vietnamwith considering the requirements of recreation of citizens*. Doctoral thesis of architecture. Institute of Architecture and Arts of Southern Federal University (2010)
25. R. Kennedy, L. Buys, E. Miller, Sustainability, **7**, 7741-7761 (2015)
26. T.L. Pham, http://www.baoxaydung.com.vn/news/vn/quy-hoach-kien-truc/cong-nghe-xanh-trong-xay-dung-giai-phap-hieu-qua-de-ung-pho-voi-bien-doi-khi-hau.html (2017)
27. A. Smith, Smith A. *Building management systems (BMS). Ceminar 1 – The basics explained*. A.G. Coombs Advisory. City of Melbourne (2017)
28. http://archive.vgbc.org.vn/index.php/pages/lotus-rating-tool (2017)
29. S.Hoang, https://kienviet.net/2014/10/07/infographic-5-tieu-chi-kien-truc-xanh-viet-nam/ (2014)
30. P.L. Nguyen, http://www.nhandan.com.vn/vanhoa/nghe-doc-xem/item/25009302-ngon-ngang-kien-truc-viet.html (2014)
31. P.C. Pham, *Maintianing and conveying the specific values of urban architecture during expanding the existing Hochiminh City centre*. Doctoral thesis of architecture. University of ArchitectureHo Chi Minh City (2015)
32. S.H.N. Nguyen, *Characteristics of traditional culture exploitation in residential architecture in big cities in Vietnam*. Doctoral thesis of architecture. University of Architecture Ho Chi Minh City (2016)

A second level of the Saint Petersburg skyline

Andrey Krasnopolsky[1] and *Sergey* Bolotin[1*]

[1]Saint Petersburg State University of Architecture and Civil Engineering, Department of Drawing, Vtoraya Krasnoarmeyskaya str. 4, Saint Petersburg, 190005, Russia
[2]Saint Petersburg State University of Architecture and Civil Engineering, Department of Basics of the Construction Organization and Management, Vtoraya Krasnoarmeyskaya str. 4, Saint Petersburg, 190005, Russia

Abstract. The article considers the history of the residential development in Saint Petersburg and states corresponding landmark dates. In recent years, changes in the altitude range of the residential development are noted, the influence of this factor on the formation of the city's silhouette is assessed. Reasons for such changes are identified. Attractiveness of high-rise residential complexes for living is assessed. Conclusions are made of tendencies in further housing construction development in terms of its altitude range. It is noted that it is possible to locate multi-storied buildings in the periphery of the city, taking into account specific visual characteristics of the construction site and silhouette of erected buildings; as for central districts, strict regulations regarding the altitude range are needed.

1 Introduction

Exceptional significance attached to architectural traditions in the life of modern Saint Petersburg distinguishes the city among other metropolitan areas of Russia. The Historic Center of Saint Petersburg and Related Groups of Monuments, included in the World Heritage List in 1990, became the first UNESCO World Heritage Site in Russia (and the USSR). Making this decision, UNESCO took into consideration the fact that "the greatness of the northern capital of Russia, its horizontal silhouette with vertical dominants, ensembles of embankments and squares — all this underlies the "imperial" image of Saint Petersburg, its genius loci" [1].

Specific "horizontal position" of the city's silhouette with a small number of vertical dominants is of special concern of Saint Petersburg citizens; it always affected and affects the multi-storied construction development in Saint Petersburg.

The study encourages to find ways for further regulation of residential development evolution, which would ensure city development, while retaining its appearance and humane environment for people.

2 Methods

[1] Corresponding author: sbolotin@mail.ru

A comprehensive approach, comprising studies and analysis of the information from literary and Internet sources on housing construction and the legislative framework related to regulations of the building height, serves as the methodological basis of the study. A full-scale survey, photographic fixation and an analysis of city's panoramas were carried out to reveal the influence of multi-storied housing construction. Landmark dates of the city development were identified.

3 Results

3.1 1703–1765–1844–1960s. Horizontal position of the Saint Petersburg silhouette

Founding the northern capital, Peter the Great considered it necessary to demonstrate the greatness and power of the renovated Russia. In 1712, he started the construction of the Peter and Paul Cathedral, and in 1722 — the construction of the Twelve Colleges building. These two buildings determined both the grand scale of the founded city and compositional features of its silhouette. The 400-meter horizontal line of the main government building rose above the trailing and swampy Vasilyevsky Island. The line continued on the Petrograd side with the elongated body of the basilica of the Peter and Paul Cathedral with the bell tower spire ascending to 112 meters over the wide Neva River. Silhouettes of two gigantic structures, having no match over hundreds of miles around, contrasted with 1–2-story wooden and clay-walled structures, huts and dugouts in the forest, in which builders lived.

Even in our time, these two buildings can be considered impressive constructions; they honorably present the image of the "new Russia" to the world. In order to assess their perception at the beginning of the 17th century, it should be noted that major gigantic construction projects of Peter the Great were implemented practically from scratch.

Urban-planning development of the capital city remained the prerogative of Russian sovereigns — heirs of Peter the Great. They personally managed the development, looked through projects of all erected buildings, established decision-making committees for the control over the execution of monarchic decisions and development of design proposals. Activities of private entrepreneurs were regulated with the help of administrative measures.

Let us mark out landmark dates of the city development.

1765 — Catherine the Great introduced rules meant to give more greatness to Saint Petersburg:

- it was prescribed to locate buildings along building lines without gaps (to ensure their adjoining) and align them by height on the front facade of blocks: "All houses to be located at the same street should be built as one solid facade with the doorsteps not coming out to the street and the same height along the building line".

-it was proposed to raise the development level, and buildings on front embankments were supposed to have the facade height of 10 sazhens (about 21.3 m), so that "the buildings along the Neva river in the least would correspond to the stone bank" [2].

Funds for the capital construction flowed in Petersburg from across Russia, nevertheless, the implementation of all these arrangements took a long time: only in the 1810–1830s, during the large-scale reconstruction of the city center under Alexander I, the height of facades on main streets reached 3–4 floors.

1844 — Nicholas I issued the decree "On restriction of high-rise construction in Saint Petersburg and limiting additional structures in existing buildings" [3].

The prohibition was caused by the appearance of 5-story houses in the city. This order is usually regarded as a concern for maintaining the "horizontal position" of the Petersburg silhouette. However, it should be taken into account that the authorities decided to limit the representative external appearance of the capital development for the first time in 140 years. The emperor's resolute reaction reminded of the subordination and limited ambitions of individuals: none of developers could erect buildings higher than the eaves of the imperial residence — the Winter Palace (23.47 m). Only bell towers and cupolas of churches ("only the God is above the tsar") could serve as exceptions. The sovereign Russian emperor followed the way which was paved by European monarchs centuries before. For example, Louis XIV, who was also a strong supporter of the principle of absolute monarchy, allocated plots for the construction in Versailles under the condition to construct buildings not higher than 18.5 meters, that is, the level of the entrance to the palace [4].

In the middle of the 19th century, the panorama of the main city space of Saint Petersburg was finally formed.

The year 1844 should be particularly noted, as it established the reached height of the development skyline in Saint Petersburg. A bit later it was cut by two new high-rise dominants. In 1868, Saint Isaac's Cathedral was consecrated, which for a long time had been surrounded by staging; its height was 101.5 m. A massive colonnade and a gleaming gold cupola rose above the standard development. The reconstruction of the spire of the Peter and Paul Cathedral, carried out in 1857–1858, which brought the total height (including the cross and angel figure) to 122.5 meters, can be considered as a response to the challenge of the erected giant. At that time, the horizontal line of the Saint Petersburg silhouette stabilized (Figure 1). Parameters of determining structures of the historical center of the city have not changed significantly to our days.

Fig. 1. Modern panorama of the Neva banks: the city development stretched along the horizontal line not exceeding the height of the Winter Palace eaves (23.47 m), the second highest dominant of the city is the Saint Isaac's Cathedral which rose up to the height of 101.5 m (photo by L. Lavrov)

In 1900s, prerequisites for adjusting the historical landscape appeared: amendments were introduced to the building code, canceling the height regulation with regard to the ridge of the eaves. Attics appeared in Petersburg immediately (Regulations SP 54.13330.2011 "Multicompartment residential buildings" defines this structural part of the building as follows: "A floor in the attic space, the facade of which is completely or partially formed by the surface (surfaces) of an inclined, kinked or curved roof" [5]). However, in the period before 1917, consequences of the radical administrative decision had no time to manifest on a large scale, and the city retained the obvious "horizontal position" of the development.

In the Soviet period, the city ceased to be the capital of the country. It was destined for the ordinary fate of a regional center, and no ambitions (in the architectural sphere as well) were permitted to manifest in provinces. The subordination was also maintained in terms of the height of buildings. In the 1930s–1950s, the silhouette of the historical center of Leningrad lost quite a lot of verticals — bell towers and cupolas of churches. It was assumed that a system of new dominants would appear instead. However, resources were sufficient only for technical objects which were functionally necessary. Then, water towers of the Meat Processing Plant and the Red Nailery plant (according to the sketch of Ya. G. Chernikhov) appeared. A tower of the Kirov District Council was built, and high-rise elements of the Industrial Cooperation House of Culture on Kirovsky Prospect and the Kirov Culture Palace on Vasilyevsky Island were significantly reduced in comparison with their project designs (Figure 2).

Fig. 2. Building of the former Industrial Cooperation House of Culture. Design drawing by architect Ye. A. Levinson and its full-scale implementation

In the late 1930s, the USSR joined the race of high-rise structures. In Moscow, the construction of a gigantic building of the Palace of the Soviets started; it was supposed to reach the height of 420 m and surpass the 381-meter American skyscraper — the Empire State Building [6]. In Leningrad, a project of the local House of the Soviets, the overall height of which corresponded to the basement level of the Moscow high-rise structure, can be considered as a response to that.

In the post-war period in Moscow, which should become a "model communist city", seven "Stalin's skyscrapers" were built of up to 200 meters high. In the capital of the Soviet Ukraine, the initiative was supported with a 73-meter house on the main street (85 m high with the spire) [7]. A lot of projects were developed based on this theme in Leningrad, but only a building on Moskovsky Prospect near the Victory Park was erected (Figure 3).

Fig. 3. Multi-storied tower-house on Moskovsky Prospect. Leningrad. Photo of 1962

During post-war restoration in the zone of the historical center, the principle of leveling the height of front facades "under the single eaves" was defining, as it was established in 1765: the missing floors were built up in the destroyed buildings, and tumbledown attic structures were replaced by capital structures in 50 cases [8].

In the 1960s, with the beginning of the Khrushchev's reforms, standard 5-story buildings got widespread, as they were considered then to be the most cost-efficient type of housing. Huge areas in periphery districts were built up with micro-districts with standard large-panel 5-story buildings. Shortly after that, related problems associated with the city transport and engineering infrastructure appeared. In the 1970s, the development of 9–12-story and, partially, 16-story houses started centrally (Figure 4).

Fig. 4. Development in the Leningrad periphery, typical for the 1980–1990s

In a short time, "high-rise" buildings (with more than 5 floors) occupied monopoly positions in the structure of housing construction in Saint Petersburg, and by 2009 they accounted for more than 2/3 of the total housing stock [9]. Enormous dimensions of erected multi-story buildings in the periphery attracted attention, as they did not correspond to the development, traditional for the city center. The living area in 551 new 16-story residential buildings, built in "bedroom districts" exceeded the living area in 33,399 1–4-story buildings, which formed the basis of the housing stock of the historical part of Saint Petersburg [9].

3.2 Preserving the "skyline". 1970–2010

Introduction of multi-story buildings in the city silhouette triggered a negative public reaction in the late 1970s. 22-story residential buildings which were erected in the peripheral district at the Victory Square were perceived then as the monumental complex at the entrance to the city. However, appearance of such buildings in the North-West of Vasilyevsky Island attracted attention of specialists immediately. Architect N. T. Vinokurova took into account their location along the embankment of the Smolenka River, in the section line of Nevsky Prospect and pointed out that, due to that, high-rise buildings would be viewed from the center of the city and, moreover, they would create a background for the spire of the Admiralty [10]. Her work can be considered the first study of the new multi-story development influence on the silhouette of the city. A speech of academician D. S. Likhachev, who intensely criticized the appearance of the lapidary silhouette of a high-rise hotel near cupolas of the Trinity Cathedral in the western part of the Fontanka River (Figure 5) and called the 18-story building "violating the skyline typical for Leningrad", can be considered as the beginning of emotional manifestations of public opinion [11].

Fig. 5. 18-story building of the hotel in the perspective of the Fontanka River

In our time, the development has turned towards the direction set by financial and economic conditions. The construction industry has undergone tremendous changes over the past 50 years. The most dynamic processes began when the country switched to the market economy. Mass privatization of apartments in the state housing stock started. The land became a commodity. Technical infrastructure enterprises turned into corporations. An official real estate market emerged; housing prices began to grow.

Large profitable projects started to be implemented. The key role is played by expanded construction enterprises and wealthy investors supported by leading banks. In the 2000s, they supported the development of modern technologies for multi-story construction, and areas for 17–25-story buildings and buildings with more stories were allocated in the peripheral districts next to 9–16-story buildings (Figure 6).

Fig. 6. Development on the northern bank of the Gulf of Finland. 22–25-story blocks constructed at present days are located near 9–12-story buildings erected in the 1990s

The strategy for the location of high-rise and multi-story buildings has changed. In the Soviet times in Leningrad, there were less than ten residential buildings with the height of more than 20 floors, and the location for each of them was determined by arrangement and urban-planning considerations. In the modern period, commercial interests became determinative. 20–25-story buildings became a characteristic element of the urban landscape in the peripheral district. Even the highest residential buildings (higher than 100 m) appeared in areas, where their construction was based on cost estimates of construction corporations.

The acute struggle with high-rise development has become one of the dynamic and popular trends of the public activity in Saint Petersburg. The struggle between developers and civil activists has been carried out for several decades with varying success. In the early 2000s, a scandal burst out with regard to the erection of the Mont Blanc multifunctional complex in the area of the historical center on the Vyborgskaya embankment of the Neva River. The construction of a 74.5-meter building was accompanied by the persistent criticism on the part of historic preservation activists. Commercial interest, desire to get more luxury apartments with beautiful views of the historical center panorama played the main role. When the construction object was topped out, it became obvious that the new building was one of the most serious recent urban-planning mistakes of Petersburg. In May 2006, the Urban Council disapproved the Mont Blanc building construction project and stated that the erected building distorted the historical and architectural panoramas of Neva banks. An attempt to introduce some adjustments without decreasing the building altitude was made, but, as the photo shows (Figure 7), it failed.

Fig. 7. High-rise volume of the Mont Blanc residential complex in the landscape of the Petrograd side

These days, the building on the bank of the Neva River is an object of the most acute critics. Complaints are also caused by the fact that "the protected panorama of Saint Petersburg was hopelessly crushed by an awkward, tactless, infinitely inappropriate massiveness of this structure" and that the developer used the remarkable potential of the panorama exclusively for commercial purposes (a unique place, where a public observation platform could appear, passed to a private luxury apartment) [12]. Information on general financial results of the sale of the dwelling with panoramic views is not reported, but the cost of 2-room apartments available in this building is 24,500,000 RUB, and the cost of 4-room apartments is 42,900,000 RUB [13].

At the present time, it is difficult to find a site in the city where construction with the same economic effect is possible; changes in laws also prevent such construction.

It is known that apartments with views in the Premier Palace residential complex, surrounding landscapes of which cannot comparable to the unique Mont Blanc situation, are more expensive than standard apartments by 20–30%. "High-rise housing is targeted, first of all, at residents with incomes much higher than average. As a rule, apartments on upper floors are more expensive than similar ones on lower floors", experts assert [14, 15].

Nowadays, commercial considerations stimulate the increase of building height and number of multi-story residential buildings, but there are no more sites available in the city center for this type of construction. At the end of the 20th century, new buildings located in blocks of the historical center posed a risk to the UNESCO protected landscape. The course of events related to the intention to build a skyscraper of 396 meters high near the baroque Smolny Cathedral, which was intended as a dominant of the Okhta Center business complex, found international resonance. It was planned to put it into operation in 2012, but the sharp reaction of citizens and architects caused intense discussion. Active counteraction led to the project failure [16].

At present, multi-story and high-rise buildings are developed in peripheral districts. The issue becomes especially acute due to the ambitious urban-planning program for the development of the "grey belt" of former industrial territories surrounding the city center in a dense semicircle. These are potential construction sites of great attractiveness for investors, as they could be used in residential construction, mainly high-rise. They are located beyond the zone where the horizontal position of the development silhouette is protected by UNESCO. Here, regulations for high-rise construction are less stringent, allowing for the construction of buildings with beautiful panoramic views of the city center. Recently, the spontaneous development of separate sites started here, resulting in that fragments of lapidary high-rise buildings situated in the "grey belt" began to take shape in the center of Saint Petersburg near cupolas of churches and spires of bell-towers (Figure 8). If the process speeds up, then individual unfortunate obstructions can turn into a continuous multi-story wall around the historical urban nucleus.

Fig. 8. High-rise buildings erected near the Smolenka river show up unexpectedly in the image of the Saint Petersburg center by Birzhevoy Bridge (Exchange Bridge) (photo by L. Lavrov)

In this regard, "standard" high-rise buildings (from 75 to 100 meters), representing a very common typological group (there were 353 such buildings as of January 15, 2015) are of particular concern. The tendency of forming a characteristic volumetric and spatial solution of such buildings causes apprehension. Building plots are very expensive, and developers counting upon a particular profit, resort to increase the number of floors. Investment calculations stimulate the output in the form of the maximum number of "apartments with views". Peripheral blocks, consisting of 10 houses of 25 floors, are already the reality [17]. Under these conditions, complexes of 25–30-story buildings, having the silhouette in the rectangular form or in the form of a dense group of towers, appear, but that does not correspond to basic compositional characteristics of the traditional Petersburg landscape. Moreover, in this case, it is possible to refer to changes in artistic and aesthetic preferences. In the early 20th century, A. N. Benois emphasized that "the rush toward the West existed from the very foundation of Petersburg" [18]. The tendency established in our years indicates "the rush toward the East" (Figure 9). Drawbacks inherent in such residential complexes are recognized by Saint Petersburg builders: "Exceeding 15–20 thousand square meters per

 https://doi.org/10.1051/e3sconf/20183301027

hectare, we get residential complexes, which are called "Shanghai". They differ by excessive crowding, lack of greenery and problems with parking lots" [19].

Fig. 9. Parnas, a new residential district in Saint Petersburg

The very limited number of residential buildings having the height of more than 100 meters can be considered as a specific feature of Saint Petersburg. Their main commercial advantage is luxury apartments with a panoramic view of the historic center of the city. The unique horizontal position of the Saint Petersburg landscape reveals many architectural dominants from the "bird's eye view". If sights of the city center can be viewed from windows of the apartment at least from afar, such apartment would cost much more. These buildings are located far enough from the city nucleus. A building providing the desired panoramic view will be viewed from the city center, posing a threat to the purity of the "Petersburg skyline". Such projects are given closer attention by both the city service for the protection of historical monuments and historic preservation activists.

It can be assumed that such choice of building parameters can be explained by ambitions of the developer and potential residents. At present, there are 8 residential buildings with the height of more than 100 meters in the city: Prince Alexander Nevsky (126 m), Bogatyrsky 2 (108 m), London Park (105 m), House at the Komendantskaya Square (105 m), Dominanta (104.5 m), Poem by Three Lakes (100 m), Twin Peaks (100 m) and Raduga (100 m). Seven of them exceed this threshold by only 5–10 meters — to such an extent that to fix the exceeded limit.

As volumes of the construction of multi-story and high-rise buildings increased, investors began to change their attitude to such structures. Problems associated with the increase in number of floors became noticeable. The demand for expensive apartments was declining. Developers report that it is quite difficult to sell apartments above the 17th floor, and high-rise residential development with 30 floors and higher (i.e. above 100 meters) is economically unprofitable [20]. In these circumstances, the city administration and leading developers began to exercise restraint and caution in opinions regarding conditions of high-rise construction in the city. Solutions restraining high-rise construction are taken at federal and regional levels. In 2006, Regional Construction Norms "Residential and Public High-rise Buildings. Saint Petersburg" entered into force in Saint Petersburg [21].

These norms cover designing of residential and public buildings with the height of up to 150 m (residential buildings with the height of more than 75 m, public buildings with the height of more than 50 m) and touch upon issues of monitoring, automation and installation of elevators, communication systems, structural as well as architectural and planning solutions, fire-fighting measures, water supply and sewerage systems, foundations and basements.

Later, administrative decisions were taken to restrain the construction of multi-story buildings. For example, in 2011, Federal Government Decree No. 207 "On requirements for the issue of certificates of admission to work on dangerous, technically challenging and unique capital construction facilities to self-regulating organizations" was issued. In 2013, the State Duma of Russia supported changes in the Urban Development Code which prohibited so-called "high-altitude deviations".

3.3 Attraction of apartments on upper floors is in question

With the advent of the real estate market, the cost of land plots intended for development became one of determinative factors in setting of characteristics of the created residential environment. Not much territories, where it was possible to conduct housing construction, were left in the city area, and they were located in peripheral zones of Saint Petersburg. High cost of land makes developers, counting upon a particular profit, to push for high density of development: firstly, to increase the number of floors in erected buildings, and, secondly, to place them as close to each other as possible. Vast peripheral territories, where development includes only high-rise buildings, has formed in Saint Petersburg. More and more blocks, consisting of 25–26-story houses, appear on the outskirts of the city [17].

According to the statistical data [9], buildings with the height of 12 floors and more amount to more than 20% of the city housing stock. At the present time, apparently, these data raise particular concerns, as more and more doubts as to the feasibility of high-rise buildings occur. Besides issues of fire safety, elevators, structure of external walls and certain household inconveniences, great attention is drawn to changes in financial conditions. Deteriorating economic indicators are of great concern. For two decades, financial attractiveness of high-rise buildings in Saint Petersburg was based on the potential of landscapes in the historical center and the view from upper floors. The silhouette of the historical center became the commodity which was sold at a good price. Apartments with views were always in demand, windows with "panoramic" views provided additional fine income. Today, only developers of residential areas in the west of Vasilyevsky Island can reckon on this potential. High-rise peripheral blocks have nothing to offer to wealthy lovers of beautiful panoramas. Landscapes of the historical center cannot be viewed from here; apartments are often located window to window with the neighboring house under construction [22]. There are no enthusiasts for upper floors here. Buyers of low-cost housing do not consider it necessary to pay extra money for panoramic views, as "views from the highest floors on the outskirts of the city are, as a rule, not as attractive as in the center, and there are a lot of concerns — an elevator can go out of order, the wind is stronger at the top" [22]. In particular, the construction of upper floors entails significant technical changes and affects the final cost of the production.

It is obvious that demand conditions make a developer pay attention to prospects of the construction of small apartments in the periphery of the city or in districts adjacent to the Leningrad Region.

3.4 Low-rise development, with the height of about 30 m

In April 2014, the city authorities paid attention to prospects of low-rise construction. It was decided to allocate sites in peripheral territories of Saint Petersburg only for low-rise blocks and provide those with infrastructure, probably, using budget funds [17].

The construction industry could not but react to administration intentions, but it was not ready to switch from the cumbersome, but well-developed production mechanism to the development of a new type of buildings, i.e. low-rise residential buildings.

Large holdings agree that they are not ready to take into account advantages of low-rise buildings, which allow for the use of more cost-efficient materials and technologies. It is stated that low-rise buildings should not exceed nine floors (conditionally, up to 30 meters high). It would be reasonable to use territories in the Leningrad Region for development, if there are roads and utilities supplied to the site. Monopolists consider the cost of 90,000 RUB per 1 square meter acceptable [23]. This means that the cost of suburban housing should increase by about 20% (at the present time, the average level is kept, depending on the location and class: 50,000–80,000 RUB per square meter) [24].

Discussions on this issue are still ongoing. Developers persistently refuse to consider 3–4-story houses as the main object of the prospective construction and focus on prospects of buildings with the "average" number of floors. Most common opinions are the following:

- residential complexes with the medium and small number of floors are more in demand among buyers than high-rise buildings;
- 5–7-story houses are attractive for potential customers who do not want to live in "warrens", but the financial effectiveness of such projects is lower than of high-rise development. A possible solution is to use 6–7-story houses with individual 25-story dominants for the development of residential complexes, such as in the Novoorlovsky residential complex;
- at the present time, mid-rise development is considered an alternative to high-rise buildings [17];
- consumer qualities of a building do not depend on the number of floors. It is difficult for a developer to ensure the economic profitability of the project in mid-rise blocks with the density of 5,000–7,000 square meters per hectare. The optimal density of development in large residential complexes is 10,000 square meters of housing per hectare. About a half from about one million square meters which are built on the outskirts of the Leningrad Region will be of mid-rise development [25].

Attempts to change the terminology instead of renewal of technologies are made:

- 10 years ago, buildings of up to 5 floors were called low-rise, now it is possible to call 9-stroey buildings low-rise [17];
- taking into account that there are already a lot of houses of 25–26 floors, it is possible to refer buildings of up to 14 floors to low-rise development;
- housing of up to 12 floors is now called "low-rise" housing [17].

At the present time, Saint Petersburg specialists refer buildings of 9 or more floors, which in the 1960s constituted a group of "high-rise buildings", to the category of "low" or "low-rise" buildings, and 9-story houses are called products of low-rise construction.

It is obvious that the planned adjustments of the typology of cost-effective dwelling in Saint Petersburg do not affect principal positions and will not affect quality parameters. The proposed changes are based on the manipulation with terms. Such unproductive approach can be explained by strong inertia of the design, construction and investment complex occupying monopoly positions.

The organizational and economic system of large holdings, covering the entire cycle of the housing production (from upstream operations to architectural and urban-planning designing), is formed with a view to the in-line construction of large multi-story buildings. Low-rise complexes will appear in the suburbs of Saint Petersburg and the Leningrad Region, but their height will vary around the level of 30 meters. Saint Petersburg low-rise buildings will become "the highest low-rise buildings in the world". Developers operating in the economy-class segment prefer to keep their process flows focused on specific features of high-rise buildings without changes, and allow only half-measures: "We used to build houses with the height of up to 27 floors. From now on, we'll build houses of 14–17 floors" [26].

4 Discussion

Retaining the image of the historical center of the city is one of the most serious problems associated with high-rise construction in Saint Petersburg. Even upon strict adherence to high-altitude regulations for central areas, high-rise buildings appearing on the border with central districts or in the so-called "grey belt" can cause irrecoverable damage to the city silhouette. Attractiveness of high-rise construction for consumers is also doubtful. Its main advantage in Saint Petersburg is the availability of apartments with panoramic views.

At present, tendencies to change the typology of housing occur; concepts relating to the number of floors in buildings are shifted. Buildings, which thirty years ago were referred to high-rise development, are now considered to be low-rise.

It is necessary to ensure strict regulation of further high-rise development in Saint Petersburg, especially on the border with historical regions.

5 Conclusions

1. The very limited number of residential buildings having the height of more than 100 meters can be considered as a specific feature of Saint Petersburg. Their main commercial advantage is luxury apartments with a panoramic view of the historic center of the city.
2. Volumes of construction and locations of high-rise buildings in Saint Petersburg are determined by the task of retaining the historical silhouette, therefore, the location of multi-story buildings is permissible in the periphery of the city, with account for specific visual characteristics of the building site and the silhouette of erected buildings.
3. High-rise buildings with the coarse rectangular silhouette, the height of which is at the limit permitted by construction regulations, inflict the greatest damage to the silhouette of Saint Petersburg. Their use should be limited in both the former industrial "grey belt" and alluvial territories in the west of Vasilyevsky Island.
4. The discussion on the feasibility of high-rise construction in Saint Petersburg is ongoing, meanwhile "the caravan moves on" and it is reported that about 30 designs of residential buildings with the height of more than 25 floors are currently under implementation in the city.

References

1. URL: http://kgiop.gov.spb.ru/media/uploads/userfiles/2015/08/24/%D0%94%D0%B5%D0%BA%D0%BB%D0%B0%D1%80 % D0% B0% D1% 86% D0% B8% D1% 8F_% D0% 92% D0% A3% D0% A6.pdf (accessed date: July 30, 2016).
2. A. V. Bunin, T. F. Savarenskaya *History of urban-planning art* (Moscow, Stroyizdat, 1979)
3. URL: http://www.nlr.ru/e-res/lawr/search.php?part=483&regim=3 (accessed date: 05.10. 2017)
4. URL: http://masterok.livejournal.com/296641.html (accessed date: 07.07.2017)
5. Regulations SP 54.13330.2011 "Multicompartment residential buildings", Moscow, 2011.
6. URL: https://www.wikipedia.org/wiki/%D0%94%D0%B2%D0%BE%D1%80%D0%B5%D1%86_%D0%A1%D0%BE%D0 % B2% D0% B5% D1% 82% D0% BE% D0% B2 (10.11.2017)
7. V. N. Stepanov *Architecture of the Soviet Ukraine*. (Moscow, 1987).
8. History of the Saint Petersburg housing stock development. URL: http://nikonovpn.spb.ru/?p=1059 (accessed date: 25.11.2017)
9. URL: http://grinkod.spb.ru/spb_house_fond.html. (accessed date: 25.11.2017)
10. N. T. Vinokurova, B. I. Pokidova, V. P. Fedorov, Silhouette: studies and preservation of the visual perception of the city. Tri Iskusstva, **1**, 4–5 (1997)
11. D. S. Likhachev Skyline of the city on the Neva. Our heritage, **1**, 8–13 (1989)
12. URL: http://spb.gdeetotdom.ru/articles/1872955-2012-09-10-zhk-monblan-oshibka-kotoruyu-ne-ispravit/ (accessed date: May 5, 2017).

13. URL: http://lifedeluxe.ru/catalog/elite/view/monblanhttp://lifedeluxe.ru/catalog/elite/view/monblan (accessed date: June 16, 2017).
14. URL: http://kommersant.ru/doc/2223512 (accessed date: September 27, 2017).
15. URL: http://rway.ru/publication/publication71-2180.aspx (accessed date: September 7, 2017).
16. URL: https://en.wikipedia.org/wiki/Okhta_Center (accessed date: November 6, 2017).
17. URL: http://gorod.spb.ru/articles/2117 (accessed date: September 27, 2017).
18. A. N. Benois Picturesque Petersburg. World of Art, **7 (1)** (1902)
19. URL: http://www.rosned.ru/mnenie_optimalnaya_plotnost_zhiloj_zastrojki_10_tys_kv_m_zhilya_na_gektar/ (accessed date: October 11, 2017).
20. URL: http://ppt.ru/news/106118 (accessed date: April 14, 2017)
21. Residential and public high-rise buildings. TSN 31-332-2006 (Regional Construction Norms). Saint Petersburg, Government of Saint Petersburg.
22. URL: https://www.kommersant.ru/doc/2830395 (accessed date: May 25, 2017).
23. URL: http://news.nb.dp.ru/a/2016/05/25/Podvodnie_kamni_malojetazhn (accessed date: May 27, 2017)
24. URL: http://rucountry.ru/note/maloetagnoi_stroitelstvo_v_sankt_peterburge_107987.html#ad (accessed date: June 23, 2017).
25. URL: http://www.fontanka.ru/2016/07/06/068/ (accessed date: September 9, 2017).
26. URL: http://www.gazeta.bn.ru/articles/2017/02/09/236998.html (accessed date: September 25, 2017).

High-rise construction in historical cities through the example of Saint Petersburg

Maria Granstrem[1*], *Milena* Zolotareva[1] and *Tatyana* Slavina[1]

[1] Saint Petersburg State University of Architecture and Civil Engineering, Department of History and Theory of Architecture, 4 Vtoraya Krasnoarmeyskaya str., Saint Petersburg, 190005, Russia

Abstract. The article sets forth results of the landscape visual analysis of the interaction of high-rise construction facilities with the environment of historical urban spaces. A toxic connection of high-rise construction facilities with the established urban landscape was analyzed and recorded. One of the latest stages of the reconstruction of historical cities, which penetrated many European countries at the end of the 20th century, also started in the beginning of the 21st century in Russia, where the reconstruction of historical facilities and territories became one of the leading trends of architectural activity. Therefore, problems of the interaction between the old city and new high-rise construction nearby historical centers are extremely relevant for Russian architects. Specific features of Russian high-rise construction within visual borders of historical cities, developed at the turn of the 20th–21st centuries, repeat past urban-planning mistakes spread in Europe in the second half of the 20th century. High-rise construction in close proximity to historical centers of cities violates an established scale and destroys a historical city silhouette.

1 Introduction

Unique historical districts of major European and Russian cities occupy only a small part of city-wide territories. In Saint Petersburg, the historical urban nucleus occupies 5% of the territory, but it is the environment of historical districts that creates a unique image of the city as the system of interrelated and interacting signs and archetypes which form stable spatial ideas about the specific historical space.

One of the latest booms of the selective reconstruction of historical cities, which penetrated many European countries at the end of the 20th century, started almost at the same time in Russia as well. It conditioned reconstruction as one of the leading trends of architectural activity [1]. Therefore, the issue of the interaction between the old and new is extremely relevant. Appearance of new high-rise dominants in so-called "gaps" of the historical center, as well as high-rise construction within visual borders of the historical development takes on special significance. Specific features of high-rise construction in historical cities, developed in Russia at the turn of the 20th–21st centuries, repeat past urban-planning mistakes spread in Europe in the second half of the 20th century. Scientifically

* Corresponding author: arch_project@bk.ru

substantiated planned reconstruction of fragments of historical territories is aimed at improving the comfort of living while retaining unique features of these fragments [2].

Almost all European capitals experienced adverse effects of high-rise construction within visual borders of historical cities. For example, the standard Parisian development, formed by rental houses in the late 19th century, began to lose its architectural and urban-planning unique character starting from the end of the 1950s in the forefront of keen enthusiasm for modern geometrized architecture "from glass and concrete". Construction of high-rise buildings also was qualified as a "modernity" feature. For example, in the late 1960s, the competition of projects for the reconstruction of the central market of Les Halles ("The Belly of Paris") represented, among others, proposals based on the "enrichment" of the historical environment with high-rise dominants [1]. For example, J. Faugeron proposed to locate five towers in the historical center, height of which would reach forty–sixty floors. Thus, the historical development was doomed to serve as a background for new high-rise buildings. The radical project of Faugeron was not approved, but, nevertheless, in 1969, the construction of the 150-meter Montparnasse tower started; it became the first dominant violating the historical architectural landscape (Figure 1).

Fig. 1. Montparnasse tower in Paris

Two years earlier in Leningrad, first high-rise facilities, i.e. Sovietskaya and Leningrad hotels, were built by the 50th anniversary of the Great October Socialist Revolution [2]; with their rectangular forms, they were the first to violate the skyline of the city's historical part (Figure 2). The construction of these facilities laid the foundation for high-rise construction both in the historical urban nucleus and in close proximity to the historical center.

Fig. 2. Sovietskaya hotel in the Fontanka River panorama

The purpose of this study is the landscape visual analysis of the interaction of high-rise construction facilities with the environment of historical urban spaces of Saint Petersburg. For this purpose, volumetric and spatial features of characteristic morphological units of the city were identified and connections between new high-rise construction facilities and the established urban landscape were analyzed.

2 Materials and methods

Central districts of Saint Petersburg represent the complex unity of the architectural and planning structure created over the period of 18th–20th centuries. It is characterized by the historical and cultural continuity, which determines the effective development of urban processes, allowing perceiving the urban planning experience and reproducing it in the light of the complication of the overall structure, functional contents and social processes [3]. Due to the established social and economic conditions, Saint Petersburg for a long time remained a city retaining its volumetric and spatial identity and its own urban-planning pattern code [4]. To carry out the analysis of the development of the system of high-rise dominants in the historical center of the city, the territory along banks of the Neva water area was determined as the most significant within the volumetric and spatial layout of the city. Urban and natural landscapes of the urban areas along embankments of the Neva, Bolshaya Nevka and Malaya Neva rivers, forming a part of the historical urban nucleus of Saint Petersburg, were studied. Spatial characteristics of this territory have the optimal balance of town-forming, local high-rise dominants with the mass of the background development and landscape, which have developed gradually in accordance with historical stages of urban planning [5]. The natural component is determined by the interaction of water spaces and green lands (parks, garden squares, squares). The Neva River serves as the center of landscape composition in this part of the city. The coherent architectural landscape of the city developed during three centuries due to the strict urban-planning discipline. Even in Petrine times, numerous imperial decrees and orders determining height restrictions, were in force in the new capital. City planners paid special attention to the silhouette of Saint Petersburg throughout several centuries; balance between the standard development and accents — urban dominants — was though through [6].

Central districts of Saint Petersburg represent the complex unity of the architectural and planning structure created over the period of 18th–20th centuries. The unique character of the city silhouette component in the layout spatial system of the city is determined by [7]:

- landscape: flat relief of the lowland area near the Neva river;
- face of the Neva river: the main composition axis of the city, forming architectural ensembles of the city center;
- compositional separation of the central and peripheral parts of the city, differing in spatial-planning qualities, including the distribution of dominants of various levels.

The historical skyline of Saint Petersburg is included into the list of 100 World Heritage sites [8] in danger; this is a way to attract the attention of the global community to problems of Saint Petersburg as a World Heritage site. Studies of the historical and urban-planning evaluation of the urban center sites and carrying out the qualitative assessment of the high-rise environment of selected areas in the city center served as the reference material for the development of the landscape visual analysis of the modern state of the altitude zoning system of the city's historical center [9]. Principles of the altitude arrangement of the territory under consideration, as well as connections of its zones and separate objects with the urban landscape, urban-forming elements of the city's central part, as well as dominants of regional and local significance were recorded and analyzed; spaces under construction representing valuable elements of the historical environment or introducing dissonance into the structure arrangement of the territory under consideration were revealed.

Thus, this study is based on revealed specific features of the modern state of urban landscapes within the boundaries of the considered territory.

3 Results

The analysis of the collected material made it possible to carry out the qualitative assessment of the Saint Petersburg center environment in order to reveal the system for the altitude zoning of its territories. As a result of the research, authors identified main types of urban-planning mistakes destroying the volumetric and spatial integrity of the city and the identity of the historical environment. The urban-planning mistake is the distortion of the historically established stable environment or unique compositional solution. Thus, such interference in the historical environment, which changes specific features and, due to new high-rise construction, levels down the role of historical dominants, can be considered as an urban-planning mistake.

In accordance with this definition, the typology of urban-planning mistakes is the following:

- Location of a new object within the visual area of a unique ensemble.
- Changes in the historical city silhouette.
- Scaling-up of new dominants relative to the scale of the historical development.
- Replication of a unique technique in new high-rise construction, leveling down the role of historical specific features or dominants.

The location of a new object in the visual area of an ensemble is often observed in historical parts of cities. In Saint Petersburg, cases of architectural and urban-planning violations, related to the introduction of high-rise dominants into the environment of historical river panoramas, are observed [10]. Here, results of the unscientific approach to solving the complex problem of combination of "new" and "old" are much in evidence. A typical example is the appearance of the Lakhta Center building within the historical river panorama of the Peter and Paul Fortress [11]. The building is located at a considerable distance from the historical part of the city, and, due to its height (462 m), it can be viewed from various points of the city center (Figure 3).

Fig. 3. Lakhta Center Building in the historical panorama of the Peter and Paul Fortress

Long before the beginning of skyscraper construction, experts developed a model according to which the skyscraper should be viewed beyond the Peter and Paul Cathedral, if an observer is on the Trinity Bridge; from this point it would be slightly smaller in height than the Grand Ducal Burial Vault. When viewed from the Palace Embankment, the skyscraper would be almost the same height as the northern rostral column [12]. No one respected the experts' opinion in 2011, and the Lakhta Center is built in 9 km from the Peter and Paul Fortress. It is already obvious that almost the entire Senate Square, a third of the Palace Square, the main part of the Palace Embankment, one third of the Peter and Paul Fortress, the whole Makarov Embankment, and Tuchkov Bridge got into its area of influenceAn example of changes in the historical silhouette with high-rise development building is shown in Figure 4. The silhouette of the classic portico of the Stock Exchange building facade at the Spit of Vasilyevsky Island is replaced by the rough outline of the administrative building built in the same visual corridor as the historical ensemble.

Fig. 4. Spit of the Vasilyevsky Island with the silhouette of the high-rise development building

The scaling-up of new dominants relative to the scale of the historical development negatively affects the volumetric and spatial integrity and characteristics of the identity of historical buildings. Smooth facades of new structures with almost no details conflict with small plastic arts of historical buildings, reducing their significance.

A typical example of such approach is the Mont Blanc residential complex, built in 2003–2008 at the Pirogovskaya embankment. A large-scale object exceeding the surrounding historical background development in height has a “spreading” silhouette, which is untypical for historical Saint Petersburg (Figure 5). Invading the panorama of the Neva embankment, it suppresses the historical silhouette line.

Fig. 5. Mont Blanc residential complex in the historical panorama of the Pirogovskaya embankment

Another example is the Aurora luxury residential complex (2005) near the building of the former School House of Peter the Great (Figure 6). The historical object is visually suppressed by large-scale forms of the high-rise structure located on the opposite bank of the Neva river.

Fig. 6. Aurora residential complex

Replication of a unique technique in new high-rise construction is presented in the architecture of the Aurora residential house. It is a technique, widespread in the modern practice, allowing to join new and old by repeating some forms of a "historical neighbor" in a new building [13]. However, this technique leads to the opposite result: the architectural significance of the historical object is reduced. The crowning tower of this large high-rise building clearly reminds the cupola-spire top of the School House (Figure 7). This large-scale "look-alike" makes the low historical building to seem even lower and the role of one of the dominants of the Neva panorama is reduced practically to zero [1].

Fig. 7. School House building

4 Discussion

The Saint Petersburg silhouette, or skyline, is unique. Silhouette lines of the Neva embankment, small rivers and canals, historical streets and squares are one-of-a-kind and distinctive, forming the image of the city [14]. The uniqueness of central districts of Saint Petersburg is based on the contrast between the city silhouette dominants and the contour line of the established residential background development, within which a unique hierarchy of city-wide and local specific features is established.

Studies of the historical and urban-planning evaluation of the urban center sites and carrying out the qualitative assessment of the high-rise environment of selected areas in the city center served as the reference material for the development of the landscape visual analysis of the modern state of the altitude zoning system of the historical urban nucleus [9]. Principles of the altitude arrangement of the territory under consideration, as well as connections of its zones and separate objects with the urban landscape, urban-forming elements of the city's central part, as well as dominants of regional and local significance were recorded and analyzed [15]. Thus, this study is based on revealed specific features of the modern state of urban landscapes within the boundaries of the considered territory.

These reference materials allowed working out a differentiated approach to the study of volumetric and spatial features of forming high-rise dominants of the historical center of Saint Petersburg.

The study allowed making the following conclusions:

- all types of high-rise construction within the historical center should not affect the perception of town-forming ensembles and unique panoramas formed by the silhouette of the background development;
- high-rise construction buildings should not result in the replacement of historically formed morphotypes of the development with newly developed ones;
- upon the construction of high-rise buildings, not only the retaining of the city's spatial arrangement as a whole should be taken into account, but also the retaining of conditions for the perception of internal spaces of historical streets.

The practical application of the study is determined by its use in the following directions:

- development of historical and urban-planning, as well as historical and architectural expert examinations of urban territories;
- activities of state heritage protection bodies, ministries and departments which control and regulate the construction activity in the territory of cities, municipal districts and regions.

Central districts of Saint Petersburg represent the complex unity of the architectural and planning structure created over the period of 18th–20th centuries. Its creation is characterized by the historical and cultural continuity, which determines the effective development of urban processes, allowing perceiving the urban planning experience and reproducing it in the light of the complication of the overall structure, functional contents and social processes.

5 Conclusions

The conducted study allowed making the following conclusions:

1. Historical patterns of the development of the high-rise dominants system in Saint-Petersburg were formed in accordance with the principle of historical continuity of the city's spatial-planning structure.
2. The high-rise arrangement of the historical center of the city was formed on the basis of natural conditions typical for Saint Petersburg.

3. The carried-out analysis showed that the arrangement of city's historical dominants is naturally connected with the Neva waterline, towards which city's festive panoramas are opening.

4. One of specific features of high-rise construction in Saint Petersburg is the fact that the highest buildings (except for the 462-meter tower of the Lakhta Center) are residential complexes, unlike high-rise buildings of functional use in other countries of the world.

5. Connections between historical dominants and objects of new high-rise construction located in overlapped air corridors are revealed.

6. Further designing of high-rise construction objects shall rely on the principle of preserving territories of the stable historical part, which implies the elimination of differences between the historical environment and modern environment, conditioned by common regularities of formation, similar proportions and, mainly, a single scale.

The authors determined future directions for studies in this field. It seems necessary to conduct the comprehensive analysis of the influence of visual landmarks — new high-rise construction objects built outside the historical center, facing directly the Neva river, as well as appearing deep in the territory and sealing the city. A hypothesis about the unfavorable influence of new and to-be-built high-rise construction objects in the formation of the image of modern Saint Petersburg and its facades in the western and eastern parts is offered.

Table 1. Influence of high-rise objects on the historical city environment (through the example of Saint Petersburg)

	Influence on the historical city			
	Distortion of the historical silhouette	Changes in the role of historical dominants	Scaling-up	Replication of techniques typical for the historical environment
1. Local dominants of 1960–1980s (Sovietskaya and Leningrad hotels)	●	●	●	
2. Residential buildings and business centers constructed in 1990–2000		●	●	●
3. Residential buildings of 2000–2017, introduced in main river panoramas of the city	●	●	●	
4. Lakhta Center (construction completion — 2018, building height — 462 m), located at the distance of about 10 km from the historical urban nucleus	●	●		

References

1. M. A. Granstrem. Are scientific methods for the reconstruction of fragments of historical cities real? Architectural Almanac, **1**, 71–78 (2016).
2. S. V. Sementsov. Stages of urban development of Saint Petersburg and typology of

master plans of 18th–20th centuries. *Reconstruction of Saint Petersburg 2003*, 93–98 (2003).

3. M. V. Zolotareva. Le Vieux Pétersbourg — L'animation d'une zone protégée. Monuments Historiques, **179**, 87–88 (1992).
4. S. P. Zavarikhin. About the silhouette and not just that. Kapitel, **1**, 34–37 (2012).
5. J. Kurbatov, V. Gorunov. The fate of the creative legacy in modern architecture in Russia. Bulletin of Civil Engineers, **23 (13)**, 203–206 (2013).
6. M. V. Zolotareva. New trends of urban development in Russia in the 18th century. Journal of Architecture and Urbanism, **41 (1)**, 71–77 (2017).
7. M. V. Zolotareva. Identification of historical patterns in the development of the high-rise dominants system in the central part of Saint Petersburg (based on works of artists and graphic artists of the 18th–early 19th centuries). *Modern science: theoretical and practical views*, 105–109 (2014).
8. M. A. Granstrem, M. V. Zolotareva. Analysis of the structure of the historical development in Saint Petersburg. Zhilishchnoe Stroitelstvo, **11**, 23–26 (2014).
9. R. T. Clement. A history of Russian architecture. Library Journal, **129: 12**, 77 (2004).
10. M. V. Zolotareva On the issue of regulation of architectural and construction process in city management structure (second half of the xix – early xx centuries). Architecture and Engineering, 1(3), 41-46 (2016). DOI: https://doi.org/10.23968/2500-0055-2016-1-3-41-46
11. Yu. I. Kurbatov. Petrograd. Leningrad. Saint Petersburg: lessons of architecture and urban-planning. (2008).
12. URL: http://kanoner.com/2011/03/14/21565/
13. S. P. Zavarikhin. Modern construction in the historical center of Saint Petersburg. *Modern Problems of History and Theory of Architecture*, 115–122 (2015).
14. D. Shvidkovsky, E. Shorban, A. Wood. Russian architecture and the West, Yale: Yale University Press (2007).
15. M. V. Zolotareva. Stages of the Saint Petersburg center formation in works of artists of the 18th–19th centuries. Bulletin of Civil Engineers, **5 (52)**, 30–39 (2015).

Structural and compositional features of high-rise buildings: experimental design in Yekaterinburg

Yulia Yankovskaya[1*], *Yuriy* Lobanov[1], and *Vladimir* Temnov[1]

[1]Saint Petersburg State University of Architecture and Civil Engineering, Department of Urban Planning, Vtoraya Krasnoarmeyskaya str. 4, Saint Petersburg, 190005, Russia

Abstract. The study looks at the specifics of high-rise development in Yekaterinburg. High-rise buildings are considered in the context of their historical development, structural features, compositional and imaginative design techniques. Experience of Yekaterinburg architects in experimental design is considered and analyzed. Main issues and prospects of high-rise development within the Yekaterinburg structure are studied. The most interesting and significant conceptual approaches to the structural and compositional arrangement of high-rise buildings are discussed.

1 Introduction

Yekaterinburg, as most of large and largest cities of Russia, transforms and acquires its new image and silhouette due to gradual high-rise development. This aspect requires timely analysis. Moreover, prospects of architectural development of the capital of the Urals shall be defined. Historically, Yekaterinburg was one of the leaders for the implementation of advanced architectural concepts developed by constructivist architects and the harbinger of avant-garde architecture in 1920–1930. Progressive trends related to intensified construction of high-rise and super-high-rise buildings, which penetrated many cities of Europe and Asia in the late 20th – early 21st centuries, affected Yekaterinburg as well. Switching of Russia to the market economy also contributed to the intensification of this process. High-rise buildings appeared in Yekaterinburg a little bit later and became a determining factor for the city development and silhouette only in the 2000s.

High-rise construction projects and the emergence of super-high-rise buildings in the 2000s posed a number of challenging issues craving resolution, i.e. new ways for the optimum introduction of high-rise buildings into the existing structure of the historical city located in unfavorable climatic conditions of the Urals region.

Historical aspects, trends and prospects of high-rise development in Yekaterinburg, conceptual issues of the formation of high-rise buildings meeting modern ideas of the comfortable architectural environment.

[1] Corresponding author: jul3203226@gmail.com

2 Materials and methods

The methodological basis of the study is a set of historical, typological, compositional-artistic approaches. To identify the origins of the formation of high-rise dominants in the city structure, the historical-genetic approach was applied. To assess existing and designed objects, the simplified typological method was used, based on two major classifications: by height and complexity of the building structure. Analysis of the role of high-rise buildings in the city structure in various historic periods and evaluation of promising trends of their formation are carried out on the basis of the compositional-artistic approach. The author conducted on-site investigations of high-rise buildings in Yekaterinburg, both constructed and under construction. The author also acted as a supervisor of research works, competitive and experimental design of high-rise buildings at the Department of Architecture [1, 2, 3, 4] of the Ural State University of Architecture and Art.

Main architectural and typological classifications used in the study are as follows: classification by height and complexity of the building structure. A summarized classification of buildings by height (with account for the interaction of the latest national and international theory and practice) is as follows [5, 6]:

- low-altitude (low-rise) buildings — up to 10 m (1–3 floors);
- medium-altitude (multi-story) buildings — from 10 to 30 m (4–9 floors);
- buildings of excess height — from 30 to 75 m (10–24 floors);
- high-rise buildings — from 75 to 200 m;
- super-high-rise buildings — more than 200 m.

Super-high-rise development can be classified by six categories:

Category 1: buildings and structures from 200 to 400 m (Aqua, Chicago)

Category 2: buildings from 400 to 600 m (Okhta Center in Saint Petersburg, design)

Category 3: buildings from 600 to 700 m (Chicago Spire, design)

Category 4: buildings from 700 to 800 m (Anara Tower, design)

Category 5: buildings from 800 to 900 m (Burj Khalifa in Dubai);

Category 6: buildings from 900 to 1000 m and more (Kingdom Tower in Saudi Arabia).

Based on this classification, we can get a good idea of the significance of an object in the high-rise development system.

Another classification used in the study describes the degree of simplicity and complexity of the structural solution in a plan and the sectional view of a building. This classification is based on the analysis of global experience in high-rise construction and national typological developments [7, 8, 9]. The classification is based on the following variables:

- rigidity/flexibility of space planning (free plan / rigid cell system / combined system / atrium system); presence/absence of central prominent structural/communicational nodes;
- penetrability / limited penetrability of vertical enclosing structures / self-sufficiency of the building enclosure;
- rigid vertical floor-by-floor separation / presence of vertical atriums.

The following summarized classification of structural features of vertically developed buildings can be presented based on those variables:

- medium-altitude buildings with a rigid planning structure;
- buildings of excess height and a rigid or flexible planning structure;
- high-rise buildings with a uniformly distributed vertical floor-by-floor structure ("buildings-stack-like structures") with a free or combined planning structure;
- high-rise buildings with vertical atriums and a free or combined planning structure;
- high-rise buildings with vertical atriums (vertical-block buildings), a free or combined planning structure and a relatively autonomous enclosure — "building – vertical city";
- super-high-rise buildings-stack-like structures with a combined planning structure;
- super high-rise vertical-block buildings with a combined planning structure.

3 Results

3.1 Historical and architectural aspect of high-rise development in Yekaterinburg

Stage 1 ("pre-revolutionary" stage) — the first half of the 18th century – the beginning of the 20th century. In general, this period is characterized by low-rise individual development. Plant management buildings and temples served as compositional dominants. Only church bell towers of those days can be classified as objects of "high-rise development". The most significant high-rise dominants of that period are as follows: the Cathedral of St. Catherine the Great Martyr (1723, height = 58 m, architect Johann Miller); Church of the Ascension (1801, height = 63.4 m, architects: M. P. Malakhov, K. G. Tursky, N. Shulayev); the Great Zlatoust Church (1876, height = 77.2 m, architect: V. E. Morgan). Such large-scale buildings as the Alexander Nevsky Cathedral (1848) and the Holy Trinity Cathedral (1839) are worth mentioning as well [10]. A complex of high-rise architectural dominants, combined by a system of visual connections, formed in the city during that period. Dominants located on the same visual axis formed systems of panoramic and spatial complexes determining silhouette characteristics of Yekaterinburg and the system of subordinated architectural area of residential and industrial buildings. Dominants of temples formed the basis for city silhouette expressiveness.

Stage 2 ("post-revolutionary" stage). 1920s–1930s. It was the period of industrialization and constructivism in the city architecture. In 1924, Yekaterinburg was renamed Sverdlovsk and acquired a new altitude range. According to the classification, in general, those were medium-altitude buildings and buildings of excess height. Constructivist buildings and complexes appeared, which largely determined the modern architectural and historical environment and cultural significance of the city: the Uralmash socialist city, Vtuzgorodok, OMM ensemble, residential complexes on Malysheva street and Lenina street, the Dinamo stadium, and a unique complex — the Chekist Town with a high-rise building of the Iset hotel (1933, height = 40m, architects: I. P. Antonov, A. M. Tumbasov, V. D. Sokolov). The aesthetic nature of the city changed. A new altitude range of Yekaterinburg–Sverdlovsk was set. A significant number of 4–6-story buildings and 8–11-story dominants appeared, such as the Central Post Office, Unipromed Institute (at the corner of Lenina street and Tolmacheva street), NKVD House (1932, height = 34m, architects: I. P. Antonov, V. D. Sokolov) [10]. Most temples determining the silhouette of the pre-revolutionary city were demolished during that period. New constructivist dominants became high-rise landmarks. A significant project of a 140-m skyscraper — a high-rise dominant in the building complex "House of Industry" — also related to the period of industrialization. Its construction started in 1931, and in 1935 it was canceled.

Stage 3 ("late Soviet" stage) — the second half of the 20th century. Only several buildings of excess height were built in Sverdlovsk in the 1940s–1960s. The Building of the Sverdlovsk City Council of People's Deputies (1930–1954, height = 61 m, architects: G. A. Golubev, M. V. Reischer), and 10-story building of the Rubin plant (1964, architect: P. D. Dyomintsev) were reconstructed [10]. In the mid-1960s, the large-scale construction of 9-story brick (and later — panel) buildings of standard series started in Sverdlovsk. In the 1970s–1990s, a television tower (1991, height = 220 m, designed by the State Union Designing Institute) became the main vertical dominant of the city. Standard 16-story buildings started to define the high-rise appearance of residential areas. Tall buildings of the constructivism period and the surviving Church of Ascension remain the dominants in that period. Three objects continued to dominate in the center of the city: the building of the Rubin plant (architect: P. Dyomintsev), grand building of the City Executive Committee with a spire (architects: G. Golubev, M. V. Reischer) and NKVD House (architects: I.

Antonov and V. Sokolov). In 1981, in the southern part of Sverdlovsk, near the bus terminal, an impressive building of the computing center, designed by architect A. Astashkin, was constructed. In the late 1970s, the construction of high-rise buildings began in the city. In 1975–1981, a 23-story high-rise building of the House of the Soviets (height = 89 m) was constructed at the initiative of Boris Yeltsin, the first secretary of the Sverdlovsk Regional Committee of the Communist Party of the Soviet Union at that period. For two decades, this building was the highest object in the city. The overall height of the large-scale urban development significantly increased during that period. However, the high-rise dominants remained within the limits set by pre-revolutionary church bell towers.

Stage 4 ("post-soviet") — the first decades of the 21st century. Generally, only this period is associated with the large-scale construction of high-rise buildings. New economic relations and changes in the social format of Russia became the basis for numerous social and cultural changes, including the widespread use of high-rise buildings in the structure of previously established cities. During that period, Yekaterinburg (renamed back in 1991) acquired particular logistical, financial and administrative, banking and business characteristics largely due to its successful location in the center of the country, at the intersection of transport and trade routes, and shifted away from its original function of the industrial city. Those trends have manifested in a new altitude range from the beginning of the 21st century to the present day. Nowadays, there are 850 buildings with the height exceeding 35 m, 40 of which have 25 floors (more than 75 m), 4 of which have 32 floors (more than 100 m) and 2 of which have the height of 200 m; Yekaterinburg ranks No. 7 in Europe and No. 3 in Russia with regard to the index of high-rise development, and in the absolute values of the altitude range in the Russian Federation it gives place to Moscow (according to the data of the Non-profit Partnership Self-regulatory Organization "Ural Builders Union" as of 2014) [11]. In 2011, a 50-story building Vysotsky (height = 188.3 m, architect: V. Grachev) was erected in Yekaterinburg, breaking world records with regard to the height of buildings [12]; it headed towards a Guinness World Record as the highest building in the world to the north of 56 degrees north latitude; in 2016, the construction of the 52-story Iset building tower was completed; it is located in the high-rise City district (height = 212.8 at the level of crown spikes, architect: Werner Sobek, LLC Dominanta) [13, 14, 15]. During the first decades of the 21st century, a number of high-rise buildings were constructed in Yekaterinburg: they distinguished by innovative architecture, unique compositional and imaginative, as well as structural solutions situated within the historically developed environment. Introduction of high-rise development, in particular, into the central (historical) part of the city, still provokes an active discussion among professional architects. It is discussed whether it is worthwhile to introduce high-rise development into the structure of Yekaterinburg or switch to the traditional architecture. In 2000–2010, trends in high-rise development, using the compositional and imaginative ideas of the Urals constructivism, emerged in Yekaterinburg. Among numerous designs and implemented solutions of that period, the following can be noted: the Summit business center in the circus area (height = 93.5 m, architect: B. Demidov, 2011), administrative building beyond the railway station (architect: V. Barantsev), high-rise retail and office building Antey-1 (height = 76 m, architect: A. Molokov, 2004), February Revolution residential complex (height = 139.6, architects: I. K. Ivanov, V. M. Ilyinykh , A. S. Bashkirov, 2010), Demidov Business House (height = 134 m (crown), designed by Arkhstroyproekt-A, 2016), partially erected complex of high-rise buildings "Yekaterinburg City", and high-rise building of the Prisma shopping mall (height = 151, architect: Chernyshev, to be completed in 2018–2019) [6]. High-rise complexes transforming and modernizing the image of the city, such as 24-story residential buildings on Radishcheva street and a complex of 5 high-rise buildings in Khimmash (Admiralsky residential complex) were built in peripheral areas and areas distant from the center of the city. If to

 https://doi.org/10.1051/e3sconf/20183301029

define the main direction of the high-rise development of that period, it may be said that it is in line with architectural and urban-planning innovations of Europe and the United States.

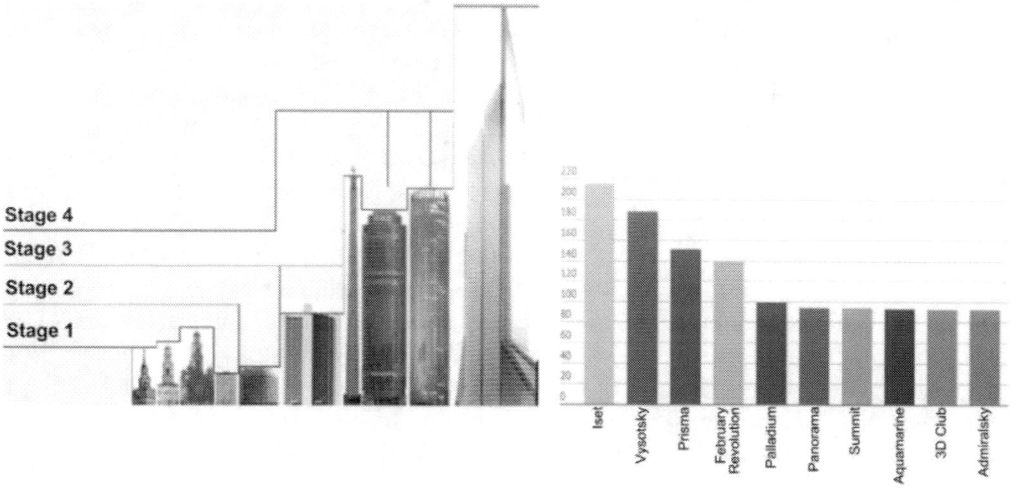

Figure 1. Historic stages of high-rise development. Altitude range diagram [16] for 10 highest modern high-rise buildings in Yekaterinburg

Taking into account the history of high-rise development in Yekaterinburg, it can be concluded that it is important to introduce buildings of excess height (rather than high-rise buildings) into the city structure; it is unlikely that super-high-rise buildings will appear in the city in the future, as they do not correspond to the large-scale structure of Yekaterinburg (as one of the most compact cities with a million-plus population in Russia) and fall out of the context of the historical part of the city (Figure 1).

Fig. 2. Two highest skyscrapers of Yekaterinburg: the Iset Tower residential complex, overall view (left), Vysotsky business center, overall view (right) [13, 8]

At the present time, mainly buildings of excess height with a rigid or flexible planning structure, as well as high-rise buildings-stack-like structures with a free or combined planning structure are erected in the city; simple vertical block atriums to arrange a part of communications spaces are used less frequently (Figure 2). Complex spatial structures with alternating vertical and horizontal atriums and autonomous enclosures are still absent in the high-rise development of Yekaterinburg. In general, complex spatial and typological structures are represented in experimental and competitive design. It can be assumed that, in the future, the high-rise development in Yekaterinburg will not head towards records with regard to the height of buildings to the north of 56 degrees north latitude, but will follow the direction of the establishment of innovative techniques for the structural organization of high-rise buildings and humanization of the internal structure of high-rise buildings.

3.2 Promising trends in high-rise development regulation in Yekaterinburg

Main directions for high-rise development regulation in Yekaterinburg were determined by the architectural community of the city based on the results of assessing the existing high-rise buildings in Yekaterinburg, public and professional discussions and analysis of construction experience in Moscow and Saint Petersburg, as well as several European cities. Let us enlist those directions.

Fig. 3. Panoramas of the center of Yekaterinburg. View of the city pond (left) and dam (right) [17]

Quantitative restrictions of high-rise development. The architectural and urban-planning practice of regulating high-rise development in Moscow gives a positive example for the introduction of temporary regulations for the limitation of high-rise development, especially in the center and historically protected area (Figure 3). The number and location of newly constructed high-rise buildings should be determined by a special plan for high-rise development, approved by the city administration.

Redistribution of high-rise buildings in the city structure. The relatively high density of high-rise buildings in the center of Yekaterinburg leads to the conclusion that it is necessary to introduce certain numerical indicators limiting the introduction of high-rise buildings into the current environment, and, possibly, optimize the **design of high-rise ensembles and complexes in the periphery areas** that have rather good connections to the center of the city with highways.

Limitation of high-rise residential development. Large-scale construction of high-rise residential complexes is not practical as Yekaterinburg has vast undeveloped territories. One of the aspects of high-rise development in Yekaterinburg is related to the overestimated land value and the desire of developers to save expenditures on landscaping and utilities, not taking into account the cost of facility operation. The second aspect is related to the fact that high-rise residential development is treated as a sort of experiment for the establishment of a new type of the comfortable living environment.

Brand function of high-rise buildings and brand positioning of the city. The obvious competition of large and largest cities with regard to high-rise development produces a certain advertising effect and can affect the economic attractiveness and tourism appeal of a non-capital city. Such effect in Yekaterinburg was achieved with the help of such high-rise objects as the Vysotsky building and Iset tower.

3.3 Creative concepts for the development of high-rise architecture in Yekaterinburg. Experimental design

Natural, climatic and environmental peculiarities of the region impose certain limitations on approaches to design. Yekaterinburg is located in the area of moderate continental climate. It is characterized by variable weather conditions and well-marked seasons of the year. Environmental conditions are unfavorable due to the large number of heavy-industry enterprises in the territory of the Urals region. The level of pollution with industrial wastes is high.

Based on environmental, natural and climatic peculiarities of the region, main tasks of designing high-rise buildings are as follows: environmental protection and compensation for adverse environmental effects; maximum use of favorable environmental factors; compactness and wind protection; establishment of optimum micro-climate inside buildings; introduction of natural components into the structure of buildings. Based on the international [18, 19] and national experience in high-rise development and considering the development prospects in Yekaterinburg, we shall mention a number of creative concepts describing prospects for the possible development of high-rise architecture, as well as structural and compositional features of high-rise buildings.

The **"green architecture" concept** of high-rise development allows taking account of local natural and climatic peculiarities, maximum use of favorable factors, optimum energy balance of the building, and introduction of the "green component" into the structure of the building. This concept involves the use of specific spatial-planning techniques, as well as methods for integrating the natural component into the structure of the building, i.e. improvement of energy-saving characteristics of the building at the expense of its shape and orientation; methods of integrating the natural component into the structure of the building (integration of the building into the natural/man-made landscape, application of "green" enclosures and "green" vertical (with necessary fire spacing) and horizontal atriums, block-garden systems for 3–4 floors, winter floor-by-floor gardens, landscaping of communications spaces, vertical landscaping of surfaces, "green" roofs, imitation of living nature, use of geo-plastics to form the underlying "basic" element of a high-rise building) (Figure 4) [4].

Fig. 4. Experimental designs of high-rise residential buildings with "green" atriums for Yekaterinburg. Projects of A. Shilkova (top) and D. Yarovenko (bottom), supervisor: A. V. Merenkov

"Building–enclosure" concept. The idea of this concept is to form a relatively autonomous internal spatial-planning structure and an external "enclosure" facade. Separation of the enclosure and internal structure is appropriate for modern high-rise buildings. The design principle is based on the development of an "environmental skin of the building" — an enclosure covering all interior spaces of the building (Figure 5) [4].

Fig. 5. Experimental design of a high-rise residential eco-building. Project of A. Grigoryeva, supervisor: A. V. Merenkov

Concept of the "architecture of information space". Its main characteristics are the following: interactivity, virtuality and adaptability. Among the methods of implementation, the following can be noted: shifting of module elements of the form in several planes; overlapping of volumetric modules and enclosures; displacements in planes; turning and changing of module proportions; deformation. Possible directions of dynamic transformations are horizontal dynamic structure of the building stylobate; centric vertical dynamic structure of the main volume; heterogeneous dynamic structure of the whole complex (Figure 6).

Fig. 6. Experimental design of a multi-functional complex. Project of P. Bartov, supervisor: A. A. Raevsky

Concept of "graffiti architecture". It is a modern direction in architecture, based on the use of graffiti in design. The following principles of graffiti introduction in architecture can be noted: packaging principle, veiling principle, screen (shield) principle, text-building principle.

Examples of the transfer of planar graffiti to architecture: "classic" planar graffiti; planar graffiti with volumetric elements; photo-realistic image of the volume of a graffiti composition, "soaring" in the air; abstract geometric compositions performed in the technique of collage with three-dimensional elements in the base; graffiti sculpture; scaleless designs of architectural compositions; human-scaled urban sculpture; conceptual designs of architectural objects (Figure 7).

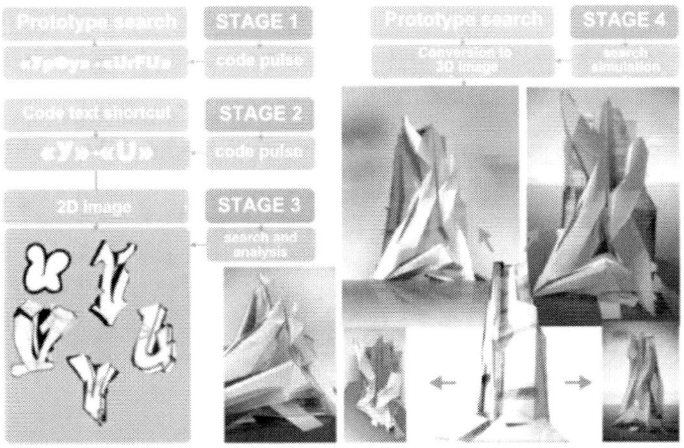

Fig. 7. Methods of shape formation in graffiti architecture. Proposed by Ye. Yevdokimova, supervisor: Yu. S. Yankovskaya

The concept of the "brand building" is especially relevant for high-rise architecture. Based on this concept, a brand building shall have the following general characteristics: awareness; uniqueness; adequate positioning (in terms of geographic location, adjacent territories, population mentality, quality of life, and time); clear and memorable image (Figure 8).

Fig. 8. Experimental design of a multi-functional high-rise business and hotel complex. Project of V. Blinov, supervisor: S. A. Dekterev

4 Discussion

The following should be noted among the obvious problems at the current stage of high-rise development in Yekaterinburg: debating issues of high-rise development arrangement in historically valuable areas of the city; lack of appropriate high-rise development regulations; low relevancy of high-rise residential development; preferences for high-rise complexes in the peripheral territories of the compact city.

At the present time, a variety of problems associated with the arrangement of high-rise buildings in the historical center and violation of the perception integrity with regard to the architectural appearance of the city need to be resolved. Let us provide the most vivid and controversial examples:

- high-rise development in the panorama of the water area of the city pond. From the city establishment until the end of the 20th century, the city pond was perceived as a significant water area. Meanwhile, masterpieces of constructivism (UNESCO monuments) — a park of the Dinamo stadium and a "ship-house" viewed from the embankment, provided the entire ensemble with a memorable view and ensured the recognition of the city panorama. At the present time, standard high-rise development prevails in this panorama, destroying the uniqueness of the architectural picture (Figure 9).

Fig. 9. Day and night views of the city pond panorama

- the building of the City Executive Committee and City Council was originally intended as a dominant on the central square of the city. The positioning of the new Palladium business center destroyed the idea of an impressive city dominant, while numerous high-rise objects completely negated the role of the building in the panorama of the city (Figure 10);

Fig. 10. Building of the City Council on the central square (left), in sight of Lenina Street (center), in the panorama of Yekaterinburg (right)

- the Vysotsky business center building in the historical center of Yekaterinburg (Figure 11)

Fig. 11. Active silhouette of the Vysotsky business center in the historical center of the city

Obvious prospects of the current stage of high-rise development in Yekaterinburg include the following: creativity of young architects, a significant role of high-rise development in the city positioning, focus on the search for new structural, compositional and imaginative solutions for modern high-rise buildings within the framework of experimental and competitive design.

5 Conclusions

1. The article provides the summarized classification of high-rise buildings by two basic characteristics: by height and nature of the building structure.
2. Four main historic stages of high-rise development in Yekaterinburg are marked out. High-altitude characteristics of large-scale development changed with time up to the end of the 20th century, while the main historically developed high-rise dominants maintained their role. Activization of high-rise development occurred at the beginning of the 21st century and resulted in a fundamental change in the city appearance, its silhouette characteristics and panoramas. Buildings of excess height prevail at the city periphery, while high-rise development prevails in the center. At the same time, as a rule, most of the constructed high-rise buildings do not have a complex spatial structure. Architectural and spatial, typological and engineering experiments are presented mainly in the field of experimental and competitive design.
3. Complex natural and climatic peculiarities, on the one hand, limit architectural experiments, and, on the other hand, give it a new interpretation within the search for innovative architectural solutions in experimental and competitive design.
4. The application of the global experience in high-rise and super-high-rise development in Yekaterinburg revealed both promising and negative aspects in the formation of a modern urban environment. In particular, this is the case of the compatibility of high-rise dominants and traditional, historically significant objects, perspectives and panoramic views.
5. Prospective directions of high-rise development in Yekaterinburg were defined: introduction of regulations for high-rise development in the city center; activization of high-rise development in the peripheral areas; the search for brand high-rise architecture requires the development of new forms and structures based on the specifics of the region and experience in experimental and competitive design.

References

1. http://www.usaaa.ru/faculties/fa/issuing/arhitektury/vysotnoe-zdani 12.12.2017
2. http://www.usaaa.ru/faculties/fa/issuing/arhitektury/mfzhk 12.12.2017
3. http://www.usaaa.ru/faculties/fa/issuing/arhitektury/diplom 12.12.2017
4. http://www.usaaa.ru/faculties/fa/issuing 12.12.2017
5. E. M Generalova, V. P Generalov, A. A. Kuznetsova. Cantilever Structure in Modern Construction. MATEC Web of Conferences. Editors: S. Jemioło, A. Zbiciak, M. Mitew-Czajewska, M. Krzemiński and M. Gajewski. 00057 (2017)
6. A. Radford, S. B. Moroc, A. Strivastava. The Elements of Modern Architecture. Understanding Contemporary Buildings, L.: Thames & Hadson (2014)
7. E. M Generalova, V. P Generalov, A. A. Kuznetsova. Innovative Solutions for Building Envelopes of Bioclimatical High-rise Buildings. *Environment. Technology. Resources Proceedings of the 11th International Scientific and Practical Conference.* 103–108 (2017)
8. C. G. Crysler, St. Cairns, H. Heynen (ed.). *The SAGE Handbook of Architectural Theory*. L.: SAGE Publication Ltd, 41–56 (2012)
9. A. Yudina. *Garden City. Supergreen Buildings, Urban Skyscapes and the New Planted Space*, L.: Thames & Hadson (2017)

10. https://ru.wikipedia.org/wiki/Список_самых_высоких_зданий_Екатеринбурга 10.12.2017
11. https://rg.ru/2014/07/08/ekaterinburg.html 12.12.2017
12. http://www.skyscrapercity.com/showthread.php?t=412073&page=316 11.12.2017
13. http://www.skyscrapercity.com/showthread.php?t=526298&page=270 12.12.2017
14. https://ru.wikipedia.org/wiki/Башня_«Исеть» 12.12.2017
15. http://img-fotki.yandex.ru/get/6712/126273919.11/0_98ed1_f8d387f1_orig 11.12.2017
16. http://ngzt.ru/news/view/02-03-2015-samye-vysokie-zdaniya-ekaterinburga-foto-video-multimedia 12.12.2017
17. https://екатеринбург.рф/жителям/город/панорама 10.12.2017
18. H. Plammer. *The Experience of Architecture*, L.: Thames & Hadson 7–27, 84–103 (2016)
19. A. Betsky. *Architecture Matters*, L.: Thames & Hadson (2017)

Evolution of high-rise construction in Leningrad – Saint Petersburg in the middle of the 20th – Early 21st centuries: projects and Implementation

Andrey Vaytens[1*], *Gennadiy* Rusanov[1] and *Pavel* Skryabin[1]

[1]Saint Petersburg State University of Architecture and Civil Engineering, Department of Urban Planning, Vtoraya Krasnoarmeyskaya str. 4, Saint Petersburg, 190005, Russia

Abstract. One of the most important issues in national urban planning is arrangement of high-rise buildings in the largest cities of Russia. This issue becomes especially acute in such cities as Saint Petersburg, which has unique architectural and urban-planning heritage preserved to a great extent. In this regard, it seems important to trace the evolution of high-rise construction development and arrangement in Leningrad – Saint Petersburg in the middle of the 20th — early 21st centuries. The goal of the article is to consider high-rise construction development regarding both public and residential buildings in comparison of project ideas and results of their implementation in the 1940s–2000s. Prerequisites and issues of high-rise construction of that period are considered. Particular attention is paid to changes in the official urban-planning policy, regulatory framework and attitude of city authorities to high-rise construction. The study was carried out with the consideration of the following historic periods differing in their urban-planning policy: the late 1940s–1950s; 1960s; 1970–1980s; 1990s; 2000s–2010s. Economical prerequisites of high-rise construction and their influence on the modern urban-planning policy during the post-Soviet period are considered. In conclusion, an attempt is made to determine tendencies of high-rise construction development in Saint Petersburg.

1 Introduction

It is known that at the present time there is no single definition for the "high-rise building". At different times this concept changed both in Russia and in the USSR according to ideological and economic attitudes. The height of multi-story buildings in the USSR ranged from 9 floors (about 30 m) up to 20 floors (about 70 m).

After the October Revolution, the construction of high-rise public and residential buildings in Petrograd (and after 1924 — in Leningrad) was not envisaged. In the first half of the 20th century, the following dominants remained in Leningrad: the spire of the Peter and Paul Fortress (122 m), the spire of the Admiralty and the dome of the St. Isaac's

[1] Corresponding author: avaytens@gmail.com

Cathedral (103 m). Before the Great Patriotic War, the height of residential and public development in Leningrad did not exceed 30 m. Only the tower of the Moskovsko-Narvsky district council and the building of the Leningrad House of the Soviets (architect N. A. Trotsky) built just before the war were exceptions: their height did not exceed 50 m (Figures 1, 2). Their height and dominating in the surrounding development were determined by ideological and urban-planning considerations.

Figures 1, 2. The tower of the Moskovsko-Narvsky district council and the building of the Leningrad House of the Soviets

2 Materials and methods

When considering the evolution of high-rise construction development in the city on the Neva river, it is proposed to use methods which are typical for architectural and urban-planning studies, i.e. analytical methods of comparison, juxtaposition and historical analogies.

In the post-war period, Leningrad architects began to pay attention to high-rise construction and participate in it only in the early 1950s.

Until that time, directions in the designing and construction of low-rise housing in Leningrad were determined by the need for renovation and reconstruction works and the restoration of military destructions.

In 1948, a Master Plan for the renovation and development of Leningrad was approved (Figure 3). In pre-war Master Plans, residential development was mainly represented by large-size city blocks, and according to the Master Plan of 1948 it was to be formed with small-size city blocks of 4–6-story buildings with all types of services. As an exception, it was allowed to increase the height of residential development up to 7 floors on main lines and squares of the city [1, p. 26].

Figure 3. Master Plan of 1948 for Leningrad renovation

In accordance with this Master Plan, in the early 1950s, it was proposed to build 6–7-story residential buildings along the main lines running from the center to new residential districts: Stalin prospect (later — Moskovsky prospect) and Stachek prospect. During this period, only buildings in the southern part of the present Moskovsky prospect (architect S. B. Speransky) and several multi-story buildings on Stachek prospect at the entrance to Leningrad from Petrodvorets and Tallinn (architect V. A. Kamensky) were constructed (Figures 4 and 5).

During Soviet times, the height control of new high-rise buildings was performed in layout designs of residential blocks and determined by urban-planning considerations: location in new districts, location of city blocks in the system of main lines, natural conditions.

Since the late 1950s and until the early 1960s, unofficial prohibition against the construction of multi-story residential buildings existed due to the large-scale construction of panel five-story buildings and development of the micro-district service system in Leningrad and other cities of the USSR (except Moscow).

 https://doi.org/10.1051/e3sconf/20183301030

Figures 4, 5. High-rise development in the southern part of Moskovsky prospect (1950s) and high-rise development in the southern part of Stachek prospect (early 1950s)

However, since the beginning of the 1960s, this prohibition had been gradually lifted in connection with the determined need for the construction of new, increasing housing volumes, and then the switch to large-scale nine-story residential development based on industrial series occurred (Figure 6).

Figure 6. Large-scale nine-story residential development (1960s)

In June 1966, a Master Plan for the development of Leningrad was approved, according to which in subsequent years significant territories were developed not only in the southern and south-western parts of Leningrad, but also in its northern and eastern parts (Kalininsky and Vyborgsky districts) (Figure 7).

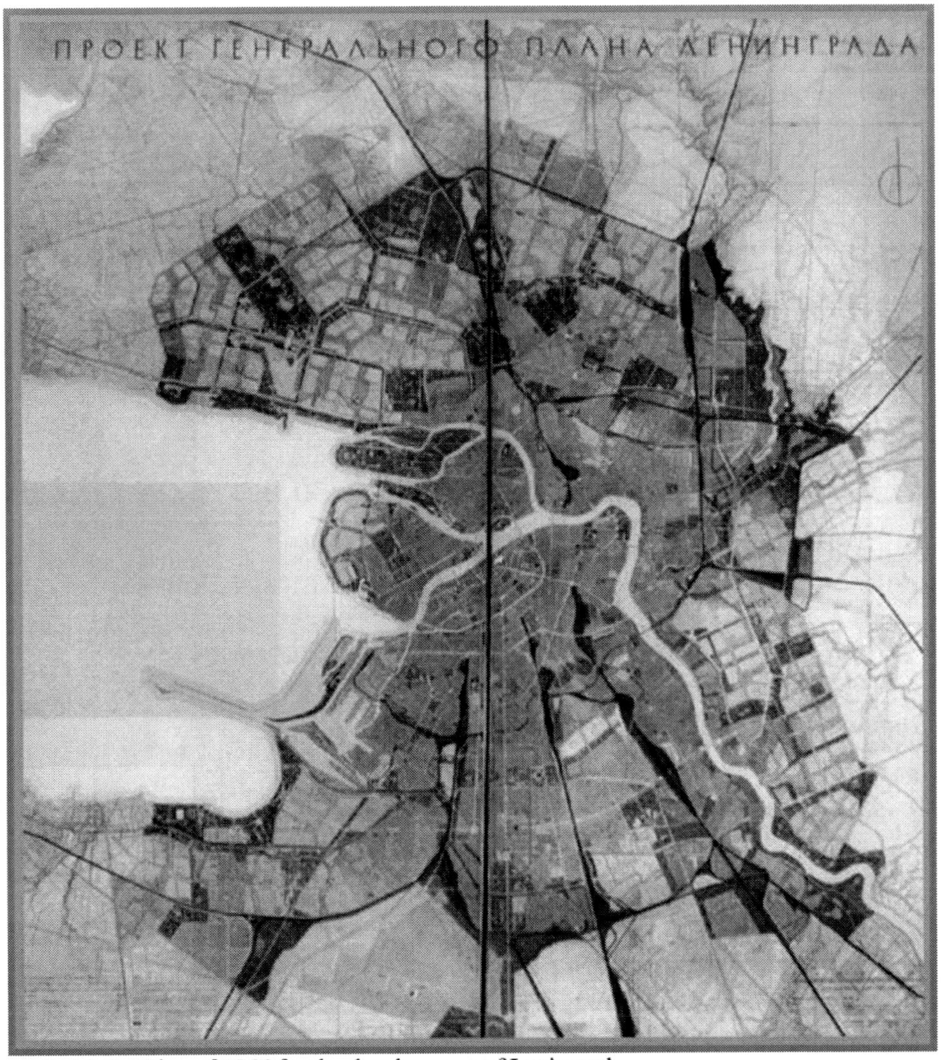

Figure 7. Master Plan of 1966 for the development of Leningrad

The development regulations included in the Master Plan of 1966 provided for the fixed ratio with regard to the arrangement of new residential housing in various districts of the city. About 45% of new housing was to be located in northern districts of Leningrad, 50% — in southern districts, including the right bank of the Neva river, 5% — in island areas. During the implementation of the Master Plan, those ratios were generally maintained [1, p. 28].

In 1969–1970s, it was planned to construct 60% of nine-story and 14% of twelve- and sixteen-story residential buildings of the total volume of housing construction in Leningrad [2, p. 7]. The average number of floors in residential buildings increased up to 10 floors in 1969–1970, according to the detailed urban-planning project for the Right Bank of the Neva river.

Residential buildings with the height of 16–18 floors were considered as integral parts of the designed large urban-planning centers of future construction areas (Figure 8). High-rise residential buildings were considered as means for the formation of city main lines. In 1973–1975, it was planned to erect eight 16-story residential buildings along the southern side of Geroyev prospect (nowadays — Leninsky prospect) [3, p. 3]. Those buildings were constructed in subsequent years. At the same time, in the northern part of Leningrad, the formation of Muzhestva square continued with high-rise residential buildings (16–18 floors) flanking the entrance to the square.

Figure 8. 16–18-story residential houses as parts of urban-planning centers in the beginning of the 1970s (models)

In the early 1970s, Leningrad architects began to consider the ratio between the new multi-story development and the historic silhouette of the city. Gradually it became clear that a significant increase in the number of floors of newly erected buildings should be considered acceptable only in areas of new residential and non-residential development, to avoid any damage to the architectural heritage. It was necessary to resolve a very important issue regarding the selection of area for the arrangement of high-rise dominants in the city. In other words, all new high-rise dominants should be included in the city plan beforehand to form a single system with high-rise compositions of the city center. In this regard, the development of a general compositional development scheme was started, which provided for the arrangement of community centers, including high-rise buildings [3, p. 5]. Unfortunately, this important urban-planning work was not completed.

In 1972, according to a special governmental decree, the size of the population of Leningrad was specified (it was expected to increase from 3.5 million to 4 million persons). Special calculations showed that the expected additional population could be completely settled on the territories stipulated by the Master Plan of 1966, provided the increase in the density of the residential development and development of new plots [4, p. 5]. As a consequence, the number of floors would increase as well. By that time, all operated buildings of standardized series were switch to the module-section basis, where at first a section, and then an apartment represented the type-design unit. Improvements affected such large-panel buildings of mass series as 1 LG-600A, 606M, 504D and others. The 137 series, based on the module-apartment method hold the key position in the list of Leningrad standard projects [5, p. 11]. Buildings of mass series had mainly 9–12 floors. In the middle of the 1970s, unique residential buildings were constructed in Leningrad, which combined 9-, 12- and 15-story module- sections of 1-LG600A series and brick 13–16 story module-inserts [5, p. 12].

Thus, the need to increase the living area in territories limited by the Master Plan, as well as urban-planning considerations with regard to the formation of large areas and main lines resulted in the increase in the residential development height up to 9–15 floors by 1966–1975. By May 1975, the 30th anniversary of Victory in the Great Patriotic War, the construction of a square with the same name was completed in the southern part of Moskovsky prospect, an ensemble at the entrance from Moscow and Kiev. Two 20-story residential buildings with the height of 70 m became vertical parts of this ensemble. At that time, they were the highest buildings in Leningrad (Figure 9).

Figure 9. Victory Square (1975) (development team under the leadership of S. B. Speransky)

In 1973, the Leningrad Design Institute was offered an assignment for designing a complex of buildings for a large R&D center engaged in advanced space development (Central Research and Development Institute of Robotics and Engineering Cybernetics). A plot in the north of Leningrad, between Polytekhnicheskaya and Akademicheskaya subway stations, was allocated for the construction. A closed tender was held, which was won by a group of architects under the leadership of architect S. V. Savin [6, p. 31]. Their high-rise complex (control and test station, 104 m) became a dominant in this part of the city (Figure 10). In 1986, the erection of this complex represented a particular stage in high-rise construction development in Leningrad. Before 1991, no residential or public buildings of such height were constructed.

Figure 10. High-rise complex of the Central Research and Development Institute of Robotics and Engineering Cybernetics (1973–1986)

In the 1990s, several attempts were made to design high-rise buildings in Saint Petersburg. Those included the unimplemented project of Peter the Great high-rise building in the western part of Vasilyevsky island (architect O. A. Kharchenko).

The late 1990s – early 2000s shall be considered as the beginning of the development and large-scale implementation of projects of high-rise buildings in Saint Petersburg (Figure 11).

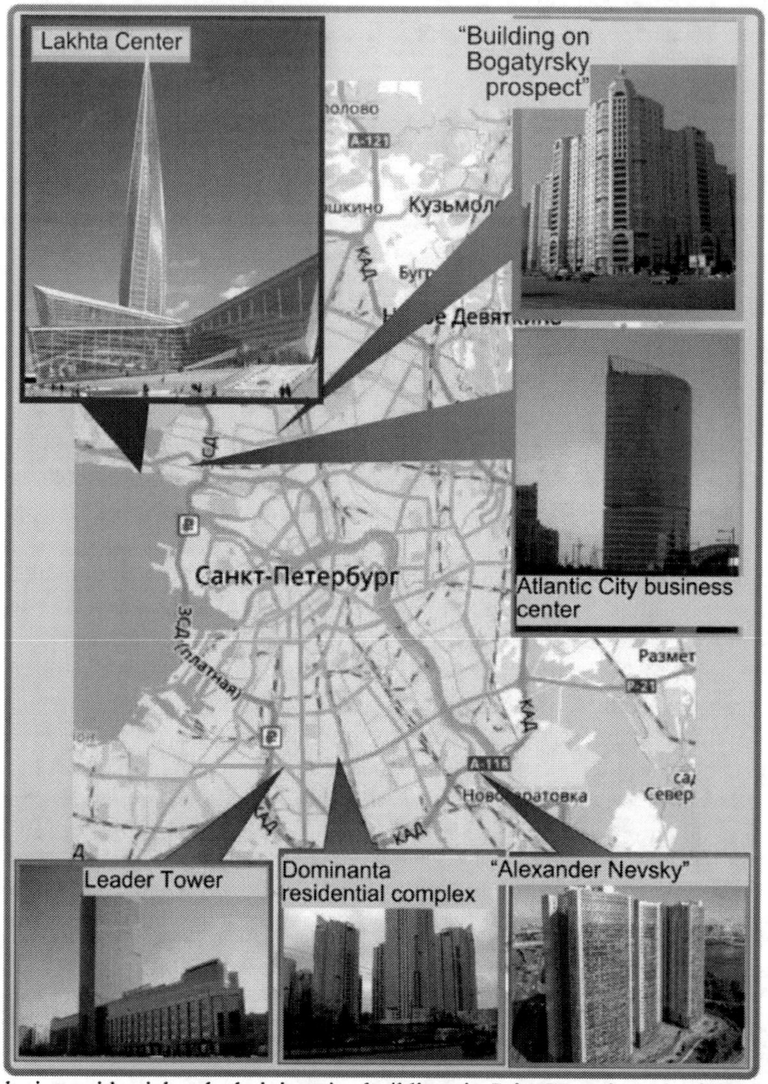

Figure 11. High-rise residential and administrative buildings in Saint Petersburg (early 2000s)

In 2000, the erection of Cloud Nine (Sedmoe Nebo) 27-story residential complex was completed, the height of which reached 101 m. This high-rise complex is located near Moskovsky prospect on Kuznetsovskaya street (Park Pobedy subway station). Another two projects were implemented in 2003: the high-rise residential "Building on Bogatyrsky prospect" (height — 108 m) and the residential complex "Poem by Three Lakes" (height — 100 m). The first building is located on Bogatyrsky prospect near the historical part of the city, the second building is located in the block of houses, limited by Engelsa prospect, Lunacharskogo prospect and Yesenina street. In 2006, a high-rise residential building was constructed on Komendantsky Prospect (height — 105 m); it was made of double multi-story plates connected with an arch. In the same year, the Mezhdunarodny residential complex (101.5 m high) located on the triangular plot, limited by Bely Kuna street, Bukharestskaya street and Turku Street, was completed. Among the buildings typical for this period, the Atlantic City business center with an extended atrium (architect S. V. Gaykovich) can be noted (Figure 12). The high-rise part of this complex (110 m) became one of dominants of the Primorsky district.

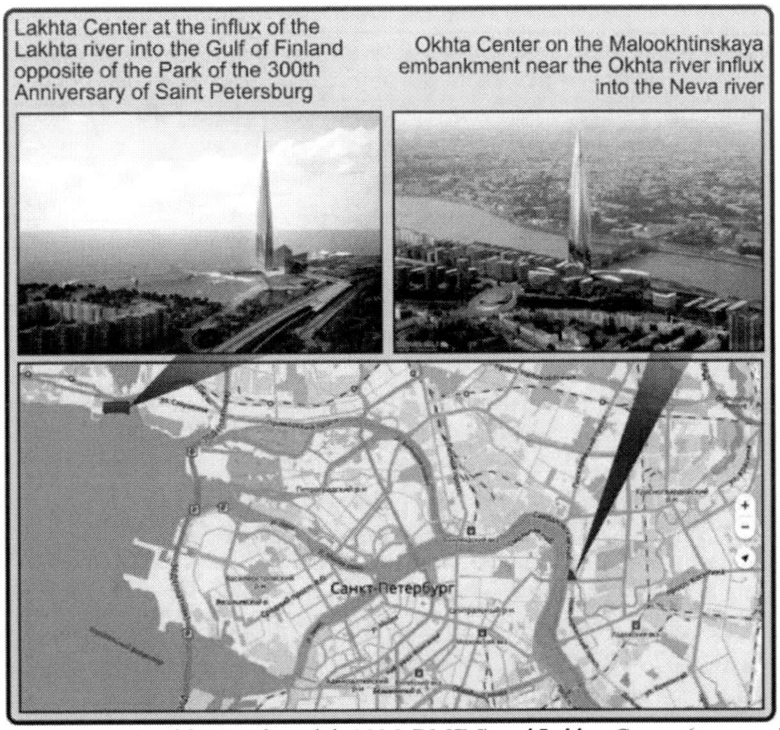

Figure 12. Okhta Center (architectural model, 2006, RMJM) and Lakhta Center (approved project, 2012)

Analysis of the results of high-rise construction in Saint Petersburg in the 2000s allowed to classify the following factors as prerequisites of its development:

- significant excess of the demand over the supply of land plots and premises (residential and administrative) near city main lines;
- development of modern technologies of frame construction;
- development of modern elevator technologies;
- development of modern engineering ventilation systems for environmental control.

In the early 2000s, the competitive designing of the Okhta Center (customer: OJSC Gazprom) became a separate area of concern. The City Administration selected the Right Bank of the Neva river for the arrangement of this complex. According to the results of the 2006 contest, RMJM London Limited was determined as the winner (Figure 13). Commissioning of the architectural dominant was expected in 2012. However, those results evoked negative responses in both specialists and the city community, and the project was denied. In 2011, OJSC Gazprom acquired the plot in another part of the city — on the bank of the Lakhta River, at a distance from the historical center. The selection of this plot was also determined by administrative decisions.

The construction of the public and business complex with the total area of more than 400,000 m2 began in 2012 (Figure 13). It was intended to house the headquarters of OJSC Gazprom. The new complex has undergone some changes in comparison with the Okhta

Center. Its extended part is divided by the atrium in the northern and southern parts; and its high-rise part (together with the spire) will be 462 m high. It is expected to complete its construction in the Q3 2018. The tower spire will be viewed from all points of the historical part of the city, as well as from the water area of the Gulf of Finland. Moreover, it will be the highest building in Europe.

The Leader-Tower business center, situated on Konstitutsii square, at the end of Novoizmaylovsky prospect, became a high-rise dominant of the southern part of the Moskovsky district (Figure 11). LLC Fregat, the initiator of this construction, acquired the corresponding land plot in 2007, obtained a site development plan and a permission for construction in 2009, and started the construction. The location for high-rise facility construction was selected well: a similar object was intended to be constructed in this territory according to the urban-planning documentation of the 1980s. However, the designed height (145 m) caused serious objections of the Committee for City Planning and Architecture of Saint Petersburg, and the plan was canceled. The customer managed to win the case in the Arbitration Court and the construction was not suspended. Eventually, the construction of this complex, which includes 52.7 thousand m2 of commercial spaces and a 5-level underground parking lot for 250 lots, was completed in 2013.

In the 2010s, the development of multi-story residential complexes in Saint Petersburg determined the silhouette of peripheral parts of the city. The selection of plots for these complexes was mainly random in nature. The lack of the necessary social infrastructure (kindergartens, schools and shops) became the main drawback of this construction.

An attempt to overcome those shortcomings can be noted in the structure of the Peter the Great and Catherine the Great (Petr Velikiy I Ekaterina Velikaya) double-tower residential complex, erected by the RosStroyInvest design and construction association near the bend of the Neva river and Rybatskoye district (Figure 14). Buildings of this complex have 32 floors (110 m) and 24 floors (80 m). A kindergarten, fitness center, cultural and educational center and cafe (much needed public infrastructure for future residents of 1,683 apartments) are designed for the construction at the site of the complex. The total residential area is 62,392 m2. RosStroyInvest company is one of the leaders in the field of high-rise residential construction in Saint Petersburg. City of Masters (Gorod Masterov), House with Chiming Clock (Dom s Kurantami), Prince Alexander Nevsky (Knyaz Aleksandr Nevsky), Two Angels (Dva Angela) and other high-rise residential complexes were also built in various parts of the city according to projects of this company.

Figure 13. Peter the Great and Catherine the Great residential complex

https://doi.org/10.1051/e3sconf/20183301030

3 Results

As a result of the consideration of implemented projects of high-rise residential and public-business complexes, it may be concluded that the location of high-rise dominants in Saint Petersburg has random nature. Occasional arrangement of high-rise buildings within limited unoccupied areas, well-established by the early 2000s, contravenes the requirements for the integrated development of urban territories. Using the imperfection of legal mechanisms for regulating urban-planning activities (and often in violation of such mechanisms), investors and owners of land plots implemented projects of high-rise buildings with disregard for the urban-planning conditions, exceeding permissible values of density and number of floors. Under such conditions, an excessive burden on the engineering and transport infrastructure occurs, while the necessary social infrastructure — schools, kindergartens, shops — is either insufficient with regard to the number of residents, or absent. Further development of these processes will lead to serious issued related to the development of Saint Petersburg, including the operation of already constructed facilities, shortage in engineering capacities and, finally, increase in social tension.

It becomes obvious that the city is still not ready for emerging changes in terms of its urban-planning development, and, in particular, for the need to develop high-rise construction. Extremely limited high-rise regulations, which are mainly adopted by fiat and without sufficient scientific substantiations, and conditioned by the compositional silhouette of the historical city and borders of protected areas of cultural heritage objects, are typical for the majority of Saint Petersburg territories.

It seems that a special approach to the formation of the modern silhouette of the city is needed, with account for the potential emergence of new high-rise dominants not violating the historical urban-planning environment. Special attention should be paid to the scientifically grounded selection of territories for the future arrangement of high-rise residential and public-business buildings. Those can be public and business centers and high-rise residential complexes. Urban-planning documentation for the arrangement of the system of high-rise dominants without violating the historical silhouette of the city is needed to be developed. Several attempts to develop such documents were already made in the urban-planning practice of Saint Petersburg. Such documentation will provide an opportunity to identify territories for future high-rise construction, suitable by their engineering and geological conditions, which are especially complicated in Saint Petersburg. For this purpose, upon the adjustment of the modern Master Plan of the city, it is necessary to determine special areas with special urban-planning regulations aimed at the altitude, density and functionality of high-rise facilities to be erected in the future.

Such approach already applied in the world practice. According to this approach, it is expedient to allocate such territories near large transport and transfer hubs, in points of the intersection of ring roads with city main lines or at a distance from historical parts of cities. Besides, direct transport connection with the historical center and all city districts shall be ensured.

The construction of the La Défense public and business center in Paris, located on a single axis with Charles de Gaulle square and having direct connection with the historical center of Paris can be considered as the best example in this field. National and international experience in designing in terms of determining and selecting territories for high-rise construction shall be considered, used and developed to achieve the sustainable development of any large city.

References

1. A. G. Vaytens Regulations in plans for urban-planning development of Petrograd – Leningrad – Saint Petersburg (1920s–2000s). Architecture and Construction, **1**, 24–31 (2007).
2. A. I. Naumov. To be constructed in 1969–1970s. Master Plan for the development of Leningrad in effect. Construction and Architecture of Leningrad, **1**, 6–8 (1969).
3. Leningrad, 1971–1975. Construction and Architecture of Leningrad, 1, 3–8 (1972).
4. G. N. Buldakov. In accordance with the Master Plan. Construction and Architecture of Leningrad, **7**, 5–10 (1976).
5. V. V. Popov. Stages of creative search. Construction and Architecture of Leningrad, **7**, 11–15 (1976).
6. V. Ivanov. *Architecture inspired by space. Image of the future in the late Soviet architecture*. Saint Petersburg, Borey Art, 252 (2017).

The importance of vertical buildings in perception and memorising the city

Rifat Alihodzic[1*], and *Domen* Zupančič[2]

[1]University of Montenegro, Faculty of Architecture, Podgorica, Montenegro
[2]University of Ljubljana, Faculty of Architecture, Slovenia

Abstract. Being aware of the surrounding we live in, among other things, means establishing of spatial relationships between oneself and the environment, equally important as relationship between oneself and others. Environment consists of facilities and space. Space, „gift by itself", is defined by terrain topography, sky horizon, plants and animals. The architecture, as a profession, is interested in space created distinctively. Perception, as elementary process of consciousness and psychological life, deals with being conscious about something. In this case, physical structures that create a city. Psychological experience of urban environment is important factor having effect on perception, memorising and orientation in urban space. Gestalt psychology of perceiving is area applying to and significant for architecture either. The importance of vertical lies in its perceiving the gravitation, forming perceiving focus, landmark, for urban units and subunits to be memorised, creating spatial hierarchy and perception logics, remembering and orientation in space. This work analyses reasons for building upright with comparative analyses in their participation in space and on human psychology. This paper's purpose is to, using fundamental facts, show the importance of vertical buildings, not as a spatial use phenomenon, but also as significant phenomenon.

1 Introduction

When perceiving any object, the perception process defines its basic features which making that particular object distinctive from others. Perceptive process determines its visual identity.

Some of the most important features of each object are its shape, size, colour, structure of material, position in space and comprehension whether the object is dinamic or static. These are some basic features, and particularities of a certain object can show some other features, especially comparative ones, which distinguish them from similar objects in vicinity. [1]

In this paper, there will be only one feature of an object (in this case, a facility) examined and that is its vertical position in space.

* Corresponding author: rifata@ac.me

Let's assume that there is a physical structure having dimensions *x,y,z*. It can be positioned in building area either horizontally or vertically. Cubic capacity of its space, its usagein other words, will be the same no matter on vertical or horizontal position.

However, participation in space in sense of perceiving, memorising and experiencing of that physical structure (form) won't be the same. That is because seeing something means giving it a place in a certain unit. [2]

Architectural facilities, in most cases, are not isolated as they seem to be in our minds when designing procedure takes place. Real space, in which they are incorporated, is always a context, no matter if it is a purely natural environment or built environment. That way, facilities are parts of a specific unit in a wider plan, but they can mostly be perceived in two ways: one of them represents a moment when the facility distinguishes because of its features in one's perception and the other one is when targeted focusing is achieved on purpose. Familiar fact, not only from Gestalt psychology, but also from experience, is that vertical facilities impose to our perception!

Important question when perceiving an object (form) basically is narrowed tothis: How to link what our perception with our memory, how to identify what we perceive and how to experience it? [3]

2 Reasons for perceiving vertical and its importance

For favouring vertical compared to horizontal in general perceiving, the gravitation is responsible, which people are accustomed to through the empathy. We should have in mind that vertical position is one of people's first experiences in space. It is not a coincidence that this is the first obligation of each designing act, when determining the way our eyes will perceive forces that the facility resists, the gravitation. Among countless directions of tridimensional space among which it is theoretically possible to move, only one direction has a gravitation as its feature: vertical. Vertical is a base and a measure for all other directions. [4]

From the geometrical point of view, there is no difference in up-down movement, but from the gravitation's point of view and designers' efforts, it surely has. The consciousness about that gives a special psychological relation towards vertical as a reference direction. What is important to perceive is that vertical position became a part of collective memory psychologically also. „Man in vertical" means the presence of moral qualities and a wide range of other positive attributes.

Let's remind ourselves that, in construction, for example, a plumb line is used and every other direction is determined according to vertical direction. That is the case when the level is used and that is done comparing to already established vertical.

Vertical is the thing that determines horizontal direction so it serves as a symmetrical axis. Unlike vertical direction, horizontal is the level of moving in every possible direction without having the „obligation" to gravitation. Horizontal level is a space of human existence and their activities. The house is assumed to be a concentrated vertical being. It rises up. The straight line dominates there. Plumb line left a seal of its sense and its balance. [5]

In favour to the comprehension of importance of vertical as a product of gravitation, there is an irresistable fact that there are no records of anyone wanting a bed or dining table on slope level. That is not only pragmatical thing, but also a thing of (dis)orientation which is obvious in stepping out of vertical. Our bodies and mind experience such situations as alarming ones. Nietzsche, who had influence of aestheitics of Hegel, Schopenhauer and Hartmann, noticed that. *„In architecture, the pride of a man, his triumph over the gravition, his will to power, assumesa visible form. Architecture is a sort of oratory power by means of form."* [6]

Vertical is „a bearer" of perceptiontaking the position as a straight line. It goes downwards, up to some horizontal to be resistant to, and that is a straight line, or, in case of architectural facility, an Earth ground. The way that contact of the Earth and facility materializes, results in specific perceiving, depending on relationship established following the principle of „visual statics", which is not the same as engineering statics, dealing with the stability of the facility.

3 Vertical facility as an urban landmark

Generally, there are two ways for better orientation in space. They are defined by K. Linch. [7] One of them represents a kind of distionction, „landmark". Other places are linked with this mark. More precise way that determines a place is a relation with some facility. It is absolutely logical that this kind of approach is more effective, especially when this is vertical. Abovementioned approaches represent fundamental elements of orientation.

Common things (facilities) do not attract our attention on regular basis, so the same applies for architecture. Psychology of perceiving teaches us that we are not able to concentrate to more things (architectural facilities) at the same time, so it is logical to focus on those that stand out, i.e. impose to our eyes. *„In architecture, as in any other culture, our sense of how some things should be, develops from our experience. Each gesture that we make means something, but the meaning depends on the culture in which gesture is understood. Architecture is a gesture made with buildings."* [8]

Without any doubt, the most significant result of vertical facility is that it becomes an orientation agent, in other words, urban mark. Urban landmarks are required as the most reliable elements of perception, memorising and orientation in space. Each vertical physical structure produces more intensive visual energy than the horizontal one because it is in position of better transmitting of vector in every direction.

Being lost in space considers panic that creates the fear of the highest order. It is known that, if a person shows disorientation and fear, people say that they are „lost". Where does it come from? *„The infant sees himself as the center of the worldsurrounding him."* [9] Their comprehension for a long time is that everything is subordinate to their needs and under their constant control. That is the thing that stays attached to humans for a long period of time and it is hard to get rid of it, so entire groups, nations and countries act from the point of „own interests" through whose centricity manifests as a feature that affects motifs and points of view. However, it is essential to observe other subjects in one's life, having their own needs that need to be respected, so that correlation necessarily grades through respecting the other „someone else's centres". At the end, it is realized that own „centre" is only one on the heap. That way, centrical tendency is being complemented with „eccentric" one. *„The eccentric spot will exhibit a dynamic tendency, probably a pull in thedirection of the centric position.."* [9]

Mutual action of two tendencies (centric and eccentric) makes the essence of human living. It is considered both in relations towards human beings and towards things that surround us. One of them, probably the most important one, is the space we live in.

Physicists of our time give us the explanation that universe is created from the energy field, so each thing, house, building, hill, represent a certain energy field. Being such, things are nothing but vector forces. In that particular sense, vectors are mentioned for the first time by psychologist K. Lewin in 1935. [10] Symmetrical frame, with vectors transmitted shaped as rays, represents typical centric composition. Establishing the relation between one's own centre and some spatial centre represents, according to previously mentioned logics, vectors transfer. These relations are multi-level ones, depending on the subject with which primary centre establishes relation. We accept that it exists in a dynamical sense, provoking perception.

From what was previously said, conclusion is that a man sees himself as a centre of everything surrounding him. During their moving, this centre is also moving with them, and the rest of the world is seen as a sum of objects eccentrically towards them. Establishing such relationship does not depend only on man, as a primary centre, transmitting their „own vectors“ towards the object, and also the objects transmit their vectors towards the beholder influencing their attention.

Vertical dimension serves as an area for visual thinking chiefly, and horizontal is the empire of activities. That moment favours vertical things as the objects of real perceiving and establishing the relationship of our „centre“ with centres of the objects around us.

Paying attention to followingcan contribute to explaining these attitudes: *„In a dynamical sense, a tower standing in an empty field or a ceramic vase on a table are centres of visual energy sending their vectors in all directions and thereby surround themselves with a tension-loaded field that spreads and gradually peters out in empty space.“* [11] So, domination of vertical in an empty space is the precondition of domination towards other objects that do not possess enough space „for transmitting vectors“ so those able to act that way impose themselves as landmarks. This way, landmarks represent accents ofcertain places, or points that they are linked to, or their parts.

Exquisite effects of landmarks can be achieved in cases when the natural landmarksmatch inherited terrain topographically and are being marked with some facility. That kind of picture is common at Italian smaller places as well as world famous cities such as Istanbul (Fig.1). This topic is frequently used, especially when building fortifications and castles, where the defense is a primary function of such facilities' locating, so the perception catches landmarks produced, whichfrequently became symbols of many towns and significant marks on the roads, because among many hills the one with a particular facility is being standing out.

Fig.1. Italian village and silhouette of Istanbul. Matching both natural and a hierarchy created by building.

It is good to mention that, matching of landmark in certain situations can be a problem for an urban structure which, because of its ruined c hierarchy, can cause some difficulties in reading the space.

4 Historical experiences with vertical buildings. Physical and spiritual landmarks

To underline the place means to mark it by vertical. From the ancient times that moment was built in inside the human collective unconsciousness because the conscious experience suggests vertical as dominant one.

Fig.2. A stone menhires and Torre Agbar in Barcelona (arch. Jean Nouvel). The same goal in space no matter on the time of origin.

If we take a look at the building history and first human experiences of space, we inevitably meet phenomenonn of vertical, instinctively perceived as a strong orientation tool. Megalith structures of menhirs, dolmens and cromlechs, still having a dominant vertical, prove that. Doubtlessly, megalith structures have a feature of identifying and matching of physical and spiritual meanings, as well as all the sacral buildings created later on. Vertical is, above all, easy to notice and, comparing to horizontal, favours itself because it is easily perceived. Quantitatively, in the initial stage of human development, it gets a saint position in a sacral. [12]

Fig.3. Simple spatial rock gate and contemporary Tori in Japan. The same idea in different temporal epochs.

If we have a glimpse at the history of certain vertical structures, we'll see its being constant in meaning and stability in form where only the envelope determines time of origin. (Fig.2,3).

Important moment in relation with vertical represents human notion of God, whose residence is connected to sky, so all sacred buildings are directed upwards and their perceiving structure has the effect of climbing. Sacral buildings, during the time, lose their prime, and are being replaced by those which, instead of God's power, manifest human power (Fig.4).

Fig.4. Mont Saint-Michel, vertical dedicated to God. Eiffel Tower, dedicated to the power of contemporary man.

From the very first marks in the space, over ziggurats and pyramides all through the total development of sacral buildings, irreplacable imortance of vertical is being sensed. So, importance of vertical was both a physical perceiving structure and a spiritual element required at sacral buildings, from historic point of view.

5 Vertical facility as a manifestation of economical power and technological culture

In human history, consuming society and economical power in its final (contemporary) phase show domination above spiritual and religious. That is not hard to notice through perceiving the city development where, up to 19th century, it was considered as perfectly normal and sacred, the rule that the highest facilities in space are those of sacral nature.

Impression created by vertical quantity estimates the height of an architectural facility and has own reasons within itself which have the notion of strong technological progress and manifestation of material power. Technological evolution is the thing that fascinates a modern man, so our civilisation considers it as almost natural and necessary process, just like biological one according to Darwin (Fig.5.).

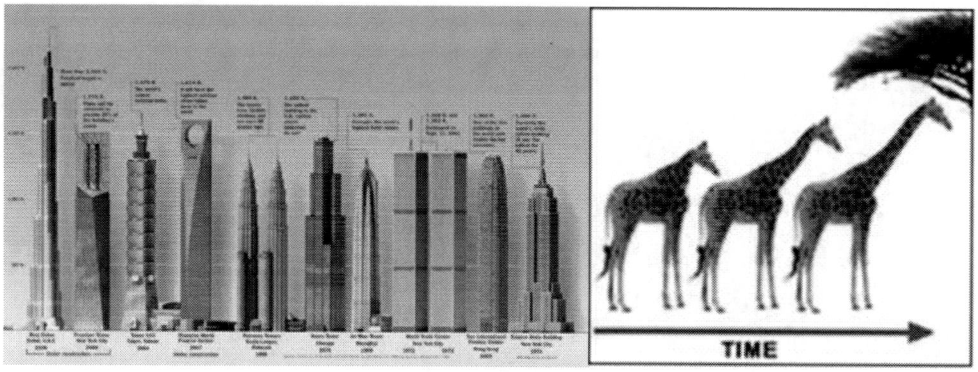

Fig.5. Each new, the highest building redefines the power scale. Similarity with Darwin's evolution theory.

Every new facility in the world, which changes scale of vertical quality, causes position redefinition, manifested in countries' power and that power is being manifested through multinational companies. If we make a comparison of the analyses based on quality of architectural facilities,by that, we can see that such system of evaluation ends only in the zone of professional circles, like Pritzker award and similar ones for instance. However, public opinion follows the fact that the new power centre will appear manifested by built vertical facility with greater curiosity and passion.

Vertical size of an object, structure and facade appearing of architectural facility are measurements of technological improvement nowadays unlike the forms from previous period. That can be interpreted as growth of new creative and available energy, which is obvious in the connection of architectural expression and technological advantages of modern and ancient civilizations.

Every new facility in the world, which changes vertical qualitative scale causes redefinition of positions showing the strength of countries and their power, manifested through multinational compositions. As a comparison, no one has made a qualitative comparison of horizontal volumes at architectural facilities so far.

6 Vertical facility as utilitarian component

If we take a look at the building history, vertical facilities made from the requirements for space quantity as a primary reason are the youngest one and in less number.

Building siloes, water towers, bridges and other engineering constructions, as well as vertical buildings represents strictly utilitarian spatial-functional category, intensively appearing with industrial society development.

Fig.6. Siloes, water towers and bridges. Beside utilitarian importance, they take the role od landmark.

Even though these facilities mostly aren't primarily accentuating the space, but came out of practical purposes, they take over a bunch of attributes connected to the meaning in space. It is particularly obvious with water towers and bridges which manifest new technological era and economical power, that they started to be significant and possible city accents in urban designer's primary purpose (Fig.6).

Building of residential buildings, for selling or renting, represents typical examples of economically justified building which find their reasons for expensive urban-construction ground. However, in terms of humanity and residential conditions, this factor is dehumanisation of the way of living so our psychological experience doesn't recognize it as the nice one. The same can be applied to student dormitories, single people accomodation and similar facilities. If our starting point is that fundamental part of urban structure of the city contains 70-85% of buildings for residential purposes, then no one can neglect the occupying of space of this kind of facilities directed upwards at the expense of humanliving (Fig.6,7).

Business buildings for renting to other companies an in better position, while and the buildings meant for the purposes of one company, having its trademark are in best position. Then, they obtain a clear and precise meaning considering power manifestation of a specific owner, company that built that facility.

Therefore, it is good to say that dehumanisation of living space in business high towers is not less significant than the one in residential areas, because a man spends most of his everyday life at his work, being awake, while is his residential space meant only for night break and sleeping.

Fig.7. High residential and business buildings make significant landmarks.

No matter the goals and reasons for building and above mentioned facilities have their role in space, theycan be caught by an eye and significantly define beholder's experience. However, categorisation of total meaning of their function does not give them significance as well as to sacral buildings and those meant for culture and sport.

No doubt that in the second half of 20th and and in 21st century, facilities of culture and sport along with business facilities dedicated to power and multinational companies, take the position over the facilities of sacral features.

7 Conclusion

Vertical and quantity of an object are the most important rules of perceiving organisation in space. [13] However, that alone is not the reason enough for completeness and perceiving the total experience. All things in process of perception carry the meaning even when those creating them do not have intention of sending a certain message. However, when designing performance aims for meaning, which the observer interprets, there can be made an effect upon it for sure. [14]

Basically, urban structure is made by landmarks built with a time residue and a requiring experience. When they establish themselves as such, they become the elements on whose basis can the whole urban picture be created. That happens when we create spatial correlation between them or we have graphically prepared structural scheme in front of them. By identification of one of such facilities can possibly reconstruct the whole urban picture. As the illustration of this claim, a part from tourist prospect of Paris was used for this purpose, where only landmarks with discreetly emphasised Seine, served the purpose to present the spatial structure of town up to unbelievable precision. The arrow directs to transversal of the city, going from Louvre, over Concord Square and Triumph Gate up to new Grande Arche de la Defense.

There are architectural facilities around the world, which represent traditional landmarks and as such they are accepted in mass culture and suggested to the individual. Paris, which many people connect with Eiffel Tower, just like Athens is connected with Parthenon, or Sydney Opera House with that city, are the first things that come to people's minds, even though they have never been there. However, when they reach Sydney for instance, they will search the famous opera house there.

If we make analyses of historical position of facility built vertically from the position of meaning in our consciousness, based on previous, we can find following reasons for their building:

- Vertical facility (or element) as the landmark in space
- Vertical facility created for praising God and aiming to the sky
- Vertical facility as a manifestation of economical and technological power
- Vertical facility as beneficial element - economical category

Fig.8. No matter the time of origin, landmarks created by vertical building have similar or the same role

What can be concluded, based on previous observations, is the fact that vertical facility is historically constant with its effect on perception and beholder's experience. Reasons for vertical building were changing throughout the history (from location in space and positioning, to religious, utilitarian, culturological, engineering etc.) and by that, facilities have various meanings but only one constant: they are landmarks and represent one of the most significant creative meand of the urban image. So, verticals are material required for successfully formation of urban picture and orientation in space (Fig.8).

An important element represents their meaning and symbolical image (created during the time) in space. That factor is one of the most important elements of cognitive processes taking place in human brain. [15] The importance of the facility in the moment of perceiving and through the history represents basic material for positioning of each object in human perception and memory. Therefore, previous information about observed significantly contribute to cognitive processes.

The abundance of meaning and respecting the beholder to urban vertical is not the same, having in mind function, meaning and the purpose of the facility.

Without any doubt, the highest value through the history of city development, had religious buildings, so cities, nations and countries identified themselves with such facilities. Through 19th, 20th and 21st century, high buildings dedicated to multinational companies are also signs of a significant place in urban, country, and even in world surrounding (Sydney Opera House, Eiffel Tower, Big Ben, etc.).

Nowadays, cultural facilities are aiming upwards, so they also become significant landmarks with one element, proved to be dominant in space: height (Fig.9).

Not only the case of Philcharmonic in Hamburg can prove that the height is an important attribute of perceiving space, adding some significance to other facilities on that particular place. That is not a wonder, because it is clear that architectures of culture and sport in 20th and 21st century surpress religious facilities.

Fig.9. Filcharmony in Hamburg. Architects: Herzog & de Meuron.

For the last 100 years, vertical facilities in contemporary historical moment and developing of the city are not in such a good position because of mutual ruthless race for space overruling to show the power of multinational companies. [16]

Fig.10. Uncontrolled race of vertical buildings destroys hierarchy of urban core of modern cities.

That speaks volumes about negative appearance manifested in disturbing, previously clear, spatial hierarchy. That is no more the church tower or the mosque tower, but a multinational company building which can be endangered by another more powerful company within a year (Fig.10).

That is why vertical facilities need to regain their position recommended by historical experience. Especially for reason of it being a powerful tool in creating spatial hierarchy and meaning carrying.

References

1. A. Geremek, M. Greenlee, S. Magnussen *Perception Beyond Gestalt*. (Psihology Press. London and New York, 2014)
2. R. Arnheimm *Art and Visual Perception.* (University of California Press. Berkley, New York, London, 1974)
3. W. R. Crozier, A. J. Chapman *Cognitive Processes in the Perception of Art.* North-Holland (Amsterdam, New York, Oxford)
4. R. Arnheim *The Dynamics of Architectural Form*. University of California Press. Berkeley, (Los Angeles, London, 1977)
5. G. Bachelard *The Poetics of Space.* Book on Demand. (2012)
6. J. D. Borstin *The Creators*. Random House. (1982)
7. K. Linch *The Image of the City.* The M:I:T Press., Massachusettc Insdtitute of Technology, Cambridge, Massachusettc. (1972)

8. A. Ballantyne, *Architecture: A Very Short Introduction.* Oxford University Press. (2005)
9. R. Arnheimm *The Power of the Center.* (University of California Press. Berkley, New York, London, 1982)
10. R. Arnheimm *The Power of the Center.* (University of California Press. Berkley, New York, London,1985)
11. K. Lewin *A Dynamic Theory of Personality.* McGraw-Hill.(1967)
12. M. Eliade *The Sacred and The Profan.* A Harvest Book. Harcourt, Inc. (1957)
13. R. Arnheimm *Visual Thinking.* University of California Press. Berkley, New York, London. (1997)
14. R. Venturi *Complexity and Contradiction in Architecture.* The Museum of Modern Art, New York (1977)
15. S. Pinker *How the Mind Works.* W. W. Norton Company. (1999)
16. R. Koolhaas, B. Mau *S, M, L, XL.* Monacelli Press (1997)

The architecture and artistic features of high-rise buildings in USSR and the United States of America during the first half of the twentieth century

Svetlana Golovina[1], and *Yurii* Oblasov[1*]

[1]Saint Petersburg State University of Architecture and Civil Engineering, Vtoraya Krasnoarmeiskaya str. 4, St. Petersburg, 190005, Russia

Abstract. Skyscraper is a significant architectural structure in the world's largest cities. The appearance of a skyscraper in the city's architectural composition enhances its status, introduces dynamics into the shape of the city, modernizes the existing environment. Its architectural structure which can have both expressive triumphal forms and ascetic ones. For a deep understanding of the architecture of high-rise buildings must be considered by several criteria. Various approaches can be found in the competitive development of high-rise buildings in Moscow and the US cities in the middle of the twentieth century In this article we will consider how and on the basis of what the architectural decisions of high-rise buildings were formed.

1 Introduction

High-rise buildings in the USSR were difficult to realize for a long time. The reasons for this were in the devastating consequences of the October Revolution and World War II. As a result, the first high-rise buildings in Moscow were erected in 1947-1957. For the Soviet Union, this became a symbolic and breakthrough event. The architecture of high-rise buildings in Moscow has become a new state symbol. This architectural style in the late Soviet architectural criticism was called "Stalin's Empire."

High-rise architecture in the US was created from the middle of the XIX century. The main centers of high-rise construction were New York and Chicago. Nevertheless, by the middle of the twentieth century high-rise buildings were being built in other large cities. The construction of skyscrapers was a phenomenon that was practically and economically profitable for the US as a developing capitalist country. Each skyscraper performed special functions and tasks according to the requirements of the customer. This influence on the expressive architectural decisions of the early skyscrapers. They had detailed facades, architectural elements and forms that represented the popular architectural styles of those years: functionalism, eclecticism, modernity and art deco [1,2].

* Corresponding author: archrick@yandex.ru

The architecture of high-rise buildings in Moscow did not appear immediately. It was required to study foreign experience to create new architectural solutions. In 1934 a group of Soviet architects led by two leaders (Iofan B.M. and Shuko V.A.) went to New York. As a result of research on US architecture and partial borrowing of fundamental foundations Soviet high-rise designing developed. At that time the architects of Soviet Union were tasked with the state importance - to build the Palace of Soviets. This symbolic building had to be erected in Moscow. However, after the Second World War, a new idea was formulated - to build not one but several high-rise buildings in the center of Moscow. The Palace of Soviets should become the main building in a single ensemble. After long design work, the Palace of Soviets was never realized. But other high-rise buildings formed the planned ensemble already by 1957 [3].

Inspired by foreign architecture Soviet architects studied construction technology, engineering and design techniques. It was important to avoid the mistakes and difficulties faced by their American counterparts. The experience of borrowing gave new opportunities for high-rise construction in the Soviet Union.

In this article seven high-rise buildings in Moscow are taken for consideration. Several buildings built in the USA during the first half of the 20th century were also selected. The article aims to compare the Moscow and American architecture of high-rise buildings.

2 Methods

The architecture of high-rise buildings was carefully studied in a professional community. One of the most famous researchers of Soviet architecture was the critic Khan-Magomedov S.O. He gave the name of the architectural style of the post-war years - "the Stalin's Empire style." Various civil constructions of that period were built in this style, but it was most clearly show in the high-rise buildings of Moscow. Creators of "the Stalin's Empire style" were architects - Iofan B.M., Shchuko V.A. In addition to practical activities, these architects also explored Soviet architecture in the process of its development from a scientific point of view. In the period when construction of high-rise buildings of that period began the most skilled Soviet architect was Oltarzhevsky V.K. His experience was based on work in the construction of high-rise buildings in the United States from 1924 to 1935. Later Oltarzhevsky V.K. became the architect of one of the seven high-rise buildings in Moscow. It was he who became one of the first architects who compared the Soviet and American high-rise buildings including from the point of view of construction technologies.

High-rise buildings and their urban value were studied in American architectural criticism in the 1920s-1930s. Some position about this was provided by the builders of skyscrapers: Farris H., Kahn A., Kahn E. It is also worth mentioning the American architect Wright F. Despite the fact that among his projects there were also high-rise buildings Wright F. claimed that the skyscraper serves only the greed of homeowners.

High-rise buildings in Moscow form the shape of the city for more than 60 years. Their place in the architecture of the city has become part of the whole ensemble. Formation of the city center is the main compositional task performed by these objects.

In this article we will consider seven high-rise buildings: the administrative building at the Red Gate, the residential building on Kotelnicheskaya Embankment, the administrative building on Smolenskaya Sennaya Square (the building of the USSR Ministry of Foreign Affairs), the building of the Moscow State University name of Lomonosov on the Vorobyovy Gory (MSU building), residential building on Kudrinskaya Square, the building of the Hotel "Leningradskaya", the building of the Hotel "Ukraine".

From the extensive list of US skyscrapers, we choose the buildings closest in architecture. In general, we will consider buildings built in New York and several in other US cities. For the analysis we choose the following buildings: the Empire State Building, the Chrysler

Building, the Wall Street 40 (The Trump Building),the Chicago Board of Trade, the Manhattan Municipal Building, the Woolworth Building, the Tudor City, the Palmolive Building.

In this article, we will set the task of analyzing buildings on three criteria: the town-planning role, the architecture of buildings and the artistic features of buildings.

As the first criterion for analysis we consider the urban planning environment - the composite space formed around architectural sites in Moscow and the United States. We analyze the connection between a skyscraper and a city. We consider the natural landscape as the base on which the city was formed and how the perception of the landscape in the process of city development has changed.

As the second criterion we analyze the construction of buildings in Moscow and the United States. As a material for analysis we consider graphical schemes. On these schemes in generalized forms we compare the compositional character of Soviet and American buildings. We analyze the compositional formation of volumes and facades. As an illustration we suggest the aforementioned schemes.

The third criterion is to consider the artistic image of the objects. We identify the features of some historical objects applied in the details and in formation of high-rise buildings. In general terms we consider the symbolism and significance of the proposed examples. The task is to reveal the artistic character of the high-rise buildings of our study. For illustration we use photographic images of historical buildings and architectural details.

The proposed analysis will allow us to consider the high-rise architecture of the USSR and the USA in the middle of the 20th century from different sides. The expected result of the analysis is the identification of pros and cons in the architecture of the objects under study. It is also interesting to consider the origins of the development of high-rise construction as a global phenomenon. For discussion we make a conceptual analysis of the development of high-altitude architecture.

3 Results

3.1 Town-planning role of objects

In the process of mastering foreign experience, a number of problems arose before Soviet architects. The main problem was that the urban planning of American cities was based on an orthogonal network of streets. In formed quarters the skyscrapers were built. The situation in Moscow was completely different. Urban spaces were formed on the basis of centuries-old conditions. This structure was close to the old cities of Europe. For example, some high-rise buildings before the Second World War were erected in Europe as well. Despite this Soviet architects had to combine the ancient town plan and modern high-rise buildings. This task caused many difficulties. But in the process of the general reconstruction of Moscow since 1935, many modern solutions became available [4].

By the beginning of the 1930s in the United States there were more than a hundred high-rise skyscrapers. In the USSR it was impossible to create a larger or equal number of high-rise buildings. The task was to correctly create even a few buildings that could organize the urban space of Moscow.

The distinctive features of US cities also included a small distance between skyscrapers. This complex situation affected the excessive blackout of the streets. This problem is tried to solve with a reduction in the size of the building with each upper level. From the total mass of the buildings the skyscrapers were tried to allocate by expressive towers: pyramidal, terraced and other more complex top. Well thought detailing of the lower floors that was close to human scale also characteristic for the architecture of skyscrapers. In the architectural decisions of the Moscow high-rise buildings, it was necessary to take into account that many elements of the structures will be clearly visible from different points of view. From the building, a special expressiveness of the architecture was required. This aspect was closely related to the reconstructed urban environment of Moscow [4, 5, 6].

The urban environment of Moscow was dependent to the natural landscape of the very foundation of the city. The structure of the city was close to the principles of low-rise building structure. The grid of the streets was also influenced by the location of the Moscow Kremlin. The shape of the city was repeating the landform in which the dominants of temple towers, bell towers and other high-rise buildings were arising.

It is important to take into account that the architecture of Moscow is not completely preserved today. Several times the city was exposed to destructive fires and other disruptions under the influence of various anthropogenic factors.

The building of the Moscow State University (further MSU) was located on a high plateau on the bank of the Moskva River.

The building of the Hotel "Ukraine", residential building on Kotelnicheskaya Embankment, was built along the water area in some the places of bends and ramifications of the Moscow River.

The building at the Red Gate and the building on Kudrinskaya Square were included in the ensemble of the Garden Ring - the main street enveloping the historical center of Moscow.

Hotel "Leningradskaya" included in the composition of the formed Area of Three Railway stations as main part of the city.

The building of the Ministry of Foreign Affairs of the USSR became dominant, built along the axis of the city that oriented to the Borodino bridge.

As we can see, three of the four buildings were located in key points of the city's natural landscape, and the rest of the buildings were included in the city plan. The renewed urban environment also repeated the historical structure of Moscow and preserved its town-planning traditions.

New high-rise buildings have changed the skyline of Moscow. After the reconstruction, the streets and other urban spaces were significantly expanded. The perspective of the streets became more dynamic and was defined by compositional accents far apart from each other. It is worth noting that the modern construction of skyscrapers in Moscow was conducted delicately. New high-rise dominants were erected in the new district of the city.

The formation of US cities was completely different. The first feature was that the cities were built from scratch and quickly built up on the existing territory. While the US cities were relatively young in a historical context, they had an increased density of buildings and population. The second feature was that the main destruction of urban buildings was caused only by fires and various natural phenomena. They have not experienced a deliberately destructive impact of anthropogenic character and have retained some primary structure. The third feature was the formation of skyscrapers which was determined by the needs of economic growth. In the middle of the twentieth century, the world financial and industrial center developed in the United States. This fact became the reason for mass building of a skyscraper as a special building for a modern type of business. The most optimal was to place the entire company in one building, rather than in a complex of buildings, as it was before [6, 7].

The natural environment of New York and Chicago have a fairly flat landscape. The architecture of skyscrapers makes the urban environment similar to mountains and canyons. High-rise buildings in the US cities completely change the landscape and dominate the earlier architecture of average and low-storey buildings. Over time this environment has become a new rapidly spreading worldwide phenomenon of urban planning.

 https://doi.org/10.1051/e3sconf/20183301032

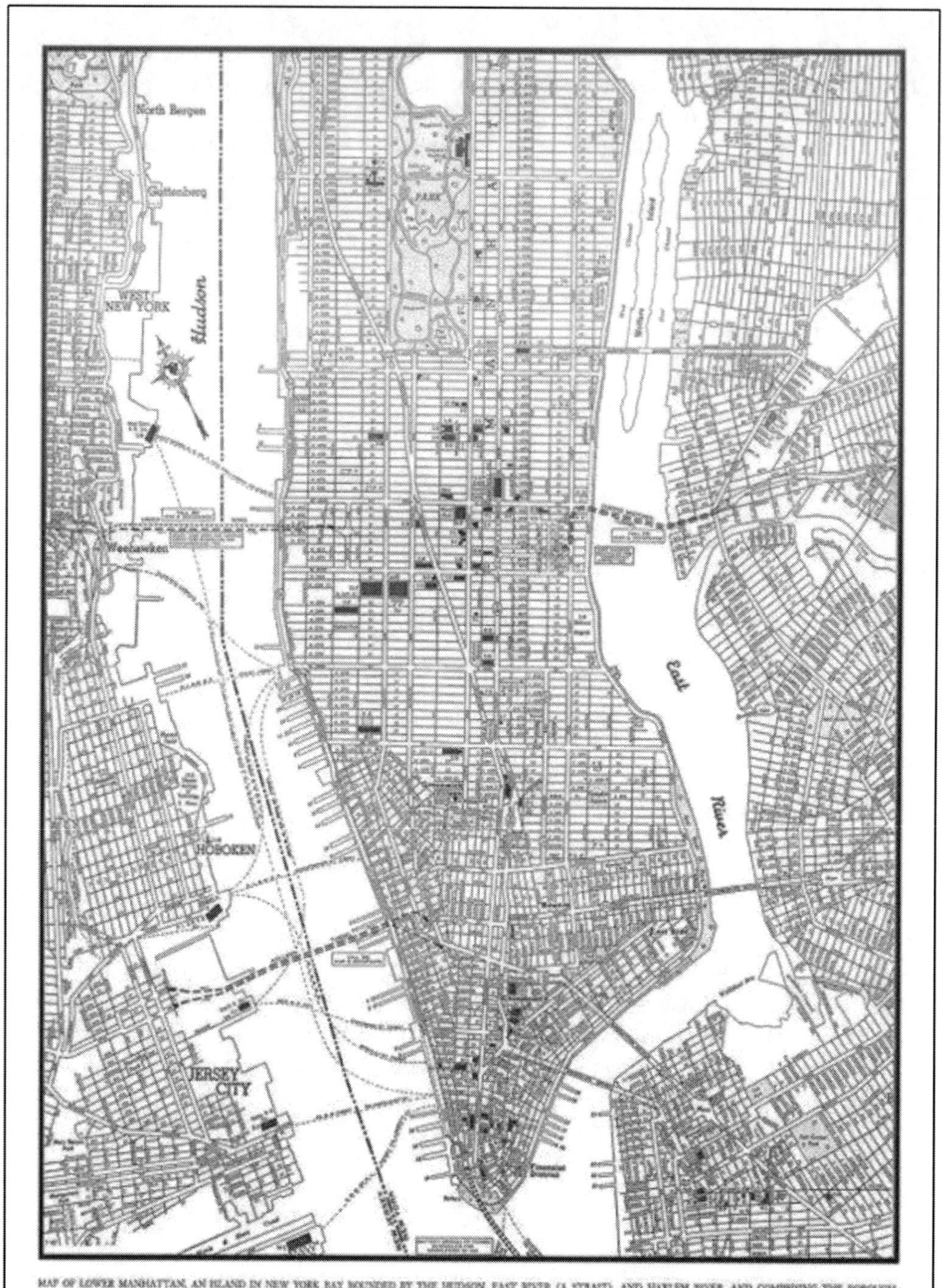

Fig. 1. New York city Manhattan street map. 1944. [21]

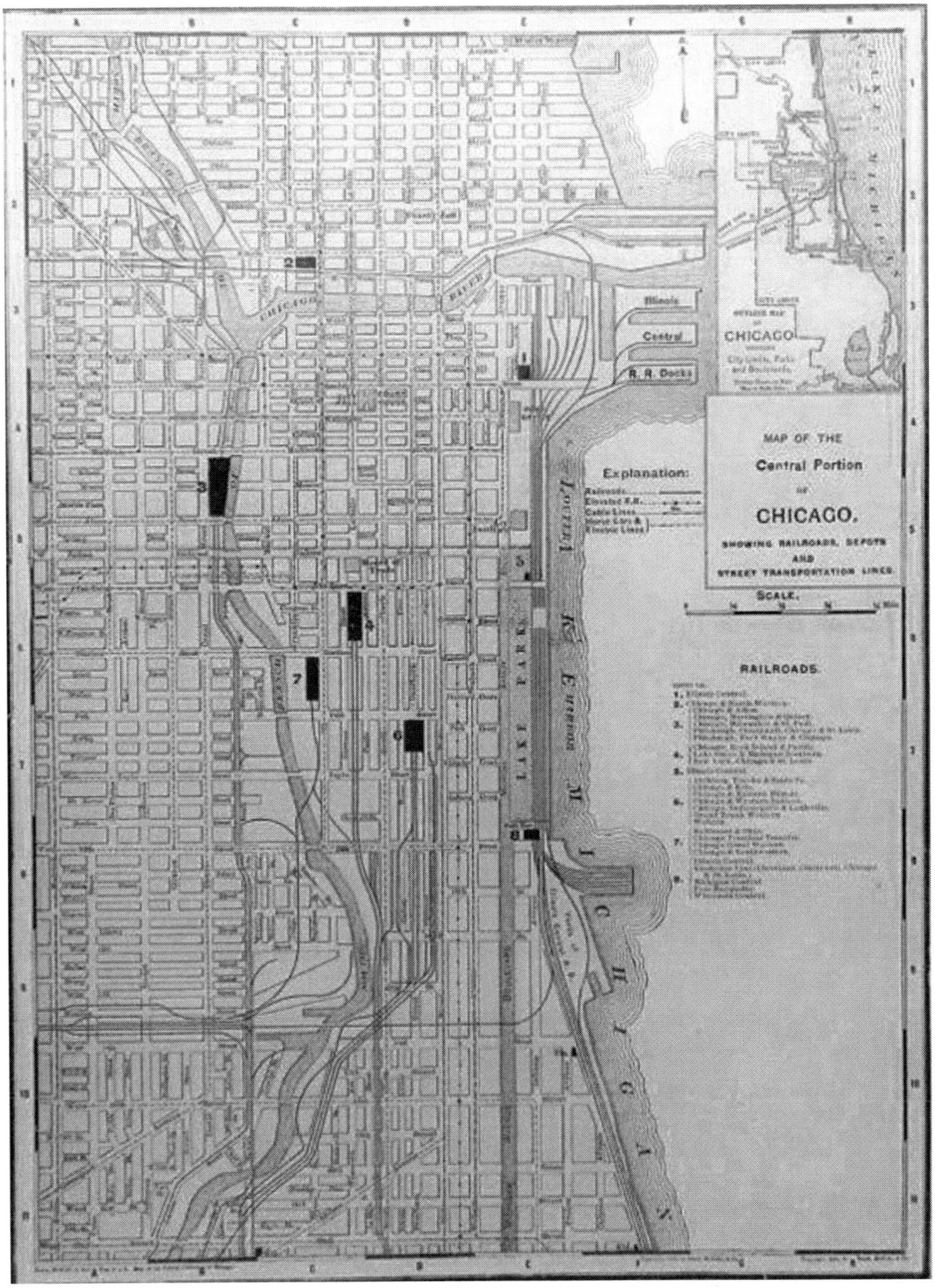

Fig. 2. Chicago city street map. 1901s. [22]

 https://doi.org/10.1051/e3sconf/20183301032

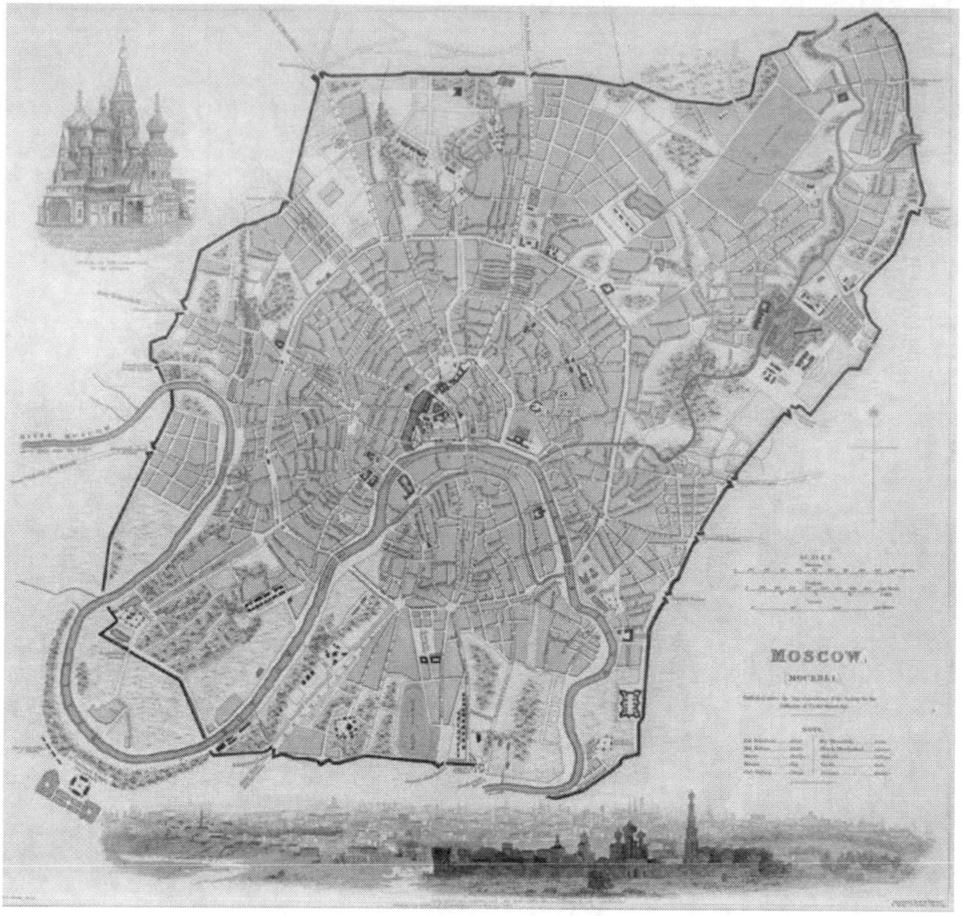

Fig. 3. The city plan of Moscow. 1836. (Maps of the Society for the Diffusion of Useful Knowledge.) [23]

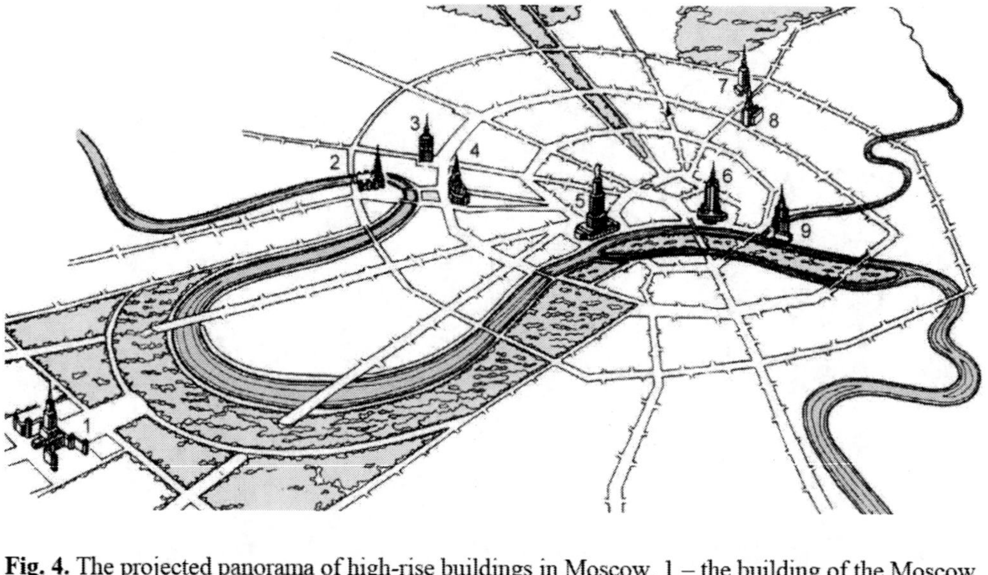

Fig. 4. The projected panorama of high-rise buildings in Moscow. 1 – the building of the Moscow State University name of Lomonosov on the Vorobyovy Gory (MSU building), 2 – the building of the Hotel "Ukraine", 3 - the residential building on Kudrinskaya Square, 4 – the administrative building

on Smolenskaya Sennaya Square (the building of the USSR Ministry of Foreign Affairs), 5 - the Palace of Soviets (not built), 6 – the administrative building in Zaryadye (not built), 7 – the building of the Hotel "Leningradskaya", 8 – the administrative building at the Red Gate, 9 – the residential building on Kotelnicheskaya Embankment. [24]

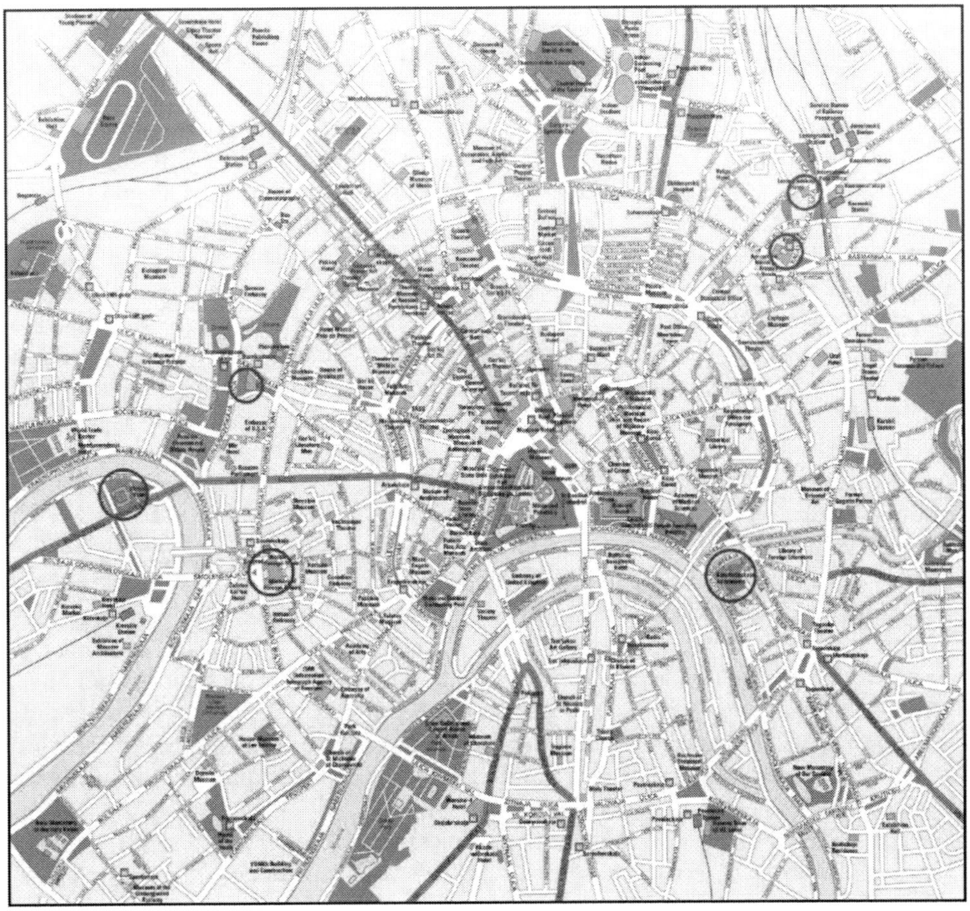

Fig. 5. Street map of downtown Moscow. Present time. High-rise buildings are marked with a red circle. The building of the Moscow State University outside the borders of the map [25]

3.2 Shape and composition of buildings

Soviet architects found out when designing high-rise buildings in Moscow that the shape of the building is very important for perceiving the object from different points of view. The experience of architects of the USA offered basically a tower type of a building. The main characteristic of tower-type skyscrapers was the reduction of the floor area with each new level. The level could included more than ten floors. This method was increasing the perspective of the building and reduced the cast shadow.

For the visual facilitation of buildings in Moscow, a multilevel construction principle was also applied. But the ledges of terraces were almost not revealed by the details on the facades in the US architecture. In architecture Moscow high-rise buildings, it was important to show the division into levels. The boundaries of the levels were richly decorated with architectural details. A clear division of volumes and facades was created. Thus an effect was obtained that softens the scale of the high-rise structure. This design of the boundaries of the levels has become characteristic of most buildings in Moscow. It should be noted that the first level was often elongated in the plan. At the same time the following levels formed the volume of a building similar to a towers of US skyscrapers [8, 9, 10].

In article we proposed schemes considered the shape of high-rise buildings in Moscow. Among them we can find the similar image. A characteristic feature is a form that remind the Kremlin towers. In Moscow houses we see a multilevel principle of volumes construction. Towers have the identical shape and similar spires.

Similarity of forms is also present in the skyscrapers of the United States. In the building of Wall Street,40 (Trump Bulding) the tower is formed from the more massive volume of the building that occupies the quartal. On the diagrams one can see that the multilevel principle of the formation of the skyscrapers towers in American practice has one more feature. In US skyscrapers, the levels are distributed mostly uniformly. In some buildings, the area of floors combined in levels does not differ significantly from the floors of subsequent levels. As an example, could be considered the building of Tudor City and the Manhattan municipal building they have the minimal difference in floor area.

The skyscrapers of the United States buildings had different top of the structures like at the Moscow objects. Often buildings were finished with multi-storeyed volumes which function as ordinary floors of the building. But the Moscow top of the buildings were rich decorated. The big problem was to see architectural details and decorations at the high levels in the close quarters of skyscrapers. The skyscrapers were perceived better on the terraces and roofs of neighboring buildings. This factor greatly influenced the changing architecture of skyscrapers during the twentieth century.

On the diagrams one can see the similarity of Moscow high-rise buildings and American skyscrapers. For example, the building of the hotel "Leningradskaya" and the building of the hotel "Ukraine" are close in shape to the towers of American skyscrapers. The principle of the volume of the building Palmolive Building and the Chicago Chamber of Commerce has in common with the building of the Ministry of Foreign Affairs and the building on Kudrinskaya Square. The shapes of the Empire State Buiding and the Woolworth Building towers that form of a dynamic axis similar to the principle of the MSU building main tower. The heavyweight composition of Tudor Building is close to the building on Kotelnicheskaya embankment.

It can be argued that many principles of building skyscrapers, which were borrowed from American practice, were largely reworked and adapted by Soviet architects. A set of architectural methods developed in the practice of building skyscrapers, found a new unique realize in the USSR.

Table 1. Graphical schemes of Moscow high-rise buildings

The administrative building at the Red Gate	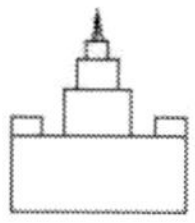
The residential building on Kotelnicheskaya Embankment	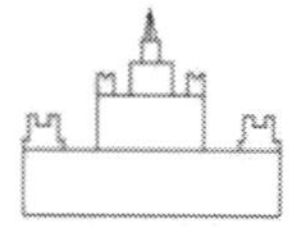
The administrative building on Smolenskaya Sennaya Square (the building of the USSR Ministry of Foreign Affairs)	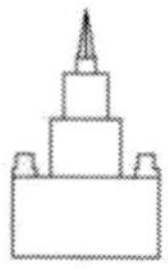

The building of the Moscow State University name of Lomonosov on the Vorobyovy Gory (MSU building)	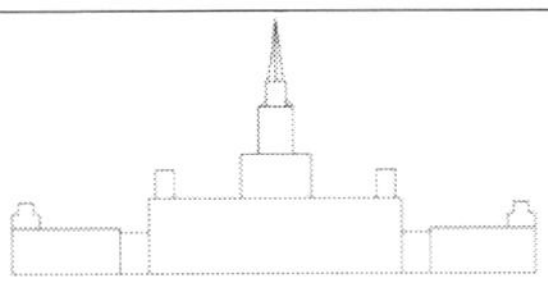
The residential building on Kudrinskaya Square	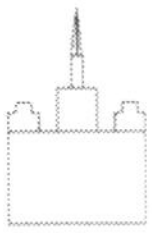
The building of the Hotel "Leningradskaya"	
The building of the Hotel "Ukraine"	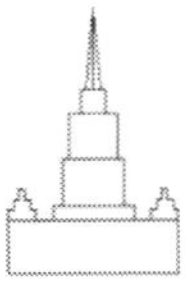

Fig. 5. The pictures of Moscow high-rise buildings

Fig. 6. The project for the construction of the Komsomol square (The square of three stations). [18]

Table 2. Graphical schemes of US high-rise buildings

the Chrysler Building	
the Empire State Building	
the Palmolive Building	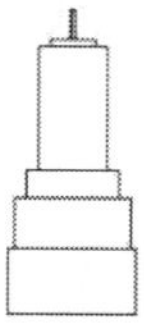
the Wall Street 40 (The Trump Building)	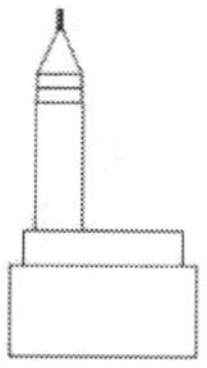
the Tudor City	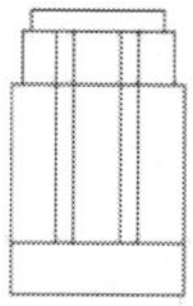
the Woolworth Building	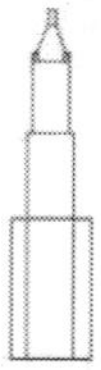
the Manhattan Municipal Building	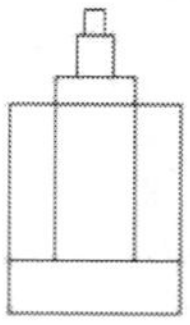
the Chicago Board of Trade	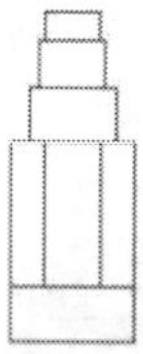

Fig. 8. The pictures of US high-rise buildings

Fig. 7. The New York: 42nd Street View. (Andreas Feininger Archive/Courtesy Zeppelin Museum Friedrichshafen). [26]

3.3 Imagery

In this section will considered the artistic features of the architecture of a high-rise building. Consider the artistic analogies of monuments from the history of architecture with high-rise buildings in America and Moscow.

Could be noted the direct parallel of the tower of the American skyscraper, with the shape of the obelisk. Obelisks were erected in memory of outstanding events and had a sacred significance. The top of the buildings in US skyscrapers is close to the obelisk, for example, the building of Wall Street 40 (Trump Bulding) has a four-sided pyramidal finish.

The image of the Gothic churches is also traced in the architecture of the buildings being studied read in the form. The main feature of the Gothic architecture the vertical ribs on the facades of the buildings. The same that rows of windows form ribs and niches on facades of skyscrapers. The architecture of the Gothic churches and towers is dynamic. At the same time Gothic architecture fulfilled a strict ideological role for the maintenance of religion and monarchical power in European countries. Many early skyscrapers in the US and Soviet high-rise buildings are related to the neo-Gothic eclectic tendencies of the early twentieth century. On the parapets of high-rise buildings in American and Soviet practice, it is easy to find a decor applied according to the principles of Gothic architecture. Among such architectural elements it is worth mentioning, arrows, pinnacles and other details [11, 12, 13].

The shape of Moscow high-rise buildings is closer to Orthodox church architecture. The principles of the formation of vertical dominants and stable horizontal composition that is typical for this architecture. As an example we can study the Church of the Intercession in Fili, Sukharev Tower and other buildings of early Moscow architecture.

The artistic image of the triumphal arch can be traced in the Moscow buildings. Arch was a majestic object much larger than the scale of human growth. A spectacular gigantic structure was created through which had passing the parades. You can find a similar principle in the architecture of high-rise buildings in Moscow. Especially the first levels of buildings that was stretched horizontally had the large scales. The lower level of the building of the Ministry of Foreign Affairs of the USSR was formed by a large-scale colonnade. There was realized principle of temple propylaeum close to the architecture of ancient Greece. Also in Soviet buildings, details of fortress architecture were used - arrows, swallowtails.

The high-rise buildings studied in this article have architectural details and forms used in the history of architecture. This phenomenon can be explained by the neoclassical and eclectic architectural trend that dominated with avant-garde in the first half of the twentieth century. However, the appeal to traditional forms gave the high-rise buildings a certain appearance. The buildings displayed an artistic image of triumph and success. The buildings with that architecture characters seems quite logical and consistent if we consider it in the complex political context of that period.

Fig. 8. The orthodox churches architecture.

https://doi.org/10.1051/e3sconf/20183301032

Fig. 9. The Egypt obelisks

Fig. 10. Forms of obelisks and ancient Russian architecture in high-rise buildings of Moscow. [18]

Fig. 11. The gothic churches architecture

Fig. 12. The examples of triumphal arks

Fig. 13. The forms of Gothic architecture in skyscrapers of New York. [27]

4 Discussion

The architecture of American skyscrapers developed during the twentieth century. This process has formed a different form of buildings. All the features of high-rise architecture gradually transformed into a new functional form. As a result, the modern appearance of skyscrapers in the cities of the USA arose under the influence of the early practice of their construction.

The building environment formed by skyscrapers has become an example of urban space contrasted with the established canons of classical architectural composition.

Moscow skyscrapers absorbed the diverse architecture of the city. But their further influence on the development of urban space did not last long. Immediately after the construction of high-rise buildings, the state architectural style changed. It was decided to abandon unnecessary details in the design of buildings. Further high-rise buildings differed from those built in "Stalin's Empire style." We can consider example of a high-rise dominant is the Council House of the RSFSR, built in 1979. This building has an ascetic architecture, but repeats the general compositional techniques of the first high-rise buildings [14].

The investigated structures were formed in opposing economic and political systems being both a visible symbol of success, status, and a rational solution. Moscow high-rise architecture has formed in the USSR the image of the capital renewed after wars and deprivations. Buildings in the center of the city presented a new Soviet state with a new ideology. It showed the capability to significant material achievements [15].

American high-rise constructions marked the newest image of a modern successful city. It became a symbol of top the era of human conflict with nature. Finally, it could be affirmed., that the natural landscape and environment completely subordinated to the genius of human thought. Arrays of skyscrapers and in the modern world are an example of the economic superiority of the city. The image of excellence and triumph in architecture is now formed not by a single pillar-like monument, but by a set of dominants that form a single complex of high-rise buildings that prevails over a human scale [16, 17].

Despite the active development and globalization of skyscrapers as an architectural unit - it is difficult to consider a skyscraper as a future architecture. In the era of depletion of resources, and heavy burden on the environment - the skyscraper acts as structure that contradicting to these tasks.

It is possible to allocate the global role of each of the considered high-rise architectural concepts according to the results of the research. American architecture marked a new scale of construction, optimized logistics costs for people working in the same system, turning a high-rise building into a commercially viable mechanism. Moscow architecture proposed a different approach. It archaized the imagery of high-rise buildings, to the imagery of medieval and ancient architecture. The skyscraper in Moscow was erected as a monument, as an image of triumph and will. A rational approach was proposed - to save resources, several such monumental buildings and temple-like buildings. It was sufficient to maintain sustainable development and high morale of the society. But widespread high-rise buildings destroy the perception of the building as an outstanding architectural monument [19, 20].

Perhaps in our time, when the construction of skyscrapers is conducted in a number of cities in the world, it makes sense to reconsider the role of high-rise buildings. It must be found a way to preserve the existing architectural landscapes. It need to not allowing them to be suppressed by dominant ones and to indicate the boundaries that are acceptable for the urban development of similar structures.

5 Conclusion

Based on the results of the article, several factors can be identified that relate and contrast the architecture of Moscow high-rise buildings and American skyscrapers. Our task was to reveal the features of buildings and also to designate the imaginative and philosophical role of high-rise constructions. To emphasize the positive aspects of the development of high-rise architecture in Moscow and the United States.

References

1 P.V. Abrosimov, A.V. Voronkov *Construction of a high-rise building Moscow State University* (Urban Management, Moscow, 1952)

2 I. Andreev *Joseph Stalin's cathedrals* (The Echo of the Planet,2013)

3 K. K. Antonov *High-rise buildings in Moscow - an outstanding achievement* (Soviet architecture and construction equipment, Moscow, 1953)

4 E. Bernasconi How rose high-rise Moscow, Echo of the planet, 2007, **35**, 40 - 42.

5 A.A. Vaskin *Architecture of "Stalin" skyscrapers. Residential house on Kudrinskaya street* (Questions of the humanities, 2007)

6 Yu. P. Volchok Multidimensional spaces of the city. High-rise buildings in Moscow, The Moscow Journal, 2011, Appendix: the history of high-rise construction in Moscow, 2 - 32.

7 A.I Kokurin, Accept the proposal of Comrade Stalin, Resolution Council of Ministers of the USSR on the construction of new buildings of the Moscow State University Lenin Hills, 1947 – 1954, Historical archive, 2004.

8 N. N. Kruzhkov *High-rise buildings in Moscow: facts from history design and construction, 1947 - 1956.* (Samara, 2007)

9 V. Lishak *Architecture of Stalin's high-rise buildings* (The World of Stamps, 2013)

10 K. P. Mikhailov, The jump in height. High-rise buildings of Moscow as creations architecture of the fourth dimension, Moscow heritage, 2008, **6**, 44 - 51.

11 V.K. Oltarzhevsky *Construction of high-rise buildings in Moscow.* (Moscow, 1953)

12 B. Rubanenko, Architecture of a high-rise building on Kotelnicheskaya embankment in Moscow, Architecture of the USSR, 1952.

13 I. Semenova *The point of architecture is to serve people!* (Moscow heritage. – 2014)

14 A. Tarkhanov Moscow skyscrapers, Kommersant, 1997.

15 N. O. Shashkova *Stalin's high-rise buildings: the symbol of the epoch and our heritage* (Architecture and construction of Russia. – 2015)

16 Christ-Janer *Christ-Janer A. Eliel Saarinen: Finnish-American Architect and Educator* (Chicago: University of Chicago Press, 1984)

17 Stern 1994 – *Stern R.A.M. New York 1930: Architecture and Urbanism between the Two World Wars* (New York: Rizzoli, 1994)

18 N. Kuleshov, A. Pozdnev. *High-rise buildings of Moscow.* (Moscow, Moscow worker, 1954).

19 https:archi.ru/russia/72752/ar-deko-i-istorizm-v-arkhitekture-moskovskikh-vysotnykh-zdanii

20 http:retrofonoteka.ru/sovarch/history/index.htm

21 https://www.etsy.com/listing/89167799/vintage-map-of-manhattan-new-york-city

22 https://redbudart.com/2012/06/25/antique-maps-of-american-cities/

23 http://www.p-porfir.ru/oldmap/w024o.htm

24 https://moscow.drugiegoroda.ru/attractions/19581-stalinskie-vysotki/

25 http://www.mappery.com/map-of/Moscow-Map

26 http://www.zeit.de/kultur/kunst/2012-06/fs-new-york-2

27 http://www.2photo.ru/en/post/30781

Architectural and town-planning reconstruction problems of the city of Voronezh

Tatyana Mikhaylova [1,*] *Dmitriy* Parshin[1], *Vitaly* Shoshinov[2], *Anatoliy* Trebukhin[2]

[1]Voronezh State Technical University, Moscow Avenue, 14, Voronezh, 394026, Russia
[2]Moscow State University of Civil Engineering, Yaroslavskoe sh. 26, Moscow, 129337, Russia

Abstract. The analysis of the state of the historically developed urban district of the city of Voronezh is made. The ways of solving the identified architectural and urban problems of reconstruction of historically developed buildings are proposed. The concept of reconstruction of a territory with historical buildings along Vaytsekhovsky Street is presented.

1 Introduction

At the present time, along with the problem of preserving and restoring historically significant buildings and architectural monuments, the number of which is steadily decreasing not only in Voronezh, but throughout Russia, there is another equally serious problem - the loss of the city's unique architectural appearance.

The situation is complicated every year by a pragmatic approach to building, in which not only new districts but also the historical center is being lined with typical high-rise apartment buildings and office centers characterized by a lack of individuality and destroying the architectural-artistic appearance of streets and the city as a whole despite the fact that the pace of construction now exceeds the needs, both in housing and in office buildings.

2 Materials and Methods

To analyze the condition of the historical development of the urban district of Voronezh, a visual inspection of the analyzed territory was carried out, and local experience of new construction, reconstruction and restoration of existing buildings, the documentation regulating the town-planning activity on the territory of the urban district of the city of Voronezh has been studied.

As the main documents and laws that served as the basis for the analysis of the historical development of the urban district of the city of Voronezh, the following documents and laws were used: the town-planning code of the Russian Federation; Federal Law of 17.11.1995 No. 169-FZ "On Architectural Activity in the Russian Federation"; federal law "On objects of cultural heritage (monuments of history and culture) of the peoples of the Russian Federation" of June 25, 2002 N 73-FZ; the resolution of the Government of the Voronezh Region "On the approval of the boundaries of the zones for the protection of objects of cultural heritage of

* Corresponding author: mtvit.1955@mail.ru

federal significance located on the territory of the urban district of the city of Voronezh, land use regimes and requirements for town planning regulations within the specified zones"; Decision of the Voronezh City Council of December 19, 2008 n 422-ii "On approval of the general plan of the urban district of the city of Voronezh"; Decision of the Voronezh City Council of December 25, 2009, No. 384-ii "On Approval of the Rules of Land Use and Development of the Urban District of the City of Voronezh"; Decision of the Voronezh City Council of 19.06.2008 №190-II "On Approval of the Rules for the Improvement of the Territories of the Urban District of the City of Voronezh", design regulations "Appearance of the facades of buildings and structures in the urban district of the city of Voronezh".

3 Results

In the process of analyzing the preserved historical built-up areas of the city of Voronezh, two conditional zones were identified: the central part of the city and its peripheral development on the slopes of the Voronezh reservoir.

When reconstructing existing built-up areas, historically important buildings and monuments of architecture, which status was not included to cultural heritage objects, often fall under demolition. At the same time, for the identified objects of cultural heritage, there is also a danger of being liquidated or modified.

Practically all the heritage buildings, representing historical and architectural value, are on the balance of cities and regions, and attraction of external funds for their maintenance is unlikely due to the peculiarities associated with their operation and economic profitability.

Due to its age, the condition of most historically significant buildings requires immediate major repairs and restoration work. Engineering systems in such buildings are either subject to complete replacement, or simply are absent, and the existing space-planning decision does not meet current requirements [1].

The adaptation of such buildings for commercial use requires re-planning or the construction of additional floors [2]. However, the possibility of such works is limited in connection with the ban on the change of the main load-bearing elements and the external appearance of the building as a whole.

Thus, the objects of cultural heritage do not represent economic value for private investors, since the profit from the construction of new multistory buildings is much greater than from the reconstruction of existing buildings. At the same time, the amount of budget funds to maintain all historical monuments in proper condition is not enough, which leads to their gradual dilapidation and destruction.

The existing provisions in the documents on protection of cultural heritage objects set forth in the Federal Law "On Objects of Cultural Heritage (Monuments of History and Culture) of the Peoples of the Russian Federation" of June 25, 2002 N 73-FZ, are not able to protect historical monuments from unauthorized alteration and liquidation. In particular, the danger to objects of cultural heritage is the allocation of any one part of the building as an object of legal protection, for example, a facade. This approach allows the elimination of the remaining unprotected part of the building, which will lead to a complete loss of its architectural and historical value.

In order to preserve and form the integral architectural-artistic appearance of the city of Voronezh, the requirement of the conformity of the architectural appearance of the new buildings and structures to the project of approved documentation was enshrined in legislation in such documents as: design regulations "Appearance of facades of buildings and structures in the urban district of the city of Voronezh"; Decision of the Voronezh City Council of 19.06.2008 № 190-II "On the approval of the rules for the urban beautification of the territory of the urban district of the city of Voronezh.". Compliance with these requirements should be monitored by representatives of the special monitoring service at all stages of construction.

In fact, completed projects can significantly differ from the approved sketches, what is excluding the possibility of achieving the stylistic unity of new buildings with existing buildings. One of the examples of the discrepancy between the finished project (Fig. 2) and the approved sketch (Fig. 1) is the residential complex "Heart of the City", located at: Voronezh, Kukolkina street, 11.

Fig. 1. Approved sketch of the residential complex "Heart of the City".

Fig. 2. Actual appearance of the residential complex "Heart of the City".

4 Discussions

It should be noted that the main value of the historical center of the city is not only its historical past, but also its architectural heritage [3].

In the conditions of the modern city, the preserved historical buildings are attractive for their color, decorativeness, decoration of facades and best proportions.

That is why the reconstruction of the historical center should assume the formation of a new and modernization of existing environment, repeating the style of architectural monuments.

As measures to create and restore the architectural expressiveness of the city, along with the current decree of the government of the Voronezh region "On the approval of the boundaries of protection zones of cultural heritage sites of federal significance located on the territory of the urban district of the city of Voronezh, land use regimes and requirements for town planning regulations within the boundaries of these zones", a plan with the division of urban area into zones should be developed within which the construction of new and restoration of existing buildings and structures will be implemented in a single architectural style, which will lead to the establishment of a single architectural-artistic appearance of streets and districts. A prerequisite is also to ensure control over the conformity of the external appearance of buildings and structures to the established architectural and compositional requirements when approving sketches, as well as at the stages of construction and operation.

To attract private investments for the restoration and exploitation of cultural heritage objects, a system of state and municipal preferences should be used that reduce the amount of rent, taxes or other conditions favorable to private investors in exchange for the restoration of heritage building.

In order to exclude the possibility of dismantling certain elements of cultural heritage objects in the process of exploitation or reconstruction, it is necessary to stop using the term "subject of protection" in the laws fixing the term, in connection with the possible misinterpretation of the present term, permitting the modification or liquidation of other structures not included in the category of protected objects. A necessary decision is also the maintenance of an inventory listing all elements and parts of the cultural heritage object for the purpose of lawful application of penalties to their owners and tenants in case of loss of individual elements of protected objects or uncoordinated changes introduced into their design.

Within the framework of the concept for the reconstruction of a zone with historical buildings developed by a student of the Voronezh State Technical University of the Department of Housing and Communal Services, it is proposed to carry out a complex of reconstructive works along Vaitsekhovsky Street.

According to Appendix No. 2 to the decision of the Voronezh City Council of December 24, 2014 No. 1694-III "Map of functional zones of the urban district of the city of Voronezh," the section of Vaytsekhovsky Street limited by Sacco and Vanzetti, Kaliayev and Tsyurupy streets refers to the zone of construction of three-floored buildings. In this area one- and three-floored stone, wooden residential and administrative buildings are situated (Fig. 3).

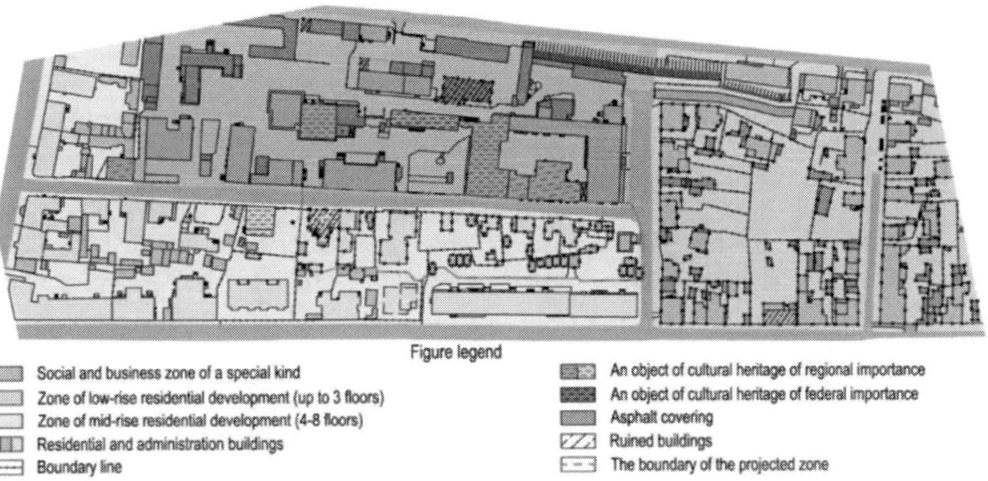

Fig. 3. Scheme of functional zoning of the Vaitsekhovsky street.

In accordance with the list of objects of cultural heritage (monuments of history and culture) of the Voronezh region, 4 cultural heritage sites of regional importance are located on the territory of the street, one of which is the "Tulinov Manor", also known as "Vigel's house", located on the territory in the quarter of the streets of Vaytsehovsky - Kalyaeva - Dostoevskogo. The building was constructed in the 18th century and executed in the Baroque and Empire style, and also includes elements of Gothic architecture (Fig.4);

Fig. 4. The current state of the house of Vigel (Tulinov manor).

Vaitsekhovsky Street is of great historical importance, both for the city of Voronezh and for the entire Voronezh region, and its historical and architectural heritage should be restored and preserved for future generations. In this regard, the first stage of the reconstruction of the territory along Vaytsehovskogo Street suggests the restoration of cultural heritage sites, with the restoration of the bearing capacity of structural elements.

In order to create a more comfortable environment and an expressive architectural background, it is proposed to assign the status of the pedestrian zone to Vaitsekhovsky Street. This solution will not only improve the conditions for acquaintance with architectural monuments and for living in the area, but also create a unity of architectural composition of the surrounding buildings and the street space.

At the same time, carrying out only restoration work under the scope of the reconstruction of the considered zone will allow solving architectural issues of cultural heritage objects, however, the issues of renewal of housing stock and landscaping will remain unresolved.

It should be noted that the majority of the housing stock in this zone has a service life of more than half a century, which implies the existence of high indicators of physical and moral deterioration of both the buildings themselves and the utility systems.

The considered site of historical development is characterized by inefficient use of the urban area in view of the predominance of low-rise individual housing and the use of significant areas for individual garages accommodating (Fig. 5).

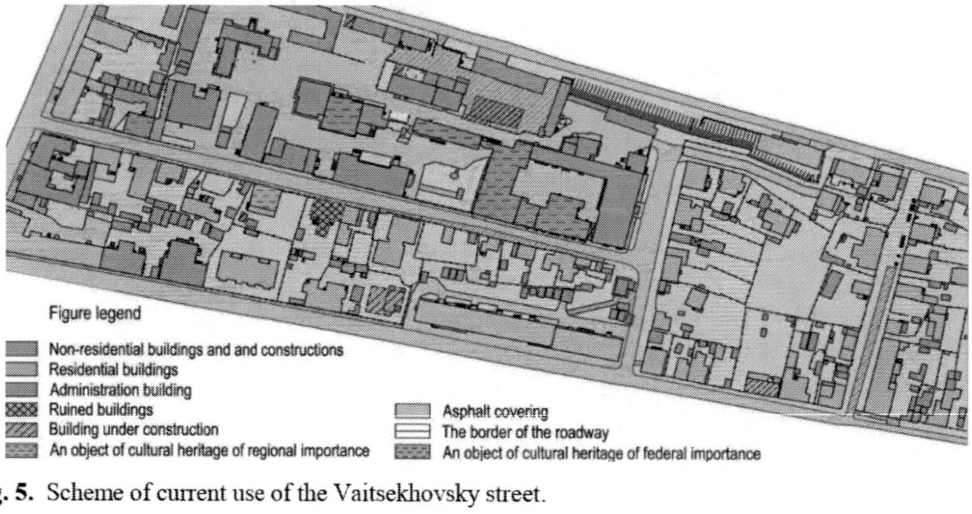

Fig. 5. Scheme of current use of the Vaitsekhovsky street.

In order to comply with architectural, planning and economic requirements, it is proposed to replace low-rise individual buildings with three-story townhouses [4], taking into account the stylistic features of existing objects of cultural heritage. The variant of the architectural execution of the facades of the townhouse, developed within the framework of the proposed concept for the reconstruction of the territory along Vaitsekhovsky Street, is shown in the Figure 6. This solution will provide additional living space within the boundaries of the reconstructed territory, equalize the total number of storeys of the building and lead to achieving a stylistic unity of the context area with the preserved architectural monuments.

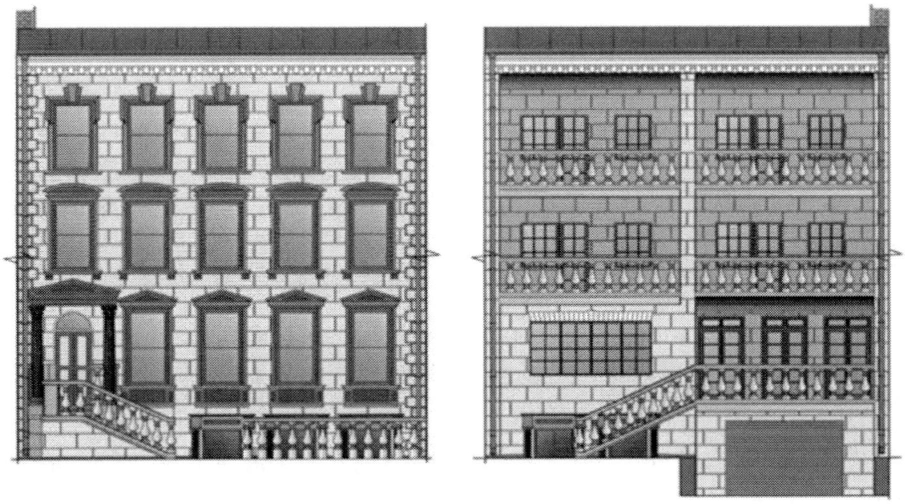

Fig. 6. Street and courtyard facade of the townhouse, developed within the framework of the proposed concept for the reconstruction of the territory along Vaitsekhovsky Street.

It should also be taken into account that the assignment of the pedestrian zone to the Vaitsekhovsky street, the restoration of cultural heritage sites, as well as the renewal and increase of the housing stock as part of its reconstruction will lead to the need to improve transport and pedestrian accessibility, namely: the expansion of the pedestrian zone, the creation of tourist and pedestrian routes and ensuring their connection with the main highways, walking and recreational areas of the city.

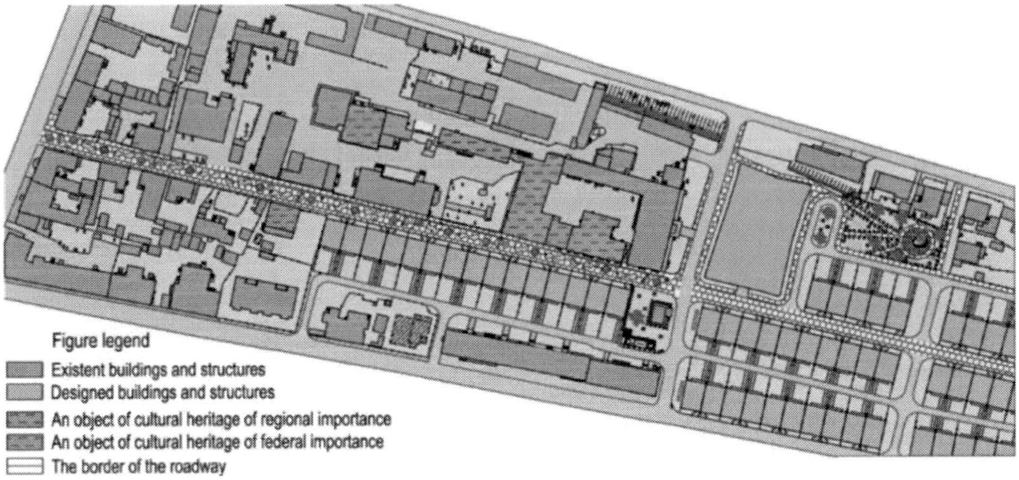

Fig. 7. Project proposal for the scheme of the planning organization of the Vaitsekhovsky street.

To solve the above problems and improve the overall transport links of the reconstructed territory, it is proposed to connect the streets of Tsyurupa and Dostoevsky by extending the Kalyaev street, as well as extending the Vaitsekhovsky street to the Massalitinov embankment (Fig. 7).

Such a solution will allow the creation of a new tourist route along Vaitsekhovsky Street, connecting the previously proposed tourist routes along Sacco and Vacnetti Street [5], Masalitinov Embankment, and Sevastyanovskiy Descent [6], into a single tourist route "Old Voronezh" on the basis of the Voronezh government decree and development program "Voronezh-tourist", making the city of Voronezh more attractive and historically valuable, both for citizens and tourists visiting the center of the Chernozem region and one of the oldest cities in the center of Russia.

5 Conclusion

As a result of the analysis of the historical development of the urban district of the city of Voronezh, it was revealed: 1) the need to improve the regulatory and legal framework governing urban planning; 2) the need to preserve the existing architectural image of the city when developing its reconstruction projects in order to preserve and recreate the artistic and architectural appearance of the city;

The proposed ways of solving the identified architectural and urban problems of reconstruction of historically developed buildings are aimed at creating and restoring the architectural expressiveness of the city, preserving and restoring the cultural heritage of Russian cities;

The proposed concept of reconstruction of the territory with historical buildings along Vaitsekhovsky Street is aimed at renewing the housing stock, creating a comfortable environment for living and acquaintance with architectural monuments, as well as achieving a stylistic unity with the preserved architectural monuments.

References

1. V. Fedorov, *Reconstruction of buildings, structures and urban buildings* (INFRA-M, Moscow, 2008)
2. D. Kincaid, *Adapting building to changing uses. Guidelines for change of use refurbishment* (London, 2002)
3. T. Bradecki, B. Uherek-Bradecka, Advanced Engineering Forum, **12** (2014)
4. I. Chereshnev, Housing construction, 11 (2007)
5. A. Guzeva, Urban planning, infrastructure, communications, 3 (2016)
6. V. Semenov, T. Mikhailova, O. Ermolenko, Urban Development, **5**, 33 (2014)

Formation and maintenance of the territorial balance in the conditions of high-rise construction in the city

Tatiana Kisel [1,*]

[1] Moscow State University of Civil Engineering, Yaroslavskoye shosse 26, Moscow, 129337, Russia

Abstract. High-rise construction results in the need of planning of infrastructure facilities, taking into account the increase in loading, as high-rise construction allows to place considerably bigger number of residents in the limited territory. For this purpose it is necessary to estimate the required and actual level of providing the population with each particular type of the facilities of social infrastructure. The compliance of required and actual level of providing can be characterized as the territorial balance, while the discrepancy acts as the territorial imbalance. The article is devoted to the development of such instruments of planning of urban development, which will allow to create the qualitative urban environment, founded on the territorial balances. Namely, it is devoted to the calculation of level of providing the population with the facilities of social infrastructure, to the determination of level of the imbalance in absolute and relative units and also to the ranging of imbalances on urgency of their elimination. The size of the imbalance is of great importance for planning and realization of managerial influences from the executive authorities, operating the city development. In order to determine the urgency of realization of actions for the construction of facilities of social infrastructure it is offered to range the imbalances according to their size, having determined the deviation size from balance, which is so insignificant that it does not demand any managerial influences (it can be characterized as balance) and also the groups of the imbalances, differing in urgency of managerial influences, directed to the decrease and elimination of the revealed imbalance.

1 Introduction

Modern megalopolises cannot be considered without high-rise construction. Projects of high-rise construction allow to use expensive city land, which is the most valuable resource, in rather compact territories with high efficiency. However, the use of high-rise construction in urban development proves the need of the use of new approaches to management of city development. One of the tasks of management of urban development is the territorial balance, which assumes maximum providing urban population with the necessary infrastructure. During planning the city development providing the population with the necessary infrastructure facilities, such as schools, kindergartens, hospitals and

* Corresponding author: doremi2@yandex.ru

clinics, facilities of culture and sport is to be considered. The existence of enough such facilities in comfortable transport availability is the necessary condition of proper activity in the city.

The modern cities and megalopolises develop mainly on the basis of high-rise construction. Among the world leaders in quantity of skyscrapers Gonk Kong, New York, Dubai, Tokyo, Shanghai and Chicago can be named [1], however, high-rise buildings are available in one and all large cities, that is certainly caused by some objective reasons. The development of technologies of construction has allowed to use with high efficiency the most valuable and limited resource – expensive city land. The territorial resource in city borders, especially in the central part or so-called "centres of gravity" (which are the territories with the high transport availability and infrastructure attracting investors, considerable human flows, etc.) is limited. Therefore the cost of such sites can be the most important element of the investments during the implementation of the investment-and-construction project.

At the same time the location and environment of the site [2], as well as the set land use mode, the mode and level of the taxation and restriction of town-planning character (for example, the instruction on style (for example, respect for historical appearance of the downtown) or the maximum height of the building) [3,4,5] can influence the value and, respectively, the cost of the land. The use of the technologies of high-rise construction allows to construct large volumes of the area in the small territory, compactly placing offices, and commercial areas and also dwellings.

However, the use of high-rise construction in urban development proves the need of the use of new approaches to the management of the city development. One of the tasks of the urban development management is the territorial balance, assuming maximum providing urban population with the necessary infrastructure. Thus, when planning construction, the need of providing the population with the necessary infrastructure facilities, such as schools, kindergartens, hospitals and clinics, facilities of culture and sport has to be considered. The existence of enough such facilities in comfortable transport availability is the necessary condition of proper activity in the city. Certainly, high-rise construction results in the need of planning of infrastructure facilities, taking into account the increase in loading as high-rise construction, allows to place considerably bigger number of residents in the limited territory. The methodical approach, allowing to form territorial balance between the residential real estate and social infrastructure in the conditions of the use of high-rise construction in urban development is offered in this article.

2 Materials and Methods

The research is based on some methodological prerequisites. So, the major and them is the use of provisions of the system approach, within which the modern city is considered as the elaborate system [6,7,8], functioning of the city as whole depends on functioning of its separate parts.

Normal functioning of the city supposes that its basic elements have to be combined: the residential real estate (that is the place of residence of citizens), production and commercial infrastructure (work places of application for residents, existence of the significant amount of places of application of work also forms the city as the main "point of attraction" for the population), transport infrastructure (it creates traficability around the city, which is the most important element of city infrastructure providing convenience of movement of citizens and also necessary for normal functioning of production and commercial infrastructure), engineering infrastructure (it is necessary for normal and comfortable functioning of the modern city and includes all the complex of engineering support by water, energy, withdrawal of drains, waste disposal, etc.), institutional infrastructure (it consists of the institutes organizations designed to provide the most effective development of economy: financial and bank structures, analytical and advice centers and also the headquarters of corporations and also the scientific and design organizations, public, political, cultural and other centers, mass media and also other economic and public institutes) and also social infrastructure (it includes objects of education, health care, trade and public catering, consumer services, sport, culture and art). As in any system, the failure in the work of one of the elements (this or that type of infrastructure) will lead to the failure in work of all the system, however in this research special attention is paid to the territorial balancing of the city in relation to the facilities of social infrastructure.

Formation of the balanced infrastructure is one of the most important tasks of management of the city. Respectively, this research is based on accounting of the main functions of management – planning, organization, coordination and control, however the main attention is focused on planning as at the stage of planning of urban development the needs for the facilities of social infrastructure have to be defined. These requirements have to be considered during the planning and placement of the facilities in the territory of the city: enough schools, clinics and hospitals, sport facilities, etc. have to be placed in availability in the territory of the city. In this research the tools, allowing to improve processes of city territorial planning for formation of territorial balance are offered.

As the tools of formation of territorial balance are offered in the research, the obligatory use of the balance method is supposed [9, 10, 11]. The balance method is one of the main methods of planning, which consists in detection of compliance (or discrepancies) and ensuring the balance in any indicators. In the conditions, when we need the solution of the main objective of economy – the satisfaction of public needs for conditions of limited resources, it is possible to consider this method to be the major. It fully belongs to the territorial planning, which main objective is the creation of conditions for the balanced and sustainable development of the city. The balance model developed on the basis of application of this method will allow to reveal discrepancy between existence of different facilities of social infrastructure and the real needs of the population, and also to determine the volume of financial resources, necessary for elimination of territorial disproportions.

The balance method (balance approach) is one of the traditional and most widespread methods of planning, used in the economy. It allows to carry out the comparison of requirements with the resources (available or perspective as well). Thus, the essence of the balance method consists in the identification and the quantitative characteristic of the relations with the parties of any activity, which counterbalance (or have to counterbalance) each other. This method is applied for the wide range of tasks, including balances of fixed assets and production capacities, work and labor, financial, transport, fuel, etc. are developed that in turn means that balances are subdivided on cost (expressed in cost indexes) and natural and grocery (expressed in physical units).

The study of theoretical questions of application of the balance method proved that, due to the universality, balance models are applicable to all the types of systems. The balance method has broad application at various levels and in the different countries [12, 13, 14, 15]. The balance method has universal character, and balance models created on its basis serve as the main instrument of maintenance of proportions in the national economy.

3 Results

As one of the most important tasks of the city development management is the formation of territorial balance, it is necessary to present its definition.

The territorial balance is understood as the compliance of qualitative and quantitative characteristics (structure and dimensional indicators) construction projects of various functional purpose within the concrete spatial fragment of the territory of the city of the established requirement. This balance represents the actual results of the spatial organization of city-planning activity and is a special case of application of the balance approach to management.

Thus, concrete expression of the territorial balance should be considered as the compliance of the actual and standard indicators on providing the population with housing-and-social facilities. In turn, the divergence of these indicators (that is, the imbalance) defines priorities of the construction activity in the concrete territory.

The distribution and practical use of model of territorial balance of the city is the relevant task, having great practical importance and allowing to realize those theoretical and methodical approaches, which conform to modern requirements to development of the urban environment.

It should be noted, that the special importance of formation of territorial balance on the basis of the facilities of social infrastructure. The standard purpose of the urban policy at the moment is formations of the comfortable environment of activity. Such environment is formed, certainly, not only at the expense of the facilities of social infrastructure, but also at the expense of recreational zones, quality of gardening, the qualitative transport system. However, the facilities of social infrastructure create the framework of the urban environment as they provide the population with the facilities, without which any normal activity of the population is impossible. Thus, it is possible to conclude, that the greatest possible providing the residents with the facilities of social infrastructure forms the basis of the comfortable living environment. At the same time the most important task of management is the decrease in the revealed territorial imbalances to their elimination (in case of existence of such opportunity):

$$L_M = \sum_{i=1}^{n} \pm\Delta_i \rightarrow \min \,, \quad (1)$$

Where L_M – criterion function of management,

ΔI - the deviation of the existing providing the population with the facilities of social infrastructure from the requirement

At the same time the deviation of the existing providing the population with the facilities of social infrastructure from the requirement is determined by some factors:

1) the structure of the population (the age structure of the population determines the requirement in number of school places and the average need for clinics, hospitals and other facilities of social infrastructure);

2) the level and structure of birth rate determines the needs for the main facilities of social infrastructure for the medium and long-term outlook;

3) when calculating the existing providing the population with the facilities of social infrastructure it is necessary to consider both the state, and private facilities of social infrastructure.

The territorial balance has to be used in 2 cases:

- when planning complex building in the territory, when there is an opportunity initially qualitatively and in the balanced way to plan the territory;

- within the developed construction when during the identification of territorial imbalances, that is excesses of need for the facilities of social infrastructure over their real quantity, there is the possibility of application of the correcting influences in the city management, that is planning and realization of the actions of urban policy directed construction new or expansion of power of the existing facilities of social infrastructure.

In the first case application of the balance method does not cause any problems regardless of that, low or multy-storied building is planned to be constructed in this territory. Absolutely other administrative situation is formed in case within the developed construction investment and construction projects of multy-storied construction which can lead to considerable change of a ratio of the residential real estate and the facilities of social infrastructure on the concrete territory are implemented. For prevention of the situation of emergence of serious imbalances, that is the shortage of the facilities of social infrastructure, during the implementation of projects of high-rise construction it is necessary to carry out the analysis of planned level of providing the population of this area with each type of social infrastructure.

Each of types of social infrastructure, generalized by expression of level of providing (LP soc.infr. (i)) expression is the following:

$$\text{LP soc.infr. (i)} = \text{Vsev. (i)} \ / \sum \text{Npop.}, \quad (2)$$

где Vserv. (i) – the actual volume of the services, provided by a concrete (i-th) type of the facilities of social infrastructure (in the corresponding measuring instruments);

$\sum$ Npop. – total population in the territory of the city or the concrete area.

Data on the actual number of these facilities and also about their power are necessary for the calculation of the indicators reflecting the level of providing with the facilities of social infrastructure (capacity). Providing level ideally has to coincide with the need for such facilities. In this case it is possible to consider the existence of territorial balance.

In case when the level of providing is not equal to the need for the facilities of social infrastructure (on each of types), it is possible to consider the existence of the territorial imbalance. At the same time the situation when the level of providing is higher than the requirement, should not lead to urgent managerial influence as in this case the situation of sufficient access, comfortable for the population, to necessary social services is provided. The shortage of the facilities of social infrastructure, that is the excess of need for the facilities of social infrastructure over the actual level of providing, on the contrary, proves the need of realization of actions of the urban policy, which are connected with the construction of new facilities of social infrastructure or expansion of already available. The lack of managerial influences can lead to some considerable decrease in the level of comfort and social service for the residents and will reduce the quality of the urban environment, that can also negatively affect the quality of the residents life.

There is a number of the editions, publishing the international ratings of the cities of the world according to the level of quality of life (for example, analytical division of the The Economist magazine, The Monocle magazine), regardless of methodology of the research and quantity of the indicators measured by them it is possible to allocate main units of indicators for the standard of living in the cities:

– political, social and economic situation;

– health care;

– educational opportunities;

– infrastructure;

– possibilities of carrying out leisure in the city (in the range from restaurants, theatres, cinemas and sporting events);

– existence of consumer goods (from food to cars);

– ecological aspects (from the number of parks to air pollution level).

Prevention and the maximum decrease in imbalances is the task of governing bodies of the city.

As the decrease in imbalances in territorial development is one of main objectives of the city management, which realization leads to formation of the comfortable and balanced territory of the city needs to be defined more precisely concepts of the territorial imbalance. Previously it is necessary to mention that the imbalance can be understood in absolute and relative expression.

The relative imbalance (Imb.rel.) is the degree of divergence of the need for providing social infrastructure with the facilities, which for the convenience (in relative expression, per 1000 people, and the corresponding actual indicators. Thus, Imb.rel. represents the deviation from the requirement, revealed on the basis of demographic data. For example, the need for providing with i-th facility of social infrastructure makes 0,15. The actual providing makes 0,11. Then there is Imb.rel. = 0,04;

The absolute imbalance (Imb.abs.) is the divergence between the need for providing and the actual indicators of providing with the facilities of social infrastructure, expressed in physical units. For example, the absolute imbalance (Imb.abs.) is determined by the ratio:

$$\text{Imb.abs.} = \text{Imb.rel.} \times \text{Qpop.} / 1000 \qquad (3),$$

where Qpop. – population of the territory (people).

As that was mentioned, the identification of imbalances within the developed building has already been told, gives information on in what territory what types of facilities to be constructed and what size (power, capacity) there have to be these facilities. However within the exfoliating building in relation to territorial balance the main and obvious objective restriction is spatial potential. That is understood as all nominal set of spatial fragments within the territory which can be used for construction. The following main components can be a part of this spatial potential:

1) a reserve of the free areas under building. This reserve in the large cities is, as a rule, extremely insignificant;

2) the areas received as the result of redevelopment of industrial territories;

3) one more component of spatial potential is implementation of actions for updating of areas of the developed building (demolition of old, functionally outdated and dangerous structures and construction in their territory of new objects)

Concerning imbalances, that should be noted the following: imbalances take place in absolute majority of real managerial situations and, depending on their size, not always demand some urgent managerial influences directed to their fastest elimination. So, for example, the situation when the imbalance does not exceed admissible level or its value so can take place a little (for example, makes about 10-3 and less) that in fact you should not consider it as the imbalance from the managerial point of view. Thus, the balance from the managerial point of view is not the exact figure characterizing full compliance between a resource and the need for him, but some interval of values allowing a deviation from balance value. This condition is expressed by a ratio:

$$\pm \Delta\, i\, (\textstyle\sum \pm \Delta\, i) \leq \varepsilon, \qquad (4)$$

Where ε is the border of managerial range of deviations.

In other words, there is some range ± Δ ε within which the managerial situation (the condition of the operated process) is recognized as the satisfactory and not demanding the correcting managerial influences. Only when deviations overstep the bounds of the specified range, there is necessary a development and adoption of the relevant managerial decisions. The considered approach is illustrated by the following scheme (figure 1).

In the drawing 1 reinforced line corresponds to such option of implementation of the operated process at which completely there are no deviations from balance. Real dynamics in the generalized view represents change of cycles of positive and negative deviations. If these deviations do not go beyond admissible limits (that is settle down in borders from - ε to +ε), then and the condition of the operated process as it was already noted, it is possible to recognize satisfactory.

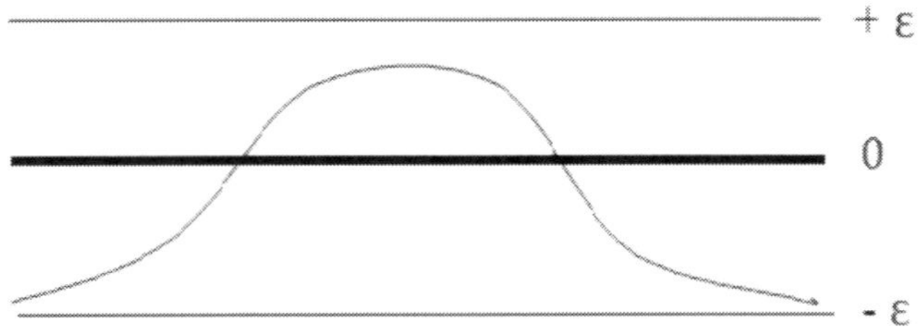

Fig. 1. Indicator tolerance from the calculated value of balance.

Calculation of balances / imbalances needs to be carried out separately for each of types of facilities of social infrastructure, having values for comfortable residence in the city:

- the number of school places;
- capacity of clinics;
- the number of places in hospitals, etc.

Formation of territorial balance (as well as identification of imbalances for their subsequent minimization and elimination) has to be provided at least at 2 levels:

1) at the level of the city in general;
2) at the level of certain areas.

In the conditions of megalopolises the second approach is especially important as the movement around the city to a facility of social infrastructure can take considerable time while necessary facilities of social infrastructure have to be available.

Comparison of indicators of need and providing with certain types of facilities of social infrastructure at the level of areas will allow to assess adequately a situation and to make the relevant managerial decisions on elimination of the available spatial disproportions (formation of the territorial balance covering, finally, all the territory of the city).

Separate task when planning urban development is determination of range $\pm \Delta \varepsilon$ within which the managerial situation is recognized as satisfactory and not demanding the correcting managerial influences. Besides, the size of an imbalance has to define urgency of measures of the city authorities for minimization and gradual elimination of territorial imbalances. Respectively, ranging of imbalances in size (deviation range between the actual level of security and requirement) is necessary. The following hierarchy of territorial balances in size (table 1) can be offered:

Table 1. Hierarchy of territorial balances.

Imbalance (Δ)	**Characteristic of the imbalance and its rank**
$\Delta \leq 2\%$	The nominal imbalance with the lowest rank, has the lowest priority when developing managerial decisions. Within the given deviation it is possible to mention the existence of territorial balance as the deviation is not considerable.
$2\% \geq \Delta \leq 10\%$	The "standard" imbalance, has the rank, next on the increase, the sequence of elimination is defined by the developed logical algorithm.
$10\% \geq \Delta \leq 25\ \%$	The pre-critical imbalance, possesses the highest rank and prime priority of elimination.
$\Delta \geq 25\%$	The critical imbalance, has the extra category rank and is the subject to extraordinary elimination.

In this case the hierarchy is made up of the imbalances of four groups (ranks) from the lowest priority in the need of managerial influences to the extra category rank of the imbalance at which need of urgent, extraordinary actions for correction of a situation and minimization of the developed territorial imbalances is supposed. Certainly, the borders between ranks can be discussed and change in each case.

4 Discussions

Calculations of the territorial balances (imbalances) allow to determine the volumes of construction works, which need to be executed for the formation of the comfortable environment of activity. It allows to solve more serious problems of planning of activity of the investment and construction sphere of the megalopolis. It is actually possible to say that on the basis of calculations of the territorial imbalance and determination of volumes of construction of facilities of social infrastructure balance models of other order can be constructed.

So, the volumes of construction allow to determine the necessary volumes of material resources which will be required to implementation of plans for minimization and gradual elimination of territorial imbalances. The need for material resources (construction materials) can be estimated in comparison with the volume and structure of production of material resources in the region. The existence of an imbalance by results of comparison of requirement and the outputs will be about the existence or lack of a material imbalance. That allows to plan purchase terms, logistics and delivery periods of material resources, necessary for construction, which shortage will be found by results of formation of imbalances.

Also it is possible to estimate the need for the manpower of different profile and qualification on the basis of data on volumes of construction. The comparison of need for the manpower and its existence in the territory of the region can be also realized within balance model and formation of labor balance. The shortage of manpower revealed, when forming labour balance proves the need of attraction of manpower from other regions for the implementation of investment and construction projects.

The highest level of balance models in relation to the realization of the city-planning actions, connected with elimination of territorial imbalances is the model of financial balance. The required volumes of construction allow to estimate in less detail volumes of financial resources which have to be spent from means of the city budget and also the amount of funds of private investors who it is necessary to attract to formation of the comfortable environment of activity and sufficient level of providing the population with the facilities of social infrastructure.

5 Conclusion

The broad use of technologies of high-rise construction in the modern cities and megalopolises both within complex development of the whole areas, and within updating of the existing building, results in the need of improvement of managerial tools on planning of the city development. The special author's attention is paid to the questions of formation of the comfortable urban environment with proper providing the population with the facilities of social infrastructure (that is compliance of level of providing with the facilities of social infrastructure and the need for them). For performance of this task use of balance models is offered and the concept of territorial balance which is understood as compliance of structure and dimensional indicators of the facilities of social infrastructure within a concrete spatial fragment of the territory of the city of the established requirement is entered. The formula of calculation of level of providing with the facilities of social infrastructure which is applied to calculation of level of security separately by each type of facilities is offered.

Separately it is necessary to notice that use of these tools is especially relevant in a situation when within the developed building investment and construction projects of multy-storied construction which can lead to considerable change of a ratio of the residential real estate and objects of social infrastructure on the concrete territory are implemented. The discrepancy of the actual level of security in objects of social infrastructure and requirement leads to formation of a territorial imbalance. The formula of calculation of a territorial imbalance both in relative, and in absolute expression is offered and proved that allows to determine finally the construction volumes necessary for elimination of territorial imbalances.

The imbalance is of great importance for planning and realization of managerial influences from the executive authorities, operating the city development. For the determination of urgency of realization of actions for construction of the facilities of social infrastructure it is offered to range imbalances in size, having determined the deviation size from balance which is so insignificant that does not demand managerial influences (it can be characterized as balance) and also the groups of the imbalances differing in urgency managerial influences, directed to the decrease and elimination of the revealed imbalance.

One can estimate the necessary volumes of construction of the facilities of social infrastructure, as well as material, labour and financial resources, which will be required for the elimination of territorial imbalances, on the basis of data on territorial imbalances.

References

1. J. Burton , Worldatlas (2017) https://www.worldatlas.com/articles/cities-with-the-most-skyscrapers-in-the-world.html
2. W.S. Mathers, Location Value https:// http://www.realmarkits.com/realestate/locationvalue.php (2013)
3. G. W. McCarthy, G. K. Ingram, S.A. Moody, Land and the City (2014)
4. V. Verkhie, The value and price of land, 40-62 (2017) https:// https://www.eolss.net/sample-chapters/C19/E1-05-03-03.pdf
5. T. Gwartney, Estimating Land Values (2018) http://www.henrygeorge.org/ted.htm
6. M. Lomedico, E.M. Bartels, A Structured Analytical Approach to Understanding and Acting in Urban Environments (2015)
7. A. Perdicoulis, The city as a system, **16** (2013)
8. G. Gardner, State of the World. The City: A System of Systems, 27-44 (2016)
9. W. Leontief, The Structure of American Economy (1919-1929)
10. W. Leontief, Input-Output Economics (1988).
11. P. Krylov, Journal Problems in Economics. The Balance Method of Planning, **11** (1968)
12. N.G. Verstina, T.N. Kisel, N.N. Taskaeva, E.M. Akimova, A.V. Fedosina, International Journal of Applied Engineering Research Managing to provide the building industry of a megalopolis with material resources on the basis of balance method (2015)
13. T.N. Kisel, M.I. Bunegin, E.M. Akimova, Economics and Entrepreneurship. Approach to planning the development of territories in conditions of ensuring a balanced development of the city, **6-3 (59-3)**, 291-295 (2015)
14. M.I. Bunegin, Kazan science, 8, 34-36 (2014)
15. A. Tlidi, International Journal of Economics and Financial Issues, **3,** 932-937 (2013)

Free space-planning solutions in the architecture of multi-storey buildings

Alexander Ibragimov [1,*] and *Alexander* Danilov[1]

[1]Moscow State University of Civil Engineering, Yaroslavskoye shosse 26, Moscow, 129337, Russia

Abstract. Here some aspects of the development of steel frame structure design from the standpoint of geometry and morphogenesis of bearing steel structures of civil engineering objects. An alternative approach to forming constructive schemes may be application of curved steel elements in the main load-bearing system. As an example, it may be circular and parabolic arches or segments of varying outline and orientation. The considered approach implies creating large internal volumes without loss in the load-bearing capacity of the frame. The basic concept makes possible a wide variety of layout and design solutions. The presence of free internal spaces of large volume in buildings of a "skyscraper" type contributes to resolving a great number of problems, including those of communicative nature.

1 Introduction

A revolutionary qualitative uplift in the construction industry in general and, in particular, in multi-storey construction occurred in connection with the expansion of reinforced concrete and steel structures. The history of construction of high-rise buildings has many centuries and even thousands of years and as such is in itself an interesting subject of research. Buildings with load-bearing walls up to ten stories tall were built in the cities of the Roman Empire. Since the subject of analysis in this paper are essentially multi-storey buildings with a steel frame, here it is necessary to confine ourselves to the nearest two past centuries. Various aspects of the development of the construction of high-rise buildings are considered in [1-3].

A turning point in the development of a constructive solution of multi-storey buildings can be considered a transition from load-bearing walls to a frame structure, that is, the implementation of one of the principles of steel design - the concentration of the load-bearing capacity of the structure. The functional separation of structural elements into load-bearing and walling appeared. The impetus for this was the appearance of new materials (reinforced concrete, cast iron and then steel). In the 1890s, concrete starts gaining place as a building material for load-bearing structures. European designers were among the first ones who disclosed the potential of reinforced concrete and used it in the construction as a load-bearing frame for high-rise construction [4].

The construction of the first skyscrapers with a metal frame was preceded by the appearance of such unique metal structures as Eiffel Tower and the Statue of Liberty

* Corresponding author: igasu_alex@mail.ru

After the devastating fire of 1871, Chicago began to build the first "skyscrapers". In 1885, the first "multi-storey" 9-storey office building appeared here. The increase in the number of storeys was due to the deficit of free areas for construction in the business center of the city and, consequently, high prices for land. The very desire to place new buildings in the center was associated with prestige and self-promotion from the very beginning.

In the middle of the XIX century, E.Otis invented an "elevator" - a mechanical means of vertical movement of cargo with a steam power unit. This allowed a dramatic increase in the number of floors, including residential buildings. Since the invention of the lift-blocking mechanism in the elevator shaft for security in case of damage (cables cut off etc.), it has become prestigious to have high-floor housing.

For the first time the metal frame was applied in England in the designs of factory buildings, but its main bearing elements were made of cast iron. The steel frame in the designs of multi-storey buildings was, apparently, first applied in Chicago in the years 1883-1885. ("Home Insurence Building", Baron Jenny), which marked the beginning of the intensive development of steel frame construction. Unfortunately, the deficit of steel products in Russia for many years has stalled the use of steel in mass construction (for example, the story of the famous Shukhov tower).

By 1902, one of the most famous high-rise buildings of Manhattan, the first Manhattan high-rise building "Flatiron Building", named so because of its unusual triangular shape, which was the result of a competent solution of the architect who had to place it at the intersection of Fifth Avenue and Broadway, crossing Manhattan diagonally. Flatiron became one of the first buildings with a steel frame, which allowed to build it with a height of 87 meters. The facade is divided into three parts: the base, lined with hewn stone, the middle main part with many levels and the upper part consisting of a penthouse with very high arches, columns and a carved eaves hanging over it. Having a three-part division with a sharp front corner, having curvature width of only two meters, Flatiron acquired the features of the antique style, reminiscent of the Renaissance palaces. Thus for the subsequent 1900-1920 years in the history of high-rise buildings-skyscrapers came the period of rampant eclecticism.

In 1913, a 59-story building, 241 m high, Woolworth Building (architect-eclecticist Cass Gilbert) was built in New York. It is called the modern cathedral and the "Temple of commerce". For the building, the architect chose the Neo-Gothic style, believing that it is best suited for a tower rushing upward. "Woolworth" became a model for many skyscrapers built later. This building kept the world record of height from 1913 to 1930, remaining the highest in the world. The facade was decorated with elements typical of the Gothic style, such as lancet arches, turrets and so on. It was with the erection of this building that a new word "skyscraper" entered the everyday life.

Since 1930, the tallest building in the world - 319 meters including the high spire above the conical roof of the building - was the skyscraper "Chrysler Building".

Soon this record was also broken. It was the building of the Empire State Building - 1931, 102 floors, 381 meter height.

Further on and up to now the construction of high-rise buildings has turned into an unrestrained races in the vertical, a kind of sports competition - who is higher, who is faster.

A hundred-meter height line was overcome in 1890. The way this race developed further [5], may be clear from Table 1, where only the multi-storey buildings are represented, which for some time have remained altitude recorders.

A kind of a long-liver among these record holders was the Empire State building, which ranked as the highest for 41 years. As for today's record holder, it is assumed that he will not remain a long-term ringleader.

At present, the number of skyscrapers has exceeded one hundred thousand. Asia is in the lead in this marathon. If initially the engine of progress in the construction of multi-storey buildings were economic considerations, then gradually the priorities shifted, and ambitions came to the fore. Governments and ruling elites are usually the promoters of unique skyscrapers construction (skyscrapers of Donald Trump). Although there is an opinion that these buildings are unprofitable objects.

Table 1. The progress in height.

Building name	Year	Number of Storeys	Hight, m
World Building	1890	20	106,4
Manhattan Insurance	1894	18	106,1
Park Row Building	1899	30	119,2
Singer Building	1908	47	186,6
Metropolitan Life Tower	1909	50	213,4
Woolworth Building	1913	57	241,4
40 Wall Street	1930	71	282,5
Chrysler Building	1930	77	318,9
Empire State Building	1931	102	443,2
World Trade Center	1972	110	526,8
Willis Tower	1974	108	527,3
Petronas Towers	1998	88	451,9
Taipei 101	2004	101	508,0
Burj Dubai	2012	160	818,0

It would be desirable to note here only, that along with rectilinear forms in some buildings arch designs were applied. However, they were not involved in the work of the basic load-bearing structures of the frameworks, but were used as supporting elements rather in auxiliary structures and as decorative elements for giving the exterior appearance of the building an architectural expressiveness and originality. The latter led to a mixture of styles and was recognized as "eclecticism".

Spatial structures in the form of domes and arches have found application mainly as roofings. For example, the 27-meter atrium is the famous palm garden of the World Financial Center (architect Caesar Pelli, 1985).

2 Types of frames of modern multi-storey buildings

Frameworks of multi-storey buildings can be classified [4] according to their design schemes:

a) frameless with parallel bearing walls;
b) core with bearing walls;
c) box-type;
d) with cantilevered flooring slabs on the level of each floor;
e) frame with non-beam slabs;
f) with height of floor consoles at the level of each second floor;
g) with suspended floors;
h) with height-to-height trusses in staggered order;
i) frame without core;
j) frame with core;
k) frame with latticed rigidity diaphragms;
l) framework with latticed triggers and latticed core;
m) double-core structure (pipe in the pipe);
n) multi-sectional box.

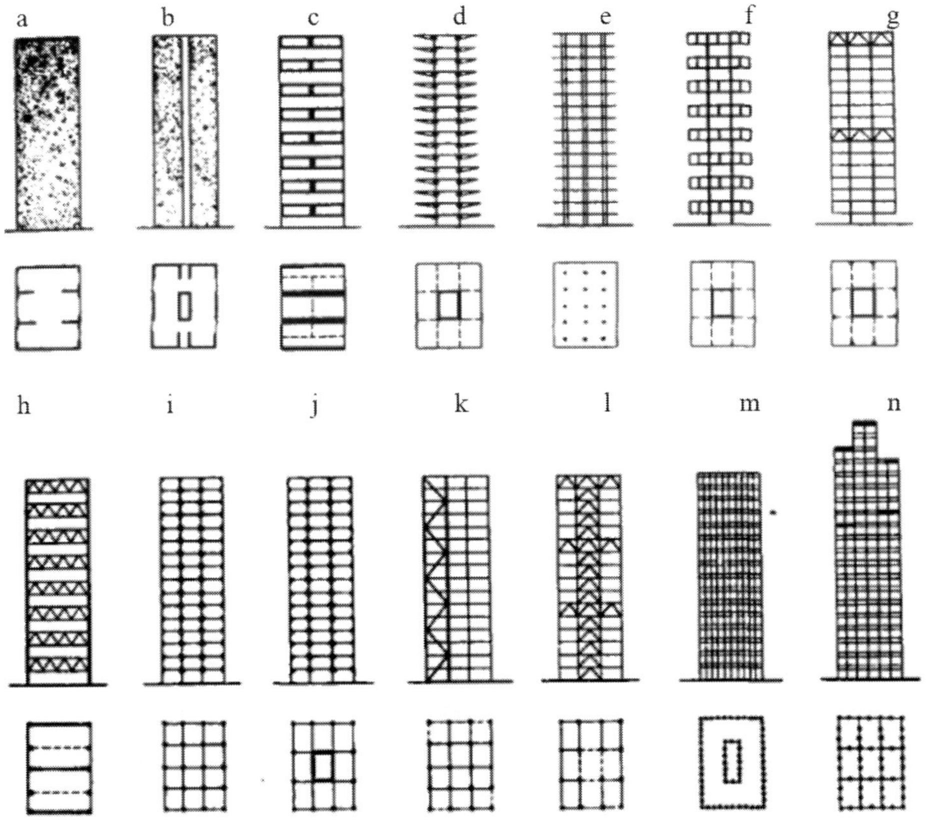

Fig. 1. Constructive schemes of high-rise buildings.

In the steel frameworks of the tallest buildings of recent decades, the last two schemes (*m* and *n*) dominate. Double-core structure (pipe in a pipe) - the outer columns and beams are located close enough to each other, and the frame of the outer walls turns into a shell with openings. The building in the whole works as a hollow tubular structure, cantilevered in the ground. The central core (pipe) increases the rigidity of the building, perceiving the horizontal loads together with the outer box (pipe). In multisectional scheme the individual boxlike sections form he complete frame according to the box scheme (bundle of pipes). External walls and intersection walls together perceive horizontal loads. In this case, an increase in the rigidity of the system is obvious. This kind of solution allows the construction of buildings of the extreme height and with a large open space of interstitial floors.

The concentration of engineering systems and, in particular, communications inside the central core, with the obvious advantages of compact placement, reduces the level of safety (viability) of the facility as a whole, and also imposes very strict restrictions on the use of the internal volume of the building. More reliable is the organization of several peripheral autonomous communication nodes, for example, if possible, along the perimeter of the circuit. It should be noted that there are high-rise buildings with two trunks (Emirates).

3 The free space concept

Large spans in frame structures can be overlapped by truss structures, which support the overlying or suspended lower structures.

Despite the variety of external forms of high-rise buildings in their load-bearing steel frameworks, however, straight lines and perpendiculars predominate, and the "old good" curve stays in the background. It is no use here to refer to, mainly, decorative arches and dome coverings.

As an alternative, it is possible to consider the use of intercepting steel elements in the form of convex (arched) or concave (rigid threads, cables) of individual flat curves or segments forming shells. Such elements should work, respectively, mainly on compression or stretching or form more effective two-chord systems. The most common case, apparently, may be the circular arch. However, constructive systems of parabolic form may prove be more effective. These same elements can effectively work in regular lattices of the "column-crossbars" without openings, redistributing force flows.

Thus, the inclusion of curvilinear elements into the steel framework allows the formation of combined structural systems of a qualitatively different action type. Of course, one should be ready to challenge the additional problems with conjunctions, mounting features, layout of the internal space, interaction with the base, wind and seismic actions, and many others. Однако, возможности создания новых конструктивных форм и компоновочных решений оправдывают дополнительные трудозатраты. However, the possibilities of creating new constructive forms and space planning solutions should justify additional expenses.

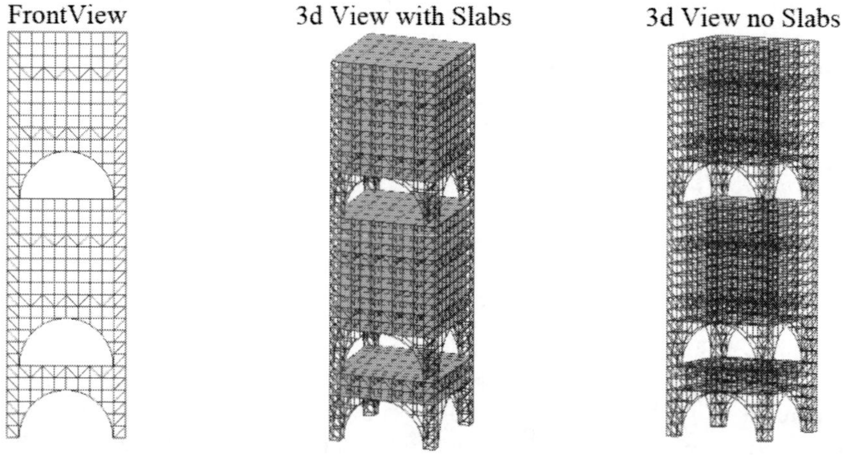

Fig. 2. Multi-storey building frame with arched supporting structures.

The wall structures are not shown here. The horizontal dimensions of the frame are 60x60 m, the height is 204 m (34 floors are 6 m or 68 by 3 m), and the radius of the semi-arch is 24 m. Communication channels are supposed to be located in the corners inside the pylons (6x6 m). The circuit is executed in axial lines.

Below are the sizes of sports grounds to be possibly located in these overall dimensions:

- Tennis: 23.78 x 8.23 m (10.9 m - for paired tennis), height to the ceiling not less than 9.14 m.
- Basketball: 28,0 x15,0 m, height to the ceiling not less than 7,0 m.
- Volleyball: the area is 18,0 x 9,0 m; a playground with adjoining zones for game 34,0x19.0 m.
- Mini football (according to FIFA requirements): length - min 20,0 / max 42,0 m; width - min 15,0 / max 25,0 m.
- Swimming pool: 50,0 x 25, 5 and 25,0 x 25,0 m - "short water".
- Curling: Length 44,5 - 45,7 m, width 4,42 -5,0 m.
- Rhythmic gymnastics, boxing, wrestling, table tennis, etc. require smaller areas.

As show the data presented, in the space of such dimension, there is a room for office space and spectators.

In this dimension, it is possible to place retail space, entertaining (restaurants, cafes), entertainment (theater, cinema, circus, planetarium, exhibition pavilions, art galleries, etc.), gardens, greenhouses, etc. All those may have the different building marks.

A special thing is parking: for cars in the lower floors of the building, for light helicopters in the upper ones.

A free three-dimensional layout of the open space is possible. Sliding gates, stained-glass windows or solid glass may close the arched openings. The creation of such interiors and their appropriate design should facilitate the solution of psychological problems of communicative nature associated with the lack of communication, conditions for relaxation, as well as cultural and everyday life. Oval lines of interiors with high ceilings, preferably with a soft natural (or imitation) insolation, in contrast to the non-characteristic in nature aggressive angular contours, it has a beneficial effect on the psychological state and facilitates communicability.

It is possible to use such constructive solutions with the preservation of the inner bearing core.

4 Conclusions

With respect to the proposed concept, we can assume the following:

1. It is not impossible to use such constructive solutions in principle
2. There are perspectives of application - functionality, an unlimited range of options for layout solutions
3. This architectural concept can be regarded as a new one
4. It opens the prospect of forming a whole class of original architectural solutions

References

1. K. Kayvani, Mechanics of Structures and Materials (ACMSM23), **1** (2014)
2. K. Kayvani, Structural Engineering Solutions for Tall Buildings in Australia (2008)
3. Le Sun, International Journal of High-Rise Buildings, **5**, 2 (2016)
4. W. Schueller, High-Rise Building Structures (1977)
5. A. Perelmuter, Essays on the history of metal structures (2015)

High-rise construction as a method for architectural development of megapolises

Vadim Kankhva [1,*]

[1] Moscow State University of Civil Engineering, Yaroslavskoye shosse 26, Moscow, 129337, Russia

Abstract. The article analyzes the current state of urban development in Moscow, there are revealed the insights into pattern of large investment projects. The regulatory framework as well as the state and the forecast of housing funds are scrutinized. A number of problems, that are related to the implementation of high-rise construction projects at all stages of the life cycles, are highlighted by the example of unique facilities, which are under construction or have already been built. Substantiation of high-rise construction by the transport hubs in megapolises is given. There are also considered main advantages of Moscow renovation project and criticism against it.

1 Introduction

Urban development of Moscow is one of the core activity areas of the construction complex of the capital. Comprehensive development is carried out in close interdependence between authorities and city investors with account of the town planning strategy. High capital intensity of urban planning investment projects as well as variability of decisions on territory development require a thorough evaluation, both from the point of view of the government policy in terms of urban development and economic efficiency of utilization of budgetary funds of Moscow [1].

In itself, the economic evaluation of investment projects involving the city budget cannot be representative, since participation in such projects is not due to the commercial interest of state structures, but to the fulfillment of obligations to the citizens.

2 Materials and Methods

Investment activity is given special attention in the legislation of the Russian Federation. In addition to determining the basic concepts of investment activity, the legislation also defines subjects of investment activity (investors, customers, contractors, users of capital investment objects and other persons). The subject of investment activity has the right to combine the functions of two or more entities, unless otherwise is stipulated by the contract and (or) the government contract, which is concluded between them.

* Corresponding author: vskanhva@mail.ru

Russian legislation considers a high-rise structure to be one that extends higher than 75 m from the lower planning level, where the road mark for fire trucks is, to the lower level of the opening in the outer wall of the upper floor.

Examples of such facilities are the Moscow International Business Center "Moscow City" and a number of multifunctional high-rise residential complexes. Meantime there are already built such objects as (Figure 1):

Fig. 1. Moscow International Business Center "Moscow City"

1. "Evolution Tower" is a multifunctional center with a height of 54 floors (255 meters), which includes commercial premises with restaurants, as well as office premises.
2. "Imepria Tower" is a business complex with a height of 60 floors (239 meters), which is a multifunctional complex that combines office space, apartments and commercial premises.
3. "City of capitals" is an architectural ensemble, symbolizing Moscow and St. Petersburg. It combines the entertainment complex, offices and luxury apartments. It consists of two towers - a 76-story Moscow and a 65-story St. Petersburg, 302 m high and 257 m high, respectively, combined with a 17-storey stylobation and a 17-story domed building with an atrium. The entire complex is located on a single podium, consisting of a 6-level underground volume and above-ground 4-level public space.
4. "Embankment Tower" is a complex intended for office premises of "A" class and includes three buildings of different number of storeys (17, 27 and 59 storeys). The common areas, located on the first level, and the trade hall unite them into a single whole.
5. "Eurasia Tower" is an office and recreational complex, comprising a 309-meter building with a total area of 207,542 m² on a three-storey podium, which houses a fitness center, entertainment centers, restaurants and shops.
6. "Federation Tower" is an office complex consisting of two triangular towers, 374 and 242 m high, respectively, located on a stylobation. There are located office space and apartments, separated by technical floors. There is also a restaurant.
7. "Mercury City Tower" is a multifunctional building with premises of residential and non-residential funds, premises for retail and cultural purposes with three underground floors.

Examples of high-rise construction of multifunctional housing estates are (Figure 2):

1. "Triumph Palace" is a housing estate with the height of 264.1 m.
2. "Mosfilm Tower" is a housing estate with a height of 208 m.
3. "Tricolor" is a housing estate with height of 194 m.

Fig. 1. Moscow multifunctional housing estates

The investment project does not necessarily require capital construction, since investments can be made in the growth of working assets, or the investor can contribute to the project a finished building or existing equipment. Significantly, investment projects, usually, are indivisible and not replicable. It is important to note that investment projects, in a certain sense, are unique. This is manifested in the fact that any project is oriented towards the use of new knowledge about nature, technosphere and society, as well as a certain combination of available resources, that are absent in the market for the purpose to carve out a certain business segment and capitalize on it [2, 3].

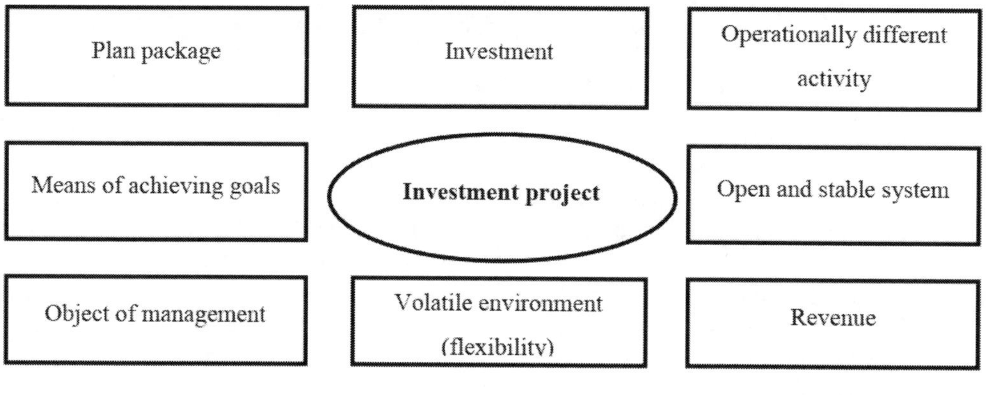

Fig. 3. Pattern of investment project

The implementation of capital investments from the Moscow city budget is carried out in accordance with the Moscow City Address Investment Program. The Address Investment Program is a document that establishes the distribution of the budgetary allocations for the implementation of investment projects for the construction, reconstruction, technical re-equipment of capital construction facilities in the city of Moscow for the next financial year and the planned period for the budget allocations for the implementation of investment projects in the city of Moscow within the arranged state programs and non-program expenses [4].

Urban development is the dominant form of spatial development of the city. Urban development is understood as the activity of public authorities, municipal bodies, legal entities and individuals in the field of architectural development of territories, taking into account the interests of public, state and civil interests, as well as national, historical, cultural, ecological, natural and other specific features of exact territories [5]. Urban planning activity assumes creation of favorable conditions of residing of the population with use of urban planning means.

The majority of municipal target programs and investment projects are implemented through urban development, so the planning and development of the territory is formed, the location of housing construction, urban and social facilities, transport and engineering communications is optimized in accordance with the zoning of the territory. The cityscape is being created and the state of the urban environment is improving.

In addition to the principles of harmonious development of the city, urban development must ensure the implementation of state, public and private interests [6].

State interests in the field of urban development are to ensure conditions for the sustainable development of settlements and inter-settlement territories, the functioning of state systems of engineering and transport infrastructure, conservation of natural resources, protection of historical, cultural and natural heritage sites.

Public interests in the field of urban development are to provide favorable living conditions, limit the harmful impact of economic and other activities on the environment, improve the environmental situation, develop the engineering, transport and social infrastructures of settlements.

Private interests in the field of urban development are the interests of citizens and legal entities, that occur in the result of the urban development.

State, public and private interests in the field of urban development are ensured by meeting the requirements of state town planning standards and rules with appropriate control. The coordination of state and public interests in the field of urban development is provided by state authorities and municipal government in accordance with the law [7].

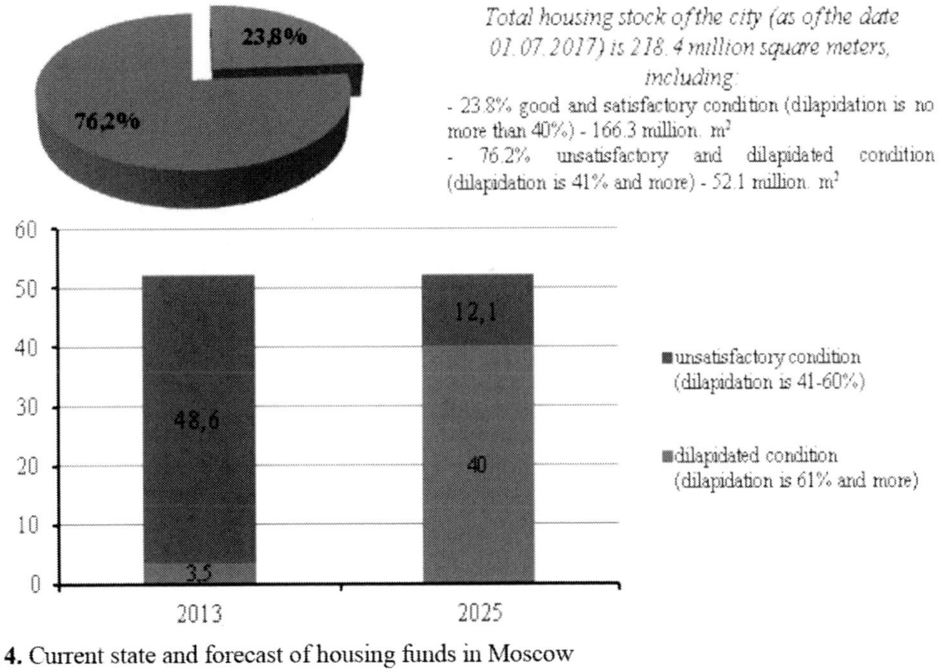

Fig. 4. Current state and forecast of housing funds in Moscow

However, urban development of the territories is necessary not only to ensure the realization of the interests of citizens and the state, but also for an objective reason - dilapidation of the housing funds. The dilapidation of the houses gradually leads them to a state of emergency. This problem is typical for the entire territory of the Russian Federation, and it is considered to be the main cause of high prices for housing. In Moscow, in order to solve this problem, there is a housebreaking program for houses of the first period of industrial construction, which is now coming to an end and a large-scale renovation program is gaining momentum. However, a fairly high rate of construction is required in order to tackle this question, taking into account the need to maintain a high level of housing per 1 person.

There are houses in the housing funds of Moscow, which major repairs are not advisable, due to the non-conformity of constructions with the modern requirements for housing (multi-storey houses with wooden ceilings, houses of the first period of industrial purpose). According to the Recorder of deeds by 2017, the residential area of such houses was 10 million square meters. Taking into account the standards for resettlement and other needs of the city for solving citywide tasks, these dilapidated funds must be compensated with a coefficient of 1.5, which indicates the need to build in Moscow 3 million square meters of housing per year. Until now, it was complicated to implement such construction volumes in the territory of Moscow, which encouraged to utilize free sites and expand the boundaries of the city [8].

3 Results

The basis for ensuring the integrated development of the city's territory is the allocation of land to investors with pre-determined technical and economic indicators of the planned development and functional purpose in accordance with the town planning regulations and the master layout plan.

Conditions for concluding an investment contract and allocating a land plot for an investment project (separation of costs for engineering support of buildings, encumbrances, concomitant development of social infrastructure facilities, etc.), as well as technical and economic pattern of development are considered at the Urban Planning and Land Commission of the City of Moscow. Within this commission, among other things, an analysis of existing investment contracts is carried out, a decision is made as to whether further implementation of such projects is advisable.

At the present moment, ensuring the conditions for the formation and development of entrepreneurship, it is necessary at the new stage of economic reforms to proceed to solving the housing problem for broader categories of citizens: young families, social workers, doctors, teachers, scientists, engineers. Moreover, there should be taken measures aimed at putting more affordable housing into operation, as well as significantly expanding the possibilities for residential lease. A number of regions have initiated pilot projects, on the basis of which various forms of public-private partnership are being carried out to ensure the functioning and sustainable development of the market for residential and commercial leasing and sales of living space, that has been built on the city budget money.

The project of renovation of housing funds in Moscow is a major example of urban development. Such an ambitious program for the development of built-up areas has its own peculiarities in terms of setting new multiple objectives. Apparently, the city budget is not able to finance this project independently and fully, the feasibility of which depends on the cooperation of private investors and city authorities.

4 Discussions

Moscow administration plans to independently set the parameters and volumes of new construction, select contractors through competitive procedures and monitor the process through a specially created State Renovation Fund. The city administration plans to abandon the involvement of private sector and investment funds, providing funding through the city budget in order to fully control the situation at the initial stage of renovation. At the start of the program, the mayor's office estimated the implementation period of the program for 10-15 years.

Despite the fact that most of Moscow's population supported the project, some experts expressed their concerns, voicing the following comments:

• Taking into account the experience of the previous demolition program for old houses under the Moscow package and private developers, the population has densified more than 3 times, with all the consequences: a heavy burden on the social and transport infrastructure, that will lead, for example, to electric "blackouts", colossal queues in kindergartens, schools and heavy traffic jams.

• According to the estimation of Greenpeace, the areas, which are going to be renovated, will lose up to 25% of green spaces, where there are especially ecologically effective zones that clean the city air.

• The expected increase in the supply of affordable housing next to the metro may lead to a price dumping and freezing of interest in secondary housing.

• The owners of mortgage apartments in five-story buildings are especially exposed to find themselves in a legal void.

5 Conclusion

Attraction of extrabudgetary funding, in order to develop urban planning environment of a city, plays a significant role in taking such decisions. Therefore, it results in necessity to create a favorable investment climate and economic conditions, that will encourage investors to cooperate with municipal authorities in order to accomplish the social tasks.

References

1. V.V. Gluhov, I.V. Ilin, Lecture Notes in Computer Science, **8638**, pp 509- 518, (2014) DOi-10.1007/978-3-319-10353-2_46
2. V.I. Churkin, O.V. Kalinina, Estimation of excess burden of labor taxation in Russia, Actual Problems of Economics, Vol **184**, IS **10**, pp 278- 282 (2016)
3. N. Safronova, E. Nezhnikova, A. Kolhidov MATEC Web of Conferences **106** 08024. (2017) doi:10.1051/matecconf/201710608024
4. R. Golov, V. Shilov, S. Silantiev ASEE International Forum, Columbus , Ohio. June 28, 2017. Paper ID #20766
5. M. Tibon, A. Borboni, M. Mor, D. Pomi 2011 Proceedings of the Institution of Mechanical Engineers. Part I: Journal of Systems and Control Engineering **225(3)** 443-451
6. N. Harmathy, J. Kontra, V. Murgul, Z. Magyar, Advances in Intelligent Systems and Computing, Vol. **692**, pp. 563 - 576 (2018) DOi- 10.1007/978-3-319-70987-1_60
7. F. Aggogeri, A.Avanzini, A. Borboni S. Pandini International Journal of Automation Technology **11(2)** pp 311-321 (2017)
8. N. Harmathy, V. Murgul, Procedia Engineering, Vol. **165**, pp. 1845 - 1852 (2016) DOi- 10.1016/j.proeng.2016.11.932

E3S Web of Conferences 33, 01037 (2018) https://doi.org/10.1051/e3sconf/20183301037
HRC 2017

The advisability of high-rise construction in the city

Natalia Sergievskaya[1]*, *Tatyana* Pokrovskaya[2] and *Natalya* Vorontsova[3]

[1] Moscow State University of Civil Engineering, Yaroslavskoe shosse, 26, Moscow,129337, Russia
[2] Moscow Polytechnic University, Bolshaya Semenovskaya str., 38, Moscow, 107023, Russia
[3] Vyatka State University, Moskovskaya str., 36, Kirov, 610000, Russia

Abstract. In this article there discusses the question of advisability high-rise construction, the reasons for its use, both positive and negative sides of it. On the one hand, a number of authors believe that it is difficult to avoid high-rise construction due to the limited areas in very large cities. On the other hand, a number of other authors draw attention to the problems associated with high-rise construction. The author of the article analyses examples of high-rise construction in several countries (UAE, Dubai "Burj Khalifa"; Japan "Tokyo Sky Tree"; United States of America, "Willis Tower"; Russia "Federation Tower") and proves the advisability of high-rise construction in the city.

1 Introduction

Nowadays high-rise construction is becoming more and more actual in the world. This is due to the fact the urbanization urban space is increasing and there is also accelerated development of construction technologies and technology of engineering support of buildings. As it was considered in USSR and then in Russia, high-rise buildings have height of 75 m that is more 25 floors. As for other countries, high-rise buildings have height from 35 to 100 metres. In the US and Europe under the skyscraper there is understood building above 150 m, and in other countries it is higher than 100 m. In Russia, the skyscrapers are considered buildings of height 75 m or more [1]. High-rise buildings play an important role in technical progress. They also promote economic development. In the modern world high-rise construction occupies leading positions in the construction industry. Skyscrapers are becoming commonplace in big cities. Thanks to high-rise construction, developing countries show their opportunities. There are following factors:

1. Economic factors. High-rise construction develops very quickly in prestigious areas in large cities thanks to the high cost of land in megapolices, etc
2. Commercial interest. It is more profitable to implement large investment projects on high-rise construction.
3. Political image. High-rise and original buildings show economic stability and prestige of countries.

* Corresponding author: doptaganka@yandex.ru

Nowadays economic factors play the most important role [2]. An international experience has shown, the buildings which height is 30 to 50 floors are most beneficial from an economic point of view. In this regard, high-rise construction is the most important direction in the modern big city. High-rise buildings may have different purposes. They may be hotels, offices, apartment houses. Usually high-rise buildings are multifunctional, that is there are parking garages, shops, offices, cinemas etc. As mentioned above high-rise construction came due to a significant shortage of territory for construction in megapolises. There is also a lack of land for office and hotel space in big cities.

The first international Symposium of the Council on tall buildings and urban environment was in 1971. The Council on Tall buildings and Urban Habitat developed and adopted international classification of skyscrapers depending on their height. The Council specifies the following types of skyscrapers (tall buildings).

1. Skyscrapers are buildings from 100 to 300 m.
2. High-rise buildings are buildings from 300 to 600 m.
3. Mega high building are buildings with the height more than 600 m.

At present time the volume of construction of skyscrapers significantly increased. High-rise construction is actively developing in the congested cities of Asia. There is trying to solve housing problem with the help the construction of high-rise buildings. Last 10 years there exist the construction of high-rise buildings which are multifunctional [3].

2 Review

Let us consider the tallest buildings in the world, in the countries; the most green buildings etc. High-rise buildings appeared in the United States. The term "skyscraper" appeared thanks to offices with a height of over 10 floors. These offices as well as the term "skyscraper" appeared in Chicago in 1891. In the following decades, the construction of skyscrapers actively took place in big cities in USA.

High-rise construction began in Europe in the late 1950 – early 1960. At first after the Second World War there was the restoration of the ruined historical centers of cities. There was also implemented mass economical housing construction of houses of the average number of storeys. After it, there began to construct high-rise buildings. As for Russia, high-rise construction began in 1913. At that time the seven-story building was built. Nowadays Russia has 38 skyscrapers and it ranks 15th in the global rankings. If we look at international experience with respect to Moscow, that it is better to master 2-3 of the construction site on small plots on the outskirts of the capital. These complexes can have both social (e.g., new jobs at the place of residence) and compositional value. If you place them on the outskirts, it will allow not to disturb the panorama of the historic part of the city [4, 5].

The rating of the International Council on tall buildings and urban habitat (ctbuh) is one of the important rankings, which is published every year. It gives the description of the 100 tallest buildings in the world. In accordance with it (the classification of CTBUH), today there are more than 3.8 thousand skyscrapers in the world. Their height is 150 m. There are 115 buildings above 300 m. There are 115 buildings above 300 m. Four Russian skyscrapers entered the hundred tallest buildings. These are "Federation Tower", Mercury City Tower and Eurasia and others.

The largest number of skyscrapers was built in 2017 year. They had 144 and they are above 200 metres. The most of the buildings were built in China (76). The following are USA (10 skyscrapers) and South Korea (7 skyscrapers). In this year, the highest constructed building has become an international financial center in China. It is 599 m and it has 115 floors. Let us consider tall buildings with the title of "the most". The tallest building, the greenest building etc.

 https://doi.org/10.1051/e3sconf/20183301037

Table 1. The tallest buildings in the world

N	The tallest building	Country	The name of the high-rise building	Height	The years of building creation
1	the first skyscraper in the world	USA, Chicago	Insurance company Home Insurance Building	42 m.	1884-1931гг.
2	the tallest building in the world	UAE Dubai	Burj Khalifa	828 m	2004-2010
	the tallest building in Asia	China, Shanghai	Shanghai tower	632 m	2008 - 2015
10	the tallest residential building	USA, New York	432 Park Avenue	425.5 m	2012 - 2015
	the largest private home built for one family	India, Mumbai		173 m	2010
11	the tallest tower in the world	Japan	Tokyo Sky Tree	634 m	2006 - 2012

There consider ten tallest skyscrapers according to "RBK-Nedvizhimost (real estate)". They are shown in Figure 1.

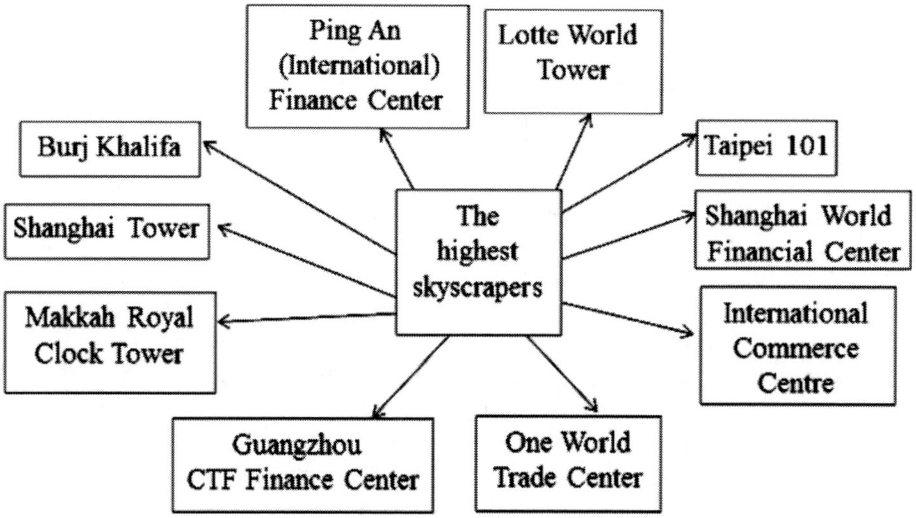

Fig. 1. The tallest skyscrapers according to "RBK-Nedvizhimost (real estate)".

Let us consider the tallest residential buildings. They are shown in Figure 2.

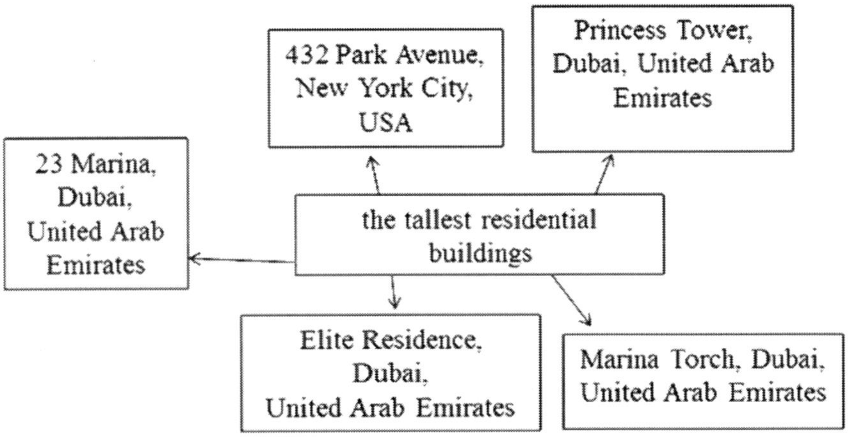

Fig. 2. The tallest residential buildings

Let us consider the tallest towers. They are shown in Figure 3.

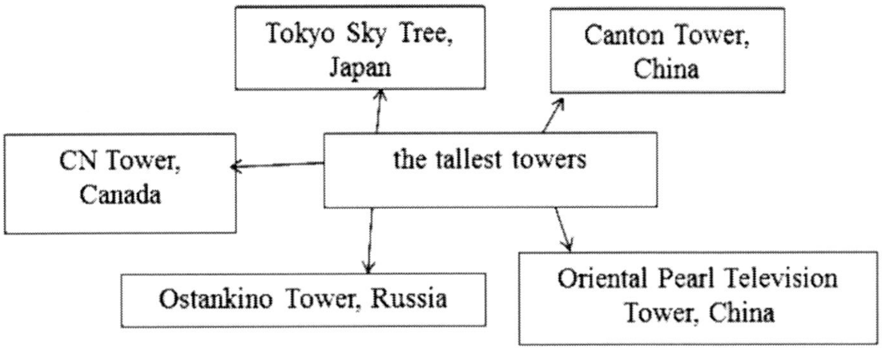

Fig. 3. The tallest towers

Table 2. The highest skyscrapers*

Country	City	The name of the skyscraper	Height	number of floors	Appointment
UAE	Dubai	Burj Khalifa	828 m	163	offices, housing and hotel
China	Shanghai	Shanghai Tower	632 m	128	hotel and offices
Saudi Arabia	Makkah	Makkah Royal Clock Tower	601 m	120	multifunctional complex
China	Shenzhe n	Ping An Finance Center	599 m	115	offices
South Korea	Seoul	Lotte World Tower	554,5 m	123	hotel and offices
USA	New York	One World Trade Center	541,3 m	94	hotel and offices
China	Guangzh ou	Guangzhou CTF Finance Center)	530 m	111	multifunctional complex
Taiwan	Taipei	Taipei 101	508 m	101	offices
China	Shanghai	Shanghai World Financial Center	492 m	101	hotels and offices
China	Hong Kong	International Commerce Centre	484 m	108	hotel and offices

*Information from "RBC- Nedvizhimost" (real estate).

There consider the most green skyscrapers.

Table 3. The greenest skyscrapers

N	Country	City	The name of the skyscraper	Feature
1	Bahrain		The Bahrain World Trade Center Towers	Three very large fans are on the front of the building. Propellers can produce up to 1100 MW of electricity per year.
2	China	Guangzhou	The Pearl River Tower	Solar batteries, photovoltaic panels reduce costs of rooms conditioning.
3	USA	New York	Bank of America Tower	The Bank is equipped with a system for electricity generation running on natural gas. It also has a system for collecting rainwater. The building has large windows. They let a lot of light. The sensors regulate the light level in the room.
4	USA	Chicago	"340 on the Park"	There is an own collection system for rainwater. It is used for technical purposes.
5	USA	Manchester	The CIS Tower	More than seven thousand panels mounted on the walls of the building. 24 powerful wind turbine installed on the roof of the tower. It provides about 10 percent of the consumption of energy for the building.

There consider the tallest hotels in the world. They are shown in Figure 4.

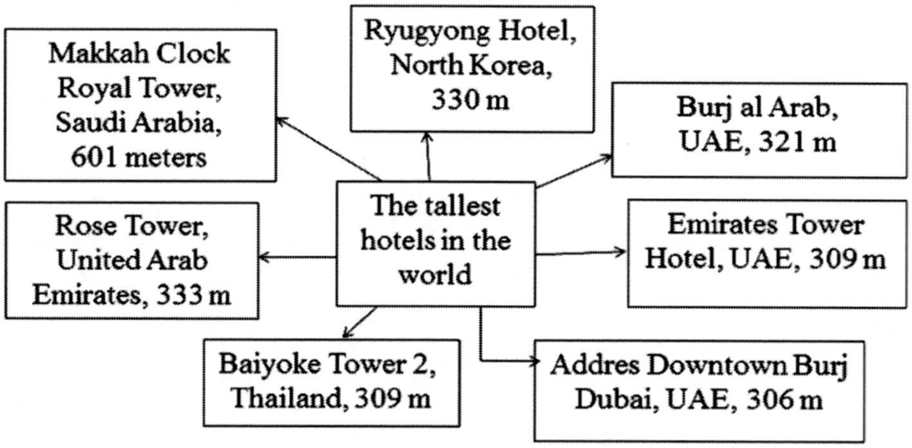

Fig. 4. The tallest hotels in the world.

3 Results

There consider some successful examples of creating high-rise buildings. These are Burj Khalifa in United Arab Emirates, Dubai; Tokyo Sky Tree in Japan; the Willis tower in Chicago and Federation Tower in Russia[6, 7] .

Table 4 Successful examples of creating high-rise buildings

N	Country	The name of high-rise buildings	Purpose	Benefits
1	UAE, Dubai	Burj Khalifa	There attract tourists	It is the tallest building in the world There is not heat up practically There is the highest observation deck There is the increase in property value There attract tourists
2	Japan	Tokyo Sky Tree	There create the most earthquake-resistant building There support antennas	There can withstand earthquakes There is the originality of design solution
3	USA	Willis tower	There located offices and business centers	There attract tourists There are heavy duty antennas. There take over lightning strikes
4	Russia	Federation Tower	There located elite apartments (residential) and modern offices	There are high speed elevators. There is automated system of ventilation and conditioning. There are energy efficient windows etc It is one of the highest buildings in Europe. There are 67 elevators There is speed super-modern escalators and elevators It was recognized the best office building in the world in 2009

1. Let us consider more detailed the called skyscrapers:

1) The first is "Burj Khalifa". This is the tallest skyscraper in the world. The project of "Burj Khalifa" was designed by American architects. American architectural Bureau Skidmore, Owings and Merrill participated in the construction of "Burj Khalifa". American architect Adrian Smith was the author of the project. The building is a work of art and marvel of engineering. "Burj Khalifa" has no equal in the world. Different states took part in design and construction of the building. Burj Khalifa was inaugurated in Downtown Dubai on 4 January 2010. Since that time it has officially became the tallest building in the world. Skyscraper has a height of 828 meters and 163 floors. This is like a miniature "city within a city". Offices and hotels, retail and residential space and own parks and boulevards are located inside the skyscraper. The first 37 floors are given to the hotel. Residential premises are located from 45 to 108 floors, and other premises are mostly offices. There are two observation decks in the skyscraper:

1. The first observation deck locates on the 125th floor, at a height of 456 meters above the ground. This gives people opportunity to see amazing panoramic views of the surroundings.
2. The second observation deck locates on the 148th floor at a height of 555 meters above the ground,. This is the highest observation deck in the world under the open sky.

Burj Khalifa is not only tourist attraction in Dubai. The skyscraper has tower with a spire, equipped with communication equipment for working of networks of cellular communication. This skyscraper is not just the tallest building in the world. Today the Burj Khalifa is called differently.

•This is the highest skyscraper

•This is the most multi-storey building
•This is the highest outdoor observation deck
•This is the high-rise building with the tallest lift
•This is the skyscraper with the high-speed elevator etc.

As a result, thanks to the skyscraper, this area and nearby areas became very prestigious, and therefore expensive. Statistics shows that the cost of such exclusive housing and commercial real estate grows, on average, by 35 percent. Moreover, it is not so important what about are in question. The cost of property near the skyscraper will be much higher than structures located away from it. Analysts believe that the rapid growth industries of tourism and trade occur in Downtown Dubai thanks to the presence of this skyscraper. In 2015 company AttractionTix declared that Burj Khalifa was the best attraction in the world. It helps to attract in Emirate of tourists from all over the world since its opening in 2010.

2. The second is "Tokyo Sky Tree" in Japan. In recent years many tall buildings have appeared. It worsened telecommunications. There were interruptions in communications. The signals were received from the old tower and they were poorly heard by last floors because of skyscrapers. Tokyo Sky Tree was built to solve this problem. In addition, the government planned to translate the country broadcasting in digital format. These and other moments have led to the need of designing the new tower. There was created the group of about 100 experienced professionals. Task was to do, not just television tower, but that it also was supposed to be earthquake resistant. "Tokyo Sky Tree" appeared in the capital by February 2012. There has emerged the world's second tallest building (634 meters) in Japan thanks to the creation of the tower. The opening of the tower was on May 22. There were involved 580 thousand people in construction. There was spent 812 million dollars. Built TV tower turned out to aseismic. Anti-seismic systems can withstand an earthquake of 7-ball force. The construction of Tokyo Sky Tree was made possible thanks to technical progress. Its height is 634 meters. This was built in 2011 and it was opened for visits of tourists in May 2012. The TV tower was built in just six years: the project was published in 2006 and it enacted in 2012. The authors of the skyscraper drew attention to three points at erection of Tokyo Sky Tree. These are the following points:

• The mixture of the traditional Japanese design with the newest modern technology.
• Tokyo Sky Tree contributed to increase capital at the expense of such grand construction
• The building was built by last word of safety regulations.

The tower must fulfill its production function. In this regard to it, its height had to exceed all existing buildings and those that might build later. The designers used new amortization technologies to protect the tower from earthquakes and hurricanes. The main goal is to support of the antennas. In addition to it, tourist use of this building was beforehand thought. Here everyone will find something to surprise oneself:

a) The tower was built in a not affluent area. Here there are many traditional Japanese buildings. So the view from the top is beautiful and varied. Tokyo Sky Tree has very high viewing platform.

b) Some people are interested in technology. They want to see unusual design solutions. They are also interested by the complexity of implementation of the anti-seismic protection.

c) Other people like exotic. They look from the height of the observation deck on buildings, which are made in traditional Japanese architecture.

3. The third example is "Willis Tower" in Chicago in USA. Each country has its own characteristics, its own symbols. Such symbol is the Willis Tower for USA. There was decided to profitably use 5000 square meters of land and to build a skyscraper in Chicago. In 1973, the tower was built. The building was erected quickly due to a number of circumstances:

1. A large number of workers worked to construct the building. These were about two and a half thousand people;

2. There were uninterrupted financing. The cost of the building is estimated at $ 150 million.

3. There were features of the design. Bruce Graham was chief architect of the building. Fazlur Khan was chief engineer. They took steel as the supporting material. Thanks to this choice, this design is able to withstand the strong blows of the Chicago winds.

The skyscraper has the height of 443 meters and 110 floors. During 25 years, the skyscraper was considered by the highest not only in USA. The tower was the highest in the world. Today the skyscraper is one of the tallest skyscrapers in the world. The inhabitants of Chicago are proud to say that they have the tallest building in the USA in their city. And they have a reason for pride:

1. Tourists come to see necessarily the observation deck. The observation deck locates on the 103th floor of the skyscraper. Every year, half a million people visit the skyscraper. During the day, 25 thousand people come here.

2. One can see 80 km around the observation deck if the weather is good.

3. The main purpose of the skyscraper is to place offices and business centers.

The skyscraper needed for broadcasting Chicago. Four heavy-duty antennas set on the roof of the building. They provide the signals this large city.

5. The above antennas avert lightning strikes approximately twice a day.

6. The fans of extreme sports and climbers are interested in the skyscraper. This building can be used as a site for films and TV programs.

4. The fourth is "Federation Tower" in Moscow in Russia. Its height is 95 floors. This skyscraper is one of the projects of Mirax Group. All decisions were thought out in advance at its construction. Much attention was paid to quality, therefore the most modern technological innovations were used. The skyscraper is located in "Moscow City". Nowadays, this building is considered one of the tallest skyscrapers in Europe. "Moscow City" has two high-rise buildings: "East" and "West". They are united by a common six-story corridor. The complex has both elite housing and modern offices.

"Federation Tower" is about 10 000 square meters, and the project is 443 000 square meters. "West" has 243 meters and 95 floors. "The East" has a height of 374 metres. The skyscraper has 67 elevators. Architect Sergei Cioban and his German colleague Peter Schweger designed the project. Thornton Tomasetti is the designer of the project. Renaissance Construction is a contractor. The facade of the tower is glazed by Yuanda. Other foreign companies also participated in the construction of the skyscraper. This is an example of international relations, international cooperation. The bookmark of the tower "West" occurred in 2006. The construction of the tower was ended in 2008. The construction of tower "the East" began in 2007. Many organizations are located in Federation Tower. Many offices of premium-class exist for entrepreneurs. There are high-speed modern elevators and escalators in the building. Offices and a shopping gallery are located in the building. Business offices are mostly located in the skyscraper. They are in the western tower from 1 to 46 floors and they are in the east on 1-60 and 63-68 floors. The total area of offices is 287 thousand square meters. Luxurious residential apartments are located on the upper floors of the skyscraper. Platinum apartments are from 90 to 95 floors in the building "East" [8, 9, 10].

4 Discussion

Different scientists have different attitudes to high-rise construction. Many recognize that this is necessary measure; some believe that it is one of the ways to exalt oneself. There exist the main reasons for the construction of high-rise buildings. The population is becoming more, and place for the erection of buildings is less. This is especially actually in large cities. The authors adhere to the two points of view about high-rise construction in Russian and foreign articles. Some people perceive high-rise construction as a benefit. Other authors refer to tall buildings warily or negatively. As an example, it is possible to consider article "Peculiarities of high-rise construction of the Central district of Yekaterinburg" by Shvets A. V. and Tarasova I. V. From our point of view in the article we can see that high-rise construction has both positive and negative sides. On the one hand, it "promotes the development of economy, business, tourism" and it is "at the forefront in the construction industry". On the other hand there are some features, which "determine negative attitude towards high-rise construction in both the public and the architectural community". For example, it may "violate historical and cultural heritage of the city" [11, 12] From the point of view of the author of this article, opinions of various authors regarding the role of high-rise construction can be very roughly divided as follows.

Table 5 The attitude of various authors [13]

Author	The title of article	+*	-*	+/-*
Shvets A. V., Tarasova I. V	Peculiarities of high-rise construction of the Central district of Yekaterinburg			+/-
Kai Chen Goh, Hui Hwang Goh, Mohd Faizal Omar, Tien Choon Toh and Abdullah Asuhaimi Mohd Zin	Accidents Preventive Practice for High-Rise Construction		-	(+/-)
O. Matejko	The history of development and modern trends in high-rise construction	+		
A.A. Aldeberky.	The influence of high-rise buildings on the environment		-	
Farzad Navaei.	An Overview of Sustainable Design Factors in High-Rise Buildings	+		
Vinay Chandwani, Dr. Vinay Agrawal, Naveen Kumar Gupta	Role of Conceptual Design in High Rise Buildings	+		
M. D. Kashin, T. B. Nabokova, V. N. Bgashev	Environmental interaction of high-rise buildings and the environment – the experience of foreign countries		-	

Positive (+)* Negative (-)* In Two Ways (+/-)*

Let us consider the advantages and disadvantages of high-rise construction based on the opinion of various authors.

Table 6 The advantages and disadvantages of high-rise construction.

Advantages	Disadvantages
• There plays a significant role in technical progress, •There promotes the development of business • There promotes the development of tourism. • There are addressed issues related to urban transport • There is a large number of private and public investments • There use labor resources actively etc.	• There may be the violation of the historic part of the city. • There are deaths and accidents • There are serious injuries • There may be air pollution, • People may feel uncomfortable on the street and in a building • There is strong wind etc.

In conclusion to the above table it can be noted that despite the fact that high-rise buildings are building more often, high-rise construction is still a high-risk area.

5 Conclusions

Thus, we see both positive and negative sides of high-rise construction. Along with the positive impact tall buildings can have serious negative effects. On the one hand high-rise construction is increased risk and great responsibility, and, on the other hand, this is an indicator of prestige and high professionalism of its participants. Much depends on the literacy development of the project and the organization implementing it. In Russia high-rise construction will be carried out due to lack of available of land for building in major cities such as Moscow, St. Petersburg and others. This can be assessed in different ways. However, as for most cities in the world, it is a forced measure on the one hand, and, on the other, a tribute to fashion and prestige.

References

1. S. Gorin, *Unique and express technologies in construction* [Digital resource] Access mode: http://www.stroinauka.ru/detailview.asp (2004)
2. W. Leontief, *The Structure of the American Economy, 1919–1929* (Cambridge: Harvard University Press, (1941) https://doi.org/10.1017/S0022050700053158 Published online: 01 February 2011
3. A. Terranova, *Skyscrapers* (New York: White Star Publishers, 2003)
4. M. Dhammy, *High rise housing. Residential Towers* [Digital resource] Access mode: https://ru.scribd.com/document/205050629/High-rise-housing
5. I. Rozhentsova et al, MATEC Web of Conferences **106** 08076, DOI: https://doi.org/10.1051/matecconf/201710608076 (2017)
6. N. Braila, C. Vahrusheva, E. Martynenko, T. Kisel, IOP: Earth and Environmental Science, **90**, 12030 (2017)
7. E. Vasilyeva, IOP (Earth and Environmental Science), **90**, 12216 (2017)
8. E. Chibisova, A. Zinatullin, Strategy of sustainable development of regions of Russia **5** 324-328 (2011)
9. I. Polyakova, E. Vasilyeva, IOP: Earth and Environmental Science **90** 012136 (2017)
10. E. Stein, N. Taskaeva, E. Chibisova, Procedia Engineering **165** 1410-1416 (2016)
11. A. Mottaeva, MATEC Web of Conferences **73** 07020 (2016)
12. A. Mottaeva , IOP Conf. Series: Earth and Environmental Science **90** 012120 doi :10.1088/1755-1315/90/1/0121209 (2017)
13. E. Petukhova ARX. The universal language of architecture **2** 108-113 (2006)

High-rise housing as a factor of the increase in Investment attractiveness of the city

Angela Mottaeva[1,2,*] and *Alexander* Zheltenkov[2]

[1] Moscow State University of Civil Engineering, Yaroslavskoe shosse, 26, Moscow,129337, Russia
[2] Moscow Regional State University , 129125 Radio str. 10, Moscow, Russia

Abstract. That is obviously for experts, irrespective of status or quality, high-rise housing is here to stay. No country is in a position to ignore or discard this conspicuous legacy of the post-war and mass housing period. Nevertheless, high-rise residential buildings are still less widespread, than offices. According to the author, there are certain problems and reasons for limitation of high-rise construction in the housing sector. The author summarizes successful experience of high-rise housing construction in the world and considers some social-and-economic prerequisites of its development in the modern city. These concrete examples, given in article, prove that the existing problems are being solved. The most effective option of construction is the combination of inhabited and commercial chambers in the high-rise building and also the creation of specific infrastructure. In that case housing high-rise estates will promote the increase in investment attractiveness and investment activity in the district and in the city as whole.

1 Introduction

That was construction of high-rise buildings for office purposes to begin at first. Thanks to that, the cities gained more modern and respectable look, and, above all - the number of the jobs (generally for "white collars"), located not only in the downtown but also out of historical areas increased. Further the turn of high-rise housing came.

Formal definitions of high-rise housing vary between countries. Many authors have focused on housing from the post-war period, which requires a lift to reach the upper floors. In general, this means high-rise is a block with a minimum of five storeys and a lift. Modern high-rise housing consist of tower (skyscraper) or point blocks reaching twenty or more storeys, or slab blocks providing hundreds of metres of separate dwellings connected by lengthy internal or external corridors. [1]

Among the architects and engineers who have made their contribution to the development of housing high-rise construction there were William Jannie, Bertrand Goldberg, Der Skatt, Dominique Perrault, Franck Lloyd Wright, Geoffry Powell, Peter "Joe" Chamberlin, Christoph Bon, Skidmore,

Owings and Merrill, Maria Rosa Cervera, Javier Pioz. In the USSR and Russia the development of high-rise housing construction is connected with the names of architects D. Chechulin and A. Rostovsky, A.N. Dushkin, B.S. Mezentsev, L.M. Polyakov and A.B.

* Corresponding author: angela-1309.m@yandex.ru

Boretsky, M.V. Posokhin, A.A. Mndoyants, A. Dobrovolsky, O. Malinovsky, P. Petrushenko, M.N. Vokhomsky (designer), engineers L.M. Gohmann, E.V. Myatlyuk, V.A. Saprykin. Modern Russian architects A. Trofimov, E. Treshchilina, V. Shteller, O. Markova, Sergey Nepomnyashchy, Anton Nadtochy and Vera Butko are also deserve mentioning.

Considering the emergence of residential skyscrapers, it is necessary to remember the term gentrification (from the English "gentry" – untitled nobility), meaning the social phenomenon which has arisen in Great Britain and the USA in the 1960th; and representing arrival to the place of poor households in the downtown of more prosperous ones. Externally gentrification was presented by reconstruction of certain buildings, quarters and the whole districts of the centres of the cities, which had the investigation substantial increase of quality in them the inhabited environment and housing prices. Evolution of this phenomenon has led to the creation of new buildings - higher and higher residential buildings.

Nowadays in the cities and megalopolises of the whole world there is a replacement of large productions out of limits of housing urban development. There is a danger of emergence of areas and the whole "ghost" towns. In order to avoid that the released territories can be effectively occupied with pointed building in the form of inhabited skyscrapers.

On the other hand, some office centres of the cities "become extinct" at the end of the working day. Construction of skyscrapers in which office and dwelling are combined, solves the problem and does the similar area more attractive. Inhabitants (households) come back to the area, founders of institutions of social service, public catering, etc. invest more actively.

The prospects of the increase in investment attractiveness of districts and whole cities at the expense of high-rise housing construction are considered in this article.

2 State-of-the-Art Review

As decades ago, at the beginning of high-rise construction, today the majority of high-rise buildings are office skyscrapers. The share of residential buildings is small – about 20%, and their height, as a rule, is much less than average skyscraper and does not exceed 140 metres, and the absolute majority of them is the share of residential buildings up to 100 metres high.

However the first European building, higher than 100 metres, was inhabited, but not administrative. It was one of seven "Stalin" high-rise buildings, created in Moscow by 1954. Those high-rise buildings have created "a ring of the high-rise dominants structuring building outside a historical part" of the city [2].

The highest inhabited building in the world is the skyscraper of Trump World Tower in New York, 262 metres high (72 storeys) is considered today. In Europe Moscow 61-storey "Triumf-Palas", which is 264 metres high (with the spikes) is in the lead. Towers of Marina City standing in Chicago (almost 250 metres high), multipurpose John Hancock Centre (343 metres high, including offices, hotel and inhabited levels at the top) are widely known. In China the "Vertical Bionic City Tower" project (a 300-storey 1228 metres high construction for 100 thousand inhabitants is being seriously discussed.

So, there is a contradiction. Numerous calculations throughout all the history of high-rise construction showed that even despite the high cost and a limited reserve of the city land (the main incentives of high-rise construction), the economic feasibility of high-rise residential buildings as mass-product comes to the end at the height of 25-30 floors. This figure also limits the vast majority of high-rise residential buildings all around the world.

This article points out the reasons of limitation of high-rise construction in the housing sector and analyzes some social-and-economic prerequisites of his development in the modern large city. The influence of high-rise housing construction on the investment activity in the city is considered.

3 Results

The majority of residential skyscrapers in the world do not exceed 75 stores.
Any further increase in number of storeys demands special measures for realization and control of constructive and fire safety, special decisions for water, heat and power supply, special complicated and expensive ventilation systems, sewerage systems and even rubbish disposal systems. Besides, construction of inhabited skyscrapers means dense population, considerable loads of social infrastructure, transport and engineering networks.

And still every day there appear more and more residential buildings, exceeding 100 metres and they are becoming higher. The matter is that construction of such residential buildings was always based first of all not on economic reasons. Except utilitarian economic purposes the high-rise ring is designed to resolve also city-planning and even political affairs.

So, for example, in case of "seven Moscow sisters" (so called "Stalin" towers), three of which are inhabited skyscrapers, they help to resolve not the housing problem: these buildings became populated "exclusively family-by-family by the citizens from nomenclature, scientific and technical, creative elite of the Soviet society" [2]. They have been designed to provide the new scale of the city, corresponding to the rank of the capital of the world power. Similarly, the construction of Marina City in Chicago has started the successful attempt to return the middle class, who was "running" to suburbs to the central regions of the city in 1964. Other projects serve to the city regulation purposes too. Such projects include rather considerable ones, such as areas La Défense in Paris and Docklands in London, the multipurpose Barbikan complexes in London, Pershing Square in Los Angeles and dozens of others projects, including great number of high-rise and superhigh-rise residential buildings. Generally all these projects have been designed to take away a certain part of business life of the city out of borders of the historic center, resolving the issue of its transport congestion and the problem of creation "the architectural education adequate to requirements of time and at the same time not breaking historical appearance of the city" [3].

Though the era of superurbanization remained in the 1970th, construction of separate multipurpose public-and-housing estates [4, 5], such as, Trump Tower for 6 thousand apartments in New York, instead the former railway depot, is conducted and today.

Large-scale projects are implemented in Southeast Asia and Australia. In the Asian region formation of similar high-rise complexes is often carried out within the programmes for reconstruction of blighted areas and also for provision of housing for the numerous population of limited territories. That is also possible to mention the "New Ring of Moscow" programme adopted in 1998. Within its framework sixty zones was allocated" in the median-and-peripheral territory of the city in 2005. That is planned to build about two hundred high-rise (30–70 storey) structures, which considerable part will be made up by the residential buildings. One can also note a local housing estate "Edelweiss", administrative-and-business buildings in the "City" zone, the 35-storey "Continental" complex, other residential buildings). The scheme of location of high-rise buildings provides several types of high-rise accents, such as: "coastal", "urbanistic", "planning accent", "entrance to the city" and so on.

What is the base for the hope, that existence of high-rise residential buildings will promote the increase in investment activity in the city?

First of all, the structure of "skyscraper" is important. In modern multilevel buildings, the top storeys are allocated to housing, below hotel rooms or offices are usually placed. Near earth level, in elevated and underground storeys of a skyscraper, shopping and entertainment centres, parkings, stations of the subway, etc. are located. One of the most widespread receptions of the architectural organization of the lower zone of similar complexes is the creation of the stilobat platform, often occupying all the ground area of the building; the total area of the chambers, which are in it can reach tens and even hundreds of thousands of meters. On such platform both one, and whole groups the of chambers of different height can accommodate. Marina City is the axiom of such organization of a complex. The first 19 storeys of the building of Marína City are occupied by the parking having design, unique for the middle of the 60th, twirled on a spiral, which is calculated on 896 parking places. On the 20th storey the laundry, the gym, warehouse, rooms for conferences, theatre, the gym, the swimming pool, an ice skating rink, several shops and restaurants are located. Storeys from 21 to 60 are given for apartments. Under one part of the platform, hanging over the river, the mooring for yachts is located [6] .

Fig. 1. Marina City (photo by Rich McNees)

Multilevel platforms of the Marina City building radically solve many functional problems, such as division of pedestrian and traffic flows, pedestrian transit and inhabited recreations, public and inhabited levels, maximal use of the urban area.

Today there are even projects of the cities-skyscrapers. In such a city inhabited and working tiers alternate with parks, technical and economic tiers. The author of two such projects is the Japanese is the well-known corporation "Takenaka". The larger one is the kilometer height of Sky City, which has to become a housing for thirty six thousand people and a workplace for one hundred thousand people. Everything is provided in such skyscraper for full-fledged life without any need to leave it: schools, parks, shops, restaurants, theaters, hospitals, offices; issues of communications and transport are resolved (figure 2). Actually such construction will become the centre of business and investment activity.

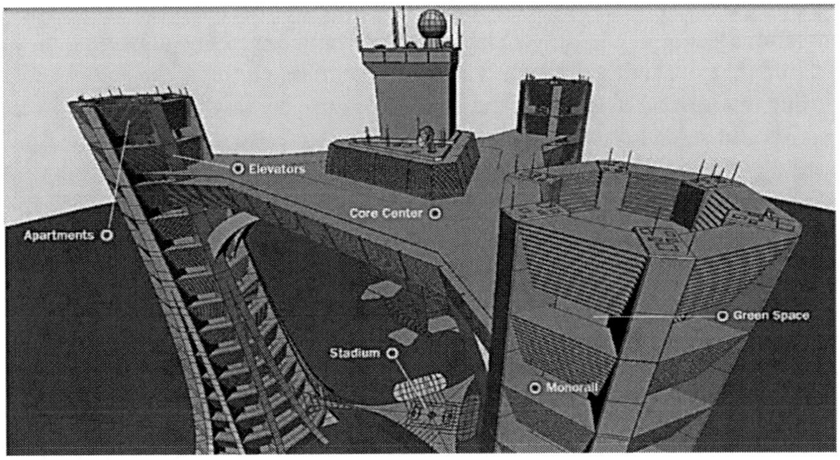

Fig. 2. Sky City project, Japan (from Web Site for Architecture & Design-Related Info)

In Russia the architect Sergey Nepomnyashchy, who offered several concepts, including the 75-storey city "Birth of Venus" and the "Pancake city", spread on a landscape in the form of a huge washer, is engaged in the projects of the cities-buildings.

It is necessary to mention significance, visibility of the high-rise building. In combination with its uniqueness the visibility of the building automatically transfers it to the category of sights of a space and social landscape of the city, makes it the subject of discussion. "The skyscraper is always a status construction" [7] not only for the owner and the inhabitant, but also for all the district and the whole city, quite often thanks to the building, becoming the subject of tourism .

In Moscow the Ukraine hotel is not only one of the best hotels of Russia, but also one of the most beautiful sights of Moscow. Leningradskaya hotel is another sight, and even when the building began to collapse, and the state had no money for reconstruction, the hotel managed to be sold rather favourably at the auction. The Sadko-Hotel network, which was engaged in reconstruction of the building, became the new owner, then the hotel opened under the new name "Hilton Moscow Leningradskaya". One more sight is the house on Kotelnicheskaya Embankment, which has its unique history, and is associated with the famous residents and surrounded with the numerous legends attracts tourists into its vicinity.

Tourism in turn becomes one more incentive of possible investments into the city economy.

4 Discussion

The natters of lication and construction of high-rise buildings and complexes in capitals and other cities, remain a subject of the closest attention of city-planners, architects, designers, engineers, ecologists, sociologists, demographers, doctors, historians, art critics and also representatives of the general city public.

For example, the US scientists completed an interesting research and found out that the chambers, which settle down not above the environmental trees (according to the American measures that means storeys from 1 to 7, according to the Russian - measures that means storeys from 1 to 10) can be recognized as the best for the inhabitants' health. According to the data of the American researchers, the people, living above the 7th storeys are subject to risk of various diseases more than those, who live below. Among "skyscrapers" persons nervous breakdowns, under immunity and cardiovascular diseases are noticed very often [7].

But despite many negative sides of life at great height, it is possible to find the particular qualities, characteristic for the high-rise dwelling, making it demanded by the consumer. Those qualities are on border between objective properties of inhabited skyscrapers as object of the social-and-economic relations and images of collective consciousness, they are "non-material", but they are rather accurately recorded in mass culture. The status and prestige of such housing, interesting design and the view from the window, existence of characteristic infrastructure and subculture in a skyscraper are among those qualities.

According to the conclusions of sociologists, the inhabitant of a skyscraper is a particular type of the citizen. That is "a rich, young, highly educated, single or married man, whose family has no more than two children" [7]. The representative behavior model, active lifestyle are inherent in it, it is rather of individualist, than a family set of values, a wide range of the formal and informal contacts, high business activity, intensive consumption of products of city culture. Such an inhabitant creates padding demand for goods and services in the area and the city in general, and in turn actively participates in their creation [8, 9, 10, 11]. Such a citizen is attracted to the city by the existence of modern status inhabited skyscrapers [12, 13, 14].

Besides medicine and psychology it is necessary to mention the following problems, arising during projection and construction of high-rise residential buildings in Moscow and demanding obligatory consideration and decision:

- justified city-planning and functional and typological need of construction;
- marginal (justified) number of storeys (height);
- the right choice of design system, the scheme and design decisions taking into account prevention of loss of the stability of the basement and the whole construction, leading to the destruction and the collapse of structures;
- inadmissibility of the deviation from the approved design solutions and change of number of storeys of buildings during construction;
- optimum containment of inhabited and non-residential, public areas (we remember the example of Marina City);
- necessary functional interaction of residential and uninhabited buildings and constructions with the transport and serving city infrastructure;
- the required containment of underground, ground and elevated parkings of the individual transport and their rational placement;
- efficient minimization of threat of external and internal danger of destruction of the building due to creation of express service of safe operation;
- the required fire and evacuation safety of the people, who are in high-rise buildings (here the towers of the World Trade Center in New York remain very effective example, though they were not residential buildings);
- rational effectiveness of the modern engineering decisions on life support and equipment of the building, to energy saving and comfort of an upkeep, etc.

The urgent matter during construction of high-rise buildings is providing them with energy. For the solution of this complexity it is necessary to consider the possibility of the use of the alternate sources of heat and energy supply which along with centralized supply will be put into maintenance . Roofs and spikes of high-rise buildings are very suitable for placement of solar batteries and wind energy installations [15, 16].

The following problem is that in view of extreme load of city highways with motor transport which number is steadily growing, close attention should be paid to transport infrastructure of the territories of high-rise complexes. But on the other hand construction of inhabited skyscrapers will create padding prerequisites for the accelerated development of transport infrastructure which in turn – is a factor of investment attractiveness and investment activity in the city. A probable solution of a transport problem are also multilevel platforms which allow to divide pedestrian and traffic flows [17].

Quite often critics of skyscrapers accuse them of violation of ecology, creation of the unnatural and unhealthy environment for life, isolation of inhabitants from the earth and nature. As the answer to this criticism there exist two Italian skyscrapers the Vertical wood (Milan, Italy) in which name the main idea of original skyscrapers is put. According to authors of the Vertical wood to return to sources and at the same time such projects will help to save residents of the megalopolis from inevitable environmental disaster.

The answer to the requirements of ecological construction is the highest building of Austria, the pollution-free skyscraper DC Tower 1, which is also known under the name Donau City Tower. This building was constructed on the bank of Danube-river in the 2013 and it at once was included in various lists of the most impressive constructions of the present. The matter is that the 250-metres building with is one of the few skyscrapers of the planet constructed according to all the standards of "green" architecture. In the building water-conservation shower, energy efficient elevators and also many other interesting counters inseparably linked with environmental issues are installed. "Life support" of the tower happens due to pollution-free energy. This creature of the French architect Dominic Perrault serves as the proof of the possibility of the solution of problems of environmental friendliness, energy efficiency and rational use of resources in the high-rise residential building.

Solution of problems, disturbing the opponents of high-rise residential building is considered in the city investment programme "New Ring of Moscow". The high-rise complexes built within the specified programme demonstrate in new approach to providing a high level of safety and comfort of accommodation. At construction of these complexes in adjacent territories the modern infrastructure is supposed: kindergartens, schools, movie theaters, shops, medical institutions and beauty shops, fitness centers, nurseries and sports grounds [18]. The territory around a complex of high-rise objects according to normative requirements has to be arranged well that surely includes also gardening. The modern city landscapes will make more comfortable social and psychological and ecological space which will attract residents and businessmen investors.

The Moscow real estate market always was and remains very attractive to investors [19, 20]. And the developed programme will help to develop and improve more actively the mechanism of interaction of authorities of the capital and various investment structures.

5 Conclusion

Great experience of high-rise construction, including housing, is accumulated in the world. The difficult and fruitful way is passed by many gifted architects.

Construction of high-rise residential buildings means certain engineering and architectural difficulties. Special responsibility lays down on administrative, executive and supervisory authorities of any city, as well as on their divisions in the field of architecture and construction, on specialized research and design institutes on creation of residential buildings and complexes, design of unique high-rise constructions. But already there exist buildings, giving the answer to these difficulties. The matter is in placing such buildings in appropriate districts of the city, without breaking architectural and esthetical integrity of areas with historical building. On the other hand, the modern European cities (and Moscow is not an exception) aspire to combine residential and office real estate. High-rise residential buildings provide necessary functional and visual and figurative concentration of inhabited function for full-fledged round-the-clock life cycle of highly urbanized fragments of the urban environment. That attracts investors and buyers of the real estate to the city.

Reference

1. R. Turkington, F. Kempen, Wassenberg Housing and Urban Policy Studies. Current trends and future prospects (Alblasserdam: OTB Research Institute for Housing, Urban and Mobility Studies) 302 (2004)
2. S. Gorin Unique and express technologies in construction [Digital resource] Access mode: http://www.stroinauka.ru/detailview.asp (2004)
3. E. Petukhova ARX. The universal language of architecture **2** 108-113 (2006)
4. J. Pridmore *Chicago Architecture and Design* (New York: Harry N. Abrams, Inc.) 296 (2005)
5. A. Terranova *Skyscrapers* (New York: White Star Publishers) 310 (2003)
6. I. Marjanović, K. Rüedi. *Marina City: Bertrand Goldberg's urban vision* (New York: Princeton Architectural Press) 175 (2010)
7. O'Sallivan A. *City economy* (Moscow: INFRAM) 706 (2002)
8. M. Dhammy *High rise housing. Residential Towers* [Digital resource] Access mode: https://ru.scribd.com/document/205050629/High-rise-housing

9. P. Dunleavy *The politics of mass housing in Britain, 1945–1975* (Oxford: Clarendon Press) 454 (1981)
10. L. Hanley *Estates: an intimate history* (London: Granta Books) 256 (2007)
11. A. Power *Property before people* (London: Allen & Unwin) 268 (1987)
12. A. Power *Estates on the edge* (UK: MacMillan) 95 (1997)
13. I. Rozhentsova et al, MATEC Web of Conferences **106** 08076, DOI: https://doi.org/10.1051/matecconf/201710608076 (2017)
14. W. Leontief *The Structure of the American Economy, 1919–1929* (Cambridge: Harvard University Press) 181 (1941) https://doi.org/10.1017/S0022050700053158 Published online: 01 February 2011
15. T. Meshcjeryakova, N. Verstina Advances in Intelligent Systems and Computing **692** 1136-1150 (2018)
16. E. Chibisova, A. Zinatullin Strategy of sustainable development of regions of Russia **5** 324-328 (2011)
17. I. Polyakova, E. Vasilyeva IOP: Earth and Environmental Science **90** 012136 (2017)
18. A. Mottaeva, MATEC Web of Conferences **73** 07020 (2016)
19. A. Mottaeva, IOP: Earth and Environmental Science **90** 12138 (2017)

National originality of the architecture of Khreshchatyk as a unique ensemble of the period of totalitarianism

Olena Oliynyk[1*]

[1]National Aviation University, Kiev, Ukraine, Kosmonavta Komarova Ave, 1, Kiev, 03680, Ukraine

Abstract. Khreschatyk is a page apart in the history of world architecture. While it has a number of distinct characteristics of totalitarian architecture, Khreschatyk is the only architectural ensemble of the period to combine national tradition with the exalted sentiment of Soviet architecture of the Stalin era. Also, it uniquely matched architecture and landscape. The façades has elements of Ukrainian baroque, which sets Khreschatyk apart from similar ensembles of the 1940s-1950s in other countries that mainly drew upon Neoclassicism or Modernism. While period architecture in other countries is typically marked by its grand scale and heavily accentuated civic spirit – complete with a denigration of the individual at the expense of the manifest greatness of Authority, Khreschatyk stand out for its pronounced harmony as an environment based on the careful preservation of old heritage, the skillful use of the landscape, and the introduction of traditional motifs, alongside an almost total lack of Soviet symbols. Unlike the grim grandness of totalitarian architecture in other countries, the facades of the residential buildings that line Khreschatyk emanate *joie de vivre* and admiration for the fertility of Ukrainian soil.

1 Introduction

The period of totalitarianism is perhaps the most dramatic not only in the history of Ukraine, but also all the countries of Eastern Europe. The tragedy experienced by the Soviet people, the physical destruction of the intellectual elite of the population, the atmosphere of constant fear and ideological pressure have affected the fact that any cultural phenomena of this period, later received a negative assessment. This primarily concerned architecture. Studies of totalitarian architecture and even today in European countries are isolated, often based not on art studies, but on political grounds. [1,2]

Yes, even in Italy, and so far, this period remains completely unexplored. In Rome, recently the funds of the former Institute of Asia and Africa were destroyed, where unique materials were collected about the Mussoliniev construction in Greece, Eritrea, Ethiopia. Consequently, not only Stalin's architecture is a symbol of tyranny, the Italian period architecture of Mussolini, too, is in the minds of Italians as a symbol of fascism, and its

* Corresponding author: archiprestig@gmail.com

research is still not prohibited, but not encouraged. Therefore, any architectural research on the architecture of this period is very relevant. [2,3]

It was at this time that majestic ensembles of the main streets of several European capitals - Minsk, Berlin, Warsaw, Kiev - were built. Most of them are built in the style of the characteristic imperial neoclassicism. Only Khreshchatyk, in contrast to the usual image of totalitarian architecture, embodies the features of Ukrainian folk architecture and the Ukrainian Baroque.

2 History of the formation

Khreschatyk, the main street of Kyiv, runs along the thalweg (the line of lowest elevation) of Khreschatyi Ravine between the Starokyivska and Pechersk hills. (Fig.1). Descending toward the street, which is located 20 to 30 meters lower that both plateaus, their terraced slopes create a multi-plane visual perception of Khreschatyk's modern buildings. Three squares – Yevropeiska, Maidan Nezalezhnosti (Independence Square) and Besarabska – were formed organically where the sloping relief recedes from the thalweg the most.

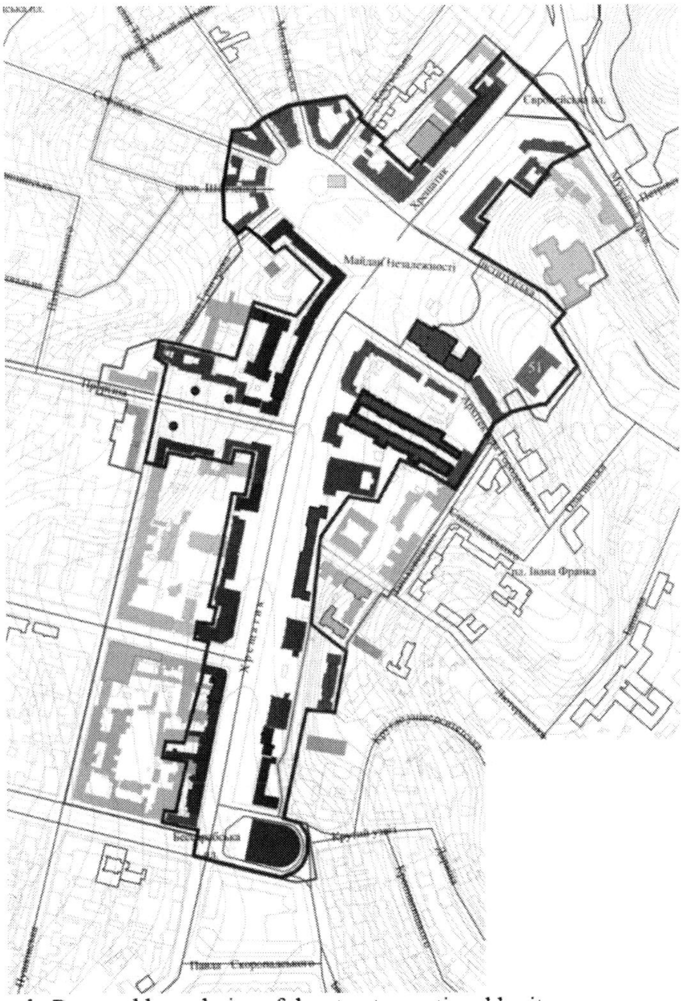

Fig.1. Khreshchatyk. Proposal boundaries of the street as national heritage.

In Kyiv Rus times, the fortifications of the Upper Town stretched across the slopes of Starokyivska (Old Kyiv) Hill above the ravine. Partly destroyed during Mongol attacks on

Kyiv in the mid-13th cent., they had all but disappeared by the early 19th cent. When the remnants had been cleared away, a square formed at the foot of Starokyivska Hill just next to Khreschatyk. That was the beginning of what we know as Maidan Nezalezhnosti today [4, 8]

The development of Khreschatyk got under way in the 1830s-1840s. It started from Yevropeiska Square and acquired its current shape in 1837 under the city's master plan (architects L. Shmihelsky, V. Beretti and L. Stanzani) to become the city's main thoroughfare. In the 1850-1860s, Khreschatyk was built up with mainly one- and two-floor buildings either to standard or individual designs by Kyiv-based architects A. Melensky, M. Samonov and L. Stanzani, which failed to constitute an integrated whole. [5]

The formative process was completed in the mid-1910s, when Khreschatyk turned into the city's most representative section. Multistoried buildings in historical and Art Nouveau styles (architects E. Bradtman, G. Schleifer, V. Nikolayev, F. Lidvall, L. Benois, V. Horodetsky, P. Andreyev and O. Schille) put Kyiv on a par with the best cities of Europe. The street housed the main government institutions (Kyiv was the seat of a large province in the Russian Empire) and boasted 17 jeweler's stores, 13 cinema theaters, seven hotels, seven major bank offices, a stock exchange, the Noble Assembly, and the town's best retail stores. Some of these pre-1917 Revolution buildings have survived to this day on the even side of the street and are of special value. These are the complex of bank and office buildings from the 1910s at the beginning of the street (##6-10) and some commercial apartments and hotels of the late 19th cent. (##40/1–52), all of them architectural landmarks.

The final stage of Khreschatyk's development came during the Soviet era. In 1941-1943, most of the street was destroyed. Khreschatyk was rebuilt from its war damage in the late 1940s-1950s to the project design of a team of Kyiv-based architects: A. Vlasov (team leader), A. Dobrovolsky, V. Yelizarov, B. Pryimak, O. Zavarov and O. Malynovsky. The centerpiece of their concept was to build up the even side of the street with a continuous front of administrative, commercial and civil structures and its odd side, with mostly residential buildings with stores, cafes and cinemas on the ground floor. The city's first subterranean passageways for pedestrians were built on Khreschatyk to facilitate traffic in 1961-1962. [6,7]

The competition for the reconstruction of Khreschatyk was announced back when the war was still on, in June 1944. It can be seen from the detailed terms and conditions of the competition that it Khreschatyk's ensemble was first supposed to play the role of a government seat and consist of civil structures only. It was recommended to push residential buildings deep inside the blocks. The level of the competition as well as the tasks set before the participants and their urban planning solutions were without parallel both in Ukraine and in all of Europe in the first postwar decade.

In January 1945, the results of the first round of the competition, in which all first-rank Soviet architects took part, were summed up. It was during the first round that the final concept formed in the main: to change the scale of the main street, give different treatment to the even and odd sides, ensure functional zoning, and create a pedestrian boulevard along the residential front on the odd side. As far as style was concerned, the designers still looked to Classicistic or Renaissance motifs, and only a few entrants dared use Ukrainian baroque forms. Three teams were selected to take part in the second round in 1945. Those were headed by A. Vlasov, V. Zabolotny and O. Tatsiy. However, V. Zabolotny did not participate in the further competition, so a duel of Tatsiy and Vlasov's teams ensued. The third round took place in 1947. Finally, Workshop #1 of the Kyivproekt Institute consisting of A. Vlasov, A. Dobrovolsky, V. Yelizarov, B. Pryimak, O. Zavarov and O. Malynovsky, with invited engineers I. Skachov and V. Repiakh, was commissioned to sum up and elaborate on the best designs of the competition.

However, the final master plan of Khreschatyk was not approved until 1949, and its realization became the task of Dobrovolsky, Yelizarov, Malynovsky and other Kyivans (A. Vlasov was appointed chief architect of Moscow in 1950). Their team was instrumental in implementing the best of the competition designs. Most importantly, they managed to lend the city a new dimension that befitted the capital of a large Union republic, underscore central Kyiv's unique natural landscape with the gentle double curve of the built-up area, and recall the traditions of Ukrainian baroque and folk architecture in the top embellishments and façade décor of the buildings.

At the implementation stage, the architects took account of almost all of the more interesting ideas put forth in the competition designs. In A. Vlasov's proposal, for example, Khreschatyk was treated as a "throughway street." It was at the implementation stage, too, that the green slopes of Khreschatyi Ravine were included in the composition (as was suggested by O. Tatsiy), the pedestrian lanes designed (O. Malyshenko), and Ukrainian baroque motifs used (V. Zabolotny).

The imagery of Khreschatyk's architecture is based on interpretations of baroque forms and methods. As the architectural heritage of the Ukrainian baroque era was not yet studied at the time, O. Zavarov, who supervised the décor part of the project, partly used drawings and measurements of some West European (primarily Spanish and German) and Latin American architecture. For the most part, however, the architects drew upon local tradition. The resulting combination lent Khreschatyk a truly unique appearance. [6,8,10]

In the process of realization, the number of civil buildings stipulated by the competition was reduced, and the odd side of the city's main thoroughfare was built up with multi-storied residential houses. (Fig.2,3)

The original conception is based on creating a "pulsating" space, which we now perceive as a logical scenario for a street with obvious landscape distinctions rather than a fruit of the architects' imagination. The short initial stretch of the street briskly flows into the broad elongated expanse of Independence Square that crosses it in a perpendicular, making it possible to engage parts of Pecherska and Starokyivska hills complete with the several tiers of buildings going up the slopes at either end of the square. Looking in the direction of the Old Town, this majestic panorama is crowned by the gold domes of Sancta Sophia of Kyiv – a WH property. On the opposite, the odd side of Khreschatyk, the square continues into a stepped rise leading to Hotel Ukraina (formerly, Moskva) with its grandstand stairs.

After its major reconstruction in 1977, the square was renamed to October Revolution Square. Its semantic purport as the city's main square, designed for holding ceremonies, gala events and festive parades, was translated into a new spatial conception. The latter was modified in 1998, when the engineering networks were reconstructed and the entire street was revamped from Yevropeiska to Besarabska Square.

Taking a gentle turn just after the Maidan, Khreschatyk runs straight as far as the diminutive Besarabska Square. At this stretch, it acquires its unmistakable asymmetrical look. While the odd side sports a festive boulevard-like pedestrian strip all along its residential front, the opposite one is a serried row of predominantly civil and administration buildings marked by a sober linear continuity of same-height principal cornices. This contrast underpins the street's artistic conception. The architecture of the even side is marked by greater laconism and a palatial restraint. On the other side, the varied profiles of the cornices, the miniature gable fronts and pendant pillars, and the ceramic faced tiles with their ornamental motifs in relief call the rich imagery of Ukrainian embroidery to mind. In combination with the original urban-planning conception, all these elements create a distinctly Ukrainian architectural ensemble.

Another complex task was to combine new construction with the historic late 19th-early 20th century buildings that had survived the war. According to A. Dobrovolsky, the team aimed at creating an integral architectural composition that would look both majestic and picturesque. The serried front of the one and the hiatuses of the other side made it possible to emphasize Kyiv's distinctive terrain. In a pioneering urban-planning technique, the pedestrian strip along the odd side united separate clusters of buildings, the hiatuses between them, and the three squares into one whole. The designers were also instrumental in seamlessly blending the triumphal scale of the project with the Romantic and ethnic motifs of the rich and variegated plastic décor as well as the visual depth of new construction on adjacent hills.

In addition to its function as the city's main thoroughfare and ceremonial public space, the visual dynamism of the combination of civil and residential buildings restored to Khreschatyk its mid-19th century scale and feel of a cozy pedestrian boulevard.

Khreschatyk is a page apart in the history of Ukrainian architecture that had defining influence on its further development. While it has a number of distinct characteristics of totalitarian architecture, Khreschatyk with its unmistakable national touch is a symbol of the country's rebirth from the tragedy of WWII and its residents and visitors' favorite public space. [9]

Khreshchatyk's new appearance had enormous influence on the development of architecture in the former Soviet Union. It still remains one of the best works of the so called totalitarian architecture. From the town planning perspective, it is an integral architectural ensemble 1200 meters long and an average of 75 meters wide, with three squares, broad sidewalks, and buildings from various periods. In my own opinion, nothing better was built in all the subsequent decades.

3 Comparative analysis with similar architectural ensembles

Its unique character was built into Khreschatyk's rebuilding program back at the inception stage. Designed as the result of several rounds of a competition to a clear-cut plan and within very stringent deadlines, it preserved all of the surviving buildings and the integrity of Kyiv's urban space as a whole. It also made full use of local vernacular tradition. Also, it was in Khreschatyk and integral part, Maidan Nezalezhnosti (Independence Square), that the dramatic events of 2004 and 2013-2014 with their strong impact on Ukrainian (and European) history and the nation's identity took place. The distinctive features of its spatial solutions and its relationship with the relief of downtown Kiev have played a central role in the organizing of mass rallies and the formation of the Ukrainian national identity.

Khreschatyk is the only architectural ensemble of the period to combine national tradition with the exalted sentiment of Soviet architecture of the Stalin era. Also, it uniquely matched architecture and landscape. The façades has elements of Ukrainian baroque, which sets Khreschatyk apart from similar ensembles of the 1940s-1950s in other countries that mainly drew upon Neoclassicism or Modernism. Khreschatyk has a unity of the imposing monumentality of proportion that befits a state capital, the lyrical intimacy of its unique terrain, the bright decorative character of Ukrainian baroque, and the comfort of a well-thought-out human environment.

While period architecture in other countries is typically marked by its grand scale and heavily accentuated civic spirit – complete with a denigration of the individual at the expense of the manifest greatness of Authority, Khreschatyk stand out for its pronounced harmony as an environment based on the careful preservation of old heritage, the skillful use of the landscape, and the introduction of traditional motifs, among them flowers, fruits, rosettes, and *rocaille,* alongside an almost total lack of Soviet symbols. Unlike the grim grandness of totalitarian architecture in other countries, the facades of the residential buildings that line Khreschatyk emanate *joie de vivre* and admiration for the fertility of Ukrainian soil. [6]

Another distinctive aspect of the Khreschatyk project is that it was designed for a territory to be rebuilt from war damage rather than a vacant space. Leaning on the landscape forms that were definitive for the historic center of Kyiv, it incorporated surviving buildings (which were painstakingly restored). The architectural and artistic treatment was different for the odd and the even part of the street, which became the key aspect of the composition.

Furthermore, landscape was not only factored in but rather given priority. Situated in a valley between two picturesque hills, Khreschatyk has numerous spatial ties to surrounding built-up areas. This can be seen from the completions of the streets connecting it to the Upper Town, the dominant high-rises on the odd side, the asymmetrical composition of the odd and even sides, the fluid turn of the street as it goes with the relief, and the hiatuses and arched driveways that create additional visual connections.

The architectural décor continues the local vernacular tradition and the distinctive Ukrainian baroque style. Colorful mosaic compositions decorating the vaults of the archways and the interiors of the Khreschatyk subway station (1960) draw upon the traditions of Mykhailo Boychuk's school of monumental art – an original if short-lived phenomenon which was almost completely eliminated in the Stalinist purges of the 1930s. All these features of Khreschatyk's appearance attest to an upsurge of the national spirit and the designers' intent to revive an artistic tradition in which monumental art and façade decoration held a central place. Specialized ceramic workshops were set up for the manufacture of tiles, ornaments and sculptural compositions at the experimental factory of the Ukrainian Academy of Architecture. Summing up, it can be said that the synthesis of arts used in the reconstruction of Khreschatyk both performed political tasks and revived national traditions in the décor, color scheme and plastic artworks of its buildings.

This interest in artistic form and intent to rethink national artistic tradition in the country's main street was a manifestation of ideological resistance and national awakening.

The ideas built into the reconstruction of Khreschatyk are topical even today. This project became a one-of-a-kind case of transforming a provincial town into a European capital that was not unlike Paris or Vienna in places. The scale and character of the street's architecture reflect the intuition and skill with which proportions needed for the city's central thoroughfare were defined and realized without detriment to the intimate, humane, and almost lyrical character this ancient city has always possessed.

Fig.2. Khreshchatyk. Even-side elevations of revival buildings

Fig.3. Khreshchatyk. Odd-side elevations

Fig.4. Khreshchatyk. Even-side elevations

Fig.5. Maidan Nezalezhnosti

Fig.6. Odd-number Side Khreshchatyk. General view.

 https://doi.org/10.1051/e3sconf/20183301039

Fig.7. Khreshchatyk. Symbols of Ukrainian baroque in architecture of facades. (Build. 13,Khreshchatyk)

Fig.8. Symbols of Ukrainian baroque in architecture of facades. (15, and 25, Khreshchatyk)

Fig.9. Architectural décor on facades. (30/1, Khreshchatyk 30/1 and 6, Prorizna)

Fig.10. Architectural décor on facades. (30/1, Khreshchatyk)

Fig.11. Architectural décor on facades. (27, Khreshchatyk 29, Khreshchatyk)

Fig.12. Architectural décor on facades (30/1, Khreshchatyk and 8, Prorizna)

Fig.13. Symbols of Ukrainian folk art in architecture of facades. (10, Prorizna)

Fig.13. Symbols of Ukrainian folk art in architecture of facades. (27, Khreshchatik)

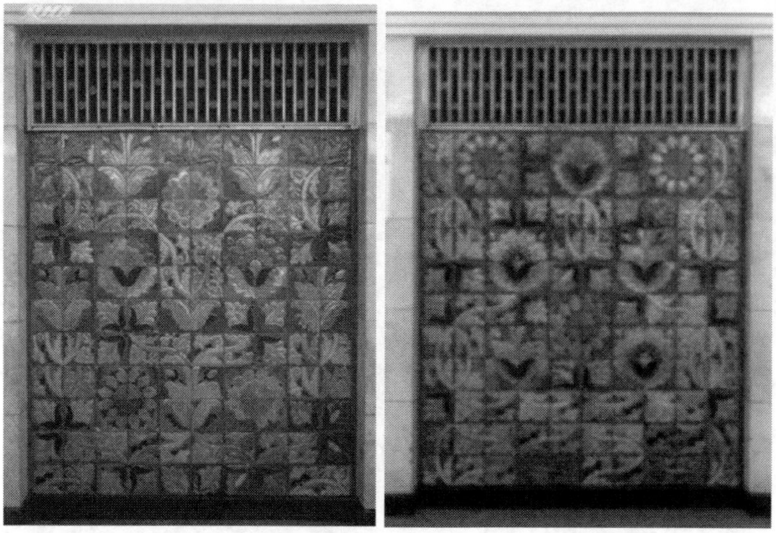

Fig.14. Mosaics in interior of metro station "Khreshchatik"

Fig.15. View of Khreshchatyk from the side of Khmelnitsky (former Lenin) street/ 1950th. Photo by G. Ugrinovich

Fig.16. Spatial ties to surrounding built-up areas. (21-23, Khreshchatik)

Fig.17. The arched driveways that create additional visual connections (24-26, Khreshchatyk)

4 Characteristic features of Khreshchatyk architecture

- Khreschatyk is an integral contemporary urban planning ensemble throughout its 1,200 m length
- Built to the winning design of Europe's first and only reconstruction competition held when WWII was still on (1944)
- Odd and even sides each has its own design reflecting unique landscape situation
- City planning methods that organically continue the scenery
- Spatial interactions with existing relief and urban structure – high top of city planning
- Incorporation of surviving buildings in the ensemble and their restoration
- The express use of elements of Ukrainian baroque and traditional art to underscore the national identity

5 Conclusions

Khreshchatyk street is an unique town-planning structure, both from the view of the general planning and space-spatial solution, as well as from the view of the architecture of buildings and small architectural forms. A feature of Khreshchatyk, which distinguishes it from other ensembles "Totalitarian architecture", is the priority of the national component. Another distinctive aspect of the Khreschatyk project is that it was designed for a territory to be rebuilt from war damage rather than a vacant space. Leaning on the landscape forms that were definitive for the historic center of Kyiv, it incorporated surviving buildings (which were painstakingly restored).

The ideas built into the reconstruction of Khreschatyk are topical even today. This project became a one-of-a-kind case of transforming a provincial town into a European capital that was not unlike Paris or Vienna in places. The scale and character of the street's architecture reflect the intuition and skill with which proportions needed for the city's central thoroughfare were defined and realized without detriment to the intimate, humane, and almost lyrical character this ancient city has always possessed.

Summing up, it can be said that the synthesis of arts used in the reconstruction of Khreschatyk both performed political tasks and revived national traditions in the décor, color scheme and plastic artworks of its buildings. This interest in artistic form and intent to rethink national artistic tradition in the country's main street was a manifestation of ideological resistance and national awakening.

References

1. X. Arendt *Dzherela totalitary`zmu*. (Kiev, Dux i litera, 2005)
2. R. Aron *Demokraty`ya y` totaly`tary`zm*. (Moscow, Tekst, 1993)
3. B. Cherkes, E. Gofer , *Arxy`tektura Ukray`nы v pauty`ne bol`shevy`zma* Arxy`tektura y` presty`zh., **2**. p7 (1996)
4. M. Berly`ns`ky`j,. *Korotky`j opy`s Ky`yeva Kratkoe opy`sany`e Ky`eva*. Repry`nt y`zdany`ya 1820g (1990)
5. O.Ye. Markova Ky`yiv: Ist. oglyad: (Karty`, ilyustraciyi, dokumenty`) Ukrayins`ky`j istory`chny`j zhurnal, **8**, pp 136–137 (1989)
6. S. Kilesso, Konkurs na proekt povoyennoyi vidbudovy` Xreshhaty`ka – poshuky` nacio-nal`noyi identy`chnosti / Arxitekturna spadshhy`na Ukrayiny` (Kiev, 1995)
7. A.V. Vlasov, Proekt zastrojky` Kreshhaty`ka. Arxy`tektura y` stroy`tel`stvo. **1**, Kiev, 1946
8. O. Oliynyk. The architectural image of Kiev's central square as a symbol of nation-al identity. Tales of Urban Lives and Spaces. De Urbanitate, Vol.3, 83-92 (2015) URL: http://sita.uauim.ro/f/sita/art/06_Oliynyk.pdf
9. B.S. Cherkes, Nacional`na identy`chnist` v arxitekturi mist. Monografiya, L`viv: Vy`davny`cztvo L`vivs`koyi politexniky`, (2008)
10. O. Oliynyk, The Maidan and Post-Soviet Evolution of Kyiv's Architecture. Current trends and methodologies on architectural design : ARChDESIGN '15, II. Ar-chitectural Design Conf., Istanbul .pp.801-824 (2015) URL: http://www.gbv.de/dms/tib-ub-hannover/862855756.pdf

Specific features of modern multifunctional high-rise building construction

Lyubov Manukhina[1*], *Natal'ja* Samosudova[1]

[1]Moscow State University of Civil Engineering, Yaroslavskoe shosse, 26, Moscow, 129337, Russia

Abstract. The article analyzes the main reasons for the development of high-rise building construction the most important of which- is a limitation of the urban areas and, consequently, the high price of land reserved for construction. New engineering and compositional solutions for the creation of new types of buildings are considered - complex technical designs of a large number of storeys completely meet the new requirements for safety and comfort. Some peculiarities of designing high-rise buildings and searching for optimal architectural and planning solutions are revealed since, with external architectural simplicity, high-rise buildings have complex structural and technological and space-planning solutions. We consider the specific features of a high-rise housing in various countries around the world, including Russia, such as the layout of the multi-storey residential buildings, depending on the climatic characteristics of the regions, assessment of the geological risk of the construction site, the choice of parameters and functional purpose of the sections of the territory of high-rise construction, location of the town-planning object for substantiating the overall dimensions of the building, assessment of changes aeration and engineering and hydrological conditions of the site. A special place in the article on the problems of improvement of the territory, the device of courtyards, landscaping, the device of playing and sports grounds. The main conclusion in the article is the following problem - when developing high-rise housing construction, the development of high-rise housing, and an increase in the population density in the territory of large cities of Russia, necessary to create a comfortable and safe level of residents living and not a decrease, but an improvement in the quality of the urban environment.

1 Introduction

The current stage of development of large cities megacities in the world is characterized by the construction of high-rise buildings and structures. Not only developed countries in various volumes develop high-rise housing construction, but almost all developing countries, with the help of high-rise housing construction, claim the mastery of technical and financial potential. The main reasons for the development of high-rise housing construction consist

* Corresponding author: 4804107@mail.ru

primarily in the limited urban areas and, consequently, the high price of land reserved for construction [1].

In our country, high-rise housing construction has developed in the Soviet period. This, so-called "Stalin's skyscrapers", which were built in the fifties of the last century. These buildings were of different functional character: institutions (the building of the Moscow State University and the Ministry of Foreign Affairs), hotels (Ukraina and Leningradskaya) and a multi-purpose residential building on Kotelnicheskaya Embankment. The peculiarity of these buildings are not only in architectural planning, but also of spatial solutions.

It should be specially noted that Moscow is the only major northern city in the world, in which there are residential high-rises. Due to the fact that weak soils predominate in Moscow, limestones with karst voids, the construction of high-rise buildings in Moscow requires particularly complex engineering and technical constructions of foundations and volume-planning solutions, in contrast to New York, where Manhattan skyscrapers stand on granites. Perhaps for this reason the construction of high-rise buildings in Moscow was suspended for a long time and had not received its development [2].

Nevertheless, the development of the construction industry, the creation of new construction technologies, as well as the lack of free areas in the city prompted a return to the decision of high-rise housing construction, including high-rise apartment buildings. In connection with the development of high-rise residential buildings, it became necessary to search for optimal architectural and planning solutions that take into account the optimization of town-planning solutions for high-rise apartment buildings not only in the central, historical part of the city, but also in other administrative districts and districts of the city. Optimization of the structure of multifunctional high-rise residential complexes becomes the main one.

Multifunctional high-rise complex it is a structure that connects into a single composition residential and public elevated areas and underground areas, house territory, and external communications [3-4]. To create a safe and comfortable environment in a person's daily life with such an unfriendly format, needed arrangement multi-storey residential development, depending on the climatic characteristics of the area, character of the relief, the conditions of insolation, the influence of the wind, the proximity of reservoirs and green areas. Especially large role in the perception of a populated place of harmonic ratios of the elements, visual connections and free spaces.

2 Experimental section

According to the current "Standards and for designing the planning of the development of sections of the territory of high-rise buildings, high-rise town-planning complexes in the city of Moscow", the section of the territory of the it's a part of the city's territory within the quarter bordered by land. Quarter size when placing a section of the territory of the complexes of high-rise buildings in it should not be more than 2.5 hectares, a section of the territory of the high-rise town-planning complexes - no more than 5.0 hectares.

Norms regulate the factors taken into account when selecting parameters and functional designation of sections of the territory of high-rise construction, including:

— Visual and landscape analysis of the location of the town-planning object for substantiating the dimensions of the building.

Fig. 1. Compositional methods of residential development of different altitude

 https://doi.org/10.1051/e3sconf/20183301040

High density of the population leads to complex compositions from vertical and extended volumes of apartment houses [5]. At engineering of city quarters of mixed number of storeys, this task is solved by compositional methods, combining low and high-rise buildings of the city.

— Analysis of the possibility of geological risk on the basis of geological survey data; is made on the basis of the "Recommendation on the assessment of geological risk in the territory of the city of Moscow."
— Calculations of the carrying capacity of the transport network, taking into account the additional load from the object with the purpose of excluding overloads of road and transport communications, are conducted on the basis of MGSN 1.01-99.
— Predictive assessment of changes in aeration conditions and engineering and hydrological conditions of the site. It is carried out in accordance with the provisions of the Federal Law "On the Protection of Atmospheric Air" and San Pin 2.1.6.1032-01.
— Light-climatic calculations of the insolation level and natural illumination, according to San Pin 2.2.1 / 2.1.1.1076-01.
— Rated duration of continuous insolation for residential and public buildings is differentiated depending on the type of flats, the functional purpose of the premises, the planning zones of the city, the geographical latitude for the central zone (58 ° N - 48 ° N), - at least 2 hours a day from 22 of March to September 22nd;

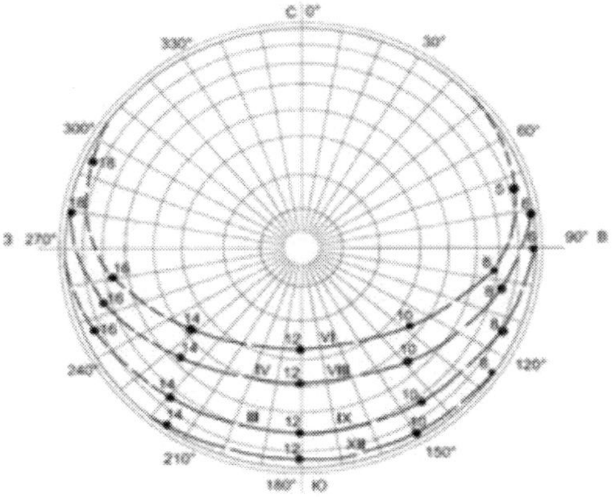

Fig.2. The solar map of Dunayev for 50 ° N. w.

Characteristics of comfort, in the same climate, may not be suitable for other climatic regions. In the northern latitudes, the buildings and territories on which they are located require as much sunlight as possible, heat; in the southern regions, conversely, they must be protected from overheating [6].

Table 1. Preferred orientation of residential development for different climatic zones

Orientation	Climatic zones			
	Tropical		Moderate	Polar
	Humid	Dry		
Solar  С Ю				
Wind	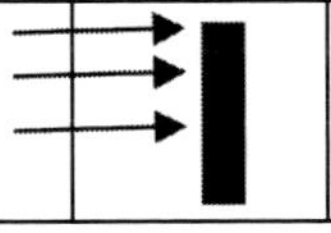	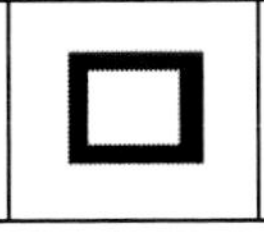		

In the territories of children's playgrounds, sports grounds of residential buildings; group sites of pre-school institutions; a sports zone, a recreation areas for general education schools and boarding schools; recreation areas of hospital facilities of stationary type, the duration of insolation should be at least 3 hours for 50% of the area regardless of geographical latitude [7].

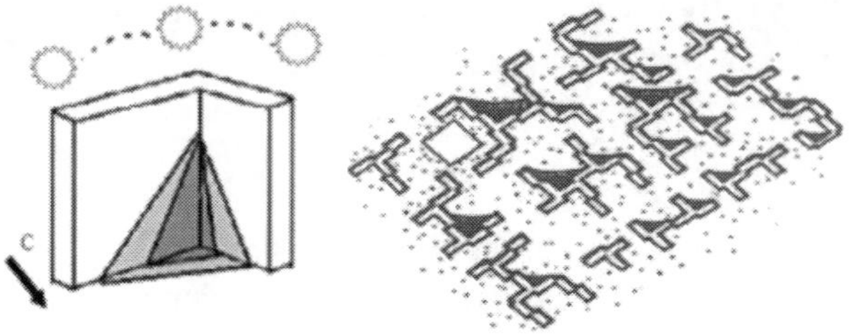

Fig.3. Nature of the darkening of territories

Result section

The minimum distance between houses is determined is determined by the number of storeys of the house, which can obscure the building, the location and orientation. With a complex orientation, special calculations by means of special software [8-9].

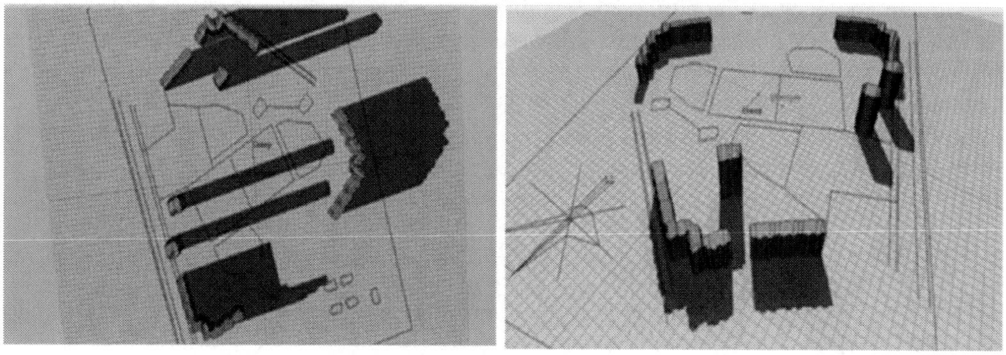

Fig.4. Calculation with the help of the program "Sitis Solaris"

High-rise buildings cast shadows, the length of which depends on the time of the year and the position of the sun, this significantly affects the microclimate of the yard area, so it is necessary to investigate the influence of cast shadows on the house area, at an early stage of design and to develop various options for the location of the building on the site.

Fig. 5. Variants of location of the complex on the site

Calculations of the population's security in the adjoining territories by gardening and public service facilities within the boundaries of the site of high-altitude education [10]. The area of gardening on sections of high-rise buildings, high-rise town-planning complexes is determined from the calculation of not less than 5.0 m^2 per inhabitant, considering landscaping of exploited roofs and special recreational spaces (winter gardens) distributed on the floors of the building. The area of the winter garden should not be less than 50 m^2.

Discussion section

Ideally, yard appears as the collective space intended solely for gaming and communication, so the yard spaces of multifunctional high-rise complexes should be clearly divided into internal and external [11]. This leads to the fact that the yards are even more separated from the street, which gives a sense of security and coziness.

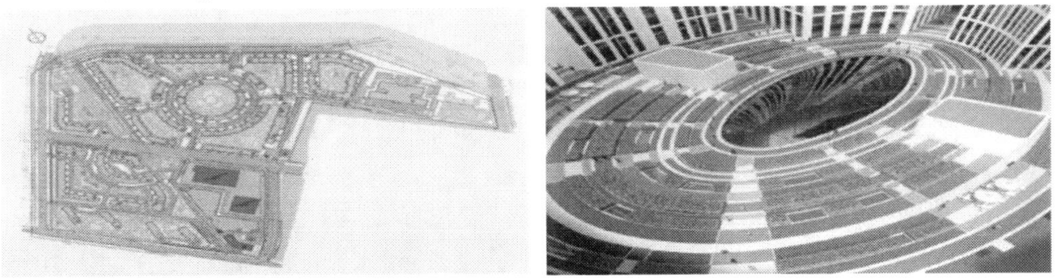

Fig.6. Options for the layout of the yard area

Landscaping involves the construction of courtyards, landscaping, the installation of playgrounds and sports grounds for children of different ages and parking [12]. Preferably, pedestrian areas, courtyards, playgrounds and boulevards are at different levels with the roadway this can be achieved with the help of a system of stylobates and pedestrian bridges connecting all the sites and pedestrian zones [13]. That is, you can get round whole residential complex with an interesting scenic route, nowhere crossing the roads.

Fig.7. Accomplishment of a residential complex (Osaka city)

Such decisions have been applied in single flagship projects of world architecture, which are being built in countries with a huge experience of high-rise construction in Singapore, the United States, and China [14]. It will raise comfort and safety to a new level and completely free the yard from the machines, because Parking is located underground, and cars leave only on the outer street of the residential complex [15].

One of the most acute and current urban development problems of high-rise buildings and complexes in Russia is the actual non-compliance existing standards and construction regulations.

References

1. L.A. Manuhina, V.A. Lukinov, Scientific journal of construction and architecture, **4**, 196-202 (2011)
2. N. Danilina, D. Vlasov, MATEC Web of Conferences 5, 05017 (2016)
3. N. Danilina, D. Vlasov Advanced Materials Research, **869-870**, 201-204 (2014).
4. R. Alidad, International Scientific Review, **17 (27)**, 45-48 (2016)
5. H.G. Jakubov, O.A. Manuhina, Real estate: economics, management, **2**, 14-17 (2015)
6. O. Kurakova, N. Khomyak, Procedia Engineering, **165**, 1221-1228 (2016)
7. P.G Grabovyj, R.V.Volkov, S.I. Beljakov, Issues of state and municipal management, **4**, 211-219 (2016)
8. P.G. Grabovy, A.K. Orlov, Procedia Engineering, **153**, 195-202 (2016)
9. D. Vlasov, E3S Web of Conferences, 01012, (2016)
10. Sherbina E.V.E., Danilina N.V.E., Vlasov D.N. International Journal of Applied Engineering Research, **22 (10)**, 43131-43138 (2015)
11. A.A. Magaj, Housing construction, **11**, 18-23 (2005)
12. N. Samosudova, Real estate: economics, management , **2**, 69-74 (2016)
13. A.A. Magaj, V.S. Zyrjanov, E.Ju. Shalygina, Housing construction, **11**, 17-20 (2015)
14. N. Samosudova, Real estate: economics, management, **1-2**. 27 (2006)
15. B.V. Prykin, L.V. Prykina, O.A. Manuhina Real estate: economics, management , 1, 1822 (2016)
16. Alshevskaya A.A., Lopatnikova Yu.A., Shkaruba N.S., Sizikov A.E., Lukinov V., Moskalev A., Sennikov S.V. Annals of the Rheumatic Diseases. **2**, 1056 (2017)

Vertical cities - the new form of high-rise construction evolution

Vera A. Akristiniy[1,*] and *Yulia* I. Boriskina[1]

[1]Moscow State University of Civil Engineering, Yaroslavskoe shosse, 26, Moscow, 129337, Russia

Abstract. The article considers the basic principles of the vertical cities formation for the creation of a comfortable urban environment in conditions of rapid population growth and limited territories. As urban growth increases, there is a need for new concepts and approaches to urban space planning through the massive introduction of high-rise construction. The authors analyzed and systematized the list of high-tech solutions for arrangement the space of vertical cities, which are an integral part of the creation of the methodology for forming a high-rise buildings. Their concept differs in scale, presence of the big areas of public spaces, tendencies to self-sufficiency and sustainability, opportunity to offer the new unique comfortable environment to the population living in them.

1 Introduction

The cities are the place of residence for a half of the population of Earth, their architecture, structure, and ecology exert considerable impact on inhabitants. This article deals with the evolution of cities, the main principles of their development and the formation of a comfortable urban environment for procuring vital activities.

Multifunctional use of urban territories risks losing its social and compositional significance and directly dependent on the economic factor. When creating new spaces and cities, architects consider past experience and offer solutions in accordance with modern requirements [1,3].

There are various opinions that cities emerged as warehouses and places of goods loading, defensive and offensive fortresses, administrative centers for the management of irrigation systems, temples. However, the ideas of planning and ordering life in the city appeared after the emergence of cities as a result of urban civilization, or the conditions for its foundation.

The form of the city is a spatial organization of the population vital activity processes and its interaction with surrounding space. Formation of space regularities are of interest: the functional zones, communications realized by structure, the basic rules of those communications. These are ideal schemes for placing functional elements.

There are many theories about the form of the city. At each stage of the humanity development, for a certain nation and terrain, it must have its own characteristics. But no matter by which path or the theory the city developed, for many centuries all over the world

*Corresponding author: 7824666@mail.ru

they share one thing: they are all characterized by low-rise buildings, with rare exceptions of religious buildings, many of which were built for decades and had non-residential functions [2,6].

This continued until the 19th century when the technology development and a number of economic factors created unique prerequisites for the development of high-rise construction. At that stage of development, the following factors can be identified [1,4]:

1. High cost of land. Cities managed to become centers of trade, administrative, political and cultural life, which led to high demand and the growth of land value. In such circumstances, each landowner sought to maximize the efficient use of their property. The only possible way to increase the usable area was to grow upwards.

2. New materials. New materials such as glass and steel acquired the increasingly wide application. The strength of the steel made it possible to shift a significant part of the load onto it, and the glass served as a good protection against the external environment instead of the walls while being several times lighter than the traditional facade.

3. Otis elevators [11]. It is difficult to overestimate the invention by Otis of the elevators safety mechanism that catches the cabin in the event of cable tear. It was thanks to him that elevators began to transport people, not just cargo, which solved the problem of inaccessibility of high-rise floors. From the once cheap, they quickly turned into the most prestigious, allowing their new owners to admire the magnificent panoramic views.

4. New constructive solutions. Until the 19th century, the increase in the height of buildings meant a thickening of the walls, which were supposed to withstand the weight of the structure. The creation of elevators and metal frames for the walls provided an opportunity for architects and engineers to design and build higher and higher buildings, increasing the number of floors. Bearing capacity was transferred to the frame of the building, the load - on the foundation.

One of the countries that concentrated all these factors was the United States, a young country open to various technologies. Not surprisingly, the world's first skyscraper, built with unique new solutions, appeared in 1885 in the American industrial city of Chicago, where the high cost of land stimulated high-rise construction. Gradually, this development led to the emergence of a large number of skyscrapers around the world. For many private corporations and developers, the construction of such erections has become a matter of prestige, between the buildings a struggle has begun for the right to be named the highest in the world. Fig.1shows the ranking of the tallest buildings as of 2018 [12].

Source: Burj Khalifa

Fig.1. World's tallest Towers (from left to right: Burj Khalifa, CN Tower, Taipei 101, Petronas Towers, Willis Tower, Jin Mao Tower, International Finance Centre, Citic Plaza, Shun Hing Square, Empire State Building, Central Plaza, Burj Al Arab)

Nevertheless, despite the progress in the field of high-rise construction of individual buildings, the horizontal growth of cities increases the number of problems [5,7]:

- Urban development occupies more and more new areas, with local flora and fauna gradually disappearing. It means there is a complete destruction of the ecosystems, which have been formed for several million years.
- The city space is not used rationally, thus creating transport and social problems. Almost in any city during rush hour, it is possible to observe congestion of all transport systems.
- The presence of industrial zones in the central part of the city. This situation stems from the fact that the factories built on the outskirts of town in the last century, today are in the central part because of the high rates of urban area proliferation. In addition to environmental pollution, the negative factor is the irrational use of industrial areas, taking into account trends in urban development.

As cities grow, there is a need for new concepts and approaches to urban space planning. High-rise construction has served as a powerful incentive for the further development of construction and related technologies necessary for the development of mega-high-rise clusters, in which people can stay for a long time with no need to leave it [4,7]. In the 21st century, there are enough prerequisites for implementing a new concept of development in the form of vertical cities (Fig. 2).

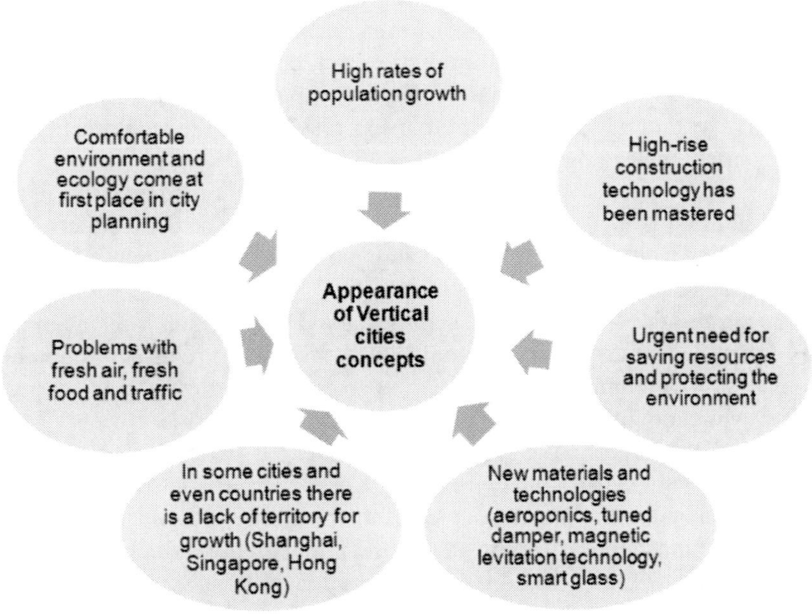

Fig.2. Prerequisites for the appearance of vertical cities

Thus, we can distinguish the following factors, which served as the basis for creating the concept of vertical cities:

1. High rates of population growth, including an increase in the proportion of the urban population, as evidenced by statistical data (Fig. 3), which must be considered in urban zoning [13].

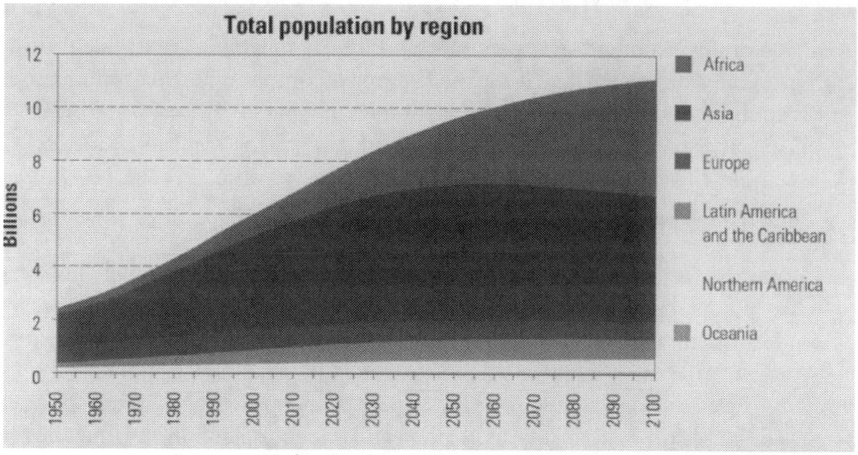

Source: Unite Nations. Department of Economic and Social Affairs

Fig. 3. World Population 2017

2. Overpopulation and deployment of a large number of industrial and transportation facilities have greatly worsened the ecological situation in the cities. A large number of harmful substances in the atmosphere, a high concentration of various types of transport, the traffic duration to the places of labor application, complicates the fresh products delivery, which grows, as a rule, outside the city. Further expansion of the city from their center will require all new communication and transport systems, the construction of which will negatively affect the ecology [8].

3. The urgent need for saving resources and protecting the environment, while the arrangement of horizontal cities does not always allow it to be done effectively, or it requires large-scale financial costs for replacing engineering communications networks or transferring objects.

4. The desire of residents to have a comfortable living, which is the main trend of the 21st century when designing city residential. For a favorable urban environment, people need parks, places for communication, a nice view from the windows and the opportunity to go in for sports, while moving as little as possible around the city [9].

5. Some of the largest cities lost the opportunity for growth and development due to the limited territory (Shanghai, Singapore, Hong Kong). This is connected to the obstacles of the physical terrain and to the inconvenience of their expansion for management and residence. China has announced to create a megacity, the population of which will exceed 130 million people. Megacity will be created by agglomeration of large Chinese cities: Beijing with a population of 22 million people, Hebei with a population of 14 million people and Tianjin, which has about 72 million people. The combination of such a large territory means the appearance of large distances, despite plans to provide the population with a convenient transport infrastructure [7,10].

2 Methods of creating a vertical city concept

Currently, there is a number of technologies that can improve the level of comfort when living in vertical cities. Special technologies are developed, parts of the structures are manufactured for special orders, and the installation process itself considers the climate features, the shape of the building, the timing of construction and other specific factors. We can safely say that high-rise construction is mastered by developers to the fullest extent and they can solve any emerging problems by applying modern design and engineering solutions using existing equipment and specialists.

The list of specialized high-tech solutions for the arrangement of vertical cities space have been systematized, serving as an integral part of creating a method of forming a high-rise cluster.

1. The newest biotechnology of aeroponics, capable of making a revolution in providing cities with food is being introduced. Aeroponics is an artificially created climate for the growth and development of plants without the use of soil and substrates [20]. With this cultivation method, the roots of the plant freely hang in the aeroponic model. The nutrient solution is supplied to the roots with micro droplets or fog, thereby creating an air-nutrient medium. The solution forms a cloud enveloping the roots, and in the pauses between spraying, aeration of the roots takes place (Fig. 4).

Source: greenhouses.ru

Fig. 4. Aeroponics

Aeroponics has many advantages [16,20], such as:

- The supply of nutrient solution and other processes controlled by the computer and correspond to the phases of plant development.
- Cultures grown on aeroponics characterized by a particularly high rate of growth and maturation.
- Active growth of roots, intensive assimilation of nutrients increases the productivity of plants, improves the quality and quantity of products.
- Accelerated period of the plant's vegetation, which allows for several harvests per year in the same area.
- Ecological compatibility, since aeroponics is a closed cycle system, there is no utilization of the nutrient solution, no use of pesticides, soil or substrates.
- The aeroponics method allows precise and quick control over all technological processes of plant growth and development.
- Low energy and material consumption of aeropon technology.

2. Technical implementation of such invention as tuned mass damper (also known as a harmonic absorber), designed to dump the high part of a building, is suggested. [14]. The safe design of high-rise buildings necessarily encounters the problem of vibrations in the upper part of the building from air currents. One of the solutions to this problem is the installation of a special high-rise tuned mass damper; an example of its effective use is the Tower of Taipei in Taiwan (Fig. 5).

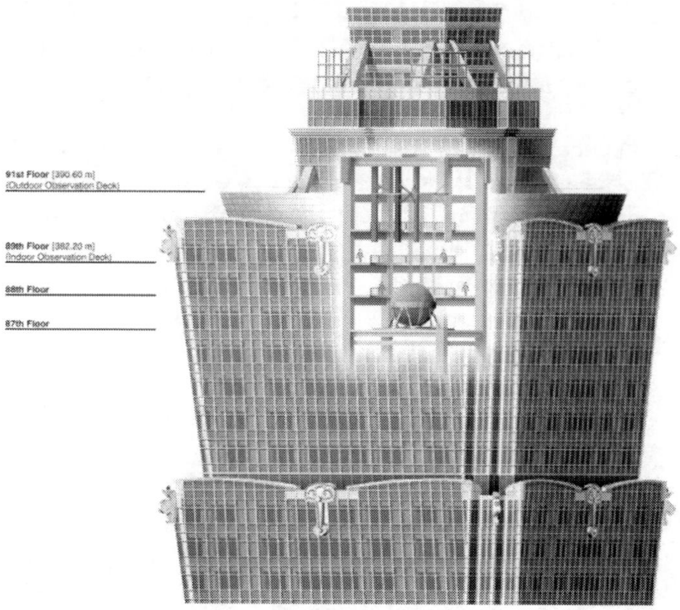

Source: en.academic.ru

Fig. 5. Scheme of the largest tuned mass damper, Tower of Taipei, Taiwan.

Firm Thornton-Tomasetti Engineers [2,23] together with the Evergreen Consulting Engineering designed 660-ton steel pendulum, which is in its operating principle an inertial vibration damper. Suspended in 87-91 floors, the pendulum oscillates, compensating for the movement of the building caused by strong gusts of wind. Its sphere, the largest in the world, consists of 41 steel plates, each 125 mm thick, which together is 5.4 m in diameter. Two other vibration dampers, each weighing 6 tons, are at the top of the spire. They soften the blows of the wind acting on the upper part of the building. The vibration energy accumulated by the dampers extinguished by a system of springs located under the dampers.

3. The practice of using smart-glass is currently widespread in the field of green building [17]. Its characteristics represent the possibility of introducing it into the glazing system of high-rise buildings. A special complex of nano-sputtering allows saving of lighting, keeping heat and preventing excessive heating from sunlight (Fig. 6,7).

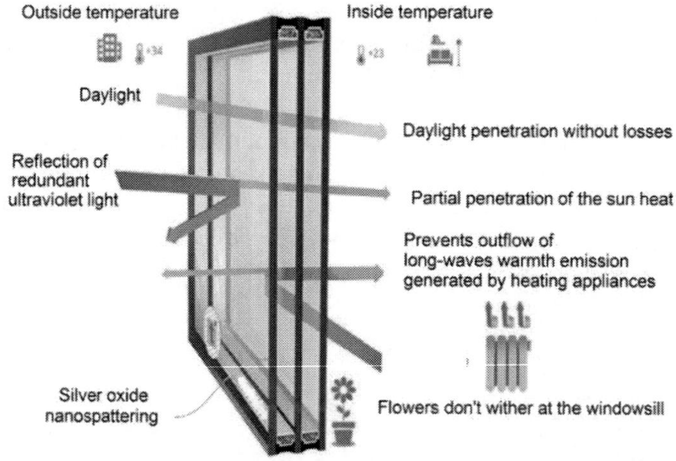

Source: almaokna.kz

Fig. 6. Smart-glass device schematic

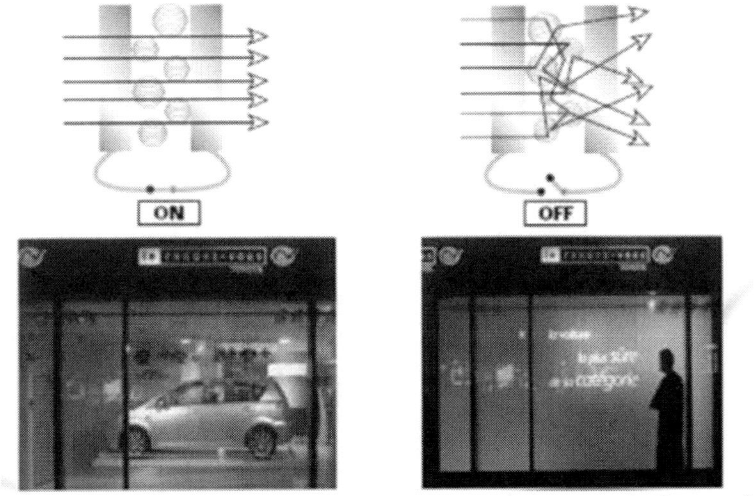

Source: kerch.biz.ua

Fig. 7. The basic principle of smart-glass

4. The MULTI Elevators (Thyssen Group, Germany) magnetic levitation technology introduced in 2017 can be an effective and high-speed method of moving in different directions of a high-rise cluster [18]. This high-tech solution without the application of wire ropes allows the use of multiple cabins in one shaft, the direction of movement of which can be vertical, horizontal or diagonal, in the case of the enclosing structures with a certain angle of inclination to the outer or inner side (Fig. 8).

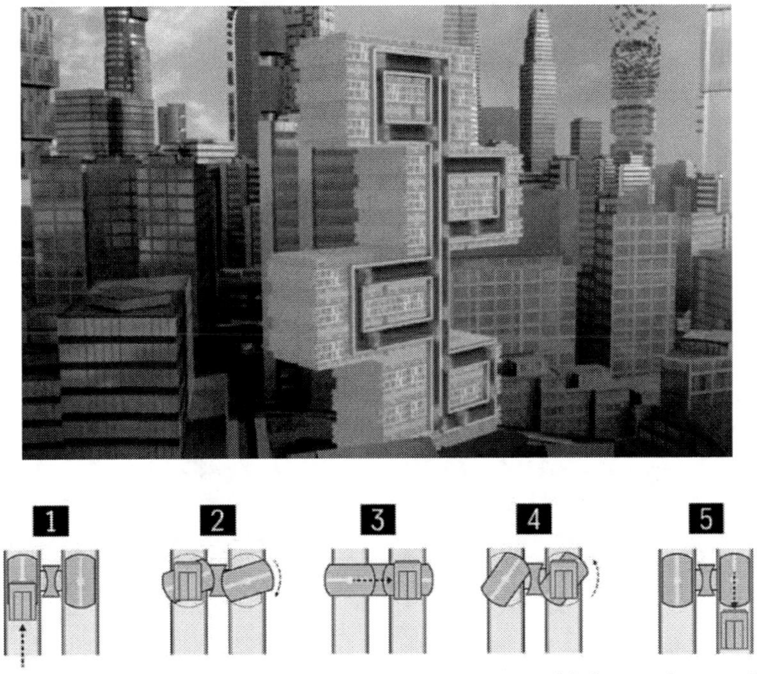

Source: multi.thyssenkrupp-elevator.com

Fig. 8. MULTI elevators (Thyssen Group, Germany)

3 Practical results

Thanks to all these accumulated technologies, a possible solution to the further growth of urban space will be emergence of vertical cities, the concepts of which architects actively propose today. One of the most interesting concepts for creating a vertical city proposed by the Russian architect S. Nepomnyaschy is the heliocluster project "City of the Sun" with the possibility of introduction in any territorial zone [6,16].

Helioclusters - the concept of unique multifunctional house-giants, comparable in scale and infrastructure with neighborhoods and even entire cities, which are as open to the natural light as possible. The concept is applicable to a variety of climatic conditions.

According to S. Nepomnyashchy, for 1 hectare of land in a heliocluster with a height of up to 35 floors, there will be 100,000 square meters of housing and 100,000 square meters of additional infrastructure, including offices, parks, traffic intersections, shopping areas and service facilities. One of the principled positions of the architect is the organization of the most compact and eco-friendly space, where openness to nature, contemplation of the natural landscape and safe interaction with the environment are the basis of human life. At the minimum area is arranged secluded, spacious accommodation with terraces and views of the forest or the park. Offices, parking lots, highways, stations - all of the city's elements are located inside one huge building (Fig. 9) [16].

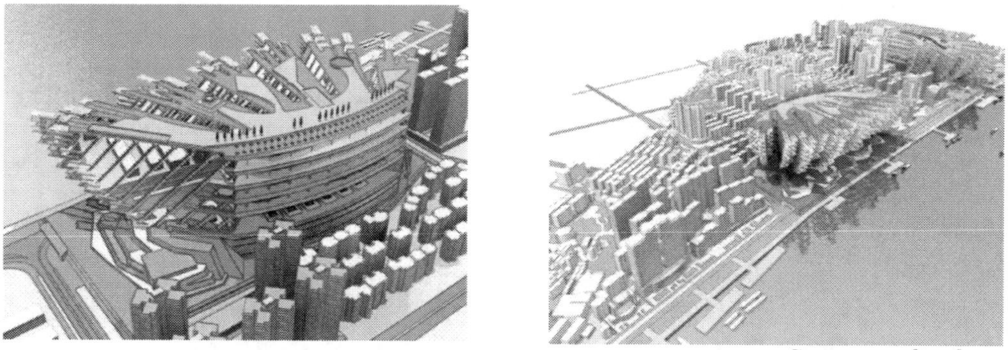

Source: realty.rbc.ru

Fig. 9. The characteristic structural elements of heliocluster

The "Heavenly Shantou gardens" heliocluster project designed for the eponymous Chinese city, introduced the developed structure of the outdoor terraces facing the sea. In the project, the main idea is to create multiple atriums with sunrooms on different levels inside the building. Light in the air will penetrate through internal channels through glass roofs or through light guides [10,15]. On the roof and at the base of residential buildings, the architect suggests dividing parks and gardens, building fountains and open pools (Fig. 10).

Fig. 10. The "Heavenly Shantou gardens" heliocluster project, China

Helioclusters projects are also developed for the Moscow suburbs. In the project "Nadmoskovie", located in the northern latitudes, the architect proposes to build closed buildings, more reminiscent of mountain massifs with facade glazing and valleys on the roof. One of their main advantages is the relatively low total heat loss through the enclosing structures and the ventilation system. This is due to the fact that such structures have a much smaller ratio of the external surface to the internal volume. Even without the use of heat recovery solutions, interior lighting, people, and household appliances produce much more energy than would be lost by such a large structure [15]. The architect proposes the construction of compact blocks with an identical number of storeys, which represent separate communities with a mixed function. They will accommodate housing, retail space and office centers. Thus, the architect lays on the basis not only the convenience and comfort of living but also the economic prospects for the development of the city or district (Fig. 11).

Fig. 11. "Nadmoskovie" heliocluster project, UNK Project.

The quintessence of the vertical cities concepts is the "Xtopia" project. Xtopia is an arcological project of a 300-story multifunctional building in Shanghai (China) or a vertical city. The planned height of the building should be approximately one mile (1609 meters) [16]. The main objective of the project is to solve the problems of overpopulation and to reduce the negative environmental impact caused by human populations (Fig. 12).

Fig. 12. The "Xtopia" heliocluster project, Shanghai, China.

Under the project, the building should accommodate about 200 000 permanent residents and another 50 000 people on a temporary basis in the building. For such a large number of people, the building must be functional and provide all the needs of the people inside it.

The shape of the building designed to minimize wind resistance, thereby reducing lateral loads. X-shaped frame located on the tripod base provides the necessary stiffness. Tuned mass damper, located in the center of the structure, will extinguish the resonant vibrations. The non-vertical shape of the building excludes the possibility of moving elevators vertically and is a typical example of the need to use elevators on a magnetic levitation system.

The building is planned to create agrarian, industrial, residential and commercial zones in order to provide all the needs of residents without a need for leaving the building. In the "tripod", there will be an industrial zone, in the center of the tower an agrarian zone, and the upper floors will occupy residential and commercial zones. The agrarian zone, according to the plan, must fully ensure the vital activity of the population living inside [19,21].

4 Discussion

At present, the concept of vertical cities is theoretical. However, many skyscrapers already carry a number of elements of such cities. For most major cities, such as Singapore and Shanghai, the creation of vertical cities is the next logical stage of development.

In their development, the cities went through a long and diversified path. At first, they arose and formed chaotically, at the intersection of trade routes, culture, providing the necessary minimum conditions for people's lives. The development of technology and the growth of the economic and social role of cities have led to an increase in the number of storeys and the height of buildings. Nonetheless, some high-rise buildings are no longer sufficient, their construction solves the problem of just increasing the areas of commercial and residential real estate, but not redistributing various functions of the city, zoning and providing it with the necessary resources [7,22].

A comprehensive approach requires new formats of high-rise construction. And the next stage of development is vertical cities. Their concept is different in scale, presence of a large number of public spaces, desire for self-sufficiency and environmental friendliness, opportunity to offer a new unique comfortable environment for its residents. Such concepts are especially relevant for a number of cities in South-East Asia, where population density is extremely high. The population of cities is growing at a rapid pace, over time, the concept of vertical cities will no longer seem utopian and, perhaps, in the near future the population will be able to spend almost the entire year in the structure of vertical cities, leaving them only for a trip outside the city or traveling to another country [15].

5 Conclusion

Regardless of form, scale and functional diversity, all cities are subject to transformation. At the same time, actual changes lag behind development needs. This leads to an imbalance and an exacerbation of the deficit of all urban spaces types, especially public spaces, open landscaped spaces, and aggravation of transport problems.

The next logical step in the development of urban planning in the right direction is the creation of a humane ecological urban environment of a new type - vertical cities [17]. Solution to ensure the required comfort characteristics of the urban environment eliminates the need for balancing on the edge between comfort and building density. High density, as an effective economic characteristic, should turn into a tool for providing comfort.

6 References

1. H. P. Cho, H. Lim, D. Lee, H. Cho, K. Kang, Technological Forecasting and Social Change, (2017)
2. J. Chen, J. Ma, S.M. Lo, Simulation Modelling Practice and Theory **74**, 99-116 (2017)

3. Yu-ting E, L. Zhou, Procedia Engineering **135**, 2016, Pages 574-577
4. J. Fu, Q. Zheng, Y. Huang, J. Wu, Q. Liu, Engineering Structures **156**, 460-471 (2018)
5. P. Lotfabadi, Renewable and Sustainable Energy Reviews **38**, 285-295 (2014)
6. P. Lotfabadi, Energy and Buildings **89**, 183-195 (2015)
7. J. X. Xiao, K. W. Michael Siu, Resources, Conservation and Recycling **131**, 172-180 (2018)
8. P. Feng, W. Xingkuan, Procedia Engineering **21**, 943-947 (2011)
9. X. Chen, H. Yang, Applied Energy **206**, 541-557 (2017)
10. L. M. Ngo, Procedia Engineering **142**, 42-47 (2016)
11. The history of Otis. Website of Otis company: http://www.otisworldwide.com / site / ru / pages / OtisHistory
12. World's Tallest Buildings 2018. http://skyscraperpage.com/diagrams/
13. World Population 2017. Unite Nations. Department of Economic and Social Affairs. Population Division. https://esa.un.org/ unpd / wpp / Publications / Files / WPP2017
14. T. Saroglou, I. A. Meir, T. Theodosiou, B. Givoni, Energy and Buildings **149**, 437-449 (2017)
15. Nadmoskovie: what Russian architects see cities of the future. Website of RBC https://realty.rbc.ru/news/577d22959a7947a78ce9162a
16. Heliolatry and not only or what the future of cities. Website of Colors Life https://www.colors.life/post/112923
17. M. Al-Chalabi, Sustainable Cities and Society **18**, 74-77 (2015)
18. R. Borck, Regional Science and Urban Economics **58**, 13-25 (2016)
19. G. Ptichnikova, Procedia Engineering **165**, 1903-1910 (2016)
20. N. Srihajong, S. Ruamrungsri, P. Terdtoon, P. Kamonpet, T. Ohyama, Applied Thermal Engineering **26**(2–3), 267-276 (2006)
21. X. Wang, X. Liu, Renewable and Sustainable Energy Reviews **47**, 241-259 (2015)
22. M. Al-Chalabi, Sustainable Cities and Society **18**, 74-77 (2015)
23. Projects of Firm Thornton-Tomasetti Engineers http://www.thorntontomasetti.com / projects/

Trend analysis of modern high-rise construction

Dmitry Radushinsky[1,*] *Andrey* Gubankov [2] and *Asiiat* Mottaeva[3]

[1]Institute for Entrepreneurship Problems LLC, 13-th line of Vassylievsky island, St. Petersburg; 192241, Russia,
[2] WEOM architectural bureau, Kolia Tomchack st., 28 St. Petersburg; 192568, Russia,
[3] Moscow State University of Civil Engineering, Yaroslavskoe shosse, 26, Moscow,129337, Russia

Abstact. The article reviews the main trends of modern high-rise construction considered a number of architectural, engineering and technological, economic and image factors that have influenced the intensification of construction of high-rise buildings in the 21st century. The key factors of modern high-rise construction are identified, which are associated with an attractive image component for businessmen and politicians, with the ability to translate current views on architecture and innovations in construction technologies and the lobbying of relevant structures, as well as the opportunity to serve as an effective driver in the development of a complex of national economy sectors with the achievement of a multiplicative effect. The estimation of the priority nature of participation of foreign architectural bureaus in the design of super-high buildings in Russia at the present stage is given. The issue of economic expediency of construction of high-rise buildings, including those with only a residential function, has been investigated. The connection between the construction of skyscrapers as an important component of the image of cities in the marketing of places and territories, the connection of the availability of a high-rise center, the City, with the possibilities of attracting a "creative class" and the features of creating a large working space for specialists on the basis of territorial proximity and density of high-rise buildings.

1 Introduction

Over the past 17 years of the XX century, the number of super-high skyscrapers in the world has quadrupled. "Superhigh buildings" or "super-high skyscrapers" are commercial or multi-functional (having both business and residential zones and sometimes other function zones) structures with a height of more than 300 m. Their number in the world in 2017 increased to 126 compared to 111 in 2016 . [1].
First of all, high rates of construction of super-high buildings are currently characteristic of the dynamically developing economies of the APR and the "monarchies of the Gulf" (oil-rich Persian Gulf Arab countries), which demonstrate their growing economic power in high-rise buildings and, in many cases, attract local and foreign investors. The most active in the construction of super-high buildings in the last two decades are China, the United Arab

* Corresponding author: D.Radoushinsky@gmail.com

Emirates, the United States, Hong Kong, Saudi Arabia, Russia, Great Britain, Taiwan, South Korea, France, Singapore, Australia and Indonesia [2-5].

In Russia, the first super-high commercial buildings (over 300 m) appeared in the 2010-s during the construction of the complex "Moscow City", in 2018-2020 will complete the construction of two extra-high buildings in St. Petersburg and Grozny. In addition to Moscow, St. Petersburg, Grozny and, probably, Ekaterinburg, construction of buildings of this height in other cities of Russia is not yet planned [6]. Therefore, despite the fact that for other cities in the country in 2000-2010-s in general, the tendency of growth of the height of the erected buildings is also typical, for the absolute majority of cases of high-altitude construction in Russia, the previously formed concept of a "high-rise building" remains relevant. In Russia and the countries of the USSR, "high-rise" are conventionally considered to be buildings taller than 75 meters or more than 25 floors [7].

In other countries of the world, high-rise buildings are considered, ranging from 35 to 100 meters. In many cities of Europe, Asia and Africa today, one of the tallest buildings remains bell towers and cathedrals / pagodas / other religious buildings built in the 19th century and earlier, for example, the Cologne Cathedral (height 157 m), the Peter and Paul Cathedral in St. Petersburg (height 122 m) and others.

According to CTBUH (Council on Tall Buildings and Urban Habitat), the world leader in skyscrapers continues to be New York, where now in the process of erecting 30 skyscrapers, and the skyscraper "Empire State Building" 443 m high remained the tallest building in the world for more than 40 years . However, the status of New York already in the 2020s. will be ready to challenge China's cities. In the last five years, 38 skyscrapers above 300 m have appeared in the People's Republic of China. In 2018, 18 super-skyscrapers will be added to them, and China's share will exceed half of the world's super-skyscrapers. Now it is 46 out of 100. Particularly impressive megaregion of the Pearl River Delta (PRD) area of approx. 40 thousand km ^{2}. This is the place where the Pearl River flows into the South China Sea and where the most densely populated and industrially developed urban area of the planet develops. In PRD there are such cities as Hong Kong, Guangzhou and Shenzhen and 20 of the hundred highest buildings. In addition to New York and PRD, the third most important region, actively developing high-rise construction, are the oil-rich Arab countries. In Dubai (UAE) is the highest skyscraper "Burj Khalifa" (828 m) and in the process of construction - 13 buildings with a height of at least 300 meters. In Mecca (Saudi Arabia), the third highest altitude at the end of 2017 from the completed skyscrapers in the world is the hotel for pilgrims "Clock Tower" (601 m).

Until the end of the 20th century, most European cities did not build tall buildings in order not to violate the historical architectural appearance (sky line). When high-rise blocks appeared in a number of capitals (the "Défense" in Paris in the 1990s, the "City" quarter in London in 2010), they were located on the outskirts of cities. The height of buildings in Europe for the first time exceeded 100 m only in the 1950s. unlike the United States, where it happened in 1913. The tallest buildings in Europe (except Russia) are in London (310 m), Paris (built 323 m), Madrid (296 m), Frankfurt (257 m), buildings higher 200 m are also available in Milan, Turin, Warsaw, Kiev, Istanbul. In addition to the Moscow skyscrapers, the highest skyscraper in Europe is the Shard London Bridge (2010), whose height is 310 m.

2 Methodological approaches and analysis

The work deals with buildings used for offices, housing, hotels in various combinations, whose height exceeds 250-300 m, TV towers, radio masts and similar technical facilities were not considered. To analyze the driving factors of building super-high buildings at the present time and the interrelationships of the image of the city, the development of occupations attributed to the creative class with the presence of high-rise buildings,

qualitative methods of analytical research have been applied. In assessing the economic feasibility of building high-rise buildings, an empirical approach is applied based on information available to authors on the payback period of a number of high-rise buildings in St. Petersburg, which in most cases is classified as a trade secret.
It should be noted that some topical works of Russian authors are devoted to high-rise construction in general [8, 9], there is also a new work published in Voronezh devoted to risk assessment, including financial risks, in the implementation of high-rise construction projects [10]. In addition to these, the main provisions and conclusions of this article are based on a number of works by foreign authors [11-15] devoted to the architectural and technological, as well as the financial and image aspects of the construction of high-rise buildings, the CTBUH database was taken into account [16], the materials of urban planning meetings councils, hearings of professional public.

2.1 Architectural and technological plus image "imperatives" of high-rise construction, the possibility of achieving a multiplicative economic effect.

It is often said that the main reason for high-rise construction in large cities, where it first appeared naturally, was the high cost of land in the center of the city, in the specific areas where concentrations of high-rise buildings began to arise. Because of the high cost of land, they tried to place the maximum possible number of useful areas (office, retail, residential), based only on existing technological and economic constraints. However, in addition to the deficit of specific land plots and areas in the city center (City), other factors play a role for modern high-rise construction:

A) The image component is the desire of the management of large corporations, the authorities to give a grandiose look to the period of their rule in the historical perspective, to form positive associations, and in the medium term to demonstrate the effectiveness and significance of their activities for both the local administrative and business community and the population, and for visitors of large national and international events, foreign tourists, representatives of the "creative class".

B) The ability to translate modern views on architecture by designers and innovative building technologies created by manufacturing companies. Well-known architectural bureaus together with the companies-producers are often lobbyists of high-rise construction. At the same time, the opponents of high-rise construction - especially in the territories of historic neighborhoods - are usually the local communities of conservative cultural workers and architects - "soil people". The latter seek to limit the height of all newly constructed buildings so that they do not conflict with existing architectural dominants, especially if they themselves do not belong to a company promoting a high-rise architectural project.

B) In the conditions of the "underheated" national economy - with signs of recession, slow growth - high-rise construction of such infrastructure can serve as an effective tool - a driver for the development of construction and a number of adjacent to the construction sectors of the national economy, which in turn , according to the views of Keynesians, institutionalists and a number of other schools of economic policy, is able to have a positive multiplier effect on the entire economic system.

Most high-rise buildings today are designed as multi-functional (with various classes of functional fire danger) and include office premises, residential premises (apartments, hotels), trade and even universities and hospitals.

Designers of most of the skyscrapers around the world are international companies, most of which are based in Great Britain and the United States. Could de mentioned the architectural bureau Skidmore, Owings & Merrill, llies and Morrison, Foster & Partners, Adrian Smith in partnership with the design bureau Thornton Tomasetti, Thom Mayne from Morphosis, which have implemented many projects around the world. To date, the "Anglo-

Saxon leadership in the architectural design of superheavy buildings" is recognized [17]. Russia is no exception, where foreign architectural bureaus are the designers of skyscrapers, while foreign companies also develop most constructive elements and parts of engineering systems of high-rise buildings, serve as contractors.

2.2 Construction features of high-rise buildings

The structural system of a high-rise building is a combination of its vertical and horizontal bearing structures that together provide strength, rigidity and stability of the structure. The main priority in the design and construction of high-rise buildings is to ensure their strength and stability, as well as rigidity, taking into account the impact of significant wind forces, overcoming the uneven shortening of vertical structures from reinforced concrete (walls and columns) under the influence of load.

In high-rise construction, materials such as steel and concrete are mainly used. It is the use of strong steel frames, which was first used in the construction of the famous New York building "Fuller Flatiron" in 1902, dramatically increased the number of storeys compared to the stone. The rigidity of the "skeleton" of construction helps to withstand the wind loads, and flexibility - to resist seismic fluctuations. In the facades of skyscrapers today, mainly steel profiles and light curtain panels of transparent and darkened glass, aluminum, and polymers are used. Popular ventilated systems, which are finished with natural or artificial stone, metal sheets, fiber concrete, i.e. concrete with fibers of metal and polymers. Innovative facade constructions are designed to increase the strength and durability, the level of protection from fire and excessive heat in the premises (in addition to the air conditioning system with high insolation), the ability to compensate for changes in loads in the internal premises, even associated with the movement of furniture. Among the newest types of cladding - ceramics with borosilicate glass (which has in its composition alkali instead of boron oxide, which makes the glass resistant to fire), panels of metal foam, nanocomposites, glass panels with a water repellent self-cleaning layer and so on.

In the construction of high-rise buildings, special requirements are advanced to the supplied elevators, air conditioning and fire extinguishing systems, to other engineering systems. In some cases, special construction technologies are used, such as the "Top-Down" technology. In Russia, the technology "Top-Down" has found wide application (in St. Petersburg - Lakhta Center, as well some facilities in the Moscow City complex). This is explained by the fact that this method allows almost simultaneously to work on the construction of the underground and ground parts of the building, which, accordingly, accelerates the construction period and the subsequent return on investment. It is also important to note that this method allows minimizing the deformations of the enclosing structures and, accordingly, the drafts of neighboring buildings. The method of construction "Top-Down" involves a device from the ground or from intermediate marks in the excavation of temporary or permanent supports inside the structure's contour, supporting the floors of the underground part of the building, which are concreted on the ground and perceive the action of the spacer from the excavation fence. The excavation work in the foundation pit is made from under the floors through the arranged technological openings. Concreting of underlying floors is carried out in sequence as the ground is removed. In the construction of the "Top-Down" technology to protect the excavation in modern conditions, the "wall in the ground" is often used, as the most versatile structure capable of absorbing vertical loads from the weight of underground floors and protection from groundwater [18, 19].

"Modular" (block) construction of high-rise buildings is also developing. Known examples are: a 30-story hotel "T30 Hotel" in Changsha, China, built in 15 days by the forces of 200 workers by technology of Broad Sustainable Building; "SOHO Apartments Darwin", 29 floors, Australia, 2014 (apartments); «Victoria Hall, Wolverhampton» 25 floors, United

Kingdom, 2009 (student hostel). Broad Sustainable Building in Changsha, China, implements a project of the construction of a super skyscraper Sky City with a height of 838 m in block technology, in early 2018 construction site preparation continues [20].

2.3 Economic feasibility of high-rise buildings

The most attractive sites for the construction of residential and office areas in many European cities, including in the cities of Russia, are in the historical center, in which buildings, as a rule, developed in the XVII-XIX. On the contrary, in cities such as New York and Chicago, in the actively built cities of China and other Southeast Asian countries, in the actively expanding cities of the "oil monarchies" (UAE, Saudi Arabia, Qatar, Kuwait), which are currently being built and planned the construction of the largest number of skyscrapers, on the contrary, it is often the new high-rise buildings that form or form the business and cultural center of the city. The problems of destroying monuments of historical architecture or nature for building on-site skyscrapers in their cases did not arise or became public domain.

A complex dilemma of combining a possible high-rise center (City) with a historical one arises in front of cities with numerous preserved architectural monuments, forming a historical center, primarily in front of European cities. As part of the solution to this problem in Paris, the high-rise quarter of Defance (the height of buildings from 180 to 323 m) is moved to the east of the city. In Moscow and Yekaterinburg, the City managed to locate in relative proximity to the historical center, which significantly increases the attractiveness of new skyscrapers for investors. [21]

In London, a different approach is applied: in the historical center, separate high-rise buildings are being constructed (for example, Shard 310 m high at London Bridge, One-Canada 235 m and others). However, the attempt to go this way with St. Petersburg was unsuccessful: it was decided to move the construction site of the tallest building in Europe and Russia to Okhta under pressure from the public to the northwest of the city to the shoreline of the village of Lakhta ..

Lahta Center is being built since 2012 in an area with exceptionally modern buildings, which greatly simplified the problem of combining architectural solutions with existing facilities, gave greater freedom of creativity to architects and designers and an opportunity to improve the transport infrastructure in the skyscraper construction area.

At the same time, high-rise buildings on the outskirts of cities are much less attractive for tenants and tenants, which makes it necessary to solve the problem of economic feasibility of such buildings. It is likely that if the Lakhta Center project were implemented not by the largest national corporation of Russia but by a local developer, then the transfer of the location of the structure from the center to the outskirts in terms of project payback could be of critical importance. The cost of renting and selling a square meter of commercial real estate is 1.5-2 times lower in new districts of the city than in the historical center.

The construction of a high-rise building in terms of payback is always a risky project. Dozens of projects for the construction of super-high and high-rise buildings in Russia in the 2010s. were delayed or postponed (see section "Undelivered and unimplemented buildings" of the resource at wikipedia.org "List_of_most_high_Buildings_in_Russia"). In particular, in St. Petersburg, the local developer Adamant in 2007-2008 planned for the construction series of office and retail buildings with a height of more than 100 m including: Warsaw Prospect (Warsaw Railway Station), Pulkovskiy (Pulkovskoye Shosse), Ladozhsky (near the Ladozhskaya metro station), Balkansky (near the station metro station "Kupchino"). Sketches of these projects, developed by the architectural workshop "Freifeld-Sedakov", were presented in the investment memorandums of the company 2005-2008. However, the high payback period for high-rise office buildings (based on 12 to 20 years or more), as well as the deterioration in the economic situation since 2008, forced Adamant to abandon these plans or implement significant less ambitious projects in existing areas.

In all cases, the most important problem remains with the transport accessibility of the area near high-rise buildings. If the transport and communal infrastructure needs to be created anew or radically reconstructed, the cost of the project may increase from 1.5 to 2.5 times. In this case, the payback period becomes the limit (50 years and more) and unattractive for investors. The current format of the multifunctional designation of super-high buildings (common everywhere except the United States and Hong Kong, where they continue to build only office skyscrapers) [22], allows the developer to sell part of the space as permanent housing, part of the office space, another part for the hotel, and in the lower floors to create an attractive shopping and entertainment center for the mass visitor.

High-rise housing construction should be considered separately. Implemented in Russia, high-rise residential projects in height do not reach office and multifunctional buildings, however, they are often huger by total area. High-rise residential construction has received a new development in St. Petersburg in 2000-2010-s with the appearance of 30-40 storey residential complexes in comparison with the record of a Soviet period 22 floors (residential buildings on Victory Square and on the Novosmolenskaya embankment). It should be noted that many experts do not see any sense in the mass erection of high-rise buildings, arguing that this has already formed the construction of cities and the availability of the opportunity to significantly expand suburban agglomeration areas due to low-rise construction [23, 24].

Most real estate developers are ready to deal with high-rise construction only with an eye to the long-term image effect. More expensive for high-rise residential buildings are both design and construction (due to complex foundation works, tougher fire regulations, increased requirements for the reliability of engineering systems), and home maintenance. "If you take two houses that are similar in terms of flatography and architecture, but different in height, the cost of a square meter for a building that is more than 75 meters tall will be 10-15% more than for a regular 25-storey building, argues director of RosStroyInvest Igor Kreslavsky. "The cost of a square meter for ultrahigh houses, from 100 meters, can increase by 50% or more because we have to use more expensive structural materials that can withstand wind loads, rent special construction equipment, etc.," the company's UNISTO Petrostal CEO Arseniy Vasilyev argues.

For high-rise construction, architects solve the problems of developing planning solutions that meet both fire and engineering standards, and are able to optimize the useful - subsequently sold or leased - area. This task is not easy, because in a skyscraper the specific area of stairs, elevators, corridors and technical rooms is higher than in typical structures. As a result, the profitability of high-rise residential projects, as a rule, is lower than "conventional" ones.

Such projects are realized for two reasons. Firstly, high-margin sales of apartments on the upper 5-7 floors (sales of "penthouses") are typical. If there are even symbolic differences from the lower floors of the building, apartments on the last floors can be sold with a coefficient of 1.2-1.5 or more from the standard price in the area for this type of housing. However, this rule works primarily for specific objects (water, historical center) and is of little use for buildings near industrial zones and within the sleeping areas. Secondly, the iconic skyscrapers are not only the "calling card" of the construction company, but also often of the district, the city, sometimes the country. In the market conditions - the construction of a unique high-rise building - a successful approach to "detuning" and differentiation among competitors.

2.4 Skyscrapers and image of cities

It should be noted that according to the existing ideas about branding of cities ("marketing of places and territories"), the success of a city in a competitive environment depends on the ability to attract investment and professional staff ("creative class" - specialists creating technological and image leadership) to its territory, flows of tourists. The creative class is the decisive factor that determines the status and functional purpose of the territory. In such global financial and business, as well as tourist centers like New York, Hong Kong, Beijing, Shanghai, London, Paris, Frankfurt, Moscow, Delhi, Dubai, Mecca, Kuala Lumpur, Chicago, Melbourne, Guangzhou and Shenzhen and others in high-rise quarters are concentrated offices of banks and insurance companies, major construction contractors and architectural bureaus, leading media, designers, trade and manufacturing companies [25-27]. It can not be said that office skyscrapers are the only place for "placing" the creative class. In low-story Silicon Valley (USA), Dublin (Ireland), Mont-Pellier (France), Ideon (Sweden) or Skolkovo (Russia) and in other world innovation clusters characterized by low-rise buildings, there are enough specialists who are referred to as the "creative class". However, it can be argued that skyscrapers today provide a convenient way and environment for the concentration of highly skilled workers and the management of significant labor resources in a territorially confined space, as well as a comfortable living environment for wealthy people with interests within the city. The unique skyscrapers on the territory of the UAE and Saudi Arabia at the same time are more focused specifically on the residence function.

3 Results

The work states three main reasons for sufficient increase of super-skyscraper erection and high-rise construction in the beginning of XXI century. The image imperatives for authorities plus the possibility of achieving a multiplicative economic effect, the lobbyism by architectural and technological companies as well as huge contractors. The new format of super-skyscraper include not only offices but multiple functions. The predominance of Anglo-Saxon specialists in the design of super-high skyscrapers and high-rise structure continues, however, there appear more and more qualified specialists and architectural bureaus from Arab countries, Asia-Pacific countries, Russia and other countries.

In the most countries in XXI the functional purpose of the super high buildings being built began to differ from the traditional office one - the world's highest hotel for pilgrims in Mecca appeared, the apartments occupy the upper part of the highest skyscraper "Burj Khalifa". In the projected skyscrapers there are hospitals, educational institutions, the public area of the lower floors becomes larger and more significant. The architecture of the projected superhigh buildings is moving farther from the usual parallipipeds and cylinders, becoming like a spiral, taking on other complex shapes.

4 Conclusion

In the paper, the following are considered consistently:

- architectural and technological and image "imperatives" of high-rise construction, the possibility of achieving a multiplicative economic effect by investing in high-rise construction;
- constructive features of construction of high-rise buildings that determine their modern appearance and provide opportunities to increase the altitude and reduce the scope of works at the facility;
- issues of economic feasibility of construction of high-rise buildings;
- the interrelation of the construction of skyscrapers with the provision of the growth of a positive image of cities within the framework of marketing of territories and attraction of representatives of the "creative class" to the cities.

References

1. Another record year for high-rise construction. – URL: https://www.bdcnetwork.com/another-record-year-high-rise-construction (publication date 20.12.2017).
2. A. Wood, Ed. Best Tall Buildings 2013: *CTBUH International Award Winning Projects*. Chicago: Council on Tall Buildings and Urban Habitat, 2014.
3. A. Wood, Ed. Best Tall Buildings 2014: *CTBUH International Award Winning Projects*. Chicago: Council on Tall Buildings and Urban Habitat, 2015.
4. A. Wood, Ed. Best Tall Buildings 2015: *CTBUH International Award Winning Projects*. Chicago: Council on Tall Buildings and Urban Habitat, 2016.
5. A. Wood, Ed. Best Tall Buildings 2016: *CTBUH International Award Winning Projects*. Chicago: Council on Tall Buildings and Urban Habitat, 2017.
6. S. Manukov. *The second skyscraper boom*. –URL: http://expert.ru/2017/12/27/neboskrebnaya-lihoradka/ (publication date 28.12.2017).
7. The Russian standard of organization STO 01422789-001-2009 "Design of high-rise buildings"; High-rise building. – URL: https://ru.wikipedia.org/wiki/High-rise_building [accessed on 28.12.2017].
8. V. P. Generalov, E.M. Generalova. *High-rise residential buildings and complexes. Singapore. Experience in the design and construction of high-rise housing*. Samara: The Book, 2013, 400 p.
9. A.I. Isakov. High-altitude construction in Russia, *Synergy of Sciences*. **6**. 442-458 (2016).
10. E.Yu. Okolelova, N.I. Trukhina. *Construction of high-rise buildings: assessing the effectiveness of projects in a risk environment*. Voronezh: Voronezh State Agricultural Academy, 2016. - 132 p.
11. M. Ventura, *Building skyscrapers. Engineering Challenges*. Mendota Heights, Minn.: North Star Editions, 2017.
12. K.Burns. *A Skyscraper Reaches Up. Be an Engineer! Designing to solve problems*. New York: Crabtee Publishing Company, 2017.
13. Parker D., Wood A. *The Tall Buildings Reference Book*. Abingdon: Routledge. 2013. 495 p.
14. C. Uffelen van, *Skyscrapers*. Salenstein: Braun Publishing, 2012.
15. Willis C. *Form Follows Finance: Skyscrapers and Skylines in New York and Chicago*. New York: Princeton Architectural Press. 1995. 219 p.
16. CTBUH Tall Building Database. The Skyscraper Center –URL: http://skyscrapercenter.com/ [accessed on 25.12.2017].
17. E. Generalova, V. Generalov. Designing High-Rise Housing: The Singapore Experience. *CTBUH Journal*. Chicago, Illinois Institute of Technology. **Issue IV,** 40-45 (2014).
18. A. Cowlard A., Bittern A., Abecassis-Empis C., Torero J. Fire Safety Design for Tall Buildings, *Procedia Engineering*, **62**, 169-181 (2013).
19. A.R. Elbakheit Why Tall Buildings? The Potential of Sustainable Technologies in Tall Buildings, *International Journal of High-rise Buildings*. **Vol. 1, № 2**, pp. 117-123 (2012).

20. Kilometer and above. 5 skyscrapers-record-holders of the future. - URL: http://www.novate.ru/blogs/150614/26693/ [accessed on 11.01.2018]
21. M. Ali, K. Al-Kodmany. Tall Buildings and Urban Habitat of the 21st Century: A Global Perspective, *Buildings,* **2**, 384-423 (2012).
22. E. Generalova, V. Generalov. Apartments in Skyscrapers: Innovations and Perspectives of their Typology Development. *Proceedings of the CTBUH 2015 International Conference «Global Interchanges: Resurgence of the Skyscraper City»*. New York, USA. 26th30th October 2015, pp. 355-362 (2015).
23. L.A. Viktorova. Vysotnye zdaniia – pliusy i minusy stroitel'stva [High - rise buildingsthe pros and cons of construction], *Ezhemesiachnyi nauchno - prakticheskii i kul'turno – prosvetitel'skii zhurnal «Arkhitektura i stroitel'stvo Rossii*, **10** (2013). – №URL: http://www.asrmag.ru/article/vysotnye-zdaniya/ [accessed on 25.12.2017].
24. L.P. Lavrov, F.V. Perov. High-rise buildings: an erroneous vector of housing construction, *Bulletin of Civil Engineers*, **5 (58),** 16-27 (2016).
25. C. Maywalda, A. Riesserb. Sustainability – The Art of Modern Architecture, *Procedia Engineering*. **155**, 238-248 (2016).
26. D. Radushinsky, A. Mottaeva, etc. The evaluation of the modernization cost of the transport infrastructure of the Northern Sea Route in the Arctic zone of the Russian Federation. *IOP Conference Series: Earth and Environmental Science*. - URL: http://iopscience.iop.org/article/10.1088/1755-1315/90/1/012137
27. D. Radushinsky. Evaluation of the integral economic effect from the implementation of Innovation projects through public-private partnership in view of their communicative component, *Bulletin of the UrFU. Series of Economics and Management*. **4 (16)**, 623–641 (2017). DOI: 10.15826/vestnik.2017.16.4.030.

Constructive systems, load-bearing and enclosing structures of high-rise buildings

Elena Anatol'evna Korol'[1*], *Yuliya Olegovna* Kustikova[1]

[1]Moscow State University of Civil Engineering, Yaroslavskoe shosse, 26, Moscow, 129337, Russia

Abstract: As the height of the building increases, loads on load-carrying structures increase dramatically, and as a result of the development of high-rise construction, several structural systems of such buildings have been developed: frame, frame-frame, cross-wall, barrel, box-type, box-to-wall ("pipe in pipe", "Trumpet in the farm"), etc. In turn, the barrel systems have their own versions: cantilever support of the ceilings on the trunk, suspension of the outer part of the overlap to the upper carrying console "hanging house" or its support by means of the walls on the lower bearing cantilever, intermediate position of the supporting cantilevers in height to the floor, from a part of floors. The object of the study are the structural solutions of high-rise buildings. The subject of the study is the layout of structural schemes of high-rise buildings, taking into account the main parameters - altitude (height), natural climatic conditions of construction, materials of structural elements and their physical and mechanical characteristics. The purpose of the study is to identify the features and systematization of structural systems of high-rise buildings and the corresponding structural elements. The results of the research make it possible, at the stage of making design decisions, to establish rational parameters for the correspondence between the structural systems of high-rise buildings and their individual elements.

1 Introduction

High-rise buildings have a specificity, significantly different from them traditional high-rise buildings and multi-storey buildings. Among the main features of high-rise buildings are:

- Significant values of both static and dynamic loads on load-bearing structures and on bases;
- high, sometimes critical value of horizontal (primarily wind) loads;
- problems of unevenness of both load values and the nature of their application;
- careful and correct selection of construction materials, excluding separate work of structural elements and ensuring homogeneity of physical and mechanical characteristics;
- Increased significance of natural (airflow, seismicity, temperature, etc.) and technogenic factors (vibration, accidents, fires, local destruction) on the safety of construction and operation;

* Corresponding author: professorkorol@mail.ru

• complex solutions of internal engineering systems and communications, accompanied by the creation of additional engineering units, which is due to the height of the building;

• Increased requirements for integrated security, including firefighting, involving the use of technical solutions of a qualitatively different level and affecting the choice as volume-planning, and constructive decisions.

These aspects must be taken into account when choosing a structural scheme of a high-rise building and designing load-bearing structures.

The publications devoted to high-rise buildings consider design, construction and operation issues based on unique experience in different countries of the world [1-7]. Despite the fact that this experience is permanently generalized by researchers in this field of construction, the appearance on the world map of each new high-rise building is of interest to a large number of specialists dealing with the problems of developing new building materials, non-traditional structural solutions for load-bearing and enclosing structures (including facades) , dealing with issues of hydrogeology, improvement of engineering systems, architects and city planners, construction monitoring services, etc. [8-19].

The unique experience in the world practice of design and construction of high-rise buildings is enriched by new objects every year, taking into account the various conditions for the construction and operation of these buildings and the trends in the development of the society's needs at the present stage in ensuring the quality of living and a comfortable environment of life activity [20-22].

2 Constructive schemes of height buildings

The choice of a concrete constructive scheme of a building depends on a complex of factors, the main of which are the height of the building, its purpose, construction conditions (characteristics of foundation grounds, seismicity, atmospheric, especially wind, impacts), architectural and planning requirements.

As the practice of high-rise construction developed in the leading countries, several structural schemes (systems) of the above-ground parts of buildings were developed, tested, improved and implemented:

• frameless with cross-bearing walls,
• frame-frame,
• frame with diaphragms of rigidity,
• barrel, frame-barrel,
• box-type (shell),
• Stem-box ("pipe in the pipe", "trumpet in the farm", "bundle of pipes").

Frameless with a cross-bearing walls scheme (Figure 1a), which for centuries has been the main solution for buildings of any purpose, in high-rise construction is applied sporadically and mainly for residential buildings and hotels, as the space-planning decisions of the buildings for this purpose are completely consistent with the requirements constructive scheme. The unpopularity of frameless wall construction scheme in the environment of architects is explained by limitations of freedom of internal space planning.

The frame-frame structural system (Fig. 1b), which served as the basis for the creation of skyscrapers at the turn of the 19th and 20th centuries, is now widely used in the construction of buildings up to 60 storeys high (in the steel version and later with the reinforced concrete frame).

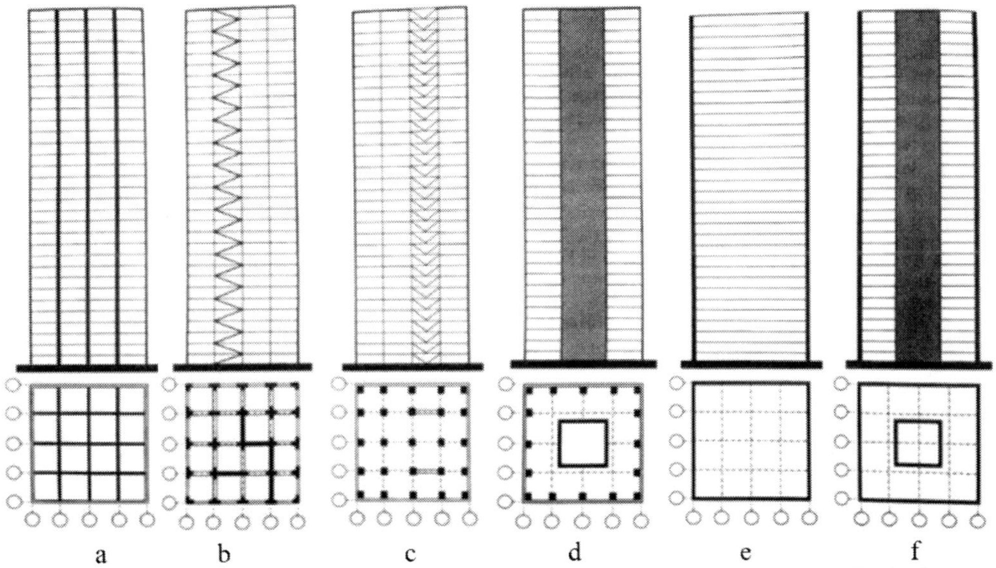

Fig. 1. Constructive schemes of high-rise buildings: a) frameless with cross-bearing walls, b) frame-frame, c) frame with diaphragms of rigidity, d) barrel, frame-barrel, e) box-shaped (shell) f) stems-box ("pipe in the pipe," "trumpet in the farm," "bundle of pipes")

However, with the increase in the number of storeys, the inevitable complication of the design of the frame units for the perception of increasing horizontal loads dictates the transition to a bonded frame with through straightened steel vertical stiffening diaphragms or with solid reinforced concrete walls - stiffening diaphragms. In connection with this, a new constructive scheme was proposed-a frame with diaphragms of rigidity (Fig. 1c).

Since the 1960s, new structural systems have been actively embarked on high-rise construction - barrel and shell systems.

The trunked structural system (Figure 1d) as the main supporting structure of the building, which receives loads and impacts, contains a vertical spatial rod - the barrel of rigidity (the "core of rigidity") to the entire height of the building.

In the stiffness of the high-rise building, which is usually located in the central part of the building, a staircase-elevator unit is located, and the structure of the stiffener itself is made of monolithic reinforced concrete, rigid steel structures or a combination thereof. In accordance with European standards, the distance from the outer walls to the core of stiffness in terms of natural light should not exceed 8 m.

The trunk system organically entered the practice of high-rise construction, as it successfully combined with the planning layout of the building. The best conditions for the spatial operation of the structures of the buildings solved according to the barrel scheme provide a strictly central location of the trunk in the plan, as well as the geometric similarity of the shapes of the building and trunk plans with the area of the "core of stiffness" about 20% of the area of the building plan. The widest distribution in the construction of buildings up to 60 storeys of various purposes (administrative, residential, hotels) was combined frame-barrel system (Figure 1d), with the frame location along the outer contour of the building. Joint horizontal movements of the frame and trunk are provided by horizontal outriggers-grillage, located through 18-20 floors. Bearing structures of barrel buildings are mainly reinforced concrete prefabricated and monolithic. The section of the walls of a monolithic trunk varies from 40-100 cm in the lower floors to 20-30 cm in the upper ones, depending on the number of storeys.

However, due to the limited dimensions of the staircase-elevator units (trunks) in the buildings of high altitude (more than 200 m), this system causes more concentrated loads on the base and can not provide the necessary rigidity, in connection with which the world practice has developed a box shell) constructive system, in which the required flexural rigidity is provided by the outer shell of the building. The shell structural system (Figure 1e) is characterized by maximum rigidity in comparison with the above schemes, since the supporting structures are located along the outer contour.

The main shell system is accompanied by two combined shell-barrel ("pipe in the pipe") (Figure 1f) and shell-diaphragm ("bundle of pipes").

As in the main system - shell, and in the combined - sheath-barrel in the center of the plan, the trunk is located with elevator shafts and halls located in its space.

The difference between these variants lies in the horizontal load distribution provided by the project: only on the shell (with the trunk only working on vertical loads from the overlaps) or on the shell and barrel. In the latter variant, the mass of the structure of the floors increases due to their inclusion in the work on horizontal impacts. Nevertheless, most of the shell-type high-rise buildings are built on a shell-and-barrel system, although individual objects have a shell structural system.

The individual specific task of designing shell buildings was the design of a bearing outer shell, combining the bearing and enclosing functions.

Over the past decades, a number of designs have been implemented:

• Spatial bezraaskasnaya multi-storey and multi-span lattice with a frequent step of the columns and floor bolts-crosspieces;

• spatial lattice macro-farm of a large module, whose braces span 10-15 floors, with a rare step of columns;

• Spatial bezraaskasnaya lattice, the rigidity of which increases the dull filling of diagonally located openings;

• lattices of diagonal rods;

• lattices of diagonal and horizontal rods;

• lattices of orthogonal and diagonal rods.

With a further increase in the height of the building, the rigidity of the shell constructions considered may not be sufficient. To this end, in projects that have not been implemented to date, it has been proposed to construct shells from cross-bar structures with the same design of horizontal outriggers-grillage. A means of increasing the hardness of the shell can also serve as a transition from the shell to the shell-diaphragm construction ("bundle of pipes").

The construction of the shell is carried out both from steel elements and from reinforced concrete. Reinforced concrete casings were made monolithic or prefabricated, with preference given to lightweight structural concrete, which made it possible to combine the load-bearing and heat-insulating functions of the wall. In recent years, shells in Europe are made predominantly monolithic from heavy concrete with subsequent insulation and external lining. For elements of steel shells, the most commonly used are rolled or welded elements of closed rectangular cross-section also with subsequent insulation and lining. At the same time, these systems do not allow the use of plastic facade solutions and require the frequent placement of load-bearing racks along the perimeter of the building.

3 Constructive solutions of the foundations

The effectiveness of the technical solution of the foundation of a high-rise building increases significantly when it is buried. The depth of foundations can be 15-25 m, and in some cases - 50 m.

At present, in the design and construction of high-rise buildings, three types of foundations have become widely used: pile, plate and pile-plate (Table 1).

Table 1. Types of foundations widely used in the design and construction of high-rise buildings

Type of foundation	Characteristic
 a) Pile foundation	The pile foundation is the most reliable and, accordingly, the most common type of foundations for high-rise buildings. This foundation construction is used for construction on bases with low bearing capacity or significant heterogeneity.
 b) Solid plate 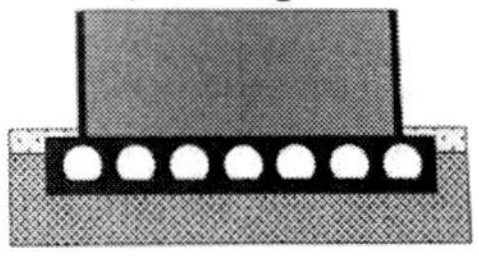 c) Box plate	Plate foundations are made either in the form of continuous monolithic reinforced concrete massifs, the thickness of which can reach 6 m, or in the form of a monolithic box structure, the internal cavities of which are used for parking lots or other technical premises.
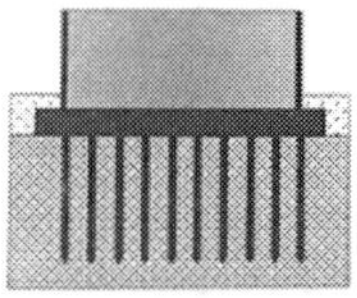 d) Pile-slab foundation  e) Disconnected pile-slab foundation	In the pile-slab foundation, up to 80% of the load from the building is perceived by piles. This type of foundation is advisable to use in weak deformable soils. The non-symmetrical (with eccentricity) application of the load can be compensated by the number and arrangement of piles in the foundation, and also by adjusting their bearing capacity by selecting the diameter and length. With a relatively small height of the building and the grounds of increased deformability, it is rational to use a disconnected pile-slab foundation, which lacks structural connections between the base plate and the piles. Piles are used in this case for differentiating the bearing capacity of the soil. The pile head is separated from the solid reinforced concrete foundation plate by a synthetic gasket, through which the load from the plate is transferred to the piles. The strength and durability of the synthetic gasket should correspond to the required design values.

Pile foundation is the most expensive type, however its application, as practice of building and operation of buildings shows, allows to minimize sediment values.

Plate foundation for sand and gravel grounds is most economical if the load is transferred without significant eccentricities

In some cases, pile-plate foundations are the most reliable and economical.

The choice of a rational design and technological solution and the design of the foundations of high-rise buildings is the following process (Fig.2).

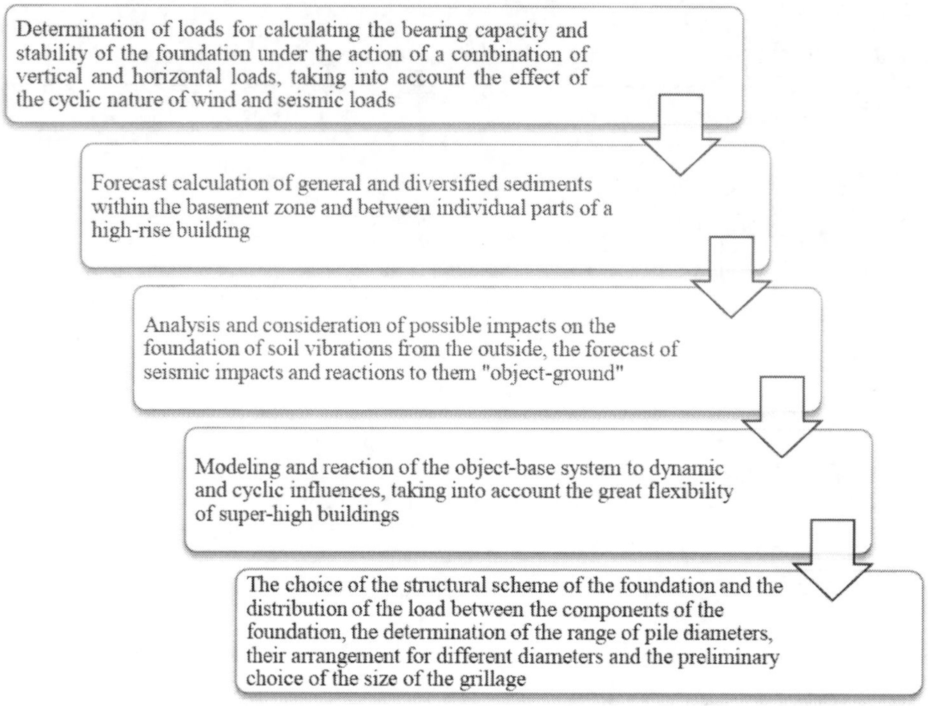

Fig. 2. The process of constructive and technological solutions and the design of foundations of high-rise buildings

3.1 Bearing structures

Load-Bearing structures of high-rise buildings at the very beginning of the development of this type of construction were mainly made of steel. And today in buildings with a height of more than 300 m, including "ultra-high", steel structures are used for load-bearing elements. Assembly of factory installation elements on the construction site is carried out with the help of bolted connections, which allows to increase the accuracy of installation by eliminating warpage of steel structures under the influence of high temperatures.

Steel structures are advisable to be used in seismic resistant regions, since steel, due to its plasticity, provides damping of the kinetic energy of seismic impacts. At the same time, steel structures require mandatory reliable protection against fire, since at 300 ° C the strength of steel is sharply reduced.

To increase the load-carrying capacity of columns with the restriction of the dimensions of its cross-section is possible due to the use of steel concrete. Internal walls, including staircase-elevator units (trunks), in recent years have been made, as a rule, from reinforced concrete or encased steel structures, which is dictated by considerations of fire safety.

In recent years, load-bearing structures of high-rise buildings are increasingly preferred to be made of reinforced concrete, since this material has a higher fire resistance and is cheaper, and its strength characteristics approach the strength of steel.

The center of research of prefabricated reinforced concrete frames was Japan. There since the 1970s, seismic resistant frames of high-strength concrete (classes B60-B100) have been developed and introduced into high-rise construction. Tests carried out by Japanese scientists confirmed the possibility of erecting high-rise buildings under conditions of high seismicity.

The structures of high-rise buildings are continuously being improved and are becoming increasingly diverse. In the last decade, pipe-concrete structures of reinforced concrete skeleton are actively used. Their high load-bearing capacity contributed to the revision of the approach developed over the past 30 years to the application in buildings above 300 m only of the shell structural system.

Horizontal load-bearing structures of high-rise buildings, as a rule, are of the same type and represent a rigid fireproof disc - reinforced concrete (monolithic, prefabricated-monolithic, prefabricated) or steel-reinforced concrete.

Reinforced concrete floors are made in the form of continuous flat slabs, including prestressed, monolithic ribbed ones with main and secondary beams, monolithic caisson slabs, prefabricated and prefabricated monolithic beam or ribbed structures. At spans of up to 8 m, non-beam ceilings made of solid monolithic plates up to a thickness of 26 cm are widely used.

Prefabricated cast-in-situ reinforced concrete floors of high-rise buildings have become widespread in Japan, since in this case the advantages of industrial manufacturing of the main volume of reinforced concrete are effectively used, which significantly increases the rate of installation of ceilings and preserves the positive qualities of monolithic structures.

A feature of overlapping of high-rise buildings is the fact that suspended ceilings and technological floors are used to accommodate numerous engineering systems (ventilation, heating, conditioning, electricity, etc.) and communications (computer, signal, video surveillance, automation, etc.). The total height of suspended ceilings and technological floors reaches 900 mm; so that at a height of 3,6 m the height of the room is 2,7 m. For the passage of engineering communications in solid beams of overlap, there are appropriate holes, the influence of which on the bearing capacity of beams should be taken into account in calculations .

According to the requirements of strength and fire resistance, all vertical and horizontal load-bearing structures of buildings with a height of more than 100 m will be made of high-strength concrete of class B60-B80 with a fire resistance rating of at least REI 240.

The structures of the coatings for the landing area for rescue helicopters should be made of the same materials and with the same fire resistance limit.

3.2 Enclosing structures

According to the constructive solution, the enclosing structures of high-rise buildings are designed either based on interstorey ceilings or suspended to them and performing only the function of protection from climatic, atmospheric factors and providing thermal insulation.

In modern high-rise buildings light profile systems with filling with panels made of aluminum, special glass, polymeric materials, ventilated systems finished with natural or artificial stone, decorative metal sheets, fiber-reinforced concrete screens and other materials are widely used.

Facade systems of high-rise buildings must be technological, air-and vapor-proof, fire-resistant, noise-proof, durable and reliable in operation, maintainable, and also have good thermal insulation properties, low coefficient of thermal expansion and low mass.

The climatic effect on the facade systems of high-rise buildings is exerted by the wind, as well as solar radiation, torrential rains, thunderstorms and smog. Facade structures should not only withstand direct wind pressure (up to 20-25 m / s), but also resist tearing forces that occur when air moves along the wall and negative pressure zones appear due to turbulence.

A special place in the varieties of enclosing structures is occupied by translucent enclosing systems. They are designed in accordance with several conditions. Profiled bearing elements for increasing strength properties and durability, as a rule, are made of steel. In translucent filling, especially strong, fire-resistant, low-emission and sun-protective glasses are used. In the practice of modern high-rise construction, double skin systems with external protective screens made of especially durable glass are widely used. They allow you to make the inner glazing partially or fully openable.

Conclusions

Each high-rise building is unique, but at the same time it allows to expand the experience of construction and operation of such buildings for further improvement. The systematization of this experience is important and can be used at the stage of developing the concept of the project and making pre-project decisions.

From the correct choice of the structural system of a high-rise building, the consumption of materials on the bearing structures and the cost of construction depends to a large extent.

Increasing the resistance of a high-rise building to wind loads can be achieved by using a rational form. As shown by numerous foreign studies and operational experience, the most suitable form for the impact of wind is the circular shape of the building. Somewhat inferior to it is an elliptical (oval) shape, and then a square one.

Load-Bearing structures of high-rise buildings distinguish high requirements for strength and deformation characteristics of materials. Traditional materials such as steel and reinforced concrete are used in combination and predetermine the development of new types of combined carrier systems in parallel with the development, research and implementation of high-strength concrete class B 100.

Requirements for facade systems designed for high-rise building construction, far exceed the requirements for the enclosing structures of conventional houses, due to the multiple increase in all types of loads - both dynamic and climatic. The facades of high-rise buildings should be air-and vapor-proof, fireproof, technological, noise-proof, durable and reliable in operation, maintainable, and also have good thermal insulation properties, low coefficient of thermal expansion and low mass.

Facade systems are constantly being improved, new technologies for fabrication and installation of structures, materials (ceramics combined with borosilicate glass, metal foam panels, nanocomposites, glass panels with a superhydrophobic self-cleaning layer, etc.) are being developed.

References

1. A. Abdelrazaq, F. Badelow, Sun-Gho-Kim, H. G.Poulos, Incheon Tower: Foundation Design in Reclamation Area. Tall Buildings **2/17,** 104-109.
2. Liu Peng, Cheng Yu, Zhu Yan-Song Constructions seismic stability Structural Design of China Zun Tower. Tall Buildings **2/17,** 110-119
3. V. Almazov, Ways and methods of counteracting the progressive destruction of high-rise buildings. Global security, 46-49 (2006)
4. Y. Granik, A. Magay, Overview of foreign construction experience in high-altitude housing construction, Unique and special technologies in construction. **1,** 20-31 (2004)
5. V. Sevostyanov, I. Mindel, B. Trifonov, Evaluation of seismic hazard for high-rise, 56-62 (2006)
6. M. Shchukina, Modern high-rise construction. Monograph. - Moscow: State Unitary Enterprise "ITC Moscomarchitecture", 440 (2007)
7. T. Maklakova, N. Senin, Architectural and constructive problems of designing high-rise buildings: scientific and educational material. (Moscow, 2009)
8. O. Prishchepa, A. Karavaev, A. Bushinskaya, Resilience of High-Rise RC Buildings to Dynamic Impacts. IOP Conference Series: Materials Science and Engineering, **262(1)** (2017)

9. O. Petrova, O. Lugchenko, M. Hammud, A. Nazhem, To the rationalization of the constructive solutions of the bridge supports. MATEC Web of Conferences, **106** (2017)
10. V. Shumeyko, The support systems of unique high-rise buildings. MATEC Web of Conferences, **106** (2017)
11. E. Generalova, V. Generalov, A. Kuznetsova, Innovative solutions for building envelopes of bioclimatical high-rise buildings. Vide. Tehnologija. Resursi - Environment, Technology, Resources **1**, 103-108 (2017)
12. Shi, S., Ren, X. Analysis of form on indoor wireless fiber network. 2nd International Conference on Artificial Intelligence, Management Science and Electronic Commerce, AIMSEC 2011 Proc. 6011402, c. 6135-6138 (2011)
13. P. Liu, C. Yin, Y. Cheng, Structural design and research of Beijing CBD Core Area Z15 Plot China Zun Tower, Building Structure, **44(24)**, 1-8. (2014)
14. Y. Cheng, P. Liu, D. Citerne, Y. Liu Structural parametric design applied in Beijing CBD Core Area Z15 Plot China Zun Tower, Building Structure, **44(24)**, 9-14 (2014)
15. P. Liu, C. Yin, K. Lee, and et al. Structural system design and study of Tianjin Goldin 117 Mega Tower, Building Structure, 42(3), 1-9. ASCE "Seismic rehabilitation of existing buildings" ASCE/SEI 41-06 (2007)
16. M. Kwok, C. Gibbons, J. Tsui, P. Liu, Y. Wang, The Structural Design of Mega Tower, China World Trade Center Phase 3, Beijing China. Paper presented at: Tall Buildings - From Engineering to Sustainability, Proc. of 6th Inter. Conf. On Tall Buildings, Hong Kong, 396-402 (2005)
17. P. Xu, Structural Design of Complex Tall Buildings, Beijing: China Architecture & Building Press (2005)
18. Ministry of Housing and Urban-Rural Development of the People's Republic of China Technical review requirement on seismic design of code-exceeding tall building structures. Jianzhi, 109. (2010)
19. M. Mikhailova, V. Dalinchuk, A. Bushmanova, L. Dobrogorskaya, Construction of unique buildings and structures. **10 (49)**, 59-74 (2016)
20. URL: http://www.antula.ru/site-rielter_271.htm (05.03.2017)
21. V. Generalov, E. Generalova, *High-rise residential buildings and complexes* (Samara, 2013)

The visual-landscape analysis during the integration of high-rise buildings within the historic urban environment

Vera A. Akristiniy[1,*] and *Elena* A. Dikova[1]

[1]Moscow State University of Civil Engineering, Yaroslavskoe shosse, 26, Moscow, 129337, Russia

Abstract. The article is devoted to one of the types of urban planning studies - the visual-landscape analysis during the integration of high-rise buildings within the historic urban environment for the purposes of providing pre-design and design studies in terms of preserving the historical urban environment and the implementation of the reconstructional resource of the area. In the article formed and systematized the stages and methods of conducting the visual-landscape analysis taking into account the influence of high-rise buildings on objects of cultural heritage and valuable historical buildings of the city. Practical application of the visual-landscape analysis provides an opportunity to assess the influence of hypothetical location of high-rise buildings on the perception of a historically developed environment and optimal building parameters. The contents of the main stages in the conduct of the visual - landscape analysis and their key aspects, concerning the construction of predicted zones of visibility of the significant historically valuable urban development objects and hypothetically planned of the high-rise buildings are revealed. The obtained data are oriented to the successive development of the planning and typological structure of the city territory and preservation of the compositional influence of valuable fragments of the historical environment in the structure of the urban landscape. On their basis, an information database is formed to determine the permissible urban development parameters of the high-rise buildings for the preservation of the compositional integrity of the urban area.

1 Introduction

Introduction of the conduct of the visual-landscape analysis practice of high-rise buildings in the system of preparation of pre-project documentation allows not only assessing the influence of hypothetically existing buildings on the perception of historically valuable objects of the urban environment. It also allows assessing the possible parameters of high-rise construction in the conditions of ensuring the continuity of the urban landscape and providing a unified urban policy [1,3].

The visual-landscape analysis is one of the tools of urban forecasting, that allows specifying the regulatory requirements and restrictions established for a particular territory.

*Corresponding author: 7824666@mail.ru

Applying the visual-landscape analysis methods has received official status in Moscow as a mandatory part of the preparation of pre-project and project documentation within the framework of the overall urban policy. The regulatory restrictions of a particular territory are directly dependent on the protection zones in which it is included, taking into account the location in the city structure, the level of historical and cultural potential, the degree of the traditional urban environment preservation, the city-forming importance and other valuable characteristics of cultural heritage objects [2].

The maintenance regimes for all zones of protection provides the optimal perception of historical and cultural monuments in the structure of the urban environment, preservation of valuable historical panoramas and types of urban landscape by adjusting the height parameters of the designed buildings in terms of the city-forming objects valuable visibility.

Methods of the visual-landscape analysis can be used to solve a variety of problems: urban planning forecast of a general nature, the output of permissible height constrictions for a particular object, computer modeling of various variants of the design solution in the system of the urban environment, a visual perception review of the designed object, revealing of the inconsistency of the realized high-rise parameters of the object to the approved project etc [4,7].

The recommended range of height, density and other parameters of the reconstruction and new construction of objects in the historical center of Moscow is determined based on general regulatory requirements and urban development restrictions of territory. Clarification of the permissible volumetric and planning characteristics of the designed buildings to be placed in the system of city panoramas, as well as compositionally interconnected with the monuments of history and culture, valuable fragments of the existing urban-planning environment, is carried out by taking into account features of the space-planning organization of the urban landscape fragment under consideration, historical and cultural studies and conclusions of the visual-landscape analysis [5].

Thus, the application of visual analysis does not contradict the perspective development of town planning regulations and allows to quickly identify inadmissible in terms of protecting cultural heritage or compositionally controversial project proposals, in the absence or inadequate elaboration of documents. The study was conducted in order to identify sectors of the interaction of designed buildings and structures with cultural heritage and analyzing peculiar overview properties of the object in the structure of urban spaces, to account for the formation of composite solutions of the designed building in the future [6,8].

2 The method of visual-landscape analysis

Visual landscape analysis implementation based on field investigations of the territory, analysis of photo fixation materials, as well as expert evaluation of the urban development situation.

The main stages of the visual-landscape analysis shown in Fig.1 [9].

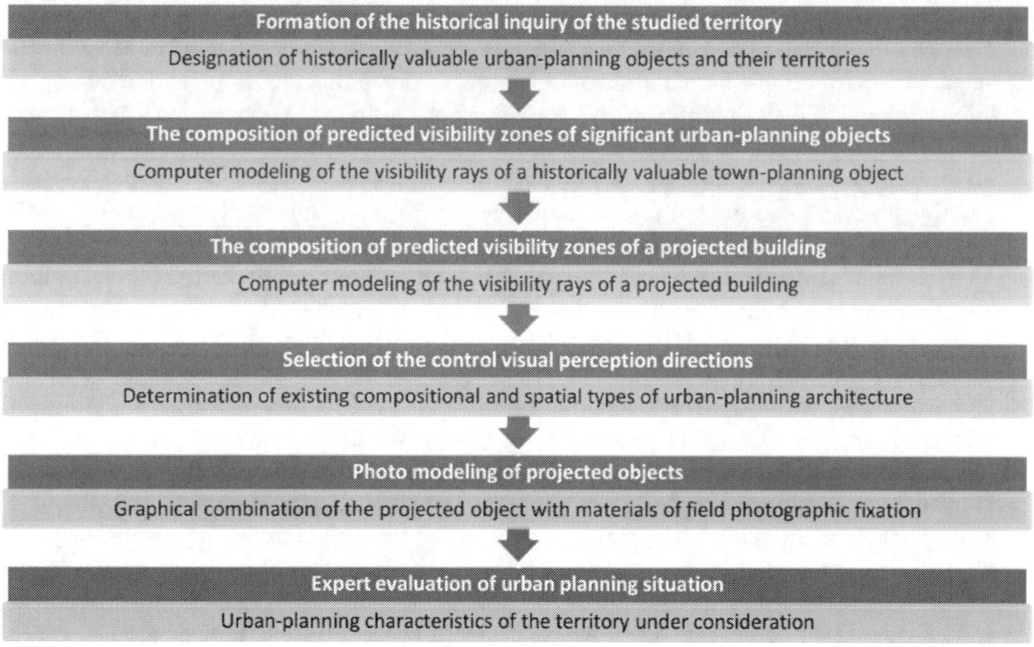

Fig.1. Contents of the main stages of visual-landscape analysis

The convincingness of the visual-landscape analysis conclusions to a certain extent depends on the exposition methods of the resulting visual sequence. The illustrative materials have a discrete character and give a fragmentary picture of the future compositional role of the new volume in the structure of the urban landscape.

New technological possibilities of materials submission allow estimating the forecasted development of the urban-planning situation by analyzing video materials. The dynamic perception of the new construction object contributes to the maximum approximation to reality, ensuring a continuous nature of the survey, and demonstrates exhaustive information about the town-planning significance of a particular type of object disclosure. The application of video makes more evident the expert evaluation of the control visual perception directions, because of which it proposed to correct the height parameters of the object [13].

The use of new visual analysis methods is promising not only for the purpose of determining high-rise restrictions in zones of interaction between new construction sites with historical and cultural heritage, but also in solving complex urban problems related to the location and expected interaction of a number of projected complexes, including new areas of the city.

The tasks of visual analysis, primarily designed to ensure the optimal perception of cultural heritage sites, also include the development of recommendations for the preservation and successive development of aesthetically expressive fragments of the urban landscape.

The documentation presented in the visual-landscape analysis differentiated according to three main sections (containing text, graphic and illustrative materials) [11]:

1. urban-planning characteristics of the territory under consideration;
2. analysis of the compositional interaction of the projected object with a valuable urban-planning environment;
3. conclusions of the visual-landscape analysis.

The urban-planning characteristic of the territory contains a generalized analysis of the most important features (geomorphological, composition and planning, scale-typological) of the valuable urban landscape fragment, as well as existing requirements and limitations of urban development of the territory, established by applicable legislative and regulatory documents. Geomorphological features include the characteristics of the territory landscape under study [12].The compositional-planning features of the territory include aspects of the urban-planning space that determine the interconnection of all its elements, which as a result create completeness and its integrity. The purpose of compositional and planning characteristics is to bring to unity the functional, technical, economic, ecological, and aesthetic requirements of the territory [14].

Within the framework of the expert evaluation of the urban-planning situation, the territory under study is analyzed on the basis of zoning schemes for the Master Plan of Moscow until 2025 (functional zoning, analysis of the reorganization areas and development zones, protection zones) [10].

The analysis of the compositional interaction of the projected object with a valuable urban-planning environment is carried out in order to identify valuable urban panoramas and types of the study area and their compositional features, determine the conditions for visual perception of particularly significant urban-planning dominants and valuable fragments of traditional buildings, as well as forecast the permissible development of compositional and visual characteristics of the territory.

Based on the materials of this section, the conditions for the correspondence of the volume-planning, including the height and compositional parameters of the projected complex to the characteristics of a valuable urban landscape are revealed (Fig. 2) [15].

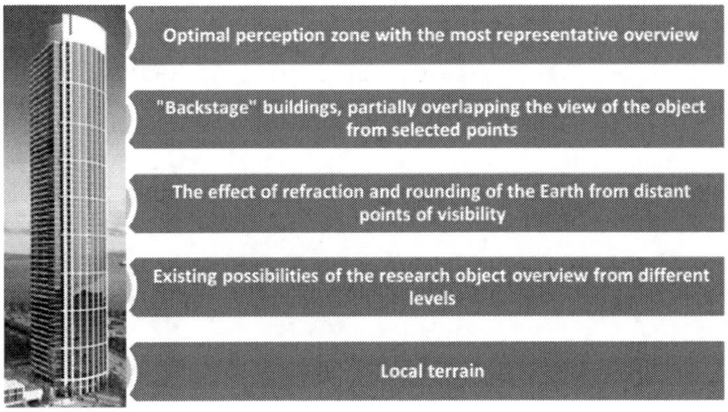

Fig.2. Scheme of the method for recording conditional visual communication lines between the viewer and the studied object

3 Practical results

The visual-landscape analysis carried out according to the presented method when integrating the high-rise building of the business center into the territory of cultural heritage objects in Moscow revealed the permissible parameters of the space-planning decisions of the projected object. According to the analysis of the studied territory of cultural heritage objects (Fig. 3), the visibility pan of the most significant cultural heritage objects (Fig. 4) and the projected high-rise buildings predicted visibility zones (Fig. 5) are determined [10].

Fig.3. Scheme of cultural heritage objects with the location designation of the high-rise object in the studied territory at the address:1/1Tishinskaya Ploshchad', Moscow,(1- Cathedral of the Immaculate Conception, 1910-1911, architect Bogdanovich F.F.; 2- Temple of the Great Martyr George the Victorious, 1788 – 1807, 1870s, 1890s; 3 - Residential house, the end of the XIX century; 4- Residential house, 1937, architect Diederichs Yu.F.; 5 - The house in which in 1902 - 1903 lived A.L. Nikolayevich; 6- Central house of cinema, late XIX - early XX centuries, architect Stamo E.M.; 7 - Garden of the Filatovskaya children's hospital; 8 - Dwelling house, 1929-1930; 9 – Firsanov's home for widows and orphans, 1883; 10 – Dwelling house, 1928, architect P.A. Andreev; 11 – Profitable house, 1905, architect E.K. Nirnsee, 1912, architect Gashinsky; 12 – Profitable house, 1908 – 1911, architect V.M. Uglichinin).

Fig.4. The visibility pan of the cultural heritage object– Simonov Monastery Ensemble (the cultural heritage object of federal significance). Address: Moscow, st. East, possession 4.

 https://doi.org/10.1051/e3sconf/20183301044

Fig.5. Predicted visibility zones of the high-rise building

At the stage of expert appraisal of the urban-planning situation, the main task of visual-landscape analysis is to verify the urban planning (volume-planning) characteristics and determine the permissible height parameters of the object under study [16,19].

In order to determine the permissible range of high-rise building parameters and to perform the visual-landscape studies, a 3D model with maximum height and volume parameters are presented (Fig. 6,7).

Fig.6. Photo modeling of capital construction object with an indication of the possible parameters of space-planning solutions. Moscow, 3rd Avtozavodskaya passage, possession 13.

Fig.7. Photo modeling reflecting the influence of high-rise building on the perception of cultural heritage objects

https://doi.org/10.1051/e3sconf/20183301044

Despite the above advantages, visual analysis is not the only factor determining the direction of the urban landscape development and the citywide panoramas formation principles. It is known that the results of the visual perception examination, with the determined degree of the object's compliance of the urban environment, are not always taken into account when making decisions on architectural councils, as well as the concept of "urban error" does not have an unambiguous interpretation, clear evaluation criteria and is subjective in its nature. Proceeding from this, determined the high probability of urban-planning errors appearance, which can be significantly reduced by the introduction of the visual-landscape analysis practice. Specific examples of the visual-landscape analysis application, as well as the consequences caused by non-compliance with the recommended height restrictions, allow us to express confidence that the positive vector prevails in the evaluation of this method [18].

Having analyzed the materials of the Committee for the state preservation of historical and cultural monuments (KGIOP) "Visual-Landscape analysis of the "Okhta Center" [17] complex high-rise building influence on the silhouette of the historical center of St. Petersburg" the ambiguity of the examination conclusions can be confirmed. An analysis of the visual perception estimated from 25 points of the city center showed that the construction of the proposed high-rise building "does not affect" only 3 of the 25 selected positions, which cannot be considered safe for the historical prospects of St. Petersburg. However, the very fact of carrying out such a study made it possible to use fragments of examination materials to justify the possibility of erecting a high-rise object in the historically developed low-rise zone based on partial presentation of materials. It is a question of the discrepancy between the space-planning parameters of the designed high-rise building for regulatory requirements and the restrictions of the territory urban development.

Fig.8. High-rise "Okhta Center" building, Saint Petersburg (a - view from the Sverdlovskaya embankment, b – view from Smolny prospect, c – view from the Proletarian dictatorship square)

The results of visual-landscape analysis recommended for accounting in the framework of urban-planning activity are oriented to the successive development of the spatial planning and typological structure of the territory, and the preservation of the compositional influence of valuable fragments of the historical environment in the urban landscape structure. Information obtained in the context of visual-landscape analysis is the basis for determining the urban-planning characteristics of the projected object and the realization of the reconstructive resource of the territory [7,10].

Visual-landscape analysis of the projected object carried out within the development framework of justifying materials for consideration of the further design issue in the selected territory. Conclusions of visual-landscape analysis considered when preparing project documentation for new construction sites, reconstruction of existing buildings and taken into account when reviewing and agreeing on project documentation.

4 Discussion

It is necessary to emphasize that conclusions and recommendations of visual-landscape analysis are not proposed to be absolutized. The degree of rigor and level of validity of the high-rise restrictions depends on many factors, such as the concretization of the urban-planning situation, the researcher's forecast about the ways of possible development of the urban landscape, previously approved project solutions.

One of the most significant problems directly related to the criteria for justifying high-rise constraints in the conduct of visual-landscape analysis is the anticipated development of composition-scale and silhouette characteristics of the urban environment. Gradual distribution of hyper-scale building complexes will lead to structural changes in citywide panoramas and the emergence of new conditions for the perception of the central part of the city. This problem requires joint development of conceptual ideas about the ways of possible urban landscape character development [14,20].

At the same time, adjusting the height parameters of the object based on the conclusions of visual-landscape analysis, optimal for the traditional urban environment, can lead to an undesirable change in the proportional structure, a decrease in the stylistic and imaginative expressiveness of the projected building, which means to promote the adoption of averaged decisions.

In this connection, it is necessary to distinguish that the recommendations made by experts in the field of visual-landscape analysis cannot be considered only true, but should be taken into account when considering project documentation on architectural boards.

In this respect, the preliminary height constraints at the pre-designed stage can be adjusted for the specific design solution. Such an approach is methodologically justified in those cases when the height constraints caused by the need to observe the compositional-spatial correspondence of the object to the structure of citywide panoramas [21].

Recommendations for preserving the composite unity of particularly valuable city-forming objects in the urban landscape structure have specific indicators, expressed in permissible absolute elevations. Conclusions of visual and landscape analysis are subject to mandatory compliance in those cases where the projected volume causes distortion of the historical and cultural monuments perception caused by a reduction in visual disclosure or a violation of background characteristics [3,21]. The regulation of the height parameters of the designed buildings in order to ensure the composition-scale compliance of the new volume with the valuable urban-planning environment is of an advisory nature. Expressiveness, quality and persuasiveness of the architectural solution of the object can substantiate the expediency of correcting the high-rise restrictions established in the framework of visual-landscape studies.

5 Conclusions

The current city regulating mechanisms combine data from general regulatory documents with detailed visual-landscape studies within the confines of a developed system of procedural approvals and examinations.

Despite the excessive construction activity, rapidly changing the face of the city, Moscow still retains especially valuable, in the true sense, peculiar fragments of the urban landscape with a high concentration of historical and cultural monuments. In connection with the fact that one of the main problems of urban-planning protection in the capital is the active reduction of the number of authentic cultural heritage objects, it is required to rethink such design and construction practice and take adequate measures corresponding to the status of the historic city [2,10]. The rehabilitation of especially valuable territories, the low level of the reconstructive resource of which compensated by the high historical and cultural significance, must be implemented within the confines of strict urban planning regulations with the predominant use of conservation methods and methods of rehabilitation.

Less valuable fragments in the historical and cultural sense of the city, including in the central part of it, may have increased the city-forming potential, allowing more free development without compromising cultural heritage. Forecasting of urban-planning activity using visual-landscape analysis methods, topical and in peripheral territories, allows to reveal the maximum reconstructive resource of a given site, and also to exclude serious composite miscalculations, by establishing the degree of conformity of the volume-planning solution and the high-rise parameters of the projected object in the structure of city panoramas [12,20].

References

1. L. Dupont, K. Ooms, M. Antrop, V. V. Etvelde, Landscape and Urban Planning **167**, 325-338 (2017)
2. S. Pardo-García, M. Mérida-Rodríguez, Computers, Environment and Urban Systems **61**(A), 56-65 (2017)
3. R. Wang, J. Zhao, Z. Liu, Urban Forestry & Urban Greening**20**, 210-217 (2016)
4. H. Jang, J. Kang, Energy and Buildings **158**(1), 656-667 (2018)
5. J.M. Lee, W. W. Braham, Ecological Modelling **363**(10), 157-171 (2017)
6. F. Peng, M. S. Wong, H. C. Ho, J. Nichol, P. W. Chan, Building and Environment **123**, 649-660 (2017)
7. M. M. Shanoer, F. M. Abed, The Egyptian Journal of Remote Sensing and Space Science **2**, (2017)
8. J. de Beer, F. Boogaard, Procedia Engineering **209**, 34-41 (2017)
9. V. Kutut, Procedia Engineering **208**, 69-75 (2017)
10. V.A. Akristiniy, E.A. Dikova, Real Estate: Economics **3**, 37-45 (2017)
11. M. Peniüa, Procedia Engineering **117**,883-890 (2015)
12. A.A. Kuzmichev, V.F. Loboyko, Procedia Engineering **150**,2095-2101 (2016)
13. V.A. Stolbova, L.I. Pavlova, Real Estate: Economics **2**, 76-84 (2015)
14. M.Kisternaya, V.Kozlov, Journal of Cultural Heritage **13**, 74-78(2012)
15. R.Vicente, T. M. Ferreira, J.A. Raimundo, M. da Silva, Journal of Cultural Heritage **16**, 1-14 (2015)
16. R. Douglas-Jones, J. J. Hughes, S. Jones, T. Yarrow, Journal of Cultural Heritage **21,** 823-833(2016)

17. All-Russian society of protection of monuments of history and culture. Saint-Petersburg city office. Analysis of materials of the landscape-visual analysis of the possible impact on the historic centre of Saint Petersburg, high-rise structures design of public and business complex Okhta-Center, (2009)
18. O.Risbol, C.Briese, M.Doneus, A.Nesbakken, Journal of Cultural Heritage **16**, 202-209 (2015)
19. N. Othman, N. Mohamed, M. H. Ariffin, M. A. Waliyuddin A. Razak, Procedia - Social and Behavioral Sciences **170**, 442-451 (2015)
20. M. Pazhouhanfar, M. Kamal M.S., Urban Forestry & Urban Greening 13(1), 145-151 (2014)
21. S. Garré, S. Meeus, H. Gulinck, Landscape and Urban Planning **92(**2), 125-135 (2009)

The Impact of High-Rise Buildings on the Living Environment

Botir Giyasov[1*], *Irina* Giyasova[2]

[1]Moscow State University of Civil Engineering, Yaroslavskoe shosse, 26, Moscow, 129337, Russia
[2]Tambov State Technical University, Sovetskaya St. Tambov. 392000. Russia

Abstract. Urbanization as a socio-economic process manifested in the concentration of the population in modern big cities contributes to the development of high-rise building construction. With the development of education and culture, changing leisure habits, city residents put forward new architectural and functional requirements to the living environment and urban infrastructure. This calls for the creation of new types and forms of residential buildings, the structure of the city and transport networks. In addition, the need to develop high-rise building construction is justified by the growing demand for residential, public and administrative buildings and the lack of free space.The paper analyzes the development of high-rise building construction in urban areas. The problem of the impact of high-rise building construction in big cities on the living environment is considered. Using analytical methods, causes and sources of pollution, such as transport and engineering infrastructure have been identified. In some urban areas, there are zones with modified thermal conditions and air exchange resulting in the formation of the "urban heat island"The qualitative and quantitative characteristics of variations in temperature and wind speed with respect to the height of the building have been calculated, using the example of the Evolution Tower of the Moscow International Business Center ("Moscow City"). Calculation and comparative analysis for the cities of Moscow, Khanty-Mansiysk and Vladivostok has made it possible to assess the variation in temperature and wind speed and their impact on the living environment under different climatic conditions.

1 Introduction

The rapid growth of the economy of the leading countries is contributing to the development of big cities. In connection with intensive internal and external migration of people to big cities, the population density is increasing. As a result, the number of modern multi-storey buildings and high-rise buildings is increasing, and the engineering and transport infrastructure is developing at an incredible pace. A modern metropolitan area is a living environment that combines a natural environment, a high-tech production environment, an intensive transport environment, a recreation environment and a high-density residential environment.

* Corresponding author: dandyr@mail.ru

The growth of modern high-rise buildings significantly affects the existing climatic conditions of the terrain and the environmental balance of the living environment. At the same time, the density of urban development, the infrastructure and transport networks play an important role in changing the living environment. Urban development as a combination of complex architectural forms significantly affects the aerodynamics of the terrain. In this regard, both industrial areas and high-density residential areas are environmentally unfavorable. Thus, high-rise buildings and structures, being an integral part of a modern city, significantly aggravate the environmental conditions of urban areas.

One of the important components of the living environment of a city is the residential environment. The unified system "man - apartment - building - neighborhood - residential area of the city" defined in the scientific literature as "residential environment" has complex features and mechanisms. A human being, interacting with the residential environment, performs non-productive activities on the territory of populated areas. In big cities, the development of the living environment, mainly consisting of multi-storey buildings, high-rise buildings and modern public spaces, causes variation in temperature and wind conditions of the terrain and aggravates the environmental situation [4].

High-density residential areas of the city, being the most important part of any metropolitan area, where comfortable living conditions are created for their residents, are exposed to the constant impact of pollution sources. The main sources of air pollution in residential areas include industrial enterprises, heating boiler houses and road transport. The economic growth and the increasing pace of construction of high-rise residential buildings, public and business centers in big and metropolitan cities, have resulted in an increase in business activity and use of motor transport. Motor transport as a mobile source of pollution is the main source of air pollution in residential areas. Due low location and close proximity to the residential area, motor vehicles create extensive and stable zones of pollution, with the concentration of pollutants in the air exceeding the permissible level by several times.

A modern city with high-rise buildings is a fairly effective system of heating the vertical surfaces of buildings and horizontal surfaces of roofs, sidewalks and roads with solar heat [8]. At the same time, the high-density construction of certain areas hinders natural aeration and causes the formation of urban heat islands that affect the pollution of the air basin. (Fig. 1) [10].

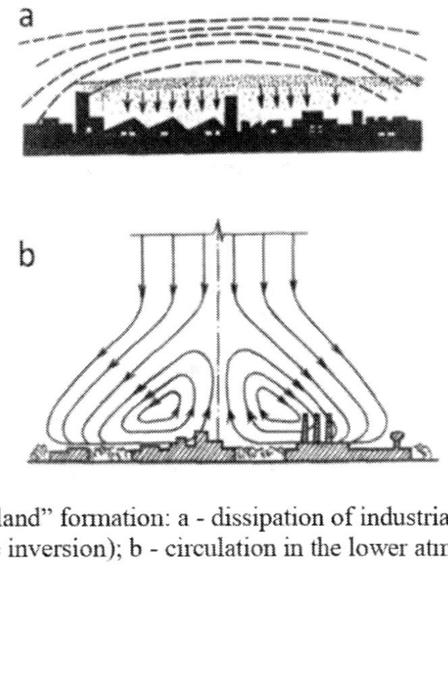

Figure. 1. "Urban heat island" formation: a - dissipation of industrial emissions in the "urban heat island" (deep surface inversion); b - circulation in the lower atmosphere over the city.

In addition, the environmental situation is negatively affected by the stagnation of air masses caused by natural and climatic factors, such as no-wind conditions, surface temperature inversion (Fig. 2) [10]. Temperature inversion in some regions can occur throughout the year.

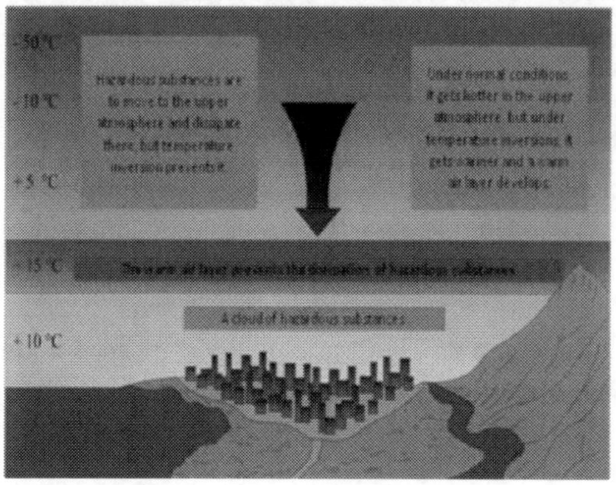

Fig. 2. Surface inversion of temperature.

Public space in residential areas is a living environment and one of the most important parts of the city's air basin. The natural wind regime in public residential areas of a modern city is affected by high density of high-rise buildings, and the air has a higher concentration of pollutants due to lack of the air exchange. Getting into the lungs of urban residents together with the inhaled air, pollutants modify gas exchange and cause oxygen starvation, thereby causing asthma or its aggravation and other lung diseases. Thus, the issue of aeration of areas with high-rise buildings is becoming increasingly important. In this regard, the most interesting are the convective currents that arise when the temperature difference between the surfaces of facades of buildings and public spaces.

In the city, the differences in the heating of open and shaded sections of streets and public space influence the local air circulation. Ascending flows concentrate above the surface of the walls exposed to light, and descending flows concentrate above the shaded walls [3]. The presence of water bodies in the cities contributes to the development of local circulation from the water body to the urban areas at daytime, and in the opposite direction at night. The speed of air flows at the outer surface of buildings, due to the difference in the temperatures of the surface of the enclosing structure and the outside air, can reach up to 10 m/s and plays an important role in the air exchange of the adjacent space (Fig. 3) [1]

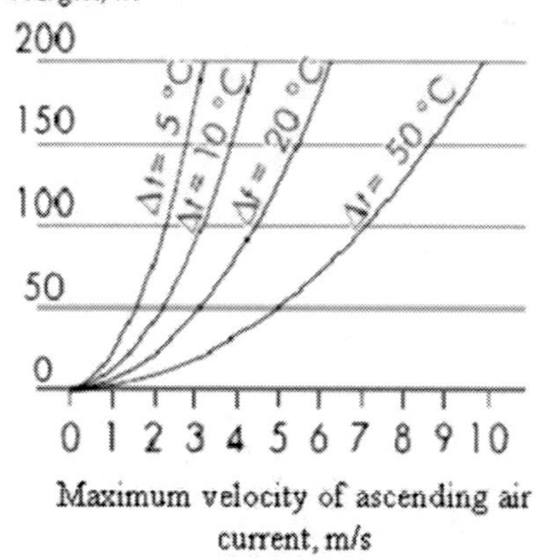

Fig. 3. Dependence of the speed of ascending flows on the height.

 https://doi.org/10.1051/e3sconf/20183301045

In the warm period of the year, the temperature of the exterior surfaces of buildings increases sharply due to irradiation with solar radiation and differs significantly from the outside air temperature. The temperature difference contributes to the formation of convective heat flow directed up the building, resulting in the emergence of the so-called near-surface (boundary) layer of heated air, (Fig. 4) [1,7,9]

The amount of solar radiation and the coefficient of solar radiation absorption by the material of the outer surface of the enclosing structures creates a temperature difference between the external surface of the building and the surrounding air.

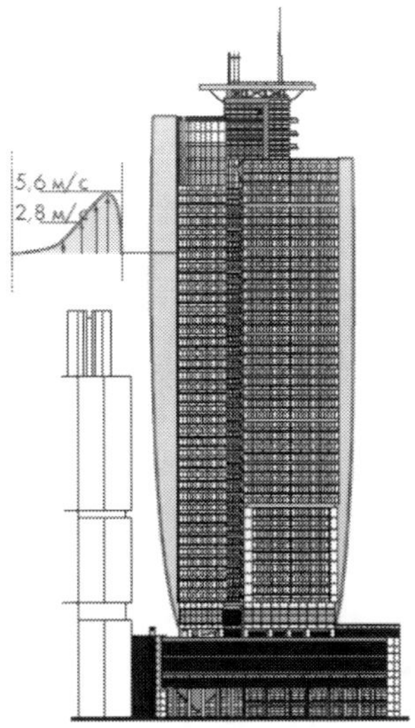

Fig. 4 An example of the velocity profile of ascending air currents at the outer surface of a high-rise building due to the effect of a temperature difference Dt = 20°C.

2 Materials and Methods

The analysis and the study of the Moscow City International Business Center revealed that compactly located towers have transformed the wind regime of the area. The rapidly developing transport and engineering infrastructure seriously exacerbates the environmental situation. Due to the density of the high-rise buildings, the Moscow City Business Center and the intensity of transport communication on the territory, there appeared extensive stable zones with stagnation of the air, where wind speeds range from 0 to 1 m/s. Thus, the territory of the Moscow City Business Center has created its own local climate, with its specific microclimatic conditions on its streets and squares. This microclimate is determined by high-rise urban development. According to our research, the formation of local microclimatic conditions is facilitated by a change in the height of buildings, air temperature and wind speed.

Using the example of the Evolution Tower, the calculation of the temperature of the outside air, atmospheric pressure and wind speed depending on the height of the building has been made.

The variation in the outside air temperature and atmospheric pressure depending on the building height can be determined by the following formulas [3]:

$$t_h = t_0 - 0{,}0065 \times h, {}^0C \tag{2.1}$$

$$p_h = p_0(1-2,25577\times10^{-5}\times h)^{5,2559}, Pa \quad (2.2)$$

where t_h, p_h is temperature, °C, and pressure, Pa, at the height h, m, respectively;

t_0, p_0 is temperature, °C, and pressure, Pa at the earth's surface, respectively.

3 Results

The calculated climatic parameters according to [4] are determined for the cities of Moscow, Khanty-Mansiysk and Vladivostok.

Using the data obtained, the graphs of the variation in the outside air temperature along the height of the building are constructed (Fig. 5,6).

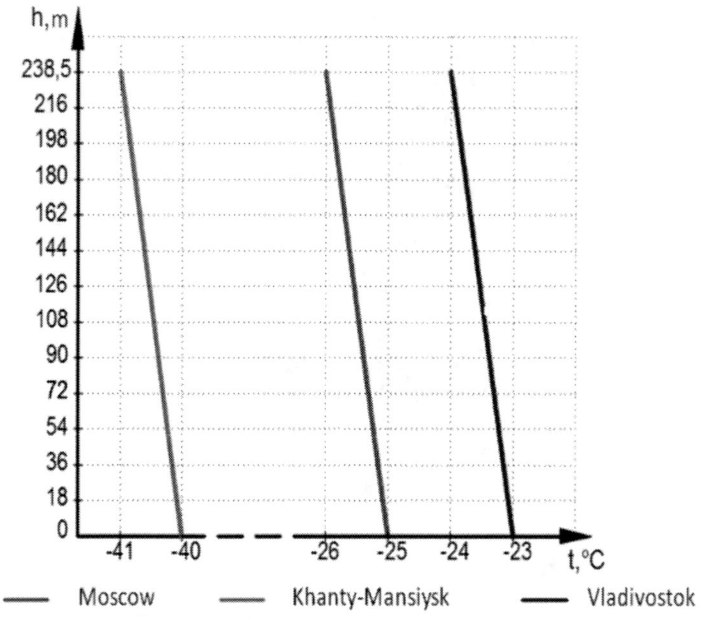

Fig. 5 - Variation in the air temperature depending on the height of the Evolution Tower during the coldest five-day period for the cities of Moscow, Khanty-Mansiysk and Vladivostok.

Various models are used to estimate the variation in the wind speed with respect to the height: the Ekman spiral, the logarithmic law, the power law. [5,6] These models estimate the wind speed V at height h if the wind speed V_0 is known at height h_0.

The power law of variation of the wind speed with respect to height has the form [5,6]:

$V_h = V_0(h/h_0)^a$, m/s (2.3)

where V_h is wind speed, m/s, at height h, m;

V_0 is wind speed, m/s, changing at the height h_0, m (wind speeds are measured at a height of 10–15 m, therefore h_0 = 10—15 m);

α is exponent depending on the type of terrain and found experimentally; it is recommended for centers of large cities to take $\alpha = 0,33$, for suburban areas α = 0,22, for open space α = 0,14 [6].

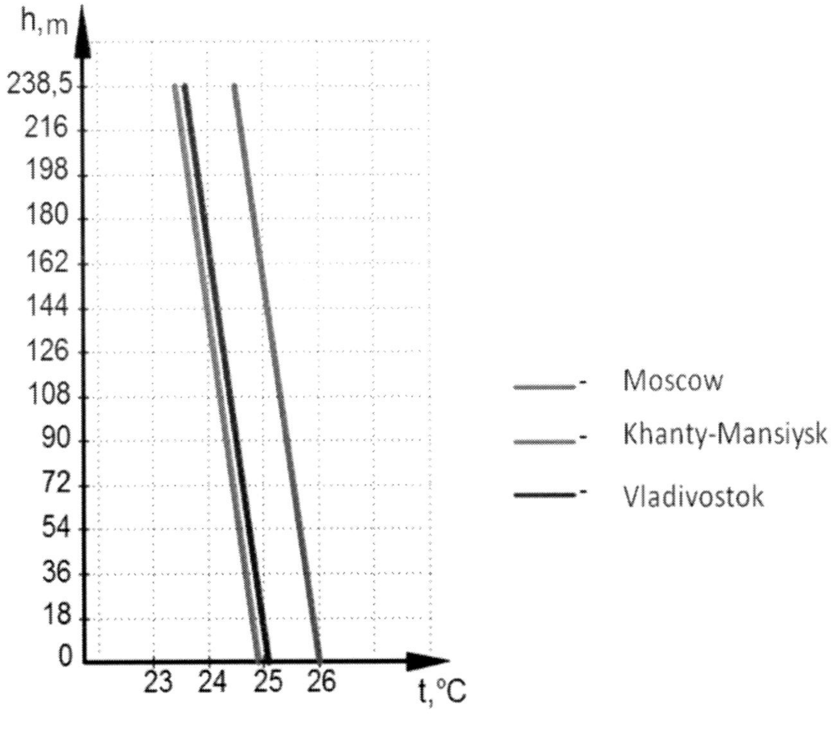

Fig. 6 Variation in the air temperature depending on the height of the Evolution Tower during the warm period for the cities of Moscow, Khanty-Mansiysk and Vladivostok.

The values of wind speed at a height of 10 m are assumed to be equal to the values from Tables 3.1 and 4.1 [4], for the city of Moscow for the warm and cold periods of the year. Based on the results of the calculation, a graph of the variation in wind speed with respect to the height of the building was constructed (Fig.7)

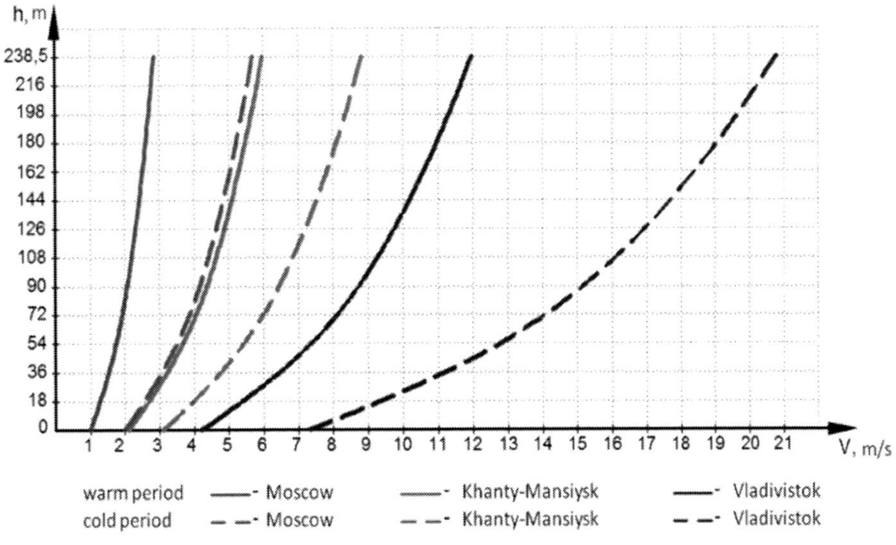

Fig. 7 Variation in the wind speed with respect to the height of the Evolution Tower the cities of Moscow, Khanty-Mansiysk and Vladivostok

As can be seen from Fig.7, with the increasing height of the building, the wind speed increases and in urban development areas this occurs more intensively.

4 Conclusions

From the findings of the study, the following conclusions can be drawn:
1. Modern urban areas with high-rise buildings have an impact both the local aerodynamics and the temperature conditions of the territory. As a result, environmentally unfavorable zones with stagnant air develop in public open spaces.
2. High-rise buildings in urban areas, being an effective system of solar heating of the vertical surfaces of buildings and horizontal surfaces of roofs, sidewalks and roads, accumulate heat and result in the formation of urban heat islands.
3. When designing high-rise buildings, it is necessary to take into account the specifics of the built-up areas, the spatial plasticity and the density of construction, which affect the aerodynamics of the terrain and the pollution of the atmospheric air.
4. The calculations and studies made it possible to evaluate the impact of high-rise buildings on the microclimate and the environmental conditions. The variation in wind speed with respect to the height of the building, the formation of ascending convective flows at the outer surface of a high-rise building are the factors that can improve the environmental situation and the microclimate of the living environment.

References

1. Ju. A. Tabunshhikov, N. V. Shilkin *Ajerodinamika vysotnyh zdanij* (AVOK, 2004)
2. B.I. Gijasov, Vestnik MGSU **3**, 13-15. (2012)
3. B.I. Gijasov , Vestnik MGSU **4**, 17-21. (2012)
4. Russian Standard SP 131.13330.2012.
5. Je.Simiu, R.Scanlon, *The impact of wind on buildings and structures* (Strojizdat, 1984)
6. ASHRAEHandbook. Fundamentals. SIEdition,1997.
7. Ju. A.Tabunshhikov, M. M.Brodach, N. V. Shilkin. Energy efficient buildings (AVOK-PRESS, 2003)
8. E.G. Maljavina, S.V.Birjukov. Calculation of the air regime of multi-storey buildings with different air temperatures inside the buildings, AVOK, 2 (2008)
9. O.D. Samarin. Zhilishhnoe stroitel'stvo, **1**, 32 – 33. (2013)
10. S.P. Hromov, M.A. Petrosjanc. *Meteorologija i klimatologija*. (Nauka, 2006)

Architecture of Stalingrad: the image of the hero city by the language of "Stalinist Empire style"

Galina Ptichnikova[1, 2*], *Alexey* Antyufeev[2]

[1]The Scientific Research Institute of the Theory and History of Architecture & City Planning, Moscow, 111024, Russia
[2]Volgograd State Technical University, Volgograd, 400005 Russia

Abstract. The article is dedicated to the peculiarities of the architecture of "Stalinist Empire style" by the example of Stalingrad architecture. This city was restored again after Stalingrad battle which took place in the period of 1942-1943, in accordance with the principals of socialist urban construction. Reconstruction of Stalingrad is considered by the authors of the article as the creation of a new type of a Soviet city – a Hero City. The authors reveal the artistic features of architectural ensembles and buildings designed according to the principles of "Stalinist Empire style".

1 Introduction

Post-war reconstruction of Stalingrad became an important stage in the architectural life, not only for the city, but for the whole country — URSS — as well. The architects were supposed to show the contours of a happy future life appearing through a harsh daily routine. Creation of visual artistic images of the city was decided by means of Soviet neo-classics, which in the post-war period obtained peculiar characteristics which allowed them to be distinguished in a style that is now referred to as "Stalinist Empire Style". The present article is dedicated to the peculiarities of form making of architecture of Stalingrad during the second half of the 1940s –1950s.

2. Search for a new style

Starting from the year of 1943, immediately after the battle of Stalingrad and liberation of the city by the Soviet forces, the Soviet Academy of architecture, the Institute of Urban development and Architectural planning of Russian Soviet Federative Socialist Republic (RSFSR) embarked on development projects of Stalingrad reconstruction.

The important ideological requirement for the establishment of an appropriate architectural and artistic image of the reborn Stalingrad was the high value of the city in the "table of ranks" in the hierarchy of Soviet cities of that time (In accordance with the Decree of the Presidium of Armed forces of the RSFSR, September 4, 1945 the city of Stalingrad was singled out into independent administrative and economic center with its own budget and is classified as a city of republican subordination ("Vedomosti" of the USSR, 1945, 30

* Corresponding author: ptichnikova_g@mail.ru

November, no. 79 (406). p. 2). Artistic image of post war mentality contributed to the idea of Stalingrad reconstruction as the creation of a new type of the city — a "Hero-City". Two main themes, influencing the search of the architectural image of a "Hero-City" were the theme of the triumph of victory in the great patriotic war and the theme of creating objects, monuments of Soviet history events. The feeling of absolute exclusiveness and importance of its universal values explains the desire of architects to capture these values in the ensembles of the city-monument and extrapolate them into the future.

Another important factor that influenced the creation of a new architectural look of the city, was that Stalingrad had been , in fact, the City – Symbol, the city of the leader, the city of the head of the State, named after him, the city with history, connected with destiny and related victories of its leader. The Chief Architect of Stalingrad P.I. Buchnev, describing the General city planning scheme of 1943, stressed the role of architecture, meant to be compatible with the name of the great Leader: "The envisaged reconstructive activities, updated residential and public development, a number of new architectural ensembles will turn Stalingrad into an exemplary well-arranged, genuinely socialist city, worthy of the great Stalin" [1]. It is worth mentioning that toponymy of Stalingrad was based on the name of the leader of the Soviet State. The main street of the city was named after I.V. Stalin, also there were Stalin Embankment and Stalin District. Among the proposals on the creation of the city center the possibility of the construction of the Palace of Stalin had been considered [2, p.41]. Many monuments to the leader were built in the city including the world's highest bronze sculpture of Stalin which was constructed in 1952 (Fig.1). Thus, the search for architectural solutions in restoration and reconstruction was made to be aimed at identifying the special status of the city.

Fig.1. Stalin's Monument at the entrance to the Volga-Don channel (dismantled). Sculptor E. Vuchetich. Photo by L. Nyman. 1950s.

The style, most in response to the decision of such kind of ideological tasks, was the style common in Soviet post-war architecture based on classical heritage which was called "Stalinist Empire style". Architecture was perceived as a mean to approval of morality, heroism, citizenship. The mission of propaganda of the deed was no less important than the various technical tasks of restoring of environmental assets. "Architecture of Stalingrad must be of high quality, the heroic spirit of the city, achievements and victories of the Soviet people under the leadership of the Lenin-Stalin party should be reflected in public buildings by means of the plastic arts", - this is the way the main artistic challenge for master plan designers was described by V.N. Simbircev, appointed the Chief Architect of Stalingrad in 1944 [3].

3 Ensembles of the center of Stalingrad

Ensemble style, monumentality and artistic integrity of buildings have become the distinguishing features of not only the center but also of other areas of Stalingrad, built in the style of post-war architecture. The idea of the architectural ensemble in those years perceived as the main in Soviet architecture. The city as an object of an artwork was built through a system of ensembles, consecutively located one after another and strung on axle of the main city arterial road — Stalin Avenue (Lenin Avenue at present). This clear system had become the foundation, which helped to bring the city arrangement together.

However, the implementation of the principle of ensemble in practice caused some difficulties. Natural development of actively recovering city did not fit into the principles of perfect ensembles statics. Therefore, no wonder that while working on the master plan of Stalingrad in 1948-1954, proposals for changing the boundaries of the city by means of its reduction had been considered [3; 4]. Due to the apparent impossibility of creating of a unified ensemble of the city which stretched almost 70 km along the Volga River by that time, the question of the reduction of the city territory and removal of the southern areas of Stalingrad from a planning "carcass" so as to form a separate city of Krasnoarmeysk was raised. Thus, the desire to create a city as a single compositional system resulted in the search for the architecture and town planning decisions concentrated on relatively small territory in comparison with the total area of the urban organism. By 1946 a drafted detailed design of the center planning of Stalingrad was developed (K.S. Aalabyan, N.H. Polyakov, D.M. Sobolev, A.E. Pozharsky, Butyagin V.A., etc) (Fig.2).

Fig.2. The project of the central part of Stalingrad.

Draft design of the central part of Stalingrad, executed under the direction of K.S. Alabyan, consisted of all the principal statements subsequently developed and fixed in the master plan of the city [5]. The central part of Stalingrad was built as a system of squares and avenues running perpendicular to the main arteries of the city. The area of the square of Fallen Fighters which was visually connected with the square of Demonstrations in front of the planned House of Councils acquired the dominant value. Through a wide Alley of Heroes, this area was connected with a new square of Victory at the Volga River Embankment. Descent to the river ended with a monumental staircase. In accordance with the principles of the ensemble in a socialist city there should necessarily be such elements as "gates to the city". The propylaea became the river gates to Stalingrad from the Volga side (architects: V.N. Simbircev, I.E. Fialko, V.S. Makarenko) (Fig.3). The image of the propylaea is similar to ancient designs. The propylaea, built in 1952, was established by two Doric porticos consisted of eight columns and arranged symmetrically along the axis of the direction of the stairs going down to the lower level of the embankment and the river. The propylaea pass through the main compositional axis perpendicular the Volga, pointing to the infinity of Volga horizon.

Fig.3. Propylaea stairs on the embankment of Volgograd. Arch. V. Simbirtsev. I. Fialko, V. Makarenko. Photo by G. Ptichnikova.

The central part of Stalingrad almost fully met strict theory of Soviet urban planning of the post-war period, which led to the conceptualization on how to build the city. For example, "seven conditions for the establishment of the city", which were listed by A.G. Mordvinov in 1945 at all-Russian conference of chief architects. [6, p.33].

The first condition is the link of the city with natural environment, revealing its beauty. In accordance with the new general plan the main artistic and logical focus of the center of ensemble was transferred to the shore of the Volga River, where the new town's main square was meant to link the city with the river [7]. River views are rhythmically opened through the transversal deep perspective. Compositionally built green embankment gave the opportunity to observe Stalingrad in all its artistic integrity from far points of the river.

The second condition was the establishment of a clear compositional center. It was executed by the main street — Stalin Ave, which through the Alley of Heroes was connected with central squares — Fallen Fighters square, newly designed "Square of Victory" and Railway Station forecourt (Fig.4, 5).

Fig.4. Railway Station forecourt. Arch. A. Kurovsky, S. Briskin. Photo by L. Nyman, 1954.

Fig. 5. Detail of the entrance of the railway station. Photo by G. Ptichnikova.

Principles of ensemble, light coloring of buildings and some other characteristics correspond to seven conditions. Unfulfilled was the only condition – establishment of the largest public building with high-rise urban composition in the main architectural node – the House of Council. After numerous competitions in 1952 L.V. Rudnev and V.O. Munz developed a project in accordance with which the House of Council was located exactly on the axis of the esplanade leading from of the Volga. The draft project was similar to the composition of Moscow high-rises and was practically "twin brother" of the House of Science and Art in Warsaw. However, due to a number of reasons the construction of the House of Council, in accordance with this project had not been initiated. Difficult searches for compositionally shaped characteristics of the main ensemble in Stalingrad were interrupted and were not completed. The symbol building, which ought to reflect the Stalin epoch, remained a monument of "paper architecture". Searches continued until the late 1980s and stopped in connection with the decision of the authorities to restore the Cathedral of Alexander Nevsky, which had been destroyed in 1932.

Stalin Avenue, stretching nearly twenty kilometers, acquired special significance as the main compositional axis, linking the city center with its suburbs. According to the development pattern the avenue was divided into several parts. The facades of the houses along the Avenue of Stalin, built in the early 1950's, served as a kind of theatrical scenery and background settings for demonstrations and festivals. The avenue development was carried out in accordance with the design of Stalingrad architects, among which one should mention A.V. Kurovskiy, M.F. Hohriakova, Z.D. Gorbushin, F.M. Lysova, D.V. Ershov, V.P. Statun, G.F. Borysenko. [7; 8].

4 "Empire" working areas

In the "Stalinist Empire style" in order to create an atmosphere of pageantry and splendor, the architecture of the Metal City (as northern areas of Stalingrad were called) had been solved. Development of streets and avenues of the working districts near "Stalingrad Tractor" plant, "Red October" plant and "Barricades" plant is characterized by the tower type, balustrades, bas-reliefs and order system. Great attention was paid to plastic art, giving the southern coloring to the developed territory by using bay windows, stanzas, verandas and balconies.

Especially highlighted are lateral streets leading from the depths of the settlements-working class areas to the main parade city road — Dzerzhinskiy St. , Shtemenko St., Metallurgists Ave, etc. (Fig.6). The houses, overlooking the crossroads, are crowned with towers, "crowns", sculptural stucco. Especially noticeable for its completed image, characterized by art deco, is a nine-storied building (93 Lenin Ave.), located at the intersection of two avenues — Stalin (Lenin) Ave. and Metallurgists Ave.

Fig.6. Prospect Metallurgists. Photo by L. Naiman.1950s.

Spatial axis of the main avenue is rhythmically counted by squares, emerging at the intersections with lateral streets, as well as squares close to plants. Dzerzhinsky square, spreading in front of the entrance of "Stalingrad Tractor" plant, is considered to be the end of the Avenue. The eastern side of the square is formed by entrance posts of tractor plant, completed in 1954 by Igor Nikolaev project and the building of plant management.

5 Sources of inspiration

The language of the archetypes was the basis of the architectural and artistic manner in post-war Soviet architecture. The search for new techniques of form making were very difficult, because external manifestation of form had to be recognizable, capable to generate historical associations already familiar to the audience , while ideologically finally oriented.

Several sources can be distinguished in the space of the archetypes of architectural forms of Stalingrad. It is, first of all, Russian classicism, which was supposed to reflect the presence of Russian national traditions in the Soviet town. There are many public buildings of that kind in the central part of the city. Among buildings, with the stylistic features of Russian classicism, an object of industrial architecture, known as the administrative building of city water supply ("urban pumping station"), should be mentioned. The building resembles rather a Russian mansion house (architect B.A. Gekker). Rotunda - pavilion with spectacular observation of the Embankment can be considered as one of the images of Russian classics (architect I.R. Arutyunova). Another powerful sources of form making were the examples of Italian Renaissance. The buildings, in which there are transversal, flexibly changing architectural themes of the Italian Renaissance, can be seen in the center of Stalingrad.

Among them – the buildings of the regional Party School, (architects V.N. Simbirzev, E.I. Levitan), the buildings of the hotel "Stalingrad" (architect A.V. Kurovskiy, V. Postoenko) and "Intourist" (architect B.G. Goldman), the administration building of Hydrostroy (architect E.I. Levitan), the building of the State Bank (architect E.I. Obuhov) (Fig.7).

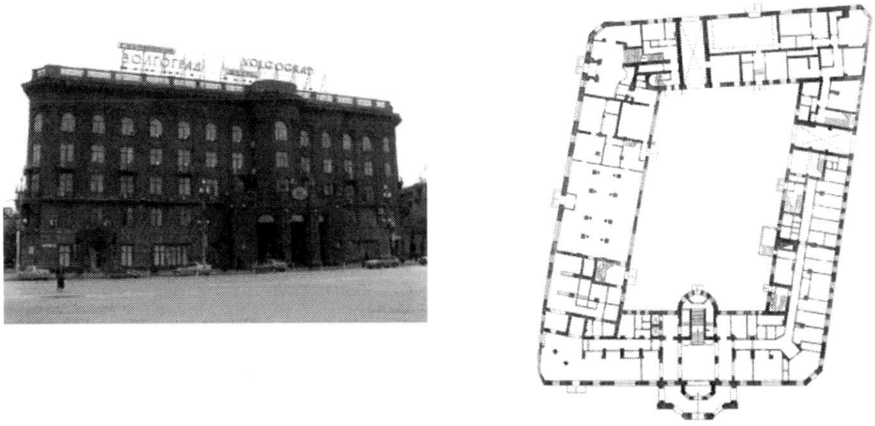

Fig. 7. Hotel "Stalingrad" (at present "Volgograd"). General view, plan., Arch. A. Kurovsky, V. Postoenko. Photo by G. Ptichnikova.

The architecture of the buildings is based on using common techniques of layers on wall façade and deep rustication on the two lower floors. Crowning of the walls in these buildings is decorated with Corinthian pilasters with elements of Soviet symbolism; moved forward cornices, balustrades on top of the buildings, arches, bas-reliefs with vases filled with bouquets of wheat and fruit, decorated components such as garlands of flowers were actively used. The building of the Regional Party School shows clear monumental composition, facades are made of a clear rhythm of pilasters, window openings and entrance stanzas with the contrast of upper smooth and low rusticated floors [9] (Fig.8). As a tribute to the tradition above the main entrance of the Regional Party School, bas-reliefs of classics Communist leaders - Marx, Engels, Lenin and Stalin were installed. The building set the tone of the entire developed area of Fallen Fighters square, defining its scope and nature (Fig.9).

Fig. 8. The building of the Regional Party School. Aarkh. Simbirtsev V.N., Levitan E.I. Photo by L. Nyman.1950-s.

Fig.9. Fallen Fighters square in Stalingrad. Photo by L. Naiman.1954.

"Great" order distinguishes buildings of the Regional Committee of the CPSU (architect E.I. Obuhov, A.S. Kulev) and the Palace of Labour (architect V.E. Maslyaev, F.M. Lysov) (Fig.10).

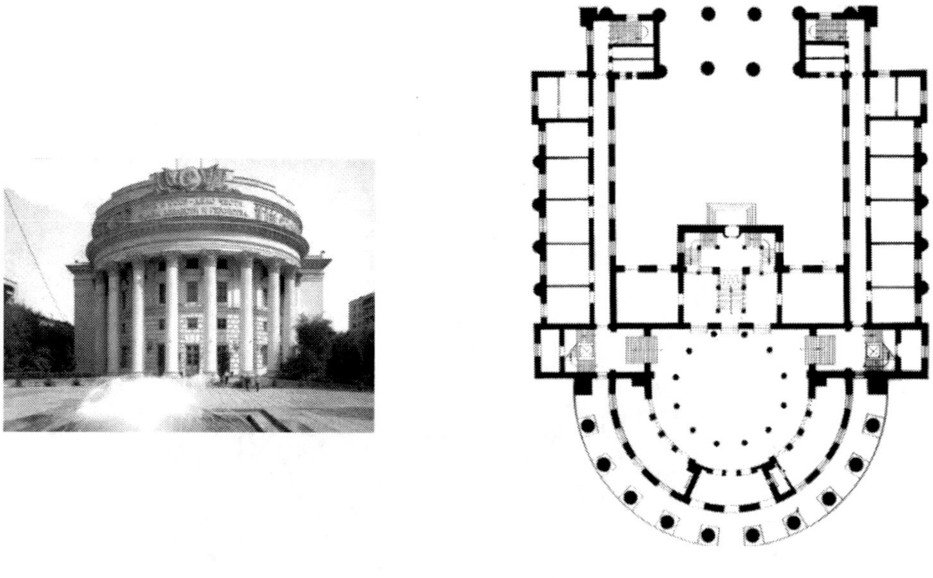

Fig. 10. Palace of Labour. General view, plan of the 1st floor. Arch. V. Maslyaev, F. Lysov Photo by N. Samsonenko, 1989.

The buildings are distinguished by the use of Corinthian order richly decorated with soviet symbols and developed entablature. The Palace of Labour (another name is the House of Trade Unions) had already been completed at the sunset of Stalin era in Stalingrad — in 1957, but on the attic of this building the words Stalin pronounced on the XVI Congress of the AUCP (b) in 1930 are still imprinted: "Labour in the USSR is a matter of honor, glory, valor and heroism". The entry section of the Palace of Labour is crowned by a sculptural composition with a facial profile of V.I. Lenin and dipped banners. Dripstones, interfloor eaves gutter, columns and pilasters of the Corinthian order are widely used in the architecture of both buildings.

"Empire shine" is visible in many buildings of post-war Stalingrad. One of the most striking examples is architecture of residential buildings of Sovetskaya Street (architects S.K. Kobelev, F.M. Lysov, G.A. Rossihin), which was basically completed in 1952-1956. The street is formed by residential buildings, which can be called without much exaggeration — Palazzo House. Large seven-storied buildings are divided into shelves and become various story, thereby simplifying large-scale external look. Facades are filled with sculptures, attics, reliefs in the shape of wheat bundles, large vases and cone firs (Fig.11). Some buildings have large two-storied arch-entrances to courtyards. The pompousness of the facades was projected on interior environment: apartments were complicated by front suites, bay windows.

Fig. 11. Sovetskaya Street, house number 28. Arch. S. Kobelev, F. Lysov. Photo by G. Ptichnikova.

A special case of using Italian motifs is a residential building No. 14 in Raboche-Krestyanskya street (architect A.V. Kurowsky) which was built in 1949. Venetian elegance and refinement are observed in its architecture. Forms of ways and shells are used in decoration of stanzas (Fig.12). The main façade of the building is decorated with bas-reliefs, the main motive of which are the symbols of the Volga region — whimsically curved sturgeon bodies.

Fig. 12. Residential building № 14, St. Raboche-Krestyanskya. Arch. A. Kurowsky. Fragment of the facade. Photo by G. Ptichnikova.

Antique designs form the basis of the development architecture of Mira street (architects K.N. Afanasyev, M.I. Sinyavsky, N.A Naumova, E.I. Levitan, A.I. Rubin, Y.N. Telenev), which became the first street of new Stalingrad, built in the post-war period. Mira Street was created as benchmark ensemble of a residential street with all its components — the beginning, the end of the composition, dominants and spatial pauses. Its length is 550 m, its width is 32, 5 m, height of buildings varies from 12 to 18 m. At the end of Mira Street, closing its perspective, in 1954, the Planetarium was built (architect V. N. Simbirtsev, M.A. Homutov) (Fig.13). The beginning of the street was completed much later (in the 1970 's.) by formation of the square near the Palace of Pioneers with picturesque asymmetric composition (architect E.I. Levitan, A.S. Leushkanov).

Fig. 13. The Planetarium. Arch. Simbirtsev. Photo by L. Nyman, 1954.

In the central part Mira Street overlooked the Demonstration square where the House of Soviets was supposed to dominate. Therefore, the street space created a ceremonial corridor for festive columns of thousands of demonstrations to approach the tribunes of the foot of the main building of the city. The main part of Mira Street development is represented by the quarters of three-five-storied houses with colonnades, open courtyards, courts of honor and fountains in the depths of the courtyards. Construction parameters, length, and number of stories in the houses; gaps between them were proportionately linked. The facades of residential buildings are formed by stanzas, galleries, sunscreens. The walls are decorated with sunflower rosettes and symbols of opulence: bundles of wheat, garlands and bouquets of flowers. The medallions depict figures of warriors and homefront workers, signs of Soviet symbolism — "hammer and sickle", five-pointed stars, as well as the attributes of military history.

Coloring of buildings has become a kind of town-planning guide in the planning structure of the Centre. Sovetskaya Street is noticeable for dark red ("Pompeian") and dark-gray colors. When building Mira Street, by contrast, the Russian classicism colors — yellow facades with white details were used. Construction of the city's main square (the Fallen Fighters square) is carried out in monochrome tones using different shades of grey or brown.

6 Conclusions

Analysis of features of reconstruction and restoration of Stalingrad for quite a short post-war period, allows following statements to be made.

Architecture in the post-war period was considered to be an effective mean of influence on mass consciousness which was performed by use of images of cultural heritage.

Experimental search of form making, associated with historic architecture and order system, was focused on the development of architectural language, stylistic means, which could express the emotional-ideological component — the pathos of the Victory of Soviet people in the Great Patriotic War and already foreseeable abundant future of the "Communism tomorrow" (Fig.14). In the post-war architecture of Stalingrad formal-aesthetic ideas and examples of Russian classicism, Italian Renaissance and ancient architecture were used as sources of inspiration.

In the post-war architecture due to the perception of the values of that time as universal, inviolable and immutable combination forms of classic past with possible forms of future, occurred.

As a rule ensemble of "Stalinist Empire style" is complete and self-contained. Implementation of one universal classic architectural and urban development model in post-war Stalingrad on a large scale predetermined certain behavior samples and preconceived mode of life of the population. Grand avenues and highways were intended to be used for demonstrations and triumphal processions, squares — for meetings, embankments- for walking. By means of "Stalinist Empire style", the "utopian" model of a Socialist City, was largely idealized and established in Stalingrad. The rigid development hierarchy of post-war neoclassic requires unconditional subordination of new objects to forged order. No wonder how painfully new objects which are introduced into urban space by the life of post-Soviet time are perceived by architectural environment of "Stalinist Empire style" [10].

Thus, the post-war architecture of Stalingrad was the complete embodiment of the ideas made of highly artistic phase of development of the domestic architecture of the 20th century. Its uniqueness lies in the creation of complex ensembles of post-war reconstruction with a unified ideological and stylistic direction as a monument to a great victory. A great advantage of these ensembles is harmonious proportion of space and height of the developed territory, its immensity to the natural landscape, the richness and expressiveness of form making tools, synthesis of architecture and sculpture.

Fig. 14. Detail of the facade of the railway station. Sculptural group "The Apotheosis of Labor". Sculptors M. Pavlovskii, N. Pavlovskaya, V. Beznukov. Photo by G. Ptichnikova.

References

1. *Characteristics of the chief architect of the mountains*. Stalingrad general layout of the city. July 29, 1943. GAVO.F.71.Op.1.D.638. L.99 (in Russian).
2. E.P. Zhorova. *Code of historical and architectural heritage of Tsaritsyn-Stalingrad-Volgograd* (Volgograd: The Panorama Publ. 2004)
3. V.N. Simbirtsev Conclusion of the chief architect "*On the main points of the new*

master plan. Stalingrad". April, 13, 1951 of the city Stalingrad. GAVO.F.71.Op.1.D.1252.L.23 (in Russian).

4. Remarks to the question "*The main provisions of the master plan. Stalingrad* "April 3, 1954. GAVO.F.71.op.1.D.1459.L.73-77. (in Russian).
5. V.M. Vodolagin, *Essays on the history of Volgograd 1589-1967.* Moscow: Nauka Publ. (1968) (in Russian).
6. Yu.L. Kosenkova, *Soviet city of the 1940s - the first half of the 1950s: from creative exploration to construction practice.* Moscow: Editorial URSS Publ. (2000) (in Russian).
7. V.I. Atopov, V.E. Maslyaev, A.F. Lypyavkin, *Volgograd.* (Moscow: Stroiizdat Publ. 1985)
8. A.M. Vyazmin, P.P. Oleinikov, *Architects of Volgograd.* (Volgograd: The Publisher Publ, 2003)
9. A.V. Zhuravlev, Yu.L. Kosenkova, *Vasily Simbirtsev.* (Moscow: Stroiizdat Publ. (1986)
10. G. Ptichnikova, Procedia Engineering (2016), **165**, pp.1903-1910.

Hybrid processing of laser scanning data

Vladimir Badenko[1,*], *Dmitry* Zotov[1] and *Alexander* Fedotov[1]

[1]Peter the Great St. Petersburg Polytechnic University, Water Resources and Hydrotechnical Engineering Department, 195251, Polytechnicheskaya, 29, St.Petersburg, Russia

Abstract. In this article the analysis of gaps in processing of raw laser scanning data and results of bridging the gaps discovered on the base of usage of laser scanning data for historic building information modeling is presented. The results of the development of a unified hybrid technology for the processing, storage, access and visualization of combined laser scanning and photography data about historical buildings are analyzed. The first result of the technology application for the historical building of St. Petersburg Polytechnic University shows reliability of the proposed approaches.

1 Introduction

Laser scanning is the most modern and actively developing type of remote sensing technologies. Nowadays the three-dimensional (3D) technologies are applied worldwide in a wide range of applications: road management, urban modeling, industry survey, heritage preservation [1]. Corresponding types of acquisition systems are the following terrestrial, mobile and airborne laser scanning [2]. Laser scanning is one of the common remote sensing methods. The principle of operation of laser scanners, regardless of their type and purpose, based on measuring of distance from the source of the laser pulse to the object. During the survey of an object, the required distance is determined according the time delay (pulse method) or the phase shift (phase method) in signal reflected by object surface [3]. The result of laser scanning is a cloud of points that is usually combined with photos [4]. Therefore for each point in the cloud user can get the following information: three coordinates, intensity of reflected signal and color (R, G, B), that accurately describe the surveyed object. Geometrical accuracy and point density of such clouds has recently significantly increased [5].

The objectives of this article are to analyze of some gaps in processing of raw laser scanning data and to present of some results of bridging the gaps discovered on the base of usage of laser scanning data for historic building information modeling. In the article results of the development of a unified hybrid technology for the processing, storage, access and visualization of combined laser scanning and photography data about historical buildings are presented.

* Corresponding author: vbadenko@gmail.com

2 Methods

An experience in architectural measurements of historical buildings using laser scanning technologies has shown that the measuring drawings of the facades and interiors of buildings most effectively created on the base of raster orthoimages, which are obtained by projecting laser point cloud on the picture plane. A similar approach has been traditionally used in photogrammetry [6]. Mobile Laser Scanning (MLS) is the most advanced and least common practice in Russian. But the number of MLS equipment in Russia has already surpassed the number of airborne laser scanners as well the amount of information per unit of time received by the MLS (up to 3-5 million points per second) several times higher than the volume produced during airborne scanning. There are numbers of promising applications that require the collection of geospatial data using MLS. However, MLS data processing are underdeveloped.

The analysis of the current state of research in this area has shown that there are a large number of publications dedicated to tasks of construction of digital models of buildings, the detection of the surface of the earth and the reconstruction of the road surface, detection and recognition of road facilities on the basis of MLS data [5, 7, 8].

Despite the existing successes, it is possible to single out the following algorithms, which require deeper study:

- for the hierarchical processing of big amount of data that provide a hybrid processing and data compression without loss of accuracy,
- for the integration of the scan data obtained from various sources,
- for the seamless linking of data processing in real time during forming orthoimage based both on a single scan, and a series of scans received from different sources for providing a reduction in the area of "dead zones".

The proposed method is a comprehensive solution to the problem by creating an algorithmic support for the hybrid system for processing of data received from all types of laser scanning (mobile, terrestrial and airborne), as well as photo. The solution proposed by authors is not the total three-dimensional modelling of urban areas or, for example, historical buildings and structures. Proposed method provides a method for semi-automatic and automatic data processing and visualization based on forming cylindrical reamers for point clouds. To solve the problems of historical building informational modelling (HBIM) [9] creation of based on laser scanning data a method for processing big volumes of data is developed. The method provides more accurate detailing of HBIM including a combination of scan data for each scale and a level of representation of survey information obtained by different scanning methods.

Factors determination and evaluation of metrics for the factors that affect the accuracy is not analyzed in the method. Special approaches for data compression without loss of accuracy and reducing the area of "dead zones" in the orthoimages are used. The problem of determining the geometry of the buildings is solved with existing methods with the usage of the integrated information obtained by various methods using interpolation and extrapolation of the collected data. However, to support the proposed method the experimental study of the effectiveness of various methods of surface modelling had been performed to ensure the acceptable level of accuracy.

3 Results

The task of the research is to develop new algorithms of cloud of points processing (big data processing) received from laser scanner. Creation of the software for the processing, visualization and vectorization of laser scanning data, as well as the presentation of data through the Internet, providing: (1) improved efficiency and accuracy of processing of laser scanning data (decoding and vectorization not less than 99.5%); (2) to provide to the customer of laser scanning process (within the engineering survey) the functionality of the post-processing of results of the scan on the basis of the primary (raw) information about the object(which is most valuable information).

Testing research for the technology was conducted to establish of 3D data for buildings of Peter the Great St. Petersburg Polytechnic University (former St. Petersburg Polytechnic Institute). St. Petersburg Polytechnic Institute was built on the model of Oxford and Cambridge as a complex with thoroughly designed and implemented infrastructure. The construction of the based buildings of the Institute was carried out after the designs of architect E. Virrich, artist architecture 1-St degree, academician of architecture since 1908, the outstanding representative of rationalist architecture, one of the pioneers in the use of reinforced concrete structures, in 1900 – 1905.

The complex of the Polytechnic Institute interesting as a coherent architectural ensemble, one of the few that have survived in a relatively unchanged form. Buildings represent the unique architectural buildings of the turn of XIX - XX centuries. The complex of buildings and Park are included in the "List of objects of historical and cultural heritage of Federal importance", approved by the decree of the President of the Russian Federation No. 176 of February 20, 1995, and are particularly valuable objects of cultural heritage of the peoples of the Russian Federation as a complex historical, cultural, architectural ensembles and structures and natural landscape monument representing material, intellectual, and decorative value.

One of the oldest constructions on the University premises is the Water Supply Tower was built in 1905 (fig. 1a), three years after the opening of the Polytechnic Institute. It is the highest point among two- and four-story institute buildings, and goes in contrast with other constructions because of its intricate and fractured forms. The Water Supply Tower is 46 meters high. Until 1953, it had been used as an engineering construction to provide the work of the water pipe system. When it was built, the Water Supply Tower has became one of the education-supporting buildings on the Institute's territory. In 1905 the annex near the Water Supply Tower the Hydraulics Laboratory was housed.

(a) (b) (c)

Fig. 1. (a) Water Supply Tower in 1905; (b) point cloud for the historical building; (c) result of semi-automatic triangulation.

There are many options for creating digital models of large objects, among the highlights of the photogrammetry and 3D laser scanning. The complexity of the research is required the use both techniques. To establish a 3D model for HBIM of Heritage Resources were used methods of 3D laser scanning and photogrammetric data processing. Creating 3D model involves the following steps: the choice of object of study; collection of visual, technical and historical information; laser scanning and photogrammetric survey to obtain data – point cloud (fig. 1b); detection and extraction of segments of dense point cloud, and the procedure of semi-automatic triangulation (fig. 1c); obtaining external color models using (fig.2a); export data in applied architectural packages (fig.2b, fig.2c).

(a) (b) (c)

Fig. 2. (a) external colour model; (b) 3D mesh for BIM; (c) 3D BIM model.

It was built dense point cloud (fig. 1a) based on image comparison using OpenCV library [10]. A dense cloud of points is 1 point per 1 cm^2, in the case of shooting large objects. In the case of this object, the point cloud was 32KK points. Processing this amount of points makes no sense and is extremely costly in terms of hardware resources. For this purpose, it is necessary to reduce the number of points. For large objects, having a complex geometry optimally reduce the number of points is in 10-fold to 3.2 KK.

4 Discussion

The survey of existing building is an activity, which is increasingly used modern laser scanning high-tech devices and equipment [11]. Appropriate methods are used for surveying of various objects, ranging from historic buildings to uninhabited territories for road construction [4, 12, 13]. Laser scanning systems collect data more completely and efficiently than traditional surveying methods [14]. Future developments of these systems continues to design a compact, lightweight and relatively inexpensive multipurpose laser scanning system that should be quick and easy to deploy, and requires a minimum amount of existing equipment for operational support [1]. The well-known features of laser scanning data are high accuracy and density and as a result of which big spatial point cloud data are created [15]. The process of obtaining point cloud for each type of laser scanning is well explored and the appropriate software have been developed by providers [16].

A significant drawback of laser scanning technologies, which significantly limits its use, is the constantly existing gap between the rapid development of the technologies and the possibilities of processing results (cloud point processing) [17]. Existing methods of geo-processing and images analysis, as well as high demands to computational resources do not allow full use of the capabilities of the technologies [18]. Today processing of laser scanning results (point clouds) are often associated with providers of equipment [19]. But there are almost no universal tools for processing and exchanging of the point clouds from different sources [9 ,20].

Laser scanning point clouds are currently collected and processed individually on specific objects and territories, according to the technological requirements (Xu et al., 2017). Automatic or semi-automatic segmentation and objects detection in 3D laser scanning data is the most resources-consuming step. There are also special software packages developed for creation of specific models types from laser scanning data such as Cyclone, RapidForm, 3Dipsos, they are all designed to work in 3D [4]. Also laser scanner providers have proposed to import point clouds into most popular CAD systems AutoCAD (Autodesk) and Microstation (Bentley). However, processing of raw 3D laser scanning data in CAD environment is inconvenient, especially if creation the 2D plans are required [21].

The complexity of historic buildings, with irregular geometry poses numerous challenges in the computer modelling and simulation of structural elements under different types of loads [22]. The responsible persons must take into account a range of possible and compatible solutions that avoid endangering the cultural significance of the historical building [9]. Nowadays, in many publications for solving the mentioned problems it has been proposed to use an approach based on HBIM. The first step of the modern HBIM usually is remote collection of survey data using a laser scanner technologies combined with digital photos [12, 22]. Therefore, there are many applications of the laser scanning technologies for data collection about historical buildings worldwide [9]. In papers has been discussed the usage of innovative laser scanning surveying technologies for modelling and monitoring of heritage buildings to forecast their behavior during the life cycle management.

5 Conclusions

Unified hybrid technology for the processing, storage and visualization of laser scanning and photography data about historical buildings based on decoding and vectorization algorithms was proposed. It is an information framework for HBIM of heritage resources in St.Petersburg, Russia. The first result of application the technology to historical building (Water Supply Tower of Peter the Great St. Petersburg Polytechnic University) shows the reliability of the proposed technology. The developed technology also can be useful for processing laser-scanning data during engineering survey of roads and existing building.

Acknowledgements

The research is carried out with the financial support of the Ministry of Education and Science of Russia within the framework of the Federal Program "Research and Development in Priority Areas for the Development of the Russian Science and Technology Complex for 2014-2020", Activity 1.2., Subsidy Agreement No. 14.575.21.0173 of 23.10.2017, unique identifier: RFMEFI57517X0173.

References

1. H. Guan, J. Li, S. Cao, Y. Yu, Int. J. of Image and Data Fusion, **7(3)**, 219-242. (2016)
2. J. Heo, S. Jeong, H.-K. Park, J. Jung, S. Han, S. Hong, H.-G. Sohn, Computers, Environment and Urban Systems, **41**, 26–38 (2013)
3. C. Wang, Y.K. Cho, Automat. Constr. **49**, 239-249 (2015)
4. P. Tang, D. Huber, B. Akinci, R Lipman, A. Lytle, Automat. Constr. **19(7)**, 829–843 (2010)
5. B. Yang, Z. Dong, ISPRS J Photogramm. **81**, 19-30 (2013)
6. O. Risbøl, C. Briese, M. Doneus, A. Nesbakken, J. Cult. Herit. **16(2)**, 202-209 (2015)
7. M. Vaaja, J. Hyyppä, A. Kukko, H. Kaartinen, H. Hyyppä, P. Alho, Remote Sens-Basel, **3(3)**, 587-600 (2011)
8. C. Cabo, C. Ordoñez, S. García-Cortés, J. Martínez, ISPRS J Photogramm. **87**, 47-56 (2014)
9. C. Dore, M. Murphy, International Archives of the Photogrammetry, Remote Sensing & Spatial Information Sciences, **42(2W5)**, 185-192 (2017)
10. V. Badenko, D. Kurtener, V. Yakushev, A. Torbert, G. Badenko, Lect. Notes Comput. Sc. **9788**, 57-69 (2016)
11. L. Barazzetti, Adv. Eng. Inform. **30(3)**, 298-311 (2016)
12. R. Volk, J. Stengel, F. Schultmann, Automat. Constr. **38**, 109-127 (2014)
13. W. Mukupa, G.W. Roberts, C.M. Hancock, K. Al-Manasir, Surv. Rev. **49 (353)**, 99-116 (2017)
14. A. Soni, S. Robson, B. Gleeson, Applied Geomatics, **7(2)**, 123-138 (2015)
15. B. Quintana, S.A Prieto., A. Adán, A.S Vázquez,. Adv. Eng. Inform. **30(4)**, 643-659 (2016)
16. V.V. Lehtola, H. Kaartinen, A. Nüchter, R. Kaijaluoto, A. Kukko, P. Litkey, E. Honkavaara, T. Rosnell, M.T. Vaaja, J.-P. Virtanen, M. Kurkela, A. El Issaoui, L. Zhu, A. Jaakkola, J. Hyyppä, Remote Sensing, **9**, 796 (2017)
17. N. Hichri, C. Stefani, L. De Luca, P. Veron, G. Hamon, International Archives of the Photogrammetry, Remote Sensing and Spatial Information Sciences - ISPRS Archives, **40(5W2)**, 343-348 (2013)
18. A. Kukko, H. Kaartinen, J. Hyyppä, Y. Chen, Sensors, **12(9)**, 11712-11733 (2012)
19. M. Lee, S. Lee, S. Kwon, S. Chin, KSCE J. Civ. Eng., **21(6)**, 2027-2036 (2017)
20. X. Xiong, A. Adan, B. Akinci, D. Huber, Automat. Constr. **31**, 325-337 (2013)
21. L. Barazzetti, F. Banfi, R. Brumana, G. Gusmeroli, M. Previtali, G. Schiantarelli, Simul. Model Pract. Th. **57**, 71-87 (2015)
22. M. Murphy, E. Mcgovern, S. Pavia, Structural Survey, **27(4)**, 311-327, 2009
23. F. Chiabrando, M. Lo Turco, F. Rinaudo, International Archives of the Photogrammetry, Remote Sensing and Spatial Information Sciences - ISPRS Archives, **42(2W5)**, 605-612 (2017)

Comprehensive assessment of the efficiency of high-rise construction projects in the form of urban blocks

Alexandr Orlov[1*], *Irina* Chubarkina[1]

[1]Moscow State University of Civil Engineering, 129337, Yaroslavskoye shosse, 26, Moscow, Russia

Abstract. The paper is dedicated to main modern trends in the area of high-rise construction. The classification of buildings and structures by height is given. Functional distribution by the height of buildings is presented. A review of positive and negative aspects of high-rise construction from the economic point of view is given. On the basis of the data obtained, it is proposed to build up residential microdistricts in the form of urban blocks. A plan of microdistricts development is presented. It takes into account urban blocks and includes their main characteristics. An economic and mathematical model was developed to carry out a comprehensive assessment of the effectiveness of high-rise construction projects.

1 Introduction

The purpose of this paper is to assess the efficiency of high-rise construction projects presented by urban blocks taking into account the social aspect of construction. Also, the need to study positive and negative factors affecting further development of the city's infrastructure and social and psychological well-being of the city's inhabitants is actualized in it.

In recent decades, a steady increase in the urban population has been observed throughout the world, which results in an increase in both the number of cities and population size. The data provided by the United Nations Organizations (UNO) confirm this trend. As a result cities and their residents face economic, social, environmental, urban planning, transport and other problems [1, 2]. And in this case the problem of urban areas rapid growth, their territorial "proliferation" directly related to an increase in the urban population size enters into the picture [3].

In this situation, due to the growth of cities, specialists are in search for new urban planning, space planning, environmental, economic, and social solutions. "Smart growth of cities" is among the trends. It means creation of so-called "compact cities", which allow to get more compact planning, to form a comfortable modern living environment without increasing the territory of a city.

* Corresponding author: alor333@gmail.com

This is primarily achieved through sound and well thought-out multipurpose skyscraper concentration. Residential bildings, offices, hotel complexes, cultural and social services system, etc. are presented in its space-planning structure [4-6].

The search for rational solutions of a "compact modern city" is conducted in various urban planning and space planning areas. Such approaches as urban densification and development of new territories are considered, and the main objective is to create compact residential areas to reduce the cost of expensive utility systems caused by their length and operation, cut the travel time from places of residence to places of employment and back [7-9].

Classification of buildings and structures by height is shown in Figure 1.

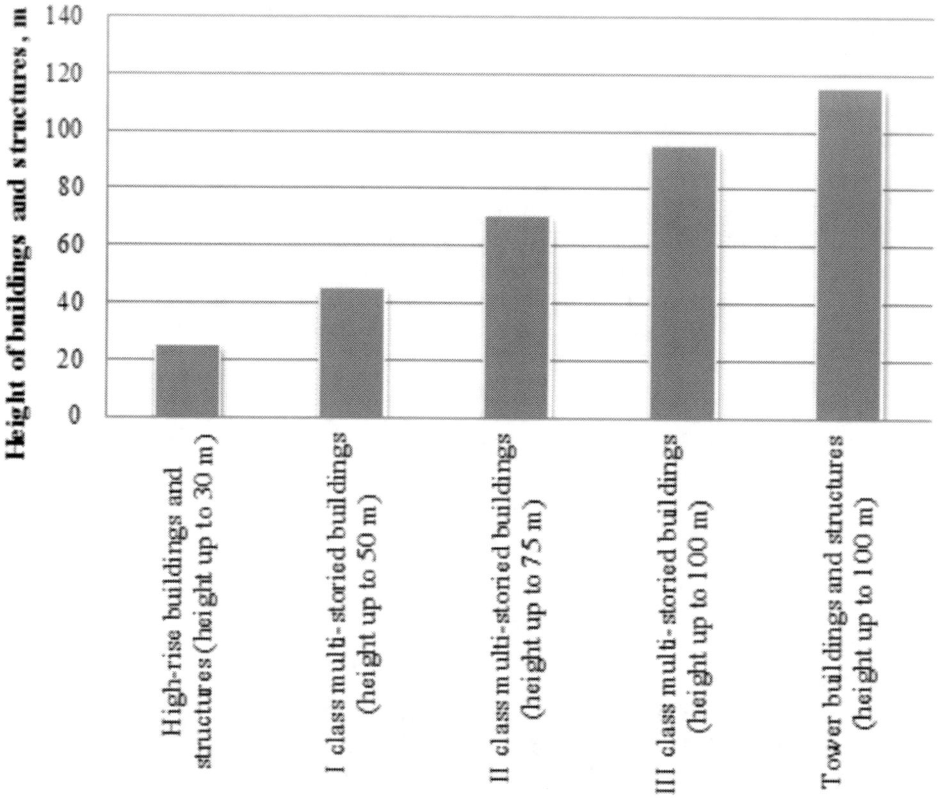

Fig. 1. Classification of buildings and structures by height.

The criterion of height in meters, but not the number of storeys, is used for the classification of skyscrapers, since storeys differ in height depending on the purpose of the building and the requirements of national design regulations.

A modern city can not be imagined without high-rise buildings for business and living. They strive to make high-rise buildings universal and expand their purpose by giving them a different form, dividing them into functional areas [10]. Multifunctionality of high-rise buildings is an economic necessity.

2 Materials and Methods

The methodological basis for the research is complex analysis, a system approach, and general scientific methods: analysis and synthesis; induction and deduction; comparison. Methods of grouping, sociological research, expert assessments and statistical processing of information were used in the course of the research.

High-rise buildings all over the world refer to objects of high level responsibility and reliability. That is why the cost of their construction is much higher as compared to conventional buildings, and it is caused by technological and constructive factors, as well as other factors ensuring safe presence of people indoors [11].

When designing high-rise buildings, it is necessary to take economically justified technical decisions without compromising the reliability of a structure [12]. Distribution of functions by height presented in Fig. 2 is the most wide-spread in the area of multifunctional high-rise buildings construction.

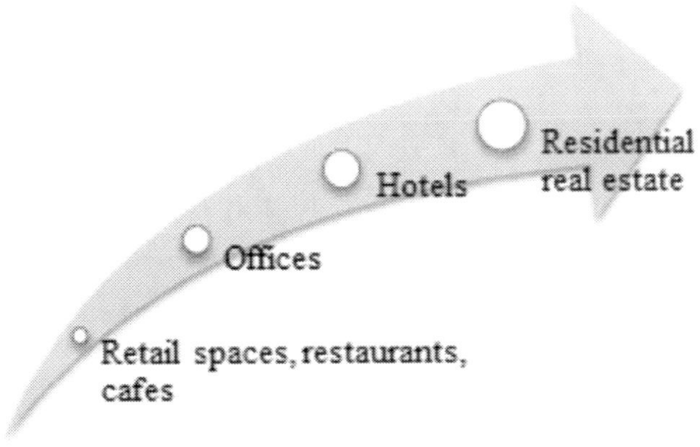

Fig. 2. Functional distribution of buildings by height.

As is seen from Figure 2, the most common type of construction objects by the functional distribution is residential real estate. However, it should be noted that at the present time consumers are not interested in purchasing apartments in residential complexes separated from social infrastructure facilities.

Social infrastructure is a system of facilities, utilities, as well as enterprises, institutions and organizations that provide social and welfare services to the population, management bodies and personnel administration, which activities are aimed at satisfying public needs of citizens that meet the established quality of life indicators. Social infrastructure unites economy, health, education, culture and art, physical education and sports, trade and public catering, personal services and contributes to the improvement of living conditions of the population. Development and effective functioning of objects that are a part of the social infrastructure, their accessibility are essential conditions for improving the standard of living and quality of life of the population [13, 14].

The main factors determining the development of the social infrastructure system are trends in socio-economic development of a settlement characterized by an increase in the population, development of the residential property market, and the services sector. Implementation of high-rise multifunctional complexes is targeted at the fulfillment of social conditions, and residents of these complexes will be provided with them by improving the quality and availability of social services (education, health, culture and social security) for all categories of residents.

Taking into account the individual approach to the design and construction of high-rise buildings, use of exclusive elements of utility systems and equipment, as well as materials that ensure the uniqueness of the building, creates significant uncertainty in the technical and economic assessment of design decisions.

Advantages and disadvantages of high-rise construction from the economic point of view are given in Table 1.

 https://doi.org/10.1051/e3sconf/20183301048

Table 1. Advantages and disadvantages of high-rise construction.

Advantages of high-rise construction	**Disadvantages of high-rise construction**
1	2
Creation of additional jobs (building maintenance personnel).	Increased requirements for the provision of integrated safety and security presuppose use of high-quality technical solutions.
Economical use of urban areas under conditions of an acute shortage of land and its high cost.	Constant monitoring of structures, as well as all utility systems shall be carried out in high-rise buildings.
High-rise buildings contribute to the development of economy, business and tourism.	Approximately 70% of the internal space is ussed in a skyscraper; the rest space is occupied by supporting structural elements, staircases and lift shafts.
Evidence of a new stage in the development of construction, the level of scientific and technological development, well-being of population	Construction and operation of skyscrapers require a much higher consumption of resources, which is associated with excessive energy consumption and colossal operating costs.

It is obvious that erection of high-rise buildings has both its advantages and disadvantages in terms of economic resources. So, there is a need to improve the efficiency of high-rise buildings. The decision on the implementation of an investment project requires carrying out of calculation and analysis of indicators of investment performance conducted within the framework of project analysis [15-17]. The basis for the assessment of the return on investment is the process of calculation and comparison of the required amount of costs and results from the implementation of projects.

3 Results

Construction of a "compact city" consisting of high-rise multifunctional buildings and service facilities creates a comfortable and convenient urban environment for the residents, and also solves problems related to the efficient use of land resources, environmental problems, etc.

Microdistricts with multifunctional high-rise buildings using a new town-building unit called "urban block" can be presented by this type of "compact city".

An urban block will be presented by multipurpose high-rise buildings (various functional groups of facilities - residential, hotel, entertainment, office, catering, trade, service, sports, etc. - will be included in the space planning structure) and a courtyard (a private area with trees and srubs, grounds for recreation and fire access roads, available primarily for residents of that particular urban block and closed for public).

A city block in this case will consist of a set of urban blocks, public areas with driveways, parking lots, recreational areas, and a block green belt.

And a microdistrict will include city blocks, streets and driveways, buildings of other functional purposes, including schools, health and fitness centers, a microdistrict center, utility facilities, recreation common areas with trees and shrubs, kindergartens, day-to-day service facilities, detached garages.

The housing project of a microdistrict in the form of urban blocks provides for a clear separation of private and public areas within a city block (Fig.3).

https://doi.org/10.1051/e3sconf/20183301048

URBAN BLOCK

PRIVATE AREAS are allocated within an urban block and represent courtyards with trees and shrubs, with playgrounds for young children and recreation; private areas are free from transport (except special vehicles)

PUBLIC AREAS are allocated within a city block and include driveways, parking lots, garden squares, boulevards; it is planned to allocate mainly utilitarian public areas (with driveways, parking lots, public utility sites) and pedestrian green areas

Fig. 3. Private and public areas within an urban block.

A microdistrict with urban blocks is shown in Figure 4.

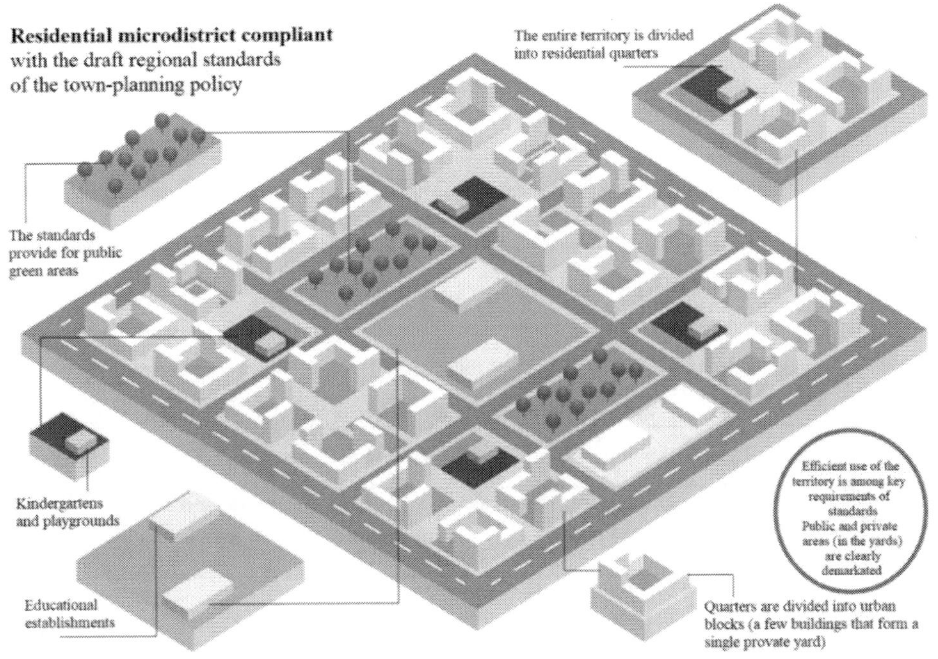

Source: Russian News (News-russia.info)

Fig. 4. A microdistrict in the form of urban blocks.

Microdistrict construction in the form of urban blocks provides for comfortable and favorable living conditions in accordance with federal urban planning, sanitary and hygienic requirements, including:

⇨ Availability of socially significant facilities (including schools and kindergartens), including their normative availability.

⇨ Creation of green courtyards free from transport (trees and shrubs).

⇨ Creation of full-fledged recreational green areas with all the necessary infrastructure (playgrounds, sports grounds, recreation areas) within a city block (public gardening) and a microdistrict (public use gardening).

⇨ Placement of trade and service facilities at street level, which allows to create a functionally saturated environment of residential streets and driveways.

⇨ Profiles of streets and driveways ensuring comfortable pedestrian traffic are worked out.

⇨ Space planning of city blocks and urban blocks is aimed at providing a single front of development, traditional for urban spaces.

The following infrastructure elements are also provided (Table 2):

Table 2. Elements of infrastructure in a microdistrict in the form of urban blocks.

Infrastructure facility	Characteristics of infrastructure facilities
Passageways	Passageways are divided into: 1. Utilitarian, i.e. those ones including access ways, parking places for motor vehicles and a public utility site; 2. Recreational, i.e. those ones including access ways and grassed and landscaped areas of shared use with recreational areas, playgrounds and sports grounds (arrangement of appropriate driveways in the form of boulevard streets with graden squares at points of their intersection)
Playgrounds	Playgrounds are intended for games and active recreation of children from three to sixteen years
Microdistrict landscaping of shared use	It is designed to provide the population with recreational landscaped areas. Microdistrict landscaped areas of shared use represent parks, garden squares, boulevards, gardens, and their borders are fixed with red lines
Landscaping of private areas (yards)	It is designed to create a comfortable environment in residential yards and ensure quiet recreation for residents
Landscaping of public areas of a city block	It is intended for the creation of convenient pedestrian connections and to ensure daily recreation for the population. Landscaping of public areas of a city block is possible in the form of boulevards, garden squares, as well as individual greenery islands
Parking areas	Parking areas are designed in accordance with the transport regulations
Dog runs	They are specially equipped and enclosed places for dogs free ranging
Recreation areas	Recreation areas are designed for quiet recreation and table games for adults
Sports grounds	Sports grounds are intended for physical exercises and designed for all age groups. Dimensions and planning configuration of a specific site depend on its specialization
Public utility sites (for dust collectors)	Public utility sites are places specially equipped for solid household waste collection

The most significant areas of social facilities development can be identified and long-term activities when creating social infrastructure facilities can be planned based on certain standards.

Assessment of efficiency from the construction of high-rise multifunctional complexes in the form of urban blocks is carried out based on achieved results of each main indicator by comparing their planned and actual values, and it is calculated by formula 1. An analysis of efficiency indicators may allow to conclude whether the implementation of an investment project is a rational decision or not. Such indicators of investment efficiency include, for example, environmental, commercial, social and budgetary efficiency, as well as enterprise costs.

$$E = N\text{actual} \,/\, N\text{planned}\; x\; 100\% \quad (1)$$

where E is the efficiency of construction of high-rise multifunctional complexes in the form of urban blocks,

N_{actual} - the actual value of indicators,

$N_{planned}$ - the planned value of indicators.

Values of indicators N are determined according to the economic and mathematical model, in which key target functions that determine the efficiency from the implementation of an investment and construction project are identified (formula 2).

$$N = f(C, E, B, K, V, Z) = \begin{cases} \text{NPV} \to \max \\ \text{IRR} \to \max \\ \text{PI} \geq 1 \\ \text{PP} \to \min \end{cases} + \begin{cases} \text{Economic efficiency (E)} \to \max \\ \text{Budgetary efficiency } (B) \to \max \\ \text{Commercial efficiency (K)} \to \max \\ \text{Social efficiency (V)} \to \max \\ \text{Enterprise costs } (Z) \to \min \end{cases} \quad (2),$$

where C is economic efficiency, a presented set of parameters such as net present value (NPV), internal rate of return (IRR), profitability index (PI), payback period (PP).

E is ecological efficiency, which is a measurable result of the environmental management system related to the control and organization of its environmental aspects, based on the environmental policy, as well as planned environmental indicators.

B is budgetary efficiency, which reflects an impact of the implementation of an investment and construction project on revenues and expenditures of the federal, regional or local budget.

K is commercial efficiency, defined as the difference between the inflow and outflow of cash from investments and real estate sales, taking into account payment of various charges and taxes.

Z is costs incurred by an enterprise, represented by a combination of resources (material, labor, financial, etc.) consumed by the construction enterprise during the reporting period.

V is social efficiency, which reflects the correspondence of costs and social results of the project unde consideration to the goals and social interests of its participants.

The main indicators of social efficiency during design and construction of high-rise multifunctional complexes in the form of urban blocks are as follows:

⇨ An increase in the life expectancy.

⇨ An increase in the birth rate.

⇨ Reduction of unemployment.

⇨ An increase in the level of health facilities availability.

⇨ An increase in the proportion of population provided with cultural facilities in accordance with the statutory values.

⇨ An increase in the proportion of population provided with sports facilities in accordance with statutory values.

⇨ An increase in the number of people systematically taking excercises and going for sports.

This calculation reflects the feasibility of project implementation and allows to hold corrective actions to be taken if unsatisfactory estimates are obtained.

4 Conclusions

Among the reasons that prevent from an increase in the volume of high-rise buildings construction is a relatively small experience in the construction of high-rise complexes and, as a consequence, the need to attract expensive foreign capacities; issues related to standartization and technical regulations for high-rise design, construction supervision, etc. remain open. However, in the long term, the demand for high-rise construction in Russia is expected to grow.

Considering the above, it can be noted that with the increase in the level of technological development and the cost of land, the disadvantages currently available may not be as big as they are now. Currently, regulatory documentation is being developed, the technology of production of some components and utility systems is being improved, and the qualification of employees is growing, which gives hope for the development of high-rise construction. In general, introduction of urban blocks can improve the return on investment in the design and construction of high-rise multifunctional complexes and enhance the competitiveness of the project developer.

References

1. A. Orlov, Yu. Zhukova, Economics and Entrepreneurship, **4-2 (81-2)**, 979-982 (2017)
2. S. Mikhaleva, A. Orlov, a collection of materials of the International Scientific and Practical Seminar "Development of the Scientific School of Real Estate Management Theory", 111-114 (2015)
3. Y. Kog, Practice Periodical on Structural Design and Construction (2018)
4. M. Rafiei, H. Adeli, Structural Design of Tall and Special Buildings (2016)
5. A. Aminmansour, K. Moon, Journal of Architectural Engineering, **16 (2)**, 47-53 (2010)
6. E. Stein, MATEC Web of Conferences, 106,01036 (2017)
7. P. Love, D. Edwards, Z. Irani, IEEE Transactions on Engineering Management, 59 (4), 6032087, 560-571 (2012)
8. E. Frolova, Sotsiologicheskie Issledovaniya, **12**, 51-58 (2014)
9. B. Semaan, J. Hemsley, ISCRAM 2015 Conference Proceedings - 12th International Conference on Information Systems for Crisis Response and Management, 321-328 (2015)
10. I. Ptuhina, T. Spiridonova, T. Musorina, Applied Mechanics and Materials, **725-726**, 153-159 (2015)
11. Z. Su, Advanced Materials Research, **1055**, 297-300 (2014)
12. P. Kostrikin, E. Buadze, Review of the Southwest State University, **1 (64)**, 83-92 (2016)
13. O. Tielietov, N. Letunovska, Actual Problems of Economics, **160 (1)**, 329-337 (2014)
14. E. Frolova, M. Vinichenko, A. Kirillov, O. Rogach, E. Kabanova, International Journal of Environmental and Science Education, **11 (15)**, 7421-7430 (2016)
15. I. Ptuhina, T. Spiridonova, T. Musorina, S. Kanyukova, A. Rezvaia, MATEC Web of Conferences, 53,01049 (2016)
16. O. Kurakova, Economics and Entrepreneurship, **4-2 (81-2)**, 648-651 (2017)
17. S. Belyakov, Construction: science and education, **2**, 2 (2015)

Shear-lag effect and its effect on the design of high-rise buildings

Bui Thanh Dat[1*], *Alexander* Traykov[2], *Marina* Traykova[3]

[1]National University of Civil Engineering, Hanoi, Vietnam
[2]University of Architecture, Civil Engineering and Geodesy, Sofia, Bulgaria
[3]University of Architecture, Civil Engineering and Geodesy, Sofia, Bulgaria

Abstract. For super high-rise buildings, the analysis and selection of suitable structural solutions are very important. The structure has not only to carry the gravity loads (self-weight, live load, etc.), but also to resist lateral loads (wind and earthquake loads). As the buildings become taller, the demand on different structural systems dramatically increases. The article considers the division of the structural systems of tall buildings into two main categories - interior structures for which the major part of the lateral load resisting system is located within the interior of the building, and exterior structures for which the major part of the lateral load resisting system is located at the building perimeter. The basic types of each of the main structural categories are described. In particular, the framed tube structures, which belong to the second main category of exterior structures, seem to be very efficient. That type of structure system allows tall buildings resist the lateral loads. However, those tube systems are affected by shear lag effect - a nonlinear distribution of stresses across the sides of the section, which is commonly found in box girders under lateral loads. Based on a numerical example, some general conclusions for the influence of the shear-lag effect on frequencies, periods, distribution and variation of the magnitude of the internal forces in the structure are presented.

1 Introduction

1.1 History of tall buildings

In the late nineteenth century, tall buildings emerged in the United States of America. Most important tall buildings were built in the USA. An official 10-story (and, after an 1890 addition, 12) tall building called Home Insurance Building located in Chicago (1885) with 55 m high is considered the world's first skyscraper [1], as shown in Figure 1.

Based on various complex factors, such as economics, aesthetics, technology, municipal regulations and politics, tall buildings also appear and rapidly increase in number in other parts of the world, especially in Asian countries, such as China, Indonesia, Japan, and United Arab Emirate. As the height of tall buildings increased and record was constantly broken, tall buildings have become a symbol of prominence. According to a data published in 2017 [2],

* Corresponding author: thanhdat55xe@gmail.com

there are more than a hundred of tall buildings above 300 m constructed in the world. Burj Khalifa (Dubai–2010) is presently the tallest building in the world, with 163 stories and 829.8 m high. By 2020, this record is expected to be broken by the appearance of Jeddah Tower-a 167-story (1000 m) building- which is under construction in Jeddah city.

Fig. 1. Home Insurance Building – Chicago [1]

Table 1. Total number of completed tall buildings in countries (2017) [2]

Country	All buildings	100 m+	200 m+	300 m+
United States	4,664	2,517	180	17
China	2,629	1,899	503	49
Canada	1,052	472	21	0
Australia	930	349	31	1
Russia	797	218	16	5
Indonesia	597	154	31	0
Japan	556	467	37	1
Brazil	532	176	1	0
United Arab Emirates	413	283	91	23
United Kingdom	368	66	5	1

1.2 Structural systems of tall buildings

A tall building is assumed as a beam cantilevering from the earth which is subjected to axial loading by gravity and to transverse loading by wind or earthquake as shown in Figure 2. As the building becomes taller, the total gravity loads increase dramatically and the building sway due to lateral loads becomes critical. The gravity loads act on floors which transfer them horizontally to the vertical walls and columns through which they are transmitted to the foundation. The magnitude of axial loading can be estimated from the slab areas, so its calculation is not usually considered to be a difficult problem. On the other hand, the response of a structure to horizontal loads is more complex because it has to carry the external shear, moment, and torque [3].

In resisting shear forces, the building must not break by shearing off and not strain beyond the limit of elastic recovery. Regarding the bending resistance, the building must not overturn from combined forces of gravity and lateral loads; it must not break by premature failure of columns; its bending deflection should not exceed the limit of elastic recovery. Torsion on a building is resisted mainly by shear in the vertical components, by the horizontal components of axial force in any diagonal bracing members, and by shear and warping torque resistance of elevator, stair, and service shafts.

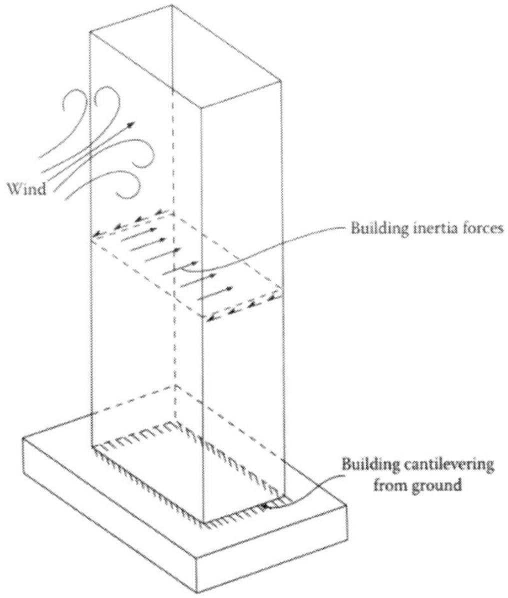

Fig. 2. Structural concept of tall building.

As discussed in a thematic report [4], structural systems of tall buildings can be classified by two main groups:

- Group of basic systems: Frame, Shear wall, Core, and tube
- Group of mixed systems: combination of two or more basic systems.

In the recent development, unlike the above classification, a system-based board classification has been proposed. In 2007, Mir M. Ali and Kyoung Sun Moon [5] divided structural system of tall buildings into two board categories:

- Interior structures: the major part of the lateral load resisting system is located within the interior of the building.
- Exterior structures: the major part of the lateral load resisting system is located at the building perimeter.

This new classification is presented based on the importance of height for tall buildings and the lateral load resisting capabilities. Therefore, it encompasses most representative tall building structural systems used today.

The category of interior structures consists of three basic types:

- The *moment - resisting frames* consist of horizontal (beam) and vertical (column) members rigidly connected together in a planar grid form. This system resists lateral deformation by joint rotation and requires high bending stiffness of columns and beams. The rigid joints are also essential for stability. However, this system is not efficient for buildings over 30 stories in height because the shear racking component of deflection caused by the bending of columns and beams causes the building to sway excessively.
- *Shear truss/shear wall – frame interaction systems* are a combination of rigid frames and reinforced concrete shear walls. This system seems to be more efficient than frame system in terms of lateral resistant for buildings up to 70 stories. The lateral rigidity of tall buildings could be increased by the combination of the approximately linear shear-typed deflected profile of frame system and the parabolic cantilever sway mode of shear wall.

- *Outrigger structures* have been historically used by sailing ships to help resist the wind forces in their sails, making the tall and slender masts stable and strong. In tall buildings where the core is considered as pure cantilever, the outriggers can reduce overturning moment in the core and transfer its reduction to the outer columns. The outrigger systems may be formed in any combination of steel, concrete and composite construction.

The category of exterior structures consists of following basic types:

- *Framed tube* is a basic tubular form which consists of closely spaced columns and deep spandrel beams rigidly connected together throughout the exterior frames. This type is reasonably efficient from 38 to 300 m in height.
- *Braced tube* is a variation of the framed tube. By using diagonal braces to connect columns at each joint, the columns can be more widely spaced and the sizes of spandrels and columns are smaller than those in framed tube. The braces also transfer gravity loads as inclined columns.
- *Bundle tube* is a combination of individual tubes. Those connected tubes act as a single unit where the three dimensional response of the structure could be improved for strength and stiffness by providing cross walls or cross frames in the building.
- Another tubular system is *tube-in-tube*. One or more tubes are located inside the perimeter tube. The floor diaphragm transfers lateral loads by connecting the core and the outer tube.
- Besides tubular form, *diagrid system* is also a type of exterior structures. It consists of multiple diagonal elements that form a diagonal grid on the face of the structure. The grid system makes the structure stable even without having any vertical columns. This is because the diagonal members in diagrid structural system can carry gravity loads as well as lateral forces due to their triangulated configuration.
- *Super-frame* is another type of lateral load resisting systems in exterior structures. This system is composed of mega-columns comprising braced frames of large dimensions at building corners, linked by multistory trusses at about every 15 to 20 stories.

The framed tube structures, which belong to the second main category of exterior structures, seem to be very efficient. That type of structural system allows tall buildings resist the lateral loads. Basically, the system consists of perimeter closely space columns connected at each floor with deep spandrel girders, thereby creating the effect of a hollow concrete tube. Framed tube structure allows fewer interior columns, and so creates more usable floor space. The system acts like a hollow cylinder, cantilevered perpendicular to the ground. However, these tube systems are affected by shear lag – a nonlinear distribution of stresses across the sides of the section, which is commonly found in box girder under lateral loads.

2 Materials and methods

2.1 Introduction to shear lag

Shear lag effect has been studied for a long time, with the observation in box girder and any hollow structure which are subjected to lateral load. According to the elementary theory of bending states (Euler-Bernoulli beam theory), a plane section remains plane before and after bending. As a result, the variation of bending stress in the cross section along flange and web panels must be varying linear. In fact, the real distribution of these axial stresses is observed not linear. Due to the shear flow developing in the section, the panels displace longitudinally in the way that the middle portion of the flange and web lag behind that of the portion closer to the corner of the box section, as discussed by Leonard [6].

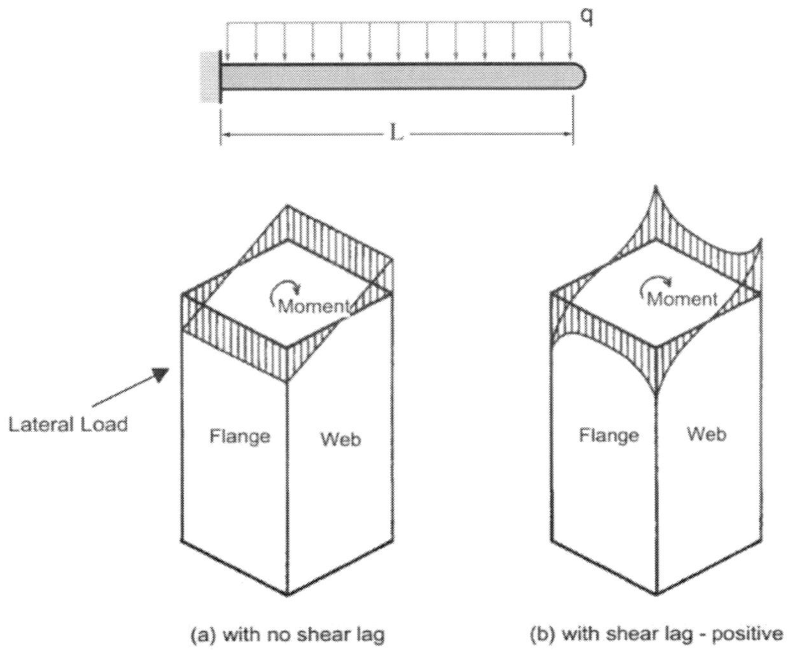

Fig. 3. Axial stress distribution in box cantilever beam [6]

Shear lag has been of interest to researchers for a long time. Shear lag in box girder was first studied by Eric Reisser in 1941. He gave applications of the fundamental equation to the actual solution of a series of shear lag problems [7]. Different aspects of the issue are considered in those researches listed in references [7-10].

2.2 Shear lag in framed tube structures

Generally, the lateral load resistance of framed tube system is highly promoted with the help of tube action. When subjected to lateral load, the tube behaves like a cantilever box beam, where the column deflects in lateral direction and beam deflects in bending. During the tube action, the shear lag occurs. This type of tubular structure also shows the real distribution of axial stresses against with the elementary theory of bending, in that the axial stresses applied in the columns of the peripheral flange are non-uniform and the stress distribution in the web panel are nonlinear. In framed tube systems, two different modes of shear lag may occur: negative shear lag and positive shear lag. As shown in Figure 4, positive shear lag in tube structures show the higher axial stress in the corner columns than that of middle columns [11]. This results in the warping of the floor slabs which leads to the consequent deflection of the secondary components.

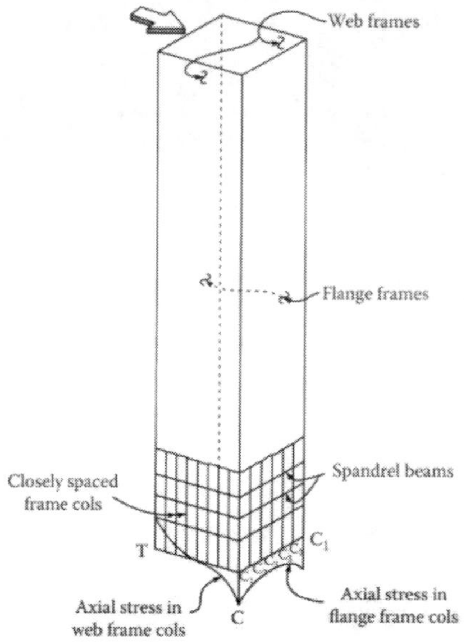

Fig. 4. Axial stress distribution in the column of framed tube building [11].

Many researchers studied shear lag effects in framed tube structures. In 1961, framed tube structural system was introduced by Fazlur Khan as one kind of exterior structures. The structure system of framed tube building was analyzed by using equivalent plane frame elements method. These studies were carried out by Coull and Ahmed (1978), Ha et al. (1978). In 1994, Kwan proposed more realistic assumption that the distribution of axial displacement of the web is independent of its in flange panel. Kwan found that shear lag is also influenced by the loads on structure and this effect at the bottom of the structure is usually higher than at the upper levels, as discussed by Leonard [6]. Y. Singh and A. K. Nagpal (1994) investigated in negative shear lag by modeling a 40-storey framed tube building. They found that positive shear lag occurs in the bottom part of the building while negative shear lag occurs at the top part, as shown in Figure 5. Shear lag effect is more significant for buildings with low ratio between the number of stories and the number of bays. They concluded that the origin of negative shear lag is positive shear lag. It is also observed that the corner columns may experience axial stresses opposite to those in the middle columns [12]. In 2000, Kang-Kun Lee, Hong Guan, and Yew-Chaye Loo used numerical method to analyze shear lag effect in framed tube structures with multiple internal tubes, as shown in figure 2-9. By investigating through a series of 40-storey buildings, they observed that the shear lag reversal points move towards the top of the structure with the increasing of shear lag and take place at a lower level in internal tubes than those in external tube [13]. Lee at el. (2002) extended a parametric study to examine the shear lag behavior of single and multiple framed tube buildings. He reported that "the axial stiffness of the columns was the most influential factor on the tube action as well as shear lag behavior, while the bending stiffness of columns and beams were a little impact perhaps due to the tube-in-tube interaction" [14]. In 2005, Kyoung Sun Moon [15] and Johan Leonard [6] discussed the behavior of diagrid system. Moon found that the optimal angle varies between 53° and 76° and this optimal angle reduces as the number of story decreases. He also found that the optimal value of "s", which is the ratio between the displacement at the top of the structure due to bending and the displacement due to shear, would be 3, 4, and 5. Leonard concluded that the optimal angle for diagrid system varies between 63.4 degree and 71.6 degree, and shear lag effect does not influence the lateral deflection of high-rise buildings.

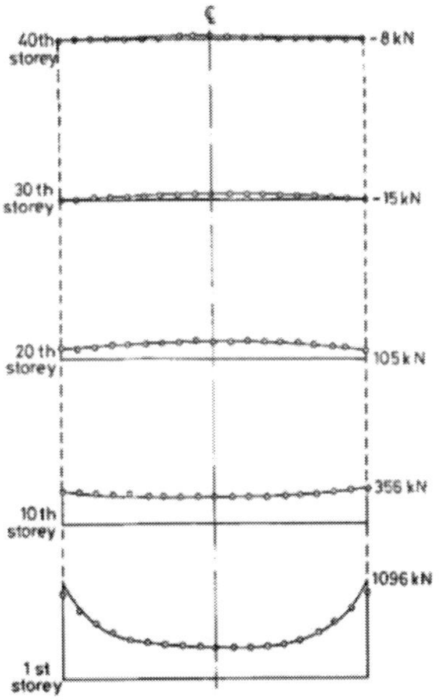

Fig. 5. Axial forces in flange – studied by Singh and Nagpal [12]

More recent studies about shear lag effect in tube structures have been conducted. In 2013, Farshid Nouri and Payam Ashtari studied shear lag in braced tube tall structures compared to framed tube tall structures as a solution to resist shear lag phenomenon. They analyzed number of factors affecting to shear lag ratio and concluded that edge columns stiffness, spandrel beam stiffness and diagrid angle are factors which play important roles on reducing shear lag phenomenon. They also recommended that it is needed to investigate separately the behavior of concrete and steel structures [16]. In 2014, D. Nagvekar and P. Hampali carried out their study on hollow structure by using Etabs software to analyze a 30-story tubular framed building. From the study, they noted that negative shear lag gets the maximum at top and occurs only after positive shear lag has occurred [17]. In 2015, Himanshu Gaur and Ravindra K. Goliya [18] conducted a study of a tube-in-tube structural system with facade bracing as a solution to mitigate shear lag. A number of different types of buildings were modeled in SAP 2000. It is found that for the heights between 120 and 48 stories, the bracing angle fluctuating between 63.43° and 45° show the least lateral deflection. Other approaches of reducing shear lag are providing additional structure, such as mage bracings or belt trusses. Those structures can be used to increase the shear stiffness of the flange and web frames of the framed tube building [19, 20].

Shear lag effect is an unexpected major phenomenon that controls the design of tall buildings using framed tube system. There are many researches with different analysis methods have been conducted to study the effect of shear lag not only in box girder, but also in tube structures. However, most of the previous studies focused on box girder, tube structures with exterior bracing system, and tube-in-tube structures. Not many studies have been attempted to describe shear lag effect and its variation along a framed tube structure.

The paper is dedicated to the modelling and the analysis of high-rise concrete building using framed tube structure with varying parameters, using ETABs software. Under the action of lateral loads, it is proposed to study the understanding of shear lag phenomenon in framed tube structure. The variables are considered in terms of building cross section and the spacing between outer columns.

2.3 Structural modeling

A 60-story framed tube concrete building with 3 m high of each story is modeled in ETABs 2015: size of columns 1x1 m2, size of beam 0.5x1 m2, slab thickness 0.25 m. There are four models with different parameters as follows:

Table 2: Different models of building

Parameter	Model 1	Model 2	Model 3	Model 4
Cross section	Square 24x24 m^2	Square 24x24 m^2	Rectangular 18x36 m^2	Rectangular 18x36 m^2
Column spacing	3 m	6 m	3 m	6 m

The building is subjected to both gravity loads and lateral load. The gravity loads are estimated as in a real office building: dead load is 4 kN/m^2 and live load is 3 kN/m^2. The earthquake load is applied according to Eurocode-8 (2004) [21], soil type C in Sofia - Bulgaria. Cross-sections of different models are presented in Figure 6.

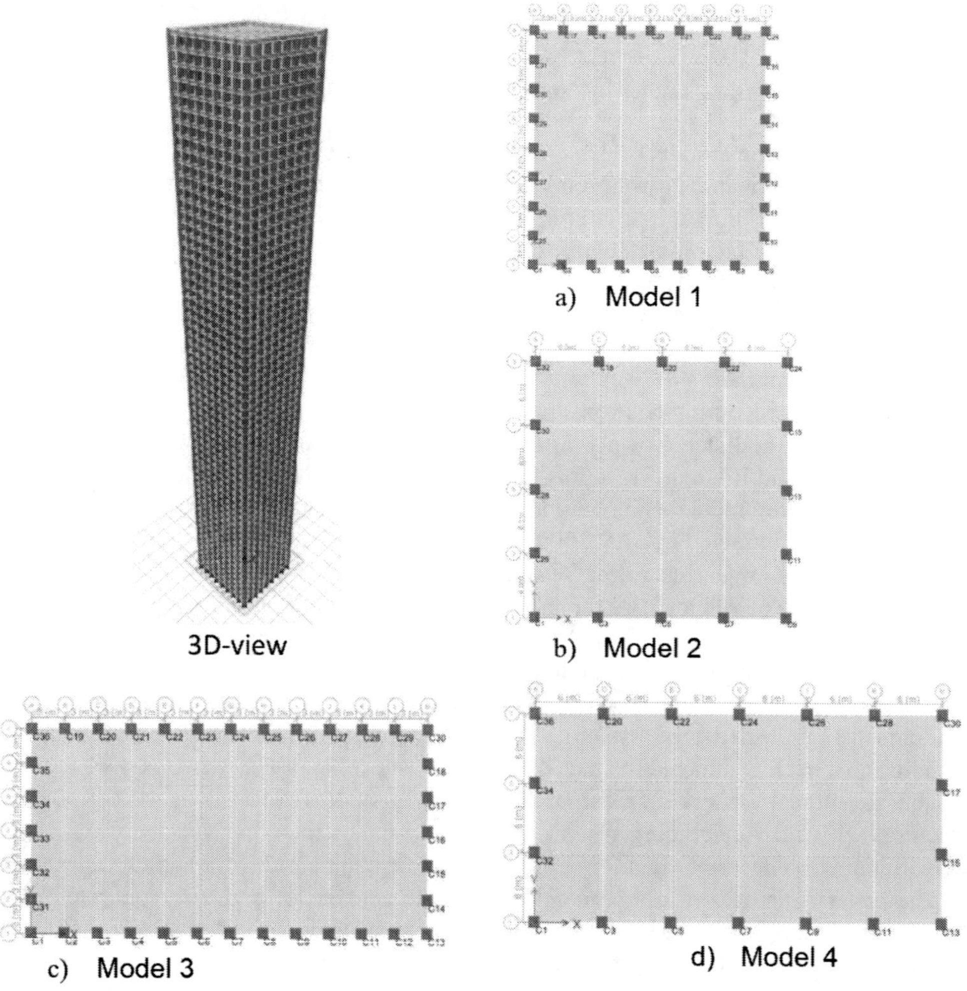

Fig. 6. Cross section of different building models

2.4 Main assumptions for analysis

The modeling and analysis approach used for parametric study are summarized as follows:

- The floor slabs in the structure are considered to be rigid diaphragms within their own plane. Thus, the high in-plane stiffness of the slabs restricts the relative lateral displacements between columns at each level.
- Joints between the spandrel beams and columns are assumed rigid.
- The building is supported by fixed supports.
- The spacing, materials, and size of the beams and columns are uniform throughout the building height.

3 Results

The following main results are obtained from the numerical modeling.

The results are presented as shear lag ratio – a non-dimensional parameter. Shear lag ratio is defined as the ratio of axial force in each column to axial force of the middle column in the same axis. In this section, the influences of building cross section and column spacing on shear lag effect are investigated.

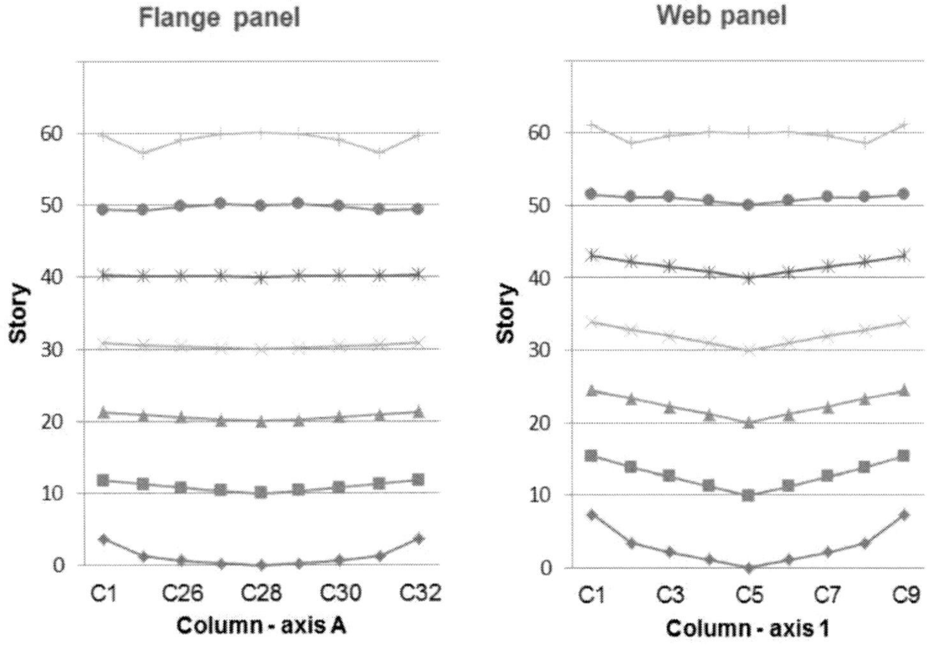

Fig. 7. Shear lag ratio distribution flange and web columns - model 1.

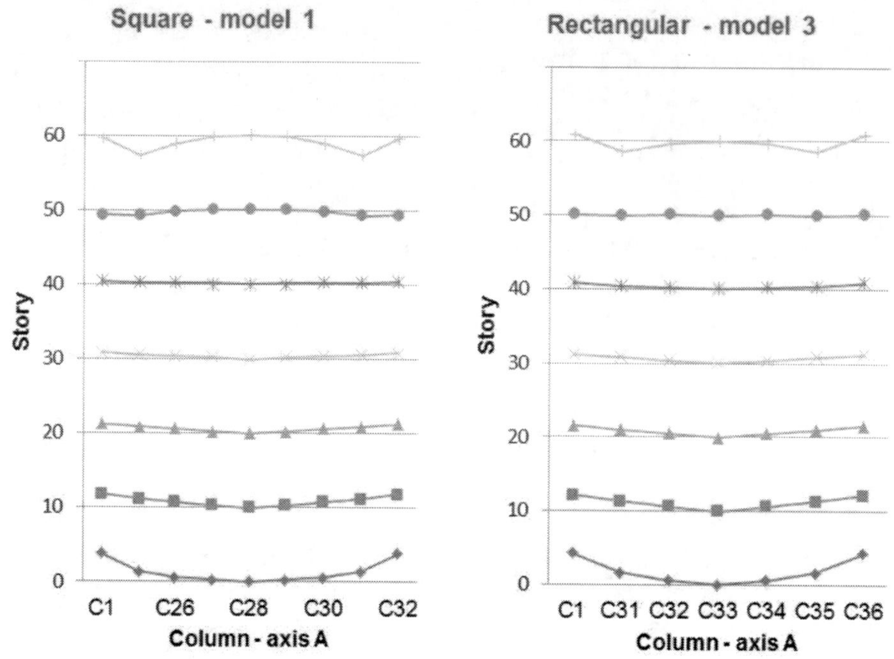

Fig. 8. Shear lag ratios distribution in building with different cross section shapes.

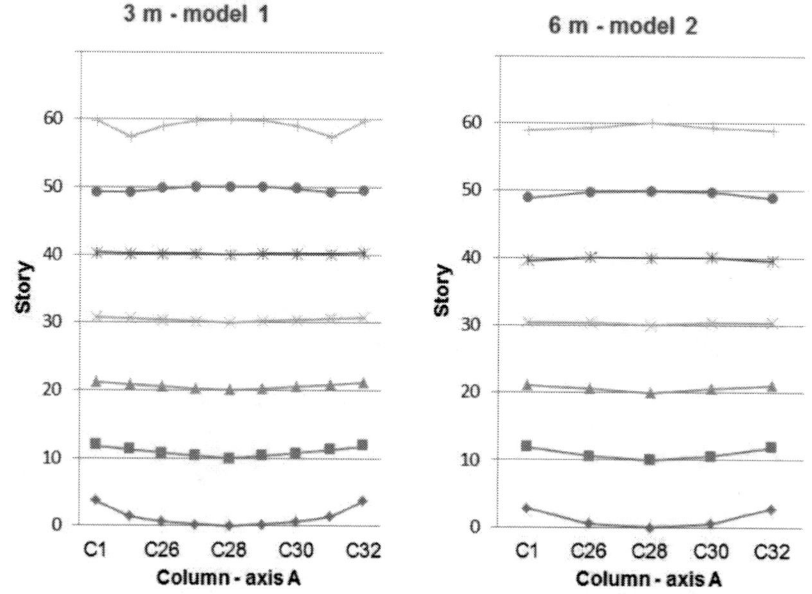

Fig. 9. Shear lag ratios distribution in square buildings with different column spaces.

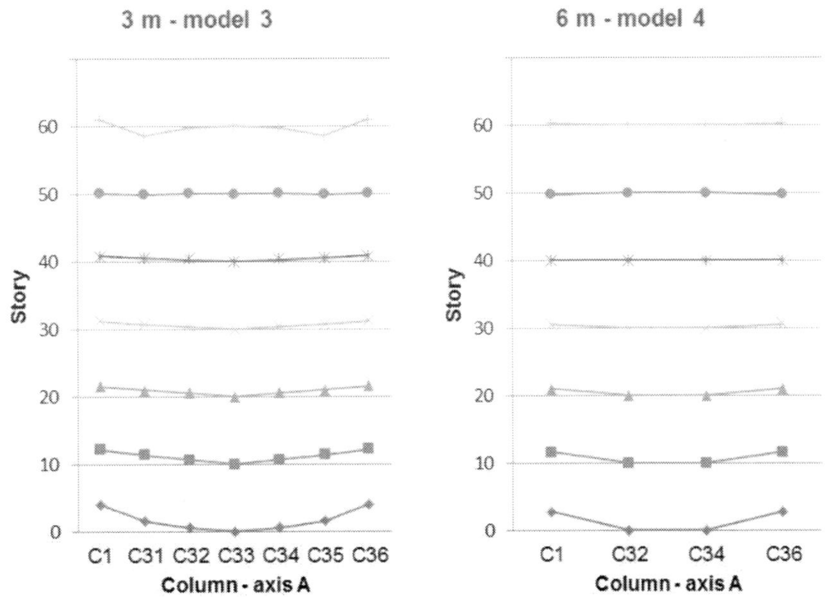

Fig. 10. Shear lag ratios distribution in rectangular buildings with different column spaces.

The description of the different models is presented in the previous section.

4 Discussion

All the results for the axial forces correspond to load combinations of vertical and horizontal loads rather than lateral loading only. That is the reason for the differences between the presented results for the axial forces in the web panels in this article and in other researches.

In Figure 7, shear lag effect can be observed in a 60-story framed tube building (model 1). The positive shear lag can be seen in the bottom part of the building, while the top part shows negative shear lag. The value of positive shear lag ratio is highest at the ground story. It decreases till it becomes zero as the increase of height, at around 40th story, then it transfers to negative shear lag. The negative shear lag ratio, on the other hand, increases from 40th story to the top of the building. The same trends are founded in the other models.

4.1 Building cross section variation

Two models with square (model 1) and rectangular cross section (model 3) are analyzed. The results of shear lag ratio in flange panel can be seen in Figure 8.

In the square section building, positive shear lag is inverted to negative shear lag at around 40th story, whereas it is changed at nearly 50th story in rectangular section building.

4.2 Column spacing variation

This section compares shear lag effect in buildings with the changing pattern of space between outer columns, from 3m (model 1 and model 3) to 6m (model 2 and model 4).

Figure 9 and Figure 10 show that the shear lag ratio in 6m-spacing buildings are smaller than those in 3m-spacing buildings. The difference can be seen clearly in the top part of the building where the negative shear lag occurs.

5 Conclusions

Based on the presented in the paper analysis, the following conclusions could be done:

- Shear- lag effect is a phenomenon typical for the framed tube structures under horizontal loads.
- The presented study shows the existence of both positive and negative shear-lag effect in the framed tube structures of the considered type.
- The positive shear-lag effect decreases with the floor levels and then it transfers to negative shear-lag effect which increases up to the top of the building.
- The shear-lag effect is not significantly influenced by the ratio of the building's dimensions in plan.
- The shear lag effect decreases when the column spacing increases. The column spacing affects the negative shear-lag effect more clearly at the top half of the building.

The present paper studies the shear-lag effect in a framed tube structure where some specific building features are not modeled. Future research shall consider the influence of the internal walls around stairs and elevators, as well as the effects of taking into account of the construction sequence.

References

1. http://www.theguardian.com. (last accessed 04.04.2017)
2. http://www.skyscrapercenter.com. (last accessed 08.04.2017)
3. B. S. Taranath, Steel, concrete, and composite design of tall buildings (New York: McGraw-Hill, 1997)
4. Thematic report: "Đề xuất lựa chọn hệ kết cấu BTCT chịu tải trọng ngang cho nhà siêu cao tầng ở Việt Nam (Solution for lateral resisting systems of reinforced concrete high-rise buildings in Vietnam)". *Division of Reinforced Concrete Buildings* (National University of Civil Engineering (NUCE), Vietnam, 2015)
5. M. M. Ali, K. S. Moon, Structural Developments in Tall Buildings: Current Trends and Future Prospects, *Architectural Science Review*, **50(3)**, 205-223 (2007)
6. J. Leonard, Investigation of Shear Lag Effect in High-rise Buildings with Diagrid System, *Master's thesis of Engineering*, (Massachusetts Institute of Technology, 2007)
7. M. H. M. Al-Sherrawi, G. A. Fadhil, Effect of stiffeners on shear lag in steel box girders, *Al-Khwarizmi Engineering Journal*, **8(2)**, 63-76 (2012)
8. V. Kristek, Z. P. Bazant, Shear lag effect and uncertainty in concrete box girder creep, *Journal of Structural Engineering*, **113(3)** (1987)
9. Y. Zhou, Analysis of the shear lag effect of cantilever box girder, *Engineering Review*, **34(3)**, 197-207 (2014)
10. R. K. Sorensen, *Evaluation of shear lag in standard H-/I-sections*, (Aalborg University Esbjerg, 2013)
11. B. S. Taranath, Structural analysis and design of tall buildings: Steel and composite construction (New York: McGraw-Hill, 1988)
12. Y. Singh, A. K. Nagpal, Negative shear lag in framed-tube buildings, *Journal of Structural Engineering*, **120(11)** (1994)
13. K.-K. Lee, H. Guan, and Y.-Ch. Loo, Simplified analysis of shear lag in framed tube structures with multiple internal tubes, *Computational Mechanics*, **26(5)**, 447-458 (2000)

14. H. A. Ghasemi, Optimal design of high-rise building bundled tube systems, *Advances in Science and Technology Research Journal*, **10(30)** (2016)
15. K. S. Moon, Dynamic relationship between technology and architecture in tall buildings, PhD thesis, (Massachusetts Institute of Technology, 2005)
16. F. Nouri, P. Ashtari, Investigation of the shear lag phenomenon and structural behavior of framed tube and braced tube tall structures, *Int. Conf. on Civil Engineering, Architecture & Urban Sustainable Development* (Tabriz, Iran, 2013)
17. Y. D. Nagvekar, M. P. Hampali, Analysis of shear lag effect in hollow structure, *International Journal of Engineering Research & Technology (IJERT)*, **3(7)** (2014)
18. H. Gaur, R. K. Goliya, Mitigating shear lag in tall buildings, *Int. Journal of Advanced Structural Engineering (IJASE)*, **7(3)**, 269-279 (2015)
19. J. C. D. Hoenderkamp, H. H. Snijder, Preliminary analysis of high-rise braced frames with façade riggers, *Journal of Structural Engineerinf*, **129(5)** (2003)
20. R. S. Nair, Belt trusses and basements as "virtual" outriggers for tall buildings, *Engineering Journal, AISC,* **35(4)** (1998)
21. Eurocode-8, Design of structures for earthquake resistance (2004)

Effect of aerated concrete blockwork joints on the heat transfer performance uniformity

Viktor Pukhkal[1] and *Vera* Murgul[2,3*]

[1]Saint Petersburg State University of Architecture and Civil Engineering, Vtoraya Krasnoarmeiskaya str. 4. St. Petersburg, 190005, Russia
[2]Peter the Great St.Petersburg Polytechnic University, Polytechnicheskaya, 29, St. Petersburg, 195251, Russia
[3]Moscow State University of Civil Engineering, Yaroslavskoe shosse, 26, Moscow, 129337, Russia

Abstract. Analysis of data on the effect of joints of the aerated concrete blocks on the heat transfer uniformity of exterior walls was carried out. It was concluded, that the values of the heat transfer performance uniformity factor in the literature sources were obtained for the regular fragment of a wall construction by approximate addition of thermal conductivities. Heat flow patterns for the aerated concrete exterior walls amid different values of the thermal conductivity factors and design ambient air temperature of -26 °C were calculated with the use of "ELCUT" software for modelling of thermal patterns by finite element method. There were defined the values for the heat transfer performance uniformity factor, reduced total thermal resistance and heat-flux density for the exterior walls. The calculated values of the heat transfer performance uniformity factors, as a function of the coefficient of thermal conductivity of aerated concrete blocks, differ from the known data by a more rigorous thermal and physical substantiation.

1 Introduction

Wall unreinforced products made of autoclaved aerated concrete have a number of advantages that have determined their wide use in Russia [1, 2]:

- availability of raw components;
- low fire hazard and high fire resistance;
- high accuracy of overall dimensions;
- Technological masonry and high productivity of work;
- low cost of products.

It is necessary to take into account the factors that affect the thermal performance of these products in order to effectively use them in the construction of buildings meeting the requirements of energy efficiency. One of these factors, characterizing the efficiency of insulation, is the heat transfer performance uniformity factor for the enclosing structures [3, 4]:

* Corresponding author: vera.murgul@mail.ru

$$r = \frac{R_0^r}{R_0^{res}}, \tag{1}$$

where R_0^r is the reduced total thermal resistance of the section of the enclosing structure, (m^2·°C)/W; R_0^{res} - thermal resistance of a section of a homogeneous enclosing structure, (m^2·°C)/W.

For example, consider the thermal imaging survey of an energy-efficient demonstration multi-apartment residential building (Building 5, Mebelnaya St., Porkhov, Pskov Region [1, 2]), which is constructed of aerated concrete blocks of D500 density and 375 mm thick. The survey revealed the influence of block joints on the uniformity of the heat flow field on the surface of the outer wall (Figure 1).

Fig. 1. Heat-conducting inclusions of aerated concrete blockworks in the form of joints

In [5-13], a significant heat transfer heterogeneity of aerated concrete blockwork is indicated. Data on the heat transfer uniformity factors are given for masonry of walls from solid-walled unreinforced products made of autoclaved aerated concrete with a blockwork size of 625x250 mm [6, 9]. Whereas, the calculation of the heat transfer uniformity factor is performed for a regular fragment of the wall structure by the addition of thermal conductivity.

The reduced total thermal resistance to heat transfer of a regular fragment of the enclosing structure is defined by the approximate dependence [3]:

$$R_0^r = \frac{A}{\sum_1^m \frac{A_i}{R_{0,i}^r}} \quad \text{(m2·°C)/W}, \qquad (2)$$

where A is the total area of the characteristic part of the enclosing structure, that equals to the sum of the areas of individual sections, m²; A_i - the area of the i-th section of the characteristic part of the enclosing structure, m²; $R_{0,i}^r$ - reduced total thermal resistance to the heat transfer of the i-th section of the characteristic part of the enclosing structure, (m²·°C)/W.

Water content of aerated concrete products proves to be higher during the operation of buildings than the design one, at which the thermophysical characteristics of the materials in [3] are given, and the values of the thermal conductivity factors for the products will be 2-3 times higher [5, 7, 8].

The use of approximate methods to calculate the heat transfer performance uniformity factor requires specification of the data. In addition, the maximum value of the thermal conductivity of aerated concrete blocks to determine the heat transfer performance uniformity factor is limited to 0.208 W/(m·°C).

2 Materials and Methods

The heat transfer uniformity factor of the enclosing structure can be defined on the basis of calculation of the heat flow field of the structure. Consider the two-dimensional heat flow field of structure of the exterior wall, which is made of aerated concrete blocks with the initial data according to [1, 2]:

- design ambient air temperature - minus 26 °C;
- design internal air temperature - plus 20 °C;
- the exterior wall is made of aerated concrete blocks of D500 density and 375 mm thick;
- the thickness of the horizontal and vertical blockwork joints is 10 mm.

A sand-cement mortar with a density of 1800 kg/m³ (thermal conductivity factor under operating conditions of the fencing B [3] - λ_m=0.93 W/(m·°C)) is used for brickwork.

The factor of thermal conductivity for the blocks (λ_b, W/(m·°C)) is assumed for the operating conditions B according to [8].

Calculation of the temperature field and heat fluxes of the external wall structure is carried out in the "ELCUT" software, where two-dimensional thermal fields are modelled.

The reduced total thermal resistance of a regular fragment of the enclosing structure is determined from the dependence:

$$R_0^r = \frac{t_{\text{int}} - t_{ext}}{q}, \text{·(m}^2\text{·°C)/W}, \qquad (3)$$

where t_{int} is the internal air temperature, °C; t_{ext} is the ambient air temperature, °C; q is the heat flux density for a regular fragment of the enclosing structure, W/m².

Design thermal resistance to heat transfer of a homogeneous enclosing structure [3] is:

$$R_0^{res} = \frac{1}{\alpha_{int}} + \sum_i \frac{\delta_i}{\lambda_i} + \frac{1}{\alpha_{ext}}, \text{ (m}^2\cdot{}^\circ\text{C)/W}, \tag{4}$$

where $\alpha_{int} = 8.7$ W/(m$^2\cdot$°C) is the heat transfer coefficient of the inner surface of the enclosing structure [3]; $\alpha_{ext} = 23$ W/(m$^2\cdot$°C) is the heat transfer coefficient of the outer surface of the enclosing structure [3]; δ_i is thickness of the i-th layer of the enclosing structure, m; λ_i is the thermal conductivity factor of the i-th layer of the enclosing structure, W/(m·°C).

3 Results

The heat flow fields of the exterior wall were calculated with the use of "ELCUT" software in the range of the thermal conductivity factor for aerated concrete blocks from 0.088 (block D300 - [1]) to 0.6 W/(m·°C). Example of calculation results is presented in Figure 2 in the form of heat flow fields and heat flux densities under design conditions.

a) heat flow field b) heat flux density

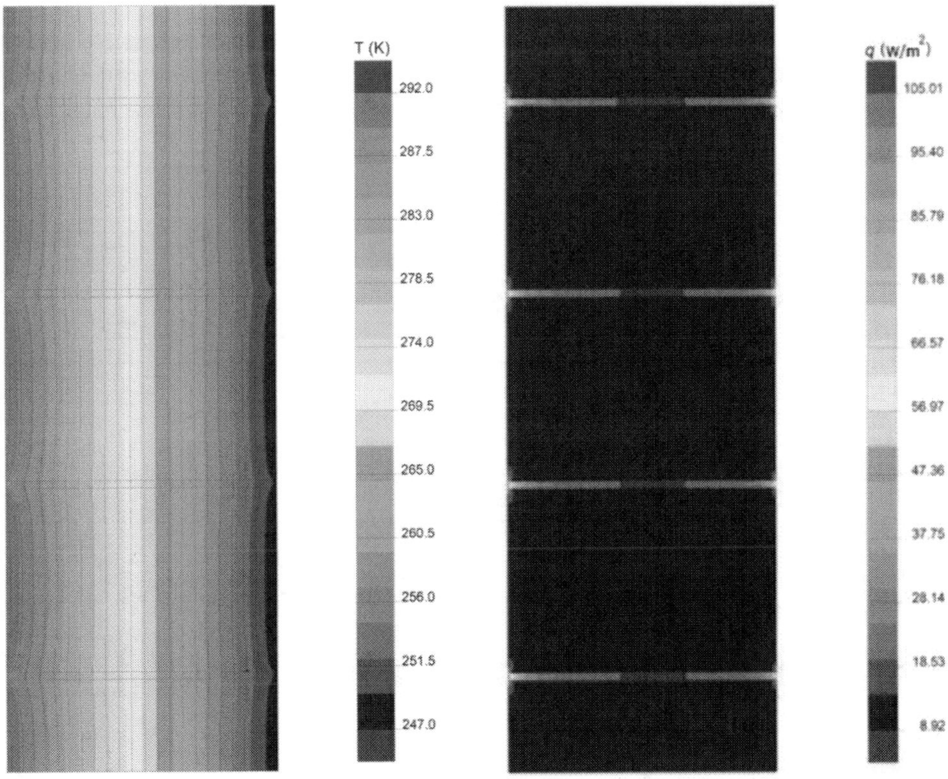

Fig. 2. Calculation example of the heat flow field and the heat flux density of the exterior wall made of aerated concrete blocks of D300 density

According to the calculations, the heat flux density through the enclosing structure and its thermal resistance are determined depending on the thermal conductivity factor of the aerated concrete blocks (Table 1). For comparison, data for a homogeneous structure are given (without taking into account the blockwork joints).

Table 1. Heat flux density through the aerated concrete blockworks and thermal resistance to heat transfer of the structure

Thermal conductivity factor of the block, λ_i, W/(m·°C)	Heat flux density, q_i, W/m²		Thermal resistance, (m²·°C)/W	
	homogeneous structure	blockwork	homogeneous structure, R_0^{res}	reduced, R_0^r
0.088	10.408	14.302	4.420	3.216
0.117	13.676	17.581	3.364	2.616
0.147	16.978	20.870	2.709	2.204
0.183	20.837	24.700	2.208	1.862
0.208	23.454	27.291	1.961	1.686
0.3	32.661	36.391	1.408	1.264
0.4	41.974	45.590	1.096	1.009
0.5	50.637	54.153	0.908	0.849
0.6	58.717	62.150	0.783	0.740

4 Discussion

According to the calculation given in Table 1, the values of the heat transfer performance uniformity factors of the enclosing structure are determined depending on the thermal conductivity factor of the aerated concrete blocks (Figure 3).

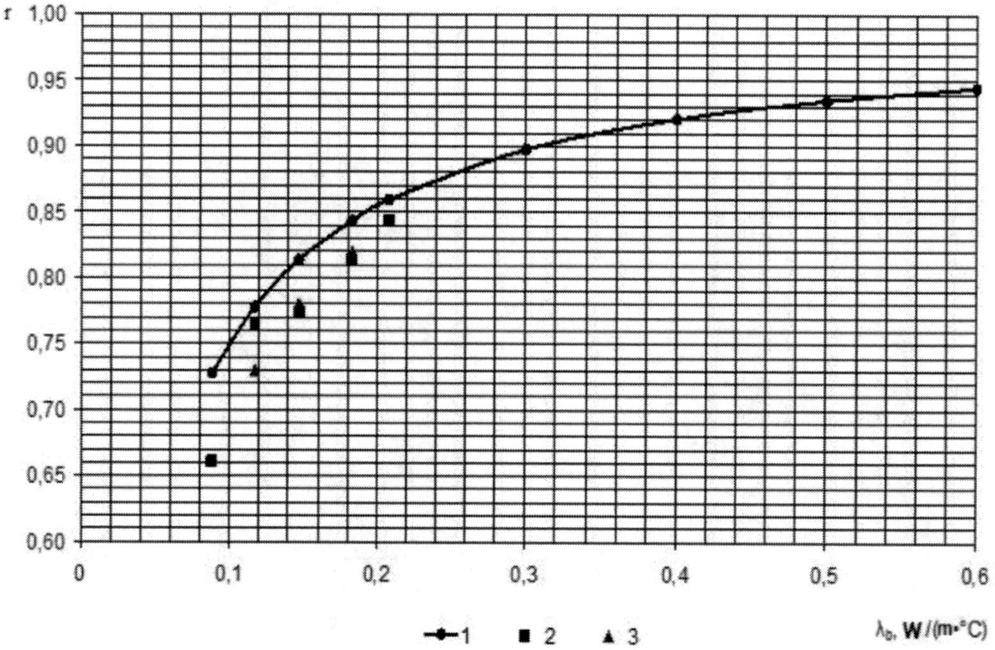

Fig. 3. Dependence of the heat transfer performance uniformity factor on the thermal conductivity factor of aerated concrete blocks: 1 – calculation results; 2 – data [9]; 3 – data [6]

As it follows from the performed calculations, an increase in the thermal conductivity factor (density) of aerated concrete blocks leads to an increase in the factor of heat transfer performance uniformity. The calculation results differ (up to 10%) from the values that are indicated in the literature [6, 9].

The relationship between the thermal conductivity factor of the joints material and aerated concrete blocks has a determining effect on the values of the heat transfer uniformity factor of the blockwork for a given thickness of the joints (Figure 4).

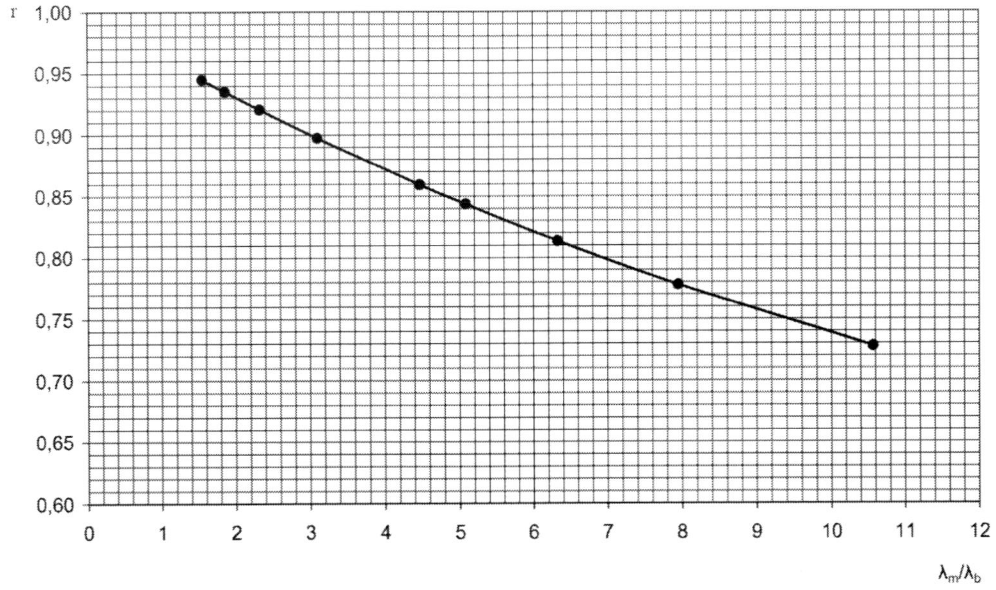

Fig. 4. Dependence of the heat transfer performance uniformity factor on the ratio of the thermal conductivity factors of the joints material and aerated concrete blocks

It is necessary to develop clamping compounds that are characterized by low values of the thermal conductivity factors in order to further improve the aerated concrete blockworks, when ensuring the required adhesion between the blocks and not impairing the performance of the enclosing structures in strength, stability, crack resistance, fire resistance, etc.

5 Conclusions

1. Design values of the heat transfer performance uniformity factor of the exterior walls of buildings made of aerated concrete can be determined using the "ELCUT" finite element modelling software, which is especially important at the stage of simulating various design solutions.
2. The calculation of the thermal state of the exterior enclosing structures out of aerated concrete and the calculated values of the heat transfer performance uniformity factor, depending on the thermal conductivity factor of the aerated concrete blocks, differ from the known data by a more rigorous thermophysical substantiation.

References

1. A. S. Gorshkov, P. A. Muravyev, A.V. Tarakin, Energosberezheniye, **8**, 30-34 (2016)
2. Muravyev P.A., Lukina-Lebedeva M.A., Tarakin A. V., Gorshkov A.S. Energoeffektivnyy demonstratsionnyy zhiloy dom, realizuyemyy po programme PROON-GEF v g. Porkhove. Inzhenernyye sistemy, № 4. 2016. S. 34-45.

3. Russian State Standard SP 50.13330.2012.
4. Russian State Standard RMD 23-16-2012
5. A.B. Krutilin, YU. A. Rykhlenok, V.V. Leshkevich, Magazine of Civil Engineering, **2(54)**, 46–55 (2015)
6. A.S. Gorshkov, A.A. Gladkikh, Magazine of Civil Engineering, **3(13)** 39–42 (2010)
7. V.A. Pinsker, V.P. Vylegzhanin, G.I. Grinfel'd, Magazine of Civil Engineering, **8(10)** 17–19 (2009)
8. N.I. Vatin, A.S. Gorshkov, S.V. Korniyenko, I.I. Pestryakov, Construction of Unique Buildings and Structures, **1(40)**, 78–101. (2016)
9. URL: http://www.hplush.ru/library
10. URL: https://elcut.ru/demo/manual.pdf
11. A. Solovyev, A. Pustovgar, A. Adamtsevich, L. Shilova, K. Degtyarev, D. Solovyev, Journal of Applied Engineering Science, **15**, 360- 370 (2017) DOI -10.5937/jaes15-14950
12. A.O. Adamtsevich, A.P. Pustovgar, S.A. Pashkevich, A.V. Eremin, Int. Journal of Applied Engineering Research, **11**, 1609 - 1611 (2016)
13. A. Solovyev, A. Pustovgar, L. Shilova, A. Adamtsevich, D. Solovev, Procedia Engineering, **165**, 1275 - 1280 (2016) DOI- 10.1016/j.proeng.2016.11.850

Energy efficiency of high-rise buildings

Anna Yu. Zhigulina[1*], Alla M. Ponomarenko [1]

[1] Samara State Technical University, Institute of Architecture and Civil Engineering, 194, Molodogvardeyskaya St., 443001, Samara, Russia

Abstract. The article is devoted to analysis of tendencies and advanced technologies in the field of energy supply and energy efficiency of tall buildings, to the history of the emergence of the concept of "efficiency" and its current interpretation. Also the article show the difference of evaluation criteria of the leading rating systems LEED and BREEAM. Authors reviewed the latest technologies applied in the construction of energy efficient buildings. Methodological approach to the design of tall buildings taking into account energy efficiency needs to include the primary energy saving; to seek the possibility of production and accumulation of alternative electric energy by converting energy from the sun and wind with the help of special technical devices; the application of regenerative technologies.

1 Introduction

High-rise construction, as the most progressive direction of the construction is particularly focused on the use of innovative technologies. For architects, designers and engineers of high-rise buildings are ideal experimental platform for testing different decisions of the construction of the tallest, the most stable, the cheapest, the lightest, the most efficient or greenest skyscrapers. The theme of this article is to analyze trends and advanced technologies in the field of energy supply and energy efficiency of tall buildings.

Energy efficient buildings owe their birth to the first global energy crisis in 1973. Humanity first began to think about the depletion of energy resources of the planet and energy conservation. At this stage it was seeking ways to reduce energy consumption for heating of buildings. The first experimental building for testing and identifying the best technical solutions for energy saving was built in Manchester (USA). It was an office building, designed by order of the General services Administration. The energy consumption of the building was reduced by the efficient use of solar radiation, double-layer fencing structures and computer control of engineering equipment of buildings [1-3]. The implementation of this project marked the beginning of the construction of energy efficient buildings around the world. Construction of the houses with low energy consumption are successfully implemented in Europe. According to various sources, Western European countries have already built about 10 thousand of such homes. Denmark, Germany and Finland are the leaders of this movement. In these countries

* Corresponding author: auzhigulina@mail.ru

the state programs of energy saving and construction of energy efficient buildings are accepted [4]. At the present stage the concept of "energy conservation" has expanded and turned into a notion of "efficiency". It implies a close intertwining of economy and ecology: people save resources and care about the environment [5]. The main principles of energy-efficient construction are maximum using of alternative renewable sources of energy such as thermal energy of earth, energy of sun, wind, and reducing the negative impact on the environment.

The presented research is part of study of modern trends of high-rise construction conducted by the Department ARPB IAC SamGTU [6-17].

2 Methods

Energy saving as the first stage of energy efficiency, involves a set of innovative architectural, structural and engineering measures to reduce consumption and the rational use of energy used for lighting, heating, ventilation, air-conditioning, operation of elevators, while maintaining a comfortable microclimate in the premises. It is known that heating, ventilation, air-conditioning of high-rise buildings require about 70 % of energy, lighting - 20%, maintenance of elevators - 10 %. The relevance of this direction is determined by the growth of energy consumption of cities in the urbanization process. According to modern statistics, about 40 % of extracted fuel in the world is spent on energy supply of buildings today.

The reduction in energy consumption can be achieved by the use of enclosing structures with optimal thermal characteristics for the climate zone of the construction site, the maximum use of reserves of natural lighting of premises with the use of larger windows. This is supported by the use of automated systems for climate control, reducing the temperature of the indoor air in the absence of people in the room; the device of night ventilation during the warm period of the year; installation of systems with variable air flow; the use of process heat to heat the cold zones and other activities [3,18].

Second, environmental, stage of efficiency – the use of alternative, renewable energy sources – solar collectors, heat pumps, wind power, thermal energy of the earth, biotechnology, etc [19].

There are several systems of rating the energy and environmental performance of buildings in the world. These systems form the standards of quality for the modern construction industry today. The most famous are the British BREEAM (Building Research Establishment Environmental Assessment Method) operating since 1990 and the us system LEED (Leadership in Energy and Enviromental Design) introduced in 1998. Both systems are rating, but the difference in approaches is that the LEED system is largely assesses the energy efficiency of buildings, and BREEAM – ecology efficiency [20-22].

The BREEAM standard differs by flexibility and adaptability to local norms and practices of construction, and LEED rigidly reglamentary using only US standards of certification and evaluation of materials. Therefore, the buildings received the BREEAM certificate may differ in the set of applied environmental elements. The buildings certified by LEED are easy to compare, since the demands of base level are standard.

BREEAM is created by the British state organization BRE Global together with the market participants. The BREEAM scheme of rating is popular not only in the UK. A version for the Gulf countries, the Netherlands, Spain, Norway, Sweden and Germany is developed now. More than 200 000 projects are certified and more than a million of registered buildings still have to go through this process today. The LEED system was developed by United States Green Building Council (USGBC) as a standard of measurement for projects of energy efficient, environmentally friendly and sustainable buildings for transition to the design, construction and operation of such buildings in the construction industry. About 278,7 million m^2 of built-up area were designed according to the LEED system today. The LEED system has expanded today, and is used as the design of new buildings and renovation of existing. The modern version of LEED – v4 (10.2013) includes 9 criteria in the field of design, construction and operation of buildings. Certification occurs at the stages of design, construction and reconstruction.

3 Results

Let`s consider several buildings, leading in the field of energy efficiency.

Skyscraper CIS Tower in Manchester (England) is lined with more than 7,000 solar panels that produce up to 180 megawatt-hours of electricity per year. The roof is equipped with wind turbines that produce about 10 % of the energy needed by the building.

Pearl River Tower, Guangzhou, China. 310-meter 71-storey building has a zero energy consumption from the external network. The South facade is supplied by double-glazed windows with ventilation between the panes to reduce heat of the building and the cost of air conditioning. Automatic blinds are rotated, depending of the angle of incidence of the sun rays, in the right direction. Tanks for rain water cleaning and recirculation are on the roof. Giant vertical wind turbines are inside the building on the technical floor. Photovoltaic panels and solar thermal collectors heat the water in the tower. According to lead architect Gordon, it is not just a building but a "high-performance tool that is generated by sun and wind." Together with architect Adrian Smith and engineer Roger Frechette Gill made a clear structure, in which the authors used virtually all of the technologies commonly used in green houses. For example, flooring is a system where circulates chilled liquid. Thus air conditioning is functioning. The engineers managed to avoid the bulk of the networks wich forced air cooling. The dehumidification system is also organized in a hidden ventilation shafts, where is situated a passive dehumidifier that removes excess moisture without the involvement of energy resources. Dry hot air will flow out of channels organized between the double glazing of the South facade. This vent glass will also reduce the overall heating of the building and, as a consequence, the cost of air conditioning.

Bank of America Tower (New York). 366-meter skyscraper made from recycled materials (used concrete with blast furnace slag). There is provided a system of accumulation and purification of rainwater, which save up to 30 million gallons of water per year.The facade is lined by the glass, which absorbes the solar radiation but provides maximum natural light. Double glazing reduces heat loss, thus reducing energy consumption. Autolighting system is regulated depending on the time of day. The cooling system produces ice reserves during the night hours to use its melting for cooling of the building in the morning and afternoon hours. The tower has a system that can produce 4.6 MW of electricity, providing part of the requirements of a skyscraper. Powerful generators eliminate losses during transport of electricity, which are characteristic for Central supply. The cost of Bank of America Tower is 1.2 billion US dollars and 3% of the total amount are given to the sustainability of the structure. Bank of America tower is the first skyscraper, which received Platinum LEED certification.

The Bahrain World Trade Centre. 240-metre complex of dual towers, located in Manama, Bahrain. It was built in 2008 by the Atkins construction firm. There are air bridges, using wind turbines in the design of these towers. The project has passed several inspections for compliance with environmental requirements. Two towers are connected by three bridges, each with a 225 kW wind turbines with a total capacity of 675 kW. Each of these turbines with a diameter of 29 meters is oriented to the North, bacause from the side of the Persian Gulf the wind blows largest number of days in a year. The towers in the shape of ships were designed in the shape of the tunnel, causing the wind passing through the tunnel, to accelerate and pass through the turbine. Beneficial use of the tunnel between the towers forming S-shaped flow was confirmed by tests showing that any wind coming at an angle of 45^{0} to each side of a Central axis will create a wind flow perpendicular to the turbines, which increases their potential for power generation. Wind turbines are designed to produce up to 1,100 MW of electricity per year, enough for functioning of all offices. This high energy efficiency is achieved by architectural and compositional form of the entire structure, allowing to create an accelerated airflow for the giant turbines. The turbines were included for the first time on 8 April 2008 and are designed to work 50 % of time daily.

Architects of Singapore Bureau WOHA, Richard Hassell and Wong Mun Summ in the course of its successful practice in the Asia-Pacific region has offered the concept of ecological breathable skyscrapers — high-rise residential buildings. WOHA architects were thinking about how to create a cottage settlement in a vertical format, and they succeeded. Located in the capital of Thailand Bangkok 69-storey residential building The Met has become a symbol of a new vision of tropical skyscrapers. Three towers with residential units are combined by bridges transitions with gardens and swimming pools every five floors. Perforated facade ignores the wind that now walks around the apartment blocks, winding through them without the aid of mechanical systems due to the bioclimatic planning of the blocks facing North and South. Structural design ensure the safety of the air tunnel on the court between the towers, for the people in case of fire or other emergency can leave the building through two adjacent towers. Shared pools in the open air between the towers, high-rise gardens, a green facade and balconies create pockets of nature even in the concrete jungle of Bangkok. Lifestyle, opened to outdoor environment and people, is possible thanks to landscape design large pool in the base unit, located there recreational areas, picnic grounds, halls for meetings or library, that allows to hold business or private meetings without leaving home from his own apartment in the green skyscraper.

Even more grandiose in area greening project implemented WOHA is located in Singapore. Residential complex Newton Suites breaks the traditional image of a polished high-rises, his shaggy paths are set at an angle sunproof metal grid, which does not block the view down. Protruding air terraces with airgardens, multi-storey vertical garden and living roof create the total area of landscaping equal to 130 % of the building footprint the horizontal area, what neutralizes its impact on the environment, carrying the function of regulation the indoor climate in premises with through ventilation. Surface Parking with natural lighting and ventilation is much more energy efficient than the underground, and the exhaust gas is absorbed by covering Parking with plants. Sports club, entertainment area and pool are situated above parking.

4 Discussion

Green technologies provide the ability to use energy resources effectively, to construct buildings more favourable from the point of view of the health of the people, to create a favourable environment and reduce operating costs. In environmentally friendly, so-called "green" buildings are applied engineering systems and technologies to increase the efficiency of resource use - energy, water and materials. At the same time technologies in these structures, are aimed at reducing the harmful effects of building on human health and the environment during the whole cycle of its existence. This is achieved by selecting the best construction, thoughtful design and construction solutions through operation, maintenance and removal of waste products. The practice of constructing high-rise buildings shows that the energy efficiency of buildings is achieved by implementation of primary energy savings and finding opportunities in the production and accumulation of alternative electric energy.
The variety and amount of creative and technological possibilities of modern architecture is so great, that allowes architects to make the right choice in favour of universal values and global objectives. Despite the limiting nature of each concept, the principles of sustainable architecture, including, of course, energy efficiency, do not stifle, but rather relieve the

imagination of the creators. In order not to be stymied in the search for aesthetic decision or the desire for basic functionality you must first imagine a better future world, and only then to pave the path towards him.

5 Conclusions

Used in the construction of high-rise buildings energy efficient strategies and technologies are evolved. Approbated at these sites optimized and effective methods need to enter into the practice of mass construction. Currently a promising vector for the development of energy efficient technologies is creating of "green" skyscrapers. This term refers to the use of construction technologies that use a variety of natural forms and resources to create healthy and comfortable conditions for human life. From the point of view of energy consumption, you must strive to create a passive (less than 45 kWh/m^2) buildings and buildings with zero energy consumption (less than 15 kWh/m^2). Innovative systems of bioclimatic design of tall buildings are designed to create more comfortable conditions for people's lives without the use of complex and costly technological solutions. The use of renewable energy sources is a symbol of developing technologies of the future.

References

1. Y. A. Tabunschikov, M. M. Brodach, N.V.Shilkin, *Energy Efficient Buildings* (Moscow, 2003)
2. Y. A. Tabunschikov, *Building materials, equipment, technologies of XXI century,* **4**, 22-23 (2002)
3. V. P. Generalov, *Features of design of high-rise buildings* (Samara, Kniga, 2007)
4. DIRECTIVE 2010/31/EU OF THE EUROPEAN PARLIAMENT AND OF THE COUNCIL of 19 May 2010 on the energy performance of buildings (recast). Official Journal of the European Union, 18.6.2010, L 153/13
5. N. A. Lekareva, *Urban Construction and Architecture*, **1**, 6-9 (Samara, Vestnik State University of Architecture and Civil Engineering, 2011)
6. E. M Generalova., V. P. Generalov, Modern trends in architecture. High-rise residential complexes as a form of mass affordable housing (for example, Hong Kong) *proceedings of Samara center of the Russian Academy of Sciences* 16, **2(2)**, 458-463 (2014)
7. E. Generalova, V. Generalov, Designing High-Rise Housing: The Singapore Experience *CTBUH Journal. Chicago, Illinois Institute of Technology*, **4**, 40-45 (2014)
8. V. P. Generalov, E. M. Generalova, Prerequisites for the formation of complexes of high-rise social housing in Hong Kong, *Traditions and innovations in construction and architecture: materials of the 71st anniversary all-Russian scientific-technical conference on the results of the NIR 2013,* 379-380 (Samara, 2014)
9. A. Yu. Zhigulina, Hygiene factors in the design of high-rise buildings *Traditions and innovations in construction and architecture: materials of the 71-th all-Russian scientific-technical conference on the results of the NIR 2013 – Samara state architectural-p.,* 384-385 (Univ. of Illinois, Samara, 2014)
10. A. Yu. Zhigulina, *Urban Construction and Architecture*, **1**, 29-30 (Samara, Vestnik State University of Architecture and Civil Engineering, 2011)
11. V. P. Generalov, E. M. Generalova, *Urban Construction and Architecture*, **1**(18), 13-18 (Samara, Vestnik State University of Architecture and Civil Engineering, 2015)
12. V. P. Generalov, E. M. Generalova, *Scientific review*, **3**, 163-167 (2015)
13. E. M. Generalova, V. P. Generalov, *Scientific review*, **11**, 46-51 (2015)

14. E. M. Generalova, V. P. Generalov, Flats in skyscrapers. Penthouse – planning features and innovations *Traditions and innovations in construction and architecture. Architecture and design: a collection of articles,* 40-45 (Samara, 2015)
15. E. Generalova, V. Generalov, Apartments in Skyscrapers: Innovations and Perspectives of their Typology Development, *Proceedings of the CTBUH 2015 International Conference «Global Interchanges: Resurgence of the Skyscraper City». New York, USA. 26th-30th October 2015*, 355-362 ISBN: 978-0-939493-44-9. (2015)
16. A. Yu. Zhigulin, A. M. Ponomarenko, the Construction of energy efficient houses in Russia *Traditions and innovations in construction and architecture. Architecture and design: a collection of articles,* 90-94 (Samara, 2016)
17. A. Yu. Zhigulina, *Urban Construction*, **2**, 84-84 (2012)
18. M. M. Brodach, Engineering equipment of high rise buildings (Moscow, 2003)
19. V. I. Ivanov, I. A. Bakhtina, T. Yu. Ivanova., S. V. Ilyin, *Urban Construction and Architecture*, **2**(19), 88-93 (Samara, Vestnik State University of Architecture and Civil Engineering, 2015)
20. A. Gutkin, *LEED rating system for energy efficient and environmentally friendly buildings,* AVOK, **6** (2008)
21. BREEM – now in Russia. *Green buildings J.*, **1** (2012)
22. A. Koryagina, *Certification System of green building. Commercial Property J.*,**2** (91) (2011)

Multi-criteria thermal evaluation of wall enclosures of high-rise buildings insulated products based on modified fibers

Alexey Pavlov[1*], *Larisa* Pavlova[1], and *Lyudmila* Pavlova[1]

[1]Samara Architectural and Civil Engineering University, 443001, Molodogvardeyskaya str., 194, Samara, Russia

Abstract. In article results of research of versions of offered types of heaters on the basis of products from the modified fibers for designing energy efficient building enclosures residential high-rise buildings are presented. Traditional building materials (reinforced concrete, brick, wood) are not able to provide the required value of thermal resistance in areas with a temperate and harsh Russia climate in a single-layered enclosing structure. It can be achieved in a multi-layered enclosing structure, where the decisive role is played by new insulating materials with high thermal properties. In general, modern design solutions for external walls are based on the use of new effective thermal insulation materials with the use of the latest technology. The relevance of the proposed topic is to research thermoinsulation properties of new mineral heaters. Theoretical researches of offered heaters from mineral wool on slime-colloidal binder, bentocolloid and microdispersed binders are carried out. In addition, theoretical studies were carried out with several types of facade systems. Comprehensive studies were conducted on the resistance to heat transfer, resistance to vapor permeation and air permeability. According to the received data, recommendations on the use of insulation types depending on the number of storeys of buildings are proposed.

1 Introduction

External walls for energy-efficient buildings are designed by multilayer in accordance with [1] with the use of effective thermal insulation materials. With the advent of new materials and systems of enclosing structures, great attention should be paid to understanding the physical processes occurring in the external walls, for their competent design and construction [2-4].

In addition to calculating the resistance to heat transfer, it is necessary to calculate the moisture state. Resistance to heat transfer, determined at the surface of the wall, is significantly different from the reduced resistance to heat transfer of the outer wall due to the presence and influence of window slopes, exterior corners of walls, partitions, ceilings, heat-conducting inclusions [5-7].

* Corresponding author: adisk63@yandex.ru

The loss of condensate in multilayered enclosing structures leads to an increase in the humidity of building and thermal insulation materials and to a deterioration in their heat-shielding properties. There is a need for calculating the moisture regime of building structures [8,9], since for an annual period of building operation there is accumulation of moisture inside the enclosing structure.

For a long time, the authors carried out theoretical studies of facade systems of multilayered wall fences [10-13], experimental studies [14-16] with various types of known insulants. The proposed work is devoted to the study of facade systems with insulation on new types of binders [17-19].

2 Materials and Methods

Theoretical studies were carried out with three options for the proposed insulation:
a) mineral wool on slime-colloidal binder;
b) mineral wool on bentocolloid binder;
c) mineral wool on microdispersed binder.

Depending on the location of the heater in the enclosure used two options for insulation:
• insulation - inside the enclosing structure;
• insulation - outside the enclosing structure.

In outside thermoinsulation two systems were used:
1 - plastering;
2 - front facing.

Advantages of the building technology of facade systems with air gap over single-walled walls have been known for a long time: air circulation and thermal radiation in the air gap provides a rapid removal of moisture from the interior of the building, supporting walls, insulation.

The multi-storey residential building in Samara has been calculate. The exterior wall constructions for theoretical substantiation have been checked in three versions, with an experimental insulation in three variants. Schemes of wall constructions and options for insulation are presented in Table 1.

Table 1. Constructions of the investigated walls and the options for insulation

The construction of the outside wall	Options for insulation	The materials of the layers
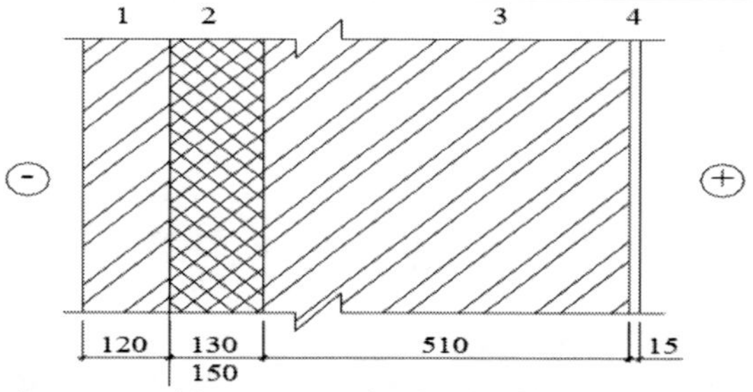 Scheme 1 - with a constructive layer of brick, thickness 0,51 m (λ_a = 0,7 W/(m°C). The insulation inside the wall. Outside – a finishing layer of ceramic bricks, thickness 0,12 m (λ_a = 0,47 W/(m°C).	a) mineral wool on slime-colloidal binder; b) mineral wool on bentocolloid binder; c) mineral wool on microdispersed binder.	1.Ceramic bricks (facing) 2.Insulation 3.Bricks 4.Lime-sand mortar

The construction of the outside wall	Options for insulation	The materials of the layers
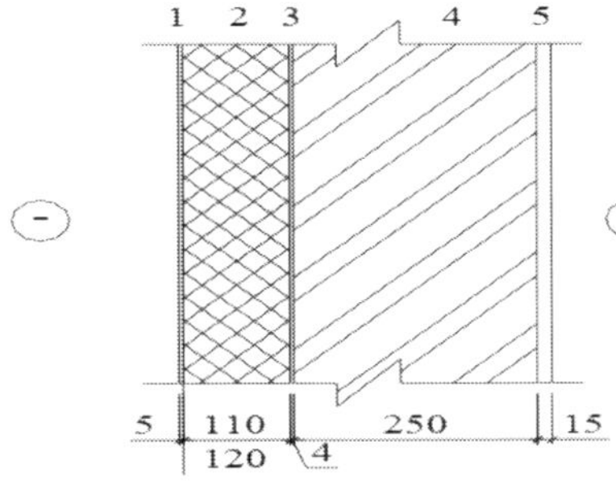 Scheme 2 - with a constructive layer of brick, thickness 0,25 m (λ_a = 0,7 W/(m°C). The outside insulation with thin layered plaster.	a) mineral wool on slime-colloidal binder; b) mineral wool on bentocolloid binder; c) mineral wool on microdispersed binder.	1. decorative plaster + plaster base with fiberglass 2.Insulation 3.Bricks 4.Lime-sand mortar
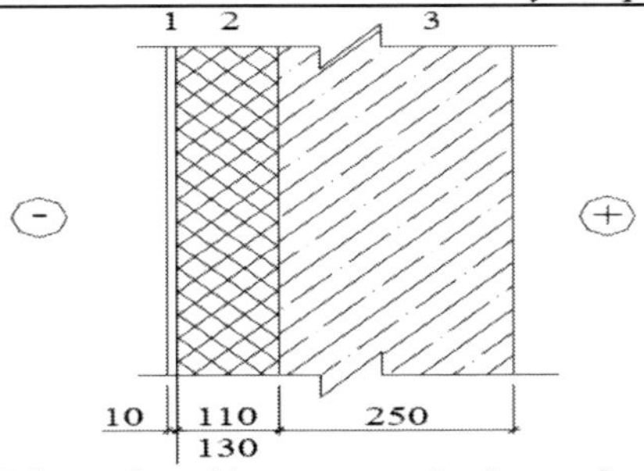 Scheme 3 - with a constructive layer of reinforced concrete, thickness 0,25 m (λ_a = 1,92 W/(m°C). The outside insulation with cement-particle board.	a) mineral wool on slime-colloidal binder; b) mineral wool on bentocolloid binder; c) mineral wool on microdispersed binder.	1. Cement-particle board 2. Insulation 3.Reinforced concrete

The technique of thermal calculation was based on the requirements [1] using the graphical method and the method of dimensionless characteristics for humidity areas. The method of calculation humidity conditions of external walls according to the method of dimensionless characteristics includes the determination of vapour pressure in the typical design cross sections for the annual cycle. If the vapor pressure exceeds the maximum elasticity of steam, in this section, the formation of condensate.

The basis of the constructive solutions of the exterior walls in determining the given resistance to heat transfer of the main fragments of the railings take the thickness of the insulation calculated in accordance with [1]. In case of insufficiency of insulation, thickness of insulation layer increases, or considering or considering the possibility of inclusion in the project energy saving measures (insulation of nodes).

The results of calculation of constructive schemes of walls 1-3 with options of insulation for residential buildings in Samara are presented in table 2.

Table 2. Results of structural analysis of the wall heat transfer resistance and water vapor

Construction scheme	Options for insulation	Insulation thickness, m	Thermal resistance R_0, $m^2{\cdot}°C/W$	Resistance vapor permeation R_n, $m^2{\cdot}h{\cdot}Pa/mg$	Required resistance vapor permeation R_n^{mp}, $м^2$·чПа/мг
1	a)	0,15	4,914	5,125	5
	b)	0,15	4,787	5,08	4,95
	c)	0,13	4,374	5,363	5,345
2	a)	0,11	3,3	2,764	0,065

	b)	0,12	3,448	2,752	0,066
	c)	0,11	3,265	3,01	0,067
3	a)	0,11	3,08	8,6	0,046
	b)	0,13	3,467	8,61	0,047
	c)	0,12	3,29	8,89	0,045

Calculations of temperature fields were carried out according to the design schemes 1-3 in the thickness of the external walls using the specialized program THERM 5.2. The results are shown in Figures 1, 2 and Table 3.

The necessity of calculation for the accumulation and condensation of moisture in building enclosures were improved.

The outer joints of the walls to the three options of heaters 1c), 2a), 3a) was chosen for thermal estimates of temperature change over the cross section of the walls.

The junction of walls with insulation inside the structure (layered masonry - see Figure 1a, 1b) does not pass heat transfer resistance (R_0=2,35 ($m^2 \cdot {}^0C$)/W < R^{req}), but passes sanitary and hygienic conditions (Δt^f=3,8 $^0C \leq \Delta t^n$ = 4 ^0C, as t_{int} = 16,9 $^0C > t_d$ = 16,2 ^0C - dew point).

Table 3. Results of calculation of temperature fields on the THERM 5.2

Construction scheme		No. of point starting from the outside surface	*T*, ^0C	*E*, Pa	*e*, Pa	Condition $e \leq E$	R_0, $m^2 \cdot {}^0C/W$	Recommendations
1	a	1	-27,6	48	720	-		Increased thickness of insulation from 0.11 m to 0.15 m, based on the calculation of resistance to vapor permeation
		2	-25,1	62	788,53	-		
		3	11,9	1323	823,77	+		
		4	19,1	2210	1273,87	+	4,914	
		5	19,3	2238	1286	+		
	b	1	-27,6	48	720	-		
		2	-25	63	789,07	-		
		3	11,7	1375	820,19	+		
		4	19	2197	1273,77	+	4,787	
		5	19,3	2238	1286	+		
	c	1	-27,6	48	720	-		
		2	-24,7	64	785,84	-		
		3	10,9	1304	841,99	+		
		4	19	2197	1274,34	+	4,374	
		5	19,2	2225	1286	+		
2	a	1	-27,4	48	720	-		The wall construction meets all requirements of heat engineering with the least thickness of insulation. The barrier structure has low vapor permeability
		2	-27	51	742,04	-		
		3	12,9	1488	794,39	+		
		4	13,3	1527	814,07	+	3,3	
		5	18,6	2142	1261,40	+		
		6	18,9	2182	1286	+		
	b	1	-27,6	48	720	-		
		2	-27,4	48	742,13	-		
		3	13	1497	792,33	+		
		4	13,6	1557	812,09	+		
		5	18,7	2156	1261,30	+	3,448	
		6	19	2197	1286	+		
	c	1	-27,4	48	720	-		
		2	-27,3	49	740,32	-		
		3	13,1	1508	832,69	+		
		4	13,3	1527	850,84	+		
		5	18,6	2142	1263,32	+	3,265	
		6	18,9	2182	1286	+		
3	a	1	-27,5	48	720	-		The wall is

Construction scheme		No. of point starting from the outside surface	T, °C	E, Pa	e, Pa	Condition $e \leq E$	R_0, $m^2 \cdot {}^0C/W$	Recommendations
		2	-27,2	49,3	736,80	-	3,08	designed to meet all the requirements of heat engineering. Non-optimal values of resistance to heat transfer
		3	16,8	1913	753,78	+		
		4	18,8	2169	1286	+		
	b	1	-27,5	48	720	-	3,467	
		2	-27,3	49	725,34	-		
		3	17,1	1949	743,25	+		
		4	19	2197	1286	+		
	c	1	-27,4	48	720	-	3,29	
		2	-27,2	49,3	725,24	-		
		3	17	1937	760,25	+		
		4	18,9	2182	1286	+		

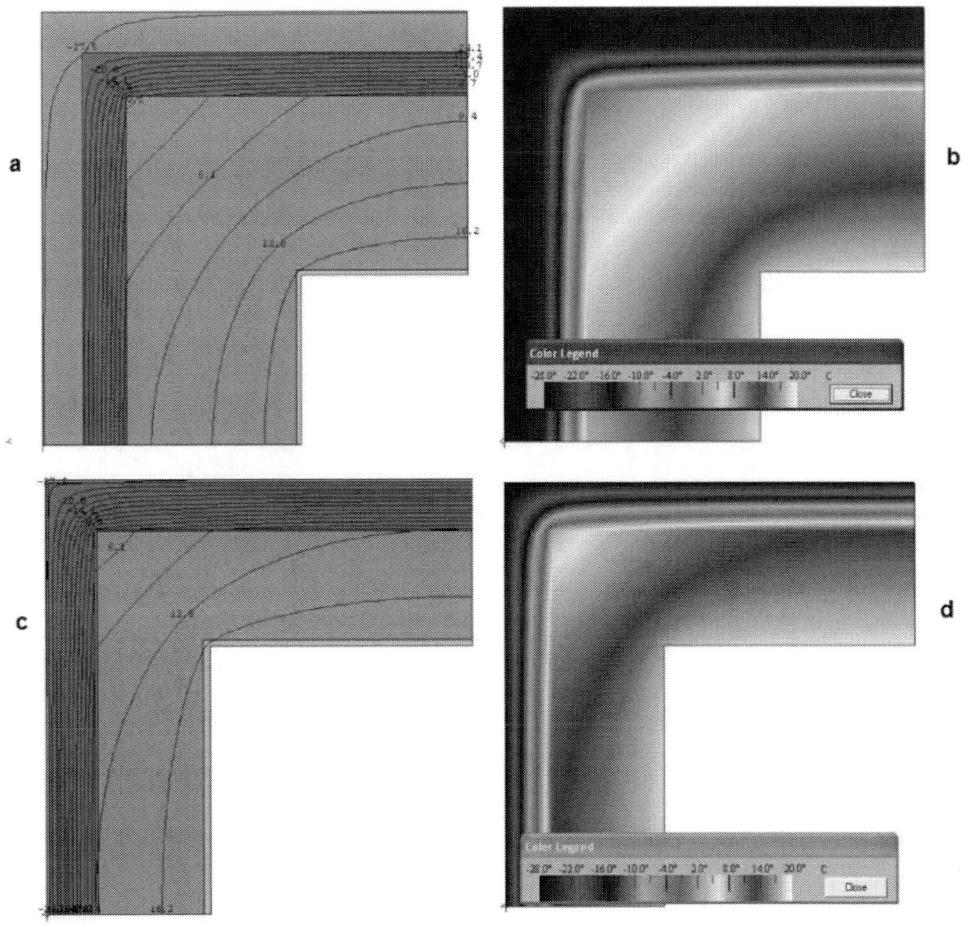

Fig. 1. Temperature fields along the cross-section of the outer joint of the walls:
a - temperature isotherms according to scheme 1c);
b - temperature distribution according to scheme 1c);
c - temperature isotherms according to scheme 2a);
d - temperature distribution according to scheme 2a)

The junction of walls, insulated on the outside (Fig. 1c, 1d) with fine plaster without the vapor barrier, the most effective (R_0=2,366 ($m^2 \cdot {}^0C$)/W, more than 2,35) and passes the sanitary and hygienic conditions (Δt^f=3,8 $^0C \leq \Delta t^n$ = 4 ^0C, as t_{int} = 16,6 $^0C > t_d$ = 16,2 ^0C).

The junction of walls, insulated on the outside (Fig.2 1a, 1b) with facing of the CPB does not pass on the resistance to heat transfer (R_0=2,333 ($m^2 \cdot {}^0C$)/W which is $\leq R^{req}$), but passes the sanitary and hygienic conditions (Δt^f=3,8 $^0C \leq \Delta t^n$ = 4 ^0C, as t_{int} = 17,2 $^0C > t_d$ = 16,2 ^0C). The insulation of the joint is required.

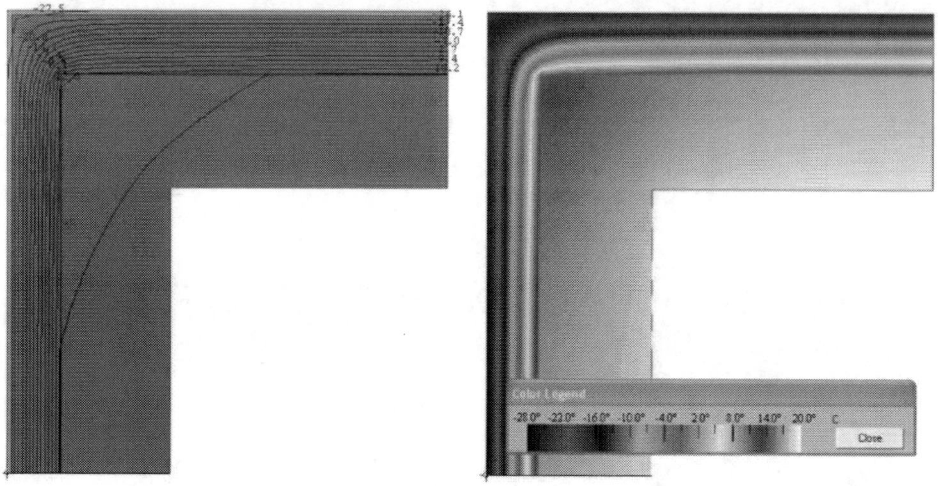

Fig. 2. Temperature fields along the cross-section of the outer joint of the walls:
a - temperature isotherms according to scheme 3a);
b - temperature distribution according to scheme 3a);

Resistance to air permeability of building envelopes and structures R_{inf}^{des} should not be less than specified resistance to air permeability R_{inf}^{req}, $m^2 \cdot h \cdot Pa/kg$, determined by the formula (1):

$$R_{\text{inf}}^{req} = \frac{\Delta p}{G_n}, \tag{1}$$

where ΔP - pressure difference of the air at the exterior and interior surfaces of building envelopes, Pa;

G_n - normalized air permeability of building envelopes, kg/($m^2 \cdot h$).

The pressure difference of the air on the outer and inner surfaces of enclosing structures, the Pa, to the windward side were determined according to [1], G_n the normalized air permeability, kg/($m^2 \cdot h$), at the table 9 [20]. Breathability G_n, kg/(m2·h), the exterior walls must be not more than 0,5.

Was determined pressure difference of the air on the outer and inner surfaces of walling on the ground floor of the building with the height of buildings 20, 40, 60, 100 m, and the normalized resistance to air permeability R_{inf}^{req}. According to the calculations on the water vapor permeability of scheme 3 based on resistance to air permeability of structures is not necessary. In the walls according to schemes 1 and 2, there are layers of plaster, which have a sufficiently high value of the resistance to air permeability.

The results of the structural analysis of walls with an experimental insulation on air permeability is presented in table 4.

Table 4. Results of the design of walls with experimental insulation for air permeability (scheme 1)

Scheme 1	H, m	$R_{inf\,turb}^{ins}$, m²·h·Pa/kg	R_{inf}^{des}, m²·h·Pa/kg	R_{inf}^{des}, m²·h·Pa/kg	Δp_i, Па	Condition $R_{inf}^{des} \geq R_{inf}^{des}$	Height, H_{onm}, m	No. of storeys
a) mineral wool on slime-colloidal binder	20 40 60 100	38,21 49,42 58,51 73,40	75,54 126,36 177,18 278,82	218,41 229,62 238,71 253,60	37,77 63,18 88,59 139,41	+ + + -	88,5	30
b) mineral wool on bentocolloid binder	20 40 60 100	22,82 29,51 34,95 43,84	75,54 126,36 177,18 278,82	203,02 209,71 215,15 224,04	37,77 63,18 88,59 139,41	+ + + -	76	25
c) mineral wool on microdispersed binder.	20 40 60 100	69,41 89,77 106,30 133,35	75,54 126,36 177,18 278,82	249,61 269,97 286,50 313,55	37,77 63,18 88,59 139,41	+ + + +	117	40

3 Discussion

1. Heat and humidity calculations showed that the investigated insulation can be applied in the system with an outer insulated layer and using them in construction with brick facing, along with basalt wool ROCKWOOL or polyester staple fiber URSA. All of the investigated insulation are non – combustible and environmentally friendly, which is one of the important factors in the structural design of the wall.

2. The design of the system with external insulation and thin plaster layer is achieved by the high thermal uniformity of the insulated exterior walls. The durability of the facade before the first overhaul is about 25 years. The comparison of the three schemes of wall constructions the wall construction with external insulation has more advantages for all types of characteristics. Dew point is shifted to the outer wall surface that prevents the accumulation of moisture in the thickness of the structure. Other advantages of this system is the lack of a vapor barrier, the possibility of realization of various architectural solutions, high maintainability.

3. Calculations on the resistance to vapor deposition and air permeability of the enclosing structures of residential buildings with insulation based on modified fibers (three schemes of wall constructions) have been shown:

- experimental insulation can be used in walls of residential buildings frameless construction system with insulation on the outside and frame system with brick facing, as they satisfy all heating requirements: resistance to heat, water vapor and lack of moisture in the enclosing structures of buildings;
- according to the resistance to air permeability, the proposed heaters are recommended for high-rise buildings. More effective from experimental insulants based on mineral fibers - using a microdispersed modifier, which can be used in buildings up to 40 storeys;
- in the construction of walls with facing brick and a thickness of the structural layer of 51 cm there is an increase of 1 cm in thickness of the proposed insulation - mineral wool

slabs with a microdispersed modifier does not require a vapor barrier layer and allows to meet the requirements of heat engineering in all respects. Other experimental heaters ensure the requirements of standards with an increase in thickness by 3 cm.

4. Studies and analysis of the results on air permeability showed that the proposed TIM on the basis of technogenic raw materials can be used in buildings up to 26 floors, and in some cases up to 40, whereas with a mineral wool insulation on a bentocolloid binder does not exceed 24 floors. The use of man-made waste as modifiers of polyfunctional action of heat-insulating materials allows to reduce the thickness of the insulation in comparison with the TIM on the bentocolloid binder.

References

1. Russian State Standard SP 50.13330.2012. Thermal performance of the buildings.
2. L. V. Pavlova, Insulation of walls and their interfaces. Abstracts/Tradition and innovation in construction and architecture: materials of the 67th scientific and technical conference on the results of the research in 2009. H.9/SGASU, 705-706 (Samara, 2010).
3. L. V. Pavlova, The Reliability and quality of thermal insulation of buildings, *Vestnik SGASU. Urban planning and architecture scientific-technical journal*, **4(12)**, 99-105 (Samara, 2013)
4. L. V. Pavlova, the Thermal protection of buildings, 40 (Samara, SGASU, 2011)
5. L. V. Pavlova, the Modern energy saving of the building envelope, Wall: textbook, 72 (Samara: SGASU, 2012) ISBN 978-5-9585-0461-9
6. L. V. Pavlova, A. A. Pavlov, Study of moisture in exterior walls//Actual problems in construction and architecture, *Education. Science. Practice/regional Materials 59 technical conference*, 407-408 (Samara, SamGASA 2002)
7. L. V. Pavlova, A. A. Pavlov, Study of moisture in the material of the exterior walls, *Modern investment processes and construction technologies/proceedings of the section "Construction" of the Russian Academy of engineering*, **2**, 211-217 (Moscow, 2002)
8. L. V. Pavlova, the Reliability and quality of thermal insulation of buildings. Abstracts, *Tradition and innovation in construction and architecture: materials of the 68th scientific and technical conference on the results of research for 2010*, **11**, 868-869 (Samara, 2011)
9. L. V. Pavlova, Investigation of thermal properties of brick walls, insulated "outside" and "inside", *Actual problems in construction and architecture, Education, Science, Practice, Materials of regional scientific 61.-technical conference on the results of the research in 2003*, **2**, 130 (SamGASA, Samara, 2003)
10. L. V. Pavlova, Prospects of development of the theory of thermal calculation of enclosures *Actual problems in construction and architecture, Education, Science, Practice, Materials 62nd all-Russian scientific-technical conference on the results of the NIR*, 2004 (SGASU, Samara, 2005)
11. L. V. Pavlova, the Defects of the exterior walls. Thermal protection, *Construction Vestnik Russia Eng. Academy, Proceedings of the section "Construction"* **6**, 270 S. K., Ed Russia Eng. Academy (Moscow, 2005)
12. L. V. Pavlova, The development of theory of heat engineering calculation of protecting designs, *Actual problems in construction and architecture, Education. Science, Practice, Materials 64 and all-Russian scientific-technical conference on the results of the research*, 564 (2005)

13. L. V. Pavlova, The Choice of insulation for exterior walls in brick buildings, *Actual problems in construction and architecture, Education. Science, Practice. The materials of the 65th scientific and technical conference on the results of NIR SamGASA* (SGASU, Samara, 2007)
14. L. V. Pavlova, The Reliability of the building envelope, *Reliability of building objects: proceedings of the Xth international scientific-technical conference* (September 2007), ed. by A. S. Lychev (SGASU, Samara, 2007)
15. L. V. Pavlova, The advantages and disadvantages of the design and operation of modern exterior walls, *Actual problems in construction and architecture. Education. Science. Practice. The materials of the 66th scientific and technical conference on the results of NIR SamGASA for 2007* (SGASU, Samara, 2008)
16. L. V. Pavlova, Study of the temperature-humidity condition of external walls of energy efficient buildings, *Actual problems in construction and architecture. Education. Science. Practice. The materials of the 66th scientific and technical conference on the results of NIR Samgasa in 2008*, **11**, 170-171 (SGASU, Samara, 2009)
17. S. F. Korenkov, Pat. 2298533 Russian Federation, IPC C04B14/46. Composition for the manufacture of hydrophobic mineral wool insulation materials / Korenkova S. F., Pavlov, A. A.; applicant and patentee VPO. No 2005121073/03; Appl. 05.07.2005; publ. 10.05.2007, bull. **13**, (2007)
18. A. A. Pavlov, Statistical analysis of chemical compositions of sludge wastes, *Building materials, equipment, technologies of XXI century*, **6**, 67 (Moscow, Publishing house LLC CSTI "Composite", 2007)
19. A. A. Pavlov, To the question about the use of a protective colloid coating for thermal insulation. materials based on mineral fibers, *Roofing and insulation materials*, **5(11)**, 26 (Moscow Publishing house LLC CSTI "Composite" 2006)

Complex analysis of energy efficiency in operated high-rise residential building: Case study

Sergey Korniyenko[1*]

[1]Volgograd State Technical University, Russia, 400005, Volgograd, Lenin avenue, 28

Abstract. Energy conservation and human thermal comfort enhancement in buildings is a topical issue of modern architecture and construction. The innovative solution of this problem makes it possible to enhance building ecological and maintenance safety, to reduce hydrocarbon fuel consumption, and to improve life standard of people. The requirements to increase of energy efficiency in buildings should be provided at all the stages of building's life cycle that is at the stage of design, construction and maintenance of buildings. The research purpose is complex analysis of energy efficiency in operated high-rise residential building. Many actions for building energy efficiency are realized according to the project; mainly it is the effective building envelope and engineering systems. Based on results of measurements the energy indicators of the building during annual period have been calculated. The main reason of increase in heat losses consists in the raised infiltration of external air in the building through a building envelope owing to the increased air permeability of windows and balcony doors (construction defects). Thermorenovation of the building based on ventilating and infiltration heat losses reduction through a building envelope allows reducing annual energy consumption. Energy efficiency assessment based on the total annual energy consumption of building, including energy indices for heating and a ventilation, hot water supply and electricity supply, in comparison with heating is more complete. The account of various components in building energy balance completely corresponds to modern direction of researches on energy conservation and thermal comfort enhancement in buildings.

1 Introduction

Energy saving and increase of energy efficiency in building is a topical issue of modern architecture and construction. The innovative solution of this problem makes it possible to enhance building ecological and maintenance safety, to reduce hydrocarbon fuel consumption, and to improve life standard of people. The requirements to increase of energy efficiency in buildings should be provided at all the stages of building's life cycle that is at the stage of design, construction and maintenance of buildings.

* Corresponding author: svkorn2009@yandex.ru

Considerable contribution to the development of the energy conception of building design in Russia was made by V.N. Bogoslovsky in the area of building thermal comfort and by Yu.A. Tabunshchikov in the area of considering the building as an entire energy system [2, 3]. This conception was developed in research works [4-20].

There are different ways of solving this problem.

Yu.A. Matrosov and I.N. Butovsky as well as other researchers [4-6] announced a new direction for the Russian national code that includes a performance requirement limiting buildings normalized annual heating energy.

As discussed by N.I. Vatin and A.S. Gorshkov [7-9] the problem of influence of building's thermal performance on heat-losses based on energy consumption modeling via heat balance method for energy performance of a building is of fundamental importance.

The author declares [10] that for increase in accuracy it is necessary to carry out a mathematic modeling of heat balance process in building envelopes with due account for thermal edge zones of building envelopes. For this purpose he developed the unique computing software [11]. As discussed elsewhere [12-14] based on computing software an analyses of thermotechnical defects of building components have been carried out.

Recently the problem of methodology of energy inspection in buildings was stated by many researchers [15-19].

Q.J. Kwong, N.M. Adam and B.B. Sahari [20] investigated thermal comfort and potential for energy efficiency enhancement in buildings.

The complex analysis of energy efficiency in residential and public buildings can be carried out based on design and measurement procedures. However, the absence of reliable and practically convenient methods of determination of energy indices in the Russian norms as well as considerable labour intensity of carrying out field measurements and processing of the obtained results complicate the solution of the above problem for operated high-rise residential buildings. Thus a topic of this original research is of importance.

The research purpose is complex analysis of energy efficiency in operated high-rise residential building.

The object of investigation is a residential multi-apartment building in Volgograd. Height of the building is 80 m. Number of floors is 23. Number of block-sections is 3. Number of flats is 234. There is heated basement in the building where there is a parking area and premises of engineering functions. There are offices on the first floor and engineering networks in the attic (Fig. 1a).

a b

Fig. 1. General views of the building: a — general view; b — individual heat supply unit.

https://doi.org/10.1051/e3sconf/20183302005

The following geometrical data are used for design object:

- Build-up area – 1093 m^2;
- Total area of the building – 25776 m^2;
- Total area of apartments – 17764 m^2;
- Envelope area – 19345 m^2;
- Heated volume — 87000 m^3;
- Glazing-to-wall ratio – 0.172;
- Building compactness ratio – 0.222 m^{-1}.

The building was built in 2007. At the moment of energy survey the life service of the building was about 5 years.

2 Materials and Methods

2.1 Climatic data

The following climatic data are used for design object:

- Location – Volgograd, Russia;
- Geographical coordinate – N 48°, E 44°;
- Height above sea level – 102 m;
- Monthly average outdoor air temperature in January – 6.9 °C below zero;
- Monthly average outdoor air temperature in July – 23.9 °C above zero;
- Design outdoor air temperature for heating – 22 °C below zero;
- Daily average outdoor air temperature during heating period – 2.3 °C below zero;
- Duration of the heating period – 176 day/year;
- Heating degree-day (HDD) – 3925 K·day/year.

2.2 Building envelope

Distinction between load-bearing and enclosing structures is a major feature of this design project. Load-bearing base of the building is a reinforced frame with precast framework with flat plates of slabs. There are perforated holes in slabs, filled with thermal inserts made of polyester foam, to decrease heat losses in the cold season along the outside contour of the slab. Heat protective cover of the building consists of different types of envelopes. Exterior walls are made of AAC 400 mm thick blocks in a form of masonry with exterior 120 mm thick brick facing. Interior surface of walls are plastered using complex mortar. Walls are supported floor-by-floor by plates of slabs. Brick masonry is connected to autoclaved aerated concrete blocks using flexible bonds. Double-glazed windows and balcony doors are made of PVC profiles (according to the design). Attic slabs and slabs over the basement are thermally insulated. Insulated bay-widows are assumed to increase level of natural lightning and to improve duration of insulation in rooms.

Thermal inserts made of efficient insulators are assumed to improve heat-conductive properties of the covers and the increase of heat protective properties. The estimation results of heat protection level and energy performance in the building and actual energy performance under field conditions are given below. Element-by-element estimation of heat protection indicators in the building are given in the Table 1.

Table 1. Element-by-element heat protection in the building based on design data.

Building envelope	Rated resistance to heat transfer R_o^{req}, $m^2 \cdot K/W$ (SP 50.13330.2012)		Estimated resistance to heat transfer R_o^{des}, $m^2 \cdot K/W$
	Minimum allowed level of heat protection	Basic level of heat protection	
Exterior walls	1.75	2.77	2.14
Windows, balcony doors	0.42	0.44	0.54
Entrance doors	0.42	0.44	0.54
Floor	3.33	4.16	3.84
Slab over basement	2.94	3.67	3.22
Slab under bay-windows	3.33	4.16	3.56

2.3 Engineering systems

There is a centralized heat supply, which source is an individual heat supply unit connected to outdoor heating networks. The individual heat supply unit is arranged in the basement. There is a double pipe heating system with bottom-pipe distribution and flat-by-flat perimeter distribution. Convectors with temperature regulators are used as heaters. Estimated parameters of a heat carrier for heating are 85/65 °C. The building is naturally ventilated. Air intake is secured through regulated window casement in living rooms and kitchens; air recovery is carried out through exhaust ventilation in kitchens and sanitary facilities. Engineering systems of buildings are equipped with metering devices for heat energy, cold and hot water and electric power (Fig. 1b).

2.4 Energy-saving technologies in the building

The following modern energy-saving technologies are realized in the building:

- Compact form of the building;
- Use of block-sections;
- Decrease in the area of windows according to requirements of natural illumination;
- Use of tambours at entrances to the building;
- Rational orientation of the building on sides of the horizon;
- Use of effective heat-insulating materials in the building envelope;
- Application of individual heat supply unit with automatic control of energy consumption;
- Use of energy efficient engineering systems and effective types of heaters.

The detailed description of the energy-saving activities applied in the project is given in article [11].

2.5 Energy efficiency estimation

The real structure of annual energy consumption in the building is complex and includes thermal energy on heating and ventilation, thermal energy on hot water supply and electrical energy supply in public place (Fig. 2).

The specific energy consumption in engineering systems during the annual period (or heating period) can be used for assessment of energy efficiency in building. The specific energy consumption is determined as ratio of the total energy consumption to the summary area of apartments. Criterion of energy efficiency in buildings is the following inequation:

$$q \le q^{bas}, \quad (1)$$

where q – design (or actual) specific energy consumption; qbas – rated specific energy consumption.

The specific energy consumption q includes the consumption on a heating and ventilation, hot water supply and electrical energy. Generally the total energy consumption Q, kWh/year, is calculated according to the formula:

$$Q = \int_{\tau_1}^{\tau_2} Q_h dz + \int_{\tau_3}^{\tau_4} Q_{hw} dz + \int_{\tau_5}^{\tau_6} Q_e dz, \quad (2)$$

where Qh, Qhw, Qe – respectively, the heat flow on a heating, hot water supply and electric power; (τ2–τ1), (τ4–τ3), (τ6–τ5) – respectively, heating, hot water supply and electrical supply periods.

The amount of thermal energy necessary for heating of the building is defined for each time point as result of summation of heat losses and heat gain, i.e. a solution of the equation of thermal balance of the building.

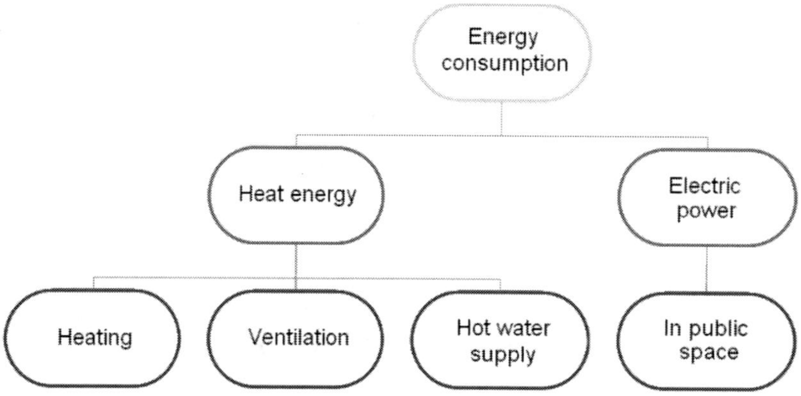

Fig. 2. Structure of annual energy consumption in the building (according to Order No. 399/Pr of Russian Ministry of Building).

Rated specific energy consumption qbas is determined for the base building which has design geometrical characteristics, minimum permissible level of building envelope thermal performance and standard engineering systems. Energy efficiency class of a building is defined on relative departure of specific energy consumption q from the rated value qbas. The computer program developed by the author makes it possible to computing building energy indices during heating period (Fig.3).

It was stated in [10] that energy consumption for heating include transmission heat loss through building envelope considering the edge zones, heat loss due to air ventilation and infiltration, indoor domestic heat generation, heat gain due to solar radiation. Preparation of initial data for computing is intuitive and easy-to-use. The calculation results in the computer program are presented in a user-friendly form. There is data base for information storage and accumulation. The computer program has modular structure that allows its further enhancement.

The measurement and registration of energy consumption indicators in the building were completed through express-method within two weeks starting on 22.12.2011 to 04.01.2012. The application of the express method allows reducing significantly the labour intensity as well as the duration of the field observation. The monitoring of the thermal energy consumption for heating and hot water supply was carried out separately on an automatic basis once a day according to the building communal heat meter (see figure 1b). The electrical energy consumption measurement was carried out on flat-by-flat basis taking into account the building communal meter reading.

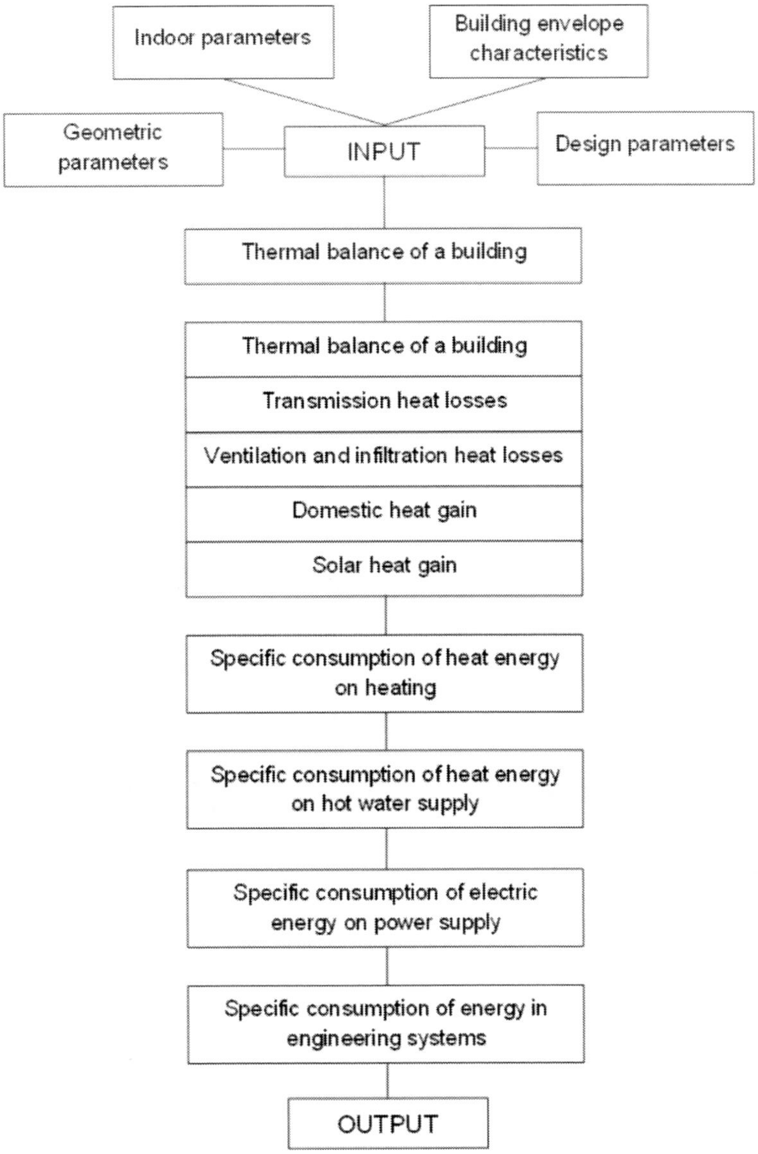

Fig. 3. Program flow chart developed by the author

Based on results of measurements the energy indicators of the building during annual period have been calculated. Dependence between a specific heating index q_h', W/(m^3·K), and a specific energy consumption q_h, kWh/(m^2·year), on heating and ventilation of a building is calculated according to the formula:

$$q_h = \frac{0.024 q_h' V_h HDD}{A_a}, \quad (3)$$

where 0.024 kWh/(W·day) – factor; V_h – heated volume, m^3; HDD – heating degree-day, K·day/year; A_a — total area of apartments, m^2.

Assessment of an energy efficiency of the building is executed by two methods:

- Using the specific heating index q_h' during the heating period;
- According to the specific energy consumption q_h on a heating and ventilation, hot water supply and electrical energy in public space.

3 Results and Discussion

The results of measuring the building heating energy consumption are given below.

- Transmission heat losses index q_{tr} = 0.149 W/(m^3·K);
- Ventilation and infiltration heat losses index q_v = 0.243 W/(m^3·K);
- Domestic heat gain index q_i = 0.076 W/(m^3·K);
- Solar heat gain index q_s = 0.037 W/(m^3·K).

The specific actual heating index q_h' = 0.279 W/(m^3·K).

The analysis of the results gives the following structure of thermal balance (Fig. 4a): transmission heat losses — 30%, ventilation and infiltration heat losses – 48%, domestic heat gain – 15%, solar heat gain – 7%. Ventilation and infiltration heat losses are the largest "specific weight" one within building thermal balance. Therefore, it is possible to increase the building energy efficiency first of all by means of reducing the ventilation and infiltration heat losses in the building.

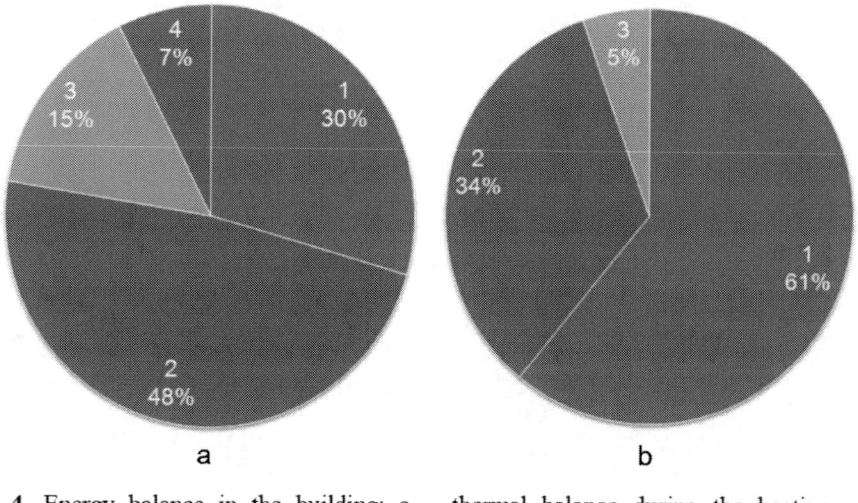

Fig. 4. Energy balance in the building: a – thermal balance during the heating period (1 – transmission heat losses; 2 – ventilation and infiltration heat losses; 3 – domestic heat gain; 4 – solar heat gain); b – energy consumption during the annual period (1 – heating; 2 – hot water supply; 3 – electrical energy supply in public place).

The main reason of increase in heat losses consists in the raised infiltration of external air in the building through a building envelope owing to the increased air permeability of windows and balcony doors. It is caused by various construction defects. The actual air change rate in the building (about 0.7 h–1) is more than design value (0.4 h–1). Minimum air change rate namin, h–1, in the building during the heating period is calculated by simplified formula:

$$n_a^{min} = \frac{30m}{\beta_v V_h}, \quad (4)$$

where m — number of persons in the building; β_v — correction index; V_h — heated volume.

According to the formula (4), using design data, the following value of minimum air change can be calculated: $n_a^{min} = 0.23\ h^{-1}$.

For ensuring the normalized air exchange at the equipment of rooms only with the exhaust shaft ventilation it is necessary to provide adjustable affluent devices in windows (thermorenovation). It is also necessary to reduce as much as possible energy consumption by external air infiltration through the building envelope.

The results of measuring the building energy indices during the annual period are given below.

- Heating index q_h = 129 kWh/(m²·year);
- Hot water supply index q_{hw} = 71.5 kWh/(m²·year);
- Electrical energy index (in public place) q_e = 11.3 kWh/(m²·year).

The total actual energy index q = 212 kWh/(m²·year).

The analysis of the results gives the following structure of energy consumption (see figure 4b): for heating — 61%, for hot water supply — 34%, for electricity supply — 5%. Heating is the "heaviest" one within building energy consumption. Therefore, it is possible to increase the building energy efficiency first of all by means of reducing the thermal energy consumption for heating.

The results of energy efficiency class assessment for the building based on heating energy consumption during the heating period are given below.

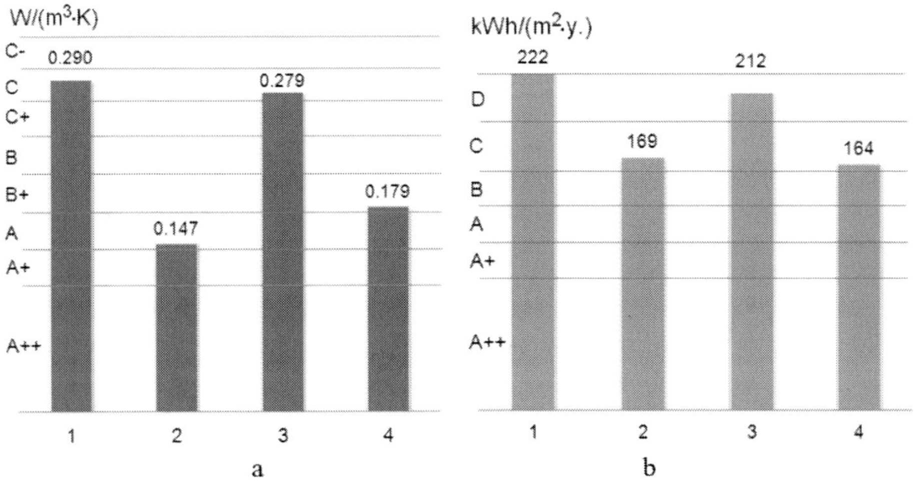

Fig. 5. Estimation of building energy efficiency class: a – heating (1 – required (standardized) level; 2 – designed level; 3 – actual level; 4 – expected level after thermorenovation; classes: A – very high; B – high; C – normal); b – total energy consumption (1 – required (standardized) level; 2 – designed level; 3 – actual level; 4 – expected level after thermorenovation; classes: A++ – highest; A+ – highest; A – very high; B – high; C – raised; D – normal).

According to the figure 5a for buildings under construction the energy efficiency class is C (normal). Use of energy saving technologies at design allows raising the building energy efficiency class to level "A" (very high). At construction various deviations from the design reduce the building energy efficiency class. Deviation of the actual level of energy consumption from design level is 90%. After thermorenovation the expected level of energy consumption approaches design level. Deviation of the expected level of energy consumption from design level is approximately 22%. The results of energy efficiency class assessment for the building based on total energy consumption during the annual year are given below.

According to the figure 5b for buildings under construction the energy efficiency class is D (normal). Use of energy saving technologies at design allows raising the building energy efficiency class to level "C" (raised). At construction various deviations from the design reduce the building energy efficiency class. Deviation of the actual level of energy consumption from design level is about 25%. After thermorenovation the expected level of energy consumption approaches design level. Deviation of the expected level of energy consumption from design level is minus 3%.

Comparing results of the building energy efficiency classes assessment based on various standards, i.e. on heating and annual total energy consumption, it is possible to note that energy consumption indices of the building will be approved with standard requirements at various stages of lifecycle of the building. Inequation (1) is fulfilled, so the building inspected conforms to the energy efficiency requirement. Thermorenovation of the building based on ventilating and infiltration heat losses reduction through a building envelope allows reducing annual energy consumption. Energy efficiency assessment based on the total annual energy consumption of building, including energy indices for heating and a ventilation, hot water supply and electricity supply, in comparison with heating is more complete. The account of various components in building energy balance completely corresponds to modern direction of researches on energy conservation and thermal comfort enhancement in buildings [3].

4 Conclusions

The research purpose is complex analysis of energy efficiency in operated high-rise residential building. Many actions for building energy efficiency are realized according to the project; mainly it is the effective building envelope and engineering systems. Based on results of measurements the energy indicators of the building during annual period have been calculated. The main reason of increase in heat losses consists in the raised infiltration of external air in the building through a building envelope owing to the increased air permeability of windows and balcony doors (construction defects). For ensuring the normalized air exchange at the equipment of rooms only with the exhaust shaft ventilation it is necessary to provide adjustable affluent devices in windows. It is also necessary to reduce as much as possible energy consumption by external air infiltration through the building envelope.

Comparing results of the building energy efficiency classes assessment based on various standards, i.e. on heating and annual total energy consumption, it is possible to note that energy consumption indices of the building will be approved with standard requirements at various stages of lifecycle of the building. Thermorenovation of the building based on ventilating and infiltration heat losses reduction through a building envelope allows reducing annual energy consumption. Energy efficiency assessment based on the total annual energy consumption of building, including energy indices for heating and a ventilation, hot water supply and electricity supply, in comparison with heating is more complete. The account of various components in building energy balance completely corresponds to modern direction of researches on energy conservation and thermal comfort enhancement in buildings [3].

References

1. V. N. Bogoslovskiy, Building Thermal Comfort, 248 (Moscow, Stroyizdat, 1979)

2. Yu. A. Tabunshchikov, M. M. Brodach, *Mathematical Simulation and Optimization of Thermal Efficiency for Buildings*, 194 (Moscow, AVOK-PRESS, 2002)
3. Yu. A. Tabunshchikov, M. M. Brodach, N. V. Shilkin, *Energy-Efficient Buildings*, 200 (Moscow, AVOK-PRESS, 2003)
4. Yu. A. Matrosov, I. N. Butovsky, R. K. Watson, Case studies of energy consumption in residential buildings in Russia's middle belt area, *Energy and Buildings*, **20(3)**, 231–241 (1994)
5. M. W. Opitz, L. K. Norford, Yu. A. Matrosov, I. N. Butovsky, Energy consumption and conservation in the Russian apartment building stock, *Energy and Buildings*, **25(1)**, 75–92 (1997)
6. Yu. A. Matrosov, L. K. Norford, M. W. Opitz, I. N. Butovsky, Standards for heating energy use in Russian buildings: a review and a report of recent progress, *Energy and Buildings*, 25(3), 207–222 (1997)
7. N. I. Vatin, D. V. Nemova, P. P. Rymkevich, A. S. Gorshkov, Influence of building envelope thermal protection on heat loss value in the building, *Magazine of Civil Engineering*, **8(34)**, 4–14 (2012)
8. N. Vatin, O. Gamayunova, Energy saving at home, *Applied Mechanics and Materials*, **672–674**, 550–553 (2014)
9. N. Harmati, Z. Jakšić, N. Vatin, Energy consumption modeling via heat balance method for energy performance of a building, *Procedia Engineering*, **117**, 924–937 (2015)
10. S. V. Korniyenko, Multifactorial forecast of thermal behavior in building envelope elements, *Magazine of Civil Engineering*, **52(8)**, 25–37 (2014)
11. S V Korniyenko, *Applied Mechanics and Materials*, **618**, 509–513 (2014)
12. S. Korniyenko, Evaluation of thermal performance of residential building envelope, *Procedia Engineering*, **117(1)**, 191–196 (2015)
13. S. V. Korniyenko, N. I. Vatin, A. S. Gorshkov, Thermophysical field testing of residential buildings made of autoclaved aerated concrete blocks, *Magazine of Civil Engineering*, **64(4)**, 10–25 (2016)
14. S. Korniyenko, Advanced hygrothermal performance of building component at reconstruction of S. Radonezhskiy temple in Volgograd, *MATEC Web of Conferences*, **53**, 01003 (2016)
15. V. Motuzienė, K. Valančius, G. Rynkun, Complex analysis of energy efficiency of public buildings: case study of VGTU, *Magazine of Civil Engineering*, **2(28)**, 9–17 (2012)
16. R. McKenna, E. Merkel, D. Fehrenbach, S. Mehne, W. Fichtner, *Building and Environment*, **62**, 77–88 (2013)
17. W. O. Brown Nils, T. Malmqvist, W. Bai, M. Molinari, *Building and Environment*, **61**, 140–148 (2013)
18. H. Sozer, Improving energy efficiency through the design of the building envelope, *Building and Environment*, **45(12)**, 2581–2593 (2010)
19. A. Roetzel, A. Tsangrassoulis, U. *Building and Environment*, **71**, 165–175 (2014)
20. Q. J. Kwong, N. M. Adam, B. B. Sahari, *Energy and Buildings*, **68**, 547–557 (2014)

Investigation the effect of outdoor air infiltration on the heat-shielding characteristics the outer walls of high-rise buildings

Yu.S. Vytchikov[1], *A. B.* Kostuganov[1], *M. E.* Saparev[1*], *I. G.* Belyakov[1]

[1]Samara State Technical University, 443100, Molodogvardeyskaya St., 244, Samara, Russia

Abstract. The presented article considers the influence of infiltrated outdoor air on the heat-shielding characteristics of the exterior walls of modern residential and public buildings. A review of the sources devoted to this problem confirmed its relevance at the present time, especially for high-rise buildings. The authors of the article analyzed the effect of longitudinal and transverse air infiltration on the heat-shielding characteristics of the outer wall of a 25-story building that was built in Samara. The results showed a significant reduction of the reduced resistance to the heat transfer of the outer wall when air is infiltrated through it. There are the results of full-scale examination of external walls to confirm the calculated data. Based on the results of the study carried out by the authors of the article, general recommendations on the internal finishing of the outer walls of high-rise buildings are given.

1 Introduction

At present, in connection with the implementation the energy saving program in the Russian Federation, significant changes have occurred in the constructive decisions of the outer walls of high-rise buildings. In the multi-storey buildings practice the most common are the exterior wall constructions shown in Figure1. The thickness of the heat- insulation material or masonry made of lightweight stones is determined based on the provision of regulatory requirements for thermal protection of buildings, guided by SP 50.13330.2012 "Thermal protection of buildings."

There are two types of air filtration in building enveloping structures - transverse and longitudinal filtration. The transverse air filtration is the movement of air in the structure, directed perpendicular to the plane of the wall surface, the longitudinal filtration -parallel to the plane of the wall. The works of Gagarin V.G., Kozlov V.G., Sadchikov A.V. [1-3, 18], Vytchikov Yu. S. [4, 6-8], Petrichenko M.R. [5] are devoted to the questions connected with research of influence of air longitudinal infiltration on heat-protective characteristics of external walls with ventilating facades [9-11, 13-17, 19-20].

Influence of longitudinal infiltration on the heat-shielding characteristics of exterior walls insulated by ventilated facades was considered in detail in work [1]. The influence of insulation density on the resistance to heat transfer of the outer wall is established on the

* Corresponding author: msx072007@yandex.ru

basis of experimental and scientific studies on the example of staple glass fiber products produced by "URSA" companies. With increasing density of staple glass fiber from 15 to 50 kg/m^3, the coefficient of thermal effect of longitudinal filtration increases from 0.77 to 0.98. By using mineral wool boards of increased density (more than 80 kg/m^3), the need in wind protection measures disappears. The purpose of this research is to study the effect of outdoor air infiltration on the thermal protection characteristics the exterior walls of modern high-rise buildings.

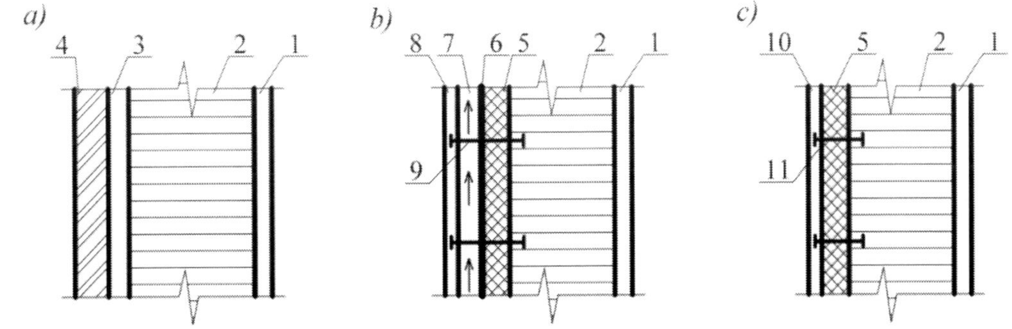

Fig.1. Constructive decisions for exterior walls of high-rise buildings: a) external wall made in the masonry's form of lightweight concrete stones; b) external wall with ventilated facade; c) external wall insulated with a facade system with thin-walled plaster: 1- internal plaster: 2- masonry from lightweight concrete stones on cement-sand or warm mortars; 3- unventilated air layer; 4- ceramic facing brick; 5- heat-insulation material; 6- windproof membrane; 7- ventilated air interlayer; 8- decorative panel; 9- metal bracket; 10- texture layer of the facade system; 11- plastic dowel.

2 Materials and Methods

Assessment of the effect the transverse filtration of outside air on the resistance to heat transfer the outer wall of a 25-storey building is made for the construction shown in Fig. 1b. The value of the heat transfer coefficient the outer wall, taking into account the infiltration of the outside air, is determined according to [6, 12] by the formula

$$k = \frac{c_p \cdot G \cdot e^{c_p \cdot G \cdot R_0^{usl.}}}{e^{c_p \cdot G \cdot R_0^{usl.}} - 1}, \text{W/(m}^2 \cdot {}^0\text{C)} \qquad (1)$$

Where c_p is the specific isobaric heat capacity of the air, J/(kg·^{0}C); G - air permeability of the outer wall, kg/(m^2·h); R_0^{usl} - resistance to heat transfer from smooth surface the outer wall, (m^2·^{0}C)/W.

According to SP 50.13330.2012 the air permeability of the outer wall is determined by the formula

$$G = \frac{\Delta P}{Ru}, \text{ kg/(m}^2\text{·h)} \qquad (2)$$

Where $Ru = \sum_{i=1}^{n} R_{ui}$ - resistance to air permeability of the outer wall, (m^2·h·Pa)/kg, R_{ui}- resistance to air permeability of the i-th layer of the outer wall, (m^2·h·Pa)/kg.

To determine the air permeability of the outer wall, it is necessary to have information on the coefficients of air permeability of the used materials, which are determined reference from the normative literature.

At present, there is no information on the air permeability of facade systems using thin-walled plasters, in the accredited Laboratory of Thermal Engineering Testing ACI SamSTU the air permeability of the most common heat insulation materials was tested on an aerodynamic stand, which was developed by the laboratory staff. The results of the investigations are presented in works [6-11].

The carried out tests of texture layers of facade systems on acrylic and silicate bases did not reveal their air permeability because of the presence of glue compositions in them, that's why the external walls, insulated with facade systems with thin-walled plaster, are practically impenetrable.

In addition to tests to determine the air permeability of building materials, the ACI SamSTU staff carried out a thermal imaging survey of the heat-shielding characteristics of external walls with ventilated facades of a 25-story building built in Samara.

Determination of heat engineering indicators of building enclosing structures was made in accordance with the requirements of Russian Standard GOST 31937-2011 «Buildings and structures. Rules for inspection and monitoring of technical condition».

The composition of works for determination of the thermal protection properties of building enclosing structures includes thermal imaging of building enveloping structures to identify places with low thermal protection indicators, performed in accordance with the requirements of Russian Standard GOST of Russia 54852-2011 "Buildings and structures. Method of thermal imaging control of insulation quality of enclosing structures".

Thermal imaging control is carried out under the close to the stationary heat transfer regime through the enclosing construction. To be surveyed surfaces should not be in the direct and reflected solar irradiation zone for 12 hours before the measurements are taken. Surfaces of enclosing structures should not be subjected to additional thermal effects from biological objects and lighting sources in the period of thermal imaging.

On the results of the thermal imaging survey, nodes of junction of enclosing structures with low heat-shielding characteristics are determined.

Thermal imager under certain temperature-humidity conditions makes it possible to determine: defects of structures connected with high humidity and leaks; presence of infiltration air flows passing through the fence due to cracks in it, poor sealing of joints, poor quality of construction and installation works, insufficient compaction of window and door structures; sections on enclosing structures, having a lower thermal resistance.

The advantages of using an infrared camera include: the speed of the survey, the contactless method of obtaining data, the absence of evacuation of people during the survey, a visual representation of the data and the presence of a sufficiently high memory (up to 200 images).

Thermal imaging is performed in infrared light. The thermal imager provides a color image of the observed object, due to this fact, it is possible to analyze in detail the state of the enclosing structures. With its help, it is possible to detect both defects in construction and installation works and design solutions of enclosing structures.

Based on the results of the thermal imaging survey of the facades of the surveyed building, the specific heat flux through the outer enclosing structures is determined by formulas

$$q = \alpha_N(\overline{\tau_N} - t_N), \text{ W/m}^2; \quad (3)$$

$$q = \alpha_B(t_B - \overline{\tau_B}), \text{ W/m}^2; \quad (4)$$

Where α_N, α_B - heat transfer coefficients on the side of the outer and inner surfaces, respectively, W/(m^2·°C);

The formula (3) is used in the case when using a thermal imager a temperature field on the outer surface of the enclosing structure is determined, formula (4) - on the inner surface.

Heat transfer coefficients $\alpha_н$ and $\alpha_в$ are determined by the results of measurements in accordance with the methodology set forth in GOST of Russia 54852-2011.

By working with a thermal imager, follow the recommendations given in GOST of Russia 54852-2011.

Resistance to heat transfer of building enclosing structures is determined in accordance with GOST of Russia 54852-2011 by formula

$$R_0 = \frac{t\,\mathrm{int}\ \ t_n}{\mathrm{q}},\ (\mathrm{m}^2{\cdot}{}^\circ\mathrm{C})/\mathrm{W}, \qquad (5)$$

where t_{int} - internal air temperature, °C; t_n, - outdoor temperature, °C; q is the specific heat flux, W/m^2.

3 Results

We estimate the influence of plaster layers on the resistance of air permeability to external walls with ventilated facades and protected with facing brick. As an example, consider a 25-storey building with a warm attic built in Samara. The external walls are made in the form of a brick ceramic masonry on a cement-sand mortar with a thickness of 380 mm. On the outside, they are insulated with basalt slabs VENTI-BATTS 150 mm thick, protected by a windproof membrane such as Isospan A and a ventilated facade made of porcelain stoneware.

As possible plasters, we will consider cement-sand and lime-sand mortars, as well as dry plaster in the form of gypsum board sheets. According to SP 50.13330.2012, the plasters mentioned above have the air permeability resistance values shown in Table 1.

Table 1. Thermal protective characteristics of finishing materials.

Denomination of the finishing material	Density in dry condition, ρ_0 kg / m^3	Layer thickness, m	Resistance to air permeability, R_u, (M2·h·Pa)/kg
Cement-sand mortar	1800	0,02	497
Lime-sand mortar	1600	0,02	189
Gypsum sheets (dry plaster)	800	0,0125	25

The location of the above materials in the outer wall is shown in Fig. 1b. Resistance of heat transfer of homogeneous construction, calculated according to the procedure set out in SP 50.13330.2012, was $R_0^{usl} = 4{,}34$ (m$^{2}\cdot{}^0$C)/W, the reduced total thermal resistance $R_0^{pr} = 3{,}47$ (m$^{2}\cdot{}^0$C)/W, which fits the regulatory requirements for residential buildings in the Samara region ($R_0^{tr} = 3{,}47$ (m$^{2}\cdot{}^0$C)/W).

Evaluation of the transverse infiltration of air to the resistance to heat transfer of the unplastered outer wall is performed using formulas (1) and (2). The results of the calculation are presented in Table 2.

Table 2. Results of calculating the resistance to air permeability of a high-rise building with a ventilated facade.

Floor Number	H, m	ΔP, Pa	G, kg/m²•h	k, W/m²·⁰C	Resistance to heat transfer under the influence of infiltration $(R_0^{usl})_{ut}$, (m²·⁰C)/W	Resistance to air permeability, (m²·h·Pa)/kg		Difference $R_u^{tr} - R_u$, (m²·h·Pa)/kg
						The required R_u^{tr}	actual R_u	
1	79	119	4,96	1,380	0,720	238,0	24	214,0
3	73	110,9	4,62	1,290	0,780	221,8	24	198,0
5	67	102,8	4,28	1,197	0,835	205,6	24	181,6
7	61	94,7	3,95	1,107	0,903	189,4	24	165,0
9	55	86,6	3,81	1,020	0,980	173,2	24	149,0
12	46	74,5	3,10	0,880	1,140	149,0	24	125,0
15	37	62,3	2,60	0,760	1,320	124,6	24	101,0
20	22	42,1	1,75	0,553	1,810	84,2	24	60,0
25	7	21,9	0,91	0,378	2,650	43,8	24	19,8

The Analyze of the influence of longitudinal infiltration on the value of the reduced total thermal resistance of the considering external wall, carried out according to the procedure described in the work [1], showed that for plates of basalt fiber with a density of 80 to 90 kg/m³, the coefficient of thermal effect of longitudinal filtration is close to 1because of a sufficiently high density.

4 Discussion

From the results of calculations presented in Table 2 it follows that when the external walls are passed without internal finishing, which is done quite often, the heat-shielding characteristics $(R_0^{usl})_{ut}$ and R_u significantly do not meet the regulatory requirements. Herewith the greatest discrepancy is observed on the lower floors.

Along with the calculated data, the negative influence of infiltration on the heat-shielding characteristics of the enclosing structures is confirmed by the results of field surveys the outer walls of high-rise buildings. Fig. 2 shows a thermogram of the inner surface of the outer wall, obtained from the thermal imaging survey of the enclosing structures of a residential apartment located on the fifth floor of a 25-storey residential building. The interior decoration of the outer walls was made with a Gypsum mortar, which does not have the necessary resistance to air permeability.

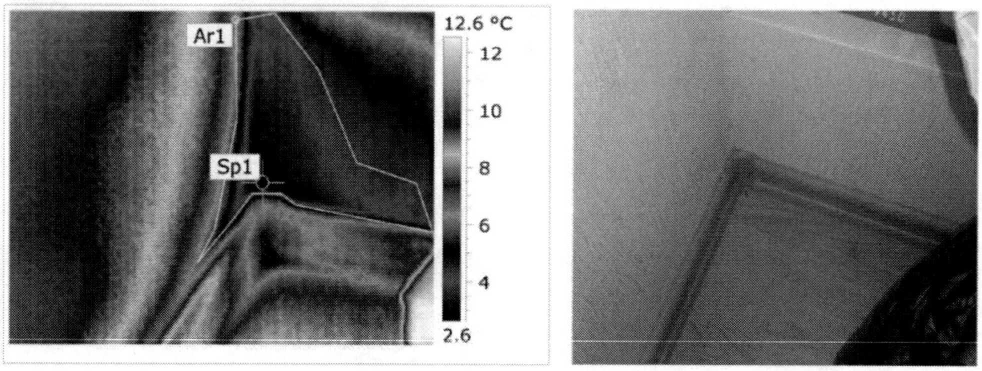

Ar1 Average Temperature 4.0 °C

Fig. 2. Thermogram of the inner surface of the outer wall

Thermal imaging of internal surfaces enclosing structures revealed areas of low temperatures in the lower part of the outer wall. At the same time, lower internal air temperatures, which do not meet regulatory requirements, were also recorded. In order to the exterior wall of the residential building to fit with regulatory requirements, it is necessary to plaster its inner surface.

5 Conclusions

Analyzing the calculated data, we can conclude that the use of cement-sand mortar as a finishing layer allows achieving the standard values of resistance to heat transfer on all floors. Lime-sand mortar can be recommended for use on the upper floors, starting with the fifth. Drywall sheets should not be used in buildings with a ventilated facade in view of the low resistance value of air permeability.

References

1. V. G. Gagarin, V. V. Kozlov, A. V. Sadchikov, Accounting of longitudinal infiltration of air at assessing the thermal wall protection of with a ventilated facade, *Industrial and civil construction*, **6**, 42-45 (2005)
2. V. G. Gagarin, V. V. Kozlov, A. V. Sadchikov, I. A. Mehnetsov, Longitudinal air filtration in modern enclosing structures, *ABOK*, **8**, 66-70 (2005)
3. V. G. Gagarin, V. V. Kozlov, A. V. Sadchikov, About the effect of longitudinal air filtration on the heat protection of walls with a ventilated facade, *Stroyprofile*, **6**, 34-35 (2005)
4. Yu. S. Vychtikov, A. Yu.Vychtikov, Investigation of the effect of longitudinal and transverse air infiltration on the heat-shielding characteristics the external walls of a production building insulated with a ventilated facade, *Improving the energy efficiency of buildings and structures*, **3**, 20-26 (2008)
5. M. R. Petrichenko, M. V. Petrochenko, Hydraulics of free convection currents in enclosing structures with an air gap, *Engineering and Construction Journal*, **8**, 51-56 (2011)
6. Yu. S. Vytchikov, A. V. Chereneva, An experimental study of air permeability of sandless claydite-concrete, *Building Materials*, **7**, 10-11 (2011)
7. Yu. S. Vytchikov, A. V. Sidorova, The organization of air exchange in modern energy-efficient buildings, *Bulletin of SGASU*
8. Yu. S. Vytchikov, A. V. Sidorova, Experimental study of enclosing structures in full-scale conditions, *Traditions and innovations in construction and architecture: materials of the 71st All-Russian Scientific and Technical Conference on the results of research work of 2013*, 284-286 (2014)
9. S. F. Korenkova, Yu. V. Sidorenko, Unfilled foam concrete, *Scientific Review*, **4**, 106-107 (2014)
10. A. S. Pershina, S. F. Korenkova, The most important components of facade compositions, *Traditions and innovations in construction and architecture: materials of the 69th All-Russian Scientific and Technical Conference on the results of research work in 2011*, 166-167 (2012)
11. A. S. Pershina, S. F. Korenkova, Decorative nano-filled cement-polymer compositions for finishing facades, *Nanotechnology in construction: a scientific online journal*, **4**, 36-46 (2011)

12. N. N. Razumov, F. V. Ushkov, I. S. Shapovalov, Resistance to heat transfer and winter design temperatures for heating engineering, *ABOK*, **5**, 78-81 (2011)
13. N. P. Umnyakova, A. A. Verkhovsky, Estimation of air permeability of the building's enclosing structures, *ABOK*, **5**, 48-53 (2013)
14. N. P. Umnyakova, Thermal protection properties of exploited hinged ventilated facade structures, *Housing construction,* **2**, 2-6 (2011)
15. P. N. Umnyakov, Infiltration and exfiltration influence on the thermal protection of enclosing structures with considering the porosity of materials, *Housing construction*, **10**, 46-50 (2014)
16. A. V. Maltsev, D. S. Sorokin, Energy saving in multi-storey residential buildings with air infiltration through the outer wall, *New University*, **3-4,** 87-90 (2015)
17. V. M. Valov, Breathable enclosing structures in a building system, *Bulletin of the Siberian State Automobile and Highway Academy*, **19**, 33-37 (2011)
18. V. G. Gagarin, S. V. Gouvernyuk, A. S. Kubenin, P. P. Pastushkov, V. V. Kozlov, To the methodology for calculating the influence of wind impact on the air conditions of buildings, News of Higher Educational Institutions, *The textile industry technology*, **4**, 234-240 (2016)
19. I. L. Shubin, A. I. Ananiev, Thermal protection properties and air permeability of ceramic Isoterex blocks in wall masonry, *Industrial and civil construction*, **3**, 57-59 (2013)
20. A. D. Zhukov, T. V. Smirnova, A. V. Chugunkov, Heat transfer in high-porous materials, *Internet-Vestnik VolgGASU*, **3**, 8 (2012)

Evaluation of punching shear strength of flat slabs supported on rectangular columns

Valery Filatov[1,*]

[1]Samara State Technical University, Institute of Architecture and Civil Engineering, Molodogvardeyskaya St., 194, Samara, 443001, Russia

Abstract. The article presents the methodology and results of an analytical study of structural parameters influence on the value of punching force for the joint of columns and flat reinforced concrete slab. This design solution is typical for monolithic reinforced concrete girderless frames, which have a wide application in the construction of high-rise buildings. As the results of earlier studies show the punching shear strength of slabs at rectangular columns can be lower than at square columns with a similar length of the control perimeter. The influence of two structural parameters on the punching strength of the plate is investigated - the ratio of the side of the column cross-section to the effective depth of slab C/d and the ratio of the sides of the rectangular column C_{max}/C_{min}. According to the results of the study, graphs of reduction the control perimeter depending on the structural parameters are presented for columns square and rectangular cross-sections. Comparison of results obtained by proposed approach and MC2010 simplified method are shown, that proposed approach gives a more conservative estimate of the influence of the structural parameters. A significant influence of the considered structural parameters on punching shear strength of reinforced concrete slabs is confirmed by the results of experimental studies. The results of the study confirm the necessity of taking into account the considered structural parameters when calculating the punching shear strength of flat reinforced concrete slabs and further development of code design methods.

1 Introduction

In the modern world one of the indicators of sustainable development of society and the state is the construction of high-rise buildings and structures. The construction and operation of high-rise buildings implies the possession of modern technology in all stages: from planning and design solutions, designing load-bearing structures of buildings and engineering systems, organization of construction works, to the monitoring of structures and engineering systems of the building to ensure failure-free operation [1–4].

The most common design solution when designing high-rise buildings is the use of monolithic gidderless slabs supported on columns-pylons [5–7]. Columns-pylons have rectangular cross-section whose height is several times greater than the width, which

* Corresponding author: vb_filatov@mail.ru

improves the lateral stiffness of the building. One of the stages of designing high-rise buildings is the calculating of flat slabs for punching. The calculation is aimed at ensuring the constructive safety of the building [8–10], including in the case of progressive collapse. Since the calculation for punching shear has an empirical basis, the carrying out of experimental studies [11–13] is a necessary condition for the improvement of computational methods [14–17]. As the results of experimental studies show [18–20], the punching strength of a slab supported on a rectangular column may be less than that of punching a square column. This effect is a consequence of the non-uniform distribution of deformations along the control perimeter of the punching of a flat reinforced concrete slab supported on a rectangular column. When punching the plate by column of circular cross-section such features of the distribution of deformation is not observed. This peculiarity for rectangular columns was observed experimentally by several researches such as Moe [21], Hawkins et al. [22], Muttoni, Sagaseta et al. [23, 24] and any others. In the experiments of these authors there is a concentration of the strains of concrete slabs in areas near the corners of the columns.

The punching shear strength of flat reinforced concrete slabs depends not only on the aspect ratio of the rectangular cross-section of the column, but also on the ratio of the side of the column cross-section to the effective depth of slab (C/d). In the articles [23, 24] it was shown that with increase of the ratio C/d punching shear strength of flat reinforced concrete slabs is reduced.

The punching shear strength of RC flat slabs at rectangular columns can be overestimated if the non-uniform distribution of strains along the control perimeter around the column is neglected in the calculations. Different approaches are used in design codes to determine the punching shear strength of a slab supported on a rectangular column. Design code ACI 318-11 [26] are used correction factor, which depends on the ratio of the sides of the column cross-section. Model Code MC2010 [27] and Swiss design code SIA 262 [28] recommend decreasing the value of the control perimeter, including for a square column, depending on the ratio C/d. There are other suggestions for improving the calculation methods [16, 17, 23–25]. Calculation methods of codes [29, 30] for punching shear strength of flat slabs the non-uniform distribution of strains along the control perimeter around the column are neglected.

2 Methods

Consider the cross-section of a flat plate I-I that passes through the body of a column (fig. 1). At the interface with the column, the slab is an absolutely rigid body in terms of flexural stiffness. The presence of such an absolutely rigid part significantly influences the plate deformation behavior and stress distribution in the interface zone of the plate with the column.

Along the perimeter of the column, a zone of constrained deformations is formed in the plate, due to the presence of a conditional absolutely rigid body in the section of the interface between the plate and the column. Suppose that for a square column the zone of constrained deformations has the form of a circle whose diameter is equal to the diagonal of the square. The maximum value of this zone is in the middle of the face of the column, at the corners of the column the zone is equal to zero. In the case of a rectangular column, the zone of constrained deformations has the shape of an ellipse. The foci of the ellipse are on the short sides of the column, so that the focal parameter is half the short side of the column section.

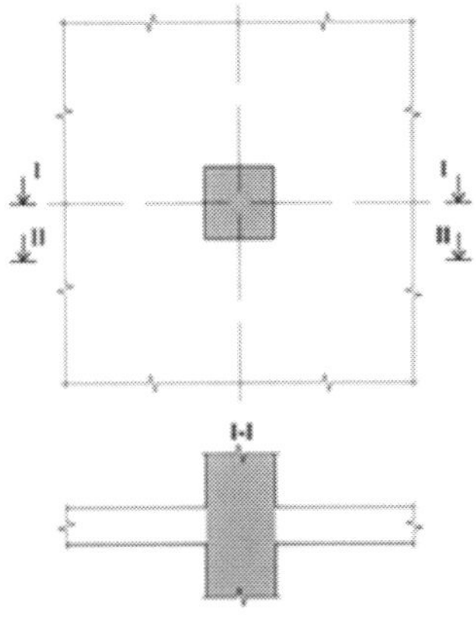

Fig. 1. Fragment of flat plate floor.

On the plate section equal to the width of the column, in the direction of the perpendicular face of the column (radial strains), the zone of constrained deformations does not significantly affect the flexural deformations of the plate. The most significant flexural deformations of the plate vary in the section of the slab along the face of the column in section II-II (fig. 1). The maximum value of strains of the plate is near the corners of the column, the minimum value is in the middle of the face of the column. Thus, the zone of constrained deformations has the greatest influence on the value of tangential strains of the plate.

To estimate the influence of the zone of constrained deformations on the flexural deformations of the plate, let us compare the flexural stiffness of two sections of the plate: from the face of the column to the control perimeter and from the face of the column to the circle denoting the zone of constrained deformations (fig. 2).

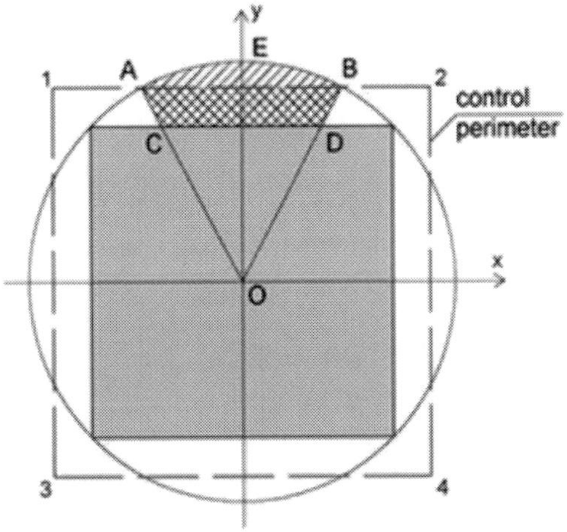

Fig. 2. Constrained deformation zone around square column.

Distance from face of column to the control perimeter will take equal to 0.5d, in accordance with [30]. Assume that the tangential strains (stresses) of the plate are reduced at the site where the circumference of the area of the constrained deformation beyond the border control perimeter (line AB). The reduction of strains proportional to the ratio of moments of inertia relative to the Y axis of area ABCD (I_{cp}) and area AEBCD (I_{cir}).

The value of the control perimeter on the side 1 – 2 will also be reduced in proportion to the ratio I_{cp}/I_{cir}. Similarly, the value of the control perimeter decreases and on the other three sides of the column section. Assuming a reduced value of the control perimeter, determine the value of the punching force according to the normative method.

The proposed approach makes it possible to determine the influence of the factor C/d on the strength of the plate under punching force (C – column face size, d – effective depth of slab).

Similarly, it is proposed to take into account the decrease in the control perimeter when the flat plate is punching by a rectangular column. The value of the control perimeter decreases in proportion to the ratio I_{cp}/I_{cir}. The values of I_{cp}/I_{cir} on the sides 1 – 2 and 3 – 4 are determined relative to the Y axis of areas ABCD и AEBCD, respectively (fig. 3).

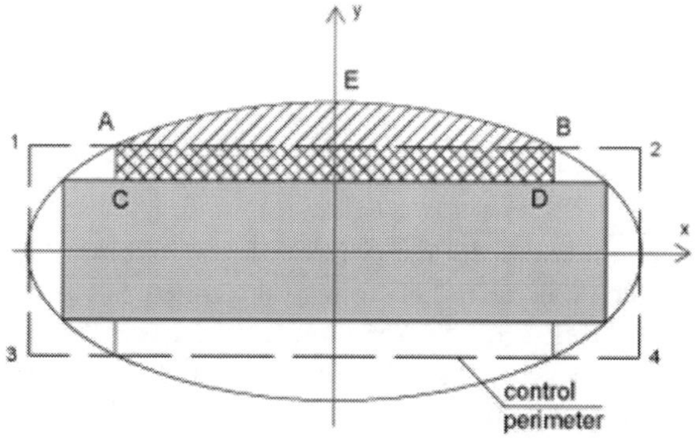

Fig. 3. Constrained deformation zone around rectangular column.

The I_{cp}/I_{cir} values on the sides 1 – 2 and 2 – 4 are determined for the respective sections (if any) relative to the X axis. In this case, when calculating the punching plate, two calculation factors are taken into account - the C/d ratio and the ratio of the sides of the rectangular cross section of the column C_{max}/C_{min}.

3 Results

In accordance with the proposed approach of considering the influence of structural parameters on the punching shear strength of flat reinforced concrete slabs was determined the dependence of strength of flat slabs under punching on the value of the factor C/d (the ratio of the sides of the column cross-section to the effective depth of slab). The change in the punching shear strength of flat slab was evaluated by adjusting the value of the control perimeter in proportion to the ratio I_{cp}/I_{cir}. The ratio of the corrected value of the control perimeter U_{red} to the initial value control perimeter U is an indicator of the reduction in punching shear strength of flat slab. The dependence of the value U_{red}/U on the calculated factor C/d for a square column is shown in fig. 4.

The strength of the plate under punching by rectangular column will depend on two structural parameters - the ratio of the sides of the column cross-section to the effective depth of slab C/d and the ratio of the sides of the rectangular cross section of the column C_{max}/C_{min}. The reduction of punching strength of flat slabs also evaluated against the adjusted and initial values of the control perimeter U_{red}/U. The dependence of U_{red}/U from the structural parameter C_{max}/C_{min} for several values of C_{min}/d (1.13, 2.7, 5) is shown in fig. 5.

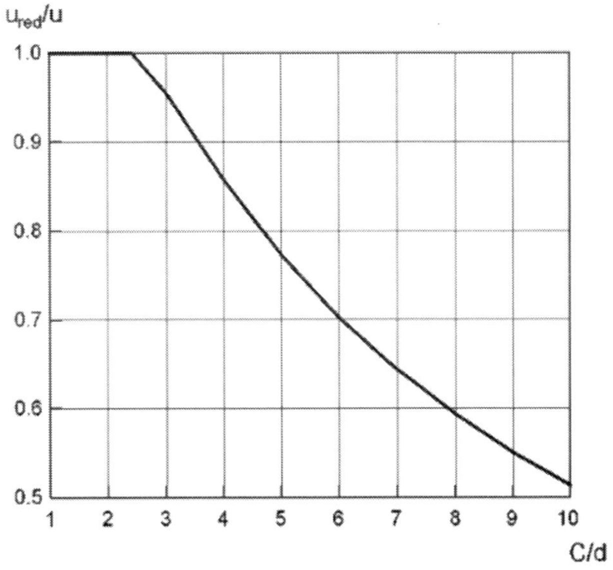

Fig. 4. Reduction of control perimeter for square column due to factor C/d

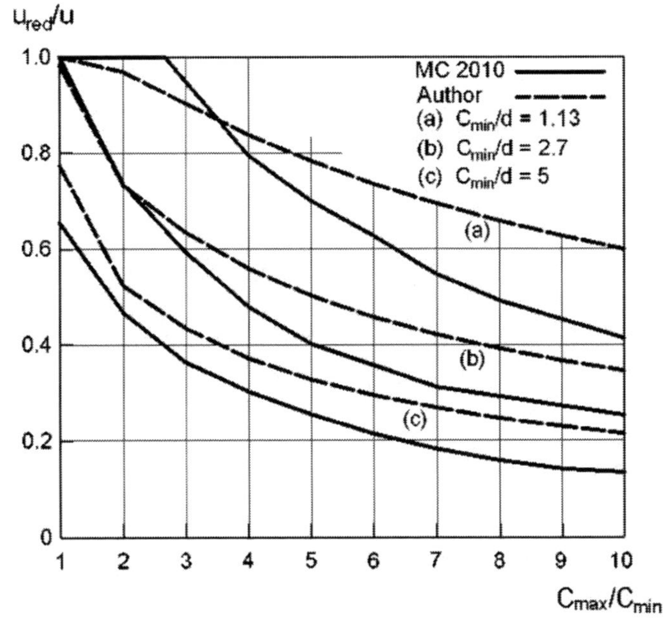

Fig. 5. Comparison of predicted shear-resisting control perimeter using MC2010 and proposed approach.

On fig. 5 also shows the dependence of U_{red}/U on the structural parameter C_{max}/C_{min} obtained in [23] using MC2010 simplified method.

4 Discussion

Analysis of the dependence in fig. 4 shows a significant influence of the parameter C/d on the punching force to flat slab. With the ratio of the sides of the column cross-section to the effective depth of slab C/d = 5 the reduction in punching shear strength is 23%, at C/d = 10 the strength reduction reaches 49%. When the ratio C/d < 2.5 the decrease in the strength of the plate punching by square column is not observed. For value of C/d = 3 MC2010 simplified method gives $U_{red}/U = 1$, the proposed approach gives the value of $U_{red}/U = 0.955$, which is a good agreement.

Analysis of the dependence on fig. 5 shows that for small values of $C_{min}/d \approx 1$, the parameter $C_{max}/C_{min} < 2.6$ does not affect the punching shear strength of a flat plate. With an increase in the ratio C_{min}/d, a sharp drop in the punching shear strength of the plate occurs. When the value of $C_{max}/C_{min} = 2$, method design code ACI 318-11 [26] does not imply a reduction in strength ($U_{red}/U = 1$), proposed approach gives the value $U_{red}/U = 0.733$ and 0.523 for $C_{min}/d = 2.7$ and 5, respectively. MC2010 simplified method gives the value $U_{red}/U = 0.74$ and 0.46 for $C_{min}/d = 2.7$ and 5, respectively. In general, the proposed approach gives more conservative results compared to the MC2010 simplified method, at $C_{max}/C_{min} = 5$ the difference is 20 - 25%.

5 Conclusions

This paper presents the results from the analytical research on punching shear of flat slabs supported by square and rectangular columns. Proposed approach was investigated to assess the shear-resisting control perimeter due to the influence of parameters C/d and C_{max}/C_{min}. The main conclusions are the following:

- The effect of parameters C/d (the ratio of the side of the column cross-section to the effective depth of slab) and C_{max}/C_{min} (the ratio of the sides of the rectangular cross section of the column) on the value of the punching force of a flat plate is significant enough and should be taken into account in design codes. Proposed approach is phenomenological and allows evaluating the effect of structural parameters on the punching strength of a flat plate on the basis of a physical model by adjusting the value of the control perimeter of the slab.
- The results from the analytical research gives more conservative results compared to the MC2010 simplified method [27] and shear field analysis or a method based on the contact pressure at the support area [23]. Analysis of research results allows identifying ranges of structural parameters, when the design codes calculation methods do not capture a possible reduction in the bearing capacity of the slab for punching that allows us to make a conclusion about the necessity of further improvement of design codes.
- The results from the analytical research on punching shear of flat slabs supported by square and rectangular columns, are confirmed the previously proposed hypothesis about the presence a zone of constrained deformation in the slab, which located along the perimeter of a square or rectangular column. The deformation features of the plate in this zone affect the distribution of strains along the control perimeter around the column and the plate resistance by the punching force, respectively. The results of the study presented in this article, as well as in [14, 18, 20, 23] confirm the necessity of taking into account the structural parameters in calculating the punching strength of flat slabs and further improving the normative calculation procedures.

References

1. E. N. Kodysh, N. N. Trekin, D. N. Trekin, Concrete and reinforced concrete – glance at future: III All-Russian (II International) *Conference on concrete and reinforced concrete*, **1**, 69–75 (2014)
2. V. I. Travush, V. I. Kolchunov, N. V. Klyueva, *Industrial and civil engineering*, **3**, 4–11 (2015)
3. V. I. Travush, S. G. Emel'yanov, V. I. Kolchunov, *Industrial and civil engineering*, **7**, 20–27 (2015)

4. A.G. Tamrazyan, *Industrial and civil engineering*, **7**, 26–27 (2012)
5. V. P. Popov, D. V. Popov, A. Yu. Davidenko, *MATEC Web of Conferences*, **86**, 01014 (2016)
6. A. O. Mateyko, *Vestnik of SSUACE*, **23(3)**, 74–78 (2016) DOI: 10.17673/Vestnik.2016.03.12
7. V. P. Generalov, E. M. Generalova, *Vestnik of SSUACE*, **10(2)**, 12–16 (2013) DOI: 10.17673/Vestnik.2013.02.2
8. N. I. Karpenko, S. N. Karpenko, *Zhilishchnoe stroitel'stvo*, **6**, 16–20 (2016)
9. V. B. Filatov, A. A. Suvorov, *Procedia Engineering*, **153**, 144–150 (2016)
10. A. G. Tamrazyan, Yu. N. Zvonov, *Industrial and civil engineering*, **7**, 24–28 (2016)
11. N. N. Trekin, D. A. Pekin, *Industrial and civil engineering*, **7**, 17–20 (2014)
12. A. N. Bolgov, A. Z. Sokurov, Concrete and reinforced concrete – glance at future: III All-Russian (II International) *Conference on concrete and reinforced concrete*, **4**, 139–149 (2014)
13. A. N. Bolgov, D. M. Yakimovich, Concrete and reinforced concrete – glance at future: III All-Russian (II International) *Conference on concrete and reinforced concrete*, **1**, 224–233 (2014)
14. A. Muttoni, *ACI Struct. J.*, **105(4)**, 440–450 (2008)
15. C. E. Broms, *ACI Struct. J.*, **87(3)**, 292–304 (1990)
16. V. B. Filatov, *Bulletin of civil engineers*, **5(40)**, 80–84 (2013)
17. V. B. Filatov, *Proceedings of the Samara scientific center of the Russian Academy of Sciences*, 16(4-3), 646–649 (2014)
18. T. Susanto, H. K. Cheong, K. L. Kuang, J. Z. Geng, *ACI Struct. J.*, **101(5)**, 678–687 (2004)
19. V. B. Filatov, E. P. Bubnov, *MATEC Web of Conferences*, **86**, 01007 (2016)
20. V. B. Filatov, E. P. Bubnov, *Industrial and civil engineering*, **2**, 86–91 (2017)
21. J. Moe, *Dev. dep. bul.* D47, PCA (1961)
22. N. M. Hawkins, H. B. Falssen, R. C. Hinojosa, *ACI Spec. Publ.*, SP-30, 127–146 (1971)
23. J. Sagaseta, L. Tassinari, M. Fernandez Ruiz, A. Muttoni, Eng. Struct., **77**, 17–33 (2014)
24. J. Sagaseta, A. Muttoni, M. Fernandez Ruiz, L. Tassinari, *Mag. Conc. Res.*, **63(6)**, 441–457 (2011)
25. V. B. Filatov, *Proceedings of the Samara scientific center of the Russian Academy of Sciences*, **16**(4-3), 642–645 (2014)
26. ACI 318-11, Building code requirements for structural concrete and commentary, (Farmington Hills, MI, ACI, 2011)
27. fib Model Code for Concrete Structures 2010, 434 (Ernst & Sohn, Berlin, 2013)
28. Code 262 for concrete structures (Zürich, SIA, 2013)
29. BS 8110, Part 1: structural use of concrete, (London, BSI, 1997)
30. SP 63.13330.2012, Concrete and reinforced concrete. Design code, (Moscow, 2012)

Construction of high-rise buildings in the Far East of Russia

Sergey Kudryavtsev[1], *Semen* Bugunov[1*], *Evgeniya* Pogulyaeva[1], *Anastasiya* Peters[1], *Zhanna* Kotenko[1] and *Danil* Grigor'yev[1]

[1]Far Eastern State Transport University, Seryshev st., 47, Khabarovsk, 680021, Russia

Abstract. The construction of high-rise buildings on plate foundation in geotechnical conditions of the Russian Far East is a complicated problem. In this respect foundation engineering becomes rather essential. In order to set a firm foundation it is necessary to take into account the pressure distribution at the structure base, in homogeneity of building deformation, which is due to collaborative geotechnical calculations complicated by a number of factors: actual over-placement of soils, the complex geometry of the building under construction, spatial work of the foundation ground with consideration for physical nonlinearity, the influence of the stiffness of the superstructure (reinforced concrete framing) upon the development of foundation deformations, foundation performance (the performance of the bed plate under the building and stairwells), the origination of internal forces in the superstructure with differential settlement. The solution of spatial problems regarding the mutual interaction between buildings and foundations with account of the factors mentioned above is fully achievable via the application of numerical modeling methodology. The work makes a review of the results of high-rise plate building numerical modeling in geotechnical conditions of the Russian Far East by way of the example of Khabarovsk city.

1 Introduction

Year by year the problem of getting new construction areas in megapolises and big cities is becoming more and more complicated because of the fact that the land price is very high. To solve this problem we have to get the max square meters quantity from each land area by means of increasing the storey number of a building.

The construction of high-storey buildings in the field of residential real estate reduces the price for a land square meter and makes it possible to decrease the price for ready housing constructions. All of the current technical difficulties are quite resolvable.

The building-up of high-storey building on plate foundation constructions in engineering-geological conditions of Khabarovsk is a complicated geotechnical goal and it is acceptable only if there is special geotechnical ground for it.

In connection with this the installation of safe foundation and base is quite urgent.

* Corresponding author: soymaa@inbox.ru

The main condition of safe foundation and base installation is the functional pressure distribution which exists in base, and also the clearance of building strain variations. The consistent assessment of these data in calculations is possible only on the base of careful geotechnical tridimensional calculations, which allow taking into consideration a lot of basic real factors:
- Soil formations; complicated geometry of a raising building;
- Space combined action of foundation soil taking into account the physical nonlinearity of an above ground construction;
- Influence of above ground constructions (concrete core) stiffness on rearrangement of base strain variations and etc.

While considering the space combined plans about building and ground interaction it makes sense to take into account the quoted factors in full with the help of numerical model method by combined finite strip-finite element method.

In the "base – foundation – high storey building" system the main role in keeping the constructional integrity is played by bases both during the building process and in the period of its exploitation. The safety of a future construction depends on resistibility of a base floor material and also on the level of deflectivity on load.

2 Materials and methods

2.1 Geotechnical conditions of Khabarovsk

Khabarovsk is "a city on three hills" and benches of the Amur. The city and the surrounding areas are situated within one of the largest trenches in The Far East – Middle-Amurskaya trench. On its territory one can find areas filled by Cenozoic poorly cemented continental buildups and also elevations made by more ancient and tough Paleozoic and Mezozoic aqueous and anogene rocks.

Khabarovsk and its surroundings are situated in severe climatic conditions with various landscape and quite diverse soil structures which you have to deal with on all the levels of building.

All rocks within the considered territory can be divided into to large groups: rocks of original base and superficial deposits.

The most important peculiarity of geological formation is the inhomogenuity of its formation, considerable variability of strength and deformational features of sedentary units. All of these factors in accordance with the normative requirements must be taken into consideration when engineering and carrying out works within the limits of the city.

Judging by the abovementioned information it makes sense to conduct the construction of buildings in the city of Khabarovsk taking into account the engineering-geological conditions, and also by using protective and preventive measures from adverse impact of different natural and industrial factors.

2.1 Research of stress strain behavior of high-rise buildings on plate foundation

The analysis of the the "base – foundation – high storey building" system was made by finite element method with the usage of bundled software "FEM Models" developed by the staff of LLC "Pi "Georeconstruction" under the charge of professor V.M. Ulitskiy, St.Petersburg. In the calculations they used elastic-plastic model of soil base in the form of volume elements with the Mor-Kulon strength condition and equivoluminar flow rule. Walls, floor structures, ground slabs are the elastic elements of a thick plank. Columns are the elastic shanks [1-3].

The analysis of the numeric model engineering results of high-storey buildings on plate foundation in Khabarovsk for different engineering-geological conditions and construction solutions has also been made:
- 23-storey single-section with crawlway domestic building located on the shore face of the Amur (plate bulk 0,8 m)
- 25-storey building of complicated configuration with the underground parking lot situated in the northern part of Khabarovsk (plate bulk 1 m)
- domestic building of variable storey number (12 and 14) with in-built shop, offices and parking lot (plate bulk 0,9 m)
- domestic building of variable storey number (10 and 12) with in-on-built shop and underground parking lot situated in the central part of Khabarovsk (plate bulk 0,7 m).

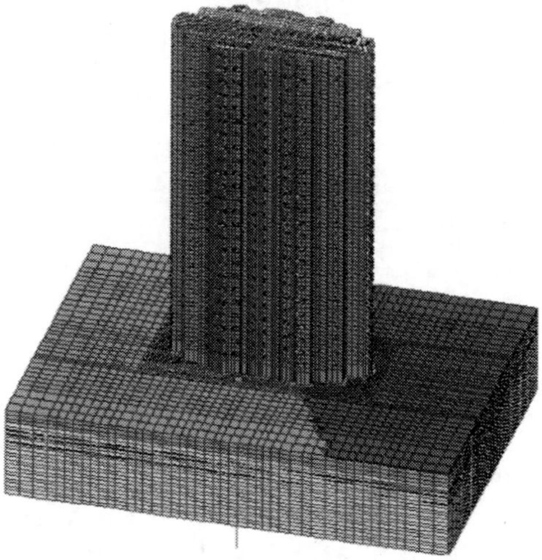

Fig. 1. Analitical model 23-storey building.

The exploration of stress strain behavior foundation of high buildings in engineering-geological conditions of Khabarovsk has shown the following:

The gradual settlement of plate foundation was got for the buildings situated on shore face of the Amur and in the northern part of Khabarovsk.

For the building situated within the limits of the first above flood-plain bench the settlement of plate foundation had the following character: 12-storey section has differential settlement (5-10 cm), and the settlement of a 14-storey section is more gradual (10-11) with clearly expressed centre. It is caused by the constructional peculiarities of the building and also by the fact that within the limits of the building area there exist filled soils and their distribution is differential (0,5-3 m).

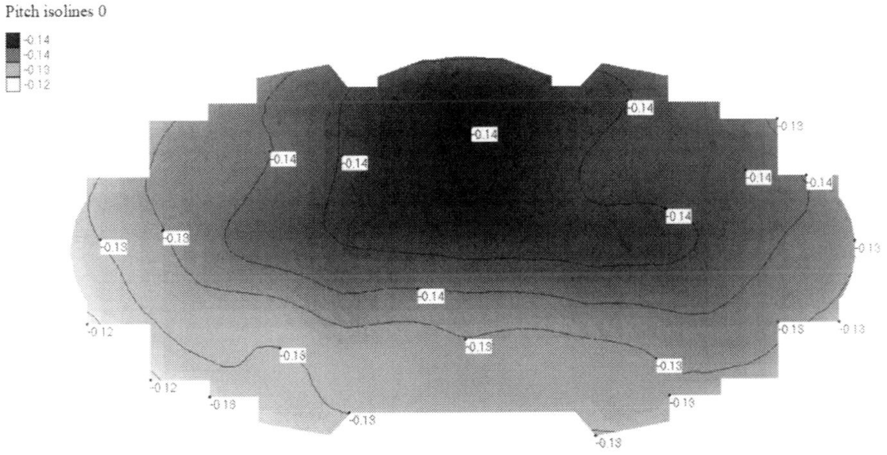

Fig. 2. Isoline of settlement plate foundation of 23-storey building.

For the building situated in the central part of Khabarovsk the settlement of foundation plate had the following character: under the whole building area the settlement was distributed differentially. The differential settlement is caused by various depth of filled soils. Under the filled soils there are plates which are practically incompressible. The settlement is given by the filled soils.

Plate foundation makes it possible to reduce the settlement variety of a foundation plate by means of its stiffness.

During the construction of high-storey buildings with quite small underground space in the urban development area in the case when there is no opportunity to expand the foundation plate square, there exist the danger of high local pressure concentration on the soil while developing building tilts. Thus there is the necessity to check the horizontal mobility.

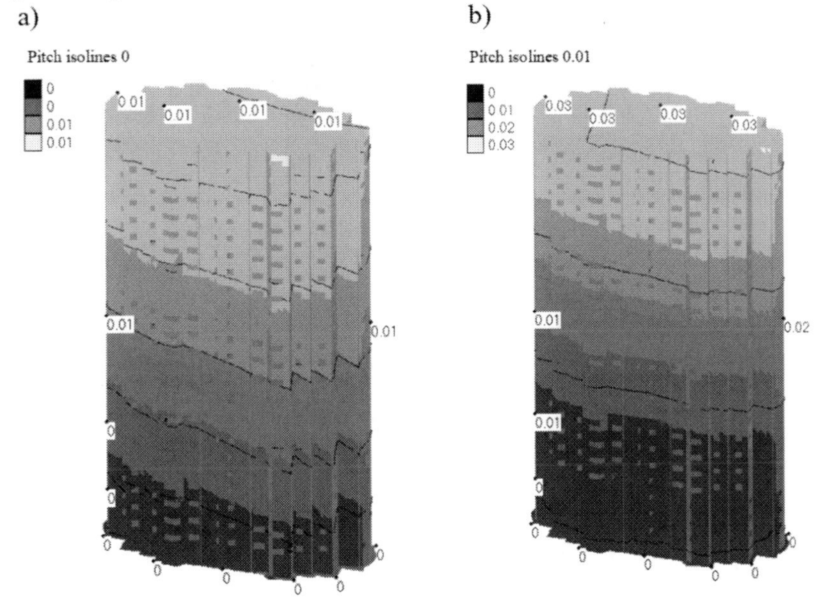

Fig. 3. Horizontal mobility of 23-storey building: a) on the X axis; b) on the Y axis.

In our cases the horizontal mobility within the combined space calculations of the "base – foundation – high storey building" system are insignificant in relation to axis. In the lower parts of the majority of the buildings (the first 5 floors) the mobility is insignificant. On the high floors the mobility on the X axis reaches 0,043 m, on the Y axis – 0,045 m, which does not exceed admissible values. The tilt i=0,001 is observed which does not exceed the admissible value i_u=0,004m.

The character of the plate foundation work of the analyzed buildings in general is moving closer to a stiff mark. The local development of plastic deformation zones which do not exceed 0,25 of the plate bulk is observed under the edges.

The moments of flection which exist in foundation plate have alternating-sign amounts, the max amounts of a flection moment M_x are: from minus 585 to plus 352 kilopascal and the one of M_y from minus 687 to plus 543kilopascal, and the flection moment of M_{xy} from minus 586 to plus 627 kilopascal. It points to the necessity of making plate reinforcement on two levels (the lower and high ones).

3 Results

3.1 Geotechnical monitoring of deformation process of plate foundation of high-storey buildings

To raise high buildings it is necessary to fulfill two procedures which provide safety while designing, building and exploitation: numeric model engineering of the "base – foundation – high storey building" system and geotechnical monitoring in the process of building and exploitation. Fulfilling of these particular procedures allows to avoid designing mistakes and also to make the suitable correction, changing and adaptation of designing and manufacturing decisions.

These procedures were held on the levels of designing and building during the construction of multicompartment 25-storey building in Khabarovsk (figure 1).

Fig. 4. Analitical model 25-storey building.

Geological lithologic section in the sphere of designing construction interaction with geological environment in the depth interval 0-26 m is differential; there are 4 engineering-geological elements in the section. Under the plate foundation the chip cab of 5 m depth is located (figure 2).

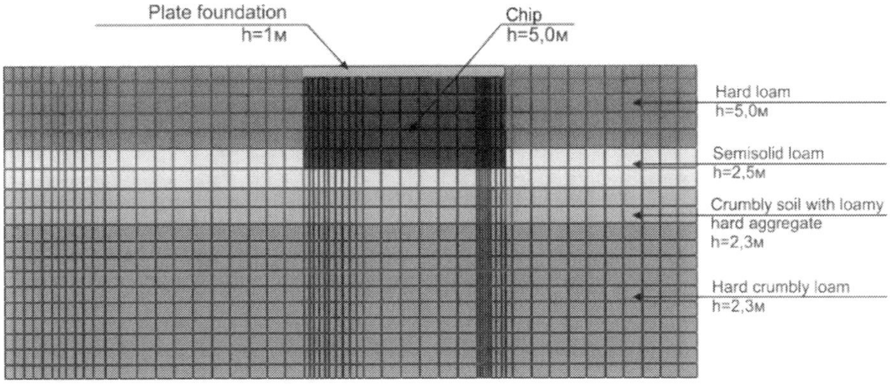

Fig. 5. Engineering-geological elements.

Before vertical displacement measuring surveying they set reference points and deformations marks which were constant during the whole observation period. 10 geodesic marks were analyzed.

The result of measuring is a summary table of sedimentary marks in each period and the accumulated settlements. Beside that we plan to work out the settlement scheme according to measuring periods.

The scheme of accumulated settlement of he analyzed marks for the period from June, 2008 to December, 2010 is on figure 6.

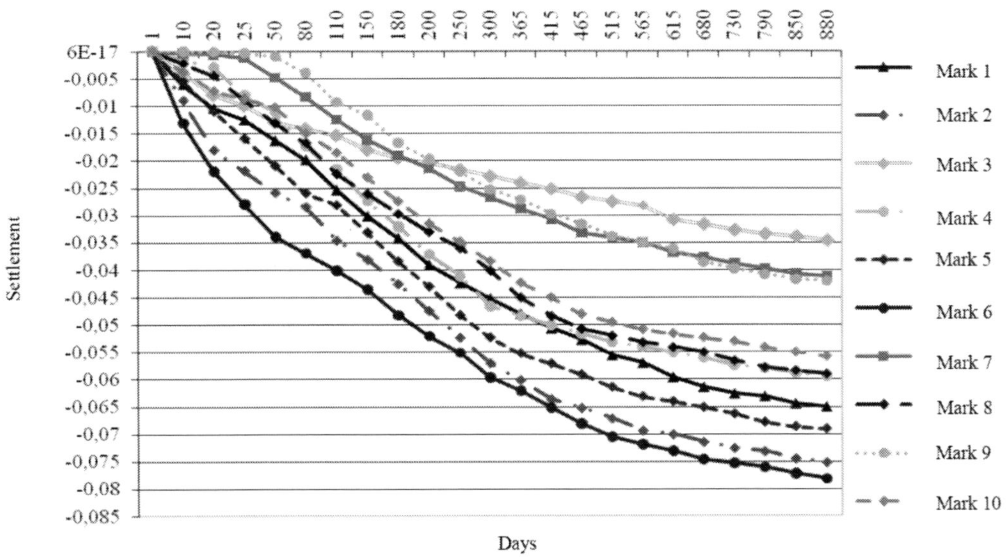

Fig. 6. Engineering-geological elements.

The settlement distribution is gradual. The max one according to the results of numeric model engineering is 87 mm, but in accordance with the geodesic monitoring it is 78 mm. The settlement amount of the raised building for December, 2010 does not exceed maximum allowable mark. It is S_u=200 m. The building tilt is insignificant and it also does not exceed the maximum allowable mark which is i_u=0,004.

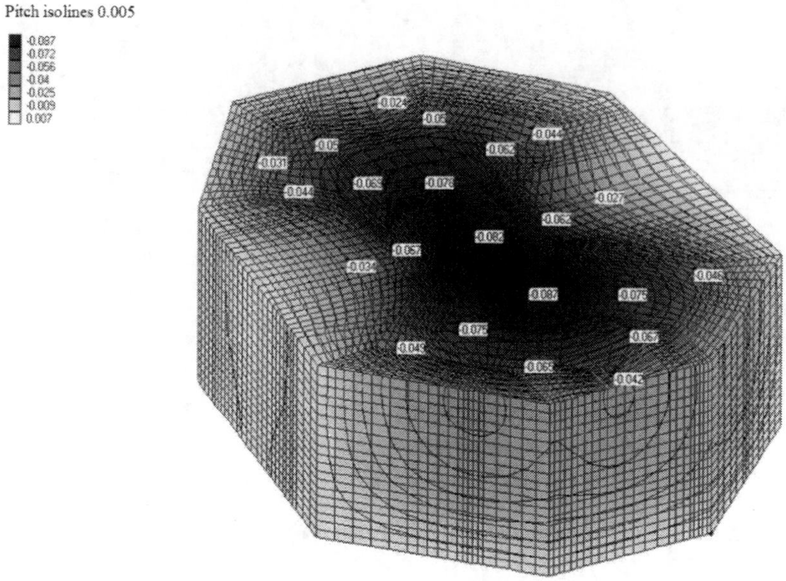

Fig. 7. Isoline of settlement plate foundation of 25-storey building

Vertical intensities in the foundation plate of a 25-storey building are distributed quite gradually. The max intensity marks are 280 kilopascal, which does not exceed the maximum allowable ones.

Engineering and numeric calculations show that during the installation of a 1 m foundation plate the building deformation is within the limits of allowable marks according to Russian norm.

4 Conclusions

The conducted explorations of the stress strain behavior of foundation plate of high-storey buildings have shown the possibility to use this particular type of underground constructions for building in engineering-geological conditions of Khabarovsk.

The settlement distribution is evenly due to plate foundation and chip's pillow under the plate. The geotechnical monitoring has shown the adequacy of the numeric model engineering in comparison with prototype observation. It makes it possible to conduct the numeric model engineering of the combined work of high-storey buildings on artificial and natural foundations in engineering-geological conditions of Khabarovsk.

References

1. S.A. Kudryavtsev, *Numerical modeling of the freezing, frost heaving, and thawing of soils. Soil Mechanics and Foundation Engineering 5,* **184** (2004)
2. S.A Kudryavtsev., V.M Ulitskii, I.I Sakharov, V.N Paramonov, *Bed – structure system analysis for soil freezing and thawing using the termoground program/ Soil mechanics and foundation engineering*: **1-7** (Springer New York Consultants . Bureau, ISSN: 0038-0741, 2015)
3. S.A. Kudryavtsev, I.V Shestakov., A.V. Kazharsky, *Monitoring and protection of the foundations of buildings of dangerous production plants on seasonally heaving ground* **191-196** (Proc. 8th Asian young geotechnical engineers conf., Astana, Kazakhstan)

Failure of structural elements made of polymer supported composite materials during the multiyear natural aging

Pavel Blinkov[1*], *Leonid* Ogorodov[1] and *Peter* Grabovyy[2]

[1]Peter the Great St.Petersburg Polytechnic University, Polytechnicheskaya, 29, St. Petersburg, 195251, Russia
[2] Moscow State University of Civil Engineering, Yaroslavskoe shosse, 26, Moscow, 129337, Russia

Abstract. Modern high-rise construction introduces a number of limitations and tasks. In addition to durability, comfort and profitability, projects should take into account energy efficiency and environmental problems. Polymer building materials are used as substitutes for materials such as brick, concrete, metal, wood and glass, and in addition to traditional materials. Plastic materials are light, can be formed into complex shapes, durable and low, and also possess a wide range of properties. Plastic materials are available in various forms, colors and textures and require minimal or no color. They are resistant to heat transfer and diffusion of moisture and do not suffer from metal corrosion or microbial attack. Polymeric materials, including thermoplastics, thermoset materials and wood-polymer composites, have many structural and non-structural applications in the construction industry. They provide unique and innovative solutions at a low cost, and their use is likely to grow in the future. A number of polymer composite materials form complex material compositions, which are applied in the construction in order to analyze the processes of damage accumulation under the conditions of complex nonstationary loading modes, and to determine the life of structural elements considering the material aging. This paper present the results of tests on short-term compression loading with a deformation rate of v = 2 mm/min using composite samples of various shapes and sizes.

1 Introduction

The mechanical properties of composite materials can be characterized using two approaches. In the first case, composite materials are considered as quasi-homogeneous (homogeneous), possessing isotropic deformation and strength properties in the case of volumetric disperse reinforcement, and possessing a certain type of anisotropy in the case of being reinforced with fibers, flat grids or fabrics. Usually models of an orthotropic or transversely isotropic bodies are adopted. Under this approach, the issue is about the

* Corresponding author: captainflint88@gmail.com

mechanical characteristics averaged in sufficiently large volumes. Another approach, which is incomparably more complex, but also more informative is to separately examine the mechanical properties of each phase with the subsequent forecasting of the properties of the entire composite. In this case, one more additional phase of the conjugation zone of the main phases is practically to be considered, for example, conjugation of matrices with reinforcing fibers. The phase boundaries are stress concentration zones, which play a particularly important role in the damage development of composite materials.

In respect to composites, which are often modeled by a homogeneous anisotropic body, the failure will be defined as the occurrence of any observed discontinuity.

Generalizations of the corresponding criteria for isotropic bodies (force, deformation and energy types and models in which the simultaneous occurrence of several damage processes is considered) were used to formulate the criteria for the failure of anisotropic materials in many works.

At present, the most common strength criteria of composites are the tensor polynomial failure criterion [2, 4] and the Goldenblat-Kopnov criterion [3].

The work [1] exposes the explanations for using the equation of force type damages in respect to orthotropic materials

$$\Pi = \frac{\sigma(\tau)}{c} + \int_0^{\tau} f[\sigma(\theta)]d\theta, \tag{1}$$

where c is a constant. This correspondence was also reproduced in [5]. In accordance with the equation (1), the value c is relevant for the "instantaneous strength of materials", which becomes the material's property under rapid (instantaneous) loading, when the process of damage accumulation does not manage to appear yet. Equation (1) leads to the definition of the law of the "instantaneous" resistance change, which decreases in time with the damage accumulation. The correspondence is as follows [1]

$$\sigma(\xi) = 1 - \int_0^{\xi} \sigma(\theta)\, d\theta, \tag{2}$$

which correlates with the functional exposed in [2], in order to determine the constricting resistance to the "instantaneous" discontinuity during the loading process. The formula (2) can be considered as a certain simple realization of this functional, at the same time the correlations (1) and (2) indicate that the method of the constricting "instantaneous" resistance and the method of kinematic equations are related in reference to the degree of damage to each other, they are alternative.

According to [3], the failure condition is as follows

$$\sum_{k,l=x,y,z} a_{kl}\sigma_{kl} + \sqrt{\sum_{k,l,r,s=x,y,z} a_{klrs}\sigma_{kl}\sigma_{rs}} = 1, \tag{3}$$

where the left-hand side can be represented by some reduced stress $\tilde{\sigma}$ in equation (1), where c = 1. In order to determine the integrand f [σ (θ)] in (1), the allowance is made according to [3], which states that the static fatigue curves coincide for the case of different stressed states in the coordinates reduced stress - time to failure. After integrating at constant stresses $\tilde{\sigma}$ and developing the failure condition Π = 1, we obtain

$$1 = \tilde{\sigma}\left(1 + \int_0^{\xi} \frac{f[\sigma(\theta)]d\theta}{\tilde{\sigma}}\right), \tag{4}$$

$$where\ \Phi(\xi) = \int_0^{\xi} \frac{f[\sigma(\theta)]d\theta}{\tilde{\sigma}} = \int_0^{\xi} \varphi(\theta)d\theta, \Phi(0) = 0.$$

Assuming that the equation of the static fatigue curve in case of uniaxial tension has the form $\tilde{\sigma} = A - Blg\xi$, the rime function can be expressed as follows

$$\Phi(\xi) = \int_0^{\xi} \frac{1}{A - Blg\xi} - 1, \varphi(\xi) = \frac{A}{2.303\xi(A - Blg\xi)^2}.$$

As a result for calculations, a kinematic equation can be written as where

$$\Pi(\tau) = \tilde{\sigma}(\tau) + \int_0^{\tau} \tilde{\sigma}(\theta)\varphi(\theta)d\theta,$$

which can be used in the case of unsteady loading. It is to be recalled that $\tilde{\sigma}(\tau)$ is the reduced stress, which is defined as

$$\tilde{\sigma} = \sum_{k,l=x,y,z} a_{kl}\sigma_{kl} + \sqrt{\sum_{k,l,r,s=x,y,z} a_{klrs}\sigma_{kl}\sigma_{rs}}, \qquad (5)$$

in the case of a linear stress state

$$\tilde{\sigma} = \left(a_{xx} + \sqrt{a_{xxxx}}\right)\sigma_{xx},$$

Coefficients at stresses for a composite are determined using the experimental data for the short-term loading under tension and compression in the directions of the main anisotropy axes, under a simple shear within these axes and axes rotated by an angle of ± 45 ° referred to reference ones.

It is possible to give more complex failure criteria, but there is value to complicate the calculation model only in case it is indicated by the experiments. A number of works [1, 2, 6] have been devoted to the research of the scattered failure stage of composites. The general destruction process of structurally heterogeneous materials can be qualitatively divided into two stages, the first of which is associated with the disseminated accumulation of stable microcracks. The appearance conditions of macrocracks from micropores and microdefects were studied in the previously mentioned papers, and also in [7, 8]. According to researches [12, 15], the process of accumulation and coarsening of cracks was considered on the basis of the concentration criterion [14]. The papers [11, 13] are devoted to the statistical, thermodynamic and kinematic aspects of the transition from microcracks to macrodestruction. A variation of the phenomenological failure theory considering the orientation of defects and a complex stress state was developed in [10].

The case of an orthotropic material under a plane stressed state was considered. The expression for the reduced stress will in this case is as follows [1]:

$$\tilde{\sigma} = a_{xx}\sigma_{xx} + a_{yy}\sigma_{yy} + \sqrt{a_{xxxx}\sigma_{xx}^2 + a_{yyyy}\sigma_{yy}^2 + a_{xyxy}\sigma_{xy}^2 + a_{xxyy}\sigma_{xx}\sigma_{yy}}, \qquad (6)$$

Coefficients at stresses for a material with sufficient brittleness under short-term loading are found using the experimental data according to the formulas:

$$a_{xx} = \frac{1}{2}\left(\frac{1}{\sigma_{uxx}^+} - \frac{1}{\sigma_{uxx}^-}\right);\ a_{xxxx} = \frac{1}{4}\left(\frac{1}{\sigma_{uxx}^+} - \frac{1}{\sigma_{uxx}^-}\right)^2;$$

$$a_{yy} = \frac{1}{2}\left(\frac{1}{\sigma_{uyy}^+} - \frac{1}{\sigma_{uyy}^-}\right);\ a_{yyyy} = \frac{1}{4}\left(\frac{1}{\sigma_{uyy}^+} - \frac{1}{\sigma_{uyy}^-}\right)^2;$$

$$a_{xyxy} = \frac{1}{4}\left(\frac{1}{\sigma_{uxy}^2}\right);\ a_{xxyy} = \frac{1}{8}\left[\left(\frac{1}{\sigma_{uxx}^+} - \frac{1}{\sigma_{uxx}^-}\right)^2 + \left(\frac{1}{\sigma_{uyy}^+} - \frac{1}{\sigma_{uyy}^-}\right)^2 - \left(\frac{1}{\sigma_{ux'y'}^+} - \frac{1}{\sigma_{ux'y'}^-}\right)^2\right];$$

where $\sigma_{uxx}^+, \sigma_{uxx}^-, \sigma_{uyy}^+, \sigma_{uyy}^-$ are the tensile and compressive strengths within main anisotropy axes; σ_{uxy} is the simple shear strength within these axes; $\sigma^{+}_{ux'y'}, \sigma^{-}_{ux'y'}$ are the strength limits for the simple shear within the axes $x'y'$, which are rotated to xy axes by an angle of + 45 ° or -45 respectively.

Further, it is necessary to have a static fatigue curve of any stress state at our disposal, for example, tension.

In the case of a biaxial stress state, the criterion can be written as

$$\frac{\sigma_1^2}{\sigma_{1P}\sigma_{1c}}+\frac{\sigma_2^2}{\sigma_{2p}\sigma_{2c}}+\sigma_1\sigma_2\left[\frac{1}{\sigma_{1P}\sigma_{2c}}+\frac{1}{\sigma_{2p}\sigma_{2c}}-\frac{1}{\tau_B}\right]+\left(\frac{1}{\sigma_{1P}}-\frac{1}{\sigma_{1c}}\right)(\sigma_1+\sigma_2)=1, \quad (7)$$

where τ_B is the shear strength, which is independent of the sign of tangential stresses (designations are retained [1]).

Considering the assumption $\sigma_{1P}=\sigma_{2P}=\sigma_P$, $\sigma_{1c}=\sigma_{2c}=\sigma_c$ criterion (7) can be written [6] as

$$\frac{\sigma_1^2}{\sigma_P\sigma_c}+\frac{\sigma_2^2}{\sigma_p\sigma_c}+\sigma_1\sigma_2\left[\frac{2}{\sigma_P\sigma_c}-\frac{1}{\tau_B^2}\right]+\left(\frac{1}{\sigma_P}-\frac{1}{\sigma_c}\right)(\sigma_1+\sigma_2)=1, \quad (8)$$

For certain correlations between the strength constants of the material, one can obtain

$$\tau_B=\sqrt{\frac{\sigma_P\sigma_c}{3}}, \tau_B=\sqrt{\frac{\sigma_P\sigma_c}{2(1+\mu)}}, \tau_B=\frac{2\sigma_P\sigma_c}{\sqrt{3}(\sigma_P+\sigma_c)},$$

which require experimental determination. In the absence of these τ_B values, the criterion of G.S. Pisarenko and A.A. Lebedeva is often used for checking

$$\frac{\sigma_P}{\sigma_c}\sigma_i+\left(1-\frac{\sigma_P}{\sigma_c}\right)\sigma_1=\sigma_P. \quad (9)$$

The theoretical contour, which was constructed according to this criterion, deviates by no more than 13% from the experimental points of multiple polymeric materials. For $\chi=\frac{\sigma_P}{\sigma_c}, 0\le\chi\le 1$, which is true for most construction materials, the limiting surface is represented as a figure, which is equally inclined to the principal axes according to equation (9).

G.S. Pisarenko offered the condition for materials with structural heterogeneity

$$\chi\sigma_i+(1-\chi)\sigma_1A^{1-J}=\sigma_P, \quad (10)$$

where $J=\frac{\sigma_1+\sigma_2+\sigma_3}{\sigma_i}$ is the stress state parameter; A is the parameter of the material structure. The value of parameter A can be determined using the torsional test data

$$A=\frac{\varphi-\chi\sqrt{3}}{1-\chi},$$

where $\varphi=\frac{\sigma_P}{\tau_B}$, τ_B is the limiting torsional stress. Condition (10) is intended for use in assessing the limiting states of fiber composites, glass-nylon composites.

2 Materials and methods

Composite materials K-211-2 and K-18-36 based on the phenol-formaldehyde resin with dispersed reinforcement consisting of wood powder particles and, in the other case, of kaolin particles, are used for structural elements, which work under conditions of different stress state types. The degree of filling of composites with particles was 30...35%. These composite materials are models of a large material class. Samples of composites were tested in their delivery condition, and a significant part of samples underwent natural aging under the conditions of a heated warehouse for 18-20 years. All tests were carried out under the same conditions.

Test materials were of the following dimensions (mm): 5x5x5; 8x8x8; 10x10x10, d/h = 10 / 17.5; d/h = 12/21. Samples were set between the plates of the presses without lubricating the ends. Tensile tests were conducted using flat samples with a bearing length of 100 mm, 25.4 mm wide and 1 mm thick. The tests of composites carried out by Yu. Y. Loginov were conducted using samples with a diameter D = 127 mm, working part length l = 190 mm, thickness t = 6 mm under conditions of a linear and plane stress states.

The sample preparation was carried out as follows: 1) cutting of a prism-shaped specimen using precision circular saw 2) grinding of the samples with an in-house built fine mechanical abrasive grinder to achieve highly parallel opposite edges of the prism sample. Cylindrical shaped samples were manufactured on a lathe. The variation in flatness along the sample surfaces was measured to be in the range of 10 μm using micrometer.

3 Results of tests

The results of tests on short-term compression loading with a deformation rate of v = 2 mm/min using K-211-2 and K-18-36 composite samples of various shapes and sizes (A is the cross-sectional area, S is the area of the free sample surface, V is the volume) are presented in Table 1 (σ_{uc} is the strength limit, ε_{uc} is the deformation corresponding to the tensile strength, ε_r is the residual deformation upon failure, E_c is the elastic modulus).

In order to forecast the mechanical characteristics of the structural materials K-211-2 and K-18-36 with larger sizes than tested ones, the results of the tests presented in Table 1 are processed in relative values depending on A, S, and V.

Table 1. Dependence of mechanical characteristics on the size of the samples.

Material	Sample dimensions, mm	A, mm^2	S, mm^2	V, mm^3	σ_{uc}, MPa	ε_{uc}	ε_r	E_c, MPa
K-211-2	5x5x5	25	100	125	220.1	0.126	0.055	3110
K-211-2	8x8x8	64	256	512	225.2	0.154	0.082	3200
K-211-2	10x10x10	100	400	100	216.0	0.144	0.077	3070
K-211-2	d/h=10/17.5	78,5	595	1374	199.6	0.082	0.041	4880
K-211-2	d/h=12/21	113	791	2374	194.6	0.072	0.027	4280
K-18-36	8x8x8	64	256	512	156.8	0.072	0.018	2860
K-18-36	d/h=12/21	113	791	2373	148.0	0.060	0.021	4080

According to the data presented in Table 1, following analytical correlations are obtained:

$$lg(\sigma_{uc}/A) = a_0 - b_0 lgA \tag{11}$$

$$lg(\sigma_{uc}/S) = a_1 - b_1 lgS \tag{12}$$

$$lg(\sigma_{uc}/V) = a_2 - b_2 lgV \tag{13}$$

$$lg(\varepsilon_{uc} \cdot 10^4/V) = a_3 - b_3 lgV \tag{14}$$

$$lg(\varepsilon_r \cdot 10^4/V) = a_4 - b_4 lgV \tag{15}$$

$$lg(E_c/V) = a_5 - b_5 lgV \tag{16}$$

where σ_{uc} and E_c values are expressed in MPa, A in mm^2, S in mm^2, V in mm^3, and the coefficients a and b for the analyzed composites at v = 2 mm/min are presented in Table 2.

Table 2. Values of coefficients (at v=2 mm/min).

Material	a_0	b_0	a_1	b_1	a_2	b_2	a_3	b_3	a_4	b_4	a_5	b_5
K-211-2	2.39	1.05	2.46	1.06	2.44	1.05	0.28	1.42	0.78	1.69	3.15	0.85
K-18-36	2.24	1.03	2.32	1.05	2.28	1.03	0.84	1.11	-	-	3.08	0.85

In order to determine the a and b coefficients for the K-18-36 composite, analytical correlations were used, which are analogous to the correlations determined for the K-211-2 composite. The mechanical characteristics related to the cross-sectional area and to the free surface area of the sample depend on their shapes, while these characteristics, which are related to the volume of the samples (for given cross-sectional area to sample height ratios), do not depend on the shape and can be sufficiently accurate described by the dependences (13) - (16).

Dependencies (11) - (16) were used to predict the mechanical characteristics of composite samples of different sizes under uniaxial compression with a deformation rate of v = 2 mm/min.

The results of the calculations are given in Table 3, where the values σ_{uc}^{A} and σ_{uc}^{S} were determined from the dependences (11) - (12), and σ_{uc}^{V}, ε_{uc}, ε_r, E_c values from (13) - (16).

Table 3. Mechanical characteristics of composite samples of different sizes.

Material	Sample dimensions, mm	σ_{uc}^{A}, MPa	σ_{uc}^{S}, MPa	σ_{uc}^{V}, MPa	ε_{uc}	ε_r	E_c, MPa
K-211-2	20x20x20	181.9	185.2	175.7	0.044	0.012	5400
K-211-2	50x50x50	166.0	166.0	153.2	0.014	0.002	8200
K-211-2	127/190/6	166.8	141.4	143.9	0.008	-	9900
K-18-36	5x5x5	157.8	166.0	162.8	0.084	-	2460
K-18-36	10x10x10	151.4	154.8	153.8	0.068	-	3390
K-18-36	d/h=10/17.5	152.4	152.4	151.6	0.064	-	3540
K-18-36	20x20x20	145.2	144.5	144.2	0.054	-	4630
K-18-36	50x50x50	137.4	131.8	132.4	0.044	-	6990
K-18-36	127/190/6	137.8	115.4	129.1	0.011	-	8420

The results of calculating the mechanical characteristics from dependences (11) - (16) and the results of K-211-2 tests are compared in Table 4.

Table 4. Mechanical characteristics of K-211-2.

Sample dimensions, mm	σ_{uc}^{A}, MPa	σ_{uc}^{S}, MPa	σ_{uc}^{exp}, MPa	ε_{uc}	ε_{uc}^{exp}	E_c, Mpa	E_c^{exp}, MPa
d/h=10/15	187.3	186.9	171.8	0.0076	0.074	3990	-
d/h=10/17.5	187.3	185.4	201.3	0.070	0.098	4070	4440

The tensile strength (σ_{ut}) of composite samples K-211-2 and K-18-36, which underwent natural aging (up to 19 years), under uniaxial tension conditions with different deformation velocities v (base velocity v = 2 mm/min) is 46.06 MPa and 37.70 MPa respectively. After the delivery, the tensile strength of tubular samples of the K-211-2 composite was 22.5 MPa and 127.6 MPa for uniaxial compression, and K-18-36 samples had 20 MPa and 85.8 MPa, respectively. After the tests of flat samples (v = 2 mm/min) of the same composites in the delivery condition, σ_{ut}= 41.89 MPa was obtained, σ_{uc} = 190.9 MPa and σ_{ut} = 35.9 MPa, σ_{uc} = 144.2 MPa. The strength of solid thin-walled samples (t = 1 mm) is approximately 1.7

times higher than the strength of tubular samples made of the same material, but larger. We should notice the results of the strength characteristics of a composite sample batch having a circular spherical surface obtained by machining (h = 21 mm, d = 12 mm, the smallest diameter of the cross section is 8.55 ÷ 9.9 mm, the width of the groove is 6.5 ÷ 7 mm). After the 19-year natural aging of the samples, the compressive strength at a deformation rate of v = 2 mm/min was σ_{uc} = 196.2 MPa for the composite K-211-2, which is substantially higher than one for the samples without machining. This fact can be explained by the peculiarity of the triaxial stress state affecting the central part of the sample.

Figures 1a and 1b represent the limiting stress values and ultimate failure lines of composite materials K-211-2 and K-18-36, respectively. The results obtained by Yu. Y. Loginov are plotted with single points (o).

The processing of the experimental data was carried out according to the stress intensity σ_i, as well as to the equivalent stresses calculated using the Pisarenko-Lebedev criterion (line 1), also experimental and calculated values were defined using the Goldenblat-Kopnov criterion. In the first quadrant σ_1>0, σ_2>0, the samples of both composites show a lower strength than the limiting ones (line 1). Higher values of σ_3 (compressive stresses) in experiments led to greater strength of the composites.

An attempt was made in order to estimate the limiting values using the Goldenblat-Kopnov criterion, where the calculation values of the limiting shear stresses were adopted according to the dependences

$$\tau_B = \sqrt{\frac{\sigma_P \sigma_c}{3}}, \tau_B = \frac{2\sigma_P \sigma_c}{\sqrt{3}(\sigma_P + \sigma_c)},$$

According to the Goldenblat-Kopnov criterion for the stresses $\sigma_1 = 2\sigma_2$ and the composite K-211-2, it was obtained 1.036 and 0.888, for the stresses $\sigma_1 = \sigma_2$ we have 1.124 and 0.928, for the composite K-18-36 it is 0.928 and 0.805, 0.860 and 0.604 in comparison with the limiting condition which equal to 1.0.

Limiting curves (2) of the tubular structural element (D = 127mm, t = 6 mm, l = 190 mm) were predicted according to Pisarenko-Lebedev criterion for a period of 20-25 years (are marked on the Figures 1a and 1b with the individual points (Δ) of the limit curve), taking into account the natural aging effect of composites and the scale factor.

Forecasting limiting curves (2) are represented according to the Pisarenko-Lebedev criterion of structural elements of smaller dimensions in the investigated scale range (in particular, A = 113 mm^2, S = 595 mm^2, V = 2373 mm^2).

Figures 2a and 2b represent the isochronous ultimate failure lines of K-211-2 and K-18-36 composites under the conditions of a plane stress state and a prolonged steady loading during the one day, ten and one hundred days.

The significant problem is the calculation of the composite damage level under nonstationary (stepwise in particular) regimes of static and cyclic loading in plane stress conditions. The equivalent (σ_{eq}) stresses (acceptable for the planning of nonstationary loading regimes) are determined according to the Pisarenko-Lebedev criterion.

For the K-211-2 it is adopted as follows:

1. $\sigma_1 = 31.02\ MPa, \sigma_2 = 15.51\ MPa, \sigma_{eq} \approx 30\ MPa;$
2. $\sigma_1 = 31.02\ MPa, \sigma_2 = 15.51\ MPa, \sigma_{eq} \approx 30\ MPa;$
3. $\sigma_1 = 31.02\ MPa, \sigma_2 = 15.51\ MPa, \sigma_{eq} \approx 30\ MPa;$
4. $\sigma_1 = 31.02\ MPa, \sigma_2 = 15.51\ MPa, \sigma_{eq} \approx 30\ MPa;$
5. $\sigma_1 = 31.02\ MPa, \sigma_2 = 15.51\ MPa, \sigma_{eq} \approx 30\ MPa;$
6. $\sigma_1 = 31.02\ MPa, \sigma_2 = 15.51\ MPa, \sigma_{eq} \approx 30\ MPa;$

For the composite K-18-36, σ_{eq} = 30 MPa is adopted. The parameter $\chi = \frac{\sigma_{ut}}{\sigma_{uc}} = \frac{\sigma_P}{\sigma_c}$ for K-211 -2 is 0.237, for the K-18-36 composite it is 0.255.

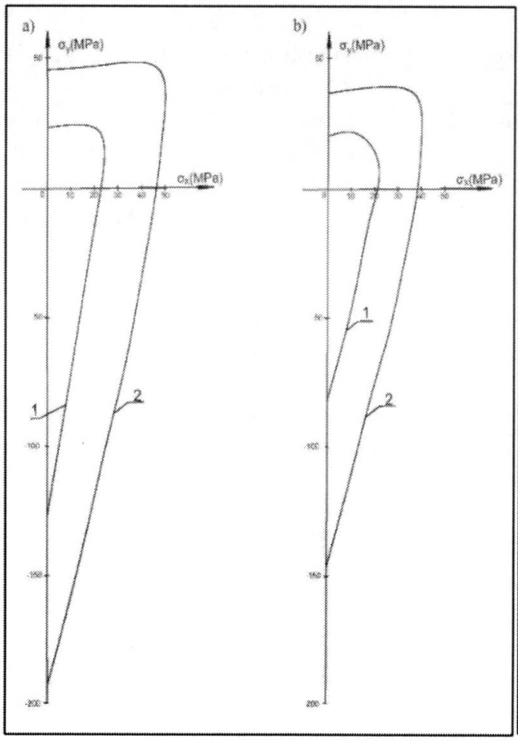

Figure 1. Limiting stress values and ultimate failure lines of composite materials K-211-2(1a) and K-18-36(1b).

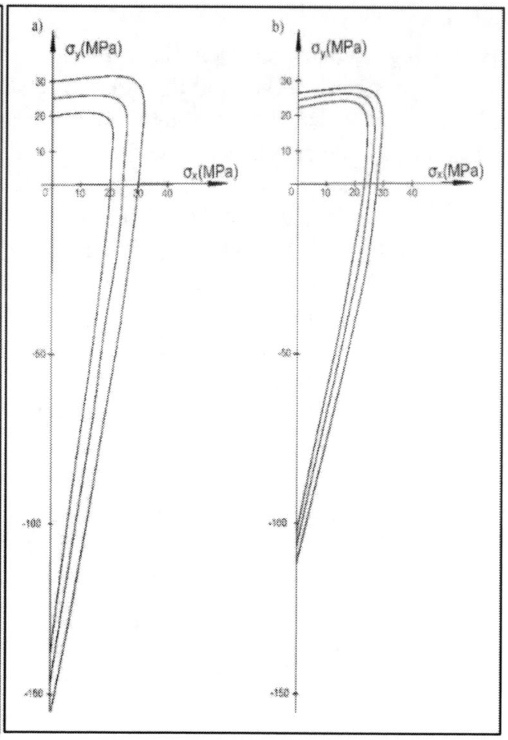

Figure 2. Isochronous ultimate failure lines of K-211-2(2a) and K-18-36(2b).

The values of stresses, loading and unloading (rest) periods, as well as calculated damage levels of composites under nonstationary loading regimes (where the order of loads (1) - (6) can be arbitrary) are exposed in Table 5 and Table 6. The version of the kinetic equation was used (4)

$$\Pi = \int_0^{\tau} \tilde{\sigma}_{eq}\, \sigma(\theta) K(\tau - \theta) \mathrm{d}\theta, \Pi(\tau) = \tilde{\sigma}(\tau) + \int_0^{\tau} \tilde{\sigma}\,(\theta) \varphi(\theta) \mathrm{d}\theta,$$

where $\tilde{\sigma}_{eq}$ is the reduced equivalent stress, $K(\tau - \theta)$ is the nucleus

Table 5. Loading regimes and the forecasted damage level of the K-211-2 composite.

Regime	Grade of load	σ_{eq}, MPa	τ, hour	τ_0, hour	Π
1	1	20	8	-	0.009
	2	30	8	16	0.017
	3	20	8	-	0.030
	4	30	8	-	0.945
					Total:1.001
2	1	20	8	-	0.001
	2	30	8	8	0.020
	3	20	8	-	0.030
	4	30	8	-	0.945
					Total:1.007
3	1	30	8	-	0.010
	2	20	8	16	0.007
	3	30	8	-	0.017
	4	20	8	16	0.014
	5	30	8	-	0.945
					Total:0.993
4	1	30	8	-	0.028

	2	20	8	-	0.030
	3	30	8	-	0.945
					Total:1.003
5	1	20	36.5	-	0.096
	2	30	4	-	0.904
					Total:1.000
6	1	30	4	-	0.003
	2	25	200	-	0.994
					Total:0.997

Table 6. Loading regimes and the forecasted damage level of the K-18-36 composite.

Regime	Grade of load	σ_{eq}, MPa	τ, hour	τ_0, hour	П
1	1	22	8	-	0.005
	2	27	8	16	0.007
	3	22	8	-	0.016
	4	27	8	-	0.969
					Total:0.997
2	1	22	8	-	0.006
	2	27	8	-	0.008
	3	22	8	-	0.016
	4	27	8	-	0.969
					Total:0.999
3	1	25	1	-	0.036
	2	30	0.25	-	0.979
					Total:1.015
4	1	30	0.5	-	0.0001
	2	25	250	-	0.996
					Total:0.996
5	1	27	8	-	0.012
	2	22	8	-	0.016
	3	27	8	-	0.969
					Total:0.997
6	1	27	8	8	0.003
	2	27	8	8	0.005
	3	27	8	8	0.007
	4	27	8	8	0.012
	5	27	8	-	0.969
					Total:0.996

4 Discussion

Manufacturing of elements made of materials and structural elements after long-term storage, re-opening, re-profiling, rejection and exploitation is economically beneficial. Recycling into secondary raw materials and then into elements requires significant expenses, and sometimes such a process is impossible. Burial or destruction of unused structural elements leads to negative environmental impacts. The use of elements made from stored and exploited materials and structural elements requires an assessment of changes of their properties, in particular of residual strength. The development of requirements for the rejection and decommissioning of elements and structures according to the residual mechanical properties of materials is a vital question. Another issue is the evaluation of the damage level and calculation of the resource of materials and structural elements under conditions of complex operating modes (loading regimes).

5 Conclusion

Unique experimental data on the mechanical characteristics of composite materials K-211-2 and K-18-36, which have undergone 18-20 years of natural aging, are obtained.

Calculated dependences, which allow to consider the effect of a scale factor on the mechanical characteristics of composites, are determined.

The results of composite tests in the plane stress state are compared with the calculations based on the Pisarenko-Lebedev and Goldenblat-Kopnov criteria, which are recommended for evaluating the strength for the flat stressed state of composites.

Failure limit lines for tubular composite structural elements are constructed, considering the scale factor in the investigated range of sizes and natural aging of materials. Nonstationary regimes of disproportionate loading of composites were developed in order to estimate the operational life of composites, as well as to estimate the damage level at the failure time under conditions of plane stress.

Acknowledgments

The authors would like to thank prof. Stolyarov O.N. for helping with the experiments.

References

1. O. A. Kurakova I. S. Sychev and M. A. Isaev 2010 *Nedvizhimost': Ekonomika, upravlenie* **1** pp 92-95
2. P. P. Zueva 2014 *Aktual'nye problem stroitel'stva, ecologii I enegosberezheniya v usloviyah zapadnoj sibiri* **1** pp 98-104
3. S. V. Sargasyan and A. A. Hromenkova 2016 *Nuchnoe obozrenie* **18** pp 36-39
4. D. N. Gotina and U. G. Tkachenko 2015 *Novye idei novogo veka: materialy mezhdunarodnogo nauchnoj konferencii FAD TOGU* **2** pp 15-20
5. I. P. Nikolaeva L. I. Ogorodov and E. L. Yakovleva 2017 *Stroitel'stvo unikal'nyh zdanij i sooruzhenij* **3** pp 8-17
6. M. N. Stepnov S. L. Chernyshev I. E. Kovalev and A. V. Zinin 2010 *Monografiya Izdatel'skij centr "Tekhnologiya mashinostroeniya" (Moskva)* 256 pp
7. V. V. Nesmelov 1993 *Fizika goreniya I vzryva* **29** pp 23-28
8. R. V. Ishchenko S. V. Avdeichik E. V. Ovchinnikov and V. A. Struk 2014 *Vesnik grodzenskaga dzyarhanaga universiteta im. Yanki Kupaly* **1** pp 63-68
9. K. Cirule 2008 *Mekhanika kompozitnyh materialov* **3** pp 463-468
10. S. K. Golyshko 2005 Vychislitel'nye tekhnologii **1** pp 134-135
11. M. I. Balzannikov and A. A. Mikhasek 2014 *Procedia Engineering* **91** pp 183-187
12. S. Chadiarakoua and P. Antoniadoub 2017 *Procedia Environmental Sciences* **38** pp 830-835
13. S. Ramakrishnan X. Wang J. Sanjayan and J. Wilson 2011 *Applied energy* **7** pp 1515-1526
14. E. A. Nekliudova A. S. Semenov B. E. Melnikov S. G.Semenov 2014 *Civil Engineering* **3** pp 25-39
15. K. B. Broberg 1982 Engineering *Fracture Mechanic* **4** pp 497 - 515

Determination of the neutralization depth of concrete under the aggressive environment influence

Anastasia Morzhukhina[1,*], *Stanislav* Nikitin[1] and *Elena* Akimova

[1]Peter the Great Saint-Petersburg Polytechnic University, 195251 St. Petersburg, Russia
[2]Moscow State University of Civil Engineering, Yaroslavskoe shosse, 26, Moscow, 129337, Russia

Abstract. Aggressive environments have a significant impact on destruction of many reinforced concrete structures, such as high-rise constructions or chemical plants. For example, some high-rise constructions are equipped with a swimming pool, so they are exposed to chloride ions in the air. Penetration of aggressive chemical substances into the body of concrete contributes to acceleration of reinforced concrete structure corrosion that in turn leads to load bearing capacity loss and destruction of the building. The article considers and analyzes the main technologies for calculating penetration depth of various aggressive substances into the body of concrete. The calculation of corrosion depth was made for 50-year service life.

1 Introduction

Exceptional durability of reinforced concrete structures is one of the main reasons for widespread use of this building material in the world.

However, material constraints, design and construction methods, as well as difficult service conditions, can lead to deterioration in quality of concrete and reinforcing bars corrosion that in turn can lead to aesthetic, functional or structural problems.

Reinforced concrete can be destroyed for various reasons, and specific damage is often the result of a combination of factors. One of the most common causes of reinforced concrete corrosion is aggressive environment penetration into the body of concrete, which subsequently causes reinforcing bars corrosion. Chemical aggression is the most violent and destructive.

Many buildings and structures are influenced by chemical attack, such as industrial buildings, tanks, chemical plants. High-rise constructions can also be affected by aggressive environments. For example, the high-rise building Velasca Tower in Milan has been exposed for a long period to carbon dioxide, which caused carbonation of concrete and corrosion of the reinforcement. Also, often in high-rise constructions design pools, in which rooms there is high moisture. In vapor and water there are chloride ions that penetrate concrete and cause chloride corrosion of the reinforced concrete structure [1-3].

* Corresponding author: a.morzhukhina@gmail.ru

Chloride corrosion

The chloride ions effect on reinforced concrete structures is one of the main reasons for premature corrosion of steel reinforcement. Chloride ions, which present in crystals of salts and sea water, penetrate the thickness of reinforced concrete elements section and cause corrosion of reinforcing bars. Corrosion occurs only with high humidity and constant exposure to oxygen. Chlorides dissolved in water penetrate through cracks in concrete and reach the steel. The risk of corrosion increases as chloride content in the concrete increases. Corrosion process occurs, when chloride content on steel surface exceeds a certain limit [4-7].

Sulfate corrosion

Sulfate aggression is the most common aggression, since sulfate ions are present in almost all types of natural and waste water. In natural form, sodium, potassium, calcium or magnesium sulfates are sometimes found in soil or dissolved in groundwater. Sulfates can attack concrete by interacting with hydroxides in cement. These reactions can cause sufficient pressure to break cement by increasing its volume, which leads to a loss of strength characteristics. Sodium sulfate reacts with calcium hydroxide and water, forming calcium sulfate dihydrate $CaSO_4 \cdot 2H_2O$ (gypsum).

Due to increase in volume of substances in the concrete, internal stresses appear. These stresses lead to cracks and destruction of its structure. It was also found that with decreasing temperature, concrete destruction under influence of sulfates is more significant [8-11].

2 Material and methods

From the experimental studies of concrete corrosion it is known that penetration of any aggressive substance into the concrete body is diffusive and causes a change in its physic and mechanical properties. Between the neutralized environment and healthy concrete, a boundary separating materials with different physical and mechanical properties can be drawn. This division is somewhat arbitrary. It is known that in a number of cases, depending on developing corrosion, the boundary between concrete layers that have been preserved in the damaged by corrosion can be traced with difficulty, and sometimes completely absent.

The results of various experimental works on reinforced concrete corrosion showed that when considering the influence of an aggressive environment on a reinforced concrete element, several factors must be considered. First, the depth of concrete neutralization is studied by an aggressive substance that penetrates into the concrete body, changes its physic and chemical properties and characterizes the protective properties of concrete in relation to the reinforcement. Secondly, consider the depth of concrete destruction, which determines reduction in the reinforced concrete element cross-section when concrete is destroyed by an aggressive substance. As shown by numerous laboratory and natural studies, the depth of concrete destruction is insignificant in comparison with the depth of neutralization for the same period [12].

The most objective quantitative characteristic of concrete corrosive process will be considered the depth of concrete neutralization by aggressive substance for a certain period of time. The rate of concrete neutralization depends on many factors, such as:

- type of aggressive environment;
- concentration of corrosive substances in contact with concrete surface;
- composition of concrete;
- observed relative humidity.

An indicator that estimates the penetration depth is proposed to be determined by a formula that looks like:

$$\delta = k(\xi) \cdot \sqrt{D^* \cdot \tau} \quad (1)$$

Here k (ξ) — coefficient; τ — time of exposure of aggressive substance to concrete; D* — effective diffusion coefficient of corrosive substance through the layer of corrosion products determined experimentally (at t = 20°C) and presented in Table 1.

For practical calculations of the corrosion depth at temperatures other than 20°C, the D * value can be calculated taking into account that an increase or decrease in temperature by 1°C gives an increase or decrease of D* by 2.6%, respectively.

Table 1 The value of effective diffusion coefficient.

Type of acid	Concentration, H	pH value	D*, sm²/h
HCl	0,0001-0,1	4-1,5	0,047
H_2 SO_4	0,0001-0,035	4-1,5	0,041

It has been suggested that k (ξ) be determined by a formula that takes into account the influence of the water-cement ratio on the chemical process [13]:

$$k(\xi) = A \cdot \sqrt{C \cdot E}$$

Here E — chemical equivalent of ratio of calcium oxide molar mass M_1 and aggressive substance molar mass M_2 . E = mM_1 / nM_2 (m, n — stoichiometric coefficients of reaction equation between acid and calcium oxide, C — concentration of aggressive substance, A — coefficient that depends on composition of concrete and is equal:

$$A = \sqrt{\frac{2\left(\frac{1}{\rho_c} + \frac{W/C}{\rho_w}\right)}{K^*}},$$

where ρ_c and ρ_w — true density of cement and water, respectively; K^* — mass content of calcium oxide in cement, W/C — water-cement ratio.

The Russian researcher Popesko proposed a modification of formula (1).

All studies on the kinetics of corrosion of concrete confirm that the depth of its neutralization is correlated with the time of action of corrosive substance. [14-15] Therefore, in the first approximation, for a mathematical description of given dependence, we can take the model in the form of a power function. Based on the results of research and experimental materials, Popesko proposed the depth concrete corrosion to be determined by a formula of the form:

$$\delta = K \cdot t^{0,5}, \quad (2)$$

where K — coefficient of aggressiveness of a particular substance to a certain class of concrete, determined experimentally; t — time of action of aggressive environment.

3 Results

The depth of neutralization of concrete depends on type of corrosive effect. We consider the effect of 1% solutions of sulfuric and hydrochloric acids. For concrete of class B25 of reduced permeability (W6) the value of coefficient A was calculated, it's equal to $5.25 \cdot 10^{-2}$. The molar masses of substances are:

$$M_{CaO} = 56{,}08 g / mole, M_{HCl} = 36{,}46 g / mole, M_{H_2SO_4} = 98{,}08 g / mole$$

The value of coefficient K for a 1% solution of hydrochloric and sulfuric acids is, respectively, 0.11 and 0.029.

The calculated values of corrosive destruction depth of low-permeability concrete (W6) under the action of 1% HCl and H_2SO_4 solutions on them are given in Table 2.

Table 2 Values of concrete neutralization depth for 50 years.

Item No.	Service life, years	Concrete neutralization depth, mm			
		By the formula 1		By the formula 2	
		1% sol. H_2SO_4	1% sol. HCl	1% sol. H_2SO_4	1% sol. HCl
1	1	2.38	5.68	1.00	3.81
2	3	4.12	9.83	1.74	6.60
3	5	5.32	12.70	2.25	8.52
4	10	7.52	17.95	3.18	12.05
5	15	9.21	21.99	3.89	14.76
6	25	11.89	28.39	5.02	19.05
7	50	16.81	40.15	7.10	26.94

4 Conclusion

In the course of this research, concrete neutralization depth was determined under influence of sulfuric and hydrochloric acid solutions on concrete for 50 years of working. When comparing the results, we see that chloride corrosion has more negative effect than sulfate corrosion under equal conditions for the same period of exposure time. Comparative analysis of two methods for calculating the depth of concrete corrosion revealed that the results are significantly different.

The expediency and effectiveness of the application of protective measures should be assessed by comparing the values of unprotected concrete corrosive destruction depth with the allowable depth of failure during the overhaul period, provided that the load bearing capacity of the structure is maintained.

If during the design service life due to corrosive effect of corrosive environment on the structure, surface layer of concrete is destroyed and the reduction in the cross-sectional area doesn't cause an increase in stresses above the maximum permissible value and there is no risk of reinforcement corrosion, then protection can be discarded.

The protective properties of concrete in a slightly aggressive acid environment can be increased by use of primary protection - by increasing the concrete density, the protective layer thickness and the structures size.

References

1. F. Lollini, E. Redaelli, L. Bertolini, CSCM **4**, 55–61 (2016)
2. M. Torres-Luquea, E. Bastidas-Arteagab, F. Schoefsb, S.-M. Sánchez, J.F. Osma, CBM **68**, 68–81 (2014)
3. N.N. Skoblinskaja, K.G. Krasilnikov, L.V. Nikitina, Varlamov, CCR 5, (1975)
4. A. Boddy, E. Bentz, M. D. A. Thomas, and R. D. Hooton, CCR **29**, 827–837 (1999)
5. X. Shia, N. Xiec, K. Fortunea, J. Gong, CBM **30**, 125-138 (2012)
6. V.V. Belov, S.E Nikitin, IOPE **4**, 142-147 (2013)

7. V.P. Selyaev, V.A. Neverov, P.V. Selyaev, E.V. Sorokin, O.A. Yudina, MCE **1**, 41-52 (2014)
8. P. N. Gospodinov, R. F. Kazandjiev, M. Mironova, CCC **18**, 401–407 (1996)
9. V. Zivica, A. Bajza, CBM **15**, 331-340 (2000)
10. P. K. Mehta, CI, 57-61 (2000)
11. J. B. Aguiar, A. Camoes, P.M. Moreira, IJCSM **2**, 21-25 (2008)
12. H. Nagano, T. Otani, Y. Sato, Y. Akiyoshi, H. Matsui, T. Mishima, PTE LTD **35**, 1-9 (2010)
13. S. Sato, Y Masuma, Y. Hasegawa, I. Natsuka, A. Aoyama, K. Yokoi, OWCS, 373-380 (2011)
14. A. Pustovgar, A. Adamtsevich, L. Shilova, Advances in Intelligent Systems and Computing, **692**, pp. 608 - 614 (2018) DOI - 10.1007/978-3-319-70987-1_64
15. A.O. Adamtsevich, A.P. Pustovgar, S.A. Pashkevich, A.V. Eremin, Calorimetric methods in analysis of composite building materials, International Journal of Applied Engineering Research, **11**, pp. 1609 - 1611 (2016)

Determination of pollutants in foundry during the manufacture of metal constructions for high buildings

Irina Ivanova [1,*] *Elena* Sushko[1], *Anna* Lyshnikova[2] and *Larisa* Prykina[3]

[1] Voronezh State Technical University, Moscow Avenue, 14, Voronezh, 394026, Russia
[2] Voronezh State Medical University named after N.N. Burdenko, Stydencheskaya st., 10, Voronezh, 394036, Russia
[3] Moscow State University of Civil Engineering, Yaroslavskoye shosse 26, Moscow, 129337, Russia

Abstract. Current developments are devoted to the environmental safety of the foundry. There is a significant amount of pollutants, according to dust, which is released in the working area, during the manufacture of metal structures for high buildings. From the point of dust extraction, the most unfavorable areas are shot blasting, sand-blasting chambers and knockout grills. The weight fraction of dust composition with diameters up to 20 µm reaches 43,8% by mass, according to experimental analysis. This kind of dust is the most dangerous to employees and also it creates problems for dust-cleaning in the air.

1 Introduction

Technological processes of manufacturing casting of metal constructions for high buildings are characterized with a huge amount of operations, in the performance of which dust, aerosols and gases are released. Dust is the main contaminant of work zone, it is formed practically on every site of the technological process of manufacturing foundry alloys:

- preparation of the molding mixture;
-molding;
- metal-smelting and metal-pouring;
- rod;
- knocking out lattices and rods, as well as cleaning castings.

A significant amount of carbon oxides, sulfur oxides, nitrogen oxides, phenol, formaldehyde, aromatic hydrocarbons, ammonia and cyanides are released into the environment in addition to dust, during the casting process. Pairs of zinc, cadmium, lead, beryllium, chlorine and chlorides, fluorides are also produced during non-ferrous metals melting. The use of organic binders in the manufacture of rods and molds leads to a significant release of toxic gases during the drying process, and especially when pouring the metal. Harmful substances such as ammonia, acetone, phenol, formaldehyde, furfural, etc. can be released into the atmosphere, depending on the class of the binder.

The most dangerous dust for employees health is that one which contains free silica - SiO_2. The poisonous of silica is not high, but with prolonged inhalation there are changes in

* Corresponding author: ivanova-eco@mail.ru

the lungs. There is a direct relationship between the weight concentration of dust in the air and the sickness rate of inhaling people. The danger of dust is also determined by its granulometric composition, i.e. the quantitative ratio of dust fractions of various sizes in it [1].

Large particles of dust linger on the mucosa of the upper respiratory tract, while particles with dimensions close to 1 μm can penetrate into the alveoli of the lungs, where the main part of them is persisted. The greatest amount of sulfur oxides is formed during the combustion of high sulfur fuel oil. Almost all sulfur is oxidized to sulfurous anhydrid $S0_2$ in emissions. Sulfuric anhydride refers to irritating gases.

At a concentration of 0.3 to 1 cm^3/m^3, it is noticed by a specific taste. At a concentration of 3 $cm^{3/}m^3$ and higher it is easily determined by its smell. At a concentration of 6 to 12 cm^3/m^3 it irritates the mucous membranes of the nose and throat, higher $20cm^3/m^3$ it also affects eyes. Inhalation $S0_2$ causes painful phenomena in lungs and upper respiratory tract. Also sometimes it causes swelling of the lungs, pharynx, as well as respiratory paralysis.

If air contains more than 50 $\mu g/m^3$ of SO_2 it can cause obstructive changes in respiratory function and it might lead to the development of chronic obstructive pulmonary disease (COPD). Sulfuric anhydride SO_3 is also an irritating gas. It is capable of causing a sensation of suffocation at a concentration of about 1 cm^3/m^3.

Nitrogen oxides emitted into the atmosphere are basically consisted of NO_2, because N0 will oxidize to $N0_2$ pretty fast. $N0_2$ has no color and smell, but it is very toxic.

Man breathing even with deadly concentration of dioxide of nitrogen, won't feel any indisposition during first hours. At the end of eight hour he will feel asphyxiation, nausea, abdominal pain, cough, fast heart rate, after which there is an violation of heart activity. Death will come in 8-48 hours. Concentrations of more than 50 cm^3/m^3, irritating the mucous membranes of the nose and eyes, should be considered as dangerous when exposed to short duration. Long exposure with nitrogen dioxide with concentrations from 25 to 100 cm^3/m^3 causes irritation of the lungs. Carbon oxides are formed in the incomplete combustion of substances containing carbon, including fuels of all types (fuel oil, coal, natural gas, etc.). It causes fast toxic effect of human's organism, if there is a huge concentration of it in the air. Some people are hypersensitive to the effects of this substance, the first symptoms of poisoning are headaches, nausea and a sense of weakness. It expels oxygen from blood's hemoglobin, which makes blood incapable of transferring oxygen from the lungs to the tissues. Due to the reduced oxygen concentration in the blood, suffocation occurs.

Phenol - possible poisoning with phenolic vapor, fine dust, formed when condensation in colder air vapors, as well as when it gets on the skin.

The lesion of 0.5-0.25 of the body surface is fatal, with the lesion of 0.25-0.17 of the surface, poisoning occurs with an increase in temperature, a violation of the functions of the nervous system, blood circulation and respiration. Formaldehyde - is a gas with a strong odor. Solutions give off formaldehyde gas even at room temperature and especially when heated. It forms explosive mixtures with air or oxygen. It is irritant gas, which also has a general toxic effect. It also has a huge influence of central nervous system. Benzopyrene – is a carcinogenic substance. It causes gene mutations and cancers. It is formed with incomplete combustion of fuel. Benzopyrene has huge chemical stability and it can easily dissolve in water. It spreads over long distances from pollution sources (sewage) and accumulates in bottom sediments.

Carbon monoxide, nitrogen dioxide, formaldehyde; sulfur dioxide and sulfuric acid aerosol; sulfur dioxide and nickel; sulfur dioxide and nitrogen dioxide; sulfur dioxide and carbon monoxide, phenol and converter dust; sulfur dioxide, carbon monoxide, nitrogen dioxide and phenol; sulfur dioxide and phenol; oxide and sulfur dioxide, ammonia and nitric oxide have a cumulative effect in the joint presence in the ambient air [2].

2 Materials and Methods

The search for highly efficient means for cleaning gas dust is largely dependent on the correct evaluation of the dispersed and chemical composition of the dust.

Experimental data shows, that dust emission in the zone of knockout lattices is 0-1% of the volume of emissions at dust concentrations from 2050 to 7850 mg/m^3.

The high requirements of regulatory documents for the purity of atmospheric air and the ambiguity of the methods for determining the scattering parameters pose the task of a complex solution in assessing the dispersion, fractional and chemical compositions of dust. The greatest dust emission is determined from the injection molding section of molds and shot blasting (Table 1).

Table 1. The concentration of dust in air removed by local exhaust systems (experimental data).

	Nature of the equipment or process	Concentration mg/m^3		Nature of dust
		Highest	Average	
1	Chipping lattices: a) bottom drainage of hopper b) full shelter grill with upper drainage	 7850 2050	 2600 650	 Humid Dry
2	Shot blasting and shot-shot cameras: a) from the camera b) from the head of the elevator	 6150 6000	 050 2200	 - -

The estimation of the true density of dust is necessary to determine the velocity of the particles of different sizes.

The estimation of the true density of dust was carried out on Le Chatelier's device. The error wasn't more then 5%. Results of the estimation are shown in table 2. The true density test was carried out by a pycnometric method.

Table 2. The true density of dust

No	Name of sample	Density, g/cm^3
Samples 1-10	Shot blasting devices	4,50
Samples 11- 20	Chipping grid of the steel casting site	2,95

Determination of the fractional composition of samples No. 1-10 and No. 11-20 was carried out on a sedimentometer by using a cathetometer with a total error of not more than 5.5%.

The results of the sedimentation analysis of the dust dispersion of the black casting shop, taken from cyclones, are showing that particles with diameters> 200 μm predominate (76% dust of shot blasting machines, and 76.7% dust from knockout gratings by mass).

The rest of the dust is represented by fractions of 200-0 μm, while the content of fine dust in the shot blasting apparatus is 2.3%.

Cyclones SK - CN of chipping grids have an IV class of a dust collector with an allowable input dust concentration of 1000 g / m3 with a hydraulic resistance of 4 kPa with a gas productivity from 2.54 to 92 thousand m^3/h with a 95% cleaning efficiency.

An evaluation of the actual efficiency of the PU established lower values of ~ 68-70%.

Table 3 shows the experimental data on the weight percent dust content of various fractions captured by a dry dust collector.

The analysis of the experimental data shows a significant presence of particles with a diameter of 15 to 125 μm coming from the fractured section and a diameter of 25 to 200 μm from knockout gratings.

Table 3. Experimental data

Dust extraction point	The weight percentage of fractions with an average particle diameter, μm						
	Less than 5	5-10	>10-20		>40-60	>60-100	More than 100
Shotcracker	8,8	9,2	30,5	18,8	12,3	12,5	7,9
Knockout gratings	6,2	8,3	30,0	18,2	11,1	12,0	14,2

Thus, the amount of fine dust reaches significant values in knockout grids - up to 44.5% (at d≤20 μm). The lack of it in the dust receptacle is confirmed by its presence of a discharge pipe and a flare discharge (H = 8 m) at the cutoff zone.

The selection of dust in the flare emission, on the contrary, shows a significant amount of fine dust, not settled in the PU. Its fractional composition is presented in Table. 4.

Table 4. Fractional composition

Dust extraction point	The weight percentage of fractions with an average particle diameter, μm				
	Less than 5	5-10	>10-15	>15-20	>25
Knockout gratings	12	16	30,2	30,5	11,3

Due to the low efficiency of dry method of dust collection, it is much more effective to carry out the preassembly of existing dust collection systems with equipment for collecting dust by wet method, for ex-ample, with a Venturi's scrubber.

Highly effective Venturi's scrubbers (class II) have higher hydraulic resistance (from 6 to 12 kPa), gas productivity from 1.7 to 84 thousand m^3/h and have the efficiency of purifying gases from dusts of any dispersed composition 96-98%.

In order to reduce dust and gas content in the working zone of the knockout grate, it is expedient to install a three-sided upper sidewall tent with two-sided drainage and compensating air [3].

The maximum allowable concentration of dust Cn, mg/m^3, in the technological and ventilation emissions that are diffused, is determined, depending on the volume of ejected air L, thousand m^3/h. With a volume of more than 15 thousand m3 / h Cn = 100R; at a volume of 15,000 m3 / h or less, Cn = K (160-4L), where K is the coefficient taken in accordance with the MPC of dust in the air of the working area of the premises in permanent workplaces:

MPC ppm, mg/m^3..............	≤ 2	>2÷4	>4÷6	>6÷10
K	0,3	0,6	0,8	1,0

Emissions of air with a concentration of dust exceeding Cn are not allowed to dissipate into the atmosphere without preliminary purification. For areas of black casting, where there is dust molding with metal, at MPC = 2 mg/m^3, Cp = 46.2 mg/m^3. The actual dust concentration at the pipe cut is 11.04 g/m^3.

The degree of risk of the technological process also depends on the purity of atmospheric air in various sections of the shop, since the technological process in all phases of production is accompanied by the release of a significant amount of dust and harmful gases.

The industrial safety of the foundry industry is associated with the risks of failures in the operation of devices and equipment, as well as the presence of causes that induce defect castings due to incompleteness of technological regimes. To a great extent, the latter is due to the strength and the clean surface of the molding mixture.

An important condition for obtaining a strong molding mixture is to ensure uniform coverage of the entire grain surface of the binder component mixture. The uniformity of its distribution is determined by the design of the mixer and the mixing time. As the agitation time increases, the compressive strength first increases, reaches a maximum (usually in 6-8 minutes), and then begins to decrease.

The resistance to the dislocation system shift in general form is expressed by the equality:

$$\frac{H}{F} = fP + C \tag{1}$$

where H - shearing force; P - specific vertical load; f is the coefficient of internal friction; C is the force of adhesion; F is the area of the cut.

The use of binders, which provide high strength of the molding mixture and new technological schemes of casting - on melted models, in thermosetting binder molds (actually plastic molds with sand filler), in forms made of liquid-glass mixtures, etc., - made it possible to obtain thin-walled nonmetallic shell shape with a high-purity surface.

The quality of castings also depends on the modes of filling the molds, the outflow of metals and their properties, external forces, cooling and hardening of the metal during casting.

In the process of filling the mold, liquid metal can destroy its walls and carry the products of destruction into different parts of the casting, forming clogs.

The number of clogs is halved in castings cast with a gating bowl, compared to castings cast with a sprue funnel, however the use of a sprue bowl causes inconvenience in a number of subsequent technological operations [4].

With prolonged exposure to a metal jet on the surface of the rod, there is burnout of organic additives, as a result of which the purity of the surface of the casting deteriorates significantly, although exfoliation does not occur.

The quality of atmospheric air in the working area of the foundry depends on the concentrations of harmful substances and their toxicity. The effectiveness of existing means of emission purification does not ensure compliance with the requirements of hygiene standards. Preliminary experimentals and calculated data show a multiple excess of the maximum surface concentrations above the maximum permissible - this excess reaches 20 times. It lasts not for long at piece production however during mass production the degree of environmental risk significantly increases.

The maximum permissible concentration of dust in the working area of MPC depends on the SiO_2 content, in particular, if $SiO_2 \leq 10\%$ MPC= 10 mg/m^3, and the actual content of SiO_2 varies in the range from SiO_2 = 35-50%, 28.2 - 46.51%, and practically all areas of the foundry have quartz. And MPC is necessary to be 2 mg/m^3. That is why the approach to assessing the health status of the work area and the effectiveness of engineering and technical measures to reduce air pollution is much harder.

The evaluation of the dispersion of dust and its true density is necessary for two reasons. First, it determines the degree of an impact on a person. In particular, the presence of dust of less than 10 microns, the so-called "pulmonary", indicates the possibility of obtaining a number of occupational diseases: dermatitis, conjunctivitisб cataract, pneumoconiosis and silicosis. Secondly, the size of suspended particles and density significantly determine the choice of dust collection devices and the optimal parameters for the emission of dusty air.

Therefore, the analysis and optimization of already operating dust-collecting complexes in combination with new promising developments and technologies for dust utilization is of particular importance. An important property of dust collecting devices is that the capture efficiency of a given particle size, known as fractional capture efficiency, increases with particle size [5]. If the capture efficiency for a given size is good (for example, 90% at 10 μm), then for particles of a larger size the efficiency will be even better. However, we have to note, that the fractional capture efficiency changes sharply with the particle size. This dependence is shown in the form of a hypothetical curve in Fig. 1, which is typical for some gas-cleaning devices. However, in the course of the experiment it was established that the system existing in the foundry shop has a low efficiency at particles less than 18 μm (Fig. 1).

In the region of small particle sizes the efficiency rapidly increases, reaching almost 100% with large dimensions it is shown on Fig. 2. The shape of this curve can vary for different types of devices and for circuit modifications within this type of device. In addition, the capture efficiency can be a function of the dust type, due to the different physical characteristics of the particles, for example, such as their shape. Figure 1 shows the dependence of the fractional efficiency of cyclones.

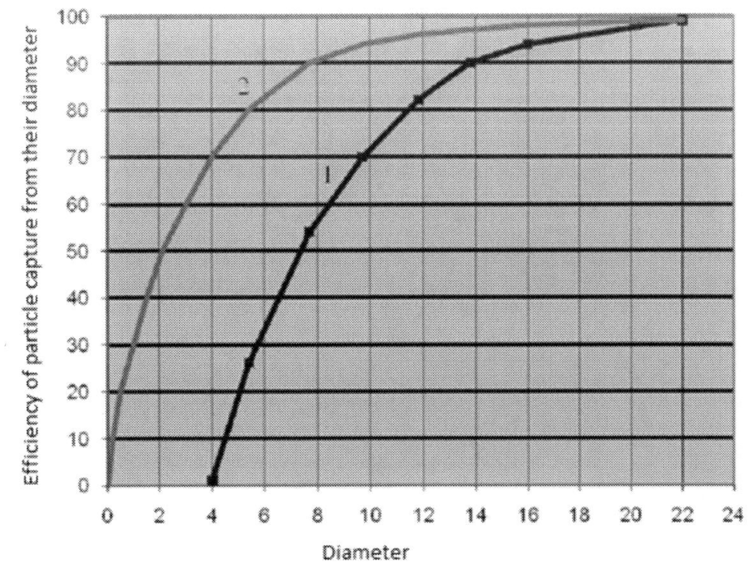

Fig.1. Experimental curve of the dependence of the efficiency of particle trapping on their diameter: 1 - experimental curve; 2 - hypothetical curve

With the joint operation of general exchange of supply and exhaust ventilation in the zone of the knockout grid, dust concentrations exceed the MAC by 7-16 times.

It is much more important to know the overall efficiency of capture than the efficiency of capturing particles of a given diameter. This means that for a particular type of dust and a certain gas cleaning device, it is necessary to have information about what percentage of the total amount of incoming dust will be removed from the stream of gas to be purified. To obtain this type of information, it is necessary to know the particle size distribution and capture efficiency as a function of the particle diameter d.

Experimental estimation of the foundry dust parameters of the working area showed that despite the piece production, there is an intense dust evolution in the areas of knockout of molds and shot blasting areas.

In the working area of shot blasting chambers, the dust content is 22-135 mg/m^3, and in knockout grates - 32-207 mg/m^3. The average values of the concentrations in the drainage, are from 650 to 2600 mg/m^3 and from 2200 to 2950 mg/m^3.

3 Results

The true density from the knockout lattices is 2.95 g/cm^3 and from the shot blasting chambers 4.5 g/m^3. This is determined by Le Chatelier's method. The dust size in the dust receptacles is 14-200 μm. Photographs of X-ray spectral microanalysis of dust captured from the shot-blasting sections of the casting and knockout gratings obtained by the scanning analyzer "Camscan S4" according to the international ASTM method, are shown on Figures 2,3.

Table 5. shows the elemental composition of dust from shot blasting machines [6].

Table 5. Elemental composition

Element	Weight, %	Atomic, %
O K	42,04	61,98
Al K	7,38	6,45
Si K	24,38	20,48
Ti K	0,29	0,14
Mn K	0,48	0,21
Fe K	25,43	10,74

Fig. 2. Photo of captured dust from shot blasting machines

Fig. 3. Photo of dust caught from the knockout gratings

The sedimentation rate of the particles is determined by the Stokes equation for particles up to 50 μm in size, since this is a laminar flow regime:

$$V_t = \frac{g \cdot d_p^2 \cdot \rho_p}{18 \cdot \mu}, \quad (2)$$

g is the acceleration due to gravity; dp is the particle diameter; ρp is the particle density; μ is the viscosity coefficient. For particle sizes not deposited in the dust receptacle, it was determined that the settling velocity of the particles is V = 3.3 cm/s (at d = 20 μm, ρ = 2.95

g/m^3) and Vt = 0.06 cm/s (at d = 1 μm, ρ = 2.95 g/m^3). The settling time of particles at a flare discharge to a height of 10 m is 303 s for d = 20 μm and 16667 s for d = 1 μm.

Table 6. Shows elemental composition of dust from knockout gratings.

Element	Weight, %	Atomic, %
O K	33,29	55,42
Na K	1,41	1,63
Mg K	0,45	0,49
Al K	15,54	15,34
Si K	4,00	3,80
S K	4,43	3,68
Ca K	0,86	0,57
Ti K	0,66	0,37
Cr K	0,43	0,22
Mn K	0,56	0,27
Fe K	37,16	17,72
Zn K	1,20	0,49

4 Discussions

The obtained data on dustiness at the industrial site and the cutoff of the flare pipe allow us to conclude that the dust cleaning system is inadequate and that more efficient removal of fine dust is required, since this dust is the most dangerous and it creates concentrations in the working zone exceeding the maximum permissible concentration.

The greatest dust content is typical for sections of shot blasting and molding out of mold. The analysis of the experimental data allows us to conclude that the flow is laminar (Re = 10-4 - 5.0). The presence in the ejection of a significant mass of dust particles of less than 20 μm was found, which determines the inadequate efficiency of its capture by the existing dust-cleaning system, as a result of which there is an excess of MPC.

5 Conclusion

Disperse dust analysis shows that particles with diameters> 200 μm predominate (76% dust of sandblasting machines, 76.7% dust from knocked out gratings). The rest of the dust is represented by fractions of 200-0 μm, while the fine dust content in the sandblasting apparatus is 2.3%, fine dust is absent from the molding portion of the mold. On the site of knocking out castings of their shape, we proposed the installation of local drainage pumps.

References

1. I.V. Ilin, O. Kalinina, O. Iliashenko, A. Levina, Procedia Engineering, **165**, pp 1683-1692 (2016) DOi- 10.1016/j.proeng.2016.11.910
2. E.C. Gini *Technology of foundry. Special types of casting* (Academy, Moscow, 2005)
3. S.V. Belov *Environmental protection* (Higher School, Moscow,1991)
4. I.A. Ivanova High technologies in ecology: materials of the 13th interregional scientific and practical conference Voronezh, pp. 23-28 (2010)
5. V.P. Novikov *The automation of foundry production. In 2 parts Part 1. Management of foundry processes*. (MGIU, Moscow, 2006)
6. O. Kalinina, O. Valebnikova, Advances in Intelligent Systems and Computing, **692**, 1315-1322 (2018) DOi - 10.1007/978-3-319-70987-1_139

The analysis of disperse and elemental composition of dust from drobestry installations of foundry at the production of construction structures for altitude buildings

Irina Ivanova [1,*] *Elena* Golovina[1], *Kirill* Kulakov[2] and *Andrey* Sorokin[3]

[1]Voronezh State Technical University, Moscow Avenue, 14, Voronezh, 394026, Russia
[2]Moscow State University of Civil Engineering, Yaroslavskoe sh. 26, Moscow, 129337, Russia
[3]Moscow Aviation Institute (National Research University), Volokolamskoe highway, 4, Moscow, 125993, Russia

Abstract. In the production of building structures for high-rise buildings, the dangerous and harmful factors of foundry production include high concentrations of dust and harmful gases released at various stages of the process. Disperse and elemental compositions of dust determine the hygienic state of the working area of the foundry. According to experts, the presence of dust of less than 10 microns in size in the air increases the danger of getting occupational diseases, such as pneumoconiosis (silicosis), bronchitis, dermatitis, conjunctivitis. The determination of the granulometric composition of the powder sample provided was carried out by the laser diffraction method realized on a laser particle analyzer Fritsch NanoTec «ANALISETTE 22».

1 Introduction

In the air environment of foundries in the production of building structures for high-rise buildings, in addition to dust, there are large numbers of carbon oxides, carbon dioxide and sulphurous gases, nitrogen and its oxides, hydrogen, aerosols saturated with iron and manganese oxides, hydrocarbon vapors and other [1]. Melting aggregates, heat treatment furnaces, dried for molds, rods and buckets, etc. are sources of contaminants. The foundry shop of machine-building production is located in the urban development. The degree of environmental safety is determined by the microclimate in the zone of shot blasting and the intercore zone of the plant (with dispersion of emissions).

When melting cast iron, in addition to dust, a significant amount of carbon oxides is produced, as well as sulfurous anhydride, hydrocarbons, and nitrogen oxide In induction furnaces in addition to the above ingredients, chlorine and hydrogen fluoride are released, when smelting aluminum. Foundry production is characterized by gas evolution under the influence of the heat of liquid metal on the molding mixtures. Depending on the type of

* Corresponding author: ivanova-eco@mail.ru

components of the molding mixtures, benzene, phenol, formaldehyde, furan, cyanide, ammonia, and other toxic substances can be released from them [2].

Dust, the main constituent of which in foundries is silica, is formed during the preparation and regeneration of molding and core mixtures, smelting casting alloys in various melting units, the release of liquid metal from the furnace, out-of-furnace processing and pouring into molds, in the knockout section of the castings, in the process stump and cleaning of casting, in the preparation and transportation of raw bulk materials.

The air condition of the working area of the foundry, as a rule, does not meet the normative hygienic requirements. This is determined by the complexity of the technological processes of production as a whole [3].

The dispersed and elemental composition of the dust also determines the hygienic state of the working zone of the foundry. According to experts, the presence of dust in the air of less than 10 microns in size increases the danger of obtaining occupational diseases: pneumoconiosis (sialicosis), bronchitis due to the low settling rate of dust and its prolonged stay in the air of the working area. In addition, fine dust and dust of less than 10 μm creates additional difficulties in dust collection [4].

2 Materials and Methods

Large particles of dust linger on the mucosa of the upper respiratory tract, particles with sizes less than 10 microns can penetrate into the alveoli of the lungs, where the main part of them is retained (Table 1). Hazard class -3, MPC = 0.5 mg/m^3.

Suspended particles can cause a toxic effect by one or more of the following three pathways [5]:

1. A particle may be toxic due to its chemical or physical characteristics.
2. A particle may interfere with one or more mechanisms by which the respiratory tract is normally cleaned.
3. A particle may be a carrier of a toxic substance absorbed thereon.

Table 1. Observed impact of suspended particles on the environment and human

Concentration, mg/m^3	**Time interval**	**Impact result**
60-180	Average annual geometric mean, in the presence of SO2 and moisture	Increased corrosion of steel and zinc plates
150	A relative humidity of less than 70%	Reducing visibility to 8 km
80-100	Simultaneous level of sulphate precipitation in 30 mg/(cm^2 per month).	Increase in mortality by 50% can occur
100-130	Concentration of S02 120 μg/m^3	Possible increase in the number of respiratory diseases in children
200	Average daily at SO2> 250 μg/m^3	Diseases of industrial workers can lead to an increase in absenteeism
300	Maximum for 24 hours with simultaneous concentrations O2>630 μg/m^3	Patients with chronic bronchitis will experience a sharp exacerbation symptoms
750	Average daily SQ2> 715 μg/m^3	Can come significant increase in diseases and deaths

Provision of safe levels of maximum permissible concentrations of pollutants in the operator's work area shot blasting area and in the wind zone of the foundry is the purpose of this work.

Due to the fact that the existing dust collection system using cyclones is not effective and there is a multiple exceeding of the maximum permissible concentration (MPC) of dust. The actual efficiency of the dust collectors in the shop is inadequate. In order to further purge the ejection into the intercorpus zone and bring the contaminants in the working area to acceptable levels, the dust collection system can be supplemented with a second "wet" stage.

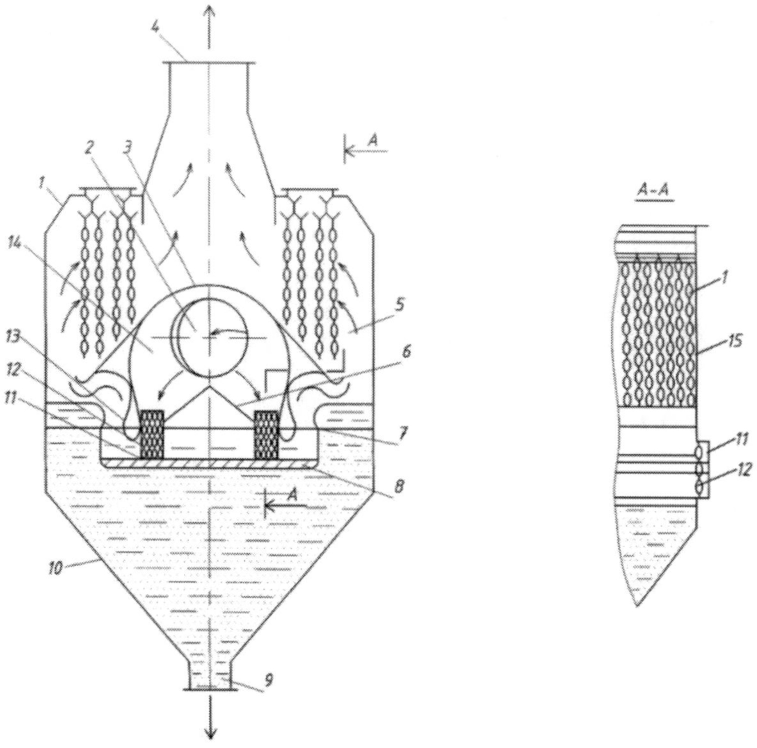

Fig. 1. Wet Dust Collector:

1 — shell; 2 — inlet branch pipe; 3 — baffle; 4 — outlet branch pipe; 5 — clean gas chamber; 6 — reflecting plates; 7 — bottom baffle; 8 — louvered grille; 9 — branch pipe for sludge removal; 10 — bunker for collection of caught dust; 11 — poket; 12 — blinded loose hanging chains; 13 — upper baffle; 14 — dusty gas chamber; 15 — drift catcher

Authors V.I. Semkin, A.V. Dumnov, N.I. Adzhiev, suggest a device for cleaning emissions, which is developed in the research and design institute for the protection of labor of building materials.

For the dust analysis, the dispersion (granulometric) analysis method was used.

Determination of the granulometric composition of the powder sample provided was carried out by the laser diffraction method implemented on the laser particle analyzer Fritsch NanoTec "ANALISETTE 22" (Fig.2, 3).

To study particles in the analyzer, the principle of comparing laser radiation from samples and the object of research is used. We used Fraungofer model. This model is used only for samples with particles larger than 0.1 μm (100 nm);

Fig. 2. General view of the laser analyzer Analysette-22 NanoTec (Fritsch, Germany)

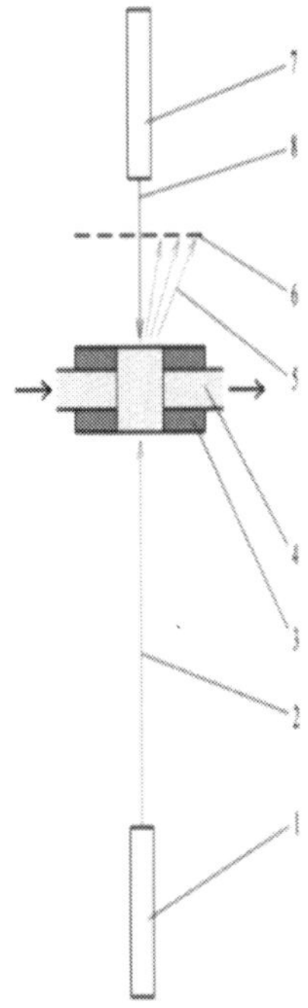

Fig. 3. Optical part of laser particle analyzer Analysette-22 NanoTec:

1 – front laser; 2 – front laser beam; 3 – measuring cell; 4 – dispersed medium containing a sample; 5 – laser-scattered sample; 6 – detector; 7 – back laser; 8 – back laser beam

3 Results

Based on the results of the analysis, the granulometric composition of the samples was determined. The results are summarized in Table 2 and are shown in Figure. 4.

Table 2. The granulometric composition of the sample, the % content of particles of a certain size.

Value,%	particle size, µm	Value,%	particle size, µm	Value,%	particle size, µm
5.0%	<= 19.285 mkm	10.0%	<=26.295mkm	15.0%	<=30.679mkm
20.0%	<=34.802mkm	25.0%	<=38.617mkm	30.0%	<=42.615mkm
35.0%	<=46.354mkm	40.0%	<=50.105mkm	45.0%	<=53.895mkm
50.0%	<=57.549mkm	55.0%	<=61.440mkm	60.0%	<=65.570mkm
65.0%	<=70.497mkm	70.0%	<=75.998mkm	75.0%	<=82.355mkm
80.0%	<=89.220mkm	85.0%	<=97.614mkm	90.0%	<=115.049mkm
95.0%	<=162.398mkm	99.0%	<=189.947mkm		

In table 1 the particle size is fixed. As a result, sample contains 90% of particles with size less or equal to 115.049µm and 10% of particles with large size. Particles of this size are dangerous for human health, as they cause pneumoconiosis.

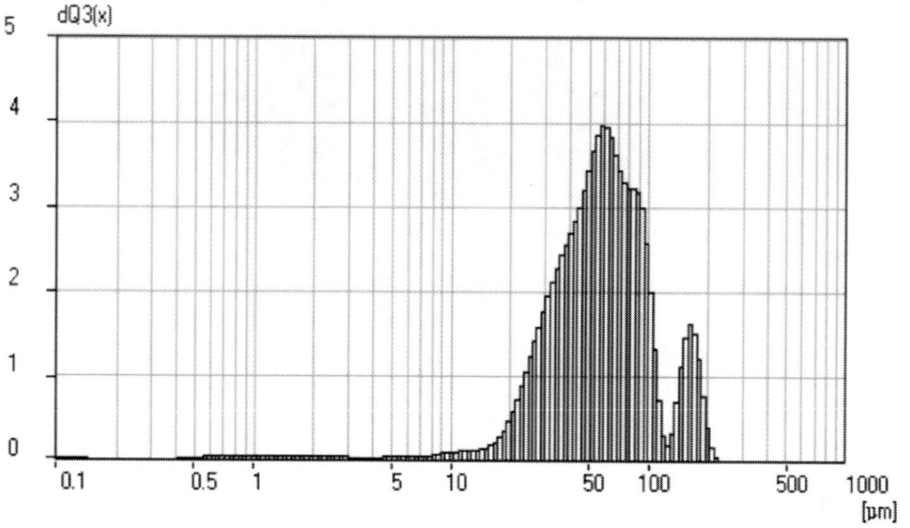

Fig. 4. Graphical results of dispersion analysis of distribution of dust particles.

The sedimentation method does not answer the question of how much particles of a certain size there are in a given system; it only shows the percentage of particles that have dimensions that are in the interval between one radius and another.

In order to estimate the dust parameters more accurately: dispersion and element composition, the X-ray phase analysis method was used, in which the following equipment was used:

1. X-ray diffractometer "DRON 04".
2. X-ray diffractometer "Radian DR-02".
3. Wave X-ray fluorescence spectrometer "Bruker S8 Tiger".

The study of the structure of dust,

The study of the structure of the dust formed during X-ray spectral analysis showed, that it, as a rule, has a form, which makes it possible to conditionally consider it spherical. When the dust particles fall, they always try to occupy the position corresponding to the greatest resistance in the air, this form contributes to their settling in the atmosphere and in inertial dust collectors. However, the presence of particles smaller than 10 µm indicates their considerable settling time and the need for highly effective post-treatment systems. As a result, 86.2% of the sample is less than 100 µm, 13.8% is heavy particles larger than 100 µm. Particles of this size are dangerous for human health, as they cause pneumoconiosis, and in the presence of a SiO2 film of silicosis [6].

A scanning electron microscope is supplemented by spectrometers of both varieties. The quantitative analysis is based on a comparison of the intensity of the characteristic X-ray radiation from the sample under study and standards of samples of known composition. The results of calculation of concentrations and absolute random errors were displayed on the display and the printer.

In the course of the experiment, the operational error control of the measurement results for each of the chemical elements was carried out.

Control is carried out by estimating the concentrations of elements in existing standards. Initially, an analysis was performed to determine the phase composition of the substance using a DRON 04 diffractometer with CoKα radiation (1.789Å) in the range of angles 2Θ (10-110) ° at U = 29kV, I = 26mA. Table 2 and Figure 3 show an X-ray spectral analysis of the dust collected from the shot blasting section of the casting.

Table 2. X-ray spectral analysis data on the «Bruker S8 Tiger» equipment.

Sample Name	Na (%)	Al (%)	Si (%)	S (%)	Cl (%)	K (%)	Ca (%)	Ti (%)	Cr (%)	Mn (%)	Fe (%)	Ni (%)	Cu (%)	Zn (%)	Ga (%)	Zr (%)	Mo (%)	Ce (%)
Dust from shot blasting plants	0.141	21.295	27.94	0.04	0.056	0.183	0.143	0.095	0.511	0.712	48.535	0.139	0.05	0.014	0.022	0.014	0.072	0.038

The manifestation of a large number of reflexes indicates the presence of several phases in the sample. To determine the phases, a diffraction measurement database "PDF-2012" was usedDue to the large number of possible elements and phases, X-ray spectral analysis was used to reduce the possible phase variants of the sample. The survey was performed twice on the Radian DR-02 diffractometer with the function of X-ray fluorescence analysis.

X-ray spectral analysis indicated the presence of a phase of iron (Fe), titanium (Ti) and chromium (Cr). The iron (Fe) phase predominates (78%). Due to the lack of the possibility of determining the "light" elements on the "Radian DR-02" diffractometer, the X-ray spectral analysis was carried out on the "Bruker S8 Tiger" wave X-ray fluorescence spectrometer at the CCU of the NO VSU.

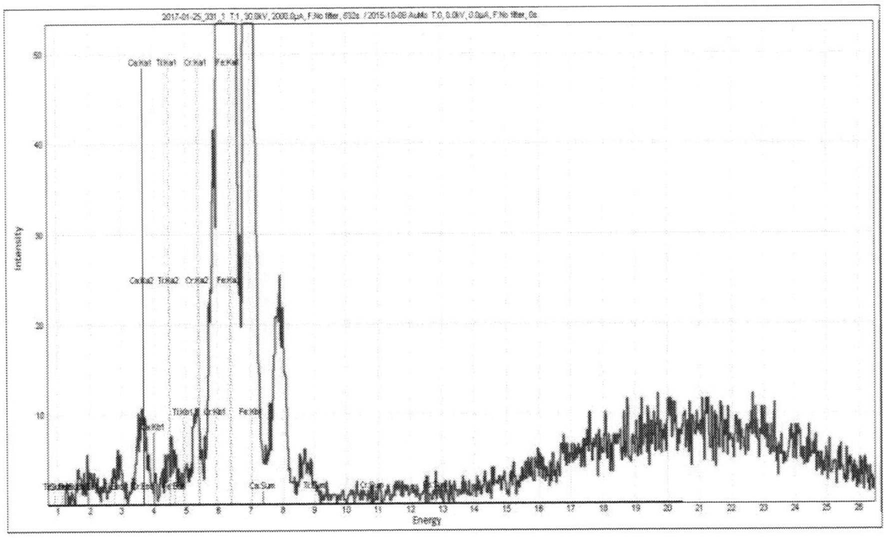

Fig. 5. X-ray spectral analysis of the sample on the diffractometer "Radian DR-02".

4 Discussions

X-ray spectral analysis showed a more accurate percentage of elements in the sample. The presence of Al, Si, Cr, Fe reduces the search time in the "PDF-2012" database for phase analysis. The phases of oxides can only be detected using the X-ray phase analysis method. The remaining elements have a small percentage value.

The results of the analysis show that in the foundries there is a presence of dispersed dust of less than 10 μm, while 78% of iron (Fe) is present in the shot blasting section in the dust. Also, the analysis of dust concentration in operator's work area revealed significant excess of the MAC of the working zone. The dispersion of the dust composition was evaluated, which showed a large amount of fine dust (less than 10 microns) of 86.2%.

In addition to dust in emissions of the foundry there are harmful gases: sulfur oxides (SO2, SO3), oxides of nitrogen (NO2, NO), carbon oxides of CO and CO2, phenol (C6H5OH), formaldehyde (HCHO), benz (a) pyrene (C20H12) [10,11]. Since the results of the parameter estimation are unstable, we had to use more accurate methods with the help of a diffractometer.

5 Conclusion

Dust structure research showed: when crushing, 86.2% of the dust produced has a fraction size of less than 100 μm, i.e. refers to medium and fine dust, having a settling rate of less than 7 cm/s.

It is established that the iron (Fe) content in the elemental composition of all dust is 78%, which happens due to the processing of castings with iron shot. The phase composition was refined in the second stage of the research with the help of the Radian DR-02 diffractometer and the diffraction measurement database "PDF-2012".

References

1. A.N. Boldin, Foundry, **3**, 33-34 (2005)
2. O. Kalinina, O. Valebnikova, Advances in Intelligent Systems and Computing, **692**, 1315-1322 (2018) DOi - 10.1007/978-3-319-70987-1_139
3. I.A. Ivanova, Bulletin of the DGTU, **2 (45)**, 200-204 (2010)
4. E.I. Golovina, I.A. Ivanova, S.A. Ivkov, Internet Journal of Science, **2 (39)**, 62 (2017)
5. V.L. Murzinov, V.Ya. Manokhin, E.I. Golovina, Polimatic network electronic scientific journal of the Kuban State Agrarian University, **130**, 155-172 (2017)
6. E.I. Golovina, I.A. Ivanova, M.V. Manokhin, Vestnik of the Don State Technical University, **2 (89)**, 141-148 (2017)

Investigation of the motion processes of wastewater in sewerage of high-rise buildings

Valentina Pomogaeva [1,*] *Lyudmila* Metechko[2], *Dmitry* Prokofiev[2], and *Tamara* Narezhnaya[3]

[1]Voronezh State Technical University, Moscow Avenue, 14, Voronezh, 394026, Russia
[2]Moscow Aviation Institute (National Research University), Volokolamskoe highway, 4, Moscow, 125993, Russia;
[3]Moscow State University of Civil Engineering, Yaroslavskoe sh. 26, Moscow, 129337, Russia

Abstract. When designing, constructing and operating sewage pipelines in high-rise buildings, issues of failure-free operation of a network arise. Investigation of the processes of wastewater moving allows identifying problem areas during operation, assessing the possibility of obstructions and breakdowns of plumbing traps on the gravity drainage sections of the pipeline. The article performs the schemes of the water outflow from the floor sewer into the riser, including the places where the riser is bent, of air delivery to the working riser under the change of the direction of drain movement with the dropout line set-up, with the installation of an automatic anti-vacuum valve, with the installation of the ventilation pipeline. Investigations of the process of sewage waste flow in a sewage riser were carried out, in order to select the appropriate structure. The authors consider structure features of some sections of sewerage in high-rise buildings. The exhaustion value in the riser is determined from the rarefactions that occur below the compressed cross-section of the riser and the loss of the air flow pressure coming from the atmosphere into the riser during the deflooding of the liquid. Preventing the formation of obstructions and breakdowns of plumbing traps is an integral part of sewage networks.

1 Introduction

Engineering networks ensure normal life activity of people living and working in high-rise buildings. Therefore, the networks should be reliable and safe. Water supply and sewage systems are the obligatory part of each building where people reside. The discharge of wastewaters into the sewage system depends on the water consumption regime. Preventing the formation of obstructions and breakdowns of plumbing traps is an integral part of sewage networks. The present work investigates the process of flow of sewage in a sewer pipe.

The sewage riser consists of a working vertical part, an exhaust part and a horizontal pipeline transporting the waste liquid to a centralized sewerage network. The exhaust part of the sewer riser is designed for air inflow from the atmosphere, when the sewage fluid moves through the riser, for normal air exchange. At the minimal wastewater removal, the sewage gases from the sewerage network emerge from the exhaust to the atmosphere. At the

* Corresponding author: pomogaeva8@mail.ru

maximum wastewater removal through the exhaust part, air is taken from the atmosphere, for normal operation not only of the sewer riser, but also of the external drainage network.

The flow of wastewater in a sewer riser is possible under different conditions: laminar, mixed or turbulent. In some parts, a water blockage can occur, covering the entire section of the pipe. In this case, a sharp pressure fluctuation arises in the riser, which can lead to a breakdown of the hydraulic shutter. The level of rarefaction will obviously depend on the speed of the falling wastewater. Since the velocity is a function that depends on the geometric length through which the volume of liquid passes during the certain time, for example, so when the liquid falls from a height of 50 m, the speed can be 4 m/s. This can lead to the occurrence of critical-value exhaustion, leading to a breakdown of the plumbing trap. As a result of analytical calculations for sewer risers not exceeding 41 m in diameter 100 mm, the throughput is 4.4-4.5 l/s, for the diameter 125 mm - 6-6.98 l/s, for the diameter 150 mm - 10 l/s, and breakage of the plumbing trap is possible at a flow rate of 9 l/s for 100 mm and 20 l/s at 200 mm.

2 Materials and Methods

When moving in a vertical sewer pipe, the sewage liquid ejects a certain volume of air (Fig. 1). The volume of air drawn into the riser first increases, then reaches a maximum value depending on the volume of water, then decreases. As a result, a water-air environment with a certain amount of sanitary waste is formed in the riser. The formation of different regimes of motion can be considered taking into account the volume gas content β and the gas saturation of the mixture α, which can be determined from the expressions:

$$\beta = \frac{u_{w0}}{u_{w0} - u_{a0}} = \frac{Q_{w0}}{Q_{w0} - Q_{a0}} \quad \text{and} \quad \alpha = \frac{1}{\omega}\sum_{i}^{n}\omega_{w,i}.$$

where u_{w0} - adjusted liquid phase velocity; u_{a0} - adjusted velocity of the gas phase; $Q_0^{''}$ - liquid discharge; $Q_0^{'}$ - attracted airflow rate; ω - liquid surface area; i - the number of sections considered in a certain period of time.

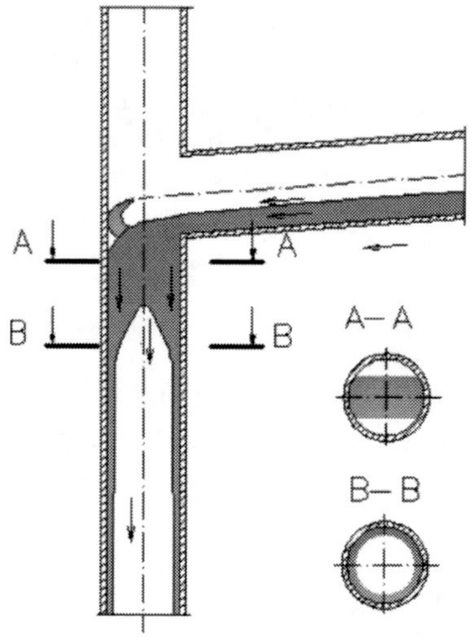

Fig. 1. Circuit of the flow of liquid from the floor sewer to the riser [1].

The differential equation for the equilibrium of forces acting in a flowing-down film has the following form:

$$\frac{d\tau}{dX} + \gamma_w = 0, \qquad (1)$$

where $\tau = \mu_w \dfrac{du_w}{dX}$ is viscous shear, Pa; γ_w - liquid specific density, N/m^3; μ_w - coefficient of dynamic viscosity of liquid, Pa·s; u_w - local velocity of liquid, m/s.

To obtain the expression describing the change in the velocity of particles in the plane of the living section of the liquid film, we integrate equation (1) under the boundary conditions $X = 0 \;\; u_w = 0$ and when $X = \delta_w \;\; \tau = 0$:

$$u_w = \frac{\gamma_w}{\mu_w}\left(\delta_w X - 0{,}5X^2\right), \tag{2}$$

where δ_w - boundary layer thickness, m.

When $X = \delta_w$, the velocity on the outer surface of the film reaches a maximum value $u_{w.max} = \dfrac{\gamma_w}{2\mu_w}\delta_w^2$, while the average velocity of the fluid in the wetted section of the film is equal to: $u_{w.cp} = \dfrac{1}{\delta_w}\int_0^{\delta_w} u_w dX.$

Taking into account equation (2), the average velocity will be:

$$u_{w.cp} = \frac{\gamma_w}{3\mu_w}\delta_w^2, \tag{3}$$

The volume flow of liquid in the film flowing down the inner surface of the pipe is determined by the equation:

$$Q_w = \pi D \delta_w u_{w.cp} = \pi D \frac{\gamma_w}{3\mu_w}\delta_w^3, \tag{4}$$

where D is the diameter of the pipe, m.

From equation (4), the thickness of the flowing fluid film is determined:

$$\delta_w = \sqrt[3]{3\frac{Q_w \mu_w}{\pi D \gamma_w}}. \tag{5}$$

From (3) and (5) the average velocity in the wetted section of the liquid film is:

$$u_{w.cp} = \sqrt[3]{\frac{1}{3}\left(\frac{Q_w}{\pi D}\right)^2 \frac{\gamma_w}{\mu_w}}, \tag{6}$$

The velocity of the gas layer at the outer boundary of the flowing liquid film in the velocity of its wetted section is not the same and is subjected to a complex dependence. At the interface between the liquid and the gas, the velocity of the fluid is equal to the velocity of the water. In the first approximation, the velocity of the gas can be assumed to change linearly; hence, the magnitude of the viscous shears is constant across the width of the layer. In this case, the equation of equilibrium of the forces acting in the gas layer has the form:

$$\mu_a \frac{d^2 u_a}{d\delta_{a.n}^2} = 0, \tag{7}$$

where μ_a - coefficient of dynamic viscosity of air, Pa.s; u_a - local velocity in the gas layer, m/s; $\delta_{a.n}$ - the distance along the normal line from the surface of the film in the cross section of the wetted section of the gas layer, m

The local velocity in the gas layer is determined from equation:
$u_a = u_{w.max}\left(1 - \delta_{a.n} / \delta_a\right)$, where δ_a - thickness of the gas layer, m.

The average velocity of motion in the plane of the live section of the gas layer, under boundary conditions $\delta_{a.n} = 0$ $u_a = u_{w.max}$ и $\delta_{a.n} = \delta_a$ $u_a = 0$ equals:

$$u_{a.cp} = \frac{1}{\delta_a}\int_0^{\delta_a} u_a d\delta_{a.n} = \frac{u_{w.max}}{2}. \qquad (8)$$

Hence, the average velocity of motion in the plane of the wetted section of the gas layer is a function of the maximum fluid velocity.

The volume flow of air entering the riser can be determined as follows:

$$Q_a = \pi D_a^{'} \delta_a u_{a.cp}, \qquad (9)$$

where $D_a^{'}$ - The diameter of the gas layer along the upper surface of the liquid film.

The thickness of the gas layer can be assumed according to the distance from the outer boundary of the liquid film, at which its velocity is maximum, that is:

$$\delta_a = 5\sqrt{\frac{L\nu_a}{u_{w.max}}}. \qquad (10)$$

where ν_a - kinematic viscosity of air ,m^2/s; 5- the Blasius coefficient.

The surface of the liquid film of the wastewater carries the air. To determine its volumetric flow, equalities (3), (6), (8), (10) are used. Taking into account that the thickness of the liquid film is much smaller than the diameter of the tube, the volume flow of air can be assumed equaling to:

$$Q_a = 2{,}5\pi D\sqrt{L\nu_a}\sqrt[3]{\frac{Q_w}{\pi D}}\sqrt[6]{\frac{\gamma_w}{\mu_w}}. \qquad (11)$$

By formula (11), it is possible to determine theoretically the ejecting capacity of a liquid when local resistances are absent.

There are local in the sewage system. For example, when wastewater leaves the floor sewer, a local hydraulic resistance arises in the riser in the form of a water-air mass acting as some kind of a piston. Consequently, there is some air deficit below the section pressed by a "piston" (Fig. 1.). In this case, the formula obtained from the Bernoulli equation can be used to determine the air flow;

$$Q_a = \mu\omega_a\sqrt{\frac{2g}{\gamma_a}\Delta p}, \qquad (12)$$

where μ - discharge coefficient; ω_a - area of the wetted section of air in the compressed section of the riser, m^2; γ_a - specific gravity of air, kg/m^3; Δp - pressure difference (Pa); g – gravitational acceleration m/s^2.

To determine the pressure difference or the exhaustion in the sewage riser, equations (11), (12) are solved relatively to the pressure difference Δp:

$$\Delta p = \frac{\left(2{,}5\pi D\sqrt{L\nu_a}\sqrt[3]{\frac{Q_w}{\pi D}}\sqrt[6]{\frac{\gamma_w}{\mu_w}}\right)^2\gamma_a}{2g\mu^2\omega_a^2} = \frac{1}{\mu^2}\frac{Q_a^2}{2g\omega_a^2}\gamma_a = \xi\frac{\upsilon^2}{2g}\gamma_a, \qquad (13)$$

where ξ - local resistance coefficient; υ - air velocity in the compressed section of the riser, m/s.

3 Results

Therefore, by analyzing the expression (13), it is possible to determine the exhaustion volume in the riser and the effect of the main parameters of liquid and air motion. The exhaustion volume will increase with an increase in the flow rate of the waste liquid. As the volume of the liquid increases, its ejecting capacity increases, and the area of the wetted section of air in the compressed section of the riser will decrease. For example, if the area of the wetted section of the air is doubled, then by (13) the pressure drop is reduced by a factor of four. Accordingly, it is necessary to increase the cross-sectional area of air in the compressed cross-section of the riser when the condition of the fluid inlet into the riser, for example, the angle of attachment of the sub-assemblies or the reduction in their diameter, changes. If we consider the option of reducing the diameter of the venting part of the riser, for example, during freezing in the winter period, then the magnitude of the discharge in the riser will tend to infinity at the area of the live cross section of air in the compressed cross section of the riser tending to zero. In this case, increasing the diameter of the venting part of the riser in comparison with the diameter of its working part is ineffective, since the airflow depends on the working diameter of the riser.

Analyzing expression (13), several conclusions can be made:

1. The exhaustion (vacuum-gauge pressure) will reach a maximum value "at the end of the length of the initial section L. Further increase in the height of the riser does not affect the volume of the exhaustion [2]".

2. The change in air parameters slightly affect the amount of exhaustions in the sewer riser and can be neglected in calculation, since the specific gravity of the air γ_a and the kinematic viscosity ν_a placed in the numerator change proportionally.

3. As the roughness of the pipeline material increases, the length of the initial section decreases, and the volume of exhaustions arising in the riser decreases.

The use of expression (13) in practical calculations is very difficult, since it is practically impossible to determine the exact values of the area of the wetted section of air ω_a in the compressed cross section of the riser, as well as and the flow coefficient μ.

Experiments conducted by A.Y. Dobromyslov [3] on sewerage systems of different diameters, heights, attachment angles, and the heights of the hydraulic seal, have shown the necessity of determining Δp. It was found that "the amount of ejection capacity increases only in the riser section equal to 90 diameters from the point of entry of liquid into it. With an increase in the working height of the riser ... the air flow rate does not change [2] ". As a result of experimental studies [1], a formula was obtained for determining the amount of exhaustions in the riser (14), which is used for practical calculations in the design of the sewage systems in Russia [6]:

$$\Delta p = \frac{366\left(\dfrac{Q_w}{(1+\cos\alpha)D^2}\right)^{1,677}}{\left(\dfrac{D}{d}\right)^{0,71}\left(\dfrac{90D}{L}\right)^{0,5}}, \qquad (14)$$

where Δp – volume of exhaustions in the riser, mm w.g.; Q_w – design flow rate of wastewater, m^3/s; α - angle of junction of the floor sewer to the riser, degr; D- design (internal) diameter of the riser, m; d - design (internal) diameter of the floor sewer, m; L - working height of the sewer riser.

Y. Dobromyslov theoretically established and experimentally confirmed that "the air flow rate coming above its compressed section is affected not by the entire height of the riser, but only by the length of its initial section equal to 90D" [4]. Thus, formula (14) will be considerably simplified when calculating the sewerage system of high-rise buildings, in which the height of the working riser is much larger than its diameter. For risers of big height, the multiplier $\sqrt{90D/L}$ is not taken into account, so the formula takes the form of

$$\Delta p = \frac{366\left(\frac{Q_w}{(1+\cos\alpha)D^2}\right)^{1.677}}{\left(\frac{D}{d}\right)^{0.71}}.$$

When $L >> 90D,\ \alpha = 90^0,\ D = d\ \ \Delta p = 366\left(\frac{Q_w}{D^2}\right)^{1.677}$.

Calculations in the design of sewage systems of buildings of various purposes, including high-rise ones are made using the formula (14). When calculating by this formula, the critical fluid flow rates $Q_{w,\,cr}$ are determined. When critical fluid flow rate exceeds, the plumbing trap can be detached. For example, with a 70 mm high plumbing trap, the critical flow rate is:
- at D = 100mm, d = 100mm and α =45^0 $Q_{w,cr}$ = 7.32 l/s, and at α =90^0 $Q_{w,cr}$ = 4.2 l/s;
- at D = 125mm, d = 100mm and α =45^0 $Q_{w,cr}$ = 12.5 l/s, and at α =90^0 $Q_{w,cr}$ = 4.3 l/s;
- at D = 150mm, d = 100mm and α =45^0 $Q_{w,cr}$ = 19.3 l/s, and at α =90^0 $Q_{w,cr}$ = 11.34 l/s.

If the height of the hydraulic shutter is 60 mm, the critical flow rate will be:
- at D = 100mm, d = 100mm and α =45^0 $Q_{w,cr}$ = 6.1 l/s, and at α =90^0 $Q_{w,cr}$ = 3.54 l/s;
- at D = 125mm, d = 100mm and α =45^0 $Q_{w,cr}$ = 10.4 l/s, and at α =90^0 $Q_{w,cr}$ = 6.12 l/s;
- at D = 150mm, d = 100mm and α =45^0 $Q_{w,cr}$ = 16.1 l/s, and at α =90^0 $Q_{w,cr}$ = 9.45 l/s.

If the height of the hydraulic shutter is 50 mm, the critical flow rate will be:
- at D = 100mm, d = 100mm and α =45^0 $Q_{w,cr}$ = 4.9 l/s, and at α =90^0 $Q_{w,cr}$ = 2.8 l/s;
- at D = 125mm, d = 100mm and α =45^0 $Q_{w,cr}$ = 8.32 l/s, and at α =90^0 $Q_{w,cr}$ = 4.9 l/s;
- at D = 150mm, d = 100mm and α =45^0 $Q_{w,cr}$ = 12.88 l/s, and at α =90^0 $Q_{w,cr}$ = 7.56 l/s.

From critical flow rates above, one can see that the higher the hydraulic shutter is, the greater the flow can pass the sewer riser and the system will operate without disruption. It has been experimentally established that the hydraulic shutter failure occurs when the air pressure in the riser becomes less than atmospheric pressure by an amount equal to or slightly higher than the height of the hydraulic shutter. The rarefaction in the riser arises from the discrepancy between the amount of the ejecting capacity of the liquid (ie, the ability of water moving down the riser to entrain air) and the value of the actual airflow entering the riser through its exhaust.

4 Discussions

Under conditions of modern construction, little attention is paid to the height of plumbing traps. This is particularly evident in the free planning of apartments, when the owners of the apartments have the opportunity to install sanitary equipment themselves. However, the minimum height of the hydraulic shutter of 50 mm will always be respected due to the design of the plumbing traps and the height of installation of sanitary devices.

In modern design of high-rise buildings, rooms of uninhabited purpose are located on floors 1-3 and the location of sanitary equipment of those floors do not coincide with the placement of sanitary conveniences of residential floors. When the location of equipment and sewage compositions of different floors are strongly differ, it is advisable to use a separate sewage system, for example using automatic vacuum valves to supply air to the lower part of the sewerage of a high-rise building. Connecting the venting part of the riser to the working riser or to a separately venting riser creates additional costs. One of the options is to unite the sewer risers from the residential part of the building and other premises, with the arrangement of indentations or overruns. In this case, it is necessary to ensure the same air pressure in the riser sections as it moves to the horizontal direction, particularly at two inflection points (point 1, 2, Fig. 2). Consider the motion of the wastewater when creating an offset.

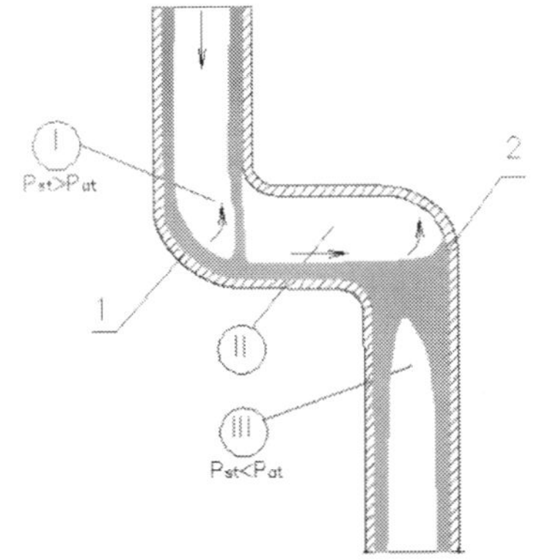

Fig. 2. Change of flow regimes of liquid and air [5]: 1, 2 – inflection points of the riser.

Since the liquid moves down the walls of the pipe in a vertical riser under the gravity, it falls in the first zone and overlap of the cross-section of the pipe surface of the horizontal section occurs (zone I, Fig. 2). As a result, a local resistance is created from the liquid film, which prevents the flow of air. The drop in the air flow rate will lead to an increase in the pressure in the sewer stand at the bend point (poit 1, Fig. 2). There are options for the appearance of a pressure that is much greater than the atmospheric pressure, in this case, if a sanitary and technical device falls into such a region, under the influence of excessive pressure, it can spill out sewage and sewage gases from the hydraulic shutter into the bowl of the device. When the pressure in the riser is stabilized, the water will fill the hydraulic shutter.

With further movement, the liquid enters the horizontal section, where the "separated" motion of the liquid and air occurs (zone II, Fig. 2). Sewage flows along the bottom of the pipe without obstructions, air moves above it. The steady-state fluid-air flow regime changes dramatically at the second inflection point (point 2, Fig. 2). Here the liquid fills the vertical section of the riser, a compressed section is formed, which prevents the movement of air. Thus, a part of the sewer riser located below the offset is cut off from normal atmospheric pressure in the following sections:

– at the point where the liquid enters the riser from the floor tap (cross-section A-A, Fig. 1);
– at the first point of inflection of the riser (point 1, zone I, Fig.2), there is a dramatic increase of the air pressure in the riser;
– under the second inflection point of the riser (point 2, zone III, Fig.2), there is a dramatic increase in the air deficit.

In this case, if the sanitary-technical device is located under the second inflection point, it is possible to break the plumbing trap of the device, since the lower part of the riser becomes unventilated at some point in time. This option is possible on the lower floors of the building. To prevent such phenomena, it is advisable to smoothly switch from a vertical section to a horizontal section of a sewer pipe using 45°, 30° or 22.5° bridles. Another option to prevent the shut-off of the plumbing trap is the installation of a bypass line (piping in the form of a loop) connecting riser sections with increased and reduced pressure (Fig. 3a) or riser equipment located below the second inflection point by a vent valve (Fig.3b, 3c) [5] .

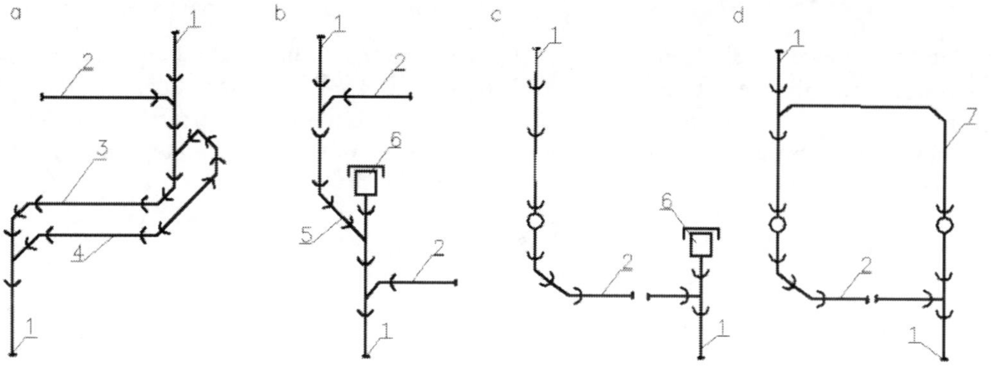

Fig. 3. Circuit of air supply to the working riser with a change in the flow direction
a) with the device of a branch line (loop-like pipeline),
b), c) with the installation of an automatic anti-vacuum valve
d) with the installation of a ventilation pipeline (bypass);
1-sewer riser; 2-floor sewer; 3 – offset; 4 - branch line; 5- overturn;
6 - automatic anti-vacuum valve; 7- - vent line (bypass)

In the premises of various purposes located on the 1-3 floors of high-rise buildings, a large number of devices can be installed on one horizontal arrangement of the sewage system. If the sanitary devices are removed far from the riser, it is advisable to install an anti-vacuum valve (Fig.3c) or a vent line (Fig. 3d) to prevent self-siphoning of the plumbing traps. The anti-vacuum valve is installed near the far device counting from the working riser (pos. 6, Fig. 3c). The ventilation pipeline also connects the sanitary equipment with the ventilation riser, connected to the horizontal pipeline and the working riser (pos.7, Fig. 3d), which is far from the working riser. Particular attention is paid to the condition that "the product of the slope of the horizontal pipeline exceeds its length by the height of the hydraulic shutter of a specific sanitary and technical device "[5].

"As the number of fixtures increases, venting needs to do as well, evolving into a venting system, with branch, circuit, and loop at the appropriate locations" [6]. In the Russian construction practice, a two-pipe system is rarely used, in contrast to the foreign experience. This is due to the increase in the cost of such a system. The diameters and design of the system are calculated in such a way that they allow the maximum possible trouble-free operation. Theoretical studies also show that the ejected air is sufficient, with one sewer riser with an exhaust port.

Thus, the article justifies the use of various design features of sewage systems in high-rise buildings, taking into account trouble-free operation in terms of plumbing traps. However, one of the significant hazards in the sewage system is the possibility of obstructions. As a result, the first floors can be flooded, since the design flow rate of effluents can reach 4-6 l/s. To prevent the flow of wastewater out of sanitary facilities, the water supply of a certain zone is turned off.

The presence of audit and cleaning device is a prerequisite for trouble-free operation of sewer risers. However, the authors assume that one of the main aspects is a smooth connection of a vertical riser to a horizontal section and the reduction in a number of offsets.

5 Conclusion

The reliability of the sewage system of high-rise buildings depends on the stability against the failure of plumbing traps and on prevention of obstructions in the gravity drainage sections of the pipeline.

The plumbing trap failure occurs when the amount of exhaustions arising during the movement of the wastewater in the riser is greater than the height of the plumbing trap. Verification calculations are performed using formula (14).

The amount of exhaustions in the riser is made up of the exhaustions that occur below the compressed cross-section of the riser and the loss of the pressure of the airflow coming from the atmosphere into the riser when the fluid flows downward.

For trouble-free operation of sewerage in high-rise buildings, the risers should be vertical without offsets.

In some cases, the two-pipe system of sewerage is not as effective as the one-pipe system, but it is considerably inferior in simplicity and cost. In each particular case, the two-pipe system of sewerage of a high-rise building should be justified by calculation.

References

1. V.P. Kamskov, I.V. Balandina, D.Y. Zemlyanushnov, Sovremennye tendentsii razvitiya nauki i tekhnologiy, **9-3**, 95-99 (2015)
2. A.S. Inosemtsev, E.V. Korolev, Stroitelnye materialy, **1-2**, 38-47 (2016)
3. A.N. Grishina, E.V. Korolev, Vestnik MGSU, **11**, 58-67 (2016)
4. A.A. Pakhratdinov, D.V. Oreshkin, Ekologiya urbanizirovannykh territoriy, **3**, 84-88 (2016)
5. A. Y. Dobromyslov, *Raschet i konstruirovanie sistem kanalizacii zdanij* (Strojizdat, Moscow, 2012)
6. D. Connelly, CPD, **27** (2007)

Design automation of load-bearing arched structures of roofs of tall buildings

Vladimir Kulikov [1,*]

[1] Moscow State University of Civil Engineering, Yaroslavskoye shosse 26, Moscow, 129337, Russia

Abstract. The article considers aspects of the possible use of arched roofs in the construction of skyscrapers. Tall buildings experience large load from various environmental factors. Skyscrapers are subject to various and complex types of deformation of its structural elements. The paper discusses issues related to the aerodynamics of various structural elements of tall buildings. The technique of solving systems of equations state method of Simpson. The article describes the optimization of geometric parameters of bearing elements of the arched roofs of skyscrapers.

1 Introduction

History of the issue. In 1976 at the international symposium adopted a classification based on building heights. Constructions up to 30 m were classified as high-rise buildings, up to 50, 75 and 100 m, respectively, I, II and III categories of high-rise buildings, over 100 m to high altitude. Within this group usually resort to additional classification, gradation of height 100 m. The number of skyscrapers over 400m worldwide in less than 20; height from 300 to 400 no more than 50, from 200 to 300 is about 150, and buildings in height from 100 to 200 m - a few thousand, and the number of such buildings is growing rapidly .

In complexity, the design of tall buildings and their construction - superior to the complexity of the construction of bridges and tunnels. This is due to the repeated predominance of height over the footprint, creating a significant load on the supporting structures. The huge height of the skyscraper results in significantly superior than typical (mid-rise) building, the degree of exposure to natural factors such as solar radiation and wind load, often exceeding the total weight of the structure. The design of tall buildings is the solution of the complex of urban, climatic, geological, architectural, constructive tasks [1, 2].

2 Materials and Methods

Aerodynamics. It is obvious that for tall buildings the impact of climate, wind, air pressure changes are extreme. The first high-rise constructed of brick have not experienced such dangerous wind impact, in contrast to the modern constructions with large spans of load-bearing structures, suspended facades, arched roofs and heights. The roof structure significantly affects the aerodynamic behavior of a skyscraper. How it is possible to reduce

* Corresponding author: kulikov-miit@mail.ru

the wind load? This is possible by: 1) the use of reliable structures with arched roofs, -2) thanks to the reliable methods of calculating their bearing capacity, 3) due to the limitation and reduction of levels of significance of factors affecting their aerodynamics. These provisions must be applied from the very beginning of designing tall buildings, with binding structure on the construction site in accordance with the wind rose, with the selection of the three-dimensional solution [3].

Aerodynamic load and the weight distribution of building structures vertically require, as a minimum, maintaining configuration throughout its height. From the point of view of sustainability tapering upwards more preferable. In this case, the structure occupies the entire plot, then the floor area is reduced. It can be a gradual change in the silhouette of a building on a sloping or curved lines or abrupt, blocky shape.

A rational spatial design of tall buildings. The most efficient forms of skyscrapers can be placed in a particular sequence, the degree of reducing the impact of air flow on their design. The absolute leader is the round plan. The lack of overhangs allows air to flow around the volume, without creating eddies that appear at the corners of rectangular buildings. Round plan involves the use of arched roofs. Second place belongs to the plans in the form of a derivative of circular, oval, shaped lenses or in the form of drops. Since the mid-twentieth century, more and more skyscrapers have similar plans, due to increase in their height, the optimal volume from the point of view of aerodynamics is not an artistic technique, but a necessity [4]. A transitional form, from the oval to the triangle with rounded corners is extremely popular and currently, due to its spatial rigidity. In third place – as common as a hundred years ago, square or diamond plans. This solution is the most popular for buildings above 60 floors, as they are more vulnerable to horizontal loads. In fourth place - high-rise buildings, coupled structurally or compositionally. They usually have a circular shape.

The roof is of arch type design curved shape in the form of an arc. With its help, often overlap houses, industrial or administrative structures to protect from rain, cold, wind, etc. Arched roofs are widely used in construction due to the following advantages:

1. Original appearance, which stands out from similar buildings with flat or gable roof.

2. Resistance to wind loads. The aerodynamic, curved shape with no sharp corners makes the arched roof immune to strong wind gusts which often lead to the breakdown of roofing from buildings and structures.

3. The reduction of snow load. Arch having a curved shape, does not delay the on-ramp of the snow mass, whereby the load-bearing design and construction is reduced.

4. The increase in the volume of the room. Due to the arched design of the roof increases the internal volume of buildings and structures.

The flow of air masses along the wall surfaces of buildings and structures occur with different speed, which depends on the efficiency of geometrical forms and their structural parameters, such as the arrow of the arch (its height), the span of the arch, forms the geometrical axis of the arch the cross-sectional size. When faced with an obstacle, the kinetic energy of wind is converted into pressure, creating wind load. This stress can be felt in any constructive element, is exposed to the air flow [5].

3 Results

The load generated by the air flow depends on the following factors: - 2) density of the air stream.(So, with high humidity, the specific gravity of air becomes greater. Therefore, increasing the value of the transferred energy); 3) the geometric shape of the stationary object (object wrapping). Based on the aspect (3), on a separate part of the building structure can influence the forces, aimed in different directions. For example: - on a vertical wall can impact a frontal force which tends to move the object. To resist this stress is the appropriate engineering design solutions, as well as the corresponding bearing capacity of the roof elements determined by a preliminary calculation. On the roof, in addition to horizontal efforts (indenting) can influence the vertical force generated from separation of the air flow when hitting obstacles. Vector vortex air flow tends to lift the roof up, trying to tear her away from load-bearing walls. The combination of all these eddy currents create wind load not only on the major elements of the building, but spreads its influence on all the elements of building structures. Power generated effort is usually proportional to the square calculation values of the wind speed [6].

Consider the method of calculation of forces from the respective air flows. The formalization of the calculation of values created by the efforts on vertical surfaces define a ratio:

$$W_m = W_0 \cdot k \cdot C \tag{1}$$

where, – W_m is the normative average value of wind stress at a height h above the ground; W_0 – normative wind pressure, depending on wind construction area, which is determined by the relevant normative document, map 3, Annex5. k-factor - coefficient of pressure fluctuations of the wind flow. C - shape factor for the façade.

So for example, for a vertical wall of a skyscraper on the ground of the construction of type (C), with the height of the object above ground level is of the order of 200m, the calculation will take the following initial data: k-factor = 0,84; (IV) area of construction, normative wind pressure W_0 = 48 kg/m^2. The shape factor for the flat facade C = 0,8. Created by stress on 1m^2 will be:

W_m = (48 kg/m^2 x 0,84 x 0,8) = 32,256 kg/m^2

At the height of the wall in the 200 meters and width 30 meters total wind load is: (200 m x 30 m x 32,256 kg/m^2) = 193536 kg or 193,536 tons, i.e. significant value. For other areas of the building, the estimated value may change in one direction or another.

Most of the damage with strong gusts of wind associated with the design of the roof. There are many good examples of how not only the individual elements of the roof, but entire roof breaks under the influence of the wind load. In a frontal direction of the wind is a collision of the wind from the front of the building and roof. Next to the vertical surface of the roof the air flow creates a vortex multidirectional vectors, in this case - there is a division on the lower, lateral and vertical components of the wind. The lower direction is the safest for the building, as all efforts are directed in the direction of the Foundation, i.e. in the direction of one of the most durable and massive parts of the structure [7, 8]. Lateral components affect the facade of the building. Vertical air flow stream at the same time - aimed directly at the eaves and creates a lifting force tending to lift the roof up and move it. The air flow is directed to the slope of the roof forms a tangential direction, sliding along the roof, enveloping the upper girder (roof ridge) and going away. This force also tends to move the roof. The perpendicular force is the normal directed inside of the roof, creating pressure, indenting the roof elements into the structure. On the leeward side of the roof, - creates an opposite force that helps create lift. Adding together all the direction of air flow, you can see that, with a high sloping roof formed efforts seeking to overturn the roof. A gentle slope helps to create large lifting forces which try to lift the design and send it in free flight. Calculation of the air load on the roof depending on the height of its location above ground level, will produce according to the formula:

$$W_p = 0{,}7 \cdot W \cdot k \cdot C \tag{2}$$

W – the standard value of the force generated by air pressure. Is determined by the maps in the Annex to the normative documents; k – coefficient showing the pressure dependence of the height above the section of the upper ground level. C – aerodynamic factor, taking into account the direction of airflow on the roof. The positive value of the aerodynamic coefficient means that the wind pushes directly on the surface. Negative values flow creates a vacuum at the surface of the roof, in other words, the "suction" of air cushion.

So for example, suppose the building is located on the shore of a large internal reservoir, and the area is of type (A). The roof is located at an altitude of *200* m, then the coefficient $k=2,45$. The prevailing winds are directed into the gable roof. The aerodynamic figure for roofs with a slope of $\alpha = 30° = -1,4$. The norm for this area of construction $W = 53\ kg/m^2$. Estimated value of the wind stress will be:

$W_p = 0,7$ x 53 kg/m^2 x 2,45 xэ (-1,4) = -127,253 kg/m^2

A negative value indicates that there is a force tending to tear the roof of the building. If the roof area of S = 120 m2, the total force will be: P = 120 m2 * (-127,253 kgf/m2) = -15270,36 kgs, about 15,253 tons..

4 Discussions

Conclusion 1. On high-rise buildings a significant and dangerous impact has the environment, atmosphere, humidity and ambient pressure. These impacts represent a number of poorly predictable alternating, and randomly alternating loads on skyscrapers. However, the article presents a modern and reliable methods to calculate their impact.

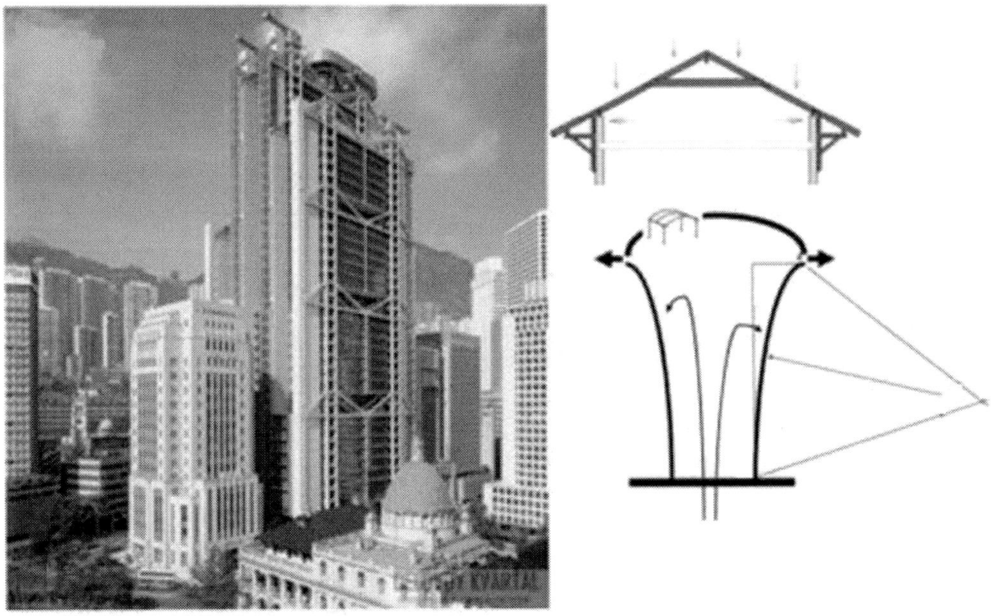

Fig.1 Possible deformation of the vertical walls of the skyscraper due to the impact of uncompensated thrust of the arched roof.

Calculate the arch of the roof of a skyscraper, is shown in Fig.1 (a) method of forces. We introduce the notation for the possible deformations of structural elements of the arched roof, fig. (b) in the following way:

$$\delta_{ij} \cdot x_i + \Delta_{ip} = 0 \tag{3}$$

Where δ_{ii} = - movement from unit loads, Δ_{ip} - movement from cargo loads.

$$\Delta_{ip} = \int_S \frac{m \cdot M_p}{E \cdot J} dS + \int_S \frac{n \cdot N_p}{E \cdot F} dS + \int_S \frac{q \cdot Q_p}{E \cdot G} dS \tag{4}$$

$$\delta_{ii} = \int_S \frac{m_i \cdot m_j}{E \cdot J} dS + \int_S \frac{n_i \cdot n_j}{E \cdot F} dS + \int_S \frac{q_i \cdot q_j}{E \cdot G} dS \tag{5}$$

Refer to fig.2, which show a virtual fragment of the arch length (*ds*) of the roof shown in fig.1 (c). Show system global ($x'\ 0\ y'$) and local ($x0y$) coordinates. Using (dx) and (dy) denote the projection (ds) to the corresponding axes $0x'$, $0y'$.

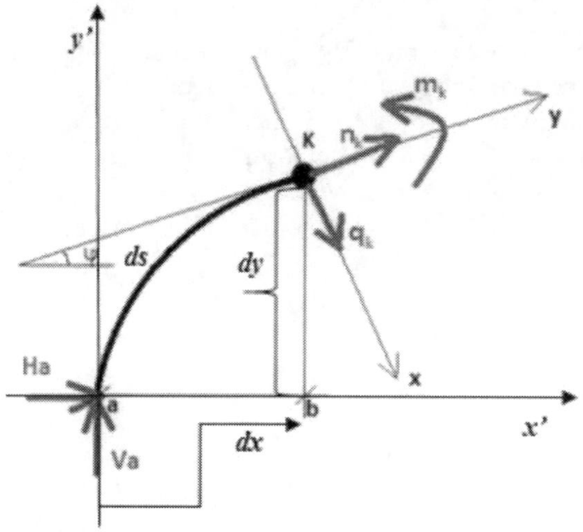

Fig.2. Detail of the arch of the roof of a skyscraper.

From fig.1, with a certain degree of accuracy, it is obvious that:

$$dS \cdot cos(\varphi) = dx \rightarrow dS = \frac{dx}{cos(\varphi)} \qquad (6)$$

Substitute (6) into (4) and move from the integration curve integral along the entire length of the arch (ds) fig.2, for integrating (dx) on certain selected sites (x1x2) of the arch considered in fig.1(c):

$$\Delta_{ip} = \int_{x1}^{x2} \frac{m \cdot M_p}{E \cdot J \cdot \cos(\varphi)} dx + \int_{x1}^{x2} \frac{m \cdot \cos(\varphi) \cdot N_p}{E \cdot F \cdot \cos(\varphi)} dx + \int_{x1}^{x2} \frac{m \cdot \sin(\varphi) \cdot Q_p}{E \cdot G \cdot \cos(\varphi)} dx \qquad (7)$$

Write the equation of the geometrical axis of the arch is calculated as $y(x) = \frac{f \cdot x(2 \cdot l - x)}{l^2}$ where (*f*) is the height of the arch, and l= (1/2) the span of the considered arch.

Bearing in mind that $m = \frac{y(x)}{f} = \frac{f \cdot x(2 \cdot l - x)}{l^2}$, we get:

$$\Delta_{ip} = \int_{A}^{B} \frac{m \cdot M_p}{E \cdot J \cdot \cos(\varphi)} dx = \int_{A}^{B} \frac{x \cdot (2 \cdot l - x) \cdot M_p}{E \cdot J \cdot \cos(\varphi) \cdot l^2} dx \qquad (8)$$

$$\delta_{11} = \int_{A}^{B} \frac{m^2 dx}{E \cdot J} = \int_{A}^{B} \frac{\left(\frac{x \cdot (2 \cdot l - x)}{l^2}\right)^2}{E \cdot J} dx = \int_{A}^{B} \frac{x^2 \cdot (2 \cdot l - x)^2}{E \cdot J \cdot l^4} dx \qquad (9)$$

Using the conversion expression (3) can be represented as follows:

$$\delta_{ij} \cdot x_i + \Delta_{ip} = 0 \rightarrow x_i = -\frac{\Delta_{ip}}{\delta_{ij}} = -\frac{\int_{A}^{B} \frac{x \cdot (2 \cdot l - x) \cdot M_p}{E \cdot J \cdot \cos(\varphi) \cdot l^2} dx}{\int_{A}^{B} \frac{x^2 \cdot (2 \cdot l - x)^2}{E \cdot J \cdot l^4} dx} \qquad (10)$$

For the taking of the integrals presented in (10), we have the formula (11) Simpson method:

$$I = \frac{h}{3} \cdot [f(x_0) + f(x_n) + 4 \cdot (\sum f(x_{odd})) + 2 \cdot (\sum f(x_{even}))] \qquad (11)$$

Conclusion 2. A significant influence of environmental factors on high-rise buildings leads to large internal stresses in various structural elements of skyscrapers. The article presents the method of calculation of the structural deformation of the arched roof of a skyscraper. This article uses the modern methods of choice of materials of construction and cross sections sufficient to prevent their destruction from extreme loads.

Formalize the notion of "Rational axis of the arch". The rational axis of the arch of a given span and a given rise. we will call such an axis in which the cross section of the arch required by the conditions of strength will be the smallest. Reducing the transverse dimensions of the arch is associated with a decrease of the actual voltage which the arc is defined by the formula:

$$\sigma = \frac{N}{F} + \frac{M}{I} \cdot y \tag{12}$$

It is obvious that the lowest value of the actual stress will be the case when the value of the bending moment in the cross section will be zero. The last statement will be true in the case that the resultant of the internal forces will pass through the center section. This condition needs to satisfy all sections of the arch, and this is equivalent to saying that the rational axis of the arch should coincide with the pressure curve. The construction of the pressure curve associated with a given distribution of the cross section of the arch. This situation greatly complicates the analytical study of matter. Therefore, the study of the question of the rational axis of the arch to produce on conventional, so-called "delicate" arch, with no real cross sizes. In this arch the curve of pressure coincides with a virtual curve for a given load. Errors that are taking place at such consideration, usually small, and their influence on the contour of the axis of the real arch can be neglected. The virtual curve depends on the size and location of the load. Position it in the arch completely determined by three points, beginning, middle, and end. But the change of its position between these points depends on the location of the load along the length of the arch.

Assume that arch is subjected to vertical external load. Mentally cutting out the arch element of length (ds), and given the fact that by combining the axis of the arch with the virtual curve, moment and shear force in its cross-section must be zero. We balance the element of two normal forces (N) and (N + dN). As mentally the cut element under the action of applied thereto of forces is in equilibrium, we can write the following equation:

$$N \cdot \sin\left(\frac{d\varphi}{2}\right) + (N + dN) \cdot \sin\left(\frac{d\varphi}{2}\right) - q_z \cdot dx \cdot \cos(\varphi) = 0 \tag{13}$$

Because of the smallness of the angle $(d\varphi)$ can be taken that $\sin\left(\frac{d\varphi}{2}\right) = \left(\frac{d\varphi}{2}\right)$. Value of $(dN\ \frac{d\varphi}{2})$ as an infinitesimal of the second order can also be neglected. $dx = ds \cdot \cos(\varphi)$. Under these conditions, equation (13) can lead to the next view

$$N \cdot d\varphi = q_z \cdot ds \cdot cos^2(\varphi) \tag{14}$$

If, under the action of external vertical loads on an arch projection on the horizontal normal force in all sections of the arch is equal to the thrust $N \cdot \cos(\varphi) = H$, equation (14) can be transformed to the following form:

$$\frac{d\varphi}{ds} = \frac{q_x \cdot cos^2(\varphi)}{H} \tag{15}$$

$$\frac{d\varphi}{ds} = \frac{1}{\rho} = \frac{y''}{\sqrt{(1+y'^2)^3}} \tag{16}$$

Using the (16) ratio, substituting the relevant expressions and $(\cos(\varphi), d\varphi, ds)$ in equation (15) and making some transformation, obtain the differential equation of the effective axis of the arch when subjected to consider the arch of the external vertical load:

$$y'' = \pm \frac{q_x}{H} \tag{17}$$

The sign in this equation should be chosen according to the location of the center of curvature along the coordinate axes. If the center of curvature is on the side of positive y, the plus sign, if negative ordinate, a minus.

$$y = \frac{q \cdot x \cdot (2 \cdot l_1 - x)}{2 \cdot H} \tag{18}$$

Conclusion 3. The article presents the method of optimization of geometrical parameters of the bearing element of the roof of a skyscraper, the arch. The use of this technique in the real design will significantly reduce the influence of external destructive factors on the structural elements of skyscrapers.

5 Conclusion

Taking into account the conclusions 1, 2, 3 can say with certainty about the possibility of using arched roofs in the design and construction of skyscrapers.

References

1. I.V. Ilin, A.I. Levina, O.Yu. Iliashenko, MATEC Web of Conf, **86**, 05028 (2016) doi - 10.1051/matecconf/20168605028
2. A. Ilin, O. Kalinina, O. Iliashenko, A. Levina, Procedia Engineering, **165**, 1673- 1682 (2016) DOi - 10.1016/j.proeng.2016.11.909
3. O. Kalinina, O. Valebnikova, Advances in Intelligent Systems and Computing, **692**, 1315-1322 (2018) DOi - 10.1007/978-3-319-70987-1_139
4. A. Zaychenko, S. Gutman, O. Kalinina, Advances in Intelligent Systems and Computing, **692**, 453-462 (2018) DOi - 10.1007/978-3-319-70987-1_48
5. T.A. Emelyanova, A.P. Denisova, Sarat. GOS. tehn. University (2013) ISBN 978-5-7433-2666-2.
6. T.G. Maklakova, *Design of civil buildings: Proc. settlement* (Stroyizdat, Moscow, 2013)
7. N.G. Vilchik, *The architecture of the buildings: the textbook* (INFRA-M, 2012)
8. V. Kankhva, Procedia Engineering, **165** (2016) doi:10.1016/j.proeng.2016.11.855

Design, construction and operation features of high-rise structures

Alexey Mylnik[1,*] *Vladimir* Mylnik[1], *Elena* Zubeeva[1], *Olga* Mukhamedzhanova[2]

[1]Moscow Aviation Institute (National Research University), Volokolamskoe highway, 4, Moscow, 125993, Russia;
[3]Moscow State University of Civil Engineering, Yaroslavskoe sh. 26, Moscow, 129337, Russia

Abstract. The article considers design, construction and operation features of high-rise facilities. The analysis of various situations, that come from improper designing, construction and operation of unique facilities, is carried out. The integrated approach is suggested, when the problems of choosing acceptable constructional solutions related to the functional purpose, architectural solutions, methods of manufacturing and installation, operating conditions for unique buildings and structures are being tackled. A number of main causes for the emergency destruction of objects under construction and operation is considered. A number of measures are proposed on the basis of factor classification in order to efficiently prevent the situations, when various negative options of design loads and emergency impacts occur.

1 Introduction

At the present moment, Russian government is developing special trends in the field of urban policy. The implementation of projects for the construction of unique buildings and structures is an urgent issue, due to the modernization and improvement of construction technologies, design, construction and operation methods of real estate, the use of modern equipment, products and materials, as well as the need to improve and enrich the existing architectural style.

In accordance with the resolution of the Government of Moscow dated 28.12.2005, a group of high-rise buildings, with a height of more than 75 m, combined with a general functional-planning and architectural-spatial solution, including a mandatory list of local service facilities is defined as a multifunctional high-rise building.

2 Materials and Methods

In terms of the great responsibility for the design of high-rise buildings, the following features should be considered:

1. Effective assessment of engineering and geological conditions.

* Corresponding author: vint215@yandex.ru

When choosing a site for the construction of a high-rise building, it is necessary to make a thorough assessment of the engineering and geological conditions, as well as the possibility of safe construction of the building, taking into account the possible effects of dangerous geological and engineering-geological processes (karst-suffosion, landslide, etc.).

2. Survey of the foundations of buildings, that are located in the zone of potential construction.

When constructing a high-rise building in a built-up area, it is necessary to perform surveys of the foundations of buildings and structures that fall within the zone of influence of high-rise construction, as well as carry out a forecast of changes in the stress-strain behavior of the soil mass and the ground water dynamics [1].

3. Consideration of design features of high-rise buildings.

3.1. Providing spatial rigidity of the structural system.

The spatial rigidity of high-rise buildings is ensured by:

- developed in plan and symmetrically located diaphragms and structural cores;
- box-shaped structural systems with load-bearing external walls along the whole perimeter of the building;
- Rigid discs' slab, combining vertical load-bearing structures and performing the functions of horizontal stiffening diaphragms under the effect of wind or seismic loads;
- rigid nodal interfaces between load-bearing structures;
- horizontal stiffener rings in the level of technical floors, providing joint work on the bend of all vertical load-bearing structures of the building.

3.2. Providing protection against the progressive collapse of load-bearing structures.

High-rise buildings should be protected from progressive collapse in case of local destruction of load-bearing structures as a result of natural and man-caused emergencies.

Natural emergencies include dangerous meteorological phenomena, the formation of karst craters and dips in the foundations of buildings [2].

Man-caused emergencies include explosions outside or inside the building, fires, accidents or significant damage to load-bearing structures due to defects in materials, poor-quality work, etc.

3.3. Additional calculations of various kinds of loads applied on the structures of high-rise buildings.

In high-rise buildings, additional short-term loads are taken into account such as:

- the load from the rescue capsule of the fire-fighting helicopter to cover;
- Load from vehicles, including fire trucks;
- load on the coverings of the stylobation (the upper part of the basement) and the underground parts of the buildings that protrude beyond the dimensions of the main volume.

In addition, a special assessment of wind and aerodynamic loads is given as part of design documentation.

It is necessary to take into account the various extra-impacts of the wind, including:

- average and pulsation components of the design wind load;
- the maximum values of the wind load acting on the enclosing structures;
- resonance vortex excitation;
- effects that excite aerodynamic unstable oscillations;
- effects that cause violations of comfort conditions of pedestrian areas, which are adjacent to the building.

Values of these effects, aerodynamic coefficients of forces, moments, internal and external pressures, evaluation of resonance vortex excitation, usually should be taken on the basis of testing models (mockups) of high-rise buildings (including existing buildings) in specialized wind tunnels [3].

It is necessary to ensure the reduction of wind flows arising at the first floors of the high-rise building and the adjacent facilities. Limiting horizontal displacements of the top of high-rise buildings and accelerating the fluctuations in overlaps of the upper floors are taken into account when the pulsation component of the wind load acts in order to ensure a comfortable stay of people in high-rise buildings.

3.4. Special technical solutions for facade structures and openings.

It is necessary to resolve issues related to the construction of hinged facades, their installation, repair and operation as well as structure and operation of non-opening window fillings of the upper floors.

Structures of windows and hinged translucent facade structures and their attachment to load-bearing structures should be calculated by strength and deformability to the effect of wind loads.

4. Calculation of seismic effect on buildings.

The calculation is carried out on the basis of the corresponding maps of seismic microzoning.

5. Thermo-technical calculations of outer walls.

Calculations of heat losses by external enclosing structures, air balance of high-rise buildings, outdoor air parameters in the places of air intake devices, etc. are performed with account of changes in the speed and temperature of the outdoor air throughout the entire height of the buildings. The parameters of outdoor air are taken under consideration of the following factors:

- decrease of the air temperature at the altitude of more than 150 m;
- increase of wind speed during the cold season;
- appearance of powerful convective heat flux affecting the facades of the building, that is irradiated by the sun.

6. Performance of additional firefighting measures.

The high-rise building has a special degree of fire resistance. It is divided into fire compartments in height. According to some experts, the division into vertical compartments is a matter of argument, since the flames on the ascending air streams bypass not only fire barrier slabs, but also technical floors, and the construction of firefighting visors at high altitudes is also a dubious solution, since the visor will be small due to snow load and rising air currents. High-rise buildings should be equipped with an automatic fire alarm system. Automatic fire detectors (or stand-alone fire detectors having an outlet to the fire alarm system) should be installed in all rooms except for premises with wet processes [4].

7. Performance of additional measures for the engineering systems and communication lines.

Engineering systems of high-rise buildings have a number of features:

- availability of at least two network inputs;
- additional technical premises;
- special systems to ensure safety of the building and premises to set them;
- special requirements for thermal protection of floors, which are located above 75 m;
- heat supply from two independent sources, and in some cases - a special boiler room;
- division of heating systems into zones to avoid excess pressure in low zones;
- additional fire pumping and spare tanks on technical floors are required for fire water supply;
- electricity is provided from two independent sources, in the presence of a third emergency source, which is a diesel power plant.

8. Performance of additional measures for the installation of garbage chutes and elevator shafts.

For high-rise buildings, there have to be special requirements for the arrangement of garbage chutes and elevator shafts. Requirements for permissible deviations of different shafts in height are stricter than for massive housing buildings.

9. Implementation of an automated building management system.

Automated building management system is required for high-rise buildings, which performs centralized monitoring, dispatching and management of engineering systems. The main functions of automated complexes, communication and information systems, which high-rise buildings are equipped with, are safety, life support of the building and provision of technological functionality of functional blocks as well as comfort of people. Technical systems must be combined into complexes that ensure the exchange of information between systems in order to eliminate redundancy and ensure the coherence of work. A flexible, freely programmable distribution system with an automated workplace, where only highly trained specialists can work, should be organized [5].

10. The arrangement of additional rooms for the placement of technological equipment for the following purposes:

- technological equipment of the departments of city emergency services
- a stationary monitoring station for load-bearing structures of a building;
- equipment for the central control point of the building security system.

3 Results

The following precedents for design, construction and operation of unique facilities can be distinguished dwelling on their specific features and development trends in this field:

1. Emergency collapse of building structures during construction and operation;
2. Damage (theft) of expensive property (products, materials, equipment) at the construction site;
3. Inconsistency of the initial permissive documentation to the legislation and norms in the field of construction of unique real estate objects;
4. Inconsistency of the construction of a unique facility to the legislation in the field of environmental safety.

Unique facilities belong to the construction objects of a high level of responsibility. Therefore, it is necessary to carry out additional measures at all stages from design to the facility abandonment in order to ensure the safe operation of the building.

A number of main causes for the emergency destruction of the unique facilities during construction and operation can be highlighted:

1. Preventive measures to avoid accidents, ensure the safety of constructed and operated facilities, taken by the executive and supervisory authorities, as well as construction and operation organizations, enterprises and associations, are insufficient. The number of accidents does not decrease, when their severity grows and the amount of victims increases;
2. The main share of accidents falls on the buildings and facilities in operation and accounts for 85% of the total number of registered accidents. The materials of the investigation show that the main causes of accidents at the operated facilities are gross violations of the rules for the technical operation of buildings and structures;
3. There is no proper control of the technical condition of buildings and structures.

The issue of more efficient control of the technical condition of a unique building or structure is one of the main issues in the system of integrated security of functioning of construction projects. Especially, it is urgent in terms of the large number of accidents and collapses that have occurred in recent years:

- the dome of the Transvaal Park in Moscow, 2004 (Fig. 1.1);
- section of the airport in France, Paris, 2005 (Fig. 1.2);
- roofing of ice rinks in Germany and Austria, 2006 (Fig. 1.3);

Fig. 1. Consequences of accidents.

The consequence of such accidents are not only economic losses, but in many cases - there are dire environmental and social consequences. A detailed analysis of the causes and consequences contributes to reducing the number of accidents.

Analysis of the circumstances of emergency destruction of structures of unique buildings allows distinguishing the following main causes:

- violation of the operation rules;
- defects at the construction stage and deviations from the design;
- violations of the technology of performance of works during construction, reconstruction and repairs;
- low quality of manufacturing of building structures.

Moreover, the following factors should be noted:

1. The development of the project and design features of unique structures, as a rule, is based on the experience and engineering foreknowledge of the designer, and the application of the acquired skills does not always give an accurate idea of the design model.
2. Imperfections, defects and damages of load-bearing elements of a structure that occur during manufacture, transportation and installation, a seemingly insignificant disparity between the final geometry of the structure and the idealized design configuration, which for traditional structures would not be significant, but for structures with unique parameters result in the occurrence of serious additional efforts and displacements.
3. The change in the mechanical characteristics of the used materials (aging, relaxation, creep, etc.), operational factors, geological conditions, that take place over time, become a serious matter due to the significant regulatory lifetime of such structures.

Thus, modern approaches to the design of unique buildings and structures provide only the initial constructional safety of the facility, i.e. design structural safety at the time of commissioning of the facility.

The constructional safety of this facility type during operation can significantly change since the design stage for the listed reasons. Firstly, due to possible deviations from design decisions during the construction of the facility. Secondly, due to operational factors, where there are changes in the object itself, in particular, as a result of material aging, corrosion of metal structures, space planning changes associated with alterations, extensions, reconstructions, etc. Thirdly, due to changes in external factors, changes in the level of surface water; physical and mechanical parameters of the foundation soil, the occurrence of dynamic impacts, the movement of soil in connection with the nearby new construction of aboveground and underground structures, etc.

Fourth, due to the "human" factor: mistakes, ignorance, negligence of the staff.

It should be borne in mind that accidents at unique buildings and structures, that are caused by their collapse, can occur according to two patterns:

- either with a gradual accumulation of stresses and deformations and subsequent collapse of load-bearing structures;
- or fleetingly (progressing, avalanche-like collapse) with possibly even a short-term but significant overload of an important supporting structural element, from the destruction of which, subsequent progressive collapse can happen. An example of this was the collapse of the World Trade Center buildings in New York City in 2001 (Fig. 1.4), where such a situation was not foreseen at the design stage.

In order to effectively prevent the first collapse pattern, it is necessary to have a regulatory framework for monitoring unique structures, that are under development.

- At the present time, protection in the second collapse pattern time can only be reliable calculation of load-bearing structural elements and appropriate structural measures to ensure the inadmissibility of a progressive collapse, since neither of the existing systems of deformation control of building structures can help at such collapse. If the process has started, even preliminary detection does not allow any action to be taken to prevent or rescue people and equipment because of its explosion-like fleetness.

4 Discussions

Diagnostics of load-bearing structures of buildings at an early stage becomes especially important for unique, technically complex high-rise buildings, since it is not easy to control the change in the stress-strain state of load-bearing structures through traditional methods of visual and instrumental inspection. In addition, in high-rise construction, a significant influence on the stress-strain state of the bearing structure is exerted by inclines and wind effects, which creates a greater dispersion of the places of accumulation of deformation stresses [6]. To date, there is no generally accepted scientifically grounded approach or the practice of designing unique structures that maintain structural integrity for various options of design loads and emergency effects.

5 Conclusion

In the context of large-scale implementation of projects for the construction of unique facilities, as well as drawbacks and a number of problems related to the regulation of the design, construction and operation of these facilities, judicial situations arise that require the expert examination of the accident which occurred with the object.

References

1. R.S. Golov, V.V. Shilov, S.A. Silantiev, ASEE International Forum (2017)
2. A. Jones, G. Fallon, R. Golov, European Business Review, **12**, 187-197 (2017) doi. 10.1108/09555340010336871
3. E. Kiseleva, M. Nekrasova, M. Mayorova, M. Rudenko, V. Kankhva, International Review of Management and Marketing, **6**, 95 (2016)
4. I. Lukmanova, M. Mishlanova, International Journal of Economics and Financial Issues, **5,** 208-216 (2012)
5. A. Konikov, G. Konikov, Procedia Engineering, **165**, 1052–1056 (2016)
6. V. Gasilov, N. Anisimova, I. Provotorov, MATEC Web of Conferences, **106,** 08035 (2017) DOI: 10.1051/ mateconf / 201710608035.

Application of tuned mass dampers in high-rise construction

Vyacheslav Teplyshev [1,*], *Alexey* Mylnik[1], *Maria* Pushkareva[1], *Murad* Agakhanov[2] and *Olga* Burova[2]

[1]Moscow Aviation Institute (National Research University), Volokolamskoe highway, 4, Moscow, 125993, Russia
[2]Moscow State University of Civil Engineering, Yaroslavskoye sh., 26, 129337 Moscow, Russia

Abstract. The article considers the use of tuned mass dampers in high-rise construction for significant acceleration and amplitude of vibrations of the upper floors under dynamic wind influences. The susceptibility of people to accelerations in high-rise buildings and possible means of reducing wind-induced fluctuations in buildings are analyzed. The statistics of application of tuned mass dampers in high-rise construction all over the world is presented. The goal of the study is to identify an economically attractive solution that allows the fullest use of the potential of building structures in high-rise construction, abandoning the need to build massive frames leading to over-consumption of materials.

1 Introduction

The construction of high-rise buildings requires the attraction of advanced achievements in many fields of civil engineering and related sciences. In order to implement ambitious construction projects, that may become symbols of cities and countries, it is necessary to conduct works on the economic substantiation of construction, architectural design, site conditions, the choice of building materials, construction design, development of building technology and its safe and durable operation.

The specialists of each of the indicated directions face unusual for their practice challenges when they deal with such projects. In this case, a large number of tests in the design of high-rise buildings fall to the share of engineers who develop building structures, whereas the situation is often aggravated by unfavorable conditions of the construction site, related to possible intense wind and seismic effects.

2 Materials and Methods

One of the effective examples of controlling the vibrations in the floors of a high-rise building, which is subject to intense wind, is the Taipei 101 skyscraper in Taipei, Taiwan. A specially designed Tuned Mass Damper (TMD) deals with damping of low-frequency vibrations and a reducing their amplitudes during typhoons, especially on the upper floors.

* Corresponding author: teplyshev@tbnenergo.com

These solutions for monitoring the amplitude-frequency response of buildings are still relatively rare, but interest in them is growing, as with a competent approach it is possible to build high-rise buildings in regions where it is usually associated with increased expenses to ensure sufficient rigidity of the load-bearing frames and stumbles upon an undesired maximum height, dictated by considerations of comfort.

The use of tuned mass dampers is advisable, including construction in areas, that are common to high-rise buildings, since it provides an opportunity to achieve qualitatively better operational and economic characteristics. However, additional research is required for the extended application of such a vibration damping system. Swayings of a high-rise structure under wind influence can quickly lead to its destruction if a resonance occurs and the amplitude reaches excessive values. However, usually the swayings are not so vigorous as to damage the structure of a skyscraper, but can cause serious discomfort to people on the upper floors.

For modern light high-rise buildings, monitoring horizontal swayings of floors to ensure sanitary requirements can be a greater challenge than the need to ensure the structural strength of buildings. Acceleration is the natural factor of all dynamic influences. The horizontal force felt by people is directly proportional to the horizontal acceleration, which is impressed upon the structure, according to Newton's first law. According to the research, the most sensitive people are able to feel acceleration from one thousandth of the acceleration due to gravity. Detailed results of the study on the susceptibility of people to accelerations in high-rise buildings are given in Table 1.

Table 1. People's susceptibility to accelerations in high-rise buildings.

	Acceleration, m/s^2	**Effect**
1	< 0.05	People do not feel the swayings
2	0.05 – 0.10	- sensitive people may notice an acceleration - dangles sway slightly
3	0.10 – 0.25	- most people feel the swayings - swayings disturb desk work - long-term exposure may cause motion sickness
4	0.25 – 0.40	- desk work becomes very difficult or unbearable - foot travel is still possible
5	0.40 – 0.50	- people strongly feel the swayings - walking becomes difficult - standing people find it difficult to balance
6	0.50 – 0.60	Most people are very uncomfortable to move; they cannot walk normally
7	0.60 – 0.70	All people are very uncomfortable to move; people cannot walk normally
8	> 0.85	Objects begin to fall; people can be injured

The requirements of standards in Russia are consistent with the results of the research described above: in the normative document [1], the maximum permissible horizontal acceleration of the floor is limited to 0.08 m/s^2 with a wind load equal to 70% of the normative value of its pulsation component.

Understanding, that it is not enough for normal operation of high-rise buildings in conditions of severe storms and hurricanes to meet the requirements of the dynamic comfort of buildings in their design, came to engineers after analysis of the operational experience of the numerous skyscrapers. However, there are only three relevant factors that can be controlled at the design stage: height, shape and mass, which affects the dynamic response of the building [2].

The limitation of excessive vibrations can be achieved in three ways: by increasing the stiffness to reduce the period of vibrations, by controlling the mass and by special design of the building's shape to improve its aerodynamic properties.

The listed above traditional approaches to design are applicable only in certain areas, behind which serious questions about the physical feasibility of projects, their cost and aesthetic qualities arise.

Thus, the attention of practicing engineers turned to various damping devices, as a means to damp the vibrations of structures under wind influences [3].

For reference, a set of possible approaches to reduce wind-induced vibrations in buildings is given in Table 2 [4, 5].

Table 2. Possible approaches to reduce wind-induced vibrations in buildings.

Approach	Type	Method and goal	Comments
Aerodynamic design	Passive	Improvement of aerodynamic properties to reduce aerodynamic coefficients	Corners with chamfers or holes; holes in the core
Structural design	Passive	Increasing the mass of the building to reduce the effect of wind pressure	Increase of cost
		Increase stiffness and natural vibration frequency	Adding bonds, increasing wall thicknesses and column stiffness
Supplementary damping	Passive	Use of materials and design joints with additional dissipative properties to increase the damping of the structure	SD, SJD, FD, VED, VD, OD
		Adding supplementary mass to increase the level of damping	TMD, TLD
	Active	Generation of inertia forces to soften dynamic reactions	AMD, HMD, AGS
		Generation of aerodynamic forces to reduce aerodynamic coefficients	Screws, turbines, aerodynamic elements
		Change in stiffness to avoid resonance	AVS
SD – steel damper; SJD – steel joint damper; FD – friction damper; VED – viscous-elastic damper; VD – viscous damper; OD – oil damper; **TMD – tuned mass damper;** TLD – tuned liquid damper; AMD – active mass damper; HMD – hybrid mass damper; AGS – active gyrostabilizer; AVS – active stiffness controller.			

The tuned mass damper is a device consisting of a mass attached to the main structure by means of a spring and a damping element. A layout view on the damper in the Taipei 101 building is shown in Figure 1.

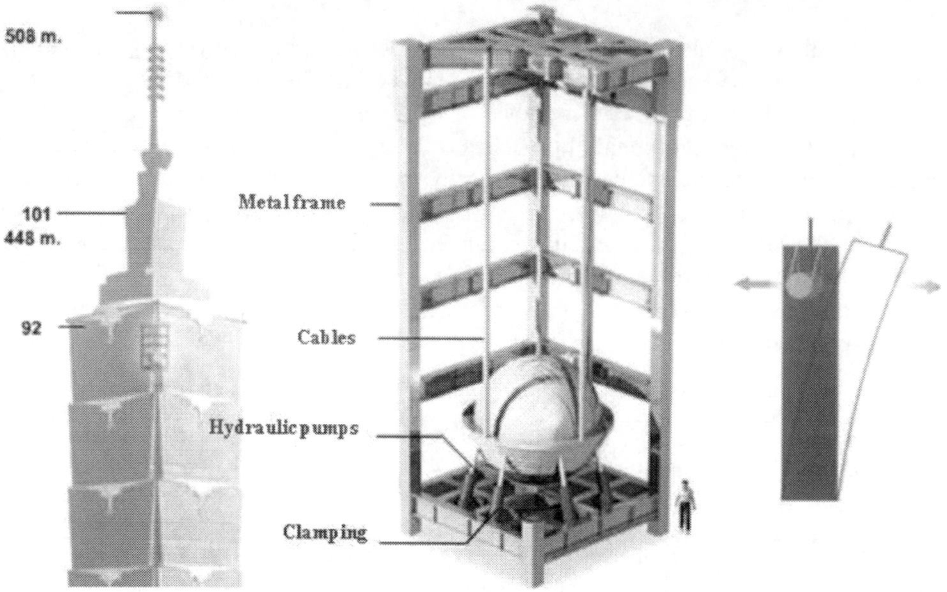

Fig. 1. Damper in Taipei 101 building.

3 Results

The spring can be performed by devices that ensure the connection of the mass with the structure of stiffness *k*. In the present work tuned mass dampers are considered, where the mount plays the role of spring, connecting the bearing structures and the mass and forming a pendulum.

In the calculations, a finite-element system is used, which is similar to the pendulum with equivalent stiffness *kd* and damping *c*.

The applicability of this analogy is justified by the following formulas.

The equation of motion in the horizontal direction corresponds the following formula:

$$T \cdot \sin b + Wd \; / g \cdot (u^{11} + ud^{11}) = 0 \quad (1)$$

where, W_d – is the weight of tuned mass damper, H; T – tension load in the mount, N.

At small values of the force *T*, the following simplification is possible:

$$T = L \cdot \sin b = L \cdot b; \qquad T = Wd \quad (2)$$

Then the equation of motion reduces to the following:

$$m_d \cdot u_d + W_d / L \cdot u_d = - m_d \cdot u^{11} \quad (3)$$

Then equivalent stiffness k_d, N/m is

$$kd = Wd \; / \; L \quad (4)$$

The length of the pendulum mount is chosen to ensure the optimal adjustment of the tuned mass damper: the vibration frequency of the mass should correspond to a certain frequency of the natural vibration of the structure. The mass should vibrate in reverse phase to the structure and damping (dispersion of the vibration energy through special damping devices) should be possible.

A special advantage of using mounts as the formation of a "spring" is the possibility of constant adjustment of the stiffness of the damper under the structure, because its dynamic characteristics can vary significantly depending on the running load imposed or changes in the stiffness of its bearing elements and node points occurred due to some external influences [6, 7].

As damping devices, the elements of viscous friction (oil buffers) are commonly used. These devises are made as cylinders with liquid, where perforated or loosely adhering pistons move. When some motion is imparted to a piston of the oil buffer, a viscous frictional force arises in the buffer, which is proportional to the speed of the piston. At the same time, the constancy of viscous properties of oils or other liquids used in buffers should be ensured. Thus, each swaying of the mass dissipates part of vibration energy of the structure and reduces the amplitudes and accelerations of its vibrations. The general dynamic design of the system with a damper is shown in Figure 2.

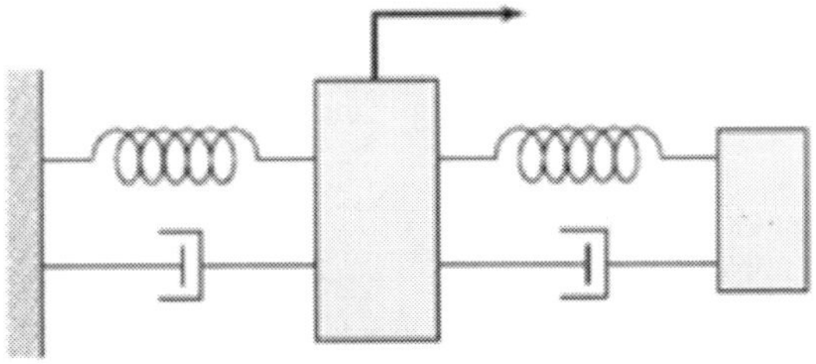

Fig. 2. General dynamic design of the system with a damper.

The mass itself usually has a relatively small mass. The use of tuned mass damper with mass equal to 1-5% of the modal mass of the structure in the form of vibrations tuned to is considered optimal.

There are several high-rise buildings with tuned mass dampers in the world. Information on some of them is summarized in Table 3.

Table 3. Several buildings with tuned mass dampers.

Name and height	**City or country**	**Number of dampers**	**Year of erection**	**Own frequency and/or mass of damper**
Taipei 101, 508 m	Taipei	1+2	2004	730 t; 2 to 4.5 t
Trump World Tower, 262 m	New York	1	2001	600 t
Canadian National Tower, 533 m	Toronto	1	1976	18 t
Akita Tower, 112 m	Akita	1	1994	0.41 Hz
Sydney Tower, 305 m	Sydney	1	1980	0.1 Hz; 0.5 Hz / 370 t
City Corp Center, 278 m	New York	1	1978	0.16 Hz / 370 t
John Hancock, 241	Boston	2	1977	0.14 Hz/ 2 300 t

4 Discussions

The structures of high-rise buildings usually have relatively high deformability in horizontal directions, which causes significant acceleration and vibration amplitude of their upper floors under dynamic wind influences.

Therefore, the significant issue of their design is to ensure the requirements for a comfortable stay of people, consisting in limiting the amplitudes and acceleration of vibration in the upper floors of structures. These requirements are provided in civil engineering standards all over the world, including Russia, Europe and the United States.

5 Conclusion

The practical importance of this study is to identify an economically attractive solution that allows the most complete use of the potential of building structures in the high-rise construction, abandoning the need to build massive frames, leading to over-consumption of materials and providing excessive strength. This work contributes to a deeper introduction of the concept of tuned mass dampers into the engineering discourse and to widening of the possible geography of their use in high-rise construction.

References

1. U.Y. Kuvshinov, *Theoretical Basics of Indoor Microclimate Providing. Scientific publication* (Moscow, Edition of Building University Association,2007)
2. V.G. Gagarin, V.A. Zemtsov, N. M. Igumnov, Roofing and Insulation Materials, **4,** 41-43 (2011)
3. A. U. Bezrukov, Microclimate of Dwelling Premises. Light Transmitting Constructions, **3** (2003)
4. P. Diachek, S. Makarevich, A. Zakharevich, D. Livansky, Vestnik MGSU, **7**, 143-147 (2011)
5. M. Jevric, M. Romanovich, Procedia Engineering, **165**, 1478 – 1482 (2016)
6. T. Simankina, M. Romanovich, O. Tsvetkov, MATEC Web of Conferences**, 53**, 01054 (2016)
7. T. Saknite, D. Serdjuks, V. Goremikins, L. Pakrastins, N.I. Vatin, Magazine of Civil Engineering, **64(4)**, 26-39 (2016)

Problems of natural lighting for deepened buildings and underground premises under screen effect of high-rise construction

Kira Larionova [1,*] and *Sergey* Stetsky[1]

[1] Moscow State University of Civil Engineering, Yaroslavskoye sh. 26, Moscow, 129337, Russia

Abstract. The main rationale and objective of the submitted research work is to create a quality lighting environment in the premises of deepened buildings and below-ground structures under screen effect of high-rise construction (high-rise buildings). It is noted, that in modern megapolises, a deficiency of vacant urban territories leads to the increased density of urban development with increased amount of high-rise construction and tendency to increase efficiency in the use of underground space. The natural lighting of premises in underground buildings and structures is the most efficient way, but it can be implemented only under use of roof lighting system in the form of roof monitors or skylights. In this case the levels of indoor natural lighting will be affected with serious screening effect of high-rise buildings in surrounding development. Such an situation is not regulated, or even considered by the contemporary building Codes and Regulations on natural lighting of interiors. The authors offered a new formula for a daylight factor calculation with roof lighting system in the described cases. The results of theoretical calculations and experimental studies showed very similar values. This proved the truth of the offered formula and elaborated method of calculation on the basis of an offered hypothesis. It prooves, that it is possible to use some factor and guide points in the daylight factors design under system of side natural lighting in the same design for a system of roof lighting.

1 Introduction

In the present time the intensive development of large cities with tendency of megapolises` creation shows the deficiency of vacant territories. This activates the high-rise construction, as well as use of an underground space.

The deepened and underground premises of buildings and structures might be used for public, industrial or even for residential functions. Unfortunately, such a premises are usually lit by artificial light. This lighting system, from contemporary point of view cannot be appraised as efficient one, due to high expense of electricity and specifics of artificial lighting effect on a human health and psych-conditions.

* Corresponding author: LarionovaKO@mgsu.ru

Thus, the creation of microclimate` environment with high quality for the premises of buildings and structures considered in such an non- ordinary conditions is quite actual. In this case it is possible to solve this problem only with use of roof natural lighting system in form of monitors, skylights or hollow tabular light guides.

But, in modern urban reality of dense city` development, the income of natural light through the elements of natural lighting system might be seriously reduced by the screening effect of surrounding high-rise buildings. For the time being, such an situation is not considered by modern codes and regulations on natural lighting and these are require to be actualize [1,2,3,4].

This determines the basic objectives of the said scientific investigations, which can be determines as "creation of qualitative natural lighting environment in premises of deepened buildings and underground structures under the screening effect of a surrounding high-rise buildings [5,6,7,8,9,10,11,12].

2 Literature review

The researches, described in this article are based on general conceptions, data and conclusions of principal works of leading researchers in the scientific field discussed, namely: A.K. Soloviev [7,13,17,20], R.Kittler [15], V.M. Slcikin [8,18], J. Mohelnikova [16], L. Brotas and M.Wilson [9,10], as well on works of the given article` authors and their postgraduates [5,6,14].

3 The researching` methods

The investigations were conducted in theoretical and practical form. Experimental studies were conducted in National Research University «Moscow State University of Civil Engineering». The studies were based on deepened block of laboratory of building physics, attached to a high rise buiding of the said university, as shown on Figure 1.

The tests were made under diffuced natural lighting, conditions which were provided by the cloudy sky, and corresponds` to the "Standard overcast skyvault of C.I.E.". During the experiments, the cloudiness percentage of the sky varies from 80 to 100%, provided the clouds were dense and low. Under the investigations of a real deepened premise all design geometric and lighting parameters were taken as a constant. The natural lighting of the interior in this case was provided with a single skylight of a pyramid shape. On the results of this stage, an working hypothesis was offered. The working hypothesis adopted says, that under universal character of an exteral natural light destribution in a space, the universal character of an internal natural light distribution in interiors must take place. Such a suggestion might be used to elaborate an universal method of a daylight factor` (D.F.) calculation, in spite of different systems of natural lighting of interiors adopted [5, 6].

On the base of this hypothesis and pilot stage of investigations, the expresion for a D.F.` design under a roof system of natural lighting with concern to screening effect of surrounding development was offered. Such a formula is based on the requirements of the modern «Codes and Regulations» [2,3,4]:

$$\mathrm{D.F.}_{\mathrm{RL}}^{\mathrm{DES}} = \left[\varepsilon_{\mathrm{RL}} \cdot \mathrm{q} + \varepsilon_{\mathrm{OB}} \cdot \mathrm{K}_{\mathrm{OB}} \cdot \mathrm{b}_{\mathrm{f}} + \varepsilon_{\mathrm{AV}}\left(\mathrm{r}_2 \cdot \mathrm{K}_{\mathrm{M,SL}} - 1\right)\right] \frac{\tau_{\mathrm{G}}}{\mathrm{K}_{\mathrm{S}}} \quad (1)$$

Where the shortenings read:

Des- designed; R.L. – roof lighting; OB- opposite building; F- façade; AV- average; M,SL – monitor, skylight; G- general; S-spare. All the factors within the formula, except ε_{OB} , K_{OB} and b_f are the factors used, according the «Codes and Regulations» in the design formula for D.F. calculation under system of side natural lighting [1-6].

4 Results

The results of experimental investigations show a pretty good matching with the data of theoretical calculations according to the offered formula (1), and this proves the truth and correctness of the concept adopted.

The results show a significant reduction of D.F. values in the situation considered, due to screening effect of opposite building.

Besides, the conclusion was made, that a complex checking of a proposed theory must be conducted, with use of a vast number of real buildings, as well as under an "artificial skyvault" dome in laboratory investigations.

All the major concepts of the work are based on the results and conclusions of former investigations, which, within the problems discussed, were carried out by domestic and foreign scientists. [11,12,15,16,19,20].

As an example of screening effect of an opposite development, determined according the results of experimental and theoretical studies, the graphs of D.F. are shown on Figure 1, and on Figure 2. Hypothetical situation of absence of the opposite building, values are represented in Table 1.

5 Conclusions and recommendations

1. The working hypothesis, used as scientific base of the investigations conducted, showed its justice. Field studies on real object installation showed a satisfactory matching of the results resieved with theoretical calculations of D.F. under system of roof natural lighting adopted, with consideration of screening effect of surrounding development, determined according the new design formula (1).

2. The work showed a possibility of correction of basic concepts of «Codes and Regulations on natural lighting», acting for the time being. In this case, the given investigations can be considered as innovative, but requiring future studies for more complex background to review the documents in question.

3. It is shown, that lighting-technique effect of opposite buildings on the D.F. levels in premises of deepened buildings and underground structures with root lighting system of a "skylight" type, has a screening and shadowing character. It is so, because the light reflection from facades of surrounding high-rise buildings, show in formula (1) as a multiplication of «K_{OB}» and «b_f» factors is much less, than «q» factor, which expresses the non-uniform luminance of a skyvault.

Table1. Results of experimental and theoretical studies.

# of points	Theoretical values of a D.F. in a real deepened premise, %	Mean Theoretical values of a D.F. in a real deepened premise, %	Experi-mental values of a D.F. in a real deepened premise, %	Mean Experi-mental values of a D.F. in a real deepened premise, %	Theoretical values of a D.F. in a real deepened premise. In a hypothetical situation of an opposite building absence, %	Mean Theoretical values of a D.F. in a real deepened premise. In a hypothetical situation of an opposite building absence, %
1	3,08	3,106	2,9	2,74	5,11	3,702
2	5,27		4,5		5,56	
3	3,89		3,7		4,11	
4	2,0		1,8		2,22	
5	1,29		0,8		1,51	

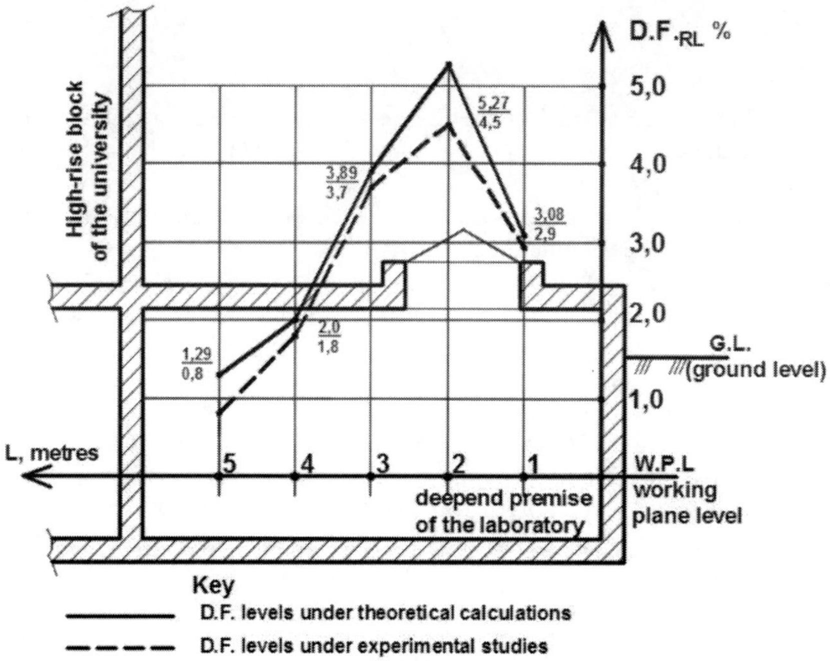

Fig. 1. Theoretical and experimental values of a D.F. in a real deepened premise. In a situation of opposite-standing high-rise building.

Note. In the case considered, the theoretical calculation of a D.F. values are conducted on the formula (1).

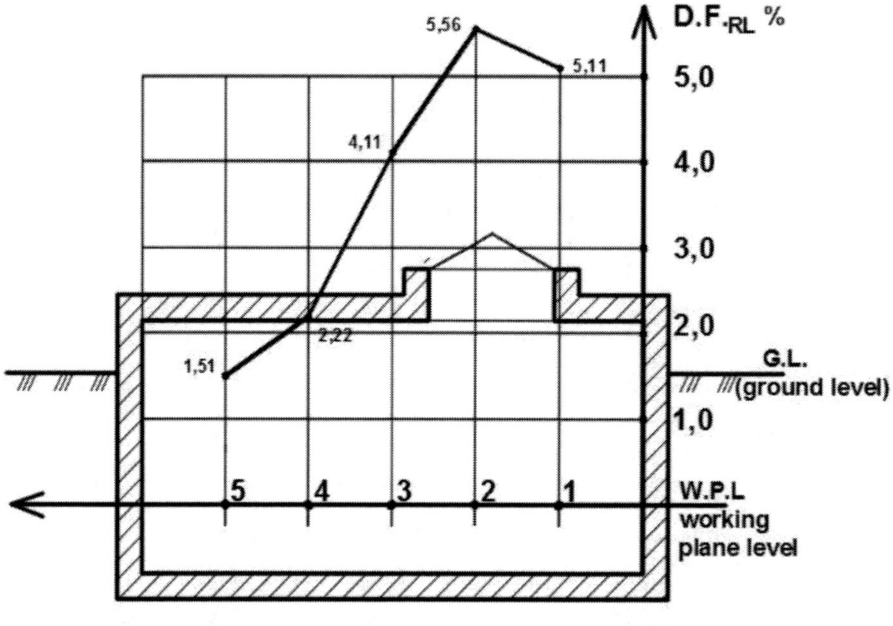

Fig. 2. Theoretical values of a D.F. in a real deepened premise under hypothetical situation of an high-rise opposite building absence.

Note. The theoretical calculation of a A. D.F. in this case is conducted on the standard formula:

$$D.F._{\cdot RL}^{DES} = \left[\varepsilon_{RL} \cdot q + \varepsilon_{AV}\left(r_2 \cdot K_{M,SL} - 1\right)\right] \frac{\tau_G}{K_S}. \quad (2)$$

References

1. Sanitary rules and standards SanPin 2.2.1/2.1.1.1278-03 (2003)
2. Buiding Codes and Regulations SNiP 23.05.95* (2005)
3. Set of Regulations SP 23-102-2003 (2005)
4. Set of Regulations SP 52.1330.2011 (2011)
5. S.V. Stetsky, K.O. Larionova, Industrial and Civil Construction, **3**, 69-73 (2015)
6. S.V. Stetsky, K.O. Larionova, Industrial and Civil Construction, **11**, 77-80 (2015)
7. A.K. Soloview, Light & Engineering, **25**, 88-93 (2017)
8. V.M. Slukin, E.S. Simakova, Academic Gerald of Ural NII proekt of RAASN, **2**, 56-60 (2010)
9. L. Brotas, M. Wilson, PLEA, 207-212 (2006)
10. L. Brotas, M. Witson, Svetotechnika, **3**, 44-47 (2008)
11. G.D. Gritsenko, Y.F. Kasyanov, Svetotechnika, **3**, 27-30 (2016)
12. A.L. Kuznetsov, E.Y. Oselesets, A.K. Solovies, M.V. Stolyarov, Svetotechnila, **6**, 4-11 (2011)
13. A.K. Soloview, Light & Engineering, **17**, 59-73 (2009)
14. S.V. Stetsky, Indusrial and civil construction, **3**, 70-72 (2014)
15. S. Darula, R. Kittler, Svetotehnika, **1**, 28-34 (2006)
16. J. Mohelnikova, Svetotehnika, **3**, 26-30 (2008)
17. A.K. Soloviev, Industrial and civil construction, **2**, 53-55 (2007)
18. V.M. Slukin, E.S. Simokova, Academic Gerald of Ural NIIproekt of RAASN, **4**, 75-77 (2011)
19. M. Knoop, The royal Danish academy of fine arts, school of architecture (2013)
20. A.K. Soloviev, Light & Engineering, **25**, 23-30 (2017)

Methods of mathematical modeling using polynomials of algebra of sets

Alexandr Kazanskiy [1,*] and *Ivan* Kochetkov [1]

[1] Moscow State University of Civil Engineering, Yaroslavskoye sh. 26, Moscow, 129337, Russia

Abstract. The article deals with the construction of discrete mathematical models for solving applied problems arising from the operation of building structures. Security issues in modern high-rise buildings are extremely serious and relevant, and there is no doubt that interest in them will only increase. The territory of the building is divided into zones for which it is necessary to observe. Zones can overlap and have different priorities. Such situations can be described using formulas algebra of sets. Formulas can be programmed, which makes it possible to work with them using computer models.

1 Introduction

At present, high-rise construction in our country is gaining wide scope. In this regard, there are also issues of safety of maintenance of such buildings. This is due not only to fire safety, but also to possible criminal and environmental threats. Proceeding from what has been said, in this article we will consider the creation of a discrete model based on methods of set theory to solve the problems of securing high-rise buildings.

The basic concepts of set theory were introduced by Dedekind in 1871 and then by his friend George Cantor. Operations on sets make it possible to form new sets of initial sets. This enables the use of sets for various practical tasks, such as the classification of any objects, analysis of various types of sociological research, analysis of data from which it is necessary to select data characterized by specific properties.

A set may be viewed as collection of objects, the elements or members of the set. We ordinarily use capital letters, A, B, C,…, to denote sets, and lowercase letters a, b, c,..., to denote elements of sets. In any application of the theory of sets, the members of all sets usually belong to some fixed large set called the universal set. A universal set is usually denoted by the symbol U. All sets under consideration at a particular time are subset of a fixed universal set U.

Choosing from the set M some elements, one can obtain a new set S that is a part of the set M then S is called a subset of M. We also say that S is contained in M or that M contains S.

If all sets are considered at a certain time and are subsets of a fixed universal set U, then we can define a universal complement, or simply a complement of the set A, denoted by As, as the set of elements that belong to U, but do not belong to A, that is,

* Corresponding author: 1940@gmail.com

$$A^c = \{x: x \in U, x \notin A\}$$

2 Methods

Sets with intersection operations, unions and complements that satisfy abstract laws [1], form an algebraic system, called the algebra of sets. This algebra is isomorphic to Boolean algebra and therefore often uses the ideas and terminology of Boolean algebra, but it should be noted that this terminology is not completely standardized, which sometimes leads to different names for the same concepts. Let us consider some concepts in more detail.

Let there be n variables, each of which defines a certain set. The expression for the algebra of sets E (or a formula) is an expression composed of these variables, connected by means of operations of union, intersection and complement, for example

$E_1 = A \cap (B^C \cap C)^C \cup (A \cap B^C \cap C^C)^C$,

$E_2 = A \cap (B^C \cap C) \cup (A \cap B^C \cap C^C)$,

$E_3 = (A \cap B^C \cap C) \cup (A \cap B^C \cap C^C)$,

$E_4 = (A^C \cup B^C) \cap (A^C \cup B^C \cup C) \cap (B^C \cup C)$,

$E_5 = (A^C \cup B \cup C) \cap (A \cup B^C \cup C) \cap (A^C \cup B^C \cup C)$

are expressions from three variables A, B and C.

Consider the following example. Let there be a list of employees of the enterprise who have access to the first object of a certain building (let's denote the list of their names as the set A). Let there also be a list of all those who have access to the second object (set B) and all those who have access to the third (set C). If now you need information about those who have access to the first and second objects simultaneously, then it will be necessary to select those who are simultaneously included in the first and second lists. For this, we need to find a new set that is the intersection of the original sets A and B, i.e. find the set $A \cap B$. But this set does not contain information about whether these employees have access to the third object. Clearly, for this it is required to find another set that is the intersection of all three sets, i.e. the set $A \cap B \cap C$. Suppose now that we need to make a list that contains the names of employees who have access to the first and second objects, but do not have a third entry. In this case it is necessary to find the set $A \cap B \cap C^c$. Considering such cases leads to the concept of fundamental products of the sets.

Suppose that there are n distinct sets $A_1, A_2, A_3, \ldots, A_n$. The fundamental product of sets is a set of the form

$$A_1^* \cap A_2^* \cap A_3^* \cap \ldots \cap A_n^*$$

where A_i^* is either A_i, or A_i^c. We also note that

(1) There are exactly 2^n such fundamental products.

(2) Any two such fundamental products do not intersect.

(3) The universal set is the union of all such fundamental products

A variable is called *a literal* or a complement to a variable, for example A, A^C, B^C, etc. A product is a literal or an intersection of two or more literals, such that no pair of them contains the same variable. For example, $B^C \cap C$, $A \cap B^C \cap C^C$, A, C^C - are products, and the expressions $A \cap B^C \cap A^C$ and $A \cap B^C \cap B^C$ - are not. We note that any intersection of literals is always reduced to either Ø (empty set) or to the product. So, for example $A \cap B^C \cap A^C = \emptyset$, because $A \cap A^C = \emptyset$ (by the law of the complement), and the intersection $A \cap B^C \cap B^C = A \cap B^C$, because $B^C \cap B^C = B^C$ (by the law of the idempotency).

If for n variables the product consists of n literals, then it is called a fundamental product (some authors call any product a fundamental product).

An expression that is the union of different products, if there is not a single work in it that is included in another work, is called the sum of products, or the normal form of the union of intersections, or a polynomial in normal form. From the previous example, E_1 is simply an expression, E_2 and E_3 are expressions in the normal form of the union of intersections.

If the expression consists of a union of fundamental products, then such an expression is called the complete normal form of the union of intersections, or a polynomial in the canonical form. The polynomial E_3, in addition to having a normal form of intersection, is also a polynomial having a complete normal form of union of intersections.

Similarly, if for n variables the union consists of n literals, then it is called a fundamental union, and if you replace the union operations with the intersection operations, and the intersection operations in the join operation, you can get the normal intersection form and the complete normal intersection form. The expression E_4 has the normal form of intersection of unions, and E_5 is the complete normal form of intersection of unions [2].

Any expression can be converted to an equivalent expression that has a normal form. One of the differences of the normal form is that in this expression the complement operation is applied only to variables. To get rid of the complements applied to the expressions, we must apply the De Morgan's laws.

3 Results

Algorithm 1 for converting the expression to the normal form of combining intersections.

Let there be an initial expression for the algebra of sets E.

Step 1. Using De Morgan's laws and involutions, we give each parenthesis to which the addition operation is applied, to the form in which the addition operation applies only to variables.

Step 2. Using the distributive law of a union with respect to the intersection, we break up the brackets containing the unions of literals with respect to the intersection operations.

Step 3. Using the associative laws, complements and idempotency, we transform every intersection of literals either in Ø or in a product.

Step 4. Using the laws of absorption and identity, we simplify the expression E and if it consists of a union of intersections, then the normal form is obtained, if not, then go to Step 2.

The laws of absorption for the algebra of sets have the form

$A \cap (A \cup B) = A$

$A \cup (A \cap B) = A$

Example 1.

We apply this algorithm to transform to the normal form of the following expression

$E = ((A \cap B^C) \cup (B \cap C^C)^C) \cap (((B \cap C) \cup (A^C \cap C))^C \cup (A \cap B))$.

Step 1. Using De Morgan's laws and involutions, we obtain

$E = ((A \cap B^C) \cup B^C \cup C) \cap ((B^C \cup C^C) \cap (A \cup C^C) \cup (A \cap B))$.

Step 2. Using the law of distributive, we expand the brackets in the right side of expression

$E = ((A \cap B^C) \cup B^C \cup C) \cap ((A \cap B^C) \cup (B^C \cap C^C) \cup (A \cap C^C) \cup (C^C \cap C^C) \cup (A \cap B))$.

Step 3. Let's transform the intersection of literals into a product

$E = ((A \cap B^C) \cup B^C \cup C) \cap ((A \cap B^C) \cup (B^C \cap C^C) \cup (A \cap C^C) \cup C^C \cup (A \cap B))$.

Step 4. Since the B^C is included in the $A \cap B^C$, then the $A \cap B^C$ is absorbed, the C^C is also included in the $B^C \cap C^C$ and $A \cap C^C$, so both these intersections are absorbed and the expression E takes the following form

$E = (B^C \cup C) \cap ((A \cap B^C) \cup C^C \cup (A \cap B))$.

Here again, we need to discard the brackets, so go to Step 2.

Step 2. Open the brackets and get

$E = (A \cap B^C \cap B^C) \cup (B^C \cap C^C) \cup (A \cap B \cap B^C) \cup (A \cap B^C \cap C) \cup (C \cap C^C) \cup (A \cap B \cap C)$.

Step 3. Let's transform the intersection of literals into a product

$E = (A \cap B^C) \cup (B^C \cap C^C) \cup \emptyset \cup (A \cap B^C \cap C) \cup \emptyset \cup (A \cap B \cap C)$.

Step 4. The intersection $A \cap B^C$ is included in $A \cap B^C \cap C$, so the latter is absorbed and the normal form for E has the form

$E = (A \cap B^C) \cup (B^C \cap C^C) \cup (A \cap B \cap C)$.

Let us consider the complete normal forms.

It should be noted that, by analogy with Boolean algebra, complete normal forms are sometimes called perfect normal forms. Consider the expression $E = E(x_1, x_2, \ldots, x_n)$, consisting of a union of products, i.e. represented in normal form. If every product consists of exactly n literals, then such an expression is called the complete normal form of the union of intersections.

Statement. Any expression of the algebra of sets can be transformed to an equivalent expression in its complete normal form, and such a representation is unique [3].

Proceeding from what was said above, one can see that any expression of the algebra of sets is transformed to an equivalent expression in normal form. Next, consider an algorithm that allows you to transform this expression into an equivalent expression in its full normal form. The idea of this algorithm is that if a product P in the expression E does not contain the i-th variable, then it can be carried into E, forming the product $P \cap (x_i \cup x_i^c)$ for $i \leq n$.

Algorithm 2 for converting the expression to the full normal form of combining intersections.

Step 1. Let there be an expression $E = E(x_1, x_2, \ldots x_n)$ represented in normal form. We find the product P in the expression E, which does not contain the i-th variable and form the product $P \cap (x_i x_i^c)$. This does not violate the equivalence of the expression, since $(x_i \cup x_i^c) = U$, and $P \cap U = P$. Remove duplicate products (this is possible since $P \cup P = P$).

Step 2. Repeat Step 1 until each product in E becomes a fundamental product, i.e. each product will not include all n variables.

Example 2.

We apply this algorithm for the expression E in the normal form obtained in Example 1 to convert it to the full normal form.

$E = (A \cap B^C) \cup (B^C \cap C^C) \cup (A \cap B \cap C)$.

Step 1. Find the product $A \cap B^C$, which does not contain the variable C and form the product $(A \cap B^C) \cap (C \cup C^C)$, we get

$E = (A \cap B^C) \cap (C \cup C^C) \cup (B^C \cap C^C) \cup (A \cap B \cap C) =$
$= (A \cap B^C \cap C) \cup (A \cap B^C \cap C^C) \cup (B^C \cap C^C) \cup (A \cap B \cap C)$.

Step 2. Since in E there is a product $B^C \cap C^C$ that does not contain the variable A, then we go back to Step 1 and form the product $(B^C \cap C^C) \cap (A \cup A^C)$,

$E = (A \cap B^C \cap C) \cup (A \cap B^C \cap C^C) \cup (B^C \cap C^C) \cap (A \cup A^C) \cup (A \cap B \cap C)$.

After the parentheses are opened in E, two identical fundamental products are formed $A \cap B^C \cap C^C$. One of them must be removed. Now E is represented in the full normal form:

$E = (A \cap B^C \cap C) \cup (A \cap B^C \cap C^C) \cup (A^C \cap B^C \cap C^C) \cup (A \cap B \cap C)$.

Thus, if there is an expression represented in normal form, then this algorithm allows algebraic transformations to bring it to the full normal form. However, this method is not the only one. For the same purpose, we can also use the Venn diagrams.

Let's consider an example. Let there be a partition of the set U shown in Fig 1. We select the set given by the formula (the normal form of the union of intersections) $A \cup (B^C \cap C)$. It defines the union of two sets: A = [4] and a set representing the intersection of the sets B^C and C, i.e. of the set $B^C \cap C = \{1, 5\}$. Therefore, the set $A \cup (B^C \cap C) = \{1, 4, 5, 6, 7\}$.This set is shaded in Fig. 1 It can be seen from the diagram that the set $A \cup (B^C \cap C)$ is given by the union of five fundamental products: to the set {4} there corresponds the fundamental product $A \cap B^C \cap C^C$, the set {6} - $A \cap B \cap C^C$, the set {7} - $A \cap B \cap C$, the set {5} - $A \cap B^C \cap C$ and the set {1} - $A^C \cap B^C \cap C$. The union of these fundamental products and gives a complete normal form

$$(A \cap B^C \cap C^C) \cup (A \cap B \cap C^C) \cup (A \cap B \cap C) \cup (A \cap B^C \cap C) \cup (A^C \cap B^C \cap C).$$

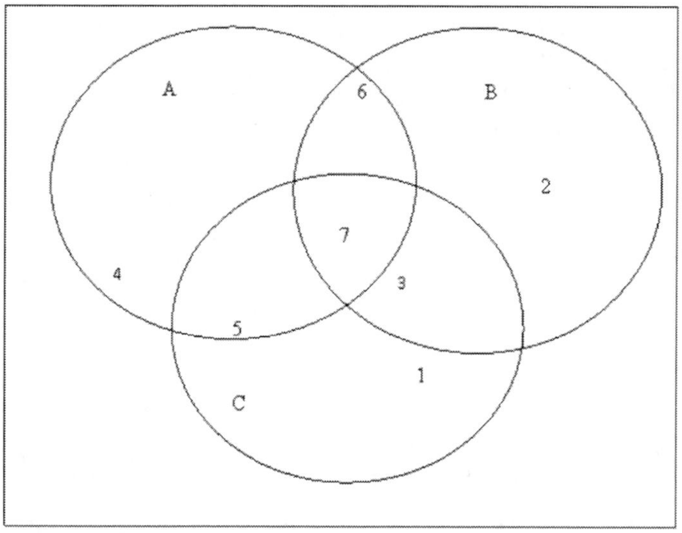

Fig. 1. Partition of the set.

We note that for any set there exists not only a single complete normal form of the union of intersections, but the only complete normal form of intersection of unions of sets. Each of these forms can be found in two ways. So for the set from the previous example $A \cup (B^C \cap C)$ we open the brackets and get an expression in the normal form of intersection of unions $(A \cup B^C) \cap (A \cup C)$. In the first bracket there is no variable C, and in the second variable B. Since the expression $(C \cap C^C) = \varnothing$, the following expression is equivalent to the original

$$((A \cup B^C) \cup (C \cap C^C)) \cap ((A \cup C) \cup (B \cap B^C)) =$$
$$= (A \cup B^C \cup C) \cap (A \cup B^C \cup C^C) \cap (A \cup B \cup C) \cap (A \cup B^C \cup C)$$
$$= (A \cup B^C \cup C) \cap (A \cup B^C \cup C^C) \cap (A \cup B \cup C).$$

The last expression is the complete normal form of intersection of unions for the original formula.

You can also find this expression using the Venn diagram. Since the elements {0, 2, 3} do not enter the initial set {1, 4, 5, 6, 7}, it is necessary to form the intersection of such unions that do not contain these three sets. The union $A \cup B \cup C$ does not contain the element {0}, the union $A \cup B^C \cup C$ does not contain the element {2} and the union $A \cup$

$B^C \cup C^C$ does not contain the element {3}. Hence, forming an intersection of them, one can obtain the complete normal form of intersection of unions of sets.

$$(A \cup B \cup C) \cap (A \cup B^C \cup C) \cap (A \cup B^C \cup C^C).$$

4 Discussion

Abstract algebra deals with the study of operations performed on certain elements. By now, the ideas of abstract algebra are used not only for mathematical methods, but also allow us to obtain practical results. The union, intersection, and complement operations performed on sets satisfy certain laws (or identities) and form an algebra of sets. Since the numerical algebra appeared earlier, the question arises which of the operations (intersection or union) is "similar" to the operation of adding numbers and which is to the operation of multiplication [5]. It is hardly possible to answer this question. For numbers, for example, only the distributivity of multiplication with addition is satisfied, and in the algebra of sets, two distributivity laws are considered: intersections with respect to union and union with respect to intersection.

Important in the execution of operations is their priority. First, the addition operation is performed, then the intersection and then the join.

The set can be given by different formulas. Although these formulas look different, but they all are equivalent in the sense that they define the same elements of a given set. For example, suppose there are two expressions in normal form

$$E_1 = (B \cap C) \cup (A^C \cap C^C),$$

$$E_2 = (B \cap C) \cup (A^C \cap B) \cup (A^C \cap B^C \cap C^C).$$

These formulas are equivalent, which is easy to verify if one transforms each of them to a complete normal form, which is the same for both E_1 and E_2

$$(A \cap B \cap C) \cup (A^C \cap B \cap C) \cup (A^C \cap B \cap C^C) \cup (A^C \cap B^C \cap C^C).$$

In order to determine which of the equivalent formulas is "simpler" we introduce the following notation. Let E be an expression in normal form and let L (E) be the number of literals in this expression (all occurrences are assumed) and F (E) is the number of products from which is formed E. For E_1, the value $L(E_1) = 2 + 2 = 4$ and $F(E_1) = 2$, and $L(E_2) = 2 + 2 + 3 = 7$ and $F(E_2) = 3$.

Let E_1 and E_2 be equivalent expressions in normal form. Then E_1 is simpler than E_2 if

$$L(E_1) < L(E_2) \text{ and } F(E_1) \leq L(E_2) \text{ or}$$

$$L(E_1) \leq L(E_2) \text{ and } F(E_1) < L(E_2).$$

The expression E, represented in normal form, is called *minimal* if there is no other equivalent expression that is simpler than E. It should be noted that there may be more than one equivalent minimal expression.

As shown in [6], for each formula there exists its minimal representation, which is very important when using similar discrete models for real applied problems. In the above article, we showed in detail how the minimization of formulas for the algebra of sets is performed using the methods of graph theory.

5 Conclusions

Proceeding from the above, one can see that the technology of using the algebra of sets formulas opens a very promising way of constructing discrete models in various directions of development in construction and economy

References

1. Buiding Codes and Regulations SNiP 23.05.95* (2005)
2. A.A. Kazansky, L.V. Larina, Informatics, **15(304)** (2001)
3. A.A. Kazansky, L.V. Larina, Informatics, **14 (255)**, (2000)
4. A. Zaychenko, S. Gutman, O. Kalinina, Advances in Intelligent Systems and Computing, **692**, 453-462 (2018) DOi - 10.1007/978-3-319-70987-1_48
5. A.A. Kazansky, Modern science success, **1**, 6 (2017)
6. O.V. Kalinina, M.V. Lopatin, Actual Problems of Economics, **182**, 392-405 (2016)

A composition for protection the stylobate in high-rise construction from the harmful effects of car exhaust gases

Irina Sokolova [1,*]

[1] Moscow State University of Civil Engineering, Yaroslavskoye shosse 26, Moscow, 129337, Russia

Abstract. In large cities, high-rise buildings are usually located along highways with heavy traffic. The study was carried out with the aim of creating a material for protection the stylobate of a high-rise building from the harmful effects of car exhaust gases. A polymer-silicate composition based on schungite and schungisite components is proposed. The composition has the properties of a wall material resistant to the corrosive environment of car exhaust gases. The results of the composition studies are presented. The possibility of increase the durability of exterior slabs for stylobate walls of high-rise buildings is substantiated, provided the proposed material is applied.

1 Introduction

High-rise buildings and complexes are being built, united by stylobates. Walls of stylobates are faced with slabs made of artificial stone, natural stone, or concrete. Increasingly, along motorways, the content of exhaust car fumes (CO, NO, NO2, CH, SO2, etc.) in air is several times higher than the maximum allowable concentrations. When interacting with air moisture, these substances form a generally acidic wet gas environment. This aggressive environment exerts a destructive effect on the material of the stylobate facing. Theoretical justification of the process of destruction of materials for finishing the facades of buildings under the influence of automobile exhausts is given in the work [1]. There is a need to protect the cladding material from corrosion.

2 The object of the study

The object of research is the material for facing the stylobate of a high-rise building. This is a polymer-silicate composition with schungite and schungisite components. Polymersilicate was chosen because this material has long been used to work in acid corrosive environments [2]. Shungite is a natural carbonaceous mineral with high acid resistance [3, 4]. Shungisite, obtained as an easy filler based on shungite rocks, is also an acid-resistant material. In addition, an important factor is the proven positive impact of schungite for the human body. Small and large fractions of schungite and shungizite were used. The

* Corresponding author: i.socolova@yandex.ru

compositions have been developed with a density of 1000 to 1200 kg / m3. The compositions differed in the size of the schungisite fractions and in the type and size of fractions of small fillings based on schungite and schungisite. The physical and mechanical properties of the developed compositions are given in the article [5, 6]. The studies were carried out according to standard methods.

This article presents the results of research on the operational properties of the material. Studies were carried out in accordance with the requirements to materials for facing facades of civil buildings The studies were carried out according to standard methods [7, 8].

3 Results

The results of testing the material to water absorption are shown in Table 1.

Table 1. Water absorption of the material by volume,%

Time, day	Compositions		
	1	2	3
2	6,0	5,5	5,8
30	10,0	9,6	9,8

It follows from Table 1 that the water absorption of the material is much lower than the water absorption of expanded clay concrete and other lightweight concretes on a cement and a polymer-silicate binder.
The results of the acid resistance tests are given in Table 2.

Table 2. The results of the acid resistance tests.

Compositions	Compressive strength of control samples, R, MPa	Concentration of sulfuric acid in s solution,%	Test period, days		
			30	270	360
1	12,8	4	0,91	0,96	0,92
		24	0,94	0,97	0,97
2	12,0	4	0,89	0,80	0,81
		24	0,99	0,82	0,84
3	16	4	0,89	0,88	0,87
		24	0,90	0,97	0,98

Figure 1 shows the results of study of moisture absorption by the material for 30 days.

All compositions have a dense structure. The results of the studies showed that the acid resistance coefficient of the proposed compositions is from 0.8 to 0.99. These indicators correspond to the requirements of regulatory documents and even exceeds these requirements. It was influenced by high acid resistance of shungizite, which is about 96%. According to Table 1, it can be concluded that the chemical resistance coefficient increases with the concentration of the sulfuric acid solution. Structure of the material becomes denser. The formation of a dark contour along the perimeter of the cross section of the sample indicates the process of compaction of the structure.

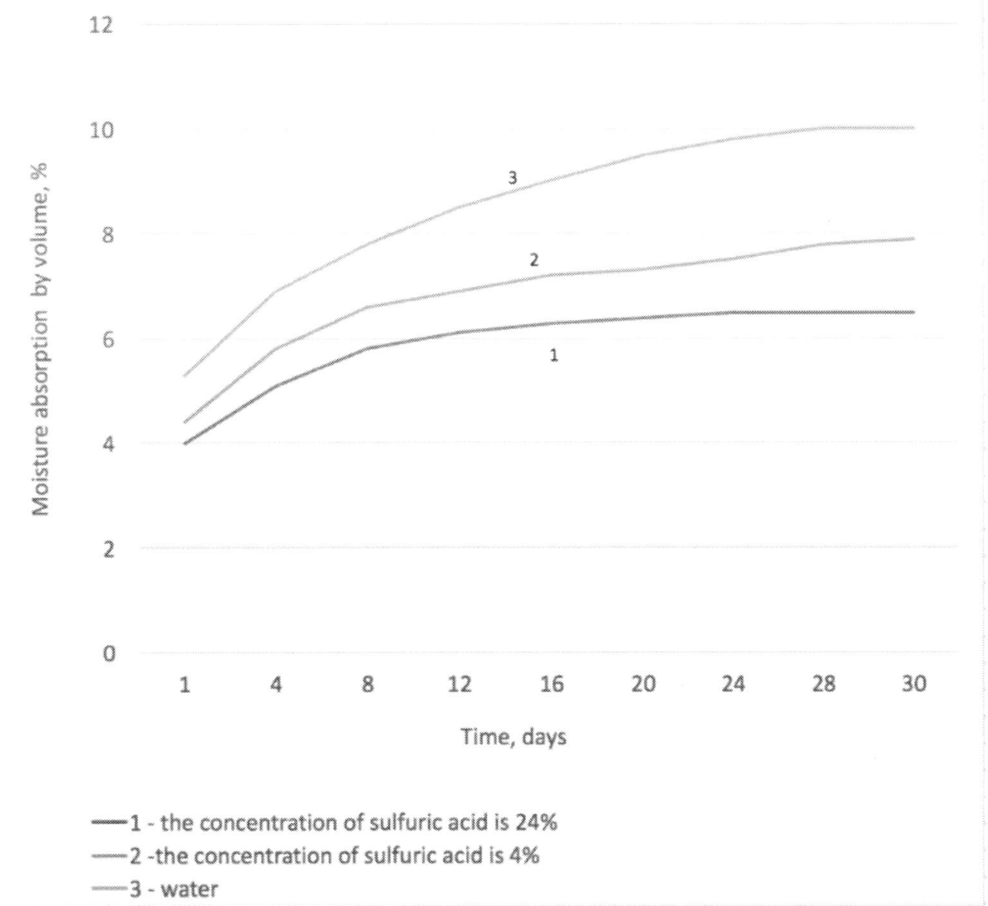

Fig. 1. The results of moisture absorption by the material for 30 days.

No changes in the samples were observed after two years of testing in a solution of sulfuric acid (cracks, chips, shells, blisters). Therefore slabs can be used without additional surface protection. According to water absorption, as an indirect indicator of penetration, the compounds meet the requirements of the standards for lightweight concrete for use in an acidic wet gas environment. The coefficient of stability of the composition in water is 0.83-0.86 at a rate of 0.8.

To conduct tests on weather resistance, the samples were installed on the site close to the highway with intensive traffic. After three years of testing, the compressive strength of the samples was determined. The coefficient of weather resistance after three years of testing was 0.91-0.96. The surface condition of the samples after three years of testing was satisfactory.

5 Conclusion

As a result of the research it was established that the polymersilicate composition with schungite and schungisite components satisfies the requirements for facing the stylobate part of the high-rise building. It is concluded that the use of the developed and investigated composition will increase the durability of the stylobate facing without an additional protective layer. Facing lits from this material are proposed for use in the construction, repair and reconstruction of stylobates of high-rise buildings.

References

1. O. Kalinina, O. Valebnikova, Advances in Intelligent Systems and Computing, **692**, 1315-1322 (2018) DOi - 10.1007/978-3-319-70987-1_139
2. A. Zaychenko, S. Gutman, O. Kalinina, Advances in Intelligent Systems and Computing, **692**, 453-462 (2018) DOi - 10.1007/978-3-319-70987-1_48
3. N.B. Yastrebova, Stroitel'nyye materialy, oborudovaniye, tekhnologii XXI veka (2003)
4. N.B. Ivashchenko, I.L. Pavlova, M.P. Kochergina, Nauchnoye obozreniye, **16**, 93-98 (2015)
5. M.M. Filippov, Gornyy zhurnal, **5** (2012)
6. I.V. Sokolova, Izvestiya VUZov, **4(370)** (2017)
7. Russian Standard STO 221 NOSTROY 2.14.132-2015
8. I.V. Ilin, A. Lepekhin, A.I. Levina, O.Yu. Iliashenko, Advances in Intelligent Systems and Computing, **692**, 1306-1314 (2018) DOi -10.1007/978-3-319-70987-1_138

Behavior of sandwich panels in a fire

Eugenia Chelekova [1,a*]

[1] National Research University "Moscow State University of Civil Engineering, Russia, 129337, Moscow, Yaroslavskoye shosse, 26

Abstract For the last decades there emerged a vast number of buildings and structures erected with the use of sandwich panels. The field of application for this construction material is manifold, especially in the construction of fire and explosion hazardous buildings. In advanced evacu-ation time calculation methods the coefficient of heat losses is defined with dire regard to fire load features, but without account to thermal and physical characteristics of building envelopes, or, to be exact, it is defined for brick and concrete walls with gross heat capacity. That is why the application of the heat loss coefficient expression obtained for buildings of sandwich panels is impossible because of different heat capacity of these panels from the heat capacities of brick and concrete building envelopes. The article conducts an analysis and calculation of the heal loss coefficient for buildings and structures of three layer sandwich panels as building envelopes.

1 Introduction

Three-layer sandwich panels are well-known in the construction of various facilities. Supermarkets, terminals, industrial buildings in the oil and gas industry are built using sandwich panels.

It is necessary to consider separately the behavior of sandwich panels in a fire. This is due to their distinctive features. If the fire appears small thermal inertia of the panels.

2 Materials and Methods

The time allotted for the evacuation of people in case of fire, largely depends on heat loss in the building envelope. Currently, these losses are determined by the ratio of heat losses [1]. The method of determining this factor is designed to brick and concrete walling [1-2] and without reason, applied to sandwich panels. In this article we attempt to determine the coefficient of heat loss from the hot gases in the sandwich panel. The study is limited to the initial stages of fire development, which lasts no more than 20 minutes. When the fire heats up only the metal sheet in contact only with hot gases. The flow of heat in the insulation at

* Corresponding author: Kiara_lion@mail.ru

the considered stage of the fire is missing. Sheet metal thickness δ=0.5-1 mm is thermally thin body and its temperature is uniform in thickness. The temperature change of the sheet in this case is described by the equation:

$$t_1 \frac{dT}{dt} = (T_s(t) - T(t) \qquad (1)$$

$t_1 = \dfrac{\delta \rho_m C_{pm}}{\alpha}$ - the characteristic time of temperature change of the sheet, δ - is the thickness of the sheet, C_{pm} and ρ_m – heat capacity the density of the sheet material, α -is coefficient of heat transfer from the gaseous medium to the structure, $T_s(t)$ - is the volumetric average temperature of the gas environment in the fire, $T(t)$ - is the temperature of the sheet is determined by:

$$T(t) = T_0 e^{-\frac{t}{t_1}} + \frac{1}{t_1}\int_0^t T_s(\tau) e^{-\frac{t-\tau}{t_1}} d\tau \qquad (2)$$

T_0 – the initial temperature of the medium and the sheet. Consider the temperature change of the environment according to the law:

$$T_c(t) = T_0\left(1 + \frac{t}{t_s}\right) \qquad (3)$$

which corresponds to the first approximation for the case of a fire with a constant burning area. [3].

Where $t_s = \dfrac{V_0 \rho_0 T_0 C_p}{Q_n \psi F_g (1-\varphi)}$ the characteristic time of measurement of the temperature of the gaseous medium in this case. V_0 - is the volume of the room, ρ_0 - is the initial density of the gaseous medium, C_{p0} – heat capacity of air, Q_n - is the heat of combustion of combustible material, ψ - specific rate of burnout, F_g – the area of the fire, φ(t) – coefficient of heat loss to the time t:

$$\varphi = \frac{q_{\Sigma(t)}}{Q_{P(t)}} \qquad (4)$$

$q_{\Sigma(t)}$ - is the amount of heat transferred from the environment in the building envelope area F_P - time t. Q_p - is the energy released by fire for time t.

$$T(t) = \left(\frac{t_1}{t_s} e^{-\frac{t}{t_1}} + 1 - \frac{t_1}{t_s} + \frac{t}{t_s}\right) T_0 \qquad (5)$$

$q_{\Sigma(t)} = Cp_m M\left(T(t) - T_0\right)$;

$M = F_P \rho_m \delta$ - is the total mass of the metallic inner coating. As a result, the ratio of heat loss is equal to:

$$\varphi = \frac{Cp_m M\left(1 + \frac{t_1}{t} e^{-\frac{t}{t_1}} - \frac{t_1}{t}\right)}{V_0 \rho_0 + C_{pm} M\left(1 + \frac{t_1}{t} e^{-\frac{t}{t_1}} - \frac{t_1}{t}\right)} \qquad (6)$$

Figure 1 presents the dependence of coefficient of heat losses for different values of the parameter: $P = \frac{C_{pm}M}{V_0\rho_0 C_p}$.

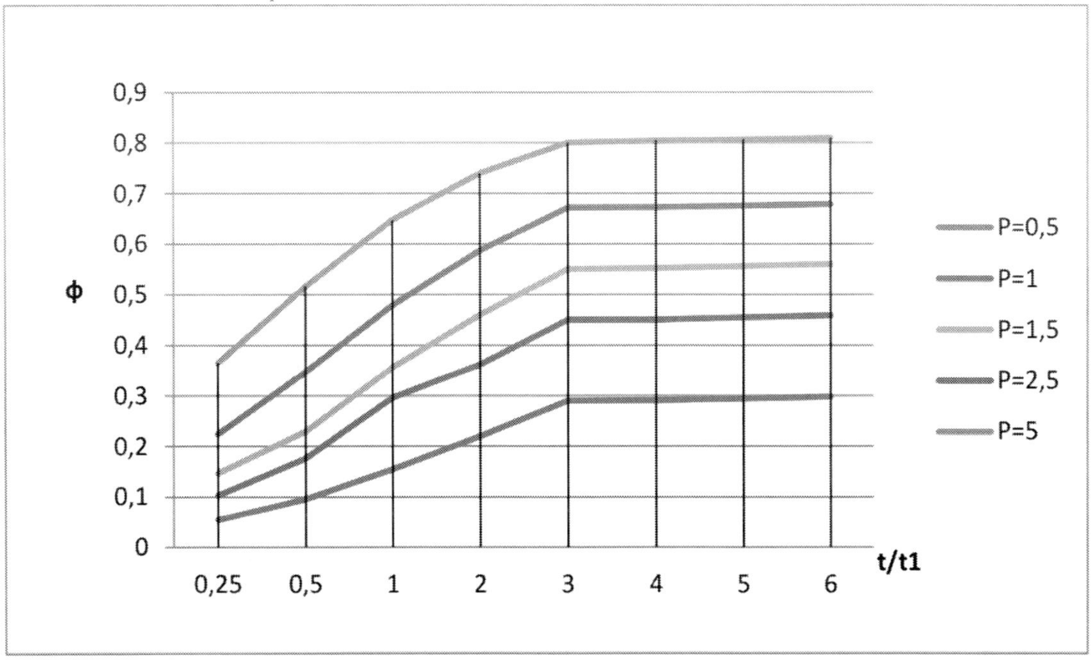

Fig. 1. The dependence of coefficient of heat losses for different values of the parameter «P».

The parameter «P» is proportional $\delta / V_0^{1/3}$.

For sheet metal: $P = \frac{3093G\delta}{V_0^{1/3}}$

Here «G» is defined as $F_p = GV_0^{2/3}$ and depends on the shape of the room.

From Fig.1 shows that with the decrease of the parameter «P» decreases and the coefficient of heat loss. Thus determined the dependence φ sheet thickness and volume of the room. It should be noted that in the approximation of a thermally thin walling coefficient of heat loss does not depend on properties of the combustible material and the heat transfer coefficient α is only in the scale of time t.

Brick and concrete wall thickness $\delta > 5*10^{-2}$ m for the initial stage of the fire $<10^3$ sec, is thermally thick solids. This means that the conditions on the opposite side do not affect the heat transfer process. For thermally thick solids, the surface temperature T_1 in the case of constant temperature T_S and with a constant heat transfer coefficient α varies with time according to [4].

$$T_1(t) - T_s = (T_s - T_0)\left[1 - \exp\left(\frac{t}{t_2}\right) erfc\left(\frac{t}{t_2}\right)^{1/2}\right] \tag{7}$$

Where $t_2 = \frac{C_p \rho \lambda}{\alpha^2}$ is the characteristic time of change of surface temperature. Brick walls with α=12 Vt/m²K, t_2 = 8680 sec. For times $t / t_2 < 0,15$ sec. the expression (7) can be replaced by the expression (5):

$$\left(T_1 - T_0\right) = \left(T_s - T_0\right)\frac{\left(\gamma\frac{t}{t_2}\right)^{1/2}}{1+\left(\gamma\frac{t}{t_2}\right)^{1/2}} \tag{8}$$

Where $\gamma = \dfrac{4}{\pi}$.

For variable temperature environment: $T_s = T_0\left(1+\dfrac{t}{t_s}\right)$ heat flux at the wall is determined.

These structures are thermally thick plates as long as the fire do not reach their critical values. For such structures the heat flow at constant temperature, the gas environment of the T_s and constant heat transfer coefficient α from the environment to the structure is equal to [3]

$$q_{\Sigma 2} = \frac{\alpha t^2 F_p T_0}{2t_s}\left[1-\frac{8}{15}\left(\gamma\frac{t}{t_2}\right)^{1/2}\right] \tag{9}$$

In this case $t_s = \dfrac{V_0\rho_0 T_0 C_p}{Q_n\psi F_g\left(1-\varphi_2\right)}$, where φ_2 – coefficient of heat loss in a brick wall for time t.

Coefficient of heat loss is equal to:

$$\varphi_2 = \frac{\alpha F_p t\left[1-\frac{8}{15}\left(\gamma\frac{t}{t_2}\right)^{1/2}\right]}{2V_0\rho_0 C_p + \alpha F_p t\left[1-\frac{8}{15}\left(\gamma\frac{t}{t_2}\right)\right]} \tag{10}$$

3 Results

Table 1 shows for comparison the dependence $\varphi_2(t)$ of (t/t_1) for different time scales t_1, that is, the brick wall is compared with sheet metal of different thickness with the same amount of space. From the comparison it is seen that the heat loss in a sandwich panel is always less than a brick wall. The maximum difference corresponds to the double and large rooms 10^4 m^3 and small thicknesses ≈0.5 mm.

Figure 2 gives the opportunity to cross-section coordinate is t/t_1 = const. to determine the dependence $\varphi_1(t)$ of the volume V_0 for different ages.

Table 1. The dependence $\varphi_2(t)$ of (t/t_1) for different time scales t_1.

t/t_1		**0,25**	**0,5**	**1**	**2**	**3**	**4**	**5**	**6**	**∞**
P/0,5	*panel*	0,0544	0,096	0,155	0,220	0,290	0,291	0,294	0,298	0,333
	brick	0,0570	0,106	0,019	0,310	0,400	0,465	0,515	0,560	-
1	*panel*	0,103	0,176	0,296	0,362	0,450	0,450	0,455	0,459	0,500
	brick	0,108	0,192	0,320	0,475	0,570	0,634	0,660	0,715	-

1,5	*panel*	0,147	0,229	0,356	0,460	0,550	0,552	0,556	0,560	0,600
	brick	0,154	0,264	0,412	0,576	0,670	0,723	0,760	0,790	1,00
2,5	*panel*	0,224	0,347	0,479	0,587	0,672	0,673	0,676	0,679	0,714
	brick	0,232	0,373	0,540	0,693	0,770	0,810	0,840	0,860	1,00
5	*panel*	0,365	0,516	0,648	0,740	0,800	0,804	0,806	0,809	0,833
	brick	0,377	0,540	0,700	0,820	0,870	0,900	0,914	0,926	1,00

4 Conclusion

The characteristic time value t_1 for the initial stage of the fire $\alpha \approx 12\text{-}16$ Vt /m^{2}K is 100-300 seconds, depending on the thickness of the sheet. By the end of the initial stage of fire when the indoor temperature reaches a critical value for the conditions of the evacuation ratio (t/t_1) is 2-6 times. In Fig.2 shows the dependence $\varphi_1(t)$ from the volume of the room for occasions: 1) $t/t_1=2$ and 2) $t/t_1=6$.

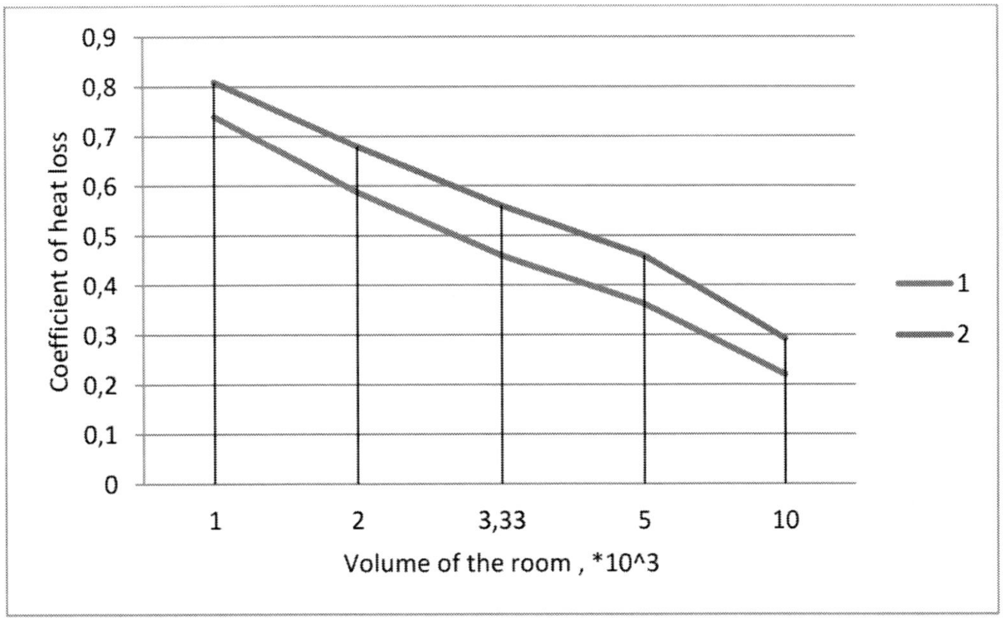

Fig. 2. The dependence $\varphi_1(t)$ from the volume of the room for occasions: 1) $t/t_1=2$ and 2) $t/t_1=6$.

The expression (9) is derived from (8) for changing the temperature of the environment using the theory of Duamel's [6]. The presented method can be used for the case of arbitrary changes in ambient temperature. The use of Duamel's theorem is due to numerical integration.

References

1. Yu. Koshmarov, Forecasting of fire hazards in the case of indoor fire. Moscow, State Fire Academy of Ministry of Interior of Russia Publ, (2000)
2. Methodology og the calculated values of fire risk in buildings, facilities and structures of various classes of functional fire hazard: Annex to the Order of the Russian Emergencies Ministry 30.06.2009 №382. Russian Newspaper, (2009)
3. Ye. Salymova, Dynamics of development of dangerous factors in buildings with enclosing structures made of sandwich panels in fires and explosions Candidate

dissertation. 05.26.03/ Salymova Ye. Yu. Moscow State University of Civil Engineering. Moscow (2015)
4. A. Lykov, Theory of thermal conductivity. Moscow, Vysshaya shkola Publ., (1967)
5. V.A.Gore, M. Fomina, Simplified calculation of heat on a flat surface. Pozharovzryvobezopasnost – Fire and Explosion Safety, 2016,vol.25, no.3, pp 5-14.
6. A.N. Tihonov, A.A. Samarskiy Equations of mathematical physics. Moscow, Nauka Publ. 1977 - 735 p.

Operational features of decorative concrete

Olga Bazhenova[1]* and *Maxim* Kotelnikov[1]

[1]Moscow State University of Civil Engineering, Yaroslavskoe shosse, 26, Moscow, 129337, Russia

Abstract. This article deals with the questions of creation and use of decorative and finishing concrete and mortar. It has been revealed that the most effective artificial rock-imitating stone materials are those made of decorative concrete with the opened internal structure of material. At the same time it is important that the particles of decorative aggregate should be distributed evenly in the concrete volume. It can be reached only at a continuous grain-size analysis of the aggregate from the given rock. The article tackles the necessity of natural stone materials imitation for the cement stone color to correspond to the color of the rock. The possibility of creation of the decorative concrete imitating rocks in the high-speed turbulent mixer is considered. Dependences of durability and frost resistance of the studied concrete on the pore size and character and also parameters characterizing crack resistance of concrete are received.

1 Introduction

In Russia and in a number of foreign countries, there are wide-ranging studies on getting new durable decorative and finishing materials that can improve a decorative and artistic look and raise operational properties of buildings and constructions.

Items of exterior finishing are placed in more difficult working conditions in comparison with other elements of designs. They are put under influence of climatic factors and mechanical tension at the same time. On the surfaces of finishing materials various defects in the process of operation are observed, which are caused by influence of fluctuations of temperature, humidity of air, impact of wind, rainfall and solar radiation, household and industrial impurity in the atmosphere (dust, soot, carbon black), impacts of static loading, thermoelastic and shrinkable tension [3, 4]

It is necessary to include sanitary and hygienic, operational and technical, esthetic requirements, durability, industrial and technological effectiveness in production and economic efficiency that can be referred to general requirements to finishing materials and items [5]. Decorative and finishing concretes and mortars have also been developed and applied a lot. At present, they are considered as structural and finishing materials used in the factory for manufacturing of concrete and reinforced concrete items with a finished front area. Thanks to the use of white and colored rocks, the palette of finishing concretes and mortars has significantly expanded. On the basis of various mineral binding and decorative

* Corresponding author: sofia.bazhenova@gmail.com

cements in combination with colored aggregates, semi-finished items are manufactured, in particular, dry mixes for decorative finishing of blocks, panels and other structural elements of buildings, and finished products with finished front area [2-6].

Properties of decorative cements in many respects depend on quality of cement clinker, and properties of the coloring tincture [7, 8]. Therefore tinctures have to be resistant to effect of alkalis, sunlight and atmospheric influences, high coloring ability and also make weak impacts on physical and mechanical properties of cements. Contents of mineral, synthetic or natural tinctures in the colored Portland cement shouldn't exceed 15 %, and organic – no more than 0.5 %.

Very strict requirements are imposed to quality of the last in these conditions. They have to suit:

- sufficient coloring ability that allows to add them in mineral quantities, providing thereby preservation of durability of concrete, not creating a considerable extra charge to its cost;
- high resistance to effect of alkalis, solar radiation and atmospheric influences;
- lack of the harmful admixtures which can negatively influence on physical and mechanical properties of the cement paste and the hardened stone or forming salt efflorescence on its surface;
-wide color tincture and reasonable price [7].

Mineral synthetic tinctures are most available and perspective for application in the cement industry, for example, ferrioxide. They can be yellow, cherry-red, brown-red and black. Ochre, saturnine red, manganese ores are most applied among natural tinctures [8].

Magmatic rocks (granite, syenite, diorite, gabbro and many others), sedimentary rocks (limestones, soapstone, sandstone and others) and metamorphic rocks (marble, gneisses) are used as aggregates for decorative concrete. It is very important to study the structure and texture of initial rocks while selecting aggregates.

Rocks are a set of minerals. Minerals are homogeneous in natural rocks in their chemical composition and physical properties. Color of minerals is of very great importance for receiving analogs of the artificial stone materials similar to natural properties according to their exterior signs. The decorative concrete received on usual technology on color cements or with tinctures and on aggregates from rocks (with the opened area) are the best decorative materials imitating natural stones. Opening of the surface of decorative concrete can be carried out by machining with a tool of refractory steels, by sand cleaning under pressure or by cutting blocks into tiles of various sizes [9].

When processing decorative concrete with tools and opening the surface layer, with the correct selection of the aggregate's grain composition, a surface is obtained that is not inferior in its properties to natural stone materials [10].

Decorative concretes and mortars, along with undoubtedly positive properties, have the same drawbacks as conventional concrete and mortar. Concrete is a capillary-porous body, the properties of which are more dependent on porosity. In the formation and subsequent hardening of concrete pores are formed as a result of chemical and physical and mechanical processes. The pore volume in dense concrete is from 9 to 20%, with increasing tincture content in decorative concrete and mortar, its porosity increases. When tinctures with a high specific surface are introduced, they are adsorbed on the grains of cement and aggregate reducing the hydration of cement grains and their adhesion to the aggregate. With an increase in the tincture content, the workability of the mixture decreases, and therefore, when the mechanical action is difficult, the trapped air is difficult to remove [11].

There are various ways to improve the performance characteristics of decorative concrete and mortars. Active mineral additives are applied, which bind calcium hydroxides to poorly soluble calcium hydrosilicates. This helps to increase the strength of cement stone due to the formation of a crystalline splice. However, the use of active mineral additives does not ensure complete binding of the excreted calcium oxide. Improvement of the properties of decorative concrete and mortars is possible with the use of surface active substances: the density of cement stone is increased, a fine-grained structure with a system of closed uniformly located pores is formed. Various plasticizing and hydrophobizing additives are used to increase the density of concrete and mortar and to reduce efflorescences on the surface of products and increase of their durability [12-13].

When determining the amount of cement, it is necessary to take into account the fact that for the color of the concrete surface, the color of the cement is of decisive importance, and not only its excess quantity per unit volume of concrete in relation to the established rate. The particular dose of cement depends on the properties of the aggregate and the technology of processing concrete. It can vary from 300 to 500 kg per 1 m3 of ready-mixed concrete. When choosing aggregates for decorative concrete, decisive importance has not only their color, but also other properties such as durability, polishability, the presence of primary and secondary impurities, color stability. In the production of aggregates for decorative concrete, the following requirements must be met:

- permanent grain-size composition of the aggregate, its careful classification, increased content of fine fractions and stone powder;
- favorable color range of the aggregate;
- the purity of the aggregate.

The content and grain-size composition of the aggregate per unit volume of concrete should be constant, as the vibrations may affect the properties of concrete, which adversely affects the color tone of the decorative concrete [8, 9].

In the process of operation, decorative materials, elements of artistic, decorative, sculptural and monumental works are subjected to aggressive influences, which are divided into two main groups:

- external, causing the destruction of the material or surface of the product;
- internal chemical and structural changes in the material that occur during the hydration of cement.

Defects of the structure of concrete are stress concentrators, as well as channels through which aggressive agents penetrate deep into the concrete. Therefore, in order to increase the durability of decorative concrete and reinforced concrete products and structures, it is necessary to increase the density and strength of surface layers of concrete, often of high porosity, due to sedimentation and water separation when compacting the concrete mix.

An effective way to increase the strength and durability of concrete is the use of silica-containing additives that interact with calcium hydroxide, increasing the density of the cement stone and improving the contact area between the cement stone and the aggregate [14-15]. The adhesion of cement stone with aggregate is provided by the quality of aggregate surface. The best results can be achieved with partial grinding of the aggregate with the improvement of the shape of the grains (approximating to the cubic shape), activation of their surface (cleaning of aggregates from clay and oxide layers). This will increase the density of concrete, deepen the process of hydration of cement and, as a consequence, reduce its specific consumption and improve the quality of decorative concrete [16-20].

Thus, the conducted analysis of production and application of decorative concrete in construction showed that the most effective artificial stone materials imitating rocks are decorative concretes with the opened inner structure of the material. In this case, it is important to distribute aggregate grains uniformly in the volume of concrete. This can be achieved only with a certain continuous granulometry of the aggregate from a given rock. In this case, the use of quartz sand as a fine aggregate is excluded, since during machining there will be either sanding out of the sand or changing the tone of the original rock. Therefore, to obtain a uniform texture of decorative concrete, it is necessary to use an aggregate of all fractions from the original rock only. To imitate natural stone materials, it is necessary that the cement stone has a color corresponding to the tone of the rock. This is achieved by using mineral or organic tinctures that are stable in alkaline environments.

2 Method

Rubble of red granite with a bulk density of 1290 kg/m^3, a density of 2.5 g/cm^3, a water requirement of 6.1 % was used in the experiment as an aggregate for decorative concretes. To reduce water consumption and improve the quality of concrete, superplasticizer C-3 (in powder form), microsilica - ferroalloy production waste with a specific surface area of 2000 m^2/kg, density 2.2 g/m^3, bulk density 0.2-0.25 g/cm^3 were used. Iron saturnine red with a specific surface of about 980 m^2/kg was used as the tinctures.

The preparation of decorative concrete mixtures was carried out in a high-speed turbulent mixer with rotor rotation at a speed of 600 rpm equipped with a propeller-like nozzle made of high-strength steel. When the components of the concrete mix are mixed, the coarse aggregate is crushed to form a continuous granulometry of the grains, the activation of the cement and an increase in the uniformity of the concrete mixture.

Systematized studies of the structure of strength, deformation properties, fracture toughness and frost resistance of decorative concretes were carried out for various cement and microsilica costs and the same mobility of the concrete mixture.

Samples of cubes were made 10 x 10 x 10 cm. The samples were tested after heat and moisture treatment in 7 days. The total and group porosity was determined in three degrees of saturation (successively over water, in water and under vacuum to constant mass), compressive strength, parametric points of the microcrack formation process under static loading, frost resistance and stress intensity factor.

3 Results and Discussion

The compositions of the concretes obtained are listed in Table 1. In all compositions, the consumption of superplasticizer C-3 in concrete was 0.63%, and aggregate (iron saturnine red) - 5% of cement mass.

Table 1. Compositions of decorative concretes

No of composition	Consumption of materials, kg m^3				W/C	Precipitation of the cone, cm
	Cement	Water	Aggregate	Microsilica		
1	500	167	1767	40	0.33	6-8
2	580	170	1683	46.4	0.29	6 – 8
3	660	173	1597	52.8	0.26	6 – 8
4	575	174	1674	46	0.3	12-16
5	650	177	1597	52	0.27	12-16
6	740	186	1542	59.2	0.25	12-16

As a result of mathematical processing of the experimental data, the strengths and frost-resistance of the examined concrete have been found to depend on the size and nature of the pores, as well as the parameters characterizing the crack resistance of concrete. A characteristic feature of the studied concrete prepared according to the developed method is that they have a different contact area between the cement stone and the aggregate compared to conventional concrete, because as a result of the preliminary treatment of the aggregate in the modified turbulent mixer, the adhesion of the cement stone to the aggregate is substantially improved and the composition of the new phases and the nature of the pores change. This is evidenced by data on the definition of group porosity in three degrees of saturation. The properties most influenced by the pores of the second group which characterize, mainly, the volume of capillary pores. The properties of decorative concretes are presented in Table 2.

Table 2. Properties of decorative concrete

No	Total porosity, % P_o	Porosity of group 2, % P_2	Strength, MPa	Frost-resistance, cycle	Stress intensity factor $MN/m^{3/2}$
1	11.1	3.4	54.5	300	3.12
2	10.5	2.6	63.7	350	3.2
3	10.0	1.6	72.4	400	3.26
4	12.0	3.5	53.4	250	3.0
5	12.2	2.7	62.9	300	2.98
6	12.4	2.0	73.2	250	2.95

The introduction of microsilica and superplasticizing additive C-3 contribute to a decrease in the content of macropores and increase strength, while maintaining high mobility of the concrete mix. The volume of macropores in concrete depends on the degree of hydration of cement in concrete, that is, on the conditions of hardening. The degree of hydration was determined by X-ray diffraction. For the studied concrete it was 0.4-0.5. By the value of macroporosity (capillary), the concrete under study should be classified as particularly dense.

The relationship between the cubic strength of concrete and the porosity of the second group (P_2) is established, which is described by the following equation:

$$R = 78.9 + 0.19P_2 - 2.17P_2^2 \quad (1)$$

The concrete cracking is characterized by a stress intensity factor (Kc), which in turn is associated with the main defect of concrete - pores, the presence of which causes the greatest concentration of stresses. Therefore, the dependence of Kc on the total porosity of concrete was established, having the form:

$$Kc = 4.96 - 0.21\,P_0 + 0.004\,P_0^2 \quad (2)$$

In frozen concrete, ice is formed first in large, and then in shallow pores, due to the different capacity of water enclosed in these pores, to pass into ice. When preparing decorative concrete mixtures in a modified turbulent mixer, the capillary porosity of concrete decreases, which leads to an increase in the frost resistance of concrete. Dependence of frost resistance on the porosity of the second group is described by the following equation:

$$F = 359 + 66.5\,P_2 - 27\,P_2^2 \quad (3)$$

The established dependencies make it possible to predict the strength and deformation properties of decorative concrete, as well as the fracture toughness and frost resistance in the process of their operation.

4 Conclusion

Analyzing the test results, and summarizing the above mentioned, we can draw the following conclusions:

- the preparation of decorative concrete mixes in a modified turbulent mixer allows to partially crush the aggregate, change its grain shape, approaching the cubic one, break down weak grains, reduce the number of microdefects, activate the aggregate surface and improve the contact area between the cement stone and aggregate;

- the influence of cement consumption and microsilica under different workability on the structure, strength and other characteristics of decorative concrete has been established;
- dependences of strength and frost resistance of concrete from the second group of porosity (capillary) are determined, and a relationship between the total porosity and the stress intensity factor is established;
- dependences of frost resistance, stress intensity factor, parameters of the microcrack formation process prepared in a modified turbulent mixer on the composition and structure characteristics are obtained.

Obtained decorative concrete, imitating various rocks, can be used for the manufacturing of pedimental slabs, paving stones, curbstone, floor slabs and decoration of interiors, as well as artistic and decorative products.

References

1. Yu. Bazhenov, V. Falikman, Build-Info, **1-2**, 289-290 (2007).
2. V. Egorov, Buil. S. of Moscow, **2**, 14-18 (2007).
3. S. Nikolaev Sec.tech. and eng. Sys., **1**, 16-20 (2006).
4. Yu. Bazhenov, E. Korolev, N. Lukutsova, S. Zavalishin, Bull. of MGSU, **6**, 73-78 (2012).
5. K. Sullivan, Cem. and its appl., **4**, 82-86 (2015).
6. O. Bazhenova, Build. Mat., equip., tech. of the XXI cen., **3**, 15 (2003).
7. B.Ya. Trofimov, L.Ya. Kramar, Vestnik of the South Ural State Univ., **15 (191)**, 36-38 (2010)
8. L. Marco *Raw materials for decorative concrete*. (Moscow: Stroyizdat, 1983).
9. V. Tserkovnikov. *Manual on the technology of obtaining artificial stone materials for works of monumental and decorative art.* (Tula: PKS, 1987).
10. O.Yu. Bazhenova, S.I. Bazhenova Adv. in mod. Sci., **6**, vol.3, 21-23 (2016).
11. P.P Gajurov, A.P. Zubehin, Cement, **7**, 28-30 (1995).
12. G.I. Berdov, N.I. Nikonenko, L.V. Ilyina, Izvestiya VUZ. Build., **12**, 25-30 (2011).
13. O.Yu. Bazhenova, V.E. Kaushansky, *Construction and decoration materials. Standards of XXI century XIII International Workshop of ATAM.* (Novosibirsk, 2006).
14. V.I. Kalashnikov, O.V. Suzdaltsev, R.A. Dryanin, G.P. Sekhposian, Izvestiya VUZ, **7 (667)**, 11-21 (2014).
15. L.A. Urkhanova, V.E. Rosina, Bull. of the Irkutsk State Tech. Univ., **10 (57)**, 97-100. (2011)
16. V.V. Loshak, S.V. Cherkasov, V.V. Vlasov // Sci. Bull. Of Voronezh State Arch. and Constr. Univ., **3-4**, 61-66 (2011).
17. D.K. Timokhin, Yu.S. Geranina, Res.-en.-eff. Tech. in the constr. complex of the region. **2(6)**, 173-176. (2015).
18. E.A. Shchetkova, G.G. Kashevarova // Bull. of the Perm Nat. Res. Polytechnic Univ. **3(27)**, 143-151 (2017).
19. S. Guziy, S. Terenchuk, ScienceRise, **2 (26)**, Vol. 9, 49-54 (2016).
20. V.I. Kalashnikov, O.V. Suzdaltsev, M.N. Moroz, V.V. Pausk, Build. Mat. **3**, 16-19 (2015)

Feasibility study tool for semi-rigid joints design of high-rise buildings steel structures

Ruslan Bagautdinov[1,*], *Daria* Monastireva[1], *Irina* Bodak[1] and *Irina* Potapova[2]

[1]Peter the Great St. Petersburg Polytechnic University, 29 Polytechnicheskaya st., St.Petersburg, 195251, Russia
[2]Astrakhan State University of Civil Engineering, 414056, 18 Tatishcheva Str. , Astrakhan, Russia

Abstract. There are many ways to consider the final cost of the high-rise building structures and to define, which of their different variations are the most effective from different points of view. The research of Jaakko Haapio is conducted in Tampere University of Technology, which aims to develop a method that allows determining the manufacturing and installation costs of steel structures already at the tender phase while taking into account their details. This paper is aimed to make the analysis of the Feature-Based Costing Method for skeletal steel structures proposed by Jaakko Haapio. The most appropriate ways to improve the tool and to implement it in the Russian circumstances for high-rise building design are derived. Presented tool can be useful not only for the designers but, also, for the steel structures manufacturing organizations, which can help to utilize BIM technologies in the organization process and controlling on the factory.

1 Introduction

One of the greatest accents in the steel structures researches is the optimization of the design model. It takes into account factors of economics, ecological influence on the environment and safety of the structure. Prediction of the structure's manufacturing costs considers as a complicated operation. The real price of the different design decisions of the building structure depends on its geometrical characteristics, material, manufacturing quality, etc. There are many ways to consider the final cost of the structures and to define, which of their different variations are the most effective from different points of view. The research of Jaakko Haapio is conducted in Tampere University of Technology, which aims to develop a method that allows determining the manufacturing and installation costs of steel structures already at the tender phase while taking into account their details. In order to take into consideration all the features of the joint fabrication process, the macros for the modelling program Tekla can be used at the tender phase. The program is based on the calculation methods developed by Jaakko Haapio in [1], which use time-based cost functions for all the process of steel structures manufacturing, transportation and installation. In the program, also, carbon dioxide emissions of the structure can be easily derived from the Building Information Model (BIM) of the steel structure.

* Corresponding author: rusish93@mail.ru

In the work Jaakko Haapio the economic analysis tool for skeletal steel structures was studied in order to make analysis and to find the most appropriate ways to optimize the design and construction processes. As a result of analysis a method was presented that allows to determine the manufacturing, transportation and installation costs of steel structures already at the tender phase while taking into account their details. This is possible by analysis of the initial building models at the tender phase. The special macros of the modelling program Tekla allows to take into consideration all the features of the joint fabrication process. Preset parameters were studied in the Jaakko Haapio time-based cost functions of all steel structures manufacturing processes at the workshop and during installations are known. The development and testing of these cost functions in Russian circumstances is the subject of the research.

1.1 Feasibility study methods

1.1.1 Feature-Based Costing Method and other cost models

At the preliminary design step the cost and emissions can be calculated by the method presented in [1]. It is based on the Feature-Based Costing Method, where feature is an attribute which affects the costs of the structure during the project.

It involves dividing the manufacturing process into cost centres of a specified floor area and height, equipment suitable for executing the required process (i.e. drilling a hole) and a certain number of workers. These resources have a fixed per minute cost, whether the process is running or not. Some cost components are related only to process time, i.e. electricity consumption of the equipment. They are called variable costs. The time required by the process is the sum of non-productive time, i.e. fixing the profile to the equipment, and process time, i.e. drilling. Total process cost is the sum of the fixed cost multiplied by total time plus the variable cost multiplied by process time. Some processes may also involve non-time related cost components. These are added to time-dependent costs.

The basis of the feature-based costing method was described in the Jaakko Haapio doctoral thesis. The material flow of a skeletal steel assembly shown in Figure 1. The workshop is divided into cost centres, where single work phases are performed. The time spent at each cost centre is converted to costs including equipment, wages, energy, rents etc.

Total cost considers as a sum of the cost derived for each cost center. The generic form of the cost for each cost centre is:

$$C_k = \frac{(T_{\mathrm{Nk}}+T_{\mathrm{Pk}})(c_{\mathrm{Lk}}+c_{\mathrm{Eqk}}+c_{\mathrm{Mk}}+c_{\mathrm{REk}}+c_{\mathrm{Sek}})}{u_{\mathrm{k}}} + T_{\mathrm{Pk}}(c_{Ck} + c_{\mathrm{Enk}}) + C_{Nk} \qquad (1)$$

where C_k = total cost of cost centre k [€]
T_{Nk} = non-productive time of cost centre k [min]
T_{Pk} = productive time of cost centre k [min]
c_{Lk} = unit labour cost of cost centre k [€/min]
c_{Eqk} = equipment installment unit cost of cost centre k [€/min]
c_{Mk} = unit cost of equipment maintenance of cost centre k [€/min]
c_{REk} = unit cost of real estate of cost centre k [€/min]
c_{Sek} = unit cost of real estate maintenance of cost centre k [€/min]
c_{Ck} = unit cost of time-related consumables needed in processing of cost centre k [€/min]
c_{Enk} = unit cost of energy needed in processing of cost centre k [€/min]
c_{Nk} = total cost of non-time-related consumables used in cost centre k [€]
u_k = utilisation ratio of cost centre k [decimal, ≤1]

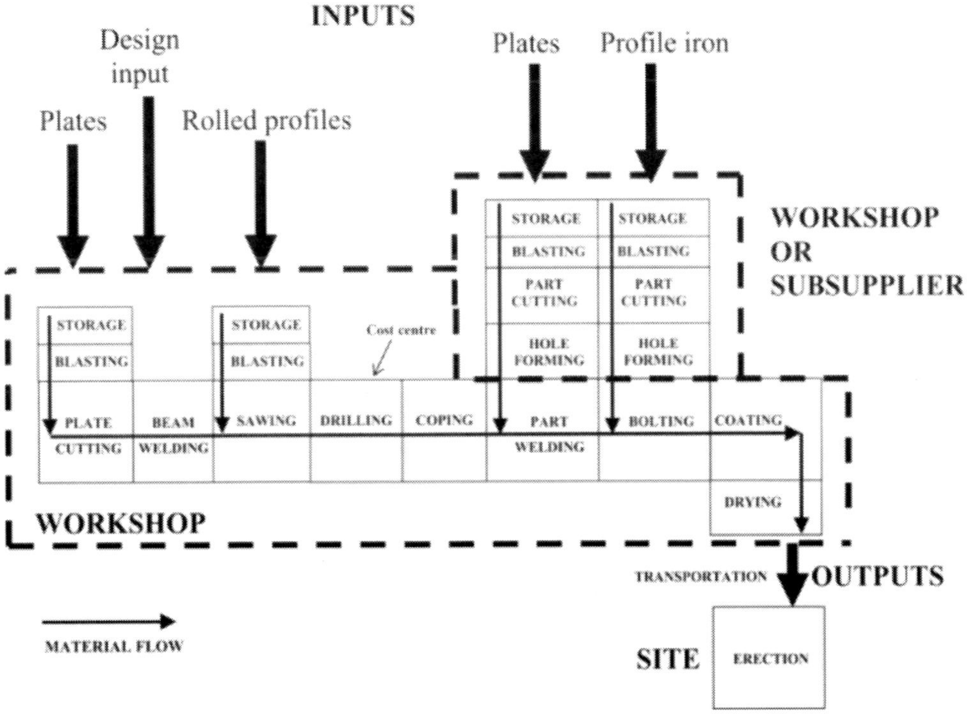

Fig. 1. Manufacturing process flows [1]

1.1.2 Feasibility study methods development in Russia

The issues of steel structures optimization process was considered in the researches of Ya.M. Lichtarnicov [2], I.S. Cholopov [3-5] and others.

The function of cost and labour intensity for the steel structures optimization process is the key function in the research of Ya.M. Lichtarnicov, that was published in 1979 y. From that time price levels, technological operations, labour intensity and other factors have changed a lot. So, today it is extremely important to find a way to deal with dynamically updated database for cost function input data. The research mostly deal with feasibility study of different design variants of space arrangements and only one chapter for the joints.

In the cost analysis method of Ya. M. Lichtarnikov the cost of steel structures estimates based on mass calculation G, that consists of two parts:

G_0 - mass of main parts,

G_s - mass of secondary parts.

Main parts mass can be determined by stress distribution and buckling resistance of the structure. Secondary parts mostly uses by any design consideration and in order to provide adequate behaviour of the structure. The mass of the secondary parts can be taken into account by construction coefficient of mass ψ. So, the formulae for structure mass:

$$G=G_s+G_0=\psi\cdot G_0 \tag{2}$$

The labour intensity can be calculated by multiplying the mass and the coefficients that Ya. M. Lichtarnikov get from the analysis of big database of different structures costs. The coefficient of seriation that can be taken into account in order to provide better accuracy of manufacturing cost calculation.

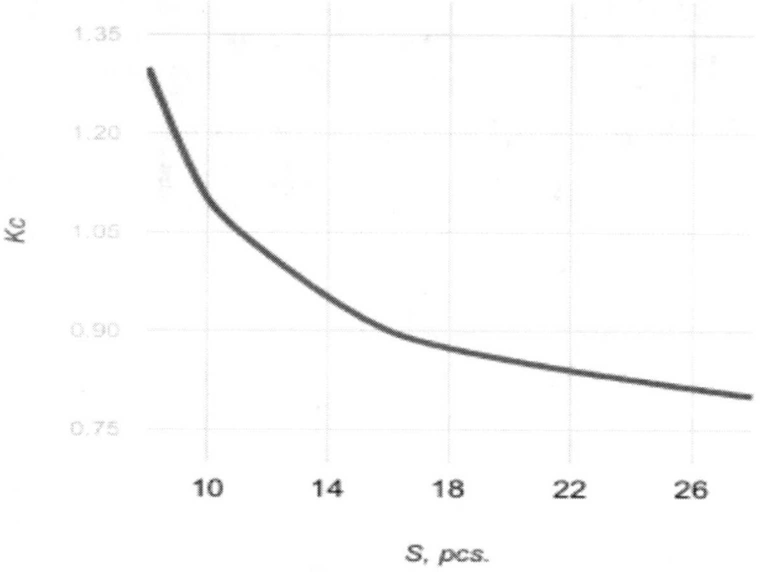

Fig. 2. The coefficient of seriation KC dependence from the quantity of equal parts in the one production run diagram [2]

For each steel grade the reduction and correction coefficients should be included in analysis that can be taken from tables 1-2. [2]

Table 1. The correction coefficient KT for different steel grades.

C38/23	C46/33	C52/40	C60/45	C70/60
1	1.5	1.18	1.22	1.36

For construction of main steel details from different steel grades the reduction coefficient α is used, that has the square root dependence for the labor intensity calculation (Table 2).

Table 2. Average value of reduction coefficient α for the effective steel classes and profiles.

	α	$\sqrt{\alpha}$
Steel type and profile	0.89	0.94
Effective steel class		
Thermally stabilised carbon steel, yield stress 290 MPa, class C44/29	0.83	0.91
Low-alloyed steel, basic bearing structure of general purpose, yield stress 330-400 MPa, class C46/33, C52/40	0.72	0.85
High-impact low-alloyed steel of general purpose, yield stress 450 MPa, class C60/45	0.69	0.82
Low-yield constructions for the structures in the northern regions, yield stress 290-330 MPa, class C44/29-C46/33	0.91	0.95
Economic hot-rolled profile		
Wide-flanged beam	0.97	0.98

The example of correlation between cost of steel structures transportation and the region of construction for columns and gantry girders can be determined from Table 3. [2]

Table 3. The enlarged average values of the transportation costs.

Type of structures	European part of USSR, besides of Komi ASSR, regions and republics of the Northern Caucasus,territories of the Urals	Territories and regions of Northern Caucasus	Ural, Western Syberia, Union republics of the Central Asia	Carelian ASSR, Eastern Syberia	Far East
Columns with mass less than 15 tons	1	1.19	1.19	1.39	4.94
Columns with mass more than 15 tons	1	1.84	1.31	1.53	5.43
Girder from hot-rolled profiles	1	1.71	1.30	1.51	5.28
Girder with the mass less than 3 tons	1	1.68	1.19	1.39	4.94
Girder with the mass 3-5 tons	1	1.86	1.38	1.57	5.10
Girder with the mass 5-15 tons	1	1.66	1.18	1.37	4.86
Girder with the mass more than 15 tons	1	1.78	1.31	1.52	5.37

While dealing with the installation cost it is necessary to include in analysis the steel grade and the location of the site. The correction coefficients for steel grade was presented in table 3. The location of construction site can be taken into considerations by the correction coefficient Kp from table 4. [2] The steel grade correction coefficient for different types of structures is given in table 5. [2] The correlation between the installation costs from the height and the cost of painting for different weights of structures is shown in the tables 5-6. [2]

Table 4. The average values of the coefficient Kp.

	The region of the structural erection	Kp
1	The European part of the USSR, besides of Komi ASSR, regions and republics of the Northern Caucasus, territories of the Ural, the piedmont of the western Urals	1
2	Udmurt ASSR, territories of the Ural, the piedmont of the western Urals, Western Syberia, republics of the Central Asia	1.1
3	Carelian ASSR, Comi ASSR Buryat ASSR, Krasnoyarsk territories and regions of the Eastern Syberia	1.13
4	Far eastern region	1.21
5	Murmansk region	1.27

Table 5. The value of the coefficient Km for different steel grades.

Type of structures	Steel class		
	C38/23	C46/33	More than C46/33
Columns with the weight less than 8 tons and framework less than 3 tons	1	1.07	1.17
Columns with the weight more than 8 tons and framework more than 3 tons	1	1	1.2
Crane beams	1	1.12	1.22

Table 6. The installation costs and the cost of painting for columns.

Constructions	Installation cost for the different heights			Cost of painting
	Less 15 m.	15-25 m.	25-40 m.	
Steel columns with weight: less than 8 t.	1	1.11	1.14	0.42
From 8 t. to 15 t.	1	1.10	1.13	0.28
More than 15 t.	1	1.06	1.07	0.25
Column connections	1	1.06	1.08	0.76
Framework structure	1	1.09	1.12	0.55

There are more different correction coefficients and more detailed cost functions, that can be derived from the research of Ya. M. Lichtarnikov [2]. It may be included in analysis in the future improvement of the method, at least, as coefficient that provide relative functions of different feature dependences, but in that thesis was excluded from calculations.

1.2 Semi-rigid joints definition by Eurocode 3

In Eurocode 3 [6] connections are classified regarding their strength and their stiffness. The stiffness classification is clear from Figure 1. It is shown (see Gomes et al. [7]) however that if plastic rotations are adopted the use of initial stiffness as a unique description parameter is not correct. In other words the initial stiffness of the connection is not enough to classify the connection properties. In this research that method of semi-rigid joints description is enough.

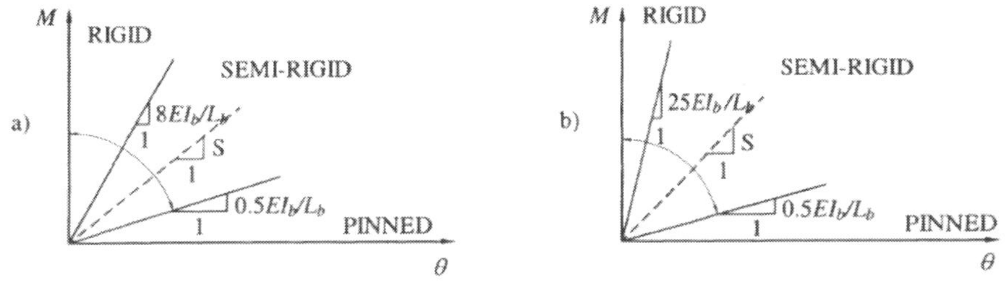

Fig. 3. Classification of beam-to-column connections by stiffness according to the Eurocode 3 - Annex J (revised): a) unbraced frames and b) braced frames. The region studied in the paper is denoted by arrow.

2 Design example

The third numerical example is a three bay, ten storey steel frame designed by Xu and Grierson [8]. Kameshki and Saka [9] as well as Foley and Schinler [10] performed weight optimization on the same example using Genetic Algorithms and evolutionary computation. The frame configuration, dimensions and loading are shown in Fig. 4. The used steel grade is S235, with a modulus of elasticity of 210 000 MPa and yield stress of 235 MPa.

2.1 Design procedure description

Basic times of different manufacturing operations are evaluated according to a fabrication cost model of joints developed at the Laboratory of the Design Optimization and Environment Engineering (LOCIE) of Polytech'Savoie (France). This cost model was first developed by Hamchaoui [11] and incorporated into a computerized module for joint design. Bel Hadj Ali updated the cost model for structural optimization with semi-rigid joints [12,13].

Design is performed considering only members as design variables while beam-to-column connections are specified to be of three type. Column bases are supposed to be rigid. Optimization variables are thus limited to 10 groups of beam members and 10 groups of column members. Beam members at each storey level are to have the same European IPE section while exterior and interior column members are to have the same European HEB section over two stories (Fig. 4).

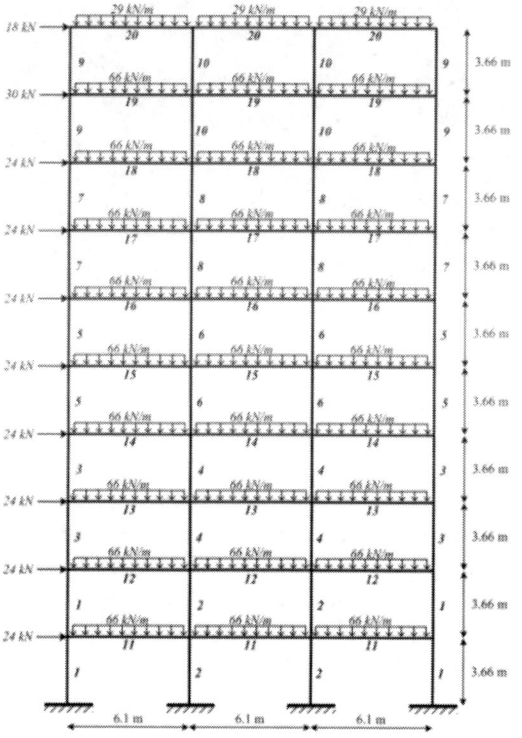

Fig. 4. Three bay, three storey frame

2.2 Results

Optimum designs obtained for frames with rigid and semi-rigid connections are presented in Tables 7, for unbraced and braced frame configurations, respectively.

Table 7. Design results for braced three bay, ten storey steel frame.

Member group	Joint type		
	Rigid	Semi-rigid	Pinned
1	HEB 300	HEB 280	HEB 260
11	IPE330	IPE330	IPE400
Total cost by original tool	82 034	66 290	75 572
Total cost by developed tool	70 970	57 632	61 214

3 Conclusions

Structural optimization has been widely studied over the last decades and extensive work has been done in the case of optimal design of steel frames [14-20]. However, engineers have few tools to approach cost optimization in a systematic manner. Here are som developments that should be made:

1. Dynamic data collection. The cost of hot-rolled steel profiles varies significantly for fairly short periods of time, thus prices monitoring for products, wages, prices for consumables and electricity can be updated by the dynamically replenished database, which will allow to change the specific coefficients of costs.

2. Localization accounting. In large countries, such as Russia, it is critically important where the construction will be arranged, where the structures should be transported. Specific costs will be strikingly different.

3. Possibility of information exchange. The designer can provide BIM models to the manufacturer, in order to plan the production process, calculate the costs, track expenses, plan purchases and reuse of production balances. In return the designer gets access to some of the information gathered by ERP and SAP systems of manufacturer, to conduct the feasibility study. So the plant can automate production planning, and the designer has the opportunity to improve the tool for feasibility studies.

Here are some of assumption made in Jaakko Haapio feasibility study tool [1]:

1. After an investment into a workshop or site facility has been made, many of the related cost factors are fixed, and the costs will run for their life time regardless of the utility rate of these factors.
2. The time used to produce a feature is essential, and a time-based approach for estimating its cost is justifiable.
3. The process consists of productive time and non-productive time. A virtual workshop was established to be able to estimate the non-productive time as well as the cost of real estate.
4. Transportation between cost centres is not considered in this thesis.
5. The aim of the used form of cost calculation is to make use of so-called deep knowledge. By using it the components of the cost function of a cost centre represent the actual, cost-causing processes and the use of statistical factors is minimised.
6. Usually, the designer does not know during the basic design phase which workshop is going to manufacture the structure, and consequently does not know which facilities will be used to manufacture it. The designer must be aware of this uncertainty and might possibly conduct a sensitivity analysis with selected parameters.
7. Typically columns are grouted to the foundations after bolting, but this process is not dealt with here. Stiffening of the frame is effected by rigid joints, steel diagonals, wall structures or a stiffening superstructure such as a concrete tower. The thesis does not deal with secondary structures, such as wall and roof claddings, stairs, handrails, floor slabs or plates, or cold formed profiles.
8. Coating is assumed to be carried out by painting, hot dip galvanising is not dealt with.
9. Columns and braces are hot rolled I-profiles (IPE, HEA, HEB), rectangular hollow sections (RHS), circular hollow sections (CHS), welded I-profiles (WI) or box profiles (WB). Beams are hot rolled or welded I-profiles or welded box beam profiles (WQ).
10. Unit costs and costs of equipment and investments are revised to correspond to the 2009 price level in Finland.

Most of them can be neglected while using above mentioned ways of method development.

References

1. J. Haapio, Feature-based Costing Method for Skeletal Structures Based on the Process Approach. PhD thesis, Tampere University of Technology, Tampere, Finland (2012)
2. Ya. M. Likhtarnikov Variantnoye proyektirovaniye i optimizatsiya stalnykh konstruktsiy. [Variant design and optimization of steel structures] Moscow: Stroyizdat, 319 p. (rus) (1979)
3. I. P. Kholopov, A. N. Popov, Mnogokriterialnaya optimizatsiya elementov metallicheskikh konstruktsiy v usloviyakh SAPR [Multiobjective optimization elements

of metal structures under CAD] // Sovremennyye stroitelnyye konstruktsii iz metalla i drevesiny. Pp. 226-234. (rus) (1999)

4. I. P. Kholopov Optimizatsiya sterzhnevykh sistem primenitelno k SAPR [Optimization rod systems in relation to CAD]. Ph.D. Dissertation Special: 05.23.17 Moscow, 39 p. (rus) (1992)
5. V. Yu. Alpatov, I. S. Kholopov, Geometrical form optimization of a spatially-rod structures // Metal Constructions Journal. №1 (15). Pp. 47-57. (rus) (2009)
6. Eurocode 3: Design of Steel Structures. Part 1.1. Revised Annex J: Joints in Building Frames, ECCS Committee TCIO - Structural Connections (1996)
7. COST CI, Control of the Semi-rigid Behaviour of Civil Engineering Structural Connections Proceedings of the International Conference, Liege, 17-19 September (1998)
8. L. Xu, D.E. Grierson, Computer-automated design of semirigid steel frameworks. J Struct. Eng.;119(6):1740–60 (1993)
9. E.S. Kameshki, M.P.Saka, Genetic algorithm based optimum design of nonlinear planar steel frames with various semi-rigid connections. J Construct. Steel Res.;59(1):109–34 (2003)
10. C.M. Foley, D. Schinler, Automated design of steel frames using advanced analysis and object-oriented evolutionary computation. J Struct Eng; 129(5):648–60 (2003)
11. M.Hamchaoui Conception économique des assemblages en construction Métallique traditionnelle. LOCIE. Chambéry, Ph.D. thesis. France: Université de Savoie; 185. (1997)
12. N. Bel Hadj Ali, J.C. Mangin, A.F. Cutting-Decelle, An overall approach to structural design of steelworks using genetic algorithms. In: Bontempi F, editor. System-based vision for strategic and creative design—Proceedings of the 2nd international conference on structural and construction engineering. Rome: Balkema; p. 481–6. (2003)
13. N. Bel Hadj Ali, Etude de la conception globale des structures en Construction Métallique—Optimisation par les Algorithmes Génétiques. LOCIE. Chambéry. Ph.D. thesis. France: Université de Savoie; 185 (2003)
14. W.MK. Tizani, D.A. Nethercot, G. Davies, N.J. Smith, T.J. McCarthy, Object-oriented fabrication cost model for the economic appraisal of tubular truss design. Adv Eng Softw;27(1–2):11–20 (1996)
15. K.B. Watson, S. Dallas, N. van der Kreek, T. Main, Costing of Steelwork from Feasibility through to Completion. Steel Construct J, AISC;30(2):2–9 (1996)
16. K. Jarmai, J. Farkas, Cost calculation and optimisation of welded steel structures. J Construct Steel Res;50:115–35 (1999)
17. U. Klansek, S. Kravanja, Cost estimation, optimization and competitiveness of different composite floor systems–Part 1: Self-manufacturing cost estimation of composite and steel structures. J Construct Steel Res;62(5):434–48 (2006)
18. L. Xu, A.N. Sherbourne, D.E. Grierson, Optimal cost design of semi-rigid, low-rise industrial frames. Eng J, AISC;32(3):87–97 (1995)
19. L.MC. Simões, Optimization of frames with semi-rigid connections. Comput & Structures;60(4):531–9 (1996)
20. L. Pavlovèiè, A. Krajnc, D. Beg, Cost function analysis in the structural optimization of steel frames. Struct Multidiscip Optim;28:286–95 (2004)

Calibration of the rutting model in HDM 4 on the highway network in Macedonia

Slobodan Ognjenovic[1,*] , *Vladimir* Okrepilov[2] , *Riste* Ristov[3], *Jovan* Papic[4]

[1]Ss. Cyril and Methodius University in Skopje, Faculty of Civil Engineering, blvd. Partizanski Odredi 24,1000 Skopje, Republic of Macedonia
[2] Federal State Budgetary Institution of Science Institute of Regional Economic Problems of the Russian Academy of Sciences, 190003, Serpuhovskaya st.,38, St. Petersburg, Russia
[3]Prostor DOO-Kumanovo, Mosa Pijade 2, 1300 Kumanovo, Republic of Macedonia
[4]Ss. Cyril and Methodius University in Skopje, Faculty of Civil Engineering, blvd. Partizanski Odredi 24,1000 Skopje, Republic of Macedonia

Abstract. The World Bank HDM 4 model is adopted in many countries worldwide. It is consisted of the developed models for almost all types of deformation on the pavement structures, but it can't be used as it is developed everywhere in the world without proper adjustments to local conditions such as traffic load, climate, construction specificities, maintenance level etc. This paper presents the results of the researches carried out in Macedonia for determining calibration coefficient of the rutting model in HDM 4.

1 Introduction

This research was carried out in the period from 2010 until 2014 on a number of sections of the highway network in Macedonia, such as Veles-Gradsko, Gradsko-Negotino and Kumanovo-Miladinovci in total length of around 100 km.

The purpose of the research was to provide initial calibration coefficient of the rutting model on the highway network in Macedonia. The data used for the research was provided by regular measurement in the period from 2002 to 2006 and from 2010 to 2014. It should be noted that no significant maintenance activities were taken in the period from 2006 to 2010. The paper presents the results of the calibration of the rutting model in HDM4 [1-8].

2 Description of rutting development model

According to HDM-4 the rutting model consists of the following components:

- Initial thickening
- Structural deformation
- Plastic deformation
- Wear and tear due to use of winter tyres

* Corresponding author: ognjenovic@gf.ukim.edu.mk

In most cases, only the first three components of the rutting development are taken into account, because there are, usually no data on the influence of winter tyres on the appearance of rutting.

3 Initial thickening

Initial thickening is given in the following equation:

$$RDO = K_{rid}\left[a_0 \cdot \left(YE4 \cdot 10^6\right)^{(a_1 + a_2 \cdot DEF)} \cdot SNP^{a_3} \cdot COMP^{a_4}\right] \quad (1)$$

RDO- rutting caused by initial thickening [mm]

K_{rid} - calibration coefficient for a certain initial thickening

YE4- average annual number of equivalent standard axes [million/lane]

DEF- average annual deflection [mm]

SNP- modified structural number of the pavement

COMP- relative compacting (%) (Table 1)

a_i - model coefficients

Table 1. Recommended values of the relative compacting level COMP

Compactness	Relative compactness COMP [%]
Total compactness in all layers	100
Total compactness in some layers	95
Adequate compactness in most layers	90
Inadequate compactness in most layers	85

What is of special importance is the relation between the advancement of the initial rutting and the standard deviation. Currently, HDM-4 enables for positive linear relation between the rutting and the resulting standard deviation of the rutting depth. It is considered that this trend can be negative in the initial stages of thickening.

Table 2. Values of the coefficient of the model of initial thickening of pavements without creation of cracks.

a_0	a_1	a_2	a_3	a_4
51740	0.09	0.0384	-0.502	-2.30

4 Structural information

Rutting is a very good indicator of the pavement structure condition. For example, on the world level, rutting is one of the key indicators of the condition that is used in the road maintenance contracts. It is supposed that the appearance of rutting in the road network shall gain importance in future, receiving increased understanding of the condition.

HDM -4 gives two forms of rutting progress on cracked and non-cracked sections:

- Structural deformation on no-cracked sections

$$RDST_{UC} = K_{rst} \cdot \left(a_0 \cdot SNP^{a_1} \cdot YE4^{a_2} \cdot COMP^{a_3}\right) \quad (2)$$

- Structural deformation after the incurrence of cracks

$$RDST_{CRK} = K_{rst} \cdot \left(a_0 \cdot SNC^{a_1} \cdot YE4^{a_2} \cdot MMP^{a_3} \cdot ACX^{a_4}\right) \quad (3)$$

ΔRDST – gradual increase of the structural deformation in the year of analysis [mm]
K_{rst} - coefficient of calibration for a structural deformation
YE4- annual number of equivalent standard axes [million/lane]
COMP- relative compacting [%]
MMP- average monthly precipitations [mm/month]
SNC- modified structure number of the pavement
ACX- area of non-indexed cracks [% from the total lane of traffic lane]
a_i model coefficients

Ако $ACX_a = 0 \Rightarrow \Delta RST = \Delta RST_{uc}$

Ако $ACX_a > 0 \Rightarrow \Delta RST = \Delta RST_{uc} + \Delta RST_{crk}$

ΔRST_{uc} - increase of rutting due to structure deformation on non-cracked pavements [mm]

ΔRST_{crk} - increase of rutting due to structure deformation on cracked pavements [mm]

Table 3. Values of the coefficients of the structure deformation model

Pavement condition	a_0	a_1	a_2	a_3	a_4
Without cracks	44950	-1.14	0.11	-2.3	
With cracks	0.0000248	-0.84	0.14	1.07	1.11

5 Plastic deformation

Due to plastic deformation, rutting is represented in HDM-4 as follows:

$$RDPD = K_{rpd} \cdot CDS^3 \cdot a_0 \cdot YE4 \cdot Sh^{a_1} HS^{a_2} \quad (4)$$

ΔRDPD- Gradual increase of plastic deformation in the year of analysis [mm]
K_{rpd} - calibration coefficient for plastic deformations
CDS- construction defect indicator
YE4- annual number of equivalent standard axes [million/lane]
Sh- velocity of heavy-duty vehicles [km/h]
HS- overall thickness of asphalt layers
a_i - model coefficients

Table 4. Values of the plastic deformation model coefficients

a_0	a_1	a_2
2.46	-0.78	0.71

6 Rutting as a result of wear and tear due to use of winter tyres

The form of the model is given by the following equation:

$$RDW = K_{rsw}\left(a_0 PASS^{a_1} W^{a_2} S^{a_3} SALT^{a_4}\right) \quad (5)$$

RDW - rutting due to the use of winter tyres in the analyzed year [mm]

$PASS$ - annual number of vehicles with winter tyres per direction
S - average velocity of traffic flow [km/h]
$SALT$ - variable depending on whether salt is used in winter maintenance ($SALT$ =1 – no salt used; $SALT$ =2 – salt usef)
K_{rsw} - coefficient of calibration of the rutting method due to use of winter tyres

The model coefficients are given on the following table.

Table 5. Values of the coefficients of the plastic deformation model

Pavement surface type	a_0	a_1	a_2	a_3	a_4
All types	0.000248	1	-0.46	1.22	0.32

7 Overall rutting size

The increase of rutting on annual level is given in the following equations:

$$\text{If } AGE4 \leq 1 \Rightarrow \Delta RDM = K_{pd}(RDO + \Delta RDPD + \Delta RDW) \quad (6)$$

$$\text{If } AGE4 > 1 \Rightarrow \Delta RDM = K_{pd}(\Delta RDST + \Delta RDPD + \Delta RDW) \quad (7)$$

$AGE4$ - age of the pavement regarding to the appearance of rutting, in years
ΔRDM - increase of the overall rutting size on annual level
K_{pd} - calibration coefficient for all rutting

The overall rutting size is determined as follows:

$$RDM = \Delta RDM + RD_a \quad (8)$$

RD_a - the overall rutting size at the beginning of the analyzed year

8 Rutting calibration methodology

Calibration of rutting progressing is a simple process of comparison of the envisaged and the existing rutting depth by establishing the relation between the geometric mean value of the measured rutting depth, ORDM and the geometric mean value of the rutting depth.

$$K_{rp} = \frac{GeometricMean[ORDM_i]}{GeometricMean[PRDM_i]} \quad (9)$$

The rutting model calibration coefficient can be calculated by the sums of the logarithms of the measured and anticipated value

$$K_{rp} = \frac{\sum(\log ORDM_i)}{\sum(\log PRDM_i)} \quad (10)$$

K_{rp} - coefficient of calibration of the rutting depth development

PRDM- rutting depth envisaged for the section
ORDM- measured rutting depth

It must be noted that, as per the HDM recommendations on calibration level 2 it is sufficient to determine the coefficient of calibration of the overall rutting depth, whereas for level 3 of calibration, it is necessary to establish all coefficients of calibration of the individual rutting components (K_{rid}, K_{rpd}, K_{rst} и K_{rsw})

9 Measuring of rutting depth

Figure 1 gives the distribution of the increase, that is, the deepening of the rutting in the periods of analysis. The figure includes all rutting measures, except on sections where subsequent measurement indicates errors. The determined precision of rutting measuring is of 1mm. The average annual increase of the rutting is of 4.2 mm (standard deviation = 1, and the changes vary from 0 to 7,4 mm/annually).

Remarks from this figure include the following:

- The annual increase of the rutting depth is within 4mm for all intervals of analysis.
- The maximal annual increase of the rutting is of 11mm, and the minimal one is of 2mm.

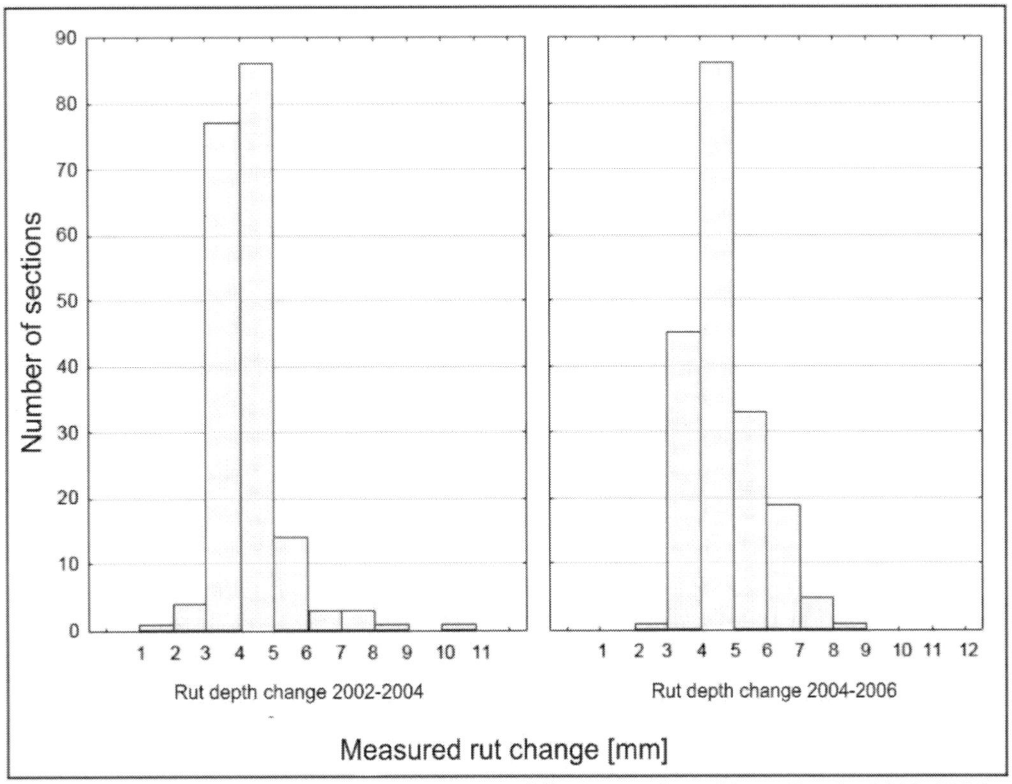

Fig. 1. Rutting increase distribution. Source: Public Enterprise of State Roads, 2002-2006

10 Calibraiton of the HDM rutting model

The format of the HDM-4 rutting model has been adopted within this research fort he establishment of the calibration coefficient. Figure 2 shows the change in the rutting depth for the non-calibrated model (the calibration coefficient being of 1).

Figures_2 and 3_indicate that there is no coincidence between the envisaged and the factual change of the rutting depth. The measured data demonstrate an annual change of 4mm, while the envisaged change of the rutting depth is of 2mm per year. On the other hand, the non-calibrated model demonstrates that the depth of the rutting on the new pavements is of 20mm, which is not appropriate (the line of the envisaged and appeared depths and changes of the rutting begins at 20mm).

The acquired calibration coefficient as pursuant to the data obtained in Macedonia is given in Table 6 (equation 9).

Table 6. Calibration coefficient for rutting depth

Type of data	Calibration coefficient (K_{pd})	Error function RMSE
All data	0.62	3657.7

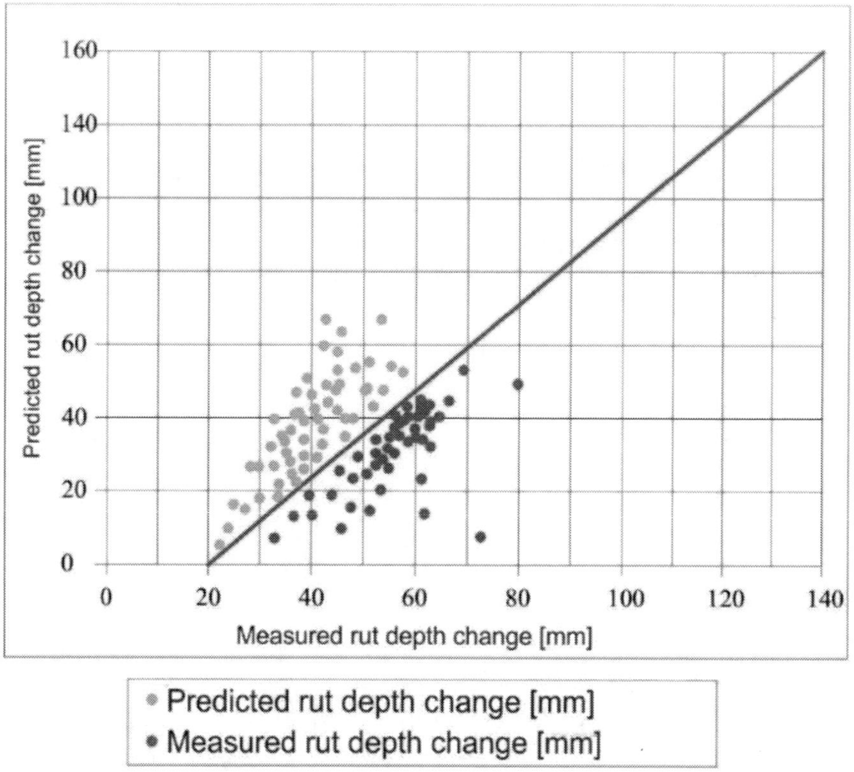

Fig. 2. Relation between the predicted and measured change of rutting depth of the non-calibrated HDM model (Default model) Source: Own research

Figure 4 demonstrates that the relation between the measured and the real change of the rutting on annual level is already established by the new value of the calibration coefficient, K_{pd}=0.62.

Figure 5 shows that the new calibration coefficient establishes a practically ideal relation between the envisaged and the measured data on the rutting depth. It leads to the conclusion that the rutting development method with the adopted calibration coefficient can be used with high level of certainty.

Note:

The RMSE error function is obtained as a square root of the difference between the envisaged and the real values of the rutting.

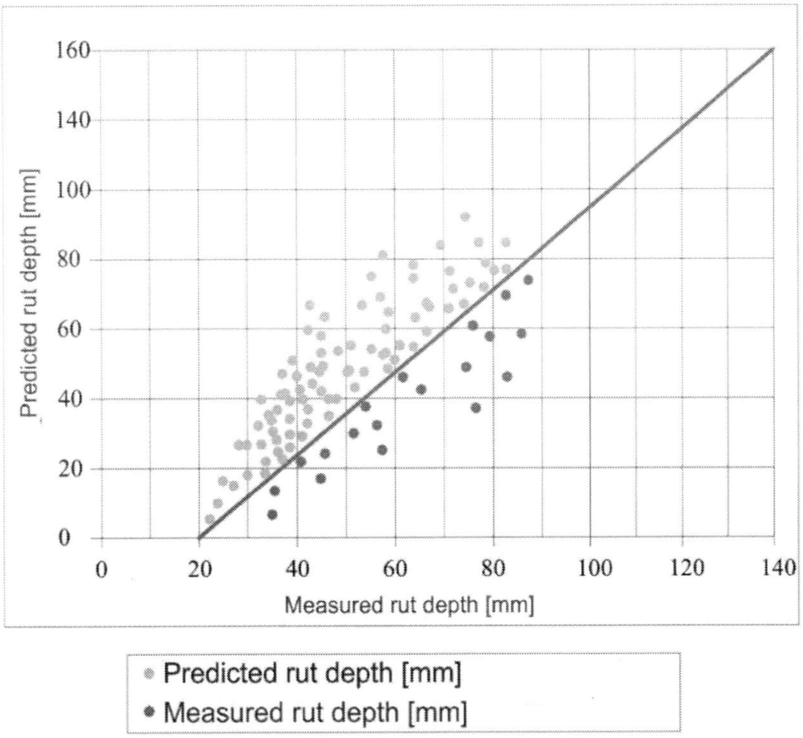

Fig. 3. Relation between the envisaged and the measured rutting depth in the non-calibrated HDM model (Default model) Source: Own research

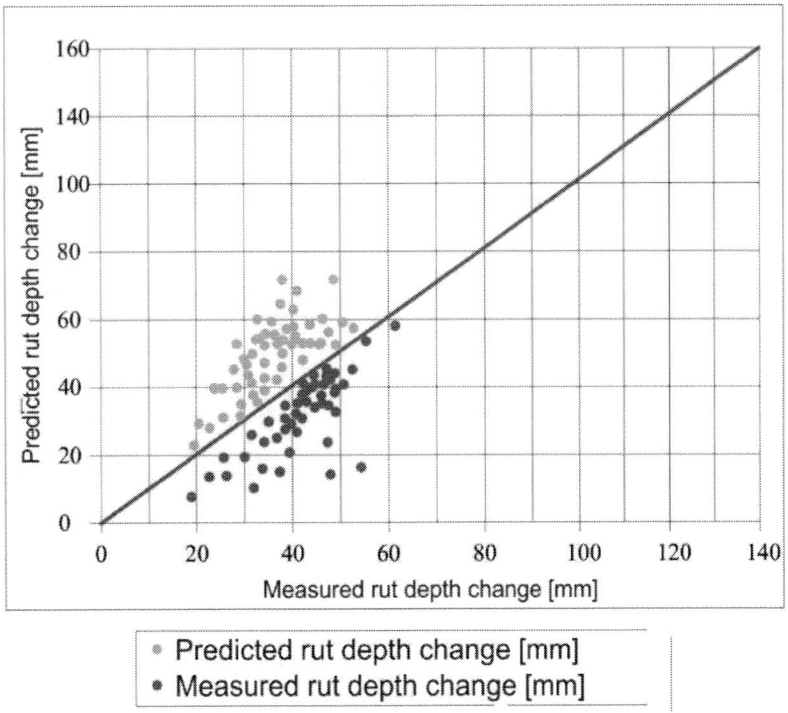

Fig.. 4. Relation between the envisaged and measured change of the rutting depth in the calibrated HDM model. Source: Own research

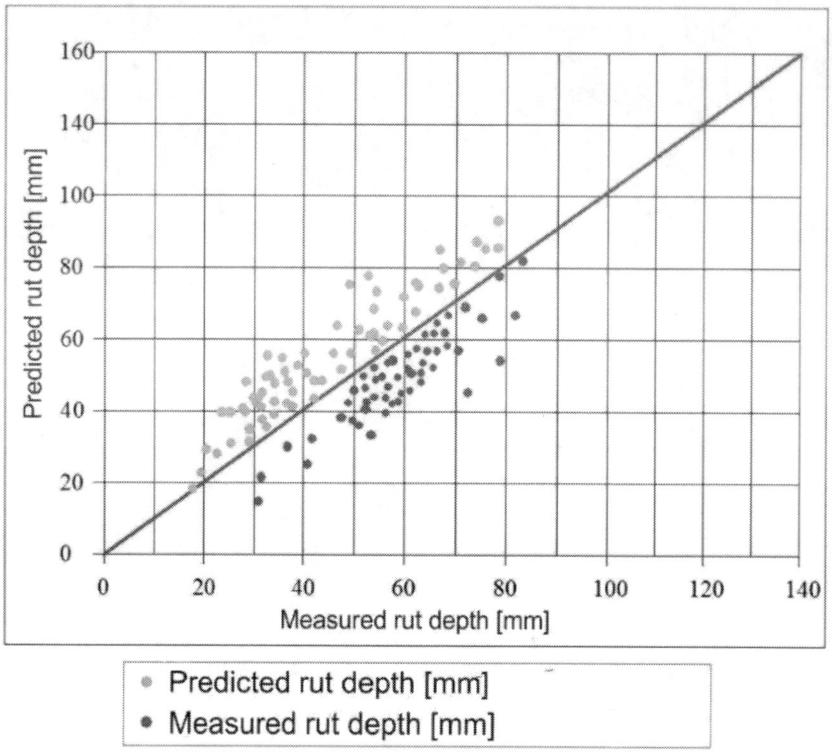

Fig. 5. Comparison between the real and the envisaged rutting depth for the HDM-4 calibrated model. Source: Own research

It must be noted that some anticipations have been made regarding certain sizes on which there are not data.

Relative compacting: $COMP = 98\%$

Crack percentage: $ACX = 3 - 17\%$ for pavements aged from 10 to 20 years

Heavy-duty vehicle velocity: $S_h = 50 km / h$

Use of salt in winter maintenance : $SALT = 2$ (salt used)

Average traffic flow velocity: $S = 80 km / h$

Average annual number of vehicles with winter tyres: $PASS = 0.6PGDS$

Conclusion

The obtained calibration coefficient yields satisfactory results in comparison with the non-calibrated model. However, it must be noted that the research lacked new data on the rutting depth.

The only available data were the ones obtained in the period of 2002-2006 and 2010-2014. Despite the incompleteness of the measures conducted from 2006-2010, they were sufficient to offer control on the rutting development in the period from 2010 to 2014.

Nevertheless, more comprehensive analyses are necessary, within the period of at least 10 years as well as possible development of another model, which in fact is calibration of level 3 according to HDM.

https://doi.org/10.1051/e3sconf/20183302023

References

1. C.R. Bennett, Paterson, W.D.O., A Guide to the Calibration and Adaptation, HDM - 4. Volume 5, The Highway Development and Management Series, pp. 48-54 (2000)
2. C.R. Bennett, Review of Experience Adapting the HDM Vehicle Operating Costs Model, Report to the International Study of Highway Development and Management Tools, pp. 78-92 (1997)
3. P.D. Cenek, Experimental Design for Determining Calibration Coefficients for HDM-III Field Deterioration Models incorporated in RIMS dTIMS, pp. 161-182 (1995)
4. W.D.O. Paterson, Scullion, T., Information Systems for Road Management: Draft Guidelines on System Design and Data Issues, World Bank, Infrastructure and Urban Development Department, pp. 202-216 (1990)
5. S. Ognjenovic, Calibration of the distress models of the pavement, Doctoral thesis, Faculty of Civil Engineering in Skopje, pp. 95-112 (2014)
6. G.T. Rohde, et. al. Henning, TFP, The Calibration and Use of HDM-4 Performance Models in a Pavement Management System, 4th International Conference on Managing Pavements. ICC Durban, (1998)
7. S., Rowshan et. al. Long-Term Pavement Performance Information Management System Data Users Guide, FHWA-RD-93-094. Federal Highway Administration. U.S. Department of Transportation. McLean, Virginia, (1993)
8. M. Y. Shahin, Pavement Management for Airports, Roads and Parking Lots, New York, (2005)

Choosing the function of mechanical properties of grounds and rock formations due to their heterogeneity

Irina Frolova[1,*] and *Murad* Agakhanov[1]

[1] Moscow State University of Civil Engineering, Yaroslavskoye sh., 26, 129337 Moscow, Russia

Abstract. The development of computing techniques to analyze underground structures, buildings in high-rise construction that would fully take account of the conditions of their design and operation, as well as the real material properties, is one of the important trends in structural mechanics. For the territory in high-rise construction it is necessary to monitor the deformations of the soil surface. When high-rise construction is recommended to take into account the rheological properties and temperature deformations of the soil, the effect of temperature on the mechanical characteristics of the surrounding massif. Similar tasks also arise in the creation and operation of underground parts of high-rise construction, which are used for various purposes. These parts of the structures are surrounded by rock massifs of various materials. The actual mechanical characteristics of such materials must be taken into account. The objective property of nearly all materials is their non-homogeneity, both natural and technological. The work addresses the matters of building nonhomogeneous media initial models based on the experimental evidence. This made it possible to approximate real dependencies and obtain the appropriate functions in a simple and convenient way.

1 Introduction

Media with mechanical properties continuous nonhomogeneity by its description can be conventionally divided into two groups to be named as media with direct and indirect heterogeneity. The first one is presented by the materials with mechanical properties that obtained various values in terms of a volume of a body under consideration during the period of manufacturing (the cementation process of the rock mass that surrounds the underground cavity, explosion loads etc.) The second group comprises materials with properties that vary during operation (temperature, radiation effects etc.) When selecting functions that describe properties change alongside the coordinates of a body from material falling into the first group, only the results of experimental investigations should be based

*Corresponding author: ifrolova51@inbox.ru

on, hence one can approximate real dependencies so that the functions obtained are relatively simple and still represent the facts.

In problems with indirect inhomogeneity the change of mechanical characteristics alongside the body was brought about not only by the dependence of properties upon a trigger (temperature, radiation) but upon distribution of this trigger in the body as well. The first part of the problem can be solved similarly as in the bodies with direct inhomogeneity, while the second part results from some equations solution, for instance, a heat conduction equation in thermoelasticity problems.

The selection of mechanical characteristics variation law in rock formations and grounds is base on ideas of rocks and grounds mechanics as well as experimental research.

The current investigations in the Earth's physics [1], rock mechanics [2] and ground mechanics [3] the natural (earthy) non-homogeneity is characterized by a medium modulus of elasticity variation by depth. Virgin ground modulus of elasticity is represented by means of exponential function of depth [1,3].

Along with this, the construction of mine workings (underground tanks) means an artificial change in the mechanical characteristics of the surrounding massif. The massif acquires the so-called technological heterogeneity due to the penetration of the mine. Small violations of physical continuity arise on the contour of the cavity and rapidly decay with fading into the interior of the massif.

As the results of existing experimental studies show [4], the resulting fracturing leads to a decrease in the modulus of elasticity. Near the hole, where the fracture is more intense, the elastic modulus is small, and as it moves away from the cavity contour into the interior of the massif, it grows asymptotically tending to the value existing before cavity penetration (Fig. 1).

In this case, the modulus of elasticity of a weighted medium with a spherical cavity is a function of the radius and can be adopted in the form [4]:

$$E(r) = E_0\left[1+(k_1-1)\left(\frac{a}{r}\right)^m\right] \tag{1}$$

where E_0 - is the modulus of elasticity in the virgin rock mass; $k_1 = \frac{E_a}{E_0}$

E_a - is the modulus of elasticity on the cavity countour, for $r = a$;

m - is an empirical coefficient.

The value of the parameters k_1 and m depends on the size of the disturbance zone and is determined experimentally [4]. The dimensions of the zone of artificial fracturing are directly dependent on the explosion power and in inverse relationship to the strength of the rocks being destroyed [4].

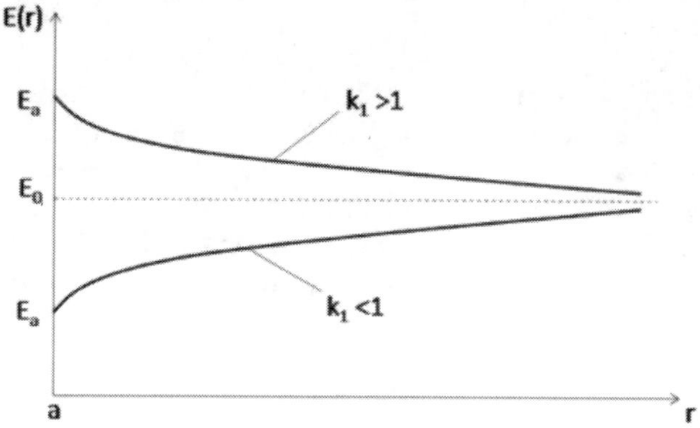

Fig. 1. Variation of rock mass modulus of elasticity near the cavity created by explosion.

2 Methodology

The ratio (1) properly describes experimental curves. Choosing the appropriate m, you can approximate a wide range of dependencies with both a slow and fast rate of change in the value of E, which corresponds to a larger or smaller area of the massif with changed properties.

Parameter $k_1 > 1$ in expression (1) corresponds to the case of rock compaction during explosion, and $k_1 < 1$ - to loosening (fracturing).

Expression (1) is successfully used by the authors of works devoted to solving problems with allowance for a one-dimensional inhomogeneity [5, 6].

When solving problems related to the determination of the stress-strain state of a rock mass under the influence of temperature effects, it becomes necessary to take into account the so-called indirect heterogeneity, while the construction of the dependence of the deformation characteristics (modulus of elasticity) on the coordinates consists of two stages.

First, the dependence of the modulus of elasticity on the temperature acting in the considered massif is determined. Thus, studies [7] of the character of the change in the Young's modulus in soils (clays) in the case of a uniaxial stress state at different temperatures have shown that with an increase in the heating temperature, this basic deformation characteristic of soils decreases (Fig. 2).

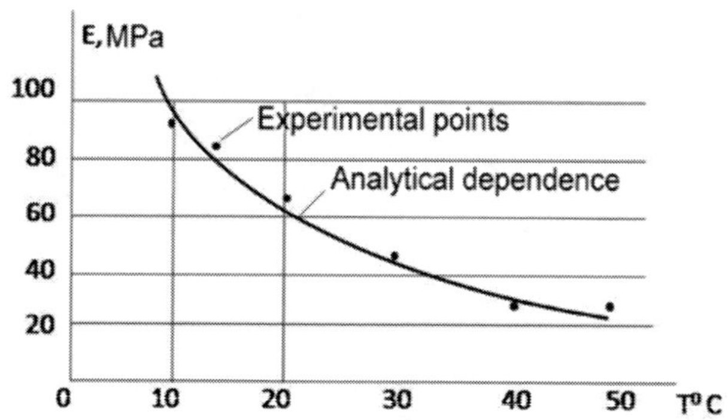

Fig. 2. Dependence of clay elasticity modulus on temperature.

Similar studies have also been carried out for rocks [2], but already for triaxial compression at much higher temperatures, since at relatively low temperatures the rocks (eg basalt) practically do not change their elastic properties (Fig. 3). Here, as in the first case, when the temperature rises, the modulus of elasticity decreases significantly (for example, in the granite, the Young's modulus at room temperature is almost three times larger than at T= 800 ° C).

The results of the obtained experimental studies can be approximated with sufficient accuracy by means of a fairly simple functional dependence of the elasticity modulus of the material of the massif on temperature:

$$E(T) = E_0 \exp(-\delta T) \tag{2}$$

Where E_0 – is the modulus of elasticity of the unheated massif; T – is the heating temperature; - δ is the empirical coefficient.

Fig. 2 and Fig. 3 show approximating relationships of the type (2). It can be seen that in some cases, the coincidence with the experimental data is quite good throughout the temperature range (Fig. 2), and on some curves the function (2) makes it possible to describe the change in the modulus of elasticity as a function of temperature only on a part of the interval (Fig.3).

For superhard rocks of the basalt type, the relation (2) can be somewhat refined:

$$E(T) = E_0 \qquad at T \le 300^\circ C$$
$$E(T) = E_0 \exp(-\delta T) \; at T > 300^\circ C$$

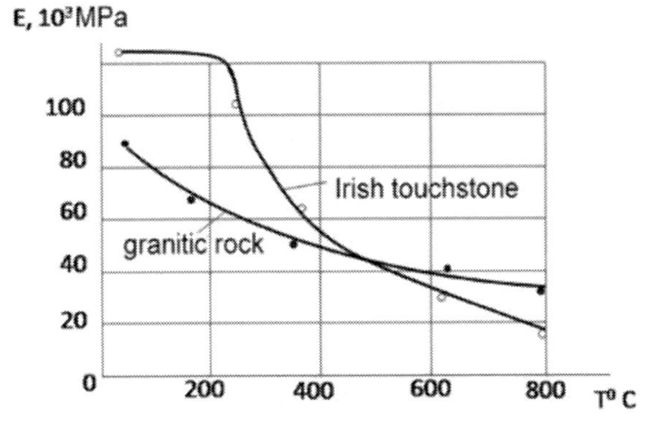

Fig. 3. Dependencies of solid rock elasticity modulus on temperature.

If the task related to the determination of temperature stresses considers a stationary heat flux, for example, caused by storage of products that tend to self-heating in the cavity (a constant temperature is maintained on the cavity surface), then the corresponding steady-state regime for a spherical cavity can be described by the dependence:

$$T(r) = \frac{(T_0 - T_1)ab}{r(b-a)} + \frac{T_1 b - T_0 a}{b-a} \tag{3}$$

where T_0 - is the temperature on the contour of the cavity; T_1 - in the massif for heating; a – is the radius of the cavity; b >> a is the conditional radius of the outer contour of the massif.

3 Results

Thus, in such tasks, taking relations (2) and (3) into account, it is easy to go from the E(T) dependence to E(T(r)) dependence.

If it is necessary to take into account both the temperature and the cracking of the rock, then the dependence of the modulus of elasticity can be written in the form [8]:

$$E(T) = E_0 \exp(-\delta T(r))\left[1 + (k_1 - 1)\left(\frac{a}{r}\right)^m\right] \tag{4}$$

4 Conclusions

One of the important issues that are considered in many studies on the calculations of inhomogeneous bodies is the question of the influence of the Poisson's ratio on the stressed state. Due to the lack of experimental data, this coefficient and the density of the material were considered to be constant in the majority of works, although, in principle, one can also assume their dependence on the coordinates of the points of the body.

References

1. F.D. Stacey, *Physics of the Earth* (Mir, Moscow, 1972)
2. G.A. Krupennikov, N.A. Filatov, *Stress distribution in rock masses* (Nedra, Moscow, 1972)
3. N.A. Tsytovich, Ground mechanics (Higher school, Moscow, 1979)
4. V.V. Rukin, K.V. Ruppeneit, Interaction mechanism of pressure tunnel lining with rock mass (Science, Moscow, 1969)
5. V.I. Andreev, Computational Methods and Experimental Measurements, **2**, 202-216 (2014)
6. V.I. Andreev, A.S. Avershyev, Procedia Engineering, **111**, 30-35 (2015)
7. S.R. Meschyan, *Grounds mechanical properties and laboratory methods of their identification* (Nedra, Moscow, 1974)
8. I.I. Frolova, Scientific review, **12**, 94-99 (2015)

Calculation of PVC windows for wind loads in high-rise buildings

Aleksandr Konstantinov [1,*] *Maya* Lambias Ratnayake[1]

[1] Moscow State University of Civil Engineering, Yaroslavskoye shosse 26, Moscow, 129337, Russia

Abstract. In the following article we examine problems faced when designing PVC windows for high-rise buildings, which are usually not considered when constructing objects for massive sites, using a high-rise residential complex as an example. We address the matters related to wind loads on windows & statistical calculation of the impact of wind loads on them. We have presented variants of installing load-bearing elements of PVC windows which accept wind loads. We conducted a laboratory experiment by simulating wind loads on the window design, which is actually used for glazing the examined high-rise building. In the course of the experiment we determined additional factors which need to be considered when constructing PVC window structures for glazing high-rise buildings. We can determine that the following calculation method for the impact of wind load on PVC windows gives higher values of the desired statistical characteristics of load-bearing elements of a window compared to the results of laboratory experiments. We provide prerequisites to improve the analytical method of calculating impact of wind loads on load-bearing elements of PVC windows.

1 Introduction

Currently, the construction volume of high-rise buildings in the world is constantly increasing, especially in the residential sector [1-5]. To reduce the cost of construction of high-rise buildings, we introduce technologies which were previously used only in mass construction sites. One such example is using PVC windows when glazing high-rise buildings instead of using traditional aluminium windows. PVC windows have lower cost & have better thermotechincal characteristics compared to aluminium windows. Nevertheless, initially PVC window systems were designed to be used low-rise & standard multistorey constructions. The use of PVC windows in high-rise buildings, which are subjected to high values of wind loads during construction & operation, needs consideration as to the technological construction features of the PVC window system, as well as the structural concept, parameters & the construction technology of the building. In the article we examine the questions of designing & statistical calculation of PVC windows using a high-rise apartment building which is 140m in height as an example.

* Corresponding author: apkonst@yandex.ru

2 Description of high-rise buildings. Description of the structural concept of windows

The considered building is a 37-storey tower shaped in the form of a rhombus in the plan and the total height is 140.3 m (figure 1). The building is located in a densely built-up urban area. For glazing the building we use large-size structures of window units from the PVC profile. The designed window unit with load area of maximum collection of wind loads is presented in figure 1.

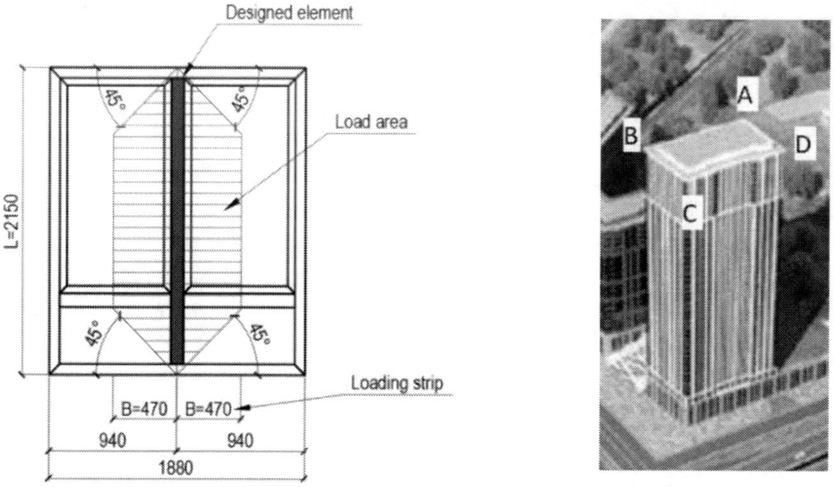

Fig. 1. Scheme of the designed window unit and General view of high-rise buildings and

3 The function of wind loads on PVC windows of high-rise buildings

The application of calculated values of wind loads of translucent structures in standard building construction sites are conducted according to valid normative documents [6,7,8]. During which, the calculated value of wind loads depend on the shape of the building, the wind area & the terrain type of construction area & it increases when the height of the building increases. For high-rise buildings using this approach is incorrect, as it does not take in to account formation features of air streams of high-rise buildings in their surrounding building area. For high-rise buildings it is essential to conduct a simulation of wind loads in laboratory conditions using an aerodynamic tube [9-12], or by using special software programs [13-16]. Research was conducted in the aerodynamic tube for the studied building to determine the actual wind load on the building. The results of simulation showed that the calculated value of wind loads on the building reach values in absolute magnitude equal to W_{mod}=1440 Pa. Meanwhile, the calculated value of wind loads obtained from valid normative documents was equal to in absolute magnitude W_{norm}=2130 Pa (i.e. 1,4 times more). W_{mod} is adopted as the calculated wind pressure on windows of the building.

4 Analytical calculation of PVC Windows for the impact of wind loads

According to the valid normative documents [8,17,18] statistical calculation for impact of wind loads on profile elements of the window unit are conducted according to conditions of inadmissibility of excessive infiltration of cold air.

Required moment of inertia of load-bearing elements of the window unit are determined by formula 1

$$J_{req} = \frac{W \cdot L^4 \cdot B}{1920 \cdot E \cdot f}\left[25 - 40\left(\frac{B}{L}\right)^2 + 16\left(\frac{B}{L}\right)^4\right] \quad (1)$$

W – the calculated value of wind load, N/mm^2;
B – width of loading strip in the stress diagram, cm;
L – working length of the load-bearing elements of the window unit , cm;
E – modulus of elasticity, Pa. For steel E = 210 000 N/mm^2;
f – maximum relative deflection of the load-bearing element of the window unit.

Maximum relative deflection of profile elements of window units are taken as equal to

$$f = \frac{1}{300}L\,. \quad (2)$$

It is considered that if the relative deflection of load-bearing elements of the window unit does not exceed the specified value, the wind does not blow through the window. For the window unit shown in figure 2, the required moment of inertia of the vertical mullion will be J_{req} = 23.15 cm^4.

5 Technological limitations of PVC window systems

The existing rules for the statistical calculation of PVC window units is based on the assumption that operational loads (wind pressure, own weight of the windows, etc.) occurs only by steel liners installed inside profile elements made of PVC.

We can provide the required characteristics of the design elements of large format PVC windows, based on the construction characteristics of the PVC window systems in two ways. In the first case, the large format window unit is divided into two parts & a front-connector is installed in between, which receives the wind load (see figure 2). In the second case the window unit is not broken into parts, but an external amplifier is installed in the vertical mullion (see figure 3). In an architectural & structural point of view, the second option is optimal because using first option significantly reduces light transmission through the window.

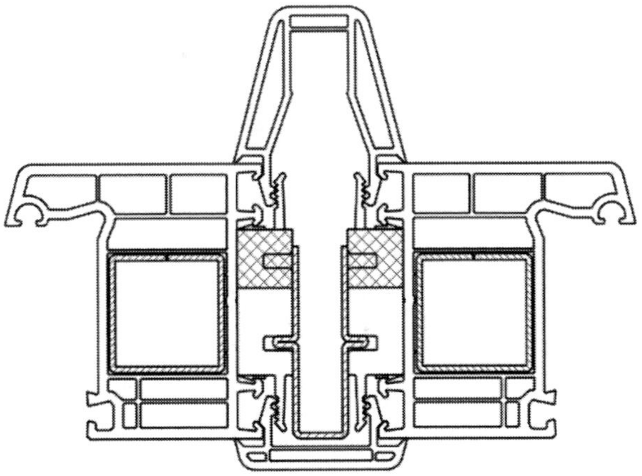

Fig. 2. Variant of strengthening PVC window units using front connectors

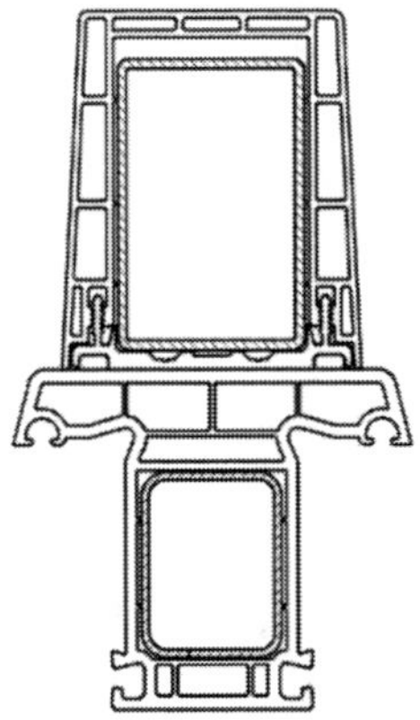

Fig. 3. Variant of strengthening PVC window units using external amplifiers.

For the considered case the actual value of the moment of inertia of a vertical mullion with an external amplifier, which is designed according to valid normative documents, is;

$$J_{req} = J_{imp} + J_{st} = 2.68 + 18.87 = 21.55\,sm^4$$

However, according to existing calculation methods when determining the actual moment of inertia of the mullion only the moment of inertia of the reinforcing liners are taken into account.

6 Laboratory experiment on simulating the impact of wind loads on PVC window

To determine the actual work of the adopted window structure under the impact of calculated values of wind pressure, a laboratory experiment was set up. The experiment was conducted on a special test stand to determine the operational performance of window units under the impact of excess pressure of air and atmospheric precipitation.

The test stand consisted of 3 basic units; test unit, control unit and computer unit. The test unit was a closed sealed chamber where the test sample is installed on one side. The control unit consisted a compressor to create positive or negative pressure inside the test unit, as well as a set of automatics and detectors to measure parameters such as temperature-humidity conditions inside the laboratory and operational factors of the test samples (permeability and resistance to wind loads). The Computer unit of the stand included a seat for the operator and the computer, which was used to determine the function of the testing program and to control the stand. To determine the relative deflections of the load-bearing elements of the window, the stand was equipped with electronic linear displacement detectors.

Before testing, the window unit was installed in a wooden frame. Fixating the window block in the frame (step and the installation of fasteners) was performed on the basis of valid standards [8,19,20]. General views of the stand with the installed test sample are shown in figures 6 and 7.

Fig. 4. General view of installed test window unit

Fig. 5. Linear displacement detectors to determine the relative deflection of mullions

7 Analysis of the results

As a result of the conducted experiment we found that when the window unit is under the impact of wind pressure equal to W_{mod}=1440 Pa & the calculated actual moment of inertia was equal to 21,55 см^4, the deflection of the vertical mullion was 5.37 mm (1/400 of the calculated length of the mullion, taken as equal to the height of the window unit). To ensure this relative deflection of the mullion, required moment of inertia of the mullion should be in accordance with the valid method of calculation (see formula 1) and equal to 30,86 см^4, which is 43% higher than the actual value. It is obvious that statistical calculation methods for PVC window units currently used by structural-designers allow the calculations to be made with redundant deflections. In order to reduce the cost of window units is necessary to optimize this technique.

It is observed that when the wind pressure is more than 600Pa, wind blows through the upper area of the hinges of window unit accompanied by a characteristic whistle. Wind blowing occurred due the contraction of upper hinges of the casement window sealing. Accordingly, in the course of the experiment we observed that – when relative deflections do not exceed normative values (1/300 L), we do not follow the basic calculation criterion of impact of wind loads on window units – thus, no wind blowing.

8 Discussions

According to the valid methodology [8], during the statistical calculation of mullions of PVC window units, only the rigidity of steel amplifier of the mullion is taken into account. The rigidity of profile of the mullion, as well as spatial work of the mullion with casement window is not included in the calculations.

It is proposed to develop a new method for calculating the impact of wind load on window units, which would take the above parameters in to consideration. We propose to conduct numerical modeling and a series of laboratory experiments in order to establish the actual work of window blocks of PVC by the impact of wind loads. It is proposed to conduct the research system for "mullion+casement window", which would take into account their work together under the impact of wind loads. The rigidity of the reinforcing liners as well as the rigidity of the translucent infill of PVC profiles should be taken in to account. On the basis of these subsequent studies, it is possible to form a refined analytical method to statistically calculate the impact of wind load on of PVC window units.

9 Conclusions

We compared the existing methods for calculating the impact of wind loads on PVC window units with the results of laboratory tests of the actual design of the window unit. It was determined that in the existing method for calculating the impact of wind loads on PVC window units includes a considerable redundancy for rigidity (40 %). It is established that when the wind pressure is above 600 Pa, while meeting regulatory requirements for deflections, we can observe wind blowing through upper hinges of window units. When installing PVC window frames in high-rise buildings we recommend the using concealed hinges.

We propose a new method of calculation to determine the impact of wind loads on window units which would take into account the collective work of the constituent elements of the window unit (reinforcing liners, window profiles with translucent infill, etc.) in order to reduce the cost of manufacturing window units by using reinforcing profiles of smaller cross section.

References

1. E. Generalova, V. Generalov, CTBUH, **4**, 40-45 (2014)
2. L. Sedegova, Vestn. Dona, **2(25)**, 123 (2013)
3. E. Generalova, V. Generalov, Social, humanitarian, medicobiological sciences, **16**, 458-463 (2014)
4. Nechaeva, Gorneva, The Fourteenth International Scentific Conference Proceedings, **1**, 268-275 (2014)
5. V. Shpara, Modern trends in design and constructing high-rise complexes, **1-2(8)**, 7-8 (2013)
6. EN 1991-1-4:2005+A1:2010
7. Russian State Standard SP 20.13330.2016
8. Russian State Standard GOST R 56926-2016
9. M. Mikhaylova , V. Dalinchuk , A. Bushmanova, L. Dobrogorskaya, Construction of Unique Buildings and Structures, **10**, 59-74 (2016)
10. Povzun, Buzun, Zimin, Construction of Unique Buildings and Structures, **3(30)**, 70-80 (2015)
11. Poddaeva, Dunichkin, Industrial and civil engineering, **4**, 42-45 (2016)
12. A. Kubilay, D. Derome, J. Carmeliet, Building and Environment, **114**, 68-81(2017)
13. A. Kubenin, A. Fedosova, Scientific Review, **8**, 130-135 (2015)
14. A. Kubenin, A. Fedosova, Scientific Review, **8**, 136-141 (2015)
15. S. Isaev, P. Baranov, Y. Zhukova, A. Tereshkin , A. Usachov, Journal of engineering physics and thermophysics, **87**, 112-123 (2014)
16. O. Poddaeva , S. Dubinsky, A. Fedosova, Industrial and civil engineering, 9, 23-27 (2014)
17. DIN 18055:2014-11
18. DIN EN 14351-1:2016-12
19. Russian State Standard GOST 30791-2012
20. Russian State Standard GOST 30674-99

Approximate method for calculating a thick-walled cylinder with rigidly clamped ends

Vladimir Andreev[1,*]

[1] Moscow State University of Civil Engineering, Yaroslavskoe shosse, 26, Moscow, 129337, Russia

Abstract. Numerous papers dealing with the calculations of cylindrical bodies [1 -8 and others] have shown that analytic and numerical-analytical solutions in both homogeneous and inhomogeneous thick-walled shells can be obtained quite simply, using expansions in Fourier series on trigonometric functions, if the ends are hinged movable (sliding support). It is much more difficult to solve the problem of calculating shells with built-in ends.

1 Introduction

As was shown in [8], the natural restriction of the method of separation of variables in the calculation of radially inhomogeneous cylinders is the impossibility of satisfying arbitrary boundary conditions at the ends. This paper considers an approximate method for solving the axisymmetric problem for a thick-walled cylinder with rigid fixed ends. The problem is solved in displacements.

2 State of the problem

In solving the axisymmetric problem will be sought displacements in the form:

$$u(r,\theta,z)=\sum_{m=0}^{\infty}\sum_{n=0}^{\infty}u_{mn}(r)\cdot\cos m\theta\cdot\cos k_n z;$$
$$v(r,\theta,z)=\sum_{m=1}^{\infty}\sum_{n=0}^{\infty}v_{mn}(r)\cdot\sin m\theta\cdot\cos k_n z; \tag{1}$$
$$w(r,\theta,z)=\sum_{m=0}^{\infty}\sum_{n=1}^{\infty}w_{mn}(r)\cdot\cos m\theta\cdot\sin k_n z,$$

From (1), the boundary conditions automatically follow

* Corresponding author: asv@mgsu.ru

$$z = 0, H; \quad w = \tau_{rz} = \tau_{\theta z} = 0 \tag{2}$$

which correspond to the hinged (sliding) support (H – cylinder height).

Below we consider an approximate calculation method cylinder with rigid fixing ends. In this case, the boundary conditions at the ends are written in the form:

$$z = 0, H; \qquad u = w = 0. \tag{3}$$

Fig. 1 shows half the cross-section of a cylindrical shell for two ways of fixing the ends.

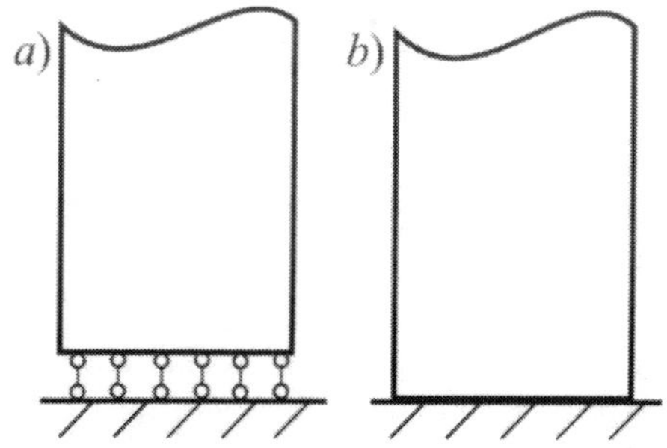

Fig 1. Thick-walled cylinder, a – hinged movable support, b – built-in (rigid fixing) end.

3 Approximate method for calculating

The proposed method is based on the algorithm for calculating cylinders for the action of surface loads, which use an expanded area, compensating loads and the collocation method. We consider the problem when the external loads are symmetric with respect to the plane $z_1=H/2$ (Fig. 2). Let us increase the height of the cylinder in both directions by an amount δ and introduce the coordinate $z=z_1+\delta$ We apply to the lateral surfaces of the cylinder in the interval $[0,\delta]$ and $[H+\delta, H+2\delta]$ such compensating loads p_a^c, p_b^c, q_a^c,q_b^c, that at the given points $r=r_1$ and $r=r_2$ at $z=\delta$ and $z=\delta+H$ (Fig. 3) the following conditions are satisfied:

$$r = r_1, r_2; \qquad u = w = 0. \tag{4}$$

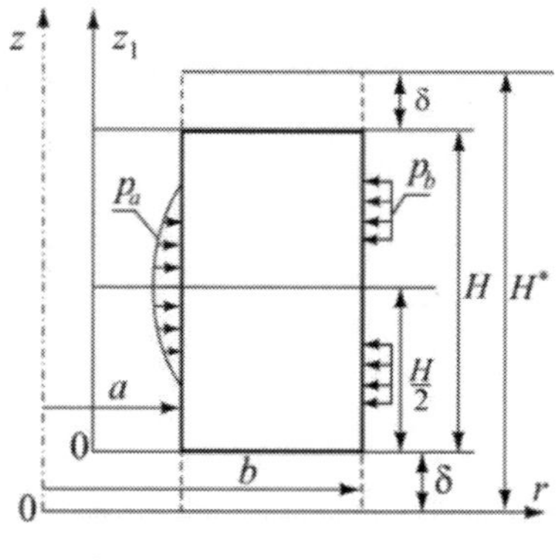

Fig. 2. Calculation scheme for a cylinder with an expanded area.

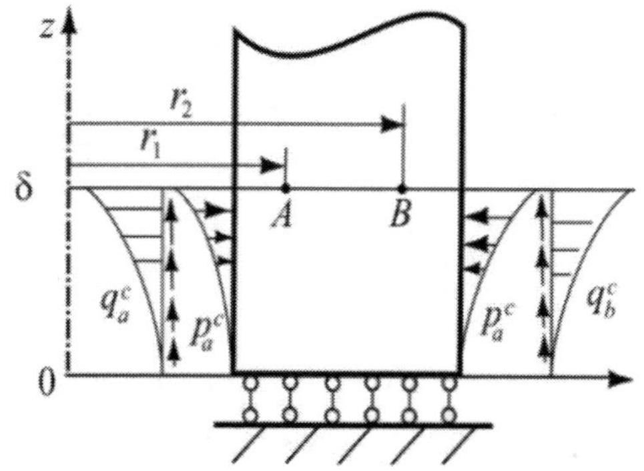

Fig. 3. Compensating loads in expanded area.

If equalities (4) are considered as conditions of real fixation, then it should be added that at the other points of the layers $z=\delta$ and $z=\delta+H$ the displacements can be either positive or negative. In this case, it is necessary to accept the hypothesis of an ideal contact between the ends of the cylinder and the support surfaces, as is done in most contact problems, for example, in the calculation of beams on an elastic foundation. Let us consider the nature of the compensating loads. If we imagine that these loads are the reaction of a semi-infinite array to the walls of a cylinder embedded in it, then as the distance from the layer $z=\delta$ to the layer $z=0$, the tangential and normal loads should tend to zero. For a description of such loads can be approximately selected functions of the form:

$$f=\begin{cases} f_0\left(\dfrac{z}{\delta}\right)^{\gamma} & 0\le z\le\delta; \\ 0 & \delta\le z\le\dfrac{H}{2}+\delta. \end{cases} \tag{5}$$

Such functions well approximate normal loads and worse – tangents, which near the layer $z=\delta$ have a peak character and when $z=\delta$ they are equal to zero. Nevertheless, if we bear in mind that when the compensating loads are decomposed into Fourier series, the peak nature of the loads q_a^c and q_b^c will also correspond quite accurately to their representations in the form of a series (Fig. 4).

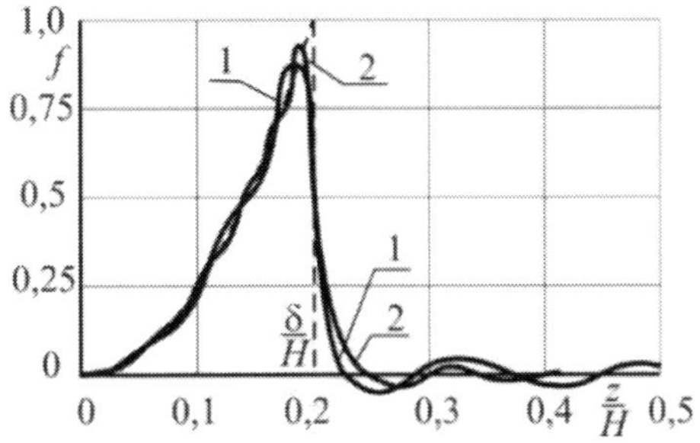

Fig. 4. Representation of function f by the Fourier series : - - - - formula (5), ——Fourier series (1 - $N=24$; 2 - $N=48$).

The essence of this method is as follows. Representing compensating loads p_a^c, p_b^c, q_a^c,q_b^c, in the form (5) and varying the parameters δ and γ, it is necessary to achieve the conditions (4) at the points *A* and *B* (see Fig. 3). It should be added that when using the collocation method, large deficiency of conditions (4) are possible at the remaining points of the interval (a,b) on the layer $z=\delta$. Therefore, the choice of the above parameters of the functions (5) must also be accompanied by an analysis of the indicated deficiency and its minimization. In this case, it is still possible to select the radii r_1 and r_2, determining the position of the points *A* and *B*.

The following sequence of solution of the problem is proposed. At the first stage, using the numerical-analytical method presented in (7, 8), the problem of the effect of external force (p_a, p_b, q_a, q_b) and temperature loads on a cylinder of length $H+2\delta$ is solved. At the same time presence of hinged (sliding) supports at the ends $z=0$ and $z=H+2\delta$, give the possibility of using the method. As a result of the solution, displacements u_p, w_p at points *A* and *B* are determined.

The index p denotes the displacements from external (active) loads. The condition of symmetry with respect to the plane $z=H/2+\delta$ allows us to restrict ourselves to the consideration of displacements on the layer $z=\delta$.

Further problems are solved successively to the action of compensating loads p_a^c, p_b^c, q_a^c and q_b^c, as defined in (5). The highest values of these loads are assumed to be equal to unity (unit loads). Defining the displacements from unit loads at points *A* and *B*, one can construct a system of algebraic equations corresponding to conditions (4):

$$U^c \cdot K + U_P = 0, \tag{6}$$

where U^c is the displacements matrix of unit compensating loads of 4 x 4:

$$U^c = \begin{pmatrix} u_{1A}^c & u_{2A}^c & u_{3A}^c & u_{4A}^c \\ w_{1A}^c & w_{2A}^c & w_{3A}^c & w_{4A}^c \\ u_{1B}^c & u_{2B}^c & u_{3B}^c & u_{4B}^c \\ w_{1e}^c & w_{2e}^c & w_{3e}^c & w_{4e}^c \end{pmatrix},$$

where the subscripts 1 ÷ 4 denote respectively the displacements from the loads p_a^c, p_b^c, q_a^c and q_b^c; $K=\{K_1,K_2,K_3,K_4\}$ is the vector of unknown coefficients; $U_P=\{u_{PA}, w_{PA}, u_{PB}, w_{PB}\}$ is the displacements vector from the active loads, including the displacements u_P, w_P at points *A* and *B*.

The coefficients $K_1 \div K_4$, found as a result of solving the system of equations (6), determine the actual values of the compensating loads at the level $z = \delta$. In other words, the compensating loads needed for the fulfillment of the conditions (4) in the section $0 \le z \le \delta$ will have the form:

$$p_a^c = K_1\left(\frac{z}{\delta}\right)^{\gamma}; \ldots q_b^c = K_4\left(\frac{z}{\delta}\right)^{\gamma} \tag{7}$$

Summarizing the solutions of five problems (one from external and four from compensating loads), we obtain the final solution of the problem that satisfies the boundary conditions.

4 Analysis of the influence of various factors on the calculation results

As noted above, the proposed method has three arbitrary factors:

- length of the fixed edge δ;
- degree of compensation load functions γ;
- choice of collocation points *A* and *B*.

Investigation of the influence of these factors on the stress-strain state of the cylinder and the determination of optimal calculation parameters was carried out in solving the problem for a homogeneous cylinder. We also considered the question of choosing the number of terms of the Fourier series *N* in the decomposition of loads and the number of steps *M* into which the integration interval is divided in the numerical solution of systems of ordinary differential equations.

On the example of the solution of the problem for a homogeneous cylinder with hinged (sliding) ends under the action of a constant external pressure $p_b = p$, let us analyze the influence of the number M on the calculation results. Obviously, in this case there is a problem of a flat deformed state, for which there is an analytic solution [4]. In particular, the axial stresses in this problem are equal to:

$$\sigma_z = -2\nu \frac{pb^2}{b^2 - a^2}. \tag{8}$$

When choosing the number M, a comparison is made of the stress values. At $M = 30$, the difference between numerical and analytical calculations did not exceed 1%. An analysis of the influence of the number of terms of the Fourier series N on the results of calculations showed that at $N = 24$ and $N = 48$, the differences in the compensating loads (see Fig. 3) at individual points reach 6%. At the same time, when calculating stresses and displacements, the difference does not exceed 3%, which is due to the Saint Venant principle. The results presented below were obtained at $N = 24$.

We now turn to an analysis of the effect of the parameters of the method of compensating loads on the value of the deficiency of the boundary conditions (4) on the interval (a,b) with the exception of the collocation points A and B, where these conditions are satisfied exactly. Let us consider the problem of calculating homogeneous cylinder loaded with an external constant pressure $p_b = p$ with rigidly clamped (built-in) ends. Note that when considering an elongated cylinder whose height is equal $H^* = H + 2\delta$, the pressure p is applied to the section $(\delta, H+\delta)$. The corresponding formulas for the coefficients of the expansion of this load in a Fourier series have the form: $p_{b,0} = pH / H^*$;

$$p_{b,n} = \frac{2p}{n\pi}\left[\sin k_n \left(H^* - \delta\right) - \sin k_n \delta\right], \text{ where } k_n = n\pi / H^*.$$

Calculations were made for the cases when the compensating loads are represented as a linear function, quadratic and cubic parabolas, to which correspond to the values of the parameter entering in (5) equal to 1, 2 and 3.

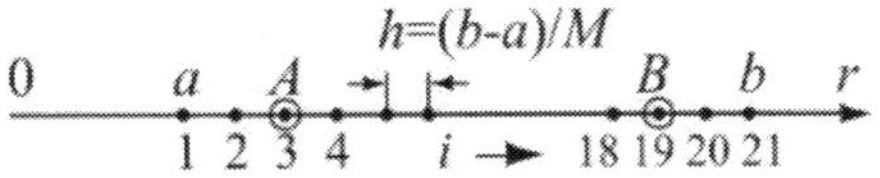

Fig. 5. Arrangement of collocation points

In addition, the depth of the "fixed edge" δ within the range of $0{,}05H \div 0{,}5H$ varied in the calculations., as well as the location of the collocation points, which were selected along the borders of the sections to which the integration interval (a,b) was divided. For $M = 20$, the number of borders of the sections is 21. Fig. 5 shows, as an example, the location of the points A and B at points with the numbers $i = 3$ and $i = 19$.

An analysis of the discrepancy between the boundary conditions on the layer $z = \delta$ was carried out by comparing the largest displacements of the compensating loads u^c and w^c on this layer with the largest displacements u^p in the middle section of the cylinder from the pressure p (the displacements w^p are negligibly small in comparison with u^p). A similar analysis was also conducted with respect to the choice of the optimal value and the depth of the "fixed edge" δ. It is established that the smallest discrepancy between the boundary conditions is reached at $\gamma = 2$. As regards the influence δ on the displacements of the points of the layer $z = \delta$, the largest displacements in this layer at $\delta > 0,2H$ remain practically constant (Fig 6). At the same time, within the range of $0.05H \div 0.5H$, the stresses σ_z change significantly. For $\delta \cong 0,5H$ the function $\bar{\sigma}_z = \sigma_z / p$ asymptotically approaches some limiting values.

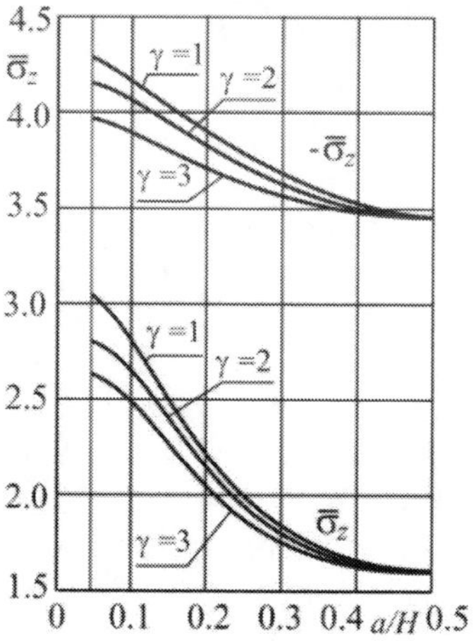

Fig. 6. Dependences stresses $\bar{\sigma}_z$ from δ at different values γ. Collocation points: 1 and 21.

Fig. 7 shows the graphs of the variation stresses $\bar{\sigma}_z$ along the axis z on the inner and outer surfaces of the cylinder, calculated with the following initial data and calculation parameters: H=2m; a=0.5m; b=1m; $E = 2 \cdot 10^4$ MPa; $\nu = 0.2$; $\nu = 0.2$; $N = 20$; $M = 24$; $\gamma = 2$; $\delta = 1$m; the collocation points are 1 and 21. In the same figure, the values of the stresses in the cylinder with the hinged (sliding) support of the ends are shown in dotted lines. In the latter case, as noted above, the problem corresponds to a flat deformed state, and the stresses σ_z are constant over the cross section (see (8)), and they are identical in all cross sections.

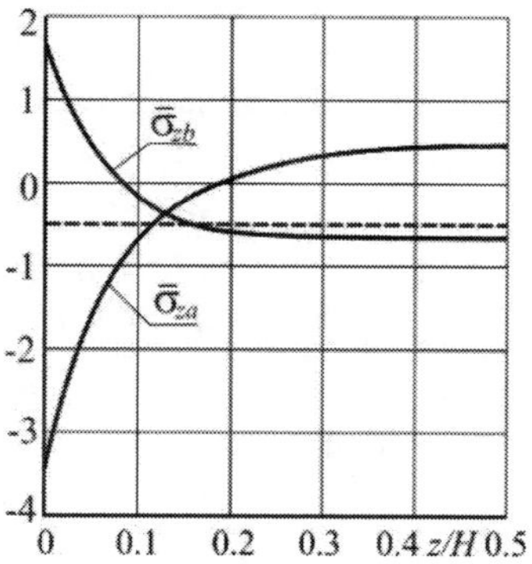

Fig. 7. Changing stresses $\bar{\sigma}_z$ along the axis z in the cylinder with: - - - - sliding support of the ends; —— rigidly clamping ends.

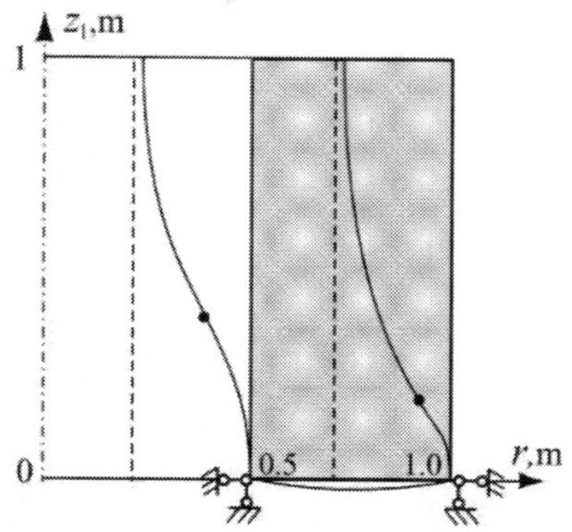

Fig. 8. Deformed state of the cylinder, - - - - sliding support, —— rigidly clamping ends, • - the inflection points.

In order to fully appreciate the difference in the results obtained in solving two problems, we consider the deformed states of cylinders with hinged support of the ends and with rigidly clamped ends (Fig. 8). At the collocation points ($r = a,b$) in Fig. 8 fixed supports are installed. Please pay attention to the following features. The deformation of the cylinder wall in the presence of a rigid clamping of the ends is qualitatively similar to the bending of a beam with clamped ends. Near the support, tensile stresses are observed from the outside, and compression is observed inside. The inflection points indicate a change in the sign of the deformations ε_z. Both recent facts agree with the stress plots in Fig. 6. Negative displacements w at the level $z_1 = 0$ are caused by the expansion of the cylinder in the axial direction due to the Poisson ratio.

4 Conclusions

Comparison of displacements and stresses in a cylinder with sliding support along the ends and in cylinder with rigidly clamped ends shows a significant influence of the type of boundary conditions on the stress-strain state of the cylinders. In sufficiently tall cylinders, this effect will act as a marginal effect, and in short cylinders and cylinders of medium height, the influence of the boundary conditions will affect the entire height [9].

This work was financially supported by the Ministry of Russian Education (7.6163.2017/ВУ)

References

1. Filon L.N.G., Phil. Trans. of the Royal Society of London, Ser. A.. V. 198. **4,** 147 (1902)
2. Prokopov V.K., PMM. V. XIII **2,** 135 (1949)
3 Volkov A.N., *Calculation of thick-walled hollow cylinders* (RUDN, Moscow, 1972)
4. Timoshenko S.P., Gudier Dj., *Theory of elasticity.* (Nauka, Moscow, 1979)
5. Pankratova N.D., Thermal stress in structural elements. **20** Kiev, 63(1980)
6. Andreev V.I., Proc. of high schools. Ser. build. and architect. **8,** 27 (1985)
7. Andreev V.I., Constr. mech. & calc. str. **2**, 8 (1989)
8. Andreev V.I., *Some problems and methods of mechanics of inhomogeneous bodies*, (ASV, Moscow, 2002)
9. Andreev V.I., Applied Mechanics and Materials. **166-169,** 354 (2012)

https://doi.org/10.1051/e3sconf/20183302027

Calculation of Prestressed Pressure Vessel Taking into Account the Concrete Temperature Inhomogeneity

Vladimir Andreev[1,*]

[1] Moscow State University of Civil Engineering, Yaroslavskoe shosse, 26, Moscow, 129337, Russia

Abstract. The paper deals with the problem of determining the stress state of the pressure vessel (PV) with considering the concrete temperature inhomogeneity. Such structures are widely used in heat power engineering, for example, in nuclear power engineering. The structures of such buildings are quite complex and a comprehensive analysis of the stress state in them can be carried out either by numerical or experimental methods. However, a number of fundamental questions can be solved on the basis of simplified models, in particular, studies of the effect on the stressed state of the inhomogeneity caused by the temperature field.

1 Introduction

In the PV which is usually constructed of reinforced concrete, under the influence of operational loads (internal pressure and temperature) there are very significant tensile stresses. In order to reduce or completely eliminate them, pre-compression in the annular and axial directions is used. Figure 1 shows a scheme of PV in which the compression in the circumferential direction is carried out by uniformly winding the metal cables along the outer surface, which leads to a constant pressure in the height [1]. Another way of pre-compression is to apply external pressure along the belts [2]. The corresponding scheme is shown in Fig.2.

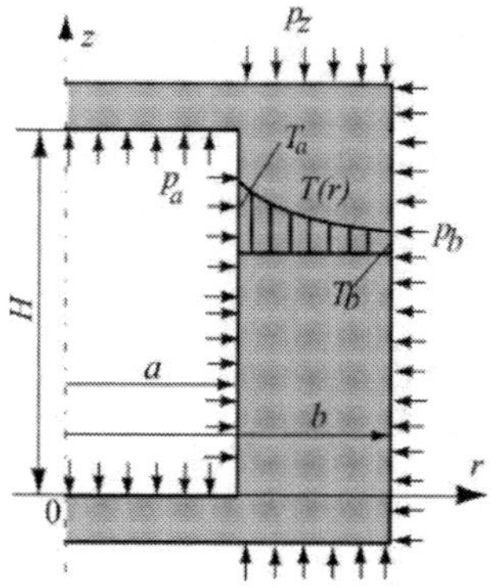

Fig. 1. Scheme pressure vessel.

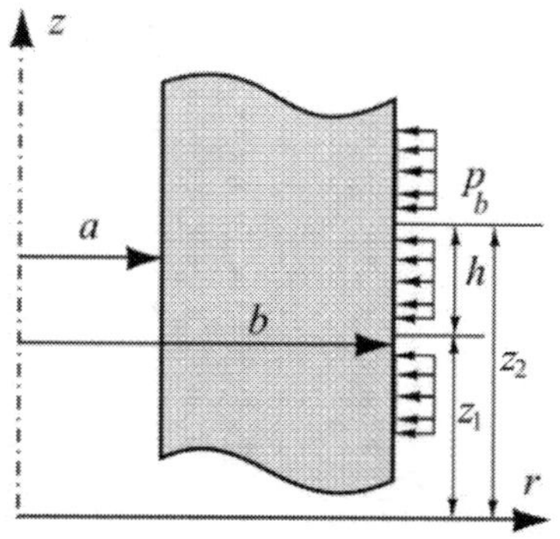

Fig. 2. Thick-walled cylinder with periodic loads. h – load period.

* Corresponding author: asv@mgsu.ru

Below are the results of calculating the stress state of the PV subjected to the action of a constant altitude temperature field $T(r)$, internal pressure p_a, pre-compression in the axial (p_z) and circumferential (p_b) directions. In this case, two methods of applying pressure p_b – uniformly along the entire height of the PV and along the belts are considered.

2 State of the problem

We shall assume approximately that the change in temperature along the radius corresponds to the solution of the heat equation for a stationary temperature field in an infinitely long cylinder:

$$T(r) = T_b + (T_a - T_b)\frac{\ln(b/r)}{\ln(b/a)} \tag{1}$$

where T_a and T_b are respectively the temperatures on the inner and outer surfaces of the vessel. The dependence of the modulus of elasticity of concrete on temperature is given in accordance with [3] and approximated by a polynomial of the fourth degree. Fig. 3 shows the inhomogeneity functions $E(r)$ for several values T_a at $T_b = 20^\circ$ C . On the basis of the analysis performed in calculating forced (temperature) deformations $\varepsilon_f = \varepsilon_T$, the coefficient of linear temperature expansion in the temperature range under consideration varies insignificantly and we will assume it to be constant: $\alpha_T = \alpha_{T0}$. Thus, the temperature deformations will be equal to $\varepsilon_T(r) = \alpha_{T0} T(r)$.

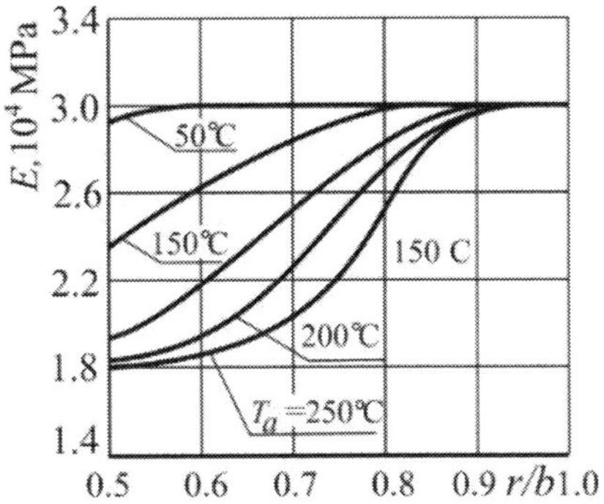

Fig. 3. Changing the modulus of elasticity in the wall of the PV.

3 Method of solution

The main task in this paper is the determination of the stress-strain state of the cylindrical part of the vessel. To calculate the thick-walled shell, with considering the inhomogeneity of the material, the numerical-analytical method described in [4, 5] was used. The solution of the axisymmetric problem is sought in displacements that are represented in the form of Fourier series:

$$u(r,z) = u_0(r) + \sum_{n=1}^{\infty} u_n(r)\cdot\cos k_n z\,; \qquad w(r,z) = \sum_{n=1}^{\infty} w_n(r)\cdot\sin k_n z\,, \tag{2}$$

where $k_n = n\pi/H$.

For a radially inhomogeneous body, the problem reduces to a system of ordinary differential equations of the second order [4]:

$$(\lambda+2\mu)\left(u_n''+\frac{u_n'}{r}-\frac{u_n}{r^2}\right)-\mu k_n^2 u_n+(\lambda+\mu)k_n w_n'+\lambda'\left[u_n'+\frac{u_n}{r}+k_n w_n\right]+2\mu' u_n'-3(Ke_n)'=0;$$
$$\mu\left(w_n''+\frac{w_n'}{r}-k_n^2 w_n\right)-(\lambda+\mu)k_n\left(u_n'+\frac{u_n}{r}+k_n w_n\right)+\mu'(w_n'-k_n u_n)+3Kk_n e_n=0. \quad (3)$$

Here, in the general case, e_n are the expansion coefficients in the Fourier series of the temperature deformations ε_T.

Similarly, surface loads are decomposed into series. Since ε_T and e_n do not depend on z, they are represented by only one term of the Fourier series, and the corresponding coefficients and functions have the form:

$$p_{a,0}=2p_a; \quad p_{a,n}=0; \quad e_0=2\varepsilon_T=2\alpha_0 T(r); \quad e_n=0(n=1,2...).$$

Pressure expansion coefficients for the case shown in Figure 2, is determined by formula

$$p_{b,0}=\frac{2p_b(z_2-z_1)}{H_1}; \quad p_{b,n}=\frac{2p_b(z_2-z_1)}{n\pi H_1}\left[\sin(n\pi z_2)-\sin(n\pi z_1)\right].$$

Here H – is the height of the layer corresponding to the load period. With uniform external lateral compression, H can be chosen arbitrarily, for example, equal to unity, and $z_1=0, z_2=H$.

As a result of calculation determined displacement u, v, stresses σ_r, σ_θ, and *preliminary* stresses $\sigma_z^{(1)}$ to which must be added the *additional* stresses σ_z^* [6], defined by formula $\sigma_z^*=2PE(r)/F^*$.

Here $F^*=2\pi\int_a^b E(r)rdr$ – is the reduced area, and P is the resultant force $P=P_1-P_2-P_3$, where $P_1=p_a\pi a^2$ is the force resulting from the effect of pressure p_a on the housing cover; $P_2=p_z\pi(b^2-a^2)$ – integral force from pre-compression in the axial direction; $P_3=2\pi\int_a^b \sigma_z^{(1)}rdr$ – is the resultant of stresses $\sigma_z^{(1)}$ determined at the *preliminary* stage of the calculation.

The solution obtained with considering the stresses σ_z^* is not exact in the whole region. In particular, the tangential stresses appearing in the contact zone of the PV wall with the cover, as well as the normal stresses caused by unbalanced bending moments from the stresses $\sigma_z^{(1)}$, remain unaccounted for. Thus, according to the principle of Saint-Venant, the results obtained will be valid in the middle part of the PV.

4 Calculation of PV without considering the temperature inhomogeneity

Fig. 4 shows the stress diagrams appearing in the PV wall from the operational loads (p_a and *T*) calculated with the following initial data: $a/b = 0.5$; $p_a = 5\,\text{MPa}$; $T_a = 100\,^\circ\text{C}$; $T_b = 20^\circ\,\text{C}$; $\alpha_{T0} = 0.12\cdot10^{-4}\,1/^\circ\text{C}$; $E_0 = 3\cdot10^4$ MPa; $\nu = 0.2$. The diagrams σ_z are constructed taking into account σ_z^* without pressure p_z. The stresses σ_z on the pressure p_z are small, so that the total diagram of σ_z is practically due only to the action of the temperature field. It can be noted that in the cylindrical part of the PV, very significant stresses arise σ_θ and σ_z from the operational loads, which must be compensated by pre-compression pressures p_b and p_z. First, it is necessary to choose the pressure p_b, since it leads to a change in the diagram of $\sigma_z^{(1)}$, then the pressure is selected, which does not affect the stresses σ_r and σ_θ.

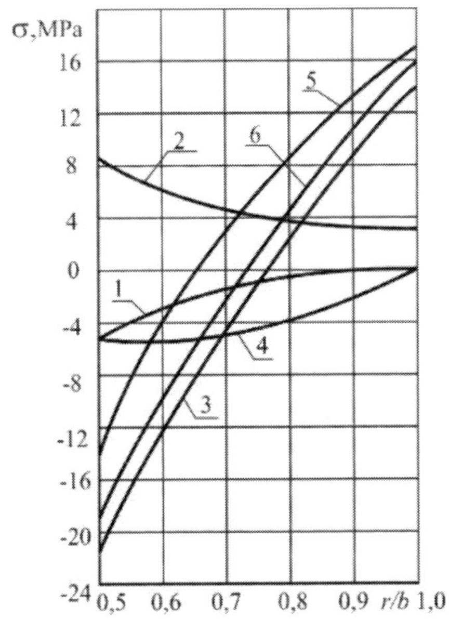

Fig. 4. Stresses in the wall of PV:
1 - $\sigma_r(p_a)$; 2 - $\sigma_\theta(p_a)$; 3 - $\sigma_z(p_a)$;
4 - $\sigma_r(p_a, T)$; 5 - $\sigma_\theta(p_a, T)$; 6 - $\sigma_z(p_a, T)$.

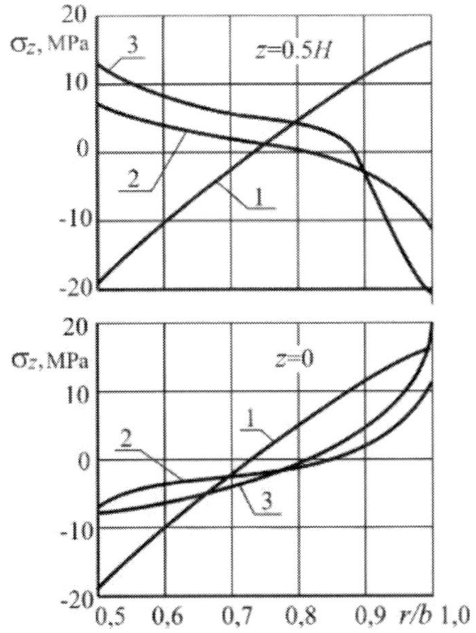

Fig. 5. Stresses in the wall of PV:
1 - from operational loads;
2, 3 – from external compression
2 - $z_1 = 0.2H$; $z_2 = 0.8H$; 3 - $z_1 = 0.4H$; $z_2 = 0.6H$.

Fig. 5 shows stress diagrams σ_z at two levels (*z*=0 and *z*=0.5*H*) from operational loads ($p_a = 5\,\text{MPa}$; $T_a = 100\,^\circ\text{C}$; $T_b = 20^\circ\,\text{C}$;) and for two cases of external compression p_b. In order for the integral value of the lateral compression to be constant the pressure value p_b was chosen inversely proportional to the width of the load belt:

$$p_b = \frac{p_b^0 H}{z_2 - z_1}.$$

The pressure p_b^0 entering into this formula corresponds to a uniform compression over the entire outer surface. From the graphs shown in Fig. 5, it can be concluded that as the width of the belt decreases, the stresses σ_z from pre-compression increase in absolute value both in the stretched and in the compressed zones. In this case, if in the section $z=0.5H$ these stresses are compensated by stresses from the operational loads, then in the sections close to the end, these stresses are added together, and for the removal of tensile stresses with a narrower load belt, greater compressive forces in the axial direction are required.

5 Calculation of PV with considering the temperature inhomogeneity of the material

Stress state of PV is due to the common action of the internal pressure p_a, the temperature field and the pre-compression pressures p_b and p_z. In order to visually observe the effect of temperature inhomogeneity on the stress state of the PV, the effect of each of these factors can be considered separately.

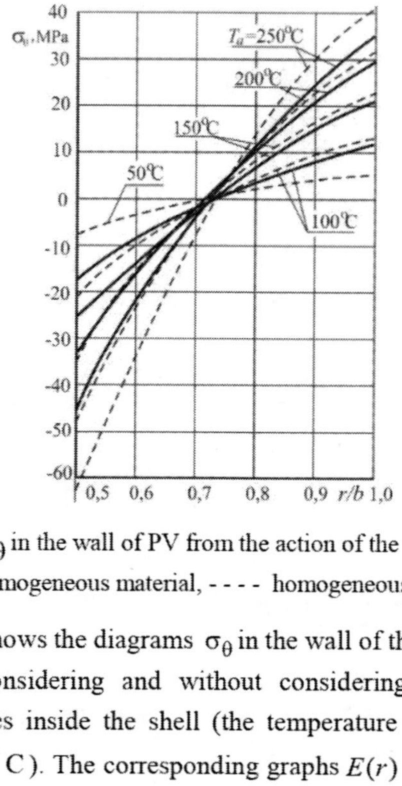

Fig.6. Stresses σ_θ in the wall of PV from the action of the temperature field:
—— inhomogeneous material, - - - - homogeneous material.

As an example, Fig. 6 shows the diagrams σ_θ in the wall of the PV from the action of the temperature field with considering and without considering the dependence $E(T)$ for different temperature values inside the shell (the temperature on the external surface is constant and equal $T_b = 20^\circ$ C). The corresponding graphs $E(r)$ are shown in Fig. 3. In the case when $T_a = 50^\circ$ C, the stress values for homogeneous and inhomogeneous materials practically coincide. As the temperature rises, the modulus of elasticity decreases significantly, which leads to a decrease in stresses. These changes are more noticeable in the compressed zone, i.e. where the temperature is higher. A similar effect is observed when calculating the stresses from internal pressure p_a.

A noticeable decrease in stresses from operational loads, obtained with considering the temperature inhomogeneity of concrete, allows reducing the amount of pre-compression p_b, which in turn leads to a decrease in axial compression. For example, it is shown that for $T_a = 200^\circ$ C and for the width of the load belt $z_2 - z_1 = 0.2H$, the pre-compression pressure p_b with considering the inhomogeneity can be reduced by almost half compared to the calculation for a homogeneous material.

Conclusions

Accounting for the inhomogeneity of materials introduces a significant correction to the stress-strain state of the structures. In the works [7, 8, etc.], the influence of various factors (radiation exposure, humidity, etc.) causing inhomogeneity of materials was considered. It is also possible to use this phenomenon for practical purposes. For example, [9, 10] describes a method for optimizing thick-walled concrete cylindrical shells by artificially changing the deformation properties of concrete.

This work was financially supported by the Ministry of Russian Education (state task #7.1524.2017/PCh)

References

1. Altshuller M.A., Vinogradov V.N. and others, Nuclear power plants. **4**, 17 (1981)
2. P.H. Margen, S. Mennon, J. of the British Nuclear Energy Soc. **14** 1, 35 (1975)
3. *Construction codes and standards* 27.13330. (2011)
4. Andreev V.I. Applied Mechanics and Materials. **752-753** , 593 (2015)
5. Andreev V.I. Applied Mechanics and Materials. **752-753**, 642 (2015)
6. Timoshenko S.P., Gudier Dj., *Theory of elasticity* (Nauka, Moscow, 1979)
7. Andreev V.I., Avershyev A. S., Int. J. for Comp. Civil and Struct Eng. **9** 3, 14 (2013)
8. Andreev V.I., Dubrovskiy I.A., Applied Mechanics and Materials. **405-408**, 1073 (2013)
9. Andreev V.I., Applied Mechanics and Materials. **166-169**, 354 (2012)
10. Andreev V.I., Potekhin I.A., Int. J. for Comp. Civil and Struct Eng.. **4** 1, 79 (2008)

Curvilinear steel elements in load-bearing structures of high-rise building spatial frames

Alexander Ibragimov [1,*], *Alexander* Danilov[1]

[1] Moscow State University of Civil Engineering, Yaroslavskoye shosse, 26, Moscow, 129337, Russia

Abstract. The application of curvilinear elements in load-bearing metal structures of high-rise buildings supposes ensuring of their bearing capacity and serviceability. There may exist a great variety of shapes and orientations of such structural elements. In particular, it may be various flat curves of an open or closed oval profile such as circular or parabolic arch or ellipse. The considered approach implies creating vast internal volumes without loss in the load-bearing capacity of the frame. The basic concept makes possible a wide variety of layout and design solutions. The presence of free internal spaces of large volume in "skyscraper" type buildings contributes to resolving a great number of problems, including those of communicative nature. The calculation results confirm the basic assumptions.

1 Introduction

The principal reasons for the steel application for high-rise buildings are:

- Steel frames are fast to erect. The construction is lightweight, particularly in comparison with traditional concrete construction.
- The elements of the framework are prefabricated. Factory conditions of manufacturing lead to established quality mounting procedures.
- The dryness of the form of construction results in less on-site activities.
- The framework is not susceptible to drying-out effects or delays due to strength gain.
- Steel frames have potential for adaptability inherent in their construction. Later modification involves relatively easy unbolting a connection; with concrete, such modifications would be expensive, and more extensive and disruptive.
- The use of steel makes possible the creation of relatively large, column-free internal spaces to divide if needed by partitions.
- With no external walls as loadbearing elements, it allows the development of large window areas incorporated in prefabricated cladding systems.

The article [1] discusses the motivation for using curvilinear steel elements (C-els in brief) in the buildings structures. In [2-3] the various aspects of load-carrying structures of low- and in [4-6] of high-rise buildings design are considered. The key issue is the volume reduction of labor-intensive routine operations requiring enhanced accuracy of fit and

* Corresponding author: igasu_alex@mail.ru

special surfacing operations at the stage of erection.

This paper presents the results of the influence of C-elements on the behavior of steel frameworks in high-rise buildings analysis.

It shows that, when properly used, such elements are capable of providing the necessary rigidity of the carcass. The vast free spaces for various purposes within the building's volume, as well as wide through openings are possible.

Obviously, with such an approach, an infinite number of variants of constructive schemes are possible. Shapes and orientation of curves far from exhausting the variety of parameters that determine the structural form of the framework. The choice of a particular variant and its justification are the prerogative of architects and design engineers.

Here it presents a relatively simple and maybe rather clumsy from the architectural point of view constructive forms. Yet it allows performing a comparative analysis and this is the main purpose of its presentation. Just from these examples it follows that the formation of a constructive form of a similar type, enough adequate to ensuring the bearing capacity, is not a routine procedure and requires the comparative structural and numerical analysis of a number of design variants and, accordingly, constructive models [7].

Allowing for a large number in tall buildings of connection joints of vertical (columns) and horizontal (crossbars, beams) steel elements with different cross sections, it is quite natural to strive for maximum simplification of such connections. The simplest and most technologically feasible solution to the junction of the beam to the column is the so-called "simple connection", where the moment transmitted by one element to the other is negligible enough. Along with the economic aspect, it should be noted here that the sensitivity of the carcass to a different kind of skew due to foundation sediment, temperature effects and other factors.

Therefore, the functioning of the frame elements differentiation takes place. Columns and crossbars provide the work of the frame on the perception of vertical loads, and bracing systems practically not involved in the perception of vertical loads perceive and transmit to the foundations horizontal impacts [8].

For curvilinear elements this functionality separation does not go. They simultaneously participate in the work of the main load-bearing elements of the framework on the perception of both vertical and horizontal loads. This, on the one hand, complicates the work of the frame and, allowing for the variety of design options, significantly complicates the calculation and design of structures. On the other hand, this opens up new opportunities for the generating of constructive and, thus, architectural forms.

2 Bearing structures analysis

Consider a flat frame (in particular, with lattice rigidity diaphragms) scheme, in which the C-elements embed in a most natural way. A 36 m long building of 24 m depth, height H = 120 m (30 floors of 4 m) has a pitch of columns 6 m in both directions.

The averaged dead (D) vertical load on the crossbars (primary beams) of the flat bearing frame is assumed to be 60 kN/m according to 10 kN/m^2 of design load on the slab. It includes here permanent like the reinforced concrete slab (200 mm) weight and, to simplify the model, all the long- and short-term (useful) variable loads.

The wind load (V) acts from left to right and, taking into account the dynamic (pulsation) component, is approximately assumed to be trapezoidal: 1.50 kN / m at ground level and 4.22 kN/m at a height of 120 m on the windward side (left) and 0.94 and 2.64 kN/m, respectively, on the opposite side (right). Load combinations [D] and [D + V] are considered.

The first case observes the effect of the curvilinear elements embedding as such into the frame system, as well as the cross-sections for these elements, on the behavior of the frame with rigid connections of the crossbars to columns. C-elements rigidly connect to columns and crossbars. The cross-section of columns is of paired I-beams 60SH2 with 90x10-angle lattice, cross-section of crossbars - 35SH2 according to STO ASCHM 20-93. Columns to foundations connections are fixed – rigid support.

The values of the maximum horizontal displacement of the middle node at the level of 120 m (frame top) and the stability factor for the first buckling mode are the comparison criteria. The stability factor (k) is the ratio of the frame global buckling load level to a given one.

The one result of the several framework variants FEA in LIRA-SAPR program is that all the frame elements have sufficient load-bearing capacity. However, as for deformability and stability of the framework system as a whole object, the results of the analysis differ significantly (Table 1).

It is easy to see that the scheme (a) does not satisfy the serviceability requirements, since the relative horizontal wind load deflections exceed the standard maximum value of 1/500. Scheme (b) with the smallest of the last three variants of the C-elements cross-section is quite acceptable. At the same time, the FEA analysis shows a number of the crossbars in the scheme (a) to be overloaded (within 10%), unlike in the schemes (b) – (c). It is clear that the scheme (a) in case of crossbars to columns hinged connections (simple connections) is practically unworkable without the diagonal stiffness elements in the frame plane.

Table 1. First buckling modes and horizontal wind displacements for the high-rise building schemes under consideration with rigid beam to column joints.

Rigid column-beam joint frames first buckling modes	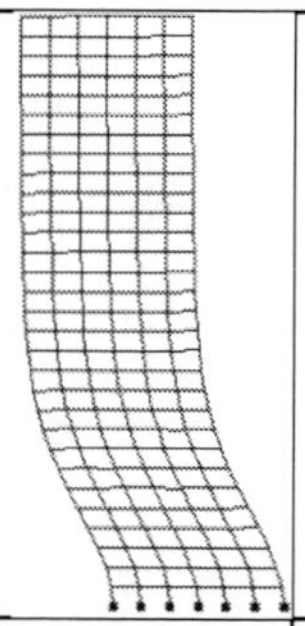	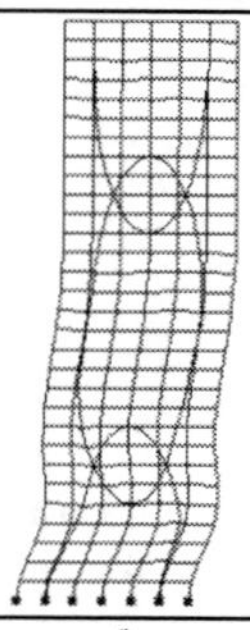	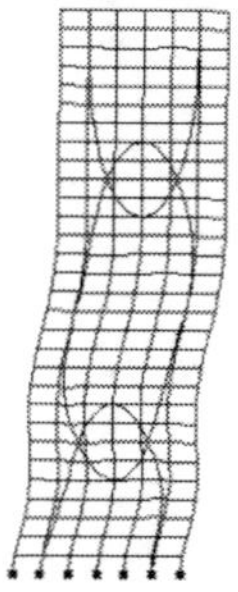	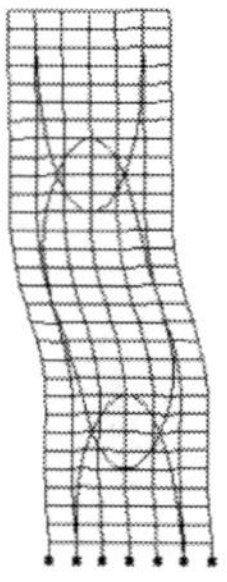
Scheme type	*a*	*b*	*c*	*d*
Details	No C-els	With C-els of paired 30SH2	C-els of paired 50SH2	C-els of paired 100SH3
Deflection dx, mm	303	148	127	104
Relative, dx/H	1/396	1/811	1/945	1/1154
Stability factor k	2.45	4.03	4.45	5.43

The Table. 2 shows that in all four cases in order to ensure an acceptable rigidity the carcass requires a stiffening diaphragms. The single chord C-elements themselves also do not provide the required stiffness of the carcass. However, one may suppose the proper shape, cross-sections and arrangement of C-elements to make it quite possible. For example, using more rigid two-chord C-elements with a diagonal grating (C-trusses).

Table 2. Horizontal displacement in frames with hinged beams

Scheme type	*a*	*b*	*c*	*d*
Deflection dx, mm	40460	739	619	346
Relative, dx/H	1/3	1/162	1/194	1/347

The stiffness diaphragms being inserted (vertical bracing of CHS 203x10 cross-section) the picture changes drastically (Table 3).

Note that in schemes (*e*)–(*h*) all horizontal elements (crossbars) have simple (no bending moments in joints) connections with columns and arches. The arch to column and column to base connections are rigid (fixed). The embedding of C-elements into the frame structure obviously several times increases the stiffness and stability of the frame (table 3) turning the "sway-sensitive" structure into "non-sway" one without any other bracing system based traditionally on a great number of diagonal elements. Scheme types (e) and (f) show the bracing system influence on the frame "sway-sensitivity".

Table 3. First buckling modes and horizontal wind displacements for the high-rise building schemes under consideration with simple joints.

Simple column-beam joint frames first buckling modes	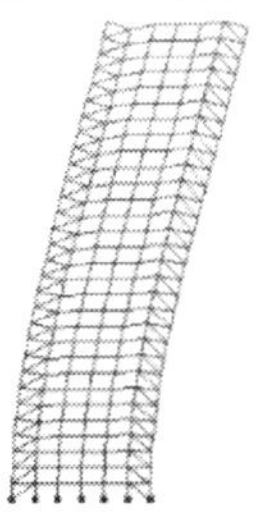	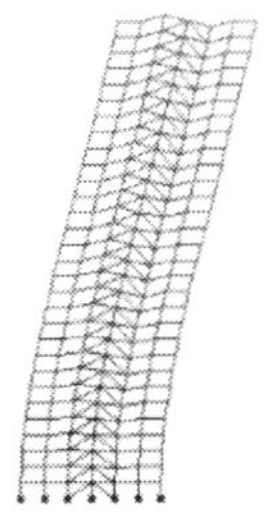	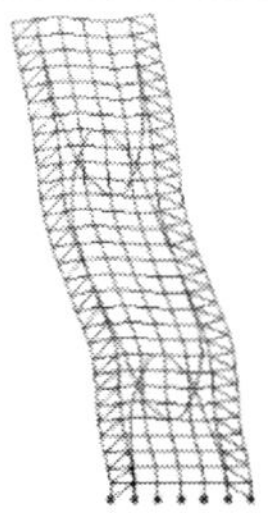	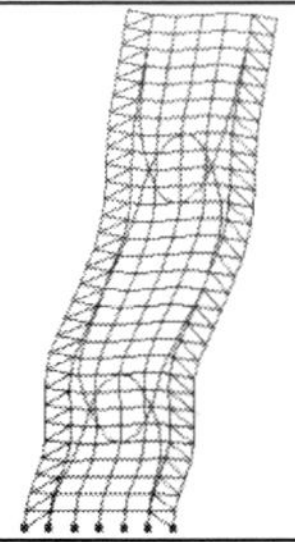
Scheme type	*e*	*f*	*g*	*h*
Details	No C-els	No C-els	C-els of paired 30SH2	C-els of paired 50SH3
Deflection *dx*, mm	504	266	106	80
Relative, *dx/H*	1/238	1/451	1/1132	1/1500
Stability factor *k*	2.22	4.16	8.77	10.11

It makes it allowable to weaken the framework by introducing some openings, say, as shown in Table. 4.

Table 4. First buckling forms and horizontal wind displacements for the high-rise building schemes under consideration with simple joints and openings.

Simple column-beam joint frames first buckling modes	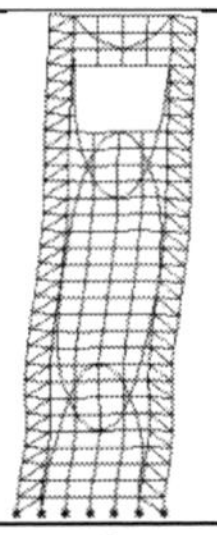	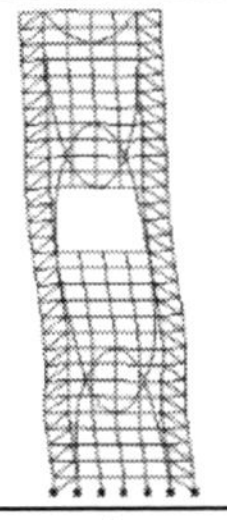	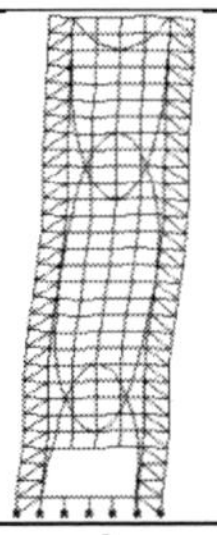	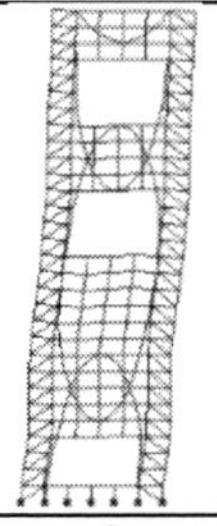
Scheme type	*i*	*j*	*k*	*l*
Details: openings	Top	Middle	Bottom	3-level
Deflection *dx*, mm	77	77	77	84
Relative, *dx/H*	1/1558	1/1558	1/1558	1/1429
Stability factor *k*	11.24	11.23	10.12	11.98

In the schemes (*i*)–(*l*) top C-elements are of paired 35SH2 cross-section and all the lower of paired 50SH2. The columns, beams and diagonals are of the same cross-sections as before: paired 60SH2, single 35SH2 and CHS 203x10 correspondently. The FEA analysis showed that in *k* and *l* cases the bottom C-elements and bottom level columns and diagonals should be somewhat enhanced. The vertical deflections on the top of frame and in the openings still meet the regulatory requirements.

The table 4 shows the significant reserve of bearing capacity of frames with openings. At least two point one should outline here: the required cross-sections of diagonals and all bearing elements but beams will decrease with increasing elevations; the presence of openings may also decrease vertical loads (see the stability factor values) and frontal wind action (reduction of impact area).

The next sample is a spatial 3d-model of a 33-storey building (H = 132 m) in the Table 5. Here the first several floors columns cross section is increased (70SH3). The reinforced concrete slabs 200 mm thick form the horizontal diaphragms at each floor level.

Table 5. 33-flor high-rise building 132 m height with C-beams and openings.

	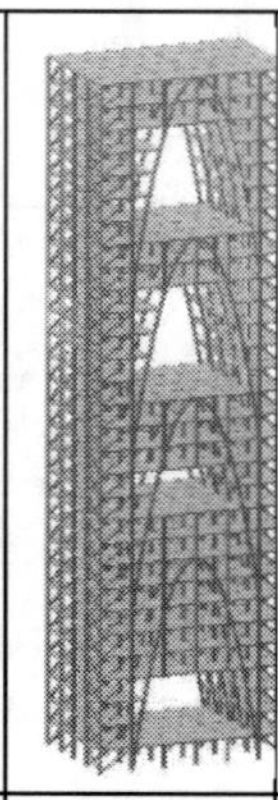	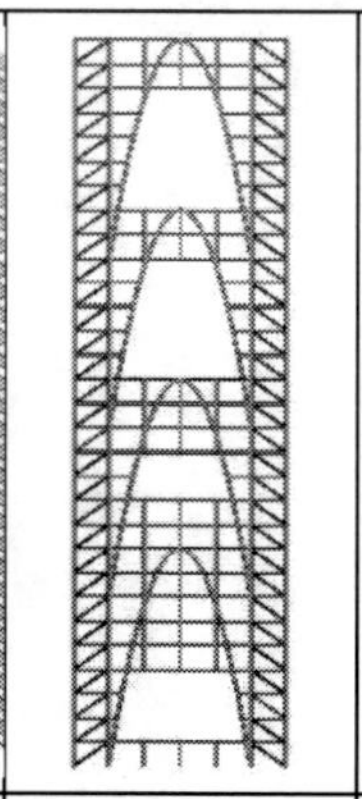	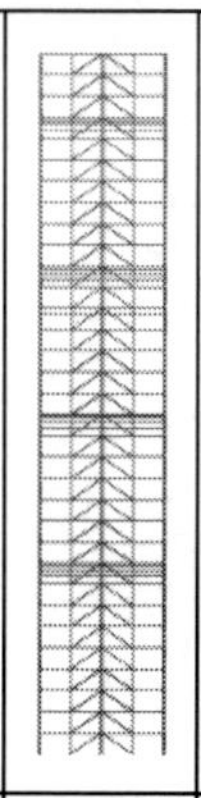	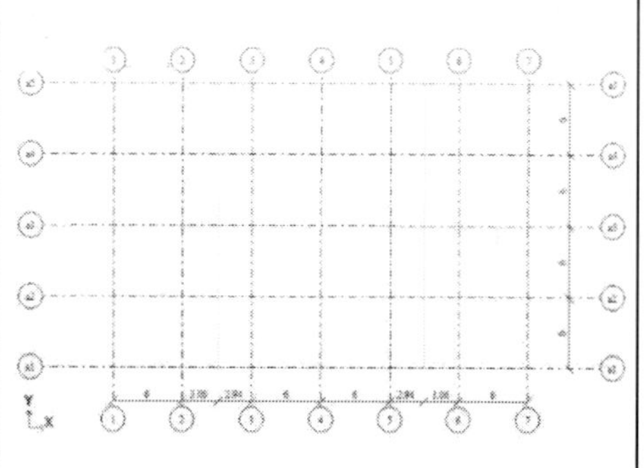
1	2	3	4	5
With front walls	Without front walls	XZ-view (no front walls)	YZ-view	XY-plane

To account for the openings in a most simple way, when setting the front and backside wind load, the flat finite elements with small shear stiffness model the enclosing structure on the front and backsides of the building design model. The vertical action is a uniformly distributed load on the floor flat finite elements of 10 kN/m^2 intensity. Lateral (X direction) and front (Y direction) wind loads have the same intensity as before. The lateral wind load applies directly to the columns (see previous models). The column-beam and arch-beam joints – simple (hinged). The reinforced concrete slab of floors connects to columns rigidly - joints with additional plated elements.

Applied loads:

1. D - Dead load
2. WX - Wind along X-axis
3. WY - Wind along Y-axis
4. SX - Seismic (5b) along X-axis
5. SY - Seismic (5b) along Y-axis

Table 6. Load combinations considered and FEA results

Combinations:		Deflections, mm			Period, s	Stability factor k
		X	Y	Z	1st mode	
1	D	0	0	-95	4.5	4.69
2	D+WX	31	0	-95		4.24
3	D+WY	0	108	-95		4.21
4	D+SX	0	231	-95		-
5	D+SY	78	2	-95		
6	0.9D+0.5WX+SX	15	231	-86		
7	0.9D+0.5WY+SY	78	55	-86		

The table 5 shows:

- deflection values lie in the admissible interval: 120000/500 = 240 (note that deflection values are obtained for design load values)

- there is sufficient margin of stability
- the first mode oscillations period may ensure the comfort of staying in the premises of the upper floors

Along with the task of ensuring the strength and stability of the bearing elements of the steel frame with increasing the height of the building and, respectively, the ratio of its vertical and horizontal dimensions, the stiffness and the stability margin of the carcass as an integral structural element become increasingly dominant factors. It extends into the rod or shell and, at some point, the provision of the load-bearing capacity of the frame elements can no longer provide the load-bearing capacity of the structural system as a whole. That is, the constructive system becomes a new quality and the effectiveness of the determining the framework stiffness bracing system acquires a special significance. All kinds of "stress concentrators" also adversely affect the behavior of the constructive system and local reinforcement is far from always save the situation.

High-rise building of 204 m height (34 floors of 6 m or 68 of 3 m, 60x60 m plan) with arched openings ([1]) is an example. One can easily design necessary element cross-sections to provide them strength and buckling resistance, minimize vertical and even horizontal deflections to meet the rules. However, in this frame the four rigid towers in the corners of the building need some additional bracing system to unite them into overall spatial structure. It is interesting to obtain in result of numerical analysis that the most problematic segments are not the ones with openings where the arches serve as bracing but the middle ones between the openings.

Table 7. High-rise building with simply supported beams and openings (48x24h m)

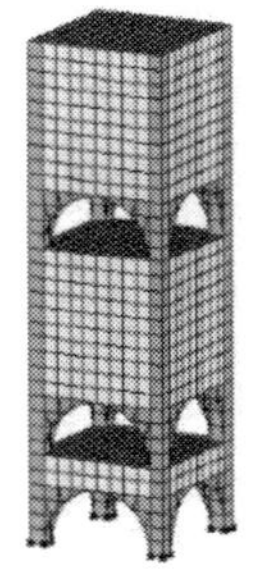	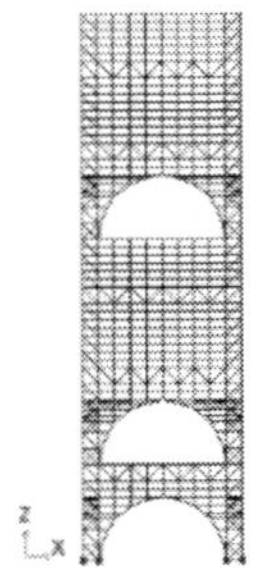	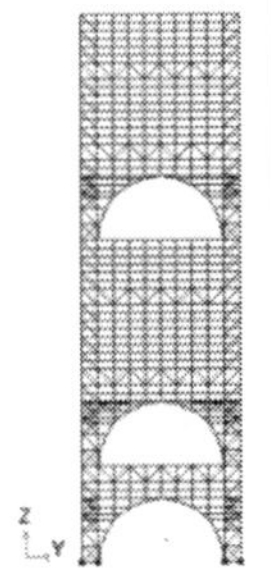	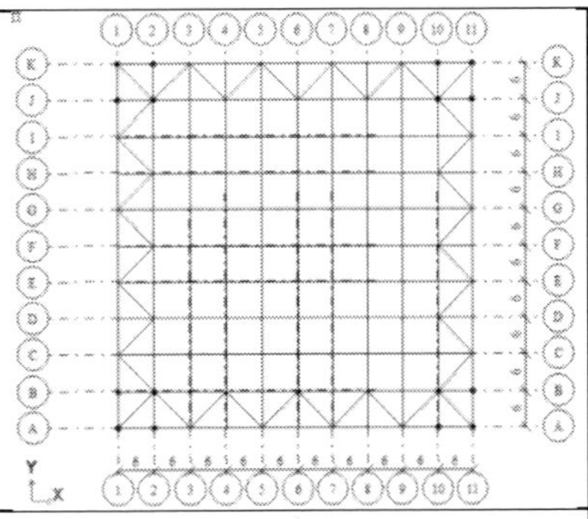	
1	2	3	4	5
Main view	XZ-view	YZ-view	Plan	1st buckling mode (k =0.64)

Introducing the considered earlier C-elements into the structure in a proper way one may possibly ensure the stability of the structure as a whole. The tables 3-5 give some examples. Various factors such as the maybe really the most critical ones like ground conditions may dictate the column grid. Long spans may be required to bridge obstructions beneath or over the ground: underground railways or archaeological remains6 buried services, etc. Generally, a widely spaced column grid is desirable since it reduces the number of foundations and increases the simplicity of construction in the ground. A steel-framed building is about 60% of the weight of a comparable reinforced concrete building resulting in smaller foundations with a consequent reduction on costs

4 Conclusions

The presented analysis results show:

1. It is possible to use C-elements in high-rise building steel frames to ensure the robustness and buckling resistance of the structure as a whole system
2. The application of such constructive solutions makes it possible to form large-span openings and vast free internal spaces
3. This kind of applications especially needs thorough numerical analysis of structure behavior in various design situations
4. The variety of shapes and combinations of C-elements enables generating various architectural configurations

References

1. A. Tusnin, Industrial and Civil Engineering, **11**, 18-22 (2017)
2. O. Tusnina, Industrial and Civil Engineering, **11**, 23-27 (2017)
3. M.F. Randolph, C.P. Wroth Analysis of deformation of vertically loaded piles. Journal of the Geotechnical Engineering Division, 1978, vol. 104, no. 12, pp. 1465-1488.
4. A. Aminmansour, K. Moon, Journal of Architectural Engineering, **16 (2)**, 47-53 (2010)
5. W. Schueller, *High-Rise Building Structures,* Wiley, p. 272 (1977)
6. V. Faghihi, K.F. Reinschmidt, J.H. Kang, Expert Systems with Applications, **16**, 7565-7578 (2014)
7. O. Kalinina, O. Valebnikova, Advances in Intelligent Systems and Computing, **692**, 1315-1322 (2018) DOi - 10.1007/978-3-319-70987-1_139
8. I.V. Ilin, A. Lepekhin, A.I. Levina, O.Yu. Iliashenko, Advances in Intelligent Systems and Computing, **692**, 1306-1314 (2018) DOi -10.1007/978-3-319-70987-1_138

Effect of Rice Husk Ash and Fly Ash on the workability of concrete mixture in the High-Rise Construction

Lam Van Tang[1,*], *Boris* Bulgakov[1], *Sofia* Bazhenova[1], *Olga* Aleksandrova[1], *Anh* Ngoc Pham[2] and *Tho* Dinh Vu[1]

[1] Moscow State University of Civil Engineering, Yaroslavskoe shosse, 26, Moscow, 129337, Russia
[2] Ruhr Universität Bochum, Bochum, Germany, IC 6-137 Universitätstr 150, 44801 Bochum

Abstract. The dense development of high-rise construction in urban areas requires a creation of new concretes with essential properties and innovative technologies for preparing concrete mixtures. Besides, it is necessary to develop new ways of presenting concrete mixture and keeping their mobility. This research uses the mathematical method of two-factors rotatable central compositional planning to imitate the effect of amount of rice husk (RHA) and fly ash of thermal power plants (FA) on the workability of high-mobility concrete mixtures. The results of this study displays regression equation of the second order dependence of the objective functions - slump cone and loss of concrete mixture mobility due to the input factors - the amounts RHA (x_1) and FA (x_2), as well as the surface expression image of these regression equations. An analysis of the regression equations also shows that the amount of RHA and FA had a significant influence on the concrete mixtures mobility. In fact, the particles of RHA and FA will play the role as peculiar "sliding bearings" between the grains of cement leading to the dispersion of cement in the concrete mixture. Therefore, it is possible to regulate the concrete mixture mobility when transporting fresh concrete to the formwork during the high-rise buildings construction in the hot and humid climate of Vietnam. Although the average value of slump test of freshly mixed concrete, measured 60 minutes later after the mixing completion, decreased from 18.2 to 10.52 cm, this value still remained within the allowable range to maintain the mixing and and the delivery of concrete mixture by pumping.

1 Introduction

In fact, modern architecture and construction would hardly be possible without the use of concrete. It becomes more and more popular material to create durable, reliable and long-lasting design, which can withstand atmospheric and mechanical influences since many years [1, 2]. If there are no problems with the manufacture or acquisition of a concrete mixtures and its issues, the supply of concrete mix (especially in case of large volumes) will be one of the most important process in the whole technological chain of organizing the construction

* Corresponding author: lamvantang@gmail.com

work. The 21st century is considered to be the century of high-rise buildings construction. It is characterized by the global cities and dense urban where are shortage of free spaces, growing demand of residential and commercial areas. These requirement will be fundamentally leading to the emergence of new town-planning concepts, including high-rise construction concept [3]. In reality, it can be said that fresh concrete mix is a rather specific material, which gradually loses its workability as hydration of the binder. Thus, this mix must be protected from sunlight, precipitation, freezing and quickly delivered to the construction site in required operating condition without any loss of quality and deterioration properties [4]. Depending on the type of buildings, the parameters of the concrete mixture and its volume are supplied by the following method [5-7]:

1. Directly applying concrete mixture to the structure from the concrete conveyor or with the help of bunker and overpasses. The main disadvantage of this approach is the limitation of application range as well as unusable for high-rise buildings.

2. Applying concrete mixture by bunker with the help of crane. These disadvantages of this method are the need to add an overload to the concrete mixture and waste time of crane installation.

3. Applying concrete mixture by belt conveyor that is stationary or mobile system based on a car or tractor. The working arrows of 10 ÷ 20 m length are used, where the conveyer belt is moving from the concrete mixture. This method is effective only for a large amount of work. It is used in the concrete structures of the zero cycle: foundations, basement walls, floors, ceilings, etc., and above-ground parts of buildings (not higher than two stories - up to 4 ÷ 6 m).

4. Applying concrete mixture by pumping. This method is used for large volumes of concrete in buildings and space constraints (Fig. 1).

Fig. 1. Pumping concrete mixture through pipelines in the high-rise constructions in Hanoi (Vietnam)

Concrete mixture with a slump cone standard of 8 to 18 cm [8-10], which corresponds to the parameters of concrete mixtures are used in the hot and humid climate of Vietnam [11], is applied directly to the concrete structure through steel pipes (diameter of 100 ÷ 200 mm). The distance and height application are up to 150 m and 40 m respectively.

Practically, in high-rise construction, depending on the building object's parameters (width, height, degree of restraint, etc.), one of the following technological schemes is being used:

- All the equipment is installed at the base of the truck mixer. This system can operate in transport mode: the concrete mixture at the plant is delivered to the facility and submitted to the installation place. It is also possible to work in a stationary mode to prepare the concrete mixture directly and apply to the installation area of the constructions.

- All equipment is mounted on the chassis of the car. The concrete mixture is delivered by other transport (dump truck or auto mixers) and loaded into the unit receiver hopper
- All equipment is mounted on the tower crane of structures.
- Using separate stationary equipment for high-rise buildings and structures construction: the receiving bunker and concrete pump are located on the ground; the boom-crane with an autonomous drive is installed on the working horizon and shifted upwards during the construction as required. The concrete mixture is applied to the vertical main pipe, which is also increased by the building erection.

The advantages of the pneumatic method in feeding a concrete mixture are high productivity, unused cranes and the possible application at any point.

This method's disadvantages are the limitations on the parameters of the mixture and the size of the aggregate (less than 40 mm), as well as the high costs to flush the pipelines.

The result of this study [12] shows that for concrete mixtures in high-rise construction, the filler selection and superplasticizers usage for ease of transportation in pumping are greatly important.

According to the studies [13, 14], a very important problem for concrete mixture, which is transported by pumping, is the necessary to ensure the stability of working ability in time. In addition, in order to provide the pipeline lubrication during the delivery of concrete pump, the concrete mixture composition must be sufficiently large number of component mortar (consisting of cement, sand, water, plasticizer and mineral filler). According to the result of this study [11], the percentage of mortar in the concrete mixture should be in the range of 40 and 44%.

Based on the above analysis, it can be concluded that using both superplasticizers and fine-graded mineral fillers (for example, fly ash of thermal power plants (FA TPP), rice husk ash (RHA), microsilica, metakaolin and so on) is an important factor in improving the pumpability and influencing the persistence of high-mobility concrete mixes.

This paper mainly focuses on the effects of rice husk ash and fly ash from TPP "Vung Ang" on the working ability of the concrete mixture and its loss for different time periods by the form of second-order regression equations.

2 Materials and Methods

2.1. Materials

- Portland Cement (PC) type SEM II 42.5 N manufactured at "Nghi Son" factory (Viet Nam). The characteristics of this Portland cement and the experimental results obtained for testing its properties are shown in Table 1.
- Standard quartz sand (QS) of Science and Technology Institute building (Vietnam): M_K = 2.85, specific weight of 2.65 g/cm^3 and the volume of compacted state is 1540 kg/m^3.
- Crushed limestone (CL) produced from the quarry "Kien Khe" (Vietnam) with the size of 5 - 10 mm, specific weight of 2.61 g/cm^3 and volume of the compacted state is 1620 kg/m^3.

Table 1. Physical and mechanical properties of Portland цемент SEM II 42.5 N "Nghi Son"

Specific weight (g/cm^3)	Fineness		Time of setting (min)		Compressive strength (MPa)		Water requirement (%)
	Retained content on sieve №009 (%)	Specific surface (cm^2/g)	Initial	Final	2 days	28 days	
3.12	3.8	3635	140	230	34.61	48.3	28.58

- Superplasticizer ACE 388 (A388) "Sure Tec" BASF (Germany). It is a new generation chemical additives based on polycarboxylate ethers with specific weight of 1.1 g/cm³ at 20 ± 5°C. The amount of A388 is 1% by the mass of Portland cement and the decrease of water content of the concrete mixture is about 30%.

- Fly Ash (FA) from TPP "Vung Ang" (Vietnam) Class C and rice husk ash (RHA). According to the results [15-17], chemical composition and physical properties of RHA and FA TЭC "Vung Ang" are presented in Table 2 and 3.

Table 2. Chemical compositions of RHA and FA TPP "Vung Ang"

Mineral additives	Average chemical composition (%)										
	SiO_2	Al_2O_3	Fe_2O_3	SO_3	K_2O	Na_2O	MgO	CaO	TiO_2	P_2O_5	LOI*
RHA	88.2	1.25	1.75	0.5	1.14	2.67	0.8	0.52	0.15	0.25	2.77
FA	54.62	25.17	7.11	0.25	1.28	0.2	1.57	1.45	1.83	1.63	2.04

Note: LOI - loss on ignition.

Table 3. Physical properties of RHA and FA TPP "Vung Ang"

Features	RHA	FA TPP "Vung Ang"
Specific weight (g/cm³)	2.15	2.22
The volume of natural porous state (kg/m³)	572	765
Specific surface area (m²/g)	11.350	14.455
Humidity (%)	2	11
Retained content on sieve with a hole size of 45 μm (%)	23.2	31
Water requirement (%)	100.7	104.1

2.2. Methods

- Calculation preliminary compositions method of concrete mixture is applied in accordance with ACI 211.4R-93 (American).

- The work ability of concrete mixture is determined by the slump of cone standard with dimensions of 100x200x300 mm.

- The mathematical modeling, depicting the dependence of RHA and FA TPP "Vung Ang" on the work ability of the concrete mixture, uses the rotatable central compositional planning method for two-factors.

3. Results and discussion

In this paper, correlation analysis made it possible to establish the mathematical dependence of standard slump cone of high-mobility concrete mixtures, which was measured at different time intervals after completion of mixing of raw materials, on the amount of RHA and FA of TPP "Vung Ang" by using Central composite rotatable design (CCRD) method for two factors [20 - 22].

CCRD method allows to obtain a more accurately mathematical description by increasing the number of experiments at the center and some special values of "star shoulder" $\alpha = \sqrt{2} = 1.414$ [23]. Matrix of the CCRD for two factors are presented in Table 4.

Table 4. Matrix of the CCRD for two factors

Sestems of experiments	Trial No.	Description of code						Y_j
		x_0	x_1	x_2	$x_1 x_2$	x^2_1	x^2_2	
Number of experiments in the core of the plan	1	+1	+1	+1	+1	+1	+1	Y_1
	2	+1	-1	+1	-1	+1	+1	Y_2
	3	+1	+1	-1	-1	+1	+1	Y_3
	4	+1	-1	-1	+1	+1	+1	Y_4
Number of experiments in "Star points"	5	+1	+1.414	0	0	2	0	Y_5
	6	+1	-1.414	0	0	2	0	Y_6
	7	+1	0	+1.414	0	0	2	Y_7
	8	+1	0	-1.414	0	0	2	Y_8
Number of experiments in the center of the plan	9	+1	0	0	0	0	0	Y_9
	10	+1	0	0	0	0	0	Y_{10}
	11	+1	0	0	0	0	0	Y_{11}
	12	+1	0	0	0	0	0	Y_{12}
	13	+1	0	0	0	0	0	Y_{13}

The determination of the preliminary composition of concrete mixture with standard slump cone from 15 to 20 cm allows to achieve concrete with the compressive strength of 70 MPa at the age of 28 days according to American standard ACI 211.4R-93. These results are shown in Table. 5.

Table 5. Preliminary compositions, properties of concrete mixture and concrete

Compositions of concrete mixture (kg/m³)							Properties of concrete mixture			Compressive strength at 28 days (MPa)
PC	RHA	FA	QS	CL	W	A388	Average density (kg/m³)	$\frac{W}{BID}$	Slump cone (mm)	
392	59	118	537	1053	159	3.92	2319.7	0.28	18.0	71.5

The water-binding ratio is constant and equal to $\frac{W}{BID} = 0.28$. The amount of superplasticizer A388 is 1% by the mass of PC.

The amounts of RHA and FA in the concrete mixture are chosen, respectively, in the range from 8% to 22% and from 20% to 40% by the mass of PC. While the contents of QS is 537 kg and CL is 1053 kg.

Description of objective functions of the empirical model:

Yi (i = 0, 30 and 60) - the slump cone of the high-mobility concrete mixtures (cm) at time of 0, 30 and 60 minutes are considered as an objective functions of the experimental model.

Input factors and their limitations are sorted as below:

- The amount of RHA is factor x_1 in the range from 0.08 to 0.22 by mass PC.
- The amount of FA is factor x_2 in the range from 0.2 to 0.4 by mass PC.

Table 6 presents the input factors and their variation intervals.

Table 6. Levels and intervals of varying factors of the CCRD

Parameter factors	Description factors	Levels varying factors					Intervals varying factors
		-1.414	-1	0	+1	+1.414	
Amount of RHA	x_1	0.0717	0.08	0.15	0.22	0.2283	0.07 and 0.0083
Amount of FA	x_2	0.2067	0.215	0.3	0.385	0.3933	0.085 and 0.0083

The number experiments of CCRD for the two factors is indicated by formula (1):

$$N = 2^k + 2\times k + m \qquad (1)$$

where: k – the number of factors, k=2;
m – the number of repeating experiments at the center of model, m = 5 according to [23].
Substitution into formula (1) obtained: $N = 2^2 + 2\times2 + 5 = 13$.

Compositions of high-mobility concrete mixtures, which were calculated by the method of central composite rotatable design for two factors, are displayed in Table 7. The values of slump cone of concrete mixture are measured in centimeter immediately after the completion of the mixing of the raw materials. These values after 30 and 60 minutes are presented in Tables 8 - 10.

Table 7. Compositions of concrete mixtures in the CCRD for two factors

Trial No.	Description of code		Parameter factors		Compositions of concrete mixtures (kg/m³)						
	x_1	x_2	RHA	FA	PC	RHA	FA	QS	CL	W	A388
1	+1	+1	0.22	0.385	348	77	134	537	1053	157	3.5
2	-1	+1	0.08	0.385	386	31	149	537	1053	158	3.9
3	+1	-1	0.22	0.215	395	87	85	537	1053	159	4
4	-1	-1	0.08	0.215	445	36	96	537	1053	162	4.5
5	+1.414	0	0.2283	0.3	368	84	110	537	1053	157	3.7
6	-1.414	0	0.0717	0.3	417	30	125	537	1053	160	4.2
7	0	+1.414	0.15	0.3933	364	55	143	537	1053	157	3.6
8	0	-1.414	0.15	0.2067	421	63	87	537	1053	160	4.2
9	0	0	0.15	0.3	391	59	117	537	1053	159	3.9
10	0	0	0.15	0.3	391	59	117	537	1053	159	3.9
11	0	0	0.15	0.3	391	59	117	537	1053	159	3.9
12	0	0	0.15	0.3	391	59	117	537	1053	159	3.9
13	0	0	0.15	0.3	391	59	117	537	1053	159	3.9

Table 8. Values of slump cone of concrete mixtures after immediately mixing completion

Trial No.	Description of code						Y_0 (cm)		$(Y_{1j}-\hat{Y}_{1j})^2$	$(Y_{01j}-\hat{Y}_{01j})^2$
	x_0	x_1	x_2	x_1x_2	x_1^2	x_2^2	Y_{1j}	$\hat{Y}_{1j}$		
1	+1	+1	+1	1	1	1	15.8	15.73	0.005	-
2	+1	-1	+1	-1	1	1	17.0	17.15	0.023	-
3	+1	+1	-1	-1	1	1	18.2	18.07	0.017	-
4	+1	-1	-1	1	1	1	20.0	19.49	0.26	-
5	+1	+1.414	0	0	2	0	16.7	16.73	0.001	-
6	+1	-1.414	0	0	2	0	18.6	18.74	0.019	-
7	+1	0	+1.414	0	0	2	16.0	15.83	0.029	-
8	+1	0	-1.414	0	0	2	18.8	19.14	0.115	-
9	+1	0	0	0	0	0	18.3	18.2	0.01	0.01
10	+1	0	0	0	0	0	18.2	18.2	0	0
11	+1	0	0	0	0	0	18.3	18.2	0.01	0.01
12	+1	0	0	0	0	0	18.0	18.2	0.04	0.04
13	+1	0	0	0	0	0	18.2	18.2	0	0
$\sum(Y_{1j}-\hat{Y}_{1j})^2 = 0.529$				$S_d^2 = 0.066$			$\sum(Y_{01j}-\hat{Y}_{01j})^2 = 0.06$		$S_{II}^2 = 0.015$	$\hat{Y}_0 = 18.20$

Table 9. Values of slump cone of concrete mixtures after 30 minutes of mixing completion

Trial No.	Description of code						Y_{30} (cm)		$(Y_{1j}-\hat{Y}_{1j})^2$	$(Y_{01j}-\hat{Y}_{01j})^2$
	x_0	x_1	x_2	x_1x_2	x_1^2	x_2^2	Y_{1j}	$\hat{Y}_{1j}$		
1	+1	+1	+1	1	1	1	13.0	12.70	0.092	-
2	+1	-1	+1	-1	1	1	13.5	14.13	0.397	-
3	+1	+1	-1	-1	1	1	15.5	15.45	0.003	-
4	+1	-1	-1	1	1	1	17.2	16.88	0.101	-
5	+1	+1.414	0	0	2	0	13.5	13.70	0.040	-
6	+1	-1.414	0	0	2	0	16.0	15.73	0.074	-
7	+1	0	+1.414	0	0	2	13.2	12.92	0.079	-
8	+1	0	-1.414	0	0	2	16.6	16.81	0.044	-
9	+1	0	0	0	0	0	14.2	14.34	0.020	0.020
10	+1	0	0	0	0	0	14.5	14.34	0.026	0.026
11	+1	0	0	0	0	0	14.4	14.34	0.004	0.004
12	+1	0	0	0	0	0	14.0	14.34	0.116	0.116
13	+1	0	0	0	0	0	14.6	14.34	0.068	0.068
$\sum(Y_{1j}-\hat{Y}_{1j})^2=1.063$				$S_d^2=0.133$			$\sum(Y_{01j}-\hat{Y}_{01j})^2=0.232$		$S_{II}^2=0.058$	$\hat{Y}_0=14.34$

Table 10. Values of slump cone of concrete mixtures after 60 minutes of mixing completion

Trial No.	Description of code						Y_{60} (cm)		$(Y_{1j}-\hat{Y}_{1j})^2$	$(Y_{01j}-\hat{Y}_{01j})^2$
	x_0	x_1	x_2	x_1x_2	x_1^2	x_2^2	Y_{1j}	$\hat{Y}_{1j}$		
1	+1	+1	+1	1	1	1	8.5	8.21	0.085	-
2	+1	-1	+1	-1	1	1	9.6	10.24	0.414	-
3	+1	+1	-1	-1	1	1	11.2	11.40	0.039	-
4	+1	-1	-1	1	1	1	14.0	13.43	0.322	-
5	+1	+1.414	0	0	2	0	9.2	9.23	0.001	-
6	+1	-1.414	0	0	2	0	12.2	12.11	0.008	-
7	+1	0	+1.414	0	0	2	9.0	8.72	0.081	-
8	+1	0	-1.414	0	0	2	13.0	13.22	0.051	-
9	+1	0	0	0	0	0	10.5	10.52	0.000	0.000
10	+1	0	0	0	0	0	10.8	10.52	0.078	0.078
11	+1	0	0	0	0	0	10.6	10.52	0.006	0.006
12	+1	0	0	0	0	0	10.4	10.52	0.014	0.014
13	+1	0	0	0	0	0	10.3	10.52	0.048	0.048
$\sum(Y_{1j}-\hat{Y}_{1j})^2=1.149$				$S_d^2=0.144$			$\sum(Y_{01j}-\hat{Y}_{01j})^2=0.148$		$S_{II}^2=0.037$	$\hat{Y}_0=10.52$

The computer program Matlab provides regression equation (2), (3) and (4):

$$y_0 = 18.2 - 0.7109x_1 - 1.17x_2 + 0.15x_1x_2 - 0.2188x_1^2 - 0.3438x_2^2 \text{ (cm)}. \quad (2)$$

$$y_{30} = 14.34 - 0.717x_1 - 1.376x_2 + 0.3x_1x_2 + 0.1988x_1^2 + 0.2738x_2^2 \text{ (cm)}. \quad (3)$$

$$y_{60} = 10.52 - 1.018x_1 - 1.595x_2 + 0.425x_1x_2 + 0.0838x_1^2 + 0.2338x_2^2 \text{ (cm)}. \quad (4)$$

By using Matlab program, the images of response surface and corresponding contour plot of the regression equations (2), (3) and (4) are shown in Fig. 2 – 4.

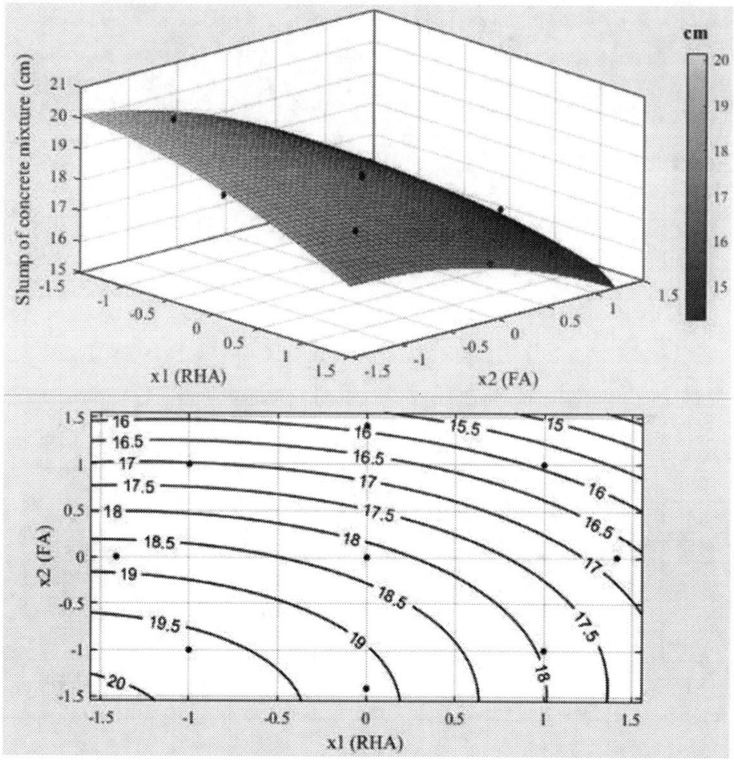

Fig. 2. The image of response surface and corresponding contour plot of the regression equations (2)

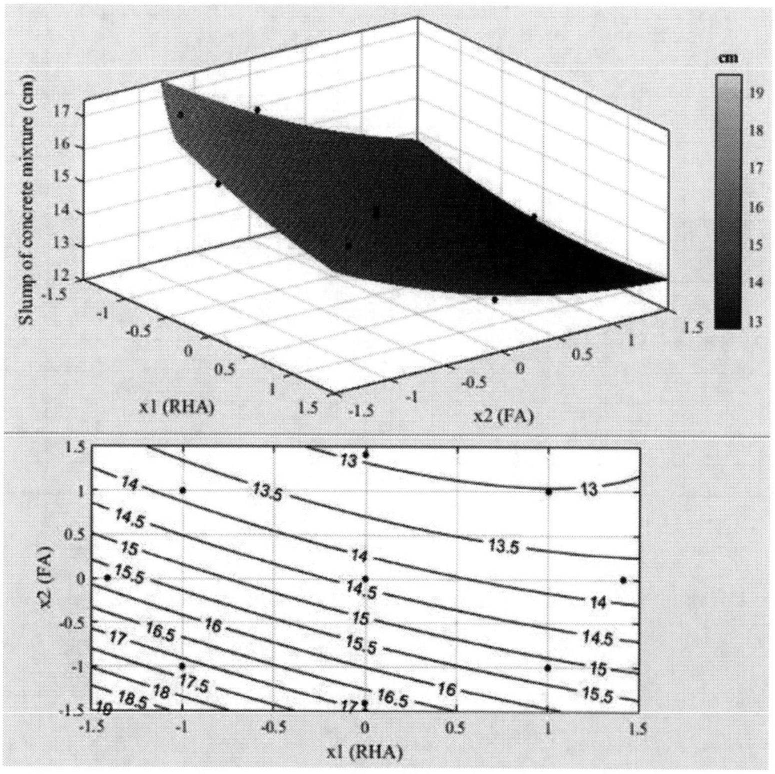

Fig. 3. The image of response surface and corresponding contour plot of the regression equations (3)

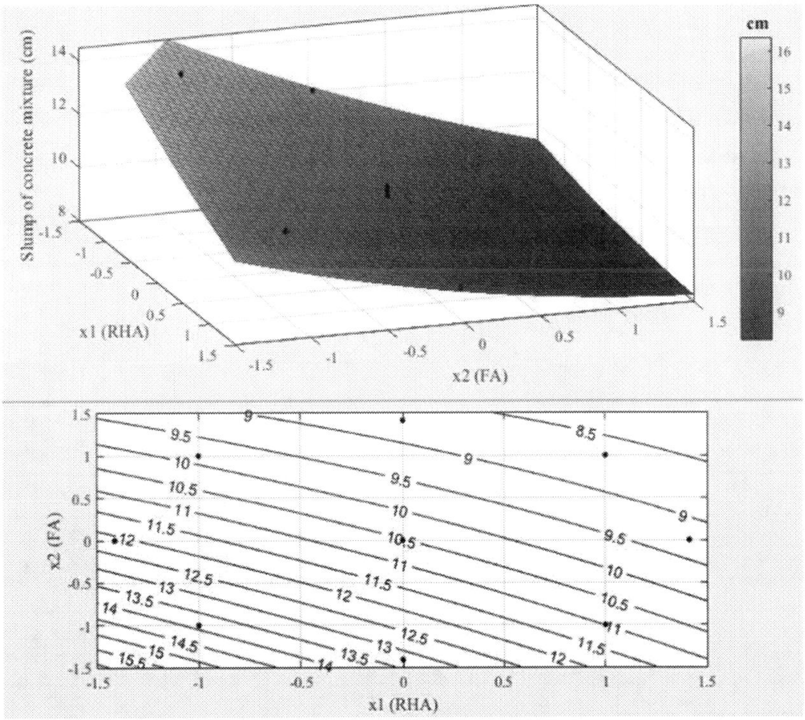

Fig. 4. The image of response surface and corresponding contour plot of the regression equations (4)

Checking the significant coefficients of regression equations

The significant coefficients of the regression equations were determined statistically by the Student's criterion ($t_\alpha(f_2)$). The coefficient b_j is considered significant if:

$$t_{bj} \geq t_\alpha(f_2), \tag{5}$$

where: $t_\alpha(f_2)$ - the value of the Student's criterion.

At the significant level $\alpha = 0.025$ and the degree of freedom $f_2 = N\times(k-1) = 8\times(3-1) = 16$ from Table 3.2 [24] obtained $\mathbf{t_{0.025}(16) = 2.7764}$.

The values of the Student's criterion t_{bj} for the coefficients b_j are calculated by the formula (6):

$$t_{bj} = \frac{|b_j|}{S_{bj}} \tag{6}$$

Estimation of dispersion of coefficients of regression equations S_{bj} are provided by the formula (7):

$$S_{bj} = \sqrt{\frac{S_{II}{}^2}{\sum_{i=1}^{N} x^2{}_{ji}}}, \tag{7}$$

where: S^2_{II} – estimation of dispersion of observation-error variance, which is determined by the formula (8):

$$S_{II}^2 = \frac{\sum_{i=1}^{m}(Y_{oj} - \hat{Y}_o)^2}{m-1}, \tag{8}$$

where: m - the number of repeated experiments at the center, $m = N_o = 5$;

Y_{oj} - the obtained value of the *i-th* experiment at the center;

$\hat{Y}_0$ - the average value of m of the experiments at the center of the experimental model, determined by the formula (9):

$$\hat{Y}_0 = \frac{1}{m}\sum_{n=1}^{m} Y_{oi} = \frac{1}{5}\sum_{i=1}^{5} Y_{oi} = 75.37 \tag{9}$$

Σx^2_{ji} – the total number of square columns x_j of the empirical matrix with two-level layout the optimum level;

$\Sigma x^2_j = N$, where: N – the number of all possible tests, N = 13.

Values of the Student's criterion of coefficients of the second-order regression equations (2), (3) and (4) are shown in Table 11.

Table 11. Values of the Student's criterion of coefficients of the second-order regression equations

j	0	1	2	3	4	5
	b_0	**b_1**	**b_2**	**b_{12}**	**b_{11}**	**b_{22}**
b_{j0}	18.2	-0.7109	-1.17	0.15	-0.2188	-0.3438
$\lvert b_{j0}\rvert$	18.2	0.7109	1.17	0.15	0.2188	0.3438
S_{bj0}	0.055	0.043	0.043	0.061	0.046	0.046
t_{bj0}	**332.285**	**16.418**	**27.020**	2.449	**4.708**	**7.397**
b_{j30}	14.34	-0.717	-1.376	0.3	0.1988	0.2738
$\lvert b_{j30}\rvert$	14.34	0.717	1.376	0.3	0.1988	0.2738
S_{bj30}	0.108	0.085	0.085	0.120	0.091	0.091
t_{bj30}	**133.144**	**8.421**	**16.160**	2.491	2.175	**2.996**
b_{j60}	10.52	-1.018	-1.595	0.425	0.0838	0.2338
$\lvert b_{j60}\rvert$	10.52	1.018	1.595	0.425	0.0838	0.2338
S_{bj60}	0.086	0.068	0.068	0.096	0.073	0.073
t_{bj60}	**122.293**	**14.969**	**23.453**	**4.419**	1.147	**3.203**

After this checking, the insignificant coefficients are discarded and the following equations are obtained as a result of the experimental model.

The regression equation for the slump cone of the concrete mixture was measured immediately after the completion of the raw materials mixing (10):

$$y_0 = 18.2 - 0.7109x_1 - 1.17x_2 - 0.2188x_1^2 - 0.3438x_2^2 \text{ (cm).} \tag{10}$$

The regression equation for the slump cone of the concrete mixture was measured after 30 minutes of mixing completion (11):

$$y_{30} = 14.34 - 0.717x_1 - 1.376x_2 + 0.2738x_2^2 \text{ (cm).} \tag{11}$$

The regression equation for the slump cone of the concrete mixture was measured after 60 minutes of mixing completion (12):

$$y_{60} = 10.52 - 1.018x_1 - 1.595x_2 + 0.425x_1x_2 + 0.2338x_2^2 \text{ (cm).} \tag{12}$$

Checking the adequacy of the experimental model

The adequacy of the second-order regression equations was checked by means of the Fisher criterion. The calculated value of which is the following ratio (13):

$$F_i = \frac{S_d^2}{S_{ll}^2}, \tag{13}$$

where: S^2_{ll}- variance estimates the reproducibility of the experiment, which is determined by formula (14):

$$S_{II}^2 = \frac{\sum_{j=1}^{m}(Y_{oj} - \hat{Y}_o)^2}{m-1}, \tag{14}$$

where: m- the number of repeated experiments at the center, m = 5;

Y_{oj} - the obtained value of the i-th experiment at the center of the experimental model;

$\bar{Y}_0$ - the average value of m of the experiments at the center of the model;

S^2_d - estimation of the dispersion of inadequacy, determined by the formula (15):

$$S_d^2 = \frac{\sum_{j=1}^{N}(Y_j - \hat{Y}_j)^2}{N-B}, \tag{15}$$

where: B - the number of coefficients of the second-order regression equation, which are significant, B = 5;

Y_j - the observed value of the i-th experiment;

$\hat{Y}_j$ - the obtained value of the experimental function in accordance with the i-th experiment;

F_α, (f_1, f_2) – the value Fisher criterion, which is obtained from table 3.5 [24], with a significant level of α = 0,05;

where: f_1- the number of degrees of freedom for residual variance, f_1 =N- B =13 - 5 = 8;

f_2- the number of degrees of freedom of an estimate of variance of observations, f_2 = m -1=5-1= 4.

Consequently: $F_{0,05}(8, 4) = 6.0410$.

The regression equation (10): $S_d^2 = 0.066$ and $S_{II}^2 = 0.015$ was obtained:

$F_0 = \frac{S_d^2}{S_{II}^2} = \frac{0.066}{0.015} = 4.40$. As $F_0 = 4.40 < F_{0.05}(8, 4) = 6.0410$, the experimental model described by regression equation (10) is adequate.

The regression equation (11): $S_d^2 = 0.133$ and $S_{II}^2 = 0.058$ was obtained:

$F_{30} = \frac{S_d^2}{S_{II}^2} = \frac{0.133}{0.058} = 2.293$. As $F_{30} = 2.293 < F_{0.05}(8, 4) = 6.0410$, the experimental model described by regression equation (11) is adequate.

The regression equation (12): $S_d^2 = 0{,}144$ and $S_{II}^2 = 0{,}037$ was obtained:

$F_{60} = \frac{S_d^2}{S_{II}^2} = \frac{0.144}{0.037} = 3.892$. As $F_{60} = 3.892 < F_{0.05}(8, 4) = 6.0410$, the experimental model described by regression equation (12) is adequate.

It can be seen in Table 8-10 and in Fig. 2-4, the results of workability determination of concrete mixtures, which are obtained in laboratory conditions, shows the usage of complex organic mineral modifiers. These mineral modifiers, contain RHA, FA TPP "Vung Ang" and superplasticizer ACE 388 in an amount of 1% of the mass of Portland cement, are relatively effective for workability of concrete mixture. However, they also lead to a decrease in their mobility. It can be explained by the fact that the water requirement of RHA (100.7%) and FA (104.1%) are greater than that of Portland cement (28.58%). In this case, the superplasticizer ACE 388 reduces the water-cement ratio, thereby increasing the density in the structure of concrete mixture. The average value of slump cone of concrete mixtures measured after 60 minutes of mixing completion are decreased from 18.2 to 10.52 cm. These values are still within acceptable range for pumping to make possible delivery of concrete mixture.

Conclusions

Based on the obtained results of the workability of concrete mixes, the following conclusions can be drawn:

1. Using the Central composite rotatable design with the help of Matlab program were obtained second-order regression equations (10), (11) and (12), as well as images of the surface response and corresponding contour plot (shown in Fig. 2 – 4). Regression equations adequately describe the dependence of slump cone concrete mixtures on variables x_1 - amount of RHA and x_2 - amount of FA, which were added to the concrete mixtures. These values of slump cone of concrete mixture are measured at various time intervals after mixing.

2. On the range of studied concentrations, the content of RHA and FA has a significant influence on the concrete mixture mobility. It can be explained by the fact that the particles of RHA and FA played an important role as peculiar "sliding bearings" between the grains of cement, leading to the dispersion of cement content in the concrete mixture. Therefore, it is possible to regulate the concrete mixture mobility when transporting fresh concrete by pumping during the high-rise construction in the hot and humid climate of Vietnam. In this case, the addition of FA TPP "Vung Ang" in concrete mixtures has a stronger effect to the slump cone as compared with the RHA.

References

1. Y.M. Bazhenov, *Technology of concrete* (Publishing. ASV, Moscow, 2011)
2. A.O. Joaquim Barros, Liberato Ferrara, Enzo Martinelli, *Recent Advances on Green Concret e for Structural Purposes* (Springer International Publishing AG, 2017)
3. S.Yu. Butrim, N.V. Gilyazidinova, *Concrete of high-rise constructions. VIII All-Russian scientific-practical conference of young scientists with international participation "Russia is young"* (Gorbachev, Kemerovo, 19-22 April, 2016)
4. Ph. D. Huu, Nguyen Ngoc Long, *High-strength and high performance concrete* (Construction Publishing, Hanoi, 2008)
5. M.V. Komarinsky, N.A. Chervova, Transport of concrete mixture in the construction of unique buildings and structures, Construction of unique buildings and structures, **1(28)**, 25-30 (2015)
6. S. Jacobsen, et. al *Pumping of Concrete and Mortar – State of the Art* (COIN Project Report 5, 2008)
7. S. Jacobsen, J.H. Mork, *Pumping of Concrete and Mortar* (Norwegian Concrete Day, SINTEF, 2007)
8. H Kosmatka Steven and L. Wilson Michelle, *Design and Control of Concrete Mixtures* (Portland Cement Association, Skokie, Illinois, USA, 2011)
9. S.A. Ambartsumian, A. Martirosyan, A.V. Galumyan, Standards performing concrete work using the bucket and the concrete pump in the high-speed monolithic housing construction, The industrial and civil engineering, **2**, 29-30 (2010)
10. M. Haist, H.S. Müller, *Optimization of the Pumpability of Self-Compacting Light Weight Concrete* (Proc. SCC, ACBM, NWU, Evanston, 2005)
11. N. N. Quy, Studying the Effect of Additives Finely Powdered Limestone and Fly ash Thermal Power to the Properties of the Concrete Mix Pump, Construction Magazine,

Hanoi, **3**, 41- 44 (2007)

12. L.M. Kolchedantsev, I.G. Osipenkova, Features of organizational and technological solutions in the construction of high-rise buildings, Scientific and technical and industrial, **11**, 17-19 (2013)
13. A.M. Neville, *Properties of concrete* (Publishing. Literature on construction, Moscow, 2003)
14. A.G. Chiknovorjan, *Technological design of prefabricated concrete and reinforced concrete production* (EBS ASV, Samara State University of Architecture and Civil Engineering, 2010)
15. N. V. Toan, Study on the Effect of Husk Ash and Superplasticizer on the Properties of Lakes, Mortar and Concrete, Journal of Science and Technology, **3+4**, 11-51 (2013)
16. Bui Danh Dai, *Effects of Active Rice Husk Ash on the Properties of Concrete Mixtures and Mortar, International Scientific Conference "Some New Achievements in Modern Building Materials Research"* (Architecture and Civil Engineering, Hanoi, 2008)
17. T. V. Lam, B.I. Bulgakov, O.V. Oleksandrova, O.A. Larsen, The possibility of using ash residues for the production of construction materials in Vietnam, Bulletin of BSTU. V.G. Shukhova, **6**, 6- 12 (2017)
18. ACI 211.4R-93, *Guide for Selecting Proportions for High-Strength Concrete with Portland Cement and Fly Ash* (USA, 2010)
19. V.M. Malhotra, P.K. Mehta, *High-Performance, Fligh-Volume Fly Ash Concrete* (Ottawa, Canada, 2005)
20. Tang Van Lam, B.I. Bulgakov, O.V. Oleksandrova, Mathematical modeling of the influence of raw components on the strength of high-quality fine-grained concrete under compression, Scientific and Engineering Journal for Construction and Architecture, **9 (108)**, 999-1009 (2017)
21. N. M. Tuyen, *Planning of the Experiment* (Publisher of Science and Technology, Hanoi, 2007)
22. L.G. Astakhova, *Lectures on the discipline "Mathematical Theory of Experiment Planning"* (Vladikavkaz, 2013)
23. T.P. Abomelik, *Methodology of experiment planning* (Collection of laboratory works for students of specialty 210201.65, Ulyanovsk, 2006)
24. L.N. Bolshev, N.V. Smirnov, *Tables of mathematical statistics* (Publishing "The science", Moscow, 2010)

Effect of rice husk ash and fly ash on the compressive strength of high performance concrete

Tang Van Lam[1,*], *Boris* Bulgakov[1], *Olga* Aleksandrova[1], *Oksana* Larsen[1], and *Pham Ngoc* Anh[2]

[1] Moscow State University of Civil Engineering, Yaroslavskoe shosse, 26, Moscow, 129337, Russia
[2]Ruhr Universität, Bochum, Bochum, Germany

Abstract. The usage of industrial and agricultural wastes for building materials production plays an important role to improve the environment and economy by preserving nature materials and land resources, reducing land, water and air pollution as well as organizing and storing waste costs. This study mainly focuses on mathematical modeling dependence of the compressive strength of high performance concrete (HPC) at the ages of 3, 7 and 28 days on the amount of rice husk ash (RHA) and fly ash (FA), which are added to the concrete mixtures by using the Central composite rotatable design. The result of this study provides the second-order regression equation of objective function, the images of the surface expression and the corresponding contours of the objective function of the regression equation, as the optimal points of HPC compressive strength. These objective functions, which are the compressive strength values of HPC at the ages of 3, 7 and 28 days, depend on two input variables as: x_1 (amount of RHA) and x_2 (amount of FA). The Maple 13 program, solving the second-order regression equation, determines the optimum composition of the concrete mixture for obtaining high performance concrete and calculates the maximum value of the HPC compressive strength at the ages of 28 days. The results contain$MaxR^{28}_{HPC}$ = 76.716 MPa when RHA = 0.1251 and FA = 0.3119 by mass of Portland cement.

1 Introduction

The concrete experts consider the last two decades of the twentieth century as decades of concrete additives. Thus, the 21st century will be the century of widespread usage of different types of high-strength concrete and high performance concrete.

The modern technology monolithic construction, used in various countries in the world as well as in Vietnam, widely applied mineral additives in the concrete mixtures. Among these additives, superplasticizers, complex of chemical organic and mineral additives are the most popular [1].

In concrete production technology, the mineral additives are separated to fine active additives and inert additives. Fine active additives helps to reduce cement content and control

* Corresponding author: lamvantang@gmail.com

the process of forming concrete structure. While the inert mineral additives in fine grinding has been used so far in order to fill and solidify the concrete structure. In which, the major role is microsilica. It is waste generated in ferroalloy and silicon production. Besides, they are more widely available due to their cheaper price, in spite of slightly less effective than imported micro silica fume. Popular inert mineral additives is fly ash, provided from thermal power plants (FA TPP), metakaolin, fine-grained slag and pumice, especially, reusing waste from brick production, gypsum-containing waste [2].

The human beings use approximately 10 billion tons of minerals and organic materials annually. At the same time, about 10% to 99% of these materials are converted into waste, causing soil, water and atmospheres pollution. For example, in coal industry, nearly 1.3 billion tons of overburden, mine rocks and about 80 million tons of wastes of coal enrichment are discharged each year [3].

The rice processing also produces a great quantity of wastes in the form of straw, bran and rice husk. In rice producing countries, one ton of grain is disposed averagely, that is more than one ton of rice straw and husk. Thus, the reuse of these wastes reasonably become high important subject [4].

Rice straw mostly uses for agricultural demand (70% for cattle food) and construction as finishing and roofing materials (about 5%). The rest of this is burnt directly in the fields (up to 15%) or simply left there, as well as natural decomposition (about 10%).

In recent times, main purpose of using rice straw is producing cellulose and its derivatives [5], while rice husk produces silica in crystalline and amorphous forms [6]. Then, amorphous silica is applied in building materials technology. Methods of utilizing agricultural residues in the world are shown in Figure 1.

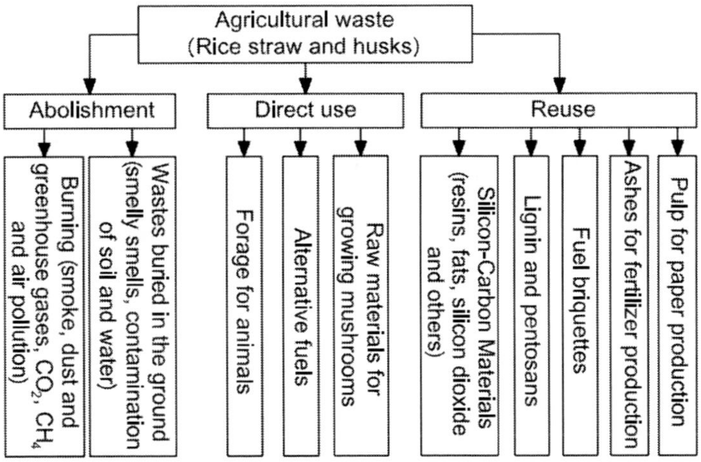

Fig. 1. Methods of disposal of rice straw and husks in the world.

In fact, rice husks have a small bulk density (about 0.1 ton/m^3) [7], thus its storage requires a large area. Moreover, this storage of waste leads to pollution of soil, water and air [8]. Therefore, processing and recycling of rice husk and straw have a great economic and social importance for the northern regions of the Vietnam, where are concentrated several rice processing companies [9].

In Vietnam, rice straw, a byproduct of rice milling, previously was only used as fuel in small ceramic kilns or for alternative fuels production - rice husk. A small amount of rice husk then was mixed with Portland cement to produce unbaked brick blocks. Presently, the ash is obtained by burning the husk, which consist of huge amorphous silicon content. This ash is widely used in Vietnam as an active mineral additives and replaced imported silica fume in concrete production with different purposes

The annual amount of industrial waste is more than 150 million tons in Vietnam. In which, metallurgical slag is about 45 ÷ 55 million tons, ash and slag TPP is nearly 50 ÷ 60 million tons and gypsum wastes is approximately 14 ÷ 16 million tons [10].

In 2016, the accumulation of solid wastes from ash and slag, particularly, waste gas and wastewater from the thermal power plant in the industrial zone "VungAng" (Vietnam), have caused seriously environmental pollution in Vietnam central provinces (Figure 2).

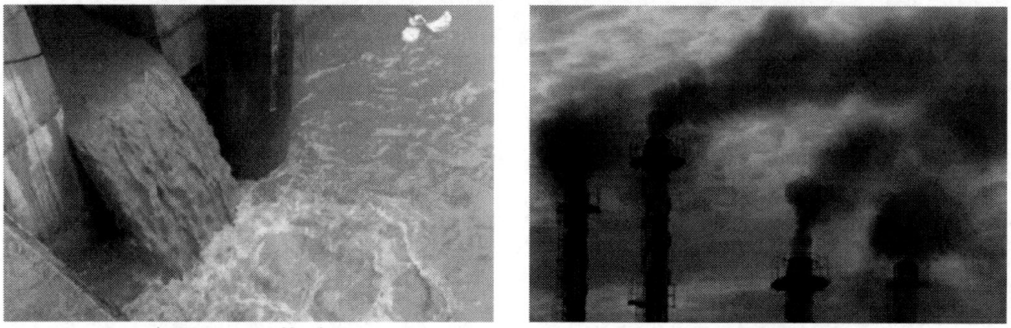

a). Water pollution b). Air pollution

Fig. 2. The contamination of the environment by industrial waste from thermal power plants in Vietnam.

TPP "VungAng" produces about 3000 tons of ash and slag waste daily. In addition, anenormous number of gaseous substances and solid particles formed as a result of solid fuel combustion enter the atmosphere through the smokestacks of this power plants [11].

The recycling of waste from industry and agriculture as mineral additives for modern concrete not only saves capital investments but also reduces production cost both in construction material producing and in industries that have so-called "by-products". It contributes to create a non-waste production cycle in the future.

The objective of this study are mathematical modeling the effect of amounts of rice husk ash and fly ash from TPP "VungAng" on the compressive strength of high performance concrete in the form of second-order regression equations. After that, the images of the surface expression and the corresponding contours of the objective function of the regression equation are received.

2 Materials and Mixture Proportioning

2.1 Cement

Portland Cement (PC) type CEM II 42.5N manufactured at "Nghi Son" factory (Vietnam), specific weight of 3.12 g/cm^3, surface area of 3635 cm^2/g. The results of chemical analysis and mineralogical composition of cement are presented in Table 1 and Table 2.

Table 1. Chemical composition of cement "Nghi Son".

Oxide composition(%)						
CaO	SiO_2	Al_2O_3	Fe_2O_3	MgO	SO_3	LOI*
62.23	22.27	5.37	4.12	2.4	2.626	0.984

Note: LOI - loss on ignition.

Table 2. Mineralogical Composition of cement "Nghi Son".

Mineral composition (%)				
C_3S	C_2S	C_3A	C_4AF	Other
56.3	23.4	4.7	12.4	3.2

2.2 Aggregates

Quartz sand (QS) originally from the golden sand of "Lo River" (Vietnam), M_K = 2.85, specific weight of 2.65 g/cm^3 and the volume of compacted state is 1540 kg/m^3.

Crushed limestone (CL) produced from the quarry "KienKhe" (Vietnam) with the size of 5 - 10 mm, specific weight of 2.61 g/cm^3 and volume of the compacted state is 1620 kg/m^3.

2.3 Chemical admixture

Superplasticizer ACE 388 (A388) "Sure Tec" BASF (Germany). It is a new generation chemical additives based on polycarboxylate ethers with specific weight of 1.1 g/cm^3 at 20 ± 5°C.

2.4 Mineral additives

Fly Ash (FA) TPP "VungAng" (Vietnam) Class C and rice husk ash (RHA) with specific weight of 2.22 and 2.15 g/cm^3, the volume of natural porous state are765 and 572 kg/m^3 and the water needs are 104.1 and 100.7% respectively. The chemical composition of the FA and RHA are shown in Table 3.

Table 3. Chemical compositions of RHA and FA TPP "VungAng"

Mineral additives	Average chemical composition(%)										
	SiO_2	Al_2O_3	Fe_2O_3	SO_3	K_2O	Na_2O	MgO	CaO	TiO_2	P_2O_5	LOI
RHA	88.2	1.25	1.75	0.5	1.14	2.67	0.8	0.52	0.15	0.25	2.77
FA TPP "VungAng"	54.62	25.17	7.11	0.25	1.28	0.2	1.57	1.45	1.83	1.63	2.04

Grain compositions of RHA and FA are determined by Laser method, presented in Table 4.

Table 4. The grain of RHA and FA ТЭС "VungAng".

Rice husk ash	% grain sizes less than D (μm)	10	25	50	75	90
		1.827	3.402	7.311	24.14	63.32
	Specific surface area (μm)	8.415				
	Specific surface area (m^2/g)	11.350				
Fly Ash	% grain sizes less than D (μm)	10	25	50	75	90
		1.025	2.812	3.873	27.65	64.64
	Specific surface area (μm)	2.588				
	Specific surface area (m^2/g)	14.455				

2.5 Methods

Method of calculation preliminary compositions of concrete mixture is applied in accordance with ACI 211.4R-93 (American). According to [12], the flowability of concrete mixture is determined by the slump flow of Abrams' cone with dimensions of 100x200x300 mm. The compressive strength of concrete is determined by a 100x100x100 mm cube pattern by Russian standard GOST 10180-2012. These cube samples are demolded after 24 hours later casting and placed in a 20°C water curing tank until the experiments. The compressive strength HPC is measured at the ages of 3, 7 and 28 days [13].

2.6 Mathematical modeling method

In this paper, the achieved second-order regression equations of the objective function is the compressive strength of HPC by using method of the Central composite rotatable design (CCRD) for two factors [14]. The method CCRD allows to obtain more accurately mathematical description by increasing the number of experiments at the center and a special value of "star shoulder" $\alpha = \sqrt{2} = 1.414$ [26].

The geometric interpretation of the CCRD for the two factors is a square, is shown in Fig. 3.

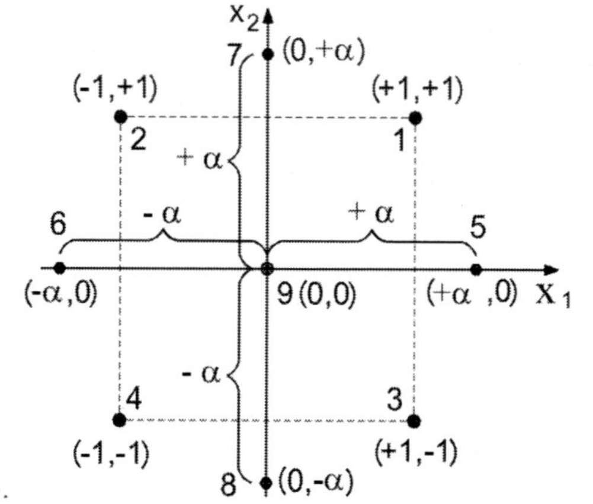

Fig. 3. Geometric interpretation of the CCRD for two factors.

3 Results and Discussion

Development of high performance concrete must possess:

- the effect of the fresh concrete with the workability is determined by the slump-flow 600 ÷ 700 mm;
- compressive strength at the age of 28 days is higher than 70 MPa.
- durability and corrosion resistance in corrosive environment.
- reduce cement consumption.

The preliminary composition of the concrete mixture was calculated according to the standard ACI 211.4R-93 [15]. The properties of the fresh concrete as well as the compressive strength of concrete areachievedand presented in Table 3.

Table 3. Preliminary compositions, properties of fresh concretes and high performance concrete.

Compositions of concrete mixture, kg/m³							Properties of fresh concretes			Compressive strength at 28 days (MPa)
PC	RHA	FA	QS	CL	W	A388	Average density (kg/m³)	$\frac{W}{BID}$	Slump flow (mm)	
392	59	118	533	1053	159	5.7	2319.7	0.28	620	72

The water/binding ratio is constant and equal to $\frac{W}{BID} = 0.28$. The amount of superplasticizer A388 is 1% by the mass of the binder. The amounts of RHA and FA in the composite binder (PC + RHA + FA) are chosen, respectively, in the range from 8% to 22% and from 20% to 40% by the mass of Portland cement. While the contents of QS is 533 kg and CL is 1053 kg.

Definition of objective functions for the description of empirical models:

Yi (i = 3, 7 and 28) - the compressive strength of the samples HPC (R_{HPC}, MPa) at the ages of 3, 7 and 28 days are considered as an objective functions of the experimental model.

Input factors and their limitationare sorted as below:

- x_1 - the amount of RHA, is in the range from 0.08 to 0.22 by mass PC;
- x_2 - the amount of FA, is in the range from 0.2 to 0.4 by mass PC.

The input factors and intervals of their variation are shown in Table 4.

Table 4. Levels and intervals of varying factors of the CCRD.

Factors		Levels varying factors					Intervals varying factors
Parameter	Description	-1.414	-1	0	+1	+1.414	
Amount of RHA	x_1	0.0717	0.08	0.15	0.22	0.2283	0.07 and 0.0083
Amount of FA	x_2	0.2067	0.215	0.3	0.385	0.3933	0.085 and 0.0083

The number experiments of CCRD for the two factors is indicatedby formula (1):

$$N = 2^k + 2 \times k + m \quad (1)$$

where: k – thenumber of factors, k=2;

m – the number of repeating the experiments at the center, m = 5 according to [28].

Consequently: $N = 2^2 + 2 \times 2 + 5 = 13$.

Compositions of high performance concrete, calculated by the method of central composite rotatable design for two factors, are displayed in Table 5. The valuesof compressive strength HPC at the ages of 3, 7 and 28 days of normal hardening are presented in Tables 6 - 8.

Table 5. Matrix of the CCRD for two factors and compositions of concrete mixtures.

Trial No.	Descriptionof code		Parameter		Compositions of concrete mixtures (kg/m³)						
	x_1	x_2	RHA	FA	PC	RHA	FA	QS	CL	W	A388
1	+1	+1	0.22	0.385	349	77	134	533	1053	157	5.6
2	-1	+1	0.08	0.385	388	31	149	533	1053	159	5.7
3	+1	-1	0.22	0.215	396	87	85	533	1053	159	5.7
4	-1	-1	0.08	0.215	447	36	96	533	1053	162	5.8
5	+1.414	0	0.2283	0.3	369	84	111	533	1053	158	5.6
6	-1.414	0	0.0717	0.3	418	30	125	533	1053	160	5.7
7	0	+1.414	0.15	0.3933	365	55	144	533	1053	158	5.6
8	0	-1.414	0.15	0.2067	423	63	87	533	1053	160	5.7
9	0	0	0.15	0.3	392	59	118	533	1053	159	5.7
10	0	0	0.15	0.3	392	59	118	533	1053	159	5.7
11	0	0	0.15	0.3	392	59	118	533	1053	159	5.7
12	0	0	0.15	0.3	392	59	118	533	1053	159	5.7
13	0	0	0.15	0.3	392	59	118	533	1053	159	5.7

Table 6. Compressive strength of HPC at 3 days.

Trial No.	RHA	FA	Description of code					$Y_3=R^3_{HPC}$(MPa)		$(Y_{3j}-\hat{Y}_{3j})$	$(Y_{03j}-\hat{Y}_{03}$
			x_1	x_2	x_1^2	x_1x_2	x_2^2	Y_{3j}	$\hat{Y}_{3j}$		
1	0.22	0.385	+1	+1	1	1	1	45.50	46.32	0.668	-
2	0.08	0.385	-1	+1	1	-1	1	48.13	52.54	19.472	-
3	0.22	0.215	+1	-1	1	-1	1	44.21	42.90	1.711	-
4	0.08	0.215	-1	-1	1	1	1	47.70	49.13	2.037	-
5	0.2283	0.3	+1.414	0	2	0	0	43.30	44.08	0.610	-
6	0.0717	0.3	-1.414	0	2	0	0	56.62	52.86	14.148	-
7	0.15	0.3933	0	+1.414	0	0	2	52.70	49.38	11.004	-
8	0.15	0.2067	0	-1.414	0	0	2	44.23	44.57	0.113	-
9	0.15	0.3	0	0	0	0	0	53.26	54.95	2.843	2.843
10	0.15	0.3	0	0	0	0	0	55.92	54.95	0.949	0.949
11	0.15	0.3	0	0	0	0	0	54.30	54.95	0.417	0.417
12	0.15	0.3	0	0	0	0	0	56.10	54.95	1.332	1.332
13	0.15	0.3	0	0	0	0	0	55.15	54.95	0.042	0.042
$\sum(Y_{3j}-\hat{Y}_{3j})^2=55.346$				$S_d^2=6.918$				$\sum(Y_{03j}-\hat{Y}_{03j})^2=5.582$		$S_{II}^2=1.395$	

Table 7. Compressive strength of HPC at 7 days.

Trial No.	RHA	FA	Description of code					$Y_7=R^7_{HPC}$(MPa)		$(Y_{7j}-\hat{Y}_{7j})^2$	$(Y_{07j}-\hat{Y}_{07j}$
			x_1	x_2	x_1^2	x_1x_2	x_2^2	Y_{7j}	$\hat{Y}_{7j}$		
1	0.22	0.385	+1	+1	1	1	1	61.35	61.43	0.006	-
2	0.08	0.385	-1	+1	1	-1	1	64.92	65.48	0.309	-
3	0.22	0.215	+1	-1	1	-1	1	60.19	58.07	4.479	-
4	0.08	0.215	-1	-1	1	1	1	63.54	62.12	2.019	-
5	0.2283	0.3	+1.414	0	2	0	0	59.25	60.19	0.888	-
6	0.0717	0.3	-1.414	0	2	0	0	65.80	65.91	0.013	-
7	0.15	0.3933	0	+1.414	0	0	2	63.82	62.87	0.901	-
8	0.15	0.2067	0	-1.414	0	0	2	56.12	58.12	4.016	-
9	0.15	0.3	0	0	0	0	0	64.30	65.47	1.364	1.364
10	0.15	0.3	0	0	0	0	0	66.21	65.47	0.551	0.551
11	0.15	0.3	0	0	0	0	0	66.30	65.47	0.692	0.692
12	0.15	0.3	0	0	0	0	0	65.10	65.47	0.135	0.135
13	0.15	0.3	0	0	0	0	0	65.43	65.47	0.001	0.001
$\sum(Y_{7j}-\hat{Y}_{7j})^2=15.375$				$S_d^2=1.922$				$\sum(Y_{07j}-\hat{Y}_{07j})^2=2.744$		$S_{II}^2=0.686$	

Table 8. Compressive strength of HPC at 28 days.

Trial No.	RHA	FA	Description of code					$Y_{28}=R^{28}_{HPC}$(MPa)		$(Y_{28j}-\hat{Y}_{28j})^2$	$(Y_{028j}-\hat{Y}_{028j})^2$
			x_1	x_2	x_1^2	x_1x_2	x_2^2	Y_{28j}	$\hat{Y}_{28j}$		
1	0.22	0.385	+1	+1	1	1	1	73.35	70.32	9.193	-
2	0.08	0.385	-1	+1	1	-1	1	75.92	74.21	2.931	-
3	0.22	0.215	+1	-1	1	-1	1	69.89	68.87	1.040	-
4	0.08	0.215	-1	-1	1	1	1	74.24	72.76	2.190	-
5	0.2283	0.3	+1.414	0	2	0	0	66.27	68.16	3.559	-
6	0.0717	0.3	-1.414	0	2	0	0	72.38	73.66	1.630	-
7	0.15	0.3933	0	+1.414	0	0	2	70.82	73.20	5.641	-
8	0.15	0.2067	0	-1.414	0	0	2	70.36	71.15	0.621	-
9	0.15	0.3	0	0	0	0	0	74.75	76.32	2.452	2.452
10	0.15	0.3	0	0	0	0	0	78.21	76.32	3.587	3.587
11	0.15	0.3	0	0	0	0	0	75.12	76.32	1.430	1.430
12	0.15	0.3	0	0	0	0	0	76.25	76.32	0.004	0.004
13	0.15	0.3	0	0	0	0	0	77.25	76.32	0.872	0.872
$\sum(Y_{28j}-\hat{Y}_{28j})^2=35.152$				$S_d^2=4.394$	$\sum(Y_{028j}-\hat{Y}_{028j})^2=8.3$					$S_{II}^2=2.087$	

Checking the adequacy of the experimental model. According to [27-29],the adequacy of the regression equations is checked by means of the Fisher criterion. The calculated value of which is the following ratio (2):

$$F_i=\frac{S_d^2}{S_{II}^2} \tag{2}$$

where: S^2_{II}- variance estimates the reproducibility of the experiment, determined by formula (3):

$$S_{II}^2=\frac{\sum_{j=1}^{m}(Y_{oj}-\hat{Y}_o)^2}{m-1} \tag{3}$$

where: m- the number of repeated experiments at the center, m = 5;

Y_{oj} - the obtained value of the i-th experiment at the center;

$\hat{Y}_0$ - the average value of m of the experiments at the center;

S^2_d - estimation of the dispersion of inadequacy, determined by the formula (4):

$$S_d^2=\frac{\sum_{j=1}^{N}(Y_j-\hat{Y}_j)^2}{N-B} \tag{4}$$

where:

B -the number of coefficients of the second-order regression equation, which are significant, B = 5;

Y_j- the observed value of the i-th experiment;

$\hat{Y}_j$-the obtained value of the experimental function in accordance with the i-th experiment;

F_α, (f_1, f_2)–the value Fisher criterion with a significant level of $\alpha = 0,05$;

where: f_1- number of degrees of freedom for residual variance, f_1=N- B =13 - 5 = 8;

f_2-number of degrees of freedom for estimating the variance of observations, f_2=m-1=5-1= 4.

Consequently: $F_{0,05}$ (8, 4) = 6,0410.

The second-order regression equations (5), (6) and (7), the values of the coefficients, the images of response surface and corresponding contour plot in Figure 4 – 6 are obtained by Matlab program.

$$Y_3 = R^3_{HPC} = 54.95 - 3.12x_1 + 1.712x_2 - 3.201x_1^2 - 3.948x_2^2 + 0.215x_1x_2 (MPa) \qquad (5)$$

$$F_3 = \frac{S_d^2}{S_{ll}^2} = \frac{6.918}{1.395} = 4.9576 .$$

As F_3= 4.9576< $F_{0,05}$(8, 4) = 6.0410, the experimental model described by equation (5) is adequate.

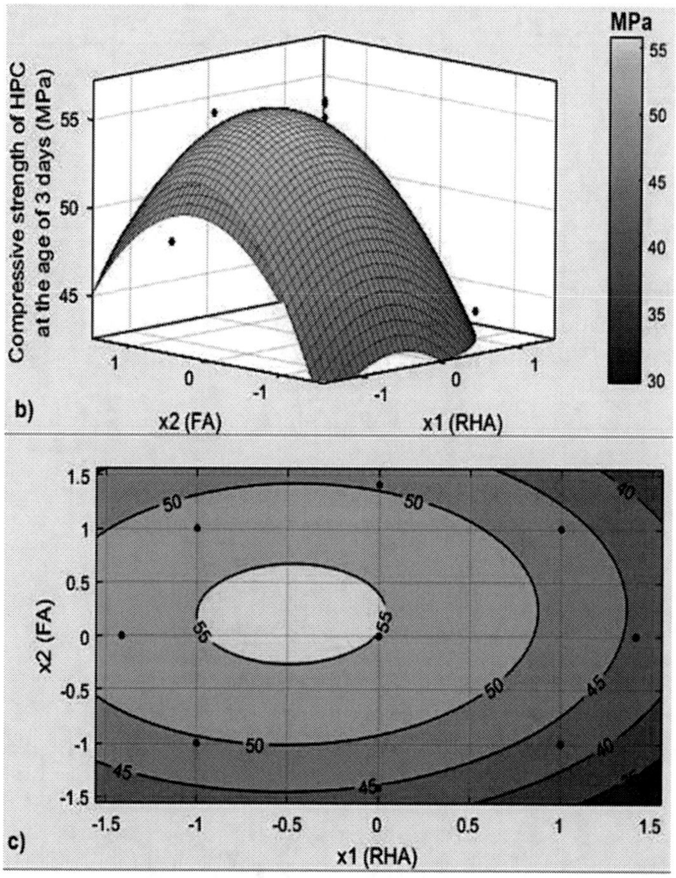

Fig. 4. The coefficients a), response surface b) and contour plot c) of the second-order regression equation (5) for compressive strength of HPC at the age of 3 days.

$$Y_7 = R^7_{HPC} = 65.47 - 2.023x_1 + 1.679x_2 - 1.158x_1^2 - 2.436x_2^2 - 0.055x_1x_2 (MPa) \qquad (6)$$

The program Maple 13 provides $MaxY_3^{opt} = 55.874$ at $x_1^{opt} = -0.4805$ and $x_2^{opt} = 0.2037$. Then, Max R^3_{HPC}=55.874MPa at RHA = 0.1164 and FA = 0.3173by mass PC, represented at point A in Fig. 7.

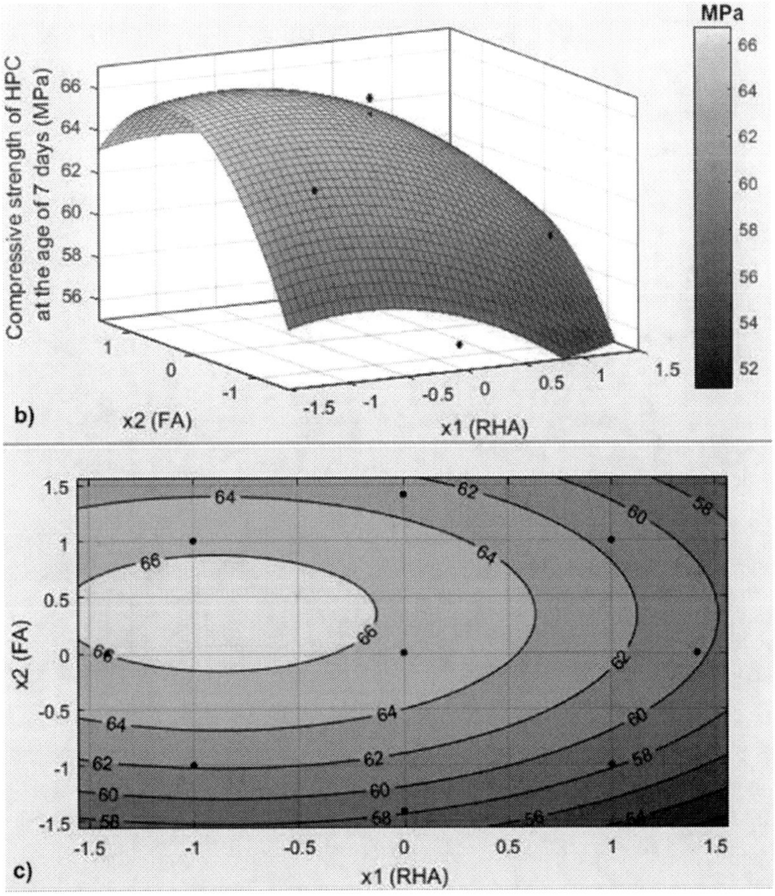

Fig. 5. The coefficients a), response surface b) and contour plot c) of the second-order regression equation (6) for compressive strength of HPC at 7 days.

$$Y_{28} = R^{28}_{HPC} = 76.32 - 1.945x_1 + 0.7239x_2 - 2.647x_1^2 - 2.015x_2^2 + 0.445x_1x_2 (MPa) \quad (7)$$

The program Maple 13 shows $MaxY_7^{opt} = 66.6597$ at $x_1^{opt} = -0.8819$ and $x_2^{opt} = 0.3546$. Then, Max R^7_{HPC}=66.6597 MPa at RHA = 0.0883 and FA =0.3301 by mass PC, represented at point B in Figure 7.

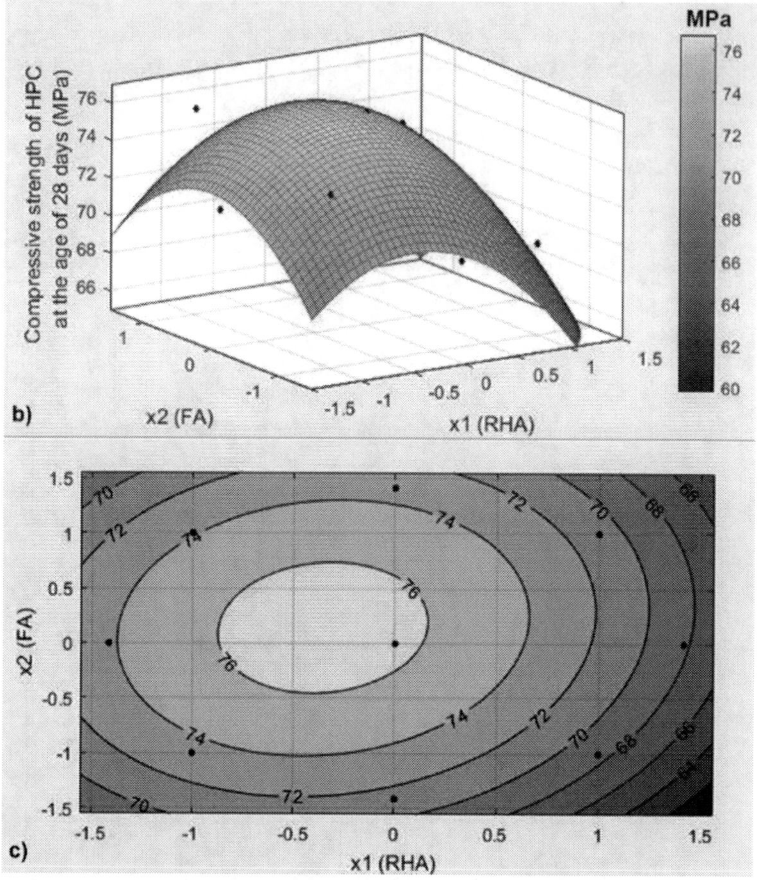

Fig. 6. The coefficients a), response surface b) and contour plot c) of the second order regression equation (7) for compressive strength of HPC at 28 days.

The program Maple 13 provides $MaxY_{28}^{opt} = 76.716$ at $x_1^{opt} = -0.3554$ and $x_2^{opt} = 0.1395$. Then, Max $R^{28}_{HPC} = 76.716$ MPa at RHA = 0.1251 and FA = 0.3119 by mass Portland cement, represented at point C in Fig. 7.

Figure 7 presents the curve of dependence of the compressive strength HPC on the amount of RHA and FA at the ages of 3, 7 and 28 days. This curve passes through points A, B and C, which are points of the optimum compressive strength of HPC at the age of 3, 7 and 28 days.

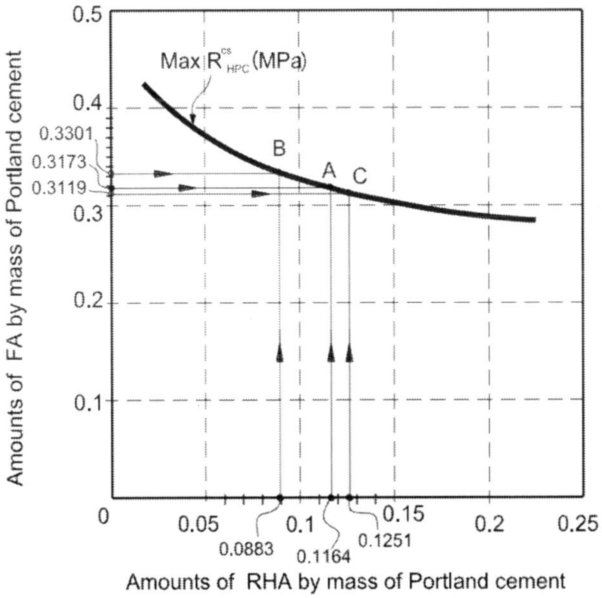

Fig. 7. Compressive strengthof HPC of amountsRHAand FA: A- optimum point of compressive strength of HPC at the age of 3 days; B - optimum point of compressive strength of HPC at the age of 7 days; C- optimum point of compressive strength of HPC at the age of 28 days.

Based on the data in Table. 6 - 8 and Figure 4 – 6, the usage of complex organic mineral modifiers, which contains RHA, FA and superplasticizer ACE 388 in an amount of 1% of the mass of the binder is gradually effective. Furthermore, the speed of concrete hardening is comparativelyhigh at the age of 3 days. It reaches 62.03% to 78.23% of its 28-day compressive strength (in Table 6 and Table 8).

4 Conclusions

1. RHA in the North of Vietnam and FA TPP "VungAng" have high pozzolanicactivity. The average content of amorphous silicon dioxide of 88.2% and 54.62%, respectively, are characterized by a low content of harmful impurities (unburned carbon, SO_3, sulfur, P_2O_5, mineral oils). Therefore, they can be considered as finely dispersed active mineral admixtures for concrete and mortar [16].

2. Using central composite rotatable design combination with Matlab program obtainsthe second-order regression equations (5), (6) and (7), as well as the images of response surface and corresponding contour plot of the objective function for these regression equations,are shown in Fig. 4 - 6. Theseequations adequately describe the dependence of the compressive strength HPC at the ages of 3, 7 and 28 days by the factors x_1 (amount of RHA) and x_2 (amount of FA).

3. The Maple 13 program determines the maximum value of the objective function-compressive strength HPC at the age of 28 days in the second-order regression equation (7), is $MaxR^{28}_{HPC}$ = 76.716 MPa at RHA = 0.1251 and FA = 0.3119 by mass of Portland cement.

References

1. E.B. Entin, L.S. Nefedova, N.V. Strzhalkovskaya, Cem. and its appl., **2** (2012)
2. N.I. Vatin, D.V. Petrosov, A.I. Kalachev, P. Lakhtinen, Mag. of Civ. Eng., **4** (2011)
3. D.F. Houston, *American Association of Cereal Chemists, St. Paul, Minnesota* (1972)

4. S.P. Shah, S.H. Ahmad, *High performance concrete: Properties and applications*. McGraw-Hill, Inc. (1994)
5. A.V. Vurasko, B.N. Driker, Ye.A Mozyreva, L.A. Zemnukhova, A.R. Galimova, N.N. Gulemina, Chem. of veg. raw mat., **4** (2006)
6. Ch.-L. Hwang, Tr.-Ph. Huynh, Constr. and Build. Mat., **101**, 1 (2015)
7. Zemnukhova, G.A. Fedorishcheva, A.G. Yegorov, V.I. Sergiyenko, Journ. of Appl. Chem, 2, **78** (2005).
8. N. Van Toan, J. of Constr. Sc. and Techn. **3, 4** (2013)
9. P.K. Mehta, V.M. Malhotra, *Adv. in Concr. Techn. (*1994)
10. B. Danh Dai. *International Scientific Conference "Some new achievements in modern building materials research"* (2006)
11. Bl. Sk. Thomas, Ren. and Sust. En. Rev., **82**, 3 (2018)
12. H Th. Le, H.-M. Ludwig, Mat. & Des., **89** (2016)
13. H. Chao-Lung, B. Le Anh-Tuan, Ch. Chun-Tsu, Constr. and Build. Mat., **25**, 9 (2011)
14. M. Tuyen Nguyen. *Planning of the experiment. Publisher of Science and Technology*. Publishing of Science and Technology (2007)
15. ACI 211.4R-93. Guide for selecting proportions for High-Strength Concrete with Portland Cement and Fly Ash (2010)
16. K.-B. Park, S.-J. Kwon, X.-Y. Wang, Constr. and Build. Mat., **105** (2016)

Experimental Research and Method for Calculation of 'Upsetting-with-Buckling' Load at the Impression-Free (Dieless) Preforming of Workpiece

Volodymir Kukhar[1], *Victor* Artiukh[2*], *Andrii* Prysiazhnyi[1] *and Andrey* Pustovgar[3]

[1]Pryazovskyi State Technical University, Universytetskaya str., 7, Mariupol, 87500, Ukraine,
[2]Peter the Great St. Petersburg Polytechnic University, Polytechnicheskaya str., 29, 195251, Saint-Petersburg, Russia,
[3]Moscow State University of Civil Engineering, Yaroslavskoe shosse, 26, Moscow, 129337, Russia

Abstract. This paper presents the results of experimental studies of load characteristic changes during the upsetting of high billets with the upsetting ratio (height to diameter ratio) from 3.0 to 6.0, which is followed by buckling. Such pass is an effective way of pre-forming the workpiece for production of forgings with a bended axis or dual forming, and belongs to impression-free (dieless) operation of bulk forming. Based on the experimental data analysis, an engineering method for calculation of workpiece pre-forming load as a maximum buckling force has been developed. The analysis of the obtained data confirmed the possibility of performing of this pre-forming operation on the main forging equipment, since the load of shaping by buckling does not exceed the load of the die-forging.

1 Introduction

Upsetting and bending are the most common operations in press-forging industry. Upsetting is mostly performed as a pressing between two flat dies, although the dies should still be regarded as a pressing deforming tool that has contoured working faces or impressions which provide the necessary shape to the product after pressing. Upsetting by uncontoured dies is more common. Its purpose is descaling, increasing of workpiece diameter for subsequent cogging, swaging, impression forging or die-forging of forgings with a round in plan view area cross section, improving the inner metal structure and growth of new fine grains [1]. Bending is used for bulk and sheet metal forming, and this operation is usually carried out using contoured dies, forming rolls or bending machines [1-3].

* Corresponding author: artiukh@mail.ru

We propose and investigate a brand new bulk forming process, such as preforming of workpiece for subsequent open die or closed die forging by upsetting of billet with large ratio of height to diameter (upsetting ratio more then 3.5), followed by buckling. Bulk forming of workpiece by upsetting with buckling is performed as a single-stroke process and one such operation combines at least three: descaling, bending and redistribution of metal along the length of a workpiece (edging, fullering or gathering – as at drop-forging). Given that it is a question of hot-forming processes, this forming operation is performed after heating the workpiece in a furnace. Buckling, accompanied by the edging and fullering of metal along the bended axis, should be attributed to one of the most effective methods of bulk dieless preforming, which has not been sufficiently studied to the present time. The practice of such operation for forming of forgings with a bended axis or dual forming [4] showed fine results in saving of metal (at least 20%), however, the load behaviour during the working stroke of 'upsetting-with-buckling' of workpieces with different upsetting ratio remain unstudied. The study of load behaviour as force modes is relevant for the right choice of appropriate production equipment, standardized gaps in the press and dies guides, angular deflection of frames and the development of combined die forging processes with the preforming and final forming of the metal product in a single die holder (sub-sow block) in automatic forging modes.

2 Review

In practice, the upsetting of workpieces with the ratio of the initial height L_0 to the diameter D_0 (upsetting ratio) $m_0 = L_0/D_0 > 3.5$ is not used to exclude buckling, which is considered a negative phenomenon. However, different researchers determined a different critical value of the upsetting ratio $m_0 = L_0/D_0$ between 2.7 and 4.3, at which the buckling of a high cylindrical workpieces occurs, which depends on the conditions at the ends of the workpiece, material type, process temperature, etc. The upper value of the possible height of the workpiece is limited by the dimensions of press frame gap. There are a lot of studies in the field of buckling of various systems, but they do not concern the practical application of the technology of bulk forming.

The strain of thin-walled shells, columns and rods has a sufficient number of modern fundamental studies [5-7], which can not be used to predict the shape changes and the behaviour of the load modes of plastic billets, which buckled during upsetting due to the discrepancy between geometric and boundary conditions. Processes which cause buckling are negative for the formation of thin-walled sheet and tubular products by the methods of sheet-metal forming, which was researched in [8] by finite element modelling and tested experimentally. It is shown that in order to predict the buckling nature, it is necessary to take into account the properties of the material and the geometric parameters of the workpiece. In addition, it is very important to exclude inaccuracies of the "press-die" system to provide quality of formed products from a sheet blanks [9]. In the paper [10], bending of thin-walled cylindrical shells that contain one welding seam was studied. The graphs of the ration of load force and bending deflection have been experimentally constructed. A comparison is made with theoretical assumptions based on Koiter's approach, relating to asymmetrically applied critical loads. The results are applicable only to the elastic region and it is impossible to estimate the variation of the load behaviour of the rods and shells experiencing plasticity buckling.

The buckling of aircraft panels under static loads to track the appearance of fatigue cracks was studied in the work [11]. Moreover, in order to increase the static load, it is proposed to introduce protective properties of the skin of the panels. Obviously, buckling strain is seen as a negative process, leading to a decrease in the durability of such panels. The results of studies [12] show that the use of workpieces with a profiled lateral surface allows to increase the value of the critical load by 20...40% in comparison with not preformed workpieces. An important factor is the shape factor, namely the effect of "side-to-thickness ratio" and for functionally structured materials for support plates production, to which specific requirements are imposed [13].

There are studies in the field of thermal buckling of silicone ribbon [14], which are produced with different pull speeds from the furnace. Among the main factors affecting the buckling behaviour, a geometric factor is also identified, i.e. shape of the ribbon profile. Positive use of the phenomenon of buckling is observed by the researchers in the simulation and experiments on the production of two-rib plates [15]. However, this effect relates to a horizontally placed workpiece in the closed die with the application of punch power load to the middle of workpiece. The flow of metal to the opposite sides leads to buckling with the tides of the metal and the beginning of the formation of two ribs on the sides of the punch.

In work [16] it was suggested to consider the preforming of workpieces by upsetting with buckling as one of the options for dieless shaping of workpieces for subsequent open die or closed die forging. Thermal preforming (thermal intensification with a given gradient of heating) of workpieces [17], and stretching with a rupture of workpieces, carried out with the localization of the heated part [18], are referred to a dieless impression-free forging technologies. Modern technologies of incremental forming without dies are carried out with a pencil-like tool with local and sequential forming of the profile on the sheet blank using CNC machines [19, 20]. Such processes, developed in Japan and spread to European countries, do not belong to bulk preforming and are carried out in a cold condition, allowing to form a sheet products in just a few minutes with utmost precision. The thermal localization of deformation allows the development of processes of dieless drawing of wires [21] and tubes [22, 23] of stainless steels and non-ferrous metals. The control of the shape-change happens due to the creation of a thermal field by a moving inductor when the required speed-related deformation modes are realized, eliminating unregulated distortions and rupture of a workpiece. This is dominated by the processes of stretching, not compression and bending, as in the case of a upsetting with buckling.

The studies on the longitudinal compression of beam beyond the elastic limit indicate an increase in the value of the load to the point of the local transition of the rod to the plastic state [24-26]. Further upsetting can be characterized by a general unloading due to an increase in the bending moment arm, which is associated, for example, with the difficulty of obtaining an increasing characteristic of ring fuses with longitudinal slots. Information about the load behaviour in the buckling of inflexible rods or beams (upsetted billets) that are completely in the plastic state with a continuous and uneven increase in the cross-sectional area is absent, which causes the scientific novelty of the problem, which in the present paper was experimentally solved.

3 The Purpose

The purpose of the present research is the experimental study of the load behaviour (force modes) at the upsetting of workpieces with the upsetting ratio $m_0 = L_0/D_0 = 3{,}0...6{,}0$ followed by bulk buckling and the development of an engineering method for calculating of the operation load to adapt to the conditions of die-forging on a particular type of equipment.

4 Material of research

The object of the study was the upsetting process of relatively high workpieces with upsetting ratio $m_0 = L_0/D_0 = 3.0...6.0$ (interval 0.5) by flat parallel dies on the universal testing machine of R-20 model (ultimate load 200 kN, indication error for loads ±1% of the measured value, starting from 10% of the scale limit value, but not below 4% of the maximum load of the machine).

The subject of the study is the load characteristics of the 'upsetting-with-buckling' process. The maximum load value (of the used scale "B") is 100 kN.

The recording of the current load was carried out continuously during the deformation by the built-in drum-type load-measuring device in the R-20 machine. Deformation tool are upsetting dies of steel 5HNV grade (C=0.5…0.6%, Si=0.1…0.4%, Mn=0.5…0.8%, Ni=1.4…1.8%, S<0.03%, P<0.03%, Cr=0.5…0.8%, W=0.4…0.7%, Cu<0.3%, Fe≈95%), heat-treated up to HRC 40-44 with roughness of the supporting surface $Ra = 6.3\mu m$.

According to the Kirpichev & Kik's basic theories of the similarity, laboratory workpieces (specimens) should be geometrically and physically similar to industrial workpieces. The workpieces with a diameter of 30 mm were made by pressing a rod made of lead-antimony alloy with a draw ratio of 7.1, thereby destroying the original cast dendritic structure. To comply with the physical similarity conditions for low-alloy structural steels classes with the presence of a single maximum on the stress-strain curve, the use of specimens from lead or antimony alloyed lead is generally accepted [27].

This is due to the fact that the deformation of lead at a room temperature is accompanied by recrystallization, as a result of which lead and heated steel detect, at different temperatures, qualitatively close dependences of the flow-stress behaviour on the strain rate and degree of deformation.

By bringing the parameters with the same dimensions to the dimensionless form the geometric similarity was observed in the subsystem, which consists of the initially independent parameters – is the dimensions of the workpieces (L_0 and D_0) and the upsetting stroke $X = (L_0 - H_K)$, where H_K is a final height of the upset-bended workpiece (that is the final distance between faces of flat upsetting dies).

The upsetting ratio $m_0 = L_0/D_0$ and the conditional degree of deformation $\varepsilon_y = (X/L_0)\times 100\%$ are the dimensionless arguments of the functions of load for a upsetting with buckling.

Workpieces (specimens) after 'upsetting-with-buckling' process are shown in Fig. 1.

The indication diagrams of the current loads (P) obtained during upsetting with buckling (stroke – X, mm) of specimens with different upsetting ratio is shown in Fig. 2.

When recalculating to dimensionless coordinates, the conditional degree of deformation of the buckled workpiece was determined as

$$\varepsilon_y = \frac{X}{L_0}(\times 100\%). \qquad (1)$$

The current specific load (p) for the 'upsetting-with-buckling' process was obtained from the value of the current load:

$$p = \frac{P}{\left(\frac{\pi D_0^2}{4}\right)}. \qquad (2)$$

The dependences $p(m_0;\ \varepsilon_y)$ are illustrated in Fig. 3.

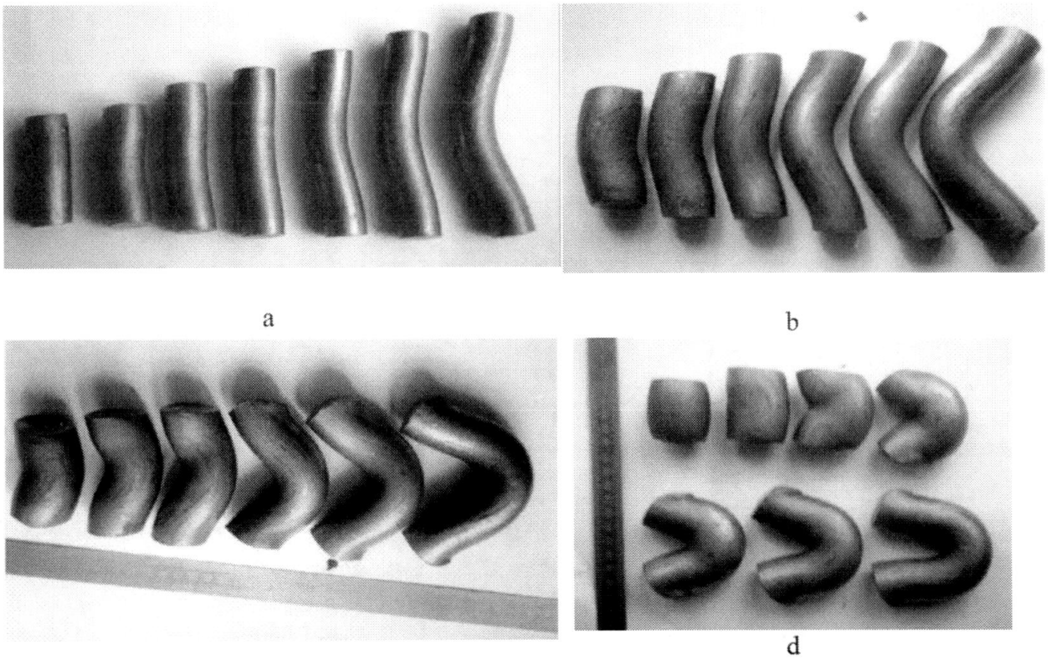

a b

c d

Fig. 1. Experimental workpieces with initial diameter D_0 = 30 mm upsetted with buckling:
a – ε_y = 13 %, m_0 = 3.0; 3.5; 4.0; 4.5; 5.0; 5.5; 6.0;
b – ε_y = 27 %, m_0 = 3.5; 4.0; 4.5; 5.0; 5.5; 6.0;
c – ε_y = 40 %, m_0 = 3.5; 4.0; 4.5; 5.0; 5.5; 6.0;
d – ε_y = 13 %, m_0 = 3.0; 3.5; 4.0; 4.5; 5.0; 5.5; 6.0

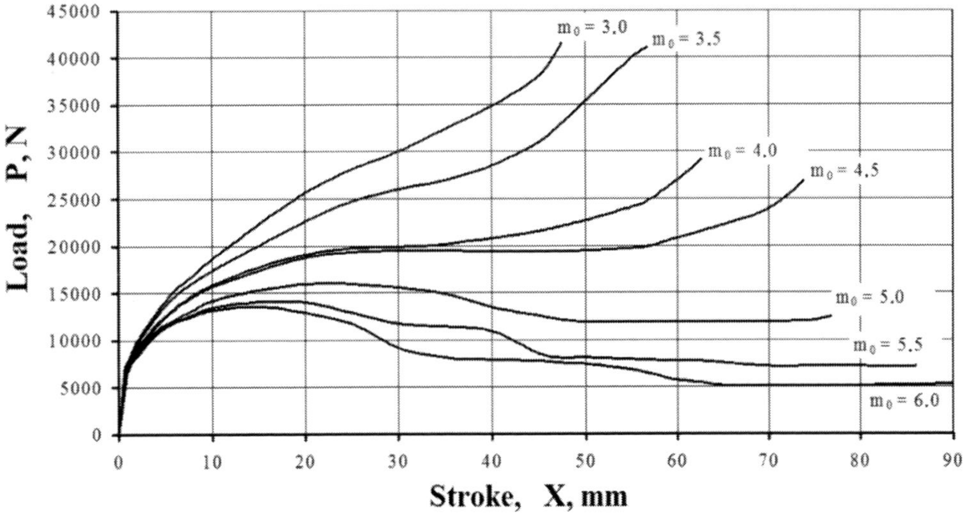

Fig. 2. Indicator diagram of the 'upsetting-with-buckling' process for antimony alloyed lead specimens

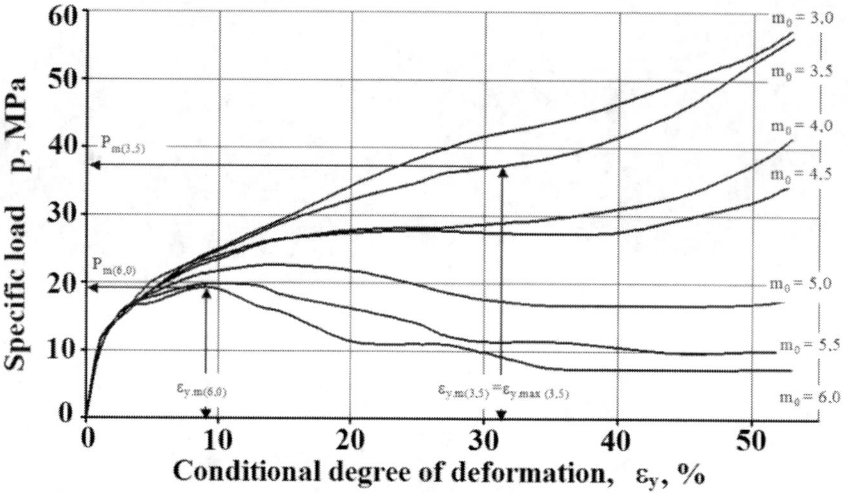

Fig. 3. Specific loads during upsetting by buckling of antimony alloyed lead specimens

The method of upsetting of cylindrical workpieces was used for obtaining a stress-strain curve $\sigma_s(\varepsilon)$, i.e. the dependence of the true flow stress (σ_s) of the specimens material on the upsetting degree – compression strain ε. A series of workpieces with $m_0 = L_0/D_0 = 1.1...2.1$ ratio were upsetted; the deforming tool is flat parallel dies of 5HNV steel grade, the roughness of the surface is $Ra = 0.63\mu m$. The deformation strain rate was of $\dot{\varepsilon} = 0.03...0.07\ s^{-1}$. The contact surfaces were lubricated with machine oil for approximate the stressed state to uniaxial compression.

The calculation of σ_s was based on the data in Fig. 4 obtained by the equations [28]: a) S.I. Gubkin & E. Siebel; b) A.F. Golovin; c) E.P. Unksov; d) K. Reisheter.

When getting the dependences of $\sigma_s(\varepsilon)$ for antimony alloyed lead (Fig. 5), it was found that the curves calculated according to the equations of Gubkin & Siebel, Golovin, and Unksov coincide qualitatively and quantitatively (a difference of less than 0.5%). The curve calculated according to the Reisheter equation is in full qualitative agreement with other graphs; however the values of σ_s are overestimated, in comparison with the previous three ones, by an average of 14% (Fig. 5).

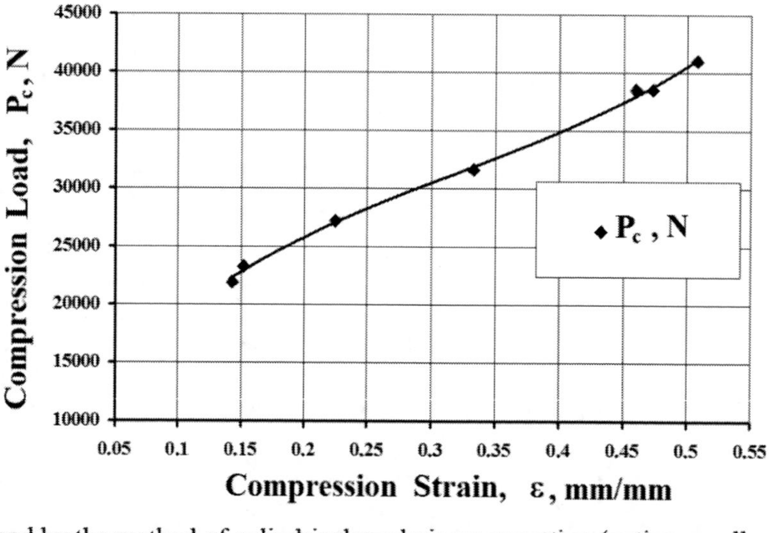

Fig. 4. Load by the method of cylindrical workpieces upsetting (antimony alloyed lead)

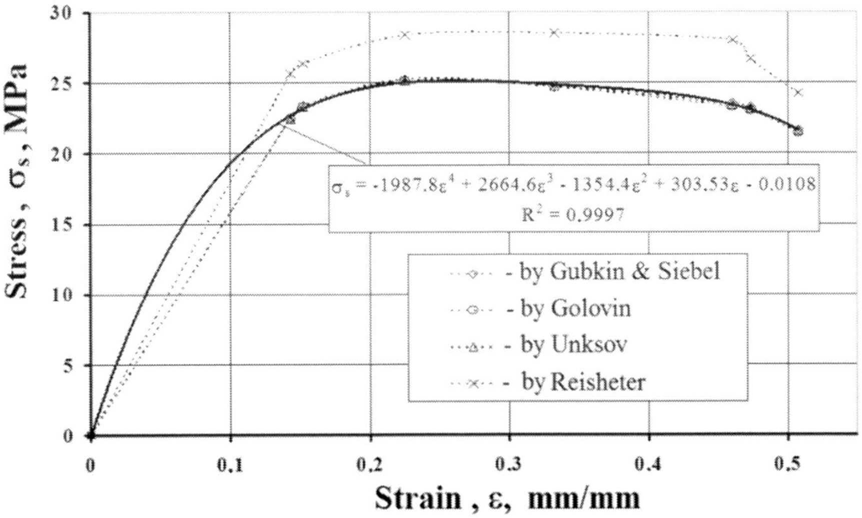

Fig. 5. The stress-strain curves $\sigma_s(\varepsilon)$ of specimens material (antimony alloyed lead) processed by different methods after the compression test

The obtained curve of the true values of the flow stress calculated by the above-mentioned Unksov's equation, were approximated (Fig. 5) with the accuracy of $R^2 = 0.9997$ by the equation:

$$\sigma_s = -1987.8\cdot\varepsilon^4 + 2664.6\cdot\varepsilon^3 - 1354.4\cdot\varepsilon^2 + 303.53\cdot\varepsilon - 0.0108. \tag{3}$$

The strain of longitudinal compression at upsetting with buckling can be fully characterized by the shortening (δ) of the axis:

$$\delta = \frac{L_0 - L_K}{L_0}(\times 100\%), \tag{4}$$

where L_K – the final length along the bended axis of the workpiece.
The shortening function of the workpiece axis is interpolated by the equation:

$$\delta\cdot 100\% = \delta(\varepsilon_y; m_0)\times 100\% = A_\delta\cdot m_0^3 + B_\delta\cdot m_0^2 + C_\delta\cdot m_0 + D_\delta, \tag{5}$$

where $A_\delta = -0.0031\cdot\varepsilon_y^2 + 0.3105\cdot\varepsilon_y - 5.2317;$

$B_\delta = 0.0454\cdot\varepsilon_y^2 - 4.4138\cdot\varepsilon_y + 72.376;$

$C_\delta = -0.212\cdot\varepsilon_y^2 + 19.908\cdot\varepsilon_y - 321.28;$

$D_\delta = 0.3127\cdot\varepsilon_y^2 - 27.574\cdot\varepsilon_y + 455.1.$

And the value of ε_y in the empirical equation (5) is substituted in percent.
The relative specific load ($\eta_\sigma = p/\sigma_s$) is possible to calculate based on the obtained results. The dimensionless quantity of η_σ is independent of the properties of the workpiece material and is a function of the process type, i.e. function of the stress-strain state, which is characteristic of this process: $\eta_\sigma = f(\mu;\ m_0;\ \varepsilon_y)$, where μ is the contact friction coefficient. Bearing

in mind when the workpiece has a ratio $m_0 > 1$ by the frictional forces are neglected [1, 5, 29], then $\mu = 0$; then $\eta_\sigma = f(m_0;\ \varepsilon_y)$.
Assuming of $\delta \equiv \varepsilon$, used values obtained in equation (5) into equation (3) in relative units. Then, using the experimental results obtained from the equation (2) the graph $\eta_\sigma(m_0;\ \varepsilon_y)$ was plotted (Fig. 6).
In the derivation of equation (5) the data of upset workpieces forming in the range of $\varepsilon_y = 13...53\%$ was used, therefore the values of η_σ for $\varepsilon_y < 13\%$ do not have a sufficient degree of reliability (in Fig. 3 they are represented by small dashed lines). The decrease in the relative specific load for workpieces with ratios $m_0 = 5.0...6.0$ in the range of $\varepsilon_y \approx 20\%$ to the value of $\eta_\sigma < 1$ is associated with the growth of the deflection (load arm), and, correspondingly, the increase of the bending moment.
The conditional degree of deformation in 'upsetting-with-buckling' is technologically limited to a certain maximum value ($\varepsilon_{y.max}$), the excess of which is corresponded with the danger of forming a "laps" or "folds" defects, therefore the graphs $\eta_\sigma(m_0;\ \varepsilon_y)$ for $\varepsilon_y > \varepsilon_{y.max}$ field are continued by a line-dotted lines (Fig. 6). Workpieces with ratio of $m_0 = 3.0$ do not buckled.
The load of the operation in the technologically permissible limits of the upsetting degree $0<\varepsilon_y \leq \varepsilon_{y.max}$ is determined by the maximum of the force (P_m) in upsetting stroke, to which corresponds a certain value of $\eta_\sigma = \eta_{\sigma.m}$ for a value $\varepsilon_y = \varepsilon_{y.m}$. In this case, for workpieces with ratios $m_0 = 3.5...4.5$, the load increases throughout the entire course of upsetting stroke (Fig. 2-3), i.e. before $\varepsilon_{y.max}$ and after this value, and consequently, the $\eta_{\sigma.m}$ value is realized at the $\varepsilon_{y.m} = \varepsilon_{y.max}$.

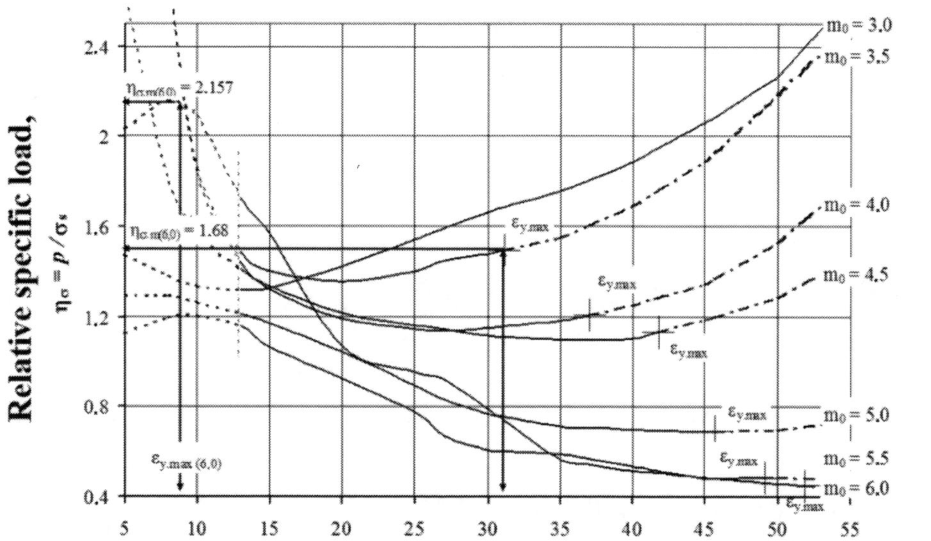

Fig. 6. Relative specific load during upsetting with buckling

The method for finding $\eta_{\sigma.m}$ for billets with ratio of $m_0 = 3.5$ ($\varepsilon_{y.m(3,5)} = \varepsilon_{y.max(3,5)} \Rightarrow \eta_{\sigma.m(3,5)}$) is shown in Fig. 3 and Fig. 6. For workpieces with $m_0 = 4.5$ a characteristic region of $\varepsilon_y = 17...40\%$ with a practically constant strain value was revealed, which is supposed to be due to the equal influence of the increasing bending moment and the increase of cross-sectional area of workpieces. When the workpieces with $m_0 = 5.0...6.0$ are deformed, the maximum load (P_m) is realized at $\varepsilon_{y.m} < \varepsilon_{y.max}$, further sediment is accompanied by a decrease in load (Fig. 2 and Fig. 3). Then the $\eta_{\sigma.m}$ value can be determined from the graph

(Fig. 6) at a known value of $\varepsilon_{y.m}$, which is found in advance in Fig. 3. The method of determining $\eta_{\sigma.m}$ for workpieces with the ratio of $m_0 = 6.0$ ($p_{m(6,0)} \Rightarrow \varepsilon_{y.m(6,0)} \Rightarrow \eta_{\sigma.m(6,0)}$) is shown in Fig. 3 and Fig. 6. The graph of the dependence $\eta_{\sigma.m}(m_0)$ is presented in Fig. 7.

Thus, the technological (maximum) load of the upsetting a workpiece of any material with buckling operation can be calculated as:

$$P_m = k \cdot \eta_{\sigma.m} \cdot \sigma_s \cdot F_0, \tag{6}$$

where $k = 1.25$ – reserve index;

F_0 – cross-section area of the workpiece;

$\sigma_s = \sigma_s(\delta)$ – the true stress, determined from the reference data [28] for $\delta \equiv \varepsilon$, after finding of δ by equation (5);

$\eta_{\sigma.m}$ – the relative specific load, determined by the curve in Fig. 7.

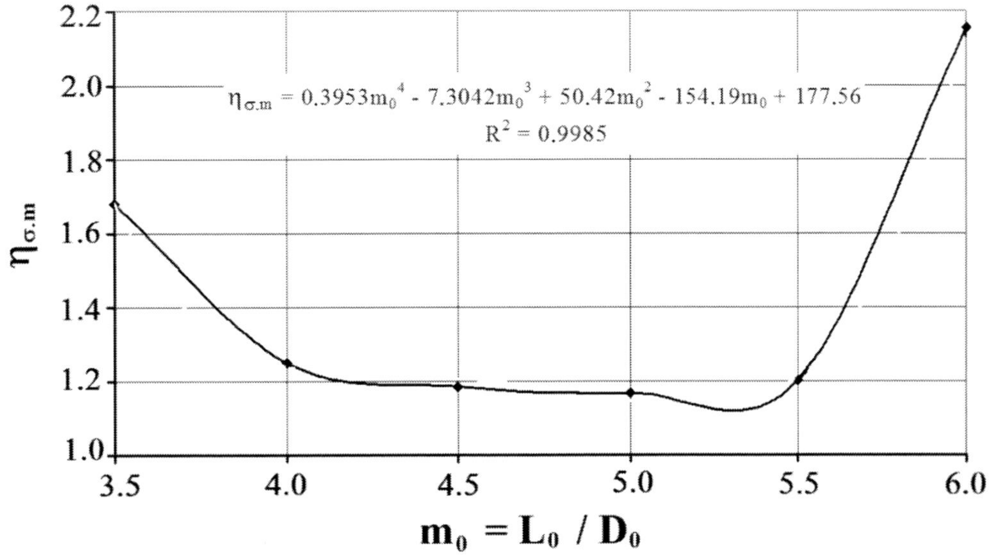

Fig. 7. The $\eta_{\sigma.m}(m_0)$ dependence

Example. Evaluate the possibility of sequential production of the preformed workpiece and closed die-forging of the "cardan fork" on a crank-press for hot bulk forming with a nominal force of $P_{pr} = 10$ MN. Material of workpiece is steel 40X grade (C=0.36…0.44%, Si=0.17…0.37%, Mn=0.5…0.8, Ni<0.3%, S<0.035%, P<0.035%,Cr=0.8…1.1%, Cu<0.3%, Fe≈97%), forging mass – 1.64 kg, upsetting temperature 1100°C. It is determined in advance: the dimensions of the workpiece are $L_0 = 216$ mm, $D_0 = 41$ mm ($m_0 = 5.27$ ratio); the required deformation degree at upsetting for preforming is $\varepsilon_y = 36\%$.

1) Calculating the axial shortening by equation (5) $\delta = 14\%$.

2) Determining $\sigma_s(\delta)$ of 40X grade steel with $\delta \equiv \varepsilon = 14\%$, $\dot{\varepsilon} = 5$ s^{-1}, according to [28]: $\sigma_s = 75$ MPa.

3) According to the graph in Fig. 7 we find $\eta_{\sigma.m} = 1.188$.

4) By the equation (6) we define $P_m = 146969$ N $= 0.147$ MN.

5) The value of the forging force for horizontal (splash) forging in the final impression is calculated from the most unfavorable conditions (chilling of the workpiece, i.e., the temperature of the closed-die forging of 966°C, the friction coefficient of $\mu = 0.5$). Taking into account the influence of the strain rate and the 25% reserve, the calculated the force for closed die-forging: $P_f = 7.35$ MN.

6) Comparing: (P_m = 0.147 MN) < (P_f = 7.35 MN) < (P_{pr} = 10 MN).
At similar calculations, the preforming loads and the forces of the closed die-forging for forgings of a suitable nomenclature [30], the first one was always considerably less than the second one.

Conclusions

Experimental studies of the load of the 'upsetting-with-buckling' process of workpiece in the range of upsetting ratio of $m_0 = L_0/D_0$ = 3.0…6.0 are performed. Analysis of the experimental data allowed to develop an engineering method for calculating of process technological load for production of the preformed (shaped) workpieces by upsetting with buckling. This preparatory operation can be carried out unimpeded on the main die-forging equipment with successive or combined forging, because the load for preforming of workpiece is always substantially less than the final forming of the forging in the die impressions or final closed dies.

References

1. V.Kukhar, V.Artiukh, A.Butyrin, A.Prysiazhnyi. Advances in Intelligent Systems and Computing, vol 692. DOI: https://doi.org/10.1007/978-3-319-70987-1_22.
2. Taylan Altan, A.E.Tekkaya (2012). Sheet Metal Forming: Processes and Applications. ASM International, OH, 400 p.
3. Metal forming handbook / Schuler GmbH (1998). Springer, Berlin Heidelberg, 563 p., DOI: 10.1007/978-3-642-58857-0.
4. V.Kukhar, V.Burko, A.Prysiazhnyi, E.Balalayeva, M.Nahnibeda (2016). East-European Journal of Enterprise Technology, 2016, 3/7(81), pp. 53–61. ISSN 1729-3774, DOI: 10.15587/1729-4061.2016.72063.
5. J.Singer, J.Arbocz, T.Weller (2007). Buckling Experiments, Basic Concepts, Columns, Beams and Plates: Vol. 1. Hoboken: Wiley, 2007.
6. J.Singer, J.Arbocz, T.Weller (1998). Buckling experiments: Experimental methods in buckling of thin-walled structures, Vol. 1, Chichester [u.a.]: Wiley, 1998.
7. J.G.Teng, J.M.Rotter (2006). Buckling of Thin Metal Shells, CRC Press, 2006, 520 p.
8. Nan Liu, He Yang, Heng Li, Siliang Yan (2016). Chinese Journal of Aeronautics, Vol. 29, Issue 1, 2016, Pages 1-14, ISSN 1000-9361, DOI: 10.1016/j.cja.2015.09.004.
9. V.Kukhar, E.Balalayeva, O.Nesterov (2017). MATEC Web of Conferences. (ICMTMTE), Vol. 129, 01041, 2017, DOI: 10.1051/matecconf/201712901041.
10. G.D.Gavrylenko (2007). Proceeding of the 13th International conference on experimental mechanics, Alexandroupolis, Greece, July 1–6, 2007, pp. 783-784.
11. D.Quinn, A.Murphy, L.Cervi (2011). Proceedings of the Institution of Mechanical Engineers, Part G: Journal of Aerospace Engineering, 225(7), pp. 791-806. DOI: 10.1177/0954410011399035.
12. J.Błachut, P.Smith (2007). Tabu search optimization of externally pressurized barrels and domes, Engineering Optimization, 39:8, pp. 899-918, DOI: 10.1080/03052150701512604.
13. B. Sidda Reddy, J. Suresh Kumar, C. Eswara Reddy, K. Vijaya Kumar Reddy. (2013). Journal of Composites, vol. 2013, Article ID 808764, 12 pages, 2013, DOI: 10.1155/2013/808764.

14. C.T.Tsai, O.W.Dillon, R.J. De Angelis (1988). Conference Record of the Twentieth IEEE Photovoltaic Specialists Conference, Las Vegas, NV, USA, 1988, Vol. 2., pp. 1545-1550, DOI: 10.1109/PVSC.1988.105968.
15. A.Dziubińska, A.Gontarz (2015). Limiting phenomena in a new forming process for two-rib plates, Metalurgija, 54(3), p. 555-558, http://hrcak.srce.hr/131621.
16. V.A.Grinkevich, V.V.Kukhar, K.K.Diamantopulo (2010). Kuznechno–shtampovochnoe proizvodstvo. Obrabotka materialov davleniem, 2010, No. 5, pp. 19–23. (in russian).
17. V.Kukhar, V.Artiukh, O.Serduik, E.Balalayeva (2016). Procedia Engineering, Vol. 165, 2016, pp. 1693–1704, DOI: 10.1016/j.proeng.2016.11.911.
18. V.V.Kukhar (2015) Metallurgical and Mining Industry, 2015, No. 6, pp. 122–132, http://www.metaljournal.com.ua/assets/Journal/MMI-6/016-Kukhar.pdf
19. M.Amino, M.Mizoguchi, Y.Terauchi, T.Maki (2015). Single Point "Dieless" Incremental Forming. In: Tekkaya A., Homberg W., Brosius A. (eds), 60 Excellent Inventions in Metal Forming. Springer Vieweg, Berlin, Heidelberg, DOI: 10.1007/978-3-662-46312-3_24
20. Anna Kochan (2001). Dieless forming, Assembly Automation, Vol. 21, Issue: 4, pp. 321-323, DOI: 10.1108/01445150110407304.
21. M.D.Naughton, P.Tiernan (2009). An Experimental Approach to Continuous Dieless Wire Drawing (Variant A) Using ELI Ti–6Al–4V Alloy, ASME. J. Eng. Mater. Technol. 2009; 131(2):021005-021005-10, DOI: 10.1115/1.3078304.
22. Y.M.Hwang, T.Y.Kuo (2013). The International Journal of Advanced Manufacturing Technology, Vol. 68, Issue 5–8, pp. 1311–1316, DOI: 10.1007/s00170-013-4922-0.
23. P.Kustra, A.Milenin, B.Płonka, T.Furushima (2016) Journal of Materials Engineering and Performance, Volume 25, Issue 6, pp. 2528–2535, DOI: 10.1007/s11665-016-2090-8.
24. S-S.Huang, I.Burgess, Z.Huang, R.Plank (2011). Proceedings of the Institution of Civil Engineers: Engineering and Computational Mechanics, Vol. 164 (2). pp. 103-119. ISSN 1755-0777, DOI: 10.1680/eacm.2011.164.2.103.
25. J.W.Hutchinson (1974). Plastic buckling, Advanced in Applied Mechanics, Vol. 14, Academic press, inc., New York, pp. 67-144.
26. Y.V.Narayana, J.B.Gunda, R.R.Pinninti, M.Ravvala (2015). World Journal of Engineering and Technology, Issue 3, pp. 185-194, DOI: 10.4236/wjet.2015.34019.
27. G.G.Shlomchack, I.Mamuzič, F.Vodopivec (1994). Rheologically similarity of metals and alloys, J. of Materials Processing Technology, Amsterdam–New-York: Elsevier, 1994, Vol. 40, № 3–4, pp. 315–325.
28. P.I.Poluhin, G.Ja.Gun, A.M.Galkin (1983). Soprotivlenie plasticheskoj deformacii metallov i splavov. Spravochnik [Resistance to plastic deformation of metals and alloys. Reference book], Moscow, Metallurgija, 1983, 352 p. (in russian).
29. V.I.Kaplanov, A.G.Prisyazhnyi (2008). Simulation of contact friction in the hot rolling of steel sheet, Steel in Translation, Vol. 38, No. 9, pp. 714–718, DOI: 10.3103/S0967091208090040.
30. A.S.Semenov, S.G.Semenov, B.E.Melnikov, V.S.Tikhonov. Magazine of Civil Engineering. 2017. No. 6. Pp. 78–101. doi: 10.18720/MCE.74.8

Quality assessment of urban areas based on neural network modeling and GIS

Olga Popova[1,*], *Julia* Glebova[1], and Andrey Pustovgar[2]

[1]Northern (Arctic) Federal University 163000, Arkhangelsk, nab. Severnaya Dvina, 17
[2]Moscow State University of Civil Engineering, Yaroslavskoe shosse, 26, Moscow, 129337, Russia

Abstract. In this article the authors carry out the research of the urban development areas structure and propose the system of its characteristics on the basis of sector affiliation of the municipal economy. The authors have developed an algorithm for quality assessment of urban development areas. The results of the research are presented on the example of several central quarters of Arkhangelsk city. The city's residential development was formed in the periods from 1900-1950, 1950-1980 and from 2002 to date. It is currently presented by low-rise wooden, homestead type residential houses and barracks-type houses; mid-rise and high-rise brick and panel buildings of typical development, buildings of large-panel housing construction. Structural SOM-analysis compiled separate quarters of Arkhangelsk into 5 groups with a high level of characteristic similarity: "Commercial", "Prospective complex development", "Sustainable development", "Perspective renovation of residential development", "Investment-unattractive". Typical development strategies for each group of quarters are determined. Most developed areas characterized by upward height. The development strategies for depressed areas is in a high-rise building, which show the economic, social and environmental benefits of upward growth of the city. Using GIS allows to visually reflect the state and assess the quality of the urban development area by the aggregate of all parameters, and also to assess the quality of the quarters for each sector.

1 Introduction

Issues of complex reproduction, reconstruction and renovation of urban development areas are considered in a great number of scientific researches [1,2,3,4,5]. However, at this moment, the planning of reproduction is fragmentary and does not have system-structural relationships. One of the most significant factors limiting the efficiency of real estate reproduction management is the lack of its monitoring support. At the same time, the information-analytical systems forming experience enables us to resolve a contradiction between the need for optimization of monitoring systems and insufficient development of theoretical and technological mechanism ensuring the fulfillment of this management task.

* Corresponding author: oly-popova@yandex.ru

Most of the existing similar methods for assessing the quality of the territory are aimed at calculating the final integral quality index with a complex internal structure of components, high degree of subjective assessment and also with a lack of optimal index value and understandable levers to change it.

This research is aimed at solving the problem of the neural network modeling and geoinformation technologies adaptation for building an intelligent monitoring system, dispatching and effective organizational and technical management of the urban development area.

2 Materials and methods

Within the framework of the research, the authors solve the problem of designing an intelligent operational management system and urban development monitoring on the basis of structural analysis and data visualization.

2.1 Structural analysis of the urban development areas quality

The area of urbanized development is a complex system and has a set of characteristics of its constituent objects. In aggregate the sectors of the municipal economy have a common task – to serve the need of city's population. That is why the development level of municipal economy sectors depends on population, its resettlement and reflects the quality of urban development areas. This research covers the analysis of urban development characteristics on the basis of sector affiliation of the municipal economy, which includes housing and utilities sector, transport sector, consumer market, construction sector, social sector, services providing public safety, management, communications, information systems and other organizations serving the needs of the city.

Each sector of the municipal economy cannot be characterized by one particular parameter. It is represented as a complex hierarchical system of parameters (characteristics) and can be described with varying degrees of detailing. The city's areas have different qualitative and quantitative indexes for each characteristic. In this research a quarter is taken for the territorial unit of urban development. An aggregate assessment of all characteristic indexes generates an overall assessment of the unit's quality (quarter).

Structural analysis of urban development multidimensional characteristics is performed by clustering the objects under study (quarters) in comparable groups with a high level of similarity of characteristics inside each group. The clustering mechanism is based on the use of neural networks of T. Kohonen's self-organizing maps (Self-Organizing Map - SOM). Self-organizing map (SOM) is an effective neural network modeling software tool for synthesizing multidimensional data [6,7].

The proposed method of mathematical modeling enables one to structure information based on material and intangible development parameters and to obtain integrated quality indexes of areas. The obtained data is used to plan the complex reproduction of urban development by forming typical management programs for objects referred to a particular group [8].

2.2 Visualization of the areas quality structural analysis

The application of geoinformation systems (GIS) involves the structuring and visual representation of data by information layers of hierarchical structure of urban development characteristics.

3 Experiment

Research results are represented on the example of separate central microdistricts of the Arkhangelsk city.

The methodology of the research included following stages:

1. Forming the databases of area characteristics for each sector of municipal economy. The list of chosen characteristics describing each sector is presented in the table 1. It should be noted that such method makes it possible to change freely both the list of studied sectors and the detailing degree of characteristics in their description, i.e. to vary the amount of characteristics describing the system in accordance with the objectives of a particular research.

Table 1. Characteristics of municipal economy sectors

Municipal economy sector		Characteristics
Housing and utilities sector	Residential development	Stage 1 - structural analysis of housing stock: - overall indexes of MDU: capital group, capacity, area of apartments, number of floors, operational lifetime, general rate of wear, replacement cost; - indexes of structural elements. In total, during the clustering of a representative sample of 316 objects 33 indexes participated in the analysis [8]. Stage 2 - Determination of the average type of residential development for the quarter according to the clustering
	Utilities	Types and length of networks in the quarter (security), network material, network wear and operational lifetime.
Transport sector		Structural analysis of the quarter's transport infrastructure quality - transport accessibility of the quarter (7 indexes for clustering 90 quarters) including quarter area, quarter form, road traffic, technical condition of roads, etc. [9]
Consumer market		Structural analysis of quarters for the supply of trade units (6 indexes for clustering 90 quarters) including distance from the quarter center to the nearest shopping mall of the city scale, number of small stores of industrial, household and grocery goods, number of public catering, etc.
Construction sector		Index for the municipality. Characterized by the number of building owners (local and/or from other regions), presence of construction sector enterprises. Used to assess the quality while comparing different municipalities within the region, country.
Social sector		Structural analysis of quarters for the supply of social infrastructure facilities (8 indexes for clustering 90 quarters) including provision of residents of the quarter with medical help, children's preschool and school institutions, the proximity and accessibility of the recreation zone, etc.
Services providing public safety		Index for the municipality. Used to assess the quality while comparing different municipalities within the region, country.
Management, communications, information systems		Index for the municipality. Used to assess the quality while comparing different municipalities within the region, country.

2. Structural analysis of urban development areas (quarters) for each sector of municipal economy. The results of the structural analysis are presented in the table 2.

The analysis was carried out by clustering objects (of housing stock [8] and separate quarters of the Arkhangelsk city) using of SOM in comparable groups with a high level of similarity of characteristics inside each group [10].

Description of the general characteristics for each group. At this stage, it is possible to define typical reproduction strategies for the objects of each residential property group, typical development strategies for each group of quarters for a particular sector.

Table 2. Clusters characteristics of municipal economy sectors in Arkhangelsk

Municipal economy sector	Results of clustering
Residential development	16 clusters, which are combined into 4 types of residential development [8]: Group 1 - New housing stock; Group 2 - Housing stock in good condition; Group 3 - Aging housing stock; Group 4 - Ramshackle and emergency housing stock.
Transport sector	4 clusters with different levels of transport infrastructure [9]: Group 1 - Good - maximum traffic and good road conditions around quarters with a shape close to square and with a large area; Groups 2, 3, 4 - Satisfactory - average traffic and satisfactory condition of roads around quarters, respectively, of rectangular shape and with a small area, a square shape and with an average area, a rectangular, strongly elongated shape and a large area; Group 5 - Poor - minimal traffic and unsatisfactory condition of roads around quarters with a form close to a square and with an average area.
Consumer market	3 clusters with different levels of consumer market development: Group 1 - Quarters, which are centers of shopping and entertainment activities of the city; Group 2 - Predominantly quarters around quarters of group 1 - public and business center of the city Group 3 - Sleeping areas of the city with a low level of consumer market development
Social sector	4 clusters with different levels of social sector development: Group 1 - insignificant removal from recreational zones and the average level of provision with social facilities; Group 2 – quarters located next to recreational zones, best provided with schools, not provided with kindergartens; Group 3 - quarters remoted from recreation zones, medium provided with schools, not provided with kindergartens Group 4 – quarters located next to recreational zones, not provided with schools and kindergartens. Selected quarters provides the same and higher level of medical and polyclinic care availability in comparison with other parts of the city.

3. The final structural analysis of urban development areas (quarters) based on the results of quality assessment of particular sectors. Structural analysis was carried out by clustering objects (particular quarters in Arkhangelsk) using SOM [10] in comparable groups with a high level of similarity of characteristics inside each group.

At this stage, an overall assessment of areas quality was performed by the totality of all parameters.

The description of the general characteristics for each group makes it possible to determine the common vector of development for them - general typical development strategies for each group of quarters (table 3). Most developed and depressed areas of the city are revealed, as well as the most investment-attractive areas for development.

The efficiency of the clusterization is shown in Fig. 1 and 2. To illustrate the dissimilarity of the city's areas the authors constructed graphs of all quarter index indices (normalized characteristics of objects) included in the representative sample (Fig. 1) [5]

Table 3. Characteristics of quarters' clusters in Arkhangelsk

Group of quarters	Housing stock	Transport infrastructure	Consumer market	Social infrastructure	Description
"Commer-cial"	Absence or a small proportion in the total development	Good - maximum traffic and good road conditions around quarters with a shape close to the square and with an average area	Concentra-tion of the main shopping and entertain-ment facilities of municipal importance	Close to recreational zones, Not provided with schools and kindergartens	Quarters, built-up with public and business facilities with a lack of or a small proportion of residential property in the central part of the city. Do not require development.
"Prospecti-ve complex develop-ment"	Large number of ramshackle and emergency housing units with a dot-building new residential development	Good - maximum traffic and good road conditions around quarters with a shape close to a square and with a large area	Average - public and business center of the city	Close to recreational zones, Not provided with schools and kindergartens	Investment-attractive for the complex construction of highly liquid real estate for residential and commercial purposes due to renovation of existing buildings
"Sustain-able develop-ment"	In good condition, partially aging, point-new	Satisfactory - average traffic and satisfactory condition of roads around quarters	Average - public and business center of the city	Average and above-average provision of social facilities	Quarters of historically sustainable development in the central part of the city with a developed social infrastructure and social and business areas. They require comprehensive rehabilitation measures: current and major repairs, spot development.
"Perspecti-ve renovation of residential develop-ment"	In good condition and aging, point-ramshackle	Bad - minimal traffic and unsatisfactory condition of roads around quarters	Low level of consumer market develop-ment	Minor distance from recreational zones and the average level of provision with social facilities	Historically developed social infrastructure. Investment-attractive for complex and point-based residential development, not attractive for commercial development on condition of the transport infrastructure development.
"Invest-ment-unattracti-ve"	In good condition and aging, lack of free areas for development	Bad - minimal traffic and unsatisfactory condition of roads around quarters	Low level of consumer market develop-ment	Close to recreational zones, Not provided with schools and kindergartens	Investment-unattractive because of the lack of free or renewable area for development. Possibility of point development. The existing and prospective development has reduced liquidity due to lack of infrastructure. Development through budget financing of infrastructure development.

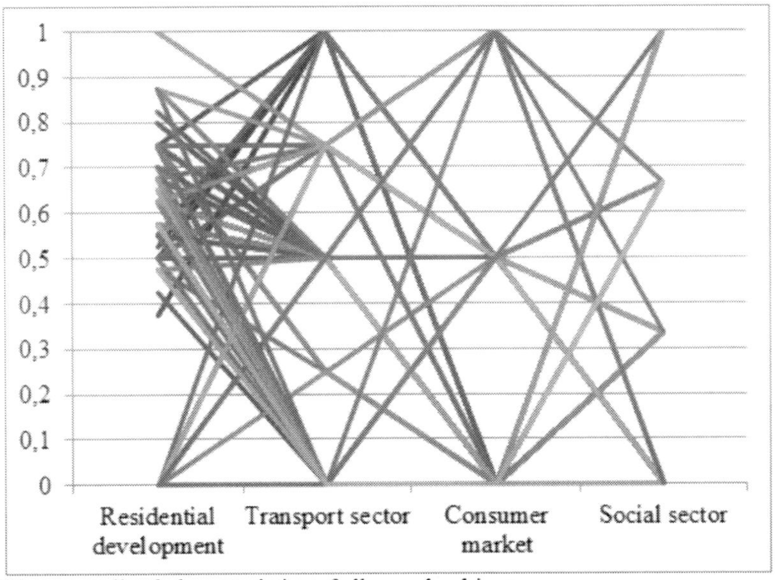

Fig. 1. Map of the normalized characteristics of all sample objects

The different situation can be seen on the graphs illustration of objects' normalized characteristics. Figure 2 shows the graphs of indices of clusters within the 1 and 3 of 5 clusters, which are described in table. 2.

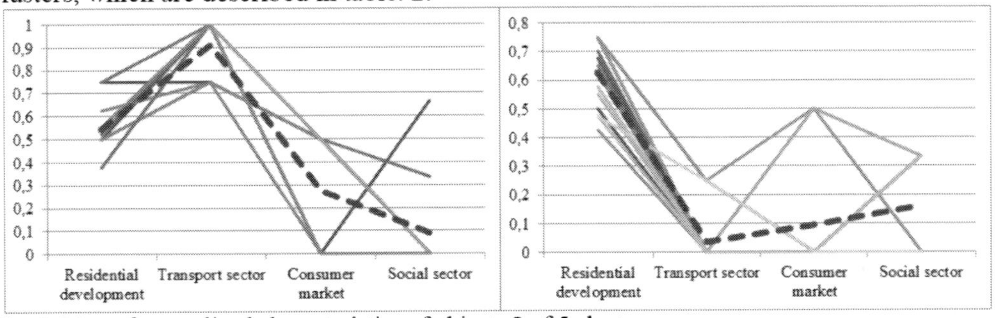

Fig. 2. Map of normalized characteristics of objects 2 of 5 clusters

4. Visualization of the obtained cluster analysis using geoinformation technologies (GIS). Using GIS enables us to visually assess the state of the urban development area, reflect both the overall areas quality assessment by the aggregate of all parameters, and the quality of the quarters for each sector. An example of visualization of the general clustering of quarters in the central part of Arkhangelsk is shown in Fig. 3.

5. The urban development state monitoring is carried out through periodic (every 1, 3 or 5 years) areas surveys and identifying the values involved in the assessment of characteristics for each sector. Periodicity of the survey is chosen on the basis of the current dynamics of changes in the state of each sector. For example, for a housing stock, it is advisable to carry out a repeated clustering every 3rd years, which corresponds to the periodicity of planning the work of capital repairs and the average time for new construction. Therefore, the state of objects during this period may change. For transport infrastructure, the periodicity of the survey should be annual, which is determined by the dynamics of road's surface deterioration and the annual repair plans.

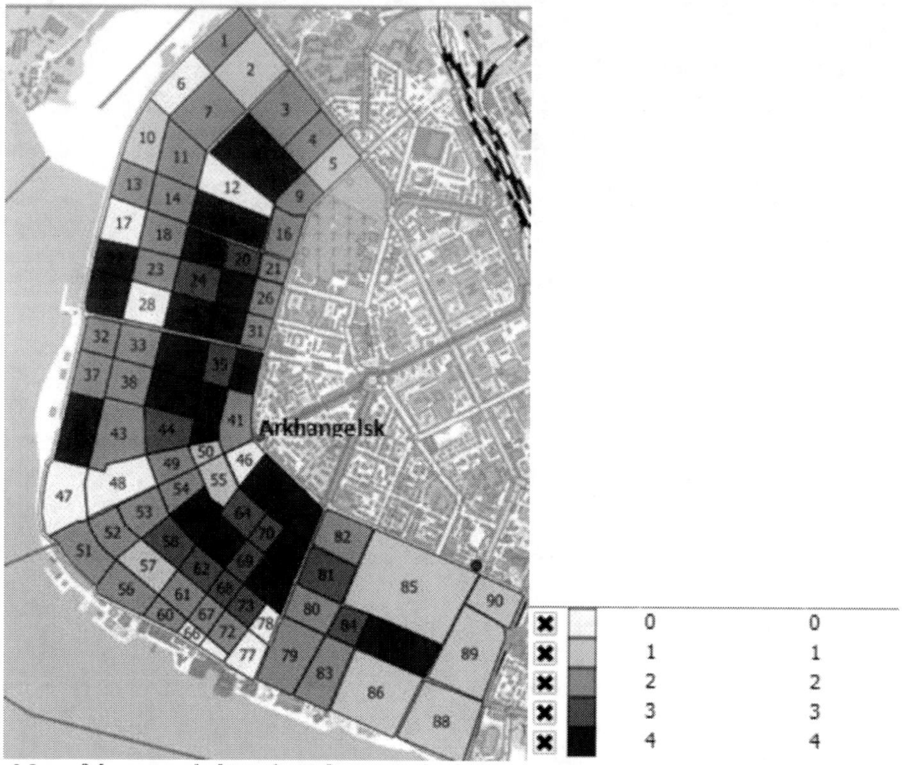

Fig. 3. Map of the general clustering of quarters in the central part of Arkhangelsk

Thus, if the overall assessment of the area quality within all parameters changes, then a quarter can get from one cluster (group) into another. At the same time the transition to a better one demonstrates the effectiveness of ongoing activities while transition to a worse quality cluster shows the lack of activities or their inefficiency and the need to take measures to improve the quality of the area (quarter).

4 Results

This article provides an algorithm for quality assessment of urban development areas. The results of the research are presented on the example of several central quarters of Arkhangelsk city. On the basis of more than 50 urban development characteristics the authors have obtained 5 groups of quarters with a high level of characteristics similarity within each group.

Thus, new computational and analytical methods for monitoring the characteristics of the hierarchical components of urban development areas were suggested. The novelty of the study is to justification for advantages of applying structural analysis for qualitative ranking of areas. The advantage of the proposed methodology is that it gives the opportunity to assess the quality of areas: to identify the most investment-attractive (unattractive) quarters for further development; to determine the strategy for its development with a list of specific activities.

It should be especially noted that the proposed method is universal, according to the authors' view. Its universality is as follows:

- the possibility of varying the list and a set of characteristics;

- the possibility of applying the method for monitoring and assessment of different areas, regardless of their geolocation and scale.

- the possibility of adapting the methodology for monitoring other processes occurring in urban areas (changes in the research topic).

The universality of such system and its "fast" processing allows one to assess the state of any urban massifs with the entire set of its capital construction objects. Systematic monitoring enables us to track the direction and development dynamics of areas. It makes it possible to respond quickly to the changes that occur.

Conclusion

The main objective of the research is to improve the efficiency of the organizational and technical management of urban development areas as complex systems of hierarchical components due to the implementation of the developed methods of monitoring and analyzing data. Application of the proposed methods will enable one:

- to form a long-term town-building strategy for the urban development area taking into account the complex comparative assessment of the individual quarters' quality
- to implement budget planning of reproduction within their investment and social efficiency.
- to accelerate the planning process and make adjustments to urban development programs in real time, to reduce the time and resources spent on monitoring and analyzing data.

Gratitude

The research was funded by the Russian Foundation for Basic Research (RFBR) according to the research project No. 16-38-00662 mol_a.

References

1. O. Popova and E. Bogacheva, AIP Conference Proceedings **1800**, 050015 (2017), doi: http://dx.doi.org/10.1063/1.4973075
2. Zivkovic M., Oliynyk O., Murgul V.A.. Construction of Unique Buildings and Structures, **1**, 102-111 (2016)
3. Aguacil, S., Lufkin, S., Rey, E., Cuchi, A. Energy and Buildings, **144**, 42-602017 (2017). (doi: 10.1016/j.enbuild.2017.03.047).
4. Wang, H., Shen, Q., Tang, B.-S. Journal of Urban Planning and Development, **141 (3)**, 05014015 (2015), (doi: 10.1061/(ASCE)UP.1943-5444.0000216)
5. Diappi L., Bolchi P., Slepukhina I. EconPapers, **11**, 1–7. (2013)
6. T. Kohonen, Self-Organizing Maps (Third Extended Edition), New York, 501 p. (2001).
7. G. Debok and T. Kohonen, Analysis of financial data using self-organizing maps, Alpina Pablisher (2001).
8. T. Simankina and O. Popova, Applied Mechanics and Materials, **584-586**, 1944–1950 (2014). https://doi.org/10.4028/www.scientific.net/AMM.584-586.1944
9. O. Popova, K. Martynov and R. Khusnutdinov, IOP Conference Series: Earth and Environmental Science, 90, 1, 012225 (2017)
10. Deductor. URL: http://www.basegroup.ru/deductor

Analysis of high-rise constructions with the using of three-dimensional models of rods in the finite element program PRINS

Vladimir Agapov*

Moscow State University of Civil Engineering, Yaroslavskoe shosse, 26, Moscow, 129337, Russia

Abstract. The necessity of new approaches to the modeling of rods in the analysis of high-rise constructions is justified. The possibility of the application of the three-dimensional superelements of rods with rectangular cross section for the static and dynamic calculation of the bar and combined structures is considered. The results of the eighteen-story spatial frame free vibrations analysis using both one-dimensional and three-dimensional models of rods are presented. A comparative analysis of the obtained results is carried out and the conclusions on the possibility of three-dimensional superelements application in static and dynamic analysis of high-rise constructions are given on its basis.

1 Introduction

High-rise buildings belong to objects with a high level of responsibility, so at the analysis of such structures by the finite element method it is necessary to use design schemes that are as close as possible to the real constructive schemes. This is especially actual for bar structures and combined systems containing rods, plates, shells and solid elements, since often in the analysis of such systems the calculation schemes are constructed using one-dimensional models of rods based on the classical theory of beam bending. The disadvantages of such approach are well known (see, for example, [1]). These include the transfer of force at a point, the inability to take into account warping and change of the shape of the cross section, the complexity of accounting for physical and geometrical nonlinearity, and a number of others. In modern software systems, such as NASTRAN [2], ANSYS [3], ABAQUS [4] et al, the combined models of rods are used. In these models, an approximation of the coordinates and displacements of points lying on the axis of the rod is specified, the coordinates and displacements of internal points are found using the hypothesis of plane sections and then, the characteristics of the finite element necessary for calculation are obtained using the three-dimensional theory. A theoretical description of such models can be found in [5-9]. However, in the final analysis, all characteristics are reduced to points lying on the axis of the rod. Thus, some of the shortcomings inherent in one-dimensional models are preserved. Completely to get rid of the drawbacks inherent in one-dimensional models, it is possible by using 3D elements to model the rods. Direct modeling of each rod by volume elements is associated with large overhead costs, since it leads to a sharp increase in the number of degrees of

*Corresponding author: agapovpb@mail.ru

freedom of the structure. The output can be found in the use of the superelement method. However, with the traditional use of this method, when all the characteristics of the modeled object are brought to the nodes lying on the surface of this object, certain problems may arise during the preparing of the input data for the calculation. Therefore, it is desirable to have special superelements, in which the necessary characteristics would be reduced to nodes lying in the end sections. This idea is implemented in the computer program PRINS[1]. To date, it has been tested and proved itself in linear and nonlinear static calculations of rod systems [10].

2 Methods

In this paper the possibility to use the previously developed superelements [10] in the calculation of high-rise construction is shown. The free vibration analysis of the spatial frame depicted in Fig. 1 and Fig. 2 was performed for this purpose. The frame was calculated by the aid of the program PRINS using two version of rod models – three-dimensional (version 1) and one-dimensional (version 2). The material used was concrete with the elasticity modulus $E = 3,2 \times 10^7$ KPa, Poisson ratio $\nu = 0,2$ and mass density $2300\ кg / м^3$. The cross-sections of all rods were assumed to be the same in the form of rectangles with side dimensions 30×30 cm. The height of one floor of the frame was assumed equal to 3.3 m. The support sections of the columns were assumed to be absolutely clamped.

Fig.1. The spatial frame

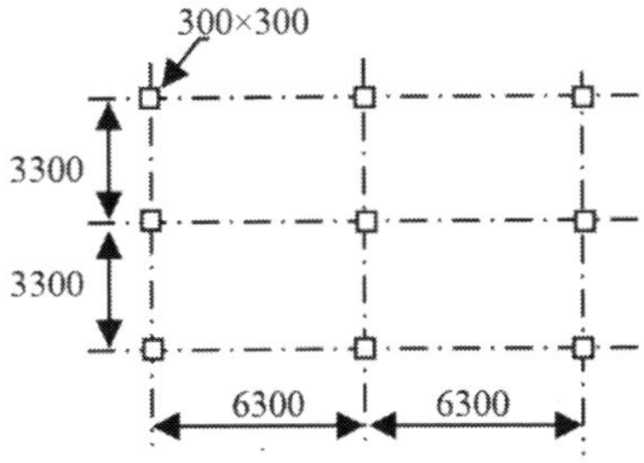

Fig.2. The mesh of columns

In the first version of the design scheme superelements of rods of rectangular cross section proposed in [10] were used. Without going into the details of the implementation, we note that the superelement is formed from three-dimensional solid 8-node finite elements with arbitrary partition by cross-section and height, as shown in Fig.3,a. From the user point of view, the superelement looks like a three-dimensional body, all characteristics of which are brought to the nodes lying in the end sections (Fig. 3,b). Intermediate nodes are excluded at the stage of the stiffness matrix, mass matrix and the load vector forming.

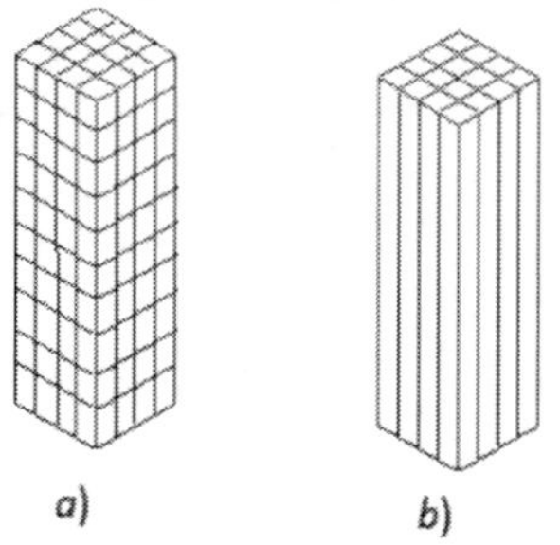

Fig.3. The superelement of the rod: *a*) partitioning of the rod on the volume finite elements; *b*) configuration of the grid after illumination of intermediate nodes

Fig. 4 shows the superelement design scheme of the first floor. Connections in the nodes were carried out as shown in Fig.5. In this case, a special volumetric module with the same mesh of nodes and made of the same material as the superelements to be connected was inserted into the node. Obviously, this way of elements connecting in the design scheme is as close as possible to the connection of these elements in a real construction. It allows you to take into account the rigidity of nodal connections, as well as the complex stress state in the connected elements in the vicinity of the nodes.

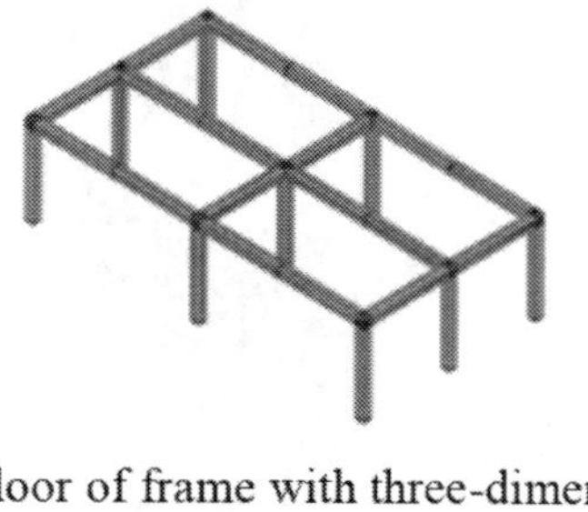

Fig.4. Design scheme of the first floor of frame with three-dimensional elements

Fig.5. Nodal connection

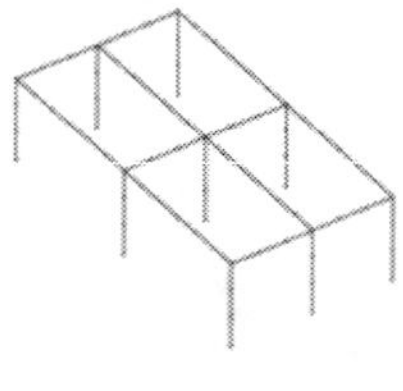

Fig.6. Design scheme of the first floor of frame with one dimensional elements

As for the design scheme using the one-dimensional elements, shown for the first floor in Fig. 6, the typical for such cases the problem of elements connecting in nodes arise. Consider, for example, a node in which two elements of a crossbar and a column are connected (see Fig. 7). In a one-dimensional scheme, the elements must be represented by their axes; the crossbars will be represented by axes 1 - 2′ and 2″ -3, and the column – by axis 2‴ - 4. But in this case the nodes 2′, 2″ and 2‴ on the design scheme will be spaced, and the elements will not be linked, which is unacceptable.

This problem can be resolved in one of two ways. In the first one, you can enter rigid inserts between the nodes 2′ and 2, 2″ and 2, 2‴ and 2, in the second - "tighten" the nodes 2′, 2″ and 2‴ to node 2 and assume that the crossbars are represented by axes 1-2 and 2-3, and the column – by the axis 2-4. Both ways are bad. In the first case, the non-existent in the construction elements are introduced into the node, which stiffen it, in the second, the calculated length of the rods increases, which increases its flexibility. In both cases, the node is considered absolutely rigid, which, as a rule, is not true.

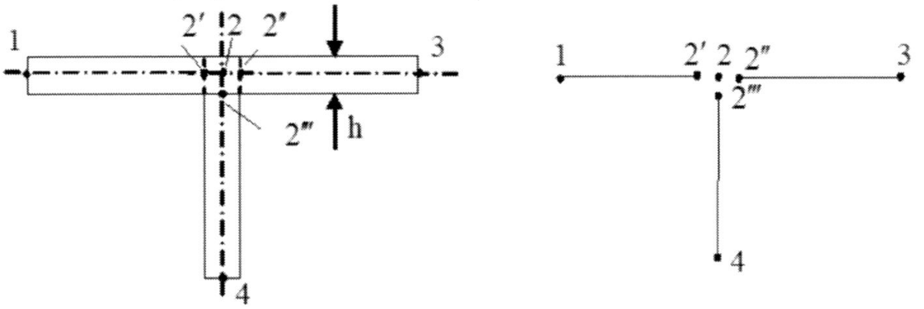

Fig.7. Connection of elements in the nodes of the design scheme with one-dimensional models of rods

In the calculations performed, the second method of connecting of one-dimensional elements in the nodes was used. This suggests that the rigidity of the construction using a design scheme with one-dimensional elements will be somewhat smaller than in the scheme with three-dimensional superelements.

3 Results

The results of the calculation in the form of natural frequencies for the first three tones are given in Table 1.

Table 1. Natural frequencies of frame

Mode number	Natural frequencies, Hz		Divergence, %
	Variant 1	Variant 2	
1	0,439	0,411	6,38
2	0,489	0,452	7,57
3	0,590	0,551	6,61

Fig.8 and 9 show the modes of natural vibrations for the first and third tones.

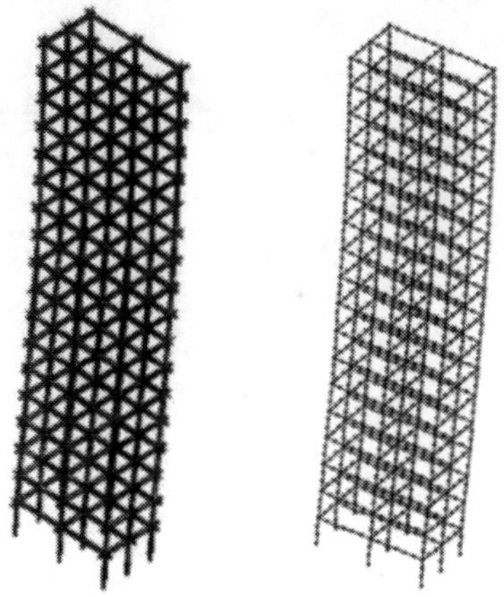

a) *b)*

Fig.8. First mode of vibration: a) version 1; b) version 2

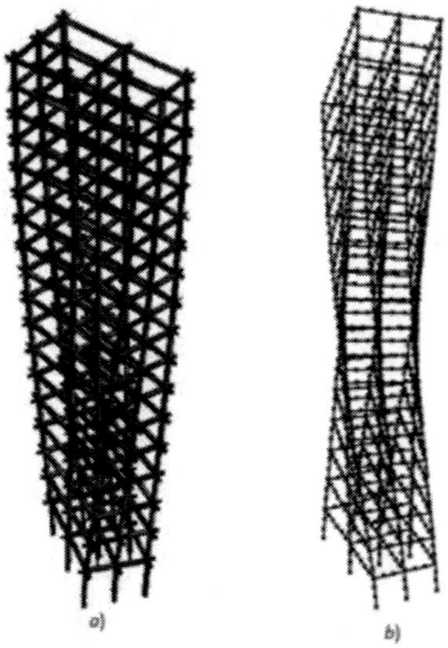

Fig.9. Third mode of vibration: a) version 1; b) version 2

4 Discussion

It can be seen from the Table 1 that the results obtained with the use of one-dimensional and three-dimensional finite elements coincide on the whole, and the frequencies in the second variant are somewhat lower than in the first. There is a simple explanation for this. When using one-dimensional elements, the rods are connected to each other at points lying on the axes of the rods without taking into account the rigidity of the nodal joints, as a result of which the calculated lengths of the rods are somewhat high and the rigidity of the structure is somewhat understated. This leads to a decrease in the frequencies of natural oscillations.

Fig. 8 and 9 show the modes of natural oscillations for the first and the third tones, obtained in the first and second variants of calculation, respectively. As can be seen from the figures, the shapes practically coincide. The number of degrees of freedom of the construction with the use of volume superelements naturally increases. However, taking into account the incomparable cost of the machine time spent for the calculation, and the cost of the construction itself, as well as the possibility to substantially approximate the design scheme to the real object, this deficiency should be considered secondary.

Conclusion

Thus, superelements of rectangular bars used in the program PRINS have adequate rigid and dynamic characteristics and can be recommended for use in static and dynamic calculations of high-rise building. This is especially important at the nonlinear calculations, when it becomes necessary to take into account the warping and changes in the shape of the cross-sections of the rods.

References

1. V. P. Agapov. Nauchnoe obozrenie (Scientific review) **8**, 79-86 (2015)
2. *MSC NASTRAN 2016. Nonlinear User's Guide. SOL 400* (MSC Software Corporation, 2016)
3. K. A. Basov. *ANSYS. Spravochnik pol'zovatelya* (*ANSYS.User Guide* "DMK-Press, 2005)
4. *ABAQUS 6.11. Theory manual*. DS Simulia (2011).
5. K.Washizu. *Variational Methods in Elastisity and Plastisity* (Pergamon Press, 1982)
6. J. C. Simo, Comp. Meth. in Appl. Mech. & Engng., **96**. 189- 200 (1992).
7. M. A. Crisfield. *Non-linear Finite Element Analysis of Solids and Structures* (John Wiley & Sons, 1997).
8. Zienkiewicz O.C., Taylor R.L. *The Finite Element Method for Solid and Structural Mechanics. Sixth edition* (McGraw-Hill, 2005)
9. K. J. Bathe. *Finite Element Procedures* (Prentice Hall, Inc., 1996)
10. V. P. Agapov, A.V.Vasiljev. Vestnik MGSU **5**, pp. 29-34 (2013)

Determining the market value of high-rise residential buildings based on evaluation of consumer properties

Svetlana Kolobova[1*]

Moscow State University of Civil Engineering, 26 Yaroslavskoye shosse, 129337, Moscow, Russia

Abstract. As you know, high-rise construction is an indicator of the practical implementation of advanced innovative technologies in the construction industry of the country. High-rise building inevitably comes to the big cities, in connection with the shortage and value of land. The life cycle of any construction project, including high-rise buildings consists of chains: of engineering survey - design-construction-operation. In the process of operation of a tall building, decisions about major repairs or reconstruction of a building are made for decision-making on further use. This article describes methods of assessing the consumer quality of high-rise residential buildings and the establishment of prices based on consumer characteristics of a tall residential building. It is proposed to assess the premises under their quality characteristics. The study was conducted to establish the influence of individual, comprehensive and integral indicators of comparable quality for effective quality living spaces. Simultaneously, there was established a relationship of quality with the consumer cost of housing, ultimately with the potential needs of owners, tenants of the home, lessor dwelling, or buyers of residential properties and other participants in the residential real estate market. This relationship further creates consumer requirements to quality standard of premises at a certain stage of socio-economic development.

1 Introduction

In Russia the leaders of high-rise housing construction are the following Moscow construction companies: "Don-Stroy", Mirax Group, Mirax Park. High-rise residential buildings are already built in residential complexes of Moscow "Triumph - Palace", "Edelweiss", "Tricolor", "Continental", "Scarlet sails", "House on Mosfilmovskaya", "House in Sokolniki". Among the economy projects in construction there is a "Northern city", "Airbus", "Volga sails" and others. The demand for housing in high-rise buildings continues to evolve. Modern high-rise apartment buildings are constructed in such a way that you have a multi-level underground Parking. On the first two floors of multi-storey residential buildings there is a shopping centre. The hotel is located above the shopping centre . Apartments are located above the hotel. The top two floors are occupied by apartments and penthouses.

* Corresponding author: KolobovaSV@mgsu.ru

Repair and reconstruction of residential development in the future will solve the issue of increasing the of consumer quality of real estate : apartments, houses, neighborhoods. Consequently, this will increase their competitiveness on the market. After the repair, residential properties have a higher assessed value compared to residential facilities, which were not repaired. It is proposed to undertake an assessment of the premises according to their quality characteristics. The study was conducted with the aim of establishing the influence of individual characteristics and indicators of quality to determine the overall (effective) quality of living spaces. Thus was established the connection of quality with the consumer cost of housing, ultimately with the potential needs of the owners, lessees, tenants or home buyers and other participants in the residential real estate market. This relationship creates a consumer demand for quality space at a certain stage of socio-economic development.

2 Literature review

The influence of consumer characteristics on the cost of housing and subsequent sales is studied by scientists of different countries. Assessment of the quality of housing in various countries are described in several scientific publications. For example, in the article "Data quality challenges in the UK social housing sector» authors Caroline Duvierac, Daniel Neagub, Crina Oltean-Dumbravaa and Dave Dickensc
explain that «The social housing sector has yet to realise the potential of high data quality» [1]. M. Norris is devoted his article to the study of the construction and management of the housing sector in the countries of Western Europe «Compared to the rest of the developed world, Western Europe is distinguished by relatively large social housing sectors».... However, institutional arrangements for the management, financing, and regulation of social housing have been commonly reformed since the 1970s. Object subsidies for the construction of new social housing have been cut and replaced by subject subsidies for tenants such as income-related housing allowances. As a result, new social housing construction has declined and the social profile of tenants has generally become more disadvantaged. In response, social landlords have devoted more attention to the social aspects of housing management and the regeneration of estates» [2]. A. Beer examines governance in the housing sector in developed economies : «Over recent decades, the management and delivery of housing assistance measures in many developed economies has become more complex as a wider range of actors have become involved in the sector» [3]. Emma R. continues the theme of the research management of the housing sector: "Through a case study of a small, affordable housing community in Sydney, Australia, it shows that housing governance can profoundly shape senses of home in aging, both contributing to and diminishing senses of home. Further, it points to a connection between housing governance and housing design with different housing typologies associated with different aging bodies and forms of management practice, with profound implications for residents' senses of home» [4].

The article discusses the influence of the quality of the workmanship and construction of new housing on sales prices and capital growth. To measure the quality of construction we use the unique situation in Singapore, where the newly completed residential projects are evaluated regardless of the quality of work in the framework of the quality evaluation system construction. «Empirical evidence suggests that apartments that are well constructed not only command a higher price for developers, but they also generate higher capital gains for homeowners and investors in the future» [5]. Other articles are devoted to the research to price discovery in the local housing market. The key question in scientific research is the quality of construction [6,7,8,9]. Russian scientists have also devoted their numerous scientific papers research housing.
[10, 11, 12, 13, 14, 15].

3 Materials and methods

The author of this article found the concept of consumer quality and cost proposed method of determining this quality of housing.

To assess current market value we introduce the concept of integrated quality. Integral quality is determined by the set of all functional and economic indicators, it is through the relationship between consumer cost and the cost of a residential facility. Structure of consumer properties of objects of housing represented in Fig.1

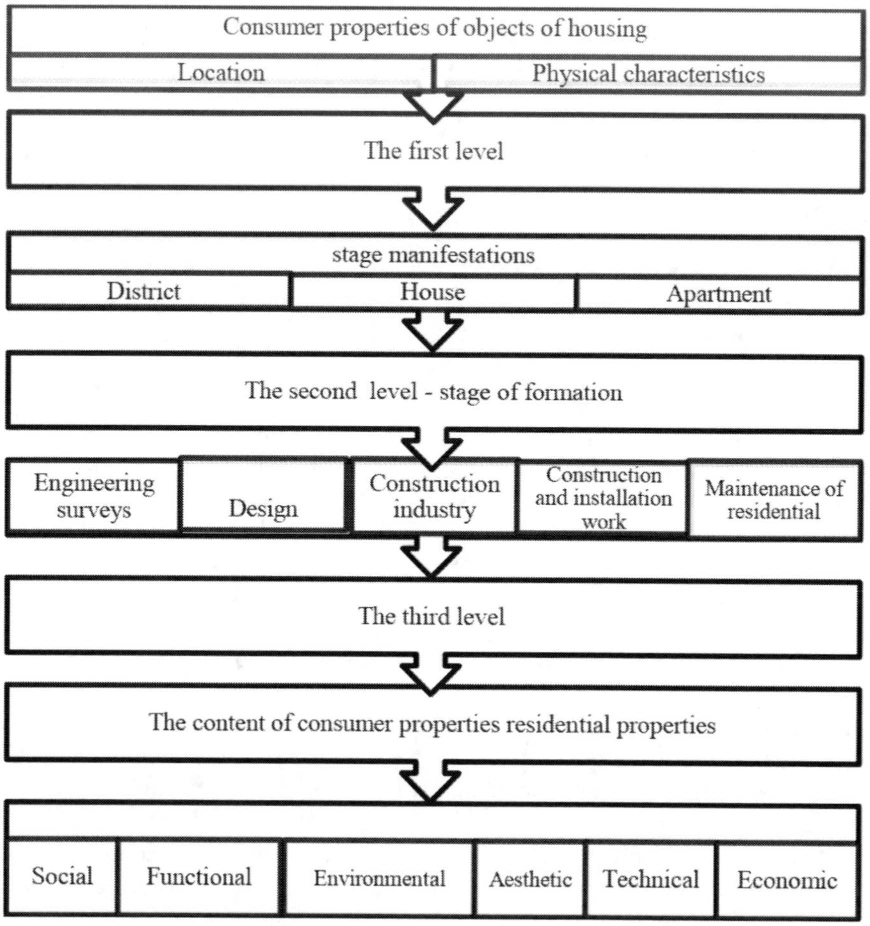

Fig.1. Structure of consumer properties of objects of housing

You must be able to measure consumer value of premises of different quality, that is, to set the level of quality. To measure the level of quality real estate by comparing it with a certain level of quality of the same object, taken as the base for comparison (the base quality). Based on this approach, the expression of quantitative relationships between the properties of a living object and the reflection of these properties in the process of operation of residential facility, namely, between the quality of construction and quality of maintenance and operation .

The number of indicators of the quality of any object, including living quarters, very large, and therefore requires the streamlining of these indicators, that is, their classification. As the basis of classification should be taken of the ability of indicators to Express the most characteristic groups of the object properties, the ability of quantitative and qualitative reflect these properties. Their set needs to determine the quality of the object. Their body needs to determine the quality of the object. For the dwelling is inherent to a certain group of properties, component quality standard: a group of environmental, functional, aesthetic, economic, technical and social properties. The system and composition of individual indicators to obtain a comprehensive characterization of the quality of the object, based on the analysis of scientific publications. The individual indicators of the quality of a residential facility must allow its assessment at three levels: district, house, apartment. Level of quality dwelling is a relative characteristic based on the comparison of the actual values of individual quality indicators (hereinafter IQI) residential facility «P_j» with the respective preset values baseline « P_j^b».

Since the properties through the relevant IQI, forming a quality residential properties, differently involved in the formation of this quality, that is, have different weight, it must be considered in determining comprehensive indicators of the quality of the estimated residential units on each of three levels of manifestation of these properties: urban district, houses, apartments. The quantitative values of the IQI of the object in each level are comparable using the weight of the IQI. The definition of weight, the IQI is produced using a introduced by the author three-point rating scale from 0 to 2 points. IQI, which have a significant effect on the formation of consumer quality of residential facility, is assigned the highest score 2, and IQI does not affect the consumer quality – 0.

Introduced coding of indicators, reflecting the level of their manifestation and the number summarizes their properties:

- consumer quality of apartments -100;
- consumer quality at house - 200;
- consumer quality of district - 300.

Since the properties through the relevant IQI, forming a quality residential facilities differ from one another by its weight, i.e., importance in the overall assessment of its level of quality, then there is a need for integrated indicators valuation for each of the three levels of manifestation of these properties : urban district, houses, apartments.

Obtaining a comprehensive measure of consumer qualities of premises associated with the approach of the calculated quantitative values of object on each level to the total index, taking into account comparability of the estimates IQI and the assigned weights of IQI.

To establish the relative weights IQI residential premises were involved 11 experts in the field of transactions with residential real estate, with practical experience in the real estate market from 6 to 15 years. The experts who participated in the determination of the functional weighting of the IQI, set out the assessment in accordance with the importance of quality. The most significant IQI was awarded a perfect score of "2", and IQI does not affect the quality of the residential object "0". In the processing of these data obtained the desired weight. Each expert was offered a questionnaire with tables. It was necessary to fill the table with given factors, "j" and IQI "Pj" to a residential facility.

An example of a questionnaire table for assessment factors given in table 1

Table 1. Evaluation of residential property.

<table>
<tr><td colspan="3">Questionnaire N ....
Mr (s) expert!
I ask You to consider tables with the factors inherent in residential facilities: apartment, house , district level .
Please top score of "2" to the factor having the most impact and "0" -factor that does not influence consumer qualities of the specified objects</td></tr>
<tr><td>Sequence number</td><td>Factors «j»</td><td>Rating</td></tr>
<tr><td></td><td>At the level of the apartment</td><td></td></tr>
</table>

1	The location of the apartment in the house	
2	Space-planning solution of the apartment	
3	The presence of the glassed balcony(s)	
4	Combined bathroom (bathroom with toilet)	
5	Improved furnish of premises under the "European standard"	
6	Sanitary state	
	At the level of the house	
1	Space-planning decision of the house	
2	The level of technical furnishing	
3	House territory improvement	
4	The proximity to public transport	
5	The building belongs to the monuments of history, culture and architecture	
6	Shopping, office space on the first floor of the house	
	At the district level	
1	The location of the location of the house relative to the center of the city	
2	The environment	
3	The improvement district	
4	The individuality of the area	

After filling in the experts table, is the processing of the obtained results. Initially it was determined the weight of characteristics at each of three levels of the house and area:

$$m_j = \frac{\sum_1^J V_{js}}{\sum_1^J \sum_1^S V_{js}} \quad \text{, in this case } \sum_1^J m_j = 1 \qquad (1)$$

Where:: m_j – the weight characteristics «j»;
Vjs – the assessment, exhibited j-th characteristic of the s-th expert;
j= 1, J – ordinal characteristics;
s= 1, S – serial number expert.

For example, the weight factor «j_1» «The location of the apartment in the house» - m_{j_1} is as follows:

$$m_{j_1} = \frac{\sum_1^1 V_{j1s}}{\sum_1^6 \sum_1^{11} V_{j1s}} = \frac{12}{66} = 0.182 \qquad (2)$$

Similarly, the calculation of the weights for other factors. Example and definition of weights for groups of factors for objects on the level are shown in table 2.

Table 2. Sample weights for groups of factors for objects on the level.

Factor j	At the level of the apartment					
Expert , S	The location of the apartment in the house	Space-planning solution of the apartment	The presence of the glassed balcony (s)	Combined bathroom (bathroom with toilet)	Improved furnish of premises under the "Euro-standard"	Sanitary state
1	1	2	1	0	0	2
2	1	2	1	1	0	0
3	2	1	1	1	0	2
4	0	2	1	1	1	1
5	2	1	2	0	0	0
6	1	1	2	2	1	0

7	2	1	1	0	1	1
8	1	2	0	1	0	2
9	0	2	1	1	0	2
10	1	1	2	1	2	0
11	1	0	1	1	0	2
$\sum_1^{11} V_{jS}$	12	15	13	9	5	12
m_j	0.182	0.227	0.197	0.136	0.076	0.182
$\sum_1^{6} m_j = 0.182 + 0.227 + 0.197 + 0.136 + 0.182 + 0.076 = 1$						

Similarly one computes the values of weights at the house and the area.
Then in the same way the weight IQI - «P_j» in each group of characteristics:

$$m_{P_j} = \frac{\sum_1^P v_{P_j S}}{\sum_1^P \sum_1^S v_{P_j S}} \text{, in this case } \sum_1^P m_{P_j} = 1 \quad (3)$$

Where m_{P_j} -- weight IQI «P_j»; $V_{P_j S}$ – the mark awarded P_j –th IQI s –th expert.
Weigh m_{P_j} is calculated by multiplying the weight $m_{P_{js}}$ on the weight characteristic «j» weight m_j . For example, $m_{P_1} = 0.182 \cdot 0.182 = 0.033$

$m_{P_2} = 0.227 \cdot 0.182 = 0.041$

$m_{P_3} = 0.197 \cdot 0.182 = 0.036$

$m_{P_4} = 0.212 \cdot 0.182 = 0.039$

$m_{P_5} = 0.182 \cdot 0.182 = 0.033$

4 Results

Example the definition of weights m_{P_j} and the average marks q_{P_j} at the level of the apartment the factors "the location of the apartment in the house" and "Space-planning solution of the apartment" are shown in table 3.

Table 3. Example the definition of weights m_{P_j} and the average marks q_{P_j} at the level of the apartment the factors "the location of the apartment in the house" and "Space-planning solution of the apartment".

Factor j	The location of the apartment in the house					Space-planning solution of the apartment					
Expert , S	Feature rooms	Floors	Material of walls, floors	The proximity to the Elevator shaft	Orientation rooms	The kitchen area of over 10 sq m	The hall area of over 10 sq m	Ceiling height	Area of residential and outbuildings	The device built-in wardrobes	Adjoining rooms in the apartment
1	1	2	1	1	2	1	2	1	0	2	0

2	1	2	1	2	0	1	2	1	1	0	0
3	2	1	1	0	2	2	1	1	1	2	0
4	0	2	1	1	1	0	2	1	1	1	1
5	2	1	2	2	0	2	1	2	0	0	0
6	1	1	2	1	0	1	1	2	2	0	1
7	2	1	1	1	1	2	1	1	0	1	1
8	1	2	0	2	2	1	2	0	1	2	0
9	0	2	1	1	2	0	2	1	1	2	0
10	1	1	2	2	0	1	1	2	1	0	2
11	1	0	1	1	2	1	0	1	1	2	0
$\sum_1^{11} V_{js}$	12	15	13	14	12	12	15	13	9	12	5
m_{P_j}	0.182	0.227	0.197	0.212	0.182	0.182	0.227	0.197	0.136	0.182	0.076
	$\sum_1^5 m_{P_{js}} = 0.182 + 0.227 + 0.197 + 0.212 + 0.182 = 1$					$\sum_1^6 m_{P_{js}} = 0.182 + 0.227 + 0.197 + 0.136 + 0.182 + 0.076 = 1$					
$q_{P_j} = \sum_1^{11} V_{js} / \sum S$	12/11=1.091	15/1=1.364	13/1=1.182	14/1=1.273	12/1=1.091	12/11=1.091	15/11=1.364	13/11=1.182	9/11= 0.818	12/11=1.091	5/11= 0.455

Table 4. An example of defining an integrated index of consumer qualities of the i-th apartment on the basis of estimates of IQI and weights

Code properties	Individual quality indicators (IQI) - (Pj)		Average score	Weight	Final score
			q_{P_j}	m_{P_j}	$q_{P_j} \cdot m_{P_j}$
100	Evaluation of consumer qualities of premises				
110	The location of the apartment in the house				
	P1	Feature rooms	1.091	0.033	0.036
	P2	Floors	1.364	0.041	0.056
	P3	Material of walls, floors	1.182	0.036	0.043
	P4	The proximity to the Elevator shaft	0.273	0.039	0.011
	P5	Orientation rooms	1.091	0.033	0.036
		Σ		0.182	0.182
120	Space-planning solution of the apartment				
	P6	The kitchen area of over 10 sq m	1.091	0.041	0.045
	P7	The hall area of over 10 sq m	1.364	0.052	0.071
	P8	Ceiling height	1.182	0.045	0.053
	P9	Area of residential and outbuildings	0.818	0.031	0.025
	P10	The device built-in wardrobes	1.091	0.041	0.045
	P11	Adjoining rooms in the apartment	0.455	0.017	0.008
		Σ		0.227	0.247
130	The presence of the glassed balcony(s) loggia (s)				
	P12	Area balcony(s) loggia (s)	0.716	0.081	0.058

	P13	Type of glazing installation, double glazing	0.534	0.081	0.043
		∑		0.197	0.101
140	Bathroom				
	P14	The presence of combined bathroom	0.678	0.042	0.028
	P15	Square bathroom	0.513	0.036	0.018
	P16	Sanitary arrangements	0.505	0.035	0.018
		∑		0.136	0.064
150	Improved furnish of premises under the "European standard"				
	P17	The need for improved finish	0.320	0.025	0.008
	P18	The equipment with modern technical devices	0.534	0.024	0.013
		∑		0.076	0.021
160	Sanitary state				
	P19	The microclimate in the apartment	1.182	0.023	0.027
	P20	Conditions of insolation and artificial illumination	1.091	0.021	0.023
	P21	The conditions of artificial lighting	1.364	0.024	0.033
	P22	Organization of space for the installation of air conditioning	0.818	0.018	0.015
	P23	The view from the window (window)	0.818	0.025	0.020
	P24	The noise level in the apartment	0.455	0.022	0.010
	P25	The need for repair	1.091	0.030	0.033
		∑		0.182	0.161
		TOTAL:		1.000	0.776

Comprehensive index of consumer qualities of the apartment:

$$C^a_{iP_j} = \sum_{1}^{25} q_{P_j} \cdot m_{P_j} = 0.776$$

Table 5. An example of defining an integrated index of consumer qualities of the i-th house on the basis of estimates of IQI and weights.

Code properties	Individual quality indicators (IQI) - (Pj)		Average score q_{P_j}	Weight m_{P_j}	Final score $q_{P_j} \cdot m_{P_j}$
200	Evaluation of the quality of consumer home				
210	Space-planning decision of the house				
	P26	The device ramps for wheelchairs and people with limited mobility	1.091	0.053	0.058
	P27	Room for bicycles, scooters, wheelchairs	1.364	0.042	0.057
	P28	The presence of a spacious hall in the hallways	1.182	0.022	0.026
	P29	The possibility of placing concierge doormanin the entrance	0.818	0.035	0.029
	P30	The presence of underground garage	0.455	0.038	0.017
	P31	Exterior facades of the house	1.091	0.046	0.050
	P32	Floors	1.091	0.046	0.050
		∑		0.282	0.287
220	The level of technical furnishing				
	P33	The availability of modern elevators	1.000	0.063	0.063
	P34	Device intercom	0.818	0.052	0.043

	P35	Device video intercom and video camera	1.343	0.032	0.043
	P36	Device automatic sprinkler	1.182	0.045	0.053
	P37	The device of systems of ventilation and conditioning	1.102	0.039	0.043
		∑		0.231	0.245
230	House territory improvement				
	P38	The possibility of using the adjacent territory	0.678	0.042	0.028
	P39	The availability of Parking machines	0.513	0.036	0.018
	P40	The playgrounds and sports facilities	0.505	0.035	0.018
	P41	The presence of fencing the local area	0.320	0.025	0.008
	P42	Gardening and landscaping	0.534	0.024	0.013
		∑		0.162	0.085
240	The proximity to public transport				
	P43	The proximity to public transport	0.678	0.042	0.028
	P44	The proximity to the metro	0.513	0.036	0.018
	P45	The proximity to the Railways stations	0.505	0.035	0.018
		∑		0.113	0.064
250	The building belongs to the monuments of history, culture and architecture				
	P46	Belonging to the monuments of history	0.320	0.026	0.008
	P47	The building belongs to the monuments of culture	0.534	0.024	0.013
	P48	The building belongs to the monuments	0.320	0.025	0.008
	P49	Finding a home in the historical buffer zone	0.534	0.024	0.013
		∑		0.099	0.042
260	The placement of household service enterprises on the ground floor				
	P50	Educational facilities	0.678	0.042	0.028
	P51	Allocation of the storage areas	0.513	0.036	0.018
	P52	The location of the office space	0.505	0.035	0.018
		∑		0.113	0.064
		TOTAL:		1.000	0.787

Comprehensive consumer house quality:

$$C_{iP_j}^{h} = \sum_{26}^{52} q_{P_j} \cdot m_{P_j} = 0.787$$

Table 6. An example of defining an integrated index of consumer qualities of the i-th district on the basis of estimates of IQI and weights.

Code properties	Individual quality indicators (IQI) - (Pj)		Average score q_{P_j}	Weight m_{P_j}	Final score $q_{P_j} \cdot m_{P_j}$
300	Evaluation of consumer qualities of the area				
310	The location of the location of the house relative to the center of the city				
	P53	The location of the area in the city	0.678	0.042	0.028
	P54	The time spent on the road	0.513	0.036	0.018
	P55	The number of available modes of transport	0.505	0.035	0.018
		∑		0.113	0.064
320	The environment				

	P56	The condition of the air environment	1.000	0.063	0.063
	P57	Background radiation	0.818	0.052	0.043
	P58	The presence of industries and businesses	1.343	0.032	0.043
	P59	Proximity to transport routes	1.182	0.045	0.053
	P60	Gardens	1.102	0.039	0.043
		∑		0.231	0.245
330	The improvement of the district				
	P61	Walking distance to the objects of trade, consumer services	1.182	0.022	0.026
	P62	Walking distance to objects of education, culture, sports, leisure	0.818	0.035	0.029
	P63	The improvement district, road surface, landscaping, lighting,	0.455	0.038	0.017
	P64	The presence of the Park, square, pond	1.091	0.046	0.050
	P65	Parking machines	1.091	0.046	0.050
		∑		0.282	0.287
340	The individuality of the district				
	P66	Historical value of residential development	1.182	0.072	0.085
	P67	Ethnic and landscape value of the territory	0.818	0.077	0.063
	P68	The presence of zones of protection of historical, cultural and architectural monuments	0.455	0.078	0.035
	P69	Density	1.091	0.063	0.069
	P70	Storeys building	1.091	0.084	0.092
		∑		0.374	0.344
		TOTAL:		1.000	0.940

Comprehensive index of consumer qualities of the district:

$$C^d_{iP_j} = \sum_{53}^{70} q_{P_j} \cdot m_{P_j} = 0.940$$

Consistency of expert opinion was verified with the value of standard deviation [16,17]. Comprehensive quality indicators "C_i" characterizing the level of consumer quality of the premises of the i-th residential facility on three levels – apartments, houses, district:

Apartment | House | District (4)

$$C_i^a = \sum_1^J q_{P_j} \cdot m_{P_j} \qquad C_i^h = \sum_1^J q_{P_j} \cdot m_{P_j} \qquad C_i^d = \sum_1^J q_{P_j} \cdot m_{P_j}$$

$$0 \leq q_{P_j} \leq 2 \qquad 0 \leq q_{P_j} \leq 2 \qquad 0 \leq q_{P_j} \leq 2$$

$$0 \leq m_{P_j} \leq 1 \qquad 0 \leq m_{P_j} \leq 1 \qquad 0 \leq m_{P_j} \leq 1$$

$$0 \leq m_j \leq 1 \qquad 0 \leq m_j \leq 1 \qquad 0 \leq m_j \leq 1$$

Где: C_i^a, C_i^h, C_i^d – comprehensive consumer quality of apartments, houses, area of the i-th residential facility;
q- average rating exposed the experts Pj –th IQI the i-th residential facility;
i = 1, n –the objects of evaluation;
j= 1, J – characteristics included in the IQI;

$P_j = \{1 \div P\}$ – the individual indicators of the quality of the IQI assessment objects.

5 Conclusions

To calculate the magnitude of the current market value of a residential facility can be used the integral index of quality of a residential facility. Methodological basis of the calculation remains unchanged , and the accuracy of calculations is high enough. For this we use relative values complex indices of consumer quality residential properties C_i^{oa}, C_i^{oh}, C_i^{od} - r_i^{oa}, r_i^{oh}, r_i^{od}, having them as the quotient of the absolute values of the complex indices on the value of the maximum of the rating scale used ($q^{max} = 2$ points):

$$r_i^{oa} = \frac{C_i^{oa}}{q^{max}}\,;\; r_i^{oh} = \frac{C_i^{oh}}{q^{max}}\,;\; r_i^{od} = \frac{C_i^{od}}{q^{max}} \quad (5)$$

Obtained three values "r" for each underlying and the estimated object at the three levels comparable because they are derived using the same methodology. The importance of these indicators in the formation of an integrated indicator of quality should be taken into account by the introduction of their weights is "y". This weight should be determined depending on the conditions that determine supply and demand for this type of residential property and the importance for the participant of the real estate market, for example, for the buyer and the seller, each of these conditions. These weights can be obtained from the analysis of the information about the cost of rent and sales of residential properties in the city or area when changed external conditions: the location, highways infrastructure and other factors.

Integral indicator of comparable quality, including estimated residential facility can be determined by the formula:

$$C_i^{int} = r_i^{oa} \cdot y^{oa} + r_i^{oh} \cdot y^{oh} + r_i^{od} \cdot y^{od} \quad (6)$$

By the proposed technique provides for the calculation C_i^{int} for a number of objects, which had previously conducted evaluation of the transaction of purchase and sale and there is information about selling price (S). These indicators will be used for the valuation of the object. Similarly being an integral indicator of the quality of a residential facility. Of the total number of objects for which received and processed information, selected object, which $S_i^{int} \approx S_0^{int}$. The presence of equality can be considered if the coefficients differ in the magnitude of the accuracy of the calculations by almost 5-7 %. If such an object is, and the price is inserted to the market of the object may be adopted same as that of the analogue.

If this is not unique, then the total number of objects selected two:
one with $S_1^{int} \leq S,$ another with $S_2^{int} \geq S_0^{int}$.

The cost is calculated based on the assumption that it changes linearly with the values S^{int} between S_1^{int}, S_2^{int}.

Then the estimated value of residential property can be determined by the formula:

$$S_0 = S_1 + \frac{(C_0^{int} - C_1^{int}) \cdot (S_2 - S_1)}{C_2^{int} - C_1^{int}} \quad (7)$$

Investors, potential buyers, tenants and other entities can calculate according to this method the market value based on an integral quality indicators, characterizing consumer properties, decides on the subsequent use of these objects: purchase, sales, rent, mortgage, etc. In the case that the objects in question do not satisfy the demand in the real estate market, the owner of a residential facility, takes the decision to repair, rebuild or overhaul to improve its consumer properties.

This technique allows to estimate the current market value of a residential facility based on the level of its consumer qualities before making a decision about capital repairs, reconstruction, and renovation of the facility.

References

1. C. Duvierac, D. Neagub, C. Oltean-Dumbravaa, D. Dickensc. ,International Journal of Information Management **38**, Issue 1, 196-200 (2018)
2. M. Norris. Social Housing Landlords: Europe International Encyclopedia of Housing and Home, 438–443 (2012)
3. A. Beer. Housing Governance, International Encyclopedia of Housing and Home, 497–501 (2012)
4. E. R. Housing Journal of Housing for the Elderly **31**, Issue 3 (2017)
5. Joseph T.L., Ooi Thao T.T., Le Nai-JiaLee. Journal of Housing Economics **26** 126-138 (2014)
6. Joseph T.L., Ooi1Thao, T.T. Le. Regional Science and Urban Economics **43**, Issue 6, 850-861, (November 2013)
7. V. Zahirovich-Herberta, K. M.Giblerb. Journal of Housing Economics **26**, 1-18 (2014)
8. E. Nezhnikova, R. Obukhov 08025 MATEC Web Conf. Volume **106** (2017)
9. V. Földváry, G. Bekö, S. Langer, K. Arrhenius, D. Petráš. Building and Environment, Volume **122**, 363-372 (2017)
10. 10 N. Safronova, E. Nezhnikova, A. Kolhidov MATEC Web Conf. Volume **106**, (2017)
11. E. A. Antipov, E. B. Pokryshevskaya. Mass appraisal of residential apartments: An application of Random forest for valuation and a CART-based approach for model diagnostics. **39**, Issue 2, 1, 1772-1778 (2012)
12. D.N. Silka, I.G. Lukmanova, A.A. Afanasev, M.A. Kasyanov, International Journal of Economics and Financial Issues. 5(3s) 121-124 (2015)
13. Kamenetsky M.I. Forecasting problems, **3**, 76-91 (2013)
14. A.A. Kokoshin, B.I. Bartenev Studies on Russian Economic Development **6**, pp. 6-18 (2016)
15. I. Lukmanova, N. Yaskova , Procedia Engineering Volume **165** , Pages 1293-1299 (2016)
16. T. Drerupa , B. Enkeb, H-M. Gaudeckerc Journal of Econometrics Volume **200**, Issue 2, Pages 378-389 (2017)
17. A. N. Koshev, V. V.Kuzina . Procedia Engineering. Volume **161**, Pages 1874-1878 (2016)

E3S Web of Conferences **33**, 02035 (2018) https://doi.org/10.1051/e3sconf/20183302035
HRC 2017

Aggregative stability of fungicidal nanomodifier based on zinc hydrosilicates

Anna Grishina[1,*] and *Evgeniy* Korolev[1]

[1]Moscow State University of Civil Engineering, Nanomaterials and Nanotechnology Research and Educational Center, Yaroslavskoye hw. 26, 129337, Moscow, Russian Federation

Abstract. Currently, there is a strong need of high performance multi functional materials in high-rise construction. Obviously, such materials should be characterized by high strength; but for interior rooms biosafety is important as well. The promising direction to obtain both high strength and maintain biosafety in buildings and structures is to manage the structure of mineral binders by means of fungicidal nanomodifier based on zinc hydrosilicates. In the present work the aggregative stability of colloidal solutions of zinc hydrosilicates after one year of storage was studied. It has been established that the concentration of iron (III) hydroxide used to prepare the precursor of zinc hydrosilicates has a significant effect on the long-term aggregative stability: as the concentration of iron (III) hydroxide increases, the resistance of the fungicidal nanomodifier increases. It was found that, despite the minimal concentration of nano-sized zinc hydrosilicates (0.028%), the colloidal solution possesses a low long-term aggregative stability; while in the initial period (not less than 14 days) the colloidal solution of the nanomodifier is aggregatively stable. It is shown that when the ratio in the colloidal solution of the amount of the substance CH_3COOH / SiO_2 = 0.43 is reached, an increase in the polymerization rate is observed, which is the main cause of low aggregative stability. Colloidal solutions containing zinc hydrosilicates synthesized at a concentration of iron (III) hydroxide used to produce a precursor equal to 0.7% have a long-term aggregative stability and do not significantly change the reduced particle. Such compositions are to be expediently used for the nanomodifying of building composites in order to control their structure formation and to create conditions that impede the development of various mycelial fungi.

1 Introduction

Currently, one of the priority areas of scientific development is the maintenance of environmental safety, including in the premises of residential buildings and public facilities [1]. This includes both the proper utilization of building materials with the expired service life, and maintenance of favorable environmental conditions in existing buildings and facilities. It is important to ensure a safe microclimate in the premises for the preservation of human health and the specified lifetime of building materials. It is known that mold fungi are

* Corresponding author: info@nocnt.ru

more dangerous compared to dust mites, bacteria and animal hair [2-5]. The formation of colonies of mycelial fungi, mainly represented by the genera Penicillium and Aspergillius [2-4, 6], is often observed on the surface and in the thickness of the enclosing structures or their finishing layers, which are capable of causing various diseases in humans [7, 8], since in most are allergenic and conditionally pathogenic; also contribute to the premature destruction of the material [9, 10] due to the chemical interaction of the products of the life activity of mold fungi with the components of the building material and with the germination of the mycelium into the thickness of the material [11-13]. Of course, the presence of mold deteriorates the decorative appearance of structures and the appearance of the premises.

The solution of these problems is the creation of modifiers that allow long-term maintenance of the fungicidal properties of the material [14-19] and be safe using [20-23]. One of the fungicidal modifiers having the listed properties is a nanomodifier based on zinc hydrosilicates, which also affects the structure of mineral binders.

However, silica-containing systems, despite the advantages of their use, have features that are their ability to polymerize, and in the presence of metal cations, hetero-coagulation is possible. These processes are accompanied by the coarsening of particles. For the modifiers represented by nanoscale particles, one of the important indicators determining the possibility of its use is the preservation of its aggregative and sedimentation stability during storage. This is necessary for the uniform distribution of the nanomodifier particles in the building material, as well as for the preservation of the efficiency of nanomodification, as assessed by the change in the properties of the resulting composite. Therefore, the purpose of the study was to establish the aggregative stability of a biocidal nanomodifier based on zinc hydrosilicates with the aim of predicting the shelf life of the zinc, allowing use in the construction industry, taking into account the delivery and storage of the nanomodifier before use. The objectives of the study were: establishment of the warranty period for storage of nanomodifier on the basis of zinc hydrosilicates and studying the processes occurring during storage of a colloidal solution of zinc hydrosilicates.

2 Materials and methods

The synthesis technology of the nanomodifier, calculation of the concentration of the precursor of silicic acid (α) was previously described in [24] and [25]. According to [26], the fungicidal properties of the material are possible even if the used fungicides have low concentrations that are not qualitatively determined. For nanomodifiers, in addition to the above, a requirement is made to preserve their aggregative stability during storage. The temperature regime during storage was the same for all systems (storage temperature was 20...25 °C), the compositions were not subjected to stirring, therefore, these factors have an equal effect on all the studied compositions. To estimate the storage time of the nanomodifier based on zinc hydrosilicates, the change in the reduced diameter of its particles after 1 year of storage by the laser diffraction method was studied using a Microtrac Zetatrac apparatus.

2 Results and Discussion

The results of studies of the change in the reduced diameter of zinc hydrosilicate particles are shown on figure 1.

https://doi.org/10.1051/e3sconf/20183302035

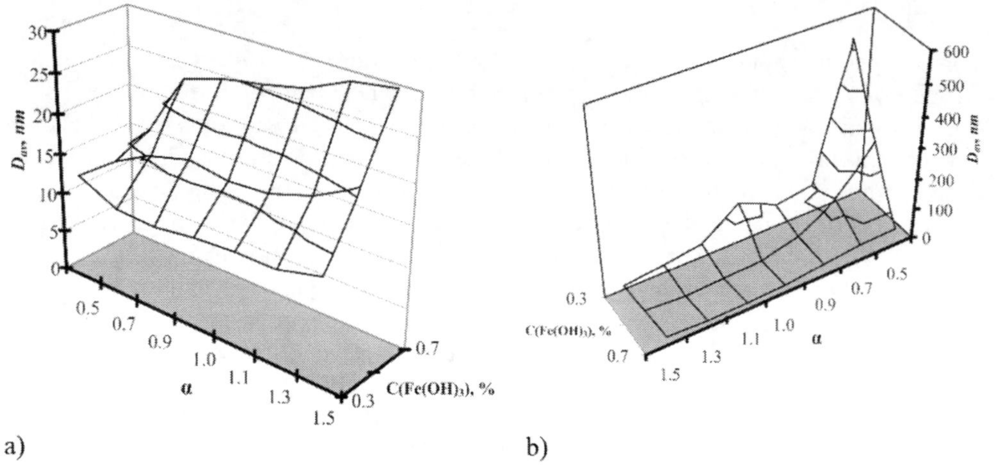

Figure 2. Size of zinc hydrosilicate nanoparticles after synthesis (a) and after 1 year of storage (b).

The analysis of figure 1 shows that the aggregate resistance of synthesized zinc hydrosilicates is influenced by the formulation. Thus, in the composition obtained with the iron (III) hydroxide sol, 0.3% concentration and the content of the precursor of silicic acid α = 0.5, a significant increase in particle size is observed - from 10.78 to 526.5 nm. In other compositions, the particle size remains in the nano-range: the particle sizes increase to 130 nm. It is known [27] that in time, depending on the conditions, both polymerization, depolymerization of silica-containing systems, and heterocoagulation are observed. To identify the processes that occur during the storage of colloidal solutions of zinc hydrosilicates, the pH of the medium has been studied (figure 2).

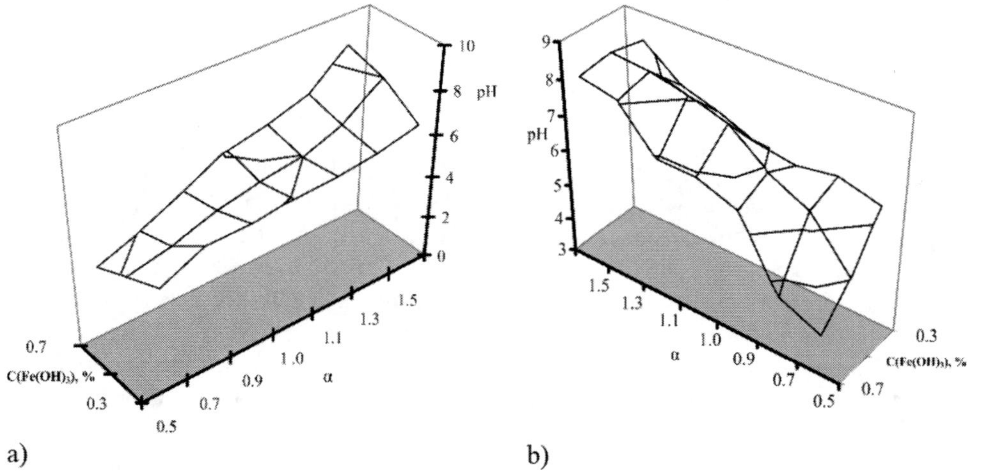

Figure 2. pH of colloidal solutions of nanosized zinc hydrosilicates after synthesis (a) and after 1 year of storage (b)

It is evident from the data presented in figure 2 that for all compositions an increase in the pH of the solutions after storage is observed, which indicates the dominance of the polymerization processes. The most significantly increased pH for compositions synthesized using a sol of iron hydroxide (III), a concentration of 0.3% and substantially less – for the compositions made using a sol of iron hydroxide (III), a concentration of 0.7%. This is consistent with the data of [27], indicating the growth of the pH of the system due to a decrease in the concentration of acid groups -SiOH.

Also, the concentration of silica and the concentration of the electrolyte, zinc acetate used for the synthesis of zinc hydrosilicates, in particular acetate ions, have a significant effect on the state of colloidal silica-containing systems. The values of the concentration of zinc hydrosilicates in the formulations and the ratio of the amounts of CH_3COOH / SiO_2 are shown in tables 1 and 2.

Table 1. Concentration of zinc hydrosilicates in colloidal solutions with different synthesis formulas, %

Concentration of $Fe(OH)_3$ sol, %	Value of α						
	0.5	0.7	0.9	1.0	1.1	1.3	1.5
0.3	0.028	0.039	0.042	0.047	0.052	0.058	0.067
0.5	0.041	0.057	0.066	0.073	0.080	0.091	0.105
0.7	0.067	0.094	0.112	0.125	0.138	0.158	0.182

Table 2. Ratio of amounts of CH_3COOH / SiO_2 substance in colloidal solutions for various synthesis formulations

Concentration of $Fe(OH)_3$ sol, %	Value of α						
	0.5	0.7	0.9	1.0	1.1	1.3	1.5
0.3	0.028	0.039	0.042	0.047	0.052	0.058	0.067
0.5	0.041	0.057	0.066	0.073	0.080	0.091	0.105
0.7	0.067	0.094	0.112	0.125	0.138	0.158	0.182

It follows from the analysis of the data of Table 1 that the concentration of zinc hydrosilicates in a colloidal solution made with the iron (III) hydroxide sol equal to 0.3% and at $\alpha = 0.5$ is the minimum of those studied. In this case, the aggregative and sedimentation stability of the specified colloidal solution does not allow it to be stored for a long time. This contradicts the known data [27] on the change in the rate of polymerization of silica-containing colloidal systems, depending on their concentration. However, the ratio of CH_3COOH / SiO_2 in the described composition is the maximum (table 2). According to [28], the influence on the stability of colloidal systems of electrolytes has a contradictory character: it was shown in [27] that with an increase in the concentration of electrolyte, the intensity of its influence on the rate of polymerization of silicic acid is most pronounced in an alkaline medium, and in [29] that electrolytes, depending on the pH of the medium, can accelerate, slow down or not affect the polymerization rate. Analysis of the data presented in table 2 shows that the relative concentration of acetate ions is the highest in the composition synthesized using the iron (III) hydroxide sol, at a concentration of 0.3%, and at $\alpha = 0.5$. Thus, the influence of alkaline medium (pH = 8.32), as well as a high concentration of acetate ions, lead to an increase in the rate of polymerization of zinc hydrosilicates. It is possible to increase the stability of this composition only by decreasing the mass of zinc acetate, which is impractical due to a decrease in the content of zinc, which is the carrier of fungicidal properties.

3 Summary

A technology has been developed for the preparation of a nanomodifier for controlling the structure formation of mineral binders and maintaining the fungicidal properties of building materials. It is shown that the guaranteed shelf life of such a modifier is at least 1 year, except the composition made using the iron (III) hydroxide sol, at a concentration of 0.3%, and at $\alpha = 0.5$. The reason for the decrease in the shelf life of the zinc hydrosilicate sol using the iron (III) hydroxide sol, at a concentration of 0.3% and at a value of $\alpha = 0.5$, is an increase in the rate of its polymerization due to the high concentration of electrolyte and alkaline solution medium. It has been established that during the storage of all the investigated colloidal solutions of zinc hydrosilicates, polymerization processes dominate the processes of formation of donor-acceptor bonds between zinc cations and active centers of the surface of silicic.

Acknowledgments

This work is supported by grant of Russian Federation President MK-8575.2016.8.

References

1. Technical regulations on safety of buildings and structures: Federal law from 30.12.2009 № 384-FZ (2009)
2. T.M. Zheltikova, A.B. Antropova, E.N. Bilanenko, V.L. Mokeeva, L.N. Chekunova, A.D. Petrova-Nikitina. Pest Management, 1 (2006)
3. O.B. Gradusova, N.E. Ivanushkina, G.A. Kochkina, S.M. Ozerskaja. Theory and practice of forensic examination 3 (2010)
4. O.A. Makanina, L.V. Gordeeva. Bulletin of Belgorod state University. Series: Natural science, 10 (2013)
5. V.M. Katola. Far Eastern journal of infectious pathology, 21 (2012)
6. T.M. Zheltikova. Problems in medical Mycology, 2 (2009)
7. V.S. Lesovoj, A.V. Lipnickij, O.M. Ochkurova. Problems of medical mycology, 2 (2004)
8. T.N. Burkutbaeva, L.K. Tastanbekova. Problems of medical mycology, 3 (2004)
9. V.T. Erofeev et al. Solid State Phenomena, 871 (2016)
10. V.T. Erofeev et al. Procedia Engineering, 165 (2016)
11. L.L. Negoda. Tradition and innovation in construction and architecture. Science and technospheric safety, 4 (2015)
12. V.F. Stroganov, V.A. Bojchuk, E.V. Sagadeev. Proceedings of the Kazan state University of architecture and construction, 2 (2014)
13. S.A. Starcev. Magazine of civil engineering, 7 (2010)
14. H. Wang, X. Xu, Y. Wang, H. Huang, Y. Li. Environmental Toxicology and Chemistry, 3 (2011)
15. F.T. Jesus, R. Oliveira, A. Silva, A.L. Catarino, A.M. Soares, A.J.A. Nogueira, I. Domingues. Ecotoxicology, 9 (2013)
16. J.A. Hingston, A. Bacon, J. Moore, C.D. Collins, R.J. Murphy, J.N. Lester. Bulletin of Environmental Contamination and Toxicology, 1 (2002)

17. A.R. Fernandez-Alba, M.D. Hernando, L. Piedra, Y. Chisti. Analytica Chimica Acta, 2 (2002)
18. K.E. Appel. Archives of Toxicology, 2 (2000)
19. N. Voulvoulis, M.D. Scrimshaw, J.N. Lester. Chemosphere, 7 (2002)
20. J.O. Morley, A.J. Oliver, M.H. Charlton. Computational and Theoretical Chemistry, 1 (1998)
21. P. Viswanathamurthi, K. Natarajan. Transition Metal Chemistry, 6 (1999)
22. J.A. Ponasik, D.J. Kassab, B. Ganem. Tetrahedron Letters, 34 (1996)
23. M. De Bolle, W.F. Broekaert, B.P.A. Cammue, S. Rees, J. Vanderleyden. Biotechnology Advances, 1 (1997)
24. A.N. Grishina, E.V. Korolev. AIP Conference Proceedings, 1722 (2016)
25. A.N. Grishina. Proc. of XIV International Conference of Students and Young Scientists "Prospects of Fundamental Sciences Development", Tomsk, 6 (2016)
26. V.F. Stroganov, E.V. Sagadeev. Introduction to the bio-construction of building materials. Kazan: KazGASU (2014)
27. N.A. Shabanova, P.D. Sarkisov. Sol-gel technology. Nanodisperse silica. Moscow: BINOM (2012)
28. T. Yokoyama, Y. Takanashi, T. Tarutani. J. Colloid Interface Sci, 2 (1991)
29. S.I Kontorovich., L.N. Sokolova, E.A. Golubeva, E.I. Kochetkova, N.P. Sokolova, E.D. Shhukin. Colloid Journal, 1 (1991)

New technology of underground structures the framework of restrained urban conditions

Mikhail Pleshko[1,2*], *Alexander* Pankratenko[1], *Alexey* Revyakin[2]; *Ekaterina* Shchekina[3] , and *Svetlana* Kholodova[3]

[1]National University of Science and Technology (MISiS), Leninskij Av., 4, Moscow, 119991, Russia,
[2]Rostov State Transport University, Rostovskogo Strelkovogo polka sq., 2, Rostov-on-Don, 344038, Russia,
[3]Don State Technical University, pl. Gagarina, 1, Rostov-on-Don, 344022, Russia

Abstract. In the paper was indicated the essentiality of large-scale underground space development and high-rise construction of cities in Russia. The basic elements of transport facilities construction effective technology without traffic restriction are developed. Unlike the well-known solutions, it offers the inclusion of an advanced lining in the construction that strengthens the soil mass. The fundamental principles of methods for determining stress in advanced support and monitoring of underground construction, providing the application of pressure sensors, strain sensors and displacement sensors are considered.

1 Introduction

The intensive city development in the Russian Federation is characterized by a steady increase in the volumes of cargo and passenger flows, an increase in the number of cars per capita, the density of urban development, growth of high-rise construction, etc. The solution of the transport problems that accompany this process should be based on an integrated approach, which provides the development of a dedicated infrastructure for public passenger transport, city extra-personal transport; closer integration into the system of express extra-modal transport of other modes of transport; construction and reconstruction of out-of-the-way pedestrian crossings; organization of a priority line for urban land passenger transport traffic on highway sections; reduction in the number of junctions; promotion of public transport and related infrastructure with the active development of the underground system, light rail transport, commuter traffic, etc. It is impossible to ensure the fulfilment of these tasks without the construction of tunnel crossings and the organization of a split-level traffic of road and rail vehicles [1,2].

In contrast, the traditional schemes for the construction and reconstruction of tunnel crossings make it necessary to significantly restrict or completely stop traffic on the existing section of the road. This leads to great difficulties and significant economic losses in the organization of traffic for the period of construction or reconstruction of the site. Illustrative examples of this problem are construction works on the construction of underground

* Corresponding author: mixail-stepan@mail.ru

transitions in the city of Rostov-on-Don on the st. Nagibina and st. Sholokhov, carried out in 2014 - 2015. Despite the high pace of work, the construction of crossings required the restriction of traffic for a period of up to 5 - 6 months, which caused the appearance of many kilometers of congestion on the streets under reconstruction.

In this regard, studies aimed at developing an effective technology of the construction of transport facilities without traffic restrictions, are very relevant.

In the foreign and domestic practice, technical and technological solutions for the construction of underground structures (tunnels, overpasses, underground passages, etc.) that minimize the effect on the earth's surface located above the construction sites have found application [3-5]. Indeed, in Germany, Japan, the USA and other countries protective screens are widely used. They are arranged by the technology of advanced lining. When applicating, the following methods are used: small diameter pipe ramming or soil sampling; horizontal drilling; slot dozing up to 3 - 4 m in length and 12 - 20 cm in height and over by means of scabbling or ditch-digging machine; shield driving of small diameter tunnel driving; microtunneling; soil solidification by chemical methods, jet grouting or freezing, etc. Since the late 90-ies of the twentieth century, similar technologies began to be actively introduced in our country. Theoretical and practical aspects of their application are considered by Alikhashkin V.A.Baklashev I.V., Bulychev N.S., Golitsynsky D.M., Garber V.A., Makovsky L.V., Merkin V.E., Frolov Yu.S., Photieva N.N. and many other scientists and specialists [3,6,7]. Their studies, as well as practical experience in the application of methods, showed that for the construction of tunnel crossingss of short length the most convenient and universal is the scheme with application advanced lining of metal pipes [8]. At the same time, the percent of its application today does not exceed 10%, and the main drawbacks of the known solutions are the lack of the possibility of effective monitoring and management of the soil mass deformation processes in the construction process which in some cases led to significant surface damage and destruction of the above-ground buildings and transport structures.

Thus, the main objective of the study is to improve the technology of tunnel crossing without restricting traffic by developing and reasoning for the technical solutions for monitoring and managing the stress-strain state of an underground facility.

2 Results

Effective building geotechnologies for the development of the underground space should be as safe as possible for the city's natural and man-made system. It should be taken into account that new objects of underground transport infrastructure are located, first of all, in the central part of the city, characterized by a high density of construction, the presence of architectural monuments and busy highways, which exclude their overlapping for a long time. In such conditions, the following basic requirements can be brought to the technologies of transport facilities construction:

- elimination of the development of excessive surface subsidence;
- exclusion of negative influence of underground objects on neighboring buildings, structures, roads and other objects of urban infrastructure;
- reduction in the volume of excavation and pumping of groundwater changing the hydrogeological situation;
- application of environmentally friendly materials and technologies for underground construction;
- Increase in technical and economic efficiency of construction [9,10].

These requirements can be met when applying combined traditional and innovative building geotechnologies, the main elements of which are:

- a technique for analyzing the stress-strain state of underground space objects in the process of construction;
- a unified system for automated management of geotechnical, engineering-geological and environmental monitoring during the construction period;
- a system of active support setting of underground structures and fixing the surrounding mass of soils on the basis of anchor systems, lining, high-strength materials and reinforcing agents;
- adaptive technology and the organization of mining operations with changing parameters of a sinking cycle, stages of opening the bottomhole, the construction of the lining;
- complex hydroprotection of underground objects with the application of new composite materials [11,12].

In view of these requirements, the technology of advanced lining using a protective screen made of pipes is realized in the following sequence.

The protective screen is constructed without stopping traffic using a microtunneling complex of the Herrenknecht type. The screen is made of steel pipes 0.8 - 1.0 m in diameter, up to 35 m long in the required quantity (Figure 1). In contrast to the known solutions, it is proposed to incorporate active control elements and anchor systems that strengthen the soil mass in the construction of the leading edge. When constructing an underground crossing, the proposed solution provides for the construction of vertical barrels on both sides of the road in places where people go into the underground space. Then, in one of the trunks, a microtunneling complex of the Herrenknecht type is installed, and in the second guide frame for the leading support.

Firstly, the pipes of the leading support of the usual construction 1 (Figure 1) of the side walls in the direction from the bottom to the top are installed. They are connected to each other with the help of special locks. The two-section pipes with internal elements of active regulation 2 in the form of hydraulic jacks are installed the last. In parallel with the installation of the shield from the pipes, the leading anchor support 3 is installed to harden the soil massif.

Under the protection of the advanced lining, the excavation is carried out under the underground crossing. During the drilling process, complex automated monitoring is organized, the main aspects of which are discussed below. When fixing the surface subsidence, it is compensated by the activation of the elements of the leading support active regulation. If necessary, the dimensions of the bottom holes, the speed of their movement and other technological parameters are corrected.

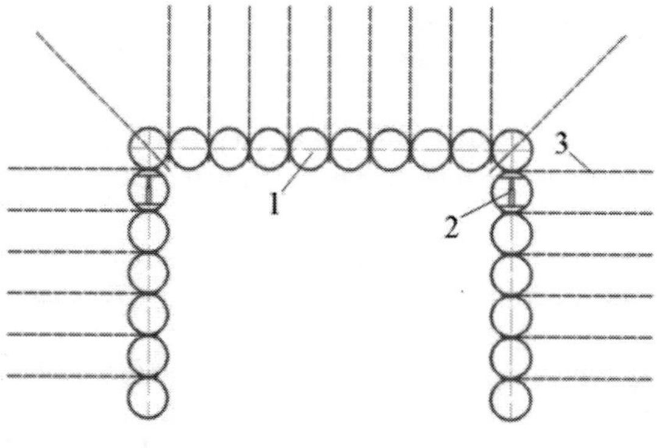

Fig. 1. The layout of the advanced lining elements.

After the sink working, it is possible to mount a membrane waterproofing from composite materials on the surface of the advanced lining and erect a constant lining of the underground crossing. In parallel, work is carried out to dismantle the microtunneling complex and to construct the portal parts of the crossing. At the final stage, decoration of the underground structure and improvement of the surrounding area is performed. The whole complex of works is done without restriction of traffic on the reconstructed part of the street.

In the process of erection of advanced lining and total development of an underground structure, a stress-strain state is monitored.

Based on the analysis of possible accident threats in the construction of an underground crossing, four groups of monitored parameters can be distinguished:

1. Parameters of the stress-strain state (SSS) of the constructions of the advanced lining: tension and deformation.

2. Indirect parameters of SSS lining: the pressure of the ground mass on the pipes of the advanced lining.

3. Technological parameters of the state of the object: mutual displacement of sections, translational displacement and roll of sections from the design position.

4. Environmental parameters: displacement and subsidence of the soil mass and the earth's surface.

The SSS of the advanced lining is determined by the loads of the surrounding soil and the geometrical parameters of the pipes. The design scheme of a pipe under the influence of external pressure is most conveniently represented in a cylindrical coordinate system with the origin at the center of the pipe section. In this case, the components of the stressed state in the section are radial and tangential stresses (Fig. 2).

It is known from the mechanics of a deformable solid body [13] that the radial stress on the outer surface of a cylindrical shell is numerically equal to the external pressure. The radial stress on the inner surface is determined by the internal pressure, and in this case they are equal to zero. Tangential stress in closed shell structures under the influence of external pressure is dominant and exeeds radial stress according to the *R*/*t* formula. Their degree is determined by the compressive forces and bending moments arising in the considered design section:

$$\sigma_\theta^N = \frac{N}{tl}; \quad \sigma_\theta^M = \frac{12My}{t^3 l}, \qquad (1)$$

N - compression strain; *M* - bending moment; *t* - the wall thickness of the pipe; $l = 1$ m - single length of the pipe of the advanced lining; *y* - the coordinate along the thickness of the pipe section with the origin at the center (see Fig. 3).

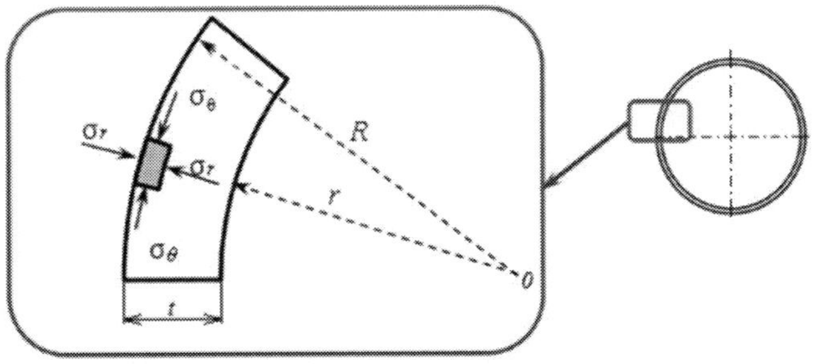

Fig. 2. Stress state in the pipe shell:

σ_r - radial stress, σ_θ - tangential stress, *R* - the outer radius of the pipe; *r* - internal radius of the pipe; *t* - the wall thickness of the pipe.

The flexural stress component reaches its maximum values at the points $y = R$ and $y = r$ (Fig. 3). At these points, the stress from bending is determined by the formula:

$$\sigma^{M}_{\theta\max} = \frac{12M\left(\pm\frac{t}{2}\right)}{t^3 l} \pm \frac{6M}{t^2 l}. \tag{2}$$

The total maximum and minimum tangential stress taking into account the formulas (1) and (2) are:

$$\sigma^{\text{tot}}_{\theta\max/\min} = \frac{1}{tl}\left(N \pm \frac{6M}{t}\right). \tag{3}$$

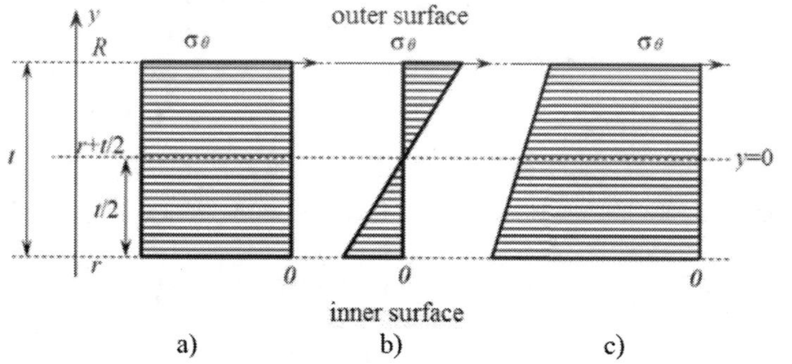

Fig. 3. Diagrams of tangential stress in the lining: a) due to the compressive force; b) due to the bending moment; c) total.

Thus, the points of the control of the parameters of the SSS of the design of the advanced lining can be determined on the basis of constructive calculations for the maximum values of tangential stress in the lining. Full control of the parameters of the lead-in SSS can be made by pressure sensors, strain sensors and displacement sensors, united in a single automated monitoring system.

In Fig. 4 as an example, the typical diagram of the distribution of the maximum tangential stress in the pipes of the advanced lining and the recommended locations for the sensors of the SSS control are shown, where the maximum values of the monitored parameters are most likely to occur.

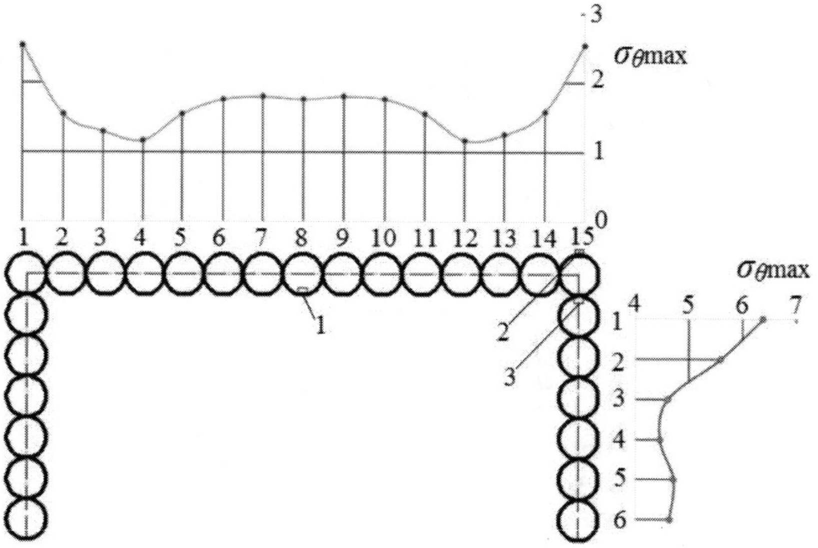

Fig. 4. Diagrams of maximum tangential stress in the advanced lining and recommended locations for the sensors of the SSS control: 1 – displacement sensors; 2 – pressure sensor; 3 – strain sensors.

3 Conclusions

The introduction of the described monitoring system and elements of active regulation ensures the effective control and management of the stressed-deformed state of the advanced lining structures and allows forecasting possible natural and man-made accident threats. This ensures high safety of construction works and creates the necessary prerequisites for large-scale construction of tunnel crossings in conditions of dense transport traffic and construction, as well as in complex engineering-geological conditions.

References

1. V. Korotaev, Gradostroitel'stvo. **6 (40).** (2015).
2. J. Denisova, Vestnik Belgorodskogo gosudarstvennogo tehnologicheskogo universiteta im. V.G. Shuhova. **11.** (2016).
3. R. Dashko, Sovremennye problemy gidrogeologii, inzhenernoj geologii i gidrogeojekologii Evrazii. Materialy Vserossijskoj konferencii s mezhdunarodnym uchastiem cjelementami nauchnoj shkoly. Nacional'nyj issledovatel'skij Tomskij politehnicheskij universitet (2015).
4. A/ Parriaux, The DEEP CITY Project: A Global Concept for a Sustainable Urban Underground Management. 11th ACUUS International Conference, Underground Space Expanding the Frontiers. 10-13 september 2007 Athens Greece. (2007).
5. Li, H., Parriaux, A. et al., The way to plan a viable Deep City: from economic and institutional aspects. The Joint HKIE-HKIP Conference on Planning and Development of Underground Space. The Hong Kong Institution of Engineers & The Hong Kong Institution of Planners. (2011).
6. Chang-Yu, Ou.: *Deep excavation theory and practice*. (London, Taylor & Francis, 2006).
7. V. Ilyichev, et al., V sbornike: Rossijskaja arhitekturno-stroitel'naja jenciklopedija. (2008).
8. B. Kartozija, Gornyj informacionno-analiticheskij bjulleten' (nauchno-tehnicheskij zhurnal). **1.** (2015).
9. E. Kulikova, Gornyj informacionno-analiticheskij bjulleten' (nauchno-tehnicheskij zhurnal). **6.** (2014).
10. S. Kaloshina, Vestnik Permskogo nacional'nogo issledovatel'skogo politehnicheskogo universiteta. Stroitel'stvo i arhitektura. **7(3).** (2016).
11. M. Bhatt, Journal of Siberian Federal University. Humanities & Social Sciences. **2(6).** (2013).
12. O. Semjonova *Gradostroitel'stvo* **3(31).** (2014).
13. A. Prokopov, Web of Conferences. **106**, 02001. (2017).
14. M. Rossinskaya, Life Science Journal 11 (SPEC. ISSUE 11). **43.** (2014).
15. M. Pleshko, ARPN Journal of Engineering and Applied Sciences. **10(1).** (2015).
16. M. Molev, ARPN Journal of Engineer and Applied Sciences. **10,** no 16. (2015).

Calculation of critical heat transfer in horizontal evaporator pipes in cooling systems of high-rise buildings

Andrey Aksenov[1], *Anna* Malysheva[1*]

[1]Moscow State University of Civil Engineering, Yaroslavskoe shosse, 26, Moscow, 129337, Russia

Abstract. An exact calculation of the heat exchange of evaporative surfaces is possible only if the physical processes of hydrodynamics of two-phase flows are considered in detail. Especially this task is relevant for the design of refrigeration supply systems for high-rise buildings, where powerful refrigeration equipment and branched networks of refrigerants are used. On the basis of experimental studies and developed mathematical model of asymmetric dispersed-annular flow of steam-water flow in horizontal steam-generating pipes, a calculation formula has been obtained for determining the boundaries of the zone of improved heat transfer and the critical value of the heat flux density. A new theoretical approach to the solution of the problem of the flow structure of a two-phase flow is proposed. The applied method of dissipative characteristics of a two-phase flow in pipes and the principle of a minimum rate of entropy increase in stabilized flows made it possible to obtain formulas that directly reflect the influence of the viscous characteristics of the gas and liquid media on their distribution in the flow. The study showed a significant effect of gravitational forces on the nature of the phase distribution in the cross section of the evaporative tubes. At a mass velocity of a two-phase flow less than 700 kg / m^{2}s, the volume content of the liquid phase near the upper outer generating lines of the tube is almost an order of magnitude lower than the lower one. The calculation of the heat transfer crisis in horizontal evaporative tubes is obtained. The calculated dependence is in good agreement with the experimental data of the author and a number of foreign researchers. The formula generalizes the experimental data for pipes with the diameter of 6-40 mm in the pressure of 2-7 MPa.

1 Introduction

Design and calculation of heat exchange surfaces with their different orientations in space, including horizontal ones, is of wide use in steam generators of thermal installations in various branches of industrial heat power engineering. Today the development of reliable and economical devices with boiling heat-transfer agents is one of the most important tasks in heat and power engineering, the successful development of which largely depends on the

* Corresponding author mani495@mail.ru

knowledge of the processes of hydrodynamics and heat transfer in two-phase working media, as well as issues related to the intensification of heat exchange and the prevention of crisis phenomena. In the present article, in order to simplify the analysis, the annular structure of a two-phase flow is the design model. The solution is based on the principle of a minimum of energy dissipation in a stabilized flow associated with irreversible losses by internal friction during the movement of real media. From the thermodynamic point of view, in the flow of a viscous liquid degradation occurs (energy depreciation), as in all other real processes taking place in nature, the characteristic feature of which is the increase in the entropy of the system. Internal nonequilibrium processes always act in a direction that causes a decrease in the rate of entropy increase. In the thermodynamics of irreversible processes, this law is formulated as follows: as the system goes over to the stationary-nonequilibrium state, the value of the incremental increase in entropy decreases, and when the stationary-nonequilibrium state is reached, it is of the smallest value compatible with external constraints [1, 2].

2 Methods

Let us consider the annular regime of a stabilized gas-liquid flow in a pipe with radius of R_0, shown in the Fig. 1. This flow will consist of a ring layer of liquid with thickness δ and a cylindrical gas core with radius $R_G = R_0 - \delta$. The mass flow of fluid G_l and gas G_g and their physical parameters (density, viscosity) $\rho_g \rho_l \mu_g \mu_l$ are assumed to be known [3,4].

In horizontal and low-inclined pipes, the lifting force (Archimedova) exerts an essential effect on the steam core:

$$P_A = \pi R_r^2 (\rho' - \rho'') g \Delta l \tag{1}$$

where R_r - radius of vapor nucleus, m; ρ' и ρ'' - liquid and gas density, kg/m³; g - projection of the acceleration of gravity on the plane of the pipe cross-section, m/s²; Δl - length of the pipe element, m.

Under this force, the co-alignment of the vapor nucleus and the annular layer of the liquid is disrupted. The thickness of the liquid layer δ becomes not uniform along the perimeter (Fig. 1). At the top of the pipe, the liquid layer becomes thinner, in the lower one, the liquid layer thickens. The vertical displacement of the flow nucleus will be accompanied by an increase in the energy dissipation intensity associated with increasing friction losses.

It is obvious that a certain amount of displacement of the steam nucleus *h* corresponding to a minimum of the second change in the energy of the system for given boundary conditions will correspond to the stabilized flow.

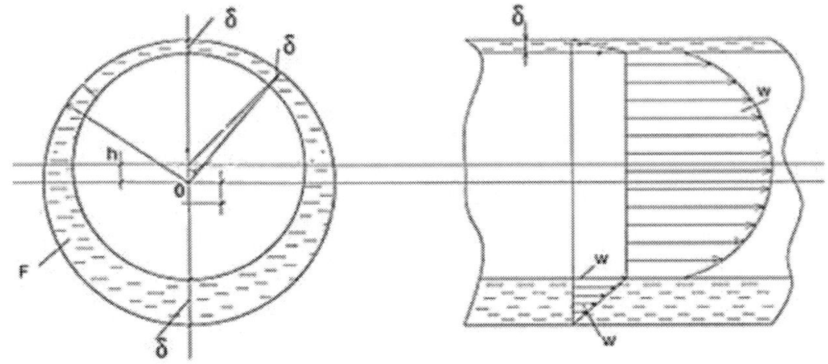

Fig. 1. The model of horizontal asymmetric annular flow of steam-water flow

We start from the assumption that the total specific work in each section of the stabilized vapor-liquid flow has a minimum [2]:

$$\sum e_i(h) = \min \quad (2)$$

where $e_i = \frac{E}{V}$;

E - secondary work of internal friction forces, W; V - the volume of the pipe Δl.
Energy flows, independent of h, are not introduced here.

Let us consider the problem of finding the magnitude of the displacement of the vapor nucleus, as well as the minimum thickness of the liquid film δ^{B}_{cr}, and of corresponding boundary heat flow density q_{cr} (W/m^2), at which a sharp deterioration of heat transfer in the region of the upper generator of the evaporation tube is observed [3].

We assume that when the vapor bucleus is displaced, its cross section is deformed insignificantly, and the absolute velocity ω_{oL} at the boundary of the vapor nucleus and the liquid layer is unchanged. In this case, the second energy loss by friction in the liquid layer$V_L = F'\Delta l$ can be expressed by the following integral.
where τ_L - frictional stress in a liquid layer, N/m^2; $\tau_L = \mu_L \cdot \frac{dw_L}{dR}$; F' - sectional area occupied by a liquid; w_L- local fluid velocity in the section under consideration, m/s.

$$dF = \Delta l R d\beta \quad (3)$$

Assuming the law of velocity distribution $w_L(R)$ in the liquid layer to be linear, we can rewrite the integral (3) in the following form

$$E_\tau = 2\Delta l \int_0^\pi d\beta \int_{R_r}^{R_r+\delta} \mu_L \frac{w_{oL}^2}{\delta^2} R dR, \quad (4)$$

where δ - thickness of the liquid layer.
Integrating over the variable R, we have:

$$E_\tau = \Delta l \int_0^\pi \frac{\mu_L w_{oL}^2}{\delta} (2R_r + \delta) d\beta. \quad (5)$$

Given that $R_r = \sqrt{\varphi} \cdot R_o$,

we receive

$$E_\tau = \Delta l \mu_L w_{oL}^2 \left(\pi + 2\sqrt{\varphi} \cdot R_0 \int_0^\pi \frac{d\beta}{\delta}\right). \quad (6)$$

According to the Fig. 1

$$\delta = \sqrt{R_0^2 + h^2 - 2R_0 h \cos\beta} - R_r, \quad (7)$$

where h - displacement of the center of the vapor nucleus relative to the axis of the pipe. With a relatively small thickness of the liquid layer, the function $\delta = f(\beta)$ can be represented as the following formal power series:

$$\frac{1}{\delta} = \text{в}_0 + \text{в}_1\beta + \text{в}_2\beta^2 + \text{в}_3\beta^3 + \cdots. \quad (8)$$

Restricting ourselves to the first four terms of the written expansion of the function $1/\delta$ in powers of β, we find the coefficients of the polynomial$\text{в}_0, \text{в}_1, \text{в}_2$ and в_3 from the following boundary conditions:

1) $\beta = 0 \quad 1/\delta = \frac{1}{\alpha - h}$;

2) $\beta = \pi \quad 1/\delta = \frac{1}{\alpha + h}$;

3; 4) $\begin{cases} \beta = 0 \\ \beta = \pi \end{cases} \frac{\partial(\frac{1}{\delta})}{\partial\beta} = 0$;

where $\alpha = R_0 - R_r = R_0(1 - \sqrt{\varphi})$.
From the first boundary condition

$$\text{B}_0 = \frac{1}{\alpha - h}. \tag{9}$$

From the second boundary condition

$$\pi \text{B}_1 + \pi^2 \text{B}_2 + \pi^3 \text{B}_3 = -\frac{2h}{\alpha^2 - h^2}. \tag{10}$$

From the third boundary condition $\text{B}_1 = 0$. (11)

From the fourth boundary condition

$$\text{B}_1 + 2\pi b_2 + 3\pi^3 \text{B}_3 = 0. \tag{12}$$

Solving jointly (10), (11), and (12), we obtain

$$\text{B}_2 = \frac{6h}{\pi^2 \cdot (\alpha^2 - h^2)}; \tag{13}$$

$$\text{B}_3 = \frac{4h}{\pi^3 \cdot (\alpha^2 - h^2)}. \tag{14}$$

Thus, the desired dependence can be represented in the following form:

$$\frac{1}{\delta} = \frac{1}{\alpha - h} - \frac{6h\beta^2}{\pi^2(\alpha^2 - h^2)} + \frac{4h\beta^3}{\pi^3(\alpha^2 - h^2)}. \tag{15}$$

Evaluate the integral (6):

$$\int_0^\pi \frac{d\beta}{\delta} = \int_0^\pi \left[\frac{1}{\alpha - h} - \frac{6h\beta^2}{\pi^2(\alpha^2 - h^2)} + \frac{4h\beta^3}{\pi^3(\alpha^2 - h^2)}\right] d\beta = \frac{\pi}{\alpha - h} - \frac{2\pi h}{\alpha^2 - h^2} + \frac{\pi h}{\alpha^2 - h^2} = \frac{\pi\alpha}{\alpha^2 - h^2}. \tag{16}$$

Taking into account (16), the expression of the second frictional losses (6) takes the following form:

$$\text{E}_\tau = \pi \Delta l \mu_L \omega_{0L}^2 \left(1 + \frac{2\sqrt{\varphi} \cdot \text{R}_0 \alpha}{\alpha^2 - h^2}\right). \tag{17}$$

Consider a second change in the energy of the selected element of the flow in the region Δl, which will be determined by the work spent against the gravitational forces on the displacement of the liquid M in the transverse direction relative to the channel:

$$\text{E}_\pi = \text{M}g(y'_c - y''_c), \tag{18}$$

where g - projection of the acceleration of gravity on the plane of the pipe cross-section, m/s^2 (for horizontal pipe $g = 9{,}81$ m/s^2); y''_c- the ordinate of the center of gravity of the vapor nuclear.
According to the earlier assumptions $y''_c = h$. (19)
The ordinate y'_c of the center of gravity of the cross section of the liquid layer when the steam nucleus is displaced by an amount h, is determined from the equation of moments of static equilibrium with respect to the horizontal axis of the section of the pipe 0-0 (Fig. 1):

$$\rho' F' g y'_c = g(F\rho' \cdot 0 - hF''\rho'); \tag{20}$$

As $F' = F(1 - \varphi)$ and $F'' = \varphi F$, then

$$F(1 - \varphi)\rho' g y'_c = -gh\varphi F\rho'. \tag{21}$$

Hence $y'_c = -\frac{\varphi h}{1 - \varphi}$. (22)

Substituting (19) and (22) into (18), we obtain

$$\text{E}_\text{п} = -\text{M}g\frac{h}{1 - \varphi}. \tag{23}$$

For the selected element with the length of Δl the second-long transverse transport of the mass M will be composed of the following flows [4]:

1) the mass of the liquid, changing into a vapor state in the amount of ΔD, kg/s

$$\Delta D = \frac{2\pi R_0 \Delta l q}{r}, \tag{24}$$

where q - heat flow density, W/m²; r - heat of vaporization, J / kg.

2) the transverse displacement of the liquid during the evaporation process, which is proportional to ΔD:

$$M_1 = C_1 \Delta D; \tag{25}$$

3) liquid particles in the result of their mechanical detachment from the inner surface of the liquid layer and entrainment into the vapor nucleus [5, 6, 7].

We assume that the flow density of dispersed particles from the surface of the liquid layer is proportional to the average mass flow rate of the liquid phase:

$$M_2 = C_2(\rho w) \cdot (1 - x) \cdot \Delta F. \tag{26}$$

In this case the coefficient of proportionality C_2 is a function of the Weber number

$$C_2 = C_2' \cdot W_e = \frac{C_2' \rho''(\overline{w''} - \overline{w'})^2 2R_0}{\sigma'}, \tag{27}$$

where $\overline{w''}$ и $\overline{w'}$ - average cross-sectional velocities of the vapor and liquid phases in the flow; σ - surface tension coefficient, N/m. As

$$\overline{w''} F'' \rho'' = (\rho w) x F_0; \tag{28}$$

$$\overline{w'} F' \rho' = (\rho w)(1 - x) F_0, \tag{29}$$

where F' и F'' - the cross-sectional area of the pipe, occupied by liquid and steam, м^2; ρw - mass flow rate, kg/m^2 · s, then

$$\overline{w''} = \frac{(\rho w) \cdot x}{\varphi \rho''}; \tag{30}$$

$$\overline{w'} = \frac{(\rho w) \cdot (1-x)}{(1-\varphi)\rho'}. \tag{31}$$

Thus, we obtain an expression for the Weber number

$$W_e = \frac{2R_0 \rho''(\rho w)^2}{\sigma'} \cdot \left[\frac{x}{\varphi \rho''} - \frac{(1-x)}{(1-\varphi)\rho'}\right]^2. \tag{32}$$

Taking into account (24), (25) and (26) the total mass flow be:

$$M = \Delta D + C_1 \cdot \Delta D + (\rho w)(1 - x) \cdot 2\pi R_0 \Delta l \left[\frac{Cq}{r} + C_2(\rho w)(1 - x)\right]. \tag{33}$$

where ΔD - mass of a liquid passing into a vapor state;

$$C = C_1 + 1.$$

The specific work against gravitational forces is as follows:

$$e_n = \frac{E_n}{V} = -\frac{2Cgh}{R_0 \cdot (1-\varphi)} \left[\frac{q}{r} + C_3 (\rho w)(1-x)\right], \tag{34}$$

where $C_3 = \frac{c_2}{c}$.

Specific work of internal friction forces:

$$e_\tau = \frac{E_\tau}{V} - \frac{\mu' w_0^2}{R_0^2}\left(1 + \frac{2\sqrt{\varphi}R_0\alpha}{a^2 - h^2}\right) \quad (35)$$

We find the value of h from the condition for the transition of the system to a stationary state. Moreover, the latter will be achieved when $\sum e(h) = \min$ [8]

$$\frac{\partial}{\partial h}(e_\tau + e_n) = 0 \quad (36)$$

Substituting expression (34) and (35) into equation (36), after differentiation, we obtain:

$$\frac{2\cdot\mu' w_0^2\sqrt{\varphi}\alpha h}{(a^2-h^2)^2} - \frac{Cg}{1-\varphi}\left[\frac{q}{r} + C_3(\rho w) - (1-x)\right] = 0 \quad (37)$$

Then

$$\frac{t_h}{(1-t_h^2)^2} = \frac{C\alpha^2 g\cdot\left[\frac{q}{r} + C_3\cdot(\rho w)\cdot(1-x)\right]}{2\mu' w_0^2\sqrt{\varphi}\cdot(1-\varphi)}, \quad (38)$$

where $t_h = \frac{h}{\alpha} = 1 - \frac{\delta^{\text{в}}}{\alpha}$ – relative displacement of the vapor nucleus.

As

$$w_0' = 2\overline{w'} = \frac{2(\rho w)(1-x)}{(1-\varphi)\rho'}, \quad (39)$$

expression (38) can be represented in the following form:

$$\frac{t_h}{(1-t_h^2)^2} = \frac{CgR_0^2\cdot(1-\sqrt{\varphi})^3\cdot(1+\sqrt{\varphi})\rho'^2\left[\frac{q}{r} + C_3\cdot(\rho w)\cdot(1-x)\right]}{8\mu'(1-x)^2(\rho w)^2\sqrt{\varphi}}. \quad (40)$$

3 Results

Based on the analysis of experimental studies of the structure of the horizontal steam-water flow by the method of "electrosounding" and generalization of the author's experimental data in the range P = 2-7MPa; ρw = 100 ÷ 1500 $\frac{\text{kg}}{\text{m}^2}\cdot s$; x> 0.2; 0.012m≤$d_0$≤0.019 m, it was obtained an expression for the reduced thickness of the liquid film at the upper pipe generator at the boundary of the zone of deteriorated heat transfer.

$$\delta_{cp}^{\text{в}} = \sqrt{\frac{k(\rho w)\mu'}{g(\rho' - \rho'')\rho'}}; \quad (41)$$

The validity of expression (41) is confirmed by experimental measurements of the effective thickness of the liquid film at the upper pipe generator at the boundary of the onset of thermal pulsations. Moreover, it is established that for smooth pipes K_α= 1,1.

From the expression (40) we obtain following expression for the boundary density of the heat flow:

$$\frac{q_{cr}}{r} = C_1\left[\frac{T}{B} - CW_e(\rho w)(1-x)\right], \quad (42)$$

$$T = \frac{t_h}{(1-t_h^2)^2} = \frac{1-\delta^{\text{в}}_{cr}/\alpha}{[1-(1-\delta^{\text{в}}_{cr}/\alpha)^2]^2}; \quad (43)$$

$$\mathrm{B} = \frac{gR_0^2(1-\sqrt{\varphi})^3(1+\sqrt{\varphi})\rho'^2}{8\mu''\cdot(1-x)^2\cdot(\rho w)^2\sqrt{\varphi}}. \quad (44)$$

Substituting the values of the numerical coefficients found from the experiments, we finally obtain the critical density of the heat flow:

$$q_{гр} = r\left[\frac{T}{B} - W_e(\rho w)(1-x)\cdot 10^{-3}\right]\cdot 1.6\cdot 10^{-4} \quad (45)$$

,

where $T = \dfrac{1-\delta^{\text{в}}_{\text{гр}}/\alpha}{\left[1-(1-\delta^{\text{в}}_{\text{гр}}/a)^2\right]^2}$; $B = \dfrac{gR_0^2(1-\sqrt{\varphi})^3\left(1+\sqrt{\varphi}\right)\rho'^2}{8\mu''\cdot(1-x)^2\cdot(\rho w)^2\sqrt{\varphi}}$.

In the Fig. 2 the dependences of the boundary heat load on the mass gas content according to the formula (45) are given. In the same place, the experimental data obtained by the author and also by other researchers are plotted [9, 10, 11].

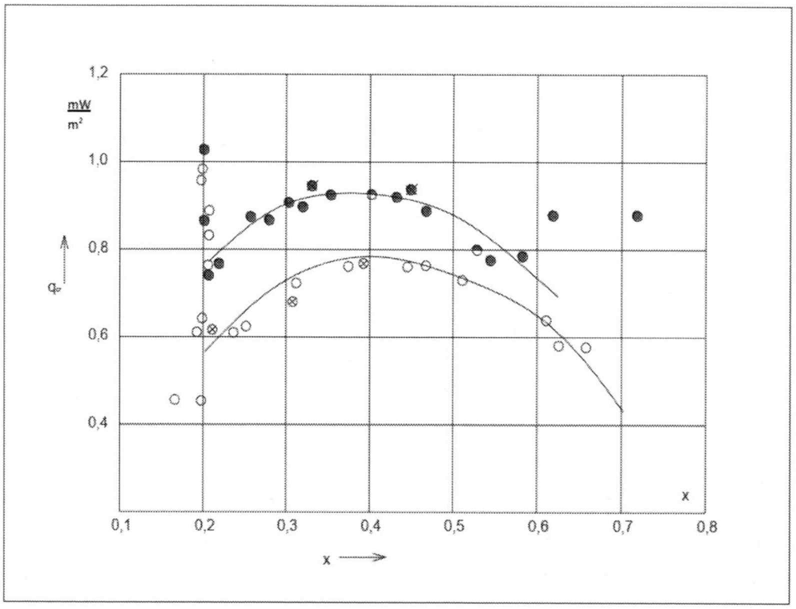

Fig. 2. Comparison of experimental data with calculation of q_{cr} for a horizontal evaporator pipe (d_0 = 0.017 m, P = 5 MPa)

4 Conclusion

The formula generalizes the experimental data obtained on steam-vaporizing pipes with an internal diameter $d_o = 0{,}012 \div 0{,}019$ in the pressure range 3-7 MPa, the range of mass flow velocities 300-1400 kg/m^2 ·s for relative enthalpies of the flow x > 0.2 with an accuracy ± 20% [12].

In the range of relative enthalpies 0.2 <x <0.3, in the experimental values q_{cr} an appreciable spread takes place, and in the range x = 0.2, the effect of the heat flow on the boundary of the zone of deteriorated heat transfer practically degenerates. The character of the dependence $q_{cr} = f(x)$ is analogous to a crisis of the second kind in vertical pipes (the experimental values obtained under these conditions in the q, x coordinates are grouped around almost vertical lines) [14, 15]. And in the normative method, it is proposed to take the degraded heat transfer zone outside the range of medium pressures for horizontal evaporative tubes at x = 0.3 for P <5.0 MPa for any q.

In this case, the dispersed-annular flow regime is only formed and has an appreciable hydrodynamic instability. At low mass velocities $\rho w < 700$ kg/m^2 ·s, the vapor clusters moving along the upper generator periodically cause "dry spots" on the channel wall, which in turn leads to the thermal pulsations of the wall near the upper generator of the pipe.

With decreasing q to a certain value $q_{\min}$, the dependence $q_{cr} = f(x)$ sharply deflects to the right and the boundary density of the heat flow q_{cr} even slightly increases. It is explained by the fact that the steam-water flow rate increases as the steam flow rate increases, which in turn reduces the asymmetry of the flow and increases the velocity of liquid droplets in the flow nucleas (i.e., the irrigation conditions of the channel wall improve) [13].

In the range of relative enthalpies 0.2 <x <0.3, the boundary heat flow density q_{cr} calculated by formula (45) corresponds to the minimum experimental values of q when the wall thermal impulses on the upper generator of the horizontal steam generating pipe were observed. A further increase in x greater than 0.4 is associated with a decrease in the thickness of the wall-mounted liquid film along the entire perimeter of the channel. At the same time, the liquid droplets in the nucleus of the stream decrease, which significantly reduces the irrigation intensity of the channel wall. It leads to a monotonic decrease q_{cr}.

In some regimes it was possible to observe simultaneously two separate zones of thermal pulsations along the length of the channel (Fig. 3): one in case of small values of x of 0.2 order and the other near the outlet section of the pipe. A similar phenomenon was also noted in [14, 15]. Two separate zones of thermal pulsations on the upper generator of a horizontal pipe can be explained on the basis of the combined influence of many factors on the nature of the dependence $q_{cr} = f(x)$, which is in complete agreement with the above physical model of the horizontal steam-water flow.

The dependence q_{cr} on the mass-velocity value for horizontal evaporative tubes is shown in the Fig. 2. It should be noted that obtained dependence has an inversion. An increase of ρw up to 1000-1500 kg/m^2 ·s leads to an increase in q_{cr}.. In the range of 1000-1500 kg/m^2 ·s an increase in the mass velocity value ofq_{cr} decreases. Moreover, at large vapor contents of the flow, the inversion point corresponds to smaller mass velocities. The observed phenomenon confirms the previously stated position that at high velocities the hydrodynamic processes in horizontal and vertical pipes are similar.

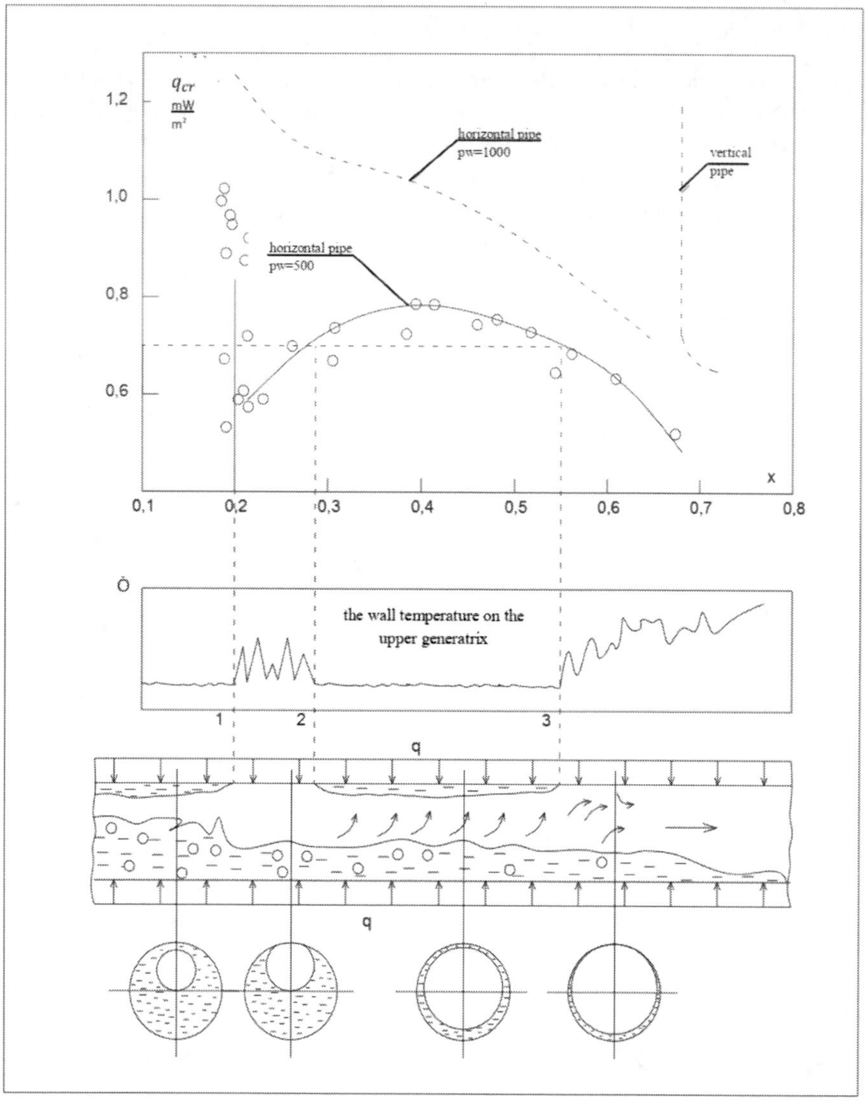

Fig. 3. Zones of deteriorated heat transfer in a horizontal evaporative tube

References

1. I.R. Prigozhin. Introduction to the thermodynamics of irreversible processes. Izhevsk, 160 (2001).
2. I.R. Prigozhin, D. Kondepudi. Modern thermodynamics. From heat engines to dissipative structures. Moscow, 44 (2002).
3. A.K. Aksenov, V.A. Bryantsev, and A.S. Selivanov. Temperature conditions of horizontal pipes during flow of steam-water flow in them. Moscow, 79 (1982). per. №1725, C.79.
4. B.Ya. kamenetskiy. Teploenergetica, **8**, 58-61 (2009).

5. B.I. Nigmatullin, V.I. Rachkov, and Yu.Z. Shugayev. Teploenergetica, **4**, 33-36 (1981).
6. E.A. Boltenko. Teploenergetica, **3**, 38-45 (2009).
7. ИA.A. Ivashkevich. Thermal physics of high temperatures, **1**, 112-117 (2012).
8. G. Nikolis. Cognition of the complex things. Introduction. Moscow, 345 (2012).
A. Klaszak. Heat and mass transfer, **37**, 443-448 (2001).
9. V. Zhylia, A. Malysheva, M. Botnar, and Ananyev A. MATEC Web of Conferences, 86 (2016).
10. O. Brukhanov, A. Rymarov, A. Malysheva, and D. Titkov. MATEC Web of Conferences, 86 (2016).
11. A.A. Malyshev, K.V. Kisser, and A.V. Zaytsev. Vestnik Mezhdunarodnoi akademii kholoda, **2**, (2017).
12. M. Goto, N. Inoue, and N. Ishiwatary. International Journal of Refrigeration, **24**, 626-638 (2001).
13. V. Soldo, T. Curko, and M. Grozdek. Transfer from Carnot to Lorenz process using zeotropic mixtures. In: IIR/IIF (Stockholm, 2002), 63-66 (2002).
14. R.F. Kelbaliev, M.Z. Iskenderov. Thermal physics of high temperatures, **43**, 456-465 (2005).

Temperature and velocity conditions of air flow in vertical channel of hinged ventilated facade of a multistory building.

Elena Statsenko[1,*], *Anastasia* Ostrovaia[1] and Andrey Pigurin[2]

[1]Moscow State University of Civil Engineering, Russia, 129337, Moscow, Yaroslavskoye shosse, 26
[2]Peter the Great St.Petersburg Polytechnic University, Polytechnicheskaya 29, 195251, St.Petersburg, Russia

Abstract. This article considers the influence of the building's tallness and the presence of mounting grooved lines on the parameters of heat transfer in the gap of a hinged ventilated facade. A numerical description of the processes occurring in a heat-gravitational flow is given. The average velocity and temperature of the heat-gravitational flow of a structure with open and sealed rusts are determined with unchanged geometric parameters of the gap. The dependence of the parameters influencing the thermomechanical characteristics of the enclosing structure is derived depending on the internal parameters of the system. Physical modeling of real multistory structures is performed by projecting actual parameters onto a reduced laboratory model (scaling).

1 Introduction

High thermal properties are considered one of the main reasons for using hinged ventilated facades in modern high-rise construction. The principle of this system is that there are heat-gravitational currents between thermal insulation and facing layer in the ventilated gap. They appear as a consequence of the action of the volume force caused by the difference in densities. There are three main types of air filtration in the designs of external fences: transverse, longitudinal and internal. All three types are possible in exterior walls with ventilated facades. The protection from air filtration is especially relevance in construction of high-rise buildings. Filtration for such buildings can wipe-out all the heat-protective characteristics of the exterior walls.

Ideally, the natural flow of air in the gap allows the wall to be permanently in a dry state, preventing the formation of condensation in the insulator layer. But the design features of each individual enclosing structure, including its technological grooved lines, affect the air velocity in the vertical channel, which also affects the temperature and humidity conditions of the enclosing structures. The article considers the effect of grooved lines and the tallness of a building on the velocity and temperature of air in a vertical ventilated gap.

* Corresponding author: staclena@mail.ru

2 Literature review

At the present times hanging ventilated facades building is relevant to many Russian and foreign scientists. A major contribution to study of free-convective current in vertical ventilated channel was made by [1-30].

The work [1] studies physical processes of free convective current and determines the conditions of cool air filtration in the gap.

The article [8] determines best hydraulic ventilated cavity of a suspended facade. The author of the article [10] estimates the average velocity of free-convective current dependence with different wall temperature.

The author of the work [11] gives the description of air-vent quarter division with different air motion modes. The work [15] determines experimentally and theoretically the average velocity and temperature profiles along the ventilated channel width.

The work [27] provides evaluation methods of thermal insulation with longitudinal air filtration of ventilated facade.

3 Goal Setting

- determining of optimal parameters for heat-mass exchange of ventilated air gap of a high-rise building
- determining the average velocity and temperature of heat-gravitational flow in air gap with open and sealed grooved lines with constant geometric parameters of the gap

4 Methods and Results

Modern high-rise construction impose strict requirements on energy efficiency of the buildings in construction. Consequently, usage of cladding structures with ventilated air cavities in the process of multi-store construction is a common practice.

The main difference between fronts of the building with ventilated air spaces and popular claddings with air cavities is that an air space has metal substructures, a thick heat-insulating layer and a facing layer that defines an architectural look of a building. While the layers with ventilated air cavities are used in low-rise buildings, the fronts with ventilated air spaces are applied in high-rise structures dozens meters high.

Primary advantages of curtain walling with ventilated air cavities are the following:
- Better heat-insulation. Consequently, this technology can help significantly decrease bills on heating system for residents of multi-store houses that have quite a broad front surface.
- Reduction of heat leakage. The system of ventilated curtain walls allows to choose an appropriate type and thickness of the insulation according to climate conditions of the region where the high-rise building is situated.
- Normalization of humidity condition of the cladding structure. This technology protects the insulation from layering and soakage what improves thermal and technical performance of a building. These advantages are provided with air flow in the air cavity. It is performed with wind and gravitational pressure. Non-saturated with moisture vapour air fills the air cavity. Circulating in the air cavity this air absorbs moisture gradually saturating with it. Moisture phase state changes depending on air temperature and cavity surfaces, as well as the mass of the incoming vapor. The increase of the partial vapor pressure changes the degree of air moisture saturation that goes up to the maximum level of saturation. Particular air humidity may cause condensate formation on the cold surface of the air cavity.

Collecting of geometric data for the air cavity such as the width of the air space and availability or absence of grooved lines provides the circulation of air in it what will not cause condensate formation on the cold surface of the air cavity. This process is also directly affected by the air velocity in the ventilated air space. The following formula for the average velocity of the thermal and gravitational flow is valid [15]:

$$v = \varphi \frac{gL}{\sqrt{RT_c}} \sqrt{\frac{1}{n} - \frac{1}{k}} \quad (1)$$

where R - individual gas constant. For air, approximately, R= 287 $m^2/(c^2K)$. In connection with simplifications made:

$$v = \varphi \frac{gL}{\sqrt{RT_c}} \sqrt{\frac{k-1}{nk}} \approx \frac{\varphi}{2} \frac{gL}{\sqrt{RT_c}} \quad (2)$$

In nondimensional form the last formula may be written in similarity criteria:

$$\frac{v}{\sqrt{gL}} = \frac{\varphi}{2} \frac{gL}{\sqrt{RT_c}} \quad (3)$$

Indeed, Froud number for measuring thermal and gravitational flow:

$$\frac{v}{\sqrt{gL}} = Fr \quad (4)$$

Mach number measuring relation of Langrangian velocity $\sqrt{gL}$ to sound velocity in cold air,$\sqrt{RT_c}$

$$\sqrt{\frac{gL}{RT_c}} = Ba \quad (5)$$

As a result:

$$\frac{Fa}{Ba} \varphi \sqrt{\frac{k-1}{k}} = \frac{\varphi}{2} \quad (6)$$

$$Fr = \varphi \sqrt{2(1 - \frac{T_c}{T_h})} \quad (7)$$

Application of similarity variables is useful for modeling of building structures on the laboratory units. We executed several laboratory tests on the stylized scheme of the ventilated air space between hot surface y = 0 (under the temperature of Th=67°C) and cold surface y = h (under the temperature of Tc=22°C). The pressure at the mark of z=0 is p_0, the pressure at the mark of z=h is p_1, provided that p_0>p_1. The average velocity and the temperature in the ventilated air space are required to be assessed.

The necessary control volume covers the air channel and extends from below starting at the zero mark and up to the surface z=L that is situated just over the output cutoff of the channel in fixed air. External air must income into the air cavity through lower weep holes. External air goes up along the ventilated air channel and goes out through the upper weep holes. [1]

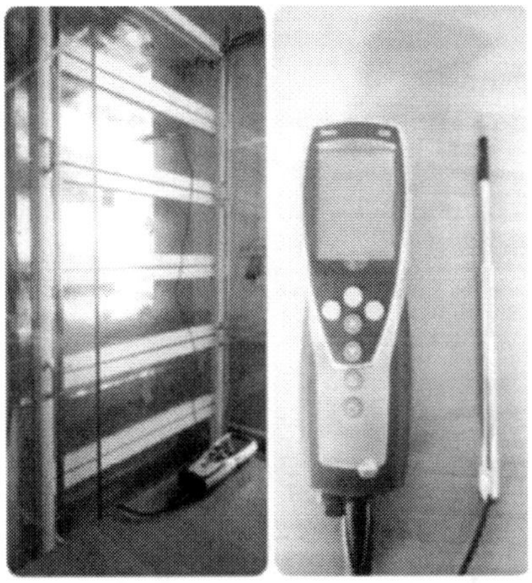

Fig. 1. Installation scheme and measuring instrument

Heating elements fixed along the height of the unit provide the heat rise. In order to distribute the heat equally, these elements are fixed to a tin sheet that has a high thermal conductivity. Consequently, Th=const. Height of the model is of L=204 cm, provided that L>>h.

When modelling a front of the building, we must take into account the temperature difference of internal and external layers explained by functioning of the heating system. Heating of the wall and heat supply from it into the air cavity are taken into account on the model. The velocity depends on the way of air supply into the unit, internal parameters of the cavity and the way of heating the hot surface. The width of the air cavity plays a significant role when building ventilated fronts on real high-rise structures. Usually the air cavity amounts to the width of 60 mm up to 150 mm. The optimal width of air cavity is defined in connection with:

- type of a facing layer;
- potential wind load;
- structure altitude.

In order to conduct a physical modeling of real multi-store structures, the actual parameters were projected onto a small laboratory model (scaling). The observed models of buildings have the following thickness of ventilated thickness ($h_{\text{of the building}}$) - 8 cm, 10 cm, 12 cm, 15 cm, 20 cm, as well as the following altitude ($L_{\text{of the building}}$):

- 5 floors ≈ 15 m.
- 9 floors ≈ 30 m.
- 22-25 floors ≈ 80 m.
- Over 30 floors ≈ 100 m.

The altitude of the laboratory unit is Lm= 2.04 meters, as a result, the thickness of the ventilated air space is (h_m) is in conversion. Then the value h_m is the following Table 1:

Table 1. The value h_m.

Building height, m	Thickness of the ventilated air space, m				
	0.08	0,1	0,12	0,15	0,2
15	0.0112	0,014	0,0168	0,021	0,028
30	0.0056	0,007	0,0084	0,0105	0,014
80	0.0021	0,003	0,0032	0,0039	0,005
100	0.0017	0,002	0,0025	0,0031	0,004

Maximum heat-shielding properties of the front design may be achieved with as low as practicable (according to the conditions of the moisture removal or for any other reasons) air space size due to the fact that the thinner the space, the faster it heats up. This is attributed to the height reduction of the temperature in the space and, as a consequence, with the increase of the average air temperature. In the executed tests the width of the space h is set to h=0,08 m=const. Radiant heat exchange occurs between the surfaces of the facing and thermal insulation, with a coefficient of radiant heat exchange depending on the temperature. Convective heat exchange occurs between the air in the space and the elements of the structure, the coefficients of convective heat exchange depend on the velocity of the air flow and air temperature and structural elements.

The velocity and temperature of the air in the space were determined with a uniform heating of the hot surface height along – Fig.2, Fig.3.

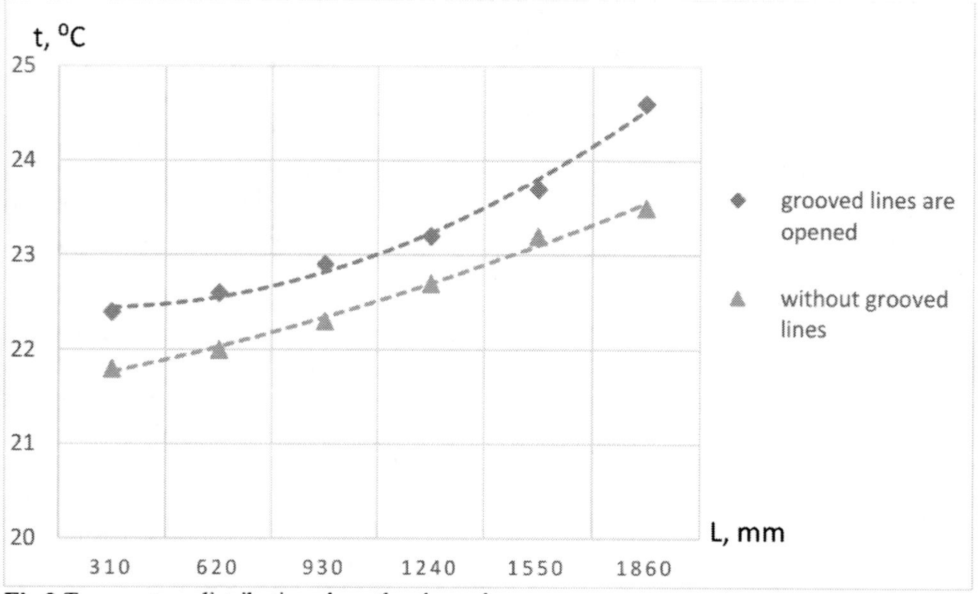

Fig.2. Temperature distribution along the channel

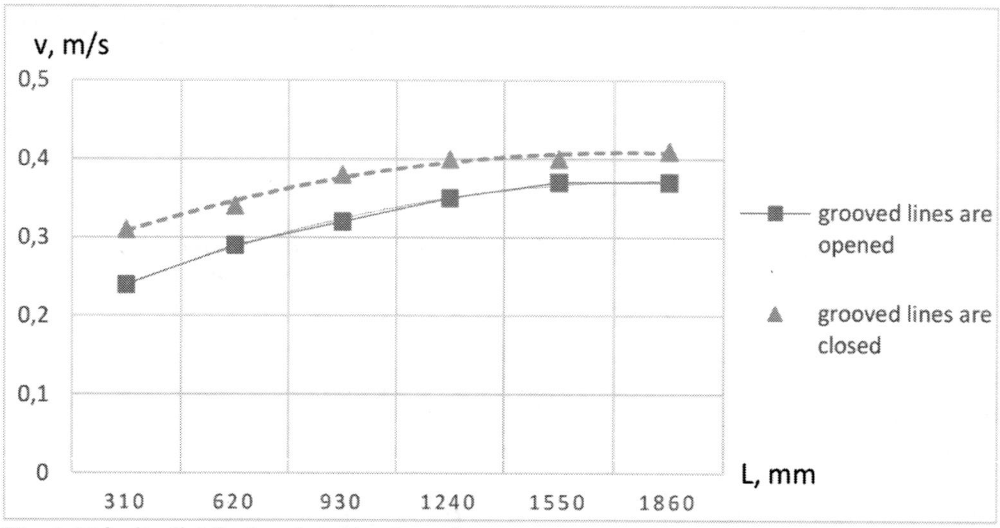

Fig. 3. Velocity distribution along the channel

The following results show that the air flow velocity in the air space is small and amounts to 1 m per sec. It depends on its average temperature. In all cases, the velocity of the air flow goes up with the increase of internal air temperature. A double increase in the height of the building front gives a small rise to the air velocity.

Due to ruptures in the facing layer, i.e. grooved lines, the temperature and air flow velocity will "tilt" and deviate from its average value for the height of the ventilated space.

5 Conclusions

Ventilated curtain walls are the most spread-out technology in the facade finishing, as well as in the multi-store construction. Advantages of ventilated fronts are obvious compared to the similar technologies.

The front of the building is sensitive to the change in number of grooved lines and the way of heating. The highest velocity is determined with sealed grooved lines (ideal channel), but this system is not yet technically possible today. The best solution is the least number of grooved lines. In this case, moisture does not stay on the external surface of the insulation, but drains through the air space. The correct collecting of geometrical data will help to prevent the condensate formation.

References

1. D. Nemova, V. Olshevskyi, D. Cejtin. Nauchno-tehnicheskie vedomosti SPBGPU, Jekonomicheskie nauki, **183**, 295–301 (2013)
2. A. Ostrovaja, E. Stacenko., V. Olshevskyi., T. Musorina,. MATEC Web of Conferences, **53**,1-5 (2016)
3. D. Nemova, V. Emel'janova, D. Miftahova, Magazine of Civil Engineering, **8**, 46-53 (2010)
4. V. Emel'janova, D. Nemova, D. Miftahova, Magazine of Civil Engineering, **6**, 67-74 (2014)
5. M. Petrichenko, Soviet research, **5 (23)**, 703-715 (1991)
6. A. Ostrovaja, M. Petrichenko, E. Stacenko, Applied Mechanics and Materials, 725-726 (2015)
7. M. Petrichenko, St. Petersburg State Polytechnical University Journal. Physics and Mathematics, 143–149 (2012)
8. M. Petrichenko, M. Petrochenko, E. Yevtushenko, Magazine of Civil Engineering, **2 (37)**, 35-40 (2013)
9. N. Vatin, M. Petrichenko, D. Nemova. Applied Mechanics and Materials,**633**, 1007-1012 (2014)
10. M. Petrichenko, M. Petrochenko, Magazine of Civil Engineering, **8 (26)**, 51-56 (2011)
11. A. Girgidov, Hydraulic engineering, 6, (2015)
12. E. Yevtushenko, Construction of Unique Buildings and Structures, **2(7)**, 55-61 (2013)
13. B. Gebhart, J. Jaluria, R. Mahajan, B. Sammakia, MIR, **5**, 678 (1991)
14. V Lapin., S. Lapin, Privolzhsky scientific journal, **2(22)**, 85-92 (2012)
15. M. Petrochenko, Osnovy gidravlicheskogo rascheta svobodnokonvektivnyh techenij v ograzhdayushchih stroitel'nyh konstrukciyah , (PhD Thesis, St. Petersburg 2012)
16. S. Kornienko, Magazine of Civil Engineering, **2**, 18-23, (2012)

17. M. Petrichenko, M. Petrochenko, Nauchno-tehnicheskie vedomosti SPBGPU, **2**, 34-39 (2012)
18. E. Yevtushenko, M. Petrochenko, Magazine of Civil Engineering, **8**, 38-45 (2012)
19. D. Nemova., Construction of Unique Buildings and Structures, 84-86 (2012)
20. V. Gagarin., V. Kozlov, Stroitel'naja fizika v XXI veke, **11**, 73-80 (2006)
21. V. Bukhartsev, M. Petrichenko, Power Technology and Engineering, **5**, 374-377 (2011)
22. J. Li, Y. Chow, Centre for Sustainable Energy Technologies (CSET), 198-203 (2014)
23. J. Hana, L. Penga, Ya. Hongxing, Energy and Buildings, **56,** 204–209 (2013)
24. A. Rosca, I. Pop. International Journal of Heat and Mass Transfer, **60**, 355-364 (2013)
25. G. Barrios, G. Huelsz, R. Rechtman, J. Rojas, Energy and Buildings, 43, 219-223 (2011)
26. H. Yang Feng, G. Xia, Q. Wan, ISHVAC 2013, 384-396 (2013)
27. L. Gaillard, S. Giroux-Julien, C. Menezo, H. Pabiou,. Chair Habitats and Energy Innovations,**7**, 54-67 (2012)
28. D. Zaborova, T. Musorina, M. Petrichenko St. P. State Polytechnical U. Journal, 231,18-26 (2017)
29. M. Petrichenko., N. Vatin, D. Nemova., V. Olshevskiy, Applied Mechanics and Materials.,**725**, 93-99 (2015)
30. A. Barabash , E. Naumova, O. Zhuvak, D. Nemova, V. Olshevskiy, MATEC Web of Conferences,**73**, (2016) DOI: 10.1051/matecconf/2016 7302006

Intumescent coatings with improved properties for high-rise construction

Andrey Ustinov[1,2*], *Olga* Zybina[1], *Leonid* Tanklevsky[1], *Vasily* Lebedev[2], and *Andrey* Andreev[3]

[1]Peter the Great St. Petersburg Polytechnic University, Higher School of Technospheric Safety, Fire Safety Department, 29 Polytechnicheskaya st., St.Petersburg, 195251, Russia
[2]Petersburg Nuclear Physics Institute, Neutron Research Department, Condensed state research division, Laboratory of neutron physico-chemical researches, PNPI, 1, Orlova roscha mcr., Gatchina, Leningrad region, 188300, Russia
[3]Peter the Great St. Petersburg Polytechnic University, Higher School of Technospheric Safety, 29 Polytechnicheskaya st., St.Petersburg, 195251, Russia

Abstract. The paper overviews the way of creating intumescent fire-protective compositions with improved properties by adding nano- and micro-sized supplements into them. Intumescent paints are inert at low temperatures, and at higher temperatures they expand and degrade to provide a charred layer of low conductivity materials. The modified intumescent paints are able to form a more stable charred layer than the classical paints. The stability of a charred layer is crucial if the fire safety in high-rise construction must be secured, because a weak charred layer will not provide a required fire endurance for steel bearing structures and they will break down in case of fire. The fire-protective properties of modified intumescent paints were estimated using an electrical furnace. Also the way of thermal decomposition of the paints was studied with thermogravimetric analysis. Results show that modified intumescent paints form a charred layer with improved fire-protective properties; it can serve as a thermal barrier for a longer period of time. Thermogravimetric analysis confirms this fact showing that the temperatures of full thermal decay in case of modified paints are higher than those of non-modified paints.

1 Introduction

Intumescent fire-protective paints are now one of the most common instruments for ensuring the fire protection of modern buildings. As was mentioned in the abstract, an intumescent coating swells as a result of heat exposure forming a charred layer which serves as a thermal barrier and does not let fire and heat to the protected surface [1]. This means that the protected construction will be able to keep fire-resistant and load-carrying properties for a longer period of time. This fact is important for high-rise buildings, large public places, airports and etc.

One of the main problems of using intumescent paints is a fire protection of steel constructions, which are commonly used in buildings mentioned in a previous paragraph. As a research [2] showed, a steel construction loses a significant part of its load carrying ability

* Corresponding author: ustinov.andrey1994@gmail.com

when the temperature in case of fire exceeds 500 ℃. The temperature of real fire is about 800–1000 ℃ and it means that the steel constructions are extremely heated and may lose their required properties in a short period of time. If a steel construction is protected by an intumescent coating, we face another circumstance: an adhesion of a charred intumescent layer to a heated steel is very low [3], and a charred layer will simply fall off from a surface. Also an intumescent layer which, in general, is a carbon-polymer matrix may burn out when the temperature of fire exceeds 1000 ℃ [4]. All these problems show the necessity of some changes in classical intumescent compositions to make them more reliable when we speak about the fire protection on high-rise and large buildings.

As the effective compositions of ingredients in intumescent paints have already been discovered [5], there is no purpose in changing these compositions. One of the ways of improving the properties of intumescent paints and the charred layers formed by them is adding different supplements into a composition. These supplements do not change the structure of the paint, but they can have a significant impact on the fire-protective properties of a forming charred layer.

The subject of a research is an intumescent paint for steel constructions produced by «Gefest» Enterprise group and based on ammonium polyphosphate, pentaerythritol and melamine. The additives used to improve the properties of this paint were fullerene-containing soot, glass microspheres and aluminosilicate microspheres.

Fullerene-containing soot is a product of the first subprocess of fullerene production; it contains of graphite's microparticles and non-purified fullerenes with different structure (C_{60}, C_{70} etc.). Such material may have a great impact on the properties of intumescent coatings [6], because a charred layer is basically a carbon matrix which can easily accept another carbon materials, such as graphite and fullerenes; the last ones have proven themselves as strengthening additives [3] which can protect a charred layer from destruction and parching [7]. Microspheres are microscopic spheres of glass, aluminosilicate and etc. which are usually between 1 and 1000 micrometers in diameter, although the sizes can range from 100 nanometers to 5 millimeters in diameter. As the microspheres are hollow their heat conductivity is low, and due to this fact they can lower the conductivity of a charred intumescent layer making it more effective and stable.

The purpose of the paper was to estimate the properties of an improved intumescent paints; the way of heat decomposition of those paints was also needed to be investigated.

2 Materials and Methods

As was mentioned in a previous section, the paper overviews the properties of an intumescent paint for steel constructions modified by fullerene-containing soot and microspheres. The paint itself is presented on the Russian market of fireproof materials. Microspheres are also a well-known product.

Fullerene-containing soot was produced in PNPI as a result of burning the graphite electrode (filled with ferrum phthalocyanine) in a voltaic arc by the method [8]. The soot is quite a cheap product and it can be used in an industrial production of modified intumescent paints. The soot used in this research contained of graphite, fullerenes C_{60}, C_{70}, and ferrum endohedral fullerenes $Fe@C_n$.

Additives were mixed with the paint; samples of modified paint were applied on steel plates (30×91 mm and 1 mm thick). Painted plates were placed in an electrical furnace and were heated with 600 ℃ by 5 min. The purpose of this procedure was to estimate the swelling coefficient of the paints as a quotient between the initial thickness of the coating and the thickness of a charred layer. This coefficient shows the efficiency of an intumescent paint in case of a fire.

Also the thermal conductivity of modified intumescent coatings was studied. Samples of paints were applied on steel tubes (15 mm in diameter). Painted tubes were also placed in an electrical furnace (600 ℃). As the tubes were hollow, the chromel-alumel thermocouples were placed in them to watch the temperature inside the tubes. The purpose of an experiment was to find the paint which forms a charred layer with the lowest thermal conductivity; it was estimated by the period of time which a painted steel tube needs to reach the critical temperature of 500 ℃; the more time it takes, the lower thermal conductivity a charred layer has, and due to this the construction's surface will start heating later.

Modified paints were also studied with thermogravimetric analysis (TGA) using a Derivatograph Q-1500D, J. Paulik, P. Paulik, and I. Erdey system, by «MOM» Hungary. The principles of this technique are described in [9]. In brief, a derivatograph simultaneously measures the parameters of a sample brought about by heating it: change in weight (thermogravimetry, TG curve), rate of the change in weight (derivative thermogravimetry, DTG curve), change in enthalpy (differential thermal analysis, DTA curve). These curves are automatically registered as a function of temperature or time. Results of thermogravimetric analysis provide qualitative and quantitative information about the processes which take place in a material while heating it.

3 Results

Table 1 represents the samples of modified intumescent paints, the type of additives and their quantity in each sample.

Table 1. Samples of modified intumescent paint

N of the sample	Additive	Additives quantity, mass.%
1	–	–
2	Fullerene-containing soot	1
3	Fullerene-containing soot	3
4	Microspheres Al-Si	10
5	Microspheres glass	10

Figures 1, 2 show the painted steel plate and tube before and after firetests. Table 2 represents the results of these tests – initial thickness of a paint layer on a steel plate, thickness of a formed charred layer, swelling coefficient, and the time of thermal barrier's existence (for tubes).

Fig. 1. Intumescent paint on steel plates before the test (left) and a charred layer formed as a result of test (right)

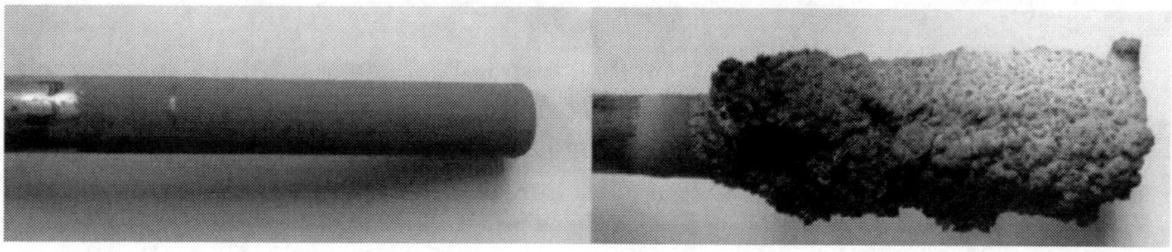

Fig. 2. Intumescent paint on steel tube before the test (left) and a charred layer formed as a result of the test (right)

Table 2. Results of firetest on steel plates and of thermal conductivity test

№ of the sample	Initial thickness, mm	Charred layer thickness, mm	Swelling coefficient	Time of thermal barrier existence, min
1	0,3	10,1	33,7	15
2	0,3	12,9	43	42
3	0,3	12,2	40,6	33
4	0,3	13,4	44,7	26
5	0,3	13,7	45,7	23

Figures 3-7 and tables 3, 4 represent the results of thermogravimetric analysis of modified paints.

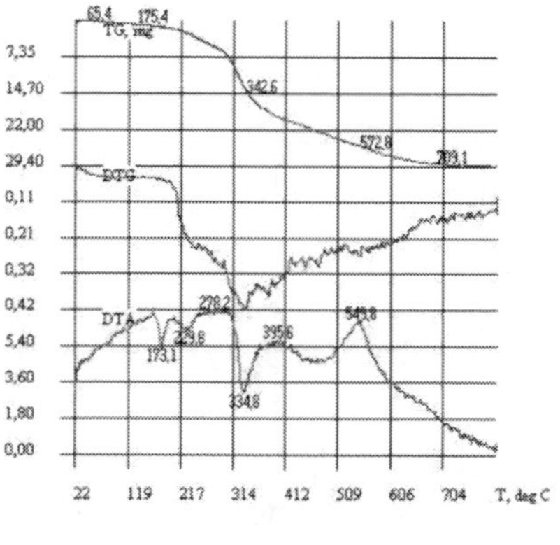

Fig. 3. Results of TGA for sample №1

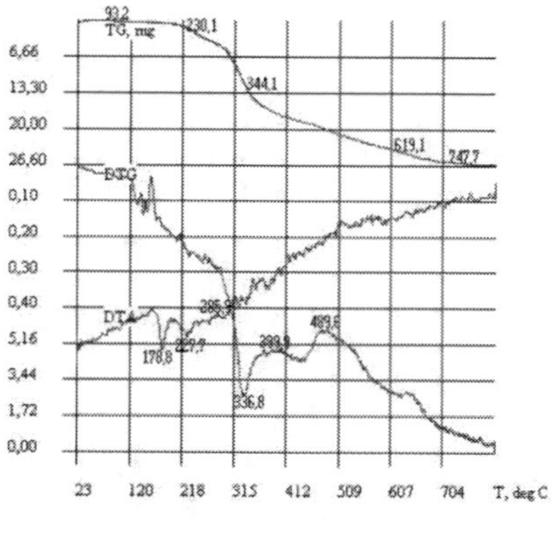

Fig. 4. Results of TGA for sample №2

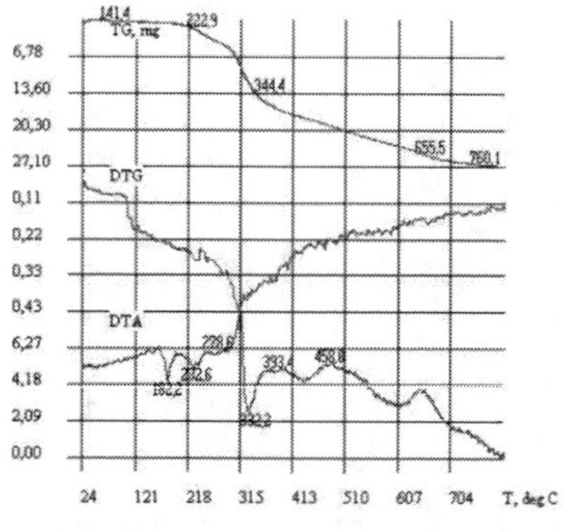

Fig. 5. Results of TGA for sample №3

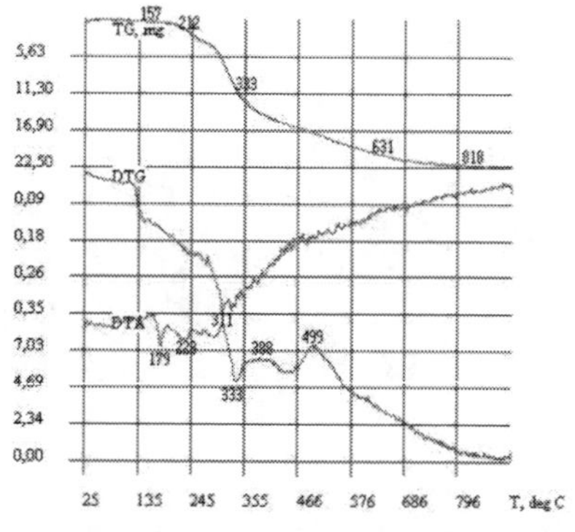

Fig. 6. Results of TGA for sample №4

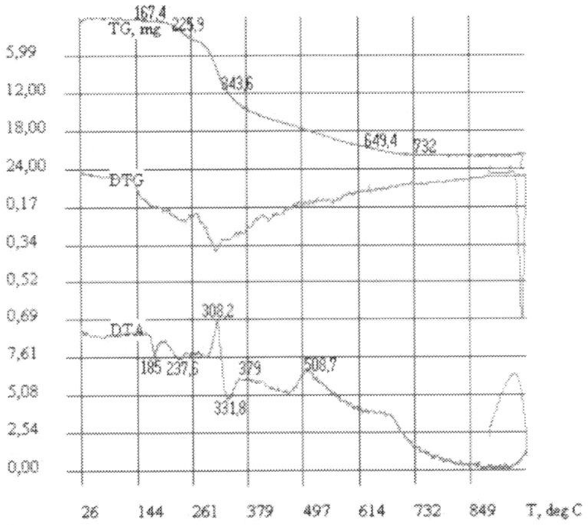

Fig. 7. Results of TGA for sample №5

Table 3. Temperatures of thermolysis stages of intumescent paints

	Temperatures, ℃				
Parameter / № of the sample	1	2	3	4	5
Start of pentaerythritol decay	173	179	182	179	185
Full pentaerythritol decay	230	228	233	223	238
Synthesis of resins	278	286	279	311	308
Ammonium polyphosphate decay	335	337	332	333	332
Hardening	396	400	393	388	379
Condensation	550	490	459	499	509

Table 4. Parameters describing the heat endurance of the paints

Parameter / № of the sample	1	2	3	4	5
T of start of sample's weight decrease, ℃	65	93	141	157	167
T of 5% of sample's weight decrease, ℃	175	230	223	212	226
T of 50% of sample's weight decrease, ℃	343	344	344	333	344
T of 90% of sample's weight decrease, ℃	573	619	656	631	649
T of full decay, ℃	709	748	760	818	732

4 Discussion

According to the results of firetests represented in table 2, all additives used for modification of an intumescent paint significantly increase its fire-protective properties. Swelling coefficient increased by 32-49%, the best results are shown by samples 4 and 5 which comprise glass and aluminosilicate microspheres. This can be explained by the fact that microspheres are hollow, and when a charred layer is forming they release an additional amount of air which increases volume of a charred layer. Fullerene-containing soot acts another way: the carbon particles help in building up the polymer matrix of a charred layer making it more solid and anisotropic [10].

As for the heat conductivity test, all modified samples of the paint show better results than initial paint; the best result was shown by the sample 2 which comprises 1 mass.% of fullerene-containing soot. This means that the carbon particles included into charred layer increase its thermal stability; the mechanism of this effect is needed to be investigated.

The data presented by tables 3 and 4 shows the way of thermal decomposition of samples during TGA. Table 3 includes temperatures of main stages of composition's thermolysis; this data shows the extent of impact of different additives on the intumescent composition's thermolysis. The most significant changes are observed on the stages «synthesis of resin» and «condensation»; in samples 4 and 5 (containing microspheres) these stages shift to the area of higher temperatures. Also all additives shift the beginning of condensation to the area of lower temperatures. These effects may be explained by strengthening the carbon matrix of a charred layer by graphite-fullerene particles and infusible microspheres.

Table 4 shows the fact that all additives increase the heat endurance of an intumescent composition; this is proven by significant shifts of temperatures of start of sample's weight decrease and of full decay to the area of higher temperatures. The best results were shown by the sample 4 (containing glass microspheres).

Conclusion

Results of this research have shown that modification of intumescent paint by different additives is an effective way of improving their properties without the necessity of changing the fundamental composition of the paint. Additives described in the paper and suggested for using in intumescent technology are quite cheap and may be found exceedingly useful in fire protection of high-rise buildings.

The data provided by thermogravimetric analysis proves the stated facts and shows that modified intumescent paints are more reliable in case of fire.

The future line of this research is to describe in details the process of modification of intumescent compositions, the interaction between paint's components and additives. Also the firetests with real steel constructions have to be done.

References

1. L. Calabrese, F. Bozzoli, G. Bochicchio, B. Tessadri , S. Rainieri, G. Pagliarini, J. of Phys.: Conf. Ser., **547**, 012005 (2014)
2. M. Gravit, V. Gumenyuk, O. Nedryshkin, Procedia Engineering **117**, 114-118 (2015)
3. O. Zybina, M. Gravit, Y. Stein, IOP Conf. Ser.: Earth and Environmental Science, **90** (1), 012227
4. M. Gravit, V. Gumenyuk, M. Sychov, O. Nedryshkin, Procedia Engineering, **117**, 119-125 (2015)
5. R. B. R.S. Oliveira, A. L. Moreno Junior, L. C. M. Vieira, IBRACON Struct. and Mater. J., **10** (1), 220-243 (2017)
6. M. Gravit, S. Antonov, O. Nedryshkin, E. Nedviga, V. Pershakov, MATEC Web of Conf, **73**, 04007 (2016)
7. M. Gravit, I. Dmitriev, A. Ishkov, IOP Conf. Ser.: Earth and Environmental Science, **90** (1), 012226, 2017

8. V. Shilin, V. Sedov, *Endohedral metallofullerenes of rare-earth elements. Part 1. The methods of their synthesis*, PNPI, 38 p. (2012)
9. J. W. Robinson, E. M. Skelly Frame, G. M. Frame II, *Undergraduate Instrumental Analysis, Sixth Edition*, CRC Press, 1079 p. (2004)
10. O. Zybina *Problemy tekhnologii koksoobrazuyushchikh ognezashchitnykh pokrytiy*, CRNS, 51 p. (2010)

The reliability of multistory buildings with the effect of non-uniform settlements of foundation

Rafik Al' Malul' [1,*], and Michail Gadzhuntsev[1]

[1]Moscow State University of Civil Engineering, Yaroslavskoe shosse, 26, Moscow, 129337, Russia

Abstract. The issue is the evaluation of reliability of construction considering the influence of the variation of the support settlement, which is changing during the lifetime of constructions due to the consolidation process of the ground. Recently, the specialists give special emphasis to the necessity to develop the methods for the estimation of reliability and durability of structures. The problem, the article considers, is the determination of the reliability of multistory buildings with non-uniform changing-in-time sediments caused by the consolidation process in soils. Failure of structures may occur before the draft reaches it`s stabilizing value, because of the violations of the conditions of normal use.

1 Introduction

The support settlement has a great effect on the distribution of stresses and deformations of the structure. The most dangerous effects on the behavior of the structure is produced by the difference of the settlement of separate parts of the structure's foundations, that difference causes some extra stresses and deformations of the structure's elements. It is known that the settlement (sagging) depends on the kind of the basement soil and the state of stress in the grounds the foundation consists of.

Let us consider the example: statically indeterminate truss which is situated in the layer of the soil.

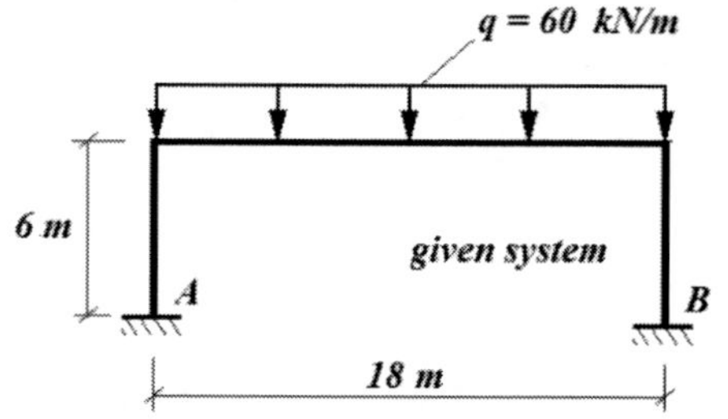

Fig.1 The truss loaded uniformly

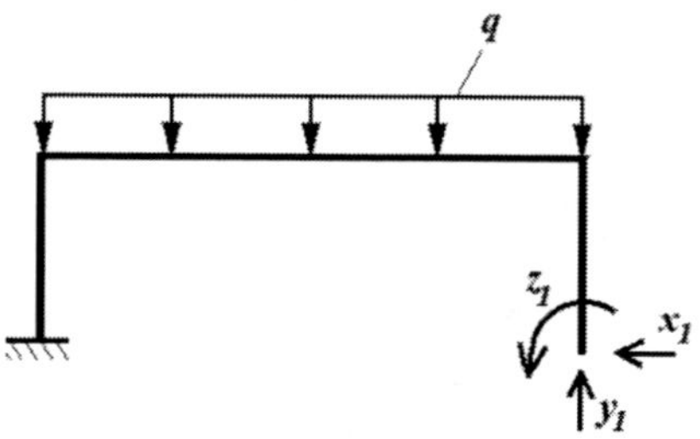

Fig.2 The main system

By the removal of one support, (B), the system is made statically determinate. Take the force of the interaction between the beam of the truss and the support (B) as the main unknown variable. The displacement of the point (B) in the main system in the

* Corresponding author: almalyul@yandex.ru

direction of the unknown force $x_1(t)$, the force was caused only by the imposed load, is denoted with Δ_{1p}, and the displacement of the same point (B) in the direction of the unit force $x_1(t)=1$, denote with δ_{11}.

The values Δ_{1p} and δ_{11} do not depend on time but on the elasticity of the system. They can be determined by the well-known methods of frame-analysis.

The strain compatibility condition for the given statically indeterminate system can be written that way:

$$\Delta_{1p} + \delta_{11}x_1(t) + s_1[x_1(t), t] = 0 \tag{1}$$

$s_1[x_1(t), t]$ — the function (yet unknown), which characterize the law of the variation of the supports` (B) settlement due to the unknown force $x_1(t)$.

The hydrodynamic stress in the soil skeleton, as we know, is determined by the integration of the equation:

$$\frac{\partial P_z}{\partial t} = c^2 \frac{\partial^2 P_z}{\partial z^2} \tag{2}$$

with the following boundary limits:

$$\begin{gathered} t = 0 \text{ и } 0 \le z \le h_0,\ P_z = 0:\ z = 0 \\ t = 0 \text{ и } z = h_0,\ P_z = P(t) = \frac{x_1(t)}{F} \end{gathered} \tag{3}$$

Solving the equation (2) with the Fourier's method, we have:

$$P_z(t) = p(t) - \frac{4}{\pi}\sum_{k=1,3,5}^{\infty}\frac{1}{k}e^{\lambda_k^2 t}\sin\frac{k\pi z}{h_0}\left[p(0) + \int_0^t p'(\xi)e^{\lambda_k^2\xi}d\xi\right] \tag{4}$$

here:

$$\lambda_k^2 = \frac{k^2\pi^2}{h^2}c^2$$

c_2 — the soil consolidation coefficient, the expression for it: $c^2 = k_0^2\frac{1+\xi_1}{\Delta\alpha}$

here:

k_0 — the coefficient of filtration, and $k_0 = const$

ξ_1 — the porosity mid-index of the soil;

α — the compacting factor under the compression of the soil;

Δ — the volume weight of water

$$H(t) = \frac{1}{\Delta}[P(t) - P_z(t)]$$

The expression for the settlement (sagging) of the soil layer is

$$s_1(t) = \frac{8k_0}{\Delta h_0}\left[p(0)\sum_{k=1,3,5}^{\infty}\frac{1 - e^{-\lambda_k^2 t}}{\lambda_k^2} + \sum_{k=1,3,5}^{\infty}\int_0^t d\tau\int_0^\tau p'(\xi)e^{\lambda_k^2(\tau-\xi)}d\xi\right] \tag{5}$$

After changing the order of integration in the last summand, we have:

$$\sum_{k=1,3,5}^{\infty}\int_0^t d\tau\int_0^\tau p'(\xi)e^{-\lambda_k^2(\tau-\xi)}d\xi = \sum_{k=1,3,5}^{\infty}\frac{1}{\lambda_k^2}\int_0^t p'(\xi)\left[1 - e^{-\lambda_k^2(t-\xi)}\right]d\xi \tag{6}$$

After partial integration of the right part:

$$\sum_{k=1,3,5}^{\infty}\int_0^t d\tau\int_0^\tau p'(\xi)e^{-\lambda_k^2(\tau-\xi)}d\xi$$
$$= -p(0)\sum_{k=1,3,5}^{\infty}\frac{1-e^{-\lambda_k^2 t}}{\lambda_k^2}+\sum_{k=1,3,5}^{\infty}e^{-\lambda_k^2(t-\xi)}p'(\xi)d\xi \tag{7}$$

Substitute (5) in (7), so we have:

$$s_1(t)=\frac{8k_0}{\Delta h_0}\int_0^t p'(\xi)\sum_{k=1,3,5}^{\infty}e^{-\lambda_k^2(t-\xi)}d\xi \tag{8}$$

Make a note that

$$P(t)=\frac{x_1(t)}{F}$$

F — the square of the foot of the support *(B)*.
Finally, we get the following expression for $s_1(t)$:

$$s_1(t)=\frac{8k_0}{\Delta h_0 F}\int_0^t x_1(\xi)\sum_{k=1,3,5}^{\infty}e^{-\lambda_k^2(t-\xi)}d\xi \tag{9}$$

The expression (9) gives us the expression for the variation of the support *(B)* settlement for our statically indeterminate system as a function of the time t and the value of reaction $x_1(t)$.
Substitute $s_1(t)$ from (9) in the strain continuity equation (1)

$$\Delta_{1p}+\delta_{11}x_1(t)+\frac{8k_0}{\Delta h_0 F}\int_0^t x_1(\xi)\sum_{k=1,3,5}^{\infty}e^{-\lambda_k^2(t-\xi)}d\xi=0 \tag{10}$$

divide the equation (10) into δ_{11}, with $t=0$, so we have:

$$x_1(0)=-\frac{\Delta_{1p}}{\delta_{11}}=x_1^0$$

$$x_1(t)=x_1^0-\frac{8k_0}{\Delta h_0 F\delta_{11}}\int_0^t x_1(\xi)\,e^{-\lambda_k^2(t-\xi)}d\xi \tag{11}$$

Let us introduce some notations:

$$\frac{8k_0}{\Delta h_0 F\delta_{11}}=\beta,\qquad \sum_{k=1,3,5}^{\infty}e^{-\lambda_k^2(t-\xi)}=k(t-\xi) \tag{12}$$

then the expression (11) turns into:

$$x_1(t)=x_1^0-\beta\int_0^t x_1(\xi)\,k(t-\xi)d\xi \tag{13}$$

So, the analysis of statically indeterminate systems in case of combined action: the load (q) and the settlement of one of it's supports, variative in time, is reduced to the solution of the integral Volterra equation, that is (13). If

$$k(t-\xi)=\sum_{k=1,3,5}^{\infty} e^{-\lambda^2\pi^2\left(\frac{c}{n_0}\right)^2(t-\xi)} \tag{14}$$

$$k(z)=\sum_{k=1,3,5}^{\infty} e^{-\lambda^2\pi^2 z} \tag{15}$$

$$z=\left(\frac{c}{n_0}\right)^2(t-\xi) \tag{16}$$

Use the integral equation (13), so

$$x_1(t)=x_1^0\left[\frac{1}{1+b}-b\sum_{k=1}^{\infty}\frac{e^{\lambda_k t}\sin 2\lambda_k}{\lambda_k(\cos^2\lambda_k+b)}\right] \quad (17) \tag{17}$$

here:

$$x_1(0)=x_1^0,\qquad b=\frac{\beta}{2}\left(\frac{n_0}{2}\right)^2,\qquad x_1(\infty)=\frac{x_1^0}{1+b},$$

$$\sum_{k=1}^{\infty}\frac{\sin 2\lambda_k}{\lambda_k(\cos^2\lambda_k+b)}=-\frac{1}{1+b}$$

Come back to our example — statically undetermined truss, shown in fig.1 (the main system is shown in fig.2). Set the physical and mechanical properties for the foundation and the construction

$F=49\ m^2$ — base area;
$EJ=9\cdot 10^3\ kg\cdot cm^2$ — the stiffness of the section;
$q=60\ kg/m$ — the load;
$k_0=2{,}2\cdot 10^7$ *cm per second* — the coefficient of filtration;
$\xi_1=1{,}368$ — the porosity mid-index of the soil;
$c^2=k_0\frac{1+\xi_1}{\Delta a}=0{,}253\cdot 10^{-2}\ cm^2/second$ — the compacting factor of the soil;
$h_0=4\ m$ — the height of the soil layer;
$\beta=6{,}24\cdot 10\text{-}7$ gramm per second;
$\Delta=5{,}17$.

Table 1 contains the values of bending moments in the truss.

Table 1. values of bending moments in the truss

M / t							
t=1 day	61,02	96,36	96,36	40,5	-49,77	-49,77	-85,11
t=7 days	86,88	89,28	89,28	-40,5	-47,52	-47,52	-49,92
t=3 months	55,51	90,85	90,85	-40,5	-44,51	-44,51	-79,85
t=1 year	54,78	90,54	90,54	-40,5	-44,1	-44,1	-79,86
t=∞	54,78	90,59	90,57	40,6	-44,3	-44,35	-79,88

The most bending moment in the framework is

$$\mathrm{M_{max}} = 162\mathrm{q}\left[\frac{1}{1+\mathrm{b}} - \mathrm{be^{-st}}\frac{\sin 2\lambda_1}{\lambda_1(\cos^2\lambda_1 + \mathrm{b})}\right] + 18{,}33\mathrm{q} - 6{,}11\mathrm{q} - 162\mathrm{q} \tag{17}$$

Let us introduce some notations:

$$M_{max} = \left(\frac{162}{1+\beta} + 18{,}33 - 6{,}11 - 162\right)q$$
$$\mathrm{D} = \frac{162\mathrm{qb}\cdot\sin 2\lambda_1}{\lambda_1(\cos^2\lambda_1 + \mathrm{b})} \tag{18}$$

The situation is that the limits of the fail-safe behavior area is determined by the condition:

$$M_{max} = M_0 - De^{s,t} \leq M_{acc} \tag{19}$$

(M_{acc} — the acceptable value of the moment)
The condition could be transformed to

$$z = \frac{M_0 - M_{\text{доп}}}{D} = e^{s,t} \tag{20}$$

Take only the first member of the raw in (17) and prelogarithmic both parts of (17), so we get:

$$ln\, z = \omega = s,t \tag{21}$$

here

$$s_1 = -\frac{\lambda_1^2}{\alpha_0^2}$$

The probability features of ω depends on the random properties of all the variables the function z depends on. The limits of the fail-safe behavior area depends on the random value $z{=}z_0$ ($\omega{=}\omega_0$), which has mathematical expectation $\overline{\omega_0}$ and the dispersion $\sigma^2(\omega_0)$. Suppose that the failure happens then $M_{max}{=}M_{acc}$, the probabilistic properties of $z_0(\omega_0)$ are defined only by the probabilistic properties of M_{max}^0 , and the probabilistic properties of s are defined by the variation of the consolidation coefficient.

The exact time of failure corresponds to the point where the stochastic process (21) and the probable limits of the no- failure behavior area cross. Suppose that then the failure happens $s{=}\omega_0/T$, the characters s and ω_0 are independent and random and that they have a truncated Gaussian distribution in the intervale $(0, \infty)$. This is so, for example, then the random value of z has a logarithmically normal distribution of the random value T, such as

$$F(t) = \alpha_{\omega_0}\alpha_s \int_0^\infty \left[\int_{\omega_0/t}^\infty f(s)ds\right] f(\omega_0)d\omega_0 \tag{22}$$

here $\alpha_{\omega0}$, α_s — normalizing multiplier factors of that distribution.

$$\alpha_n = \left[0{,}5 + \Phi\left(\frac{m_n}{\sigma_n}\right)\right]^{-1} \tag{23}$$

$$n = \omega_0, s$$

The expression for the density of distribution for no-failure behavior is that:

$$f(t) = \frac{dF(t)}{dt} \tag{24}$$

Then, in accordance with (4), it can be written down:

$$\frac{d}{dt}\int_{\omega_0/t}^{\infty} f(s)ds = \frac{\omega_0}{\sigma_{(s)}t^2\sqrt{2\pi}} exp\left[\frac{1}{2}\left(\frac{\omega_0}{\sigma_{(s)}t} - \frac{ms}{\sigma_{(s)}}\right)^2\right] \tag{25}$$

Substitute (22) into (24) and take into account (25), then we get the expression for the density distribution for the time of no-failure behavior of the construction:

$$f(t) = \frac{\alpha_{\omega 0}\alpha_s exp[0{,}5(\beta^2/\sigma + \alpha^2)]}{2\pi(t^2+\sigma^2)} + \frac{\alpha_{\omega 0}\alpha_s(\beta t + \alpha\sigma^2)}{\sqrt{2\pi(t^2+\sigma^2)}} exp\left[-\frac{(\beta - \alpha t)^2}{2(t^2+\sigma^2)}\right]\left[0{,}5 + \Phi\frac{\beta t + \alpha\sigma^2}{\sigma\sqrt{t^2+\sigma^2}}\right] \tag{26}$$

$$\sigma = \sigma\frac{\omega_0}{\sigma(s)};\quad \beta = \frac{\overline{\omega_0}}{\sigma(s)};\quad \alpha = \frac{\bar{s}}{\sigma(s)}$$

The reliability function of the system is determined by the following expression:

$$P(t) = 1 - \int_0^t f(t)dt \tag{27}$$

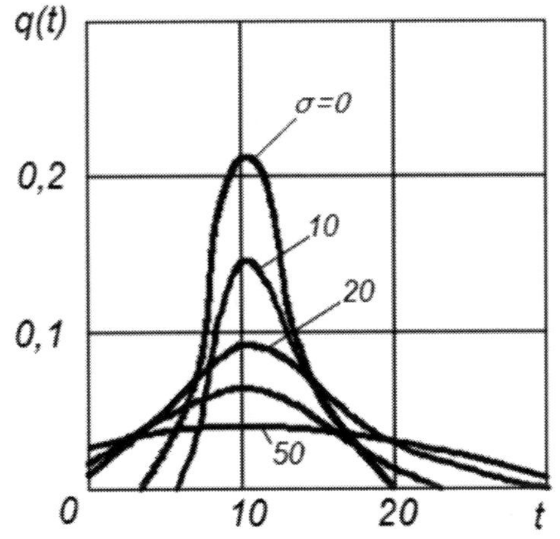

Fig.3 The distribution density of non-failure work

Conclusions

The variation of the support settlement has a significant effect on the distribution of moments in the left vertical pier and in the beam of the truss. The resultant expression (26) for the density distribution of the time of no-failure behavior corresponds to the physical meaning of the task. The rate of accumulation of settlement decreases over time. We can find from (27) that the rate of the decrease of the reliability function is the most of all at the start moment and decreases gradually. For example, let the coefficient of variation, the consolidation coefficient be 0,2 and β be 50, then we get, — for various values of σ in accordance with (26), — different curves describing the density distribution for the time of no-failure behavior.

We should make a note that $t = 0,\ f(t) \neq 0$.

Fig.3 shows that the density distribution of the time of no-failure behavior increases from the constant value $t=t_i$ and, with increasing value of t, the density distribution decreases.

References

1. Al Malul R.M., PhD Thesis
2. N.H Arutyunyan *Some problems of the creep theory* (Moscow, 2006)
3. A.R. Rzhanitsyn The theory of analysis for the reliability of engeneering constructions. M.; StroyIzdat, 1978, 240 c.
4. V.D. Raizer, The reliability theory of constructions. Association of the Civil Engineering Universities, Moscow, 2010, 382 p.
5. V.D. Raizer Reliability Optimization for Pipelines to Corrosion Wear, Proc.of ASCE conference Advances in Pipeline Engineering And Construction. Pipelines 2001. — San-Diego, CA, US, 2001.
6. Russian State Standard SNIP 2.03.01.84

E3S Web of Conferences **33**, 02041 (2018) https://doi.org/10.1051/e3sconf/20183302041
HRC 2017

Complex assessment of urban housing energy sustainability

Olga Popova[1,*], *Julia* Glebova[1], and *Irina* Karakozova[2]

[1]Northern (Arctic) Federal University, 163000 Arkhangelsk, Russia
[2]Moscow State University of Civil Engineering, Yaroslavskoe shosse, 26, Moscow, 129337, Russia

Abstract. The article presents the results of a complex experimental-analytical research of residential development energy parameters - survey of construction sites and determination of calculated energy parameters (resistance to heat transfer) considering their technical condition. The authors suggest a methodology for assessing residential development energy parameters on the basis of construction project's structural analysis with the use of advanced intelligent collection systems, processing (self-organizing maps - SOM) and data visualization (geo-informational systems - GIS). SOM clustering permitted to divide the housing stock (on the example of Arkhangelsk city) into groups with similar technical-operational and energy parameters. It is also possible to measure energy parameters of construction project of each cluster by comparing them with reference (normative) measures and also with each other. The authors propose mechanisms for increasing the area's energy stability level by implementing a set of reproduction activities for residential development of various groups. The analysis showed that modern multilevel and high-rise construction buildings have the least heat losses. At present, however, ow-rise wood buildings is the dominant styles of buildings of Arkhangelsk city. Data visualisation on the created heat map showed that such housing stock covers the largest urban area. The development strategies for depressed areas is in a high-rise building, which show the economic, social and environmental benefits of upward growth of the city. An urban regeneration programme for severely rundown urban housing estates is in a high-rise construction building, which show the economic, social and environmental benefits of upward growth of the city.

Introduction

A large number of scientific researches on the issues of resource conservation are related to construction technology, studies of the material energy characteristics products and structures [1-5], research of separate construction projects and utilities energy parameters. In its order urbanized areas are a complex system of interrelated elements and presume the need for a comprehensive study of the energy systems parameters.

* Corresponding author: oly-popova@yandex.ru

1 Materials and methods

The research carried out by the authors has made it possible to identify and clarify the complex characteristics of the urbanized development area energy system. The set of energy security and energy intensity parameters along with the dynamic indexes of energy efficiency of all area energy system components determines its integrated assessment - the energy sustainability of the area.

Energy parameters of development are an important component of the area energy sustainability in general since it is one of the main consumers of energy resources.

The methodology for assessing the energy parameters of the development area represents itself the detection of the energy parameters list (formation of factor space), complex assessment of construction project by selected parameters, determination of mechanisms for increasing the level of area's energy sustainability. The methodology includes:

1. Experimental-analytical research of urban development energy parameters.

Experimental-analytical research of energy parameters lays in the examining of existing developments (definition of designation, technical-operational, energy parameters and other characteristics of construction projects) and forecasting of its complex reproduction (new construction, capital repairs, modernization, etc.).

Technical-operational and energy parameters of the existing development are characterized by the number and structure of consumers (construction projects), constructive solutions, technical condition, the level of obsolescence.

The possibility for new construction depends on the capacities of existing energy sources or alternative ones (the technical access to utilities, engineering communications, the availability of free heating energy). The possibility of reproduction activities for existing development (capital repairs, modernization, etc.) is mainly determined by technological capabilities and economic efficiency of projects.

The result of the first stage is the formation of the factor space and database.

2. Structural analysis of urban development area's energy parameters via clustering with the use of self-organizing maps (SOM).

The created datebase, which includes multidimensional data of the development projects' characteristics, serves as a basis for structural analysis with the usage of Deductor Studio Academic software product, the platform – Kohonen's self-organizing maps (SOM). Buildings and constructions of different designation are grouped by the highest similarity of chosen characteristics. The resulting groups of projects (clusters) have a high level of characteristic similarity within one group.

3. Designing the strategy for increasing the level of development energy sustainability for each group of projects.

Each group of projects can be described by the group average parameter values. After that, a development strategy for projects of each group is formed including the required list and amount of reproductive activities for increasing the energy sustainability of buildings considering the amount and efficiency of capital investments.

4. Data visualization using geo-informational systems (GIS)

The urban development projects under investigation are displayed on a geo-informational map. Data visualization by means of GIS (Software product QGIS 2.18) allows one to present the urbanized development projects with their description and geo-referencing. The opportunity of forming thematic heat cards gives a chance to assess the overall state of an area's energy sustainability.

2 Experiment

The housing stock of the city has a long history of formation and represents a set of construction projects for individual and commercial developments of different periods. Each period is characterized by features of technology, materials, design and construction norms. By the current moment of time, each urban development has formed its specific identity. This is expressed in a single factor space for any area (the same list of characteristics), but in a different values of those factors for each area in particular.

The authors carried out the research of residential development energy parameters on the example of the Arkhangelsk city. It is one of the largest seaports in the North-West of Russia. The climate is maritime subarctic with long winters and short cool summers. Climatic features of the territory determine the design and energy characteristics of the development.

The main part of the city's residential development was formed in the periods from 1900-1950 and 1950-1980. It is currently presented by wooden individual houses and barracks-type houses, brick and panel buildings of typical development, buildings of large-panel housing construction. This type of development is typical for most of the existing urbanized areas of the Arctic zone.

2.1 Experimental-analytical research of residential development energy parameters

As a result of the research the authors have formed a data array containing up-to-date information on the residential development characteristics in the number of 2196 units located in the central part of the city.

Energy parameters of residential development are determined by the most essential elements of comparison:

1. Characteristics of the existing housing stock:
- special arrangements (type, house series, number of storeys, total building area);
- the design resistance to heat transfer of the building's main cladding is determined by the calculation method on the basis of typical design decisions considering materials used in the construction. General characteristics of the housing stock of different construction periods are structured and presented in the Table 1;
- year of construction;
- technical condition (level of physical deterioration);
- moral obsolescence (availability of in-house engineering communications – heat and water supply, water disposal, electrification).

2. Characteristics of the new construction area:
- availability of development free allotments or already built-up ones that requires renovation through demolition and new construction;
- availability of free energy capacities, which is determined by the possibility of connecting new construction projects to supporting utilities. This research uses the values of the heat load increase (Gcal / h) for the period from 2018-2027 and the load level of power centers by city districts.

2.2 SOM clustering of residential development projects

The clustering made it possible to identify homogeneous groups of projects with a high similarity in the technical-operational and energy characteristics within the same group shown in Table 2.

As a result of the research, it was revealed that the main part of capital construction projects does not meet the requirem of energy efficiency being in unsatisfactory technical condition with signs of obsolescence.

The design resistance to heat transfer of the residential development cladding in the central part of the Arkhangelsk is R0sd =1,79 m^2×°C/W, which is twice lower than the normative value (Rreq =3,54 m^2×°C/W).

Table 1. Enlarged structuring of housing stock in Arkhangelsk.

Con-struc-tion year	Series, type of house	Wall Material	R_{0sd}, m^2×°C/W[25]	Technical condition [27] (physical deterioration), %	Obsolescence
1930-1960	Wooden multidwelling houses (2 and 3 blocks)	Wood, beam 150*150 mm.	1,43	Unsatisfactory – 45-60 %; Dilapidated – 61-80%	Basically, the lack of connection to one or more of the city's utilities(central heating, water supply, water disposal)
1950-1960	Construction projects of the postwar development	Ceramic brick with a thickness of 2 - 2.5 bricks	0,91	Unsatisfactory – 45-55 %	Not found
1958-1965	Typical residential buildings of 1-464A series	Single-layer expanded-clay concrete panels	0,77	Unsatisfactory – 41-50 %	small kitchen areas (4-6 m2); connected sanitation units; incapability of capital redesign; low sound insulation
1960-1970	Typical residential buildings of 1-335 series	Double-layer reinforced concrete panels	0,43	Unsatisfactory – 41-50 %	small kitchen areas (4-6 m2); low ceilings; narrow corridors and small staircases; connected sanitation units; incapability of capital redesign; low sound insulation
1965-1975	Typical residential buildings of 1-335-AK-11 series	Double-layer reinforced concrete panels	0,43	Satisfactory – 35-40 %; Unsatisfactory – 41-45 %	small kitchen areas (4-6 m2); incapability of capital redesign; low sound insulation
1960-1980	Typical residential buildings of 1-447; 1-447C-47; 1-447C-48 series	Calcium silicate brick with a thickness of 2 - 2.5 bricks	0,68	Satisfactory – 35-40 %; Unsatisfactory – 41-45 %	small kitchen areas (4-6 m2)
1985-1998	Typical residential buildings of 111-93 series	Three-layer Reinforced concrete panels	3,55	Satisfactory – 31-40 %	Not found
since 1990	Individual cottage houses	Wooden, mixed construction with the use of energy-efficient insulation	4,06	Good – up to 20%; Satisfactory – 31-35 %	Not found
since 2000	Typical residential buildings of 93 series with improved cladding	Reinforced concrete panels, effective insulation, facing brick	3,56	Good – up to 20%; Satisfactory – 31-35 %	Not found
Since 1990	Residential buildings with individual planning	Ceramic brick with a thickness of 2 - 2.5 bricks, effective insulation	4,33	Good – up to 20%	Not found

Table 2. Clustering of housing stock in the central part of Arkhangelsk.

N of clu-ster	Ener-gy facili-ties	Con-struc-tion year	Material of enclosing structures, project type	Technical condition (physical deterioration) [27]	Utilities	$R_0^{d\ av}$, $m^2 \times °C/W$	Comp-liance with regula-tions	Energy capacities
0	29	1995-2015	wood, effective thermal insulation - individual residential houses	good and satisfactory (10-35%)	partly	4,2	Yes	sufficiently
1	151	1980-2014	brick with heat-insulating materials, large-panel buildings, multilevel and high-rise construction buildings	good and satisfactory (20-40%)	fully	4,07	Yes	sufficiently
2	138	1960-2014	brick with heat-insulating materials, large-panel building, multilevel and high-rise construction buildings	good and satisfactory (20-40%)	fully	4,04	Yes	limited
3	198	1940-1995	silicate brick - massive series; ceramic brick - postwar period, reinforced concrete - panel of the Soviet period	satisfactory, unsatisfactory (30-65%)	fully	0,61	No	sufficiently
4	307	1920-1985	ceramic brick - postwar period, reinforced concrete - panel of the Soviet period	satisfactory, unsatisfactory (30-65%)	fully	0,64	No	limited
5	626	1900-1980	Wood, 2-3 block multi-dwelling apartments of barrack type, individual houses	satisfactory, unsatisfactory, dilapidated (40-90%); partly with overhauling	partly (electrification)	1,43	No	sufficiently
6	747	1900-1980	Wood, 2-3 block multi-dwelling apartments of barrack type, individual houses	satisfactory, unsatisfactory, dilapidated (40-90%); partly with overhauling	partly (electrification)	1,43	No	limited

2.3 Designing the strategy for increasing the level of development energy-resistance for each group of construction projects

Comparative characteristics' analysis of the project groups with reference values (normative) and with each other allows one to define common development strategies for groups including a list of possible reproductive activities shown in Table 3.

Table 3. Housing stock of cluster characteristics in the central part of Arkhangelsk.

Cluster's N	Characteristics of the cluster / Description of recommended reproduction activities
0 1 2	**Construction projects with heat losses complies with standards, in good condition suitable for further operation**: - fulfilment of current and capital repairs according the normative periods of its conduction; - modernization of existing utilities (full complex of utilities improvement for residential development); - new construction on the development free allotments with free energy capacities (cluster 0 and 1)
3 4	**Construction projects with the largest heat losses, in good condition suitable for further operation**: - conduction of current and capital repairs with increasing resistance to the heat transfer of the enclosing structures (external walls, floor, coverage, window and door openings) - energy-efficient repairs; - repair and / or replacement of in-house sanitary equipment; - new construction on the development free allotments with free energy capacities (cluster 3).
5 6	**Construction projects with high heat losses in unsatisfactory condition**: - No economic effect from energy-efficient repairs – conduction of reproductive activities with a supporting character until demolition; - Demolition, new construction considering the financial costs of the planning period and the availability of energy capacities (cluster 5); - modernization of existing utilities; - creation of the new available energy capacities for renovation of the development

2.4 Data visualization using GIS

Urban development projects forming a cluster with a certain set of energy parameters characteristics are displayed on a geo-information basis - a map (Fig. 1). The created heat map displays design reduced resistance to heat transfer of constriction project walls in the central part of the Arkhangelsk. Gradient red marked projects have below the normative value of the parameter. Gradient blue marked projects have above the normative value of the parameter.

As a result of the research, it was revealed that the main part of capital construction projects does not meet the requirements of energy efficiency. The cluster analysis data shows that such projects belong to clusters 3-6. Those are projects in satisfactory, unsatisfactory, dilapidated and emergency condition and requiring different amount of reproduction activities up to capital repairs, modernization, demolition and new construction. The total project capacity of 3-6 cluster is 1878 pc., which is 85% of the existing residential development.

At the moment, in view of the available financial resources for clusters 3,4 - brick and panel construction of post-war and Soviet periods (22% of the residential development in the central part of the city) – it would be mostly productive to carry out a complex of energy-efficient repairs aimed at reducing the consumption of resources considering possible construction technologies, the amount and efficiency of capital investments. The technical condition of the projects (the average physical deterioration - 30-65%) will allow them to operate until the next inter-repair period.

Projects of clusters 5,6 - wooden 2-3 blocks multi-dwelling barrack type houses and individual houses partially equipped with utilities (63% of residential development in the central part of the city) – do not meet the heat engineering standards, while in unsatisfactory and dilapidated condition, not suitable for further operation. Demolition and new construction are seen more cost-effective and investment-attractive. An urban regeneration programme for severely rundown urban housing estates is in a high-rise construction building, which show the economic, social and environmental benefits of upward growth of the city.

Fig. 1. Heat Map - clustering of housing stock in the central part of the Arkhangelsk by energy parameters.

3 Results

As the main component of the city's built-up area housing stock characteristics lead to a conclusion about the high energy capacity of the city territory.

Clustering of residential development by energy parameters allows us to identify groups of construction projects requiring updating and propose a typal list and scope of reproduction activities for each group. On the basis of the proposed strategies it is possible to develop targeted development programs for individual areas (for example quarters) taking into consideration the characteristics of the formed residential development aggregate. Targeted programs should be developed in light of detailed project surveys (visual, instrumental) and regulatory requirements of operation along with possible dynamics of residential development wear. Construction and installation work plan will reduce its terms and cost.

The development strategy for each cluster assumes the implementation of reproductive activities to increase the energy-resistance of buildings, improvement of urban area and financing mechanisms for investment-construction projects.

Periodic study of the development energy parameters and changes in the heat map of the project characteristics will give an opportunity not only to monitor the indexes of energy sustainability, but visualize them on the geo-basis and assess the transformations in the city's energy system with reference to the particular area.

Conclusions

The proposed methodology is unified for any type of development and can be adapted for implementing on any territory. Such approach considers not only the required reduction of energy consumption for individual projects, but also a comprehensive study, an assessment of the city's energy system, long-term reproduction and utilities planning (including new construction, reconstruction, modernization, capital and current repairs) in municipalities and regions.

References

1. A.N. Belous, et al Construction of Unique Buildings and Structures **11 (50)** (2016) (doi: 10.18720/CUBS.50.1)
2. A.I. Kopylova, et al Construction of Unique Buildings and Structures **10 (49)** (2016)
3. S.V Korniyenko, et. Al. Construction of Unique Buildings and Structures **6 (45)** (2016)
4. H. U. Rehman. Experimental performance evaluation of solid concrete and dry insulation materials for passive buildings in hot and humid climatic conditions. Applied Energy **185** (2017) (doi: 10.1016/j.apenergy.2016.01.026).
5. S.Yu. Tarabukina, T.L. Simankina, et. Al. Construction of Unique Buildings and Structures **3 (54)** (2017) (doi: 10.18720/CUBS.54.4).

Resistance to the destruction of concrete in constructions of height buildings at dynamic loads

Mikhail Berlinov[1] *Marina* Berlinova[1*], and *Alexandr* Tvorogov[1]

[1]Moscow State University of Civil Engineering, Yaroslavskoe shosse, 26, Moscow, 129337, Russia

Abstract. The analysis of the criterion of strength of concrete in structures of high-rise buildings under vibration and shock impacts is presented. The idea of an energy approach to ensuring the strength of concrete and the durability of building structures from reinforced concrete under the influence of shock impacts on the life of such structures is presented in a high-rise construction. A method for determining the strength and durability of concrete in load-bearing building structures made of reinforced concrete for irreversible thermodynamic processes has been developed. Dependences that determine the behavior of concrete in reinforced concrete structures of high stores on the load-bearing structures of a building under the influence of damped oscillations from the operation of air transport on the landing site are determined, taking into account the impact arising from its landing.

1 Introduction

Today, effective influence on the structure formation of concrete is a feature of new technologies for making structures. As the investigations of various authors [2, 7] show, all known strength theories are phenomenologically based on experimental data. Determination of the criteria for the service life of concrete in building structures [4] is based on the limiting state and the beginning of the process of material destruction [5]. The energy theory of strength for concrete is acceptable with a slightly different formulation, namely: destruction occurs when the energy necessary for its destruction reaches its limiting value [6].

In the scientific and technical literature [1], the problem of stable and unstable force resistance of concrete was demolished depending on the level of the stressed state under hysteresis energy dissipation of deformed reinforced concrete elements. Within the framework of the energy approach, a method of compliance was used, which determines the intensity of energy release during the experimental evaluation of the development of cracks in concrete under the condition of dynamic load.

A stable thermodynamic equilibrium state is an infinitesimal small effect and causes an infinitesimal change in the system, but if an infinitesimal effect can cause a finite change in the thermodynamic state of the system, then it is in an unstable equilibrium state.

* Corresponding author: marina.tvor@mail.ru

All equilibrium processes are thermodynamically reversible, which without external influence returns the system to its original state, which cannot be said of thermodynamically irreversible processes. So, especially for concrete, deformation is thermodynamically irreversible process.

2 Results section

According to the first law of thermodynamics, the transition of the system to a certain final state "2" from the initial "1", the sum of the work W, the heat Q, and the energy of mass transfer Z does not depend on the manner in which the transition occurs. The character of the internal state of the system determines the value of the internal energy of the system U, determined by the difference in the values of these states

$$U2\text{-}U1 = A + Q + Z \tag{1}$$

For the case of thermodynamically reversible deformation processes, Z = 0, this parameter (the work of inertial forces) can be neglected. Then the elementary work done over the system

$$\delta A = \sum_{k=1}^{n} X_k \delta x_k, \tag{2}$$

here X1, ..., Xn are generalized coordinates for the system of external forces. The work done by these forces (for example, from zero to some final state values) will be equal to

$$A = \int_0^{(X_1,\dots X_n)} \sum_{k=1}^{n} X_k \delta x_k.. \tag{3}$$

The curvilinear integral (3) depends on the path of integration, since the elementary work (2) is not a complete differential. Then the recording of the total differential will reflect the increment of the internal energy of the system when work done above it is δA and the heat input δQ

$$dU = \delta A + \delta Q, \tag{4}$$

that with reference to the process of deformation of concrete it is possible to write down

$$\delta A = \sigma\delta\varepsilon \tag{5}$$

At an equilibrium transition from a given state to an infinitely close one, the increment is determined in terms of the internal energy density U

$$dU = \frac{\partial U}{\partial \varepsilon}\delta\varepsilon + \frac{\partial U}{\partial T}\delta T \tag{6}$$

Here, taking account of (4) and (5), the record will have the form

$$\frac{\partial U}{\partial T}\delta T + \frac{\partial U}{\partial \varepsilon}\delta\varepsilon = \sigma\delta\varepsilon + \delta Q \tag{7}$$

Proceeding from the above, the total differential is the differential of energy, reflecting its change between two states:

$$U_2 - U_1 \int_1^2 dU. \tag{8}$$

The relationship between stresses, strains and temperature is determined by the equation of state:

$$\sigma = \sigma(\varepsilon, T) \text{ or } \varepsilon = \varepsilon(\sigma, T).$$

According to the second law of thermodynamics, it is impossible to extract heat from a source and turn it into work at a constant temperature without making changes in the given system and the surrounding space. Then for reversible (equilibrium) processes the entropy of the system:

$$dS = \delta Q / T. \tag{9}$$

And, accordingly, for irreversible (nonequilibrium) processes

$$dS > \delta Q / T. \tag{10}$$

Having written (7) with allowance for T, we obtain

$$\frac{\delta Q}{T} = \frac{1}{T}\frac{\partial U}{\partial \varepsilon}\delta\varepsilon + \frac{1}{T}\frac{\partial U}{\partial T}\delta T + \frac{1}{T}\sigma\delta\varepsilon, \tag{11}$$

here for reversible processes δQ / T is a total differential:

$$\left(\frac{\partial U}{\partial \varepsilon}\right)_T = \sigma - T\left(\frac{\partial \sigma}{\partial T}\right)_\varepsilon \tag{12}$$

and then the entropy difference for two states of the system for reversible processes will have the form

$$S_1 - S_2 = \int_1^2 \frac{\delta Q}{T} \tag{13}$$

and for irreversible

$$S_1 - S_2 > \int_1^2 \frac{\delta Q}{T}. \tag{14}$$

Since the parameters of temperature and stress tensor, etc. can be measured by conventional instruments, as in the case of nonequilibrium processes, a natural phenomenological definition is obtained. Accordingly, when developing thermodynamic methods for determining the strength and durability of concrete in building constructions for irreversible processes, it can be assumed:

- for the irreversible processes, the principle of local equilibrium (the second law of thermodynamics) remains valid both for reversible processes (4) and (9);
- thermodynamic functions of coordinates and time do not depend, because in each local volume of a concrete body are functions of the same parameters as in the case of equilibrium;

gradients of velocities, temperatures, stresses, etc. in the system under consideration are sufficiently small;

- in the individual elements of the system, the total change in energy and entropy add additively from the changes in the thermodynamic functions.

As can be seen from the analysis, the space of the thermodynamic state of any system is three-dimensional and is determined by parameters characterizing its internal state, when the two parameters are independent, and the third one can always be calculated through them (1- absolute body temperature T, 2-specific body volume V, 3- the specific pressure ρ or the internal energy density U, or the entropy density S, or the enthalpy density i, etc.). The change in internal energy as a function of states "1" and "2" is determined by the difference (U2-U1) of the corresponding values and is completely independent of the nature of the process, then dU is the total differential

$$\int_1^2 dU = U_2 - U_1 \text{ and } \Phi dU = 0. \tag{15}$$

Considering the processes of deformation of concrete in building structures, along a reversible path, for V and ρ equivalents are the stresses σ and relative deformations ε.

When operating high-rise buildings, the problem of their maintenance arises, including in emergency situations, when emergency evacuation of people on the upper floors is necessary, which makes it expedient to use air transport-helicopters. However, when using them, it becomes necessary to provide in the building special landing areas, which are located on the cover. In this case, it is necessary to take into account the impact of additional shock effects arising from the action of the helicopter.

The load from the operation of air transport is represented in the form of damped harmonic oscillations.

Damping occurs as a result of damping of the oscillations due to the features of the structure of the concrete, which is mathematically reflected by adding to the oscillation equation the recording of a damped time function [1]: $y = A \cdot e^{-\frac{\delta t}{T}} \cdot \sin(\omega t + \varepsilon)$, where $e^{-\frac{\delta t}{T}}$ is the damped function of time.

One cycle of oscillations can be described using the "loading-unloading" diagram presented in [8] (Fig. 1).

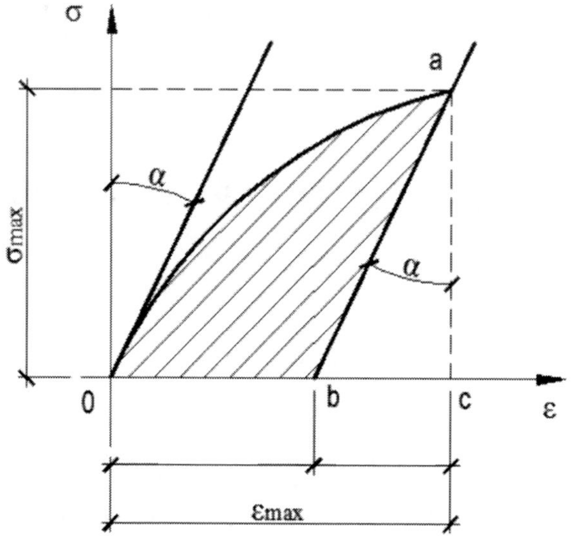

Fig. 1. The diagram "loading-unloading" of a concrete sample: $\varepsilon_{Ob.}$ - reversible deformations when unloading to the instant t; $\varepsilon_{H.Ob.}$ - irreversible deformations; figure 0ab - hysteresis loop.

In this case, the amplitude of the oscillations will correspond to the maximum stresses in each cycle. It is known that the total deformations correspond to the energy expended for complete deformation in one "loading-unloading" cycle, which can be defined as the area of figure 0ac in Fig. 1:

$$W = \int_0^{\varepsilon_{\max}} \sigma d\varepsilon = \sigma_{\max}\varepsilon_{\max} - \int_0^{\sigma_{\max}} \varepsilon d\sigma \qquad (16)$$

Then for n the number of oscillations, the total deformations can be defined as the sum of the total deformations of each load-unload cycle:

$$\varepsilon_{full} = \sum_{i=1}^{n} \int_0^{\varepsilon_{\max}} \sigma d\varepsilon = \sum_{i=1}^{n} \sigma_{\max}\varepsilon_{\max} - \int_0^{\sigma_{\max}} \varepsilon d\sigma \qquad (17)$$

Thus, with the help of (17), complete deformations from the operation of air transport on the structure of a high-rise building can be determined. The properties of concrete vary from age, as well as from the influence of external influences and the regime of the previous loading.

The influence of impact when landing air transport on the cover-area can be reflected in the form of a diagram of stress-strain state of concrete with preloading at time t (Fig. 2).

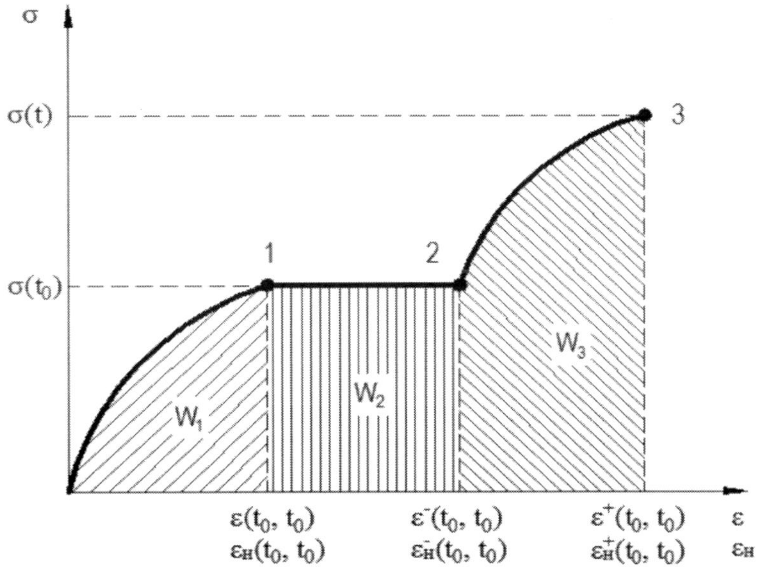

Fig. 2. The diagram of deformation of compressed concrete at its preloading at time t.

Here $\sigma(t_0)$ - the stress in the concrete at the time t_0; $\varepsilon(t_0,t_0)$, $\varepsilon_d\left(t_0,t_0\right)$ - respectively, the relative and irreversible deformation of concrete under the action of stress; the area of the figure W_1 - the production of entropy when loaded at a time t_0; section 1-2 corresponds to simple loading at a stress $\sigma(t_0)$ for a time interval t_0 from t; the area of the figure W_2 - the increment in the entropy production density over a period of time t_0 from t; t - the moment of time, $\sigma(t)$ - the stress at the moment of time t, in which the overload occurs; $\varepsilon^+(t_0,t_0)$, $\varepsilon_d^+\left(t_0,t_0\right)$ - respectively, the relative and irreversible deformation of concrete under the action of stress $\sigma(t)$; the area of the figure W_3 - the increment in the density of entropy production during preloading.

For the process of deformation of real bodies along an irreversible path, in particular concrete, which has the properties of nonlinearity and nonequilibrium, the stresses σ and deformation ε are functionals and cannot simultaneously serve as state parameters, since they strongly depend on deformation.

The total increment in the entropy production density will be equal to the sum of the incremental increments, then at the moment of failure under impact:

$$\Delta S\times Te=\left\{\frac{V_k}{E_b(t)}\left[\frac{\sigma(t_0)}{R(t_0)}\right]^{mk}+\left\{1-k_{ob.k}+V_c\left[\frac{\sigma(t_0)}{R(t\ )}\right]^{mc}\right\}C^*(t,t_0)\right\}\sigma^2(t_0)-$$

$$\frac{V_k}{E_b(t_0)}\frac{\sigma^{mk+1}}{R^{mk}(t_0)}d\sigma-\int_0^{\sigma(t_0)}\left\{1-k_{ob.c}+V_c\left[\frac{\sigma}{R(t_0)}\right]^{mc}\right\}\sigma C^*(t_0,t_0)d\sigma+$$

$$-\int_0^{\sigma(t_0)}\begin{aligned}&+\left\{\frac{V_k}{E_b(t)}\left\{\left[\frac{\sigma(t)}{R(t)}\right]^{mk}\sigma(t)-\left[\frac{\sigma(t_0)}{R(t)}\right]^{mk}\sigma(t_0)\right\}\right\}+\left\{1-k_{ob.k}+V_c\left[\frac{\sigma(t)}{R(t)}\right]^{mc}\right\}\sigma(t)-\\&-\left\{1-k_{ob.c}+V_c\left[\frac{\sigma(t_0)}{R(t)}\right]^{mc}\right\}\sigma(t_0)\ \}C^*(t,t)\ \}\sigma(t)-\end{aligned} \tag{18}$$

$$-\int_{\sigma(t_0)}^{\sigma(t)}\frac{V_k}{E_b(t)}\frac{\sigma^{mk+1}}{R_b^{mk}(t)}d\sigma-\int_{\sigma(t_0)}^{\sigma(t)}\left\{1-k_{ob.c}+V_c\left[\frac{\sigma}{R_b(t)}\right]^{mc}\right\}\sigma C^*(t,t)d\sigma$$

Where $C^*(t,t)$, $C^*(t,t_0)$, $C^*(t_0,t_0)$ - a measure of creep; m - the parameter of the nonlinearity of the deformation, depending on the loading regime; k - material constant; $E_b(t)$ - instantaneous strain module. On the basis of the analysis given in [8], the derivation of the formula of long-term strength (19) follows, which applicable in the case under consideration:

$$\frac{\Delta ST}{R_{dl}(t)}=\frac{V_k}{E_b(t)}\left[\beta^{mk}\lambda^2+\frac{m_k+1}{m_k+2}(1-\lambda^{mk+1})\right]+\left(1-k_{ob.c}+V_c\beta^{mc}\right)\lambda^2C^*(t,t_0)-$$

$$-\frac{1}{E_b(t_0)}\frac{V_k}{m_k+2}\beta^{mk}\lambda^2-\left(\frac{1-k_{ob.c}}{2}+\frac{V_c}{m_c+2}\beta^{mc}\right)\lambda^2C^*(t_0,t_0)+\left[\left(1-k_{ob.c}\right)\frac{(1-\lambda)^2}{2}+\right.$$

$$+V_c\eta_R^{mc}\frac{m_c+1-\lambda^{mc+1}(m_c+2)+\lambda^{mc+2}}{m_c+2}\ \Big]C^*(t,t) \tag{19}$$

Conclusions

Thus, it can be seen that on the basis of the assumption introduced in [1] on the significant effect of the dissipation of the force resistance due to the complex influence of the medium and the absorption of the deformation energy, it is possible to determine the influence of vibrational influences on the structure of the building from the operation of air transport when landing on the roof of a high-rise buildings.

Based on the analysis of the second law of thermodynamics with respect to the process of deformation of concrete carried out in [3], and also taking into account the modified record of the first law of thermodynamics with respect to the deformation process, it becomes possible to use the entropic strength criterion for analysis of deformation of structures, in particular, from shock effects in the considered the case.

The application of such criteria as the effect of dissipation of force resistance, or entropic criterion of strength in the analysis of deformation of structures of buildings and structures will allow to take into account the real working conditions of structures that will arise during operation, including in extreme situations.

References

1. V.M. Bondarenko Report on research work on the topic: Fundamentals of applied dissipative theory of survivability of structures. Moscow. 2008.
2. M.V. Berlinov Calculation of the structure of the frame of buildings under dynamic influences from industrial equipment // Industrial and civil construction. 2004. № 6. Pp. 48-49.
3. M.N Berlinova., M.V. Berlinov, A.V. Tvorogov To the question of ensuring the strength of concrete by the method of thermodynamics // Scientific review. 2015 No. 22. Pp. 148-153.
4. M.N. Berlinova, M.V. Berlinov, A.V. Tvorogov Entropy criterion of strength of concrete in building structures // Scientific Review. 2015 No. 22. pp. 162-165.
5. M.V. Berlinov, E.A. Makarenkov Calculation of reinforced concrete structures by the f inite element method taking into account a real description of the actual physical processes // Vestnik MGSU. 2013. № 11. pp. 26-33.
6. M.V. Berlinov, E.A. Makarenkov Criterion of strength of concrete in conditions of a triaxial stress state under dynamic influences // Scientific Review. 2014. No. 7-1. Pp. 197-200.
7. M.N. Berlinova, A.V. Tvorogov The regime strength of concrete in building structures // Natural and technical sciences. 2015. № 6 (84). Pp. 530-532.
8. M.N. Tvorogova Resistance to deformation and destruction of reinforced concrete structures taking into account nonlinear and nonequilibrium properties and loading regimes // Thesis of the candidate of technical sciences. Moscow. 2006.

Partitions for high-rise construction using phosphogypsum

Sergey Zolotukhin[1], *Olga* Kukina[1,*], and *Anatoly* Abramenko[1]

[1]Voronezh State Technical University, Voronezh, Russia

Abstract. Gypsum blocks are usually used to make partitions in high-rise construction. Reducing the cost of materials used in high-rise construction is an urgent task of modern material science. Phosphogypsum dihydrate, which has binding properties, is one of the large-tonnage waste. The authors have proved that, after years of storage in heaps, water-soluble phosphates, fluorides and other additives included in the structure of fresh phosphogypsum dissolved in water due to weathering (humidification-drying, freezing-thawing in a water-saturated state), and the calcium hydro- and dihydrogen phosphates ingressed in the lattice underwent complete hydrolysis and disintegration, thereby changing the physicochemical properties of phosphogypsum. The data obtained by the authors on the absence of water-soluble compounds of phosphorus, fluorine in stale phosphogypsum indicate its ecological purity and the possibility of application in housing construction. Having analyzed the data of modern methods of differential scanning calorimetry and scanning electron microscopy, the authors predicted and proved by the energy of dehydration of phosphogypsum dihydrate, lime, sandy loam, the possibility of obtaining non-flammable materials with sufficient strength for wall materials. Understanding the processes occurring in water films (the thickness of the water film, the pressure, the temperature and the pH of the aqueous extract of the mixture, the drying of the materials produced), made it possible to develop a technology for obtaining wall products from lime-sandy phosphogypsum material using typical silicate brick production equipment and vibropresses for key-cog blocks production.

1 Introduction

In mass construction of high-rise buildings, partitions made of gypsum blocks are widely used. This is due to the fact that the key-cog blocks are produced with a clear geometry, a smooth surface that does not need additional processing. The quality of the surface meets the requirements for finishing decoration – paperring, coating – which can greatly reduce the cost of finishing works. Inter-apartment partitions should have good sound insulation. The sound insulation of hollow gypsum key-cog blocks is 43.0 dB. According to SniP 23-03-

* Corresponding author: lgkkn@rambler.ru

2003, chapter "Noise isolation", soundproofing of interior partitions should be 41 dB. A partition can be double or single. In the case of a double partition, there must be an air gap of 40 mm. Electrical wiring and communications can be safely hiden into partitions made of gypsum blocks. In fact, this material is an alternative to traditional gypsum plasterboard. It also allows to build partitions much faster, since it does not require installation of a frame, but has large dimensions itself. In addition, the gypsum blocks are rather narrow, which saves space in the room.

Phosphogypsum dihydrate, which has binding properties, is one of the large-tonnage waste. The annual volume of its production in the world is 225 million tons per year, while no more than 5% is used, forming heaps, which accumulate over decades, occupying huge areas. In the literature there are no data on investigation of the properties of stale phosphogypsum.

Because of high content of gypsum dihydrate in phosphogypsum dihydrate, numerous studies on its processing have been aimed at obtaining gypsum using classical technologies: burning, cooking in gypsum-making boilers, autoclaving.

However, the use of phosphogypsum dihydrate as a basic material for processing by traditional technologies did not lead to a wide use of phosphogypsum due to high power inputs, which are 2-3 times higher than the ones for processing ordinary gypsum stone. Available developments in mechano-hydrochemical activation, pressing of materials based on gypsum binders enable the creation of energy-efficient and cheap materials using phosphogypsum dihydrate. The factors holding back their application include the instability of the properties of waste materials and the lack of knowledge in the theory of condensation hardening of these systems. The development of the technology using standard equipment for the production of wall materials will simplify the wider use of phosphogypsum in construction.

In view of the above, the objectives of this work are:

- to conduct research on the properties of stale phosphogypsum;
- to develop the theory of structure formation of composites;
- to obtain energy-efficient low-cost materials used in high-rise construction.

In the structure formation of composite building materials, the most important are the electrical interactions, which depend on the masses, the magnitude of the charges of their constituent particles, the distance between them, and the penetrability of the interaction medium. The dependences, proposed by B.V. Deryagin and L.D. Landau (Fig. 1), allow us to assume that these interactions are realized by alternating fields of attraction and repulsion formed around the particles that make up the structure of composite building materials.

A special place in the formation of the structure of composite building materials belongs to the water. Here and further, "water" means ions and colloidal particles of materials forming the composite dissolved in water. Water is a medium that sharply changes the dielectric constant of the medium, predetermining the interaction forces between the particles.

When the components of raw mixtures are mixed, water is adsorbed as films on hydrophilic particles. The maximum strength characteristics of the obtained materials are reached at thicknesses of water films $1\cdot10^{-7}$ m> Δ> $1\cdot10^{-9}$ m.

At the temperature 65…70°C and the change in the pH of water extracts (Fig. 2), water films change their thickness dramatically. At the same temperatures, the penetrating power of water sharply increases.

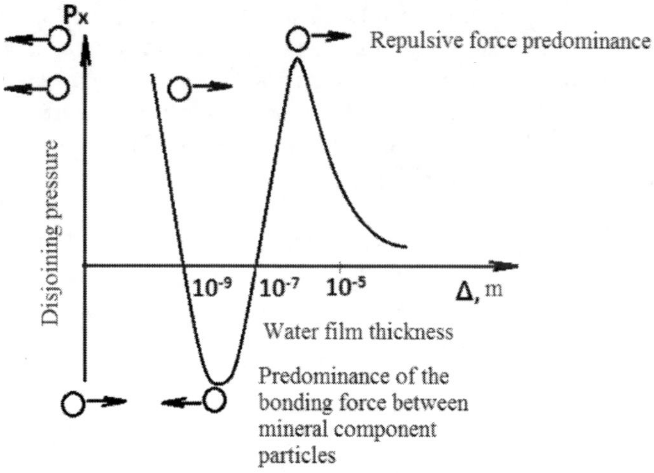

Fig. 1. The change in the disjoining pressure as a function of the thickness of the water film for hydrophilic particles

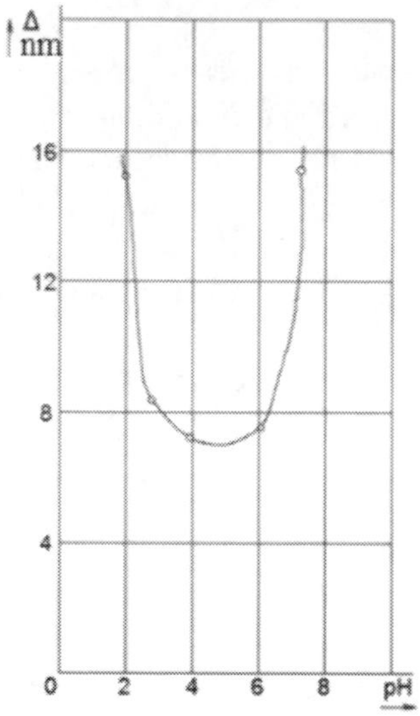

Fig. 2. Dependence of the thickness of the boundary layer on quartz particles on the pH

2 Materials, equipment and methods

The following materials were used as raw mix components:

- construction quicklime (manufacturer: Rossosh, "Pridonkhimstroyizvest'", prepackaged: Voronezh, "Stroytorgservis");

- stale phosphogypsum dihydrate of the Uvarov Chemical Plant, liquidated in 2000 due to bankruptcy. The area of the heaps is 6.4 km^2, the volume reaches about 35 million tons;

- sandy loam from the quarry of the Tambov region of the Uvarov district, located in the immediate vicinity of the phosphogypsum deposits.

The data of granulometric, microchemical, roentgen-phase, microscopic, and thermal analyzes are given in [1].

The physical and mechanical properties of the materials were determined in accordance with the requirements of GOST 23789-79, GOST 8736-93, GOST 5802-86, GOST 310.4.81, GOST 22688-77, GOST 125-79 on the universal electromechanical testing system Instron 5982 with the error in the load ±0.5%. X-ray phase analysis was performed on the automatic diffractometer PANalytical EMPYREAN. The differential scanning calorimetry (DSC) was performed on the synchronous thermal analysis device STA 449 F5 A-0082-M (NETSCH, Germany) with the NETSCH Proteus software with a built-in, reliable, monolithic weighing system and high-sensitivity DSC and DTA sensors in the modern device STA 449 ***F3*** *Jupiter*®. Investigation of the microstructure of the materials was carried out using the scanning electron microscope JSM-6380LV. The images demonstrate the microstructure of the cleavages surface of the samples. The micron marker present in the images makes it possible to estimate the grain sizes and intergranular thicknesses of the water films.

Based on the results of the pH analysis of the water extracts of phosphogypsum dihydrate, data of microchemical, microscopic, X-ray phase and thermal analysis, it can be concluded that with age (from 1998 to 2017), the phosphogypsum changed its physicochemical parameters. Phosphoric acid, water-soluble phosphorus compounds, fluorine in the form of mechanical inclusions or compounds of the phosogypsum dihydrate crystalline lattice in the form of solid solutions have not been detected.

Analyzing the changes in the habit of phosphogypsum dihydrate, which are clearly visible when comparing the photomicrographs (Fig. 3, 4) of the stale phosphogypsum dihydrate, it can be concluded that over the years of storage in heaps, the water-soluble phosphates, fluorides and other impurities usually presented in the structure of fresh phosphogypsum dissolved in water due to weathering (humidification - drying, freezing - thawing in a saturated water state), and the calcium hydro- and dihydrogenphosphates ingressed in the crystal lattice underwent complete hydrolysis and disintegration under the action of the water molecules, thereby modifying the physicochemical properties of phosphogypsum.

Fig. 3. A photomicrograph of phosphogypsum dihydrate at 300:1 magnification, made in 1998.

Fig. 4. A photomicrograph of the phosphogypsum dihydrate (×350), made in 2017.

The obtained data on the absence of water-soluble compounds of phosphorus, fluorine in stale phosphogypsum indicate its ecological purity and the possibility of application in housing construction, which is confirmed by the data of the "Center for Hygiene and Epidemiology in the Voronezh Region" [2].

The analysis of the results of the differential scanning calorimetry of materials used in the study showed the following:

- the energy of dehydration of the sandy loam of the Tambov region of the Uvarov district is 1.29 J/g, which indicates the absence of binding properties of this component of the lime-sandy phosphogypsum composite system (Fig. 5, 6);

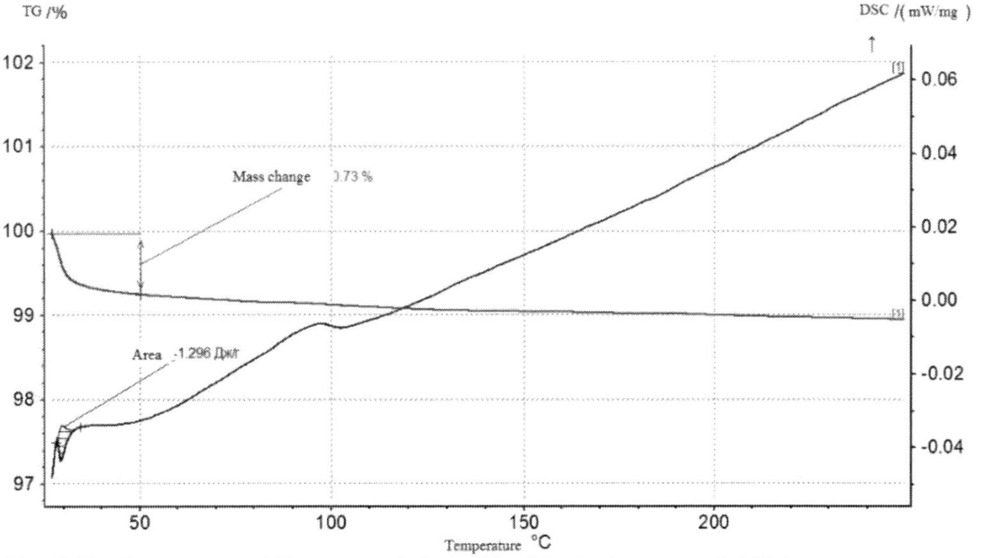

Fig. 5. The thermogram of disperse sandy loam (the dehydration energy 1.3 J/g)

- the energy of dehydration in quicklime is more than 600 J/g. On the first day of hardening of hydrated lime, pressed at the pressure of 5 MPa, an endoeffect with the dehydration energy of 295.9 J/g and the change in mass of 24.52% was observed. The 3 day of hardening of lime showed the endoeffect with the dehydration energy of 126.9 J/g and the change in mass of 9.52%. And by the 24th day of hardening of hydrated lime, the endoeffect had the dehydration energy of 10.21 J/g and the change in mass of 3.97%. Such phase changes in lime confirm its binding properties only in the initial period of its hydration. However, in the case of the hydrated lime, the energy of dehydration decreases sharply with age, which indicates a decrease in binding properties in time due to the carbonization of its surface (Fig. 6-10). However, the high dispersion of lime allows to increase the number of crystallization contacts, which ultimately leads to an increase in the strength of the resulting materials;

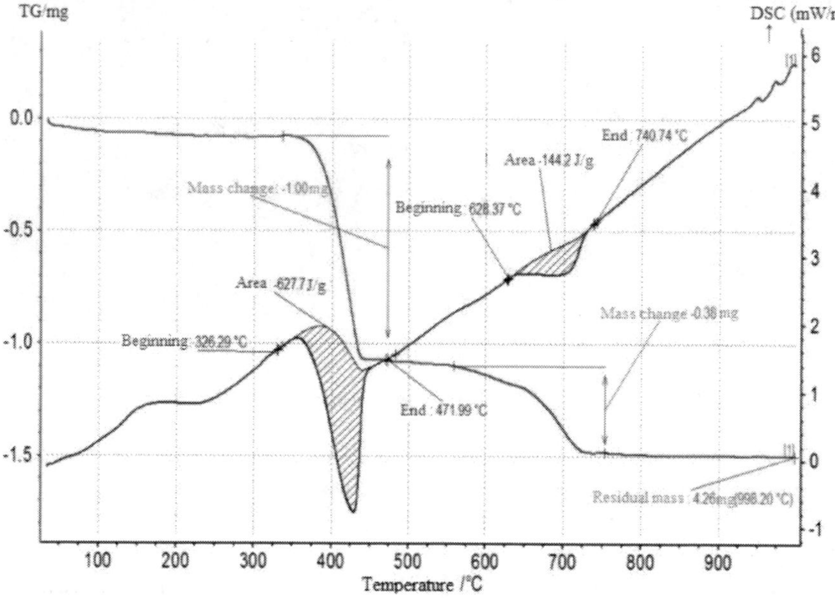

Fig. 6. The thermogram of a disperse sample of quicklime (the dehydration energy 772 J/g)

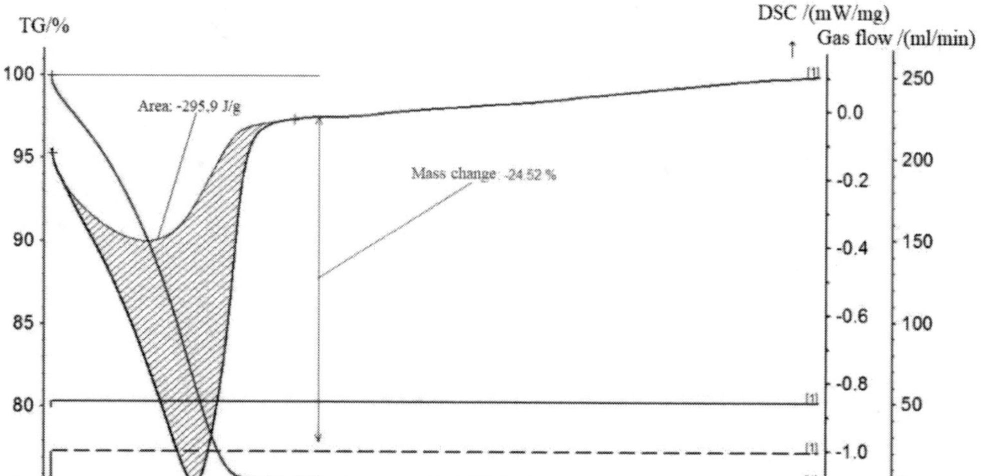

Fig. 7. The thermogram of a sample of hydrated lime pressed at 5 MPa on the 1st day of hardening (the energy of dehydration 296 J/g)

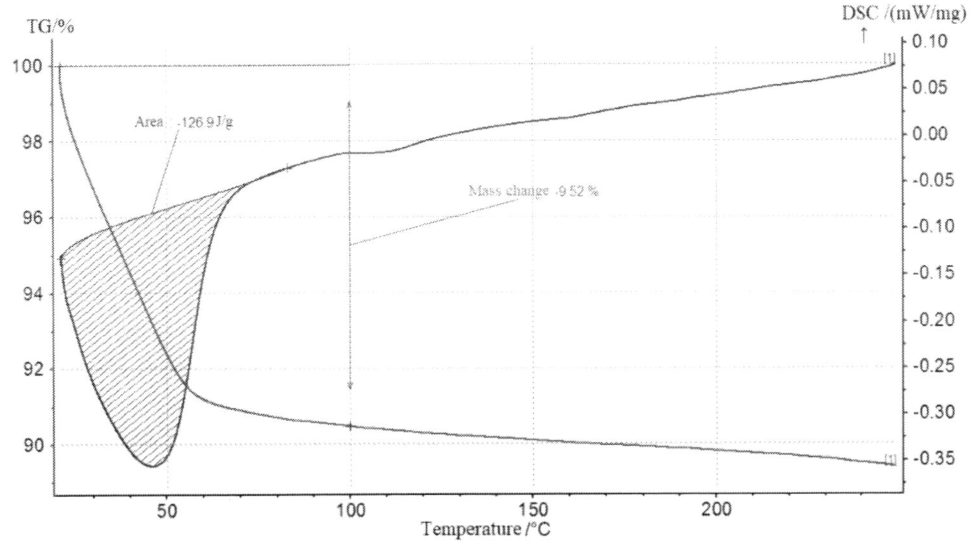

Fig. 8. The thermogram of lime pressed at 5 MPa on the 3rd day of hardening

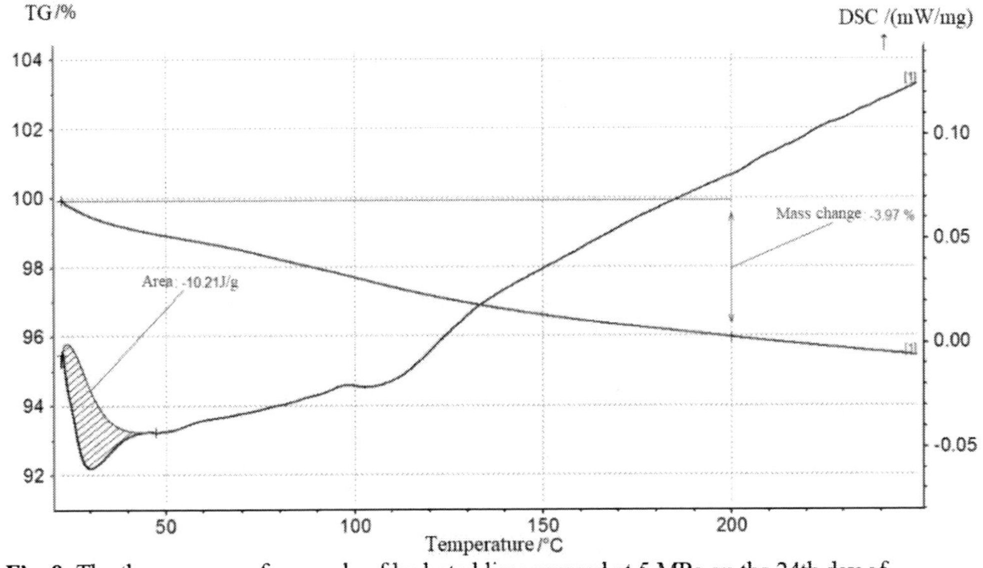

Fig. 9. The thermogram of a sample of hydrated lime pressed at 5 MPa on the 24th day of hardening (the energy of dehydration 10 J/g)

https://doi.org/10.1051/e3sconf/20183302043

- the energy of dehydration of the stale phosphogypsum in 1998 (Fig. 11) and 2017 (Fig. 12-15), which show that over the years of storage in heaps, it has changed its binding properties. According to the thermographic studies of phosphogypsum dihydrate (Fig. 12), performed in 1998, the thermograms of phosphogypsum have a pronounced double endothermic effect of dehydration of the gypsum dihydrate, which is revealed in the interval 100-180°C [3]. By 2017, the phosphogypsum has lost the pronounced double endothermic effect of dehydration of gypsum dihydrate. The onset of the first endoeffect of phosphogypsum appears at 100°C, reaching the maximum at 145°C. Basing on the thermograms of phosphogypsum pressed at 5 MPa at the age of 1, 3, 24 days, it can be concluded that when the samples are compacted, an increase in dehydration energy is observed. The methods of differential scanning calorimetry register the constant changes in binding properties that are taking place in phosphogypsum for decades. While the thermograms show that the first endoeffects have a dehydration energy of more than 200 J/g and do not disappear in time, which indicates that phosphogypsum shows binding properties at all times.

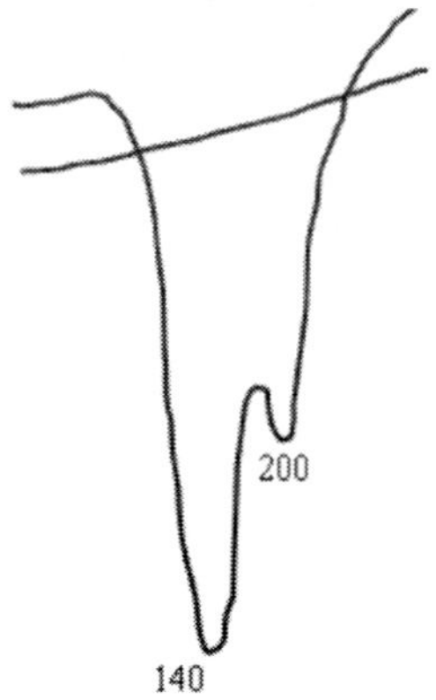

Fig. 10. The thermogram of phosphogypsum dihydrate (1998)

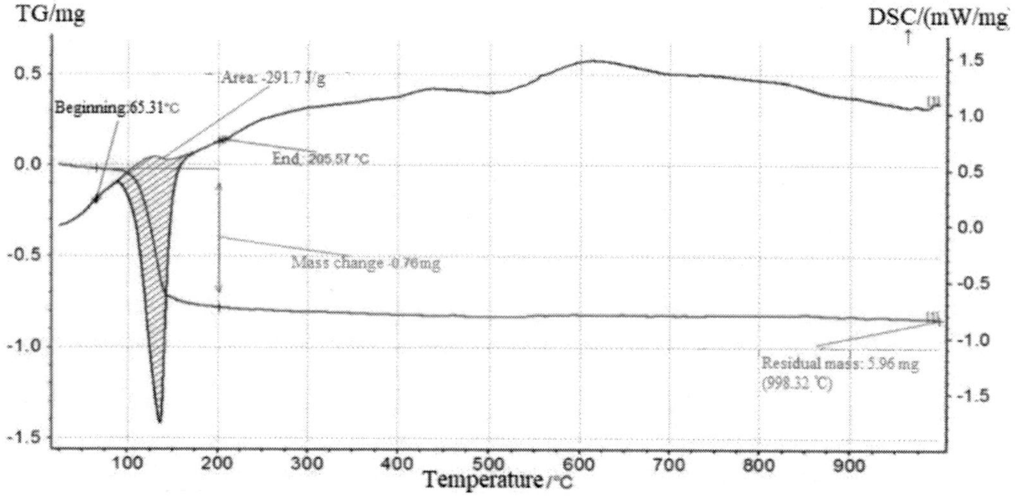

Fig. 11. The thermogram of phosphogypsum dihydrate (2017) (the energy of dehydration 291.7 J/g)

The results of the differential-scanning calorimetry of pressed samples of the stale phosphogypsum dihydrate at the pressure of 5 MPa are shown in Fig. 13 - 14.

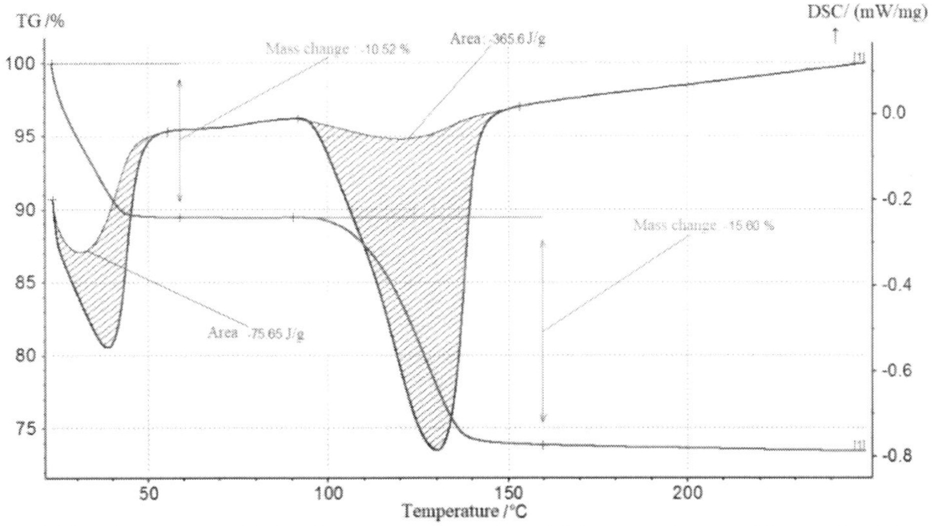

Fig. 12. The thermogram of phosphogypsum pressed at 5 MPa on the 1st day of hardening (the dehydration energy 441 J/g)

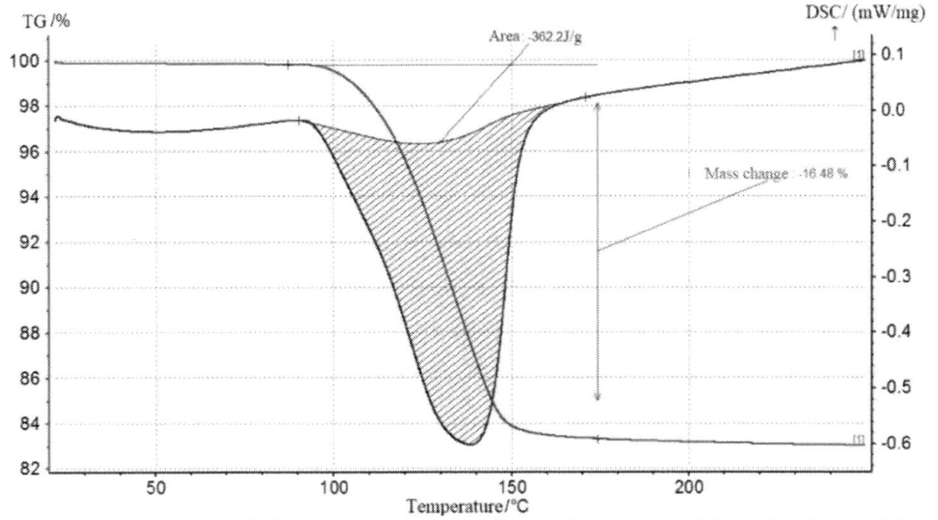

Fig. 13. The thermogram of phosphogypsum pressed at 5 MPa on the 3rd day of hardening (the energy of dehydration 362 J/g)

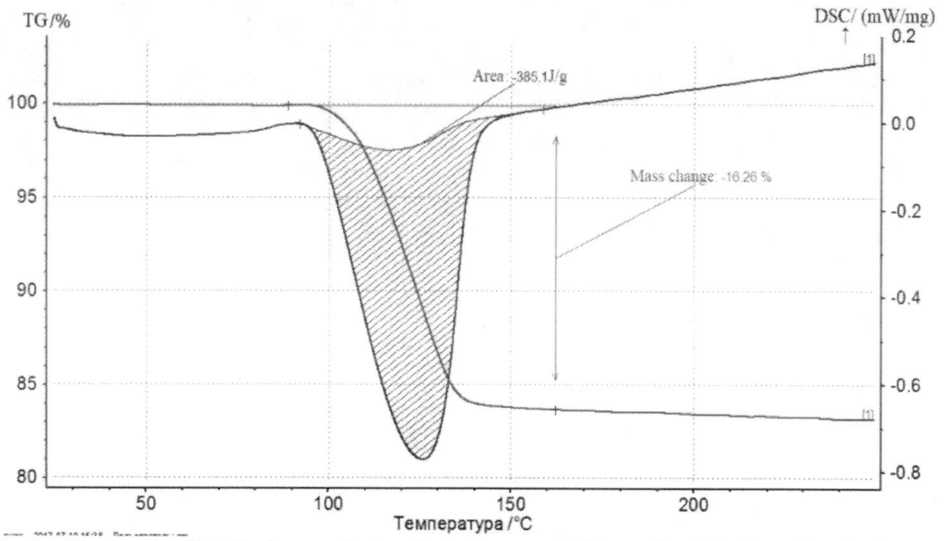

Fig. 14. The result of DSC of pressed phosphogypsum at 5 MPa on the 24th day of hardening (dehydration energy 385 J/g)

Previous studies [4] have shown that water films, self-organizing according to the usual technologies of preparation of mortars, stabilize at the thicknesses equal to $1 \cdot 10^{-7}$ m, and the maximum strength parameters in the obtained materials can be reached at the thicknesses $1 \cdot 10^{-7}$ m>Δ>$1 \cdot 10^{-9}$ m, so to reduce the thickness of the water films, it is necessary to bring the mineral components closer, which can be achieved by pressing or drying the resulting composites.

Understanding the contribution of each of the materials described above and the directional effect of the parameters of technological processes (thickness of the water film, pressure, temperature and pH of the aqueous extract of the mixture, drying of the produced materials) [5-21] allowed to develop a technology for obtaining wall products from lime-sandy phosphogypsum material on a typical equipment for the production of silicate bricks and vibropress plants for the production of key-cog blocks.

The raw meal for the production of building material included stale phosphogypsum dihydrate, hot water t = 90-100°C, dispersed sandy loam with the specific surface area of 4,000-4,500 cm^2/g and quick lime, at W/S = 0.3-0.4 [22].

The technical result showed the increase of strength of the material (Table 1), simplification and cheapening of the technology and improvement of the product quality.

Table 1. Dependence of the strength of lime-sandy phosphogypsum material on the pressure

no	Pressure, MPa	Strength, MPa
1	2,5	6,74
2	5	8,06
3	10	11,50
4	20	18,06
5	25	19,72
6	40	30,1

Studies of the strength parameters of pressed lime-sandy phosphogypsum material have shown the possibility of direct production of building blocks with the strength of 5-20 MPa with a softening factor of 0.5-0.9.

The cost of production of 1 m^3 of wall gypsum materials, depending on the capacity of the equipment, varies from 900 to 1950 rubles/m^3, which is 2 to 3 times cheaper than the analogues.

Thus, the developed technology of lime-sandy phosphogypsum material molding will allow to produce cheap effective gypsum products that can be used in production of structural materials for low- and medium-quality wall materials for internal works.

3 Conclusions

1. After years of storage in heaps, the water-soluble phosphates, fluorides and other additives included in the structure of fresh phosphogypsum dissolved in water due to weathering (humidification-drying, freezing-thawing in a water-saturated state), and the calcium hydro- and dihydrogen phosphates ingressed in the lattice underwent complete hydrolysis and disintegration, thereby changing the physicochemical properties of phosphogypsum.
2. Methods of differential scanning calorimetry according to the first endoeffects allow to predict binding properties of dispersed materials.
3. It has been registered that the binding properties of phosphogypsum dihydrate are changing with time. The thermograms show that the first endoeffects have the dehydration energy of more than 200 J/g and do not disappear in time, which indicates that phosphogypsum shows binding properties constantly.
4. The contribution of highly dispersed sandy loam and hydrated lime materials to the strength of the resulting materials consists in the structuring of water films and the significant growth of crystallization contacts.
5. Water films, self-organizing according to the usual technologies of preparation of mortars, stabilize at thicknesses equal to $1 \cdot 10^{-7}$ m, and the maximum strength parameters in the obtained materials can be reached at thicknesses $1 \cdot 10^{-7}$ m$>\Delta>1 \cdot 10^{-9}$ m, so to reduce the thickness of the water films, it is necessary to bring the mineral components closer, which can be achieved by pressing or drying the resulting composites.
6. Understanding the processes and the influence of each of the components of lime-sandy phosphogypsum material made it possible to develop the technology for the production of wall materials.
7. Studies of the strength parameters of pressed lime-sandy phosphogypsum material have shown the possibility of direct production of building blocks with the strength of 5-20 MPa with a softening factor of 0.5-0.9.
8. The cost of production of 1 m^3 of wall gypsum materials, depending on the capacity of the equipment, varies from 900 to 1950 rubles/m^3, which is 2 to 3 times cheaper than the analogues.
9. The developed technology of molding of lime-sandy phosphogypsum material will allow to produce cheap effective gypsum products that can be used in production of structural materials for low- and medium-quality wall materials for internal works.
10. The results of this study allow to directly form the structures of composite building materials with any dispersed systems.

References

1. S.N. Zolotukhin, O.B. Kukina, A.A. Abramenko, E.A. Soloveva, E.A. Savenkova, *IOP Conf. Ser.: Earth Environ. Sci.* 90 012088 (2017)
2. V.N. Semenov, PhD Thesis (2002)
3. G.D. Shmelev, PhD Thesis (1998)

4. S.N. Zolotukhin, Dr. tech. sci. diss. (1990)
5. R.N. Mirsayev, V.V. Babkov, S.S.Yunusova, L.K. Kuznetsov, I.V. Nedoseko, A.I. Gabitov, *Phosphogypsum waste of the chemical industry in the production of wall products* (Moscow, 2004)
6. R.N. Mirsayev, V.V. Babkov, I.V. Nedoseko, S.S. Yunusova, I.I. Akhmadulina, U.Sh. Shayakhmetov, Const. mat. **6**, 6-9, (2009)
7. I.V. Nedoseko, Full Dr. tech. sci. diss. (2002)
8. I.A. Gar'kina, A.M. Danilov, Ye.V. Korolev, Const. **3-4**, 30-37 (2009)
9. E.V. Korolev, Nanotech. in constr.: Sci. Int. J. **1**, 66-79 (2009)
10. A.N. Bormotov, I.A. Proshin, Ye.V. Korolev, *Mathematical modeling and multicriteria synthesis of composite materials* (Penza, 2011)
11. E.V. Korolev, V.A. Smirnov, Adv. Mat. Res. **746**, 277-280 (2013)
12. A.F. Bur'yanov, Const. Mat. **8**, 30-33 (2008)
13. G. Yakovlev, I. Polyanskikh, A. Gordina, G. Fedorova, A.B. Buryanov, *7th Sci.-Tech. Conf. on Mat. Prob. in Civ. Eng., MATBUD*, 13-21 (2015)
14. Alaa M. Rashad, J. of Clean. Prod. **87**, 717-725 (2015)
15. Liu Laibao, Zhang Yunsheng, Tan Kefeng, Adv. in Cem. Res. **27-10,** 567-570 (2015)
16. Hua Sudong, Wang Kejin, Yao Xiao, Const. and Buil. Mat. **121,** 290-299 (2016)
17. Yoon Seongjin, Mun Kyoungju, Hyung Wongil, J. of As. Arch. and Buil. Eng. **14-1**, 189-195 (2015)
18. Shen Yan, Qian Jueshi, Huang Yongbo, Cem. & Conc. Comp. **63**, 67-75 (2015)
19. Shen Yan, Qian Jueshi, Adv. in Cem. Res. **27**, 515-525 (2015)
20. B.Lamia, R.M. Jamel, Ch. Mohamed, Rom. J. of Mat. **47-1**, 106-111 (2017)
21. Hua Sudong, Wang Kejin, Yao Xiao, Cem. & Conc. Comp. **72**, 299-308 (2016)
22. S.N. Zolotukhin, F. Ibragim, E.A. Savenkova, E.A. Soloveva, A.S. Lobosok, A.A. Abramenko, A.A. Drapalyuk, Yu.B. Potapov, Patent of the RF 2584018, (2016)

E3S Web of Conferences 33, 02044 (2018) https://doi.org/10.1051/e3sconf/20183302044
HRC 2017

Research on Chinese Intelligent Guanglianda in High-rise Construction

Marina Romanovich[1], *Peiyu* Shi[1,2], *Ziyue* Wang[1,3], *Liang* Zhao[4,*], and *Vera* Gerasimova[5]

[1]Peter the Great St. Petersburg Polytechnic University, 195251, Polytechnicheskaya, St.Petersburg, 29, Russia
[2]Institute of Civil Engineering, China University of Mining and Technology, No.269, Liberation south road, Xuzhou, Jiangsu, China
[3]Institute of Civil Engineering, Southwest Jiaotong University, No. 111, North 1st Section of Second Ring Road, Jinniu District, Chengdu, Sichuan, China
[4]Civil Engineering Department, Xi`an Jiaotong University, 28 Xianning West Road, Xi`an, Shaanxi, China
[5]Moscow State University of Civil Engineering, Yaroslavskoe shosse, 26, Moscow,129337, Russia

Abstract. Since the reform and opening up in China, the rapid development of China's construction industry, especially after 2011, BIM technology in China has a rapid development and depth of application. This article mainly discusses the process of development of China's construction industry management and the process of development of BIM technology in China and, the use of software in China, the main use of revit in 3D modeling (design stage), the use of Guanglianda in the whole process of construction management. This article researches on Guanglianda and the applications were described, for example, in the integrated project of management information, labor management information, construction video surveillance, material management, material site acceptance. This paper also elaborates on the cooperation of Revit and Guanglianda in different stages. This paper also elaborates on a construction case in China, and makes a description of the problems of crane, facts and time, and then explores the reasons behind the rapid construction of China. And the feasibility of China's construction management approach to the introduction of Russia.

1 Introduction

The introduction and development of BIM technology in China has gone through three stages. The first stage is the concept introduction stage (1998-2005). This stage is mainly due to the introduction of IFC standard. In 1998, Chinese industry researchers began to contact and study IFC Standard, and in 2002 China Academy of Architecture hosted the "IFC standard technology seminar", China's construction industry according to their own situation on the IFC standard in the practical application of the project were studied and expanded[1]. The second stage is the theoretical research and preliminary application stage (2006-2010),

* Corresponding author: luke1996@stu.xjtu.edu.cn

BIM at the stage of the concept has been a preliminary understanding of scientific research institutions for BIM technology becoming a theoretical study, in 2010,
Tsinghua University and Autodesk joint development BIM technology has also begun to pilot applications in high-end complex demonstration projects, such as the Shanghai World Expo project, the Shanghai Bund SOHO project, etc. BIM technology in the "China BIM standard framework research", proposed the establishment of China Building Information Model Standards, CBIMS. China in the design stage from the application to the construction process for the whole process of expansion of the trend, in 2008 the world's third high-rise Shanghai Center project in the BIM technology: curtain wall and steel structure design, professional collision inspection, engineering calculation, construction simulation aspects of the application, but the software is still dominated by foreign software, such as revit. The development of BIM in china shown in fig1.

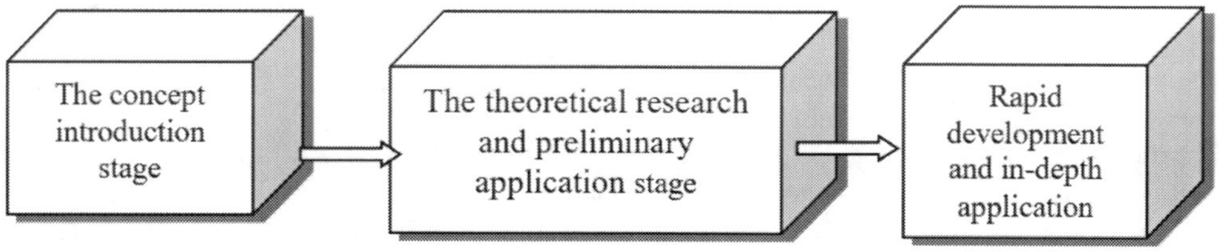

Fig. 1. Development of BIM in China.

(2011 - present), China's national policy clearly put forward the development of BIM to support, in 2011 China's Ministry of Housing and Urban and Rural Development issued (2011-2015) construction information Development plan, the BIM included in the "second five" key promotion technology, China's governments at all levels are also actively promoting the application of BIM, BIM's technical research work is also included in the national subject, the relevant national standards have entered the preparation stage, the current BIM technology in the project mainly presents the following characteristics: firstly, enterprise application BIM technology project scope is expanding, 43.3% of enterprises in the project has started using BIM technology[2,3] . BIM application is more and more, and gradually achieve the whole professional, the whole process of BIM collaborative design. BIM technology is applied not only to the design stage of use, but also widely applied in the engineering calculation, construction simulation, depth design, professional coordination and progress control, based on 4D (3D + cost)[4], 5D (3D + cost + progress) of the overall construction project management solutions are also implemented in a number of large projects[5]. China's own BIM application software is also a large number of practical projects, such as Guanglianda, luban and so on. In general, BIM technology is the depth of the application of the typical "BIM +", that is, BIM technology as the core of the three-dimensional information management system[6-7].

The Construction organization plan is the product of economic system. Before the 1970 s, Chinese construction enterprises are under the planned economy system, they were insufficient recognition of the construction organization plan, they took construction organization plan as a technical document with the single function. In the 1980s, with the development of market economy, the construction organization plan was constantly changing its character, construction organization plan including the project management functions. Nowadays, with China joining into the WTO, the Guanglianda which is based on BIM become a tendency of future design and construction development, it puts forward new requirements for the optimization of construction organization plan. The development process of construction organization plan is shown in Table 1.

Table 1. Development process of construction organization plan.

Time	Event
1951	Chinese first batch of key industrial projects
1953	Ministry of Water Resources promulgates "Notice on Construction Organization Plan"
1970s	Construction organization plan is a technical document
1980s	Construction organization design includes project management function
2001	China joins WTO, optimization of construction organization plan
2011	Guanglianda becomes the future development trend of construction design

After 2011, with the use of Guanglianda, the construction organization plan has been further optimizing, Chinese construction industry ushers in a faster development. According to the data of "Ministry of Housing and Urban-Rural Development of the People's Republic of China" and "National Bureau of Statistics of the People's Republic of China", we can get the development situation of construction industry in China. Fig2 is the development of built-up area in China and fig3 is Gross output value of construction in China.

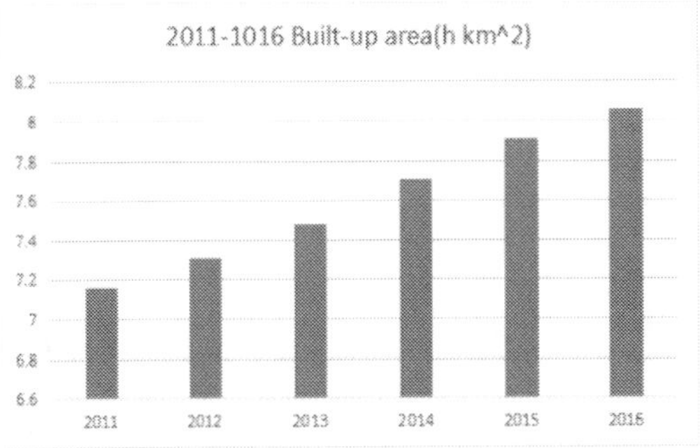

Fig. 2. 2011-2016 The development of built-up.

Fig. 3. 2011-2016 Gross output value of construction in China.

2 Methods

2.1 PM methods

According to China (unified standard for buiding information modeling)[8], BIM5D technology covers from the design, cost, construction, project operations and other aspects to achieve all aspects of the precise cost management. And the definition of BIM5D, 1D is hand drawing, 2D can be understood as electronic drawing, 3D is 3D simulation, 4D is in 3D on the basis of the simulation, we add the time dimension. In order to plan the feasibility of the project and optimize the construction task, the 5D is the cost factor added on the basis of 4D, and the calculation process is done for the construction of the project. Cost, improve the quality, shorten the duration. On June 16, 2015, China was promulgated by the Ministry of Housing and Construction (on the guidance of advancing the application of building information model). The construction of the project was marked by laws and regulations, and the technology was marked as a bureau, and the application of BIM was popularized and deepened in the construction field. Construction management cover mainly in the following aspects:

1) Coordination management

A project construction often has a lot of units involved, so in coordination of various units, improving the ability to exchange information is particularly important, BIM5D technology for all information are integrated and digital management, optimizing the site construction environment and allocating resources, reducing the construction site of the participants.

2) Cost control

The best performance of BIM5D technology in the cost control is the dynamic cost management, according to the progress of changes in cost control tracking and reviewing, the rapid formation of the project cost plan, in the construction simulation according to the construction schedule which can find the corresponding cost percentage, if expected cost is found more then that of the total cost, you can immediately modify the schedule and construction mode, the loss will be reduced to a minimum stage.

3) Progress control

BIM5D technology is based on BIM construction model, including integrated management of participants and professional schedule, comprehensive and dynamic grasp of construction progress, solving the conflict and contradiction between construction and resource allocation, based on the three main factors of resources, cost, material, in advance, things, and after the control, making the construction schedule to optimize the process, to ensure the completion of the task. 3D Module of typical floor with Guanglianda(GCL) is shown in fig4 .

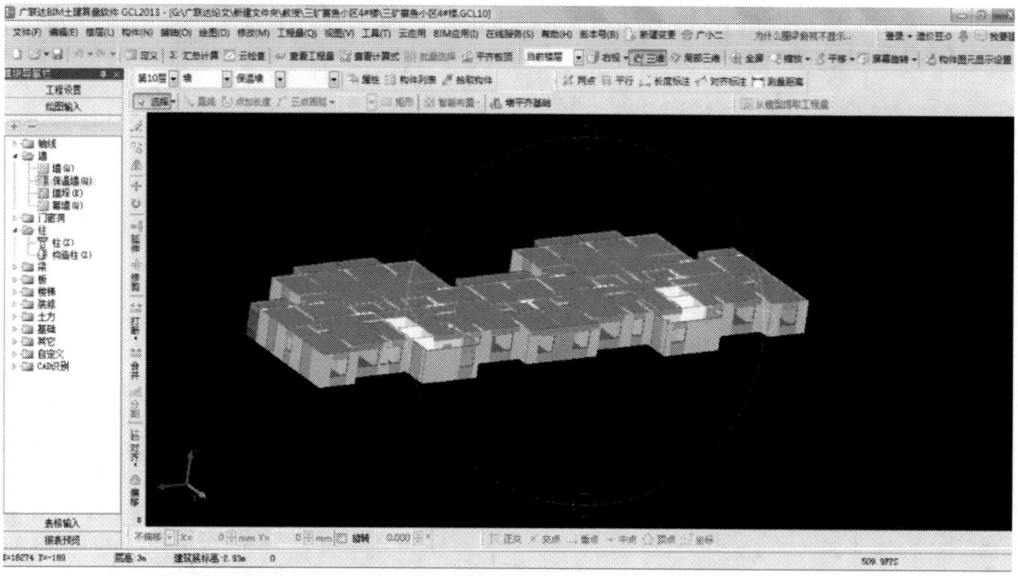

Fig. 4. 3D Module of typical floor with Guanglianda (GCL)

4) Quality control

Quality control follows the principle of PDCA cycle through the planning, implementation, inspection, processing to start the control, for the complex construction process of digital simulation, to achieve three aims of lofting, positioning and detection, to achieve remote quality testing which can carry out construction, equipment and other professional and pipeline in the construction of the various stages of the collision detection, analysis and simulation, good pre-control, reduction of reworking.

5) Document management

The construction of each data is the basis of post-query. BIM completion model can include engineering start, project planning, project implementation, engineering control, project accomplishments of the whole process of information data which increases the data accuracy and provides the decision-makers with the guiding role during the late operation and management.

2.2 Navisworks and Guanglianda

Autodesk Navisworks software can combine the data of AutoCAD or Revit design with the geometry and information from other design tools, we usually consider it as the whole three-dimensional project it is reviewed by a variety of real-time formats. Navisworks software products can help all related project to be a whole, so as to optimize design from the decision-making, implementation, performance prediction and planning to the facility management and operation stage[14].

Autodesk Navisworks software includes three kinds of products, which can help us to strengthen the control of the project, it can also improve work efficiency even in the most complex projects to ensure the engineering quality.

1. Autodesk Navisworks Manage software is a comprehensive review solution, it is used by design and construction management professionals to ensure the smooth progress of the project.

2. Autodesk Navisworks Simulate software can accurately reproduce the design intention, make accurate four-dimensional construction schedule, and realize the visualization of construction projects,

3. Autodesk Navisworks Freedom is a browser, it is free to view all the engineering drawings of Navisworks Review, Navisworks Simulate or Navisworks Manage. Navisworks Freedom provides design professionals with efficient communication methods to enable them to review the NWD format's project files in a convenient, secure and smooth manner.the comparison between Navisworks and Guanlianda shown in table2.

Table 2. Comparison between Navisworks and Guanglianda.

Name	Navisworks	Guanglianda
Features and Speciality	·NWD and 3D DWF Publishing ·Collaboration Toolkit ·Photorealistic Visualization ·Object Animation ·4D Scheduling ·Clash and Interference Detection	Different functional parts: GLM: ·Advanced Technological Platform ·Advanced product management system letting ·Strong customization capability GJK: ·Labor management network ·Real-time management GLK: ·Automatic risk warning and risk management ·Multidimensional analysis and problem tracing system

	·Clash and Interference Management ·Real-Time Navigation	GBQ: ·Efficient and comprehensive pricing ·Convenient group price; ·Fast report processing; ·Automatic bid management

Guanglianda`s application functions mainly include: GLM, GJK, GLK, GBQ, the following table 3 to explain these basic functions, but also a brief overview of the core advantages.

Table 3. Basic function and core-advantages of Guanglianda.

Function	Description	Core-advantages
GEPS	• Guanglianda of construction enterprises project management information solution GEPS is designed for the business input to the construction process of the whole process of management as the main line. • Focusing on the construction business core value chain; • Running through the upstream and downstream industry chain of management product platform to help enterprises to enhance the intensive management and project management level.	• Technology platform: a strong organizational structure, workflow, statements, instant messaging, integration and other capabilities; • Advanced product management system letting customers enjoy the benefits of continuous product upgrades; • Providing the overall solution for project management, all the project management products can be a perfect integration. • A strong customization capability, customers can choose the most efficient, most stable, the highest quality customization service.
GLM	• Guanglianda GLM labor management system at construction site. • Focusing on the construction enterprises to enhance the project management capability; • Committing to the enterprise management centralized and standardized construction.	• The use of internet technology connects the independent components. • Composing of labor management management of the neural network where enterprise as a hub. • Playing a monitoring and regulation-making role
GJK	• GJK system is a comprehensive utilization of video, network, communications and other advanced technology to achieve a control of scientific and technological means. • Its aim is to realize the real-time management and supervision of the construction site. • GJK standardizes management behavior, improves management efficiency and improves the quality of the entire construction process management.	• Real-time video monitoring; • Automatic risk warning and risk management,; • Multidimensional analysis and problem tracing system
GLK	• GLK aims to build an efficient and collaborative management information platform which consists of six subsystems: communication management, document management, information dissemination,	• Providing the various information and dealing with real-time management on smart phones at anytime, anywhere; • Readily access to information, mobile collaboration,;

	administrative affairs, job management, mobile office. GLK is supported by strong technical platform which consists of micro-portal platform	• Group mutual trust mechanism achieving distributed deployment and messages, documents, data transmission exchange
GBQ	• Guanglianda pricing function GBQ is the core of the whole solution of the construction budget management. • It mainly adopts the three modules of bidding management and inventory pricing to realize the bidding business electronically.	• Efficient and comprehensive pricing. ; • Convenient group price; • Simple and fast report processing; • Automatic bid management

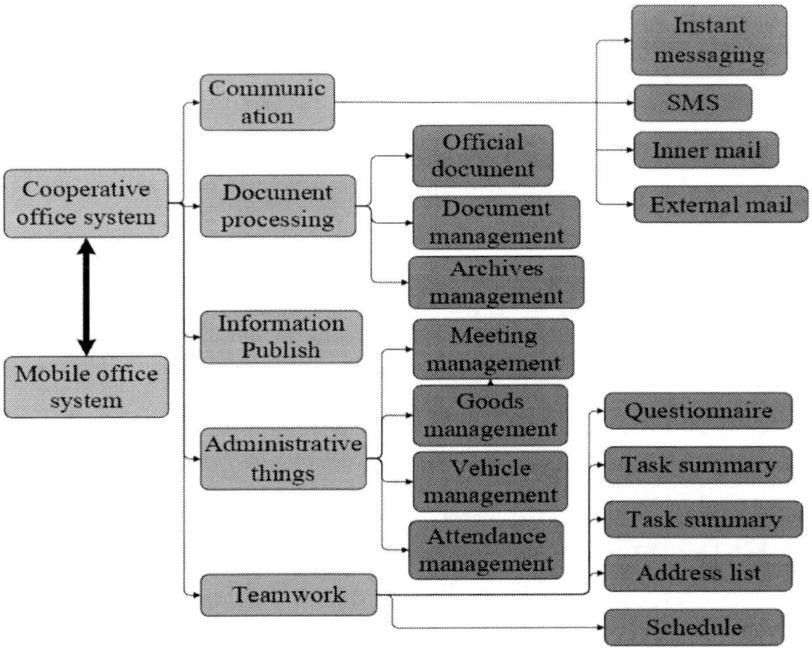

Fig. 5. Function module diagram.

2.3 Case Analysis

Taking the typical floor of a residential building in China as a research object, the floor plan is divided into two bays. The duration of each event is calculated with RSMeans (further-RSMeans, United States of America, USA), and then the total construction period is determined. Use the Gunaglianda to calculate the volume of project.Specific The project volume is exported with the table form. The corresponding cost has also been derived table form.

The volume and the cost generated by the software(GBQ) of the tables, in order to make the research results more detailed, respectively, in accordance with concrete and steel; masonry; template category statistics. The volume and cost of concrete and steel show in table 6. The volume and cost of masonry show in table 7. The volume and cost of template show in table 8.

Table 6. The volume and cost of Concrete and steel with Guanglianda (GBQ)

Line number	material	unit	The volume of 10-stories	The integrated unit price (Euro)	Total cost (Euro)
10504001004	Bearing wall(200mm thick,C30)	Cubic meter	251.3	49.17	12356.92
10504001004	Internal wall(200mm thick,C30)	Cubic meters	322.41	49.17	15853.65
10503002003	Concrete beam(C30)	Cubic meters	8.81	49.55	436.56
10503002003	Concrete frame beam(C30)	Cubic meters	10.16	49.55	503.46
10505003009	Plates:concrete-C30	Cubic meters	88.39	49.8	4401.42
10502002004	Structure column:C25concrete	Cubic meters	5.84	49.87	291.2
10506001003	Stairs:concrete-C30	Cubic meters	16.4	60.52	992.63
10515001021	Steel:HPB300	ton	1.639	546.83	896.25
10515001017	Steel:HRB400	ton	27.5	583.98	16059

Table7. The volume and cost of masonry with Gaunglianda(GBQ)

Line number	material	unit	The volume of 10-stories	The integrated unit price(Euro)	Total cost (Euro)
10402001001	Bearing wall: 200 thick aerated concrete block	Cubic meters	104.73	36.45	3817.22 4
10401003010	Bearing wall:120 thick gangue brick	Cubic meters	77.41	47.03	3640.49
10402001001	Internal wall:200 thick aerated concrete block	Cubic meters	285.87	36.45	10419
10402001001	Internal wall:120 thick aerated concrete block	Cubic meters	63.12	36.45	2300.57
10401003010	Internal wall:120 thick gangue brick	Cubic meters	104.91	47.03	4933.82

Table 8. The volume and costs of template with Guanglianda(GBQ)

Line number	material	unit	The volume of 10-stories	The integrated unit price(Euro)	Total cost(Euro)
11702011001	Bearing wall(200mm thick,C30)	square meters	518	2.95	1526.84
11702011001	Internal wall(200mm thick,C30)	square meters	641.31	2.95	1890.22
11702011001	Concrete beam(C30)	square meters	106.56	2.95	314.04
11702006001	Concrete frame beam(C30)	square meters	126	7.74	975.25
11702016001	Plates:concrete -C30	square meters	737.5	6.78	5000.83
11702003001	Structure column:C25	square meters	85.2	11.30	962.37

Table 9. Table of comprehensive unit price analysis with Guanglianda(GBQ)

Quota Number	Quota Project Name	Quota Unit	Quantity	Unit Price				Total Price			
				X1	X2	X3	X4	X1	X2	X3	X4
A4-206	Pump Pre-mixed Concrete Steel Template Wall	Cubic meters	0.1	15	390	2	38	2	39	0.2	4
A4-198	Pump Pre-mixed Concrete Single Beam	Cubic meters	0.009	14	393	2	39	1	39	0.2	4
A4-210	Pump Pre-mixed Concrete plate	Cubic meters	0.1	16	393	2	39	2	39	0.2	4
A4-195	Pump Pre-mixed Concrete structural columns	Cubic meters	0.1	44	367	2	40	4	37	0.2	4
A4-215	Mixed Concrete straight stairs	square meters	0.1	10	97	1	10	1	10	0.1	1
A4-415	Cast-in-situ components round bar	ton	1	140	334	8	70	140	334	8	70
A4-418 R*1.1	Cast-in-place components rebar	ton	1	164	340	10	76	164	340	10	76
A3-49	Fly ash Areated concrete block wall	Cubic meters	0.1	71	259	4	34	7	26	0.4	3
A3-20	Plasted brick	Cubic meters	0.1	141	286	6	42	14	29	0.6	4
A12-62	Template of cast-in-situ concrete steel template wall	square meters	0.01	98	117	54	29	1	1.2	0.5	0.3
A12-46	Template of cast-in-situ concrete	square meters	0.01	218	435	64	65	2	4	0.6	0.6

X1-Labort cost; X2-Material Cost; X3-Mechanical Cost; X4-Management Cost and Profit.

Then, the construction schedule was constructed with the critical path method [15]. the schedule will show in Guanglianda.

https://doi.org/10.1051/e3sconf/20183302044

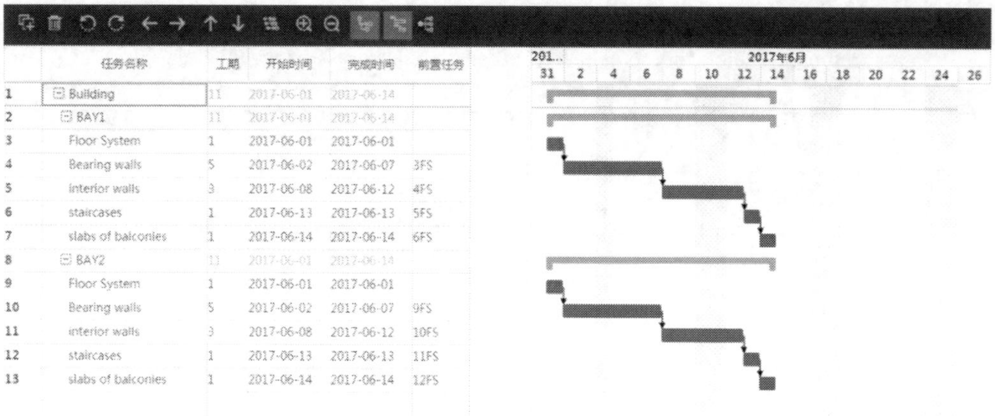

Fig.6. Schedule of construction (one typical floor of a building) and technical complex with the application of the RSMeans framework created by Guanglianda

Input a residential typical floor plan in Guanglianda (GCL), through the choice of building materials, could calculate the volume of the work.

3 Results

There are numerous problems facing today`s construction process management. Many of these are a result of inappropriate pre-preparation issues including budget estimation, workforce considerations, safety, etc. Successful construction process management insures the completion of the construction project in time, within budget, and to do the project specifications[2].In addition, the planning for labors, construction materials, machinery, including cranes, and others should be made effectively, safely and economically[16]. Process management is the core of a successful construction project, including management model propose, implementation, construction site inspection and supervision in the process[3]. If one part of the whole chain falls, the project could be largely influenced even be suspended. The development of the process management is the vital part of the every single project. However, in actual construction, construction units that will always be perfunctory[3], fail to take it seriously and not realize that the progress of the implementation plan is the vital core of the plan, which results in ineffective implementation of the situation[17]. In addition, expenditures of time, money and resources, including human and material, are wasted each year because ineffective or failed balance of time, cost and quality[6]. In the construction process management, the time management can effectively be advanced if the budget is estimated high between the time and cost[18]. To improve speed of the construction, in another word, to improve time efficiency, the quality side is likely and largely to be influenced and eliminated. Quality and cost, as long as problems of the quality arise, you need funds to remedy; the quality will decline if funds shrink[3]. So the relationship between the three needs proper management to control in a process management model. The comparison between Nevisworks and Guanglianda shows that Guanglianda is more efficient and economic in construction management[20]. Guanglianda`s competence of three-dimension, real-time and intelligent management provides companies accurate reference for decision-making. According to the 2012-2018 The Precast Analysis Report of Chinese Construction Management Industry, the number of enterprises that use software of Guanglianda is already nearly 60%, some of non-Guanglianda users around 40% of construction companies in China are positive about the future of Guanglianda due to its high efficiency and commercial benefits[21].

4 Conclusion

This paper compared the main use of Revit in 3D modeling (design stage), the use of Guanglianda in the whole process of construction management. This paper then proposed possible ways of application of Guanglianda and Chinese project management methods into Russia after the researches on the different functions of Guanglianda.

1.It is confirmed that the most effective, economic and fast way of management and construction is the combine method of BIM technology, such as Revit, and Chinese Guanglianda platform.

2.We have found that the functional parts of Guanglianda are the contributing facts that Chinese construction has been conducted in an unprecedented speed.

3.When researching on the development of BIM technology and Guanglianda construction, we still find there are some restraints considering statics, construction cases, etc. In order to apply Chinese Guanglianda platform and project management methods into European construction process, this is an open area for further research.

References

1. China City Press, China construction industry information development report, (2004)
2. J. Xu, Research on application of BIM 5D Technology in central grand project, school of management and engineering. Journal of Zhejiang University. (2016)
3. Z. Xinsheng, using BIM techology to carry out life cycle application to enhance he quality of project [J], Focusing on information, 31(6): 20-24 (2013)
4. J. Zhang, Construction resource dynamic management and real-time monitoring based on 4D-BIM [J], *Construction Technology*.(2011)
5. L. Yong, Construction project BIM prediction method research [D], Journal of Wuhan University of Technology, (2014)
6. LIU xian-wei, GAO Hong-gang, WANG Fu-sheng, Application value and implement of BIM in construction field [J], *Construction Technology,* (2012)
7. LIU Zhao-qiu,WANG fu-lei, LI-Yun-gui, BIM construction and its Value Analysis in design and Construction [J], *Acta Scientiae Scientia Sinica,* (2014)
8. Unified standard for building information modeling, (2017)
9. L. S. Chin, et al, The practice of time management on construction project, the 5th International Conference of Euro Asia Civil Engineering Forum (EACEF-5)
10. X. Li, et al, Research on Construction Schedule Management Based on BIM Technology, 13th Global Congress on Manufacturing and Management, GCMM (2016)
11. A. Izmailov, et al, Project management using the buffers of time and resources, 12th International Strategic Management Conference, ISMC, (2016)
12. K. Zhou, Autodesk Revit Building, *Architecture Journal,* (2017)
13. H. Mallawaarachchi, S.Senaratne, Importance of Quality for Construction Project Success, 6th International Conference on Structural Engineering and Construction Management (2015)
14. CHAU K W, ANSON M, ZHANG J P. 4D dynamic construction management and visualization software, *Automation in Construction*, (2005)

15. M. Romanovich, et al, *International Scientific Conference Energy Management of Municipal Transportation Facilities and Transport EMMFT 2017* pp265 - 272. (2017)

16. KYMMELL W. Building information modeling: planning and managing construction projects with 4D CAD and simulations [M]. New York: McGraw-Hill, (2008)

17. LI H, CHAN N, HUANG T, et al. Optimizing construction planning schedules by virtual prototyping enable resource analysis. *Automation in Construction*, (2009)

18. GOEDERT J D, MEADATI P. Integrating construction process documentation into building information modeling. *Journal of Construction Engineering and Management*, (2008)

19. E. C. Teicholz et al,. BIM Handbook: A Guide to Building Information Modeling for Owners, Managers, Designers, *Engineers and Contractors* [M]. NY John Wiley and Sons, (2008)

20. Computer Integrated Construction Research Group, Building Information Modeling Execution Planning Guide Version 2, (2010)

21. R.Jardim - Goncalves. Building information modeling and inter operation. *Automation in Construction*, (2010)

Humidity Distributions in Multilayered Walls of High-rise Buildings

Olga Gamayunova[1,*], *Tatiana* Musorina[1] and *Alexander* Ishkov[2]
[1]Peter the Great St. Petersburg Polytechnic University, 29 Politechnicheskaya Str., St. Petersburg, 195251, Russia
[2]Moscow State University of Civil Engineering, Yaroslavskoe shosse, 26, Moscow,129337, Russia

Abstract. The limitation of free territories in large cities is the main reason for the active development of high-rise construction. Given the large-scale projects of high-rise buildings in recent years in Russia and abroad and their huge energy consumption, one of the fundamental principles in the design and reconstruction is the use of energy-efficient technologies. The main heat loss in buildings occurs through enclosing structures. However, not always the heat-resistant wall will be energy-efficient and dry at the same time (perhaps waterlogging). Temperature and humidity distributions in multilayer walls were studied in the paper, and the interrelation of other thermophysical characteristics was analyzed.

1 Introduction

Thermal insulators are structural elements that reduce the heat transfer process and act as the main thermal resistance in the structure. Thermal insulators have low thermal conductivity, low strength and are porous media with increased hygroscopicity, so the heat insulator should alternate with a strong wall. Consequently, the operability of the construction of a multilayered wall is determined by the distribution of temperature and the distribution of moisture across the layers.

In this article we want to show that even in independent processes of thermal conductivity and diffusion, the temperature and humidity distributions are related: the wall temperature sets the pressure of saturated water vapor and thereby determines the moisture level.

Is the "dry" wall energy-efficient? We will try to prove that this is so. Is the heat-resistant wall energy-efficient and dry at the same time? We will try to prove that these conditions are not fulfilled simultaneously [1-4].

2 Materials and Methods

To assess the humidity of the wall, we need to know the average wall temperature, which determines the average moisture level. The average temperature of the wall is determined by the construction of the wall, that is, by alternating layers [3,5]. The best combination of layers, providing a maximum of average temperatures, is achieved with a monotonic decrease

* Corresponding author: ganayunova@inbox.ru

in the thermal conductivity coefficient along the course of the heat flux, that is, achieved with the average temperature that is the least different from the maximum wall temperature [10, 12-14].

Quantitative analysis of the distribution of temperature and humidity will be carried out on the basis of a structure consisting of 10 layers ($i=\overline{1,10}$); the thickness of each layer is 0.05 m. The values of the thermal conductivity coefficients are given by the following conditions:

$$\lambda_x^a = 0.1 + 0.1(i-1)$$
$$\lambda_x^g = 1 - 0.1^{\frac{1}{2(i-1)}} \tag{1}$$

The temperature inside the room (T_h) is taken as 18 °C, outside (T_c) – 0 °C.

The distribution of temperature across the wall is defined as follows:

$$\theta_x = \frac{T_h - T_x}{T_h - T_c} = \frac{\sum_{i=0}^{x} \frac{\delta_i}{\lambda_i}}{\sum_{i=0}^{N} \frac{\delta_i}{\lambda_i}}, \tag{2}$$

θ - dimensionless temperature difference;
T_x – the temperature of the layer x.

From the formula (2) it is possible to determine the temperature in the layer x:

$$\theta_x = \frac{T_h - T_x}{T_h - T_c} \rightarrow T_x = T_h - \theta_x(T_h - T_c) \tag{3}$$

The diffusion coefficient (D) is given in relative units in the form of an arithmetic progression with a difference of 0.1 in the range [1,2]:

$$D_x = 1 + 0.1\cdot(i-1), \quad D \in [1, 2], \quad i = \overline{1,10} \tag{4}$$

Using equations (5) and (6), we find the relative humidity (φ_x) and the reduced relative humidity (ω_x).

$$\varphi_x = \omega_x(\varphi_c - \varphi_h) + \varphi_h \tag{5}$$

$$\omega_x = \frac{\varphi_x - \varphi_h}{\varphi_c - \varphi_h} = \frac{\sum_{i=1}^{x} \frac{\delta_i}{D_i}}{\sum_{i=1}^{N} \frac{\delta_i}{D_i}} \tag{6}$$

The value of the relative humidity inside the room (φ_h) is taken equal to 50% (in accordance with the average normative values). Relative humidity outside the room (on the cold side) (φ_c) is determined by the temperature-humidity regime of the air. The air can be dry to saturation $\varphi_c \leq 1$ or waterlogged (water is in a wet steam state). Then $\varphi_c > 1$. From the point of view of the humidity regime of the wall, the second case is dangerous. Therefore, further calculations and analysis will be carried out with respect to the case for $\varphi_c > 1$. The values calculated using formulas (1 - 6) are summarized in Table 1.

Table 1. The values of thermophysical characteristics in the alternation of layers 1-10.

	Layer thickness, m	Coefficient of thermal conductivity		Temperature, °C		The diffusion coefficient	Reduced relative humidity	Relative humidity
x	λ	λ^a	λ^g	T^a	T^g	D	ω	φ
1	0.05	0.1	1.00	11.85	17.60	1	0.139	0.583
2	0.05	0.2	0.68	8.78	17.01	1.1	0.266	0.659
3	0.05	0.3	0.44	6.73	16.09	1.2	0.382	0.729
4	0.05	0.4	0.32	5.20	14.84	1.3	0.489	0.793
5	0.05	0.5	0.25	3.97	13.23	1.4	0.588	0.853
6	0.05	0.6	0.21	2.94	11.28	1.5	0.681	0.908
7	0.05	0.7	0.17	2.07	8.98	1.6	0.768	0.961
8	0.05	0.8	0.15	1.30	6.34	1.7	0.849	1.010
9	0.05	0.9	0.13	0.61	3.34	1.8	0.927	1.056
10	0.05	1	0.12	0.00	0.00	1.9	1.000	1.100

3 Results and Discussion

For a more detailed analysis of moisture distributions in multilayered walls of high-rise buildings, we will additionally consider how the temperature, relative humidity and order of alternation of layers affect the vapor pressure (absolute humidity).

Absolute humidity (or vapor pressure) is determined by the formula:

$$p_{n_x} = \varphi_x \cdot p_{s.s._x} = \varphi_x \cdot \left(\frac{T_x}{100}\right)^4 \tag{7}$$

$p_{s.s.}$ - pressure of saturated steam;

The found values are reduced to the Table 2.

Table 2. The values of thermophysical characteristics in the alternation of layers 1-10

	Layer thickness, m	Coefficient of thermal conductivity		The diffusion coefficient	Temperature, °C	Steam pressure, 10^{-6}, Pa	Temperature, °C	Steam pressure, 10^{-6}, Pa
x	λ	λ^a	λ^g	D	T^a	p^a	T^g	p^g
1	0.05	0.1	1.00	1	11.85	115.227	17.60	269.332
2	0.05	0.2	0.68	1.1	8.78	39.214	17.01	265.749
3	0.05	0.3	0.44	1.2	6.73	14.982	16.09	235.387
4	0.05	0.4	0.32	1.3	5.20	5.785	14.84	184.900
5	0.05	0.5	0.25	1.4	3.97	2.114	13.23	125.773
6	0.05	0.6	0.21	1.5	2.94	0.682	11.28	70.778
7	0.05	0.7	0.17	1.6	2.07	0.175	8.98	30.084
8	0.05	0.8	0.15	1.7	1.30	0.029	6.34	7.830
9	0.05	0.9	0.13	1.8	0.61	0.002	3.34	0.634
10	0.05	1	0.12	1.9	0.00	0.000	0.00	0.000

For a monotonically increasing sequence of coefficients of thermal conductivity and diffusion, and for an ordered alternation of layers (from 1 to 10), the dependence of the vapor pressure and temperature will have the form (Fig. 1).

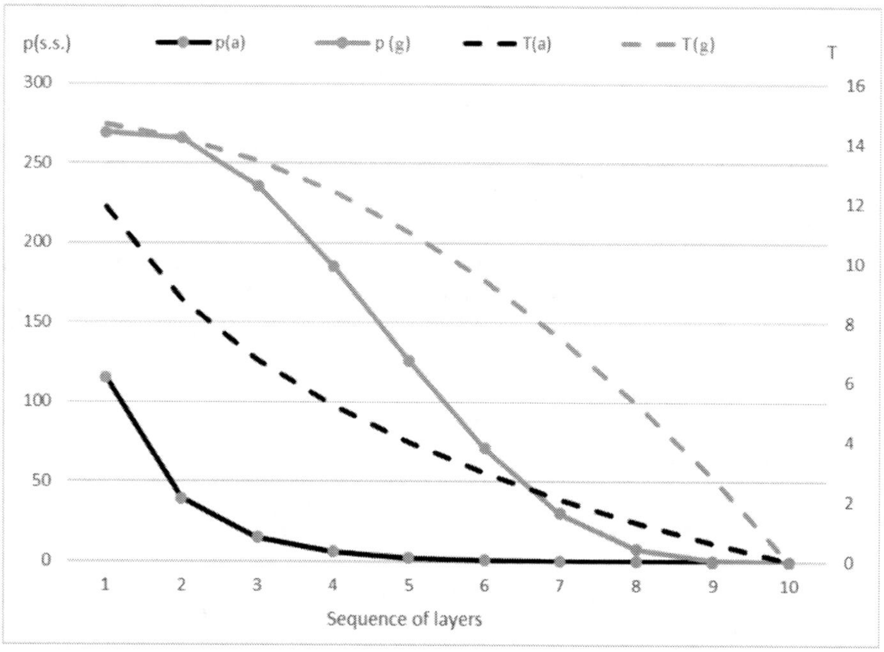

Fig. 1. Distribution of vapor pressure and temperature in a multilayer structure

Average values of temperature and humidity serve as representative values. As analysis of the values obtained with successive permutation of layers shows, they depend on the alternation of the transfer coefficients (that is, on the alternation of layers).

The best option is a design in which the average vapor pressure (absolute humidity) will be less, because this guarantees the preservation of the thermal characteristics of the structure.

Limit distributions of vapor pressure along the layers (worst and best) for $\lambda_x^a = 0.1 + 0.1(i-1)$ and $\lambda_x^g = 1 - 0.1^{\frac{1}{2(i-1)}}$ shown on Fig.2.

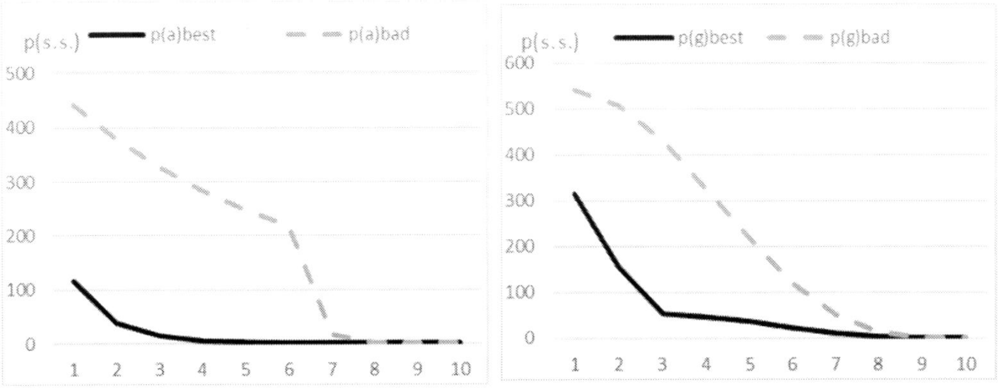

Fig.2. Limit distributions of vapor pressure along the layers (worst and best) for $\lambda_x^a = 0.1 + 0.1(i-1)$ and $\lambda_x^g = 1 - 0.1^{\frac{1}{2(i-1)}}$

The best design from the point of view of moisture distributions will be achieved by alternating the layers (1,2, ..., 9,10) for $\lambda_x^a = 0.1 + 0.1(i-1)$, as well as alternating layers (9,10,1,2, ..., 7,8) for $\lambda_x^g = 1 - 0.1^{\frac{1}{2(i-1)}}$. In this case, the diffusion coefficient varies along the layers, increasing monotonically from 1 to 1.9 in steps of 0.1. At the same time, absolute humidity naturally decreases with decreasing wall temperature.

Energy efficiency requirements is the low thermal conductivity and, consequently, low average layer temperatures. In this case, the absolute humidity decreases with the same value of relative humidity.

Simply put, an energy-efficient wall is drier than a wall with high thermal stability and a high average temperature, leading to a high absolute moisture content of the layers.

The moisture state of the enclosing structures, including high-rise buildings, determines their heat-shielding and sanitary-hygienic qualities, has a significant effect on the durability of the structure as a whole. The reduction in the heat-shielding properties of the enclosing structures during their wetting is associated with an increase in the thermal conductivity of moistened materials, that is, a decrease in the resistance to heat transfer of the entire structure. The increase in the thermal conductivity of materials occurs due to replacement of air in the pore space with water, the thermal conductivity of which is much higher than the thermal conductivity of the air.

To keep the heat in the room, it is necessary to have a coefficient of thermal resistance not less than the required, i.e. the normative value of the reduced resistance to heat transfer should be taken at least normalized values. To fulfill this requirement, multi-layer wall structures with heat insulators are necessary.

Having a sequence of n-layers with different coefficients of thermal conductivity and diffusion, we can construct 2n! Combinations of layers. It is quite obvious that alternating layers by even transpositions, we have 2n-combinations of layers, of which for small n (for example, 4, 5, ...), it is easy to choose the best combination that ensures low humidity while maintaining thermal resistance. This was done. The best options are shown in Fig.2. These options simultaneously demonstrate minimum absolute humidity and high thermal resistance.

4 Conclusions

Energy efficiency requirements are low coefficients of thermal conductivity and, as a consequence, low mean layer temperatures. In this case, the absolute humidity decreases with the same value of relative humidity.

Simply put, an energy-efficient wall will be drier than a wall with high thermal stability and a high average temperature, leading to a high absolute moisture content of the layers.

The article showed that even in independent processes of thermal conductivity and diffusion, the temperature and humidity distributions are related: the wall temperature sets the saturated vapor pressure of water and thereby determines the moisture level.

It is shown that the "energy-efficient" is a "dry" wall. In this case, the heat-resistant wall is not both energy-efficient and "dry" at the same time.

References

1. D. D. Zaborova, M. I. Kukolev, T. A. Musorina, M. R. Petrichenko St. Petersburg State Polytechnical University Journal **4**, 28-33 (2016).
2. D. D. Zaborova, T. A. Musorina, M. R. Petrichenko St. Petersburg State Polytechnical University Journal **1 (23),** 18-26 (2017).
3. D. Zaborova, M. Petrochenko, et al, MATEC Web of Conferences **73,** 2-7 (2016).
4. M. Petrichenko N. Vatin, D. Nemova, N. Kharkov, A. Korsun Applied Mech. and Mat. 672-674, 1903-1908 (2014).
5. M. Petrichenko, A. Ostrovaia, E. Statsenko Applied Mech.and Mat. **725-726**, 87-92 (2015).
6. M. Petrichenko, N. Vatin, D. Nemova, N. Kharkov, A. Staritcyna Applied Mech. and Mat. 627, 297-303 (2014).
7. M. Petrichenko, N. Vatin, D. Nemova, V. Olshevskiy Applied Mech. and Mat. **725-726**, 93-99 (2015).
8. M. R. Petrichenko Examples of fundamental and applied research XXV International Scientific and Practical Conference. 31-35 (2016).
9. M. V. Petrochenko, K. I. Strelets, M. R. Petrichenko, E. B. Yavtushenko Applied Mech.and Mat. **672-674**, 567-570 (2014).
10. M.R. Petrichenko, N.S. Kharkov St. P. State Polytechnical U. Journal **63**, 237-242 (2008).
11. V. Murgul, MATEC Web of Conferences, **106**, 06002 (2017) DOi-10.1051/matecconf/201710606002
12. A. Gorshkov, V. Murgul, Advances in Intelligent Systems and Computing, **692**, 577 - 591 (2018) DOi - 10.1007/978-3-319-70987-1_61
13. V. Pukhkal, V. Murgul, IOP Conference Series: Earth and Environmental Science, **90**, 012211 (2017) DOi - 10.1088/1755-1315/90/1/012211
14. V. Murgul, MATEC Web of Conferences, **73**, 02001 (2016) DOi-10.1051/matecconf/20167302001

Smart glass as the method of improving the energy efficiency of high-rise buildings

Olga Gamayunova[1], *Eliza* Gumerova[1,*] and *Nadezda* Miloradova[2]
[1]Peter the Great St. Petersburg Polytechnic University, 29 Politechnicheskaya St., St. Petersburg, 195251, Russia
[2] Moscow State University of Civil Engineering, Yaroslavskoye shosse, 26, Moscow, 129337, Russia

Abstract. The question that has to be answered in high-rise building is glazing and its service life conditions. Contemporary market offers several types of window units, for instance, wooden, aluminum, PVC and combined models. Wooden and PVC windows become the most widespread and competitive between each other. In recent times design engineers choose smart glass. In this article, the advantages and drawbacks of all types of windows are reviewed, and the recommendations are given according to choice of window type in order to improve energy efficiency of buildings.

1 Introduction

Today high-rise buildings are erected more often than it was even 10 years ago. It can be said that this fact becomes global. Claims on facade glazing for these houses differ markedly from requirements to windows of other. In view of some specific conditions unique window constructions will be required for upper zones of skyscrapers.

Furthermore, we should give consideration to energy consumption and conservation. Pride of place goes to windows since current state of its thermal protection is inferior to heat shield of wall envelopes [2].

In accordance with data of energy survey one of the reasons of energy loss is glazing in old wooden frames, which is installed mostly in budgetary organizations (kindergartens, schools, etc.). Obviously, these windows are subjected to be replaced. Energy and economic effectiveness depend on choice of type of window construction.

Heat losses through the window occur for several places: window units, air conduction losses and radiation.

It is clear that value of heat losses through the window unit directly depends on window construction, materials and manufacturing quality. In reality it is about 10% of whole heat losses [6-8].

In case of constant rate increase on electricity and heat, improvement of thermotechnical characteristics is consider to be one of the most effective ways to decide the problem of energy consumption.

* Corresponding author: eliza_gumerova@mail.ru

2 Main types of modern window structures

The days, when windows were only wooden and have ordinary glazing, are gone. Innovative technologies and building materials improve significantly fabrication. Window products change their usual design and become variable, more functional and durable.

At the present time basic materials for window production in Russia are wood, aluminum and PVC. Due to various estimates, percent of plastic windows ranges from 60 to 80% of all translucent structures.

Each of these types of windows works both ways (table 1) and particular consumer [1]:

- PVC windows – the most economic glazing for industrial, public and residential buildings;
- aluminum windows come into operation gradually in residential houses, not only in industrial and public buildings;
- modern wooden windows refer to elite translucent structures and are settled mainly in premium-class flats and villas. Traditional windows are used in weekend cottages [9].

We consider in more details the main properties of PVC and wooden windows. In this article we will not deal with aluminum windows for its primary use in industrial buildings. Furthermore, since there is sufficiently frigid climate in Russia, the value of quantity of settled aluminum windows is small and cannot have an impact on energy effectiveness.

Table 1. Advantages and drawbacks of wooden, PVC and aluminum windows.

Window	Advantages	Drawbacks
Wooden	• high strength of timber, low density provide the high quality; • high heat and sound conductivity; • high freezing resistance; • environmental friendliness; • appearance of natural material; • repair capability	• presence of cracks and knots; • putrescibility; • hygroscopity; • flammability; • requirement for periodic servicing (painting, etc.); • necessity of ventilation; • very high demands to qualification of manufacturers and windows fitters
Aluminum	• almost infinite life; • high strength, low density; • corrosion, deformation and environmental hazard resistance; • lack of necessity to servicing; • ability to manufacture windows of any forms and dimensions	• aluminum smelting is energy-consuming and capital intensive process; • possibility of electrochemical reactions; • high heat conductivity; • necessity of ventilation
PVC	• durability, weather sustainability • impermeability of all joints • lack of necessity to servicing; • high heat and sound conductivity	• necessity of ventilation; • very high demands to qualification of manufacturers and windows fitters

3 Energy-conservating PVC windows

Fast moving progress in production of energy-saving glazing can be one more argument for buildings with facade glazing. Energy-conserving coverings are made with using of nanotechnologies [10].

Energy efficient k- or i-glass are used in energy-saving windows. I-glass has better properties than k-glass. Differences between i- and k-glasses are as in the manufacturing technology as in the emissive coefficient. Application of window units with i-glass not only saves energy, but also increases the comfort of being in apartments.

Drawback of i-glass is a low abrasion resistance that presents some particular difficulties during transportation. However, taking into account the fact that energy-conserving covering of i-glass is inside window unit, this disadvantage does not impact on operational characteristics of i-glass.

Advantages of energy-saving glass:

- one-chamber glass with k- or i-glass has more value than triple-pane window with traditional glass;
- one-chamber glass with energy-saving glass is lighter than triple-pane window, that provides extended operational period and decreases the load on window hardware;
- double-pane glass with low emissivity has more light transmission than two-chambered with customary glass;
- thanks to mass production the cost of one-chamber glass with i- or k-glass is nearly the same as of two-chambered with traditional glass [11].

4 Smart-glass

In recent years hi-tech solutions, for example "smart glass" with changeable transparence, are all the rage. Thanks to thin liquid-crystal film window transparency can be regulated by remote control panel. (Fig. 1)

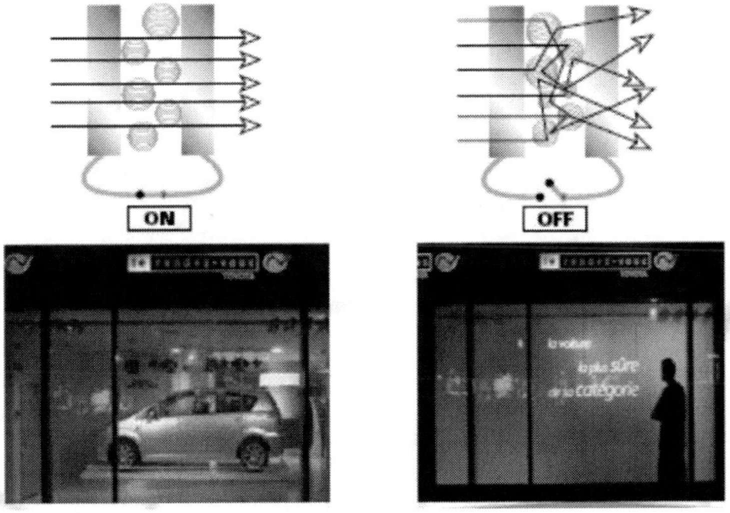

Fig. 1. Smart-glass [4]

Film with changeable matte appearance can be used in combination with various types of glass. The most appropriate variant of matte glass production is chosen due to some peculiarities:

Table 2. Characteristics of ways of making glass with variable mat.

Variant of production of glass with changeable matte surface		Appendix
Triplex		Film between two glasses for installation into frame (dimension to 3x1.5 m)
On glass		Film is applied on glass (for big dimensions or all-glass execution)
In window unit		Film is applied on glass inside the window unit for sound- or heat-insulated structures (for installation into frame)

Analysts note that smart-glass allows to decrease heat losses, cut down expenses on ventilation and lightning, and answers for alternative to louvre shades and curtains. Smart-glass in clear state does not hold violet rays.

5 Features of glazing in high-rise buildings

It is not so long ago as the glazing of buildings was implemented by exceptionally standard windows and cases, but nowadays the following trends are visible distinctly in modern architecture [5]:

- increasing area of glazing (up to 80% of area of facade);
- enlargement of number of storeys in buildings;
- extensive use of window structures inside the buildings;
- increase of dimensions of applying elements of glazing;
- functional extension of glazing, which decreases consequences of negative factors of external environment [12-18].

It is necessary to execute glazing works taking into account architectural and planning concept of building or structure, environmental factors of building site, functional peculiarities and anticipating operating conditions.

There are several key features of glazing in high-rise buildings [3].

Glazing ought not to be a threat to human life and health as indoors, as outdoors during installation and maintenance works. Collapsed glazing and possible fall and wounds can be illustration of hazardous situations which are associated with glazing. Degree of risk is rising with increase of building height. Thus, it is needed to apply only the safety glass (harden, multi-layered, reinforced) or install protection obstacles.

Furthermore, the requirements to fire-resistance of glazing in multi-storey buildings are more strict because of more intensive air draft and long evacuation routes. As a result, air flow rises sharply in place of initial fire during the glass failure. In this case, combustion process intensifies abruptly and fire begins to dpread rapidly indoor. Fire-resistance of glazing is to have be sufficient in order to provide the safe escape time.

It is essential to define loads and actions, which will act on it taking into account that they can be easily changed due to height above ground.

It should be considered that wind load and solar radiation intensity enlarge, temperature and air pressure decrease with increase of height.

As for the optical features, special attention is given to distortions in reflected light and color of applied glass. Daylight area enlarges and in this situation, the exterior of building is worsen markedly by the optical distortions and color inhomogeneity of glazing.

Glazing should maintain its operation abilities while meeting the operation and maintenance requirements during the full life of building. Glazing with the service life smaller, than it was expected, is ought to be replaceable or repairable. This question is relevant in high-rise buildings because maintenance operation, repair and replacement of outside glazing is complicated engineering problem.

Using of glass with self-cleaning coating can be recommended in order to increase operating life and reduce the time of maintenance operation.

6 Conclusions

To increase the energy efficiency, the priority should be given up to modernized structures of energy-saving windows, which can decrease energy consumption by 30% with high profitability of investments without wall insulation.

PVC window units gained popularity in Russia. Nevertheless, wooden windows are environmentally friendly, there is interchange of air in it, but they less resistant to moisture. That is why air will be fresher in rooms with wooden window units than in premises with PVC windows.

Applying of smart glass can solve the problem of safety and energy-saving glazing in high-rise buildings. It cuts down consumption by 40%. According to forecasts, the volume of using the smart glass will be increased by 6 times, and operating income will be annually raised for 21.2% by 2022 year.

Finally, it is highly recommended to take account into increase of risks, that causes the necessity of more strict requirements to safety, complicacy of operation conditions of glazing due to increasing value of loads, growth of aesthetic requirements for whole daylight area.

References

1. Online resource. URL: http://www.techart.ru/files/publications/44-45.pdf
2. Online resource. URL: https://www.msesupplies.com/blogs/news/34196932-global-smart-glass-market-to-grow-to-5-8-billion-usd-by-2020
3. Online resource. URL: http://www.oknamedia.ru/novosti/smart-steklo-novaya-realnost-v-stekolnoy-industrii-45174
4. Online resource. URL: http://tybet.ru/content/articles/ index.php?SECTION_ID=371 &ELEMENT_ID=29320
5. URL: glassinfo.ru/articles/ 2006_01_osobenosti_proektirovania_osteklenia_visotn_zdani.pdf
6. Vatin N., Gamayunova O. *Energy saving at home*. Applied Mechanics and Materials, **672-674**, 550-553 (2014)
7. Vatin N., Gamayunova O. *Energy efficiency and energy audit: the experience of the russian federation and the republic of Belarus*. Advanced Materials Research, **1065-1069**, 2159 (2016)
8. Vatin N., Gamayunova O. *Choosing the right type of windows to improve energy efficiency of buildings*. Applied Mechanics and Materials, **633-634**, 972-976 (2014)

9. Vatin N., Gamayunova O. *The role of the state and citizens in improving energy efficiency*. Applied Mechanics and Materials, **725**, 1493 (2015)

10. Vatin N., Gamayunova O., Nemova D. *An energy audit of kindergartens to improve their energy efficiency*. Advances in Civil Engineering and Building Materials IV - Selected and Peer Reviewed Papers from the 2014 4th International Conference on Civil Engineering and Building Materials, CEBM 2014 4th, 305-308 (2015)

11. Gumerova E., Gamayunova O., Shilova L. *The optimal decision of insulation in cladding structures for energy efficient buildings*. MATEC Web of Conferences, **06020** (2017)

12. Vatin N.I., Nemova D.V., Tarasova D.S., Staritcyna A.A. *Increase of energy efficiency for educational institution building*. Advanced Materials Research, **953-954**, 854-870 (2014)

13. Vatin N.I., Gorshkov A.S., et. Al.*The energy-efficient heat insulation thickness for systems of hinged ventilated facades*. Advanced Materials Research, **941-944**, 905-920 (2014)

14. Kiryudcheva A.E., Shishkina V.V et. al. *Energy efficiency of building envelopes for public buildings*. Construction of Unique Buildings and Structures, **5 (44)**, 19-30 (2016)

15. V. Pukhkal, V. Murgul, IOP Conference Series: Earth and Environmental Science, **90**, 012211 (2017) doi :10.1088/1755-1315/90/1/012211

16. N. Harmathy, V. Murgul, Procedia Engineering, **165**, 1845 - 1852 (2016) DOi-10.1016/j.proeng.2016.11.932

17. A Gorshkov, V. Murgul, Advances in Intelligent Systems and Computing, **692**, 577 - 591 (2018) DOi - 10.1007/978-3-319-70987-1_61

18. N. Harmathy, J. Kontra, V. Murgul, Z. Magyar, Advances in Intelligent Systems and Computing, **692**, 563 - 576 (2018) DOi- 10.1007/978-3-319-70987-1_60

AUTHOR INDEX

AUTHOR INDEX

AUTHOR INDEX

AUTHOR INDEX

AUTHOR INDEX

EDP Sciences – Web of Conferences
17, Avenue du Hoggar
Parc d'Activité de Courtabouf
BP 112
F-91944 Les Ulis Cedex A
France

ISBN 978-1-5108-5945-6

9 781510 859456

CURRAN ASSOCIATES INC.
proceedings
.com

International Scientific Conference "High-Rise Construction 2017" (HRC 2017)

E3S Web of Conferences Volume 33 (2018)

Samara, Russia
4 – 8 September 2017

Part 2 of 2

edp sciences

International Scientific Conference "High-Rise Construction 2017" (HRC 2017)

E3S Web of Conferences Volume 33 (2018)

Samara, Russia
4 – 8 September 2017

Part 2 of 2

Editors:

D. Safarik
Y. Tabunschikov
V. Murgul

ISBN: 978-1-5108-5945-6

Printed from e-media with permission by:

Curran Associates, Inc.
57 Morehouse Lane
Red Hook, NY 12571

Some format issues inherent in the e-media version may also appear in this print version.

Printed by Curran Associates, Inc. (2018)

For additional information, please contact EDP Sciences – Web of Conferences at the address below.

EDP Sciences – Web of Conferences
17, Avenue du Hoggar
Parc d'Activité de Courtabœuf
BP 112
F-91944 Les Ulis Cedex A
France

Phone: +33 (0) 1 69 18 75 75
Fax: +33 (0) 1 69 28 84 91

contact-edps@webofconferences.org

Additional copies of this publication are available from:

Curran Associates, Inc.
57 Morehouse Lane
Red Hook, NY 12571 USA
Phone: 845-758-0400
Fax: 845-758-2633
Email: curran@proceedings.com
Web: www.proceedings.com

TABLE OF CONTENTS

PART 1

PART 2

Modern energy efficient technologies of high-rise construction

Inessa Lukmanova [1,*] and *Roman* Golov [2]

[1]Moscow State University of Civil Engineering; 26, Yaroslavskoe Shosse, Moscow, 129337, Russia
[2]Moscow Aviation Institute (National Research University), Volokolamskoe highway, 4, Moscow, 125993, Russia

Abstract. The paper analyzes modern energy-efficient technologies, both being applied, and only introduced into the application in the construction of high-rise residential buildings. All technologies are systematized by the authors as part of a unified model of "Arrows of Energy-Efficient Technologies", which imply performing energy-saving measures in the design, construction and operation of buildings.

1 Introduction

In conditions of legislative introduction of new standards in the field of energy efficiency of multi-apartment buildings under construction (hereinafter referred to as MAB), one of the important criteria in the field of high-rise construction is the use of modern energy-efficient materials, equipment and technologies in construction of new buildings. The development of appropriate regulations and rules has allowed to establish the necessary legislative base [1-4].

2 Methods

In accordance with the developed documents, a growing number of construction companies are using more energy-efficient solutions. One of the reasons for this attention to the level of energy efficiency of MAB is their prevailing role in the total energy consumption of the entire urban economy shown by the example of Moscow in the following diagram (Fig. 1) [5].

In accordance with the foregoing, it can be concluded that energy-efficient construction in the housing sector is one of the promising areas of high-rise construction. In practice, it is aimed not only at achieving compliance of buildings under construction with legislative norms, but also at increasing demand for such real estate from buyers. Certain attention is paid in the domestic science to the studies in the field of energy saving technologies in housing and communal services, but, as practice shows, existing developments, unfortunately, are not of a systemic nature and are devoted to selective analysis of certain energy-efficient technologies.

* Corresponding author: lukmanova@mgsu.ru

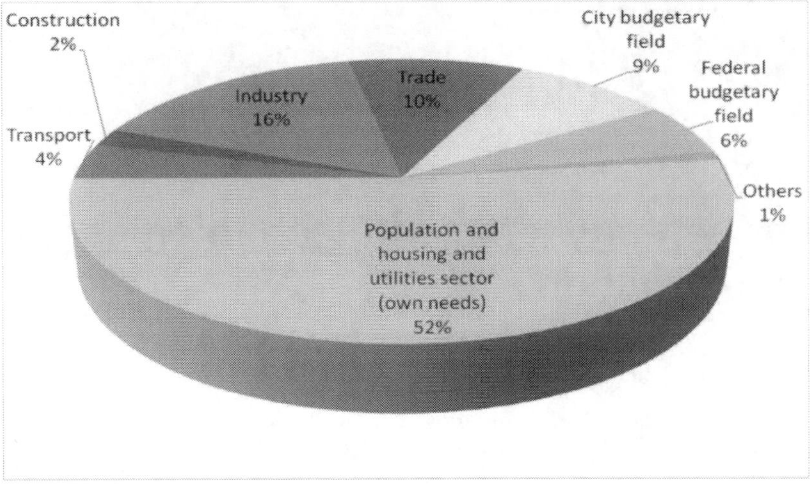

Fig. 1. Distribution of energy consumption in Moscow.

As part of the study, the authors developed a unified model called the "Arrow of Energy-Efficient Technologies", which combines energy-saving technologies implemented in all structural and engineering subsystems of MAB, as well as those solutions that should be applied at the building design stage in order to pre-select and analyze individual technologies and options for their integration into a single project. Systematicity in the issue of selecting and implementing energy-efficient technologies, according to the authors, is one of the most important factors for success in MAB construction. Visual representation of the proposed model is shown in the diagram (Fig. 2).

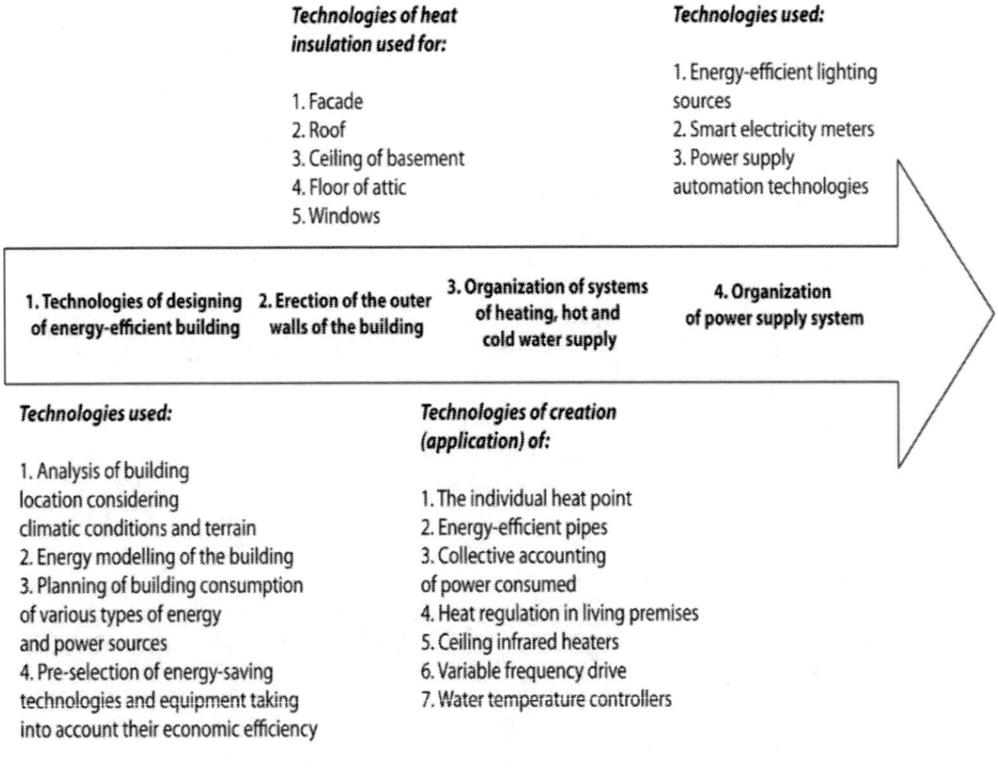

Fig. 2. the "Arrow of Energy-Efficient Technologies".

The model developed by the authors is based on previous studies in the field of energy saving in the form of three components: for training of personnel (a complex educational model of the "Arrow of Energy-Efficient Competencies" was developed [6]); as part of developing a unified approach to automation of energy saving processes (the foundations of an integrated information and analytical system that implements all the functions of energy-efficiency management of buildings based on data from metering devices were formulated, while developing a methodical approach to improving energy efficiency, recommended for use in various industries [7].

3 Results

The implemented system has become one of the most popular and successful in the Russian market, allowing to analyze the energy efficiency of heating subsystems, hot and cold water supply, power supply subsystems in automatic mode, calculating the current level of energy consumption, detecting violations of qualitative and quantitative parameters of energy resources received by the building, managing its energy installations, and issuing recommendations on the introduction of additional energy-saving technologies.

Let's consider separate, most significant components of the mentioned blocks of the "Arrow of Energy-Efficient Technologies". Within the first block, when analyzing the building location, taking into account climatic conditions and terrain, it is also necessary to take into account the environmental requirements for MAB construction [8].

The software complexes used for energy modeling allow forming, calculating and analyzing heat, lighting and aerodynamic processes: heat exchange between the facade of the building and the external environment, heating loads, heat generation, etc. The most popular software systems in the field of energy modeling include DOE 2.2, Equest, Trane TRACE 700, Carrier HAP, which are widely used in Europe and the US and have enough modules to calculate various energy and engineering parameters of the buildings under construction.

The Project Expert software package proved to be suitable to pre-select energy-efficient technologies and equipment with the calculation of their economic efficiency.

In the second block, the most important role is played by the technology of thermal insulation of the building facade, since it is through the walls that the building loses up to 40% of heat. Energy-efficient technologies in this area are: a suspended ventilated facade that contributes to the building's strength and provides excellent thermal insulation, fire safety, protecting the walls from negative weather effects, liquid ceramic insulants (TSM KERAMIK), providing energy-saving effect 10 times higher than traditional materials, and extruded polystyrene foam.

One of the promising directions in the development of insulation for roofing is the technology of incorporating natural components into mineral wool insulation materials, which ensure a higher level of their environmental compatibility and safety, for example, the incombustible insulation Ekos developed by Knauf. For the insulation of the ceiling of basement - extruded polystyrene foam, as well as sprayed insulation materials from polyurethane foam Ecotermix Bio produced by the Russian company Ecotermix.

When insulating the floor of attic, it is advisable to use reflective insulation materials: penofol, armofol, liquid foil, etc., which can reduce the thermal conductivity of insulation materials up to 10 times.

Currently, there are several types of energy-saving double-glazed windows. A type of such products quite common in the market is double-glazed windows with a heat-reflecting film with a layer of oxide of metal, silver or titanium oxide, with chambers filled with gas: argon, krypton or xenon). A key area of energy saving in heating and hot water supply systems in buildings (where losses reach up to 70% of the total energy losses of the entire building) is the installation of an individual heat point (IHP), which automates and optimizes the process of supplying the heat carrier to consumers. Such control units are produced by many Russian manufacturers; of the largest foreign structures selling IHP in the domestic market is Danfoss [9-10].

4 Discussion

Experts recommend metal pipe pipes, which have higher strength characteristics and provide long service life, as the most durable version of pipes.

One of the leading Russian manufacturers of common building metering devices is TBN Energoservis company. Its products are characterized by high quality, durability and protection of its devices from external mechanical influences and attempts to electronically hack them. One of the trends in energy saving in the housing and utilities sector is the per-apartment implementation of energy-efficient technologies, allowing residents to individually create the most comfortable microclimate with thermoregulators and thermostats (the most famous are Honeywell, Genius, Danfoss, Drayton, Hive, etc.).

Using ceiling infrared heaters, which heat not the air itself, but furniture and furnishings that give off heat, oxygen is not burned in the rooms, warm air practically does not accumulate under the ceiling, and heating itself is more uniform and the heat energy is used more efficiently (Almac, Ballu, Timberk, Friro, Ecoline and others).

It is recommended to install a variable frequency drive to increase the efficiency of the asynchronous motor, achieving energy saving in comparison with the throttle mechanism up to 60%, energy carrier transfer up to 25%, reducing the failure rate of heating and water supply networks, and avoiding water-hammer effects.

As an effective tool for optimizing the temperature regime of hot water and coolant, temperature controllers should be used to control it in an automatic mode, providing a comfortable room temperature for residents. Within the power supply system, it is recommended to use LED light sources: General Electric, Gauss, LG, Philips, Osram, Cosmos, Navigator, etc., smart power meters transferring data via various types of controllers (GPRS controller, LPWAN controller, or Wi-Fi controller). At the same time, the latter is the most optimal, capable of maintaining a stable connection and being fault tolerant.

As part of the automation of the lighting system, various types of sensors have proven themselves: a photocell, a digital timer with a weekly or annual program, an astronomical timer, a motion sensor providing savings up to 70% (Toyama, Schneider Electric, Lutron Electronics, Vantage Controls, etc.).

5 Conclusions

In modern conditions, the energy efficiency improvement of multi-apartment buildings has become an integral part of high-rise construction. The development of legislation in the field of energy saving in Russia contributes to the development of this area in our country. The research of the authors allowed to consolidate the most effective of the applied technologies as part of a single model. The authors believe that only the systemic and integrated introduction of energy-efficient technologies at all stages of the life cycle of the Moscow Ring Road provides the maximum energy effect, which brings tangible economic benefits to apartment owners, making their housing one of the most demanded in the modern real estate market.

References

1. D.N. Silka, M.A. Ivanova, Proceedings of Moscow State University of Civil

Engineering, **12**, 5(104), (2017) doi: 10.22227/1997-0935.2017.5.572-58
2. I.N. Aleksandrovich, Int. J. Appl. En. Res., **10**, 21, 42415-42418 (2015)
3. N. Ivanov, Proc. Eng., **153**, 228–231 (2016) http://dx.doi.org/10.1016/j.proeng.2016.08.107
4. A.A. Pakhratdinov, D.V. Oreshkin, Ekologiya urbanizirovannykh territoriy, **3**, 84-88 (2016)
5. A. Y. Dobromyslov, *Raschet i konstruirovanie sistem kanalizacii zdanij* (Strojizdat, Moscow, 2012)
6. D. Connelly, CPD, **27** (2007)
7. N.Yu. Yaskova, Bulletin of the Irkutsk State Technical University, **1 (60)**, 178-186 (2007)
8. I.G. Lukmanova, N.Yu. Yaskova, Economics of construction, **4 (28)**, 13-19 (2014)
9. N. Harmathy, J. Kontra, V. Murgul, Z. Magyar, Advances in Intelligent Systems and Computing, **692**, 563 - 576 (2018) DOi- 10.1007/978-3-319-70987-1_60
10. N. Harmathy, V. Murgul, Procedia Engineering, **165**, 1845-1852 (2016) DOi- 10.1016/j.proeng.2016.11.932

Comparison of Thermal Stability of Dry High-strength Concrete and Wet High-strength Concrete

Tatiana Musorina[1], *Aleksandr Katcay*[1,*], *Anna Selezneva*[1] *and Victor* Kamskov[2]
[1] Peter the Great St. Petersburg Polytechnic University, 29 Politechnicheskaya St., St. Petersburg, 195251, Russia
[2] Moscow State University of Civil Engineering; 26, Yaroslavskoe sh., Moscow, 129337, Russia

Abstract. High-strength concrete is a modern material, which occupies it`s own niche on the construction material market. It is applicable in a large-scale high-rise construction, particularly an underground construction is a frequently used solution for a space saving. Usually underground structure is related to a wet usage environment. Though not all properties of the high-strength concrete are investigated to the full extent. Under adverse climatic conditions of the Russian Federation one of the most important properties for constructional materials is a thermal capacity. Therefore, the main purpose of the paper is to compare a thermal capacity of the high-strength concrete in humid conditions and a thermal capacity of the high-strength concrete in dry operational condition. During the study dependency between thermal capacity and design wall thickness and ambient humidity has to be proven with two experiments. As a result the theoretical relation between thermal capacity characteristic – thermal inertia and wall thickness and ambient humidity was confirmed by the experimental data. The thermal capacity of a building is in direct ratio to the construction thickness. It follows from the experiments and calculations that wet high-strength concrete has less thermal stability.

1 Introduction

Nowadays the construction of a high-rise constructions is highly developed. From this perspective the industry requires high-strength materials, in particular concrete. High-strength concrete is a modern material, which occupies unique niche on the world construction market.

High-strength concrete of classes over B60 is made of a Portland cement based binder [1]. High-strength concrete during hardening cures much quicker than common concrete. This is related to the low water-cement ratio, which is achieved through the use of a plasticizer and high heat of setting due to quick humidification and high level of a binder content [2].

Since most of Russian cities are located in a climate with low temperature scenario, it is important to consider to what extent a material used in the construction possesses characteristics necessary for comfortable indoor climate, such as: thermogravitational

* Corresponding author: zenit67ru@mail.ru

convection[3], energy efficiency[4-9], thermal stability. High-strength concrete has to be strength, frost resistance and water penetration resistance tested [10,11], however under Russian harsh climatic conditions thermal capacity is an extremely important characteristic of a building envelope. The accumulative capacity of wall structures determines the amount of heat absorbed and passed by the structure and used to maintain a required temperature level of the room and enclosing wall when the temperature of the external heat source changes [12]. According to [13], thermal stability is one of the most important topics in the building design, as it measures how quickly a material gains or loses heat to the environment. The thermal stability is property of the thermal conductivity coefficient, density of the material, specific heat and heat transfer resistance [14]. Consequently a thermal stability calculation may be summarized as estimation of an internal surface temperature swing or as measurement of damping of temperature oscillations in building structures. [15]. In superstructure building construction with underground and semi-basement floors, basements it is important to consider moisture-holding capacities of rooms with building structures and external environment.

The method of thermal stability estimation for building structures is based on determination of a range of temperature on internal surface of a building structure [16]. Stationary method based on the regularities of stationary thermal stability, as shown by theoretical and experimental studies, provides greater accuracy [17]. The application of heat-resistant materials reduces energy consumption of heating or wall thickness. According to [18] the thermal capacity characteristic is a thermal lag D

$$D = R \cdot s \tag{1}$$

$$R = \frac{\delta}{\lambda} \tag{2}$$

$$D = \frac{\delta}{\lambda} \cdot s = \frac{s}{\lambda} \cdot \delta \tag{3}$$

Nomenclature

D	thermal lag
s	estimated coefficient of heat absorption
R	thermal resistance of certain
δ	thickness
λ	thermal conductivity

According to app. [18] in dry operational conditions for crushed-stone concrete estimated coefficient of heat absorption is 16.77, thermal conductivity is 1.74, for wet operational conditions 17.88 and 1.86 subsequently.

Then in dry operational conditions:

$$\frac{s}{\lambda} = \frac{16.77}{1.74} = 9.64$$

In wet operational conditions:

$$\frac{s}{\lambda} = \frac{17.88}{1.86} = 9.61$$

Hence with equal thickness of structure the thermal lag D in wet operational conditions is less than it is in dry operational conditions.

The thermal stability of construction depends on the following factors [19]:

1) physical condition and structure, which are determined by the phase state of matter;

2) a degree of crystallization and size of crystals;

3) anisotropy of the thermal conductivity of crystals and the direction of heat flow;

4) a volume of porosity of the material and the characteristics of the porous structure;

5) chemical composition and presence of impurities (the latter especially affects the conductivity of crystalline bodies);

6) maintenance conditions, depending on temperature, pressure, humidity of the material;

7) thickness of the structure;
8) sequence of layers.

Dry air enclosed in small closed pores, which is practically not possible convective heat transfer, has the minimum conductivity. In this case, the thermal conductivity of air is minimal and counts 0.023 W/(m·°C). Therefore, the increase in porosity of the material is the main way to reduce thermal conductivity.

The thermal conductivity of the materials increases significantly with hydration. This is because the moisture that falls into the pores and partly replaces the air and increases the thermal conductivity of the material, because the thermal conductivity of water λ = 0.58 W/(m·°C) greater in 25 times than the thermal conductivity of air. Thermal lag is in inverse ratio to thermal conductivity, hence with thermal conductivity decrease thermal stability increases. Estimated coefficient of heat absorption depends on material and humidity under operating conditions. Whereas thermal lag is in inverse ratio to estimated coefficient of heat absorption, which is bigger in the wet operational conditions than in normal or dry ones.

Nevertheless, there are a few articles on the issue of a thermal stability of a high-strength concrete. Because of this, the purpose of this article is to discover the characteristics of thermal stability for high-strength concrete.

1.1 Goal

From the foregoing it follows that in order to reasonably conclude about the thermal features of wet high-strength heavy concrete, it is necessary to examine it and see how its thermal features differ from the properties of a dry high-strength heavy concrete and from which it depends.

1.2 Tasks

To achieve the objectives of the article it is required:

- to compare the thermal stability of the structure depending on its thickness;
- to carry out experiments to study the thermophysical features of materials;
- to compare properties of wet high-strength heavy concrete and dry high-strength heavy concrete;
- graphically see the difference between the characteristics of wet high-strength heavy concrete and dry high-strength heavy concrete.

2 Materials and Methods

2.1 Materials

For the experiment there were made 2 samples of wet high-strength heavy concrete and 1 sample of dry high-strength heavy concrete with the following composition:

Table 1. Composition of concrete.

№	Concrete	Cement, [kg]	Slag, [kg]	Sand, [kg]	Crushed stone, [kg]		Water, [kg]	Water-cement ratio	Plasticizer PF 3196 0,5%, [kg]	Density, [kg/m^3]	Strength, [MPa]
					5-10 mm, [kg]	10-20 mm, [kg]					
1	high-strength heavy concrete	380	76	798.4	273	728.5	175	0.38 (0.47)	1.90	2428	70.1

https://doi.org/10.1051/e3sconf/20183302048

2.2 Methods

The experiment consists on simulating the real conditions of typical structures in cold climates, it is by drastic temperature changes in few hours during the same day. It has built a box (imitation of a closed room) for dry high-strength concrete simulating the indoor conditions. It consists of the following: a wooden board, a heat insulator, a concrete sample. It is represented in fig. 1. Also it has built a 2 cubes for wet high-strength concrete. It consist of the following: a heat insulator, a concrete sample. It is represented in fig. 2.

Installing the appropriate sensors and equipment inside it is the way to be in accordance with real characteristics of buildings [20-24]. The initial temperature of the air inside the chamber was 18°C, internal environment. The initial temperature of the air inside the chamber is 30°C, followed by a drastic drop down to 0°C and then up to the initial temperature. The temperature was measured with the help of 3 sensors: on the outer surface, inside the model in the middle, inside the model at the edge. In this way it is able to assess the behavior which really apply in real life conditions for giving future conclusions about the material. As explained before and in accordance to the aim of this paper, the idea is to simulate with all possible details all conditions of environments in similar climatic conditions tendencies.

Fig. 1. Dry high-strength concrete.

Fig. 2. Wet high-strength concrete.

The sample was placed in a chamber (initial temperature inside the structure was approximately 30°C). The sample was at 30°C degrees for 60 minutes. Then followed an decrease of temperature in the chamber to 0°C degrees. The sample was heated until the sensor inside the structure began to register an increase in temperature. Due to the fact that the material is thermally stable, it continues to cool down, even if the chamber has at plus temperature. Then the material was cooled until the temperature of the sensor began to fall.

3 Results

The results of the experiment are shown at fig.3 where presented changes in temperature of wet high-strength concrete that are depended on temperature in samples through time.

 https://doi.org/10.1051/e3sconf/20183302048

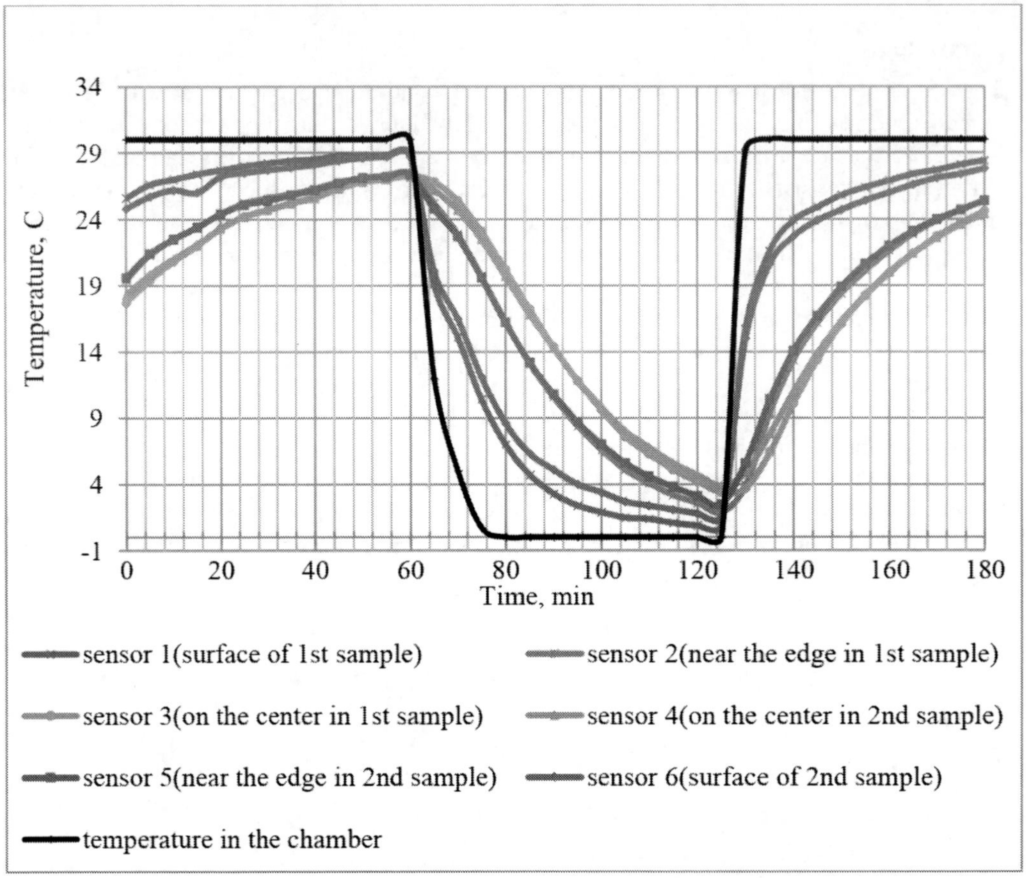

Fig. 3. Graph of temperature of concrete change from temperature through time.

The dependence of a temperature lag from material thickness is confirmed by the diagram Fig.3: sensors 2, 5, located close to the edges of models, shows more abrupt change of temperature during a switch of a heat regime inside the chamber than sensors 3, 4, located inside the models in the middle. Hence with increase of material thickness the thermal stability of a building construction increases.

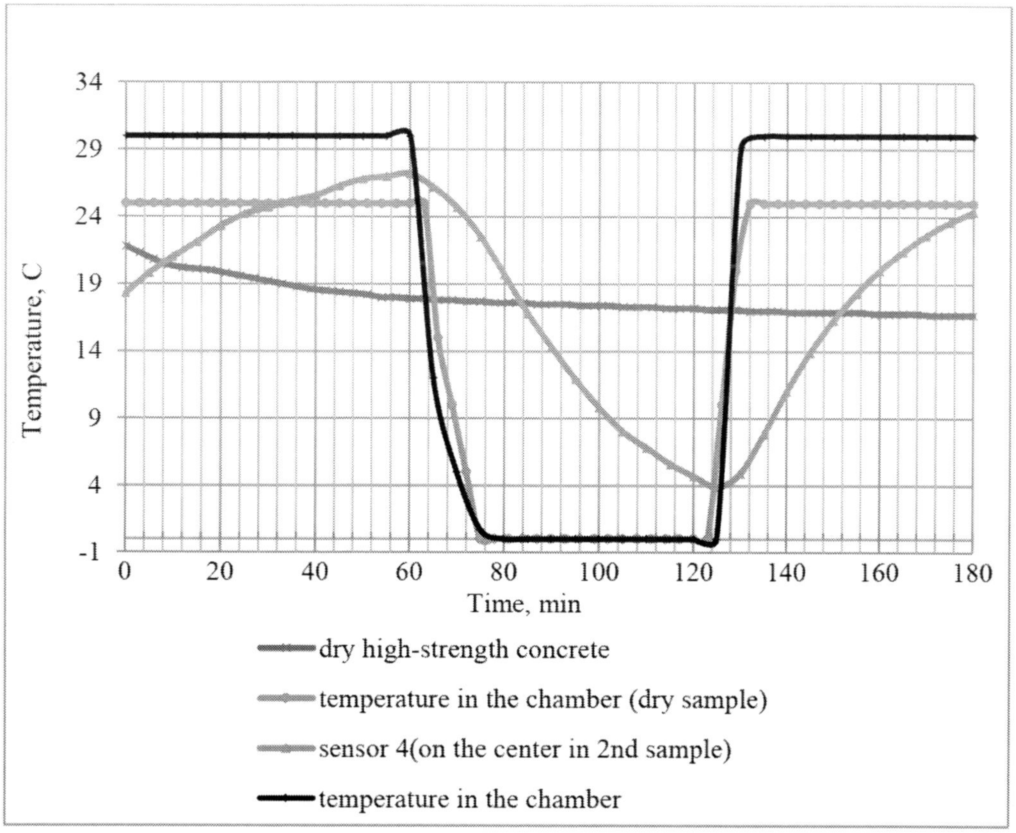

Fig. 4. Graph of temperature of wet and dry concretes change from temperature in chamber through time.

At fig. 4 it can be seen that in wet high-strength concrete the temperature swing is significantly more than one in dry high-strength concrete. That confirms accuracy of the above theoretical comparison of thermal stability of concrete under wet and dry conditions.

4 Discussion

In underground construction it should be considered, that thermal stability of wet high-strength concrete is less than thermal stability of dry high-strength concrete, since alternatively thermal properties of the basement areas will differ considerably from thermal properties of the building superstructure.

For more accurate research of thermal properties of high-strength concrete for usage in underground construction a frost resistance experiment should be conducted.

Conclusions

1. It was experimentally proved, that thermal capacity of a building is in direct ratio to the construction thickness;
2. It follows from the experiments and calculations that wet high-strength concrete has less thermal stability;

Acknowledgement

Thanks to the laboratories of Institute of civil engineering from Peter the Great St. Petersburg Polytechnic University which provided all equipment and materials and helped to do experiments.

References

1. O. M. Smirnova Mag. of Civil Eng. **6**, 12-22 (2016).
2. M. M. Batdalov, H. N. Maiev, D. K. S. Batayev, S. G. K. Bata Safety of struct. **1**, 45-47 (2013).
3. N. Vatin, M. Petrichenko, D. Nemova: Applied Mech. and Mat. **633-634**, pp. 1007-1012 (2014).
4. N. Vatin, O. Gamayunova.: Advanced Mat. Research. **1065–1069**, pp. 2159–2162 (2015).
5. V. Kostenko, N. Gafiyatullina, A. Semchuk, M. Kukolev: Mag. of Civil Eng. **68, No 8**, pp. 18-25 (2016).
6. D. D. Zaborova, T. A. Musorina, M. R. Petrichenko St. P. State Polytechnical U. Journal **231**,18-26 (2017).
7. D. Zaborova, M. Petrochenko, L. Chernenkaya MATEC Web of Conferences **73**, 2-7 (2016).
8. S. Korniyenko Procedia Eng. **117**, 191–196 (2015).
9. M. Petrichenko, N. Vatin, D. Nemova, N. Kharkov, A. Staritcyna Applied Mech. and Mat. **627**, 297-303 (2014).
10. State Standart 31914 *High-strength heavy-weight and fine-grane concretes for situ-casting struct. Rules for control and quality assessment* (2012).
11. EN 1992 Eurocode **2**. *Design of concrete structures* (1992).
12. D. Zaborova, T. Musorina, A. Selezneva, A. Butyrin, Therm. Resist. and accum. of Heat by the Wall Constr. (International Scientific Conference Energy Management of Municipal Transportation Facilities and Transport EMMFT), 473 (2017).
13. D. D. Zaborova, T.A. Musorina, M.R. Petrichenko: St. P. State Polytechnical U. Journal **23. 1**, 18-26 (2017).
14. Y.G. Barabanshikov, S.V. Belyaeva, M. Antonova, S.V. Vasyutina: Procedia Eng., 41-49 (2015).
15. The Ministry of construction of Ukraine DBN **V.2.6**. *Thermal insulation of buildings*, 64 – 65(2006).
16. State Standart 26253 *Buildings and struct. Method for determining the building struct ability to maintain* (1984).
17. M. Y. Leshchinsky, Testing of concrete: Right allowance (1980).
18. SP 50.13330 *Thermal protection of buildings* (2012).
19. Y. P. Gorlov and etc. The techn. of thermal insulation mat. (1980).
20. A. Keyvanfar, A. Shafaghat, M. Zaimi A. Majid, B. L. Hasanuddin, M. W. Hussin, Kherun Nita Binti Ali, A. D. Saad Journal of Renewable & Sustainable Energy Reviews **39**, 277-293 (2014).
21. B. Nagy, S. G. Nehme, D. Szagri, Energy Procedia **78**, 2742 – 2747 (2015).
22. V.Borovkov, A. Kalyutik, V. Sergeev, Heat and mass transfer equipment. Publishing house of Polytechnic University, 231 (2008)
23. V. Pukhkal, V. Murgul, IOP Conference Series: Earth and Environmental Science, **90**, 012211 (2017) DOi - 10.1088/1755-1315/90/1/012211
24. N. Harmathy, J. Kontra, V. Murgul, Z. Magyar, Advances in Intelligent Systems and Computing, **692**, 563 - 576 (2018) DOi- 10.1007/978-3-319-70987-1_60

Strength resistance of reinforced concrete elements of high-rise buildings under dynamic loads

Mikhail Berlinov[1*]

[1] Moscow State Construction University (National Research University), 129337, Moscow, Yaroslavl highway, d. 26, Russian Federation

Abstract. A new method for calculating reinforced concrete constructions of high-rise buildings under dynamic loads from wind, seismic, transport and equipment based on the initial assumptions of the modern phenomenological theory of a nonlinearly deformable elastic-creeping body is proposed. In the article examined the influence of reinforcement on the work of concrete in the conditions of triaxial stress-strain state, based on the compatibility of the deformation of concrete and reinforcement. Mathematical phenomenological equations have been obtained that make it possible to calculate the reinforced concrete elements working without and with cracks. A method for linearizing of these equations based on integral estimates is proposed, which provides the fixation of the vibro-creep processes in the considered period of time. Application of such a technique using the finite-difference method, step method and successive approximations will allow to find a numerical solution of the problem. Such an approach in the design of reinforced concrete constructions will allow not only more fully to take into account the real conditions of their work, revealing additional reserves of load capacity, but also to open additional opportunities for analysis and forecasting their functioning at various stages of operation.

1 Introduction

Increasing efficiency and quality of high-rise construction under conditions of dynamic loads from wind, seismic, traffic and process loads requires the use of more economical and reliable constructions, which becomes possible if the advanced methods of calculation are used in the design [1]. The creation of such methods is associated with the need to more fully reflect the real properties of materials accounting their actual behavior under the influence of external loads. Development of such methods of calculation is closely linked with the consideration of the nonlinear and rheological properties of materials [2].

It is known that the technique becomes all the more perfect, the more it links the object with the surrounding real world, but it is known from philosophy that the number of such

* Corresponding author: berlinov2010@mail.ru

links is infinite, but allowing for a sufficient majority of the basic relationships will help to solve this problem [3].

Obviously, the improvement of computational methods lies in the ways of their increasing complexity due to the necessity of using a very complex mathematical apparatus, since it is this factor that can realize the aforementioned variety of connections [4], and the use at the present stage of the development of society. And the use of such a powerful tool as modern computers and their software allows you to implement fairly complex mathematical solutions with the help of numerical methods and the necessary linearization of nonlinear dependencies [5].

2 Results section

For dynamic calculation, as a rule, well-known equations of dynamic equilibrium are used [6] , which have the following form:

$$\left.\begin{aligned}\frac{\partial\sigma_x}{\partial x}+\frac{\partial\tau_{xy}}{\partial y}+\frac{\partial\tau_{xz}}{\partial z}+X=\rho\frac{\partial^2 u}{\partial t^2}\\ \frac{\partial\tau_{yx}}{\partial x}+\frac{\partial\sigma_y}{\partial y}+\frac{\partial\tau_{yz}}{\partial z}+Y=\rho\frac{\partial^2 v}{\partial t^2}\\ \frac{\partial\tau_{zx}}{\partial x}+\frac{\partial\tau_{zy}}{\partial y}+\frac{\partial\sigma_z}{\partial z}+Z=\rho\frac{\partial^2 w}{\partial t^2}\end{aligned}\right\} \qquad (1)$$

where: X, Y, and Z are the projections of the volume forces on the corresponding axes; ρ is the mass of the material; u, v, and w displacement along the Cartesian coordinate axes, t-time.

The proposed calculation technique is based on the main assumptions of the modern phenomenological theory of a nonlinearly deformable elastic-creeping body, and takes into account the geometric nonlinearity expressed in nonlinear dependencies between strains and stresses (the Cauchy conjecture is regarded as usually fair). Therefore, the resolution of these equations, taking into account all these factors, will be substantially complicated.

For the derivation of resolving mathematical dependencies, we use a well-known technique that allows us to obtain resolving equations in stresses:

$$\Theta(t)=\frac{1-2\nu}{E_0(t)}S_m(F)F(t)-(1-2\nu)S_p(F)F(t)K(F,\omega,\omega_{\min},t)C(t,t_0)$$
$$-(1-2\nu)\int_{t_0}^{t}S_n(F)F(\tau)K(F,\omega,\omega_{\min,}\tau)\frac{\partial}{\partial d\tau}C(t.\tau)d\tau \qquad (2)$$

here: v - coefficient of relative transverse deformations; E_0 (t) is the modulus of elastic-instantaneous deformations; $\Theta = \varepsilon_x (t) + \varepsilon_y (t) + \varepsilon_z (t)$ is the relative volume deformation;

K (σ_x, ω, ω_{min}, τ) is the vibro-creep coefficient; C (t, t_0) is a measure of creep strains.

The nonlinearity functions of elastic-instantaneous deformations and creep strains have the form:

$$\left.\begin{aligned}&S_m(F)F(t)=\sigma_x(t)S_m[\sigma_x(t)/R(\rho,t)]+\sigma_y(t)S_m[\sigma_y(t)/R(\rho,t)]+\sigma_z(t)\,S_m[\sigma_z(t)/R(\rho,t)];\\ &S_p(F)F(\tau)K(F,\omega,\omega_{min},\tau)=\sigma_x(\tau)S_p[\sigma_x(\tau)/R(\rho,\tau)]K(\sigma_x,\omega,\omega_{min},\tau)+\\ &+\sigma_y(\tau)S_p[\sigma_y(\tau)/R(\rho,\tau)]K(\sigma_y,\omega,\omega_{min},\tau)+\sigma_z(\tau)S_p[\sigma_z(\tau)/R(\rho,\tau)]K(\sigma_z,\omega,\omega_{min},\tau)\end{aligned}\right\} \qquad (3)$$

here: S_m [σ_x (t) / R (ρ, t)] is the nonlinearity function for uniaxial deformation; R (ρ, t) - concrete strength under uniaxial dynamic deformation; ρ - coefficient of asymmetry of the cycle.

If the connection between stresses and strains is taken in a known inverse form, which in the case of a uniaxial stress-strain state has the form:

$$\sigma_x(t)=E_0(t)\varepsilon_x(t)\varphi_m[\varepsilon_x(t)]+Q(t,t_0)\varepsilon_x(t)\varphi_p[\varepsilon_x(t)]-\int_{t_0}^{t}\frac{\varepsilon_x(\tau)\varphi_p[\varepsilon_x(\tau)]}{K(\sigma_x,\omega_{\min},\omega,\tau)}\frac{\partial}{\partial\tau}Q(t,\tau)d\tau \quad (4)$$

where: $\varphi_m=1+\eta_m\left(\frac{\varepsilon}{\varepsilon_R}\right)^{a_m}$ and $\varphi_p=1+\eta_p\left(\frac{\varepsilon}{\varepsilon_R}\right)^{a_p}$ (5)

- functions of nonlinearity of deformations; Q (t, t_0) - the hereditary function of the second kind

here: ε_R - is the limiting relative deformation of a given class of concrete, at the moment immediately preceding the failure, with acceptable accuracy, determined from experimental data or analytically by the formula:

$$\varepsilon_R=\frac{R(\rho,t)}{E(t)}, \quad (6)$$

here: E (t) is the elasto-plastic modulus of concrete deformations

It should be noted that the coefficients of dependences (5) can be obtained directly from the real diagram (σ-ε) or by solving logarithmic equations systems mathematically.

For tangential stresses, the formula takes the form:

$$\tau_{xy}(t)=\frac{E^0_m(t)}{2(1+\nu)}\gamma_{xy}(t)\varphi_m[\gamma_{xy}(t)]+\frac{Q(t,t_0)}{(1+\nu)}\gamma_{xy}(t)\varphi_p\times$$
$$\times[\gamma_{xy}(t)]-\frac{1}{2(1+\nu)}\int_{t_0}^{t}\frac{\gamma_{xy}(\tau)\varphi_p[\gamma_{xy}(\tau)]}{K(\sigma_x,\omega_{\min},\omega,\tau)}\frac{\partial}{\partial\tau}Q(t,\tau)d\tau \quad (7)$$

The values of the characteristics entering into equation (7) can be found similarly to (4)

The hereditary function of the second kind in the dependences (4) and (7) can be found analytically by the generally accepted method from the solution of the following integral equation:

$$L(t,\tau)-Q(t,\tau)=\int_{t_0}^{t}L(\xi,\tau)Q(t,\xi)d\xi \quad \} \quad (8)$$

where: $L(t,\tau)=-E(\tau)\frac{\partial}{\partial\tau}\left[\frac{1}{E(\tau)}+C(t,\tau)\right].$

The relationship between stresses and deformations for an elastic-creeping body is established by considering this expression together with equation (2), from which we obtain:

$$\sigma_x(t)=\frac{\nu E_0(t)}{(1+\nu)(1-2\nu)}\varphi_m(\Theta)\Theta(t)+\frac{\nu Q(t,t_0)}{(1+\nu)(1-2\nu)}\varphi_p(\Theta)\Theta(t)-\frac{\nu}{(1+\nu)(1-2\nu)}\int_{t_0}^{t}\frac{\varphi_n(\Theta)\Theta(\tau)}{K(F,\omega_{\min},\omega,\tau)}\cdot$$
$$\cdot\frac{\partial}{\partial\tau}Q(t,\tau)d\tau+\frac{E_0(t)}{2(1+\nu)}\varepsilon_x(t)\varphi_m(\varepsilon_x)-\frac{1}{2(1+\nu)}\int_{t_0}^{t}\frac{\varepsilon_x(\tau)\varphi_n(\varepsilon_x)}{K(\sigma_x,\omega_{\min},\omega,\tau)}\frac{\partial}{\partial\tau}Q(t,\tau)d\tau \quad (9)$$

where: analogous to (3), φ_m (Θ) is a unified function of nonlinearity determined on the basis of experimental data or analytically:

$$\varphi_m(\Theta)\Theta(t)=\varepsilon_x(t)\varphi_m[\varepsilon_x(t)/\varepsilon_R(t)]+\varepsilon_y(t)\varphi_m[\varepsilon_y(t)/\varepsilon_R(t)]+\varepsilon_z(t)\varphi_m[\varepsilon_z(t)/\varepsilon_R(t)] \quad (10)$$

For the case of quasilinear creep, formula (9) can be somewhat simplified to a more acceptable form:

$$\sigma_x(t) = \Theta\left[\frac{\varepsilon(t)}{\varepsilon_R(t)}\right]\Big\{ \frac{\nu E_0(t)}{(1+\nu)(1-2\nu)}\Theta(t) + \frac{\nu Q(t,t_0)\Theta(t)}{(1+\nu)(1-2\nu)} - \frac{\nu}{(1+\nu)(1-2\nu)}\int_{t_0}^{t} \frac{\Theta(\tau)}{K(F,\omega_{\min},\omega,\tau)} \cdot$$

$$\cdot \frac{\partial}{\partial \tau} Q(t,\tau)d\tau + \frac{E_0(t)}{2(1+\nu)}\varepsilon_x(t) - \frac{1}{2(1+\nu)}\int_{t_0}^{t}\frac{\varepsilon_x(\tau)}{K(\sigma_x,\omega_{\min},\omega,\tau)}\frac{\partial}{\partial\tau}Q(t,\tau)d\tau \Big\} \quad (11)$$

To reduce the writing, equations (9) and (11) can be represented as:

$$\sigma_x(t) = \tilde{\lambda}(t)\Theta(t) + 2\tilde{\mu}(t)\varepsilon_x(t) \quad (12)$$

where: $\tilde{\lambda}(t)$ and $\tilde{\mu}(t)$ are respectively non-linear integral operators.

The dependences for tangential stresses taking into account (8) take the form:

$$\tau_{xy} = \frac{E_0(t)}{2(1+\nu)}\gamma_{xy}(t)S_m\left[\frac{\gamma_{xy}(t)}{\gamma_R(\rho,t)}\right] + \frac{Q(t,t_0)}{2(1+\nu)}\gamma_{xy}S_p\left[\frac{\gamma_{xy}(t)}{\gamma_R(\rho,t)}\right] -$$

$$- \frac{1}{2(1+\nu)}\int_{t_0}^{t}\gamma_{xy}(\tau)S_n\left[\frac{\gamma_{xy}(\tau)}{\gamma_R(\rho,\tau)}\right]\frac{1}{K(\tau_{xy},\omega_{\min},\omega,\tau)}\frac{\partial}{\partial\tau}Q(t,\tau)d\tau \Big\} \quad (13)$$

By resorting to similar arguments, one can obtain analogous relations for σ_y (t), σ_z (t), τ_{yz} (t) and τ_{zx} (t), which can finally be represented in the following form:

$$\begin{aligned} &\sigma_x(t) = \tilde{\lambda}(t)\Theta(t) + 2\tilde{\mu}(t)\varepsilon_x(t); \qquad \tau_{xy} = \tilde{\mu}(t)\gamma_{xy}(t) \\ &\sigma_y(t) = \tilde{\lambda}(t)\Theta(t) + 2\tilde{\mu}(t)\varepsilon_y(t); \qquad \tau_{yz} = \tilde{\mu}(t)\gamma_{yz}(t) \quad (14) \\ &\sigma_z(t) = \tilde{\lambda}(t)\Theta(t) + 2\tilde{\mu}(t)\varepsilon_z(t); \qquad \tau_{zx} = \tilde{\mu}(t)\gamma_{zx}(t) \end{aligned}$$

Analyzing equations (14), we can conclude that the rejection of nonlinearity and rheology of deformation will turn them into mathematical expressions of the inverse form of Hooke's law.

Considering the compatibility of deformation of concrete and reinforcement and carrying out similar reasoning, it is possible to easily obtain solving equations for reinforced elements in orthotropic reinforcement (Figure 1.a), working both with cracks and without them. For example, for elements operating without cracks, the integral operators will have a slightly different form than in (14), since they will take into account the reinforcement of the element, namely:

$$\hat{\lambda}_i(t) = \frac{\nu E_m^0(t)(1-\mu_i)(1+\alpha\mu_i)}{(1+\nu)(1-2\nu)}\cdots -$$

$$- \frac{\nu(1-\mu_i)}{(1+\nu)(1-2\nu)}\int_{t_0}^{t}\frac{\cdots}{1+\mu_i E_s K(F,\omega_{\min},\omega,\tau)}\frac{\partial}{\partial\tau}Q(t,\tau)d\tau \Big] \quad (15)$$

$$\hat{\mu}_i(t) = \frac{E_m^0(t)(1+\alpha\mu_i)}{2(1+\nu)}\cdots - \frac{1}{2(1+\nu)}\int_{t_0}^{t}\frac{\cdots}{1+\mu_i E_s K(\sigma_x,\omega_{\min},\omega,\tau)}\frac{\partial}{\partial\tau}Q(t,\tau)d\tau$$

The proposed method is, in principle, allows to consider any direction of reinforcement, such as an inclined A_i (Figure 1,b). In this case, the reinforcement ratio: $\mu_{si} = A_{si}/a_1a_2$; a_1

and a_2 - the distance between the individual rods with oblique reinforcement. Note that in this expression ratio $\alpha_c = K(\sigma_{bp}, \omega_{min},\omega,\tau)C(t,\tau)/ K_s(\sigma_{sp},\omega_{min},\omega,\tau)\, C_s(t,\tau)$ can be considered as reinforcement reduction coefficient for concrete, interpreted with respect to the creep deformation in contrast to the values of $\alpha = E_s / E^0_m(t)$ valid for elastic-instantaneous deformations.

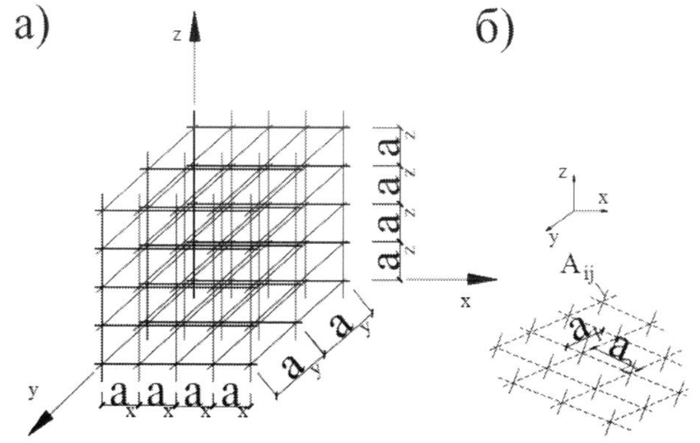

Fig.1. Directions of reinforcement; a) - is orthotropic; b) – inclined.

Omitting even more obvious arguments for elements working with cracks without taking into account the nonlinearity and rheology of deformation of reinforcing steel in all three directions of the coordinate axes, we obtain:

$$\hat{\lambda}_i(t)=\frac{\nu E_s^0(t)\psi_{si}\mu_i}{(1+\nu_s)(1-2\nu_s)}\cdots-\frac{\nu\psi_{si}\mu_i}{(1+\nu_s)(1-2\nu_s)}\int_{t_0}^{t}\cdots\frac{\partial}{\partial\tau}Q_s(t,\tau)d\tau \;]$$
$$\hat{\mu}_i(t)=\frac{E_s^0(t)\psi_{si}n_{\tau i}\mu_{si}}{2(1+\nu)}\cdots-\frac{\psi_{si}n_{\tau i}\mu_{si}}{2(1+\nu)}\int_{t_0}^{t}\cdots\frac{\partial}{\partial\tau}Q_s(t,\tau)d\tau \qquad (16)$$

Where the integral operators turn into physical constants for the deformation of the reinforcement, really taking into account the work of concrete on the sections between the cracks. Those, in the final analysis, we obtain three resolving systems of equations differing only in the values of integral operators, which take into account the reinforcement and appearance of cracks.

Obviously, taking into account the dependences (14), (15) and (16) when searching for a numerical solution, with the aid of some calculating apparatus (the finite element method, for example), will encounter insurmountable mathematical difficulties, since it will require closed integration of the resolving equations; therefore, in order to obtain engineering solutions in numerical form, it is necessary to linearize the problem at least in the physical area. This linearization is possible on the basis of the well-known method of integral estimates.

Discussion

Calculation of reinforced concrete constructions in the operational state according to the accepted design standards is usually performed with the help of the well-known linear apparatus of the theory of elasticity or structural mechanics, and in the classical theory nonlinearity, rheology and anisotropy of deformation are ignored. These imperfections of materials are estimated only by multiplying the calculated characteristics by some empirical coefficients that do not depend on the modes of action of external loads and, what is especially important, the duration of their action. More advanced engineering methods that have become popular recently take into account the so-called ideal nonlinearity of deformation, allowing to obtain some additional opportunities for economical design of building constructions. At the same time, they do not give an opportunity, unfortunately, to investigate the stress-strain state of reinforced concrete constructions in the operational stage under time-varying external load regimes, as well as to reveal the peculiarities of the redistribution of internal forces from the more loaded sections and the cross-section components to less loaded ones.

Modern norms and design practice allow for the consideration of rheology and nonlinearity of deformation at the recommendatory level and mainly for static calculations for planar and three-dimensional structures. Although there are numerous methods and proposals that allow taking into account these factors, it is true, basically, again for static loading conditions.

The current methodology for assessing the adaptability and redistribution of forces under dynamic effects on reinforced concrete constructions is inaccurate and, in fact, empirical. It does not allow us to seek a complex solution of the problem, taking into account the nonlinearity and duration of deformation, the regime and duration of loading, the features of energy absorption, vibro-creep, etc. and their influence on the process of redistribution of efforts.

To answer the above questions, the integral estimation method allows. In this case, it is necessary to consider the following prerequisites as valid: the geometric scheme of the constructions does not change even at the moment immediately preceding the failure. This is due to the fact that at that moment the load-bearing capacity of the cross-sections exists, with at least a minimal, but still finite rigidity. And in dangerous sections, the dangerous development of forces under the influence of an increasing external load is inhibited by the process of redistribution of forces and reaches its maximum at the time of destruction; the assumption of smallness of deformations with a finite value of displacements remains; the hypothesis on the additivity of causes and the consequences of deformation, which is otherwise called the principle of the independence of the action of forces, is considered fair. The mechanical interpretation of the law of conservation of energy in modern mechanics of a deformable solid body, expressed by the principle of possible displacements, which asserts the equality of the work of external and internal forces on every possible infinitesimal displacement of points of the system, is applicable both for ideally elastic and for nonlinearly, non-equilibrium and anisotropic- deformable systems. The application of this principle to perfectly elastic systems assumes that the entire work of internal forces is spent only on the formation of deformations and accumulates in the form of potential energy. Its use in systems with imperfections of deformation requires taking into account possible irreversible energy consumption for hysteresis losses, residual deformation, overcoming of internal friction and heat generation.

In connection with the fact that the energy expenditure during the deformation of real bodies depends both on the regimes and the duration of the action of external loads, and on the level of internal forces formed, then in the energy balance equations the irreversible energy losses mentioned above, must be taken into account, in addition, and equations of equilibrium and energy balance equations must be written for an infinitesimal time interval.

The assumption of smallness of deformation is applicable to reinforced concrete constructions, since their absolute dimensions are always much greater than their absolute deformations, and this assumption is valid regardless of the type of the equation of the mechanical state of the materials used.

The additivity hypothesis, formulated in mechanics in the form of the principle of independence of the action of forces, is of fundamental importance in the methods of calculating building constructions. However, for calculation of non-linearly deformable systems it is, strictly speaking, not applicable, but for linear but nonequilibrium-deformable systems it is suitable only for synchronous changes in time of external loads and mechanical properties of materials. At the same time, the additivity hypothesis is not limited to the principle of the independence of the action of forces, it can be interpreted as a feature of the mathematical form of recording the resolving equations in displacements or in stresses (15).

If we assume that the displacements of the system must reflect the overall level of the stress-strain state of the system, then they become dependent on this level of the stressed state, but the form of recording the equations does not change. However, the implementation of such a mathematical apparatus is possible only through the application of the method of integral estimates in combination with a system of successive approximations that clarifies the stress state with allowance for all kinds of redistribution of forces and stresses in the cross sections. The hypothesis of additivity is expedient to use only when the linearizing sense of the integral estimator method is taken into account.

Conclusions

In the method of integral estimates, process of successive approximations is not just a mathematical technique used instead of closed solution, but a method of integral refinement, based on an iterative process, internal forces and stresses, transforming over time due to their redistribution from more stressed sections and components of the cross sections on the less loaded .

Following the adoption of such an assumption, derived above, integral-differential equations apply to non-linear differential permitting decisions based on the finite difference method or finite element, which in turn transforms them into a system of nonlinear algebraic equations.

Under these conditions, it is necessary to take into account that the coefficients of the system of algebraic equations themselves become nonlinear and depend on the level of the stress state and require constant refinement in the course of the solution based on the method of successive approximations. It is necessary to use special techniques that improve the convergence of the iterative process.

Considering the process of dynamic deformation of an elastically creeping body, it is possible to construct a calculation technique under dynamic load conditions of any type, and the case of harmonic effects will be the simplest.

To realize the solution of the problem in the non-linear formulation, which provides the dependence of the physic-mechanical characteristics of the materials on the level of the stressed state of the material, it is required to find the magnitude of the acting stress on each of the faces of each volumetric finite element at each stage of the iteration process. To this end, we should use the well-known dependences (14), bearing in mind that instead of the constants adopted in the theory of elasticity, we must consider nonlinear integral operators (15), (16).

It is necessary to bear in mind too one significant drawback inherent in the finite element method, which consists in the fact that the displacement field in a deformable body divided into finite elements is continuous, since for the chosen law of change of displacements for the entire system the conditions of continuity of displacements are automatically satisfied. As for the field of stresses and deformations, while remaining constant (under the linear law of displacement) in each finite element, they undergo discontinuities at the boundaries between the elements, i.e. on the lines of division of the continuum into finite elements.

Thus, the above reasoning makes it possible to find a numerical solution of the posed problem for a triaxial nonlinear and elasto-creeping deformed body under the conditions of the dynamic action of external loads using the method of integral estimates based on the stepping method and successive approximations. The implementation of such method in high-rise construction will more fully reflect the real properties of materials accounting their actual behavior under the influence of dynamic loads that will make it possible to save material resources.

References

1. V.M. Bondarenko, V.I. Rimshin. The theory of dissipative power resistance of reinforced concrete, 287 (2015)
2. E.M. Zverjaev, M.V. Berlinov, M.N. Berlinova. The integral method of definition of basic tension condition of anisotropic shell.International Journal of Applied Engineering Research, 11(8), 5811 (2016)
3. E.A. Korol, M.V. Berlinov, M.N. Berlinova, Kinetics of the strength of concrete in constructions.Procedia Engineering, 165, 292 (2016)
4. E.A. Korol, Yu.A. Kharkin, *Book of Reports of XX Russian-Polish-Slovak Workshop "Theoretical Fundamentals of Construction"* 401 (2011)
5. V.M. Bondarenko,V.I.Kolchunov. Calculation models of the power resistance of reinforced concrete, 112 (2004)
6. A.G. Tamrazyan, S. G. Yesayan. Mechanics of creep of the concrete, 490 (2012)

Estimation of erosion-accumulative processes at the Inia River's mouth near high-rise construction zones.

Natalya Sineeva [1,*]

[1]Novosibirsk State University of Architecture and Civil Engineering, 630099 Str. Leningradskaya 113, Novosibirsk, Russia

Abstract. Our study relevance is due to the increasing man-made impact on water bodies and associated land resources within the urban areas, as a consequence, by a change in the morphology and dynamics of Rivers' canals. This leads to the need to predict the development of erosion-accumulation processes, especially within the built-up urban areas. Purpose of the study is to develop programs on the assessment of erosion-accumulation processes at a water body, a mouth area of the Inia River, in the of perspective high-rise construction zone of a residential microdistrict, the place, where floodplain-channel complex is intensively expected to develop. Results of the study: Within the velocities of the water flow comparing, full-scale measured conditions, and calculated from the model, a slight discrepancy was recorded. This allows us to say that the numerical model reliably describes the physical processes developing in the River. The carried out calculations to assess the direction and intensity of the channel re-formations, made us possible to conclude, there was an insignificant predominance of erosion processes over the accumulative ones on the undeveloped part of the Inia River (the processes activity is noticeable only in certain areas (by the coasts and the island)). Importance of the study: The study on the erosion-accumulation processes evaluation can be used in design decisions for the future high-rise construction of this territory, which will increase their economic efficiency.

1 Introduction

Small water objects of Novosibirsk lost their former importance. Over the past 40-50 years Novosibirsk water objects no longer are used take water for household needs, drinking, and their fishery value is practically lost. The Inia River is not an exception in this sense, since it flows through a territory subjected to anthropogenic influences. The indirect source of influence on the water flow and sediment in the estuary area of the Inia River, located within Novosibirsk, is the Novosibirsk hydroelectric complex, the natural course of water levels in the section of the Inia River is distorted by influence of the Obi River backwater. The increased source, in comparison with the "natural", sediment inputs in the Inia River within the city of Novosibirsk are HPS-5, Borok quarry, Sibtransmash plant. All of them contribute finely-dispersed material in the River, which has a high adsorption capacity, immobilizes various elements, organic compounds, oil products, and in its turn there is the

* Corresponding author: ief@sibstrin.ru

source of secondary River pollution, for example, under condition of the project's implementation for the River canal and coastal areas improving. The project of the small city River improvement may include clearing the channel, relief geo-plasticity, bank protection structures, retaining hydraulic structures, etc. In any case, this contributes to a change in the geometry of the River channel, for example, it's narrowing and, as a consequence, channel processes activation, such as deep or lateral erosion.

In this regard, it is necessary to improve hydrological and ecological monitoring, which includes not only a system of observations, but also an assessment, and a forecast of changes in the water bodies state, erosion-accumulation processes, the direction and intensity of channel reorganization. Carrying out of such works involves modern computer systems adaptation.

The present work was carried out on a grant of the Novosibirsk City Mayoralty. The main goal of the work is the numerical model adaptation for the assessment of the direction and intensity of channel deformations for the River mouth in the zone of perspective high-rise build-up of a residential microdistrict, developed by the Institute of Water and Environmental Problems of the SB RAS, by Dr. phis.-mat. sciences, V.A. Shlychkov. As an initial data, an explanatory note to the project for the planning of the residential, public-business and recreational area (the Inya River mouth) was send to the city mayor's office

In order to achieve this goal, three main tasks are formulated, the first two of which are aimed at obtaining the initial data for the mathematical model adaptation:

1. Analytical study of bottom sediments of Inya mouth region, which allows obtaining data on their composition and properties, including the study of granulometric, mineral compositions and organic constituents.

The density and moisture capacity of bottom sediments, which determine the ability of soils to water exchange, depend on the granulometric composition. Water exchange affects the rate of chemical reactions, physico-chemical exchange processes, and a number of biological processes in soils. Many of their physical, physical - chemical and chemical properties were associated with the mineral constituent of the bottom sediments. Obtaining data on the organic matter amount is important, since organic matter is involved in the formation of compounds complexes with finely dispersed silty or colloid fractions of inorganic components that are suspended or in bottom sediments (metals, clay particles, etc.). Organic acids (acetic, oily, etc.), carbohydrates, proteins, fats, amino acids, humic substances, esters, enzymes were found in natural waters.

The whole complex of organic substances is continuously transformed, some substances disintegrate, and others are formed. Organic carbon is a reliable indicator of the total content of organic matter in natural waters. Humic and fulvic acids make up a significant part of the surface waters' organic matter.

Thus, bottom sediments are considered as a multiphase heterogeneous system of an open type, exchanging substances and energy with other media.

2. Hydrometric survey of a water body with the channel reception survey of the Inya River mouth's speed structure, turbidity of the water, water flow and slope of the water surface.

3. Performing process on a numerical experiment to assess the direction and intensity of channel re-formations, which includes the following main stages: analysis of natural-natural factors and hydraulic features of the River section; selection of source materials containing full-scale hydrometric and morphometric information; the formation of a digital terrain model in the nodes of a regular grid with a given spatial resolution; a curvilinear finite-difference grid generating, taking into account the geometry of the solution domain and the requirements for spatial detail, which provides the required accuracy of calculation; the formation of initial fields and boundary conditions for the numerical model; study the of the model sensitivity and calibration of parameters based on actual information; the carrying out of working calculations, analysis and interpretation of the results.

2 Materials and Methods

The numerical experiment is a leading method to evaluate the direction and intensity of channel deformations based on two-dimensional numerical models developed by the Institute of Water and Environmental Problems of the SB RAS, by Dr. phis.-mat. sciences, V.A. Shlychkov. With a purpose to obtain the initial data for the purpose of mathematical models adaptation, bottom sediments were studied at different points of the River section to determine the granulometric, mineral composition and organic component by laboratory methods, according to N.A. Kachinsky works; quantitative X-ray phase analysis; I.V. Tyurin (in the modification of V.V. Ponomareva and T.A. Plotnikova). The hydrometric survey of the water body was carried out in accordance with the adopted recommendations.

1. Determination of the granulometric composition.

Within the present framework, the bottom sediments' granulometric composition of River silts (as well as the mineral composition and content of organic substances) was studied by selecting a number of samples in the River canal, followed by their laboratory treatment. The study of the bottom sediments' granulometric composition was carried out in the soil biogeochemistry laboratory of the Institute of Soil Science and Agrochemistry of the SB RAS using N.A. Kachinsky method, based on the relationship between particle sedimentation rate in an aqueous medium and their diameter (O.B Kuznetsova. et al, 2014).

2. Determination of mineral composition.

On the basis of the Geochemistry Laboratory of Rare Elements and Ecogeochemistry and the X-ray structural analysis laboratory of the Joint Institute of Geology, Geophysics and Mineralogy of the SB RAS, an analysis of the bottom sediments' mineral composition of the object was made. This study was carried out using quantitative X-ray phase analysis with a help of the DRON-3 diffractometer (Fetisov GV, 2007), which made it possible to determine the intensity dependence of the phase diffraction peaks being determined from its content in the sample.

3. Determination of organic constituents.

The study of organic constituents was carried out at the Institute of Soil Science and Agrochemistry of the SB RAS using the laboratory method of chemical analysis according to I.V. Tyurin scheme (modified by V.V. Ponomareva and T.A .Plotnikova) (O.B. Kuznetsova et al, 2014) content of organic carbon; brown humic acids carbon; fulvic acids carbon was defined.

4. Hydrometric survey.

Hydrometric surveys were carried out at the mouth of the Inia River near the motor bridge (Fig 2). In order to obtain channel survey, investigations were carried out using the Garmin GPS 178c measuring complex, which allows depth measurements with an accuracy of up to 0.1 m and determine positions in geographic coordinates, depending on the number of satellites with an accuracy of 1 m (Lyakhin Yu.S. & Perepelitsa D.I., 2009). Also, the positions of the left and right coast water cuts were recorded. Measurements of flow velocities by a hydro-metric turntable of the GR-21M mark were made in the measuring section at the site in question. Determination of water consumption and the discharge of suspended sediments was carried out according to recommendations of (Bespalova L.A. & Berdnikov S.V., 2016). The water flow rate is calculated by analytical ($Q = \int_F V df$) and grapho-analytical ($Q = \int_B q db$) by the ways, where F – flow area, m^2; V – speed, m/s; q – specific consumption, m^2/s; B – width of the River, m. The rate of suspended silts was determined analytically. The water level fluctuations were recorded on the water gauge rail. To determine the turbidity of the water flow (mass concentration of suspended sediments), water samples were taken by the bottles of GR-16M with subsequent processing of the selected samples.

3 Equations and mathematics

Mathematical modeling to assess the direction and intensity of channel deformations was carried out on the basis of two-dimensional numerical models, developed by developed by the Institute of Water and Environmental Problems of the SB RAS, by Dr. phis.-mat. sciences, V.A. Shlychkov. Numerical experiment is based on two models having the following blocks in this research paper:

1. двумерная, вертикально осредненная модель плановых течений в водоемах суши (гидродинамическая составляющая);
2. модель транспорта наносов и деформации русла (деформационная составляющая).

1. two-dimensional, vertically averaged model of planned currents in the land's water bodies (hydrodynamic component);
2. model of the silts transport and deformation of the River canal (deformation component).

Из всего многообразия формул для расхода наносов в качестве базы использовалась хорошо апробированная и широко применяемая в расчетах зависимость Бэгнольда, не предусматривающая деление наносов на взвешенные и донные. Формула Бэгнольда базируется на анализе мощности, расходуемой потоком на перемещение твердых частиц (Боровков В.С., 1989). Соотношение для насыщающих значений c_* концентрации выглядит как

Of all the formulas' varieties for sediment discharge, a well-tested and well-known Bagnold correlation was used as a base, which did not involve the division of silts into suspended and bottom sediments. Bagnold's formula is based on the analysis of the power consumed by the solid particles flow (Borovkov V.S., 1989). The ratio for saturating concentration c_* values looks like

$$c_* = \sigma \frac{u_*^2}{g\,h}\left(\frac{e}{tg\,\varphi} + 0.01\frac{|\mathbf{u}|}{w_g}\right) \tag{1}$$

where ρ, ρ_c – density of water and solid slurry material, $\sigma = \rho_c / \rho$ – relative density of alluvium, u_* – friction speed, e – coefficient that takes into account the energy losses for sediment transport, $\varphi \approx 32^\circ$ – angle of internal friction of sediments in water, w_g – hydraulic size. Variable-free expression (1) sets the total critical concentration of suspended and attracted sediment; the first term in parenthesis (1) determines the velocity of sediment transported, the second is suspended. The total sediment consumption by Bagnold is written as

$$\mathbf{q}_b = h\,\mathbf{u}\,c_* . \tag{2}$$

We now state the boundary conditions..In the input line, the total flow of the River is known Q_1. In the cross-section of the output line, the level of the free surface is recalculated into depths h .

The hydrodynamic problem statement closes the initial conditions on the velocity components $u = v = 0$ and spatial distribution of depths h in moment The required parameters of the flow were obtained by integrating the equations of Saint-Venant's planned currents in time before reaching the steady-state regime (Shlychkov V.A., 2008).

The main numerical experiments were carried out under steady-state flow conditions with a water flow rate $Q =$ 34,01 m^3/s. One of the numerical model components for determining the rate of deformation is the digital model of the River channel relief, which was built on the basis of the field surveys' results. The problem was solved in two stages. First, a vector velocity field was calculated u, v and the scalar field of the stream depths h, corresponding to a given flow rate Q. At the second stage, the trends of the basic deformation processes were determined on the basis of the previously calculated velocity field characteristics.

4 Results

1. The granulometric composition curve of the bottom sediments is shown in Fig. 1. Investigation of the granulometric composition showed that the bottom sediments of the Inia River mouth area. In accordance with the accepted classification (Kramarenko V.V., 2016) is represented by sand, dust and mud.

2. X-ray analysis of the bottom sediments' mineral constituent showed the presence a large part of rock minerals of massive crystalline magmatic origin (quartz and acid plagioclase (in the sum of ≈ 60-70%). And also the presence of dense and loose sedimentary rocks: carbonates are represented by calcite; clay minerals are present in a subordinate amount: illite, chlorite, smectite; trace amounts (no more than 10% of the total) marked potassium feldspar and hematite

3. The results of the determination of the organics' fractional composition (humic acids, fulvic acids, insoluble residue of sediments) are expressed as a percentage of the total organic carbon content of more than 10%. The value of the sediments' insoluble residue (53.4% of the total organic carbon of the sample) characterized the fixing strength of substances with the clay fraction and the degree of the organic matter humification.

4. The total length of the channel survey was 500 m (Figure 2). Based on the research results and geodetic work, a plan of the channel section was constructed in the isobaths - in 0.25 m (Fig. 3) at a given working water surface level, and also the morphometric characteristics of the channel in the adopted section were determined: the area of the water section is 131.12 m^2; the River width along the working level of the water surface is 108.4 m; average depth is 1.2 m; The maximum depth was 2.0 m. The velocity of the water flow was measured on 12 high-speed verticals in a detailed way in a real gauging gate. The values 131,12 $м^2$;of the average velocities on the verticals are in the range 0.192 ... 0.346 m/s. The flow rate of the water flow, calculated analytically, is 33.89 m^3/s, and graphoanalytical is 34.13 m^3 / s. The calculated flow of suspended sediment is 2612,002 g/s.

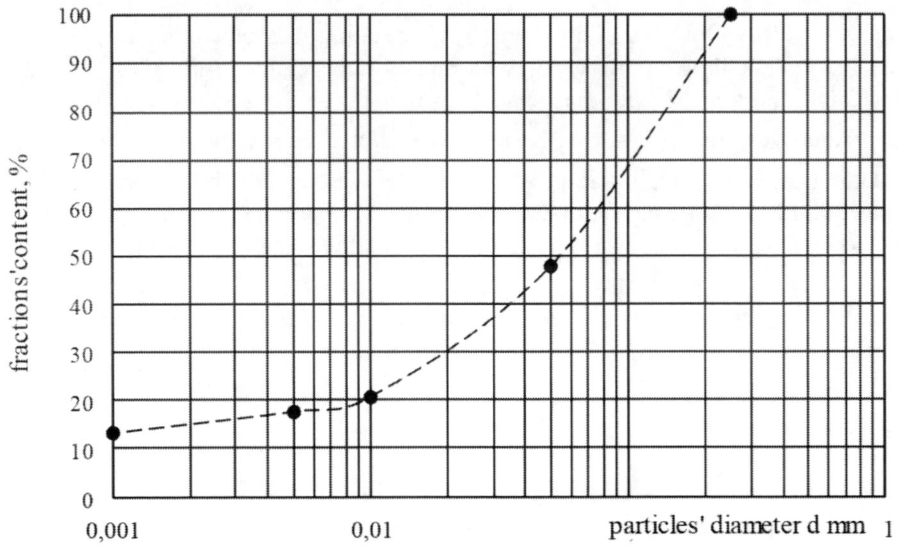

Fig 1. Curve granulometric composition of the bottom sediments in the Inia River's estuary region

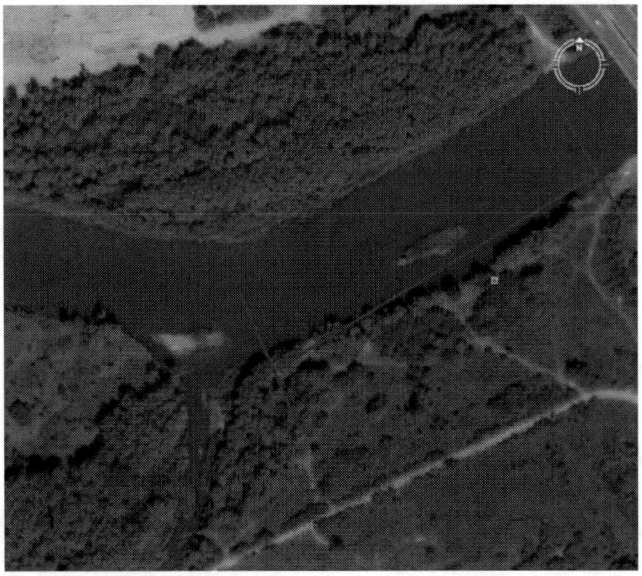

Fig 2. Space shot of the channel section

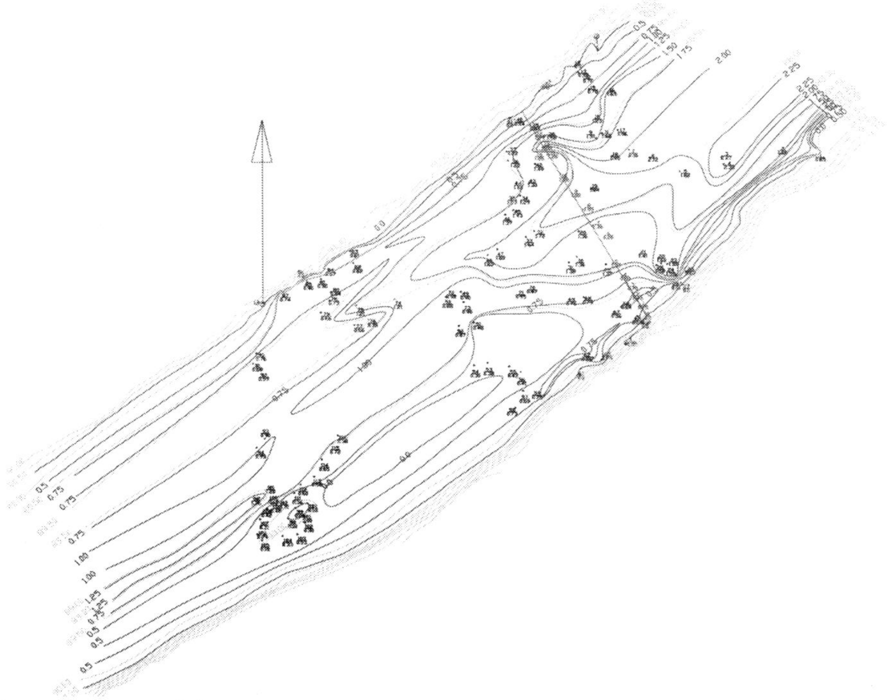

Fig 3 Channel survey of the area.

5. Results of mathematical modeling.

The purpose of the calculations was to study the features of erosion and sediments' accumulation in the selected area of the Inia River in "domestic" (natural) conditions. Undoubtedly, its further urbanization strengthening will make significant changes in channel processes.

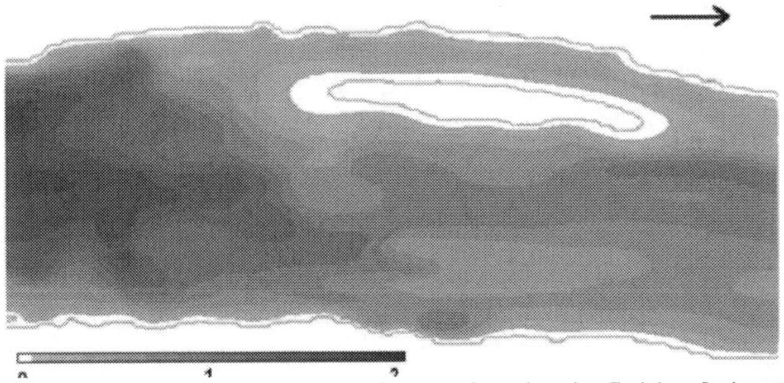

Fig 4 Depth field the Inia River' area. It shows the depth field of the River section, constructed using the planned numerical model used

As a result of calculations in the Inia River' area, the prevalence of erosive processes over accumulative ones was revealed; there is a tendency to erosion along the right bank along the River, where the highest erosion values (up to 1.4 sm / day) are observed in the area 100-200 m from the origin, as well as in the middle of the left side and the most of the left (small) (Fig 5). The average erosion values in the area under consideration are 0.12 sm/day, accumulation is 0.09 sm/day, a pronounced erosion process occurs only near the coast and the island.

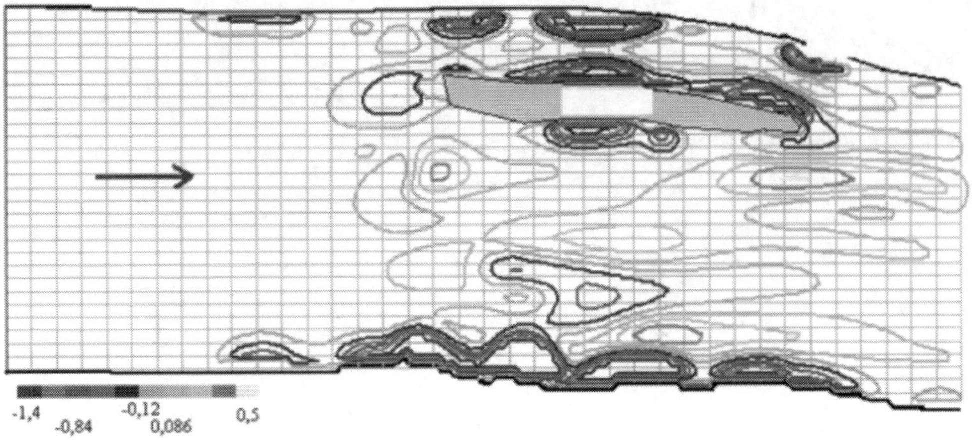

Fig 5. Tendencies of erosion and accumulation processes in the River section

The process of accumulation is clearly expressed on the right side of the island (yellow color in Fig 5), the strain rate is 0.5 sm/day. The same intensity of deformations occurs in places of flow velocities decreasing, which causes the acceleration of sedimentation processes and the gradual accumulation of sediments. In the area, which is near the motor bridge and to the island in the area in question (0-100 m from the origin), the intensity of erosion-accumulation processes is an order of magnitude lower than in the narrowing of the water flow (about 100 meters further from the origin).

5 Discussion

1. The granulometric, mineral compositions and of the bottom sediments' organic constituents of the Inia estuary region were studied. The obtained data are presented in the form of a granulometric graph and a descriptive part and used in the adaptation of a numerical model.

2. In order to obtain the initial information on the main water flow hydraulic characteristics, a hydrometric survey was carried out for the mathematical calculations of channel deformations in the River region of its mouth area. It is no more than 500 m long in the zone of direct residential microdistrict's construction.

3. The performed calculations on the evaluation of the direction and intensity of the channel reorganizations made it possible to adapt the mathematical model in terms of orientation and intensity (Shlychkov V.A., 2008). Also, the results made it possible to conclude that there is an insignificant predominance of erosion processes over accumulative processes in the area in question (mainly, in some areas, near the coast and the island). In general, we can talk about the approximate state of the dynamic equilibrium, characteristic of natural conditions.

6 Conclusion

Strengthening the urbanization factors will contribute to pollution of the River channel, regime violation of the levels and flow rates, channel regime, which can lead to the development of irreversible channel changes that can not be eliminated without a directed engineering impact. The task of regulating channel processes is to provide the necessary hydraulic, operational, ecological indicators of the River channel as a whole. In this regard, the hydrological-ecological monitoring is rather important, its main tasks of are: regular observations of the natural characteristics' current state; identification of factors and patterns of technogenic and natural changes in ecosystems in time and space; these changes assessment; modeling and forecasting of changes in water systems under anthropogenic impact; development of recommendations for nature management processes.

References

1. O.B. Kuznetsova, O.I. Shirikova, Z.V. Kireeva, Korobeinikova, *Chemical methods of studying the state of the environment (*Vologda, 2014)
2. G.V. Fetisov, *Synchrotron radiation. Methods for studying the structure of substance-* : (OOO Publishing company "Physics and Mathematics ", Moscow, 2007)
3. Yu.S. Lyakhin, D.I. Perepelitsa, *Experience of using combined echo sounders with GPS and GIS-technologies for constructing cart diagrams of water objects. Water management in Russia: problems, technologies, management* (Russian Research Institute of Integrated Use and Protection of Water Resources, Ekaterinburg, 2009)
4. L.A. Bespalova, S.V. Berdnikov, *Workshop on hydrology staging and conducting observations on the Rivers: educational-methodical manual* (Southern Federal University, Rostov-on-Don, 2016)
5. V.S. Borovkov, *Channel processes and dynamics of River flows in urban areas* (Gidrometeoizdat, Leningrad, 1989)
6. V.A. Shlychkov, Numerical modeling of River flows with allowance for vortex generation at the bed-floodplain boundary, Water resources. - **U. 35**, 546-553 (2008).
7. V.V. Kramarenko, *Ground science: a textbook,(* Yurayt Publishing House, Moscow, 2016)

The assessment of waters ecological state of the Crimea coastal near high-rise construction zones

Natalya Vetrova[1,*], *Tatyana* Ivanenko[1] , and *Emran* Mannanov[1]

[1]*Academy of Civil Engineering and Architecture of the Crimean Vernadsky Federal University*, 295493 Simferopol, *Republic of Crimea*

Abstract. The relevance of our study is determined by the significant level of coastal sea waters pollution by sewage near high-rise construction zones, which determines the violation of the sanitary and hygienic of sea waters 'characteristics and limits the possibilities for organizing recreational activities. The purpose of this study is to identify the ecological state of the marine aquatic area by the example of the Western Crimea near high-rise construction zones. The studies confirmed that the recreational and coastal area wastewater is intensely mixed with seawater, as a result, the pollution in the coastal strip of the sea in the area of deep water discharges sharply decrease. This happens because of water rapid rise to the surface and under the influence of the continuous movement of sea water huge masses with deep-water discharge, fresh wastewater is actively mixed with sea water. However, with no doubt, it is inadmissible to discharge sewage into the sea directly from the shore, but only at the estimated distance from the coast. The materials of the article can be useful for the management bodies and organizations involved in monitoring the quality of the coastal zone of the sea, teachers and students of higher educational institutions when assessing the ecological situation of the territories.

1 Introduction

The sea is an important resort and recreational resource of the Crimea. In the coastal zone of the Western Crimea territory is a strip of relative shallow water (depths up to 20 m), at the same time, the coastline length from the city of Sevastopol to the city of Evpatoria is about 75 km, the beaches stretch almost a continuous strip.

The use of the sea coastal part and the beach is the main curative factor that attracts vacationers, it should be provided with a level of sanitary parameters. The identification of the marine area ecological state of the Western Crimea near high-rise construction zone was the study purpose.

Modern theoretical and applied researches to study problems of the Crimea coastal zones' ecological safety are carried out by scientists of the Marine Hydrophysical Institute V.A.

* Corresponding author: sapronovat@mail.ru

Ivanov, Yu.N. Goryachkin, V.V. Fomin, V.F. Udovik (Goryachkin, 2010), design works in the coastal zone are carried out by the TsNTU "Inzhzashchita" under the guidance of M.N. Ryzhyi (2015).

2 Materials and methods

The development of coastal recreational zones is associated with the use of considerable space, and the tourism uncontrolled development causes damage to the environment: the intensive development of coastal recreational zones significantly changes the natural landscape, leads to pollution with their waste, pollution of the air as a result of the increase in the number of road transport.

The greatest damage to coastal waters is caused by local sources of pollution. The most vulnerable ones are the water areas of bays and gulfs with weak water exchange, which requires the development of measures to improve the ecological state of the sea coastal zone (Aibulatov, 2005).

During the research, the experimental (route inspection, sea water coastal sampling according to existing methods), analytical (analysis of selected samples by different chemical and physicochemical methods, sanitary-hygienic assessment of the environment under study), and the method of statistical generalization (determination of parameters zones of pollution of the coastal zone of the sea) were successfully used.

The condition of the coastal recreational zone ecosystem is determined through an assessment of its natural and technogenic factors and indicators (parameters). Such an ecosystem condition assessment is carried out on the basis of a criteria limited number that ensure a confident qualification of its condition in a joint examination. This requires a unified approach to assessing the state of both the ecosystem and its constituent components.

The territory of coastal recreational zones is a set of different functional zones, united by multiple connections into a complex dynamic system. Town-planning scheme for the construction of settlements is largely determined by the relief (the valleys of Belbek rivers (Lyubimovka village), Kacha (Orlovka village), Alma (Peschanoe village)).

The western coastal recreational zone of Crimea is an example of a linear type resort agglomeration. Currently, the planning structure of the Western Crimea settlements is in the process of formation. Due to the intensive construction of sanatorium, health resorts and recreational facilities, as well as spontaneously emerged areas of individual construction, individual settlements joined into a single urbanized strip. This phenomenon significantly worsens the quality of the habitat.

The territory under consideration is covered by a combined system for the removal of surface waters, a closed local stormwater network and open rain storms, located mainly in the health resorts and boarding houses. Pollution of surface runoff by heavy metals occurs during flushing with roadway plums.

Then, the surface runoff falls without preliminary purification into water bodies through the system of drainage trays and culverts.

In order to assess the ecological condition of water resources and the level of its disturbance in the coastal recreational zone of the Western Crimea, natural and analytical studies of the sea water quality in the coastal zone were conducted and the results of observations for 1998-2010 were processed. The results of the research are grouped in the following areas:

- level of pollution by the coastal drains;
- sanitary and hygienic characteristics of sea water for bathing.

To clarify the factors that formed the coastal zone ecological condition, the existing information on the ecological situation of the territory was analyzed. The maps of ecological and engineering-geological surveys were drawn up for some Western Crimea territories and settlements, with the following aspects: functional zoning of territories' development; density of recreational facilities development; objects (enterprises, facilities) operating on the TSO; organized and unorganized sources of pollutant emissions; discharge sources of sewage and sewage; maps of engineering-geological conditions. Full-scale geochemical investigations with water sampling and analysis in the laboratory using accredited methods were performed.

The A, B, C, D classes standards of the European Quality Movement for the resort cities "Blue Flag of Europe" were used, in order to assess the quality of wastewater treatment, the sea water. During the study, an experimental (route survey, sea water sampling, according to existing methods, analytical analysis of selected samples by various methods of chemical engineering, chemical and physicochemical methods, sanitary and hygienic assessment of the environment under study, zone of the sea) were used

3 Results

Sewage in the Western Crimea is fed to the Evpatoria, Saki and Kachin treatment facilities through a sewage pumping station system. After treatment, sewage is discharged into the sea through deep-water discharges at a distance of 1600 m from the shore (Evpatoria CBS), 1800 m - Saki CBS, 1250 m - Kachin CBS (Figure 1).

Table 1. Results of experimental studies of the coastal waters' quality of Crimea Western (in the area of releases) (August 1998-2010) *

N	Water composition indicators	Pollutants in the sea water in Saki area CBS, mg/dm3			The content of pollutants in the sea water in Saki area CBS, mg / dm3			The content of pollutants in the seawater near the Kachinsky area CBS		MPC mg/dm3	Exceeding the MPC C, mg / dm3 MPC (n times)		
		1998 y	2004 y	2010 y	1998 y	2004 y	2010 y	1998 y	2010 y		Evpatoria	Saks	Kachinsky
1	BOD complet	-	17,36	17,455	-	30,6	30,6	-	-	17,46	0,99	1,7	
	BOD5	8,8	-	-	19,9			6,8	1,0	3,00	2,9	6,6	2,2
2	Suspended substances	9	14,81	14,90	20	30,0	30,0	8,0	1,0	0,75	0,99	2,0	0,5
3	Ammonia nitrogen	2,97	31,2	31,13	22,09	13,86	13,86	2,2	0,01	2,0	15,6	11,0	1,1
4	Nitrite	0,071	6,68	6,735	0,495	2,31	2,31	0,031	0,001	3,3	2,04	0,7	0,009
5	Phosphates	-	693	642,5	-	7,64	7,64		-	3,5	198	2,1	-
6	Nitrates	1,3	0,881	0,894	0,6	10,3	10,3	0,7	н/о	45,0	0,01	0,2	0,015
7	Petroleum	0,02	0,018	0,0179	0,08	-	-	0,02	н/о	0,3	0,06	0,2	0,06

	products												
8	Iron	0,05	0,099	0,099	0,01	-	-	0,02	-	0,3	0,3	0,03	0,06
9	Copper	-	0,018	0,0179	-	-	-	-	-	1,0	0,01	-	-
10	Chromium	-	н/о	н/о	-	-	-	-	-	0,05	-	-	-
11	SPAC	0,046	0,202	0,202	0,247	1,06	1,06	0,025	н/о	0,23	0,8	4,6	0,1
12	Zinc	-	0,048	0,0479	-	-	-	-	-	-	1	-	-

Analysis of the study results showed that the sewerage networks of cities and towns in the coastal recreational zone of the Western Crimea have significant physical wear (62-85%) and insufficient capacity, the sewage treatment system is created and operates without taking into account storm water.

It is established by (Ivanenko et al., 2011) in the period of rainfall precipitation anthropogenic pollution (household and industrial waste, oil products, fertilizers, etc.) coming from the entire catchment to the sea with surface runoff

4 Discussion

As a result of the natural and analytical studies conducted for the coastal recreational zone of the Western Crimea, zones of environmental disturbances in populated areas were identified (Table 2): pollution of the sea coastal zone by deepwater and emergency discharges (production of insufficiently treated sewage with CBS, emergency releases of sanatoriums, storm sewage); unsatisfactory technical condition of bank protection structures.

Due to the lack of storm sewerage development in the new construction zone of resort facilities, the surface runoff falls directly into the coastal zone of the sea, contaminating it with heavy metals, oil products, organic compounds.

In addition, there are local characteristics of both the formation and discharge of effluents.

There are no treatment facilities in the village Novofedorivka, so the natural pond is turned into a reservoir-settler of the village sewage.

Table 2. Violations of the marine waters ecological state in the coastal recreational zone of the Western Crimea

In a settlements	Category of purification grade	Method of sewage treatment	Violations of the marine waters' ecological state
Lubovka village	Normally – limited purified	Mechanical and biological ways	- pollution of the sea coastal zone by insufficiently purified (for biological treatment) discharges; - exceeding the maximum permissible concentration limit for sea water quality indicators BOD, BOD5, suspended solids, ammonium nitrogen; - emergency condition of sludge treatment facilities.

In a settlement s	Category of purification grade	Method of sewage treatment	Violations of the marine waters' ecological state
Orlovka village	Normally – limited purified	biological way	- pollution of the sea coastal zone by insufficiently purified (for biological treatment) discharges due to insufficient capacity of the WWTP for channeling recreational facilities; - excess of MPC for sea water quality indicators BOD5, suspended substances, nitrites, phosphates, ammonium nitrogen; - the absence of the second (reserve) collector from the SPS to the CBS.
Andreevka village	Normally – limited purified	biological way	- pollution of the sea coastal zone by insufficiently purified (for biological treatment) discharges in the emergency condition of structures; - discharge of insufficiently treated sewage by the release of 15 l.m. from the water's edge in the beach area; - insufficient CBS capacity for development of the recreation zone
Kacha Town	Normally – limited purified	biological way	- pollution of the sea coastal zone by insufficiently purified (by biological treatment) discharges - the absence of the second (reserve) collector from the SPS to the CBS.

The large number of organic substances entering in the pond of a given locality colormates a few natural drains in sand interlayer and limits the possibilities of filtration from the pond. This leads to its overflow and an increase in the level of groundwater and nearby area flooding. As a result, an emergency discharge through the trenches to the Black Sea is carried out annually, at the end of the holiday season.

An important source of bacterial pollution of the sea, as it was established by the authors, can be insufficiently purified sewage from sewage treatment plants located in Yevpatoria, Saki,. Nikolaevka, Peschanoe, Andreevka and Kacha town. The quality of water cleaning does not meet sanitary requirements (sewer networks are worn out, high accident rate of their work). Only in the Nikolaevka water from sewage treatment plants enter the fields, but not into the sea.

As a result of the pollution transfer by currents, the entire marine area is contaminated with ammonium nitrogen, nitrites, phosphates, and surfactants. It should be emphasized that the beach strip is also polluted.

The highest pollution levels are observed in the Yevpatoriya Sea Port. The content of suspended solids at the outlet from the WWTP is reduced by 90-95%, BOD - by 80-85%, COD by 80-98%, etc. According to these indicators, the cleaned sewage of the WWTP of the Crimea West Coast complies with the norms for discharging substances from urban treatment facilities according to the "European Directive". At the same time, in addition to the normatively, clean waters that have already been cleared at sewage treatment plants (CBS), insufficiently purified water and water are discharged into the sea without purification.

In general, the quality of sea water of the Western Crimea coast in terms of microbiological and physicochemical indicators (in terms of BOD5 MPC is exceeded in 2.2-6.6 times, in terms of Ammonium nitrogen in 1.1-15.6 times) can be attributed to class C (periodic pollution).

At the same time, the most promising class A water quality is located in areas with a minimum recreational load (Uglovoye village, Beregovoe village, Kacha village).

In general, the problems of ensuring the environmentally safe state of the Western Crimea marine waters require further research to prepare engineering and technical measures that, in the complex, will allow them to be solved in the current period and prevent future development

5 Conclusion

In accordance with the study objectives, the analysis of the sea waters ecological state of the Western Crimea coastal was made. It was found that the inadequate sewer networks coverage and structures, their unsatisfactory condition and insufficient capacities in the zone, contribute to the pollution of certain coastal territory sections and the sea by organic compounds, oil products, heavy metals new construction of resort facilities, unorganized surface runoff to the sea coastal zone. Justifying technical solutions to this problem, it is advisable to take into account that in order to maintain the coastal strip of the sea in an uncontaminated state, it is necessary to have a gap between the coastline and the boundary of the urbanization zone of 200-300 m.

References

1. Yu.N. Goryachkin, Current state of the Black Sea coast of the Crimea, Reports NAS of U, **No. 10,** 78-82 (2010)
2. M.N. Ryzhy, Problems of engineering protection of the Crimean sea coasts and ways of their solution in modern conditions Sevastopol, 13-15 (2015)
3. N.A. Aibulatov, *Russia's activities in the coastal zone of the sea and the problems of ecology*, Science, Moscow, 304 (2005)
4. T.A. Ivanenko, Ecological condition of big Yalta water resources Building and technogenic security, Crimea, **Vol. 40** (2011)

Technology of surface wastewater purification, including high-rise construction areas

Anna Tsyba [1,*], *Yury* Skolubovich[1]

[1]Novosibirsk State University of Architecture and Civil Engineering, 630099, Leningradskaya str. 113, Novosibirsk , Russia

Abstract. Despite on the improvements in the quality of high-rise construction areas and industrial wastewater treatment, the pollution of water bodies continues to increase. This is due to the organized and unorganized surface untreated sewage entry into the reservoirs. The qualitative analysis of some cities' surface sewage composition is carried out in the work. Based on the published literature review, the characteristic contamination present in surface wastewater was identified. The paper proposes a new technology for the treatment of surface sewage and presents the results of preliminary studies.

1 Introduction

Based on the literature review [1-5], it was revealed that one of the most important hydrosphere pollution factors are surface sewage of high-rise construction areas, which composition, in recent period, increasingly depends on the results of human activities. So far, surface sewage was considered conditionally clean and not dangerous for water bodies. However, it has now been established that surface wastewater discharged from the territories of cities and industrial enterprises is significantly polluted and can not be discharged into water bodies without preliminary treatment [5]. Therefore, an organized diversion with further surface sewage purification is an important task for improving the ecological state of water bodies. In order to determine the qualitative characteristics of the surface sewage contaminants, a theoretical analysis of some cities' surface wastewaters were carried out.

The nature and degree of surface sewage pollution of high-rise construction areas, from cities' territories and industrial enterprises, are different. Their qualitative and quantitative characteristics depend on the level of a city municipal improvement, sanitary state of water catch basin area, surface atmosphere, as well as hydrometeorological parameters of precipitation: intensity and duration of rains, period of dry weather, and the intensity of snow melting process. The amount of surface runoffs pollutants depends on the population density, level of landscaping, type of surface cover, frequency of streets cleaning, traffic intensity, presence of industrial enterprises and amount of atmosphere emissions . [6].

* Corresponding author: michmacha@mail.ru

The recommendation on the system calculation for collecting, diversion and cleaning of surface runoff from residential areas showed that the main sources of surface runoff pollution occurring in urban areas and industrial sites are as a rule of:

- road surface wastes, products of road paving;
- products of soil surface erosion; emissions of suspended substances in the atmosphere by industrial enterprises, motor transport, heating systems;
- spills of oil products on the pavement surface; loss of bulk and liquid products, raw materials, semi-finished products;
- places for domestic garbage.

The highest level of surface runoff pollution is observed in the territories of large shopping centers, highways with intensive traffic, areas of industrial and trucking enterprises, and unorganized construction sites.

Pollution of surface runoff can be classified as follows [6-8]:

- organic and mineral impurities of natural origin, formed as a result of gases absorption from the atmosphere and erosion of the soil (sand particles, clay humus);
- substances of technogenic origin (oil products, synthetic surfactants, heavy metal compounds, etc.);
- bacterial contamination entering the drainage system with unsatisfactory sanitary and technical condition of the territories and sewerage networks.

Surface runoff from industrial sites has, as a rule, a more complex composition, and the concentration of pollution in it is higher than in urban runoff. The content of suspended solids can reach tens of grams per liter. High biochemical oxygen demand values are observed in the surface runoff from light and food industry enterprises, meat processing plants, tanneries and dairies, textile enterprises. From the ash dumps of the HPS, precipitation is washed out by phenols, which content in the surface runoff can reach hundreds of milligrams per liter. The bottom sediments forming in the zones of surface runoff discharges become secondary sources of water pollution by heavy metal compounds, which determine the increased background content in water of the corresponding ingredients. Forming hardly soluble complexes, heavy metal compounds practically do not linger on traditional treatment facilities of water pipes [9].

The oil products present in the surface runoff, getting into the water bodies, significantly affect the oxygen regime. At high concentrations of suspended solids, partial precipitation occurs in the discharge section of surface wastewater and downstream. That leads to water bodies' siltation and prevents the normal course of biological processes at the bottom of the reservoir. The organic part of suspended solids is oxidized by dissolved oxygen in water.

Present in the surface runoff oil products, when ingested, significantly affect the oxygen regime. First, oxidizing, they consume dissolved oxygen in the water and, secondly, a great damage from oil products is applied to microorganisms and fish farming. Fish are very sensitive to changes in the water chemical composition and the presence of petroleum products, especially, during the embryonic period. Oil products, getting into a reservoir, also lead to plankton death, an important component of fish fodder base. The presence of petroleum products in reservoirs and bottom sediments leads to the decrease of dissolved oxygen and microorganisms in water, as a result the processes of natural self-purification of water bodies are sharply slowed down [7].

Based on the theoretical analysis of the surface sewage composition in some Russian cities, their characteristics were determined (Table 1), which allowed giving some average quantitative characteristic of the surface urban runoff contamination. The content of suspended solids is 1.5-6 g / l, BOD5 40-120 mg O2 / l, COD 400-750 mg O2 / L, petroleum products - 20-25 mg / l. Contain of heavy metal compounds of asbestos fibers in the surface runoff from urban areas can be found with great consistency. The amount of surface runoff is 10-100 times higher than in domestic wastewater. The amount of urban run-offs pollution exceeds the MAC values for suspended substances, BOD, petroleum products, phosphates and heavy metals. Surface runoff accounts for approximately 75% of suspended solids, 20% of organic substances (BOD), 68% of petroleum products, and 65% of heavy metals entering reservoirs [2].

Table 1. Characteristics of the surface runoff quality of some Russian cities' territories

Indexes	Samara [8]	Saint Petersburg [1]	Volgograd [1]	nternational Review [14]	MPC p.x.[2]
Suspended substances, mg / l	50-1450	300-600	420-1250	150	+0,25 background
Iron total, mg / l	0,03-10,7	2-12	-	-	0,1
BOD, mg / l	5,2-316	20-50 (BOD_5)	39,2-118,5	-	3
Petroleum products, mg / l	0,125-475	7-12	0,75-3	-	0,05
Sulphates, mg / l	63,4-792	-	126,8-216	-	100
Nitrogen ammonium, mg / l	3,8-11,2	8-10	0,45-2,08	2,10	0,39
pH	7-8	-	7,41-7,	-	6,5-8,5
Zinc, mg / l	0-0,035	-	0,039	0,25	0,01
Copper, mg / l	0-0,58	-	0,22	-	0,001
Aluminum, mg / l	0-0,1	1-6	-	-	0,04
Cadmium, mg / l	0-0,05	-	-	0,0045	0,005
Chromium, mg / l	0-0,065	-	-	-	0,001
Nickel, mg / l	0	-	-	0,03	0,01

2 Method

In order to purify the surface wastewater, the following cleaning methods are used: mechanical, chemical, physicochemical and biological.

Mechanical cleaning of surface sewage is carried out by means of: sedimentation, filtration, flotation. The classical scheme of sewage treatment plants includes grids, sand catchers, settling tanks. Filters (sand, coalescent, sorption) and biological ponds are used for post-treatment of surface sewage from oil products. Flotators can be used for purification of industrial surface wastewater with high concentrations of petroleum products instead of sedimentation tanks.

The biological method of cleaning surface wastewater is based on the use of aerobic microorganisms' vital activity. Biological purification is advisable to use for the removal of dissolved organic compounds (COD and BOD) from the surface runoff, as well as to reduce the content of synthetic surfactants, nitrogen and phosphorus compounds.

New wastewater treatment technology was developed using a new filtering facility of the reactor-clarifier in NGASU (Sibstrin), in order to improve the efficiency and the facilities for wastewater treatment development (Fig. 2) [10]. According to the technological scheme, averaging of sewage in the settler-regulator is assumed. Then the wastewater gets the mixer-air separator, where it is freed from air bubbles and mixed with the flocculant. Further, the wastewater is purified in the clarifier reactors.

Purification of water in the clarifier reactor occurs in the bottom upwards direction with the rate of the upward flow 1.5 times higher than the filtration rate in conventional contact clarifiers under normal operation conditions [11]. At the same time, the load increased by 10%, this was necessary for the grains separation and its stagnant zones elimination. The loading grains were in a stationary state relative to each other. Grains separation leads to their working surface increasing, the loading layer porosity and decreasing the hydraulic resistance to the motion of the clarified liquid in it. As contact loading, quartz sand or any other fine-grained material that meets the requirements for chemical resistance and mechanical strength is used. As a result, the clarifier reactor combines advantages of granular filters and clarifiers with suspended sediment, has high productivity, reliable operation and low operating cost [12-15].

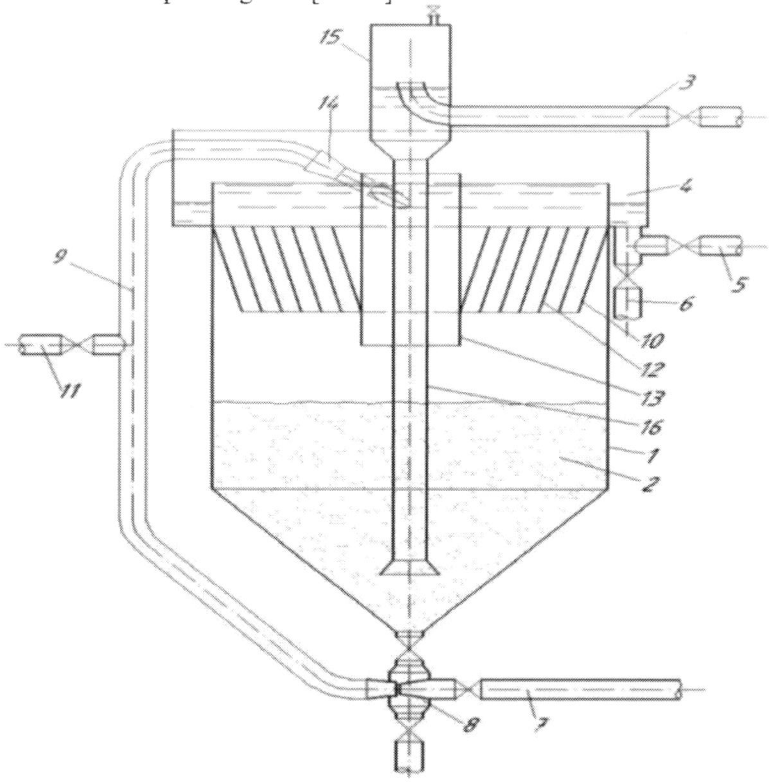

Fig. 1. Reactor-clarifier
1 - housing; 2 - contact loading; 3 - pipeline of source water; 4 - collecting trough; 5 - the pipeline of clarified water; 6.7 - pipelines for tapping and supplying wash water; 8 - hydroelevator; 9 - pulp transport pipeline; 10 - conical diaphragm; 11 - pulp discharge pipeline; 12 - thin-layer module; 13 - semisubmissive cylinder; 14 - air ejector; 15 - air separator; 16 - lower distribution tube.

3 Results

In order to confirm the efficiency of wastewater treatment, a technological scheme for cleaning with an experimental model of the clarifier reactor loaded with burnt fuels with 0.8-1.25 mm fractional composition and a height of the loading layer of 0.96 m was collected in the NSASU (Sibstrin) laboratory. Laboratory tests of the reactor-clarifier on surface sewage waters of Kemerovo region were carried out. The characteristics of laboratory tests are presented in Table 2.

During the experiment, the ascending flow rate was 8.6 m / h. High degree of water purification by suspended matter was observed up to 10 mg / l at the exit from the plant for 10 hours period, the field of manifestation of breakthroughs with values up to 30 mg / l was stopped. The results of the studies are summarized in Table 2.

Table 2 Investigation results of the reactor-clarifier efficiency

N	Indexes	Unit. measurements	Contamination values in source water	The values of contamination in purified water
1	chromaticity	degrees	186	5,0
2	feculence	EFM mg/dm^3	386,0 224	1,0
3	pH	unit.pH	7,3	7,6
4	Oxidizing property, permanganate	mgO/dm^3	5,0	3,0
5	Alkalinity	$mmol/dm^3$	6,8	7,4
6	Rigidity	F	8,3	7,5
7	Calcium	mg/dm^3	50,1	42
8	Magnesium	mg/dm^3	70,5	65,7
9	Ammonium ion	mg/dm^3	0,4	0,2
10	Ammonium ion	mg/dm^3	0,05	0,04
11	Ammonium ion	mg/dm^3	2,1	1,5
12	Iron	mg/dm^3	0,2	0,1
13	Manganese	mg/dm^3	0,1	0,1

Preliminary results of investigations in the purification technological scheme of clarifier reactor using showed a high degree of water clarification.

On the basis of the analysis, it can be proclaimed that surface wastewater causes significant harm to water bodies, polluting them with suspended substances, petroleum products, heavy metals and require cleaning. Technological scheme for the surface wastewater purification was adopted on the basis of carried out research

References

1. V.I. Kichigin, P.G. Bykova, Water supply and sanitary engineering. Issue. **1**. 28-32 (2002)
2. M.A. Gridneva, *Improvement of the disposal and purification of surface sewage from urban areas: the author's abstract.* PhD Thesis, Samara. (2004)
3. A.G. Melekhin, I.S. Shchukin, Construction and architecture. **1,** 122-132 (2012)
4. I.N. Sukhanov, *Analysis of ecological safety of the city of Novosibirsk.* Novosibirsk, **215** (2012)
5. A.A.Volchek, I.V. Bul'skaya, Bulletin of the Brest State Technical University. Issue **2**, 41-43 (2012)
6. M.I. Alekseev, A.M. Kurganov, *The organization of surface (rain and thawed) drainage from urbanized territories. Tutorial.* (Moscow, Publishing House of the ASV, 2000)
7. *Recommendations on the calculation of systems for collecting, diversion and cleaning of surface runoff from residential areas, sites of enterprises and determining the conditions for its release into water bodies*, Moscow, JSC "NII VODGO" (2014)
8. V.S. Dikarevsky, A.M. Kurganov, A.P. Nechaev, M.I. *Alekseev Removal and cleaning of surface sewage. Tutorial. L .* (Stroiizdat, 1990)

9. L.P. Ignatieva, Sanitary protection of water objects: textbook, MO Potapova; Ministry of Health. Russia, Department of Communal Hygiene and Hygiene of Children and Adolescents. Irkutsk. IGMU (2016)
10. E.L. Voitov, Yu.L. Skolubovich, Patent 23070754, the Russian Federation, MKI C02F 1/52. The device for water treatment, **27**. Publ. 27.09.2007
11. H.P. Duncan, *Urban Stormwater Quality: A Statistical Overview*, Cooperative Research Center for Catchment Hydrology, Melbourne, Australia (1999)
12. E.L. Voitov, *Preparation of drinking water from surface sources with increased natural and anthropogenic pollution*, Novosibirsk. NGASU (Sibstrin), (2010)
13. A.E. Krutkov, E.L. Voitov, Yu.L. Skulubovich, A.A. Tsyba, Efficiency of wastewater treatment of coal industry enterprises, Novosibirsk State University of Architecture and Civil Engineering (Sibstrin). Issue **1 (64),** (2017)
14. Yu.L. Skolubovich, E.L. Voitov, A.E. Krutkov, Increasing the efficiency of wastewater treatment in coal mining enterprises, News of Higher Educational Establishments. Construction works. **4** (2015)
15. Yu.L. Skolubovich, E.L. Voitov, A.Yu. Skolubovich, Preparation of drinking water from surface sources in ecologically unfavorable areas, the Academy of Sciences' Reports of the Higher School, Russia, Novosibirsk State Technical University, Issue 1 (2008)

Ecologically Safe Geothermal Energy Resources in Western Siberia near high-rise construction zones

Alexandr Shevchenko [1,*] and *Olga* Shiganova [2]

[1] Novosibirsk State University of Architecture and Civil Engineering, 630008 Novosibirsk, Russia
[2]Siberian Research Institute of Geology, Geophysics and Mineral Resources, 630008 Novosibirsk, Russia

Abstract. The development of geothermal energy in combination with other renewable energy sources (the sun, the wind) will help to solve the problem of heat supply and electrification in near high-rise construction zones of the country, especially in sparsely populated parts, where centralized energy and heat supply is economically unacceptable, and will improve the ecological situation. The aim of the research is to analyze the geothermal resources of the main aquifers in Western Siberia and to develop recommendations for further study and use of heat and power resources of this territory. The article gives retrospective of state research programs and potential use of hydrothermal resources of administrative units geographically entering the territory under consideration. It is noted that by now such programs have been curtailed for various reasons, although there are examples of their successful and effective use in various fields of industry and agriculture. According to the decision of the Supreme Ecological Council of the State Duma Committee of the Russian Federation adopted in 2014 on the beginning of the development of federal targeted programs for the use of heat power water as a source of electricity and heat supply, the Ministry of Natural Resources and Ecology of the Russian Federation made proposals for further research and use of hydrothermal waters in Western Siberia. Implementation of the programs proposed by the authors, alongside with other positive aspects, will solve the problems of heat supply in remote territories and improve the environmental situation in the region.

1 Introduction

Nowadays the energy industry of Russia mainly uses traditional fossil fuels, such as oil products, natural gas and coal. Inexhaustibility and environmental cleanliness of renewable energy sources make it necessary to carry out intensive study of the prospects and ways of their usage.

* Corresponding author: michmacha@mail.ru

Western Siberia is a huge, mostly sparsely populated territory, occupying about 15% of the territory of Russia. Its subsoil, along with organic fuel resources, contains significant reserves of fresh and mineral waters, waters suitable for industrial extraction of valuable chemical elements and compounds, as well as thermal and subthermal waters (fig.1).

At present the energy industry of Russia mainly uses traditional non-renewable sources, such as oil products, natural gas and coal. However, in the coming decades depletion of hydrocarbon reserves is projected. According to some estimates, taking into account world oil consumption, these reserves will last for 28 years and in Russia - only for 15 years [2].

Inexhaustibility and environmental cleanliness of renewable energy sources make it necessary to research prospects and ways of using them. The main renewable source is hydropower. The geothermal energy market has tremendous potential, since this kind of energy source is much more stable if compared to other renewable energy resources.

Today the majority of countries use geothermal heat to a certain extent [3]. Depending on its temperature, the geothermal energy heat transfer agent is used for power generation, heating of buildings, greenhouses, swimming pools, fisheries, as well as for balneological purposes [4]. There are several reasons for poor development of geothermal energy in Russia. Reserves of the majority of the country's geothermal deposits have low and medium temperatures, which currently do not allow them to be competitive with traditional energy sources.

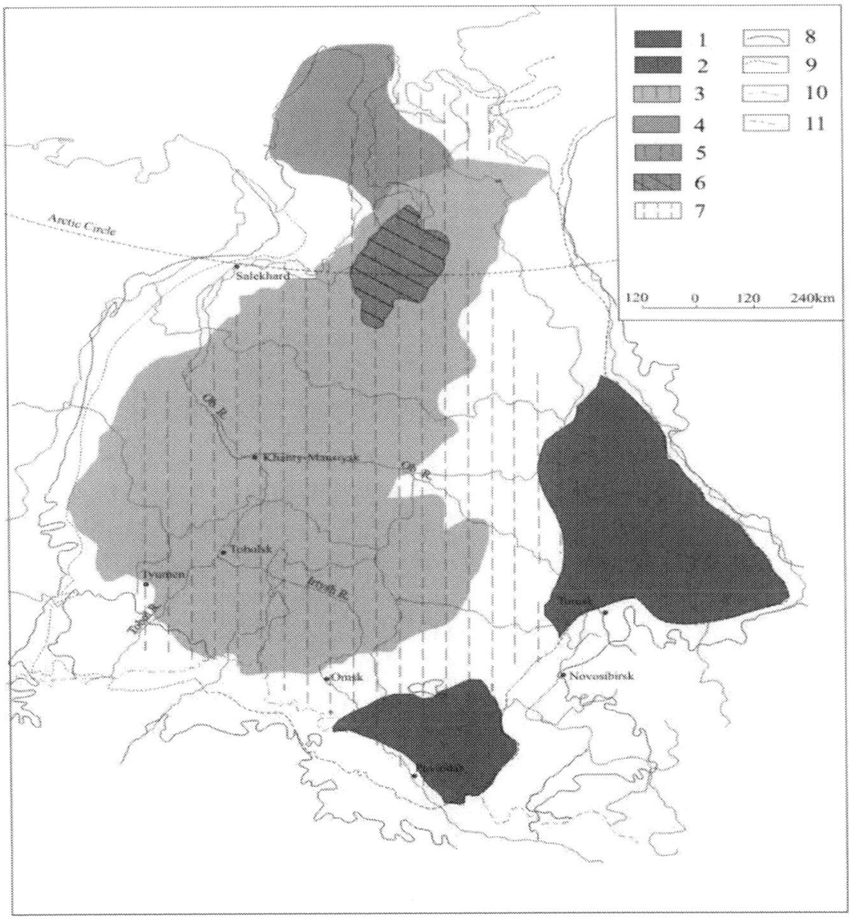

Fig. 1. Zonation of West Siberian Artesian Basin Aquifer Complex (K_{1-2} Cenomanian-Aptian Pokur Formation and its Analogues)[1].

Area of Water Distribution:1-Fresh (Up to 1 g/l); 2-Fresh Subthermal (20-50°C) ; 3-Iodine-bromine Mineral and Thermal; 4-Iodine Mineral; 5-Iodine Mineral and Thermal; 6-Industrial Iodine and Thermal; 7- Mineral Subthermal. *Boundaries:* 8-West Siberian Artesian Basin; 9-Aquifer Complex; 10-RF Administrative Regions, 11- Russia.

Other reasons are lack of methodological, technological and regulatory support for the extraction and the use of geothermal resources under Russian climatic conditions, lack of attention paid to environmental aspects of energy production.

In 2014 according to the decision of the Supreme Ecological Council of the State Duma Committee of Russia, the Ministry of Natural Resources and Ecology of the Russian Federation was given the task to initiate the development of federal targeted programs for the use of heat-power water as a source of electricity and heat [5]. In this regard, the analysis of the existing programs and potential use of geothermal energy in administrative units geographically located on the Western Siberia territory, as well as the experience of other regions of Russia, is particularly important. The analysis of their implementation experience will help to develop recommendations for the programs planned.

Energy potential of thermal waters of Western Siberia

Despite the huge heat potential of groundwater in different regions of Russia, their resources were not considered as the primary sources of geothermal energy even in geographical areas not provided by traditional sources. The history of studying thermal energy resources of the region's underground waters is considered in [6]. For northern and central regions of Western Siberia, the studies were carried out on the materials of testing deep wells drilled for oil and gas. In southern regions, which were not objects of large-scale study of subsurface resources for hydrocarbons, data on wells for domestic and drinking water supply were mainly used. From the hydrogeological point of view, the territory of Western Siberia is a huge artesian basin with several hydrogeological complexes (Fig.2). This basin is a huge reservoir of groundwater, which can fully meet the needs of domestic and drinking water supply, balneology, chemical industry and serve as a source of geothermal energy [8,9].According to

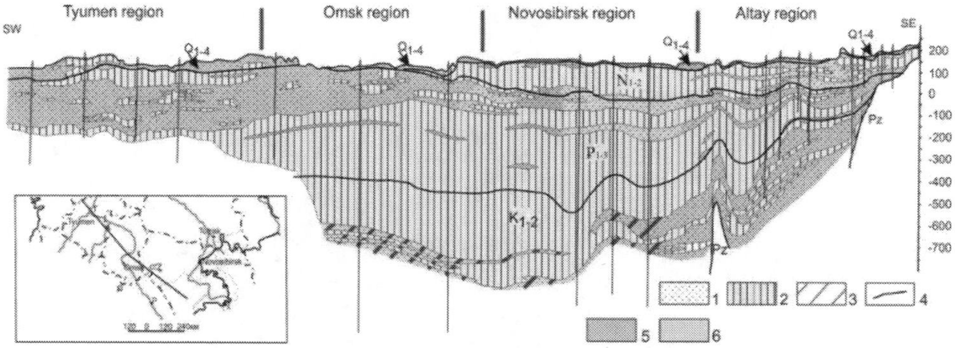

Fig.2 Cretaceous – Quaternary Hydrogeological Section of West Siberian Artesian Basin [5]

1. Water-bearing Deposits; 2. Water-proof Deposits; 3.Thermal and Subthermal (20-50°C) Waters; 4.Stratigraphic Boundaries.5.Fresh Water; 6.Mineral Water. [9], central regions of Western Siberia are regarded as "promising" with water temperatures at the mouth of wells from 40 to 120 °C, areas on the periphery of the basin are "with limited prospects" with water temperatures at the mouth of the wells from 20 to 40°C. Maximum temperatures up to 100 °C and more are observed in internal depressions. The Cenomanian water-bearing complex (K_{1-2}) has the best hydro-geological parameters (depth, rate, mineralization, temperature) for geothermal raw materials for Western Siberia. Its permeable part is confined to the Pokur formation and its analogs. The depth of the

complex roof is from 150 to 1300 m, the total thickness is from 200 to 800-1000 m. The water pressure in most of the basin is 20-40 m above the drill mouth, the water temperature is + 30-43 0C, self-discharge rate is 10-25 l / s [1]. Such subthermal waters cannot be directly used in hot water supplying and heating systems. The use of geothermal systems makes it possible to expand the prospects of the subsoil thermal power potential.

Geothermal systems include a wide variety of designs and operations(see, for example, [10]). If groundwater is considered as a source of energy for heating a structure they are usually called groundwater-source heat pumps (GSHP). These days GSHP systems are one of the most widespread "green" technologies for heating and cooling of buildings. An overview of the current state-of-the-art, latest developments and trends in GSHP system technology can be found at [11]. Unfortunately, for a number of reasons, as it will be shown later, in Western Siberia and Russia in general, the use of geothermal systems is extremely limited.

2 Analysis of geothermal energy use in Western Siberia

In Western Siberia the direct use of thermal waters (with the formation temperature of 40-100 ° C and mineralization of 1-5 g / l) is practiced in the Tyumen, Tomsk, Novosibirsk and Omsk regions for balnearies. In agriculture these waters are used in fish farming (wintering ponds) and greenhouse production.

Since the 1980s regional programs for assessing the resources of hydrothermal waters have been operating in a number of administrative areas in the southern part of Western Siberia. In the Novosibirsk region the regional target program "Introduction of heat pumps at fuel and energy complex facilities on the territory of the Novosibirsk Region for 1999 - 2002" was adopted. Based on this program "perspective" and "very promising" objects were chosen in 26 settlements of the region where heat pumps could be installed.As of 01.01.2003, heat pumps operated in 6 settlements using groundwater resources of the Cretaceous Aquiferous Complex (K_{1-2}). In Karasuktown a separately located building with the total square of 6,000 m^2(secondary school) has been heated by artesian water from a well with a temperature of 24 °C (at the mouth) for 25 years. The heat pump system consists of two GSHP systems NKT-300 manufactured by the Russian company "Energia"[12]. Its total heat output is 700kW. The facility was launched in 1992. The cost of geothermal heat is about 40% of the cost of heat produced by coal boilers in the region; the payback period was 1.1 years [13]. Unfortunately, in the region these examples of heat pumps application are unique.

In the neighboring Omsk region the "Program for Development of Heat Power Resources in the Omsk Region for 2005 -2010" was developed and implemented. The aim of the Program was a balanced development of the mineral and raw materials base to meet the needs of the region in thermal energy of underground water resources. The main tasks were: to meet the needs of the region's population in thermal water and formation of mechanisms of state regulation in the sphere of research, reproduction, and use of the ground waters. The Program was financed both by the federal budget and by subsoil users. Unfortunately, because of not enough funding those very important tasks were reduced mainly to evaluation and inventory of geothermal water resources in three administrative districts of the region. Legal and economic aspects of the geothermal energy use have not been developed and implemented. The effectiveness of introducing geothermal systems into the Russian market can be demonstrated by the example of the Chuvash Republic. This administrative region is located in the European part of Russia with a temperate continental climate similar to the climate of the southern and central parts of Western Siberia. Favorable conditions for the introduction of heat pumps are created there. In 2002 for the first time in Russia the Cabinet of Ministers of the Chuvash Republic brought into effect separate energy tariffs. For energy generated by heat pumps the tariff is approximately 2.15 times lower than for energy produced by traditional heat supply systems. At the same time the use of heat pumps gives more than 70% fuel savings for heat supply of industrial buildings and public facilities [14]. Thus, even insignificant support from the state through tariffs regulation will allow the geothermal systems technology to take the firm place of gas and coal boilers on the market of heat-producing equipment.

3 Conclusion

The analysis of the thermal waters resource base in Western Siberia, makes it possible to draw the following conclusions. Geothermal energy of underground waters of Western Siberia, as a renewable ecologically clean natural resource, has been used successfully, but in a very limited scale, for the purposes of heat supply of social, agricultural objects and residential buildings, depending on the temperature of a heat source directly or with the use of GSHP systems.

Speedy development of the thermal power of groundwater must be achieved through the development and implementation of Federal targeted regional programs for the use of geothermal energy. These programs should be successors of regional programs that existed in some administrative regions of Western Siberia in recent decades. The main developers should be the Ministry of Natural Resources and Ecology of the Russian Federation, scientific and research institutes, R&D firms. Financing of programs is possible partly by the federal budget, partly by subsoil users (public-private partnership).

Along with other positive moments, these measures will allow to solve the problems of sustainable heat supply of remote territories and improve their ecological situation.

References

1. O.V. Shiganova, Water: Ecology and Technology, Moscow, Russia, **Vol.4** (2000)
2. V.P. Gavrilov, Forecast for Potential Trends of Russian and Global Energy Complex Development, *Oil and Gas Geology*, May 2016 (In Russian).
3. V. Svalova, K. Potapov, Geothermal Energy Use in Russia, Country Update for 2010-2015, *World Geothermal Congress*, Australia (2015)
4. O.V. Shiganjva, Y.Y. Marchenko, A.A. Shevchenko, Underground Waters of the South of Western Siberia (Russia) as a Source of Renewable Energy, Symposium on Thermal Science and Engineering in Serbia,Sokobanja,17-20, **Vol.18th** (2017)
5. Hydrogeology of the USSR, **Vol. XVI**, 1971 (In Russian).
6. A.E. Kontorovich, V.S. Surkov, West Siberia. Geology and Mineral Resources of Russia. Saint-Petersburg, **Vol.2** (2000)
7. URL: http://www.gshp.org.uk/ground_source_heat_pumps.html
8. J.D. Spitler, Latest Developments and Trends in Ground-Source Heat Pump Technology, European Geothermal Congress Strasbourg, France, 19-24 http://teplo.in/index.php/ob-ekty (2016)
9. Ju. M. Petin The Experience of a Heat Pumps Manufacturing Decade in CJSC (ZAO) "Energia", *The Energy Policy*, **Vol.3**, (2001)
10. V.P. Prochenko, V.G. Gorshkov, S.V.Osipovich, About the Experience of Heating Pumps Using in the Chuvash Republic, available online at: http://www.rosteplo.ru/tech_stat/stat_shablon.php?id=499.

Scanning the welded joints of aluminium alloys using subminiature eddy-current transducers

Sergey Dmitriev[1,*], *Alexey* Ishkov[2], *Vladimir* Malikov[1], and *Anatoly* Sagalakov[1]

[1]Altai State University, 656049 Lenina Av., Barnaul, Russia
[2]Altai State Agricultural University, 656049 Krasnoarmeyskiy Av., Barnaul, Russia

Abstract. Aluminium has a reputation for ease of use, strength and durability. In addition to its exceptional aesthetic properties, solid aluminium does not burn. As architects, contractors, consultants and real estate owners look to meet stringent safety requirements in the construction and refurbishment of high-rise constructions for both residential and commercial uses, aluminium cladding provides an alternative that is not only safe but that is also durable and attractive. One of the ways to connect elements into a aluminium construction is welding. friction stir welding is one of the most efficient. The authors developed a measuring system based on subminiaturized eddy-current transducers aimed at examining locally the defects of welded joints in aluminium-magnesium alloy plates connected by means of friction stir welding. The authors made a modification of the Delyann filter, which allowed them to increase considerably the signal-noise relations. The dependency of the eddy-current transducer response on defects was provided, i.e. concealed cuts and openings inside the welded joint, at the frequencies of 100-10000 Hz of the exciting winding.

1 Introduction

Constructions made of aluminium alloys have got a frequent practice in current technology. Application of these alloys in such spheres as aircraft engineering, high-rise constructions, automobile construction, spacecraft engineering, demands higher standards of the structure and mechanical characteristics of the material. One of the ways to connect elements into a construction is welding. However, common types of fusion welding are not always applicable for some aluminium-based alloys. In order to join elements of such alloys, one has to use special methods and processing technologies, among which, friction stir welding is one of the most efficient [1,2].

Due to significant temperature and deformity gradients, obtained welded joints are characterized by high inhomogeneity and deficiency of the structure. Affected by external mechanical impact, structural defects of the welded joints cause early crack formation and destruction. As soon as the obtained constructions can be used in extreme operation conditions, the mentioned circumstance requires the performance of welded joints thorough diagnosis. Currently, friction stir welding (FSW) is widely applied in the production of body frame elements made of aluminium alloys. This method allows us to get permanent

* Corresponding author: osys@me.com

qualitative joints even made of those alloys that are not subject to electric-arc welding or argon-arc welding. The ultrasonic technique with the use of phased array antennas demonstrates capability of detecting defects within the whole volume of the weld pass. The obtained results are highly valid. However, except for defects, located within the joint volume, surface defects, such as bridging, interstices and cracks in the surface layer and others, influence the joint strength substantially. One of the most highly-efficient and exact methods, which allows detecting defects located in the surface or in the thin subsurface layer, is the eddy-current control method.

Some authors have been made some researches enforce to find reliability NDT techniques to detect and sizing the micro root defects on FSW. Nevertheless, in conductivity materials, when high reliability is required as mentioned before, eddy current is the most adequate technique. However, eddy current (EC) flow can change not only due to defects, but also due to the variations of the probe lift-off. The high-sensitive lift-off effect of conventional EC probes introduces noises which mask the signal of small defects, making their detection difficult or even impossible. Moreover, the low sensitivity of common EC probes as well as other electronic issues may not enable to distinguish between defect and non defective conditions.

Catalin Mandache et al. [3] applied PEC in FSW, but even in the presence of significant size root defects the inspection accuracy is low. PEC variant has the capability to detect deeper defects, nevertheless micro-defects in FSW can hardly be detected.. Neil Goldfine et al. [4] have developed a MWM® probe which is able to detect some FSW defects, but only quite large (typically above 150 micron). Lamarre et al. [5] were among the first to apply phased array ultrasonic on FSW, and P. H. Johnston [6] highlighted difficulties in detecting the oxides alignment defects on the nugget of the FSW using this NDT technique.

M. Moles et al [7] used ultrasonic and eddy currents arrays for inspection of friction stir welds in aluminum. L .S. Rosado et al [8] used eddy current probe to detect the imperfections in the friction stir welding of metals. Li et al [9] have reported on multiple nondestructive testing methods on the FSW of AA 2219-T6. Hu et al [10] also employed a high-precision magnetic sensor to detect the weld defects in aluminium Friction stir welds.

In [11] a new NDT system is presented. The system is composed by a planar eddy currents probe (IOnic probe), electronic devices for signal generation, conditioning and conversion, automated mechanized scanning and analysis software. The IOnic system was developed mostly to be used for the micro-defects detection on aluminum solid state processed alloys by FSW. The experimental results in AA2024 welded by FSW clearly show that this system is able to detect imperfections around 50 μm, which contribute to increase the reliability on NDT of micro imperfections.

Therefore, the task of developing an eddy-current transducer capable of examining welded joints obtained by means of friction stir welding and the creation of eddy-current measuring system capable of detecting flaws in such materials seems to be crucial.

The purpose of this work is the detection of defects in welded FSW joints of aluminium alloys with the application of the eddy-current transducer and the development of the eddy-current control procedure for such joints.

2 Materials and methods

The authors have developed a subminiature eddy-current transformer for local monitoring of the physical parameters in titanium-alloy plates and weld seams [11]. In contrast to existing sensors, it permits local measurements with sections measuring a few microns, to depths of the order of 5 mm. The electrical conductivity of the alloy is directly measurable, and its distribution over the sample surface and thickness may be readily established. The eddy-current method is based on the dependence of the current magnitude and distribution on the geometric and electromagnetic parameters of the sample and on the mutual position of the sensor and the sample. The basic informative parameter β_0 is a generalized characteristic of the object, the eddy-current sensor, and the frequency of the electromagnetic field [12]. We need to develop an adequate model of the response of a eddy-current transformers of plate type, which are sensitive to many variables and permit reproduction of the voltage hodograph at small values of β_0. We have plotted hodographs illustrating the influence of various parameters on the induced voltage on the basis of the proposed model [13,14].

The eddy-current transformer is connected to the sound board of a computer with special software that controls the voltage in the transformers exciting winding and also reads the voltage at the measuring winding (in arbitrary units). On the basis of preliminary calibration, these readings are converted to values of the electrical conductivity. The exciting winding of the transducer consists of ten turns; its diameter is 0.12–0.13 mm. The measuring winding consists of 130 turns and has a diameter of 0.05–0.08 mm. To minimize the influence of the exciting winding on the recording signal, the circuit contains a compensation winding that is connected to the measuring winding in accordance with the well-known differential circuit. This consists of 20 turns. A copper wire with a 5 μm thickness is used for winding turns. The turns are wound around a pyramidal core. The proposed shape of the core is favorable for the area of the magnetic field. The core is made of ferrite with an initial magnetic permeability of 500.

Different transducers that are based on cores that have the same ratio of the base diagonal (400 μm) and edge length (1 mm) were calibrated using samples with a well-known electric conductivity. The characteristics of the designed transducers allow one to efficiently localize the magnetic field within 2500 μm^2 and provide penetration of the magnetic field into the studied object at a depth of up to 5 mm.

The eddy-current transducer (**Fig. 1.**) is a transformer with measuring (1), exciting (2), and compensation (3) windings and a magnetic circuit 4, which is located inside the cylindrical platform 5 with tracks that are cut on the external side for windings. The platform is impregnated with a compound 6 at a temperature of 200°C to prevent the disintegration of the windings when the ferrite screen 7, which is intended for the localization of the electromagnetic field on the tested object, is put in place. From the outside the transducer is contained in a corundum washer 8, which protects the core 4 from contacting the tested object.

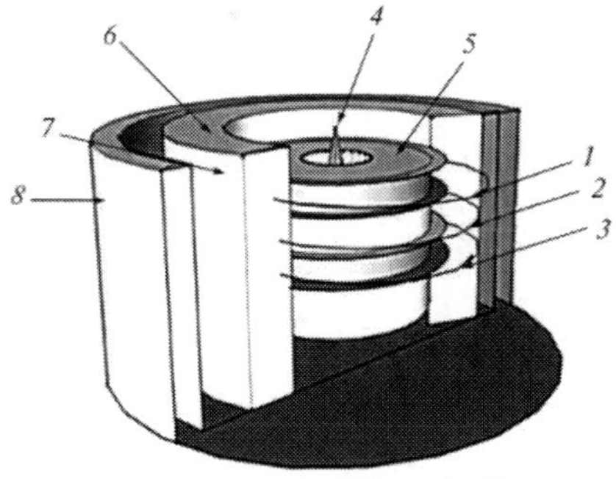

Fig. 1. A block diagram of the eddy-current transducer.

The measuring system, which is based on a miniature eddy-current transducer, operates as follows. The software of the personal computer controls the operation of the generator, which produces a train of rectangular voltage pulses with the repetition rate f_1 that is necessary for the operation of the eddy-current transducers. The voltage pulses are transmitted from the generator output to two series integrators. They are then directed to the input of the power amplifier. From the amplifier output the voltage pulses arrive at the exciting inductance coils of the eddy-current transducers. The difference of the output voltages of the measuring coils of the transducers contains information on the structural heterogeneities of the tested object that is located in the effective area of the eddy-current transducers. It is detected and amplified in a special microphone amplifier. The signal arrives at the amplitude detector after the transmission through two series high-quality low frequency filters and two series selective amplifiers. The signal is then transmitted through an analog-to-digital converter to a personal computer. Due to the simultaneous control of the generated signal frequency at the exciting coil and the cut-off frequency of the filtering system and the selective amplification, the useful signal, which contains information on the electric conductivity distribution inside the object, in particular, on possible flaws of the object, is detected. The control program allows one to change the operating frequency of the measuring system so that the signal that is received from the measuring winding is reliably recorded.

3 Result and discussion

Defectoscopy was performed with the help of a transformer eddy current probe of the overhead type, field intensity of generating winding - 800 A/m. The measured characteristic is introduced voltage induced by the field of eddy currents arising in monitored object. The sensor calibration was performed before the beginning of the measuring operations, the calibration itself consisting in the definition of the input voltage from the defect-free section. The calibration was performed at the frequency of 700 Hz. After that, the detector was moving over the scanned defective area, while the voltage input in the measuring winding of the transducer was being registered. The authors tried to determine the frequency that provided for the most input voltage deviation from the voltage value obtained at the defect-free section of the specimen. The voltage value corresponding to this frequency was viewed as a parameter, which behavior allowed inferring the presence of a defect. Scanning was carried out by moving of sensor across welds or across deforming area.

Sample No. 1: two aluminum-magnesium alloy slats connected by friction welding. Thickness of the slats was 8.3 mm. Width of the weld was 2 cm. Scanning was carried out by moving the sensor perpendicular to the joint line, measuring the distance from the calibration area to the measurement area.

a b

Fig. 2. Photo of Sample No. 1 and Sample No. 2 surface with marked scanned areas.

In the experiments with sample No. 1(**Fig.2.a.**), it was found that slats welded together by weld, give different values of introduced voltage that, obviously, indicates different materials of slats. The results of the experiment are presented in **Fig.3.** The boundaries of weld correspond to drop of the introduced voltage. At the boundaries of weld, it is possible to see sections of significant drop in the introduced voltage, corresponding to the boundaries of the weld.

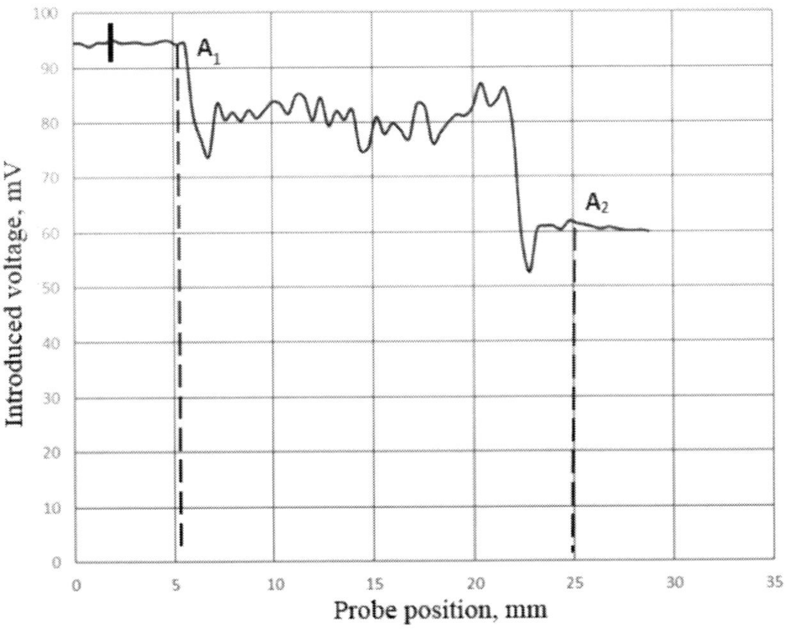

Fig. 3. Amount of the voltage introduced to measuring winding of the transducer in the area of a weld seam. Scanning frequency is 700 Hz. A_1-A_2 are welding edges

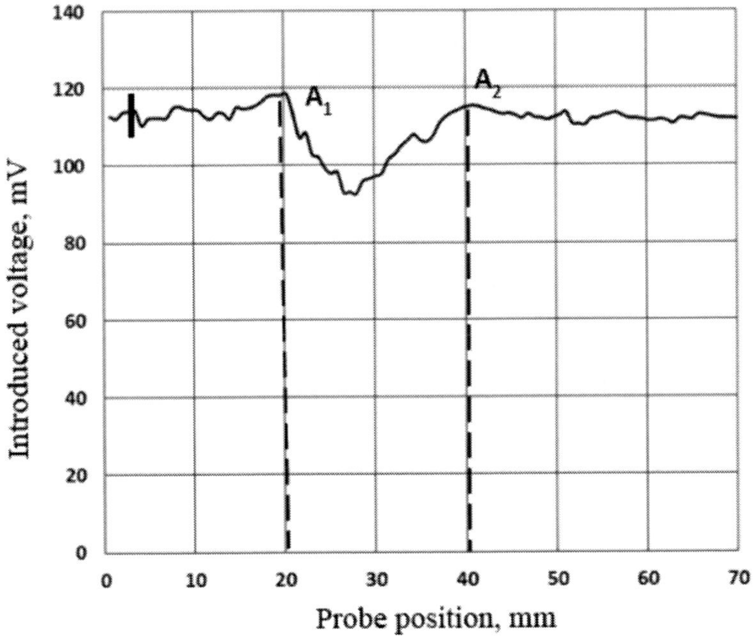

Fig. 4. a. Measurement results for the defective weld seam, Area 1. A_1-A_2 are welding edges, C is the location of the crack

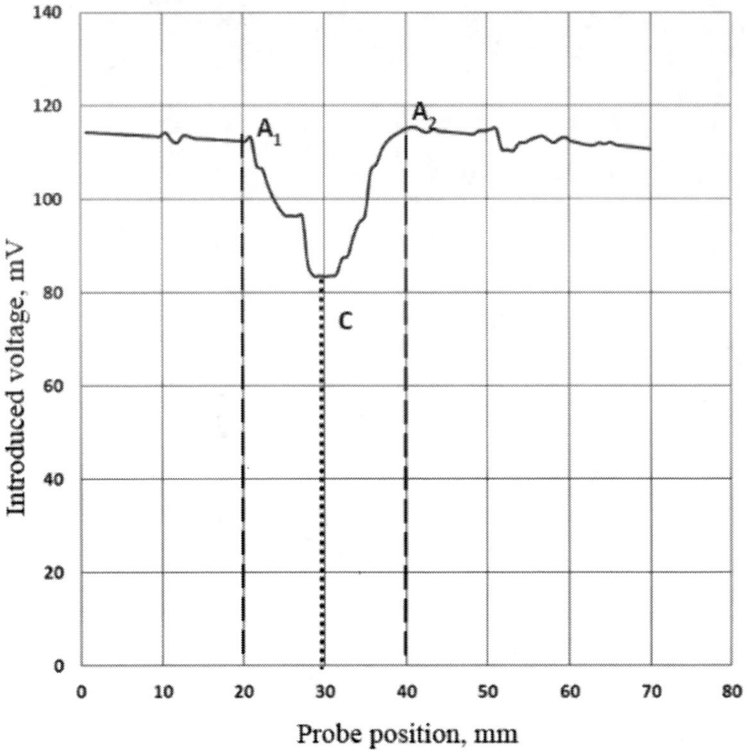

Fig. 4. b. Measurement results for the defective weld seam, Area 2. A_1-A_2 are welding edges, C is the location of the crack

Sample No. 2 (**Fig. 2.b.**) is a 5-mm-thick plate having technological flaws in the area of the welded joint (scanned sections are labelled with numbers 1, 2, 3, 4, 5, 6, 7, 8 and are presented in the picture). Scanning was performed by moving the sensor perpendicularly to the joint line, measuring the distance from the calibration area to the measurement area. In this sample, the scanning was performed by moving the sensor over the area of weld with the defects contained therein.

In the area 1, inside the sample, there was a cavity with diameter 2 mm, the depth of occurrence from the scanning surface was 3 mm. The cavity was fixed during scanning with the use of an eddy current probe operating at a frequency of 700 Hz (**Fig. 4.a.**). The boundaries of weld correspond to the drop in introduced voltage. The dotted line indicates the defect area. The drop of voltage across the measuring winding of the probe in the defect area was 25 mV compared to the calibration area and 7 mV compared to the weld area

In areas 2 and 3, there was a surface crack on the welding joint. The width of the crack was 0.2 mm. The surface crack was easily fixed by the probe at practically all frequencies: the change in the applied voltage was about 35 mV compared to the calibration area and about 15 mV compared to the weld area (**Fig 4.b., 5.a**.).

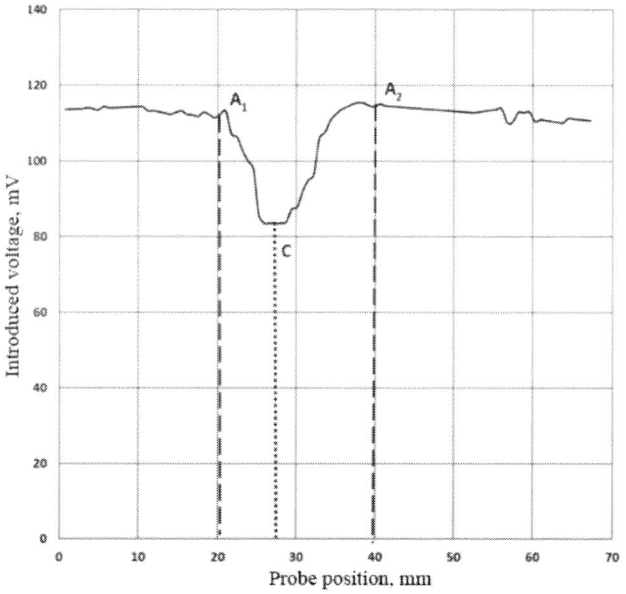

Fig. 5. a. Measurement results for the defective weld seam, Area 3. A_1-A_2 are welding edges, C is the location of the crack, B_1-B_2 are the cavity edges.

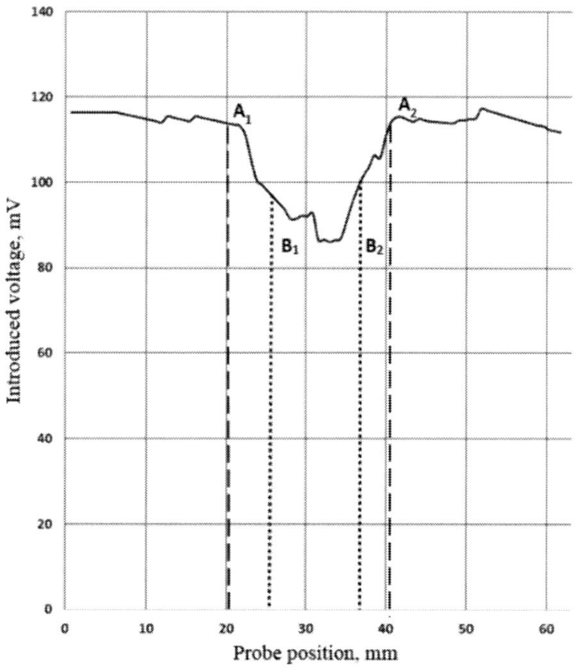

Fig. 5.b. Measurement results for the defective weld seam, Area 4. A_1-A_2 are welding edges, C is the location of the crack, B_1-B_2 are the cavity edges.

Area 4 is a milled internal cylindrical cavity located on the back of the slat, 4 mm in depth, 15 mm in diameter, with depth of 1 mm from the scanning surface, which is clearly visible on the probe (**Fig. 5.b.**). The change of introduced voltage in the area of the cavity is 30 mV compared to the calibration area and 10 mV compared to the weld area.

Areas 5-8 were presented as a weld area, having internal cavities (holes) marked with characteristic dips at the edges of the joint. (**Fig. 6.a-d**). The drop in introduced voltage across the gaps was 5-10 mV compared to the neighbouring defect-free area of the sample. Since a similar pattern was observed in a sample No. 1 (**Fig. 3.**), it can be concluded that this feature is inherent to this type of welding.

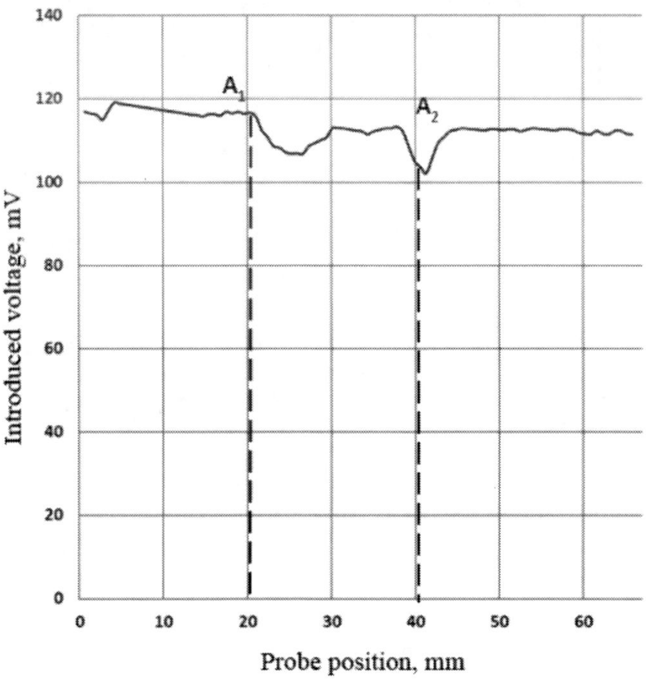

Fig. 6.a. Measurement results for the defective weld seam, Area 5. A_1-A_2 are welding edges

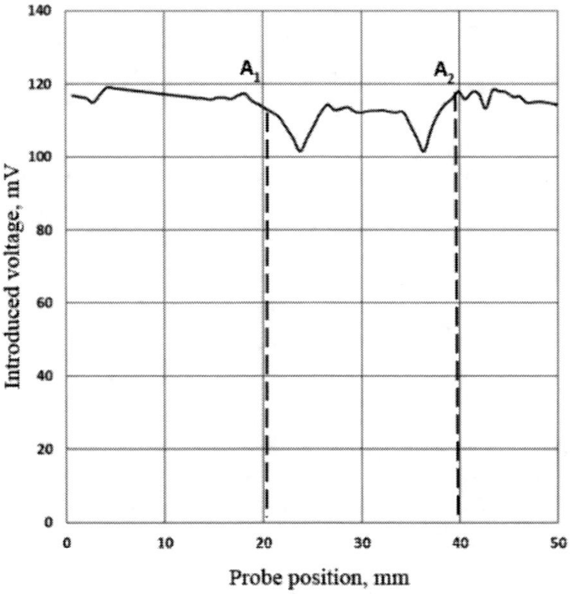

Fig. 6.b. Measurement results for the defective weld seam, Area 6. A_1-A_2 are welding edges

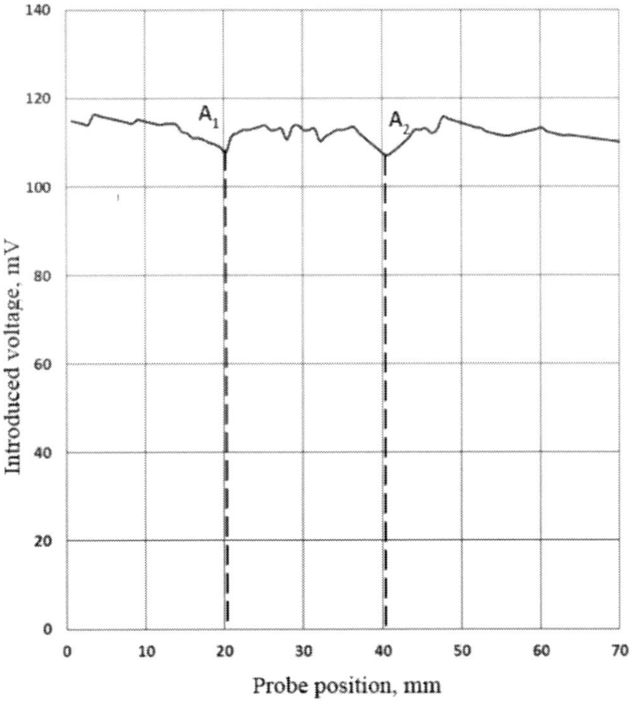

Fig. 6.c. Measurement results for the defective weld seam, Area 7. A_1-A_2 are welding edges

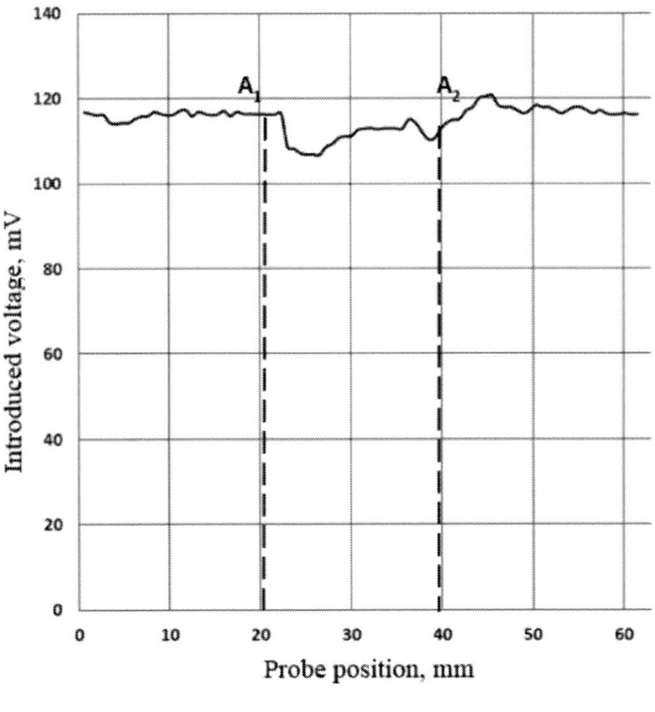

Fig. 6.d. Measurement results for the defective weld seam, Areas 8. A_1-A_2 are welding edges

4 Conclusion

We worked out a measuring system aimed at the examination of plates made of aluminium-magnesium alloys joint with the help of friction stir welding method. The conducted research demonstrates the potential of the eddy-current control method applied to search for midget flaws in the welded joints. Subminiaturized eddy-current transducers, constructed on the basis of pyramid-shaped cores, allow effectively detecting cracks and cavities having the diameter from 0.2 to 2 mm located at the depth of up to 3 mm. In addition, we demonstrated the possibility to control the areas of weld seam edges.

References

1. S. Mironov, Physical Mesomechanics, **17**, 103 (2014)
2. V Rubtsov, E Kolubaev, S Tarasov, P Vasilyev, V Bakshaev, AIP Conf. Proc, **1623**, 535 (2014)
3. C Mandache, L Dubourg A Merati, Materials Evaluation, **4**, 382 (2008)
4. D Grundy, V Zilberstein, N Goldfine, *Proc. of 7th International Conference on Trends in Welding Research,* **1**, 227 (2005)
5. A Lamarre, M Michael, AIP Conference Proceedings, **509**, 1333 (2000)
6. P Johnston, *35th Annual Review of Progress in Quantitative Nondestructive Evaluation*, **1**, 1 (2008)
7. M Moles, A Lamarre, O Dupuis, *16th World Conference on NDT*, **84**, 50 (2004)
8. L Rosado, T Santos, M Piedade, P Ramos, P Vilaca, Measurement: Journal of the International Measurement Confederation, **43(8)**, 1021 (2010)
9. B Li, Y Shen, W Hu, Materials and Design, **32(4)**, 2073 (2011)
10. B Hu, R Yu, H Zou, NDT&E International, **47**, 66 (2012)
11. V Pedro, G Telmo, R Luís, M Rosa, *The 12th International Conference of the Slovenian Society for Non-Destructive Testing »Application of Contemporary Non-Destructive Testing in Engineering»*, **1**, 329 (2013)
12. S F Dmitriev, V N Malikov, A M Sagalakov, M A Davydchenko, *Proceedings of 2015 International Conference on Mechanical Engineering, Automation and Control Systems, MEACS 2015,* **1**, 1 (2016)
13. V N Malikov, S F Dmitiev, Russ. Eng. Res., **8**, 32 (2016)
14. S F Dmitriev, V N Malikov, A M Sagalakov, Welding International, **31(8)**, 608 (2017)

 https://doi.org/10.1051/e3sconf/20183302055

Design of solar systems in high-rise buildings

Alexander Kolosov[1*], *Dmitry* Chudinov[1], *Sergey* Yaremenko[1]

[1]Voronezh State Technical University, Moscow Avenue, 14, Voronezh, 394026, Russia

Abstract. Nowadays, the renovation program is being implemented in the megapolises of Russia. Innovative high-rise buildings are built instead of morally and physically obsolete houses, where non-traditional renewable energy sources are used to the fullest extent, under the effect of which they are located. The possibility to use solar systems with variation of their design parameters is considered. It is established that solar systems have high technical potential. The share of heat load, that is provided by using solar energy, varies from 4 to 84% depending on the time of the year. Economic indicators restrain the use of such panels. The payback period is about 8 years at the current cost for thermal energy.

1 Introduction

Currently in Russia, comfortable high-rise buildings are being erected within the renovation program in order to substitute morally and physically obsolete dwellings. Apartments on the last floors are often two-level. The area of apartments is 200 to 300 m^2, which is comparable with the area of a cottage. It is possible to use alternative resources for energy supply in order to increase energy efficiency, reduce the operating costs of similar apartments and buildings in general.

The use of solar energy to heat water can be practically significant and promising. Solar power plants have a number of advantages: saving fuel and energy resources; low operating costs; simplicity of design and reliability in work; minimal maintenance; durability; complete safety. The profitability of the use of such systems for different regions of the country is determined by the gross, technical and economic potential of solar energy [1-4].

The gross potential of solar radiation is determined by its intensity, which depends on many factors: latitude, longitude, air conditions, hydrometeorological conditions of the region, time of year and day. The existing formulas for determining solar radiation do not always give an accurate estimate [5]. The authors suggest using the data of long-term actual observations when calculating the gross potential of solar radiation.

The data on solar radiation, presented in the reference books in the form of tables, should be used in the form of correlations, determined on the basis of the approximation of these tabular data in order to intensify large-scale thermal engineering calculations.

Analytical dependencies that allow determining the values of the total solar radiation on a horizontal and vertical surface amid a cloudless sky for the corresponding latitude were obtained.

---

* Corresponding author: kolossn@yandex.ru

$$H_{jan} = -614.08 \cdot \ln(\varphi) + 2586.09 \quad (1)$$

...

$$H_{june} = 0.00005571 \cdot \varphi^5 - 0.01486526 \cdot \varphi^4 + 1.57018502 \cdot \varphi^3 - 82.03553868 \cdot \varphi^2 + \\ + 2117.99506113 \cdot \varphi - 20707.95629264 \quad (2)$$

...

$$H_{dec.} = -623.75 \cdot \ln(\varphi) + 2596.38 \quad (3)$$

In order to take into account the cloud cover, the values obtained from these correlations must be multiplied by the corresponding correction factor (ψ), which is the ratio of the value of the total solar radiation under the actual cloudiness conditions (I_{Σ}^{O}) to the total solar radiation amid a cloudless sky (I_{Σ}^{Y}).

$$\psi = \frac{I_{\Sigma}^{O}}{I_{\Sigma}^{Y}}. \quad (4)$$

The replacement factor (the share of the heat load provided by solar energy) is applied to determine the technical potential of solar energy in heat supply and hot water supply systems.

The estimate of the dependence of the replacement factor on the variation in the design parameters of the solar heat supply and hot water supply systems is given in the article. The estimate of economic potential at a variation of cost for thermal energy is also given.

2 Materials and Methods

Among the various possible aspects of the use of solar energy, the most technically and economically prepared field is the utilization of solar energy for the heat supply of housing-and-utility and agricultural facilities. The obtained in this case ecological effect is determined by the consumption of the replaced organic fuel and is directly dependent on the efficiency of solar heat supply systems.

Solar heating systems operate on a technically simple principle. The core element of the solar heating system, where solar radiation is converted into thermal energy, is the solar collector. A solar thermal collector collects heat by absorbing sunlight. A collector is a device for capturing solar radiation. Solar radiation is energy in the form of electromagnetic radiation from the infrared (long) to the ultraviolet (short) wavelengths.

The term "solar collector" commonly refers to solar hot water panels, but may refer to installations such as solar parabolic troughs and solar towers; or basic installations such as solar air heaters. Solar collectors can be used both with concentration and without concentration of solar radiation. A surface in flat collectors that receives solar radiation is also a surface that absorbs radiation. Focusing collectors, which usually have concave reflectors, concentrate radiation incident on their entire surface onto a heat exchanger with a smaller surface area, thereby increasing the energy flux density. The advantages of flat collectors include the possibility to use both direct and scattered solar radiation, they do not require tracking the sun and daily maintenance. In a constructional way, they are simpler than a system with concentrating reflectors, absorbing surfaces and tracking mechanisms. At the present moment, flat collectors are used in solar heating systems.

The solar energy, that is converted by the solar collector into heat, is accumulated in the storage tank due to the heat capacity of the liquid and is used when necessary to provide the heat load of the heating and hot water supply. Water is used as the heat carrier in the heating system and hot water supply system and antifreeze is used in the solar circuit to prevent freezing during the cold period of the year. A conventional boiler is used as a backup source of energy, which provides heating in those cases when the storage bank is empty. The set of solar heating system equipment includes regulators, safety valves, pumps and pipelines.

3 Results

The following values of the replacement factors were obtained in [5]: 21.2 to 33.7% (19.5 to 34.7 are experimental values) with a variation in the area of the solar collector from 5 to 20 m^2; 23.9 to 31.6% (22.4 to 34.2 are experimental values) with a variation in the tank volume of the battery from 0.25 to 1 m^3 (Figure 1, 2).

The values of the replacement factor grow with the increase of the area of the solar collector, however, an increase of the solar collector area over 15 m^2 does not contribute to a significant increase in it. The optimal tank volume of the battery is in the range from 0.75 to 1 m^3.

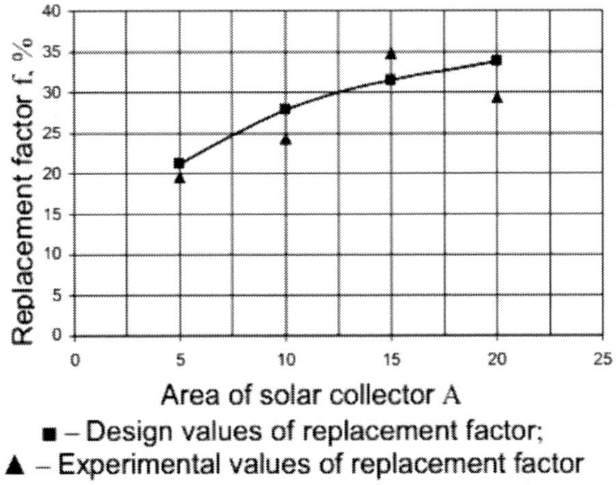

Fig. 1. Dependence of the replacement factor on the area of the solar collector (the tank volume of the battery is 0.75 m^3).

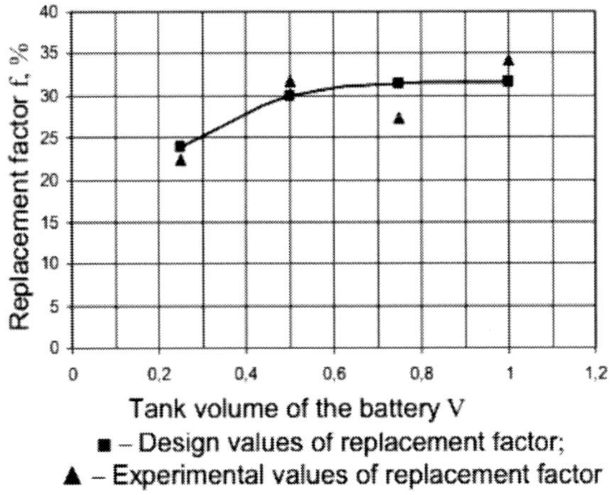

Fig. 2. Dependence of the replacement factor on the tank volume of the battery (solar collector area is 15 m^2).

4 Discussions

The evaluation of the efficiency of investments in solar energy systems can be defined in accordance with the Recommendations on the Assessment of the Economic Efficiency of the Heat Supply Investment Project developed by the NP "ABOK" . There were identified the following criteria: a) additional discounted expenses (DE) for solar heating and hot water supply systems over the accounting period; b) the payback period (Flow) of additional expenses for solar heating hot water supply systems.

A residential house (cottage) was considered as an object to evaluate its economic efficiency, which can be compared with multi-level apartments in high-rise buildings. The need for thermal energy is carried out either by the traditional system of power supply (I option), or by a combination of solar and traditional systems (II option). The system is equipped with solar collectors with a total area of 12 m^2, the tank volume of the heat-insulated battery is 600 liters. There is antifreeze in the solar collector circuit. The battery tank has a built-in heat exchanger. The system is equipped with a device for coolant circulation and automation. The replacement factor of thermal load was defined for the calendar year as a whole according to the method presented in [6].

The additional discounted expenses (DE) for the compared options are defined from the expressions [7]:

$$DE_I = \left(K_I + \sum_{t=1}^{Tsl} \frac{E_{It} - E_{Iat}}{(1+r)^t} \right), \tag{5}$$

$$DE_{II} = \left(K_{II} + \sum_{t=1}^{T_{sl}} \frac{E_{IIt} - E_{IIat}}{(1+r)^t} \right), \tag{6}$$

where (according to the options): K - capital investment; E_t, E_{at} - operating expenses and depreciation charges per year t; r - interest rate (discount rate), rel. units [ ]; T_{sl} - time horizon (acounting period).

The obtained on the basis of a number of performed calculations results are illustrated in Figures 3 and 4.

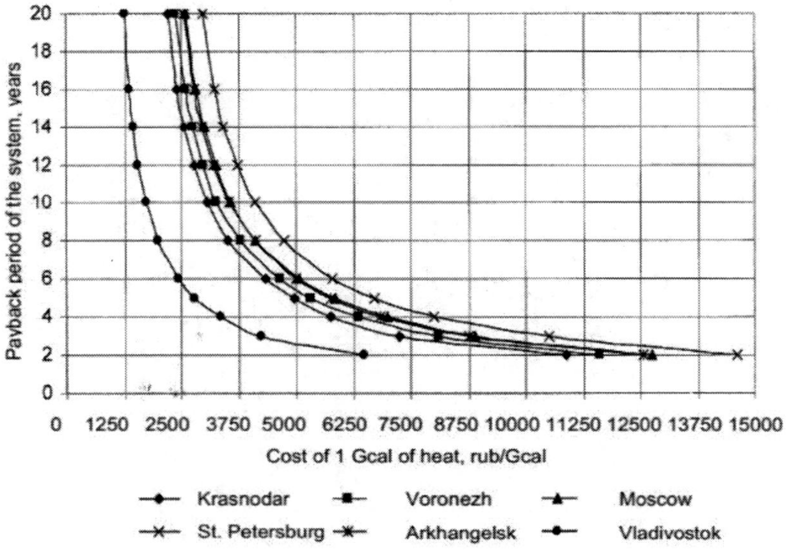

Fig. 3. Dependence of the payback period of the solar energy system on the cost of heat.

https://doi.org/10.1051/e3sconf/20183302055

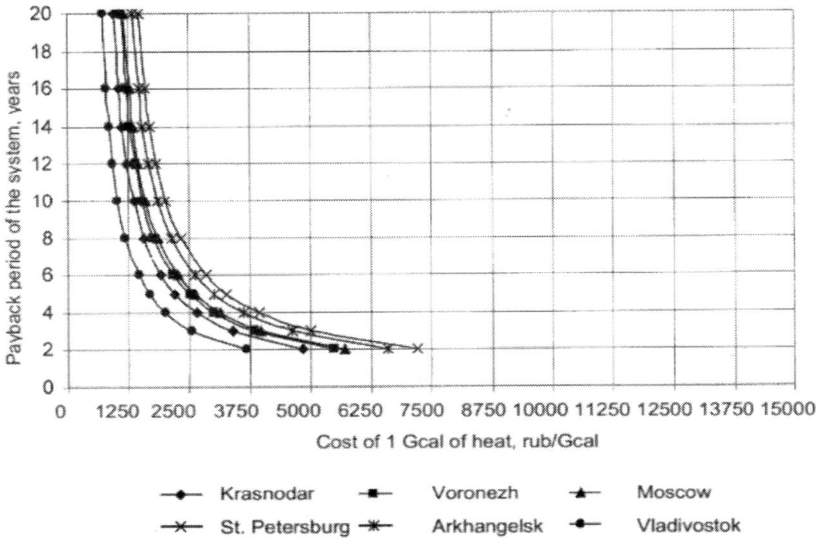

Fig. 4. Dependence of the payback period of the solar hot water supply system on the cost of heat.

5 Conclusion

The scale of solar systems use in Russia is determined by a number of factors: climatic conditions, economic reasons and technical capabilities.

It is possible to increase the energy efficiency of buildings and reduce operating expenses through using solar systems for the needs of heat supply and hot water supply within the renovation program. Solar power plants are among of the most satisfactory technologies of heat supply.

Mutual efforts of researchers, developers, manufacturers and the state are required in order to successfully introduce solar systems for partial energy supply of high-rise buildings. It is necessary to increase their competitiveness by applying improved technical solutions (combined generation of thermal and electric energy) and reducing cost parameters.

The payback period of solar systems (according to the assumed initial data), depending on the region where they are applied, varies from 3 years (hot water supply systems) to unprofitable value (heat supply systems).

It is apparent that an increase in the cost of heat will lead to a reduction in the payback period of solar plants.

References

1. E. Nezhnikova, IOP Conf. Series: Earth and Environmental Science, **90** (2017) DOI:10.1088/1755-1315/90/1/012161
2. V. Kankhva, IOP Conf. Series: Earth and Environmental Science, **90** (2017) DOI:10.1088/1755-1315/90/1/012175
3. O. Sotnikova, Vestnik Voronezhskogo gosudarstvennogo tekhnicheskogo universiteta, **6**, 56-63 (2005)
4. D. Chudinov, T. Shchukina, O. Sotnikova, Promyshlennaya ehnergetika, **9**, 44-46 (2008)
5. D. Chudinov, Energosberezhenie, **7**, 64-80 (2009)

6. A. Borboni, D. De Santis, Meccanica, **49**, 1327-1336 (2014) DOI: 10.1007/s11012-014-9895-z
7. A. Borboni, F. Aggogeri, R. Faglia, International Journal of Advanced Robotic Systems, **10**, 149 (2013) DOI: 10.5772/55539

Possibilities of using aluminate cements in high-rise construction

Maria Kaddo[1]*

[1]Moscow State University of Civil Engineering, Yaroslavskoe shosse, 26, Moscow, 29337, Russia

Abstract. The article describes preferable ways of usage of alternative binders for high-rise construction based on aluminate cements. Possible areas of rational use of aluminate cements with the purpose of increasing the service life of materials and the adequacy of the durability of materials with the required durability of the building are analyzed. The results of the structure, shrinkage and physical and mechanical properties of concrete obtained from dry mixes on the base of aluminate cements for self-leveling floors are presented. To study the shrinkage mechanism of curing binders and to evaluate the role of evaporation of water in the development of shrinkage was undertaken experiment with simple unfilled systems: gypsum binder, portland cement and «corrosion resistant high alumina cement + gypsum». Principle possibility of binder with compensated shrinkage based on aluminate cement, gypsum and modern superplasticizers was defined, as well as cracking resistance and corrosion resistance provide durability of the composition.

1 Introduction

Construction is one of the main branches of economy. A huge number of diverse resources is required when new buildings and structures are erected.

In particular, in the European Union (EU) the construction sector accounts for approximately 40% of energy consumption, more than 50% of recoverable materials, 30% of water and waste consumption and 35% of greenhouse gas emissions. This estimate takes into account all the stages of the life cycle of the buildings, including extraction of raw materials for the production of building materials and products, construction, operation and maintenance of buildings, their demolition and waste management. At the same time, the construction materials industry uses byproducts and waste from other industries (metallurgical slag, TPP ashes, woodworking waste etc.) as raw materials, contributing to the solution of environmental problems [1].

High-rise buildings have different purposes. The share of skyscrapers intended only for housing is about 30%. Among the buildings with a height of more than 200 meters, which were constructed in recent years, the share of office buildings has slightly decreased. But the majority of high-rise buildings is multifunctional. Apart from the general purpose rooms, they are fitted with parking lots, shops, offices, cinemas, etc. As more and more

* Corresponding author: m.kaddo@yandex.ru

multifunctional facilities appear on the market, the demand for a combination of different types of building materials within a single building is growing. The invariable change of tenants means that offices are constantly being repaired using a large number of different finishing materials, leading to a large amount of waste. Waste produced during the demolition of buildings accounts for a large proportion of all solid waste generated worldwide; more than 50% of the waste comes from the construction industry. Reducing the loss of materials at construction sites is one of the main problems the industry is faced with. During the construction, operation and repair of high-rise buildings different types of waste are formed. These include that of concrete, reinforced concrete structures, steel, aluminum, ceramic materials, wood, plastics, glass, etc. Therefore, it is important to use durable materials even when decorating and repairing rooms of high-rise buildings with a significant working lifespan. In the 1970s the main structural material for high-rise construction was steel. Nowadays the share of all-metal structures is only 3%, and high-strength reinforced concrete structures are used everywhere. The main binder for the production of high-strength concrete is Portland cement and its varieties. Other binders are used extremely rarely. Without challenging the position of Portland cement in concrete structures building, we would like to propose the areas of usage for other binders (aluminous cement, construction and high-strength gypsum) in the construction and finishing of certain special-purpose rooms of high-rise buildings.

In the modern high-rise buildings, the share of auxiliary and technical premises (cars parks, sports halls, premises for accommodation of engineering systems of life support of buildings, climatic equipment, etc.) is high. Requirements for the floors of such premises are more technical than aesthetical. High strength, wear- and abrasion resistance and durability are required. Repair and replacement of such coatings requires significant investment. It is important that materials used are technological. From this point of view, self-leveling floors on the base of aluminate cement are the most interesting ones [2-9].

The production of aluminate cements causes less damage to the environment. Moreover, the use of aluminate cements eliminates losses caused by corrosion of cement stone. Corrosion is probable after a certain period of service in the underground and basement technical rooms of high-rise buildings.

The appearance of cracks on the surface of the floors is the main type of defects. Cracking is caused by deformation of the material. It causes stress which exceeds the tensile strength and is not compensated by relaxation processes. Such deformations arise during curing and exploitation and are of different nature (have different causes). The main types are deformation caused by physical and chemical processes in the curing binder (contraction, expansion), and shrinkage caused by the evaporation of moisture from the cement material both immediately after its placement (plastic shrinkage) and in the process of curing (drying shrinkage). Drying shrinkage is the main reason for cracking.

Shrinkage of two kinds, contractual and humid, can occur during the curing of concrete. The contraction has almost no effect on the volume of construction, and leads to an increase in porosity of the concrete. Humid shrinkage is associated with the evaporation of water from the curing concrete and exceeds the contraction by 5-10 times [10, 11].

The problem of cracking resistance of thin-film coatings of cement compositions can be solved by binders with controlled shrinkage (expandable binders, whose shrinkage is compensated by the expansion). Resistance to cracking would be higher in the case when expansion and shrinkage are close not only in size, but also in the development time.

For shrinkage compensated concrete can be used organic expanding components in an amount of 0.5-2% [12-15]. More widespread has expanding mineral supplements, the application of which produce products of hydration and cause the expansion of the system and increase the strength of the cement stone. Using the expanding additive requires a high production standards. Expansion effect depends on many factors and is not stable [16-18].

The most promising method is the use of binders based on aluminate cements, which expansion is provided by the formation of ettringite $3CaO \cdot Al_2O_3 \cdot 3CaSO_4 \cdot 31\text{-}32H_2O$ (CASH) [19].

2 Materials, methods and mixture proportioning

For the experiment were analyzed compositions and properties of the produced aluminate cements. The data on the chemical composition of aluminate cements are given in Table 1.

Table 1. The chemical composition of aluminate concretes, (%).

Composition	Al_2O_3	CaO	Fe_2O_3	SiO_2	SO_3	R_2O
CRHAC -75-0,5 (Russia)	no less than 75	no less than 20	no more than 0.5	no more than 1.3	no more than 0.15	no more than 0.35
CRHAC -70-1 (Russia)	no less than 70	no less than 20	no more than 1	no more than 2	no more than 0.15	no more than 0.5
ISTRA 40 (Germany)	39 –42	37 – 40	14 – 17	2 – 5	0,4	MgO 1.2
SECAR 51 (France)	50.8 – 54.2	35.9 – 38.9	1.5 – 2.5	4.0 – 5.5		MgO < 1,0 TiO_2 < 4.0 K_2O+Na_2O < 0.5

Of the proposed cements for study was selected corrosion resistant high alumina cement (CRHAC), because it contains at least 70% Al_2O_3. The chemical composition of used cement: Al_2O_3 – 70.5%, CaO – 28.1%, SiO_2 – 0.87%, Fe_2O_3 – 0.3%.

The mineral composition of used cement: calcium aluminate (CA and CA_2) in the ratio 1:1. As a result of x-ray curing products principally consist of CAH_{10}. In the age of 3 days hydration degree of CA is about 80-85%, CA_2 - about 50%. Selected cement is characterized by high strength.

Curing speed is less high compared with other aluminate cements, that corresponds to the task (Table. 2). As a second component of binder necessary for expansion, was used natural gypsum ($CaSO_4 \cdot 2H_2O$) and hemihydrate gypsum ($CaSO_4 \cdot 0,5H_2O$).

Table 2. Technical characteristics of the material.

Value	CRHAC -75-0,5	CRHAC -70-1	ISTRA 40	SECAR 51
Specific surface area, m^2/kg	no less than 450	no less than 450	300 – 340	375 – 425
Curing time: the beginnig –the end, hour	not earlier than 2, not later than 12	not earlier than 2, not later than 12	not earlier than 1, not later than 5	not earlier than 2, not later than 5
Compressive resistance, MPa	after 3 days of curing no less than 50	after 3 days of curing no less than 50	after a day of curing 65–75	after a day of curing 55–85

3 Results and Discussion

For experimental verification of the component composition of binder and the type of gypsum were conducted studies in a wide range of ratio of the components at a fixed value of the plasticity of the mixture (Table 3).

Table 3. Test results of two-component binders of different composition

CRHAC, %	100	80	70	60	50	40	0
gypsum, %	0	20	30	40	50	60	100
$CaSO_4 \cdot 2H_2O$							
W/C	0.50	0.50	0.50	0.50	0.50	–	–
compression strength, MPa	54.5	50.2	53.1	48.4	34.6	–	–

Composition contents 60 ... 80% of CRHAC (the amount of gypsum 40 ... 20%) have a high compression strength. Both in terms of strength and economic reasons more efficient use of natural gypsum.

Composition was fixed after 3, 7, 24 and 72 hours. In first hours of hydration was observed rapid formation of CASH, decaying after 24 hours (Table 4).

Table 4. Changes in the phase composition of the system «CRHAC + gypsum», depending on time of hydration.

Hydration time, hour	Relative intensity of peaks, %		
	ettringite		
	9.83Å	5.65Å	3.89Å
0	0	0	0
3	50	50	50
7	83	75	83
24	90	90	95
72	100	100	100

To study the shrinkage mechanism of curing binders and to evaluate the role of evaporation of water in the development of shrinkage was undertaken experiment with simple unfilled systems: gypsum binder, portland cement and «CRHAC + gypsum» (70% + 30%). Selection of the first two systems was defined as sufficient of their study, and the fact that in the first is observed a rapid formation of a solid skeleton formed of crystalline neoplasms ($CaSO_4 \cdot 2H_2O$), and in the second - a solid skeleton is formed of several phases, a large share of which is occupied by phase of gel. So, on the example of these two systems are easier to specify the new features of binder deformation «CRHAC+gypsum» (Figure 1).

Deformation of curing two-component (CRHAC + gypsum = 70 : 30) binder developed as a result of the superposition of two competing processes: the expansion due to the formation of CASH and formation on its basis of the cement stone structure and shrinkage caused by the evaporation of water from the curing composition.
As in the initial period is very active the formation of cement stone structure, we observe the rapid expansion, reaching after 4...5 hours – 0.6 mm/m.

As a result, the deceleration rate of hydration, sealing, but not the extension of curing system due to hydration products (indicated by a rapid increase of the strength of the composition after 4...5 hours after mixing) and the evaporation of water from the cured material, we observe a sharp decline in the rate of expansion, and then the resulting transition effect process after 5...6 hours from expansion to shrinkage. Although at this time, continues the process of expansion by hydration of the remaining share of CA and, in particular, CA_2. After 3…4 days the rate of shrinkage is stabilized and the initial expansion is compensated by shrinkage. The shrinkage of the composition is terminated after 10...12 days at a value of shrinkage of 0.8...0.9 mm/m. (to compare according to our measurements of Portland cement, it is about 8...10 times higher).

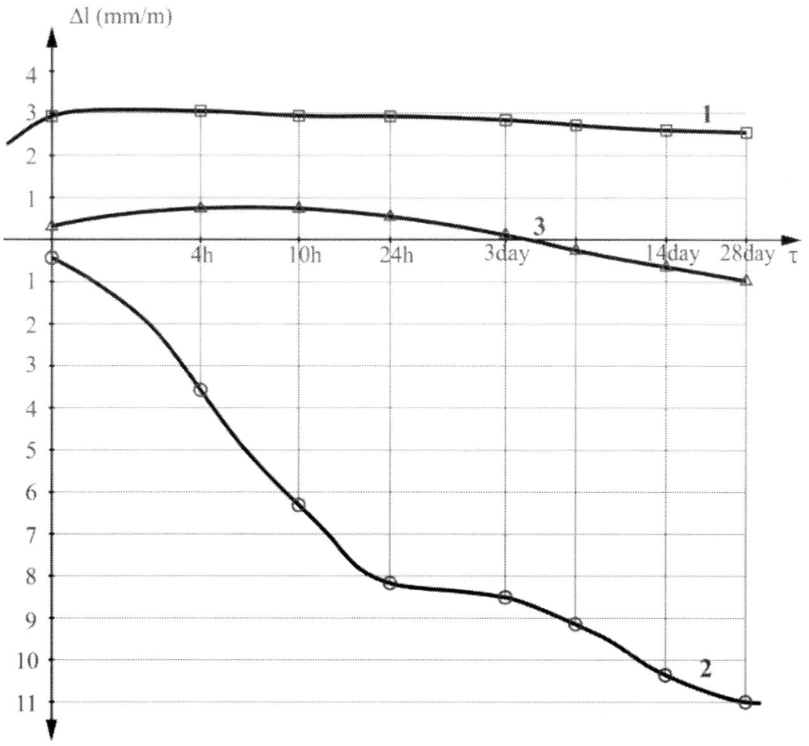

Fig. 1. Deformations of various binders (mm/m): 1 – gypsum; 2 – Portland cement; 3 – two-component binder

Using aluminate cement as a main component for dry mixtures for self-leveling floors requires an adequate selection of the superplasticizer.

Commercially available additives for use with aluminate cement to produce dry mixtures for floors were analyzed. Supplements of three main groups were selected for study: polymeric sulfomelamin (Peramin SMF20 Perstorp Construction Chemicals Inc., Sweden) modified with polyethylene glycol (Melflux PP100F, Melflux PP200F, SKV Polymers GmbH, Germany) and polycarboxylate (Sika Viscocrete 105P, Swiss). Evaluating of the effectiveness of the superplasticizer joint venture was carried out in accordance with GOST 30459-2008 "Admixtures for concretes and mortars. Determination and estimate of efficiency". (EN 934-6:2002 Admixtures for concrete, mortars and grout – Part 6 Sampling, conformity control and evaluation of conformity). All the properties were determined on the samples with dimensions of 4x4x16 cm according to ASTM and GOST for the tests. Water-reducing effect is 57...63%. The study works were done by the equipment of MSUCE.

It is known that using aluminate cements mixtures with superplasticizers a rapidly lose flowability. Cause of this were made studies for saving flowability of mixture after 20, 40 and 60 minutes after mixing. Additives were introduced in the amounts recommended by the manufacturer. Saving flowability (decrease no more than 10% in 60 minutes) showed only superplasticizers based on polycarboxylate.

The two-component composition based on a two-component binder (CRHAC + gypsum), filled with quartz sand (Binder:Sand. = 1:1.5) and modified by superplasticizers based on polycarboxylates (Sika Viscocrete 105P, Swiss) characterized by the following technological and operational properties:

- average density - 1750 ... 1850 kg/m^3;
- saving of self-leveling for 30…60 minutes. after mixing;
- setting time: beginning - 2 h 00 m ... 2 h 20 m, end - 3 h 00 m… 3 h 15 m;

- strength of composition: compressive strength after 7 hours is about 5 ... 6 MPa, after 1 day. - 23 ... 25 MPa, after 3 days. - 42 ... 44 MPa;
- flexural strength after 3 days. - 11...12 MPa.

4 Conclusion

During the construction of high-rise buildings, it is possible and rational to use binders alternative to Portland cement for some special purpose rooms. For construction of floors for technical premises (cars parks, sports halls, premises for accommodation of life support engineering systems of buildings, climatic equipment, etc.) is expediently to use mixes based on aluminate cements. The production of aluminate cements causes less damage to the environment. Moreover, the use of aluminate cements eliminates losses caused by corrosion of cement stone. Corrosion is probable after a certain period of service in the underground and basement technical rooms of high-rise buildings.

Producing dry mixtures for self-leveling floors became possible after the creation of modern superplasticizers, which provide a shear strength. Analysis of the results shows that for dry mixtures (with the use of portland cement) modern superplasticizers based on polycarboxylates work effectively. At the same time, there is no consensus about the factors that have influence on the compatibility of superplasticizers with portland cement, aluminate cement, gypsum and mixtures of these materials.

References

1. A. Kylili, P.A. Fokaides, SCC, **35**, 280 (2017)
2. U. Klansek, S. Kravanja, Int. J. of Conc. Str. and Mat., **62**, 434 (2006)
3. M. Zajac, J. Skocek, F. Bullerjahn, M.B. Haha, CCR, **84**, 62 (2016)
4. F. Winnefeld, B. Lothenbach, CCR, **40**, 1239 (2010)
5. M. Garcia-Mate, I. Santacruz, Angeles G. De la Torre, L. Leon-Reina, Miguel A.G., CCC, **34**, 684 (2012)
6. L. Pelletier-Chaignat, F. Winnefeld, B. Lothenbach, G. LeSaout, C.J. Muller, C. Famy, CCC, **33**, 551 (2011)
7. F.P. Glasser, L. Zhang, CCR, **31**, 1881 (2001)
8. J. Pera, J. Ambroise, CCR, **34**, 671 (2004)
9. Q. Zhou, N.B. Milestone, M. Hayes, JHM, **136**, 120 (2006)
10. T. Zhang, P. Gao, R. Luo, Y. Guo, J. Wei, Q. Yu, CBM, 662 (2013)
11. G. Bernardo, A. Telesca, G.L. Valenti, CCR, **36**, 1042 (2006)
12. T. Le-Bihan, J.F. Georgin, M. Michel, J. Ambroise, F. Morestin, CCR, **42**, 1055 (2012)
13. M. Collepardi, A. Borsoi, S. Collepardi, J.J. Olagot, R. Troli, CCC, **27**, 704 (2005)
14. F. Rajabipour, G. Sant, J. Weiss, CCR, **38**, 606 (2008)
15. B. Rongbing, S. Jian, CCR, **35**, 445 (2005)
16. Q. Cao, Zhongguo John Ma, CBM, **75**, 450 (2015)
17. K. Sisomphon, O.Copuroglu, E.A.B. Koenders, CBM, **42**, 217 (2013)
18. C. Maltese, C. Pistolesi, A. Lolli, A. Bravo, T. Cerulli, D. Salvioni, CCR, **35**, 2244 (2005)
19. M. Kaddo, MATEC Web Conf., **106**, 6 (2017)

Construction of high-rise building with underground parking in Moscow

Vyacheslav Ilyichev[1,2] , *Nadezhda* Nikiforova[2,3*] and *Artem* Konnov [2,3]

[1]Russian Academy of Architecture and Construction Sciences, 107031, 24/1, Bol'shaya Dmitrovka, Moscow, Russia
[2]Research Institute of Building Physics, 127238, 21, Lokomotivnyj proezd, Moscow, Russia
[3]Moscow State University of Civil Engineering, Yaroslavskoe shosse, 26, Moscow, 129337, Russia

Abstract. Paper presents results of scientific support to construction of unique residential building 108 m high with one storey underground part under high-rise section and 3-storey underground parking connected by underground passage. On-site soils included anthropogenic soil, clayey soils soft-stiff, saturated sands of varied grain coarseness. Design of retaining structure and support system for high-rise part excavation was developed. It suggested installation of steel pipes and struts. Construction of adjacent 3-storey underground parking by "Moscow method" is described in the paper. This method involves implementation of retaining wall consisted of prefabricated panels, truss structures (used as struts) and reinforced concrete slabs. Also design and construction technology is provided for foundations consisted of bored piles 800 мм in diameter joined by slab with base widening diameter of 1500 мм. Experiment results of static and dynamic load testing (ELDY method) are considered. Geotechnical monitoring data of adjacent building and utility systems settlement caused by construction of presented high-rise building were compared to numerical modelling results, predicted and permissible values.

1 Introduction

In 2014, on Leninsky Avenue in Moscow, BESTCON company (construction project by Metropolis LLC and ZEST LTD.) began the construction of a high-rise building with an attached three-storey parking. The building has class KC-3 according to [1] and has unique status because its height is 108 m. In compliance with the requirements of Russian standards [2] design and construction of the project was conducted with scientific support under the guidance of V. A. Ilyichev (academician of Russian Academy of Architecture and Construction Sciences (RAASN)) and carried out by the Research Institute of Building Physics of the RAASN (NIISF RAASN).

In the course of work on scientific support the following issues were resolved: design approval of the excavation retaining structures for one-storey underground level and three-storey underground car park, review of pile testing protocols, monitoring of concrete strength in above-ground and underground structures, the settlement prediction of the high-rise

* Corresponding author: n.s.nikiforova@mail.ru

building, the adjacent school building and utility systems, testing of window units and piers on the upper floors of the building subjected to wind load, prediction of the impact on surrounding buildings and the necessity of mitigation measures, analysis of the geotechnical monitoring results and etc.

2 Project Details

2.1 Geology of the site

Geological engineering cross-section of the construction site from the top to down is represented by the following soils (in brackets are the thickness of the layers): anthropogenic soils, predominantly lean clays (0.4 to 9.1 m), quaternary deposits - firm-stiff clays (0.7-3.6 m), lean clays soft-to-stiff (4-20 m), firm-stiff clays (0.5-2.2 m), very soft-stiff silty clays (0.5 to 4.9 m) covering early cretaceous sands from silt to medium (up to 59.2 m). Physical properties and strength of soil are given in table 1.

In the soil mass there are perched water table at a depth of 1.0-5.1 m, and two water levels: at a depth of 4.5-6.7 m, with a hydraulic head of 0.6-2.6 m and above jurassic deposits at a depth of 22.8-24.0 m. The level of surface is 196,300 m [3].

Table 1. Physical properties and soil strength.

Type of soil	Thickness, m	*E, MPa*	C_n, *kPa*	φ_n, °
anthropogenic soil: lean clay firm-stiff	0.4-9.1	-	-	-
clay firm-stiff	0.7-3.6	15	0.28	18
lean clay soft-firm	1.3-8.1	17	0.21	17
lean clay firm-stiff	1.0-5.2	11	0.27	18
lean clay stiff	1.0-3.7	22	0.43	19
clay firm-stiff	0.5-2.2	17	0.49	18
silty clay very soft-stiff	0.5-4.9	19	0.21	23
sand fine, from medium dense to dense	up to 59.2 m	40-50	0.18-0.08	35-36

2.2 Structures of high-rise building and underground parking

Design of above-ground and underground building structures was carried out in accordance with the requirements of Russian standards for high-rise [4,5] and underground construction [2,6,7]. High-rise construction is a research subject for Russian ([8], etc.) and foreign authors ([9], etc.).

Experience of underground space development of Russian cities, reflecting the problems of mitigation measures application for surrounding buildings and structures, evaluation of technological settlement, is described in the papers [10-17].

2.2.1 Structural design of high-rise building

Residential 33-storey building with built-in 1-storey underground level (Fig.1) has an attached underground 3-storey car parking with single storey aboveground pavilion, which is the entry to the underground parking. One-storey underground level under the high- rise part of the building connected by an underground passage in the first underground floor at level -5,200 with a 3-storey underground parking. Dimensions of high-rise part are 71.5x29.6 m, attached underground part – 103.3x35.0 m. The retaining structure of a one-storey underground level excavation was designed from metal pipes 426x10 mm 13 m long with metal wale beam and wooden lagging, with corner struts and raker struts of pipes with a diameter 530x10 and 630x10 mm, respectively (Fig.2). Prime excavation was pre-arranged with depth of 4.0 m. The load bearing structures of the building are reinforced concrete walls and columns.

Fig. 1. Construction of the high-rise building at Leninsky Avenue street in Moscow with attached underground parking.

Foundations under high-rise part of the building consist of bored piles joined by slab 2.0 m of thickness. From the bottom of the excavation, having a depth of 7.45 m, length of the piles is 23.5 m with a base widening of 1.5m and a height of 1.5 m. The relative level of the low pile end is -29, 450 m. Bored piles have a diameter of 800 mm, the spacing is 3.5 and 2.5 m. Under the end of piles there are the cretaceous fine sands. Piles are drilled in this stratum to not less than 1 m. Absolute level of the pile end is 166.85 m. The piles are drilled from a ground surface, although they are casted up to the bottom of the pit. Above the pit shafts are filled to the surface with gravel.

Fig. 2. The retaining structure for the excavation and pile drilling.

The base widening of the piles was carried out by mechanical means. Fig. 3 shows the used widening tool.

Fig. 3. The widening tool.

2.2.2 Structural design of 3-storey underground parking

The depth of the pit was 12.20 m (absolute level of the bottom was 182.595 m). Diaphragm wall, thickness of 600 mm – precast-monolithic, load bearing, buried into confining bed, buried depth below the bottom of the pit is 5.0 m (abs. level 177.60 m). Permeability without block wall secured through the use of prefabricated elements of high water-resistance (W12) concrete, which are arranged on the internal side and provide no filtering of water through the section of the diaphragm wall (Fig. 4).

Fig. 4. Precasted panel for diaphragm wall of 3-storey parking.

The construction of underground parking is realized by “Moscow method”, which is an improvement of the "top-down" method and includes retaining technology for excavations executed by two major slabs, "truss" bars between them and the two cable anchorages at the corners of the pit (Fig.5). This excavation retaining method is protected by Russian patent No. 2220258 [18] etc.

Fig. 5. Construction of 3-storey underground parking. Slab supported by truss structures.

3 Results of scientific support

3.1 Evaluating of piles bearing capacity on the site

The bearing capacity of the bored piles was determined on-site by ELGADTOP company. Two methods were used: static load (two piles) according to [19] and dynamic load by the ELDY method (two piles) according to the test programme, developed by ELGADTOP [20] and proved by NIISF RAASN. While testing pile No. 66 (Fig.6) with static load the supplement piles No. 36,38,96,98 were used as an anchors. The load was applied in five steps. The settlement of pile reached 18.9 mm. Bearing capacity of the pile estimated F_d = 885 ts.

Fig. 6. Pile No. 66 static test.

The image of the pile after testing by ELDY method (Fig.7) indicates that the pile is not damaged.

Fig. 7. Pile No. 237 after ELDY test.

Fig.8 depicts a computer monitor that records the process and parameters of the test.

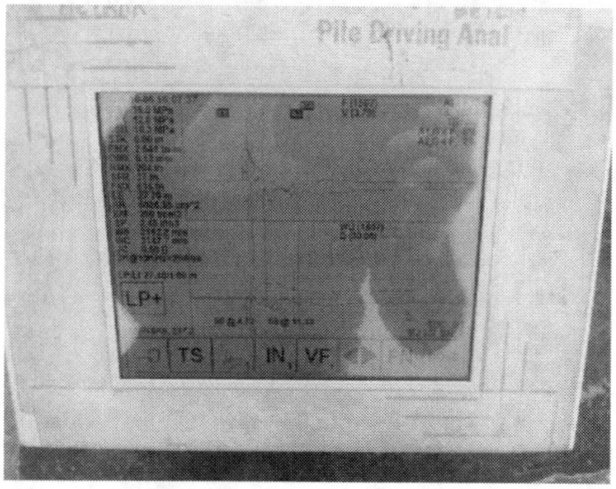

Fig. 8. Computer monitor recording the ELDY test process of pile No. 237.

The design length of pile No. 145 – 29.48 m; pile No. 237 – 29.62 m. Class of concrete is B40 , 21 days passed after casting. Achieved settlements: pile No. 145 – 18 mm, pile No. 237 – 16 mm. Bearing capacity of the pile according to the test (F_d): No. 237 – 1113,7 tf, pile No. 145 – 1082,4 tf.

It's necessary to admit that the piles were casted and tested from the surface but would work from the depth of ~8m taking into account underground part of the building. To determine the bearing capacity of piles F_d it's required to subtract the skin friction carrying capacity to a depth of 8.0 m from the surface, which is ~ 33,6 tf for the pile No. 237, and ~ 38,5 tf for the pile No. 145 and the pile No. 66 according to the test data collected by ELGADTOP Co. Thus, F_d of piles No. 237, No. 145, No. 66 after completion of underground part would be 1080,1044 and 847 tf , respectively.

In accordance with clause 7.1.11 [6] the estimated load (N) transmitted to the pile is 739,3 tf (No. 237), 714,6 tf (No. 145) and 676,4 tf (No. 66). The least of three values was chosen for the project – N≤676 tf.

3.2 Geotechnical forecast and monitoring results

The school building and utility systems were in the zone of the construction impact. Geotechnical forecast of high-rise building settlements and settlements of the surrounding buildings and structures was provided by PODZEMPROEKT Co. using the software package Midas GTS NX in three dimensional model [3]. An improved model of Mohr-Coulomb was used for these calculations.

Geotechnical monitoring included control of foundation slab displacements under the high-rise part of the building (45 cycles of measurements), monitoring of diaphragm wall around attached underground parking, 3-storey school building, fencing of the Embassy of Germany territory, electric pole and existing underground utilities in the zone of construction impact. Settlements of the high-rise building and the surrounding building and structures were measured at 124 geodetic marks in accordance with [21].

Foundation slab settlements under the stylobate part of the building reached -7.5... -15,7 mm , and under high-rise part of the building -11,4 mm... -22,3 mm and did not exceed calculated values of settlement, equal to 65 mm (Fig. 9) [22, 23].

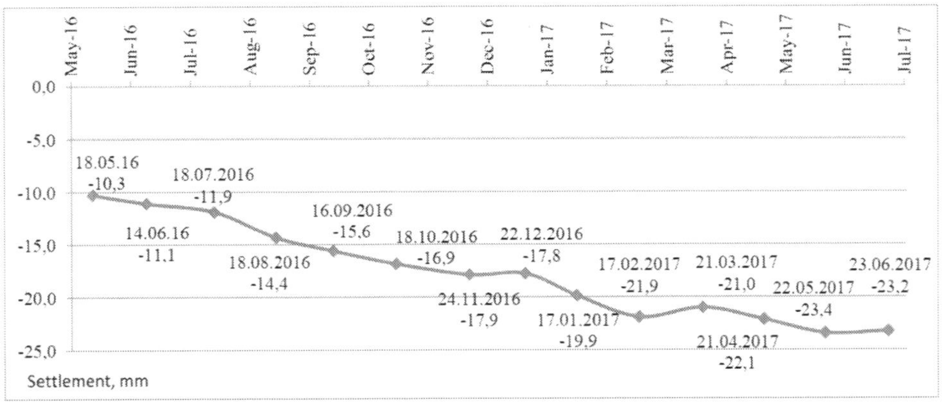

Fig. 9. Maximum settlement of foundation slabs under high-rise and stylobate part of the building.

Displacements of buildings and utilities didn't exceed the predicted values. Mitigation measures for the surrounding buildings and structures weren't required.

Horizontal displacements of the diaphragm wall were measured by engineer's microlevel NE-3e (developed by the Institute of Physics of the Earth (IPE) RAS) at 60 microlevel marks on -1, -2 and -3 storeys of the underground parking (20 marks on each floor). 37 measurement cycles were made (Fig.10).

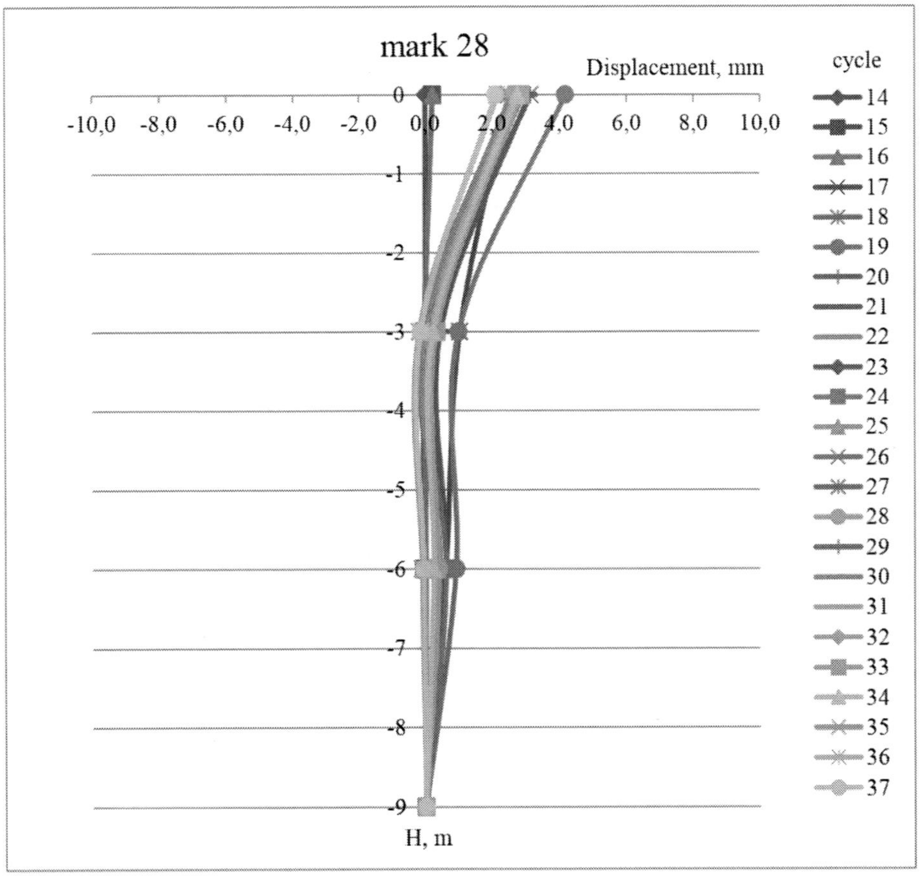

Fig. 10. Schedule change of a horizontal wall position during the construction of parking at cross-section of the observation mark No. 28.

3.3 Additional works, executed within scientific support for the design and construction of the high-rise building.

On the basis of the verification calculations done by PODZEMPROEKT Co. the length of the retaining metal piles was reduced by 1...2m on three sides of the pit. Also a design of the support system was changed. Raker struts based at foundation slab were changed to the longitudinal strut resting on a midway support which saved metal and accelerated the works on-site.

During concrete strength control (class B55) in the structures of high-rise building problem pylons have been revealed at level -0,250. A concrete in these pylons at the time of testing has strength no more than 90% of the design value. For this reason verification calculations have been done establishing that the margin of safety in them ranged from 9 to 34 %.

In connection with a partial change of the design solutions verification calculations was performed concerning reinforcement of interwindow piers on the influence of the wind load specified in the recommendations given by CNIISK V.A. Kucherenko. These calculations were followed by on-site tests of interwindow piers on wind load, showed the adequacy of reinforcement.

Also tests of window units with valves were conducted to determine the resistance to heat transfer, air and water permeability and also soundproofing. Received results confirmed the usability of chosen window units.

4 Conclusions

Scientific support was able to solve a number of issues during the design and construction of high-rise building, for example, potential necessity of mitigation measures for utilities, need for a reliable forecast of high-rise building settlement, confirmation of the "Moscow method" effectiveness for retaining and strutting of underground car park structures, etc.

Acknowledgments

The authors would like to thank BESTCON Co., ZHILKAPSTROY Co., PODZEMPROEKT Co. for given project documentation and cooperative work at the site.

References

1. Russian Standard GOST 27751-2014. Reliability for constructions and foundations. General principles.
2. Russian Standard SP 22.13330.2016 Soil bases of buildings and structures.
3. The calculation of the impact on existing utilities and the school building during the construction of a multifunctional residential complex at the address: Moscow, Leninsky prospect, 97-99, as well as development of recommendations for mitigation measures for existing utilities and school building, scientific and technical report PODZEMPROEKT Co., (2014)
4. Russian Standard SP 267.1325800.2016 High rise buildings and complexes. Design rules.
5. Russian Standard MGSN 4.19-05.Multifunctional high-rise buldings and complexes, vol. 1. Moscow standard document.
6. Russian Standard SP 24.13330.2011 Pile foundations.
7. Russian Standard SP 45.13330.2017 Earthworks, Grounds and Footings.

8. O.A. Shulyatev, *Osnovaniya i fundamenty vysotnyh zdanij* (ACB, Moscow, 2016)
9. H.G. Poulos, Razvitie gorodov i geotekhnicheskoe stroitel'stvo, **1**, 1 (2011)
10. V.A. Ilyichev, R.A. Mangushev, N.S. Nikiforova, Osnovaniya, fundamenty i mekhanika gruntov, **2,** 17 (2012)
11. *Geotechnical Engineer's Handbook. Footings, foundations, underground structures. 2-nd edition.* (edited by V.A. Ilyichev, R.A. Mangushev) (ACB, Moscow,2016)
12. V.A. Ilyichev, N.S. Nikiforova, A.V. Konnov, *Proc. of 19th ICSMGE*, 1789 (Seul, Korea, 2017)
13. V.A. Ilyichev, R.A. Mangushev, N.S. Nikiforova, A.V. Konnov, BST Byulleten' stroitel'noj tekhniki, **6**, 68 (2017)
14. R.A. Mangushev, N.S. Nikiforova, *Tekhnologicheskie osadki v zone vliyaniya podzemnogo stroitel'stva* (ACB, Moscow, 2017)
15. N.S. Nikiforova, Obespechenie sohrannosti zdanij v zone vliyaniya podzemnogo stroitel'stva (MGSU, Moscow, 2016)
16. N.S. Nikiforova, *The Proc. of the First U.S.A.- Russia Geotechnical Engineering Workshop*, 72 (Oakland, California, 2012)
17. O.A. Shulyatev, O.A. Mozgacheva, V.S. Pospekhov, *Osvoenie podzemnogo prostranstva gorodov* (ACB, Moscow, 2017)
18. C.O. Zege, I.A. Zege, N.C. Zege, Sposob vozvedeniya mnogoehtazhnogo podzemnogo sooruzheniya. Russian Federation patent №2220258 (2003)
19. GOST 5686-2012 Soils. Field test methods by piles. Russian standard document.
20. Programme of field pile test using impact load (ELDY method) and static load. ELGADTOP Co. (2014)
21. Russian Standard GOST 24846-2012 Soils. Methods of measuring the strains of structure and building bases.
22. V.A. Ilyichev, N.S. Nikiforova, A.V. Konnov, V.R. Irtuganova, ZHilishchnoe stroitel'stvo, **6**, 29 (2016)
23. Report for the scientific and technical work. "Geotechnical monitoring for the construction project "Multifunctional residential complex with underground parking at the address: Moscow, Leninsky prospect, 97-99". Step 11. Geotechnical monitoring at the construction site (April-June 2017)". NIISF RAASN Research institute. (2017)

Improving of the monolithic foamconcrete quality for used in the high-rise constructions

Anastasia Sychova[1,*], *Andrey* Solomahin[1], *Vitaliy* Kotovich[1], *Larisa* Svatovskaya[2], and *Yuriy* Kamenev[1]

[1]Military Space Academy named after A.F.Mozhaisky, 197198 Zhdanovskaya st.13, St.Petersburg, Russia
[4]Emperor Alexander I St. Petersburg State Transport University, 190031 Moskovsky pr.9, St.Petersburg, Russia

Abstract. The article is devoted to the problem of obtaining the desired rheological characteristics of the foamconcrete when transporting it by the feeding pipelines to the place of installation in high-rise construction. The dependence of the density change of foamconcrete on the height to which it is transported is shown. It is stated that the use of stabilizers for the foam, as well as certain mechanisms of stabilization results in the concrete mix of the required density. To assess the rheological properties of the concrete mix the coefficient of volumetric relaxation is proposed.

1 Introduction

Interest in the use of monolithic non-autoclave foam concrete in walling and partitions of high-rise buildings is due to the fact that its application allows you:

-to simplify and reduce the cost of construction due to the production of the foam concrete on the construction site;

-to work year-round (in winter at temperatures down to -15°C without the additional care of the concrete; the foamconcrete can accumulate heat due to the implementation of the thermos effect);

-to increase the volume of work more than ten times, in comparison with the laying of aerated concrete and foam concrete blocks, with a significant reduction in the number of workers;

-to minimize the use of crane and lifting equipment (the supply of monolithic foamconcrete is made from ground mark);

-to abandon the use of adhesive mortar and lintels, as in the case of manufacturing gas - concrete masonry (to eliminate the “cold bridge”);

-to reduce significantly the removal of debris from the construction site (breakage of gas-aerated concrete blocks);

-to reduce transport expenses and the area of the zones of materials storage;

-to get the extra dwelling space by reducing the thickness of the thermal insulation of external walls by using foam concrete of low densities [1, 2].

[*]Corresponding author: amsychova@yandex.ru

Practice shows that in constructing a sixteen-storey building with the perimeter of 400-500 meters an additional dwelling space of 500 m^2 is formed. It takes place due to the use of monolithic foam concrete with medium density of 300 kg/m^3 filled between the outer cladding walling in a half-brick and its inner lining-water-resistant gypsum board (Fig.1).

The possible applications of heat-insulating of monolithic foam concrete with medium density of 300 kg/m3 in high-rise building are shown in Fig.1-3.

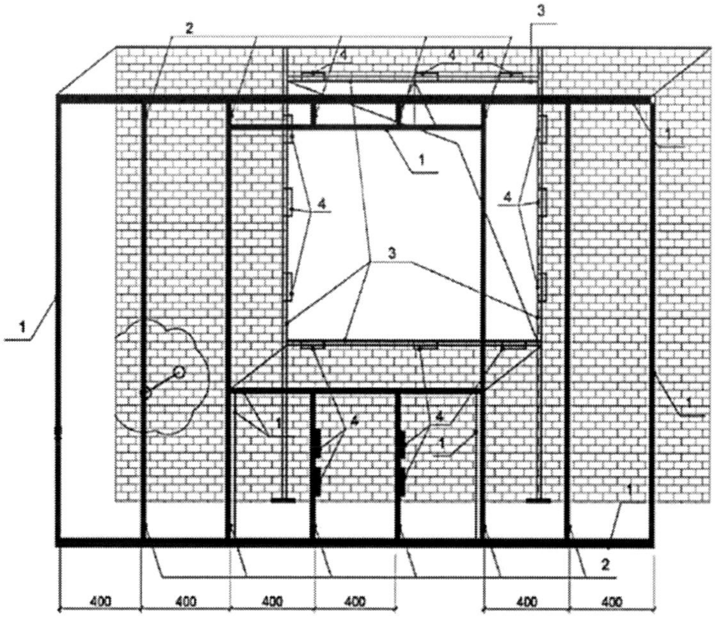

Fig.1. The construction of the outer walls of frame-monolithic building, in which the monolithic foam concrete is placed: 1 - guide profile 30x27мм; 2 - profile 60x27мм; 3 - profile 66x30мм; 4 - packing wood treated with anti-septic.

The construction of the outer wall of frame-monolithic building, in which the monolithic foam concrete is placed is shown in Fig.1.

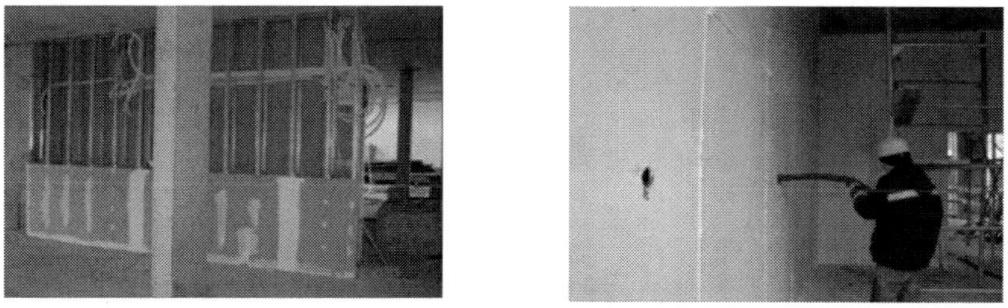

a b

Fig.2. Internal partitions of the building with monolithic foamconcrete filler: a –frame construction of the interior walls, b – a variant of the layer-by-layer pouring of foam concrete into the fixed formwork of the partition

The possible internal partitions of the building with using the monolithic foam concrete is shown in Fig.2.

Fig. 3. The protecting constructions of the high-rise building with filling a foamconcrete into the fixed formwork.

Figure 3 shows a high-rise building, in which the foam concrete is poured into the fixed formwork, consisting both from the outer and inner side of the frame on the basis of light galvanized steel profiles of different cross-section. It can be sheathed from the outside, for example with cement-bonded plates with the subsequent finishing of the outer surface and with moisture-proof gypsum cardboard sheets inside.

In all the proposed schemes heat-insulating foamconcrete with its layer-by-layer placing into the structure is used. The layer height is up to 600 mm.

The main feature of these technical solution application is the feeding technology of monolithic foam concrete from zero design level up to 70 meters vertically and up to 200 meters horizontally, in full compliance with the required design and rheological characteristics.

It is possible to realize it under the following conditions:

-first, only in the separate method of producing foam and water-cement-sand mixture with their further mixing in the common mixer;

- second, at the altitudes of above 20 meters, you must perform additional steps to increase foam stability;

-thirdly, due to the change in the scheme of foam supply to the height of over 20 meters without additional operations of improving the quality of the foam.

The change of feeding scheme means the transfer of foam concrete plant to the place of pouring, i.e. the breakdown of the building on the catches of the certain height takes place.

For monolithic foam concrete, the study of rheological properties is of great practical interest because they determine the behavior of foam-cement systems under stress when supplying to the place of pouring, its deformability and its ability to restore the original pore structure after load removal [3].

Nowadays the investigations of the problems of the rheological composition of the concrete mix change during transportation over long distances in the vertical were not undertaken. There are no systematic data about the critical height of feeding the foam concrete mixes of different rheological composition by using foam feeding equipment with different principles of work. Receiving, during transportation, the foam concrete with stable physico-mechanical characteristics is rather difficult.

Table 1 shows the values of changes in the density of foam concrete mix (for the average density D300 in a dry state) when feeding to the height from 0 to 70 meters.

The table shows that when feeding the foamconcrete to a height its density increases up to 50%, and as a result, the foamconcrete does not meet the design brand and has a higher heat conductivity and lower thermal characteristics.

Table 1. The change in the density of the foam concrete average density D300 depending on the delivery height

The type of foamconcrete	Feeding height, m							
	0	10	20	30	40	50	60	70
	The density, kg/m^3							
Damp foamconcrete mixture	420	490	525	560	575	600	626	646
Foamconcrete in the project age	300	370	400	440	455	480	500	520

To understand the processes of the concrete mix rheology change let us consider the scheme of its feeding by gerotor pump to the height of H and the resulting hydraulic pressure in the feeding line, Fig.4.

A column of the mixture P_{mix} (depending on the feeding height) as well as the starting pressure of the pump P_{pump} press on the bottom layer ΔH of the foamconcrete mix.

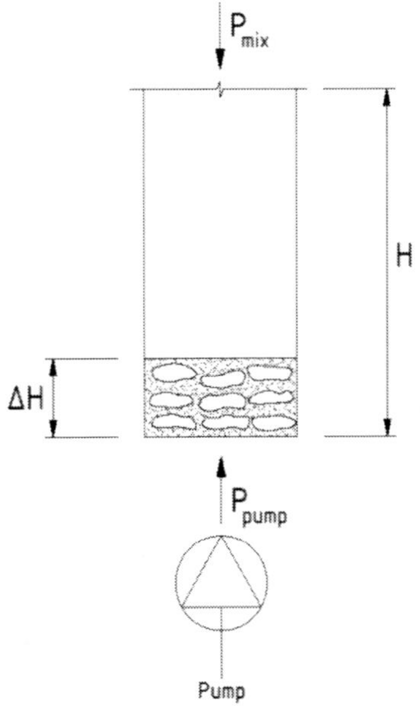

Fig. 4. Diagram of the impact of different types of load on the bottom layers of foamconcrete during the vertical transportation

2 Materials and Methods

To assess the rheological stability of the concrete mix when transporting it to the building site at a considerable vertical distance, it was proposed to introduce the coefficient of volumetric relaxation of the foamconcrete mass, *Kv.r.* The value of this coefficient is proposed to be assessed as the ratio of the volume mass of the foam concrete mix at the outlet of the foamconcrete plant, Vm_1 to the weight mass of the foamconcrete mixture at the outlet of the pipeline in the place pouring, Vm_2, the formula 1:

$$Kv.r.=Vm_1/Vm_2 \quad (2)$$

From the formula we see that the higher *Kv.r.*, the higher is the ability of the compound to perceive the various loads during the transportation without destroying. The dependence of the coefficient of volume relaxation upon the height of lifting the foamconcrete mix of the average density D300 is shown in Fig.5.

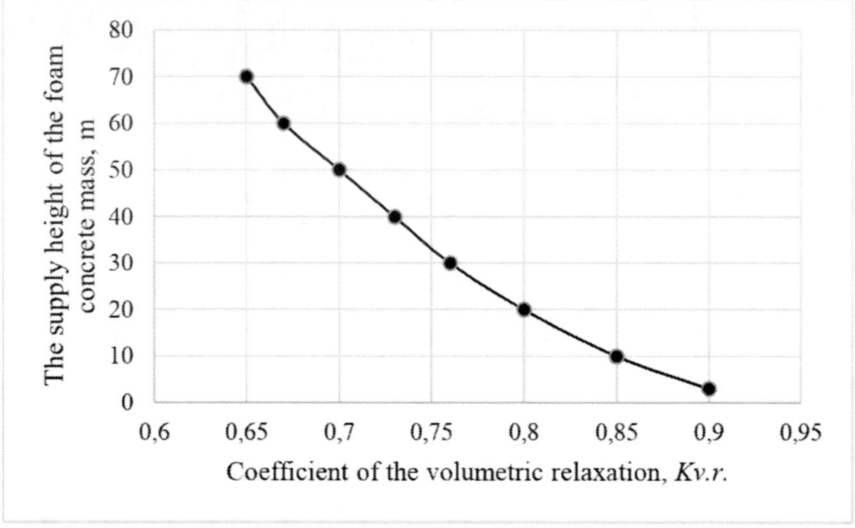

Fig. 5. The dependence of the coefficient of volumetric relaxation of the mixture upon the lifting height of the foam concrete mass

3 Results

The main idea of this work is the following assumption as the most "fragile element" in the concrete mix is a construction foam, its strengthening and stabilization should increase the Kv.r. of the foam concrete mix. This step should increase the stability of the concrete mix under different loads determined by the pressure in the pipeline and the height to which the mixture is lifted.

For experimental research a protein foaming agent was used, because it is the best from the point of view of its influence on the hydration of cement [4, 5, 6].

The stabilizers providing a different mechanism for the stabilization of the foam were used. It's strength was estimated at the time of the stability in comparison with the control unstabilized foam [7]. Stability of the control foam was 40 minutes. The following agents were used as stabilizers: polyvinyl acetate; latex, $Fe_4[Fe(CN)_6]_3$ and $K_3[Fe(CN)_6]$ [8, 9]. Sols are colloid solutions that contain particles with a size 1-100 nm as a disperse phase [10, 11].

Experimental studies showed Fig. 6 that all the selected substances have a positive effect and increase the time of the foam stability up to 4.8 times. The best effect is when $K_3[Fe(CN)_6]$ was used.

 https://doi.org/10.1051/e3sconf/20183302058

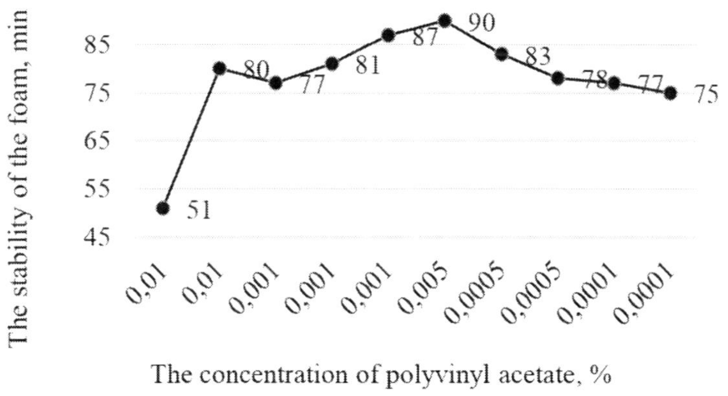

a

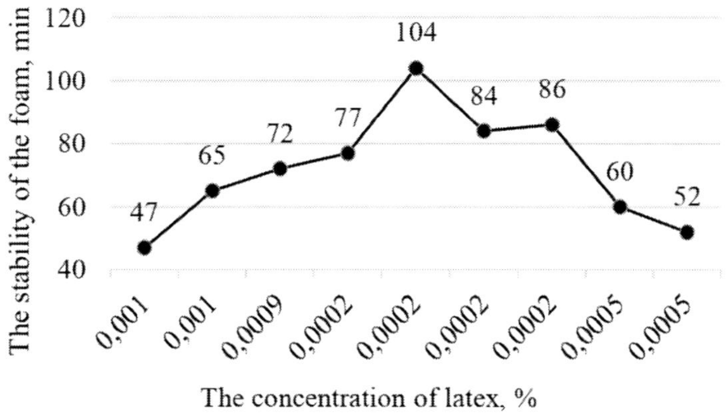

b

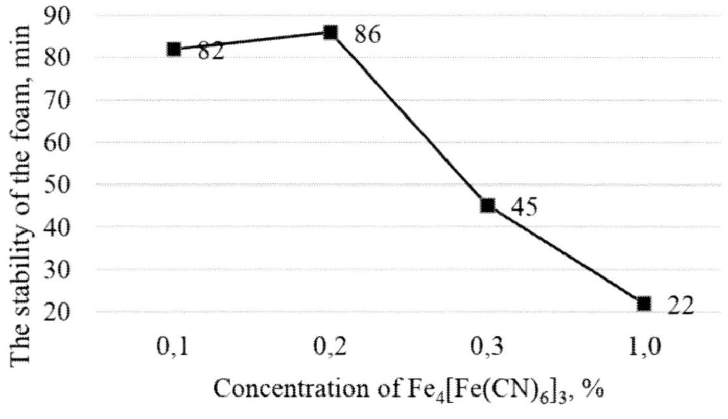

c

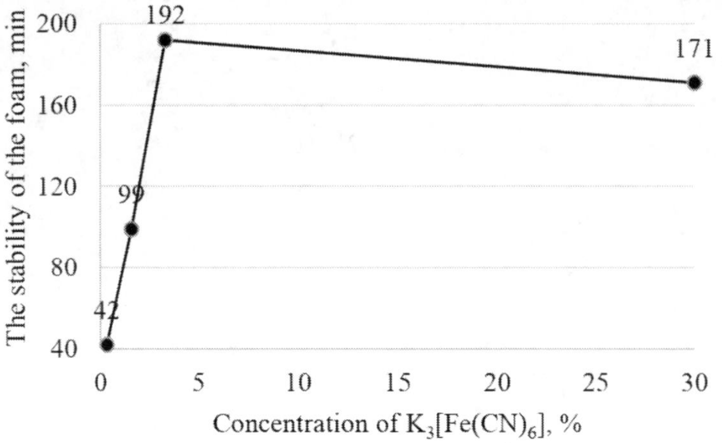

d

Fig. 6. Stability of the foam prepared on the basis of various stabilizers: a – polyvinyl acetate; b – latex; c – $Fe_4[Fe(CN)_6]_3$; d – $K_3[Fe(CN)_6]$.

4 Discussion

The effect of the foam stabilization can be connected with the manifestation of two mechanisms. In the case of using polyvinyl acetate and latex the viscosity of the fluid foaming agent [12] increases, and in the case of $Fe_4[Fe(CN)_6]_3$ and $K_3[Fe(CN)_6]$ containing the ions Fe^{2+} and Fe^{3+}, its interaction with protein molecules of the foaming agent is possible. This interaction formed a strong complex compounds, increasing the thickness of foam films [13]. In both cases, the stabilization effect results in the decrease in leakage speed of liquid from foam films under the action of gravity and increase the time of its sustainability.

It was further suggested that to maintain the rheology of the concrete mix when supplying it to the place of laying vertically not only the stability time, but also its structural and mechanical properties are important. Preliminary experiments showed that with the introduction of stabilizers based on iron ions strong, but fragile foam film is formed. It can't withstand hydraulic pressure in the pipeline during supply to the high altitude. In the case of using polymers that increase the viscosity of the liquid from foam films, the stability time of the foam is lower, but the foam has greater ductility, i.e. the ability to restore the shape of the bubble, and to relax after the action of the pressure in the pipeline [14, 15].

Fig. 7 shows the coefficient of volumetric relaxation of foamconcrete mixture prepared on the basis of the construction foam, stabilized with latex. From the figure it follows that the presence of the stabilizer increases the value of the *Kv.r.* up to 38% when supplying the foam to the height of 70 m.

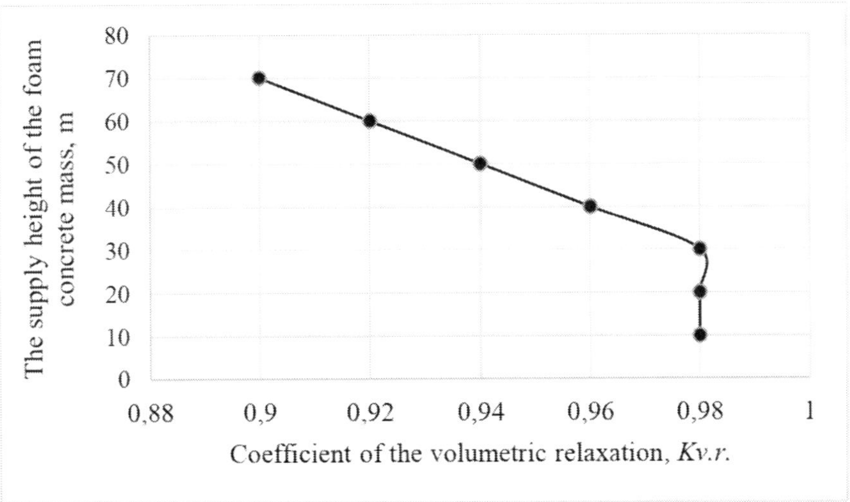

Fig. 7. The dependence of the coefficient of volumetric relaxation of the lifting height of the foamconcrete mass prepared on the basis of the foam stabilized with latex.

When the height of a building is more than 70 meters it is recommended to use the method of laying foamconcrete by separating the building on the sections with a step of up to 70 meters and transfer the foam plant to the place of pouring. This method can provide the pouring of foam concrete mass in the process of buildings construction of any height.

5 Conclusions

1. It is propossed to assess the stability of wet foam concrete mass during transportation to the place of laying to the height of 70 meters by using the coefficient of volumetric relaxation, as the relationship of the density of the foam at the outlet from the foam plant to the density of the foam at the outlet from the pipeline.
2. For the first time it is proposed to increase the level of rheological properties of the of foam concrete during its transportation to the place of laying into the construction through the stabilization of foam and increase the strength of the foam bubble.
3. Foam stabilizers of different nature are considered and it is shown that the of foam stability in their presence increases up to 4.8 times; mechanisms of the stabilizing effect and their influence on structural and mechanical properties of the foam are proposed.
4. It is shown that by using foam stabilization, coefficient of volume relaxation of the foam concrete mass during its transport vertically up to 70 meters increases up to 38%.
5. In the construction of buildings more than 70 meters height it is proposed to use the method of laying foamconcrete by separating the building on the section with a step of up to 70 meters and transfer the foam plant to the place of pouring.

References

1. V. Mikulski, V. Kupriyanov, *Building materials* (ASV, Moscow, 2000).
2. I. Rebjev, *Building materials* (Higher school, Moscow, 2003).
3. L. Shakhova, *Technology of foam concrete. Theory and practice* (Publishing house Association building universities, Moscow, 2010).

4. L. Svatovskaya, M.Shershneva, M. Baidarashvili, N. Yakimova, A. Khitrov, Proceedings of the International Conference on Sustainable Waste Management and Recycling: Construction Demolition Waste, 199-203 **(2004).**
5. A. Sychova, M. Sychov, E. Rusanova, Procedia Engineering, **189,** 681-687 **(2017),** DOI: 10.1016/j.proeng.2017.05.108
6. L. Svatovskaya, A. Sakharova, M. Baidarashvilly, A. Petriaev, Proceedings of the 14th International Conference of International Association for Computer Methods and Recent Advances in Geomechanics, 152 **(2014)**.
7. A. Sychova, A. Solomahin, A. Hitrov, Procedia Engineering, **189,** 688-694 **(2017),** DOI: 10.1016/j.proeng.2017.05.109
8. L. Svatovskaya, A. Kabanov, M. Sychov, IOP Conference Series: Earth and Environmental Science, **90**, **(2017)**, DOI: 10.1088/1755-1315/90/1/012010
9. A. Sychova, L. Svatovskaya, S. Mjakin, I. Vasiljeva, Electron Beam Modification of Solids: Mechanisms, Common Features and Promising Applications, 35-37, **(2009)**.
10. A. Sychova, L. Svatovskaya, S. Mjakin, I. Vasiljeva, Electron Beam Modification of Solids: Mechanisms, Common Features and Promising Applications, 39-47, **(2009)**.
11. A. Zimon, *Colloid and surface chemistry* (Agar, Moscow, 2007).
12. B. Summ D. *Foundations of colloid chemistry* (Academy, Moscow, 2007).
13. L. Shahova, V. Balyasnikov, *Foam for cellular concrete* (Belgorod, 2002).
14. N. Shabanova, *Fundamentals of Sol-gel technology of nano silicon oxide* (Akadembook, Moscow, 2004).
15. M. Gel'fman, N. Kirsanova, O. Kovalevich, L.Holohanowa, N. Rosalino, *Workshop on colloid chemistry* (Lan', Saint-Petersburg, 2005).

The method of abstraction in the design of databases and the interoperability

Nikolay Yakovlev[1*]

[1]Moscow State University of Civil Engineering, Yaroslavskoe shosse, 26, Moscow, 129337, Russia

Abstract. When designing the database structure oriented to the contents of indicators presented in the documents and communications subject area. First, the method of abstraction is applied by expansion of the indices of new, artificially constructed abstract concepts. The use of abstract concepts allows to avoid registration of relations many-to-many. For this reason, when built using abstract concepts, demonstrate greater stability in the processes. The example abstract concepts to address structure – a unique house number. Second, the method of abstraction can be used in the transformation of concepts by omitting some attributes that are unnecessary for solving certain classes of problems. Data processing associated with the amended concepts is more simple without losing the possibility of solving the considered classes of problems. For example, the concept "street" loses the binding to the land. The content of the modified concept of "street" are only the relations of the houses to the declared name. For most accounting tasks and ensure communication is enough.

1 Introduction

Still J. Forrester [1] defined the city as urbanized territory, i.e. a system in which interact with different types of business, residential and non-residential people. A detailed description of these components and check the related facts and events are executed in the units of the authorities and city management. An integral part of the functioning of these units is now becoming the operation of automated information systems (AIS) or the automated control systems (management information system). Information support of automated systems, database structures are created on the basis of a set of indicators appearing in the documents. At the beginning of creation of the automated systems in the city of Moscow was made and in some cases specifically stipulated by the customer, when you build to describe the logical structure of the respective databases (at the level of global logical description [2]) most accurately reflect ideas about the subject area, formed in the non-automated control system and reflected in the relevant documents. It can be shown that this approach often leads to difficulties of identification and subsequent integration. These difficulties are due to the natural active role of people in processes of functioning of the city and the nature of these, reflecting the complex activity.

* Corresponding author: yakovlev50@inbox.ru

The data about real objects are often characterized by a plurality of representations. For example, the address of a building in the city of Moscow can consist of the following parts: street; Number and index of the house; Number and index of the hull; Number of the building. Historically, the city was a collection of land plots - estates with buildings on them. The numbering of buildings on the land plot clockwise still takes place in the internal documentation of the Moscow City Bureau of Technical Inventory. From there, the numbering of buildings sometimes penetrates into real addresses. The concept of "hull" is more associated with the processes of mass construction, is not tied to the land and reflects the sequence of construction of nearby buildings. Buildings later built have a larger hull number. In some areas of the city, for example in Zelenograd, often there is no house number, it is replaced by the number of the hull, and part of the hull number is the number of the neighborhood. In addition, the structure can have addresses on several streets, often two for corner buildings at intersections. The streets are subject to renaming. Most often this happens when you replace "building" addresses with "mail" addresses. Depending on the age of the document, the street name may correspond to the position before or after renaming.

In these conditions, the addresses of the same structure in different documents could differ quite significantly. And if the information system required that information on the same structure should be combined, and the data on the structure came from different sources, then the solution of such tasks involves a significant amount of identification work.

Therefore, the phenomenon that is required to deal, lies in the multiplicity of representations of the locations virtually any structures in the city. The building has multiple addresses. And if we have some address yet to be identified, then this address will have to compare with multiple sets of addresses each set of addresses belongs to a certain building.

Traditional development of information support involves the use of objects and their attributes from the subject area [3,4,5,6,7,8,9]. While implicitly it is assumed that for the successful operation of the future system this material enough. Nothing beyond existing information to add not required. The identification procedure can be quite complicated [10]

The second phenomenon that needs to deal with is a varied territorial division of the city, allowing not less than a variety of ways of its representation in information security. The division of the city into spheres of influence of the territorial units of the urban structures is very diverse. Describe the parts of the territorial division can be performed in a verbal description of boundaries or a depiction of the boundaries on paper or electronic maps and charts, or, listing the address lists of the serviced buildings. The situation with borders is sometimes compounded by the fact that not all land within the city is divided into local area. View the land through the list of addresses relevant to the plot of the buildings is complicated by the multiplicity of addresses for each building.

The choice of method of representation of concepts depends on the nature of the tasks we are going to solve with the use of this concept. For most accounts, search tasks and identification tasks, the unit of territorial division on the map is not important. Since the activities of legal entities and individuals in the city is concentrated in buildings, and it's relation to particular structures to unit territorial division. A lot of problems to be addressed in an automated way connected with the use within a single city service information prepared in another city service with the features of territorial division within each service. The translation task pairing such information into language representations on the territorial division of the city leads to the processing of data about land plots, their borders, and mutual complementarity. Such processing is characteristic of geographic information systems.

2 Methods

In order to cope with the variety of locations, designing the corresponding concept applied a variant of the method of abstraction based on the following features. On the one hand, the applied abstraction to the maximum technically possible. On the other hand, after you build the required concepts are applied the most complete Supplement of this concept to those items that were discarded in the initial abstraction. A new concept has been called "Unique house number" (UDOM) and treated any individual structure. When you build this concept applied principles:

1) primary code structure needs to be abstract, i.e., do not depend on any physical parameters, location and similar characteristics of relations of the structure on site

2) filling of the database must ensure that the maximum technically possible completeness of data on all buildings in the city

3) filling of the database must ensure that the maximum technically possible completeness of data on all addresses of each of the buildings

For solving problems related to the diversity of territorial division of the city, in order to avoid the necessity of processing data on land parcels, used the method of abstraction. In this case, abstracting themselves from the land which divided the territory of the city. After this abstraction remains such an attitude: "the name territorial unit" - the list of unique house numbers belonging to a given territorial unit. In terms of the logical data description is most convenient to represent these relations using certain facets assigned to unique house numbers. Each facet corresponds to a certain territorial division of the city. Since the activities of legal persons connected with the placement in the buildings, with addresses, the idea of productively when considering the territorial aspects, especially if analyzing, and processing the exposed crossing of the territorial divisions.

3 Results

3.1. Unique house numbers and addresses

In 1987-1989 on the basis of hardware and software of automated system for accounting and distribution of residential area in Moscow (management information system "KURS") was established on the basis of a future Database, "The Housing and uninhabited Fund" (for the Moscow city Bureau of technical inventory (MosgorBTI)). In the description of the data in the Statistical card for the building was applied the first two postulates.

The implementation of the first postulate was ensured by the automatic generation (the first time you enter the corresponding statistical card) the so-called "unique house number" (UDOM), which is then not changed throughout the life cycle of the buildings. That is, even after the demolition of buildings and construction on this place another house archival data on demolished structure in the database were contacted by former unique house number.

The second postulate was provided by enter in the database information about all included in MosgorBTI buildings in the city of Moscow and the further maintenance of the database. Note that the entry in the database " The Housing and uninhabited Fund " information about all apartments in all buildings in the city and about all areas all the apartments were made in 1990-1993. Thus, in the machine information base, information about a particular structure is associated with the corresponding unique house number. The ability to restore the original form of the document, from where a specific address came, is ensured by storing in a special table the connection between the unique house number and the entire set of real addresses associated with this building.

The transition from the usual address, which occurs in documents, to the unique number of the house is carried out through identification procedures. These procedures are performed either interactively with direct participation of the operator, or in batch mode, when processing of an array of data with addresses received from another information system is required. If it is not possible to uniquely identify interactively, the document where the address is found with multiple interpretations is returned to the source for clarification. This situation arises if the database has two buildings with one house number and buildings 1 and 2, in the address for identification there is only the house number, the hull number and the building number are missing altogether.

When identifying in batch mode, man-machine procedures are used with sequential refinement of SQL queries and a gradual increase in the percentage of identified addresses. The implementation of the second principle is very useful and allows increasing the share of the identified addresses. It is important to note that the identification for the address data is performed once - the first time they are entered into the system. In further transformations of data, a unique house number participates. Thus, the principle of one-time data entry into the system is realized. The third principle is used when expanding the use of unique numbers to other automated systems, and ensuring interoperability. The result of the application of this principle increases the efficiency of identification processes.

3.2. The interaction of automated system

The subject of interaction between systems is quite extensive. We will discuss only the question of the use in one system of information about a certain object or phenomenon from another system in those cases when the concepts of the same subject or phenomenon in different systems are different. For example, let it be information about apartments in the same house, whose addresses in interacting systems are represented in different ways: in one system this will be the mailing address, and in another it will be the address of the technical inventory office. In this case, the addresses of the same structure in different systems may not coincide. Large companies that deal with interoperability issues, in particular, offer mutual agreement of interfaces, data formats and exchange protocols. In our case, this is not enough, since the content difference in addresses will not disappear anywhere. Staying at the level of universal methods of ensuring interaction before taking into account specific differences and to ways of overcoming them will not reach you.

Inclusion in the interaction process of abstract concepts that unambiguously represent the subject of interaction is one of the ways to actually ensure the interaction of systems that have multiple ambiguous data about the same subject. We give a number of examples of the implementation of this proposal. For the first time, unique house numbers were generated and applied in the database "Housing and Uninhabited Fund", created for the Moscow City Bureau of Technical Inventory.

The easiest from the point of view of the effectiveness of data exchange, in which information about buildings is involved, would be an option when both systems have a binding to unique house numbers. And such approaches have been organized. When launching an automated social security management system for Moscow residents, information was provided on addresses linked to unique house numbers. In the future, this system provided data to the tax authorities with information about pensioners who were entitled to property tax relief. Since the data was provided with unique house numbers, there were no problems with the identification of addresses.

For the tax authorities in the city of Moscow in 1999 - 2004 was created an automated information system (AIS) "Tax 2 Moscow". For the subsystems of this system, related to the property tax, a part of the database containing information about the buildings was based on unique house numbers. When transferring to the tax authorities from MosgorBTI assessments of apartments for tax purposes, which also contain unique house numbers, again there were no problems with identification. The importance of obtaining "good", tied data with the appraisals of apartments is due to their application both at the annual tax assessment and throughout the year with the direct appeal of taxpayers to the inspection.

For the postal service of the city of Moscow (Glavmospochtamp), a binding of their directory with information on the buildings on the Database "Housing and Uninhabited Fund" with the use of unique house numbers was made. Along the way, the guide itself was refined, which was helped by the realized postulate of the completeness of the data. Within the AIS "Tax 2 Moscow" it was required to solve the problem of mass printing and dispatch of notices on property taxes in such a way that by mail the notices came within the allotted time. Transferred to the tax authorities from Glavmospochtampa information about the buildings contained a unique house number and postal code. As a result, there were no problems with identification and the postal service began receiving in mass order from the tax authorities letters with documents on property taxes, issued by all postal rules and with correct indices.

Feature of the city of Moscow is a strong departmental disunity. Each information system is created taking into account the established traditions and categorically does not want to change itself to facilitate interaction with other systems. For those cases when the direct inclusion of unique house numbers in the operating information system did not result in the creation of an external service that received data from the sending system, processing them with identification and binding to unique house numbers and subsequent transfer of replenished data to the user system. This situation arose with data on the deceased, processed in the information system of the registry office. The fact is that it is impossible to send documents on property taxes to deceased people, every mistake in such a case is fraught with a scandal, and there were no unique house numbers at the addresses of registration in the database of the registrar's offices. This external processing of data from the registry offices solved the identification problem.

3.3 Abstract concepts in the tasks of territorial division of the city.

When describing the units of the territorial division of the city, we will abstract from the configuration of the boundaries and topology of parts of the land. Then for the representation of a specific territorial unit there is only a list of unique house numbers pertaining to the territorial unit in question. And this representation is quite enough for solving a wide range of tasks of automated systems. The implementation of this approach is fairly simple due to the high level of abstraction. At the logical level, a flat table is defined, containing a unique house number and signs of belonging to the corresponding units of territorial division. Let us give some examples.

A street is a collection of buildings belonging to a street. In this case, the structure can belong to different streets. This is the case for corner houses, taking into account the renaming of streets, etc.

The district (according to the old division of the city before perestroika) is a set of buildings that form the district.

The administrative district is a set of buildings forming the administrative district.

The municipal district is the totality of the buildings forming the municipal district.

The administrative district is a set of buildings forming the administrative district.

The municipal district is the totality of the buildings forming the municipal district.

The area of responsibility of the police stations is the totality of the buildings to which the responsibility of the relevant branch is covered.

And so on.

"Layout" of buildings with codes, for example, territorial tax inspections is a routine work, which is sometimes greatly facilitated by knowledge of specific features. So the territory of the "marginal" territorial tax inspectorates (Nos. 11 and on) basically coincided with the old district division of the city, which allowed to reduce the time for the primary marking of the buildings.

We will show the application of this approach for the tasks of processing data on taxes on the property of individuals. During the transfer of data to the tax authorities on the availability of real property subject to taxation from physical persons, identification of addresses of buildings in which there are owned apartments is performed, and binding of addresses to unique house numbers. After processing the data on the owners together with the information on pensioners for accounting for tax benefits, then together with information about the deceased and information on the cost of apartments for taxation, it is required to transfer the data to the territorial tax inspectorates. It is not difficult to isolate the data for each inspection, since for each unique number of the house there is its relation to the corresponding territorial tax inspection. Further, having received the documents with assessed taxes in the territorial inspection, they must be printed and packed in envelopes with the indication of postal codes, which is provided by the presence of a unique home number connection with the postal code of the liaison office. This same connection allows you to solve the problem of possibly even loading the post offices delivery of tax notices in order to meet the temporary restrictions on the transfer of documents.

4 Discussion

The application of the above abstraction is not necessary, but ignoring them can lead to undesirable consequences. When created in 1991-1992 the automated system of the present Department of Property of the City of Moscow developers, it was decided to consider in the database ' structures solely as they appear in the input documents. It was motivated by the fact that the relevant legislation requires that documents in the information system be stored in an identical paper document with the volume and composition of information. As a result, for three months of operation of the system in the database, building addresses accumulated, more than twice the number of all buildings in the city. Under these conditions, the task of identifying data to collect information even for one structure, for example, for resettlement before demolition or reconstruction, often led to the need to refer to paper archives. In a few years, the Department entered into a contract to bind the accumulated addresses with unique house numbers and identification was performed.

5 Conclusions

In the case when information about the same object have many views, how many different addresses of the same structure, it is advisable by abstraction to create a new concept. Implementation in the information system of this concept must be associated with the maximum achievable diversity of the cast when abstracting data. Design and use abstract concepts in problems of functioning of automated information systems and problems of interoperability of information systems in the context of multiple representations of the same objects showed its effectiveness.

Not all data is formally associated with complex objects, for example, land must be handled with appropriate difficulty. Application of the method of abstraction allows you to simplify the data representation and processing. Simplicity, in turn, reduces the likelihood of errors. Of course, this is true for a certain group of tasks. The method is tested in the interoperability of automated systems.

References

1. J. Forrester. Dynamics of the city development. - M, Progress, 1974
2. J. Martin. Organization of databases in computer systems. -M, World, 1980

3. C.J. Date Introduction to Database Systems An 8 Williams (2016)
4. Devacademy.ru (http://devacademy.ru/posts/sqlite-vs-mysql-vs-postgresql)
5. M. A. Poltavtseva Storing complex data structures in relational databases. PhD Thesis. (Tver 2008)
6. S.V. Mescherjakov, V. M. Ivanov Effective technology of information systems (SPb: Publishing house Politekhnika, 2005) http://www.polytechnics.ni/
7. S.V. Mescherjakov, V. M. Ivanov Methods of optimal designing of databases of production equipment - SPb: Publishing house Polytechnic University,2012. http://gpupress.ru/
8. T. Halpin, *Information Modeling and Relational Datadases: from conceptual analysis to logical design* (Morgan-Kaufmann, San Francisco, 2001)
9. Data Modeling and Database Design (Course Technology, 2007)
10. N. S. Nikolaeva, A. A. Shilnikova, M. V. Yurova Automation, mechatronics and information technologies: materials of the I International scientific and technical, Omsk: Omsk State Technical University (2010)

The strength of compressed structures with CFRP materials reinforcement when exceeding the cross-section size

Petr Polskoy[1], *Dmitry* Mailyan[1], *Sergey* Georgiev[1], *Viktor* Muradyan[1*]

[1]Don State Technical University, Rostov-on-Don

Abstract. The increase of high-rise construction volume or «High-Rise Construction» requires the use of high-strength concrete and that leads to the reduction in section size of structures and to the decrease in material consumption. First of all, it refers to the compressed elements for which, when the transverse dimensions are reduced, their flexibility and deformation increase but the load bearing capacity decreases. Growth in construction also leads to the increase of repair and restoration works or to the strengthening of structures. The most effective method of their strengthening in buildings of «High-Rise Construction» is the use of composite materials which reduces the weight of reinforcement elements and labour costs on execution of works. In this article the results of experimental research on strength and deformation of short compressed reinforced concrete structures, reinforced with external carbon fiber reinforcement, are presented. Their flexibility is $\lambda h=10$, and the cross-section dimensions ratio b/h is 2, that is 1,5 times more, than recommended by standards in Russia. The following research was being done for three kinds of strained and deformed conditions with different variants of composite reinforcement. The results of the experiment proved the real efficiency of composite reinforcement of the compressed elements with sides ratio equal to 2, increasing the bearing capacity of pillars till 1,5 times. These results can be used for designing the buildings of different number of storeys.

1 Introduction

The growth of capital construction in Russia and abroad extremely requires the use of new modern building materials and structures. Primarily it should be directed to the increase of efficiency and durability of construction structures.

A new push in development of the construction industry and science is come from the «High-Rise Construction» and also from the need of using new kinds of constructive decisions. The more there is an increase of construction, the more there are repair and reconstruction costs of existing and newly erected buildings. Structures restoration is impossible without their strenthening, that is why the composite materials have been included into new and promising materials.

* Corresponding author: muradyan2007@yandex.ru

Taking into account the above mentioned, the Department of Reinforced Concrete and Stone Structures at Don State Technical University (DSTU) has been investigating and developing new innovative and energy saving technologies in the construction and the improvement of culculations for more than 10 years.

Among such scientific works are: the high-strength concrete and their design features study [1,2]; effective structures with the use of self-sealing high-strength and light concretes on porous aggregates [3,4]; prestressed elements with variable prestressing by element length [5,6]; reinforced concrete structures study, strengthened by composite materials with different kinds of tensely-deformed conditions.

The latest works are the strengh study of normal and oblique sections of bending elements at different percentages of external composite reinforcement [7,8]. The third way is the determination of strength and stiffness of compressed reinforced concrete elements with different variants of external longitudinal and transverse reinforcement, and also with different eccentricity of load applications.

The Set of Rules, published in Russia in 2004, about strengthening of reinforced concrete structures with composite materials made the design issues legal. However, the first publication of this Set of Rules couldn't solve many other problems, because according to [9,10], many of its provisions are based on foreign authors' researches, for example, on Aprile's results of research and others [11], concerning the strength issues of oblique sections on the transverse force. The same happens with strengthening of compressed reinforced concrete elements. It is connected with the dimensions ratio of elements sections of rectangular profile (b/h), which cannot be more than 1,5. The flexibility value of strengthened elements $\lambda_i=l_0/i$ is limited and cannot exceed 50. However, a large number of structures or their elements, existed in Russia, still exceeds these constraints. This refers, for example, to compressed pillars and braces in reinforced concrete girders, and also deadbolts of girders' structures and others.

There is not any advice on calculating the compressed elements for which the eccentricity of load application is out of inequality limits $0{,}3h \leq e_0 \leq 0{,}1h$. However, it is this variant of a strained deformed state which occurs during the construction of high-rise buildings. The following article is of an experimental nature and deals with test results of bearing capacity of short pillars ($\lambda_h=10$) with sides ratio (b/h)=2,0 with three values of eccentricity of load application ($e_0=0$; $e_0=0{,}16h$; $e_0=0{,}32h$) and with different ways and variants of their strengthening with the usage of carbon fiber reinforced plastic.

The final purpose of the work is the proposal development of reliable normative base creation for strenghtened structures calculations using non-traditional materials at any kinds of strained-deformed state.

2 Materials and methods

The following investigations were carried out using rectangular test pieces made of heavy concrete of B30-35 class which is often used in the construction nowadays.

For getting comparable results of testing, all the short pillars had similar cross-section 250x125 (h) and length 1200mm. The least of the section size was taken as the height of the pillars (h) to guarantee knowingly a planned destruction including a case of stability loss.

For heavy concrete of B30 project class, the unfraction crushed stone made of dense limestone 5-20mm, of 800 brand was used. For concrete of B35 class the granite crushed stone of the same fraction of 1000 brand was used.

As a working armature in cross-section of pillars, 4Ø12A500 was used. Knitted cross reinforcement, set with a pace of 180 mm, was accepted with a diameter of 6mm made of steel of B 500 class. To prevent concrete crushing under the steel head, on the end sections of the carcasses additionally with a pace 50 mm 6 constructive grids made B 500 with a diameter of 4 mm. The size of the cells 50x40mm.

For elements strengthening two types of external composite reinforcement – transverse and longitudinal were used. For transverse reinforcement three-layer holders or holders with gaps – clamps made from carbon fiber reinforced cloth with unidirectional fibers of cold hardening (MBRACE FIB CF 300/4900.300g/5.100m). The longitudinal reinforcement is represented by laminate – the stripes of unidirectional fibers of hot hardening (MBRACE®LAM CF210/2400.50x1,4.100m). Both types of carbon fiber reinforced plastic are made in Germany.

By test results according to the State Standard 25.601-80 the test strength indicators for three-layer carbon fiber reinforced cloth made $\sigma_{f,u}$ 2888,8 MPa and coincides with the standardized technical characteristics of the manufacturer's factory.

All the pillars strengthened by the carbon fiber reinforced plastic were tested at the age of 680-720 days. After strenthening with composite materials the sample before the moment of their testing were kept during 10-15 days.

3 Results

As it was mentioned above, at the first stage of studying the compressed reinforced concrete elements strengthened by composite materials, 16 test short (λh=10) pieces were investigated. In the process of testing three eccentricity values of load application (e0=0; e0=2,0; e0=4,0 cm) and nine variants of external reinforcement were varied. The structure of all strengthening variants is represented in picture 1. Here you also can see the cross sections and the pace of transverse clamps.

It is necessary to mention separately, that all the pillars near the metal heads had design three-layer clamps at 100 mm wide. For elements with external transverse composite reinforcement these clamps played the role of anker increasing the reliability of its anchoring.

The results of the conducted experiment on the strength of the test pieces are represented in Table 1. For easy using there is an index number of a piece together with its pressmark. Besides, having the natural inaccuracy of concrete strength, the results of the experiment were analyzed taking into account the strength of samples but not the destructive efforts (Column 4). Their bearing capacity $N_{s,f}^{red}$ was obtained by means of multiplication the test destructive load by the coefficient $k_{s,f}^{red}$, equal to concrete classes of standard and strengthened samples.

Simultaneously for increasing the reliability of the experiment's results, separate strengthened samples which showed little strength in each series of tests were included into the standard samples. These samples in Table 4 are marked with *. This value is indicated in the denominator of column 6.

The method of direct comparison of strengthening the reinforced samples – $N_{s,f}^{red}$, and standard samples – $\overline{N}_{s,f}^{red}$ was taken as the main method of the efficiency assessment of different variants of strengthening. Their ratio is named the coefficient of strength – $k_{s,f}^{red}$.

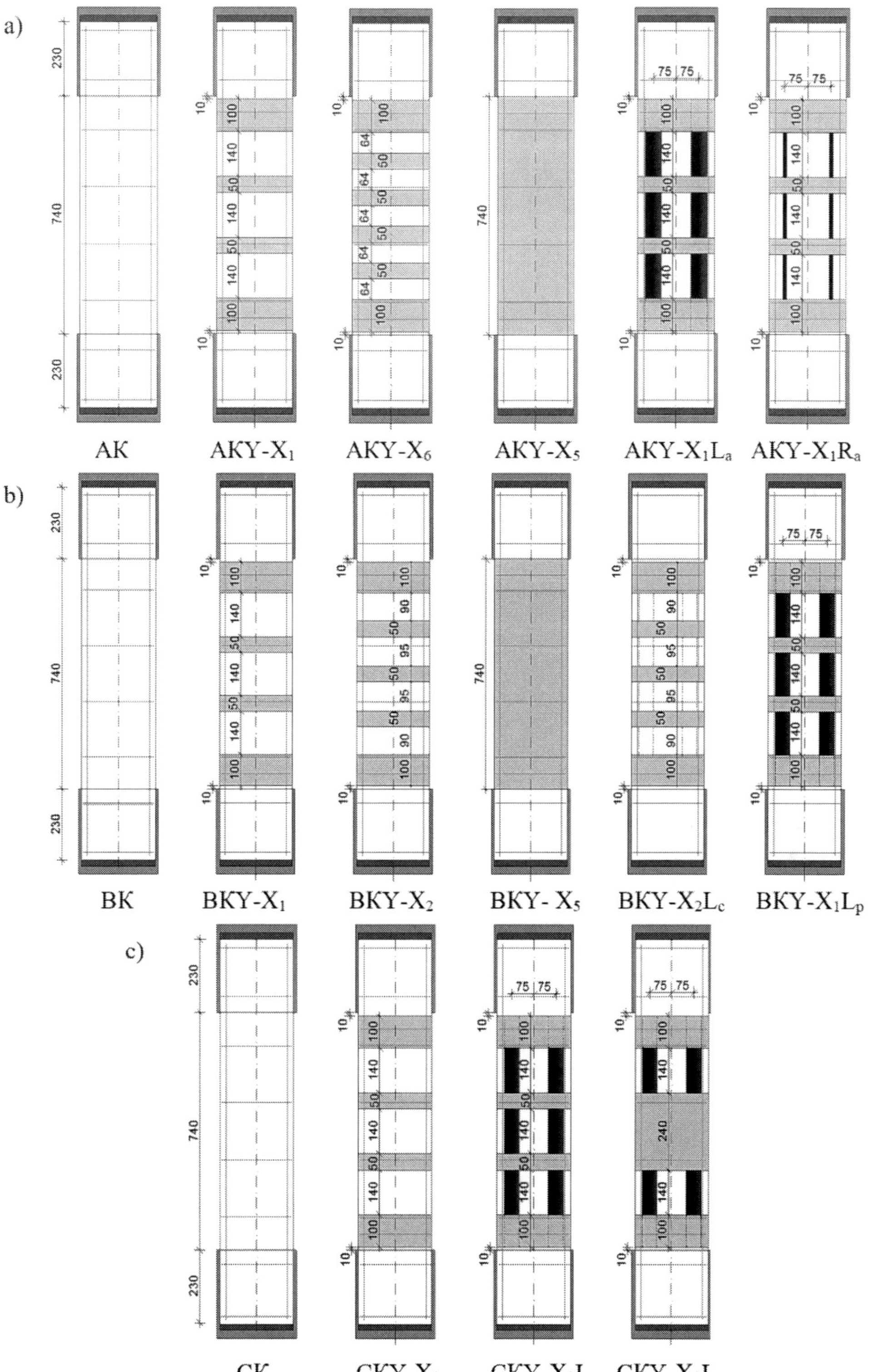

Fig. 1. Variants of strengthening the short (λ_h=10) pillars of external transverse and longitudinal reinforcement with eccentricity of load application e_0=0 – (a); e_0=2cm – (b); e_0=4– (c).

4 Analysis

Data analysis in Columns 6-9 of Table 1showed the following:

The pillars stregthened by reinforced three-layer holder with the variant X_5, tested with eccentricity $e_0=0$, showed maximum bearing capacity. Specifically it can be seen by its actual value of load growth $\Delta N_{s,f}^{red}$ for strengthened pillars. With eccentricity $e_0=2{,}0$cm (0,16h) the efficiency of the entire holder reduced 2 times.

The second effective variant is X_1R_u, in which the strenthening is done with circle bars Ø8mm, pasted into the groove, cut in the protective layer of the concrete. The same picture is in the studies [12], where instead of circle bars the stripes of 25 mm wide and of 1,4 mm thick were put into the groove.

The clamps of transverse reinforcement of w_f=50mm wide had three different paces in axis - S_f=190; 140; 125mm. Evaluating the influence of the gap size equal to 140; 95; 65mm respectively (Picture 2), it can be noted that there was no strong influence on the change of bearing capacity of pillars, strengthened by these variants (X_1, X_2, X_6).

With eccentricity $e_0=2{,}0$ cm (0,16h) in our experiments the clamps X_1 and X_2 showed a load growth for short pillars, comparable with an effect of the entire holder.

The clamps in variant X_3 had an increasing effect only in the complex variant of strengthening together with external longitudinal complex reinforcement.

Table 1. Comparison of the following test strength of standard and reinforced short pillars by changing the variants of external composite reinforcement.

№ of pillars	Pressmark of pillars	Class of concrete B, MPa	Test load (кN):		The given values:			
					Strength (кN):		Load growth (кN) $\Delta N_{s,f}^{red}$	Coefficient of strength $k_{s,f}^{red}=\frac{N_f^{red}}{N_s^{red}}$
			At destruction N_s, N_{sf}	At maximum permissible deformation $N_{\Delta,ult}$	At destruction $N_s; N_{sf}^{red}$	At maximum permissible deformation $N_{\Delta,ult}^{red}$		
1	**2**	**3**	**4**	**5**	**6**	**7**	**8**	**9**
Series A. Axial eccentricity $e_0=0$								
1	AK	33,2	1150	1500,0	1150,0	1150,0	-	1,0
2*	AKY-X_1	30,0	1190,5	1190,5	1317,5	1317,5	112,2	1,093
5*	AKY-X_1L_a	31,8	1100,0	1100,0	1148,4 1205,3	1148,4 1205,3	-	1,0
3	AKY-X_6	39,3	1600,0	1600,0	1351,6	1351,6	146,3	1,12
4	AKY-X_5	30,1	1625,0	1625,0	1792,3	1792,3	587,0	1,487
1	**2**	**3**	**4**	**5**	**6**	**7**	**8**	**9**
6	AKY-X_1R_a	30,1	1379,0	1379,0	1521,0	1521,0	315,7	1,262
Series B. Axial eccentricity e_0=2cm(0,16h)								
7	BK	30,3	592,5	545,0	592,5	545,0	-	1,0
12*	BKY-X_1L_p	31,8	700,0	630,0	667,06 629,7	600,3 572,65	37,3	1,06
8	BKY-X_1	30,0	778,9	685,0	786,7	691,85	157,0	1,249

9	BKY-X_2	30,3	794,7	700,0	794,7	700,0	165,0	1,262
10	BKY- X_5	32,4	844,0	750,0	789,3	701,4	159,6	1,253
11	BKY-X_2L_c	32,4	800,0	730,0	748,1	682,7	118,4	1,188
Series C. Axial eccentricity e_0=4cm(0,32h)								
13	CK	33,2	422,2	335,0	422,2	335,0	-	1,0
14*	CKY-X_1	33,2	482,5	380,0	482,5 452,3	380,0 357,5	330,2	1,067
15	CKY-X_1L_p	33,2	530,0	435,0	530,0	435,0	77,7	1,172
16	CKY-X_3L_p	39,4	608,0	455,0	512,3	383,4	60,0	1,133

It should be noted, that the efficiency of transverse clamps influence, pasted on the longitudinal elements of external reinforcement, **reduces** and appears regardless the eccentricity value.

The longitudinal external reinforcement of short pillars with the usage of laminate at e_0=0 and 2,0 cm did not influence on their bearing capacity because this armature went wrong when loaded within (0,75-0,8) N_{ult}. The similar results were got in the work [13]. With eccentricity e_0=4,0 cm the laminate (stripes), pasted on the extended face, on the contrary, showed the positive effect. It is well conformed with the regulations [14,15].

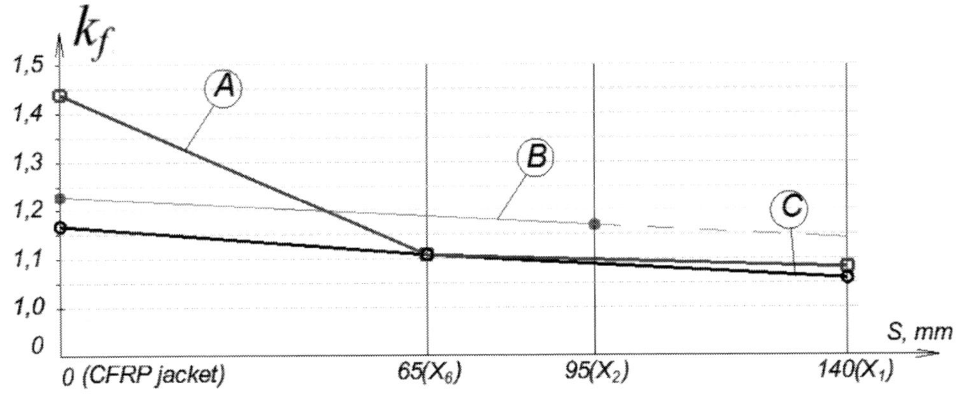

Fig. 2. The influence of the gap value between clamps (S) on the efficiency of three-layer transverse reinforcement of short pillars (λ_h=10) with eccentricity of load application e_0=0 (A); e_0=2,0cm (B); e_0=4,0cm (C).

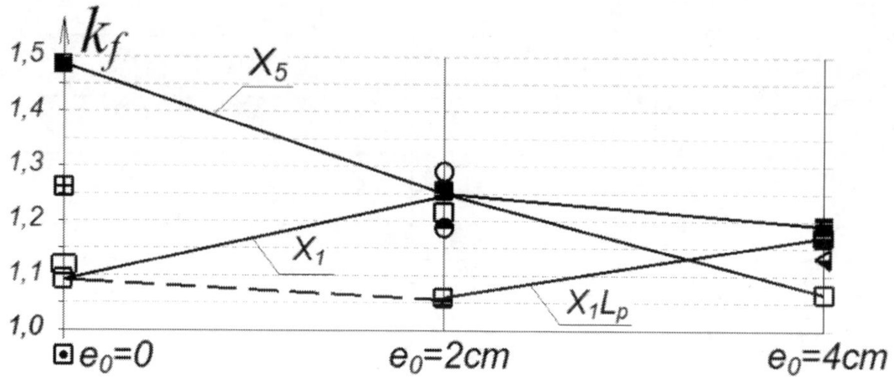

Fig. 3. The influence of eccentricity of load application on the change of the coefficient of composite strength (k_f^{red}) of short pillars. Conventions are done in Table 2.

Table 2. Conventions for Picture 2 and Picture 3.

Variant of strengthening	standard column	$Y\text{-}X_6$	$Y\text{-}X_5$	$Y\text{-}X_1L_a$	$Y\text{-}X_2L_c$	$Y\text{-}X_1L_p$	$Y\text{-}X_3L_p$	$Y\text{-}X_1R_a$
Conventions of variants of the strengthening		□	■	⊡	◓	⬓	◀	⊞

All mentioned above about the influence of variants of strengthening on the bearing capacity of test samples can be seen if we analyze the graphics in Picture 3, where the change of test values of the composite strength coefficient k_{sf}^{red} is shown, depending on the eccentricity of the load application.

It is important to note again that according to our experiments the efficiency of transverse reinforcement reduces a bit if the clamps are pasted on the longitudinal stripes – laminates of external strengthening. Apparently, it can be explained by the fact that the cocrete can easily be deformed in the transverse direction in the intersection places between longitidinal and transverse composite armatures. It also happens, in our opinion, because of less tight fit of clamps to concrete and absence of full adhesion in the places of clamps intersection (made of cloth) and elements of longitudinal reinforcement (from stripes of laminates) because different pasted compositions are used for these materials. Besides, it is necessary to put extra repair composition to prevent height difference at the intersection of strengthening elements.

5 Conclusions

1) The conducted experimental research of compressed reinforced concrete elements with the sides ratio b/h=2 with the flexibility λ_h=10, strengthened with carbon fiber reinforced plastic in transverse and longitudinal directions and tested with axial eccentricity of load application e_0=0; e_0=0,16(h) и e_0=0,32(h), proved not only the possibility but also the high enough efficiency of composite reinforcement.

2) It was established that the strengthening effect in compressed elements by composite materials is achieved not only with eccentricity of load application $e_0 \leq 0{,}1h$ and $e_0 > 0{,}3h$, but

also for compressed elements working with little eccentricity e_0=0,16h. The latest are not taken into account.

3) The most reliable for elements loaded at $e_0 \leq 0,16h$, is strengthening by complete holders. However, it should be noted that efficiency of composite reinforcement at the similar percent of reinforcement reduces because tension in the elements of strengthening does not achieve calculated resistance and that requires improvement of calculating apparatus.

4) The efficiency of external transverse reinforcement by clamps in the presense of external longitudinal reinforcement at e_0=0 reduces at 6-12%. For elements working with big eccentricity (e_0=0,32h), transverse reinforcement is functional because it increases the reliability of adhesion for external longitudinal composite reinforcement. Its pace must not exceed the value $S_{fw} \leq 15d_s$.

5) Symmetric longitudinal reinforcement of a short pillar of internal circle carbon fiber reinforced plastic consisted of 4 bars with diameter of 8 mm (μ_f=0,64 %), pasted in propylene within the thickness of a protective layer of concrete led to the increase of bearing capacity of a test pillar at 31% with preservation of rigidity similar to non-strengthened samples, i.e standard.

References

1. A.M. Mkrtchyan, D.R. Mailyan, *Nauchnoe obozrenie.* **11**, 72-76 (2013)
2. D.R. Mailyan, V.N. Aksenov, N.B. Aksenov, *International scientific conference energy management of municipal transpiration facilities and transport EMMFT* (2017)
3. A.N. Davidyuk, D.R. Mailyan, G.V. Nesvetaev, *Technologies of concrete.* **1-2**, 57-59 (2011)
4. A. V. Belyaev, G. V. Nesvetaev, D.R. Mailyan, *MATEC Web of Conferences.* **106**, (2017)
5. D.R. Mailyan, R.I. Mailyan, V.H. Khuranov, *News of higher education institutions. Construction.* **5**, 4-11 (2004)
6. V.A. Muradyan, D.R. Mailyan, A.A. Lyapin, V.E. Chubarov, Earth and environmental science **90** (2017)
7. P.P. Polskoy, D.R. Mailyan, D.A. Dedukh, S.V. Georgiev, *Global Journal of Pure and Applied Mathematics.* **12.2**, 1767-1786 (2016)
8. P.P. Polskoy, D.R. Mailyan, A.A. Shilov, Z.A. Meretukov, *New technologies.* **4**, 44-48 (2015)
9. T.A. Mukhamediev, D.V. Kuzevanov, *Concrete and reinforced concrete.* **2**, 18-20 (2014)
10. T.A. Mukhamediev, D.V. Kuzevanov, *Concrete and reinforced concrete.* **6**, 20-24 (2013)
11. A. Aprile, A. Benedetti, *Proceedings off First Course on FRP applications for R/C elements rehabilitation*, Bologna, Italy (2001)
12. N. Chikh, M. Gahmous, R. Benzaid, *Proceedings of the World Congress on Engineering,* **4 - 6**, (2012)
13. S. Matthys, PhD, Ghent University, 345 p. (2000)

14. ACI 440.2R-02, Guide for the Design and Construction of Externally Bonded FRP Systems for Strengthening Concrete Structures. *American Concrete Institute* (ACI), Committee 440 (2008)
15. CNR-DT 200/2004, Guide for the Design and Construction of Externally Bonded FRP Systems for Strengthening Existing Structures, Rome (2004)

On the design of high-rise buildings with a specified level of reliability

Andrey Dolganov[1], and *Pavel* Kagan[2,*]

[1] Ltd "Sev. R. Development", Syromyatnicheskaya Nizh., 10 p. 2, Moscow, 105120, Russia
[2]Moscow State University of Civil Engineering, Yaroslavskoe shosse, 26, Moscow, 129337, Russia

Abstract. High-rise buildings have a specificity, which significantly distinguishes them from traditional buildings of high-rise and multi-storey buildings. Steel structures in high-rise buildings are advisable to be used in earthquake-proof regions, since steel, due to its plasticity, provides damping of the kinetic energy of seismic impacts. These aspects should be taken into account when choosing a structural scheme of a high-rise building and designing load-bearing structures. Currently, modern regulatory documents do not quantify the reliability of structures. Although the problem of assigning an optimal level of reliability has existed for a long time. The article shows the possibility of designing metal structures of high-rise buildings with specified reliability. Currently, modern regulatory documents do not quantify the reliability of high-rise buildings. Although the problem of assigning an optimal level of reliability has existed for a long time. It is proposed to establish the value of reliability *0.99865* (3σ) for constructions of buildings and structures of a normal level of responsibility in calculations for the first group of limiting states. For increased (construction of high-rise buildings) and reduced levels of responsibility for the provision of load-bearing capacity, it is proposed to assign respectively *0.99997* (4σ) and *0.97725* (2σ). The coefficients of the use of the cross section of a metal beam for different levels of security are given.

1 Introduction

High-rise buildings have a specificity, which significantly distinguishes them from traditional buildings of high-rise and multi-storey buildings. Among the main features of high-rise buildings are:

- Significant values of both static and dynamic loads on load-bearing structures and on bases;
- high, sometimes critical value of horizontal (first of all, wind) loads;
- problems of unevenness, both the magnitude of the loads, and the nature of their application;
- careful and correct selection of construction materials, which excludes separate operation of structural elements and ensures homogeneity of physical and mechanical characteristics.

These aspects should be taken into account when choosing a structural scheme of a high-rise building and designing load-bearing structures.

* Corresponding author: kagan@mgsu.ru

Bearing structures of high-rise buildings in the initial period of development of this type of construction were mainly made of steel. And today in buildings with a height of more than 300 m, including "ultra-high" buildings, steel structures are used for bearing elements. Assembly of factory installation elements on the construction site is carried out with the help of bolted connections, which allows to increase the accuracy of installation by eliminating warpage of steel structures under the influence of high temperatures.

Steel structures are advisable to be used in seismic resistant regions, since steel, due to its plasticity, provides damping of the kinetic energy of seismic impacts. At the same time, steel structures require mandatory reliable protection against fire, since at 300° C the strength of steel is sharply reduced [1-13].

2 Materials and Methods

Currently, modern regulatory documents do not quantify the reliability of structures. Although the problem of assigning an optimal level of reliability has existed for a long time. Thus, in 1890, the Russian version of Hütte 's handbook [11] with a developed system of reliability coefficients was published. In 1926, prof. M. Maier published a paper [12] in which he criticizes the calculation of the allowed voltages and proposes to calculate the designs from the account of an unprofitable combination of loads and material resistances. In 1929 N.F. Khotsialov [10] developed the ideas of M. Mayer. He, referring to the stochastic variability of mechanical and geometric parameters of structures, instead of the slogan "Build without destruction, by all means" proposed a new one - "Construction with an expedient number of destructions." Design, according to N.F. Khotsialov, should be conducted taking into account both capital costs and probable "defective deviations" and the amount of losses caused to the state by an accident.

In 1945, in connection with the development of new norms for calculation and design, the Commission for the unification of calculation methods, organized by the Narkomtyazhstroy adopted a conditional scheme of design coefficients proposed by I.I. Gol'denblat, M.G. Kostyukovsky, A.M. Popov. According to this scheme, the overall safety factor depended on the coefficients of homogeneity, overloads and operating conditions of the structure. In the future, the proposed scheme was included in the calculation method for limiting states. It was accepted that the designs must meet the requirements for them with a reasonable degree of risk.

Abroad [1], the formula for the normalization of the probability of failure $Q(t)$ became very popular:

$$Q(t) = 10^{-5}\, \xi_S\, T / L, \qquad (1)$$

where ξ_S - coefficient of social significance (Table 1); T - estimated service life in years; L is the average number of people inside the building or in its immediate vicinity during the period for which the risk is assessed.

Table 1. Coefficient of social significance ξ_S

Type of construction	ξ_S
Places of meeting of people, dams	0.005
Residential, office, commercial and industrial buildings	0.05
Bridges	0.5
Towers, masts, structures on the sea shelf	5

For buildings of a normal level of responsibility, the required reliability, $R(t)$, according to (1) will be:

$$R(t) = 1 - Q(t) = 1 - 10^{-5} \times 0.5 \times 50 / 50 = 0.999995 \text{ or } 0.9^{5}5. \quad (2)$$

Professor Ryush (table 2) proposes to normalize the reliability of *P(t)* constructions in the probability of their failure *Q(t)*, where *Q(t) = 1 - P(t)*.

Table 2. Rationing of reliability of reinforced concrete structures

Type and characteristics of failure	*Q(t)*
Failure without preliminary signals (brittle failure, loss of stability, etc.)	10^{-7} ... 10^{-5}
Loss of load capacity with a preliminary signal	10^{-4}
Inability to normal operation without loss of load capacity (similar to the 2nd group of limit states)	10^{-3} ... 10^{-2}

In the Russian Federation, the reliability of the mechanical characteristics of materials for reinforced concrete and metal structures is regulated by GOST 5781-82 *, GOST R 52544-2006 for rod reinforcement, GOST 18105-2010 for concrete, and also by the building rules of SP 63.13330-2012 "Concrete and reinforced concrete structures" and SP 16.13330-2016 "Steel structures". The reliability of loads is regulated by SP 20.13330.2016 "Loads and impacts".

According to these documents, the reliability (security) of the material's standard resistances is assumed to be *0.95 (1.64σ)*, and the availability of design resistances is approaching *0.99865 (3σ)*: the standard resistances are divided by the reliability coefficients for materials that are greater than 1. Constant and long-term normative loads are defined as mathematical expectations (mean values) of physical characteristics of materials. Normative time loads are the mathematical expectations of extreme technological or natural processes. For calculations on the first group of limiting states, the standard loads are multiplied by the reliability factor, which is usually greater than 1.

Thus, in the calculations for the first group of limiting states of the resistance of materials, they have a security of about *0.99865 (3σ)*, and the load security varies from 0.6 to 1. In the calculations for the second group, the limiting states of the resistance of materials have a *0.95 (1.645 σ)*, and the load is approximately 0.5.

Therefore, the value of reliability *0.99865 (3σ)* in calculations for load-bearing capacity and should be assigned to structures of buildings and structures of a normal level of responsibility. For increased and reduced levels of responsibility [3], the load-bearing capacity should be assigned with respectively *0.99997 (4σ)* and *0.97725 (2σ)*.

3 Results

To prove the rigidity of structures, consider a simple example. In calculations for the second group of limiting states, for example, the calculated deflection f we compare with the permissible [f]. That is, we achieve the fulfillment of the condition:

$$f \leq [f] \quad (3)$$

In formula (3), the calculated parameters are represented by their mathematical expectations. It follows that the results obtained by formula (3) have a security close to 0.5.

Table 3 summarizes the formulated proposals for designating the level of reliability of building structures, depending on the group of limit states (PS).

Table 3. Proposals for designation of the level of reliability of building structures

Level of responsibility of structures	**Provision for the group I of limit states**	**Provision for group II of limit states**
Increased	0.99997 (4σ)	0.95 (1.64σ)
Normal	0.99865 (3σ)	0.50 (0×σ)
Reduced	0.97725 (2σ)	Not regulated (if condition (3) is fulfilled for constant and long-term loads)

Let us consider an example of designing a metal beam with a specified reliability of *0.97725 (2σ)* in terms of bearing capacity for constructing a reduced level of responsibility [3]. The span of the beam is *10 m*. The standard and design loads are given in Table 4.

Table 4. Normative and design loads (example)

Name of the load	**Normative, kN/m²**	**Coefficient reliability, γ_f**	**Estimated, kN/m²**
Net weight of the beam	0.14	1.05	0.15
Mechanisms of airing	0.02	1.10	0.02
Coating flooring	0.65	1.10	0.72
Snow	1.00	1.40	1.40
Connections horizontal	0.06	1.05	0.06
Total:	1.87	—	2.34

For the loads given in Table 4, we determine the required moment of resistance of the section of the beam, m³, from the condition for the strength of normal sections:

$$w = M / (R_y\gamma_c) = 2.34 \times 2 \times 10^2 / 8 / (245000 \times 0.9) = 0.000265.$$

According to the assortment we take the I-beam 26B1, $w = 0.000312$ m³; $J = 0.00004024$ m⁴. The coefficient of section usage will be:

$$k = qL^2 / (8R_y\gamma_c w) = 2.34 \times 2 \times 10^2 / (8 \times 245000 \times 0.9 \times 0.000312) = 0.852. \quad (4)$$

Let us check the rigidity of the beam. The allowable deflection is:
[f] = L / 250 = 10000/250 = 40 mm.
The calculated deflection is:

$$f = 5/384 \times qL^4 / (EJ) = 5/384 \times 1.87 \times 2 \times 10^4 / (200000000 \times 0.00004024) = 0.048 \text{ m}. \quad (5)$$

As can be seen from (5), the calculated deflection exceeds the allowable deflection by 1.2 times.

We estimate the reliability of the beam, taking into account the variability of the random variables. Random resistance is the steel resistance and the total value of the loads.

To calculate the reliability of the cross sections, we use the linearization method (6) - (10). This method assumes a formula dependence of the function of random arguments, and this dependence must be differentiable.

$$m_Y = \varphi(m_{X_1}, m_{X_2}, \ldots, m_{X_n}), \tag{6}$$

$$D_Y = \sum_{i=1}^{n}\left(\frac{\partial\varphi}{\partial x_i}\right)_m^2 \cdot D_{x_i} + 2\sum_{i<j}^{n}\left(\frac{\partial\varphi}{\partial x_i}\right)_m\left(\frac{\partial\varphi}{\partial x_j}\right)_m K_{ij}, \tag{7}$$

or through the mean square deviations:

$$\sigma_Y^2 = \sum_{i=1}^{n}\left(\frac{\partial\varphi}{\partial x_i}\right)_m^2 \sigma_{x_i}^2 + 2\sum_{i<j}^{n}\left(\frac{\partial\varphi}{\partial x_i}\right)_m\left(\frac{\partial\varphi}{\partial x_j}\right)_m r_{ij}\sigma_{x_i}\sigma_{xj} \tag{8}$$

$$r_{ij} = K_{ij} / (\sigma_i \cdot \sigma_j). \tag{9}$$

where r_{ij}, σ_{x_i}, σ_{x_j} – respectively, the correlation coefficient and the mean square deviations of the random variables X_i, X_j; K_{ij} – is the correlation moment of the random variables X_i, X_j:

$$K_{ij} = M[(X_i - m_{Xi})(X_j - m_{Xj})]. \tag{10}$$

The average values, q_m, and the load variation coefficient, v_q, are determined from Table 3. The average value of steel resistance, R_{ym}, is determined with the coefficient of variation, V_s, equal to 0.04.

In this way,

q_m = 1.87 kN/m^2;

V_q = [(2.34 – 1.87) / 1.64] / 1.87 = 0.156;

R_{ym} = 250000 / (1–1.64×0.04) = 267550 kN/m^2.

The average square deviations of the resistance of steel and loads are respectively equal, kN/m^2: σ_s = 267550 × 0.04 = 10702; σ_q = 1.87 × 0.156 = 0.291.

We take the partial derivatives in (4) with respect to R_y and q, and substitute R_y and q R_{ym} and q_m. in the expressions obtained. When calculating by formulas (11) – (12) γ_c is equal to 1; the load distributed per 1 m^2 is reduced to the running load, multiplying by a step the beams, equal to 2 m. We obtain

$$\partial Y_k / \partial R_y = -q_m l^2 / (8R_{ym}^2 \gamma_c w) = -1.87\times10^2 / (8\times267550^2\times0.000312) = 2.09\times10^{-6}, \tag{11}$$

$$\partial Y_k/\partial q_m = l^2 / (8R_{ym} \gamma_c w) = 10^2 / (8\times267550\times0.000312) = 0.299. \tag{12}$$

Using formula (8), taking into account (4), we determine the standard deviation of the coefficient of cross-section use, MPa:

$$\sigma_k = [(2.09\times10^{-6})^2\times10702^2 + 0.299^2\times0.291^2]^{1/2} = 0.09.$$

We define the reliability index t for k = 0.852:

$$t = (1 - k) / \sigma_k = (1 - 0.852) / 0.09 = 1.648. \tag{13}$$

For the normal distribution law, the bearing capacity of the beam under consideration P (t) is 0.95, which is somewhat lower than the required value of 0.97725 according to Table 3.

We define the derivative with respect to q of the deflection function (5), m^3/kH:

$$\partial Y_k/\partial q_m = 5/384 \times B \times L^4 / (EJ) = 5/384 \times 2 \times 10^4 / (200000000 \times 0.00004024) = 0.03. \tag{14}$$

By (8), we determine the standard deviation of the deflection function, m:

$$\sigma_f = (0.03^2 \times 0.291^2)^{1/2} = 0.0094.$$

We define the reliability index t for the actual deflection f = 0.048:

$$t = ([f] - f) / \sigma_f = (0.040 - 0.048) / 0.0094 = -0.883. \tag{15}$$

4 Discussion

For the normal distribution law, the security of the deflections (rigidity of the beam under consideration) P(t) is 0.1887. The probability of exceeding the deflections from the load limits calculated for the second group is 0.8113.

Define the design parameters for which the load carrying capacity and rigidity of the structure would correspond to the specified reliability values. The required security according to the carrying capacity according to Table 3 is 0.97725 (2σ), and for normal operation it is not regulated.

The coefficient of cross-section use for the security of 0.97725 (2σ) is:

$$[k] = 1 - 2\sigma_k = 1 - 2 \times 0.09 = 0.82. \tag{16}$$

It is sufficient that the cross-section of the middle part of the beam with a length of *3 m* is symmetrically reinforced with plates *0.7 mm* thick, so that the coefficient of cross-section usage becomes *0.82*.

For technological reasons we weld plates with a thickness of 3 mm. Reliability by the criterion of bearing capacity has become *0.999232 (3.168σ)*. For stiffness - *0.264653*.

Thus, section *26B1* reinforced with two plates *3 mm* thick with a length of *3 m*, installed in the middle of the beam, meets the requirements of reliability and safety for building a reduced level of responsibility. The coefficient of use of the cross section for the case under consideration became *0.754*.

References

1. G. Augusti, A. Baratta, F. Kashiati, *Veroyatnostnye metody v stroitel'nom proektirovanii* (Stroyizdat, Moscow, 1988)
2. A. Gemmerling On the Reliability of Mass Structures, Stroitel'naya mekhanika i raschet sooruzheniy, **5**, 69-73 (1974)
3. GOST 27751–2014. Nadezhnost' stroitel'nykh konstruktsiy i osnovaniy (Standartinform, Moscow, 2015)
4. V. Telichenko , E. Korol , P. Kagan , S. Komissarov , S. Arutyunov , Constructive solutions for high-rise buildings, Vysotnyye zdaniya: zhurnal vysotnykh tekhnologiy, **4**, 102-109 (2008)

5. A. Dolganov, *Reliability of core reinforced concrete structures* (JSC "MAOBTI", Magadan, 2001)
6. A. Dolganov, *Optimization of reinforced concrete structures with regard to reliability criteria and minimum cost* (Izdatel'stvo Severnogo mezhdunarodnogo universiteta, Magadan, 2002)
7. A. Dolganov, Evaluation of the reliability of rafters F-1 warehouse terminal "Belaya Dacha Market", Sbornik nauchnykh trudov MGSU, 75–77 (2002)
8. A. Rzhanitsyn, The economic principle of calculation for safety, Stroitel'naya mekhanika i raschet sooruzheniy, **3**, 3-5 (1973)
9. Rusian State Standard SP 20.13330.2016
10. N. Khotsialov, Stocks of strength, Stroitel'naya promyshlennost, **10**, 840 (1929)
11. *"Hütte". A reference book for engineers, architects, mechanics and students. The ninth edition.* (T-vo skoropechati A.A. Levensona, Moscow, 1916).
12. M. Maier, *Die Sicherheit der Bauwerke und ihre Berechnung nach Grenzkräften anstatt nach zulässigen Spannungen* (Springer, Berlin, 1926)
13. A. Dolganov, Estimation of the Reliability of the Monolithic Multistory Buildings, Promyshlennoe i grazhdanskoe stroitel'stvo, **8**, 50-51 (2010)

Industrial waste utilization in the panels production for high buildings facade and socle facing

Irina Vitkalova[1] , *Anastasiya* Torlova[1] , *Evgeniy* Pikalov[1,*], and *Oleg* Selivanov[1]

[1] Vladimir State University named after A.G. and N.G. Stoletovs, 600000 Vladimir, Russia

Abstract. The research presents comprehensive utilization of such industrial waste as galvanic sludge, broken window glass as functional additives for producing ceramics for facade and socle paneling in high-rise construction. The basic charge component is low-plasticity clay, which does not allow producing high-quality products if used without any functional additives. The application of the mentioned above components broadens the resource base, reduces production cost and the mass of the products in comparison with the currently used facing ceramics. The decrease of product mass helps to reduce the load on the basement and to use ceramic material in high-rise construction more effectively. Additional advantage of the developed composition is the reducing of production energy intensity due to comparatively low pressing pressure and firing temperature thus reducing the overall production cost. The research demonstrates the experimental results of determining density, compressive strength, water absorption, porosity and frost resistance of the produced ceramic material. These characteristics prove that the material can be applied for high buildings outdoor paneling. Additional research results prove ecologic safety of the produced ceramic material.

1 Introduction

Buildings facing is well known to be used for protecting facades and socles against negative environmental impacts such as temperature drops, snow and rain falls. Besides buildings paneling is significant for creating beautiful design and presentable appearance.

Nowadays a great variety of materials is used for buildings facing. The choice of materials depends on operating conditions, shape and building height, design solutions and facing materials price.

Ceramics is one of the most ancient known materials, but today it still remains very popular. Its principle advantages are its durability, reliable protection against environmental impact and ecological safety. Moreover ceramics is characterized by a wide assortment of color and texture, fire resistance and safety, it does not fade with the time.

Building facades and socles facing with ceramics is widely and successfully applied both in low and high-rise construction. Speaking about tall building facing, ceramics

* Corresponding author: evgeniy-pikalov@mail.ru

advantages include its stable resistance towards temperature drops and almost any wind pressure, which gets stronger with the buildings height. The most effective way is to produce large area panels because it simplifies and hastens their mounting. In case with socles it is better to use special adhesive compositions, and for facades facing it is advisable to mount them on the purlins according to the ventilated facade system. Like any other material ceramics has several disadvantages including comparatively big mass, high price, fragility and brittleness. The last two features cause the requirement of careful mounting of the panels.

The research objective is to develop the charge composition for ceramic facing material production providing lower production cost and mass in comparison with the existing analogues (terracotta tiles, clinker and porcelain stoneware) and at the same time retaining its strength and other performance characteristics for using the material in the facing of high buildings construction.

The objective is suggested to be reached by using low cost raw materials and decrease of pressing pressure and firing temperature. Most facing ceramic materials are produced using high plasticity clays. Our research suggests using low plasticity clay and additionally secondary raw materials serving as functional additives but not as cheap fillers.

Basing on the previous experiments [1] the authors stated that ecologically safe ceramic materials can be produced adding 2.5mass % of galvanic sludge and 2 mass % of boric acid. Moreover boric acid stipulates the formation of vitreous phase at firing increasing the strength properties and complicating the migration of heavy metals, contained in the galvanic sludge, into the environment. In this connection a special interest ranks the boric acid replacement in the charge composition directly by the cullit, which is also a waste material.

Using secondary raw materials will stipulate cost reduction and quality improvement of the ceramic facing panels, on the one hand and the possibility of comprehensive waste utilization on the other.

2 Materials and Methods

The experiments during the research process were carried out using raw materials available in the Vladimir region and regarding regional peculiarities.

The basic component of the developed charge was the clay from Suvorotskoe field in Vladimir region with the following composition (mass %): SiO_2 = 67,5; Al_2O_3 = 10,75; Fe_2O_3 = 5,85; CaO = 2,8; MgO = 1,7; K_2O = 2,4; Na_2O = 0,7. The presence of aluminum, calcium and magnesium oxides testified of the low clay plasticity [1], plasticity rate amounted 5.2, determined by the standard method, and hence in compliance with GOST 9169-75 the researched clay referred to low plasticity group. Consequently the production of high quality products from this clay without introducing any functional additives is impossible though its application may help developing regional raw materials base and reduce production cost.

The second component of the charge composition is galvanic sludge (moisture 60-70% mass %) received after chemical treatment of electroplating effluents at machine building enterprises [1]. Galvanic sludge contained the following chemicals (mass %): $Zn(OH)_2 \approx$ 11,3%; $SiO_2 \approx$ 7,08%; $Ca(OH)_2 \approx$ 16,52%; $Cr(OH)_3 \approx$ 9,31%; $(Fe^{2+})Cr_2S_4 \approx$ 4,17%; $CaCO_3$ ≈40,25 %; CaO ≈ 3,45%; ZnO≈ 2,41%; $Cu(OH)_2 \approx$ 2,38%; $Ni(OH)_2 \approx$ 2,62%; $Mn(OH)_2 \approx$ 0,64%; $Pb(OH)_2 \approx$ 0,14%. The relatively big number of compounds, containing zinc, chromium, nickel and copper as hydroxides, prove of galvanic sludge toxicity [1]. According to the public data [1, 2] nowadays no technology provides the problem solution of galvanic sludge accumulation in the Vladimir region, therefore the development of their utilization method is very vital. The application of galvanic sludge in the construction materials production including building ceramics might become a real solution of the problem.

Another charge component is window glass cullit of the following composition (mass %): SiO_2 = 73,5; CaO = 7,4; MgO = 1,9; Na_2O = 11,1; K_2O = 5,2; Al_2O_3 = 0,9. Glass cullit in Vladimir region is a result of local glass production enterprises operation and household consumption, which is accumulated in the region. This waste is not completely processed by the enterprises because of the composition diversity [3] and sources, therefore it is piled in big amounts (up to 10% of the total waste) in storage and waste disposals [4]. Consequently its utilization deserves special attention.

Boric acid of B type, 2-nd class (GOST 18704-78) was introduced into the charge as an additional component which increases ceramic material strength and prevents heavy metals migration into the environment [1, 2].

The samples were made according to the following technology: clay and galvanic sludge were dried to constant mass and then grinded. Further research was conducted using fraction with particles size less than 0.63 mm. Afterwards raw components were mixed, moistened for reaching mold moisture of 8 mass %. The cube samples with 50 mm side were formed under the pressing pressure of 15 MPa from the received charge and then fired at the temperature of 1050 °C. The samples of the researched compositions were made by series of 3 samples each.

To assess the impact additives composition of the charge in the samples according to standard methods for ceramic material the following basic characteristics were determined: density (ρ, kg/m^3) compressive strength (σ_{cmp}, MPa), porosity (P, %), water absorption (WA, %) and frost resistance (FR, cycles).

As the used galvanic sludge refers to 2-3 hazard category [1] additional research was carried out to prove ecologic safety of the produced ceramic material using the method of Daphnia magna Straus under the impact of toxic substances in the daily water extraction of the researched samples [5].

Scanning electron microscope REM Quanta 200 3D was used to the analysis and assessment of the samples structure.

3 Results

At the first stage of the research it was decided to increase galvanic sludge amount in the previously developed composition of the charge [1] from 2.5 to 5 mass %, to exclude boric acid and to introduce from 10 to 30 mass % of glass cullit in every 5 mass %. The results of the ceramic properties determination based on the researched compositions are demonstrated in figures 1 and 2.

The presented data demonstrates that the strength properties and frost resistance of the ceramic material increase but porosity and water absorption decrease with the enhanced content of glass cullit amount in the charge composition. It can be described by the increased amount of vitreous phase at firing, which not only fills the interstices between the ceramic particles in the material, but also serves as a binder between them thus forming a firm frame, which is proved by the images made by REM and demonstrated in figure 3. However during samples toxicity tests over 50 % Daphnias deaths (Daphnia magna Straus) were observed in all the cases during first 50-75 hours. That is why it was decided to introduce boric acid into the charge in the amount of 2.5 and 5 mass %, thus decreasing ceramic material toxicity with galvanic sludge [1].

https://doi.org/10.1051/e3sconf/20183302062

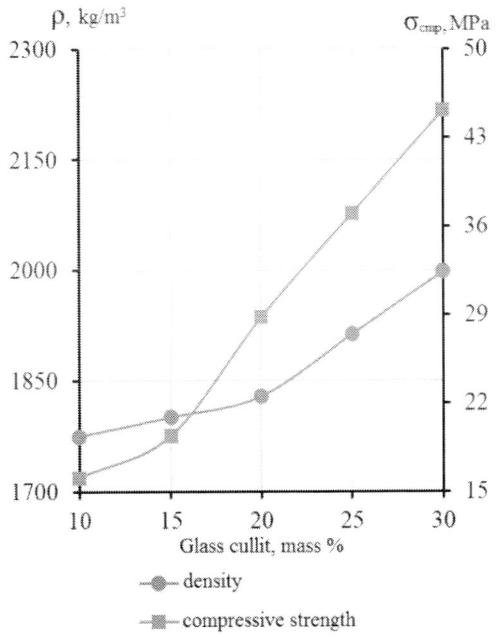

Fig. 1. The researched samples density and compressive strength.

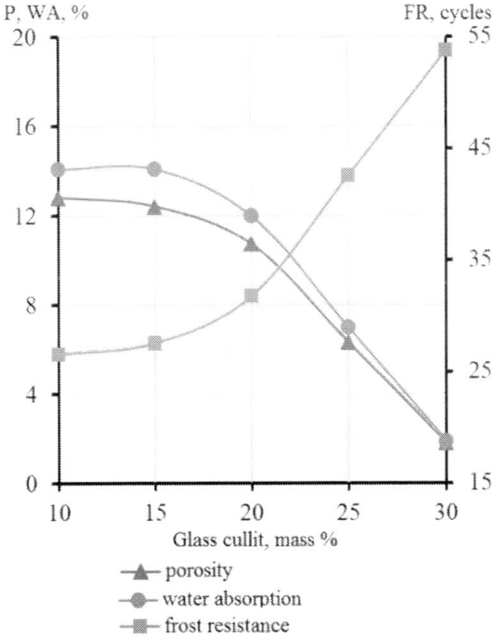

Fig. 2. The researched samples porosity, water absorption and frost resistance.

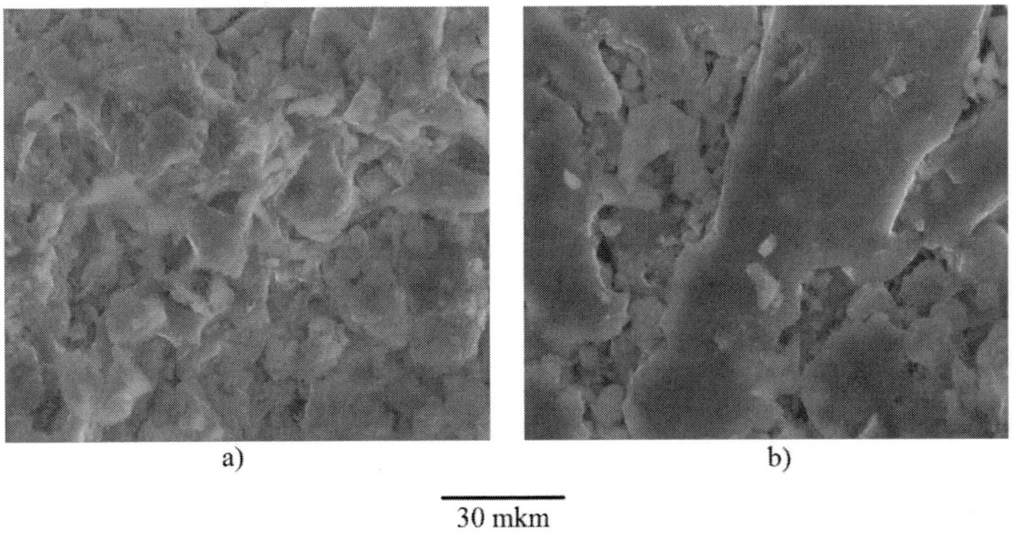

Fig. 3. Ceramic material structure at the introduction of 10 (a) and 30 (b) mass % of glass cullit in the charge composition (2000 times increase)

Samples toxicity test results when boric acid was introduced into the charge are shown in figure 4. Thus the sample received when 5 mass % of galvanic sludge (GS), 30 mass % of glass cullit (GC) and 2.5 mass % of boric acid (BA) were introduced into the clay can be considered ecologically safe. The results of determining physical-mechanical properties of the samples, received on the basis of compositions containing boric acid in comparison with the composition without additives are presented in the table.

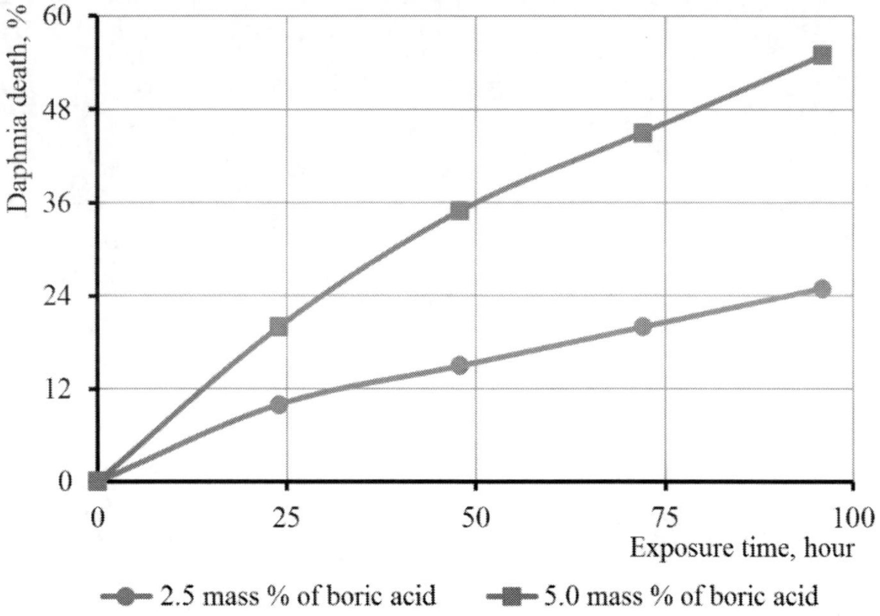

Fig. 4. Daphnia deaths dynamics in aqueous extract with various content of boric acid in the charge composition.

Table. Comparative characteristic of ceramic samples properties.

Additives content, mass %			ρ, kg/m³	σ_{cmp}, MPa	P, %	WA, %	FR, cycles
GS	GC	BA					
-	-	-	2099,4	14,3	6,9	7,5	41
5	30	2,5	1687,0	29,4	3,0	3,1	51
5	30	5	1865,6	17,7	2,8	2,9	51

According to the table data, additives introduction stipulates the enhancing of ceramic material strength characteristics and its frost resistance and also the reduction of porosity and water absorption in comparison with ceramics produced in compliance with the similar technology but without additives introduction. It should be stated that, when glass cullit and boric acid were introduced into the charge composition, the self glazing effect is observed on the surface and in depth of the ceramic material thus additionally reducing water absorption and increasing frost resistance.

4 Conclusions

On the basis of the research results the conclusion can be made that the developed charge composition allows producing durable and ecologically safe ceramic material from low quality raw materials by introducing 2 types of waste including toxic galvanic sludge. Herewith considerably large amount of waste (30 mass % of glass and 5 mass % of galvanic sludge) is added to the charge thus stipulating its comprehensive utilization.

The production of items on the basis of the developed charge composition will allow receiving quality products under lower pressing pressure and firing temperature: 15 MPa and 1050 °C for the developed composition in the comparison with currently produced facing ceramics under ≥ 30 MPa and 1000 – 1450 °C.

The new ceramic material density is lower than of the existing analogues (1900 – 2200 kg/m³ for terracotta and clinker; 2400 – 2600 kg/m³ for porcelain stoneware) thus it decreases the mass of facades and socles, paneled with that type of material, reducing the load on the building basement. It means that the developed material can be effectively applied in high-rise construction.

All the rest properties of the developed ceramic material allow using it for outdoor buildings facing. The glazer layer formed on the ceramic material surface is a self-cleaning surface, i.e. cleaned by the rain that can save money on washing facades and socles. It is very important especially for high building construction with large paneled areas.

References

1. M.A. Sukharnikova, E.S. Pikalov, O.G. Selivanov, V.Y. Chukhlanov, Glass and Ceramics **73 (3-4)**, 246-248 (2016)
2. Environment condition and population health in the Vladimir region in 2014: annual report. Issue 22 Vladimir Region Administration, environment Dpt. Vladimir: Transit-X, p.120 (2015)
3. Y.I. Wiseman, A.A. Ketov, Vestnik PNIPU. Urbanistics **4**, 78-75 (2011)
4. Toxicity determination method of water and water extractions from soil, waste water sediments, waste by Daphnia death and fertility change. Federal registry FR.ФР.1.39.2007.03222. Moskow: Aquaros, p.41 (2007)
5. E.S. Pikalov, O.G. Selivanov, V.Y. Chukhlanov, N.V. Chukhlanova, Ecology and industry in Russia **6**, 24-29 (2017)

Variation principle in calculating the flow of a two-phase mixture in the pipes of the cooling systems in high-rise buildings

Andrey Aksenov[1], and *Anna* Malysheva[1*]

[1]Moscow State University of Civil Engineering, 12933, Moscow, Yaroslavskoe shosse, 26, Russia

Abstract. The analytical solution of one of the urgent problems of modern hydromechanics and heat engineering about the distribution of gas and liquid phases along the channel cross-section, the thickness of the annular layer and their connection with the mass content of the gas phase in the gas-liquid flow is given in the paper.The analytical method is based on the fundamental laws of theoretical mechanics and thermophysics on the minimum of energy dissipation and the minimum rate of increase in the system entropy, which determine the stability of stationary states and processes. Obtained dependencies disclose the physical laws of the motion of two-phase media and can be used in hydraulic calculations during the design and operation of refrigeration and air conditioning systems.

1 Introduction

Questions related to the distribution of a two-phase medium in a stream, i.e. with the true volumetric gas content (the relative cross-sectional area occupied by the gaseous medium) in the gas-liquid flow, and also with the slip coefficient (ratio of the average phase velocities) are key in the hydrodynamics of two-phase flows, for example, in the study of steam flow in air conditioning systems of industrial civil tall buildings .

Among the numerous studies in this area, experimental methods predominate, and in the calculations, empirical formulas and nomograms are mainly used.

When analyzing a two-phase flow the Fauske, Zivi relation can be used [1,2]:

$$\frac{\overrightarrow{w_g}}{\overrightarrow{w_l}} = \left(\frac{\rho_l}{\rho_g}\right)^{\frac{1}{2}},$$

$$\frac{\overrightarrow{w_g}}{\overrightarrow{w_l}} = \left(\frac{\rho_l}{\rho_g}\right)^{\frac{1}{3}},$$

where w_g - gas phase velocity; w_l - liquid phase velocity; ρ_l – liquid density; ρ_g – gas density.

The first of these formulas is derived from the minimum of the total amount of motion of the liquid and gas phases in a two-phase flow. These formulas do not reflect the influence on the slip coefficient of either the mass content of gas or the physical properties of the liquid

* Corresponding author: mani495@mail.ru

and gaseous media (except for the ratio of their densities), which limits the possibility of their application.

In the present work, it is made an attempt to solve the problem of the volume gas content and the phase slip coefficient applied to two-phase annual flows on the basis of the principle of minimum energy dissipation in a stabilized flow associated with irreversible losses by internal friction during the motion of real media.

To simplify the analysis, the annular structure of the adiabatic two-phase flow is adopted as the design model. The solution of the problem of the relative area occupied by the gas nucleus φ and the corresponding phase slip coefficient $s = (\overline{w_g} \,/\, \overline{w_l})$ in a symmetric horizontal annular flow, depending on the determining parameters.

From the thermodynamic point of view, in the flow of a viscous liquid degradation occurs (energy depreciation), as in all other real processes taking place in nature, the characteristic feature of which is the increase in the entropy of the system. Internal nonequilibrium processes always act in a direction that causes a decrease in the rate of entropy increase [3, 4]. In the thermodynamics of irreversible processes, this law is formulated as follows: as the system goes over to the stationary-nonequilibrium state, the value of the incremental increase in entropy decreases, and when the stationary-nonequilibrium state is reached, it is of the smallest value compatible with external constraints [5].

2 Methods

Let us consider the annular regime of a stabilized gas-liquid flow in a pipe with radius of R_0, shown in the Fig. 1. This flow will consist of a ring layer of liquid with thickness δ and a cylindrical gas core with radius $R_G = R_0 - \delta$. The mass flow of fluid G_l and gas G_g and their physical parameters (density, viscosity) $\rho_g \rho_l \mu_g \mu_l$ are assumed to be known

Let us select a cylindrical element of radius R of unit length in the course of the flow. On the surface of the selected element, the force of internal friction acts. It is equal to

$$\mathrm{T} = \tau 2\pi R_1, \tag{1}$$

where τ- tangential friction stress, N/m^2.

The second work of the internal friction force is equal to the product of the value of τ by the rate of shear deformation on this axis:

$$dE = 2\pi R\tau \frac{dw}{dR} dR \ . \tag{2}$$

The second work of internal friction along the entire section of the gas-liquid flow is expressed by the sum of the integrals:

$$E = 2\pi \int_0^{R_0-\delta} \tau_\Gamma \cdot R \frac{dw_g}{dR} \cdot dR + 2\pi \int_{R_0-\delta}^{R_0} \tau_l \cdot R \frac{dw_l}{dR} \cdot dR. \tag{3}$$

In this case the first term E_g refers to the gas nucleus, and the second one E_l to the annular layer of the liquid; $E = E_g + E_l$; τ_g и τ_l – tangential friction constraint in the gas and liquid phases of the flow; N/m^2; w_g и w_l - velocity of the gas and liquid phases on the surface of radius R in the cross section, m/s.

The value E expresses the second amount of dissipation in the cross section of the flow over a section of unit length.

Taking into account the law on the minimum rate of entropy increase in stationary irreversible processes or the variation principle of least action known in classical mechanics, we can conclude that the distribution of the gas and liquid phases along the cross section will be such that for any given mass flow gas content and other defining flow parameters, the second work of internal friction forces is minimal. It ensures the stability of stationary flows. Proceeding from this, it is possible to establish the relative cross-sectional area occupied by the gas phase $\varphi_K = \frac{F_g}{F}$ (true volume gas content) and the ratio between the average velocities of the gas and liquid phases $\overline{w_g}\ /\ \overline{w_l}$. [6]

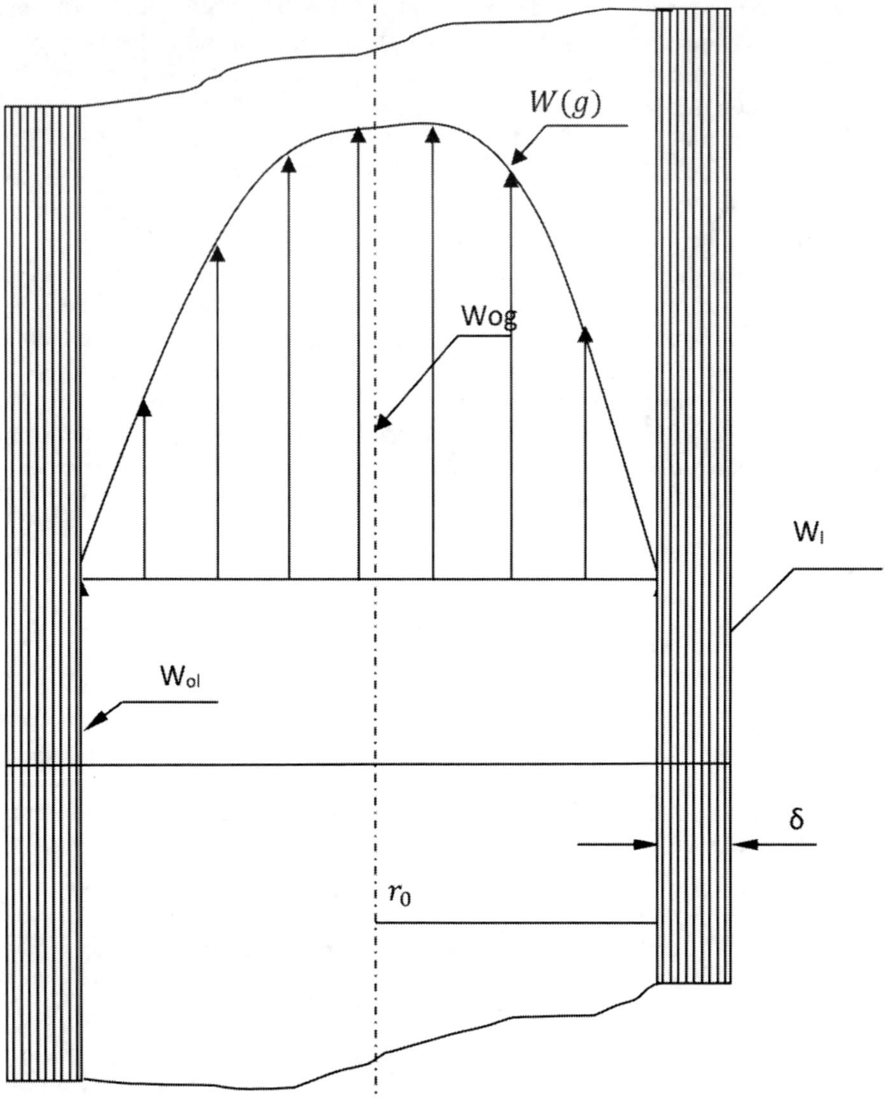

Fig. 1. Calculation model for the flow of a two-phase annular gas-liquid flow

We consider the problem posed for various cases of the annular regime of the cross section.

2.1 Laminar flow in the gas nucleus and in the liquid layer

For laminar flow, the magnitude of the tangential frictional stress is expressed by Newton's law:

$$\tau = \mu \frac{dw}{dy},$$

where $y = R_0 - R$.

Consequently, the integral (3) can be rewritten in the following form:

$$E = 2\pi\mu_g \int_0^{R_0-\delta} R \cdot (\frac{dw_g}{dR})^2 \cdot dR + 2\pi\mu_l \int_{R_0-\delta}^{R_0} R \cdot \left(\frac{dw_l}{dR}\right)^2 \cdot dR. \quad (4)$$

where μ_g и μ_l - dynamic viscosity coefficients of gas and liquid media, Pa·s.
The law of distribution of the phase velocities over the cross section of the flow will be assumed to be parabolic in the gas nucleus and linear in the liquid ring layer:

$$w_g = w_{0g}\left[1 - (\frac{R}{R_0-\delta})^2\right] + w_{0l}, \tag{5}$$

where ω_{0g} - relative velocity on the axis of the gas nucleus, m/s;

$$0 \le R \le R_g;\ R_g = R_0 - \delta;$$

$$w_l = w_{0l}\frac{R_0-R}{\delta}, \tag{6}$$

where $R_g \le R \le R_0$.

According to the formula (5)

$$(\frac{dw_g}{dR})^2 = \frac{4w_{0g}^2 \cdot R^2}{(R_0-\delta)^4}.$$

Consequently

$$E = \frac{2\pi\mu_g - w_{0g}^2}{(R_0-\delta)^4}\int_0^{R_0-\delta} R^3 \cdot dR = 2\pi\mu_g w_{0g}^2. \tag{7}$$

Since the relative velocity on the ω_{0r} axis is related to the average relative velocity in the nucleus by the relation $w_{0g} = 2\overline{w_g}$, then

$$E_g = 8\pi\mu_g w_g^2, \tag{8}$$

where

$$\overline{w_g} = \overline{w_{g\,(abs)}} - \omega_{0l}.$$

Let us find the relationship between the maximum w_{0l} and the average velocities in the section of the liquid phase. To do this, we calculate the integral

$$\overline{w_l} = \frac{2\pi\int_{R_0-\delta}^{R_0}\rho_l R w_l dR}{\pi\rho_l[R_0^2-(R_0-\delta)^2]} = \frac{2}{(2R_0-\delta)\delta}\int_{R_0-\delta}^{R_0} R w_l dR. \tag{9}$$

Taking into account relation (6), we will obtain

$$\overline{w_l} = \frac{2w_{0l}}{(2R_0-\delta)\delta^2}\int_{R_0-\delta}^{R_0} R(R_0 - R)dR = \frac{w_{0l}(3-2\eta)}{3(2-\eta)}, \tag{10}$$

where η – the relative thickness of the annular liquid layer; $\eta = \frac{\delta}{R_0}$.

Consequently

$$w_{0l} = 3\overline{w_l}\frac{2-\eta}{3-2\eta}. \tag{11}$$

Taking into account continuity:

$$w_{0l} = \frac{3G_l}{\pi\rho_l R_0^2(3-2\eta)}. \tag{12}$$

Average absolute velocity of the gaseous medium

$$\overline{w_{g(abs)}} = \frac{G_g}{\pi\rho_g R_0^2(1-\eta)^2}, \tag{13}$$

where G_lи G_g - mass flow of fluid and gas in the flow, kg/s.

Taking into account formulas (12) and (13), expression (8) can be rewritten in the following form:

$$E_g = \frac{8\mu_g G^2(1-x)^2}{\pi R_0^4 \rho_l^2} \cdot \left[\frac{A}{(1-\eta)^2} - \frac{3}{\eta(2-3\eta)}\right]^2, \quad (14)$$

$$\text{where } A = \frac{x}{1-x} \cdot \left(\frac{\rho_l}{\rho_g}\right); \quad G = G_g + G_l; \quad (15)$$

x - mass flow rate of the gas phase in the flow.

Let us find the second term of the expression (4). According to the formula (6)

$$E_l = 2\pi\mu_l \int_{R_0-\delta}^{R_0} R\left(\frac{w_{0l}}{\delta}\right)^2 dR = \pi\mu_l \omega_{0l}^2 \frac{(2-\eta)}{\eta}. \quad (16)$$

Taking into account formula (12)

$$E_l = \frac{9\mu_l G^2(1-x)^2(2-\eta)}{\pi\rho_l^2 R_0^4 \eta^3 (3-2\eta)^2}. \quad (17)$$

Thus, the sum (14) and (17) will be equal to:

$$E_l = \frac{8\mu_g G^2(1-x)^2}{\pi R_0^4 \rho_l^2} \cdot \left\{\left[\frac{A}{(1-\eta)^2} - \frac{3}{\eta(2-3\eta)}\right]^2 + \frac{9\mu_l G^2(1-x)^2(2-\eta)}{\pi\rho_l^2 R_0^4 \eta^3 (3-2\eta)^2}\right\}. \quad (18)$$

It is easy to show that $(1-\eta)^2 = \frac{F_g}{F} = \varphi$;

$$G = \pi R_0^2 (\rho w),$$

where ρw – mass velocity of the gas-liquid mixture, kg/m^2·s.

Therefore, expression (18) can be written in the following form:

$$E = \frac{8\pi\mu_g (\rho w)^2 (1-x)^2}{\rho_l} \left\{\left[\frac{A}{\varphi} - \frac{3}{(1-\sqrt{\varphi})(1+2\sqrt{\varphi})}\right]^2 + \frac{9\mu_l(1+\sqrt{\varphi})}{8\mu_g(1-\sqrt{\varphi})^3(1+2\sqrt{\varphi})^2}\right\}. \quad (19)$$

Let us find the required value of the true volume gas content φ_{CT} and the relative thickness of the liquid annual layer η (in the stationary flow) from the condition of a minimum of the value E as a function of the argument φ (19) or the argument η (18).

The necessary condition for the extremum:

$$\left(\frac{\partial E}{\partial \varphi}\right)_{\varphi=\varphi_{st}} = 0, \text{ or } \left(\frac{\partial E}{\partial \eta}\right)_{\eta=\eta_{st}} = 0.$$

The nature of the functions makes it possible not to justify sufficient conditions for a minimum.

For the sake of simplicity, we do not write the index "st" for η and φ in the sequel.

Let us find the derivative:

$$\frac{\partial}{\partial\eta}\left\{\left[\frac{A}{(1-\eta)^2} - \frac{3}{\eta(3-2\eta)}\right] - \frac{9\mu_l(2-\eta)}{8\mu_g \eta^3 (3-2\eta)^2}\right\} = 2\left[\frac{A}{(1-\eta)^2} - \frac{3}{\eta(3-2\eta)}\right] \cdot \left[\frac{2A}{(1-\eta)^3} - \frac{3(3-4\eta)}{\eta^2(3-2\eta)^2}\right] - \frac{9\mu_l[\eta(3-2\eta)+(2-\eta)(9-10\eta)]}{8\eta^4(3-2\eta)^3}. \quad (20)$$

Equating the resulting expression to zero, after a simple transformation we obtain the following expression:

$$[A\eta(3-2\eta)-3(1-\eta)^2]\cdot[2A\cdot\eta^2(3-2\eta)^2+3(3-4\eta)(1-\eta)]\cdot\eta-\frac{9}{8}\cdot\frac{\mu_l}{\mu_g}[(3-2\eta)^2-\eta](1-\eta)^5=0. \quad (21)$$

This equation expresses the analytical relationship between the true volumetric gas content $\varphi=\frac{F_g}{F}=(1-\eta)^2=0$ and the mass flow rate vapor content

$$x=G_r/G.$$

Concerning the variable φ, equation (21) will look as follows:

$$[A\cdot(1-\sqrt{\varphi})(1+3\sqrt{\varphi})-3\varphi][2A\cdot(1-\sqrt{\varphi})^2\cdot(1+2\sqrt{\varphi})^2+3\sqrt{\varphi}\cdot(4\sqrt{\varphi}-1)](1-\sqrt{\varphi})-\frac{9}{8}\cdot\frac{\mu_l}{\mu_g}[(1+2\sqrt{\varphi})^2-(1-\sqrt{\varphi})]\cdot\varphi^2\sqrt{\varphi}=0. \quad (22)$$

We solve the problem for a parabolic velocity profile in the annual liquid layer and in a gas nucleus. In this case

$$w_l=\frac{w_{0l}}{\delta\cdot(2R_0-\delta)}\cdot(R_0^2-R^2); \quad (23)$$

$$\frac{dw_l}{dR}=-\frac{2w_{0l}\cdot R}{\delta\cdot(2R_0-\delta)}, \quad (24)$$

where $R_0-\delta\leq R\leq R_0$.

The average velocity of fluid in the annual layer of the liquid is equal to

$$w=\frac{1}{2}w_{0l},$$

$$w_{0l}=\frac{2G_l}{\pi\rho_l\cdot[R_0^2-(R_0-\delta)^2]}=\frac{2G_l}{\pi\rho_l\cdot R_0^2\cdot[1-(1-\eta)^2]}. \quad (25)$$

Second work of internal friction forces in liquid annular layer of unit length

$$E_l=2\pi\mu_l\int_{R_0-\delta}^{R_0}R\left(\frac{dw_l}{dR}\right)^2dR=\frac{8\pi\mu_l w_{0l}^2}{\delta^2\cdot(2R_0-\delta)^2}\cdot\int_{R_0-\delta}^{R_0}R^3dR. \quad (26)$$

After calculating the integral (26), and taking into account the relation(20), we obtain

$$E_l=\frac{8\pi\mu_l w_l^2\cdot[1+(1-\eta)^2]}{1-(1-\eta)^2}. \quad (27)$$

Using the continuity equation

$$E_l=\frac{8\pi\mu_l G^2(1-x)^2\cdot[1+(1-\eta)^2]}{\pi\rho_l^2R_0^4\eta^3(2-\eta)^3}. \quad (28)$$

According to formula (8) the second work of internal friction forces in a gas nucleus of unit length

$$E_g=8\pi\mu_g(\overline{w_{g(\text{абс})}}-w_{0l})^2=\frac{8\pi\mu_g G^2(1-x)^2}{\pi\rho_l^2R_0^4}\cdot\left[\frac{A}{(1-\eta)^2}-\frac{2}{\eta(2-\eta)}\right]^2.. \quad (29)$$

Consequently

$$E_g=\frac{\delta\mu_g G^2(1-x)^2}{\pi\rho_l^2R_0^4}\cdot\left\{\left[\frac{A}{(1-\eta)^2}-\frac{2}{\eta(2-\eta)}\right]+\frac{\mu_l[1+(1-\eta)^2]}{\mu_g\eta^3(2-\eta)^3}\right\}. \quad (30)$$

In the expression (30) we replace the variable η with another one: $(1-\eta)^2$, i.e. the value of the true volumetric gas content - φ. Then:

$$E_g = \frac{\delta\mu_g G^2(1-x)^2}{\pi\rho_l^2 R_0^4}\cdot\left[(\frac{A}{\varphi}-\frac{2}{1-\varphi})^2+\frac{\mu_l}{\mu_g}\cdot\frac{(1+\varphi)}{(1-\varphi)^3}\right]. \quad (31)$$

We differentiate the value E with respect to the variable φ.

Equating the first derivative to zero

$$\frac{\partial}{\partial\varphi}\left[(\frac{A}{\varphi}-\frac{2}{1-\varphi})^2+\frac{\mu_l}{\mu_g}\cdot\frac{(1+\varphi)}{(1-\varphi)^3}\right]=2\left(\frac{A}{\varphi}-\frac{2}{1-\varphi}\right)\cdot\left[-\frac{A}{\varphi^2}-\frac{2}{(1-\varphi)^2}\right]+$$

$$+\frac{\mu_l\cdot(1-\varphi)^3+3(1-\varphi)^2(1+\varphi)}{\mu_g\cdot(1-\varphi)^4}=\frac{2\mu_l(2+\varphi)}{\mu_g\cdot(1-\varphi)^4}-2\left[\frac{A}{\varphi^2}+\frac{2}{(1-\varphi)^2}\right]\cdot(\frac{A}{\varphi}-\frac{2}{1-\varphi}),$$

we obtain the following equation expressing the minimum condition for the resultant second work of internal friction forces:

$$\left[\frac{A}{\varphi^2}+\frac{2}{(1-\varphi)^2}\right]\left(\frac{A}{\varphi}-\frac{2}{1-\varphi}\right)-\frac{\mu_l}{\mu_g}\cdot\frac{(2+\varphi)}{(1-\varphi)^4}=0; \quad (32)$$

$$(0<\varphi<I).$$

Analysis of the obtained solutions (22) and (32) shows their differences. It confirms the advantage of the method of integral relations used in the problem under consideration. The method is based on one of the fundamental laws of mechanics. The solutions can also be used for the wave flow of a liquid film if the value η means relative thickness of the liquid layer. It is also to introduce the so-called effective viscosity coefficient as a parameter that takes into account wave nature of the interface between the gas nucleus and the liquid layer [7,8].

2.2 Turbulent flow in the gas nucleus of two-phase flow, laminar flow in the annular liquid layer

The velocity distribution over the cross section for a turbulent flow of a medium in a cylindrical channel is expressed by the formula:

$$\frac{w_g}{w_{0g}}=1-\frac{Arth\sqrt{mt}-\text{arctg}\sqrt{mt}}{0.805\cdot\text{lg}Re_g}, \quad (33)$$

where t- relative distance from the nucleus axis; $t=\frac{R}{R_0-\delta}=\frac{R}{R_g}$; $Arth\sqrt{m}-\text{arctg}\sqrt{m}=0{,}805lgRe_g$.

For $Re_g>400$ the value m is very close to one.

The choice of this law of velocity distribution is due not only to the fact that it very well reflects the experimental data, given the deformation of the velocity profile with a change in the Re number, and primarily because the simpler distribution laws (power law and logarithmic) used in hydrodynamics have a significant drawback in the fact that the derivative on the wall turns to infinity, which limits the possibilities of solving this and other boundary problems.

The stress of turbulent friction in accordance with the Prandtl-Karman theory is expressed here by the formula

$$\tau_g=x^2\rho_g(R_0-\delta)^2\cdot(1-m^2t^2)^2\cdot(\frac{dw_g}{dR})^2.$$

Differentiating expression (33) with respect to the variable R, we have

$$\frac{dw_g}{dR}=\frac{dw_g}{dt}\cdot\frac{dt}{dR}=-\frac{w_{0g}m\sqrt{mt}}{0{,}805lgRe_g\cdot(1-m^2t^2)\cdot(R_0-\delta)}. \quad (34)$$

Consequently, for the tangential frictional stress in the gas nucleus, we obtain the following expression:

$$\tau=\frac{\rho_g x^2 m^3 w_{0g}^2 t}{(0{,}805lgRe_g)^2}. \quad (35)$$

The average flow rate $\overline{w}_g$ is determined by the following formula:

$$\overline{w_g} = \frac{2\pi \int_0^{R_0-\delta} \rho_g R w_g dR}{\pi \rho_g (R_0-\delta)^2}. \quad (36)$$

After substituting expression (33) and calculating the integral (36), the relationship between the average and maximum flow velocities takes the following form:

$$\frac{\overline{w_g}}{w_{0g}} = \frac{\lg Re_g - 0{,}83}{\lg Re_g}. \quad (37)$$

Consequently, the expression for the tangential frictional stress (35) can be rewritten in the following form:

$$\tau = \frac{x^2 \rho_g m^3 w_g^2 t}{0{,}805^2 (\lg Re_g - 0{,}83)^2} = ct. \quad (38)$$

Second work of internal friction forces in the gas nucleus is expressed by the integral

$$\mathrm{E}_g = 2\pi 0 \int_0^{Rg} \tau_g R \frac{dw}{dR} dR. \quad (39)$$

According to formulas (34) and (37)

$$\left|\frac{dw_g}{dR}\right| = \frac{c_2\sqrt{mt}}{1-m^2t^2}, \quad (40)$$

$$\text{where } \mathrm{C}_2 = \frac{2\pi c_1 c_2 R_g^2}{m^3} \int_0^m \frac{(mt)^2\sqrt{mt}}{1-(mt)^2} d(mt). \quad (41)$$

To calculate the integral in expression (41), we introduce the notation $\sqrt{mt} = z$.

Then $d(mt) = 2zdz$.

The required integral can be rewritten in the following form:

$$2\int_0^{\sqrt{m}} \frac{z^6}{1-z} dz = 2\int_0^{\sqrt{m}} \left(\frac{z^2}{1-z^4} - z^2\right) dz = 2\int_0^{\sqrt{m}} \left(\frac{1}{4(1+z)} + \frac{1}{4(1-z)} - \frac{1}{2(1+z^2)} - z^2\right) dz =$$
$$Arth\sqrt{m} - arctg\sqrt{m} - \frac{2}{3}(\sqrt{m})^3 \approx 0.805(\lg Re_g - 0{,}83); \quad (42)$$

$$Re_g > 400.$$

Substituting the calculated value of the integral in expression (41), we obtain

$$E_g = \frac{2\pi x^2 \rho_g \overline{w_g^3} R_g}{0{,}805^2 (\lg Re_g - 0{,}83)^2}. \quad (43)$$

The average velocity in the gas nucleus relative to the liquid film is less than the absolute average velocity $w_{g(\text{абс})}$ by the value w_{0l}:

$$\overline{w_g} = \overline{w_{g(\text{абс})}} - w_{0l}.$$

According to formulas (13) and (25)

$$\overline{w_{g(\text{абс})}} = \frac{G_r}{\pi \rho_g R_0^2 \varphi} = \frac{Gx}{\pi \rho_g R_0^2 \varphi}; \quad (44)$$

$$w_{0l} = 2\overline{w_l} = \frac{2G_l}{\pi \rho_l R_0^2 (1-\varphi)} = \frac{2G_l(1-x)}{\pi \rho_l R_0^2 (1-\varphi)} \quad . \quad (45)$$

That is why

$$\overline{w_g} = \frac{G(1-x)}{\pi \cdot R_0^2 \rho_l} \cdot \left(\frac{\mathrm{A}}{\varphi} - \frac{2}{1-\varphi}\right). \quad (46)$$

Substituting the obtained expression of the velocity (46) into the formula (43), we obtain

$$E_g = \frac{2x^2 \rho_g G_l^3 \sqrt{\varphi}}{0{,}805^2 \cdot \pi^2 R_0^5 \rho_l^3 (lgRe_g - 0{,}83)^2} \cdot \left(\frac{A}{\varphi} - \frac{2}{1-\varphi}\right)^3. \tag{47}$$

According to the formula (28)

$$E_l = \frac{8\mu_l G^2{}_l (1+\varphi)}{\pi \rho_l^2 R_0^4 (1-\varphi)^3}. \tag{48}$$

Thus, the resulting dissipative characteristic of the gas-liquid flow for the studied case will be

$$E_l = \frac{8\mu_l G^2 (1-\chi)^2}{\pi \rho_l^2 R_0^4} \cdot \left[\frac{C_3 \sqrt{\varphi}}{(lgRe_g - 0{,}83)^2} \cdot \left(\frac{A}{\varphi} - \frac{2}{1-\varphi}\right)^3 + \frac{1+\varphi}{(1-\varphi)^3}\right], \tag{49}$$

where $C_3 = \frac{0{,}386 \cdot x^2 \rho_g G (1-\chi)}{\pi \rho_l \mu_l R_0}$.

3 Results

From the minimum condition of this function ($\frac{\partial E}{\partial \varphi} = 0$), we obtain the following analytical relationship between the mass flow (x) and the true volumetric (φ) gas content and the two-phase flow or the corresponding phase slip coefficient

$$\left(\frac{A}{\varphi} - \frac{2}{1-\varphi}\right)\left\{\frac{B+0{,}87}{2\sqrt{\varphi}B} - 3\sqrt{\varphi}\left[\frac{A}{\varphi^2} - \frac{2}{(1-\varphi)^2}\right]\right\} + \frac{\rho_l B^2 (2+\varphi)}{\rho_g Re_{lg} (1-x)(1-\varphi)^4} = 0; \tag{50}$$

$$0 < \varphi < 1; \quad 0 < x < 1;.$$

In this case

$$B = \lg\left(\frac{Re_{g(0)} x}{\sqrt{\varphi}}\right) - 0{,}83,$$

$Re_{гr} = \frac{(\rho w) \cdot d_0}{\mu_g}$ – the reduced Reynolds number for the gas; $Re_{lg} = \frac{(\rho w) \cdot d_0}{\mu_l}$ – the reduced Reynolds number for the liquid phase; d_0 – inner diameter of the pipe, m; (ρw) – mass rate of gas-liquid flow, kg/m^2·s.

4 Discussion

The calculated values of φ according to the formula (50), found from the condition of a minimum of the dissipative characteristic of the two-phase "*E*" flow, are in good agreement with the experimental data of various authors [9-15] and are very close to the values calculated of the known dependences for the steam-water flow of Z.L. Miropol'skiy. It confirms the reliability of the physical model underlying the proposed theoretical solution and points to a new approach to this problem different from the traditional one.

A comparative analysis of the experimental data according to φ in heated and unheated flows showed that in the range $x > 0{,}1$ the thermal flow practically does not affect the value of the true volumetric vapor content.

Equation (50) is somewhat cumbersome for engineering calculations. However, when using a computer, the establishment between the values of φ and x for different values of other parameters of the gas-liquid flow is not difficult.

The performed mathematical analysis showed a very small effect of the logarithmic functions entering into the equation (50). It is explained by the fact that the coefficient $(\lg Re_g - 0{,}83)^2$ entering the initial formula (49), which reflects the deformation of the velocity profile in the gas nucleus, changes insignificantly in the turbulent flow.

If, when differentiating function (49), the indicated coefficient is considered to be independent of φ, then we have

$$\frac{\partial}{\partial\varphi}\left[c\cdot\sqrt{\varphi}\left(\frac{A}{\varphi}-\frac{2}{1-\varphi}\right)^3+\frac{1+\varphi}{(1-\varphi)^3}\right]=0. \quad (51)$$

Or

$$\frac{2}{2\sqrt{\varphi}}\left(\frac{A}{\varphi}-\frac{2}{1-\varphi}\right)^2\left[\frac{A}{\varphi}-\frac{2}{1-\varphi}-6\varphi\left(\frac{A}{\varphi^2}-\frac{2}{(1-\varphi)^2}\right)\pm\frac{2(2+\varphi)}{(1-\varphi)^4}\right]=0; \quad (52)$$
$$0<\varphi<1.$$

5 Conclusion

Ffter simple transformations we obtain the following equation:

$$\mathrm{C}[\mathrm{A}(1-\varphi)-2\varphi]^2[5\mathrm{A}(1-\varphi)^2+2\varphi(1+\varphi)]-4\varphi^3\sqrt{\varphi}(2+\varphi)=0. \quad (53)$$

In this case

$$\mathrm{C}=\frac{0{,}193\aleph^2\rho_g Re_{lg}(1-x)}{\pi\rho_l(lgRe_g-0{,}83)}. \quad (54)$$

The disadvantage of equations (32), (50) and (53) is that they express the dependence in implicit form.

$$\varphi=f(x,{}^{\rho_l}/_{\rho_g},{}^{\mu_l}/_{\mu_g},\rho w)\cdot$$

For engineering calculations, these equations can be substantially simplified. Since the velocity at the boundary between the gas and liquid phases for the values $x>0{,}1$ is much less than the absolute velocity of the gas phase, then with sufficient accuracy for practical calculations it can be assumed that

$$w_{0g}=k_1\overline{w_g}, \quad (55)$$

where k_1- averaged coefficient of proportionality.

$$w_{0g}=\frac{k_1\cdot G\cdot x}{\varphi\,\rho_g}=\frac{k_1(\rho w)x}{\varphi\rho_g}. \quad (56)$$

Therefore, the expression for the dissipation energy (31) for a two-phase flow can be written in the following form:

$$E=\frac{8\pi\mu_g(\rho w)^2(1-x)^2}{\rho_l^2}\left[\frac{k_2A^2}{\varphi^2}+\frac{\mu_l(1+\varphi)}{\mu_g(1-\varphi)^3}\right]. \quad (57)$$

From the condition for the minimum of this function (with respect to the variable φ), we obtain the following expression:

$$\frac{(2+\varphi)\varphi^3}{(1-\varphi)^4}=\frac{k_2A\mu_g}{\mu_l}. \quad (58)$$

The left side of equation (58) is well approximated by expression

$$7{,}1\cdot\left(\frac{\varphi}{1-\varphi}\right)^{3{,}6}.$$

Therefore, we can write

$$\frac{\varphi}{1-\varphi}=\left(\frac{k_2A\mu_g}{7{,}1\mu_l}\right)^{0{,}278}. \quad (59)$$

$$\text{Then } \varphi=\frac{1}{1+k\cdot\left(\frac{\mu_l}{A^2\mu_g}\right)^{0{,}278}}, \quad (60)$$

where $\mathrm{A}=\frac{x}{1-x}\cdot\frac{\rho_l}{\rho_g}$.

By choosing the appropriate value of the coefficient k, we can take into account not only the character of the velocity distribution over the cross section, but also the change in the parameters ρ_l/ρ_g and μ_l/μ_g in the result of mechanical carryover of liquid particles into the gas nucleus, and also "loosening" of the liquid annular layer with intensive vaporization in steam generating pipes.

Equation (60) satisfactorily generalizes the experimental data on the steam-water flow at a value of $k = 0.5$.

In accordance with formulas (44), (45) and (60), we obtain the following expression for the slip coefficient:

$$S = k\left[\frac{x\rho_l}{(1-x)\rho_g}\right]^{0,444}\left(\frac{\mu_l}{\mu_g}\right)^{0,278}. \qquad (61)$$

The problem of the distribution of the phases of a gas-liquid flow in plane channels is very interesting. The theory considered allows us to investigate this question.

Since the main purpose of this work was to study the flow of two-phase flows in pipes, we give only two formulas obtained for determining the true gas volume content φ and the slip coefficient in flat channels

$$\varphi = \frac{1}{1+\frac{k(1-x)\rho_g}{x\rho_l}\sqrt{\frac{\mu_l}{\mu_g}}}; \qquad (62)$$

$$s = k\sqrt{\frac{\mu_l}{\mu_g}};$$

$k = 0,5$.

The advantage of the obtained equations and formulas for determining the true volumetric vapor content, in contrast to the known ones, is that they reflect the influence of the viscous characteristics of the gas or liquid media and can be used to calculate the vapor-liquid and gas-liquid flows of various components.

The latter is especially important for the study of many technological processes taking place in various apparatuses of the chemical industry, refrigeration and heat power engineering.

References

1. H. G., Vassilicos Ch. (eds.) Prediction of Turbulent Flows. (Cambridge University Press, 2005)
2. G.A. Dreytser. Efficiency of the use of flow swirls for intensification of flow in tubular heat exchangers. Moscow (2005).
3. I.R. Prigozhin. Introduction to the thermodynamics of irreversible processes. Izhevsk, 160 (2001).
4. I.R. Prigozhin, D. Kondepudi. Modern thermodynamics. From heat engines to dissipative structures. Moscow, 44 (2002).
5. G. Nikolis. Cognition of complex things. Moscow, 345 (2012).
6. B. Palm Heat transfer in heat exchangers designer for minimum charge. In: IIR//IIF. Stockholm, **5**, 84-89 (2002).
7. S.E. Tarasevich, A.B. Yakovlev. *On the boundary of the dispersed-annular flow regime in channels with continuous twisting*. (Moscow; Novosibirsk, 2004).
8. Bandara Filho Enio P., Saiz Jabardo Jose. Int. J. Refrig., **8**, 895-903 (2004).
9. V. Soldo, T. Curko, and M. Grozdek. Transfer from Carnot to Lorenz process using zeotropic mixtures. In: IIR/IIF. Stockholm, 63-66 (2002).
10. G.A. Dreytser. Efficiency of the use of flow swirls for intensification of flow in tubular heat exchangers. Moscow (2005).

11. Yu.M. Brodov, A.Yu. Ryabchikov, and K.E. Aronson. Investigation of a number of methods for intensifying heat transfer in energy heat exchangers, **6,** 49 (2002).
12. V.V. Yagov. Heat transfer and crises during boiling in swirling flows. Moscow (2005).
13. A. Klaszak Heat and mass transfer, **37,** 443-448 (2001).
14. Yu.I. Shanin. *Heat transfer when the flow is twisted in a square channel.* (Moscow, 2005).
15. V.G. Bukin, A.Yu. Kuzmin. Refrigeration machines operating on non-azeotropic refrigerant mixtures (Astrakhan, 2003).

The method of selecting an integrated development territory for the high-rise unique constructions

Svetlana Sheina[1*], *Elina* Shevtsova[1], *Alexander* Sukhinin[1] and *Elena* Priss[2]

[1]Academy of Construction and Architecture of Don State Technical University, Rostov-on-Don, Russia.
[2]Don State Technical University, sq. Gagarina, 1, Rostov-on-Don, 344010, Russia

Abstract. On the basis of data provided by the Department of architecture and urban planning of the city of Rostov-on-don, the problem of the choice of the territory for complex development that will be in priority for the construction of high-rise and unique buildings is solved. The objective of the study was the development of a methodology for selection of the area and the implementation of the proposed method on the example of evaluation of four-territories complex development. The developed method along with standard indicators of complex evaluation considers additional indicators that assess the territory from the position of high-rise unique building. The final result of the study is the rankings of the functional priority areas that takes into account the construction of both residential and public and business objects of unique high-rise construction. The use of the developed methodology will allow investors and customers to assess the investment attractiveness of the future unique construction project on the proposed site.

1 Introduction

In the XXI century on the territory of large cities a top priority of urban activity is the reconstruction of urban development. It happens so because every year the need for new construction of residential, public and commercial buildings increases to meet the needs of the population, but suitable areas for new construction in the city are almost gone. The maximum density of urban development is achieved in the Central part of the city, and the remaining areas that suitable for new construction are too small. In this regard, increases the popularity of high-rise unique building being an effective solution for construction in the dense urban areas and a small area of land.

The Government of the Russian Federation faced the problem of search of new territories, meeting the demands of investors, developers and the population in general. The program of complex development of territories was developed by the Government of the Russian Federation, which is based on the optimum economic development of territories due to the demolition of dilapidated housing and low-value buildings, removal of industrial enterprises outside the city and reconstruction of existing buildings. Integrated development of territories

* Corresponding author: rgsu-gsh@mail.ru

is possible on the initiative of the right holders of land plots and (or) located on them objects of real estate [1], or on the initiative of local authorities [2]. Under this program the Department of Architecture and Urban planning of the city of Rostov-on-Don marked four areas within which a decision may be taken on the complex development on the initiative of local authorities.

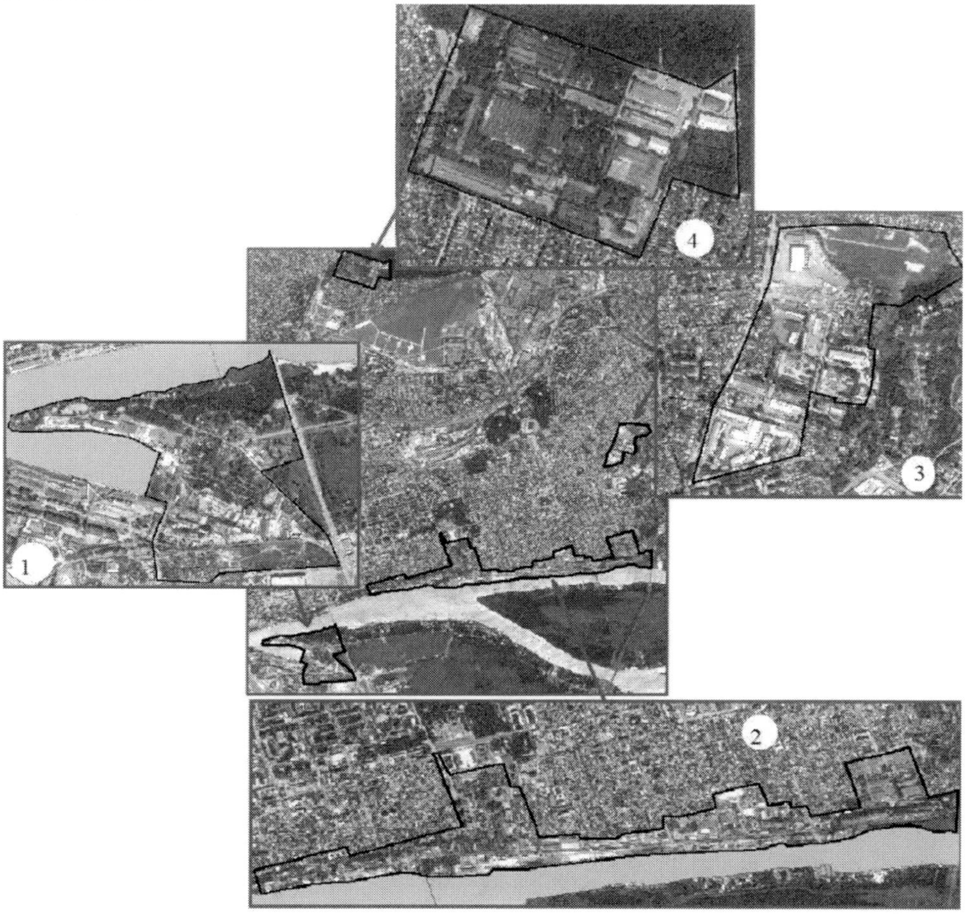

Fig. 1.A fragment of the electronic map of the city of Rostov-on-Don (1,2,3,4 – areas of integrated development)

The study of the areas for complex development helped to identify the problem of assessing their suitability for high-rise unique building. The objective of the study was the development of a methodology of selection of priority areas and the implementation of the proposed method on the example of evaluation of four territories complex development. The choice of the most priority areas is based on the techniques of a complex assessment of the territory, taking into account additional factors, evaluating the territory from a position of allocation of a unique high-rise buildings.

2 Materials and methods

To solve this problem, we developed a methodology of choice of the territories for integrated development of unique high-rise buildings. The developed methodology includes the following metrics (figure 2):

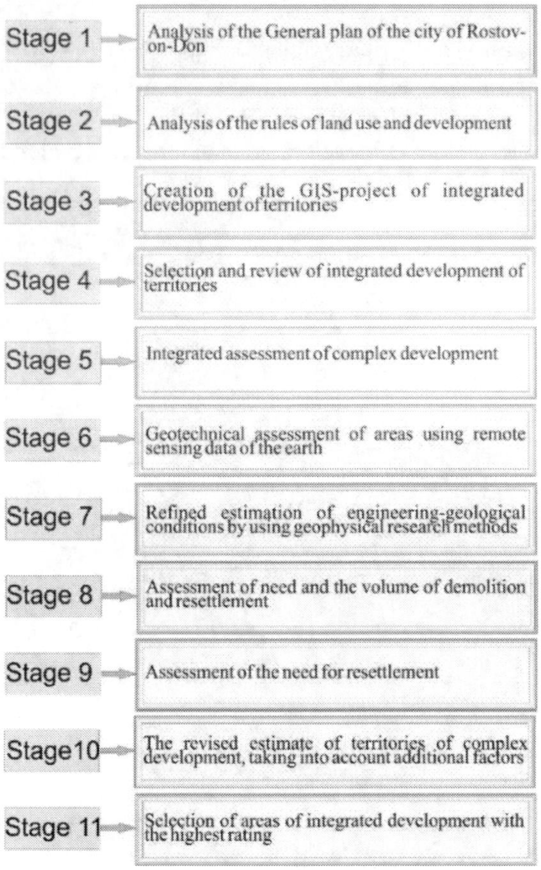

Fig. 2. The method of selection of areas of integrated development for high-rise and unique buildings

Stage 1: Analysis of the General plan of the city of Rostov-on-Don

Analysis the General plan of the city is based on the electronic map of the city of Rostov-on-Don in PC ArcGis 10.1

Stage 2: Analysis of the rules of land use and development

Analysis of the rules of land use and development is based on the electronic map of the city of Rostov-on-Don in PC ArcGis 10.1

Fig. 3. Map of the city of Rostov-on-Don

Stage 3: Creation of the GIS-project of integrated development of territories

In ArcMap 10.1 GIS database of the territories complex development of the city of Rostov-on-Don is designed and developed. In the electronic map of Rostov-on-Don the above described territory was applied, which is divided into four modules.

The first module contains the general characteristics of the areas (location areas, perimeter, area, compliance with the rules of land use and development). The second module contains information about engineering-geological conditions of territories (rate of rise of the groundwater, the depth of the groundwater, the presence of landslides, type of soil, flooding floods and geological hazards). The third module is the data on the ecological risk of territory and density of noise pollution. The fourth module of the GIS database contains information about the investment attractiveness of the territory (zones of protection of monuments of history, the degree of accessibility to the city center, security engineering equipment and the data of rating of functional territory priority depending on the type of the placed object)

Stage 4: Selection and review of integrated development of territories

Using the GIS-project of integrated development of territories, at this stage the general idea of the territories of complex development, their location on the city map, the land area, the positive and negative aspects of each territory.

Stage 5: Integrated assessment of complex development

Complex assessment of the territory is formed on the result of comparing the components' estimates from the point of view of the usage in various purposes: industrial construction, agriculture and forestry, for the formation of residential zones and zones of mass rest of the population. The goal of a complex site assessment is determining the degree of its suitability for different uses.

The territory of Rostov-on-Don was divided into pieces of land for evaluation, occupied by various land use categories. For the boundaries of the assessment sites were taken natural boundaries, boundaries of land occupied by large engineering structures, boundaries of enterprises, agricultural and forestry land.

In this study as the indicators of a complex assessment of the territory there were taken the factors of the relative values used in the assessment of land settlements [3]:

- Availability of public business, the administrative center;
- The level of engineering arrangement (communications, transport);
- The level of development of the sphere of cultural-and-everyday service of the population;
- Existence of monuments of history and culture;
- State of the environment;
- Engineering-geological conditions affecting the construction;
- Landscape and recreational terms.

Based on these factors, the calculation of the coefficients of relative values of the area are made. The numerical values of the coefficients for each of the assessment sites for each of the 7 groups of factors are summarized using the following formula:

$$K=1+\kappa_1+\kappa_2+\kappa_3+\kappa_4+\kappa_5+\kappa_6+\kappa_7; \quad (1)$$

K -is the total value of the coefficient of relative values of the assessment area;
K1–K7 -is a sum of the values of the coefficients of the factors (from 1 to 7 gr.).

The methods of selecting of the priority areas for the complex development imply a solution of the problem of choosing the evaluation criteria of these territories from the point of view of their attractiveness for certain types of their priority functional use. For solution of this task the determination of the coefficients of the desirability of evaluation factors of each function is fulfilled. Further the calculation of the integral index of investment attractiveness of the given area for a given function takes place [3].

To assess the degree of influence of the i – evaluation factor on the investment attractiveness of the area for the specified function the coefficient of desirability is entered, whose values are established on the basis of expert recommendations (table 1).

Table 1. Scale of values of the coefficients of the desirability

Rang	Evaluation	K_g
1	Critically necessary	4
2	highly desirable	2
3	preferably	1
4	doesn't matter	0
5	undesirable	- 1
6	highly undesirable	- 2
7	critically undesirable	- 4

Calculation of ratings is performed according to the formula:

$$A_i = K_{gi} * K_i \qquad (2)$$

A_i -is the rating-dependent valuation factors
K_{gi} - the coefficient of the desirability of the i- evaluation factor
K_i - is the coefficient of relative values of the i- station.
The final integrated ranking of the estimates are obtained by the formula:

$$A_j = \sum_{i=1}^{n} A_i + C \qquad (3)$$

A_j- -integrated rating score, C =10.
Thus, based on the study of the methodology of integrated assessment the rating of the functional priority of the territories complex development with the help of IAS "a simulation model of the master plan" that is available on the Department of urban construction and economy of ACA DSTU, which automatically calculates the rating depending on the specific evaluation of the site and of its functional use.
The final rating assessment of the territory was calculated by the formula:

$$B = (A_1 + A_2 + \dots + A_n)/n \qquad (4)$$

B is the final rating assessment of the territory,
A_n-the rating of the n- section,
n is the number of plots on site.
The results of the calculations are presented in table 2.

Table 2. Rating of the functional priority of the territories of complex development

N	The rating depending on the function	
	Residential area	*Public-business zone*
Area No 1	12.84	11.63
Area No 2	14.7	12.95
Area No 3	11.25	11.06
Area No 4	12.12	11.73

According to the functional priority of the highest priority for the construction of residential, public and business buildings is the area No2.

https://doi.org/10.1051/e3sconf/20183302064

Step 6: Geotechnical assessment of areas using remote sensing data of the earth

Earth remote sensing (ERS) – obtaining information about the Earth's surface and objects on it by contactless methods, in which a registered device is deleted from object of research to a considerable distance. The method consists of interpreting the results of measurement of electromagnetic radiation reflected or emitted by the object and is registered in some remote from it space.

Remote sensing techniques are based on the use of sensors that are placed on the spacecraft and record the electromagnetic radiation adapted for digital processing formats.

Radar interferometry - processing method of radar data based on the allocation of the phase difference of signals reflected from different parts of the area. It allows to calculate the path traversed by radio waves to the Earth's surface and, consequently, to obtain high-precision information about the absolute elevation of the terrain and the displacement of the surface caused by different factors.

The use of interferometric radar gives the ability to monitor:
- landslides;
- dynamics of process of wetting of the soil massifs;
- subsidence and suffosion deformations;
vertical displacements of buildings, structures, pavements, etc.

To monitor the city of Rostov-on-Don was used remote sensing data of the radar satellites COSMO-SkyMed-1-4 for the period from may 2011 to November 2013. The image processing performed by the software ENVI and SARscape Intererometric Stacking. As a result of processing of images of the array 3000-30000 points on m^2 and evaluated the dynamics of the displacements of points in millimeters.

Fig. 4. Pockets of vertical offsets of the city of Rostov-on-Don, obtained on the basis of data of interferometric photography. Red dots indicate significant subsidence of the earth's surface

On the basis of remote sensing it is possible to draw the conclusions about the possibility of further building development and feasibility of construction in the area based on the information of the vertical displacements of the earth surface. Comparative analysis of the territories complex development of the city of Rostov-on-Don on the basis of interferometric photography is presented in table 3.

https://doi.org/10.1051/e3sconf/20183302064

Table 3. Analysis of data of remote sensing for integrated development of territories

The name of the area	Interferometric data shooting	Conclusion on the basis of interferometric data
Territory No 1 (Leninsky district)		There are intensive vertical displacements of the earth's surface, buildings and structures
Territory No 2 (Kirovskiy/ Proletarian district)	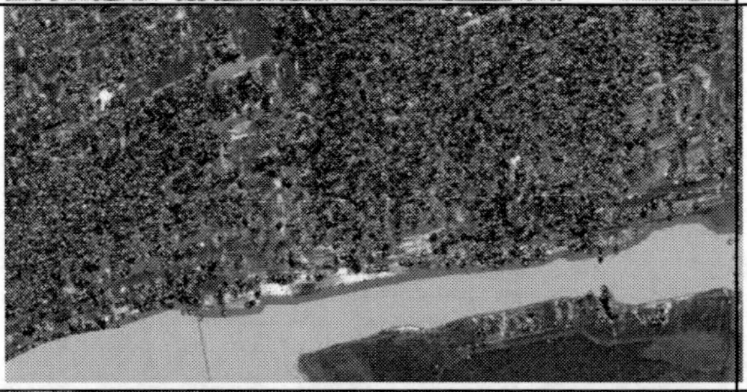	There are vertical displacements of the earth's surface, buildings and constructions of different degrees of intensity.
Territory No 3 (Voroshilovskiy district)		Vertical offset discovered only on the territory of industrial enterprises, planned for the suburb.

Territory No 4 (Proletarian district)		Vertical offset discovered in the territory of an economic building and a residential building.

Step 7: Refined estimation of engineering-geological conditions by using geophysical research methods

For an updated assessment of engineering-geological conditions of territories of complex development, one should choose methods that are not accompanied by high noise or vibration impacts. These methods include ground-penetrating radar and seismic low-power oscillators. The use of these methods does not require space for the deployment of equipment, and can be used in dense urban areas. These methods provide the opportunity to assess the volume status of the soil, its properties and structure, to determine the representativity of the drilling and to compare the results of sampling soil and spot methods.

GPR method is based on the emission of pulses of electromagnetic waves and detecting signals reflected from different objects in the sensed environment. The result of the GPR studies is the radargram. The horizontal axis of radargram — axis profile (in meters), the vertical axis of radargram — time axis beginning at the time of sending the probe pulse (figure 5).

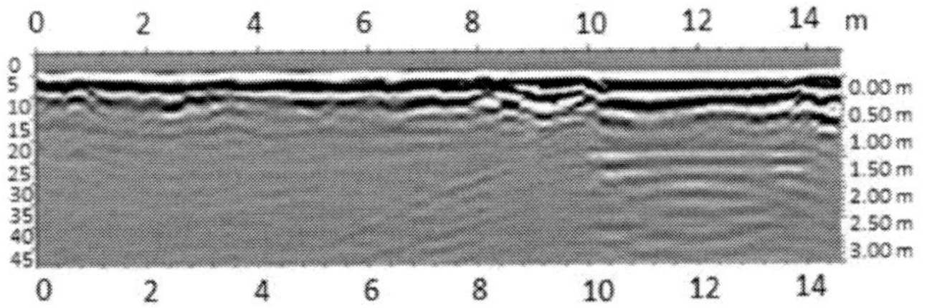

Fig. 5. Radargram

The principle of operation of the seismic low-power oscillators is based on the discharge of cargo from a height of 6 – 8 meters. The moment of contact of the load surface of the soil is marked by a special device, and is transmitted as marking the beginning of the excitation on the seismic station. Seismic exploration allows to evaluate the physical properties of the soil, which can then confirm the verification by drilling.

In the territories of the integrated development was used both geophysical method of investigation, which resulted in the specification of the engineering geological conditions of territories complex development and definition of the boundary of occurrence of loess forest subsidence of soils of different types. After receipt of finalized results were built GIS electronic map in ArcGIS 10.1, which will be applied as the basis for selecting the priority areas of complex development.

Stage 8-9: Assessment of need and the volume of demolition and resettlement

For assessing the need for relocation and demolition in the territories of complex development it is necessary to estimate the number of existing buildings and their functionality. For the evaluation based on electronic maps and satellite images of Rostov-on-Don, which revealed the following:

Table 4. Number of buildings of various functional purposes in the territories of the integrated development

	Territory No1	Territory No2	Territory No3	Territory No4
Residential building	0	>100 buildings	0	0
Industrial buildings	30-40 buildings	20-30 buildings	32 buildings	23 buildings
Administrative building	15 buildings	0	2 buildings	18 buildings

Thus, the greatest levels of relocation and demolition are required on territory No. 2 due to the large number of residential buildings (100 buildings).

Stage 10: The revised estimate of territories of complex development, taking into account additional factors

As additional factors to the revised site assessment, taking into account the unique construction of high-rise buildings, the following additional indicators were pointed out:

• Analysis of the need for resettlement and demolition;

• Analysis of the geotechnical situation;

• Availability of residential complexes on the territories of integral development;

•The possibility of building of high-rise complex and unique buildings;

At stages 6-9 the analysis of the necessity of relocation and demolition, geotechnical situation was fulfilled.

To analyze of the availability of residential complexes it is required the usage of electronic cards of Rostov-on-Don, as well as assessing the real estate market.

Table 5. Analysis of the availability of residential complexes on the territories of the integrated development

	Territory №1	Territory №2	Territory №3	Territory №4
The number of residential complexes (2017)	0	3 "Aquamarin", "On the Sedov street","Quiet Don"	2 "Izmail Park", "Horizon "	2 "Cascade, New town "

Thus, the greatest number of residential complexes identified in No. 2.

To assess the possibility of building in the territories of integrated development of complex high-rise buildings is necessary from the point of view of the feasibility of placing the building of high-rise buildings in the area. When selecting property one must take into account the architectural appearance of buildings and structures. The construction of a detached high-rise unique building could disrupt the architectural integrity of the urban composition, in this connection, it is expedient to build a complex of high-rise residential buildings. During the construction of the unique complex of high-rise buildings should consider the idea of creating a demonstration zone high-tech, zone of superior comfort or a whole district of high-rise and unique buildings.

For the final updated assessment with additional factors applied to the same sequence of actions as in the comprehensive assessment areas, using the main factors. As the evaluation factors are 4 additional indicator, and then calculates successively the formulas (1), (2), (3) and (4).

The results of the calculation are summarized in table 6.

Table 6. Rating of functional priority territories for integrated development high-rise and unique construction

№п/п	The rating of high-rise and unique buildings depending on the function	
	Residential area	*Public-business zone*
Territory No 1	10.82	10.90
Territory No 2	11.10	11.69
Territory No 3	13.12	12.27
Territory No 4	12.00	11.31

3 Results

Stage 11: Selection of areas of integrated development with the highest rating
According to the functional priority for building unique buildings, calculated in stage 10, the highest priority was the area No. 3.

Fig. 6. The selected area of comprehensive development

Territory No. 3 is located in the Voroshilov district, the area of 608 thousand m2. This area is located in the area of less dangerous geological and environmental risk, low noise, no landslides and flooding.

4 Conclusions

The developed method has allowed to solve the problem of choosing the most suitable for the construction of high-rise and unique buildings of the territory integrated development. Deciding on the best location for the building has always been an acute problem, both investors and customers in connection with a large number of factors that must be considered when choosing a site for the future construction project. On the other hand, the development of the construction industry and the emergence of new types of buildings require review of standard procedures for the selection of site for facility construction in connection with the introduction of new building technologies, new materials, new requirements of modern facilities. Unique high-rise building, like any new branch of the construction industry, has its own characteristics, norms and requirements. The right choice of site for the construction of tall unique buildings should be based on consideration of factors that are of importance for this type of buildings.

Discussed in the article additional factors were presented on the base of the unique characteristics of high-rise construction in general. The final rating score assigned to the territories of complex development, can be used as justification for selecting this site, and may change depending on user settings. Methodology choice of site complex development, designed for high-rise and unique construction will allow investors and customers to assess not only environmental, geological and other standard risk factors, but also take into account the investment attractiveness of the future of the unique object to the proposed site.

References

1. The town-planning code of the Russian Federation of 29.12.2004 N 190-FL (ed. from 29.07.2017). Article 46.10. Integrated development on the initiative of local authorities. –FL of 03.07.2016 N 373-FL.
2. The city-planning code of the Russian Federation of 29.12.2004 N 190-FL (ed. from 29.07.2017). Article 46.9. Integrated development on the initiative of the authorities land owners and (or) located on them of objects of real estate. –FL of 03.07.2016 N 373-FL.
3. S. G. Sheina, L. L. Babenko, R. B. Matveyko, A. A. Kamalova. *Complex assessment of the territory in urban construction* (Rostov State University of Civil Engineering, 2015)
4. S. G. Sheina, L. V. Gyria. *Organizational and technological support of the reconstruction of urban development with consideration of ecological monitoring of the territory* (Rostov State University of Civil Engineering, Rostov n/D, 2012)
5. S. S. Bachurina. *Metropolis: methods and models for process control of complex reconstruction of existing buildings* (SINTEG, Moscow, 2004)
6. L. B. Zelentsov, S. G. Sheina. *The control system of quality of reconstruction of urban development* (Rostov State University of Civil Engineering, Rostov n/D, 2005)
7. V. F. Kasyanov. *Reconstruction of the residential development of cities*. Textbook. (Association of building Universities, Moscow, 2005)
8. Russian Standard SNiP 2.07.01-89* Urban development planning and construction of urban and rural settlements.
9. N. P. Shepelev, M. S. Shumilov. *The reconstruction of urban development*. Textbook. (Highest school, Moscow, 2003)
10. M. A. Gyria. *Organizational and technological support of the reconstruction of urban development on the basis of state monitoring of the geological environment*. PhD Thesis, Rostov n/D, (2007)
11. A. V. Gridnevsky. *Methods of collecting and processing information for monitoring of geodynamic hazards of urban areas*. (Rostov State University of Civil Engineering, 2016)
12. A. V. Grinevskii, S. G. Sheina. *Monitoring and assessment of geological hazards of the territories of the Rostov region* (Rostov State University of Civil Engineering (2015)

13. A. V. Gridnevsky. Materials of year session of scientific Council of RAS on problems of geoecology, engineering geology and hydrogeology, **14** (2012)
14. A. V. Gridnevsky. Sergeev readings: materials of annual session of RAS Scientific Council on problems of Geoecology, engineering Geology and hydrogeology (2014)
15. A. V. Gridnevsky, S. G. Sheina. Possibilities of radar imagery of the earth's surface for monitoring of geodynamic hazards of urban areas (Rostov n/D, Rostov State University of Civil Engineering, 2015)

The use of nanomodified concrete in construction of high-rise buildings

Sergei Prokhorov[1*]

[1]Vladimir State University named after A. and G. Stoletovs, Gorkiy street, 87, Vladimir, 600000, Russia

Abstract. Construction is one of the leading economy sectors. Currently, concrete is the basis of most of the structural elements, without which it is impossible to imagine the construction of a single building or facility. Their strength, reinforcement and the period of concrete lifetime are determined at the design stage,taking into account long-term operation. However, in real life, the number of impacts that affects the structural strength is pretty high. In some cases, they are random and do not have standardized values. This is especially true in the construction and exploitation of high-rise buildings and structures. Unlike the multi-storey buildings, they experience significant loads already at the stage of erection, as they support load-lifting mechanisms, formwork systems, workers, etc. The purpose of the presented article is to develop a methodology for estimating the internal fatigue of concrete structures based on changes in their electrical conductivity.

1 Introduction

The main requirements for reinforced concrete structures during construction and operation period are strength and resilience. [1]. When the concreting works, hardening rate of concrete and its real bearing capacity can be determined by various methods, including non-destructive. However, it is well known that the physical and mechanical properties of building structures changes under the action of various atmospheric, climatic, technological, human and other factors. Nondestructive testing methods are most often used to assess these changes and determine the actual strength [2,3,4]

However, these techniques have a number of disadvantages:

1. Inability of constant control over the state of structures.

Many constructions are based in inaccessible places or hidden under the finish coatings, which significantly complicates the assessment of their technical condition and makes it impossible to use the equipment;

2. Lack of opportunity to track changes in the fatigue state and the integrity of the structures at the micro level. In high-rise buildings high-strength concrete and steel are applied, however it should be noted that an increase in the strength of concrete increases its fragility and reduces its plastic-deformation properties, which entails an almost instantaneous destruction of the material when it reaches the limit state. [5,6] Therefore, the problem of determining (revealing) critical fatigue at an early stage is very acute.

* Corresponding author: oc204@bk.ru

3. Some of the methods requires special preparation of the surface;

4. Methods have a local character of strength determination. [7]

The engineering systems of high-rise buildings widely use remote control of equipment and its condition, based on the collection and transfer of information to the dispatch center. This allows to diagnose malfunctions at an early stage and enables the use of preventive measures or let to prepare for the repair more carefully. [8,9]

The determination of internal fatigue according to the change in electrical resistivity of concrete, can become an alternative to existing methods.

Attempts to determine the internal fatigue in the concrete on the basis of changes in conductivity have been made long time ago. Adding modifying additives which fill the voids at the molecular level and turn the concrete from, practically, the dielectric into an electroconductive structure.[10,11] Among the most available, there can be distinguished method based on the determination of the electrical conductivity according to the transmission of electric current through the body of the structure, via electrodes located on one side of the structure. [12,13] When using this method it's rather difficult to make an adequate estimate of the structure's state, due to the low convergence of the results. Furthermore it was found that the electrical resistance is strongly dependent on the water saturation of concrete, and as a result of current transmission, polarization of free ions occur, which leads to a distortion of results.

Another method of investigating the stress-strain state of concrete is tomography of electrical resistance, when 2 or 4 w electrodes through which voltage is applied connected to the sample, and fix changes in electrical resistance under load. The method is based on measurement of the internal fields due to the piezoresistance effect. [14,15]

2 Materials and methods

The researches made on the base on Vladimir State University showed that the results of the change in the electrical conductivity of concrete samples depend on the concentration of the modifier (carbon pipes), electrode material, their configuration (there were investigated plate, perforated, mesh electrodes), and the depth of immersion.

Tests of fine-grained concrete were conducted on samples of cubes with edge size of 7 cm and beams with dimensions 4 × 4 × 16 cm in accordance with GOST 10180 - 2013. Portland cement PC-TO-500-H manufactured by JSC "Mordovcement" was used as binder, and Quarry quartz sand was used as fine aggregate for obtaining samples. To prepare the mixture, distilled water was used in accordance with GOST 6709-72 in the volume of water-cement ratio 0.5.Multilayered carbon nanotubes were used as a nano-modifying additive. Except increasing the strength of elements, the use of nano additives makes it possible to increase the electrical conductivity of concrete and to use it for internal fatigue estimation. At the stage of preparing carbon tubes 2 mkn size are added to the a concrete. The preparation of the dispersion was carried out on a SONOPULS HD 3200 unit by ultrasonic treatment with a flat titanium probe TT 13, 13 mm in diameter.The treatment was carried out for 5-30 minutes at amplitude values of 10, 50 and 100%. The process was carried out in a glass beaker with constant cooling.
The temperature of the treated dispersion was controlled by means of a thermocouple entering.To uniformly distribute the nanomodified additives in the suspension, polycarboxylate hyper-plasticizer P-11 was used in the amount of 0.25% of the cement mass.

At the first stage of the research there were conducted experiments to select the qualitative and quantitative composition of the concrete mixture. The fine-grained concrete with a water-cement ratio of 0.6 based on portland cement M 500 served as a standard. Samples were prepared with a percentage of carbon pipes 2 mkn in length 0.01-0.13% by weight of cement.

During the experiment, the strength of the samples and the electrical resistance at the age of 1, 3, 7, 14 and 28 days were determined.

The next step was to analyze the change in the electrical conductivity of concrete samples, depending on the material and method of the electrodes installation.

Three types of electrodes were tested:

lamellar solid (1 type)

grid with cells 0.5 × 0.5mm (2 type)

perforated with a cell size of 2x2mm (type 3)

Before installation, the electrodes were stripped and embedded in the samples at the manufacturing stage.Indications were taken before the tests and during the loading to failure.The results were recorded with help of multimeter Mastech MS8040 and video fixation (Fig.1)Compression and bending tests was carried out on a hydraulic press BM-3,5.

Fig. 1. Testing of samples

3 Results

The results of the effect of the percentage of nano modifying additive on strength and electrical resistance, which were produced during the research, are given in Table 1.

Table 1. The change in the strength characteristics and electrical resistivity of concrete, depending on the content of the nano modifying additive

Content of nano modifying additive,%	Storage duration					
	1day		day 14		28 days	
	Rc, MPa	R, MOhm	Rc, MPa	R, MOhm	Rc, MPa	R, MOhm
0 (reference)	22	1,5	49	4,5	55	7,3
0.01	26	1,5	65	4,5	73	7,1
0.02	38	1,5	68	4,3	76	7,1
0.03	31	1,3	71	4,3	77	7
0.05	33	1,2	72	4,2	79	6,8
0.07	34	1,2	73	4,0	82	6,7
0.10	36	1,1	78	3,8	88	6,5
0.13	38	1,1	71	3,8	76	6,5

As can be seen from the table adding of carbon tubes not only improves the strength of concrete, but also increases its conductivity. This is explained by the creation of a more accurate pore size distribution and replacement of voids by nanomodified additives, which leads to a decrease in the resistivity. However, it should be noted that by increasing the concentration above 0.1% we get the decrease in strength of laboratory samples. Subsequently, tests were performed on all samples with a content of 0.1% carbon nanotubes by weight of cement.

Investigations of the effect of materials and electrode types on the electrical resistivity of concrete were carried out under a load of compression and bending.

The dependence of the change in the electrical resistivity of the control sample during bending is shown in Fig. 2.

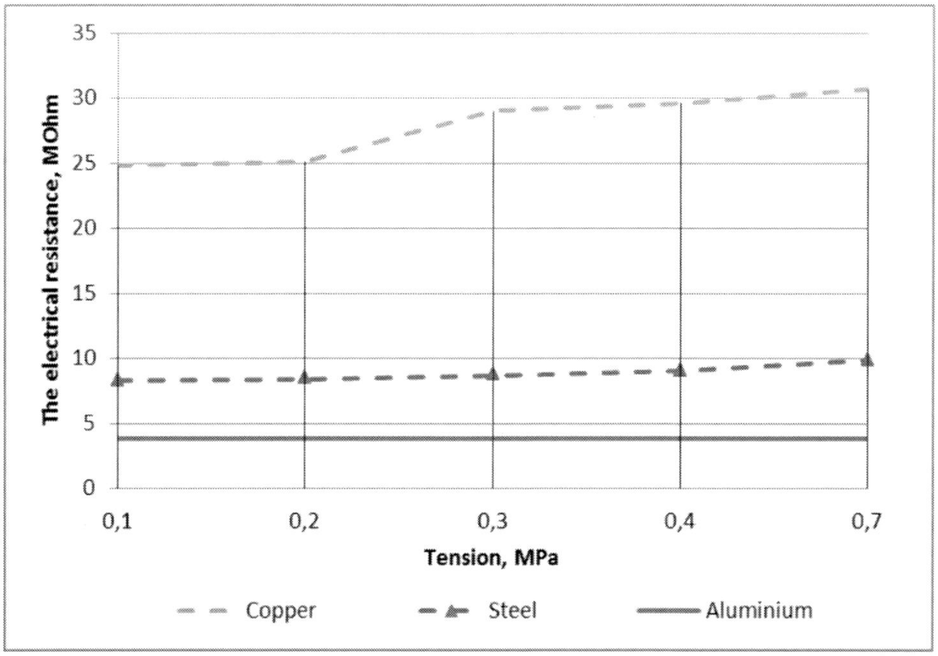

Fig. 2. The dependence of the change in the electrical resistivity of the control sample during bending

As can be seen from the graph (Fig. 2), the greatest change in electrical conductivity occurs on electrodes of type 3, when bending.This is connected to the smaller specific surface area of the electrode located in the compressed zone, and when the microcracks appear, the electrical resistivity begins to increase. At the same time, when using the electrodes of type 2, the change in electrical resistivity is insignificant due to better integration into the body of the sample, whereby the electrode moves together with the concrete without breaking the circuit.

In the graph of conductivity analysis using solid electrodes (type 1), it can be noted that the increase in electrical resistance occurs at voltages close to the limit. This is due to loss of a contact interface between the concrete and electrode, and the appearance of microcracks in the contact zone.

During the compression test of the sample, the overall picture of changes in the electrical conductivity is retained, but it takes a distinct character (Figure 3).

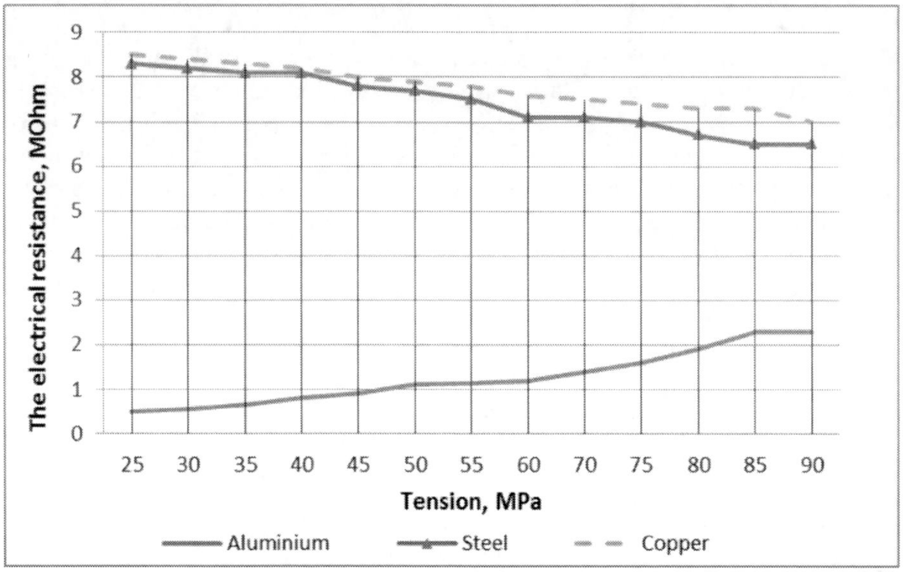

Fig. 3. Dependence of the change in the electrical resistivity of a control sample during compression.

In particular, type 1 and type 3 electrodes behave almost identically, as when microcracks appear in the compression zone, they quickly lose contact with the concrete, which leads to a sharp increase in the electrical resistivity. Electrodes of the second type show a smooth increase in electrical resistance, depending on the load, due to more deep integration into the body of the sample. In this case, the initial resistance is less than the resistance at destruction in 4 times, and for electrodes of the second type in 5 times.

The results showed that the specific resistance of concrete depends significantly on the resistance of the contact itself and the density of its adhesion to concrete.

If there is a gap between the electrode and the concrete, the contact area decreases and the resistance increases. Furthermore, there is a polarization of ions around the electrode, which entails a concrete polarization. This leads to an underestimation of the actual value.

When analyzing the change in the total electrical conductivity, depending on the period of concrete hardening , it can be noted that it decreases in the range of 40-150%. This is explained by the contraction of free ions in the pores of K +, Na +, O- and less in Ca2 + and SO2.

When testing samples by cyclic loading, nano modified concrete shows the repeatability of electrical resistance values, while ordinary concrete has a significant spread after the load is removed (partially or fully), followed by loading again.

4 Discussions

In general, the researches have shown that the introduction of nano additives into concrete makes it possible to study the change in the stress-strain state of structures on the basis of electrical resistivity, more fully. During the hardening, microvoids in cement stone are filled with carbon pipes, which leads to an increase in strength by 10% on one hand, and on the other, to the formation of stable bonds for the passage of electric current along the body of the structure.The change in electrical resistance occurs more evenly.

The obtained results allow us to define the internal voltage in the sample with an accuracy of 10-15%.The work on the adjustment of the composition of the concrete mix and the development of software for the interpretation of the results is in progress, to improve the accuracy. The researches about how reinforcing percentages and the type of reinforcement effects on the change of the electrical conductivity of concrete are also in progress. At the present moment, work is underway to assess the effect of nano additives on electrical resistivity changes in reinforced structures using steel and fiberglass reinforcement.

Conclusions

The researched have shown that the shape of the electrodes and the percentage of nano modifying additives affect the change in the electrical conductivity of concrete. By analyzing these data, it is possible to estimate the change in internal fatigue in the body of a structure in real time. Currently, works are underway to analyze the influence of reinforcement on the change in electrical conductivity and to develop software modules for processing data to determine the magnitude of the fatigue on electrical resistance.

References

1. 1.*GOST 22690-88 Concretes. Determination of strength by mechanical methods of nondestructive testing* (State Construction Committee USSR, 1988)
2. D.Y.Snezhkov, S.N. Leonovich, Construction and reconstruction, **2 (58)**, 152-160 (2015)
3. H. Wiggenhauser, H. Azar,i Infrastruct Syst [Internet], **23(4)** (2017)
4. A.V. Ulybin, Civil Engineering Magazine, **4**, 10-15 (2011)
5. A.V.Petukhov, M.O. Korovkin, N.A. Eroshkina, Modern scientific research and innovation. **№ 3** (2017) [Electronic resource]. URL: http://web.snauka.ru/issues/2017/03/79416 (reference date: 11.12.2017).
6. V.R. Kulkarni Special Issue 20, **5**, 26-33 (2016)
7. V. Stengel, Insp. of build. and struct.: probl. and ways of their solut. Proceed. of the VI Internat. Scienti. and Pract. Conf., 189-195 (2015)
8. E.V. Moskvin, A.S. Glazyrin, S.V. Gorodilov, A.V. Putov, Proceedings of the ETU ETU, **6,** 49-53 (2016)
9. A.V. Demenev, N.L. Pavlov, Sciences of Europe, **6-2 (6)**, 23-27 (2016)
10. A.V. Rizhenko; V.H. Rizhenko, S.V. Lankin International Research Journal, **8-3 (50)**, 86-89 (2016)
11. L.A. Urhanova, S.A. Lhasaranov, S.P. Bardakhanov, Building materials, **8**, 52-55 (2014)
12. L.Courard, F. Michel, Cement and Concrete Composites,**45**, 111-116 (2014)
13. I.You, D .Yoo, & G. Zi, Sensors (Switzerland), **17** (2017)
14. K. Kawaai, I.Ujike & H.Okuno, ISEC 2017, (2017)
15. A. Asgharzadeh, K. Reichling, M. Raupach, Concrete Solutions,181-186 (2016)

Prospects for development of heat supply systems in high-rise districts

Viktor Zhila[1], *Elena* Solovyeva[1*]

[1]Moscow State University of Civil Engineering, Yaroslavskoe shosse, 26, Moscow, 129337, Russia

Abstract. The article analyzes the main advantages and disadvantages of centralized and decentralized heat supply systems in high-rise districts. The main schemes of centralized heat supply systems are considered. They include centralized heat supply from boiler houses, centralized heat supply from autonomous heat sources, heat supply from roof boiler houses and door-to-door heating supply. For each of these variant, the gas distribution systems are considered and analyzed. These systems vary depending on the heat source location. For each of these systems, technical and economic indicators are taken into account, the analysis of which allows choosing the best option for districts where high-rise buildings predominate.

1 Introduction

The orientation of the Russian energy sector to district heating and centralized heat supply as the main way to meet the heat needs of cities and industrial centers did not justify itself technically and economically. Performance of district heating and heating systems is improper, there are a lot of unsuccessful technical solutions, unused reserves, which reduce the economy and reliability of these systems. Especially it is necessary to pay attention during heat supply of high-rise buildings.

The industrial character of the structure of district heating systems (DHS) with combined heat and power plant (CHP) and boiler houses, the groundlessness of the scale of onsumers connection and the practical uncontrollability of the operating modes of the DHS (sources - heating networks - consumers) have largely devalued the advantages of centralized heat supply. If the sources of thermal energy are still comparable with the world level, the analysis of the DHS shows the following

- The technical equipment and the level of technological solutions for the construction of heat networks correspond to the state of the 1960s, while the radii of heat supply increased sharply, and a change to the new standard sizes of pipe diameters took place;
- The quality of the metal heat pipes, heat insulation, shutoff and control valves, construction is much inferior to foreign analogues, which leads to large losses of thermal energy in the networks;

* Corresponding author solovyva@mail.ru

- Poor thermal and water insulation of pipelines and heating network channels increased the damage of underground heat pipes, which led to serious problems of replacing the equipment of heating networks;
- The domestic equipment of large CHP plants corresponds to the average foreign level of the 1980s, and at present steam turbine CHP plants are characterized by high fault rate, as almost half of the installed capacity of turbines has developed a calculated resource;
- There are no flue gas cleaning systems from NOx and SOx at the operating coal-fired CHP plants, and the efficiency of participate trapping often does not reach the required values. These environmental problems affect the living conditions of people who live in high-rise buildings;
- At the present stage competitiveness of the CHP can be ensured only by new technical solutions, both in the structure of systems, and in the schemes, as well as by supplement of energy sources and heat networks.
- It should be mentioned that traditional modes of operation of district heating in practice have the following disadvantages:
- Problems with regulation of heat supply for heating of buildings during transitional periods, when wind, solar radiation, domestic heat release exert a particularly great influence on the thermal conditions of heated rooms;
- Overrun of fuel and over-heating of buildings in the warm periods of the heating season;
- Large heat losses during its transportation (10% and even more);
- Irrational consumption of electricity for pumping heat carrier, due to the very principle of central quality regulation;
- Long-term operation of the supply pipes of the heating network in an unfavorable temperature regime, characterized by an increase in corrosion processes, etc.

The advantage of centralized heat supply lies in the combined generation of thermal and electrical energy at the CHP plant. The thing is that the increase in the number of autonomous boiler houses will definitely not lead to a decrease in fuel consumption at the CHP plants (provided that the power generation is unchanged) [1]. This indicates that fuel consumption is increasing in the city in general, and the level of air pollution is increasing.

2 Methods of research

The modern system of decentralized heat supply represents a complex set of functionally interconnected equipment, including an autonomous heat generating installation and building engineering systems (hot water, heating and ventilation systems).

Recently, many Russian regions are interested in the energy-efficient technology for door-to-door heat supply of multistorey buildings, that is a kind of decentralized heat supply, when each apartment in a multi-family house is equipped with an autonomous system for heat and hot water providing. The main elements of the apartment heating system are the heating boiler, heating appliances, air supply and flue gas systems. Arrangement of pipe-lines is performed using steel pipes or modern piping systems – plastic or metal-plastic.

Objective prerequisites for the use of autonomous (decentralized) heat supply systems are:

- The absence in some cases of free capacities on centralized sources;
- Increase in number of housing facilities;
- a significant part of the construction is in the area with undeveloped engineering infrastructure;
- Lower capital investment and the possibility of phased thermal loads;

- Comfortable conditions in the apartment at one's own will, which in turn is more attractive in comparison with the centralized heat supply, where temperature depends on the directions on the beginning and the end of the heating period;

- The appearance of a large number of different modifications of domestic and imported heat generators of low power on the market.

Heat generators can be placed in the kitchen, in a separate room on any floor (including attic or basement) or in anaddition.

The most common scheme of autonomous (decentralized) heat supply includes: a single-circuit or dual-circuit boiler, circulation pumps for heating and hot water supply, check valves, closed expansion tanks, safety valves. In the case of a single-circuit boiler, a capacitive boiler or plate heat exchanger is used for the preparation of hot water [2].

Advantages of decentralized heat supply are as follows:

– There is no need in land allocation for heating networks and boiler houses;
– Reduction in heat losses due to lack of external heat networks, in network water losses, as well as the reduction in expenses for water treatment;
– Significant reduction in expenses for repair and maintenance of equipment;
– Full automation of consumption modes.

In autonomous heat supply systems it is not recommended to use unprepared water from the water pipe because of its aggressive impact on the boiler elements, which necessitates the use of filters and other water treatment devices.

Increasing the economy of such a system is achieved by using the heat released during the combustion of natural gas as a source of energy.

Door-to-door heat supply is popular in many European and Asian countries. In Russia multistorey buildings with autonomous heat supply have already been built in Smolensk, Serpukhov, Bryansk, St. Petersburg.

Among the experimental buildings built in the Russian regions, there are both high-rise elite and mass building houses. Their price is higher than of houses with centralized heat supply, but the level of comfort gives advantages in the real estate market, because owners of such houses can decide for themselves on the quantity of heat and hot water.

Decentralized systems of any kind allow excluding energy losses during its transportation (as a result, the cost of heat for the end user is reduced), increasing the reliability of heating and hot water supply systems.

Despite all advantages of decentralized heat supply, there are also negative sides. In small boiler houses, including "roof" ones, the height of the chimneys, as a rule, is much lower than for large ones. With the total equal of heat power, the emission values do not change, but the dispersion conditions deteriorate sharply. In addition, small boiler houses are located, as a rule, near a residential area.

However, the impact of large boiler and CHP emissions on different floors of residential buildings is almost the same; because of concentrations leveling in height occurs depending on the remoteness of the pollution source. The height of the pipes of small boiler houses and the height of apartment houses are commensurable, and therefore emissions are carried out practically at the level of the upper floors.

In the fuel balance of municipal heating, natural gas is about 45%. Because of obsolete technologies, gas combustion is approximately 1.5 times greater than required [3, 4].

Combustion of natural gas as an environmentally friendly fuel should be as close as possible to the consumer that makes it possible to reduce losses in the thermal energy transport. This can be achieved by autonomous and quarterly heat supply and by use of fundamentally new system of quantitative and qualitative regulation, ensuring the coincidence of the modes of heat generation and consumption with the least losses and costs.

Autonomous heat supply can be successfully implemented in the construction of new areas, as well as existing areas of mass construction.

In a number of Russian cities (Smolensk, Serpukhov, Nevinnomyssk, Cheboksary, Yakutsk, Belgorod) apartment heating systems are being introduced. But it leads to some problems [5,6]:

- heating of stairwells and cellars;
- allocation, construction and material of chimneys arrangement of gas meters;
- Location of gas pipelines.

The main disadvantages of centralized heat supply systems:

- poor quality of heat networks leads to heat losses (20...25%);
- high level of costs for operation of heating networks (about 50% of all costs in heat supply systems);
- high degree of deterioration of heating networks and excess of critical level of failure rate;
- overheating of buildings during the transition period;
- impossibility of consumers (population) to determine the quantity, quality and price of service;
- significant overestimation of heat consumption in residential buildings in comparison with the actual;
- low coverage of residential buildings with heat meters;
- low heat insulation characteristics of residential buildings.

In fact, the population is not a buyer of heat, but only a consumer, it can not control the volume, quality and price of heat energy. It should be noted that the heat supply in Russia have been developed gearing to district heating introduction (combined heat and power generation). Very often, residential areas located near a large industrial hub were supplied with heat from a centralized source. The economic crisis of the last 15 years has led to an industrial recession.

Advantages of autonomous heat supply systems are as follows:

- there is no need to alienate land for construction of a heat source and networks;
- reduction of heat losses due to lack of external heat networks, reduction of network water losses, as well as of costs for water treatment;
- a significant reduction in the cost for equipment repair and maintenance;
- full automation of consumption modes.

3 Results of research

The heat supply system includes a source, heat networks, switchgears (IHP-individual heat point, CHC – central heating center) and a consumer.

One of the variants for decentralized heat supply is boiler stations located on the roof. In this case there are no heat networks, no problems with additional land allocation and installation of a high chimney. In the boiler station located on the roof, lightweight structures with aluminum heat exchangers or copper finned tubes are used. It is possible because of the favorable hydraulic operating conditions of the equipment, because of static pressure lack.

Roof boiler stations provide maintenance of a necessary temperature mode for heating of an individual building.

The technology of the "floating floor" of the boiler room and the installation of vibration compensators on the supply and discharge pipelines, makes it possible to provide noise control and reduce vibration forced by operating equipment.

Hot-water boilers operating on natural low pressure (5 kPa) gas with a coolant temperature up to 115 ° C is used in roof boiler stations.

Constant presence of maintenance personnel is not needed in boiler stations because of high level of automation.

At buildings heating from roof boiler stations it is possible to support necessary comfortable temperature of air by selection of the individual schedule of temperatures of the water submitted in a heating network.

The roof boiler station works in a mode of 70…75% of the rated capacity in the heating season and in a mode of 25% in summer. This makes it possible to carry out preventive and repair work with a partial shutdown of thermal capacities in any period [7].

The time for preventive maintenance of roof boiler stations is no more than 3 days. At the same time, with centralized heat supply, the system switches off for 3 weeks.

To reduce the extent of heat networks, autonomous boiler stations are used in the attached construction (directly adjacent to the building).

Discussion

After the collection of all necessary information, a comparison of four heat supply systems was carried out on the example of the Kurkino district of Moscow. In this case, all apartments are equipped with electric cookers.

Variant 1 – centralized heating from boiler stations.

Variant 2 - centralized heating from autonomous heat sources.

Variant 3 - decentralized heating from boiler stations located on the roof.

Variant 4 – door-to-door heating.

In the variant 1, a district heating system is developed, where the heat source is a boiler station, from which two pipes are provided for heating networks to the central heating station, and after the central heating station there are four pipes for heating and hot water supply. In this case, the gas is supplied to the boiler station.

In the variant 4, a local heat source is installed in the apartment, which ensures the supply of heat to the heating and hot water supply systems.

In this scheme, a two-stage gas supply system is proposed:

first stage – medium-pressure gas pipeline, which is laid in the residential district; in each house a cabinet control point is installed;

second stage - low-pressure domestic gas pipelines. Gas is supplied only to a local source of heat.

Variants 2 and 3 are intermediate between variants 1 and 4. In the second case, autonomous heat sources are used as a source of heat, from which two pipes are laid from autonomous heat sources to individual heat point, and there are four pipes for heating and hot water supply from individual heat point [8].

In this case, gas is supplied to autonomous heat sources via medium-pressure gas pipelines. In the variant 3, a relatively low power (from 300 to 1000 kW) boiler station located on the roof is used as a heat source.

The gas pipeline to the boiler station is laid along the outer wall of the building, in places convenient for servicing and excluding the possibility of its damage [9].

Variants of heat supply systems are presented in the Figure 1.

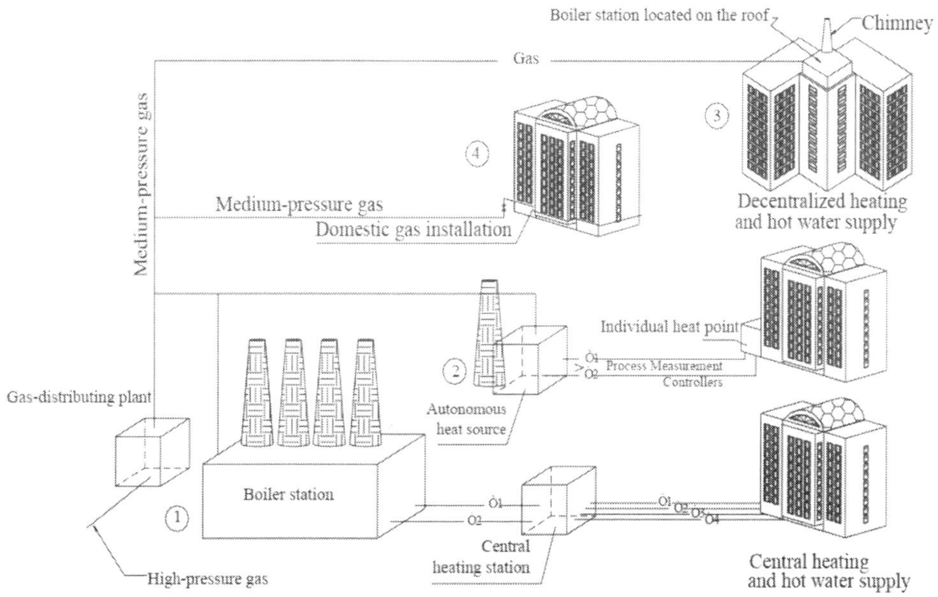

Fig. 1. Schematic diagrams of heat supply systems

Technical solutions for heat supply according to the several variants should be made on the basis of technical and economic calculations. The best variant can be determined by comparing possible solutions [5].

The most expensive variant of heat supply is the variant 1 - centralized heat supply from the boiler station. In this case, a large part of the costs are accounted for heat networks taking into account the central heating station, which is 63.8% of the total cost of the whole system. Among them, 84.5% account for the laying of only heat networks. The cost of the heat source is 34.7%, of gas networks accounting for gas-distributing plant and cabinet-type gas control point is 1.6% of the total amount for the system [7,10].

Variant 4 is only 4.2% cheaper than the variant 1 (Figure 2). Hence, they can be accepted as interchangeable. If in the variant 1 the most part of the costs are heat networks, then in the variant 4 it belongs to the heat source, i.e. wall boilers (62.14% of the total cost of the whole system) [11].

It is worth paying attention to the other two variants: roof boiler stations and autonomous heat sources.

Concerning economics, the variant 2 is more profitable, that is, centralized heat supply from autonomous heat sources. In this variant, the most part of the costs is accounted for heat networks with regard to individual heating station, which is 67.3% of the total cost of the system. Among them, the heat networks account for 20.3%, 79.7% - individual heating stations. The cost of the heat source is 26%, the share of gas networks, taking into account gas-distributing plant and cabinet-type gas control point is 6.7% of the total amount of the system.

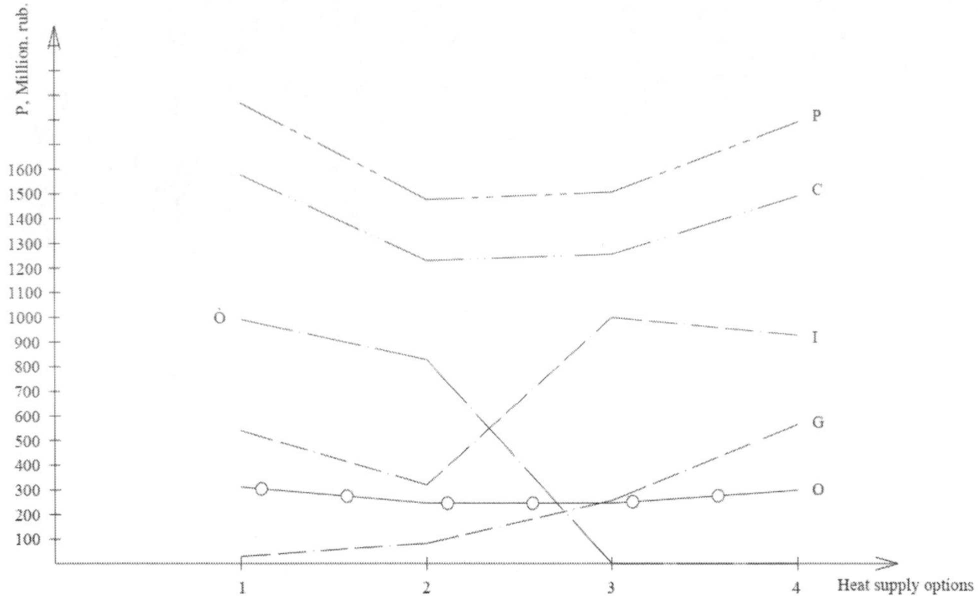

Fig. 2. The schedule of the resulted expenses:
P - resulted expenses; K – capital cost; I - capital cost for heat sources; T – price of heat networks taking into account central heating station/ individual heating station; G - capital investments in gas network taking into account the price of gas-distributing plant/cabinet-type gas control point

The cost of laying the pipelines of the heat supply system depends on the extent of the heat networks. Consequently, the approach of a heat source that runs on gas to the consumer by installing attached, built-in, roof and individual heat generators will significantly reduce the cost of the system. In addition, statistics show that most of the failures of centralized heating system accrue to heating networks, which means that the reduction of the length of heat networks will increase the reliability of the heat supply system [12].

Since heat supply in Russia is of great social importance, increasing its reliability, quality and economy is the most important task. Any disruptions in providing the population and other consumers with heat have a negative impact on the country's economy and increase social tension.

In the current tense situation, it is necessary to implement resource-saving technologies.

In addition, in order to improve the reliability of the pipelines, pre-insulated pipelines of the non-channel gasket with polyurethane foam insulation in the polyethylene jacket (pipe in the pipe) must be used [13, 14].

The essence of the housing and communal services reform should be not the increase in tariffs, but the regulation of the rights and obligations of the consumer and the heat producer. It is necessary to harmonize the legal and regulatory issues and to develop the basis for technological regulation. All conditions of economic attractiveness for investments should be created.

4 Conclusion

The analysis of the presented results on the study of heat supply systems makes it possible to choose the optimal option for high-rise buildings, which allow introducing resource-saving measures and creating conditions for attracting investments.

References

1. P.A. Khavanov, N.A. Kharlamova, Ye.B. Solov'yeva. Principal moments in the development of thermal mechanical parts of autonomous heat generating plants. In: Nauchnoe obozrenie. **5**, 297-300 (2012).
2. Energy, Technology, Perspectives 2008. Scenarios and strategies to 2050. International energy agency. OECD/IEA (2008).
3. P.A. Khavanov *Sources of heat of autonomous heat supply systems* (Moscow, 2014).
4. URL: http://www.minstroyrf.ru/ (25.04.2017)
5. Ye.Ye. Novgorodskiy et al. Integrated energy-technological use of gas and protection of the air. Moscow, 368 (1997).
6. URL: http://www.minstroyrf.ru/ (25.04.2017)
7. URL: http://base.garant.ru/ (25.04.2017)
8. D. Butterworth Simplified Methods for Condensation on a Vertical Surface with Vapour Shear. UKAEA Rept. AERE-R9683 (1981).
9. URL: http://www.consultant.ru/ (25.04.2017)
10. V.A. Yakovlev Report at the Government of the Russian Federation Meeting. In Heat Supply News, **6** (2006).
11. L.D. Boguslavskiy, A.A., Simonova M.F. Mitin Economy of heat and gas supply and ventilation. Moscow, 351 (1988).
12. Ye.B. Solov'yeva, A.A. Malysheva. Selection of the optimal trace of the distribution systems of heat and gas supply. In: Agrarian Scientific Journal. **11**. 48-50 (2015)
13. N.A., Kharlamova T.S. Pirogova. The process of heat spreading in pipelines as the basis of the "heat wave" method. In. Nauchnoe obozrenie. **10**. 53-56 (2015).
14. N.A. Kharlamova, T.S. Pirogova. The "heat wave" method as a method of determining the actual heat losses. In: Nauchnoe obozrenie. **10**. 66-69 (2015).
15. A. Ya. Sharipov. Autonomous Heat Supply for Highrise Buildings and Complexes. ABOK. **3** (2016)

Conducting experimental investigations of wind influence on high-rise constructions

Olga I. Poddaeva[1], *Anastasia N.* Fedosova[1], *Pavel S.* Churin[1] and *Julia S.* Gribach[1,*]

[1]Moscow State University of Civil Engineering, Yaroslavskoe shosse, 26, Moscow, 129337, Russia

Abstract. The design of buildings with a height of more than 100 meters is accompanied by strict control in determining the external loads and the subsequent calculation of building structures, which is due to the uniqueness of these facilities. An important factor, the impact of which must be carefully studied at the stage of development of project documentation, is the wind. This work is devoted to the problem of studying the wind impact on buildings above 100 meters. In the article the technique of carrying out of experimental researches of wind influence on high-rise buildings and constructions, developed in the Educational-research-and-production laboratory on aerodynamic and aeroacoustic tests of building designs of NRU MGSU is presented. The publication contains a description of the main stages of the implementation of wind tunnel tests. The article presents the approbation of the methodology, based on the presented algorithm, on the example of a high-rise building under construction. This paper reflects the key requirements that are established at different stages of performing wind impact studies, as well as the results obtained, including the average values of the aerodynamic pressure coefficients, total forces and aerodynamic drag coefficients. Based on the results of the work, conclusions are presented.

1 Introduction

The construction of high-rise complexes is one of the priority directions in the field of construction in the modern world. Designing skyscrapers allows you to competently form the surrounding space, save territorial resources. The development and progression of technologies makes it possible to create unique architectural masterpieces (Figure 1), which have no analogues in the world [1-5].

However, the ratio of objects to the class of increased responsibility requires a competent approach to ensuring the reliability and safety of construction projects in the development of project documentation. One of the most influential external factors in the construction of high-rise buildings is the wind. Air flows often have an adverse effect on skyscrapers, which entails deformation of the structures [2,6-9]. In the current regulatory documents (SP 20.13330.2016 "Loads and impacts") it is established that in order to ensure the reliability and safety of high-rise buildings, a special assessment of the wind regime and

* Corresponding author: js-995@mail.ru

aerodynamic parameters should be included in the design documentation, including the average and pulsation components of the estimated wind load, maximum values of wind load, resonance vortex excitation, etc. However, the determination of these values is possible only based on testing the layout in high-rise objects in wind tunnels of architectural and construction type [3,4,10].

Fig. 1. High-rise buildings in Moscow (Moscow-city)

The purpose of this work is to describe the methodology for carrying out experimental studies of wind impact on high-rise buildings and structures to ensure their reliability and safety of objects, developed at the Research and Production Laboratory for Aerodynamic and Aeroacoustic Testing of Building Structures of the NRU MGSU [1-3], and its approbation on the example of a projected residential located in the city of Tyumen.

2 Methods of research

In this paper, a technique is presented for carrying out experimental studies of wind effects on high-rise buildings, developed and tested in the Educational and Research Laboratory for Aerodynamic and Aeroacoustic Testing of Building Structures of the Moscow State University of Civil Engineering.

In carrying out experimental studies of wind effects on buildings higher than 100 meters is to be executed the following steps:

• Climatic analysis of the territory, including the analysis of topographic and meteorological data: to assess wind loads, the nature of the variation in the average wind speed in height, the structure of atmospheric turbulence, the dependence of average velocities and turbulence on the roughness of the underlying surface of the terrain.

• Design and creation of the layout on a reduced scale, including an assessment of the geometric characteristics of the construction site, installation of measuring equipment for determining wind characteristics.

• Experimental research in a specialized wind tunnel and processing of the obtained results.

• Analysis of the received data, and development of conclusions and recommendations for ensuring the reliability and safety of structures.

Experimental studies are carried out on the basis of NRU MGSU. As a test bench was used a large gradient wind tunnel (Figure 2) of an architectural and construction type with a working zone length of 18.9 m, which makes it possible to model the surface layer of the atmosphere. Figures and tables, as originals of good quality and well contrasted, are to be in their final form, ready for reproduction, pasted in the appropriate place in the text. Try to ensure that the size of the text in your figures is approximately the same size as the main text (10 point). Try to ensure that lines are no thinner than 0.25 point.

Fig. 2. Large Gradient Wind Tunnel NRU MGSU

3 Approbation

Approbation of this technique was carried out on the example of a projected multifunctional business center, located in Tyumen (Figure 3, a)). The building of the business center is a tower with a height of 110 meters. In terms of construction, the structure has the form of a rectangle with a rounded side with dimensions in the plan of 33.6 and 32.6 meters.

As already mentioned above, the climate analysis of the area is primarily carried out. Based on the climatic analysis, the type of terrain is selected in accordance with the requirements of SP 20.13330.2016, and the corresponding incident velocity profile is calculated. In this case, the type of development site corresponds to type "B".

a) b)

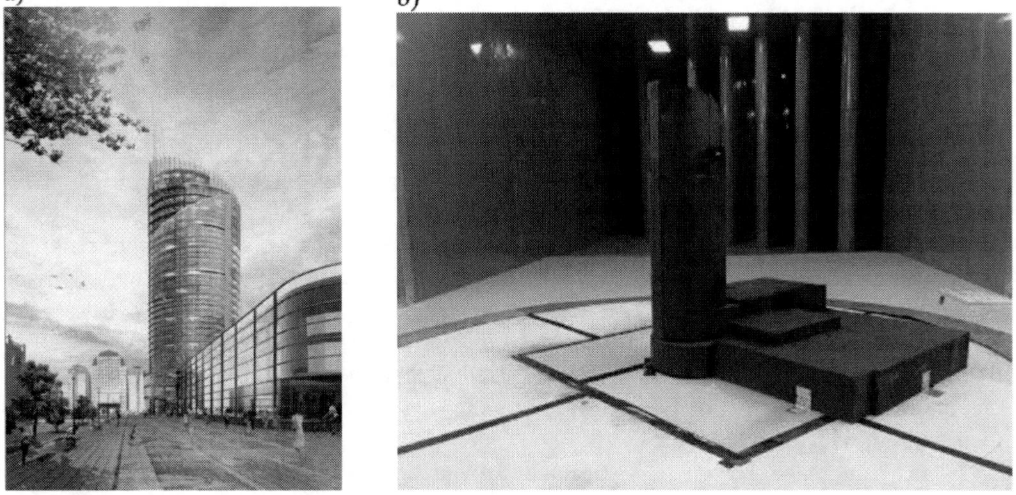

Fig. 3. a) Multifunctional business center, b) Model of a multifunctional business center

Design and production of the model on a reduced scale for experimental studies is performed. The scale of the layout is chosen from the compliance conditions established for the given tunnel of restrictions on the degree of cluttering of the cross section of the working part (in this case, M 1: 100) (Figure 3, b)) (GOST R 56728-2015) Buildings and structures Method for determining wind loads on the enclosing construction). To measure the flow velocity and pressure at characteristic points, a drainage system is installed (Fig. 4). Through the holes, the pressure is transferred through copper and silicone tubes to differential pressure sensors MPXV7002dp, the error in measuring pressure by these sensors, after their individual calibration, was 0.15%.

The model of the investigated building was installed on a specialized turntable in the working area of the wind tunnel. After that, the installation was started and the air flow was fed to the model. The characteristic velocity of the air flow in the pipe is chosen to be sufficient to ensure self-similarity in the Reynolds number, when a further increase in speed does not lead to a significant change in the averaged values of the aerodynamic coefficients at the control points on the model (GOST R 56728-2015). During this series of tests, the wind speed in the working area was 15 m / s.

Fig. 4. Drainage system installation

3 Results and discussion

The result of the tests is the pattern of pressure distribution over the surface of the layout. Based on the obtained data, average values of aerodynamic pressure coefficients for various angles of attack of the wind from 0 ° to 360 ° are calculated with a definite step according to formula (1).

$$C_p=(2\times p)/(\rho\times\vartheta\hat{}2\ ) \quad (1)$$

where Cp – aerodynamic pressure coefficient

p – pressure (obtained by the experimental method)

ρ – air density (at temperature +20° ρ=1.2041 kg/m3)

υ – air speed (15 m/s)

Similarly, the peak values of the aerodynamic pressure coefficients needed to calculate the peak component of the wind load on the enclosing structures and facade systems are calculated.

Aerodynamic drag coefficients relative to the X, Y and Z axes - Cx, Cy and CMz are calculated by means of six-component tensors and ATI DAQ F/T software (aerodynamic drag coefficients relative to the X, Y and Z axes), graph of the dependence of the aerodynamic coefficients Cx, Cy, Cmz is plotted with a change in the angle of attack from 0 ° to 345 ° (Figure 5).

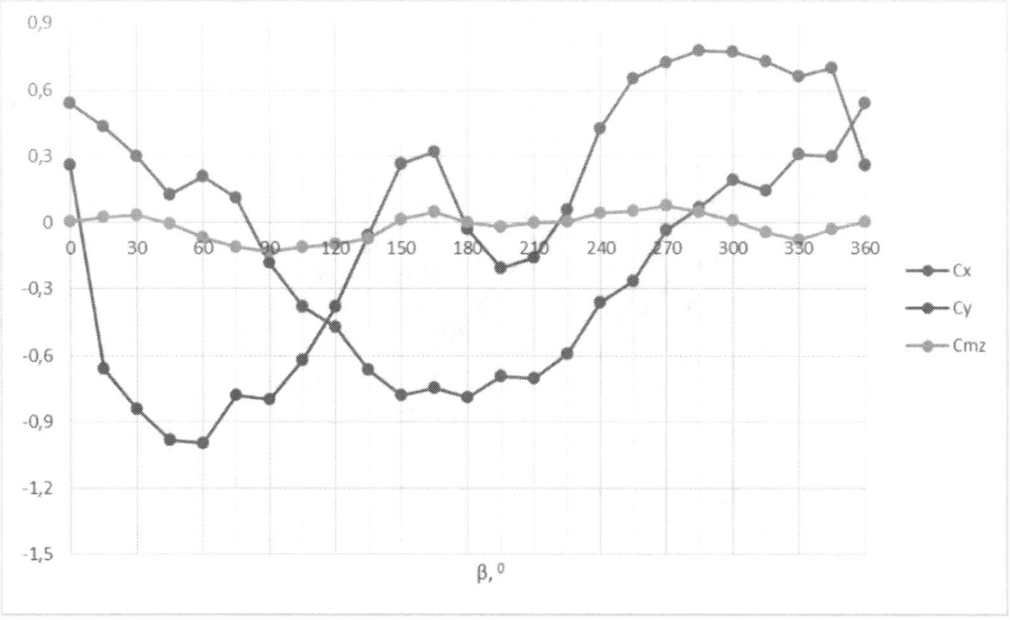

Fig. 5. Graph of the dependence of the aerodynamic coefficients Cx, Cy, Cmz with the angle of attack varying from 0° to 345°

5 Conclusions

Based on the obtained results, it can be argued that the average and peak values of the aerodynamic coefficient Cp at control points on the surface of the layout, while the average values are in the range from 1.5 to -2.5, and the peak values in the range from 2.5 to -4. Zones with the maximum values of the peak aerodynamic coefficients are located on the lateral (in relation to the incoming wind) areas of the facades, in the characteristic zones of stalling the vortices from the structure.

The values of the dimensionless aerodynamic coefficients Cx, Cy, Cmz are determined with respect to the center of the base of the investigated object, while the maximum values of the coefficient Cx arise at the blowing angles 150° and 180° (Cx = -0.78), the maximum values of the coefficient Cy arise at the blowing angles 45° and 60° (Cx = -0.99). In general, the values obtained are fairly well correlated with the values of the aerodynamic coefficients given in SP 20.13330.2016, not exceeding them, which eliminates the need to develop specialized recommendations for structural modification.

Acknowledgments

The work was financial supported by the Ministry of Education and Science of the Russian Federation within the framework of the state #7.6075.2017/BCh, Project «Investigation of

the phenomena of aerodynamic instability of building structures in aeroelastic statement, including the development of an innovative methodology for analysing meteorologica data to refine the parameters of the wind load».
All tests were carried out using research equipment of The Head Regional Shared Research Facilities of the Moscow State University of Civil Engineering.

References

1. O.I. Poddaeva, J.S. Buslaeva, D.S. Gribach *Physical model testing of wind effect on the high-rise*, AMR, **1082**, 246-249 (2015)
2. O.I. Poddaeva, J.S. Buslaeva, D.S. Gribach Experimental study of wind loads on a multifunctional high-rise residential complex, Vestnik Belgorodskogo gosudarstvennogo tehnologicheskogo universiteta im. VG Shuhova, **6**, 58-62 (2015)
3. O.O. Egorychev, P.S. Churin Experimental study of wind loads on high-rise buildings, Zhilischnoe stroitelstvo, **6**, 20 (2015)
4. Á.F. Saemundsson *Wind effects on high rise buildi*ngs (2007)
5. J. D. Holmes *Wind loading of structures*, CRC press. (2015)
6. D. Čehel'ová, M. Janák, B. Bielek The impact of relative position of the high - rise buildings on the wind flow, AMM, **820** (2016)
7. Quan Yong, Gu Ming, Tamura Yukio Experimental evaluation of aerodynamic damping of square super high-rise buildings, WS, **8**, 5, 309-324 (2005)
8. Huang Dongmei, Zhu Ledong,Ding Quanshun, Zhu Xue, Chen Wei Journal of Fluids and Structures, **69**, 355-381 (2017)
9. M. Iancovici, M.A. Riley, F. Sadek, E. Simiu *Wind effects on high-rise buildings: database-assisted design versus the high-frequency force-balance technique,* ICWE (2017)
10. B. Yan, Q.S. Li Wind tunnel study of interference effects between twin super-tall buildings with aerodynamic modifications, JWEIA, **156**, 129-145 (2016)

Technological aspects of lift-slab method in high-rise-building construction.

Pavel V. Gaidukov[1,*], and *Evgeny* M. Pugach[2]

[1]Moscow State University of Civil Engineering, Yaroslavskoe shosse, 26, Moscow, 129337, Russia

Abstract. The utilization efficiency of slab lifting technology for high-rise-building construction is regarded in the present article. The main problem of the article is organizing technology abilities indication, which proves the method application possibility. There is the comparing of lifting technologies and sequential concrete-frame extension, as follows: the first one: the parameters are defined, and the second one: the organizational model is executed. This model defines borders of the usage methods, as well. There is the mathematic model creating, which describes boundary conditions of the present technologies usage. This model allows to predict construction efficiency for different stored-number buildings.

1 Introduction

1.1 Demand and high-rise construction development

The process of the population urbanization has started in the middle of 20th century, which has led to the agglomeration cities with demanded infrastructure situated. This has made it possible to provide comfortable existence for the million population. However, the increase of the townscape areas has led to the sharp appreciation of the land properties cost for the major construction work objects in the investment appealing districts. This had effect on the square meters cost of the buildings. For example, 1m2 cost of the one-roomed flat in Moscow has been changed from 600$ to 4200$ for the period from 2000 to 2017, agreeably. There is tendency on the infill areas organization by the mixed orientation high-rise-complexes on the back of land value appreciation. These high-rise-complexes include as accommodation apartments as office spaces. All these factors has led to the high-rise-construction demand on the territories that had been previously planless for such projects.

1.2 Common problems of high-rise-building construction

The main arresting factor of the high-rise-building construction is high-produced organization problem of the work performance [1]. Such type of building construction represents chain of cyclic processes held in the space on one typical floor territory. In its turn, the confined space does not allow to employ maximum quantity of the labor capacity

* karmodel@gmail.com

and does not allow to make good use of building equipment, as well. Concrete works represent key technology operation on high-rise-building construction, which efficiency is expressed as form reuse. Mr. Hisham Abou Ibrahim and Mr. Farook Hamzeh in the article named as «Role of Formwork Systems in High-Rise Construction» note two problems of works organization, include the following:

The first one is: standard solutions represented as panel formwork that is held on the tower crane work, and thus the performance efficiency is decreased.

The second one is: the building height and organization complexity of lifting equipment additional installation points (except in the stiffening core) restrict no ability to launch any additional workflow for construction productivity increase connecting with the vertical construction transfer. The usage of the panel formwork is led to concrete defects precipitation. All of this demands to the additional resources mobilization for the defects isolation, as well. There is the proposed solution as self-climbing formwork set in the above article. [1.2] However it also has weaknesses such as necessity for horizontal form installation to concrete floor slabs. All of this turns us back to the question of lifting tackle usage.

The third one is: concrete mixture handling to the providing works level. Powerful concrete pump stations or organized interception points on the compounded floors, both are used for handling concrete to the high-levels of buildings. The usage of the way named as crane-bucket significantly reduces the concrete works performance.

Another side of the problem is hospitable working environment creation for labors. The necessity of the safety practices implementation during labors working on the height is led to the stuff movement restriction and to the checking performance increase by the technical customer, as well.

1.3 Lift slab technology as high-rise-building construction problem solving

In the present Article, the lift slab technology is proposed for above high-rise-building construction problem solving, which was first proposed by French engineer Lafarge. However, this solution could not be realized for the reason of the corresponding elevation equipment absence to lift such heavy constructions. The first multistory building was built in 1951 in the USA by the lift slab technology applying. Hereafter the same method was used during constructing buildings in Japan, Europe and the USSR. Summary of this technology is consisted in the enlarged block production, such as slab or a floor in general (slab and the higher walls) at the zero mark. In the following, they are lifted up along the piers, and excluding any horizontal moves, they are locked at the project mark, and then there is further vertical bearing support installation.

The present type preferences are the following

The first one is: flexibility of the bearing support material (rolled iron, prefabricated reinforced concrete, and monolith);

The second one is: flexibility of the space-planning decision;

The third one is: a great number of process operations that are performed at height absence; and

The forth one is: requirement for tower cranes installation, as well.

The lift method consists of the slab package production and moving up, in the following. General information about the building production technology is indicated in the article named as "New developments of lift slabs" by FRANK A. RANDALL, JR. The purpose of the present article is organizing technology abilities searching that prove the lift method utilization efficiency in the high-rise-building construction The present technology foundation is production of the precast flat slab conduit in the edge form at the zero mark and the preparation of the separation layer between several package slabs. Herewith the package slab must be capable of the main conditions requiring, as follows:

The first one: to exclude the bond above the contact slabs surface of different floors;

The second one: to possess abrasion and weather resistance. It means that it should not fail in the workers movements during reinforcing overlying slab, and lose its characteristics from the environmental factors;

The third one: to provide the production works continuity;

The forth one: to provide the constructions and concrete surfaces quality, as well.

These problems solution could be used separating layers preparation made of polyurethane film former lacquers [1.3]. Concreting and reinforcing of the construction are the same corresponding processes as using during building constructions by the standard methods. Exception to this applies collar installations into reinforcing cage, which are needful to lock slabs at the project marks. The sliding formwork usage is a rational way for the building stiffening core construction in real life of modern construction practice. Such type of methodic excepts any requirement in tower crane exploitation and interrupts for the concrete form panels movements [3].

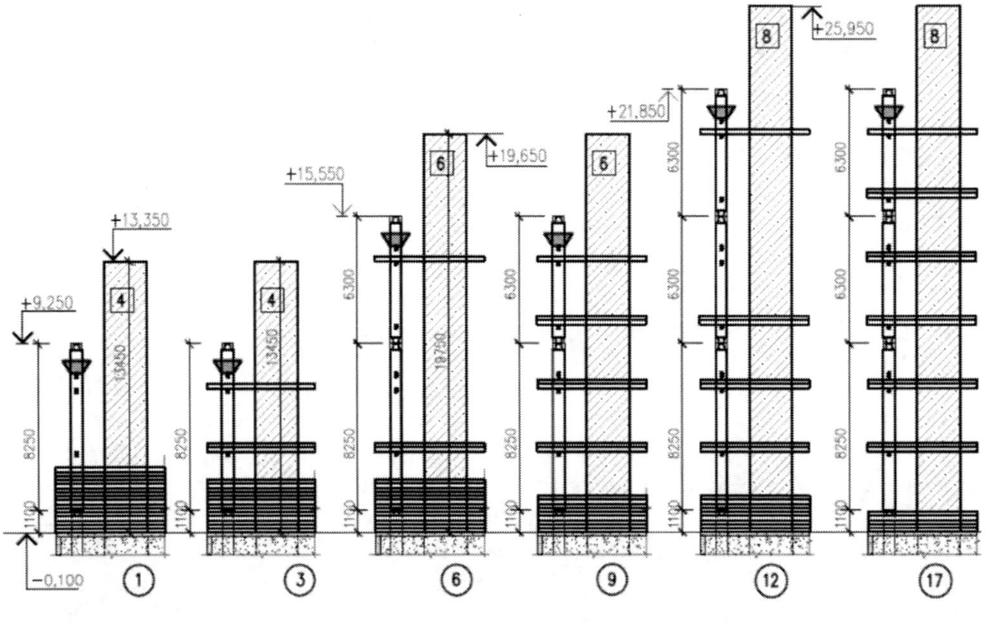

Fig. 1. Lifting model.

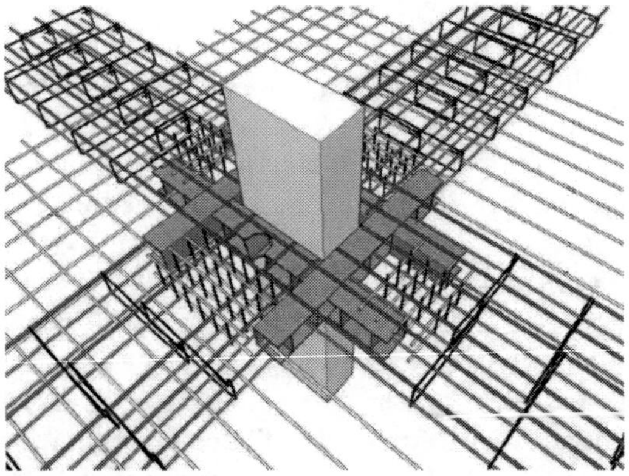

Fig. 2. Collar in the reinforcement. [3]

2 Materials and methods

The analytic method to prove high efficiency of the high-rise-building construction is used in the present article. The reinforce concreate high-rise-building frame installation was comparing by slab lifting technology and sequential extension method, as well. Numerical investigations were conducted on basis of labor and machine time costs available data [4]. 30-storied building was confirmed as searching model, and had the following parameters: 96 meters in height and its typical floor square was 784 m2.

3 Results

The labor input calculation and machining time cost, both, were represented for the propose of organizing model construction for lift-slab method and sequential extension technology (both are listed in table 1). There are main technology processes, which include in scope of works on building construction, set in the article. The technology operations, which are representative of both construction types indicating labor costs, are listed in table 1.

Table 1. Labor costs of the technology operations.

Processes name	Unit of measure	Scope of works		Providors	Schedule time, hour	
		Slab lifting	Sequential extension		Slab lifting	Sequential extension
Vertical structures construction						
Colomn and stiffening core building	item	1140			465	476
Stiffening core building	m^3	1582.5			540	540
Slab construction						
Framework handling	100 tonnes	0.38	7.254		3	130
Form work installation	m^2	765.20	24180		112	2440
Reinforse handling	100 tonnes	10.41			10	18
Slab reinforcement	tonne	1010.72			2646	2650
Collars installation	item	2280			200	-
Concrete mixture handling	m^3	16.33			36	40
Pouring concrete	m^3	1608.75			272	280
Separation layer preparation	100 m^2	241.80	-		168	-
Sanding surfaces	100 m^2	241.80	-		360	-
Electromechanical elevator installation	item	1260	-		88	-
Package lifting	cycle	158	-		32	-
Total					4932.00	5558.00

Analyzing data, it is possible to notice that the following type of works make the main difference in labor costs:
the first one is: floor shuttering construction works (2,328 hours);
the second one is: framework handling (127 hours); and
the third one is: the works which are not included in building frame extension works scope.

For illustration, the cyclogram of works performance is purposed (pic.3). This cyclogram measures both, as sequence of providing work, as rationality of one or another method usage. Picture 1 illustrates the line graphs of works cross at the level of 22 floor. The slab lifting is ahead of time the building frame works extension.

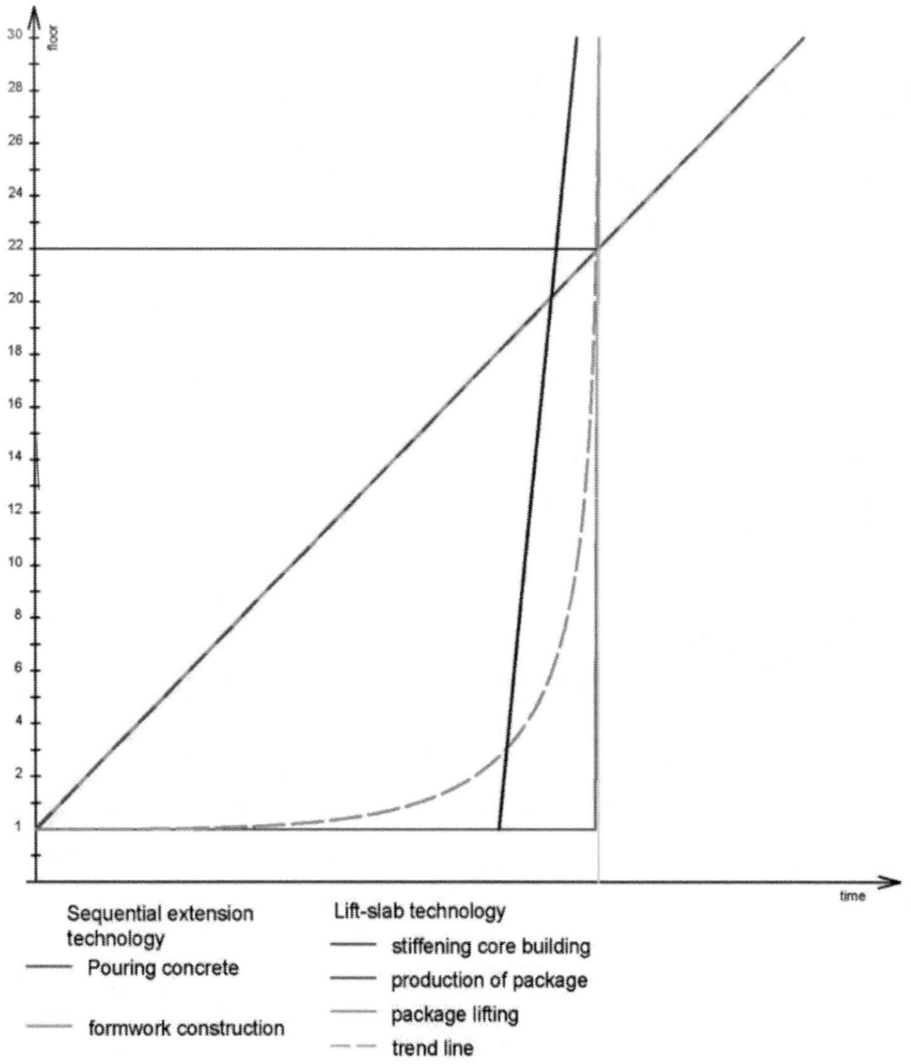

Fig. 3. Cyclogram of works performance.

This fact and the determined crossing point affords ground for function researching to prove taking the construction method. Approximation of the found functions allows us to assimilate the graph of lift-slab method made in system floors / time to the function 1, set as: *y=e^(x-0.01*a)*. This also allows us to assimilate the building frame extension method to the function 2, set as: *y= x/(0.01*a)*. (pic. 4).

https://doi.org/10.1051/e3sconf/20183302068

There are the following shown the graph the
X (axis of abscissae) is shown as axis of time base;
Y (axis of ordinate) is shown as axis of floors;
A is shown as parameter equals to the labor effort for the associated building construction works (labor-days).

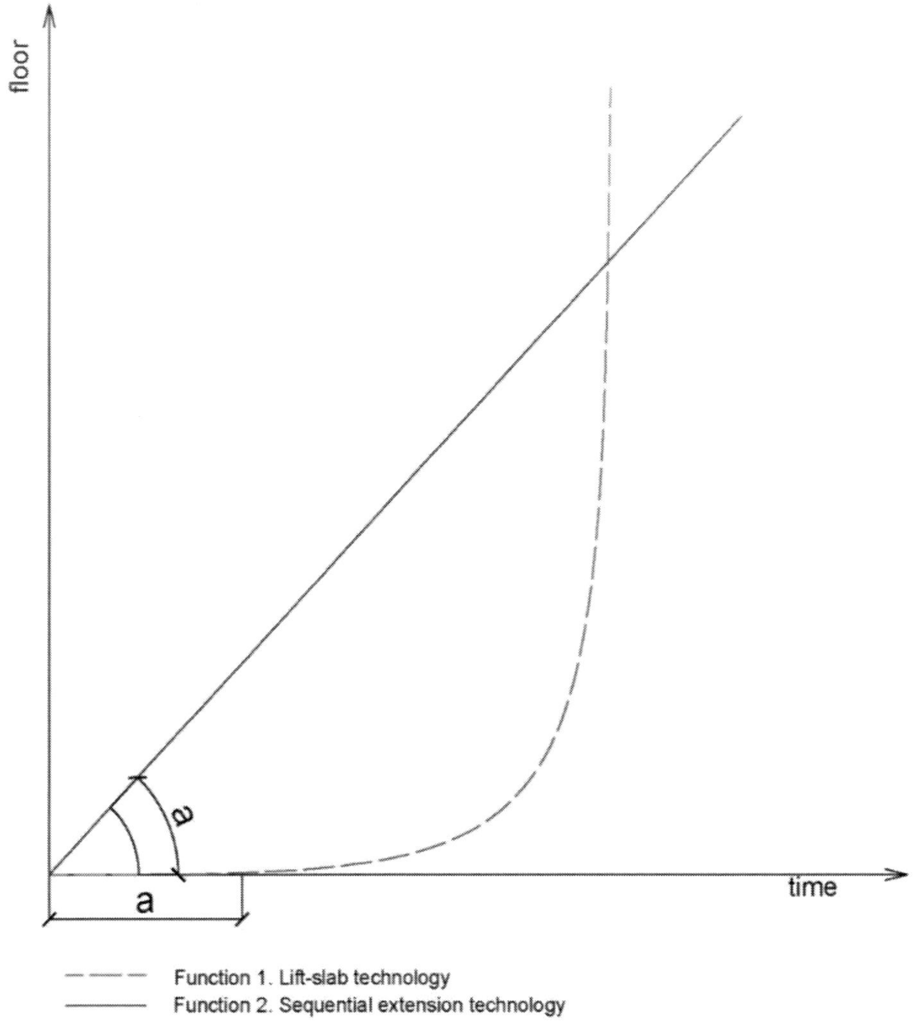

Fig. 4. Graph of the functions 1 and 2.

4 Conclusions

The crossing point founded by this way allows us to take stock the necessity of one or another method usage.
The choice problem of one or the other technologies is very subjective and it depends not only on the labor costs calculating. For each technology additional and external factors are very important. These factors are such as the following:
The first one is: construction site square;
The second one is: the number of work providers; and
The third one is: the organization possibility of safe working practices at height.

The lift-slab method except the hazard factor for workers, which is connected with high-rise operations and other associated works. However, this method does not allow to combine at the same time the following processes, such as finishing works, or engineering networks installation.

References

1. A. Hisham Abou Ibrahim1,2 and Farook R. Hamzeh11 Department of Civil and Environmental Eng ineering, American University of Beirut, Lebanon Role of Formwork Systems in High-Rise Construction. Available from: Role of Formwork Systems in High-Rise Construction [accessed Jan 15 2018].
2. A. Frank, J. Randall, *Structural Engineer Concrete Construction*, (1981)
3. A.O. Saakyan, R.O. Saakyan, S.Kh. *Shakhnazaryan Vozvedenie zdaniy i sooruzheniy metodom pod"ema*. (Stroyizdat, Moscow, 1985)
4. Edinye Normy i rastsenki na stroitel'no-montazhnye raboty. Body E-1-7, E-4. (Glavstroy Ministerstva stroitel'stva SSSR 1972)
5. J.I. Potapova, Method Of Lifting Of Overlappings – Town-Planning Accent To Low Budget Habitation, (The Rostov State Building University, Rostov-on-Don)
6. M. Ashraf et al Adapting Lift-Slab Technology To Construct Submerged Pile Caps, Journal Of Construction Engineering And Management (2000)

Effect of changing of the parameters of the cable network of monitoring systems of high-rise buildings on the basis of string converters on their operability

Nikolay Gusev[1], *Larisa* Svatovskaya[2], and *Alexandr* Kucherenko [1,*]

[1]Military Space Academy named after A.F. Mozhaysky, Zhdanovskaya st.13, St. Petersburg, 197198, Russia
[2]Emperor Alexander I St. Petersburg State Transport University (PGUPS), Moskovsky pr.9, St. Petersburg, 190031, Russia

Abstract. The article is devoted to the problem of improving the reliability of monitoring systems for the technical conditions of high-rise buildings. The improvement is based on string sensors with an impulsed excitation method ensuring the maximum signal-to-noise ratio at their output. The influence of the parameters of the monitoring system on the shape of the excitation impulses of the string, and, consequently, on the amplitude of the string vibration of the string converter is also considered in the article. It has been experimentally proved that the parameters of the excitation impulses of the string converters. The article presents the results of the experiments showing the effect of the fronts duration of the excitation impulses on the amplitude of the oscillations of the strings. The influence of the fronts duration of the excitation impulse with the frontal lengths up to 0.5 ms is studied at the excitation impulse duration not exceeding 0.5 times the duration of natural oscillation periods of the string. The experimental data are compared with the theoretical ones and hypotheses explaining their difference are advanced. The article suggests some methods of reducing the influence of the cable-switching equipment system parameters on the amplitude of string oscillations. The possibilities of improving the reliability of the systems developed on the basis of string sensors with an impulsed excitation method and used for monitoring the technical conditions of the high-rise buildings are proposed.

1 Introduction

In the practice of monitoring the stress-strain state of building structures and soil bases of buildings and construction throughout the whole period of their operation the long-term monitoring systems [1-3], and the systems for monitoring the state of high-rise buildings [4-10], have been widely applied.

* Corresponding author: kucher_90@bk.ru

In the terminology of the Ministry of Emergency Situations of the Ministry of Emergency Situations [11-13], such systems were called SMES (structured systems for monitoring and managing the engineering systems of buildings and structures). In hydraulic engineering such systems have received the name ASDC - automated systems of the diagnostic control [14-17].

ASDC and SMES systems, as a rule, are developed on the basis of the string converters. In addition to primary string converters, the SMES and ASDC systems include: switching and secondary equipment (specialized period-meters), as well as the cable network [17].

The method of measurement by using string converters is based on the dependence of the frequency ω (or period T) of the frequency resonator (string) oscillation on the string extension stress which in each particular string converter is proportional to the measured parameter P (linear displacement, pressure, temperature, etc.).

To excite the string vibrations of the string measuring transducers with impulse excitation (STMS) electromagnetic coils are used. When reeding an excitation impulse of a rectangular shape of standard amplitude and duration to the terminals, an excitation impulse with a certain amplitude F and duration τ_u is generated in the coil. The force impulse excites the damped oscillations of the string y by "pinch" that induce e, the frequency ω, or the period T in the electromagnetic coil e, which can be measured up to some preassigned value of the amplitude determined by interference level corresponding to the minimum value y_{min} string oscillation amplitude. The maximum value of the amplitude of the string oscillations y_{Max} is limited by the design gap between the string in the unexcited state and the core of the electromagnetic coil of excitation δ. The value of the gap for commercially available STMS lies in the range 0.35 ÷ 0.50 mm [17].

Each STMS has an individual calibration, as shown in **Fig. (1)**, produced with the excitation of string vibrations by standard rectangular impulses, their parameters being defined in control documents [17]. The stated parameters of the excitation impulses should provide the value of the amplitude of the damped oscillations of the string in the range from y_{Max} to y_{min} for the entire measurement interval ($t_{изм}$). These conditions ensure stable oscillation of the STMS strings and reliable operation of the SMES and ASDC systems.

From the theory of oscillations [18-22] it is known that when an oscillating system is acted upon by a impulse the duration of which is equal to or multiple the system oscillation period, there appears the probability that the system oscillations will not arise or will have a negligible amplitude. Comparative analysis of the excitation impulses duration (τ_u) of serially produced secondary equipment for STMS [23], the values of which lie in the range from 300 to 700 mks, with the duration of the oscillations periods of the STMS (T) strings in the operating range of the measured parameters, 400 to 2000 mks [23], showed that the range of durations τ_u by 25 ÷ 30% overlaps the operating range of the values of the oscillation periods of the strings T, as shown in **Fig. (1)**.

This means that in the mentioned range of the oscillation period of the STMS string, the values of its amplitude and, consequently, the EMF excitation induced in the electromagnetic coil will fall, which will lead to the decrease in the "signal/noise" ratio at the output of the STMS and the instability of its operation.

 https://doi.org/10.1051/e3sconf/20183302069

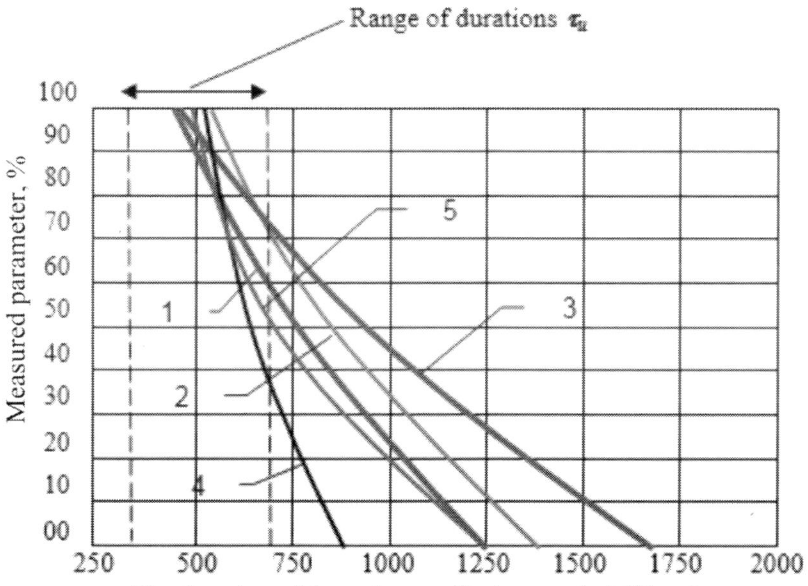

Fig. 1 Dependence of measured quantities on the period of string vibration for string converters with impulsed excitation of string vibrations.

The studies have shown [22] that the parameters of the vibration of the STMS string depend on the amplitude $\boldsymbol{F}$ and the duration of the excitation impulse of the string. The dependence of the string vibration amplitude on the duration of the excitation impulse for various types of STMS is presented in **Fig. (2)**.

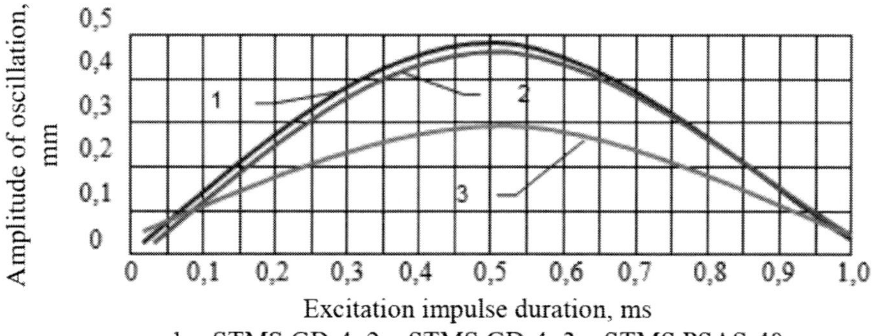

Fig. 2 Dependence of the vibration amplitude of the STMS string on the duration of the excitation impulse.

The invariance of the (standard) values of the parameters of the amplitude $\boldsymbol{F}$ and the duration τ_u of the string excitation impulse with a change in the frequency $\boldsymbol{T}$ of oscillations $\boldsymbol{y}$ of the string in the operating frequency range of the STMS results in the sharp decrease of the string oscillations amplitude. This defect was eliminated in subsequent developments of the secondary equipment with automatic correction of the parameters of the amplitude $\boldsymbol{F}$ and the duration τ_u of the string oscillations impulse as a function of the value of $\boldsymbol{T}$ [22].

Experience of the SMES exploitation showed that the string vibration amplitude of each specific standard sensor in the systems of the high-rise buildings monitoring, even when adjusting the parameters of the amplitude $\boldsymbol{F}$ and the duration τ_u of the excitation impulse, differed, and these differences tended to change with time.

The authors put forward the hypothesis that the excitation impulse parameters change while passing through communication cables and switching equipment.

When calibrating the STMS, the secondary equipment is connected directly to the transducer and of rectangular form is performed with a rectangular excitation calibration [24]. However, on real objects in the SMES or ASDC STMS systems, string converters are connected to the secondary equipment by means of cables, for example, CRHD 2x1.5 (TC 16.505.065-75) and switching equipment, as it is regulated in the normative and technical documentation for STMS. The cable CRHD 2x1.5 has a capacitance $\boldsymbol{C}$ (the capacitance of 1 m of cable is not more than 150 pF) and the electrical resistance of the conductors $\boldsymbol{R}$ (no more than 15 Om per 1 km of cable length). Consequently it is an integrating circuit. When passing through it a rectangular excitation impulse distorts its shape, which is expressed in the formation of the front and back fronts with a duration of τ_f, caused by the time constant of the chain $t = RC$. For normal operation conditions of the CRHD cable 2x1.5, the value of the chain time constant on the real objects does not exceed 0.1 ÷ 2.5 mks, which is incommensurably small in comparison with the excitation impulse durations (from 300 to 700 mks) and does not significantly change the process of string oscillations excitation.

However, during operation, the normalized values of $\boldsymbol{C}$ and $\boldsymbol{R}$ of the cable and the contact resistance R_{nks} of the switching equipment change as a result of aging or external influences and can increase tens or even hundreds times, which results in the significant changes in the shape of the excitation impulse.

The main factors affecting the capacity of the cable are: electrical aging, oxidation, thermal aging and humidification. The investigations carried showed that, with a constant resistance of the cable veins, the formation of the excitation impulse fronts, with the duration τ_f, commensurable with the duration τ_u, takes place with the increase of the cable capacity CRHD 2x1.5 tens times, which is quite probable in practice under the influence of the actual parameters of the influencing factors.

The resistance of the cable line consists of the resistance of the cable cores and the contact resistance of the switching equipment of the SMES system.

The most significant contribution to the total resistance of the system "cable line – switching equipment" is made by the contact resistance of the commuting equipment, which can increase tens or even hundreds times during corrosion or oxidation of contacts of switching equipment. Thus, during the operation of the cable network and the switching equipment of the SMES systems, their parameters change within the limits that significantly affect the time constant of the circuit and the formation of the fronts of the excitation impulses that approximate its shape to the trapezoidal.

The purpose of this investigation was to state the regularities of the string vibration amplitude change of the string converters used in the systems for high-rise building monitoring. The change is caused by the distortion of the excitation impulse shape. The influence of the process an the efficiency of the monitoring systems has also been investigated.

2 Materials and methods

The amplitude y_n of the oscillation of the n'th string harmonic excited by a trapezoidal impulse depends on the coefficient of the n'th harmonic (ξ_n). The coefficient ξ_n depends on the duration of impulse excitation τ_u, the duration of its fronts τ_f, as well as the circular frequency of the nth harmonic ω_n, and has the form [22]:

$$\xi_n = \frac{2}{\omega_n \tau_f} \sin\frac{\omega_n \tau_f}{2} \sin\frac{\omega_n \tau_u}{2} \quad (1)$$

Fig. (3). shows the theoretical dependence of the amplitude y_1 of the first string harmonic vibration on τ_u and τ_f STMS (PLDS -150) at T_1 = 1.0 ms.

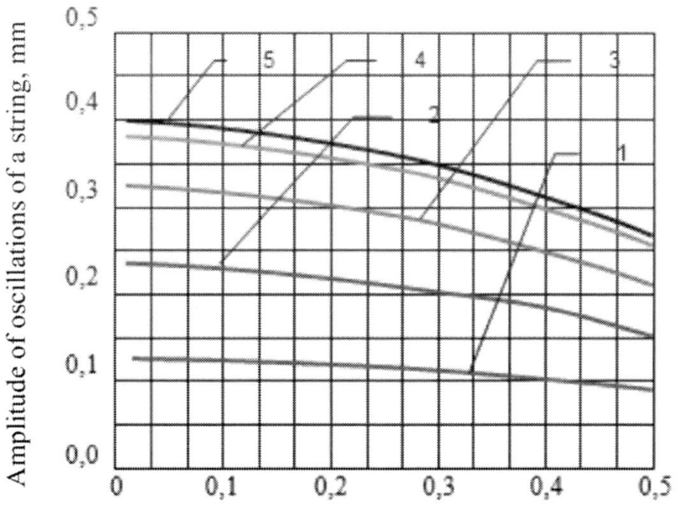

Fig. (3). – Dependence of the first harmonic amplitude (n = 1) of the string vibration on the duration of the excitation impulse front τ_f for the STMS (PLDS -150) for various values of τ_u at T_1 = 1.0 ms.

The investigations were carried out by using standard sensors of the PLDS-150 type with the parameters specified in the standards [14, 17], and according to the standard procedure [23, 25]. The devices for recording static and dynamic displacements (DRSDFD) as the secondary equipment were used. They have the following main parameters:

The parameters of the request impulse (excitation) with a smooth adjustment:

voltage	135 ÷ 165 V,
durability	0,25 ÷ 1,00 ms,
repetition period	not less than 2 s.

Parameters of STMS (PLDS-150):

constructive gap δ	0,4 mm
output period range	450 ÷ 1250 mks

Experimental conditions:

output cable	such as CRHD 2x1,5
ambient temperature	+20^{0}C
recording equipment	oscilloscope S1-96M

The experiment was carried out at five values of the impulse duration in excitation τ_u in the range from 0.1 T_1 to 0.5 T_1 in steps of 0.1 T_1 and with a front duration change $\tau_ф$ in the range from 0.0 to 0.5 T_1 in steps of 0.1 T_1 at T_1 = 1.0 ms. The amplitude of the excitation impulse ensured the maximum value of the string vibration amplitude $y_{Max} \leq \delta$ at a value of T_1 = 1.0 ms at an impulse voltage of 145.5 V.

The shape of the excitation impulse obtained during the experiment is shown in **Fig. 4**

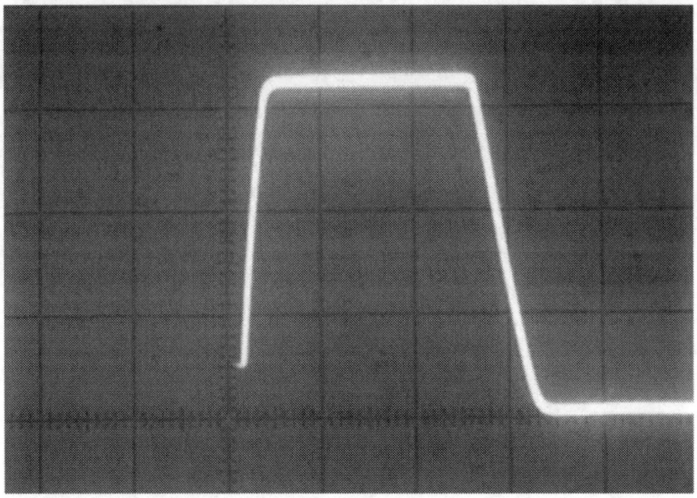

Fig. 4 The shape of the excitation impulse at $\tau_u = 0{,}5T_1$ and $\tau_f = 0{,}1T_1$.

3 Results

The results of the experiment are presented on **Fig. 5**

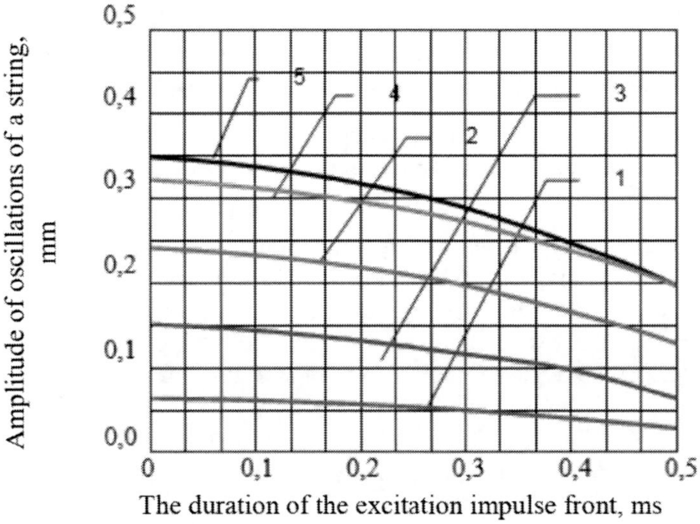

1-τ_u=0,1T$_1$; 2-τ_u=0,2T$_1$; 3-τ_u=0,3T$_1$; 4-τ_u=0,4 T$_1$; 5-τ_u=0,5T$_1$

Fig. 5 Experimental dependences of the amplitude of the first harmonic (n=1) of the string vibration on the duration of the excitation impulse front τ_f for STMS (PLDS-150) at various values of τ_u.

The materials presented in **Fig. 5** show that:

1) The amplitude of the oscillations of the string has a nonlinear dependence on the duration of the excitation impulse front τ_u and its fronts duration τ_f.

2) As the duration of the excitation impulse front increases, the amplitude of the vibration of the string y decreases. So for $\tau_f < 0.5T_1$, i.e. at the maximum amplitude of the excitation impulse and the preservation of its trapezoidal form, the string oscillation amplitude drops by 45-50%.

4 Discussion

The experimental data of the dependence of the amplitude of the string vibrations on the duration of the fronts of the excitation impulse τ_f, presented in **Fig. (5)**, differ from the theoretical ones presented in **Fig. (3)**. Thus, the values of the amplitude of the string vibrations obtained in the course of the experiment are lower than the theoretical values presented in **Fig. (3)**, up to 26% at $\tau_u = 0.5$ and $\tau_f = 0.5T$ and up to 45% at $\tau_u = 0.2T_1$ and $\tau_f = 0.5T_1$.

This phenomenon is obviously due to at least three circumstances:

1) The theoretical dependence (1) was obtained for the case of equality of the values of the front and back fronts, i.e. symmetric shape of the excitation impulse. The actual shape of the excitation impulse during the experiment was different from the symmetric one, as shown in **Fig. (4)**, namely, the leading front is 25-35% shorter than the trailing edge.

2) Oscillation amplitude of only the first harmonic y1 without taking into account highest harmonics (n>1) was fixed while the energy of the excitation impulse was distributed between them.

3) During the experiment, it was not possible to fix the effect of the string oscillation precession, in the presence of which the oscillations amplitude will always be lower than the oscillations in one plane.

The experiment confirmed the general tendency of the string oscillation amplitude decrease when the duration of the excitation impulse front increases in the range up to $\tau_\phi < 0.5T_1$. The decrease of the string vibration amplitude and, consequently, of the excitation of the EMF ***e*** induced in the electromagnetic coil results in the decrease of the "signal/noise" ratio at the output of the STMS and the instability of the whole monitoring system.

5 Conclusions

During the experimental investigations it was proved that in order to maintain the maximum amplitude of the string oscillations y_{Max} (provision the maximum "signal/noise" ratio) over the entire measurement interval (t_{izm}), the amplitude of the excitation impulse of the STMS strings must be corrected in proportion to the duration of the excitation impulse fronts τ_f, i.e. to the change of the parameters ***C*** and ***R*** of the «cable-switching equipment» system. These conditions ensure the stable oscillation of the SIPI strings and reliable operation of the STMS systems built on their basis. If these conditions are not fulfilled, STMS begins to give unstable readings soon after the introduction of the SMES systems into operation (from 1 to 5 years). So, they are excluded from the measurements as coming out of service and the informative basis for making decisions on the technical state of the high-rise buildings is reduced.

The experimental results obtained allow us to formulate future directions of the research:

- investigation of the influence of different durations of the front and back fronts of excitation impulses on the string oscillations amplitude;
- investigation of the influence of the precession of the string oscillation and the highest harmonics on the string oscillations amplitude;
- investigation of the ways of reducing the values of the parameters ***C*** and ***R*** of the «cable-switching equipment» system or their changes in time.

References

1. A. Zhivaev, G. Bolyrev, A System for Static and Dynamic Monitoring of Ice Sport Arena Text, IWSHM, 8th International Workshop on Structural Health Monitoring (Stanford, USA, September 2011)
2. L. Olson, Dynamic Bridge Substructure Evaluation and Monitoring, USA: US Department of Transportation, Federal Highway Administration, 219 (2005)
3. N. Sokolova, Sensors follows the safety. Monitoring as a pledge of quiet. St. Petersburg.: Building and City economy in St. Petersburg and region, **88** (2006)
4. D. Lysov, V. Guryev, V. Dorofeev, N. Nazjmov, Structural mechanics of engineering buildings and constructions, **2**, 55-61 (2011)
5. D. Zavjalo, Engineering systems of high-rise buildings, High-rise buildings, **1**, 107-109 (2013)
6. A. Abdelrazak, High-rise buildings, **1**, 96-103 (2016)
7. V. Gurje, V. Dorofeev, Unique and special technologies in buildings, **1** (2004)
8. TSN 31-332-2006 Residential and public high-rise buildings
9. MGSN 4.19-2005 Time standard and projecting principles of multifunctional high-rise buildings and residential complexes in Moscow
10. N. Kapustin, V. Tarakanovsky, A. Voznyuk, A. Klimov, Preventing of building and construction crash: Collected scientific articles. Issue, **9**, 192-206 (2010)
11. CODE R 22.1.13-2013 "Safety in the state of emergency. Measures concerning civil defense, measures concerning prevention emergency state of natural and techno genic character. Structurizational monitoring system and control of the engineering systems of buildings and constructions. Requirements to the requence of building and exploitation"
12. N. Gusev, Significant problems of protection of inhabitants and territories from fires and catastrophes, (2006)
13. N. Gusev, V. Golovin, Sh. Iskhakov, V. Kishik, E. Pavlov, Scientific journal "Ber Board", **6** (2006)
14. STO 17330282.27.140.021-2008: Control-measuring systems and equipment of hydro-technical constructions of hydro-electric power stations. Arrangement of explaitation and technical maintenance. Standarts and requirements
15. Bernd Hiller, I. Sukhov, V. Lee, Automatized system of deformation monitoring (ASDM) at Sayano-Shushenskaya hydro-electric power stations, St. Petersburg: Hydrotechnique, **2**, 12-15 (2015)
16. N. Gusev, Regional Informatics - 2006. Materials of the jubilee X International Conference. St. Petersburg, (2006)
17. SO 34.21.546-2003 (RD 153-34.2-21.546-2003) Regulations of creating and conducting natural observations and investigations at the dams out of soil materials, (2003)
18. L. Etkin, Vibro-frequency sensors. Theory and practice, (2004)
19. N. Gusev, Collection scientific articles. SPb, 21-34 (2007)
20. R. Zinoviev, Collected materials RUDN. Series "Engineering investigations", **2**, 64-66 (2005)
21. N. Gusev, Materials of international scientific conference. St. Petersburg, (2006)
22. N. Gusev, Journal "Life and Safety", **3-4**, 46-50 (2004)
23. STO 70238424.27.140.004-2008 Control-measuring systems and equipment of hydro-technical constructions. Regulations of development. Standards and requirements
24. STO 70238424.27.140.025-2009: Hydro-electric power stations. Control-measuring systems and equipment of hydro-technical constructions. Metrological equipment and estimation of technical state and efficiency. Standards and requirements
25. R RSK 002-06 General requirements to the calibration methods used in Russian system of calibration

Use sulfoferritic cements in construction

Svetlana V. Samchenko[1] and *Dmitriy A. Zorin*[1,*]

[1]Moscow State University of Civil Engineering, Yaroslavskoe shosse, 26, Moscow, 129337, Russia

Abstract. Currently, high-rise construction has received increasing attention around the world. In the big cities under construction is less space and one solution is the high-rise construction. However, high-rise buildings use special requirements, such as strength, thermal insulation, wind load and others. When concrete is exposed to continuous loads by wind or to mechanical loads, it undergoes abrasion. Resistance to this process depends on the characteristics of materials that the concrete and finishing seams are made of. Research on increasing impact and abrasion resistance of calcium sulfoferrite-based cement stone from the perspective of formation of cement stone structure will be instrumental in developing durable materials for application in high-rise construction.

1 Introduction

To ensure permanent integrity of a structure is an engineer's primary goal. Therefore, a high-rise construction is constructed taking into account wind loads and the way they influence buildings integrity, wind noise, thermal insulation and also the materials used in this construction.

Wind load is viewed as the most critical load among temporary loads on high-rise construction. This is due to the fact that using steel and reinforced concrete elements with light-weight re-bars and fillers in multistory construction has resulted in reduced buildings weight and has eliminated restrictions for structures height. Implementing new materials and construction methods has led to a significant reduction in structural stiffness, hence today wind loads have started to be of vital importance in construction [1].

Wind loads imply not only exposure to strong winds but also take into account the movement of air masses with different moisture and temperature carrying large amounts of solid particles. That is why the choice of building materials for high rise construction is limited not only by their technological features, but also by their ability to withstand chemical and physical exposure factors [2].

Erosion, or weathering, is caused by wind, water, and icing, and results in loss of material off the concrete surface and chipping. As concrete is a brittle material, the edges of joints and seams are fractured by impacts. When concrete is exposed to continuous loads by solid particles or to mechanical loads, it undergoes abrasion. Resistance to this process depends on the characteristics of materials that the concrete and finishing seams are made of.

* Corresponding author: dim-z@yandex.ru

In constructions exposed to weather carbon dioxide results in formation of calcium carbonate. In hydraulic structures, its presence in the composition of water produces leaching that mainly affects binding materials [3,4].

Concrete leaching is a similar process, with the exception that it develops in the presence of moisture and consists in removal of cement stone. Degradation intensifies under the impact of water that contains carbon dioxide, or a sulphuric acid of organic origin [5].

Physical factors of concrete degradation include its shrinkage, both plastic and drying shrinkage, that results in formation of micro-cracks, cracks or warping. This is especially true for materials used to seal joints and seams. At present, there is a record of using expanding admixtures for sealing joints during metro tunneling where these materials are used for joint packing and manufacturing of tubings [6-8].

Much attention is devoted to the problem of concrete corrosion damage. Researchers explore opportunities to increase resistance of cement formulations by using various types of binding materials, as well as by introducing corrosive-resistance aggregates and fillers into their composition [9-12].. Researches have been completed on corrosion resistance of cement stone with sulfate-bearing clinkers [13-16]. Hence, research on increasing impact and abrasion resistance of calcium sulfoferrite-based cement stone from the perspective of formation of cement stone structure will be instrumental in developing durable materials for application in high-rise construction.

Calcium sulfoferrite-based cements are capable of not only compensating for shrinkage but also of leading to system expansion. Besides, these cements are characterized by high strength properties [8,17,18].

This study is aimed at obtaining a calcium sulfoferrite-based binding material having non-shrinking properties and resistant to abrasion and impacts.

2 Materials and methods

In the study, binding material was obtained using sulfoferrite clinker (SFC) and Portland cement clinker (OPC). Clinkers physical properties and chemical analysis are given in Table 1. Mineralogical composition of cements is listed in Table 2.

Table 1. Physical and chemical properties of cements and sulfoferrite clinker.

	OPC	SFC
Specific gravity	3.16	2.96
Blaine fineness, cm^2/g	2700 (3500)	4230
Loss of ignition, %	0.21	0.01
SiO_2, %	22.03	12.23
Al_2O_3, %	5.15	3.09
CaO, %	65.41	51.83
MgO, %	1.20	1.92
SO_3, %	0.34	5.19
Fe_2O_3, %	4.86	23.79

Binding material was obtained by replacing a part of Portland cement clinker with sulfoferrite clinker powder. Portland cement clinkers were ground to Blaine specific surface area of 2700cm^2/g and 3500cm^2/g. Calcium sulfate dihydrate was ground separately to specific surface area of 3500cm^2/g. Sulfoferrite clinker powder was ground to fractions of: >80μm, 63-80μm, 45-63μm, 28-45μm and less than 28μm. The mixtures were prepared by mixing components together in the following ratio: PC clinker – 80%, expanding admixture (of a certain fraction) – 10%, and gypsum – 10%. Finished cements were mixed with water at the w/c ratio = 0.4. Cement paste was formed into specimen hardened under normal conditions for 1, 3, 7, 14 and 28 days. Then the samples were tested for strength and expansion, as well as went through physical and chemical analysis. Strength tests were conducted in compliance with GOST 30744-2001 [21]. Abrasion tests were conducted in compliance with GOST 13087-81 [22]. Formation of cement stone structure of binding material was studied using scanning electron microscopy (SEM).

Table 2. Mineralogical composition of cements

	OPC	**SFC**
Alite	67	-
Belite	8	25
Tricalcium aluminate	5	-
Calcium aluminoferrite	13	-
Calcium sulfoferrite	-	75

The study focused on how expanding component dispersity affects the morphology of ferrous ettringite formation and the properties of binding material. To eliminate the influence of gypsum dispersity, the samples were mixed with saturated gypsum solution. After that they were hardened for 6, 12 and 24 hours, as well as for 3, 7, 14 and 28 days, and were studied using a variety of physical and chemical methods. The quantity of formed ferrous ettringite was determined using the chemical method described in the monograph [19]. Simultaneously, studies were conducted on microslides using optical microscopy and X-ray phase analysis.

3 Results

3.1 Effect of expanding component dispersity on the morphology of ferrous ettringite

The effect of expanding component dispersity on the morphology of ferrous ettringite formation was studied using the minerals of sulfate-bearing clinkers $2CaOFe_2O_30.8CaSO_4$. For the study, it was ground to fractions of: >80μm, 63-80μm, 45-63μm, 28-45μm and less than 28μm. To eliminate the influence of gypsum dispersity, the samples were mixed with saturated gypsum solution. After that they were hardened for 6, 12 and 24 hours, as well as for 3, 7, 14 and 28 days, and were studied using a variety of physical and chemical methods.

The results of experiments revealed that the smaller the size of initial hydrating grains of the $2CaOFe_2O_30.8CaSO_4$ mineral, the faster its hydration occurs, which is evidenced by the growth of hydration degree as shown by the data of X-ray phase analysis. Measurement results are given in Table 3.

Calcium sulfoferrite $2CaOFe_2O_30.8CaSO_4$ is characterized by low hydration activity which is confirmed by slow liquid phase saturation with calcium ions and slow growth of medium pH factor. For hydration of the less than 28μm fraction, the maximum value of pH = 13.5 is reached in 24 hours (Fig. 1). At the same time, a higher amount of formed crystals is observed in the same fraction (Fig. 2).

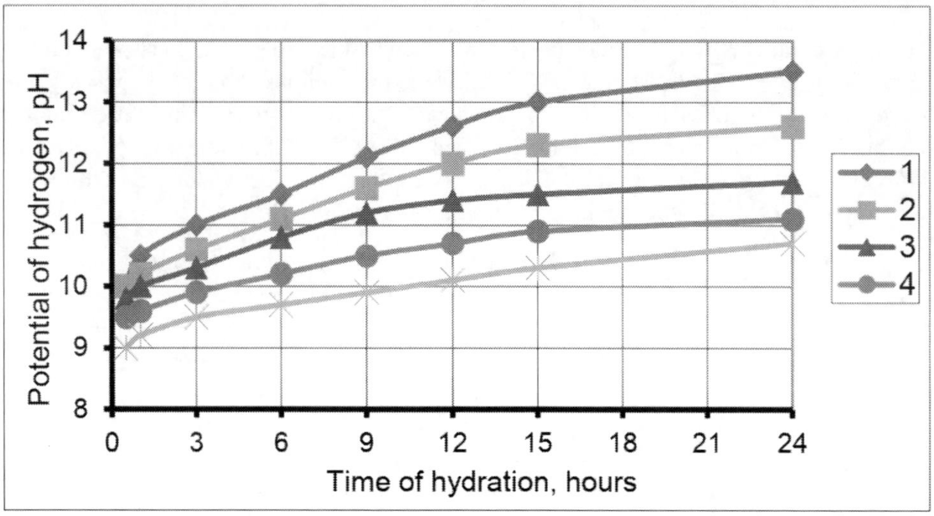

Fig. 1. Medium pH variation during hydration of 2CaOFe2O30.8CaSO4different fractions (1 – fractions less than 28μm,; 2 – 28-45μm; 3 – 45-63μm; 4 – 63-80μm; 5 – >80μm).

Table 3. The characteristic of the process of $2CaOFe_2O_30.8CaSO_4$ mineral hydration for different fractions.

Grain size, μm	Time of hydration, days	Degree of hydration (X-ray), %	Amount of ferrous ettringite
< 28	0,25	12,3	14,8
	0,5	23,2	25,3
	1	26,8	28,2
	3	29,8	36,9
28-45	0,25	11,6	13,5
	0,5	21,6	23,9
	1	25,6	29,5
	3	28,8	36,1
45-63	0,5	18,4	13,1
	1	21,3	24,1
	3	30,5	37,3
63-80	0,5	14,8	11,5
	1	18,9	12,5
	3	24,2	32,2
	7	30,8	37,7
> 80	1	7,1	7,7
	3	8,3	9,2
	7	15,0	12,9
	14	20,3	16,5
	28	26,8	30,1

The study of hydration processes of calcium sulfoferrite mineral with the composition of $2CaOFe_2O_30.8CaSO_4$ in microslides demonstrated that hydration of small-size mineral fractions (less than 28μm) occurs at the time of its partial dissolution along with crystallization of short prismatic crystals (l=50-60μm, d=5-6μm). Moreover, in small-size fractions (28-45 and less than 28μm) formation of such crystals around initial mineral grains begins as early as in 24 hours, in 45-63 fractions it occurs at the age of 3 days, whereas in large-size fractions (63-80 and >80μm) it occurs only by the 14th day.

Formation of ferrous ettringite prismatic crystals always occurs regardless of the fraction of mineral $2CaOFe_2O_30.8CaSO_4$. The mineral's dispersity affects only its hydration activity. Smaller-size fractions hydrate faster, while medium-size and coarse fractions are very slow to hydrate due to low hydration activity. Formation of ferrous ettringite prismatic crystals is observed only at the age of 7-14 days.

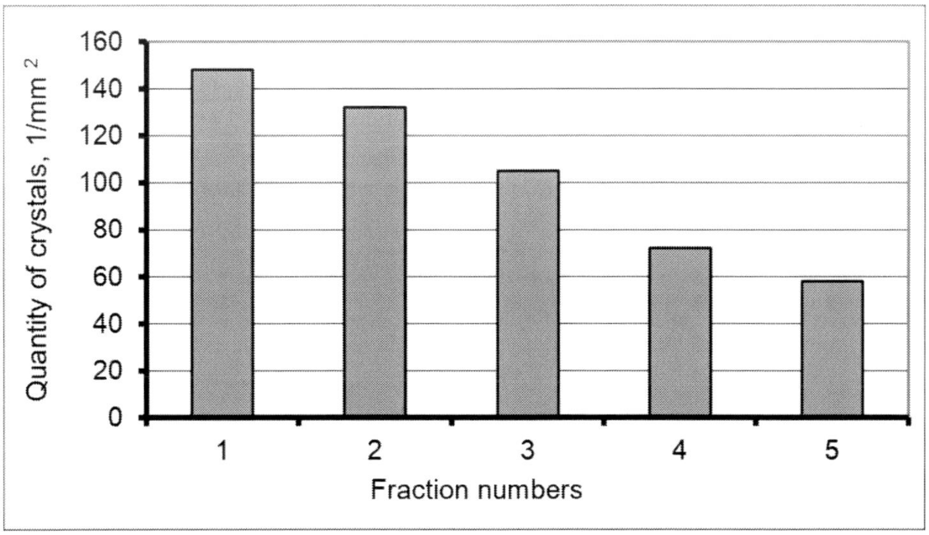

Fig. 2. Initially formed crystals per 1mm^2 of microscope slide for hydration within 24 hours of $2CaOFe_2O_30.8CaSO_4$ different fractions (fraction numbers as indicated in Fig. 1).

As obtaining a mono-sized composition of expanding admixture is hardly possible in the field conditions, experiments were conducted to study the influence of poly-sized composition of $2CaOFe_2O_30.8CaSO_4$ mineral on the morphology of formed ettringite crystals.

Poly-sized mineral compositions were prepared by mixing of fine and coarse fractions with the ratio of 1:1. Less than 28μm and 28-45μm fractions of the $2CaOFe_2O_30.8CaSO_4$ mineral were selected, as these fractions demonstrate the most intensive hydration within the first 24 hours. This choice of fractions was made because in the first hours of hydration small-size fractions saturate the liquid phase very quickly, contributing to formation of large numbers of crystallization centers. On the contrary, coarse fractions gradually saturate the liquid phase, and its ions should be used in crystals formation.

The obtained data revealed that the quantity of ferrous ettringite in the first hours and days of hydration is somewhat lower than in small-size fractions, but still higher that in large-size fractions, if compared to the quantity formed in mono-sized compositions.

3.2 Effect of expanding component dispersity on formation of cement stone structure.

Cement stone structure mainly depends on the quantity of hydrates and porosity. It widely varies depending on composition and crystallinity degree of crystalline hydrates which, in turn, depends on factors such as impurities presence in minerals, hardening conditions, as well as hardening system dispersity [16,20].

To study the influence of expanding component dispersity on cement stone formation, Portland cement was mixed with different fractions of calcium sulfoferrite expanding component (less than 28μm, 28-45, 45-63, 63-80 and over 80μm fractions).

Portland cement clinker was ground to Blaine specifics surface areas of 2700 and 3500cm2/g. Calcium sulfate dihydrate was ground separately to specific surface area of 3500cm2/g. The mixtures were prepared by mixing components together in the following ratio: PC clinker – 80%, expanding admixture (of a certain fraction) – 10%, and gypsum – 10%. Finished cements were mixed with water at the w/c ratio = 0.4. Cement paste was formed into specimen that fastened under normal conditions for 1, 3, 7, 14 and 28 days. Then the samples were tested for strength and expansion. Obtained results are given in Table 3 and Figure 3.

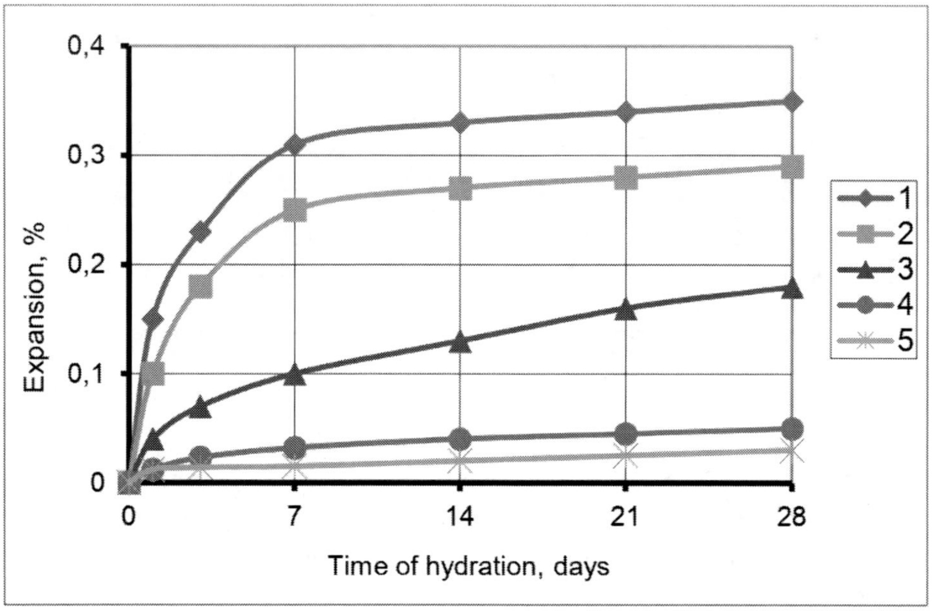

Fig. 3. Expansion of produced cement with expansive admixture and gypsum. fraction numbers as indicated in Fig. 1

Table 4. Strength and self-stressing of mixed cements for different fractions.

Grain size, μm	**Mixed cement with S = 2700 cm²/g**				**Mixed cement with S = 3500 cm²/g**			
	Strength, MPa		self-stressing, MPa		Strength, MPa		self-stressing, MPa	
	1 day	28 days	1 day	28 days	1 day	28 days	1 day	28 days
<28	15	58	0,48	1,5	18	61	0,6	1,8
28-45	14,8	56	0,44	1,3	17	59	0,55	1,6
45-63	14,2	53	0,41	1,12	15	57	0,5	1,5
63-80	12,5	51	0,32	0,96	13	54	0,4	1,3
>80	10,6	44	0,08	0,9	11	48	0,1	1,2

The strongest expansion was observed in samples that contain small-size fractions of SFC minerals, as their hydration occurs more intensely that hydration of coarse fractions of the minerals. Expansion relationships in ORC-based cements with different specific surface areas (2700 and 3500cm²/g) were almost identical (Fig. 3). Furthermore, a less porous stone structure is formed that ensures high strength of samples, and if expansion is restricted, it also ensures self-stressing of the stone. Notably, the best results are obtained in combination with Portland cement clinker ground to specific surface area of S = 3500cm^2/g. (Table 4).

According to the data of electron microscopy studies, formation of cement stone structure with finely-ground Portland cement clinker and fine fraction of calcium sulfoferrite is caused by formation of ferrous ettringite prismatic crystals and by finely-crystalline gelatine mass of calcium silicate hydrates.

Conducted studies reveal that grinding fineness and particle size distribution of cement produce a strong impact on formation of cement stone structure. Furthermore, dispersity of both the Portland cement component, and the expanding component is significant.

Increasing the fineness of Portland cement grinding with the permanent fraction of expanding component results in higher cement strength. The effect of expansion depends both on the fineness of expanding component, and on its type.

In case of calcium sulfoferrite-based expanding component, the strongest expansion effect is produced when the component is introduced into cement in small-size fractions (<28μm and 28-45μm). Their hydration accelerates both due to increased reacting surface of particles, and their activity. Higher hydration activity of small-size particles ensures that formation of cement stone non-shrinking structure accelerates, while fast hydration of finely-ground Portland cement component contributes to compaction of cement stone structure and increase of its strength.

3.3 Resistance of calcium sulfoferrite-based cement stone to abrasion

The above-mentioned studies demonstrate how a calcium sulfoferrite-based binding material with non-shrinking properties can be obtained. When hardened, this cement forms a strong and dense cement stone, and can ensure its resistance to abrasion if used for sealing joints and seams.

To confirm this assumption, tests were conducted on samples made from cements that contained calcium sulfoferrites. Obtained results are given in Table 5.

Table 5. Resistance to abrasion of mixed cement with calcium sulfoferrite.

Consumption of cement, kg/m^3	400	425	450
Water–cement ratio	0.45	0.42	0.40
Weight loss (kg per 1 m^2 of surface abrasion for 1 hour) water-saturated samples	0.55	0.48	0.30

4 Discussion

The study of hydration in poly-sized compositions of $2CaOFe_2O_30.8CaSO_4$ mineral demonstrated that small-size fractions ensure the formation of crystallization centers, while the less than 45μm particles, constantly interacting with the liquid phase, provide for gradual growth of crystals.

Conducted studies allow to conclude that for calcium sulfoferrite-based expanding admixtures it is preferable to use a poly-sized composition that combines only fine fractions (<28μm and 28-45μm).

To produce expanding cements based on calcium sulfoferrite admixtures the best practice is to introduce these minerals in fine fractions. To obtain dense strong cement stone, Portland cement component should be ground to specific surface area of at least 3000cm^2/g.

The results demonstrate that samples made from cements containing calcium sulfoferrites are characterized by resistance to abrasion. Hence, they can be recommended for using in structures exposed to increased physical impacts.

Conclusion

As a result of conducted studies, the following conclusions can be made:

1. Formation of ferrous ettringite prismatic crystals always occurs regardless of the fraction of mineral $2CaOFe_2O_30.8CaSO_4$. The mineral's dispersity affects only its hydration activity. Smaller-size fractions hydrate faster, while medium-size and coarse fractions are very slow to hydrate due to low hydration activity. Formation of ferrous ettringite prismatic crystals is observed only at the age of 7-14 days.
2. During hydration in poly-sized compositions of $2CaOFe_2O_30.8CaSO_4$ mineral small-size fractions of less than 28μm ensure formation of crystallization centers, while particles of 28 to 45μm, constantly interacting with the liquid phase, provide for gradual growth of crystals.
3. To produce calcium sulfoferrite-based expanding additives it is preferable to use a poly-sized composition that combines only fine fractions (<28μm and 28-45μm).
4. Formation of cement stone structure with finely-ground Portland cement clinker and fine fraction of calcium sulfoferrite is caused by formation of ferrous ettringite prismatic crystals and by finely-crystalline gelatine mass of calcium silicate hydrates.
5. Calcium sulfoferrite-based binding materials, having a high strength and an ability to expand, as well as to form cement stone resistant to abrasion, may be used in high-rise construction as a material for joints and seams sealing.

References

1. Osobennosti otsenki vetrovogo vozdeystviya na vysotnye zdaniya, Youth science forum: Technical and mathematical Sciences, Moscow, **1 (30)** (2016).
2. S.V. Samchenko, Sul'foalyumoferritnye tsementy, svoystva i primenenie/ Beton i zhelezobeton, Sbornyy zhelezobeton, **1 (10)** 18-21 (2014)
3. S.V. Samchenko, T. V. Kouznetsova, Resistance of the calcium sulphoaluminate phases to carbonation, Cement, Wapno, Beton. **5**, 317-322 (2014)
4. T. V. Kouznetsova, et al., Carbonation of the constituents of hydrated Portland cement, aluminate and sulphoaluminate cements, 13 Int. Baustofftagung, Ibausil, Weimar, Bundesrepublik Deutschland, 2, 2-0543-2-0546.
5. P. Hou, R. Zhang, Y. Cai, X. Cheng, S. P. Shah Construction and Building Materials, **113**, 890-896 (2016)
6. T.V.Kouznetsova, et al., The use shrinkage-compensating cements in metro building/ 12th International Congress on the Chemistry of Cement (ICCC). Montreal, Canada, pst. 5.07 (2007)
7. T. V. Kouznetsova, et al., Modifitsirovanie portlandtsementa dlya ustraneniya usadki betona, Tsement i ego primenenie, 14-15 (2007)
8. S.V. Samchenko, *Sul'fatirovannye alyumoferrity kal'tsiya i tsementy na ikh osnove*, (Moscow, 2004)
9. A.S. Brykov, et al., Sul'fatostoykost' portlandtsementnogo kamnya s alyumosoderzhashchimi uskoritelyami skhvatyvaniya, Tsement i ego primenenie, **5**, 59-63 (2013)

10. G.S. Royak, et al., Shlakoportlandtsement dlya predotvrashcheniya korrozii betona, Tsement i ego primenenie, **3**, 104-106 (2011)
11. K.B. Safarov, V.F. Stepanova, Regulirovanie reaktsionnoy sposobnosti zapolniteley i povyshenie sul'fatostoykosti betonov putem sovmestnogo primeneniya nizkokal'tsievoy zoly-unosa i vysokoaktivnogo metakaolina Stroitel'nye materialy. **5**. 70-73 (2016)
12. L.Ya. Kramar et al., Betony vysokoy sul'fatostoykosti, Tsement i ego primenenie. **4**. 127-131 (2011)
13. S.V. Samchenko, O.V. Zemskova, D.A. Zorin Corrosion resistance of sulfated cements in carbonate and in carbonate-sulfate mediums, MATEC Web of Conferences **106**, 03014 (2017).
14. S.V. Samchenko, Yu. R. Krivoborodov, Vlijanie dispersnosti special'nogo cementa na strukturu tverdejushhego kamnja, Vestnik Belgorodskogo gosudarstvennogo tehnologicheskogo universiteta im. V.G. Shuhova. **2/5**. 238-240 (2003)
15. A.P. Osokin, Yu. R. Krivoborodov, Svojstva rasshirjajushhihsja cementov i ih primenenie, Cement i ego primenenie. **6**. 43 (2004)
16. S.V. Samchenko, et al., Influence of fineness of expansive components of cement properties, Cement-Wapno-Beton, XIII/LXXV, **5**, 254-257 (2008)
17. Yu. R. Krivoborodov, S.V. Samchenko, *Fiziko-khimicheskie svoystva sul'fatirovannykh klinkerov* (Tsementnaya promyshlennost', Moscow, 1991)
18. I.N. Borisov, et al., Sintez sul'foferritnogo klinkera dlja proizvodstva bezusadochnyh i rasshirjajushhihsja cementov. Sovremennye problemy nauki i obrazovanija, **2**, 269 (2012).
19. S.V. Samchenko, *Rol' jettringita v formirovanii i genezise struktury kamnja special'nyh cementov* (RHTU im D.I. Mendeleeva, Moscow, 2005)
20. S.V. Samchenko, *Formirovanie i genezis struktury tsementnogo kamnya* (Moscow, 2016)
21. Russian Standard GOST 30744-2001 CEMENTS Methods of testing with using polyfraction standard san.
22. Russian Standard GOST 13087-81 Concretes. Methods for determination of abrasion.

On certain development aspects of an ipsas-based system-target approach to evaluation of net asset sustainability level projects in high-rise construction

Ruben Kazaryan[1,*]

[1]Moscow State University of Civil Engineering, 26 Yaroslavskoe shosse, Moscow, 129337, Russia

Abstract. Problems of accounting and reporting of net assets and the procedure of their formation taking into account the specifics of the economic and legal status of property of a non-commercial autonomous institution are some of the most controversial in the accounting for entities of the public sector. The study focuses on justification of accounting rules for net assets of public sector entities. The methods used in the study are as follows: comparison, synthesis, analysis, logical approach, and system approach. The article examines legal aspects and specifics of recognition of assets of public sector entities in accordance with IPSAS standards (International Public Sector Accounting Standards are a set of accounting standards issued by IPSASB (Council for International Financial Reporting Standards for Public Sector Organizations) used by state-owned enterprises worldwide in preparation of financial statements as of the 31st of August, 2015. The most crucial factor in the modeling of key performance indicators of the system-target approach to estimation of the sustainability level of net assets on the basis of IPSAS is a multicriterial evaluation of the basic management strategy for quality system elements used in operational and strategic planning projects operations in high-rise construction. We offer an alternative evaluation of assets due to be returned to the right holder (the state controller) in the event of liquidation of a public sector entity.

1 Introduction

The concept of the public sector efficiency, forming the basis of public administration reforms in this area, faces a number of outstanding methodological and organizational problems. In particular, such problems include the absence of a generally accepted notion of public sector's economic efficiency in scientific literature and practice; lack of established economic efficiency criteria and indicators. Methodological approaches and principles of the economic efficiency determination in public and commercial sectors must be uniform. We believe that this approach is untenable in view of fundamental differences of the principles and objectives of the non-profit sector, and its deep-rooted economic and social distinctions, and requires some additional research in the field of financial accounting and reporting.

*Corresponding author: r.kazarian@mail.ru

Acritical indicator of economic efficiency is "net assets". The key purposes of assessment of a non-profit organization should include not only the achievement of maximum social impact, but also the optimization of budget resources and economic performance. In this regard, special significance is acquired by the "net assets" category, which allows, on the one hand, a review of the investment process in the public sector, and, on the other hand, an estimation of increase in the total value of a non-profit enterprise. This indicator can also be used traditionally as a measure of growth of retained earnings from chargeable operations of a non-profit organization.

In the conditions of risk and uncertainty, there is a need for continuous management of economic processes in the public sector. Changes of the last decades have shifted the focus financial and managerial accounting from cost management and financial flows onto economic processes' management (financial standing, risks, backup enterprise system, reorganization processes, value added control), based on the use of accounting engineering tools (monitoring, financial, hedging, or other derivative reports). The net assets indicator, in conjunction with net liabilities, represents one of the most important measures of assessing economic processes, efficiency, and sustainable development of a nonprofit enterprise. However, as concerns businesses that are not focused on profit-making or satisfaction of public demand for services as the outcome of investment, this indicator is difficult to determine. Therefore, a lot of proven models and methods of accounting for net assets cannot be directly applied in this area. This situation leads to a mismatch between the urgent need for scientific methodology in the field of information and analytical support of management and evaluation of the efficiency by using the net assets indicator. A significant enhancement of the financial information quality and the formation of adequate information support for control and management processes to a wider audience, improving the decision-making process regarding the allocation of resources, ensuring greater transparency and accountability of decision-makers (Bellanca&Vandernoot, 2014). Application of these standards allows an improved control and supervision of the budget as well as development of communication tools for promoting dialogue and synchronizing the work of state institutions of different countries. (European Commission, 2012b; European Commission, 2012a). IPSAS have actually become international benchmarks for evaluation of accounting practices in the public sector throughout the world. For these reasons, IPSAS deserves attention of accounting policy makers, as well as practitioners and researchers (KPMG, 2013). Today, the development of national financial accounting and reporting standards is based exactly on this dynamic group of international standards applied worldwide [1-5].

This gives rise to a problem associated with the need to develop new national and international accounting models and tools for net assets in the non-profit sector, and to adapt the existing ones, as well as to develop effective methodologies based thereon.

2 Methods

The study uses logical system, synthesis, analysis, and analogy methods, as well as the comparative analysis method. The purpose of analysis is to identify differences between IPSAS and the public sector standards published by the Ministry of Finance of Russia. The purpose also includes a system review of draft standards and draft resolutions on the basis of published information and opinions, surveys of the Ministry of Finance and the parties concerned (public sector executives, accounting and state control personnel). The surveys have been conducted in the form of a questionnaire. Of special interest are the respondents' views on the expedience of the use of IPSAS in the Russian Federation. We appreciate the loyalty of respondents to the prospects of using IPSAS as the IPSAS basis. We also research in the possibility of the accounting tools' synthesis based on IPSAS and of organization of state and public supervision, including that based on the "Electronic Budget" technology.

The synthesis method is used to categorize positive and negative consequences of the IPSAS introduction as the basis of Russian accounting standards in the public sector. The analysis is based on expert examination of publications and documentary materials of the main regulatory authorities (reports, regulations, conference proceedings and discussions).

3 Results

3.1 The influence of theoretical and legal features of operations of public sector entities on interpretation of the "Net assets" category

Public sector entities are characterized by features that pre-determine the peculiar characteristics of the composition and disclosure of information about the accounting items. As defined in IFRS (International Financial Reporting Standards) 22 "Disclosure of Financial Information about the General Government Sector", a non-profit organization is a legal entity or an organization of other forms, which that are created with the purpose of production or distribution of goods and services, but which do not generate financial benefits for regulatory bodies [2-3]. Thus, their distinctive feature is also the complete absence, or limitation, of the right of ownership to property and other resources administered by a non-profit institution, which are publicly owned and controlled by the state authorities.

Distinctive features of non-profit entities include:
- lack of generation of economic benefits as the main goal of activities.
- Predominance of a non-market way of organizing activities.
- Production, distribution, and consumption of public goods.
- Lack/restriction of the ownership right to property and other resources administered by a non-profit institution, which are publicly owned and controlled by the state authorities.
- Ensuring an economic equilibrium between demand and supply of public goods through state mechanisms (social institutions, infrastructure, and resources).
- The ability, within acceptable limits, to carry out activities aimed at deriving additional economic benefit while maintaining the objective functions of the public goods distribution and achievement of goals of the state in meeting social needs.

Table 1 summarizes key distinctive features of non-profit institutions and determines the influence on recognition of and accounting for net assets/equity of such institutions.

The particular features of interpretation of net assets of a non-profit institution shown in the table are approached differently in the methodology of national and international financial reporting standards. Application of international standards and accounting principles is based on the following groups of documents:
- For commercial organizations – International Accounting Standards (IAS) and International Financial Reporting Standards (IFRS) developed by the Council for IASB (IASB).
- For public sector – International Public Sector Accounting Standards (IPSAS) developed by the Council for IASB OS (IPSASB) under the International Federation of Accountants (IFAC). Definition of the targets and application of both groups of standards is based on the principle of prevalence of economic substance over legal form. This allows qualification of a commercial entity for the purpose of applying the IFRS OS solely on the basis of the objectives of the enterprise and its functions to be implemented in the economic and social environment, regardless of the legal form of ownership and the structure of investment by the state. Therefore, recognition of assets and funding sources in the paradigm of IFRS is based solely on the criteria of their use for achievement of the main goals of the enterprise [2-5].

Table 1. Influence of distinctive features of non-profit institutions on recognition of and accounting for net assets.

Description	Distinctive features of recognition and accounting in Capital/Net Assets categories
Lack of generation of economic benefits as the main goal of activities	Eliminates the possibility of using the term "Capital" for financial evaluation of investment into the operational and financial cycle for the purpose of deriving economic benefits
Predominance of a non-market way of organizing activities,ensuring an economic equilibrium between demand and supply of public goods through state mechanisms (social institutions, infrastructure, and resources)	There is no traditional profit-making and accumulation mechanism characteristic of commercial organizations, with no accumulation of net assets over time
Lack/restriction of the ownership right to property and other resources, administered by a non-profit institution, which are publicly owned and controlled by the state power authorities	Eliminates the recognition of assets and capital in terms of control over economic benefits/ opportunities of beneficial use.
The ability to, within acceptable limits, carry out activities aimed at deriving additional economic benefit while maintaining the objective functions of the public goods distribution and achievement of goals of the state in meeting social needs	Creates a traditional business model of the resources' flow, introduces the concepts of "capital", "revenue", "profit", thus determining a mixed character of the net assets category (containing both a non-commercial component, a financial assessment of the resource transferred by the control entity (the state), and own assets recognized in terms of control of future economic benefits and derived profit)

Pursuant to paragraph 17 of IFRS OS 1 "Presentation of financial statements", statement of financial position should as a minimum disclose the value of net assets/equity. Pursuant to paragraph 95 of IFRS OS, if an organization does not have any share capital, it should disclose its net assets/ equity either in the financial statements or in the Notes, with breakdown for:
- Contributed capital meaning the aggregate amount of contributions by owners less payments to owners as of the reporting date,
- Accumulated surplus or deficit.
- Reserves with description of the nature and purpose of each reserve in the assets/ equity.
- Minority interest.

Review of the net assets of a non-profit institution should take into account particular features of functioning of these enterprises. For instance, net assets of a non-profit enterprise should be understood as a difference between its assets and liabilities. However, this category is not homogeneous from the point of view of its economic and legal content. The proposed composition of the items does not reflect all the specifics of a non-profit organization's activity; thus, this grouping as implemented within the concept of international accounting standards does not allow for a full analysis or presentation of reliable and relevant information about net assets to the users.

3.2 The problem of recognizing the property and assets of public entities controlled by the state.

"Absence/restriction of the right of ownership to the property and other resources administered by a non-profit institution that are publicly owned property belonging to state authorities", identified earlier deserves special attention. In connection with this restriction, there is a problem of IFRS OS application for recognition of the property of a public entity as assets and for reflection thereof within its net assets and sources. There is a problem of recognition of such property within assets, its share in the value of net assets of the company, as well as the order of recognition of settlements with the founders as sources of funding as the organization's capital or other type of liability. Legal and scientific literature offers different approaches to solving this problem. However, in the light of the IFRS OS positions and reforms of accounting on the basis laid down in the specified standards, these issues remain unsolved and are of a disputable nature.

3.3 Categories of the assets of a public entity in terms of "Control of assets by the state"

We believe that property and its sources should be categorized as follows:
- The property of the founder as the holder of the right of ownership and other types of property, which may not be foreclosed upon liquidation of the legal entity.
- The property created and controlled by the non-profit organization at the expense of its own and third-party private sources, which may be foreclosed upon liquidation of the legal entity.

We propose the following categories of property for subsequent recognition within the assets (a share in the net assets) (Table 2).

Table 2. Categories of the assets of a public entity in terms of "Control of assets by the state".

<table>
<tr><th>Type of asset</th><th colspan="2">Subtype of asset</th><th>Conditions of the asset's return to the state controller</th></tr>
<tr><td rowspan="2">The property of the founder as the holder of the right of ownership and other similar types of property</td><td>The right vested in the public entity</td><td>Acquired at the expense of the funds allocated by the subject as the holder of the right of property</td><td rowspan="2">Demand of return to the founder or another owner in case of liquidation/ termination</td></tr>
<tr><td colspan="2">Most valuable property as reflected by separate accounting. Can be withdrawn as redundant or used other than as intended.</td></tr>
<tr><td rowspan="2">The property created and controlled by a non-profit organization at the expense of its own and third-party private sources, which may not be foreclosed upon liquidation of the legal entity, etc.</td><td colspan="2">Other state-owned property that may be administered independently</td><td rowspan="2">May foreclosed upon liquidation of the legal entity.</td></tr>
<tr><td colspan="2">The property acquired at the expense of income-generating activities, the property received gratuitously from the third parties and other similar types of property</td></tr>
</table>

The following problems of accounting for and assessing the value of net assets of a non-profit organization are of special urgency:
- The conditions and criteria for recognizing the aforesaid property as assets on the IFRS OS basis, including the concepts of control, risks, and benefits.
- The conditions and criteria for recognizing settlements with the owners and property controllers, as well as proceeds thereof within the liability side of the balance-sheet (Capital, Liability).
- Accounting for the aforesaid types of property in the assessment of the net assets of a non-profit autonomous institution from the point of view of serving the different groups of the users' interests (paying capacity, liquidity, investment attractiveness, sustainable development, etc.).

- The criteria for recognizing the revenue and the accrued financial result.

According to IPSAS, a public sector entity has control of its assets, if it can use them for receiving future economic benefits or useful potential in the achievement of its goals and can exclude or otherwise regulate the access of other entities to such economic benefits or the useful potential [1-3]. The useful potential contained in the assets is their ability, independently or jointly with other assets, to be used for providing the state (municipal) services (paid services, works, manufacture of products) and/or performance of the state (municipal) and other functions in compliance with the entity's purposes, without generating any cash flow (cash equivalents) for the entity [1-3]. In other words, the benefits generated by the asset cannon be evaluated in cash terms, but do exist and have a social or another other useful potential for the society, a feature quite typical for non-profit organizations and their non-market functioning principles.

Pursuant to para 7 (IPSAS) 23 "Income from non-exchangeable operations (taxes and transfers)", control of an asset arises when the organization can use the asset or otherwise derive benefit therefrom, pursuing its goals, as well as close or otherwise regulate the access to this benefit for other parties. The said criterion of the entity's control over its economic benefits is strictly observed, and all property mentioned in Table 2 can be recognized as assets of a non-profit organization in accordance with IPSAS. In this regard, the right of ownership to the transferred property should be neglected in compliance with the principle of priority of the economic content before the legal form. The property is fully reflected in the assets of the balance sheet / statement of financial position.

3.4 The problems of recognizing the value of assets to be returned to the right holder (state controller) upon liquidation of a public sector entity. Evaluation of net assets

There is an out standing problem of the conditions and criteria of recognition of settlements with owners and property control entities in the liability side of the balance-sheet (Capital, Liability). Delineation between the funding sources in the liability side of the statement of financial position and their qualification as capital or revenue is contained in IFRS OS 23 "Revenue from non-exchange transactions". It provides for distinction between contributions of the owner to the institution's net assets and valuables received in non-exchange transactions (i.e. for no equal valuable consideration) [3,4].

The owner's contribution is qualified according to IFRS OS 1 and means future economic benefits (including the possibility of beneficial use) from third parties giving no rise to liabilities for the organization and forming a share in the organization's net assets /capital which:

- Gives the right to payments allocable by the organization, both as 1) the future economic benefits or a possibility of beneficial use for the period of its existence at the discretion of the owners or their representatives, and 2) any excess of assets over liabilities in the event of the organization's liquidation.
- May be sold, exchanged, transferred, or redeemed.

Considering the correspondence of the highlighted assets in the composition of the organization with the specified definitions, we should put an increased focus on the property assigned to the institution and purchased at the expense of the funds that have been allocated by the subject who is the holder of an ownership right [4,5].

There arises a contradiction between the economic and legal substance of this indicator and the definition of the contribution of the owner, forming the capital/net assets of the institution. Since the requirement of the ability to distribute any excess of the assets over liabilities in the event of liquidation of the organization is not implemented to the full extent. Analysis of IPSAS 1 and IPSAS 23 shows that IPSAS does not contain any detailed instructions on how to reflect the specificity of the settlements accounting with state property owners. The definition of "the owner's contribution" is contrary to the economic and legal content of the category of settlements with the founders of a non-profit institution, as it does not take into account the need for the return of property in case of liquidation. According to IPSAS, the paid-up capital in the public sector can be confirmed by the transfer of resources between the parties[3,4,5].The issue of equities during the transfer of resources is not a necessary condition for matching the transfer to the definition of payables from owners. The transfer of resources which rewards a share in the net assets/ capital of an organization is distinguished from the transfer of other resources if it can be confirmed by the following:

• By an official confirmation of the contractual parties on the transfer of resources (or on a type of such transfer) as an integral part of the net assets/ capital, either before or during the implementation of the contribution. For example, when creating a new organization, the budgetary authority at the Department of Finance may perceive the initial transfer of resources to the organization as a share in form of the net assets/ capital, but not as providing funds to meet current needs,

• By a formal agreement in respect of the resources' transfer, creating a new share or increasing the existing one in the net assets/ capital of the organization that may be sold, transferred, or redeemed.

In the scientific literature, there is the authors' position that the settlements with the founding shareholders represent in their initial assessment a current long-term debt of the company, which is based on the approach that the institution must return the received assets during liquidation. As criteria of recognition of the current liability, the authors cite such arguments as occurrence of the event in the past (transfer of property) and the eventual outflow of future economic benefits (in the event of liquidation of the organization).

We believe that each approach has its advantages and disadvantages, but does not fully solve the problem of preparation of information on investments of the owner in a part of the secured property and the property created in the process of the enterprise's functioning, as well as recognition in the financial statements.

The net assets of a non-profit institution are formed as an increase of economic benefits arising in the operation process, of the result of changes in the value of assets. However, disputable remains the question of recognition of value of the investments of the owner directly in the composition of net assets, which may be seized without the consent of the institution at the time of liquidation of the organization or other cases stipulated by the law.

3.5 Estimation of the net assets on the basis of value adjustment of the assets subject to return to the right holder (state supervisor) during the liquidation of the public sector entity

The composition and value of the net assets represents significant information for the users of external reporting of a non-profit organization. The non-profit organization submits public accountability paperwork based on the IPSAS principles, primarily with the goal of providing users with information about the financial standing both to ensure control and to attract alternative sources of funding. Quite relevant for such institutions there seems to be the investment attractiveness, in connection with additional features of such organizations to provide services and perform projects and works for third parties on a commercial basis. The amount of net assets under this approach characterizes both the financial security, development stability, and the investment component of non-profit institution activities (both from the part of the owner of state property, and directly by the non-profit institution itself).

At the same time, the net assets of the company traditionally serve as a tool for assessing solvency and resource endowment from the point of view of their possible distribution among the creditors in case of liquidation of the enterprise, as well as in its current activity. In this case, the principle of conservatism requires a precautionary approach to the recognition of formation sources of assets directly in the capital funds, as well as the inclusion of the assets within the net assets taking into account their feasibility and use to settle claims of the contractors. Despite the fact that the resources transfers can be executed by means of confirmation or official agreement, the organization determines the nature of such transfers on the basis of their economic substance and not merely their legal form[6,7,8].

To achieve the aforesaid purposes of providing the financial statements' users with the relevant and reliable information, the current IPSAS approach described above is not able to provide any universal method for the formation of the net assets of a nonprofit organization.

4 Discussion

4.1 The methodology for the recognition and qualification of assets subject to return to the right holder (the state supervisor) during the liquidation of the public sector entity

Assets subject to return to the right holder (the state supervisor) during the liquidation of the public sector entity may be perceived as the initial transfer of resources to the organization as a share in net assets/ capital, and not as provision of funds to meet current needs. However, based on the definition of the owner's contribution into IFRS OS 1 "Submission of financial statements", there is no fulfillment of the requirement on possibility of distribution of any excessive assets over liabilities in the event of liquidation of the organization. As already noted, this property may be withdrawn and will not increase the assets in case of liquidation of the organization. We believe that classifying such investments of founders as "the owner's contribution" in the statement of financial standing is not appropriate, and does not meet the requirements of IFRS OS.

The property to be transferred can be fully depleted in the course of the institution activity throughout the useful economic life and physical ability to participate in the operational and financial cycle of the organization. Loss, damage, the asset's loss of its physical and moral characteristics due to the use does not entail the obligation to repay to the owner the original value of the property or to represent property, similar in its characteristics. The return of the asset to the right holder (the state supervisor) during the liquidation of the public sector entity involves the return of a physical object, not a money equivalent. Thus, no requirements arise in respect of the compulsory equivalent value of the founder's investments, subject to returned at the time of liquidation and equal to the value of the initial gross proceeds.

The transfer of such asset is the symptom of a non-exchange transaction [6-10].Therefore:
- The property is settled on a non-profit organization for the implementation of its statutory activities (extraction of economic benefits or opportunities for beneficial use) with the priority objective to fully utilize throughout the useful economic life/ tangible existence.
- The property shall be returned in the event of liquidation of the organization as a physical object (right) on the date of liquidation and not as a money equivalent.
- The property shall be returned in the same condition and with the same physical and moral characteristics, which it has had on the date of liquidation of the organization.

That is, there are two features of the property return in liquidation:
- The property will be returned not in full (only that property, which exists physically on the date of the liquidation).
- The property returns in terms of moral and physical wear and tear, damage, and partial loss of properties.
- The property returns after an indeterminate period of time, which affects its cost measurements.

The requirement of IPSAS to reflect the value of such property at fair value, i.e. the amount by which it is possible to exchange an asset or settle a liability, when a transaction is being made between the well-informed parties, wishing to make such a transaction, and independent from each other. In our opinion, it is necessary to use the rating recommended by IPSAS, as it most accurately reflects the value of the property as contributions of the founder to the organization, provides the required control of evaluation objectivity and allows a more reliable estimation of net asset amount[8-12].

The fair value is, in fact, is an alternative rating of the governmental investments to creating an institution, as a means of implementing the functions of distribution and redistribution of public goods, which is the "effect" of such investments. Thus, when evaluating gross revenue in the part of the fixed assets, we should, as far as possible, be guided by the approach of market evaluation.

4.2 Alternative evaluation of the assets subject to return to the right holder (the state supervisor) during the liquidation of the public sector entity

Based on the above analysis, we can conclude that the asset to be returned to the right holder (the state supervisor) during the liquidation of the public sector entity corresponds to the two sources of formation. The property on the date of investments is the gross inflow of economic benefits in their fair assessment and reflects the investment of the state property owner that implements its objectives through a non-profit institution. At the same time, it is impossible to ignore the fact of the pre-emptive right for the withdrawal of such property during the liquidation of the organization in physical terms and as of the date of the institution's liquidation. The pre-emptive right of liquidation, of course, creates a present obligation for the enterprise, since the following requirements are met at the same time:

- The property shall be returned in the event of liquidation of the organization as a physical object (right) on the date of liquidation and not as a money equivalent.
- The property shall be returned in the same condition and with the same physical and moral characteristics, which it has had on the date of liquidation of the organization.

However, as noted above, the obligation arises only in relation to a physical object in the volume and condition of wear on the liquidation date. The amounts of accumulated depreciation and residual value of this object does not, in our opinion, reflect the true assessment of the property. As depreciation, although they may be correlated with a tendency of loss of physical and moral characteristics by the object in some cases, as a whole, are predominantly a tool to compare the income and expenses of the organization and the resulting residual value cannot give a reliable valuation of the asset at the current date [9-13].

In its economic content, the property object assigned to the institution represents potential investments of the owner into alternative projects at the time of return (it can be transferred to public auctions and privatization, in front of the other institution, used in private activities, etc.). From this point of view, the cost of the property return is a recoverable amount. The most accurate assessment is, in our opinion, the liquidation value of the asset. According to article 13 IFRS OS 17 "fixed assets", liquidation value is the estimated amount that the organization would receive currently from the disposal of the asset after deducting the expected costs of disposal, as if the asset was already of age and in the condition expected at the end of the beneficial use [10-14]. Thus, the author of the dissertation research suggests recognizing the obligation of the pre-emptive right for the recovery of property assigned to a public non-profit institution from the part of the owner in the amount of the asset's liquidation value, and to exclude the specified amount from the composition of the net assets of the organization. The scheme of the liquidation value calculation is as follows:

- Corrected (revaluated) cost of the asset.

- Sum of current costs associated with the liquidation (the costs of maintaining assets prior to their sale, management costs, etc.).
- The value of all obligations associated with the liquidation.

At the same time, on the liabilities side of the statement of financial standing there must be shown the gross inflows of economic benefits at the fair value of the property assigned to a state non-profit institution from the part of the owner that will result in an increase of the net assets. In our opinion, when calculating the net assets from the gross revenue of economic benefits in fair evaluation on the date of receipt, we must exclude the amount of the liabilities, calculated based on the liquidation value of the assets subject to exemption on the date of liquidation. Thus, the cost excess of the gross income from the founder over the amount of the specified liabilities will amount to the proportion of the net assets that are "owned" by the institution and implemented in the process of activity throughout its existence. Also, it is proposed to eliminate the excess of liabilities amount over the initial carrying value of assets in terms of property assigned to the institution. In this case, when evaluating the liability, the author of the dissertation research recommends to use the lower of the two ratings:
- The liquidation value of the property intended to be seized.
- The carrying value of the property intended for seizure, recognized at the time of investment by the owner.

This is due to the principle of interaction between the property owner and the public non-profit institution, according to which the investments sent by the state property owner do not have either the purpose of extracting economic benefits from the use of the transferred assets, or otherwise gaining additional economic benefits. Thus, the owner may not recover economic benefits more than he has invested during the formation and functioning of the non-profit institution in accordance with the principle of non-market interaction [12-20].

5 Conclusions

According to the recommended method, at the time of formation or subsequent investment by the founder it is reflected as the gross inflows of economic benefits/opportunities for beneficial use directly in an increase of the net assets of the organization (see the Section "Net assets/Capital" in accordance with IFRS OS 1 or the Section "Other liabilities" in accordance with Federal Accounting Standards in the public administration sector "Presentation of financial statements"). On the subsequent reporting date of the financial statements, it is recommended to reclassify the above economic category within the liabilities, thus ensuring the exclusion of the property value subject to seizure from the net asset This will ensure adequate assessment and information disclosure:
- On the obligations in front of the founder for the property, on which there is a preferential right of withdrawal upon liquidation in the most reliable (liquidation) assessment, reflecting the actual characteristics of the property on the date of withdrawal.
- On the share of the gross income from the founder forming the net assets of an institution and "functioning" in the organization.

References

1. M. Bisogno, S. Santis, A. Tommasetti, Int. Journal of Public Administration, Public-Sector Consolidated Financial Statements: An Analysis of the Comment Letters on IPSASB's Exposure Draft, 49.,**38 (4)**, 311-324 (2015)
2. S. Bellanca, &J. Vandernoot, Journal of Modern Accounting and Auditing International Public Sector Accounting Standards (IPSAS) Implementation in the European Union (EU) Member States, **3**, 257-269 (2014)

3. P. Brown, Some observations on Research on the benefits of Nations adopting IFRS. The Japanese Accounting Review. (2013) Doi:10.11640/tjar3-2013.01
4. J. Christiaens, J. Rommel, Accrual accounting reforms: Only for business-like (parts of) government. Financial Accountability and Management, 24 (1), 59-75 (2008)
5. Ernst and Young (2012). IPSAS Explained: A summary of International Public Sector Accounting Stanards. John Wiley and Sons, Ltd
6. URL:http://www.epp.eurostat.ec.europa.eu/portal/page/portal/public_consultations/documents/Privacy_statement_feb_2012_EN.pdf (02.04.2017)
7. URL:http://www.epp.eurostat.ec.europa.eu/portal/page/portal/public_consultations/documents/IPSAS_stakeholders_consultation_paper.pdf (02.04.2017)
8. M. Jesus, S. Jorge Governmental budgetary reporting systems in the European Union: is the accounting basis relevant for the deficit reliability? International Review of Administrative Siences, **81 (1)** 110-133 (2015)
9. URL: https://www.ifac.org/sites/default/files/uploads/IPSASB/IPSASB-Project-Timetable-Template-November-2014.pdf (02.04.2017)
10. KPMG, IPSAS, Africa Implementation survey Price Waterhouse coopers, Towards a new era in government accounting reporting (2013)
11. E. S. Kosonogova, *The Interpretation of financial statements as an object of studies/ Accounting: a view from the past into the future. Abstracts of the international scientific conference Sokolov reading II* (SPb State University, 2013)
12. Order of the Ministry of Finance of the Russian Federation On Approval of the program for federal accounting standards development for public sector organizations of April 10, 2015 № 64н. (In Russ.)
13. Order of the Ministry of Finance of the Russian Federation On Approval of Rules for preparation and refinement of the program for Federal accounting standards development for public sector organizations of March 23, 2015 № 45н. (In Russ.)
14. Order of the Ministry of Finance of the Russian Federation On Approval of the program for federal accounting standards development for public sector organizations of April 10, 2015 № 64н. (In Russ.)
15. B.A. Lyovin, R.R. Kazaryan et. al. *Infographics of anthropotechnical management. Vol.1 Infographicmodeling in the mental activity philosophy* (Moscow, 2016)
16. B.A. Lyovin, R.R. Kazaryan et. al. *Infographics of anthropotechnical management. Vol.2 Conception of advanced development of anthropotechnical security of functioning and life quality* (Moscow, 2016)
17. R.R. Kazaryan et. al. *Infographics of anthropotechnical management. Vol.3 Anthropotechnical management as a means of provision of activities service Lyovin* (Moscow, 2016)
18. R.R. Kazaryan, I.A. Bun'kina, Regarding some aspects of economic basis of products and services certification system development. Nautchnoye obozrenie [Science review] **7** (2016)
19. A.A.Volkov, V.O.Chulkov, R.R.Kazaryan, M.A.Fachratov, E.K.Rahmonov, Applied Mechanics and Materials 584-586 2387-2390 (2014) doi: 10.4028/www.scientific.net/AMM.584-586.2387.
20. A.A.Volkov, V.O.Chulkov, R.R.Kazaryan, M.A.Fachratov, E.K.Rahmonov, Applied Mechanics and Materials, **584-586** 2391-2394 (2014) doi: 10.4028/www.scientific.net/AMM.584-586.2391

On the of neural modeling of some dynamic parameters of earthquakes and fire safety in high-rise construction

Larisa Haritonova[1,*]

[1] Volgograd State Technical University, 400005, Volgograd, Lenin Avenue, 28, Russia

Abstract. The recent change in the correlation of the number of man-made and natural catastrophes is presented in the paper. Some recommendations are proposed to increase the firefighting efficiency in the high-rise buildings. The article analyzes the methodology of modeling seismic effects. The prospectivity of applying the neural modeling and artificial neural networks to analyze a such dynamic parameters of the earthquake foci as the value of dislocation (or the average rupture slip) is shown. The following two input signals were used: the power class and the number of earthquakes. The regression analysis has been carried out for the predicted results and the target outputs. The equations of the regression for the outputs and target are presented in the work as well as the correlation coefficients in training, validation, testing, and the total (All) for the network structure 2-5-5-1for the average rupture slip. The application of the results obtained in the article for the seismic design for the newly constructed buildings and structures and the given recommendations will provide the additional protection from fire and earthquake risks, reduction of their negative economic and environmental consequences.

1 Introduction

On of the construction priorities in the world today is the high-rise construction in big cities. It is explained by the lack of vacant territories OR their high price. The high-rise construction boom in Russia is yet to come. The high-rise buildings' construction in cities or in the country enhances their status unofficially. However, it requires to secure safe functioning processes in these structures. The UNO data shows that by the number of those who suffered or were killed among various modern day catastrophes, the first place is taken by hydro-meteorological disasters (tsunami, floods, etc.), the second place take the geological ones (volcanic eruptions, earthquakes, etc.), the third place is taken by man-made disasters. In which case, in recent years we can observe the regularity of increasing number of man-made disasters at the simultaneous increasing of the natural calamities [1]. More than half of the accidents registered by the Major Accidents Reporting System MARS (functioning under the authority of the European Committee – Italy) at the United Research Center were caused by the following factors: inefficiency of monitoring the environment safety systems and the low

* Corresponding author: haritonova410@yandex.ru

level of occupational safety [2]. The high-rise structures have a higher degree of potential danger in comparison with the common ones. Fires in these buildings can result in more injured people and material damage. The firefighting is hampered by the structure height, difficulties in fire extinguishing instruments delivery to the fire outbreak level and hydro-meteorological conditions, for instance, strong wind speed fluctuations that usually increases with the building's height [3-5].

Along with the climatic conditions, the fire situation in the high-rise buildings is influenced also by social factors; part of them is given in the paper [5]. The article [4] analyzes fires in high-rise buildings in the Russian Federation and abroad, and the recommendations are given to increase the efficiency of firefighting systems. Some critical factors are shown for high-rise buildings:

-the blocking of escape routes inside the structure;
-difficulties in the supply of firefighting means and rescue efforts;
- a very rapid smoke and fire spread.

The fire consequences in high-rise buildings over the recent five years are analyzed in the article [6]. It is stated that in fire guard garrisons of the Russian Emergency Ministry (especially in the provinces but not in the capitals), there are often no life saving equipment in the high-rise buildings. And even if there are some cranked elevators or rung ladders, the firefighting brigades cannot quickly find a place for regrouping or even simply to install the needed rescue equipment. The work [7] shows some means of warning and controlling the fire evacuation which function on the basis of the network technologies. The solution of the firefighting problem is often seen in providing the fire brigades with very expensive turntable ladders and elevators [4]. However, their exploitation is not always effective at the building's height of more than 60 meters.

2 Recommendations on the enhancement of fire extinguishing efficiency in the high-rise building

The number of fire safety rules failures increases significantly over a large time period of using high-rise buildings. For example, the fire-alarm and systems of smoke removal get out of order (resulting from an improper use). Very often, the trapdoors between the outer ladders stairways are blocked or there are no means of individual protection.

It seems advisable
-to cut the circle radius of fire stations;
-to equip the high-rise buildings with special elevators (not less than one for a fire compartment) for moving fire platoons;
-it is extremely important to foresee zones with the increased fire resistance period and equipped with fire cranes;
-to use high fire resistance materials in construction and finishing the high-rise buildings.

It is less expensive but efficient, to rehearse periodically after the appropriate training with the residents or staff members working in the high-rise structures. It is also an effective the installation of video observation systems and the fire extinguishing systems that can work automatically.

3 Use of the neural simulation method for some dynamic parameters of earthquake sources

The urbanization of territories put forward demands to securing the reliability of buildings and structures in order to reduce the risk of living in seismic danger regions. This is especially important due to the fact that we can observe the process of global seismic activization.

Under these circumstances, if in the years 2003-2006 the number of man-made catastrophes exceeded those of natural origin by several times, already in the years 2007-2010 the number of natural disasters significantly exceeded the man-made ones(by several times) [1]. A very intensive construction of high-rise buildings and structures is being observed today not only in big cities but also on the south coast of Crimea which is a seismically dangerous area. The seismic activities of magnitude up to 9 are observed here. Nowadays, the various methods of simulating processes are widely spread [8-17].

When designing and constructing high-rise buildings, it is insufficient to study results of only macro-seismic activity for predicting the possible seismic impact [18-20].

The paper [18] obtained the equations of connections (empirical dependences) with the help of the orthogonal regression method by the average interval values for various dynamic parameters of earthquake sources upon an energetic class. These dependences which this work calls "long-term ones" were obtained by the data of analog seismic recording systems as a result of generalization of 512 pairs of dynamic parameters in the Crimean earthquake sources within the years 1955-2007.

The author has compared the graphically presented above-mentioned empirical dependences with the values of dynamic parameters determined by the data obtained at the digital seismic stations. This comparison showed that the value of the U parameter (the dislocation value(or the average rupture slip) found from the determining the seismic moment (the power characteristic of the earthquake source) by the data of the digital seismic stations were lower in general than were assumed by the long-term dependences (by 1/5 order of magnitude). Therefore, the given article proposes to simulate some dynamic parameters with the help of the Neural Network simulation. Were use the statistic data [18] for the average earthquake dynamic parameter analysis at the foci of 58 Crimean earthquakes in the years 2007-2011. These data are restored by the 306 amplitude spectrums of the digital seismic stations' records.

The neural network architecture included the two nodes in the input layer. The following two input signals were used: the power class $K_П$ and the number of earthquakes N. With the two input variables and two hidden layers, the number of neurons in each of the hidden layers will be taken as1; 2; 4; 5. Let us take the network structures in the form: (2-5-5-1). The U-parameter, the dislocation value (or average rupture slip) was taken as the target data. A 2 x 58 - matrix was utilized. The Levenberg-Marquardt algorithm has been chosen for the network training in this study. As the Performance function - the Mean Squared Error was used. As the transfer function, the hyperbolic tangent was applied.

4 Results

The neural model geometry and some applied algorithms are given in Fig. 1, and some results of the experiments are shown in Fig. 2-3.

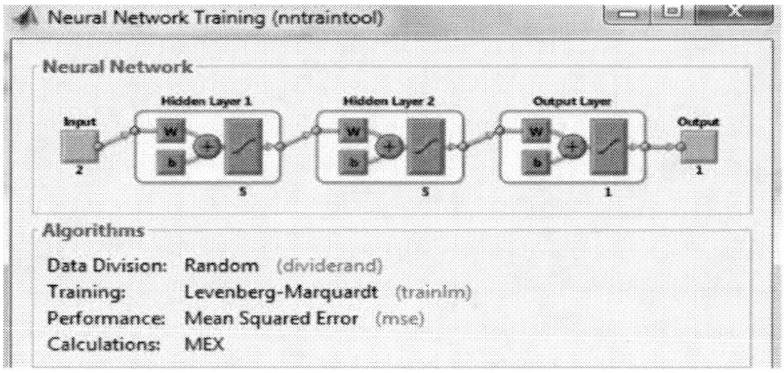

Fig. 1. The example of the network structure 2-5-5-1 and algorithms.

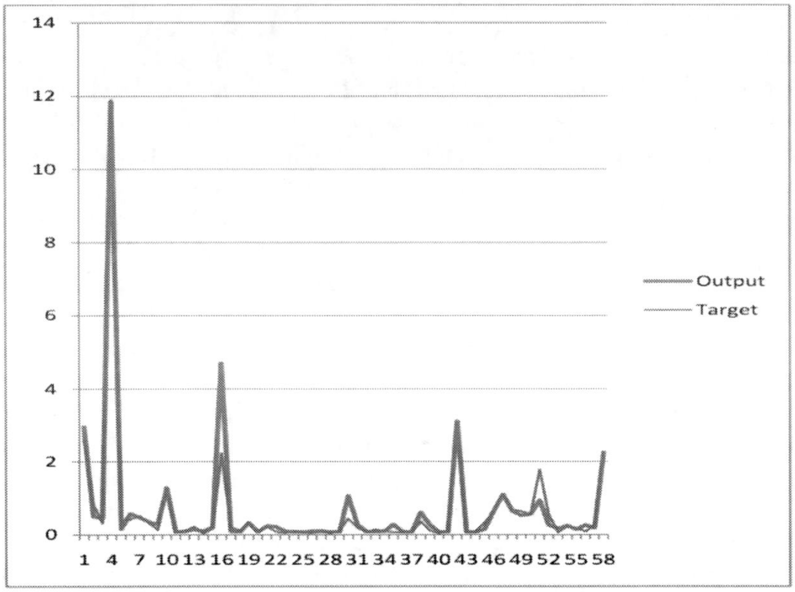

Fig. 2. The graph of the comparison target data for the average rupture slip from the publication [18]) with the output values (calculated with the help of neural network) for the network structure 2-5-5-1.

The equations of the regression obtained in this study for the outputs and target, and the correlation coefficients in training, validation, testing, and the total (All) for the network structure 2-5-5-1 for the average rupture slip are presented in Fig.3.

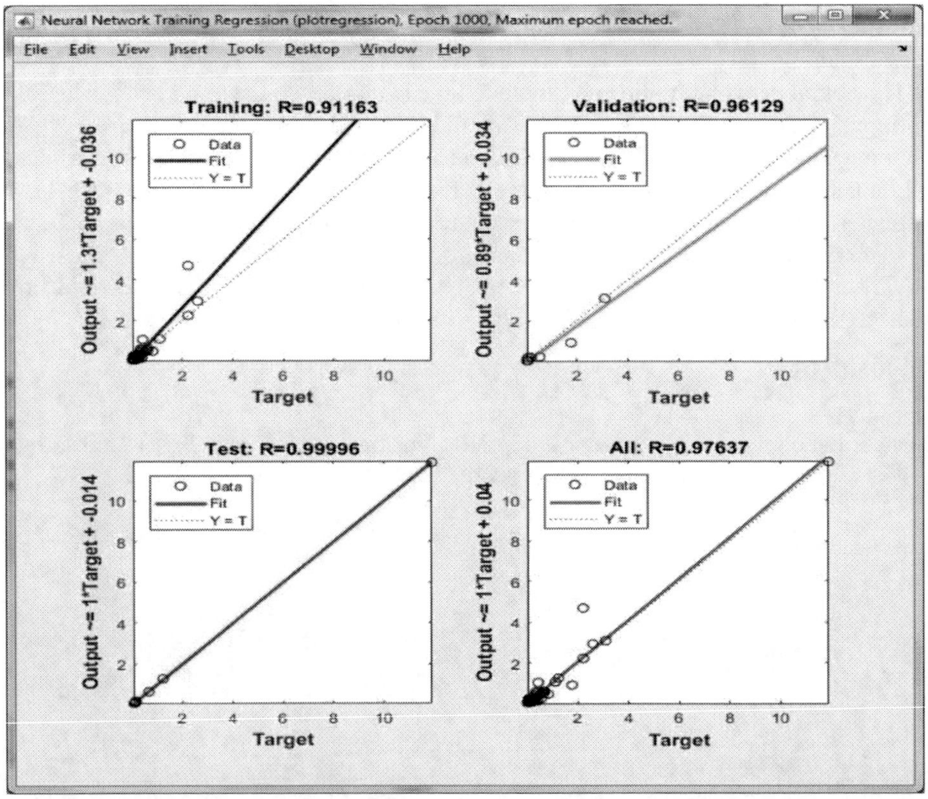

Fig. 3. Equations for the target outputs, target and correlation coefficients in training, validation, testing, and the total (All) for the network structure 2-5-5-1 for the average rupture slip.

The comparison target data for the average rupture slip from the publication [18]) with the output values (calculated with the help of neural network) for the network structure 2-5-5-1 (the graph 2) show a good agreement.

5 Conclusions

In the future it is supposed to continue the study of using the neural network modeling for other dynamic parameters of earthquakes foci to solve practical problems in the engineering seismology. The application of the obtained results in the seismic design for the plotted and constructed buildings and structures as well as the presented recommendations will result in the additional protection from possible fire and earthquake risks, the reduction of their negative environmental and economic consequences.

References

1. Y.V. Sychev, Internet Zhurnal Tekhnologii Tekhnosfernoy Bezopasnosti, **1** (**41**) (2012)
2. W.G.Garrison, *Large property damage losses in the hydrocarbon-chemical industries. A thirty-year review* (New York, 1998)
3. E.E. Kirukhancev, V.N. Ivanov, Internet Zhurnal Tekhnologii Tekhnosfernoy Bezopasnosti, **4**(**50**) (2013)
4. E.E. Kirukhancev, V.N. Ivanov, Internet Zhurnal Tekhnologii Tekhnosfernoy Bezopasnosti, **5**(**51**) (2013)
5. Tien Minh Nguyen, E.E. Kiryukhantsev, Internet Zhurnal Tekhnologii Tekhnosfernoy Bezopasnosti, **2** (**72**) (2017)
6. S.V. Sharhuv, Technospere Safety, **4** (**9**), 37- 42 (2017)
7. S.V. Sharhun, Journal «Pozharovzryvobezopasnost. **22(2)**, 60-64 (2013)
8. L. Haritonova, Advances in Intelligent Systems and Computing. Springer, Cham., **692** (2018)
9. L. Haritonova, Procedia Engineering, **206**, 1002-1008 (2017)
10. L.P. Haritonova, International Scientific and Practical Conf. World science, **8** (**12**), 47-55 (2016)
11. L.P. Haritonova, MATEC Web Conf. **106**, 06006 (2017)
12. L.P. Haritonova, International research journal, **8–3**(**50**), 155–158 (2016)
13. L.P. Haritonova, Proceedings of the Second International Scientific and Practical Conferences "*The Top Actual Researches in Modern Science (28–29 July 2016 Ajman UAE)*" International Periodical Scientific Edition: International Scientific and Practical Conferences "World science", **8** (**12**), (2016)
14. S.V. Novikova, *Neural network simulation for solving the monitoring problems under conditions of incomplete fuzzy information (by an example of ecological monitoring)* (DrSc thesis, Kazan State Technical University) (2013)
15. L.P. Haritonova, Int. Sci. J. (ISJ) Theor. Appl. Sci. **34(2)**, 164–171 (2016)
16. L.P. Haritonova, International research journal, **3(45)**, P **2** 91-93 (2016)
17. L.P. Haritonova, Intern. periodical scientific edition: Intern. Sci. and Practical Conf. World science (Proc. of the II Intern. Scientific and Practical Conf. "*Modern Scientific Achievements and Their Practical Application (Dubai UAE)*"), **3** (**3**), 54-56 (2015)

18. B.G. Pustovitenko, E. A. Merger, A. A. Pustovitenko, Geofizicheskiy Zhurnal, **35** (**5**), 172-186 (2013)
19. B.G. Pustovitenko, E. A. Merger, Geodynamics, **1** (**14**), 124–128 (2013)
20. B.G. Pustovitenko, V.E. Kulchitskiy, I.V. Kalinyuk, E.A. Merger, Seismostoykoye stroitelstvo. Bezopasnost sooruzheniy, **6**, 20-23 (2012)

The improvement of thermal characteristics of autoclave aerated concrete for energy efficient high-rise buildings application

Pavel Khavanov[1], *Ekaterina* Fomina[2*], and *Natalia* Kozhukhova[2]

[1]Moscow State University of Civil Engineering, Yaroslavskoe shosse, 26, Moscow, 129337, Russia
[2]Belgorod State Technological University named after V.G. Shoukhov, 308012, Belgorod, Kostukova Street, 46, Russia.

Abstract. Nowadays, the problem of energy saving is very relevant. One of the ways to reduction energy consumption in construction materials production and construction of civil and industrial high-rise buildings is the application of claddings with heat-insulating performance. The concept of energy efficiency of high-rise buildings is closely related to environmental aspect and sustainability of applied construction materials; reducing service costs; energy saving and microclimate comfortability. A complexity of architectural and structural design as well as aesthetic characteristics of construction materials are also should be considered. The high interest focused on materials with combined properties. This work is oriented on the study of energy efficiency of buildings by improving heat-insulation and strength performance of autoclave aerated concrete. The applied method of sulfate activation of lime allows monitoring phase and structure formation in aerated concrete. The optimal mix design of aerated concrete with the compressive strength up to 8.5 MPa and decreased density up to 760 kg/m^3 was proposed. Analysis of structure at macro- and microscale was performed as well as the criteria of an optimal porosity formation was considered a number, size, shape of pore and density of interior partition. SEM analysis and BET method were performed in this research work. The research results demonstrated the correlation between structure and vapor permeability resistance, also it was found that the increase of strength can lead to reduction of thermal conductivity.

1 Introduction

The modern regulations require construction materials to meet the strict requirements for thermal and strength performance for construction of polyfunctional energy-saving high-rise buildings and residential complexes. Special requirement are made for materials applied in high-rise buildings. These materials should provide the required level of comfortability and optimal microclimate for human life and activities as well as required heat insulation [1]. Due to a significant environmental impact of construction materials production the energy

* Corresponding author: fomina.katerina@mail.ru

efficiency of buildings become a problem number one in the construction sector [2]. A "Zero-energy" house is one of the effective approaches in construction was introduced and become very popular in Europe [3].

The new generation composites are the functional materials with improved performance and combine several characteristics such as structural, decorative, thermal-insulation and sound insulation etc. [4]. The application of autoclave aerated concrete as building envelopes is a good way to realize the concept. Aerated concrete performs well in the constructions with steel reinforced-concrete and high-rise monolithic structures [5]. Nowadays, autoclave aerated concrete is a multi-purpose construction material, produced using «green» technology [6]. The aerated concrete is a light-weight structural material with high thermal- and sound- insulation, reduced labor intensity; reduced transportation and others associated costs. Die to its light-weight and plasticity, aerated concrete can be easily cut in simple and more complex architectural elements, which allows to expand the applicability of the material and be a good fit in high-rise buildings and others complex structures construction where the use of high-density elements is not possible. Moreover, aerated concrete is characterized by low thermal-conductivity providing an optimal air condition control and microclimate.

Development of innovative approaches and energy-efficient and eco-friendly technologies for aerated concrete production with improved thermal characteristics is based on modification of cellular structure saving the strength performance [7–9]. Smart concrete is among of the innovative construction materials. It can work in extreme climate and service conditions [10] considering social and aesthetic aspects [11].

The reduction of density with saving and increasing of strength performance is the key factor for efficiency of production and application of aerated concrete. An effective method to achieve an excellent thermal-insulating characteristics in aerated concrete is the control of hydration process. Often, a low quality autoclave material is associated with uncompleted chemical interaction between lime and quartz components in the binder system. Therefore, intensity and completeness of lime slaking as well as chemical reactivity of $Ca(OH)_2$ are the key parameters at first stage of hydration in order to produce a high quality material.

However, at industrial level of production, it is difficult to precisely control the hydration process. The previous studies [12] showed that slaking of high-reactive lime (reactivity is 86 %, water demand is 32–64%) at the temperature range of 160– 190°C at the presence of 0.05–0.25 % gypsum, the gypsum dehydration takes place leading to formation of β-hemihydrate of calcium sulfate and $Ca(OH)_2$-particles with higher dispersity. Incorporation of high-dispersed $Ca(OH)_2$ and β-hemihydrate calcium sulfate into the aerated concrete mixture causes the change in hydration kinetics, setting time and rheological performance [13–14]. In addition it helps to combine the periods of intensive structure formation and gas release (blow out of aerated concrete). That provides high-performance matrix structure of the final autoclave composite.

2 Materials and Methods

The mix of aerated concrete was designed to be applicable for a small-sized blocks cutting technology production in order to provide density of 800 kg/m^3 and compressive strength of 2.5–7.5 MPa [15]. The sulfate activation of lime was performed during binder specimens casting process. The β-hemihydrate calcium sulfate in content of 0.15 % was added to lime. The obtained mixture was slaked. Water/ solid ratio was 0.32. The obtained slaked binder was added into the aerated concrete mixture consisting of cement, aluminum paste and quartz sand. It was observed that the most intensive gas release was achieved with providing the certain thermal conditions during exothermic reaction caused by lime slaking process when. Therefore, only about 20 % of the lime-gypsum mixture needs to be slaked before adding into the concrete mix. The mix of aerated concrete was poured in cubic metal forms of 10×10×10 cm using castable technology. The components of the aerated concrete mixture was in the following proportion: lime – 1; quartz sand – 3; cement – 1; aluminum paste – 0.0048.

The poured specimens were cured in ambient conditions for 3 hours followed by autoclave treatment under the following curing regime: air-flush of the specimens in the curing chamber for 40 min; after that pressure was buildup by up 10 atmospheres and temperature up to 180°C during 1 hour; curing process was performed at a constant 10 atmospheres and 180°C for 5 hours followed by pressure release during 2 hours. The portion of β-hemihydrate calcium sulfate in the binder was determined using cation-exchange method [16]. Microstructure of the aerated concrete was investigated using high-resolution scanning-electron microscope Hitachi S-800 and the microporosity of concrete specimens was measured by BET method using SoftSorbi–II ver.1.0. machine.

Thermal-conductivity of the autoclaved specimens was determined by the cylindrical probe method using ITP-MG4 Zond apparatus. Vapor permeability was measured according to the procedure described in the Russian Standard GOST 31359-2007 [18].

3 Results

The mix design of aerated autoclaved concrete was developed using design of experiment method considering the following parameters: slaked product content, water/solid ratio and content of β-hemihydrate calcium sulfate (%). The aerated concrete with density of 800 kg/m^3 and compressive strength of 4.5 MPa was used as a reference.

In order to describe the binder system response by variation of the given parameters, adequate regression equations and yield surfaces were developed as depicted in Figure 1.

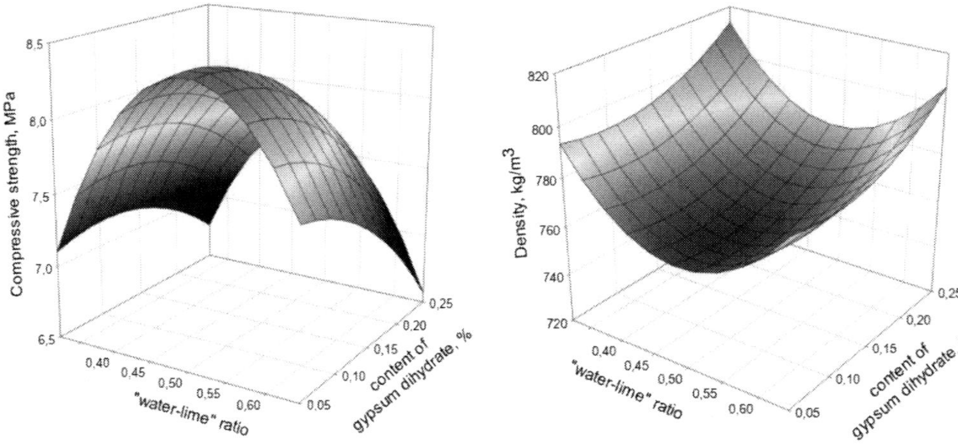

Fig. 1. The effect of water/solid ratio and content of β-hemihydrate calcium sulfate on density and compressive strength of aerated concrete.

Figure 1 demonstrated that the incorporation of previously slaked gypsum-lime binder into the aerated mixture enables to improve compressive strength performance and reduce density and water demand from 0.5 to 0.48 of the specimens.

The aerated concrete after autoclave treatment demonstrated the absence of SO_4^{2-} ions, confirming the reactivity of β-hemihydrate calcium sulfate during structure formation. Therefore, the optimal mix composition of aerated concrete contented 0.15% of β-hemihydrate calcium sulfate with 0.4 water/solid ratio, resulting in a significant increase of compressive strength from 4.5 to 8.5 MPa, and density reduction from 800 to 760 kg/m^3.

Sulfate activation positively effected pore formation in concrete matrix. Structural skeleton of the optimal composition demonstrated porosity increase from 70 to 75% as well as homogeneous pore distribution (Figure 2 *b*). Pores were spherical with 0.3–0.9 mm in diameter.

a

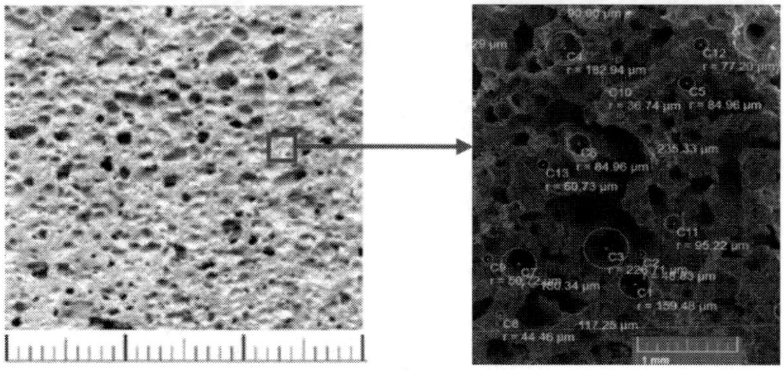

b

Fig. 2. SEM images of the aerated concrete after autoclave treatment: a – composition 1; b – composition 2 (Table 1).

According to the surface area BET data presented in Table 1, the volume of nanopores in concrete specimens was changed, where the volume of pores less than 94.6 nm was 0.007 cm^3/g for reference specimen, and the volume of those for the experimental specimen was 0.018 cm^3/g. Formation of proper porous structure is the key parameter to improve thermal characteristics of concrete, because density reduction causes the decrease of thermal conductivity from 0.19 to 0.17 W/(m·C°) (Table 1), improving thermal characteristics of aerated concrete.

Table 1. Characteristics of the aerated concrete after autoclave treatment.

Mix	Content of gypsum dihydrate in previously slaked binder, %	Water/solid ratio	Density, kg/m^3	Compressive strength, MPa	Heat conductivity, W/(m·C°)	Vapor permeability, mg/(m·h·Pa)	Porosity, %	Pore space with pore of 94.6 nm (cm^3/g)
1	0	0.50	800	4.5	0.19	0.14	70	0.007

2	0.15	0.48	760	8.5	0.17	0.16	75	0.018
According to Russian Standard 31359-2007	–	–	800	2.5–7.5	0.19	0.14 at least	–	–

Mix 2 of the produced aerated concrete specimen meets the requirements of the Russian Standard GOST 31359-2007 [19] in terms of vapor permeability, which was 12 % higher than that for reference mix 1 (Table 1). That means it can provide the "breathing wall" effect and keep in-door microclimate optimal.

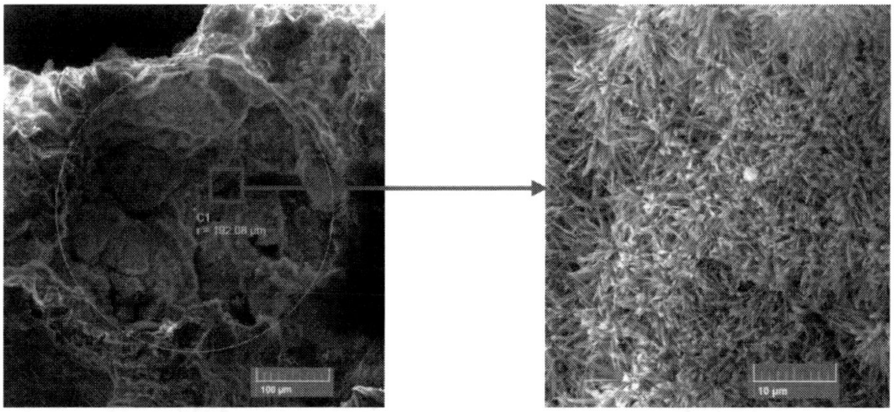

Fig. 3. SEM images of the mix 2 (Table 1).

Microstructure of an interporous zone as the most important zone in aerated concrete matrix responsible for with density and strength performance demonstrated the increase of structural packing density (Figure 3). Hydrothermal curing conditions are favorable for needle-like and stem-like crystal formations, which are typical for 11Å-tobermorite that serves as a reinforcing element at micro level structure strengthening structure of aerated concrete.

Conclusions

This work demonstrated the opportunity to control thermal- and sound-insulating performance of aerated concrete due to homogeneous pore distribution, required density of interporous zone, required size and shape of particles. It can be achieved by controlling of hydration process in the binder system. Using method of lime sulfate activation allowed maximally intensify lime reactivity and stimulate the chemical interaction between components in the changing and rheological characteristics of fresh concrete mixture and mineral composition of aerated concrete. Increasing porosity, formation of multudispersed pores and their homogeneous distribution led to density reduction of the aerated concrete up to 5 %. The formation of needle-like crystals helps to reinforce matrix structure improving density of interporous zone, reduce structure imperfection and double the compressive strength of aerated concrete.

The developed aerated concrete is characterized by complex of improved thermal characteristics including heat conductivity reduction from 0.19 to 0.17 W/(m·°C), vapor permeability increase from 0.14 to 0.16 mg/(m·h·Pa) and that meets the requirements of modern standards of thermal insulation, energy saving and microclimate parameters. The application of developed concrete is good approach to produce environmentally friendly, sustainable and energy efficient structure at the stage of design, production and operation of high-rise buildings in today's megapolises.

Acknowledgements

The work is realized in the framework of the Program of flagship university development on the base of the Belgorod State Technological University named after V.G. Shoukhov, using equipment of High Technology Center at BSTU named after V.G. Shoukhov.

References

1. Yu.A. Matrosov, V.N., Yarmakovsky, Construction technologies, **6(40)**, 24–28 (2005).
2. S.G. Abramyan, T.A. Matviichuk, Engineering Journal of Don, **1 (44),** 74 (2017).
3. Yu.A. Tabunshchikov, M.M. Brodach, ABOK: Ventilation, heating, air-conditioning, heat supply system and construction thermal physics, **1**, 14–21 (2013).
4. V.S. Lesovik, Vestnik MGSU, Vol. 12, **1 (100)**, 9–16 (2017).
5. N. I. Vatin, A.S. Gorshkov, S. V. Kornienko, I. I. Pestryakov, Roofing and insulating materials, **2**, 22–31 (2016).
6. R. Lesovik, Y. Degtev, M. Shakarna and Levchenko A. Modern Applied Science, **9 (1)**, 45–50 (2015).
7. A. E. Knyazev, Technologies of technosphere safety, **3 (55)**, 35 (2014).
8. A.S. Inozemtcev, E.V. Korolev, Smirnov V.A. Structural Concrete. T. 18, **1**, 67–74 (2017).
9. V.S. Lesovik, M.Yu. Elistratkin, M.V. Absimetov, E.V. Kogut, Regional architecture and engineering, **3 (32)**, 11–20 (2017).
10. V.S. Lesovik, E.V. Fomina, O.V. Khotomchenko, Investigation of application conditions for construction materials taking into account a growth of environmental impact (Energy- and source-saving, environmentally-friendly chemical-engineering processes of environmental protection. Proceedings of International scientific and technical conference of BSTU named after V.G. Shoukhov, Belgorod, 2015) 409–414.
11. V.S. Lesovik, Belgorod: BSTU, pp: 287 (2016).
12. E.V. Fomina, N.P. Kudeyarova, Izvestia Vusov. Severo-Kavkazsky region. Technical sciences, **6**, 17–19 (2006).
13. E.V. Fomina, N.I. Altynnik, V.V. Strokova, V.V. Nelubova, A.B. Bukhalo, Construction materials J., **9**, 33–35 (2011).
14. E.V. Fomina, Bulletin of Belgorod State Technological University named after V.G. Shoukhov J., **4**, 7–11 (2013).
15. Russian standard 25485-89 «Cellular concretes. Specifications».
16. Russian standard 30256–94 «Building materials and products. Method of thermal conductivity determination by cylindrical probe».
17. Russian standard 25898–83 «Building materials and products. Methods of steam-tightness determination».
18. Russian standard 5382-91 «Cements and materials for cement production. Chemical analysis methods».
19. Russian standard 31359-2007 «Cellular autoclave curing concretes. Specifications».

Bending of an Infinite beam on a base with two parameters in the absence of a part of the base

Maxim Aleksandrovskiy[1,*] and *Lidiya* Zaharova[1]

[1]Moscow State University of Civil Engineering, Yaroslavskoe shosse, 26, Moscow, 129337, Russia

Abstract. Currently, in connection with the rapid development of high-rise construction and the improvement of joint operation of high-rise structures and bases models, the questions connected with the use of various calculation methods become topical. The rigor of analytical methods is capable of more detailed and accurate characterization of the structures behavior, which will affect the reliability of objects and can lead to a reduction in their cost. In the article, a model with two parameters is used as a computational model of the base that can effectively take into account the distributive properties of the base by varying the coefficient reflecting the shift parameter. The paper constructs the effective analytical solution of the problem of a beam of infinite length interacting with a two-parameter voided base. Using the Fourier integral equations, the original differential equation is reduced to the Fredholm integral equation of the second kind with a degenerate kernel, and all the integrals are solved analytically and explicitly, which leads to an increase in the accuracy of the computations in comparison with the approximate methods. The paper consider the solution of the problem of a beam loaded with a concentrated force applied at the point of origin with a fixed value of the length of the dip section. The paper gives the analysis of the obtained results values for various parameters of coefficient taking into account cohesion of the ground.

1 Introduction

Modern design and construction of high-rise buildings and structures are unthinkable without a clear understanding of the features of the work as well as a number of constructive elements of the future facility. At the same time, the issues related to the refinement and improvement of the parameters of the design schemes reflecting the interaction of structures with the base become urgent. The design schemes of unique objects, among which, first of all, are modern skyscrapers, meet the most stringent requirements in terms of their strength, reliability and durability. The uniqueness of high-rise buildings requires the improvement of calculation methods, taking into account the numerous factors of operational, technological and constructive nature. A significant part of the calculations related to the constructions of such objects is connected with the specification of efforts in the elements interacting with the base, and any refinement of calculations significantly affects the cost. This formulation of the problem fully relates to the tape foundations of high-rise buildings and structures, industrial

* Corresponding author: AleksandrovskiyMV@mgsu.ru

and civil facilities, airfield and road surfaces, the operating conditions of which are associated with possible subsidence of the base and the formation of voids. When designing strip foundations located on karst rocks and subsidence soils in the form of loess deposits, which, when soaked, give a large draft, it is necessary to take into account the possibility of formation of depressions, dips, etc. under the foundations. The absence of contact of the beam with the base can occur at various sites, in particular, at the center or near the edge, which can be reduced to different statements of problems, for example, if the load and dip in the ground are in the central part of the strip foundation, then it can be calculated according to the scheme of infinite beams.

Despite the diversity of the computational models of bases and methods for calculating them, these methods are usually not interrelated and are used to solve particular problems. Among the elastic base models using two parameters, one can note the results presented in the works of M.M. Filonenko-Borodich [1], V.Z. Vlasova and N.N. Leontiev [2] and P.L. Pasternak [3]. The main advantage of such models was the ability to take into account the distributive capacity of the base. In the two-parameter model of the elastic base, the first coefficient characterizes the rigidity of the base by compression, and the second, by the shear. Despite the various ideas proposed for the calculation of beams on an elastic base, the authors come to the same kind of differential equation

$$\frac{d^4w}{dx^4} - 2r^2\frac{d^2w}{dx^2} + s^4w = \frac{qa^4}{E_1J} \tag{1}$$

where r^2 and s^2 are the generalized elastic dimensionless quantities, a is a half-length of the beam.

Despite the availability of various techniques for determining the values of coefficients [4, 5, 6] in the reports on engineering and geological surveys, as a rule, information on the above coefficients is absent. In SCAD Office in the Pasternak satellite program, the method proposed by M.I. Gorbunov-Posadov [7] is used as one of the possible.

In this formulation of the problem, the model proposed by V.Z. Vlasov, N.N. Leontiev [2] is used as a base, and the calculated values of the shear coefficient according to [8] are: $k = 0$, $k = 0.5$ и $k = 0.9$.

The method of solving a differential equation of this kind was considered both on the basis of rigorous analytical methods, for example, [9, 10], and with the widely used, approximate numerical methods [11, 12].

2 Materials and Methods

To find the solution of the original differential equation, the method of generalized solutions proposed by V.I. Travush [10] is used. Later, this method was used to obtain an analytical solution to the bending of an infinite beam on a Winkler elastic foundation in the absence of a base under the beam part [13]. However, the main drawback of such a model of the basis is the lack of possibility to take into account its distributive capacity. The general form of the analytical solution for equation (1) was presented in [14], however, in our opinion, the authors made a number of inaccuracies, which in the case of implementing the above algorithm should lead to incorrect results.

Consider an infinite beam lying on an elastic base with two bed coefficients [2], and there is no base under the central part of the beam (Fig. 1.).

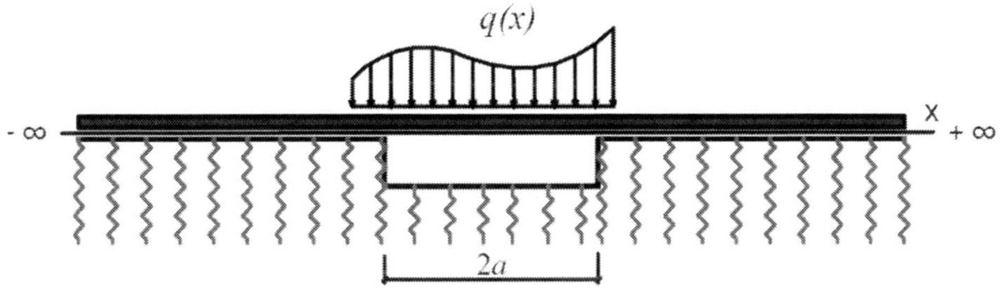

Fig.1. The scheme of an infinite beam lying on the base with two parameters in the absence of a base on a given section.

The arbitrary load $q(x)$, applied to the beam can be decomposed into symmetrical and inverse symmetric components, relative to the vertical axis of the beam. The differential equation describing the deflection of a beam for a given condition in the treatment of [2] takes the form

$$EJy^{IV}(\xi) - 2t\theta(\xi - a_1)\cdot y''(\xi) + k\theta(\xi - a_1)y(\xi) = q(\xi) \tag{2}$$

In the treatment [8], (10), the authors do not reflect the allowance for the absence of tangential stresses under that part of the beam where there is no base (see (13) [8]);

The equation (2) can be reduced to dimensionless coordinates by introducing the commonly used change of variables:

$$x = \xi\beta;\ y(x) = y(\xi\beta);\ \theta(x-a) = \theta(\xi\beta - a_1\beta);\ k_1 = \frac{2t}{EJ\beta^2} = \frac{4t}{\sqrt{kEJ}};$$

$$\frac{k}{EJ} = 4\beta^4;\ \beta = \left(\frac{k}{4EJ}\right)^{1/4};\ q_0(x) = \frac{1}{EJ}q(x). \tag{3}$$

Here E is the modulus of elasticity of the beam material; J is its moment of inertia; t and k are the parameters of the base model; k_1 is the second coefficient of the bed; $\theta(x-a)$ is the Heaviside function

$$\theta(x-a) = \begin{cases} 0, & x < a \\ 1. & x \geq a \end{cases}$$

The differential equation for the deflection of a beam in dimensionless coordinates and functions, taking (2) into account, takes the form of

$$y^{IV}(x) - k_1\theta(x-a)y''(x) + 4\theta(x-a)y(x) = 4k^{-1}q(x) \tag{4}$$

To solve equation (4), we can use the Fourier transform to solve problems with an infinite domain, when the function $y(x)$ and its first three derivatives have no discontinuities in this area [15].

Under the action of the symmetric load component, applying the direct cosine of the Fourier transform separately for each term of equation (2), we obtain:

$$\sqrt{\frac{2}{\pi}}\int_0^\infty y^{IV}(x)\cos\lambda x dx - k_1\sqrt{\frac{2}{\pi}}(\int_0^\infty y''(x)\cos\lambda x dx - \int_0^a y''(t)\cos\lambda t dt) +$$
$$+\sqrt{\frac{2}{\pi}}4(\int_0^\infty y(x)\cos\lambda x dx - \int_0^a y(t)\cos\lambda t dt) = \sqrt{\frac{2}{\pi}}\frac{4}{k}\int_0^\infty q(x)\cos\lambda x dx \tag{5}$$

Introducing the corresponding notation of Fourier transform:

$$\sqrt{\frac{2}{\pi}}\int_0^\infty y^{IV}(x)\cos\lambda x dx = \lambda^4\sqrt{\frac{2}{\pi}}\int_0^\infty y(x)\cos\lambda x dx = \lambda^4 Y(\lambda) \tag{5a}$$

$$\sqrt{\frac{2}{\pi}}\int_0^\infty y''(x)\cos\lambda x dx = -\lambda^2\sqrt{\frac{2}{\pi}}\int_0^\infty y(x)\cos\lambda x dx = -\lambda^2 Y(\lambda) \tag{5b}$$

$$\sqrt{\frac{2}{\pi}}\int_0^\infty q(x)\cos\lambda x dx = \sqrt{\frac{2}{\pi}}Q_0(\lambda) \tag{5c}$$

The general form of the equation with respect to a new dimensionless coordinate is

$$Y(\lambda) = \frac{4}{k}\sqrt{\frac{2}{\pi}}\frac{Q_0(\lambda)}{(\lambda^4 + k_1\lambda^2 + 4)} + \frac{(k_1\lambda^2 + 4)}{(\lambda^4 + k_1\lambda^2 + 4)}\sqrt{\frac{2}{\pi}}\int_0^a y(t)\cos\lambda t dt \tag{6}$$

In contrast to [8], here we take into account the change in the form of the cargo term on the right-hand side, due to the consideration of the distributive properties of the base.

Applying the inverse cosine of the Fourier transform to the resulting equation (6), we obtain an expression for the desired deflection function.

$$y(x) = \frac{2}{\pi}\frac{4}{k}\int_0^\infty \frac{Q_0(\lambda)}{(\lambda^4 + k_1\lambda^2 + 4)}\cos\lambda x d\lambda + \frac{2}{\pi}\int_0^\infty \frac{(k_1\lambda^2 + 4)\cos\lambda t\cos\lambda x}{(\lambda^4 + k_1\lambda^2 + 4)}d\lambda\int_0^a y(t)dt \tag{7}$$

Thus, applying the forward and inverse cosine of the Fourier transform to equation (2) allows us to reduce the differential equation (2) to the integral equation with respect to the sought deflection function of the beam $y(x)$.

$$y(x) - \int_0^a y(t)K^c(x,t)dt = y(x)^c_\infty \tag{8}$$

In this equation, the right-hand side $y(x)^c_\infty$ is the deflection of an infinite beam loaded with a given load and completely in contact with the two-parameter base along its entire length.

The general expression for the deflection function of an infinite beam for symmetric components $y(x)^c_\infty$,using [16] is written in the following form

$$\begin{aligned} y(x)^c_\infty = {} & \frac{2}{\pi}\frac{4}{k}\frac{P}{2}\beta\frac{1}{2}\{\frac{\pi}{2}\frac{2}{\sqrt{16-k_1^2}}\frac{1}{2}e^{-|x_0-x|\sqrt{1+\frac{k_1}{4}}}(\sqrt{1-\frac{k_1}{4}}\cos((x_0-x)\sqrt{1-\frac{k_1}{4}})+ \\ & +\sqrt{1+\frac{k_1}{4}}\sin(|x_0-x|\sqrt{1-\frac{k_1}{4}}))+\frac{\pi}{2}\frac{2}{\sqrt{16-k_1^2}}\frac{1}{2}e^{-|x_0+x|\sqrt{1+\frac{k_1}{4}}}(\sqrt{1-\frac{k_1}{4}}\cos((x_0+x)\sqrt{1-\frac{k_1}{4}})+ \\ & +\sqrt{1+\frac{k_1}{4}}\sin((x_0+x)\sqrt{1-\frac{k_1}{4}}))\} \end{aligned} \tag{9}$$

For the special case, when the load in the form of a concentrated force is applied in the center, i.e. $x_0 = 0$ the above expression takes the form of

$$y(x)^c_\infty = \frac{2P}{k}\beta\frac{1}{\sqrt{16-k_1^2}}e^{-|x|\sqrt{1+\frac{k_1}{4}}}(\sqrt{1-\frac{k_1}{4}}\cos(x\sqrt{1-\frac{k_1}{4}})+\sqrt{1+\frac{k_1}{4}}\sin(|x|\sqrt{1-\frac{k_1}{4}})) \tag{10}$$

Introduce the notations:

$$1-\frac{k_1}{4} = A;\ 1+\frac{k_1}{4} = B;\ (4-\frac{k_1^2}{2}) = C;\ -2\sqrt{A}(4+k_1) = D;\ 2\sqrt{B}(4-k_1) = F;$$

$$2k_1\sqrt{A}\cdot\sqrt{B} = G;\ \frac{2\pi}{D^2+F^2} = H;\ CF-GD = S;\ [CD+GF] = -L. \tag{11}$$

The use of the theory of residues makes it possible to express the kernel of the integral equation (8) $K^c(x,t)$ through a combination of trigonometric and hyperbolic functions:

$$K^c(x,t) = \int_0^\infty \frac{(k_1\lambda^2 + 4)\cos\lambda x\cos\lambda t}{(\lambda^4 + k_1\lambda^2 + 4)}d\lambda = \begin{cases} H(\varphi(x)_{1>}\psi(t)_{1>} + \varphi(x)_{2>}\psi(t)_{2>}) & (x \ge t) \\ H(\varphi(x)_{1<}\psi(t)_{1<} + \varphi(x)_{2<}\psi(t)_{2<}) & (x \le t) \end{cases} \tag{12}$$

where

$$\varphi(x)_{1>} = e^{-x\sqrt{B}}\cos(x\sqrt{A}),\ \varphi(x)_{2>} = e^{-x\sqrt{B}}\sin(x\sqrt{A}),$$
$$\psi(t)_{1>} = S\cos(t\sqrt{A})ch(t\sqrt{B}) - L\sin(t\sqrt{A})sh(t\sqrt{B}),$$
$$\psi(t)_{2>} = S\sin(t\sqrt{A})sh(t\sqrt{B}) + L\cos(t\sqrt{A})ch(t\sqrt{B}),$$
$$\varphi(x)_{1<} = ch(x\sqrt{B})\cos(x\sqrt{A}),\ \varphi(x)_{2<} = sh(x\sqrt{B})\sin(x\sqrt{A}),$$
$$\psi(t)_{1<} = (S\cos(t\sqrt{A}) + L\sin(t\sqrt{A}))e^{-t\sqrt{B}},\ \psi(t)_{2<} = (S\sin(t\sqrt{A}) - L\cos(t\sqrt{A}))e^{-t\sqrt{B}}. \quad (13)$$

It follows from (12) that the kernel of the integral equation (8) – $y(x)^c_\infty$ is degenerate and that, when solving it, we must consider two cases. The first of them assumes that the point at which the deflection is sought is in that part of the beam where there is a base beneath it. Obviously, in this case it follows from (8) and (12), an expression for determining the deflection function of a beam is

$$y(x)^c = y(x)^c_\infty + H\sum_{i=1}^{2} c_{i>}\varphi(x)_{i>} \quad (14)$$

here, the introduced notation is

$$c_{i>} = \int_0^a y(t)^c\psi(t)_{i>}dt \quad (i = 1,2) \quad (15)$$

The coefficients $c_{i>}$ can be determined as a result of the standard procedure for solving integral equations with a degenerate kernel [5]. To this end, both sides of equation (14) are multiplied consecutively by $\psi(x)_{1>}$ and $\psi(x)_{2>}$, and then integrate from 0 to a.

The functions $\Phi_{i>k>}$ appearing in these expressions are integrals of combinations of functions $\varphi_{i>}$ and $\psi_{k>}$, the values of which can be expressed in explicit form.

$$\Phi_{1>1>} = \int_0^a e^{-x\sqrt{B}}\cos(x\sqrt{A})(S\cos(x\sqrt{A})ch(x\sqrt{B}) - L\sin(x\sqrt{A})sh(x\sqrt{B}))dx =$$
$$= \frac{1}{2}\{S[\frac{\sin(2a\sqrt{A})}{4\sqrt{A}} + \frac{a}{2} + \frac{-e^{-2a\sqrt{B}}(-\sqrt{A}\sqrt{B}\sin(2a\sqrt{A}) + B\cos(2a\sqrt{A}) + A + B) + A + 2B}{4\sqrt{B}(A+B)}] -$$
$$-L[\frac{\sin^2(a\sqrt{A})}{2\sqrt{A}} - \frac{\sqrt{A} - e^{-2a\sqrt{B}}(\sqrt{B}\sin(2a\sqrt{A}) + \sqrt{A}\cos(2a\sqrt{A}))}{4(A+B)}]\}$$

$$\Phi_{1>2>} = \int_0^a e^{-x\sqrt{B}}\sin(x\sqrt{A})(S\cos(x\sqrt{A})ch(x\sqrt{B}) - L\sin(x\sqrt{A})sh(x\sqrt{B}))dx =$$
$$= \frac{1}{2}\{S[\frac{\sin^2(a\sqrt{A})}{2\sqrt{A}} + \frac{\sqrt{A} - e^{-2a\sqrt{B}}(\sqrt{B}\sin(2a\sqrt{A}) + \sqrt{A}\cos(2a\sqrt{A}))}{4(A+B)}] -$$
$$-L[\frac{a}{2} - \frac{\sin(2a\sqrt{A})}{4\sqrt{A}} - \frac{e^{-2a\sqrt{B}}(Ae^{2a\sqrt{B}} - \sqrt{A}\sqrt{B}\sin(2a\sqrt{A}) + B\cos(2a\sqrt{A}) - A - B)}{4\sqrt{B}(A+B)}]\}$$

$$\Phi_{2>1>} = \int_0^a e^{-x\sqrt{B}}\cos(x\sqrt{A})(S\sin(x\sqrt{A})sh(x\sqrt{B}) + L\cos(x\sqrt{A})ch(x\sqrt{B}))dx =$$
$$= \frac{1}{2}\{S[\frac{\sin^2(a\sqrt{A})}{2\sqrt{A}} - \frac{\sqrt{A} - e^{-2a\sqrt{B}}(\sqrt{B}\sin(2a\sqrt{A}) + \sqrt{A}\cos(2a\sqrt{A}))}{4(A+B)}] +$$
$$+L[\frac{\sin(2a\sqrt{A})}{4\sqrt{A}} + \frac{a}{2} + \frac{-e^{-2a\sqrt{B}}(-\sqrt{A}\sqrt{B}\sin(2a\sqrt{A}) + B\cos(2a\sqrt{A}) + A + B) + A + 2B}{4\sqrt{B}(A+B)}$$

$$\Phi_{2>2>} = \int_0^a e^{-x\sqrt{B}} \sin(x\sqrt{A})(S\sin(x\sqrt{A})sh(x\sqrt{B}) + L\cos(x\sqrt{A})ch(x\sqrt{B}))dx =$$

$$= \frac{1}{2}\{S[\frac{a}{2} - \frac{\sin(2a\sqrt{A})}{4\sqrt{A}} - \frac{e^{-2a\sqrt{B}}(Ae^{2a\sqrt{B}} - \sqrt{A}\sqrt{B}\sin(2a\sqrt{A}) + B\cos(2a\sqrt{A}) - A - B)}{4\sqrt{B}(A+B)}] +$$

$$+L[\frac{\sin^2(a\sqrt{A})}{2\sqrt{A}} + \frac{\sqrt{A} - e^{-2a\sqrt{B}}(\sqrt{B}\sin(2a\sqrt{A}) + \sqrt{A}\cos(2a\sqrt{A}))}{4(A+B)}]\}$$

For the case $x_0 = 0$ with allowance for (10) and (11) we obtain:

$$\Phi_{1P>} = \frac{2P}{k}\beta\frac{1}{\sqrt{16-k_1^2}}\int_0^a e^{-|x|\sqrt{B}}(\sqrt{A}\cos(x\sqrt{A}) + \sqrt{B}\sin(|x|\sqrt{A}))(S\cos(x\sqrt{A})ch(x\sqrt{B}) - L\sin(x\sqrt{A})sh(x\sqrt{B}))dx =$$

$$= \frac{P}{k}\beta\frac{2}{\sqrt{16-k_1^2}}(-\frac{\sqrt{A}(A+B)(-\sqrt{B}L - \sqrt{A}S) + \sqrt{B}(2\sqrt{A}BL - 3A\sqrt{B}S - B^{3/2}S)}{8\sqrt{A}\sqrt{B}(A+B)} +$$

$$+\frac{1}{8\sqrt{A}\sqrt{B}(A+B)}e^{-4\sqrt{B}}(\sqrt{A}(A+B)(-4Be^{4\sqrt{B}}L - \sqrt{A}S - \sqrt{B}(L - 4\sqrt{A}e^{4\sqrt{B}}S)) +$$

$$+\sqrt{B}(A^{3/2}(-1+e^{4\sqrt{B}})L + \sqrt{A}B(1+e^{4\sqrt{B}})L - B^{3/2}e^{4\sqrt{B}}S - A\sqrt{B}(2+e^{4\sqrt{B}})S)\cos(4\sqrt{A}) +$$

$$+\sqrt{B}(B^{3/2}e^{4\sqrt{B}}L + A\sqrt{B}(-2+e^{4\sqrt{B}})L + A\sqrt{B}(-1+e^{4\sqrt{B}})S + A^{3/2}(1+e^{4\sqrt{B}})S\sin(4\sqrt{A}))$$

$$\Phi_{2P>} = \frac{2P}{k}\beta\frac{1}{\sqrt{16-k_1^2}}\int_0^a e^{-|x|\sqrt{B}}(\sqrt{A}\cos(x\sqrt{A}) + \sqrt{B}\sin(|x|\sqrt{A}))(S\sin(x\sqrt{A})sh(x\sqrt{B}) + L\cos(x\sqrt{A})ch(x\sqrt{B}))dx =$$

$$= \frac{P}{k}\beta\frac{2}{\sqrt{16-k_1^2}}(-\frac{\sqrt{A}(A+B)(-\sqrt{A}L + \sqrt{B}S) - \sqrt{B}(3A\sqrt{B}L + 2\sqrt{A}BS + B^{3/2}L)}{8\sqrt{A}\sqrt{B}(A+B)} +$$

$$+\frac{1}{8\sqrt{A}\sqrt{B}(A+B)}e^{-4\sqrt{B}}(\sqrt{A}(A+B)(\sqrt{A}(-1+4\sqrt{B}e^{4\sqrt{B}})L + \sqrt{B}S + 4Be^{4\sqrt{B}}S) - \sqrt{B}(B^{3/2}e^{4\sqrt{B}}L + \tag{16}$$

$$+A\sqrt{B}(2+e^{4\sqrt{B}})L + A^{3/2}(-1+e^{4\sqrt{B}})S + \sqrt{A}B(1+e^{4\sqrt{B}})S)\cos(4\sqrt{A}) + \sqrt{B}(\sqrt{A}B(-1+e^{4\sqrt{B}})L +$$

$$+A^{3/2}(1+e^{4\sqrt{B}})L - B^{3/2}e^{4\sqrt{B}}S - A\sqrt{B}(-2+e^{4\sqrt{B}})S)\sin(4\sqrt{A})))$$

After determining the coefficients $c_{i>}$, the deflection function is known, hence the expressions for the angles of rotation, bending moments and transverse forces on the section of the beam in question are also known. We give their real values for the case of a concentrated force applied at the center of the beam:

$$y(x)^c = \frac{P}{2EJ\beta^3}\frac{1}{\sqrt{16-k_1^2}}[e^{-|x|\sqrt{B}}(\sqrt{A}\cos(x\sqrt{B}) + \sqrt{B}\sin(|x|\sqrt{A})) +$$

$$+C_{1>}e^{-x\sqrt{B}}\cos(x\sqrt{A}) + C_{2>}e^{-x\sqrt{B}}\sin(x\sqrt{A})];$$

$$\varphi(x)^c = \frac{P}{2EJ\beta^2}\frac{1}{\sqrt{16-k_1^2}}[-\sqrt{B}C_{1>}e^{-x\sqrt{B}}\cos(x\sqrt{A}) + \sqrt{A}C_{2>}e^{-x\sqrt{B}}\cos(x\sqrt{A}) -$$

$$-\sqrt{A}C_{1>}e^{-x\sqrt{B}}\sin(x\sqrt{A}) - \sqrt{B}C_{2>}e^{-x\sqrt{B}}\sin(x\sqrt{A}) + e^{-x\sqrt{B}}(\sqrt{A}\sqrt{B}\cos(x\sqrt{A}) -$$

$$-A\sin(x\sqrt{A})) - \sqrt{B}e^{-x\sqrt{B}}(\sqrt{A}\cos(x\sqrt{A}) + \sqrt{B}\sin(x\sqrt{A}))];$$

$$M(x)^c = -\frac{P}{2\beta}\frac{1}{\sqrt{16-k_1^2}}[(A+B)e^{-x\sqrt{B}}(-\sqrt{A}\cos(x\sqrt{A}) + \sqrt{B}\sin(x\sqrt{A})) +$$

$$+C_{1>}e^{-x\sqrt{B}}[(-A+B)\cos(x\sqrt{A}) + 2\sqrt{A}\sqrt{B}\sin(x\sqrt{A})] +$$

$$+C_{2>}e^{-x\sqrt{B}}[-2\sqrt{A}\sqrt{B}\cos(x\sqrt{A}) + (-A+B)\sin(x\sqrt{A})]];$$

$$
\begin{aligned}
Q(x)^c = &-\frac{P}{2}\frac{1}{\sqrt{16-k_1^2}}[3A\sqrt{B}C_{1>}e^{-x\sqrt{B}}\cos(x\sqrt{A}) - B^{3/2}C_{1>}e^{-x\sqrt{B}}\cos(x\sqrt{A}) - \\
&-A^{3/2}C_{2>}e^{-x\sqrt{B}}\cos(x\sqrt{A}) + 3\sqrt{A}BC_{2>}e^{-x\sqrt{B}}\cos(x\sqrt{A}) + A^{3/2}C_{1>}e^{-x\sqrt{B}}\sin(x\sqrt{A}) - \\
&-3\sqrt{A}BC_{1>}e^{-x\sqrt{B}}\sin(x\sqrt{A}) + 3A\sqrt{B}C_{2>}e^{-x\sqrt{B}}\sin(x\sqrt{A}) - B^{3/2}C_{2>}e^{-x\sqrt{B}}\sin(x\sqrt{A}) + \\
&+3Be^{-x\sqrt{B}}(\sqrt{A}\sqrt{B}\cos(x\sqrt{A}) - A\sin(x\sqrt{A})) + e^{-x\sqrt{B}}(-A^{3/2}\sqrt{B}\cos(x\sqrt{A}) + \\
&+A^2\sin(x\sqrt{A})) - B^{3/2}e^{-x\sqrt{B}}(\sqrt{A}\cos(x\sqrt{A}) + \sqrt{B}\sin(x\sqrt{A})) - \\
&-3\sqrt{B}e^{-x\sqrt{B}}(-A^{3/2}\cos(x\sqrt{A}) - A\sqrt{B}\sin(x\sqrt{A}))].
\end{aligned} \tag{17}
$$

It is more difficult to determine the deflection function in the beam portion where there is no base, i. e. $-a \le x \le a$. In this case we obtain from (8) and (12)

$$
y(x)^c = y(x)^c_\infty + H\sum_{i=1}^{2}[\varphi(x)_{i>}\int_0^x y(t)^c\psi(t)_{i>}dt + \varphi(x)_{i<}\int_x^a y(t)^c\psi(t)_{i<}dt] \tag{18}
$$

The solution of this equation can be obtained, for example, by the method of successive approximations; the corresponding integrals are calculated in explicit form.

$$
\begin{aligned}
y(x)^c &= y(x)^c_\infty + H\sum_{i=1}^{2}[\varphi(x)_{i>}\int_0^x y(t)^c\psi(t)_{i>}dt + \varphi(x)_{i<}\int_x^a y(t)^c\psi(t)_{i<}dt] = \\
&= H[\varphi(x)_{1>}\int_0^x y(t)^c\psi(t)_{1>}dt + \varphi(x)_{1<}\int_x^a y(t)^c\psi(t)_{1<}dt + \varphi(x)_{2>}\int_0^x y(t)^c\psi(t)_{2>}dt + \varphi(x)_{2<}\int_x^a y(t)^c\psi(t)_{2<}dt] + y(x)^c_\infty
\end{aligned} \tag{19}
$$

$$
C_1 = \int_0^x y(t)^c\psi(t)_{1>}dt;\quad C_2 = \int_x^a y(t)^c\psi(t)_{1<}dt;\quad C_3 = \int_0^x y(t)^c\psi(t)_{2>}dt;\quad C_4 = \int_x^a y(t)^c\psi(t)_{2<}dt \tag{20}
$$

$$
y(x)^c = C_1\varphi(x)_{1>} + C_2\varphi(x)_{1<} + C_3\varphi(x)_{2>} + C_4\varphi(x)_{2<} + y(x)^c_\infty \tag{21}
$$

Substitute (21) in (20)

$$
\begin{aligned}
C_1 &= \int_0^x y(t)^c\psi(t)_{1>}dt = \int_0^x (C_1\varphi(t)_{1>} + C_2\varphi(t)_{1<} + C_3\varphi(t)_{2>} + C_4\varphi(t)_{2<} + y(t)^c_\infty)\psi(t)_{1>}dt; \\
C_2 &= \int_x^a y(t)^c\psi(t)_{1<}dt = \int_x^a (C_1\varphi(t)_{1>} + C_2\varphi(t)_{1<} + C_3\varphi(t)_{2>} + C_4\varphi(t)_{2<} + y(t)^c_\infty)\psi(t)_{1<}dt; \\
C_3 &= \int_0^x y(t)^c\psi(t)_{2>}dt = \int_0^x (C_1\varphi(t)_{1>} + C_2\varphi(t)_{1<} + C_3\varphi(t)_{2>} + C_4\varphi(t)_{2<} + y(t)^c_\infty)\psi(t)_{2>}dt; \\
C_4 &= \int_x^a y(t)^c\psi(t)_{2<}dt = \int_x^a (C_1\varphi(t)_{1>} + C_2\varphi(t)_{1<} + C_3\varphi(t)_{2>} + C_4\varphi(t)_{2<} + y(t)^c_\infty)\psi(t)_{2<}dt.
\end{aligned} \tag{22}
$$

We give the form of the calculated values of the first four integrals representing the corresponding coefficients:

$$
\begin{aligned}
f_{11} = H\int_0^x \varphi(t)_{1>}\psi(t)_{1>}dt = &\frac{H}{4\pi}[S(\frac{1}{\sqrt{B}} + \frac{\sqrt{B}}{A+B} - \frac{e^{-2\sqrt{B}x}}{\sqrt{B}} + 2x + \frac{e^{-2\sqrt{B}x}(-\sqrt{A}\sqrt{B}\cos(2\sqrt{A}x) + (A + Ae^{2\sqrt{B}x} + Be^{2\sqrt{B}x})\sin(2\sqrt{A}x))}{\sqrt{A}(A+B)}) - \\
&-L(\frac{(-1 + \frac{Ae^{-2\sqrt{B}x}}{A+B})\cos(2\sqrt{A}x)}{\sqrt{A}} + \frac{\frac{B}{\sqrt{A}} + \sqrt{B}e^{-2\sqrt{B}x}\sin(2\sqrt{A}x)}{A+B})]
\end{aligned}
$$

$$
\begin{aligned}
f_{12} = H\int_0^x \varphi(t)_{1<}\psi(t)_{1>}dt = &\frac{H}{4\sqrt{A}\sqrt{B}(A+B)\pi}[\sqrt{A}\sqrt{B}(-L\sqrt{B} + S\sqrt{A})ch[2\sqrt{B}x]\sin(2\sqrt{A}x) + \sqrt{B}(A+B)S(2\sqrt{A}x + \sin(2\sqrt{A}x)) + \\
&+\sqrt{A}((A+B)S + (\sqrt{A}\sqrt{B}L + SB)\cos(2\sqrt{A}x)sh[2\sqrt{B}x]]
\end{aligned}
$$

$$
\begin{aligned}
f_{13} = H\int_0^x \varphi(t)_{2>}\psi(t)_{1>}dt = &\frac{H}{4\pi}[S(\frac{1}{\sqrt{A}} + \frac{\sqrt{A}}{A+B} - \frac{\cos(2\sqrt{A}x)}{\sqrt{A}} - \frac{e^{-2\sqrt{B}x}(\sqrt{A}\cos(2\sqrt{A}x) + B\sin(2\sqrt{A}x))}{(A+B)}) - \\
&-L(-\frac{1}{\sqrt{B}} + \frac{\sqrt{B}}{A+B} + \frac{e^{-2\sqrt{B}x}}{\sqrt{B}} + 2x - \frac{e^{-2\sqrt{B}x}(\sqrt{A}\sqrt{B}\cos(2\sqrt{A}x) + (Be^{2\sqrt{B}x} + A(-1 + e^{2\sqrt{B}x}))\sin(2\sqrt{A}x))}{\sqrt{A}(A+B)})]
\end{aligned}
$$

$$f_{14} = H\int_0^x \varphi(t)_{1<}\psi(t)_{1>}dt = \frac{H}{4\sqrt{A}\sqrt{B}(A+B)\pi}[\sqrt{B}(A+B)L(2\sqrt{A}x - \sin(2\sqrt{A}x)) +$$
$$+\sqrt{A}\sqrt{B}(L\sqrt{A} + S\sqrt{B})ch[2\sqrt{B}x]\sin(2\sqrt{A}x) - \sqrt{A}((A+B)L + (-BL + \sqrt{A}\sqrt{B}S)\cos(2\sqrt{A}x))sh[2\sqrt{B}x]]$$

$$f_{1p} = \int_0^x y(t)_\infty^c \psi(t)_{1>}dt = \frac{1}{16\sqrt{A}\sqrt{B}(A+B)\sqrt{16-k^2}}[-\sqrt{A}(A+B)(-L\sqrt{B} - S\sqrt{A}) +$$
$$+\sqrt{B}(2\sqrt{A}BL - 3A\sqrt{B}S - B^{3/2}S) + e^{-2\sqrt{B}x}(\sqrt{A}(A+B)(-S\sqrt{A} - 2Be^{2\sqrt{B}x}Lx - \sqrt{B}(L - 2\sqrt{A}e^{2\sqrt{B}x}Sx)) + \tag{23}$$
$$\sqrt{B}(A^{3/2}(-1+e^{2\sqrt{B}x})L + \sqrt{A}B(1+e^{2\sqrt{B}x})L - B^{3/2}e^{2\sqrt{B}x}S - A\sqrt{B}(2+e^{2\sqrt{B}x})S)\cos(2\sqrt{A}x) +$$
$$+\sqrt{B}(B^{3/2}e^{2\sqrt{B}x}L + \sqrt{A}B(-2+e^{2\sqrt{B}x})L + \sqrt{A}B(-1+e^{2\sqrt{B}x})S + A^{3/2}(1+e^{2\sqrt{B}x})S)\sin(2\sqrt{A}x))]$$

The expression determining the values of the bending moment at the section where the base under the beam is absent will take the following form

$$M(x)^c = -\frac{P}{2\beta}\frac{1}{\sqrt{16-k_1^2}}[(A+B)e^{-x\sqrt{B}}(-\sqrt{A}\cos(x\sqrt{A}) + \sqrt{B}\sin(x\sqrt{A}))] +$$
$$+C_{1<}e^{-x\sqrt{B}}[(-A+B)\cos(x\sqrt{A}) + 2\sqrt{A}\sqrt{B}\sin(x\sqrt{A})] - C_{2<}[(A-B)\cos(x\sqrt{A})ch(x\sqrt{B}) + \tag{24}$$
$$+2\sqrt{A}\sqrt{B}\sin(x\sqrt{A})sh(x\sqrt{B})] + C_{3<}e^{-x\sqrt{B}}[-2\sqrt{A}\sqrt{B}\cos(x\sqrt{A}) + (-A+B)\sin(x\sqrt{A})] +$$
$$+C_{4<}[2\sqrt{A}\sqrt{B}\cos(x\sqrt{A})ch(x\sqrt{B}) + (-A+B)\sin(x\sqrt{A})sh(x\sqrt{B})]$$

3 Results

As an example of using the obtained dependences, we give the calculation of an infinite beam, under the central part of which there is no base for the given section $2a = 1$. The beam is loaded in the center by concentrated force P. The diagrams of these deflections and bending moments are shown in Figure 2 and Figure 3, for different values of the coefficient of adhesion. In the same place, the corresponding diagrams for an infinite beam everywhere contacting the base are shown in dashed lines. Actual values of deformations and forces can be obtained using (17), (18), and (24).

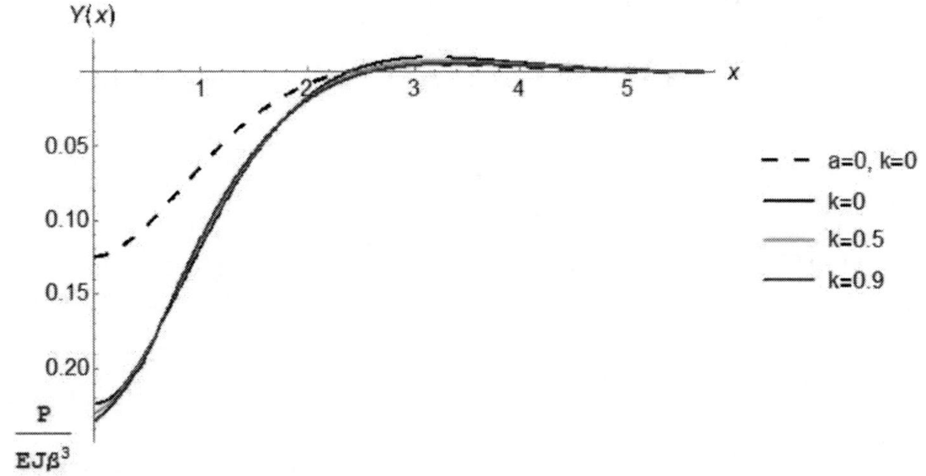

Fig.2. Diagrams of an infinite beam deflections with $a = 0.5$ and $k = 0$, $k = 0.5$, $k = 0.9$.

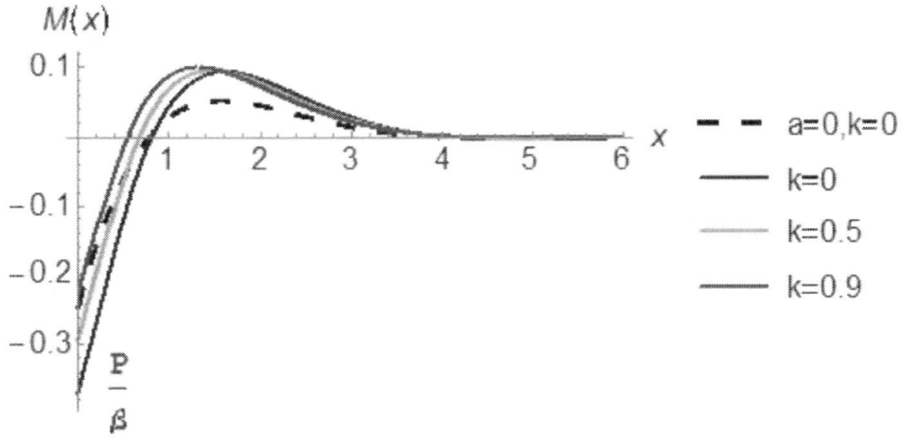

Fig.3. The bending moment diagrams of an infinite beam with $a = 0.5$ and $k = 0$, $k = 0.5$, $k = 0.9$.

When a skew-symmetric load acts on the beam, the application of the sine-transformation to differential equation (4) makes it possible to reduce it to an integral equation of the type (12), but the right side of it will be the deflection function of an infinite beam from a given skew-symmetric load $y(x)^s_\infty$, and the kernel of the equation is the function

$$K^s(x,t) = \int_0^\infty \frac{(k_1\lambda^2 + 4)\sin\lambda x \sin\lambda t}{(\lambda^4 + k_1\lambda^2 + 4)} d\lambda = \begin{cases} H\{\varphi(x)_{1>}\psi(t)_{1>} + \varphi(x)_{2>}\psi(t)_{2>}\} & (x \ge t) \\ H\{\varphi(x)_{1<}\psi(t)_{1<} + \varphi(x)_{2<}\psi(t)_{2<}\} & (x \le t) \end{cases} \quad (25)$$

Here the functions:

$$\varphi(x)_{1>} = e^{-x\sqrt{B}}\cos(x\sqrt{A}),\ \varphi(x)_{2>} = e^{-x\sqrt{B}}\sin(x\sqrt{A}),$$

$$\psi(t)_{1>} = S\cos(t\sqrt{A})sh(t\sqrt{B}) - L\sin(t\sqrt{A})ch(t\sqrt{B});$$

$$\psi(t)_{2>} = S\sin(t\sqrt{A})ch(t\sqrt{B}) + L\cos(t\sqrt{A})sh(t\sqrt{B}),$$

$$\varphi(x)_{1<} = sh(x\sqrt{B})\cos(x\sqrt{A}),\ \varphi(x)_{2<} = ch(x\sqrt{B})\sin(x\sqrt{A}),$$

$$\psi(t)_{1<} = (S\cos(t\sqrt{A}) + L\sin(t\sqrt{A}))e^{-t\sqrt{B}},\ \psi(t)_{2<} = (S\sin(t\sqrt{A}) - L\cos(t\sqrt{A}))e^{-t\sqrt{B}}. \quad (26)$$

Conclusions

The perfection of the behavior simulation of an infinite beam lying on a base determined by two parameters and having a dip under the loaded part is of great importance for the design of high-rise buildings and structures. Taking into account the dips in the ground significantly changes the values of the calculated deformations and forces in the beam. The study can be used to analyze the strength and practical design, for example, of strip foundations. The analysis of the results obtained using different values of the coefficients showed that the proposed calculation method and the model of the base under consideration are simple and convenient for practical calculations and can constitute an alternative to other engineering methods. The proposed method of calculation can be easily extended to cases where the dip occurs not under one, but under several sections of the beam simultaneously.

References

1. M. Filonenko-Borodich Collection of papers MEMIIT, Issue 53. (1945)

2. V. Vlasov, et al. *Beams, Plates and Shells on Elastic Foundation.* - Moscow: Fizmatgiz, p. 75 (1960)
3. P. Pasternak *Fundamentals of a New Method for Calculating Foundations on Elastic Foundation with the Help of Two Bed Coefficients.* - M. - L .: Gosstroyizdat, p. 56 (1964)
4. A. Kerr, *On the formal development of elastic foundation models,* Ing.-Arc., (1984)
5. S. Kolyasina , et al. The electronic scientific publication "Scientists' Notes PNU", **5**, 2, pp. 21 - 34, (2014)
6. C. Vallabhan, , et al. *An improved model for beams on elastic foundations*, Elastic-Plastic Failure Modelling of Structures with Applications Presented at the (1988)
7. M. Gorbunov-Posadov , et al. *Calculation of Structures on an Elastic Foundation* / M.I. Gorbunov-Posadov, (Moscow, Stroyizdat, 1994)
8. V. Travush , et al. *Bending of Uninsulated Rectangular Slabs Lying on a Two-parameter Base* (Izv. universities. Stroitelstvo, 1998)
9. D. Dobromir Engineering Mechanics, **19**, No. 6, p. 381–392, (2012)
10. V. Travush V.I. *The Method of Generalized Solutions in the Problems of Bending of Slabs on a Linearly Deformable Bed* / Stroitelnaya Mechanica i Raschet Sooruzheniy. 1, 24-28 (1982)
11. G. Wei , *A New Algorithm for Solving some Mechanical Problems, Computer Methods in Applied Mechanics and Engineering*, 190,2017-2030, (2001)
12. G. Wei, *Discrete Singular Convolution and its Application to the Analysis of Plates with Internal Supports.* Part 1: Theory and algorithm International Journal for Numerical Methods in Engineering, 55,913-946, (2002)
13. V. Travush II Republican Scientific and Technical Conference. Scientific and technological progress and ecology (Aktau, 1992)
14. A. Marufiy, Vestnik KSUCTA **1** pp.252-256 (2016)
15. I. Geldfand I.M., Shilov G.E. *Generalized Functions and Actions over Them* (Moscow, Dobrosvet (2007)
16. I. Gradshtein *Tables of Integrals, Series and Products* (St. Petersburg, BHV-Petersburg,2011)

Preparation and properties of hybrid materials for high-rise constructions

Tatyana Matseevich[1*]

[1]Moscow State University of Civil Engineering, Yaroslavskoe sh. 26, Moscow, 129337 Russia

Abstract. The theme of the research is important because it allows to use hybrid materials as finishing in the high-rise constructions. The aim of the study was the development of producing coloured hybrid materials based on liquid glass, a polyisocyanate, epoxy resin and 2.4-toluylenediisocyanate. The detailed study of the process of stress relaxation at different temperatures in the range of 20-100°C was provided. The study found that the obtained materials are subject to the simplified technology. The materials easy to turn different colors, and dyes (e.g. Sudan blue G) are the catalysts for the curing process of the polymeric precursors. The materials have improved mechanical relaxation properties, possess different color and presentable, can be easily combined with inorganic base (concrete, metal). The limit of compressive strength varies from 32 to 17.5 MPa at a temperature of 20 to 100°C. The values σ_∞ are from 20.4 to 7.7 MPa within the temperature range from 20 to 100°C. The physical parameters of materials were evaluated basing on the data of stress relaxation: the initial stress σ_0, which occurs at the end of the deformation to a predetermined value; quasi-equilibrium stress σ_∞, which persists for a long time relaxation process. Obtained master curves provide prediction relaxation behavior for large durations of relaxation. The study obtained new results. So, the addition of epoxy resin in the composition of the precursor improves the properties of hybrid materials. By the method of IR spectroscopy identified chemical transformations in the course of obtaining the hybrid material. Evaluated mechanical performance of these materials is long-time. Applied modern physically-based memory functions, which perfectly describe the stress relaxation process.

1 Introduction

The purpose of this study was to develop hybrid materials that combine the properties of polymeric and inorganic substances which may serve as materials for high-rise constructions. These materials must be produced in one stage, have good mechanical properties, easy to be painted and connect with an inorganic base (concrete etc.). Because polymeric materials are important relaxation properties, in this work, we conducted experiments on the stress relaxation in the temperature range from 20 to 100°C at the constant deformation equal to 3 %.

* Corresponding author: MatseevichTA@mgsu.ru

2 The review of the literature

A promising direction of improvement of technological, physical-mechanical, thermo-physical and other properties of composite materials is the development and use of complex binders, among which of particular interest are hybrid organic-mineral composition [1-3]. One of the ways of producing of hybrid composites is the introduction into a polymer matrix of inorganic particles with the surface modified by the grafted organic molecules [4-5]. This allows obtain nanocomposites possessing unusual combination of properties.

Structure and properties of metal-polymer composites are described in [6]. In a review article [7] analyzed the structure and properties of polymer nanocomposites, in which matrix polymers have different chemical structure. Among them polymer-layered silicates, as well as composites in which the dispersed phase is nano-diamonds. Considered hybrid materials based on liquid crystal polymers. Significant attention is paid to chemical modification of nanoparticles and intermolecular interactions between the nanoparticles and the matrix polymer. In the other review article [8] presents the results of a structural study of hybrid nanomaterials. Among them the polymer-layered silicates, for example, nanodiamonds are described. The methods of modification of nanoparticles and their dispersion in the polymer matrix are discussed. The role of interactions between polymer matrix and fillers and the morphology of the nanoparticles occurring during processing, has also received considerable attention.

The series of works [9-12] on preparing and research of properties of hybrid materials, the modification of organic silicon compounds polyurethanes [9-10] and polyetherurethanes [11-12] have been published.

In our work, was used as inorganic component, an aqueous solution of silicates of alkali metals containing significant amounts of water. In this regard, as modifiers to the silicate compositions used isocyanat-containing chemical compounds (ICC), which have a high reactivity towards water. In the case of using aqueous solutions of silicates of alkali metals in combination with ICC manifests a chemical interaction between the components with the formation of covalent bonds.

3 Materials and methods

As objects of research we used the PIC (brand MELLIONATE MR-200), 2.4-TDI, epoxy oligomer (technical brand ED-20), LG.

To evaluate the heat resistance of the prepared compositions was determined by thermomechanical curves on the instrument TMA Q400 (TA Instruments).

To conduct micro-photographing used polarizing microscope POLAM-312P.

To evaluate the heat resistance of the obtained compositions was determined by thermomechanical curves (dependence of deformation on temperature) on the instrument TMA Q400 (TA Instruments).

The compression curves were measured on samples of size 4×4×6 mm at a deformation rate 4.68 mm/min.

IR spectra were measured on a Fourier transform infrared spectrometer Tensor 37 (Bruker) with a resolution of 2 cm^{-1}. The investigated composition was placed between the windows of KBr. FTIR spectra with a resolution of 4 cm^{-1} were measured on Fourier spectrometer Vertex 70v with an attachment for FTIR GladiATR with diamond work item. Correction of FTIR spectra was performed using OMNIC programs.

As a research objects were used aqueous solution of sodium silicate (liquid glass LG), polyisocyanate (PIC), epoxy resin technical brand ED-20 and 2.4-toluenediisocyanate (2,4-TDI). In the process of producing samples of polymer materials were also used fat-soluble dye (Sudan blue G). This dye is a derivative of aminoanthraquinone. In synthesized systems are used as a catalyst. Also the dye of the phthalocyanine green was used.

The samples of polymer materials were produced by fractional sequential addition of the original components. Mechanical mixing was carried out using a laboratory stirrer vane type. Initially the mixture of LG and PIC in the ratio of 32.9÷26.3 wt. % was prepared. Under mechanical stirring for 3-4 min observed segregation of the mixture. Total processing time was 5-6 min to achieve homogeneity of the system. In order to remove the gaseous products of the resulting mixture was vacuumed at 40°C and 2 mm Hg. art. for 30-40 min. Then to reaction mixture were injected ED-20 in the amount of 34.2 wt. %. In this case, the process of mixing in the reaction mixture of segregation has not been observed and the process time was about 3-4 minutes. Then, just as in the previous case, the resulting mixture was vacuumed under the same conditions.

The final stage consisted in the addition into the reaction mixture of an excess of 2.4-TDI in the ratio of 4.7 wt. %. Here the mixing process was accompanied by intensive increase of the viscosity of the mixture, and the duration of the process was 2-3 min; the reaction mixture was vacuumed for 20 to 25 minutes at a temperature of 40°C. The amount of dye in the polymer mixtures was 1.9 wt. %.

The prepared polymer mass was a homogeneous viscous liquid. The curing process of the mixture (precursor) was carried out by stepwise heating from 40 to 70°C; the total curing time was 5-6 hours.

4 The measurements conducting and the approximation of stress relaxation

To conduct the stress relaxation in the uni-axial compression at the device for micromechanical test structures Dubov-Regel were prepared samples with a size of 4.0×4.0×6.0 mm. The measurements were performed at deformation of 3% at different temperatures (20-100°C). The rate of deformation was $1.87 \cdot 10^{-1}$ mm/min.

The stress relaxation curves were approximated in the linear region of mechanical behavior using the Boltzmann equation:

$$\sigma = \sigma_0 \left[1 - \int_0^t T(\tau) d\tau \right], \tag{1}$$

where σ_0 is the initial stress appearing in the output of the job "instantaneous" deformation; $T(\tau)$ is the memory function; τ is the current time, which runs through values from 0 to t; t is the end time.

For describing experimental data on stress relaxation used the memory function based on consideration of thermodynamic functions and their changes during the process [13-15]. The main assumption made in [13-15], is that the process of stress relaxation occurs as a result of interaction and diffusion of kinetic units (called often relaxers). The relaxers can be the groups of atoms arranged in repeating units, large fragments of macromolecules and their segments. The individual elements of the free volume (in this case, the "empty volume"), i.e., microcavity, stress raisers, etc. are also often. Together, these microcavities can merge, restructure and diffuse into the polymeric material during the stress relaxation, forming a structure that provides reduced stress relaxing. The process of restructuring has been proven by direct experiments for the study of the annihilation of positrons with the help of a special device that allows the relaxation process directly in the measuring chamber [16-17].

The emergence of two types of kinetic elements (relaxers and non-relaxers) and their diffusion in the material leads to the production of entropy of the system, which increases in the course of stress relaxation.

The entropy production (or the rate of change of entropy) is determined by the following relationship: $\frac{dS}{dt}\frac{1}{v}$ where S is the entropy; t is the time; v is the volume of the system. Considering the entropy of mixing of two types of kinetic units (relaxers and non-relaxers), in the works [13-15] obtained the General form of the memory function:

$$T(\tau) = -\frac{S_0}{k_B m}\left[\frac{1}{\alpha \ln \alpha + (1-\alpha)\ln(1-\alpha)} - \frac{1}{\ln 0.5}\right], \tag{2}$$

where $m = m^* \int_0^\infty T^*(\tau)d\tau$; k_B is the Boltzmann constant; m^* is the total number of kinetic units (in this case relaxers and non-relaxers per unit volume); α is the share relaxers of the total kinetic units, $T^*(\tau)$ is the variable part of the memory function; S_0 is the initial entropy of the system (specimen).

If the change in α over time τ due to the interaction between relaxers and their transition to non-relaxers material, the following kinetic equation takes place:

$$\frac{dc}{d\tau} = kC^n, \tag{3}$$

where k is the constant of reaction rate; C is the concentration.
The result:

$$C = \frac{C_0}{\left[1 + C_0^{n-1}(n-1)k\tau\right]^{\frac{1}{n-1}}}, \tag{4}$$

where C_0 is the initial concentration of any type of relaxers (for simplicity assumed that these concentrations are equal for different types of relaxers). Then:

$$\alpha = \frac{C}{C_0} = \frac{1}{(1 + k^*\tau/\beta)^\beta}, \tag{5}$$

where $k^* = kC_0^{n-1}$; $\beta = \frac{1}{n-1}$; n is the reaction order.

If the relaxation process is limited by the rate of interaction of relaxers, there is the memory function of $T_1(\tau)$, which is obtained by substituting (5) into (2).

If the relaxation process is limited by the diffusion rate of non-relaxers, the value α is determined from the relation [18]:

$$(1 - \alpha) = a\tau^{b/2}, \tag{6}$$

where $0 < b < 1$ and a is a constant.
It is the memory function of $T_2(\tau)$, which is obtained by substitution of (6) into equation (2):

$$T_2(\tau) = -\frac{S_0}{k_B m_2}\left[\frac{1}{a\tau^\gamma \ln a\tau^\gamma + (1 - a\tau^\gamma)\ln(1 - a\tau^\gamma)} - \frac{1}{\ln 0.5}\right] \tag{7}$$

where $\gamma = b/2$.

Approximation curves of the stress-relaxation was performed using a personal computer, into the software which was pre-recorded tabulated values of the variable parts of the memory functions which are given in [13-15] were incorporated. Automatically the selection of those parameter values for which the sum of squares of deviations of experimental values of the stress relaxing from the settlement was minimal and the correlation coefficient is maximum.

5 The results of the study

The study by the method of IR spectroscopy shows that the obtained polymer has a complex structure. Isocyanate groups remain in the final product in very small quantities, turning into NH-groups of the polyurethane. The formation of isocyanurate cycles in significant quantities is not observed.

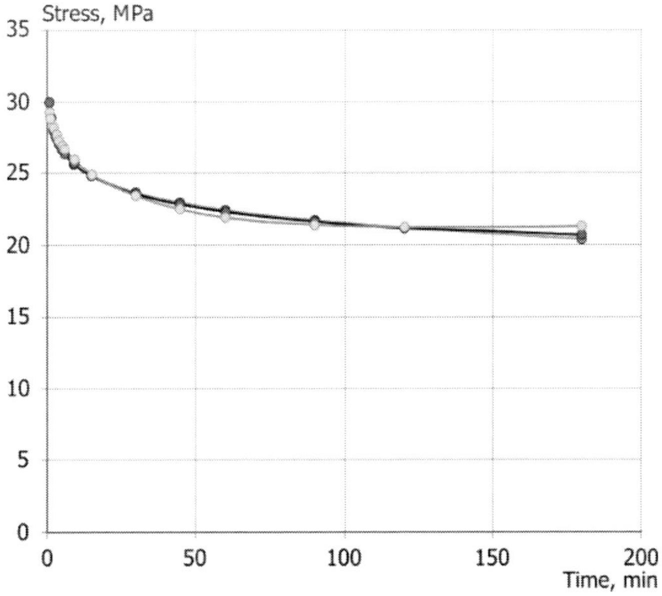

Fig. 1. The approximation the stress relaxation curve. Temperature is 20°C.

Figure 1 shows an example of experimental curve of stress relaxation for painted sample and the calculated curves obtained using the memory functions $T_1(\tau)$ and $T_2(\tau)$. Figure 1 is taken from the monitor of the computer program "Cascade" (INEOS RAS). It is seen that the approximation provides a good convergence of calculated and experimental data. In all cases the coefficients of correlation r close to 1.

Values of r for most samples slightly higher when using the memory function $T_1(\tau)$ than when using the memory function $T_2(\tau)$. In the first case, this means that the limiting stage of the process is the rate of interaction of relaxers and, and in the second case, the rate of diffusion of the non-relaxers. The rate constant of the interaction of relaxers for almost all samples is equal to 0.001 min^{-1}.

The values of initial and final stresses and the compressive strength are shown in Table 1. It is clear that our materials have high strength at different temperatures and relatively little stress relaxation.

Table 1. The properties of the painted samples.

Temperature, °C	Tensile strength at compression, MPa	The initial stress, σ_0, MPa	The equilibrium (final) tension, σ_∞, MPa
18	32.0	30.0	20.4
50	24.0	26.0	16.2
85	18.0	16.7	8.8
100	17.5	16.1	7.7

6 Conclusion

The painted materials based on sodium LG, 2,4-TDI and PIC, modified ED-20, have been prepared and investigated. Curing in the mild conditions at temperatures of 60-70°C leads to the formation of network structures. Monitoring method IR spectroscopy showed that the final product remains the only band of stretching vibrations of NH-groups. The IR spectrum corresponds to mainly the spectrum of the epoxide, and the difference spectrum matches the spectrum of the polyurethane. Chemical interaction is via the terminal groups of epoxy oligomer, i.e. for oxiranyl cycles.

Found that the modification of the initial mixtures of ED-20 results in more uniform grain structure of the composites and improving their mechanical properties. The detailed study of the process of stress relaxation at different temperatures in the range 20-100°C is provided. The physical parameters of the memory functions are evaluated. The initial relaxation modulus reached values of 1000 MPa at 20°C, and 540 MPa at 100°C. For non-modified epoxide starting materials of the relaxation module was 730 MPa at 20°C and at 100°C to 520. Obtained master curves provide prediction relaxation behavior for large durations of relaxation. The greatest relaxation module has material containing an increased amount of PIC.

Availability of simple technology for producing hybrid materials based on LG, their excellent mechanical properties and the ability to be painted to turn enables their use as decoration materials for the high-rise constructions.

References

1. M.A. Meador, E.F. Fabrizio, F. Ilhan, A. Dass, G. Zhang, P. Vassilarias, J.C. Johnston, N. Leventis, J.Chem. Mater., **17**, 1085 (2005).
2. D.J. Boday, D.A. Loy, Polymer Preprints, **50**, 282 (2009).
3. D.M. Delozier, R.A. Orwoll, J.F. Cahoon, J.S. Ladislaw, J.G. Smith, J.W. Connell, Polymer, **44**, 8, 2231 (2003).
4. V.I. Roldugin, O.A. Serenko, E.V. Getmanova, N.A. Karmishina, S.N. Chvalun, A.M. Muzafarov, Dokl. Akad. N., **449**, 5, 552 (2013).
5. A.S. Zhil'tsov, I.B. Meshkov, T.S. Kurkin, O.B. Gorbatsevich, V.V. Kazakova, A.A. Askadskii, O.A. Serenko, A.N. Ozerin, A.M. Muzafarov, Ros. Nanotekhn., **8**, 9-10, 46 (2015).
6. A.D. Pomogaylo, Ros. Khim. Zh., **XLVI**, 5, 64 (2002).
7. V.A. Gerasin, E.M. Antipov, V.V. Karbushev, V.G. Kulichikhin, G.P. Karpacheva, R.V. Tal'roze, Ya.V. Kudryavtsev, Usp. Khim., **82**, 4, 303 (2013).
8. A.V. Povolotskaya, A.V. Povolotskiy, A.A. Man'shina, Usp. Khim., **84**, 6, 579 (2015).
9. H. Sardon, L. Irusta, M.J. Fernandez-Berridi, M. Lansalot, E. Bourgeat-Lami, Polymer, **51**, 5051 (2010).
10. K. Szwarc-Rzepka, T. Szatkowski, F. Ciesielczyk, T. Jesionowski, Physicochem. Probl. of Mineral Proc., **49**, 2, 673 (2013).
11. A. Rekondo, M.J. Fernarndez-Berridi, L. Irusta, Europ. Polymer J., **42**, 2069 (2006).
12. M.S. Shabalina, E.A. Antipova, G.A. Khlebov, Kauchuk i rezina, **5**, 26 (2014).
13. 13 A.A. Askadskii, *Computational Materials Science of Polymers* (CISP, 2003)
14. A.A. Askadskii, V.I. Kondrashchenko, *Komp'yuternoe materialovedenie polimerov*. (Nauch. Mir, 1999).

15. A.A. Askadskii, A.R. Khokhlov, *Vvedenie v fiziko-khimiyu polimerov* (Nauch. Mir. 2009).
16. A.A. Askadskii, S.A. Tishin, V.V. Kazantseva, O.V. Kovriga, Vysokomolek. Soed., **A32**, 12, 2437 (1990).
17. A.A. Askadskii, S.A. Tishin, M.I. Tsapovetskiy, V.V. Kazantseva, O.V. Kovriga, V.A. Tishin , Vysokomolek. Soed., **A34**, 1, 62 (1992).
18. R.J. Gaylord, B. Joss, J.T. Bendler, E.A. Di Marzio, Brit. Polymer J., **17**, 2, 126 (1985).

Modelling of energy consumption at construction of high-rise buildings

Elena Korol[1*] and *Oleg* Korol[1]

[1]Moscow State University of Civil Engineering, Yaroslavskoe shosse, 26, Moscow, 129337, Russia

Abstract. High-rise building structures in the course of its erection suppose primary use of methods provided for erection, concrete and external finishing works. Erection works do not differ significantly from usual ones: traditional equipment, accessories and techniques are used which are based on erection of structures in project position using a crane. Structures to be assembled in building frame include steel columns and beams, wall panels, form elements of columns, walls and floor structures. We can note heightened attention to operational control for quality of erection, but it is attributable to all works in the course of high-rise construction. During high-rise erection by means of cast in-situ reinforced concrete all formworks to be used do not have any special differences except systems specially designed for high-rise erection using sliding formwork or vertical traveling forms. In these systems special attention is paid to safety of elevated works. Working methods of placement and curing of concrete and structures as a whole remain traditional – the requirements for controlling such operations become toughened. The most evident differences in high-rise erection with regard to equipment, machinery and accessories used are in means provided for load transportation and safety of works at heights. Particularity of internal finishing works which are also obligatory during construction of skyscrapers allows not considering them in as technological differences from usual construction as far as the «height» of its execution is limited by height of particular floor and determined by price and building class.

1 Introduction

High-rise erection process consists of execution of well-known technological processes on installation and erection of building structures: foundations, walls, columns etc. However particular features of such structures and factor of high altitude at which works should be performed predetermine several differences in methods of work execution expressed in appearance of new methods, mechanical aids, types of work and design documentation.

The construction site of High-rise erection is characterized by the integration of various types of energy consumers. The main ones are machinery and equipment, technological processes, as well as infrastructure of the construction site.

* Corresponding author: KafedraGKK@mgsu.ru

2 Mechanical aids and equipment to be used within high-rise erection

2.1 Cranes designated for high-rise erection

According to current data the use of traditional column cranes within building construction is limited by load-lifting height at 70– 80 m (in case of larger height the ratio of parameters «safety, lifting capacity, weight, cost» becomes nonoptimal). Lifting of loads at 130– 140 m is provided by attachment crane; its vertical tower can be detached on erected structures of building under construction. Integrated use of column cranes (erection phase at height up to 50–60 m) and attachment cranes (higher part of building) is used rather often when erecting buildings of height up to 130 m. Optimal ratio of parameters «load capacity, weight, cost» of attachment cranes is limited in case of building height of 130 m and more; however some cranes of this type are able to provide lifting of load up to 150 m in height. This is the field of application of selfclimbing cranes that do not have limitations in load lifting height. These cranes are fixed on hard stocks of stiffening cores or on external outline of building and its height of tower and attachment and movement structure provide crane operation in levels 30–40 m. After erection of building all cranes will be demounted and get down in parts using winches. According to practices of construction a technique is used when cranes of such type are preserved and remain on the roof for subsequent use during major repairs.

2.2 Equipment designated for transportation of concrete mixture to floors

Buckets and stationary concrete pumps are primarily used for transportation of concrete mixture to height. The use of buckets is defined by small volumes of monolithic works in precast frames. The use of concrete pumps is more specific for solid-cast frames. In this case the majority of concrete pumps of this type provide effective pumping of mixture at 40–50 m in vertical line, thus mixture supply at larger height is provided in stages using several pumps and intermediate containers. Placement of concrete mixture directly into cast formwork is performed using concrete placement booms installed on self-lifting or vertical traveling pillar supports of height 30–40 m. Cycles of vertical displacement of such devices correspond to erection rate of 3–4 floors. Primarily the supply of concrete mixture to working levels of high-rise buildings currently do not cause a problem and technically it is well cared.

2.3 Equipment designated for lifting of loads, personnel and for works on face of building

When erecting high-rise buildings the traditional problem of lifting small loads at the stage of finishing works is added by question of obligatory lifting of construction personnel at the stage of building frame erection. Special cargo-and-personnel lifting machines with lifting capacity up to 3 tones and personnel capacity up to 15–20 men with average lifting height up to 300 m are used for these purposes. Installation of a lifting machine is performed after erection of 5–10 floors of superstructure; the machines are provided for both finishing works and lifting of personnel to floors located near working levels of erection or concrete works. Number and type of lifting machines is defined on the base of building configuration and requirements for arrangement of construction works on site.

3 Technological features of foundation work

High-rise buildings are widely erected with developed embedded substructure designated for bearing vertical and tipping loads of building tower. The substructure area as a rule is larger than superstructure area of a building in 1.5–2 times and depth of foundation is 10–20 m.

3.1 Stage of excavation works

Heavy increase of depth of foundation in comparison with traditional depth of 3–4 m has led to inability to perform common excavation of high-rise building foundation using natural slopes of excavations that enlarge area of building foundation in 2–3 times. Substructure development methods «diaphragm wall» and methods of land cofferdam and fender have been actively used in order to solve these problems. It should be noted that these methods require special equipment and accessories.

Technological problems of deep foundation installation with vertical fencing of slopes are based on provision of stable release of walls enclosing ditch for foundation and ground as a whole up to the moment of creation of spacely rigid structure of building substructure. In order to solve this problem vertical fencing of slopes are mostly combined with external walls of building substructure using various methods for stiffening fences and hidden excavation works by means of small construction equipment together with installation of bearing piles in the course of excavation. These questions come up sharply when erecting highrise structures in fully developed areas.

Excavation works in the course of substructure installation are usually performed using a crane-type excavator equipped with front or bottom-dump bucket. In this case excavation is performed by levels in 3–4 m. Loaders and mini-excavators can be used in case of using methods of hidden excavation; earth excavation from trench can be performed by cranes. Excavation works as a rule are performed by specialized construction organizations equipped with respective excavation and transportation means.

Stage of excavation works in high=rise construction is additionally accompanied by special activities related to water drawdown and protection from ground water. Technology of execution of such works requires use of special equipment.

3.2 Stage of foundation slab installation in high-rise buildings

Foundation slab installation in the course of works on development of high-rise building substructure represents complicated and rather long individual stage. Rather often structure of foundation of high-rise buildings is designed in the form of pile-and-slab foundations that introduces installation of foundation slabs by works on installation of cast in-situ piles.

Foundation slabs of high-rise buildings are widely made of cast in-situ reinforced concrete and have rather complex basic geometry in plan with thickness 2–4 m and total volume of 3000–4000 m3. Practice of domestic construction allows specifying the following main technological problems related to its installation:

1. Reinforcement cage of foundation slabs represents heavy lower and upper multilayer nets resting on vertical reinforcement bars with system step 0.5–1 m and with local additional reinforcement in area of bearing columns and pillars. Due to that the reinforcement cage becomes an independent structure that requires solution on provision of stability and safety at all stages of its installation.

2. Provided high-strength concrete (usually class B40 and higher with low-cement content and high consistency of concrete mix) significantly increases requirements for processes of production, transportation and placement of mixture.

3. Requirements for continuity of foundation slab concreting in combination with its large thickness and area of concreting zones lead to:

– Necessity of mixture retardation after placement to 12–20 hours;

– Use of concrete placers of higher performance in combination with use of greater quantity of transportation means;

– Use of greater number of vibrators and thorough provision of mixture vibration for all local volumes of placement.

4. Holding of massive foundation slabs after placement of mixture is performed in compliance with extremely strict temperature requirements for rates of concrete heating-cooling and temperature differences between central and edge zones.

3.3 Stage of installation of vertical and horizontal structures of substructure

Structures of walls and floors of building substructure made of cast in-situ reinforced concrete do not have any significant differences within execution of formworks, reinforcement or finishing works. They are primarily characterized by high solidity (columns, pylons, bearing walls and beams), intensive reinforcement and use of high-strength concrete. Intensity of concreting is greatly reduced in comparison with foundation slabs within its installation; method of concrete works is within the traditional frameworks of using automobile or stationary concrete pumps with placer masts or it is performed using method «crane-bucket». Respectively there is no problem when using cranes of any types for provision of formworks or concrete works.

3.4 Stage of erection of superstructure, enclosing structures, finishing works

Erection of superstructure is performed using well-known technologies and technical means. Vertical enclosing structures (external walls of building) shall be installed either within the stage of floor erection (erection of wall panels), or as an individual set of works following works of installation of building bearing structures. In the latter case such works will be performed in levels with delay from works on structure erection in 5 or more floors.

Finishing works to be performed during high-rise construction shall be executed in levels in the course of its construction. Such level structure of finishing works is directly related to presence of crawl floors in a building so far as these floors are designated for solving matters of installation of roofing system and provision of temporary supply of finishing work areas with water, heat and energy.

4 Results

The construction site is characterized by the integration of various types of energy consumers. The main ones are machinery and equipment, technological processes, as well as infrastructure of the construction site. Therefore, integrated schedule for the consumption of fuel and energy resources can be presented in the form of a diagram of the daily fuel consumption by all types of energy consumers in unified units of measurement (kg of fuel equivalent), the general view is shown in Figure 1.

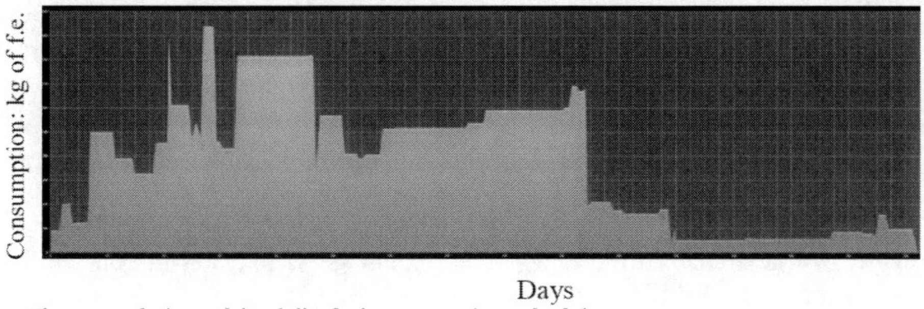

Fig. 1. The general view of the daily fuel consumption schedule

The calculation of fuel and energy resources consumption by different consumers occurs with the calendar factor to their hourly expenditure and the degree of employment during working hours. In general, the formula (1) for calculating the consumption of fuel and energy resources by various consumers is:

$$W\text{consumer} = Q\text{hour} * k * T\text{day} * T\text{work} * B, \quad (1)$$

where Wconsumer – the consumption of the fuel and energy resources consumer,
Qhour – the hourly consumption of fuel and energy resources by the consumer,
k – the calendar factor for the hourly expenditure of fuel and energy resources,
Tday – the duration of the working day,
Twork – the duration of the works,
B – the level of employment of fuel and energy resources consumers.

Modeling of energy and fuel resources consumption for **high-rise erection** monolithic housing facilities, as an example, revealed that the variation in the timing of the construction (for the same duration of works) significantly affects the integral consumption of energy resources during the production of the construction site. Thus, calculations will decrease the overall consumption of energy resources by 27% at the start of work in April compared with the beginning of October (Figure 2).

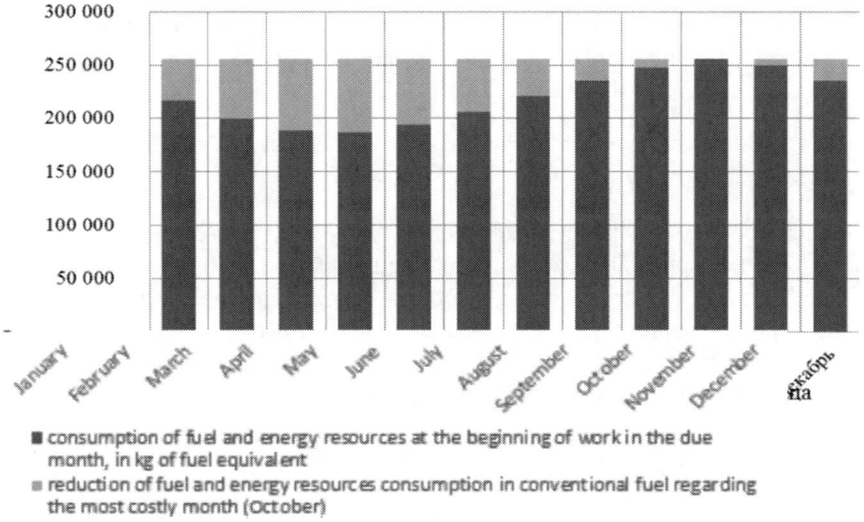

Fig. 2. Diagram of fuel and energy resources consumption changes depending on selection the month of beginning of the construction works

5 Discussion

The carried-out analysis of a consumption of energy resources in a generalized view on the integrated groups of consumers – machines, mechanisms and the equipment; technological processes and infrastructure of the building site I have revealed the following features.

Variation of terms of the beginning of construction at preservation of identical duration of works has revealed essential differences in a consumption of energy resources for separate groups of consumers. Consumption of energy resources by the "Machines, Mechanisms and Equipment" group at the beginning of construction in October is 13,7% higher, than at postponement of the beginning of construction for April.

The made comparative analysis of consumption of energy resources is submitted by all types of cars, mechanisms and equipment in table 1. The biggest energy consumers in this group are the tower crane (21%) and the stationary concrete pump (19%), the smallest - painting (0,03%) and plaster stations (0,11%) at the most adverse beginning of works in October.

Table 1. Consumption of Fuel and energy resources (FER) by cars, mechanisms and the equipment at various terms of the beginning of construction

Machines, mechanisms and equipment	FER expense in %	
	october	april
Bulldozer	0,43	0,44
Mobile crane	6,7	6,5
Tower crane	21	23
Mini-loader	3,2	3,5
Excavator	8,3	8,1
Vibropogruzhatel	1,4	1,4
Vibrotamper	0,26	0,23
Welding transformer	6,76	6,5
Concrete pump	3,2	3,07
Avtobetonosmesiteli	4,5	4,3
Stationary concrete pump	19,4	17,1
Construction elevator	0,98	1,08
Plaster station	0,11	0,15
Compressor	8,78	8,85
Painting station	0,03	0,04
Electric tool	7,5	8,5
Electric winch	6,4	6,12
Table for reinforcing works	0,73	0,73
Total	100	100

Consumption of energy resources by the Technological Processes group at the beginning of construction in October in 8 above, than at postponement of the beginning of construction for April. It is explained, first of all, by the fact that the summer period of construction heat treatment of newly-laid concrete mix is carried out only for the purpose of an intensification of set of durability of concrete and reduction of terms of construction of the facility. By production of monolithic works during the winter period it is necessary to provide a set of critical durability of concrete in the conditions of the negative temperature of external air that is connected with increase in duration and power of warming up of reinforced concrete designs of walls, columns, heating of overlappings and also concrete preparation and a base plate.

The made comparative analysis of consumption of energy resources by the Technological Processes group is presented in table 2. At the most adverse term the beginning of works in October the biggest energy consumers to this group are processes of electrowarming up of walls and columns (50%) and an electrical heating of overlappings (31%), the smallest - heating of concrete preparation and a base plate (1%) and was warmed soil vertical electrodes (6%).

Table 2. Consumption of FER by the Technological Processes group at various terms of the beginning of construction

Technological processes	FER expense in %	
	october	october
Electrowarming up of walls and columns	49,6	33,8
Electrical heating of overlappings	31,2	2,5
Heating of concrete preparation and base plate	0,9	0
Having warmed soil vertical electrodes	5,6	20,4
Heating of fittings electroheaters	12,6	43,3
Total	100	100

Consumption of energy resources by the Infrastructure of the Building Site group at the beginning of construction in October is 60% higher, than at postponement of the beginning of construction for April, t. to costs of heating of points of heating of working and household rooms, and in lighting – due to reduction of duration of light day increase.

The made comparative analysis of consumption of energy resources by the Infrastructure of the Building Site group is presented in table 2 At the most adverse term the beginning of works in October by the biggest consumers of energy resources in this group infrastructure elements of the building site - heating of household rooms (34%) and points of heating of workers (25%) are. Calculation and comparison of a consumption of energy resources on publicizing of external works at the beginning of a stroitelsv in October increases by 85%, and the electricity consumption on emergency and security lighting changes slightly. The comparative consumption of energy resources on basic elements of infrastructure is presented to the building site in table 3.

Table 3. Consumption of FER by the Infrastructure of the Building Site group at various terms of the beginning of construction

Infrastructure of the building site	FER expense in %	
	october	october
Point of heating of workers	25,2	2,02
Heating of household rooms and points of protection	34,1	41,3
Lighting of household rooms and points of protection	2,9	4,8
Security lighting	15,09	23,6
Emergency lighting	10,9	17,9
Publicizing of external works searchlights of 2 kW (working day 24 hours)	11,7	10,3
Total	100	100

Conclusion

The development of a methodological base for organizational and technological modeling in the part of building diagrams for the consumption of energy resources during the erection of high-rise building structures was proposed. The construction of energy consumption diagrams in the period of the complete building construction cycle allows planning and flexibly adjusting the consumption of energy resources both for individual types of consumers and in general.

Varying the timing of the beginning and completion of construction in the construction of a diagram of energy consumption allows you to achieve a reduction in energy consumption during the erection of high-rise building structures to 20-40%.

References

1. G.I. Abdullayev, V.Z. Velichkin, T.N. Soldatenko, Magazine of Civil Engineering, **3,** 43-50 (2013)
2. P.G. Grabovyi, E.A. Gusakov, A.M. Krygina, *Real Estate: Economics, Management.* **2.** Pp. 14-19 (2013)
3. P.G. Grabovyi, L.A. Manukhina, Real Estate: Economics, Management. **No 1-2.** Pp. 6-8. (2014)
4. G. Keshavarz, M., E.K. Zavadskas, Z. Turskis, J. Antucheviciene, Computers and Industrial Engineering. **112**, pp. 156-174 (2017)
5. K. Kianpour, A. Jusoh, A. Mardani, K.M. Nor, E.K. Zavadskas, Sustainability (Switzerland). **9,** iss.9. (2017)
6. E.A. Korol, *Real estate: economics, management.* **2.** Pp.32-35 (2013)
7. E.A. Korol, P.B. Kagan, *Collection of materials of the VI International Scientific and Practical Confuel and energy resources* Pp. 100-107 (2014)
8. E.A. Korol, N.S. Shushunova, Increase the level of environmental safety of a metropolis in the construction of "green" buildings Scientific Review. **No 7.** Pp.144-147 (2014)
9. E.A. Korol, N.S. Shushunova, Technologies of eco-building of exploited roofing coatings Scientific Review. No. 8. Pp.42-45 (2015)
10. E.A. Korol, N.S. Shushunova, Modern technologies of reconstruction of roofing coatings in eco-building, Construction and reconstruction **No 3** (65). Pp. 114-118. (rus) (2016)
11. P.P. Oleinik, A.L. Bolsherotov, Industrial and civil construction. **No. 10**. Pp. 20-21. (rus) (1987)
12. P.P. Oleinik, V.I. Brodsky, Industrial and civil construction**. No 12.** Pp. 30-32. (2012)
13. P.P. Oleinik, V.I. Brodsky, Basic requirements for the composition and content of the project for the production of works *Technology and organization of construction*. No. 3 (4). Pp. 35-38 (2013)
14. T. Saati T. Decision-making. The method of analyzing hierarchies. Moscow.: Radio and Communication, P. 278. (1993)
15. Zh.G. Mogilyuk, S.I. Zavalishin, M.S. Khlystunov, Scientific review, **No. 7-1**, 241-244 (2014)
16. Zh.G. Mogilyuk, M.S. Khlystunov, V.I. Prokopiev, International Journal for Computational Civil and Structural Engineering, **10, 2,** 120-129 (2014)

17. S.I. Zavalishin, Zh.G. Mogilyuk, M.S. Khlystunov, Scientific review, **7,** 245-248 (2014)
18. S.I. Zavalishin, M.S. Khlystunov, Zh.G. Mogilyuk, Ecology of urbanized territories, **4,** 35-41 (2015)
19. V.V. Poduvaltsev, M.S. Khlystunov, Zh.G. Mogilyuk, Safety in the technosphere, **3,** **2**, 14-17 (2014)
20. V.I. Prokopiev, M.S. Khlystunov, Zh.G. Mogilyuk, International Journal for Computational Civil and Structural Engineering, **2**, 105-110 (2014)
21. M.S. Khlystunov, V.I. Prokopiev, Zh.G. Mogilyuk, Procedia Engineering, **111,** 672 (2015)
22. M.S. Khlystunov, Zh.G. Mogilyuk, Safety of buildings, **3,** 16 (2003)

Numerical verification of composite rods theory on multi-story buildings analysis

Alaa El-Din Mansour[1,*], *Vladimir* Filatov[1], *Michael* Gandzhuntsev[1] and *Nikita* Ryasny[1]

[1]Moscow State University of Civil Engineering, Yaroslavskoe shosse, 26, Moscow, 129337, Russia

Abstract. In the article, a verification proposal of the composite rods theory on the structural analysis of skeletons for high-rise buildings. A testing design model been formed on which horizontal elements been represented by a multilayer cantilever beam operates on transverse bending on which slabs are connected with a moment-non-transferring connections and a multilayer columns represents the vertical elements. Those connections are sufficiently enough to form a shearing action can be approximated by a certain shear forces function, the thing which significantly reduces the overall static indeterminacy degree of the structural model. A system of differential equations describe the operation mechanism of the multilayer rods that solved using the numerical approach of successive approximations method. The proposed methodology to be used while preliminary calculations for the sake of determining the rigidity characteristics of the structure; are needed. In addition, for a qualitative assessment of the results obtained by other methods when performing calculations with the verification aims.

1 Introduction

Nowadays, the most common and demanded structures analysis and calculation method is the finite element method. In parallel, other methods (both numerical and numeric-analytical) are also under development. Further development and improvement of design methods is mainly carrying out to ensure the reliability of design decisions, especially when studying unique buildings and structures.

One of the methods are mainly functioned to analyses multi-story skeletons through calculating the configuring elements with the aid of Composite Rods Theory [1] (for its improver †A.R. Rzhanitsyn). One of those who first applied this was P.F. Drozdov [2]. Further later; the theory of composite rods was also functioned to calculate building frames for seismic actions [3].

†Alexey Rufovich Rzhanitsyn (1911 – 1987) born to a family of a Latin language teacher in one of Moscow high schools. After graduation, worked for some time in manufacturing. Then, studied at Moscow construction research institute (MICI) [Now; Moscow State University of Civil Engineering (MGSU)], which he graduated in 1936, and work in the Central Research Institute of the Construction Materials Industry. In 1945, Rzhanitsyn defended his doctoral dissertation thesis while in 1952, became the head of the materials strength laboratory at the central research institute where he worked. In parallel, taught at the MICI, where in 1972 been promoted to be a professor at structural mechanics department.

* Corresponding author: alaa_for_all@hotmail.com

Reinforced concrete structures [4-6], perforated steel beams [7, 8] and structures made of timber [9, 10].

In such an approach, the calculation design model of the structure is imposed as multilayer composite cantilever beams, where layers are acting as vertical supporting structural elements. Slabs act as elastic diaphragms with neglected local deformation. Thus, those individual beams are absolute rigidly connected by cross-links and elastic shear bonds. Due to such rigid linking, the resulting system is equivalent to (n+1) joining bars along the same bending axis (Fig. 1), where (n) is the number of joints.

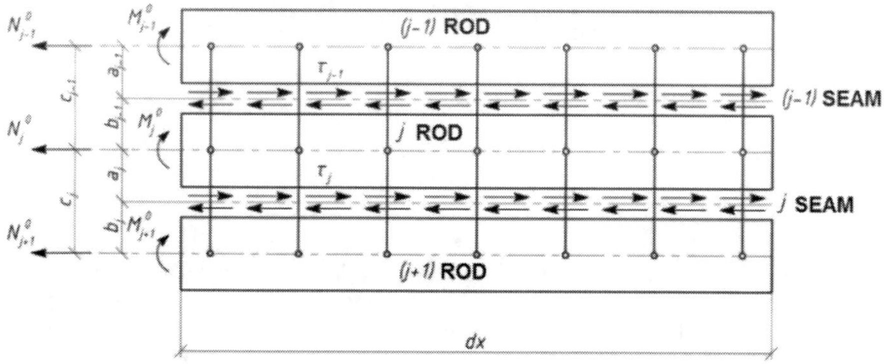

Fig. 1. Rigid linking in between individual beams.

When the composite rod is deformed, shearing forces arise in the shear connections of each connecting point. In order to simplify the model, discrete connections are replaced by a shearing forces function on seam per unit length as shown above. The stress-strain state of a compound rod is generally determined by solving a system of (***n***+***2***) differential equations of the 2nd degree.

Below, A numerical calculation technique for a rod consisting of two identical end (*n* = *1*).

2 Calculation Methodology

The differential equations system of solution take its dimension-less form through

$$\frac{d^2m}{d\psi^2}=-p; \quad (1)$$

$$\frac{d^2w}{d\psi^2}=-\frac{1}{2}\left(m-\tilde{t}\right); \quad (2)$$

$$\frac{d^2\tilde{t}}{d\psi^2}=-k\left(m-s\tilde{t}\right), \quad (3)$$

Where

$$\psi=\frac{x}{l};\tilde{t}=T\frac{cl}{EI};m=M\frac{l}{EI};w=\frac{y}{l};p=\frac{ql^3}{EI};s=\frac{4I}{Fc^2}+1;k=\frac{\xi}{2}.\frac{c^2l^2}{EI};$$

x - Is the coordinate measured along the axis of the composite rod; **E**, **I**, **F** - modulus of elasticity, moment of inertia, cross-sectional area of a single end of the composite rod respectively; **q** - A distributed load intensity; **M** - The total bending moment over the section of the composite rod devoid of shear connection from the action of the external load; **T** - shear forces accumulated along the length of the rod from its origin to the section under consideration; **l** - span of the rod; **y** - deflections of the composite rod (identical for each end due to the absolute rigidity of the cross-links); **c** - the axial distance between rod sides; ξ - the shear stiffness of the seam.

When solving the above shown system of differential equations, boundary conditions given in [11] should be considered As per (Fig .2), In the case of free shear by the end of a composite rod In both cases: with the hinged support of the composite beam, at a free edge; at a rigidly embedded edge, where, (φ-friction coefficient) are specified, in particular, straining actions (bending moment & transverse force), deformations [draft (slippage) & rotation] a zero values.

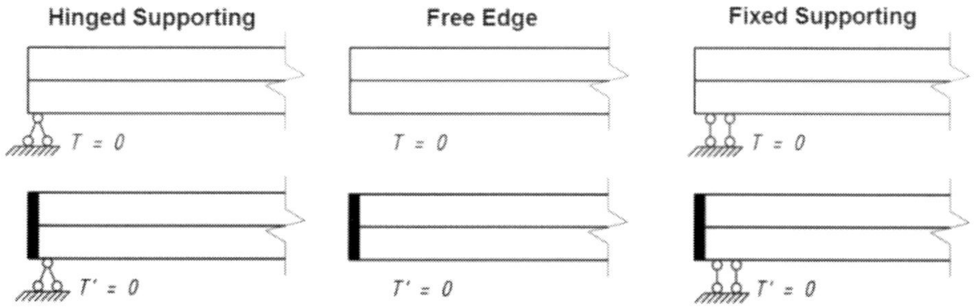

Fig. 2. Different boundary conditions.

So that to solve the differential equations system of EQ.(1), EQ.(2) & EQ.(3), the successive approximations method (MPA) developed in [11]. This method has high accuracy specially when considering the finite continuity of applied function, its first and second derivatives, and discontinuities of the right-hand sides of the differential equations without resorting to the use of generalized functions. The MPA was early successfully verified on plates, shells calculation on both rigid supports and elastic foundations. Also, on compressed-curved monolithic beams subjected to static linear & nonlinear actions, dynamic loads and on stability [11].

Now, the difference equation of MPA approximating EQ.(1) on a uniform mesh with step h at a regular point

$$m_{i-1}^R - 2m_i^L + m_{i+1}^L + \Delta m_i + h \cdot \Delta m'_i =$$

$$= -\frac{1}{12}h^2(p_{i-1}^R + 10p_i^L + p_{i+1}^L) + \frac{5}{12}h^2 \cdot \Delta p_i + \frac{1}{12}h^3 \cdot \Delta p'_i, \quad (4)$$

Where $\Delta m_j = {}^L m_j - {}^R m_j$; $\Delta m'_j = {}^L m'_j - {}^R m'_j$; ${}^L m'_j = m'_{j-0}$; ${}^R m'_j = m'_{j+0}$;

$m' = \dfrac{dm}{d\psi}$; The remaining quantities of this type have a similar meaning. Approximation (1) at the left boundary node

$$hm'_i + m_i - m_{i+1} = \frac{1}{12}h^2(hp'_i + 5p_i + p_{i+1}) \quad (5)$$

The difference approximation of both EQ.(2) & EQ.(3) can be performed by analogy also with the use of [11]. The resulting difference equations to be written successively for all calculation nodes of the model.

3 Numerical example

To illustrate the previous explained methodology, down here calculation results of a multi-story reinforced concrete frame that shown in (Fig. 3) on four different proposals, is presented

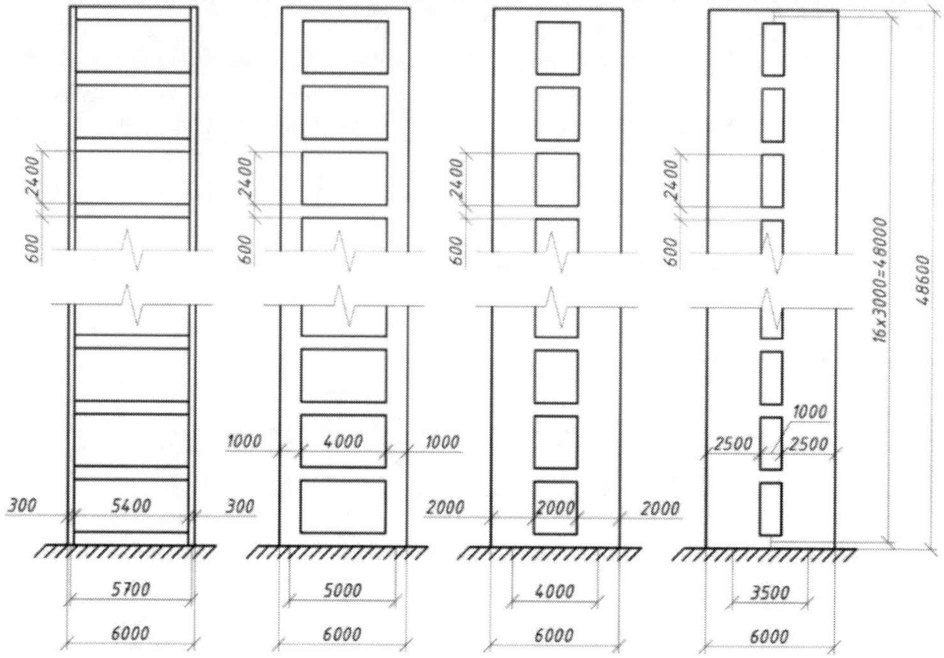

Fig. 3. A multi-story frame with its different proposals.

4 Initial Data

Modulus of elasticity of reinforced concrete E = 32.5·10³ MPa; the structural initial thickness is 300 mm, the remaining geometric dimensions in millimeters are shown in (Fig. 3); four versions of the frame under study different from each only in the width of vertical structural elements such as 300, 1000, 2000 and 2500 mm. respectively with a distributed load of intensity q = 1 kN / m is applied over the entire height from one side of the frame.

The shear seam stiffness is determined from [12] as

$$\xi = \frac{24E}{Bc^2\left(\frac{2c}{I_{п}}+\frac{B}{I_{В}}\right)}, \tag{6}$$

Where I_R - the horizontal bridges moment of inertia (horizontal elements); I_B - the (vertical elements) moment of inertia; B – axial dimension in between bridges (height of the floor is 3000 mm).

5 Results

Table 1 shows the values of horizontal displacements (drifts) at 48 meters for four different calculations by the theory of composite rods and in parallel via LIRA-CAD (LIRA-SAPR) program that based on finite element method (FEM).

The finite element results were obtained using elements modeled as normal (usual non-special) frame elements.

For the aim of solution convergence evaluation using composite rods theory, several numerical resulting values been carried out on bases of meshes different in dimensions by dividing the rod along the length by 4 segments (h_1 = 1/4), by 8 segments (h_2 = 1/8), by 16 segments (h_3 = 1 / 16), by 32 segments (h = 1/32) and by 256 segments as well.

Table 1. Horizontal frames drifting at elev. 48 meter from base on mm

Calculated Solution No.	Numerical solution using Composite Rods Theory					FEM solution (usual frame elemental model)
	h=1/4	h=1/8	h=1/16	h=1/32	h=1/256	
1	*39.81*	*41.11*	*41.58*	*41.69*	*41.72*	*42.34*
2	*12.55*	*12.71*	*12.73*	*12.73*	*12.74*	*12.72*
3	*7.94*	*7.96*	*7.96*	*7.96*	*7.96*	*7.91*
4	*6.57*	*6.58*	*6.58*	*6.58*	*6.58*	*6.53*

For calculated solutions (2), (3) & (4), an additional fineness been added to computations by FEM via using additive shell elements sized (100 x 100) mm that resulted out with values shown down in Table 2

Table 2. Horizontal frames drifting at elev. 48 meter from base on mm

Calculated Solution No.	Composite Rods Theory	FEM solution (frame elements model)	FEM solution (shell elements model)
2	*12.73*	*12.72*	*10.97*
3	*7.96*	*7.91*	*5.33*
4	*6.58*	*6.53*	*4.34*

Conclusions

The demonstrated theory shows an effective numerical approach for calculating composite rods which accuracy is high and converges the solutions rapidly considering discontinuous load functions.

This technique extremely suits the calculation of structures that lead to the composite rod models: brace-less trusses, Virendel girders [12], perforated beams and skeletons of high-rise buildings.it also suits those structures that behavior simulate rod structures.in such cases, Composite rods theory always give satisfying results on various sparse grids (with a small number of breakdowns of the structure along its length into the calculated sections) and as usual, so that we can obtain the required parameters (forces, displacements) at any point on a cross section, this point should be a node of the computational mesh. The technique procedure for determining forces in a section of a composite rod we described in details in [13]. Finally, when it is desired to calculate skeletons of multi-story buildings with considerable cross sections and rigid shear walls, the proposed method can just be used to perform preliminary or estimation calculations.

References

1. A.R. Rzhanitsin. *Composite rods and plates*. Moscow: Stroiizdat, 316 p. (1986)
2. P.F. Drozdov, M.I. Dodonov, L.L. Panshin, R.L. Surukhanyan. *Designing and calculation multi-storey civil buildings and their elements*. Moscow: Stroiizdat, 351 p. (1986)
3. A.N. Kuleshov. Bulletin of the Adyghe State University. Series 4: Natural-mathematical and technical sciences, **4**, pp. 148-154. (2008)
4. K.Z. Bashirov ACADEMIA. Architecture and construction, **2**, p.125-128. (2013)
5. V.S. Fedorov, K.Z. Bashirov, V.I. Kolchunov. ACADEMIA. Architecture and construction, **2**, p.116-118. (2014)

6. R.F. Fardiev, A.K. Ashrapov, Kazan State University of Architecture and Civil Engineering, **4**, pp. 363-369. (2015)
7. A.I. Pritykin, News of Institutions of Higher Education. Construction, **10**, pp. 110-116. (2009)
8. A.I. Pritykin, Bulletin of MGSU, **4**, pp. 177-181. (2009)
9. V.I. Linkov, Construction mechanics and calculation of structures, **5**. pp. 30-35. (2011)
10. Y.V. Krasnoschekov, Bulletin of the Siberian State Automobile and Highway Academy, **14**. pp.28-32. (2009)
11. R.F. Gabbasov, A.R. Gabbasov, V.V. Filatov, *Numerical construction of discontinuous solutions to the problems of structural mechanics.* Moscow: Publisher ASV, 280 p. (2008)
12. V. Shuller, Construction of tall buildings, Moscow, Stroyizdat Publ., 248 p. (1979)
13. V.V. Filatov, Calculation of through trusses using theory of composite rods A.R. Rzhanitsyna, Bulletin of MGSU, **9**, pp. 23-31. (2013)

Proposed strategies for designing sustainable high-rise apartment buildings in Ho Chi Minh City responding to critical urban issues

Nguyen Hoang Long Truong [1,*], Ngoc Huan Giang [1] and Trong Binh Duong [1]

[1]University of Architecture Ho Chi Minh City, 196 Pasteur str., Ward 6, District 3, Ho Chi Minh City, Vietnam

Abstract. This paper aims at finding practical strategies for designing sustainable high-rise apartment buildings in Ho Chi Minh City responding to varied municipal issues. Two steps are made. Step-1 identifies the critical issues of Ho Chi Minh City which are associated with high-rise apartment building projects. Step-2 finds potential and applicable strategies which are solutions for the critical issues in Step-1 with reference of seven selected assessment methods. The study finds the set of 58 strategies applicable to designing sustainable high-rise apartment buildings in Ho Chi Minh City.

1 Introduction

The emerging term 'sustainable development' proclaimed initially in the 1987 World Commission on Environment and Development has caught widespread concerns and become the global tendency responding to the critical environmental matters of greenhouse gas (GHG), global warming, and climate change [1]. The concept of sustainable development is acknowledged to be an intersection of the environmental, social and economic dimensions which are interrelated and interdependent to form the comprehensive approach of an issue [2]. Regarding built environment, sustainable design is a three-dimensional thinking of problem-solving. The strategies for sustainable design in planning, design, and construction should be studied in a local situation and diverse contexts in order to achieve large effects of sustainability on the global scale.

Vietnam government has profoundly proclaimed that sustainable development is a crucial future of the country. Vietnam early affirmed the sustainable development at the 1992 Rio Earth Summit and again concretized that affirmation at the 2002 Johannesburg Earth Summit as launching the National Agenda 21 which was domestically introduced in the Decision No. 153/2004/QD-TTg, namely, 'The Strategic Orientation for Sustainable Development in Vietnam.'

Ho Chi Minh City (HCMC), the most concentration of population and development in Vietnam, recently has a remarkable increase in developing high-rise apartment buildings (HAB) for meeting the great accommodating demands of citizens. High-rise dwellings are determined as the only solution for solving the housing demands in the coming decades in

* Corresponding author: long.truongnguyenhoang@uah.edu.vn

the urban area of HCMC. Recent official proclamation intensively insists that developing apartment housing must be the principal priority for all new residential projects developed in the major cities of Hanoi and HCMC. It particularly targets that the ratios of apartment buildings in new residential projects in HCMC must reach 80% in 2015 and 90% in 2020 respectively [3, 4]. HCMC today has been in the progress of upgrading and transforming urban housing from low-rise townhouses to HAB. It is expected that high-rise residences development is the potential solution for reserving land for infrastructures, public facilities, and green spaces, as well as facilitate municipal management and public transportation development [5]. However, most of built and ongoing HAB projects were profoundly carried out in the profit-oriented driven rather than the ecological and social establishment [6]. In addition, although sustainable high-rise residence has become a majority in the prospective period, the relative supports for HAB design are insufficient. The demand for practical design strategies thus is greatly appreciated.

Although there are multiple of green building assessment tool dedicated for Vietnam such as LOTUS rating system of Vietnam Green Building Council (VGBC), EDGE Resource Efficiency Certifications in Vietnam of International Finance Corporation (IFC) supporting HAB, the contents of criteria specify the general aspects of environmental issues and comprehensive Vietnam context rather than HCMC urban critical issues typically. Aside from that, the available rating systems are constructed with a huge set of criteria dedicated for rating and pointing the integral building life cycle but for guiding design procedures. They are therefore complicated for common planners, investors, or designers. This problem is the gap for this paper to fill in. This paper focuses on the practical strategies for designing sustainable HAB in HCMC, which are driven by the presently critical issues of the city.

2 Literature review

Many scholars have conducted studies on the designing sustainable HAB in HCMC and brought about multiple aspects of literature. Environmental issues of HAB is greatly studied in the thesis of Giang [7] and Le [8]. Giang [7]'s master thesis works on design methods for HAB in HCMC towards micro-climatic comforts and climate-driven energy-efficiency with analyzing traditional/vernacular houses and drawing lessons of climatically responsive design methods. However, this thesis merely investigates the climatically responsive aspects of HAB rather than a comprehensive approach to HAB as a whole. In another aspect, Le [8] suggests models of HAB in Vietnam based on lessons of passive and unique characteristics learned from Vietnamese vernacular houses. In her doctoral dissertation, two stages were carried out, analyzing case studies of vernacular houses and townhouses for collecting passive design strategies and suggesting typical models for HAB applicable to Vietnamese condition. Although this dissertation conducts research to detailed models of HAB but the solutions presented in the proposed models are hard to be applied in the actual construction industry.

Recently, there are noticeable two studies on sustainable HAB in HCMC, including Truong [9] and Giang [10] doctoral dissertation. The former provides a guidelines framework for HAB in HCMC and the latter provides criteria system of sustainable architecture for designing HAB in Ho Chi Minh City. The both studies consider the sustainability terms in integral three dimensions. However, Giang [10] attempts to provide comprehensive rating system with environmental, social and economic dimensions which result in the criteria system huge and complicated to be applied. Meanwhile, Truong [9] covers the urban and general sustainability issues to establish a framework of criteria dedicated for composing guidelines.

In general, despite multiple studies on sustainable HAB in HCMC, practical strategies easily applicable to designing sustainable HAB in HCMC are still opaque. This shortage of knowledge is an opportunity that paper aims to tackle.

3 Materials and methods

The strategies of sustainability applicable to HAB in HCMC should stem from the critical issues of HCMC urban and match with the local construction industry, designing and constructing techniques and living context. Local context, thus, should be attached to the process of study and become a key axis throughout the study. This study is designed to conduct in two steps, including Step-1 and Step-2. Step-1 identifies the critical issues of HCMC urban which are associated with HAB projects. Step-2 finds potential and applicable strategies built as solutions for the critical issues in Step-1 with reference to criteria of seven selected green building assessment methods. The procedure of the study is summarized in below diagram (Fig. 1).

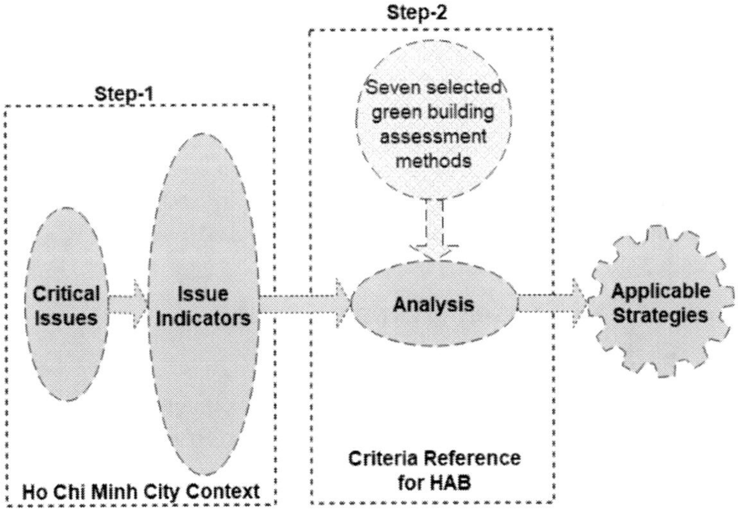

Fig. 1. Research diagram

Step-1 works on secondary data which is mostly used for documentary analysis. The data derives from various relevant sources, including books, narratives, reports, journals, magazines, archival data, and websites with intent to figure out a scenario of critical issues and its indicators. The result of Step-1 is the issue indicators of all critical urban issues.

Step-2 works on finding solutions for solving the issue indicators found in Step-1. Reference sources for strategies are seven selected green building assessment methods prevalent in the international, regional and domestic scale, all of which support designing HAB. Analyzing is the basic method used in this process. The proposed strategies are built on an integral reference and actual HAB built environment in HCMC. The documents of seven selected assessment methods are popular and available for downloading from organizational websites (Table 1).

Table 1. Seven selected green building assessment methods

No.	Sign	Assessment Method Title	Version	Publishing Year	Country	Global	Regional
1	①	Leadership in Energy and Environmental Design	LEED-NC V.3	2009	USA	✓	
2	②	Code for Sustainable Homes	-	2010	UK	✓	
3	③	The Comprehensive Assessment System for Building Environmental Efficiency (CASBEE)	CASBEE-NC	2014	Japan	✓	
4	❹	BCA Green Mark Scheme	4.1	2012	Singapore		✓
5	❺	Green Building Index (GBI)	GBI-RNC 3.1	2014	Malaysia		✓
6	❻	Building for Ecologically Responsive Design Excellence (BERDE)	BERDE-NC-VRD 1.1.0	2013	The Philippines		✓
7	❼	LOTUS	LOTUS-R EP	2013	Vietnam		✓

4 Results

4.1 Step-1: Identifying critical issues of HCMC

As common megacities in emerging countries, HCMC urban area today are inevitably struggling with various chronic problems in the urban living environment which are threatening the living quality and environment of HCMC.

4.1.1 Climate and Foreseen Climate Change

a. Urban Climate

Located in the tropical equatorial monsoon, the climate of HCMC is typically hot and humid with two distinctive seasons: the rainy season, from May to November, and the dry season, from December to April. Generally, HCMC has high air temperature and humidity; absorbs a great amount of solar radiation; and receives monsoons all year round [11, 12].

Air Temperature

Since HCMC is located in the tropical equatorial climate zone, the air temperature is quite high. The city is reported to receive 140 Kcal/cm² per year with insolation time of 5-9 hours per day and 160-270 hours per month. The annually average temperature is as high as 27°C and 330 days per year have the average temperature of 25-28°C. April is the hottest month with the highest mean temperature (28.8°C) while the average temperature is lowest (25.7°C) between December and January. The absolute highest and lowest temperatures are 40°C and 13.8°C [11, 12].

Humidity

The humidity is slightly high in HCMC. The mean relative humidity is 79,5% in annual, particularly 80% in the rainy season and 74.5% in the dry season. The absolute highest humidity is 100% while the lowest is 20% [11, 12].

Rainfall

The rain intensity in HCMC region is significantly high. The annually average rainfall reaches 1,949 mm and approximately 90% of which happens in the rainy season from May to November. The average number of rainy days in annual is 159 days. June and September are the two months recorded the highest rainfall while January, February, and March having no significant rainfall events [11, 12].

Prevailing Wind Directions

Two principal prevailing monsoons are corresponding to the two seasons, including the west-southwest monsoon flowing in the rainy season, from May to November, and north-northeast monsoon flowing in the dry season, from December to April. The west-southwest monsoon comes from the Indian Ocean with the average speed of 3.6 m/s and reaching the highest speed of 4.5 m/s in August. The north-northeast monsoon comes from the East Sea with the average speed of 2.4 m/s. Aside from the two main monsoons, the south-southeast monsoon occurs between March to May with the average speed of 3.7 m/s. Specifically, HCMC is located out of typhoon zone while the typhoon often strikes from the North to South Central Coast of Vietnam [11, 12].

b. Foreseen Climate Change

The impacts of climate change have evidently acknowledged in Vietnam. Vietnam was enumerated into the most vulnerable countries suffering the consequences of climate change seriously [13]. Specifically, HCMC locates in the high-effect zone of this influence. Recent researchers show evidence that there are relations of climate change and unfavorable changes in weather factors in HCMC, such as increasing air temperature, rainfalls, and sea-level rise [14]. The air temperature in HCMC tends to increase significantly in the coming decades. The average temperature is projected to increase 1°C by 2020 and 2.6°C by 2100 [15]. Besides, the rainfall tends to decline during the dry season and increase during the rainy season [15]. Statistical documents of rainfall show an increasing trend of the intensity of annual highest intensity rainfalls with an average rate of about 0.8mm per year and frequency of high intensity (greater than 100mm) rainfalls is clearly enhanced (

Fig. 2 and Table 2). In addition, sea levels are predicted to rise 30 cm by 2050 and 65-100 cm by 2100 compared with average sea level during the period 1980-1999 [15]. Consequently, the increasing frequency of heavy rains higher than 100 mm combined with more serious tidal flooding due to sea-level rise contributes critical urban inundation in HCMC urban area.

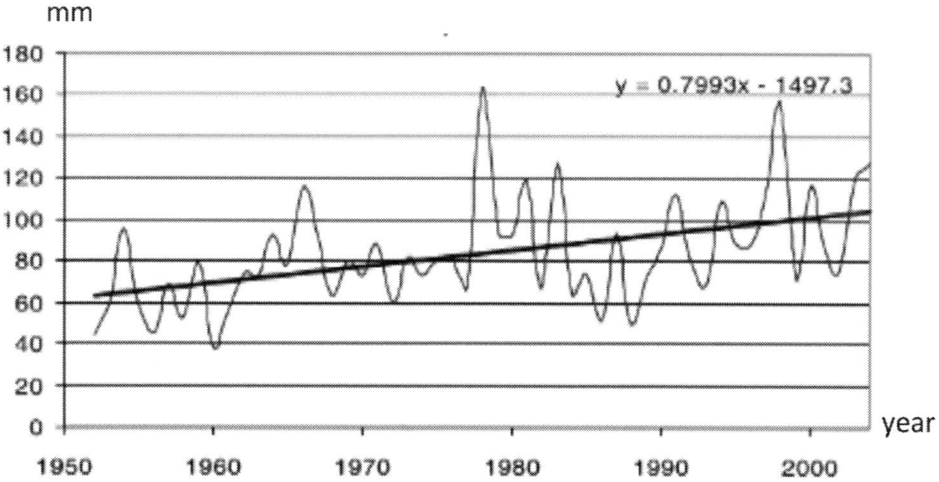

Fig. 2 Trend of highest intensity rain in HCMC [16]

Table 2 The frequency of high-intensity rainfall (greater than 100 mm) lasting 180 minutes [16]

Period	1952-1961	1962-1971	1972-1981	1982-1991	1992-2002

Occurred Times	0	1	2	2	4

4.1.2 Environmental Risks

a. Energy Consumption and Greenhouse Gas Emission

Located in the tropical belt, Vietnam possesses rich sources of renewable energy including solar energy, wind energy, and biomass. Despite having high potentials of renewable energy, majority energy consumed in HCMC is fossil fuel based sources and the exploitation of renewable energy is modest. The records show that HCMC consumes one-third of nationwide commercial energy, daily consumes 40 million kWh and annually consume 14.6 billion kWh [5]. As a result, the greenhouse gas emission in HCMC is extremely increasing since the volume of greenhouse gas emitted is proportional to the amount of fossil fuel energy consumed. HCMC is responsible for one fourth of the nationwide greenhouse emission originated from oil and coal consumption [5].

Due to climatic characteristics of HCMC, a large portion of electricity is consumed in residential buildings for maintaining a thermally comfortable indoor environment. Dehumidification and refrigerated cooling are primary strategies to alleviate overheat and humid air. With the development of mechanical cooling technologies, designers, builders, and owners misused the HVAC equipment without considering the energy efficient systems or indoor thermal comfort management. The rising household incomes accelerates the demands of using air conditioning for cooling interiors, in both office and residential buildings [17].

Besides, the greenhouse emission is also directly from material processing and transportation. A large portion of private motorcycle vehicles is the principal contributor to air pollution, noise pollution, and $CO2$ emission.

b. Emissions into Land and Water

The construction activities are evidently taking a large part in land and water sources degradation in the HCMC. The expansion of the construction site to the vegetated areas, river and canal banks during urbanization process has reduced significantly the greenery area and narrowed the water surface. Due to the high demands of housing, a considerable greenery areas are cleared for construction purposes without recovering after finishing construction. HCMC urban area possesses multiple rivers and canals for regulating the water flows. However, the recent urban expansion shows that many large apartment projects lie at banks of the rivers and canals and sometimes even encroach on the rivers and canals. These actions have accelerated the flow alteration, river bank erosion, and urban inundation. Besides, the construction sites are often backfilled, leveled and remained empty long before the actual constructing activities to be carried out, which brings about erosion, land slips under heavy rains, and causing critical pollution in land and water sources.

Moreover, HCMC construction activities contribute to solid waste with an essential proportion disposed into landfills. Reducing wastes is seldom paid attention in planning and design process. Most of the contractors do not consider seriously for classifying waste at construction sites and disposing of their construction waste in landfills. Only a small number of large contractors ready to pay more for waste management and disposition, the medium and small builders, however, often ignore treating waste properly. Moreover, most of the landfill sites are open to nature and lack of essential safety environmental protection methods to stop the solid waste, especially hazardous elements from penetrating into the soil, groundwater, river and the sea. These facts result from the unawareness of environmental protection, lack of public responsibility toward general safety, imbalance between profits and environmental preservation.

c. Urban Inundation

Urban inundation has been a critical problem in HCMC. Many factors contribute to the flooding in HCMC, from natural elements such as low elevation, heavy rainfall, flood tide, and sea level rises to improper human activities in urban planning and construction activities. Vietnam belongs to the top ten countries that their land and population is situated in low-lying coastal areas, 10 m above mean sea level [18]. HCMC is situated mostly on low-lying and marshy land in the north-eastern plain of Mekong Delta, the most vulnerable to flooding due to its low-level plain, population concentration, and economic activities. HCMC metropolitan area has a quite low level when 55% of the urban area is situated on one meter above mean sea level and only 28% is placed on two-meter level.

Sea level rise due to climate change is also a threatened potential since HCMC is 50 km from the sea. The sea level is reported to rise at the rate of 2-4 cm per decade and could rise 30 cm by 2050, 75 cm by 2100 when 10% of the area of 204 km² will be inundated.

The mixture of factors including flood tide, sea level rise, and rainfall accelerates urban flooding in HCMC. The records show that if the occurrence of heavy rainfall over 100 mm, coinciding with the flood tide, the long spell of extensive flood might occur [19]. The flooding has occurred in most districts and seriously affected the health, living activities, and production of citizens. The prolonged heavy rains often cause various roads of districts, e.g. District 6, 8, 11, Tan Phu, and Binh Tan, immersed in water, even in the new road and districts. This fact drives the city to face various side problems such as traffic congestion, environmental pollution, and public diseases.

Aside from the natural factors, improper human activities are also the reasons for the urban flooding, for example, the weak management in water conservation projects. To cope with the urban inundation, the city has carried out multiple steps for controlling flooding through many projects for renewing the capacity of the basin and drainage of canals such as Nhieu Loc – Thi Nghe, Tau Hu – Ben Nghe, Tan Hoa – Lo Gom. However, the implemented solutions partially solve the flooding in central areas, but new areas appear to be a brand new inundated place, especially in suburban areas. Due to the weak management of drainage infrastructure deployed projects, most of 'anti-flood' projects, however, have not been met the requirements and progress schedules.

4.1.3 Socio-Economic Issues

a. Shortage of Greenery and Open Space

Standards for the public greenery area and open space has been early prescribed in Vietnam planning and construction standards. The detailed ratio of greenery land for public use is defined in Decision 01/2006/QD-BXD with particular requirements as Table 3. Accordingly, HCMC, an extraordinary urban, must adopt the public greenery land as high as 12-15 m² per person. However, the city's greenery and open space for the public purposes, in reality, has been modest and unevenly distributed. Even though actual greenery area average per capita citywide reaches 13.74 m², the inner city attains only 1.95 m² per capita while suburban achieves an enormous number, 473.6 m² per capita [5]. The picture seems to get worse when the ratio of the greenery urban area tends to decrease in the context of increasing demands for city's infrastructure expansion and population concentration.

Table 3 Standard of area greenery land for public use [20]

Urban Type	Standard of greenery land for public use (m²/person)	Standard of green park land (m²/person)	Standard of gardens plants land (m²/person)	Standard street trees land (m²/person)

Extraordinary	12 - 15	7 - 9	3 – 3.6	1.7 – 2
I & II	10 - 12	6 – 7.5	2.5 – 2.8	1.9 – 2.2
III & IV	9 - 11	5 - 7	2 – 2.2	2 – 2.3
V	8 - 10	4 - 6	1.6 – 1.8	2 – 2.5

Responding to this situation, the city government has implemented many actions. In 2000, a project of HCMC's green park planning until 2010 was approved to set a short-term target for the city [21]. The greenery urban park must reach an average of 6-7 m^2 per capita by 2010, not including street trees, isolated trees in the industrial park, and private housing trees, of which, the existing urban is 3-4 m^2 per capita and new urban is 8-10 m^2 per capita. However, the above project was finally failed. The statistical data carried out at the end of 2011 showed an unexpected urban green park area achieved merely less than one m^2 per capita [22]. HCMC administrators also made an effort to bring about greenery spaces by renovating existing urban systems through projects in 2009 with strategies learned from cases of Singapore. Accordingly, 28,000 m^2 of sidewalk in the old districts (District 1,3,5, and 10) was recovered with greenery area of grasses and plants. Besides, there were approximately 13,000 m^2 of land was covered with the green landscape in the overhead bridge sites such as An Suong, Tan Thoi Hiep, Nga Tu Ga and Tan Thuan overhead bridges [22].

Although taking many efforts to fill the gap of greenery area, HCMC government impossibly manages the ratio as the standard required. The principal reason is that many residential projects did not comply with greenery area as planned. Much of the land reserved for developing greenery space is used for another purpose due to economic issues. Meanwhile, the official managers did not fulfill their monitoring functions.

b. Transportation Issues

The city planning is still imperfect, and the public transportation facilities are limited. The urban rail systems, e.g. subway, are under construction while the bus systems are not favorable to public use. As a result, the majority of people prefer using the private vehicles, e.g. motorcycle and car. Accompanying with the predominant townhouses in HCMC, the motorcycle has been the most popular vehicles serving transportation demands of people. Highly relying on private motorcycles leads to a number of side effects in urban area such as greenhouse gas emission, air and noise pollution, traffic congestion, accidents, and disorder city residence. The city environment is significantly polluted with dust, noise and polluted air. Traffic congestion has been a critical issue in HCMC when the growth of traffic volumes is increasing over the existing road networks' capacity.

For reducing congestion and CO_2 emission, in recent decades, the government has released policies to promote using bus systems in the city center with the efforts reducing the number of private motorcycles, however, the effects of those policies remain modest. In addition, the long-term mass transit systems, e.g. subways and sky trains, have been planned and constructed with the high expectation to be a good solution for solving the traffic issues.

c. Living Habits & Culture

Lack of public interests on sustainability

The majority of public in HCMC have been indifferent to the issues of sustainability due to the diverse reasons. This results from both managers and citizens' sides. From managers' side, the term sustainable development and its related themes have not been evidenced yet in the system of national laws, policies, and standards. The municipal planning has not been consistently connected with long-term plans on socio-economic development towards environment-friendly [23]. The city administrators are spending time for abundant social problems rather than issues of sustainability. On the public side, the environmental sustainability is likely not an urgent or mandatory issue rather than their basic living demands. Also, the environmentally friendly design or construction is often thought to be more expensive than the common method. People often believe that green building is accompanied with the huge additional fees. As a result, organizations and businesses have no interest in issues of green building due to lack of financial motives.

Particularly, the limited knowledge and technology have narrowed the boundary of strategies for energy efficiency, reuse and recycle materials. The shortage of typically sustainable design also diminishes the encouragement for sustainable design. Weak financial foundation, short-term business and backward constructing fashions have hindered the research and implement the sustainable design and construction. The materials and equipment associated with sustainable construction are often imported that increases cost.

Living habits

The traditional Vietnamese village contributes intimate community connectivity among villagers through the close interaction and cooperation among residents in the common daily activities. Since traditional village acts as an independent administrative and economic unit, the order and relationship among residents are defined by the village customs that are sometimes more powerful than the remote king's rule as a Vietnamese proverb said: 'The will of king yields to the people's customs.' In this sense, community connectivity has become distinctively cultural identities of the Vietnamese community [24]. However, the community sense in municipal modern living has greatly changed which results in various drawbacks on social life.

Besides, the common living habits of Vietnamese have specific manners of space arrangement and living which are blurred in the urban living, especially in HAB. All the critical issues of HCMC are summarized in Table 4 with the detailed issue indicators.

Table 4 Summary of critical issues of HCMC and issue indicators

Critical issue	Sub-issue	Issue indicator
A. Climatic issues	A1 Harsh climate & climate change effects	A1.1 Overheating
		A1.2 High rain intensity
B. Environmental risks	B1 GHG emission	B1.1 Low rate of renewable energy use
		B1.2 High demand of air-conditioning system
		B1.3 Materials processing
	B2 Emission into land and water	B2.1 Land erosion and river/canal pollution
		B2.2 High amount of solid waste
	B3 Urban inundation	B3.1 Low level land
		B3.2 Weak water conservation infrastructure
C. Socio-Economic issues	C1 Shortage of greenery and open space	C1.1 Lack of public open space
		C1.2 Lack of greenery land
	C2 Transportation	C2.1 Overloaded private vehicles
		C2.2 Unattractive public/bus transportation
		C2.3 Noise, polluted air, accidents
		C2.4 Traffic congestion
	C3 Living habits & culture	C3.1 Lack of community connectivity
		C3.2 Low-rise/Townhouse living habits

4.2 Step-2: Finding applicable strategies

Step-2 finds applicable strategies built as solutions for the critical issues in Step-1 with reference to criteria of seven selected green building assessment methods. A strategy could be the solution for multiple issue indicators, of which, a primary issue indicator is the main issue-driven strategy. Table 5 represents all outcomes of the analysis in Step-2. The far right column represents the reference sources that are referred in the certain strategy.

Table 5 Proposed strategies built from solutions of issue indicators

Primary issue indicator	No.	Proposed strategy as solution	Secondary issue indicator	Reference sources
A1.1 Overheating	01	Optimizing the building orientation (facing south)	B1.1	❼
	02	Incorporate the buffer space (balcony or loggia)	B1.1	-
	03	Provide buffer spaces (staircases, lifts, store, toilets, etc.) on majority of the west wall	B1.1	-
	04	Incorporating exterior shading devices; vertical greenery system	B1.1	①, ③, ❹, ❺, ❻, ❼
	05	Provision of thermal insulation on building envelope; vegetated roof	B1.1	③, ❹, ❺, ❻, ❼
	06	Light color textures and rough textures for building external facades, roofing and hardscapes	B1.1	①, ③, ❺, ❻, ❼
	07	Provide shade from a tree canopy, solar panels structures or solar reflective architectural structures	B1.1	①, ❺, ❻, ❼
	08	Interior sunlight control elements (e.g., curtain, drapes, blinds, adjustable louvers)	B1.1	-
	09	Selecting appropriate glazed size for daylighting and avoiding excessive illumination levels inside the building	B1.1	❺
	10	Use appropriate double-glazed units with Low-e glass, or glass coatings	B1.1	❺
A1.2 High rain intensity	11	Design drainage systems that adequately dispose of rainwater and prevent flooding from excessive runoff	B3.2	-
	12	Use permeable/pervious paving materials or open grid pavement system for hardscapes	B3.2	①, ②, ③, ❺, ❼
B1.1 Low rate of renewable energy use	13	Install on-site renewable energy generation systems	-	①, ②, ③, ❹, ❺, ❻, ❼
	14	Use solar water heating system	A1.1	③, ❼
	15	Use geothermal systems, cogeneration, or other alternatives for water heating and space cooling	A1.1	③
	16	Incorporate skywell for daylighting and air movement	B1.2, A1.1	-
	17	Provide daylighting, cross and/or stack natural ventilation for all public, circulation spaces and large portion of habitable rooms	B1.2, A1.1	③, ❹, ❺, ❻, ❼
	18	Arrange apartment unit's openings in opposite sides and interior doors to facilitate the cross ventilation	B1.2, A1.1	-
	19	Enhancing the penetration of daylight by light color interior surfaces, light shelves	-	-

Primary issue indicator	No.	Proposed strategy as solution	Secondary issue indicator	Reference sources
	20	Increase floor-to-ceiling heights and decrease distance of habitable spaces from windows	-	-
	21	Planning interior partitions and furniture that avoids blocking daylight	-	-
B1.2 High demand of air-conditioning system	22	Use operable windows in combination with mechanical ventilation systems	B1.1, C2.3	-
	23	Use flexible occupant-controlled openings	B1.1, C2.3	①, ❻
B1.3 Materials processing	24	Use certified wood/timber	B2.2, B2.1	①, ❺, ❻, ❼
	25	Minimize the use of highly processed materials	B2.2, C2.3	-
	26	Use non-baked materials for non-structural walls	B2.2, C2.3	❼
B2.1 Land erosion and river/canal pollution	27	Recycle/compost landscape and/or organic waste and incorporate compost to promote healthy topsoil in landscaping	B2.2	❺
	28	Preserve the site with existing and adapted vegetation	B3.2, C2.3, A1.1	①, ②, ③, ❹, ❺, ❻, ❼
B2.2 High amount of solid waste	29	Use local/regional materials; natural/organic materials	B1.3, B2.1	①, ②, ③, ❺, ❻, ❼
	30	Reuse/renovate an old building; reuse building nonstructural components	B1.3, B2.1	①, ②, ③, ❺, ❻, ❼
	31	Use reusable, recyclable materials	B1.3, B2.1	③, ❻
	32	Use the reused, refurbished, and recycled materials	B1.3, B2.1	①, ②, ③, ❹, ❺, ❻, ❼
	33	Use materials with appropriate durability, self-evident care and maintenance	B2.1	③
	34	Provide a centralized and easily accessible space for collection and storage of recyclables	B2.1	①, ❺, ❻, ❼
	35	Design for flexibility and adaptability to future changes in building use (e.g., layout, building structures)	C1.2	②, ③
B3.1 Low level land	36	Locate the building on elevated position	A1.2	②, ❻, ❼
	37	Elevated ground floor	A1.2	-
B3.2 Weak water conservation infrastructure	38	Use water-conserving plumbing fixtures and appliances	B2.1	①, ②, ③, ❹, ❺, ❻, ❼
	39	Use site-reclaimed water (non-potable water) for landscape irrigation, or toilet flushing, or cooling tower	B2.1	❺, ❻, ❼
	40	Use ecology-based treatment processes for sewage treatment	B2.1	①, ②, ③, ❺, ❼
	41	Grey/black water treatment, recycling and reuse for non-portable purposes	B2.1	①, ②, ❺, ❻, ❼
C1.1 Lack of public open space	42	Enhance vegetated open space and biodiversity	C2.3, C3.1	①, ②, ③, ❻, ❼
	43	In site selection, avoid building on environmentally, culturally, and socially sensitive sites.	C3.1	①, ②, ③, ❺, ❻, ❼
C1.2 Lack of greenery land	44	Optimize the layout and reduce the development footprint to save land and natural resources, and preserve open space	B2.1, C2.3	①, ②
	45	Rehabilitate a brownfield or previously occupied site	B2.1, C2.3	①, ②, ❺, ❻, ❼
	46	Provide limited parking capacity to meet but not exceed minimum local zoning requirements	C2.2	①, ❻

Primary issue indicator	No.	Proposed strategy as solution	Secondary issue indicator	Reference sources
C2.1 Overloaded private vehicles	47	Provide an amount of preferred parking spots for buses, sharing vehicles, and high-performance hybrid vehicles	C2.2, C2.4	①, ❻
	48	Locate preferred parking, bicycle parking, pick-up areas, and covered waiting spaces close to main building entrances	C2.4	❻
C2.2 Unattractive public/bus transportation	49	Locate a site in walking distance to public transit stops	C2.4	①, ❺, ❻, ❼
	50	Provide public transport information and assistance for residents	C2.4	❼
C2.3 Noise, polluted air, accidents	51	Design buildings for noise pollution reducing	-	②, ③, ❹, ❺, ❼
	52	Take into account security, noise and dust concerns when providing natural ventilation	A1.1; B1.1	-
	53	Positive soundscapes	-	-
C2.4 Traffic congestion	54	Provide favorable public or private walkway	C2.2, C2.3	-
	55	Provide favorable bicycle lanes and secure and easily accessible bicycle storage/parking on the site	C2.3	①, ❺, ❻
	56	Locate a site in walking distance to many community services and amenities	C3.1	①, ❺, ❻, ❼
C3.1 Lack of community connectivity	57	Incorporate dedicated community facilities flexible for varied activities (e.g. recreational facilities, meeting room, childcare facilities, gardens)	C1.1	❼
C3.2 Low-rise/Townhouse living habits	58	Proper layout of apartment unit (living, kitchen and bedroom space)	-	-

5 Discussion

Presently, the architectural designing activities in HCMC has no much concern over the terms of sustainability. The numbers of 'green' buildings designed recently bring about merely formal appearance rather than the real sustainable performance of the building. In addition, various sustainable strategies applied are general strategies rather than straightly contribute to the real issues of HCMC. The result of this research with 58 strategies stemming from the critical issues of HCMC urban is an up-to-date guidance responsible to the actual context of society. The set of strategies is formed on the thinking of applicability, thus it is practical for the designing contemporary HAB in HCMC toward sustainability.

Since the starting points of this strategies building are the critical issues of HCMC, they certainly bypass some aspects of sustainability such as energy consumption issues and VOC emission issues which are the limitation of the above strategies. This research implementation is not the copy of available strategies and criteria but filtering out what is necessary and facilitating to apply to the actual condition. The research contributes highly practical values to HAB built environment in HCMC.

6 Conclusions

This study is conducted on the critical issues of HCMC urban and developing the practical set of strategies as solutions to those problems based on the reference of various sustainable criteria. The set of 58 strategies is ready to apply in the actual design process on HAB in HCMC.

References

1. G. Brundtland, "Our common future: The world commission on environment and development," Oxford University Press, 1987.
2. ISO, "ISO/DIS 15392, Sustainability in building construction – general principles," International Organization for Standardization, 2006.
3. Phê duyệt chiến lược phát triển bền vững Việt Nam giai đoạn 2011-2020," P. Minister, ed., 2012.
4. Approval of national housing development strategy to 2020 and vision to 2030 [in Vietnamese]," P. Minister, ed., 2011.

A. D. Tran. "Phát triển nhà ở cao tầng Thành phố Hồ Chí Minh theo hướng xanh - hiện đại - bản sắc," October 15, 2014; http://118.70.241.18/english3/news/?41939/Phat-trien-nha-o-cao-tang-Thanh-pho-Ho-Chi-Minh-theo-huong-xanh---hien-dai---ban-sac.htm.

5. "Gỡ vướng trong phát triển nhà ở tại TP Hồ Chí Minh," October 15, 2014; http://cand.com.vn/dia-oc/Go-vuong-trong-phat-trien-nha-o-tai-TP-Ho-Chi-Minh-372833/.
6. N. H. Giang, "Giải pháp thiết kế nhà ở cao tầng tại TP.Hồ Chí Minh theo hướng đảm bảo điều kiện tiện nghi vi khí hậu & sử dụng năng lượng có hiệu quả [Design Methods for High-Rise Housing in Ho Chi Minh towards Micro-Climatic Comforts and Energy Efficiency]," Master thesis, University of Architecture Ho Chi Minh City, Ho Chi Minh, 2007.
7. T. H. N. Le, "An Analysis of Passive Design and Unique Spatial Characteristics Inherent in Vietnamese Indigenous Housing and Their Applications to Contemporary High-Rise Housing in Vietnam," Doctoral dissertation, Inha University, Incheon, 2011.
8. N. H. L. Truong, "Developing Design Guidelines Framework for Sustainable High-Rise Apartment Buildings in Ho Chi Minh City," Doctoral dissertation, Seoul National University, 2016.
9. N. H. Giang, "Hệ thống tiêu chí Kiến trúc bền vững áp dụng cho thiết kế nhà ở cao tầng tại Thành phố Hồ Chí Minh [Criteria system of sustainable architecture for designing high-rise housing in Ho Chi Minh City]," Doctoral dissertation, University of Architecture Ho Chi Minh City, 2016.
10. Vietnam Ministry of Construction, QCVN 02: 2009/BXD Vietnam building code - Natural Physical & Climatic Data for Construction, Hanoi: Vietnam Ministry of Construction, 2009.
11. "Khí hậu, thời tiết," October 15, 2015; http://www.hochiminhcity.gov.vn/thongtinthanhpho/gioithieu/Lists/Posts/Post.aspx?List=9efd7faa-f6be-4c91-9140-e2bd40710c29&ID=5497&Web=9d294a7f-caf2-456d-8ca0-36b393b8c052.
12. S. Dasgupta, B. Laplante, C. Meisner et al., "The impact of sea level rise on developing countries: a comparative analysis," World Bank policy research working paper, no. 4136, 2007.
13. Asian Development Bank, Viet Nam : environment and climate change assessment, 2013.
14. "Biến đổi khí hậu và Hiệu ứng đảo nhiệt đô thị ở TPHCM," June 15, 2015; http://quanlidothi.com/xem/1089/bien-doi-khi-hau-va-hieu-ung-dao-nhiet-do-thi-o-tphcm.html.
15. H. Long Phi, "Inundation and Stormwater Drainage in Ho Chi Minh City."

16. K. Nihon Enerugi Keizai, APEC energy demand and supply outlook, Tokyo: Asia Pacific Energy Research Centre, Institute of Energy Economics, Japan, 2013.
17. G. McGranahan, D. Balk, and B. Anderson, "The rising tide: assessing the risks of climate change and human settlements in low elevation coastal zones," Environment and Urbanization, vol. 19, no. 1, pp. 17-37, 2007.
18. Ministry of Natural Resources and Environment, Climate Change, Sea Level Rise Scenarios for Vietnam, Hanoi, 2009.
19. "TCXDVN 362-2005 Quy hoạch cây xanh sử dụng công cộng trong các đô thị - Tiêu chuẩn thiết kế," M. o. Construction, ed., 2006.
20. "Phê duyệt dự án quy hoạch công viên cây xanh Thành phố Hồ Chí Minh," P. s. C. o. HCMC, ed., 2000.
21. Van Nam. "TPHCM: Diện tích công viên giảm gần 50%," October 15, 2014; http://www.thesaigontimes.vn/Home/dothi/hatang/27565/.
22. "Định hướng chiến lược phát triển bền vững ở Việt Nam (Chương trình Nghị Sự 21 của Việt Nam)," P. Minister, ed., 2004.
23. H. L. Phan, C. Tu, D. N. Nguyen et al., The Traditional Village in Vietnam, Hanoi: The Gioi, 1993.

Ensuring living condition for ageing population by public–private partnership (PPP)

Miha Konjar[1,*], *Matej* Nikšič[2], *Janez Peter* Grom[1], *Sabina* Mujkić[2], and *Alenka* Fikfak[1]

[1]University of Ljubljana, Faculty of architecture, 1000 Zoisova 12, Ljubljana, Slovenia
[2]Urban Planning Institute of the Republic of Slovenia, 1000 Trnovski pristan 2, Ljubljana, Slovenia

Abstract. Lack of financial resources has become one of the main issues in fulfilling social and physical needs in urban development. The declining levels of public resources make the collaboration between public and private investors necessary. When facing the challenges of ageing population, shared investment may contribute to the appropriate development of sheltered housing to meet the goals of spatial planning as well as certain standards at the level of urban design. By ensuring appropriate living conditions for all generations such urban PPP projects may contribute to the fulfilment of the public interest. The paper presents practice of PPP implementation in Ljubljana, Slovenia, where local authority with the collaboration of private partners ensured more than 400 sheltered apartments in the last years. Examples show the extension of the idea from the 70s onwards in finding new models of housing for the aging population. The development of new models can be a good example of strengthening the cooperation between public and private partners in the field of urban development practice.

1 Introduction

Today's apparently "chaotic" settlement pattern in Slovenian suburbia and countryside is a consequence of political, social and economic conditions and changes that have been taking place in the post-WWII period for the last fifty years. The ideological orientation of the communist movement gave absolute priority to collective society and interests. A higher level of collectivism was to be achieved more by restriction rather than considering individual interests. In physical planning social housing construction was favoured. Uniform models and standards were predominant both for large urban agglomerations as well as for rural areas. Restrictions and directed construction together with regulated planning were supposed to give a complete overview of everything that was taking place in space [1]. Decline of policies on detached housing was one of the reasons why organized and planned forms of housing cooperatives were not developed. At the same time the priority given to social housing served as an efficient instrument to achieve quality and certain aesthetic level of organized settlement pattern.

The need for a new type of dwellings has occurred and was left to new processes of pre-structuration in agrarian economy accompanied with the social and demographic

* Corresponding author: miha.konjar@fa.uni-lj.si

restructuring (abandonment of agrarian land and employment in industry) [2]. The first turning point in the development of legislation were the laws adopted in 1967 about urban planning [3] and a regional spatial planning law [4]. This legislation was first used for urban planning at the level of municipalities. It was meant as a guideline for development in urban areas, while avoiding spatial dispersion of single-family houses in extra-urban area (including urban periphery). The changes in the production system of rural space based on mixing and complementing of agrarian with non-agrarian functions, influenced the change of traditional settlement patterns and of their constituent parts – "homes" – by introducing new typologies. The countryside was left to develop by itself. Facilities were built individually, mostly in the form of replacement construction for the old farms. The design and construction were still based on the typology of social housing as a "free-standing single-family residential house, positioned in the middle of the plot". This process exploded in the 1970's and drastically changed the settlement patterns in Slovenian.

The conducted research deals with the given topic comprehensively and interdisciplinary, with the integration of practical knowledge from different fields, such as: Urbanism, Architecture, Landscape Architecture, Spatial planning, Administration, Law and Economy. The aim of this research was to identify the connection between problems of housing for ageing population and the development of models for implementation in a PPP manner. Special attention was given to the realization of the public interest in such projects, especially from the point of view of urban design and spatial planning goals to ensure quality living environments. Comparable projects, national as well as international, have been taken into consideration while trying to determine the current level of PPPs in Slovenia. The working method has been oriented towards a review of sites developed by PPP and their urban dimensions. The research was focused on 11 Slovenian city-municipalities where semi structured interviews in the form of informal discussions were conducted. The paper presents a case study of shelter housing development as PPP in Ljubljana, Slovenia.

When exploring contemporary housing issues in relation to the issue of the aging population in Slovenia and also elsewhere some important questions have to be raised.

- How to grant the older generation decent living conditions?
- What is happening with the housing fund built during the 70's in which elderly live today?
- Do these buildings offer suitable conditions for the elderly?
- How does the housing policy reflect different aspects of public interest?
- How do public-private partnerships (PPPs) address the development of housing?
- Do sheltered apartments respond to the context of a socially useful "public interest" and allow and support a long-term investment for and to the society and its social environment?

2 Ageing in Slovenia

The aging of the population is a general process in Slovenia and elsewhere in developed word that cannot be avoided. When we are talking about life of elderly today, we are no longer talking about a few years after retirement, but about a life period of 20-30 years. The elderly maintain their independent life longer and longer, independent of care and help. Of course, their needs increase over the years, but in general the old population of today is "rejuvenated", people remain active and vital after retirement for a long time [5]. As a result, the old population is becoming more and more heterogeneous, and therefore the boundary between middle age and old age cannot be precisely defined. The paradox of today's society of the third life period is that older people have more personal freedom but fewer possibilities for achieving their goals [6].

The share of the population aged 65 increased from 10.6% to 16.5% in the period from 1989 to 2009. In the same period, the share of the population aged 0-14 decreased from 20.9% to 14.0% [7]. Such a trend is projected to continue; in 2029, at least 65 years old residents would be 24.8%, and in 2059 33.5% [8]. Only 12.8% are supposed to be 15 years old at that time. The population of Slovenia will not only be much lower but it will also be significantly older according to these assumptions. According to the forecasts, the number of those aged 65 or more will represent one third (33.4%) of the entire population by the end of 2060 [7]. Another matter are the living conditions of the elderly and their spatial exclusion. Studies show that in Slovenia around 20.8% of the elderly live in unfit conditions (with at least one issue – overcrowding, humidity, lack of space, no bathroom or toilet), while the spatial exclusion is even much higher. The problem of spatially excluded older people, that do not have basic services in their vicinity, is extremely problematic as they represent 49.2% in this demographic group [9].

Demographic studies predict large changes in the structure of the population in the next decades, due to the already mentioned negative demographic trends. As we are still not able to comprehend such a social phenomenon, we are still not able to cope with it neither in a cultural or political sense. As a result, if these unpredicted consequences are not entirely dealt with, the old age will make the ageing population progressively more excluded and isolated without guaranteed medical care.

3 Housing in Slovenia: social environment and financial mechanisms

Marta and France Ivanšek, pioneers in research of Slovenian housing, work had extreme importance and impact for Slovenian post-war architecture, urban design and planning, and society in general. Several key movements were promoted by their actions, which were advanced and unique for Slovenian space. They dealt with housing systematically and comprehensively. The major influence on their vision, attitude and reflections had a five-year work period spent in Sweden. The efforts of the Swedes for a Highly-Living Culture made a great impression on them. Economical attitude to space, retained design and social ideas typical of Scandinavia have become the main motive of their work. On the basis of these experiences and their own vision they established the Department for Housing Research at the Urban Planning Institute where they carried out the first residential research for a higher standard of living. Their work in this field is regarded as pioneering, since they were the first in Slovenia, that systematically monitored the state and development of housing construction, investigated the quality and utility value of apartments and living habits as well as the desires of people. On the basis of the results of this research, they established further orientations of their work, both planning and awareness-raising by educating users about their own living environment. Their work and efforts were based on three main approaches: research of urban and architectural arrangements of settlements (housing research), interior design, and education of people, that was intended to raise the awareness of people in the field of housing. The goal of their work was to improve the living environment of all people in everyday life, to create beautiful and comfortable homes for everyone and to raise the living culture of the Slovenian population. Special dedication was given to understand the one-family house, the free-standing house which gives the designer and the tenant relative maximum individual freedom. However, when such freedom leads to the fact that from someone's kitchen the only possible view is to the neighbour's bedroom, the view from the cooking corner to the neighbour's living room and with the terrace that does not allow any privacy, all the relative advantages of this individual freedom quickly disappear and such situation can quickly turn into an absolute failure [10].

The Ivanšek were convinced that a quality residential space is one of the basic needs of every human being, particularly the elderly. The transition to creating a better living environment for the elderly was therefore an understandable step in their careers. Initial research in this field produced the results at an early stage. They created special interior design elements and furniture (Retirement Centre Tabor, Ljubljana, 1965) which lead them to the design of designated housing for older people (Pensioners' House in Poljan, 1973-75 and Kolezija, 1979-82 in Ljubljana). Still today Marta and France Ivanšek's principles can be applied when solving contemporary housing problems to ensure the appropriate quality of living.

Slovenian planning legislation states that the main objective of spatial planning is to enable coherent spatial development by the consideration and coordination of different development needs and interests and assuring public benefits, especially in the areas of environmental protection, the conservation of nature and cultural heritage, etc. As it can be seen, from one of the basic principles of spatial planning, the identification and consideration of the public interest is one of the key components of spatial planning in general. One of the main principles in spatial planning is the dominance of the public interest, which requires state and municipal authorities to take into account both public and private interests, carefully consider them both in accordance with the objectives of spatial planning and ensure that private interests do not negatively affect the public one [11]. Gantar [12] points out that local or state institution do not have the legitimacy for interventions in space, which can by their nature restrict certain civil rights, if such an intervention is not justified by the public interest. More than that, this public interest must be legally codified through specific forms of legal acts (spatial plans of various forms) [13]. Alexander [14] lists three roles of public interest in connection with spatial planning:

- the legitimization of the state's functioning, which is related to the planning,
- sets a norm for planning practice, and
- serves as a criterion for the evaluation of planning and its products (plans, projects and policies).

Planners encounter conflicting interests of individual actors (users of space, local communities, investors, representatives of other professions, etc.). When this happen, the spatial planner should act as a neutral negotiator among the participant stakeholders [15]. Gantar [12], however, draws attention to the possibility of a discrepancy between the planner's professional conviction and the public interest. He highlights the existence of a dilemma: "Should the planner simply accept the definition of a public interest in relation to a certain intervention in a space as it is, or whether it should be actively involved in its design and therefore a certain degree of politicization of his operation" [12].

Between different public interests the quality of life of all generations is equally important, as it is the life span. Therefore, the aging population should be equally enabled to live the same full, active, healthy, safe and satisfied life as integral part of the society. In the foreground comes the phrase "active age", which emphasizes that this is not a passive period of life, but the period in when people should come to life with new forms of socially beneficial activity.

3.1 Overview of housing financing in 20th century in Slovenia

One of the essential factors for planned construction is also the economic scheme or the provision of financial resources, which is conditioned by the relevant legislation [16]. This area has changed together with the socio-political system in Slovenia. We can define eight major time-bound sets that have been conditioned by different types of financing from the beginning of the 20th century to this day:

1) [Before the Second World War] – Controlled development of housing can be found only in cities. In addition to privately built and owned houses, there was also a fund of profitable apartments, owned by large and small private investors, intended for renting. Municipalities and private legal entities owned a certain proportion of social non-profit apartments intended for socially endangered population. Additionally, a special type of rental

housing intended for employed and residential "colonies" existed too, owned by large companies, usually located in a vicinity of operating industrial facilities.

2) [1945-1956 Restoration years] – With large scale industrialisation programmes the influx of new residents into the urbanised centres became increasingly strong. The National Committees established offices that had oversight over the entire housing stock at the local level. During this period, housing was an essential instrument of social policy – minimum living conditions were guaranteed to every citizen. There were some changes introduced between 1947 and 1956. On one hand, the housing right and the resulting rights to the unhindered use of the dwelling and the prohibition of silence and displacement of the inhabitants were permanently entrenched; and on the other hand the financing of housing construction was settled through the state budget. Collective housing and the first neighbourhoods were built with budgetary funds. The socially-owned housing was mainly intended for political and professional staff. The social function of housing policy has declined; instead, housing policy has become part of human resources policy.

3) [1956-1965] – In 1956, municipal housing construction funds were formed from the compulsory contribution; from the income of employees. This created a relatively broad economic basis for more extensive construction. Individuals were able to get a loan from municipal funds for the construction of single-family houses. Newly established funds were the basis for the development of housing construction and related industry. The law on the nationalization of apartments for rent and construction land entered into force in 1958, bringing new developments in the field of housing policy: 1) The law enforced a class attitude towards renting buildings and building land; and 2) It was estimated that official underdeveloped construction operative bodies would not be able to cope with the existing funds intended for housing construction. The principle of land acquisition through expropriation was too slow and nationalization took place. Agricultural land was mainly nationalized in the vicinity of cities, which caused their rapid spread. During this period, the growth of individual buildings, mainly one family houses has caused the extensive urban sprawl.

4) [1965-1972] – An economic reform came into act in 1965. With the resolution on the further development of the housing economy system, the whole mechanism of the housing economy moved into the economic field. The society has supported the development of construction companies in housing and residential expansion. This way the housing policy was deprived of the social component. Deformations occurred in the housing construction model: 1) The price-structure was not public and the producers behaved in a monopoly way, consequently the prices rose; and 2) There were no norms on the size and structure of the dwellings (the price of m2 of housing was the same for one- or two-room dwelling). Since housing policy had no longer a social component, many of the working people could not solve housing basic situation. These reasons led to the construction of solidarity apartments, which were financed only by current buyers and construction companies, and not by the society as a whole.

5) [1972-1990] – Housing projects were financed by companies and others so called users of social assets. Funds were collected in several ways: 1) Income from solidarity (non-refundable) pooling – for solidarity housing construction; 2) Funds from net income for reciprocal pooling – for solving housing issues for company employees, for the construction and renovation of private homes, single-family homes, etc.; and 3) Funds from net income, disbursed after the final account – for solving the housing problems of employees, providing loans to employees, for earmarking in banks, etc. Until 1987, the volume of housing construction was relatively high, and in the early 1980s, more than 10,000 dwellings were built each year. In the second half of the 1980s, the share of housing construction in the social sector began to decline compared to the private sector. In 1989, real interest rates were

introduced. With the elimination of self-management socialism, the then existing housing system was abolished.

6) [1991-1999] – The Housing Act [17] brought two essential novelties in the housing field: 1) Social ownership in the residential sector was abolished. The main driving force was a process of privatization, which was based on a concept of purchase under the very advantageous conditions, which led to the divestment of social housing. 2) The Housing Fund of the Republic of Slovenia (1991) was established, which was supposed to finance the national housing program, encourage housing construction, renovation and maintenance of apartments and houses. The Housing Act is based on the principle that the concern for resolving the housing problem lies on every individual, with social corrections when the state provides for those who cannot solve it alone. Instead of the principle of "providing housing" that was used in the past, the term "facilitating the acquisition of housing" is introduced. In this way, the housing area was de-bureaucratized, and the administrative allocation of housing was replaced by the procedures of the organized housing market and by supporting private initiatives. The most important provider of housing loans with a favourable interest rate is the Housing Fund of the Republic of Slovenia. The main source of financing for the fund was a 20% share of the purchase price during the privatization of housing. Under the provisions of the Housing Act, housing funds may also be established by municipalities and non-profit housing organizations. However, funds are lacking in particular for lending to social and non-profit housing. The law categorized types of housing: owned, non-profit, social, office and profit housing.

7) [1999-the present time] – The national housing savings scheme, adopted by the government in 1999, increased the volume of favourable housing loans, as it encourages long-term savings. The national savings scheme is implemented through the Housing Fund of the Republic of Slovenia and the banks that are selected by the public tender. The attractiveness of such savings is provided in the form of attribution of a certain amount of the premium after the expiry of each saving year. At the same time, savers are offered comfortable housing loans. Long-term savings in the country are expected to provide a sufficient amount of funds in the future for favourable long-term housing loans. In 1998, the Housing Fund issued the second, in 2001 the thirds, in 2002 the fourth bonds issue for legal entities, intended for recapitalization of the Fund for the implementation of its legally defined activities. The construction and renovation should provide at least 10,000 apartments annually in Slovenia. According to the National Housing Program, about 2,500 non-profit rental apartments, 2,000 social rental apartments, 5,000 own homes and 500 profitable apartments.

8) [present time] – The period of Public-private partnerships (PPP), which play an important role as an instrument of planning and implementation of public infrastructural projects, including urban projects that affect the dynamics of urban development. Even if in Slovenia PPP sector is still partly underdeveloped (the Public-Private Partnership Act [18] was for example adopted only in 2007), the PPP principle shows high potential for achieving spatial planning and urban development goals. When used correctly the PPP tool could enable the successful realization of different urban projects, which would be otherwise often hindered or even not realized due to the financial obstacles.

4 Public-private partnership (PPP)

Well defined PPPs are often mentioned in the context of privatization, but it is necessary to distinguish between the two concepts. Unlike privatization, PPPs operate on the basis of a joint decision-making process where public and private partner's interests have been interpreted and defined in a contract or some other form of agreement [19]. In the Green Paper [20], the European Commission defines partnerships as: "cooperation between public authority and the business world, which is intended to provide financing, construction, reconstruction, management, maintenance of infrastructure or the provision of services".

PPPs, on the principle of »people in the first place«, ensure that people are key stakeholders. They focus on improving the quality of life in communities, especially in poor environments, and create local sustainable jobs that address food and health issues, strive for gender equality, regulate access to drinking water, energy, transport and education for all, encourage social cohesion and fair regulation, and try to eliminate all kinds of racial, ethnic, religious and cultural discrimination. Public-private partnerships are being promoted and expanded, as they give people access to better services at more favourable prices.

The good governance of the public sector is the public interest. The public sector thus works in the public interest, and the limits of orientation towards the user are determined by the public interest, which means that the public interest also directs the amount and method of public activity in a particular field. The public concern is more complex, driven by legislation, regulation and authorities, political opinion, democratic decision making, minimizing the risk and maximizing the social value [21]. PPPs are generally developed for public interest missions: roads, railways, ports, airports, water and wastewater, waste, energy, health, security and prisons, to give some examples. The government, the ultimate guardian for the provision of these services, entrusts the private sector with the responsibility of designing, financing, building and operating the infrastructure and/or service [22].

4.1 PPP tool in Slovenian spatial development

The PPP tool has been introduced into Slovenian spatial development practice rather recently. Most often different improvement of the quality of the living environment are achieved by the implementation of PPP, which also contributes to a more controlled creation of urban space and increases the flexibility of the traditionally rather ossified planning process in Slovenia. The lack of regional administrative level and due to a division of Slovenia to 212 small municipalities at a local level, Slovenia has a specific pattern when implementing PPP practice. In general, there are few PPP projects that exceed the threshold value to apply for some EU founds available for example in the new EU Investment Plan.

The Slovenian legislation defines the PPP as »... the ratio of private investment to public projects and / or public co-financing of private projects in the public interest and cooperation between public and private partners in connection with the construction, maintenance and management of public infrastructure or other projects of public interest ...« [18]. Such a public-private partnership necessarily includes at least the provision of services representing a public service or service otherwise carried out in the public interest, the construction of facilities and facilities necessary for the provision of such services (public infrastructure) or, as a general rule, both the construction of public infrastructure and the provision of public services. The essential element of public-private partnership is the public interest, and this element should not be neglected by the State and the local community, but they must undoubtedly address it in this relationship.

4.2 The role of the "public interest" in PPP in Slovenia

The role of "public interest" is crucial in the implementation of PPP. In the draft of the new Act on Investment Incentives [23], the public interest is defined as follows: "The public interest is the general interest of the organized wider community or the welfare of the general public, according to its value criteria of its benefit, from the increase in capital, the capacity to increase production capabilities and technological progress towards the direct and indirect creation of new jobs, with the transfer of knowledge and technologies, and cooperation with educational institutions that are above the interest or the benefit of the individual. The implementation of a strategic investment is what is considered to be in the public benefit, which, due to spatial, technological and logistical needs, is suitable for insertion in space and fosters economic development."

The National Housing Program [24] envisages the next strategy in relation to the public interest, PPP and building development. The program specifies housing cooperatives as an additional source of funding for housing construction: "An additional source of funding for housing construction will also be housing cooperatives, which are a form of housing self-help for citizens in partnership with the public-sector" [24]. Housing cooperatives have a form of housing self-help for citizens in partnership with the public sector. As these will be public works, housing cooperatives will receive various benefits from the public sector, in particular favourable loans or subsidies for interest and guarantees for loans and land at favourable prices, for sale or long-term lease. The value of the dwelling will be determined by the actual costs of the construction of the dwelling or housing. To ensure and enforce that the system will work in favour of the tenants the investor will have to provide low building expenses. This will have to be achieved without compromising on the safety and other building standards, but rather be achieved by building on cheaper land and with thoughtful, modest building details and finalization. The investor will find cheaper land in cooperation with the municipality by public-private partnership.

5 Housing, sustainable urban strategies and the use of PPPs in the city of Ljubljana

Ljubljana offers a vast distinction of different types of housing which were recognized as one of the cultural values of the city (Fig. 1).

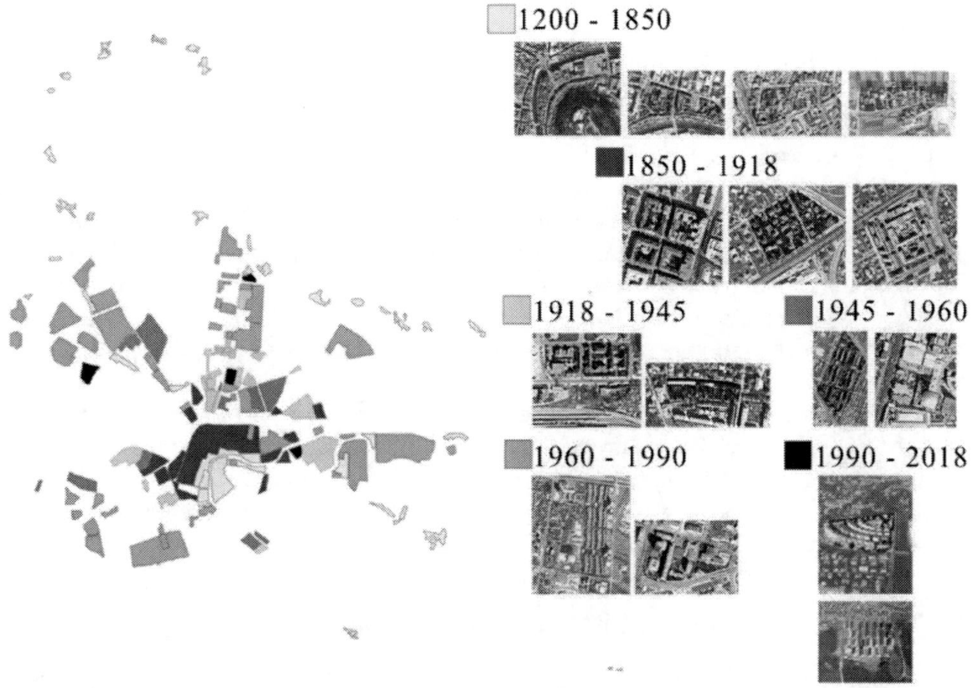

Fig. 1. Characteristic areas in Ljubljana [25].

The general contents of the recently adopted Sustainable Urban Strategies (SUS) for the city municipality of Ljubljana (MOL) that entitles different goals and directions for future urban development of the city are:

- Sustainable (co-natural) development addressing more sustainable form and functioning of cities and their influential areas.
- Economic development with the goal to ensure good conditions for preservation and creation of new jobs and opening of new business opportunities.
- Rise of living conditions by providing a high quality living environment and ensuring high environmental standards, quality green and open public spaces, quality renovation of residential and other parts of the city, preservation of cultural heritage, etc.
- Development of urban character connected to the development of (new) urban functions, social infrastructure, public spaces of urban character, revitalization of degraded areas and cultural heritage with city functions, development of cultural activities, revitalization of neighbourhoods in areas of good urban life standard etc.
- New forms of management such us new forms of decision-making and city management.

The directions mentioned in the SUS MOL [26] are important also from the perspective of the implementation of PPP projects. Additionally, SUS defines the mayor suitable locations for urban project placement. Identified degraded areas are some of these locations. The current success of PPP projects in Ljubljana can be also attributed to the comprehensive survey and study of degraded areas available and suitable for different urban projects [27-29]. To reuse such degraded areas is also one of the strategic and at the same time main objectives of managing the territory of Ljubljana according to the strategic spatial development of Ljubljana [30] and the implementation of Municipal Spatial Plan of MOL [31]. As Burgar [29] points out, the deteriorated areas of a city are valuable urban development areas, where the City of Ljubljana should challenge itself and enforce public interest in order to reinvent beyond the speculative logic of the market. Two examples of renovation of brownfields, which were carried out as PPP projects are the construction of a multi-purpose football stadium, a sports hall and an accompanying building with the service infrastructure in the Stožice district, and the Center for Contemporary Arts "Rog".

The overview of newsletters published in the Official Gazette of the Republic of Slovenia [32] reveals a multitude of acts on the topic of the implementation of PPP projects adopted in MOL: Ilirija - implementation of the construction of the Ilirija Swimming and Gymnastics Center; Introduction of a wireless network in the area of the City of Ljubljana; Underground car-park under Plečnik's market and the extension of the "Mahrova" house; renovation of Vodnik's square and the underground car-park under the "Krekov" square with its renovation; Sports Park "Črnuče"; Revitalization of the sports center "Savsko naselje"; National Gymnastics Center; Sheltered apartments "Šiška"; Old Airport Ljubljana; Football Hall in "Štepanjsko" district; energy renovation of municipality owned buildings, etc.

6 Sheltered housing projects in the City of Ljubljana, as a model of PPP

The projects in the field of housing and housing construction is less present. The exception are the interventions in the urban space in the form of sheltered apartments carried out with the private partner Mijaks in Rožna dolina, Koseze, Murgle and Dravlje-Šiška districts. Local authorities were able, with the collaboration of private partner, to ensure more than 400 sheltered apartments in the last years.

6.1 Sheltered apartments

The sheltered apartments are intended for older people who are still able to live independently and are not interested in being accommodated in nursing homes. The service is carried out in the functionally connected environment with the dwellings that are adapted to the needs of elderly. They differ from the usual dwellings in that they are adapted for the ageing people: the buildings must stand in a well-regulated and peaceful environment, the transport infrastructure must be properly regulated and accessible, access to facilities must be good, the interior spaces of the building must be designed for the motorically less capable people in mind, medical and social care must be provided 24 hours a day (technical aids, home care, home help, organized nutrition, personal hygiene and care, protection and organization of personal contacts, transport, accommodation of emergency call devices, security of housing). An important attribute of sheltered apartments is to maintain all the advantages of conventional living standard, especially by assuring autonomy and privacy, while at the same time provide residents with all institutional care services.

Residents of sheltered dwellings generally have the option of choosing individual services or service packages depending on the amount of assistance that an individual needs, according to their state of health. Despite the fact that the basic concept of sheltered housing has already been exceeded in more developed countries, as is this next step in the development of more modern forms of living for the elderly, sheltered housing in Slovenia still does not have the effects desired. Because of the high proportion of private owned dwellings by the elderly and the lack of understanding what sheltered apartments really are and what they offer, elderly in Slovenia are still quite reserved when deciding about leaving their privately owned homes and move to such "institution". So far, there are relatively few sheltered apartments buildings in the country, but the construction trend is increasing, due to the aging of population and related necessity of care services. It is expected that this concept as a form of living for the elderly will largely replace the traditional homes for the elderly that offer(ed) only institutionalized stays.

6.2 Case study of sheltered housing in Šiška Dravlje, Ljubljana

The presented case study of the urban project carried out by means of a PPP, consists of the establishment and construction of two buildings intended to provide sheltered housing for elderly. It includes all the services needed in relation to the management and maintenance of the sheltered apartments throughout the whole duration of the partnership. The private partner had two years to prepare the project documentation, obtain a building permit, build the building and transfer the dwellings agreed upon to the public partner. This is an example of one of the projects carried out by the Mijaks company in partnership with the public partner Public Housing Fund of the city municipality of Ljubljana [33]. The project is called "Sheltered housing Mijaks - Ljubljana Šiška Dravlje. 54 residential units were built in 2016 with a total gross area of 4,477.76 m2. The success of the project is also that all the apartments were already sold in the same year as they were built. Beside the construction of sheltered apartments, the contract also predicted a provision of 46 outdoor parking spaces and the design of the surrounding area, as well as the provision of maintenance and management for the entire project. The adopted PPP model was DBFTO (design-build-finance-transfer-operate) for the part transferred to the public partner while it was a DBFO (design-build-finance-operate) for the part kept in private ownership. The criteria for selecting a private partner were focused on: finding the largest total floor space of sheltered housing offered to the public partner, finding the highest share of rented apartments for the public partner within the private part of the project, the offered market rent amount of sheltered apartments in the private part of the project and the selling price of the supplied apartments in the private part of the project.

6.3 The method of activating publicly owned land with private capital

For the purpose of the implementation of the project, the public partner transferred the building right to the private partner, for the construction period of the project. The scope and duration of the given building right has been agreed upon in the process of selecting a private partner. The owner of the land was the Municipality of Ljubljana. The Municipality invested in the project an area of 4,045 m2 intended exclusively for the construction of sheltered apartments and additional financial founds for the acquisition of the building right and other municipal contributions. The total stake of the public partner amounts to € 1,153,900 (excl. VAT). For the future needs of the residents from the new sheltered apartments, the public partner also enabled the joint use of the nearby park, that is part of the already existing nursing home Šiška. Within the project, the public partner had to establish mechanisms, to ensure effective public interest protection through the whole period of the public-private partnership.

6.4 Gains in city urban development

The private investor found a "niche market" and successfully established cooperation with the public sector in order to improve the conditions and quality of life for the older generation in Ljubljana. In the case of sheltered housing projects, the main gain for the city municipality was to obtain a working concept of sheltered apartments, adapted for people in their third life period, providing them with a new, higher level of living quality. This reduces the pressure on nursing homes for the elderly and ensures a wider range of activities and living possibilities for the aging population. Sheltered housing is one of the important aspects of improving the quality of life which can be achieved through the combination of public sector and private funds and inputs. The construction of sheltered apartments in Ljubljana, implemented through the PPP mechanism, is a good example of a dynamic, controlled development of the city.

However, in the process that lead to the presented project there were also some mistakes made. Only by addressing these mistakes the new generation of such projects can improve. For example, the process of preparation and implementation of the project did not take into account some of the main principles of modern and sustainable spatial development such as: public participation, integration with local residents, verification of the project within the context of the existing urban plan. The definition of different urban parameters, that were supposed to regulate the project, was not done according to the existing urban design of that area, but was determined only based on the criteria and general provisions that apply according to the planned land use in the regulatory plans which are too general. One of such general provisions is also the guideline from the Sustainable urban strategy that envisage the concentration of urbanisation. This is often carried out by constructing on quality vacant plots within the urban structure, resulting in the increased build-up factors, higher population density, increased traffic which results in rising of air pollution and noise pollution. If the norms and regulations are not strictly defined, there is a high possibility that such project might bring disapproval and dissatisfaction among the public, together with some other negative effects on urban space. With the combination of over rationalization in the public sector, with the goal to achieve the planned results as soon as possible and the private sector, which is mainly driven by the possible profit, the care for the quality of life of all citizens it easily neglected. It can often happen that such combinations might affect the successful realization of the PPP project. Even more, the realization of such urban project might consequently effect or even deny the basic role and mandate of PPP that should seek for higher "public good".

7 Conclusions

The awareness of the importance of the existence of social housing is relatively high in the sphere of spatial development in Slovenia, where these issues are mainly addressed within the urban environments. This includes granting housing to young families, homes for the elderly, student dormitories, etc. Nevertheless, there are not many social housing projects carried out due to the lack of the legislative and financial incentives.

These types of projects may be better implemented through mutual rewards offered in the PPP processes, although they are often lacking verification from the point of view of spatial planning. These projects mostly strive for the maximization of the gross built surfaces to find positive results for both public and private partner to justify the project, but by doing so they ignore contemporary principles of urban planning and urban design, including sustainability and resilience principles. Usually synonyms of these contemporary principles, e.g. participatory urban planning and the verification of expert opinions to enforce public interest, are left out of the negotiation and planning processes. This comes partly due to the fact that the urban planners themselves lack the expertise in the field of PPP. In order to make a better use of PPP in providing more and better social housing in Slovenia some additional endeavours must be made in terms of educating public administration and decision makers on how to manage the PPP processes. On the other hand, the awareness raising is needed at the side of private investors too so that they would understand the need for a well thought and designed urban spaces which will provide a high quality urban environments and will thus enhance their funds initially invested in a long term too.

Acknowledgments

The research presented is part of the ongoing project called Analysis of options for implementing of urban projects using public-private partnership conducted by three partner institutions: two form University of Ljubljana, Faculty of Law and Faculty of Architecture and the Urban Planning Institute of the Republic of Slovenia. The project is financed by Slovenian Research Agency and Ministry of Spatial Planning of Slovenia.

References

1. S. Kosanović, A. Fikfak, Energy Building, **115**, 69-77 (2016)
2. A. Fikfak, *Naselbinska kultura slovenskega podeželja - Goriška brda* (University of Ljubljana, Faculty of architecture, Ljubljana, 2008)
3. *Zakon o urbanističnem planiranju (ZUP)*. Official Gazette of the RS, no. 16/67 (1967)
4. *Zakon o regionalnem prostorskem planiranju (ZRPP)*. Official Gazette of the RS, no. 16/67 (1967)
5. R. Sendi, B. Černič Mali, A. Jakoš, M. Filipovič Hrast, *Stanovanjske potrebe upokojencev in drugih starejših oseb* (Urban planning Institute of the RS, Ljubljana, 2003)
6. A. Krajnc, AS, **3/99**, 5-18 (1999)
7. *Eurostatove projekcije prebivalstva za Slovenijo, 2008-2060, EUROPOP2008, konvergenčni scenarij* (SORS, Ljubljana, 2008)
8. N. Vertot, *Starejše prebivalstvo v Sloveniji* (SORS, Ljubljana, 2010)
9. V. Hlebec, M. Kavčič, G. Ogulin Počrvina, *Staranje, izziv za izobraževanje in medgeneracijsko sodelovanje* (Zveza ljudskih univerz Slovenije, Ljubljana, 2013)
10. F. Ivanšek, *Enodružinska hiša: od prosto stoječe hiše k nizki zgoščeni zazidavi* (Ambijent, Ljubljana, 1988)

11. *Zakon o prostorskem načrtovanju (ZPNačrt)*, Official Gazette of the RS, no. 33/07 (2007)
12. P. Gantar, Geodetski vestnik, **51 (2)**, 246-254 (2007)
13. T. Križman, *Vloga civilne iniciative v slovenskem prostorskem načrtovanju, diplomska naloga* (University of Ljubljana, Biotechnical Faculty, Ljubljana, 2012)
14. E. R. Alexander, Planning Theory, **1 (3)**, 226-249 (2002)
15. P. Bradwell, I. Johar, C. Maguire, P. Miner, *Future planners, propositions for the next age of planning* (Demos publications, London, 2007)
16. V. Pust, AB, **68/69**, 23-24 (1984)
17. *Stanovanjski zakon (SZ)*, Official Gazette of the RS, no. 18/91 (1991)
18. *Zakon o javno-zasebnem partnerstvu (ZJZP)*, Official Gazette of the RS, no. 127/06 (2006)
19. L. B. Sagalyn, Journal of the American Planning Association, **73 (1)**, 7-22 (2007)
20. *Green paper on public-private partnerships and Community law on public contracts and concessions* (Commission of the European Communities, Brussels, 2008)
21. B. D. Jones, *Reconceiving decision-making in democratic politics: attention, choice, and public policy* (The University of Chicago Press, London, 1994)
22. C. O. Cruz, R. C. Marquez, *Infrastructure Public-Private Partnerships. Decision, Management and Development* (Springer-Verlag Berlin Heidelberg, 2013)
23. *Osnutek zakona o spodbujanju investicij (ZSI)* (Ministry of Economic Development and Technology of the RS, Ljubljana, 2017)
24. *Nacionalni stanovanjski program 2013-2022 (NSP)* (Ministry of the Environment and Spatial Planning of the RS, Ljubljana, 2013)
25. M. Šašek-Divjak, K. Dimitrovska Andrews, et al., *Občinski prostorski načrt MOL: strateški del. Mapa 1, Odlok o Strateškem načrtu Mestne občine Ljubljana* (Urban planning Institute of the RS, Ljubljana, 2009)
26. *Trajnostna urbana strategija Mestne občine Ljubljana 2014-2020 (SUS MOL)* (Municipality of Ljubljana, Ljubljana, 2015)
27. J. Koželj, L. Ažman Momirski, T. Maligoj, B. Omerzu, D. Flere, *Degradirana urbana območja* (Ministry of Environment and Spatial Planning of the RS, Ljubljana, 1998)
28. J. Koželj, P. Filipič, P. Hočevar, K. Strle, D. Kušar, *Merila in kriteriji za določitev degradiranih urbanih območij (DUO)* (University of Ljubljana, Ljubljana, 2016)
29. B. Burgar, *Vključevanje mesta v prenovo degradiranih območij in ponovno uporabo obstoječih objektov* (Municipality of Ljubljana, Ljubljana, 2003)
30. M. Šašek Divjak, Urbani izziv, **19 (1)**, 11-24 (2008)
31. *Municipal Spatial Plan of the City Municipality of Ljubljana*, Official Gazette of the RS, no. 87/2010 (2010)
32. *Official Slovenian Newsletter*, https://www.uradni-list.si/glasilo-uradni-list-rs/vsebina/ (accessed on 17 March 2017)
33. *Project presentation. Sheltered housing Mijaks Šiška Dravlje* (Mijaks, Ljubljana, 2017) http://mijaks.si/slovensko/oskrbovana-stanovanja-mijaks/dravlje-siska/ (accessed on 16 March 2017)

Model and Algorithm for Substantiating Solutions for Organization of High-Rise Construction Project

Vladimir Anisimov[1,*], *Evgeniy* Anisimov[2] and *Anatoliy* Chernysh[1]

[1]Peter the Great St. Petersburg polytechnic University, Institute of Industrial Economics, Management and Trade, 195251 St. Petersburg, Russia
[2]Peoples' friendship University of Russia, 117198, Moscow, Russia

Abstract. In the paper the models and the algorithm for the optimal plan formation for the organization of the material and logistical processes of the high-rise construction project and their financial support are developed. The model is based on the representation of the optimization procedure in the form of a non-linear problem of discrete programming, which consists in minimizing the execution time of a set of interrelated works by a limited number of partially interchangeable performers while limiting the total cost of performing the work. The proposed model and algorithm are the basis for creating specific organization management methodologies for the high-rise construction project.

1 Introduction

Quality management projects for the construction, modernization, technical re-equipment, reconstruction or major repair of buildings and structures for various purposes, as well as their complexes is inextricably linked with the adoption of sound management decisions on the organization of material technological and logistics processes of the construction project and their financial support. Complexity of projects and significant costs, caused by possible miscalculations in the organization of these processes, lead to the fact that the relevant decisions should be based not only on experience and intuition, but also on objective scientific justifications. The tool for such justifications can be the corresponding mathematical models and algorithms [1-4].

A wide class of problems of justifying decisions on the organization of material technological and logistical processes of a construction project and their financial support can be formalized in the form of a model and algorithm for minimizing the execution time of a complex of interrelated works, with restrictions on the number of performers, their interchangeability and the cost of attracting certain work [5]. The development of such an algorithm is the purpose of this article.

* Corresponding author: an-33@yandex.ru

2 Model Description

When building a model, the composition and interrelation of the works reflecting the corresponding material technological and logistical processes of the construction project should be displayed in the form of a network [6-8].

$$G = \{(i, j)\}, \quad i, j = 0,1,\dots,m, \quad i < j, \tag{1}$$

Authors should use the forms shown in Table 3 in the final reference list.
where i, j are network node numbers;
$(m+1)$ are total number of nodes.

Each task in the network (1) is associated with an arc (i, j), connecting the i and j nodes. The node $i = 0$ corresponds to the event of the beginning of the execution of the tasks package represented by the network (1). Nodes $i = 1,2,\dots,m$ the nodes correspond to the events consisting in the completion of all the tasks described by the arcs included in each of them. The total number of tasks (arcs) is N.

The sequence of task obeys the logistic rule: the task corresponding to the arc emerging from any node can not be started until all tasks corresponding to the arcs in this node is completed.

Each task (i, j) is characterized by the necessary quantity $n(i,j)$ of handlers and duration $\tau(i, j)$.

A lot of handlers involved in the implementation of the set of tasks will be designated.

$$R = \{1, 2,\dots,k,\dots,K\} \tag{2}$$

(k – conditional numerical order (identifier) of handler, K – number of handlers).
The interchangeability of handlers in a formalized form is represented by a matrix

$$\Delta = \left\|\delta_k(i, j)\right\|, k = 1,2,\dots,K, \quad (i, j) \in G \tag{3}$$

Where

$$\delta_k(i,j) = \begin{cases} 1, if\ the\ performer\ can\ be\ involved \\ \quad to\ the\ carrying\ out\ (i,j)\ task, \\ 0\ otherwise. \end{cases} \tag{4}$$

The cost attracts handlers to the task execution as a component of the vector

$$\mathrm{C} = \|\mathrm{c}_k(i,j)\|,\ k = 1,2,\dots,\ K,\ (i,j) \in G, \tag{5}$$

where $\mathrm{c}_k(i, j)$ is a cost per time unit of engaging k handler to execution of (i, j) task.

The logistic plan for the implementation of the set of tasks is determined by the set of

$$Y = \{x_Y(i, j), r_Y(i, j) | (i, j) \in G, r_Y(i, j) \subset R\}, \tag{6}$$

where $x_Y(i, j)$ determined by the plan Y timestamp corresponding to the beginning of (i, j) task.

$r_Y(i, j)$ is variety of handlers, involved to (i, j) task execution in accordance with plan Y.

We will assume that the interruption of each started $(i, j) \in G$ task is not allowed and the composition of the marked handlers is not changed.

Taking into account the accepted designations, the cost $\Omega_Y(i, j)$ of the task execution during the Y plan realization is determined by the relation

$$\Omega_Y(i,j) = \sum_{k \in r_Y(i,j)} c_k(i,j)\tau(i,j) \tag{7}$$

Let us denote by the G_L the set of all paths of the network (1) connecting its initial and final vertices.

The time for executing the entire set of tasks for the implementation of the plan (6) will be equal to the maximum $L \in G_L$ path T_L length from the initial vertex $i = 0$ of the network (1) to the final one $j = m$.

Taking into account the accepted designations, the task of forming a logistic plan (6) that minimizes the time of execution of a set of tasks (1) with restrictions on the number, interchangeability and cost of handlers in a formalized form is reduced to the following problem of mathematical programming:

define a logistics plan

$$Y^* = \{x^*_Y(i, j), r^*_Y(i, j) \mid (i, j) \in G, r_Y(i, j) \subseteq R\} \tag{8}$$

of the executing the set of tasks, ensuring compliance with the condition

$$T^* = T(Y^*) = \min_Y \max_{L \in G_L} T_L(Y) \tag{9}$$

under constraints

$$X_Y(i,j) \geq \max_{l,i}\{x_Y(l,i) + \tau(l,i)\}, \quad (i,j) \in G; \tag{10}$$

$$\sum_{(i,j) \in r_Y(i,j)} \delta_k(i,j) = n(i,j); \tag{11}$$

$$\sum_{(i,j) \in F_Y(t)} n(i,j) \leq K; \tag{12}$$

$$\sum_{k=1}^{K} \delta_k(i,j) \geq n(i,j); \tag{13}$$

$$\sum_{(i,j) \in G} \Omega_Y(i,j) \leq \Omega^*, \tag{14}$$

where $F_Y(t)$ is variety of the complex`s tasks, performed at each timestamp t when implementing the logistics plan Y;

Ω* is the maximum permissible cost of the completing set of (1) tasks.

In the task (8) - (14) condition (9) formalizes the desire to minimize the time of the set of tasks execution.

Restriction (10) reflects the logistic rule that task originating from any network node (1) can begin only after the completion of all task included in this node.

Restriction (11) formalizes the requirement to allocate for each task a specified number of handlers.

Restriction (12) reflects the natural condition that the number of handlers simultaneously attracted to the task execution can not exceed their total number.

Restriction (13) formalizes the requirement that the number and interchangeability of handlers must ensure the execution of each complex`s task (1).

Restriction (14) means that the total cost of performing the set of tasks (1) can not exceed the established permissible level.

2 Algorithm Description

Tasks (8) – (14) refers to nonlinear problems of discrete inhomogeneous resources distribution of an arbitrary network. It is an NP- complex object of discrete programming [9]. Precise methods for solving problems of this class were first proposed in [2]. However, in the models considered in these studies only the handlers and their productivity were taken into account. At the same time, when creating plans for realizing real material technological and logistical processes of construction production, along with renewable resources (executors), it is necessary to take into account also available non-renewable (for example, financial) resources. In tasks (8) - (14), such resources are the cost Ω_Y (i, j) of performing each task when the plan is implemented. The limitations of these resources formally represents the ratio (14). The model (8) - (14) obtained as a result of introducing this relation is a further generalization of the model considered in [2] model. Exact algorithms for solving the problem (8) - (14) are currently absent. At the same time, the need for management practice for the implementation of complex projects requires their development[10]. One of the purposes of this scientific article is the construction of such an algorithm. For the existence of solution of tasks (8) - (14) it is necessary and sufficient that:

the composition and interchangeability of the handlers ensured the possibility of executing the entire set of tasks (1);

the established level Ω * of the allowable cost made it possible to carry out a set of tasks (1).

In the formalized form, the fulfillment of the first of these requirements consists in the fulfillment of the constraint (12). It means that from the staffing you can select specialists capable of performing any complex`s task (1).

In order to verify the second requirement is satisfied, can be used the set

$$R^* = \{r^q(i,j) \mid (i,j) \in G, q = 1,2,\ldots\} \quad (15)$$

of all possible options for assigning resources to the relevant tasks. There are many associated with this set

$$\Omega = \{\Omega^q(i,j), q = 1,2,\ldots, (i,j) \in G\} \quad (16)$$

of values Ωq (i, j) of the tasks execution corresponding tasks is associated with the corresponding options of resource assignment. Elements of the set (15) are defined by

$$\Omega^q(i,j) = \tau(i,j) \sum_{k \in r^q(i,j)} c_k(i,j). \quad (17)$$

In view of (16), the second of the requirements ensuring the existence of a solution of the task under consideration (7) - (13), in a formalized form, is represented by the relation

$$\sum_{(i,j) \in G} \min_q \Omega^q(i,j) \leq \Omega^*, \ q = 1,2,\ldots \ . \quad (18)$$

It is expedient to carry out a verification of the feasibility of conditions (13) and (18) before the procedure for finding the solution to problem (8) - (14) is initiated. If they are not satisfied, then the solution of problem (8) - (14) does not exist. If they are carried out, the procedure for optimizing the logistic plan for performing the set of tasks (1) can be implemented [11].

The basis of this procedure (the algorithm for the formation of an optimal plan) for searching can be based on the approach first proposed in [2].

It is based on the following constructions [2]:

representation of a set of $V = \{S\}$ admissible by restrictions of fragments S of the logistics plan Y in the form of a tree of subsets (threadinging);

calculation of the lower boundary of the objective function (8) for the threads of the tree (selected subsets);

searching for feasible options for the logistics plan;

checking the established permissible options for optimality.

The procedure proposed in [2] allows to define the logistics plan (8) for the implementation of a set of tasks (1) that satisfies conditions (10) - (13).

Distinctive feature of this algorithm considered in this paper is the necessity to account, on every stage, for branching and the condition (14). If condition is broken, the execution of current thread is no longer possible and we proceed to the next thread [12].

As in the algorithm [2], branching in the proposed algorithm is carried out on the basis of the dichotomous scheme. When implemented, each vertex vs of the S-th thread of the variational tree is an element of the logistics plan. Moreover, if the task (i, j) corresponding to this element starts at the moment of time $x_S(i,j)$ for $r_S(i,j)$-th variant of handler assignment, then let us set

$$v_S = \{x_S(i,j), r_S(i,j)\}. \tag{19}$$

If task (i, j) does not begin at the time $x_S(i,j)$ for $r_S(i,j)$-th variant of handler assignment, then let us set

$$v_S = 0 \tag{20}$$

For each thread $S \in V$ the values $x_S(i,j), \quad (i,j) \in G$, (initialization timestaps) must be selected from ascending sequence that corresponds to such thread $t_S = \{t_S^n\}, \quad n = 1,2,...$.

Herewith $t_S^1 = 0$, and the subsequent moments $t_S^n, \quad n = 2,3,...$ are determined on the basis of relation

$$t_S^n = \min_{(i,j)\in F(t_S^{n-1})} \{x_S(i,j) + \tau(i,j)\}, \tag{21}$$

where $F_S(t_S^{n-1})$ is variety of tasks (i,j), previously included in the S-th thread and unfinished by the time point t_S^{n-1}, that is

$$F_S(t_S^{n-1}) = \{(i,j) \mid (i,j) \in G, x_S(i,j) \le t_S^{n-1} \le x_S(i,j) + \tau(i,j)\} \tag{22}$$

Thus, it is a sequence of moments of time, in which task is completed, included in the next thread of the tree and releasing the respective handlers.

The condition $t_S^1 = 0$ reflects the fact that all variants of the logistics plan begin at the time $t = 0$.

Let us prove that the assignment of the dates for the commencement of task, not in accordance with the sequences t_S, does not allow to shorten the total execution time of the set of tasks under consideration (1).

Indeed, for any plan Y containing a fragment S, the early term of the beginning of any task $(i,j) \in S$ by definition belongs to the sequence t_S. Consequently, the late dates of the beginning of the task lying on the path critical for plan Y also belong to this sequence. For tasks that do not belong to critical paths, it is possible to vary the timing of the start within the appropriate time reserves. At the same time, the boundaries of these reserves also belong to this sequence, and the variation within the boundaries does not change the time of execution of the set of tasks as a whole [13]. Therefore, the task starting and completion dates for the optimal logistics plan (9) according to criterion (9) must belong to the sequence corresponding to this plan t_S.

In the interest of realizing the adopted dichotomic threading scheme, we introduce the set associated with (15)

$$D = \{d_q \mid q = 1,2,\ldots\} \tag{23}$$

in which $d_q = 1$ if the resource assignment option $r^q(i,j)$ is used to execute the (i,j) task, or $d_q = 0$ otherwise.

In this case, the serial number q of the element $d_q = 1$ of the set D characterizes both the task performed, the version of the resource assignment, and the cost of the resources involved in its implementation.

Taking into account (22), the threading process in the interests of compiling the optimal logistic plan (7) consists in choosing for each next time point t_S^n the admissible variables $d_q \in D$ and establishing their values, that is, the relation (18) ($v_S = \{t_S^n, d_q = 1\}$) exists if the corresponding d_q task $(i,j) \in G$ starts at the timestamp $x_S(i,j) = t_S^n$ with $r^q(i,j)$ the resource allocation option or (19) ($v_S = \{t_S^n, d_q = 0\}$), if the specified work does not begin with the considered resource assignment option at the moment t_S^n.

The set P_S^n of variables $d_q \in D$ that can be included in the S-th fragment of the logistics plan (7) at the timestamp t_S^n, contains quantities $d_q \in D$, corresponding to the tasks $(i,j) \in G$ not previously included in the thread S under consideration, satisfying the conditions:

$$x(l,i) + \tau(l,i) \le t_S^n, \quad (l,i) \in G \tag{24}$$

$$r^q(i,j) \subseteq R_S^n, \tag{25}$$

$$\Omega^* - \sum_{m \in D_S(t_S^{n-1})} d_m \Omega_m + \Omega_q \ge 0, \; q = 1,2,\ldots, \tag{26}$$

where R_S^n is variety of uncommitted resources for S-th fragment of the plan at the timestamp t_S^n;

$D_S(t_S^{n-1})$ is a set of variables d_q included in the thread S of the variants tree to the moment of time t_S^{n-1}.

Condition (24) identifies tasks for which all the preceding ones have been executed. Condition (25) identifies tasks for which there are permissible orders of free handlers, and (26) are the cost-admissible orders of free handlers.

As an estimate of the lower bound of the objective function (9), for each fragment of the calendar plan, the maximum path length from the initial vertex of the graph G to the final one, determined without taking into account the resource constraints (12), (14) for tasks which are not included in S., can be adopted. In this case, if at the next thread step corresponding to the moment, it is established, then the following is supposed to be defined similarly to [2]:

a) task $(l,i) \in G$ that was previously included in the S-th fragment of the plan (the task for which $x_S(l,i) < t_S n$, begins at the appropriate moments $x_S(l,i)$ and ends in the moments $x_S(l,i) + \tau(l,i)$);

b) for the task $(i,j) \in G$ corresponding to the variable $d_q = P_S^n$ and, therefore, included in the considered step in the S-th thread of the tree version - $x_S(i,j) = t_S^n$;

в) for the task $(e,h) \in G$, corresponding to the variables $d_u \in P_S^n,\quad u \neq q$, which according to the resource constraint (2.9) at the moment t_S^n cannot be included in the plan at the same time as the (i,j), start time is t_S^{n+1}, and the duration is determined by the relation $t_S^{n+1} + \tau(e,h)$.

If the thread is set at the considered step $d_q = 0$, then to determine $W_S(d_q = 0)$ in addition it's important following:

a) tasks $(l,i) \in G$, previously included in the S-th fragment of the logistics plan (the task for which $x_S(l,i) < t_S^n$), begin at the appropriate moments $x_S(l,i)$ and end in moments $x_S(l,i) + \tau\ (l,i)$;

б) the task (i,j), corresponding to the variable $d_q \in P_S^n$, begins at the appropriate moments t_S^{n+1} and ends in the moments $t_S^{n+1} + \tau\ (l,i)$;

в) other tasks $(e,h) \in G$, corresponding to variables $d_u \in P_S^n,\quad u \neq q$, begin in the moment t_S^n and end in the moment $t_S^n + \tau\ (e,h)$.

An important element of the algorithm for solving the problem (7) - (13), which significantly affects its convergence, is the method of choosing the next task and the option of assigning resources to it. Formally, it consists in the choice of variables $d_q \in P_s^n$ to include the S-th thread at t_S^n time. In the proposed algorithm d_q, the selection at the next thread step is carried out in two stages: the first one selects task, and the second - the resource assignment option. The selection of the next task is carried out in accordance with the following sequence of preferences:

$$\min T_j^{(n)} \rightarrow \max \tau(i,j) \rightarrow \min i \rightarrow \min j,$$

i.e., the first in the plan includes task, which corresponds to a smaller late completion date $T_j^{(n)}$. If there are several such tasks, then the maximum volume is selected from them. If there are several such tasks, then tasks with the smallest numbers (i, j). In this case, late completion dates should be determined taking into account the considered fragment of the S logistic plan.

Scenario $r(i, j)$ to identify the handler for the selected task (i, j) is determined from the condition of the minimum value

$$\sum_{k \in r(i,j)} \sum_{(l,h) \notin S} \delta_k(l,h),$$

i.e. the handlers who are least universal for the remaining tasks $(l,h) \notin S$ are appointed

The task selected in this way (i, j) and scenario $r(i, j)$ the assignments of the handlers uniquely determine the next variable $d_q \in P_S^n$, included in S-th thread of the tree of the plan's variants at the moment t_S^n.

The skip of the tree is organized in accordance to the rule "go to the right". That allows storing in the memory of a computer only the current fragment of the logistics plan, the smallest of the previously obtained values of the objective function and the corresponding acceptable version of the plan.

This rule, in combination with the method of selecting tasks and handlers, is an approximate algorithm for solving the problem (8) - (14), which allows obtaining the first feasible solution in a finite number of steps equal to the number of N tasks in the network (1).

Each S thread ends if it includes all N tasks, that is, an acceptable logistic plan Y has been received, or if

$$W_s \geq T^0(1-\mu), \quad 0 \leq \mu \leq 1, \tag{27}$$

where T^0 - the least value of the objective function for the previously obtained allowable logistic plans (record);

μ - specified deviation of the objective function from the optimal (optimization accuracy).

The executive of condition (27) means that on the thread under consideration it is impossible to improve the previously obtained record by more than 100μ% and its continuation within the specified optimization accuracy does not make sense.

The procedure for finding the solution ends if all the remaining threads satisfy the condition (27). When the method for traversing the variants tree of such a situation is established, the second return to the root vertex corresponds. At the same time, the last record is the desired value of the objective function (8), and the corresponding feasible plan for performing the set of operations is the optimal logistic plan[14-15]. The use of the considered algorithm provides analysis of all possible variants of the plan and excludes repetitions when viewing them.

4 Conclusion

In general, the proposed algorithm for solving the problem of forming an optimal plan for the implementation of a set of interrelated works provides the possibility of obtaining both exact and approximate solutions. They can be relatively simply integrated into specific decision support systems, since the requirements for the efficiency indicator and the limitations of the model (8) - (14) are sufficiently general that allows them to form a wide range of specific methodologies for justifying managerial decisions on the organization of material technological and logistical processes of the construction project and their financial support.

References

1. V.G. Anisimov, YE.G. Anisimov, A branch-and-bound algorithm for one class of scheduling problem, The magazine of computational mathematics and mathematical physics, 32 - 2, 2000-2005, (1992)
2. V.G., Anisimov, E.G. Anisimov, *The magazine of computational mathematics and mathematical physics*, 37-1, 54-60, (1997)
3. V.G. Anisimov, E.G. Anisimov, T.N. Saurenko, M.A. Sonkin, , *Journal of Physics: Conference Series*, **803, 1**, 012006, (2017)
4. V. Anisimov, E. Anisimov, M.A. Sonkin, *International Journal of Applied Engineering Research,* **10(17)**, 38127-32, (2015)
5. A.R. Bril, O.V. Kalinina, I.V. Ilin, A.S. Dubgorn, O.Y. Iliashenko, Forecasting the turnover growth in the risk management system as management decisions support, Proc. of 2017 20th IEEE Inter. Conf. on Soft Computing and Measurements, SCM 2017, 7970692, 692-693, (2017)
6. A. Kozlov, S. Gutman, I. Zaychenko, E. Rytova, The formation of regional strategy of innovation-industrial development. Advances in Intelligent Systems and Computing, 37th Int. Conf. Information Systems Architecture and Technology, 15-126 (2017)
7. A. V. Izotov, O. V. Rostova, Development of a system of sectoral investment priorities, Proceedings of the 29th International Business Information Management Association Conference – Sustainable Economic Growth, Education Excellence, and Innovation Management through Vision 2020, pp. 1812-1822, (2017)
8. A.R. Bril, O.V. Kalinina, I.V. Ilin Small innovative company's valuation within venture capital financing of projects in the construction industry, MATEC Web of Conferences, 106, 08010, (2017)
9. A. O. Alekseyev, O. G. Alekseyev, V. G. Anisimov, Ye. G. Anisimov, N. I. Yachkula, Application of Markov chains in estimating the computational complexity of the simplex method, *Soviet journal of computer and systems sciences*, **5**, 130-134, (1988)
10. O.Yu. Iliashenko, A.I. Levina, A.D. Borremans, The Approach to the Formal Specification of Static Structure of the System: mapping UML to EXPRESS-G, Proc. of the 29th IBIMA Conference – Sustainable Economic Growth, Education Excellence, and Innovation Management through Vision 2020, 1177-1186 (2017)

11. G.Y. Silkina, Information and Communication Technologies in Ensuring of Innovative Development, Proc. of the 29th IBIMA Conf. – Sustainable Economic Growth, Education Excellence, and Innovation Management through Vision 2020, 1165-1176, (2017)
12. I.V. Ilin, O.V. Kalinina, A.I. Levina, O.Yu. Iliashenko, Approach to Organizational Structure Modelling in Construction Companies, MATEC Web of Conferences, 05028, (2016)
13. I.V. Ilin, A.I. Levina, O.Yu. Iliashenko, Enterprise architecture approach to mining companies engineering, MATEC Web of Conferences, **106**, 0806 (2017)
14. A.B. Anisiforov, A.S. Dubgorn, Proc. of the 29th IBIMA Conf. - Education Excellence and Innovation Management through Vision 2020: From Regional Development Sustainability to Global Economic Growth, 2920-2930, (2017)
15. B. Ozorhon, K. Oral, Drivers of Innovation in Construction Projects, Journal of Construction Engineering and Management, **143(4)**, 04016118, (2017)

Two-stage commercial evaluation of engineering systems production projects for high-rise buildings

Aleksander Bril[1], *Olga* Kalinina[1*], *Anastasia* Levina[1]

[1]Peter the Great St. Petersburg Polytechnic University, Institute of Industrial management, Economy and Trade, 195251, Polytechnicheskaya, 29, St.Petersburg, Russia

Abstract. The paper is devoted to the current and debatable problem of methodology of choosing the effective innovative enterprises for venture financing. A two-stage system of commercial innovation evaluation based on the UNIDO methodology is proposed. Engineering systems account for 25 to 40% of the cost of high-rise residential buildings. This proportion increases with the use of new construction technologies. Analysis of the construction market in Russia showed that the production of internal engineering systems elements based on innovative technologies has a growth trend. The production of simple elements is organized in small enterprises on the basis of new technologies. The most attractive for development is the use of venture financing of small innovative business. To improve the efficiency of these operations, the paper proposes a methodology for a two-stage evaluation of small business development projects. A two-stage system of commercial evaluation of innovative projects allows creating an information base for informed and coordinated decision-making on venture financing of enterprises that produce engineering systems elements for the construction business.

1 Introduction

Internal engineering systems and networks are the basis of the infrastructure of any construction site. Elements of these systems include units and blocks of sewerage systems, heating and power supply, automatic water fire extinguishing installations, fire water supply, details of "smart houses" and so on. The cost of construction is largely determined by the technical characteristics and cost of these elements.

Details and blocks of engineering systems are produced in small enterprises, as well as in divisions of medium and large companies in the construction business. There is a high level of competition, new production technologies and innovative projects of products with improved operational characteristics are actively developing.

Financing of development of such enterprises is most expedient due to venture business. At present, it is considered that venture business is one of the most effective ways of financing innovative projects. Venture funds invest in little-known, promising innovative organizations

* Corresponding author: olgakalinina@bk.ru

at its earliest stages of development, when it is unknown and has not yet completed the implementation of the "business idea" underlying its business. The investment period is from three to seven years. Venture fund in exchange for its investments receives a stake in the authorized capital of an innovative enterprise and after a certain period of time comes out of it by "buyback": it sells its stake in the open market or to a strategic investor. [1-9]

Some of these projects are unprofitable or low-profitable, but some of them bring significant profits and even over-profit. In this regard, the average rate of return of venture funds is about 25-30%, which significantly exceeds the incomes of conventional investment projects with an indicator of 5-10%.

In Russia, venture funds appeared in the early 90's. Restructuring of the economy and privatization processes have caused a sharp increase in interest in the market of direct investment in Russia from various investors and, as a result, reflected in the more than dynamic start of the venture business. This activity is supported by some large companies, and venture funds and innovation development networks are being created [1-4, 7-10]. The main tasks of these organizations are to search for new effective ideas - innovations and create conditions for their transformation into highly profitable commercial enterprises.

The most common mechanism for venture financing of business is the system of organizing a joint venture with a "buyback". To implement this methodology, it is necessary to find a prospective enterprise - "start-up", determine its real value, the amount of the authorized capital, the value of the joint venture at the beginning of financing and at the time of the agreed redemption period. At the stage of choosing effective "start-ups" dozens and hundreds of proposals are eliminated. At the same time, venture investors often use their own expert assessments [7,8]. It is at this stage that most of the disagreements between investors and the founders of the company arise, and at this stage a significant part of the deals is broken. The literature suggests a sufficiently large number of methods for assessing the effectiveness of innovation projects and calculating the company's value at various stages of development [11-15]. Many methods are rather complicated and require initial information, which at the stage of "start-up" can not be sufficiently reliable.

The article proposes a technique for a two-stage commercial evaluation of innovative projects at the "start-up" stage. The methodology is intended for use in the process of joint venture fund operation and an innovative enterprise (due diligence). It involves the evaluation of an innovative project from two sides: from the economic efficiency of the project as a whole and from the financial situation and the level of risks of the manufacturer. The proposed methodology is based on UNIDO methodology [16] and allows, first of all, to more rationally approach the choice of an innovative enterprise for financing from a venture fund. The second most important issue that allows us to solve the two-stage method of calculating economic and financial indicators is the possibility of creating the initial information for calculating the company's value at various stages of its development, including, at the time of the "redemption".

2 Methods

Venture financing is a long-term direct investment in small innovative enterprises with a high or relatively high level of risk in order to generate significant profits in the period from 3 to 7 years. There are external and internal venture business. The development of domestic venture business is most often carried out in large corporations at the expense of the resources of the initiators of the idea. Financing external ventures is by raising funds from external sources: investment, insurance, charitable, pension funds, state resources, private investors. Organizationally, this is done either through the creation of venture funds or through the individual activities of entrepreneurs, called business angels. At the same time, if you do not go into the sectoral details, three types of financing objects have become most widespread:

applied research (the stage of "sowing" according to the US-European classification of the stages of innovation development), start-ups in the same classification and small innovative enterprises with a lack of financial resources. The main types of external venture business are presented in Fig. 1.

External Venture Business		
Private investors (business angels)		Specialized companies (venture funds)
Business idea ("seeding" stage)	Completed research (start-up phase)	A small enterprise with a deficit of finance and qualified personnel

Fig. 1. The enlarged scheme of varieties of external venture business, compiled by the authors

Venture investments for decision-making on financing require a feasibility study. Two main points must be taken into account. The first is the availability of information for planned calculations, which is quite different for such, for example, financing objects as "sowing" or operating a small enterprise. And the second is the rules for making decisions about financing from venture funds, industrial companies or business angels. If a business angel - a private investor decides to participate in innovative projects mainly on the basis of his own experience and risk assessment [8], venture funds have specially developed standards for selecting projects for investment. In this paper, the methodology for assessing the effectiveness of future investments only for venture funds is considered.

The proposed system of settlements is aimed at ensuring their effective development, creating conditions for operational management and decision-making at various stages of development. It should be noted that the literature offers a lot of classifications of stages and stages of development of innovative projects [11,12,13].

Here, the authors maintain a classification that includes 4 stages:

- preparation for the opening of the enterprise, initial financing and organizational development (start-up),
- rapid initial growth in sales volumes, significant investment in production, possible negative cash flow,
- achievement of significant economic results, financial payback, positive cash flow,
- a decrease in revenues as a result of increased competition, the search for new ideas and options for the development of the company, possible negative cash flow.

For effective implementation of the project, a specific system of financial and economic calculations is appropriate at each stage.

The first is a feasibility study based on a methodology for commercial project evaluation, and the development of a business plan. On the second - a regular assessment of the financial condition of the enterprise and monitoring the implementation of financial and economic indicators of the business plan. The third is to monitor the financial condition of the company and determine the value of the enterprise to prepare for the transition to the fourth stage. On the fourth - again, the valuation of the company, the development of new ideas and business planning. To effectively manage projects and make rational decisions, a correlation between calculation methods and decision criteria at each stage is necessary (Table 1).

The basis for making initial decisions on the project is its feasibility study and the development of a business plan. The business plan is a clearly structured by sections section of the work and operations necessary for the implementation of the project. At the same time, a project is understood as a plan for investing money and capital in order to achieve previously planned benefits.

The main objective of developing a business plan is to justify the costs of the project and, accordingly, the rationale for the sources of its financing. The business plan is one of the main documents in their general list for presentation to investors and creditors. Initially, the development of business plans was carried out only with respect to investment projects. Accordingly, for all works and operations included in this document, costs and results - incomes should be justified and calculated.

Table 1. Stages of the project and methods of financial and economic calculations.

Stages of multi-project	Basic methods of financial and economic calculations	Decision Criteria
1 Preparation for the opening of the enterprise, initial financing	Feasibility study and business plan development	Net present value, profitability and operating leverage
2 Rapid initial growth in sales, possible negative cash flow	Regular assessment of the financial condition of the enterprise and control over the implementation of business plan indicators	Special system of financial indicators
3 Financial payback, positive cash flow	Monitoring the financial condition of the company and determining the value of the company	System of financial indicators, company value
4 The aggravation of competition, the search for new ideas and options for development	Again, valuing the company, developing new ideas and business planning	The company's value, profitability and operating leverage

These calculations are information for decision-making by investors or creditors: to finance the project or to refuse financing. The decision on financing is made on the basis of the size of such characteristics as the payback period and the simple rate of return for the project, as well as discounted indicators, the net present value of NPV and the internal rate of return of the IRR.

With the development of business planning, this system of calculations showed significant shortcomings associated with the fact that the above decision criteria are not included in the indicators characterizing the effectiveness of enterprises.

The financial condition of the enterprise and the effectiveness of its development is characterized by a system of indicators of business activity and financial stability. The most important characteristics here are the profitability of sales, profitability of assets and equity, as well as such characteristics of financial stability as the share of equity in assets, financial leverage (lever), the coefficients of general and urgent liquidity, solvency ratios. For effective project management, it becomes necessary to combine these project-specific and enterprise-wide evaluations into one business planning system.

Based on research and numerous calculations by the authors, it is recommended to follow the two approaches in the economic evaluation of innovative projects:

- two-stage system of financial and economic calculations for investment projects;
- one-stage settlement system for operational projects.

The methodology proposed by the authors for evaluating innovative projects for the production of engineering systems for a high-rise construction business is based on the first approach and includes a two-stage calculation system [14]:

- stage 1- determination of commercial efficiency on the basis of payback calculations and discounted indicators;
- stage 2 - study of the impact of project implementation on the growth of the company's profits and labor productivity, profitability indicators and level of operational (operational) risks.

Below is a list of the main financial and economic indicators for describing the planned results at both stages of the project evaluation.

Stage 1. Definition of economic efficiency on the basis of calculations of a recoupment and the discounted indicators;

INDICATORS

- Payback period of investments r
- Simple rate of return on investment i
- Net present value of NPV
- Internal rate of return IRR

Stage 2. Study of the impact of project implementation on the growth of the company's profits and labor productivity, profitability and level of operational risks.

INDICATORS

- Profit increase from sales at the enterprise
- Increase in labor productivity (sales per employee per month)
- Profitability of sales of ROS
- Operational (production) leverage of ODA

It should be noted that the list of indicators for the second stage can be significantly expanded in accordance with the features of each company implementing an innovative project and considered as an object of venture financing. The choice of additional indicators is based on the methodology of financial diagnostics of enterprises, developed by the authors within the system of economic and financial calculations for innovative projects. [14].

In this presentation, business planning at the first stage of the innovation project allows solving two main tasks:

- provide a business case for the venture financing of an innovative project,

- to create an information base for planning the development of a new enterprise, calculating its initial value and the future value of the "buyback", taking into account the quantitative assessment of the existing financial and production risks.

3 Results

The income of venture companies is formed due to the system of "buyback", i.e. sale of the share of the venture company in a joint innovation enterprise after its release in 3-7 years of operation. To forecast the value of this income, it is necessary to determine the value of the applicant's company, the amount of the authorized capital, the cost of the joint venture at the beginning of financing and at the time of the agreed repurchase period. A two-stage system of commercial evaluation of innovative projects allows creating an information base for determining the value of an innovative enterprise. We will dwell in more detail on the methodology for estimating the value of the company, which is necessary at the third and fourth stages of the implementation of innovative projects for the production of engineering systems for construction (Table 1).

In accordance with the two-stage settlement system, the authors propose a set of relatively simple evaluation methods that are based on the results of the second stage of calculations and logically fit into the overall system of financial and economic calculations of the planned innovation enterprise. The theory of business valuation offers three approaches to solve this problem: costly, profitable and comparative. The author's methodology suggests estimating the company's value based on various development options and the results of the company's financial diagnostics in the second and third stages of development.

To determine the initial value of the company at the third stage of the innovation project implementation, it is necessary to consider at least two development options:

- the first one - without additional investment of the project; here it is supposed to be further implemented in accordance with the indicators of the previously developed business plan with a two-stage commercial evaluation system,

- The second is to attract investments in accordance with the leadership strategy; here in the near future, competition, commercial and financial risks are expected to grow, and the prices for company services will decline.

For both variants of development, the company's value at the time of Stom's assessment is determined on the basis of a cost approach, after conducting an audit and diagnosing the financial condition of the operating enterprise. To expand the information on the value of the enterprise at the time of valuation, it is also necessary to use the income approach methods: "profit capitalization", "50% of sales" and "fivefold profit from sales." If there is information about the purchase and sale of similar enterprises, it will significantly supplement the information on the company's value. Based on the agreement of the results obtained, the actual value.

St Mo F this characteristic is the basis for further calculations of the company's value in the planned period of time. Table 2 shows the results of the valuation of the innovative enterprise implementing the project for the production of elements and units for the equipment assembly in the construction business.

Table 2. Results of the initial valuation of the company at the third stage of the multi-project implementation

Company valuation method	Sum, mio RUR
Cost	14,0
Capitalization of profit	11,0
"50% of sales"	14,5
"Fivefold profit on sales,	16,0
Approved St Mo F	13,5
The value of the company after investing in assets, taking into account the appreciation factor St In	18,0

After determining the actual value of the company at the time of valuation, it is necessary to establish a planned period for calculating changes in the value of the company. Its magnitude depends on the specifics of the project, but should not exceed three years.

For the second variant of development it is necessary to determine the company's value at the planned moment of attracting additional investments St in. Here, the rise in the cost of the enterprise due to the attraction of additional capital should be taken into account. When determining the coefficient of appreciation, the key factors are the turnover of the assets of the enterprise (the number of assets turnover per year), and the profitability of ROA assets. These are the indicators determined at the 2nd stage of the commercial evaluation of the innovation project. After the establishment of the St in, one can proceed to the calculation of the company's value for the anticipated planning period. Here you can apply several methods of evaluation. In the author's methodology, a venture approach and a method of future value are recommended. Thus, the two-stage system of commercial project appraisal allows creating an information base for clarifying and coordinating the actual and planned cost of the enterprise, adjusting the project implementation process, solving organizational, production and financial issues.

4 Discussion

The most difficult issues for the venture business are the procedure for selecting innovations for financing and evaluating the company's future value for "buyback". The effective organization of this process involves accelerated screenings and presentation to fund the best and most promising ideas.

The system of "buyback" is realized through three main methods:

- Management buy-out (sale of the share of the venture company to existing shareholders and management);
- external redemption (sale of a share of a venture company to an outside investor);
- sale of the share of the venture company through the public offering of IPO shares.

The literature suggests and uses a large number of methods and approaches to determine the value of a company and attract venture investors [6-7]. The task is to systematize these methods in relation to different stages of venture financing and, most importantly, in relation to different levels of innovation. Radical innovations associated with new products in the whole world or in Russia, and processes, are more risky and, accordingly, their results are more difficult to predict. For such innovations, it is advisable to conduct planned calculations of cost indicators on the basis of simple aggregated methods.

For simple innovations related to significant, and sometimes, and with minor changes in known processes and products, more detailed methods can be used. Most innovations in the field of engineering systems for high-storey construction are just such. In accordance with this, it can be reasonably recommended to apply the two-stage system of commercial evaluation of innovative projects for the organization of venture financing in the construction business.

In the process of this work, a revised and agreed business plan for the development of the joint venture should be developed, on the basis of which further sources of financing, order and price of the buyback are determined. The main task of the two-stage feasibility study and analysis of the company's value is precisely to achieve this consent.

5 Conclusions

1. Two-stage system of commercial evaluation of innovative projects allows creating an information base for informed and coordinated decision-making on venture financing of enterprises that produce elements of engineering systems for the construction business.

2. Based on the results of the calculation of indicators for the second stage of the commercial evaluation, financial diagnostics and valuation of the company for the entire planned period of its development are carried out. This creates a fairly broad base of financial and economic indicators for agreeing and clarifying operational plans for further implementation of the venture financing of the project.

3. Relatively simple algorithms recommended by the authors of the calculation methods and available software products allow you to concentrate when monitoring the development of the project to ensure the reliability of the original data.

References

1. URL: http://www.rusventure.ru/ru/programm/analytics (13.03.2017)
2. Popov, P. Roosenboom, Journal on Banking and Finance, **37,** Issue **12**, 4695–4710 (2013)
3. J.L. Woolley, A.V. Bruno, E.D. Carlson, Journal of Management for Global Sustainability, **2**, 7–30 (2013)
4. J.P. Killing, Strategies for Joint Venture Success. Routledge Library Editions: International Business, **22**, 132 (2013)
5. E.M. Rogova, M.A. Sapozhnikova, Problems of Value Russian Innovative Companies Evaluation at Early Stages of Development by Venture Investors. Scientific and technical sheets of the St.-Petersburg State Polytechnic university. Economic sciences, **6–1 (185)**, 150–158 (2013)

6. L.V. Nikolova, J.J. Kuporov, D.G. Rodionov, International Journal of Economics and Financial Issues, **5 (3S)**, 68–72 (2015)
7. A.O. Adamtsevich, A.P. Pustovgar, S.A. Pashkevich, A.V. Eremin, International Journal of Applied Engineering Research, Vol. 11, pp. 1609 - 1611 (2016)
8. V. Kankhva, IOP Conference Series: Earth and Environmental Science, Vol 90, 012175 (2017) DOI - 10.1088/1755-1315/90/1/012175
9. G. Gogoberidze et al, Indicator method of estimation of human impact assessment for coastal local municipalities. Measuring and Modeling of Multi-Scale Interactions in the Marine Environment - IEEE/OES Baltic International Symposium, BALTIC 2014. 6887840
10. A.V. Kozlov, E.V. Rytova, et al, The valuing of the indicator of a regional industrial development: The fuzzy logic approach. Proceedings of the 19th International Conference on Soft Computing and Measurements, SCM 2016. Paper number 7519842, 545-547 (2016)
11. Rice, A. L. (Ed.). (2013). The enterprise and its environment: A system theory of management organization (Vol. 10). Routledge.
12. A.B. Anisiforov, A.S. Dubgorn, Organization of enterprise architecture information monitoring. Proc. of the 29th IBIMA Conf. - Education Excellence and Innovation Management through Vision 2020: From Regional Development Sustainability to Global Economic Growth, pp 2920-2930 (2017)
13. V. Berens, P.M. Havranek, *Guidelines for assessing the effectiveness of investments: translate Engl.* (Moscow, Interexpert, Infra-M, 1995)
14. A. Pustovgar, L.A. Shilova, A.O. Adamtsevich, L.A. Shilov, International Journal of Applied Engineering Research, Vol. 10, pp. 40988 - 40990 (2015)
15. V. Izotov, O. V. Rostova, Development of a system of sectoral investment priorities, Proc. of the 29th IBIMA Conference – Sustainable Economic Growth, Education Excellence, and Innovation Management through Vision 2020. Pp. 1812-1822.

Project Management Life Cycle Models to Improve Management in High-rise Construction

Andrey Burmistrov[1*], *Maria* Siniavina[1] and *Oksana* Iliashenko[1]

[1]Peter the Great St.Petersburg Polytechnic University, Polytechnicheskaya, 29, St. Petersburg, 195251, Russia

Abstract. The paper describes a possibility to improve project management in high-rise buildings construction through the use of various Project Management Life Cycle Models (PMLC models) based on traditional and agile project management approaches. Moreover, the paper describes, how the split the whole large-scale project to the "project chain" will create the factor for better manageability of the large-scale buildings project and increase the efficiency of the activities of all participants in such projects.

1 Introduction

In the paper are presented some factors affecting the management of high-rise building projects and other projects, unique features of managing a complex project in terms of uncertainty at different stages of its life cycle, proposals that improve the quality of project management.

The objective of the current study was to demonstrate a possibility to improve project management in construction high-rise buildings through the use of various Project Management Life Cycle Models (PMLC models) based on traditional and agile project management approaches.

1.1 Importance

Currently, urbanization is global trend and in Russia as well - for example, federal government is planning some "organizational, regulatory and institutional measures are required to streamline and accelerate the development of agglomerations as a basic condition for the development of the postindustrial economy in Russia in the medium term (for 6 years)" [1], in particular, it is considered that priority will be focused not only on Moscow and St. Petersburg, but also on 36 Russian agglomerations with a population more than 500 thousand people. [2] Because of constantly increasing of the cost of land in large cities and growing of population, it is obvious that the construction of high-rise buildings will also increase. (At the same time, we do not consider the construction of skyscrapers, which for the purposes of our paper can also be attributed to high-rise buildings). Consequently, the

* Corresponding author: olgakalinina@bk.ru

relevance of the project management of exactly high-rise buildings construction will increase for construction companies.

1.2 Specificity of high-rise buildings as an element of large cities

At present, not only in Russia there are no single criteria for concepts, but also in the world: "multi-storey building", "high-rise building" and others. According to Emporis Standards, "A high-rise building is a structure whose architectural height is between 35 and 100 meters. A structure is automatically listed as a high-rise when it has a minimum of 12 floors, whether or not the height is known. If it has fewer than 40 floors and the height is unknown, it is also classified automatically as a high-rise. A multi-story structure between 35-100 meters tall, or a building of unknown height from 12-39 floors. Example buildings: The Lloyd's Building (London, United Kingdom), Lever House (New York City, U.S.A.), Hansa-Hochhaus (Cologne, Germany)" [3]. In Russia, various definitions of the term "high-rise building" are used, for example, experts indicate that high-rise building is a building with hight more than 75 m (more than 25 floors) [4], according to [5] tall buildings are considered to be more than 75 meters high and up to 400 meters, and according to [6] high-rise residential and public buildings up to 150 meters high are considered residential buildings with a height of more than 75 m, public buildings - more than 50 m). Different definitions of high-rise buildings in the Russian Federation, see in [7].

In comparison with buildings of lower number of floors, high-rise buildings have the following features: as part of the system of the "city" system, they also touch issues of aesthetics, traffic management and in a broader sense sustainable urban development. From the point of view of the internal arrangement, higher demands are placed on high-rise buildings in terms of reliability, resistance to wind loads, and so on. The processes of creation and operation of high-rise buildings are also subject to higher requirements: engineering surveys, design, construction, etc.

In general, it can be certified that a high-rise building as a unique engineering and organizational system has a much higher complexity than low-rise buildings, this complexity is characteristic both for the system itself (the high-rise building) and for all the processes associated with the creation of this systems, including those with higher responsibilities, risks, investments, etc.

Due to this technology, the management of high-rise building projects requires the use of methods and tools that are designed to overcome the ever-increasing uncertainty and complexity of the life cycle processes of high-rise buildings.

2 Methods

In this paper, the data obtained from the scientific and periodicals, which indicate the quality status of project management in Russia and in the world, and some methods that are available to improve project management are being considered.

According to Lean Construction Institute (LCI), 70% of all construction projects are over budget & delivered late. Construction labor efficiency and productivity has decreased, while all other non-farming labor efficiency has doubled or more since the 1960s. Currently, 70% of projects are over budget and delivered late. [8]

According to PMI research, organizations lose $ 109 million for every $ 1 billion which was invested in projects and programs [9]. The KPMG study provides data that only 31% of all respondents' projects came within 10% of budget in the past 3 years, and just 25% of projects came within 10% of their original deadlines in the past 3 years. [10]

The reasons for the lack of success of projects are actively studied all over the world, and tools for improving project management are also offered. In particular, the implementation of BIM is proposed; Lean Construction [11]; Project Definition Rating Index (PDRI) for Building Projects [12], [13], "project implementation process based on the passage of strictly defined stages and milestones" (Stage-Gate Process,), "complex project monitoring system" (Control Tower) [14] and many other tools.

3 Results

Practitioners and researchers estimate the lack of elaboration elaboration of project requirements at the initial stages is the main factor of the project's failure. As indicated in [13], industry practitioners are perceiving that early planning efforts in the project life cycle have a greater impact on the success of planning delivery. [15] McKinsey experts also consider that it is necessary to implement a system of a phased project management system that provides the suspension and completion of inefficient projects in the early stages. [14]

The research [16] indicates that the top five causes of troubled projects were:

1. Requirements: Unclear, lack of agreement, lack of priority, contradictory, ambiguous, imprecise.

2. Resources: Lack of resources, resource conflicts, turnover of key resources, poor planning.

3. Schedules: Too tight, unrealistic, overly optimistic.

4. Planning: Based on insufficient data, missing items, insufficient details, poor estimates.

5. Risks: Unidentified or assumed, not managed.

Similar views are expressed by Russian practitioners: "the main factors of project's failure:

1. Mistaken or incorrect definition of the main parameters at the start of the project (cost, timing).

2. Abnormality of the technology of project management, design and construction.

3. Formation of false goals ("technological death").

The cause of theseis defects in the project management system. "[17]

Besides the tools for the improvement described above, , the experience of the industry which learns new methods and models of project management – IT industry faster that others can be used to improve the system of construction project management (including construction of high-rise buildings). The answer to the growing uncertainty and complexity of software development was the development of Agile based technologies.

However, in construction as opposed to IT industry, as a rule, project management models historically based on the PMI model are applied. This is a linear ("waterfall") model, which is the most efficient, but less flexible (is non-tolerant to changes). [18,19] This is understandable and appropriate, since buildings can not be reconstructed by "sprints", as in Scrum. Therefore, it seems that Flexible Project Lifecycle Management Models (models of project life cycle management) have no place in construction projects. However, it just seems. Firstly, stages of construction projects are heterogeneous (shown in the Table 1 below), and, secondly, there are several flexible (agile) project management models.

3.1 Heterogeneity of stages of construction projects

Let us consider the Influence and Expenditure Curve for the Project Life Cycle model (see Figure 1), which identifies the conceptual relationship between influence and expenditure in a project life cycle.

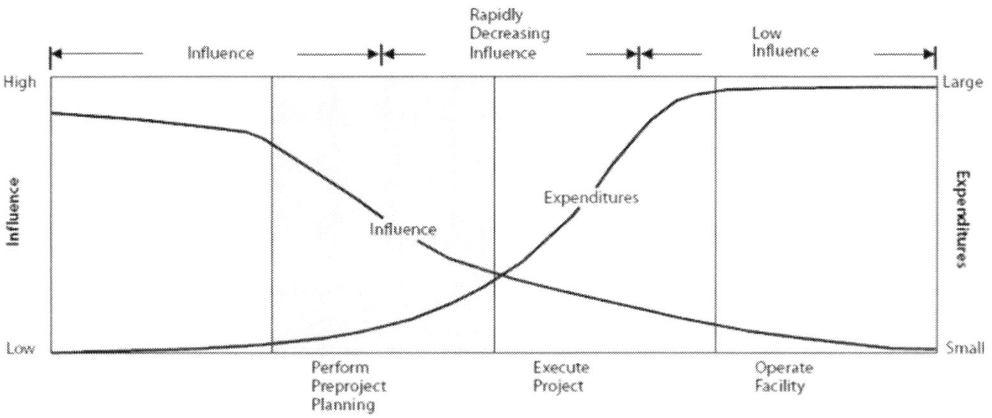

Fig. 1. Influence and Expenditure Curve for the Project Life Cycle

The curve labeled "influence" in Figure 1 reflects a company's ability to affect the outcome of a project during various stages of a project. The diagram illustrates that it is much easier to influence a project's outcome during the project planning stage when expenditures are relatively minimal than it is to affect the outcome during project execution or operation of the facility when expenditures are more significant. [20]

As we see that the key characteristics that affect project management (costs and possibility of influencing the project) vary with stage of the life cycle in opposite directions, which drastically changes the requirements for methods and models of project management.

3.2 Project Management Life Cycle Models (PMLC models)

Robert K. Wysocki highlights following PMLC models (see Figures 2 and 3):

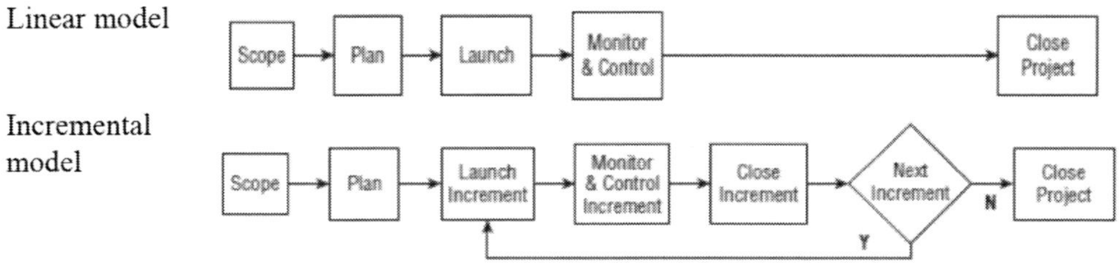

Fig. 2. Two models based on *Traditional Project Management Approaches*

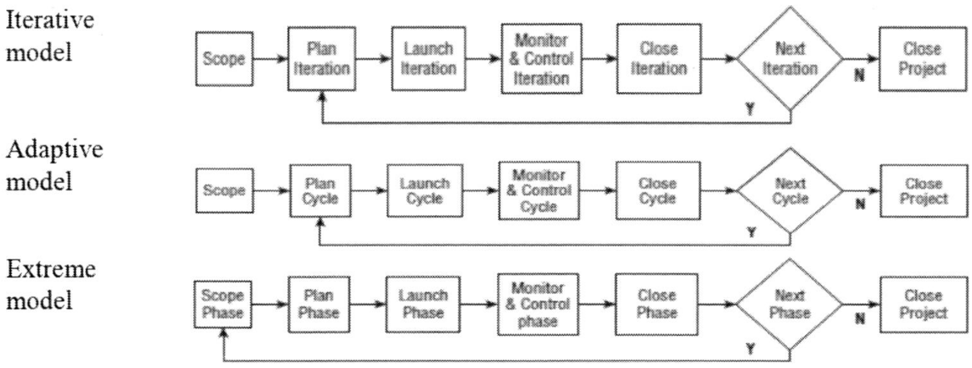

Fig. 3. Three models based on Agile Project Management Approaches, Source: [21]. In Russian, see the review of these models in [22].

3.3 High-rise construction project can be viewed a sort of project chain

High-rise construction project can be viewed a sort of project chain, at least, on starting position. Separate links are separate stages of the overall project. This follows from the fact that, in the case of increased attention to the early stages, the identified unpromising projects can be terminated at the Concept or Design stage (in order to avoid higher costs in the future). In addition, from Table 1 you can clearly see that the different Lifecycle Stage for a complex project is very different. Thus, we can consider the Concept and Design stages as separate projects (entering the chain) and use for them different models of project lifecycle management.

Table 1. Characteristics of different Stage of Project Lifecycle in Development (e.g. for high-rise buildings)

Project Lifecycle Stage	Work volume and costs	Possibility to influence the project	Cost of making changes	Goals, Requirements, Solution	Management Life Cycle Model
Concept	Low	Very High	Low (x 1)**	Unclear	Extreme or Adaptive model
Design & Planning	Medium	High	Medium (x 5)	Partially Clear	Iterative model*
Procurement and Construction	High	Medium	High (x 12)	Clear	Incremental or Linear model
Start up	Medium	Low	Very High (x 40)	Very Clear	Linear model

Note:

*- We might notice that on this stage BIM is used quite similar to software or production prototype.

**- Quantitative estimates from INCOSE (The International Council on Systems Engineering (INCOSE) is a not-for-profit membership organization founded to develop and disseminate the interdisciplinary principles and practices that enable the realization of successful systems. http://www.incose.org/about) data for all kinds of complex projects. Here is given as a possible assessment for construction projects.

General approaches to the selection of the Management Life Cycle Model, which is recommended in [21] is shown in Figure 4.

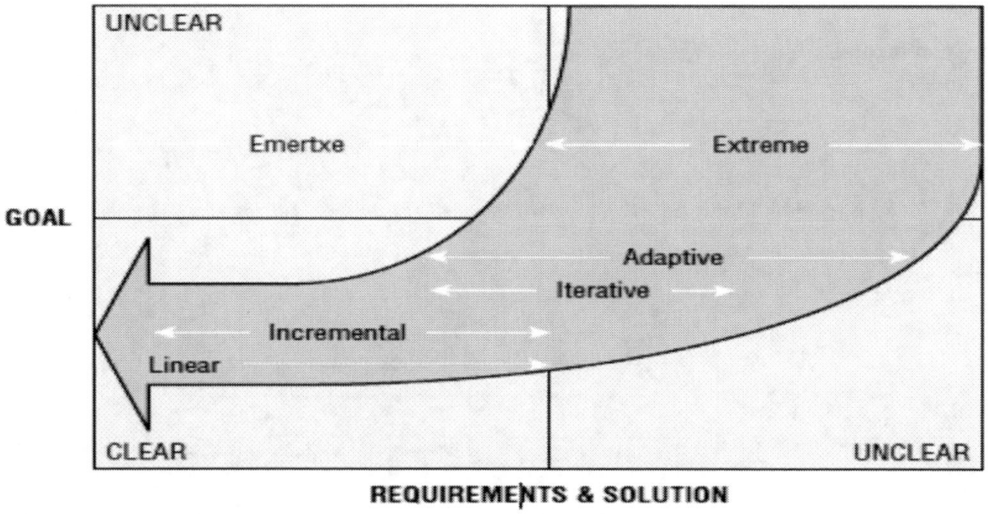

Fig. 4. PMLC approaches

As is shown above, these approaches can be used to manage complex construction projects.

4 Discussion

The paper shows that an important improvement in the management of complex construction projects, including for high-rise buildings, is the increased attention to the initial stages of the life cycle of such projects. The characteristics of different stages of projects are considered, and a complex project of high-rise buildings with a high degree of uncertainty is proposed, not as a single project, but as a chain of projects.

The main link of such a chain should be a relatively independent project corresponding to the "Concept" stage, which is managed by the Extreme or Adaptive Management Life Cycle Model. These models have the greatest flexibility and the ability to change all the important elements of the "main" project at the very beginning. In this case, the participants in this project have more opportunities for an objective review of the project and, if it is found inexpedient or unprofitable - for rejection. As shown in paper [14], the more inefficient projects are rejected at an early stage, the more successful the remaining projects are.

If the construction is deemed expedient, the next project "Design & Planning" is started, which is managed on the basis of the Iterative Management Life Cycle Model, which allows gradually getting more and more complete "prototypes" of the building under construction. The criteria for the success of the completion of this project should be the recognition by all participants of the fact that all important circumstances have been taken into account and that the uncertainty has been reduced to the level at which the next project - "Procurement and Construction", with the largest volumes of work and cost, can be managed by the least adapted to the changes, but the most economical model is the Incremental or Linear Management Life Cycle Model. If, in the course of the project "Design & Planning", circumstances are revealed that it is inexpedient or unprofitable to continue the project, the "Design & Planning" project also ceases, which also saves resources for all participants.

Also very important issues are ways to harmonize the interests of different stakeholders in the course of the "Concept" and "Design & Planning" projects. These topics include, in particular, such components as the formation of a «common language», comprehension and formulation of shared values and on their basis shared goals by all participants, and implementing of systems engineering methods. These issues are critically important, but they are beyond the scope of this paper. They were considered by us, in particular, in [23] (on an example of projects on creation of transport systems in agglomerations).

5 Conclusions

In order to respond more flexibly to the increasing uncertainty and complexity of the buildings project, it is useful to decompose the whole large scale project and to split it into the "project chain". In this chain, the former stages of "Concept", and "Design & Planning" become separate projects, executed sequentially. Then, for managing these projects, which are very different in their characteristics (and where the uncertainty is much higher than for the subsequent project "Procurement and Construction"), it will be possible to apply special Management Life Cycle Models based on Agile Project Management Approaches that are specifically designed for these conditions. In case of revealing inexpediency or ineffectiveness of the continuation of the work, it is easier to stop the "small" projects than the large-scale project. At the same time, for the project "Procurement and Construction", when the goals, requirements and solutions are already defined, it is advisable to use the traditional Incremental or Linear Model Management Life Cycle Model.

In summary, the split the whole large-scale project to the "project chain" will create the factor for better manageability of the large-scale buildings project and increase the efficiency of the activities of all participants in such projects.

References

1. The plan of measures ("road map") "Development of agglomerations in the Russian Federation: The Economic Development Ministry of the Russian Federation. 19.11.2015.
2. Priority project passport "Safe and quality roads" (approved by the Presidium of the presidential Council of the Russian Federation for strategic development and priority projects (Protocol of 21 November 2016. № 10.
3. URL: https://www.emporis.com/building/standard/3/high-rise-building
4. Yu. Granik. Energy saving, 2, 92 (2004)
5. URL: http://docs.cntd.ru/document/1200042296
6. Russian State Standard TSN 31-332-2006
7. Standards of urban design, Moscow region (app. the resolution of the Government of Moscow region from August 17, 2015 N 713/30).
8. URL: http://blog.trestlescs.com/an-overview-of-lean-construction
9. Project Management Institute: Pulse of the Profession 2014 – The High Cost of Low Performance/
10. URL: https://assets.kpmg.com/content/dam/kpmg/pdf/2015/04/Global-Construction-Survey.pdf
11. URL: http://blog.trestlescs.com/an-overview-of-lean-construction
12. The Construction Industry Institute. (1999). Project Definition Rating Index (PDRI)—Building Projects. Implementation Resource 155-2, Austin, TX.
13. Wang, Y.-R. & Gibson, G. E. (2002). Using PDRI for project risk management: an empirical study of current practices. Paper presented at PMI® Research Conference 2002: Frontiers of Project Management Research and Applications, Seattle, Washington. Newtown Square, PA: Project Management Institute.
14. URL: http://vestnikmckinsey.ru/transport-infrastructure-and-logistics/krupnyhe-proektyh-klyuchevyhe-faktoryh-uspekha
15. I. Ilin., Kalinina O.V., Levina A.I., Iliashenko O.Yu. MATEC Web of Conferences, **83**, 2016. Article number 05028.
16. URL: http://www.pmsolutions.com/audio/Strategies_for_Project_Recovery_Research_Report.pdf
17. URL: http://proestate.pro/images/proestate/2017/prezentacii/are1/Gennadij-Kirkin_Irbis.pdf
18. V. Glukhov et al Lecture Notes in Computer Science. 2014. **8638** LNCS. Pp. 509-518.
19. I. Ilin, Anisiforov A.B. WSEAS Transactions on Business and Economics. 2014. 11. Pp. 757-764.
20. The Construction Industry Institute. (1995). Pre-Project Planning Handbook. Special Publication 39-2, Austin, TX.
21. R. Wysocki. *Effective Project Management: Traditional, Agile, Extreme.* Seventh Edition. (John Wiley & Sons, Inc., 2014)
22. D. Butin, A. Burmistrov. The life cycle model of the project: comparative analysis / Collection: Week of science SPbSPU materials of scientific conference with international participation. Polytechnic University of Peter the Great, 2016. P.310-312.

23. V. Schwetsov, A. Burmistrov, V. Morozov. Stages of Modernization an Effective Transport Management System in Russian Agglomerations: Invitation to Discussion , Collection: Transport planning and modelling: II International scientific-practical conference; SPbGASU. – SPb2017. – 341 p. P.298-307

Calculation of high-rise construction limitations for non-resident housing fund in megacities

Oksana Iliashenko[1], *Sergey* Krasnov[2] , and *Sergey* Sergeev [1]

[1] Peter the Great St.Petersburg Polytechnic University, Russia, 195251, St.Petersburg, Polytechnicheskaya, 29 ilyashenko_oyu@spbstu.ru

Abstract. The paper is devoted to topical issues of urban planning in terms of high-rise construction of a non-resident housing stock in relation to megacities. We consider this issue taking into account the limitations of natural, communal and social resources. The problem is especially acute for the overwhelming majority of the state capitals, as well as cities with historical heritage that are of great interest due to the rapid development of tourism and the high mobility of the population in the world. The growth of the population of many states capitals led to the use of high-rise buildings as a non-resident housing stock. However, there are a number of restrictions on the high-rise construction of non-resident housing stock in megacities. The authors formalize the problem of determining the optimal ratio of the volume of urban buildings belonging to the high-rise buildings types and intended for non-residents to a common housing fund. We conduct economic calculations to determine the quantitative indicators. It can be used as the basis for administrative measures aimed at limiting the people flow arriving with the intention of temporarily deploying in megacities.

1 Introduction

Urban planning, as a rule, covers a significant time horizon. At the same time, a set of tasks related to practical functions is solved, where the arguments are economic indicators [1, 2] demography, ecology and limited environmental resources intended for common use.

Indicator's analysis such as the population size, the volume of housing construction carried out among the major cities of the world shows that the list of cities from Alpha to Alpha ++, coincides with the participants in the ranking of the most attractive places for tourists. For comparison, according to the data of the Globalization and World Cities Research Network, the Global city table as the most important nodes of the world economy, ranked in descending order Alpha indicator and the List of the most visited cities ranked according to the data provided by the research company Euromonitor. For such megacities is characterized by the presence of an increased ratio of the volume of residential premises (hotels, hostels, tourist hostels, etc.) for non-residents to the total area of the apartment fund. In addition, over the past few years, through the penetration of the concept of Web 3.0., the world's B-corporation networks have emerged, concentrating on online sites such as Airbnb, CouchSurfing, a powerful attraction potential for travelers. They take advantage of the integration with social networks such as facebook, as well as a wide range of mobile applications for smartphones on common platforms iOS and Android. At the same time, a global information and technology service field is implemented to simplify travel and accommodation in any city in the world. [3] In the number of persons temporarily staying in the selected territory, it is also necessary to include the category of citizens migrating for the period of temporary earnings (Table 1)

Table 1. Comparison of data on global cities

NN	Global City	Attractiveness
1	New York*	Hong Kong*
2	London*	Bangkok
3	Singapore	London*
4	Hong Kong*	Singapore
5	Paris*	Paris*
6	Beijing	Dubai*
7	Tokyo*	New York*
8	Dubai*	Kuala Lumpur
9	Shanghai*	Rome
10	Sydney	Taipei
11	São Paulo	Seoul*
12	Milan	Tokyo*
13	Chicago	Prague
14	Mexico City*	Delhi
15	Mumbai*	Vienna
16	Moscow*	Berlin
17	Frankfort	Madrid
18	Madrid	Moscow*
19	Warsaw	Beijing
20	Johannesburg	Athens
21	Toronto	Budapest
22	Seoul*	Lisbon
23	Saint Petersburg	Saint Petersburg

Thus, it becomes possible to formulate the research task. It is necessary to model [4] the use of the limited potential available to the city. At the same time, its composition can be both natural (environment, air, water), and technological (utilities, transport), as well as social (medicine, culture, law and order). As a result, it is necessary to find the maximum number of the total population, both permanently resident and non-residents, based on the potential limits of the loaded resources [5] of public use.

1 Materials and Methods

The solution of the problem will be sought using the methods of mathematical modeling [6], using the economic applications of game theory (Game theory), a series of results of Queuing theory and auxiliary formulas for describing stochastic processes from the economic point of view.

We introduce a number of necessary formalisms, which make it possible to uniquely identify the parameters [7] of the investigated processes. We define common resource as CR_i. We set $i = 1...\Omega$, where Ω – the amount of common resources concentrated in the territory under consideration. First of all, high-rise buildings that accumulate the main stream of non-resident population and transport are included here. High-rise buildings can have different purposes: hotels, offices, residential buildings. Most often, the high-rise building is multifunctional, in it, in addition to the main premises, parking lots, shops, offices, cinemas, etc. are located. Most of the non-resident housing stock is concentrated in high-rise buildings [8]. Considering transport, it is necessary to separate communications with external and internal environment. The category of transport that implements external communications includes railway, air communication, passenger port, radial roads. The second type is represented by urban traffic, metro, public transport.

The next group CR includes utilities such as sewage, electricity [9], water, fire, medical facilities, garbage collection. The socially significant resources should also include natural, characterized by indicators of air pollution, water bodies, the presence of space in parks, the density of the population in the city. Another resource is telecommunications. An advanced telecommunications network is one of the most important indicators of the quality of services provided to visitors to megacities [10].

It is necessary to supplement these data with the following specific types of resources considered in the mathematical model [11]. All the cities listed in Table 1 are characterized by a high concentration of tourist attractions, which are concentrated in a limited area. Such places of attraction are called POI (place of interest). Usually it is either a historical center of the city, which is typical for most European cities, or cultural objects (theaters, museums) or places of world-wide events (olympiads, military battles). We introduce the following notation:

λ_i - intensity of user flow $i - CR$, where $i = 1...\Omega$

μ_i - capacity indicator $i - CR$, где $i = 1...\Omega$

Then, using Kendall's symbols for the model parameters in the form: $M/M/n/Q$ we can calculate the average time spent in finding and using CR_i:

$$\overline{T} = \overline{R}_i / \lambda_i, \text{ where } \overline{R}_i = \frac{\rho_i^2(1-(M+1-M\rho_i)\rho_i^M)}{(1-\rho_i)(1-\rho_i^{M+2})},\ \rho_i = \frac{\lambda_i}{\mu_i},\ i = 1...\Omega$$

M – acceptable queue length for this resource. For this, we used the well-known solution of the Erlang equations.

Now formally imagine the city as a set of resource users. For the issues considered in this problem, it is convenient to present the representation in a consolidated form, as a set of houses, hotels, hostels. We denote their number by n. Denote the load of each user as l_i where $i = 1..n$, then in vector form we can write down:

$$\overline{\Lambda} = (l_1, l_2, ..., l_n) \tag{1}$$

while the total load L on the resource is:

$$L = \sum_{i=1}^{n} l_i$$

Next, we will establish the value of the weighted average variable costs U, and a function $g(l)$. The function reflects the value of the revenue per unit of the load of each user. we suppose that with the total load starting from a certain number $L > L_0$, the profitability of the resource falls both in connection with the activities of competitors, and because of the saturation of the market. This can be expressed by the relation:

$$g'(l) < 0 \tag{2}$$

Further, for the same reasons, we add factors related to the negative component of the impact of the activities of these n users on the resource and its natural limitations. It follows that the second derivative:

$$g''(l) < 0.$$

We write the ratio for calculating the i-th user's winnings as:

$$R_i = l_i\, g(l_1 + l_2 + ... + l_n) - U\, l_i = l_i\, g(L) - U\, l_i \tag{3}$$

Then, assuming the existence of Nash equilibrium [11], there must exist a load value of the i-th user under which the value of expression (3) reaches a maximum with the remaining components of the vector (1):

$$\overline{\Lambda}_i^*(l_1^*, l_2^*, ..., l_{i-1}^*, l_{i+1}^*, ..., l_n^*)$$

We shall seek the solution of the problem by taking the partial derivatives $\frac{\partial R_i}{\partial l_i}$ from the extremum condition:

$$\frac{\partial R_i}{\partial l_i} = 0 \text{ for } i = 1...n.$$

Further, denoting as $l_{-i}^* = \sum_{k \neq i} l_k^*$ we obtain the following expression:

$$g(l_i + l_{-i}^*) + l_i g'(l_i + l_{-i}^*) - U = 0 \text{ for } i = 1...n.$$

Summing up the values at the desired equilibrium point L^*, we obtain the equation:

$$L^* = n\frac{U - g(L^*)}{g'(L^*)}.$$

If we compare this expression with the search for a maximum from the point of view of the level L_0 - the optimal use of the resource by the city, then the extremum of the equation will be achieved if the following relation is satisfied:

$$L_0 = \frac{U - g(L_0)}{g'(L_0)}$$

Taking into account (2) and the fact $n > 1$, it's obvious that $L^* > L_0$, а с учетом (2) и того, что $n > 1$, ясно, что $L^* > L_0$. The meaning of the results you can see in calculations on real data.

2 Results

To illustrate the above procedure, we will calculate the above formulas using specific values obtained from the analysis of the municipal economy of St. Petersburg. From Russian cities, this is one of the most sought after points for travel of domestic and foreign tourists.

This is confirmed by a survey of UNESCO and the WTO, according to which the northern capital is among the most attractive cities in the world in terms of its potential. Such a flow of visitors imposes a number of requirements on the organization of city services and provides a wide field for the development of optimal [12] solutions for the planning and implementation of all types of business associated with temporary stay in the city. In the economic theory, this problem has been developed insufficiently both in the general methodological aspect and in the field of practical implementation.

To conduct analysis and formulate the commercial policy of the city administration on an extended horizon, it is necessary to apply well-developed mathematical methods and models [13].

When compiling them, it is necessary to take into account a number of factors:

- uneven demand for this type of services, not only related to annual seasonality, but also such periods as school holidays, traditions, public holidays;
- provide for the coordination of simulation results with GDS (Global Distribution System) systems, CRS (computer reservations system);
- business planning must comply with USALI (The Uniform System of Accounts for the Lodging Industry) standards;
- modeling should provide for the implementation of services not only for individual customers, but also wholesale consumers, as well as in the MICE (Meetings, Incentives, Conferences, Exhibitions) and TMC (Tourism Marketing Concepts) segments.

Let's consider separately each of the listed items.

The study of issues related to the modeling of commercial processes should be carried out, on the one hand, by visual methods, which implies the use of simplified schemes. On the other hand, it is necessary to use the capabilities of computer technology and the application of tools of mathematics and statistics. As in any commercial segment, planning for the reception of visitors to the city is impossible without reliable estimates of future market needs. First of all, this is due to the fact that such an organization as, for example, a hotel, is closing in a long chain consisting of advertising, travel companies and intermediaries, each link being characterized by its own lag. Additionally, in such a business there is the influence of seasonality and, as already mentioned, there is still a correlation with socially significant events, vacations and other known market factors.

Mathematics provides a number of ways to predict future values of a process based on extrapolation of known results. The simplest way is regression and trend analysis [14], but the polynomial trend realized in mass office products does not allow to build analytical models. Next, it should be noted that the cycle for any type of service is not necessarily a one-off (maybe several seasons a year). Intervals between cycles within a year are also not necessarily the same, that is, there is usually a non-uniform frequency characteristic of an asymmetric shape [15]. The initial data will be considered indicators of the number of arriving citizens for a fairly long period of time. It is desirable that it includes at least two complete annual cycles. For certainty, we will assume that the data is received monthly, which is consistent with the tax accounting. Using the possibilities of programming on a computer, one can obtain by simple means a function that describes the data on the number of visitors for the previous period.

It is known that a periodic function can be represented as a Fourier series, then, adding it to a sufficient planning horizon, it is possible to continue calculations based on the data already obtained. Formally, it can be represented:

$$F(x) = A_0 + \sum_{k=1}^{+\infty} A_k Cos(2\pi \frac{k}{T} x + \theta_k),$$

where $F(x)$ - function, to calculate the number of visitors, A_0 - constant characterizing the displacement, A_k - amplitude of the k-th harmonic oscillation, $2\pi \frac{k}{T}$ - circular frequency of harmonic oscillation, θ_k - initial phase of the k-th oscillation. Our task is to determine the unknown coefficients of the Fourier expansion A_0, A_k, θ_k, T.

Since a non-linear model is needed for the solution, it is necessary to add the corresponding Solver for Nonlinear Programming [9]. In order to find a solution with data on the volume of guest flow by months, and the period can naturally be taken to be one year, then, for example, the expression for the first harmonic of the Fourier expansion looks as follows:

$$F(m) = A_0 + A_1 Cos(2\pi \frac{1}{12}(m + \theta_1)),$$

where m – month number (tabular, not calendar). In addition, the target value and optimization criterion will be needed [16]. As a criterion, the minimum standard deviation of the calculated data is taken. The target value is also calculated using Solver for Nonlinear Programming. When choosing either DEPS Evolutionary Algorithm or SCO Evolutionary Algorithm, we proceed from the fact that DEPS (Differential Evolution and Particle Swarm Optimization) consists of two independent algorithms and represents each vector of the solution as coordinates in the dimension space of the arguments, simultaneously in each iteration choosing one of the two strategies and applying to the current solution vector. SCO (Social Cognitive Optimization) mimics the behavior of learning and sharing information. At each iteration searches for the solution, comparing it with the data already received. As a result, we already obtain in the analytical form a formalized expression for the intensity of resources. The coefficients A_0, A_k, θ_k, T will be used for further calculations.

Using data on the tourist flow to St. Petersburg, the capacity of the POI, as a result of calculations on a computer, we obtain data on the intensity of the load, as well as a forecast for saturation of the city's capabilities as a visiting site from the condition of loss of attractiveness of its sights.

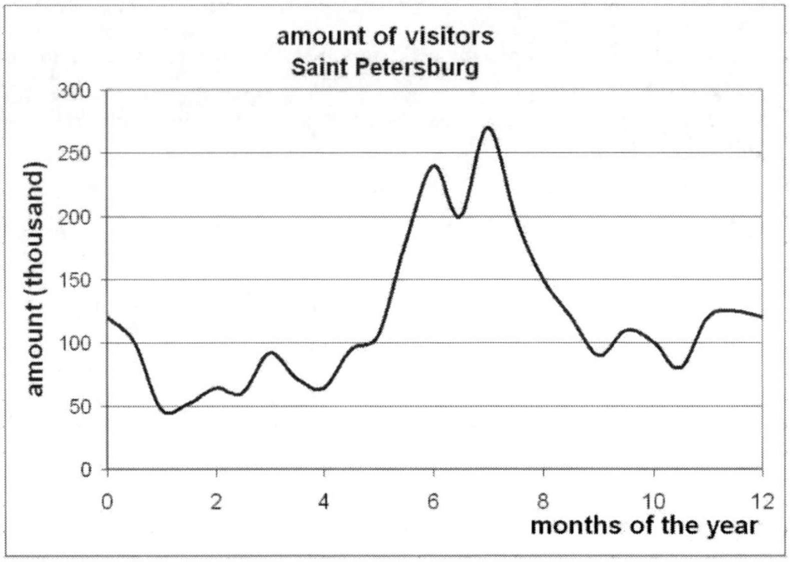

Fig. 1. Calculating the intensity of visiting the city.

Figure 2 shows data on the degree of satisfaction of requests. Its assessment is carried out as an indicator of how many POIs in reality managed to visit a visitor. The graph shows the percentage of the actual value of the POI from the planned one. Note that the data are averaged, for specific dates it is necessary to take into account the unevenness of the graph of Figure 1, with an inversely proportional dependence. Figure 2 also indicates the current level of ratio of visitors to the resident population.

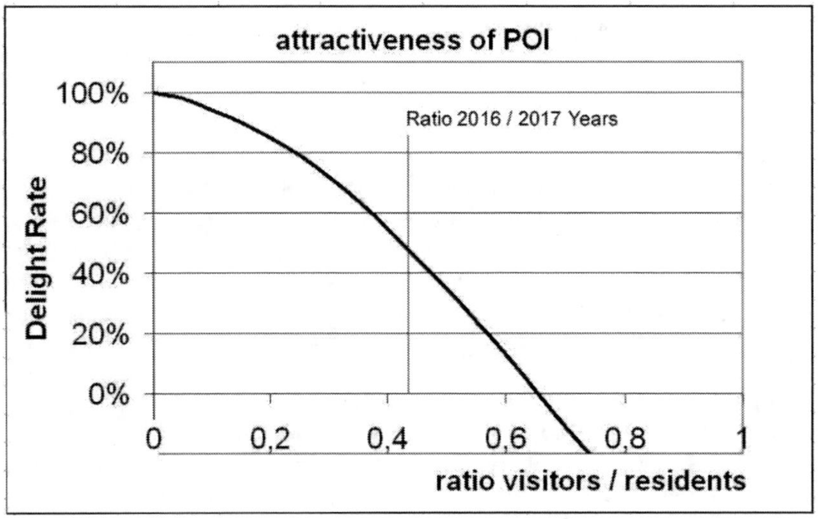

Fig 2. Level of satisfaction with requests.

The final result of the calculation should be a set of recommendations for the optimal reception of visitors. Turning to world experience, it can be concluded that the most common approach is to apply restrictive measures. It is known that there is usually no technical solution here, and it is within the competence of local authorities to establish rules of the game, administrative regulation. In this case, mathematical modeling serves to make recommendations based on the forecast for the adoption of these legal or other measures that will be made during the next phase of business in the hospitality industry by constraints formalized in the compilation of dynamic equations.

3 Discussion

In Table 1, an asterisk indicates cities in which, according to the information of GaWC, there is a maximum number of foreigners, the need to increase the share of non-resident housing stock, the most intense passenger traffic by main types of transport and the largest population. At the same time, their authorities were the first to introduce restrictions.

There were many interrelated reasons for their initiation. The restriction was introduced to tourists visiting historical sites in order to prevent their destruction, as well as restrict access to a number of unique natural sites. Another serious limitation was the lack of hotels in demanded seasonal visit periods. Another circumstance was transport problems, traffic jams, and a decrease in the average speed of traffic. The listed cities are at the same time the most important transport hubs, a similar capacity limit is experienced by seaports due to cruise ships and air harbors. One of the most tangible results was a sharp increase in air pollution in these densely populated cities. But the most significant reason for the authorities is the discontent of resident residents, the creation of hotbeds of social tension due to the limited resource of public order services.

As a result, the paradigm for improving the quality of life of visitors was formulated. An effective solution was the introduction of an environmental tax on visitors. Its size is differentiated depending on the level of the hotel. In order to exclude unaccounted visitors, there are large penalties for using resources such as Airbnb, CouchSurfing. In addition, the city authorities impose restrictions on the construction of new hotels, located in separate buildings. As an alternative, we propose to place hotels in high-rise buildings with a developed infrastructure.

Figure 3 shows the calculation of the optimal (averaged) Ecotax. The graph represents the calculations as theoretical, from the formulas based on the model considered in this paper, and obtained from real data.

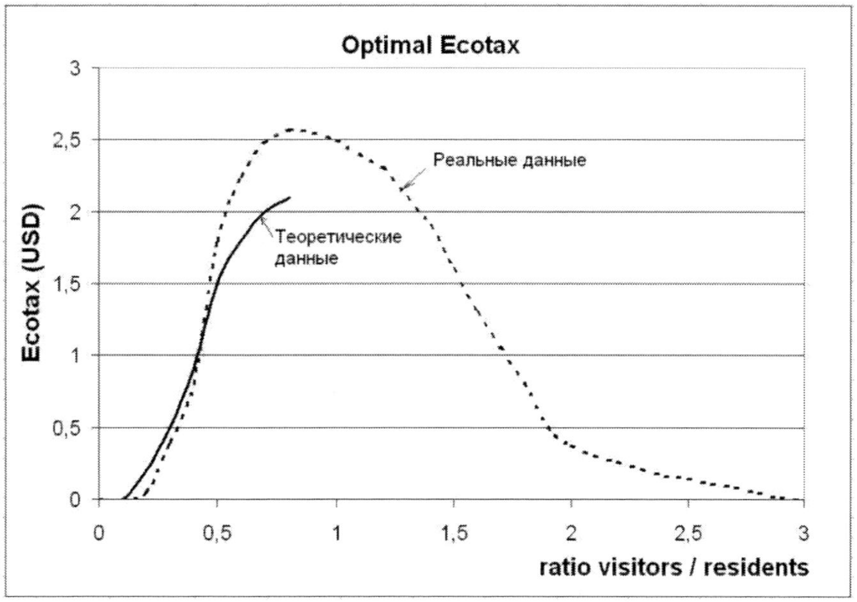

Fig. 3. Calculation of the optimal value Ecotax.

Note that the theoretical calculation covers only part of the possible ratio of the number of visitors to residents. On the contrary, real data show a strongly different character at large values of this ratio. This is explained by the reasons that can't be taken into account in the model and are of a social nature. First of all, this refers to the regions in which business on tourists is the main share of the economy. For their populations, a maximum influx of visitors is required during the high season, since the rest of the year most residents will not have other earnings.

4 Conclusions

In the conduct of economic activity, most often competing in one territory between business entities, over time factors appearing relating to the sphere of universal values. Since hotels in attractive cities are concentrated around historic centers with attractions, the problem of using this resource is very acute in Madrid, Barcelona, Rome, Venice, Tallinn, St. Petersburg and dozens of other landmark places for travelers.

Theoretical and methodological provisions for the formation of a forecasting system that interacts with the current and prospective management of hotel enterprises in a single complex that determines the target orientation and ensures the reality of achieving the necessary economic and social results in the development of hotels have not been adequately developed in the economic literature.

From the standpoint of the business analyst, who forms theoretical models, multidirectional processes must be taken into account in the problem under consideration. This is an increase in the burden on resources and the resulting decline in the quality of these types of common resources, while simultaneously assessing the negative component of the impact of business processes.

Various socio-economic factors necessitate the search for various measures to address the issues of regulating the flow of tourists in megacities.

1. Administrative regulation of this business segment is necessary. Many cities are already solving this problem. One of the ways to solve the problems is to develop mechanisms for regulating the volumes of non-resident housing stock. One such mechanism is the restriction in the construction of new buildings for hotels and hotel accommodation in high-rise buildings. At the same time, it is necessary to take into account a number of restrictions on high-rise construction of non-resident housing stock in megacities. Another mechanism is the introduction of an environmental fee or tourist tax. This process is everywhere, in many cities and countries, and the figures vary significantly, but the general principle remains the same - the amount of tax should balance between the decrease in the number of tourists and the damage from the increase in the intensity of the flow of guests. The methods outlined in this paper make it possible to scientifically justify the dynamics of these multidirectional processes and simulate the main parameters of such interaction on the planning horizon in 2-3 years.
2. The usual static modeling and stationary solution for forecasting is not enough. It is necessary to use computer calculations on dynamic models with the use of scientific methods based on serious mathematical methods.

References

1. I.V. Ilin, A.I. Levina, O.Yu. Iliashenko, MATEC Web of Conf, **86**, 05028 (2016) DOi - 10.1051/matecconf/20168605028
2. I.V. Ilin, A.I. Levina, O.Yu. Iliashenko, MATEC Web of Conf., **106**, 08066 (2017) DOi-10.1051/matecconf/201710608066

3. A.N. Petrov, News of the St. Petersburg State Economic University. (2016). **No. 2 (98).** - p. 74-80.
4. V. Kankhva, IOP Conference Series: Earth and Environmental Science, Vol 90, 012175 (2017) DOI - 10.1088/1755-1315/90/1/012175
5. R. Gilemkhanov, R. Bagautdinov, V. Kankhva, Advances in Intelligent Systems and Computing, Vol. 692, pp. 1036 - 1045 (2018) DOI- 10.1007/978-3-319-70987-1_112
6. A.Pustovgar, A. Adamtsevich, L. Shilova, Advances in Intelligent Systems and Computing, **692**, pp. 608 - 614 (2018) DOI - 10.1007/978-3-319-70987-1_64
7. A.O. Adamtsevich, A.P. Pustovgar, S.A. Pashkevich, A.V. Eremin, International Journal of Applied Engineering Research, Vol. 11, pp. 1609 - 1611 (2016)
8. G. Binder. *Sky High Living: contemporary high-rise apartment and mixed-use building*. Tall Buildings. (2006).
9. V. Glukhov, Ilin I.V. Lecture Notes in Computer Science. (2014). **Volume 8638** LNCS. Pp. 509-518.
10. I.Ilin, Levina, A.I., Iliashenko, O.Yu. *Enterprise architecture approach to mining companies engineering*. (2017). MATEC Web of Conferences, **106**, 0806
11. E. Istomin E. P., Sokolov A. G., Abramov V. M., Gogoberidze G. G., Popov N. N. (2015) International Multidisciplinary Scientific GeoConference Surveying Geology and Mining Ecology Management, SGEM..
12. A. Anisiforov Proceedings of the 29th International Business Information Management Association Conference - Education Excellence and Innovation Management through Vision 2020: From Regional Development Sustainability to Global Economic Growth.
13. S.M. Sergeev Economics. Society: Selected Papers of the International Scientific School "Paradigma" (Summer-2015, Varna, Bulgaria) Compiling Editor Dr.Sc., Prof. E.Sibirskaya. Yelm, WA, USA, 2015. p. 38-41.
14. S. Sergeev Selected Papers of the International Scientific School "Paradigma" Winter-2016 (Varna, Bulgaria) Compiling Editor Dr.Sc., Prof. O.Ja. Kravets. Yelm, WA, USA, 2016. p. 95-102.
15. S. Sergeev International Scientific School "Paradigma" Summer-2016 Selected Papers. Yelm, WA, USA, 2016. p. 92-97.
16. S. Sergeev Proceedings of the XXIII-th International Open Science Conference (Yelm, WA, USA, January 2018). Science Book Publishing House. Yelm, WA, USA (2018). p.169-175.

IT Solution concept development for tracking and analyzing the labor effectiveness of employees

Igor Ilin , *Svetlana* Shirokova, *Aleksandr* Lepekhin

Peter the Great St. Petersburg Polytechnic University, Institute of Industrial management, Economy and Trade, 195251, Polytechnicheskaya, 29, St.Petersburg, Russia

Abstract. Labor efficiency and productivity of employees is an important aspect for the environment within any type of organization. This is particularly crucial factor for the companies, if which operations are associated with physical labor, such as construction companies. Productivity and efficiency are both very complicated concepts and a huge variety of methods and approaches to its analysis can be implemented within the organization. Despite that, it is important to choose the methods, which not only analyze the key performance indicators of employee, but take into account personal indicators, which might affect performance even more than professional skills. For this complicated analysis task it is important to build IT solution for tracking and analyzing of the labor effectiveness. The concept for designing this IT solution is proposed in the current research.

1 Introduction

All around the world, large companies are trying to provide an appropriate combination of approaches for increasing the effectiveness of staff (for example, by increasing labor productivity) and reducing injuries of employees.

In high-rise construction, this problem is most acute, as they appear to be complex systems with different types of internal challenges [1]. On one hand, It is necessary to secure the employees, to predict the state of their ailment. On another hand, it is important to increase the effectiveness of their activities, interactions and day-to-day operations.

It is not surprising that currently this topic is especially relevant, because increasing the efficiency of workers means new opportunities for the company, while the competitive environment puts considerable pressure on the company's development [2]. Thus, it is necessary to propose different approaches to the analysis of efficiency and productivity of labor in a direct relationship with the continuous analysis of the worker's physical condition and compare these indicators.

This research paper addresses:

• Different approaches to the analysis of the efficiency and productivity of working process,

• Comparative description of the approaches and the proposal of necessity of developing an IT solution using the most appropriate approach,

• The necessity of implementation of a monitoring module for analysis of the physical condition of the employee.

The goal of the research is the development of an integrated approach as the basis for an IT solution that combines an analysis of approaches to the evaluation of productivity in combination with methods for assessing the capabilities and physical condition of staff [3-8]. The research structure is the following. The second chapter addresses the analysis of different approaches to evaluation of labor productivity. The third chapter analyses the applicability of Biological feedback approach to tracking and improving the labor productivity. The Discussion and Conclusion parts summarize the results of the current research.

2 Materials and methods

Current chapter presents an analysis of approaches to the evaluation of labor productivity, which aims to increase the efficiency and productivity of daily activities of employees to further create an IT solution that uses the most appropriate method.

The research methodology consists of:

• identifying of various existing approaches to the evaluation of labor productivity;
• conducting a direct analysis with the identification of key features;
• providing the choice of approach to assessing the performance of intellectual work.

When choosing approaches to solving scientific and technical problems, one of the main criteria is the ability of one or another approach to create value for consumers of products by increasing labor productivity [3-5].

Another criterion is the theoretical basis of the approach, the existence of a practical experience confirming the practical feasibility of the method.

A number of studies have been carried out, on the basis of which it is possible to optimize labor productivity. Labor productivity analysis can be performed using several approaches, such as:

• Economic approach,
• Biometric approach,
• Analysis of external factors (manifestations of human activities),
• Biological feedback approach (BFD).

Figure 1 demonstrates the reasons for choosing an approach to analyzing labor productivity [6].

Each approach was studied and its concept was disclosed.

The economic essence of labor productivity lies in the fact that it is represented as a function of real labor, which means that it is an indicator of evaluation and a criterion for the efficiency of production. The growth of labor productivity allows increasing production volumes and creating conditions for the further development of the economy [3].

Balanced Scorecard (BSC) is one of the most popular and universally accepted concept of managing the implementation of the strategy developed by the professors of Harvard University D. Norton and R. Kaplan (USA). BSC - is a mechanism for the consistent delivery of strategic success factors to the staff, the company's goals and monitoring of their achievement through key performance indicators.

KPI is an indicator of the attainability of the goals set, as well as the characteristics of the effectiveness of business processes and the work of each individual employee. KPI and staff motivation make it possible to implement the company's strategy, described and formalized using of BSC [10].

This approach involves measuring the economic performance of the organization, while labor productivity is treated as an economic indicator that characterizes the participation of an individual employee in the company's financial performance.

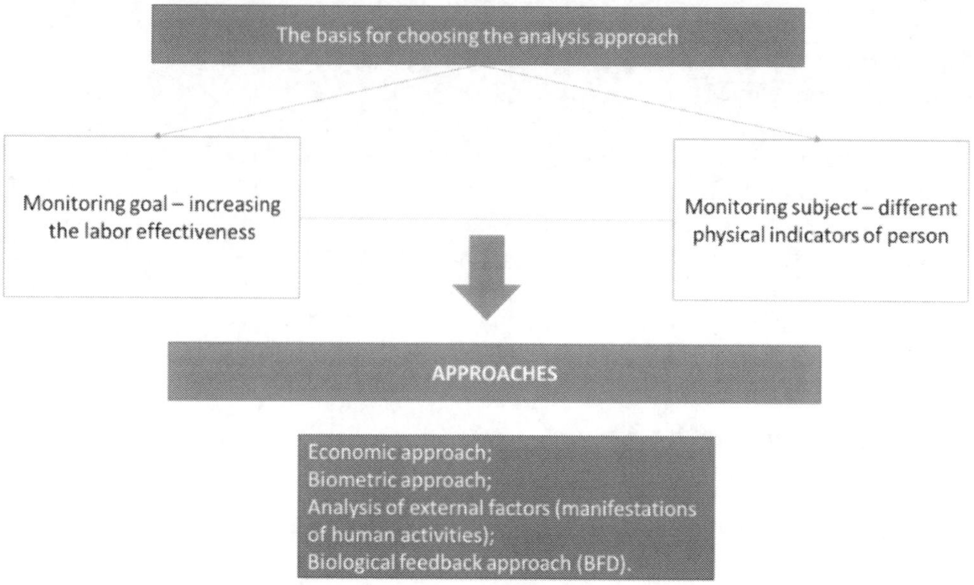

Fig. 1. Approaches to the analysis of the labor productivity.

However, this approach does not allow to optimize the individual daily activities of each employee. KPIs are used at a higher level of detail, both in terms of the organizational hierarchy and in terms of the planning horizon.

To track the individual contribution of the employee and monitor the personal effectiveness, measurements of a number of biometric parameters are used. The characteristics of biometric indicators are:
• Measurability,
• Resilience to the environment,
• Resistance to forgery,
• Accuracy of recognition.

The sources of information for this approach could be:
• Electroencephalogram,
• Electrical activity of the skin,
• Thermometry (surface temperature of the body),
• Photoplethysmogram and other.

Having obtained all the necessary data from the research, it will be necessary to create an information system that processes the data obtained and issues appropriate recommendations. For the procedure of complex examination, it is possible to use methods of organizing complex examinations [7,10,11].

The problem is the need to compare the requirements for increasing the efficiency of labor, on the one hand, and the possibility and physical condition of the employee, on the other. These problems are associated with the need to analyze the effectiveness of approaches to assessing performance in conjunction with the assessment of physical state. The quantitative evaluation criteria are normally heterogeneous, so the problem of comparability of criteria arises.

As a result, there arises the need to create a system for organizing a complex examination based on the use of structuring methods that divide a large initial uncertainty into more visible parts. The information approach can be used to obtain estimates of the degree of influence of the approach to the performance evaluation taking into account a specific situation and to bring heterogeneous criteria (quantitative and qualitative) to unified information units for further comparison and analysis.

The Figure 2 illustrates an example of the assessment of the requirements for improving productivity, using different approaches, taking into account the physical characteristics of employees.

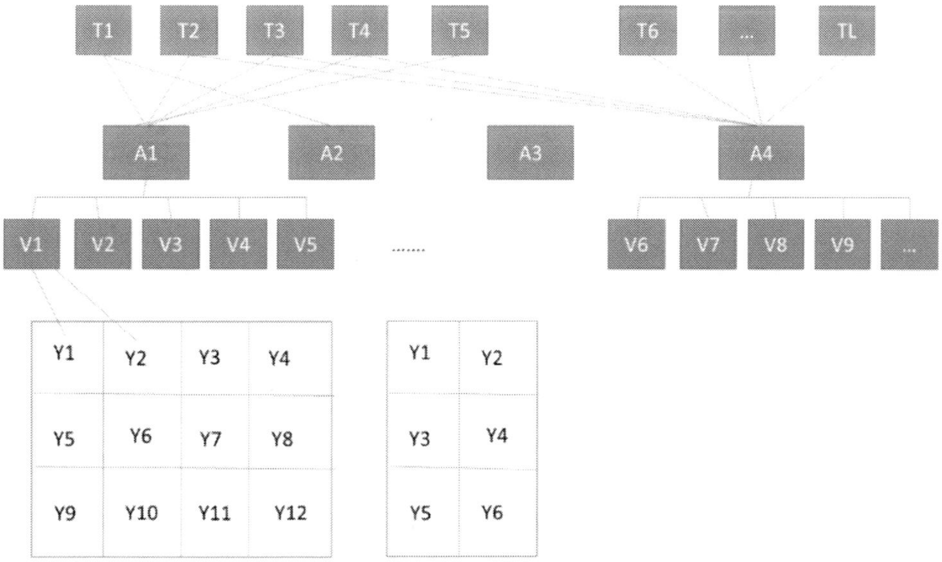

Legend: A1—A4 – approaches to the evaluation of labor productivity
T1---TL – parameters for assessing the physical condition of the employee - quantitative and qualitative;
V1---VZ - groups used in the approach to assessing performance
Y1---Yw components (a set of parameters) for assessing the physical state

Fig. 2. The method of organizing a complex examination, based on the information approach.

For the evaluation, you the quantitative criteria given in the upper part of Figure 2 can be used. But the qualitative characteristics are important, they can only be quantified only partially.

When assessing by qualitative criteria, pi' is determined as a degree of influence of i-parameter of evaluation of a person's condition or taking into account the use of the i-parameter when assessing the physical state in a situation.

3 Results

Analysis based on external factors involves a set of methods and tools that allow you to register and process data from the external manifestation of human activity. Data can be used:

- Movement activities of the user,
- Electromyograms,
- Oculograms,
- And other.

The investigated parameters can be used for analysis of mental activity, emotional stress, employee fatigue.

Audit of user activity is possible with the help of specially developed software applications, which have the following functions:

1. Register a record of user actions in the storage of space-time data of the system;
2. Form arbitrary reporting based on accumulated data

Biological feedback is a method of changing the psychophysiological state of a person. It is based on an individual approach, the choice of an individual standard of the optimal psychophysiological state and on this basis the definition and current correction of the tactics of its change [7].

Biological feedback aims to track and analyze the following indicators as well as their influence the overall labor intensity and effectiveness:

• Heart rate,

• Breathing rate,

• Skin galvanic reaction.

Table 1 demonstrates the summary analysis of different approaches.

Table 1. Strong and weak points for labor intensity analysis.

	Economic approach	Biometric approach	Analysis of external factors	Biological feedback
Strong points	KPI implementation	Monitors personal effectiveness	A set of methods and tools that allow processing the data of the external manifestation of human activity is used.	The active participation of a person in his training in the management of the functions of his own organism.
Weak points	It does not allow to optimize the individual daily activities of each employee;	Only physical indicators of employees are taken into account, all other aspects are ignored	There may be a feeling of "excessive observation"	Not detected

Inclusion of these models in the procedure of integrated assessment made it possible to formulate a procedure for determining the effectiveness of activities taking into account the physical condition of a person.

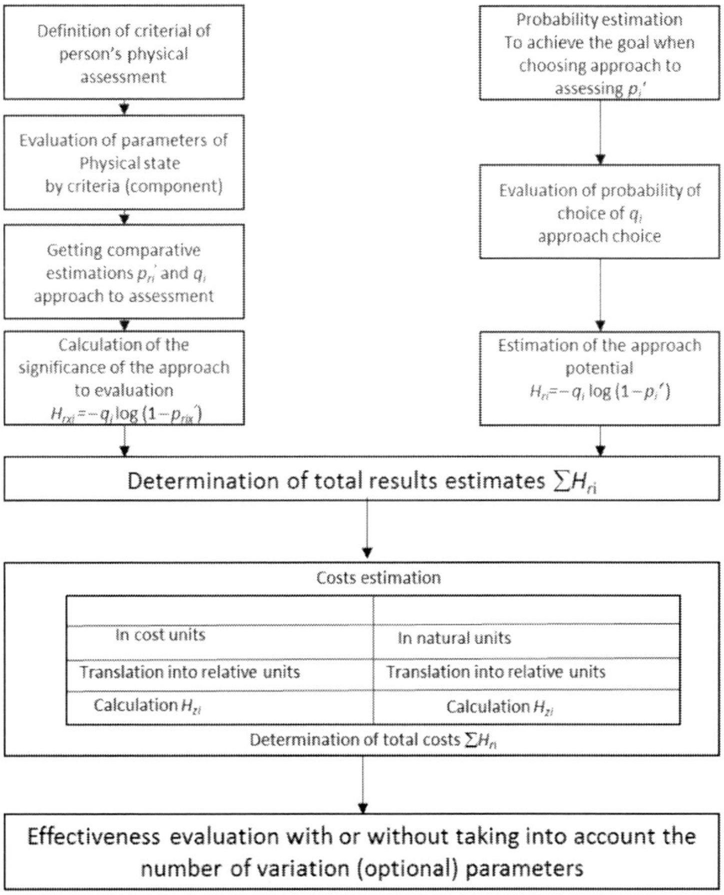

Fig. 3. The algorithm of the integrated assessment procedure.

The proposed algorithm of Hi estimation based on pi results in the following advantages of the proposed approach:

• Not only p_i' (degree of influence) can be taken into account, but also q_i – The probability of taking this parameter into account in the complex estimation,

• Simplification to obtain generalized estimates of the impact of options or their components on the implementation of sub-goals.

The developed integrated assessment procedure is the basis for developing an IT solution. Joint use of methods of organizing complex examinations and BFD technologies allows one to effectively approach the solution of the problem of increasing the efficiency of labor taking into account the physical condition of the employee in this situation. This allows you to develop a software and hardware system.

4 Discussion

Evaluating the information presented in different studies and taking into account the problem nature, it can be concluded that the convenient approach to track personal labor intensity is the Biological feedback approach. The method provides an instrument for the disclosure and use of body reserves, organically fits into any scheme for improving the quality of life and improving health. The use of Biological feedback technologies is a promising way to increase labor productivity, since it allows to regulate the cognitive load and the psycho-emotional state of a person. This aspect is rarely addressed in the existing literature yet. Few research paper proposed an importance of personal indicators analysis and their influence of work performance and, as a result, on productivity and organizational environment [12]. Some studies even focus on relation between health risks and work productivity, which means that there is potential for further research and analysis [13]. The systematic assessment, proposed in this paper, could enable a comprehensive assessment of productivity and efficiency, taking into account the physical state and using biological feedback. This approach could also complement different approaches to enterprise development, such as enterprise architecture projects, in terms of human resources development [14].

The development of IT solutions using this approach might improve the efficiency and productivity of work. The design and development of IT services for tracking personal indicators could become the direction for future research.

5 Conclusions

The research paper addressed the problem of labor productivity and efficiency, which is specifically important for construction companies because of high labor intensity among employees. It was proposed that in order to increase the efficiency and productivity, it is particularly important to track and analyze different indicators of employee's performance. Different methods were studied and the concept of algorithm of the integrated assessment procedure was proposed. This concept summarizes different personal indicators and methods of their tracking and analysis. The proposed ideas could further become the basis for IT services development, which would have potential for tracking, analyzing and increasing labor efficiency.

References

1. I. Ilin I.V., et al, MATEC Web of Conferences. **83**. 05028 (2016)
2. A.Bril A.R., Kalinina O.V., Ilin I.V. (2017) MATEC Web of Conferences, 106, 08010, International Science Conference SPbWOSCE-2016 “SMART City”.
3. M.Koch, M. J., & McGrath, R. G. (1996). Improving labor productivity: Human resource management policies do matter. Strategic management journal, 335-354.
4. V. Volkova et al. Proceedings of the 19th International Conference on Soft Computing and Measurements, SCM 2016 2016. C. 470-473
5. S. Brody, Biofeedback and SelfRegulation, Vol. 19, 1-13 (1999)
6. A.Rice, A. L. (Ed.). (2013). The enterprise and its environment: A system theory of management organization (Vol. 10). Routledge.
7. J. Freeman, J. A. (1975). U.S. Patent No. 3,890,957. Washington, DC: U.S. Patent and Trademark Office.
8. M. Shvartsman, M., Srivastava, V., & Cohen, J. D. (2015). A theory of decision making under dynamic context. In Advances in neural information processing systems (pp. 2485-2493)..
9. S. Alter, S. (2015). Work system theory as a platform: response to a research perspective article by Niederman and March. Journal of the Association for Information Systems, 16(6), 485.
10. T. Alesinskaya, et al (2017). Conception Bsc for Investment Support of Port and Industrial Complexes. Academy of Strategic Management Journal, 16, 9A.

11. A. Bril et al Proceedings of 2017 20th IEEE International Conference on Soft Computing and Measurements, SCM 2017. 6 July 2017, Paper number 7970692, Pages 692-693
12. J. Collins, et al. (2005). The assessment of chronic health conditions on work performance, absence, and total economic impact for employers. Journal of Occupational and Environmental Medicine, 47(6), 547-557.
13. G. Johns, G. (2010). Presenteeism in the workplace: A review and research agenda. Journal of Organizational Behavior, 31(4), 519-542.
14. I. Ilin, I.V., Levina, A.I., Iliashenko, O.Yu. Enterprise architecture approach to mining companies engineering. 2017 MATEC Web of Conferences, 106, 0806

The Application of the Real Options Method for the Evaluation of High-Rise Construction Projects

Aleksandr Izotov[1] , *Olga* Rostova[2] , and *Alissa* Dubgorn[1,*]

[1]Peter the Great St. Petersburg Polytechnic University, Institute of Industrial Economics, Management and Trade, 195251 St. Petersburg, Russia

Abstract. The paper is devoted to the problem of evaluation of high-rise construction projects in a rapidly changing environment. The authors proposed an algorithm for constructing and embedding real options in high-rise construction projects, which makes it possible to increase the flexibility of managing multi-stage projects that have the ability to adapt to changing conditions of implementation.

1 Introduction

The development of an investment and construction complex based on application of new technologies and equipment, the introduction of progressive forms of organization of construction production, the use of modern materials and products cannot be sufficiently effective without using the theory of construction projects investment management [1].

High-rise construction projects are subject to a number of risks, which, if implemented, can lead to significant deviations in the actual results of project implementation from the planned. The most significant are the following risks: the technologies used may not be viable due to engineering mistakes, planned technical parameters are not met, the quality of the constructed facilities does not meet the requirements of consumers, rapid and frequent innovations can lead to obsolescence of the technology even before the end of the project. The risk of incorrect choice of the project site, its inconsistency with the surrounding buildings, errors in constructive solutions, incorrect selection of materials and inaccurate definition of the layout of the facility should also be noted [2,3]. The end of the project may coincide with a slowdown in the development of the regional economy or unfavorable changes in market conditions, which may lead to a lack of demand for a new facility or a lower demand compared to the planned one. Features of analysis and classification of risks of the construction industry are considered in [4,5].

There is a problem that having assessed the risk of implementing a high-rise construction project as high, companies refuse to invest. However, if you take into account the possibility of making a decision on its further implementation at certain stages of the project, when analyzing the project, its investment attractiveness can significantly increase.

* Corresponding author: dubgorn@spbstu.ru

Most often, the cash flow analysis method (DCF method) and the definition of its net present value (NPV) for the current period [6] are used to assess the commercial attractiveness of high-rise construction projects. But the longer the project, the more innovative it is, the more difficult it is to assess both the cash flows themselves and their possible deviations.

In various sources, it is recommended to use various modifications of the DCF method to assess project risks, in particular, analysis of possible scenarios, the Monte Carlo method, fuzzy-set theory methods, statistical methods [7,8]. But all these approaches are either too lengthy or require a very large amount of initial information. In addition, these methods do not have sufficient flexibility in a rapidly changing environment. They do not take into account the possibility to change the decision on the further implementation of the project at its certain stages. Therefore, when projects that allow step-by-step implementation or give an opportunity to exit the project before its completion are considered, it is expedient to use the real options method [9,10]. Unlike the cash flow method, this method evaluates not the project parameters, but the value of the investor's right to make flexible investment decisions within the project in case of changing conditions in the real estate market.

2 Materials and Methods

The research proposes an algorithm for constructing and building real options into high-rise projects that allow an investor, on the one hand, to protect against the consequences of unfavorable events, and on the other hand, not to miss the potential opportunities that may appear during the course of implementation.

The evaluation of investment projects using the real options method is based on the assumption that any investment opportunity for the company can be considered as an option, that is, the company has the right, and not the obligation to create or acquire assets for some time [11,12].

However, if the company does not have the opportunity to implement the project step-by-step or in case of failure to leave the project before its completion, minimizing losses, then the company faces the choice of investing now or not, which does not contain real options.

The use of this estimation algorithm assumes at the first stage the evaluation of the project by applying traditional methods of investment analysis. The indicators of commercial efficiency of the project are calculated, risk analysis is carried out using the sensitivity analysis method, the scenario approach and the Monte Carlo method. In the second stage of the evaluation, the real options method is used, scenarios of variability in making investment decisions in the project implementation process are identified.

When analyzing a project, it is necessary to identify what its flexibility is and what real options can be present in it [13]. It is also necessary to assess the cost of implementing options, in which cases it is possible and what benefits it will bring.

This method is based not on the study of the most probable or averaged scenario (like DCF analysis), but on the study of project milestones in which several development options are possible. It is directed not only to detection, but also to the purposeful creation of such points. The optional characteristics are determined at the design and development stages of the project and can be used or not used, depending on the current state of the project and fluctuations in the external environment at the stages of project implementation and completion [14-15].

3 Results

For approbation of proposed method, the project of building a high-rise business center (BC) was considered.

The project implementation period is 7 years. The construction phase takes 2 years, the lease of premises is carried out within the next 5 years. At the end of the 5th year of operation, the BC is being sold.

The required volume of the project financing is 28 260 thousand dollars, including own investments of 10 900 thousand dollars. The cost of the land plot is 1 110 thousand dollars.

The length of the construction is 24 months. It is assumed that all the approvals, the development of the architectural concept, the project documentation will be completed within 6 months, and the construction itself will be completed within 1.5 years. The discount rate is 16.74%.

Initially, the project was analyzed on the basis of the cash flow discounting method. The main financial indicators were obtained (for a realistic scenario):

- Discounted payback period of investments (DPBP) (DPP) = 7 years,
- Net Present Value (NPV) = 4 300 thousand dollars,
- Internal Rate of Return (IRR) = 21.6% per annum,
- Investment profitability (PI) = 1.2.

Analysis of these indicators shows that the project is acceptable for implementation:

To analyze the risks of this project, a simulation method was applied. As a result of 10,000 experiments, it was found that the probability of failure of this project is 6.5%. However, the coefficients of variation of the project parameters are great. So for the NPV coefficient of variation is 65.7%, which indicates a high uncertainty.

Consider the real options embedded in the project: option to delay; option to expand; option to refuse, option for stage-by-stage project implementation (stage option).

3.1 Option to delay

After buying a land plot, you can postpone the commencement of construction until the market conditions are favorable. In accordance with the obligations of the developer, the construction can be postponed only for 3 years. The basic asset is the current value of expected revenues from the project, the strike price is equal to the investment size.

We form the tree of possible values of the underlying asset for three years (Figure 3). The measure of uncertainty obtained by means of simulation modeling makes 17%. It is assumed that during each period of time the value of the underlying asset can either increase with a coefficient of 1,185, or decrease with a coefficient of 0.84.

At the second stage, the option value tree is built (Figure 2). Calculations begin with the rightmost nodes. If the cost of income from the project in the node exceeds the amount of investment in the project, then at this point it is advisable to invest and get a difference between these values, otherwise it is worth refusing the project.

In the intermediate nodes the most favorable way is chosen: to execute the option immediately or delay. The valuation is performed by constructing an equivalent portfolio, which consists of a basic asset and risk-free bonds and completely repeats the yield of the option. Figure 3 shows the calculation of the equivalent portfolio parameters for the top node of the second period.

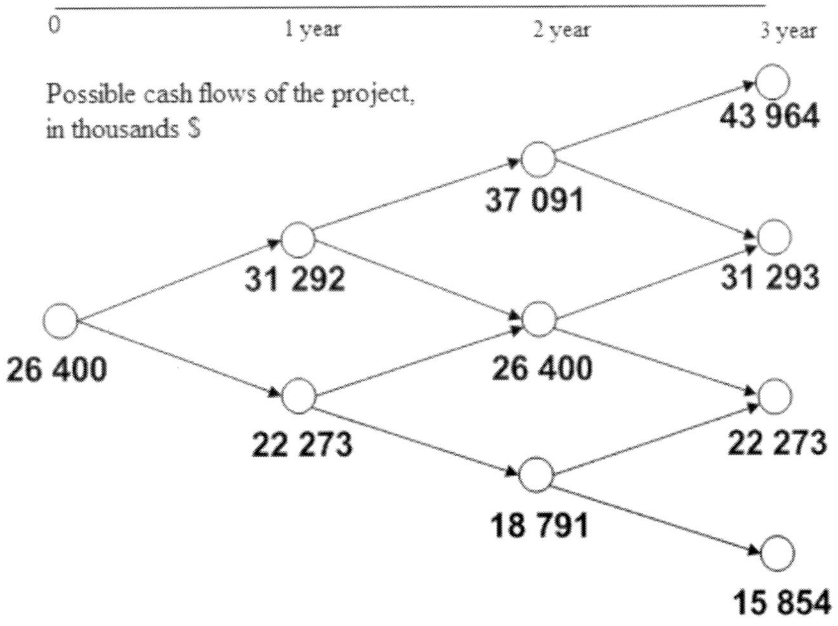

Fig. 1. The tree of possible values of the basic asset.

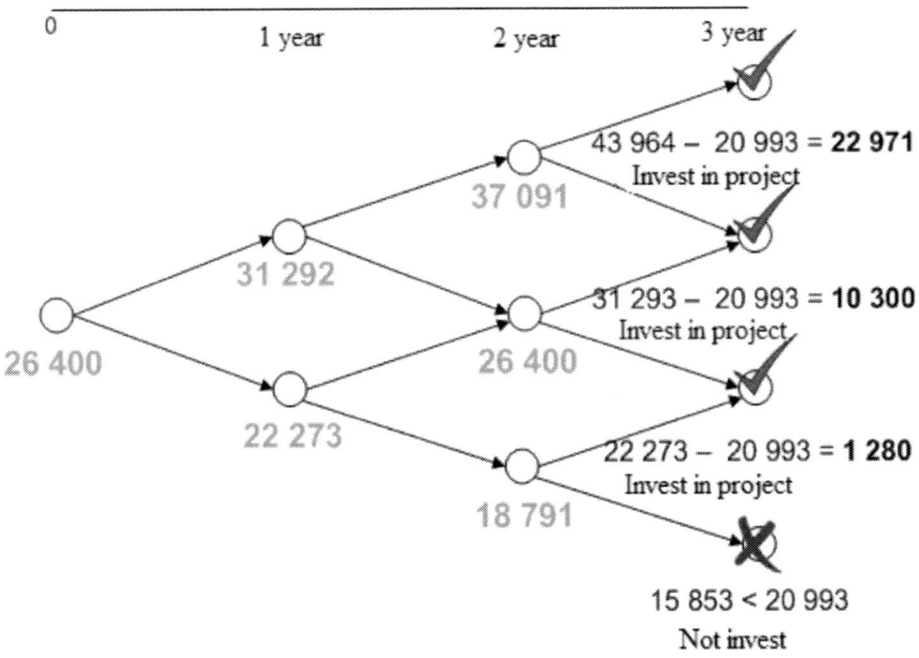

Fig. 2. The tree of option value.

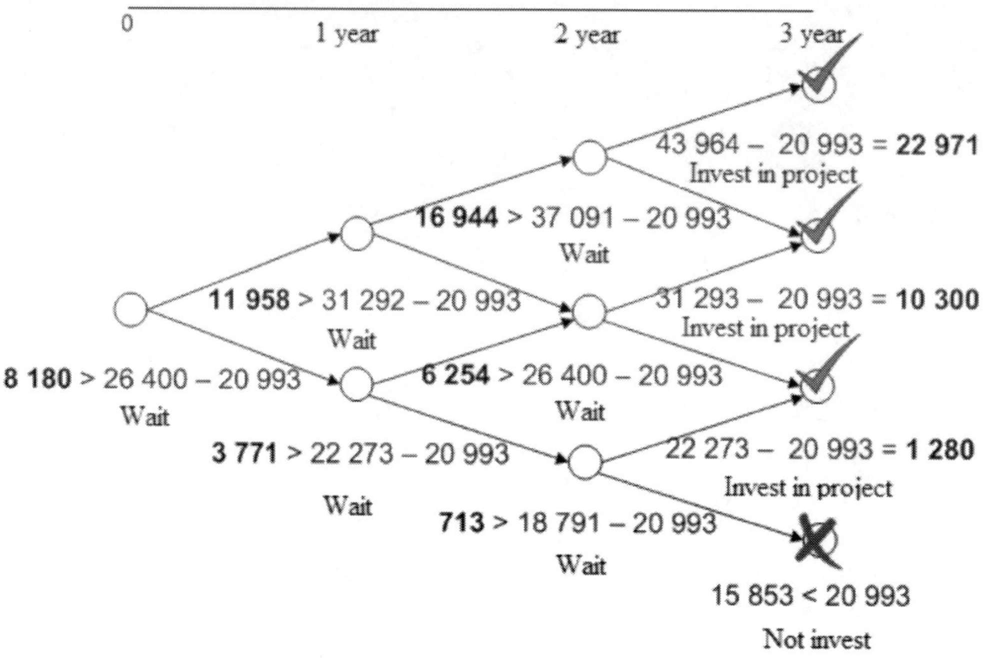

Fig. 3. The tree of option value.

Folding the model from right to left, we find the desired value - the value of the option at time t0. Taking into account the cost of acquiring a land plot, the cost of the project when choosing the option to delay will be $ 7 million. The net present value has increased approximately 1.5 times.

3.2 Option to refuse

It is assumed that within the first two years after the start of the project, the developer can sell the site to a third party buyer with a partially constructed object. The price of sales is defined as the sum of the capital costs at the previous stages.

The project cost tree, taking into account the built-in refuse option, is shown in Figure 6. It may be profitable for a developer to leave the project following the results of the second year if the market develops according to a pessimistic scenario. This real option raises NPV by 1 million dollars.

https://doi.org/10.1051/e3sconf/20183303008

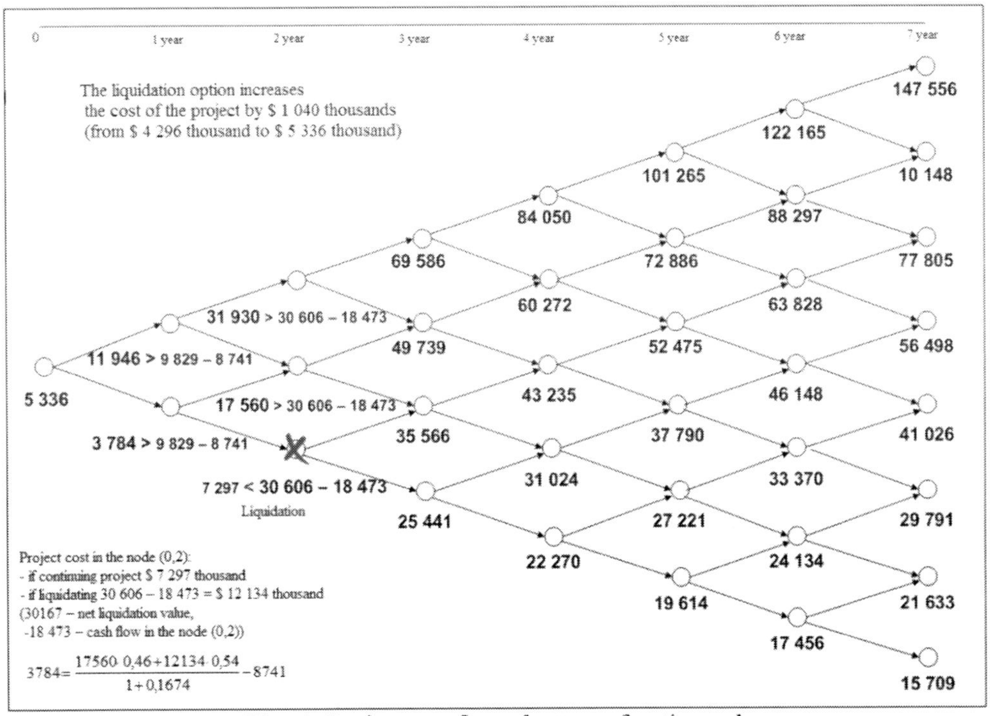

Fig. 4. Option to refuse: the tree of option value.

3.3 Option to expand

A real option to expand the scale of the project exists at the stage of development of project documentation. The real option to increase the office center area by 10%, valid for 6 months was considered. Flexibility in this case raises the project cost by 600 thousand dollars. Figure 5 shows the decision tree for this option.

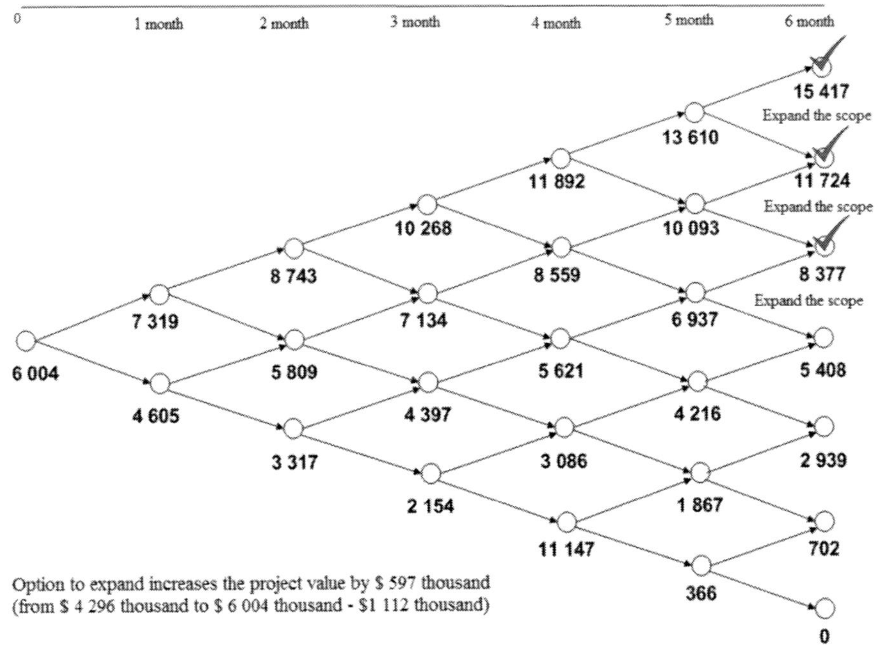

Fig. 5. Option to expand: the tree of option value.

3.4 Stage option

Since the construction of a large facility at once is rather risky, the construction project can be split into two stages. This is an opportunity to gradually enter the market, as well as take into account new information in the implementation of subsequent phases of the project. The option for phased implementation raises the project cost by $ 2.4 million. The decision tree of this option is shown in Figure 6.

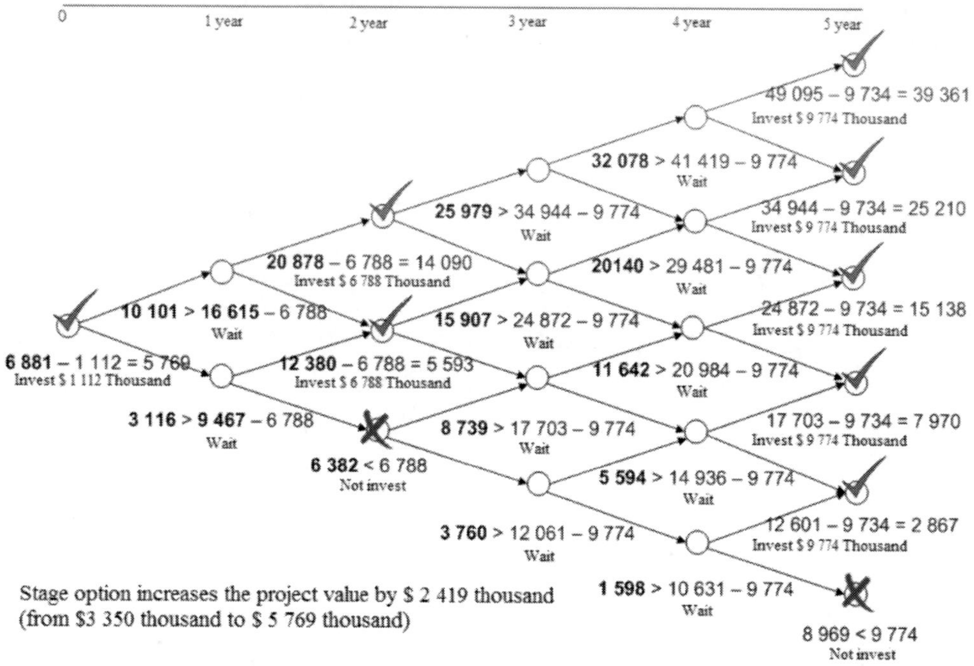

Fig. 6. Stage option: the tree of option value

The table 1 shows main characteristics of the real options discussed above.

Table 1. Summary of discussed real options.

Option	Type	Basic asset	Strike price	Option duration	Option value, thousand $
Stage option	compound	Cash flow of two-stage project $ 20 984 thousand	The amount of investment in the first stage of $ 6,788 thousand, in the second stage $ 9 734 thousand.	2 years for the first stage and 5 years for the second stage	2 419
Option to refuse	put	Cash flow of project $26 400 thousand	Project liquidation price	2 years	1 040
Option to delay	call	Cash flow of project $26 400 thousand	The amount of investment in the project (construction cost) - $ 20 992 thousand.	3 years	2 773
Option to expand	call	Cash flow of initial project $26 400 thousand, expanded project $ 29 427 thousand	The cost of construction of the initial object - $ 20 992 thousand and the expanded project - $ 24 085 thousand.	6 months	597

4 Discussion

Analysis of the project from the position of real options has allowed not only to identify additional opportunities hidden in the project, allowing the investor to feel more confident in a risky environment, but also to quantify them.

The possibility of influencing the course of the investment process has a certain value, which is taken into account when assessing the effectiveness of the project. Therefore, the proposed scheme for designing optional characteristics of an innovative project at the pre-investment stage of its implementation allows the investor to avoid underestimation of NPV cash flows in conditions of uncertainty of the external environment. The method of real options allows to include in the project a number of additional opportunities that can reduce investors' uncertainty about the future cash flow, as well as assess the contribution of these opportunities to the overall value of the project even at the stage of pre-project preparation.

It is important not only to calculate the value of the project, but also to plan for future flexible solutions in the draft. If the cash flows exceed expectations, then the project can be expanded; if there are less than expected, the project can be reduced or stopped altogether. Using the real options method allows management to focus not on absolutely accurate forecasts, but on determining alternative ways of project development, the ability to flexibly respond to changing external conditions.

5 Conclusion

The valuation of options provides important additional information and thus contributes to the reasonableness of the decisions made about the feasibility of implementing innovative projects, especially when projects are considered that have significant opportunities for adaptation to changing conditions, as well as multi-stage projects. The evaluation of real options should not be considered as an alternative, but as an addition to the DCF method, refining the estimates obtained by traditional methods.

References

1. D. Chubrik, Introduction of BIM: Myths and Reality, High-rise buildings, No. 4, pp. 139-147, (2014)
2. A.E. Plahin, A.Yu.Kokovihin, E.S. Ogorodnikova, S.A. Suslov, Assessment of influence of infrastructure projects on development of the territory, NGIEI bulletin, No. 11, pp. 139-147, (2017)
3. M.N. Shchukina, Modern high-rise construction, Moscow: GUP "ITC of Moscomarchitecture", 401 p., (2007)
4. T. F. Morozova, L. A. Kinayat, A. Zh. Kinayat, Assessment of risks in construction, Construction of Unique Buildings and Structures, No. 5 (10), pp. 68-76, (2013)
5. B. S. Seytzhanov, The risks of the construction industry – classification and characteristics analysis, URL: http://www.be5.biz/ekonomika1/r2011/00585.html (20.12.2017)
6. A. Kalashnikov, N. I. Vatin, Organization, management and planning in construction. Basic principles and fundamentals of organization of investment and construction projects, SPb.: Publishing house of Polytechnic University, (2011)
7. A.V. Izotov, O. V. Rostova. Development of a system of sectoral investment priorities, Proceedings of the 29th International Business Information Management Association Conference 3-4 May 2017 Vienna Austria, pp.1822-1832, (2017)

8. S. Ptuhina, A. A. Liskov, I. A. Ptuhin, Development of total cost management in construction, Construction of unique buildings and structures, No. 5. pp. 17-20, (2012)
9. I.V. Ilyin, O.V. Rostova, Investment Management, SPb.: Publishing house of Polytechnic University, (2017)
10. T.A. Khudyakova, A.V. Shmidt, Uncertainty of External Environment during the Construction of the Controlling System of Sustainability, Proceedings of The 26th International Business Information Management Association Conference, pp. 1607–1611, (2015)
11. I.S. Sokolova, E.V. Gubanova, S.V. Solovyeva, The use of financial instruments in the formation of an effective securities portfolio, NGIEI bulletin, No. 9 (64), pp. 123-137, (2016)
12. L. E. Brandao, J. S. Dyer, W. J. Hahn, Using Binomial Decision Trees to Solve Real-Option Valuation Problems, Decision Analysis, Volume 2, Issue 2, pp. 69-88, (2005)
13. I.V. Ilin, O.V. Kalinina, A.I. Levina, O.Yu. Iliashenko, Approach to Organizational Structure Modelling in Construction Companies, MATEC Web of Conferences. 5th International Scientific Conference on Integration, Partnership and Innovation in Construction Science and Education, IPICSE 2016, article number 05028, (2016)
14. A.R. Bril, O.V. Kalinina, I.V. Ilin, Small innovative company's valuation within venture capital financing of projects in the construction industry, MATEC Web of Conferences, Volume 106, 08010, International Science Conference SPbWOSCE-2016 "SMART City", (2017)
15. V.V. Glukhov, I.V. Ilin, Project Portfolio Structure in a Telecommunications Company, Lecture Notes in Computer Science, Volume 8638, pp. 509-518, (2014)

IT Requirements Integration in High-Rise Construction Design Projects

Anastasia Levina[1*], *Igor* Ilin[1], *Rustam* Esedulaev[1]

[1]Peter the Great St. Petersburg Polytechnic University, Institute of Industrial management, Economy and Trade, 195251, Polytechnicheskaya, 29, St.Petersburg, Russia

Abstract. The paper discusses the growing role of IT support for the operation of modern high-rise buildings, due to the complexity of managing engineering systems of buildings and the requirements of consumers for the IT infrastructure. The existing regulatory framework for the development of design documentation for construction, including high-rise buildings, is analyzed, and the lack of coherence in the development of this documentation with the requirements for the creation of an automated management system and the corresponding IT infrastructure is stated. The lack of integration between these areas is the cause of delays and inefficiencies both at the design stage and at the stage of putting the building into operation. The paper proposes an approach to coordinate the requirements of the IT infrastructure of high-rise buildings and design documentation for construction. The solution to this problem is possible within the framework of the enterprise architecture concept by coordinating the requirements of the IT and technological layers at the design stage of the construction.

1 Introduction

The rapid development of large cities in Russia led investor-builders and local authorities to a solution of the issue of scarcity of territories and the high cost of land, which is the construction of high-rise buildings, both residential and public. The high-rise building is, on the one hand, the point of economic growth and development center of the given territory, on the other - the challenge to the local infrastructure, the city authorities and the whole system of state building. Effective activities to create high-rise buildings require a certain maturity in the following aspects:

• technological: professionally trained architects, engineers and workers are required; a certain level of development of the building materials industry is needed;

• infrastructure: the development of engineering and social infrastructure in the territory under consideration, good logistics of traffic flows;

• economic: the availability of investors, the willingness of local authorities to contribute to the solution of issues of territorial planning, construction and commissioning of the facility;

* Corresponding author: olgakalinina@bk.ru

• regulatory and legal: the existence of a legislative framework that allows to regulate the design, construction, expertise, operation of constuctions.

Regarding the last point, the construction of high-rise buildings requires special conditions and approaches, while in practice it turned out that normative documents in the field of high-rise construction until recently (until 2017) in Russia were just absent.

The buildings and structures in modern cities are not just a set of building structures. Modern buildings, providing an effective environment for life, must be equipped with a variety of engineering systems, the list of which depends on the functional purpose of buildings and that are the more difficult, the more complex constructed facilities are. High-rise buildings, as a complex technical object, require extra attention to engineering systems. For example, some fire safety requirements and the resulting requirements to ensure the evacuation of people in case of fire at high-rise buildings determine the existence of complex automated systems of devices for the collection and transmission of data on the detection of a fire, that must be transmitted to a single center for processing and decision making. Obviously, the issue of managing the design, creation and subsequent operation of these systems should be payed as much attention as it is payed to creation of the building itself, since without the effective functioning of these systems in the modern world, the building will be just an empty box. Management of the whole set of engineering subsystems of modern buildings is impossible without information technologies. The control system for engineering subsystems of high-rise buildings should provide the ability to monitor the status of various objects, collect, transmit, process information in a single information space. [1] Because of that, the task of the need to regulate the development of an integrated automated control system for engineering systems of high-rise buildings, as well as the development of requirements for the IT infrastructure of the erected object and their accounting when creating project documentation for construction arise. At present, there are no requirements and norms for designing, creating and implementing such a system as part of the high-rise building project.

In view of the lack of a unified concept for the creation of an automated management system and a clear procedure for managing its implementation, approved at the early stages of the construction of a new high-rise building, the following problems may arise [2]:

- spontaneity of the implementation of various subsystems of the information system at different time periods;
- inconsistency of the decisions of various developers and, as a consequence, the lack of the possibility of subsequent integration of individual subsystems into a single information management system;
- increase the cost of implementing individual subsystems;
- achievement of local optimisations at the expense of the effectiveness of the entire management system;
- lack of replicability of expanding the functionality of automated control systems as the enterprise develops;
- usage of solutions that place the customer in dependence on specific performers and manufacturers during the operation phase;
- increase in operating costs.

At the construction phase, all this entails an increased risk of delay in putting the building into operation because of the lack of consideration for the requirements of the IT infrastructure in the construction design. On the operating object the disjointedness of the automated systems leads to [3]:

- subjectivity and significant influence of the human factor on the collection and interpretation of information;
- significant costs for monitoring the reliability of information;
- making decisions based on distorted information;

- propensity to spontaneously manage deviations;
- permanent overstatement of planned costs for all components of the building management process.

The application of a systematic approach to the construction of a management system that includes as a key component an integrated automated information system and its corresponding IT infrastructure allows to avoid the problems discussed above and minimize the risks associated with them. The implementation of the automated management system and IT infrastructure, coordinated with the stages of construction of the high-rise facility, will allow to eliminate in advance many of the problems that arise in the creation, integration and development of automated systems, as well as reduce future expenses for the modernization and maintenance of the building.

The purpose of this article is to formulate requirements for the IT infrastructure of high-rise buildings, which must be taken into consideration at the stage of designing high-rise buildings.

2 Materials and methods

To achieve this goal, the following tasks were consistently solved in the article:

- identification of the existing regulatory framework governing the construction of facilities in general and high-rise buildings in particular, also regulating the creation of automated control systems;
- analysis of the existing regulatory framework in the field of building high-rise buildings and the creation of IT infrastructure for the presence of indications of meaningful and chronological harmonization of these standards when creating new facilities;
- search and analysis of publications on the problem of harmonizing the requirements of regulatory documents for construction and the creation of automated management systems and IT infrastructure;
- forming proposals to integrate the requirements for the IT infrastructure into the high-rise buildings design phase.

As part of the study, publications on the use of information technologies in construction were analyzed. For example, in [4] it is said that the development of information technology affects construction in general. However, the authors focus not on how information technologies should be taken into account in future buildings, but how the process of modeling, planning, and erecting buildings has changed due to the emergence of information systems of certain classes. These information systems can reduce the labor costs associated with modeling and preliminary tests of models. The same aspect is analyzed in [5].

Forbes in [6] notes that planning is the stage of the project, which should consider the life of future owners of the building, but attention is paid only to the fact that it is necessary to plan the building so that future tenants can live comfortably in it. It is proposed to conduct a survey of future owners.

The following databases were searched to find research or case studies about the issue: Google Scholar, Springer, ScienceDirect, IEEE Explore, Wiley InterScience. As a result, ready templates and techniques were not found in order to have the most efficient and convenient for future users the future structure of information technologies, which will be used by end-users.

In addition to the analysis of normative documents and publications on the problem of the study, during the period from 2015 to 2017 consultations were held with specialists in the field of civil and industrial facilities and related areas (construction companies, industry design institutes, providing companies). All experts confirm the absence of a statutory requirement to harmonize the stages of designing and creating buildings and structures and the stages of designing and developing an automated control system for the building object.

At the same time, respondents emphasize the need for such an agreement with a view to in-time solvation for the problems of integrating construction sites and elements of an automated management system.

To reconcile the requirements of the design documentation for the construction and IT infrastructure, this problem was considered point of enterprise architecture. Enterprise architecture is the commonly used framework of management science in the field of integrated design of business management systems and in a broader context - socio-economic systems. [7-11] The approach to integrated enterprise design based on the enterprise architecture concept is of a universal nature, i.e. applicable for all types of businesses. However, the methodology of designing individual layers and elements of architecture depends on the specific industry.

Within the framework of the enterprise architecture as a complex model of enterprise management, a number of elements, united in the so-called layers. The layers and their elements are interconnected and interdependent: one defines requirements to another one. Layered structure of the enterprise architecture model specifies the relationships between core components of the system. TOGAF [8] defines the following layers of the enterprise architecture: Business, Application and Technology layer. The last one traditionally refers to IT-infrastructure, which is defined by the structure of the Application layer and it requirements.[12] In the meantime, in a broader context Technology layer includes not only IT, but all the physical infrastructure of the enterprise or a system.[13] From this point of view, the physical facilities of the building and its IT-infrastructure has to be designed in parallel and taking into account the requirements of the future IT system of the building.

3 Results

It is difficult to overestimate the role of the design stage when creating any object. It is at this stage that all the principles of the functioning of the created system are stated, at this stage it is possible to introduce the required corrections with minimal costs in comparison with the subsequent stages. The cost of fixing errors embedded in the project increases with each next stage of object creation.

After analysis of the approaches of the creation of enterprises and management systems, documented both as regulatory documents and as collections of best practices, three main points of view on this task were identified:

1. point of view of construction;
2. point of view of automated control systems (IT);
3. point of view of the architecture of the enterprise.

The order of enumeration of points of view corresponds to the degree of theoretical elaboration of the approach and the degree of its practicability in practice in descending order. It is advisable to use an integrated approach to the design of the enterprises being created, and the approaches related to the enterprise architecture concept should play an integrating role. [14] Since the constructing of buildings and structures has a much longer history, the construction industry has implemented a full cycle of introducing best practices and technologies into practice: the technologies are scientifically substantiated, fixed in the form of standards, in accordance with which the design of buildings is realized and the examination of finished objects is carried out. The design of the IT component of buildings and structures is a new area of knowledge and also it is dynamically changing. In this regard, despite the existence of state standards for the design of automated systems (GOST 34), there is no unified approach to design, and therefore, and examination of information systems. The development of integrated management systems, including on the basis of the enterprise architecture concept, despite the growing popularity of this area among managers and developers of information systems for business, as well as the existence of standards and collections of best practices, this concept is not used in the design of buildings.

Since 2017, the construction and design of high-rise buildings is regulated additionally by the Code of Rules SP 267.1325800.2016 Buildings and high-rise buildings. Design rules [15]. Additional documents are being developed, but none of them (according to official sources) is concerned about the integrated automated management system of high rise buildings. In this document, in addition to the features of the building design of high-rise buildings, it is pointed out that it is necessary to take into consideration in the project numerous engineering subsystems that ensure the operation of such constructions. Thus, the engineering infrastructure of high-rise buildings includes:

- Heating
- Ventilation and air conditioning
- Cold supply
- Water supply, sewerage and watercourses
- Power supply, power electrical equipment and electric lighting
- Communication systems, signaling, automation and dispatching systems (including telephone communication, radio, television, Internet, fire protection system, security system, etc.)
- Vertical transport (including elevators, escalators, etc.)
- Garbage removal
- Safety of functioning and operation of engineering systems

Despite the fact that in the regulatory document SP 267.1325800.2016 Buildings and high-rise complexes. Design rules denote the importance of designing these engineering systems and provides links to the regulatory documents governing the design of such systems, there is no indication of the need to link the content and time of works on the design and implementation of high-rise buildings and engineering systems. In addition, there is no indication of the need to develop and implement a common automated control system for engineering systems of high-rise buildings. Thus, the current regulatory documentation on the design of high-rise buildings regulates the creation of a "material shell" for the future building, but completely misses the issue of creating a "neural network" of the future system, thereby inducing the imperfection of the system being built at the project stage.

It is necessary the structure of the project for the creation of buildings and structures. A typical project consists of 5 stages:

1) Preparatory stage - drawing up a draft of the house, drawing up and coordination with the authorities of documentation permitting construction
2) Zero cycle - excavation, foundation laying and building of the ground floor, laying of communications under the foundation, construction of a common grounding bus
3) Major construction works - building walls and floors, installation of internal partitions and staircases, roofs, exterior finishing
4) Carrying out communications - conducting electricity, running water, sewerage, gas, as well as ventilation and security-fire alarms
5) Interior finishing - installation of interior doors, wall treatment, doing makeover.

In any project activity, the boundary stage, without which the project can not continue, is the planning stage of the project. In the construction projects at this stage is drawn up and the plan for future construction. The plan is necessary in order to understand what result is necessary to obtain at the end of the project.

Because of in the construction after the planning the following stages must follow the plan (this is due to the fact that it is necessary to deliver the object within a specified period, and the slightest deviation from the plan entails an increasing deviation in the final product), the Waterfall model must be used. Features of this model are the following:

1) each stage must be completed before the next one begins, one can not return to the previous stage (for example, you can not start building walls before foundation is completed)
2) the requirements for the final product are clearly defined (at the beginning of the work, the exact parameters of the building should already be determined)
3) there is a fixed cost of work (the cost of the object is determined even before the construction of it)
4) the cost of the error made in the previous stage is extremely high.

At the same time, the stage of writing a plan can take several cycles, which means using the Agile methodology.

For these reasons, it is necessary that a place for future hardware provision be provided at the planning stage.The most common hardware IT services are the following:

- Servers;
- Wiring for power;
- Routers;
- Ethernet wiring;
- Monitors, keyboards and mouse;

It is necessary that the correct placement, that is conducive to convenient and long-term operation, should be provided at the planning stage. The authors compiled a table, which is presented below, in which the main components in the hardware part of any information system are given, with recommendations for planning (Table 1).

Table 1. Objects of IT infrastructure and their place at the design stage of the constructioning

Name of equipment	Equipment place	Notes
Servers	Well-ventilated rooms with a low degree of humidity	It is recommended when planning to allocate rooms of the ground floor for hosting the server room
Wiring for power	Since the server requires a large number of wirings, it is necessary to provide for deep trenches in the walls for it	Trenches must be drilled in order to further increase the capacity of servers, which means that you need to add wiring
Ethernet wiring	In case if the object is a residential house, it is necessary to provide Ethernet wiring into each apartment	It is recommended to conduct cables to the room where Wi-Fi routers will be located for further connection with them
Routers	Routers should be located in such a way that the signal coming from them reaches each room of the apartment	Hardware must be installed in such place where there are no reflective grids and wiring between the router and workstations
Monitors, keyboards and mouse	These accessories do not need special care, but it is necessary that it be installed so that the user's eyes are not less than 50 cm (3) (4)	Since it is necessary to have a minimum distance, while from the user's point of view, it is necessary to save space, it is desirable that these components are located in a corner or near a wall.

Usage of the above recommendations will achieve the following results:

1) The object at the stage of delivery will have an additional competitive advantage, which will make it more attractive to buyers, as a result - the time of sale / rental of premises will decrease;
2) Reducing the costs of future users for the deployment of IT infrastructure due to the lack of costs for preparing premises for servers and components;
3) Increase in the duration of the operational period of the equipment due to the creation of a comfort environment for it.

Since planners and builders do not have extensive knowledge in the field of information technology, it is recommended to involve people with deep knowledge of network and system administration in the process of planning buildings. This concept of creating a plan has a flaw: since more resources are attracted to the planning stage, the cost of the project is increased. These costs can be transferred to the cost of the final product. However, prices in the real estate sector are relatively large, for this reason, the value in relative terms does not increase much, and this does not repel potential buyers.

In conclusion, it should be noted that with the growth of the pace of construction and dissemination of information technologies, the urgency of the problem of integration of the two spheres will increase. Using the tips given in this article, you can avoid a lot of bottlenecks. This will benefit both owners of construction companies who will increase their sales and profits, and end users who will reduce the cost of installing computer equipment.

4 Discussion

In this paper, an approach is proposed to take into account the requirements of hardware IT support for high-rise buildings when developing design documentation for construction. The solution of this problem is possible within the framework of the enterprise architecture concept by coordinating the requirements of the IT and technological layers at the design stage of the facility. Consideration of the problem in a broader context presupposes the need to develop an approach not only to harmonize the stages of designing the material objects of buildings and structures and their IT infrastructure, but also to agree on the stages of designing an integrated object management system that includes an information system.

The solution of such a problem is the subject of further research within the framework of the development of the topic touched upon in this article. Such a study requires an in-depth analysis of the regulatory framework for the development of management systems and its alignment with project design documents for construction, as well as studying best practices in the integration of construction standards and the creation of integrated management systems.

5 Conclusions

The article highlights the importance of automated management systems and the corresponding IT infrastructure for modern high-rise buildings. The growing role of IT support is due, on the one hand, to the complexity of managing the engineering systems of high-rise buildings, on the other hand, to the requirements of future users of the building infrastructure. Based on the analysis of the existing regulatory framework in the field of creating design documents for construction, including high-rise buildings, the article states the lack of consistency in the development of this documentation with the requirements for the creation of an automated management system and the corresponding IT infrastructure. The lack of integration between these areas is the cause of delays and inefficiencies both at the design stage and at the stage of putting the building into operation. The solution of this problem is possible within the framework of the enterprise architecture concept by coordinating the requirements of the IT and technological layers at the design stage of the facility. The article identified the main groups of IT infrastructure objects and proposed a format for accounting for the requirements of these facilities at the design stage of the building.

References

1. E. Istomin et al International Multidisciplinary Scientific GeoConference Surveying Geology and Mining Ecology Management, SGEM. Volume 1, Issue 2, 2015, Pages 607-614
2. I. Ilin, et al. *Enterprise architecture approach to mining companies engineering*. MATEC Web of Conferences, **106**, 0806. (2017)
3. A.Bril A.R., Kalinina O.V., Ilin I.V. (2017) *Small innovative company's valuation within venture capital financing of projects in the construction industry* // MATEC Web of Conferences, Volume 106, 08010, International Science Conference SPbWOSCE-2016 "SMART City".
4. URL: http://www.tdx.cat/bitstream/handle/10803/6160/03Nfm03de12.pdf?sequence=3
5. URL: http://www.tandfonline.com/doi/abs/10.1080/01446199400000016
6. URL:https://www.forbes.com/sites/houzz/2013/11/27/6-steps-to-planning-a-successful-building-project/#4fda8d083b45
7. M. Lankhorst M. Enterprise Architecture at Work. Modelling, Communication, Analysis (Springer-Verlag, 2013)
8. The Open Group 2009 TOGAF Version 9. The Open Group Architecture Framework (TOGAF). London: TSO
9. URL: http://www.zachman.com/
10. URL: http://cisr.mit.edu/research/research-overview/classic-topics/enterprise-architecture
11. URL: http://www.gartner.com/it-glossary/
12. A.Anisiforov et al Proceedings of the 29th International Business Information Management Association Conference - Education Excellence and Innovation Management through Vision 2020: From Regional Development Sustainability to Global Economic Growth. 2017, 2920-293 (2017)
13. R.Giachetti R.E. (2012) *A Flexible Approach to Realize an Enterprise Architecture*. Procedia Computer Science 147-152.
14. A.Bril et al Proceedings of 2017 20th IEEE International Conference on Soft Computing and Measurements, SCM 2017. 6 July 2017, Paper number 7970692, Pages 692-693.
15. Russian Standard SP 267.1325800.2016

Analysis of administrative barriers in the industry of the high-rise construction in Russian Federation

Irina Zaychenko[1], *Alexandra* Borremans[1*] and *Svetlana* Gutman[1]

[1] Peter the Great St. Petersburg Polytechnic University, Institute of Industrial management, Economy and Trade, 195251, Polytechnicheskaya, 29, St.Petersburg, Russia

Abstract. The article describes the concept and types of administrative barriers encountered in various areas of the enterprise. The particularities of the Russian high-rise construction industry are described and a comparative analysis of administrative barriers in this sector is performed. The main stages and administrative procedures when the developers implement investment and construction projects in the field of high-rise construction are determined. The regulatory and legal framework for the implementation of investment and project activities in the high-rise construction industry has been studied and conclusions have been drawn on its low level of precision in the issue of the formation of competitive and efficient high-rise construction markets. The average number of administrative procedures for the implementation of the investment and construction project in the field of high-rise construction is determined. The factors preventing the reduction of administrative barriers in the high-rise construction industry are revealed.

1 Introduction

Administrative barriers in construction are one of the most significant problems of the industry. Because of these issues the problems inherent in the Russian construction market are constantly self-reproducing: monopolization, corruption and as the result of such conditions the ineffectiveness of the market mechanism. Construction organizations face insurmountable administrative obstacles that may adversely affect the socially important task - the realization of affordable and comfortable housing for Russian citizens.

The relevance of this study is determined by the need to improve the efficiency of realization of investment and construction projects in the field of high-rise construction by optimizing administrative procedures [1]. It is necessary to develop and implement new measures to reduce administrative barriers in high-rise construction.

The aim of the work is to analyze the administrative barriers in the high-rise building industry and to identify the problems that arise when eliminating them.

* Corresponding author: alexandra.borremans@mail.ru

2 Materials and Methods

The construction sector, and in particular the high-rise construction sector, is the object of close attention of public authorities. This is primarily due to the need of addressing at the state level such issue as the turnover of land, preserving the architectural and functional integrity of the urban infrastructure and the technical safety of the constructed facilities.

However, excessive state regulation of the sector can lead to a business activity restriction of organizations, and the availability of an administrative resource can promote the monopolization of the construction market. According to different estimates, the GDP losses caused by the overestimated construction cost due to competition restrictions in this sector are annually from 0.21% to 0.54% [2].

Moreover, it can be assumed that administrative barriers restrain the development of competition in the industry, reduce the employment of the population, increase the corruption of business conditions, increase the cost and time required to complete a concrete transaction. Weak competition in the construction market hinders the development of this sector, blocks the improvement of construction quality, and also leads to an overestimated cost per square meter of housing, resulting in its low availability for the population. The reasons for weak competition in the construction sector, nevertheless, are linked not only to the policy of constructors, but also to the situation in the related markets and the presence of an excessive number of administrative barriers.

Administrative barriers are the rules established by decisions of state and municipal authorities, observance of which is a prerequisite for business in the area, which significantly increases the cost and time of passing bureaucratic procedures.

The analysis of previous researches by types of administrative barriers [3-6], allows making the following classification, presented in tab. 1.

Table 1. Classification of administrative barriers.

Scope of the administrative barrier	Barrier type
Obtaining the right to conduct business	Company registration
	Agreement of project documentation with government agencies
	Obtaining licenses
	Obtaining the right to purchase (rent) the premises
	Access to loans, equipment leasing
Access to markets for goods	Quoting volumes of purchases and deliveries
	Certification of goods and services
	Branding of goods
	Registration of goods
	Technical examination and identification of goods
Implementation of economic activities	Control and supervisory activities of government agencies
	Establishment of forms of mandatory reporting

	Inspection activity of authorized government agencies
	Introduction of norms of mandatory approvals and registrations of decisions by government agencies
Administrative barriers of a general nature	Contradiction of current legislation
	Imperfection of the current legislation
	Violation of legal guarantees for business entities

These types of administrative barriers are common. However, as mentioned above, entrepreneurial activities within the urban development sector are the most susceptible to the influence of administrative barriers. Thus, 74% of surveyed construction companies assess the administrative barriers in the market as "high" or "prohibitively high" [7].

In addition, the level of administrative barriers in the largest Russian cities varies significantly. In Moscow, for example, the number of all reconciliation procedures averages 54, takes 704 days and 2612.8% of per capita income, while, for example, in Rostov-on-Don, only 22 procedures need to be completed, spending 194 days and 272.9% of per capita income [8]. For business related to high-rise construction, such regional differences in the difficulty of obtaining approvals and permits for construction in the implementation of investment projects are one of the determining factors when choosing an object for investment.

High entry barriers to local high-rise markets limit the access of constructors to such markets, which leads to monopolization of the industry. Imperfect competition in the markets significantly limits the potential for increasing the volume of high-rise construction and, ultimately, leads to a reduction in the affordability of housing for the population.

Analysis of the regulatory framework [9-12] shows that there is formally a certain legal and regulatory framework for the implementation of investment and project activities in the high-rise construction industry, but in practice this is not applied. There is no institutional environment conducive to the formation of competitive and efficient markets for high-rise construction. The main reason for this is the presence of redundant administrative barriers, namely, numerous conciliation and licensing procedures that are mandatory for the developer in the implementation of the investment and construction project.

In order to identify administrative barriers in the high-rise construction industry, the main stages and main administrative procedures of investment and construction projects implementation in this area by developers were determined. Thus, when implementing an investment and construction project in the field of high-rise construction, it is common to distinguish five main stages [13]:

1. Land-building and urban planning.
2. Preparation of project documentation.
3. Construction.
4. Commissioning of the facility.
5. State cadastral registration and state registration.

The tab. 2 presents the main administrative barriers at each of these stages.

Table 2. Problems in the passage of administrative procedures at each stage of the investment and construction project in the field of housing construction.

Stage of the investment and construction project in the field of high-rise construction	Issues in passing administrative procedures

Land-building and urban planning	Issues associated with obtaining a town planning plan for a land plot.
	Issues associated with obtaining technical conditions for connection to networks of engineering infrastructure.
	Issues associated with the holding of auctions for the provision of land for housing construction.
Preparation of project documentation	Duration and high cost of passing the state expertise of project documentation and engineering survey results.
	A large number of approvals of project documentation in local government agencies, city services, organizations operating engineering and maintenance networks.
	High cost of connection to the networks of engineering support.
	Lack of qualified personnel in the design.
Construction	The length of time for obtaining a building permit.
	The length of time for obtaining a warrant for earthworks.
	A large number of inspections by state construction supervision agencies.
Commissioning	Long terms of Rostekhnadzor (the Federal Service for Ecological, Technological and Nuclear Supervision) procedures for commissioning the facility.
	Long time for obtaining permission to put the facility into exploitation.
	A large number of approvals with organizations operating engineering and technical support networks and other organizations necessary to obtain a permit to put the facility into exploitation.
	The requirement to conclude contracts for connection to the engineering support network in the number of documents required to obtain a permit for commissioning the facility.
	Double technical inventory of the object - before and after putting the facility into exploitation.
State cadastral registration and state registration of rights	Long term registration of ownership of real estate.

Thereby, the main administrative barriers in the high-rise construction industry are: excessive number of administrative procedures; long terms of passage of administrative procedures; high cost of passing administrative procedures.

For realization of the investment and construction project in the field of high-rise construction, the developer needs a certain regulation that would reflect administrative procedures step-by-step, which he must pass with a mandatory indication of the time, amount and cost. Nowadays, the industry has not established a clear, transparent regulation with an established list of administrative procedures for the developer to implement the investment and construction project in the field of housing construction, which strongly provokes the problem of excessive administrative barriers.

At the stage of land management and town planning training, if the right to land is not obtained by the developer at the auction, it is necessary to pass about 2 times more procedures than foreign practices [14]. In this case, if it is necessary to change the permitted type of land plot use, this requires especially a lot of time - an average of 229 days [8].

One of the most common ways to obtain the right to a land plot is to acquire a property right or lease a land plot at an auction. In this case, the period of passing the procedures at the stage of land management and urban planning is almost halved (to 136 days). At the same time experts pointed out such problems as the absence or a small number of auctions for the provision of land for housing construction, the lack of information on the conduct of such auctions, the high price of land, a small number of land plots. A serious problem in the administrative procedures at the stage of land management and urban planning in many cities is the long period of obtaining a town-planning plan for a land plot. For instant, it is on average 54 days in case of obtaining rights to a land plot at auction and 166 days in case the developer has rights to a land plot, received not at auction. This is related also with the fact that the developer must first obtain technical conditions for connecting to the networks of engineering and technical services [15].

At the design stage, one of the main problems is a large number of approvals for project documentation in local government agencies, city services and organizations operating engineering and technical support networks - on average, 18 procedures are required, spending an average of 144 days on this. At the same time, in some cities the number of procedures at the design stage reaches 26. It is important to note that the federal legislation on urban development does not require any coordination of project documentation, except for state expertise. Unjustifiably long and expensive are the procedures associated with the passage of state expertise of project documentation and engineering survey results, requiring an average of 66 days and 711 thousand rubles. The cost of connection to the networks of engineering and technical support and the coordination of technical connection conditions at this stage (on average 20 million rubles) are also estimated as high.

During the construction phase, as the main problems, experts noted the long terms of obtaining a building permit and a large number of inspections by state construction supervision authorities. For the passage of procedures related to obtaining a building permit and other documents required for construction work at this stage, an average of 140 days is required. In the process of construction, on average, 11 inspections should be conducted by the state construction supervision authorities. However, their number can reach 43. It should be noted that the total number and regularity of inspections of construction by state construction supervision authorities in federal legislation are not defined.

The stage of commissioning the facility is the third in terms of the duration of the passage of all administrative procedures in the process of realization an investment and construction project in the field of housing construction. On average, at this stage it is necessary to pass 35 procedures and spend 189 days, 70% of this time is required for passing procedures related to the confirmation of the readiness for operation of public utility systems and the conclusion of contracts for resource supply, which is a significant problem in the process of implementing the construction project. Passage of administrative procedures at this stage is estimated at an average of 621 thousand rubles. Averages of 86% are the costs associated with the procedures for confirming the readiness for the operation of public service systems and with the conclusion of contracts for resource supply.

At the stage of state cadastral registration and state registration of property rights for real estate, the main problem is a long period of state registration. At this stage, it is necessary to undergo an average of 3 procedures, the total average cost of which is 3090 thousand rubles. The passage of administrative procedures at this stage of the project takes an average of 59 days; however, in some cities it reaches 120 days. The procedure for registering ownership of real estate is regulated by federal legislation and is carried out by the territorial bodies of the Federal Service for State Registration, Cadastre and Cartography. The duration of this procedure is from 3 to 60 days - whereas on average it takes 23 days.

Fig. 1 shows the cumulative average number of procedures for the implementation of investment and construction project in the field of high-rise construction in the following options:

- Obtaining ownership rights to a land plot or the right to conclude a lease contract for a land plot at an auction (Option 1);
- The land plot is owned or leased from the developer for other reasons and does not require a change in the permitted type of use (Option 2 without considering the change in the permitted use);

- The land plot is owned or leased from the developer for other reasons and requires a change in the permitted type of use (Option 2, taking into account the registration of the permitted type of use).

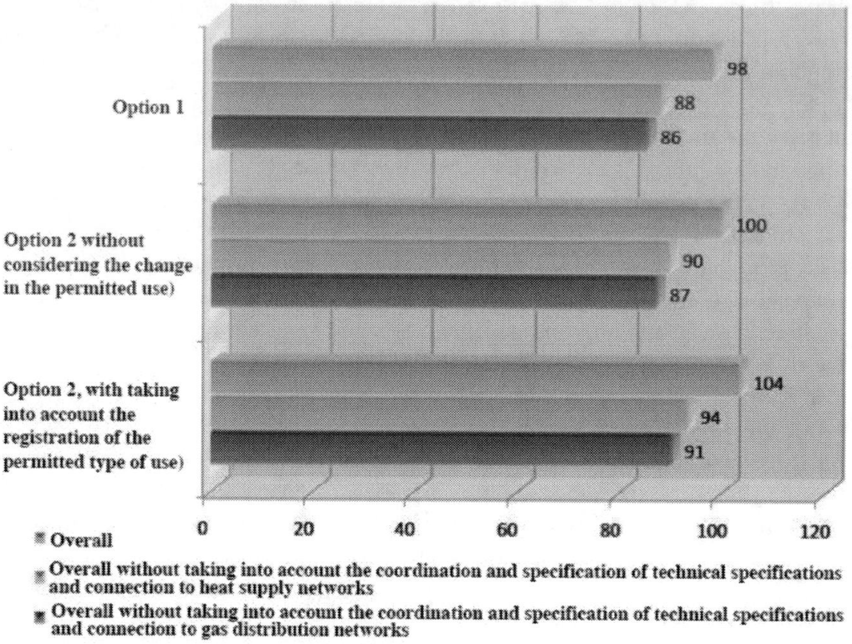

Fig. 1. Aggregate average number of all administrative procedures for the implementation of investment and construction project in the field of high-rise construction.

In general, for the entire period of implementation of the investment and construction project in the field of housing construction, the average number of procedures is approximately the same for all three indicated options for obtaining or having a land plot and ranges from 98 to 104 procedures.

However, the number of administrative procedures is significantly changing depending on the construction phase. Fig.2 shows the minimum, maximum and average values of the number of administrative procedures at particular stages of the investment project in the field of high-rise construction for the case of the developer obtaining rights to the land plot at the auction.

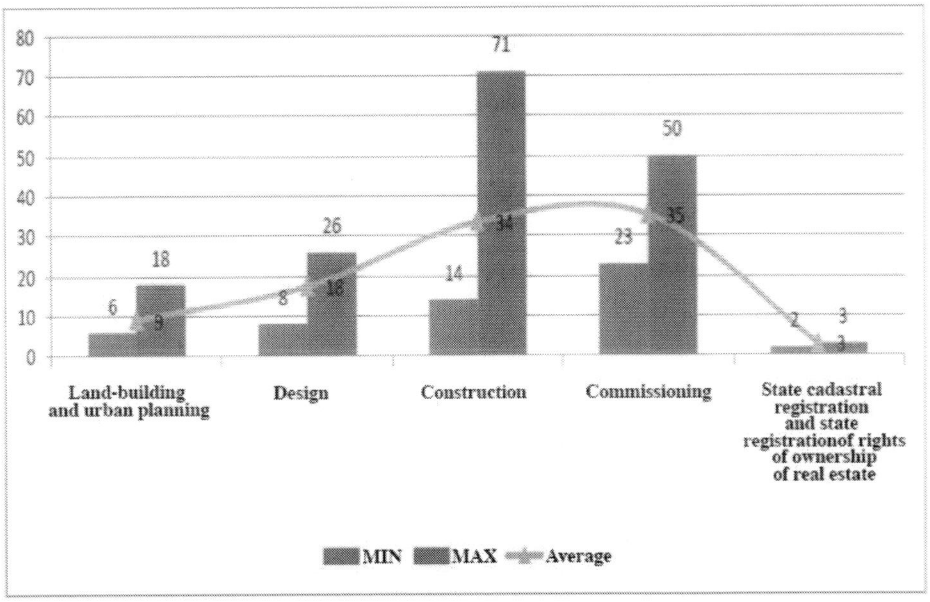

Fig. 1. The average, minimum and maximum number of administrative procedures at particular stages of implementation of the investment and construction project in the field of high-rise construction.

On average, when implementing an investment and construction project in the field of high-rise construction, constructors pass 100 procedures, spending about 3 years and about 25 million rubles [16]. Most of these costs are the costs of connection to the networks of engineering and technical support, including electrical, thermal, gas distribution networks and water supply and sanitation networks.

3 Results

Based on the conducted analysis, the following conclusions can be highlighted regarding the number of administrative procedures for the implementation of the investment and construction project in the field of high-rise construction:

- An average developer needs to complete about 100 procedures. The maximum number of procedures reaches 143 in the case when the land plot that the developer has not received at the auction requires a change in the permitted type of use. The minimum number of procedures is 73 and is achieved in the case when the developer receives rights to the land plot at the auction;
- The general list of all possible procedures includes 182 and 180 procedures respectively in the option of obtaining rights to the land plot at the auction and the option of the developer's rights to the land plot obtained for other reasons;
- Most procedures refer to the stage of commissioning the facility (an average of 34 procedures) and the construction phase (an average of 34 procedures);
- There are significant differences between cities in the composition of administrative procedures;
- Of the total list of all possible procedures common to almost all cities (required in 80-100% of cases) are only about a third of the procedures - 56 procedures in case of obtaining rights to the land plot at auction and 61 in the event the developer has rights to the land plot, received not at auction;

- The greatest differences in cities are noted in the procedures related to the period before the start of construction (stages of land management and urban planning, as well as design);
- The smallest differences are noted in the procedures related to the period after the completion of construction (the stages of commissioning the facility, as well as state cadastral registration and state registration of property rights for real estate objects).

4 Discussion

The study of world (foreign) experience in the implementation of projects in the high-rise construction industry shows [17-22] that excessive administrative barriers lead to a decrease in the efficiency of construction business, significant losses for society. The results of the sub-national study "Doing Business" conducted by the World Bank and the International Finance Corporation [23] showed that in the ranking of countries, according to the criteria for passing permitting procedures in the construction industry, the Russian Federation ranks 115 out of the represented 190 countries, which may indicate a sufficiently high level of administrative barriers. Nowadays in Russia there is a radical transition to the system of self-regulation in the field of urban development, namely, the All-Russian non-governmental non-profit organization "National Association of Self-Regulating Organizations, based on the membership of persons engaged in construction" (NOSTROY) was established. This organization is actively conducting research in the field of construction of the Institute of Urban Economics [24], work is beginning to translate the service for issuing construction permits into electronic form [25]. This is the most significant reform to reduce administrative barriers in the era of digitalization of the economy, but still there is a lack of comprehensive measures to reduce administrative barriers in the high-rise construction industry. Moreover, this does not allow the construction community to take an active position in addressing the problem of reducing excessive administrative barriers, including issues of monitoring and evaluating the effectiveness of administrative procedures in the implementation of investment and construction projects. However, there are factors that prevent the reduction of administrative barriers in the construction of high-rise buildings:

1. Imperfection of the legislation regulating activities in this industry, for instance, there is no clear transparent list of administrative procedures when the developer passes all phases of the investment and construction project and the lack of detailed, approved regulatory acts that reduce administrative barriers at all stages of the investment- building project.
2. Lack of mechanisms for strict control and measures of responsibility for the implementation of legislative procedures.
3. Lack of consideration of regional peculiarities of construction of high-rise buildings and structures.
4. Practically complete absence of "feedback" and coordinated work of federal and municipal authorities.
5. The absence of a unified policy on pricing in the construction, reconstruction and capital repairs of municipal facilities at the expense of budgetary funds.
6. Lack of well-established cooperation with banks on the restructuring of debt on loans received by construction companies in Russian credit institutions, in the implementation of investment and construction projects.
7. Conflict of interests of state authorities and developers.
8. The complexity of procedures for securing the capital construction site with engineering and technical infrastructure, including electrical, thermal, gas distribution networks and water supply and sanitation networks.

Conclusions

Thereby, in order to improve the efficiency of investment and construction projects in the high-rise construction industry, it is necessary to make changes in the legislation to reduce the impact of factors that hinder the reduction of administrative barriers at all stages of project development and implementation, and also provide developers with the opportunity to pass the relevant administrative procedures via digital services.

References

1. A. Bril, O. Kalinina, I. Ilin, *In: MATEC Web of Conferences. EDP Sciences* 8010 (2017).
2. Bjulleten' o razvitii konkurencii [*Bulletin on the development of competition*], (2013), Vypusk **1,** URL: http://ac.gov.ru/files/publication/a/509.pdf. (cited: 19.12.2017).
3. A.V. Anisimov, Jekonomika i biznes: teorija i praktika, 2-11 (2016).
4. A.O. Sitnikov, Nauchnyj al'manah, **7-1**(33), 52-57 (2017).
5. R. I. Malikov, K. E. Grishin, Vestnik UGUJeS. Nauka, obrazovanie, jekonomika, **4** (6), 55-65 (2013).
6. D.V. Manushin, Aktual'nye problemy jekonomiki i prava, **4** (32), 52-60 (2014).
7. M.V. Podbivalova, Zhurnal pravovyh i jekonomicheskih issledovanij, **4,** 264-267 (2013).
8. Portal. Nacional'noe obedinenie stroitelej. NOSTROJ. [*Portal. National Association of Builders. NONSTROY*] – URL: http://nostroy.ru/department/sektor-reestra/monitoring-deyatelnosti-sro/. (cited: 14.12.2017).
9. Gradostroitel'nyj kodeks Rossijskoj Federacii ot 29.12.2004 g. № 190-FZ [*Urban Development Code of the Russian Federation of 29.12.2004 № 190-FZ*] (v red. 31.12.2017. – URL: http://www.consultant.ru/document/cons_doc_LAW_51040/. (cited 10.12.2017).
10. Rasporjazhenie Pravitel'stva RF. № 1336-r [*Order of the Government of the Russian Federation* № 1336-r] «Ob utverzhdenii plana meroprijatij («dorozhnoj karty») «Sovershenstvovanie pravovogo regulirovanija gradostroitel'noj dejatel'nosti i uluchshenie predprinimatel'skogo klimata v sfere stroitel'stva» ot 29 ijulja 2013 g.
11. Postanovlenie Pravitel'stva RF № 403 *[Resolution of the Government of the Russian Federation No. 403*] «Ob ischerpyvajushhem perechne procedur v sfere zhilishhnogo stroitel'stva» ot 30.04.2014.
12. Rasporjazhenie Pravitel'stva RF № 1554-r [*Order of the Government of the Russian Federation No. 1554-r*] «Plan meroprijatij po podgotovke proektov federal'nyh zakonov, aktov Pravitel'stva RF i vedomstvennyh aktov, neobhodimyh dlja otmeny izbytochnyh i (ili) dublirujushhih procedur, a takzhe sovershenstvovanija realizacii procedur, vkljuchennyh v ischerpyvajushhij perechen', utv. Postanovleniem Pravitel'stva RF ot 30.04.2014 g. № 40313.08.2015 g.» ot 13.08.2015.
13. I.V. Ilin, O.V. Kalinina, A.I. Levina, O.Yu. Iliashenko, *In: MATEC Web of Conferences* 5028 (2016).
14. N. Kosareva, A. Puzanov, Fast Forward: City Planning in a Hyper Dynamic Age. ISOCARP Review. (2012).

15. V.V. Gluhov, I.V. Ilin, *In: International Conference on Next Generation Wired/Wireless Networking. Springer* 509–518 (2014).
16. T.G. Polidi, Voprosy imushhestvennoj politiki, **2**(149) (2014).
17. Y. Mahgoub, B. Abbara, Procedia-Social and Behavioral Sciences, **36**, 640-649 (2012).
18. A. Collins, S. Watts, M. McAlister, CTBUH 8th World Congress, 175-185 (2008).
19. V. Antoniucci, C. D'Alpaos, G. Marella, International Journal for Housing Science & Its Applications, **39**. – №. 2 (2015).
20. P. De Jong, H. Wamelink, CTBUH 8th World Congress, 3-5 (2008).
21. E. L. Glaeser, B. A. Ward, Journal of Urban Economics, **T. 65.** №. 3, 265-278 (2009).
22. D. Parker, A. Wood (ed.). The Tall Buildings Reference Book. – Routledge, 512 (2013).
23. Portal Doing Business. Ocenka biznes regulirovanija. [*Portal Doing Business. Evaluation of business regulation*] – URL: http://russian.doingbusiness.org/rankings. – (cited: 20.12.2017).
24. Portal. Institut jekonomiki goroda [*Portal. Institute of Urban Economics*]. – URL: http://www.urbaneconomics.ru/research. (cited: 22.12.2017).
25. Agentstvo novostej «Stroitel'nyj biznes» [*News Agency "Construction business"*] - URL: http://ancb.ru/news/read/4617. (cited: 27.12.2017).

Normative and economic foundations of high-rise construction in the city of Samara

Olga V. Didkovskaya[1], *Aleksey Y.* Bocharov[1], *Marina V.* Ilyina[1] and *Olga A.* Mamaeva[1]

[1]Samara State Technical University, 443001 Molodogvardeyskaya St., Samara, Russia

Abstract. Every year the number of free land plots for construction of buildings is steadily decreasing in cities. In this regard, the participants in the investment and construction process are increasingly seeking to maximize the use of land. A logical way for this is to use high-rise construction. However, until recently, builders encountered serious obstacles in the form of lack of special norms and rules for the design of high-rise buildings. It led to the need for individual coordination of each high-rise facility, the development and approval of special technical conditions, the passage of numerous administrative approvals. Thus, investment activity regarding the construction of high-rise buildings in the Russian Federation is reduced. In 2016, there were regulatory changes that substantially alleviated these difficulties. In this article, the authors analyze the features of the town-planning normative-legal field of high-rise construction and its development, track the interrelations between the regulatory regulation of the construction of similar facilities, with the real need, the technical feasibility and economic feasibility of their construction in the urban district of Samara. Conclusions and suggestions are also based on the analysis of the norms of urban zoning, the residential real estate market and the value of land plots.

1 Introduction

The regulatory vacuum in the system of town-planning regulation in the Russian Federation with respect to the dynamically developing segment - construction of high-rise buildings - is finally overcome by the adoption in 2016 of codes of practice (SP): JV 253.1325800.2016 "Engineering systems of high-rise buildings" (introduced from 04.02.2017), SP 267.1325800.2016 "High-rise buildings and complexes" (introduced from 01.07.2017) [**Ошибка! Источник ссылки не найден.**,**Ошибка! Источник ссылки не найден.**]. With the introduction of these standards, all participants in the investment and construction process [**Ошибка! Источник ссылки не найден.**,**Ошибка! Источник ссылки не найден.**] will have the opportunity to act uniformly throughout the country in the construction of skyscrapers. Consequently, we can conclude that the Ministry of Construction of Russia is going to streamline the construction of high-rise buildings.

In this article, analytical, statistical and expert methods solve the problems of identifying the category of "high-rise buildings", studying the features of their design and

[1] Corresponding author: kafedra_cen@mail.ru

construction, establishing dependencies between the technical and economic indicators of high-rise construction.

2. Methods

Despite the approved JV, regulating the rules for the design of high-rise buildings and their engineering systems, the main question to which the unequivocal answer has not been received is "what is the height of the building that will be referred to the category of high-rise buildings?".

For this purpose, the current system of regulatory legal regulation of the issue was monitored and analyzed. It was found out that the height of buildings in various normative documents is determined in different ways depending on the tasks for which the content of the norm or the law is directed:

– SP 42.13330.2011 "Urban planning. Planning and development of urban and rural settlements"- meters (the maximum parameter -" objects height from the surface of the earth 50 m and more ") [**Ошибка! Источник ссылки не найден.**];

– SP 54.13330.2011 "Residential apartment buildings" - meters (the maximum parameter - "this code of rules applies to the design and construction of newly constructed and reconstructed multi-apartment residential buildings up to 75 m high") [**Ошибка! Источник ссылки не найден.**];

– SP 14.13330.2014 "Construction in seismic regions" - meters (the maximum parameter - "designing of buildings with a height of more than 75 m should be carried out with the support of a competent organization") [**Ошибка! Источник ссылки не найден.**];

– JV 118.13330.2012 "Public buildings and structures" - meters (maximum parameter - "this code of rules applies to the design of new, reconstructed and overhauled public buildings and structures up to 55 m high") [**Ошибка! Источник ссылки не найден.**];

– SP 253.1325800.2016 "Engineering systems of high-rise buildings" - meters (maximum parameter - "this set of rules applies to the design of engineering systems of newly constructed and reconstructed public buildings of more than 55 m in height and residential buildings of more than 75 m in height, including multifunctional buildings and buildings of one Functional purpose ") [**Ошибка! Источник ссылки не найден.**];

– SP 267.1325800.2016 "High-rise buildings and complexes. Design rules »- meters (the maximum parameter -" this code of rules can be used in the design and construction of public buildings above 50 m, as well as multifunctional buildings in which public spaces are located at an altitude of more than 50 m ") [**Ошибка! Источник ссылки не найден.**];

– SP 160.1325800.2014 "Buildings and complexes multifunctional. Design rules »- meters (maximum parameter -" this code of rules applies to the design and construction of new, reconstructed and overhauled multifunctional buildings up to 75 m high with the placement of public premises on floors not higher than 55 m ") [**Ошибка! Источник ссылки не найден.**];

– Federal Law No. 123-FZ "Technical Regulations on fire safety requirements" - floors and meters (the maximum parameter is "more than 17 floors or 50 meters" [**Ошибка! Источник ссылки не найден.**]);

– Urban Development Code of the Russian Federation - meters (the maximum parameter - unique elevations include "a height of more than 100 meters") [**Ошибка! Источник ссылки не найден.**].

Thus, from the above analysis of existing regulations, which refer to the height or number of storeys, we can conclude that the notion of a high-rise building is ambiguous.

Even in newly adopted joint ventures, the difference in determining the altitude is 25 m-50 m and 75 m.

Graphical interpretation of the variability of the regulatory requirements for altitude is shown in Fig. 1.

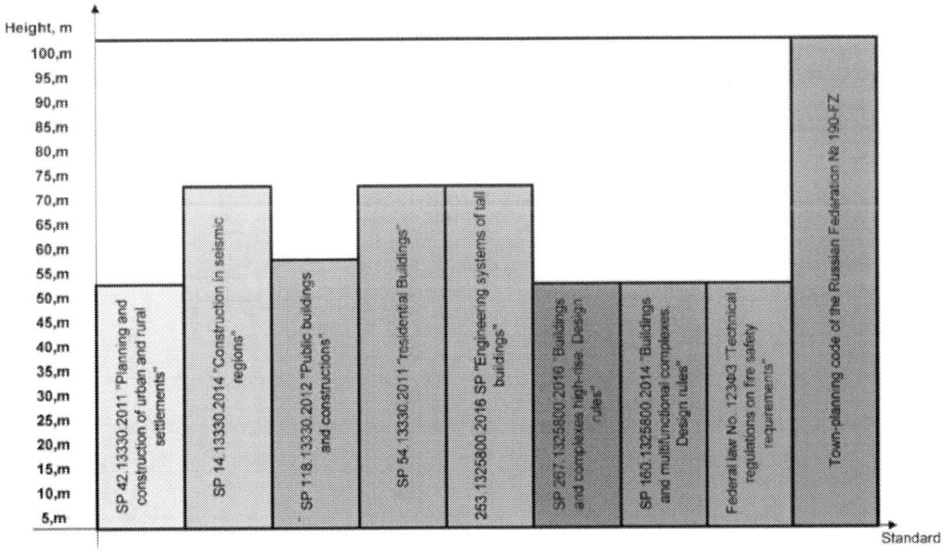

Fig. 1. Regulations for high-rise construction

Based on the results of the analysis of the new joint venture, a number of the following most important provisions can be distinguished when designing high-rise buildings:

- The JV contains stringent requirements agreed with the experts of the Ministry for Emergency Situations of the Russian Federation regarding the safety of construction for buildings with a height of more than 50 meters;
- the need to use structures with an increased fire resistance limit is fixed;
- Provide backup power supply;
- the need is fixed for all floors of equipment of isolated security zones for inactive citizens, where they could wait for firefighters;
- provides for the presence of a refractory service elevator for rescuers, and also takes into account the requirements and regulations for the smoke removal system.

The above requirements significantly increase the unit cost of construction in high-rise buildings (cost per square meter of the total area). At the same time, additional costs arising in comparison with other capital construction objects are made to ensure the safety of users, therefore owners (tenants, users) will demand for such areas.

The interconnected urban planning requirements for the height of buildings the size of the land plots on which they are located are contained in Chapter 4 "Urban Planning Zoning" of the Urban Development Code of the Russian Federation. In particular Art. 38 defines "the maximum (minimum and (or) maximum) sizes of land plots and the maximum parameters of permitted construction, reconstruction of capital construction objects", namely "the maximum number of floors or the maximum height of buildings, structures, structures." Sub-legal requirements are defined in the "Rules of land use and development in the city of Samara" - a document of town-planning zoning, which is approved by the regulatory legal acts of local governments, regulatory legal acts of the state authorities of the constituent entities of the Russian Federation and establishes territorial zones, town planning regulations [**Ошибка! Источник ссылки не найден.**].

On the basis of this standard, Table 1 is compiled reflecting the parameters of the maximum permissible number of storeys (altitude) of residential and non-residential buildings in the city of Samara.

Table 1. The maximum allowable number of storeys (high-rise) residential and non-residential buildings in the g. o. Samara

Types of territorial zones	Maximum height of buildings	
	The main permitted uses of land	Ypes of real estate use that require special approval
Zones of business and service functions		
C-1 Zone of business and commercial enterprises within the boundaries of the historical part of the central district	Multi-apartment houses (MCD) up to 3 floors	MCD above 3 floors
C-2 Zone of regional and city centers	MCD over 4 floors	MCD up to 5 floors
C-3 Service area of local (district) population	MCD above 3 floors	Residential buildings (railway) up to 3 floors
C-4c Zone of specialized commercial use of sports and entertainment facilities	-	-
C-4т Zone of specialized commercial use of public transport facilities	-	-
C-5м Zone of specialized non-commercial use of medical facilities	-	-
C-5н Zone of specialized non-commercial use of educational and scientific facilities	-	-
C-5к Zone of specialized non-commercial use of objects related to the worship	-	-
Residential areas		
G-1 Zone of low-rise buildings with individual houses	Railway station up to 3 floors / 12 m	Railway station up to 4 floors / 12 m
G-2 Low-rise mixed residential development zone	Railway station up to 4 floors / 12 m, MKD / 15 m	-
G-3 Medium-rise residential area of 3-6 floors	MKD up to 6 floors / 30 m	MKD above 6 floors / 51 m
G-4 Zone of multi-storey residential development of 5-16 floors	MKD up to 9 floors in the historical part of the city / 30 m or MKD above 5 floors outside the historical part of the city / 51 m	MKD above 9 floors in the historical part of the city / 51 m
G-5 Zone of planned housing	-	-
Recreation Area		
P-1 Zone of the center of recreational areas	-	-
P-2 Zone of parks, boulevards, embankments	-	-
P-3 Zone of natural landscapes	-	-
P-4 Zones of Specially Protected Natural Areas	-	-
P-5 Area of horticultural sites and collective gardens	-	-
Industrial and communal areas		
PC-1 Zone of enterprises and warehouses of V-IV hazard classes (sanitary protection zones - up to 100 m)	-	-
PK-2 Zone of enterprises and warehouses of the third class of harmfulness (sanitary protection zones - up to 300 m)	-	-
PC-3 Zone of enterprises and warehouses of II-I hazard classes (sanitary protection zones - up to 500 m and more)	-	-
Rzv Reserve areas	-	-

Based on the analysis of Table. 1, we can conclude that the data on the maximum parameter of buildings are limited to 51 meters. Therefore, given that this type of document regulates the parameter of permitted building of ordinary buildings, we can assume that the height of 51 m is the minimum height with which high-rise buildings begin. In this case, buildings taller than 100 meters in accordance with the Town Planning Code of the Russian Federation are unique.

3 Results

Analysis of the regulatory legal framework of the issue under consideration allows us to define the category of "high-rise buildings" for the conditions of their construction in the territory of the city. Samara with the following parameters:

- height - more than 50 meters;
- storeys - more than 17 floors;
- accommodation in the residential development zone Zh-3, Zh-4;
- possible placement in the area of business and service functions.

The assumption of the prospects for the development of high-rise construction in the city of Samara was made by the authors on the basis of the analysis of the map "Rules of building and land use in the city of Samara"[11].

From the analysis of the map of the "Rules of development and land use in the city of Samara" we can conclude that the potential location of skyscrapers is possible in an area of up to 10% of the city. This is a high index in comparison with the majority of cities in both the Volga Federal District [**Ошибка! Источник ссылки не найден.**] and the Russian Federation as a whole.

Next, we will analyze the housing market in the city. Samara for March - April 2017 in the context of proposals for the sale of apartments in buildings with a distribution by altitude. According to the number of storeys, the apartment houses are conditionally divided into five groups [**Ошибка! Источник ссылки не найден.**]:

- low-rise - with a height of 1-3 floors (up to 9 meters);
- medium storeys - 4-5 floors (up to 15 meters);
- increased number of storeys - 6-9 floors (up to 27 meters);
- multi-storey - 10-16 floors (up to 48 meters);
- high-rise - more than 17 floors (more than 51 meters).

This division formed the basis for the analysis of the market of apartments in multi-apartment houses (hereinafter referred to as MCD) (see Table 2).

Table 2. Distribution of the offer of the market of apartments in inhabited multi-apartment houses by number of storeys

No. in order	MKD group of storeys	Number of apartments, pcs.	Share of total, %
1	Low-rise	133	9,72
2	Average number of storeys	477	34,87
3	Elevated number of storeys	314	22,95
4	Multi-storey	336	24,56
5	**High-altitude - more than 17 floors (more than 51 meters)**	**108**	**7,89**
	Total	**1368**	**100,00**

Analysis of Table. 2, shows that the segment "high-rise buildings" in the residential real estate market. Samara is represented by 108 offers for sale, which is almost 8% of the total market of apartments. In Table. 3 presents the price analysis of the market for the supply of apartments in the MKD in the context of their accepted classification by number of storeys.

Table. 3 shows that the dependence of the cost of an apartment on the height of the building on the market is present, although it is substantially smoothed as the number of storeys increases. Further, the market of apartments located in high-rise buildings was investigated (Table 4).

Table 3. Analysis of the average cost of apartments in residential apartment buildings, depending on the number of storeys

No. in order	The group of number of storeys	The average cost of 1 square meter, thousand rubles.
1	Low-rise	47,07
2	Average number of storeys	59,50
3	Elevated number of storeys	65,33
4	Multi-storey	66,52
5	**High-altitude - more than 17 floors (more than 51 meters)**	68,46

Table 4. Distribution of apartments in high-altitude MKD by districts of the city. Samara

No. in order	Districts of the city	Number of apartments, pcs.	Share of total,%
1	Railway district	23	21,30
2	Kirovsky district	5	4,63
3	Leninsky district	11	10,19
4	Oktyabrsky district	50	46,30
5	Industrial area	16	14,81
6	Samara Region	1	0,93
7	Sovetskiy district	2	1,85
	Total:	**108**	**100,00**

Analysis of Table 4 showed that more than 92% of all "high-rise" apartments are located in four of the seven districts of the city. Kuybyshevsky and Krasnoglinsky districts do not have objects related to the "high-rise buildings" segment on their territory, therefore they are not presented in Table 4.

4 Discussion

For the discussion in this article, the following research results and suggestions of the authors are made:

- the assumption of the development of high-rise construction in the territory of the city of. Samara, made on the basis of an analysis of the rules of land use and development (Table 1, Figure 2);
- dependence of the cost of the apartment on the building's height, established as a result of the analysis of the residential property supply market (Table 3, Figure 3);
- dependence of the supply of apartments in high-altitude MKD on the value of the ratio of the value of 1 sq.m. Of the land to the cost of 1 sq.m. Apartments (Table 5, Figure 4).

The figure presented below. 3 reflects the form of the statistical dependence of the average cost of flats (1 sq.m.) on the number of floors in which they are located, detected analytically (Table 3). The logarithmic function in this case is the most reliable (reliability of approximation (R2 = 0.959)) for constructing the trend. This contributed to several factors: first, the area of the apartment, and secondly the quality of the finish. These factors were not taken into account in this study.

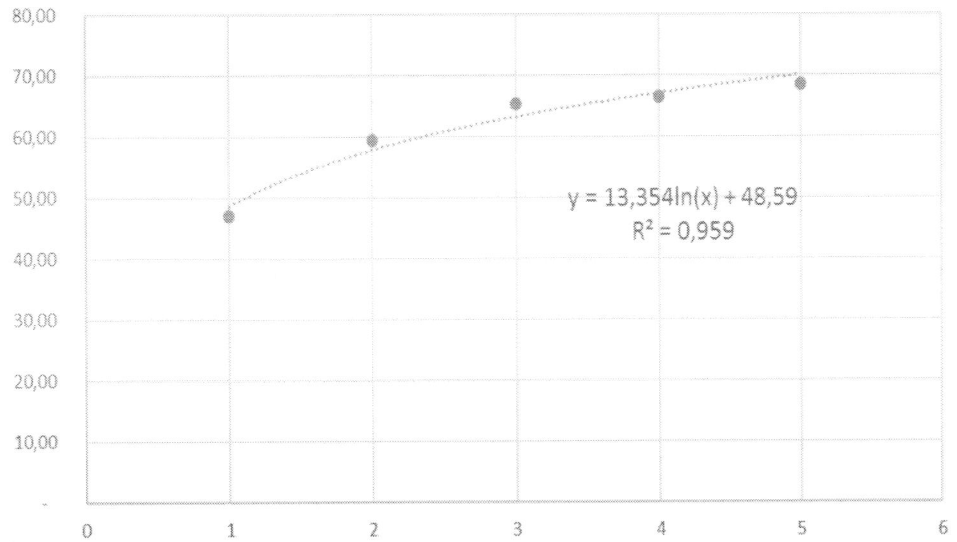

Fig. 3. The logarithmic dependence of the average cost of apartments (1 sq.m.) on the number of storeys

Table. 5 and Fig. 4 illustrate the relationship between the cost of land and apartments in high-altitude MKD.

Table 5. Analysis of the average cost of apartments in high-altitude MKD and the value of land plots by districts of the city. Samara

No. in order	Districts of the city	The average cost of 1 square meter, thousand rubles.	The average cost of 1 sq.m. of the land plot, thousand rub.	The cost ratio of 1 sq.m. of the land to the cost of 1 sq.m. apartments, %
1	Leninsky district	86,06	11,90	14%
2	Oktyabrsky district	64,75	6,38	10%
3	Railway district	67,80	5,98	9%
4	Industrial area	67,11	5,90	9%
5	Sovetskiy district	73,02	4,17	6%
6	Kirovsky district	72,39	3,30	5%
7	Samara Region	130,47	7,00	5%

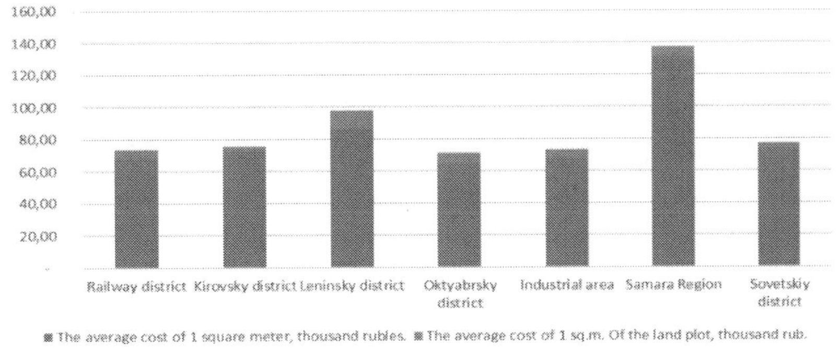

Fig. 4. Share of the cost of 1 sq.m. of the land plot to the cost of 1 sq.m. apartments

From the analysis of Table 5, it can be assumed that there is a dependence of the supply volume in the segment of the sale of apartments in high-rise apartment blocks of multi-apartment buildings on the ratio of the average value of land plots and apartments. Those. From Table 5, it can be seen that when the indicator "Value ratio of 1 sq.m. Of the land to the cost of 1 sq.m. Apartments "the share of the land plot in the cost of apartments in high-rise apartment buildings is reduced (Figure 4).

This conclusion is made on the basis of a short-term study of the local market and requires detailed comprehensive study. Lonely this circumstance is quite logical, i.e. With growing demand for the final construction product - flat (in the case under consideration), calculation of demand for land in a certain location (area) [14 - **Ошибка! Источник ссылки не найден.**]. And this circumstance in turn prompts the builder to increase the number of storeys, thereby reducing the share of the land plot in the cost of the apartment.

5 Conclusion

The construction of high-rise buildings, the number of floors of which has already reached a hundred, is a promising area in urban development. The ability to locate on a relatively small plot of land a huge amount of residential and commercial space makes the construction of skyscrapers very attractive for developers. In the Russian Federation in general and in the Samara region in particular, high-rise construction was accompanied and accompanied by many difficulties. In this connection, this type of construction does not have a high investment attractiveness.

Strengthening the legitimacy of design and construction standards in this area, in the opinion of the authors, will contribute to the orientation of the construction complex of the region for capital investments in the construction of high-rise buildings of various functional purposes [0-**Ошибка! Источник ссылки не найден.**].

The comprehensive analysis of town-planning norms, land and real estate markets and the dependencies established on their basis show the developers economic and other prospects for building high-rise buildings in the capital of the region

References

1. http://docs.cntd.ru/document/456044284 (date of circulation: 04.24.2017).
2. http://docs.cntd.ru/document/1200139948.
3. http://docs.cntd.ru/document/1200084712 (reference date: April 24, 2017).
4. http://docs.cntd.ru/document/1200084096 (reference date: April 24, 2017).
5. http://docs.cntd.ru/document/1200111003 (the date of the application: 04.24.2017).
6. http://docs.cntd.ru/document/1200092705 (reference date: April 24, 2017).
7. http://docs.cntd.ru/document/1200113272 (reference date: April 24, 2017).
8. http://docs.cntd.ru/document/902111644 (date of circulation: 04.24.2017).
9. http://docs.cntd.ru/document/901919338 (reference date: April 24, 2017).
10. http://docs.cntd.ru/document/945005955 (reference date: April 24, 2017).
11. http://map.samadm.ru/pzz/ (date of circulation: 04.24.2017).
12. http://pfo.gov.ru (reference date: 04.24.2017).
13. I. A. Sinyansky, Typology of buildings: a textbook for stud. Establishments of environments. Prof. Education I. A. Sinyansky, N. I. Maneshina, 288 (Moscow, Publishing Center "Academy", 2014) ISBN 978-5-4468-0595-2

14. M. I. Balzannikov, A. A. Mikhasek, V. P. Popov, *Study Results Assessment of Quick-hardening Liquids Penetration into Soil Elements of Hydraulic Engineering Constructions, Procedia Engineering*
15. E. E. Kireeva, A. D. Kornilova, *Research of Instruments for Financing of Innovation and Investment Construction Projects, Procedia Engineering*
16. O. Mamaeva, A. Bocharov, Construction audit as a modern instrument of investment and construction process management, *MATEC Web of Conferences* 86, 05007
17. O. V. Didkovskaya, O. G. Mamayeva, Development of Cost Engineering System in Construction, *Procedia Engineering*
18. S. V. Domnina, E. V. Savoskina, N. V. Shekhova, On Innovative Decisions in the Investment-construction Cycle, *Procedia Engineering*.
19. E. M. Generalova, V. P. Generalov, A. A. Kuznetsova, Modular Buildings in Modern Construction, *Procedia Engineering*.
20. Y. V. Sidorenko, S. F. Korenkova, Self-organization processes in silicate and cement construction materials, *Solid State Phenomena*.
21. T. Y. Vavilova, N. D. Potienko, I. V. Zhdanova, On Modernization of Capital Construction Projects in the Context of Sustainable Development of Social Sphere, *Procedia Engineering*.
22. D. V. Zelentsov, K. L. Chertes, O. V. Tupicina, Theoretical Basis and Experimental Study of the Aeration Characteristics of the Composting Mixtures during the Design and Construction of the Aeration System of the Oily Waste Biodegradation Complex, Procedia Engineering, **153**, 903-908
23. A. Bocharov, I. G. Burtsev, Cost Engineering Investment-construction project Economic aspects of management of the building complex in modern conditions: a collection of articles ed. M.I. Balzannikova, K.S. Galitskova, N.V. Shekhovoy; SGAU, 224-229 (Samara, 2016)
24. Yu. P. Sologubov, T. E. Gordeeva, Analysis of planning solutions for residential buildings for energy efficiency, *Urban planning and architecture*, **4(21)**, 104-107 (2015)

 https://doi.org/10.1051/e3sconf/20183303012

Specificity of high-rise construction and real estate markets in the regional economy: an analysis of Russian practice (example of St. Petersburg)

Viktoriya Vilken , *Olga* Kalinina, *Alissa* Dubgorn*

Peter the Great St. Petersburg Polytechnic University, Institute of Industrial management, Economy and Trade, 195251, Polytechnicheskaya, 29, St.Petersburg, Russia

Abstract. In paper features of the regional markets of construction the commercial and residential real estate on the example of St. Petersburg are defined. The current situation is analysed, the specific features of branch are revealed. The major factors influencing investors making decisions on construction of various types of objects are defined. The main methods of advance of real estate objects are considered.

1 Introduction

Among the elements of a market economy, a special place is occupied by real estate and construction, which act as means of production and objects of consumption (land, houses, dachas, apartments, garages). Real estate is the basis of personal existence for citizens and serves as a base for economic activity and development of enterprises and organizations of all forms of ownership. In Russia there is an active development of the real estate market and an increasing number of citizens, businesses and organizations are involved in real estate transactions. Real estate is the main subject of discussion in the privatization of state and municipal property, when renting non-residential premises, when buying and selling residential premises.

In turn, construction is one of the key, fund-forming industries that largely determine the pace of development of the economy of the region and the country as a whole, as well as the solution of the most important social and economic tasks.

The level of competition in the North-West region of Russia is very high, and in order to withstand competition, it is necessary to know the market, the strengths and weaknesses of competitors, to be superior in terms of professionalism in the chosen field.

The purpose of this research is to identify the main and specific features that characterize the markets of construction of commercial real estate and residential housing construction, as well as the real estate market in the field of rent in the Russian practice on the example of St. Petersburg.

* Corresponding author: alissa.dubgorn@gmail.com

2 Analysis of real estate market of St. Petersburg

The commercial real estate market is oligopolistic, the following barriers are established at the entrance to the industry: first of all, state regulation and strict standards that need to be met, the scale economy factor is also important.

The competition between already functioning companies in the market is centered around the struggle for the client. Also, companies compete for areas and construction sites, in terms of price, service levels and additional services provided to customers.

The logic of the differentiation strategy requires that the company puts behind the basis of differentiation such attributes of the product that would distinguish it from the similar product of competing companies. Indeed, there are very few multifunctional structures in the field of office real estate, because the threat of incorrect positioning is very high and the costs for differentiation are also high [1]. Unlike the strategy of cost leadership, implementation of the differentiation strategy does not require the presence of only one leader in the industry - in this case there may be several companies that successfully implement the differentiation strategy [2].

The heavy deficit of hotel rooms in St. Petersburg, as well as the declared preferences from the city authorities for developers of the hotel sector, led to an increase in interest in multi-profile facilities that combine hotel and other functions.

According to the forecast for the development of the market for high-quality office real estate, the demand for quality office space will grow by the time the facility is put into operation, with more attention paid by tenants to the availability of services, of which the most significant are: the required number of parking spaces and availability of a hotel in the nearest zone [3-4]. Due to the fact that rental rates in business centers of class "A" are in the highest price segment, the tenants of the business center will be companies for which the image component of their location and the level and scope of services are important [5-6].

Tenants of class "A" office space are most often represented by international companies, large Russian companies and foreign representational offices.

By type of activity, the most represented companies are in the wholesale and retail trade (21%), as well as in the services sector (11.5%).

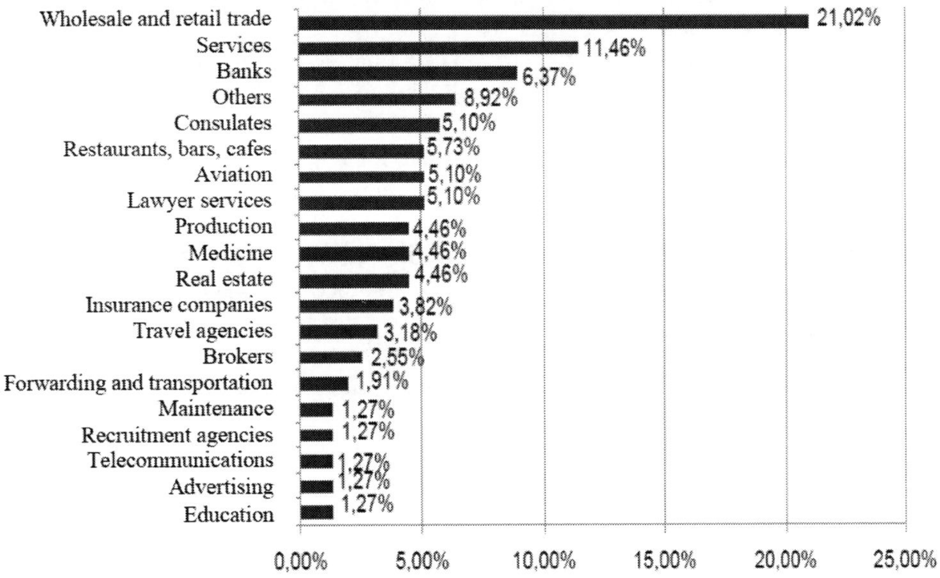

Fig. 1. Distribution of tenants by business lines in St. Petersburg, 2017

The greatest demand in the market as a whole is for offices with an area of up to 100 square meters, as well as 100-300 square meters. Tenants of such areas are Russian companies and representations of foreign companies. The demand for premises of 300-500 square meters is reduced. High demand is the premises of 501-600 square meters. Tenants are represented by international companies, large Russian companies and foreign representation offices. Annually before the crisis there was an increase in demand for large areas - from 1,000 square meters. Tenants of such areas were financial, insurance, legal, consulting companies. Most likely, this trend will resume after the end of the economic recession. Such companies conclude their contracts for six months or more before the end of the construction of the business center.

Is it expedient to choose the city of St. Petersburg for the implementation of a project? What is the basic information about this segment of the market? According to official statistics, the St. Petersburg economy employs approximately 2,445,200 people. Tenants of business centers are mainly companies working in the field of computer technology, finance, services, real estate, communications.

The share of population employed in these industries is 586 200 people, or 24% of the total number of workers. One employee requires about 7 square meters of office space.

Thus, the market capacity, according to approximate calculations, is 4 103 400 square meters. With an available supply volume of 1,644,700 square meters, the deficit reaches 2,458,700 square meters.

According to the company Praktis CB, the provision of Petersburg with high-quality office real estate is 359 square meters for 1 thousand people, which is 8 times less than the similar average in Europe. The average In Europe the average account of office space per person is between 2.5 and 4.5 square meters. In particular, in Warsaw, the availability of business areas is 1 thousand 450 square meters per one thousand citizens, in Paris - 5 thousand 600 square meters, in Stockholm - 5 thousand 700 square meters, in London - 7 thousand 100 square meters. It is also worth noting that in Moscow this indicator is 540 square meters per one thousand citizens. The total market potential in 2017 was 14.6 million square meters. Based on the average European supply of office space, the capacity of the business real estate market in St. Petersburg will be 15.97 million square meters in the next 10-15 years.

Due to the current post-crisis situation, the market for business centers is unstable, but some positive trends are to be noticed.

The volume of input of real estate in first quarter of 2016 was 34.2 thousand square meters, which is comparable to the average value of input in the first quarters of 2009-2014. The commissioning in 2017 amounted to 210-220 thousand square meters, in 2018 this indicator will be about 160-170 thousand square meters.

Dynamic growth of high-quality supply is continuing from 2009 to the present. There is a redistribution of the structure of business centers in the direction of increasing the shares of classes A and B and reducing the share of the class C.

At the moment when choosing the class of an object it is necessary to take into account the number of objects being under construction, as well as the number of frozen and suspended objects, the input of which can significantly affect the dynamics of supply and demand in the industry. The number of business centers of class B being under construction is 57%.

LCMC specialists have revealed that for the period from 2010 to 2017 the construction of 24 commercial real estate projects (including shopping complexes, warehouses and hotels) with a total area of 10.8 million square meters was suspended for an indefinite period. Most of them belong to the class B. This means that no matter how high the demand for office space of B class, it will be satisfied at the expense of frozen and suspended objects.

Accordingly, it can be concluded that the development of investment projects in Class A is purposefully.

In the process of designing a new multifunctional center, it is necessary to solve the problem of the future offices floor-space. Offices should be spacious enough, but not too much so that the tenant does not get the impression that he pays for extra meters.

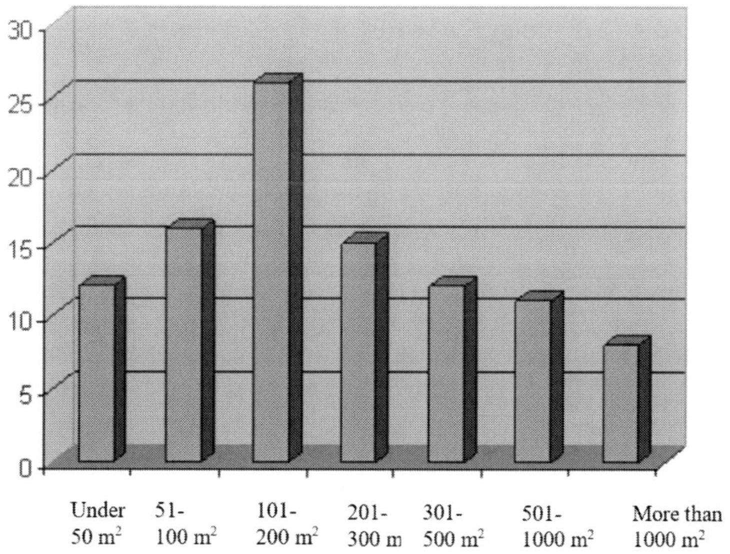

Fig. 2. The structure of demand in the business centers of class A by the size of floor-space in St. Petersburg, 2017

As can be seen from Fig. 2. it is advisable to offer offices from 51 to 300 square meters. In the pre-crisis period, the market had a tendency to increase the demand for offices more than 1000 square meters. Most likely, by the time the objects are completed, this trend will resume.

It is necessary to clearly distinguish the profiling function of the object. Depending on the main purpose of the object, one of the functions is the main one, and the second is the "satellite".

The combination of hotel and office functions is not so common as hotel-shopping. Hospitality is the most complicated function in terms of implementation and management. A fairly successful concept of the facility will help to ensure not only the accommodation (in the hotel) of business travellers working in the office part of the multifunctional center, but also to accommodate numerous non-resident participants of the events carried out in the same center.

Thus, more than half of the declared office space of class "A" are situated in the multifunctional centers.

The infrastructure objects in business centers of class "A" are located primarily on the ground floors. Food facilities can be located on the upper floors. Large infrastructure facilities are located separately from the main building. The class "A" business center usually includes the following infrastructure objects: food facilities (cafes, restaurants); services (small shops, bank office, ATM, currency exchange, laundry, medical center, dry cleaning, travel agency); business infrastructure (conference hall, meeting room, hotel); sports and recreational infrastructure (sauna, fitness center, sports hall, tennis court); entertainment (theater, nightclub, casino, children's entertainment center).

Based on all of the above, we can say about the conclusion about the greater preference on the part of investors for the construction of multifunctional centers.

This will help to diversify risks to some extent, but at the same time increase investment costs.

Generally, there are two groups of methods for promoting real estate objects:

1. Methods based on the key role of the marketing department.

These methods include placing advertising in the media and other types of advertising media, and then receiving and processing calls from potential customers. This method uses outdoor advertising, print media and, of course, the Internet. In recent years, the Internet has become one of the best means of advertising. One of the most effective ways to promote real estate is contextual advertising on the websites.

2. Methods based on the key role of the brokerage department.

The active methods of attracting tenants include: contacts on the client base, personal connections. The effectiveness of the promotion method increases, if it is combined with a positive image of the company in the market. "Cold" calls, i.e. contact by phone with the offer of their goods and services are also a method based on the key role of the brokerage department.

A costly investment project needs to be assessed in terms of the effectiveness of investing money. For complex technology and long-term investment projects, it is more expedient to use a systematic approach and formalized methods of economic evaluation of the expected return on investment. This was discovered already in the second half of the last century by engineers designing a completely new technology and technology for the time: power lines, bridges and railways [7]. They needed to estimate the cost recovery for additional railway branches or additional copper consumption for wires with better conductivity and lower thermal losses.

Within the engineering economics, new directions have appeared already in the 20th century, with a large share of mathematical and statistical calculations (operation research, optimization of technical solutions in conditions of uncertainty), and more applied technical and economic ones (the method of reduced costs, functional and cost analysis). Based on these methods, as well as the development of the theory of financial management and the analysis of cash flows, the methodology of commercial project appraisal develops by the mid of 1970s.

In the USSR, at the end of the 20-ies in the metallurgy and power engineering and other branches of the developing heavy industry, the methodology of recoupment as a method of reduced costs begins to be used for a formalized evaluation of projects. Since the mid of 30s, the share of systematized estimates has fallen and is almost completely replaced by an intuitive approach (discretionary decisions). Among the systematized assessments of numerous at that time projects, the analysis of only the cost of production of future production becomes the most common (without taking into account the initial capital expenditures – investments). Practical decision-making on the criterion of the minimum cost price becomes a prerequisite for the creation of a new methodology of functional and cost analysis in the 40s of the 20th century.

It should be noted that the method of reduced costs is still used by some developers in Russia to evaluate the project and usually leads to unsuccessful attraction of investors. However, this method has in the current situation a somewhat different and rather narrow scope. It is effective for comparing various technological alternatives within a single enterprise, while investing own funds in the development and re-equipment of existing production with new equipment.

The method of recouping costs in a market economy was transformed by the mid of 1970s (based on the use of new methods of financial management) in the methodology of commercial project evaluation.

The variety of methods used by various companies and state bodies necessitated their systematization, which was what the intergovernmental organization at the United Nations did. The so-called UNIDO methodology was created, on which the most widely used international standards for evaluating investment projects and business planning are based now.

The use of formalized methods of economic evaluation, even with significant uncertainty of projects, allows decision-makers to better understand the risk factors in investment alternatives, to obtain information for a more objective choice of economical options [8]. The application of systematic methods of economic evaluation makes it possible to exclude strong-willed decisions and demonstration of executives extravagance as a basis for choosing investment decisions [9-10].

However, when choosing the method of commercial evaluation of real estate construction projects, it is necessary to dwell on the general criteria for making decisions by investors whose choice is currently debatable, namely:

• Completeness and reliability of the initial information on the project (for example, in addition to data on planned costs and revenues, ownership rights of all project participants, public opinion and many other, often non-financial and non-economic information);

• Reliability and credibility of prospective project managers and developers, assessment of the possibility of solving financial and organizational issues, their financial and technological viability;

• Reliability and credibility of analysts and experts in the commercial evaluation of the project.

3 Conclusion

As a result of the analysis of the development of construction and real estate markets in St. Petersburg, the following conclusions can be drawn:

1. The most attractive target is the target segment of industrial construction, it is the most profitable and high-tech segment in terms of the use of global and innovative technical solutions, demanding the appropriate technical specialists in the company. Also, the entry barrier for competitors is the availability of the necessary documents confirming their qualifications and financial stability, preventing them from starting work with minimal financing or by confirming guarantees, by pledging assets. All these factors do not allow medium and small companies to enter this segment, leaving only strong players who lead the consolidation and are struggling to increase the share of the target segment in the industry, not by lowering the price, but by offering additional value to the consumer.

2. The segment of construction of commercial real estate is the most competitive segment of the construction industry in the field of engineering; it contains most of the competing companies that do not have a clearly pronounced competitive advantage in relation to each other. Moreover, the consumer in this segment is not ready to pay for the advantage and the price of services that generates price wars comes to the fore.

3. The segment of residential housing construction is the least profitable segment due to the primitiveness of engineering systems. Intermediate consumers of this market (general contractors) follow the path of direct integration, creating within their structure assembly units that perform these works for the end user. Through the effect of scale, general contractors create serious competition for engineering companies, reducing profitability to a minimum or zero in this segment. The main criterion of the consumer in this segment is the price. Authors should use the forms shown in Table 3 in the final reference list.

https://doi.org/10.1051/e3sconf/20183303012

References

1. D.G. Proverbs, Holt, G.D., Olomolaiye, P.O. International Journal of Project Management. Vol. **17**, Issue 3, pp. 195–204 (1999)
2. V. Ireland, Construction Management and Economics, Vol. **3**, Issue 1, pp. 59–87 (1985)
3. R. Sacks, Journal of Construction Engineering and Management. Vol. **133**, Issue 5, pp. 374–384 (2007)
4. Y.-M. Cheng, International Journal of Project Management. Vol. **32**, Issue 5, pp. 850–860 (2014)
5. R. Turkington, High-rise housing in Europe. Current trends and future prospects. DUP Science. Delft University Press. The Netherlands.
6. G. Finkel, *The economics of the construction industry. Routledge*. (Taylor and Francis Group. New York. USA, 2015)
7. I. Ilin, et al, Model of asset portfolio improvement in structured investment products. Life Science Journal, Vol. **11**, Issue 11, pp. 265–269 (2014)
8. L.V. Nikolova, D.G. Rodionov, N.V. Afanasyeva, European Research Studies Journal, Vol. **20**, Issue 2, pp. 396–410 (2017)
9. V.V. Glukhov, I.V. Ilin, Lecture Notes in Computer Science, Vol. **8638 LNCS**, pp. 509–518 (2014)
10. I.V. Ilin, A.B. Anisiforov, WSEAS Transactions on Business and Economics, Vol. **11**, pp. 757–764 (2014)

Life quality and living standards in big cities under conditions of high-rise construction development

Elena Avdeeva[1,*] *Tatiana* Averina[1], *Larisa* Kochetova[2]

[1]Voronezh State Technical University, Moscow Avenue, 14, Voronezh, 394026, Russia
[2]Russian University of Justice (Central branch), 20-letya Oktyabrya 95, Voronezh, 394006, Russia

Abstract. Modern urbanization processes occurring on a global scale inevitably lead to an increase in population density in large cities. People assess the state of life quality and living standards of megalopolises under conditions of high-rise construction development ambiguously. Using SWOT analysis, the authors distinguished positive and negative aspects of high-rise construction, highlighted threats to its development and its opportunities. The article considers the model of development of the city's industry and infrastructure, which enables determining the optimal volume of production by sectors and branches of city economy in order to increase its innovative, production and economic potential and business activity.

1 Introduction

Due to the increase in the number of people wishing to live in major centers of a country and the world, there is a sustainable tendency of forming megalopolises with a population of more than 1 million people and urban agglomerations that count more than 10 million people. This process began in the second half of the last century. In those years, the population of New York was more than 10 million people. In 1975, there were 5 cities of such size, in 2000 the number was 19, and according to the forecast 23 cities will overcome this line by 2025. The total number of high-rise buildings in the seven years has tripled. Leadership in the construction of skyscrapers belongs to Asia and the Middle East.

The top five global cities of GCI-2015 included New York, London, Paris, Tokyo and Hong Kong. Moscow was on the 14th place. As the evaluation criteria were used: political interaction, cultural and information exchange, human capital and economic activity.

According to Wikipedia, the megalopolis is the largest form of settlement, formed when a number of urban agglomerations merged. For the first time the term megalopolis appeared in 1915 in the book "Cities in Development" by the Scottish sociologist Patrick Geddes. Later, in 1938, American historian and philosopher of technology L. Mumford described the megalopolis as the first stage of urban overdevelopment and social decay. He made the conclusion about the initiating role of the subject in the placement of industry; considered hierarchical civilizations as megamachines and concluded that technology become an active subject of reality, transforming a person. In 1945 french geographer Jean Gottmann used this term to denote a continuous urban development with a length of more than 1000 km

* Corresponding author: avdeeva_ea@mail.ru

and a width of up to 200 km along the Atlantic coast of the United States, which interconnected agglomerations of Boston, New York, Philadelphia, Baltimore, Washington. The population there amounted to 40 million people. He determined the mechanism of folding regions, states, civilizations as a natural interaction of close and related local "ways of life" .

The aim of this work is to study and evaluate the world trend of the formation of megalopolises and the rapid growth of high-rise construction, to analyze the impact of increasing the density of the population on the quality and standard of living in them. With the help of modeling the feasibility and economic attractiveness of the large city development was shown.

Large cities are leaders of regional and national development. Megalopolis is a weakly structured economic system with a large number of internal and external links. This economic and social system ensures the processes of current consumption, the reproduction of resources, the realization of future needs [1].

It attracts flows of labor resources and entrepreneurs, also being a large consumer market, where consumer and industrial goods and services are sold on a large scale. V. V. Ivanov believes that this is "the spatial structure of the post-industrial economy, the development mechanisms of which require special study. They are concentrated huge economic potential "[2].

2 Materials and Methods

For the analysis, statistical methods are used and SWOT analysis is applied. Any modern city is distinguished by the presence of high-rise buildings, characterizing it as highly developed, since this implies a high development of construction technologies, the availability of highly qualified specialists of architects, designers, developers, etc. Increasing the population density of large cities requires qualitatively new approaches to ensuring the safety of life and the use of new technologies as a factor of sustainable growth. The experience of European countries shows that the main increase in GDP is created in megacities [3]. These are the so-called "global cities" - a city that acts as the main driving force of progress in all areas of society, both in the scale of the national economy and the world system (Tables 1 and 2).

Table 1. 10 countries with a high share of the world population, density of their people and GDP per head.

№	**Country**	**Population, people as of 10/07/2017**	**Share of world population, % as of 10.07.2017**	**Density of populatio n, people per km2, as of 10.07.201 7**	**Average life expectancy , years (according to 2015)**	**GDP (nominal) per capita, $ (according to IMF data, 2016)**
1	China	1 384 160 000	18.4%	648	76.1	8113
2	India	1 318 270 000	17.5%	357	68.3	1723
3	USA	325 375 000	4.33%	32	79.3	57436
4	Indonesia	261 600 000	3.48%	127	69.1	3604
5	Brazil	207 709 000	2.76%	23.6	75	8727
6	Pakistan	197 750 000	2.63%	229	67.4	1468
7	Nigeria	188 500 000	2.51%	165	54.5	2211
8	Bangladesh	162 752 000	2.17%	1154	71.8	1411
9	Russia	146 400 000	1.95%	8.56	70.5	8928
10	Mexico	129 100 000	1.69%	57	76.7	8555

About 4.3 billion people or about 58% of the world's population live in these ten countries.

Table 2. The largest cities in the world in terms of population density (according to the data for August 2016).

№	Cities with high-rise buildings	Population	Total area (km^2)	Population density (people/km^2)	Country
1	Shanghai	24 256 800	6 340.50	3 826	China
2	Karachi	23 500 000	3 527.00	6 663	Pakistan
3	Beijing	21 516 000	16 410.54	1 311	China
4	Delhi	16 349 831	1 483.00	11 025	India
5	Lagos	16 060 303	1 171.28	13 712	Nigeria
6	Tianjin	15 200 000	11 760.00	1 293	China
7	Istanbul	14 160 467	5 461.00	2 593	Turkey
8	Tokyo	13 513 734	2 191.00	6 168	Japan
9	Guangzhou	13 080 500	7 434.40	1 759	China
10	Mumbai	12 442 373	603.40	20 620	India
11	Moscow	12 197 596	2 510.12	4 859	Russia
12	São Paulo	11 895 893	1 521.11	7 821	Brazil
13	Shenzhen	10 467 400	1 991.64	5 256	China

Shanghai is the largest and most highly developed city in mainland China. The population density is almost 6 times higher than the average for China. The average life expectancy is also greater: in 2017 it is 83.37 years . GDP per head in 2016 was $ 17 548. This is 2.2 times more than China's GDP per head. 2.74% of residents age 15 and older were illiterate in 2010.

Karachi is the main economic center and seaport of Pakistan. The density of the city's population is 29 times higher than the density in the country. A similar situation is observed in Delhi: the population density is 31 times higher than the population of India.

The total population density of Russia was 8.56 people/km^2 as of January 1, 2017.

The population is disposed extremely unevenly: 68.3% of Russians live in the European part of Russia, which comprises 20.82% of the territory. The density of the population of European Russia is 27 people/km^2, when the population of Asian part of the country is 3 people/km^2. The urban population is 74.27% (2017). Moscow is the leading city by the number and density of the population counting 4834.31 people/km^2.

Moscow is the only city of Russia, which is considered a megalopolis with high-rise buildings. The tallest building in Moscow is the "Tower of Moscow", which is part of the "City of Capitals" complex. The height of the Moscow skyscraper built in 2009 is 301 meters. In the 76-storey building there are residential apartments, as well as a business center. Currently, the "Tower of Moscow" has become the tallest building in Europe and the CIS countries. In Moscow, there are 4.5 thousand buildings and structures with a height of more than 50 m. Moscow leads in all indicators of socio-economic development. Per head income in 2016 was 59 567.0 rubles per month, and in the Moscow region - 38 395.9 rubles per month, which is significantly higher than the average for Russia - 30 448.1 rubles per month. In absolute terms, the GRP of Moscow in 2016 amounted to 13.9194 trillion rubles.

The success of the development of large cities is one of the factors in the transition to an innovative economy and the competitiveness of the Russian economy. The problem of providing citizens with comfortable, affordable housing is still relevant. There is a structural housing deficit in the country, 75-80% of housing stock is equipped with basic communal services, the proportion of dilapidated and emergency housing is sufficiently high. For many cities in Russia, high-rise buildings will help solve the above problems. A number of cities began to implement programs of high-rise construction, for instance Novosibirsk and Yekaterinburg [4, 5].

3 Results

The conducted analysis of the strengths and weaknesses of high-rise buildings is presented in Table 3.

Table 3. SWOT matrix of high-rise development analysis

Strengths	**Weaknesses**
The political weight, determined by the number of projects in the field of international cooperation, the number of foreign embassies and international missions Saving of land High level of business activity Innovative scientific and educational potential National and world scale of commodity and stock markets Concentration of large financial structures The presence of a large number of well-paid jobs Active information exchange through the media of global importance Highly developed information and communication infrastructure High cultural level availability of a large number of theaters, museums, concert and exhibition halls of the world level, large flows of foreign tourists The level and quality of life is much higher than other territories, the presence of social infrastructure	High concentration of population Limited area Very high environmental load Expensive home exploitation Lack of own natural resources High environmental load, huge waste streams and emissions of road transport
Opportunities	**Threats**
Dispelling fears of people about life in high-rise buildings Reducing the cost of operation (after 20 floors each next 3 floors add an additional 10% to the 150% of the cost of exploitation, which already exists). Demonstration of uniqueness (view from windows, infrastructure in the house and possible work place) Possibility of development in the city center The use of vertical landscaping of buildings as a means of decoration Safety of elevators - the use of water poles supporting special metal structures floating in evacuation mines, their carrying capacity is up to thousand people Equipping a helicopter landing site in case of evacuation of people Using BIM technologis	The complex residential structure, attracting an unfavorable social climate, vandalism, crime Risks related to quality, safety - seismic hazard, strong wind, draft high cost of high-rise housing propaganda of low-rise buildings

4 Discussions

Even by the number of positions in each column, one can talk about the expediency and prospects of high-rise buildings of large cities. This can be justified by statistical data. The Human Development Index (HDI) in 2014 was 0.946 in Moscow, compared with 0.874 for Russia as a whole. This is a high factor, which is comparable to Poland, the Czech Republic, and the Baltic countries. It takes into account the expected life expectancy, duration of citizenship education and national income per capita. Among the countries of the world, Russia ranks 55th [6,7]. Therefore, it is important for Russia to promote the development of large cities.

In order to determine the directions of the megacity development, it is necessary to allocate the following groups of territories: industrial, agricultural, special purpose, engineering infrastructure, transport infrastructure, public-business, recreational, natural, water areas. Investing in any of these areas involves getting as much income as possible on invested capital. Not only investors and enterprises are interested in this, but budgets of all levels need them. Expansion of investments in the economy of a particular city increases its production and economic potential and business activity. There is an increase in the employment of the population, an increase in the collection of taxes, the social sphere is supported. The main indicator of efficiency in the long run is net discounted income, the guarantee yield index. As Marshall wrote, "the time factor underlies the main difficulties in solving almost any economic problem." The development of the program for the development of the functioning of the industrial-infrastructural complex is carried out using modeling based on the model of V.V. Glukhova and M.E. Oseevsky.

The task is to determine by years the planned period of capacity, output by sector of the economy and infrastructure sectors. It is necessary to take into account own consumption in the branches, inter-branch relations, external deliveries, attracted number of employees [8].

Introduce the unknown planned values of x and y, then the volume of production is determined by the need of the city's economy:

$$y_n(t) = y_n^{\,0}(t) + y_n^{\,1}(t) + \sum_m a_{mn}(t)x_{mn}(t) \quad (1)$$

where $x_{mn}(t)$ is a product consumption of n th industry in m-th industrial sector in a year t

$a_{mn}(t)$ is a specific consumption of products of n-th service industry in m-th industrial sector in a year t

$y_n^{\,0}(t)$ - internal consumption of the product of n-th industrial sector in a year t

$y_n^{\,1}(t)$ - other consumption of the product of n-th service industry transfer out of the territory of a megalopolis, consumption in the housing and communal services, in the non-production sphere in the year t.

Capabilities of infrastructure are limited by its capacities:

$$y_n^{\max}(t) \geq y_n(t) \quad (2)$$

where $y_n^{\max}(t)$ – capacity of n-th service industry

$y_n(t)$ – determined product demand of n-th industry in a year t.

The possible volume of production of the industrial complex is determined by external need and internal cooperation:

$$x_m(t) = x_m^{\,0}(t) + \sum_n x_{mn}(t) + \sum_m \sum_l z_{ml}(t) \quad (3)$$

Where $x_m^{\,0}(t)$ – own consumption in m-th industrial sector in a year t

$x_{mn}(t)$ – consumption of industrial sector products in n-th service industry in a year t

z_{m1} - product consumption of m -th industrial sector in the first sector of industry in a year t.

Capabilities of industry sectors are limited by their capacity

$$x_m^{max}(t) \geq x_m(t) \tag{4}$$

where $x_m^{max}(t)$ - capacity of m -th industrial sector

$x_m(t)$ - planned consumption in m -th industrial sector in a year t. The number of workers is one of the limiting resources, which is described by the conditions:

$$\begin{gathered} N_n(t) = b_n y_n(t), \\ N_m(t) = b_m y_m(t), \\ N(t) = N_n(t) + N_m(t), \\ N^{max}(t) \geq N(t), \end{gathered} \tag{5}$$

where $N(t)$ – total demand in employees

$N_m(t)$ - demand in workers of m -th industrial sector

b_m - need for workers per unit of output in m -th industrial sector

$N_n(t)$ - need for workers in n -th service industry

b_n - need for workers per unit of output in n - th service industry

$N^{max}(t)$ - the maximum permissible labor resource.

Wages of workers depend on the expected average payments:

$$W(t) = \sum c_n(t) N_n(t) + \sum c_m(t) N_m(t) \tag{6}$$

where $W(t)$ is total wage

$c_n(t)$ и $c_m(t)$ - average wages in m -th industrial sector and n -th service industry respectively.

The form of production process and the degree of availability of products depends on the volume of materials, components and tools delivered:

$$B_s = \sum_n b_{sn} y_{sn} + \sum_m b_{sm} x_m \tag{7}$$

where B_s – the total amount of incoming materials by S -th flow entering the city, b_{sn} and b_{sm} - specific demand for materials in m -th sector of industry and n -th service industry, received by s -th flow entering the city

Material flows determine the capacity of transport highways:

$$B_s^{max} \geq B_s \tag{8}$$

where B_s^{max} – maximum capacity of s -th flow entering the city

B_s - total planned amount of incoming materials by s -th flow entering the city

Tax revenues to the city budget are determined by the normative base adopted in the city, tax benefits granted by enterprises, production volumes:

$$F_k(t) = \sum_m f_{km}(t) x_m(t) + \sum_n f_{kn}(t) y_n(t) \tag{9}$$

where $F_k(t)$ – amount of k -th form of tax revenue to the city budget

f_{km} and f_{kn} - specific amount of taxation of k -th form of m -th product in industrial sector and of n -th product of service industry.

The integral indicator, the minimum of which must be ensured, is the sum of deviations from the desired amount of tax revenues to the budget. In this case, deviations are taken into account only in the form of a shortfall of the income amount

$$\min J_1 = \min \sum_t g_t \sum_k e_k F_k(t) \quad (10)$$

Maximum amount of tax revenues to the budget can be an alternative indicator:

$$\max J_2 = \max \sum_t g_t \sum_k e_k F_k(t) \quad (11)$$

where $F_k^{\,0}(t)$ – desired tax revenue path of k-th form to the city budget

$F_k(t)$ - amount of k-th form of tax revenues to the city budget in a year t

e_k - validity of k-th form of tax for the budget

g_t - validity of t-th year for the total planned period.

Thus, when elaborating development scenarios, it is necessary to focus on the development of capacities, production volumes, the supply of labor resources, and the capacity of the infrastructure sectors. The application of this model is possible to megalopolises and urban agglomerations.

5 Conclusion

The regions of Russia are distinguished by a strong heterogeneity in socio-economic development, different natural and resource potential, climatic conditions, transport infrastructures, etc. If the control arms are concentrated in a certain region, it allows attracting financial resources, highly skilled specialists, implementing innovative projects in a number of industries, financing budgetary areas. All listed above enhances the quality and standard of living of the governor's region, as well as the manufacturing regions, as investment flows are directed into industrial production.

References

1. A. Larionov, E. Nezhnikova, International Journal of Applied Engineering Research, **6**, 4433-4439 (2016)
2. E. Nezhnikova, IOP Conf. Series: Earth and Environmental Science, **90,** 012161 (2017) doi:10.1088/1755-1315/90/1/012161
3. V. Kankhva, IOP Conf. Series: Earth and Environmental Science, **90,** 012175 (2017) doi:10.1088/1755-1315/90/1/012175
4. V.V. Ivanov, Nauka (2015)
5. S.A. Kolesnikov, Vestnik MGSU, **1**, 7—15 (2015)
6. E.A. Avdeeva, T.E. Davydova, Ekonomika i predprinimatel'stvo, **10-1** (2017)
7. T.A. Averina, Sistemy upravleniya i informacionnye tekhnologii, **45** (2011)
8. I.V. Ilin, A.B. Anisiforov, WSEAS Transactions on Business and Economics, **11**, 757-764 (2014)

Model of investment appraisal of high-rise construction with account of cost of land resources

Ella Okolelova [1,*] *Marina* Shibaeva[1], and *Natalya* Trukhina[1]

[1] Voronezh State Technical University, Moscow Avenue, 14, Voronezh, 394026, Russia

Abstract. The article considers problems and potential of high-rise construction as a global urbanization. The results of theoretical and practical studies on the appraisal of investments in high-rise construction are provided. High-rise construction has a number of apparent upsides in modern terms of development of megapolises and primarily it is economically efficient. Amid serious lack of construction sites, skyscrapers successfully deal with the need of manufacturing, office and living premises. Nevertheless, there are plenty issues, which are related with high-rise construction, and only thorough scrutiny of them allow to estimate the real economic efficiency of this branch. The article focuses on the question of economic efficiency of high-rise construction. The suggested model allows adjusting the parameters of a facility under construction, setting the tone for market value as well as the coefficient for appreciation of the construction net cost, that depends on the number of storey's, in the form of function or discrete values.

1 Introduction

At present moment, special attention is paid to high-rise construction, as the most promising from the point of view of rational use of land resources, the introduction of new construction technologies, the creation of a modern image of cities. High-rise construction is revving up around the world, what is due not only to the high degree of urbanization, population growth, the deficiency of suitable territories for construction in many countries of the world, but also to the desire to make cities more modern, original, comfortable, and creative, using the latest innovative solutions and technologies in the field of construction.

At the very beginning of its development, high-rise construction took into account, mainly, a lack of land resources, what is a common problem for many countries. Nowadays, not only the shortage of land forces architects and builders to begin a kind of "vertical race". This is a tribute to the modern architectural fashion, the desire to modernize the cityscape, to make them comfortable for living and working.

Each high-rise building is an intellectual product. Its construction should be approached in a comprehensive way: innovative architectural and planning solutions must be supplemented with modern engineering infrastructure, which provides the best solution for creating a reliable, safe, comfortable building, allowing its owner to achieve maximum economic efficiency.

* Corresponding author: Ella.o2011@yandex.ru

A lot of scientists, including Russian ones: Alferov V.I., Balabanov I.T., Graboviy P.G., Panibratov Y. P., Maklakova T.G., Schukina N.M. - and their foreign colleagues: J. Friedman, N. Ordway, Pham H.X. studied the construction of high-rise buildings, its economic efficiency and relevance. So, there were stated main principles and methods of high-rise design, considered the questions of energy efficiency, complex security, fire protection, supervision over the facility and advanced building technologies and materials [1].

In the work of Pham H. X. "Innovative methods for evaluating the effectiveness of design solutions in the field of high-rise construction" [2], methods for appraising the economic efficiency of design solutions for the implementation of high-rise buildings are considered.

Despite the diversity of aspects of assessing the prospects for the development of high-rise construction in various countries of the world, the problem, in our view, is not sufficiently covered. In connection with the foregoing, the aim of the work is to review and analyze the factors, that determine the prospects for high-rise construction in various countries, evaluate the negative factors as project risks, and calculate the optimal number of storeys of the building in terms of maximum economic efficiency.

The solution of this task starts with an analysis of the process of high-rise construction in the aggregate of interrelated problems, and the break down of factors, that determine the trends and growth rates of high-rise construction as a modern global strategy for the construction industry. Moreover, the risks that accompany the erection and subsequent operation of high-rise buildings should be estimated.

2 Materials and Methods

The theoretical methods of research, which are used in the work, are presented by theoretical analysis and generalization of the scientific literature, as well as by studying the world experience in high-rise construction. The process of investment appraisal of the construction of a high-rise building can be classified as methods of experimental and theoretical level.

High-rise buildings are complex architectural and engineering structures. The problem in this case is not only a significant rise in the cost of construction. It is necessary to take into account many factors, for example, geographical ones. As it is known, there are many seismically active zones on the Earth, on the territory of which the erection of buildings, even of standard storeys, requires additional expenses to use constructional solutions that increase the seismic resistance of buildings. Another problem of operating high-rise buildings is the wind load. Therefore, various complex architectural forms are used, based on the larger size of the building's footing relative to its top. In addition, construction of high-rise buildings must take into account the architectural landscape of the adjacent territories from the point of view of their insolation [3].

The issue of the subsequent operation of the high-rise building also requires attention. First of all, complex utility systems are used in this type of buildings, which are necessary to run the building and ensure its functionality. Unfortunately, this also leads to a significant increase in the cost of the facility.

The increase in the cost of high-rise buildings should be compensated by the use of high-tech materials and structures, modern technologies for erecting buildings and organizing production. It is the systematic approach to the construction of high-rise buildings that makes it possible to assert that this branch is the most promising to develop not only in the countries with limited land resources, but also in the countries where this issue is not so burning.

 https://doi.org/10.1051/e3sconf/20183303014

High-rise construction allows increasing the economic impact from the sale of the object manyfold by increasing the total area of the building while reducing the costs of allocating (purchasing) a land plot for construction. At the same time, high-rise construction has many problems. It is possible to single out the positive factors that determine the profitability of the project, and negative factors that affect the growth of costs during its construction and operation [4].

Table 1. Factors that determine the efficiency of constructing high-rise buildings

Positive factors	Negative factors
Social	
Housing (residential construction)	Architectural and aesthetic assimilation (the need to harmonize high-rise buildings with old and historic ones)
The growth of retail, office and other type of spaces (construction of commercial real estate)	Excessive urbanization of the country
Increase in the number of jobs	Labor migration from other sectors of the national economy
Comfortable housing and high quality residential and commercial property	Increase in social differentiation
Change to the modern cityscape, reaching the world standards	
Dynamic development of the construction industry and related fields	
Economic	
Reducing the cost of purchasing land for construction	Shift in consumer preferences and decrease in consumer demand
Growth of the investor's income due to upgraded amenities of buildings, modern equipment, high quality of service, etc.	The level of income of the population, consumption possibilities
Attracting foreign investment, increasing capital inflows into the country	Increase in costs when erecting high-rise buildings
Development of intraeconomic relationships with other industries	Increase in the cost of sites for construction
Development of urban infrastructure	
Technological	
Use of modern technologies in construction	Complicated to construct
Modern building materials	Complex engineering support and subsequent maintenance and operation
Development and implementation of innovative technologies in construction, including buildings engineering	Use of sturdier structures
Geographical	
Rational use of land resources	Possible landscape change due to the industrial development of the country
Environmental	
Use of new technologies for treatment plants and waste processing	Worsen of the environment situation due to the growth of manufactures, decrease in natural areas and recreational zones, decrease in insolation of adjacent territories.

The complexity of high-rise construction, due to the need for large capital investments, has led to a specific form of construction organization.

The objective of initiators of such projects is a deep analysis and comparison of costs for the construction of a complex high-rise facility with the results of economic profitability from its operation. The next step to solve the problem is to construct a model for calculating the optimal height of a high-rise object from the point of view of its economic efficiency.

The cost of land resources determines the architectural and planning decisions in many ways. The issue to appraise the economic efficiency of an investor when constructing high-rise buildings and their subsequent implementation remains one of the most urgent in modern urban planning. To date, there is no universal algorithm that allows to solve with certain accuracy the issue of the optimal number of storeys and the height of the building [5].

The main objective of the investor is to solve the economic problem of finding the balance between saving the cost of acquiring land for construction, the possibility to utilize a larger number of areas on the one hand, and the cost of construction of high-rise buildings on the other hand. The increase in the cost of constructing a high-rise facility is due to the use of high-strength structures, taking into account wind effect and seismic resistance in some countries. In addition, the complexity of creating utility systems in high-rise buildings is many times more expensive than creating the same systems in buildings of low and medium height. The analysis of the factors that allow assessing the strengths and weaknesses of the construction of high-rise buildings is presented in Table 2.

Table 2. Economic impact and additional expenses.

Positive economic impact	Additional expenses
Reduction of the building area, saving of land resources, reduction of costs to acquire a land plot for construction	Increase in the cost of design works
Reduction of perimeter of load-bearing wall structures	Creation of more complex utility systems, lift maintenance
Reduction of costs for roofing and materials	Use of high-strength structures
Reduction of material costs for the floor structures	Complexity of installation, use of lifting machines and mechanisms, supply of materials and products
Localization of utility systems (water supply, heating, sewerage), minimization of expences for urban utility systems	Strengthening the load-bearing structures and foundation of buildings
Increase in the amount of sold area	Complexity to supply energy resources to the upper floors of the building

In order for the investor to assess the efficiency of investments in high-rise construction, it is necessary to determine which factors will have a greater impact on the financial result of the project. For this purpose, a model has been developed that makes it possible to evaluate the economic efficiency of investments in the construction of a high-rise building [6].

Assume the cost of a small storey building be equal to:

$$C_1 = S_1 \times n_1 \times z_1, \tag{1}$$

where S_1 is the area of the 1st floor of a low-rise building, m^2; n_1 is the number of floors of a low-rise building; z_1 is the primecost of 1 m^2.

The cost of a high-rise building will be:

$$C_2 = S_2 \times n_2 \times z_2, \tag{2}$$

where S_2 is the area of the 1st floor of a multi-storey building, m^2; n_2 is the number of floors of a multi-storey building; z_2 is the primecost of 1 m^2.

Assume the total area of buildings is the same, that is:

$$S_{total} = S_1 \times n_1 = S_2 \times n_2 = const \Rightarrow \frac{S_1}{S_2} = \frac{n_2}{n_1}, \quad (3)$$

The area of the land plot intended for building is reduced in proportion to the increase in the number of storeys of the building. Consider the cost of a land plot for two construction options.

$$L_1 = S_1 \times l, \quad L_2 = S_2 \times l, \quad (4)$$

where L_1 and L_2 are the cost to acquire land for construction for low-rise buildings and high-rise ones, respectively; l is the price of 1m2 of a land plot.

The coefficient of savings on land allocation is proportional to the growth of the number of storeys (provided the total area of the building is constant) and is equal to:

$$m = \frac{L_1}{L_2} = \frac{S_1}{S_2} = \frac{n_2}{n_1}. \quad (5)$$

It should be taken into account that the primecost of 1 m^2 increases due to the use of high-strength structures, construction of utility systems, the complexity of installation works, etc. with increasing the number of storeys of the building. Consequently,

$$z_2 = kz_1, \quad (6)$$

where k is the coefficient of increase in the cost of construction with the increase in the number of storeys.

The total project expenses function is defined as following:

$$y(x) = l(x) + z(x) + r(x), \quad (7)$$

where l is the cost of purchasing a land plot, which is intended for building, \$/m^2; z is construction primecost, \$/m^2; r are expenses to neutralize risks, \$; x is the number of storeys.

The function of total project expenses is shown in Figure 1.

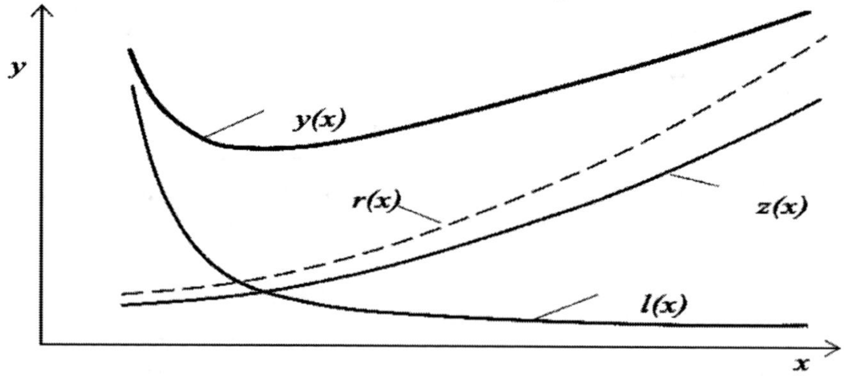

Fig. 1. Graphic definition of total expenses.

3 Results

The work of the model at a real high-rise facility is presented as the results of the study. As an example, the construction of a 35-storey building with certain parameters of the area, height of the building, the primecost of construction, selling price is considered. The initial data is the following:

- number of floors is 35;

- the area of one floor is 1500 m^2 (therefore, the total area of the building is 52.5 thousand m^2);
- Construction primecost of 1 m^2 is 650 \$/m^2;
- selling price of 1 m^2 is 1800 \$/m^2;
- coefficient of increase in price of construction is k = 1.03.

The coefficient of increase in price k according to the calculated data represents an exponential dependence of the form $k = 0{,}9707e^{0.0296\,n}$, where n is the number of floors (Fig. 2).

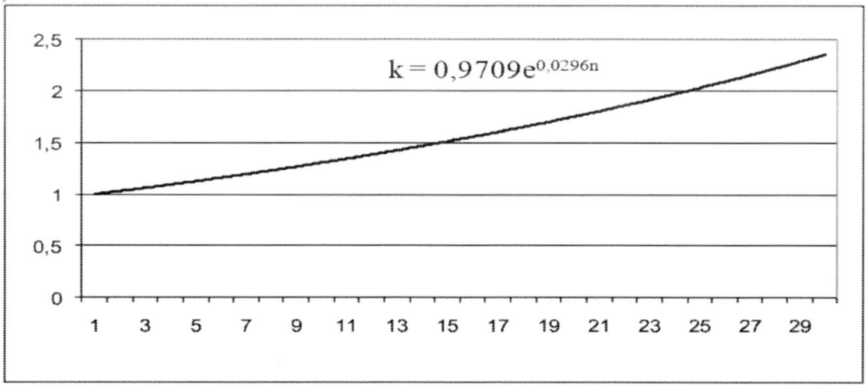

Fig. 2. Dynamics of the coefficient of increase in the primecost of construction, depending on the number of storeys of the building

Define the economic impact from the accomplishment of the building (Table 3).

Table 3. Calculation of economic efficiency from the accomplishment of a multi-storey facility depending on the number of storeys

Coefficient of increase in price of structures	Number of storeys	Primecost, thousand \$	Revenue, thousand \$	Profit, thousand \$
1	1	975.0	2700	1725.0
1.06	2	2070.6	5400	3329.4
1.09	3	3200.5	8100	4899.5
1.13	4	4397.2	10800	6402.8
1.16	5	5663.9	13500	7836.1
1.20	6	7003.7	16200	9196.3
1.23	7	8419.9	18900	10480.1
1.27	8	9915.7	21600	11684.3
1.31	9	1149.9	24300	12805.1
1.35	10	13161.1	27000	13838.9
1.39	11	14918.1	29700	14781.9
1.43	12	16770.0	32400	15630.0
1.48	13	18720.7	35100	16379.3
1.52	14	20774.8	37800	17025.2
1.57	15	22936.6	40500	17563.4
1.62	16	25210.8	43200	17989.2
1.67	17	27602.2	45900	18297.8
1.72	18	30115.9	48600	18484.1
1.77	19	32757.1	51300	18542.9

Coefficient of increase in price of structures	Number of storeys	Primecost, thousand $	Revenue, thousand $	Profit, thousand $
1.82	20	35531.3	54000	18468.7
1.88	21	38444.1	56700	18255.9
1.93	22	41501.3	59400	17898.7
1.99	23	44709.1	62100	17390.9
2.05	24	48073.7	64800	16726.3
2.12	25	51601.9	67500	15898.1
2.18	26	55300.3	70200	14899.7
2.25	27	59176.2	72900	13723.8
2.32	28	63236.8	75600	12363.2
2.39	29	67489.9	78300	10810.1
2.46	30	71943.4	81000	9056.6
2.53	31	76605.5	83700	7094.5
2.61	32	81484.9	86400	4915.1
2.69	33	86590.5	89100	2509.5
2.77	34	91931.4	91800	-131.4
2.86	35	97517.3	94500	-3017.3

The profit graph is shown in Figure 3.

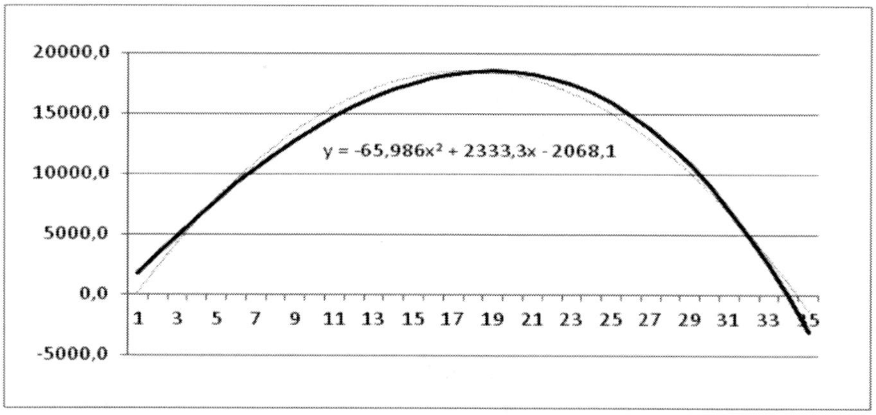

Fig. 3. The profit graph from the accomplishemnt of the facility depending on the number of storeys

Graphic interpretation of the model allows to visually evaluate the optimal number of floors. The maximum possible profit is achieved under the given conditions if the number of storeys is 19. For a more accurate calculation, the profit function should be scrutinized, which is approximated by a second-order polynomial [7]. The first-order derivative of the function, which is represented in the graph, will be the following:

$$y' = -131{,}97x + 2333{,}3$$

The optimal number of floors $x \approx 18$ is obtained by equating the first derivative to zero and solving the resulting equation, i.e. the construction of an 18-storey building under the given conditions will bring the developer maximum profit. Further increase in the number of storeys will gradually reduce the profitability of the project. Profit is almost zero at $x \approx 34$, further increase in the number of storeys can run up the expenses on the project so drastically, that they will exceed the revenue from the sale of the facility.

Dynamics of the k coefficient in further studies showed that the efficiency of increasing the number of storeys of the building is not infinite, and has a limit (Fig. 4).

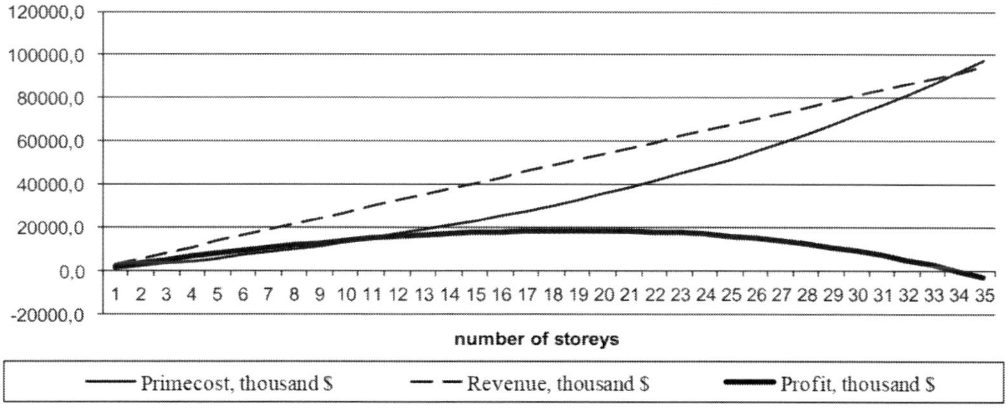

Fig.4. Graphic definition of the optimal number of storeys and the maximum profit for the investor.

Based on the preformed calculations, the optimal number of storeys of the building can be determined, which will bring the maximum profit to the investor. In this example (Figure 4), profit reaches its maximum value with an 18-storey structure. This point can be called the point of the optimal number of storeys for given parameters S_1 and S_2, as well as the dynamics of growth of the coefficient of primecost increase, which was determined by calculation.

4 Discussions

It is hard to call a building of 18 floors high-rise, but in this case it was required to prove the work of the model on some illustrative example. Verification of the model allows to conclude that it can be applied on a real and larger project. The suggested model allows to construct various options of architectural solutions, while linking them into a single mechanism with economic solutions [8, 9].

The considered example does not take into account the possibility of appreciation of the market value of 1m^2 of the area of a high-rise building, that is built with the use of new technologies and materials and has modern architecture, engineering, comfort of living, etc. In this case, the revenues from the accomplishment of this facility can significantly increase, especially from the sale of commercial space located on the first floor of the building. But at the same time, expenses will grow.

The suggested model allows to adjust all parameters and set, for example, the dynamics of the market value of 1m^2 or the coefficient of increase in the construction primecost, depending on the number of storeys, either in the form of functional dependence or in the form of discrete values.

For example, the coefficient of increase in primecost, depending on the number of storeys, is given by an exponential dependence of the form $k=e^{0,03n}$, where n is the number of storeys. Whereas, the cost of selling residential and commercial spaces of the facility increases depending on the design and comfort of the building, as well as depending on the number of storeys.

Assume that the real estate market appraisal showed the following results of appreciation of 1m^2 depending on the prestige, architecture, safficiency of the utility infrastructure and other factors: the market value of 1m^2 of the building of 10-14 floors is increased by 5%; 15-19 floors - by 10%; 20-24 floors - by 15%; 25-30 floors - by 20%. For the practical implementation of the model, both functional dependencies and discrete values can be used.

5 Conclusion

The model allows combining the architectural and cost characteristics of a building into a single system, consider the real estate market conditions, and also allows evaluating the investment project of high-rise construction for the inclusion of modern smart control systems of climate, energy resources and other elements of comfort, since the market value of such an facility is growing drastically.

Factors of increasing and descreasing character for more accurate appraisal of efficiency of high-rise construction are considered and systematized. Especially important in efficiency appraising are descreasing factors, which are defined as project risks.

References

1. E. Nezhnikova, Procedia Engineering, **165,** 1300-1304 (2016) doi: 10.1016/j.proeng.2016.11.854
2. S. Uvarova, V. Kanhva, M. Bovsunovskaya, Economika i predprinimatel'stvo, **9,** 829-831 (2014)
3. K. Gumba, O. Belyantseva, L. Kochetova, Procedia Engineering, **165,** 1323-1327 (2016)
4. V. Murgul, MATEC Web of Conferences, 06002 (2017)
5. V. Kankhva, S. Uvarova, S. Belyaeva, Procedia Engineering, **165,** 1046-1051 (2016)
6. N.I. Vatin, V.N. Chechevichkin, A.V. Chechevichkin, Y.S. Shilova, Magazine of Civil Engineering, **37(2)** (2013) DOI: 10.5862/MCE.37.12
7. A. Borboni, M. Lancini, Journal of Vibration and Acoustics **137(3)** (2015)
8. M. Ageeva, Bulletin of the University, **8,** 119-118 (2011)
9. V. Kankhva, MATEC Web of Conferences, **106,** 08027 doi:10.1051/matecconf/201710608027

Development of innovative methods for risk assessment in high-rise construction based on clustering of risk factors

Ella Okolelova [1,*] *Marina* Shibaeva[1], *Oleg* Shalnev[1]

[1] Voronezh State Technical University, Moscow Avenue, 14, Voronezh, 394026, Russia

Abstract. The article analyses risks in high-rise construction in terms of investment value with account of the maximum probable loss in case of risk event. The authors scrutinized the risks of high-rise construction in regions with various geographic, climatic and socio-economic conditions that may influence the project environment. Risk classification is presented in general terms, that includes aggregated characteristics of risks being common for many regions. Cluster analysis tools, that allow considering generalized groups of risk depending on their qualitative and quantitative features, were used in order to model the influence of the risk factors on the implementation of investment project. For convenience of further calculations, each type of risk is assigned a separate code with the number of the cluster and the subtype of risk. This approach and the coding of risk factors makes it possible to build a risk matrix, which greatly facilitates the task of determining the degree of impact of risks. The authors clarified and expanded the concept of the price risk, which is defined as the expected value of the event, 105 which extends the capabilities of the model, allows estimating an interval of the probability of occurrence and also using other probabilistic methods of calculation.

1 Introduction

The development of high-rise construction as a global process of urbanization is happening at a high pace in many countries. The shortage of land resources, the aspiration to change the cityscape, make it more modern and attractive not only for residents but also for tourists, and also the objective to increase the area of construction for trade, office and residential facilities make architects strive to search for new constructional forms and engineering solutions.

Many countries actively develop high-rise construction as a new branch of the construction industry with a view to the most rational use of land resources. At the same time with the erection of buildings that impressively rise above the ground, investors do not forget about its depths and build, if not underground cities, a very significant underground structure. All this allows initiators of high-rise construction projects to carry out the principle of obtaining maximum profit from a limited building area.

* Corresponding author: Ella.o2011@yandex.ru

Undoubtedly, from the point of view of builders and architects, the decoration of cities with modern buildings is not only a tribute to urban fashion, the beauty of the cityscape and its more modern look, but also convenience, new residential and commercial spaces, new jobs, a new lifestyle.

From an investor's point of view, any project is a risk. High-rise construction is a special category of facilities, which carries an even higher level of risks. Much depends on the geographical features of the construction region, the level of social development, the mentality of the population and many other factors, that are identified as project risks.

In countries, which experience a lack of construction sites, high-rise construction is a prerequisite not only for the dynamic development of the construction industry, but also for ensuring the necessary standard of living and the activities of society as a whole. At the same time, each country has its own peculiarities, which determine the risk values and the probability of losses for the investor within the implementation of the high-rise construction project. The cost of construction depends largely on the types and levels of country and project risks.

For instance, the important factors, that determine the cost of construction of high-rise buildings, are the climatic conditions. Ground movements, the influence of the waters, wind load complicate the project and cause a significant rise in price not only at the design stage of the object, but also during construction. The geography of the country largely determines the level of risks and potential natural and climatic threats. Nevertheless, risks associated with the socio-political and economic level of development should be taken into account.

Thus, speaking of various kinds of risks from the point of view of investment costs, it is necessary to understand that reducing the impact of risks is, first of all, a rise in the construction cost. For example, it is required to apply new technologies and modern methods, when building a life support system for a high-rise facility, as well as specialists who have the necessary skill level to install and subsequently operate it. All this causes a significant rise in the construction cost of facilities, despite of construction of high-rise buildings in climatically dangerous regions, including seismically active and coast zones, mountainous areas, etc. High-rise buildings constructed in such zones require to use structures of a different type in order to preserve the resistibility of buildings during the adverse impact of natural and climatic factors.

Therefore, the problem of risks can be considered as a price factor that affects the investment cost of constructiont o a certain extent. The most likely and common types of risks are considered in order to quantitatively evaluate the risks as a practical tool for assessing the level of danger for a high-rise facility.

A lot of works of domestic and foreign scientists, such as Maklakova T.G., Almazov V.O., Granik Y.G., Burkov V.N., Trukhina N.I., Pham H. X., are focused on the problem of risk evaluation. These works provide a risk evaluation, including analysis of the construction industry in terms of various options for implementing investment projects. The issues of classification, analysis and assessment of risks of high-rise construction require separate consideration due to their specifics, what has determined the relevance of the work.

The goal of the study is to develop a risk analysis model that allows determining the investment value of risk taking into account the probability of a risk event and possible consequences.

2 Materials and Methods

Statistical data of observations, statistical analysis, which allows to evaluate systemic and non-systemic risks of high-rise construction in different regions, are applied in the work as an empirical method of research. The method of mathematical modeling, dwelled on cluster analysis, allows assessing the degree of impact of risks and the possible consequences in case of risk events [1].

Risks are differentiated by characteristics and are united into clusters by these characteristics. For convenience of further calculations, each type of risk is assigned a separate code with the number of the cluster and the subtype of the risk.

Table 1. Risk classification in high-rise construction

Types of risks in terms of impact and consequences		Code
Impact	*Possible consequences*	
Cluster 1 (Geological, geographical and climatic risks)		
Seismic activity	Destruction of buildings and structures as a result of earthquakes of various strengths	101
Impact of waters, including seas and oceans	Possibility of hurricanes, typhoons, tsunami, floodings, etc.	102
Factor of a mountainous area	Possibility of diastrophic activity, avalanches and mudflows	103
Cluster 2 (Country risks)		
1. Political		
Legal support	Absence of the necessary regulatory framework to develop the high-rise construction (land ownership and real estate issues, protection of investor interests, financial institutions, credit facilities, etc.)	211
Legislative framework	Absence of the necessary normative acts for the high-rise construction (issues of land notification, ecology, urban development planning, construction regulations, etc.)	212
Criminality rate	Threat of terrorist attacks, protests of the population, etc.	213
2. Economic		
Inflation rate	Devaluation of the national currency, drop of the exchange rate, rise in the cost of materials and resources.	221
Level of economic development	Negative consequences of the general economic crisis, possible economy contraction, a decline in incomes of population, absense of a developed real estate market, a decrease in effective demand, insufficient development of construction and related industries, a lack of material and technical base to develop high-rise construction, a lack of foreign investment, access to credit resources, taxation, etc.	222
3. Social		
Unemployment rate	Lack of trained specialists to develop high-rise construction in the region	231
Social differentiation	Decline in income per capita, increase in the number of people living below the living minimum wage, etc.)	232
Local mentality	The discontent and protests of the population against the active urbanization of the country.	233
Cluster 3 (Investment risks)		
Supply of capital	Lack of the necessary amount of equity capital, the difficulty of attracting foreign investors, insufficiently developed credit mechanisms, high loan interest rates, etc.	301
Growth of investment expenditures	Land notification; administrative and legal decisions; rise in price of materials; the complexity of object construction and the use of high-strength structures; expenses for innovation (new materials, technologies, methods of production and management); utility systems; object maintenance; green issues, etc.	302
Market risks	The difficulty to enter new markets for construction products,	303

Types of risks in terms of impact and consequences		Code
	reducing the effective demand of consumers, the dynamics of the value of real estate, etc.	
Inflation risks	The duration of the construction of the facility, long terms of positioning, the complexity to anticipate profit, etc.	304
Cluster 4 (Technological risks)		
Project risks	Design failures of a high-rise building (geological surveys, account of natural and climatic conditions, design features, etc.)	401
Complexity of erection	Insufficient provision of machinery, mechanisms and structural components necessary for high-rise construction	402
Complexity of service	Lack of profile services to ensure constant and trouble-free operation of utility systems, lack of a control system for high-rise facilities	403
Cluster 5 (Operational risks)		
Failures of life support systems	Time eliminate the emergency and restore the operation of the systems; impossibility to restore the working capacity of technical systems during the normalized recovery time	501
Fire risks	Absence or malfunction of the fire safety system, the difficulty of evacuating people in emergency cases; local destruction of structural elements of buildings, etc.	502
Explosion risks	Detection of chemically hazardous substances in the building; The threat of destruction of the structure of a building partially or completely due to the presence of explosive chemicals	503
Cluster 6 (Environmental risks)		
Technological risks	Increased costs for the use of modern technologies to create life support systems (energy-saving technologies, water use and treatment systems, sewage filtration system, etc.)	601
Natural risks	Possibility to offset the use of the country's natural resources (deforestation, water pollution, landscape modification, violation of the natural and ecological balance, etc.)	602

Risks are evaluated by the likelihood of occurrence and the degree of impact. The likelihood of risk occurrence is determined on the basis of statistical observations and anticipations on the probability of events. The probability of an event can be evaluated as the frequency of its occurrence in practice or the modeling of the situation. The idea of the suggested method is to assess the likelihood of risk occurrence, as well as to evaluate the frequency and the scale of possible damage from the risk event [2].

Thus, the price risk can be defined as the amount of probable loss in the event of adverse effects. If the probability of occurrence of i-th risk is p_i and the value of possible losses is V_i , the price risk is defined by:

$$R = \sum_{i=1}^{n} p_i V_i \qquad (1)$$

The price risks are losses with account of the probability of its occurrence. This statement allows us to construct a risk matrix where the loss value is defined as a percentage of the maximum possible damage. The price risk can be determined in terms of value, but in this case it is more correct to use relative values. Thus, the price risk is calculated as a relative indicator and is presented as a percentage of the cost of construction. In accordance with the classification, the probabilities of events and the value of the maximum loss are determined. Identification of risks by qualitative and quantitative characteristics was carried out on the basis of cluster analysis, which makes it possible to group risks not only by sources of their occurrence and likelihood of occurrence, but also by the value of damage [3]. As a method of quantitative risk analysis, it is suggested to allocate risks in terms of significance, which assumes the maximum possible loss when implementing an investment project. Table 2 presents the results of assessing the likelihood of occurrence of risks and the value of the maximum loss according to the impact.

Table 2. Quantitative risk analysis

Code	Likelihood of occurrence	Maximum loss,% of the construction cost	Price risk,% of the construction cost
101	0.1	90	9.00
102	0.1	90	9.00
103	0.1	80	8.00
211	0.15	10	1.50
212	0.15	10	1.50
213	0.2	10	2.00
221	0.5	20	10.00
222	0.4	30	12.00
231	0.2	15	3.00
232	0.2	10	2.00
233	0.1	5	0.50
301	0.5	20	10.00
302	0.3	25	7.50
303	0.6	15	9.00
304	0.8	20	16.00
401	0.7	25	17.50
402	0.5	25	12.50
403	0.3	20	6.00
501	0.3	60	18.00
502	0.3	60	18.00
503	0.3	50	15.00
601	0.75	30	22.50
602	0.4	10	4.00

Based on the exact price risk, a cluster grouping was conducted. The complexity of the problem is in the fact that certain types of risk according to the conditions of empirical observations have a very low probability to occur. These risks, for example, may include man-made disasters. At the same time, the damage will be very great in case of their event. And vice versa: risks, that have a high probability to occur, do minor damage to investment project. The objective of risk analysis is to detect the most active group of risks with relatively high likelihood of occurence and value of damage.

Grouping of objects by a number of characteristics is performed on the basis of methods of cluster analysis. The parameter of clustering is the price risk. The range of price risk is [0.5; 22.5]. Clusters are formed on the basis of the price risk in the order of increasing of the specified parameter (Table 3).

Table 3. Clustering by price risk.

Code	Likelihood of occurrence	Maximum loss,% of the construction cost	Price risk,% of the construction cost
233	0.1	5	0.5
211	0.15	10	1.5
212	0.15	10	1.5
213	0.2	10	2
232	0.2	10	2
231	0.2	15	3

602	0.4	10	4
403	0.3	20	6
302	0.3	25	7.5
103	0.1	80	8
101	0.1	90	9
102	0.1	90	9
303	0.6	15	9
221	0.5	20	10
301	0.5	20	10
222	0.4	30	12
402	0.5	25	12.5
503	0.3	50	15
304	0.8	20	16
401	0.7	25	17.5
501	0.3	60	18
502	0.3	60	18
601	0.75	30	22.5

Three main clusters of risks were formed and a cluster classification was introduced based on the qualitative and quantitative analysis (Table 4).

Table 4. Risk clusters

Risk cluster	**Description**	**Range of price risk**
Conservative risks	Risks that have a low probability of occurrence and a minor value of damage in case of event	[0.5; 7.5]
Moderate risks	Risks that have either an average probability of occurrence or an average value of damage in case of event	[8; 15]
Aggressive risks	Risks that have either a high probability of occurrence and a minor value of damage or a low probability of occurrence, but a significant value of damage	[16; 22.5]

The following presents a risk matrix, which displays the risk clusters that pose the greatest threat to the project (Fig. 1).

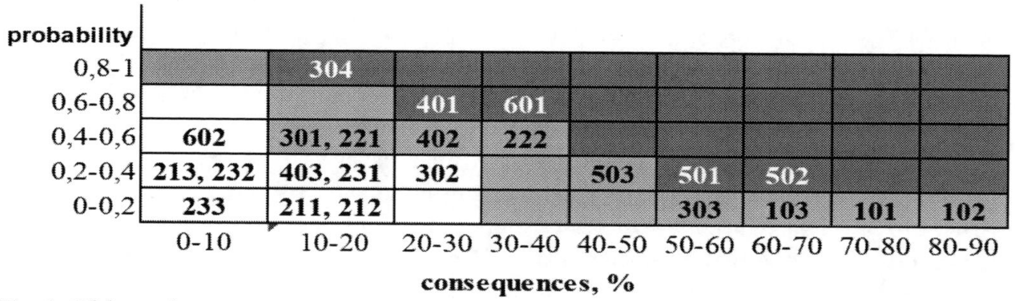

Fig. 1. Risk matrix

Note: the risk clusters are highlighted in Figure 1 with the following colors:

(1) - - conservative risks

(2) - - moderate risks

(3) - - aggressive risks

All kinds of possible risks are clustered according to certain characteristics. The objective is to determine the impact of each risk cluster on the construction of high-rise buildings.

Therefore, it is necessary to determine the extent of the impact of risks, i.e. it is required to estimate the value of losses in case of unfavorable events for the project. It can be man-made disasters, harsh climate conditions, inflation, political protests in the society and many other factors that can ultimately affect the profitability of the project [4].

As it is shown in Figure 1, three risk clusters have been formed. Each cluster includes risks identical to the price risk, regardless of the nature of the event, the likelihood of occurrence, etc. The range of the price risk was adopted as a basis to form the cluster, which made it possible to identify and group risk factors.

The statement that the factor weight of a group can be represented as a cluster distance between points in a multidimensional factor space was adopted as the working hypothesis of risk analysis. Consequently, the factor influence of each group is calculated as the Euclidean distance:

$$d_{ij} = \sqrt{\sum_{t=1}^{n} (x_{it} - x_{jt})^2}, \quad I,j = 1, \ldots, n. \tag{2}$$

where d_{ij} is the Euclidean distance between the i-th and j-th objects; x_{it}, is the value of the i-th parameter for the i-th object.

A square matrix of distances D is obtained with a dimension nxn (by the number of objects); This matrix is symmetric, i.e. $d_{ij} = d_{ij}\ (i,j=1,\ldots,n)$.

$$D = \begin{pmatrix} d_{11} & d_{12} & \ldots & d_{1n} \\ d_{21} & d_{22} & \ldots & d_{2n} \\ \ldots & \ldots & \ldots & \ldots \\ d_{n1} & d_{n2} & \ldots & d_{nn} \end{pmatrix} \tag{3}$$

As a result, a cluster of risks has been formed that most actively affect the anticipation of the project profitability [5]

3 Results

The price risk is defined as the expected value of the event on the basis of the risk study, cluster analysis of risk factors and the determination of mechanisms for calculating the impact of risks on the design and construction of a high-rise facility. This evaluation option is the most accurate. Thus, the considered model allows estimating the interval probability. In this case, all other parameters will also have a certain range of values. Therefore, the developed risk evaluation model allows using various calculation methods.

A further calculation method is presented in the example studied earlier. The distance between clusters is determined in accordance with formula (2), and a distance matrix is constructed (Table 5).

Table 5. Matrix of cluster distances

Cluster name	Conservative	Moderate	Aggressive
Conservative (1)	0	23.2	34.5
Moderate (2)	23.2	0	24.6
Aggressive (3)	34.5	24.6	0

A "risk triangle" is obtained, which is graphically presented in Figure 2.

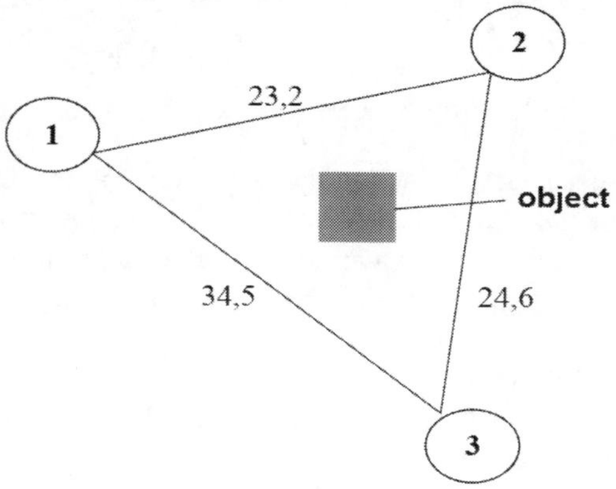

Fig.2. Geometric model "Risk triangle".

Geometric interpretation: The object is in the center of the triangle, i.e. it is at the point of intersection of the medians of the triangle. The degree of impact of each risk cluster (vertixes of the triangle) is determined geometrically as 2/3 of the median [6].

Risk clusters are formed on the basis of the price risks, what is the value of expected losses in case of the risk event with account of the probability and maximum damage. The distance between clusters, respectively, is also expressed in terms of the price risk.

If the distance between the vertixes of the triangle increases, so the price risk will grow. It is necessary to determine the lengths of the medians of the triangle in order to determine the degree of influence of the risk clusters on the object (Figure 3).

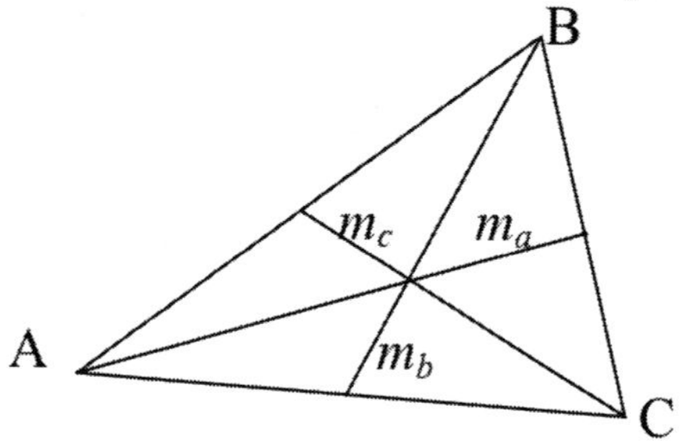

Fig. 3. Geometric interpretation of risks.

The median lengths of the triangle are calculated as follows:

$$m_a = \frac{1}{2}\sqrt{2b^2 + 2c^2 - a^2}\ ; \quad m_b = \frac{1}{2}\sqrt{2a^2 + 2c^2 - b^2}\ ; \quad m_c = \frac{1}{2}\sqrt{2a^2 + 2b^2 - c^2} \tag{4}$$

Distances for the risk cluster:

$$m_1 = \frac{1}{2}\sqrt{2 \cdot 34{,}5^2 + 2 \cdot 23{,}2^2 - 24{,}6^2} = 26{,}7$$

$$m_2 = \frac{1}{2}\sqrt{2 \cdot 24{,}6^2 + 2 \cdot 23{,}2^2 - 34{,}5^2} = 16{,}6$$

$$m_3 = \frac{1}{2}\sqrt{2 \cdot 34{,}5^2 + 2 \cdot 24{,}6^2 - 23{,}2^2} = 27{,}6$$

The center of the triangle is at the intersection of medians, with medians at this point being related as 2:1. Consequently, the degree of influence of the R_{ki} risk cluster on the object can be defined as 2/3 of the median length of the corresponding cluster vertex:

$R_{k1} = 2 \cdot 26.7/3 = 17.8$; $R_{k2} = 2 \cdot 16.6/3 = 11.04$; $R_{k3} = 2 \cdot 27.6/3 = 18.4$.

5 Conclusion

The price of cluster risks is determined and the degree of their influence on the implemetation of the project is investigated. Despite the fact that clusters represent three groups of risks (conservative, moderate and aggressive), the degrees of influence differ slightly. This is due to the fact that the probability of occurrence of a risk with a high level of damage is usually much lower than the probability of occurrence of conservative risks with minor possible damage. Consequently, there is a "leveling" of the price risk, and as a result, risk clusters have almost equal impact on the facility.

Risk evaluation allows not only to assess the value of possible damage with the highest probability, but also to take measures in order to neutralize risks. For example, buildings and facilities are built using more sturdy structures in seismically active zones, where the probability of earthquakes is high.

Based on the suggested model, risks were identified, clustered in order to highlight the most significant ones, and also their impact on the project environment was evaluated.

References

1. H.K. Fam, Ekonomika i menedzhment sistem upravleniya, 1(7), 111-118 (2013)
2. E.Yr. Okolelova, H.K. Fam, Ekonomika, 7, 23-27 (2011)
3. V.N. Burkov, H.K. Fam, Moscow, 51-56 (2013)
4. I.V. Ilin, A. Lepekhin, A.I. Levina, O.Yu. Iliashenko, Advances in Intelligent Systems and Computing, **692**, 1306-1314 (2018) DOi -10.1007/978-3-319-70987-1_138
5. O.V. Kalinina, M.V. Lopatin, Actual Problems of Economics, **182**, 392-405 (2016)
6. V. Kankhva, MATEC Web of Conferences, **106**, 08027 doi:10.1051/matecconf/201710608027

Methodical bases of selection and evaluation of the effectiveness of the projects of the urban territory renovation

Evgeniya Sizova [1,*] *Evgeniya* Zhutaeva [1], and *Andrei* Chugunov[1]

[1] Voronezh State Technical University, Moscow Avenue, 14, Voronezh, 394026, Russia

Abstract. The article highlights features of processes of urban territory renovation from the perspective of a commercial entity participating in the implementation of a project. The requirements of high-rise construction projects to the entities, that carry out them, are considered. The advantages of large enterprises as participants in renovation projects are systematized, contributing to their most efficient implementation. The factors, which influence the success of the renovation projects, are presented. A method for selecting projects for implementation based on criteria grouped by qualitative characteristics and contributing to the most complete and comprehensive evaluation of the project is suggested. Patterns to prioritize and harmonize renovation projects in terms of multi-project activity of the enterprise are considered.

1 Introduction

Renovation of urban areas as a complex integrated process is designed to adapt the existing territory and facilities to modern needs in order to increase economic, social, environmental efficiency and rational use of the territory as a whole. At present moment, the renovation is given special attention both by the state and by market actors. Companies are interested in implementing projects that are characterized by economic profitability and achievable planned results with optimal capacity utilization.

Problems of renovation of urban areas are discussed in [1, 2]. The authors note that renovation is a multi-faceted process, requiring the participation of a large number of subjects, and that is characterized by high value of the results. Based on the works [3, 4], one can draw a conclusion about the main prerequisites for changing the current urban environment:

- Loss of objects of their functional qualities or initially inefficient functional specialization of buildings or territories;
- Lack of territory for new construction;
- a change in the entire urban environment, leading to the need for changes within specific territories;
- obsolescence and dilapidation of the existing funds;
- the need to preserve the territory of the historical center, objects of cultural heritage.

* Corresponding author: sizova@vgasu.vrn.ru

The questions on economic rationale of high-rise construction were considered in [5]. There was noted that the specifics of constructional solutions, as well as strict requirements for reliability and safety lead to a significant increase in project costs. However, the resulting increase in the rationality of the use of territories in terms of their limitation is an important aspect that makes high-rise construction in demand. Criteria for selection of investment projects to implement and evaluation of economic efficiency were considered in [6-10]. Based on the results of the literature review, it can be concluded that the question of selecting criteria for the evaluation of investment projects is one of the most important, but it has not been adequately reflected in the studies in terms of projects for urban renovation.

2 Experimental part

According to the established renovation procedure, an investment project is formed, before the procedure begins, which defines all types of the planned works. The analysis of the ways to select renovation projects precedes the identification of features, on which the composition of the project participants will depend, the coordination patterns, the sources and procedure for financing, and many other features.

The specific features of renovation projects, which must be taken into account when evaluating the efficiency of investment, include the following:

1. The presence of additional difficulties in the "working" objects in an already existing environment with the least damage (especially in densely built-up conditions) and the need to ensure the functional consistency of objects.

2. Uncertainty of the targets at the initial stages of the project.

3. Difficulties caused by the definition of the planning horizon for the project.

4. Uncertainty in the technical and economic evaluation of design decisions, caused by the unique nature of the objects and the non-reproducibility of the conditions, as well as the susceptibility to the influence of internal and external factors.

5. Strongly expressed public importance of projects.

6.There is a need to coordinate the interests of a significant number of participants, that are involved in the project, directly or indirectly (state and municipal authorities, owners, enterprises engaged in the renovation process, etc.)

In addition, renovation projects are characterized by restrictions on the timing of implementation, content, resources, and also ones require significant investment. Renovation projects, including the construction of high-rise, large-span and other unique buildings and structures or work of a large scale, bring special requirements on participating in projects organizations. Only large enterprises can fully meet these requirements. They have a set of characteristics that make them the most preferred subjects to implement renovation projects (Fig. 1).

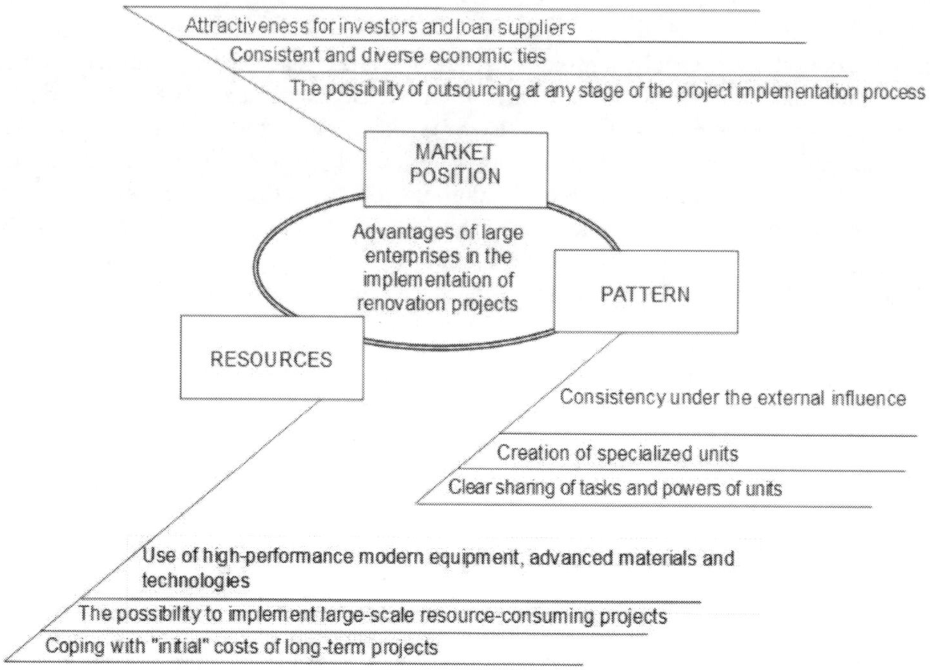

Fig. 1. Advantages of large enterprises in the implementation of renovation projects.

Large enterprises have significant resources to implement large-scale, high-risk, long-term projects. The significant expenses at the initial stages is offset by financial reserves and incomes in other branches, unlike small enterprises, which may not have reserves and capacities for similar projects. Significant revenue of large companies leads to an increased demand for scientific and technical developments, which allows achieving significant rates of return on the costs of their implementation

3 Results

Selection of projects to implement is an important stage for the enterprise, requiring a set of criteria that allow to comprehensively evaluate the project, while maintaining a certain level of materiality (priority) of the criteria. Based on the identified features of renovation projects and the factors that affect the efficiency of project implementation, with account of the advantages of large enterprises as participants in the implementation of renovation projects, and the peculiarities of projects for the construction of high-rise, large-span and other unique buildings and structures, the following system of criteria selection and evaluation of the effectiveness of renovation projects is suggested. The system of criteria is presented in general terms in Fig. 2.

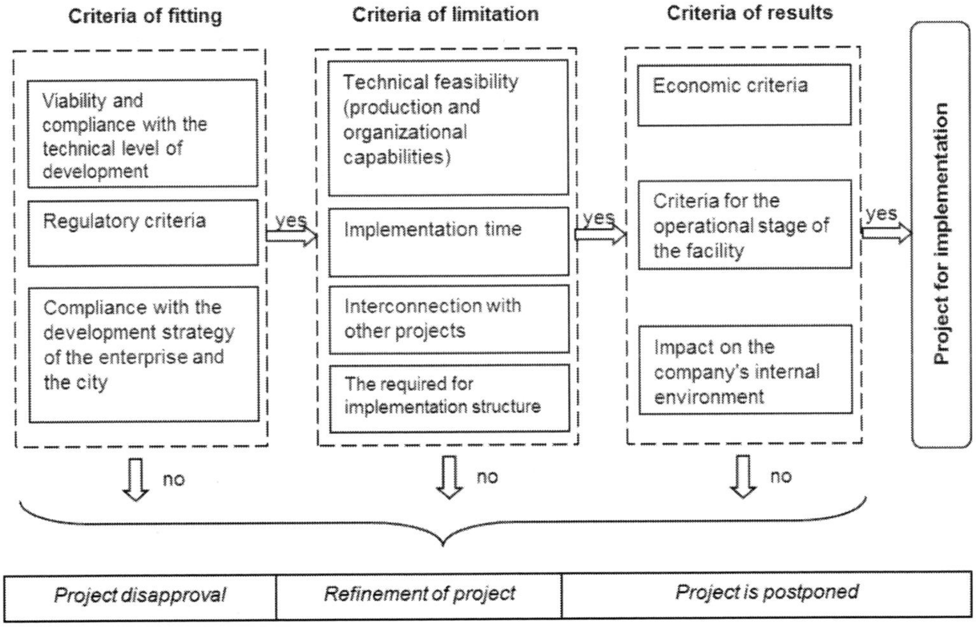

Fig. 2. Criteria for selection of projects.

The criteria of fitting limit the range of renovation projects to those that satisfying the basic development vectors at the macro, meso and micro levels. The compliance of the project with the progressivity ratio at the global and national levels of technology development is checked, as well as the prospects and necessity for the city and for the enterprise itself. Compliance with regulatory requirements is evaluated. Methods for harmonizing existing and required ratio are considered, alternative regulatory sources are analyzed, and scales of forthcoming adjustments are evaluated since the implementation of projects with innovative technology solutions or individual elements is usually not envisaged in the existing normative documentation at the time of the project analysis.

The criteria of limitation relate to the conditions necessary for the implementation of the project. Since any enterprise has a limited potential for objective reasons, the project should be implemented within the available resources, as well as the correlations with projects that are already implemented for consistency, degree of dependence and the possibility of harmonization. The preliminary number of the entities, which are planned to be involved in the project implementation, are considered, including the possibility, if necessary, to attract external actors, their willingness to participate is evaluated, and the cooperation between them are contemplated. One of the most important criteria for limitation is the period of implementation. The fitting of the set deadlines is determined individually by the economic entity.

The criteria for the results relate to the set and objectively anticipated results of the project implementation. From the economic indicators, the following are traditionally calculated:

1. Payback period.
2. Net present value.
3. Project profitability index.

In addition, a plan of financial flows is drawn up, absolute costs, net profit, additional financing, interest payments and other indicators characterizing the results of the project in terms of money are estimated.

Moreover, the qualitative indicators of the operational stage of the facility are relevant to the result criteria, and the possible consequences for the internal environment of the enterprise that may entail the implementation of the project are anticipated. Such consequences include changes in the structure of the enterprise, the impact on the position and status of the enterprise in the market, the benefits that are obtained in the form of replenishing the knowledge base, increasing experience, etc.

The features of large construction companies allow implementing several renovation projects simultaneously. Multi-project workflow requires a special approach, since the company's objective is to optimally distribute its resources and time the deadlines with stages of implementation.

A lot of the projects under consideration on the principle of priority can be divided into priority and non-priority ones. Prioritizing (importance for a particular company) projects is a difficult task. The prioritizing process is exacerbated in large enterprises, that involved in the implementation of renovation projects, by the multidimensionality and diversification of their activities, as well as the significance of the project for municipality. Since, based on inherent features, the projects of any company are characterized by a variety of tasks, directions, characteristics, parameters, it is difficult to comprehensively describe its weight out of the context that determines the characteristics of the company. At the same time, prioritizing of projects can not be dwelled solely on complying with the list of priority vectors in accordance with the development strategy. Therefore, the company should independently be guided by the system of goals, objectives, development priorities at the time of the consideration of the project in order to set the priority. The prioritizing process for the renovation project can be described as follows (Fig. 3).

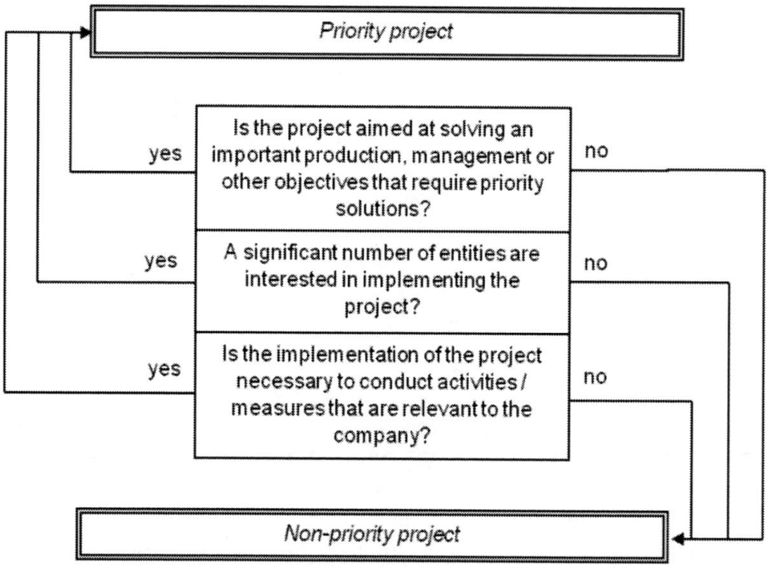

Fig. 3. Prioritizing the project.

The inconsistency and contradictions in the renovation projects implemented by the enterprise can lead to non-compliance with the terms of their implementation, budget overruns, results that do not meet the anticipations, as well as project closing. It is necessary to evaluate in advance the level and nature of the relationships between all projects under consideration, implemented and postponed ones in order to avoid these consequences,

From the authors' point of view, the dependence of renovation projects can be manifested in three aspects: resources, result and pattern.

Dependence on the outcome occurs in the case when projects should be implemented sequentially in accordance with their internal logic and characteristics, i.e. without executing one project (or its specific stage), there is no way to begin the implementation of the next one. This type of dependence results in two ways: firstly, the start and end dates for the implementation of the priority project affect the timing of the beginning of the subsequent interrelated one. Secondly, the qualitative and quantitative results, which are directly obtained at the output, can influence the requirements for the subsequent project or affect its principal components. Dependence on the structure is the requirements for the participants involved in the implementation of the renovation project. Dependence of projects on resources (labor, material, technical and other resources that are necessary for different projects) is coordinated through the implementation of a number of traditional activities for project management (identification of needs, scheduling, assessment of labor reserves, etc.) using tools for production, distribution and financial logistics, as well as methods of managing human capital assets

4 Discussion

Despite the adopted at the enterprise patterns for selection renovation projects, there are a number of factors that affect the success of the project. From the position of the commercial entity, the following factors can be distinguished:

1. Nature of interaction with state and municipal authorities.
2. The degree of influence on the implementation of the project of existing regulatory framework and regulatory documentation in the field of renovation and construction.
3. Presence and possibility to overcome administrative barriers in project implementation.
4. Reliability and stability of funding sources.
5. Organizational structure of the enterprise, as well as individual features of economic activity (methods of work, qualification of personnel, etc.).
6. The level and nature of cooperation and integration with other participants in the life cycle (general contractors/subcontractors, suppliers of building materials, designers, etc.).
7. Previous experience in the implementation of renovation projects.

It is also necessary to take into account the complexity of accurate prediction in the process of selection and implementation of projects in a number of cases, since renovation projects are characterized by long terms of implementation.

5 Conclusion

To date, the specifics of activities for the implementation of urban renovation projects is determined not only by industry aspects, but also by the scale of activities. Based on the capital intensity and the need to overcome barriers, the most preferred participants among the business sector will be large enterprises that have an optimal, well-designed and coordinated pattern for selection of projects with account of its own features. Successfully implemented projects on the renovation of the territory will encourage to use the urban development potential fully and increase the utility level of territories and enhance the overall competitiveness of the municipal.

References

1. E.A. Karaulanova, I.V. Kukina, Molodyozh' i nauka, 1-3 (2011)
2. A.S. Vilkova, D.V. Marennikova, Mezhdunarodnyj nauchno-issledovatel'skij zhurnal, **6(6)**, 155-159 (2017)

3. L.V. Tolstoj, R.G. Abakumov, Innovacionnaya nauka, **1**,100-103 (2017)
4. A.K. Orlov, G.A. Smirnov, Ekonomika i predprinimatel'stvo**, 10-3 (75-3)**, 510-513 (2016)
5. D.H. Ajbazova, Ekonomicheskie nauki, **138**, 36-39 (2016)
6. S. Uvarova, S. Belyaeva, L. Myshovskaya, IOP Conference Series: Earth and Environmental Science, **90(1),** 012169 (2017)
7. K. Gumba, S. Belyaeva, IOP Conference Series: Earth and Environmental Science, **90(1)**, 012163 (2017)
8. S. Belyaeva, D. Voronov, S. Erypalov, MATEC Web of Conferences, **106**, 08033 (2017)
9. K. Gumba, S. Uvarova, S. Belyaeva, S. Revunova, MATEC Web of Conferences, **106**, 08023 (2017)
10. A. Bukreev, O. Vasilyeva, V. Vlasenko, E. Sizova, MATEC Web of Conferences, **106** (2017)

To the issue of the environmental development of housing construction in the integrated land use management

Violetta Politi [1,*]

[1]Moscow State University of Civil Engineering; 26, Yaroslavskoe sh., Moscow, 129337, Russia

Abstract. The paper considers environmental problems arising during the implementation of housing programs for the integrated development of territories by high-rise buildings. The need to assess the quality of the environment in the functional zoning of territories is identified and a calculation of the index of the ecological potential of the territories is proposed. The main stages of organization of construction of environmental real estate and the introduction of energy-saving and resource-saving technologies at the regional level are formulated.

1 Introduction

Currently, in accordance with the concepts of development and general organization of urban and rural settlements, the territory for the development of cities must be chosen taking into account the possibility of its rational functional use. The basis for the choice of the functional use of the territory is comparison of various options for architectural and planning solutions and the assessment of technical and economic, and sanitary and hygienic indicators, accounting for fuel, energy, water and all territorial resources. In addition, it is necessary to assess the state of the environment taking into account the forecast of future changes in natural and other conditions, while considering the state of the environment and its impact on the living and health conditions of the population [1].

2 Materials and methods

The emergence of environmental problems is associated with the intensification of anthropogenic processes, which do not take into account the natural potential to compensate for the negative consequences of the results of the activities of all mankind. Also, in part, the emergence of environmental problems is associated with the orientation to social and economic priorities of the habitat of the population of urban and rural areas. In addition, people began to forget that they are part of the natural system, and influencing their habitat, they directly influence their health. This has become the reason for special attention to the problems of environmental and economic nature and to formation of the so-called "green" economy, which allows to significantly reduce risks to the environment and prevent its degradation [2]. The emergence of industrial zones, polluting the environment was the result of such an uncontrolled human impact on nature. Reorganization of such zones by

* Corresponding author: polity_violca@list.ru

turning it into a safe residential area is a new task for both state authorities and large business.

Unfortunately, with the functional zoning of the planned development site, the only criterion for safety for the health of the population today is to ensure that all factors outside the sanitary protection zone are reduced to the required hygienic standards, which are the maximum one-time MPCs. Special studies are needed to assess the health risk [3] to determine the possibilities of using the territories for specific purposes.

The theory and practice of urban development solves problems associated with the elaboration of territorial development plans. They include both architectural and artistic, construction and engineering, economic issues, as well as sanitary and hygienic issues. However, proper attention is not given to the latter. Even at the pre-project stages of the investment study, with integrated development of the territory, it is necessary to pay attention to the study of all sources of pollution, assessment of the current state of atmospheric air, taking into account the aeration mode, the terrain and microclimatic conditions of the territory.

The application of the concept of functional zoning of the industrial territory to be reorganized, using the methodology for assessing the health risk to the population, will allow allocating only a part of the territory for perspective residential development; the rest of the territory, depending on its characteristics, was divided into functional zones of a different purpose: public-business, and communal. As part of traditional approaches to modeling and decision-making on the basis of compliance with MPC of pollutants, the entire territory would be adopted to accommodate a promising residential development [4]. Particular attention should be paid to territories with existing buildings, especially those located in the zone of negative impact of industrial facilities.

3 Results

When assessing the quality of the environment, with integrated development of the territory, an environmental risk indicator that estimates the probability of occurrence of ecologically caused diseases in the population may be chosen as the determining criterion [6]. The *US EPA* methodology (*developed by the Erisman Center for Hygiene*) makes it possible to determine the quantitative values of this risk. Using the values of the total environmental risk of this methodology, it is possible to determine the overall environmental risk:

$$\mathrm{ER} = \frac{\mathrm{HI}}{\mathrm{N}}, \qquad (1)$$

where *ER* - is the environmental risk; *HI* - is the total risk; *N* - is the number of pollutants (chemical pollution of anthropogenic origin).

Depending on the value of *ER*, you we can introduce the index of the environmental potential of the territory, taking into account the number of points depending on the magnitude of the risk component (Table 1).

Table 1. The value of the index of the environmental potential of urban territory

Total environmental risk (ER)	Level of the territorial potential (Ip)	Points
Over 10	*Catastrophic potential*	1,0
7 – 8	*Critical potential*	2,0
4 –6	*Minimum permissible potential*	3,5
1 –3	*Average potential*	5,0
0,6 – 1,0	*Increased potential*	7,0
0,4 – 0,6	*High potential*	9,0
Less than 0,4	*Optimum potential*	10,0

Studies show that a new high-yield segment of the housing market has been established in Russia related to the complex development of new urban areas. This segment accounts for an average of 45% of all construction work in the residential real estate market performed by large developer companies. It should be noted that there is a tendency for increasing this share.

Based on the analysis of the practice of developing new territories, it should be noted that the investment and construction complex, showing high rates of growth in housing construction, has accumulated sufficient experience in the construction of "spot" development in residential neighborhoods, and block development in the territories of former industrial zones, borders of the city, on the outskirts (the peripheral belt of the city) [5]. This led to urbanization of a residential zone of urban territories, namely, to an increase in the density of residence, and, consequently, to a decrease in the comfort of the habitat (Table 1). The higher the concentration of real estate per unit area of the urbanized zone, the more diverse and complex the problems that arise in different fields, (see Table 2).

Table 2. Consequences of spot and block development of the territory in the residential and former industrial districts of the city

Pos. No.	Factor name	Result of the change
1	Technical and technological factors	Increase of the load on engineering networks (power networks, water supply, water disposal)
		The emergence of waste disposal and pollution problems; problems with cleaning of the areas
		Increasing infrasound sources in the urban environment
		Increasing sources of low-frequency electromagnetic fields
2	Socio-cultural factors	Pollution of the socio-cultural environment
		Acoustic (noise) pollution
		Visual (aesthetic) pollution
		Decrease in the degree of insolation of residential premises
		No parking spaces, places for walking and rest; places for sports
		Increased burden on schools and kindergartens
3	Environmental factors of the city [5, 6]	Increasing content of anthropogenic pollutants in the city's natural environment (atmosphere, soil, drinking water)
		The presence of geochemical soil cover anomalies
		Environmental hazards and environmental risks (in Moscow): - occurrence of karst danger (local subsidence and dips of the earth's surface); - erosion and landslide hazards on the banks of the rivers Moscow and Skhodnya; - suffosion hazard and suffosion risk; - hazard and risk from frost heaving leading to a rise in the earth's surface, squeezing out of piles, pillars and foundations.
		Decreasing area of green plantations
		Pollution of atmosphere by gas exhausts of motor transport
		Negative technological impact on the subsoil, underground and subsurface
4.	Result: deterioration of quality of life and psychosomatic state of the population	

For example, according to the town planning SNIiP, the following is supposed to account for 1 resident of the house: 6 sq.m. of green plantations; 0,1 sq.m. for the rest of the adult population; 0,7 sq.m. for children's playgrounds; 2 sq.m. for sports; 0,3 sq.m. for economic purposes; 0.8 sq.m. for parking of motor vehicles. Therefore, the area of the adjacent territory per 1 resident should be 9.9 sq.m. However, the house under construction is not provided with these norms [5].

In addition, spot development leads to oversaturation of the residential area with personal vehicles, and, consequently, to deterioration in the environmental quality of the human habitat, including increased burden on public transport, worn out engineering networks of the city, and educational institutions. Even with the development of large areas of former industrial zones, the investor-developer does not burden himself with the obligation to erect public and social facilities - schools and kindergartens [6].

Solution of these problems is in the scope of the interests of professional development. There are such new activities in the traditional schemes of development as the development of individual land plots, land-development for integrated land use management. It is the development of land plots for integrated use of territories is an innovative form of professional development.

It should be noted that land development in Russia is of a one-sided investment nature. However, adopting the experience and the mechanism for implementing investment and construction projects implemented in Western countries, the Russian land development, according to experts, will take new innovative forms that will successfully implement large-scale projects. Over the past 5 years, regulatory acts have been developed that make it possible to implement the activities of land developers on a qualitatively new level. Modern forms of development can allow changing the habitat of people in accordance with modern requirements of environmental development. Namely, instead of a quantitative growth of the urban environment, including spot and block development of the territory, land development is able to implement large-scale projects for the integrated development of the territory, combining the necessary set of residential and social facilities that meet the requirements of "green construction" [7].

To promote environmental development in the construction sector it is necessary to change the current Russian model of management. Without developing appropriate concepts of effective management, which should be offered specific scientific and practical approaches to real overcoming the inconsistencies of the existing Russian model of governance requirements of the new economy, it is impossible to solve this problem.

The results of the study of organizational culture at the enterprises of the construction industry confirms the conclusion about the significant influence of authoritarian style leaders of regional businesses because of the dominance of "vertical culture" with a strong cult of unity of command. A multidimensional empirical analysis of the level of implementation of effective management tool in enterprises of the construction sector of the economy indicates a significant excess number of barriers to development of effective management over its resources (Fig. 1).

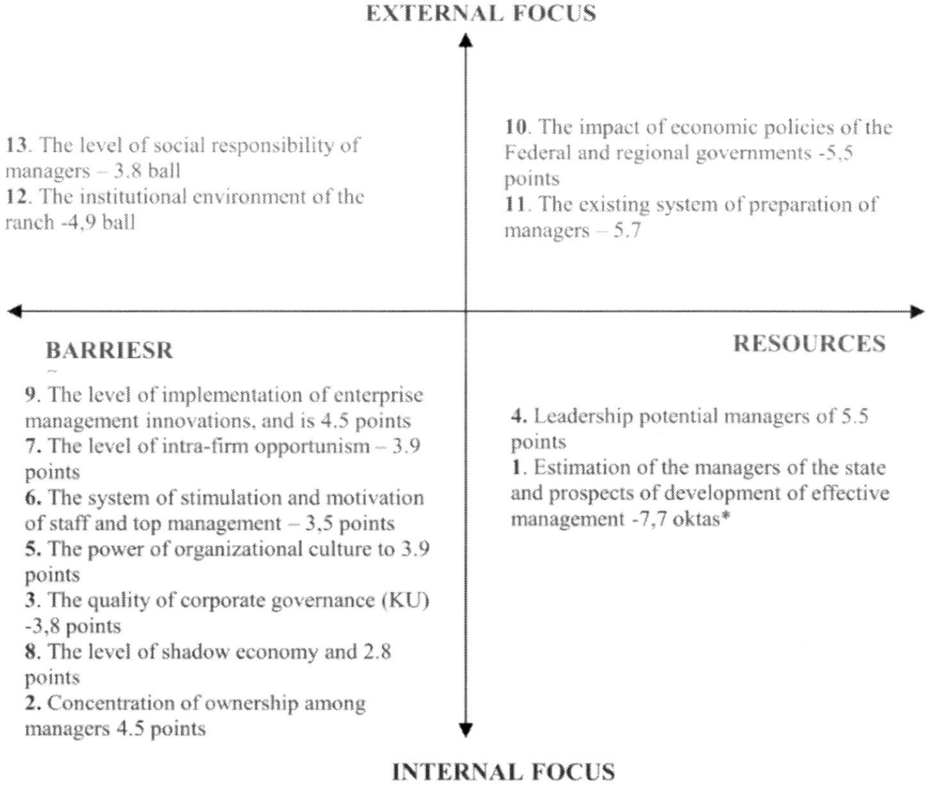

Fig. 1. Matrix resources and barriers to the development of effective management in the construction enterprises (*internal factors: 1-9, external: 10-13)

In recent years, the main regulatory acts have been adopted that allow implementation of activities related to land development, in accordance with established market relations. In particular, modern forms of development are aimed both at a complex, but also in a qualitatively different change in the human habitat. This implies the rejection of the quantitative growth of the city, block, and even more spot construction, and the creation of modern complex buildings that combine residential, public and social real estate, and that comply with the requirements for the environmentally friendly habitat, that is, the requirements of "green" standards of construction and operation of facilities. Land development is intended to ensure a proportional and balanced development of built-up, including residential, public and industrial areas, and undeveloped territories, including natural, green areas and territories of the street and road network [8].

4 Discussion

Land development is intended to choose the types of use of land plots and architectural and town-planning solutions within the framework of town-planning requirements and restrictions established by law. The solution of this task is directly related to the ecologization of the economy, which implies the introduction of innovative safe technologies, resource-saving and energy-saving industries, including in the investment and construction sector and the housing and communal services that make it possible to create an environment for human life activities. Organization of construction of eco-real estate and the introduction of energy-efficient and resource-saving technologies is implemented through the introduction of standards regulating green construction, the development of innovative solutions for the introduction of saving technologies, and other necessary measures (Table 3).

Table 3. Stages of organization of construction of eco-real estate and introduction of energy- and resource-saving technologies at the regional level

STAGE 1.	Ecological support of construction facilities that ensure compliance with the requirements for urban territory
STAGE 2.	The introduction of international standards (ISO-14000, ISO-4001) and the standards regulating the construction of "green buildings" to the construction sector companies
STAGE 3.	Designing of the construction of eco-real estate facilities creating a comfortable internal living environment
STAGE 4.	Development of innovative organizational and technological solutions for the implementation of energy- and resource-saving technologies

The investment attractiveness of territorial entities is determined by a set of interrelated indicators, among which we can single out the pace of housing construction, the level of development of regional real estate markets, improving the comfort of the population and safety of business functioning.

5 Conclusions

At present, sufficient scientific and practical potential has been created in assessing the resource efficiency and energy efficiency of residential real estate, as well as some experience was accumulated in analyzing the mutual influence of the level of development of the regional economy and the pace of housing construction. At the same time, the environmental component and its assessment remain aside. A systematic approach to the solution of this issue is at the stage of formation, and the existing regulatory framework is fragmentary. In connection with the above issue, it is necessary to highlight the leading conceptual approaches to the solution of the problem of improving the ecology of residential real estate.

The first direction was created by the system of normative documents in the field of compliance with the requirements for the ecology of housing during construction and operation of buildings. The direction is represented by a certain structure of normative and technical documentation, however, its application is not an obligatory condition for conducting construction operations. Eco-development is at the stage of formation in Russia, and the "green" code for construction and operation of buildings is mandatory only for environmentally literate entrepreneurs.

The second approach is based on the world experience of "green" construction and building certification in accordance with the requirements of the standards of national systems for voluntary standardization of capital construction projects (in the *UK in BREEFM*, in the *US in LEED*, etc., 32 systems in 24 countries).

The third approach is the concept of biosphere compatibility proposed by the Russian Academy of Architecture and Building Sciences (supervisor – *academician V.A. Ilyichev*). In accordance with this concept, the degradation of the natural environment in the process of inefficient production and economic activity reduces the assimilation potential of the biosphere, hinders further development and reduces the potential of future generations. In this context, the self-sustaining development of housing construction requires the coordinated development of the scientific and technological field and the evolution of the consumption structure.

The approaches considered above are logically connected with the concept of sustainable development of the country's economy. The development of this concept resulted in the emergence of a new model of social and economic development, for which the priority vector for modernizing the economy is to improve the living conditions of the current generation of people and to provide favorable conditions for future generations.

References

1. A. Orlov, I. Chubarkina MATEC Web of Conferences **106** 08015 (2017)
2. R. S. Golov, V. V. Shilov, S.A. ASEE International Forum, Columbus , Ohio. June 28, 2017. Paper ID #20766.
3. V. Kankhva, Justification of system of assessment of ecological safety degree of housing IOP Conf. Series: Earth and Environmental Science **90** 012175 (2017) doi:10.1088/1755-1315/90/1/012175
4. I.V. Ilin, V.I. Koposov, A.I. Levina, Model of asset portfolio improvement in structured investment products, Life Science Journal, **11**, pp 265-269 (2014)
5. O.V. Kalinina, Universal approach to building the progressive scale for income taxation, Actual Problems of Economics, Vol **176**, IS **2**, pp 387-400 (2016)
6. A. Mottaeva, IOP Conf. Series: Earth and Environmental Science **90**, 012120 doi:10.1088/1755-1315/90/1/0121209 (2017)
7. A. Pimenova, S. Kuzmina, N. Morozova, A. Mottaeva, MATEC Web of Conferences, **73**, 07018 (2016) DOI: https://doi.org/10.1051/matecconf/20167307018
8. E. Nezhnikova, IOP Conf. Series: Earth and Environmental Science **90** 012161 (2016) doi:10.1088/1755-1315/90/1/012161

E3S Web of Conferences 33, 03018 (2018) https://doi.org/10.1051/e3sconf/20183303018
HRC 2017

High-rise construction in Russia as one of the objectives of improving the management mechanism for living standards of the population

Svetlana Sergeeva [1,*] *Lyudmila* Komyshova[1], *Natalia* Nagibina[1]

[1] Voronezh State Technical University, Moscow Avenue, 14, Voronezh, 394026, Russia

Abstract. The article is aimed at determining the significance of solving the problem of high-rise construction in Russia as one of the tasks of improving the mechanism of managing the living standards of the population. To determine the effectiveness of measures aimed at improving the quality of life management mechanism, the expert evaluation method is used. The paper presents the measures to improve the quality of life and consequently the migration attractiveness of the region. Despite a number of actual difficulties and lack of sufficient experience, there is an interest in the development of high-rise construction. The result of the analysis of the general demographic dynamics for the last fifteen years is shown. Migration is the most mobile tool for managing short-term demographic processes; therefore, great attention should be paid to this component. Moreover, it is necessary to take into account other consequences of migration.

1 Introduction

Under the current conditions of the actual economy in Russia, improving the quality of living standards of the population is one of the priority directions of social and economic policy. The living standard of the population being some kind of touchstone reflects the state of the economy. Many demographic indicators depend on it, for instance, the migration growth. People always choose a place where it is better to live. The economy is created by people and in for people's well-being. The connections are simply direct: sustainable economic development ensures employment of the population. The implementation of activities aimed at improving the living conditions of the population is one of the main groups of activities aimed at improving the standard of living and infrastructure security.

2 Materials and Methods

To determine the effectiveness of measures aimed at improving the mechanism for managing the quality of life of the population, the expert evaluation method is used.

* Corresponding author: sergeevas2@yandex.ru

The expert evaluation method is based on the use of indirect and incomplete information, experience of experts and intuition. Persons involved as experts should be highly qualified specialists in the complying knowledge areas, have a great professional and practical experience.

3 Results

The expert is provided with the maximum possible amount of objective data related to the problem under analysis. The qualification of experts is tried to be improved through special tuition and training. In addition, the expert should be informed enough about the sources of the problem and ways of solving similar problems in the past. In order to ensure the independence of assessments, the possible interinfluence of experts is eliminated and the impact of extraneous factors is reduced. To determine the possibility of appearing of the solution outcomes, the method of estimate coordination is used, which consists in assessments given by experts independently from each other, and then in combining these assessments into one generalized [1]. The method of the analytical note is a variation of this method, which indicates the possibility of the solution outcome appearance and an explanation on this possibility. To coordinate the estimates, the weighted average probability value is calculated taking into account the validity of the expert.

Today, in the conditions of a "demographic winter", activities that attract migrants play a special role [2, 3]. For this purpose, specific measures for the Voronezh region are determined (Table 1).

Table 1. Measures to improve the quality of life in terms of migration attractiveness

№	Measure	Costs, mln. rub	Effect, %	Efficiency, %
3.	Migration attractiveness of the region.	0,78	5,1	6,5
3.1	Ensure the access of migrants to the social services: health, social, educational and pension services.	0,1	0,7	7
3.2	Simplify the registration system in the living quarters.	0,05	0,9	18
3.3	Develop a network of personnel and recruiting agencies.	0,05	0,1	2
3.4	Develop a public service for the vocational guidance of compatriots living abroad.	0,05	0,3	6
3.5	Direct compatriots living abroad on professional trainings.	0,1	0,8	8
3.6	Improve the system of financial and material assistance to compatriots living abroad.	0,05	0,2	4
3.7	Enlarge the membership of potential migrants for the targeted program of resettlement assistance.	0,1	0,1	1
3.8	Improved information package and information materials on the state program of resettlement.	0,05	0,7	14
3.9	Improve regulatory and legal acts that consolidate the rights and obligations of compatriots living abroad (amend the regulatory documents to shorten the deadline, establish a special procedure for issuing permits for temporary residence and residence permits for compatriots living abroad).	0,01	0,1	10
3.10	Analysis of the implementation of regional resettlement programs, evaluation of their effectiveness.	0,01	0,2	20
3.11	Development of measures to address housing problems of	0,01	0,1	10

	compatriots living abroad.			
3.12	Improvement of the system of state support for compatriots living abroad.	0,2	0,9	4,5

The diagram is used for clarity of understanding (Fig. 1).

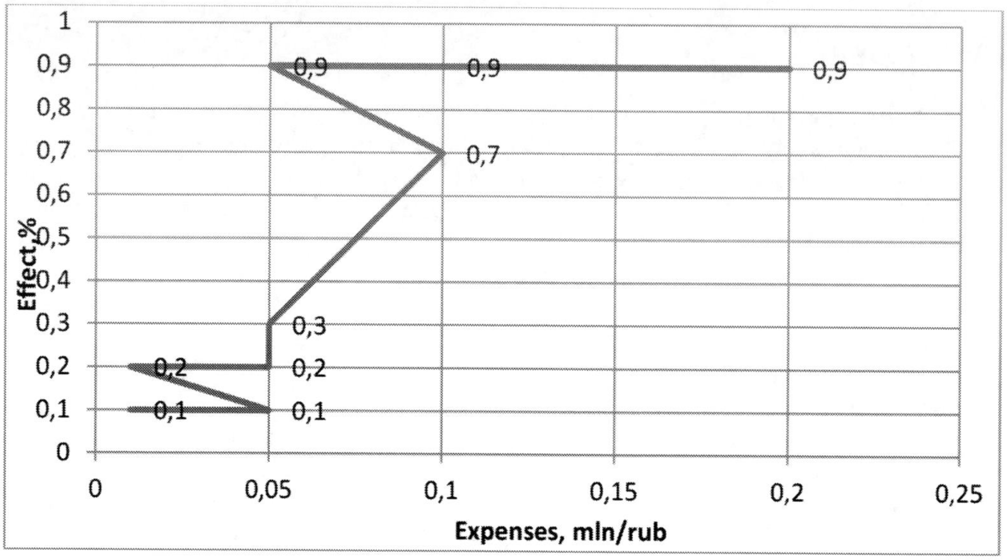

Fig. 1. Diagram of dependency between the effect (%) and the expenses (million rubles) in terms of migration attractiveness of the region.

We use artificial neural networks to predict time series [4].

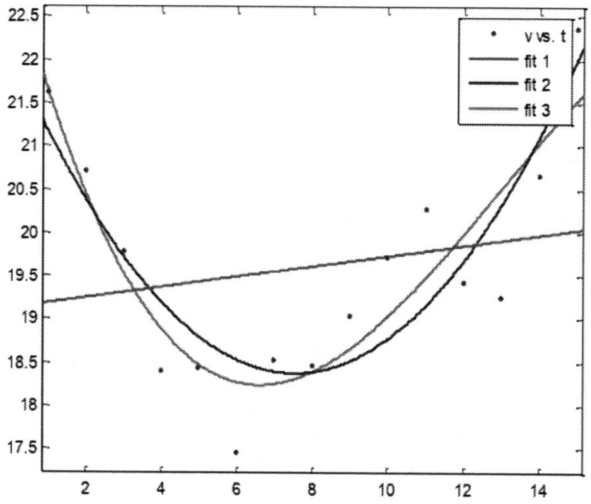

Fig. 2. Approximation of fertility data by linear - 1, quadratic - 2 and cubic - 3 polynomials

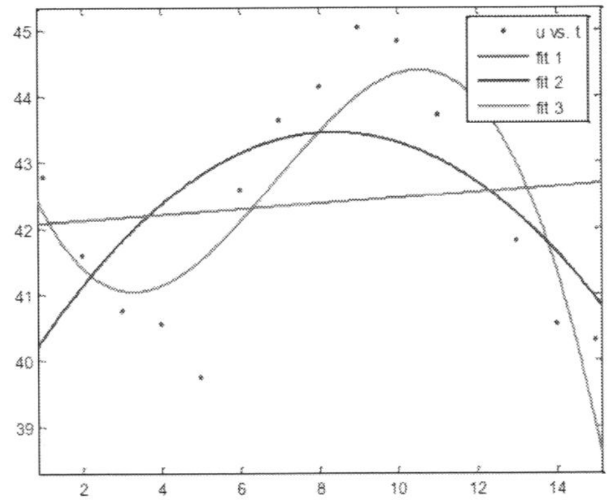

Fig. 3. Approximation of mortality data by linear - 1, quadratic - 2 and cubic - by 3 polynomials

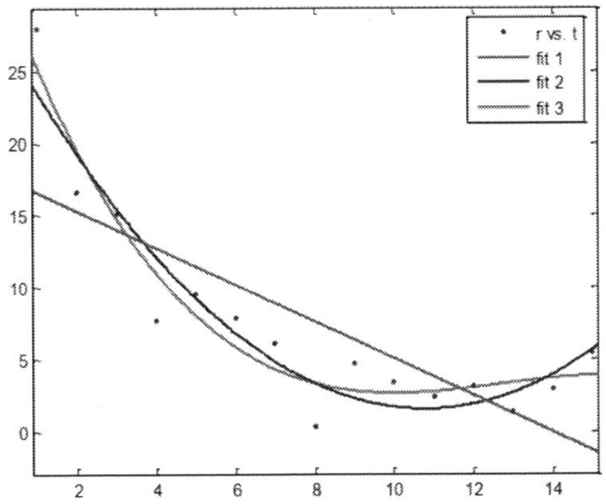

Fig. 4. Approximation of migration data by linear - 1, quadratic - 2 and cubic - by 3 polynomials

"From the analysis of figures 2-4 it follows that mortality and fertility will not change very much and in the first approximation they can be considered even constant. In addition, it can be assumed that there is an anticorrelation between them, which should be further verified. The most dynamic component was the migration one. It can be successfully approximated in the section of the parabolic curve under consideration"[5].

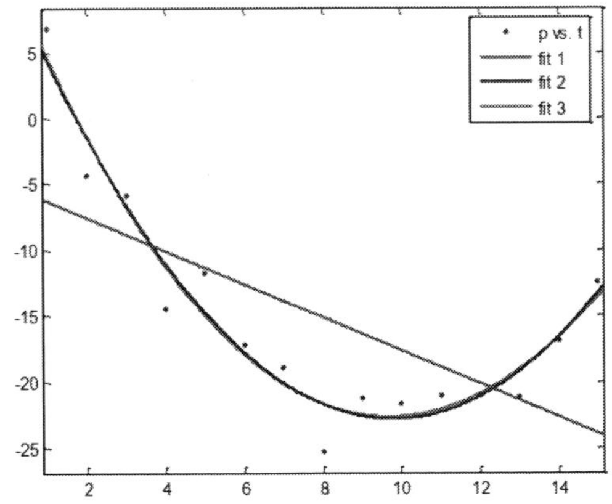

Fig. 5. Approximation of the general demographic dynamics by linear - 1, quadratic - 2 and cubic - 3 polynomials

Figure 5 shows the result of an analysis of the overall demographic dynamics over the past fifteen years. The parabolic law is seen to reflect the dependence quite well. Estimating the accuracy of the approximation using the R-square criterion gives a value of 0.94. It is worth mentioning that the quality of the approximation is determined by the closeness of the value to one, which means that the result obtained is good enough. In general, the integral population change curve has a simple parabolic shape, and we will simulate it in more detail a little later. It can be seen that migration is the most mobile tool for managing short-term demographic processes, so attention to this component should be quite large. Moreover, it is necessary to take into account other consequences of migration.

4 Discussion

"Perhaps the most important issue of modern housing construction along with the problem of increasing the volume of housing commissioning, is the question: "What should be built?". The development of market relations has led to an "explosion" in the housing construction area, which is typological although barely noticeable. Along with the usual large-panel housing construction, houses with different planning and constructive structures have appeared, the construction of single-family and low-rise buildings with high-density indices has become more active. The phenomenon of elite housing has been identified. New construction technologies are being introduced. So the sharper the question arises: "What of things that can be built is really need to be built?" [6]. Despite a number of objective difficulties and lack of sufficient experience, there is an interest in the development of high-rise construction. High-rise buildings can have different purposes: hotels, offices, residential buildings. A high-rise building is most often multifunctional and besides the main premises contains parking lots, shops, offices, cinemas, etc. The implementation of high-rise construction is dictated by the real shortage of construction sites in the largest cities, partly by the deficit of office and hotel areas, which are rational to place in high-rise buildings, as international experience shows. In the short term, particularly this direction in the development of skyscrapers should be expected, with the refusal to place residential premises there. Finally, the people's psychological attraction to the victory over the height should be always considered. The most valuable experience for domestic practice is the experience of city planners in European capitals, first including:

- consistent concentration of forces on a very limited number of sites.
- subordination of the design of the building to the principles of integrated urbanism with the complexity of development and the deployment of transport networks at several levels;
- ensuring the complexity of the development due to the combination of objects of different functional purposes in buildings, the space-planning solution, which most harmoniously meets their functions, which first of all means the combination of various buildings for various purposes in complex development;
- combination of a wide range of buildings in housing development (offices, dwellings, hotels, public services, educational institutions, trade, entertainment and sports) in order to create a wide range of workplaces for the majority of the population of the complex and its full service;

In modern conditions, the study of demographic processes in regulating and planning the socio-economic development of the regions is becoming increasingly important. The stage of the demographic crisis has finished: the "demographic dividend" has been exhausted, so Russia is expected to continue the natural decline of the population at the new stage of development, but now it will pass within a period of unfavorable changes in age composition.

Without taking into account the demographic factors, it is impossible to implement in practice and scientifically justify medium-term and long-term regional economic development programs. Being the main productive force of society, a person is the creator of social wealth; therefore, the number of people and, especially, the working population with its qualifications is a factor that influences the opportunities for further social and economic development.

Investigations and works of many economists were dedicated to the issues of raising the standard of living. The works of K. Marx played a significant role in the development of their methodological foundations. A significant contribution to the study of the standard of living of the population, taking into account the peculiarities of the Russian economy, was made by the following domestic economists: Gerasimov, V.V. Drobysheva, V.V. Zharikov, O.A. Zhuravleva, M.A. Kolomeitseva, S.I. Sergeeva.

It is worth mentioning the studies of the All-Russian Center for Living Standards, conducted under the leadership of VN. Bobkova. Definition of the quality of life according to V. N. Bobkov.: "The quality of life is a variety of abilities and satisfied needs of the individual, social groups and society as a whole, predetermining their development and well-being" [7].

The above-mentioned and other scientists made a significant contribution to the study of the standard of living of the population. "This refers to the definition of this concept, indicators of measurement, characterization of the state and the search for ways to grow in different countries. The standard of living is one of the most significant categories describing the social and economic state of society. The well-being of the population is the level of income, material and housing security, which is an important indicator of the quality of life of the population. Housing security is one of the main indicators of well-being "[8].

The aim of the study is to find solutions of the problems of life quality, using advantages of the high-rise construction.

As methods for predicting demographic processes in the Voronezh region, the method of expert assessments, the method of statistical analysis, the method of mathematical modeling, neural networks are used.

The main priority in the design and construction of high-rise buildings is to ensure their strength and stability, as well as rigidity, taking into account the impact of considerable wind forces.

This requires the search for new technical and technological solutions.

5 Conclusion

When forecasting the demographic indicators of the Voronezh region, it should be noted that in our version of the demographic burden, the negative component predominates, but the migration processes will enlarge the population. Providing the population with quality housing leads to improving the quality of life of the population.

In conclusion, a few words should be said about the economic side of the issue. The cost of high-rise buildings is much higher than of the objects of mass construction, and is determined not only by specific design solutions, but also life support systems and complex security requirements. When designing high-rise buildings, economically justified technical solutions surely should be made, but they should not reduce the reliability of the structure and make it a source of increased danger to people and the environment. Only under these conditions high-rise buildings will become a kind of visiting card of the state, speak for its economic well-being and achievements of scientific and technical progress in the construction industry [9].

The world practice of high-rise construction has revealed that uneven reduction in length under the load is one of the most difficult aspects for vertical bearing structures of reinforced concrete.

This requires the search for new technical and technological solutions.

References

1. P.P. Oleynik, T.K. Kuzmina, Technology and organization of construction industry, **2(3)**, 18-20 (2013)
2. A. Purnus, C.-N. Bodea, Engineering Procedia, **164**, 98-105 (2016)
3. N. Grau. Social and Behavioral Sciences, **74**, 10-20 (2013)
4. V. I. Popov, Scientific problems of design, construction and exploitation of bridges and tunnels, Moscow, 11-20 (2014)
5. I.G. Lukmanova, M.Y. Mishlanova, International Journal of Economics and Financial Issues, **5**, 208-216 (2015)
6. V.N. Bobkov, vestnik vgu, **2**, 26-36 (2009)
7. B.F. Bylov, Nauka (1966)
8. S.I Sergeeva, Voronezh, 116-119 (2010)
9. S.I. Sergeeva, L.N. Komyshova, Voronezh, 247-253 (2010)

Problems of increased transport load as a result of implementation of projects of high-rise constructions

Ivan Provotorov [1,*] *Valentin* Gasilov[1], and *Nadezhda* Anisimova[1]

[1]Voronezh State Technical University, Moscow Avenue, 14, Voronezh, 394026, Russia

Abstract. The structure of problems of high-rise construction us suggested, which includes the impact on environment, design solutions, transportation problems, financial costs for construction and operation, and others. Positive and negative aspects of high-rise construction are considered. One of the basic problems of high-rise construction is the problem of increased transport load. Construction of the subway on the basis of the concession mechanism, with the use of unmanned control of rolling stock is proposed as the most expedient solution. An evaluation of the effectiveness of this project is presented, it shows quite high performance indicators for a private investor. Main problems that the project implementation may face in conditions of lack of scientific and methodological support are outlined.

1 Introduction

Scientists note that the increase in the number of urban population and the need to ensure comfortable living conditions in the face of a shortage of land resources (especially in central parts of cities) leads to the need for building high-rise buildings around the world [1]. In scientific and industrial community there are two different opinions about the expediency of construction of high-rise buildings in Russia. According to the first one, positive aspects of such construction are noted. High-rise buildings contribute to the development of the economy, business and tourism. [2]. Second one Second one states that there are significant land resources in Russia and there is no need for the erecting high-rise buildings [3]. Authors of this study Without taking unequivocally both positions, the authors of this study hold the view that the construction of high-rise buildings is an extremely complex area, which, of course, has both positive and negative sides. A similar approach is used in a number of scientific papers [4]. At the same time, the tendencies occurring in foreign countries show an increase in the volume of such construction, which indicates that the construction of high-rise buildings is a trend of the current stage of development. Russian reality demonstrates a significant interest in the problems of high-rise construction, which is indicated, in particular, by the regular holding of the international high-rise and unique construction forum 100+ Forum Russia, for example in 2017 it was visited by 6,112 participants, including representatives of 97 Russian cities, 22 official delegations from cities and regions of Russia [5].

* Corresponding author: ivanprovotorov@yandex.ru

2 Materials and Methods

Problems arising during the development and implementation of high-rise construction projects require systematization and the search for optimal methods for overcoming them. Problem of high-rise construction is quite common in scientific literature, basically it concerns design, technical, technological, financial and economic aspects. But specific directions are also considered, for example, psychological aspects that need to be considered in the design of high-rise buildings [6, 7]. Along with it, they do not emphasize the search for comprehensive ways to solve the problem.

The analysis of scientific papers, sectoral studies and practice of implementing high-rise construction projects allows us to highlight the problematic field of such projects. Table. 1 presents systematized problems of the implementation of high-rise construction projects.

Table 1. Problems of high-rise construction projects implementation

Positive aspects	Problem	Negative aspects
In projects of high-rise construction, modern ecological materials and structures are often used. There is an opportunity to use technologies based on renewable energy sources.	Impact on the environment	Increase the burden on the environment, taking into account the increase in population density.
Development of building science and practice. The use of advanced materials and structures.	Project solutions	Design solutions become more complicated due to the complexity of the construction task and emerging risks.
The construction of complex high-rise buildings that combine leisure, business, residential, commercial areas reduces the impact on the transport network.	Transport problem	Increased load on the road network, passenger vehicles. Increase in traffic congestion, increase in travel time. Additional costs for transport infrastructure improving.
Minimization of the land component in the cost of construction. Reducing the cost of obtaining permits (in comparison with the construction of several objects). Reducing the cost of relocation of workers, machines and mechanisms. In the central parts of the cities, in most cases, there are few land resources left.	Financial costs of construction and operation	The increase in construction costs due to the complexity of the construction process, the uniqueness of buildings, the need to use modern expensive materials and design solutions. A significant part of the internal volume of the high-rise building is not used. The requirements for structures and their cost are increasing. High cost of repair and maintenance
High-rise construction allows the use of energy-efficient technologies aimed at saving resources.	Impact on engineering infrastructure	Increased load of the engineering infrastructure, calculated for a certain number of users. Accelerated wear of equipment under the influence of increased load. Exploitation requires more resources
Living in high-rise buildings is associated with success, prestige. The presence in the building of commercial and business facilities increases the comfort of living.	Psychological aspects, comfort of living	Living in high-rise buildings can lead to the emergence and development of various phobias, including fear of height or confined space. There may be a different kind of psychological discomfort.
Modern solutions aimed at creating a comfortable environment in high-rise buildings can positively influence the health of residents (for example, air purification). Reduced noise for residents of the upper floors. Cleaner air on the upper floors.	Impact on the health of the resident of population	There are studies showing the occurrence of negative health effects from living in high-rise buildings. Deterioration of bacteriological protection of the population. Strong wind on the upper floors.
The impact can be less than the combined effect of erecting several objects of a smaller number of storeys.	Impact in the process of construction	negative impact of high-rise construction projects on adjacent territories during the construction period is possible. It will consist in the deterioration of the living environment for the population living near the construction site
Modern video monitoring and security systems allow you to minimize the risks of living.	Security of residence	Increased fire risk, increased risks of terrorist attacks and illegal actions. Complexity of evacuation in case of emergency.

Aspects of problems of high-rise buildings construction above are not exhaustive, but they allow us to estimate positives and negatives of erecting such object, and to develop measures aiming at minimizing negative impacts. In this paper, we consider the transport aspect, as one of the most important.

We will consider problems of high-rise construction and ways of solving transport problems on Voronezh-city example (Russia). It is situated in European part of Russia, in 515 km to the south-south-east of Moscow. Population is above one million people. There are few high-rise buildings in the city. In the coming years, it is planned to put into exploitation several high-rise residential buildings, including a 100-meter high residential building. Along with it, last years are characterized by residential buildings construction, maximum height of which is near to 75 m (74,16 m and 74,56 m). This is largely due to the reluctance of developers to exceed the height of 75 m, which, in accordance with domestic regulations, is the criterion for classifying a building as a high-rise. The reasons are increased The reasons are an increase in the requirements for this kind of facilities, as well as the unavailability of fire departments to eliminate the consequences of fires in high altitude facilities. Fire safety of high-rise buildings is one of the most important problems, it is noted that for the development of high-rise construction, financial support should be provided to improve fire-fighting and rescue equipment.

Along with it, the main problem, which At the same time, the main problem hampering the development of high-rise construction in Voronezh should be recognized the problem of the transport sector, which has been operating in the regime of significant overload over the past decade. At the beginning of 2017 the city of Voronezh took the third place in Russia in terms of motorization of the population, which amounted to 308 vehicles per thousand people. This indicator shows both positive aspects - growth of living standards and mobility of the population, as well as negative ones - increasing the negative impact from motor transport, and increased load on transport infrastructure.

Since the beginning of 21-st century motorization level in Russia increased highly, and motor transport market became one of the most dynamically developed in Europe. Despite the fact that the period from 2014 to 2017 became pretty heavy (for example, in 2014, the level of cars sale fell by about 10%), we should expect the increase in motorization level of population. This can be concluded on the basis of comparison with foreign countries figure 1.

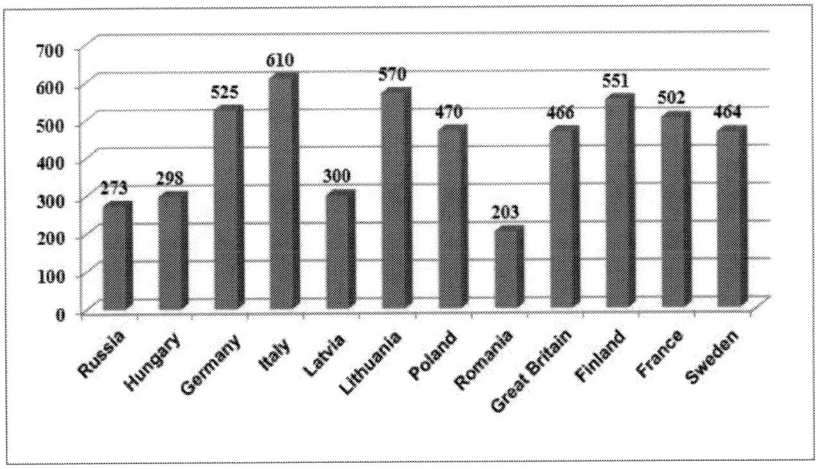

Fig.1. Motorization level in Russian Federation and different world states, passenger cars on one thousand people of population [8]

At the same time, it should be noticed that a significant lag of motorization level from developed foreign countries allows us to say, that in near years after the end of economic crisis, the significant increase in fleet of vehicles will take place (due to deferred demand). This will lead to increase in the intensity of traffic and will cause an even greater burden on the road network.

The volume of cars of all types production in Russia is decreasing under low demand level influence. The decrease in the incomes of the population also reduces the traffic volumes. As a result, transport problem is not taken serious in last years. This creates prerequisites for "explosive" increase in traffic intensity in the period of overcoming the crisis phenomena because of vehicle acquisition due to delayed demand. Thus it is necessary to develop the set of measures to overcome these inevitable phenomena that could have a significant impact on transport complex [9].

One of the most actual areas for transport problem solving in Voronezh is a subway construction project. Unfortunately, lack of funds does not provide an opportunity to implement such project. A new round of development of the subway construction project in Voronezh began in April 2017, during the visit of the Japanese delegation to Russia. During meeting, various issues related to the development of the urban environment and the participation of Japanese architects and developers were discussed, including the construction of the metro. After 2 months, an open tender was announced for work on the feasibility study for the creation of an urban high-speed rail passenger transport system.

As a result of the feasibility study, the following work was carried out by the following schedule:

- I stage until 31 July of 2017;
- II stage until 31 August of 2017;
- III stage until 20 December of 2017.

During the substantiation of the economic part of this project, the authors of this study evaluated the effectiveness of the investment project. Calculations show that the total amount of investment for the construction of the first stage of the metro, taking into account the rise in cost of construction and rolling stock, will be about 48.344 billion rubles (about 0.7 billion euros). As a result of the calculations, the problem of insufficient indicators of the project's effectiveness is revealed, which does not allow to return the invested funds at the desired time. The feasibility of integrated automation of metro lines is being studied, allowing unmanned rolling stock management, which can be one of measures aimed at improving the efficiency of the project. This is consulted with general designer. This measure allows you to reduce labor costs, which should lead to increased profitability. At the same time, complex automation causes additional costs for rolling stock and the creation of integrated automated control systems.

In view of the complexity of implementing the project only at the expense of budget funds, the basic option is the scheme of infrastructure concession. According to the Federal Law "On concession agreements," objects of the concession agreement may be subways and other public transport [10, 11].

3 Results

Based on the comparison of various options for the implementation of the subway project on a concession basis, a financing option for the project was chosen when the state finances the project at the initial stage (the first 4 years), and for 5-th year the private investor will make all investments in the amount of 19762.655 million rubles (about 286 million euro), which will amount to 40.879% of the total investment under the project. To determine the feasibility of the project, the project's cash flows for the concessionaire were modeled - figure 2.

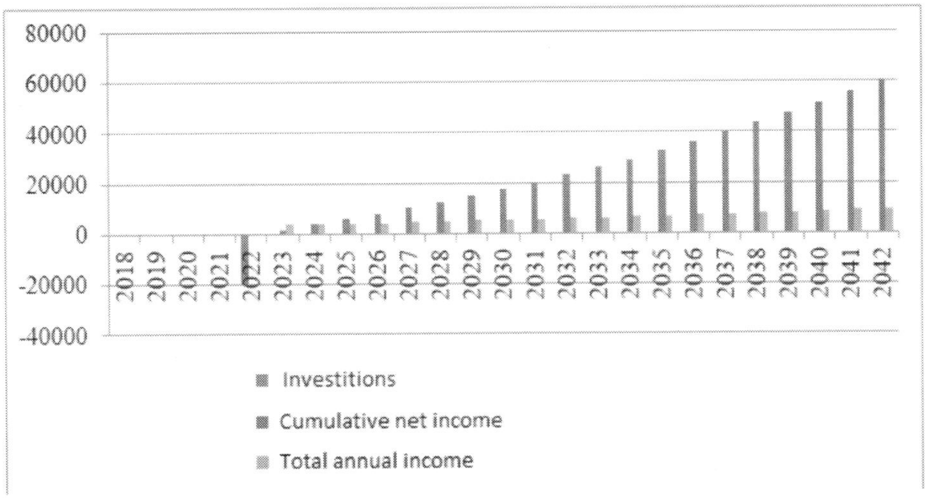

Fig. 2. Modeling of cash flows of a subway construction project for a concessionaire

The calculations show sufficient performance indicators for the concessionaire:

- IRR = 12%;
- DPP = 15,76 year;
- NPV = 3094,852 million rubles (about 45 million euros);
- DPI = 1,172.

4 Discussions

The problem of ensuring the efficiency of the transport network for the development of high-rise construction requires more attention from the scientific community, since it is one of the basic problems. The proposed project of creating high-speed rail passenger transport will involve certain risks. This is connected to the complex technical task of constructing the most complex facility, financial and economic risks, and social risks associated with the implementation of the project on the existing transport and residential system of the city. An assessment of the readiness of the theoretical and methodological framework for implementing such projects on the basis of work [12, 13] shows a rather low preparedness of Russian reality for the implementation of such projects. One of the main problems in the implementation of the subway construction project is the difficulty of providing the required passenger traffic (300 thousand people a day on average in 2023). In addition, a significant amount of investment in the project in the conditions of economic instability in the country's economy is required. Currently, it is necessary to develop approaches to improving the effectiveness of such projects [14].

5 Conclusion

Solving the basic problems associated with the construction of high-rise buildings is the main condition for the development of this direction. The approaches to solving the problem presented in this paper can be relevant for a number of cities, faced with the difficulties of providing favorable transport conditions for the development of high-rise construction.

References

1. V.P. Generalov, E.M. Generalova, Bulletin of SASU, **1 (18)**, 13-18 (2015)
2. L.V. Tolstoy, R.G. Abakunov, Innovative Science, **1-1**, 100-103 (2017)
3. E. Kalcheva, A. Takia, Y. Hadi, Social and Behavioral
4. I.V. Ilin, A.B. Anisiforov, WSEAS Transactions on Business and Economics, **11**, 757-764 (2014)
5. V.V. Gluhov, I.V. Ilin, Lecture Notes in Computer Science, **8638**, 509- 518, (2014) DOi-10.1007/978-3-319-10353-2_46
6. K. Olga, Intern. Journal of Economics and Financial Issues, **6**, 995- 1002 (2016)
7. O.V. Kalinina, M.V. Lopatin, Actual Problems of Economics, **182**, 392-405 (2016)\DOI: 10.1016/j.sbspro.2015.12.093.
8. Yu.S. Dvirnik, Scientific almanac, **4-3(30)**, 52-54 (2017) DOI: 10.17117/na.2017.04.03.052.
9. J. Tamošaitienėa, E. Gaudutisb, M. Kračkaa, Procedia Engineering, 57, 1151–1155 (2013) DOI: 10.1016/j.proeng.2013.04.145
10. C. Haitaoa, L.Leileic, Q.Jiuzid, Procedia Engineering, **43**, 23–27 (2012) DOI: 10.1016/j.proeng.2012.08.005
11. P. Fenga, W. Xingkuan, Procedia Engineering, **21,** 943–947 (2011) DOI: 10.1016/j.proeng.2011.11.2098
12. V.A. Kazarnovskiy, L.A. Manuhina, L.V. Katina, Economics, Management, **1**, 44-47 (2015)
13. K.A. Karelina, APRIORI, **1**, 1-11 (2016)
14. V. Gasilov, N. Anisimova, I. Provotorov, MATEC Web of Conferences, **106,** 08035 (2017) DOI: 10.1051/ mateconf / 201710608035

Evaluation of possibility to increasing sustainability of high-rise buildings through use university intellectual property

Igor Potekhin[1,*], *Valeryi* Mischenko[1], *Angela* Mottaeva [2,3], *Alexander* Zheltenkov[3]

[1]Voronezh State Technical University, Moscow Avenue, 14, Voronezh, 394026, Russia
[2]Moscow State University of Civil Engineering, Yaroslavskoe sh. 26, Moscow, 129337, Russia
[3]Moscow Regional State University , 129125, Radio str. 10, Moscow, Russia

Abstract. In this article explained approach of valuation of intellectual property of Voronezh State Technical University, as her usefulness to increasing the sustainability and ecological safety of high-rise building. High-rise building is main type of buildings in modern cities. They include large volume of material mass, high volume of energy using and high volume of emissions. Using innovation solutions to improving ecology safety of high-rise buildings has large potential to city in whole. Explained in the article methods of calculation of effects helps to value sustainable solutions of present and future generations. Thus usefulness of patents express through usefulness regarding to high-rise building, including for sustainable development.

1 Introduction

The purpose of this section is defining method of patents evaluation. In case of uncertainty methodic of its evaluation, known of modern methodic. Supporting the development process for building products by the use of research portfolio analysis: A case study for wood plastics composite materials «Assuming that R&D projects are based on an anticipated potential analysis, our study is the first to examine the use of project databases as an information pool about trends and future developments in WPC with respect to the construction industry. We derived from it a project-based future matrix by explaining current WPC developments in the light of the *Technological S-Curve*concept to Foster (McKinsey). This method predicts whether a specific research area is still in the phase of generic research or already close to commercialisation. Findings are in the interest of the bio-plastics industry and particularly of the medium sized European WPC companies which will be enabled to easily gather significant input for their corporate innovation roadmap» [1].

«Although today's WPC research shows activities in both technical and ecological aspects, our portfolio did not undermine a potential trend in the combination of these, such as an eco-innovation approach. Obviously, manufacturers still focus on mechanical material and product properties rather than on ecologically driven sales arguments, i.e. cascading or

* Corresponding author: potekhin_300587@mail.ru

sustainable disposal. Research in this field is solely driven by public institutions and is therefore in the generic state similar to green-composites but with orientation to applications»

2 Materials and Methods

We will use matrix and patents of portfolio, but in our case we will seek direct benefit from patents, application to high-rise buildings and technological process of its construction.

Table 1. Patents of Voronezh state technical university

Name of patent Registration of patent	Technological result				
	Raw material economy	Energy economy	Decreasing labor intensity	Increasing an accuracy	Increasing an ecology
1	2	3	4	5	6
Raw material mix for making construction products by non-firing technology RU 2584018		+			+
Bearing support part of bridge RU 2547531				+	
Method of making coating of sport playground RU 225094	+				+
Node joining of column with monolithic concrete slab RU 2547035	+		+	+	
Nano-structure binder for composite building materials RU 2408552	+			+	
Method of manufacturing composite organic-mineral modificator of concrete RU 2454381	+				
Method of manufacturing foam decoration RU 2276659	+				+
Firesafety intumescent material RU 2317274					+
Combined pneumatic formwork for making monolithic concrete span structures RU 2478158	+		+		
Unit for bio-cleaning of waste sewage water RU 2114793		+			+
Method of manufacturing hollow filler RU 2186047	+				+
High-regenerative level meter RU 2418270			+	+	
High-frequency volume meter RU 2393434			+	+	
Construction of road pavement RU 2340720				+	
Polymer-concrete mix			+	+	

RU 2394786					
Composition for making building material RU 2345968	+	+			+
System of definition of coordinates of underground pipeline RU 2437127	+			+	
Complex additive for concrete mix RU 2443648		+	+		
Sun heat collector RU 2444682		+			+
Accumulation heat exchanger RU 2333433	+	+			
Sun shielding window RU 2327847	+	+			
Grained filter RU 2116117	+				+
Sound absorbing cell panel RU 2206458		+			+
Compound blade of land grader RU 2498022	+			+	
Vehicle power generator RU 2521742		+			
Stator of windpower unit RU 2517168		+			+
Rotor of windpower unit RU 2515577		+			+
Method of vibroshield of mobile construction machines RU 2310781				+	+
Method of prevention of fatigue crack on base of magnetodipole RU 2353909		+		+	
System of automatic managing of work tool of earthmoving machine RU 2327010		+		+	
Method of manufacturing granite-cement products RU 2500655	+	+			+
Noise silencer RU 2338964					+

To evaluation usefulness the innovation of VSTU, we use definition High-rise building. A building is usually considered a “high-rise” if its height exceeds 23 m (75 ft) [2]. In this book is written about main types of structures and in weight mass of each structure.

In our case we choose more distributed structure for city building in Voronezh for building with 18 to 25 floors. This structure is made of reinforced concrete. Its weighted weight is: 32 kg/m2 of rebar and 0,34 m3/m2 of concrete [2].

 https://doi.org/10.1051/e3sconf/20183303020

3 Modeling

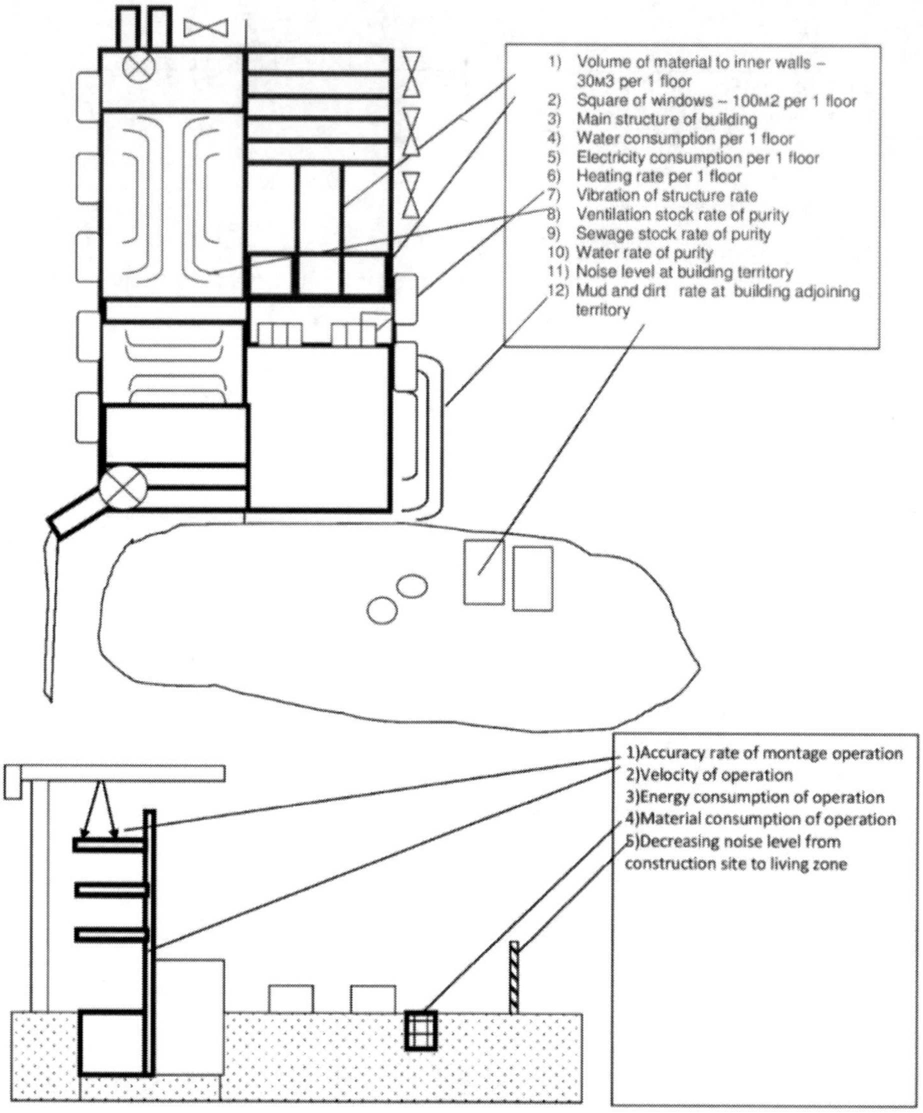

Fig. 1. Application of university patents at high-rise building construction structure and at site

Requirements to ecology and sustainability of city housing is written in the book «Reconstruction and renovation of current city housing».

«Development cities is accompanied with increasing a density of housing, especially in central heritage districts. By this is forgotten, that presenting open spaces improves temperature condition, insolation, aeration, giving conditions of dispersions of harmful exhausts in atmosphere, provide conditions to formulation ecological structure of settling. At present time, investors and city administration don't take into account these ecological criterions».

Developed measures of protection from noise include urban planning, architecture, construction activities – installation of shielding bands of forest plants and special fences, increasing sound insolation, using special noise protection screen-bariers between noise issues and objects, which need to be protected. System of water sewage have to take into account possibility of receiving drainage waters from different issues and common collectors. By technical possibility and coordination with nature protection administration, it could be use to feed garden plants by these sewage and drainage water through private pipeline.

Pathogenic zones could be technogenic or anthropogenic, which conditions is caused by industry impact, transport impact, power plants impact, garbage dumps, geochemical wastes. In this zones, the first system in human, which threated is nervous system – peoples feels irritability, headache, mood drop, fatigue [3].

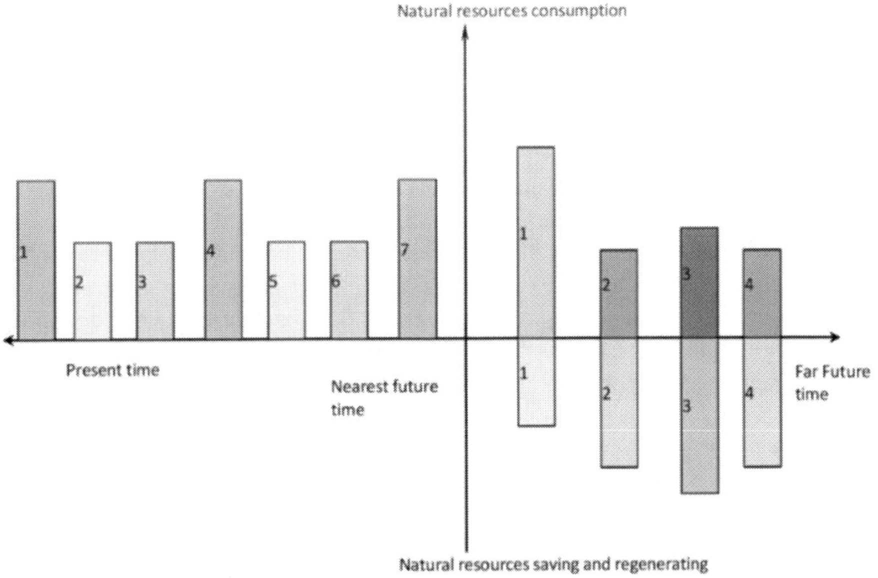

Fig. 2. Meaning of patents application at high-rise building to the sustainability of city

4 Results

Table 2. Explanation of sustainability of high-rise building

Present time	Future time
1.Usage of energy	1.Economy of natural raw materials
2.Simplicity of building structure and construction	2.Saving a biodiversity on building territory
3.Usage of raw materials	3. Saving comfort for people on building territory
4.Payment only for natural raw resources	4. New job specialties
5.Fee of building usage	
6.Clean around a building	
7.Pure exhaust from building	

Our consideration of using some most meaningful patents to increasing sustainability of building.

Based on the plastic and hardened test results on of 14 types of concrete mixes, Mix 4 from Group M (360 kg total binder with, 80% GGBFS, 0.35 w/c) and Mix 14 from Group L (300 kg, 60% GGBFS, 0.38 w/c) are proposed for use in the construction of Masdar City in the UAE. These mixes meet the strength, durability, setting times, and workability requirements of Masdar City as well as they are cost effective compared to MC baseline mix. Although Mix 10 with 50% GGBFS and 30% FA achieved good durability and shrinkage performance as well as low carbon emissions it could not meet the workability requirements of Masdar City [4].

Production and transportation of many engineering construction materials requires high amounts of energy and has high levels of GHG (greenhouse gas) emissions associated with it. This can have a detrimental impact on the environment especially with the recent realization of the severity of climate change and global warming. Concrete is one of the most widely used construction materials and has CO_2emissions associated not just with the manufacturing process of cement, but also transport of ingredients over long distances. One of the solutions to reduce the environmental impact of concrete is to use more environmentally friendly ingredients and reduce the amount of transportation required in shipping these ingredients and/or the finished material. One of the building materials, Rammed Earth (RE) also known as "Pisé de terre" or simply "Pisé" (Anderson, 2000) is the material that Ecosol Design & Construction (ED&C) Ltd and the builder members of the North American Rammed Earth Builders Association (NAREBA) have been using for construction in North Western Washington State, USA; and Southern Alberta and British Columbia, Canada. The material typically used consists of locally available sand, soil, or gravel and is stabilized using nominal quantities of cement. The author was approached was approached by the Cement Association of Canada (CAC) to undertake a research project to study mechanical properties of RE. In the recent years, RE walls construction has become popular [5].

Natural sand has been the conventional fine aggregate in concrete production for many decades. However, there has been extensive research into alternative materials suitable to replace sand in concrete. The stress-strain curves (SSC) for concretes with quarry dust and river sand as fine aggregate showed similar behaviour. The SSCs for concrete with quarry dust showed higher maximum strains, 15% higher than those of conventional concrete. Based on the characteristics of the SSCs, it can be concluded that the structural design process of concrete with quarry dust within the elastic and inelastic range can be carried out normally without the need for modifications.

The modulus of elasticity of concrete with quarry dust was 8.6% lower than concrete with sand. However, concrete with 25% sand replacement gave the highest (7.9% higher than concrete with no sand replacement) modulus of elastic of concrete. The equations to predict modulus of elasticity of quarry dust concrete based on the compressive strength derived in this study varies greatly from those proposed in standard codes of practice [6].

This study investigates the first known application of light polyurethane foam (LPF) in construction of an extension to existing noise barriers. An experimental program was carried out to determine the LPF mechanical properties as well as the lateral resistance of the entire wall system. A fatigue test was also performed on a flexi-wall in order to evaluate its long-term performance under a repetitive loading. According to the results of this study, the following conclusions can be drawn:

The construction of a flexi-wall is significantly faster and more economic than the conventional method for the investigated application. The outcomes of the experiments showed that the structural performance of the wall for accelerated construction is completely acceptable. The mechanical performance of the flexi-wall is mainly governed by the steel rebars and as LPF is lighter and less expensive than cementitious material, a flexi-wall can be a good alternative to other types of extension for the investigated application.

Even though not examined in the current study, it is expected that applying a layer of shotcrete on the wall surfaces could enhance the flexural resistance, surface resistance and fatigue characteristics of the wall. Noise barriers or sound walls are usually constructed along the roadways to mitigate the airborne noise emanating from vehicles. Most of the provinces across Canada have established a "noise barrier retrofit program", which mainly involves the extension and renewal of the existing sound walls in urban environments, to mitigate noise pollution and minimize its impacts on public health.

In comparison with conventional masonry walls, a flexi-wall is more noise-absorbing and can be built significantly faster. The construction of flexi-walls along roadways is also less obstructive since there is no need for construction vehicles and mobile cranes, which usually block or constrict the roads during roadside construction. This accelerated technique also reduces traffic congestion, construction noise and risk of road accidents during the construction period which results in lower cost of the project [7].

It was proven that as the concrete compressive strength increases, the effect of the construction increases and causes more loss in the bending capacity of the structural element. A procedure to compensate for the reduction in the moment capacity due to the existence of a construction joint is presented [8].

There has been a drastic increase in the world population in recent years, new megacities are born and existing megacities become more populated. Besides new economical, managerial and social challenges associated with growing cities, a deformed energy budget pulls them toward a warmer climatic condition, known as urban heat island (UHI). Highly populated areas mandate cities to develop either vertically or horizontally, resulting in more released anthropogenic heat, a higher blockage effect against urban ventilation, a higher absorption of solar radiation due to the implementation of artificial materials, and eventually a reduced long-wave emission to sky due to the blockage effect of buildings. The UHI models are diverse in terms of scale with respect to the aim of a study, changing from building-scale for investigation of the impact of the UHI on thermal comfort of a pedestrian to urban-scale for exploring the effect of synoptic wind on urban ventilation.

Table 3. Cost valuation of non-financial effects of sustainable development, using a patents VSTU

Effect	Mechanism of realization	Recipient of benefit	Financial analog of services	Calculating effect on 1 year
Economy of nature wealth (mineral deposits)	-saving nature deposits and landscape -decreasing necessity in fossil fuel consumption, generation energy at power station	-majority of city	1) decreasing fees for garbage deposition 2) decreasing fees for energy consumption	1) 100 000 RUB 2) 30 000 RUB
Saving a biodiversity at buildings adjoining territory	-alive plants -different kinds of animals -saving earth from erosion	-majority of city -buildings owner	1) decreasing costs on advertisment 2) снижение расходов на озеленение территории 2) decreasing costs on gardening a territory	1) 60 000 RUB 2) 40 000 RUB 3) 300 000 RUB 4) 30 000 RUB
	Water quality		building 3) increasing service life of building foundation 4) decreasing costs on water	
Saving comfort people around a building	-clear air -low noise -clean	-buildings owner	1) decreasing costs on advertising 2) decreasing costs on tiding up 3) decreasing illness of habitants	1) 60 000 RUB. 2) 30 000 RUB. 3) 20 000 RUB.
New job speciality	Education new skills and speciality	-majority of city -buildings owner	1) dducating new specialities	1) 40 000 RUB.

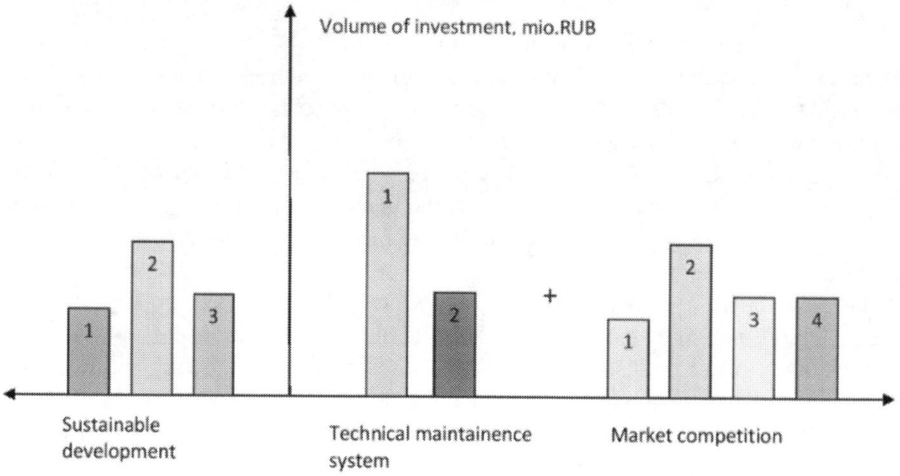

Fig. 3. Distribution of investments to development of construction company

Table 4. Explanation of distribution investments to development of construction company

Investments in sustainable development	Imperative investments	
	Technical maintainence system	**Market competition**
1. Prototype creation	1.Renovation of equipment and tools	1.Development new models
2. Testing and modification of prototype	2.Repair of equipment	2.Purchase new equipment for making a new models
3. Commercial production		3.Testing and improving new models
		4.Commercial production of new models

4 Conversation

The influence on sustainable development of city with intellectual property of Voronezh State Technical University. The evaluation of useful potential of intellectual property. Modern requirements to ecological is regulated by LEED and BREAM, which most famous in the world. About the standards and another ecological sustainable construction is showed in book Kubba. Handbook of green building design and construction [9].

«Sustainable building materials also reduce landfill waste of which the IgCC codes mandate a minimum of 50% of construction waste must be diverted from landfills, and at least 55% of building materials must be salvaged, recycled-content, recyclable, biobased, or indigenous. The IgCC also mandates that buildings must be designed to span for a minimum of 60 years of life, and must show a service plan that justifies that. The following aspects should be considered when choosing building materials for a project:

• Choose sustainable construction materials and products whenever possible. Their sustainability can be measured by several characteristics such as recycled content, reusability, minimum off gassing of harmful chemicals, zero or low toxicity, durability, sustainably harvested materials, high recyclability, and local production. Use of such products promotes resource conservation and efficiency, minimizes the adverse impact on the environment and helps to harmonize with its surroundings.

• Employing dimensional planning and other material efficiency strategies reduce the amount of building materials needed and cut construction costs. For example, the design of rooms to 4-foot multiples minimizes waste by conforming to standard-sized wallboard and plywood sheets.

• If possible, reuse and recycle construction and demolition materials. Using recycled-content products also cuts costs and assists in the development of markets for recycled materials that are being diverted from landfills, an example of which is the use of inert demolition materials as a base course for a parking lot.

• Allocate adequate space to facilitate recycling collection and to incorporate a solid waste management program that reduces waste generation.

• Require waste management plans for managing materials through deconstruction, demolition, and construction.

Employing recycled/reused materials helps to ensure the sustainability of resources. If building projects use only virgin raw materials these materials will gradually be exhausted. As the availability of raw materials become scarce, prices will rise and before long the materials will no longer be obtainable.

This trend has already started to impact the availability of certain raw materials which are either no longer available or have become very scarce, and can only be obtained recycled from existing projects.

Improved technology is making it much easier and more cost-effective for designers and engineering professionals to incorporate sustainability into their high performance design strategies. Likewise, there are many recommended practices that can reduce the environmental and resource impacts of buildings, and enhance the health and satisfaction of their occupants. The most prominent strategies that come to mind include:

1. Using less to achieve more: The most effective green design solutions are able to address a number of needs with only a few elements. For example, a concrete floor may be simply finished with a colored sealant that reflects daylight for better illumination, and eliminates air pollutant emissions from floor coverings. The floor can also be used to store daytime heat and nighttime cold to provide occupant comfort. Thus a carefully designed element serves as structure, and finished surface, distributes daylight, and stores heat and cold, thus saving materials, energy resources, capital and operating costs.

2. Incorporate design flexibility and durability: Buildings that are designed with the flexibility to adapt to changing functions over long useful lives reduce life-cycle resource consumption. Durable sustainable structural elements that contain generous service space and are able to accommodate movable partitions can last for many decades, instead of being demolished because they are incapable of adapting to changing building functions. Durable envelope assemblies reduce life-cycle maintenance and energy costs and improve comfort.

3. To achieve maximum effectiveness combinations of design strategies must be carefully considered: Green buildings are incorporating increasingly complex systems of interacting and interrelated elements. Intelligent green design must consider the impact of these elements and systems on each other, and on the building as a whole. As an example, the need for mechanical and electrical systems is greatly affected by building form and envelope design. Combining strategies like daylighting, solar load control, and natural cooling and ventilation can all work together to reduce lighting, heating and cooling loads.

4. Take advantage of site conditions: Buildings are usually considered more sustainable when they respond to local microclimate, topography, vegetation and water resources; they are also usually more comfortable and efficient than conventional designs that rely on technological fixes and ignore their surroundings. As an example, Santa Monica in California has exemplary solar and wind resources for passive solar heating, natural cooling, ventilation and daylighting, but has meager local water supplies (some of which have recently been polluted). Taking advantage of such free natural resources, and conserving scarce high-priced commodities are appropriate approaches to reduce costs and connect occupants to their surroundings.

5. Adopt preventive maintenance, not repairing after the fact: Addressing potential problems from the beginning by applying preventive maintenance is both practical and

economically prudent. For example, using low-toxicity building materials and installation practices is more effective than diluting indoor air pollution from toxic sources by employing large quantities of ventilation air.

Also it is necessary to take into account the involvement stakeholders and owners in ecological and sustainable investments.

What are motivations for stakeholder willingness to participate? For instance, it is found that perception of environmental risks and benefits influence willingness to participate and that stakeholders willingness to participation with effort, money and time differ with reference to whether a city region is new or already established. Willingness to participate was analysed by means of statistical analyses in Egypt, which seemed to vary across cities and neighbourhoods across the country, although it was concluded that action approaches were needed for stakeholder participation to facilitate dynamic transformation towards increased sustainable landscape. In order to identify motivations and barriers for collaboration between stakeholders in urban logistics in Singapore, a study showed that environmental benefits and risks influence motivations to participate. Barriers to stakeholder willingness in land use in Hong Kong include difficulties in urban renewal, inherent shortage of land, lengthy time involved in coordination and responsiveness to stakeholders, and overcomplicated administrative processes.

5 Conclusion

Urban areas account for 70% of carbon emissions, and are likely to be the locus of attention to reduce future emissions in developing countries. It is expected that, by 2030, about 2 gigatons CO_2eq of GHG emissions should be cut per annum in order to reach the 1.5 ° level by 2050 (CAT, 2015). To make that happen, cities must take concrete steps towards defining and implementing various low-carbon projects and mitigation policies. Moving from business as usual (BAU) to a low carbon future requires an annual investment of between US$4.5 and 5.4 trillion globally (CCFLA, 2015).

For instance, Asian Development Bank (ADB) places greater attention on financing projects that offer an array of GHG mitigations, energy efficiency, reduced accidents and other social engagement potentials.

An illustrative comparison between the market and nonmarket benefits of low-carbon investment projects in Nairobi, Colombo and Balikpapan. The vertical axis denotes the overall nonmarket benefits in millions of US dollars and horizontal axis represents the projects' market return. This diagram shows four different situations on which low-carbon investments can be judged:

(1)Win-win: This will happen when both of the market and nonmarket NPVs are above zero. This creates a desirable situation for both investors and the public to implement a low-carbon project

(2)Win-lose: This occurs when a project offers large nonmarket benefits but the market benefits are negative (i.e., the project is financially infeasible). In this case, interventions may be required in case the project results in significant benefits to the society. This situation applies to all of the projects in this study.

(3)Lose-lose: This is when a project offers both negative nonmarket and market benefits. This is not desirable for either investors or the societies receiving the benefits of the investments.

(4)Lose-win: This is a scenario in which a project results in a positive market NPV but negative nonmarket NPV. In this case, project implementation needs to be avoided from a public policy perspective.

References

1. N. Nagib, C. Gerges , A. Issa, S. Fawaz, Case Studies in Construction Materials, **5**, 112-123 (2016) https://doi.org/10.1016/j.cscm.2016.09.004
2. K. Rashidi, M. Stadelmann, A. Patt. Sustainable Cities and Society, **34**, 69-78 (2017) https://doi.org/10.1016/j.scs.2017.06.003
3. K. Soma, M.W.C.Dijkshoorn-Dekker, N.B.P.Polman, Sustainable Cities and Society, **37**, 438-450 (2018)
4. M. Elchalakani, T. Aly, E. Abu-Aisheh, Case Studies in Construction Materials, **1**, 10-24 (2014) https://doi.org/10.1016/j.cscm.2013.11.001
5. R. Gupta, Case Studies in Construction Materials, **1**, 60-68 (2014) https://doi.org/10.1016/j.cscm.2014.04.002
6. A. Rozhentsova, A. Mottaeva, MATEC Web of Conferences, 08076, DOI: https://doi.org/10.1051/matecconf/201710608076 (2017)
7. A. Mottaeva, IOP Conf. Series: Earth and Environmental Science, **90**, 012120 doi:10.1088/1755-1315/90/1/0121209 (2017)
8. A. Pimenova, S. Kuzmina , N. Morozova, A. Mottaeva, MATEC Web of Conferences, **73**, 07018 (2016) DOI: https://doi.org/10.1051/matecconf/20167307018
9. S. Kubba, Elsevier (2017)

Methodical foundations of the preparation of highly valued personnel for high-rise construction

Svetlana Belyaeva [1,*], *Oksana* Belyantseva[1], *Nataliya* Safonova[2], *Olga* Vasilyeva[3]

[1]Voronezh State Technical University, Moscow Avenue, 14, Voronezh, 394026, Russia
[2]State University of Marine and River Fleet in the name of admiral S.O. Makarov (Voronezh branch), Leninsky pr., 174 "L", Voronezh, 394033, Russia
[3]Moscow State University of Civil Engineering, Yaroslavskoe sh. 26, Moscow, 129337, Russia

Abstract. When carrying out design and survey and construction and installation works for such an innovative type of activity as high-rise construction, the problem of personnel qualification becomes urgent. The article poses a research problem, identifies the main reasons for the need for training highly qualified specialists in construction, and suggests areas for improving training. The expediency of development of mentoring system was proved, the corresponding model of interaction between educational institutions and construction enterprises was offered, key interaction effects were evaluated.

1 Introduction

Globalization and unification of the world economy, the active formation of the market system of management in Russia, the specificity of the current stage of scientific and technical progress [1] make it necessary to pay special attention to the problem of improving the workforce and ensuring its compliance with modern social and economic requirements.

Improvement of higher education, maintenance of its quantitative and qualitative characteristics at a level that meets all modern requirements of building science and practice is a necessary condition for high-quality and safe performance of works, an important factor in ensuring safety, environmental friendliness, energy efficiency and intellectuality of high-rise buildings.

On the basis of data of the amount, quality, structure of highly skilled specialists in construction and their use, it is necessary to determine the real need for them for the type of activity, and hence the scale of their reproduction [2].

The problem of improving the training of specialists in the system of Russian higher education in the construction sector is of great importance, at least for 4 reasons:

1) to improve the quality of education in general;

2) to improve the quantitative, structural and qualitative characteristics of the workforce;

3) for the successful and civilized development of the construction industry;

4) to ensure economic stability and economic growth.

* Corresponding author: belyaeva-sv@mail.ru

Note that all these moments are interrelated and interdependent, and the urgency of the problem of increasing the efficiency of reproduction of workers with higher education is not only great but also obvious, predetermined by the very content of the category "highly qualified specialists", by the huge role that they play in the economy[3] .

This problem has two interrelated aspects - quantitative and qualitative [4]. But the improvement of the quality of the workforce in accordance with the current level of scientific, technical and cultural achievements in the society, the trends in their development, and changes in the socioeconomic nature is of particular relevance [5].

Many studies confirm that the insufficiently high quality of the labor force, firstly, leads to its accelerated physical and moral aging, and secondly, causes crisis and regressive conditions in all spheres of society, but primarily in the economy [6]. Thus, the improvement of the quality and structure of the workforce should become an important part of the current stage of development of the Russian economy.

2 Materials and Methods

The main direction of the improvement of the quality of workforce, in our view, is the improvement of its educational and professional training, which includes the development of the following elements:

- a system of general and vocational education aimed at training new workforce;
- systems for in-house training of personnel, including workplaces, as well as "pointed", "targeted" training of interns (for example, graduates of higher education institutions);
- Self-study and distance learning systems with appropriate conditions;
- a continuous education system aimed at preserving and improving the workforce;
- systems of vocational guidance and psychological support for young people, unemployed citizens, as well as those who are employed, but who wish to change the work profile;
- services of the employment service for training unemployed citizens, as well as a mechanism for supporting and organizing advanced training for workers at risk of dismissal.

Let us dwell on some of these problems in more detail. First of all, it should be noted that the Russian multifunctional system of general and professional education, retraining and advanced training of personnel is going through a difficult period of reform. At the same time, changes in the education system are largely inadequate to changes in the labor market [7]. State educational structures, which have a rich experience of teaching adolescents, often can not organize effective adult training.

Thus, changes in the system of vocational education should be directed to meet the needs of the population for professional educational services, but also they should take into account the corresponding needs of regional economies and the development trends of regional labor markets [8].

According to the survey of employers, it was found that the average period of adaptation to work in a team of a graduate of a technical college is 1-3 years (the main reasons are a low level of development of "flexible skills" - the ability to work in a team, negotiate, find an approach to people , persuade, make decisions independently, select and use relevant information, etc., as well as "lag" in technical knowledge and corresponding "hard skills" from the real level of innovation development (because from the moment of introduction of innovation at a particular enterprise before it is included in the relevant normative and technical documents, and, accordingly, in lecture courses, takes place, in particular, in construction, 1-2 years).

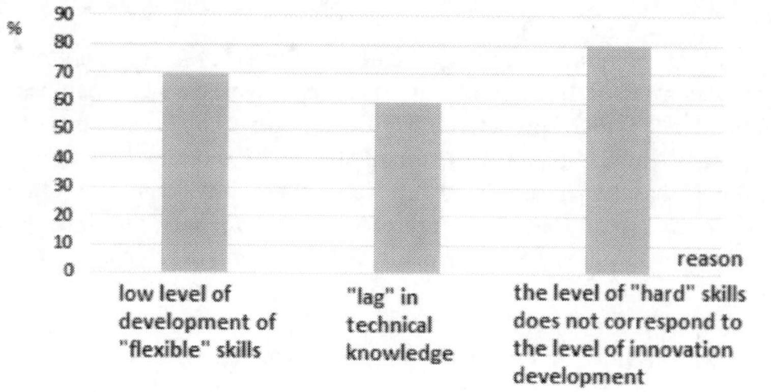

Fig. 1. Results of the survey of employers on the reasons for the duration of adaptation of graduates of higher education institutions in the collective of the enterprise

As a solution to existing problems, we consider it expedient to introduce a mentoring mechanism in the university. Mentoring is an activity for the individualization of education aimed at clarifying the educational motives, professional interests and competences of the student, choosing educational resources for creating an individual educational trajectory, working on professional orientation, forming the educational and professional reflection of the student.

An experienced employee of an educational organization or an employee of a partner enterprise can be a mentor [9]. The subject of mentoring is the formation and implementation of an individual educational trajectory of students, starting with the diagnosis of the basic competences of talented people and including the process of support after employment.

The basis of the proposed model of the mentoring system is the construction of the goal of the educational trajectory (work with tutors and staff of partner enterprises) and the individualization of the educational trajectory by including services for the development of "flexible" skills (coaching, training of emotional intelligence, business simulation, open lectures, webinars, etc.) and the rapid development of innovative professional competencies and the specifics of more narrowly professional activities (mentoring of an employee as a rule, university graduates) on the basis of a preliminary assessment center for identifying "strong" and "weak" parties to the participants, and conducting gaming sessions to introduce into the profession with the final formation and promotion of the resume (or the support of a business project).

In general, the model of the mentoring system in the educational organization is presented in Figure 2.

We also consider it expedient to create a single information and communication platform on the website of an educational organization that must contain the following necessary services and accumulate them in one place:

- portfolio of participants;
- database of planned contests, grants, conferences;
- database of mentors (describing their career success and the skills that they needed);
- database of winners of competitions, participants in projects of an educational organization, etc. (associated with the portfolio, infographics of quantitative indicators);
- database summary "Gold reserve personnel";
- database of enterprises-partners, employers and enterprises inviting students for practice;

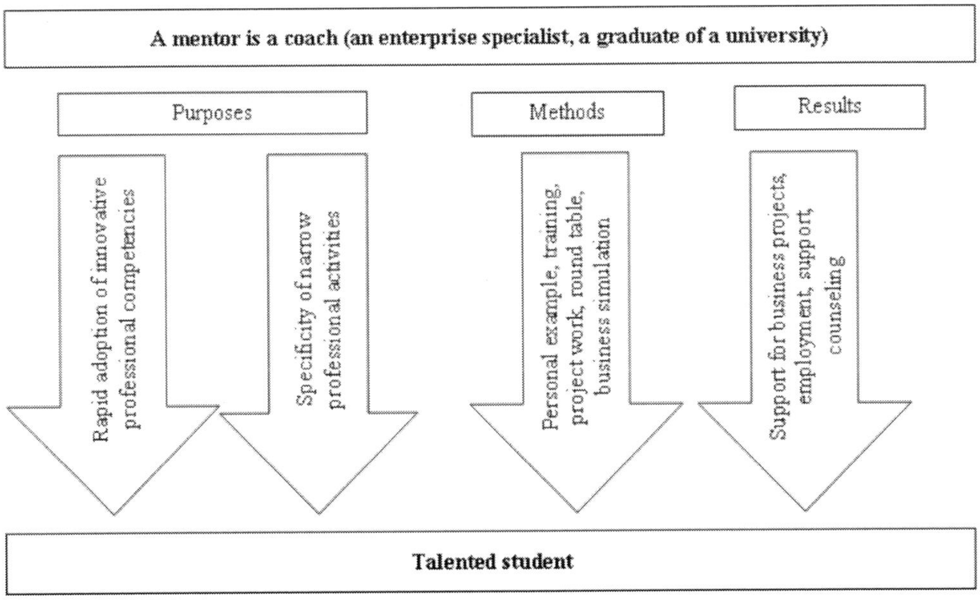

Fig. 2. Conceptual model of the mentoring system in the educational organization of the construction profile

- chat rooms for participants and mentors;
- e-learning resources and links;
- database of projects implemented by the educational organization and its partners;
- information from mentors;
- access to the service of distance educational technologies.

Mentor system consists of following. The tutor (specialist of the educational organization) helps to develop an individual trajectory of learning based on the planned goal and the results of individual evaluation (with the exception of the formation of an individual curriculum that includes participation in certain activities to form the necessary "soft" and "tough" skills, conferences, etc., obtaining additional education or professional retraining, upgrading skills for certain programs) and adjust it for the participants that are involved in the project (taking into account the impact of the participant is determined by results of participation in project activities and business modeling).

The mentor-coach (the specialist of the enterprise-partner, as a rule, the graduate of the educational organization) carries out certain measures for the formation and improvement of innovative professional competencies (personal example, open lectures, round tables, master classes, etc.), introduction to the specifics of activities enterprise, conducts project activities with participants of the relevant focus, accompanies the participant after employment for this enterprise, conducts consultations, may initiate topics , assignments for the implementation of project activities of participants. By the decision of the mentor the participant can be employed at the enterprise even before the end of the training.

In general, a block diagram of the algorithm for implementing the mentoring system in an educational organization is presented in Figure 3.

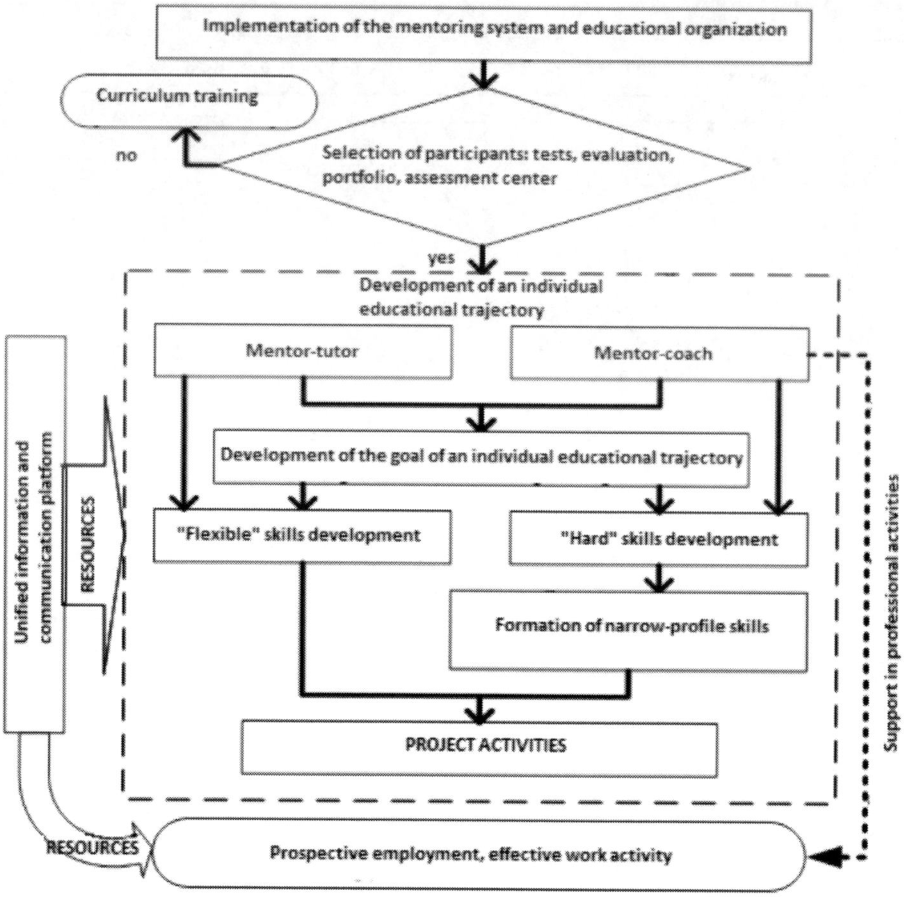

Fig. 3. Block diagram of algorithm for implementing the mentoring system in educational organization

From the consideration of the problem of improving the quality of training of workers in educational institutions, we turn to the problem of improving the workforce at enterprises and organizations.

It is known that the qualitative level of employees of Russian enterprises and organizations is significantly inferior to the requirements imposed on the international labor market.

Most of trainde workers in the process of finding work face the need to either upgrade their existing qualifications or acquire a new specialty. However, employers who are experiencing financial difficulties do not pay attention and do not show interest in upgrading their employees' skills. Volumes of in-house training are constantly shrinking.

Training programs for employees at enterprises are focused mainly on obtaining primary qualifications, there are practically no training programs for highly skilled workers.

Insufficient attention is paid to motivation of employees to improve their professional and qualification level, there are practically no managers and other specialists dealing with issues of in-house training and having for this purpose special training.

Practice shows that the minimum costs necessary for a simple reproduction of the professional potential of the enterprise must be at least 1.5% of the wages paid [10]. However, at present they are 0.5-0.7%, that is 2-3 times less than the minimum.

From what has been said, it is necessary to restore and develop the system of in-house training with the appropriate state support. A number of measures can be proposed in this regard. Let's name some of them.

First, relying on the positive experience of developed countries, it would be necessary to introduce a system of certification of workers in the Russian Federation, which, in our view, would significantly improve the quality and competitiveness of the workforce [11]. Such practice, based on the system of certification of employees of enterprises and organizations, exists in Russia, but requires significant changes and additions in accordance with global trends and modern features of the domestic economy.

Secondly, the experience of the mentoring system should be extended to the organization of training in the workplace.

Third, the role of system of professional development and professional retraining should be strengthened.

Fourth, a system for monitoring the level of quality of the workforce should be introduced. In this regard, there are a number of indicators for the characterization of highly skilled specialists. However, a harmonious system of interrelated quantitative and qualitative indicators, full and versatile characterizing specialists, has not yet evolved. In addition, based on the existing realities, conditioned by the cardinal social and economic transformations in the country and the constant development of scientific and technical progress, the available indicators have to be adjusted and supplemented. At the same time, special attention should be given to the indicators characterizing the level of training of specialists, as well as unemployment and migration among them. Note also that the existing statistical information is extremely poor, irregular, the possibilities of current statistics are limited

3 Results

So, the contribution of a specialist with higher education to production and economics is important and very multifaceted: scientific and technical programs and developments, new technology, advanced technologies (including organizational, informational, humanitarian, social), project analysis, etc.

We emphasize that the production function, which characterizes the production process of an individual firm and expresses the patterns and efficiency of allocating its resources, contains indicators of the quantity and quality of highly qualified specialists as the most important characteristics of the labor resources used.

In addition, as a result of implementing the proposed system, both the educational organization and the partner firm receive a number of effects, systematized conceptually in Figure 4.

Accordingly, students participating in the system acquire the necessary "flexible" and "tough" skills, knowledge of innovative technologies, work skills (and, correspondingly, almost 100% probability of employment) at the mentor's enterprise, the possibility of online communication with participants and mentors (networking), access to the database of information and communication platform services (database of competitions, grants, video recordings of events, links, portfolios and other resources), creating a professional resume and its promotion in HR-agencies of tht region, held at educational conferences Organization of Employers ("Gold Fund staff"), developed by the implementation of a business project in its compliance criteria on the basis of grant support or creation of small innovative enterprises.

Educational organization is empowered to attract employers, to create a community of graduates and strengthen cooperation with the enterprises that they represent, to increase the share of employed graduates, to strengthen the link between science and production, to promote the organization's brand in the HR community, and to develop new educational resources.

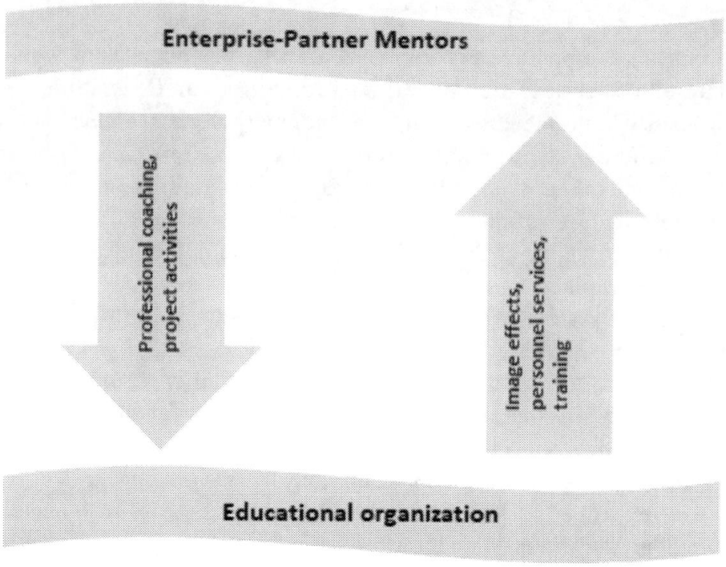

Fig. 4. Effects of the educational organization and the partner enterprise as a result of the introduction of the mentoring system

The effect for the enterprise is to shorten the period of adaptation of new employees (increase in labor productivity), to obtain image effects, to train mentors (including obtaining the necessary skills for mentoring at the workplace), implementation of projects initiated by the enterprise (with cost savings).

As a result, following the fractal construction of the management system for the investment and construction complex, an effect is achieved at the meso level, which consists of the growth of the share of highly qualified graduates remaining at construction enterprises, increase in labor productivity, and increase in competitiveness of construction companies in the region

4 Discussions

Taking into account the specifics of modern conditions, we propose to use the following system of indicators to characterize specialists.

First of all, indicators characterizing highly skilled specialists should be divided into two groups:

1) quantitative,

2) qualitative.

In the case of highly qualified specialists, quantitative indicators are indicators of the number of specialists grouped according to a certain trait, and qualitative indicators reflect indicators of the specific set and amount of knowledge and skills acquired by specialists in higher education institutions and then replenished in the course of their professional activities.

Within both groups of indicators (quantitative and qualitative) it is expedient to distinguish three subgroups:

1) indicators characterizing persons with higher education;

2) indicators characterizing highly skilled specialists engaged in the economy;

3) indicators characterizing non-working professionals registered with employment services.

Note that not all non-working professionals are included in last subgroup, but only those registered with employment services, it is dictated by the need to assess the potential of specialists in the country which actively seek for work and claim the relevant jobs. Information about all non-working professionals can be obtained as a difference between the corresponding indicators of the first and second subgroups.

When studying the quality of specialists at micro level, first of all within the framework of individual enterprises, their associations, sub-sectors, along with the above-mentioned interests, indicators such as length of service: general and specialty; number and structure of specialists by category, etc.

In addition, unemployed, but previously working specialists can be characterized by such an indicator as the average score obtained on the basis of the latest attestation. For a microanalysis of the quality of unemployed professionals who previously worked, such indicator as the length of work: general and specialty can be used.

It should be emphasized that the analysis of the quantitative and qualitative characteristics of specialists can not be conducted in isolation from the analysis of the state and development of higher education, especially in the context.

5 Conclusion

It should be noted that in general, the system of indicators proposed for the characterization of highly qualified specialists contains a number of such indicators that reflect not only the availability, but also the use of specialists, which in the first place belongs to the second subgroup of indicators-those characterizing specialists engaged in construction.

Finally, it should be noted that an indicator of the effectiveness of specialists' activity is an original synthesizing indicator reflecting the availability and use of highly qualified personnel, their quantitative and qualitative characteristics.

In order to increase labor productivity, reflecting, among other things, the effectiveness of specialists, it is necessary to introduce a mentoring mechanism in educational organizations and further in the production process. The introduction of mentoring in the educational organization will allow graduates to adapt more quickly and painlessly to the team of the enterprise, to master narrower, close to the specifics of the company's competences, to reduce the gap between the speed of introduction of innovations into practice and coverage in the training process, will reveal the "secrets" of the profession and avoid a lot of mistakes, sometimes fatal, to young specialists.

References

1. B. Karatop, C. Kubat, Ö. Uygun. Computers & Industrial Engineering, **86**, pp 127-136 (2015)
2. E. E. Makarius, Mahesh Srinivasan Business Horizons, **60 (4)**, pp 495-505 (2017)
3. H.N. Folz, Tracy L. Sprunger, Amy H. Sheehan, Josephine Aranda, Jasmine D. Gonzalvo. Currents in Pharmacy Teaching and Learning, In press, corrected proof, Available online 12 January 2018
4. V. Kankhva. IOP Conf. Series: Earth and Environmental Science **90** 012175 (2017) doi:10.1088/1755-1315/90/1/012175
5. L.N. Komyshova, S.I. Sergeeva Tasks and functions of educational systems in conditions of reforming the education system. Modern education: content, technology, quality. pp 162-163 (2011)
6. M.C. Meyers, M. Woerkom. Journal of World Business, **49(2)**, April 2014, Pages 192-203

7. P.R. Sparrow, H. Makram Human Resource Management Review, **25(3)**, (2015)
8. E. Nezhnikova Procedia Engineering **165** 1300-1304 (2016) doi 10.1016/j.proeng.2016.11.854
9. S.S. Gutman, I.M. Zaychenko, O.V. Kalinina, Selection of strategy implementation tool for shipbuilding cluster of Arkhangelsk Oblast Proc. of the 29th IBIMA Conf. Education Excellence and Innovation Management through Vision 2020, pp 1430- 1438 (2017)
10. I.V. Ilin, O. Kalinina, O. Iliashenko, A. Levina, Procedia Engineering, **165**, pp 1683-1692 (2016) DOi- 10.1016/j.proeng.2016.11.910
11. V.I. Churkin, O.V. Kalinina, Estimation of excess burden of labor taxation in Russia, Actual Problems of Economics, **184**, IS **10**, pp 278- 282 (2016)

Formation of the perspective innovation policy of the enterprise in the conditions of organizational-economic changes

Svetlana Uvarova [1,*], *Vyacheslav* Vlasenko[1], *Anatoly* Bukreev[2], *Ludmila* Myshovskaya[1], *Olga* Kuzina[3]

[1]Voronezh State Technical University, Moscow Avenue, 14, Voronezh, 394026, Russia
[2]Department of Economic development of Voronezh region, Lenin Square 1, Voronezh, 394018, Russia
[3]Moscow State University of Civil Engineering, Yaroslavskoye shosse 26, Moscow, 129337, Russia

Abstract. This article is based on the analysis of modern condition and dynamics of innovational development of high-rise buildings construction. A number of cardinal organizational and economic changes in management at the macro and meso-levels is taken into the account. Principal scheme of development of the methodology of formation of perspective innovation politics in high-ruse buildings construction based on inculcation of modern methods of strategic control of innovational activity is suggested in this article.

1 Introduction

The radical reorganization of the economic management system associated with the transition to self-regulation and innovative development is one of the most important , - one of the most important directions of reforms carried out in our country [1]. This problem is of particular importance at the level of the construction organization, whose position in the market economy is radically changing.

The The process of building production is formed under the influence of a large number of organizational factors. Transformation of the management system and its improvement means, first of all, a change in organizational relations and, accordingly, organizational forms of management.

Often, these changes lead to the reorganization and restructuring of enterprises, to the search for new network interactions that are more or less effective than already existing ones. The effectiveness of organizational changes in the activities of the construction organization is correlated with the conditions of the external environment, fluctuations in the economic cycles of business activity, realized and spontaneous organizational and economic changes, communication system of the enterprise [2].

From this point, the study of the specifics of both strategy and formation of perspective innovational politics of construction organization in modern economy in Russia is an urgent research topic that requires a serious scientific approach. The development of the problem of

* Corresponding author: uvarova_s.s@mail.ru

the creation of the strategy of the effective management and the formation of perspective innovation politics of the development of the construction organization considering cyclicality of economic development, already realized and also planned changes (organized, as well as institutional) on managements macro level, basing on the realization of the potential of the development and innovational infrastructure institutes, due to the improvement of regional investment and construction complexes in the regions is a relatively new scientific direction.

2 Materials and Methods

Effective economical politics in innovation activity of construction companies management should be dedicated not only to maintain positive development trends for the development of the investment and construction complex, but to remove restrictions that prevent their continuation [3].

First trend is to form the "raw-innovation" contour. Since innovations relate, first of all, to resource saving [4], a market is formed for innovations in the investment and construction complex, namely, in the building materials industry. The logically arising task of modernization of transport and energy infrastructure is also connected with the construction industry. Thus, the general economic trend of the innovative development of the commodity sector is transformed into a development trend of the investment and construction complex, which consists in the formation of the need for innovative development of construction technology and technology for resource conservation.

Second important trend is the formation of a high-quality consumer market. This is due to the sociological trend of forming a "new middle class". The trend that has arisen in the service market has spread to the housing construction sector, and subsequently to the commercial real estate market, which reflects the idea of creating a new quality human environment [5].

Thus, it is necessary to have management, which will provide resonant effect on systems elements to develop trends further, which is the first imperative condition for achieving innovative development of the investment and construction complex.

However, as the research results and their own analytical calculations of the authors show, there are a number of restrictions for further innovation for sustainable economic development, including investment and construction complex. The nature of the emergence of restrictions can be either spontaneous (spontaneous organizational and economic, institutional, technical and technological, natural changes in both the external and internal environment can lead to the formation of constraints), and initiated (initiated changes made without assessing the consequences, incorrect methods, or without agreement with the system's own movements can also lead to the formation of restrictions). The effect of restrictions can lead the system of investment and construction complex to the choice of an unstable development trajectory, or to a change in the region of attraction of the chosen attractor, which leads to a change in the "goals" of the system, and, ultimately, to the choice of an unstable development trajectory [6].

To overcome the limitations and ensure a qualitatively new development of the investment and construction complex, it is necessary to qualitatively change the regulatory role of the state associated with the necessary adjustment of the market mechanism and its regulators, based on coordination of the activities of state institutions and public structures, their de-bureaucratization, and the development of a national development trend. Realization of tasks which are necessary to achieve innovative development must be accompanied by certain managerial decisions of power structures and actions that must be taken ahead of time in conditions of risk and uncertainty [7]. Therefore, the state management of the investment and construction complex should be based on the analysis of worldwide and Russian development trends, public and business community views, on the presentation of future forms of organization of inter-industry interactions, especially in connection with the processes of globalization.

The public administration of the investment and construction complex should take into account the already accomplished spontaneous and initiated changes.

Accordingly, there is a need to develop conceptual and methodological foundations for the strategy and perspective innovative policy of construction enterprises, taking into account changes, as well as planning the reforms that have already been taken to ensure their systemic and successful implementation, which is appropriate for a program-targeted approach.

The mechanism and methodological basis for the implementation of the innovation strategy is the innovation process and its innovative projects. The methodological basis of the innovation strategy (based on the thesis on the two-level nature of the concept of strategy) is the concept of the overall management of the enterprise and its functional subsystems.

Thus, innovational strategy is a conception of interaction between company and external environment, based on the methodology of anticipating the implementation of technological, organizational, economic changes (representing a portfolio of innovative projects that implement the innovation process) with the aim of adapting the internal environment and shaping market development trends in the form of sustainable competitive advantages.

Proceeding from the foregoing, an objective need arises to justify a new concept for the formation of a promising innovation policy and the development of a strategy of a construction organization, allowing to significantly improve its investment and construction activities by stimulating innovative processes, introducing modern methods of strategic management of regional zoning and creating "points of innovation growth" that take into account the institutional, economic, organizational, information, analytical and regional aspects of governance.

Proceeding from the complexity and multidimensionality of the positioned problemp, there is a need to solve a number of interrelated but quite diverse tasks from the point of view of both their hierarchy (macro-, meso- and microlevel) and the direction (institutional, organizational, economic, regional aspect), as well as the methodology used (econometric methods, methods the theory of graphs, the methodology of territorial zoning, etc.). In general, a basic scheme for the development of a methodology for the formation of a promising innovation policy in construction on the basis of the introduction of modern methods of strategic management of innovation activity is presented in Figure 1.

At the first stage, in order to determine the current state (initial coordinates) and problems of innovation development of construction organizations, as well as the structure of construction management, taking into account the implemented transformations, we consider it necessary to assess the statics and dynamics of innovation in construction, to analyze the impact on its dynamics and the effectiveness of state and regional policy [8]; to assess the effectiveness of the functioning of development institutions and the correlation with the results of the activity of construction enterprises; to carry out the analysis of innovation strategies of large construction enterprises; to determine the specifics of innovation activity of large, medium and small enterprises; assess the dependence of innovative activity of enterprises on the pricing and audit of investment and construction projects; to formulate the target blocks of problems and the factor space of innovation in construction.

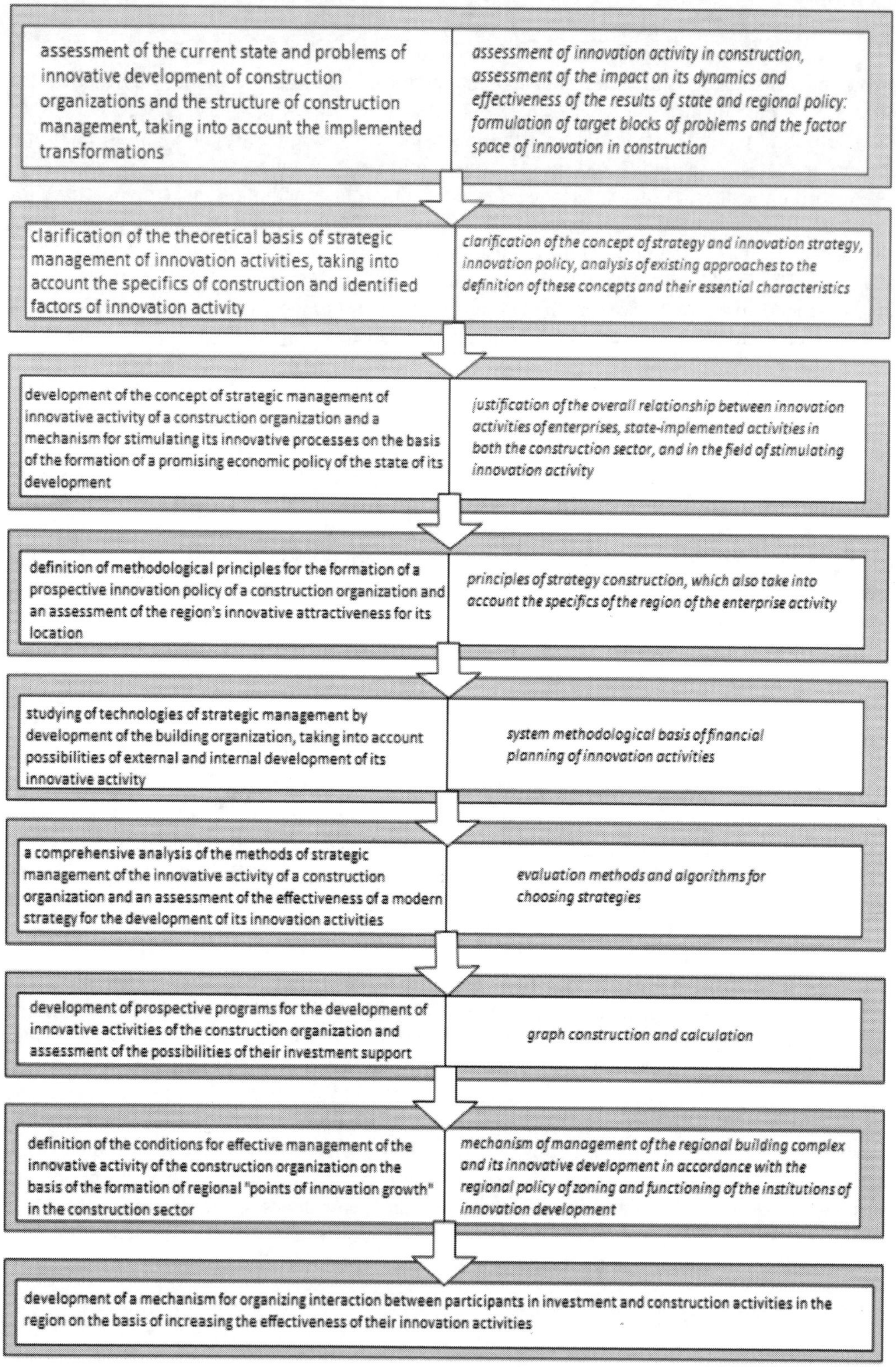

Fig. 1. Principal scheme for the development of a methodology for the formation of a promising innovation policy in construction on the basis of the introduction of modern methods of strategic management of innovation activities

Further, in order to unify the conceptual apparatus and to increase the effectiveness and coherence of institutional transformations, it is necessary to clarify the theoretical basis for the strategic management of innovation activities, taking into account the specifics of construction and identified factors of innovation activity. Within the framework of the solution of this complex of problems, we analyzed the concept of an innovation strategy, defined significant features of the concept, which will be sent methodological principles, as well as methodological tools developed for a large construction enterprise.

On the basis of the concept, the principles for building a strategy are formulated, which also take into account the specific nature of the region of the enterprise's activity (in this aspect, we consider it necessary to take into account, among other things, the specific state zoning policy at the regional level).

Further, within the framework of a comprehensive analysis of the methods of strategic management of the innovative activity of the construction organization and evaluation of the effectiveness of the modern strategy for the development of its innovative activity, we developed evaluation methods and algorithms for selecting optimal strategies. When forming a strategy, a sustainable development of an enterprise is a prerequisite, in which it develops at any stage of the business cycle, returning to the trajectory of its development after the crisis.

The most important stage in the development of the methodology for the formation of a promising innovation policy is to determine the conditions for effective management of the innovative activity of the construction organization on the basis of the formation of regional "points of innovation growth" in the construction sector. The mechanism of management of the regional construction complex and its innovative development should be formed in accordance with the regional policy of zoning and functioning of innovation development institutions: it is construction that should give impetus to innovative development, uniting enterprises of various industries in the operation of technoparks and special economic zones, development of the infrastructure of the city and the region.

3 Results

As a result, when implementing the proposed recommendations, especially at the meso-level of management, we consider it necessary to make an assessment of the quality of management at each stage of the life cycle when planning and implementing the initiated changes in the formation of a promising innovation policy.

It is worth noting the difference in the understanding and criteria for assessing the quality of management, depending on the purpose of the assessment, the scale of the problem and the policy being implemented, but ultimately the quality of public administration manifests itself in its impact on the socio-economic development of the country, region, complex, industry. In general, the following scheme for assessing the quality of public administration can be cited (Figure 2).

Management cost should provide not only actual reduce of administrative expenses for companies and regional investment and construction complex, but also reduction initial wastes on regulation which can be provided by continuity of institutional changes.

When determining the public administration effect on each type of measures, it is necessary to take into account the prospects for the development of technologies and methods that ensure the economy of natural resources, environmental safety of production activities, improve the level and quality of life of the population, modernize the country's production base, preserve and grow the competitiveness of the enterprise under WTO conditions.

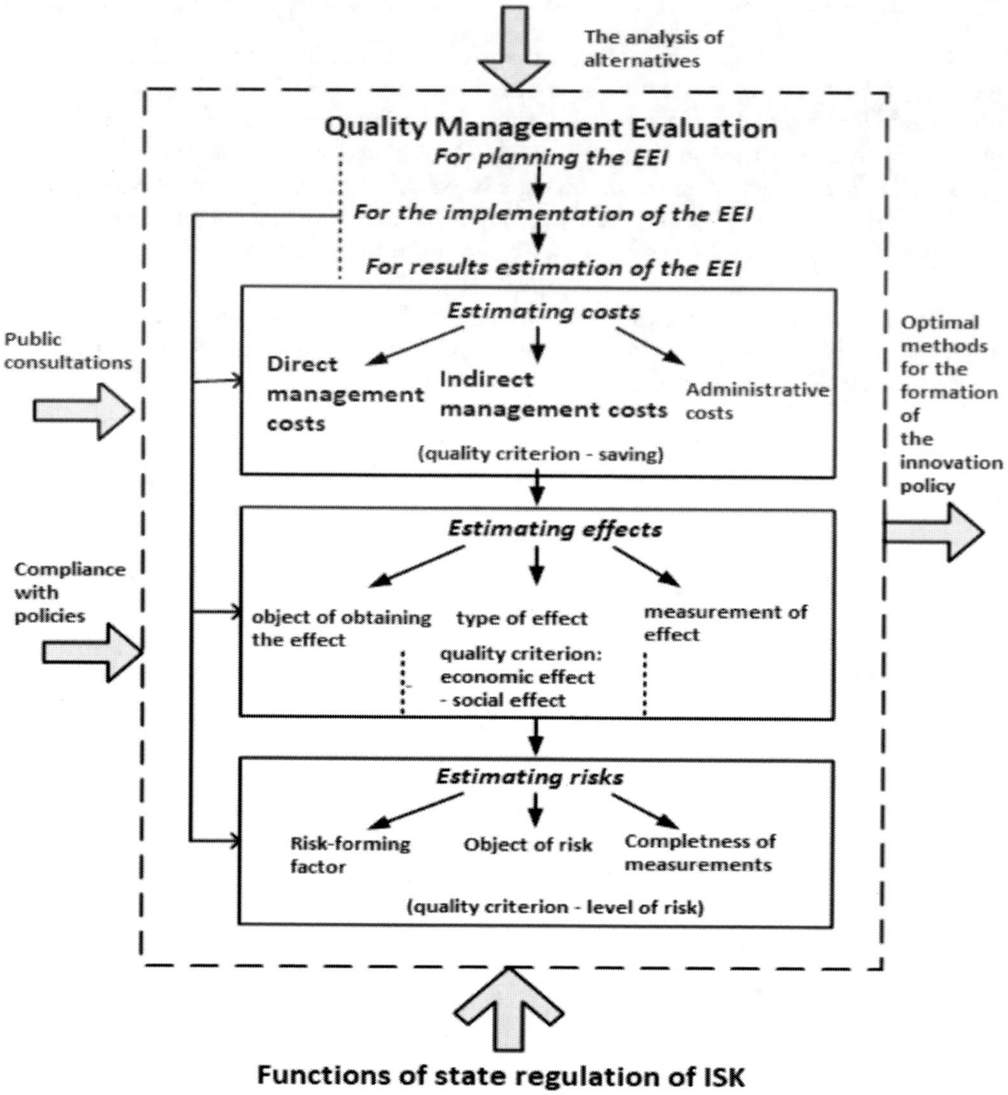

Fig. 2. Management quality assessment scheme

Reducing the risk degree of management of an enterprise or a regional investment and construction complex implies the need to take into account the already accomplished spontaneous and initiated changes, which ensures the implementation of the principles of optimality and complexity. In addition, when planning the formation of innovative policy, it is necessary to build on the analysis of global and Russian development trends, the views of the public and the business community, on the presentation of future forms of organization of inter-industry interactions, especially in connection with the processes of globalization

4 Discussions

It should be noted that innovations (including technological ones) are often a source or a catalyst of organizational and economic changes.

Analyzing the scientific approaches to the content of the notion of innovation, it is possible to identify a number of authors, that distinguish an essential feature of innovation as a change (replacing the old with a new one). Thus, innovations can be interpreted as changes, therefore, applying the methodology of planning and implementing organizational and economic changes to the planning and implementation of innovations, it can be concluded that innovation is constantly introduced by all elements of the managed subsystem of the investment and construction complex. In addition, in the previous paragraphs of the study it is shown that each element of the managed subsystem of the investment and construction complex is in the process of constant updating on the basis of economic changes characterized by signs of innovation (the production and economic subsystem), institutional changes (institutional subsystem), organizational changes (social subsystem), which has the character of self-development. Self-development is caused by the tendency to self-regulation and constant learning in the social subsystem, by the formation of legislative initiatives by the business community in the institutional subsystem, by the search for financing mechanisms and by the interaction in the implementation of projects in the production and economic subsystem. The aspiration of all the classification elements of a managed subsystem to continuous improvement and renewal is their integrative characteristic. In accordance with the integrative characteristic, the principle of innovation is put forward: technical and technological innovations are one of the main driving forces and reasons for organizational and economic changes (organizational innovations).

This thesis confirms and justifies the necessity and timeliness of developing the methodology and methodological foundations for the formation of a promising innovation policy in the context of the already implemented and planned organizational and economic changes in the management system of the investment and construction complex.

5 Conclusion

The proposed methodological foundations can be used to improve the mechanisms of strategic management and effective development of innovative activities of the construction organization in the modern economy of Russia, as well as to increase the efficiency of the regional investment and construction complex by building a promising innovation policy in accordance with the implemented organizational and economic changes.

References

1. A. Zehavi, D. Breznitz. Distribution sensitive innovation policies: Conceptualization and empirical examples. Research Policy, Volume **46(1)**, p. 327-336 (2017)
2. K.M. Gumba, M.I. Mamaev, Conceptual foundations for the formation of sustainable competitive advantages. News of the Irkutsk State Economic Academy (Baikal State University of Economics and Law). **1.** pp 11 (2014)
3. K.M. Gumba, S.V. Rodionova Substantiation of the essence and effectiveness of managerial innovations in the enterprise. Economics and Entrepreneurship. **11-3 (52-3)**. P. 645-647 (2014)
4. K.M. Gumba, S. Belyaeva IOP Conference Series: Earth and Environmental Science **90**. 012163. (2017)
5. V. Kankhva Procedia Engineering **165** 1300-1304. (2016) doi:10.1016/j.proeng.2016.11.855
6. V.S. Kankhva, S.S. Uvarova, S.V. Belyaeva Procedia Engineering. **165.** p. 1046-1051. (2016)
7. L. Grilli, M. Mazzucato, M.e Meoli, G. Scellato. Technological Forecasting and Social Change, **127**, Pages 1-7, (2018)
8. E. Nezhnikova IOP Conf. Series: Earth and Environmental Science **90** (2017) 012162. doi :10.1088/1755-1315/90/1/012162

The substantiation of methodical instrumentation to increase the tempo of high-rise construction in region

Svetlana Belyaeva [1,*], *Tatyana* Makeeva[1], *Andrei* Chugunov[1] and *Peraskovya* Andreeva[2]

[1]Voronezh State Technical University, Moscow Avenue, 14, Voronezh, 394026, Russia
[2] Moscow State University of Civil Engineering, Yaroslavskoye shosse 26, Moscow, 129337, Russia

Abstract. One of the important conditions of effective renovation of accommodation in region on the base of realization of high-rise construction projects is attraction of investments by forming favorable investment climate, as well as reduction if administrative barriers in construction and update of main funds of housing and communal services. The article proposes methodological bases for assessing the state of the investment climate in the region, as well as the methodology for the formation and evaluation of the investment program of the housing and communal services enterprise. The proposed methodologies are tested on the example of the Voronezh region. Authors also showed the necessity and expediency of using the consulting mechanism in the development of state and non-state investment projects and programs.

1 Introduction

As the results of statistical studies show, the provision of housing for the population in the regions of Russia remains at an extremely low level, despite a fairly large volume of construction [1]. In addition, housing renovation programs planned and implemented in a number of cities and regions, including in the Voronezh region, are aimed not only at providing affordable housing for the population, but also to ensure that housing meets the requirements of innovation, environmental friendliness, and energy efficiency [2]. Compliance with these requirements, along with ensuring the efficiency of urban development, postulates the need for growth in high-rise construction in the residential structure. Investments attraction at the same time can only be carried out with the formation of a favorable investment climate in the region [3].

The rapid erection and efficient functioning of high-rise buildings depends on many factors and, in particular, on the structure and the degree of renewal of the basic production assets of utilities, which affects not only the possibility of connecting the building to networks [4].

The analysis of the current state of the industry Indicates that the achievement of the conceptual tasks of reforming to improve the effectiveness of this sphere and to ensure the required quality of the services provided is impossible without a well-founded approach to

* Corresponding author: belyaeva-sv@mail.ru

the problem of searching for and purposeful use of financial resources for the technical re-equipment of enterprises in this sphere.

2 Materials and Methods

According to representatives of the construction complex of the Voronezh region, one of the main reasons restraining the tempo of high-rise housing construction is the excessive length of the period for issuing initial permits for construction. Summarizing the results of analyzing the perception of the relative size of administrative barriers by the subjects of the construction market, we can arrange them for operating firms in the following order: problems with real estate (rent and purchase of land and buildings), legal system, regulation (Figure 1).

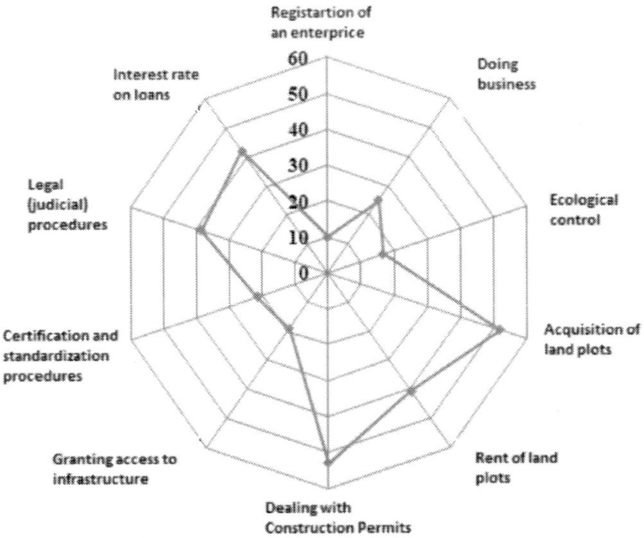

Fig. 1. Assessment of the importance of administrative barriers for the development of construction business

In general, the costs of overcoming administrative barriers by the construction organizations of the Voronezh region are proposed to be determined as follows (Figure 2).

$$Z_{vr}+Z_n$$

$$C_{ab} = Z_r + Z_e$$

$$Z_l+Z_p+Z_{reg}+Z_{ra}$$

Fig. 2. The costs of overcoming administrative barriers by construction organizations in the Voronezh region,

where Z_{vr} - costs which are connected with the loss of management time to resolve issues in connection with the implementation of mandatory requirements and procedures; Z_r – costs which are connected with tax accounting (expenses for the organization of tax accounting, including reporting); C_{ab} - total costs for administrative barriers; Z_r – regular costs; Z_b – one-off costs; Z_l – costs connected with participation in CPO; Z_p – cost connected with verifications (labor costs + fines and penalties); Z_{reg}- registration costs (official and unofficial payments); Z_{ra} – access costs (costs in connection with land and real estate permits)

Based on the calculations carried out, the costs of overcoming administrative barriers on average in the organizations of the construction complex in the Voronezh Region have decreased over the past few years, but still amounted to about 164.12 thousand rubles.

Thus, despite the existence of state program-targeted measures to provide housing for the population, renovation of housing, the mechanism for implementing appropriate investment programs, both at the federal and regional levels, needs to be clarified and adjusted to overcome the identified shortcomings.

Second key moment is stimulating of investment activity in the region. Certain measures of influence on business environment should be taken to stimulate investment activity in order to form favorable investment climate [5].

We identified key factors of investment potential and risk. In fact, based on the results of the analysis, one can answer the question of what is included in the concept of the investment climate in the region. For a detailed study, two ratings of the region's investment attractiveness were selected: ratings based on the rank of the potential and the rank of risk. These rating contain enough data for economic analysis [6].

Data on factorial loads allow us to formulate conclusions about the relative weight of an individual trait in the structure of the resulting factor. In turn, data on factor weights determine the ranking of objects for each factor.

After the calculations, the objective functions will have the following form:

The objective function of the risk of the investment policy of the region:

$$F = 0{,}287Q_Z + 0{,}004Q_P + 0{,}375Q_E + 0{,}474Q_F + 0{,}600Q_S + 0{,}373Q_K + 0{,}006Q_{EK} \quad (1)$$

where Q_z -- the coefficient of optimum or the rank of legislative risk; Q_p – the coefficient of optimum of political risk; Q_e – the coefficient of optimum of economic risk; Q_f – the coefficient of optimum of financial risk; Q_s– the coefficient of optimum of social risk; Q – the coefficient of optimum of criminal risk; Q_{ek} – the coefficient of optimum of ecological risk; *a, b, c, d, e, f,h* – measures of the influence of determining factors on the resultant, or the magnitude of factor loads.

The objective function of the investment policy potential of the region

$$F = 0{,}851Q_T + 0{,}916Q_P + 0{,}922Q_E + 0{,}925Q_F + 0{,}888Q_I + 0{,}778Q_K + 0{,}135Q_{IN} + 0{,}149Q_{PR} \quad (2)$$

where Q_t – the coefficient of optimum or rank of labor potential; Q_p – the coefficient of optimum of consumer potential; Q_e – the coefficient of optimum of production potential; Q_f – the coefficient of optimum of financial potential; Q_i – the coefficient of optimum of institutional potential; Q_k – the coefficient of optimum of innovation potential; Q_{in} – the coefficient of optimum of infrastructure potential; Q_{pr} – the coefficient of optimum of natural resource potential.

In our opinion, some of the indicators are highly correlated and, therefore, are interdependent and replaceable. Therefore, in order to optimize the calculations for a potential investor, we consider it expedient to modify the methodology of RA «Expert» in the following way:

1. We suggest to calculate the investment risk analysis by three components, the tightness of which is maximized with the resulting indicator (social risk, financial risk, economic risk).
2. It is also expedient to calculate the investment potential for three main factors that have the greatest impact on the final potential (financial potential, production potential, consumer potential).
3. It is necessary to add third section to analyzed indicators “"Legislative and Legal Potential of the Region".

Having received a modified methodology for assessing the investment potential of the region and making a recalculation of all indicators, the values of the Spearman correlation coefficient and the determination coefficient remained unchanged. The calculation of the concordance coefficient by components has changed, but the generalizing characteristic has remained unchanged. We can assume that the optimization of the method is possible, since the results of calculations are close.

In this case, the objective functions acquired the following form:

Function of risk of investment policy of the region:

$$F = 0{,}375Q_E + 0{,}474Q_F + 0{,}600Q_S \tag{3}$$

Function of the investment policy potential of the region:

$$F = 0{,}916Q_P + 0{,}922Q_E + 0{,}925Q_F \tag{4}$$

The third key point in reviving the tempo of high-rise housing construction is the need to update the fixed assets of housing and communal services [6]. In order to ensure a sufficient level of quality of services, to maintain productive capacity in working order and to ensure innovative development, the utility company, along with the production program, has the right to form an investment program [7, 8].

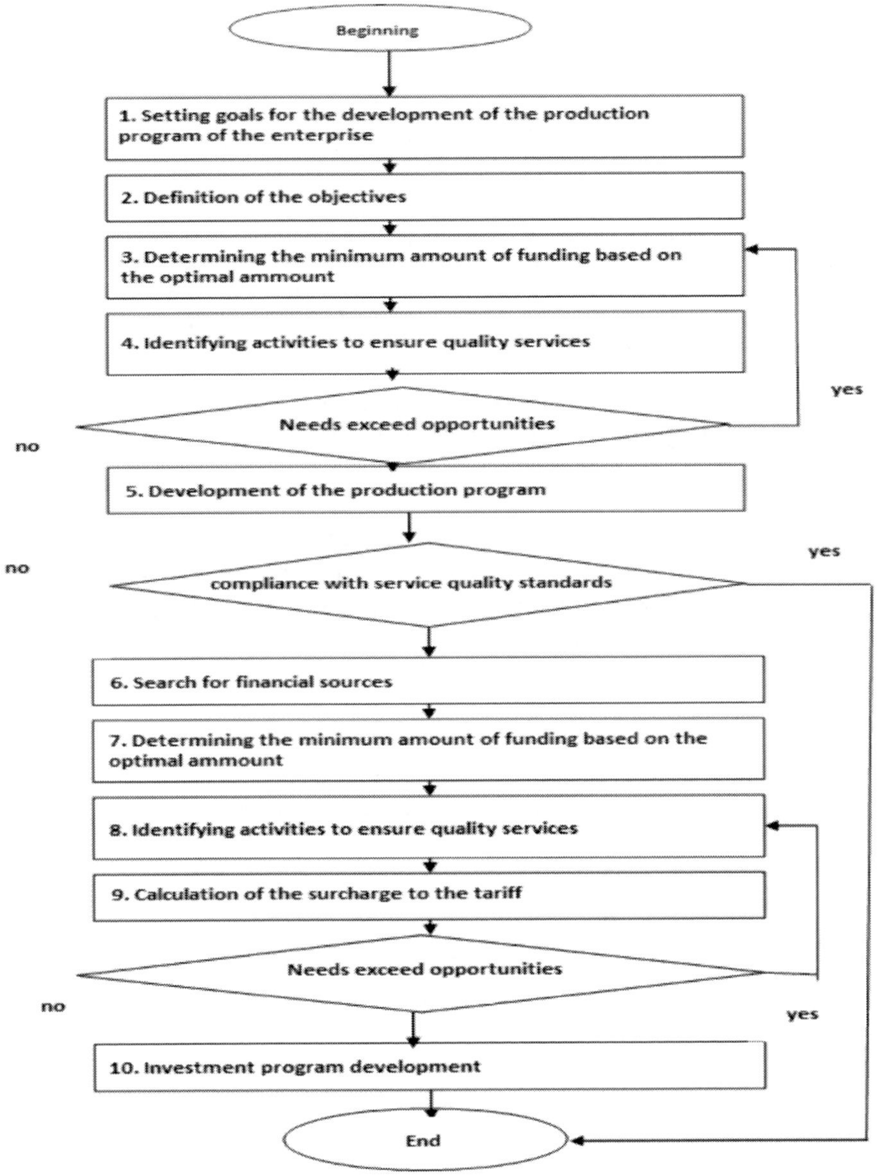

Fig. 3. Investment program development algorithm

There are no practical recommendations on the formation of investment programs. In this regard, it seems appropriate to take into account regional specifics, to specify the set of requirements, to recommend an algorithm and procedure of the formation of an investment program (Figure 3, 4).

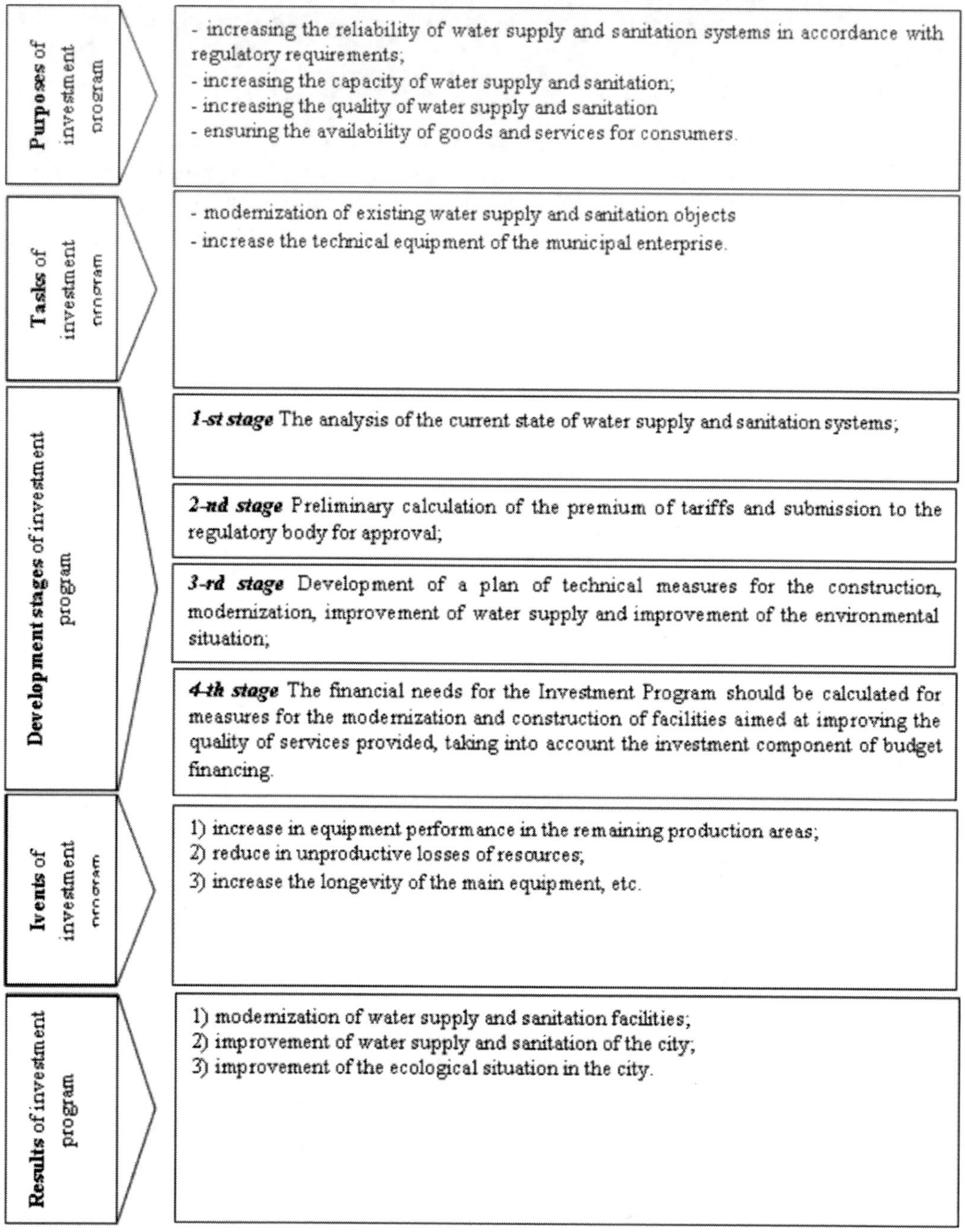

Fig. 4. The procedure for the formation and content of the investment program

To assess the effectiveness of formed investment program, a system of indicators characterizing the increase in the availability of services and the trouble-free operation of fixed assets achieved as a result of the program implementation should be used [9, 10]. Such indicators should be a characteristic of condition and development of communal infrastructure systems, operating conditions and parameters that are accessible to observation and measurement, providing the required level of quality. They can be used not only to estimate the reliability, balance of systems of communal infrastructure, the availability of services for customers, but also to provide ecological requirements and the quality of services provided.

3 Results

Based on the obtained methodology, an analysis of the investment attractiveness of the regions of the Central Federal District was made. Among the regions of the Central Federal District Voronezh takes the third place by the availability of investment potential and the sixteenth (from 18 regions) by the riskiness of investments in this area. As a result, the overall rank of investment attractiveness is quite high.

With regard to assessing the effectiveness of investment programs of housing and communal enterprises, we propose the following system of indicators (Table 1).

Table 1. Recommended composition and characteristics of the assessment of the investment program of the utility (fragment)

Indicator	Characteristics and calculation procedure
1. RELIABILITY	
1.1 Increasing the reliability of fixed assets	Reduction in the number of consumer complaints registered during the year.
1.2 Accident rate	The ratio of the number of accidents recorded during the year to the total length of the network.
1.3 Average time to eliminate accidents	The ratio of the total duration (time) expended in eliminating accidents during the year to their number.
1.4 Resource volume	The dynamics of the resource volume is shown in terms of the following parameters: useful leave (including the population and other consumers), losses, resource consumption for own needs.
1.5 Movement of fixed assets	Indicators of the movement of fixed assets: the coefficients of renewal, wear, fitness, retirement, and the average wear factor should be estimated for all fixed assets and by groups of funds, including separately for networks

Formation of investment programs based on the proposed provisions will increase the efficiency of municipal enterprises and improve the quality of services.

4 Discussions

In our opinion, one of the possible methods of activating the tempos of high-rise housing construction, which contributes to the optimization of state and non-state investment programs, is the use of management consulting as construction enterprises and housing and communal services enterprises (as a rule, on the examination of operational and tactical business improvements, which are less costly, but they give visible results in a relatively short time), and government authorities (on territories, on which there are no shortage of funds, as well as problematic municipalities, which are trying to form a long-term plan to overcome the crisis). With the use of consulting tools, it is possible to develop program measures to improve the investment climate in the region, as well as investment programs to improve the efficiency of the housing and communal services in the region, which, in our opinion, will contribute to a more dynamic development of high-rise housing construction in the region.

The close cooperation of a team of consultants, expert analysts and officials on projects of state and municipal importance, including public-private partnership, will further promote the improvement of life in regions and municipalities, and will also promote the development of consulting itself. Particularly important areas of such consulting in the future will be: the development and examination of concepts, strategies, programs of sectoral, municipal and regional social and economic development, territorial marketing and regional branding, structural reorganization of complexes, industries and enterprises, preparation of projects to justify attracting private investments in programs development of housing and communal services and high-rise construction [11].

5 Conclusion

To improve the investment climate further, it is necessary to influence the identified key components of investment potential and risk in the region, in particular, the formation of the opportunity to implement projects in the high-rise construction segment, reduce administrative barriers, grant additional tax benefits and preferences.

There is a number of problem areas in the implementation of program activities, including insufficient attraction of extra budgetary sources of funding, lack of concentration of allocations for key problems, shortcomings in consolidating budgetary allocations at various levels, and in the system of goal-setting and responsibility. The development of measures aimed at overcoming shortcomings contributes to an even more dynamic growth in the effectiveness of state and non-state investment programs at all levels, including unstable economic environment. We consider it appropriate to develop such events, with the involvement of professional consultants in this area of consulting.

References

1. A. Ohotina, O.Lavrinenko. Procedia - Social and Behavioral Sciences, **174**, (2015)
2. N.A. Anisimova, V.G. Popov, T.I. Smotrova Planning and evaluation of the effectiveness of measures for advertising and promotion in the residential real estate market. Modern economy: problems and solutions. 2 (86). P. 96-106. (2017)
3. P.M. Belmiro Duarte, M.C. Nuno Oliveira, O. Lino Santos Dynamics of quality improvement programs. Optimal investment policies. Computers & Industrial Engineering, **91**, (2016)
4. D. Dollar, M. Hallward-Driemeier, T. Mengistae. World Development, **34(9)**, (2006)
5. V.S. Kankhva, S.S. Uvarova, S.V. Belyaeva Procedia Engineering. **165**. (2016)
6. Y Kolyhaeva Procedia - Social and Behavioral Sciences, Volume **166,** (2015)
7. A. Larionov, E. Nezhnikova Role of energy efficiency in improving the quality of housing projects 2016 International Journal of Applied Engineering Research. - №6.
8. N. Bayraktar, Procedia Economics and Finance, **5**, (2013)
9. T. Hayes, F. Murtinho, H. Wolff World Development, **93**, (2017)
10. R. Bril, O. Kalinina, I. Ilin, MATEC Web of Conf. **106**, 08010 (2017) DOi - 10.1051/matecconf/201710608010
11. I.V. Ilin, A.B. Anisiforov, Improving the efficiency of projects of industrial cluster innovative development based on enterprise architecture model, WSEAS Transactions on Business and Economics, **11**, pp 757-764 (2014)

Formation of the portfolio of high-rise construction projects on the basis of optimization of «risk-return» rate

Svetlana Uvarova [1,*], *Olga* Kutsygina[1], *Elena* Smorodina[1], *Khuta* Gumba[2]

[1]Voronezh State Technical University, Moscow Avenue, 14, Voronezh, 394026, Russia
[2]Abkhazian State University, Universitetskaya Avenue, 1, Sukhum, 384904, Abkhazia

Abstract. The effectiveness and sustainability of an enterprise are based on the effectiveness and sustainability of its portfolio of projects. When creating a production program for a construction company based on a portfolio of projects and related to the planning and implementation of initiated organizational and economic changes, the problem of finding the optimal "risk-return" ratio of the program (portfolio of projects) is solved. The article proposes and approves the methodology of forming a portfolio of enterprise projects on the basis of the correspondence principle. Optimization of the portfolio of projects on the criterion of "risk-return" also contributes to the company's sustainability.

1 Introduction

When managing production and economic activities of the enterprise, it is necessary to respect the principle of conformity, which states that it is necessary that for sustainable development the system needs to ensure that the goals are consistent with the available resources, and that with efficiency - the costs incurred [1]. The conformity principle is realized in the management by introducing the proposed methodology for the formation of the portfolio of high-rise construction projects on the basis of optimizing the "risk-return" ratio, and is assessed by determining the efficiency of high-rise construction companies. Compliance with this principle at the enterprise management level is correlated with other levels of the management hierarchy through the fractality of management systems at all levels. That means, that the proposed methodology can be applied when planning and implementing a portfolio of projects of changes at the level of state or sectoral management, taking into account the modification of methods for calculating costs, effect and risk.

2 Materials and Methods

Treating the production program of the enterprise as a portfolio of projects it realizes, we consider it expedient to apply elements of the methodology of project and portfolio management.

* Corresponding author: uvarova_s.s@mail.ru

Formation of the portfolio of projects is usually done taking into account diversification, by which we understand not only the expansion of the scope of the firm's activity and the development of new segments of the market of contract works in construction [2], but also the typology of activities for a number of characteristics. In portfolio management, portfolio diversification is treated as one of the methods to reduce the risk of a portfolio by placing financial assets in more than one type of asset whose prices or yields are weakly correlated [2]. Carrying out the typology of the elements of the enterprise's portfolio of projects on various attributes, corresponding to the terminology of the portfolio management, the portfolio of projects is diversified.

Portfolio management theory distinguishes several basic approaches to solve the optimization problem (Table 1).

Table 1. Possible approaches to optimize the "risk-return" relationship in the formation of a portfolio (compiled on the basis of [3])

Approach	Substantiation of the essence of the approach	Formalization of the criterion
«Maximum revenue»	select the portfolio that provides the greatest result (income, discounted income, profit) while providing an acceptable level of risk for the investor	$D \to max$ $r \leq r_{pr}$
«Optimum probability»	Select a portfolio that provides an acceptable income for the investor, taking into account the probability	$M(D) \to max$
«Optimum variability»	Select a portfolio that has the least variability of income when the conditions change, that is, with a minimum variation	$CV(D) \to min$
«Minimal risks»	Select a portfolio that ensures minimization of risk while keeping income at a given level	$D = D_{pr}$ $r \to min$

The of investment portfolios is based on the method of "medium-variance analysis" proposed by G. Markowitz. The essence of the average variance analysis is the formation of the portfolio, based on the requirements of the investor on the basis of the minimum risk approach or maximum profitability [4]. This approach allows you to take into account the level of risk at all stages of the typology of the enterprise portfolio.

In general, similar to the portfolio approach, the methodology for forming a portfolio of projects is of a probabilistic nature.

Then we denote the set of states of the market environment as **Ω**. Vector D=(D1, D2,…Dm) represents a set of alternative elements of the portfolio for each feature of the typology T=(T1,…Tn). That means, that the entire portfolio of projects of the enterprise (as well as the investor's investments) can be represented in the form of a matrix of dimension m×n, where the columns show signs of typology, and the lines - available alternatives. The methodology calculates one attribute of the typology.

The effect indicators corresponding to each alternative element of the project portfolio (revenue, revenue or profit, depending on the alternative) are a vector E=(E1, E2,…Em) of random values, which are defined in the set **Ω**, and performance indicators form a vector φ of the same dimension and characteristics [5].

The production program is a portfolio of projects whose permissible aggregate is defined as S. Then the enterprise forms such m-dimensional vector of the portfolio P, which characterizes the distribution of the resource Z, representing the volume of the production program, into m types of activities (or CIW, or projects) according to the shares d_{ij}.

The parts are determined in the case of the optimal probability approach according to the competitive mechanism, in other cases - according to the Markowitz model.

It should be noticed, that in last case, it is necessary to observe the condition of non-negativity of parts (in contrast to the task of classical portfolio management). It is also necessary to calculate the covariance matrix of the efficiency of alternative elements of the production program $C = \{C_{jk}\}_{j,k=1}^{m}$,which is determined positively and has inverse matrix.

It should also be noted that operating the average values in this case is mathematically incorrect, therefore, in calculations for the account of the stochastic nature of the data, it is necessary to operate with probability values, that is, mathematical expectations of the values [6].

Proceeding from the foregoing, we proposed a methodology for the formation of a portfolio of enterprise projects on the basis of optimizing the "risk-return" ratio (Figure 1).

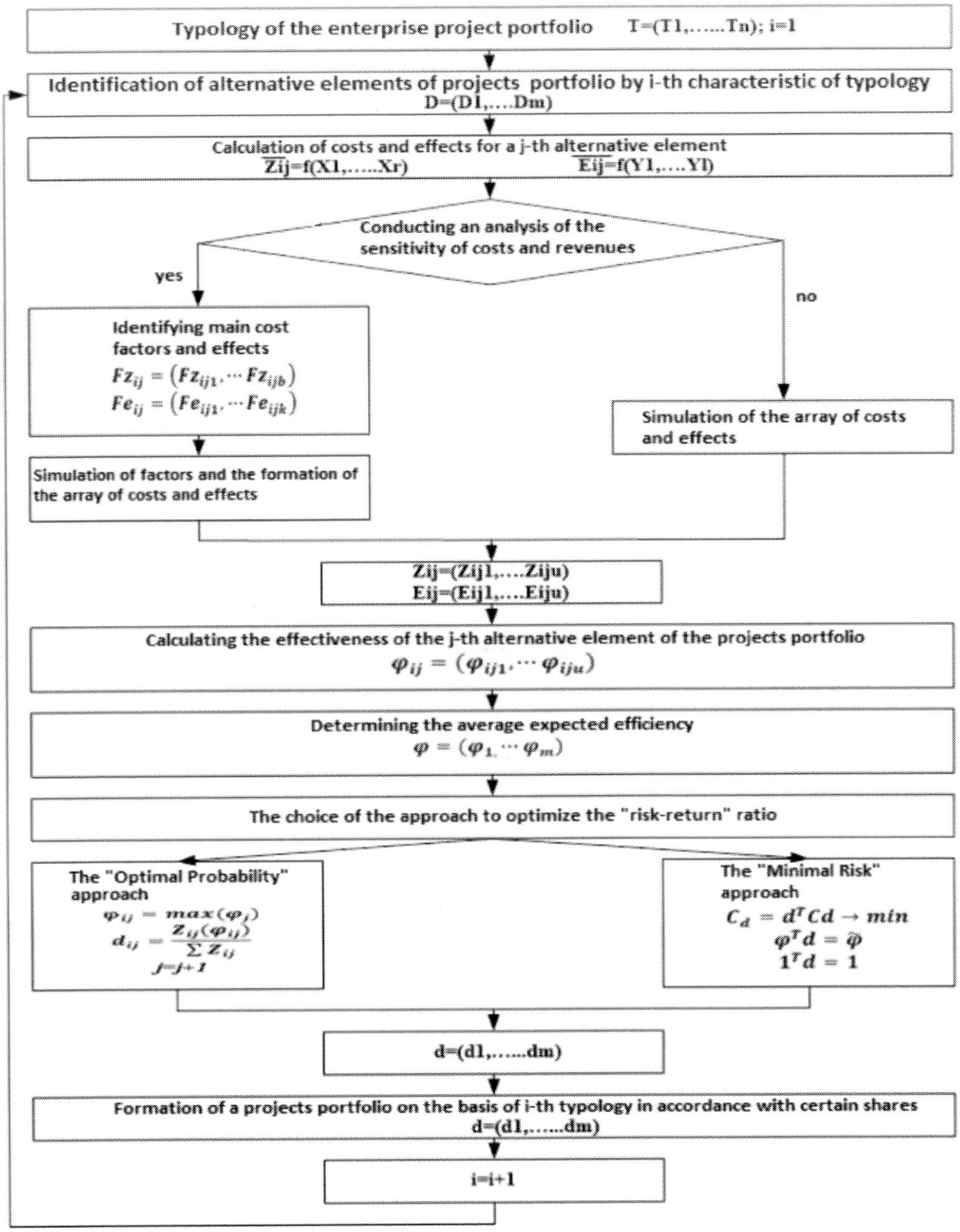

Fig. 1. Block diagram of the algorithm of the methodology for forming the portfolio of projects of the enterprise on the basis of optimization of the ratio "risk-return".

3 Results

We will site the algorithm for the formation of a portfolio of projects according to the proposed methodology using the example of the "Souzspetsstroy" enterprise.

The enterprise performs general construction and special construction works, performs the functions of general contractor and subcontractor, carries out work on state and non-state orders. The typology of projects portfolio is shown in Table 2 (is not exhaustive).

Table 2. The typology of projects portfolio of "Souzspetsstroy" enterprise

Characteristics of typology	Alternative elements
Form of ownership of the customer	Participation in the projects of a state customer Participation in the projects of a non-state customer Participation in PPP projects
Performed functions	Gencontracting Subcontracting
Phases of the life-cycle of the project	Laboratory quality research Construction and installation work
Types of civil and erection works according to Rosstat typology	45.1 Preparation of the construction site 45.2 Construction of buildings and structures 45.4. Finishing work 45.5 Renting of construction machinery and equipment
Types of construction works	45.21.1 Executing of common building works for erection of buildings 45.21.2 Executing of common building works for erection of bridges, elevated highways, tunnels and underground roads 45.21.7 Mounting of buildings and structures

As part of the calculation of the proposed methodology, the classification can be aggregated depending on the set of characteristics.

The initial data for the calculation based on the typology "Types of general construction works" are given in Table 3. It should be noticed, that calculations are made, depending on three alternatives (m=3) as part of one of typological characteristics. Then the profit values for each alternative form vectors E1, E2, E3, the cost values are vectors Z1, Z2, Z3, and the efficiency values are vectors φ of the same dimensions and characteristics.

Table 3. Cost and proceeds from the construction of residential high-rise buildings, highways and mounting the prefabricated structures, planned on the basis of actual data (thousand rubles accrued) and the primary design characteristics of data sets

Index	Quarter				Average	Dispersion	Standard deviation	The coefficient of variation
	1	2	3	4				
Road construction								
quarterly revenue	10413	22987	34446	47399	11849,75	1319544,92	1148,71	0,10
cost per quarter	9854	20619	30383	41357	10339,25	383516,92	619,29	0,06
efficiency	1,06	1,11	1,13	1,15	1,14	0,01	0,06	0,05
Civil high-rise construction								
quarterly revenue	36582	66443	127494	180152	45038,00	205441944,67	14333,25	0,32
cost per quarter	29415	62026	119692	164998	41249,50	166878579,00	12918,15	0,31
efficiency	1,24	1,07	1,07	1,09	1,10	0,02	0,14	0,13
Mounting of prefabricated structures								

quarterly revenue	17377	32158	53719	71572	17377	7802208,00	2793,24	0,16
cost per quarter	17676	31476	51435	64733	17676	10163046,25	3187,95	0,20
efficiency	0,98	1,07	1,08	1,34	1,12	0,02	0,16	0,14

In order to take risks into account in assessing the effectiveness of each type of work, we performed simulation of costs and effects, then we calculated the arrays of values of the mathematical expectation of efficiency. In the presence of cost and effect dependencies on external and internal factors, analogous calculations can be made by sensitivity analysis, which is reflected in collective monograph. Results of simulation modeling are partially given in Table 4.

Table 4. Results of simulation modeling of the arrays of costs and effects, calculation of the efficiency

	Costs for the implementation of road construction, thousand rubles	Proceeds from the implementation of road construction, thousand rubles	Efficiency of road construction
1	10792,05	8409,611	0,67073
2	11002,73	13268,68	1,262933
3	9857,095	12583,42	1,179416
4	11328,16	11974,03	1,105146
5	10740,19	13320,62	1,269264
etc. 500 immitations	…	…	…

The affiliation of the obtained data sets to the normal distribution of the quantities is confirmed by the calculation of the asymmetry error and kurtosis error criteria [7].

Histograms of efficiency are presented in Figures 2 – 4.

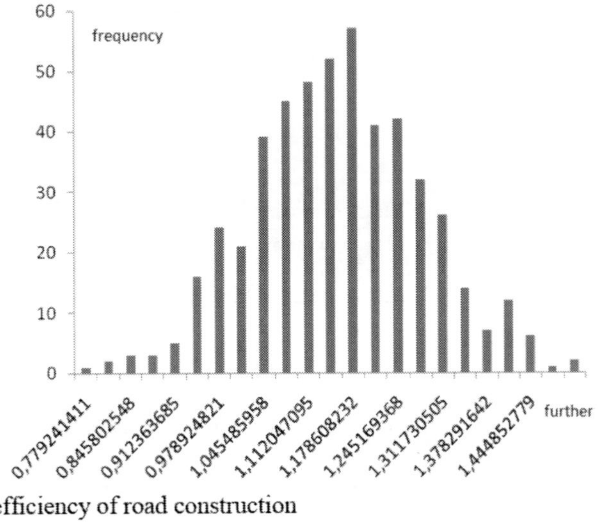

Fig. 2. Histogram of efficiency of road construction

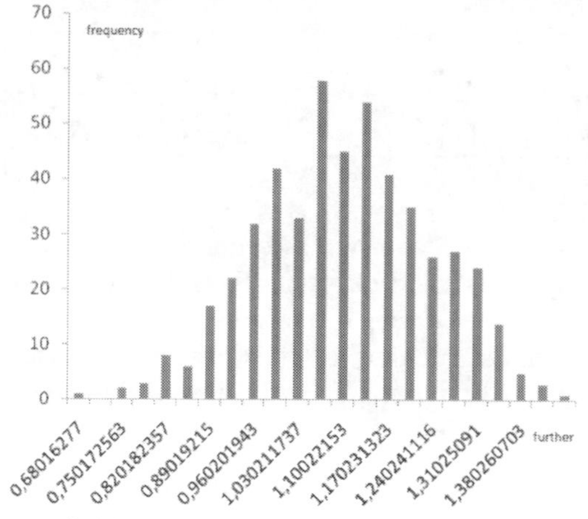

Fig. 3. Histogram of efficiency of civil high-rise construction

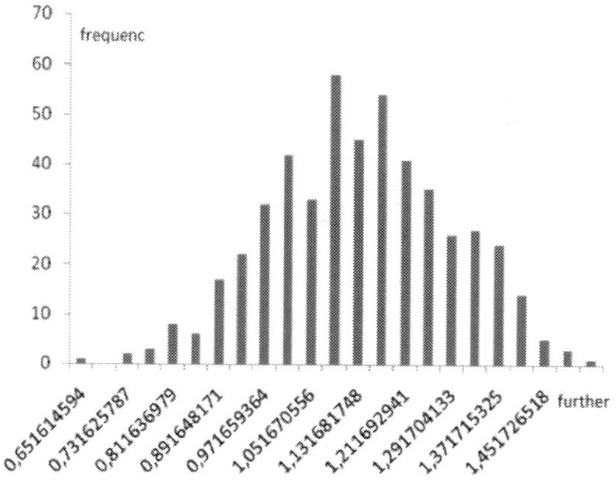

Fig. 4. Histogram of efficiency of mounting of prefabricated structures

The results of calculating the effectiveness and risk of the elements of the project portfolio are summarized in Table 5.

Table 5. Values of the effectiveness and risk of elements of the project portfolio

Alternative elements of projects portfolio	Average effectiveness expected	Risk (standard deviation)
Road construction	1,14	0,13
Civil high-rise construction	1,10	0,14
Mounting of prefabricated structures	1,12	0,15

According to the competitive mechanism of resource allocation, construction works on construction of highways will be included as elements of the project portfolio; in case of the presence of resources - work on the mounting of structures from prefabricated structures, in the case of the availability of resources - construction works for the construction of high-rise civil buildings. We will compile a portfolio of projects on the criterion of risk minimization. For this, it is necessary to calculate the covariance matrix of efficiencies (Table 6).

Solving the optimization problem in the SPSS program, we obtained the following results. Formation of the production program based on the ratio of 30% - performing general construction works on the construction of highways, 70% - performing civil works for the construction of high-rise civil buildings will ensure the efficiency of the activity at the level of 1.11 at a risk level of 0.016 (the share of elements of the production program amounted to 0.27 and 0,72 respectively).

Table 6. Covariance matrix of efficiencies

	Efficiency of high-rise civil construction	Efficiency of road construction	Efficiency of mounting of prefabricated structures
Efficiency of high-rise civil construction	0,018666	0,014628	0,021333
Efficiency of road construction	0,014628	0,016125	0,016718
Efficiency of mounting of prefabricated structures	0,021333	0,016718	0,02438

Calculation of the model according to the "maximum effect" approach gives the following result: the efficiency of the enterprise at the level of 1.14 is provided by performing works on the construction of highways, the risk at the same time will be 0.018. The Soyuzspetsstroy management entities have chosen to implement the project portfolio option corresponding to the "minimum risk" approach.

4 Discussions

Implementation of the proposed methodology in the practice of construction companies, in our opinion, contributes to the competitiveness of enterprises, effective structural changes in the industry, which becomes particularly relevant in the context of Russia's membership in the WTO. As part of the proof of the truth of the correspondence principle, we calculated the coefficient of pair correlation of indicators and performance factors of enterprises of the construction industry of Russian Federation for 2009-2016. The results of the correlation analysis are given in Table 7.

Table 7. Results of the correlation analysis of performance enterprises of the construction industry of Russian Federation

	Presence of basic funds	Commissioning basic funds	Depreciation of basic funds	Growth rates of construction and installation works	Price Index	Number of employees	Salary	Financial results	Profitability of goods	Profitability of actives	Number of profitable organizations
Availability of fixed assets, billion rubles	1,00	-0,77	0,73	-0,82	-0,82	-0,71	0,83	-0,95	-0,95	-0,94	-0,98
Commissioning of fixed assets, %	-0,77	1,00	-0,12	0,92	0,86	0,42	-0,55	0,55	0,54	0,73	0,67
Degree of depreciation of fixed assets,%	0,73	-0,12	1,00	-0,23	-0,33	-0,59	0,75	-0,90	-0,90	-0,71	-0,81
The index of growth rates of construction and installation works	-0,80	0,92	-0,23	1,00	0,99	0,72	-0,40	0,60	0,61	0,64	0,67
Consolidated index of construction products prices, %	-0,82	0,86	-0,33	0,99	1,00	0,82	-0,40	0,66	0,67	0,63	0,69
Average number of employees, thousand people	-0,71	0,42	-0,59	0,72	0,82	1,00	-0,26	0,69	0,72	0,44	0,61
Average monthly salary, rubles	0,83	-0,55	0,75	-0,40	-0,40	-0,26	1,00	-0,87	-0,84	-0,96	-0,92
Balanced financial result, billion rubles	-0,95	0,55	-0,90	0,60	0,66	0,69	-0,87	1,00	1,00	0,92	0,98
Profitability of the sold	-0,95	0,54	-0,90	0,61	0,67	0,72	-0,84	1,00	1,00	0,90	0,97

goods,%											
Return on assets,%	-0,91	0,73	-0,71	0,64	0,63	0,44	-0,96	0,92	0,90	1,00	0,98
Number of profitable organizations, units	-0,98	0,67	-0,81	0,67	0,69	0,61	-0,92	0,98	0,97	0,98	1,00
Number of unprofitable organizations, units	0,86	-0,95	0,30	-0,99	-0,97	-0,68	0,52	-0,67	-0,68	-0,74	-0,75
Percentage of costs, in% of the cost of construction and installation work	0.99	-0,98	0,80	-0,97	-0,93	-0,60	0,73	-1,00	-0,99	-0,92	-0,99

The number of profitable organizations, as well as the balanced financial result of organizations' activities, are in close relationship with the profitability (efficiency) of sales, the price index of construction products [8], the cost of production, which indicates the need to maintain the optimal correspondence between costs and efficiency of the enterprise to achieve the main goal of entrepreneurial activity - maximizing profit.

5 Conclusion

When forming the portfolio of projects, the methods of portfolio management (mean-variance analysis), probability theory, simulation modeling, econometric methods, sensitivity analysis, methods of the theory of operations research, the theory of active systems are used according to the proposed methodology.

The implementation of the proposed methodology will yield the following results:

– decrease in the overall risk of the construction company in an unstable environment;

– acceleration of adaptation of enterprises to organizational and economic changes, initiated at macro and mezzo levels;

– implementation of organizational and economic changes at the micro level with response to the development of the system;

– provision of synchronization of investment and construction complex subsystems at the micro, meso and macro levels;

– making scientifically based decisions on the diversification of production.

Optimization by the criterion of the structure of projects by the form of ownership of investment sources will allow to increase the efficiency of work with the state and non-state customers, as well as determine the feasibility of participation in projects based on PPP. According to the proposed methodology, it is possible to determine the optimal structure of investment.

References

1. O. Belyantseva, A. Panenkov, N. Safonova, IOP Conference Series: Earth and Environmental Science, **90(1)**, 012167, (2017)
2. I. Pan, A. Korre, S. Durucan Computers & Chemical Engineering, **89**, pp 84-105 (2016)
3. K. Gumba, O. Belyantseva, L. Kochetova Procedia Engineering. **165.** P. 1323-1327 (2016)
4. S. Dedu, F. Şerban Procedia Economics and Finance, **32**, (2015)
5. Wins, R. The *Mathematics of Money Management* Publisher Alpina Publisher. 2013.
6. X. Han, H. Wu, Q. Yang, J. Shang International Journal of Production Economics, **182**, (2016)
7. I.V. Ilin, O.Yu. Iliashenko, A.I. Levina, Proc. of the 28th IBIMA Conf. - Vision 2020, pp 768-781 (2016)
8. O. Kalinina, O. Valebnikova, Advances in Intelligent Systems and Computing, **692**, pp 1315-1322 (2018) DOi - 10.1007/978-3-319-70987-1_139

Strategic advantages of high-rise construction

Natalya Yaskova [1,*]

[1] Moscow State University of Civil Engineering, Yaroslavskoye sh. 26, Moscow, 129337, Russia

Abstract. Traditional methods to assess the competitiveness of different types of real estate in the context of huge changes of new technological way of life don't provide building solutions that would be correct from a strategic perspective. There are many challenges due to changes in the consumers' behavior in the housing area. A multiplicity of life models, a variety of opportunities and priorities, traditions and new trends in construction should be assessed in terms of prospective benefits in the environment of the emerging new world order. At the same time, the mane discourse of high-rise construction mainly relates to its design features, technical innovations, and architectural accents. We need to clarify the criteria for economic evaluation of high-rise construction in order to provide decisions with clear and quantifiable contexts. The suggested approach to assessing the strategic advantage of high-rise construction and the prospects for capitalization of high-rise buildings poses new challenges for the economy to identify adequate quantitative assessment methods of the high-rise buildings economic efficiency, taking into account all stages of their life cycle.

1 Introduction

At all times, people tried to subordinate space and time to the demand to meet their growing needs. Progress manifested itself in the inevitable effect of the objective law of growing labor productivity, reducing time spent on the production of goods and requiring a structural arrangement of the living space. Consuming resources to structure the space, people created capital assets with various degrees of efficiency, solved the problems of accommodation and ensured functioning of industrial, infrastructural, social, and ecological frameworks of vital activity. During periods of economic growth, the rate of living space arrangement was accelerated, whereas the cost of the allocated resources often played a secondary role. The scale of goals determined the rate, not always (for example, in a socialist economy) putting cost-effectiveness at the forefront. Until recently, regardless of socio-economic structures, environmental consequences and quality of living space were not taken into account in achieving the goals. At present, it is impossible to ignore these factors. It is quite obvious that deteriorating living environment critically affects the

* Corresponding author: mcua3@yandex.ru

standard of living, slows down the rate of development, hinders business activity [1, 2], therefore, creating conditions for crises.

A period of crises in all cases makes us reassess values, restructure natural and economic environment. This is particularly relevant for real estate, creating a framework of life and responsible for the quality of housing, transport, and common industrial infrastructure. Not only properties of real estate are subject to change, but also the processes of their formation and replacement of obsolete capital construction facilities [3]. At the same time, the laws of evolution dictate the need to create a steady trend of reducing resource consumption, distributed over all stages of life cycle, not only for manufactured products and provided services, i.e. the sphere of production, but also that of consumption, seeking to save operating costs and to increase the value of consumer goods, especially long-term ones [4].

2 Materials and Methods

One of the economic axioms is the existence of a direct connection between the duration of real estate life cycle and the need for giving it the maximum perspective advantage in relation to operational cost savings. This is especially relevant due to active formation of new megatrends in development. Drastic changes literally pervade everything: natural environment, including biosphere and land resources; production activities and services; research and educational activities; corporate standards and business models; organizational and economic management models; transport and supply systems; social sphere, including health care; public administration; financial and product markets; legal framework and existing institutions; socio-economic relations in the system "nature – society – man" etc.; personal needs for nutrition, recreation, communication, self-expression, entertainment, etc. New technological trends concern all sectors and spheres of national economy, creating a new picture of physical, biosphere, and digital world [5]. At the same time, convergence of new technologies occurs at all stages of the real estate life cycle. Thus, housing becomes not only "smart", replacing a number of traditional human functions, but also self-developing, adaptive. Many experts [6] argue that modern computers are capable of self-development, which makes it difficult to predict which PC applications and when will arise. For instance, unmanned vehicles and drones, virtual assistants and translators, doctors, medicines and entertainment activities, foodstuffs, forms of communication, vertical cities and many other things that technical experts call "the surrounding mind". Laws of new wave of production and consumption development will inevitably affect the change of Man with its inherent new structure of needs and quality of life.

One of the most stable trends in modern social development is the growing urban population. Together with the population, the height of buildings also increases. Only in the last 15 years, the height of skyscrapers has increased from 270 to 350 m. In this connection, Russia is by no means an exception, the country is a full participant in high-rise construction. Among 135 skyscrapers higher than 300 m 5 buildings are located in the Russian Federation. Historically, the leading Russian cities' verticals consisted of belfries, churches, minarets, etc. As early as in the 18th century, Russia was a leader in high-rise buildings. In Saint Petersburg, the spire of the Peter and Paul Fortress was 122 m height, being the highest one in Europe. The form of modern Kremlin should have changed significantly, as the height of the Palace of Soviets, according to the original conception, should have been 400 m. The construction plans were disrupted by war, but seven high-rise buildings were established. Thus, contrary to the prevailing illusion of Russia's chronic underdevelopment in high-rise construction, the country has always participated in the race for high-rises [7].

In terms of design feature and space-planning solutions, high-rise construction has been sufficiently well investigated [8]. It is obvious that high-rise buildings must be safe, smart, and energy-efficient. In terms of economic evaluation, these buildings are traditionally considered to be much more expensive as compared to traditional ones. It is true for the most part, but it is necessary to take into consideration other aspects that can alter such unambiguous conclusions.

First of all, a consumer behavior is changing. A new generation of buyers enters the real estate market, who have grown up in a new information environment on the one hand, one the other - they impart much greater importance to design, information and communication quality of life environment; and also they have certain ambitions and propensity to experiment. This creates an increased demand for accommodation in high-rise buildings.

Secondly, to assess the economic efficiency of high-rise buildings and any other real estate, it is necessary to take into account their whole life cycle. The modern technologies of real estate construction and operation allow for saving to 30% of the life cycle cost component [9]. However, lower operational costs save not only money, but also the time of owners for making utility payments, telemetrically reading the data of water and electricity meters.

Thirdly, technological innovations solve critical mass of technical problems associated with high-rise buildings, such as the problem of vertical transport, water supply, energy and heat supply, parking spaces, wind load, etc.; with complication of tasks, their price in most cases decreases. Thus, for instance, technology of horizontal and vertical movement of people in buildings using a levitation method (tested in a German city of Rottweil). The system does not use cables, does not require designing a core for elevators in the center of high-rise building, which ensures a much more maneuverable approach to flat arrangement, allowing different buildings linking and creating a new space design.

Fourthly, improving environmental friendliness of future cities is a reality observed all over the world. The cost estimate of environmental friendliness is not only possible, but it is already widespread, for example, by housing buyers. They “vote with a ruble”, buying real estate for a price of 20–25% more expensive, where there are park areas and clean air. In this regard, it must be noted that, unlike traditional buildings, high-rise development provides more free space for green plantations, leaving the land for parks and public areas, etc.

Fifthly, when planning a modern life framework, one should think about the perspective; the today’s needs may change drastically in the future. For instance, the answer to the question about large-scale parking necessity is rather negative. With the development of car-sharing, unmanned vehicles, the number of parking spaces can be reduced significantly. The approach to retail, public and other spaces can also change. If in the first case, online shopping will reduce the functional shopping areas, in the second case, the demand will clearly increase.

All these factors should be taken into account when assessing high-rise buildings’ economic efficiency. Meanwhile, a modern developer is aimed at getting quick profit. As a result, they continue using outdated building format with a low standard of life, which, surely, contradicts the interests of State and society. Moreover, many examples of economically successful skyscrapers worldwide demonstrate a very high system-dynamic building effect. For instance, Burj Khalifa in Dubai. The tower is capital intensive, and taking into consideration the fact that the final construction stage took place in global crisis, the real estate turned to be unprofitable. At the same time, the tower became part of a large-scale construction project, consisting of many surrounding objects of various functional purposes. Regardless of the real estate type – commercial, banking, sports & entertainment or residential – the view of the future tower ensured positive dynamics of efficiency. The mega-tower of Burj Khalifa has paid off, and the area around it has made and continues to make profit for its developer – EMAAR. The mentioned example also emphasizes that criterion for assessing economic efficiency of high-rise buildings in modern environment must "capture" the comprehensive effects during the entire life cycle of high-rise buildings.

3 Results

The main theoretical premise for assigning the integrated economic efficiency of high-rise construction is identifying the strategic orientation of construction. It means that we cannot neglect how the urban environment will change or what the effect of renovation will be, what innovations will be implemented or how the urban environment will react, etc. Bearing in mind that, according to F. Nikandrov, the Chief Architect of GORPROEKT, "high-rise construction is always a challenge" (financial, constructive, organizational, etc. challenges), its strategic focusing becomes the main condition for launching the project as early as at a project stage. It is necessary to emphasize that the competitive advantage, fixing the economic efficiency of a high-rise construction project at a given time, not only poorly reflects the outlook, but is also based on numerous assumptions in its calculation part, which does not meet the requirements of new social development context. In view of the above mentioned, from the author's point of view, the *strategic advantage* is a concept reflecting the contemporary contexts of high-rise construction as discussed above. The latter is expedient, if it corresponds to the trend of dynamically changing social needs and preferences in terms of the configuration and structure of life framework. Note that strategic interests of creating a comfortable urban environment form the priority with regard to assessment judgments on high-rise real estate economic efficiency.

As for the tools to provide the strategic advantage from the point of view of optimal adaptation of skyscrapers to the urban development priorities, let us mention that they represent a combination of mechanisms, models, and technologies for assessing real estate economic efficiency at all stages of its life cycle.

From the economic point of view, common approach to real estate life cycle cost evaluation [9] is based on the understanding of aggregate costs of real estate owner. With reference to high-rise buildings, in general, these are cost of their designing, construction, operation, and disposal. Note that the costs at the operation stage are up to 75% of the total. Those of the construction stage sometimes are up to 20%. Accordingly, designing and disposal will make from 5 to 10% of the total real estate costs. A case study showed that, in high-rise construction, possible savings at designing stage, even using information technology, are insignificant. Construction stage also features a high capital intensity and technological discipline. As a result, the main cost savings reserve lies in operational costs.

The modern innovative context presupposes the construction of energy-efficient and environmentally friendly buildings. The studies showed that the average resource economy when using modern technology reaches 30%. In this case, everything else being equal, absolute value of economic efficiency (E^{a}_{lc}) of a modern high-rise building can be estimated as

$$E^{a}_{lc} = \sum z^{б}_{lc} - \sum z^{in}_{lc} \quad (1)$$

where:

$\sum z^{б}_{lc}$ are the cumulative life cycle costs of a high-rise building, constructed without the use of modern technology;

$\sum z^{in}_{lc}$ are the cumulative life cycle costs of a high-rise building, constructed with the use of modern technology.

In general, savings will be more than 20% in the current structure of life cycle costs. Thus, for a property owner, the effect is translated in saving total costs. But, as shown by the study, modern ownership contexts should not be limited only to expense component. Market valuation of high-rise buildings is important, i.e. its capitalization (over time, it becomes increasingly important). Capitalization of high-rise buildings is an increase in their value over time. Our study of capitalization processes showed that they have inherent rules, as shown in the Fig. 1.

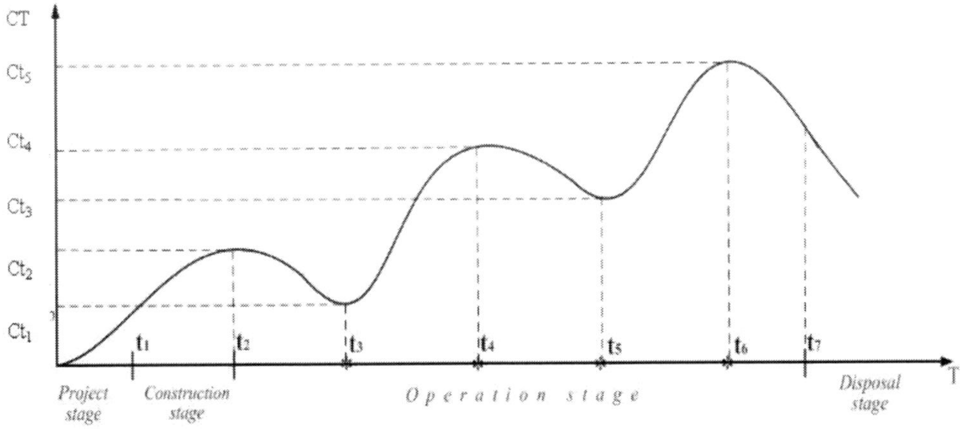

Fig. 1. High-rise buildings decapitalization rule

CT is high-rise real estate market price;
Ct_1 is actual price of property at the moment (t2) of putting into service;
Ct_2 is actual price of property at the moment (t3) of launching the overhaul;
Ct_3 is actual price of property at the moment (t4) of the end of the overhaul;
Ct_4 is actual price of property at the moment (t5) of renovation / restructuring;
Ct_5 is actual price of property at the moment (t6) of the end of;
(t2 ÷ t3), (t4 ÷ t5), (t6 ÷ t7) are periods of decapitalization.

As follows from the diagram, a strategically focused life cycle must include some stages in property operational phase. The following relationships characterize them:

$$Ct_1 < Ct_2 < Ct_3 < Ct_4 < Ct_5 \quad (2)$$

5 Conclusion

Thus, high-rise objects capitalization is the result of ensuring their normal operation, as well as real estate overhaul and restructuring. For its owner, it is of fundamental importance that additional costs for capitalization activities were lower than the increment in value at all phases and stages of its life cycle. In this case, economic benefits of saving life cycle costs are supported by high-rise real estate capitalization factors. Moreover, the owner should, already at a startup stage, evaluate economic potential of overhaul and reconstruction, thus ensuring the strategic advantage of a high-rise building in the long term.

References

1. V.V. Gluhov, I.V. Ilin, V.I. Koposov, A.I. Levina, Asian Social Science, **10 (24)**, pp 209- 216 (2014) DOi- 10.5539/ass.v10n24p209
2. N. Yaskova, Modern formats of the strategy of urban real estate development. Scientific review. **7-1**. p. 392-396 (2014)
3. N. Yaskova. Evolution of the development of investment and construction activities. Bulletin of the Irkutsk State Technical University. **1 (60)**. P. 178-186. (2012)

4. I.G. Lukmanova, N. Yaskova. Development of the scientific foundations of the evolutionary economy in the current conditions of investment and construction activity. Economics of construction. **4 (28)**. p. 13-19. (2014)
5. K. Schwab. *The Fourth Industrial Revolution*. M., Eksmo, 208 p. (2016)
6. V.S. Kazeikina, S.A. Baronina The cost of owning residential real estate at the total cost in the life cycles of reproduction as the basis for energy efficiency management / under the general. Ed.. Astana, 212 p. (2015)
7. V.V. Peshkov, N. Yaskova. Myths and reality of modern administration. Ekonomichny chasopis-XXI. **3-4.** - P. 32-36. (2015)

Use of formwork systems in high-rise construction

Oksana Kurakova [1,*]

[1] Moscow State University of Civil Engineering, Yaroslavskoye sh. 26, Moscow, 129337, Russia

Abstract. Erection of high quality buildings and structures within a reasonable time frame is the crucial factor for the competitiveness of any construction organization. The main material used in high-rise construction is insitu reinforced concrete. The technology of its use is directly related to the use of formwork systems. Formwork systems and formwork technologies basically determine the speed of construction and labor intensity of concreting operations. Therefore, it is also possible to achieve the goal of reducing the construction time and labor intensity of works performed by improving the technology of formwork systems use. Currently there are unresolved issues in the area of implementation of monolithic technology projects, and problems related to the selection of a formwork technology, high labor intensity of works, poor quality of materials and structures, etc. are the main ones. The article presents organizational and technological measures, by means of which introduction it is possible to shorten the duration of construction. A comparison of operations performed during formwork installation according to the conventional technology and taking into account the implemented organizational and technological measures is presented. The results of a comparative analysis of economic efficiency assessments are also presented on the example of a specific construction project before and after the implementation of the above mentioned measures. The study showed that introduction of the proposed organizational and technological model taking into account optimization of reinforcing and concreting works significantly improves the efficiency of a high-rise construction project. And further improvement of technologies for the use of insitu reinforced concrete is a promising direction in the construction of high-rise buildings.

1 Introduction

Today erection of high quality buildings and structures within a reasonable time frame is the crucial factor for the competitiveness of any construction organization. It is possible to solve this problem, first of all, by means of introduction of modern technologies and optimization of existing organizational and technological solutions.

* Corresponding author: ks-home@mail.ru

For example, hundreds of well-known skyscrapers were built on the basis of a reinforced concrete skeleton, including world record holders - the Burj Dubai tower (818 m high) and the Petronas Towers in Malaysia (432 m high). The potential of mix-in-situ concrete is not used to the fullest extent in Russia. Obviously, extention of the scope of its application in high-rise construction is facilitated by the development of new technologies, including creation and implementation of modern formwork systems. It is formwork systems and formwork technologies that basically determine the speed of construction and labor intensity of concreting operations [1, 2].

To date there are unresolved issues in the area of monolithic technology projects and the main ones are as follows:

- Difficulties in choosing the most optimal formwork system for a specific development project.
- Insufficient informativity with respect to the rules and terms of concrete curing and further form stripping.
- High labour intensity of major types of construction and installation works.
- Low quality of finished structures.
- Low level of control over construction processes related to an improper regulatory framework.
- Low quality of building materials of home manufacture.

An integrated approach to the organization and technology of in-situ concreting is needed to solve the above mentioned problems.

- Improvement of applicable regulatory inactments.
- Improvement of solutions for the performance of construction and installation works.

Development of approaches to solve organizational and technological issues affecting the economic efficiency and overall profitability of project implementation.

2 Materials and Methods

A formwork system is a temporary or permanent construction form, which includes formwork and elements that ensure its rigidity and stability - supporting structures, scaffolds, and fasteners. There is a wide variety of formwork systems (Fig. 1).

Hydraulically operated travelling self-climing forms are used in the construction of external walls of buildings, which height exceeds 30 storeys. They represent a formwork module consisting of outer and inner formwork panels, working scaffolds and form anchors. It should be noted that sometimes it is impossible to use creanes at an altitude exceeding 100 m because of winds and fogs. They can be used 4-5 days per week, and this time is enough to build at least 1 storey. Self-climbing forms solve problems of shuttering and mechanical formwork removal, formwork mechanical movement by height, ensuring of safe working conditions and maximum protection from wind. Formworks differ in nature; they are designed and manufactured for a specific project. Special projects involving formwork movement by height, a hydraulic distribution arm and individual cranes placed on the carcass under construction are developed for extremely complex high-rise buildings [3, 4].

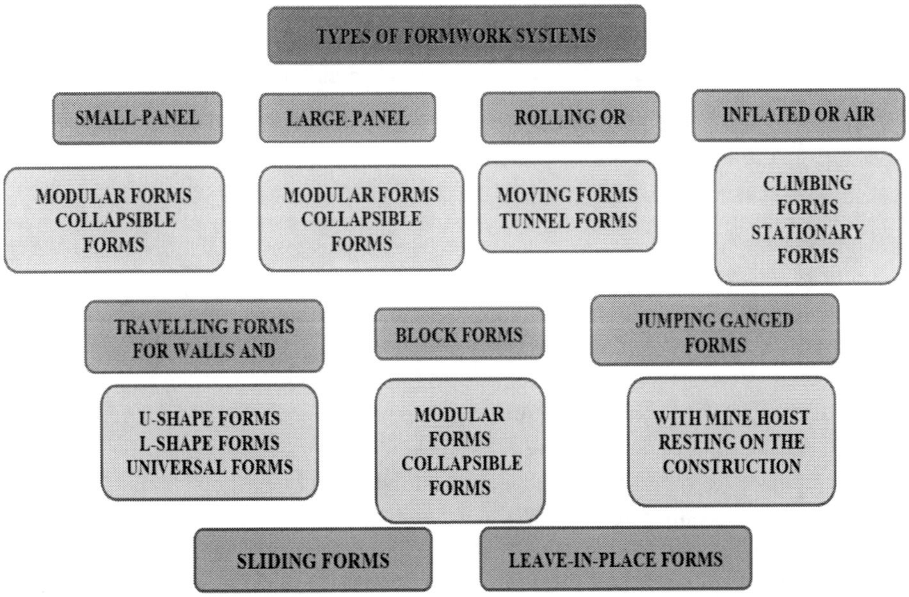

Fig. 1. Types of formwork systems.

The number of in-situ concreting operations performed, as well as their sequence, in which these types of formwork systems are used, is changed in the process of development of organizational and technological solutions for the improvement of formwork.

3 Results

When studying shuttering work processes, a number of operations, which optimization will reduce the duration of construction and, consequently, increase project efficiency were identified.

Improvement of methods of formwork installation:

- rational choice of formwork, including enlargement of shuttering panels into reusable cards with no need to demount them and application of shaft formwork, which shorten the duration of shuttering works
- application of coatings eliminating adhesion and the need to lubricate formwork
- reduction of the distance of formwork movement by a crane

A reasonable approach to choosing a formwork system is an important factor for reducing the period of construction of buildings and structures. When choosing formwork it is necessary to take into account characteristics and parameters of a future building, dimensions of structures under concreting, as well as climatic conditions. Use of the latest formwork systems for the implementation of monolithic technology projects contributes to improving the use of technology and construction quality [5].

A formwork system is chosen at the design stage, when the number of panels and the sequence of works on erection (dismantling) are selected. Main ideas and solutions for works optimization are considered at this stage.

A development project is divided into assembly areas, tiers, bays and a working project is drawn up for formwork system selection. Currently, there are many specialized software systems that allow carrying out a detailed calculation of the amount of formwork, its layout for horizontal and vertical structures. Introduction of optimization solutions allows to reduce the area of the shuttering system used and achieve the maximum possible reduction in the number of panels.

Depending on the configuration of the structures being erected, the number of formwork panels is selected. By means of additional swivel and hinge elements formwork panels can be put longitudinally, transversally, perpendicularly. Formwork auxiliary elements are as follows [6]:

- Leveling beams.
- Scaffolds brackets.
- Struts.
- Clamps.

A competent calculation of formwork elements affects the reliability of the system, which ensures the following:

- Strength and rigidity for various types of loads.
- A possibility to use automatic means during installation.
- Humidity conditions that are required for concrete hardening and curing.
- Fast installation and dismantling.
- Convenience for repair and replacement of individual elements.
- Minimization of labor and material costs.

The maximum number of formwork elements is determined taking into account the technology of works execution and the sequence of construction processes in the course of a development project erection. Formwork panels for the construction of vertical and horizontal structures are selected based on the number of bays, 2 largest or 3 respectively. It should be borne in mind that one system is used for reinforcement works, the second one - for concrete works and the third one - for installation [7].

In the course of the study it was found out that three sets on the floor for horizontal structures are not always enough in case of reduction of the time of works execution up to two days per floor. It is required to take into account one additional formwork set per bay. The formwork is removed from the structures when concrete cures and support racks are installed. Intermediate support racks are required only if concrete strength is 45-85% [3].

To calculate the number of oversupporting racks, it is required to take into account several factors, such as:

1) Climatic parameters, i.e. air temperature in the area of the development project construction.
2) Number of floors.
3) Speed of works execution.
4) Concrete hardening time.

In order to calculate the number of oversupporting racks, it is necessary to calculate coefficient K. As a rule, oversupporting racks are calculated per floor, per bay or per one set of forms for floors. In practice, K coefficient per floor is calculated by the following formula for multi-storey development projects:

$$K^e = \frac{\sum_{i=1}^{i=n}(m_i * n_i)}{k} \quad (1),$$

K coefficient per bay or per workform system set will be calculated as follows:

Kz=Ke*Kbay (2),

where

mi - the percentage of oversupporting racks for a given level of intermediate strength of a floor used for concreting divided by 100%.

ni - the time of relevancy of this percentage of racks determined by the schedule of concrete strength development in days.

k - the average speed of floor erection, days/floor.

Kbay - the number of bays on the floor.

Kfw - the number of bays, for which a formwork set is designed.

The following list of works presented in Table 1 is recommended to optimize the execution of shuttering works of vertical structures.

Table 1. Comparative analysis of operations during shuttering works of vertical structures.

According to the current Russian standards	Optimized option
• Marking of places for formwork installation along laying out axes. • Installation of panels. • Panels bolting. • Formwork alignment. • Formwork fastening by struts, ledgers, braces and ties.	• Formwork sorting. • Assembling into cards according to the working project. • Installation of scaffolds on the formwork. • Installation of scaffolds for external walls. • Marking of installation points on the concrete. • Slinging and installation of one side of the formwork, removal of slings. • Installation of plastic cones and tubes, installation of elements forming openings. • Slinging and installation of the second side of the formwork. • Installation of leveling beams. • Formwork fastening with ledgers and nuts, fixation of struts, closing of ends, closure of gaps between the formwork and concreted structures. • Alignment of the formwork prior to concreting, final alignment of the formwork after concrete placement.

To optimize works on walls formwork removal, let's compare the option of operations that is applied today according to the current standards with an optimized option for the performance of this type of works.

Table 2. Comparative analysis of operations during wall wormwork removal.

According to the current regulations	Optimized option
• Removal of struts. • Loosening of bolted connections of panels and removal of formwork anchors. • Removal of the formwork from the surface of concrete. • Removal of panels and fixtures and their transfer to the place of storage. • Formwork cleaning. • Formwork lubrication.	• Removal of struts and nuts from the formwork. • Release of struts. • Removal of ends fixtures. • Warehousing of fasteners. • Panels removal from the concrete surface. • Panels transfer to the intermediate site, panels cleaning from concrete and their lubrication. • Removal of end cutoffs and elements forming openings. • Removal of plastic cones.

For the installation and dismantling of forms for floors, a comparison of the operations, which are used today in accordance with the current regulations, with an optimized option for this type of works was carried out.

Table 3. Comparative analysis of operations during installation and dismantling of forms for floors.

According to the current regulations	Optimized option
• Installation of sub-curve pieces and fastening. • Installation of curve pieces. • Installation of panels.	• Formwork sorting. • Supply of formwork elements to the bay. • Installation of racks with tripods. • Installation of universal forks.

According to the current regulations	Optimized option
• Formwork alignment. • Installation of frame boards and fastening.	• Installation of longitudinal beams. • Installation of intermediate racks without tripods. • Installation of transverse beams. • Initial formwork alignment. • Plywood installation. • Closure of seams between plywood sheets. • Installation of elements forming openings (casings). • Final formwork alignment.

The proposed organizational and technological measures for the execution of shuttering works should ensure a reduction in the duration of construction and labor intensity.

4 Conclusion

The technical and economic assessment of the proposed organizational and technological model for the implementation of investment and construction monolithic technology projects is based on a comparison of options. The comparison was carried out on the example of implementation of a 30-storey multifunctional residential complex with an underground parking. Construction is carried out using monolithic technology [8, 9].

• In the first option the project is implemented taking into account the accepted organizational and technological solutions, where the average speed of storey erection is 3 days.

• In the second option the project is implemented in the ordinary and usual course (traditional technology), where the average speed of storey erection is 6 days.

The average time for the execution of works of a single cycle is 30 calendar days, taking into account delivery of the required resources/materials. It is necessary to understand that there may be unplanned downtimes at the construction site for various reasons, so it should be taken into account when planning work days. As practice shows, most often a downtime does not exceed 10-15 days. As a result it takes 90 working days to build the carcass.

As per calculation, about 95 m of concrete is poured per day: walls 0.22 m thick - 38 m3, horizontal structures 0.18 m thick - 57 m3.

According to the calculation, about 420 m2 of formwork is required for concreting of 1 bay of vertical structures. Carrying out simple calculations that take into account the volume of the concrete mix per day, it was found that 11m of vertical formwork is required for 1m3 of concrete mix. At least 840m2 of formwork is required for two bays to ensure uninterrupted works execution.

Table 3. Comparison of technical and economic indicators of projects before and after the implementation of organizational and technological measures.

Economic efficiency indicators	Option 1 (taking into account the organizational and technological decisions)	Option 2 (using the conventional technology)
NPV	384,436,335.53 rubles	370,956,858.09 rubles
IRR	64.55%	46.20%
PI	1.27	1.1
T	8 city blocks	9 city blocks

It was stipulated earlier that the technology of execution of formwork installation works to a greater extent depends on the chosen formwork. Formwork temporary indicators of "PERI ELPOS" and "TIPOS DOKA" systems are used for the implementation of the development project.

It should be noted that organizational and technological solutions were implemented not only for formwork operations, but also for the execution of reinforcing and concrete works.

References

1. Y. Shin, D.-W. Kim, S.-W. Yang, H.-H. Cho, K.-I. Kang, ISARC 2008 - Proceedings from the 25th International Symposium on Automation and Robotics in Construction (2008).
2. Y. Kog, Practice Periodical on Structural Design and Construction (2008)
3. M. Rafiei, H. Adeli, Structural Design of Tall and Special Buildings (2016)
4. A. Aminmansour, K. Moon, Journal of Architectural Engineering, **16(2)**, 47-53 (2010)
5. P. Love, D. Edwards, Z. Irani, IEEE Transactions on Engineering Management, 59(4),6032087, 560-571 (2012)
6. E. Frolova, Sotsiologicheskie Issledovaniya, **12**, 51-58 (2014)
7. B. Semaan, J. Hemsley, ISCRAM 2015 Conference Proceedings - 12th International Conference on Information Systems for Crisis Response and Management, 321-328 (2015)
8. I. Ptuhina, T. Spiridonova, T. Musorina, Applied Mechanics and Materials, **725-726**, 153-159 (2015)
9. A.K. Orlov, MATEC Web of Conferences **106**, 08013, (2017)

Economic efficiency of high-rise construction in the Moscow program of renovation of housing stock

Andrey Misailovov [1,*]

[1] Moscow State University of Civil Engineering, Yaroslavskoye shosse 26, Moscow, 129337, Russia

Abstract. The article considers a new initiative of the regional authorities of updating the housing stock designated as the renovation of housing. Its main aspects are analyzed, including the nature of program, economic efficiency of its implementation due to high-rise construction and the regulatory and legislative framework, the procedure for implementing the program, and the time frame for its implementation. The role of the program for regions in which high depreciation of the housing stock is combined with a limited number of sites for a new housing construction is disclosed. The high-rise construction in the renovation program is presented as a variant of a successful solution not only of the tasks of renovating the housing stock, but also of filling the regional budget. The social and economic orientation of the high-rise construction and the involvement of residents in the process of making town-planning decisions in the field of high-rise construction at all stages of implementing the program are shown.

1 Introduction

Economic efficiency of investments in high-rise construction is a guarantee of successful solution of social tasks and maintaining the stability of the region. In the Moscow region the development of the construction industry has not changed the number of facilities, budget financing, contracts price. Invested in the high-rise construction Finance provide significant economic impact and are considered to be an important anti-crisis measure. The correct policy of investment in high-rise housing construction helps the city administration to increase annually the volume of budgetary funding and to implement more large-scale entrepreneurial projects.

The high level of business development is supported by investments into creation projects of qualitatively new housing. It basis on the progressive innovative technical and technology solutions, designs, materials and the equipment, normative, technical, scientific, methodical documentation. These projects have successfully passed the analysis of economic efficiency (including budget expenses calculated with the use of the base prices of material resources, labor input of works, costs of use of cars and mechanisms, overhead costs, wholesale prices for production) in comparison with applied earlier analogs. Positive results of using new projects in high-rise housing construction have allowed to finish complex technical and material work and since the end of 2014 to notify all house-building

* Corresponding author: stanu@yandex.ru

plants of the Moscow region about transition of construction to the houses of new series. It is important to note that Moscow from the city budget supported those organizations which according to the new project modernized their production in order to recieve in 2016-2017 full transition to new housing series [1].

The model of development of business in high-rise housing construction considered complex relationship between the state and private business which cornerstone the aspiration to get profit from rivalry and interaction.

One of the features of modern construction business is in combining efforts of the companies for production and receiving profit thanks to more effective mechanisms of sales. As show examples of business cooperation, the modern construction market isn't any more the place only of competitive fight. In fact it is possible to speak about formation of new type of the housing construction market. The awareness of need of strategic alliance for acquisition and support of competitive advantage and a possibility of simultaneous competition and cooperation with other construction companies is one of its characteristics [2]. Appeared model of interaction between the state and private business allowed to develop the program of renovation of the Moscow housing stock directed to solve urgent social tasks and which is, in fact, the favorable entrepreneurial project.

2 Materials and Methods

The Moscow city law dealing with the program of demolition five-floor houses adopted in 1998 nowadays is in a conflict with a large amount of the federal norms appeared the last years. The lack of necessary precepts in the law does not allow execution of the program of renovation in a reasonable time. In particular, existing rules don't provide demolition of shabby housing (like five-floor houses) and concern only emergency housing. The existing standards of the town-planning, land and budgetary legislation are not so suitable for renovation of the existing urban development. As a result demolition of each residential district of five-storey buildings drags on for years, and the whole program – for decades. It is considered that this law has completely sputtered out, and the solution of problems of the accelerated renovation becomes possible with adoption of the special federal law. It will give the start to large-scale implementation of the program of renovation of housing stock. The program of renovation is designed to update housing stock of Moscow without waiting five-storey buildings finally turn into the hazardous dwelling. Thanks to the program of renovation the Muscovites living in shabby five-floor houses will be provided with modern well-planned apartments.

The program of renovation of housing stock represents demolition of shabby houses (often five-floor) and construction of modern residential multi-storey quarters. The program is socially oriented and as much as possible meet the interests of Muscovites. The inhabitants of the demolished five-storey buildings are free of charge and will receive equivalent well-planned apartments in new high-rise houses in the area of residence. The broad information support of the program is provided. Reference information can be obtained on the official Internet portal of the mayor and the Government of Moscow (mos.ru), in information centers of justices of areas and prefectures of administrative districts, in the offices of the state services "My Documents". The problem of information maintenance includes creation such conditions which allow inhabitants of shabby houses make a conscious choice about inclusion or non-inclusion their houses in the program of renovation. So, the offices of justices of areas and prefectures of administrative districts have to provide the broadest informing of citizens about the program [3].

The need of carrying out renovations is connected with that fact that five-floor houses of the first period of industrial housing construction in Moscow (from 1957 to 1975) have saved up the sufficient level of wear. Besides, in the city there are five-floor houses built earlier and there are houses two-four floors high, technical characteristics of which are similar to five-floor houses. The majority of these houses according to calculations should operate within 25–50 years. Since 1988, the Government of Moscow worked with the program of resettlement and demolition of five-floor houses to help over 160 thousand Moscow families to receive new apartments. In the 2000th years the method of so called "wave resettlement" was practiced. Inhabitants of the demolished houses moved to new houses. Only after that demolition of old and construction in the freed territory of new residential high-rise buildings was carried out. At the same time, in Moscow there are about eight thousands five-floor houses of the first period of industrial housing construction carried to the so-called not taken down series, and as have shown technical inspections considerable part of which is also in an unsatisfactory state, and conditions of accommodation of inhabitants don't meet modern safety requirements and comfort.

Since February 2017, active work was carried out to prepare the necessary laws and regulations. The Government of Moscow has set a number of the tasks which solution is necessary to start implementation of the program, among them:

- to adopt the federal law on renovation and city norms and rules providing its execution;
- to find out opinion of inhabitants on inclusion of their house in the program of renovation and then on the basis of the received decision define the lists of the five-floor houses to demolish;
- to pick up free sites for construction of starting (first) houses, including residential high-rise buildings;
- to allocate budgetary funds for the start of the program;
- to define mechanisms of investments because of the lack of budgetary funds for implementation of the program of renovation.

The main terms of granting new apartments according to the Moscow standard and to the program of renovation are the following:

- granting the equivalent apartment in the new house with identical number of rooms, living space not less then in the apartment in the five-floor house which is subjected to demolition, more spacious rooms of general use (kitchen, a hall, a corridor, a bathroom, a toilet);
- granting the new apartment in the area of residence in any district in the city, except central district, Zelenograd, Troitskiy district and Novomoskovskiy district in which housing will be provided within the administrative district;
- granting new apartments in property, and also (at the request of employers of housing in the demolished houses) under the contract of social hiring;
- granting new apartments to people in the waiting list with simultaneous improvement of living conditions and space;
- new apartments are provided with comfort class finishing;
- in quarters of renovation high-quality improvement is carried out.

It is emphasized that despite the increase in the area of new apartments (first of all, at the expense of more spacious rooms of general use - kitchens, halls, corridors, bathtubs, toilets) it will cost nothing to people.

For resettlement of inhabitants of five-floor houses, new high-rise houses will be built of modern materials (a monolith or the panel of new generation) and using modern projects - with elevators and spacious entrances. Ceilings in new apartments will be higher, and sound insulation will be much better, than in panel five-storey buildings. Also double-glazed windows are envisaged in apartments. In apartments, as a rule, will become possible carrying out re-planning. Entrances and lift halls will be at one level therefore the disabled persons or parents with a baby carriage will be able to come freely into the house and to rise at any floor.

Due to the announced initiatives, the appearance of Moscow will be improved by bright and non-standard facades of the new high-rise houses built on the places of the demolished five-storey buildings. The predicted term of their service will be at least 100 years, and in case of appropriate service and repairing much longer. Along with the fact that it is guaranteed to inhabitants of five-floor houses that the new apartment will be in the same area (except for central district, Zelenograd, Troitskiy district and Novomoskovskiy district), inhabitants also have an opportunity to chose apartments provided according to the program of renovation in the other desirable free area.

Since the beginning of 2017 the Moscow City Architecture Committee and prefectures of administrative districts conduct active work on selection of launch pads in all areas where it is planned to begin the program of renovation. The actual list will be approved after adoption of the program. Renovation supposes demolition of quarters of five-floor houses and construction new high-rise buildings, so called "point" demolition is possible as an exception for separate five-storey houses. For an additional fee residents of the moved houses have an opportunity to get the apartment with a large number of rooms and the bigger space, than apartment which is free of charge according to the program of renovation. The cost of "square meter" will be market in case of acquisition the bigger apartment, however it is important to note that at the same time various flexible systems of discounts and payments are provided.

The program is especially favorable to a large number of the Muscovites standing in a queue on improvement of living conditions. First, expectation terms in a turn will be considerably reduced, and Muscovites will need only one moving. Secondly, within the program of renovation they will get new apartments with simultaneous improvement of living conditions on norms of granting living space depending on family composition. Great social value belongs to mechanisms of resettlement of families, especially in not privatized apartments where divorced spouses and the grown-up children live. Muscovites can refuse to get equivalent apartment in the new house and choose one of resettlement options. Their resettlement in the area of residence will become possible on condition of additional payment, and in case of unwillingness to incur material inputs, such Muscovites will be able to be settled to the area with lower cost of apartments. It should be noted, an opportunity to choose the way of resettlement combines both options. Resettlement of owners of communal flats will be carried out to different apartments according to the law of the city of Moscow of May 31, 2006 No. 21 "About ensuring the housing rights of citizens at resettlement and release of premises (houses) in the city of Moscow" which forbids granting rooms to citizens in communal flats.

According to the program of renovation the homeowners in the five-floor house registered in other apartment will be able to receive the equivalent apartment, or receive monetary compensation in a natural form instead of the apartment. At the same time the amount of compensation will be calculated, proceeding from the market value of the old apartment. Considering that the cost of the new apartment will be 20-30 percent higher then old, supposed that receiving compensation is economically unprofitable for inhabitants of five-storey buildings. Besides, the rules of receiving compensation will have to guarantee respect for the housing rights of socially vulnerable categories of citizens – first of all, minor children and incapacitated persons [4].

Positive social factor of implementation of the program is planning of the work with preferential categories of citizens - lonely pensioners and families of pensioners who will be given help not only in moving, but also arrangement in the new apartment. Resettlement of inhabitants to new apartments will be carried out regardless of existence their debts on payment of housing and communal services in old apartments, as moving to the new apartment won't exempt the debtor from a duty to pay the debt. No special measures will be taken to debtors in connection with implementation of the program of renovation, will be provided to them just the same apartments, as well as to inhabitants who have no debt on housing and communal services.

3 Results

Options of work with apartments in five-floor houses which are in pledge at bank are worked out in the program. The terms "equivalent apartment" and "equal price apartment" are specified. The equivalent apartment is understood as the new apartment which according to the consumer characteristics (the apartment's area, number of rooms, district) is same (or the best) in comparison with old. The equal price apartment is understood as the new apartment which market value is equal to the market value of the old apartment. It is considered that receiving the equivalent apartment is much more favorable to inhabitants, than receiving equal in price apartment because the cost of square meter in new houses will be much higher, than in five-floor houses. In quarters of renovation on the place of the demolished houses the modern standards of improvement (including creation of local parks, bicycle tracks, public sports, children's and leisure infrastructure) will be applied, at the same time the area of green plantings will be kept or increased. It is additionally noted that projects of new high-rise quarters will be developed with participation of the best domestic and world urban architects, experts in the field of transport, design and urban environment. Warranties and the accounting of interests of owners of non-residential premises in the demolished houses are important for preservation of infrastructure in areas of building. At own choice people will be able to receive either compensation of market value of the withdrawn real estate, or equivalent non-residential premises to continue business activity.

It is defined that for inclusion the house in the program of renovation, it has to satisfy two main conditions:

- the five-floor house has to be in unsatisfactory technical condition;
- most of inhabitants (owners, employers of apartments under the contract of social hiring, constantly registered citizens) have to support inclusion the house in the program.

Houses which have been built till 1957, but with the same characteristics can be involved in the program of renovation with the consent of inhabitants. Comfortable five-floor houses which are in satisfactory condition won't be involved in the program of renovation, and problems of such houses will be solved within the program of capital repairs. On the basis of the analysis of the received results of polls, the preliminary list of houses involved in the program of renovation has been formed. Development of the program of renovation could not be possible without serious analytical researches among which:

- the carried-out analysis of more than 250 thousand appeals of inhabitants of the five-floor houses sent to Moscow authorities in recent years;
- initiatives in March-April, 2017 of prefectures of administrative districts about carrying out consultations with deputies of local government and seniors of houses about inclusion five-storey buildings in the program of renovation;
- the conducted telephone survey (in the second half of April, 2017) of inhabitants of five-floor houses to clarify their preliminary opinion of desirability/undesirability in participation in the program of renovation;
- analyzing technical information of condition of houses.

The final list of the houses involved in the program of renovation will be defined according to the results of vote of inhabitants of five-storey buildings taken from May 15 to July 1, 2017. Inhabitants of the five-storey buildings included in the preliminary list will receive notices (via message boards near their entrances, e-mail) with the offer to take part in vote. Some inhabitants will be notified by SMS mailing. The categories of inhabitants of the five-floor houses included in the preliminary list will be the following: owners of apartments (including all owners of shares), the citizens having contracts of social hiring and the citizens having constant registration. From each apartment several people will be able to vote, at the same time each inhabitant has only one voice. The tenants of apartments and those people who have temporary registration with successors and relatives of inhabitants of five-storey buildings have no right to take part in vote.

Discussion of the program has revealed a wide range of supporters of renovation of whose houses which were not included into the preliminary list. Therefore the inhabitants of the five-floor houses which aren't included in the preliminary list will be able to fill up the application for participation in the program of renovation in the electronic system "Active Citizen" or in the multipurpose center "My Documents". If the house gets a significant amount of applications, then additional vote with participation of all inhabitants will be taken. In case of positive result of vote the house will be included in the program of renovation. At the same time it is noted that during discussion an influence on decision-making is allowed in any honest ways, without violating the standard rules of conduct. In case of existence of the town-planning opportunity, first of all will be resettled those houses which inhabitants most actively participate in vote and have collected the maximum poll for participation in the program of renovation [5].

Other houses can also be included in the program of renovation, for example, panel nine-floor buildings, but at observance of a number of conditions. First, according to technical and constructive characteristics nine-floor or any other houses has to be same as houses of the first period of industrial housing construction. Secondly, technical condition of the house has to be unsatisfactory. Thirdly, the house has to be in borders of quarter of renovation (that is near five-floor houses). Besides, inhabitants have to suggest including the house in the program of renovation and by vote to collect a majority of votes «for renovation». If the house meets the first three requirements, then its inhabitants can fill up the application for participation in the program of renovation in the "My Documents" center or using the electronic system "Active Citizen". It is expected that renovation will be carried out at the same time across all territory of the city as in perspective areas, and in less prestigious areas [6]. The sequence of resettlement and demolition is put into dependency on readiness of inhabitants to participate in the program, existence of free areas for building and from some other conditions.

4 Discussions

Projects of the organization of high-rise construction will be made to cause as little inconveniences to inhabitants of nearby houses as possible. In this regard terms of construction will be reduced. Federal and city laws on quality control of construction works and respect for silence on the days off and at night fully extend to the program of renovation. Projects of the organization of construction of new high-rise houses will be made to exclude any damage to nearby building, so neither strengthening of the bases, nor any other preventive actions for the houses adjoining on quarters under construction will be required.

5 Conclusion

Thus, the program of renovation of housing stock is an entrepreneurial initiative of the Moscow authorities allowing to solve social problems: to reach comfort class in living conditions for people on the waiting list and inhabitants of not prestigious five-storey buildings on suburbs of the city, to apply resettlement of communal flats, to start improvement in quarters of accommodation, to get an "alignment" of an external architectural form of all districts in relation to the center. Also some economic problems can be solved: financing of the project from non-budgetary funds, gaining on an economic efficiency due to increase in number of storeys, guaranteeing payments of owners to the city budget for a prolonged period of time.

References

1. P.G. Grabovy, A.K. Orlov. Procedia Engineering, **153**, 195-202 (2016)
2. V.S. Kankhva. Immovable property: economy, management, **3**, 17-19 (2009)
3. S.V. Domnina, E.V. Savoskina, N.V. Shekhova. Procedia Engineering, **153**, 741-746 (2016)
4. A. Larionov, E. Nezhnikova. International Journal of Applied Engineering Research, **6 (11)**, 4433-4439, (2016)
5. A. Larionov, E. Nezhnikova. ARPN Journal of Engineering and Applied Sciences, **3 (11)**, 2023-2029, (2016)

High-rise housing construction as a way of solving the problem of providing people with comfortable habitation

Andrey Misailovov [1,*]

[1] Moscow State University of Civil Engineering, Yaroslavskoye shosse 26, Moscow, 129337, Russia

Abstract. The article analyzes the role of high-rise construction in solving the problem of providing people with comfortable habitation. High-rise construction is considered as a part of urban environment of big cities, a way of effective land use and development of entrepreneurship, including small and medium-sized enterprises. The economic efficiency of high-rise construction, an increase in budgetary financing and the number of introduced innovations are discussed.

1 Introduction

Sanctions of some countries helped Russia to begin active development of the construction business. Minister M. Men presented a development strategy and prioritized projects in the construction industry. In this way housing construction has become one of the priority projects and Russian population meet with enthusiasm this project. A new legal framework for housing construction was developed. Agreements are concluded with all the entities that participate in the implementation of projects. Particular attention was paid to high-rise construction. Houses with more than 25 floors and a height more than 75 meters helped to solve the problem of providing housing for people in multi-populated cities.

At least 80 million square meters of housing was created in 2015, 2016, 2017. The amount of funds for the improvement of the urban environment reached 27.2 billion rubles in the year 2017. 50 million rubles was allocated for the training of employees responsible for the implementation of the housing project. The main architects of cities provide the beauty and convenience of urban areas. Their work has been improved, as the State Council of the Russian Federation made a decision on the direct subordination of the chief architects to the governors of cities [1].

High-rise buildings becoming gradually a part of the urban environment actively involve residents in the process of improvement of the adjoining territory - squares, embankments, parks. Municipalities conduct polls and meetings concerning projects of improvement. Public and environmental commissions actively work; only 500 million rubles was allocated for the renewal of parks.

* Corresponding author: stanu@yandex.ru

2 Materials and Methods

The construction of high-rise buildings helps to use effectively federal, regional and municipal land in cities. A new methodological and regulatory framework for the construction of high-rise buildings has been developed. It is aimed at the integrated development of the urban environment and the creation of comfortable places for mass recreation. The construction of high-rise buildings has long been considered the prerogative of large enterprises. At present, the Ministry of Construction guides a work on the large-scale development of small and medium-sized businesses. Small and medium businesses will receive large state orders.

Throughout most of the twentieth century it was believed that greater economic efficiency have a large enterprise. This opinion was based on the Ford ideas - concept of specialization of workers and a strict sequence of manufacturing operations are possible only in large-scale production. The work of the major companies was connected with the development of innovative processes, scientific and technological progress. Small and medium-sized companies were given niche providing services in the industries producing consumer goods. However, at the end of the century, small and medium businesses have proven the ability to generate innovations and to develop new technologies that allow the economy to make a strong leap forward. An example is the economic growth of Vietnam, Malaysia, South Korea and especially China and India, where the competent investment in the area of software and computer support gave high economic results [2].

In the construction industry over the last two or three decades formed the active medium business sector that possesses the actual qualities as ease of management, flexibility and mobility. In addition, in contrast to the "clumsy" major construction partners focused on mass production, it took advantage of the flexibility of its production system to improve the quality of implementation of individual orders and the expansion of a range of prestigious services and manufactured goods. Medium business is ready to participate in the construction of multi-storey high-rise buildings.

In this process not only the fact of invention was important, but also the process of monitoring the business profitability in the market of production and services where innovations appeared. It should be noted that in the early stages innovations were often associated with the entrepreneurial spirit and creative ingenuity of entrepreneurs in traditional architectural solutions and projects, construction technologies and materials. By varying and combining in new ways traditional solutions, entrepreneurs of small and medium enterprises offered the market a lot of new construction services and products.

The process of diffusion of construction innovations creatively varied – adapted to the new conditions and the wishes of the customer (for example, in accordance with regional, cultural and traditional high-rise construction requirements), supplemented technological findings (based on natural conditions and construction materials). So, numerous innovations in the field of cottage and low-rise construction were widely circulated and led to a massive spread of innovations. It allowed for a short period of time to transform completely the architecture of many Russian country areas. Now there are beautiful, comfortable, high-quality residential complexes. Gradually the balance of competition and cooperation in the relations between construction firms was formed. In many respects the construction sector of small and medium businesses contributed to the creation of the conditions for the improvement of the economy: the development of a competitive environment, the increase in new jobs and expansion of the consumer sector.

Despite the difficulties of transition to market relations and small time period, by the end of the first decade of the XXI century, the contribution of small and medium enterprises in the gross domestic product of the country amounted to more than 20% (in developed countries - more than 50%). This was facilitated by a deliberate process when gradually at the state level were established Federal and regional programs to support small and medium enterprises. The experience of foreign financial and credit funds of small business support in the USA, France, Japan was taken into account.

In 2008 in 29 subjects of the Russian Federation programs of support the export-oriented small enterprises were successfully implemented; in 35 subjects the Federal budget supported programs for small business development; in 31 regions acted program of development loans and leasing for small business. The Russian government plans to increase the share of medium-sized businesses to 60% in economy of the country. Although some economists are skeptical about this, it should be noted that targeting this sector of entrepreneurship to receive large orders will help to increase and expand its development. In this way, the business will be able to participate in high-rise housing construction [3].

At present there are a lot of tax incentives on earnings. For small and medium enterprises are provided:

- special tax regimes, simplified rules of tax accounting, tax returns for individual taxes and levies;
- simplified methods of accounting and reporting;
- simplified procedure for preparing statistical reports;
- peculiarities of privatization of state and municipal property;
- participation in the procedures of placing of orders for delivery of goods, performance of works, rendering of services for state and municipal needs;
- measures to ensure the rights and legitimate interests in the implementation of state control (supervision);
- financial support measures;
- measures for the development of support small and medium enterprises.

More and more entrepreneurs in the Russian Federation use the basic principles of the state policy on development of small and medium entrepreneurship enshrined in the law:

- responsibility for the support of small and medium enterprises between the Federal bodies of state power, bodies of state power of subjects of the Russian Federation and bodies of local self-government;
- liability of public authorities of the Russian Federation, bodies of state power of subjects of the Russian Federation, bodies of local self-government for providing enabling environment for the development of small and medium enterprises;
- participation representatives of small and medium enterprises, associations of entrepreneurs, other non-profit business organizations, including chambers of Commerce and the Association of employers in the development of state and municipal policies for the development of small and medium enterprises and expert examination of draft normative legal acts on the issues of development of subjects of small and average business;
- equal access of small and medium-sized enterprises to support [4].

In the year 2015 for small and medium business there were made tax holidays, which covers the period up to 2020. In all regions were created favorable conditions for entrepreneurial activity in housing construction to increase the competitiveness of the economy. In Moscow, the volume of construction of high-rise buildings grows every year. High-rise construction has become an important part of the project of providing people with habitation. Since 2011 there was adopted and implemented a program "Stimulation of economic activity", the part of which was the sub-program "Development of small and medium enterprises". This sub-program aims to increase the competitiveness of the economy of Moscow and the welfare of its population by the creation of favorable conditions for entrepreneurial activity and sustainable development of small and medium enterprises, including micro enterprises and individual entrepreneurs. In this regard, it is important to mention the significance of training of entrepreneurship in universities. In Moscow state university of civil engineering students study various aspects of urban construction, analyze the economic indicators of the feasibility of construction (including altitude), the calculation of estimated costs with basic prices for tangible and intangible resources.

The successful implementation of the program is connected with the following indicators:

• reduced and simplified administrative procedures for implementation of investment projects and starting business;

• stimulated business activity and business support;

• created competitive environment;

• improved infrastructure conducive to business development;

• training, retraining and re-employment resources in favor of high-performance industries [5].

These changes have created favorable conditions for development of entrepreneurship of small and medium construction companies: annual sustainable business development was reached, contribution to the economy of the city was increased, the number of employed people in small and medium construction business was increased, transparency of the activity of enterprises was increased. The activities of small and medium construction enterprises by turnover Finance, number of actors, the share of the employed in small and medium businesses of the total population in the industry are annually monitored. Some medium-sized enterprises already have experience of participation in the construction of high-rise residential buildings. It is supposed to determine the position of Russia in the "Doing Business" ranking during the evaluation of the sub-program dedicated to the development of enterprises.

In general, in the process of implementation of the sub-program it is planned to increase the following indicators:

- the turnover of small and medium-sized businesses with 6870 billion rubles to 8360 billion;

- the number of actors from 360 to 420 thousand;

- to increase the share of the employed in small and medium businesses up to 60%;

However, the unfinished tasks of a General nature on the formation favorable environment for the development of small and medium enterprises can be highlighted: Federal authorities are still configured to support mostly big business and small and medium enterprises are supported by regional and local authorities that often lack the administrative resources and economic potential; significantly reduce the development of medium business high cost loans, and the inability to pay a Deposit or provide a guarantor for the loan while the unions of small enterprises and special funds currently do not act as guarantors for loans; there are no special banks to service small and medium businesses [6].

3 Results

At present, the construction industry solves a huge task. The national project – the program "Housing for Russian family" was ready since 2015 for implementation. It was prepared by the Ministry of construction and housing and coordinated with all departments. Despite the fact that housing provision Russians is gradually changing for the better, as evidenced by the growth in mortgage lending and housing construction, the request for the new, comfortable and affordable housing is still high. Almost 2/3 of citizens wants to improve their housing conditions, but do not have the capability. The program "Housing for Russian family" will allow 460 thousand Russian families to improve the living conditions. New housing program is designed to 2017 inclusive. 36 regions got applications for participation in the program. Until the end of 2017 year in these regions it is planned to build at least 25 million square meters of economy class housing with the entire social and engineering infrastructure. Those families who are unable to buy apartments on the market conditions can take advantage of this opportunity. In accordance with the social orientation of the program the cost of one square meter of housing will not exceed 30 thousand rubles. It is planned to build 5 million square meters in 2015 year, 6 million in 2016 and in 2017 14 million. In large cities, high-rise construction has become an important part of the project of providing people with comfortable housing.

Families in which the housing for one of its members does not exceed the maximum amount set in a specific region, but not more than 18 square meters per person can participate in the program. In addition, in these apartments could be settled citizens, living in substandard housing, families with two or more minor children. The total income of such families and different financing mechanisms – mortgage loan, the parent capital and other forms of state support should allow them to buy an apartment. Thus, people can choose apartments in high-rise buildings both expensive and cheap [6].

The government hopes that citizens will mostly use mortgage lending opportunities. The sum of issued credits in Russia in 2015 was more than 2.8 trillion rubles. Despite the large magnitude of the sum, the credits are too expensive, largely due to insufficient development of the mortgage market. According to the Agency for housing mortgage lending, the share of mortgage loans in 2016 was about 9%. In 2017, the law was amended to improve the protection of the rights of participants in the shared construction of high-rise buildings. However it should be noted the increase in the number of mortgage agreements in 2017.

It is important to stress that program "Housing for Russian family" is planned to implement at the expense of developers and funds of the parent capital in public infrastructure provision on the plots. It is considered that total volume of long-term borrowings will be 100 billion rubles for the construction of infrastructure, more than half of which were previously allocated funds of Vnesheconombank, and the rest will be brought on the market.

Currently in the high-rise construction is actively growing entrepreneurial activity, rapid deployment and commercialization of results of scientific research and development connected with extension, renewal of product range and the quality improvement of construction materials and services; also there is an improvement in their production technology and appearance of effective finds in the system of the sale of goods and services. In the new project of a high-rise housing construction can be used entrepreneurial innovations that improve the quality of building.

Among the innovations it should be noted that some experimental objects of high-rise housing are ready to be built in Moscow with the use of nanotechnology, they may be the prototype of the housing of the new generation for the whole country. About thirty different kinds of innovative technological solutions will be applied in high-rise construction. Among them - air filters which clean the air in rooms and disinfect it with ultraviolet light, the separate water filtration systems, enriching water with minerals and giving the opportunity to clean it up to drinking. In the high-rise construction will be applied innovative technologies allowing to reduce operating costs of the buildings and to increase the intervals overhaul and current repairs. Such technologies were mainly used in the construction of commercial buildings and now it is planned to apply them in the construction of municipal buildings and social facilities. It will give the opportunity to determine the pros of used innovative approaches and to adjust the vector of development of technologies to meet new challenges for the high-rise construction of the city's facilities. In the high-rise construction of residential areas (especially in the implementation to mass high-rise construction) is interesting the technology that allows personalizing the look of the high-rise buildings. Data analysis of consumer demand in the property market show that people prefer high-rise residential complexes that are designed with the latest innovative technology, gradually moving away from the format of the purchase of a typical apartment in a typical house. So, the development of standards for the implementation of such projects is of importance.

Important findings are related to the production of environmental technologies that do not harm the nature. It is known that the enterprises for manufacture of building materials must meet severe requirements to comply with environmental safety. Using environmentally friendly building materials solves the problem of the health of descendants. This primarily refers to synthetic materials. They have adverse effects on the environment, require more energy expenditure, harmfully recycled and are not capable to quick natural self-destruction.

Usually, preference is given to the organization of space in a way that it meets the requirements of comfort for both inside the apartment and in public areas. Typically, innovative high-rise construction uses a high quality products, different first-class characteristics and interesting design. When choosing material the number of functions it performs is counted (for example, could be glazed with the option of a balanced light distribution). When planning the adjoining areas thought out landscaping, convenient location, public areas, safety of residents. Much attention is paid to the presence of playgrounds and sports grounds, walkways and recreational areas, Jogging and Cycling trails. In addition to high-rise residential buildings in the infrastructure are included kindergartens, schools, sports facilities, shopping facilities. An important direction in modern high-rise construction is individualization, in which the layout of the apartments provides a variety of solutions. For example, in one house can be represented by more than 20 variants of layouts, among which multi-level apartment, apartment with separate entrance, panoramic apartment, Studio apartments, one-roomed flat.

4 Discussions

However, it should be noted that with the increase of the possibilities of innovations and their number in high-rise residential construction the price of apartment in a building designed by the latest technology, increases. Under these conditions, the economy of high-rise construction becomes an important problem of optimizing the cost of construction. It is important that during construction, the objective was to introduce the maximum amount of eco-friendly solutions at every stage of construction and implement them with minimal cost. The account of only some aspects of the construction helps to form a conclusion about the possibility of solving such problems.

Economy advantages in high-rise buildings can give the use of energy-saving system. The increase in the thickness of the insulation of external walls, replacement of inefficient ventilation system and windows provides savings of thermal energy by 25-27%. The use of new building materials or an innovative combination of traditional building materials is economically reasonable. For example, the reduction in the number of fixtures by increasing the amount of concrete in those places where the size of the reinforced concrete structure does not matter; good results are obtained by using fiberglass reinforcement instead of steel (rebar at equal load-bearing characteristics is thinner, lighter, more corrosion-resistant).

5 Conclusion

The construction of monolithic buildings, in which is used cellular concrete instead of the standard concrete can be cost-effective. This kind of building material while maintaining the strength improves thermal insulation and facilitates the design of the house and this solves the problem of the quality of mobile communication that occurs in the monolithic structures.

Thus, the implementation of large-scale housing projects in modern high-rise construction requires a combination of innovative technologies that allow creating a comfortable environment for consumers and a wide range of capabilities for developers and they will receive not only the gratitude from consumers, but also the financial gain. The construction of high-rise buildings allows attracting more than 80% of extrabudgetary financing, solving the social problem of improving the living conditions of citizens and promises profit to investors.

References

1. Z.I. Ivanova, O.V. Yudenkova, A.D. Ishkov, E.A. Shnyrenkov, International Education Studies. **5**, 232-239 (2015)
2. A.D. Ishkov, T.N. Magera. Procedia Engineering, **117**, 148-153 (2015)
3. M.I. Kamenetskii, N.Y. Yaskova, Studies on Russian Economic Development, **26(2)**, 124-131 (2015)
4. M. Aladdess. Engineering Systems and Structures, **3**(16), 38-43, (2014)
5. A. Vorotynseva, M. Aladess. Finances, Economics, Strategy, **6**, 48-54, (2016).

Working capital management in the process of financial support of investment and construction projects and of the construction material industry

Nadezhda Danilochkina[1,*] *Inessa* Lukmanova[2], *Olga* Roshchina [3], *Nikolay* Voytolovskiy[4]

[1]Moscow Aviation Institute (National Research University), Volokolamskoe highway, 4, Moscow, 125993, Russia
[2]Moscow State University of Civil Engineering, Yaroslavskoye sh., 26, 129337 Moscow, Russia
[3] Metropolitan Financial and Humanitarian Academy, Shosseinaya str., 90/17, Moscow, 117623, Russia
[4]Sankt-Petersburg State Economic University, Sadovaya street, 21, St. Petersburg, 191023, Russia

Abstract. The article presents the analysis of working capital in the process of financial support of high-rise construction investment projects. The factors influencing the choice of the working capital management model were analyzed, the reasons of the change in the requirement for the values of current assets in the process of construction of high-rise facilities were determined. The author has developed the scheme of interrelation between production, operational and financial activity cycles of enterprises implementing investment projects of unique buildings and structures and made a comparative description of their financing sources.

1 Introduction

In order to provide an uninterrupted production process of the company, it is necessary to purchase primary products, materials, containers, etc. For the permanence of the output realization, the business entity should accumulate certain volumes of end construction products. Calculations for finished products extract cash from the turnover, sometimes creating accounts receivable. In order to maintain liquidity, the company can temporarily place free cash into various short-term investments, which should have high liquidity and be able to be freely converted back into monetary assets. The own working capital of the company, its composition and structure, the turnover speed and efficiency of working capital use predetermine the status of the company and the stability of its position in the financial market [1]. J. Brigham and L. Gapenski, the leading experts in the field of financial management, note that short-term financial decisions imply management of cash flows expected in the nearest future (within the current year). The origin of most of these cash flows is associated with transactions relating to working capital and short-term liabilities.

2 Methods

* Corresponding author: nadanilochkina@yandex.ru

Financing of the current activity of the company is related to the financing of current assets and the development of a perfect mechanism for managing the company's current assets. The effective application of such a mechanism in practice is an extremely urgent problem today, as the effective formation and regulation of the volume of current assets helps maintaining a preferable level of liquidity, ensures the promptness of the production and financial activity cycles, and as the result sufficiently high paying capacity and financial stability of companies.

This explains the importance of the working capital's role in the financial provision of the activity of the company.

The definition of current assets formulated by I.A. Blank also reveals the essence and importance of working capital in ensuring the current activity of the organization. Under current assets, he implies assets characterizing the complex of property values of the company that serve the current production and commercial (operational) activities and are fully consumed within one production and commercial cycle.

E. M. Shakhbazyan distinguishes several differences between working capital and current assets (Figure 1).

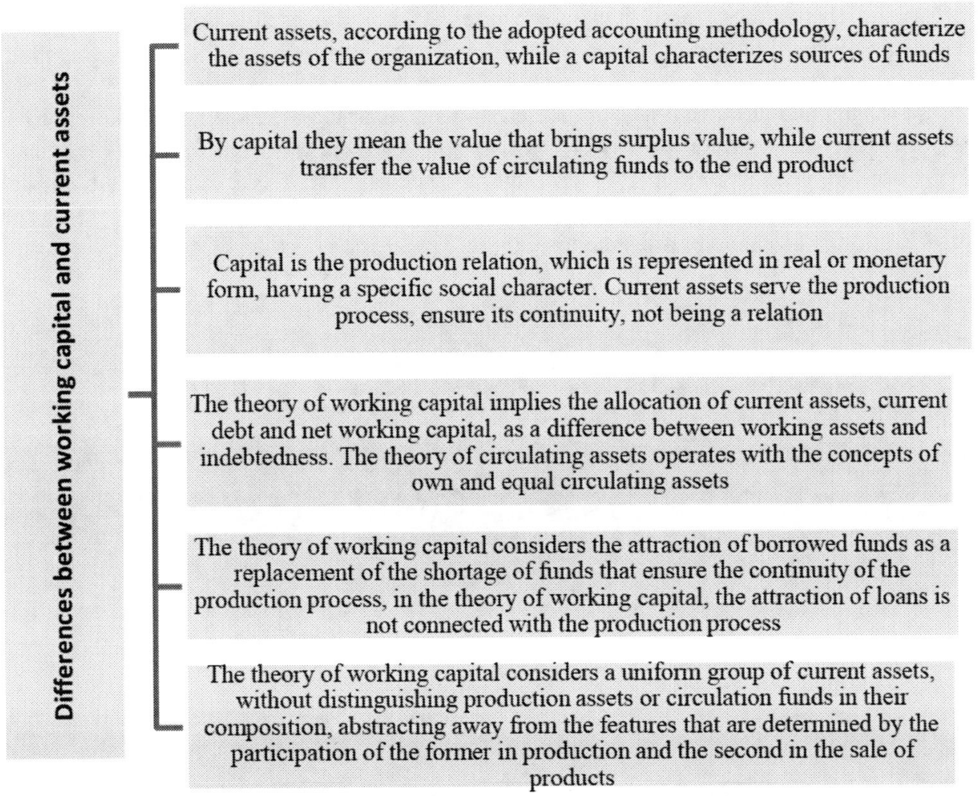

Fig. 1. Differences between working capital and current assets

The current assets' management policy and current assets themselves are important primarily in terms of ensuring the continuity and effectiveness of the current activities of the company. As V.V. Kovalev assumes, the change in current assets is accompanied by a change in short-term obligations, therefore both objects of accounting should be considered together in terms of the policy of net working capital management.

Current asset management includes the work of the administration on the development of the main direction of the policy concerning current assets, and specifically the ways to finance them [2].

By the policy of management of current assets, I. A. Blank means part of the overall financial strategy of the company, consisting in the formation of the necessary volume and composition of current assets, rationalization and optimization of the structure of their sources of financing.

3 Results

In order to justify the choice of the volume of short-term loans for financing investment construction projects, S. Ross, R. Westerfield and B. Jordan suggest analyzing the following factors:

– monetary reserves. A flexible financial policy implies a surplus of money and small short-term loans. This policy reduces the chance that the company will experience a financial famine.

– insurance of urgency. Most organizations aim to reconcile the urgency of assets and liabilities. They finance stocks through short-term bank loans, and fixed assets through long-term financing. Organizations try to avoid financing long-term assets through short-term loans. Such a mixture of urgency of payments makes frequent refinancing necessary and is risky, as short-term interest rates are more volatile than long-term rates.

– relative interest rates. Short-term interest rates are usually lower than long-term ones. This means that in most cases it is less profitable to resort to long-term loans than to short-term loans.

Management of current assets of the company is associated with specific features of the formation of its operational cycle, which characterizes the time interval between the acquisition of production reserves and the receipt of cash from sales of the products produced of these reserves. The constant process of circulation of the assets under consideration is graphically presented in Figure 2 [3].

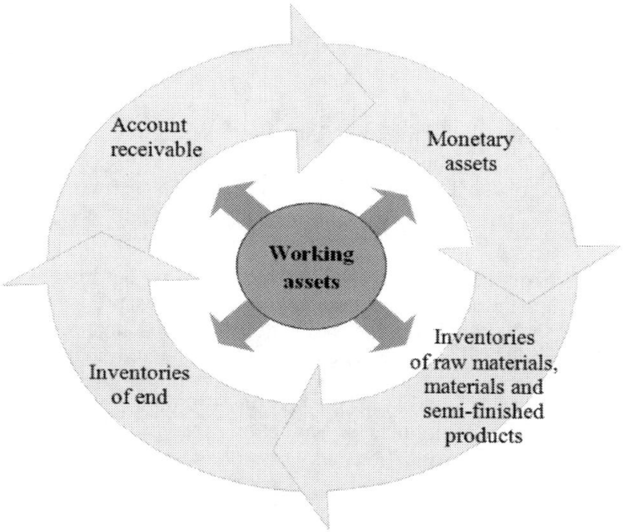

Fig. 2. Characteristics of the movement of current assets within the process of their circulation

The flow of the current assets of the organization goes through four main stages within the process of their circulation, consistently changing their forms [4]:

1. Stage 1 - monetary assets are used to purchase raw materials;
2. Stage 2 - reserves of tangible current assets are converted into finished products due to production activities;
3. Stage 3 - finished products are sold to consumers and converted into current receivables before the payment
4. Stage 4 - current receivables are re-converted into cash assets.

Developing the position formulated by J. Brigham and L. Gapenski, V.V. Kovalev draws attention to the fact that the definition of the volume and structure of working capital, the sources of their coverage and the relationship between them is the objective of working capital management. He assumes that this will be sufficient to ensure the long-term productive and efficient financial activities of the company.

The stated objective is strategic. The subject matter of the same importance is the maintenance of working capital of the amount optimizing the management of current activities. In these terms, its liquidity is the most important financial and economic attribute of the company [5].

In this regard, the importance of liquidity management along with the management of circulating assets arises. According to J. Brigham and L. Gapensky, liquidity management includes planning for the entry and use of liquid resources in such a way as to be able to pay off their short-term obligations on time (Figure 3).

From all of the above, it follows that liquidity management is a broader concept than working capital management, since it involves planning sales of permanent assets, issuing long-term securities and evaluating any other long-term projects that affect the organization's liquidity.

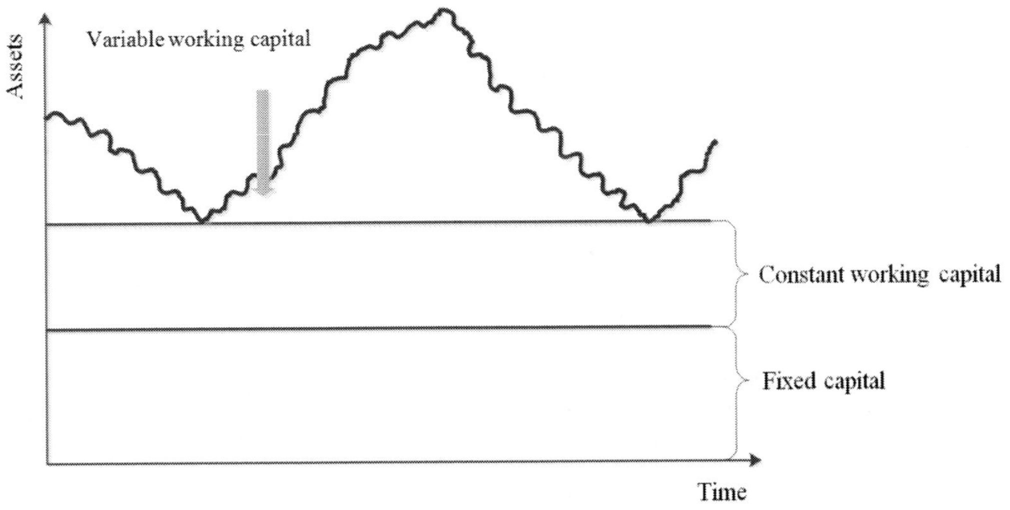

Fig. 3. Constant and variable working capital as components of a dynamic range of assets of the organization

Working capital is usually divided according to the period of the functioning of current assets into a constant (or system) capital and variable capital [6].

The first category of obligations relates to the management of both current assets and liquidity, while the second relates only to managing liquidity.

V.V. Kovalev assumes that the task-oriented management of current assets of the organization determines the need for their prior classification. In terms of financial management, this classification of current assets is based on the following main features, which are graphically presented in Figure 4.

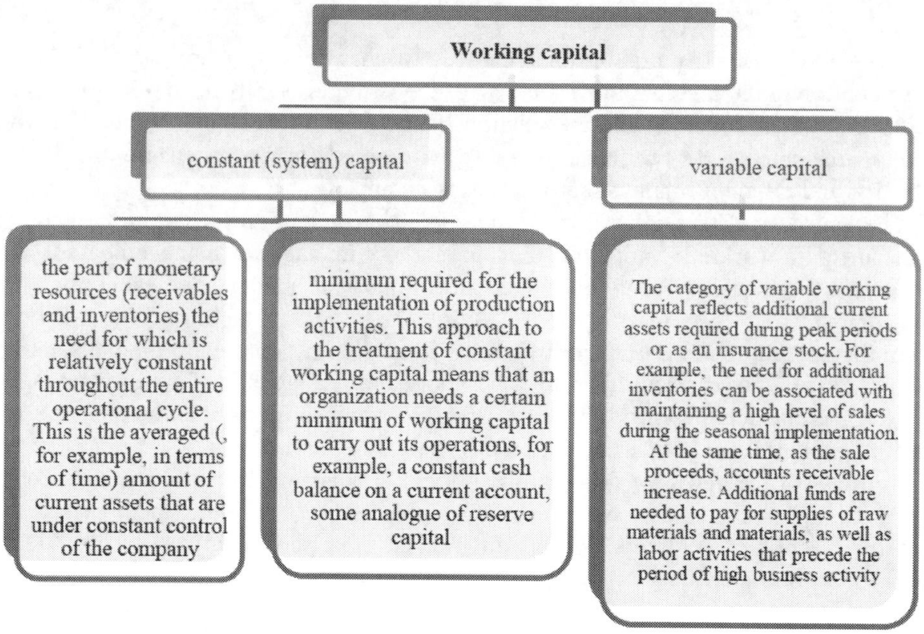

Fig. 4. Working capital breakdown

The level of working capital is directly related to the risk of liquidity loss and the amount of profit. Due to this, the working capital management policy should provide finding a compromise between the risk of liquidity loss and efficiency of work. It comes down to solving two important problems [7]:

1) Ensuring solvency. There is no solvency, when a company is not able to fulfill obligations and can possibly declare bankruptcy. A company that does not have sufficient working capital can face insolvency risk.

2) Ensuring an acceptable volume, structure and profitability of assets. Because different current assets have different effects on profits, a high level of reserves will require significant expenses. In turn, end products can help increase sales volumes and increase revenues in the future. Decisions related to the determination of the level of different groups of current assets should be considered both in terms of the profitability of this type of assets and in terms of m the position of their optimal structure.

In work [2] a clear sequence of stages of implementation of the management policy for current assets is formed, which is graphically presented in Figure 5.

Taking into account the considered features of management of current assets, it should be mentioned that each of the components of current assets has its own features [8]:

- 1. the justification of the reserves should be based on the calculation of the optimal delivery schedule and the average daily balance, taking into account an effective control system for their movement
- 2. management of accounts receivable implies not only an analysis of the dynamics of its state, its specific gravity, composition and structure for the previous period, but also the formation of a credit policy with respect to purchasers of products, a credit system, and systematic monitoring of debtors
- 3. cash management involves not only monitoring the level of absolute liquidity, but also optimizing the average balance of all cash on the basis of calculations of operational, insurance, compensation and investment reserves

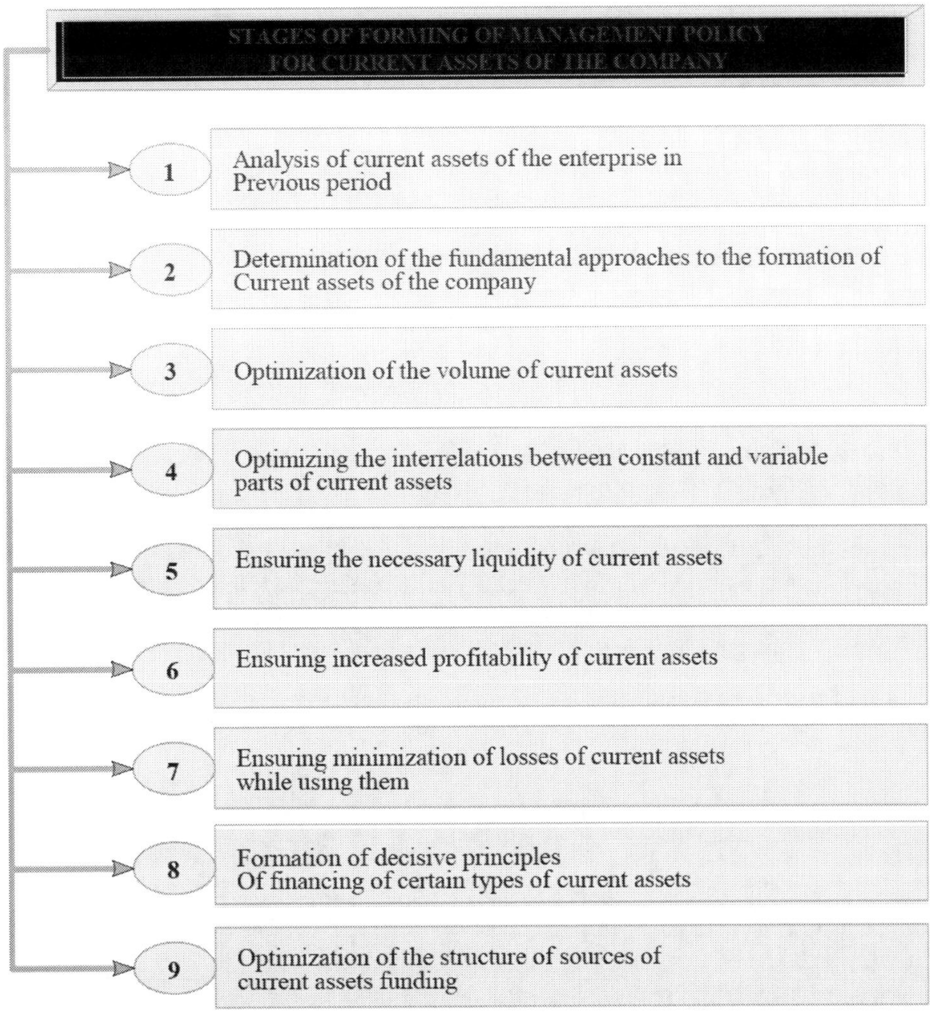

Fig. 5. The main stages of the formation of policy of managing current assets of the company

4 Discussions

Short-term financial decisions are made within the short-term financial planning, the main task of which is to provide and maintain the liquidity of the company. A liquid company is able to carry out its current activities. That means that it has sufficient funds to make the necessary payments to employees, suppliers and investors, pay taxes, etc. Conversely, a company that is not liquid cannot find the funds needed for these payments in a short time, which leads to a disruption of current operations. Current assets are used to create funds that are not aimed at a certain period. The description of this process is more complete, because they directly provide the procedure of processing and conversion, selling products, as well as the formation of monetary resources and their expenditure.

5 Conclusion

Thus, financial stability management is an important aspect of the financial and economic departments of a company and includes a number of organizational activities covering planning, operational management and the creation of a flexible coordination structure for the management of the entire company and its departments. Particular attention is paid to the effective use of working capital, which takes a great part in ensuring the normalization of the company's work.

References

1. R. S. Golov, V. V. Shilov, S. A. Silantiev ASEE International Forum, Columbus , Ohio. June 28, Paper ID #20766. (2017)
2. E. Nezhnikova Procedia Engineering **165** 1300-1304 (2016) doi: 10.1016/j.proeng.2016.11.854
3. A. Jones, G. Fallon, R. Golov European Business Review, **12-4**, pp.187-197, doi. 10.1108/09555340010336871
4. I.V. Ilin, A.I. Levina, O.Yu. Iliashenko, MATEC Web of Conf, **86**, 05028 (2016) DOi - 10.1051/matecconf/20168605028
5. I.V. Ilin, O. Kalinina, O. Iliashenko, A. Levina, Procedia Engineering, **165**, pp 1673-1682 (2016) DOi - 10.1016/j.proeng.2016.11.909
6. I. Zaychenko, S. Gutman, O. Kalinina, Advances in Intelligent Systems and Computing, **692**, pp 453-462 (2018) DOi - 10.1007/978-3-319-70987-1_48
7. A. Borboni F. Aggogeri A. Merlo, N. Pellegrini, CAmici 2015 International Journal of Advanced Robotic Systems, **12**, art. no. 42. DOI: 10.5772/60052
8. V. Pukhkal, V. Murgul, M. Garifullin Procedia Engineering **117** 624-627 (2015)

Methodical approaches to value assessment and determination of the capitalization level of high-rise construction

Vitaly Smirnov [1,*] *Leonid* Dashkov[2], *Roman* Gorshkov[3], *Olga* Burova[3], and *Alina* Romanova[4]

[1]Moscow Aviation Institute (National Research University), Volokolamskoe highway, 4, Moscow, 125993, Russia;
[2]Russian University of Cooperation Moscow region, Mytischi, ul. V.Voloshina, 12/30, 141014, Russia
[3]Moscow State University of Civil Engineering, Yaroslavskoe sh. 26, Moscow, 129337, Russia
[4]Moscow humanitarian-economic University, Leninsky highway, 8/16, Moscow, 119049, Russia

Abstract. The article presents the analysis of the methodological approaches to cost estimation and determination of the capitalization level of high-rise construction objects. Factors determining the value of real estate were considered, three main approaches for estimating the value of real estate objects are given. The main methods of capitalization estimation were analyzed, the most reasonable method for determining the level of capitalization of high-rise buildings was proposed. In order to increase the value of real estate objects, the author proposes measures that enable to increase significantly the capitalization of the enterprise through more efficient use of intangible assets and goodwill.

1 Introduction

Under conditions of Russian economy modernization, the top priority is given to the development of methodological approaches to the parameters of the capitalization process of real estate objects. Capitalization ensures their balanced value based on reaching a consensus of economic interests of the main participants of investment and construction activities. Russian scientists assume that while implementing social programs, there is a need to reach a consensus among the main participants of investment and construction activities simultaneously, which will help to create a balanced value of housing construction projects by justifying the actual growth characteristics being the foundation of capitalization.

2 Materials and Methods

The cost of real estate, as well as the cost of the company is affected by a number of factors presented in Fig. 1

* Corresponding author: svgvgy@mail.ru

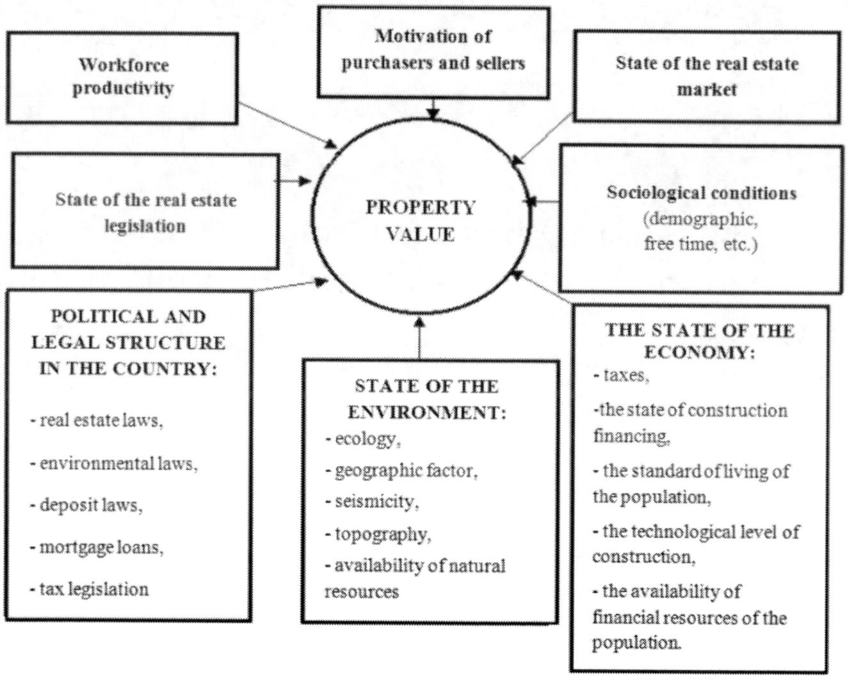

Fig. 1. Factors determining the value of real estate.

Today there is no a comprehensive methodology for assessing the capitalization of high-rise construction objects and of the entire enterprise. Theoretical and methodological issues of evaluation, which are presented in current economical scientific literature and theses, are systematized and shown in Table 1 [1].

Table 1. Methods for assessing the capitalization of enterprises and objects of real estate

Method name	Type (variant)	Brief description	Formula
Estimation of capitalization of joint-stock companies			
Stock method	-	Capitalization (RK) is determined through the exchange value of shares (P) and their quantity issued by the issuer (N).	$RK = P x N$
V.V. Piven's method	The multiplier of market capitalization	Shows the amount of changes in the capitalization (ΔMS) resulting from the purchase and sale of stocks.	$MMS = \Delta MS / S$
Estimation of capitalization of other enterprises			
The multiplier method		The algorithm of the method is reduced to two stages: • at the first stage, the multiplier coefficient (k) is calculated, characterizing the dependence of capitalization on profit, sales volume, assets and other parameters of financial and economic activity; • at the second stage, the capitalization (C) of the evaluated enterprise is calculated on the basis of its parameters (FP) for profit, sales volume, assets and multiplier coefficient	$C = k * FP$
	P/E ratio	P/E ratio is the quotient of dividing the capitalization of the enterprise by their net profit. The capitalization of the i-th enterprise is the product of the P/E ratio by its net profit.	$C = k * P_{net}$
	P/R ratio	P/R ratio is the quotient of dividing the capitalization	$C = k *$

		of enterprises by their sales volume. The capitalization of the i-th enterprise is the product of the P/R ratio by its sales volume (V) or proceeds from the sale of goods, works, services	V
	P/NAV ratio	P/NAV ratio is the quotient of dividing the capitalization of enterprises by their net assets. The capitalization of the i-th enterprise is the product of the P/NAV coefficient on its net assets (A_{net}) or proceeds from the sale of goods, works, services	$C = k * A_{net}$
	P/BV ratio	The P/BV ratio is the quotient of dividing the capitalization of enterprises by their balance currency (book value). The capitalization of the i-th enterprise is the product of the P/BV ratio at its book value (C_{bal})	$C = k * C_{bal}$
	P/BVA ratio	P/BVA ratio is the quotient of dividing the capitalization of enterprises by their balance currency (book value). The capitalization of the i-th enterprise is the product of the P/BVA ratio on its balance currency (B_{bal})	$C = k * B_{bal}$
Method of capitali-zation	A.G. Berberyan's method	The reference value of the P/E ratio is taken equal to 20. Then the capitalization of the enterprise is equal to 20 volumes of its net profit with an adjustment for the coefficient of variation (k_{var}), which is determined by the set of counted elements estimated in points	$C = 20*P_{net} * *k_{var}$
Ratio of capitali-zation	A.V. Bubnova's method	The coefficient of capitalization is a quotient of the division of the market value of equity capital to the book value of borrowed capital	$C_{cap} = EC_{mark} / BC_{bal}$
Methodo-logy of rating evaluation	Y.S. Karpuzov's method	The procedure includes 7 stages: 1. construction of a dynamic series of absolute indicators; 2. express diagnostics of absolute indicators, 3. Calculation of relative indicators of the dynamics of capitalization; 4. comparison of the results with the criterial values of dynamic capitalization; 5. rating assignment by the dynamics of accumulated capitalization; 6. rating description; 7. ranking of enterprises by the dynamics of capitalization	-
Goodwill	N.S. Pogorelova's method	The method of surplus profits, which is based on the fact that they bring intangible assets unreflected in the balance sheet, which provide yield on assets and on equity capital above the industry average. 1. Calculation of the value of all company assets 2. Adjustment to obtain a normalized net profit. 3. Determination of the average branch index of return on assets 4. Definition of expected profit. 5. Calculation of surplus profits, which include the impact of goodwill. 6. Calculation of the value of goodwill.	-

Capitalization of a real estate object reflects an increase in its value, therefore the concept of capitalization is directly related to the concept of value.

3 Results

The value management of real estate objects and construction companies implies the creation of positive cash flows. This management is most effective when the following management methods are combined:

1) financial, which are based on the discounted cash flow model (DCF);

2) non-financial (they include the balanced scorecard model (BSC)).

There are three main approaches for estimating the value of real estate objects and enterprise for determination of the level of the company capitalization:

1. Income approach

A. Discounted cash flow method

Today at almost all enterprises, the analysis of new investments (real investments) is based on the discounted cash flow method, which fully corresponds to the valuation of stocks, which is equal to the discounted value of future cash flows expected by stockholders at any given moment. The DCF method is reasonable to apply for a large enterprise, because it enables taking into account the risks related to the uncertainty of the activities. This method is also suitable for those enterprises where the financial situation depends on the contracts completed [2].

The discounted cash flow method consists of several stages:

1. Choosing the cash flow model. When assessing the value of an enterprise, one of two models of cash flow is used: for net worth and total invested capital.

2. The retrospective analysis and the forecast of gross proceeds from sales require detailed consideration of the following factors: production volumes, enterprise growth rates, available production capacities, prospects and possible consequences of capital investments, the general situation in the economy, the share of the estimated enterprise on the market, long-term growth rates of the enterprise [3].

3. Analysis and forecast of costs, which can be divided into production (operating) costs and capital costs (investment costs, which are aimed at extensive and intensive development of production). Accounting for production costs and capital expenditures shows their impact on the value of the enterprise at various stages of the LC [4].

4. Analysis and forecast of investments, including analysis of own working capital, analysis of capital investments and analysis of the need for financing

5. Calculation of the cash flow for each year of the forecast period. There are two main methods: the direct method, which is based on the analysis of cash flow by items of income and expenditure, i.e. on accounting accounts; the indirect method, which analyzes the cash flows by the directions of business.

6. Determination of the discount rate.

Due to the risk associated with obtaining future income, the discount rate should exceed the rate of risk-free investments and provide additional income for the entire period of risk.

7. Calculation of the value of business in the post-forecast period. In the post-forecast period, the calculation of the discount rate should be made using the Gardon method, which is based on the forecast of obtaining stable incomes in the residual period and implies the equality of the amount of depreciation and capital investments [5].

$$PV = D_n * \frac{1+g}{(R-g)}, \quad (1)$$

were R_v = PV is for residual value of the enterprise (reversion) in the post-forecast period, $D_{n+1} = D_n * (1 + g)$ is for income that can be gained within the first post-forecast year (D_{n+1}), g is for expected annual growth rates of income, R is the coefficient of capitalization.

9. The calculation of the present value of future cash flows and the value within the post-forecast period is the summation of the present value of periodic cash flows and the current value in the post-forecast period expected in the future.

10. Final amendments. After determining the preliminary value of the enterprise cost, there is a need to make final adjustments to the value of the non-performing assets and adjust the value of the working capital in order to obtain the final market value [6].

B. Income capitalization method

It is applied when insignificant fluctuation of the future incomes is expected. According to this method, the value is determined by dividing the results of activity (profit or cash flow) for the base period by the capitalization ratio.

2. Cost approach

A. Method of net assets.

The general formula for estimating a company using the net assets method is:

$$NAV = \Sigma\ VAi - D + G,\ i=1, \quad (2)$$

Where, NAV is the value of the company's net assets; VAi is the market value of ith asset; D is market value of obligating documents; G is goodwill. As it can be seen from the formula, firstly, the company's assets disclosed in the balance sheet are corrected for their market valuation, then the market value of obligating documents is deducted and the value of goodwill is added.

B. Residual value method

Liquidation value is the value that is obtained when the enterprise is liquidated and a separate sale of assets is possible. The evaluation formula is:

$$LV = PV\ (\ \Sigma\ AVi) - D - CL\ ,\ i=1 \quad (3)$$

where LV is the liquidation value of the company; AVi is the value of the ith asset; D is market value of obligating documents; CL is the cost of liquidating the company.

3. Comparative (market) approach

A. Capital market method

This method is based on market prices of stocks of comparable companies that are listed on the stock market.

B. The method of transactions

This method is based on the analysis of prices, the acquisition of control stocks of comparable enterprises or the analysis of the acquisition prices of enterprises in whole.

C. Sectoral coefficient method

• This method is used for an approximate valuation of enterprises. This methos consists in analysis of large statistical data of specific transactions on sales of enterprises

4 Discussions

Particular attention in assessing the value of real estate objects and an enterprise should be paid to modern approaches to assessing the business reputation (goodwill) of the company. A professional valuation of goodwill will help to reveal the real value of the business and, in particular, sell it at a much higher price.

Today there are no decisive methods to calculate accurately reputational assets. Studies made by domestic scientists show that a change in reputation by 1% leads to a change in market value of 3% [7].

For construction companies, the reputation is especially important due to industry specific features: high production costs, multi-level risks, duration of the technological cycle, etc. Buyers usually carefully choose real estate. They study the market, the quality of products, the reputation of a company. In this context, the reputation of the construction company is the key to its success, so the management should pay special attention to the key factors shaping the reputation: the success and experience of a company, the openness and transparency of business operations and financial matter, private trust to the first persons of a company, social responsibility and others.

5 Conclusion

First of all, the reputation is developed by the fulfillment of the obligations assumed: the deadlines for delivery of objects and the quality of construction. Elaboration of houses and proper planning of apartments, which ensures comfortable living, also play an important role. The conclusion from the above is that capitalization is a complex process of increasing the value of enterprises through various sources: retained income, revaluation of assets, increase in the value of stocks, etc. Today construction companies pay insufficient attention to issues related to their capitalization, especially through intangible assets and goodwill.

References

1. V. Pukhkal, V. Murgul, M. Garifullin Procedia Engineering **117** 624-627 (2015)
2. R. Golov, V. Shilov, S. Silantiev ASEE International Forum, Columbus , Ohio. Paper ID #20766. (2017)
3. V. Kankhva, Procedia Engineering **165** 1300-1304 (2016) doi:10.1016/j.proeng.2016.11.855
4. A. Jones, G. Fallon, R. Golov European Business Review, **12-4**, pp.187-197, doi. 10.1108/09555340010336871
5. A. Bril, O. Kalinina, O. Valebnikova, Lecture Notes in Computer Science, **9870**, pp 766-775 (2016) DOi -10.1007/978-3-319-46301-8_67
6. A.K. Orlov, I.Y. Chubarkina MATEC Web of Conferences **106** 08015 (2017)
7. I.V. Ilin, A. Lepekhin, A.I. Levina, O.Yu. Iliashenko, Advances in Intelligent Systems and Computing, **692**, pp 1306-1314 (2018) DOi -10.1007/978-3-319-70987-1_138

Methodology for determining the investment attractiveness of construction of high-rise buildings

Ekaterina Nezhnikova [1,*] *Valentin* Kashirin[2], *Yana* Davydova[3], and *Svetlana* Kazakova[3]

[1]Peoples' Friendship University of Russia (RUDN University), Miklukho-Maklaya Street, 6, Moscow, 117198, Russia
(The publication was prepared with the support of the "RUDN University Program 5-100")
[2]Moscow Aviation Institute (National Research University), Volokolamskoe highway, 4, Moscow, 125993, Russia
[3]Moscow humanitarian-economic University, Leninsky highway, 8/16, Moscow, 119049, Russia

Abstract. The article presents the analysis of the existing methods for assessing the investment attractiveness of high-rise construction. The authors determined and justified the primary choice of objects and territories that are the most attractive for the development of high-rise construction. A system of risk indicators has been developed that allow making a quantitative adjustment for a particular project in the evaluation of the efficiency of investment projects. The study is aimed at developing basic methodological concepts for a comparative evaluation of the prospects of construction of high-rise facilities that allow to take into consideration the features of investment in construction and to enable quantitative evaluation of the investment effectiveness in high-rise construction.

1 Introduction

Although there is a number of different methods for assessing investment attractiveness and the riskiness of certain projects, it is not always possible to use them because of their "general" nature. The industry specificity of a particular business is negligibly taken into account, the types of objects are not considered, there is no linkage to specific strategic priorities and tasks. Thus, traditionally used methods for assessing the investment attractiveness of high-rise construction are based on the macro level analysis.

Investors evaluate the attractiveness of projects in comparison with alternative investment options. Estimating investment potential, the investor brings his subjective view in its description, considering attractive and prospective the one that best meets his ideas and expectations. As a result, the concept of investment attractiveness can be formulated as an investor's subjective assessment of the region, formed on the basis of an analysis of the objective characteristics of the investment climate [1].

* Corresponding author: katnej@mail.ru

2 Materials and Methods

The study of such approaches to analyzing investment potential as financial, resultant, resource, etc., has shown that investment potential is a complex of investment opportunities that is formed due to integration and synergy of infrastructure, resource and macroeconomic potentials, limited by the level of investment risks [2].

The total investment potential of the system includes the following components:

– Resource and raw component: provision of the territory with balance reserves of the main types of natural resources;

– Production component: the aggregate result of the economic activity of the population in the region;

– Consumer component: aggregate purchasing power of the population of the region;

– Infrastructural component: economic and geographic location of the region and its infrastructural saturation;

– Innovative component: the level of science development, the introduction of scientific and technological progress and the level of manufacture modernization in the region.

– Labor component: availability of professionally trained labor and qualified engineering staff.

– Institutional component: providence of the development of the region with the necessary institutions of a market economy.

– Financial component: the volume of the tax base and the profitability of enterprises.

The authors suppose, that the structural approach should be complemented with institutional components. As practice shows, a favorable investment climate is not sufficient for the attractiveness of the project without taking into account measures that stimulate investment activity. Therefore, when assessing, due attention should be paid to the state participation in the investment climate development. A generalized scheme of the relationship between the elements of the investment process is shown in Fig. 1.

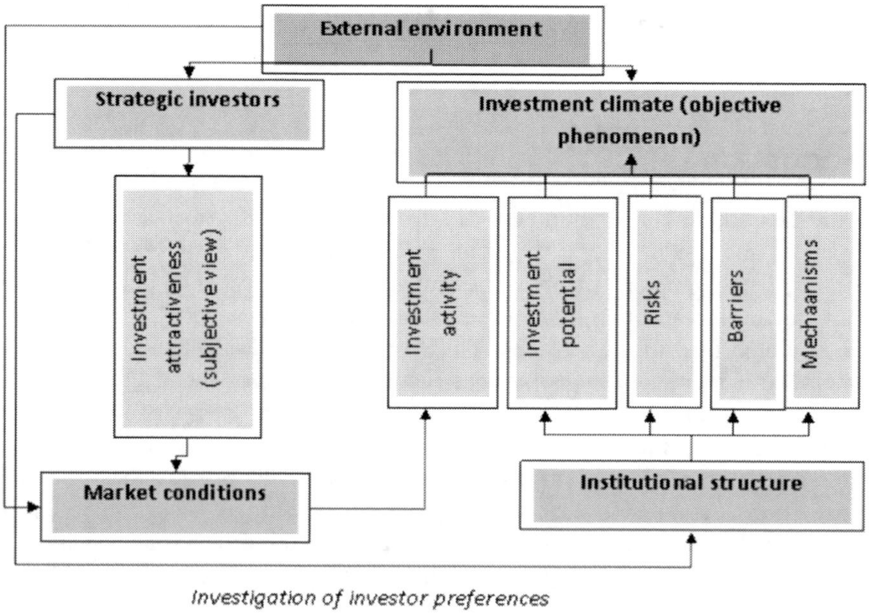

Fig. 1. Interrelation of elements of investment process

A large number of studies and publications are aimed at the issues of investment climate assessment. Each method satisfies the preferences of a certain "consumer". Consequently, in each case it determines its own set of basic factors and indicators, as well as the approach to the organization of the research process, etc. As a result of the analysis of these sources, a classification of methods for assessing the investment attractiveness of economic systems was made (Figure 2).

The assessment of the investment climate in terms of methodological approach is the most commonly used classification. There are three most distinctive approaches [3]:

1. Constricted approach: the calculation uses the indicator of the level of profitability of the used assets;

2. Factor approach: the identification of a set of influencing factors and the determination of the weighted average estimate;

3. Risk approach: risks are assessed in terms of the possibility of investment losses and income on them.

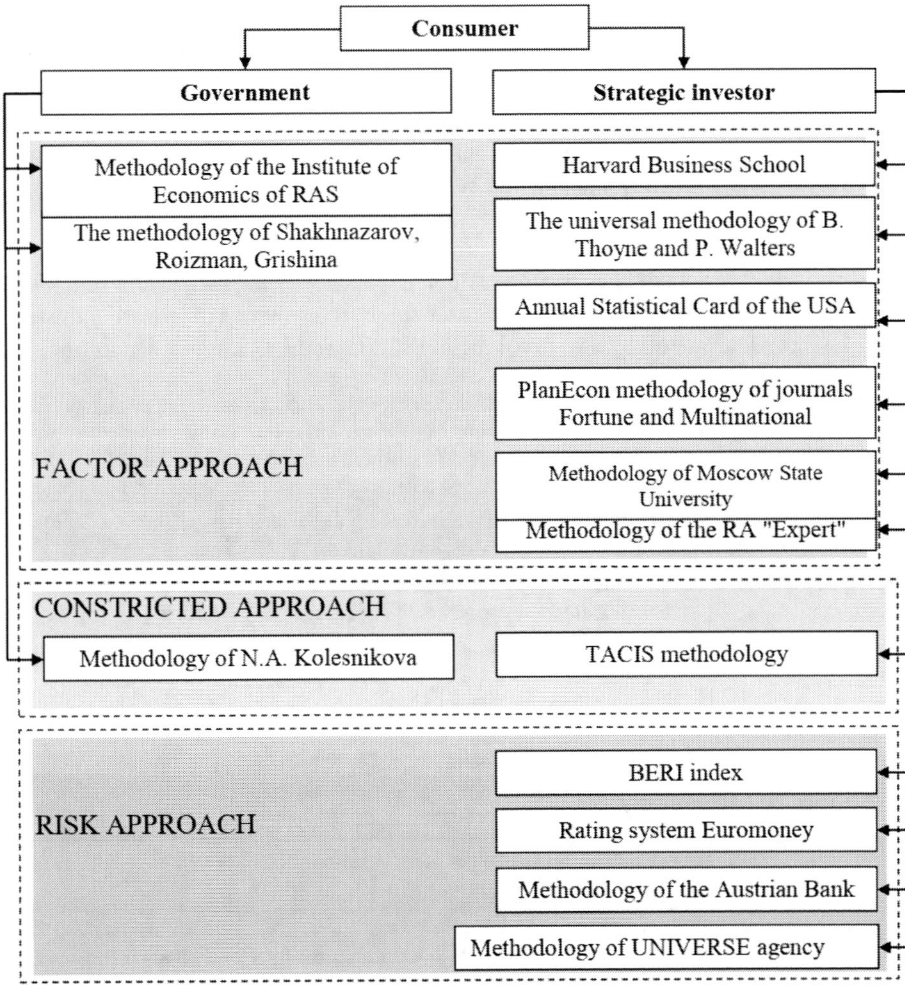

Fig. 2. Classification of methodologies for assessing the investment attractiveness of economic systems

All methods imply consideration of a set of various factors that influence the prospects of investment projects. The number of these factors, their specifics and structure, the methods of evaluation, the methods of measurement, the weights of the factors vary considerably and represent positions of certain methodologies and the certain specifics of application.

However, the following conclusions can be made in general:

1. There is no unified international or even national standard for a comparative assessment of the investment attractiveness of high-rise construction projects.

2. There is a limited set of indicators to be considered, which does not take into account the differentiation of the regions of the Russian Federation.

3. The generalized estimate does not take into account the significant factors of attractiveness of high-rise construction projects in a particular case.

4. There is no account of the legislative conditions for investment, especially in the part of regional and local legislation.

5. There is no clear justification for the principles of aggregation of a set of valuation indicators.

6. There is uncertainty about the period of research.

7. Often the methods are of a general nature, they do not take into account the sectorial specifics and priorities of certain groups of investors.

3 Results

Within the research the authors offer methodical concepts of an estimation of investment attractiveness of high-rise construction projects for choosing a building companies' development directions. The elaborated approach includes both qualitative and quantitative analysis for a number of key factors that reflect the specifics of the industry and specific markets [4].

The methodology assumes a two-dimensional assessment: at the first stage, the investment attractiveness of the territory is determined, which enabled the initial selection of areas of territorial expansion; on the second stage investment criteria are adjusted to the degree of risks. The methodology involves eight consecutive stages, which are shown in Fig. 3.

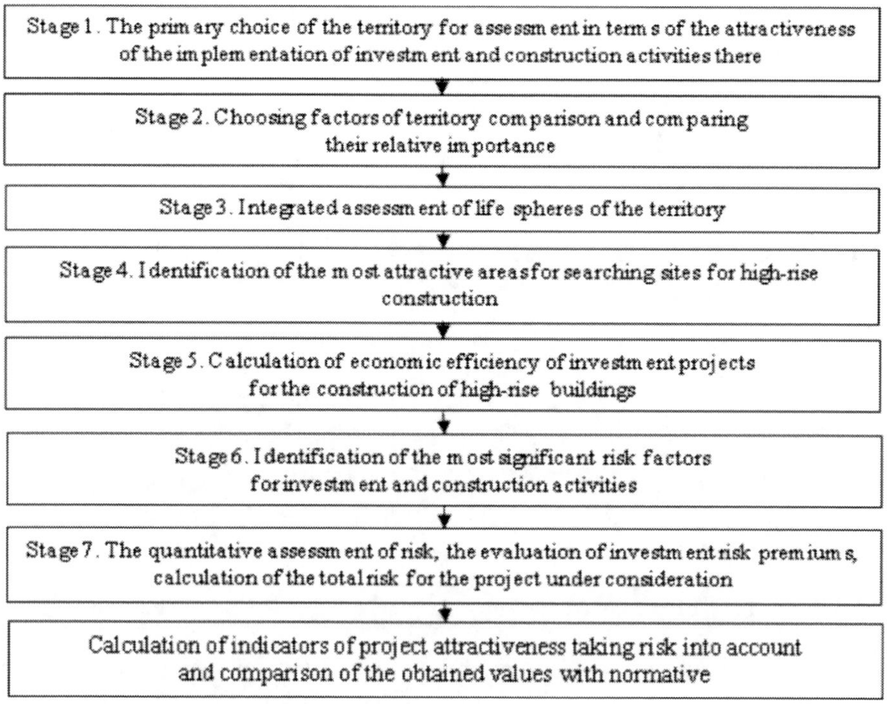

Fig. 3. Algorithm for evaluating the investment attractiveness of high-rise construction projects

Some of these stages are obvious and standard, others need special explanations [5].

Stage 1. The first stage implies consideration of areas with the greatest concentration of quantitative housing demand.

Stage 2. This stage implies the use of the method of deducing 3 indexes: the index of the non-movable market level, indexes of the socio-economic and socio-demographic spheres of the regions' life-activity.

Stage 3. In the context of further evaluation, indicators that have a negative value, such as "unemployment rate", were given negative scores, because these indicators are disincentives of investment activity.

When creating scoring, in addition to the initial data on the values of the indicators, a scale was set to estimate each indicator. To calculate the scoring using the continuous scale, the following formulas were used:

For driver indexes:

$$b_{ij} = b_{\min j} + (b_{\min j} - b_{\max j}) * (x_{ij} - x_{\min j}) / (x_{\min j} - x_{\max j}) \tag{1}$$

For disincentive indexes:

$$b_{ij} = b_{\max j} - (b_{\min j} - b_{\max j}) * (x_{ij} - x_{\min j}) / (x_{\min j} - x_{\max j}) \tag{2}$$

where $b_{max\ j}$ and $b_{min\ j}$ are respectively the maximum and minimum possible scores for the j-th index according to the scale accepted for it (in our case, from 1 to 10); $x_{max\ j}$ and $x_{min\ j}$ - respectively, the maximum and minimum value of the j-th index when assessing the sphere of the i-th regions.

The complex estimate (index) of the investigated spheres was calculated by the formula (3):

$$I_i = \sum_{j=1}^{n} D_j * b_{ij} \tag{3}$$

where I_i is the complex cumulative score (index) of the i-th region in the investigated sphere of life; b_{ij} is the score of the region in the j-th indicator; D_j is the specific weight of the j-th index in the integral estimation of the investigated sphere; j ... .n- number of indicators.

Based on calculated complex comparative assessments, each region is assigned a certain place in the field under study.

Stage 4. Based on the indicators obtained at the previous stage, the territories are ranked, which makes it possible to select the most prospective for the deeper comparison from the previously proposed list.

Stage 5. The calculation of the economic efficiency of investment projects for the construction of high-rise buildings in the regions selected earlier is performed on the basis of standard investment analysis methods and involves the estimation of NPV and IRR.

Stage 6. In the developed methodology, an important role is played by risk factors that are associated with the economic, political and social situation of individual administrative or geographical regions. Their classification, taking into account the specifics of the construction industry, is given in Table. 1.

Table 1. List of main regional risks for the construction company

Group of risks	Type of risk	Brief description
Socio-economic risks	Change in income of the population	In different periods of time, the population may not be able to pay.
	Using mortgage services	High interest rate on a mortgage loan, impossible for most of the population.
Risks of the construction industry	Change in demand for housing	The competitiveness of high-rise buildings is determined by the surrounding environment and the specificity of individual preferences of buyers.

(sector-specific risks)	Rivalry	The main content of rivalry is the struggle for complete satisfaction of the needs of the consumer.
	Availability of land plots	The scarcity of land is connected with the lack of or poorly prepared infrastructure for new construction
	State of engineering infrastructure	The scarcity of land is connected with the lack of or poorly prepared infrastructure for new construction
	Legal and regulatory framework	The risk of increasing the time for obtaining all necessary town planning documentation for design and construction.
	Human Resources	Lack of qualified personnel in the region. Low proportion of the able-bodied population. Deficiency of workers for construction
	Development of the construction industry	High cost of building materials. Shortage and deterioration of machines and mechanisms. Absence of "good" contractors for design and construction works
	Types of construction	The predominance of low-rise buildings (low profit for construction companies). Risk of non-viability of the project and risk of non-completion of construction.
Financial risks	Bank interest rates	High rates for bank loans, with the use of borrowed funds.
	Tax rates	High tax rates, incl. rent for land, etc.
	The rate of inflation	High inflation rate
Macroeconomic risks	Geographical location	Unfavorable or least profitable economic and geographical position of the region
	Climat	The harsh climate leads to an increase in the cost of construction. The longevity of the house under construction depends on the quality of the soil research in the construction site.

Stage 7. A methodology similar to the country risk expert assessment, widely used to determine the discount rate in the valuation of a business, was offered to be used as a quantitative tool for assessing regional risks that could be used in further calculations.

For risk assessment (C), individual risk factors are suggested to be assessed in the range from 1% (minimum risk) to 10%. Evaluation of the importance of this or that factor is carried out by the expert way and gives an opportunity to focus experts' attention on different factors of regional differences that are most interesting for a particular enterprise in a particular situation [6].

The total value of regional risk is determined by the formula:

$$C = \frac{}{r * e} \sum_{i=1}^{10} Ni * Ri \quad (4)$$

where N_i is the number of observations of the *i*-th degree of risk; R_i is the value of the degree of risk (from 1% to 10%); r is a number of risk factors; e is a number of experts.

As a basic region in the methodology, it is proposed to use the region of primary placement of the company planning territorial expansion of its activities.

Stage 8. To assess the attractiveness of the region, it is proposed to use indicators of net present value and internal rate of return and to adjust them to the degree of investment risk.

The total risk adjustment is calculated using the following formula:

$$R = B + C \quad (5)$$

where B is a corporate business risk, a risk premium for investing in a particular company; C is a- regional risk.

The index of attractiveness of the investment project is calculated according the following formula:

$$A^* = \frac{A}{(1+R)} \quad (6)$$

where A is the selected economic indicator; R is the total amount of risk.

The values of A^* obtained must be greater than or equal to the corresponding normative values adopted in the company, taking into account the experience and existing practice in implementing investment projects for high-rise construction.

4 Discussions

The attractiveness and economic reasonability of high-rise construction is determined by the development level of the territory where the project is planned to be implemented. The complexity here is to choose an alternative; which means to choose the territory more preferred for investing in a high-rise construction project [7]. Strengthening competition and the gradual saturation of demand in the construction market of megacities around the world make the largest construction corporations diversify their business, directing development to certain regional market sectors.

5 Conclusion

The need for methodological support of territory development is becoming increasingly important under the condition of continuing consolidation and growth of business, as it occurs in the construction industry. The development of high-rise construction requires new assessment approaches that make this development the most effective and sustainable for each particular construction company, allowing to assess adequately the possible risks and neutralize them.

References

1. I.V. Ilin, A.I. Levina, O.Yu. Iliashenko, MATEC Web of Conf., **106**, 08066 (2017) DOi-10.1051/matecconf/201710608066
2. V. Kankhva, Procedia Engineering **165** 1300-1304 (2016) doi:10.1016/j.proeng.2016.11.855
3. A. Jones, G. Fallon, R. Golov European Business Review, **12-4**, pp.187-197, doi. 10.1108/09555340010336871
4. A. Bril, O. Kalinina, O. Valebnikova, Lecture Notes in Computer Science, **9870**, pp 766-775 (2016) DOi -10.1007/978-3-319-46301-8_67
5. A.K. Orlov I.Y. Chubarkina *MATEC Web of Conferences* 08015 106 pp (2017)
6. R Golov, V. Shilov, S. Silantiev ASEE International Forum, Columbus , Ohio. June 28, 2017. Paper ID #20766.
7. E. Nezhnikova. IOP Conf. Series: Earth and Environmental Science **90** 012161 (2017) doi:10.1088/1755-1315/90/1/012161

Models of resource planning during formation of calendar construction plans for erection of high-rise buildings

Irina Pocebneva [1,*] *Vadim* Belousov, and *Irina* Fateeva[1]

[1] Voronezh State Technical University, Moscow Avenue, 14, Voronezh, 394026, Russia

Abstract. This article provides a methodical description of resource-time analysis for a wide range of requirements imposed for resource consumption processes in scheduling tasks during the construction of high-rise buildings and facilities. The core of the proposed approach and is the resource models being determined. The generalized network models are the elements of those models, the amount of which can be too large to carry out the analysis of each element. Therefore, the problem is to approximate the original resource model by simpler time models, when their amount is not very large.

1 Introduction

Generalized network models (GNM) are a convenient tool that allows using mathematical methods efficiently in calendar planning and construction management when erecting high-rise buildings and structures. However, network methods do not allow a comparative analysis of the alternatives that arise when assigning resources that ensure the performance of these specific works. A mathematically correct approach to solving resource allocation problems is the combination of resource and time analysis in the process of plan formation. The problem of using the branch and bound method for the case, when the selection of workflows with indivisible resources is demanded, became the main topic of many works. However, the necessity to consider resource alternatives imposes the necessity to abandon the network setting of tasks. At the same time, a turning back to network models and their generalizations at a qualitatively new level allows the use of network methods [1]. This paper presents the resource-time analysis as a combination of resource analysis (i.e. the implementation of a sequence of RM transformations, including branching for a partial resource assignment) and time analysis (i.e., obtaining estimates for selecting the most promising branch node).

2 Materials and Methods

Assume that two finite sets X, Z are given, where $X \neq t$, $X \cap Z \neq t$.

The space of functions of dimension $|X| + |Z|$ is denoted by Θ. The functions are defined

* Corresponding author: ipocebneva@vgasu.vrn.ru

on the set $X \bigcup Z$. The arbitrary function T Θ will be called a plan.

Each x element of the X set is interpreted as an identifier of an event, which can imply the beginning, the end of the work or its parts, the construction stages, etc. For the T plan, the T_x coordinate is the time for the event x to occur. Each element $z \in Z$ is understood as a resource identifier with a continuous intensity scale. The T_z number, which will be further called "tempo", is the reciprocal of the intensity. It is assumed that in the process of performing the work, the tempo of T_z cannot be changed, although it can be selected from this range $[l_z, L_z]$ in the planning phase [2]. The duration of work (x, y) is equal to bT_z where b is the volume of resource consumption z at work (x, y). If the resource usage rate T_z is determined, the duration of the work that consumes this resource is automatically determined.

Sometimes it is convenient to understand the plan T as a vector of dimension $|X| + |Z|$ with the coordinates $\{T_x\}_{x \in X \bigcup Z}$.

Definition 1. The following system of inequalities is called a resource scale:

$$\{l_z \le T_z \le L_z\}_{z \in Z}, \tag{1}$$

for the plans T, where $\{l_z\}_{z \in Z}$, and $\{L_z\}_{z \in Z}$, answer the correlations $0 \le l_z \le L_z$.

Definition 2. The temporary model (BM) S is given when the resource scale (1) is defined as well as the finite system of conditions of the following form:

$$T_x \le L \quad (x \in X) \tag{2}$$

$$-T_x \le -1 \quad (x \in X) \tag{3}$$

$$T_x - T_y \le a \quad (x, y \in X, x \ne y), \tag{4}$$

$$T_x - T_y \le bT_z \quad (x, y \in X, x \ne y, z \in Z), \tag{5}$$

where on the right sides are arbitrary numbers L, 1, a, b, with b>0.

Definition 3. For a BM S, a plan T is said to be S-admissible if answers all relations occurring in S of the form (1) - (5). The set of all S-admissible plans is denoted by D(S).

Using the S temporary model, a wide range of scheduling tasks can be formulated and solved. Note that all relations (1) - (5) are linear. This allows to use linear programming methods to solve problems with linear criteria $F = \sum_{x \in X \bigcup Z} \lambda_x T_x$ relatively to $\{T_x\}_{x \in X \bigcup Z}$.

The optimized criterion can be vector. The following problem is the example a vector minimization problem:

$$\{T_x\}_{x \in X \bigcup Z} \to \min;$$

where the time limits of the events are minimized and the intensity of work is maximized. In certain cases, when solving a multicriteria problem, the plans obtained are not only answering the Pareto condition, but also not being improved by any of the criteria [3]. An example is the problem of finding early plans on a computer equivalent to a GNM in the following formulation:

$$\{T_x\}_{x \in X} \to \min;$$

Definition 4. A resource model (RM) is a finite aggregate of TM:

$$R = \{S_\alpha\}_{\alpha \in \sigma}, \tag{6}$$

where σ* is a finite set of resource alternatives.

Definition 5. R* is the initial resource model, when

$$R^* = \{S^*_\alpha\}_{\alpha \in \sigma^*},$$

where σ * is the set of variants with the full assignment of resources.

Definition 6. For PM (6) we denote by

$$D(R) = \bigcup_{\alpha\in\sigma}(S_\alpha)$$

Plans $T \in D(R)$ will be called R-admissible.

A number of scheduling tasks can be set on the resource model. The example of an optimization task that is solved under the scheduling of work complexes is the following task: for a resource model (RM) R, R-admissible plan should be found that provides an optimum for some vector criterion F.

Direct analysis of the set D(R) is often not possible because of the large number of alternatives for resources, and possibly for the intensity of their use.

3 Results

One of the approaches to solving such problems is connected with the application of the scheme of branches and boundaries [4]. For the effective use of it is necessary to specify the methods of branching (dividing the set into subsets), the methods for obtaining estimates on subsets in accordance with F criterion. The way to divide RM into parts and approximate them by temporal and generalized network models for obtaining estimates by means of a sequence of corresponding transformations is shown below.

Definition 7. Assume that RM R' depends on the resource model R(R'<R) when D(R') $\subset$ D(R). When herewith D(R')$\neq$D(R), then the model R' will be called more rigid than the model R. When D(R)`= D(R), then the models R and R' are called equivalent .

The concept of subordination is obviously a partial order relation on the RM family (the properties of transitivity, reflexivity and antisymmetry are fulfilled).

The set D(R*), where R* is the original model, is divided into a sufficiently large finite number of subsets D ($S_{\alpha*}$) satisfying the systems of linear inequalities. Using some properties of the model (for example, the partial assignment of resources), R* can be decomposed into subsets for which D(R_1), ..., D (R_k) are such that D(R*) contains each of D(R_i) and each D ($S_{\alpha*}$) belongs to some D(R_i). Therefore, R_i is a RM which is intermediate between the original RM and TM $S_{\alpha*}$. The R_i sets are constructed in such a way that the prospects of the plans included in D(R_i) can be estimated, for example, on the basis of the time analysis R_i sufficiently accurate. When performing this analysis, R_i is replaced by S which is less rigid than or equivalent to TM. Such a replacement can be carried out gradually by elementary operations by constructing the sequence R^i <R` <R`` <... <S.

Note that each constraint (1) - (4) is determined by the linear form f(T) on the left side and the value on the right side. There are not more than 2 of such forms (|X| + |Z|) + |X| (|X| -1).

Suppose the linear form f(T) and RM R are given. Consider the variants of the occurrence of the form f(T) in $S_\alpha \in R$. S_α can be contained in one or more inequalities of one of the forms (1)-(4), in which there is f(T). Of the two such inequalities (T) <A, f (T)<A', that one for which the right side is larger can be discarded (when A'$\geq$A, the second inequality is needless). In some cases (considering the further transformation of the RM), it is advisable to leave the second inequality in order to discard the first, moving to a less rigid RM. It is worth noting that the set of occurrences of the form f(T) in S_α may turn out to be empty.

4 Discussion

There are various ways to change over to a simpler and yet equivalent RMs.

Thus, if for some $z \in Z$ we have $1_z = L_z$, then (5) can be replaced by two inequalities of the form (4):

$$\begin{cases} T_x - T_y \le bL_z, \\ T_y - T_x \le -bl_z, \end{cases} \quad (7)$$

which facilitates conducting temporary analysis.

Operation of a fictitious extension of the scope of use of the resource.

Suppose that the resources z, z 'are used in BM S at work (x, y). Then Tx-Ty = b1Tz and Tx-Ty = b2Tz. In addition, let z 'be used at work (x', y '): Tx` - Tu` = b3Tz ,. We add the equality Tx, -Ty, = b4Tz, where b4 = b1b3 / b2. We obtain a BM equivalent to S.

The operation of matching the tempo range. Let the resources z, z 'be used in the BM at work (x, y).

It means that $l \le T_z \le L$, $l` \le T_{z`} \le L`$, $T_x - T_y = bT_z$, $T_x - T_y = b`T_{z`}$.

It is reasonable to consider the intersection of tempo ranges. Instead of the inequality $l \le T_z \le L$ while working with models, the following inequality should be taken:

$$\max\left\{l, \frac{b`l}{b}\right\} \le T_z \le \min\left\{L, \frac{b`L}{b}\right\}.$$

TM which is equivalent to the original one is obtained similarly for $T_{z`}$.

Operation of fragmentation of the tempo range.

Suppose that for $S_\alpha \in R$ when some $z \in Z$, there is the strict inequality $l_z < L_z$. Consider $l \in (l_z\ L_z)$ and replace TM S_α by two BMs $S_{\alpha`}$ and $S_{\alpha``}$ that differ from $S_\alpha$ in that the inequality (1) is replaced by $l_z < T_z < l$ and in $S_{\alpha``}$ by $l < T_z < L_z$ under a condition of given z in $S_{\alpha`}$.

The operation of fragmentation enables obtaining RM R', which is the equivalent to R model, for using it later in the branch and boundary analysis. In the limiting case, a change of TM to an infinite set of GNM $s_{\alpha`}$ with a fixed rate of resource consumption is obtained, i.e. with equalities $l_z = L_z$ and inequalities (7).

The following operations make it possible to obtain less rigid RMs.

Generalization operation.

Suppose that PM (6) and the form f are given. The inequality $f(T) \le A$ is assumed to be general for R if it occurs in all Sα. Other inequalities containing f are assumed to be special. We define a new RM R'>R by discarding all the partial inequalities containing f from R. This procedure will be called the operation of generalization. If some S_α does not contain an inequality with f form, then R ' also does not contain any of such inequality. When each S_α contains at least one such inequality, then it is useful to include the inequality $f(T) \le M$ in each S_α, where M is the maximum of right sides in all occurrences of f into R. In this case, all $S_{\alpha`}$ will contain this inequality, wherein they will not contain any other inequalities of f form.

The operation of merging the tempo range.

This operation is a special case of a generalization operation with preliminary maximization of the right sides. It concerns inequalities (1) for a fixed z. Each $S_{\alpha`} \in R$ ' is obtained of S_α by replacing inequality (1) by $lz^* < Tz^* < Lz^*$, where $l_{z*} = \min_\alpha l_z$, $L_{z*} = \max_\alpha L_z$.

Operation of transition to a variable rate.

In TM S, equalities (5) for fixed T_z are replaced by two inequalities (8):

$$\begin{cases} T_x - T_y \le bL_z, \\ T_y - T_x \le -bl_z, \end{cases} \quad (8)$$

If the z resource served only one work (x, y), the new TM S' would be equivalent to S, since the tempo can be arbitrary. However, if there are several works serviced by the resource z, then the inequalities (8) mean that the rate of use of the resource z is set for each work. Thus, we pass to a less rigid TM. The usefulness may consist in that when we make a variable rate for every resource, there will not be any equality of type (5) in the TM, and it will turn into an GNM, which is more convenient in terms of time analysis.

Until now, we have considered the ways to reduce rigidity in order to obtain an "optimistic" assessment of the RM. The reasonability of further branching is also determined by the "pessimistic" assessment. If these two assessments are close, then there is not much sense in further branching indeed. A "pessimistic" assessment can be obtained by moving to a more rigid RM and then by the sequent time analysis of this RM. One of the methods for obtaining a more rigid RM is the resource assignment operation.

Assignment operation of z resource.

We take an arbitrary value from the rate T_z in the TM S within the limits defined by inequality (1), and replace (1) by $c \le T_z \le c$. Now inequalities (3) can be replaced by a pair of inequalities $T_z \le c$ and $-T_z \le -c$.

Assigning all the resources $z \in Z$, we obtain the GNM, for which it is easy to conduct a time analysis and, if feasible plans exist, get a "pessimistic" estimate.

Since each $S_\alpha \in R$, which is considered as RM, is more rigid than R, the rigidifying of resources in S_α will result in obtaining the GNM which is also more rigid than R. This is sufficient for a pessimistic assessment of R.

Consider the peculiarities that arise when different types of resource constraints are taken into account both at the initial stage of determining the initial RM and at the stages of its partitioning into TM subsets.

Assume that resource z is used at several works (or their parts)

$$(x_1, y_1),\quad (x_2, y_2),\quad \dots,\quad (x_n, y_n) \tag{9}$$

and can have any intensity from the segment $[I_0, I_1]$ of the positive semi axis. This means that the resource can be consumed at any rate from the interval [1, L], where $1 = 1\backslash I_1$, $L = 1\backslash I_0$.

However, if the tempo is chosen, then it is considered unchanged within the process of executing the serviced z resources of works.

Assume that $b_1, b_2, \dots, b_n$ are the volumes of resource consumption in works (9). Using the resource at the tempo of $T_z \in [l, L]$ leads to the fact that $b_1T_z, b_2T_z, \dots, b_nT_z$, of time units will be expended on works respectively (9), which is described by inequalities (1) and (5).

In fact, there are no alternatives, if we are satisfied with TM (1)-(5). They will occur only if in the time analysis there will arise a need to get rid of non-networked equalities (5).

With the operation of moving to a variable rate, one can get a less rigid GNM and analyze it. If this estimate is preliminary, the fragmentation of the tempo range should be made by dividing the range [l, L] into two segments, for example $\left[l, \frac{L+l}{2}\right]$ and $\left[\frac{L+l}{2}, L\right]$. The time analysis of each new more rigid TM will undoubtedly be more accurate. The division can be continued until the boundaries [l`, L] become almost indistinguishable.

The case of a discrete set of intensities. Suppose that instead of the range [1, L], a finite set of permissible usage rates of the resource $\Lambda = \{\lambda_1, \dots, \lambda_m\}$, $0 < \lambda_1 < \dots < \lambda_m$ is and any $T_z \in \Lambda$ can be chosen. After such a choice, there will be equality of a network character:

$$T_{xi} - T_{yi} = b_i T_z.$$

As a result, m GNM is obtained for each z. The m number can be found to be too large. In this case, the TM with the inequality (1) of the following form can be investigated similarly to the way it was done in the previous section:

$$\lambda_1 \le T_z \le \lambda_m. \tag{10}$$

By allowing T_z to take continuous values (10) instead of discrete values, we get a linear problem instead of a search problem. In order to obtain a more accurate estimate, the fragmentation operation can be replaced by considering more rigid TMs corresponding to the two tempo intervals $[\lambda_1, \lambda_i]$ and $[\lambda_{i+1}, \lambda_m]$.

Suppose that there is an indivisible z resource that serves a number of works or their parts (x_1, y_1), ..., (x_n, y_n), wherein the order of their execution is not defined in advance. Suppose π is a permutation of numbers 1, ..., n from n. In order to require the execution of work in the order

$$(x_{\pi_1}, y_{\pi_1}),\dots,\ (x_{\pi_n}, y_{\pi_n}), \tag{11}$$

it is enough to set the constraints of transport type:

$$T_{y_{\pi_1}} \le T_{x_{\pi_2}}, T_{y_{\pi_2}} \le T_{x_{\pi_3}}, \dots, T_{y_{\pi_{n-1}}} \le T_{x_{\pi_n}} \tag{12}$$

The number of interchanges grows rapidly with increasing n. Therefore, there is a need for a gradual fragmentation of the set of interchanges during the investigation of the RM for a given n. This can be done indirectly by the sequent including of inequalities (12).

At the first step suppose dividing $R=\{S_\alpha\}$ into two classes R' and R''. We assign all the S_α for which $T_{y1}<T_{x2}$, to R', to the second class we assign all the S_α for which $T_{y2}<T_{x1}$,.

Continuing the branching in the R' class, we add one of the inequalities $T_{y1}<T_{x3}$, or $T_{y3}\le T_{x2}$, to $T_{y1}<T_{x2}$. If $T_{y1}<T_{x3}$ is chosen, in the next step we consider the inequalities $T_{y2}<T_{x3}$ or $T_{y3}<T_{x2}$. After choosing the second of these inequalities, the inequality $T_{y1}<T_{x2}$ becomes needless, so it can be discarded. Obviously, a binary branch can be used to view the entire set of interchanges.

In cases when the downtime of a resource is inadmissible, along with the definition of the order of servicing of works (11), it is worth pointing out that the next work should begin immediately after the end of the previous work. Formally, this means that inequalities (12) should be supplemented by the "opposite" inequalities:

$$T_{x_{\pi_2}} \le T_{y_{\pi_1}}, T_{x_{\pi_3}} \le T_{y_{\pi_2}}, \dots, T_{x_{\pi_n}} \le T_{y_{\pi_{n-1}}} \tag{13}$$

Here, with the gradual assignment of inequalities (12), (13), a more subtle approach is required. Suppose, for example, that $T_{y1} \le T_{x2}$ is chosen at the first step. Then, choosing $T_{x2}\le T_{y1}$ at the second step, we clearly indicate that the second work should be performed immediately after the end of the first work. An alternative to this does not have a simple expression by the formulas (4), that is why one has to use a strict inequality, meaning to insert obligatorily some other work between the first and second work in the future.

One of the approaches consists in firstly fixing the work that will be performed first, then the work that will be the second, etc. The most effective way is determined by the specific features of the problem. For the task of determining the order of work at construction sites with continuous work of the performers, a detailed algorithm and some results of computer calculations are described in [5, 6]. It should be noted that several resources can be used on the same work. There are no obstacles to the theoretical apparatus proposed here. Moreover, the same resource can belong to two types (P, Q), (P = 3.1, 3.2, Q = 3.3, 3.4 - section numbers). This means that there is a need to determine both the rate of resource consumption, and the order of maintenance work. Such a resource can be treated as two different resources. When using the scheme of branches and boundaries, instead of full detailing, first one resource, then another, etc., in a number of problems, it is advisable to alternate a partial assignment to different resources [7].

5 Conclusion

The paper considers ways of transforming the scheduling tasks: decomposition, aggregation, etc. The modern theory of structured programming and the principles of developing software complexes largely use the ideas of Descartes. The same ideas comprise the core of the method of branches and boundaries. This paper considers methods of transformation for approximating the scheduling tasks by linear and network tasks. The present work shows that the branching (decomposition) within the formation of calendar schedules for erecting high-rise buildings and structures should be considered in close conjunction with aggregation which is the reverse operation. Resource-and-time analysis is presented as a flexible process, involving a dialogue, in one respect, with consumers (specialists in the field of organization, technology and production management of unique works with the aim of adjusting resource and network constraints), and, in other respect, with a computer while conducting resource and time analysis of the RM, its parts or their transformations.

References

1. V.V. Gluhov, I.V. Ilin, Lecture Notes in Computer Science, **8638**, 509- 518 (2014) DOi-10.1007/978-3-319-10353-2_46
2. I.V. Ilin, A.I. Levina, O.Yu. Iliashenko, MATEC Web of Conf, **86**, 05028 (2016) DOi - 10.1051/matecconf/20168605028
3. S.A. Barkalov, Nguen Van ZHang, Nguen Than' Zhang, Sistemy upravleniya i informacionnye tekhnologii, **3.1(53)**, 116-119 (2013)
4. V.E. Belousov, Ekonomika i menedzhment sistem upravleniya, **3.2(17)**, 227-232 (2015)
5. V.E. Belousov, Sistemy upravleniya i informacionnye tekhnologii, **4(62)**, 31-33 (2015)
6. R. Bril, O.V. Kalinina, I.V. Ilin, A.S. Dubgorn, O.Yu. Iliashenko, SCM, 692- 693 (2017) DOi - 10.1109/SCM.2017.7970692
7. E. Nezhnikova, Procedia Engineering, **165,** 1300-1304 (2016) doi 10.1016/j.proeng.2016.11.854

Economic efficiency of application of innovative materials and structures in high-rise construction

Roman Golov [1,*], *Varvara* Dikareva[2], *Roman* Gorshkov[2], *Anatoly* Agarkov[3]

[1]Moscow Aviation Institute (National Research University), Volokolamskoe highway, 4, Moscow, 125993, Russia
[2]Moscow State University of Civil Engineering, Yaroslavskoye sh., 26, 129337 Moscow, Russia
[3]Moscow Financial and Economic Institute, Novomoskovskaya street, 15a, Moscow, 129075, Russia

Abstract. The article is devoted to the analysis of technical and economic efficiency of application of tube confined concrete structures in high-rise construction. The study of comparative costs of materials with the use of different supporting columns was carried out. The main design, operational, technological and economic advantages of the tube confined concrete technology were evaluated, conclusions were drawn about the high strength and deformation properties of axial compression of steel tubes filled with high-strength concrete. The efficiency of the tube confined concrete use is substantiated, which depends mainly on the scale factor and percentage of reinforcement affecting its load-bearing capacity.

1 Introduction

The key objective of high-rise construction is to reduce the weight of buildings, their material consumption, reduce the volume of building structures and labor costs. The most suitable system for such buildings is a frame or frame-core structure. In this system, bending moments arising from horizontal loads are accepted primarily by monolithic trunks or stiffening cores. The objectives, that are set by traditional methods and materials, cannot be solved. The questions of creating effective designs of high reliability and minimum weight are posed.

These requirements are met by building structures from tube confined concrete. Tube confined concrete is a complex structure, consisting of a steel shell and a concrete core, working together. Tube confined concrete elements, that perform little flexibility and small eccentricities of longitudinal force application (which is typical for vertical load-bearing elements of high-rise building frames), have an exceptionally high load-bearing capacity at relatively small cross sections, being an example of a successful combination of valuable properties of metal and concrete. This gives significant savings in steel and concrete, leads to a reduction in the cross-sectional dimensions of the elements, their mass and transport

* Corresponding author: roman_golov@rambler.ru

costs, as well as the preservation of all the advantages of metal structures in terms of installation.

This is facilitated by the confinement effect, which creates a steel shell for the concrete core, performing simultaneously the functions of both longitudinal and transverse reinforcement. This effect is especially manifested in columns of circular cross section. The lateral pressure of the pipe prevents the intensive development of microcracks in the fracture in the concrete core, which, under conditions of comprehensive crimping, withstands stresses that greatly exceed the prismatic strength. The steel casing, in turn, is protected from loss of local and general stability due to the favorable effect of the internal pressure of the solid medium.

2 Materials and Methods

The main point in the construction of high-rise buildings, taking into account the use of tube confined concrete elements, is to ensure the joint work of the concrete core and the steel shell under operational loads, which is an essential design drawback of these structures. In terms of the difference in the initial coefficients of transverse deformation of concrete and steel (vb $\approx$ 0.18, vs $\approx$ 0.3), during the gradual loading, the core and the casing work jointly only in the initial loading period. Then, the outer shell tends to tear away from the surface of the concrete, because of the indicated difference in deformation properties and low strength of the bonding of concrete to steel, contributing to the occurrence of radial tensile stresses in it, which leads to a disbonding. At this moment, there is no natural transverse compression of concrete in the tube, and concrete works under uniaxial compression conditions, whereas the tube - as a longitudinal reinforcement. Due to this drawback, the issue of modeling the tube confined concrete element in simple computing systems (Scad, Lira, etc.), which takes into account the properties of the materials in the complex, has not been resolved.

When constructing high-rise buildings with tube confined concrete structures, it is necessary to take into account the structure of butt joints of columns in height and overlap, which requires additional study of the main structural interfaces to other constructions of the building and the development of fundamentally new structural and technological solutions. An industrial and high-performance method is required to fill the tubes with concrete, which ensures high strength and uniformity of the concrete core, when using tube confined concrete structures

Fig. 1. The "Canton Tower" TV Tower with the height of 600 m (Guangzhou). a - the general view of the tower; b - course of construction, c - shell of tube confined concrete elements.

In 2010, one of the world's tallest structures was built in Guangzhou. This is the 600m TV tower "Canton Tower" (Figure 1), which has a unique architectural cone-shape.

The monolithic reinforced concrete trunk is closed by a shell of a triangular lattice consisting of tube confined concrete elements. This design allows wind flows to pass through the structure without creating swirls, which guarantees the stability of the tower during the typhoons [1]. The new constructional system "SWMB", named after the company that developed it, "Skilling Word Magnusson Berkshire Inc." in the 1970s, contributed to the construction of even more than a dozen high-rise buildings in the US that demonstrated the efficiency of tube confined concrete. A special feature of the system is the use of tube confined concrete structures as columns of ultra-high-strength concrete [2].

Specialists also note that the use of tube confined concrete columns with high-strength concrete provides seismic stability of the skyscrapers. Low material intensity, high fire resistance of tube confined concrete, the possibility of combination with various structures of overlapping is noted by French specialists who develop frame-type structural systems of buildings with supporting columns of tube confined concrete.

3 Results

The efficiency of these or other building structures is determined by the consumption of necessary for their production materials, labor efforts and, finally, the cost of these structures. As experience of design construction of structures from tube confined concrete shows, saving of materials, labor and cost is achieved, when using tube confined concrete. For example, in comparison with a metal span structure, the actual cost of a tube confined concrete bridge, which was built by V.A. Rosnovsky, turned out to be 20% less. At the same time, savings of metal amounted to 52%. When replacing reinforced concrete columns with tube confined concrete ones at the Semiluki plant, the expenses were almost halved. In addition, the tube confined concrete pillar weighs only about 2 tons comparing with a design weight of a reinforced concrete column of more than 13 tons. A great reduction in the cost of structures has been achieved. High economic efficiency of tube confined concrete structures has been proved in various works long time ago. An important advantage of tube confined concrete structures in comparison with regular reinforced concrete is that they do not need any embedded parts and formwork, since the casing and the fittings are replaced by a casing tube. Many embedded parts of various purposes and types are used in reinforced concrete columns. Today, there is a huge number of standard sizes of embedded parts, which does not allow to ensure the rhythm of their configuration. According to our research, tube confined concrete has an exceptionally high load-bearing capacity with small cross sections of columns, being an excellent example of the optimal combination of outstanding metal and concrete capabilities (Table 1). At the same time, steel tubes act as a fixed formwork when concreting, providing both longitudinal and transverse reinforcement of concrete. It takes loads in all directions and at any angle. Concrete in the tube confined concrete structures is under conditions of all-round compression and it withstands stress, which substantially exceeds its prismatic strength [3].

Experience in the construction of buildings with tube confined concrete structures showed the advantages of using such structures, which is confirmed by numerous studies in different countries. It is possible to compare the technical and economic efficiency of tube confined concrete structures with steel and reinforced concrete structures using the example of designed industrial and civil structures. The economics and expediency of the tube confined concrete structures of the skelp mill-300 of the Krivoy Rog metallurgical plant is reflected by the data: direct costs for building structures are decreased by 40%, labor effort - by 38%.

Table 1. Comparative costs of materials on supporting columns (1500 tons load)

Column material	Cross-sectional area of the column, m^2	Column diameter, m	Area of metal, m^2	Area of concrete, m^2	Metal consumption,%	Concrete consumption, %
Reinforced concrete	0.405	0.670	0.023	0.382	127	118
Metal	0.059	1.000	0.059	-	304	-
Tube confined concrete	0.321	0.630	0.019	0.302	100	100

The main advantages of tube confined concrete technology:

1. Structural and operational

- High load-bearing capacity of tube confined concrete columns
- Efficiency of the steel casing - tubes instead of fittings
- Increase of strength and durability parameters of concrete in the tube
- Three-axial compression of concrete in the tube
- Reduction of the weight of the structural frame of the building
- Increase of fire resistance of steel frame structures
- High building resistance to seismic effects, explosions, extreme loads and impacts

2. Technological

- Steel tube execution of the role of the primary frame of the building and fixed formwork for concrete
- Durability during the winter season
- High speed of erection of frames out of tube confined concrete, what is 3-4 times superior to that of classic reinforced concrete
- Reduction of welding works is 2-3 times

3. Economic

- Reduction of the consumption of metal for the erection of the building frames is 1,8-2 times
- Reduction of the construction time of cases is 1.5-2 times
- Reduction of the cost of cases by 25-30%

Parameters for metal and reinforced concrete columns were taken from the projects of the constructed facilities. Tube confined concrete columns were calculated from the loads affecting the similar steel and reinforced concrete columns. These costs include installation, painting, capital investments in related construction industries, capital investments in fixed assets, as well as operating costs. Joints are simplified in tube confined concrete columns and metal consumption is reduced. The structures of the cantilevers of the tube confined concrete columns are welded and can be solved in different ways taking into account the architectural and construction requirements. The manufacture of the column is feasible both in the conditions of factories and in the conditions of the construction site. Analysis shows that tube confined concrete elements are more efficient than reinforced concrete and steel structures. When replacing reinforced concrete structures with tube confined concrete, the consumption of concrete and metal is significantly reduced due to embedded parts, and labor costs are almost halved. The weight of the structures decreases. The use of tube confined concrete structures is particularly efficient at high loads in centrally compressed and eccentrically compressed elements with small eccentricities. Moreover, the use of tube confined concrete definitely requires additional study of the main interfaces with other structures [4].

The choice of constructional solutions during the design stage should be carried out in conjunction with the methods of manufacturing and erection adopted in the project, as well as the requirements for the economic operation of structures [5].

Tube confined concrete structures should be designed primarily welded, with a wide application of automatic and semi-automatic welding. Welded joints of tube confined concrete structures should be designed so that they can be carried out by simple means with the obligatory satisfaction of the requirements for quality control in the conditions of construction. The sizes of tube confined concrete structures should be assigned with account of the requirements of standardization, modularity and unification. It is necessary that the number of diameters and wall thicknesses used in one design should be kept to a minimum. Separate structural elements should be enlarged as much as the lifting capacity of the assembly mechanisms, dimensions, as well as the conditions for manufacturing and transportation allow. When designing buildings and structures with load-bearing tube confined concrete structures, clear structural schemes must be adopted to ensure the necessary strength, overall stability, and also the spatial constancy of the building or structure. Strength and stability of tube confined concrete structures should be provided both in operating conditions, and during transportation and installation. The necessary strength, rigidity and stability of tube confined concrete structures, as well as individual elements and connections at all stages of operation and erection are determined by calculation; while the design schemes must meet the accepted constructional ones. Special attention should be paid to the strength, stiffness and durability of the joints of the elements (Fig. 2) when designing the tube confined concrete building structures, ensuring reliable transfer of forces. The constructional solutions of the joints can have different variants. Pairings of tube confined concrete elements are made by means of a figured cut, which is welded along the contour without any additional fasteners.

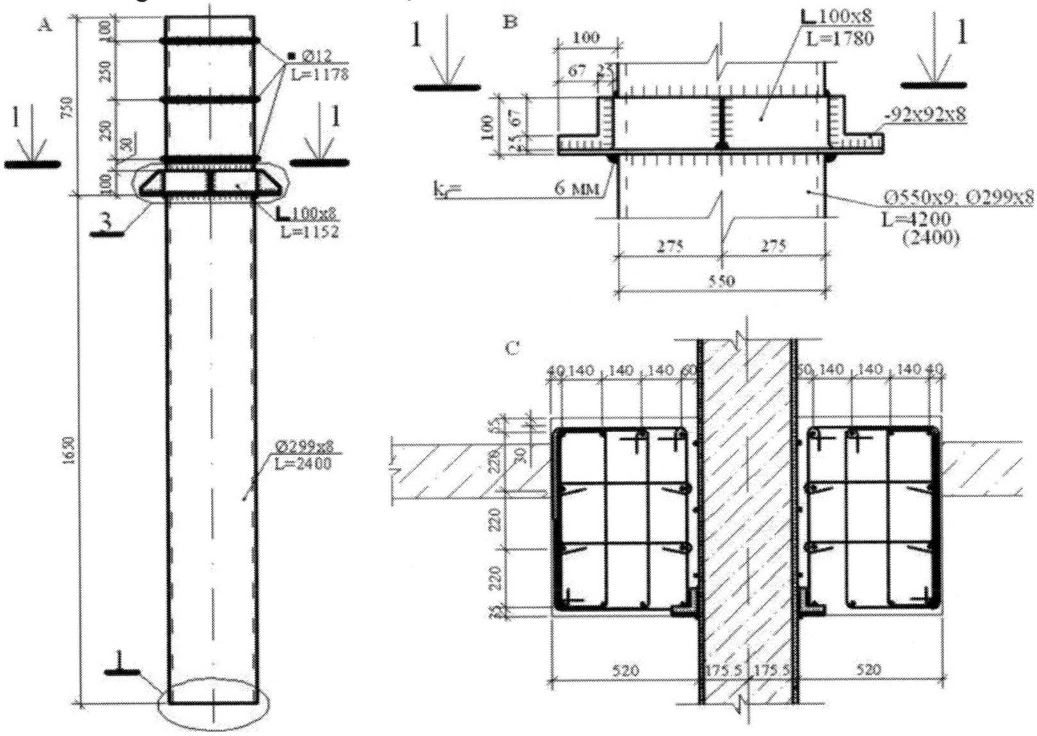

Fig. 2. A - General view of tube confined concrete columns B - Joints in general view C - Cut of tube confined concrete through section

The processing of the tube ends in this case is carried out on automatic universal installations for figured cutting of tubes. When designing tube confined concrete structures, there should be provided measures to protect them from corrosion. They should provide extra thorough coloring or protection with other coatings. Sinuses and troughs should be avoided at the design stage, which hinder moisture or airing. It is necessary that all parts of tube confined concrete structures are accessible for observation, cleaning and painting during operation of the facility [6].

4 Discussions

In the architectural and construction business, tube confined concrete structures have been being used for more than 50 years. In the Soviet Union, Europe, the United States, Japan and other industrialized countries, serious attention was paid to tube confined concrete structures 80 years ago and they were used in industrial, urban multistorey housing and high-rise buildings, multi-trestle road junctions and various special facilities.

The wide application of tube confined building structures in Russia is hindered by the absence of domestic regulatory documents for their calculation and design, not only for seismic regions, but also for non-seismic regions. Although the strength of tube confined concrete has been studied by many researchers for dozens of years. The existing methods of calculation differ significantly from each other. They do or do not take into account the material properties not in a complex, the main features and specificity of the resistance of tube confined concrete to deformation are incompletely reflected depending on the nature of the affecting load. In addition, the prospect of using high-strength concretes in tube confined concrete columns opens up, which will also make it possible to significantly reduce the cross-sectional dimensions of load-bearing structures, and hence the overall construction costs.

5 Conclusion

Tube confined concrete structures create conditions that reduce the phenomenon of progressive collapse due to man-caused and other impacts. As it is known, the destruction of buildings and structures can occur because of a strong earthquake. And, of course, not only from horizontal seismic forces, as it was believed up to now, but also from gravitational forces. This is the weight of the structure and the vertical component of the earthquake, affecting structures that already have seismic damages and horizontal displacements. Therefore, the advantage of tube confined concrete columns is their ability to withstand seismic impacts in the horizontal and vertical directions without being destroyed. Tube confined concrete structures are very reliable in operation. In the limiting state, they do not lose their load-bearing capacity instantly, as reinforced concrete, and for a long time are able to withstand the load, even getting large deformations, the tube confined concrete rod is able to keep carrying a significant load further on.

References

1. V.P. Kamskov,, I.V.Balandina, D.Y.Zemlyanushnov, *Sovremennye tendentsii razvitiya nauki i tekhnologiy*, **9-3**, 95-99 (2015)
2. A.N. Grishina, E.V.Korolev, AIP Conference Proceedings, **20-26** (2016)
3. A.S. Inosemtsev, E.V. Korolev, Stroitelnye materialy, **1-2**, 38-47 (2016)
4. E.V. Tkach, MATEC Web of Conferences, **86**, Moscow (2016)
5. N.Y. Yaskova, D.N. Silka, V.S. Kankhva, I.G. Lukmanova. *Finance and credit in construction.* College textbook for the students. Moscow (2011)
6. A.D. Ishkov, T.N. Magera. Procedia Engineering, **117**, 148-153 (2015)

 https://doi.org/10.1051/e3sconf/20183303034

Algorithm for evaluating the effectiveness of a high-rise development project based on current yield

Elena Soboleva [1,*]

[1] Moscow State University of Civil Engineering, Yaroslavskoye sh. 26, Moscow, 129337, Russia

Abstract. The article is aimed at the issues of operational evaluation of development project efficiency in high-rise construction under the current economic conditions in Russia. The author touches the following issues: problems of implementing development projects, the influence of the operational evaluation quality of high-rise construction projects on general efficiency, assessing the influence of the project's external environment on the effectiveness of project activities under crisis conditions and the quality of project management. The article proposes the algorithm and the methodological approach to the quality management of the developer project efficiency based on operational evaluation of the current yield efficiency. The methodology for calculating the current efficiency of a development project for high-rise construction has been updated.

1 Introduction

To solve the problem of operational evaluation of the current efficiency of the development project for high-rise construction and further management, an algorithm is proposed based on the current yield of a project as a tool for monitoring the implementation of the project. To determine the effectiveness of a development project, it is necessary to carry out a performance analysis, which is defined as the ratio of the achieved result to costs spent to obtain it. The use of performance metrics has allowed to develop a certain efficiency management algorithm [1].

2 Methods

Periodic updating of the economic and financial model of a development project reveals a change in pre-defined indicators, which requires an improvement in the evaluation of current efficiency. To solve the problem of the operational management mechanism for evaluating the effectiveness of the high-rise building project, it is proposed to consider the main implementation factors and efficiency factors.

The first stage involved drawing up a chain of economic and organizational ties for the implementation of the development project. in accordance with a comprehensive

* Corresponding author: alenas_06@mail.ru

analysis of project practices, the tasks of the development project are structured (Table 1) to determine the share in the formation of the factors for the condition of aim achievement.

Table 1. The objectives of a high-rise construction development project

The name of objective of the high-rise construction development project	Designation
Professional support of the project	DPO-1
Professional definition of the project strategy and development of the technical project	DPO -2
Providing control over the development of the project and the schedule for the construction of real estate	DPO -3
Formation of the budget of the development project	DPO -4
Ensuring control over the implementation, optimization and minimization of the budget of the development project	DPO -5
Maximum reduction in construction terms	DPO -6
Ensuring control over maintaining the quality of construction of facilities	DPO -7
Achieving optimal project efficiency	DPO -8
Protecting the interests of the investor to subcontractors and government agencies	DPO -9
Preventing the situation of increasing cost and reduction of quality	DPO -10

The factors were defined that in one respect are based on the main objectives of the development project, and in other respect are the basis for determining the factors of the project's effectiveness. The pyramid of forming the tasks of a high-rise development project is made. An analysis of the conducted survey of developers enabled to express the effectiveness of the tasks of the project.

$$E_{DPT} = f\left[\left(R_{dp}; Q_{dp}; T_{dp}\right) * (\alpha; \beta; \gamma)\right] \Rightarrow \max \qquad (1)$$

where: R_{dp} is obtaining greater results at lower costs,, Q_{dp} is creating a quality product, specifically the construction facility, T_{dp} is obtaining the maximum planned result on time;; E_{DPT} is effectiveness of the tasks of the development project; α, β, γ are coefficients of execution of components at all stages of the life cycle of the development project [2].

At the third stage, the author proposes to determine the factors for the implementation of the development project, which are the driving force behind the effectiveness of the project. The author defines the factors for the implementation of the development project as the fundamental driving characteristics of the organization and functioning of the project, which establish its life cycle and the effectiveness of the results (Table 2).

Table 2. Implementation factors for the high-rise construction development project

N		Factor name	Influence on the result of the project	Impact share, %
1	Fp	Creating an effective team of professionals for project management	Ensuring the successful implementation of all stages of the development project and quality control of the project being implemented.	3
2	Fr	Level of detailed real estate market research	Allows determining the existing supply-demand ratio in different market sectors.	20
3	Ft	Full accounting of cycles when choosing the time of implementation of the development project	Development projects are long-term, therefore it is necessary to take into account the patterns in the development of the real estate market – cyclicality: the beginning of the project should accrue to activity decay, the finishing of the	5

			project - to the peak of activity.	
4	Fl	Qualitative choice of the location of the future development project	The location factor is the key in the formation of the sale price and the speed of implementation.	4
5	Ff	Reasonable choice of the way of financing of the development project	Observe the proportion between own and borrowed funds, where own at the start-up stage should not be less than 10%.	10
6	Fa	Application of the competitive approach at all stages of the development project	Executors for each stage of the development project should be carefully selected on a competitive basis	1
7	Fs	Systematic control over compliance with the implementation deadlines	Affects the project payback and cost over time	7
8	Fq	Availability of a quality management system	Provides the final result of the project and removes reputational risks of the developer	15
9	Fv	Qualitative calculation and control of investment costs and sale values of a production unit	Directly affects the overall objective of the project	25
10	Fp	Timeliness of preparation of the subsequent stages	All stages of the development project should be carefully calculated and planned	10

To determine the effectiveness of the development project, the method of multi-criteria evaluation of alternatives is applied using a procedure of ranking by several indicators. Based on the survey conducted and the experimental data of the research of development projects, the factors are ranked by the value of the necessary impact on the result of the project. There were obtained the following key factors determining the efficiency of the development project: realization price (Fi) and the investment cost (Fc) of the square meter of the sold area of the facility being constructed under the development project [3].

3 Results

The main tools of project performance management are methodologically expressed through the design ratio - the current yield index of the development project. When evaluating the effectiveness of projects, the index of investment yield is used, which is determined as the ratio of the sum of the cash flow elements from operating activities to the absolute value of the sum of the elements of the cash flow from investment activity. Initial planned and further actual data on the project in the process of its implementation are recorded in the economic and financial model of the high-rise development project [4].

The yield index is an indicator of the return of investments, defined relatively to the total indicators of net cash flow and investments for the economic life of their investments. Estimation of the current yield Y_i is the momentary evaluation of the profitability of the development project in a specific period of time through the indicators: $V_{real(i)}$ - the price or market value of the facility under construction, $P_{act(i)}$ - the actual investment prime cost of the facility under construction, $C_{ness(i)}$ - the remaining costs at the current time *(i)* necessary to complete the construction and installation works on the development project facility under construction. A graphic model for assessing the current efficiency of a development project within the management time-frame is proposed (Fig. 1)

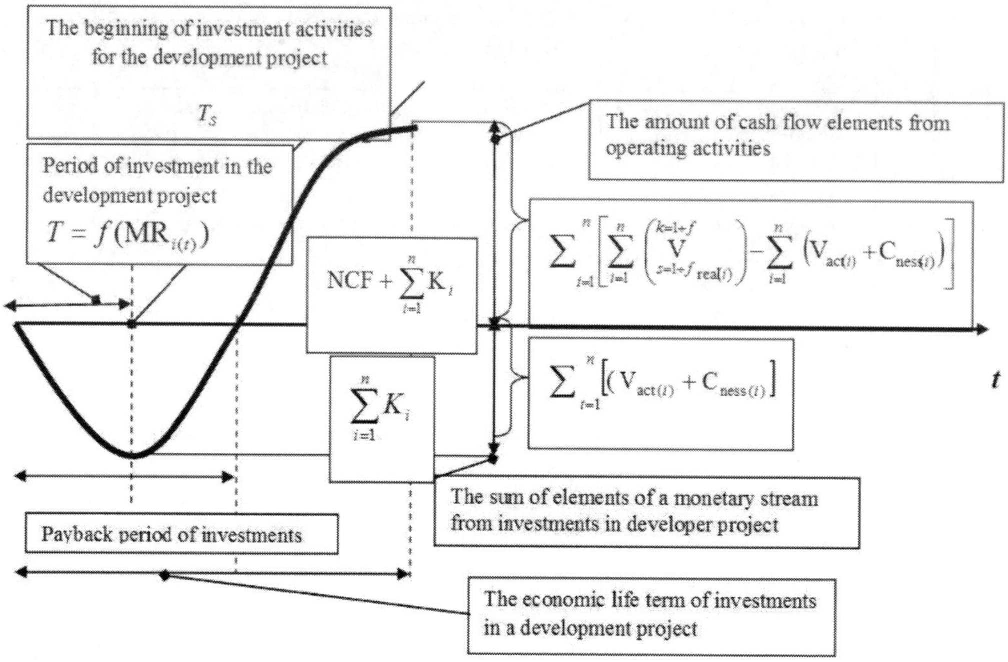

Fig. 1. Graphical model for assessing the current efficiency of a high-rise construction project during the life-cycle management time-frame in a development

The start of the investment activity T_S on the development project depends on the investment period:

$$T = f(\mathrm{YP}_{i(t)}) \tag{2}$$

where: t is the period of implementation of the development project, which includes all stages of the life cycle, taking into account the time-frame of operational management of the development project. Money resources entering the development project:

$$\sum_{i=1}^{n} \mathrm{MR}_{i(t)} = \sum_{i=1}^{n} \mathop{\mathrm{V}}^{ki}_{si}{}_{\mathrm{real}(i)} + \sum_{i=1}^{n} \mathrm{K}_i \tag{3}$$

where: K is the volume of investment of the development project. The author offers the conditions for determining and forming the price for the implementation of the development project facility under construction [4].

$$\mathrm{V}_{\mathrm{real}(f-g(i))} = \sum_{i=1}^{n}\Big(\sum\big(\mathop{\mathrm{V}}^{k1}_{s1}{}_{\mathrm{real}(f(i))} + \mathop{\mathrm{V}}^{k1}_{s2}{}_{\mathrm{real}(j(i))} + \mathop{\mathrm{V}}^{k1}_{s3}{}_{\mathrm{real}(g(i))}\big) + \sum\big(\mathop{\mathrm{V}}^{k2}_{s1}{}_{\mathrm{real}(f(i))} + \mathop{\mathrm{V}}^{k2}_{s2}{}_{\mathrm{real}(j(i))} + \mathop{\mathrm{V}}^{k2}_{s3}{}_{\mathrm{real}(g(i))}\big) + \ldots \sum\big(\mathop{\mathrm{V}}^{kf}_{s1}{}_{\mathrm{real}(f(i))} + \mathop{\mathrm{V}}^{kf}_{s2}{}_{\mathrm{real}(j(i))} + \mathop{\mathrm{V}}^{kf}_{s3}{}_{\mathrm{real}(g(i))}\big) \tag{4}$$

where: **k** is the object class. Several classes of real estate objects are possible from 1 to f (economy, economy-plus, average, business); **s** is the object status; **sf** is the number of statuses in accordance with the number of classes of real estate objects; **s1** is the status of the object at the state of readiness of the real estate development project facility 35-60%; **s2** is the status of the object at the state of readiness 61-80%; **s3** is the status of the object at the state of readiness 81-100%; $\mathbf{V}_{\mathbf{real(f(i))}}$ is the implementation value of the real estate development project facility under construction at the state of readiness 35-60%; $\mathbf{V}_{\mathbf{real(j(i))}}$ is the implementation value of the real estate development project facility under construction at the state of readiness 61-80%; $\mathbf{V}_{\mathbf{real(g(i))}}$ is the implementation value of the real estate development project facility under construction at the stage of readiness 81-100% [5].

The solution of the problem of an operational evaluation of the high-rise development project current efficiency involves using a generalized indicator, the investment cost. It designates a unit of production - a square meter of the sold area of the real estate object, which represents the total costs expended at all stages of the life cycle of the development project. The author proposes the method of forming and controlling the change in the investment cost of the real estate object under construction within the framework of the project.

$$P_{invest(i)} = \frac{P_{act(i)} + C_{ness(i)}}{\sum_{i=1}^{n} \sum_{si}^{ki} S_{real(i)}} =$$

$$= \frac{\sum_{i=1}^{n} ((P^{act}_{con\&inst(i)} + C^{ness}_{con\&inst(i)}) + (P^{act}_{techcust(i)} + C^{ness}_{techcust(i)}) + (P^{act}_{overh(i)} + C^{ness}_{overh(i)}) + (P^{act}_{real(i)} + C^{ness}_{real(i)}) + + (P^{act}_{design(i)} + C^{ness}_{design(i)}) + (P^{act}_{dev(i)} + C^{ness}_{dev(i0}) + (P^{act}_{loan(i)} + C^{ness}_{loan(i)}))}{\sum_{i=1}^{n} \sum_{si}^{ki} S_{real(i)}} \quad (5)$$

where: S_{real} is area for the sale of real estate buildings under construction; $C_{con\&inst}$ is costs for construction and installation work; $C_{techcust}$ is direct costs of the construction technical customer; C_{overh} - overhead costs of the construction technical customer; C_{sale} - costs for the sale of the real estate buildings under construction including advertising and marketing; C_{design} the costs of designing real estate buildings under construction; C_{dev} - developer management costs; C_{loan} - costs for the loan debt of the project. Then the index of current profitability is determined by this formula:

$$Y_i = \frac{V_i - (P_{act(i)} + C_{ness(i)})}{(P_{act(i)} + C_{ness(i)})} = \frac{V_i}{(P_{act(i)} + C_{ness(i)})} - 1 \quad (6)$$

The dynamics of the current yield index and the interrelation of the parameters $P_{act(i)}$, $C_{ness(i)}$, $V_{real(i)}$ are analyzed and presented graphically by the example of a development project of the complex development of a territory with the construction of residential real estate in area 200 000 m^2, commercial premises and social infrastructure [6].

4 Discussions

Thus, the development project of high-rise construction is considered as sustainable and effective within the implementation process assuming that the inequality is fulfilled: $Y_i > Y^*_{(i)}$. Therefore, the index of the current yield of the development project should be above zero. In case of inequality: $Y_i < Y^*_{(i)}$ at the state of the real estate object construction readiness 80-100%, a program is being developed to stabilize the development project through the project's managed parameters: the sale price and residual project costs. In case of a stable situation, it is recommended to conduct similar temporary monitoring of the development project before the end of its final stage. The control calculation with unstable financing of the development project should be performed monthly, with stable financing and execution of the terms - quarterly.

5 Conclusion

The management of high-rise development project efficiency involves a set of management processes: planning, execution arranging, monitoring and analysis. These processes strategically determining the mission enable the developer to assess and manage the project implementation activities fast in order to achieve the goals under appropriate usage of available resources. The management system is based on the principles of cost management.

References

1. E. Nezhnikova R. Obukhov MATEC Web of Conferences **106** 08025 (2017) doi: 10.1051/matecconf/201710608025
2. I.V. Ilin, V.I. Koposov, A.I. Levina, Model of asset portfolio improvement in structured investment products, Life Science Journal, **11**, pp 265-269 (2014)
3. K. Olga, An innovative approach to the formation of a progressive taxation probabilistic model on personal incomes, Intern. Journal of Economics and Financial Issues, Vol **6**, IS **3**, pp 995- 1002 (2016)
4. A.K. Orlov, Arrangement and planning of developer's activities in the course of construction megaprojects implementation MATEC Web of Conferences **106**, (2016)
5. V. Kankhva, Procedia Engineering **165** 1300-1304 (2016), doi:10.1016/j.proeng.2016.11.855
6. E. Nezhnikova IOP Conf. Series: Earth and Environmental Science **90** 012161 (2017) doi:10.1088/1755-1315/90/1/012161

Methodology of project management at implementation of projects of high-rise construction

Oksana Papelniuk [1,*]

[1] Moscow State University of Civil Engineering, Yaroslavskoye shosse 26, Moscow, 129337, Russia

Abstract. High-rise construction is the perspective direction in urban development. An opportunity to arrange on rather small land plot a huge number of the living and commercial space makes high-rise construction very attractive for developers. However investment projects of high-rise buildings' construction are very expensive and complex that sets a task of effective management of such projects for the company builder. The best tool in this area today is the methodology of project management, which becomes a key factor of efficiency.

1 Introduction

Today it is impossible to underestimate influence of high-rise construction on formation of a modern city landscape and also importance of the innovative technologies which are applied at their construction. The conventional criteria in the world according to which the building can carry the proud name "skyscraper" have been established by Council for high-rise buildings and the urban environment. According to classification of the organization, a skyscraper — the building not less than 150 meters high, and multi-storey buildings higher than 300 meters it is accepted to carry to ultrahigh.

By the end of 2015 on the planet there was the 3214th skyscraper, from them 56 ultrahigh, construction of 300 more similar buildings among which 90 will rise by height more than 300 meters was also conducted.

However implementation of similar investment and construction projects is very expensive and difficult, hiding in itself a lot of calls for construction company. Construction even of a usual, not ultrahigh skyscraper costs the company builder in the sum not less than 80 million dollars. And in case of construction of champions on height can reach 700-800 million dollars. Expensive researches begin even prior to construction — development, careful check and the approval of the project of future building, part in which is taken only by highly skilled engineers and architects, studying of the place of alleged construction taking into account "wind loadings", features of soil, a seismological situation in the area demand considerable expenses. Then even more essential expenses on laying of the strongest base, operation and the maintenance of the construction equipment and also compensation to thousands of workers begin.

* Corresponding author: oksana-17@mail.ru

 https://doi.org/10.1051/e3sconf/20183303035

The complexity and cost intensity of such projects sets a task of use of the most effective methods of management therefore use of tools of project management gains ground at implementation of similar projects for the company builder. Today the methodology of project management is universal, and can be successfully used in various spheres and branches, including high-rise construction.

2 Materials and Methods

In methodology of project management the project is considered as a system of the purposes and mechanisms of realization in the form of technological, technical and organizational documentation, regulation processes by material, technical, labor and other resources for achievement of goals.

The concept of project management is mainly described by means of the following main structural and functional components:

- context of project management;
- processes of project management;
- main functions of project management;
- methods of project management;
- practice and effective management of the project.

Disclosure of a being of these components of structure can be tracked by the analysis of structure of knowledge and components of project management which are submitted in figure 1.

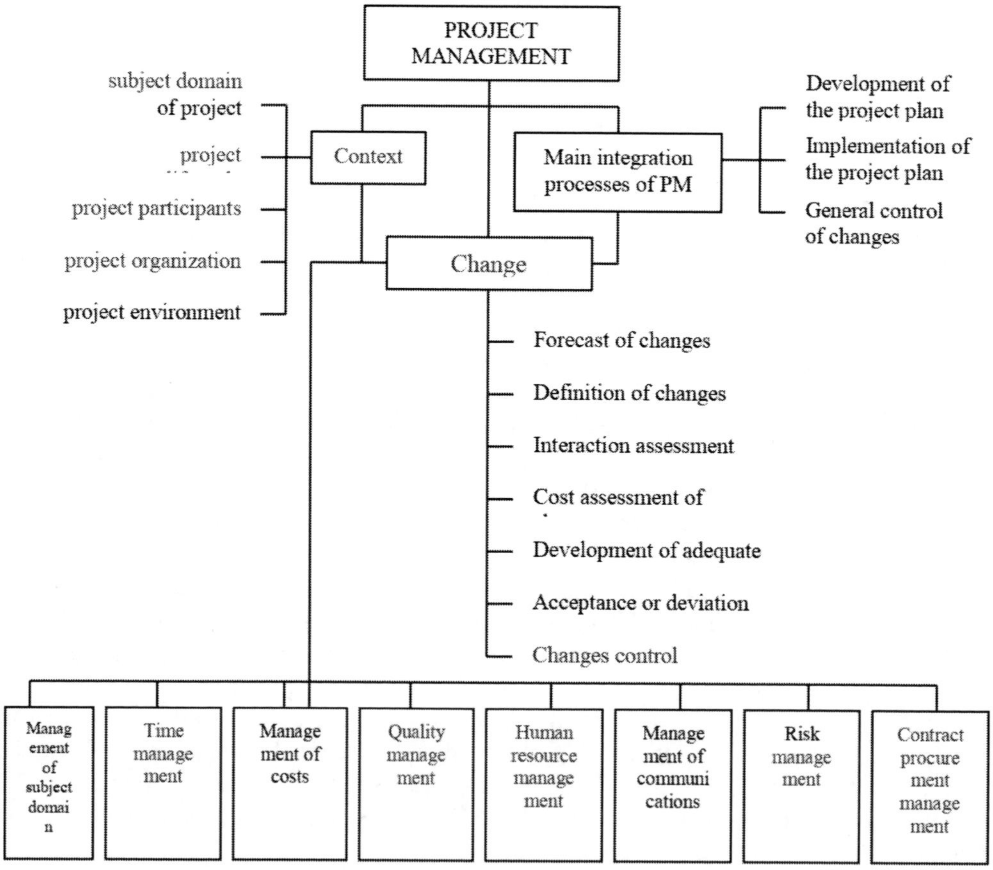

Fig. 1. Structure of knowledge and components of project management

3 Results

The structure of project management given above is universal and can be used at implementation of projects of high-rise construction in which development and realization many participants of the project take part. Their structure is presented on figure 2 . But all of them, naturally, can't be presented in each specific project. Very often one participant can assume several functions of certain participants which are given figure 2.

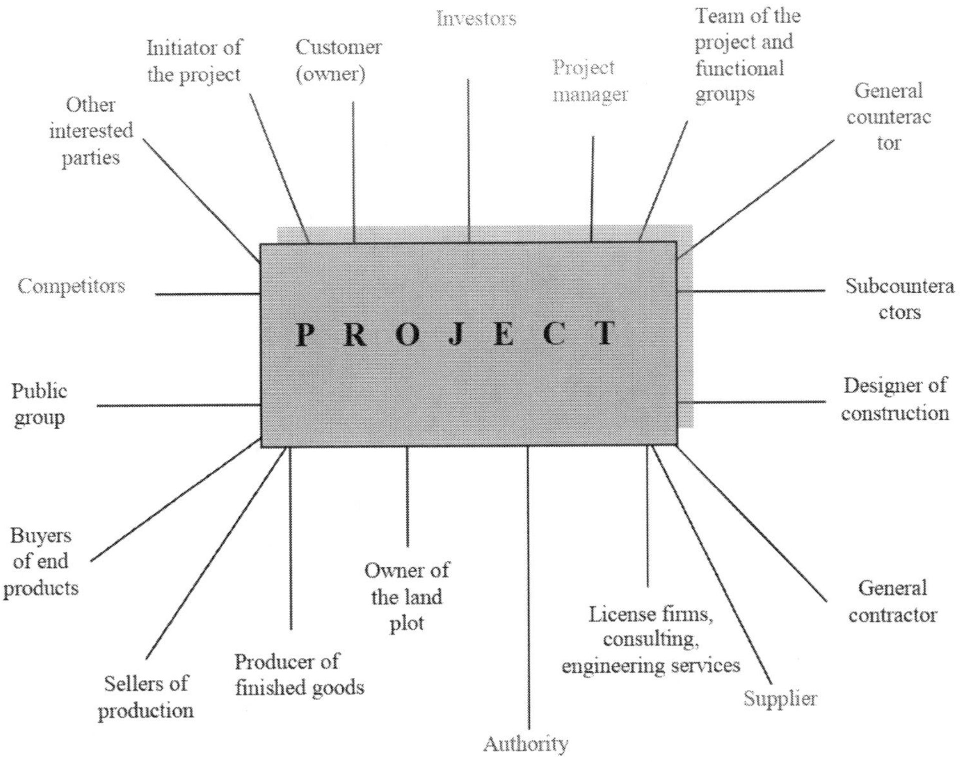

Fig. 2. Schematic diagram of participants of the project

In case of development of the system of project management its surrounding is considered. It is a set of the external and internal factors influencing project implementation [1, 2]. In a general view factors and other internal variables of any organization (subjects of managing) can be traced according to figure 3 (figure 3).

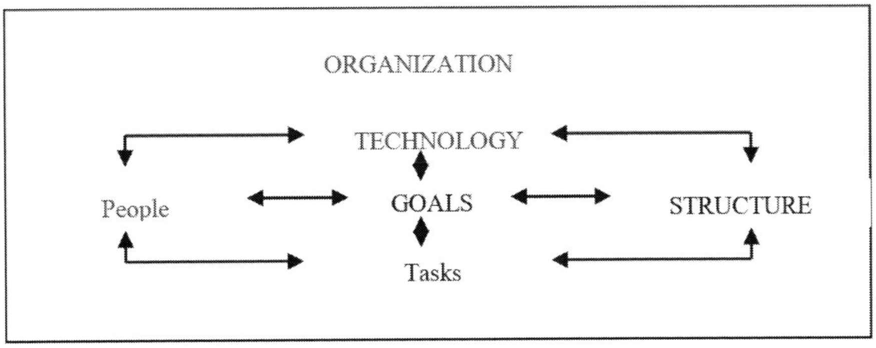

Fig. 3. Interrelation of internal variables of the organization

The purpose is an end result. The object can be set by the management of the organization, as well as other factors of the internal environment of the organization realizing the project. All other components of the internal environment as a taxonomy to requirements of the purpose and the tasks following from this [3, 4].

4 Discussions

There is a question whether it is necessary to recognize the provided structure of the project management system of high-rise construction accepted for creation. According to some experts it is necessary to consider also such variable as financial resources [5]. And experts divide structure of factors of an external environment for the organization and a project management system into two components: direct and indirect influence. Composition of these two units, it is possible to see in figure 4 [6, 7].

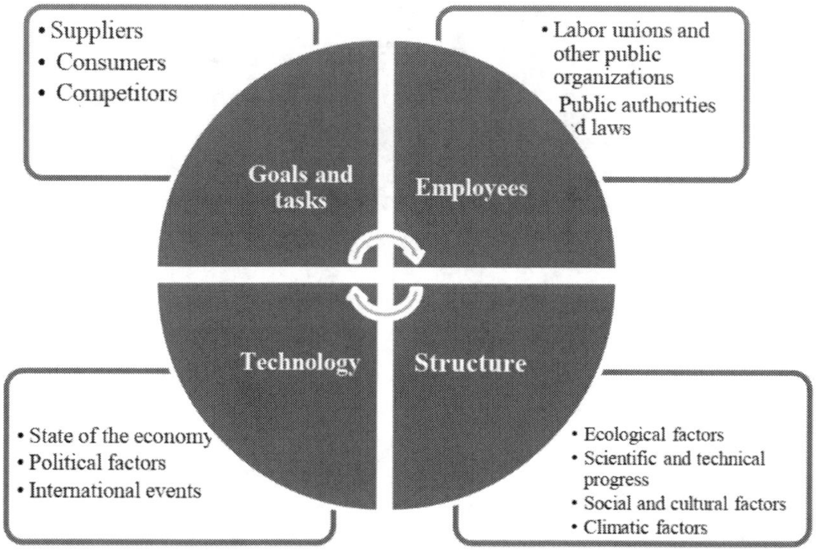

Fig. 4. Factors of the internal and external environment

5 Conclusion

Thus, high-rise construction is the direction of construction in which use of new technologies, introductions of progressive forms of the organization of production, use of modern materials is most intensively carried out. All this can't be rather effective without use of methodology management of investment and construction projects. The prevalence of investment and construction projects around the world imposes increased requirements to knowledge of management of them.

Construction is not only the first branch in which began to use provisions of the theory of project management, but also subject to introduction of new techniques and instruments of project management. Now project management is a key factor of efficiency and at implementation of projects of high-rise construction.

References

1. I.G. Lukmanova, A.G. Korolyov, E.V. Nezhnikova, *Project Management,* MGSU, (2013)
2. E. Nezhnikova. Procedia Engineering **165** 1300-1304 (2016) doi 10.1016/j.proeng.2016.11.854
3. N. Safronova, E. Nezhnikova , A. Kolhidov MATEC Web of Conferences **106** 08024 (2017) doi. 10.1051/matecconf/201710608024
4. E. Nezhnikova, R. Obukhov MATEC Web of Conferences **106** 08025 (2017) doi: 10.1051/matecconf/201710608025
5. A. Larionov, E. Nezhnikova ARPN Journal of Engineering and Applied Sciences **3** 2023-2029 (2016)
6. A. Larionov, E. Nezhnikova International Journal of Applied Engineering Research **6** 4433-4439 (2016)
7. A. Jones, G. Fallon, R. Golov, European Business Review **12 (4)** doi:10.1108/09555340010336871

Formation of costs of high-rise objects of housing and civil purpose based on enlarged norms

Anna Vorotyntseva [1,*] *Andrei* Ovsiannikov[1], and *Vladimir* Bolgov[1]

[1] Voronezh State Technical University, Moscow Avenue, 14, Voronezh, 394026, Russia

Abstract. When determining the cost of capital construction objects, for purposes of pre-design workings out and purposes of initial maximum initial price determination on tenders, construction price norms are used (CPNs). Modern CPNs are not designed to determine the value of high-rise buildings. It is necessary to adapt modern CPNs to get opportunity for the possibility to take into account special cost factors in determining the cost of high-rise buildings. The main ways can be: selection of new representative objects or application of additional correction factors.

1 Introduction

International experience of building modern cities shows that, because of high land cost and high density of construction the most justified from an economic point of view are buildings ranging in height between 30 to 50 floors. That is why high-rise building construction is a priority in modern urban environment. Implementation of high-rise construction in large cities is dictated by a high deficit of the territory for construction, including a deficit of office and hotel areas, which it is more reasonable to place in high-rise buildings. In Russia, high-rise construction has not yet become as widespread as in the west, but nevertheless is developing actively. This is confirmed by a number of completed projects in Russian cities [1, 2].

Construction of high-rise buildings for housing and civil purposes is a huge multi-stage process, in which, at the planning stage of future work, it is often necessary to know how much the construction of the planned project will cost. At present, the task of developing and introducing aggregated cost indicators of construction of buildings and structures to the construction industry is more acute than ever. The need to use such indicators is dictated by emerging economic issues at the stages of developing an investment project. The most convenient option in this situation is the compilation of an estimate for the aggregated standards for the price of constructive solutions (SPCS) and construction price norms (CPNs), which allows the maximum construction cost to be determined as accurately as possible without making a detailed estimate [3]

What is the difference between SPCS and CPNs, and on what stages are they necessary?

* Corresponding author: allslavin@mail.ru

2 Materials and Methods

Standards for the price of constructive solutions (SPCS) – show the amount of funds to build separate construction elements of the object calculated per unit of measurement. SPCS are used at the design stage as a rule.

CPNs at its turn – show the amount of funds necessary and sufficient to build capital building object, calculated per unit of measurement. CPNs are used in the planning, investment and design stages. In this article we will consider the purpose and technology of compiling budget estimates with the help of CPNs for building construction as a whole. Also we will find out why is necessary to use it.

CPNs compilations are used for:

- investment planning for capital investments;
- assessment of the effectiveness of the use of funds directed to capital investments;
- preparation of technical and economical significates in the designing task;
- investment project design
- justification of the cost of construction;
- minimization of subjective indicators in the valuation of construction objects.

CPNs were developed and approved on the basis of design estimates for facilities were developed and approved on the basis of design estimates for facilities designed and built in the Russian Federation in the period from 1995-2009. Primary information was collected on a large number of objects of various functional purposes, and then a thorough analysis of design estimates for these facilities was carried out. As a result, the basis of CPNs development is the data that had passed the expertise and met the town-planning and space-planning requirements, imposed on modern construction complexes and facilities. It resulted in adoption of the order of Ministry of Regional Development of Russian No. 187 at 22.04.2011 «About approval of enlarged standards for the price of construction of various types of capital building projects for non-production purposes and engineering infrastructure" with prices for January 1, 2011». Later these norms documents were updated and supplemented. Next norms were approved by the order of the Ministry of the Region of Russian Federation No. 643 at 30 December of 2011. They contain prices for January 1, 2012. CPNs - 2014, approved by the order of Ministry of Construction Housing and Communal Services of Russian Federation at 28.08.2014 No.506. Currently, there are norms approved by Ministry of Construction at 13 June 2017 No. 867. At present, the computation of the estimate for CPNs is made in prices for 2017. It should also be borne in mind that all prices in the enlarged norms are taken for the base (Moscow) region, respectively, for calculations on other territories, it is necessary to apply the transition coefficient from the prices of the base region to the price of the subjects of the RF [4, 5].

Depending on the functional purpose all CPNs are combined in collections:

CPNs 81-02-01-2017 «Apartment buildings»
CPNs 81-02-02-2017 «Administrative buildings»
CPNs 81-02-03-2017 «Objects of public education »
CPNs 81-02-04-2017 «Health facilities»
CPNs 81-02-05-2017 «Sports buildings and facilities»
CPNs 81-02-06-2017 «Objects of culture»
CPNs 81-02-07-2017 «Railways»
CPNs 81-02-08-2017 «Highways»
CPNs 81-02-09-2017 «Bridges and overpasses»
CPNs 81-02-10-2017 «Metro objects»
CPNs 81-02-11-2017 «Outdoor communication networks»
CPNs 81-02-12-2017 «Outdoor electrical networks»

CPNs 81-02-13-2017 «Outdoor thermal networks»
CPNs 81-02-14-2017 «Water supply and sewerage networks»
CPNs 81-02-15-2017 «Gas supply networks»
CPNs 81-02-16-2017 «Small architectural forms»
CPNs 81-02-17-2017 «Planting of greenery»
CPNs 81-02-18-2017 «Civil aviation objects»
CPNs 81-02-19-2017 «Urban Infrastructure Buildings and Structures»
CPNs 81-02-20-2017 «Objects of sea and river transport»
CPNs 81-02-21-2017 «Energy facilities (except for linear ones)».

3 Results

The use of compilations and the compilation of calculations for NDCs is regulated by the MDS 81-02-12-2011 "Methodological recommendations on the application of state estimate standards - enlarged standards for the price of construction of various types of capital construction facilities for non-production purposes and engineering infrastructure" approved by the order of the Ministry of Regional Development of the Russian Federation of October 4, 2011 No. 481.

When using CPNs it is necessary to keep in mind what costs it includes [6]:

- costs for the construction of capital building projects that meet the urban planning and space planning requirements for modern facilities, as well as the costs of construction of individual buildings and structures designed using standard (reusable) design solutions;
- costs provided by the current normative documents in the field of pricing for the performance of work in the construction of the facility in normal (standard) conditions, not complicated by external factors;
- costs for the purchase of building materials and equipment, costs for labor remuneration for workers and maintenance of construction machinery (mechanisms), overheads and estimated profits, costs for the construction of temporary buildings and structures, additional costs for the production of work in winter, costs connected with receipt by the customer and project organization of initial data and specifications for design, carrying out the necessary approvals for project solutions, expenses for insurance (including construction risks); costs for design and survey work and project expertise, customer service content and construction control, reserve funds for unforeseen work and costs.

It is necessary to keep in mind that cost parameters of CPNs are not taken into account, and, if necessary, should be taken into account separately:

- works and costs associated with the allocation of land for construction, travel expenses of workers, transportation of workers, construction and maintenance costs of shift camps, land tax payments during construction, fees for connection to external engineering networks;
- additional costs, arising under special construction conditions (in remote settlements from the existing infrastructure (additional transportation costs), cramped working conditions), which should be taken into account additionally. Special conditions for the construction of the facility are taken into account by the coefficients provided in the technical parts of the collections of the CPNs. Additional transportation costs are accounted with the use of zonal coefficients for changing the cost of construction in context of the

subject of Russian Federation specified in Appendix No. 2 to this Methodological Recommendations.

To calculate the cost of construction of high-rise buildings for civil purposes, it is necessary to take into account structural features of buildings. High-rise buildings have different purposes: hotels, offices, residential buildings. Most often high-rise building is multifunctional, in addition to premises of main purpose it also has parking lots, shops, offices and cinemas. High-rise building also has its own structural features of the main bearing elements, which depend on the specific weight of the cost of structures and types of work.

How can we calculate the cost of planned for construction object, using CPNs? Recommended for calculations formula is contained in MDS :

$$C_{fi} = \left[\left(\sum_{i=1}^{n} CPN_i \cdot P \cdot K_C \cdot K_{tr} \cdot K_{reg} \cdot K_{zone}\right) + Z_p\right] \cdot I_{fi} + VAT \quad (1)$$

where:

CPN_i – used index of the state estimate norm - the enlarged norm of the construction price for a specific facility for the base district (Moscow region) at the price level at the beginning of the current year;

N – total number of used indexes of the state estimate norms - an aggregate construction price norm for a specific facility for the base district (Moscow region) in the price level at the beginning of the current year;

P – capacity of the object planned for construction (total area, number of places, length, etc.);

I_{fi}– forecast index, it is determined basing on indexes of manufacturers prices by types of economic activity in line «Capital Investments» which are used to forecast social and economic development of Russian Federation;

K_{tr}– coefficient of transition from the prices of the base district (Moscow region) to level of prices of subjects of Russian Federation, it is used to calculate planned cost of construction of facilities which are financed by federal budget funds, also it is determined basing on state estimated norms – Construction Price Norms. The value of this transition coefficients is set annually by the orders of Ministry of Regional Development of Russia;

K_{reg}– coefficient, which considers climatic conditions of region of construction (differences in constructive decisions) in regions of Russian Federation comparing to basic region;

K_C– coefficient, which characterizes rise in cost of construction in seismic regions of Russian Federation;

K_{zone}– zoning coefficient, which considers difference in resources cost in region;

Z_p – additional costs, which are considered as separate calculation, in order, which is conditioned by construction production cost determination method on Russian Federation's territory MDS 81-35.2004, , approved by the Resolution State Committee of Russian Federation on 5 March 2004;

Z_p – additional costs, which are considered as separate calculation, in order, which is conditioned by construction production cost determination method on Russian Federation's territory MDS 81-35.2004, , approved by the Resolution State Committee of Russian Federation on 5 March 2004, 15/1;

VAT – value-added tax, currently 18%.

Existing approach does not allow to calculate the cost of high-rise buildings of civil purpose, due to the lack of developed standards for construction prices.

There are different types of high-rise residential multi-apartment buildings:

- Residential multi-apartment high-rise building (more than 16 floors) bricked, mono carcassed;

- Residential multi-apartment high-rise building (more than 16 floors) paneled [7].

For object of other types there is no calculation of altitude. It is necessary to develop CPNs for most common types of objects:

- residential houses;
- administrative buildings;
- trade objects – shopping centers;
- combined objects of different functional orientation.

It is necessary to develop CPNs in one of directions:

1. Definition of representative-objects from high-rise ones.
2. To use coefficient method of cost accounting.

In case of using first method it will not have any differences from existing ones. It is only necessary to demarcate the gradation of high-rise objects correctly – to combine representative-objects in different high-rise groups and update current CPNs.

To calculate the altitude of objects with second method following approach is proposed [8, 9, 10].

The cost of base object of construction is calculated with next formula:

$$C_{obj} = (\sum_{i=1}^{n} CPN_i \cdot S_i) \cdot K^{alt} \quad (2)$$

where:

CPN_i - used index of the state estimate norm - the enlarged norm of the construction price for every type of object (Residential buildings, Health facilities, Sports buildings and structures, Cultural objects), included in the object;

S_i – specific weight of the i-th type of object in total volume of construction of the facility;

K^{alt} – coefficient of «altitude» of object, which considers the ratio of the building area and the total area of the building. This coefficient is important to This coefficient is necessary to exclude the individual "duplicated" types of work (foundations) considered in each i-th kind of object.

5 Conclusion

In conclusion it will not be wrong to say that calculations made with CPNs collections are very important on planning stage, design stage and investing of high-rise buildings of civil purpose stage.

When considering issues of construction residential buildings, health and cultural facilities, external engineering networks, roads, etc. customer can estimate how much it will cost to erect a particular building and structure, by using the collections of CPNs, which is an actual task in the conditions of limited budget funds.

References

1. Mir M. Ali. Evolution of Concrete Skyscrapers: from Ingalls to Jin mao Electronic Journal of Structural Engineering. **1(1)** (2001)
2. J. Orgill-Meyer, M. Jeuland, J. Nathan Cutler., Ecological Economics,Volume **146**, Pages 250-264, (2018)
3. A. Vorotynseva A. Ovsiannikov V. Bolgov MATEC Web of Conferences, **106** (2017)

4. O.V. Didkovskaya, O.A. Mamayeva, M.V. Ilyina. Procedia Engineering, **153**, pp 131-135. (2016)
5. T. Ahmad, A. Aibinu, M. Jamaluddin Thaheem. Procedia Engineering, **180**, pp 1695-1704. (2017)
6. V. Zahirovich-Herbert, K. M. Gibler. Journal of Housing Economics, **26**, P.1-18 (2014)
7. S. Uvarova, S. Belyaeva, L. Myshovskaya, IOP Conference Series: Earth and Environmental Science. **90(1)**, 012169. 2017
8. S. Belyaeva, D. Voronov, S. Erypalov. MATEC Web of Conferences. **106**, 08033. – (2017)
9. N. Safronova, E. Nezhnikova, A. Kolhidov MATEC Web of Conferences **106** 08024 (2017) doi: 10.1051/matecconf/201710608024
10. V. Kankhva, Procedia Engineering **165** 1300-1304 (2016) doi:10.1016/j.proeng.2016.11.855

 https://doi.org/10.1051/e3sconf/20183303037

Justification of directions of technological and price audit systems changes for the purpose of high-rise construction innovating

Yana Rogacheva [1,*], *Andrey* Panenkov[1], *Zinaida* Petrikova[1] and *Ekaterina* Nezhnikova[2]

[1]Voronezh State Technical University, Moscow Avenue, 14, Voronezh, 394026, Russia
[2]Moscow State University of Civil Engineering, Yaroslavskoe sh. 26, Moscow, 129337, Russia

Abstract. Improving the quality of high-rise buildings under modern conditions should be based not only on compliance with the norms of technical regulations, but also on ensuring energy efficiency, environmental friendliness, and intellectuality, which can be achieved only through the introduction of innovations at all stages of the life cycle of the investment project. Authors of this article justified the need for a mechanism of technological and price audit of projects. They also suggested the model of life cycle of organizational and economic changes, connected with implantation of the mechanism of projects audit. They showed innovation character of ecological high-rise construction for the whole life cycle. Authors also made proposals to change the audit system for high-rise construction projects in the focus of its environmental friendliness.

1 Introduction

The analysis of statistical data, as well as the results of expert studies on this issue [1] indicate a rather low innovation activity of enterprises in the construction industry, not only in Russia, but also in the world. Nowadays, a technological and price (project) audit serves as the instrument that promotes the introduction of innovations in the construction of high-rise buildings, the purpose of which is to innovate real estate objects according to the criteria of innovation, energy efficiency, environmental safety, minimizing anthropogenic impact on the environment, and economy. Thus, the key aspect in the projection of innovative construction is the quality of construction products, and according to the opinion of leading scientists, this is the quality of both final and intermediate products. It should be noted that the life cycle of construction products is quite long, which leads to the conclusion that it is necessary to take into account the quality of exploitation of the constructed real estate object [2]. Moreover, in our opinion, due to its duration, it is the operational phase of the life cycle that has an impact on the welfare of future generations. Therefore, today the problem of energy intensity and energy efficiency, ecological compatibility of construction products is sharply raised. The consolidation of all these requirements is achieved when implementing the concept of "green" construction.

* Corresponding author: jarogacheva@mail.ru

2 Materials and Methods

In the process of analyzing the structure of commissioning high-rise apartment houses based on wall materials and the number of rooms, we can assume the development of two trends: first, the development of the trend "the formation of a high-quality consumer market" [3], which is confirmed by the structure of housing commissioning on wall materials (the predominance of monolithic and brick structures) and by the number of apartments (stable dynamics of three-room apartments and more); secondly, the formation of a trend of saving current costs, which is confirmed by stabilization and a slight decrease in the average size of apartments, as well as a slight decrease in the share of entering apartments with a large number of rooms. The second trend is also confirmed by the analysis of public spending on housing and utilities and fuel.

In the conditions of forming trends in the development of the real estate market, therefore, aspects of improving the quality of real estate and reducing current operating costs are becoming topical, which is in the focus of the problem of "green" construction. All researchers of this problem note their innovativeness as one of the key properties of the objects of "green" construction [4]. In "green" construction, all types of innovations are used. In addition, it is very important to "pass through" the introduction of innovations at each stage of the life cycle of the "green" construction.

Considering the certification as the main criterion for classifying the building as "green", it should be noted that at each stage of the investment and construction project, as well as the life cycle of the building, an environmental assessment is carried out (Figure 1) - from the section of the EIA of projects documentation and environmental expertise in the examination of design estimates to assess compliance with environmental requirements in accordance with GOST R 54964-2012 at each stage of the life cycle of the property and its certification for one of the "green" standards. It should also be noted that if the certification of the facility as a "green" at the time of its commissioning for is a mandatory, then at the stage of exploitation and, especially, the recycling of such a system of confirmation of the "green" status is not a mandatory, which, in our opinion, can provoke a change in the "green" status of the object during its physical life, especially after major overhaul or reconstruction. It is also necessary to comply with the principles and criteria of "greening" at the stage of liquidation of the facility, which is one of the directions for improving the system of standardization and certification of "green" construction.

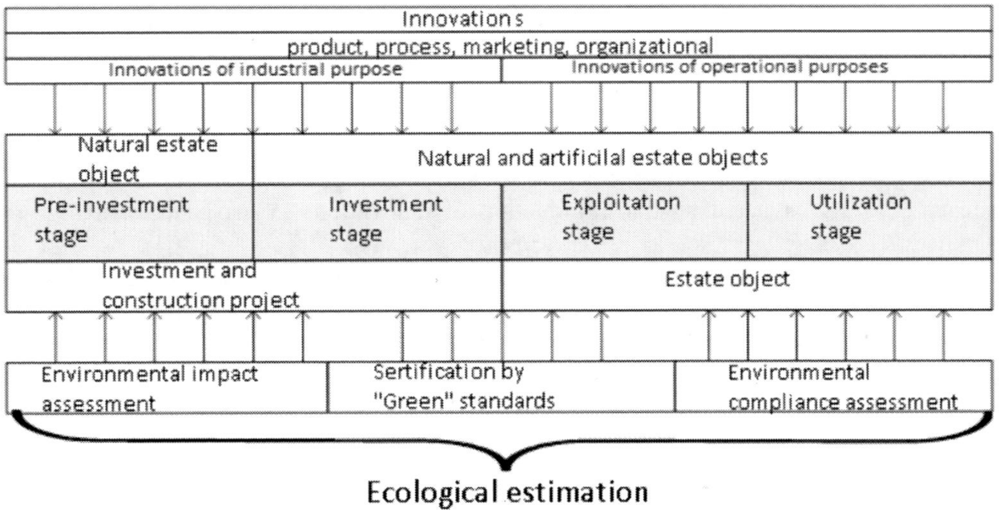

Fig. 1. Innovative character of "green" high-rise construction throughout the life cycle of the facility

The implementation of the principles of "green" construction, the improving of quality parameters of high-rise buildings, the introduction of innovations are associated with increased costs. Therefore, given the insufficiently high level of solvent demand in Russia and the high degree of competition with conventional construction projects, the demand for adequate economic profitability of the final and intermediate products of "green" high-rise construction is revealed. It should be noted that the indicator of the efficiency in this case can not be considered as the listed costs. Only discounted flows of funds, combining capital and current costs, taking into account the material expression of all types of effect, can serve as a criterion for the effectiveness of "green" high-rise construction.

Nowadays, a mechanism providing a comprehensive evaluation of construction investment projects and allowing to ensure the implementation of effective innovative solutions both at pre-investment and investment stages is a public technological and price audit implemented for projects with state participation. The implementation of the mechanism of public technological and price audit allows to optimize the price-quality ratio of both investment projects [5] and the resulting objects, based on the requirements of the "green" standards. It should be noted especially, the mechanism of public technological and price audit covers all stages of the life cycle of the project, which fully corresponds to the approach we adopted above (Figure 2).

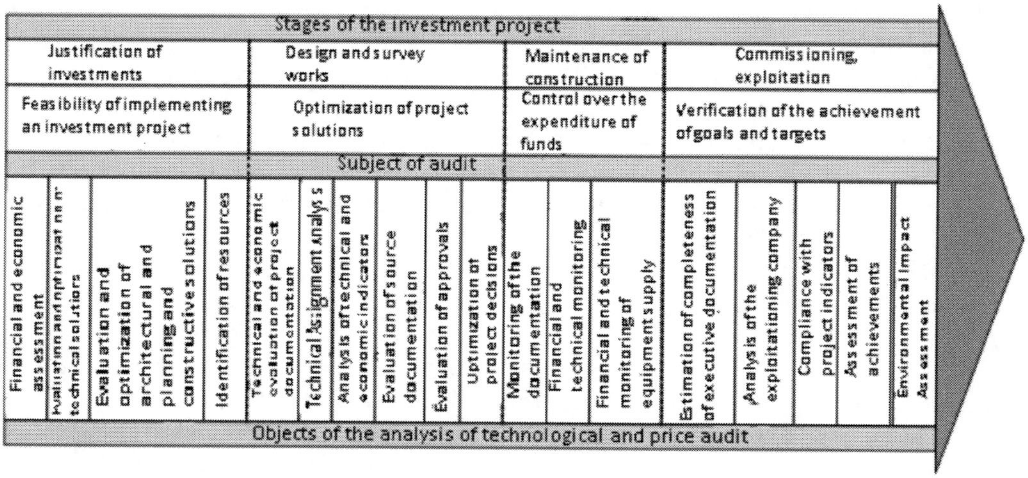

Fig.2. Substantiation of the subject-object content of the public technological and price audit throughout the life cycle of the project

The introduction of a public technological and price audit can be interpreted as changes, therefore, applying the methodology of planning and implementing organizational and economic changes to the planning and implementation of the audit. At the same time, it is necessary to take into account the innovative nature of the audit mechanism, as well as the fact that the need to implement changes aimed at the creation and operation of this mechanism is also dictated by the introduction of innovations in the process of high-rise construction.

Planning and implementation of organizational and economic changes, including the mechanism of technological and price audit, on the basis of compliance with the life cycle of organizational and economic changes contributes to the solution of the task of ensuring the complexity of planned reforms.

The mechanism of public technological and price audit, ensuring the formation of the property of "innovation" among its participants, also launches the mechanism of interaction of innovations in the creation of construction products, stimulating demand for technological innovations.

Despite all of the advantages of current mechanism of technological and price audit, its one-sided focus on major investment projects with state participation should be noted. The need to ensure the implementation of innovative design and construction solutions in high-rise construction in accordance with the requirements of "green" standards leads to the expediency proposed in [6] projecting this mechanism for the implementation of any construction investment projects through the implementation of project audit. Consequently, there is a need to build a methodology for implementing a project audit in high-rise construction based on the correspondence to the life cycle of organizational and economic changes (Figure 3).

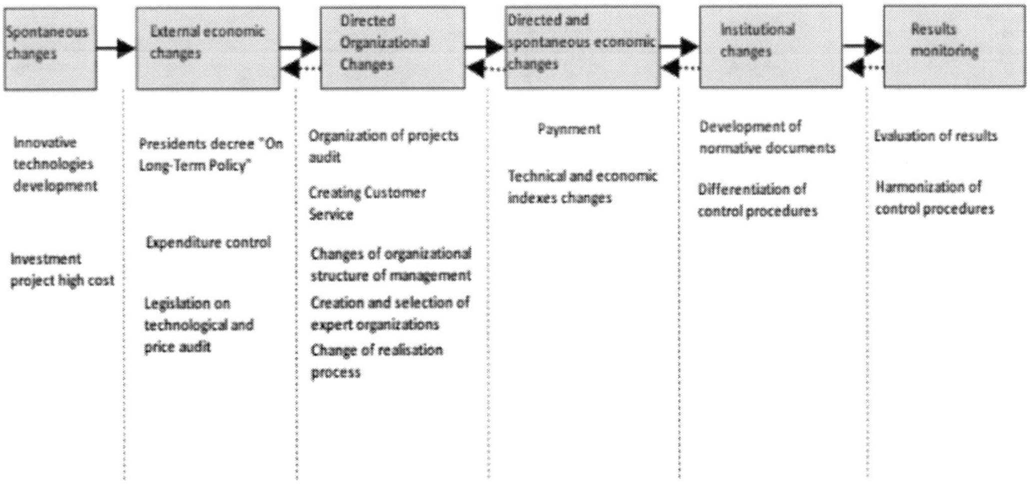

Fig. 3. Life cycle of organizational and economic changes of projects

The proposed model will make it possible to implement the necessary transformations on the basis of the developmental spiral

3 Results

Project audit of investment projects in high-rise construction is a tool for confirming proposed innovative design solutions, selecting optimal technological solutions for new investment projects, optimizing the cost and timing of their implementation, and introducing the requirements of "green" standards.

The procedure for project audit, the subject of which is the evaluation of the selection of optimal innovative design solutions that correspond to the "green" standards, and the evaluation of the cost parameters of the investment and construction project, ensures their systematic and consistent implementation. The implementation of project audit at the enterprise of the investment and construction complex begins with the adoption of a strategic decision on its implementation and it is developed in accordance with the life cycle of organizational and economic changes (Figure 4).

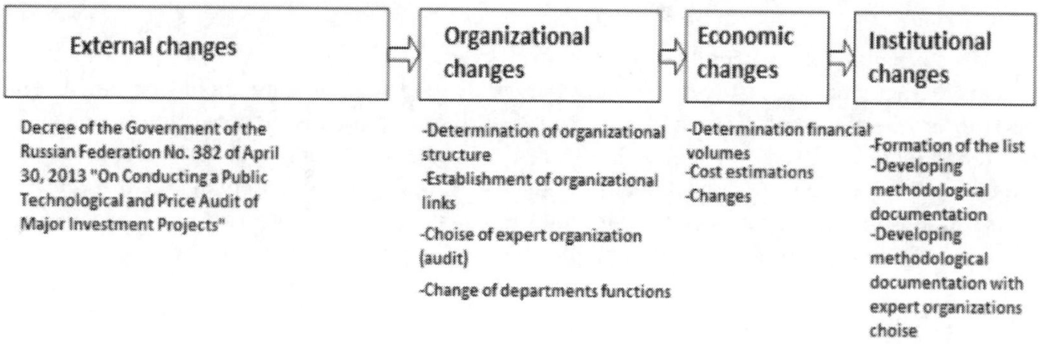

Fig. 4. Life cycle of organizational and economic changes of introduction of projects audit on the enterprise

Summarizing the conclusions above, we consider it expedient to present high-rise construction from the point of view of the system approach as a feedback system in which modern problems and requirements for construction products are realized in effects through the innovation of the processes of creating and operating real estate objects until the criteria of "greenness" project audit (Figure 5).

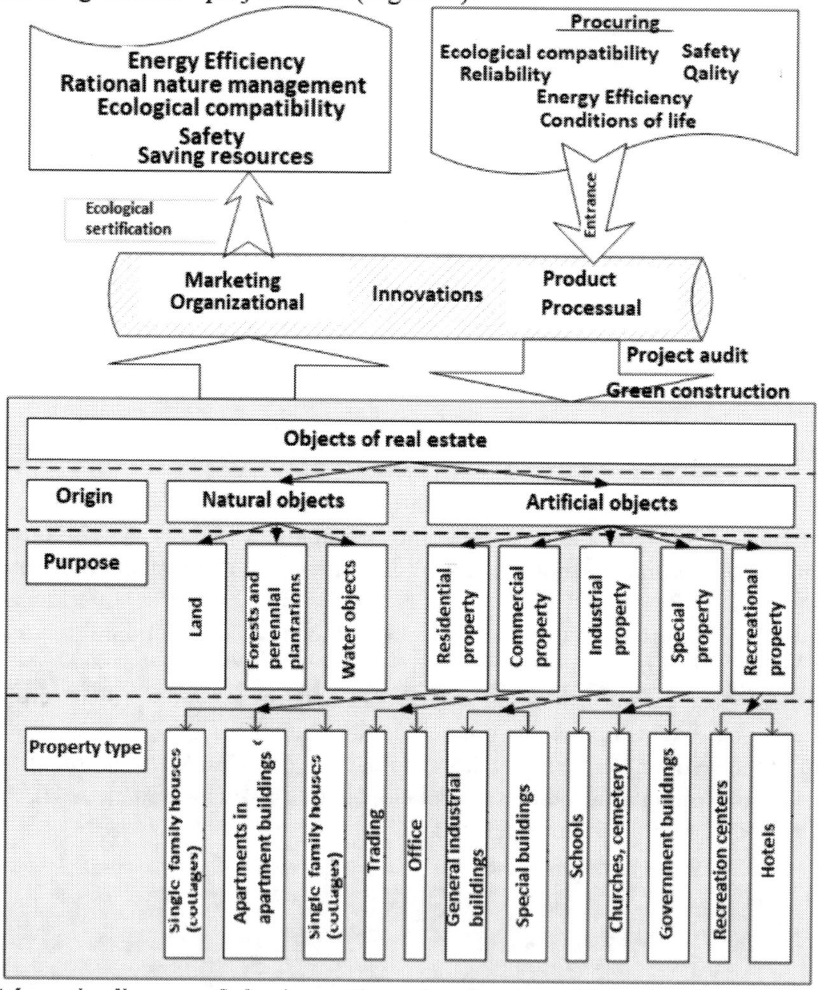

Fig. 5. Schematic diagram of the innovation of real estate objects by the criteria of "green" construction

4 Discussions

Considering the existing state level requirements for energy efficiency and energy saving, as well as environmental compatibility, it is advisable to develop innovative construction technologies in this segment. This is the reason why constant updating of the system of construction norms and quality standards on the basis of monitoring of innovative technologies is necessary. In our opinion, the system of estimations of the effectiveness of project taking into account operating costs based on energy-saving technologies should be specified. It is also necessary to take into account the environmental effect when assessing the effectiveness of all projects, taking into account the deferred costs associated with environmental impact. These findings should be implemented in the practice of expert organizations that implement mechanisms of public technology and price, as well as project audit.

5 Conclusion

As a result of conducted research, we consider it necessary: at first, implementation of the requirement of compliance with energy efficiency and environmental standards in the system of evaluation of investment projects for objects of public technological and price, and project audit; secondly, the feasibility of implementing a project audit with the relevant requirements for the evaluation and selection of high-rise construction projects; third, the introduction at the enterprises of the projects audit mechanism on the basis of compliance with the life cycle of organizational and economic changes to ensure compliance with their requirements of completeness and complexity; fourth, harmonization of existing mechanisms for monitoring investment projects based on innovation in construction.

We believe that adherence to the above principles will allow to increase not only the innovative activity of construction production and the level of innovation in construction products, but also to ensure, first of all, the quality of construction projects, while observing the condition of a certain balance of economic and non-economic (environmental, social) efficiency, especially production.

References

1. I.V. Ilin, A. Lepekhin, A.I. Levina, O.Yu. Iliashenko, Advances in Intelligent Systems and Computing, **692**, pp 1306-1314 (2018) DOi -10.1007/978-3-319-70987-1_138
2. O. Kalinina, O. Valebnikova, Advances in Intelligent Systems and Computing, **692**, pp 1315-1322 (2018) DOi - 10.1007/978-3-319-70987-1_139
3. V. Kankhva, IOP Conf. Series: Earth and Environmental Science **90** 012175 (2017) doi:10.1088/1755-1315/90/1/012175
4. E. Nezhnikova, IOP Conf. Series: Earth and Environmental Science **90** 012162 (2017) doi :10.1088/1755-1315/90/1/012162.
5. S.N. Shipova Peculiarities of adaptation of public technological and price audit in modern economic conditions. Economics and Entrepreneurship. **2-1 (67-1)**. P. 628–631 (2016)
6. S.S. Uvarova, I.G. Lukmanova The mechanism for ensuring effective control of investment projects on the basis of system integration of existing institutions. Izvestiya Vysshikh Uchebnykh Zavedenii. Technology of the textile industry. **3 (369)**. P. 29-33. (2017)

To the problem of ensuring stability of activities of construction companies engaged in high-rise building construction

Nikolay Ivanov[1,*], and *Ahmed* Safe Aldeen[1]

[1]Moscow State University of Civil Engineering; 26, Yaroslavskoe Shosse, Moscow, 129337, Russia

Abstract. Recently, more and more attention in scientific literature has been drawn to improving the sustainability of organization. The growth in the volume of high-rise construction in Russia makes the task of assessing and ensuring the sustainability of organizations and enterprises leading this type of construction very relevant. The article considers the approach to assessing the sustainability of the organization's activities in the context of functioning of quality management system (QMS). It puts forward the hypothesis that assessment of sustainability of an organization that has a real and efficient functioning quality management system can be based on the results of assessing the effectiveness of the QMS. The article describes in sufficient detail the sequence of actions to form a list of criteria for assessing the effectiveness of the QMS and sustainability of the organization, and to evaluate both characteristics on the basis of these criteria. For a clear interpretation of the results obtained, the authors use so-called petal diagrams. It suggests an original approach to their creation and analysis. Based on the results of the study, the authors conclude that in order to assess the sustainability of enterprises and organizations analysis of the dynamics of changes in the basic sustainability factors is mandatory.

1 Introduction

The current period of development of construction in Russian Federation is characterized by the creation of unique structures and architectural and building complexes, forming increasingly complex construction and operation systems. High-rise construction, original planning and structural solutions of structures based on the achievements of high technologies and modern engineering solutions, represent the progressive tendencies of the modern construction complex.

In these context enterprises and organizations of the sector pay great attention to the quality of housing construction [1, 2], to the improvement of the efficiency of management decisions and sustainable development [3, 4, 5], to the mitigation of risks in construction [6-8]. These issues are especially relevant for design and construction organizations of high-rise construction that have implemented the quality management system (QMS). As noted in [8], the existence of an existing system of quality management and a certificate for

* Corresponding author: IvanovNA@mgsu.ru

it indicates that the company can be trusted, that the risks in the company are predictable, and the identified nonconformities are eliminated along with the causes for their occurrence. The basis of any effectively functioning QMS according to the ISO 9001 standard is the principle of continuous improvement. The organization in which the QMS is functioning effectively is in a state of continuous improvement of its activities, and such an organization will always be sustainable. That is why to assess the sustainability of the organization that has implemented the QMS, it is necessary and sufficient to assess the effectiveness of the QMS of this organization. Assessment is proposed to do in the light of the approach outlined below.

2 Methods

At the first stage, it is necessary to form a list of indicators on the basis of which an assessment of the effectiveness of the QMS of the organization will be conducted. The actions of this stage are described in detail in a series of publications [9, 10]. The choice of the initial set of indicators is proposed to be carried out on the basis of the cause-effect "Fishbone" (Ishikawa) diagram [11]. The set of indicators is formed on the basis of the 5M rule that is taking into account the five main factors: machines, materials, man, methods and measurements [12]. For each factor, a set of risks is defined, the manifestation of which is possible during the functioning of the organization, as well as a set of criteria that allows assessing the effectiveness of measures to reduce the impact of these risks on the functioning of the QMS and, as a consequence, on the sustainability of a organization.

The base of the second stage is the application of an expert approach for ranking a set of indicators for assessing the sustainability of the organization and selecting the most significant indicators. A group consisting of K experts, using the group assessment method (for example, Delphi method or the "Brainstorm" method), forms the final set of criteria for assessing the effectiveness of the QMS functioning for each factor. This set can contain N indicators, each of which, according to experts' general opinion, reflects in some way the effectiveness of the QMS. Then each expert is invited to order the indicators included in one of the five factors in terms of importance, ranking the indicators of each factor in the range from 1 to N. In this case, the value 1 will correspond to the most significant indicator, and the value of N will be the least significant. Work with each expert at this step is carried out separately, which allows experts to express their opinions regardless of the opinions of colleagues. As a result of this step, a matrix of expert assessments of the criteria is formed, the rows of which correspond to the criteria being evaluated, and the columns to the experts making the assessment. The matrix element Oij (Table 1) is the numerical value assigned to the i-th indicator by the j-th expert.

Table 1. Matrix of expert estimates of criteria.

	Experts					
Criteria	1	2	…	j	…	K
1						
2						
…						
i				Oij		
…						
N						

Columns are added to the matrix, which reflect the total score of each of the indicators, as well as the average value of each indicator, taking into account the views of all experts. When ranking indicators, experts should take into account both the sectoral features of construction and the characteristics of enterprises of high-rise design and construction. An example of a matrix of expert assessments is presented in Table 2.

Table 2. Ranking of indicators within an individual factor.

	E1	E2	E3	E4	E5	E6	E7	E8	E9	E10	Sum	Ave
P11	3	2	2	3	3	2	3	3	3	2	26	2,6
P12	7	8	7	8	8	8	7	6	8	8	75	7,5
P13	4	4	5	4	4	4	4	4	5	4	42	4,2
P14	8	7	8	7	7	7	8	8	7	7	74	7,4
P15	5	6	4	6	5	6	5	5	4	5	51	5,1
P16	1	3	1	1	2	3	1	2	2	3	19	1,9
P17	6	5	6	5	6	5	6	7	6	6	58	5,8
P18	2	1	3	2	1	1	2	1	1	1	15	1,5

To judge the degree of consistency of the opinions of experts participating in the ranking of indicators, it is necessary to calculate the coefficient of variation of the estimates of Vi and the concordance coefficient W. The variation coefficient Vi is calculated for each i-th index and reflects the degree of agreement of experts' opinions about the relative importance of this indicator:

$$V_i = \frac{\sigma_i}{C_i}, \qquad (1)$$

where $\sigma_i = \sqrt{D_i}$ is the standard deviation of the estimates of each of the indicator and $D_i = \frac{1}{k}\sum_{j=1}^{k}\left(C_{ij} - C_i\right)^2$ is the variance of the estimates.

Estimating expression (1), we can conclude that the smaller the value of Vi, the level of consistency of experts' opinions is higher.

To determine the degree of consistency of the group opinion of experts as a whole for the aggregate of indicators within a certain factor, the concordance coefficient is used:

$$W_i = \frac{12\Delta S^2}{k^2(n^3 - n)}, \qquad (2)$$

where ΔS2 - a measure of the degree of consistency of judgments of experts, which is determined by the formula:

$$\Delta S^2 = k^2 \sum_{i=1}^{n} \left[r_i - \frac{n+1}{2}\right]^2 \qquad (3)$$

Substituting ΔS2 in (2), we obtain:

$$W_i = \frac{12\sum_{i=1}^{n}\left[r_i - \frac{n+1}{2}\right]^2}{n^3 - n} \qquad (4)$$

The concordance coefficient can take values from 0 to 1. For W = 0, the experts' opinions practically do not coincide. The value W = 1 shows that the experts practically rank the indicators in terms of their importance. The dynamic variation of W from 0 to 1 reflects an increase in the degree of consensus among experts.

Then the experts again jointly discuss the results and determine the "feature line". By "feature line" we mean the sum of all estimates of one indicator, which is still considered by experts to be worthy of consideration. Indicators with a total score greater than the "feature line" are excluded from further processing. It is proposed to leave six of each factor as significant indicators, thus, the feature line will be the total score of 60 points.

The next step is to process the resulting sample of significant criteria. For each criterion, average expert judgment is determined, and then the rank of the criterion in points is calculated. It is assumed that the sum of the quantities inversely proportional to the average estimates of each criterion corresponds to 100 points, and the rank of each criterion in points is proportional to the value inversely proportional to the average criterion estimate:

$$Ri = \frac{Bij}{100} * \sum_{j=1}^{K} Bij, \quad (5)$$

where $ij = \frac{K}{\sum_{j}^{K} Oij}$.

The rank of each criterion is rounded to an integer, subject to the restriction that the sum of the ranks of all criteria is 100. The total score of all the criteria identified by experts, equal to 100, corresponds to the "ideal" state of an effectively functioning quality management system. This condition assumes the fulfilment of all the requirements of ISO 9001: 2015.
Finally, the experts again jointly discuss the rules for determining the actual indicators for meeting the requirements of the standard for the selected QMS functioning criteria. For example, for the criterion "Number of claims received from consumers", these rules are presented in Table 3.

Table 3. Rules for adjusting the value of the rank of the criterion.

Criteria name Value range	The range of values	Points
Number of claims from consumers	No claims Less than 3 claims 3 to 5 claims More than 5 claims	Maximum points Minus 3 points Minus 6 points Minus 8 points

In the third stage, the sum of the scores of the significant criteria calculated for the current year is compared with the lower limit of the score for the quality management system, the functioning of which can be considered effective. As a rule, according to existing methods of assessing the effectiveness of QMS functioning, the boundary is usually set in the 72-74 points area. If the calculated sum of points of the current year is less than the specified limit, the functioning of the QMS of the construction industry organization is considered ineffective and it is recommended to develop a set of corrective actions aimed at eliminating the causes that reduce the effectiveness of the QMS. Absolute stability of the organization is achieved if all factors have a value of at least 74 points.

3 Results

If at least one factor goes beyond the stability line, then it is considered that the organization loses its stability. To assess the general state of the organization, it is necessary to analyze the state of the factor level with the help of a petal diagram, the general form of which is shown in Fig. 1.

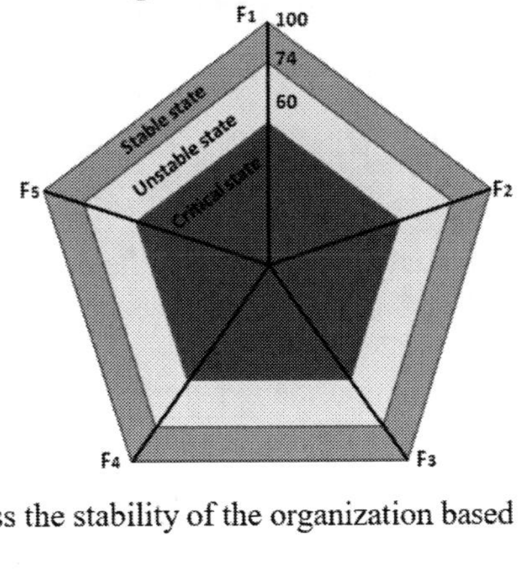

Fig.1. A petal diagram to assess the stability of the organization based on the 5-factor model.

To interpret the results of assessing the values of factors reflecting the sustainability of the activity of a organization in the construction industry, we introduce three regions of stability: a stable state (green zone), an unstable state (yellow zone), and a critical state (red zone). Define the boundary values for each of these zones. So in the green zone there are factors whose values can range from 100 to 74 points. In the yellow zone, the values of factors are determined from 74 to 60. The red zone reflects the values of factors from 60 to 0. Depending on the values adopted, the factors can be unevenly distributed within these three zones. Let's consider the basic variants of an arrangement of values of factors which can arise in the course of an estimation of stability of activity of the organization of high-rise construction. Suppose that all factors considered in the current period have values that fall within the range of values of the green zone. In this case, you can call the state of the organization sustainable. This state is the most preferable for any organization, since it speaks of an effectively organized activity. However, in this case the position of the factors within the zone itself is important: factors that are located close to the lower boundary of the zone require special attention from the management of the organization. Otherwise, there is a high probability of an organization moving from a stable state to an unstable state or even a critical state. It is these factors that require targeted impacts through activities aimed at reducing the likelihood of risk occurrence or its complete elimination.

In the conditions of high competition and scarcity in resources that have developed recently in the construction market, the absolutely stable state of the organization for organizations in the construction industry is not a frequent phenomenon. The state where one or more factors are outside the green area in the yellow zone is called unstable. The main task in such a situation is to identify these factors and develop corrective actions for overcoming the unstable state and transition from an unstable situation to a sustainable situation. The selection of optimal measures is important, as a result of which, the organization will not be in an even worse position.

There is a situation in which one or more factors are in the critical red zone. Other factors may be in the yellow or even green zone. In this case, the organization is in a critical, near-bankruptcy state. Such organizations require a special strategy to overcome the crisis situation.

Assessing the stability of an organization at any point in time does not provide exhaustive information about the development of the organization [13]. That is why it is advisable to assess the state of the factors of sustainability in the dynamics: if the value of the factor has a positive dynamic, we can speak about the effectiveness of measures to increase sustainability within this factor. If the value of the factor acquires a border stable state, then one should also evaluate the dynamics, specifying whether it is positive or negative. Particular attention should be paid to those factors, the dynamics of the values of which have a negative value, which corresponds to the tendency to deterioration of stability.

The stability of the organization is proposed to be evaluated by comparing the values of the factors for the current year and the previous year. An example of a diagram reflecting the dynamics of factor values for 2 years (previous and current) is shown in Fig 2. Analysis of the diagram shows that the values of factors F1, F4 and F5 tend to increase, the value of factor F2 decreases, the value of factor F3 remained unchanged. At the same time, despite the increase in the value of factor F4, the stability of the organization within this factor is considered not achieved, in contrast to the values of the factors F1 and F5, within which the organization has taken a stable position. From the results of the analysis it follows that special attention is paid to factors F2 and F4.

On the basis of the obtained data on the dynamics of the efficiency factors of the QMS of a organization, in addition to evaluating each factor separately, it is possible, if necessary, to calculate the integral indicator of the organization's stability.

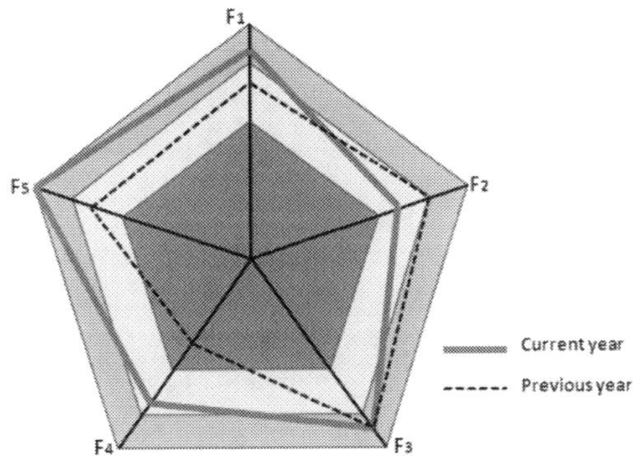

Fig. 2. Dynamics of values of factors reflecting the stability of the organization.

4 Discussion

As noted in [14], in the general case, the complex stability of the organization of the construction industry can be described by the following function:

$$y(t) = f\{F_1(t); F_2(t); F_3(t); F_4(t); F_5(t)\}, \quad (6)$$

where $F_1(t), F_2(t), \ldots, F(t)$ are the factors influencing the complex stability of the organization.
The interpretation of the values of the complex stability function is presented in Table 4.

Table 4. Interpretation of the values of the complex stability function

Function Value y(t)	Level of stability
$0 \le y(t) \le 0{,}6$	Critical state of organization
$0{,}61 \le y(t) \le 0{,}73$	Unstable state of the organization
$0{,}74 \le y(t) \le 1$	Stable state of organization

The basis of calculations of the evaluation is to be based on the degree of compliance with real radar chart of an ideal chart, which represents the absolute stability in all factors. Geometrically, the value of the integral indicator for an ideal system can be compared to the area of the figure, which is a regular polygon (in the presented version - a pentagon) with a diagonal of 100 (Fig. 3).

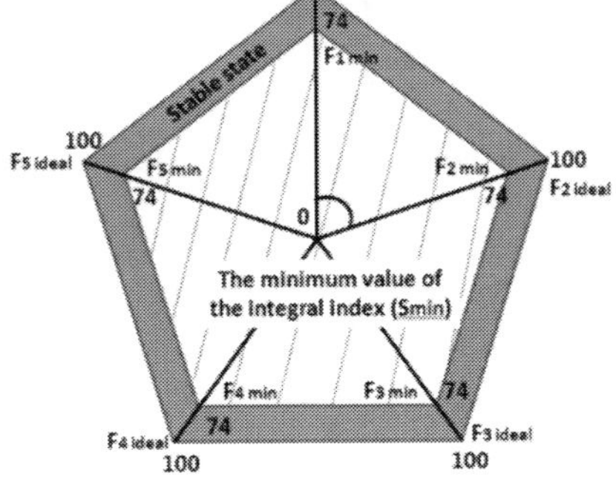

Fig. 3. Dynamics of factor values, reflecting the stability of the organization.

The lower boundary of the stability zone will correspond to the area of the polygon with the diagonal $F_{i\,min}$. The minimum value of the integral indicator of stability of the organization is calculated by the following formula:

$$S_{min} = \frac{1}{2} F_{i\,min}^{2} * \sin\frac{360}{n} * n, \tag{7}$$

where $F_{i\,min}$ is the minimum permissible value of the i-th factor (not less than 0.74), n is the number of factors.

When determining the integral indicator of sustainability, the basic condition of the complex stability of the organization must be fulfilled - all factors entering the integral indicator should have a value of at least 0.74. Only in this case it makes sense to evaluate the dynamics of the integral indicator. When assessing the sustainability of the organization, it is advisable to assess the state of complex sustainability in dynamics, which allows us to evaluate the correctness of management decisions, as well as monitor the dynamics of the organization's sustainable development:

$$\Delta S = S_{current} - S_{previous}, \tag{8}$$

where $S_{current}$ - the value of integrated sustainability in the current year,

$S_{previous}$ - the value of complex sustainability in the previous year.

If the value $\Delta S > 0$, then the activity of the organization under consideration is effective and has a sustainable development.

Thus, assessing the results of the current period and comparing them with the results of the previous period, we can conclude on the level of sustainability of the activities of the organization, as well as the effectiveness of management decisions.

Conclusions

The approach proposed in the article can be seen as a useful tool for practical assessing and ensuring the sustainability of construction organizations and enterprises and as tool for theoretically predicting the future's needs along with the success of a business model.

References

1. E.V. Nezhnikova, The Economics of Construction Journal, **5(35)**, 14-21 (2015)
2. E. Nezhnikova, R. Obukhov, MATEC Web of Conferences, **106**, 08025, (2017) https://doi.org/10.1051/matecconf /201710608025
3. D.N. Silka, M.A. Ivanova, Proceedings of Moscow State University of Civil Engineering, **12**, 5(104), (2017)ndoi: 10.22227/1997-0935.2017.5.572-58
4. N.nOertwig, Life Cycle Engineering and Management (2017)
5. E. Nezhnikova, Proc.Eng, **165,** 1300-1304 (2016) https://doi.org/10.1016/j. proeng.2016.11.854
6. H.M. Bernstein, M.A. Russo, D. Laquidara-Carr, Smart market report (2011) http://www.pepperlaw.com/resource/5452/0C0
7. N. Ivanov, MATEC Web of Conferences, **86,** 05004 (2016) https://doi.org/10.1051/matecconf /20168605004
8. I.N. Aleksandrovich, Int. J. Appl. En. Res., **10**, 42415-42418 (2015)
9. N. Ivanov, Proc. Eng., **153**, 228–231 (2016) http://dx.doi.org/10.1016/j .proeng.2016.08.107
10. H. Allen Hu, S.H. Chen, C.W. Hsu, Int. J. Environ. Sci. Technol., **9-2,** 343-354 (2012) https://doi.org/10.1007/s13762-012-0037-7

11. K. Ishikawa, Productivity Press Taylor & Francis (1990)
12. http://asq.org/learn-about-quality/cause-analysis-tools/overview/fish bone.html
13. V.N. Goncharov, E.A. Kolosova, Journal "Organizer of production, **3(58)**, 56-59 (2013)
14. I.G. Lukmanova, E.V. Nezhnikova, Proceedings of Moscow State University of Civil Engineering, **6**, 158-164 (2014) https://doi.org/10.22227/1997-0935.2014.6.158-164

Problems and solutions in the field of improving control and supervision activities in the construction sector

Varvara Dikareva [1,*] and *Anna* Mostovaya [1]

[1] Moscow State University of Civil Engineering, Yaroslavskoye shosse 26, Moscow, 129337, Russia

Abstract. The improvement of control and supervision activities in high-rise construction is an important goal of the state. In the paper, the author considered the initiatives of the state in terms of improving the control activity. He shows the main approaches that will be applied by the state. The application of information and communication technologies in control activities is described in detail. It was noted that the quality of the regulation of various procedures remains a serious issue. At the same time, modern innovations in the field of control should take into account the increasing volumes of data and information about various economic processes.

1 Introduction

Control and supervision activities in the construction sector performs an important function. The state exercises control and pursues several goals. First, it is necessary to ensure correctness of drafting of all documentation in order to exclude material, physical and other damage during construction of the facility and its commissioning. Second, it is necessary to assess the effectiveness of the overall design activity in construction in order to adjust the system of measures of sectoral management. The main function in control and supervision activities is performed by the Ministry of Construction of Russia and various controlling state bodies, including the Accounts Chamber of the Russian Federation. Also, control is exercised independently by business entities. The basic requirements of control and supervision activities of the Ministry of Construction of Russia were established on the basis of the federal program "Reform of Control and Supervision Activities" approved on December 21, 2016 by the Presidium of the Council under the President of the Russian Federation for Strategic Development and Priority Projects. The implementation period is until 2025. Directly the Ministry of Construction of Russia, as a federal executive body which exercises normative and legal regulation in the areas of exercising control and supervision functions in terms of regional types of control (supervision), is a participant in this program. The totality of measures contained in this federal program has not only a large quantitative composition. Achievements of four levels are also provided. Analysis of the situation in practice shows that various violations continue to manifest themselves in the

* Corresponding author: dikareva-va@rambler.ru

construction industry [1, 2, 3]. At the same time, active actions on the part of the authorities can have a positive result if the proposed solutions are subjected to a serious expert assessment and fully comply with the trends of the time.

2 Materials and Methods

Many levels of state administration of the industry, as well as the complexity of the projects themselves, their specific diversity create barriers to the creation and adequate application of effective management methods, regulation of all works, etc. [4]. Regulations, instructions, guidelines, methodical documents remain the main instrument of the state when exercising control and supervision activities. For business, the poor elaboration of government regulatory documents means the emergence of unforeseen temporary and financial costs. The authorities at the level of the constituent entities of the Russian Federation, as well as at the federal level, can not directly receive qualitative feedback for assessing their own managerial functions. The totality of regulatory and legal documents at the federal level comprehensively enough disclose the content of managerial functions and tools. At the same time, certain practical implementation at the level of the subjects of the Russian Federation and municipal bodies is significantly out of sync. As a result, formation of new regulatory and legal documents at the federal level is also subsequently implemented with obviously lower efficiency than expected. Fig. 1 shows the problematic issues in terms of control and supervision activities, and solutions that are implemented at the federal level, but which, among other things, will be aimed at solving problems in the construction sector.

This concept aims to increase the level of awareness of the executive authorities and their officials about the objects of control, their condition, and ensure effective planning of resources for performing control and supervision measures. For business representatives, it will be possible to quickly and in the medium term see the composition of control measures addressed to them, decisions made by state bodies on the basis of inspection findings and other information. At the same time, the industry still has a set of systemic problems that will impede the introduction of new control and management mechanisms.

Based on the foregoing, it is possible to state systemic limitations on the effectiveness of investment activities as part of administrative and market regulatory mechanisms. At the state level, they are as follows:

- Firstly, the low level of organization of work at the level of executive bodies;
- Secondly, the lack of a unified system for planning the activities of the authorities in terms of control and supervision;
- Thirdly, low efficiency of the feedback system, inconsistency of the system of normative legal documents being developed with tasks that require priority solving.

For business it is:

- Firstly, the lack of exhaustive regulations on interaction with authorities;
- Secondly, high risks of conducting supervised activities without a clear understanding of the rules and requirements;
- Thirdly, the lack of feedback regarding evaluation of the effectiveness of state control measures;
- Fourthly, low responsibility of state bodies for actions restricting entrepreneurial activity without availability of reasons.

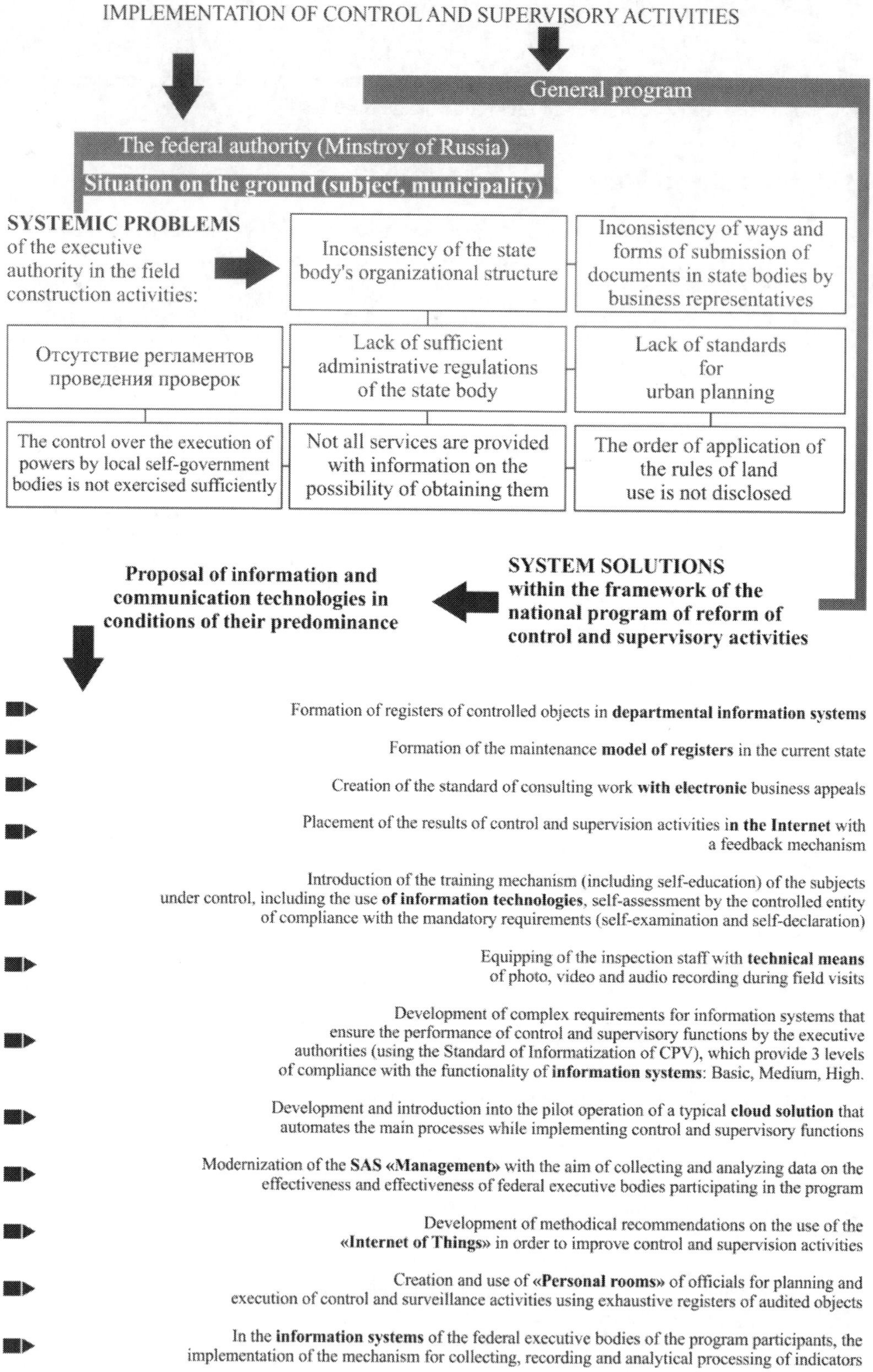

Fig. 1. Problems and solutions in the field of control and supervision activities.

Taking into account the information in Fig. 1, it should be noted that in the future the state sees the use of information technology tools as the basis for control and supervision activities. This is an invariant trend of time. In the future, effectiveness of investment activities will depend on how much the aggregate of new instruments will be effectively integrated into the economic processes of government and business entities.

3 Results

An analysis of the directions for improving control and supervision activities shows active implementation of new generation management systems. They are mainly proposed at the state level. Here is the need to improve the efficiency of capital investment development, provided that the sources of funds are extremely centralized, and work performance and management of the investment cycle are in contrast decentralized. The scale of such dispersion is unparalleled at the level of individual industries and fields of activity. State investments are being appropriated in every constituent entity of the Russian Federation and the municipality, and hundreds of developers and customer services, technical supervision services, etc. are involved. No major corporation has such a large scale of activity. The exception may be transnational corporations (TNCs), but in fact there are very few such forms of doing business in our country.

Due to the logical connection of management systems in TNCs and the state mechanism of capital investments, where the Ministry of Construction and Housing and Communal Services of the Russian Federation currently plays a key role, we note the following. The subject and municipal authorities should have a good command of the communication system in order to ensure a complete understanding of all information flows between the federal level (the Ministry of Construction of Russia, etc.) and budget administrators. An analysis of the directions of the reform of control and supervision activities shows that a lot of work is being performed in this field, and this is consistent with the principles of organizing large business.

So, the results obtained on the basis of the author's analysis have shown that modern and prospective fields of control and supervision activities should be based on a serious analytical basis. Now there are really new opportunities for collecting and studying various information and data generated due to the active introduction of information technologies in the system of public administration. The scheme shown in Fig. 1 clearly indicates that the Internet, various information technologies are now envisaged in almost all measures in the field of control activities.

4 Discussions

The issues of increasing the efficiency of investment and construction activities are still extremely urgent. Many researchers at the same time point to the need to develop special measures precisely at the level of the bodies of sectoral management [5, 6]. In this regard, the development of communication mechanisms at the state level is indeed in demand. Communication of this level allows managing the efficiency of the construction industry. At the same time, it becomes possible to move on to a new generation of control tools. From the point of view of the debatable aspects, there remains the issue of improving the quality of the regulation of various procedures controlled by the state. New communication technologies do not improve the quality of control and do not prevent violations themselves. Participants in construction activities should function within a certain system of rules. In this regard, it is the improvement of the order that really remains a debatable issue, when the main focus of state programs is on various information technologies. At the same time, organization of work to increase the availability of various information, its systematization, ensuring the convenience of working with it will be the key to the success of the reform of control and supervision activities.

5 Conclusion

Control and supervision activities in the construction industry require many improvements. At the disposal of the industry, technologies emerge that allow instantaneous reporting of information to various participants, and at the same time, to monitor the current situation at a very high speed. Such opportunities create a strong information flow and complicate the management of the industry. Systematization of information in terms of regulations, rules, procedures will be the most important field of activity of state management bodies in construction.

References

1. V.S. Kankhva, S.S. Uvarova, S.V. Belyaeva, Procedia Engineering, **165**, 1046-1051 (2016)
2. V. Kankhva, D. Silka, Procedia Engineering, **165**, 1337-1342 (2016)
3. D.N. Silka, MATEC Web of Conferences, 07012 (2016)
4. N. Yaskova, Bulletin of the Irkutsk State Technical University, **1(60)**, 178-186 (2012)
5. R.S. Golov, V.V. Shilov, S.A. Silantiev, ASEE International Forum, 20766 (2017)
6. N. Yaskova, T. Alexeeva, MATEC Web of Conferences, 07025 (2016)

Methods of erection of high-rise buildings

Nadezhda Cherednichenko [1,*] and *Pavel* Oleinik[1]

[1] Moscow State University of Civil Engineering, Yaroslavskoye shosse 26, Moscow, 129337, Russia

Abstract. The article contains the factors determining the choice of methods for organizing the construction and production of construction and installation work for the construction of high-rise buildings. There are also indicated specific features of their underground parts, characterized by powerful slab-pile foundations, large volumes of earthworks, reinforced bases and foundations for assembly cranes. The work cycle is considered when using reinforced concrete, steel and combined skeletons of high-rise buildings; the areas of application of flow, separate and complex methods are being disclosed. The main conditions for the erection of high-rise buildings and their components are singled out: the choice of formwork systems, delivery and lifting of concrete mixes, installation of reinforcement, the formation of lifting and transporting and auxiliary equipment. The article prescribes the reserves of reduction in the duration of construction due to the creation of: complex mechanized technologies for the efficient construction of foundations in various soil conditions, including in the heaving, swelling, hindered, subsidence, bulk, water-saturated forms; complex mechanized technologies for the erection of monolithic reinforced concrete structures, taking into account the winter conditions of production and the use of mobile concrete-laying complexes and new generation machines; modular formwork systems, distinguished by their versatility, ease, simplicity in operation suitable for complex high-rise construction; more perfect methodology and the development of a set of progressive organizational and technological solutions that ensure a rational relationship between the processes of production and their maximum overlap in time and space.

1 Introduction

The construction of high-rise buildings has a long history. The birthplace for them is the United States. For the first time in urban planning there were created skyscrapers that determined the business centers of New York, Chicago and other major cities. In the second half of the last century, the active urbanization of European cities caused a variety of high-rise buildings in London, Paris, Frankfurt-on-Main, etc. Particularly should be highlighted Défense multifunctional complex (France), which is distinguished by the originality of design solutions and high unification of structural systems of high-rise buildings. Later, the rapid economic growth ensured the construction of record high buildings in East and South-East Asia: Singapore, Hong Kong, Shanghai, Taipei, etc. [1].

* Corresponding author: CherednichenkoND@mgsu.ru

Our country has also got a certain experience of high-rise construction: erected buildings on the Smolenskaya Square, near Krasnye Vorota area, the construction of the New Arbat district, the construction of the buildings of the RF Government and CMEA (Council for Mutual Economic Assistance) headquarters, etc. [2].

Thus, the construction has gained a vast experience in the design and construction of high-rise buildings in various natural and climatic conditions, different design systems and complexity. This experience has been studied and widely used in our country, which has now begun a new stage in the construction of high-rise buildings in the cities of Moscow, Nizhny Novgorod, Krasnoyarsk, Kemerovo, St. Petersburg, Volgograd and other.

2 Materials and Methods

In our country the high-rise buildings are considered to be the structures above 100 m high. In accordance with the Urban Planning Code of the Russian Federation they are also considered to be unique buildings. The buildings with a height of 30 to 100 m are assigned to I, II and III category of storey number. As a rule, in Russia the priority is given to the construction of multi-functional complexes, which include not only office premises, but also residential floors, shops and cinemas. But at the same time, the low-capacity complexes are being built, for example, "Tricolor" and "North Park" residential complexes, offices of "Gazprom", "Sberbank" companies, etc.

The forms of a tower type of high-rise buildings are used, as a rule, in connection with large wind loads, but rather complex structural systems such as skeleton-frame, wall, barrel, shell, combined are used at the same time. Each of these constructive systems has its own rational area of use. For example, the wall system is used for residential buildings, the frame system is used when building up to 60 floors, the barrel system is functionally and constructively effective due to the load-bearing core, and the shell system provides the possibility of erecting buildings with a height of over 200 m [3,4,5,6].

There are also quite complex elements engineering systems for high-rise buildings as follows: plumbing, heating, ventilation, power supply, hot water supply, sewerage, drains, automatic fire extinguishing systems, automation and dispatching systems, electrical equipment, low-current systems, Internet networks, etc. For example, electrical equipment consists of emergency lighting systems, smoke removal, power supply of fire protection systems, security and fire alarm systems, and the water supply system, in addition to the traditional elements, includes a number of technical premises in 10-12 floors with an optional mounted pumps. Or, for example, the power supply system can be autonomous, which is actual in case of switching off the central power supply, the operation of elevators, supply of water and electricity to the apartments.

High-grade concrete, a wide range of metal structures and translucent materials, various construction equipment and equipment are used for the construction of high-rise buildings. A large number of organizations is directly involved in the creation of high-rise buildings: construction, assembly, specialized, design and research, transport, supervisory, as well as a number of organizations that perform the functions of testing, coordination, etc.

The described factors, as well as the construction conditions, including the influence of natural and climatic factors, ultimately determine the choice of methods for organizing construction and production of construction and installation works.

3 Methods of organization and production of works

The construction of the underground part of high-rise buildings is associated with the need to build powerful pile-slab foundations of high-strength concrete. Constructive solutions of such foundations, for example, in Moscow, are dictated by the ground conditions and the technology of their arrangement is similar to foundations of the multi-storey buildings. But at the same time, the installation of foundations and bases under assembly crane that stand the large ground loads. In addition, large volumes of excavation work are typical with the burial of the basement into several tiers for building the underground part of high-rise buildings.

The above-ground part of high-rise buildings is most often single-sectioned and compact enough; it is less often performed as multi-section. The structural basis of buildings is usually a steel, reinforced concrete or combined frame with a hard core or with flat diaphragms. The load-bearing core is arranged mainly in a monolithic variant in sliding or shuttering formwork. There are two schemes for building up the structures. In the first scheme, the concreting of the load-bearing core outruns the mounting of the frame by 1-2 tiers, and in the second scheme, the mounting of the load-bearing core lags behind the carcass assembly. In this case, it is recommended that the maximum height of the backlog must be no more than 8 storeys, but we must also take into account the obligatory welding and concreting of the structures [7, 8, 9].

When applying a reinforced concrete frame, we can use prefabricated structures (columns, crossbars, slabs), monolithic structures, complex structures (monolithic frame, prefabricated slabs). Then, the technology of erecting a high-rise building includes the following cycles: concreting the load-bearing core, assembling prefabricated structures or erecting a monolithic frame, general building work on a tier (partitioning, welding, preparing for laying utility networks, etc.), special building work (building of engineering systems, laying of networks, installation of equipment, etc.) and finishing works [10, 11].

Decisions on the organization of construction should be based on the principles of in-line production of construction and installation works, providing the fragmentation of the work process into the tiers and construction staging. A tier consists of one floor, and the number of construction staging is from 2 to 4 per section. Each cycle of work is performed by a corresponding specialized brigade (unit) or several brigades in the case of a large amount of work. The leading flow is the installation of the framework and therefore the rhythm of the performance of other works is linked to it according to temporal and spatial parameters [12, 13].

The main condition for the erection of high-rise buildings from reinforced concrete is compliance with the requirements for achieving the design strength of concrete, including in the nodes and joints of bearing structures.

In the case of arrangement of a steel or composite frame, there are used columns and crossbars (joined in two directions by rigid welded assemblies into frame systems). The columns are made of standard rolling profiles and must be concreted to increase their fire resistance, and the crossbars are made of welded I-sections. The slabs of inter-storey floors are laid on their broadened lower shelf. The load-bearing core can be either reinforced concrete monolithic, or in the form of a closed shaft of steel structures; all elements of the frame are attached to steel structures and, together with the slabs, form a hard disk.

Fastening of crossbars is made immediately after installation and alignment of the whole cell.

The construction of a high-rise building with a steel or combined frame can be carried out in separate or complex method. With a separate method, the steel frame is mounted to its full height and only then the general construction, special and finishing works are performed. The application of the integrated method provides for simultaneous execution of all types of work on approximately 10 floors. On the upper 2-4 floors, the steel structures of the framework are assembled according to the scheme "installation - reconciliation - connection", below 2-3 floors the assembly of prefabricated or monolithic ceilings is carried out, the framing of the frame is provided on the lower floors, and the special and finishing work is organized at the very bottom. Here we should take into account that the installation of steel structures is necessarily carried out tier by tier. At the same time, the first work is performed on the central load-bearing core of building and the installation of other structures is proceeded only after reconciling all of its elements. Stability of the columns is achieved with the help of conductors or temporary braces before they are being fixed. The fastening of crossbars is made after installation and alignment of the entire cell.

The main condition for the construction of high-rise buildings with a steel frame is to ensure the strength and stability of each tier.

4 Formation of organizational and technological solutions

When building high-rise buildings in Russia, there is a tendency for a widespread use of monolithic reinforced concrete structures. In addition to reducing the weight of the building, compared to buildings with a steel frame, the economic effect is also provided by the absence of the need for additional fire protection of load-bearing structures. The development in the native practice of monolithic housing construction is promoted by a diverse range of formwork systems capable of performing the most sophisticated production technology works.

The choice of formwork system is determined not only by the complexity and scope of work, but also by climatic conditions. Strong winds, fogs, precipitation make it difficult to work at height, and in some cases they cause the halt. Therefore, self-elevating shuttering systems with hydraulic drives are getting serious development. Such systems include shuttering panels, anchors, locks, scaffolding elements, etc. For their application, it is advisable to develop projects that include the layout of the shuttering boards and their movement along the height.

To ensure a high rate of erection of monolithic high-rise buildings, we require a clear organization of continuous production of concrete in large quantities and serve it over long distances, both vertically and horizontally. The concrete shall be presented fairly stringent requirements for durability, frost resistance, resistance to aggressive chemicals. According to experts, the most optimal is the concrete of S60 - S80 classes [14, 15].

Delivery of concrete mix is carried out either by truck mixers from the district and central factories of ready-mixed concrete or from an automated concrete unit located directly on the site. The second technological scheme of delivery of a concrete mix is certainly more effective.

Feeding of the concrete mixture to the object is carried out with the help of powerful concrete pump installations in the form of truck-mounted concrete pumps and stationary concrete pumps. The truck-mounted concrete pumps are used mainly for the erection of the underground part of the building and its lower floors. For this purpose, the truck is equipped with a distributing boom. For continuous supply of concrete mix to the upper floors, stationary concrete pumps are used, which have concrete conveying pipes connected to a hydraulic distributor boom. Such a boom is mounted on the monolithic structures of the corresponding catch and carries out the supply and distribution of the concrete mixture to the mounted formwork.

Additional requirements are also applied to the installation of reinforcement. In view of the fact that welding of reinforcement for high-rise buildings is unacceptable, therefore, couplings or wire bindings are used to connect it.

A set of lifting- transport and auxiliary equipment is also significantly different from the traditional construction. For example, the tower cranes are not advisable when building buildings above 60-70 m high. Therefore, for work up to a height of 130-140 m, it is necessary to use attachable tower cranes, which are attached to the structure of the building., Self-lifting assembly cranes attached to the central load-bearing core of building are used when erecting a building over 130 m. Such cranes provide work on a level with a height of 30 to 40 m.

Special cargo-passenger lifts are used for the supply of other cargoes (finishing materials, sanitary and electrotechnical products, small-piece loads) to the tiers, as well as lifting of workers; the number and type of them are selected depending on the organization of work. The lifts with a capacity of up to 20 people and a lifting capacity of up to 3 tons are most commonly used. As a rule, lifts are installed after the erection of 8-9 floors.

The solutions for the installation of facade systems are also important in the construction of high-rise buildings. As practice shows, it is most expedient to perform facade works in combination with the erection of a frame of a building with a gap of 5-7 floors. To date, a number of effective facades has already been created to meet modern requirements for thermal properties, noise insulation, air permeability, water permeability and fire resistance. For example, the facade series of the "TATPROF" system was highly appreciated. In particular, the facades of the series EK-40V and EK-69V are made in the form of multicellular structures made of frame elements that allow the glazing of almost all types of facades - straight, vertical, hinged, built-in, etc. The glazing of the building is made indoors and can be applied for different climatic zones.

The efficiency of building of high-rise buildings depends to a large extent on the qualitative development of organizational and technological solutions as part of construction projects and projects for the production of works. To the present moment, the speed rate of construction of high-rise buildings is characterized as not less than 4-5 floor construction per month. The available reserves for reducing the duration of construction can be used by creating:

- complex mechanized technologies for the efficient construction of foundations in various soil conditions, including in the heaving, swelling, hindered, subsidence, bulk, water-saturated, etc.;
- complex mechanized technologies for the construction of load-bearing monolithic reinforced concrete structures, taking into account the winter conditions of work and using mobile concrete batching complexes and new generation machines;
- modular formwork systems, distinguished by their versatility, ease, comfort of operation, taking into account the features and complexities of high-rise construction;
- more perfect methodology and the development of a set of progressive organizational and technological solutions that ensure rational relationship between the processes of production and their maximum combination in time and space.

5 Conclusions

At present, a new stage in the construction of high-rise buildings of different design systems and degree of complexity is being implemented in Russia, based on the use of progressive foreign and native experience in their design and construction in various natural and climatic conditions.

In the native practice, both single-section and multi-sectional high-rise buildings are built. The structural basis of buildings, as a rule, is a steel, reinforced concrete or combined frame with a hard core or flat diaphragms. The construction of such buildings is carried out according to two basic schemes - the advanced creation of the central load-bearing core and the advanced assembly of the frame.

When using a reinforced concrete frame, the technology of erecting a building includes cycles of concreting the load-bearing core, assembling prefabricated structures or building a monolithic frame, performing general, special and finishing works. The leading flow is the installation of the frame, with the rhythm of which the performance of other types of work on temporal and spatial parameters is coordinated.

In the case of arrangement of a steel or composite frame, all work is performed in separate or complex methods. With a separate method, the steel frame is mounted to its full height and only after that other work is performed. The application of the integrated method provides the simultaneous execution of all types of work on approximately 10 floors.

The rate of construction of high-rise buildings provides construction of no less than 4-5 floors per month. The available reserves for reducing the duration of construction include the creation, of complex mechanized technologies for the erection of load-bearing monolithic reinforced concrete structures, modular shuttering systems and improved methodology of interconnection and maximum combination of technological processes.

References

1. http://library.stroit.ru/articles/verhstroy/
2. http: // fasad-rus.ru/vysotnoe-stroitelstvo-v-rossie-article 466.htme
3. J.S. Volkov, Construction expert,14 (2003)
4. M.F. Markovsky, Architecture and construction, **1,** 44-47 (2007)
5. V. Ali. Mir, Electronic journal of Structural Engineering, **1,** 2-14 (2001)
6. P.P. Oleynik, T.K. Kuzmina, Mates Web of Conferences 1177,00129 (2017)
7. M.F. Markovsky, New opportunities in building technologies (2008)
8. V.I. Telychenko, Highrise building, **2** (2008)
9. T.K. Kuzmina, N. Cherednichenko, MATEC Web of Conferences **5**, 05012 (2016)
10. https://ais.by/story/12614
11. https://infopedia.su/17 y 4dc5.html
12. Construction" National Research Center, MDS 12 -23.2006 (2006)
13. P.P. Oleinik, T.K. Kuzmina, V. Zenov, MATEC Web of Conferences, **86,** 05019 (2016)
14. Rules for construction, TKP 45-1.03-109-2008 (2008)
15. B. Tour, M.F. Markovsky, A. Shherbach, Architecture and construction, **2,** 72-81 (2008)

Installation of ventilated facades without scaffolding in high-rise buildings

Lyubov Gnedina [1,*] *Arina* Muchkina[1], *Alexander* Labutin [2]

[1] Moscow State University of Civil Engineering, Yaroslavskoye shosse 26, Moscow, 129337, Russia
[2] LTD "Polystyrene concrete", Ivanovo, Russia.

Abstract. This article consider the use of polystyrene concrete blocks during assembling enclosing structure of ventilated facades in high-rise monolithic housing construction. Comparing with traditional technology devices hinged ventilated facade the main advantage of the proposed design is an exception of using scaffold, that leads to a cheapening of the enclosing structure. Proposed solutions are confirmed by patents of the Russian Federation.

1 Introduction

Scientific and industrial cooperation of the Department of Construction materials Ivanovo State Architectural and Construction Academy and Labutin & Co. started in 1996. The results of this cooperation are: the creation of "LIKO" raw mix for the production of filler heat insulation [1], a number of publications in the press [2,3,4,10], check recording of calculation programs in sector funds of algorithms and programs [5 and 6], successful implemented developments in construction (sinus filling of grouted masonry of a 5-storey residential building in Pedagogichesky Lane and the assembling of heat-insulating layer on the roof of the baby food plant in Ivanovo).

2 Materials and Methods

Prefabricated blocks, wall panels (Fig. 2), beams and lintels made from custom (poured) heat insulation (Fig. 2). They passed a comprehensive certification test in the research, design and technological institute of concrete and reinforced concrete - NIIZhB named after A.A.Gvozdev and the Central Research Institute of Building Constructions named after VA Kucherenko - CNIISK and are widely used as load-bearing structure and enclosing structures. Monolithic high-rise housing construction as enclosing structures, as well as in the construction of public. Material and products made of the mixture proved its viability and included in the Code of Rules 32-101-2004. Fig. 3-5

* Corresponding author: gnedinalyu@mgsu.ru

Fig. 1. a) The polystyrene concrete block with facing layer, b) The polystyrene concrete wall panels.

Fig. 2. The polystyrene concrete lintel.

Fig. 3. Construction of the cottage with block

Fig. 4. Construction of a multistory building.

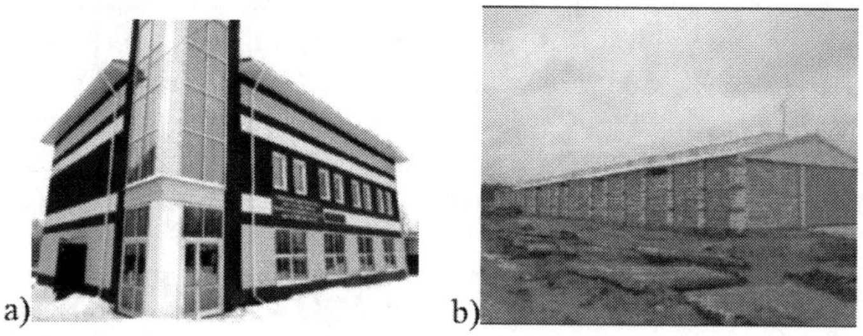

Fig. 5. Public building made of polystyrene concrete blocks and Industrial building for agricultural purposes.

The advantages of polystyrene concrete over other slubby and poured heat insulations:
– The availability of raw materials;
– Ease of manufacture (does not require special high-tech equipment), minimum energy consumption for manufacture and relatively low cost;
– Adhesion of the filler material to the building materials and construction is comparable with the adhesion of construction masonry mortar;
– High thermal performance and saving them for a long time (for example, for the moderate climate Russia is not more than 200 mm in the thickness of the desired block of polystyrene grade 600 density);
– Massive foundations aren't required for low-rise buildings;
– Reducing the load on foundations in multi-storey and high-rise buildings;
– Heavy lifting equipment aren't required for the construction of prefabricated buildings;
– Ecological and biological stability is comparable to concrete and reinforced concrete;
– High maintainability;
– The possibility of forming on any templates;
– The possibility of ventilated facades without scaffolding

The disadvantages are low impact strength and the need for exterior and interior decoration of walls.

The last drawback is eliminated by the device inside polystyrene concrete blocks embedded details made of roll-formed sheet shape (Fig.10 and 11) [7 and 8]. Guides for the gypsum covering constructions (indoors) or fasteners for mounting tiles (outdoors) are attached to them. It is also possible to install ventilated facade that helps to remove condensation from the fence body.

3 Results

A construction of multi-storey buildings with ventilated facades is widespread in Russian cities at the present time. Conventional (classical) scheme of the ventilated facades installation on frame buildings is:

1. A filling of openings (with brick, gas silicate, foam concrete blocks, etc.);
2. An assembling of scaffolds;
3. An additional insulation with slabs of increased rigidity;
4. An installation of sufficiently strong metal frame;
5. Cladding with porcelain tile or composite panels;

This technology of ventilated facade installation is quite expensive and time consuming.

An undoubted advantage over the classical scheme has the 400 mm wall construction made of D-300-400kg / m3 polystyrene blocks with an installation of ventilated facade without scaffolding and additional heat insulation [9].

The erection of this enclosing structure is carried out after the erection of load-bearing structures of walls, columns and monolithic inter-floor reinforced concrete floor (Fig. 9 and 10).

 https://doi.org/10.1051/e3sconf/20183303041

Fig. 6. The construction of the exterior walls using technology [9].

Fig. 7. The construction of the exterior walls using technology [9].

On the solid-cast overlap (1) (Fig.6 and 7) aligning profiles (2) are installed with metal brackets (6) in steps equal to the width of ceramic granite (3) or other finishing material. Thereafter, the cleats (4) are fixed on leveling profiles (2) and the porcelain tile (3) is installed from the ceiling to other ceiling. After cladding process polystyrene concrete blocks (5) with hardened metal padding (7) are fitted inside. Using the metal bracket (6), the unit is attached to the leveling profile (2).

So the construction of the enclosing wall is carried out without scaffolding directly from the overlap.

A heat insulation of such wall is not necessary, since the thickness of the polystyrene blocks is 400 mm and provides the required regulatory heat-shielding properties.

Taking into account the foregoing, the erection of a walling structure with a ventilated façade from interfloor floors has the following advantages:

1. There are no costs for the installation of scaffolding;
2. There are no costs for insulation, hydro and wind-proof membrane;
3. The reduction of metal construction costs;
4. The reliability of construction by direct attachment to the porcelain stoneware wall with a ventilated gap of 30-40 mm (in the classical device ventilated facade porcelain stoneware is at a distance of 200-250 mm from the wall);
5. The reduction of the time of construction and installation works and ensuring the continuity of the process sequence of the construction of the enclosing structure (simultaneously conducted facing and filling of openings);
6. The total reduction in the cost of construction and installation works in 2 times.

Cost of 1 m^2 of the construction (construction and installation work + materials + transport) is 5500-6000 rubles. The above technology was developed by the Director of

OOO Polystyrene Concrete (Labutin and Co) by Alexander N. Labutin and patented [9]. (phone for communication: +7 903 878 06 40). This technology has already been worked out in practice. Currently, the company carries out in the city of Ivanovo construction of multi-storey buildings with a ventilated facades without scaffolding and additional insulation. All works are carried out with interfloor overlappings.

4 Conclusions

Summarizing the above we can conclude that it became possible to abandon the multi-layered enclosing structures with the internal location of the heat-insulating layer, and to return to proven and well-established design solutions of almost single-layer walls made of polystyrene concrete on a new qualitative level. This method will significantly reduce one-time and operating costs. In terms of heat loss such enclosing structure will have only one "narrow" place - contiguity place of window and door blocks to the body of the masonry or panel, so in this direction further scientific and engineering developments should be implemented.

References

1. S.V. Fedosov, A.M. Ibragimov, L.Yu. Gnedina, A.N. Labutin, *Syr'evaya smes' dlya izgotovleniya zalivnogo uteplitelya* (1998)
2. S.V. Fedosov, L.Yu. Gnedina, *Zalivnoy uteplitel' na osnove penopolistirola* (1999)
3. L.Yu. Gnedina, *Optimal'noe mestopolozhenie uteplitelya v mnogosloynykh ograzh-dayushchikh konstruktsiyakh* (2000)
4. L.Yu. Gnedina, *Zalivnoy uteplitel' «LIKO» v trekhsloynykh ograzhdayushchikh paneyakh* (2006)
5. S.V. Fedosov, A.M. Ibragimov, L.N. Aksakovskaya, *Raschet temperaturnykh poley raspredeleniya potentsiala perenosa massy v trekhsloynoy stenovoy paneli* (2003)
6. S.V. Fedosov, A.M. Ibragimov, L.Yu. Gnedina, S.A. Ignat'ev, *Raschet tolshchiny teploizolyatsionnogo (srednego) sloya trekhsloynykh stenovykh paneley (stena 2)* (2005)
7. Pat. № 74142 RF. 2007138288/22 15.10.2007 (2008)
8. Pat. № 143759 RF. zayavle-no 2014109155/03 11.03.2014 (2014)
9. Pat. № 164519 RF. 2016109385/03 15.03.2016 (2016)
10. L.Yu. Gnedina, A.M. Ibragimov, A.A. Titunin, A.N. Labutin, Building materials, **10** 21-23 (2016)

Risk management at the stage of design of high-rise construction facilities

Violetta Politi [1,*]

[1]Moscow State University of Civil Engineering; 26, Yaroslavskoe sh., Moscow, 129337, Russia

Abstract. This paper describes the assessment of the probabilistic risk of an accident formed in the process of designing a technically complex facility. It considers values of conditional probabilities of the compliance of load-bearing structures with safety requirements, provides an approximate list of significant errors of the designer and analyzes the relationship between the degree of compliance and the level of danger of errors. It describes and proposes for implementation the regulated procedures related to the assessment of the safety level of constructive solutions and the reliability of the construction process participants.

1 Introduction

High-rise multi-profile buildings should be classified as technically complex construction projects. As we know, with the increase in the number of characteristics and elements of any facility, the number of probabilities of failures in the operation of the system elements also increases. Traditionally, the reliability of load-bearing structures is the main indicator ensuring safety of operation at any stage of the life cycle for the construction facility. However, for a complex construction facility, safety must also be ensured by the reliability of engineering systems and equipment. Also, special attention is paid to the foundation, which bears a colossal load. The most important factors for ensuring the safety of people inside a high-rise building include fire protection. Therefore, it is necessary to solve the problems of ensuring reliability of both the structural part and the operation of the building, strive to reduce the material consumption, optimize the ventilation and hydraulic regime, and also the energy consumption regime at the design stage of a complex facility. It is possible to manage the risk of erecting and operating a technically complex facility on the basis of the development and implementation of regulated procedures related to the assessment of the safety level of constructive solutions and the reliability of both engineering systems and all construction process participants.

2 Materials and methods

Distinctive features of high-rise construction are high load on load-bearing structures, in particular, on bases and foundations, the increased importance of the impact of natural and man-made factors, such as seismic impacts, solar radiation, aerodynamics, vibrations, and uneven load of structural elements [1]. These features require increased attention of

* Corresponding author: polity_violca@list.ru

designers to the selection of design and space-planning solutions. It should be noted that for high-rise buildings, complicated in the architectural and planning terms and with a large extension of engineering networks, the most important factors of trouble-free operation are the correct designer's choice of automatic heating, ventilation, gas supply, electricity, and water supply systems.

All of the above factors indicate the complexity of designing and operating high-rise buildings and, as a consequence, the high probability of errors in the design, construction and operation of a complex construction facility.

The risk of an accident is a vector quantity consisting of an adjustable part, for example, a facility risk, and an unadjustable part generated by the external environment.

A facility risk is the risk that can be subject to the targeted effect, and which is formed by errors of designers, builders, supervisors and controlling organizations.

In most cases, accidents of buildings and structures, as well as the failure of engineering systems supporting the life of the consumer, occur as a result of a collision of two independent events - the appearance of an external unintended impact and effects determined by a set of human errors that led to a reduction in the structural safety of the facility [3, 4].

The forecast of an accident risk of construction facilities on the basis of the classical probabilistic approach is impossible for two reasons. First, accidents of buildings and structures are very rare events, and second, unpredictable human errors allowed in the process of designing, erecting and operating a construction facility, introduce significant uncertainty in the response (behavior) of the supporting frame of the facility to external influence. Since the risk of an accident and the degree of uncertainty of the technical state of the facility's structural framework are closely related concepts, the accident risk assessment should be based on a logic-probability approach based on probability theory theorems, the methods of the theory of fuzzy sets, fuzzy logic techniques and decision-making methods under uncertainty [5].

The values of the actual risk of an accident, physical (structural) wear and safe operating life of buildings and structures are also interrelated. To determine these values, the accident risk distribution law should be used, which is an integral indicator of the level of structural safety of a construction facility.

The probability of occurrence of the risk of an accident at a complex construction facility should be carried out on the basis of functioning of an expert system, which is the result of the interaction of the expertise of specialists, the mathematical apparatus and information technology.

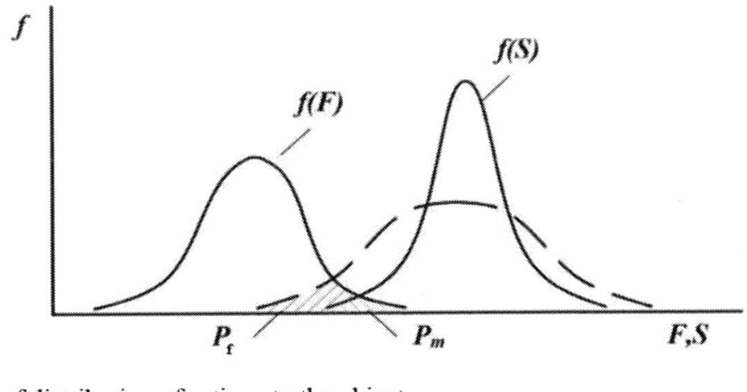

Fig. 1. The law of distribution of actions to the object.

(F - the force of the impact; S - the resistance force; P_F – the fact of the likelihood; P_T-the theoretical probability of an accident; «point line» – the influence of human errors on the operation of the law)

2.1 About the Accident Risk Distribution Law

So, it is known from the practical experience that the determining factors for the occurrence of emergency situations are the errors of the construction participants. It is also known that by the end of the facility construction, the actual probability of an accident *(P_f)* in comparison with the theoretical *(P_t)* increases several times. Further, the actual probability of an accident on an unlimited number of new buildings can be represented as:

$$P_f = P_t + P_d \tag{1}$$

P_d – additional probability generated due to errors of the construction process participants (designers, builders, experts, supervisory and controlling organizations).

To determine P_d, the hypothesis theorem (Bayes formula) is used, which allows one to recalculate the a priori probabilities in terms of information about the errors committed. Two opposite events are introduced: B – errors in erecting load-bearing frameworks of buildings exist, $B*$ – no errors. Events B and B^* form a complete group of incompatible events, are identified in the construction process, and before the construction starts, the set C is empty.

The following designations are accepted:

P(B*)=v – the probability that there are no errors in the constructed building,

P(B) = (1–v) – the probability of the opposite event.

A priori (before construction) probabilities of occurrence and non-occurrence of an accident of buildings (structures) are known and are respectively equal to P_m and $(1–P_m)$. The ratio follows from the Bayes formula:

$$P_d = \frac{P_m P\left(\frac{C}{A}\right)}{P_m P\left(\frac{C}{A}\right) + (1-P_m) P\left(\frac{C}{A^*}\right)} \tag{2.1}$$

where P(C/A) is the probability of the event C, provided that the accident occurs; P(C/A*) is the probability that the event C will occur.

Based on the use of fuzzy logic techniques, we can adopt the following relations: P(C/A*) = v, in turn, P(C/A) = 1 - v.

Indeed, if an accident does not occur during the construction or operation of the facility, then the event $B*$ occurs, in the opposite case, the event B occurs.

After substituting these equalities in the Bayesian relation, we have:

$$P_d = \frac{P_m(1-v)}{v} \tag{2.2}$$

The addition of two probabilities gives the following relationship: $\frac{P_d}{P_m} = \frac{1}{v}$, where v is the structural reliability of the building.

It is important to understand that the probability P_t exists and can hypothetically arise if the project requirements will not be violated in the design and erection of the bearing framework of the building. However, in practice, such violations always occur, and then the ratio of the two quantities will always be greater than one, namely $(P_f \,/\, P_t) > 1)$. This ratio shows how many times the actual risk of an accident is higher than the theoretical risk taken into account in the building design. Consequently, it can be adopted as an integral indicator of the technical state of the load-bearing framework of a high-rise construction facility. Therefore, the result of calculating this ratio can be correlated with the magnitude of the risk assessing the probability of an accident at the construction facility r :

$$r = P_f/P_t = 1/v \tag{3}$$

However, there are no statistical data on the possible values of r. Therefore, in order to find the distribution law for the probability density of the occurrence of a random variable r on an unbounded set of new non-typical high-rise buildings, the author sets forth the following axioms based on the laws of logic and practice in the construction industry:

Axiom 1. The probabilities of $r \le 1$ are equal to zero.

Axiom 2. The distribution curve is strictly asymmetric; the mode of the random variable r is shifted to the left of the mean value.

The one-parameter Rayleigh distribution, which has the following form, corresponds to the above axioms:

$$f(x) = (r-1)/\sigma\,\sigma^2\,exp[-(r-1)2/2\sigma^2] \tag{4}$$

It is known that in the distributions the parameter σ is related to the mathematical expectation (average value) of the accident risk R as follows:

$$R = 1 + 1{,}25\sigma \tag{5}$$

To determine the mean value of the risk R, one should use the following relation derived from (3):

$$R = 1/Mv, \tag{6}$$

where Mv - is the mean value (mathematical expectation) of a random variable v which numerical values are in the range from 0 to 1.

3 Results

On the Management of Building Accident Risk at the Design Stage

Below is a list of comments most often encountered when reviewing project documentation in the section "Constructive solutions and space-planning solutions" in terms of design solutions:

- Design tasks and design documentation do not specify or incorrectly specify the identification characteristics of the buildings (structures) being designed (reconstructed) (including the belonging to hazardous production facilities specifying the hazard class, the level of responsibility of buildings and structures, the general seismic zoning map adopted to determine design seismicity in the design, etc.);
- Calculation substantiation of design solutions for two groups of limiting states and calculations of the influence of new construction on the structures of the surrounding development are not provided;
- Calculations for an emergency situation (calculations of resistance to a progressive collapse) are not provided for buildings and structures with an increased level of responsibility and for public and industrial buildings with a normal level of responsibility;
- Information is not provided on the strength and deformation characteristics of the soil at the base of the foundations, as well as drawings with characteristic geological sections with the foundations marked;
- The results of determining aerodynamic coefficients based on the results of purging of models of structures in wind tunnels or on recommendations developed by specialized organizations are not provided for buildings and structures with an increased level of responsibility;
- There is no provision for scientific and technical support in the design, manufacture and installation of structures, as well as their technical monitoring during erection and operation for buildings and structures with an increased level of responsibility;
- There are no construction tasks for the arrangement of process equipment with loads and references;
- The results of a survey of the technical condition of the erected buildings and structures necessary for the development of project documentation are not performed and are not provided.

Obviously, to increase the safe resource of the constructed facility, it is necessary to develop procedures for regulating the possible risk of an accident planned for the facility erection. The initial stage of regulation is to predict the probability of occurrence of an accident risk at the initial, design stage. The information base for predicting can be an analysis of the effectiveness of quality management systems, which applies to each construction process participant. According to practical observations, a high level of quality management systems ensures a high level of safety of the designed facility. It is possible to predict the probability of occurrence of an accident risk using equation (6), and the average reliability levels *Mp* of the groups of the same type of bearing framework structures are determined according to the results of the examination of the efficiency of the quality systems used by potential construction process participants [6, 7, 8].

Table 1. Requirements of ISO 9001 for elements of quality management systems.

Pos. No.	Elements of quality management systems
1.	Availability of quality policy and quality work motivation system
2.	Correspondence of engineering and workers' professions to the production profile
3.	Availability of actualized programs for upgrading the skills of different categories of employees
4.	Compliance of mechanisms and equipment with the requirements of the quality of technological operations
5.	Availability of a maintenance system and a system for documenting the procedures of performing technological operations
6.	Regular internal audit conducted to assess the effectiveness of the quality management system
7.	Availability of programs, techniques, experts with definition of their responsibility, forms of documentation and registration of incoming and production quality control of products
8.	Availability of equipment, metrological and laboratory support of procedures for identifying product quality

The average reliability level *Mp* of the structure group can be estimated from the probability *P(B)* of its compliance with the requirements of regulatory documents in terms of ensuring strength, rigidity and stability. Indeed, it follows from the equation of the total probability that

$$Mp = P(B)=\sum P(B/qi)\ P(qi), \tag{7}$$

where q = {qi} = {MCD, M*BD, MB*D, MBD*, M*B*D, M*BD*, MB*D*, M*B*D*} is the complete group of events including the following independent events: M – there are no errors of suppliers of materials (products); B – there are no errors of builders; D – there are no errors of designers; M*, B* and D* are events opposite to events M, B and D.

The conditional probabilities *P(B/qi)* in this equation are invariants that have the meaning of the probability of the correspondence of load-bearing structures to safety requirements while observing certain conditions (Table 2).

Table 2. Values of the conditional probabilities of compliance of load-bearing structures with safety requirements.

Pos. No.	Probability value	Comment
1.	P(B/q1) = 1	- no errors of all construction participants
2.	P(B/q2) = a	- there are only suppliers' errors
3.	P(B/q3) = b	- there are only builders' errors
4.	P(B/q4) = c	- there are only designers' errors
5.	P(B/q5) ≅ ab	- no errors of designers only
6.	P(B/q6) ≅ ac	- no errors of builders only
7.	P(B/q7) ≅ bc	- no errors of suppliers only
8.	P(B/q8) ≅ abc	- there are errors of all construction participants

Invariants in aggregate represent a priori information that can be compiled from an analysis of the causes of accidents.

In the Russian Federation, ~80% of accidents occur due to errors made by the construction process participants, including ~20% due to the fault of suppliers, ~50% due to the fault of builders, and ~10% due to the fault of designers. Thus, a =0,8, b = 0,5, c = 0,9.

If we designate the probabilities of the independent events *M, B* and *D* as $P(M)=\mu.m$, $P(B)=\mu.b$, $P(D)=\mu.d$, then the following mathematical model follows from the total probability equation:

$$Mp = \mu.m\mu.b\mu.d+0,8(1-\mu.m)\ \mu.b\mu.d\ +0,5\mu.m(1-\mu.b)\ \mu.d+\ 0,9\mu.m\mu.b(1-\mu.d)+0,4(1-\mu m)(1-\mu.b)\ \mu.d++0,72(1-\mu.m)\ \mu.b(1-\mu.d)+0,45\mu.m\ (1-\mu.b)(1-\mu.d)+0,36(1-\mu.m)(1-\mu.b)(1-\mu.d) \quad (8)$$

In the equation, the values $\mu.d$, $\mu.m$ *and* $\mu.$ bhave the meaning of the degrees of compliance of load-bearing structures with the requirements of regulatory documents. The equation allows to predict the average level of reliability *Mp* of groups of structures of the bearing framework, both at the stage of the building design, and at the stage when the facility project already exists

4 Discussion

Let's consider the case of declaring the level of structural safety planned for the erection of a building (structure) when the facility project already exists. When determining $\mu.d$, the expert's tasks are:

1) to track the presence of significant (weighty) errors in the project (Table 3);
2) to assess the level of danger of errors found in the project;
3) to take measures to eliminate the errors found.

Table 3. Sample list of significant designer's errors.

Error No.	Characteristic of possible errors in the design of structural elements
1.	The physical and mechanical characteristics of soils adopted in the project do not correspond to the actual state of the foundation. There are no laboratory tests of soil samples with undisturbed structure
2.	When choosing the foundation calculation model, the geological engineering surveys did not reveal or take into account the characteristic dependences of the soil deformation under load
3.	Dimensions of the foundation and the position of the masses on the facility plan do not ensure the uniformity of settling
4.	The design scheme of the supporting frame does not correspond to its actual work under load
5.	When entering the initial data in the software, errors in dimensions and values of loads, resistances, and rigidity were made. The calculation is not duplicated according to another software
6.	Non-compliance of the designed nodal connections of the framework elements with node rigidity estimation scheme adopted in the design
7.	Bonded structures do not provide the required rigidity of the supporting frame of the facility

The level of danger of gross errors found in the project is estimated according to a special rule (Table 4).

Table 4. Relationship of the degree of compliance with the level of danger of errors made.

Error danger level	Degree of effect of errors found on safety in a linguistic form	Project reliability level
First	Virtually no effect	0.939
Second	Imperceptible effect	0.987

Third	Insignificant effect	0.686
Fourth	Significant effect	0.568
Fifth	Very significant effect	0.828
Sixth	Extremely high effect	0.500

The danger levels and the degree of compliance are determined for each significant error found in the project. Together they form a fuzzy set of project quality assessments. In accordance with the algebra of fuzzy sets, the final numerical estimate of the project is determined as μd= min { (μ.d)i }.

To determine μ.m and μ.b, a hypothesis is introduced about the direct dependence of the safety level of constructive solutions of afacility on the effectiveness of the functioning of quality systems of organizations participating in the investment construction process.

The task of the expert is to assign the degrees of correspondence (μ.m)i and (μ.b)i for all elements of the quality system (Table 1).

By analogy with μd, the final numerical estimates of the quality systems of the supplier organization and the contractor organization are determined as: μm = min {(μm)i }, μb = min {(μ.b) i }.

The technology of declaring the safety of constructive solutions of a facility at the stage of design development consists of the following operations:

1. A "tree" of the states of the facility is formed, which is a hierarchical sequence of the erection of groups of the same type of structures of the supporting framework.
2. A numerical estimate of the project μ.d. is determined.
3. Based on the results of diagnostics of quality systems, μ.m and μ.b are assigned for all construction process participants.
4. According to the equation (8), the values of the average reliability levels Mp of all groups of structures of the supporting framework of the facility are determined.
5. According to the equation (8), the actual risk of an accident R after the facility erection is determined, and this risk is compared with the normal risk of an accident Rn.
6. If R>Rn, a diagram is built for the average predicted reliability levels Mp in groups of the same type of structures of the supporting framework of the facility, specifying the standard normal Rn level of the structural reliability.
7. The diagram identifies the construction participants which potentially will bring the greatest risk of the accident to the facility, and an administrative decision is made based on the information in p. 3.

Conclusions

The methodology for calculating the risk of an accident and the safe residual life of buildings and structures are closely related to procedures for assessing and regulating the level of structural safety of both buildings under construction and in operation.

The application of this methodology makes it possible to determine the "contribution" of each group of structures of the supporting framework to the value of the accident risk of the facility under study; calculate the safe residual life of the facility and make a prediction of the period of operation, after which measures should be taken at the facility to reduce the risk of an accident.

References

1. I. Ilin, O. Iliashenko, A. Levina, Vision 2020, 768-781 (2016)
2. I. Ilin, A. Anisiforov, WSEAS Transactions on Business and Economics, **11**, 757-764 (2014)

3. O. Kalinina, O. Valebnikova, Advances in Intelligent Systems and Computing, **692**, 1315-1322 (2018) DOI - 10.1007/978-3-319-70987-1_139
4. A. Zaychenko, S. Gutman, O. Kalinina, Advances in Intelligent Systems and Computing, **692**, 453-462 (2018) DOi - 10.1007/978-3-319-70987-1_48
5. V. Kankhva, Procedia Engineering, **165,** 1300-1304 (2016) DOI:10.1016/j.proeng.2016.11.855
6. V. Kankhva, MATEC Web of Conferences, **106**, 08027 DOI: 10.1051/matecconf/201710608027
7. E. Nezhnikova, IOP Conf. Series: Earth and Environmental Science, **90,** 012161 (2017) DOI:10.1088/1755-1315/90/1/012161
8. A. Jones, G. Fallon, R. Golov, European Business Review, **12**, 187-197 https://doi.org/10.1108/09555340010336871

Student satisfaction as an element of education quality monitoring in innovative higher education institution

Elena Razinkina[1], *Ludmila* Pankova[2], *Irina* Trostinskaya[3], *Elena* Pozdeeva[3], *Lidiya* Evseeva[3], and *Anna* Tanova[3,*]

[1]Peter the Great St. Petersburg Polytechnic University, Rector's Office, 195251 St.-Petersburg, Russia
[2]Peter the Great St. Petersburg Polytechnic University, Basic Educational Programmes Department, 195251 St.-Petersburg, Russia
[3]Peter the Great St. Petersburg Polytechnic University, Advertising and Public Relations Department, 195251 St.-Petersburg, Russia

Abstract. Topicality of the research is confirmed by increasing student involvement into the educational process, when not only the academic staff and administration participate in the improvement of higher education institution's activity, but also education customers – students. This adds a new dimension to the issue of monitoring education quality and student satisfaction with higher education. This issue echoes the ideas of M. Weber about the relationship between such components as cognitive motivation, personal development and student satisfaction with higher education. Besides, it is essential to focus on the approach of R. Barnet to defining the quality of education with the emphasis on a priority of development of an educational institution as the system that meets customers' needs. Monitoring student satisfaction with education quality has become an integral part of the educational process not only in a number of European universities, which have used this monitoring for decades, but also in Russian universities, which are interested in education quality improvement. Leading universities in Russia, including Peter the Great St. Petersburg Polytechnic University, are implementing policies targeted at increasing student satisfaction with higher education quality. Education quality monitoring as a key element in the system of providing feedback to students contributes greatly to this process.

1 Introduction

Global civilizational changes increase the value of the scientific and technological potential in the society. Information, knowledge, scientific breakthroughs become a significant resource in the unifying medium that bonds people and technology. Therefore, great importance is attached to the issues related to transformations of roles common for subjects in the educational space and an increase in the professional potential of university graduates. All the above-mentioned considers the quality of professional education as the priority trend

* Corresponding author: tanovaann@mail.ru

for competitive professional training and makes educational institutions actual subjects of the market, capable of accumulating their competitive advantages to improve the education quality.

Expansion of the higher education system in Russia, changes in its structure and content are caused by its increasing contribution into the economic progress, by the improvement of living standards, and by the rising demand for qualified staff. The society development trends, determined by the fourth industrial revolution, imply modification of educational strategies and building a new cultural and educational model. All these prerequisites make monitoring in the education quality management, including student satisfaction monitoring, relevant to the today's situation.

The object of this research is students, as they have always been the most responsive part of the young generation. They are the social community capable of bringing educational innovations into the economy. Both student responsiveness and good education lead to further social activity. Integrated into the social structure of the society, the student community shows their aspiration for personal and professional identity, creativity, involvement into socio-economic and cultural reforms.

It is worth noting that global acceleration and extensive accumulation of new information and communication products surge the demand for specialists who would be knowledgeable, broad-minded, flexible, capable of adapting their occupational fields to market needs and to the necessity to change their personal or professional preferences, who would be skilled at dealing with information and knowledge, adopting high-tech products to achieve their life goals. Therefore modern workers should continuously enhance quality, their personal and professional skills. Global information environments ensure emergence of new knowledge, relevant to challenges of the technogenic civilization.

According to M. Weber, the goal of a higher education institution is to form and to anchor in the student perception the so-called "philosophy of life", their future role as researchers and lecturers, which was the key role for M. Weber at the turn of the century. He saw the supreme purpose in serving the science, being committed to the common cause, aspiring to acquire knowledge. All this is correlated with the student satisfaction with higher education and issues of student personal development [1].

2 Problem setting

As early as 1970s European universities were aware of the necessity to analyze student satisfaction with their academic program to achieve their educational goals [2]. At present there has been accumulated great experience of satisfaction monitoring studies: in Europe they amount to 30 years, in UK – 20 years [3]. The monitoring research has contributed to the modern trend and set the course for new research. However, the Russian experience in the area of education research is insufficient.

Satisfaction monitoring is based on studying student satisfaction feedback, which is a sign of successful performance for a higher educational institution [4]. *Student satisfaction feedback* is defined as opinions of students about the service they receive as students. This may include perceptions about the learning and teaching, organization of the educational process, the learning support facilities, the learning environment [5]. International researchers agree that the student feedback, which is assiduously collected in higher education institutions, will contribute to continuous education quality improvement. Along with public information data, the research helps upgrade education management. Public awareness also matters: it forms a new student culture based on their participation in management. Students have to be sure that changes in their universities are caused by the feedback they show in the polls. Criteria for measuring satisfaction depend on the university profile. For instance, researchers at the School of Business and Law, Liverpool John Moores University, provided 60 variables that influence student satisfaction [6].

In Russia opinion polls illustrate that university studies discourage a number of students and make them disappointed at their chosen occupational field [7]. This fact forces major Russian universities to do comprehensive research into factors and indicators of student satisfaction with education quality.

Satisfaction with the quality of engineering education is significant in preparing a qualified innovation-oriented engineer who possesses the main professional competence – ability to create "human-dimensional" systems and apply "organizational creativity" [8].

Building a satisfaction model is based on the methodology of empirical interpretation, which implies searching for empirical elements [9]. Objectification of the "satisfaction" category means differentiation between satisfaction as the state of consciousness and satisfaction as the behavioral reaction. Moreover, satisfaction is defined as a derivative of a set of factors that reflect the synthesis of the value of a measured object and the quality of a measurer, i.e. a student.

The satisfaction model combines the results of satisfaction measurement in accordance with various parameters of the educational process and situation, behavioral aspects showing students' life strategies and their trajectories to fit into the labor market, and institutionally determined factors resulting from the trends of the modern educational paradigm. Student satisfaction with higher education and education quality measurement depend on student values and attitudes to education – incentives to study, value of education and knowledge.

The ultimate value of education can be defined by a number of the following meanings: ability development, cognitive needs, general literacy. Besides, the attitude to education is explained by the factors: belonging to different social groups, level of income, education of parents, modernization in the system of education. In total these factors determine the socio-situational aspect of the student attitude to education [10].

Satisfaction with higher education is a parameter characterizing the professional socialization process of students as subjects of the educational activity. The education satisfaction model consists of the following main elements: satisfaction with social status; satisfaction with the profession/area of expertise students are trained for; satisfaction with the educational process [11]. In addition, satisfaction with the curriculum may be viewed as an integrative indicator marking the degree of fulfillment of student social expectations from educational activities, formed during their professional socialization.

The importance of higher education satisfaction in the modern legal framework makes it urgent to measure the quality of educational services on a regular basis, which is possible when combining the statistical and sociological approaches implemented in the form of monitoring. Monitoring satisfaction as part of sociological research in the system of education quality management enables getting the full picture of student attitude to education and characterizing the educational environment, putting special emphasis on such aspects as: satisfaction with student life and the selected higher education institution; with the educational process; participation in extra-curricular activities; the productivity of awareness-raising activities [12].

The practice of conducting education satisfaction surveys at engineering universities and colleges shows that satisfaction is considered as comprehensive assessment primarily based on the motivational aspect, whose positive parameters correspond with satisfaction indicators (surveys carried out at engineering universities and colleges in the Tyumen region, The Republic of Bashkortostan and the Republic of Komi) [13].

According to the survey, first year students of Siberian State Technological University assess the peculiarities of studying at university in comparison with school quite adequately and they also see the special features of university education, especially in terms of greater independence. They also note the need for free creative activities both during the classes and after. It is evident that freedom is the most important prerequisite for the development of a self-actualizing creative person. Thus, ensuring freedom to obtain satisfaction with the educational process, society can, on the one hand, help individuals to satisfy their needs, and on the other – provide students with an opportunity for self-development and responsibility enhancement [14].

Quality, as one of the key indicators of a successful higher education institution, is viewed here as a number of characteristics of an educational service, which result in the necessary skills and expertise gained by university graduates. In this respect, the quality of educational services is the major concern of the market, which is defined by competitiveness of the educational establishment and is summarized in the form of academic rankings [15]. According to R. Barnet, three main approaches, characterizing quality, are agreed upon today: the objectivist, relativist and developing [16]. Integrating these approaches helps to describe the activities of an educational establishment as a system (its systemic measurable qualities), it enables weighing the achievements in a multi-tasking environment and indicating the strategic area of development towards the utmost satisfaction of consumer needs.

When measuring quality, customer satisfaction with educational services serves as the core element, as establishments depend on their consumers, and they have to take into account their needs and interests, fulfill their requirements and try to exceed their expectations [17]. The most significant areas of consumer monitoring, used to study satisfaction and quality assessment, lie in defining the indicators connected with external consumers (school-leavers, parents, employers); measuring the quality of the teaching process; defining the indicators reflecting students' progress.

At Peter the Great St. Petersburg Polytechnic University (SPbPU) student surveys are frequently carried out as online and paper questionnaires, which enables monitoring student satisfaction with education. The survey aimed to analyze the quality of educational programs offered at SPbPU was conducted in May and June 2017. The comprehensive assessment was based on the concept of satisfaction with education, in which satisfaction is formed as a result of systemic interaction between the following parameters:

- Correlation between the education received and the expectations of a student;
- Intentions of students in terms of further employment with regard to their field of studies;
- Opportunities for creativity and self-fulfillment offered by the University while studying, gaining skills and knowledge relevant for the current market situation;
- Wide and adequate information support, as well as support from the teaching staff and other university employees;
- Sufficient facilities;
- Satisfaction with the formal procedures accompanying the studying process;
- Comfortable environment for communication.

The following important factors were also taken into consideration in the survey: the high profile of the University, scientific focus of education, the opportunity to study foreign languages, the possibility to meet interesting people and take part in various projects, which is a top priority for leading universities and is especially welcomed by young people hoping to build up successful careers. In the questionnaire a lot of attention was drawn to the student awareness indicator, which does not only show the degree of involvement in the educational process, but it also defines the quality of work done by academic bodies and the supervisors of educational programs.

The given research was based on the main hypothesis that the quality assessment of a certain educational program directly depends on the fact how well a student is informed about the opportunities offered by various academic bodies of the University.

The following assumptions acted as complimentary hypotheses:

- Student satisfaction is influenced by insufficient awareness of certain separate elements of academic programs, which have been introduced only recently or those that have undergone innovative changes, such as project-based learning, distance learning, optional subjects;
- Being aimed at finding employment connected with their area of studies, students put a special emphasis on a practice-oriented approach to education.

3 Monitoring results

4040 students took part in the survey, representing all the institutes which SPbPU comprises.

First of all, it should be mentioned that, on the whole, the results of the survey demonstrate quite a high level of student satisfaction with education, i.e. 66% of the surveyed students (see fig.1). There is interconnection between the indicators on which the parameters of the concept of student satisfaction are based. Thus, 65% of the students replied that education lived up to their expectations, 68% of students had a clear vision of their future jobs, 64% of respondents were going to look for a job closely connected with their area of studies and 82% confirmed the fact that they had made an informed choice to study a certain field. These data show that more than 2/3 of students are implementing the strategy of practice-oriented training, gaining skills and knowledge relevant for the current market situation.

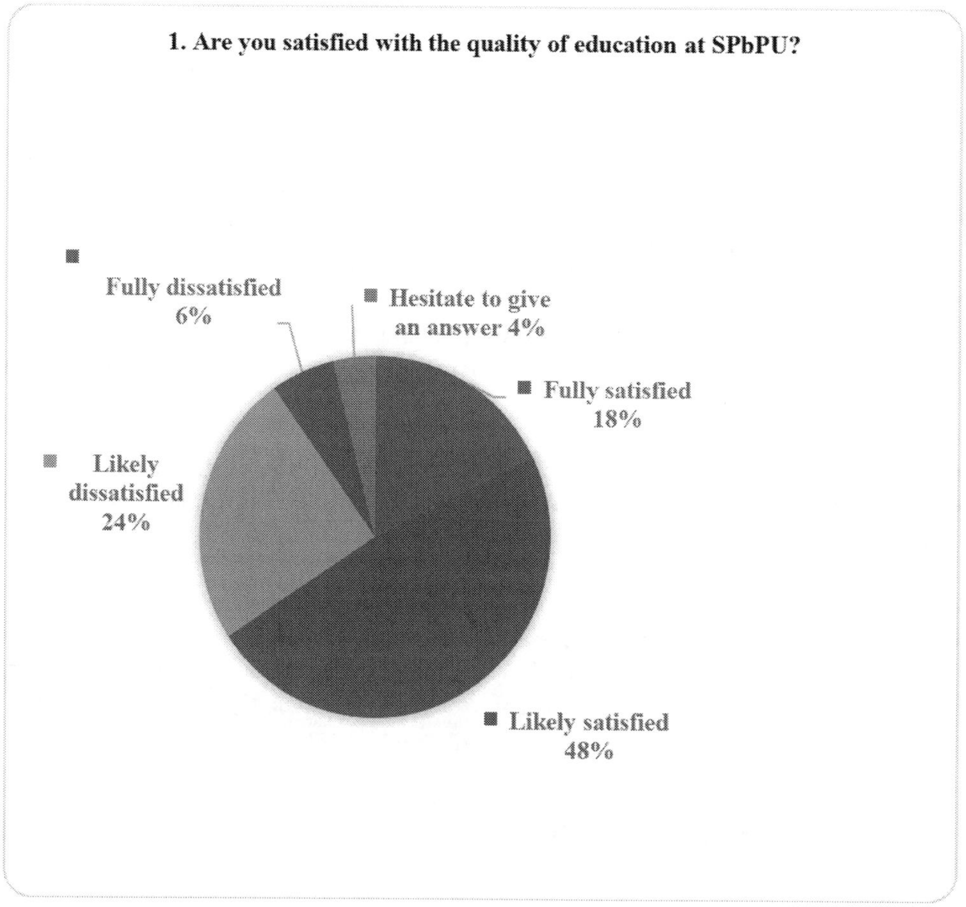

Fig. 1. Satisfaction with the quality of education at SPbPU

The practice-oriented approach to education is confirmed by the fact that students do various internships while studying at the Polytechnic University. Thus, only 8% of the students said that they were not doing an internship, while 92% of respondents replied that establishments where students could do an internship were chosen either by the University (12%), or by the students themselves (22%), or both ways were possible (52%) (see fig.2).

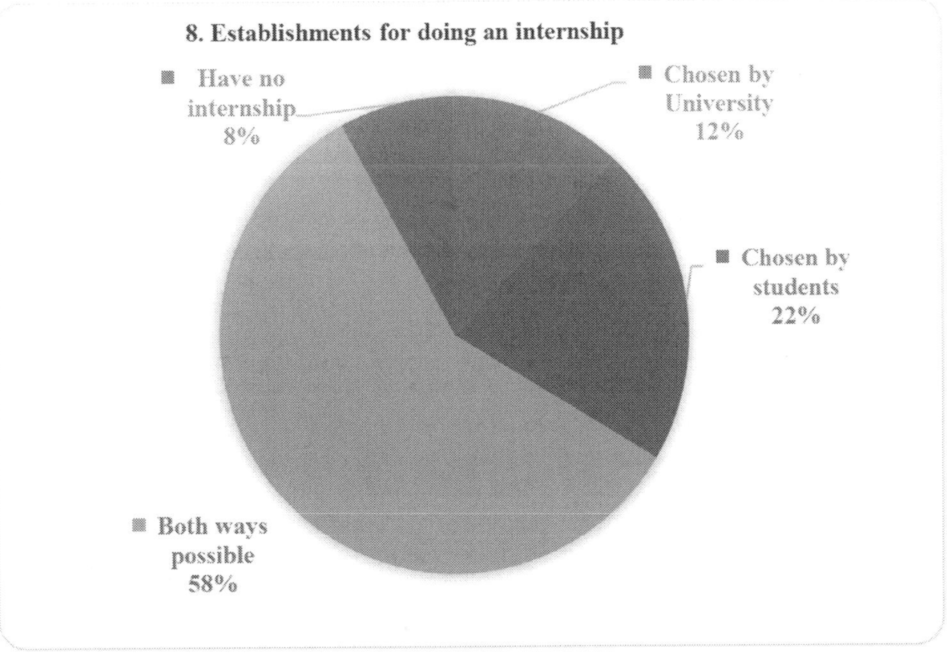

Fig. 2. Establishments for doing an internship.

Efficiency of an internship regarding the future employment connected with their current field of studies was mentioned by 68% of the students.

Extra-curricular activities also serve as a significant indicator of student satisfaction with education. Research has shown that on the whole 36% of the students are satisfied with the way leisure activities are organized (they gave 4-5 points), 34% awarded 2-3 points, and 30% of respondents gave 1 point and less (see Table 1). The results of numerous surveys demonstrate the fact that, as a rule, about one third of students take an active part in extra-curricular activities, while the rest of the students consider the way leisure activities are organized not as participants, but as spectators, attending the events held at the University.

When assessing the university facilities, the focus was on the following indicators: the condition of classrooms, infrastructure for social and cultural events, sporting facilities and equipment, the availability of various facilities and equipment (see Table 1). The results of the survey show that the availability of various facilities and equipment is not highly evaluated by the students, most of them giving 2 or 3 points to this parameter. This testifies the fact that, enjoying an outstanding reputation among students (80% and 58% of the respondents respectively replied that the high profile of the University and the education received there are worth 4-5 points), SPbPU still has room for improvement and that it should concentrate on developing its facilities.

Table 1. Assessment of Various Aspects of Educational Activities at the University.

	0 - 1 points	2 – 3 points	4 -5 points
The condition of the infrastructure for social and cultural events	25	33	43
The condition of gyms and other sporting facilities	27	41	32
The opportunity to use gyms and other sporting facilities	31	38	31
The condition of classrooms	18	49	33
The availability of various facilities and equipment	16	45	39
Leisure activities for students	30	34	36

Another area for further development may be found in the field of innovative educational technologies, including distance learning, project-based learning and optional subjects within a mobility module, as well as information support for these processes. To illustrate this statement with the help of some research results, it can be mentioned that 26% of the respondents approve of distance learning technologies, 40% do not approve and 34% are undecided. 79% of the respondents are unaware of the opportunity to select certain subjects within a mobility module (and 21% are, on the contrary, informed about it), 81% know nothing about project-based learning (19% claim the opposite).

4 Conclusions

The general conclusions based on the given research make it possible to admit that the sociological monitoring has justified expectations, as it has shown student satisfaction with higher education received at Peter the Great St. Petersburg Polytechnic University, and it has also highlighted some problem areas, as well as areas for further development. Further quality enhancement is primarily connected with the systemic development of facilities and equipment as a significant component of the educational process, increasing flexibility and individualization of the educational process by means of electronic and distance-learning technologies, offering more opportunities for practice-oriented learning, more active information support covering the advantages of innovations introduced, opportunities for self-development resulting from project-based learning, extra-curricular and leisure activities for students.

To sum it up, it should be mentioned that the hypotheses (the main one and the complimentary ones) proposed at the beginning of the given research were primarily confirmed when analyzing the results of the survey, therefore, the purpose and the tasks of the research have been fulfilled, which confirms its value and scientific prospects, as well as the fact that surveying can be further applied as an efficient tool for monitoring the quality of education.

References

1. M. Weber, *Self-Consciousness of European Culture of the 20th Century* (Politizdat, Moscow, 1991)

2. B. R. Morstain, Journal of Higher Education, **1,** 1 (1977)
3. V. A. Zotova, Journal of Creative economics, **11 (3),** 303 (2017)
4. M. Shah, C. Nair, J. Richardson, *Measuring and enhancing the student experience* (Chandos Publishing, Cambridge, 2017)
5. L. Harvey, Journal of Quality in Higher Education, **9 (1),** 3 (2003)
6. J. Douglas, A. Douglas, B. Barnes, Journal of Quality Assurance in Education, **14 (3)**, 251 (2006)
7. D. L. Konstantinovsky, *Education and science in Russia: state and development potential* (Center for Sociological Research, Moscow, 2016)
8. A. L. Andreev, *Education and science in Russia: state and development potential* (Center for Sociological Research, Moscow, 2016)
9. V. A. Yadov, *Strategy of sociological research. Description, explanation, understanding of social reality* (Dobrosvet, Book house University, Moscow, 1998)
10. Yu. A. Zubok, V.I. Chuprov, Journal of Sociological research, **8,** 103 (2012)
11. A. S. Spassky, Journal of Innovations in education, **5,** 63 (2003)
12. N. P. Makarkin, *Monitoring of students' satisfaction with educational process* (Publishing house Mordovian University, Saransk, 2007)
13. S. Yu. Frolov, Journal of Theory and practice of social development, **4** (2015)
14. A. V. Andrienko, Journal of Siberian Medical Sciences, **1** (2007)
15. N. A. Bonyushko, A. A. Semchenko, Journal of Omsk Scientific Bulletin. Series "Society. History. Modernity", **3 (129),** 31 (2014)
16. R. Barnet, *Improving higher education: total quality care* (SRHE and Open University Press, 1992)
17. S. S. Kotova, I. I. Hasanova, Journal of Education and Science, **9,** 43 (2016)

4D modeling in high-rise construction

Anastasiya Balakina[1], *Tatyana* Simankina[2*] and *Vitaly* Lukinov [3]

[1,2] Peter the Great St. Petersburg Polytechnic University, Polytechnicheskaya, 29, St. Petersburg, 195251, Russia
[3] Moscow State University of Civil Engineering, Yaroslavskoe shosse, 26, Moscow, 129337, Russia

Abstract. High-rise construction is a complex construction process, requiring the use of more perfected and sophisticated tools for design, planning and construction management. The use of BIM-technologies allows minimizing the risks associated with design errors and errors that occur during construction. This article discusses a visual planning method using the 4D model, which allows the project team to create an accurate and complete construction plan, which is much more difficult to achieve with the help of traditional planning methods. The use of the 4D model in the construction of a 70-story building allowed to detect spatial and temporal errors before the start of construction work. In addition to identifying design errors, 4D modeling has allowed to optimize the construction, as follows: to optimize the operation of cranes, the placement of building structures and materials at various stages of construction, to optimize the organization of work performance, as well as to monitor the activities related to the preparation of the construction site for compliance with labor protection and safety requirements, which resulted in saving money and time.

1 Introduction

With the complication of construction projects, there is a need to use more perfected and sophisticated tools for planning and construction management. Risk assessment at the design and construction stage is included in the main stages of the business plan development. The price of incorrect accounting of the risk component is the underfunding of the investment project. Therefore, when developing a business plan, the risk of underfunding, the likelihood of its occurrence, and the severity of its consequences are taken into account.

In order to be competitive and effective both in the construction market and in the design market, companies need to introduce new progressive methods of organizing work. Because of it, there is a need of systems that work more efficiently (save time and money, demand fewer resources and don't compromise quality), provide better coordination and

*Corresponding author: talesim@mail.ru

communication between the participants of the project team, and also transfer all necessary information about the project to all Interested parties involved in the life cycle of the project.

At present time, BIM-technologies allow minimizing the risks associated with design errors and errors that occur during construction. Which means that BIM-technology can act not only as a tool for qualitative design, but also as a process of organizing this design itself. An important role is also played by the strategic management of the enterprise, which is engaged in designing. Today, experts talk about BIM-technologies, as an instrument that allows to reduce design time, to implement the project in a high-quality and fast manner, and also as an information base that allows reliably to operate the constructed objects.

Creation of the BIM-model is a process of collective creation of the model by all participants of the investment and construction process. Each specialist (engineer-designer) works in his field and forms a model of his section, i.e. all specialists work in a single information field, forming a 3D model of the project. There are many programs for project management, which are widely used in the construction industry (Primavera P3, Microsoft Project, etc.). However, such schedules force users to visualize and interpret the sequence of actions in their minds. Interpretation of the schedule may range depending on the level of knowledge and experience. 4D models are created to solve this problem [1].

2 Methods

2.1 4D planning

Software products like Bentley Systems Inc, Autodesk, SmartPlant 3D are useful for creating a BIM-model. They could also support BIM-technology. During the use of traditional design tools, buildings are usually found in the finished state [2]. At the same time, as a result, one of the main limitations of 3D models is their inability to display the exact state of the progress of construction [3]. In addition, traditional construction planning tools, such as schedules and diagrams, don't facilitate the visualization of the process and require the creation of a mental representation of the construction [4].

The addition of the time attribute to a 3D (x, y, z) environment results in what it is broadly known as 4D (x, y, z, t) environment. This extra feature provides the model with more dynamism in terms of representing the behaviour of the building elements along time, extending in this way its usage for other purposes. [5-6].

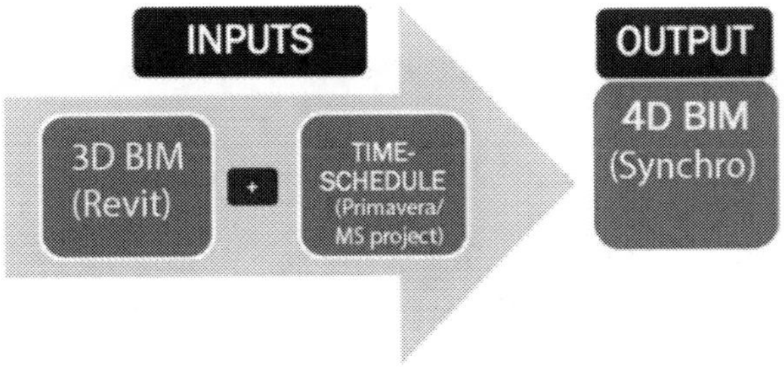

Fig. 1. From conventional BIM model to 4D BIM model

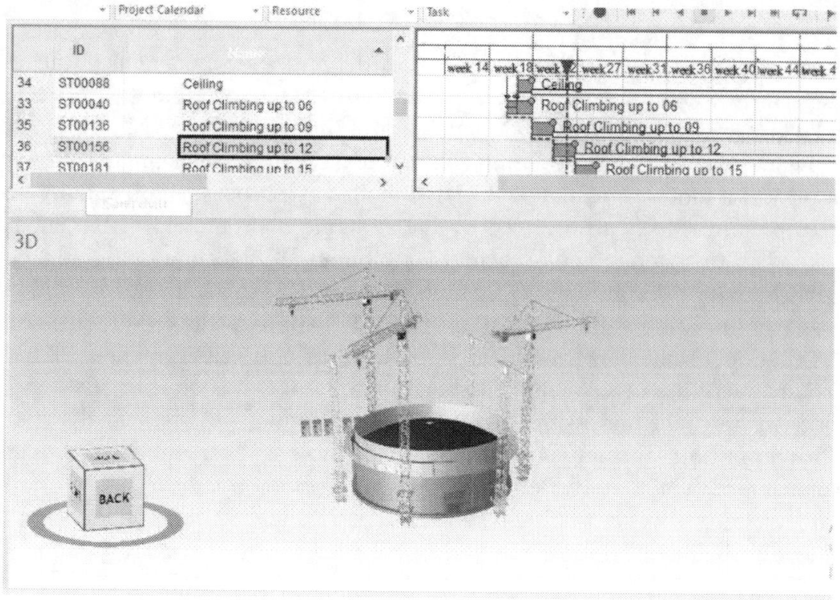

Fig. 2. Illustration of 4D model

Regardless of the project phase, there is no doubt that these 4D models enhance project understanding, particularly in those large-scale projects with especially high complexity. This makes it easier to understand and detect errors before the construction starts, which leads to saving money and time. It is commonly known that the detection of problems during construction, which had to be solved at the planning stage, leads to time and financial resources.

It should be noted that the grouping of objects in the building model should be carried out in accordance with the stages of construction and is associated with the relevant activities in the construction schedule. For example, if a concrete slab needs to be built in three stages, then the slab model should be divided into three sections so that the sequence can be effectively planned and illustrated. In addition, temporary works, such as scaffolding and tower cranes, should also be included in the building model (Figure 2).

Knowledge of the contractor is very important when building a 4D model for the planning process. If the model is built during the design phase of the building, the contractor can give feedback on constructiveness, the approximate cost of construction and consistency. Modeling 4D for the planning process serves as a communication tool for identifying potential bottlenecks and as a method of improving cooperation between different project teams [7].

2.2 Study subject

Let us consider an example of using 4D modeling in the construction of a 70-storey building (Fig. 2-a).

To construct a constructive part of the building, in the program Primavera P3 was created a schedule of works, which includes 700 works (Figure 2-b). Accordingly, the number of working areas and the work included in them change.

At the first review of the schedule and 2D drawings it seems difficult to understand the construction logic, as well as to find any inconsistencies and errors.

To create the 4D model, the following actions were performed: a 3D model was created from 2D drawings, which was integrated with the construction schedule in the Synchro PRO program.

This model allowed to visualize the sequence of construction works, and also facilitated the verification of the calendar schedule. Re-examination of the sequence of construction works using the 4D model made it easier to understand the construction logic and determine the presence of inconsistencies and errors. 4D model was thoroughly checked in order to ensure the authenticity of the schedule and compliance with its construction process.

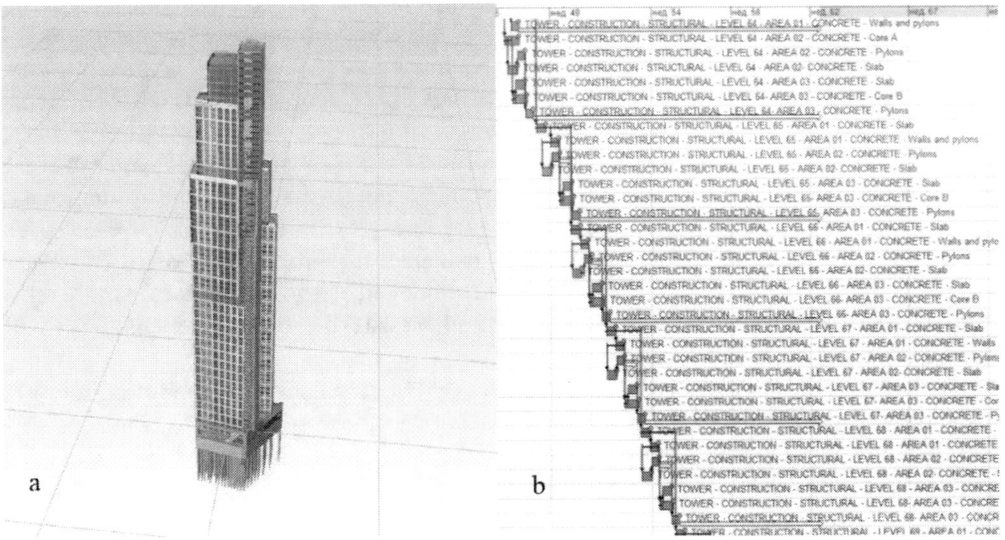

Fig. 3. a - 70 storey building, b – construction schedule

3 Results

During the analysis of the created model, several errors in the construction schedule were revealed (figure 3):

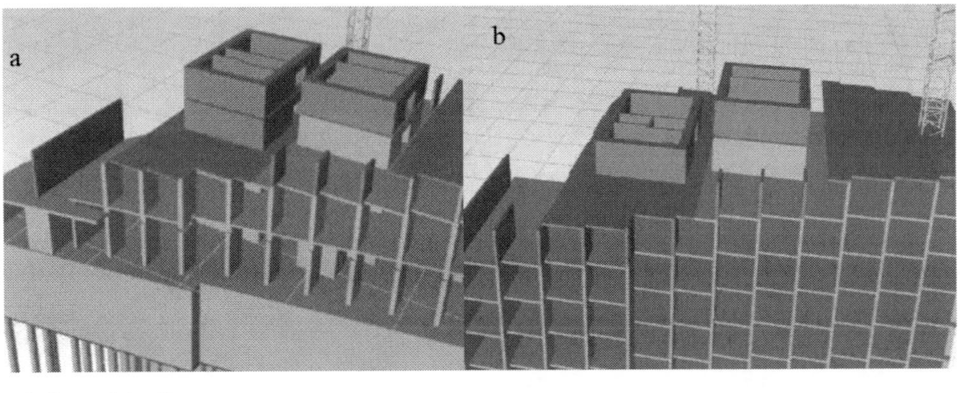

Fig. 4. Error detection

1. On figure 4-a we can see that the stiffening core is erected on the 3rd floor, but the construction of the stiffening core on the 4th floor has begun. However, such a situation can't be realized.
2. On figure 4-b shows that the walls are being erected, but install of the slab is started too, which does not correspond to the construction technology.

In addition to identifying design errors, 4D modeling has allowed to optimize the construction, as follow: to optimize the operation of cranes, the placement of building structures and materials at different stages of construction, to optimize the organization of work, and to monitor the activities related to the preparation of the construction site for compliance labor protection and safety engineering [8].

4 Conclusions

In the process of analyzing the process of creating a BIM-model in comparison with a flat design, there were identified such advantages as: minimization of risks and mistakes between different sections of the project documentation, the ability to provide technical, aesthetic, economic, environmental, social and other requirements for the future facility, as well as space-planning solutions according to the specifics of the adopted norms, check the model on collisions and constructive errors.

With the help of visualization and communication, the planners, project team can achieve a better understanding of the project scope and objectives, which can improve the construction planning and execution process significantly leading to the project success. Implementing 4D modeling allows planners to detect the problems prior to construction phase which lead to reduction in the amount of rework and clashes. Therefore, a more reliable and detailed work plan can be obtained which assists the project to complete within prescribed time and budget.

Effective planning of construction of high-rise buildings plays an important role in the creation and maintenance of the project. One of the most significant advantages of 4D modeling is the provision of visualization of construction work, which can't be achieved with 2D drawings and schedule.

The 4D model is the best way to simulate the construction process before actually starting the construction of the project, as well as providing a basis for identifying potential problems in space and time that are difficult to take into account in the schedule.

Reference

1. I. V. Nesterov SAPR I GIS avtomobil'nih dorog, **2,** 33-36 (2004)
2. M. McKinney, K. Fischer, Automation in Construction, **6**, 433-447 (1998)
3. H.J. Wang, J.P. Zhang, K.W. Chau, M. Anson, Automation in Construction, **13**, 5, 575-589 (2004)
4. B. Koo, M. Fischer, Journal of Construction Engineering and Management, **126**, 4, 251-260 (2000)
5. U. V. Alberto, Assessment of 4D BIM applications for project management functions, **34**, pp 108 (2014)
6. K.W. Chau, M. Anson, J.P. Zhang, Journal of Construction Engineering and Management, **130**, 4, 598-606 (2004)
7. C. Eastman, P. Teicholz, R. Sacks, K. Liston, BIM Handbook: A Guide to Building Information Modeling for Owners, Managers, Designers, Engineers and Contractors (2011)
8. V.V Sharmanov, T.L Simankina, A.E. Mamaev Magazine of Civil Engineering, **1**, 69, pp. 77-78 (2017)

Elaboration of technology organizational models of constructing high-rise buildings in plans of construction organization

Irina Osipenkova[1], *Tatyana* Simankina[2*], *Taisiia* Syrygina[3], and *Vitaliy* Lukinov[4]

[1] Saint-Petersburg State University of Architecture and Civil Engineering, 2-nd Krasnoarmeiskaya St., 4, St. Petersburg, 190005, Russia
[2, 3] Peter the Great St. Petersburg Polytechnic University, Polytechnicheskaya, 29, St. Petersburg, 195251, Russia
[4] Moscow State University of Civil Engineering, Yaroslavskoe shosse, 26, Moscow, 129337, Russia

Abstract. This article represents features of the elaboration of technology organizational models of high-rise building construction in technology organizational documentation on the example of the plan of construction organization. Some examples of enhancing the effectiveness of high-rise building construction based on developments of several options of the organizational and technological plan are examined. Qualitative technology organizational documentation allows to increase the competitiveness of construction companies and provides prime cost of construction and assembly works reductions. Emphasis is placed on the necessity to comply with the principle of comprehensiveness of engineering, scientific and research works, development activities and scientific and technical support.

1 Introduction

In recent decades Russian construction companies have accumulated enormous experience of high-rise and top buildings construction. They take into account design characteristics of tall buildings that have a significant impact on technological and organizational solutions of high-rise construction.

This article represents certain aspects of technological and organizational solutions for the design and construction of high-rise buildings of monolithic reinforced concrete. In most cases, both abroad and in Russia load-bearing framework of high-rises are built of monolithic reinforced concrete [1].

The main feature of high-rise construction that affects the technology organizational model of its erection is a concentration of large amounts of monolithic concrete in construction footprint [2]. If in a multi-storey (25-30 floors) building there are 9-11 m^3 of reinforced concrete per m^2 of construction footprint whereas in high-rise buildings with 100-200 floors this parameter reaches 50-60 m^3/m^2.

The vertical orientation of a high-altitude house, limited size of floor working area, large amount of floors require minimizing the amount of time wasted on construction of one floor.

* Corresponding author: talesim@mail.ru

Besides, significant increases in wind speed and intermittent wind loading mode that occur from an increase in building's size are a very important circumstance that influences the speed if high-rises construction.

It follows that in plans of construction organization (PCO) that are a component of project documentation many factors affecting the technology organizational model of constructing high-rises should be envisaged. Most important challenges of PCO include making decisions that allow to provide the possibility to implement a row of effective organizational and technological plans, because PCO is the main document of strategic planning of the construction process which defines policy decisions on technology and organization of production that are detailed in further documents.

2 Methods

The elaboration of the technology organizational model of high-rise construction assumes a whole range of effective organizational and technological measures that promote to shorten the timescales and reduce labour intensity input at the same level of quality.

Linkage between the requirements of project/working documentation, production technology, the conditions of construction area, definition of technical feasibility, safety, economy and ecology is undertaken by the method of formulating of a particular technical task and the detailed reasoning on the way to address it that meets the indicated requirements. Organizational and technological plans merged in the settlement pattern are the result of elaboration of technological and organizational solutions.

Technology organizational model links substantiation, methods and rules that apply to the linkage between the requirements of the project documentation, technology and safety of production, specific conditions for construction and the environment during erection of certain object's constructions or its parts in specific conditions.

Technology organizational model is a primary core for establishing the costs of the construction and for its security assessment. This understanding makes it possible to evaluate consolidated design, technological and organizational solutions tailored to high-rise building. In technology organizational schemes it is important to define the most effective construction methods and the sequence of high-rises building choosing appropriate machinery, mechanisms, equipment and devices for the accomplishment of the main types of work [3]. Placing concrete batching plant (portable factory) at a construction site is an example of the effective organizational and technological solution for high-rises building [4]. It's possible to smoothly provide the facility to be constructed with ready-mixed concrete as the transportation of concrete mixture to the place of its preparation does not depend, for example, on heavy traffic.

Fig. 1. Portable concrete batching plant

Reloading-free supply of concrete mixture is possible while concreting constructions using «crane - hopper» method. In this case concrete mixture goes from a concrete mixer to concreting hoppers installed in the body of a drop-side track which delivers them to the coverage of the assembly crane. With a crane the mixture is delivered to the working horizon and poured into the formwork. Thus, apart from productivity growth, we reach concreting cost reduction as the cost of an auto concrete mixer's machine-shift is 2,5-3 times bigger than the cost of the machine-shift of a drop-side truck. That plan of concreting organization is relevant in concreting of load-bearing structures (columns, cross bars, floor slabs etc.) of the lower floors (for example, from 1 to 20) of high-rise buildings. Justification of the scope of this scheme of concreting is needed in a case-by-case basis.

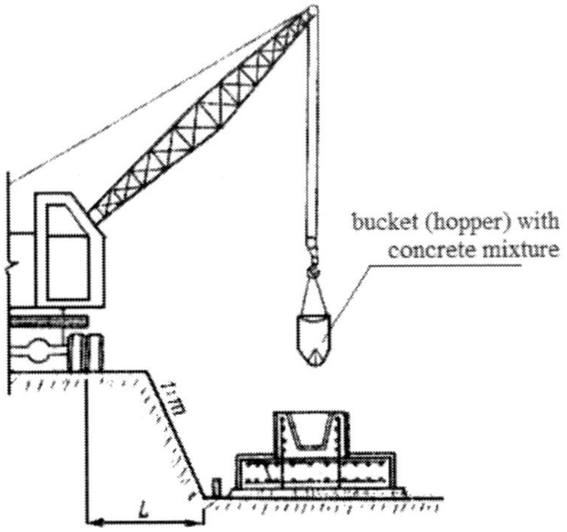

Fig. 2. «Crane - hopper» method in concreting the part of the structure

It is possible to organize transportation of concrete in borders of building site from the mixing plant to the area that needs to be concreted with stationary concrete pumps. That plan of concreting can be relevant in high-rise buildings' monolithic foundation and carcassing.

Fig. 3. Stationary concrete pump

Depth of laying a foundation varies from 3-4 to 8-10 m. [1, 5]. In so doing, most of concrete is placed lower than the surface level. In this case concrete pipe of concrete pump placed in the area of batching plant is made of steel links of concrete pump, whereas supplying of concrete mixture to concrete block foundation is accomplished via a flexible concrete pipe.

While concreting building's upper stories the system for concrete supplying consists of three parts: horizontal - from a batching plant to the placement of vertical concrete pipe; vertical concrete pipe; stationary boom. Depending on technical parameters of used concrete pumping equipment, in junctions of stated parts and throughout its length and (or) heights are set auxiliary stations of concrete pumping [6].

In contrast to linear type facilities that provide during its construction the suitable conditions for simultaneous execution of rebar, formwork activities and concreting in different work zones in borders of one floor, it high-rise building the area of a floor is smaller and the challenge of acceleration of increasing of stripping strength of concrete are pressing. The placement of batching plant near the high-rise building on a construction site creates favorable conditions for intensification of concrete pouring in hardening of concrete.

One of the directions of the accelerating of load-bearing structure building is an expansion of the borders of applicability of the thermos method. A set of rules 70.13330 "Load-carrying and cladding structures" allows the heating of concrete mixtures with Portland cement to +35°C but not more. This temperature restriction of concrete mixture heating in the process of its preparation is due to the requirement of maintaining of concrete workability for a time from its preparation to its pouring. In normal conditions of the massive construction the distance from the factory that supplies custom concrete can reach dozens of kilometers and transport time of concrete mixture - 1,5-2 hours. If a concrete production plant is placed on a construction site then transportation takes 10-20 min.

3 Results

The study of the setting time of concrete in increased temperature conditions (provided on [7]) has revealed that, for example, Portland cement M500 starts to set in 1,2 - 1,3 hours even at a temperature of 50°C. Consequently, with experimental justification load-bearing structures are recommended to pour with mixtures heated to 40-50°C at a batching plant and to use hot-thermos method. The hardening time of concrete will be comparable to heating methods with minimized energy consumption and labour input. Moreover, the quality of concrete, prepared of preheated mixture is better that the quality of concrete that is heated, for example, with heating cable. This is because both in the hot-thermos method and in the method of electric preheating of concrete mixture, heat is delivered to concrete mixture before its pouring and setting. It follows that in both these methods destructive phenomena caused by uneven volumetric expansion of concrete mixture components are minimized.

It should be mentioned that the hot-thermos method with its increased temperature of concrete mixture heating that provides shortening of time frames, reduction of energy consumption and labour input, improving the quality of concrete does not apply to the concreting of massive monolithic foundations. Applied to the foundations of high-rises, appears the opposite problem of diminishing the influence of cement's exothermic heat on the concrete's state of tension to avoid cracks. Directions and ways of solving this problem are not examined in this article.

Usage of high-strength sorts of concrete in high-rise construction in the building of a load-bearing framework of overground building's parts are a strong trend. Experience of designing and construction of top buildings indicates that in construction of supporting structures of the same building different classes of concrete(from B120-B100 to B60-B40) are applied basing on the dependency of the pressure on elevation [6].

4 Conclusion

Increased strength characteristics of high-performance types of concrete is provided by rational choice of its components, including highly-effective modified supplements. Most ready-mixed concrete factories are targeted at the production large deployments concrete, including high-quality (but not high-strength). That is another argument bearing witness to the feasibility of placing a concrete batching plant at a construction site capable of preparing concrete mixtures, taking into account specificities of high-rise building including a row of high-strength sorts of concrete.

In addition, it's impossible to avoid research and development related to implementation of separate project, organizational and technological decisions that reflects the specificity of high-rise construction.

It is also necessary to provide scientific and technical support at all stages of design, construction and exploitation of high-rises [8].

It this article were examined some examples of improving the effectiveness of high-rise construction on the basis of development of several options of technology organizational schemes. Qualitative technology organizational documentation allows to increase the competitiveness of construction companies and to reduce prime cost of construction and assembly works.

References

1. M.F. Markovskiy, Arkhitektura i stroitel'stvo, **2**, 220 (2011)
2. M.A. Romanovich, Increase of technology organizational reliability in cast-in-place construction based on model engineering of time schedule parameters] (PhD Thesis), Saint-Petersburg: SPSUACE (2015)
3. L.M. Kolchedantsev, S.V. Volkov, Zhilishchnoe Stroitelstvo, **11**, 21-26 (2015)
4. L.M. Kolchedantsev, I.G. Osipenkova, Zhilishchnoe Stroitelstvo, **11**, 17-19 (2013)
5. I.G. Osipenkova, T.L. Simankina, R.R. Nurgalina, *Fundamentals of organization and management in construction* (Saint-Petersburg, SPSUACE, 2013)
6. M.F. Markovskiy, A.V. Shcherbach, Arkhitektura i stroitel'stvo, **1**, 82–89 (2013)
7. L.M. Kolchedantsev, Intensification of concrete works on the base of the thermovibrating processing of concrete mixture (PhD Thesis), Saint-Petersburg: SPSUACE (2002)
8. Russian Regional guidance document 31-04-2008 «Recommendations for the construction of residential and public buildings»

Non-price competition in the regional high-rise construction market

Elena Ganebnykh[1,*] *Tatyana* Burtseva[1] *Ekaterina* Gurova[1] and *Irina* Polyakova[2]

[1]Vyatka State University, Moskovskaya str., 36, Kirov, 610000, Russia
[2]Moscow State University of Civil Engineering, Yaroslavskoye Shosse, 26, Moscow, 129337, Russia

Abstract. The article analyzes the market of high-rise residential construction in the city of Kirov (Russia). A minimal significance of price factors has been revealed in the process of the market analysis. This suggests that a lower price does not guarantee an increase in consumer demand. Thus, factors of non-price competition are of great importance in the market in question. The expert survey has identified the factors of non-price competition which influence consumer perceptions. A perceptual map has been constructed on the basis of the identified factors by means of the factor analysis to determine the positioning of each high-rise building relative to the consumer requirements. None of the high-rise residential buildings in the market in question meets the consumers' expectations of an "ideal facility".

1 Introduction

In Russia, a high-rise building is traditionally considered to be a structure of 75 meters high or having more than 25 floors [1]. In foreign countries, structures between 35-100 meters high or having 12-39 floors are referred to as high buildings according to Emporis Standards [2]. Buildings of over 100 meters high are classified by Emporis as skyscrapers.

The history of high-rise construction in Russia has a long history, but it was mainly applied in the construction of churches and other religious buildings (such as bell towers, monasteries). Mass construction of high-rise civil structures began in Russia in the middle of XX century, and it later spread to industrial and engineering construction [3]. Most high-rise buildings were erected mainly in Moscow and St. Petersburg for many years without being built in other regions of Russia. This trend is partly explained by urbanization processes, population growth and built-up density. Since the 90s of XX century, the fashion coming from abroad after "opening the borders" after the collapse of the USSR has become an additional factor in the rapid growth of high-rise construction. So construction of high-rise buildings in cities with relatively small population is mainly determined by fashion. Although, urban density of such cities also often results in infill construction of high-rise buildings in the city center, which is explained by an increased demand for housing in these parts of a city.

The city of Kirov is the center of the Kirov region; its population is 500 000 people. Since the city has a history of 640 years, its central part has low-rise buildings, which is typical of

* Corresponding author: ganebnykh@mail.ru

old Russian and European cities. Urban construction, constrained from the east by the Vyatka River, has gradually developed into the south and south-west, but the business center is still the old part of the city (Figure 1).

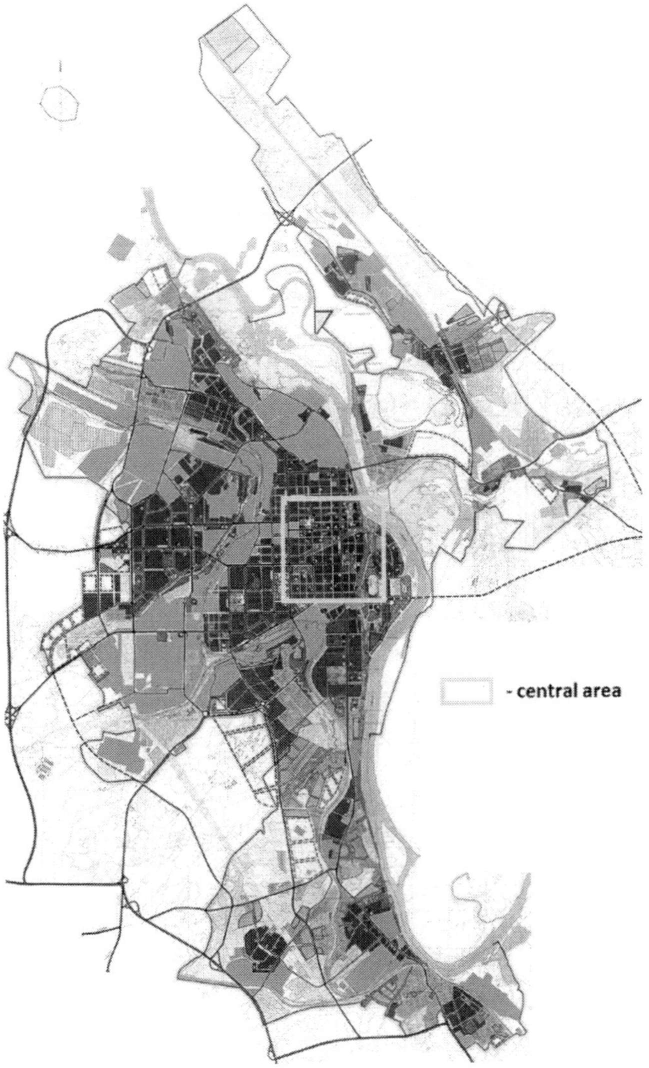

Fig. 1. Map of Kirov districts.

The following restrictions were in force in high-rise construction in Kirov in 2017:
- land plot area between 0.5-1.6 ha;
- the maximum number of above ground floors is 17;
- building factor not exceeding 1.6.

At the same time, it is allowed to increase the built-up density by 30% when reconstruction of the existing buildings takes place [4].

In 2014, the residential real estate market in Russia entered the hypersupply phase [5], characterized by an increase in the number of unoccupied buildings and also an increase in the number of new facilities non-demanded by the consumer. In fact, this kind of market is the buyer's market. Developers should make a lot of effort in terms of marketing and seeking financial support to implement the ongoing projects. In such conditions of development of the regional market of high-rise residential real estate, competition factors become of special importance as they allow developers to improve their position in their struggle for scarce resources (the consumer).

A set of instruments of competition between the entrepreneurial businesses constitutes competition methods [6]. Any marketing strategy takes advantage of both price and/or non-price factors of the selling proposition.

Price competition, as the name implies, is associated with strategies aimed at reducing prices. Such competition is acceptable when the goods or services offered are similar so that the price is their only difference [7]. However, a characteristic feature of the real estate market is a close link to the geographical location (location of the building), which is of very important, almost crucial, significance for the consumer. Therefore, developers also have to resort to factors of non-price competition in their struggle for the consumer. Non-price competition is based on offering a more reliable product or service; which is of a higher quality, having a longer service life and having other improved consumer properties [8].

2 Methods

At first, it is necessary to make a strategic analysis of the target market to assess the factors of price and non-price competition in the market for high-rise construction (Figure 2).

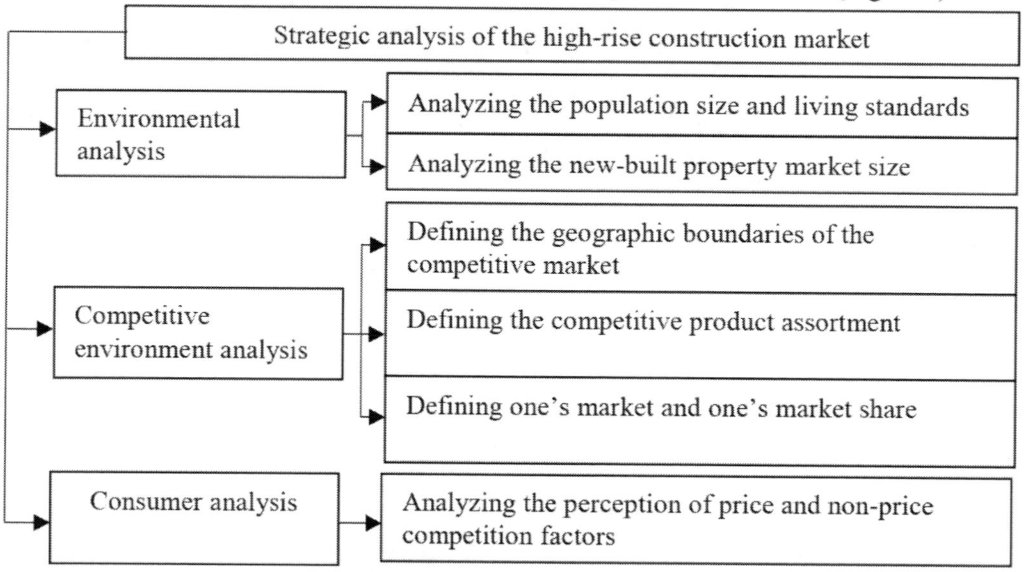

Fig. 2. Scheme of the high rise construction market analysis.

First of all, it is necessary to describe target consumers in terms of perceiving these factors to make a comprehensive assessment of the real level of price and non-price competition. It is necessary to determine the main characteristics of those buyers who are perceptive to price competition methods, as well as those who are perceptive to non-price competition methods. So it will be possible to define the target segment and the number of the target segment consumers to assess their perceptions.

Thus, a strategic analysis of the high-rise construction market will be carried out. This analysis can become the basis for developing the marketing strategy of the participants of this market.

3 Results

Marketing agencies, as well as real estate agencies of Kirov regularly publish the results of their market research, based on the statistical data concerning the changes in supply and demand in the market. However, their results are not of analytical nature and their aim is only to describe the dynamics of property prices in the region. In addition, the research data summarizes the results for the entire real estate market, without segmenting it; they do not describe the high-rise construction market separately. Thus, a comprehensive strategic analysis including a number of steps is necessary to develop the marketing strategy for the construction companies operating in the high-rise market.

3.1 Environmental Analysis

The environmental analysis is usually considered to be an initial stage of strategic management [9]. This analysis allows us to identify factors that contribute to or hinder the commercial success of an enterprise. As the dynamics of the external environment increase their influence, such analysis becomes a vital factor for the enterprise, ensuring its survival. The elements of the external environment are not of the same value for an enterprise, so an important part of the analysis is to identify those ones which play the most important and significant role.

3.1.1 Analyzing the population size and living standards

Kirov is the regional center of the Volga Federal District; it was founded in 1374. The territory of the city is 169.7 km^2 with the population density of 1954.5 people per square kilometer. The size of the permanent population of Kirov is steadily growing; there are some variations in the growth, but on the whole, they coincide with the demographic changes in Russia. In 2017, the city's population was 500 thousand people. Migratory flows are mainly directed towards other regions (mostly Moscow, St. Petersburg, Kazan), but the urban population remains replenished thanks to urbanization of the Kirov region as a whole, i.e. movement of the rural population to the center of the region to settle there permanently.

Comparing income and expenditure of the population gives an understanding of the living standards in the city. The state statistics bodies do not have a separate database for each municipality, so statistics indicators are based on the averaged data for the Kirov region as a whole.

Table 1. Indicators of living standards in the region.

Indicator	2012	2013	2014	2015	2016
Average monthly salary, thousand rubles per month.	16.9	19.3	21.0	22.1	23.4
Subsistence level, thousand rubles per month.	6.2	6.9	7.6	9.3	9.3
Proportion of the population with below the subsistence level income, %	12.4	13.6	12.7	14.9	no data
Expenditure pattern, %					
- food products;	32.5	34.7	33.8	38.9	36.2
- non-food products;	44.8	41.3	43.5	36.9	40.3
- service sector payments.	22.7	24.0	22.7	24.2	23.4

Analyzing the indicators of the living standards of the Kirov region population, presented in Table 1, one can see a noticeable increase in the average monthly salary, however, the average salary in the region is significantly lower than average figures in Russia. The average monthly salary in the Kirov region was 65% of the national average in 2016. The population income grew almost in proportion to the subsistence minimum (an increase by 1.4-1.5 times in the last 5 years), but the expenditure pattern changed as well: spending on non-food products was reduced and the expenditure shifted towards food products. When most of the income is spent to buy food, it means a decrease in the real income of the population. Real estate refers to luxury goods; therefore growth in its sales is always conditioned by the population income growth and availability of loan funds. In Kirov, the decrease in the population expenditure coincided with the hypersupply phase in the real estate market, which puts developers in an extremely tough situation.

3.1.2 Analyzing the new-built property market size

The new-built property market size is estimated by the number of buildings (square meters of buildings) put into operation for a particular period of time. The dynamics of changes in the size of the new-built property market is shown in Figure 3.

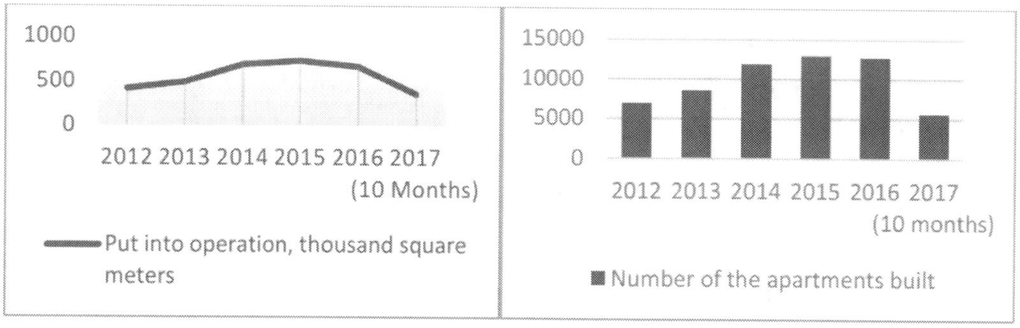

Fig. 3. Indicators of the new-built property market size.

The infographic in Figure 3 is indicative of the hypersupply phase when the supply exceeds the demand. Starting from 2015, construction activity has been slowing down, so the market size is going to decrease until there is lack of supply.

The indicators in Figure 3 reflect the general situation in the residential real estate market, while this study is aimed at the high-rise construction market. In 2017, there were only 4 high-rise residential buildings under construction in Kirov, 2 of them are located in the central district and 2 in the south-western district of the city

The central district:

- Iceberg Residential Complex– 27 floors;
- Alye Parusa Residential Complex– 25 floors.

The south-western district:

- Malakhit Residential Complex– 27 floors;
- Olimp Residential Complex– 27 floors.

All residential complexes are of premium class; they are multi-sectional and are commissioned in a phased manner.

3.2 Competitive environment analysis

If we define competition as a struggle for customers, then all aspects of the competitive environment and competitive advantages are the principal indicators which distinguish one firm from another and allow the firm to attract the attention of customers [10]. Therefore, it is necessary to make a comparative analysis of the competitive environment in order to project the market situation.

3.2.1 Defining the geographic boundaries of the competitive market

The number and density of the population, built-up density and low demand for high-rise construction in the Kirov region have defined clear boundaries of this market. In fact, there are only 3 more high-rise buildings in the city in addition to the four high-rise complexes mentioned above: Edelweiss Residential Complex (25-storey building, commissioned in 2012), Studenchesky Residential Complex (24-storey building, commissioned in 2015) and Prestige House Residential Complex (25-storey building, commissioned in 2014). The geographic boundaries of the market under study lie within the boundaries of the Kirov municipality. In fact, they are located in the central and south-western districts of the city. The south-western part of the city is essentially a bedroom district, but it has a developed infrastructure and good transport accessibility.

3.2.2 Defining the competitive product assortment

The product assortment in the market for high-rise construction is conditioned both by general characteristics of a residential complex and by the characteristics of apartments. The characteristics of the four high-rise residential complexes are compared in Table 2.

Table 2. Characteristics of high rise residential complexes.

Indicator	**Iceberg**	**Alye Parusa**	**Malakhit**	**Olimp**
General characteristics of the residential complex				
Location district	center	center	south-west	south-west
Number of floors	27	25	27	27
Number of apartments	136	414	192	751
Wall material	brick and monolith	brick and monolith	brick and monolith	brick and monolith
Class of the residential property	premium class	premium class	premium class	premium class
Infrastructure	highly developed	highly developed	semi-developed	under developed
Parking lot	underground and ground level	underground and ground level	ground level	ground level
Children's playground	in the yard	in the park	no	in the yard
Characteristics of apartments				
Average metric area of a 1-room apartment	44.9	47.7	46.9	42.5
Average metric area of a 2-room apartment	74.6	67.0	59.6	57.0
Average metric area of a 3-room apartment	97.6	99.6	67.7	74.6
Average metric area of a 4-room apartment	112.0	-	-	93.7
Finishing	shell and core	shell and core	fine finish	shell and core

The comparative analysis of the product assortment in the high-rise construction market in Kirov has shown that Iceberg Residential Complex and Alye Parusa Residential Complex have very similar characteristics. Alye Parusa differs only with its location. It is located at the intersection of the city's central streets, which causes quite a noisy environment and an unfavorable ecological situation with regard to air pollution caused by exhaust gases. The construction declaration does not provide additional noise insulation of the premises, which means the consumer has to bear extra costs connected with it.

The characteristics of Malakhit Residential Complex are different from the other high-rise buildings as it is located in the district which previously was not inhabited, and, accordingly, it does not have a well-developed social infrastructure. However, the complex is located in one the central city streets, which provides good transport accessibility. At the same time, there are two shopping centers and a hotel complex being built within a radius of 0.5 km, which means the infrastructure is going to develop in the short term.

Olimp Residential Complex is located on the border of the residential and industrial zones of the south-western district, so the infrastructure there is quite underdeveloped (there is even no public transportation). The complex is a large construction facility, consisting of 9 sections and 751 apartments in total. The number of future tenants will certainly become a catalyst for the development of social infrastructure in this part of the city. The design documentation for Olimp Residential Complex specifies non-residential premises which are likely to provide the residents with essential commodities.

3.2.3 Defining the market share of the market players

There are 39 construction developers in the housing construction market in Kirov (according to the data of 2017), but only 3 of them are engaged in high-rise construction: OOO SK DomInvest, OOO Kirovsky SSK и OOO Stroisoyuz. The market shares of the business entities in this market are considered to be a part of the resources of the market. Depending on the size of the market share, the developer can be a leader or an outsider in the market having a strong or weak competitive position. Since construction of high-rise buildings is, in fact, the first experience for all three market participants, one can not define the market share using such indicators as the number of the constructed facilities or square meters put into operation.

The market share is calculated according to the formula:

$$M_{sh} = \frac{Q_P}{Q_c} \text{x } 100\%, \tag{1}$$

where:
M_{sh} – market share, %;
Q_P – the volume of the area in high-rise buildings to be commissioned, square meter;
Q_c – the total volume of the area in high-rise buildings to be commissioned, square meter.

Thus the market shares of the developers are as follows:

6 % - Ltd SK DomInvest (11 751 square meters);
49 % - Ltd Kirovsky SSK (95 694 square meters);
45 % - Ltd Stroisoyuz (86 725 square meters).

When calculations were made, the total area, including non-residential real estate, was taken into account, therefore the market share of Ltd Kirovsky SSK is bigger thanks to the shopping center located on the ground floor of Alye Parusa Residential Complex (46 586 square meters).

It is impossible to estimate an increase in the share because the high-rise buildings in question are these developers' first experience of high-rise construction in Kirov. It is interesting to note that the developers of all three previously commissioned high-rise buildings are no longer working (they are closed or undergoing bankruptcy proceedings).

To assess the efficiency of the price factor in the market, the price pressure coefficient (C_{pp}), which characterizes the variation of prices in space and time, is used:

$$C_{pp} = \frac{\sigma}{\overline{P}}, \quad (2)$$

where:
C_{pp} – price pressure coefficient;
σ – standard deviation of product prices;
$\overline{P}$ – average price of the product.

$$\sigma = \sqrt{\frac{1}{n}\sum(P_i - \bar{P})^2}, \quad (3)$$

where:
n – number of developers in the high-rise construction market.

Thus, statistical information on the prices for the facilities in question is necessary to assess the effect of the price factor (Table 3).

Table 3. Prices for the real estate units in the high-rise construction market in Kirov.

Indicator	Iceberg	Alye Parusa	Malakhit	Olimp
Cost of 1 m² of 1-room apartment, rubles.	48 000	58 000	49 000	48 000
Cost of 1 m² of 2-room apartment, rubles.	46 000	52 000	49 000	44 000
Cost of 1 m² of 3-room apartment, rubles.	45 000	55 000	49 000	38 000
Average cost of 1 m² of the real estate unit, rubles.	46 333	55 000	49 000	43 333

Thus, the price pressure coefficient is 0.08. The closer C_{pp} is to zero, the lower the price competition level is. Thus, it can be said that price competition in the market of high-rise construction is of minimal significance. This suggests that a lower price will not guarantee an increase in consumer demand. Therefore, non-price factors of competition become more important.

3.3 Consumer Analysis

Despite the fact that the goal of the consumer analysis is usually to identify segments or groups of people having similar requirements that can be directly targeted in marketing activities [11], the method of an expert survey was used in this study. Experts were realtors, as well as buyers of the residential real estate.

To analyze the perception of non-price competition of real estate units in the high-rise construction market in Kirov, all non-price competition factors influencing the consumer perception have been revealed:

- location;
- social infrastructure;
- transport accessibility;
- convenient access way, parking lots;
- construction quality;
- exterior of the building, building design;
- metric area of apartments;
- quality of common facilities;
- apartment planning;
- reputation of the developer.

As a result of the factor importance ranking, four factors were identified as having the greatest importance for consumer perception: location, social infrastructure, construction quality, and reputation of the developer. Further, the main components of non-price competition in this market were assessed by experts on a 10-point scale; the average figures are given in Table 4.

Table 4. Non-price competition factors in the high-rise construction market.

Real estate unit	Location	Social infrastructure	Construction quality	Developer's reputation
Iceberg Residential Complex	6.6	7.8	8.4	8.2
Alye Parusa Residential Complex	8.4	7.8	8.4	7.8
Malakhit Residential Complex	8.0	6.8	7.2	7.8
Olimp Residential Complex	5.8	6.6	8.2	7.4

The factor analysis [12] made with the Minitab Statistical Software 14 is used to process the data presented in Table 4; the result of this analysis is the product perceptual map. The analysis involves another object, the so-called "ideal point", i.e. an object that the consumer would prefer to any other, including the objects that can be displayed in space but do not really exist. The ideal point allows one to determine the position of each object in relation to the consumers' requirements, and also allows one to assess the position of objects relative to each other.

The results of the statistical data processing were interpreted to identify the axes of the perceptual map. The statistical indicator of factor loads was used as a basis [13]. Factor loads allow one to determine the correlation of characteristics of an object with the map axes, to determine which characteristics the factors strongly correlate with, and accordingly, give names to the map axes (Table 5).

Table 5. Factor loads of the perceptual map axes.

No	Main Factors	Axis 1: Social infrastructure and construction quality	Axis 2: Location and developer's reputation
X_1	Location	0.323	0.940
X_2	Social infrastructure	0.813	0.573
X_3	Construction quality	0.955	0.252
X_4	Developer's reputation	0.771	0.593

Analyzing the correlation of factor loads with the perceptual map axes, one can conclude that high correlation values for Axis 1 are observed between the variables of X_2 (social infrastructure) and X_3 (construction quality). As for Factor 2, high correlation values are observed between the variables of X_1 (location) and X_4 (developer's reputation). Consequently, it is reasonable to analyze the results relative to Axis 1 which is "Social infrastructure and quality of construction" and Axis 2 which is "Location and reputation of the developer".

 https://doi.org/10.1051/e3sconf/20183303046

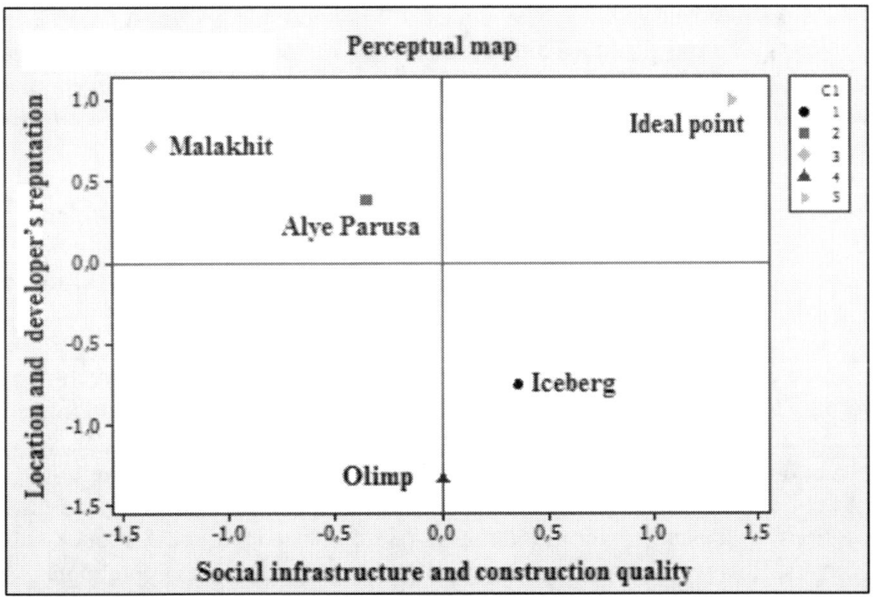

Fig. 4. Perceptual map of non-price competition of high-rise residential buildings.

The map can be interpreted by analyzing the coordinates and the relative positioning of objects with respect to the ideal point and to the axes of the perceptual map. The objects which are close to each other compete more rigidly. As one can see, these are Malakhit Residential Complex and Alye Parusa Residential Complex. The isolated objects have their own unique image, which is not always beneficial for the developer. The gaps on the map may indicate prospective opportunities to promote and position one's products in the real estate market.

The objects most distant from the axis are stronger in this characteristic. As one can see, Malakhit and Alye Parusa have a very strong position in regards to location and reputation of the developer, but the position in regards to available social infrastructure facilities and construction quality is quite weak. Iceberg, on the contrary, is quite competitive in terms of available social infrastructure facilities and construction quality, but it has a weak position in regards to location and the developer's reputation. The position of Olimp Residential Complex requires purposeful work aimed at both creating a positive perception of the location and the developer's reputation, as well as social infrastructure and construction quality.

4 Discussion

Thus, the perceptual map can show the positioning of real estate units in the marketplace and preferences of consumers in regards to efficiency of non-price competition factors. In general, the analysis has shown that all the objects under study are not able to compete in terms of perception of non-price competition; their position on the map is quite distant from the "ideal point".

Analyzing the ranks of factors of non-price competition, one can see that the location of a high-rise building is more important for the consumer. At the same time, this factor correlates with the social infrastructure factor weakly. It means selecting an attractive land plot for construction is a matter of high priority for the developer, and the infrastructure can be developed later, at the stage of construction or even after the facility has been commissioned. However, an important restriction for developers is the lack of land in attractive districts of the city.

Another problem of Kirov and other cities having a long history is the historic buildings in the central part of the city, which often have cultural and historical value. Modern high-rise buildings do not fit into the urban ensemble of the low-rise historical center. In this regard, the problem of finding an attractive location can be solved by demolishing old buildings. The construction of Iceberg Residential Complex and Alye Parusa Residential Complex began in this way.

An alternative solution may be a new microdistrict, which is more costly in terms of infrastructure development, but it gives a lot of scope to implement different architectural concepts. One can solve the problem in this way only during the Expansion phase when the demand for the residential real estate is growing. It is due to a large metric area of high-rise buildings, the construction of which should be supported by real demand.

Further analysis of consumer perceptions of the development quality with regards to high-rise construction has revealed some bias against monolithic construction technology. According to consumers, this technology has not been used in Russia until quite recently, so there are no statistical data on the quality of such construction. More than that, Russian construction companies often do not have enough experience in high-rise residential construction, but a violation of construction technology is very dangerous in this case. Such perceptions of consumers, despite the high-rise construction appeal, reduce the number of real buyers. However, a positive thing about it is that this factor is going to have less impact as the high-rise construction market is developed.

Conclusions

The conducted research allows us to draw a conclusion about the development of the high-rise construction market in Russia. The crisis of 2014 and the subsequent recession in the real estate market have also affected the high-rise construction sector.

On the whole, the environmental analysis of Kirov has shown that the residential high-rise buildings are of an infill construction type, which is explained by the demand for such facilities in the central districts of the city and lack of available land there. At the moment, 4 high-rise residential buildings are being built in the city of Kirov.

However, assessment of the price factor influence has shown minimal values. This fact suggests that lowering prices will not guarantee an increase in consumer demand. Therefore, non-price factors of competition become important.

The most significant factors of non-price competition that have an influence on consumer perceptions have been revealed to analyze the perception of non-price competition of real estate in the high-rise construction market in Kirov. They are location, social infrastructure, construction quality and the developer's reputation. The factor analysis procedure has been used to construct a perceptual map to determine the positioning of each building relative to the consumers' requirements, as well as to assess the positioning of the buildings relative to each other. On the whole, none of the buildings in the market meets the consumers' expectations of an "ideal facility". Each of the high-rise buildings has its advantages, but each of them is also noticeably distant from the "ideal point".

References

1. Yu. Granik, Zhilishchnoe Stroitel`stvo Journal, Vol. 2, pp. 20-22 (2008.)
2. URL: https://www.emporis.com/building/standard/3/high-rise-buildin
3. URL: https://sciencepop.ru/kratkaya-istoriya-vysotnogo-stroitelstva-v-rossii
4. E. Stein, N. Taskaeva, E. Chibisova, Procedia Engineering 165, 1410-1416 (2016)

5. V. Formigle, The Four Phases of the Real Estate Cycle (CrowdStreet, 2016)
6. I.Rozhentsova et al, MATEC Web of Conferences, 08076, DOI: https://doi.org/10.1051/matecconf/201710608076 (2017)
7. URL: http://www.stanford.edu/~jdlevin/papers/matching.pdf.
8. S.L. Brue and C.R. McConnell *Economics–Principles, Problems and Policies* (15th edition). (Boston: Irvin/McGraw-Hill, 2002).
9. O. Soboleva, T. Burtseva, E.Barmina, E. Ganebnykh. and P.Tokareva, Overcoming Uncertainty of Institutional Environment as a Tool of Global Crisis Management, Contribution to Economics / Ed. by Popkova E.G. Springer International Publishing AG, pp. 565-574 (2017)
10. D. Aaaker, Strategic Market Management, 10th edition. Wiley, (2013)
11. S. Silbiger, The Ten-Day MBA: A Step-by-Step Guide to Mastering the Skills Taught In America's Top Business Schools. HarperCollins (1999)
12. A. Pimenova et al, MATEC Web of Conferences, 73, 07018 DOI: https://doi.org/10.1051/matecconf/20167307018 (2016)
13. Mottaeva, MATEC Web of Conferences, 08072, DOI: https://doi.org/10.1051/matecconf/20167308072 (2017)

 https://doi.org/10.1051/e3sconf/20183303047

High-rise construction in the Russian economy: modeling of management decisions

Tatyana Miroshnikova [1] and *Natalia* Taskaeva[2*]

[1] Vladivostok State University of Economy and Service, Gogolya str., 41, Vladivostok, 690014, Russia,
[2] Moscow State University of Civil Engineering, Yaroslavskoye Shosse, 26, Moscow, 129337, Russia

Abstract: The growth in the building industry, particularly in residential high-rise construction, is having considerable influence on the country's economic development. The scientific hypothesis of the research is that the execution of town-planning programs of high-rise construction depends to a large extent on the management of the provision of material resources for the construction of a millionth city, while the balance model is the most important tool for establishing and determining the ratio between supply and demand for material resources. In order to improve the efficiency of high-rise building management, it is proposed to develop a methodology for managing the provision of construction of large cities with material resources.

1 Introduction

At the present time in Russia there is a sharp change in the priorities of the practice of town planning towards the construction of high-rise public buildings and residential complexes, which can be regarded as a positive trend. As the world practice shows, in large cities the construction of high-rise buildings is getting more and more developed, which is due to the high cost of land, limited urban areas, intensive population growth and other reasons. The accumulated experience abroad shows that, taking into account the cost of land, 30-50-storey building objects are optimal for economic indicators. High-rise buildings form the modern urban landscape of the cities of the USA, Canada, China, and Europe. As the world practice shows, the large number of storeys of buildings with a high population density leads to an improvement in the quality of life. Residential objects and objects of social sphere, located not far from each other, make it possible to eliminate the difficulties associated with overcoming unnecessarily large land areas by residents. Construction is unique in stimulating the growth of other sectors (Hillebrandt 1985) and this contribution varies according to the nature of the society in which construction takes place (Okpala and Aniekwu 1988) [1].

[*] Corresponding author: natalia.taskaeva@yandex.ru

According to the scientist Didenko V.G. the statistics of developed countries indicate a direct relationship between the density of the urban population and the cost of energy conservation. In cities with higher density, achieved due to high-rise construction, energy saving costs are significantly lower, due to the smaller length of thermal and electric networks [2].

The great age of high-rise buildings began after the end of World War II, when the world economy and population again expanded. It was an optimistic time with declining energy costs, and architects embraced the concept of the tall building as a glass prism. This idea had been put forward by the architects Le Corbusier and Ludwig Mies van der Rohe in their visionary projects of the 1920s.

As environmental control systems increased in cost, economic pressures worked to produce more efficient structures. In 1961 the 60-story Chase Manhattan Bank Building, designed by Skidmore, Owings & Merrill, had a standard steel frame with rigid portal wind bracing, which required 275 kilograms of steel per square metre, nearly the same as the Empire State Building of 30 years earlier. The same firm demonstrated economy of structure in tall buildings only nine years later in the John Hancock Building in Chicago

Building construction has settled into a period of relative calm after the explosive innovations of the 19th century. Steel, concrete, and timber have become mature technologies, but there are other materials—such as fibre composites—that may yet play a major role in building [3].

2 Methodological approaches and analysis

The Russian practice of high-rise construction boils down to construction, which in most cases refers to the compacting of existing buildings, which does not correspond to world practice. This in turn leads to poor ventilation of the space between the buildings, weakening of the bearing capacity of the soil, the removal of gardening, the lack of places for recreation and the creation of a visually aggressive environment.

High-rise buildings, especially mixed use, are rational because they can accommodate a large number of people in a small piece of land. This avoids the development of agricultural land and reduces energy consumption, as well as carbon emissions to the environment associated with pendulum migration (people moving from suburbs to work and vice versa). These facilities offer effective systems of vertical and horizontal movement, contributing to the creation of cities for walking.

High-rise buildings allow saving the city space and the municipality's expenses for infrastructure: pulling the network to a skyscraper per square meter of area is much more expedient than to a low-rise building or a cottage. The second aspect is environmental: the inhabitants of the upper floors, are not affected by the pollution and noise of the metropolis. Another positive moment is the image: unique high-rise buildings attract world attention (examples of the Empire State Building (449 m) in New York and the Burj Khalifa building (828 m) in Dubai are very indicative in this respect). Investigating the problems of high-rise construction should take into account such factors as the state of the air environment; the need for mechanical ventilation systems; the characteristic of aerodynamics; quality of the air environment and the need for engineering and environmental training.

With all the obvious economic feasibility, high-rise construction in large Russian cities is not carried out in sufficient volume. There are several objective reasons for this. Among the first, one should mention the absence at the federal level of technical regulations for the design and construction of high-rise complexes and the lack of experience in their construction. The issues of construction supervision over high-rise facilities and the interaction of federal and city structures in this area remain open.

Issues related to the placement and erection of high-rise apartment buildings and complexes in large cities of Russia should be the subject of the closest attention of city planners, architects, designers, engineers, ecologists, sociologists, demographers. It is obvious that high-rise housing construction is at the forefront of the construction sector today. The presence of skyscrapers is distinguished by any modern city, testifies to the high development of construction technologies, the intellectual capabilities of designers, the readiness of the corresponding material and technical base. The annual conference of the Council on High-Rise Buildings and Urban Environment CTBUH collects all the necessary information and statistics on skyscrapers and compiles the altitude ratings. According to NEDO (1988), timely completion of projects is the most important indicator of an efficient construction industry. Time overrun is the main cause of cost overrun and it affects both clients and contractors.

Even though government spending is increasing, it is not nearly enough to solve the housing problems and, therefore, over the last 10–12 years, private companies have stepped in to initiate real estate business through the construction and sale of middle to high-rise apartment buildings. The main reasons for high-rise rather than low-rise construction are scarcity of land and land prices (Figure 1).

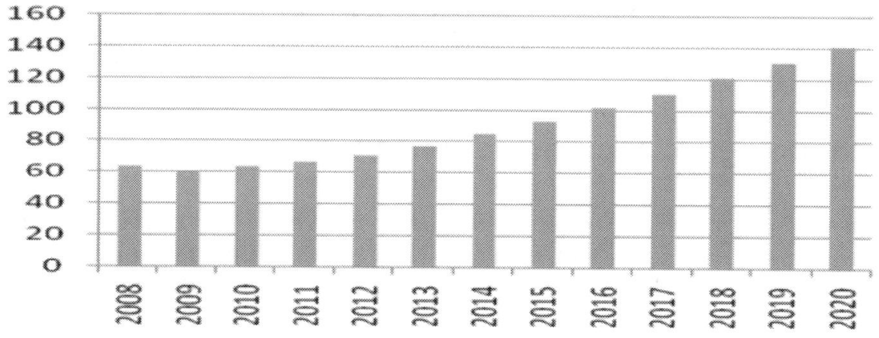

Fig. 1. Forecast of housing construction in the Russian Federation until 2020, million square meters [4]

In 2016, the volume of housing construction in Russia decreased, as compared to 2015, by 6.5%. A total of 1156,500 apartments with a total area of 79.8 million square meters were put into operation in the country in a year. meters - by 5.5 million square meters. meters less than in 2015 [5].

Leaders in terms of the commissioned housing in addition to the Moscow region and the Krasnodar Territory were Moscow (3361.8 thousand square meters), St. Petersburg (3116.3 thousand square meters), the Republic of Bashkortostan (2698.5 thousand square meters. m) and Tatarstan (2406.5 thousand square meters).

In the first quarter of 2017, 13.1 million square meters were commissioned. meters of residential (Figure 2).

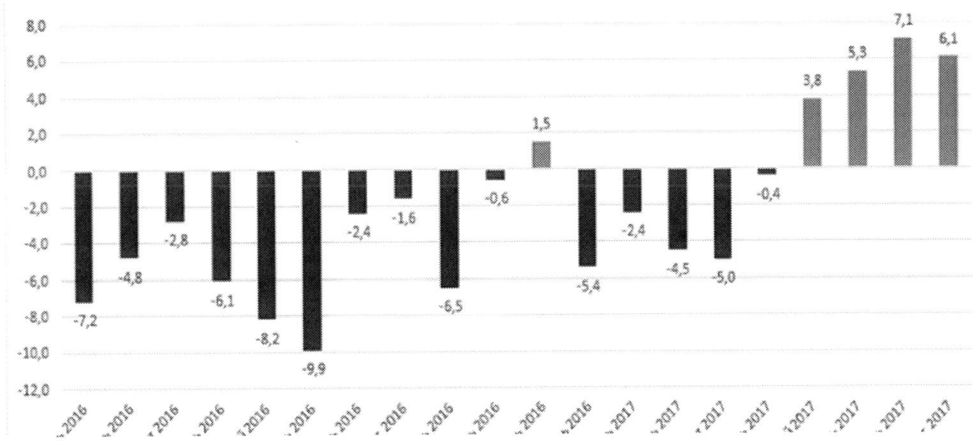

Fig. 2. The dynamics of the volume of work performed by type of activity construction in percent [6]

3 Results

Because of the relatively small experience of building high-rise complexes in Russia, not all of the above factors of high-rise construction have been thoroughly studied to date. The most acute issue in the construction of high-rise buildings and complexes is the provision of their power capacities. In view of the current electricity deficit in the city, it is necessary to consider the possibility of using alternative sources of heat and power supply. The following additional questions require further study: questions of engineering support; emergency situations; analysis of geoecological assessment; preliminary check of visual-landscape perception of high-altitude objects; Enlarged evaluation of the possibilities of transport services for high-rise facilities [7].

Among the problems arising in the design and construction of high-rise residential buildings in a big city and requiring mandatory accounting, review and decision, include the following [8]:

- the justified town-planning and functional-typological necessity of erection;
-limitable permissible height (height);
- the correct choice of the structural system, scheme and design solutions, taking into account the prevention of loss of stability of the base and the structure itself, leading to the destruction and collapse of structures;
- inadmissibility of deviation from the approved design decisions and changes in the number of storeys of structures in the process of construction;
- Optimum capacity of residential and non-residential, public premises;
- the necessary functional interaction of residential and non-residential buildings and structures with the transport and service infrastructure of the city;
- Required capacity of underground, land and aboveground parking lots of personal transport and their rational location;
- effective minimization of the threat of external and internal danger of building destruction by creating a special service for safe operation;
- the required fire and evacuation safety of people in high-rise buildings;
- rational efficiency of modern engineering solutions for the life-support and equipment of the building, energy saving and comfort of service.

In order to improve the efficiency of high-rise building management, it is proposed to develop a methodology for managing the provision of construction of large cities with material resources .[9]. It should be aimed at ensuring the possibility of identifying an imbalance between supply and demand by groups of material resources, as well as the formation on this basis of management decisions related to the increase in the capacity of existing enterprises or the construction of new enterprises for the production of building materials, products and structures, with a shortage of material resources for purposes of high-rise construction.

4 Discussion

In general, the history of management is among the most elaborate. The primary basis of classical administrative theory was laid by A. Fayol in the beginning of the 20th century. To create a scientific theory of management, scientists such as F. Taylor and G. Ford worked with him. Theoretical questions of the structural organization of various control systems were developed in the works of M. Weber [10].

Socio-economic and managerial aspects of urban development were investigated and described in the works of foreign and domestic researchers, to which, in the first place, include J. Vickery, V.E. Volkova, V.L. Glazychev, G. Simmel, A.N. Kabatskova, OL Leibovich, V.A. Lefevre, S.V. Pirogova, N.E. Pushkov, J. Forrester, N.V. Shushkov.

Modern economic processes are characterized by uncertainty and high risk. This makes it necessary to make managerial decisions based on variation calculations of the comparison of demand and supply of material resources for the construction of a metropolis in the implementation of urban development programs. The balance model allows making well-founded management decisions aimed at increasing the level of provision of high-rise construction with material resources.

The scientific hypothesis of the research is that the execution of town-planning programs of high-rise construction depends to a large extent on the management of the provision of material resources for the construction of a millionth city, while the balance model is the most important tool for establishing and determining the ratio between supply and demand for material resources [11].

The effectiveness of any actions to provide the building complex with material resources, including balanced balancing, depends on the availability of a single methodological toolkit. In this case, such a toolkit is understood as a rational composition of interrelated tasks of determining the balance, the result of which is a comprehensively grounded conclusion on the adoption of a concrete version of the balanced provision of high-rise construction with material resources for practical implementation.

The magnitude of the need for material resources (MRreq.) Is presented in the form of a uniquely defined relationship and is understood as the existence of a balance:

$$\sum \text{MRreq.} = \text{H MR (уд.)} \times \sum Q, \tag{1}$$

where H MR (ud.) - specific standards for the consumption of material resources;

ΣQ - the total volume of construction and installation works (the same as the total number of units of construction objects).

The following systematization of indicators of the material balance of high-rise construction is associated with the definition of the proposed amount of material resources of the construction complex of a large city. It can be determined on the basis of the total capacity of the construction enterprises, which should be divided into specific groups of material resources. Such an approach will allow developing managerial decisions on the rational distribution of orders for the production of building materials, taking into account the focus and capabilities of specific construction enterprises.

 https://doi.org/10.1051/e3sconf/20183303047

With reference to the balance of material resources, the main and obvious objective restriction of the supply of material resources by the construction complex of a metropolis can be the aggregate production capacity of enterprises that are part of the construction complex of a metropolis. If the relationship takes the form:

$$\sum \text{Buld.enterpr.} < \sum \text{VMR}, \tag{2}$$

This situation means that the enterprises belonging to the construction complex of a megacity of the enterprise are not able to provide the production of the necessary amount of material resources for all the nomenclature used. The balanced provision of the construction complex of the metropolis with material resources for the four groups of the enlarged nomenclature should be based on the following ratio:

$$\sum_{Z=1}^{Z} \text{MR}_{\text{kz(Buld.enterpr.)}} \geq \sum_{d=1}^{D} \text{MR}_{\text{dk(resours)}}, \tag{3}$$

where $\text{MP}_{\text{kz(Buld.enterpr.)}}$ – material resources of the **k**-th type produced by the **z**-th enterprise;

$\text{MR}_{\text{dk(resours)}}$ - material resources of the **k**-th type, consumed in the production of resources at the **d**-th building site;

k – the specific nomenclature group of products of construction enterprises (the total number of nomenclature groups k = 1, .., 4);

z - the enterprise of a building complex (the total number of enterprises z = 1, .., Z);

d – construction objects, where material resources are consumed (d = 1, ..., D).

The balanced provision of the construction complex of the metropolis with material resources for the four groups of the enlarged nomenclature should be based on the following ratio:

$$\sum_{z=1}^{Z} \text{MR}_{\text{kz(Buld.enterpr)}} \geq \sum_{d=1}^{D} \text{MR}_{\text{dk(resours)}}, \tag{4}$$

where $\text{MP}_{\text{kz(Buld.enterpr)}}$ – material resources of the **k**-th type produced by the **z**-th enterprise of the construction complex;

$\text{MR}_{\text{dk(resours)}}$ - material resources of the **k**-th type, consumed in the production of construction and installation work on the **d**-th building site;

k – the specific nomenclature group of products of construction enterprises (the total number of nomenclature groups k = 1, .., 4);

z - enterprise of the construction industry (total number of enterprises Buld.enterpr z = 1, .., Z);

d – construction objects where material resources are consumed (d = 1, ..., D).

The excess of the values $\sum_{z=1}^{Z} MR_{\text{kz(Buld.enterpr)}}$ над $\sum_{d=1}^{D} \text{MR}_{\text{kd(resours)}}$ is explained by the objective necessity of having certain stocks (at least in the size of the insurance stock), which is an indispensable condition for rational management of the movement and use of material resources and should be taken into account in the balance sheet being formed. Quantitatively, this excess is determined by:

$$\text{M}R_{\text{k}} = \sum_{z=1}^{Z} \text{MR}_{\text{kz(Buld.enterpr)}} - \sum_{d=1}^{D} \text{MR}_{\text{kd(resours)}} \tag{5}$$

When forming a balance model, the fact that the consumption of material resources is carried out in the context of specific types of objects under construction is also taken into account. This means that the following ratio should be observed:

$$V_{MR\,(k)} \geq \sum_{d=1}^{D} V_{MR\,(kd)} \tag{6}$$

This ratio means that the total demand for material resources included in the k-th nomenclature group should slightly exceed (not less than the amount of the insurance stock) the amount of consumption of material resources included in this nomenclature group for all created objects, where this type material resources are used.
The balance sheet model of the material resources of the construction complex must be based on ensuring the balance of material resources in the context of specific enterprises that are part of the construction complex. This condition of balance has the form:

$$\sum_{z=1}^{Z} \Pi_{Buld.enterprK_{(z)}} \geq \sum_{k=1}^{K} MR_{k(Buld.enterpr)}, \tag{7}$$

where $\sum_{z=1}^{Z} \Pi_{Buld.enterprK_{(z)}}$ – total volume of production of building materials and structures for all enterprises;

Z– the total number of enterprises in the construction sector (z = 1, .Z).

At that, in the composition of each share of the total volume of production (Buld. $enterprK_{(z)}$ includes a proportional part of the insurance stock.

5 Conclusion

The adopted managerial decision entails the need to form a new material balance of the construction complex of high-rise construction of a large city. It should take into account the changes envisaged by management decisions, and, accordingly, the adjustment of all indicators of the material balance. There must be defined a new result that characterizes the presence or absence of a balance of supply and demand for the material resources of the construction complex of high-rise construction, and the basis for adopting a new management decision is created.

In general, the balance sheet model of material resources of the construction complex of high-rise construction should provide for the possibility of adjusting the indicators as when changing the standards for the consumption of material resources, the volume of construction and production capacity of enterprises in different periods of time, and when adjusting the planned indicators of the high-rise building complex.

Thus, the developed conceptual model of the material balance of the construction complex of high-altitude construction covers all stages of the development of a methodology for managing the provision of material resources for the construction of a metropolis from the stage of setting management objectives to the stage of practical implementation.

The main results include the following developments:

1. The conceptual model of the material balance of the building complex of high-rise construction, covering all stages of developing methodological tools for providing the building complex of large cities with material resources from the stage of setting management objectives to the stage of practical implementation.
2. The system of indicators that determine the balance of supply and demand for the material resources of the construction complex and take into account the change in the price index for them.
3. Model of material balance in high-altitude construction, allowing to detect the magnitude of imbalance in various groups of building materials, products and structures in the enlarged nomenclature.

4. Methodological provisions for the justification of management decisions to ensure the construction of material resources on the basis of the material balance of the construction complex, decisions that differ in their variability and are adopted depending on the presence or absence of a balance of material resources and the possibility of eliminating imbalances.

References

1. S. Salam, et al Hughes, W (Ed.), 15th Annual ARCOM Conference, Liverpool John Moores University (1999)
2. J.R. Baldwin, Association of Researchers in Construction Management, Vol. 1, p. 23-32(1999)
3. J.R. Baldwin, J.M. Manthei, Causes of delay in the construction industry. Journal of Construction Division, ASCE, 97(2), p.177–187 (1971)
4. V. Didenko, High-rise construction: problems and perspectives/ sociology of management 1, p.73(2012)
5. K. Rozhentsova et al, MATEC Web of Conferences, 08076, DOI: https://doi.org/10.1051/matecconf/201710608076, (2017)
6. E. Stein, Procedia Engineering, Volume 165, Pages 1410-1416 (2016)
7. A.Mottaeva , IOP Conf. Series: Earth and Environmental Science 90 012120 doi :10.1088/1755-1315/90/1/0121209(2017)
8. A.Mottaeva, MATEC Web of Conferences, 73, 07020, DOI: https://doi.org/10.1051/matecconf/20167307020, (2016)
9. M.Rahman, World Habitat Day '96. Ministry of Housing and Public Works, Bangladesh, (1996)
10. O. Ogunlana, K. Promkuntong, Construction delays in a fastgrowing country: comparing Thailand with other economics. International journal of project management, 14(1), p. 37–45(1996)
11. W.Trochim, The research methods knowledge base. 2ed. Ithaca, NY: Cornell Custom Publishing, (1999)
12. D. Okpala,. Causes of delay and cost overrun in the construction industry, CENSER. Seminar series, (1986)
13. Kwakye, Construction project administration in Practice. Essex: Longman, (1997)

Insurance of risks as the instrument of protection of investments into high-rise construction

Elena Vasilyeva[1,*] and *Vladimir* Okrepilov[2]

[1] Moscow State University of Civil Engineering, 129337, Yaroslavskoe shosse, 26, Moscow, Russian Federation

[2]Federal State Budgetary Institution of Science Institute of Regional Economic Problems of the Russian Academy of Sciences, 190003, Serpuhovskaya st.,38, St. Petersburg, Russia

Abstract. Insurance now is one of the important sectors for both world, and national financial systems. The mechanism of insurance of risks plays an essential role in the process of construction of high-rise buildings and as well as other unique objects. The author argues that insurance is categorically necessary because of the following reasons: huge scale and complexity of high-rise projects and great expenses, the use of unsafe sites for construction, great possible losses in case of the project, important role for the city infrastructure, numerous organizations, involved into the process of construction and maintenance, multypurpose use of the building. The features of insurance in the field of high-rise construction is also considered in the article. The prerequisites, confirming the loss occurrence are named.

1 Introduction

Construction industry as whole and high-rise construction particularly is subject to more risk and uncertainty than many other industries.

The development of a high-rise construction project from inception to completion takes a long time and involves many phases. It brings workers with different skills and interests together and involves the use of large and diverse set of equipments. All of these complex requirements have to be handled with proper co-ordination to provide a smooth flow of activities. It is necessary to identify and analyse the risks that may appear during project risks means the control of the project itself .Insurance is, of course, only one means of managing risks associated with such projects. It needs to be put into context and understood that not every risk can be insured against adequately or insured for a price that is acceptable [1].

Speaking about insurance of high-rise buildings, we have to consider first of all the whole complex of various self-contained types of insurance created in last years which are bound to the construction and maintenance of such objects. Besides, it is necessary to think about

* Corresponding author: elena.chibisova_metr@mail.ru

the features and padding risks distinguishing insurance of high-rise buildings upon insurance of other constructions.

The article aims in conducting an investigation on how insurance is used as a risk transfer tool in the construction industry and examine the interaction between construction players and insurance companies with respect to risk management. The attempt to present interaction of main types of insurance of structural risks, their temporary sequence and transitions between them is also made in the article.

2 Methods

How insurance of investments into construction of high-rise buildings is carried out? In the figure 1 the main steps which need to be undertaken are shown [2].

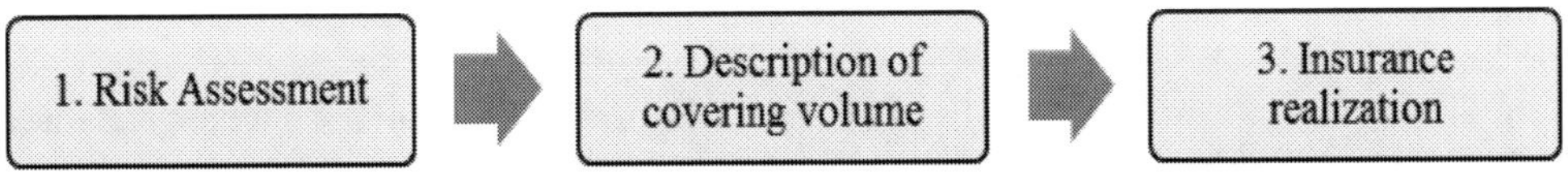

Fig. 1. Stages of insurance of investments into high-rise construction.

1. Risk assessment is the first stage. On its basis the insurer has to define whether he or she is ready to insure this risk in a way desirable to the client. And, besides, on the basis of risk assessment the insurer decides in what volume he intends to participate in him. Assessment of risk is carried out on the basis of definition of the insured sum and the level of insurance premiums which is based on her. Quite often during construction of the building the cost of construction works considerably increases: wrong assessment, the arisen difficulties, inflation, etc. can be the reasons. It is recommended to keep track constantly of dynamics of expenses and in due time to coordinate the correcting measures with the insurer.

Also it is recommended to coordinate a higher insured sum, an extra charge to an award, in order to avoid a possible underinsurance at the stage of signing of the contract. Upon the termination of insurance term the difference comes back to awards to the insurer.

2. Along with the definition of the insured sum the description of volume of a covering is formed. Types of coverings are concretized in reservations of the contract, certain risks are excluded. As for high-rise projects, that can be:

- special conditions of fire safety on building sites;
- special conditions at a construction of the base of deep laying;
- responsibility to the third parties - vibration, removal or weakening of the bearing elements;
- special conditions of laying of underground cable lines, pipelines and him similar constructions.

These provisions are in writing fixed in the policy.

3. The third step is the insurance realization. Insurers are partners of the insurer here. However they also are economically interested in fast, effective implementation of the project properly. For this purpose they conduct examination of the building site [3, 4].

The insurer first of all is interested in the following aspects [5, 6]:

- management, control;
- the updated job schedule;
- work flow;
- storage conditions of materials;
- measures of fire safety;
- administrative work;
- special conditions on a building site;
- unprofitability;

• construction documentation (photo of the site);
• recommendations about improvement of the organization of construction and security measures;
• control of dates of performance, results;
• the plan of measures on prevention of damage from a storm or a flood.

3 Results

The need and economic validity of insurance of installation and construction works and also the subsequent insurance of property of buildings are so not obvious anywhere, as in the field of a construction and operation of high-rise objects. For this purpose there are several reasons:

1. High-rise projects represented and represent huge investment projects. Costs of construction at the level of 1 billion US dollars presently is not a rarity any more.

In most cases these means will be mobilized by means of the credits. The investor, naturally, wants to insure himself from technical, natural and human risks which can threaten his investments into construction time [7].

Decrease in risks allows to reduce the cost of the credits.

2. High-rise buildings play an important role in city infrastructure. As a rule, they are built in the center of the large cities and influence other elements of infrastructure. It assumes exact observance of the provided terms of carrying out construction works, etc.

3. Modern high-rise objects become more and more complex systems. Hundreds of the organizations are often involved in their construction.

4. Modern high-rise buildings are often intended for complex use: individual housing, restaurants, hotels, offices and administration premises, halls for holding various actions, the movie theaters, theaters, parkings, etc. belonging to various operating organizations getting or renting parts of the building.

5. Quite often they are constructed in territories where natural cataclysms are possible: in deserts, seismo-dangerous zones, on unsafe sites of the coast, in the sea, etc. They are under construction in areas with unreliable soil and in zones of storm influence.

6. Their construction is quite often conducted on a limit of technical capabilities of the time.

7. Modern high-rise objects (and not only they) represent today not so much buildings how many some kind of "machine for accommodation". That increases the importance of the machinery and equipment, and respectively, and insurance of technical risks.

What contents and features of insurance of construction risks of high-rise objects?

If to one looks at property insurance in a chronological order, then it is necessary to begin with insurance of construction risks which actually begins with with delivery of the first construction materials and the equipment. And it comes to the end with the termination of a construction of an object and its transfer to operation.

The purpose of insurance is in protecting construction (it can be both capital repairs, and reconstruction of the building, and installation of constructions and the equipment) from all usually expected technical and natural risks, to provide to the contractor or to the customer an opportunity to eliminate with the fastest image the arisen damage to avoid more long breaks in production and idle times which can arise for the called reasons.

Insurance of construction risks for high-rise buildings should be carried out as CAR insurance (Construction All Risks – all construction risks) within technical insurance. It covers the risks threatening construction of the building or installation of equipment. Insurance "from all risks" means that all risks, except specified in Exceptions are insured. Risks of the fire, explosion, destruction or collapse of parts of already built design and also, as a rule, natural disasters, such as earthquakes, storms, floods, etc. and also thefts of construction materials and the construction equipment, etc. become covered.

Perhaps also combined design insurance of the high-rise building (Comprehensive Project Insurance) includes all basic elements today. Such policy represents individual option (tailor made) corresponding to uniqueness of the majority of high-rise buildings.

If we consider a complex of the corresponding types of insurance in an approximate chronological order, then at the beginning of a chain there is an insurance of responsibility of architects, designers and engineers which subject is civil responsibility. This type of insurance allows to cover risk of emergence of damage to the third parties at design and construction which results from mistakes or omissions of architects or design engineers .

He is followed by the insurance of the construction risks (CR) which is carried out by us in the form of "all construction risks" and "all assembly risks". At the same time it is about a covering on the basis of insurance "from all risks".

This type of insurance uses terms checked in the policy (named perils insurance – insurance of the named risks). It allows to indemnify the material damage which has arisen during operation of the building and can carry the name "insurance of operational risks" (operational cover) also [8].

In parallel the operating organization (organizations) or various users of an object quite often sign contracts of insurance of civil liability before the third parties in connection with damage which they can incur as a result of operation of this building.

All above-named phases of implementation of the project can be followed by insurance of losses from a delay of the beginning of production activity (e.g., in case of failure to deliver for rent because of untimely commissioning), has arisen as the instrument of protection of investments for a covering of material damage which can arise in the course of a construction of an object and installation of equipment. In most cases as the insurer here as a rule the contractor (contractor) in coordination with the customer (principal) acts.

Before construction and during a construction of a high-rise object on the building site materials and quite often very expensive structural elements of the building, the equipment have to be delivered.

At this stage in process of the growing volume and cost of the transported elements the expediency and importance of signing of the contract of transport insurance increases. These risks can be included also in the form of additional reservations in the contract of insurance of SMR.

In parallel, and it is frequent within the insurance program, integrated with CAR, the contract of insurance of responsibility of civil engineering firm which protects the insurer (the customer, the contractor, subcontractor, etc.) from risks of presentation of the financial requirements following from civil liability for the loss caused to the third parties in the course of construction of this object is signed. The civil liability of civil engineering firms gains the growing value especially when the part of the building even in the course of construction is put into operation.

Now trade and the congress centers quite often begin to function on first floors of the building before completion of construction of the building that can cause the necessity of insurance of electronic devices [9].

There is also a number of special insurance policies which can be acquired depending on purpose of the building, etc.

Here it is possible to carry also types of personal insurance, such as insurance of civil liability of the employer.

The decision on what types of insurance to choose to the owner operating the organizations, to the customer of construction works, the contractor, the architect (each party having insurance interest) are accepted first of all by the customer and the contractor in common with the insurer on the basis of the project documentation. She contains technical parameters of the ready high-rise building, a basis of his integration into the existing platform infrastructure.

It, naturally, increases the cost of the policy of assurance of civil liability.

A few years ago the Munich reinsurance society has put into operation more modern insurance, combined in many sections of projects, called – Comprehensive Project Insurance (CPI). It contains up to four sections and can include all above-stated segments (insurance of transport, construction-and-installation works, construction equipment and civil liability) which can be also combined with insurance of losses from the delays of the beginning of a production activity which have resulted from suspension of installation and construction works or supplies of equipment as a result of problems with transport.

After transfer of an object to the owner or the operating organization the insurance described above is succeeded by annual renewable property insurance. The builder, the owner or the operating organization (operator) becomes the insurer now.

This insurance includes compensation of material damage and some other types of a covering.

The most important of all types of insurance of high-rise buildings, undoubtedly, is insurance upon the fire. It is caused by assessment of scale of possible losses, or PML (Probable Maximum Loss – the greatest possible loss). The speech in this case goes about insurance of property which covers the risks stipulated in the policy.

4 Discussion

Sceptics can object that receiving compensation from insurance company – a difficult procedure and that the company can have occasions to refuse to the insured person payment. To disprove these fears, it is necessary to discuss in what case payments for insurance are carried out?

Really, the refusal of compensation payment in principle is possible. But two prerequisites which existence authentically confirms the loss occurrence are always important and guarantees compensation payment:

1) There has to be a sudden, unforeseen event, which came irrespective of the insurer and the insurer and led to a loss.

2) An object to which damage is caused has to be insured from risks, i.e., insurance validity has to extend to it.

Besides, it is necessary to learn precisely who in a concrete case is an insurer as cases when they mistakenly address not to that firm are frequent.

The question is subject to examination whether there was a loss, for example, in the course of construction and whether he has been caused by one of the hidden dangers or it is about a loss, responsibility for which is born by the third organization [10, 11].

Reinsurance plays important role in insurance of installation and construction works.

In effect overcautious persons are insurers of insurers. They make available to insurers financial means for a covering of large risks.

From the aforesaid becomes clear that reinsurance has to play especially important role in insurance of installation and construction works on high-rise buildings. It is caused by the following factors [12]:

– The considerable sums and limits of insurance which follow from high investment expenses and the greatest possible loss.

– Danger of accumulation of losses on different types and categories of insurance (for example, installation and construction works, responsibility of builders, personal insurance).

– Accumulation of risks turns out and as a result skyscrapers for infrastructure and town-planning reasons are quite often built at small distance from each other. As it was already mentioned above, they are structurally connected to other buildings and infrastructure facilities or are under construction in close proximity to them.

– Now skyscrapers are built in the areas subject to natural risks: to storms, earthquakes, tsunami, etc.

5 Conclusion

It is possible to prove that if there is no insurance and effective work of the insurance companies, the majority of high-rise objects would not be under construction today, they were not under maintenance and were not on sale.

Large investment projects on construction of high-rise buildings can turn back in case of failure huge risks for the investor. That itself proves need of fears,

It is clear, that insurance in the sphere high-rise constructions has the features. Large investment projects on construction of high-rise buildings in most cases are unique, sometimes not having analogs at the national level. Insurance companies have to accumulate the international experience of a construction of similar objects. They have to have qualified personnel that will give them the chance to give support to clients in assessment of risks, underwriting, formation of insurance programs and in settlement of losses.

It is important that insurers chose financially steady insurance companies having the sufficient capital which in case of damage could pay the share on a loss. As each insurer can and has to assume only a certain share of risk (from the insured sum or limits), one or several overcautious persons are necessary for him. Re-insurers also have to limit the responsibility.

References

1. A. Babu, S. Kanchana, International Journal of Civil Engineering and Technology, **5 12**, 206-210 (2014)
2. A. O'Sallivan, *City economy* (Moscow, INFRAM, 2002)
3. N. Bunni, *Risk and insurance in construction* (Spon, London, 2003)
4. D. Heidenhain, Geneva Papers on Risk and Insurance - Issues and Practice, **26(2)**, 268-276 (2001)
5. M. Rahman, M. Kumaraswamy, Engineering, Construction and Architectural Management, **9 (2)**, 131-151 (2002)
6. E. Rendell, D. Yablonsky, *Risk manager's insurance guide* (Department of Community and Economic Development, Harrisburg, 2003)
7. V.P. Generalov, *Features of design of high-rise buildings Samara* (Samara State architecture-and-construction University, 2009)
8. W. Leontief, *The Structure of the American Economy, 1919–1929* (Cambridge: Harvard University Press, 1941), https://doi.org/10.1017/S0022050700053158 Published online: 01 February 2011
9. J. Pridmore, *Chicago Architecture and Design* (New York: Harry N. Abrams, Inc., 2005)
10. A. Mottaeva, MATEC Web of Conferences, **73,** 07020 (2016)
11. A. Mottaeva, IOP (Earth and Environmental Science), **90,** 12138 (2017)
12. *Prospects of construction and maintenance of skyscrapers in the 21st century* [Electronic resource] – Access mode: https://regrazvitie.ru/perspektivy-stroitelstva-i-ekspluatatsii-neboskrebov-v-21-veke/ (2018)

https://doi.org/10.1051/e3sconf/20183303049

Conversion of wastelands into state ownership for the needs of high-rise construction

Elena Ganebnykh[1,*]

[1]Vyatka State University, Moskovskaya str., 36, Kirov, 610000, Russia

Abstract. High-rise construction in big cities faces the problem of land shortage in downtown areas. Audit of economic complexes showed a large volume of wastelands. The conversion of wastelands into state and municipal ownership helps in part to solve the problem of the lack of space for high-rise construction in the urban area in the format of infill construction. The article investigates the problem of the conversion of wastelands into state and municipal ownership. The research revealed no clear algorithm for converting wastelands into state and municipal ownership. To form a unified system for identifying such plots, a universal algorithm was developed to identify and convert ownerless immovable property into state or municipal ownership.

1 Introduction

Empirical observations have shown that high-rise residential construction is mainly distributed in the metropolitan central districts with a high density of construction. This phenomenon can be explained by the fact that traditionally downtown areas are more prestigious for living, therefore, the increased demand for housing should be provided by an appropriate proposal from the developers. However, a serious limiting factor here is the land shortage for this type of housing development. At the same time there is a large number of wastelands in cities, which is confirmed by the results of the city's economic complexes audit initiated by municipal bodies. Only in Moscow there were found 25 million square meters of wasteland [1].

The problem of wastelands is partly associated with "losses" during the restructuring of government bodies, so most often it concerns abandoned roads and green areas. A very important question is adequate consideration of the assignment of these wastelands use, since from the ecological point of view green areas should be improved and converted into park areas. However, partially abandoned wastelands can be undoubtedly given for the construction of housing stock. If such plots are located in an acceptable zone, then this will help develop high-rise construction in cities.

Though the conversion of wastelands into state or municipal ownership does not completely solve the problem of lack of land for high-rise construction in downtown areas,

* Corresponding author: ganebnykh@mail.ru

it can partially provide for infill construction. At the same time, the conversion of wastelands into state (municipal) ownership with subsequent making plots over for construction is a way of the municipal budget replenishing.

The conversion of land plots into state and municipal ownership is one of the most pressing issues of land and property relations. The process of converting wasteland plots into state and municipal ownership has a number of difficulties related to the lack of a full-fledged system for identifying abandoned property - therefore, not every state and (or) municipal authority considers such work to be a wise measure.

Historically, after the "Decree on Land" in 1917, all land in Russia was automatically converted free of charge into state ownership, that is, it was seized from the owners voluntarily and expropriated, and then was included in the administrative circulation through redistribution between citizens and legal entities only based on decisions of the relevant bodies. At that time, the fundamental principles were the principles of labor land use (cultivation of land only in-house) and equalization of land in accordance with consumer and labor standards. Any transactions with land plots for their alienation (sale, lease and (or) security) were prohibited. As a result of these transformations, the State became the largest and the monopolist owner, and the leading forms of land use, corresponding to the interests of the state and society, were exclusively social (communes, partnerships, and later collective and state farms) [2]. Thus, the Soviet period ruled out the possibility of wastelands in legal terms, but in fact, wastelands always exist, especially on the territory of such a huge country.

After issuing in 1990 the RSFSR law "On Land Reform", a number of economic reforms began, which resulted in forming free market relations in Russia. The main directions of land reform in the 1990s in the Russian Federation include: reforming the legal framework; administrative redistribution of agricultural land; introduction of an economic regulation mechanism in the land and property sphere (land purchase, land payments); formation of governing bodies in land and property sphere; information support of land reform (land cadaster, land monitoring) [3].

By the end of a long ten-year period of transformation of land relations, it became clear that there was an urgent need for more rigorous codification of all changes (adoption of a new Land Code [4]), more active participation of the State as a correcting and guaranteeing institution and "generally carrying out state policies in the field of land relations targeted not at buying and selling land and its speculating, but at rational use of the land, ensuring an increase in soil fertility and yield of agricultural crop, environmental protection" [5]. The main reasons for the existence of wastelands at present stem from the processes that took place in Russia in the 1990s, namely: transition to the market economy, establishing private property, privatization of land, and formation of a new largely imperfect legislation.

2 Materials and Methods

To analyze the state of the problem of wastelands it is necessary to conduct a comprehensive study including the following stages:

1. Legal provisions for the conversion of wasteland plots into state and municipal ownership:
- studying the unity existing in legal understanding and terminology concerning wastelands;
- studying availability and completeness of the legislative framework ensuring the identification and circulation of wasteland;
- studying procedures for converting ownerless immovable facilities into the state ownership.

2. Analysis of procedures for registration of wasteland plots in State cadastral records.

3. Assessment of capabilities and ways of disposing wasteland plots by state and municipal authorities:
- analyzing formats for granting land to own or dispose by private individuals or legal entities;
- studying the process for granting land plots to own or dispose by private individuals or legal entities;
- studying price formation when selling land plots to own or dispose by private individuals or legal entities;
4. Analysis of the practice of converting wasteland into state and municipal ownership on the example of the City of Kirov municipal entity administration:
- identifying weak points in the practice of converting wasteland plots into state and municipal ownership;
- searching for advantages for state and municipal authorities from the conversion of wastelands into their ownership.

The results of the research made it possible to work out a number of recommendations to improve the system for converting wasteland plots into state and municipal ownership, and to form an algorithm for identifying and converting ownerless property into state or municipal ownership.

3 Results

The research investigated and analyzed theoretical aspects of the system of converting wasteland plots into state and municipal ownership and ways of disposing such land plots after converting them into state and municipal ownership. If all idle lands on the territory of our country are put into civil circulation, this will positively affect the country economy as a whole. According to economic calculations, if all the land in Russia is either leased out or transferred into ownership, the GDP will increase by 10% per year, that is, it will be higher than that of China - through land resources alone [6].

3.1 Legal provision for converting wasteland plots into state and municipal ownership

According to the RF legislation [7], objects can be recognized as ownerless in 3 cases: objects do not have an owner, the owner of the object is unknown or the owner refused the object. It is important to understand that wasteland plots are considered to be only those land plots that have no owner or which owner is unknown. At the same time, land plots, where the owner refused from the right of ownership, are not classified as wasteland plots.

In most municipalities, the process for identifying ownerless immovable property is approved at the level of municipal legal acts. For example, the City of Kirov municipal entity has the Resolution of the Administration of the City of Kirov of 30.08.2017 No 3202-p "On the approval of the Regulations on the process of registration of ownerless immovable property into the ownership of the municipal entity "City of Kirov". And this act regulates the process of converting not only wasteland plots, but all the ownerless property located on the territory of the city.

The main stages in converting wasteland plots into the state and municipal ownership include the identification of wasteland, their state registration as ownerless, as well as declaring state or municipal ownership on the wasteland plot in a judicial proceeding.

There is no definite procedure codified by law to identify ownerless property, therefore each municipality is compelled to develop such a procedure independently, based on territory peculiarities and (or) activities of the property management committees.

Practice shows that wasteland plots are identified as a result of:

- information sent by individuals and legal entities to the territorial departments of the City or District Administration;
- notifications from the State cadastral registration authorities to the Administration about land plots without registered rights;
- incoming owner's abandonment of the land plot ownership to the local government body at the location of the land plot;
- identification of ownerless property in the course of auditing the immovable property use on the territory of a municipality conducted within the framework of municipal control.

Thus, in the course of conducting such municipal land control, the Administrations of municipal entities work to ensure compliance with the requirements of land legislation in the use of land plots in accordance with their designation, category, permitted use, as well as identify previously unrecorded land plots, and list them. The main body providing land control is a specially formed Commission.

After identifying wasteland plots and their listing, the Commission examines them directly and draws up an inspection report, which is sent to the Office of the Federal Service for State Registration, Cadaster and Cartography of the region to take essential measures. The identified wasteland plots are subject to mandatory state cadastral registration, which is carried out at the request of the local government body on which territory such plots are located.

Within a year after registering the wasteland plot, it can be returned to the owner or transferred to another person under the rule of acquisitive prescription [7]. If during this time nobody declares their rights to this land plot, then in this case it is possible to declare state or municipal ownership of the wasteland plot in a judicial proceeding on the application of state and (or) municipal authorities. Such applications are examined primarily by courts of general jurisdiction or arbitration courts.

After the court decision on recognition of the right of municipal ownership of the city, district on the land plot becomes effective, the Committee applies for state registration of the right of municipal ownership of the immovable property. The Certificate of the municipal title is issued in a month's time, and then the resolution of the City Mayor, the Head of Administration simultaneously includes the land plot in the Unified Register of the City (District) Municipal Property and excludes it from the Register of Ownerless Immovable Facilities.

3.2 Analysis of procedures for registering wasteland plots in state cadastral records

The process of cadastral registration of wasteland plots is regulated by the Order of the Ministry of Economic Development of Russia of 10.12.2015 N 931 "On Establishing the Procedure for Registering Ownerless Immovable Property". Governing bodies in municipal territorial entities (city and district) have the right to adopt their own statutory acts regulating issues relating to registration of ownerless immovable property, including land plots, unless otherwise stipulated by the legislation.

Thus, having analyzed the legal framework, the following process was identified for the registration of ownerless immovable property:
- taking an application for registering a land plot as wasteland with the essential documents attached;
- verifying the documents authenticity and credibility of information reflected in them;
- making entries in the Unified State Register of Immoveable Property on the registration of wasteland plots.

Depending on the territorial location of the land plot, the relevant bodies have the right to file an application:

- if the land plot is located on the territory of an urban, rural settlement or city district, then the application is submitted respectively by local government bodies of urban, rural settlements and city districts;
- if the land plot is located in inter-settlement territories, then the application shall be submitted by the local government body of the municipal districts on which territories the land plot is located;
- if the land plot is located on the territories of cities with federal status, then the application is submitted by the executive State government bodies of these cities with federal status;
- if the land plot is located on the territory of more than one municipal entity, then the application is submitted by any local government body with reference in this case to the name of another municipal entity on which territory the immovable property object is also located.

The decision to register a land plot is expressed in making the entry of registration in the Unified State Register of Immovable Property, after which the applicant receives notification of the object registration.

3.3 Assessment of the capabilities and ways of disposing wasteland plots by state and municipal authorities

Land plots in state or municipal ownership are one of the main sources for replenishing the budgets of the State, region, and municipality. The main way to draw income from such land plots is to dispose them skillfully, which is carried out through using the land plot in one's own interests or by assigning land plots to use of individuals or legal entities. The general procedure for granting land plots is shown in Figure 1.

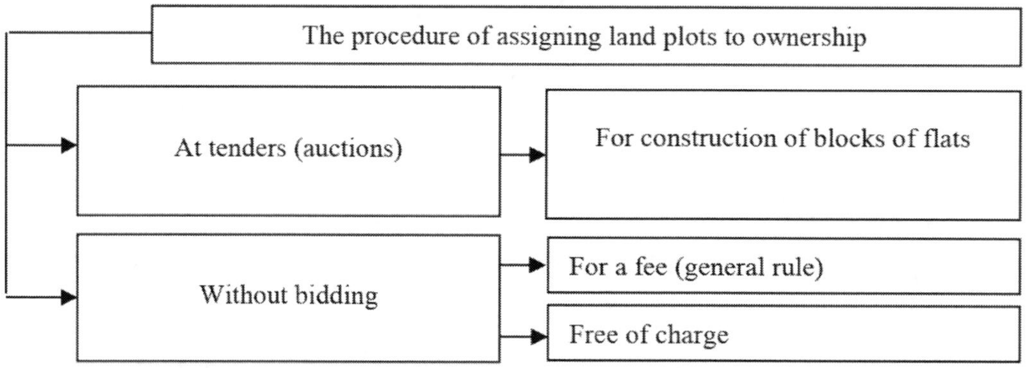

Fig. 1. General procedure of assigning land plots to ownership

According to the provisions of the Land Code of the Russian Federation, the powers to assign land plots from the state and municipal property, depending on the specific form of the land plot ownership, are given to the relevant executive bodies of state power and local government bodies.

Free assignment of land plots from the state and municipal property is carried out in the following cases:
- assigning a land plot to religious organizations;
- assigning a land plot to a non-profit organization of citizens established for gardening, truck farming and related to communal facilities;
- assigning a land plot to a citizen who has used it in accordance with the established permitted use for the right of uncompensated use for 5 years;
- assigning a land plot to citizens with 3 or more children;
- assigning a land plot to certain categories of citizens.

In other cases, land plots are provided on a paid basis. Therefore, forming land prices is of particular importance, since the receipts are revenues of budgets of the corresponding levels. The formation of the transaction price of the land plot of the state or municipal ownership is shown in Figure 2.

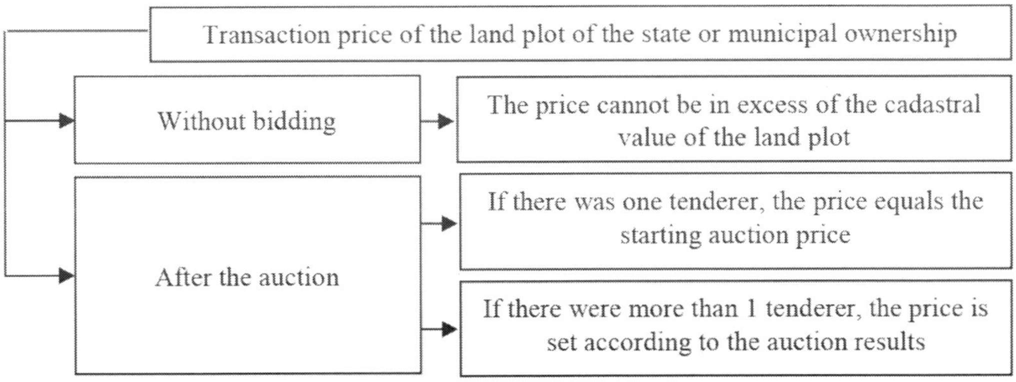

Fig. 2. Price formation of land plots of the state or municipal ownership

The procedure for auctioning the ownership of a land plot of the state or municipal ownership depends on the initiator of the auction (a party concerned or authorities) and on the auction organizer (the authorized government body or a specialized organization acting on the basis of a contract with the authorized government body).

Thus, Russian legislation establishes and regulates many ways to manage state and municipal lands. Choosing the way of disposal lies entirely with the authorized state and municipal authorities.

3.4 Analysis of the practice of converting wasteland plots into state and municipal ownership on the example of the City of Kirov municipal entity administration

Analyzing data available from the official websites of the Administration of the City of Kirov, it becomes evident that the process of converting ownerless property into state and municipal ownership in the Kirov region is poorly developed. Local government bodies are mainly engaged in converting into municipal ownership such ownerless immovable objects as ownerless flats, ownerless houses, as well as ownerless engineering communications. Conversion of wasteland plots into municipal ownership is quite a rare phenomenon. This situation is typical for most municipalities due to a number of reasons.

1. The complexity of the procedure for converting wasteland plots. As a result of constant updating of the regulatory and legal framework, which, on the one hand, improves and regulates certain issues more clearly, but, on the other hand, changes too often, all processes slow down because specialists need time to understand and produce a system for applying these norms in practice.

2. Lack of clear delimitation of duties of government bodies in this issue. The law does not stipulate that it is the duty of municipalities to register rights to the ownerless property, therefore, using this fact local government bodies do not undertake such duties. On the other hand, the law says that the only authorities that have the right to file an application for registering ownerless property are local government bodies, since only they have the right to apply to the registration authority to register such property in cadastral records [8].

3. Absence of a clear algorithm of actions to convert wastelands into state and municipal ownership. Currently, there is no general provision regulating the procedure for converting wasteland plots into state and municipal property. Therefore, as a rule, local government bodies establish this procedure.

4. Lack of a clear system for identifying wasteland plots. The procedure for identifying ownerless property is not regulated by any general statutory act either. The legislative base is all the same Decisions and Regulations of local government bodies, which do not provide a clear procedure for identifying ownerless property [9].

5. Weak interest of the authorities in converting wastelands into state and municipal ownership. This process can be described as costly and protracted, as, firstly, to obtain documents with the description of the land plot, it is necessary to complete cadastral work. Only a qualified cadaster engineer has the right to conduct such work, so his work must be paid by the customer. The duration of this process is stipulated by the legislation, which states that it must take at least one year from the moment of registering a wasteland plot in the cadastral records till the moment it is recognized as municipal or state property. This time allows the possibility that the owner of this land may show up. In this case, the municipality will not be able to register the ownership of this plot.

Thus, analyzing the possible unfavorable development of the situation, the authorities do not consider this procedure viable. In practice, in order to avoid a protracted process of converting wastelands, heirless property, or unclaimed land shares into state and municipal ownership, as well as extra costs, local government bodies often write official papers stating that this land plot is not ownerless but is in the municipal ownership or that the data on the rights to the land plot, the owner of which has died, are registered in the Unified State Register of Real Estate by mistake, since authorized bodies haven't issued any certificate on the right to life (inherited) possession or the right to permanent (unlimited) use [10].

Nevertheless, the advantages of converting ownerless property into state and municipal property are also evident. Firstly, the number of municipal or state immovable property objects is growing for a minimal fee, which can be used for their own purposes[11]. Secondly, it becomes possible to expand and replenish the tax base due to ownerless immovable property, which leads to an increase in taxes incoming to the state budget.

Thus, having analyzed the system of converting wasteland plots into state and municipal property, there was revealed a large number of shortcomings and defects. At the same time, the advantages of the system are revealed in the possibilities of disposing converted land plots, this system is particularly favorable for housing construction lands, as it incurs significant revenues to the budget.

4 Discussion

As a result of the research, the following important aspects of the system of converting wasteland plots into state and municipal property were revealed:

- need to secure duties of implementing and monitoring the process of identifying ownerless property, respectively, to state authorities or local government bodies (delineation of levels of responsibility);
- need to allocate responsibility for dealing with ownerless property, located on the municipal territory, to the Head of administration of this municipality (personal responsibility);
- availability of a clearly-defined algorithm of actions to identify and transfer ownerless property into state or municipal ownership (uniform procedure).

The results of the research made it possible to formulate a number of recommendations on improving the system of converting wasteland plots into state and municipal property.

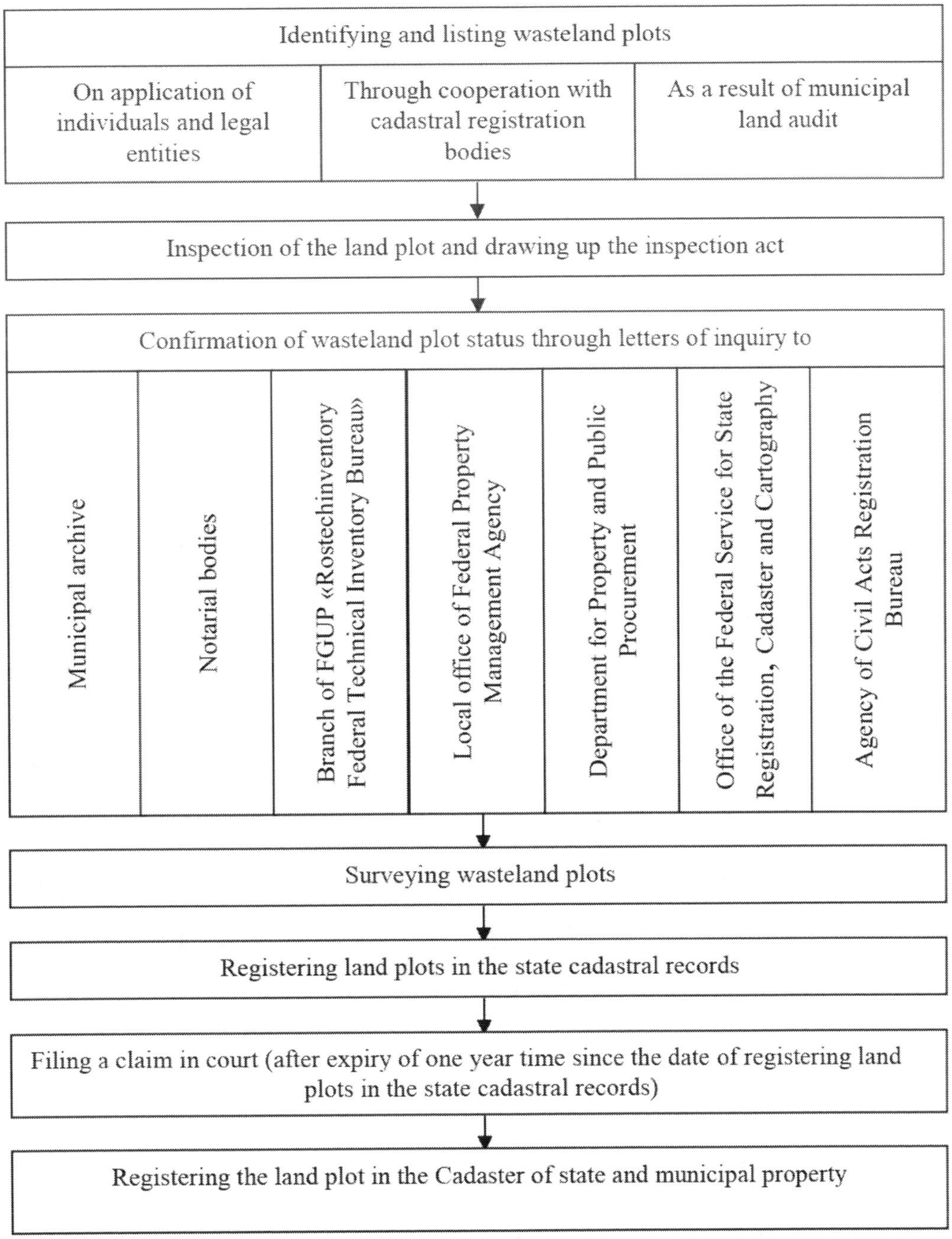

Fig. 3. The uniform algorithm of actions for identifying and converting ownerless property into state or municipal ownership

Firstly, it is necessary to impose the responsibility for converting ownerless property on the Heads of Administrations of the respective municipalities by securing this item in the regulatory framework. This measure will stimulate local authorities to implement these activities. Secondly, it is necessary to create and legislatively fix a permanent Commission, which will be engaged in identifying previously unregistered land plots, as well as their listing. Thirdly, it is necessary to oblige local authorities to report regularly on the progress in identifying, examining and converting wasteland plots of the municipalities in question. This measure will help to systematize the results of the work performed, as well as monitor the process. Fourthly, it is necessary to introduce a uniform algorithm for identifying and converting ownerless property into state or municipal property. The algorithm proposed for implementation is shown in Figure 3.

This algorithm clearly shows the options for identifying wastelands. It is proposed to add to the generally accepted procedure of converting wastelands a compulsory a survey of the land plot and to draw up the inspection report. This measure will help to study the land plot and its condition, determine its potential purpose, and also make sure that this land plot is not really used and in the process of its conversion there will be no person claiming this land on the basis of acquisitive prescription. Among other things, the algorithm determines all possible organizations, which may possess information about rights, or the lack of rights to the surveyed land. Responses to letters of inquiry to these organizations are basic documents confirming the lack of registered rights to this land plot, or lack of data on such rights. These documents will form the evidence base used in court. The other stages correspond to the generally accepted system.

Conclusions

After studying the history of land and property relations, it became obvious that the problem of wastelands can be traced back for a long historical period. The wasteland of our time is a direct consequence of the events that took place during the collapse of the USSR and the formation of the Russian Federation. Until now, in the legislation there is no clear regulation of converting wastelands, which causes the complexity, high labor input and cost of this process. The research studied and analyzed theoretical aspects of the system of converting wasteland plots into state and municipal ownership and ways of disposing such land plots after converting them into state and municipal ownership.

In general, there are active attempts to form a regulatory and legal framework that, on the one hand, is improved and regulates certain issues more clearly, on the other hand, changes too often. In this connection, all processes slow down because specialists need time to understand and develop a system for applying such norms in practice.

Analysis of the practice of converting wastelands on the example of the city of Kirov municipal entity administration showed that in fact there is no clear delimitation of duties of government bodies in this issue, and there is also no clear algorithm for converting wastelands into state and municipal property. To form a unified system for identifying wasteland plots, a universal algorithm was developed to identify and convert ownerless property into state or municipal property.

The results of the research made it possible to formulate a number of recommendations for improving the system of converting wasteland plots into state and municipal property, as there was revealed a weak interest of the authorities in solving this problem.

Despite the fact that the conversion of wastelands into state or municipal property does not completely solve the problem of lack of land for high-rise construction in downtown areas, it can partially help in the format of infill construction. At the same time, the conversion of wastelands into state (municipal) ownership with subsequent making plots over for construction is a way of the municipal budget replenishing. In this regard, it will be relevant to study price formation on land plots allocated for high-rise construction.

References

1. S. Ryazanov,. Mayor fights with ghosts // Arguments of the week. **24** (2011)
2. E.Smirnova, E.A. Actual problems of the humanities and natural sciences, **4**,. 158-162 (2016)

3. V. Zasyad-Volk, Administrative consulting, **4**, pp. 107-117(2011)
4. The Land Code of the Russian Federation of 25.10.2001.
5. Volkov, S.N. Land relations in Russia: materials of the round table // Economics of Agriculture, **10**, 10-11(1998).
6. URL: https://www.ikirov.ru/news/29780-chto-nam-nesyot-zemelnaya-reforma (reference date: 1.11.2017).
7. Civil Code of the Russian Federation. from 30.11.1994.
8. Yu. Kulagina, Municipal Economy, **2**, 90-83 (2016)
9. A. Mottaeva et al , MATEC Web of Conferences, 07018,DOI: https://doi.org/10.1051/matecconf/20167307018 (2016)
10. A.Pimenova, et al, MATEC Web of Conferences, 73, 07018 DOI: https://doi.org/10.1051/matecconf/20167307018 (2016)
11. A.Mottaeva, N.Gritsuk, MATEC Web of Conferences, 08083, DOI: https://doi.org/10.1051/matecconf/20167308083 (2017)

Simulation-based modeling of building complexes construction management

Aleksandr Shepelev[1,*], *Galina* Severova[1] and *Irina* Potashova[2]

[1]Northern (Arctic) Federal University named after M.V. Lomonosov, Department of Construction, 163002 Northern Dvina Embankment, 17, Arkhangelsk, Russia
[2] Sochi State University, Departmen of Finance, credit and world economy chair, 354000, Sovetskaya St., 26A, Sochi, Russia

Abstract. The study reported here examines the experience in the development and implementation of business simulation games based on network planning and management of high-rise construction. Appropriate network models of different types and levels of detail have been developed; a simulation model including 51 blocks (11 stages combined in 4 units) is proposed.

1 Introduction

In modern conditions great interest for all participants in the town-planning process is represented by effective methods of organizing and producing works for the construction of buildings of different types and floors, especially high-rise residential and multifunctional buildings and complexes built with the use of advanced technologies [1-3]. We have accumulated sufficient experience in erecting high-rise buildings and complexes in Moscow, St. Petersburg and a number of regional centres of Russia [4, 5]. As the domestic and world practice shows, an increase in the volume of such construction is inevitable.

At the same time, the existing federal state standards of higher education require a very high level of competencies of future construction managers: specialists whose major is "Construction of unique buildings and structures" and graduates holding a bachelor or a master degree and majoring in Construction. Achievement of these requirements is largely facilitated by business simulation games, as evidenced by the research of a number of universities and institutes [3, 6], including more than 35 years of experience in the department of construction at Northern (Arctic) Federal University (NArFU, formerly ALTI-ASTU) [7, 8].

The business simulation game (SG) "Network Planning and Construction Management of a Complex" (NPCMC), developed at the Department of Economics and Organization of Construction of the Kiev Civil Engineering Institute under the guidance of Professor V.I. Rybal'skiy, has become one of the most widely used in our country games for the training of civil engineers in recent decades. It is based on simulation of planning and management of the on-line construction of complexes of similar industrial buildings. The development of new modifications of SG "Network" and "Network-2" [7, 8] was the result of extensive and

* Corresponding author: a.shepelev@narfu.ru

creative use and improvement of this SG in NArFU. Many years of experience and research have shown that one of the main advantages of business games is that they simulate situations that can arise (and they do arise) on real construction sites, solutions for emerging problems can be considered in advance, a team successfully trained for advanced methods of planning and management being formed [6-8].

Thus, it follows from the above that it is timely, relevant and objectively useful to develop a version of the business game based on the existing ones, which contributes to the improvement of the training of students majoring in Construction in the field of organization, planning and management of construction production and is based on network modeling of high-rise buildings complexes erection.

2 Development of network models for construction of high-rise buildings

It is known that the relevant organizational and technological documentation (OTD) of various temporal and hierarchical levels is being developed. At the same time, the theory and practice of the construction industry [1, 2, 4, 9] show that one of the most important elements in the OTD are calendar plans for construction (production of works) that are developed and displayed on the basis of various organizational and technological models. Network models (NM) of various types are the most effective tool for solving scheduling problems at different stages of preparation and subsequent construction management [1, 4].

At the same time, a variety of design solutions for buildings (including high-rise buildings), contractor and subcontracting construction and assembly organizations, situations and organizational and technological conditions and solutions, the possibilities for combining the stages of construction and types of work lead to the need for a "typical" modeling of the construction of facilities.

Based on experience in planning of works organization and long-term studies carried out at the Department of Construction Production and summarized in [9], a system of generalized deterministic time-based NM (conventional NMs have been known since the 1970s) reflecting the real relationships between construction works in more detail was proposed as the main tool for the development of calendar plans at various levels.

As an example, Figure 1 shows an enlarged network model for the construction of a high-rise residential or multi-functional building (for a combined method of organizing work). It has 15 enlarged works: 1 - Preparatory works; 2 – Excavation of a foundation pit; 3 – Laying the foundations; 4 - Construction of the underground part of the building; 5 - Waterproofing, backfilling; 6 - Erection of the elevated part of the building; 7 - Roof installation; 8 - Special works, 1 cycle; 9 - Plastering and cladding works; 10 - Carpentry and painting, 1 cycle; 11 - Special works, 2 cycle; 12 - Carpentry and painting, 2 cycle; 13 - Installation of external networks; 14 - Landscaping, gardening and small forms; 15 - Preparation for facility commissioning.

This model uses elements of generalized network models - front links of the type "not earlier", allowing to take a variety of options for organizational and technological solutions for the construction of objects of this type (different number of floors, work volumes, labor, intensity and timing of work and stages, their terms, resource allocation, etc.) into account.

The available experience shows the possibility of successful implementation of such network models for scheduling the erection of buildings and complexes as part of construction projects, and with greater detail (up to 30-50 names of activities) - for the development of work schedules, operational planning and management of their construction.

At the same time, calendar plans for specific facilities, conditions and organizational and technological decisions are developed and displayed in the form of network graphs (respectively, enlarged or detailed) based on the simplest or multi-purpose deterministic temporary network models. They are used in the process of preparation and implementation of SG for the planning and operational management of the construction of buildings.

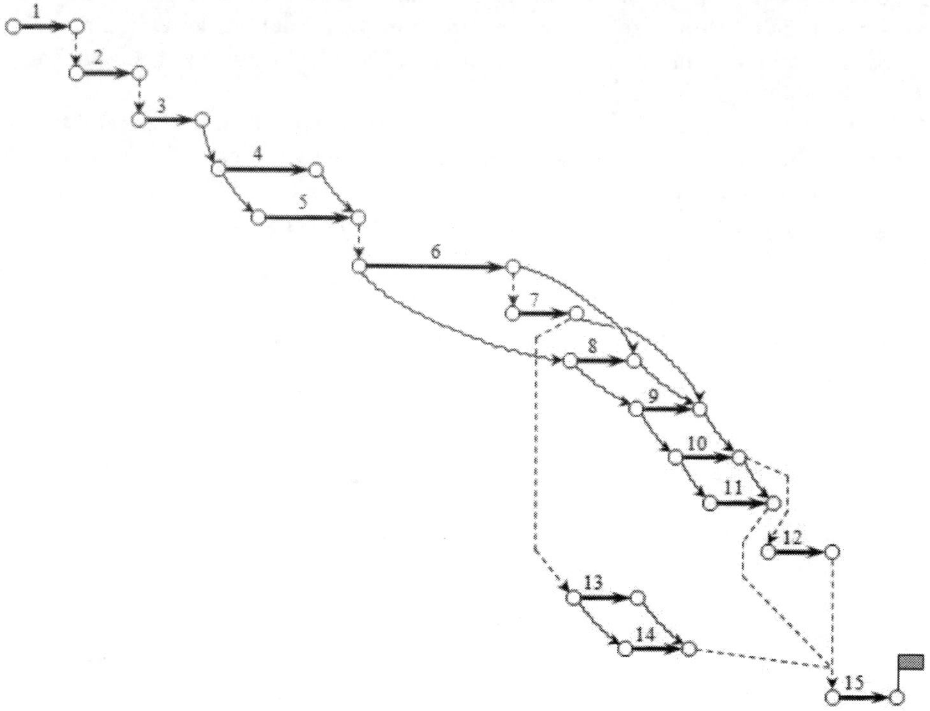

Fig. 1. A generalized network model for the construction of a high-rise building.

3 Results of the modeling of the business game related to the organization of construction of the complex

As a result of the long-term use and improvement of the SG for the organization and management of construction (NPCMC, "Network", etc.) [7, 8] we proposed a new modification of the SG - "Network-3". In the course of this SG, the functioning of the system of network planning and management (NPM) of the construction of a complex of buildings (including high-rise residential or multifunctional) is simulated. Multipurpose network models (developed on the basis of generalized network models) are used, rational use of resources is taken into account, and the main stages are "Initial planning" and "Operational management".

In view of the above, during the modeling of the SG "Network-3", developed, documented and successfully applied: the necessary input data, instructions to participants, forms and examples of documentation, reference and graphic materials. The block diagram of this SG includes 51 blocks in 11 stages, combined in 4 units. The enlarged model of the SG "Network-3" is presented in Table 1.

An important result of this SG was the students' awareness of the organizational and technological reliability of the decisions they made with the planned intensity of work and other conditions of construction that are affected by destabilizing factors arising in random combinations [10-12].

During the implementation of business games in the teaching and learning process, anonymous questioning of participants (students, teachers, representatives of construction organizations) was regularly conducted, the feedback questionnaires being developed by our department.

Table 1. Model of simulation game "Network-3".

SG stages	SG blocks	SG participants
Unit 1. Preparation for the game		
1. Preparation	1.1. Making decision on what the complexes should consist of and how inter-related their parts should be	SG instructor (SGI), arbitration (A)
	1.2. Setting up the length of construction periods and bonus accumulation	SGI, A
	1.3. Election (appointment) of complexes managers, release of construction documents and initial data	SGI, A
	1.4. Appointment of NPM staff and executives (E), release of construction documents and initial data	Complexes managers (CM)
	1.5. Reviewing of instructions and documents	CM
	1.6. Reviewing of instructions, documents and initial data	NPM staff (NPMS)
	1.7. Reviewing of instructions and initial data	E
Unit 2. Initial planning of the construction of buildings and complexes		
2. Tasks description distribution	2.1. Preparation and distribution of tasks description	SGI, A
	2.2. Development of construction execution strategy	CM
3. Initial network models development	3.1. Development of tasks on initial network models (INMs) development and distribution of work sheets for them among the executives	NPMS
	3.2. Development of INMs and work sheets for them	E
	3.3. INMs and work sheets testing	NPMS
	3.4. INMs and work sheets testing control	CM
4. Development and engineering analysis of complex network models	4.1. Development of complex network model (CNM)	NPMS
	4.2. Engineering analysis of time on CNM	NPMS
	4.3. CNM testing	CM
	4.4. CNM testing	SGI, A
5. Making CNM parameters consistent with the set up lengths of construction periods	5.1. Engineering analysis evaluation, making proposals on reduction of construction periods lengths	NPMS
	5.2. Engineering analysis evaluation, preparation for operational meeting	CM
	5.3. Making proposals on reduction of construction periods lengths	E
	5.4. Participation in the meeting, correction of parameters due to decisions made at the meeting	E, NPMS
	5.5. Participation in the meeting, coordination of proposals	CM
	5.6. Monitoring of the conduct of the meeting	SGI, A
	5.7. Correction and recalculation of CNM	NPMS
6. CNM approval and analysis of initial planning results	6.1. Making proposals on approval of redeveloped CNMs	NPMS, CM
	6.2. CNM approval and/or correction of time restrictions	SGI, A
	6.3. Scoring	E
	6.4. Scoring and monitoring of executives	NPMS
	6.5. Scoring and monitoring of NPMS	CM
	6.6. Monitoring the CNM and scoring	SGI, A
Unit 3. Operational management of construction works (3 cycles)		
7. Obtaining operational data about the condition of complexes and its processing	7.1. Delivering information about the time of obtaining data and changing of set up construction periods	SGI, A
	7.2. Processing of data on the performance of construction works and adjustment of estimation of coming period	E
8. Renewal of models, data processing and	8.1. Analysis of input operational information; fixation of construction works performance on CNM; data processing and CNM (re)calculation	NPMS
	8.2. (Re)calculation of CNM of complex (stage)	NPMS

CNM (re)calculation	8.3. CNM (re)calculation monitoring	CM
	8.4. Construction works performance evaluation and disclosure of delays	NPMS
9. Factual report on condition of complexes and optimization of CNM parameters	9.1. Making proposals on reduction of work of executives	CM, NPMS
	9.2. Making proposals on elimination of delays	E
	9.3. Participation in the meeting; correction of parameters due to decisions made at the meeting	E, NPMS
	9.4. Discussion and making decisions on further performance of construction works	CM
	9.5. Monitoring of the conduct of the meeting	SGI, A
	9.6. Correction and recalculation of CNM	NPMS
10. Analysis of operational management cycle outcomes	10.1. Scoring	E
	10.2. Scoring and monitoring of executives	NPMS
	10.3. Scoring and monitoring of NPMS	CM
	10.4. Monitoring of CNM and scoring	SGI, A
Unit 4. Analysis and implications of simulation game results		
11. Follow-up activities	11.1. Final analysis of the conduct and outcomes of performance, distribution of feedback questionnaires among the SG participants	SGI, A, CM
	11.2. Announcement of the results and analysis of the conduct of the simulation game	SGI, A
	11.3. Completion of feedback questionnaires	CM, NPMS, E, A
	11.4. Feedback questionnaires review; conclusions and recommendations	SGI, A
	11.5. Awarding of winners	SGI

All participants noted the originality of this type of practical training, excitement, competition of units and complexes, the opportunity to make collective and independent decisions, efficiency for studying relevant disciplines - that is, the applicability and necessity of business games in the teaching and learning process, as well as in real planning and construction management.

In the process of conducting business games and based on their results, interesting and useful suggestions on the content of SG, improvement of criteria, scoring and forms of documentation, additional participants, etc. were received. The findings indicate good prospects for further development and improvement of simulation exercise in construction, the fact that game methods are an integrated part of education and belong to the best and most effective ways of training specialists.

4 Results

Based on the foregoing, it can be concluded that the goal of the work has been achieved and modeling of the construction of building complexes, including high-rise buildings, using technologies and methods of conducting business (management) games is quite effective and has good prospects.

5 Conclusion

To recapitulate the findings of the present research: preparation, improvement and modeling of the business simulation games make the teaching and learning process at universities' department of construction more effective. Business simulation games are applicable in real network planning and management of construction, including construction of high-rise buildings and complexes.

The research highlighted the need for further development in this field, aiming at expanding the scope of simulation games application and achieving greater adequacy of games, used in real construction activity.

Acknowledgements

The authors are grateful to all the teachers and students who have been participating in the preparation, implementation and improvement of business simulation games at the NArFU Construction Department for more than 35 years, as well as to the 6th year students majoring in "Construction of unique buildings and structures" for joint development of network models for the construction of a group of high-rise buildings in 2017.

References

1. L.V. Kievskiy, *City development: Collection of research papers 2006-2014* (2014)
2. P.P. Oleynik, V.I. Brodskiy, Vestnik MGSU, **6**, 119-125 (2012)
3. V.I. Rybal'skiy, Industr. and civil eng., **9**, 8-11 (2003)
4. *Temporary recommendations on technology and organization of construction of multifunctional high-rise buildings and building complexes in Moscow. MDC 12-23.2006* (2006)
5. *Recommendations on the construction of residential and public high-rise buildings. RMD 31-04-2008 St. Petersburg* (2008)
6. S.A. Barkalov, V.F. Babkin, A.V. Shchepkin, *Business simulation games in organization and management* (2003)
7. A.N. Fedotov, A.L. Shepelev, V.I. Rakovskiy, *Contribution of ASTU scientists to the development of science and technology: Collection of research papers*, 232-239 (1999)
8. A.L. Shepelev, G.V. Severova, V.I. Rakovskiy, *Theoretical foundations of construction: Collection of reports of the XIX Slovak-Polish-Russian seminar*, 451–458 (2010)
9. A.L. Shepelev, Industr. and civil eng., **10**, 41-44 (2012)
10. A.A. Gusakov, A.V. Ginzburg, S.A. Veremeenko, etc, *Organizational and technological reliability of construction* (1994)
11. S.E. Klimov, *Methodology of scheduling of construction production in the harsh conditions of the Far North* (2005)
12. K.C. Crandall, J. of the Constr. Division, **102**, 415-423 (1976)

Current-day matters of administration and law in the field of high-rise construction

Elena Voskresenskaya[1,1], *Vitaly* Snetkov [1] and *Alexander* Tebryaev[1]

[1]St. PetersbPeter the Great Saint - Petersburg Polytechnic University, 195251, St. Petersburg, Polytechnicheskaya str., 29, Russia

Abstract. The article touches upon main reasons for high-rise construction: increase in energy consumption and limited availability of site in the big cities of Russia. Increase in energy consumption is related with construction, transportation and applying of ventilation and air conditioning systems. Nowadays, there are developed a lot of design and engineer solutions, that include autonomous systems as well as passive methods with low energy consumption rate, which are interrelated with local climate conditions. Certain architectural solutions contribute to energy consumption decrease: building orientation with respect to the cardinal directions, taking into account the prevailing cold wind directions, maximum glazing of the southern facades and minimum glazing of the northern ones, what plays a big role in hard climate conditions. Limited availability of site for construction in the big cities resulted in rapid development of the high-rise construction, which today prevails in terms of quantitative indicators of civil engineering.

1 Introduction

A need of regulatory framework arose in the face of development of the high-rise construction in Russia, as the result of the growth of the construction base, investment efficiency, deficiency of territories in a number of cities, as well as its prestige. In accordance with the adopted Act 'On Technical regulation', it is planned to develop technical standards 'On High-Rise Building Safety', a set of rules and Moscow urban construction code for high-rise construction. These documents should reflect those requirements, that are specific for high-rise construction, but absent in the current regulatory framework.

The only regulatory document up to date, that regulates the legislation in this field, is 'Temporary regulations and design rules for multiuse high-rise buildings and complexes in the cities of Moscow and Saint-Petersburg', whereas all existing high-rise construction projects are carried out on the basis of this regulatory document. Despite the status of the document, it doesn't fully meet the needs in terms of high-rise construction design, though 16 design and research institutions were involved to develop these set of regulations. Nowadays, the process of training personnel has been stepped up, the educational programs for "Construction of high-rise buildings and structures" have been opened in leading

[1] Corresponding author: elenvoskr@mail.ru

universities of Russia, which allows us to hope to get qualified specialists in the near future. As the result, it will push the process of greater participation of domestic companies in the construction of such objects. That is why the issue of the modern regulatory framework becomes particularly urgent, since the use of foreign experience alone hinders domestic developments in this direction. It should be noted that the role of foreign influence in the regulatory framework for the construction of high-rise buildings is significant. All the fundamental issues related to classifications of objects by their standard functional support, etc., are resolved at meetings of international public organizations of engineers and architects - IABCE - ASCE and CIB. Russia also takes part in the work of these organizations and coordinates its regulatory framework with the requirements of world practices.

2 Methods

Current-day matters of administration and law in the field of high-rise construction were studied by the authors of this article [1] and in the works of other authors: Mir V. Ali [2], Potapova Y. I. [3], Markov M. F. [4], Tour V., Markov M., Shcherbach A. [5], Telichenko V.V. [6].

3 Results

The main provisions of the energy policy of the Russian Federation are aimed at designing energy-saving comfortable buildings, in which rational architectural and technical solutions are required. Currently, about 40% of all fuel produced in the country is spent on heating and cooling of buildings, but at the same time, stocks of traditional natural fuels (coal, oil, gas) are gradually depleted both in Russian and throughout the world.

One of the most important problems to solve in high-rise buildings and complexes is their energy efficiency. The energy efficiency of a high-rise building is influenced by such factors as the location of the object, its orientation with respect to the cardinal directions, the functional purpose, the three-dimensional and constructive solution, the applied engineering systems and equipment. The concept of energy efficiency of high-rise buildings is to consider the problem as a single system, including the functioning of buildings and the environment, their mutual influence and interdependence on each other and finding a common, rational path of development.

The limited availability of site for construction in the major cities of Russia, and especially in Moscow and St. Petersburg, was the reason for the rapid development of high-rise construction, which today prevails in terms of quantitative indicators of construction. Multistorey structures are preferably built not only in the segment of residential real estate. The overwhelming majority of business centers today also aspire to the sky with an impressive number of storeys. The big news of the industry come from the sphere of commercial real estate. So, the business world and ordinary people closely follow the development of high-rise business complexes (Moscow City, Lakhta-Center), that significantly affect the urban landscape.

Moscow from a small frontier military point on the site of the village of Kuchkovo for 870 years has turned into the largest in Europe agglomeration area of 13 000 to 26 000 square km by 01.01.2017, whereas St. Petersburg from the five historical areas has grown up to 18. The process of growth of urbanized areas is rapid. According to the UN report, as early as the 1950s, only 746 million people lived in cities, now it is 3.5 billion. In the next 30 years, the figure will almost double - they predict the number to increase by 2.5 billion. The number of megapolises with a population of more than 10 million people rose from 10 in 1990 to 28 in 2014, and in the following 13 years there will be 41 of them.

The construction of skyscrapers, that is conducted in the bedroom suburbs, is mostly welcomed, since it positively influences the development of the uptown districts of the cities. Nevertheless, the construction of large-scale multistorey facilities in downtown of megapolises is ripe for disputes. Thus, the adherents of the traditional architectural style, that prevails in the historical center of the above-mentioned settlements, are skeptical about the construction of such objects and doubt that skyscrapers will harmoniously fit into the cityscape. The defenders of the traditional cityscape of St. Petersburg played a major role in cancellation of the construction of the controversial project "Okhta-Center" on the right bank of the Neva river.

High-rise construction returns in large cities after a 30-year break in the construction of high-rise, mostly administrative, structures. Debates unfold around the new high-rise projects, e.g. the Okhta Center in St. Petersburg. Solution of the problems in construction of specialized (offices, hotels) and multiuse high-rise buildings have to be started almost from scratch. The situation is complicated by the absence of new domestic regulatory framework for the design and construction of high-rise facilities.

In most European countries, building is considered as high-rise, if its height is above 100 m., whereas in the USA - above 150 m. In Russia, buildings above 100 m are defined high-rise, regardless of their purpose.

Skyscrapers have a distinctive feature - a decrease in unit cost per square meter due to the large number of storeys on one foundation. The more storeys are there in the building, the cheaper construction of each subsequent storey is. About 30-40% of capital investment in the construction of a skyscraper accrue to the zero cycle (ground works), and the first 6-7 floors of the building. But in all cases we are talking about buildings, which height does not exceed 150-200 meters. When building taller buildings, the cost of construction increases: the more storeys are there in the building, the more expensive the construction of each subsequent floor will be. The scope of work performed by type of economic activity "Construction" in Russia is presented in Table 1.

Table 1. The scope of work performed by type of economic activity "Construction" in Russia

Year	million. rub. (1992 year – tln. rub., in actual prices)	In % to comparable prices	
		To the previous year	to 1990
1992	1.5	64.0	63.0
2000	203.8	113.5	36.1
2005	1754.4	113.2	57.7
2010	4454.1	105.0	82.8
2012	5714.1	102.5	89.2
2013	6019.5	100.1	89.3
2014	6125.2	97.7	87.3
2015	6148.4	95.2	83.1
2016	6184.4	95.7	79.5

The optimal height of the building, in terms of its payback, varies for different countries, and largely depends on the design of the building, its purpose, climatic conditions and taxation in the country.

The average number of floors in multifamily houses under construction is approaching to 16. By August 1, 2017, the average number of storeys of multifamily buildings being built in the Russian Federation amounted to 15.9 floors. The leaders in the average height of construction are Moscow, St. Petersburg and the Ryazan Region. Since November 2015, the average number of storeys of multifamily buildings being built in Russia has increased from 15.5 to 15.9 floors per square meter of the house.

The leader in terms of the height of construction is still Moscow with an index of 21.3 floors on average per square meter under construction.
From July 2017 to October 2017 the following changes occurred in the TOP-10 regions with the highest average height of buildings under construction. The Ryazan Region, where the average height of the houses under construction was 19.1 floors (in July this figure reached 18.5 floors) was the second largest. St. Petersburg moved from second place to third. Rostov region with an index of 17.8 floors rose from the seventh to the fifth place (the number of storeys in July was 17.7). Perm krai also added some storeys to the average index (it was 17.6, and became 17.8 floors), what resulted in the seventh place among the regions, having risen up a notch. Voronezh region, although added 0.5 floors to the average number of storeys (it was 17.2, became 17.7 floors), retained the ninth position. For the first time the Krasnoyarsk krai came in the top with the average number of storeys in multifamily buildings of 16.9 floors. Lipetsk region left the top. Statistical data on the average number of storeys by region are presented in Table 2.

Table 2. The average number of storeys of building in the top-10 regions of Russia

Region	Average number of storeys per m^2
Moscow	21.3
Ryazan region	19.1
Saint-Petersburg	19.1
Sverdlovsk region	18.7
Rostov region	17.8
Primorsky krai	17.8
Perm krai	17.8
Novosibirsk region	17.7
Voronezh region	17.7
Krasnoyarsk krai	16.9

The largest share of housing construction in Russia today accrue to houses with a height of 13-17 floors: they make up 27% of the total area of housing under construction. The share of high-rise (25 and more floors) residential buildings in the Russian Federation by October 1, 2017 is 14.3%. Data on the number of storeys of residential buildings are presented in Table 3.

Table 3. The distribution of the number of storeys in the houses under construction and commissioned for operation

Number of storeys	Under construction		Dwelling units		S, total of dwelling units	
	units	%	units	%	units	%
1-3	3763	25.1	96451	4.0	4915552	4.1
4-8	2862	19.1	238058	9.8	11724627	9.8
9-12	3116	20.8	499984	20.5	24619745	20.5
13-17	2631	17.2	657493	27.0	32411811	27.0
18-24	1802	12.0	593128	24.3	29112724	24.3
25 and more	830	5.5	352510	14.5	17197663	14.3
Total	15004	100	2437624	100	119982112	100

The smallest average number of storeys of multifamily buildings being built is fixed in the Jewish Autonomous Region - 3.0 floors. Multifamily buildings with an average number of storeys of 7 floors are being built in 4 more regions.

The data are presented in Table 4.

Table 4. The smallest number of storeys of buildings in the regions of Russia

Region	Average number of storeys per m^2
Kamchatka krai	6.6
the Yamalo-Nenets Autonomous District	6.5
Republic of North Ossetia-Alania	6.1
Nents Autonomous District	5.5
Jewish Autonomous Region	3.0

According to state construction supervision, in the first half of 2017 supervision was carried out for the construction of 60 unique buildings and structures with a total area of 6 million 101 thousand square meters. In accordance with the Town Planning Code of the Russian Federation the object of capital construction acquires the status of 'unique' in the event if at least one of the following characteristics is provided in its design documentation: a height of more than 100 m; width of spans over 100 m; the presence of a console over 20 m; the penetration of the underground part below the planning mark of the land by more than 15 m. One of the high-rise buildings of a unique residential complex is problematic. It is a 32-storey "Academ-Palace", located at 78, Vernadsky Prospekt. The complex has been being built since 2005. Construction stopped in 2014 because of litigation between the developer "OSC" and the owner of the rights to land - the Moscow State Institute of Radio Engineering, Electronics and Automation. If the developer does not resume the construction of the facility, the Committee may use its right to apply to the arbitration court for liquidation of the company. The most remarkable unique non-residential objects being built in Moscow are the Luzhniki sports arena, the reconstructed Dynamo stadium, the spire of the Russian Foreign Ministry building on Smolenskaya Square, the Moscow City MIBC building complex. The core of the Lakhta-Center tower passed for 380 meters. This means that the St. Petersburg supertall has become the tallest building in Europe. The predecessor is the Moscow East Federation (374 m). In 2018, the gap will be 88 meters.

Stage construction was started in late August 2015, the first major milestone happened less than in a year - in June 27, 2016. The core overcame 147-meter mark and the Lakhta Center became the tallest building in St. Petersburg. A couple of weeks later, the skyscraper overtook the highest building in Belgium Tour du Midi, which is located in Brussels. It is 150 meters and was commissioned for operation in 1967. In the same July, the building became taller than the Maastoren - a high-rise building of 164.75 m in Rotterdam, the Netherlands, which was built in 2010. The tallest skyscraper of Switzerland - Roche Turm Bau 1, 178 m, which was built in 2015, the building of the Lakhta Center exceeded in August 2016. Then the tallest Swedish Tower Turning Torso of 190 m was left behind. It is also a newbuild.It was commissioned for operation in 2005. Italy's tallest building, UniCredit Tower in Milan, height - 217.7 m. built in 2012, was outstripped in October 2016. At that time the Tour First - the tallest skyscraper of France after being renovated in 2011, when 72 meters were added to the existing 159 - was overtaken. The Warsaw Palace of Culture and Science was overtaken in November 2016. The constructive height of the palace is 230.7 m, according to the Emporis Skyscraper Award. In January 2017 the Spanish the Torre de Cristal with the height of 250 meters, that was built in 2008, was outstripped. April - the German Commerzbank Tower - 259 m., Built in 1997, was overtaken. And world-known English 'The Shard', the 306-meter tower, that was built in 2013, was left behind. The issues of quality of production of construction and installation works and applied materials, products and structures in the construction of high-rise buildings are crucial in ensuring the reliability and durability of structures and the integrated safety of buildings in general.

4 Discussion section

Modern construction in large cities is characterized by a wide application of new, advanced design solutions, materials, structures and technologies, as well as better qualification of performers and the associated large number of participants in the construction process. Under these conditions, effective control over the timely introduction of technical innovations, strict observance of technological discipline and all requirements for the quality of materials and structures, that are used at work sites, becomes crucial along with competent organization of the process and well-coordinated interactions of performers. In the Committee of State Construction Supervision in Moscow, a special Office for Supervision of the Construction of Unique Buildings and Structures was established. In the first half of 2017, the Department already conducted 281 inspections of the Moscow facilities, that are under construction and were recognized as unique, and 253 tests were carried out on the implemented structures and materials. 1152 violations were found in the course of inspections. Quality control at construction sites of the city is continuously improved. A major contribution to the quality improvement of construction is made by comprehensive inspections, when not only representatives of self-regulating organizations in the construction sector took part, but also experts from research organizations and other supervisory bodies. Data on the availability of fixed assets of construction organizations are presented in Table 5.

Table 5. The fixed assets of organizations, conducting construction activities, as% of the total

	2005	2010	2012	2013	2014	2015
Fixed assets – total,	100	100	100	100	100	100
Which include:						
Buildings and constructions	36.0	39.9	35.5	36.3	37.0	40.2
Machinery and equipment	42.1	41.9	43.0	40.9	40.5	38.6
Means of transport	18.4	18.6	18.9	20.2	19.8	17.9
Other types of fixed assets	3.5	3.6	2.6	2.6	2.2	3.2

The results of comprehensive inspections revealed a number of typical violations that occur in the construction of high-rise buildings. Such violations, in particular, include the use of obsolete or worn out equipment. The state of equipment by January 1, 2017 is displayed in Table 6.

Table 6. The presence and condition of the main construction machinery in construction organizations

	Total, thd. units	Of the total units of machinery, %	
		Machinery with expired service life	Machinery of foreign manufacture
Single-bucket excavators	11.8	33.4	71.9
Scrapers	0.3	79.2	53.4
Bulldozers	8.9	48.1	41.7
Tower cranes	3.7	46.2	32.1
Vehicle-mounted cranes	7.2	36.1	27.6
Pneumatic wheel-mounted cranes	1.4	43.2	58.8
Crawler-mounted cranes	1.2	63.1	39.4
Autograders	4.1	45.5	30.1

5 Conclusions

The development of high-rise construction in Russia highlighted a number of problems that require mandatory consideration and solution: imperfection of the regulatory framework; lack of experience in the design and construction of high-rise complexes; lack of qualified civil engineers; justified town-planning and functional-typological necessity of construction; maximum permissible number of storeys (building height); the correct choice of the structural system, scheme and design solutions, taking into account the prevention of loss of stability of the structure foundation and the structure itself, that leads to collapse of structures; inadmissibility of deviation from the approved design and changes in the number of storeys during the construction process; necessary functional interaction of residential and non-residential buildings and structures with the transport and service infrastructure of the city; required capacity of underground, land and aboveground parking lots for personal transport and their rational location; effective minimization of the threat of external and internal danger of building collapse by creating a special service for safe operation; the required fire and evacuation safety of people in high-rise buildings; rational efficiency of modern engineering solutions for life support and equipment of the building, energy saving and comfort of service.

6 References

1. Elena Voskresenskaya Vitaly Snetkov Alexander Tebryaev and Zokhidjon Askarov V 2017 *MATEC Web of Conferences* **106** 08055
2. Mir V Ali V 2001 E*lectronic Journal of Structural Engineering* **1** 2–14 pp
3. Potapova Y I V 2012 *Progress of modern natural sciences* **6** 14-16 pp
4. Markov M F V 2007 *Architecture and construction* **1** 44-47 pp
5. Tour V Markov M Shcherbach A V 2008 *Architecture and construction* **2** 72-81 pp
6. Telichenko V V 2008 *High-rise construction* 2008 **2** 112-122 pp

Development of national standards related to the integrated safety and security of high-rise buildings

Elena Voskresenskaya[1*] and *Lubov* Vorona-Slivinskaya[2]

[1]Peter the Great Saint - Petersburg Polytechnic University, 195251, St. Petersburg, Polytechnicheskaya str., 29, Russia
[2]Federal State Budgetary Educational Institution of Higher Education «Saint Petersburg State University of Architecture and Civil Engineering» (SPSUACE), St. Petersburg, 190005, Russia

Abstract. The article considers the issues of developing national standards for high-rise construction. The system of standards should provide industrial, operational, economic and terrorist safety of high-rise buildings and facilities. Modern standards of high-rise construction should set the rules for designing engineering systems of high-rise buildings, which will ensure the integrated security of buildings, increase their energy efficiency and reduce the consumption of resources in construction and operation.

1 Introduction

The Federal Law № 116-FZ of 21.07.1997 "On Industrial Safety of Hazardous Production Facilities" outlines the main tasks of the state policy in the field of ensuring the population safety and the protection of hazardous facilities. These tasks cover the following issues: development and legislative consolidation of technical regulations on safe operation and security of hazardous facilities; development of fundamental and applied science, technology and technics, including development and implementation in practice of information and forecasting and analytical systems; systems, means and methods of technical diagnostics of the installed service life of hazardous equipment; creation of systems for monitoring the status of hazardous facilities for assessment and operational prediction of the consequences of their destruction (damage); improvement of systems and facilities for the protection of hazardous facilities; ensuring the stable and safe operation of information and telecommunications systems of hazardous facilities; ensuring continuous monitoring of the condition of power supply systems and standby power sources at hazardous facilities. Since January 1, 2014, in accordance with clause 3 of Section 11 of the Federal Law of July 21, 1997 № 116-FZ, "organizations operating hazardous industrial facilities of a certain hazard class are required to create systems for managing the industrial safety of facilities and ensure their operation.

Current-day matters of administration and law in the field of high-rise construction were studied by the authors of this article [1] and in the works of other authors: Mir V. Ali [2],

* Corresponding author: elenvoskr@mail.ru

Potapova Y. I. [3], Markov M. F. [4], Tour V., Markov M., Shcherbach A. [5], Telichenko V.V. [6].

2 Experimental section

A modern building or facility is a complex system, which includes a system of structures, engineering life support systems and systems of implementing resource and energy conservation processes, maintaining comfort and ensuring safety. All the systems that make up the building (facility) interact with each other and the environment. Safety-related buildings or facilities of the system (SRBF systems), interacting with the structures and volumes of the premises of the facility, with the equipment of engineering systems, the environment perform safety functions that lead to a reduction in the risk of harm to people, property and the environment. SRBF systems operate in a specific environment in interrelation with other systems. It is possible to assess the risks associated with the operation of these systems or to assess and confirm compliance only in each particular building (facility) with the connection to the terrain and taking into account local conditions. By this, systems related to the safety of buildings differ from systems, for example, related to the safety of machinery or vehicles.

SRBF systems include systems for monitoring the condition of bearing and enclosing structures of a building (facility), the ground of its foundation, the condition of the equipment of engineering systems, the state of the environment in the building (facility) and its surroundings; fire alarm systems, smoke and heat removal systems, burglar alarms, access control and management, television surveillance, sound systems for alarm purposes and other systems. These systems, being integrated into a single complex security system, simultaneously counteract many dangers arising from internal and external hazardous impacts of natural or man-caused nature on a building (facility) and its systems.

In accordance with the provisions of paragraphs 5.1 and 5.2 of the Manual International Organization for Standardization (ISO)/International Electrotechnical Commission (IEC) 51 and B.2 of IEC 61508-5, there is no absolute security, since after taking protective measures, residual risk remains. Considering security measures, it is possible only to talk about reducing the risk to a level that is acceptable to the population at a certain time in the certain country, taking into account the economic, social political situation, the existing traditions, and the benefits from the use of products. An acceptable risk is established resulting from finding a compromise between the interests of those who generate risks, those who are at risk, and the authorities that regulate relations between legal subjects and manage the resources of the country.

Bearing and enclosing structures and engineering equipment of a building (facility) create a favorable environment for people's life and activities, for the implementation of processes, and act as protection from the environment. At the same time, there remains some risk of harm caused by the behavior of structures and engineering systems under the influence of external and internal impacts of natural or man-caused nature. In order to reduce the risk degree, compensatory measures are taken with the use of electric, electronic, programmable electronic SRBF-systems. In case of not achieving the acceptable risk using SRBF systems, external risk reducing facilities can be additionally used. Among them, there are architectural forms, fencing or landscape solutions, which prevent, for example, unauthorized approach for vehicles to a building.

In modern SRBF systems, programmable electronic (PE) means are being increasingly used, the operation of which is controlled by the software and becomes an integral part of the SRBF systems.

3 Results

High-rise buildings are new town-planning and functional-planning formations, formed along the vertical axis. However, a complex of town-planning requirements (solving transportation problems, providing objects of public service, gardening, etc.) is preserved, which should be met taking into account the spatial specifics of these objects. The Town Planning Code of the Russian Federation does not contain the term of high-rise building. There is the concept of a unique building, to which the objects of capital construction are referred with height of more than 100 m according to the design documentation.

Currently, the Worldwide Academy of Sciences of Complex Security and the University of Comprehensive Systems of Security and Engineering are developing seven basic national standards GOST R 53195 under the general title "Safety of functional systems associated with the safety of buildings and structures." The standard describes the measures (methods and/or means) for risk reduction and assessment methods established in GOST R 53195.3 and GOST R 53195.4 for SRBF systems and their components.

These standards developed for the first time cover all the necessary requirements for the full life cycle of a wide class of systems of any technological complexity used to ensure the safety of buildings and structures. Objects of standardization are means of automation and control systems, including fire-extinguishing means, firefighting equipment, other aspects of fire protection; alarm and warning systems; low-voltage switched equipment and control equipment, other switching equipment and control equipment.

A series of standards being developed will serve as a basis for newly developed and updated group standards and standards for individual safety systems and systems harmonized with international ISO and IEC standards.

The standards being developed do not have direct foreign analogs. For their development, the provisions of the IEC 61508-1 - IEC 61508-7, ISO 9000, ISO 9001, ISO 9004 standards are used.

The national standards under developing along with other national standards of this issue will enable providing a regulatory framework for the creation and application of modern security systems for a wide class of real estate, including highly dangerous, technically complex and unique objects; carrying out risk analysis and assessment, evaluating the completeness of safety and confirming compliance with the requirements of the technical regulations for the safety of buildings and structures; promoting the development of applied science, technology of information and forecasting and analytical systems and methods for diagnosing equipment of engineering systems and structures of buildings and facilities; promoting the improvement of systems and means of protection of buildings and facilities; ensuring the stable and safe functioning of information, telecommunication and control systems of protected objects; ensuring continuous monitoring of the condition of all systems of protected objects, including life supporting systems, energy and resource saving, process implementation, security and other systems; reducing the risk of harm to the life and health of citizens of the Russian Federation, property and the environment and/or reducing the severity of consequences in case of occurrence of hazardous events and reducing the costs of eliminating the consequences of emergencies; reducing technical barriers to international trade, including real estate trade; increasing the competitiveness of the security system products produced by domestic manufacturers on the Russian and international market.

4 Discussion section

Complex security of construction projects includes economic security. Certain indicators of economic security in construction should be considered. The dynamics of construction activity and the distribution of work scope in the construction sector in Russia are presented in Table 1.

Table 1. Distribution of work scope performed by type of activity "Construction" by federal districts, as % of the total

	2015	2016
The Russian Federation	**100**	**100**
Central Federal District	27.6	25.8
North-West Federal District	13.1	14.5
Southern Federal District	8.4	7.7[1)]
North-Caucasian Federal District	4.3	4.4
Volga Federal District	20.0	19.6
Ural Federal District	11.6	11.6
Siberian Federal District	9.1	9.9
Far Eastern Federal District	5.8	6.5
Crimean Federal District *	0.1	-

** In accordance with the executive order of the President of the Russian Federation № 375 "On the Southern Federal District" of July 28, 2016, the Southern and Crimean Federal Districts were reorganized into the Southern Federal District.*

Over half of the construction work in 2016 was carried out in the following constituent territories of the Russian Federation: Moscow (11.5% of the total work in Russia), St. Petersburg (6.7%), Tatarstan (4.7%), Bashkortostan (3.1%), Krasnodar territory (3.7%), Krasnoyarsk territory (3.2%), Moscow Region (5.4%), Nizhny Novgorod and Samara Regions (2.5% each), Tyumen (without autonomous regions) (2.6%), the Yamalo-Nenets Autonomous District (2.9%) and the Khanty-Mansiysk Autonomous Region Ugra (2.8%).

In the IV quarter of 2016, the results of surveys of business activity of construction companies indicate an unstable economic situation in the construction sector. The index of entrepreneurial confidence in the fourth quarter of 2016 was -21%, which is 8 percentage points lower than in the fourth quarter of 2015 and 5 percentage points lower than in the third quarter of 2016.

The increase in business activity in the IV quarter of 2016 was observed in 13 constituent entities of the Russian Federation. The results of the survey show that in 24 out of 82 subjects of the Russian Federation participating in the survey, the index of entrepreneurial confidence is higher than the average for Russia. Positive value of this indicator was noted in the Kaliningrad and Tver Regions. The lowest value of the index of entrepreneurial confidence (-60%) is recorded in the Khabarovsk Territory.

The volume of investment in fixed capital for the development of construction activities in 2016 amounted to 445.0 billion rubles or 3.0% of the total volume of investments in fixed capital, which is 0.1% more than in 2015.

In the total volume of investments aimed at developing their own base of construction organizations, the share of small businesses in 2016 was 28.2% compared to 29.9% in 2015.

At the end of 2016, the full accounting value of fixed assets in construction amounted to 2233.9 billion rubles, or 1.3% of the value of fixed assets of all types of economic activity.

The commissioning of fixed assets by type of activity "Construction" amounted to 176.2 billion rubles, or 1, 6% of the total volume of fixed assets introduced in the economy as a whole. Information on the level of depreciation of fixed assets in construction is presented in Table 2.

Table 2. The level of deterioration, the coefficients of renewal and retirement of fixed assets in construction, %

	2014	2015	2016
Coefficient of renewal of fixed assets (in comparable prices)	3.8	3.4	3.2
The coefficient of liquidation of fixed assets (in comparable prices)	1.1	1.0	1.0
Deterioration rate of fixed assets (in current prices, at the end of the year)	51.2	50.4	51.8

The renewal coefficient of fixed assets in type of activity "Construction", calculated in comparable prices, in 2016 was 3.2% and decreased by 0.2 percentage points in comparison with 2015. The coefficient of retirement of fixed assets in construction in 2016 was 1.0% and did not change in comparison with 2015. In the last two years, the renewal coefficient of fixed assets in construction was slightly lower than in the economy as a whole, the retirement ratio was slightly higher, and the level of depreciation remains higher than the average for the economy. At the end of 2016 the degree of deterioration decreased by 1.4 percentage points as compared to 2015. The state of equipment as of 01.01.2016 is given in Table 3.

Table 3. The presence and condition of the main construction machinery in construction companies

	Total amount, thousand units	Of total machinery amount, %	
		Machines with expired service life	Machines of foreign manufacture
One-bucket excavators	12.3	31.2	70.8
Tournapulls	0.3	74.2	51.1
Vehicle-mounted cranes	2.4	63.6	39.6
Pneumatic wheel-mounted cranes	7.7	36.3	27.5
Caterpillar-tracked cranes	1.5	42.8	53.1
Tower cranes	4.1	46.6	26.1
Land levellers	4.1	46.3	29.7
Tractors	7.0	50.1	39.5
Bulldozer earthmovers	9.3	46.7	40.9

In 2016, the financial result of profitable organizations exceeded the financial result of unprofitable organizations by 130.6 billion rubles. In 2015, the financial result of profitable organizations was lower than the financial result of unprofitable organizations by 54.3 billion rubles. Indexes of financial performance of construction organizations are presented (Tab. 4)

Table 4. Financial result of construction companies

	2015	2016
Balanced financial result (profit minus loss), million rubles	-54329	130560
In % to the corresponding period of the previous year	-	139.5
The share of unprofitable organizations relative to the total number of organizations,%	32.8	24.1
Amount of loss, million rubles	320030	114424
Profitability of sold goods, products, works and services, %	3.8	5.5

In 2016, about 900 construction companies were unprofitable. In 2016 the share of unprofitable organizations in the total number of construction organizations decreased by 8.7

percentage points in comparison with 2015. In 2016 the profitability of sold goods, products, works and services increased by 1.7 percentage points in comparison with 2015.
At the end of 2016 in organizations involved in construction activities, accounts payable exceeded receivables by 1195.4 billion rubles, or 43.6% (at the end of 2015 - by 1317.3 billion rubles or 52,4%). At the end of 2016, the share of overdue receivables in the total amount of accounts receivable in organizations engaged in construction activities was 4.5% and increased by 0.5 percentage points in comparison with the end of 2015.
At the end of 2016, the share of overdue accounts payable in the total volume of accounts payable of these organizations was 2.3% and remained at the level of 2015.
By end of 2016, overdue accounts payable of organizations engaged in construction amounted to 91.3 billion rubles, increased by 2.1% in comparison with the corresponding period of 2015, and overdue accounts receivable amounted 123.0 billion rubles, and increased by 22.2%, respectively.

5 Conclusions

The Federal Autonomous Institution "Federal Center for Standardization, Technical Standardization and Compliance in Construction" has developed a set of rules for high-rise construction. It entered into force on February 2, 2017. The new set of rules "Engineering systems of high-rise buildings" covers the design of engineering systems for construction and reconstruction of public buildings higher than 55 m and residential buildings higher than 75 meters. The document sets the rules for the design of engineering systems of high-rise buildings, which will provide comprehensive security of buildings, increase their energy efficiency and reduce the consumption of non-renewable resources in construction and operation. Before the adoption of new rules, in order to build a facility above 100 m, it was necessary to develop and adjust special regulations (SR), which restrained the development of high-rise construction. For 2017, the Ministry of Construction of Russia planned to develop a set of rules “High-rise buildings and complexes. Fire safety requirements”.
Basic documents, firstly sets of rules for high-rise construction, began to function in 2017, but the complete number of documents that enable minimizing elaboration of various special regulations, will be developed within the period of 2017-2019.

6 References

1. E. Voskresenskaya, V. Snetkov Alexander Tebryaev and Zokhidjon Askarov V 2017 *MATEC Web of Conferences* **106** 08055
2. Mir V Ali V 2001 *Electronic Journal of Structural Engineering* **1** 2–14 pp
3. Y I Potapova 2012 *Progress of modern natural sciences* **6** 14-16 pp
4. M F Markov 2007 *Architecture and construction* **1** 44-47 pp
5. T. Markov M Shcherbach A *Architecture and construction,* **2** 72-81 pp (2008)
6. V. Telichenko *High-rise construction,* **2** 112-122 (2008)

Normative regulation of material incentives for workers in the sphere of high-rise construction

Anna Kopytova[1*], *Elena* Matys [1], *Natalia* Zotkina[1], *Irina* Reshetnikova[1], *Natalia* Meller[1], *Inna* Nekrasova[1]

[1]Tyumen Industrial University, Volodarskogo str., 38, Tyumen, 625000, Russia

Abstract. The article is devoted to the problem of normative and legal regulation of incentives for workers of the building companies. The company considered is engaged in high-rise construction. The part of the document "Regulations for the incentives of employees of the enterprise" is presented. This document is introduced into the practical activities of the Tyumen enterprise. The presented part of the document regulates the issues of material incentives for employees of the enterprise. The document "Regulations for the incentives of employees of the enterprise" is developed in accordance with the approach proposed by author. The document was developed, after the authors had carried out an analysis of the scheme of incentives at the enterprise.

1 Introduction

In a highly competitive environment, trade organizations need to look for reserves to improve the efficiency of their activities. Successful work in the service sector assumes that the staff of the enterprise has certain incentives, which, in turn, is impossible without an appropriate effective incentive system.

Material incentives for workers in a construction enterprise are the most important motivational unit that affects the effectiveness of the organization. [1- 10]

2 Methodology

When creating a system of incentives for workers in a construction organization, a rational incentive system was used (Fig. 1). [4, 6, 7]

An important feature of a rational incentive system is that it makes possible self-determination by the organizations (according to the needs of workers as well as their own capabilities, available resources and the existing limitations) of the structure of their incentive system [8]. The proposed model provides a choice (like the menu in the cafeteria) by the organization needed and possible (rational for it) components of the incentive system.

For the Tyumen construction company, we developed a document "Regulations for the incentives of employees of the enterprise." The article provides an overview of a part of this document, namely, material incentives for employees.

*Corresponding author: a.copytowa@yandex.ru

 https://doi.org/10.1051/e3sconf/20183303053

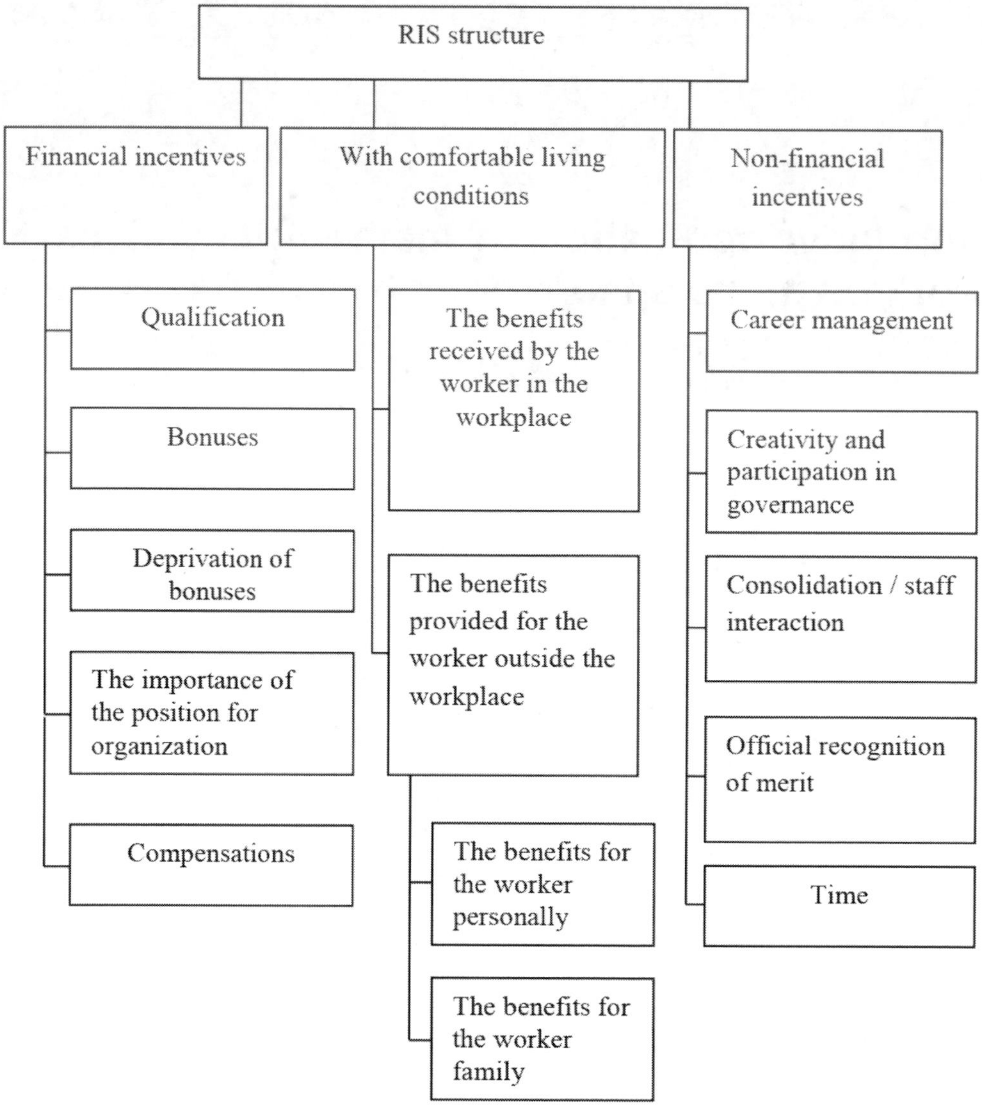

Fig. 1. Model and structure of the rational incentive system.

3 Results

Material incentive is one of three types of incentives (material, non-material and stimulating with comfortable living conditions) and includes five components: qualification, bonuses, deprivation of bonuses, importance of the post for organization, compensations. Taking into account these five elements, the material incentive unit, which is part of the "Regulations of incentives for employees of the enterprise," was developed at the enterprise.

The first element is qualificationю. it provides for the assessment of the employee in four areas: the level of education, the compliance of the employee's education with the position held, the length of service in the position held and the length of service in the organization.

1. The level of education. Salary increases in accordance with the assessment scale presented in Table. 1.

Table 1. Remuneration rate according with the employee's education level

№	Education	Coefficient to the salary
1	General, secondary, primary vocational	–
2	Secondary vocational	1,02
3	Higher professional education	1,05
4	Few University diplomas	1,1

Thus, an employee with a secondary vocational education receives a bonus of 2% of salary monthly, and an employee with a higher education has bonus of 5% of salary.

2. For the compliance of the education of an employee with the position held by him, an additional payment of 5% to the salary is provided. The enterprise under consideration is a trade organization with a small number of personnel, most of the employees have the right to receive such a payment.

3. Length of service in the position held. The estimated scale is presented in Table. 2.

Table 2. Coefficients of remuneration for length of service in a particular profession

№	Length of service in the position held	Coefficient to the salary
1	Up to 3 years	-
2	From 3 to 5 years	1,03
3	From 5 to 10 years	1,05
4	More than 10 years	1,1

4. Length of service in the organization. Remuneration of employees is carried out in accordance with Table. 3.

Table 3. Coefficients of remuneration for length of service in the organization

№	Length of service in the organization	Coefficient to the salary
1	Up to 3 years	-
2	From 3 to 7 years	1,07
3	More than 7 years	1,12

The component "Bonuses" includes two elements:

1. Fulfillment of planned targets. Planned targets for the positions of employees are established taking into account the functional duties approved in the job descriptions of the organization and are accepted in accordance with Table. 4.

Table 4. Efficiency indicators for employees' positions

Profession	Indicator	Remuneration
Accountant	Absence of sanctions by tax authorities	15% to the salary
	Compliance of work with the established terms (timely submission of accounts, calculation of wages, absence of overdue receivables, etc.)	15% to the salary
	Optimization of taxation	20% of the identified benefits to the organization
Sales Manager	Number of new customers	0,01% to the salary
	Customer satisfaction (no claims, complaints, litigation, etc.)	15% to the salary
	Sales level (volume, growth rate)	2% of manager sales; 5% of sales in excess of the plan
	Quality of documents (no comments)	10% to the salary

Profession	Indicator	Remuneration
Warehouse employees	The percentage of losses in the total volume (one need to strive to reduce the breackage, damage of goods during the picking, storage, movement in the warehouse)	15% to salary, if the established limits are observed
	Material flow of the warehouse (workload, implementation of the plan)	15% of the salary when the plan is implemented
Commercial Director	Increase of profits through marketing activities	20% of the profit received through marketing activities
	Introduction of innovations (allocation of additional spheres of activity - filling the market, partial change of service, etc.)	20% of the benefit of the enterprise through the introduction of innovations
	Revenues from sales	2% of sales, 4% of revenue in excess of the plan
Lawyer	Faultless execution of documents	20% to the salary
	The amount of work (the number of vessels won, the number of inspections, etc.)	15% to the salary when the plan is implemented
	Knowledge of the legislative framework	15% to salary in the absence of cases of identification of non-competence of a lawyer in the legislative framework of different areas

2. Compliance with the working time regime (labor discipline). Payments occur once a quarter. Absence of admissions (considered on the basis of the schedule approved at the beginning of each month) for sickness, schooling and other valid reasons is rewarded with a bonus of 7% of the official salary.[11, 12, 13, 14]

The component "Deprivation of bonuses" provides for two elements:

1. Violation of the workflow. Personnel actions that entailed material losses (sanctions on the part of counterparties, etc.), changes in the quality of products sold, inappropriate execution of documents, worsening of the opinion of contact groups on the quality of the enterprise, litigation of the enterprise with various subjects, etc. can lead to the deprivation of the employee of the entire stimulating part of the payment. [15, 16, 17, 18]

2. Violation of discipline (the presence of delays, appearance in the workplace in an unacceptable form, etc.) may entail the deprivation of the employee of the entire stimulating part of the payment.

The component "The importance of the post for the organization" implies an additional payment for those professions that are recognized as the most significant for the enterprise. Coefficients of surcharge for the importance of the post for the Tyumen enterprise under consideration are presented in Table 5.

Table 5. Coefficients of additional payments for the importance of individual posts

№	Post	Coefficient to the salary
1	Commercial Director	1,1
2	Accountant	1,08
3	Sales managers who fulfilled the plan for the previous 3 months	1,1

Payments for the "Compensation" component are made when recognizing the presence of harmful labor conditions for certain posts in accordance with the current legislation of the Russian Federation:

— if the time for rest is decreased (work at holidays, weekends);

— if there is the impact on biorhythms of sleep and wakefulness of a person;
— if there are high physical and mental loads, etc.

4 Discussion

Material incentives for workers in a construction organization are the powerful tool for influencing the work activity of workers. [11, 19, 20, 21, 22, 23, 24, 25] Material incentives include: wages, bonuses, allowances, surcharges, compensations. This grouping can be considered in a different ways, which may be a direction for further research.

Wages are characterized by remuneration for work, depending on the qualifications of the employee, the complexity, quantity, quality and conditions of the work performed.

Bonuses are used to stimulate the quality of performance and overfulfilment of the plans (tasks) put. Thanks to such an incentive element as bonus, the economic interrelation between the growth of the effectiveness of the organization and the increase in wages is manifested. Bonuses can be of a systematic or one-time nature. The incentive bonuses may not depend on the results of work (paid for the holidays).

Supplements are used to stimulate conscientious attitude to work, increase the efficiency of workers, auxiliary processes.

Stimulating surcharges give impetus to the development of skills. If the employee has sufficient knowledge and skills to perform work of higher qualification than required by the position held, the employer can entrust him with this work with an appropriate surcharge.

Compensation is used at enterprises with harmful working conditions; with a decrease in the time for rest (working at holidays, weekends); in conditions of influence on biorhythms of sleep and wakefulness of the person; in the conditions of high physical and mental loads, etc. Such payments can be used for employees of laboratories, lifters, security guards, etc.

Conclusions

Thus, the material incentive system developed for the Tyumen Construction Company within the framework of the regulations for the stimulation of employees currently operates and extends to all employees of the organization. And it provides five components: qualification, bonuses, deprivation of bonuses, the importance of the post for the organization, compensation.

References

1. M. Froli, F. Laccone, Journal of Architectural Engineering **24(1)**, 04017030 (2018) DOI: 10.1061/(ASCE)AE.1943-5568.0000285
2. J.M. Stritch, R.K. Christensen, International Journal of Manpower **37-5**, 840-858 (2016) DOI: 10.1108/IJM-12-2014-0249
3. M. Gravit, O. Zybina, A. Vaititckii, A. Kopytova, Advances in Intelligent Systems and Computing **692**, 1093-1101 (2018) DOI: 10.1007/978-3-319-70987-1_118
4. N. Zotkina, M. Gusarova, A. Kopytova, Advances in Intelligent Systems and Computing **692**, 1204-1213 (2018) DOI: 10.1007/978-3-319-70987-1_129
5. M. TepePublic Management Review **18-4**, 508-538 (2016) DOI: 10.1080/14719037.2015.1014396
6. A. Kopytova, Procedia Engineering **165**, 1132 (2016) DOI: 10.1016/j.proeng.2016.11.830

7. A. Kopytova, MATEC Web of Conferences, **106**, 08056 (2017) DOI: 10.1051/matecconf/201710608056
8. N. Zotkina, S. Bardasov, M. Gusarova, A. Kopytova, MATEC Web of Conferences, **106**, 08050 (2017) DOI: 10.1051/matecconf/201710608050
9. K. Pykhtin, T. Simankina, V. Sharmanov, A. Kopytova, IOP Conference Series: Earth and Environmental Science, **90 (1)**, 012065 (2017) DOI: 10.1088/1755-1315/90/1/012065
10. N.S. Zotkina, Finances and Credit **48 (480)**, 21-28 (2011)
11. V. Lezier, M. Gusarova, A. Kopytova, IOP Conference Series: Earth and Environmental Science, **90 (1)**, 012034 (2017) DOI: 10.1088/1755-1315/90/1/012034
12. A. Minnullina, R. Abdrazakov, Advances in Intelligent Systems and Computing **692**, 1224-1233 (2018) DOI: 10.1007/978-3-319-70987-1_131
13. A.I. Kazannik, Bulletin of Omsk University **4 (33)**, 35-43 (2012)
14. A.V. Kopytova, Exchange of intellectual property **3 (XIV),** 31–37 (2015)
15. B. Liu, J.L. Perry, Review of Public Personnel Administration **36-1**, 4-30 (2016) DOI: 10.1177/0734371X14549672
16. A. Minnullina, R. Abdrazakov, IOP Conference Series: Earth and Environmental Science, **90 (1)**, 012089 (2017) DOI: 10.1088/1755-1315/90/1/012089
17. J. Yeo, International Journal of Public Administration **39-3**, 216-225 (2016) DOI: 10.1080/01900692.2015.1004083
18. M.S. Gusarova, A.V. Kopytova, Bulletin of Omsk University **4**,14–22 (2014)
19. N.S. Zotkina, A.V. Kopytova, Economics and entrepreneurship **12–2(53–2)**,714–719 (2014)
20. V. Lezier, M. Gusarova, A. Kopytova, IOP Conference Series: Earth and Environmental Science, **90 (1)**, 012069 (2017) DOI: 10.1088/1755-1315/90/1/012069
21. A. Mottaeva, A. Minnullina, IOP Conference Series: Earth and Environmental Science, **90 (1)**, 012123 (2017) DOI: 10.1088/1755-1315/90/1/012123
22. M. Gravit, O. Zybina, A. Vaititckii, A. Kopytova, IOP Conference Series: Earth and Environmental Science, **90 (1)**, 012103 (2017) DOI: 10.1088/1755-1315/90/1/012103
23. E.A. Mitrofanova, Personnel and intellectual resources management in Russia **2**, 36-41 (2014)
24. N. Zotkina, A. Kopytova, M. Zenkina, O. Zhigunova, MATEC Web of Conferences, **106**, 08058 (2017) DOI: 10.1051/matecconf/201710608058
25. A. Minnullina, MATEC Web of Conferences, **106**, 08067 (2017) DOI: 10.1051/matecconf/201710608067

Strategic and tactical aspect of the relations between the participants of ICP in high-rise construction

Maksim *Fedorov* [1], Elena *Matys* [1], and Anna *Kopytova*[1*],

[1]Tyumen Industrial University, Volodarskogo str., 38, Tyumen, 625000, Russia

Abstract. Outlined the main trends in the development of the construction sector in the Russian Federation at the present stage of its development are identified problems and shortcomings of current investment in the construction industry. It is proved that to consider the investment and construction process should be in terms of the strategic and tactical aspects. It was established that, on the basis of the spirit and the participants of investment and construction process, the strategic aspect it should be focused primarily on the formation of financial resources (through various mechanisms) and to attract investors. A tactical aspect of the investment and construction process is focused on the choice of organizational form of the project and the rational use of resources generated by the investment in construction.

1 Introduction

Construction industry, closely connected with investment processes, has decisive importance for modern Russian economy, which suffers in recent years several problems, connected with sanctions, lack of financial resources and unstable currency rate. Construction is one of the industries with prolonged cycle of production, and capital investment problems may affect for a long-time period. The distinctive characteristic of investment and construction process is that construction companies act as accumulative organizations, which encourage investment from different sources in order to complete investment and construction project.

It is impossible to over valuate the importance of the construction industry for every single citizen of the country, as this industry brings not only construction results, but also working places, forms GDP, provides pumping up the budget on different levels. Economic and political reforms brought profound changes in the functioning of domestic investment and construction industry. Economic reform in the construction sphere became a fundamental change in optimality criteria in the construction and investment industry. New guidelines led to changes in both structure and measures in the activities of a construction company. List of participants changed and expanded, principle of mutual interaction became more complex, guidelines and objectives also changed. .

*Corresponding author: a.copytowa@yandex.ru

In the new post-crisis ownership conditions, efficiency of the creation process of final construction products mostly depends on the concerted efforts and interaction between many participants of the investment and construction process: investors, customers, project-making and construction companies, material and equipment suppliers etc. The success of reaching the main goals – creation of final construction products and improvement of the cost, time and quality parameters, depends on effective organization and regulation of the interactions between stated above participants.

2 Analysis of latest publications on the topic of research

The problem of interaction between the investment and construction process participants is reviewed in works of specialists of a quite wide circle: economists, financiers, engineers, logistics experts, managers. Research on the chosen topic requires close focus on the works of the following scientists: V. Berens, I.A. Blank, A.H. Zaveryuha, A.V. Martynov, O.A. Menshikova, E.V. Ulianov, A.I. Ostashevskaya, E.A. Tkachenko, R.N. Holt, V.D. Shapiro

In is necessary to underline the fact that in recent years there are a lot of publications concerning construction problems, financial supply of the construction, realization risks of the investment projects in the sphere of finance, raise of the investment activity of the construction process participants. At the same time, the problem of specification of strategic and tactical directions of interaction between the investment process participants requires a more detailed review due to lack of publications on the given topic.

3 Research goal

On the basis of the description of the chosen topic and evaluation of the amount of works on the reviewed problem, it is possible to state the research goal: compilation of both theoretical and practical recommendations oriented on the formation of strategic and tactical directions of interaction between investment and construction process participants.

4 Presentation of basic material of the research

At the end of 2016, Russian economy shows several revival characteristics, though in general remains in the state of depression (Fig. 1).

According to The Ministry of Economic Development and Trade, GDP decrease in the first quarter of 2016 slowed down to 2.5 %, but, if seasonality is excluded, showed monthly decrease equal to 0.1 %. The GDP dynamics remained unchanged during the second quarter (- 0.6 % annually) (Residential building and real estate market in the period of economic decrease (2016)). Positive development is shown by the third quarter, though GDP quantum index gives more information (comparison with the same quarter of the preceding year as %), and it has not shown 100 % and above since the end of 2014. Therefore, GDP increase in real prices is counter-balanced by inflationary processes.

Taking into account the fact that GDP index is the basic indicator of national economic development and mostly reflects the tendencies of most industries development, it is necessary to move forward to the analysis of the construction industry state in Russian Federation on the modern stage of development of the domestic economy. The construction sphere in Russia entered the crisis back in the second half of 2013, when stagnation of innovative process became evident in general. More than a half of all investments in Russia fall to housing (15 % of all the investments in 2015) and non-residential buildings (41 %), and the housing share have been increasing starting from 2010.

https://doi.org/10.1051/e3sconf/20183303054

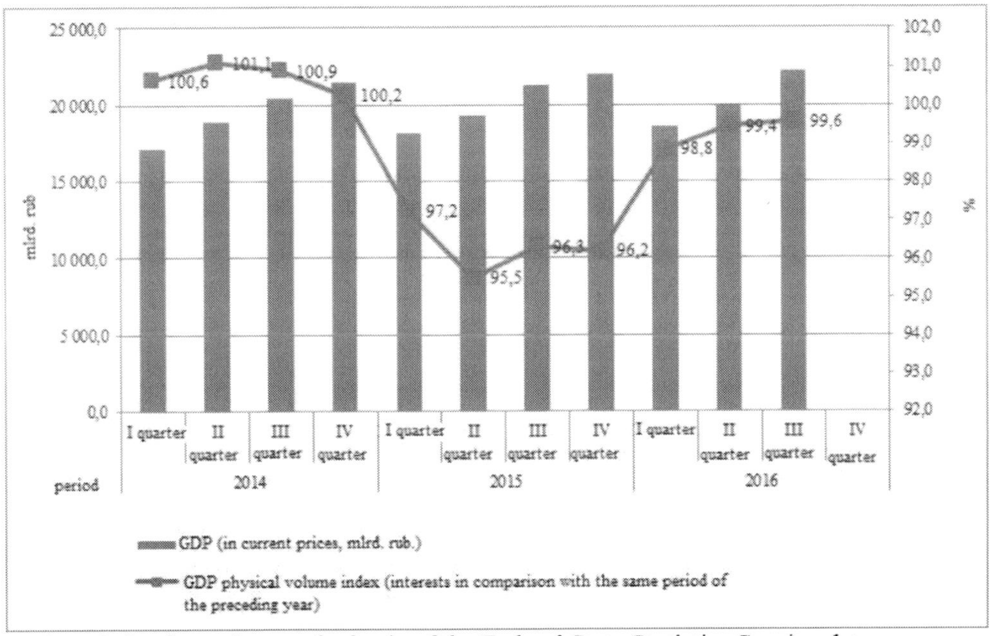

* made by the author on the basis of the Federal State Statistics Service data

Fig. 1 GDP dynamics in current prices and in interests compared with the same quarter of the preceding year

At the same time, according to the Federal State Statistics Service, overall investment volume in all economic spheres keeps on decreasing, and investments in capital stock (in other words, in construction), also decrease, though they are more unstable (pic. 2). Despite the fact that investments in construction increased in 2013, there is a significant decrease of this parameter since 2014.

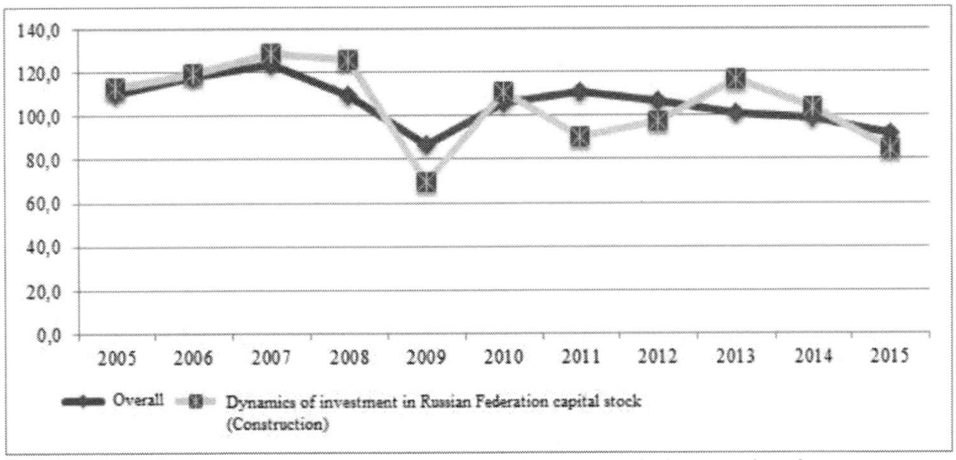

* made by the author on the basis of the Federal State Statistics Service data

Fig. 2 Comparison of investment in Russian Federation capital stock dynamics in all types of economic activity, in construction in particular

The first stage of overall investment decrease, including investments in the construction, happened in the period of world financial crisis in 2008-2009, when investment increase fell from 120 % to 70%, consequently affecting the domestic economy in general. As a result of smart policy of National Bank and profile ministries, the index raised back by 2010, though it didn't reach the pre-crisis figures. The crisis of 2014-15 influenced at the investment processes greatly. As a result, many foreign investors left the market, and domestic ones are not active due to unstable economy and national currency rate. Construction as a direction for long-time investments nowadays considered is a high-risk one.

It is considerable to evaluate volume and dynamics of the investments in the construction sphere particularly, which has been increasing in real prices since 2005, but index of physical volume of capital investments proves that the whole picture is not so bright (pic. 3). The problems can be traced back to 2012, when pace of capital investment growth slowed down. During last two years, experts note the tendency of slow capital investment decrease.

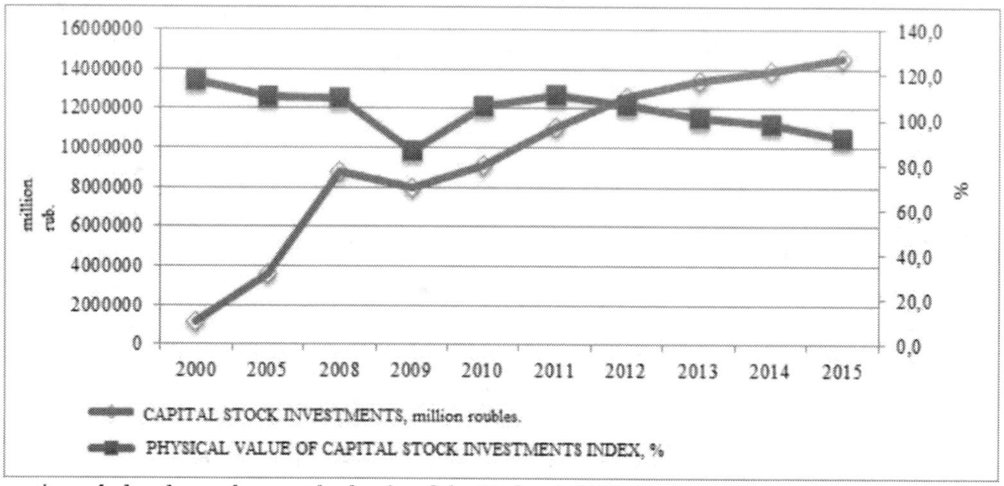

* made by the author on the basis of the Federal State Statistics Service data

Fig. 3. Dynamics of absolute and comparative values of capital stock investments.

Statistic data prove that it is necessary to develop effective mechanisms of investment improvement, especially in the construction sphere, as it is vital for residential, industrial and infrastructural construction, in other words, it reflects development of all branches of economic activity.

Systemic changes in the construction sphere should be started with development of strategic directions of investment activity, so it is necessary to note what influence on the relations between participants of the investment and construction process.

Contents of investment and construction processes as a single process come from the term "investment and construction complex", which has the following meaning:

Integration of construction and investment processes into one single complex;

Interaction between two activity types and consolidation of construction market participants and real investments;

Interaction or consolidation of constructive and real investment resources.

It is considerable to account the investment and construction process in theoretical-economic view as a system of relations between investment and construction cycle participants, concerning resource development and creation of material values. In business aspect, all participants of the investment and construction process fall into several groups:

Design organizations;

Enterprises;

Enterprises in the sphere of construction materials (brick factories, homebuilders, trusts etc.);

Infrastructure (transportation, energetics etc.).

Special place in investment and construction process have the elements, the activity of those is connected with investment attraction. These are investment funds, investment banks, facilitating agencies and other structures. On a peculiar stage of economic development appears necessity of close analysis not of special organizations, but their specific combinations, which serve as inter-industry and branch complexes. In other words, investment and construction process is a peculiar set of relations between participants of investment activity, project, construction and installation works, concerning mobilization and rational use of resources. On the basis of conducted analysis, it is possible to distinguish the meaning of investment and construction process in strategic aspect. It turns out to be a set of all participants of investment and construction activity, oriented on the achievement of common goals

Investment and construction activity is practical (financial, organizational, productive, economic) activity of the government, incorporated and unincorporated businesses, connected with accumulation and achievement of financial resources in order to invest them in fixed assets and provide their effective usage. The amount of constructed and put in place fixed assets (buildings, constructions, objects) is defined by investors' abilities. These abilities are evaluated by size of capital investments, which investors are able to use in order to create fixed assets. Capital investment volume depends on the following factors:

Assets (owned capital) of investors, earned from business activity – profit, amortization, profit from capital issues etc.;

Opportunity and conditions of access to bank credits and other borrowed funds (funded loans, capital issues etc.), as well as use of population funds, attracted funds of labour collectives and incorporated businesses.

Economic rationality and ability of capital asset consolidation in within pool centers, finance and production groups, international companies; presence of investment potential of institutional structures such as pension funds and insurance companies.

Investment abilities of governmental and local budgets. Investment and construction complex should be considered as a set of economically, organizationally and technologically interconnected owned subjects, which common activity is oriented on putting in work capital funds of both productive and non-productive activity and satisfaction of own economic needs within the investment project realization.

Consequently, on the basis of meaning and participants of investment and construction process, it should be strategically oriented of financial resources formation (using various mechanisms) and investors attraction. The investors attraction problem should be solved not only by construction companies themselves, but also by the government by creation of favorable investment climate, attraction of institutional investors in the construction sphere and realization of domestic investors' stimulation mechanisms.

The problem of realization of investment policy is observed in T.R. Bareev's work (T.R. Bareev, 2015), which is devoted to the fact, that on the basis of investment policy's goals and taking into account structural priorities of economic development, it is possible to highlight its' main goals: increase of investment resources of the region on the basis of budget and non-budget sources of all levels of governmental management, financial resources of private domestic and foreign capital and population assets; concentration of investment resources on high-priority economic spheres and solution of the most actual social development problems by creation of favorable investment conditions; development and putting into practice market mechanism of investment process regulation; specified distribution of grace investments among chosen projects, economic sectors and territories; increase of investment supportability of depressing territories and stagnated branches (T.R. Bareev (2015)). Generalizing and systemizing all strategic approaches to the participants' of investment and construction process relations, it is possible to work out a scheme of the relations described above (Fig. 4).

After all the strategic goals and directions of investment and construction process realization are defined, it is possible to move forward to development of tactical directions of this process' organization. On tactical level it is necessary to solve a larger amount of different problems, which arise on every stage of investment and construction project's realization.

According to the author, the primary direction within process of tactical problem's solution, which needs close attention, is choosing the form of organization of a construction project. The problems of realization of the given investment and construction process will be solved within the chosen form of organization

In Russian scientific research organizational form of investment and construction process' control is sometimes equal to organizational structure of the company. This approach is not correct. Participants of investment and construction process include many organizations of different specializations. It is a rare situation, when one company acts as customer, designer and contractor at the same time. Also one company is rarely able to provide the whole project with investment resources.

It is reasonable to take attention at the analysis of effectiveness of three main forms of investment and construction process: traditional, key ready and professional construction management. In organizational form of management in traditional contractual way the object's construction is made by specialized construction organizations (contractors), who sign an agreement with the customer.

Traditional form is connected with fixed price. It guarantees the project accomplishment in time within fixed (estimated) price, which is a high-risk one for the contractor, as he needs to forecast all possible problems, which may arise in the construction process at the stage of signing the agreement.

The experience of usage of traditional organizational form of investment and construction process organization shows for several years already that customers have to face such disadvantages as estimated price overrun and violation of estimated time of putting object into action. Moreover, in traditional form of construction control organization the customer has to execute many additional high-risk functions, not connected with his main activity profile. In other words, in traditional form of investment and construction process organization there is no function diversification depending on the final result of a given sub-process. Within this form, the customer signs a design project with one general contractor, who supplies the directors with main construction process phases: design and construction. It helps to speed up decision making and realization and, as a result, speeds up the construction and lowers the costs. General contractors are usually well-known design and construction organizations.

General contractor fully operates the resources and controls the construction process itself. As an alternative to traditional one, in foreign practice, key-ready organizational form of construction is considered as a mark of high professional skills level of all construction company personnel, a criterion of effective management system and solid advantage on investment and construction market. At the same time, it is necessary to understand the importance of shift to more progressive forms of participant relations organization on the basis of design and construction agreements, within which new demands to all investment process participants, contractors in particular, are set.

In modern countries in the latest decade rapidly develops the form of professional construction management (Fesenko T.G., Minaev D.M. (2016)). This form is an evolution of the key-ready form. Integration of design and construction stages, object complexity, necessity of further speeding up the construction, resulted as an introduction of new management type – professional construction management and its' particular operators – professional managers.

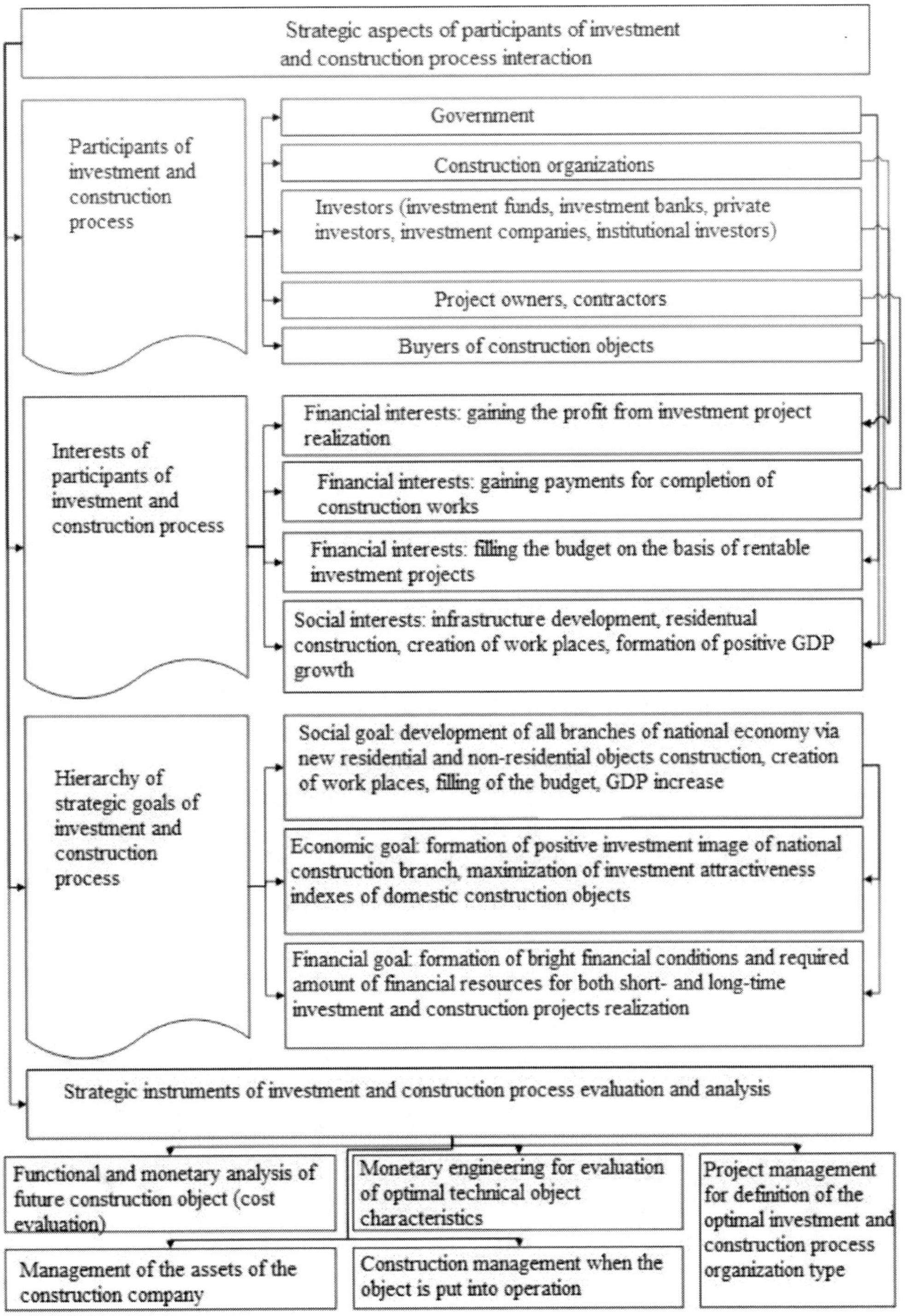

Fig. 4 Strategic aspects of participants of investment and construction process interaction (developed by the author)

In unstable conditions of construction organizations environment, both customer's and investor's main goal is effective investment into new, perspective type of activity, development of new productions. One of the main criteria is quick investment refund. In this situation the customer gives control of the whole investment cycle to professional managers. While in modern countries the professional construction management and project management market is already formed, it is a still developing process for Russian construction sector, what is proven by the fact that traditional contractual customer-operator relations still prevail in the realization of the investment and construction process. The peculiarity of organizational form of professional construction management is that the customer, acting as both incorporated business and investor, signs all contract types with design and construction organizations, while planning, coordination and control over their realization is made by a professional manager. The same manager checks the quality of work and ensures rational usage of investor's financial resources.

Usage of professional construction management organizational form gives the customer all advantages of the key-ready form: opportunities to shorten the whole investment cycle by interaction of design and construction stages. The customer has an opportunity to control contractors' actions via construction manager, what decreases the risk of investment assets overrun, which form from investment resources.

For formalization of advantages of the choice of an organizational form within realization of tactical aspects of the investment and construction process we offer a methodical approach, which includes basic stages of choice of organizational form in a given situation (pic. 5).

Rationality and reasonability of usage of a chosen organizational form depends on objective factors. If financial resources are limited (and the company designed the project and has construction experience), it is not reasonable for the construction company to look for design and construction organization or hire a professional manager. In such situation it is reasonable to conduct a tender and find a contractor, who will do the work responsibly and for minimal price.

Along with design and construction companies attraction in the key-ready organizational form, it is possible to reach an effective work process and connect design and construction stages by hiring development companies, which provide construction within professional construction management form. Usage of a development company is oriented on effective organization of a construction process and profit increase from further object exploitation.

Developer is a businessman, who makes profit from construction object creation, where he acts as:

a) author of a project (what to create and where);

b) buyer of a land property for construction;

c) organizer of an object project process (made by licensed designers), employer of the customer, general contractor, brokers for object consignment, residential managers (if necessary) – gives financial resources himself or finds investors, if necessary.

There are two main development types within tactical realization of investment and construction process:

1) Developer doesn't cover financial risks and has a fixed fee (fee-development);

2) Developer creates commercial buildings, acting as a single organizer of the project.

Actually, developer completes structure and realization of financial scheme of the project. Developer may add into the project his own resources, which are an axis of future financial scheme, acting as investor at the same time. This development time is known as speculative development. The financial scheme of large development projects is usually a complex combination of developer's own resources, attracted investments, bank loans and payments from future residential owners

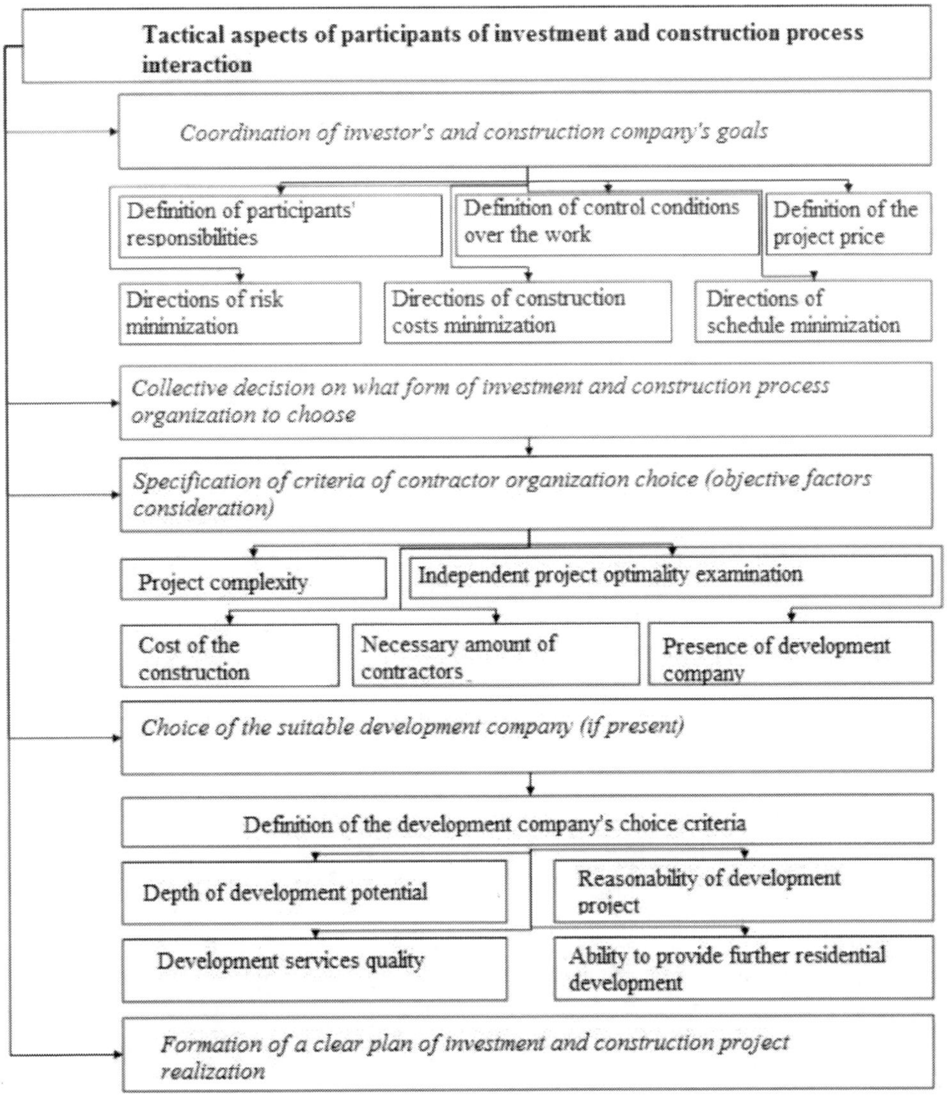

Fig. 5 Tactical aspects of participants of investment and construction process interaction (developed by the author)

Consequently, when company moves from traditional contractor form to professional construction management and key-ready forms, as well as reorganization of general contractor companies, significant part of responsibility is moved from highest level to the middle one.

The most responsible people, in modern understanding of investment and construction process, are situated on the level of contractors or developers, depending on the type of organizational form used. That is why general goals of the director of the construction in the process of design and construction process realization, along with action coordination of builders, subcontractors, material and equipment suppliers, are expanded by interaction with designer and investor.

The director of construction project implements the following functions:

Planning of design works according to the construction schedule;

Examination and evaluation of different design solutions on the stage of initial design, choice of the best one for customer's confirmation;

Definition of the order of project documentation sets distribution;
Coordination of execution of design and construction schedule;
Registration and control of project changes in the construction process.

In is necessary to underline the fact that in recent years there are a lot of publications concerning construction problems, financial supply of the construction, realization risks of the investment projects in the sphere of finance, raise of the investment activity of the construction process participants. At the same time, the problem of specification of strategic and tactical directions of interaction between the investment process participants requires a more detailed review due to lack of publications on the given topic.

Conclusion

As a result, it is reasonable to highlight, that it is necessary for the company to use various methods, instruments and ways of construction activity organization in both tactical and strategic aspects in order to achieve successful realization of investment and construction process.

At strategic level it is necessary to coordinate interests of innovation and construction process participants, develop approaches of governmental stimulation of investors in construction sphere, provide maximization of investment attractiveness. The government is interested in investment and construction process activation, as it help to fill the budget, form GDP, develop infrastructure at different levels.

At tactical level the main goal is to choose the organizational form of investment and construction process support. As a rule, every of the forms described, provides the director of the construction with information related to a particular sphere. In conditions of dynamic foreign market, it is necessary to have a systematic approach to development of efficiency of construction production management, built on the interacted analysis of problem complex of various methods implementation.

It is necessary to highlight that with increase of the number of the investment and construction process participants and of interaction complexity, it is vital to develop management methods. That is why for successful realization of design and construction contracts, apart from organizational changes in management structure, it is necessary to have investor activity mechanisms and mechanisms of increase of national economy investment attractiveness.

References

1. M. Froli, F. Laccone, Journal of Architectural Engineering **24(1)**, 04017030 (2018) DOI: 10.1061/(ASCE)AE.1943-5568.0000285
2. J.M. Stritch, R.K. Christensen, International Journal of Manpower **37-5**, 840-858 (2016) DOI: 10.1108/IJM-12-2014-0249
3. M. Gravit, O. Zybina, A. Vaititckii, A. Kopytova, Advances in Intelligent Systems and Computing **692**, 1093-1101 (2018) DOI: 10.1007/978-3-319-70987-1_118
4. N. Zotkina, M. Gusarova, A. Kopytova, Advances in Intelligent Systems and Computing **692**, 1204-1213 (2018) DOI: 10.1007/978-3-319-70987-1_129
5. M. TepePublic Management Review **18-4**, 508-538 (2016) DOI: 10.1080/14719037.2015.1014396
6. A. Kopytova, Procedia Engineering **165**, 1132 (2016) DOI: 10.1016/j.proeng.2016.11.830

7. A. Kopytova, MATEC Web of Conferences, **106**, 08056 (2017) DOI: 10.1051/matecconf/201710608056
8. N. Zotkina, S. Bardasov, M. Gusarova, A. Kopytova, MATEC Web of Conferences, **106**, 08050 (2017) DOI: 10.1051/matecconf/201710608050
9. K. Pykhtin, T. Simankina, V. Sharmanov, A. Kopytova, IOP Conference Series: Earth and Environmental Science, **90 (1)**, 012065 (2017) DOI: 10.1088/1755-1315/90/1/012065
10. N.S. Zotkina, Finances and Credit **48 (480)**, 21-28 (2011)
11. V. Lezier, M. Gusarova, A. Kopytova, IOP Conference Series: Earth and Environmental Science, **90 (1)**, 012034 (2017) DOI: 10.1088/1755-1315/90/1/012034
12. A. Minnullina, R. Abdrazakov, Advances in Intelligent Systems and Computing **692**, 1224-1233 (2018) DOI: 10.1007/978-3-319-70987-1_131
13. A.I. Kazannik, Bulletin of Omsk University **4 (33)**, 35-43 (2012)
14. A.V. Kopytova, Exchange of intellectual property **3 (XIV),** 31–37 (2015)
15. B. Liu, J.L. Perry, Review of Public Personnel Administration **36-1**, 4-30 (2016) DOI: 10.1177/0734371X14549672
16. A. Minnullina, R. Abdrazakov, IOP Conference Series: Earth and Environmental Science, **90 (1)**, 012089 (2017) DOI: 10.1088/1755-1315/90/1/012089
17. J. Yeo, International Journal of Public Administration **39-3**, 216-225 (2016) DOI: 10.1080/01900692.2015.1004083
18. M.S. Gusarova, A.V. Kopytova, Bulletin of Omsk University **4**,14–22 (2014)
19. N.S. Zotkina, A.V. Kopytova, Economics and entrepreneurship **12–2(53–2)**,714–719 (2014)
20. V. Lezier, M. Gusarova, A. Kopytova, IOP Conference Series: Earth and Environmental Science, **90 (1)**, 012069 (2017) DOI: 10.1088/1755-1315/90/1/012069
21. A. Mottaeva, A. Minnullina, IOP Conference Series: Earth and Environmental Science, **90 (1)**, 012123 (2017) DOI: 10.1088/1755-1315/90/1/012123
22. M. Gravit, O. Zybina, A. Vaititckii, A. Kopytova, IOP Conference Series: Earth and Environmental Science, **90 (1)**, 012103 (2017) DOI: 10.1088/1755-1315/90/1/012103
23. E.A. Mitrofanova, Personnel and intellectual resources management in Russia **2**, 36-41 (2014)
24. N. Zotkina, A. Kopytova, M. Zenkina, O. Zhigunova, MATEC Web of Conferences, **106**, 08058 (2017) DOI: 10.1051/matecconf/201710608058
25. A. Minnullina, MATEC Web of Conferences, **106**, 08067 (2017) DOI: 10.1051/matecconf/201710608067

Determination of sustainable values for the parameters of the construction of residential buildings

Larisa Grigoreva[1]* *and Vladimir* Grigoryev[2]

[1]Moscow State University of Civil Engineering (MGSU), 129337, Moscow, Russia
[2]JSC "Atomtechenergo", Department of Information and Analytical Systems, 101000 Moscow, Russia

Abstract. For the formation of programs for housing construction and planning of capital investments, when developing the strategic planning companies by construction companies, the norms or calculated indicators of the duration of the construction of high-rise residential buildings and multifunctional complexes are mandatory. Determination of stable values of the parameters for the high-rise construction residential buildings provides an opportunity to establish a reasonable duration of construction at the planning and design stages of residential complexes, taking into account the influence of market conditions factors. The concept of the formation of enlarged models for the high-rise construction residential buildings is based on a real mapping in time and space of the most significant redistribution with their organizational and technological interconnection - the preparatory period, the underground part, the above-ground part, external engineering networks, landscaping. The total duration of the construction of a residential building, depending on the duration of each redistribution and the degree of their overlapping, can be determined by one of the proposed four options. At the same time, a unified approach to determining the overall duration of construction on the basis of the provisions of a streamlined construction organization with the testing of results on the example of high-rise residential buildings of the typical I-155B series was developed, and the coefficients for combining the work and the main redevelopment of the building were determined.

1 Introduction

In Russia, the practice of multi-storey mass housing construction and design standards was previously focused on the height of buildings up to 75 m. At present, high-rise buildings are considered to be above 75 m. According to the number of floors, the existing classifications are rather conventional. Residential buildings are considered high-rise if there are at least 25 floors.

In works [1, 2], the authors consider the trends in the development of high-rise construction in Russia. In works [3-5], the authors study the international experience of high-

* pehel@yandex.ru

rise construction, identify the features of building technologies and the role of high-rise buildings in the urban space.

The erection of high-rise buildings implies the development of a single regulatory and legal framework [6, 7]. In February 2017, a set of rules «Engineering systems of high-rise buildings» came into effect, which provide for the integrated security of buildings, will improve energy efficiency and reduce the consumption of non-renewable resources in construction and operation. However, these standards do not regulate the duration of construction.

For the organizational and technological design of high-rise construction, there is a need for a system that includes a variety of organizational, technological, technical, economic and time factors and provides the opportunity to regulate the performance indicators [8, 9].

The purpose of the work is to study and evaluate the indicators of the duration of construction of residential buildings, the establishment of sustainable indicators of the duration of construction of major stages.

2 Methods

To compare and study the parameters of the construction of residential buildings, a statistical array of initial information was collected for typical series of wide application for more than a decade of their construction in Moscow and the Moscow region. The main parameters of the construction of residential buildings are, first of all, indicators of the duration of the construction of the building as a whole and its individual redistribution (stages) - the preparatory period, the main period, including the erection of underground and aboveground parts.

Calculated indicators of the duration of construction of facilities can be determined on the basis of organizational and technological documentation, the degree of object readiness, statistical evaluation of durations [10-12]. In [13], the authors showed that, in general, the reconstruction of high-rise buildings is similar to the reconstruction of other buildings and structures. At the same time, similar steps are used when drawing up a package of design estimates, previous and accompanying work. The main difference is in the choice of schemes for mechanization of dismantling and assembly works of reconstruction and further application of organizational solutions.

3 Results

The main indicators of development and adoption of optimization solutions are:
total duration of construction, T;
duration of the preparatory period, T_1;
duration of the erection of the underground part, T_2;
duration of erection of the above-ground part, T_3;
duration of laying of external engineering networks, T_4;
duration of improvement, T_5.

There are various options for determining the total duration of construction of a residential building. We have identified 4 main options:

Option 1.

$$T = T_1 + T_2 + T_3 + (1 - \alpha)\, T_4 + (1 - \beta)\, T_5 \quad (1)$$

$$\text{at } T_4^{/} > T_3^{/} \text{ and } T_5^{/} > T_3^{/}$$

where T is the total duration of the construction of a residential building, days;
T_1 - duration of the preparatory period, days;
T_2, T_3, T_4, T_5 - duration of erection (device), respectively, of underground, aboveground parts of buildings, external engineering networks, improvement, days;

$T_3^{/}, T_4^{/}, T_5^{/}$ - date of completion of works, respectively, for erection of the above-ground part, laying of external engineering networks and improvement;
α - the coefficient of overlapping in time of work on the laying of external engineering networks with the erection of the elevated part of the building;
β - coefficient of time alignment of works on the improvement of the territory with the erection of the elevated part of the building.

Option 2.

$$T = T_1 + T_2 + T_3 \quad (2)$$
$$\text{at } T_4^{/} \le T_3^{/} \text{ and } T_5^{/} \le T_3^{/}$$

Option 3.

$$T = T_1 + T_2 + T_3 + (1 - \alpha)\, T_4 \quad (3)$$
$$\text{at } T_4^{/} > T_3^{/} \text{ and } T_5^{/} \le T_3^{/}$$

Option 4.

$$T = T_1 + T_2 + T_3 + (1 - \beta)\, T_5 \quad (4)$$
$$\text{at } T_4^{/} \le T_3^{/} \text{ and } T_5^{/} > T_3^{/}$$

More demanded is Option 2, while the work on laying external engineering networks and landscaping are carried out simultaneously with the erection of the elevated part of the residential building. For each option set the duration of each stage, taking into account the identification of rational duration of work and their maximum overlap. For this purpose, for the set of works (Figure 1), the duration of their implementation can be defined as:

$$t = t_n + \sum_{i=1}^{n-1} (t_i - \psi_t t_i) \quad (5)$$

where t – the duration of the whole set of works $i = \overline{1, n}$;

t_n – the duration of the last n-th work;

t_i – the duration of the i-th work, $i = \overline{1, n}$;

Ψ_i – the coefficient of combination in time i and (i + 1) works, equal to

$$\psi_t = \frac{t_{i+1}^c}{t_i} \quad (6)$$

where t_{i+1}^c – duration of the (i + 1) work, of combination with i-th works, $t_{i+1}^c \le t_{i+1}$, substitute the expression for Ψ_i in (3):

$$t = t_n + \sum_{i=1}^{n-1} (t_i - t_{i+1}^c) \quad (7)$$

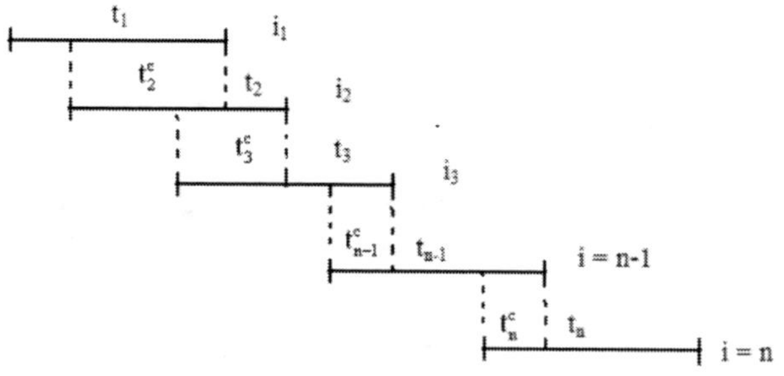

Fig. 1. Fragment of the planned schedule of works

Residential building series I-155B is a 25 storey single-section tower. Residential buildings of this series can be attributed to high-rise buildings. The duration of the preparatory period (T_1) for both normative documents and actual data is one month [14, 15].

The underground part. Terms for the erection of the underground part are made up of the length of the excavation excavation and the device in the combination of a monolithic plate and the zero cycle (Figure 2):

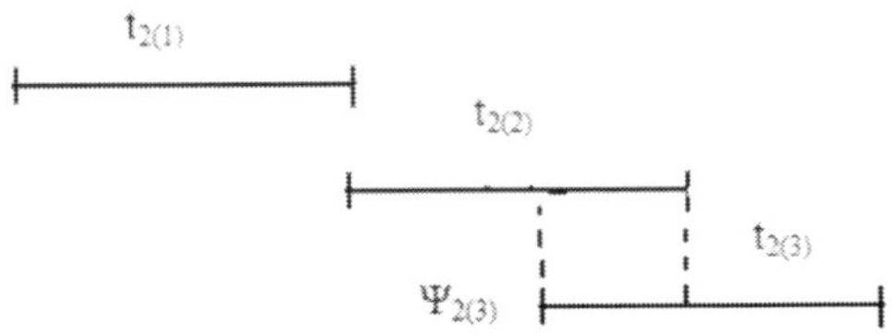

(8)

Fig. 2. Fragment of the calendar plan for the erection of the underground part of the residential building I-155B

$$T_2 = t_{(2)3} + t_{(2)1} + t_{(2)2} - \Psi_{2(3)} \cdot t_{(2)2}$$

$t_{(2)1}$, $t_{(2)2}$, $t_{(2)3}$ – the length of the excavation, the device of a monolithic slab, the erection of the technical underground;

$\Psi_{2(3)}$ – coefficient of combination of works on the erection of the technical underground with the device of a monolithic plate, varying within:

$$\Psi_{2(3)} = \begin{Bmatrix} \frac{5}{24} \div \frac{5}{20} = 0,21 \div 0,25 \\ \frac{10}{24} \div \frac{10}{20} = 0,42 \div 0,50 \end{Bmatrix} \qquad 0,21 \le \Psi_{2(3)} \le 0,50 \tag{9}$$

at the device of a monolithic plate with pile base:

$$\Psi_{2(3)} = \begin{Bmatrix} \frac{5}{30} \div \frac{5}{25} = 0,17 \div 0,20 \\ \frac{10}{30} \div \frac{10}{25} = 0,33 \div 0,40 \end{Bmatrix} \qquad 0,17 \le \Psi_{2(3)} \le 0,40 \tag{10}$$

Installation of the tower crane is carried out in parallel with the excavation of the excavation and therefore does not stand out as a separate type of work.

Thus, the duration of the erection of the underground part of the residential building of the I-155B series will be:

at the device of a monolithic plate without a pile foundation

$$\min T_2^{/} = 25 + 15 + 20 - 0,21 \cdot 20 = 55,8 \text{ days}$$

$$\min T_2^{/} = 35 + 20 + 24 - 0,5 \cdot 20 = 50,0 \text{ days}$$

$$\min \overline{T_2^{/}} = 52,9 \text{ days}$$

$$\max T_2^{/} = 35 + 20 + 24 - 0,21 \cdot 24 = 74,0 \text{ days}$$

$$\max T_2^{/} = 35 + 20 + 24 - 0,5 \cdot 24 = 67,0 \text{ days}$$

$$\max \overline{T_2^{/}} = 70,5 \text{ days}$$

at the device of a monolithic plate with pile base:

$$\min T_2^{//} = 25 + 15 + 25 - 0,17 \cdot 25 = 60,8 \text{ days}$$

$$\min T_2^{//} = 25 + 15 + 25 - 0,4 \cdot 25 = 55,0 \text{ days}$$

$$\min \overline{T_2^{II}} = 57{,}9\text{дн.}$$

$$\max T_2^{II} = 35 + 20 + 30 - 0{,}17 \cdot 30 = 79{,}9 \text{ days}$$

$$\max T_2^{II} = 35 + 20 + 30 - 0{,}4 \cdot 30 = 73{,}0 \text{ days}$$

$$\max \overline{T_2^{II}} = 76{,}5 \text{ days}$$

$\min T_2^{I}$, $\max T_2^{I}$ – duration of the erection of a subterranean part during the construction of a monolithic slab without a pile foundation with values $\Psi = 0{,}21$ и $\Psi = 0{,}50$;

$\min T_2^{II}$, $\max T_2^{II}$ – duration of the erection of a subterranean part when constructing a monolithic slab with a pile base at values $\Psi = 0{,}17$ и $\Psi = 0{,}40$;

$\min \overline{T_2^{I}}$, $\max \overline{T_2^{I}}$ – minimum and maximum average duration of the erection of a subterranean part when constructing a monolithic slab without a pile foundation;

$\min \overline{T_2^{II}}$, $\max \overline{T_2^{II}}$ – minimum and maximum average duration of the erection of a subterranean part when constructing a monolithic slab with a pile foundation.

The aboveground part. The erection of the aboveground part includes the following main works: installation of prefabricated reinforced concrete structures, construction, plumbing, electrical and finishing works.

The parameters of assembly of prefabricated reinforced concrete structures have a wide range of values both for the duration of installation and for the number of floors to be mounted (Table 1). At the same time, the maximum duration of assembly of prefabricated reinforced concrete structures is typical for variants with strongly pronounced unevenness of the assembled floors (for example, 1-3-6-7-7-1, 3-4-5-6-6-1).

Table 1. Variants of erecting floors by time periods (fragment)

Cipher (number)	Number of installed floors by month							
	1	2	3	4	5	6	7	8
69	4	4	6	5	6			
71	1	6	7	7	4			
72	1	4	4	8	8			
73	1	1	4	10	9			
74	5	9	8	3	-			
75	1	4	8	8	4			
77	3	8	7	7	-			
79	2	4	3	2	5	4	4	1
81	1	3	6	7	7	1		
82	3	4	5	6	6	1		
83	2	4	6	4	6	3		
Total number of floors	24	51	64	67	55	9	4	1
Average value of the floors being erected	2	4	5	6	6	2	-	-

Analysis of the options for erecting floors by time periods showed that such a situation arose as a result of the discrepancy between the available labor and material and technical resources to the number of simultaneously erected objects. So, for example, during 2016, the installation of the aboveground part was carried out at 28 sites. At the same time, the maximum volume of installed floors falls on November and December, respectively, at 63 and 56 floors. In January, only 29 floors were erected, and in the remaining months the number of mounted floors ranged from 35 to 50.

To obtain statistical values of the duration of installation of prefabricated reinforced concrete structures of the above-ground part of the residential building I-155B, an array of initial information is formed, in which $\underline{t}$ – the duration of montage of n-floors (months), t_0 – the duration of one-storey installation (month), n – the number floors (Table 2).

Table 2. Statistics on the installation of residential buildings I-155B

n	3	2	8	16	22	23	25	24	25	25	25
$\underline{t}$	1	1	2	4	6	6	5	5	5	5	5
t_0	0.33	0.50	0.25	0.25	0.27	0.26	0.26	0.20	0.21	0.20	0.20
n	25	25	15	25	17	25	13	25	25	25	
$\underline{t}$	4	5	3	4	3	8	2	6	6	6	
t_0	0.16	0.20	0.20	0.16	0.18	0.32	0.15	0.24	0.24	0.24	

After processing of the specified data array, the following equations of communication were obtained between the duration of installation of prefabricated reinforced concrete structures and the number of installed floors:

linear $t = 0{,}25 + 0{,}21 \cdot n$

quadratic $t = 0{,}57 + 0{,}14 \cdot n + 0{,}22 \cdot 10^{-2} \cdot n^2$

logarithmic $t = 1{,}49 + 2{,}07 \cdot \log n.$

Calculation of linear dependence: the regression $t = f(n)$ has the form $t = k \cdot n + b$, where k, b are coefficients.

$$k = \frac{\overline{t \cdot n} - \bar{t} \cdot \bar{n}}{D_n} = \frac{98{,}667 - 19{,}905 \cdot 4{,}381}{55{,}23} = 0{,}207$$

$\bar{t} = 4{,}381$ – mathematical expectation t;

$\bar{n} = 19{,}905$ – mathematical expectation n;

$\overline{t \cdot n} = 98{,}667$ – mathematical expectation $t \cdot n$;

D_n – dispersion n, $D_n = 55{,}23$; $b = \bar{t} - k \cdot \bar{n} = 4{,}381 - 0{,}207 \cdot 19{,}905 = 0{,}249$

Thus, the equation of linear dependence has the form $t = 0{,}25 + 0{,}21 \cdot n$.

To determine the relationship, the correlation coefficient is calculated by the following formula:

$$r = \frac{\overline{t \cdot n} - \bar{t} \cdot \bar{n}}{S_n \cdot S_t}$$

S_n – standard deviation n; S_t – standard deviation t;

$$r = \frac{98{,}667 - 4{,}381 \cdot 19{,}905}{7{,}432 \cdot 1{,}785} = 0{,}864$$

Similar calculations were made for the quadratic and logarithmic functions.

Linear regression is better than quadratic and logarithmic approximates the relationship between the duration of installation of prefabricated reinforced concrete structures and the number of installed floors: Correlation coefficient for linear dependence – 0.864; Correlation coefficient for quadratic dependence – 0,833; Correlation coefficient for the logarithmic dependence – 0,649.

Based on the analysis of design and practical solutions, the following important circumstance is revealed: the duration of erection of the aboveground part can be defined as the arithmetical sum of the installation time of prefabricated reinforced concrete structures, the incompatible part of the plumbing work, the incompatible part of the finishing works.

All other types of work are carried out within the specified total duration of erection of the aboveground part.

In the case of different lengths of performance of plumbing and electrical work, the maximum duration is assumed.

Thus, the duration of erection of the above-ground part is represented as

$$T_3 = t_{3(1)} + t_{3(2)} + t_{3(3)} - \Psi_{3(2)} \cdot t_{3(2)} - \Psi_{3(3)} \cdot t_{3(3)}$$

$t_{3(1)}$, $t_{3(2)}$, $t_{3(3)}$ – duration, respectively, installation of prefabricated reinforced concrete structures, performance of sanitary (or electrical), finishing works;

$\Psi_{3(2)}$ – coefficient of combination of plumbing works with the installation of prefabricated reinforced concrete structures;

$\Psi_{3(3)}$ – coefficient of combination of finishing and plumbing (or electrical) work.

The indicators $\Psi_{3(2)}$, $\Psi_{3(3)}$) will be applied within the following limits:

$$\Psi_{3(2)} = \begin{cases} \Psi'_{3(2)} = \frac{15}{165} \div \frac{15}{120} = 0,09 \div 0,13 \\ \Psi''_{3(2)} = \frac{30}{165} \div \frac{30}{120} = 0,18 \div 0,25 \end{cases}$$

$$0,09 \le \Psi_{3(2)} \le 0,25$$

$$\Psi_{3(3)} = \begin{cases} \Psi'_{3(3)} = \frac{80}{180} \div \frac{80}{150} = 0,44 \div 0,53 \\ \Psi''_{3(3)} = \frac{100}{180} \div \frac{100}{150} = 0,56 \div 0,67 \end{cases}$$

$$0,44 \le \Psi_{3(3)} \le 0,67$$

The duration of erection of the above-ground part is multivariant, therefore it can be represented as an interval of values between the minimum and maximum limits, i.e.

$$\min T_3 \le T_3 < \max T_3$$

In Table 3, only the minimum and maximum performance indicators are presented and therefore all intermediate values of T_3 will be in the interval 265.5 days $\le T_3 \le$ 393.2 days.

Table 3. Indicators for the duration of erection of the aboveground part depending on the duration and degree of overlap

Key performance indicators, days					Duration indicator erection of the aboveground part, days
$t_{3(1)}$	$t_{3(2)}$	$t_{3(3)}$	$\Psi_{3(2)}$	$\Psi_{3(3)}$	T_3
1	2	3	4	5	6
120	150	100	0,09	0,44	312,5
120	150	100	0,25	0,44	288,5
120	150	100	0,09	0,67	289,5
120	150	100	0,25	0,67	265,5
165	180	115	0,09	0,44	393,2
165	180	115	0,25	0,44	364,4
165	180	115	0,09	0,67	366,8
165	180	115	0,25	0,67	338,0

Conclusions

The total duration of the construction of a residential building, depending on the duration of each redistribution and the degree of their overlapping, can be determined from one of the identified four options.

Approaches are proposed for determining the duration of the erection of underground and aboveground parts of residential buildings, taking into account the wide spread, both the duration of their work, and the degree of their overlap with the indication of the minimum and maximum values and conditions for their operation. So, for example, for residential buildings I-155B the coefficients of combination of works during the device of technical underground vary in the range from 0.17 to 0.50 with the duration of erection of the underground part from 52.9 days. up to 76.5 days, and when erecting the above-ground part, the coefficients of work combination vary from 0.09 to 0.67 with the duration of this redistribution from 265.5 days. up to 393.2 days.

When forecasting and planning the residential development of neighborhoods and neighborhoods, a simplified approach can be used to determine the total duration of construction of residential buildings, based on the statistical relationship between the duration of construction and the total area.

References

1. A. Magay, Housing construction, **2,** 9-12 (2008)
2. O. Zhuchkov, E. Marinich, S.Turmov, Mod. soc.: probl., ideas, innov., **4,** 89-93(2015)
3. T. Maklakova, High-rise buildings. Urban planning and architectural and constructive design problems (Moscow, Publisher ASV, 2008)
4. A. Matejko, Vestnik SGASU, **3**, 74 (2016)
5. A. Kozhas, M. Rakhimov, A. Kasimov, S. Kozhasov, Proc. of the University, **3** (60), (2015)
6. E. Generalova, K. Galstyan, Traditions and innovations in construction and architecture. Architecture and design (Samara, 2015)
7. A. Bimagambetov Bull. of scien. conferences, **4-1** (8), (2016)
8. T. Kravchunovska, E. Zajats, S. Yepifantseva, Prydniprovs'ka State Academy of Civil Engineering and Architecture, **11** (176), 4-8 (2012)
9. V. Bolshakov, T. Kravchunovska, E. Zayats, S. Yepifantseva, Prydniprovs'ka State Academy of Civil Engineering and Architecture, **10** (175), 4-9 (2012)
10. P. Panchenko, T. Bobrova, The assessment of the duration of the installation work in the construction of buildings on the basis of the valuation technological operations (Omsk, 2016)
11. A. Bronevitsky, S. Bronevitsky, Prydniprovs'ka State Academy of Civil Engineering and Architecture, 11-12 (164-165), 40-45 (2011)
12. S.A. Bolotin, A.Kh. Dadar, I.S. Ptuhina, Engin. and constr. magazine, **7**(25), 82-86 (2011)
13. YU. Kontorchyk, B. Zakorko, Prydniprovs'ka State Academy of Civil Engineering and Architecture, **4** (205), 55-61 (2015)
14. SNiP 1.04.03-85. The norms of the duration of construction and the backlog in the construction of enterprises, buildings and structures.

https://doi.org/10.1051/e3sconf/20183303055

15. Regional standards for the duration of construction of buildings and structures in the city of Moscow (SUE "NIAC", 2007)

Economics of high-rise construction: the feasibility of skyscrapers building in the Russian cities

Artur Petrov[1*] and *Daria* Petrova[2]

[1]Tyumen Industrial University, Melnikaite street, 72-221, Tyumen, Russia
[2]Federal University named after the first President of Russia B.N. Yeltsin, Ekaterinburg, Chapaev street, 16a-511, Ekaterinburg, Russia

Abstract. The article considers the economic aspects of constructing high-rise buildings in the world and in Russia. Data on the number of high-rise buildings in Russian cities with a million population are presented. It is proved that interest in high-rise construction in Russia has been formed only in Moscow and partly in St. Petersburg and Yekaterinburg. The analysis showed that the reason for this is the expensiveness of high-rise construction. According to the enlarged macro-calculation, the cost of building 1 m^2 of the area of the Federation Towers complex (Moscow City) is about 2710 \$/m^2. Practically a possibility of return on investments in the foreseeable time interval exists only in Moscow. For the regions of Russia this task is rather complicated. Population density in regional Russian cities is quite low, business entities do not have the necessary financial resources for investing in high-rise construction, and investments from abroad absent.

1 Introduction

High-rise construction has become fashionable in the last 30-50 years practically all over the world. Projects of buildings with a height of more than 1000 m are no longer surprising. According to [1] today in the world there are 8 non-technical buildings with a height of more than 500 m; 23 buildings with a height of more than 400 m; 236 buildings with a height of more than 300 m; 1317 buildings with a height of more than 200 m; 4238 buildings with a height of more than 150 m. These facts form an opinion on the practical and economic expediency of building skyscrapers. However, the Russian experience does not confirm this conclusion. In fact, by the end of 2017, only 40 buildings with a height of more than 150 m (less than 1 % of analogues in the world) have been built or are in the active phase of construction in Russian cities. What are the reasons of this situation? As a working hypothesis of the study, the results of which are presented in the article, the economic entities' lack of economic interest in the high-rise construction in Russian cities is considered.

2 Methods

* Corresponding author: artigpetrov@yandex.ru

2.1 Initial assumptions

The construction of skyscrapers and urban population density are weakly connected with each other. It is known that the most densely populated city in the world is Manila (Philippines). The average population density in this city is 41511 people/km^2, and in Manila City (9th district) this indicator reached 71263 people/km^2 in 2015 [2]. In the world there are a number of cities with an average population density of more than 20,000 people/km^2 (Mumbai, Dhaka, Kathmandu, etc.), but there are practically no skyscrapers in these cities. At the same time, there are cities in the world that have many high-rise buildings and the population density is very high in all these cities.

2.2 Factors contributing to the choice of high-rise building strategy in the city

The works [3] clearly indicate that the most important reason for constructing high-rise buildings in the city is the desire of developers to attract «hot money» of investors to the city. So it was in American cities at the dawn of high-rise construction. Now the same situation is repeated in the cities of South-East Asia, Near East and Middle East. Another significant factor is the rapid growth of the economy, which requires huge volumes of office and residential construction, preferably of high quality and with a high level of individuality [4]. High-rise construction meets these basic requirements.

2.3 High-rise construction in Russia

As at December 2017, in Russia only 40 buildings with a height of more than 150 m have been constructed or are in the active construction phase [5] (36 of them in Moscow, 3 in Ekaterinburg and 1 in St. Petersburg). Moscow and partly St. Petersburg and Yekaterinburg are leading in high-rise construction in the country (Table 1).

Table 1. Statistics of high-rise buildings (more than 20 floors) in Russian cities with a population of more than 1 million people [5]

City	Number of buildings with more than 20 floors	The total number of floors in buildings with more than 20 floors
Moscow	1039	25712
St. Petersburg	553	13124
Yekaterinburg	217	5411
Novosibirsk	117	2818
Samara	75	1775
Perm	70	1640
Krasnoyarsk	67	1593
Rostov-on-Don	58	1280
Kazan	41	921
Ufa	41	920
Voronezh	34	809
Chelyabinsk	29	658
Nizhny Novgorod	24	556
Volgograd	16	385
Omsk	2	43

2.4 Establishment of a statistical relationship between the height of tall buildings and the fiscal capacity of Russian cities

In order to form a general idea of the relationship between the city's economy and the demand for high-rise construction, we will use regression analysis methods of the connection between the predictor and the resultant (Table 2). As a predictor (influence factor) we used the indicator of fiscal capacity; as a resultant - the characteristics presented in Table 1.

Table 2. Fiscal capacity of Russian cities with a population of more than 1 million people

City	Populatio n, people	Population density, people/km^2	Municipal budget (2017), bln. rub.		Specific budget expendi-tures (2017), rub./pers.
			Incomes	Outcomes	
Moscow	12380664	4833	1996.0	2211.0	178.58
St. Petersburg	5281579	3764	502.87	572.69	108.43
Novosibirsk	1602915	3170	36.13	36.92	23.03
Yekaterinburg	1455904	3111	38.10	39.10	26.86
Nizhny Novgorod	1264075	3078	29.65	32.72	25.88
Kazan	1231878	1915	20.28	20.97	17.02
Chelyabinsk	1198858	2262	30.24	30.64	25.56
Omsk	1178391	2079	15.43	16.23	13.77
Samara	1169719	2160	26.13	27.97	23.91
Rostov-on-Don	1125299	3229	34.91	37.42	33.25
Ufa	1115560	1576	26.11	26.25	23.53
Krasnoyarsk	1082933	2765	28.73	29.35	27.10
Perm	1048005	1310	24.50	25.50	24.33
Voronezh	1039801	1743	19.02	21.06	20.25
Volgograd	1015586	1182	20.37	20.57	20.25

Regression models constructed according to the data of Table 1 and 2 are shown in Fig. 1. The statistical relationship exists ($R^2 = 0.911$ for Fig. 1A and $R^2 = 0.902$ for Fig. 1B).

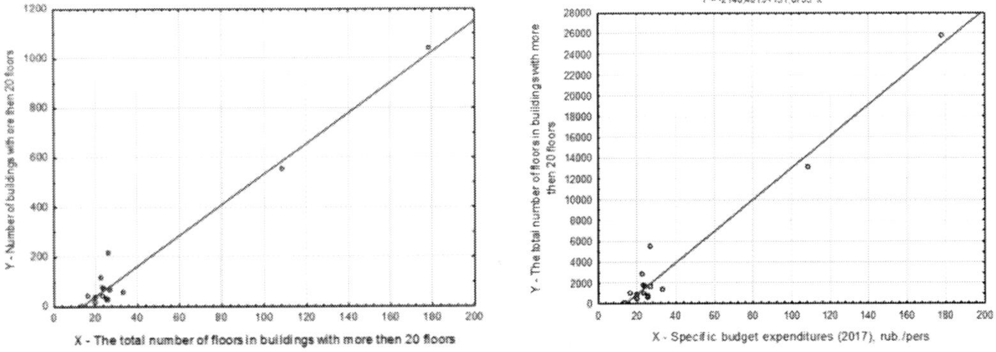

a) Number of buildings with more than 20 floors in city

b) The total number of floors in buildings with more than 20 floors in city

Fig. 1. Influence of fiscal capacity on the investment attractiveness of Russian cities with a population of one million people from the position of high-rise construction

It can be argued that high-rise constructing has chances for development in cities, fiscal capacity of which is relatively high. However, this is not a final, but only an auxiliary conclusion, leading to the main idea of research.

2.5 Ideology and research tasks

Using the example of Moscow International Business Center (Moscow City) [6], it is necessary to consider the ideology of formation of investment attractiveness of building brand-new high-rise constructions cluster. The business center management sources told that $ 12 bln. have already invested in MIBS [7]. According to the chief architect of Moscow S. Kuznetsov, to end the building $ 3 bln. more should be invested. The final cost can make up to $ 15 bln. How comparable are these costs to global analogs and are there any chances to recoup investments within a reasonable time?

For the purpose of proving the hypothesis of the lack of economic entities' economic interest in the management of high-rise construction in the Russian cities, three tasks were solved:

- Ascertainment of the structure of the cost of construction the 1 m^2 area in different countries of the world;
- Comparison of the results of the enlarged macrocalculation of the cost of the construction 1 m^2 area of the MIBC with world analogues;
- Assessment of the possibility of the MIBS construction project recoupment by the attracted business.

3 Results

3.1 Ascertainment of the structure of the cost of construction the 1 m2 area in different countries of the world

Fig. 2 presents data on the existing cost ranges for the construction of 1 m^2 in high-rise buildings in London [8], the cities of the UAE, Shanghai, Melbourne, New York. Note that the regional gap in the cost of construction can be threefold [9]. Fig. 3 shows the structure of the cost of high-rise construction of 1 m^2 area in London (4100 $/$m^2$), Riyadh (2900 $/$m^2$), Shanghai (1300 $/$m^2$).

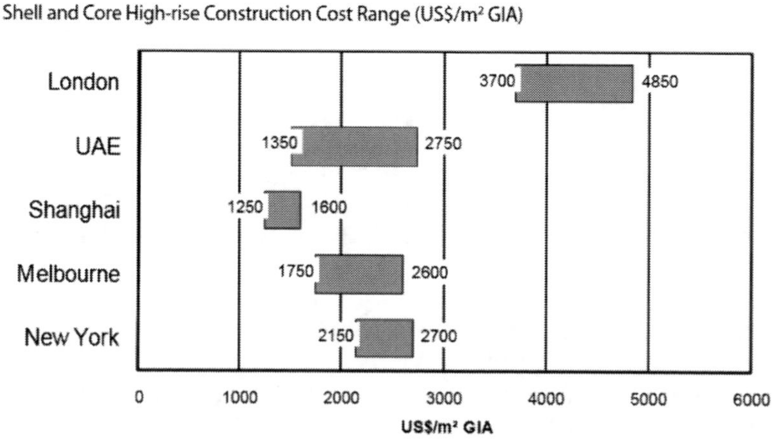

Fig. 2. Differences in the cost of high-rise construction in the cities of the world [9]

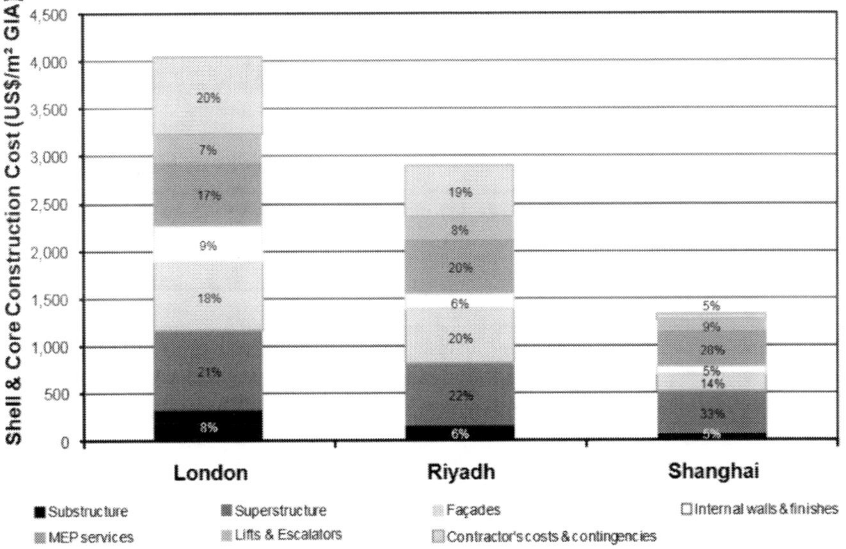

Fig. 3. Typical Elemental Build-up of Shell & Core Construction Costs for landmark High-rise Office Buildings in Europe, Middle East & Far East [9]

3.2 Comparison of the results of the enlarged macro calculation of the cost of the construction 1 m2 area of the MIBC with world analogues

The most expensive project in «Moscow City» was the two towers of the Federation Towers complex, whose total cost of construction amounted to 1.2 billion dollars. The total area of Federation Towers is 442915 m^2. The estimated cost of 1 m^2 Federation Towers is $ 1.2 bln. / 442915 m^2 = 2710 $/$m^2$. Table 3 presents the results of a comparison of the complex Federation Towers construction economy with analogues in the UAE, Taiwan, USA. Analysis of Table 3 shows that the estimated cost of building 1 m^2 area of Federation Towers (Moscow) is close to the upper limit of the cost range for the analogues in the UAE and the USA.

Table 3. Comparison of the estimated cost of building 1 m^2 area of Federation Towers (Moscow) with the cost of building 1 m^2 area of analogous high-rise buildings in the UAE, Taiwan, USA

Country, city, high-rise building	Height to the roof, m. (without spire)	Floors amount	Total cost of building, $ mln. (year)	Total area of skyscraper premises, m^2	Estimated cost of construction 1 m^2 area, $/$m^2$
Russia, Moscow, «Federation Towers»	373.8	93	≈ 1200 (2016)	442915	≈ 2710
UAE, Dubai, «Burj Khalifa»,	584.5	163	≈ 1500 (2010)	309473	≈ 4847
Taiwan, Taipei, «Taipei 101»	438.0	101	≈ 1700 (2004)	357719	≈ 4752
USA, New York, «Empire State Building»	373.1	102	40.949 (1931) or 499.577 (2017) considering $\$1_{1925}$ = = $\$12.20_{2017}$	208879	196.1 (2031) or 2391.2 (2017) considering $\$1_{1925}$ = = $\$12.20_{2017}$

3.3 Assessment of the possibility of the MIBS construction project recoupment by the attracted business

Table 4 presents some known data [10] describing the business of the owners of Moscow International Business Center (MIBC) high-rise buildings.

Table 4. Information about some objects of MIBC

Data	Some high-rise buildings, included in MIBC			
	Federation Towers	OKO	Mercury City Tower	City of Capitals
Owner	AEON (R. Trotsenko)	Capital Group (V. Doronin, P. Te, E. Berman)	Mercury Development (I. Kesaev)	Capital Group (V. Doronin, P. Te, E. Berman)
Estimated cost	$ 1,2 bln.	$ 0,7 bln.	$ 0,7-1,3 bln.	No data
Total area, thousands m^2	443	250	174	288
Office area, thousands m^2	287	115	89	72
Residential area, thousands m^2	78	135	22	101
Number of apartments	212	588	137	470
Rental rate	$ 400-650	$ 540-560	$ 400-430	$ 650-940
Vacancy	8 %	70 %	55 %	7 %
Estimated cost of Sale, $/$m^2$	No data	360 thousands rub.	$ 8000	$ 13000
Purchasers	VTB Bank, Sberbank of Russia, Mercury	Moscow City Hall	JTI, NorNickel	Renaissance Credit, Capital Group

Taking into account the data from table 4, the following conclusions can be drawn:

- The various buildings of MIBC have different investment attractiveness for the Moscow business.
- Having an estimated construction cost of 1 m^2 area of about 2700-2800 $/$m^2$ and selling areas at $ 6000-13000 $/$m^2$, the owners have the opportunity not only to compensate the investment expenses, but also to make a profit.
- At the same time, information about vacancies leads to the conclusion that the refund of investments will be long and difficult.

4 Discussion

The article [11] provides information on the structure of the foreign direct investment distribution in the Russian economy (Table 5). The data of the Central Bank of Russia illustrate the thesis that only Moscow is an object of investors' interest (including Russian investors, who return finances from offshores). The main focus of investment interest, that also includes the field of high-rise construction (up to 68 % of the total investment in Russia), is concentrated in Moscow. Foreign investors are significantly less interested in St. Petersburg (5-8 %). The Sverdlovsk oblast (the regional center - Yekaterinburg) received in 2011-2015 only 0.55-3.21 % of foreign direct investments. And this despite the fact that Yekaterinburg is the third city in the country by the number of high-rise buildings. Probably, this fact confirms the thesis that there is practically no economic interest of economic entities in the management of high-rise construction in Russian cities. It is also important that the density of people in Russian cities is low enough (table 2), and investors are not interested in investing in projects that are attractive from a position of marketing, but ineffective in terms of capital productivity.

Table 5. Regional structure of foreign direct investments inflow (part of the table [11])

Federal Districts (FD) and Federal Subjects of Russian Federation	2011	2012	2013	07.2014… …06.2015
	Regions shares in foreign direct investments inflow, %			
Russian Federation	100.0	100.0	100.0	100.0
Central FD	73.44	59.21	62.51	64.51
Moscow	68.01	51.27	55.82	53.98
……………………………………………				
North-West FD	6.91	12.89	10.66	11.47
St. Petersburg	4.48	7.84	6.74	5.63
……………………………………………				
Ural FD	8.94	14.27	8.13	7.42
Sverdlovsk oblast	0.55	3.21	0.56	0.69

5 Conclusions

According to the results above we can formulate the following conclusions.

1. By the end of 2017, only 40 buildings with a height of more than 150 m (less than 1% of analogues in the world) have been built or are in the active phase of construction in Russian cities; most of them in Moscow, St. Petersburg and Yekaterinburg because of the high level of fiscal capacity.

2. The cost structure of building 1 m^2 area in different countries of the world significantly varies. The Russian experience of high-rise construction (on the example of Federation Towers in MIBC) is close to analogues in the UAE and Saudi Arabia.

3. The estimated cost of building 1 m^2 of Federation Towers in MIBC is 2710 US $\$/m^2$.

4. The possibility of recoupment of MIBC construction project by the involved business exists in theory, but is very difficult in practice.

5. Regional high-rise construction projects in Russia have few chances of development. The reason is a large volume and a continuous refund of the required investment.

References

1. http://www.skyscrapercenter.com/ (Accessed 31.10.2017)
2. http://psa.gov.ph/content/philippine-population-density-based-2015-census-population (Accessed 2.11.2017)
3. http://global.ctbuh.org/resources/papers/download/1269-building-cost-and-eco-cost-aspects-of-tall-buildings.pdf (Accessed 7.11.2017)
4. P. de Jong, and S. Van Oss. Hight Rise Ability. In *14th Annual European Real Estate Society Conference*. ERES: Conference. London, UK, (2007).
5. http://skyscraperpage.com/diagrams/?cityID=14 (Accessed 13.11.2017)
6. http://citymoscow.ru/ (Accessed 14.12.2017)
7. https://www.bfm.ru/news/290715 (Accessed 14.12.2017)
8. D. Langdon, S. Watts. Building, **6** (2002)
9. S. Watts, D. Langdon, CTBUH Journal, **3** (2010)
10. http://www.forbes.ru/kompanii/infrastruktura/infographics/311921-gorod-ambitsii-komu-prinadlezhit-moskva-siti (Accessed 4.01.2018)
11. Kuznetsova O. International processes. Issue 14. № 3 (in Russian)

Leasing instruments of high-rise construction financing

Olga Aleksandrova[1,*], *Elena* Ivleva[2] , *Viktoria* Sukhacheva[2], and *Anna* Rumyantseva[2]

[1] Military Educational-Scientific Center of the Air Force of "The Air Force Academy named after Professor N.E. Zhukovsky and Yu.A. Gagarin", 1, Marshala Zhukova str., Syzran, Russia
[2] Private Educational Institution "St. Petersburg University of Management Technologies and Economics", 44, Lermontovsky Prospekt, St. Petersburg, Russia

Abstract. The leasing sector of the business economics is expanding. Leasing instruments for high-rise construction financing allow to determine the best business behaviour in the leasing economy sector, not only in the sphere of transactions with equipment and vehicles. Investments in high-rise construction have a multiplicative effect. It initiates an active search and leasing instruments use in the economic behaviour of construction organizations. The study of the high-rise construction sector in the structure of the leasing market participants significantly expands the leasing system framework. The scheme of internal and external leasing process factors influence on the result formation in the leasing sector of economy is offered.

1 Introduction

The replacement of fixed assets is required to ensure sustainable economic growth. There is a problem of investments accessibility for financing the enterprises of the real economy. Considering that investments in high-rise construction, along with investments in the development of the industrial sector and the transport system, have a multiplicative effect, the solution of this problem is also associated with the growth of the final economic effect (result).

The most effective way to finance a growing economy is leasing. The current policy of import substitution enables the leasing sector to supply equipment for the producing units retooling, which stabilizes both the leasing sector and the economy as a whole.

Today we can state: leasing financing is successfully developing and increasing the power. In 2016, the amount of new contracts increased by 39% and reached a volume of 1.2 tn rub. [6]. Leasing is gaining increasing recognition in the regions. In 2016, regional transactions increased by 33% and amounted to 375 bln. rub.; leasing transactions are becoming larger. The average transaction amount increased compared to 2015 from 6.7 to 8.2 mln. rub. The share of leasing business in GDP reached 1.2% that corresponds to the pre-crisis level. Leasing business has a steady tendency to increase its portfolio, compared with investments in fixed assets, which have a negative dynamics (-3%) according to the Ministry of Economic Development [12].

* Corresponding author: post_graduate@mail.ru

Despite the positive dynamics in the leasing sector, there are bottlenecks; the share of bad debt is one of them [11]. In 2016, it was 10%, which is 2% lower than the previous year. The result of debt reduction was achieved through the withdrawal of equipment from insolvent debtors that leads to new problems for the leasing company (loss of financial assets, deterioration in the quality of assets, the problem of the seized equipment sale, etc.). In 2016, the volume of restructured debt increased to 20% [13].

Methodology and theory

There is still no special regulation of the leasing services market. Leasing has a direct impact on investment and national wealth, which determines the importance of reform for economic development. By the end of 2017 it is expected to adopt a law on the regulation of leasing companies that should take effect next year. The Central Bank is ready to develop guidance on the methodology for calculating capital, industry standards of accounting that are associated with amendments to the law on leasing [8]. The main activities are designed to transform management standards in leasing, improve the risk management and operational environment of leasing companies, which will raise investor confidence and improve the business environment in the leasing sector. It is expected that the reform will positively affect the volume of the leasing market and will bring leasing to a new level of development. Full and transparent regulatory norms will improve the efficiency of the leasing sector and ensure economic mutual benefits for all participants of the leasing process [5].

The effectiveness of the leasing economy sector reform largely depends on the synchronization of innovations. A systematic approach when implementing reform measures will achieve the best result. Management in the leasing economy sector is advisable to consider on the basis of a systematic approach that allows to review the situation comprehensively, to study not only individual elements but their totality. The system approach allows to find the source of problems both inside the system and in the external environment. Existing links in the system allow to transfer the properties of an individual element entering the system to the rest of its elements. The effective functioning of the system depends on the effective interaction of its elements, which consists in achieving the stated goal and result, both of each element of the system, and of the system as a whole [4].

The main goal, which is set in the system design, is the resolution of emerging problems related to its functioning or the influence of the external environment on it.

The leasing process is supposed to be viewed as a set of economic entities engaged in a certain type of business and joined with transaction links. The communication of the participants is quite complex and constantly forms new relationships. Business in the leasing economy sector is characterized by a state of emergence, when the overall effect of the leasing process does not correspond to the sum of the effects of its participants [1].

The leasing process is based on the division of labor, which leads to the dispersal of entrepreneurial functions and the complication of the process itself. It is expressed in the autonomy of the leasing process participants, who are authorized to make decisions independently, carry out entrepreneurial activities within the established framework that affects the efficiency of not only individual participants, but also the entire leasing economy sector. To improve the performance of the leasing economy sector is possible when integrating the goals of the leasing process participants, taking into account the fact that the substantive side of the participants' activities does not change.

To implement the system approach in the management of the leasing economy sector, including the field of high-rise construction, it is necessary to clearly represent the elements of the system, the relationships between them, their goals, objectives and principles of operation [3]. Currently, the leasing system is formed spontaneously, but in the process of its functioning, in search of effective options for interaction between the leasing transaction participants, a new paradigm is created. The goal of building a leasing system is to increase the effectiveness of each participant of leasing process and leasing business in general, by reducing the risks and uncertainties of the process.

Leasing business is oriented towards providing services to economic entities and involves many participants to its implementation. Leasing business is defined as a set of independent organizations interacting with each other, engaged in various entrepreneurial activities, including leasing, investment, trade relations, and placed under the obligations for the purposeful use of financial resources for the acquisition of the leased asset with the aim of making a profit [3].

The system of leasing business is complex, because each participant is an independent economic unit engaged in its own kind of entrepreneurship; each has its own goals that can be contradictory, its tasks, and each one achieves them in different ways. A mismatch of interests can lead to devastating consequences. The defining condition for the leasing system development is the unity of business objectives. The actions of entrepreneurs are aimed at meeting market demand. The profit that entrepreneurs expect to receive from their activities is an indicator of the resource use effectiveness.

The leasing system is an economic system that has adaptive properties in the process of performing a complex of leasing operations in transport, industry and construction, which is formed from several subsystems with a high degree of interaction between themselves and the external environment [2].

Management performs various functions in the continuous process of entrepreneurial activities. We have identified the following functions in the process of managing leasing activities:

1. Organization of leasing process. Shows principles of the leasing process participants interaction as elements of the system and determines the order of their operation.

2. Planning and forecasting of the leasing process. Determines the sequence of actions to achieve the desired and possible development options. This function is based on increasing equity capital, building up the norm of profitability, coordinating the volume of operations.

3. Regulation. The measures that determine the leasing process framework due to compliance with the laws and regulations as a rule.

4. Motivation of activities. The basis is the argumentation in favor of a decision that motivates the leasing participants to certain actions to achieve the goal, both personal and general.

5. Stimulating the leasing business development. We have in mind the existence of reasons that would lead to actions resulting in the leasing economy sector development. External (macrolevel) reasons caused by long-term events and not dependent on the decision of the leasing transaction individual participant can be identified, as well as internal (microlevel) ones that have arisen inside the leasing system, i.e. between the participants as a result of their actions.

6. Accounting and control. It reflects the results of the leasing transaction participants activities in a documentary manner and verifies the correctness of the use of the relevant provisions and statutory acts provided for by law.

The leasing process management is connected with the final financial and economic result of the transaction sites activities, which depends on a variety of particular factors [1].

Defining the strategy for managing leasing business, it is necessary to strive to achieve the following goal: the leasing process implementation should be aimed not only at creating additional capital, but also at preserving the base capital of all leasing participants.

To make a decision in conditions of uncertainty, the following rules should be followed:

- not to risk more than own capital can allow, guided by the framework of the capital structure theory;
- always think about the chances and consequences of risk;

- a positive decision is taken only in the absence of doubt;
- not to throw a whale to catch a sprat;
- always evaluate the options [9].

The effective functioning of the leasing sector largely depends on the environment in which the leasing process takes place. Carrying out a leasing deal, participants are exposed to both internal and external factors, which has a certain effect on the outcome of both the individual leasing transaction participants and the leasing economy sector.

We propose a scheme (Figure 1) of the influence of external and internal factors of each economic subject (participant) of the leasing process on the formation of the result in the leasing sector.

$B^{n.c}$ - the amount of effective demand.

$B^{h.m}$ - state of the scientific and technical products market.

B^{κ} - growing competition.

$B^{p.l}$ - leasing development opportunities.

$B^{f.r}$ - the financial market state.

$B^{\kappa.n}$ - credit policy.

$B^{e.p}$ - economic and political situation in the country.

$B^{n.p}$ - legal regulation.

$B^{y.u}$ - rate of inflation.

$B^{h.n}$ - tax policy.

B^{u} - infrastructure development.

$P^{h.f}$ - availability of financial resources.

$P^{n.o}$ - the production equipment state.

$P^{n.n}$ - product parameters.

$P^{m.p}$ - methods of business regulation.

$P^{f.n}$ - financial potential of the company.

$P^{d.n}$ - business policy and management of the organization.

$P^{p.p}$ - efficient allocation of resources.

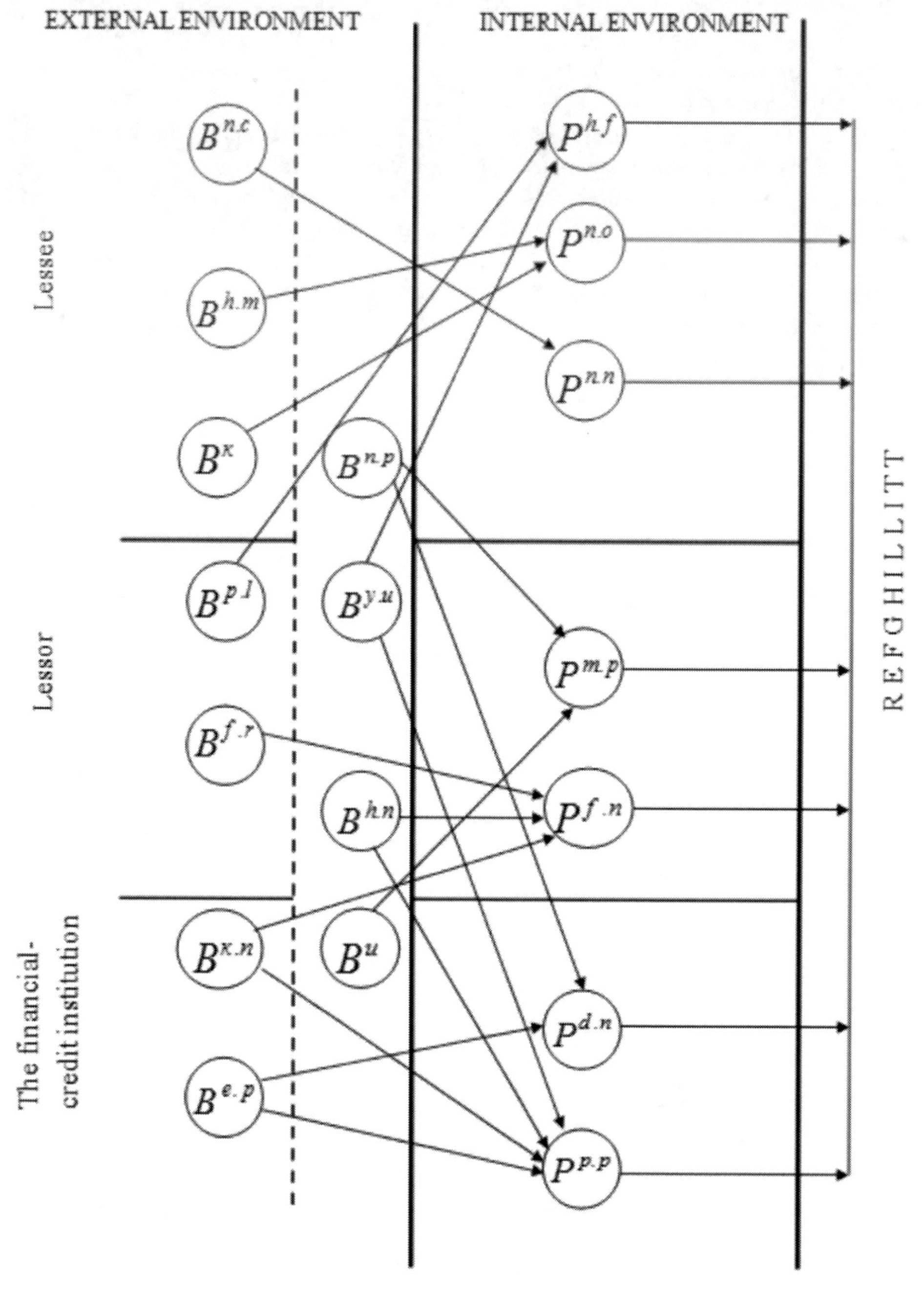

Fig. 1 Factors of external and internal environment of the leasing business participants, affecting the result.
Prepared by the authors.

The formation of internal factors of the leasing process participants and their financial results are influenced by external factors [15]. The leasing participants feel the influence of external factors as well as factors of an individual nature. The internal environment factors are determined by the entrepreneur on the basis of an external environment analysis and are used to establish the boundaries of the business structure functioning. Connections between the leasing system participants exist; they can be material (relations with respect to the leasing object), monetary (settlements between participants), informational (receipt, exchange of information). Internal links between the leasing participants have a stronger manifestation than external ones, they are of a cyclic nature, as the flows move progressively. The links between the leasing process participants can be direct, reverse, indirect.

The effectiveness of the leasing market functioning depends on the management system of both a separate leasing participant and the leasing process as a whole. Reducing the impact of external and internal factors to the leasing process functioning is aimed at solving two problems:

- ensuring of leasing business effective development;
- reduction of entrepreneurial activity participants risks [11].

To reduce financial losses, leasing companies are increasingly turning to risk management. The quality of risk management in the leasing market is low. Risk management is based on the lessee analysis, assessment of the subject of leasing and control over the process of transferring leased property. The advance amount is at the heart of reducing the possible risk. The leasing process participants need to know the actual cost of the risk, which is expressed in the actual loss of the lease participants or the costs of reducing, compensating for the losses that have occurred [10].

The leasing process participants act in concert in time, interact on a single vector, i.e. the effect caused by the coordinated actions in space and time of various economic entities leading to qualitative changes refers to the synergistic effect, which should lead to a qualitative breakthrough in the functioning of the system.

The leasing system is dynamic, which is confirmed by its change in time and space and is characterized by the following properties: stimulation, functionality; variability in time; existence in a changing environment.

The leasing system is exposed to the external environment, i.e. there are certain processes at the entrance to the system that affect it and lead to a change in the behavior of participants in the system. This property is called stimulation.

Functionality of the leasing system is considered as its behavior in the external environment and the dependence of processes on the output of the system. The functions of the leasing system are determined by the multiplicity of exits. The leasing participants use the system for their own purposes, evaluate its functions and arrange them in relation to their needs, as a result of which the functions are of importance: from the main to the secondary.

The main property of the dynamic system is its variability in time. As a rule, leasing deals are tied up not for one year, besides there is a tendency to lengthen the term of the deal. The terms of the leasing deal are the boundaries of the leasing system. Therefore, when organizing the leasing process, it is necessary to take into account possible changes during the validity of the leasing contract that affect the functioning of the leasing system and respond to them in a timely manner [3].

Changes that can occur at different rates affect the internal parameters of the system first of all. Positive changes cause the leasing system development. The emergence of a new participant in the leasing system leads to a change in the structure of the system, which affects its properties and leads to the growth of the system. So consideration of construction in the structure of the leasing market participants essentially expands the scope of the system.

Discussion. Financing leasing activities in the field of high-rise construction, studying the best economic practices in this area expands the scope of research on leasing activities. A special direction in this sphere was identified by V. Gazman – securitization of leasing assets in the conditions of reducing the volume of leasing financing by banks and complicating the conditions for crediting lessors [7].

The system exists in a continuously changing environment, which leads to a change in the system itself. To survive in such conditions, the system must adapt to external changes. Internal changes must be more active than external ones. And here it is necessary to compare the rate of change within the system with the rate of change in the external environment. In the leasing system, the rate of internal changes should be higher than the rate of external changes, which will reduce their negative development and increase the adaptability of the system.

The processes at the entrance to the leasing system have a stimulating character when it is organized and developed. They facilitate the integration of economic entities into a system on the basis of a leasing agreement. Most of the movement occurs within the system itself between its participants.

The greater the dynamic complexity of the system, the longer it takes for the signal to pass through the structure. One delay can lead to a significant delay in the signal [8]. Therefore, in order to change the situation in the right direction, it is necessary to take into account the discrepancies in the time of receipt of the signal and the result. To be able to change the situation, it is necessary to clearly understand the processes occurring in the leasing system.

The system growth and development lead to the emergence of new links between the elements. Additional relationships are formed as a result of the appearance of new elements in the system. Such connections can grow exponentially, i.e. each subsequent element increases the number of links to a greater extent than the previous one [8]. The appearance of a new system member leads to a change in the requirements for the final state of the system, which is not compatible with the previous one.

The conclusion

The connection between the elements (participants) of the system forms the system structure and serves as a guarantee of its stability. When the stability of the leasing system is violated, there are divergent processes that cannot be controlled and lead to the disintegration of the system.

Any system has a growth limit. When the optimal value is reached, when the system functions with maximum efficiency, the subsequent growth can lead to a decrease in efficiency. An uneven load of the system participants can provoke the occurrence of such a situation. At the point of effective functioning, it is necessary to determine the limitations the system can meet and take measures to avoid the growth limit. The applied system approach to the leasing process makes it possible to consider the management of interrelated and interdependent participants, which influence the efficiency of leasing business in the field of high-rise construction.

References

1. O. Alexandrova, O. Burgonov, E. Ivleva, Leasing tools of business economics. Russian scientific journal. Economics and management **5**, 45-48 (2016)
2. Alexandrova O., Ivleva E., N. Shashina, Leasing mechanisms of business economics. Coll. of scient. papers of II International Research and Practical Conference Modernization of the Russian economy. Forecasts and Reality, 14-19 (2016)
3. O. Alexandrova, O. Burgonov, E. Ivleva, N. Shashina, Management of developing the leasing sector of entrepreneurial economy. Procedia Engineering **165**, 980 – 989 (2016)
4. O. Aleksandrova, E. Ivleva, A. Kirdyashkin, N. Shashina, Infrastructural development factors of leasing entrepreneurship in real sector of economy. IOP Conf. Series: Earth and Environmental Science, **90** (2017)

5. O. Adamchuk, The Central Bank mitigated the reform of leasing. Newspaper Vedomosti, (2017)
6. R. Bulatov, Leasing in Russia: restoration after stress. Leasing Revue **3**, 2-15 (2017)
7. V. Gazman, Overcoming of stereotypes in leasing. Questions of economy **2**, 136 (2017)
8. I. Drogobytsky, System analysis in the economy. 2nd ed. UNITY-DANA Publ., 423 (2011)
9. S. Moiseev, Reform in leasing: how to fix an important mechanism for the economy. RBC, 79 (2017)
10. A. Shapkin, V.Shapkin, Theory of risk and modeling of risk situations: Textbook. Moscow, Dashkov and K Publ., 880 (2005)
11. D. Shperber, Organization of the leasing company management. Leasing business technology **3**, 7-12 (2017)
12. The first leasing broker [Electronic source]. Available at: http://1lizing.ru (accessed 15.10.2017)
13. Federal Service of State Statistics. Available at: http://www.gks.ru/free_doc/doc_2015/invest.pdf (accessed 15.10.2017)
14. "Expert" Rating Agency. Available at: http://static2.banki.ru/ugc/4f/c4/05/59/Rynok_lizinga_po_itogam_2016_goda.pdf (accessed 22.10.2017)
15. A.Rumyantseva, Problems of financing the activities of small enterprises. Transactions of St. Petersburg University of Management Technologies and Economics **2**, 41-52 (2015)

 https://doi.org/10.1051/e3sconf/20183303058

Procedure of the preparatory works execution during the remediation of high-rise buildings debris due to the disaster

Denis Sinitsyn[1*]

[1] Moscow State University of Civil Engineering, Yaroslavskoe shosse, 26, Moscow, 129337, Russia

Abstract. The article covers one of the preparatory works types, in particular – the debris clearance. The types of machines and machinery, the purposes of the debris clearance, as well as the tasks for decrease of debris mass and scope are specified herein. The purpose of the article is to show the current level of the domestic machinery, as well as to share the experience of preparatory works. The article describes one of the preparatory works types, in particular – the debris clearance during the high-rise buildings demolition. Given are the equipment and machinery types, covered are the purposes and procedure of debris clearance defined are the tasks on reduction of debris mass and sizes. Specified are the types and methods of the preparatory works execution. The purpose of this article is to demonstrate the domestic equipment current state, and to share the experience in preparatory works execution.

1 Introduction

Nowadays the main task of preparatory works, under the conditions of new facility construction, is the development of required and safe conditions for the main works execution. To execute the preparatory works, upon management decision in some cases, developed is the project documentation concerning their execution procedure. The following shall be reflected in the work execution design: types of the preparatory works and their order of priority; the shortest and the most safe traffic routs through the facility territory, access ways to the certain buildings, facilities as well as the areas of sites development for the machinery arrangement; methods of walls demolition or dismantling, of the debris moving; technical solutions on the temporary fastening arrangement of the certain structures and elements, threatening to collapse failure; technical solutions on the equipment protection [1,2,3,4,5,6,7].

The preparatory works are divided at the site of their execution into the off-site works executed beyond the company territory, and the intrasite ones executed at the company territory. The preparatory works are divided into the following types:

- debris clearance;
- demolition of buildings and theirs parts;

* Corresponding author: slavinaay@mgsu.ru

- temporary fastening of structures, threatening to collapse failure;
- equipment protection;
- use of materials from the debris clearance and storage of reusable materials and structures.

The debris clearance is executed for purposes of the site preparation for further reconstruction works at the facility, of remaining equipment withdrawal, as well as for assessment of damage and the opportunity of use of remaining structures, foundations, channels and utilities.

2 Materials and Methods

Depending on the task in hand the debris is cleared in part or in whole. The partial clearance is executed during the development of access ways and areas for the machinery arrangement and manipulation, as well as during the essential industrial equipment withdrawal.

The debris clearance in whole is executed during the territory clearing for further reconstruction. The general principle during the debris clearance is the works execution downwards and in all directions, where it is possible to clear the debris; if there is the insufficient quantity of such directions it is required to develop them intentionally.

It is recommended to clear the debris by straight-line flow method, which allows putting into effect promptly the available workforce and means, and using them in the most complete and smooth manner for the duration of works, as well as widely applying for overall mechanisation. To execute the debris clearance successfully along the separate directions, it is reasonable to develop the complex groups equipped with lifting machines, bulldozers, transport means, machines and apparatus for breakdown of shatters of structures and equipment, auxiliary and service machines. The group accepts the tasks and limits of works per shift, per day or several days. When executing the works in shifts, it is set the task for a day shift to clear the debris and dispose the structures, and for the night shift-to develop the working area for the day shift [8,9,10].

The development of access ways and areas is executed for the purpose of free access directly to the work areas and for arrangement of mechanical means.

The access ways are arranged as single-line ones of 3-3,5 m in width, with siding lines every 200-250 m as the areas of 2,5-3,0 m in width and of 15-20 m in length.

The clearance is executed by means of the shatters displacing out of the access way using the heavy hydraulic-motored bulldozer of work equipment. The operation of the single bulldozer is possible only in the debris consisting mainly of brick shatters. The making of access ways in the debris of mixed type with high content of reinforced concrete components is possible only by applying the complex of machines. The complex contains 3-4 machines, for example, one bulldozer with the capacity of 180-220 kW with ripper or with means for separation of the debris components, and two or three bulldozers with the capacity of 70-80 kW.

3 Results

The lead machine of the complex is the heavy bulldozer: it displaces, withdraws and removes the large-size shatters; the other bulldozers spread the access way for wheeled machines. In case of insufficient tractive force of even the heavy bulldozer, the heavy elements in debris are blasted or broken by the pneumatic tool. For these purposes, as a rule, the teams of construction machines are added with three or four persons with means for cutting of reinforcement and steel structures, as well as groups for works execution.

The debris clearance for purpose of the territory complete clearing shall be executed as follows:

- Firstly, withdrawn and removed are the certain beams, columns, roof slabs and other long-length elements raised from the debris;
- Large remnants of demolished reinforced concrete structures (beams, trusses, etc) are preliminarily free from the other elements of debris, are broken into the smaller parts and then removed from the debris;
- Large brick or concrete blocks are removed in whole, if it is allowed by a lifting capacity dimensions of construction machines and transport means; if it is impossible, then they are preliminarily broken and separated into the smaller parts;
- Small shatters and material debris are used for filling and planning of the territory or carted away beyond the facility territory and bulked in the specially designated area [11].

The machine sets for the debris clearance shall consist of the bulldozers, автокранов, excavators, tractor loaders, compressor station, pneumatic tool, machines with diamond tool, hydraulic wedges, etc.

As a first approximation, the addition of construction machines into complexes may be performed according to Table 1 (where given are the four groups of buildings and the appropriate machines complexes). These machines may be replaced by machines of equivalent types, if required and depending on their availability in the train.

The content of groups of industrial buildings and the list of machines, which may be used for the debris clearance, are given below.

Table 1. The four groups of buildings and the appropriate machines complexes

No. of group	Data on the building structure	Machines for selection of complexes for the debris clearance
1	Industrial single storey buildings with the height of 10-15 m Industrial multi-storey brick buildings with number of storeys: 2, 3, 4, 5	Bulldozers of traction category 3-4t: DZ-29, DZ-101A, DZ-42G, etc.
		Tractor loaders TO-18, TO-11
		Hydraulic shovels, 3-4 grade, with the shovel capacity of 0,5-0,65m^3: EO-4322A (MRO), EO-4121, EO-4221B
		Truck cranes with a lifting capacity of 4-10 t (the short-baseline ones are preferable): KS-3662A, KS-3571
		Tipping lorries (KRAZ-256B) with the lifting capacity of 5-7 t
2	Industrial single storey prefabricated buildings with bays of 12,18 m, with the height to the bottom of framework structures of 3,6-9,6 m, with mass of elements of 5-9 t, with the columns step of 6-12 m Industrial multi-storey prefabricated buildings with a storey height of 3,6; 4,8; 6 m, the bay width of 2, 3, 4 m, mass of elements of 4, 7, 12 t	Bulldozers of traction category 10-15 t: DZ-27(160 h.p.), D-35 (180 h.p.) with widener of blade
		Tractor loaders: TO-21, TO-24
		Tipping lorries with the lifting capacity of 7-16 t
		Hydraulic shovels, 3-4 grade, with the shovel capacity up to 1m^3: EO-4121, EO-3322A; EO-4125
		Truck cranes with a lifting capacity of 10-16 t: KS-5263, MKG-10A, MKT-25
		Tipping lorries with the lifting capacity of 7-16 t

3	Industrial single storey prefabricated buildings with bays of 18, 24, 30 m, with the height to the bottom of framework structures of 6-15 m, mass of elements of 8-16 t, with the columns step of 12 m Industrial multi-storey prefabricated buildings with the mass 7-12 t	Bulldozers of traction category 15-25 t: D-34, D-59, D-384, D-685, D-439A Tractor loaders: TO-21, TO-24 Hydraulic shovels, 4-5 grade: EO-4121, EO -5122, EO-5124, EO-5015A Truck cranes with the lifting capacity of 25-40 t: KS-5363, KS-5122, KS-6371 short-baseline, MKG-25, SKS-40 Tipping lorries with the lifting capacity of 9-25 t
4	Industrial single storey prefabricated buildings with bays of 24, 30, 36 m, with the height to the bottom of framework structures of 12-18 m, mass of elements of 15-35 t, with the columns step of 12 m Industrial multi-storey heavy-duty buildings of mixed type with the height of 15-35 m, mass of elements of 15-25 t	Bulldozers of traction category 25 t: DZ-34 Tractor loaders: TO-21, TO-24 Hydraulic shovels, 5 grade: EO-5122, EO-5124, EO-5015A Truck cranes with the lifting capacity of 40 t, 63 t Tracked jib cranes 40-100 t: SKG-40, SKG-100 Tipping lorries with the lifting capacity of 16-45 t

The calculation on number of machines and machinery for the debris clearance is based on the following:

- actual scope of works – mass of debris in tons or its volume in m^3;

-fixed time limits of clearance, under the specified shiftwork and shifts duration;

- production capacity of available mechanical means;

While calculating the actual scope of works it is recommended to use Table 2; the production capacities of mechanical means on the debris clearance are given in Table 3.

The volume weight of debris, its content and structure during the demolition of some types of buildings.

Table 2. The actual scope of works

Type of buildings	Volume weight of debris t/m^3	Content of shatters, weight %				
		Brick blocks up to 1m^3 brick rubble	Shatters of			Material debris
			Reinforce concrete and concrete structures up to 0,8m^3	Wooden structures	Steel structures (including the machining equipment)	
Brick production buildings	1,514-1,297	25	55	3	10	7
Brick residential buildings	1,087-0,931	50	15	15	5	15
Wooden buildings	0,478-0,399	10	-	80	-	10

Large-panel buldings	0,982-0,819	-	75	18	2	5

Table 3. Production capacities of mechanical means on the debris clearance

Name of works	Unit of meas.	Production capacity per working hour
Debris clearance by excavator with loading to transport means, with a shovel capacity, m^3: 0,3 0,5 1,0	m^3	12 18 40
Debris clearance with displacing of mass by tractor-bulldozers, type C-100 and T-141 at the distance of, m: 10 20 30	m^3	50 35 30
Debris clearance of access ways by bulldozers, with the width of access way of 3,5 m, with a medium height of debris, m Up to 0,3 m Up to 0,5 m	m	170-200 100-150
Transportation of blocks and material debris from the debris clearance by using the tipping trucks, with loading by excavator at the distance of, km: 1 2 3 5	m^3	5-8 4-6 3-5 2-3
Loading of blocks to transport means by mobile cranes. Demolition by tractor of buildings walls with the height of storeys: Up to 3 More	m^2	35 20

During the clearance of debris made by the brick buildings, the hydraulic shovel with convertible attachments (back or front shovel, pneumatic hammer or jack hammer) will take the lead in the complex. The shovel is used for clearance and loading of debris structure to the tipping trucks and for breakage of large brick blocks. The debris elements for loading will be displaced and grouped using by applying the bulldozer. In case of clearance of mixed type debris, consisting of demolished reinforced concrete and steel structures, at the first process stage the crane with the lifting capacity of 25-100 t will be held the key place, with the radius of working range up to 30-40 m. Such crane shall operate together with machine and equipment for cut through the reinforcement of reinforced concrete structures and steel structures. To decrease a mass and dimensions of debris shatters used are the methods equivalent to ones of buildings demolition or applied are the demolition means, specified in Table 4. The grinding is executed in situ or at the specific areas. The demolition means of the construction structure materials recommended during the preparatory works execution.

Table 4. The demolition means

Demolition means	Production capacity	Effect of demolition mean to a material	Process technology	Recommended complex of machines and machinery	Possibility of application indoor
Hydraulic-wedged plants	0,45-20 m^3/hr per one breaking device (hydraulic-wedged)	Breaking of concrete and reinforced concrete masses to pieces	Preparation to works: blast-hole drilling; arrangement of hydraulic-wedged breaking device in a blast hole. Breaking. Clearance of broken parts	Pneumatic drill IP-4701 or PR-20L, compressor ZIF SVE 4,8/1,0SHR, CMJ-83 (ISP-400), hydraulic-wedged breaking device (it is possible to use several breaking devices)	Possible
Jack hammers and pneumatic hammers	1,5-3,5 m^3/hr	Complete demolition	Preparation to works: material demolition. Disposal of demolished concrete from the area of demolition.	Base excavator EO-4121, jack hammer SP-62	Possible provided that the height of premises is more than 8,4 m
Diamond drilling plant	1,2-4,8 m^3/hr (opening/hr)	Making of opening, formation of slots, severing of reinforced concrete, concrete and brick structures	Preparation to clearance. Drilling of openings. Disposal of demolished material	Diamond drilling machine, set of diamond drills, compressor ZIF SVE 4,8/1,0SHR	Possible
Pneumatic and electric demolition hammers	0,25-0,6 m^3 per shift	Breaking of concrete	Preparation. Breaking. Disposal of demolished material	Demolition hammer ZUBR ZM-40-1700 K	Possible
Flame jet burner	-	Severing of concrete and reinforced concrete made in any position	Preparation to severing. Severing. Disposal of separated material	Set. Oxygen containers	Possible provided the safety and ventilation procedure compliance
Oxygen lance. Powder-oxygen cutting plant	1,5-5,1 m/hr	Oxygen lance drilling. Cutting	Preparation to works. Cutting to the depth of 4m. Clearance of demolished material	Metal tube 3-5 m in length. Oxygen containers. Plant, type UPKR-2	Possible provided the overventilation
Blasting materials	0,55-0,59 m^3/hr. Thickness of the material to be demolished - any	Breaking to pieces in any position of a structure	Preparation to works: blast-hole drilling. Blast. Disposal of products after the blast	Blasting materials	Limited due to seismic effect and shatters distribution
Blasting generator VN-2	45-150 m^3/hr, number of blasts per 1 minute 80-1500.	Breakage in any position of a structure	Preparation to works. Breakage	Blasting generator VN-2. Hydraulic shovel, 3grade, or machine KrAZ-257K	Limited due to large dimensions

	Thickness - any				

After the withdrawal and clearance of large components of the debris started working the universal hydraulic shovels with the shovel's capacity of 0,65-1,25 m^3, and the wheeled tractor shovels. The final process of the debris clearance is the loading and disposal of damaged structures shatters and material debris to the waste tips. When selecting the transport diagram of the debris clearance, it should be considered that the favourable ratio of capacity of the excavator shovel and the tipping lorry body is the ratio 1:4 or 1:5.

The preliminary lifting capacity of transport means depending on the excavator shovel capacity is accepted as per Table 5.

Table 5. The preliminary lifting capacity of transport means

Capacity of the loading machines shovels [m^3]	Lifting capacity of the tipping lorry [t]
0,3-04	3,5-10
0,5-1	7-10
1,25-2	10-25
3-4	25-40

In Fig. 1. given are the process flow diagrams of the lumped debris clearance (characterized by the side lengths ratio as 1:1) and the extended debris clearance (characterized by the side lengths ratio as more than 1:2) [12].

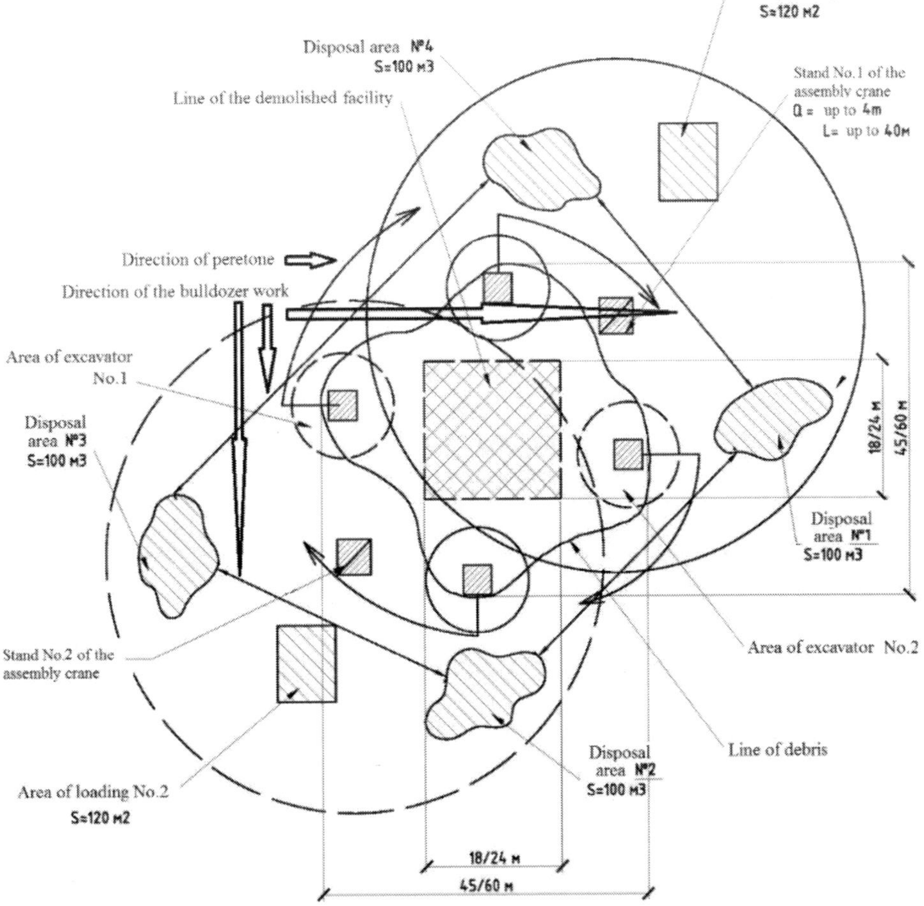

Fig. 1. Debris clearance of the lumped facility.

4 Conclusions

The upgrading and use of new machinery allow applying the new design and process solutions. The types and purposes of the debris clearance, the selection of machines and machinery, tasks for decrease of the debris mass and dimensions, selection of a certain type of motor transport depend on the technical-and-economic characteristics, characteristics of the machines, duration and length of haul. Generally, it is preferable to use the specific types of transport for the debris clearance – the hydraulic-motored bulldozers and excavators.

References

1. P.P. Oleynik *Construction organization. Conceptual frameworks. Models and methods. Information-utility systems.* (2001)
2. B.V. Zhadanovskiy *Machining of concrete and reinforced concrete by the diamond tool, Ind. and civ. constr.* **4** (2017)

3. M.S. Vainshtein, B.V. Zhadanovskiy, S.A. Sinenko, A.A. Afanas'yev *et al Selection of methods for construction and installation works mechanization*, Sc. Rev., **13**, 123-127 (2015)
4. T.V. Zolina *Application for assessment methods of operation life of industrial building with cranage throughout its life cycle*, Astrakhan, AISI (2014)
5. T.V. Zolina, Vestnik MGSU, **14**, 98-108 (2014)
6. B. V. Zhadanovsky *Organizational and technological solutions for labor safety in work production* (Moscow, ASV Publ., 2015)
7. A. Ginzburg, S Kachanov International Journal of Applied Engineering Research ISSN 0973-4562 **10, (20)**, pp 40869-40872 (2015)
8. B.V. Zhadanovskiy, Sinenko S.A., Kuzin F. Journal of Technology and organization of construction production. **1(6),** pp.38 (2014)
9. B.F. Shirshikov, et al *Organizational-process solutions on the labour safety in the work execution designs* (2015)
10. T K. Kuz'mina, Sinenko S A *Basic forms of implementation of the customer functions upon the major investment-construction projects organization and management Science review* 7 pp 222-227 (2016)
11. M.N. Ershov, Lapidus A.A. *Modern technologies of civil buildings reconstruction*, (Moscow, ACB, 2014)
12. SV. Shatov News of PUBA.. **7-8 (172-173)**. pp.39-44 (2012)

Organizational-economic model of formation of socio-commercial multifunctional complex in the construction of high-rise buildings

Ariadna Kirillova[1*] *and Oksana* O. Prytkova[2]

[1] Moscow state university of civil engineering, 129337, 26, Yaroslavskoye Shosse, Moscow, Russia
[2] LLC "AVStroy",16a, Selskohozyaystvennaya street, Moscow, Russia

Abstract. The article is devoted to the features of the formation of the organizational and economic model of the construction of a socio-commercial multifunctional complex for high-rise construction. Authors have given examples of high-altitude multifunctional complexes in Moscow, analyzed the advantages and disadvantages in the implementation of multifunctional complexes, stressed the need for a holistic strategic approach, allowing to take into account the prospects for the development of the city and the creation of a comfortable living environment. Based on the analysis of multifunctional complexes features, a matrix of SWOT analysis was compiled. For the development of cities and improving the quality of life of the population, it is proposed to implement a new type of multifunctional complexes of a joint social and commercial direction, including, along with the implementation of office areas - schools, polyclinics, various sports facilities and cultural and leisure centers (theatrical, dance, studio, etc.). The approach proposed in the article for developing the model is based on a comparative evaluation of the multifunctional complex project of a social and commercial direction implemented at the expense of public-private partnership in the form of a concession agreement and a commercial multifunctional complex being built at the expense of the investor. It has been proved by calculations that the obtained indicators satisfy the conditions of expediency of the proposed organizational-economic model and the project of the social and commercial multifunctional complex is effective.

1 Introduction

High-rise construction is an important factor in the development and perspective direction of the Russian real estate market. Among the many advantages of high-rise construction, the most important is the effective use of urban areas, due to the high cost of land, its scarcity and demand for residential and commercial real estate.

The vast foreign experience accumulated in America, Japan, Korea and other countries is widely studied and applied in Russian practice. Moscow continues the trend for the development of high-rise construction. Currently in the capital there are several dozens of

* Corresponding author: kirillova_an@mail.ru

modern residential complexes with a height of more than 100 meters, among them such projects as "Triumph Palace", "Tricolor", "Edelweiss", skyscrapers in MIBC "Moscow City". At the end of 2016 has began the construction of a new complex of residential skyscrapers "Neskuchny Home & SPA". Five towers will have from 70 to 90 floors. The project is located on Leninsky Prospekt within walking distance of the Neskuchny Garden. In the stylobate part of the residential complex there will be shopping galleries, a fitness club, author's cuisine restaurants, a cafe, a medical center, a bakery, a fresh-water shop, an underground parking lot for more than 3.000 cars and much more. A kindergarten for 250 children, a 24-hour children's club and a center for the development of children's creativity will be organized for children. The main attraction is a SPA-complex with a panoramic swimming pool, located in the transition between the two towers at a level of 130 meters above the ground. At the initial stage of construction, prices for apartments start from $ 167 200). In this case, the most expensive lot - penthouse area of 532 square meters - costs $ 6.8 million High-rise construction is determined by the multicomponent, high degree of risk and capital intensity, complexity and multifunctionality. The construction of such multifunctional complexes (MFC) is aimed at the complex solution of important urban tasks for the development of housing stock, the creation of office centers, business parks, leisure facilities for residents and recreational areas, redevelopment of territories for the withdrawal of industrial zones from the city center, etc. [1, 3, 4, 5, 11].

There are also drawbacks in the implementation of MFC, which are often due to inaccurate assessment of their potential, scale, insufficient validity of investment and credit decisions, lack of competitiveness of offered goods and services, and experience in effective operation. The effectiveness of MFC projects is also adversely affected by errors that are allowed when choosing the optimal ratio of the functional composition of the MFC and determining the optimal balance of areas in the complex, which leads to inconsistencies in various segments of real estate [12].

A holistic strategic approach is needed to take into account the development prospects of the city and create a comfortable living environment. At the same time, the strategic development of the MFC of the social and commercial direction should comply with the following principles:

-promote the development of the city in the long term and positively influence the development of specific urban areas and the city as a whole, ensure the formation of a favorable urban environment and improve the living conditions of citizens;

-ensure the development of urban areas by combining the efforts of private investors and the state;

-use specific mechanisms to regulate the problems that arise in the state and investors in the joint implementation of MFC projects [8].

At present, Russian experts refer to multifunctional complexes, mainly, projects with two or more functional components, each of which forms independent demand. Foreign experience on this topic shows that the MFC should consist of at least three income-producing components that generate revenue and have independent demand. Specialists note that these functions in the MFC are united by a single space or building in order to create the most comfortable environment, where the basic functions of the city's life will be realized. In the domestic practice there are two types of IFC: residential and commercial [9].

Residential MFC is a building or complex of residential buildings united by a common infrastructure and a single architectural concept. Such complexes fulfill the housing and social function. The public structure appears on the first floors, and the residential structure on the next floors.

Commercial IFC is a real estate object combining such categories as office, retail, hotel, etc. The most common in the Russian practice is the combination of office and retail real estate, with any ratio of areas. Also, recently, an increasing number of multifunctional complexes include a hotel function [10].

Based on the analysis of the characteristics of the MFC was compiled a matrix of SWOT analysis (Table 1). It analyzes the strengths and weaknesses of the MFC, as well as their capabilities and threats.

The investment attractiveness of multifunctional complexes directly depends on the economic efficiency of interconnected structures, premises, institutions into a single whole, a single real estate object that is limited to a specific territory.

Table 1. SWOT analysis of MFC

Strengths of MFC	**Weaknesses of MFC**
The maximum efficiency from the use of land.	The emergence of difficulties in the design and implementation of the project.
Reducing the cost of resources.	The possibility of a collision of functions with an erroneous zoning and concept.
Reducing the unit cost of creating an object due to its scale.	The most complex promotion of the object.
Multifunctionality of the complex increases attendance.	Specificity, greater complexity, cost of operation and management of the object.
Synergetic effect due to the attendant functions, which can complement each other with visitor flows.	When zoning the territory there may be restrictions on the legislation.
The possibility of repeated use of free areas of the IFC.	The need to maintain an object class.
Multifunctionality of the object increases the investment attractiveness of the project, because there is a reduction in risks due to investing in different types of real estate.	If there are apartussments in the MFC that lack the social infrastructure necessary for human life.
The ability to re-profile the object in a variety of adverse situations.	
Capabilities	**Threats**
Reorganization of industrial areas with a lack of free space.	The general state of the Russian economy.
The emergence of legislative acts that regulate the implementation of MFC projects.	Instability of the US dollar, euro.
Expansion of types of MFC (implementation of the MFC socio-commercial direction).	Lack of free land in large cities.
	Market oversaturation, lack of demand.

The quality of projects in the MFC in Russia lags far behind their number, this indicates their relative immaturity. It will take several years for the MFC segment to acquire the features of a civilized market. It is necessary to develop organizational and economic models for the implementation of MFC projects, expand the list of types of MFC, including through the appearance of the MFC socio-commercial function.

2 Method and algorithm

For the development of cities and improving the quality of life of the population, it is proposed to implement a new type of multifunctional complexes of a joint social and commercial direction, including, along with the implementation of office areas - schools, polyclinics, various sports facilities and cultural and leisure centers (theatrical, dance, studio, etc.). Earlier in the Russian practice, projects of this type were practically not implemented.

The essence of this proposal is that in one complex of buildings will be combined: school, polyclinic, sports and leisure facilities, so children who will study at this school will be able to visit sports (pool, sports grounds, arenas, gyms, etc.) and leisure institutions (theatrical, dance, art, music studios) without spending too much time as it usually happens now. Parents, in order to educate and diversify their children, lead them to various education centers. In terms of the magnitude of megacities, it needs a huge time and resources.

The combination of these functions in a single high-rise building will increase the economic efficiency of the use of space by combining the premises of various activities [10].

Organizational models of MFC project implementation are formed on the basis of the project management model simultaneously in several directions: design, construction management, operation, including facility management (operation organization, relationships with related organizations and suppliers, all types of work with tenants), repair of building structures, engineering systems of facilities (inspections, monitoring, preparation for seasonal operation, current and major repairs), the sanitary maintenance (cleaning of places of the general using, cleaning of adjacent territories, care of green plantations) [8.9].

Economic models of MFC implementation are formed on the basis of project financing methods. It refers to the financing of investment projects, which is characterized by a special way to ensure return on investment, which is based on the investment quality of the project itself, the revenues received from the company being created or reconstructed in the future. A special mechanism for project financing includes an analysis of the technical and economic characteristics of the investment project and an assessment of the risks that are associated with it, and the return on investment is based on project revenues that remain after covering all financial expenditures [13, 14]. Financing of real estate projects is carried out by the developer's own funds or borrowed. Using a developer or builder borrowed funds in the Russian market, namely, a bank loan in practice is associated with: insufficiency of the developer's own funds, need to provide the bank with sufficient security, lack of opportunities for Russian banks to lend investment and construction projects in sufficient quantities and for a specific time frame. In these circumstances, when constricrion and operating the MFC, it is advisable to apply the mechanisms of public-private and municipal-private partnership [2, 6, 15]. One of the main priorities of state policy is the construction of socially important facilities. The proposed MFC project, being socially significant, will have the following functions: educational, sports, cultural, wellness and office. The economic model will be formed by combining state and private financing, i.e. a model of public-private partnership (PPP) is applied. The proposed type of MFC is presented in the following scheme (Figure 1).

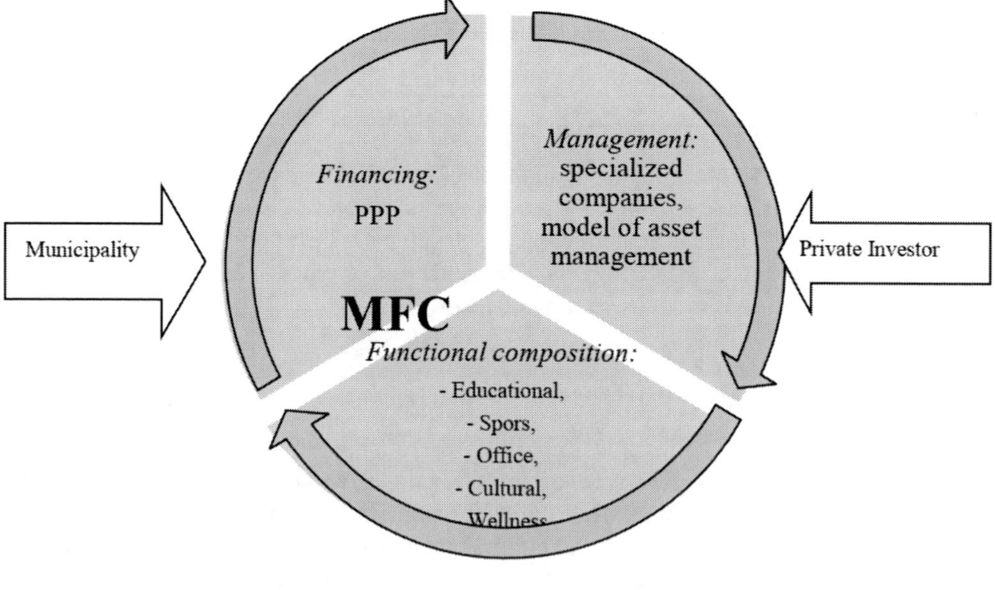

Fig. 1. Methodological approaches to the organizational-economic model of MFC

School and polyclinic in this MFC will be public (free), sports and cultural facilities - paid. Accordingly, the school and polyclinic should be financed by the municipality, sports and cultural facilities - by private investors. Since all these blocks should function interconnected in one real estate object, the author is invited to implement the MFC of the social and commercial direction (Table 2) using the following organizational and economic model.

Table 2. Functional components of the socio-commercial MFC

Functions	Function blocks	Management	Financing
Educational	Educational institution (school)	Municipality	Means of the municipality
Sports	Paid sports facilities (swimming pool, sports grounds, playpens, gyms, etc.)	Private owner	Private and loan investments
Cultural	Paid cultural and leisure centers (theater, dance, art, music studios)	Private owner	Private and loan investments
Wellness	Polyclinic	Municipality	Means of the municipality
Office	Office area in rent	Private owner	Private and loan investments

3 Results

To select from the possible alternative combinations of functional modules of the most optimal, socially significant and profitable MFC project that will meet the investor and city requirements, an algorithm for modeling the MFC project of a social and commercial direction in Moscow is proposed, as well as a comparison with the concept of implementing a typical MFC project. This algorithm will make it possible to clearly identify the best and most effective solutions for both the investor and the municipality. The algorithm is shown in Fig. 3.

The proposed approach to the development of the model is based on a comparative evaluation of the MFC project of the social and commercial direction implemented through PPP in the form of a concession agreement and a commercial MFC being built at the expense of the investor. The question of the legal relationship to the land plot will be decided in completely different ways. When implementing a commercial MFC, the land plot will be leased, which means that the item "Lease of the land plot" will appear in the cost structure. When the MFC realizes a social and commercial direction, a concession agreement will be concluded between the investor and the municipality, according to which the land plot will be granted to the investor by the municipality for free use for a certain period, which will correspond to the payback period of the project and the receipt of profit for the investor. Therefore, in the costs of implementing the socio-commercial MFC, there are no costs associated with leasing or buying a land plot.

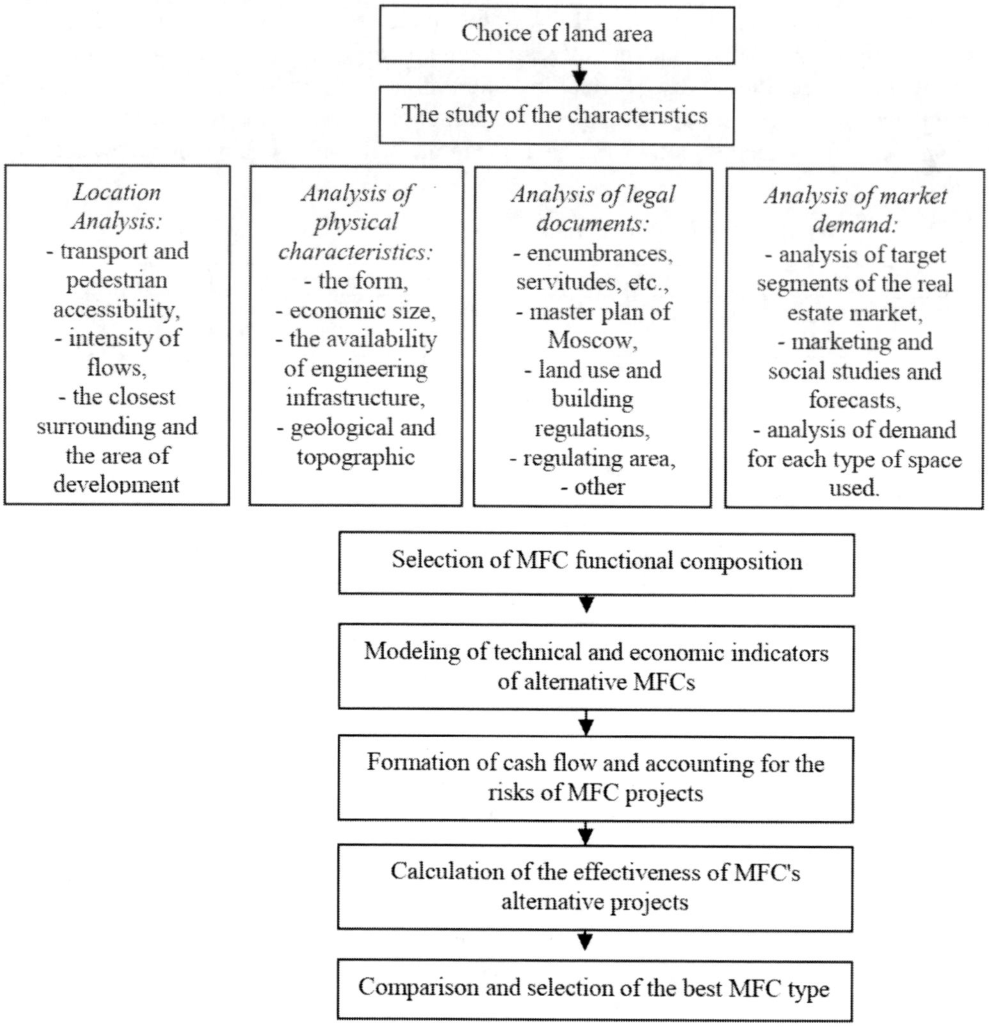

Fig. 3. Algorithm for modeling the MFC project in the city

The main income of a commercial MFC consists of leased space, apartments, parking spaces, and sales areas. In the social and commercial MFC, revenues are derived from the rental of offices, areas for the sports and cultural and leisure sector, as well as rental of parking spaces. Table 3 compares the obtained indicators of commercial efficiency of two projects.

Table 3. Comparative performance indicators of MFC

Indicator	Commercial MFC	Socio-commercial MFC
Internal rate of return (IRR)	3.9%	2.4%
Internal rate of return (IRR), % per annum	46.2%	28.8%
Present value of cash flows, USD	22 134 316 604	9 787 739 698
Net present value (NPV), USD	220 488 311	201 236 861
Profitability index (PI)	1.84	1.54
Payback period (PBP), months.	70	70
Discounted payback period (DPBP), months.	72	73

From the results obtained, it follows that the calculated indicators of the commercial MFC are significantly higher than the socio-commercial MFC. This is due to the fact that in the commercial MFC revenues come from the entire constructed area of the MFC, unlike another MFC, where part of the building area is given to a school and a budgetary polyclinic, and the rest of the premises are profitable. All received indicators satisfy the condition of the proposed model and the social and commercial MFC is effective.

4 Conclusions

One of the effective tools for improving the spatial structure of cities can be high-rise multifunctional complexes that play the role of an effective management tool for the territorial development of real estate. The advantages of the project implementation are the effectiveness of children's education, transport unloading, walking accessibility of services and increasing the investment attractiveness of the surrounding buildings. The implementation of this MFC is not only a means of profit for the developer, but also has a significant impact on the city budget, economic and social relations, which will certainly improve the living standards of citizens, improve the quality and diversity of the urban environment.

References

1. *Urban Development Code of the Russian Federation. Official text* (Moscow, Prospekt, Omega-L) 104 (2009)
2. *Federal Law of the Russian Federation of July* 21, 2005 No. 115-FZ "On Concession Agreements" (2005)
3. Federal Law of 25.02.1999 No. 39-FZ "*On investment activities in the Russian Federation, implemented in the form of capital investments*" (1999)
4. Law of the City of Moscow of March 3, 2004 No. 13 *"On the Basics of Urban Development in the City of Moscow"* (2004)
5. *Moscow City Building Codes 4.04-94 "Multifunctional buildings and complexes"* (1994)
6. O.Yu. Atkov, *Options for business and state cooperation for the development and improvement of social infrastructure: the prospects for using PPP mechanisms* (Moscow) (2008)
7. N.F. Vecher, A.A. Olkhlvsky, *Investments in commercial real estate. Life cycle of the facility (*Moscow: Business project) (2005)
8. T.A. Osipova, *Formation of a strategy for the development of multifunctional complexes in a large city,* PhD Thesis (2012)
9. P.G. Graboviy, S.I. Beliakov, A.N. Kirillova & other, *Surveying: organization, expertise, management. Part two. Expertise of real estate and construction control* (Moscow, ASV, Prosvetitel, 2015)
10. P.G. Graboviy, S.I. Beliakov & other, *Economics and real estate management* (Moscow, Prospekt, 2013)
11. V.M. Ostretsov *The proceedings of the annual city conference-exhibition"Unique and special technologies in construction of UST-Build 2004* (Moscow) (2004).
12. R.I. Ibragimov, E.P. Kotelnikova, M.A. Morozov, A.A. Sutugin, Main problems of construction of height buildings, Scientific and Practical Electronic Journal of the Alley of Science **9** (2017)

13. D.L. Whitley, *Genetic Algorithms and Evolutionary Computing* (Van Nostrand's Scientific Encyclopedia) (2002)
14. I. Chamey Igal, *The conditions for capital investment in the real estate, University of Toronto,* PhD Thesis (2000)
15. G. Ian. *Thrall, Business geography and new real estate market analysis (*NC, USA, Oxford University Press, Incorporated) (2002)

Determining the Supply of Material Resources for High-Rise Construction: Scenario Approach

Anna Minnullina[1,*] and *Vladimir* Vasiliev[2]

[1]State Budget Educational Institution of Higher Education «Industrial University of Tyumen», Volodarskogo str., 38, Tyumen, 625000, Russia
[2]State Budget Educational Institution of Higher Education «Industrial University of Tyumen», Volodarskogo str., 38, Tyumen, 625000, Russia

Abstract. This article presents a multi-criteria approach to determining the supply of material resources for high-rise construction under certain and uncertain conditions, which enables integrating a number of existing models into a fairly compact generalised economic and mathematical model developed for two extreme scenarios.

1 Introduction

Every year the number of high-rise buildings and their floors is significantly increasing. Constructions around the world become complex, technical and engineering systems, which results in increased technical, environmental, psychological, aesthetic, architectural and other requirements. High-rise erection is closely connected with the active use of intellectual and physical human resources, as well as material resources; pollution and human isolation from the environment. By all means, the above mentioned and many other problems of high-rise construction require a careful study and implementation of the positive results achieved in the course of this research.

Thus, high-rise construction is among the most cost-intensive sectors of economy, with the largest expenses in the sphere of logistics. Due to serious investment in high-rise construction, it is important to correctly determine the supply of material resources. To solve this problem, we have decided to set a multi-criteria problem for determining the economic order quantity of material resources presented for two extreme scenarios.

Nowadays there is a large number of studies and papers related to these problems [1-4], but all these works and approaches examine the objectives in a predominantly logistic format under certain conditions with a simplified justification of the criterion of optimality.

The modern Russian literature on economy often uses the Harris-Wilson model which is related to dynamic deterministic models of inventory management in order to define the economic order quantity (1) [5-8,16,17]. This simple classical model helps to understand the movement of stocks and in many cases enables to regulate and control their level.

We calculate the optimal size of supply using the following formula (1):

* Corresponding author: minnullinaay@yandex.ru

$$\overline{N} = \sqrt{\frac{2\alpha Q}{\beta T}} \tag{1}$$

where α is shipping and handling expenses as per a delivery lot of material resources;

T is a time period when material resources are used;

Q is a gross requirement of material resources during T time;

β is variable expenses for handling a resource unit per unit time.

However, this model is based on certain conditions, i.e. supposed resources are used evenly and as soon as the size of the current stock reaches zero, new supplies are instantly delivered. Demand is regarded as constant in time, and the replenishment of stocks is instant. It is assumed that there is no deficit, therefore only the current stock is considered, the level of which ranges from the maximum value equal to the whole lot at the moment of its receipt to the minimum value equal to zero.

In the classical model of the economic batch quantity, a deficit is not provided but in some cases deficit losses are comparable to the costs of stock maintaining. If there is a deficit, the economic batch quantity model is as follows (2):

$$\overline{N'} = \sqrt{\frac{2\alpha Q}{\beta}} \times \sqrt{\frac{\beta+\gamma}{\gamma}}, \tag{2}$$

where γ is shortage costs of the stock of resource units per unit time.

2 Methods

The multi-criteria approach for certain and uncertain conditions we have developed is of a systematic nature and allows integrating a number of existing models into a fairly compact generalised economic and mathematical model presented for two extreme scenarios.

We introduce the following notations for the formal representation of this model:

$l = \overline{1, L}$ is the number of a material resource supplier;

$j = \overline{1, m}$ is a type of the resources used;

B_j is a company's demand for the *j* resource bought;

A_{lj} is the aggregate supply of *j* resource by *l* supplier;

$L_0^{(j)} = (l/A_{lj} > 0, l = \overline{1, L})$ is the subset of supplier numbers prioritised for *j* resource;

a_{lj} is the maximum possible schedule line quantity of *j* resource by *l* supplier;

α_{lj} is delivery expenses for the lot of *j* resources by *l* supplier;

β_j, γ_j are handling and deficit expenses of a *j* resource unit per unit time (if a deficit is no lower than $\gamma_j = +\infty$);

T_{lj} is the schedule date of *j* resource supplies by *l* supplier;

C_{lj} is the procurement price of a *j* resource unit realised by *l* supplier;

X_{lj} is the supply scope of *j* resource by *l* supplier.

Scenario 1

All the components of supply (α), storage (β), deficit (γ) and delivery period (T) costs are deterministic. In this case, we represent the multi-criteria definition of the economic order quantity of material resources delivered by suppliers to some construction organisation as the following cluster of $F(x)$ optimality criterion and S_1 constraint system:

$$F_1(x) = \sum_{j=1}^{m} \left| \sqrt{\frac{\gamma_j}{\beta_j + \gamma_j}} \right| \sum_{l \in L_0^{(j)}} \sqrt{2\alpha_{lj}\beta_j X_{lj} T_{lj}} \to \min; \quad X \in S_1 \tag{3}$$

$$F_3(x) = -\sum_{j=1}^{m}\sum_{l\in L_0^{(j)}} \frac{X_{lj}}{\sum_{j\in L_o^{(j)}} X_{lj}} \log_2 \frac{X_{lj}}{\sum_{j\in L_0^{(j)}} X_{lj}} \rightarrow \min; \quad X \in S_1 \tag{4}$$

$$F_4(x) = \max_{\substack{1\le j\le m \\ l\in L_0^{(i)}}} \left| \sqrt{\frac{2\alpha_{lj}T_{lj}}{\beta_j X_{lj}}\left(\frac{\beta_j+\gamma_j}{\gamma_j}\right)} \right| \rightarrow \min; \quad X \in S_1 \tag{5}$$

$$F_5(x) = \max_{\substack{l,K\in L_0^{(j)} \\ l\neq K}} \left\| \sqrt{\frac{2\alpha_{lj}X_{lj}}{\beta_j T_{lj}}\left(\frac{\beta_j+\gamma_j}{\gamma_j}\right)} \right| - \left| \sqrt{\frac{2\alpha_{kj}X_{kj}}{\beta_j T_{kj}}\left(\frac{\beta_j+\gamma_j}{\gamma_j}\right)} \right\| \rightarrow \min;\ j=\overline{1,m}. \quad X\in S_1$$

$$S_1: \begin{cases} \sum_{l\in L_0^{(j)}} X_{lj} \ge B_j, j=\overline{1,m}; \\ \sum_{l\in L_0^{(j)}} X_{lj} \ge \sum_{l\in L_0^{(j)}} A_{lj}, j=\overline{1,m}; \\ \left| \sqrt{\frac{2\alpha_{lj}X_{lj}}{\beta_j T_{lj}}\left(\frac{\beta_{j+\gamma_J}}{\gamma_j}\right)} \right| \le a_{lj}, l\in L_0^{(j)}, j=\overline{1,m}; \\ \sum_{l\in L_0^{(j)}} \frac{a_{lj}T_{lj}}{\left|\sqrt{\frac{2\alpha_{lj}T_{lj}}{\beta_j X_{lj}}\left(\frac{\beta_{j+\gamma_J}}{\gamma_j}\right)}\right|} \ge \sum_{l\in L_0^{(j)}} A_{lj}, j=\overline{1,m}; \\ X\ge 0\ (or\ X\ge 0\ and\ integer-valued). \end{cases} \tag{6}$$

Scenario 2

As a rule, supply (α), storage (β), deficit (γ) and delivery period (T) expenses are undetermined and can be represented as the following quantities:

$$\alpha \in S_\alpha : \{\alpha_{lj}^{(1)} \le \alpha_{lj} \le \alpha_{lj}^{(2)}, j=\overline{1,m}\}$$

$$\beta \in S_\beta : \{\beta_j^{(1)} \le \beta_j \le \beta_j^{(2)}, j=\overline{1,m}\}$$

$$\gamma \in S_\gamma : \{\gamma_j^{(1)} \le \gamma_j \le \gamma_j^{(2)}, j=\overline{1,m}\}$$

$$T \in S_T : \{T_{lj}^{(1)} \le T_{lj} \le T_{lj}^{(2)}, j=\overline{1,m}\}$$

where (α^1,α^2), (β^1,β^2), (γ^1,γ^2), (T^1,T^2) are correspondingly the maximum and minimum values of expense vectors/matrices and resource delivery time. The value changes of a, A, C matrices and B vector are considered to be irrelevant. Otherwise, the system is expanded by $S_1, S_A, S_{c,} S_B$ quantities.

The optimality criterion and constraint system are as follows:

$$F_1(x) = \max_{\substack{\alpha\in S_\alpha \\ \beta\in S_\beta \\ \gamma\in S_\gamma \\ T\in S_T}} \sum_{j=1}^{m} \left| \sqrt{\frac{\gamma_j}{\beta_j+\gamma_j}} \right| \sum_{i=1}^{n}\sum_{l\in L_0^{(j)}} \left|\sqrt{2\alpha_{lj}\beta_j X_{lj}T_{lj}}\right| \rightarrow \min; \quad X\in S_2 \tag{7}$$

https://doi.org/10.1051/e3sconf/20183303060

$$F_3(x) = -\sum_{j=1}^{m}\sum_{l\in L_0^{(j)}} \frac{X_{lj}}{\sum_{j\in L_0^{(j)}} X_{lj}} \log_2 \frac{X_{lj}}{\sum_{j\in L_0^{(j)}} X_{lj}} \to \min; \quad X \in S_2 \tag{7}$$

$$F_4(x) = \max_{\substack{1\le j\le m \\ i\in L_0^{(j)} \\ \alpha\in S_\alpha \\ \beta\in S_\beta \\ \gamma\in S_\gamma \\ T\in S_T}} \left| \sqrt{\frac{2\alpha_{lj}T_{lj}}{\beta_j X_{lj}}\left(\frac{\beta_j+\gamma_j}{\gamma_j}\right)} \right| \to \min; \quad X \in S_2 \tag{8}$$

$$F_5(x) = \max_{\substack{l,K\in L_0^{(j)} \\ l\ne K \\ \alpha\in S_\alpha \\ \beta\in S_\beta \\ \gamma\in S_\gamma \\ T\in S_T}} \left\| \sqrt{\frac{2\alpha_{lj}X_{lj}}{\beta_j T_{lj}}\left(\frac{\beta_j+\gamma_j}{\gamma_j}\right)} \right| - \left| \sqrt{\frac{2\alpha_{lj}X_{lj}}{\beta_j T_{lj}}\left(\frac{\beta_j+\gamma_j}{\gamma_j}\right)} \right\| \to \min; \; j=\overline{1,m}. \quad X \in S_2 \tag{9}$$

$$S_2: \begin{cases} \sum_{l\in L_0^{(j)}} X_{lj} \ge B_j, j=\overline{1,m}; \\ \sum_{l\in L_0^{(j)}} X_{lj} \ge \sum_{l\in L_0^{(j)}} A_{lj}, j=\overline{1,m}; \\ \max_{\substack{\alpha\in S_\alpha \\ \beta\in S_\beta \\ \gamma\in S_\gamma \\ T\in S_T}} \left| \sqrt{\frac{2\alpha_{lj}T_{lj}}{\beta_j X_{lj}}\left(\frac{\beta_j+\gamma_j}{\gamma_j}\right)} \right| \le a_{lj}, l\in L_0^{(j)}, j=\overline{1,m}; \\ \min_{\substack{\alpha\in S_\alpha \\ \beta\in S_\beta \\ \gamma\in S_\gamma \\ T\in S_T}} \sum_{l\in L_0^{(j)}} \frac{a_{lj}T_{lj}}{\left| \sqrt{\frac{2\alpha_{lj}T_{lj}}{\beta_j X_{lj}}\left(\frac{\beta_j+\gamma_j}{\gamma_j}\right)} \right|} \ge \sum_{l\in L_0^{(j)}} A_{lj}, j=\overline{1,m}; \\ X \ge 0 \; (or\; X \ge 0 \; and\; integer-valued). \end{cases} \tag{10}$$

In both scenarios the criteria of optimality represent the following aspects:
$F_1(x)$ minimises total production, handling and deficit expenses related to resources;
$F_2(x)$ minimises the gross price of the resources bought;
$F_3(x)$ minimises the entropy, i.e. the uncertainty of supply quantities;
$F_4(x)$ minimises the maximum interval between resource deliveries;
$F_5(x)$ minimises the maximum resource delivery tolerance $j=\overline{1,m}$ by various suppliers.
Limitations made by S_1 constraint region should meet the following criteria:
– demand for each type of resources should be satisfied;
– resource delivery cannot exceed the total offer by all possible suppliers;

– one delivery lot by suppliers cannot exceed their maximum values;
– the total value of all possible resource deliveries by all possible suppliers is no less than the total offer;
– non-negativity and/or integrality of supplies.

Conclusions

In case of scenario 2, the meaning of the optimality criteria and S_2 constraint system remains the same but the implementation is conducted in the format of the "guaranteed result".

The two model scenarios we presented solve the following problems: realise a multi-criteria setting under certain and uncertain conditions; represent an elegant combination of classic models (the transport and Harris-Wilson models); enable to use the well-known mathematical methods for determining the economic order quantity of material resources; allow further development and improvement in various modified forms; have a higher level of adequacy in relation to real-life conditions of construction organisation activities.

Therefore, we can assert that determining the economic order quantity of material resources in high-rise construction should skip the initial data on procurement in the form of criteria for optimal system functioning that limit production conditions and indicators through a kind of filter represented by an economic and mathematical model for making managerial decisions and implementing purchases.

References

1. A.G. Butrin, Yu.G. Amerhanova, Integratedlogistics, **4**, 20 (2009)
2. V.P. Fedko, V.A. Bondarenko, Commerciallogistics, **304** (2006)
3. V.V. Scherbakov, Logistics and Supply Chain Management, **582** (2015)
4. A.A. Kanke, I.P. Koshevaya, Logistics bases, **576** (2013)
5. Yu.G. Lebedev, Logistics harmonized supply chain, **448** (2005)
6. A.P. Tyapukhin, Design of distribution systems based on logistics, **240** (2007)
7. L.B. Mirotin, A.K. Pokrovsky, Fundamentals of Logistics, **192** (2013)
8. M. Linders, F. Johnson, A. Flynn, H. Fearon, Procurement and Supply Management, **752** (2014)
9. D. Bauersoks, D. Kloss,Logistika: integrated supply chain,**640**(2010)
10. Yu.A. Borodina, O.M. Syardova. Thestrategyofsustainabledevelop**18**, 56 (2013)
11. A.V. Sysolyatin, Electronic scientific journal**13**, 1-7(2015)
12. V.S. Lukinskiy, V.V. Lukinskiy, N.G. Pletnev, Logistics and SCM, **360** (2016)
13. V.I. Sergeev, Corporate Logistics, **976** (2008)
14. V.I. Sergeev, I.P. Elyaschevich. SupplyLogistics, **398** (2016)
15. A.Kopytova, Procedia Engineering**165**, 1132 (2016)
16. M.N. Grigoryev, V.V.Tkach, S.A. Uvarov,Commerciallogistics, **490** (2016)
17. A.Yu. Minnullina,Economics and Entrepreneurship**5-2**, 734(2014)
18. A.V. Kemenov,Logistics building complex,**320** (2012)

Increase in Population Density and Aggravation of Social and Psychological Problems in Areas with High-Rise Construction

Elena Romanova1,*

[1] Moscow State University of Civil Engineering, 26 Yaroslavskoye shosse, 129337, Moscow, Russia

Abstract. High-rise apartment houses have technical and economic advantages in areas with dense population. Their placement in the central part of the city allows increasing the number of living space in the limited territory, to bring population to the place of employment and reduce pendular migration. But increase in population density leads to psychological problems: level of a stress, fatigue increases, the number of phobias grows, infectious diseases extend quicker. These problems can be solved at resettlement of inhabitants to the suburb. However such decision leads to aggravation of a transport problem and the pulsing increase in population density in the downtown and on its suburb. To solve a transport problem, it is necessary not to increase the square of the cities. Therefore in the suburbs is also used high-rise construction. But high-rise residential districts on the suburb of the city get own social problems which are capable to destroy all advantages of high-rise construction.

1 Introduction

High-rise apartment buildings have technical and economic advantages in areas of high population density. They have become a distinctive feature of housing accommodation in virtually all densely populated urban areas around the world. In contrast with low-rise and single-family houses, apartment blocks accommodate more inhabitants per unit of land area and decrease the cost of municipal infrastructure. A means of assessing population and building densities at the same time is often used measure dwellings per hectare. It is a useful proxy for comparing housing projects.

A concept “density” is borrowed from physics where the meaning is clear – mass divided by volume. Yet when it transferred to the city, nothing is so simple. Concentrations of people are generally measured as residents per hectare based on census data. But it's wrong. Population densities cannot be based on residents alone, since the numbers of people in a given neighbourhood at a given time include those who work there or are visiting. In a mixed-use neighbourhood, residents may be a small proportion of the population density.

Also there is the distinction between gross and net densities. For this assessment we also incorporate water bodies, freeways and unbuildable sites, so the average density

* Corresponding author: romanova_e_v@mail.ru

diminishes. The gross density is always lower than net density and it is the one that matters in debates over urban density.

This indicator in Russia is regulated Construction Norms and Regulations 2.07.01-89*. The recommended density (gross) for the cities with various number of inhabitants (from less than 20 000 – to over 1 000 000): high – 130-220 persons/hectare, average – 190-210 persons/hectare, low – 70-180 persons/hectare. For territories of various town-planning value she can increase or decrease: "… population density, as a rule, shouldn't exceed the 450 persons/hectare" and to be "not less 40 persons/hectare" [1]. But even in large megalopolises and the regional centers these norms aren't carried out. So the most densely populated region of Russia on density - the city of Moscow has density of 4 834,31 people/km2 that is made by 48,3 persons/hectare. It is slightly higher than the minimum border! For comparison: before accession of the new territory population density in Moscow was 105,8 persons/hectare.

With the greatest population density except Moscow have entered the five territorial subjects of the Russian Federation: St. Petersburg – 37,6 persons/hectare, Sevastopol – 4,9 persons/hectare, the Moscow region – 1,7 persons/hectare, the Republic of Ingushetia - 1,3 persons/hectare. In other cities and regions population density has made less than one person on hectare [2]. The chief architect and the head of the Council on development of city infrastructure from Barcelona, professor José Asebilyo Marin, considers that low population density is the main problem of the Russian cities: "It is necessary to remember that the cities with low density are inefficient both economically and ecologically. The high density and compactness of the cities are positive lines from the town-planning point of view "[3].

Population density is not only the number of residents but fluctuates over time and with functional mix. There are also population density rhythms, as people move from place to place throughout the day and week. The same urban precinct can be densely populated during work hours and empty on weekends.

A distinction should be made between "internal" and "external" densities – the numbers of people in a room or apartment versus those in an urban precinct. At the other extreme, internal crowding largely defines a slum [4]. If you look at new high-rise housing in the evening, you can find many rooms and even apartments unoccupied.

High-rise construction reduces internal density and increases external at the same time. It promotes strengthening of individualization, autonomy of the person, estrangement from the outside world and society. The disturbing tendencies in behavior of people connected with growth of number of storeys and population density in residential zones have been described in the Soviet researchers works in the 70-80th years. Scientific interest in this problem doesn't weaken also in the 21st century [5, 6, etc.] Researches of the English and Singapore scientists [7] show negative correlation between the level of happiness and population density. If the person is the whole day among a large number of other people, he needs the place and time for a privacy. He finds such place in the apartment. So at citizens idea of the dwelling as about the sovereign territory where the entrance is forbidden strangers was gradually created. So culture, spirituality, traditions have changed.

At last, there is a question of density streetlife – concentration people in public space [8]. Here it is almost impossible to determine density. We can measure the number of pedestrians in a minute or on square meter. It doesn't give us understanding of population density and necessary density of building. This density - the fact that we often call city "noise" or "intensity" - provides the greatest contribution to social and economic development of the city together with negative elements, such as social and psychological overload.

Objective of this research is the description and the analysis impact of high-rise buildings on mental health and social wellbeing of their inhabitants.

During the research the following problems were solved:
- To classify high-rise buildings by a way of their use.
- To estimate the relation to high-rise construction of respondents in different age groups (students and teachers).
- To reveal and describe psychological and social phenomena which can arise at the person during stay in high-rise residential buildings and residential districts of high-rise building.
- To analyze the prospects of using architectural planning solutions of high-rise construction for removal psychological and social tension of the population.

2 Methods

For identification psychological phenomena which arise at the person during stay in high-rise buildings information on results of earlier conducted researches and also stories by builders-spidermen, inhabitants of multi-storey buildings and residential districts which are in free access with high-rise building have been analyzed.

The questionnaire "Your relation to high-rise construction" has helped to estimate the relation to high-rise construction in different age groups of people. The questionnaire includes a series of questions the opened and closed type and reveals opinions of the respondent high-rise construction objects in general, his preferences concerning the residence and fears connected with accommodation in the high-rise building.

The research was conducted at the Moscow State University of Civil Engineering (MGSU) in 2017. 119 teachers and students aged from 17 up to 69 years with various social statuses, income level, constantly living in Moscow and its suburb and also arrived from other cities of Russia have participated in a research.

3 Results

In practice of design and construction the most various views of buildings therefore it is possible to classify them by a large amount of signs meet. The most important in this research is classification by the following signs: to destination and on number of storeys. Here there is a set of various classifications too.

On number of storeys we can allocate low-rise buildings – 1–2 storeys, average-rise buildings – 3–5 storeys, multi-storey buildings – 6–10 storeys, high-rise buildings – 11–25 storeys, skyscrapers – more than 25 storeys [9]. As height of floors in different buildings and the different countries is various, on CIB symposium passing in 1976 in Moscow the general classification of buildings by their height in meters has been accepted.

The exact height above which a particular building is deemed a high-rise is specified by fire and building codes for the country, region, state, or city where the building is located. In Russia buildings higher than 75 m (more than 25 floors) carry to high-rise.

In book High-Rise Security and Fire Life Safety [10] addresses the following types of high-rise buildings:
- office buildings;
- hotel buildings;
- residential and apartment buildings;
- mixed-use buildings.

A mixed-use building may contain offices, apartments, residences, and hotel rooms in separate sections of the same building. Hotel-residences are another type of mixed-use occupancy.

We have asked students and teachers of MGSU about the relation to these types of high-rise buildings. 67% of respondents (generally students) are positive to construction of high-rise buildings (higher than 11 floors). From them 43% approve construction of skyscrapers (more than 25 floors). Skyscrapers attract respondents with the positive sides: beautiful appearance, prestigiousness, an opportunity to place a large number of people in densely built up city. Other respondents have shown negative attitude to high-rise construction. They noted shortcomings: discrepancy to historical appearance of the city, high fire danger, big population density. Now 3% of respondents live in low-rise buildings, 7% in average-rise buildings, 12% in multi-storey buildings, 76% in high-rise buildings, 2% in skyscrapers. Results of the relation to different types of high-rise buildings (more than 25 floors) in Table 1.

Table 1. Relation to high-rise buildings.

Comments	Building type							
	high-rise buildings				skyscrapers			
	Office building	Hotel building	Residential and apartment building	Mixed-use building	Office building	Hotel building	Residential and apartment building	Mixed-use building
in high-rise buildings have to be …	88%	93%	95%	37%	87%	52%	35%	63%
I visited several times	2%	22%	100%	41%	53%	12%	5%	23%
I regularly visited *	70%	-	87%	100%	1%	-	2%	32%
I live constantly*	-	-	76%	-	-	-	2%	-
I would like to live in	-	-	18%	6%	-	1%	27%	28%
I want to work in	53%	-	-	10%	31%	-	-	51%
in Moscow it isn't necessary to build high-rise buildings …	-	-	2%	1%	2%	5%	60%	-

* All hostels of the university and the office building belong to the high-rise buildings group.

Results of poll show that students and teachers of MGSU treat high-rise buildings up to 25 floors high more tolerantly. These buildings are widespread in Moscow; they are habitual for perception and operation. Other relation to skyscrapers, it alerted and partly negative. Those respondents who wish to live in high-rise residential buildings have specified that buildings have to be in the downtown. Nobody wants to live in the high-rise apartment houses located on the suburb. The reason for that are the psychological and social problems arising at residents of high-rise residential districts. The experts studying influence of the urban environment on the person around the world meet in the conclusions that high-rise housing is a concentration of negative factors.

Residential high-rise buildings increase population density in this quarter. In certain cases it isn't bad. High-rise buildings in the center are occupied by rich people. It is good for the area and it is prestigious for residents. On the megalopolis suburb absolutely other tenor of life.

Apartments in the multi-storey building on the suburb of the city have lower price. They are got by people with other level of prosperity, education and culture. Also to these areas occupy people according to social programs: orphans, needy, immigrants [11]. Comfortable cohabitation requires a certain level of culture of all inhabitants of the house. It most often is impossible. Residents of different social groups form numerous subcultures which are at enmity with each other. Therefore multistoried housing and dense population in suburban areas negatively influence moral atmosphere of life. Here it isn't necessary to wait for rapprochement and improvement. Hostile faceless environment of panel houses in which there live Russians today, makes a certain impact on their behavioral installations. She forms at them total estrangement and individualism in the worst sense. The environment of skyscrapers doesn't promote communication to people. Residents of multi-storey buildings often don't know even the neighbors in a staircase. The yards in such houses don't perform the main function of space of socialization [12], and serve only as an intermediate zone between the apartment and the street. Reduction of contacts between inhabitants, growth of feeling of the inhabited environment anonymity, leads to growth of household vandalism and crime, in particular among teenagers.

Modern new buildings situated near Moscow are multistoried same buildings in the middle of the huge yards waste grounds. Such building isn't proportional to the person. The more storeys in the building, the more the person feels small and as though "pressed to the earth". Similar "pressure" of skyscrapers, though isn't realized, but can create an additional stress that is expressed usually by feeling of fatigue, irritability and desire to be fenced off from people around. It is difficult to master huge empty spaces around the house psychologically. The microspace of communication makes about 10 m [13]. The space out of these borders remains to strangers and turns into the huge and uncomfortable waste grounds. The dense population provides to the person less than 10% of necessary personal space that causes discomfort, fear and aggression. Everything together it promotes growth of crime and suicide.

Results of poll have shown that accommodation on the top floors of high-rise buildings even in prestigious districts of the city also not too attracts people. The reason for that is care of physical and mental health.

The statistics of diseases shows that at inhabitants above the fifth floor, the frequency diseases of respiratory organs grow. At those who live above the ninth floor this indicator increases several times. Pathogenic bacteria are carried by air which climbs from first floors staircases, mines of elevators. These air inhabitants of the top floors breathe.

Advantage of clean air at height is practically not felt. At many skyscrapers on the top floors of a window don't open. Because of the high wind speed outside the window it is always noisy and is very cold. At a height wind can reach such force that will constitute real danger. The woman, for example, can just not hold a shutter of a window in hands and receive blow by a window frame. It is from the windward. And with leeward side so low pressure that can exhaust the person literally outside.

Need to use the liftleads to reduction of time spent in the fresh air. Children from low houses can walk one. Parents control their behavior through a window. Children from skyscrapers are deprived of such opportunity; they are forbidden to be used the liftindependently. Also parents can't control and even see the child from a window of the 20th floor. The elevator creates obstacles for the accompanying adult's child: inconvenience using a carriage, need to control the child during the trip, etc. Statistically children from multi-storey buildings lag behind in physical development the peers living in other conditions. The same treats also health of elderly people, many of which are subject to fears when using the lift.

Accommodation in skyscrapers can cause mental health problems. At height at the person visual touch with the earth is lost. Living on the 40th floor sometimes feels as on the plane. Psychologists also note that inhabitants of the top floors can be subject to height fear – an acrophobia, fear of the closed and opened spaces – claustrophobia and an agoraphobia respectively, besides — aero acrophobia — fear of open spaces at height. If they are summarized with fears which accompany the person at stay underground (for example, in an underground parking) [14], the level of a stress becomes hazardous to health.

Besides a skyscraper inhabitants are very vulnerable at technical malfunctions. The most terrible is the fire. Therefore fire safety of skyscrapers is given especially large number of researches [15 - 17, etc.]. One more feature of skyscrapers is the raised pressure in all pipes. Otherwise not to lift water on the top floors [18]. Because of it breaks of pipes in skyscrapers - the phenomenon constant, quite often fills in on several floors. Hard it is necessary to the inhabitant of a skyscraper and at power outage. Elevators will stop - and the person will be locked in the apartment.

According to poll of Institute of comparative social researches, preferences of city dwellers now on the party of low low-density building [19]. The modern resident prefers to walk in the parks and squares, to see green plantings [20].

The most part of respondents have noted in questionnaires that they would prefer to live in houses from 3 or 4 floors. Slightly less respondents have shown willingness to live in seven - or the nine-floor house. At the same time nobody has agreed to acquisition of housing in the 22-storey building, even on condition of fine transport infrastructure in the area. It means that with growth of welfare inhabitants will leave residential districts with skyscrapers and big population density to more comfortable areas. And in multi-storey quarters there will be only outcasts who give nothing to the city but only consume his resources.

Survey conducted within the real research has yielded a bit different results. 51% of respondents (generally teachers and nonresident students) have preferred to live in low areas; 22% - in houses from 5 to 25 floors high; 27% - in the high-rise buildings located in the downtown. About preferences of the last group of respondents already it has been told above.

4 Discussions

Multi-storey and high-rise housing has arisen in the large cities as the answer to land platforms deficiency. People knew about it in Ancient Rome. Gradually multistoried housing became "standard" for a city landscape, and today most of the population on the Earth lives in apartment houses. And since the end of the 19th century as deification of an engineering thought also the first skyscrapers have appeared. Now in the world there is no megalopolis in which there would be no "highest" building.

However single skyscrapers in the downtown and inhabited residential districts on the suburb have absolutely different level of expediency comfort [21].

Esthetics, convenience, comfort, ecology of the majority the Russian multistoried new buildings very low. Increase in density of building leads to such pollution that the territory gradually turns in unsuitable for life not only in respect of comfort, but also from the medical point of view. The more the number of floors in the house and population density, the are higher probability that someone from residents will be an outcast and will damage the lift, will draw at an entrance, etc. And further the condition of housing will only worsen [22]. Faceless, depressive and in many respects the anti-human environment of multistoried residential districts really produces the outcasts and attracts strangers, promotes formation a criminogenic situation. In the West have faced it long ago. For example, in the USA in 1972 in the city of St. Louis have demolished a housing estate of Pruitt-Aygou. In the beginning it was considered as a modernism masterpiece, but has gradually turned into the most real ghetto. The American authorities then have understood that such social housing brings more problems, than advantage. In Russia there is a similar situation now.

The cities have to develop so that the quality of the urban environment increased, the soil and air were cleaned, population density and traffic flow decreased, access to social infrastructure was facilitated. Houses at most up to 3 floors high, without elevators, with underground parking is the most optimum at the price and comfort option. 1-2-3-storey lodges with the second inhabited floor, with profitable business (shop, a pothouse, etc.) on the first floor, with delivery of rooms on the attic and a warehouse in the cellar are correspond to traditions of the European and Russian architecture [23]. And domination of multistoried boxes over surrounding space puts the psychological pressure upon consciousness an individual.

However to find the land plots for low housing construction it is real only on suburbs of the cities. And it is most important "minus" of such building with which also other problems – transport try to keep step, infrastructure and also social – everything that is connected with absence in suburban settlements of kindergartens, schools and hospitals with policlinics [24]. Therefore resettlement people to remote suburbs doesn't solve a problem. It leads only to increase in number of motor transport and number of accidents, growth of CO2 emissions and increase in level of a stress from constant trips and traffic to the place of study, work of rest and back [25]. The remoteness from social infrastructure is one of the most important reasons which forces able-bodied population to be flown down to the cities with their uncomfortable multistoried housing and dense population, to raise and raise children in these conditions, to form a certain social environment, to accumulate psychological and medical problems.

5 Conclusions

Equations should be centred and should be numbered with the number on the right-hand side. High-rise housing construction is on the advanced positions in the construction sphere today. Existence of skyscrapers distinguishes any modern city, and ability to build them speaks about high development of construction technologies, intellectual designers' opportunities, readiness of the corresponding material and technical resources.

Skyscrapers (especially buildings of the mixed use) are rational in essence [26]. In them a large number of people on a small ground can be placed. It allows to save agricultural grounds from building and reduces the energy consumption and emissions of carbon in the environment connected with pendular migration (i.e. with moving of people, say, from suburbs, by work and back).

Skyscraper is also the integral element of prestige for the city and even the whole country. It is no wonder what is under 85% of the largest skyscrapers in the world constructing in Asia where special attention is paid to this aspect.

However high-rise housing construction in Russia has the restrictions. Most of the population in the cities are immigrants from rural areas in the first or second generation. They have still very strong requirement of a unification with the nature, requirement of close direct contacts with relatives, friends. Most of citizens tolerantly belong to buildings up to 25 floors. Skyscrapers cause the alerted relation. As the residence most of people prefer low areas or skyscrapers in the downtown. Multi-storey residential quarters in the suburbs with sad appearance are least attractive to housing. At the same time they remain popular because of low cost. The uncomfortable yards, long distances from the house to the house and to places of social infrastructure negatively influence mental and social wellbeing of inhabitants. People come to the big city behind happy life, but get depression and isolation. High-rise construction in Russia doesn't promote the solution of social problems though has indisputable pluses before low building. Rather it lays the foundation for new problems which consequences are coming to us.

References

1. Russian State Standard SNiP 2.07.01-89*

2. URL : http://www.statdata.ru/nasel_regions
3. URL : http://capitalgroup.ru/press/media/2012/7879
4. I. Ptuhina, T. Spiridonova, T. Musorina, S. Kanyukova, A. Rezvaia, MATEC Web of Conferences, **53**, 01049 (2016)
5. A.Y. Barkovskaya, M.P. Nazarova, Izvestiya VolGTU, **16 (5)**, 37-42 (2014)
6. N. V. Anisimov, E. A. Ivashechkina, Molodoy uchenyy, **11**, 1840-1842 (2015)
7. C. Ingraham, The Washington Post, March 18 (2016)
8. E. Pafka - http://hdl.handle.net/11343/52749
9. A. Yugov, A. Tymoshko, Modern industrial and civil construction, **10(4)**, 217–224 (2014)
10. A. Zaiter, High-Rise Building Definition, Development, and Use. *High-Rise Security and Fire Life Safety* (Elsevier Inc., 2009)
11. N. Petković-Grozdanović, B. Stoiljković1, M. Shubenkov, MATEC Web of Conferences, **73**, 06001 (2016)
12. A.A. Jamaludin, H. Hussein, K. Md. Tahir, MATEC Web of Conferences, **66**, 00098 (2016)
13. Z. Ivanova, I. Pryadko, MATEC Web of Conferences, **106**, 01012 (2017)
14. E. Romanova, Procedia Engineering, **165**, 1176 – 1183 (2016)
15. V. Pershakov, A. Bieliatynskyi, I. Popovych, K. Lysnytska, V. Krasheninnikov, MATEC Web of Conferences, **73**, 01001 (2016)
16. X. Zhixiang, T. Yong, Procedia Engineering, **45**, 705 – 709 (2012)
17. M. Qianlia , G. Weib, Procedia Engineering, **45**, 685 – 689 (2012)
18. URL : http://tinybop.com/Tinybop-EL07-Skyscrapers-Handbook-EN.pdf
19. A.S. Kurbatova URL : http://terraplan.ru/pdf/ecoreal_17_razvitie.pdf
20. I.I. Akulova, Gradostroitelstvo. Infrastruktura. Kommunikacii, **1(1)**, 21-28 (2015)
21. D. Korolchenko, V. Kholshchevnikov, MATEC Web of Conferences, **106**, 01038 (2017)
22. I.P. Pryadko, International Journal of Applied Engineering Research, **21**, 42147-42152 (2015)
23. V. Kasyanov, O. Landysheva, MATEC Web of Conferences, **86**, 04060 (2016)
24. A. Babayeva, E. Zakablukovskiy, T. Smetanina, N. Shmeleva, N. Shilovskaia, MATEC Web of Conferences, **106**, 01003 (2017)
25. URL : https://www.econstor.eu/ERSA2014_01736.pdf
26. S Abd-Wahab, A.I. Che-Ani, A. Sairi, N.M. Tawil, M. Abd-Razak, MATEC Web of Conferences, **66**, 00103 (2016)

Measures to reduce construction time of high-rise buildings

Leonid Kolchedantsev[1], *Aleksey* Adamtsevich[2], *Olga* Stupakova[1*] and *Alexander* Drozdov [1]

[1]Saint Petersburg State University of Architecture and Civil Engineering, 2-ya Krasnoarmeiskaya st., 4 St. Petersburg, Russia, 190005
[2]Moscow State University of Civil Engineering, Yaroslavskoye shosse, 26, Moscow, 129337, Russia

Abstract. The organizational and technological solutions for high-rise buildings construction efficiency increase are considered, primarily – decrease of typical floor construction time and improvement of bearing structures concrete quality. The essence of offered technology is: a concrete mixing station and a polygon mainly for load-bearing wall panels with starter bars casting are located on the building site; for reinforced concrete components manufacturing and butt joints grouting the warmed-up concrete mixtures are used. The results of researches and elaborations carried out by the SPSUACE in area of a preliminary warming-up of concrete mixtures are presented. The possibility and feasibility of their usage in high-rise buildings and of excess height buildings construction including cast-in-place and precast execution are shown. The essence of heat-vibro treating of concrete mixture is revealed as a kind of prior electroresistive curing, and the achieved results are: accelerated concrete strength gain, power inputs decrease, concrete quality improvement. It is shown that the location of a concrete mixing station on the building site enables to broaden possibilities of the "thermos" method use and to avoid concrete mixtures warming up in medium-mass structures erection (columns, girders) during the high-rise buildings construction. It is experimentally proved that the splice between precast elements encased with warmed-up concrete mixture is equal with conjugated elements in strength.

1 Introduction

The most important constructional and technological peculiarity of buildings of excess height (25-30 storeys) and high-rise buildings (100-150 and more storeys) is its significant height excess on dimensions in plan. From this it follows that the construction time of its substructure massively depends on typical floor construction time. 5-6 working days per storey is considered to be acceptable result according to literature [1, 2]. From this it follows

* Corresponding author: olgan_70@mail.ru

that the only substructure will be under construction about three years. Thus, a typical floor construction time decrease is quite an important task.

Because of limited sizes in plan decrease of typical floor construction time by means of compacting time schedules of different work types complete with traditional technologies is of little promise. The most typical floor construction decrease time effect may be got due to realization of organizational and technological conclusions being the point at issue of this article which are: location of a concrete mixing station and casting yard (mainly for wall panels manufacturing) on the building site; the warmed-up concrete mixtures usage for reinforced concrete components on site casting and butt joints grouting; concrete mixture preparation with 35^0C - 45^0C temperature at the polygon for medium mass structures concrete pouring and their subsequent maturing by using the "thermos" method.

There are substantiations of each listed organizational and technological solution below. It is important to note that only combined application of them enables to get the desired result that is typical floor construction time decrease of 1,5 - 2 days.

2 Methods

It is well known that the extra-strong concrete classes (B50-B80) are used for lower storeys construction and high-strength concrete classes – for upper storeys construction in high-rises building. Mentioned parameters of concrete physical and mechanical properties are estimated relatively (relative compressive, tensile, bending strength, etc.), what gives the opportunity to apply the lab and field tests results to concretes of different classes.

In addition to traditional (Gostovskaya) techniques, for example, GOST 10180-90 Concrete. Methods of determining strength on test specimens, for test of joint of precast-monolithic buildings was designed by a special technique. On active area of the "old" concrete which are the ends of combined panels the vertical inclined chases 4 mm wide and 6 mm depth were made. The concrete mixture of «new» concrete was warmed-up just before the placing with alternate electric current of industrial frequency (50 Hz) to + 50^0 C and + 70^0C temperature and then placed with vibration in splice cavity. For «old» and «new» concretes samples shear testing postgraduate student Mustafin R.R. created special method excludes the appearance of the torsional moment about «old» and «new» concretes contact plane [3].

The essence of concrete mixture thermal treatment that is a kind of a preliminary warming-up, its rational application field, used materials, and the achieved effect are presented in writing [4]. It is important to note here that only application of warmed-up concrete mixtures made at site polygon enables to get concrete transport strength in minimum time with minimum power inputs and required concrete quality provided. The grouting of butt joints between precast elements with warmed-up mixture provides the equality in strength of the joint and conjugated elements [5, 6].

The feasibility of a concrete mixing station location on a site of a high-rise building construction, its parameters are substantiated in writings [7, 8].

3 Results

3.1. The mobile concrete mixing station (CMS) location on high-rise building site.

The mobile concrete mixing station (CMS) location on high-rise building site enables to provide:
- uninterrupted concrete mixture supply with no depending on the road traffic;

- delivery cost minimizing for concrete mix and prefab reinforced concrete fabricated at the site polygon;
- costs reducing for concrete on-site transportation, due to partly piping, for example.

3.2. The production of reinforced concrete products on the on-site test.

The production of reinforced concrete products on the on-site test site using heated mixtures is schematically shown in Figure 1. The daily turnover of the molds is provided by the heat introduced into the mixture when it passes through the HVTCM device and the exothermic heat of the cement.

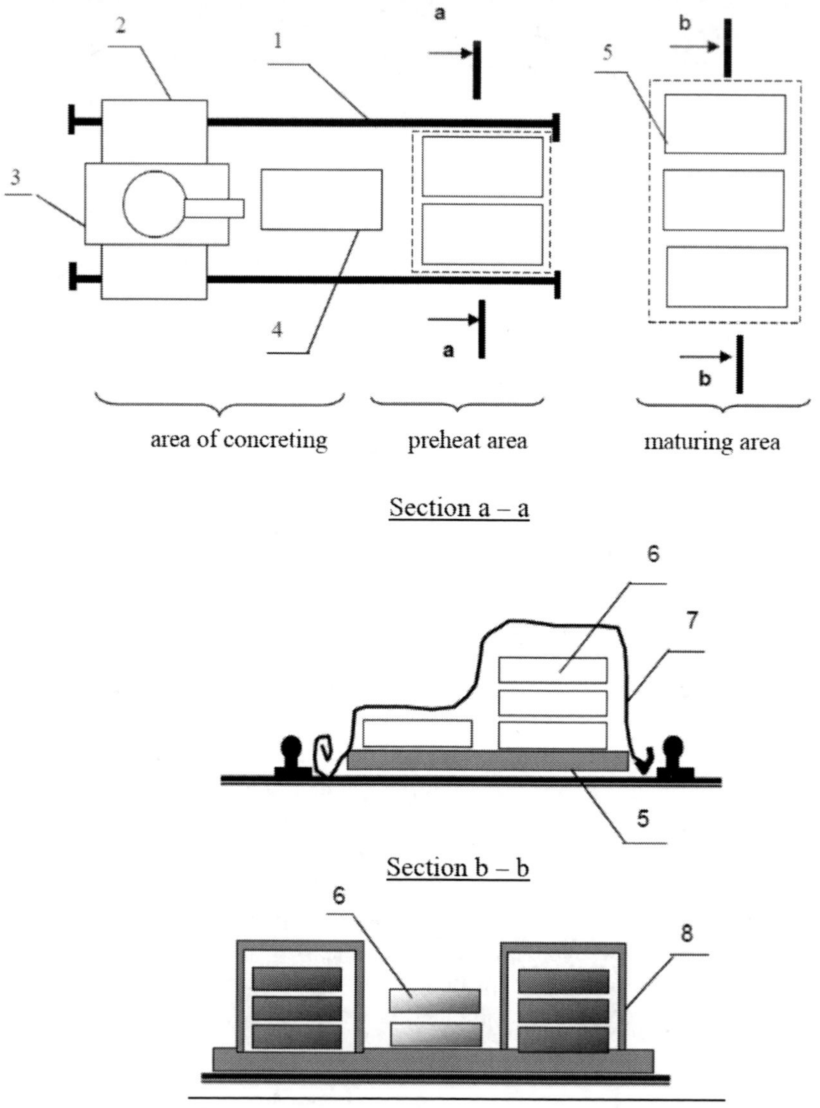

Fig. 1. Site casting yard scheme for combination of HVTCM and thermoreactive curing:
1 - rough track; 2 - self-moving portal; 3 – HVTCM plant; 4 - structure under concreting; 5 - thermoactive footing; 6 - molded product; 7 - flexible heat seal packaging; 8 - thermoactive cap.

3.3. Heat-vibro treating of concrete mixtures.

Heat-vibro treating of concrete mixture (HVTCM) is kind of preheating. Its essence is in continuous and overload (1 - 3 minutes) electrical curing of concrete mix along with vibration, extreme pressure and vapour exposure. Heatup and vibration are the main so the method of concrete mixture treatment and required equipment are called so. At that, vibration is used primarily as a way of concrete mix transportation during its treatment. Activating vibration property, vapour, excess pressure and electrostatic field action during electrical curing exerts in concrete characteristics improving, such as its strength against pure heating of concrete mix.

HVTCM plant general form, which is recommended to use, is presented in Figure 2. System's critical components are: concreting hopper; vertical heating tube; inclined heating tube with catch; gate-operating mechanism and electrical box (OFC).

On internal surface of vertical tube current-collecting devices are electrically secured assigned, which connected to alternating current system with voltage of 380/220 V. Electrodes have a triangular cross section, at that adjacent electrodes are parallel. Inclined pipe body is neutrally assigned, electrically insulated central rod with circular current-collecting device are co-axial assigned inside the case. The plant works as follow. Traditionally made and transported to site concrete mix fills into hopper with fully closed gate. For the sake of internal volume filling with concrete mix and close contact between mix and electrodes procuring the vibrator turns on for 3-5 seconds. Then voltage is applied on electrode cells. After concrete mix temperature required value realization (near 70 - 80°C), after 1,5-2 minutes traps open and vibrator turns on. Further processing of the mixture proceeds in steady state. Since in such a short heating time the concrete mix to this temperature, the fluid body liquid phase is heated, and cement and sand grains from it, the coarse aggregate is heated later due to the thermal conductivity from the warm-up components of the mixture, so that after exiting the plant the average mix temperature falls by 5-20°C (depends on the aggregate size and its amount).

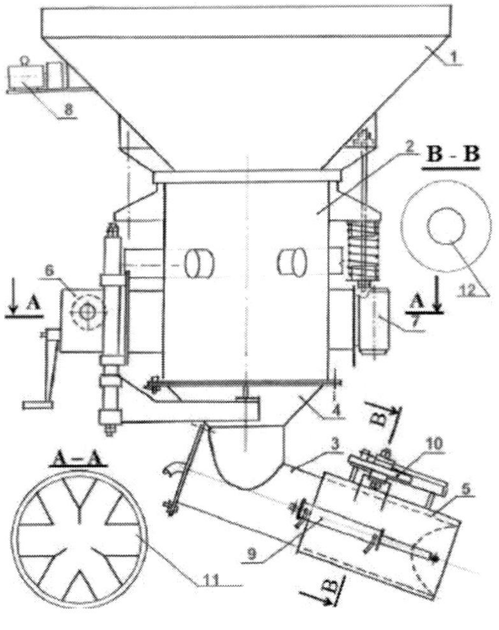

Fig. 2. HVTCM plant general form. 1 - concreting hopper; 2 – vertical heating tube; 3 – inclined heating tube with reducing pipe 4; 5 - catch; 6 - elevator; 7 - vibrator; 8 - pumping set; 9 - hydrocylinder; 10 - synchronising linkage; 11 – triangular electrodes; 12 – circular electrodes.

Processes during heat-vibro treating of concrete mixture may be described this way. According to the Joule-Lenz law, when a current passes through a concrete mixture with an electrical resistance, heat is released in it. High temperature increases reaction of cement hydration. Vibration leads to deflocculating of cement grains, partly its dispergation, which is complemented by chemical dispergation. Concrete mix preheating is accompanied by partly vapour formation; and since treating passes in closed volume, this leads to excess pressure formation near 0.01 MPa. Water viscosity reducing with high temperature and vapour existence, which penetrability is greater than water, in combination with excess pressure keep to moisture penetration in cement grains. The combination of these effects intensifies chemical reactions, leads to an increase of cement mass, involved in the interaction with water.

The usage of heat-vibro treating concrete mixtures enables:

- strength gain increasing (40 – 50% of R28 after 6 – 8 hours, 70 – 100% of R28 after 24 hours) with 2 °C/hour cool-down rate (Figure 3);
- energy input minimizing (up to 50 kWh /m3);
- improvement of concrete quality for warm-up mixes against thermal curing for density, strength, concrete-to-steel bond;
- deletion of irretrievable electrodes losing, heating wire, which is unescapable during concrete curing;
- man-hour reducing for concrete maturing.

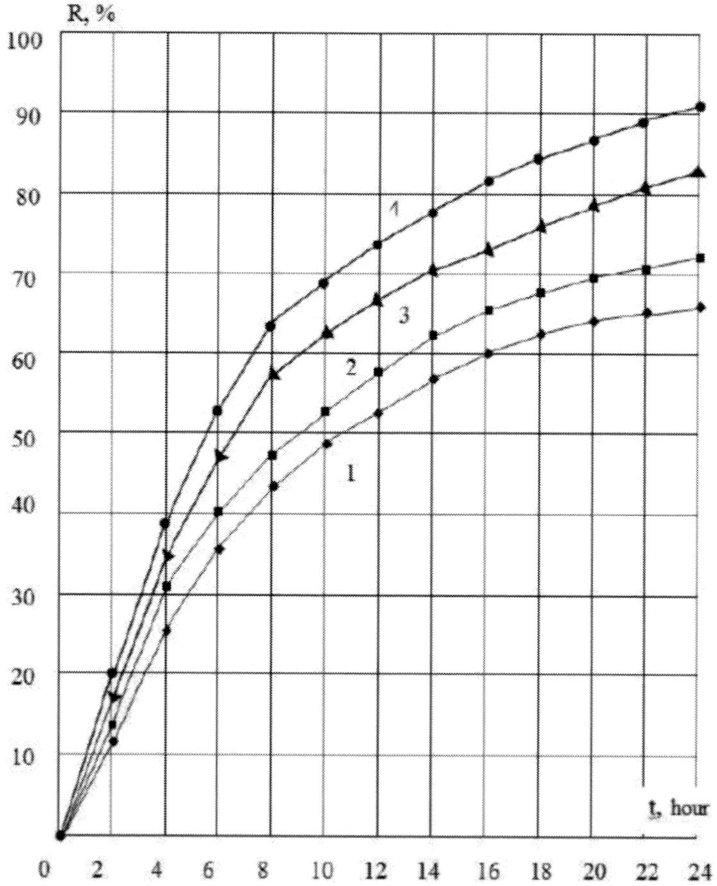

Fig. 3. Strength development for fresh concrete:
1 – heat-vibro treating with 50^0 C; 2 – heat-vibro treating with 60^0 C; 3 – heat-vibro treating with 70^0 C; 4 – heat-vibro treating with 80^0 C.

https://doi.org/10.1051/e3sconf/20183303062

3.4. Embedment of joints between wall panels and floor slabs

Embedment of joints of reinforced concrete wall panels and slabs with overlap of the rebar, heating the concrete mix allows you to:

- providing the pace of cast-in-place and precast construction comparable to panel buildings;

- arc welding operations minimizing during construction period of residential unit;

- erection of prefabricated buildings, which model of structure is similar to cast-in-place buildings, that is safety and lifetime rising for dynamic loads or man-triggered actions;

To ensure the rate of concrete cooldown in the splice no more than 2 ° C per hour, the splice formwork must be thermally active. It was experimentally proved that combination of concrete mixture pre-heating before its laying in the splice, treating the active surface of "old" concrete and heating formwork, providing a cooling rate of concrete at a splice no more than 2°C per hour, It makes it possible to obtain splices of prefabricated-monolithic buildings with equal strengths with monolithic concrete with the shear works structure [3].

Butt joint grouting technology with warm-up concrete mixes for cast-in-place and precast construction, developed by Kolchedantsev L.M., Stupakova O.G. and Mustafin R.R., is following (Figure 4).

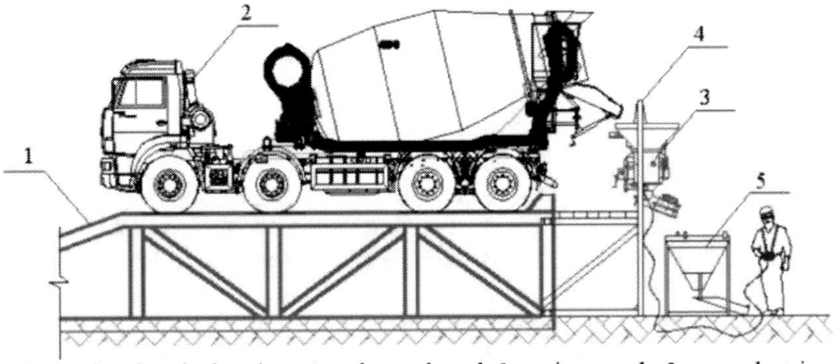

a) organization scheme for electrical curing. 1 – elevated road; 2 – mixer truck; 3 – pre-electric curing device for concrete mix; 4 –supporting frame; 5 – distributing hopper

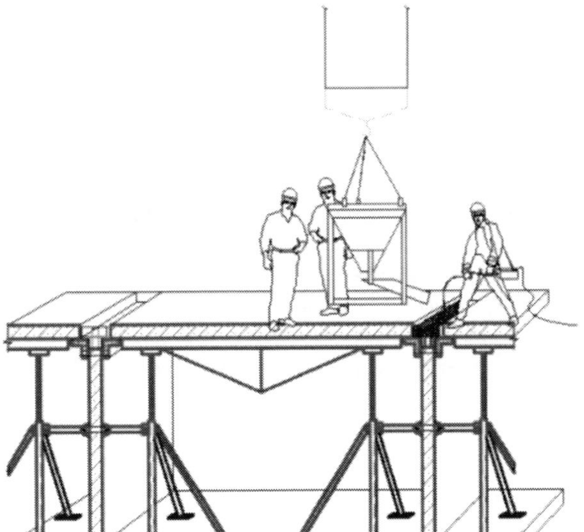

b) concrete placing and poly-roll

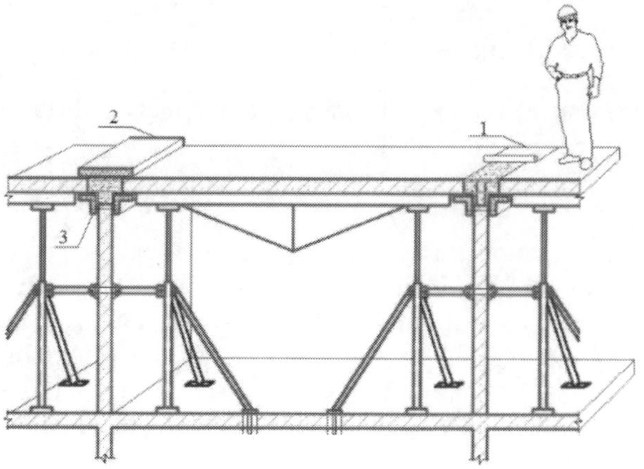

c) splice hiding PF and TFC: 1 – polyfilm; 2 – thermoactive flexible coating; 3 – thermoactive coating (TAC)

Fig. 4. Key phases for splice making with warm-up concrete mixes.

After the wall panels rigging up the formwork with heating elements is being installed and with its help the butt joint heats to 15-20°C temperature. On the elevated road in assembly crane action area the HVTCM plant with under 2 m^3 per hour productivity is being set. The device may work with energy from diesel generator with 100-150 kVA power in case of freely installed electric power absence. Concrete mixture with plasticizing agents increasing time of warmed-up concrete mixture placeability maintenance added in making period is being uploaded from concrete mixer truck to the HVTCM plant receiving bunker.

There are three bunkers 0,5- 0,75 m^3 volume located in HVTCM plant location area below its deep pipes with batch gate. The bunkers are set on trolley frame that can move relative to the HVTCM plant. During the winter time the bunkers are insulated and equipped with hinged covers. After being filled with heat-vibro treated concrete mixture first bunker is transferred by crane to working horizon for grouting the butt joints. At the same time the concrete mixture is put from the HVTCM plant to the bunker №2. The bunker №3 is backup in case of some deviation from the continuous-cyclical schedule correlating the butt joints grouting process and HVTCM plant work.

The concrete mixture after heat-vibro treating is densifying with pervibrator in butt joints grouting process. After joint grouting its open surface is being covered with film. The required concrete maturing mode in joint is maintained with formwork with heating elements.

Floor butt joints are grouted the same way.

3.5. The setting of stiffening cores columns and girders provided spatial rigidity and overall stability of a high-rise building

The formwork and reinforcement setting of these structures is performed with traditional technology. Considering these structures belong to medium or low-mass type the «thermos» method is recommended to apply for their grouting and maturing.

Considering the general orientation of this work, namely, the reduction of the time for the construction of high-rise buildings, and taking into account that the concrete-mixing unit is located directly on the construction site, the temperature of the concrete mixture output from the mixer may exceed the maximum permissible values specified in the Code of Practice SP 70.13330.2012 Bearing and enclosing structures, the updated version of SNiP 3.03.01-87. However, this recommendation can be implemented only after an experimental check of the timing of setting the cement at elevated temperatures.

4 Conclusions

Thus, the total effect of known technological solutions (concrete mixing station and polygons for floor and wall panels production location on a building site) in combination with a preliminary concrete mixtures warming-up applying for reinforced concrete elements production and grouting of the butt joints between them after their rigging up, and also the «thermos» method applying extension in medium and low-mass structures construction is in following: about 65-70% of high-rise building structures should be built not in the cast-in-place but in the precast and cast-in-place method that enables to decrease a typical floor construction time by 1,5-2 days.

Reference

1. Concrete placing and finishing equipment, Concrete construction, **5**, 4-11, (2004).
2. Concrete-making materials and accessories, Concrete construction, **5**, 17-24, (2004).
3. L. Kolchedantsev, R. Mustafin, O. Stupakova, Building Materials, **4**, 17-19 (2012)
4. L. Kolchedantsev, Intensification of the technology of concrete work on the basis of thermal vibration treatment of concrete mixtures (SPbGASU, St. Petersburg,. 2001.)
5. 5. L. Kolchedantsev, A. Vasin, I. Osipenkova, O. Stupakova , *Technological bases of monolithic concrete. Winter concreting: a monograph (under the editorship of L. Kolchedantseva)*, (publishing house "Lan", St. Petersburg, 2016)
6. R. Mustafin L. Kolchedantsev, Patent 2468158, The method of concreting structures with the use of permanent reinforced concrete and (or) reinforced cement formwork (2012)
7. L. Kolchedantsev, S. Volkov, A. Drozdov, Housing construction, **2**, 38-43 (2015)
8. L. Kolchedantsev, I. Osipenkova, Housing construction, **11**, 17 – 19 (2013)

 https://doi.org/10.1051/e3sconf/20183303063

Development of high-rise buildings: digitalization of life cycle management

Elena Gusakova[1,*]

[1]Moscow State University of Civil Engineering, Yaroslavskoe shosse, 26, Moscow, 129337, Russia

Abstract. The analysis of the accumulated long-term experience in the construction and operation of high-rise buildings reveals not only the engineering specificity of such projects, but also systemic problems in the field of project management. Most of the project decisions are made by the developer and the investor in the early stages of the life cycle - from the acquisition of the site to the start of operation, so most of the participants in the construction and operation of the high-rise building are far from the strategic life-cycle management of the project. The solution of these tasks due to the informatization of management has largely exhausted its efficiency resource. This is due to the fact that the applied IT-systems automated traditional "inherited" processes and management structures, and, in addition, they were focused on informatization of the activities of the construction company, rather than the construction project. Therefore, in the development of high-rise buildings, the tasks of researching approaches and methods for managing the full life cycle of projects that will improve their competitiveness become topical. For this purpose, the article substantiates the most promising approaches and methods of informational modeling of high-rise construction as a basis for managing the full life cycle of this project. Reengineering of information interaction schemes for project participants is considered; formation of a unified digital environment for the life cycle of the project; the development of systems for integrating data management and project management.

1 Introduction

Construction and operation of high-rise buildings is one of the most complex types of real estate development. Project ideas of high-rise construction are always unique and always connected with the development trend of megacities. To date, sufficient experience has been gained not only in the design and construction of such buildings, but also the data of their long-term operation and monitoring in Russia and many other countries [1, 2]. Bright examples of the implementation of such projects in Moscow are seven buildings of "Stalin" skyscrapers, the construction of the New Arbat, a complex of high-rise residential, hotel and office complexes of the business center "Moscow City".

Analysis of the experience of high-rise construction shows that the more high-rise buildings are built, the longer buildings are used, the more their engineering and

* Corresponding author: e_gusakova@inbox.ru

organizational-technological specifics are revealed. It is associated with large volumes of long-term investments for new construction, for the preparation of the territory, and for the coordination of relevant legal, regulatory procedures, financial relations and legal and urban issues [3]. Specificity is associated with the complexities and risks that almost all participants in high-rise construction projects - customers, investors, designers, builders, operators, expect [4]. The customer and the investor are certain risks of long-term forecasting of future costs and project management, designers have difficulty in accounting for specific loads (sediment, wind, terrorist attacks, fire, etc.), builders - each time the development of new organizational and technological solutions . At the operational stage - the complexity of the management company in ensuring the reliable operation of engineering networks and the safety of the facility. In a high-rise building of a heating, ventilation, air-conditioning, heat supply and cooling system and other engineering systems - this is a very complex set of interconnected functional systems of equipment and utilities. In addition, the operation processes of each high-altitude object are unique. This is due to the fact that high-rise buildings are multifunctional, they have functional areas in various combinations: residential, administrative-office, hotel, car parks, various technical rooms, shops, restaurants, cafes, entertainment public spaces for meetings, exhibitions, etc.

Meanwhile, the unique risks associated with high-rise projects lie not only in solving engineering and technical problems. System gaps and uncertainties in management are found, primarily in the life cycle management concepts projects. And this is quite understandable. All decisions that determine the purpose of the high-rise construction project, the functionality of the building, profitability, partners, organizational and economic mechanisms for selling and disposing of property are accepted by the customer and the investor at the early stages of the life cycle - from the acquisition of the site to the sale of the high-rise building [5]. Therefore, initially the majority of participants in the construction and operation of a high-rise building are far from the strategic management of the project life cycle.

2 Methods

It is obvious that high-rise construction projects require the development of a strategy for organizing and coordinating the stages and stages of construction and maintenance. The forecast of the life cycle of a high-rise development project is far from simple. This is a large amount of information on the project, for processing which, as experience shows, not enough computer technology alone. These issues are a separate, specific area of research that requires developers and architects, designers, engineers, ecologists, sociologists, demographers, representatives of different layers of the urban community to carry out a multifaceted and professional analysis of the project life cycle.

It should be understood that the approach to creating a digital environment for building projects was evolving. First, tools and systems for automating the production of paper-based documentation were created and widely introduced: computer-aided design systems for the production of drawings, specifications, and technological documentation. Then - the automated production management system for creating production plans and reports on its progress, office systems - for the preparation of text and table documents, etc. And it was natural, because for many decades the paper documentation was the generally accepted form of presenting the results of intellectual activity and the instrument of information interaction. Most of the existing management processes, as well as organizational, production and management structures were formed based on it. Until now, thousands of designers, engineers, designers, technologists, manufacturers, managers, employees of management companies and commercial structures are engaged in the development, distribution, storage and actualization of out-of-mass information files. Many participants of the project are brought up on standards, norms, rules and schemes, which are based on case management using paper documentation [6].

In the case of high-rise construction projects, as with most complex projects, the traditional approach to informatization has practically exhausted its efficiency resource to date. Restriction is the "inherited" project management structures that use only a small fraction of the potential of information and network technologies [7]. Practice has shown that the automation of the production of traditional paper documents does not solve not only the management problems, but also the more simple tasks of information exchange between numerous participants in the life cycle of a complex high-rise construction project (customers, designers, builders, operating and maintenance organizations, etc.). Different systems "speak different languages" and do not understand each other well. Information structures and systems were developed both from above and from below, on various principles of data structuring and software, which creates difficulties for the exchange of electronic data, complicates the integration of the information environment. When transferring data from one information system to another, it takes a lot of labor and time to unify data structures and re-encode, which fundamentally reduces the effectiveness of IT systems. Problems of digitalization of project life-cycle management are also related to the fact that the development of IT-systems is focused on informatization of the activities of the construction company, rather than the project. Features of the life cycle of the project are associated with a large number of project participants, the presence of highly complex, multidirectional and multi-criteria relationships: the investor - the customer of construction - the general contractor - subcontractors; specificity of cash flow associated with procurement of materials and equipment, hiring of labor, permanent relocation of workers from one construction site to another, etc. [8]. For this reason, it is difficult to accumulate and ensure the continuity of data on the stages and subsystems of the life cycle of a high-rise building. The developer, acting to achieve tactical and strategic design goals(developing the organizational and economic mechanism for the implementation of the project, planning the work and the timing of their implementation, the need for labor, resources, resources and resources, coordinating the scope of work with the customer and the contractor and other participants, etc.), constantly faces the uncertainty of the life cycle of the project and conflicting management tasks. At the same time, he must solve them in the absence of an organized information space for the project and decision support systems [9].

3 Results

Thus, the potential for improving the life cycle management of high-rise construction projects is largely exhausted, due to two problems:

• Methodological approach to computerization, which automates the activities of project participants and the processes of organizational, production and management structures, which until then were formed in the context of traditional workflow, is outdated;

• Specifics of construction, where the goal is a project (building, construction, real estate in general). Most software systems do not take this into account to the extent necessary, since they are developed by analogy with other industries for the needs and business processes of a manufacturing enterprise, construction firm or company. Those informatization in construction is not project-oriented [11].

4 Discussion

To solve these problems, new concepts of control digitalization, based on:

1. the formation of a single digital environment for the full life cycle of a high-rise construction project, which should reveal the potential for strategic management, taking into account the features of the facilities.
2. organizational reengineering of schemes of information interaction of project participants.

The progressive approach to informatization, to the structuring of data, their storage, archiving and updating, is based on constant information support of the life cycle of the project. All transformations of a real building object in time must be accompanied by changes in the data about it. Technical support of information systems, including means of collection, registration, accumulation, processing, transmission, display, output of information; devices for automatic information retrieval, communication and computer networks allow it to be carried out practically in real time. Ideally: the real building object is accompanied by its digital copy or virtual building object, which continues to exist and is updated at the stage of building operation. It can be realized in an integrated digital environment of the project, which ensures, under the condition of observing common standards, the sharing and continuity of information at all stages of the project's full life cycle [12].

The interplay of high-rise construction participants can also become more effective if it is implemented in the integrated digital environment of the project life cycle. First, during the construction phase, intensive exchange of results between project, survey and construction organizations, including the customer, the general contractor, subcontractors, suppliers and other project participants, often geographically separated from each other and using incompatible computer platforms and software, is needed. Then - at the stage of operation of the high-rise building. The life cycle of high-rise buildings has been realized for many decades, including possible upgrades, reconstruction or redevelopment, and, if necessary, the liquidation of a project that has exhausted its potential. Almost all projects are subject to repeated changes in the concept of the project and redevelopment. Changes are inevitable in the implementation of long-term projects. The need for changes arises during the life cycle of a building repeatedly and practically at any stage. Accordingly, there is a need for technology change management. They include long-term planning of the investment and construction process, consideration of the cyclic patterns of macroeconomics in the high-rise building, much more complex project analysis and consolidation of the stages and participants of the project. In practice, setting and solving such problems can only be achieved by large specialized development companies. But as a result, the adaptability of the project increases, which makes it easier to manage changes and is determined directly by the extent to which the most probable changes and directions of development of high-rise building project subsystems were provided for in the design decisions.

At each stage the project can be reflected by a completely unambiguous information object and model. During the life cycle, the initial model is sequentially detailed, new subsystems appear, the amount of data increases. Accordingly, the full life cycle of a high-rise building project can be represented as a boundedly increasing sequence of interconnected information objects. The tree of the project objectives and the direction of development are determined by its participants. At any stage of the life cycle, the information object or model of the project is an integrated environment for the formation of the composition of the work of this stage (where it is determined what exactly it is necessary to design, produce, mount, repair, re-equipment, what is the laboriousness of the work, performers, costs, etc.).

Management of changes from this point of view is methodologically close to genesis as a method of research. Genesis focuses on the future of the project, considers it as a developing system, all the characteristics of which are due to cause-effect relationships and are adequate to changing external conditions and factors of influence. Substantiation of solutions by methods of organizational and technological genesis is the initial condition of the project's fitness for changes, its adaptability [13]. Dynamic classes of models can be used for this. Traditional project analysis can be supplemented by taking into research feedbacks and information flows obtained on the basis of predicting the regularities and parameters of the life cycle of a project at pre-investment stages.

The analysis of the software of construction companies and developers shows that the creation of a integrated information model of the life cycle of a high-rise building is possible on the basis of integration and subsequent consolidation of data management systems and project management, i.e. PDM (Product Data Management) and PM (Project Management) classes. The functions of the PDM-system are to ensure the reception of data and provide tools at each stage to form the composition of the works, the functions of the PM system - to formulate a plan of these works at each stage. At the same time, the process of data exchange between PDM and PM, including the transformation of production problems into informational ones, should be convenient for work in an iterative mode for all project participants.

The analysis of the accumulated experience shows that the BIM (Building Information Modeling) technologies and CALS (Continuous Acquisition and Life Cycle Support) technologies play a special role in solving the problems considered.

BIM-technology can be used as a basis for creating software-compatible formats for data modeling of high-rise buildings. In practice, BIM is used for the most part to visualize design solutions and present them to the customer. Its use with respect to the full life cycle is limited to a low level of development of the classification of objects, works and resources in construction and, as a result, low compatibility of data formats.

The idea of information support of projects and information integration of the stages of the product life cycle formed the basis of CALS technologies (Continuous Acquisition and Life Cycle Support). This idea consists in abandoning the "paper environment", which traditionally reflects the development and sale of products, and the transition to an integrated information environment covering all stages of its life cycle. Information integration is the result of the fact that all automated systems used at different stages of the project life cycle do not operate with traditional documents (for example, scanned drawings) or even with their electronic displays, but with formalized information models describing the product, its production technologies and use. Models exist in an integrated information environment in the form of information objects. As necessary, those systems that need specific information objects for work can extract them, process them, create new objects, and place their work results in the same environment. It is a collection of distributed databases in which there are unified, standard rules for storage, updating, retrieval, and transmission of information through which paperless information interaction takes place between all participants in the product life cycle.

The integrated digital project life cycle includes a full set of indicators, documents, classifiers and codifiers of information, files, databases, knowledge bases, methods of using them in the subject activity, as well as the methods of representation, accumulation, storage, transformation, a specific system to meet the information needs of users in the right form and at the required time. At the same time, information, once created, is stored in a single digital lifecycle environment, is not duplicated, does not require any transcoding in the exchange process, preserves its relevance and integrity. In order to make all this possible, information models and corresponding information objects should be standardized, including modeling tools, exchange file formats, program interfaces for access to shared data, application protocols - product data models.

For the leading industries (mechanical engineering, aerospace industry, which produce the same type of products), the international community has developed standards STEP (from the English STandard for Exchange of Product model data. The content and organization of information objects is determined by the features of the domain and the information needs of users of the automated system, their goals and objectives. At the same time in high-rise construction each object is unique, which significantly complicates and increases the cost of creating compatible data formats.

5 Conclusions

Thus, managing the life cycle of a high-rise development project is an urgent task that requires serious scientific and project research. Traditional schemes for managing the life cycle of the project and the interaction of all participants in the high-rise construction project should be substantially modernized in the context of Building Information Modeling and Continuous Acquisition and Life Cycle Support. At the same time, the development of systems for integrating data management and project management will ensure a multiple increase in the efficiency and competitiveness of a high-rise building project at all stages of its life cycle.

References

1. E. Gusakova International Scientific and Technical Journal "Real Estate: Economics, Management", **2**, 61-66 (2009)
2. A. Orlov, Economics and Entrepreneurship, **6-3 (59-3)**, 545-548 (2015)
3. P. Grabovyi, E. Gusakova, S. Lunyakov International scientific and technical journal "Real Estate: Economics, Management", **3-4**, 74-79 (2014)
4. S. Baronin, A. Porshakova Journal of Izvestia of the Penza State Pedagogical University, **24** (2011)
5. E. Gusakova, Yu. Ushakova International Scientific and Technical Journal "Real Estate: Economics, Management", **1**, 54-60 (2015)
6. T. Kuz'mina, S. Sinenko, A. Slavin Industrial and civil engineering, **6**, 71-75 (2016)
7. A. Pavlov, E. Gusakova *Fundamentals of organization and management in construction* (Publishing house URAIT, 2016)
8. A. Volkov,V. Chulkov, D. Korotkov Advanced Materials Research, 2577-2580 (2014)
9. A. Khripushin, A. Ginzburg Vestnik MGSU. **3**, 207-211 (2013)
10. A. Volkov , P. Chelyshkov D. Lysenko Procedia Engineering **153**, 833–837 (2016)
11. A. Ginzburg Industrial and civil engineering, **9**, 61-65 (2016)
12. E. Gusakova, A. Krugina Real Estate: Economics, Management", **32**, 84-87 (2012)
13. A. Ginsburg Information Resources of Russia, *5,* 28-31 (2016)

Economic justification of costs at inspection of industrial safety of high-rise marine structures

Pavel Garibin[1], *Evgeniy* Ol'khovik[1,*] and *Igor* Rastorguev[1]

[1]Admiral Makarov State University of Maritime and Inland Shipping, Hydraulics Structures Department, 198035, Saint-Petersburg, Russia

Abstract. The task of technical and economic regulation within mutual international recognition of testing laboratories are considered. Codes and procedures within requirements of international ISO/IEC standards of a series 17000 for elimination of non-tariff barriers and interlaboratory exchange of experts in the field of high-rise marine construction are considered. In paper, the methods of assessment of formation of economically justified cost of works at inspection and monitoring of technical condition of high-rise marine wharf engineering port structure based on settlement and actual labor input were applied. For the countries of EU, data on the average cost of works of testing laboratory within a week have been taken as a basis. Such approach will be objective as considers only expenses on obligatory actions in the course of inspection of technical condition of port engineering constructions. The analysis of public results of financial activities of the accredited organizations allowed to calculate the main indicators of the size of necessary profit and overheads at observance of all requirements imposed to test laboratories including taking into account their future technical development. The offered practice corresponds to the general direction by mutual international recognition of independent testing laboratories and can be use in the future.

1 Introduction

Ensuring of industrial safety is an actual problem as it affects the most important aspects of human life. Modern system researches of economy of industrial safety [1] and industrial accidents [2] are known since the beginning of the 80th years. In article [3] the economic measures which belong to designing of industrial processes and ensuring of their safety are considered. The quantitative criteria based on consequences simulation and in total with the economic indicators with use of multi-purpose optimization of programming are offered. The NSGA-II algorithm [4] it can be obtained as the procedure of optimization for which the amalgamation of the emergency costs calculation for various constructive schemes included a high-rise construction and marine overhead road is offered.

In work [6] authors consider economic and technical solutions for a possibility of prolongation of service life of safety of crucial components when they reach the termination of life term. The qualified decision about continuous work outside the service life is offered

* Corresponding author: OlhovikEO@gumrf.ru

to be carried out based on estimates on two indexes: instrument assessment of technical condition and economic assessment of benefit.

In paper [7] authors on the example of Shenzhen's port consider the questions of assessment of the economic losses in the industry related to destruction of marine port, recommendations on risk management based on a statistical approach and a comprehensive analysis of Petri Net are offered.

There is no uniform and centralized system for technical safety assessment for the maritime Port of Seattle in article [8] (the Pier and wharf structures belong to different owners), that leads to some problems. So, in case when one part of port pier is in unsatisfactory technical condition it naturally influences industrial safety of the adjacent constructions and all infrastructure of port in general. Financial consequences of remedial measures for business and engineering practice were offered.

The similar idea was checked in work [9] where the damping properties of the wharf 375 meters long were investigated. Although the wharf was designed as a number of independent sections, the research had shown that the construction part and pipelines located on the top part of the wharf connected the adjacent sections together and had significant effect on dynamic behaviour, especially in the longitudinal direction. In current time in Russia there are about 10 thousand port hydraulic engineering constructions (on river and at marine), at the same time about 30% from them are in a critical condition and another 30% have restriction at operation. The similar situation takes place in other countries, where the main trend is the decommissioning (closing) of small port objects and transfer of freights to large port terminals, which are hubs and overload several million tons of freight per year.

2 Methods

As the main method for determination of initial maximum cost of works in Russia uses the resource-indexed method or a way of the open auction (an auction with decrease), that we for the analysis used open data on tenders for the last five years. For the analysis of pricing in the European Union we used cost data handbooks of the base prices "RS Means" and "Spon's" which give the average prices for the last time period on the actual performed works.

All wharf structures are objects of technical regulation, their condition has to be checked with the set periodicity - examination procedure or confirmation of conformity. Carrying out of these procedures is an mandatory condition and it is under the supervision of federal or local authorities, but for objectivity the inspection is carried out by the test centres (laboratories) having accreditation according to the international standard ISO/IEC 17025:2005 "General requirements for the competence of testing and calibration laboratories" (in Russia - State Standard ISO/MEK 17025-2009).

Some states gradually reduce the responsibility for maintenance of transport safety by attraction to examination of private companies, and then they are responsible for complete responsibility for checks and cannot reduce standards on qualification of the specialists who are carrying out inspections. The similar tendency exists also in Russia, the requirements to experts are constantly supplemented, the assessment of qualification is partially brought out of control of self-regulatory organizations, checks by expert community become stricter, a part of examinations was transferred from regional to federal level. Moreover, federal and local authorities conduct work on introduction of economic interest for owners of hazardous production facilities.

In 2015 the European commission based on offers of national authorities of five countries created the Unified register of certification agencies and test laboratories of the EU where 950 certification agencies and 2300 test laboratories are working according to 34 Technical regulations. They issued 450 thousand certificates of conformity and registered 1.5 million conformity Declarations.

The "Rapprochement of the technical regulation systems of and standardization of the EU and the Russian Federation" project (EuropeAid/132827/C/SER/RU) recommends to limit the scope of the technical regulation of the Customs union concerning buildings and constructions by only the basic requirements to buildings and constructions which will be the foundation for establishment of essential characteristics for construction materials and products in technical standards to these regulations, and also forms and schemes of conformity confirmation of construction materials and products, requirements to the state supervision, etc. The similar system of technical regulation in construction branch has been operating in Europe since 1989 (in Russia from 2009 in updated version).

Objective calculation of works cost shall include all costs incurred at implementation of port constructions inspection and besides according to the technological scheme of works performance accepted at the entity (described in the Quality Guidelines) because different types of works require various personnel qualification and use of the equipment. Our researches on change of works cost on examination and monitoring of marine and inland wharf structures for the last five years have shown the following:

- there is an increase in number of the organization's (test laboratories) rendering services in the field of examination of industrial safety that in turn increases the competition;
- the initial contractual cost of works on inspection of technical condition remains at one invariable level for a long time;
- price reduction (underbidding) is gradually increased during the tender procedures.

Phase 1: Preparation;

- the works connected with the conclusion of contractual commitments and ensuring of quality management.

Phase 2: Initial operations on inspection of technical condition;

- analysis of technical documentation;
- development of the program of inspection technical condition.

Phase 3: carrying out of visual inspection, measurement, surveying;

Phase 4: carrying out of instrumental inspection, test materials by non-destructive testing method;

Phase 5: cameral data processing and drawing up of reports;

Phase 6: registration of measurements protocols, delivery of work results to the customer, closing of the agreement.

Works on agreement of the accounting with the public supervisory authorities [10] and execution of conformity Declarations (currently these works are performed by the owner of construction or the organization which carries out exploitation).

The works on safety concerning the environment and work on automation of monitoring of port hydraulic engineering constructions [11] or the water area of approach channels [12] can be included as a part of the general works at carrying out inspection and monitoring according to the Russian acts of technical regulations (No. 384, 620, 623) and the European Regulation (EU) No 305/2011.

For the purpose of an economic justification of expenses on examination of industrial safety of marine wharf structures it is offered to take as a basis the technological regulations of works by means of which to calculate specific labor input for standard objects and average cost of experts wage depending on their qualification. Such approach will be the most objective as considers only direct expenses on obligatory actions in the course of inspection of technical condition of marine wharf structures. At the same time all works can be estimate as percentage according to the offered partition on Phases that also provides objectivity as the part of works cannot be performed because of features of construction, i.e. it can be dispensable.

The statistical data of the state authority responsible for observations in the field of work and social protection were use for determination of the average level of the salary. For more objective assessment search was carried out for concrete regions. The labor market research, i.e. the analysis of open vacancies was carried out in addition.

In calculations there was used the following scheme which is direct calculation of salary only for the performed works on monitoring of industrial safety by group of specialists with separation into individual works:

$$SL = \sum_{i=1}^{n} T_{ij} \cdot C_j \qquad (1)$$

where SL – personnel salary, T_{ij} – labor input of i operation performance by j specialist, C_j – average hourly earnings of j specialist (expert).

At the same time the following scheme was applied to the certain experts having various competences different the requirements imposed to the expert:

– labor input of operation performance by the expert was estimated with coefficient 1.0;

– labor input of operation performance by the engineer (in the field of tests or cameral processing) was estimated with coefficient 1.2;

– labor input of operation performance by the expert with considerable experience (the senior expert, the head of department, the chief engineer, etc.) was estimated with coefficient 0.8.

Thus, all performed works were carried out conditionally by conventional "expert" that facilitated a possibility of calculation of labor input for other conditions, other object or other region.

As a result, the basic cost of works can be calculate on the following formula:

$$C = \frac{T \times C_j}{K_C} \times (1 + Kp) \qquad (2)$$

where C – basic cost of works on monitoring of technical condition (examination of industrial safety); T – standard (specific) labor input, the amount of labor input of works on inspection in [persons]•[hour]; C_j – average hourly earnings of j specialist (expert); K_C – the coefficient establishing a wages share in total costs for examination of industrial safety (0.35÷0.5); K_P – the coefficient establishing a profit share in the cost of works – profitability (0.1÷0.15).

For the accounting of special factors which complicate (simplify) the process of work performance on inspection of high-rise construction or marine overhead port engineering constructions, the additional coefficients can be applied to a formula (2), for example, depth of construction, weather conditions, the adverse period of year, etc.

3 Results

The offered concept for justification of works cost on confirmation of technical condition of constructions is more flexible and corresponds to modern recommendations as European commission, and norms which are applied to testing laboratories in Russia.

Our research confirmed the results of work [13], which was established strong correlation between the safety budget and the industrial accident rate enabled modelling of an industrial safety system with these variables as the input and the output, respectively.

The made analysis of practice for the last 3 years on several organizations has shown that the size of the overheads connected with the providing of quality of laboratory management and metrological requirements constitutes up to 25%, and profit constitutes to 10%. Such economic indicators provide high quality of works performance and scientific and technical development of enterprise, including training and retraining of personnel. Level of the salary fund depends on a similar indicator in the chosen region and can be adjuste by various coefficients.

Always there is some interrelation between the level of industrial safety and the budgetary costs of monitoring. It is confirm as our researches of emergency cases for the last five years and the international practice – [14]. For maintenance of high level of safety of designs use of special systems of management [15] and project management, including and risk analysis is possible. Expenses on these actions have to be shifted to the owner of a design or the organization, which exercises control and operation. We compared the methods that were suggested in this paper with methods from well-known international agencies – «RS Means»[16] and «Spon's»[17], who publish the Engineering Price Book (or Cost Handbook) and obtained the correct correlation.

4 Conclusions

Offers on basic structure of works on examination of industrial safety of marine wharf structures are developed. The economic justification of costs is based on the acts of technological regulations of works performance and average compensation of experts and test laboratory. It is revealed that the actual cost of services of the test labs for inspection of technical condition of buildings and constructions can significantly differ depending on the country of stay. At the same time, the volume of the performed works according to technical regulations is approximately identical in the different countries. Different approach when determining and forming cost for basic works is the main reason for such distinction [18-20].

The essence of settlement determination of the price of technical inspection depends on the level of the salary of the experts who are directly carrying out technical observation. The problem of the international rapprochement of testing laboratories consists in creation of equally acceptable working conditions of experts, including and in compensation. This method has been used in this paper for what the analysis of a large number of works from inspection of technical condition of marine wharf engineering constructions has been made.

The proposed practice corresponds to the direction of mutual international recognition of independent testing laboratories.

References

1. Y. Walter On the economics of industrial safety. Law and Contemporary Problems. **38**, 669 (1974)
2. J. Chelius The control of industrial accidents: Economic theory and empirical evidence. Law and Contemporary Problems. **38**, 700 (1974)
3. S. Eini, H. Shahhosseini, M. Javidi, M. Sharifzadeh. *Process Safety and Environmental Protection*. **104**, 254 (2016) doi: 10.1016/j.psep.2016.09.010
4. L. Pan, K. Li, W. Xue Multi-objective optimization for building performance design considering thermal comfort and energy consumption, *35th Chinese Control Conference (CCC),* (2016) doi: 10.1109/ChiCC.2016.7553789
5. S. Eini, B. Abdolhamidzadeh, G. Reniers, D. Rashtchian Optimization procedure to select an inherently safer design scheme. *Process Safety and Environmental Protection,* **93** (2015) doi: 10.1016/j.psep.2014.05.002
6. M. Shafiee, I. Animah, N. Journal of Loss Prevention in the Process Industries. **44**, 299 (2016) doi: 10.1016/j.jlp.2016.09.013
7. X. Tan, J. Lam, Y. Zhang Economic impact of port disruptions on industry clusters: A case study of Shenzhen. *International Conference on Transportation Information and Safety (ICTIS),* (2015) doi: 10.1109/ICTIS.2015.7232069

8. C. Caudill, J. Hingle, S. Asavareungchai, A. Sidler The Execution of a Comprehensive Dock Condition Survey. *Proceedings of Ports '13: 13th Triennial International Conference,* (2013) doi: 10.1061/9780784413067.060
9. R. Boroschek, H. Baesler, C. Vega Experimental Evaluation of the Dynamic Properties of a Wharf Structure. Engineering Structures. **33**, 344 (2011) doi: 10.1016/j.engstruct.2010.10.014
10. A. Nicholson Why Regular Inspection of Transportation Facilities is Critical to Public Safety. *Structures Congress.* (2006) doi: 10.1061/40889(201)161
11. P. Garibin, E. Ol'khovik, S. Shabanov The methodology of organization of continuous monitoring of technical state of hydraulic structures of water transport. News of higher educational institutions. Construction. **674**, 58 (2015)
12. M. Champiri, S. Sajjadi, S. Mousavizadegan, F. Moodi Journal of Intelligent and Fuzzy Systems. **32**, 1945 (2017) doi: 10.3233/JIFS-161411
13. A. Afonin, E. Ol'khovik, A. Tezikov Development of the assessment methods of anadromous depths on the northern sea route depending on the detail of survey of the bottom relief. Vestnik Gosudarstvennogo universiteta morskogo i rechnogo flota imeni admirala S.O. Makarova. **38**, 62 (2016) doi: 10.21821/2309-5180-2016-8-4-62-68
14. G. Choi, B. Loh Control of Industrial Safety Based on Dynamic Characteristics of a Safety Budget-Industrial Accident Rate Model in Republic of Korea. Safety and Health at Work. (2016) doi:10.1016/j.shaw.2016.11.002
15. S. Bovteev, S. Kanyukova Development of methodology for time management of construction projects. Magazine of Civil Engineering. **62**, 102 (2016) doi: 10.5862/MCE.62.10
16. *Cost Planning and Estimating for Facilities Maintenance*, RSMeans Data. (2016)
17. J. Smith, D. Jaggar, P. Love. *Building cost planning for the design team*. (2016)
18. C. Ghani, N. Zaimah, N. Myeda, A. Shah Ali. Operations and Maintenance Cost for Stratified Buildings: A Critical Review, *MATEC Web of Conferences*. **66**, (20160
19. H. Krstić, S. Marenjak Analysis of buildings operation and maintenance costs. *Građevinar*. **64**, 293 (2012)
20. A. Suffian Some common maintenance problems and building defects: Our experiences. *Procedia Engineering,* **54**, 101 (2013)

Socioecological Aspects of High-rise Construction

Michael Eichner[1*] and *Zinaida* Ivanova[2]

[1]German University in Cairo – GUC, New Cairo City – Main Entrance of Al Tagamoa Al Khames, Egypt
[2]Moscow State University of Civil Engineering, Yaroslavskoe shosse, 26, Moscow, 129337, Russia

Abstract. In this article, the authors consider the socioecological problems that arise in the construction and operation of high-rise buildings. They study different points of view on high-rise construction and note that the approaches to this problem are very different. They also analyse projects of modern architects and which attempts are made to overcome negative impacts on nature and mankind. The article contains materials of sociological research, confirming the ambivalent attitude of urban population to high-rise buildings. In addition, one of the author's sociological survey reveals the level of environmental preparedness of the university students, studying in the field of "Construction of unique buildings and structures", raising the question of how future specialists are ready to take into account socioecological problems. Conclusion of the authors: the construction of high-rise buildings is associated with huge social and environmental risks, negative impact on the biosphere and human health. This requires deepened skills about sustainable design methods and environmental friendly construction technologies of future specialists. Professor M. Eichner presents in the article his case study project results on implementation of holistic eco-sustainable construction principles for mixed-use high-rise building in the metropolis of Cairo.

1 Introduction

Social and environmental problems of high-rise construction are more and more discussed in relation with the growing number of high-rise buildings and the expansion of their functional purpose and urban presence. In different countries there are different approaches to determining the number of floors of high-rise buildings. According to Russian building codes, buildings that rise from 75 meters (public objects from 50) to 120-150 meters, with the number of floors not exceeding 30, are classified as high-rise buildings. All buildings higher than this are usually called super-high buildings or skyscrapers. Tower buildings are mostly multifunctional, accommodating offices (International Business Centre, Seul), banks, stock exchanges (Taipei Financial Centre, Taiwan), hotels and casinos (Las-Vegas), apartments (Moscow City), educational buildings (Moscow, MGU). There are high-rise apartment buildings, such as the well-known residential complexes "House on

*Corresponding authors: ivanovazi@mail.ru

Mosfilmovskaya" in Moscow, the "Aqua Tower" in Chicago or the "Bosco Verticale" in Milano. If high-rise constructions were considered to contribute economic and innovative value, they are also responsible for many social and urban problems, changing the faces and atmosphere of cities and can have negative impact on the public live, city infrastructure, the environment and living quality in cities. During last decades these implications became more acute, as the number of storeys of buildings grew, their number increased, with over densification of city centres and over forming former low-rise districts of established quarters with out of scale point constructions.

Today, the attitude towards high-rise construction is ambivalent. There are supporters and opponents of tall buildings, both among specialists, architects, builders and residents.

2 Literature Review

Works devoted to the development of architectural typology and the history of high-rise buildings are manifold. The first works appeared in the second half of the twentieth century. In particular most works, published at that time, the history of the formation of various types of high-rise buildings was examined [1] and their constructive and technological evolution [2]. The dispute between supporters of high-rise constructions and opponents was controversial discussed in the press, scientific conferences and public hearings during the second half of the twentieth century. Some authors believed, that multi-storey construction is the bright future of buildings, helping to solve problems that are associated with the growth of the population of cities. Today, more than half of the world's population lives in cities. This figure will soon approach in Europe 85% and in Russia 74% [3]. Land is becoming more expensive in cities, with one square meter of land in the historical centre of Moscow of average 21,000 rubbles. This price varies from 15,800 to 18,836 rubbles in different parts of the capital [4]. The rise in land prices is the main reason for the construction of high-rise buildings, according to some authors. [5]. Other researchers object this argument and call the rise in land prices a secondary factor. They put forward the desire of architects, town planners, city authorities to build something that stands out, which would become a symbol of the city and surpasses the height of buildings in other cities. In other words, the problems of the development of high-rise buildings and their architecture can not only be seen in relation to social or technical needs, but also in the auspicious images of high-rise towers and their possibility to create an accent in cities, raising their status among the world's cities by displaying advances in architecture and technology [6,7]. Tall buildings become symbols of technological progress, prestige and define the face of cities.

Equally important is the economic side of high-rise construction: the majority of high-rise buildings are commercial with high economic and profit advantages, as more square meters can be built on a smaller site areas, generating higher leased for offices [6]. The subsequent high costs of operation will be compensated by higher rents, achieved.

The attitude of different authors to the complex problems of designing and maintaining technical systems like communication, ventilation, technical operation, fire safety of high building structures is also ambiguous. Problems like aerodynamics, big ventilation shaft areas per floor, complicated heating and control systems inevitably arise during construction and operation and require high planning and construction effort. From this point of view, high-rise buildings should be "smart" and "low-tech", minimizing necessary technical systems and following the idea of "bioclimatic architecture" in areas with complex climatic conditions [8,9]. Social and environmental aspects of high-rise constructions are the most controversial. According to some Russian authors, "today, high-rise construction is perceived and evaluated by the population extremely negatively, rejected as a disharmonious and ugly creation of a technocratic civilization, that is aggressive and hostile to the centuries-old historic cities" [5]. Other Russian authors believe, that in the next 10 years skyscrapers will account for up to

7% of the total annually newly built area of Moscow's apartments, since many Muscovites would like to live in high-rise buildings [10]. The academic of the Russian Academy of Architecture and Building Sciences Yu.P. Grigoriev, former deputy chief architect of Moscow holds a special point of view. In his opinion, high-rise construction should not at all be residential (residential development should be within 9-12 floors), despite the fact that it is profitable to build high-rise residential buildings [11].

3 Research problem and objectives

The purpose of this article is to identify socio-ecologic problems of high-rise architecture, problems that affect physical and psychological conditions including the well being of people, giving them a feeling of comfort or discomfort. These problems were examined by interviewing inhabitants of high-rises to find out, which negative consequences follow the construction of high-rises. The article uses the results of a sociological survey of Muscovites and polls, conducted at the National Research Moscow State University of Civil Engineering. Respondents were students, studying in the field of "Construction of unique buildings and structures" (high-rise construction). The surveys were conducted in November 2016 and November 2017. In addition, methodological analysis of scientific conference publications and public hearing materials was conducted, including interviews with experts, government officials to analyse opinions of various social groups. The second author of the article, Professor M. Eichner carried out a case study research project on the possibilities to enhance high rise architecture of mixed-use housing complexes by a comprehensive implementation of sustainable housing standards during the design phase of projects, to find out, whether in capital cities of developing countries these instruments can minimize harmful influences on the biosphere and enhance habitat condition of inhabitants of tall buildings in metropole cities.

4 Results

4.1 Socio-ecological problems of high-rise architecture

Socio-ecologic aspects of high-rise buildings and vertical districts characterize the level of comfort and safety of people in buildings and the quality of the surrounding environment.

4.1.1 Ensuring the safety of high-rise building users

Increased effort is necessary to ensure convenient and save inside space conditions, balanced room temperature, air quality, acoustics, hygienic factors, water supply and sanitation in high-rise buildings for residential, hotel, office purposes. Uninterrupted operation of elevators is essential in order to ensure the movement of passengers "up and down" in a regulatory regime, limiting the waiting time of elevators according norms, with waiting times not interrupting the functional rhythm of the building. High-rise buildings are often equipped with high-speed elevators, which have reached a high technological level but are contraindicated in some cases for passengers with certain diseases.

Security problems are manifold and also related to engineering systems. Modern high-rise buildings are equipped with mechanical ventilation systems, which have to be protected of fire and smoke distribution but at the same time against terrorist acts with poisonous substances at air intake devices. The probability of death of a large number of people is higher in comparison to low rise buildings in case of natural catastrophes. The authorities of Macon County (North Carolina, USA) announced in December 2006 an 11-month moratorium on issuing permits for high-rise constructions. They argued this prohibition by the fact, that high-rise buildings have a larger population density and, consequently, a greater risk of death and injury of people during natural disasters [12].

Urban integration of tall buildings and security issues are directly related. Modern cities, especially metropole cities, are highly equipped with underground communication systems and infrastructure, different geological formations, landslide threats, which creates the danger of subsidence and destruction of buildings. In addition, the absence of strong winds and seismic danger, minimum humidity and good load-bearing capacity of the soil are required, raising issues of choosing suitable natural, climatic, soil and location quality conditions [13].

4.1.2 Fire safety as a complex and specific high rise problem

Building codes in all countries require increased fire safety conditions and fire-resistant load-bearing structures for tall buildings with specially designed details for load-bearing walls, fixing elements and columns, including the use of fire-resistant building materials.

Tall building structures require the presence of fire compartments, which divided the building in different sections, to limit the spread of fire. Special requirements are also imposed on the number of staircases as main escape routes in case of fire. According to Yu. Gluhovenko (the head of the Centre for Complex Security of the Central Research Institute of the Ministry of Construction of Russia), it is necessary to use fire lifts in high-rise buildings to evacuate people in emergency situations. If only stairs are used for this purpose, evacuation can last several hours. Such elevators should have specialized cabins and lift shafts [14].

4.1.3 Ecological impact and use of natural lighting in high-rise buildings

Of particular concern is the negative impact of high-rise buildings on the biosphere and human health [15,16]. According to insolation standards, high-rise buildings require certain distances between buildings of at least 100 meters, to minimize shading of lower floors even if all norms are observed [17]. In residential buildings, apartments on lower floors are uncomfortable, their economic quality and market value is low. According to modern building practice, ground floors are usually used for shops, commercial use or offices, to provide working places. Second and third floors also mostly lack natural light and are as a consequence of lower living quality. Neighbourhoods and the rights of neighbouring property owners of tall buildings may face violation of its rights for reasonable natural light and visual quality. In Bahrain, environmentalists have demanded, that all planned high-rise construction projects undergo during the planning permission phase an additional three-stage procedure for environmental, social and health assessment. They stated, "skyscrapers should be part of the social and physical urban environment, as well as to add to the value of the local community." Low air quality and floor heights lead to the syndrome of the "sick building" (overwork, convulsions, joint pain, dry and itchy skin, exhaustion, headaches, palpitation, cough, sore throat, inability to concentrate, hypersensitivity to smells and dyspnoea), causing a drop of productivity and higher fluctuation of people [18].

4.1.4 Psychological discomfort of people in tall buildings

For thousands of years people are living on earth and built houses on its ground, traditionally dwellings of 1-2 floors. People live nearly never under ground and high above grounds, because both underground structures and high-rise buildings can cause unpleasant living conditions and apprehensions. Modern technologies ensure the safety of living even in such places, but it must be taken into account that a considerable number of people have a fear of heights. According to safety standards, glazing of balconies and loggias is mandatory, however, glass does not create a sense of security. It is necessary to use railing (fencing) to reduce psychological discomfort. Above the 20th - 22nd floor open private outside areas like balconies, loggias and terraces are usually not realized, reducing the access of fresh air for rooms and inhabitants [19]. This problem can be partly solved by designing winter gardens to create psychological comfort and improve natural room ventilation. As studies show, people prefer to live in low-rise houses. In many cities, which became pioneers in the construction of high-rise buildings, a slowdown in the pace of tall building construction can be recognized. As an example for this can be mentioned the decreasing high-rise construction in Dubai, a world known city of dynamic high-rise building activity [5]. In various cities of Russia and abroad, there is also a clear tendency to abandon the construction of high-rise buildings

4.1.5 Visual quality of high-rise buildings

A person can feel discomfort not only on high floors, but also on the street next to high-rise buildings. Tall buildings and skyscrapers of low architectural quality can put psychological pressure on people, creating a sense of threat. People feel "tired", eye irritation, discontent with themselves and their environment in areas with monotonous architecture (economic high-rises). On the contrary, single high-rise buildings or a well-designed ensemble of towers has the potential of representing the city as a symbol (cathedral, iconic construction) and becoming a matter of pride ("and we are not worse, we also have our own skyscraper"). There are few such architectural masterpieces in the world, defining the "sensitiveness" of high-rise architecture design. The erection of high-rise buildings in city centres can violate the mental map of the urban inhabitants and create a sense of loss of their historic city [20].

4.2. Sustainable standards as planning method for multifunctional high rise buildings

The rapid growth of cities is a global tendency, which can be observed in nearly every cultural and political context today. Not only city centres are undergoing density transformations but potentially every kind of urban territory in medium and large size cities is exposed to economic development pressure to becoming a vertical city with manifold social, economical and technical challenges for inhabitants, the way, how people live together and the ecologic environment. High rise buildings and vertical city districts with groups of tall buildings are in most of the cases designed and built like low rise buildings but with a multiple number of floors, not considering the increased requirement of green spaces for recreation and communication and increased requirements for individual and public transport infrastructure. Successful cities were always open cities with socially orientated building typologies, balanced in functions and architectural styles, with enough open accessible green spaces and every day services.

Fig.1. Vertical public high-rise development in Cairo, Prof. M. Eichner, R. Maher 2017

4.2.1 Sustainable standards for high rise structures

Providing inhabitants of vertical building structures with a convenient and socially balanced living environment, using minimum energy and other natural resources and at the same time integrating the high rise structure into the urban context, represent the big challenges today. To promote social synergies and a conflict minimized urban atmosphere, facilities like urban district councils, childcare, community stations and public accessible recreation and communication spaces of sufficient areas should be provided. Modern sustainable building standards for housing buildings require from planners and developers to integrate within their development public accessible outside areas for communication, recreation, kids playing and youth activities as barrier-free landscape spaces. A sufficient size is around 10 m² per person (synergy effects are allowed, for example the use of fire access roads or similar). In low rise housing districts, public spaces of significant size and high landscape quality are usually integrated in the form of squares, courtyards of pocket parks. Providing in high-rise buildings social valuable open spaces with pleasant and ecologic landscape design for different social groups and generations is widely neglected. One of the authors explored the advantages of an application of sustainable standards in multifunctional high-rise designs for central districts in downtown Cairo.

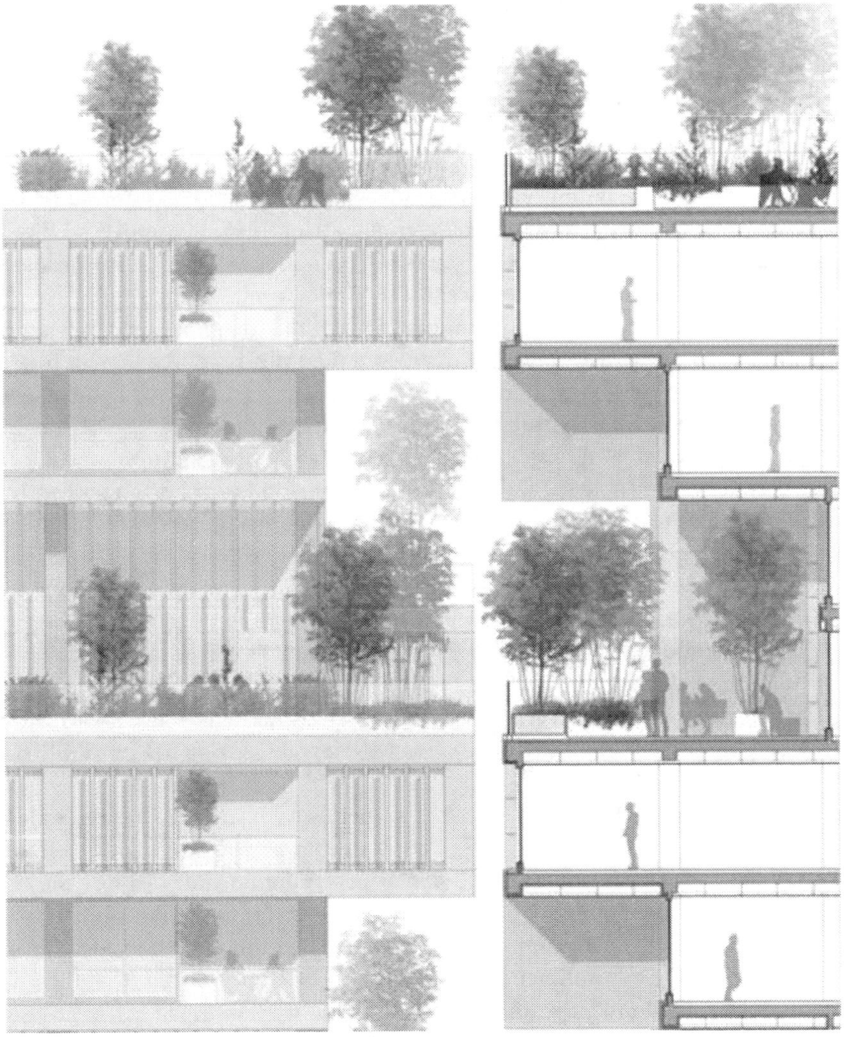

Fig.2. Vertical living city Cairo, detail section, Prof. M. Eichner, R. Anwar, GUC 2017

https://doi.org/10.1051/e3sconf/20183303065

The aim of this case study design project was to find out, whether it is possible today, within the economical and cultural context of the rapidly transforming Capital of Egypt to initiate a paradigm shift from socially isolated high-rise objects to socio-cultural and resource-efficient vertical cities structures, providing similar living quality like low rise housing districts. The question today is not whether high-rise buildings are preferable to low-rise buildings but how to design multi-storey buildings and districts, which meet all important sustainable and social criteria in an economical and balanced way.

Fig.3. Vertical garden tower, Cairo, detail section, Prof. M. Eichner, H. Hamdy, GUC 2017

4.3 Sociological surveys as indicator for needs of inhabitants of high rises

City authorities should comprehensively study needs and opinions of their urban inhabitants prior to realization of high-rise buildings. Town planners should learn how residents feel while living in upper floors and physicians should study the physical and mental state of people, living or working in high-rise buildings. Such systematic studies are not conducted in Russia, but information on individual surveys can be found. The Internet based "Catalog of local home networks of the city of St. Petersburg" for example published the results of a sociological poll, how Moscow residents are thinking about skyscrapers and the possibility of living in them. The results of the study showed, that a significant part of Muscovites would refuse to live in a residential high-rise building. Out of the total number of respondents, 58% answered negatively, the remaining 42% expressed their desire to live in a high-rise building. The interviewed Muscovites, who did not agree to live in high-rise apartment buildings, explained their fundamental refusal for living in skyscrapers with fear of height, possible problems with technical building systems and elevators. In addition, most of them are confident, that there is a greater likelihood of accidents and terrorist attacks in buildings of this type. This allows to conclude, that low-rise buildings are preferred by many people. The smaller part of the people, who participated in the survey in Moscow (42% of respondents) like modern high-rise buildings. They relate them to beauty and comfort and consider them as the solution of the housing shortage problem in the capital. In addition, the representatives of this group mentioned the advantage of apartment layouts in skyscrapers and the availability of garages. They also noted the opportunity to enjoy the beautiful scenery and get the full spectre of emotions [21]. These results show, that Muscovites differ in their opinions with a significant number of people in Moscow, assessing life in a high-rise building positively.

The authors of the article conducted a sociological survey at the Moscow State University of Civil Engineering in November 2017, asking questions about the influence of high-rise buildings on the urban ecology. Sixty students were interviewed: 66% of the respondents answered, that high-rise construction negatively affects the environment. 6% of the respondents concluded, that the impact of Russia's contemporary construction industry is even devastating for the environment. At the same time, students said, that they do not have enough environmental knowledge to judge professionally the impact of high-rise buildings on the ecology of urban territory. 28.3% of respondents do not know what "green standards" or “sustainable construction standards” are. The results of this survey indicate that in Russia to date little attention is paid to the environmental education of engineers, planners and builders. Students lack knowledge about modern eco-sustainable technologies for the construction of unique buildings and structures.

5 Conclusions

The study conducted by the authors of the article allows concluding, that the socio-ecological impact of high-rise buildings has not been sufficiently studied. On the one hand, their negative impact on the biosphere and people is obvious. On the other hand, there are social groups interested in high-rise construction, among which are, city authorities, developers, architects and a part of the urban population. Each of these groups has its own interests: economic, financial, urban development, creative or political. The opinion of the authors of the article: despite the slowdown in the tendency of high-rise construction in some countries, it will continue and even grow. It is impossible and not useful to build barriers for a creative search of new construction technologies for tall buildings, innovative high-rise development strategies and expressions skyscraper architecture. However, the interests of the biosphere, mankind and society should become the main goal of high-rise construction. Each project should be correlated with people's needs, a humanitarian balance and the low impact on the biosphere. This requires a wider and more creative application of sustainable planning solutions and technologies during the design and construction of high-rise buildings, briefly presented in this article with projects of Professor M. Eichner.

References

1. G. Binder, Tall buildings of Europe, the Middle East and Africa (Sydney: Images Publishing, 2006).
2. T.G Maklakova, High-rise buildings. Urban planning and architectural-constructive design problems. (ASV, Moscow, 2008).
3. URL: https://www.golosagorodov.info/istoricheskaja-spravka/goroda/skolko-naseleniya-zhivet-v-gorodakh-sovremennoy-rossii.html
4. Evaluation news. Moscow: the cost per square meter of land. URL: https://proocenka.ru/otrasl-news/1476-moskva-stoimost-kvadratnogo-metra-zemli.
5. A. V. Korotich, Akademicheskij Vestnik UralNIIproektRAASN 2, 51-55 (2016).
6. A. V. Korotich, Akademicheskij Vestnik UralNIIproektRAASN 3,43-46 (2011).
7. K. Fender, Asia & Australasia: A Selection of Written Works on the World's Tall Building Forefront, Conference Proceeding, 154-163 (2015).
8. C. Stalker, Beyond the Podium: Urban Spaces for Tall Buildings in a Subtropical City. Asia & Australasia: A Selection of Written Works on the World's Tall Building Forefront, 172-183 (2015).
9. K. Wan, G. Cheung & V. Cheng, CTBUH Journal, Issue III, 20-25 (2015).
10. URL : http://refy.ru/8/13532-obespechenie-kompleksnoy-bezopasnosti-pri-stroitelstve-vysotnyh-zdaniy.html
11. URL: https://www.abok.ru/for_spec/articles.php?nid=1387
12. URL: http://www.maconnc.org/MORATORIUM%20ON%20%20HIGH-Rise%20construction%20Adopted.pdf.

13. C. Kopczynski & M. Whiteley, CTBUH Journal, Issue III, 20-26, (2016).
14. URL: www.https//andexpert.ru
15. URL: http://ecom.su/city_building/index.php?id=969;
16. Ecological threat threatens Almaty due to high-rise construction URL: https://www.kn.kz/news/47984/
17. URL: http://global.ctbuh.org/resources/papers/download/2404-integrated-and-intelligent-buildings-an-imperative-to-people-the-planet-and-the-bottom-line.pdf
18. URL : http://www.gulf-daily-news.com/Story.asp?Article=199234&Sn=BNEW&IssueID=30232.
19. URL: https://www.abok.ru/for_spec/articles.php?nid=2444
20. URL : http://www.international.icomos.org/risk/2006/istanbul2006.htm
21. The popularity of skyscrapers among Muscovites: a sociological survey. URL: http://spb-nets.ru/articles/na.php?id=22

Formation of production structural units within a construction company using the systemic integrated method when implementing high-rise development projects

Azary Lapidus[1] and *Ivan* Abramov[1,*]

[1]Moscow State University of Civil Engineering, 26 Yaroslavskoe shosse, Moscow, 129337, Russia

Abstract. Development of efficient algorithms for designing future operations is a vital element in construction business. This paper studies various aspects of a methodology required to determine the integration index for construction crews performing various process-related jobs. The main objective of the study outlined in this paper is to define the notion of integration in respect to a construction crew that performs complete cycles of construction and assembly works in order to find the optimal organizational solutions, using the integrated crew algorithm built specifically for that purpose. As seen in the sequence of algorithm elements, it was designed to focus on the key factors affecting the level of integration of a construction crew depending on the value of each of those elements. The multifactor modelling approach is used to assess the KPI of integrated construction crews involved in large-sale high-rise construction projects. The purpose of this study is to develop a theoretical recommendation and a scientific methodological provision of organizational and technological nature to ensure qualitative formation of integrated construction crews to increase their productivity during integrated implementation of multi-task construction phases. The key difference of the proposed solution from the already existing ones is that it requires identification of the degree of impact of each factor, including the change in the qualification level, on the integration index of each separate element in the organizational and technological system in construction (integrated construction crew).

1 Introduction

In the conditions of increasing requirements [1] to the quality of works and given the current shortage of skilled resources in the labor market, quality labor has turned into a key element in construction operations.

Developers [2] who employ significant numbers of skilled workers carrying out integrated construction and assembly operations have a clear competitive edge vs. all other construction market participants. Construction business is set up and grows in an environment where standard operational requirements are applied. All structural subdivisions

* Corresponding author: ivan2193@yandex.ru

of such an organization, including construction crews are often operating with their labor force strength and qualification at the levels lower than what is required for optimal operations (in work zones and sub-areas).

The proposed comprehensive and systemic approach to contracting market includes the following requirements to standard construction teams which must have qualified workers with a large number of skills capable of performing multi-task operations.

A critical task of this study is to define the concept of universality for production units, including construction teams and explore the impact of a variety of destabilizing factors on their operations [3].

Part of that job is to define the notion of integration applied to a construction crew implementing complete cycles of construction and installation works in order to find optimal organizational solutions [3,4].

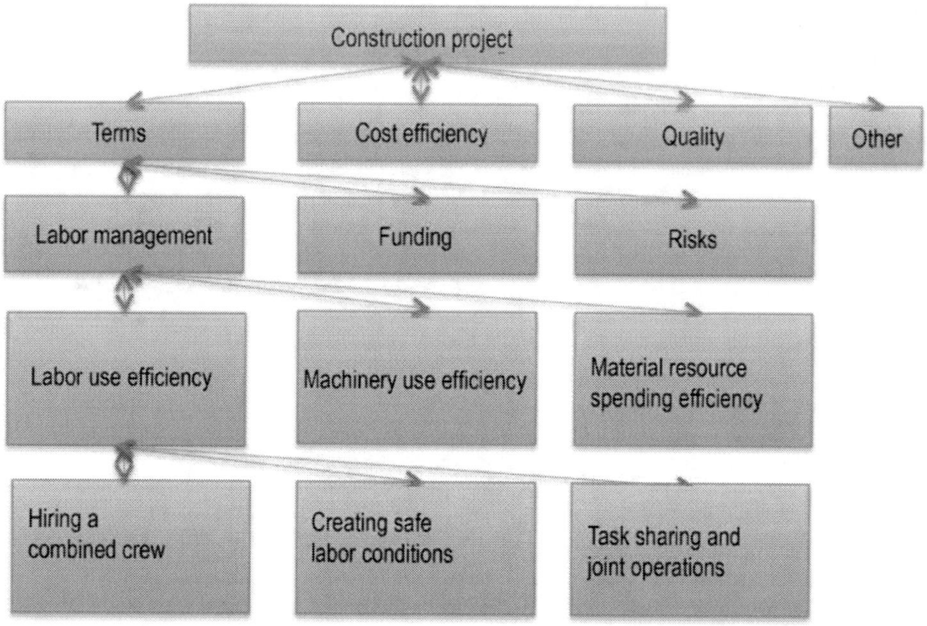

Fig.1. Key factors affecting optimal implementation of a construction project.

Our study was based on works associated with design of organizational structures of large-scale construction processes using a wide range of implemented solutions, including the study of the integral efficiency potential model [4,5] with regard to organizational and technological solutions of construction sites, allowing to take into account the impact of organizational, technological and management decisions in the implementation of a construction project.

We have also studied several papers [6,7,8] on how to approach the task of optimizing allocation of labor during the planning phase of construction and installation works. Labor resources are made up of workers organized as construction crews, and their number and workload is one of the key factors in the planning process affecting the overall schedule of construction operations.

Another paper [9] is quite interesting in the sense that it proposes a methodology for a quantitative assessment of the following key factors affecting the quality and efficiency of masonry operations: compatibility, compliance and craft.

There is another study [10] which argues that productivity growth does not depend on performing a maximum number of tasks regardless of the calendar schedule, workload increase or the number of hours worked. Performance improves when the work schedule becomes more predictable, making it possible to align the workload with the labor resources in the most optimal way.

Following our analysis and research, the first steps were taken to develop a model that allows one to theoretically develop and practically define the integration index for an integrated production crew. The model is based on the key factors affecting the end result of the overall construction process, including vocational training; level of skills; labor productivity; labor management; absence (presence) of violations of labor discipline, and the degree of integration of production processes performed by construction crews.

As noted earlier, studies that dealt with similar tasks and labor resources (construction crews) [7] were carried out prior to our study. In our work, however, we have tried to approach the problem in a significantly more profound way and focus on the methodology for determining the integration index for integrated construction crews engaged in construction operations. Our methodology was developed on systemic and integration approach and principles in order to produce the required construction project deliverables.

2 Methods

One of the main aspects in solving the problem of optimizing the systemic integrated method of performing a full cycle of construction and installation works is the need to create an algorithm in which an integrated construction crew will play a significant role. This structural unit may usually be formed right at the initial stage of the project, with clearly formulated tasks primarily focusing on labor productivity, quality and deliverable deadlines [11].

We have developed an algorithm for achieving the set task, which can be graphically described in the following way:

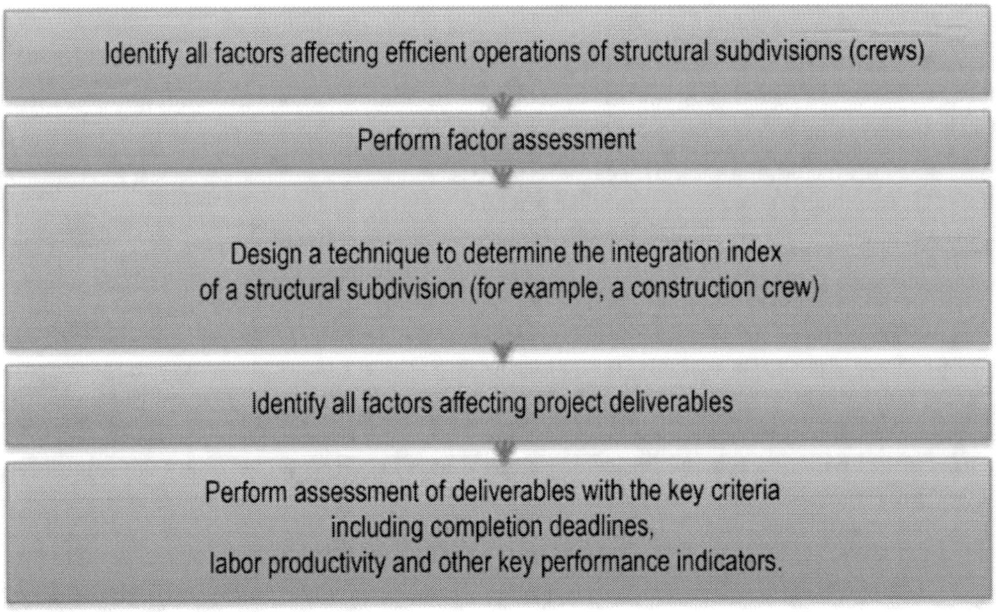

Fig. 2. Algorithm for optimizing the systemic integrated method.

o identify all factors affecting efficient operations of structural subdivisions;
o perform factor assessment;
o design a technique to determine the integration index of a structural subdivision (for example, a construction crew);
o identify all factors affecting project deliverables;

○ perform assessment of deliverables with the key criteria including completion deadlines, labor productivity and other key performance indicators.

On-site studies carried out at a number of construction operations have shown that an integrated construction crew is the most effective operational unit. The advantage of an integrated crew comparted to any individual forms of performing construction works is explained by the team work model which makes it possible to deliver all required final results in a construction project [11,12]. From the point of view of construction system engineering, an integrated construction crew is assumed as a subsystem within a system, the latter being a large-scale construction project. Let's review the key factors of variable significance affecting the subsystem whose variability depends on the behavior of the subsystem and the system as a whole and grouped into the external and internal categories.

The external factors include:

- Project funding;
- Site location;
- Climate conditions.

The internal factors, with their variability affecting quality deliverables, include:

- qualification level;
- labor efficiency of unit crews inside an integrated construction crew;
- effective compliance with performance quotas set for specific climate conditions;
- organization of labor within construction crews;
- material incentives;
- loss of working time when distributing work scopes among crews, units and other performers of specific jobs;
- loss of working time due to injuries or illnesses;
- loss of working time due to violation of labor discipline;
- number of crews performing similar jobs;
- degree of overlapping jobs performed by integrated construction crews, units and other contractors;
- projected non-attendance.

All these factors have a probabilistic nature.

Following analysis, all key factors are identified and assessed.

Factor assessment needs to be carried out via expert assessment method, prior job deliverable assessment method, and performance tests applied to a specific crew.

As an example, let us take some conventional deliverables obtained with the application of the above-specified methods. In the mathematical assessment of the integration index of a crew working in a large-scale construction project, we would suggest applying the multi-factor system modeling method [11,12].

To start with, we need to assess the significance of each factor affecting the result (FAR).

- high qualification level (FAR = + 1)
- low qualification level (FAR = -1)
- high level of experience (FAR = + 1)
- low level of experience (FAR = -1)
- high crew unit performance (FAR = + 1)
- low crew unit performance (FAR = -1)
- high labor organization within a construction crew (FAR = + 1)
- low labor organization within a construction crew (FAR = -1)
- high loss of working time during transfer of work scopes among crews, units and other contractors (FAR = + 1)

- low loss of working time during transfer of work scopes among crews, units and other contractors (FAR = -1)
- high loss of working time due to violation of labor discipline (FAR = + 1)
- low loss of working time due to violation of labor discipline (FAR = -1)
- large number of crew units performing the same processes (FAR = + 1)
- small number of crew units performing the same processes (FAR = -1)
- large number of overlapping jobs performed by construction crews (FAR = + 1)
- small number of overlapping jobs performed by construction crews (FAR = -1)
- large number of overlapping jobs performed by construction crew units (FAR = + 1)
- small number of overlapping jobs performed by construction crew units (FAR = -1)

In order to build a mathematical model of the studied processes, we need to add another factor describing former employer recommendations. That recommendation "k" assumes a "0" value if the assessment of performance of works is extremely negative and is backed with documents and relevant acts, while factor k = 1 means the crew has performed the previous scopes of work with no complaints regarding its performance and no failures due to their fault. As a result, the information obtained and required for participation in the selection process and characterizing the significance of each factor affecting the result (FAR) must be formalized in a general mathematical model. The integration index for a production crew, Ccc (integrated construction crews), is calculated by the formula:

$$C_{cc} = k\sum_{i=1}^{n} V_i = k(V_1 + V_2 + \ldots + V_n), \tag{1}$$

Where V_i is the aggregate FAR and "k" is the factor of former employer recommendations.

Job applicant assessment:

-9≤ Ccc ≤0 – crew is excluded from participation in the project;
0< Ccc ≤4 – crew is a weak candidate for participation in the project;
4< Ccc ≤7 – crew is an average-level applicant for participation in the project;
7< Ccc ≤9 – crew is the highest-level applicant given preference for participating in the project.

After the analysis, these factors are evaluated for their integration levels. Crew performance indicators are used in further assessment to obtain the results that will be used for project implementation status timeline estimates.

Identification of the all required parameters and their evaluation with the help of the algorithm and mathematical models will make it possible to justify the need for combining labor resources in a multipurpose integrated construction crew.

3 Conclusions

The proposed solution of the problem is essentially different from all modern solutions, because the need to identify the level of impact of each factor, including changes in qualification levels, on the integration index of a separate element of the organizational and technological system in construction (integrated construction crew).

We have developed theoretical recommendations and scientific and methodological provisions of organizational and technological nature critical for efficient formation of integrated construction crews with the aim of increasing their productivity in the integrated implementation of all phases of high-rise construction projects.

References

1. Rulebook SP 48.13330.2011, Organization of Construction. The Updated Edition of Russian Construction Rules SNiP 12-01-2004, 14-19 (2011).

2. A. Lapidus, I. Abramov, Organizational and process-related design of private low-rise construction projects when developing a calendar plan, Science Bulletin, 4 (2017).
3. A. Lapidus, I. Abramov, Calculating a construction company's resource potential as a bidder, Science Bulletin, **9**, 6-9 (**2017**).
4. A. Lapidus, Efficiency Potential of Organizational and Process-Related Solutions of the Construction Project, Bulletin of MGSU, **1**, 175-180 (2014).
5. A. Lapidus. A. Makarov, Shaping up organizational and process-related potential of manufacturing roofing structures for high-rise residential buildings, Bulletin of MGSU, **8**, 150-160 (**2015**).
6. V. Mishchenko, D. Yemelyanov, A. Tikhonenko, Developing a method for optimized resource distribution in calendar planning based on genetic algorithms, VSUACE, 76-78 (2013).
7. V. Mishchenko, D. Yemelyanov, A. Tikhonenko, Industrial and Civil Engineering, **10**, 69-71 (2013).
8. V. Mishchenko, E. Gorbaneva, Y. Rithy, F. Lin, Application of the flow method in low-rise urban residential development in hot climates, Ho Chi Minh City University of Architecture, 28-38.
9. L. Floreza, J. Cortissoz, Procedia Engineering, **164**, 42-48 ((2016).
10. M. Liu, G. Ballard, Proc. for the 16th Conf. of the Int. Group for Lean Construction, Planning and Control, 657-666 (2008).
11. I. Abramov, T. Poznakhirko, A. Sergeev, The analysis of the functionality of modern systems, methods and scheduling tools, MATEC Web of Conferences, **86**, (2016)
12. V. Chulkov, R. Ghazaryan, O. Kuzina, Proc. of the Int. Sci. and Prac. Conf., **5-6**, 82-94 (2014).
13. A. Fini, T. Rashidi, Journal of Construction Engineering and Management, **5**, 142 (2015).
14. P. Oleynik, Modeling of investment process term reduction, Natural and Technical Sciences Journal, **10**, 412-414 (2015).
15. Department of the Army, U.S. Army Corps of Engineers, Grouting Technology, **1110-2-3506** (2017).
16. The Government of Western Australia, Impact of new building techniques and technologies on the residential housing sector of the construction industry, Construction Training Fund (**2015**).
17. E. Heuvelhof, E. Hardoy, I. Mitlin, J. Satterthwaite, Environmental Problems in Third World Cities, (1992).
18. P. Cooke, Introduction: regional innovation systems - an evolutionary approach, Regional Innovation Systems (2004).
19. S. Bolotin, A. Dadar, I. Ptukhina, Engineering and Construction Journal, **7**, 82-86 (2011).

Modeling work of the dispatching service of high-rise building as queuing system

Marina Dement'eva[1*], and *Anastasiya* Dement'eva [1]

[1]Moscow State University of Civil Engineering, Yaroslavskoe shosse, 26, Moscow, 129337, Russia

Abstract. The article presents the results of calculating the performance indicators of the dispatcher service of a high-rise building as a queuing system with an unlimited queue. The calculation was carried out for three models: with a single control room and brigade of service, with a single control room and a specialized service, with several dispatch centers and specialized services. The aim of the work was to investigate the influence of the structural scheme of the organization of the dispatcher service of a high-rise building on the amount of operating costs and the time of processing and fulfilling applications. The problems of high-rise construction and their impact on the complication of exploitation are analyzed. The composition of exploitation activities of high-rise buildings is analyzed. The relevance of the study is justified by the need to review the role of dispatch services in the structure of management of the quality of buildings. Dispatching service from the lower level of management of individual engineering systems becomes the main link in the centralized automated management of the exploitation of high-rise buildings. With the transition to market relations, the criterion of profitability at the organization of the dispatching service becomes one of the main parameters of the effectiveness of its work. A mathematical model for assessing the efficiency of the dispatching service on a set of quality of service indicators is proposed. The structure of operating costs is presented. The algorithm of decision-making is given when choosing the optimal structural scheme of the dispatching service of a high-rise building.

1 Introduction

High-rise buildings have become a common phenomenon in the world practice of construction. They are a factor of scientific and technological progress, an indicator of the economic power of the state, determine the prestige of large construction holdings and management companies. Modern cities need renovation. The construction of high-rise buildings solves the problems of the development of territories, raises the diversity of the urban environment, improves the usability. The planning solutions of high-rise buildings are both highly specialized and multifunctional, ideally representing the so-called "city in the city" mode [1-4]. This approach allows us to realize the greatest number of functions in

* Corresponding author: 7dem@mail.ru

one construction site and the most economical and ergonomic use of urban space. The Shanghai Center (China), Time Warner Center in Manhattan (New York, USA), Moscow-City (Russia) [5] can serve as vivid examples of such objects.

However, high-rise construction has also negative aspects: the complication of architectural, planning and constructive solutions; more energy consumption, more complicated engineering systems, difficulties in evacuating people, environmental problems due to high density of construction, transport and population [6-9]. The multi functionality of high-rise buildings leads to additional problems: the complexity of the distribution of human flows, the construction of independent engineering systems increases the number of equipment serviced, the uneven use of equipment during the day complicates its management. Despite the fact that the listed problems are solved at the design stage, the complexity of the object leads to the complexity of its further operation. Failures of systems, equipment, structures of high-rise buildings can have more catastrophic consequences, which justifies the increased requirements to the quality of work of managers and exploitation organizations.

Studies show that, according to regulatory requirements [10], the composition of exploitation activities of high-rise buildings is similar to the exploitation of mass-building buildings and includes:

– management of facilities (organization of operation, relationships with related organizations and suppliers, all types of work with tenants and tenants);

– maintenance and repair of building structures and engineering systems of facilities (inspections, monitoring, preparation for seasonal operation, current and major repairs);

– sanitary maintenance (cleaning of common areas, cleaning of houses, care of green areas).

From the point of view of organizing the exploitation of any building, all the works can be divided into two groups according to the volumes, types and terms of implementation: permanent and probabilistic. Permanent works include the sanitary maintenance of buildings and territories, maintenance, scheduled maintenance, technical inspections. Since the types and terms of these works are determined by regulatory documents, and their volumes depend on the building parameters, the structural diagram of the management and operational service makes it possible to determine the relationship between the units, the area of responsibility of each employee and thereby ensure the requirements for the quality of operation. Probabilistic work can occur at random times, because they are emergency. Consequently, the time, volume and nature of probabilistic works are not defined. Only on the basis of a long collection of information on failure statistics can a different degree of probability be judged on the possibility of an accident. For example, the probability of failure of the wall panel is much lower than the failure of the tap in the water supply system. The refusal in the hot water supply system is higher than in the cold water supply system. The presence of probabilistic works complicates the maintenance and requires the optimization of the structural scheme of the managing organization on the basis of the theory of mass service [11-21].

Traditionally, to monitor the reliability and safety of the functioning of the building, timely detection and elimination of emergency situations, the dispatching service integrated into the united dispatching service of the district and the city is provided. However, high-rise buildings are unique objects that require specialized work. Consequently, the dispatch service should be part of the structure of the management company serving the high-rise building. At present, the dispatching service of high-rise complexes is a centralized point of automated operation control, using its own specialized software. Consequently, this subdivision from the from auxiliary passes into the category of primary, without which it is impossible to effectively manage the infrastructure of a high-rise building.

The increase in the number of assigned functions complicates the structure of the dispatching service, which entails the problem of reducing the quality of its work [22-27]. The purpose of the work is to develop an algorithm for making decisions on the selection of the optimal structural scheme of the dispatching service of a high-rise building, taking into account the minimization of operating costs and costs for the maintenance of the dispatching service, and also minimizing processing time and fulfilling orders. To solve the set goal, the following tasks were solved:

– analysis of the cost structure in the modeling of various schemes for organizing the work of the dispatching service;

– analysis of the structure of applications entering the dispatch service at failures of engineering system elements in various functional segments of a high-rise building;

– calculation of the performance indicators of the dispatching service as a queuing system.

2 Materials and Methods

The object of research in this article is the organization of the work of the dispatcher service of a high-rise building as a queuing system. The subject of the study is the application of queuing theory in the modeling of various schemes for organizing the work of the dispatching service of a multi-purpose high-rise building and assessing the change in the parameters of its performance.

Statistical analysis of data on failures, functional-cost analysis of operating costs, and system analysis were used to analyze and summarize the data. The performance indicators of the dispatching service were determined on the basis of queuing theory.

Since the design standards in Russia are oriented to the height of buildings up to 75 m, the trend has been to refer objects to high-rise buildings above 75 m. The paper considers a 25-storey building with a maximum mark of the parapet of the roof of 83.5 m, built in 2008. The building has a zone of apartments, office space, hotel, underground parking. The area of the functional area of the apartments is 40000 m^2, office and hotel complex is 45000 m^2.

The quality of the work of the dispatching service as a single control center for the operation of a high-rise building can be estimated by the following quantitative indicators:

1. The average number of detected defects during regular inspections - characterizes the quality of monitoring, the qualification of personnel.

2. The average idle time of engineering systems and building elements in anticipation of maintenance and in the process of maintenance - characterizes the operational efficiency of the operation service.

3. The coefficient of use of working time as a ratio of the average time of employment of workers to the total time for performing operations for the technical operation of the building.

4. Economic exploitation - the expedient use of material and labor resources.

5. The number of justified complaints of the population on the quality of work, the operational efficiency of the units - characterizes the creation of maximum amenities for the residents.

6. The probability of failure-free operation for a specific time interval and the duration of operation of the elements between failures - characterizes operational reliability.

It is impossible to assess the effectiveness of the organization of the work of the dispatching service for any one of the listed indicators, since the values of other indicators may be inadmissible with the best value of one indicator. For example, low material and labor costs may result from a low level of organization of operation, which may be the cause of unacceptable operational efficiency. Therefore, in the practice of operation, it is advisable to apply the following approach to the assessment of the dispatching service: the main indicator of the efficiency of management of the exploitation process is adopted, for it the extreme value is determined. The remaining indicators impose restrictions, determined by the standards, technological features, market conditions. As indicators of efficiency, it is advisable to take the profit of the management company and the time for performing repair work with the constraint imposed. Then the mathematical model for assessing the efficiency of the dispatching service can be presented in the following form:

$$\begin{cases} S_{pr} = P(S_{in}; S_{cost}) \rightarrow \max \\ Z_{sist} = P(Z_q; Z_{serv}) \rightarrow \beta \end{cases} \quad (1)$$

Where S_{pr} – profit in the course of work on troubleshooting; S_{in} – income from the activities of dispatching services; S_{cost} – expenditure from the activity of the dispatching service; Z_{sist} – the processing time of the application (being in the system); Z_q– the time that the application is waiting for service (waiting in line); Z_{serv} – time of the application by the technical specialist (being in service); P – the likelihood that service performance indicators will not exceed the allowable value; β – the restriction imposed on the time indicator equal to the duration of the shift.

The average work time can be determined as follows:

$$Z_{sist} = Z_q + k \cdot (Z_{el} + Z_m + Z_w) \quad (2)$$

Where Z_{el} – time of movement in special equipment (movement in the elevator); Z_m– travel time without special equipment; Z_w – time of work execution; k – time loss factor for technical and technological breaks (receipt and storage of materials, elimination of several faults in one application, etc.).

The average profit from troubleshooting activities can be determined as follows:

$$S_{pr} = S_{in} \cdot (1 - P_0) \cdot \lambda - \left[S_m \cdot (1 - P_0) \cdot \lambda + S_d + S_w + S_q \cdot Z_q \cdot n_q + S_{id} \cdot P_0 \cdot \lambda \right] \quad (3)$$

Where P_0 – probability of idle technical personnel; λ – the intensity of incoming applications from the dispatcher to the technical staff; S_m – cost of materials for troubleshooting; S_d – the cost of organization and maintenance of the dispatching service (personnel costs, rent, communication services, depreciation of fixed assets, software, consumables, etc.); S_w – the cost of the organization and maintenance of repair teams; S_q – cost of costs when the application is in the queue (for example, due to loss of resources in the event of an accident); n_q – average number of requests in the queue; S_{id} – cost of costs associated with idle workers.

The probability of loading the dispatching service regardless of the structural scheme of the organization:

$$p = \lambda_a \cdot (Z_{serv} + Z_d) \quad (4)$$

Where λ_a – the number of incoming applications to the dispatch service; Z_d – time of transfer of the application from the dispatcher to the executor of works.

Regardless of the structural schemes considered, the dispatch service is staffed with the necessary number of workers of the required specialty for the most rapid elimination of possible malfunctions of the elements of the high-rise building.

3 Results

Three models of the organization of work of the dispatching service were considered. The first model is a single dispatch center for collecting and registering applications from the entire building, regardless of the functional purpose, dividing the engineering systems into zones vertically or horizontally (Figure 1). The processed applications are transferred to the brigades, who also carry out work at any point of the building. Such an arrangement is most optimal if there is one owner, tenant or one management company in the building. As a rule, this situation is not typical for high-rise buildings.

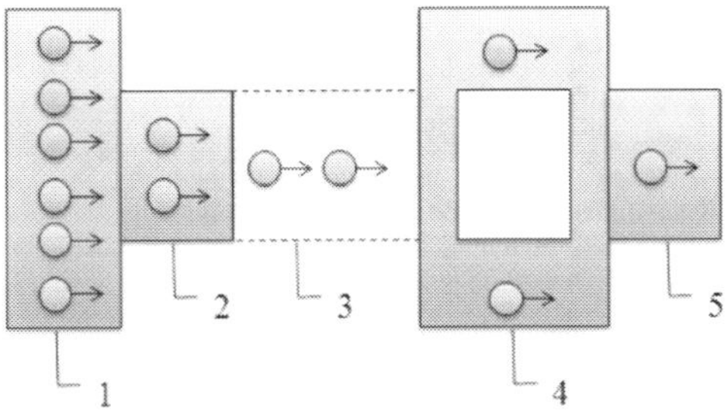

Fig. 1. The model of the organization of a single dispatch center with a common brigade service (D1): 1 – flow of incoming applications; 2 – reception and processing of applications by the dispatching service; 3 – order execution queue; 4 – application service; 5 – throughput of technical personnel.

Performance indicators are defined as for a queuing systems one-channel with an unlimited queue. Probability of downtime of technical specialists:

$$P_0 = 1 - p \tag{5}$$

Average number of requests in the queue:

$$n_q = \frac{p^2}{1-p} \tag{6}$$

Average time of the application in the queue:

$$Z_q = \frac{p^2}{\lambda_a \cdot (1-p)} \tag{7}$$

Average time of the application in the system:

$$Z_{sist} = \frac{1}{\lambda \cdot (1-p)} \tag{8}$$

The second model is also a single dispatch center for the collection and registration of applications. However, processed applications are transferred to teams that have fixed service areas (Figure 2). The distribution of service areas can be carried out according to the functional purpose, height of the building, etc.

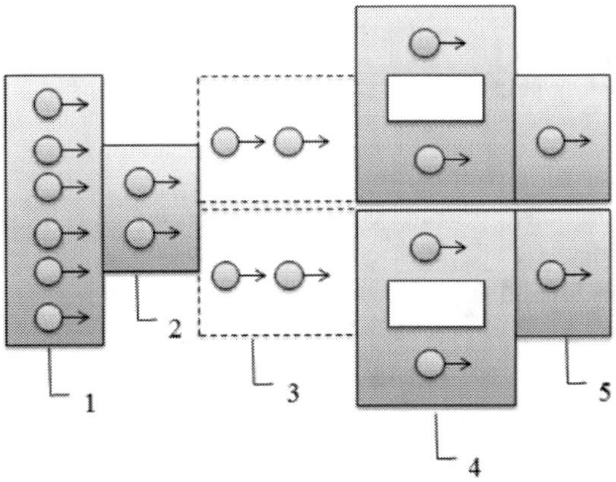

Fig. 2. The model of the organization of a single dispatch center with a specialized brigade service (D2).

This variant of the organization of work is advisable in the presence of different owners, different management companies in the same building. Then the organization of work of repair crews can be dealt with by each managing company separately. Performance indicators are defined as for a multi-channel queuing system with an unlimited queue. Probability of downtime of technical specialists:

$$P_0 = \left(1 + p + \frac{p^2}{2!} + \frac{p^3}{2!(2-p)}\right)^{-1} \tag{9}$$

Average number of requests in the queue:

$$n_q = \frac{p^3}{2 \cdot 2!\left(1 - \frac{p}{2}\right)^2} \cdot P_0 \tag{10}$$

Average time of the application in the queue:

$$Z_q = \frac{p^2}{\lambda_a \cdot 2 \cdot 2!\left(1 - \frac{p}{2}\right)^2} \cdot P_0 \tag{11}$$

Average time of the application in the system:

$$Z_{sist} = Z_q + \frac{1}{\lambda} \tag{12}$$

The third model consists of several dispatch center accepting applications from fixed objects (premises, engineering systems, equipment, owners, etc.) and transferring them to strictly repaired repair teams (Fig. 3).

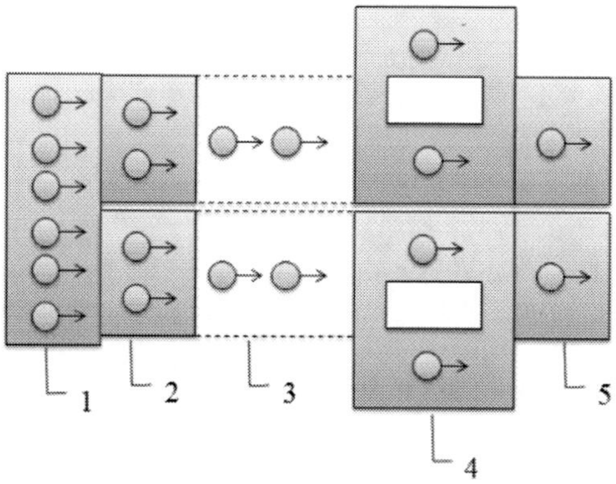

Fig. 3. The model of the organization of several dispatch centers with specialized brigade service (D3).

Such an option for organization of work is expedient if it is necessary to provide the maximum level of reliability, but it is characterized by significant operating costs. Performance indicators are defined as for a single-channel queuing system with an unlimited queue using formulas (5-8).

With a decrease in the number of incoming orders to the dispatch center, up to 40-50% of the calculated value, the management company will incur losses (Figure 4). In further calculations, it is advisable to consider a decrease in the number of incoming orders to 60% of the calculated value as critical.

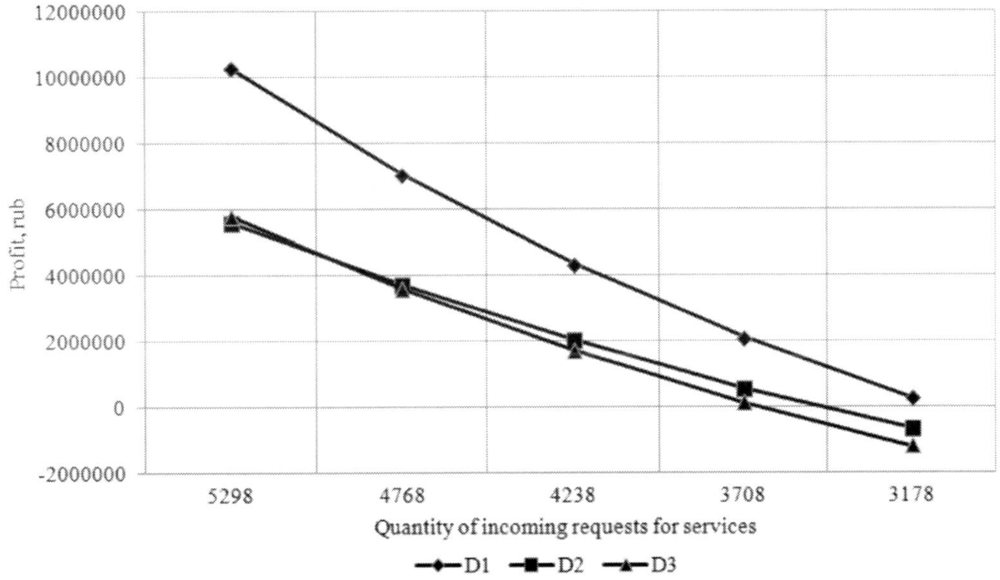

Fig. 4. Indicators of profit depending on the number of incoming applications.

Taking into account the maximum admissible value of the time of the application in the system equal to eight hours, all dispatch services are applicable (Figure 5).

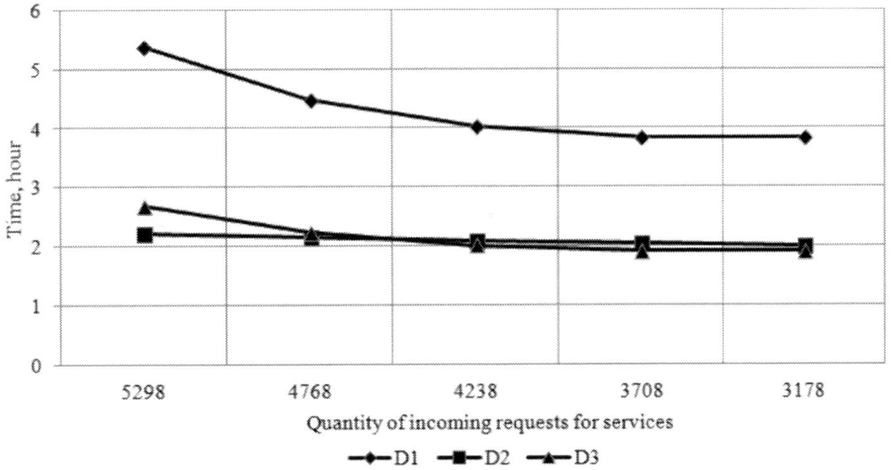

Fig. 5. Indicators of the time of application in the system depending on the number of incoming applications.

The profit indicator is the best when the normative intensity of the processing of applications by technical experts approaches the number of incoming applications to the control center as much as possible. With the increase in productivity, the costs associated with the idle time of specialists increase, and with a decrease, the costs associated with increasing the queue length and losses when the application is in the system are increased (Figure 6).

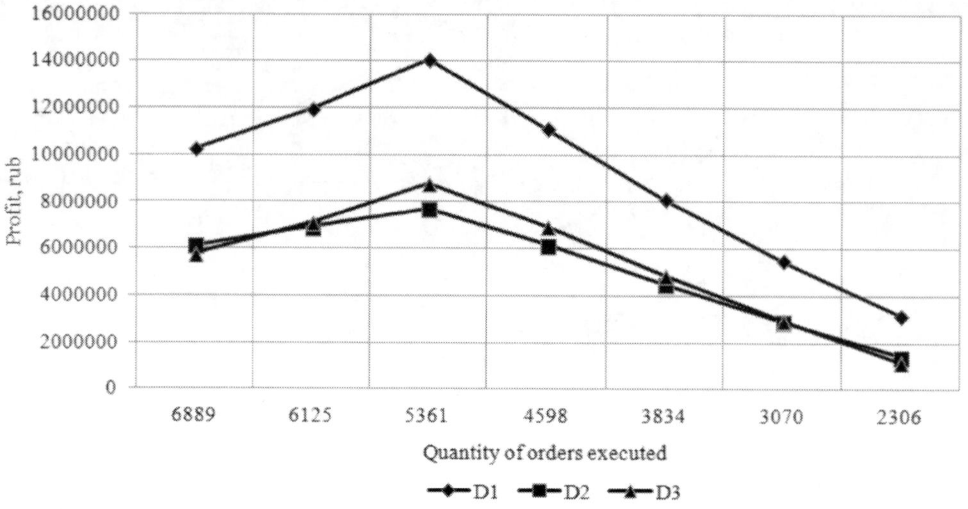

Fig. 6. Indicators of profit, depending on the intensity of work performed by technical specialists.

The time spent in the system is significantly increased in single-channel systems with a 10% or more reduction in performance (Figure 7).

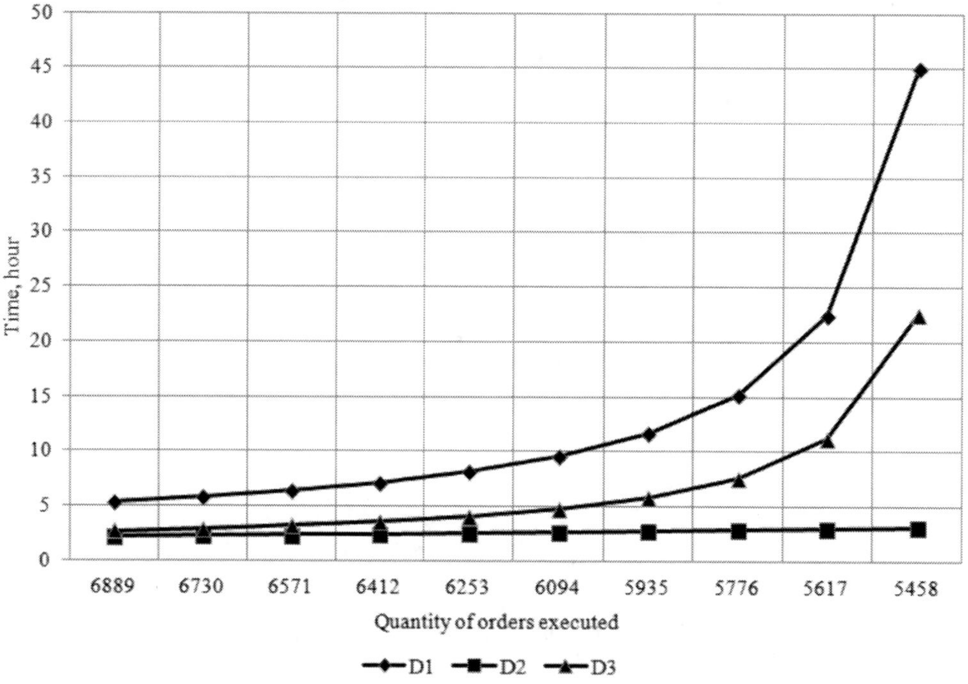

Fig. 7. Indicators of the time of application in the system, depending on the intensity of work performed by technical specialists.

4 Discussion

The carried out researches have allowed to define problem situations at the organization of work of dispatching service of a high-rise building. The novelty of the research is that analytically and graphically the dependences of the profit amount were revealed during the dispatching service operation from the costs of organization of the dispatch center, the cost of organizing the work of repair teams, the costs of materials for troubleshooting in engineering systems, the costs associated with the queue and the regulatory intensity execution of applications by technical specialists, as well as the time of finding the application on the network and the time of elimination of the failure from the normative intensity technical experts.

It was found that the maximum revenue is associated with the highest execution time of applications. In addition to the flow of incoming applications, it is necessary to optimize the performance of teams. The second and the third model (D2, D3) are close in terms of their performance, despite the different number of throughput channels.

The practical significance of the study is justified by the developed algorithm for making decisions on the selection of the most effective dispatching service scheme (Figure 8). The algorithm is designed to automate decision making, develop a computer program to help workers of exploitation services. The algorithm presents the initial data influencing the change in the efficiency indicators of the dispatching services, the calculated indicators and the decision-making mechanism.

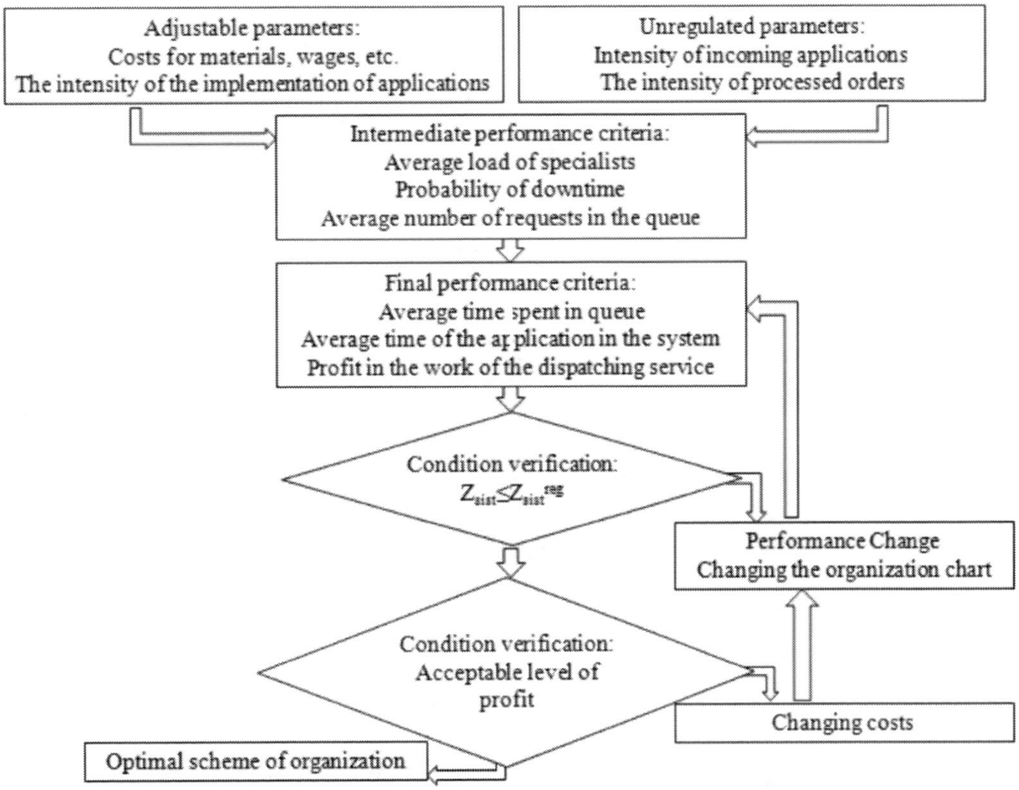

Fig. 8. Algorithm for choosing the optimal scheme for the organization of the dispatching service.

Conclusions

To solve the task set in the work, the following tasks were solved:

– the analysis of expediency of construction of high-rise buildings in conditions of existing city building is executed;

–the specifics of the organization of the dispatching service in the structure of the managing company of the high-rise building are analyzed;

–fulfilled a review of the main exploitation measures and their classification into two groups of work: constant and probabilistic;

–proposed a time loss factor for technical and technological breaks, preparation of a specialist for the processing of an application, acceptance and storage of consumables in a warehouse, the elimination of several faults in one application, affecting the processing time of an application by a technical specialist;

–used the methodology used to calculate the performance indicators of the dispatching service based on the theory of mass service;

– dependencies of the influence of the changed indicators of the dispatching service on the time and profit criteria are established;

– developed an algorithm for making decisions on the selection of the optimal scheme for the organization of the dispatching service.

References

1. V.V. Aurov, Works of the Moscow Architectural Institute, **1**, 185-187 (2015)
2. P.P. Zueva, Int. Conf. Actual problems of construction, ecology and energy conservation in Western Siberia, **3**, 98-104 (2014)
3. P.P. Zueva, Architecture and construction of Moscow, **553**, 5, 42-54 (2010)
4. D.A. Efremyan, A.Yu. Sidorenko, T.O. Kondratenko 4. Science yesterday, today, tomorrow, **9 (43)**, 57-62 (2017)
5. URL: http://mingitau.livejournal.com/112278.html (2018)
6. M. Pecio, Bezpieczenstwo i Technika Pozarnicza, **42**, 159-165 (2016)
7. S.Sh. Sairiddinov, Urban planning and architecture, **7**, 2 (27), 38-47 (2017)
8. A.A. Raevskaya, A.O. Bykov, A.I. Platitsyna, World science: problems and innovations, **1**, 161-163 (2017)
9. Yu.A. Matrosov, Construction Materials, **10**, 90-92 (2005)
10. PM-2957 Recommendations for the operation of multifunctional high-rise buildings and complexes, Moscow (2006)
11. A.P. Kirpichnikov, D.B. Flax, K.N. Galyamova, Bulletin of Kazan Technological University, **19**, 11, 122-125 (2016)
12. A.P. Kirpichnikov, D.B. Flax, Bulletin of Kazan Technological University, **17,** 24, 242-245 (2014)
13. L.N. Polyakova, V.V. Karelin, V.M. Bure G.M. Hitrov, Bulletin of St. Petersburg University, **1**, 75-82 (2015)
14. M.I. Busarev, A.P. Kirpichnikov, D.B. Flax, Bulletin of Kazan Technological University, 22, 155-161 (2011)
15. I.M. Yakimov, A.P. Kirpichnikov, G.R. Zainullina, Yu.G Isayeva, Z.T. Yakhina Bulletin of Kazan Technological University, **19,** 5, 141-145 (2016)
16. S.B. Malkov, News of SFU. Technical science, 7 (42), 94-101 (2004)
17. I.N. Valeev, A.P. Kirpichnikov, Bulletin of Kazan Technological University, 4, 75-79 (2006)
18. A.P. Kirpichnikov, I.N. Valeev, Survey of Applied and Industrial Mathematics, **14**, 5, 891-893 (2007)
19. Yu.E. Malashenko I.A. Nazarova, Informatics and Applications, **8**, 1, 12-20 (2014)
20. T.V. Zakharova, Bulletin of Moscow University, **15**, 4, 32-37 (2007)
21. E.V. Larkin, V.V. Kotov, A.N. Ivutin, A.N. Privalov, Journal of Computational and Engineering Mathematics, **3**, 2, 14-24 (2016)

22. G.D. Drozdov, Innovations, **11,** 115-119 (2011)
23. A.N. Loshakov, Plumbing, Heating, Conditioning, **3 (147**), 78-79 (2014)
24. M.E. Dement'eva, A.D. Vishnevetskiy Proceedings of Moscow State University of Civil Engineering, **2,** 203-209 (2014)
25. A.G. Retivin, A.I. Pestryakov, K.A. Pavlychev, The Bulletin of the NGIEI, 4 (35), 107-112 (2014)
26. Yu.K. Khabibullin, O.B. Barysheva, News of Kazan State University of Architecture and Civil Engineering**, 2 (40),** 198-205, (2017)
27. A.I.Frolov, Information systems and technologies, **1-4**, 226-230 (2006)

Scale factor management in the studies of affine models of shockproof garment elements

Oleg Denisov[1], *Mikhail* Pleshko[1,2*], *Irina* Ponomareva[3] and *Vitaliy* Merenyashev[1]

[1]Don State Technical University, Gagarina sq.1, Rostov-on -Don, 344010, Russia
[2]Rostov State Transport University, Rostovskogo Strelkovogo polka narodnogo opolcheniya sq., 2, Rostov-on-Don, 344038, Russia
[3]Southern Federal University, Bolshaya Sadovaya st.105/42, Rostov-on-Don, 344006, Russia

Abstract. New samples of protective garment for performing construction work at height require numerous tests in conditions close to real conditions of extreme vital activity. The article presents some results of shockproof garment element studies and a description of a patented prototype. The tests were carried out on a model which geometric dimensions were convenient for manufacturing it in a limited batch. In addition, the used laboratory equipment (for example, a unique power pendulum), blanks made of a titanium-nickel alloy with a shape memory effect also imposed their limitations. The problem of the adequacy of the obtained experimental results transfer to mass-produced products was solved using tools of the classical similarity theory. Scale factor management influence in the affine modeling of the shockproof element, studied on the basis of the equiatomic titanium-nickel alloy with the shape memory effect, allowed us to assume, with a sufficient degree of reliability, the technical possibility of extrapolating the results of experimental studies to full-scale objects for the formation of the initial data of the mathematical model of shockproof garment dynamics elastoplastic deformation (while observing the similarity of the features of external loading).

1 Introduction

The Russian Federation is currently implementing the national standard GOST P 54934 - 2012 "Occupational safety and health management systems", however, statistics highlight some increase in occupational morbidity. For occupations characterized by an increased risk of injury, it is possible to get a large joint injury at the workplace. It is described that when jumping and falling at the moment of landing, forces, that exceed its weight by almost five times, affect on the body. Incorrect landing, especially in combination with twisting in the knee joint, as a rule, leads to micro- and macro-injuries of the joint, which causes further development of pathology, including a decrease in the quality of life and disability. The construction works at a height are of particular concern [1- 7, 14].

A sufficient number of different protective aids are used for the prevention of injuries, the most convenient of which are the knee pads [8]. However, at the present time, the use of

* Corresponding author: mixail-stepan@mail.ru

such protective devices is not widespread for common use (except for athletes), in addition, there are practically no means to protect large joints in occupations that are characterized by an increased risk of injury, innovative technical solutions and materials of increased protection and effectiveness are rarely used [3, 4, 17].

When analyzing modern technical solutions that underlie the domestic means of personal protection of personnel [13, 16], it was found that there is a large number of them. Most of them describe elastic-plastic elements, such as elastic bandages, lamellar materials, soft fittings. All of them, as a rule, are made in the form of an elastic deformable element based on soft polyurethane and / or gel. However, all existing samples can be further developed for more optimal use towards the efficiency of shock absorption increase, the energy capacity increase for example, due to the operation of the damping element of the material used, not only in the elastic but also in the plastic deformation regions, the inclusion of rational energy-absorbing elements in the protective devices. along with the implementation of tasks to reduce the overall size of the model.

The solution of this problem calls for the determination of methods and technical devices that facilitate a convenient and rapid change of fixation points on garment, as well as the need to investigate the possibility of transferring the results of field experiments and theoretical calculations to real industrial samples of shockproof elements.

2 Initial data

Let us consider some results of experimental studies to determine the power characteristics of the shock protection elements made on the basis of elastic-plastic inserts made of the alloy with shape memory effect.

The test of the prototype model of the shockproof device was carried out on the drop crusher (GOST 10708-82), specially modified for the present studies at different temperature regimes, accompanying work in various climat and production conditions. During the experiment, at a given temperature, the rotation moment and the magnitude of the shift strain were measured. The temperature of the helix of the model was subjected to controlled changes by application of an electric contact heater, measured with a digital thermometer (GOST P 8.625-2006). The signal was transferred from a thermocouple, the junction was fixed according to a conventional technique [12].

Peculiarities of the experimental procedure: a model of a shockproof device matched to the size, was worn on the basis, imitating the model of the knee joint of a human. As a test model, the following adaptation was used.

The developed scale model of a shockproof garment element (Fig. 1) was an overlay fastened with a velcro fastener and a protective and decorative valve. The model included a base in the form of spirals from an equiatomic titanium-nickel alloy with shape memory effect. The base was made in the form of wire helices of a flat section with risks along the entire length and included in an elastic deformable element of silicone, which made it possible to perform a relative spatial fixation of the spiral in the base, to provide the required distance between the turns of the spiral, and also to participate in the absorption of the impact energy. The base on both sides was covered with a perforated heat-conducting tape based on a copper alloy, with C-shaped or U-shaped "petals" and on one side equipped with a textile velcro, and on the other side the fabric had a slotted punching.

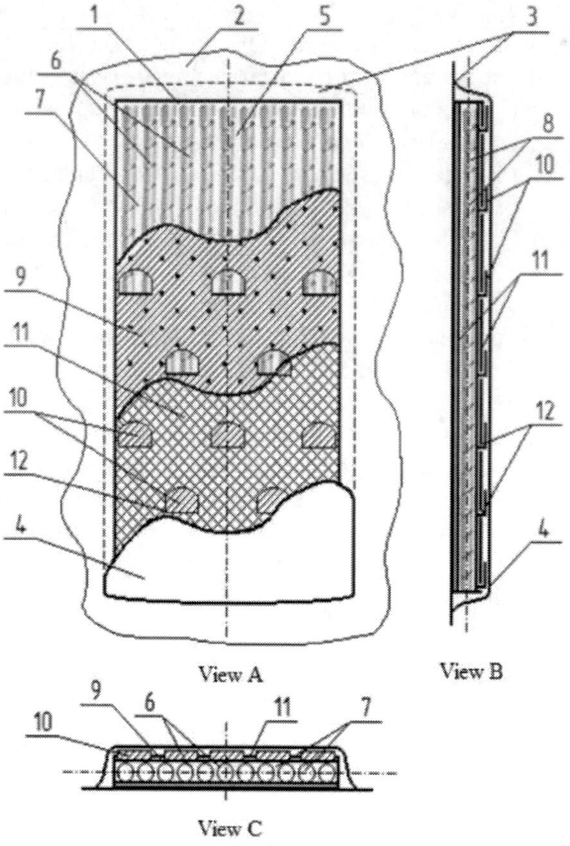

Fig. 1. Shockproof garment element.

Legend: 1 - shockproof element, 2 - garment, 3 - velcro, 4 - fabric decorative and safety valve with velcro, 5 - base of the shockproof garment element, 6 - wire helices of flat section with risks along the entire length of the equiatomic titanium nickel alloy with shape memory effect, 7 - elastic deformable element of silicone, 8 - spiral turn, 9 - thermal conductive tape based on copper alloy perforated with micro holes and C or U-shaped holes, 10 - "petals" of heat-conducting tape based oncopper alloy C- or U-shaped, 11 - fabric with slotted holes, 12 - slotted punching of fabric.

The experimental design was developed in accordance with the standard method for evaluating the energy absorption for plastic-elastic elements of similar systems [4, 10, 11].

It was experimentally established that when the impact of a drop crusher with a working area of $S_y - 1 \cdot 10^{-3} m^2$ (1000 mm^2), about 150 ... 200 turns were deformed in the elastoplastic region, absorbing in total about 150 ... 180 J, and taking into account the deformation of the elastic filling of the fabric support base - up to 200 ... 240 J of impact energy.

In order to determine the characteristics of energy absorption at different temperatures, one of the stages of the study included testing with the measurement of dynamic characteristics in correlation with heating of the deformed elements.

The main results of the experiment:

1. The energy-absorbing capacity of given plastic elements with shape memory effect to a large extent correlates with temperature (Fig. 2).

2. The actual energy absorption is in the range of more than 100 J.

3. The nature of the curves is practically repeated for various shock loads up to critical ones.

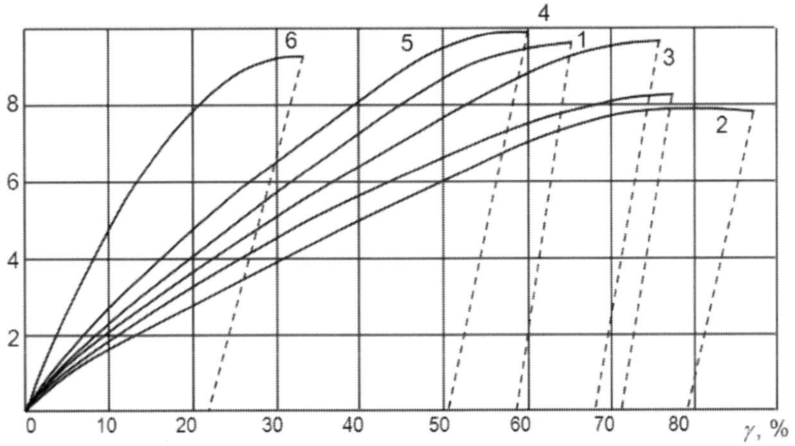

Fig. 2. Conditional stress-strain diagrams of the model based on the TiNi-1 alloy under shift (torsion) and high. Legends: 1 - 300 K; 2 - 320 K; 3 - 325 K; 4-335 K; 5 - 355 K; 6 - 385 K (dashed: elastic discharge)

It is known [4, 14] that an energy impact of more than 100 J can become dangerous for a human, therefore, a capacity for energy absorption of about 150-200 J is nominal for a technical protective device.

Based on the results of the field experiment, the features of the expected operation of the real device are determined. In case of shock, at first, deformations of the model elements occur in the elastic region, ensuring the absorption and dispersion of the shock energy. When the shock effect exceeds the level of elastic deformation of the protection elements, the outer layers of the model collapse to the limiting states of elastic hysteresis, and the spiral wires of the device are deformed elastically, and then plastically. It is established that the deformations of spiral wires are accomplished due to twisting of single turns and elastoplastic bending and torsion at a nonlinear impact site. In this case, the elastoplastic deformation of the site results in the greatest absorption of the shock energy [8, 9, 15, 16].

It is shown that with an increase in the number of cycles of shock deformation, a gradual decrease in the degree of restoration of the original dimensions of the model is observed, up to a certain stabilization of the value. As the shift strain amplitude is increased, the degree of reduction of the shape is reduced, and more cycles are required to achieve stabilization. In practical application, single strokes are typical, under which, obviously, there will be a stable recovery of the sample. It is possible to explain this behavior of the TiNi elastic-plastic element by non-dislocation deformation mechanisms, which correlates with the literature data and the results of studies of the mechanical memory effect [3, 9, 10].

3 The calculation results

The processing of the results of experiments of scale models showed the principal possibility of transferring the results of the experiment to real shockproof elements based on the use of a system of dimensionless complexes.

To estimate the influence of the scale factor, the results of the model tests were presented in the form of relations between the similarity criteria [6]. The linear dimensions and other characteristics are considered as characteristic parameters: the length of the section subjected to the effect l, the radius of the twisting of the spirals R, the value of the external force (per unit length) of the force action F, the relative linear strain of the experimental model of the shockproof device φ, the relative linear deformation of the model of the shockproof device in the elastic zone φ_{Π}, the integral ultimate strength of the materials of the model of the shockproof device in the zone of elastic deformation *G*, the integral model of the tensile strength of materials in the elastoplastic deformation zone shockproof device *Φ*. As a result, the following main parameters adopted:

$$F,\ l,\ R,\ G,\ \Phi,\ \phi,\ \phi_{\Pi}, \tag{1}$$

which dimensions are represented on the basis of the similarity theory [6] in the form of a matrix:

$$\begin{Vmatrix} 2 & 1 & 1 & -1 & -1 \\ 1 & 0 & 0 & 1 & 1 \\ -2 & 0 & 0 & -2 & -2 \end{Vmatrix}, \tag{2}$$

and the dimensionless parameters φ and φ_{Π} are not included in this determinant.

It is important to determine the number of independent dimensionless dependencies associated with a given matrix. It is equal to $n-r=5-2=3$, where r is the rank of the matrix (2).

To create a system of dimensionless complexes, the matrix is transformed into the following form:

$$\begin{Vmatrix} 1 & 0 & -1 & 0 & -3 \\ 0 & 1 & 0 & 0 & -1 \\ 0 & 0 & 1 & -1 & 0 \end{Vmatrix}. \tag{3}$$

Using the matrix (3) presented above and angular quantities φ и φ_{Π}, a system of dimensionless complexes is now compiled in the form of similarity criteria:

$$\Pi_1 = \frac{F}{GR^3},\ \Pi_2 = \frac{l}{R},\ \Pi_3 = \frac{G}{\Phi},\ \Pi_4 = \varphi,\ \Pi_5 = \varphi_{\Pi}. \tag{4}$$

In this case, the first dimensionless complex was considered as a known similarity law, analogous to the formulation of Kirpichev-Barb-Kik [12].

As the main characteristic size, the radius of the spirals *R* is adopted, and the length of the linear section subjected to the influence l of the experimental model 1 and the real object 2 is expressed in terms of the ratio:

$$L_i = K_i R_i,\quad i=1,\ 2. \tag{5}$$

Then:

$$\frac{l_1}{R_1} = \frac{l_2}{R_2} = K_1 = K_2 = K. \tag{6}$$

Using (6), we present a system of equations for elastoplastic deformation of models in the form:

$$F_i = \frac{\pi}{2K} D_i,\quad i = 1,\ 2,\ \ldots$$

$$D_i = \begin{cases} G_i R_i^3 \varphi_i,\ \varphi_i < \varphi_{ni} \\ G_i R_i^3 \left[\frac{4}{3}\left(\varphi_{ni} - \frac{4\varphi_{ni}^4}{\varphi_i^3} \right) + \right. \end{cases}$$

$$\left. + \frac{\Phi_i}{G_i}(\varphi_i - \varphi_{ni})\left(1 - \frac{\varphi_{ni}^4}{\varphi_i^3}\right) + \frac{\varphi_{ni}^4}{\varphi_i^3} \right],\ \varphi_i \geq \varphi_{ni}. \tag{7}$$

For geometrically similar samples 1 and 2 scales of spiral radii R_0, relative linear strain φ_0, integral strength limits of the materials of the shockproof device model in the zones of elastic and elastoplastic deformation G_0 and Φ_0 were introduced:

$$R_0 = \frac{R_1}{R_2};\ \varphi_0 = \frac{\varphi_1}{\varphi_2};\ G_0 = \frac{G_1}{G_2};$$
$$\Phi_0 = \frac{\Phi_1}{\Phi_2};\ \varphi_{\text{по}} = \frac{\varphi_{\text{п}1}}{\varphi_{\text{п}2}}, \tag{8}$$

In this case we consider the conditions of external loading models, in which the relationships remain constant applied external linear power effects:

$$\frac{F_1}{F_2} = F_0 = const. \tag{9}$$

Using the scales (8), the condition of simple loading of torsions (9), by geometric similarity, transformed (7) into a dimensionless form:

$$\frac{F_0 F_2}{G_0 R_0^3 G_2 R_2^3} = \frac{\pi \phi_0 \phi_2}{2K}\ \text{при}\ \phi_0\phi_2 < \phi_{\text{по}}\phi_{\text{п}2}. \tag{10}$$

$$\frac{F_0 F_2}{G_0 R_0^3 G_2 R_2^3} = \frac{\pi}{2K}\left[\begin{array}{c}\frac{4}{3}\left(\phi_{\text{по}}\phi_{\text{п}2} - \frac{(\phi_{\text{по}}\phi_{\text{п}2})^4}{4(\phi_{\text{по}}\phi_{\text{п}2})^3}\right)+ \\ +\frac{\Phi_0\Phi_2}{G_0 G_2}\times(\phi_0\phi_2 - \phi_{\text{по}}\phi_{\text{п}2})\times \\ \times\left(1 - \frac{(\phi_{\text{по}}\phi_{\text{п}2})^4}{4(\phi_{\text{по}}\phi_{\text{п}2})^4} + \frac{(\phi_{\text{по}}\phi_{\text{п}2})^4}{(\phi_{\text{по}}\phi_{\text{п}2})^3}\right)\end{array}\right] \tag{11}$$

при $\varphi_0\varphi_2 \geq \varphi_{\text{по}}\varphi_{\text{п}2}$;

$$\frac{F_2}{G_2 R_2^3} = \frac{\pi \phi_2}{2K}\ \text{при}\ \phi_2 < \phi_{\text{п}2}; \tag{12}$$

$$\frac{F_2}{G_2 R_2^3} = \frac{\pi}{2K}\left[\begin{array}{c}\frac{4}{3}\left(\phi_{\text{п}2} - \frac{\phi_{\text{п}2}^4}{4\phi_2^3}\right) + \frac{\Phi_2}{G_2}(\phi_2 - \phi_{\text{п}2})\times \\ \times\left(1 - \frac{\phi_{\text{п}2}^4}{\phi_2^4}\right) + \frac{\phi_{\text{п}2}^4}{\phi_2^3}\end{array}\right] \tag{13}$$

where $\varphi_2 \geq \varphi_{\text{п}2}$.

In experiments with models of prototypes of shockproof devices [6], the second dimensionless complex was changed. The remaining similarity criteria remained constant. In Figure 3, for the fixed values of the similarity criteria from the results of several experiments, a simple criterion equation was created $\frac{F}{GR^3} = f\left(\frac{l}{R}\right)$.

As the value of the second dimensionless complex decreases, the nonlinear effect of plastic hardening degenerates.

During the test of the models, the change in the heating temperature of the samples can also significantly affect the mechanical properties of the power elements of the shockproof device.

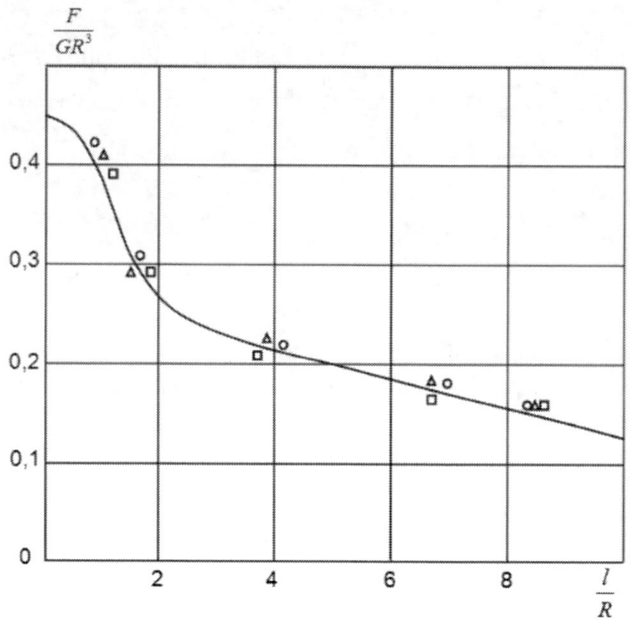

Fig.3. Experimental values of dimensionless power characteristics of models in the criterial form.

The experiments were processed in a wide temperature range in the criterial form $\frac{F}{GR^3} = f\left(\frac{l}{R}, \varphi_{\Pi}\right)$.

An assessment of the scale factor influence on the possibility of transferring the results of experimental studies to real objects showed a good agreement of the criteria in the given region of deformation. In this case, the discrepancies for model samples amounted to no more than 15-20% with a confidence coefficient of 0.95, which can be explained by the difference in the nonlinearity of hardening in elastoplastic deformations in models based on an alloy with phenomenal properties. The nonlinearity of the criterial dependence in this case is a consequence of the special and random nature of the external forces applied to the experimental models.

4 Conclusions

Thus, the evaluation of the scale factor effect in the simulation of a shockproof garment element based on the alloy with shape memory effect makes it possible to extrapolate the results of experimental studies to other objects to form the initial data of the mathematical model of the dynamics of elastoplastic deformation of shockproof devices while observing the similarity of the external loading characteristics.

An integrated approach to the investigation of individual shockproof elements placed on garnet can contain such key points as:

- study of information on similar samples used in production, sports, life;
- development of sketches and approval of the pilot design of the device;
- selection of structural materials with the definition of the range of major suppliers;
- production of prototypes;
- carrying out full-scale experiments;
- an assessment of the possibilities to transfer the experimental results to real objects;
- marketing research of the industrial production possibility.

References

1. A. E. Astvatsaturov et al. *Obespechenie bezopasnosti pri tehnogennyh i prirodnyh chrezvychajnyh situacijah*. (Rostov n/D: DSTU, 2004).
2. O. V. Denisov et al. Inzhenernyj vestnik Dona: Jelektronnyj nauchnyj zhurnal. **2.** (2017).
3. O. V. Denisov et al. IOP Conference Series: Earth and Environmental Science. "Ecology and Safety in the Technosphere: Current Problems and Solutions". **012044.** (2017).
4. O. V. Denisov et al. Sbornik nauchnyh trudov nauchno-metodicheskoj konferencii, posvjashhennoj 85-letiju DSTU (Rus) [Innovative technologies in science and education: Collection of scientific works scientific-methodical conference]. Rostov on/D (2015).
5. O. V. Denisov et al. *Kompleksnaja bezopasnost' naselenija i territorij v chrezvychajnyh situacijah. Problemy i reshenija*. (Rostov on/D: DSTU, 2016).
6. O. V. Denisov et al. Izvestija vuzov. Estestvennye nauki. **4**. (1999).
7. O. V. Denisov et al. Materialy III Regional'noj nauchno-prakticheskoj konferencii Juzhnogo federal'nogo okruga. Krasnodar (2015).
8. I. I. Eremin, O. V. Denisov Inzhenernyj vestnik Dona: Jelektronnyj nauchnyj zhurnal. **36. (2-2).** (2015).
9. Ya. B. Fridman, *Mehanicheskie svojstva metallov. Ch.1. Deformacija i razrushenie* (M.: Mashinostroenie, 1974).
10. A. I. Kostoglotov et al. Izvestija vuzov. Estestvennye nauki. **4. (**1999).
11. A. A. Kozlov, et al. Vestnik Sankt-Peterburgskogo gosudarstvennogo universiteta tehnologii i dizajna. Serija 1: Estestvennye i tehnicheskie nauki. **2.** (2016).
12. G.B. Iosilevich *Prikladnaya mekhanika* (Moskva: Vysshaya shkola, 1989).
13. Protivoudarnoe prisposoblenie: patent na izobretenie [Shockproof device: patent for invention]. B. Ch. Meshi, O. V. Denisov, Yu. I. Bulygin, A. E. Ponomarev, I. A. Ponomareva. No RUS 2578997. (08.2014).
14. B. A. Rabinovich Zhurnal avtomobil'nyh inzhenerov. **6** (2016).
15. A. G. Tihomirov,, et al. Izvestiya vysshih uchebnyh zavedenij. Severo-Kavkazskij region. Seriya: Estestvennye nauki (Rus). **3.** (2004).
16. I. V. Topilin, I. A. Ponomareva, Inzhenernyj vestnik Dona: Jelektronnyj nauchnyj zhurnal. **36 (2-2) (**2015).
17. V. V. Vershinin,, A.A. Loktev, Vestnik Permskogo nacional'nogo issledovatel'skogo politehnicheskogo universiteta. Mehanika. **1** (2011)

High-Rise Construction in Densely Dwelled Cities: Requirements for Premises Insolation and Consequences of their Violation in Russian Law and Jurisprudence

Boris Gongalo[1], *Lubov* Gudovicheva[1], *Anna* Gubareva[1] and *Larisa* Dobrynina[1*]

[1]Ural State Legal University, 21, Komsomolskay st., Ekaterinburg, 620137, Russia

Abstract: The issues of constructing high-rise, primarily residential, buildings have a great social significance. Not every plot of land, acquired in the Russian Federation is suitable for high-rise construction. Therefore, every construction company that plans to erect a multi-apartment building, a high-rise office building, or a skyscraper must take into account not only technical norms but as well sanitary legislation regulations that set obligatory requirements about insolation of apartments. The article includes a short study of several norms in the Russian legislation regarding insolation of dwellings; analises the problems of judicial interpretation of the statutory limitations. In this aspect it researches the debatable questions arising in practice of state arbitration courts dealing with the lawsuits on allocation of land-plots by the local administration. The analysis of the judicial practice is followed by description of the difficulties facing the developers of land-plots, concerning the project and territorial planning documentation.

1 Introduction

The cities differentiate between each other in many aspects, starting from their geographical position and ending with their area and population. The cities with a high density dwelling are standing out, because their planning, architecture and height of building have to be correlated with the so-called "infill development". The city of Ekaterinburg, situated in the Sverdlovsk region of Russia is one of them. The intensive construction in the city leads to some unavoidable issues of providing comfortable and safe living and working there. The construction of high-rise buildings is stimulated by the rocky ground underneath Ekaterinburg. High-rise buildings become, up to a point, a "visiting card" of the city and will continue to be built in the future.

Every type of multi-apartment house satisfies one or another requirements of population and has its own advantages and disadvantages. Therefore, they have their certain shares in the structure of urban dwelling. The structure of municipal housing fund is formed by the centuries of the city's history and depends on traditions, mentality, economic and political realities, existing legislation etc. [1] At the same time, arises the question of the ways to

* Corresponding author: dovre@yandex.ru

observe all rights of citizens, living in the buildings next to skyscrapers, including their right for necessary insolation.

2 Methods

The current research uses the methods of observation, comparative legal studies, formal logic, description and interpretation. The goal of research is to show that the problems of insolation have systemic nature and can not be isolated from the issues of urban development, economics of high-rise building and personal intangible interests of population. The research aims at finding solutions for problems of high-rise construction.

Insolation is the solar irradiance by direct sunbeams of different objects, including developed territories, fencing constructions and rooms of the building. Placing and orientation of residential and public buildings must provide continuous insolation for apartments - at least 2,5 hours per day between April 22 and August 22 for the Northern geographical zone (to the north of 58− of northern latitude), or 2 hours per day between April 22 and August 22 for the Central geographical zone (between 58− and 48− of northern latitude), or at least 1,5 hours per day between February 22 and October 22 for the Southern geographical zone (to the south of 48− of northern latitude) [2]. It may seem that buildings with panoramic windows may be the solution for this problem. For the first time this solution was proposed by the French Architect Le Corbusier in his article "The Five Points of the New Architecture" published in the journal "L'Esprit Nouveau" in the 20-ies of the last century. As the fourth point of the new architecture he proposed elongate shape window. Pileses with the intermediate beams form the rectangular opening in a facade, through that light and air pass into the room anliberal amount. A window stretches from a bar to the bar, becoming, thus, by the extended window. An apartment is identically lighted up in all places - from a wall to the wall. It is well-proven, what an apartment is illuminated in 8 times more intensive, than the same apartment with vertical windows. All history of architecture runs around exceptionally round the window opening. And the reinforced concrete opens possibility of maximal illumination through the extended windows" [3].

The sunlight is vital for the human being's feeling and mood, it acts as an antiseptic, keeps rooms dry etc. The right for the sunlight is one of inalienable rights of a person, which can be limited only under federal law (like in case of imprisonment) or insurmountable circumstances (like polar night). The personal unproperty relations are folded concerning the non-material blessing that is inherent to any individuals from birth (life, health, favourable natural environment) or acquired by virtue of pointing of law (name, inviolability of place of inhabitation etc.). In obedience to Russian legal tradition, the non-material (spiritual) blessing are the object of the personal unproperty rights. Most in detail the scientifically reasonably personal unproperty rights, including, right on a health, are considered by L.O. Krasavchikova [4] and M.N. Maleina [5].

According to Charter of Worldwide organization of health protection, "ahealth is the state of complete physical, heartfelt and social prosperity, rather than absence of illnesses or physical defects" [6]. In accordance with the first point of the 150 articles of the Civil code of Russian Federation (further on text - civil code), a right on a health is attributed to the personal unproperty rights for a citizen (individuals). This right belongs to all individuals regardless of age, sex, citizenship and other differences. Studying a current legislation, it is possible to distinguish the basic types of health of individual that it is possible in this context to quality a man, such, as: physical and psychical health. There are factors, that influence on a physical health: terms of residence, quality of feed, drinking-water, air, condition of work and other. It is necessary to include some factors? That qualify the psychical health of man:

absence of stress situations, quality of mutual relations with other people, correct estimation of itself in the society, etc.

At the same time, there are such violations of life quality, that simultaneously influence on physical and psychical health of man. To such harmful factors it is necessary to take violation of terms of insolation of dwellings apartments.

In obedience to the article 22 of the Federal law on 30.12.2009 No. 384-FZ "Technical regulation about safety of building and structures" [7], (point 1) building must be projected so that sufficient duration of insolation or sun protection was provided in dwellings apartments, for creation of safe terms of residence, regardless of his term; (point 2) implementation of these requirements, it must be provided with measures on the orientation of dwellings apartments on compass point, and also by the measures of structural and plan character, including on equipping with modern amenities of adherent territory. In accordance with Article 23 of the Federal Law from 30.03.1999 No. 52-FZ (as amended) "On Sanitary and Epidemiological Welfare of the Population" [8], (point 1) dwellings apartments (area of a room planning, luminosity, insolation, microclimate, ventilation, sound-levels, vibration, ionizing and unionizing radiations) must comply with sanitary and epidemiological requirements in order to ensure safe and harmless terms of residence regardless of his term; (point 2) settling of the dwellings apartments confessed in accordance with the sanitary legislation of Russian Federation useless for a residence is equal, as well as grant to the citizens for the permanent or temporal residence of non-residential premises, shut out.

The requirements for insolation are also determined at the level of by-laws. Point 14.21 of the JV on 42.1333.2011. Set of rules. Town planning. Planning and construction of urban and rural settlements, updated version SNiP 2.07.01-89 * [9] envisaged, that placing and orientation of dwellings and public building must provide duration of insolation of apartments and territories in accordance with the Sanitary norms and rules the "Hygienically requirements to insolation apartments of dwellings and public building and territories. SanPiN 2.2.1 / 2.1.1.1076-01 ratified by the Main state health-officer on 19.10.2001 [10].

In accordance with a point 2.1 of requirements, to irradiation of surfaces and spaces by direct sunlight (insolation) are made when placing objects, in planning and building projects for neighborhoods and quarters, for the construction and reconstruction of individual buildings and structures, and for supervising the construction and operating facilities. Point 2.2 of this Regulations is set that implementation of requirements of norms of insolation is arrived at by placing and orientation of building on compass point, and also by volume of-plan decisions.

In a point 2.3 this act underline, that insolation is an important factor rendering revitalizing influence on the habitat of man, and must be used in dwellings, public building and on territory of dwelling building. By a point 3.1 duration of insolation in dwellings building must be provided no less then in one room of 1 - 3 room apartments and no less thenin two rooms of 4th and more room apartments, and point 3.3 talks that irregularity of duration of insolation at that one of periods must be no less than 1,0 hour is assumed. Setting of norms of insolation for every zone is set in a division 2 normative acts indicated higher.

Some authors suggest to simplify exposition of this concept, adopting analogues from a pre-revolution right: the designed building ignores belittling light, if distance from him to the nearby estate there is a more height of the designed structure "on a compass point, considering from a top to the terrene" or if "meantime et al the street of the first or second digit of legal width"(Article 603 of the Civil Code of the Zurich Canton.

Edition of the Editorial Commission for the Drafting of Civil Code. printing house, 1887. P. 108) [11]. More short suggestion: a person, who has a right on light, "cannot oppose, if access of light will be covered no more than ½ of the height of direct corner, considering from a window-sill to up" [12].

3 Results

Duration of insolation is regulated by normative acts only as it applies to next structures: to dwellings building (to the apartments); to child's preschool establishments; to educational establishments of general, primary, middle, additional and professional education, to boarding-schools, child's houses and other; to medical and preventive, sanatory-health and resort establishments; to establishments of public welfare (boarding houses for the disabled and the elderly, hospices, etc.).

A failure to observe of requirements to insolation of dwellings apartments can bring to confession of dwelling apartment unfit for a residence (see "Regulations on the recognition of a dwelling, a dwelling unfit for living and an apartment building as emergency and subject to demolition or reconstruction", approved by RF Government Decree on28.01.2006 No. 47) [13].

The most common consequence of the violation of insolation, educed at planning of dwellings or unoccupied building is a violation of insolation in next dwellings doors as a result of realization of building. In this case, a positive opinion of the state construction expertise will not be obtained, that will lead to the refusal to issue a building permits (Articles 49, 51 of the Town Planning Code of the Russian Federation [14]).

In one of the cases, building company appealed to the court with a lawsuit about nullification negative conclusion of state examination on project documentation of object of capital building. The position of a court is that company" sproject documentation on the object of capital building "dwelling-house with built-in non-residential premises on a ground floor and underground garage down Roshinskaya Street in city Ekaterinburg" falls short of to the requirements of technical regulations and normative technical documents.

In obedience to project documentation, after placing of the designed dwelling-house on a contiguous area with an already present house, it will result in the complete exception of insolation of north-eastern facade of new house (Decree of the Federal Antimonopoly Service of the Urals District on February 24, 2014 No. Ф09-186/14 in case No. А60-18408/2013) [15].

There can be another situations, when a high-rise building can not be located on a land plot provided for construction, because of parameters of the permitted construction do not correspond to the actual construction parameters, and the land plot is not suitable for the construction of a residential house, taking into account the requirements of all regulatory and legal acts, compliance with which is necessary for the design, construction and commissioning of this building. In another case, the construction company appealed to the court with the lawsuit about nullification the contract with the city district administration about the termination of the land lease for the design and construction of a high-rise 10-storey residential building and recovery of damages in the amount of 1,117,200 rubles. Lawsuit requirements were satisfied. The court found that on the basis of the resolution of the administration, an auction was held to sell the rights to conclude lease contracts for land plots, for housing needs for the design and construction of a residential building, for a period of 3 years. When the preliminary design was coordinated with the management of the city administration architecture, it was established that it was impossible to construct a residential building above four floors on the given land plot, due to the lack of space for children's playgrounds, parking for cars, and the insolation of the kindergarten adjacent to the northern side to the disputed land. Guilt of building company is absent.

In case of assumption of errors at planning, when them were not found out by examination, the construction of a building or structure that violates insolation in the existing dwelling – house, building can be forbidden, the given permission on building is confessed by invalid in the judicial order (Article 1065 of the Civil Code), and also another methods of protection of rights (Article 12 of the Civil Code of the Russian Federation) are applied.

In case of impossibility of demolition of the constructed building, the developer may be obliged to move citizens of flats that do not receive sufficient sunlight to other living quarters or with their consent to apply other compensatory measures, for example, to install on the facade of a building under construction a fluorescent lamp, apply mirror facades, etc. At the same time, all these requirements deal with the proprietors of dwellings apartments only, and also row of public building, for example, of sanatoriums. It is suggested to enter the concept of insolation for non-residential premises, as many people during a light day are at work, and, a sunlight is also required them for maintenance of the health. In this connection, it is suggested also to forbid using apartments in that people spend a lot of time, if in such apartments there are not windows or corresponding compensative events are not conducted, in particular, as setting of lamps of daily illumination of high quality.

Suggestion about the necessity of distribution of requirement to insolation on non-residential premises is coused by the circumstance, that citizens can live in such apartments that in accordance with the current Russian legislation is not subsumed dwellings, they are unoccupied. In the number of such apartments it is necessary to include the suites of rooms, lofts [16], apartments in dormitories, office apartments.

4 Discussion

World practice of planning and construction of high-rise buildings in cities with dense building, and also in cities with dense building of center applies requirements to insolation to the dwellings apartments. Practice of application of the same norms to the apartments, where citizens, probably, live temporally, for example, in hotels, is not uniform.

Authors of researches of the legal mode of suites of apartments, loft-apartments, studios are convinced, that the norms of insolation do not spread to these apartments. Does developer can to ignore the norms of insolation of apartments in existing next to the new construction of the building if he knows this circumstance?

And in theory, and practically, on this question it is necessary to answer in the affirmative. In addition, it be possible to say, that physical persons, acquiring a studio or loft for a permanent residence, already beforehand show the unreasonableness of the behavior, as they must suppose that many requirements that is produced to the dwellings apartments (to the individual dwelling-houses, apartments in condominiums, apartment dwellings houses) do not spread to such apartments.

In particular, one of such situations was considered the arbitration (state) courts of the Ural region in regard to requirements to the building company (Decree of the Federal Arbitration Court of the Urals District on14.05.2008 Case No. A60-28553/2007) [17]. To present tense business data are considered the "classics" of permission of dispute on insolation.

This building company got the positive conclusion of state examination, and then permission on building of two high-rise dwelling-houses in city Ekaterinburg (the Aston Plaza project).

But later, in connection with the appeal of deputy of Regional legislative Body and citizens, resident in a next door, public supervisory body, pulled out to person building a requirement about stopping of building in order of the article 1065 Civil Code, and also brought a building company to the administrative account. Violations of legislation were educed in area of providing of sanitary and epidemiology prosperity of the population expressed in violation of operating sanitary rules and hygienically norms.

That is, actions of building company, on building of new housing estate worsened, and in a row case fully eliminated insolation and natural luminosity of apartments of north and north-western orientation of already existent dwelling-house, that violates a current legislation. A building company appealed the actions of public body in an arbitration court.

The dramatic effect of situation consisted of requirements of public body, to settl out of lodgers of 30 apartments from home in that citizens lived in the own apartments that located next to new building, in accordance with the Russian legislation. Building company, understanding, what losses it draws, contested bringing to administrative responsibility, because this act was set his guilt as a developer. And guilt is one of four necessary elements of bringing to responsibility as reimbursement of losses. If person building had to resettle the tenants of lodgers 30 apartments, then he, firstly, would so rise the price building of the high-rise residential buildings, that hardly acquisition of such apartments was expedient from the side of potential customers, and, secondly, would tighten a building process on long years, that would do building unprofitable, and person building forced to appeal to the court with a requirement about confession by his bankrupt.

It would be desirable to underline that a requirement to insolation is so meaningful and considerable, that can result in the most different economic and legal unprofitable consequences for person building, projector and even for the state because of the lack of the proper supervision after planning during realization of state examination of building project.

Originally, courts of first and second instances decided that the requirements of declarant were not subject satisfaction. It was set that reason of delivery of positive calculations and conclusions on insolation, executed in 2006 by organization by a designer, was wrong establishment of orientation of apartments (a dwelling-house has a not north and north-western orientation).

At establishment of guilt of building company, a court came to the conclusion, company realized possibility of violation of legislation about health and epidemiological prosperity of population at building of new dwelling-house and foresaw the offensive of harmful consequences as worsening of terms of insolation of apartments of north and north-western orientation of already existent dwelling-house. By virtue of parts 11, 15 of articles 48 of the Town Planning Code of the Russian Federation, a building company is a person responsible for authenticity of project documentation.

The court of the third instance abolished previous decisions, specifying that an already existent dwelling-house, insolation in that could be broken, was originally built and used as a house with apartments, having status of official. The house was considered originally not suitable for a permanent residence the result was built not three high-rise dwelling-houses and only two. This was the cost of failure to observe of requirements of insolation already on the stage of choice of lot land for building and at the subsequent planning.

The contradiction of right and practice, related to all non-residential premises, is that status of house confesses primary, and secondary is a fact of permanent residence in him physical persons constantly, by families, with children. It would be desirable to underline that the main culprit of such losses is local self-government body, giving such problem area for building, and also public state body giving out positive conclusions without realization of the real deep researches. To date reimbursement of such losses maybe on the basis of articles 15, 1064, 1069 of the Civil Code of the Russian Federation.

5 Conclusions

In place of the indicated person building there can be any other, including foreign. On another-strange individual and legal persons the national mode, that is expressed not only in the grant of rights but also in an incumbency, spreads, including, on reimbursement of losses.

It is necessary to mark the following. Judicial practice of Russian Federation is oriented to satisfaction of requirements in regard to dwellings apartments, but not always. Such requirements that touch are mainly satisfied, so-called, neighborly spores, when the question is not about erection of sky-scraper in the courtyard of ordinary "Khrushchev" five-storey houses. Examining the lawsuits of lodgers of such houses, court, as a rule, acknowledge them not interested in the serve of such lawsuits. These facts are marked in literature. In publications on this question, a legislation over is brought and practice of other countries, acknowledging lodgers resident in apartments in a house, located with the site area of the high-rise building interested in challenging of documents accompanying new building. In particular, experience of the USA is analyzed, where disputes about building of new building next to already existing always requires the presence of locals as the third persons [18]. The same position must accept the Russian courts that try to find a necessary parity between development of city on the whole, the construction of new high-rise buildings that will decorate a city and "save" many hectares, and interests of physical persons, in particular, by their right on the necessary amount of sunlight in apartments, where they constantly or temporally live and work, after those exceptions that must be set forth at the level of law, equal as and a self right on a sunlight must get a due settlement.

The questions considered in the real research are actual for world practice of building high-rise buildings and other meaningful objects that confirms a followins fact.

When this article was about ready, many world mass medias passed the report of BBC's radio and television that family from Great Britain appealed to the court with a requirement to forbid building of stadium of Stamford Bridge for FC "Chelsea" in London because of fears, that a sunlight will cease to fall in the windows of their house. [19].

References

1. S.N. Glazunov, V.S. Samoshin. *Housing issue in Russia: problems and prospects.* (Omega-L Publishing House, Moscow, 2008)
2. R.T. Nagaev. *Real estate: encyclopaedic dictionary (land management and cadaster of real estate, urban planning and architecture, real estate economy and land law)* (Publishing House Economics, Moscow, 2008)
3. URL: http://arx.novosibdom.ru/node/547 (03.04.2017)
4. L.O. Krasavchikova. *The concept and system of personal non-property rights of citizens (individuals) in the civil law of the Russian Federation*. (Ekaterinburg, 2004)
5. MN Maleina. *Personal non-property rights of citizens: concept, implementation, protection* (Moscow, M3 Press, 2000)
6. URL: https://www.dostupnigorod.ru/wp-content/uploads/2015/11/sp-42.13330.2011.pdf (03.04.2017)
7. URL: http://sudact.ru/law/?law-txt=&law-date_from=&law-date_to (03.04.2017)
8. D.A. Monakhov Vestnik civil law, **1** (2012)
9. K.P. Zmillov. Journal of the Ministry of Justice, **11** (1895)
10. URL: http://www.consultant.ru (03.04.2017)
11. M.M. Monastyrev. Legislation and Economy, **84** (2015)
12. E.D. Polydi, T.K. Baykova, E.V. Abbots. Property Relations in the Russian Federation, **4** (2016)
13. S.V. Savina. Turnover of apartments as housing: directions for improving legislation // Law, **4** (2015) (http://www.consultant.ru)
14. URL: http://sudrf.kodeks.ru/rospravo/document/770462219 (03.04.2017)
15. B.R.D. Alarie, A.J. Green. Osgood Hall Law Journal, **48** pp 386 - 387 (2010)
16. URL: http://www.footballtop.ru/news/smi-semya-iz-britanii-mozhet-pomeshat-abramovichu-postroit-novyy-stadion-chelsi (03.04.2017)

Principles and methods for ensuring safe operation of high-rise buildings

Oleg Korol [1*] and *Yuliya* Kustikova[1]

[1] Moscow State University of Civil Engineering, Yaroslavskoe shosse, 26, Moscow, 129337, Russia

Abstract. The purpose of monitoring the technical condition of high-rise buildings is to prevent possible negative situations leading to significant socio-economic losses by timely warning of the emergence of such situations. To achieve this goal, it is necessary to solve the following main tasks, such as: identifying the time and place of origin and development of negative processes that lead to the emergence of an emergency situation; analysis of the possible development of the situation in time; development of management decisions; formation and submission of warning signals; obtaining new knowledge about the operation of the object, the factors of influence on this object, the speed of development of destructive processes. When solving the above problems, an important role is played by constructing an adequate mathematical model of the object, the parameters of which should be calibrated according to the current monitoring results.

1 Introduction

Damages and defects in building structures, as practice shows, occur 50% or more at the construction stage, 20% at the operational stage and 30% due to errors in geological research and design. Avoiding these problems will help to monitor the condition of soils, foundations, structures during construction and operation, as well as the use of safety systems.

As for high-rise buildings, security issues are of crucial importance here. Each such building is a complex constructive system with a large number of engineering communications. The increased number of buildings and, consequently, the presence in them of a significant number of people with limited opportunities for their evacuation require the designers to solve additional problems. In particular, the inclusion in the draft measures to prevent, detect and eliminate emergencies, evacuation and rescue people.

2 Literature review

Until recently, in the world of construction science, there was a problem of constructing theoretical models and software for project security of buildings and structures, taking into account the degradation of the systemic static and dynamic stability of load-bearing structures caused by thin geodynamic, geological and geophysical and other natural accumulative processes of deferred action [1-7].

* Corresponding author: mrkorol.oleg@gmail.com

This problem was particularly acute and contradictory in seismic-passive regions of urbanized territories with bases that, on the one hand, are not affected by accidentally dangerous seismic loads and impacts, and on the other, have a fundamentally heterogeneous thin structure of the geological section that is inconsistent with a rigorous design assessment, down to the depths crystalline basement [8-10].

The results of the measurements show that under the influence of external sources the building performs rather complex resonant oscillations of at least two types - the oscillations of the building as a single system of the "object-base" type and the oscillation of individual elements or parts of the building [11-13].

As a result of the full-scale instrumental studies of geodynamic safety of the responsible building objects, it became clear that the initial cause of the launch of local geodeformation mechanisms is the previously unknown and therefore poorly studied processes of cyclic "fatigue" of base soils, assembly junctions and building materials. In turn, the experimental data obtained and the analytical identification of cause-effect relationships have made it possible to reliably establish microseismic and microgravitational mechanisms for the activation of such phenomena [14].

The obtained statistical and experimental data and identification of resonant cause-effect relationships allowed to establish microseismic and microgravitational mechanisms of activation of such phenomena reliably [15-17].

According to the results of the research it was also established that they are geoecological manifestations of the effect of long-term accumulation of nonlinear effects of technogenic excitation of resonant microseisms in the bases of these objects. Despite the significant difference in the scale and nature of the evolution of cosmic and terrestrial catastrophic processes, they are united by the common physical nature of the gravidynamic cause-effect relations of the launching of dangerous global meteorological, climatic, geophysical, ecological, seismic, geodeformation and other terrestrial processes [18-20].

3 Methodology

The main reasons for the collapse of high-rise buildings are:

– exposure to natural factors that lead to aging and corrosion of construction materials and reduce their physico-mechanical characteristics: air environment, atmospheric moisture, groundwater, subsidence ground, negative air temperature, etc;

– natural disasters causing destruction: hurricanes, storms, tornados, tsunamis, floods, earthquakes, landslides, mudflows, snow slips, etc;

– design and production defects of structures and technical systems: errors in exploration and design, poor quality of construction or building materials and structures;

– the impact of technological processes on materials and structures: additional loads, high temperatures, vibrations, oxidants;

– violation of the rules for the operation of structures, technical systems and resulting fires, explosions of fumes of gasoline, chemicals, gas, etc.

The basic principles of monitoring the technical condition include:

- continuity of the monitoring process;
- reliability of the monitoring system;
- reliability of the results;
- the effectiveness of the system.

In practice, in order to consistently comply with the principle of continuity, it is necessary to build systems for monitoring the technical state in such a way as to provide the minimum possible intervals for obtaining information on the characteristics of the object under given economic constraints. This is due to the need to comply with the following principle, namely, the principle of reliability.

Table 1. Shows the ways and forms of implementation of these principles.

Principle	Content
Continuity of the monitoring process	Organization of a monitoring system for the technical condition of high-rise buildings and structures in which information about the state of an object or its elements is recorded at intervals not exceeding the characteristic times of events associated with the impact on the object or its ability to resist the impact.
Reliability of the monitoring system	Integrates the reliability of the equipment - and the reliability of determining the parameters of the impact and resistance to the impact, taking into account the probabilistic nature of these parameters. Such accounting is extremely important for the choice of time and amplitude characteristics of the elements of the monitoring system, including such as temporal resolution, degree of sampling, dynamic range and the like.
Reliability of the results	The definition of the parameter under consideration by two or more alternative methods sharply raises the reliability of measurements and the quality of the monitoring system as a whole.
Efficiency of the system	The ability of the system to register, carry out data transformation and analysis, and generalize the results obtained to achieve the set goals and solve the tasks.

The monitoring system should be designed in such a way as to provide the required probability of registering the process essential for the facility.

A significant number of methods used in the construction of monitoring systems, including methods for determining the stress-strain state, are indirect. At the same time, the result of measurements is influenced by a large number of factors that are difficult to take into account, which reduces the reliability of the results obtained. In this connection, it seems necessary to use methods based on different physical principles for measuring the parameters of a structure, comparing and analyzing the results obtained. Given the high responsibility of high-rise buildings and structures, it is necessary to install several independent monitoring systems for the same physical parameters.

An important aspect of the efficiency principle is the economic aspect. So, for example, when registering changes in the stress-strain state by measuring the relative elongations in concrete, economic costs are sharply reduced with increasing the base of the relative strain sensor. To achieve the same reliability in the case of using sensors with a short base, much more of them are required, as well as the application of mathematical averaging procedures. Similar tasks arise in the design of other monitoring subsystems.

Analysis of the structural features of high-rise buildings allows us to formulate the following proposals for the organization of a monitoring system: (Table 2).

In practice, in order to consistently comply with the principle of continuity, it is necessary to build systems for monitoring the technical state in such a way as to provide the minimum possible intervals for obtaining information on the characteristics of the object under given economic constraints. This is due to the need to comply with the following principle, namely, the principle of reliability. This phenomenon is well studied in mechanical engineering as a multi-cycle and low-cycle fatigue under a constant load cycle. These processes can explain the numerous accidents occurring when there are no significant impacts on the structure. Constant monitoring of the nature of the change in impact allows us to conclude that the degree of damage accumulation, and thus the residual resource of the structure, its reliability.

Table 2. Methods and methods for providing controlled parameters.

Controlled measures and parameters	Methods
Integrity control and detection of defects in foundation slabs and slabs and covering of the stylobate part of the structure	• acoustic, based on the sensing elements installed in the body of the plates; • fiber-optic, based on reflectometry in fiber optic lightguides, embedded in the body of plates near their surface
Determination of vertical deformations and bends of the foundation slab, revealing the unevenness of the draft	• geodetic leveling; • hydrostatic leveling, installation of inclinometers
Measurement of relative strains (stresses) of the main structural elements, the list of which is determined on the basis of numerical modeling and expert evaluation, in comparison with reinforcement maps. The most likely locations for the installation of sensors are: slabs of the stylobate part, support bearing structures of the hulls, their spans and consoles	• independent strain-gauge systems based on fiber-optic Bragg sensors (VOD), string and electromechanical sensors, and also on the basis of acoustic methods (including acoustic emission)
Reconstruction of spatial spaces of structural elements of a structure	• measurements by means of a complex of robotic total stations; • digital photogrammetry; • laser scanning
Observation of the amplitude-frequency characteristics of elements and structures of a structure	• installation of triaxial accelerometers in controlled areas of the facility, installation of seismometers
Climatic observations, including temperature and humidity, wind and snow loads	• methods and means of climate control
Situational monitoring of parameters determined on the basis of development of scenarios of adverse impacts and events in their probable aggregate	• all of the above methods applied to the design zones defined by scenarios developed on the basis of interdisciplinary expert assessments

Taking into account that the parameters of the reliability of construction objects depend to a large extent on the geodynamic stability of the object-foundation systems, the solution of the safety problem begins with a pre-project geotechnical justification for the stability of the fine structure of the geological section of the bases, followed by their own geodynamic resonances, geodeformation and geoecological risks at the design stage and concludes with regular monitoring of the evolution of the banks and sediment of buildings and structures during the construction period operation and maintenance.

According to the results of the research it was also found that, especially in megacities, unexpected super-project geodesic information bases of engineering structures and communications are in fact geological manifestations of the effect of long-term accumulation of non-linear consequences of technogenic microcyclic excitation of vibroseis in the zone of laying or placement of these objects. In other words, there is a technogenic violation or acceleration of the natural evolution of geological and hydrogeological processes in base soils, i.e. according to already established terminology there is a technogenic violation of natural geoecological conditions in the zone of construction sites, and especially in the territories of megacities and areas of mass development.

The latter statement is due to the fact that it is megacities that are distinguished by increased technogenic vibroseismic activity, as well as heterogeneity of the fine structure of the geological section and uneven static and dynamic load in the distributed "object-base" system, which significantly enhances the geoecological efficiency of natural and technogenic microseisms on the grounds of the calculated volume grounds and the building object as a whole.

Thus, it can be concluded that if in geospatial zones the geodynamic stability of geological bases of building objects is disturbed by earthquakes, then in seismic-active zones - by ultra-low-frequency geodeformation processes responsible for the latent long-term accumulation of hydrogeological and physicomechanical inhomogeneities in geological sections.

However, geo-information processes under the foundations of building objects and engineering structures in seismic-passive regions can be significantly strengthened or "accelerated" as a result of natural and technogenic excitation of vibroseismic and gravis-seismic resonances of objects and their geological bases.

The work on designing and creating a monitoring system is carried out in several stages (Fig. 1.).

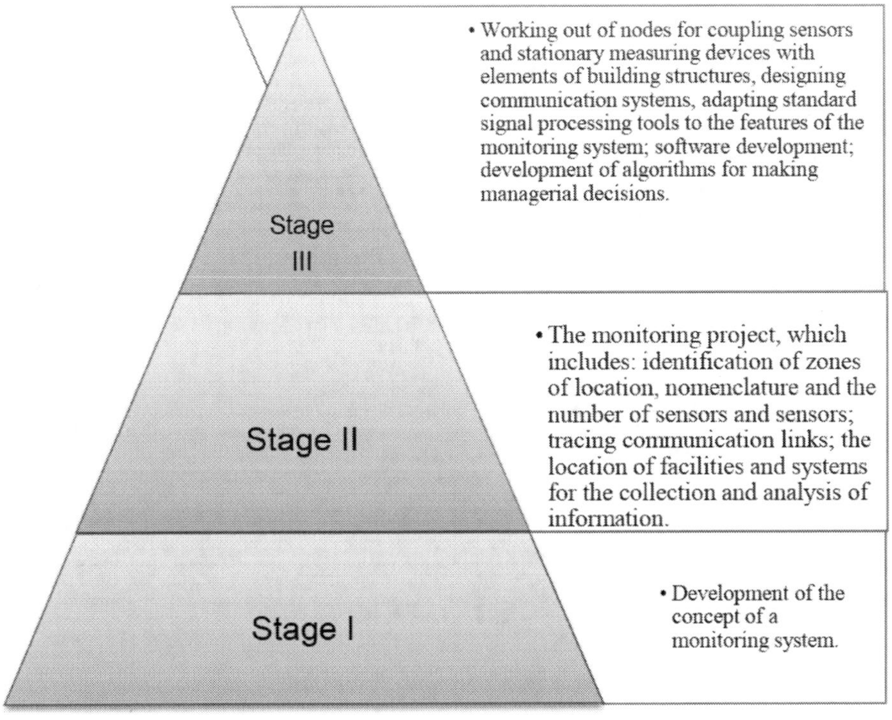

Fig. 1. Stages of development of the monitoring system.

It should be noted that the installation of sensors of the monitoring system should be carried out at the earliest possible stages of construction as the structure is erected. This requirement is extremely important, as it provides information on the "zero" loading of the structure and allows you to track the change in the parameters of structures at all stages of construction, assessing their compliance with design values.

However, the progressive development of the building science, along with new advances in the field of information technology, allows us to successfully overcome a number of such limitations. Among them, in the first place, we should mention essentially new results of research and development in recent years in the field of project modeling and monitoring of the dynamic stress-strain state of essentially heterogeneous bases of building objects [3]. These works open up new possibilities for project modeling of cause and effect relationships for the implementation of previously poorly studied phenomena and the effects of a difficultly controlled smooth decrease in the stability of object-base type systems and long-term operational reliability of building objects throughout their life cycle.

To assess the various types of monitoring of buildings from the point of view of the most effective solution, their multicriteria analysis was carried out. The following were chosen as the main criteria:

C_1 - the range of diversity of structural elements and the building as a whole
C_2 - is the latitude of the spectrum of the measured parameters
C_3 - accuracy of measurements of specified parameters
C_4 - real-time range
C_5 - speed of monitoring the current state of structural elements (and buildings)
C_6 - the complexity of installing the system
C_7 - level of automation of the monitoring process
C_8 - minimization of investment and operating costs

The monitoring methods discussed above are ranked taking into account the integrated assessment of the monitored parameters for ensuring the complex safety of high-rise buildings.

Table 3. Ranking of methods for monitoring high-rise buildings.

Types of monitoring		Rank of importance
M_1	Monitoring of the state of objects by point sensors	4
M_2	Georadiolocation	3
M_3	Fiber-optic monitoring	1
M_4	Geophysical monitoring	2

We determine the coefficient of importance by the formula:

$$C_{Mn} = \frac{2(n-i+1)}{n(n+1)} \cdot 100 \quad (1)$$

Table 4. Matrix of paired comparisons of criteria by measures.

$\Omega_{1,1}=1$	$\Omega_{1,2}=0,5$	$\Omega_{1,3}=0,25$	$\Omega_{1,4}=0,3$
$\Omega_{2,1}=2$	$\Omega_{2,2}=1$	$\Omega_{2,3}=0,5$	$\Omega_{2,4}=0,6$
$\Omega_{3,1}=4$	$\Omega_{3,2}=2$	$\Omega_{3,3}=1$	$\Omega_{3,4}=1,3$
$\Omega_{4,1}=3$	$\Omega_{4,2}=1,5$	$\Omega_{4,3}=0,75$	$\Omega_{4,4}=1$

$$N_i = \left(\Omega_{1.1} \times \Omega_{1.2} \times \Omega_{1.3}\right)^{1/3} \quad (2)$$

Then we normalize the numbers obtained by the formula:

$$\alpha_i = N_i \sum N_i n_i = 1 \quad (3)$$

Table 5. Ranking of evaluation criteria for types of monitoring.

Criteria	Ranks / Indicators of importance by type of monitoring			
	M_1	M_2	M_3	M_4
C_1	4/10	3/20	2/30	1/40
C_2	3/20	4/10	1/40	2/30
C_3	3/20	4/10	1/40	2/30
C_4	4/10	2/30	1/40	3/20
C_5	4/20	3/10	2/30	1/40
C_6	4/10	3/20	1/40	2/30
C_7	4/10	3/20	1/40	2/30
C_8	1/40	3/20	2/30	3/20

Table 6. Matrix of paired comparisons of criteria by types of monitoring and priority vectors.

Matrix of paired comparisons								Vectors of priorities	Normalization
Event M_1									
$\Omega_{1,1}=1$	$\Omega_{1,2}=0,5$	$\Omega_{1,3}=0,5$	$\Omega_{1,4}=1$	$\Omega_{1,5}=0,5$	$\Omega_{1,6}=1$	$\Omega_{1,7}=1$	$\Omega_{1,8}=0,25$	1,548	0,111
$\Omega_{2,1}=2$	$\Omega_{2,2}=1$	$\Omega_{2,3}=1$	$\Omega_{2,4}=2$	$\Omega_{2,5}=1$	$\Omega_{2,6}=2$	$\Omega_{2,7}=2$	$\Omega_{2,8}=0,5$	1,842	0,132
$\Omega_{3,1}=2$	$\Omega_{3,2}=1$	$\Omega_{3,3}=1$	$\Omega_{3,4}=2$	$\Omega_{3,5}=1$	$\Omega_{3,6}=2$	$\Omega_{3,7}=2$	$\Omega_{3,8}=0,5$	1,842	0,132
$\Omega_{4,1}=1$	$\Omega_{4,2}=0,5$	$\Omega_{4,3}=0,5$	$\Omega_{4,4}=1$	$\Omega_{4,5}=0,5$	$\Omega_{4,6}=1$	$\Omega_{4,7}=1$	$\Omega_{4,8}=0,25$	1,548	0,111
$\Omega_{5,1}=2$	$\Omega_{5,2}=1$	$\Omega_{5,3}=1$	$\Omega_{5,4}=2$	$\Omega_{5,5}=1$	$\Omega_{5,6}=2$	$\Omega_{5,7}=2$	$\Omega_{5,8}=0,5$	1,842	0,132
$\Omega_{6,1}=1$	$\Omega_{6,2}=0,5$	$\Omega_{6,3}=0,5$	$\Omega_{6,4}=1$	$\Omega_{6,5}=0,5$	$\Omega_{6,6}=1$	$\Omega_{6,7}=1$	$\Omega_{6,8}=0,25$	1,548	0,111
$\Omega_{7,1}=1$	$\Omega_{7,2}=0,5$	$\Omega_{7,3}=0,5$	$\Omega_{7,4}=1$	$\Omega_{7,5}=0,5$	$\Omega_{7,6}=1$	$\Omega_{7,7}=1$	$\Omega_{7,8}=0,25$	1,548	0,111
$\Omega_{8,1}=4$	$\Omega_{8,2}=2$	$\Omega_{8,3}=2$	$\Omega_{8,4}=4$	$\Omega_{8,5}=2$	$\Omega_{8,6}=4$	$\Omega_{8,7}=4$	$\Omega_{8,8}=1$	2,190	0,157
Event M_2									
$\Omega_{1,1}=1$	$\Omega_{1,2}=2$	$\Omega_{1,3}=2$	$\Omega_{1,4}=0,7$	$\Omega_{1,5}=2$	$\Omega_{1,6}=1$	$\Omega_{1,7}=1$	$\Omega_{1,8}=1$	1,345	0,128
$\Omega_{2,1}=0,5$	$\Omega_{2,2}=1$	$\Omega_{2,3}=1$	$\Omega_{2,4}=0,3$	$\Omega_{2,5}=1$	$\Omega_{2,6}=0,5$	$\Omega_{2,7}=0,5$	$\Omega_{2,8}=0,5$	1,232	0,117
$\Omega_{3,1}=0,5$	$\Omega_{3,2}=1$	$\Omega_{3,3}=1$	$\Omega_{3,4}=0,3$	$\Omega_{3,5}=1$	$\Omega_{3,6}=0,5$	$\Omega_{3,7}=0,5$	$\Omega_{3,8}=0,5$	1,232	0,117
$\Omega_{4,1}=1,5$	$\Omega_{4,2}=3$	$\Omega_{4,3}=3$	$\Omega_{4,4}=1$	$\Omega_{4,5}=3$	$\Omega_{4,6}=1,5$	$\Omega_{4,7}=1,5$	$\Omega_{4,8}=1,5$	1,414	0,135
$\Omega_{5,1}=0,5$	$\Omega_{5,2}=1$	$\Omega_{5,3}=1$	$\Omega_{5,4}=0,3$	$\Omega_{5,5}=1$	$\Omega_{5,6}=0,5$	$\Omega_{5,7}=0,5$	$\Omega_{5,8}=0,5$	1,232	0,117
$\Omega_{6,1}=1$	$\Omega_{6,2}=2$	$\Omega_{6,3}=2$	$\Omega_{6,4}=0,7$	$\Omega_{6,5}=2$	$\Omega_{6,6}=1$	$\Omega_{6,7}=1$	$\Omega_{6,8}=1$	1,345	0,128
$\Omega_{7,1}=1$	$\Omega_{7,2}=2$	$\Omega_{7,3}=2$	$\Omega_{7,4}=0,7$	$\Omega_{7,5}=2$	$\Omega_{7,6}=1$	$\Omega_{7,7}=1$	$\Omega_{7,8}=1$	1,345	0,128
$\Omega_{8,1}=1$	$\Omega_{8,2}=2$	$\Omega_{8,3}=2$	$\Omega_{8,4}=0,7$	$\Omega_{8,5}=2$	$\Omega_{8,6}=1$	$\Omega_{8,7}=1$	$\Omega_{8,8}=1$	1,345	0,128
Event M_3									
$\Omega_{1,1}=1$	$\Omega_{1,2}=0,75$	$\Omega_{1,3}=0,75$	$\Omega_{1,4}=0,75$	$\Omega_{1,5}=1$	$\Omega_{1,6}=0,75$	$\Omega_{1,7}=0,75$	$\Omega_{1,8}=1$	1,612	0,141
$\Omega_{2,1}=1,3$	$\Omega_{2,2}=1$	$\Omega_{2,3}=1$	$\Omega_{2,4}=1$	$\Omega_{2,5}=1,3$	$\Omega_{2,6}=1$	$\Omega_{2,7}=1$	$\Omega_{2,8}=1,3$	1,314	0,115
$\Omega_{3,1}=1,3$	$\Omega_{3,2}=1$	$\Omega_{3,3}=1$	$\Omega_{3,4}=1$	$\Omega_{3,5}=1,3$	$\Omega_{3,6}=1$	$\Omega_{3,7}=1$	$\Omega_{3,8}=1,3$	1,314	0,115
$\Omega_{4,1}=1,3$	$\Omega_{4,2}=1$	$\Omega_{4,3}=1$	$\Omega_{4,4}=1$	$\Omega_{4,5}=1,3$	$\Omega_{4,6}=1$	$\Omega_{4,7}=1$	$\Omega_{4,8}=1,3$	1,314	0,115
$\Omega_{5,1}=1$	$\Omega_{5,2}=0,75$	$\Omega_{5,3}=0,75$	$\Omega_{5,4}=0,75$	$\Omega_{5,5}=1$	$\Omega_{5,6}=0,75$	$\Omega_{5,7}=0,75$	$\Omega_{5,8}=1$	1,612	0,141
$\Omega_{6,1}=1,3$	$\Omega_{6,2}=1$	$\Omega_{6,3}=1$	$\Omega_{6,4}=1$	$\Omega_{6,5}=1,3$	$\Omega_{6,6}=1$	$\Omega_{6,7}=1$	$\Omega_{6,8}=1,3$	1,314	0,115
$\Omega_{7,1}=1,3$	$\Omega_{7,2}=1$	$\Omega_{7,3}=1$	$\Omega_{7,4}=1$	$\Omega_{7,5}=1,3$	$\Omega_{7,6}=1$	$\Omega_{7,7}=1$	$\Omega_{7,8}=1,3$	1,314	0,115
$\Omega_{8,1}=1$	$\Omega_{8,2}=0,75$	$\Omega_{8,3}=0,75$	$\Omega_{8,4}=0,75$	$\Omega_{8,5}=1$	$\Omega_{8,6}=0,75$	$\Omega_{8,7}=0,75$	$\Omega_{8,8}=1$	1,612	0,141
Event M_4									
$\Omega_{1,1}=1$	$\Omega_{1,2}=1,3$	$\Omega_{1,3}=1,3$	$\Omega_{1,4}=2$	$\Omega_{1,5}=1$	$\Omega_{1,6}=1,3$	$\Omega_{1,7}=1,3$	$\Omega_{1,8}=2$	1,352	0,130
$\Omega_{2,1}=0,75$	$\Omega_{2,2}=1$	$\Omega_{2,3}=1$	$\Omega_{2,4}=1,5$	$\Omega_{2,5}=0,75$	$\Omega_{2,6}=1$	$\Omega_{2,7}=1$	$\Omega_{2,8}=1,5$	1,307	0,125
$\Omega_{3,1}=0,75$	$\Omega_{3,2}=1$	$\Omega_{3,3}=1$	$\Omega_{3,4}=1,5$	$\Omega_{3,5}=0,75$	$\Omega_{3,6}=1$	$\Omega_{3,7}=1$	$\Omega_{3,8}=1,5$	1,307	0,125
$\Omega_{4,1}=0,5$	$\Omega_{4,2}=0,7$	$\Omega_{4,3}=0,7$	$\Omega_{4,4}=1$	$\Omega_{4,5}=0,5$	$\Omega_{4,6}=0,7$	$\Omega_{4,7}=0,7$	$\Omega_{4,8}=1$	1,246	0,119
$\Omega_{5,1}=1$	$\Omega_{5,2}=1,3$	$\Omega_{5,3}=1,3$	$\Omega_{5,4}=2$	$\Omega_{5,5}=1$	$\Omega_{5,6}=1,3$	$\Omega_{5,7}=1,3$	$\Omega_{5,8}=2$	1,352	0,130
$\Omega_{6,1}=0,75$	$\Omega_{6,2}=1$	$\Omega_{6,3}=1$	$\Omega_{6,4}=1,5$	$\Omega_{6,5}=0,75$	$\Omega_{6,6}=1$	$\Omega_{6,7}=1$	$\Omega_{6,8}=1,5$	1,307	0,125
$\Omega_{7,1}=0,75$	$\Omega_{7,2}=1$	$\Omega_{7,3}=1$	$\Omega_{7,4}=1,5$	$\Omega_{7,5}=0,75$	$\Omega_{7,6}=1$	$\Omega_{7,7}=1$	$\Omega_{7,8}=1,5$	1,307	0,125
$\Omega_{8,1}=0,5$	$\Omega_{8,2}=0,7$	$\Omega_{8,3}=0,7$	$\Omega_{8,4}=1$	$\Omega_{8,5}=0,5$	$\Omega_{8,6}=0,7$	$\Omega_{8,7}=0,7$	$\Omega_{8,8}=1$	1,246	0,119

Table 7. Baseline data for assessing the significance of the criteria.

Significance of events		Significance of evaluation criteria							
		C_1	C_2	C_3	C_4	C_5	C_6	C_7	C_8
M_1	α_1=0,200	α_{1M1}=0,111	α_{2M1}=0,132	α_{3M1}=0,132	α_{4M1}=0,111	α_{5M1}=0,132	α_{6M1}=0,111	α_{7M1}=0,111	α_{8M1}=0,157
M_2	α_2=0,175	α_{1M2}=0,128	α_{2M2}=0,117	α_{3M2}=0,117	α_{4M2}=0,135	α_{5M2}=0,117	α_{6M2}=0,128	α_{7M2}=0,128	α_{8M2}=0,128
M_3	α_3=0,356	α_{1M3}=0,141	α_{3M3}=0,115	α_{3M3}=0,115	α_{4M3}=0,115	α_{5M3}=0,141	α_{6M3}=0,115	α_{7M3}=0,115	α_{8M3}=0,141
M_4	α_4=0,269	α_{1M4}=0,130	α_{4M4}=0,125	α_{3M4}=0,125	α_{4M4}=0,119	α_{5M4}=0,130	α_{6M4}=0,125	α_{7M4}=0,125	α_{8M4}=0,119

We will evaluate the significance of the criteria for all the developed measures.
Criterion C_1: 0,200x0,111+0,175x0,128+0,356x0,141+0,269x0,130=0,130
Criterion C_2: 0,200x0,132+0,175x0,117+0,356x0,115+0,269x0,125=0,121
Criterion C3: 0,200x0,132+0,175x0,117+0,356x0,115+0,269x0,125=0,121
Criterion C4: 0,200x0,111+0,175x0,135+0,356x0,115+0,269x0,119=0,119
Criterion C_5: 0,200x0,132+0,175x0,117+0,356x0,141+0,269x0,130=0,132
Criterion C_6: 0,200x0,111+0,175x0,128+0,356x0,115+0, 269x0,125=0,119
Criterion C_7: 0,200x0,111+0,175x0,128+0,356x0,115+0, 269x0,125=0,119
Criterion C_8: 0,200x0,157+0,175x0,128+0,356x0,141+0,269x0,119=0,136

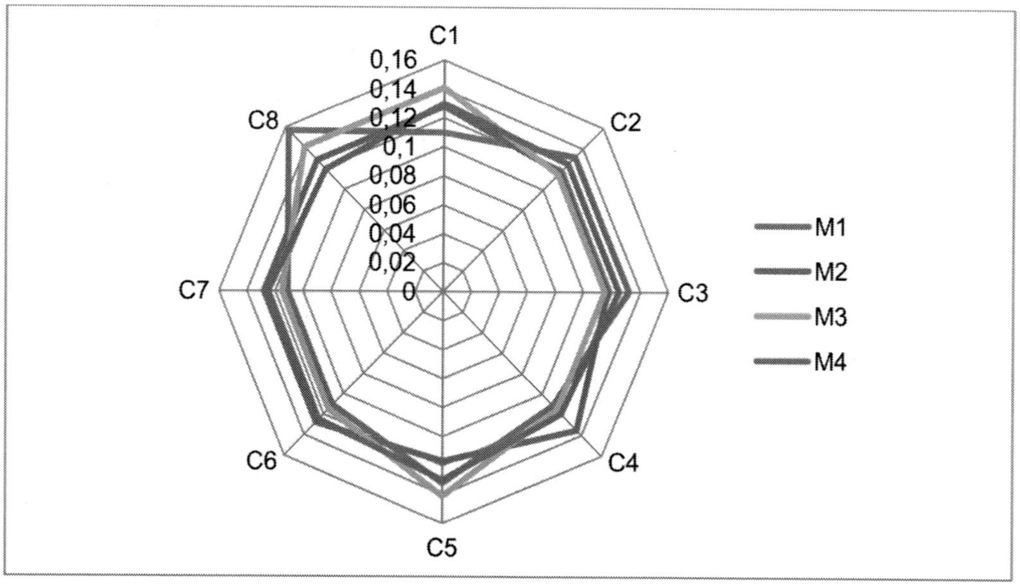

Fig. 2. Indicators of the importance of evaluation criteria for monitoring tall buildings.

The above monitoring systems will allow to track the integral indicators of the state of individual elements and structures in general. On the basis of integral indicators, it is possible to make timely managerial decisions: on calling specialists and conducting instrumental surveys of certain structural elements, making changes to the design decisions at the construction stage, limiting operational loads, evacuating personnel, prohibiting access to certain areas and the like.

Given the high responsibility of high-rise buildings and structures (significant material and human losses in the event of an emergency, and even more so when the structure is destroyed), several independent monitoring systems should be installed for the same physical parameters, i.e. organization of a multi-loop or complex monitoring system.

In particular, the following actual private tasks of intellectual monitoring were solved:

- predictive automated analysis of the evolution of geodynamic characteristics and resonances of object-base type systems;
- Assessment of the effect of seismoacoustic and physicomechanical inhomogeneities in the calculated volume of grounds on the accelerated degradation of geodynamic stability of systems of the "object-base" type of high-rise buildings and structures;
- Assessment of the impact of geodynamic resonances of buildings and structures on the growth of risks of stability loss of the fine structure of the geological section in the estimated base volume;
- Evaluation of the long-term impact of microcircular gravis-seismic loads on the risks of uneven sediment in high-rise buildings and structures and the degradation of their systemic stability.

Conclusion

1. Existing technical solutions and principles of monitoring the state of building structures are intended mainly for monitoring linear characteristics, vibroseismic loads and transfer functions of buildings and structures. At the same time, the processes of weakening and the dynamic characteristics that reflect them are fundamentally nonlinear in their physical essence.

2. The "linear" approach to solving the problem of controlling the evolution of the residual resource of stability and reliability of buildings and structures allows modeling linear dynamic properties of building structures.

3. Widely distributed calculation software for interpreting monitoring data based on various modifications of the "finite element" method are correct in applying only quasistatic stability for design simulation, but they do not have the necessary mathematical sufficiency for calculating and predicting the modeling of nonlinear processes representing the physical nature of the wear processes and the degradation of the safety of building materials, continuously distributed stiffness's and masses is difficult constructed mechanical oscillation systems.

4. The results of theoretical and field studies of domestic and foreign scientists, including specialists in the field of engineering geophysics, confirm the presence of infra-low-frequency and high-quality geodynamic resonances of high-rise buildings and structures. Reliable control over the non-linear evolution of parameters requires a significant expansion and tightening of the requirements for instrumental-methodical monitoring tools, including on the frequency band and dynamic error in the infra-low-frequency range of dynamic loads.

References

1. Zh.G. Mogilyuk, S.I. Zavalishin, M.S. Khlystunov, Scientific review, **7-1**, 241-244 (2014)
2. Zh.G. Mogilyuk, M.S. Khlystunov, V.I. Prokopiev, International Journal for Computational Civil and Structural Engineering, **10(2),** 120-129 (2014)
3. S.I. Zavalishin, Zh.G. Mogilyuk, M.S. Khlystunov, Scientific review, **7,** 245-248 (2014)
4. S.I. Zavalishin, M.S. Khlystunov, Zh.G. Mogilyuk, Ecology of urbanized territories, **4,** 35-41 (2015)
5. V.V. Poduvaltsev, M.S. Khlystunov, Zh.G. Mogilyuk, Safety in the technosphere, **3(2),** 14-17 (2014)

6. V.I. Prokopiev, M.S. Khlystunov, Zh.G. Mogilyuk, International Journal for Computational Civil and Structural Engineering, **2**, 105-110 (2014)
7. M.S. Khlystunov, V.I. Prokopiev, Zh.G. Mogilyuk, Procedia Engineering, **111,** 672 (2015)
8. M.S. Khlystunov, Zh.G. Mogilyuk, Safety of buildings, **3,** 16 (2003)
9. Zh.G. Mogilyuk, M.S. Khlystunov, V.I. Prokopiev, Procedia Engineering, **111,** 672 – 678 (2015)
10. V.I. Prokopiev, M.S. Khlystunov, Zh.G. Mogilyuk, Procedia Engineering, **111,** 666 – 671 (2015)
11. S.I. Zavalishin, Zh.G. Mogilyuk, M.S. Khlystunov, Ecology of urbanized territories, **2,** 45-50 (2006)
12. Zh.G. Mogilyuk, M.S. Khlystunov, Scientific review, **13,** 90-93 (2015)
13. V.V. Poduval'tsev, M.S. Khlystunov, Zh.G. Mogilyuk, Safety in the technosphere, **26 (45),** 33-37 (2013)
14. V.I. Telichenko, E.A. Korol, M.S. Khlystunov, High-rise buildings, **1**, 5-7 (2008)
15. M.S. Khlystunov, V.I. Prokopiev, Zh.G. Mogilyuk, Procedia Engineering, **111**, 297 – 301 (2015)
16. M.S. Khlystunov, Zh.G. Mogilyuk, V.V. Poduval'tsev, Measurement Techniques, **58 10**, 1139-1143 (2016)
17. V.I. Prokopiev, M.S. Khlystunov, Zh.G. Mogilyuk, Procedia Engineering, **111**, 672 – 678 (2015)
18. E.A. Korol, I.V. Rubtsov, A.V. Kukhta, N.P. Chetverik, High-rise buildings, **5,** 123-125 (2008)
19. V.I. Telichenko, E.A. Korol, P.B, S.V. Komissarov, S.G. Arutyunov, A.A. Afanasiev, *Management of programs and projects for the erection of high-rise buildings: Scientific publication*. (Moscow: ASV Publishing House, 2010)
20. V.I. Telichenko, M.S. Khlystunov, V.I. Prokopiev, Zh.G. Mogilyuk, Ecology of urbanized territories, **2,** 13-21 (2011)

One of the criteria for selecting a contractor for high-rise construction

Zalina Tuskaeva[1,*] and *Timur* Tagirov[2]

[1]FSBO HE North-Caucasian Institute of Mining and Metallurgy (State Technological University) 44 Nikolaeva Str., Vladikavkaz, 362021, Russia
[2] Moscow State University of Civil Engineering, Yaroslavskoe shosse, 26, Moscow, 129337, Russia

Abstract: The mechanisms for management of the building complex used and proposed to date do not always provide the required result in the assessment of the construction organization facilities. Therefore, the development of new effective methods for such an assessment is an urgent task especially in questions related to high-rise construction. The article formally sets the task of assessing the technical facilities of a construction organization. Due to the use of expert methods, the weighted values of the coefficients of local indicators for technical facilities are identified.

1 Introduction

High-rise buildings are an integral part of modern city-planning policy of megacities. They reflect financial possibilities of a state, a particular region or city, t construction technology and technical equipment level, availability of the required specialists in design and construction [1-5].

According to the data of the Council on Tall Buildings and Urban Habitat 101 high-rise buildings had been built in the RF by 2014 [6].

During the hundred years of high-rise buildings design and construction, the world practice has accumulated relevant theoretical and practical experience, identified the key problems of their design, constru.ction and maintenance. The development of high-rise construction in the RF identified a number of serious problems to be solved, and namely:

- Disadvantages in legal framework;
- lack of necessary experience in design and construction;
- lack of relevant personnel;
- inadequate equipment of contractors [4].

Although nowadays the required minimum legal and regulatory framework for choosing a contractor has been created, many issues remain insufficiently developed. Applicants are often selected with no scientific and methodological regard of their technical and financial status and ability to manage the project. A proper systematic evaluation is not carried out, resulting in errors in the selection of the contractor and failure of contractual obligations.

The intensification of problems associated with high-rise construction requires an assessment of contractors' reliability in terms of technical compliance or equipment.

* Corresponding author: tuskaevazalina@yandex.ru

To solve this problem, it is necessary to develop a methodology for assessing technical facilities of organizations committed to win the contract for high-rise construction. This will limit access of uncompetitive, technically inadequately equipped contractors to high-rise construction.

Construction machinery is one of the most important criteria for assessing the level of the construction organization technical equipment, since modern construction industry uses a large variety and number of labor saving devices [7-18].

Increase in the machinery and its maintenance cost raise the questions of optimization of technical and economic performance indicators, which ensure an increase in productivity and a reduction in the cost of construction. The management mechanisms used and proposed today do not provide the desired result in the assessment of the level of construction works mechanization [17]. Therefore, the development of methods for assessing the overall technical facilities of a construction organization claiming to win a contract for a high-rise building is one of the key tasks to solve the problems associated with the construction of skyscrapers.

2 Methodology

Many modern methods of assessing the activities of organizations are based mainly on parametric one-criterion estimates. Such an approach does not allow fully reflecting the peculiarities of the problem in question and taking into account the mutual influence of a set of indicators [19].

In general, the task of assessing the technical facilities of a construction organization can be represented as follows:

$$y_{tp} = \sum_{i=1}^{k}(d_i \cdot X_i) / \sum_{i=1}^{k} d_i \quad , \tag{1}$$

Where d_i – weighted coefficient of i-factor of technical facilities;

k – number of factors examined;

X_i – discreet values of i-factor.

This problem is divided into two components:

1. Quantitative estimation of weight coefficients of technical facilities factors;
2. Development of a technique for calculating the discrete values of each factor.

This article is devoted to the solution of the first part of the problem-the determination of the quantitative values of d_i.

The procedure for assessing technical facilities is necessary to analyze the capacities of the construction organization in high-rise building construction included into the category of unique buildings and structures. To develop this idea, we will immediately formulate the problems: "How to assess these capacities?", "Which of the performance indicators of the contractor organization are the most significant and important?", "How to assess comprehensively?". The entire list of issues under consideration requires a detailed account of the actual capabilities of the contractor.

The possibilities of applied mathematics for finding effective solutions are limited. To solve such problems, heuristic methods are widely used, based on the simulation of the decision-making process by a person who has the necessary information and possesses sufficient experience. The solutions chosen in this case are the best ones that can be obtained with the help of the developed algorithm.

A combination of heuristic methods for the random search of successive approximations with quantitative methods for solving optimization problems is

Indicators reflecting certain areas of the state of contractor technical facilities are presented in the work by particular quantitative indicators. To summarize the assessment, it is advisable to use a composite set of indicators, called the integral indicator of the technical potential level.

The study and choice of indicators will be performed consistently in several stages:

1. Drawing up a list of indicators that objectively reflect the entire level of technical equipment.
2. Creating a group of experts necessary to identify the most significant indicators.
3. Conducting an expert survey and processing the results in order to identify significant indicators.

3 Results

The process of forming or selecting the most significant indicators was carried out by the expert evaluation method, since it is impossible to assign quantitative values to a number of factors included into the investigated problem, using other methods [19]. The required number of experts was defined according to the simplified Wentzel formula [20]:

$$N = k \cdot P_d / (P_{o.shc} + 0{,}5) = 11 \cdot 0.95 / (0.05 + 0.5) = 19 \qquad (2)$$

A group of 22experts were offered the following indicators for assessing the level of technical facilities (see table 1).

Table 1. Indicators of technical facilities

Item N	Indicators	Symbol
	Indicators characterizing the material component of technical facilities I. Indicators characterizing the material component of technical facilities (availability)	
1	Level of basic machines facilities	X1
2	Level of completeness of construction equipment	X2
3	Proportion of machines exceeding service life	X3
4	Proportion of equipment produced overseas	X4
	II. Indicators characterizing the material component of technical facilities (management)	
5	Basic machines availability coefficient	X5
6	Coefficient of the intensity of construction machinery use	X6
7	The level of compliance with environmental requirements for construction equipment	X7
	III. Indicators characterizing the material component of technical facilities (economic characteristics)	
8	The percentage (level) of use of depreciation charges by the organization	X8
	IV. Indicators characterizing the labor component of technical facilities	
9	Coefficient of operators sufficiency	X9
10	Coefficient of operators qualifications	X10
11	Level of use of progressive forms of payment	X11

The results of expert assessment of technical facilities indicators are presented in table 2.

Table 2. Results of expert assessment

Indicators	expert mark										
	1	2	3	4	5	6	7	8	9	10	11
X1	10	10	9	10	8	9	9	10	8	10	9
X2	8	6	8	7	8	7	9	10	8	9	7
X3	5	7	7	8	7	7	6	5	6	6	6
X4	8	9	7	9	10	8	6	10	6	9	10
X5	9	10	9	10	10	10	10	10	9	9	10
X6	10	9	10	10	10	10	10	10	10	9	8
X7	6	5	5	6	7	7	6	5	4	8	5
X8	8	7	9	9	6	6	4	7	6	6	7
X9	7	10	9	7	10	9	9	10	10	9	10
X10	10	10	9	8	10	9	9	10	9	10	8
X11	9	7	9	8	10	10	7	10	9	8	9

Table 2. Results of expert assessment. Continued

Indicators	expert mark											Average mark	Varation coefficient
	12	13	14	15	16	17	18	19	20	21	22	23	24
X1	8	9	9	8	10	9	10	9	10	8	9	9,14	0,084
X2	9	6	7	6	7	6	10	9	7	7	7	7,64	0,168
X3	9	9	7	6	7	6	6	8	8	7	7	6,82	0,153
X4	6	6	8	5	8	9	6	6	8	6	8	7,64	0,209
X5	9	9	9	9	10	9	10	9	8	8	9	9,32	0,071
X6	10	10	9	10	10	10	10	10	9	10	10	9,73	0,058
X7	6	5	6	4	7	4	7	7	3	4	3	5,45	0,263
X8	7	5	5	5	6	6	7	5	4	3	6	6,09	0,244
X9	10	10	9	10	10	9	10	9	10	10	9	9,36	0,080
X10	9	10	10	8	9	10	9	10	8	10	9	9,27	0,083
X11	9	7	9	6	9	10	10	7	7	6	8	8,36	0,162

The average mark is defined as the quotient of dividing the sum of the proposal marks by the number of experts. To establish the degree of reliability of the study conducted, the average proposal mark, variance and standard deviation were determined.

The average mark for indicator i is defined according to the formula:

$$\overline{C_{ij}} = \frac{1}{n}\sum_{j=1}^{m} C_{ij}, \quad (3)$$

Where n– is the number of experts;

m– number of evaluation indicators;

C_{ij}– evaluation of indicator i by expert j;

$\overline{C_{ij}}$– average mark for indicator i.

The variance of marks is determined by formula:

$$D_i = \frac{1}{n-1}\sum_{j=1}^{m}\left(C_{ij} - \overline{C_{ij}}\right), \quad (4)$$

where D_i– variance of marks.

Variation coefficient is calculated by formula:

$$\vartheta_i = \frac{\sqrt{D_i}}{\overline{C_{ij}}} = \frac{\partial_i}{\overline{C_{ij}}}, \quad (5)$$

Wher ϑ_i– variation coefficient;

∂_i– standard deviation.

Since the variation coefficient in all cases was within $0 < \vartheta_i < 0{,}3$, – the consistency of experts is considered sufficient.

For the convenience of the comparative analysis, the marks are converted to ranks. Activities that score the minimum amount of ranks are considered to be the priority (table 3)

Table 3. Results of expert evaluation ranking

Indicators	Expert mark										
	1	2	3	4	5	6	7	8	9	10	11
X1	2,0	2,5	4,5	2,0	7,5	5,0	4,5	4,5	6,5	1,5	4,5
X2	7,0	10,0	8,0	9,5	7,5	9,0	4,5	4,5	6,5	5,0	8,5
X3	11,0	8,0	9,5	7,0	9,5	9,0	9,0	4,5	9,0	10,5	10,0
X4	7,0	5,5	9,5	4,5	3,5	7,0	9,0	10,5	9,0	5,0	2,0
X5	4,5	2,5	4,5	2,0	3,5	2,0	1,5	4,5	4,0	5,0	2,0
X6	2,0	5,5	1,0	2,0	3,5	2,0	1,5	4,5	1,5	5,0	6,5
X7	10,0	11,0	11,0	11,0	9,5	9,0	9,0	10,5	11,0	8,5	11,0
X8	7,0	8,0	4,5	4,5	11,0	11,0	11,0	9,0	9,0	10,5	8,5
X9	9,0	2,5	4,5	9,5	3,5	5,0	4,5	4,5	1,5	5,0	2,0
X10	2,0	2,5	4,5	7,0	3,5	5,0	4,5	4,5	4,0	1,5	6,5
X11	4,5	8,0	4,5	7,0	3,5	2,0	7,0	4,5	4,0	8,5	4,5

Table 3. Results of expert evaluation ranking. Continued

Indicators	Expert mark										Average rank
	12	13	14	15	16	17	18	19	20	21	22	23
X1	8,0	5,0	4,0	4,5	2,5	5,5	3,5	4,5	1,5	4,5	3,5	4
X2	5,0	8,5	8,5	7,0	9,0	9,0	3,5	4,5	8,5	6,5	8,5	7
X3	5,0	5,0	8,5	7,0	9,0	9,0	10,5	7,0	5,5	6,5	8,5	8
X4	10,5	8,5	7,0	9,5	7,0	5,5	10,.5	10,0	5,5	8,5	6,5	7
X5	5,0	5,0	4,0	3,0	2,5	5,5	3,5	4,5	5,5	4,5	3,5	4
X6	1,5	2,0	4,0	1,5	2,5	2,0	3,5	1,5	3,0	2,0	1,0	3
X7	10,5	10,5	10,0	11,0	9,0	11,0	8,5	8,5	11,0	10,0	11,0	10
X8	9,0	10,5	11,0	9,5	11,0	9,0	8,5	11,0	10,0	11,0	10,0	9
X9	1,5	2,0	4,0	1,5	2,5	5,5	7,0	4,5	1,5	2,0	3,5	4
X10	5,0	2,0	1,0	4,5	5,5	2,0	3,5	1,5	5,5	2,0	3,5	4
X11	5,0	7,0	4,0	7,0	5,5	2,0	3,5	8,5	8,5	8,5	6,5	6

The reliability of the study is estimated with the help of ranks.:

$$W = \frac{12\sum_{i=1}^{m}(S_i-\bar{S})^2}{n^2(m^3-m)} \tag{6}$$

where Si – sum of ranks for object i;

$\bar{S}$– average rank.

For the total ranks $\bar{S}$ = 131. According to formula (6), W= 0,57. This value of the general consistency coefficient indicates a sufficient level of consistency and, therefore, sufficient accuracy of the findings of the expert survey. The final results of ranking the indicators of the construction organization technical facilities are shown in figure 1.

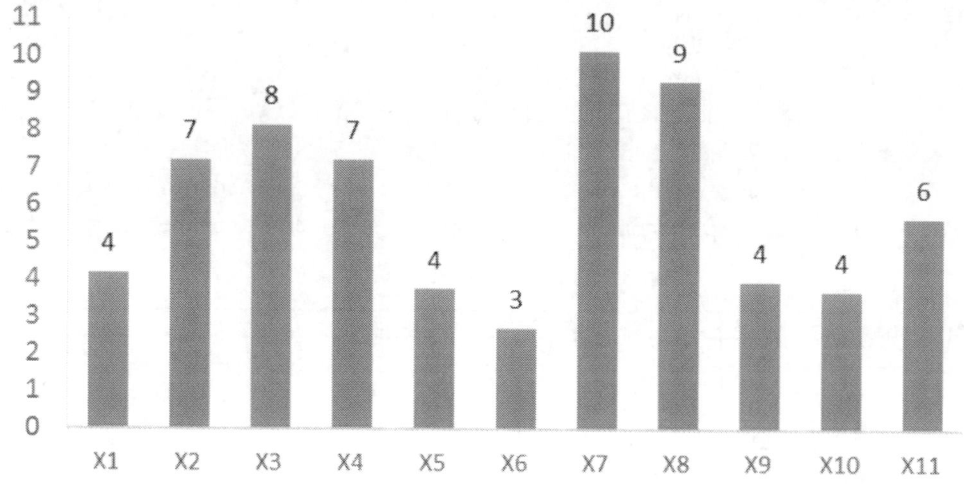

Fig. 1. Ranking the indicators of the construction organization technical facilities

The value obtained is divided by the number of indicators studied and multiplied by the average score of each indicator: To obtain the quantitative values of the weight coefficients from equation (1), we determine the sum of all the average estimates (graph 23, table 2) $\sum_{j=1}^{m}\overline{C_{ij}}$= 88,8. The reported value is divided by the number of indicators studied and multiplied by the average score of each indicator:

$$d_i = \frac{1}{n}\sum_{j=1}^{m}\overline{C_{ij}} \cdot \overline{C_{ij}} \quad (7)$$

As a result of calculating the weight coefficients of the technical facilities indicators according to equation (7), the general formulation of problem (1) is as follows::

$$У_{tp} = 0{,}103 \cdot X_1 + 0{,}086 \cdot X_2 + 0{,}077 \cdot X_3 + 0{,}086 \cdot X_4 + 0{,}105 \cdot X_5 + \\ + 0{,}110 \cdot X_6 + 0{,}061 \cdot X_7 + 0{,}069 \cdot X_8 + 0{,}105 \cdot X_9 + 0{,}104 \cdot X_{10} + 0{,}094 \cdot X_{11} \ (8) \quad (8)$$

where X_i – discreet value of i- factor.

Discussion and conclusion

Equation analysis (8) enables to identify the most significant indicators of technical facilities, such as (in descending order): coefficient of intensity of the construction equipment use, basic machines availability coefficient, coefficient of operators sufficiency, coefficient of operators qualification, ratio of the availability of machinists, coefficient of classroom machinists, level of equipment of basic machines, level of basic machines facilities.

Recommendations

Considering that high-rise construction faces a rather large number of tasks, the problem of the technical facilities of domestic construction companies requires a compulsory solution. Compliance with the level of equipment for the construction of high-rise buildings, which are classified as unique, can encourage domestic companies to participate in the construction of similar structures.

Today, high-rise construction in Russia is going through rough times; nevertheless, interest in high-rise construction continues to grow.

References

1. L.A. Koklugina, A.V Koklugin. *Technology and organization of construction of high-rise multifunctional buildings: Textbook.* (Kazan: Publishing house of Kazan state architectural university, 2016)
2. L. K. Petrenko, S.E. Mangilevskaya Scientific review, **7**, 2014;
3. URL: https://regrazvitie.ru/perspektivy-stroitelstva-i-ekspluatatsii-neboskrebov-v-21-veke;
4. URL: https://www.natural-sciences.ru/ru/article/view? id=30334;
5. URL: http://www.vseon.com/i/themes/sovremennaya-strojka/item/samye-vysokie-zdaniya-rossii-2.html
6. Rules and regulations 267.1325800.2016 «Buildings and high-rise complexes. Rules of design"
7. V.V. Asaul. Economics of Construction. **1**, 14-25 (2005)
8. A.N Asaul, M.K Starovoitov, R.A Faltinskiy. *Cost management in construction* (S.Petersburg: IPEV, 2009)
9. D.G. Babaeva Analysis of the state of fixed productive assets in industry // Bulletin of DGINH: collection of scientific works. Issue. X. Makhachkala, 2006. p. 76-81.
10. S.M. Buttaeva. Bulletin of the Russian state social university, **2(54)**. p.119-130 (2007)
11. D.P. Volkov, S.N. Nikolaev. *Reliability of construction machinery and equipment.* (Moscow, Vyshaya shkola,1979)
12. D.A. Dvizov, N.V. Skidanov, X Interuniversity scientific conference of young scientists and students of Volzhskiy, p.4-5 (2004)
13. V.N. Ivanov, *R.F. Salikhov.* Omsk scientific bulletin. Omsk: OmGTU, 1. 92 – 94 (2007)
14. M.I. Kamenetskiy, M.F. Kostetskiy. Economics of construction, **4**, 17–22 (2010)
15. E.P. Pankratov, O.E. Pankratova. Economics of construction. **3(33)**, 4–17 (2015)
16. S.V. Repin, A.V.Savelyev. Construction technique 31-35 (2006)
17. Z. R. Tuskaeva. International Journal of Applied Engineering Research, 6, 4369–4376 (2016)
18. Z.R. Tuskaeva. Software Procedia Engineering, 165, 1184–1191 (2016)
19. S.S.Uvarova, S.V. Belyaeva, V.S. Kankhva. *Economic sustainability of construction companies and projects.* (Scientific library of MSEU, Moscow, 2011)
20. E.S. Ventzel. Theory of probability: Textbook. 6th edition. (Vyshaya shkola, Moscow, 1999)

Principles for the formation of an effective concept of multifunctional high-rise construction investment projects

Sergei Beliakov[1,*]

[1]Moscow State University of Civil Engineering, Yaroslavskoe shosse, 26, Moscow, 129337, Russia

Abstract. Investment projects of high-rise construction have a number of features that determine specific risks and additional opportunities that require analysis and accounting in the formation of an effective project concept. The most significant features of high-rise construction include long construction time, complexity of technical and technological solutions, complexity of decisions on the organization of construction and operation, high cost of construction and operation, complexity in determining the ratio of areas designed to accommodate different functional areas, when organizing and coordinating the operation of the facility, with internal zoning. Taking into account the specificity of high-rise construction, among the factors determining the effectiveness of projects, it is advisable to consider as key factors: organizational, technological and investment factors. Within the framework of the article, the author singled out key particular functions for each group of factors under consideration, and also developed a system of principles for the formation of an effective concept of multifunctional high-rise construction investment projects, including the principle of logistic efficiency, the principle of optimal functional zoning, the principle of efficiency of equipment use, the principle of optimizing technological processes, the principle maximization of income, the principle of fund management, the principle of risk management . The model of formation of an effective concept of investment projects of multifunctional high-rise construction developed by the author can contribute to the development of methodological tools in the field of managing the implementation of high-rise construction projects, taking into account their specificity in the current economic conditions.

1 Introduction

In the conditions of high cost and shortage of land intended for construction in the large cities of Russia in recent decades, dynamic development has received high-rise construction, which allows to maximize the technical and economic parameters of investment projects in relation to the small size of the built-up area [1].

Typically, for the classification of high-rise buildings, the height criterion is used in meters, rather than in floors, because the heights of the floors are adopted different depending

* Corresponding author: serj-bel@yandex.ru

on the purpose of the building and the requirements of national design standards. In Russia high-rise buildings are considered buildings with a height of more than 75 meters.

High-rise buildings can have different purposes: hotels, offices, residential buildings, however, given the large scale, most often high-rise buildings are designed as multifunctional, and in addition to premises for main purposes, parking lots, service facilities, etc. are located.

2 Materials and Methods

Investment projects of high-rise construction have a number of features that determine specific risks and additional opportunities that require analysis and accounting in the formation of an effective project concept. The most significant features of high-rise construction include:

• ***long terms of construction,***

Investing in high-rise construction projects is of a long-term nature, since the construction time of high-rise buildings is substantially higher than for similar projects of low and medium-story level. At the same time, the effectiveness of projects of high-rise construction is significantly influenced by changes occurring in the economic environment. Therefore, the use of effective methods of forecasting, investment management and risk control becomes the most important factors determining the effectiveness of high-rise investment projects.

• ***complexity of technical and technological solutions,***

The construction of high-rise buildings is associated with the need to develop unique technical and technological solutions, including those aimed at ensuring the acceptable level of safety at various stages of construction, as well as during operation and repair works [2].

• ***complexity of decisions on the organization of construction and operation,***

The organization of construction of high-rise facilities, as a rule, requires the development of complex plans that ensure the rational performance of material support processes, the performance of construction work in the conditions of a limited construction site, environmental protection, security in construction, etc. Also, the solution of the problem of efficient operation of high-altitude objects, in particular, combining several functions is required.

• ***high cost of construction and operation,***

The cost of construction and operating high-rise buildings is significantly higher than that of mass construction, and is due not only to specific design solutions, but also life support systems and complex security requirements.

• ***complexity in determining the ratio of areas designed to accommodate different functional areas, when organizing and coordinating the operation of the facility, with internal zoning.***

The formation of the functional concept of a high-rise building is always connected with the need for a comprehensive analysis of target risk segments, long-term forecasting of the economic environment, complex analysis of the competitive environment, solving architectural and commercial tasks of the internal zoning of the facility, etc.

Thus, taking into account the specificity of high-rise construction, the issues of conceptual planning of the project are of particular importance, since at this stage prerequisites are formed for all further stages of project implementation and its risks and effectiveness are predetermined.

Depending on the nature of the factors that affect the efficiency of investment construction projects, it is customary to distinguish [3-6]:

• economic factors,
• social factors,
• administration factors,

- environmental factors,
- organizational factors,
- technological factors,
- investment factors.

In accordance with this classification of factors, the objective function of the efficiency of the multifunctional high-rise construction investment project may be presented in the following form:

$$E_{project} = f\left(\sum_{i=0}^{n} \begin{pmatrix} F^i_{economic}; F^i_{social}; F^i_{administration}; F^i_{environmental}; \\ F^i_{organizational}; F^i_{technological}; F^i_{investment} \end{pmatrix}\right) \rightarrow max \quad (1)$$

Taking into account the specificity of high-rise construction, among these factors, it is advisable to consider as key factors: organizational, technological and investment factors. This choice is due to the fact that these groups of factors have the most significant impact on the cost, risks, timing of the works under the project, these factors are amenable to change and control within the project management processes. The degree of influence of these factors depends on the characteristics of a particular project [5, 7].

Thus, the objective function (1) may be reduced to the following form:

$$E_{project} = f\begin{cases} \text{Organizational efficiency} \rightarrow max \\ \text{Technological efficiency} \rightarrow max \\ \text{Investment efficiency} \rightarrow max \end{cases} \quad (2)$$

Within the framework of this article, the author singled out key particular functions for each group of factors under consideration:

$$Organizational\ efficiency = f\begin{cases} logistics\ costs \rightarrow min \\ combination\ of\ consumer\ flows \rightarrow max \end{cases} \quad (3)$$

$$Technological\ efficiency = f\begin{cases} total\ cost\ of\ use\ of\ equipment \rightarrow min \\ technological\ sustainability \rightarrow max \end{cases} \quad (4)$$

$$Investment\ efficiency = f\begin{cases} commercial\ profitability \rightarrow max \\ cost\ of\ capital\ raised \rightarrow min \\ ratio\ of\ profitability\ and\ risks \rightarrow max \end{cases} \quad (5)$$

3 Results

Based on the identified groups of factors and key particular functions, author developed a system of principles for the formation of an effective concept for multifunctional high-rise construction investment projects and a corresponding principal model (Figure 1).

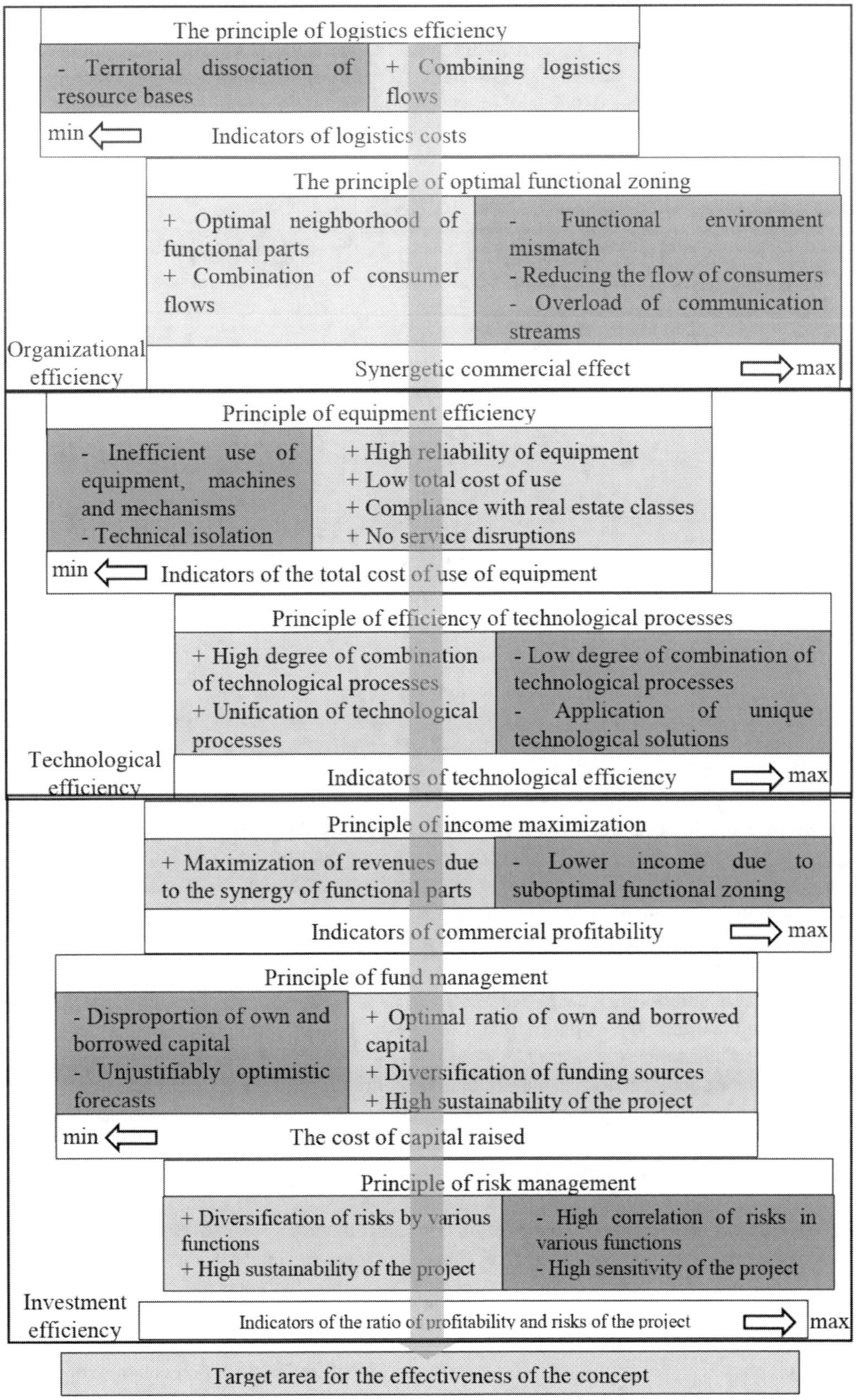

Fig. 1. Model of the formation of an effective concept of investment projects of high-rise construction.

In this model, the target area for the effectiveness of the concept of the multifunctional high-rise construction investment project is represented as an axis, and the blocks intersecting the axis contain the principles of efficiency for each of the key groups of factors and the target striving for maximum or minimum. It should be noted that the factors under consideration are not fully independent of each other and, accordingly, the control impacts on individual factors and groups of factors cause the emergence of synergistic effects in the context of the impact on the overall effectiveness of the project. Thus, the proposed principles for the formation of an effective concept of investment projects should be considered comprehensively within the framework of a unified management system.

4 Discussion

The principle of logistics efficiency.

When forming the concept of the project, it is expedient to plan the flows of resource support for all technological processes (both during construction and operation of the facility), which, when combined, reduce logistics costs [8]. Within the framework of logistic networks, it is allowed to increase costs for individual flows, provided that this will lead to an increase in the efficiency of the project as a whole.

The principle of optimal functional zoning.

When forming the concept of spatial zoning of a multifunctional high-rise object, it is necessary to consider the influence on the efficiency of the object's operation of such factors as the optimal neighborhood of functional elements, the combination of consumer flows, preferences for location within a high-rise building [9].

Principle of equipment efficiency.

Since the cost of equipment is a significant part of the total cost of the high-rise construction project, it is advisable to plan types and types of equipment at the stage of concept formation and further design, which are characterized by an optimal ratio of reliability and the volume of total costs throughout the life of the operation [10-12].

It is necessary to conduct an analysis of the conformity of the proposed equipment with the criteria for determining the classes of commercial real estate in accordance with certain types of functional purpose. It is also advisable to consider the possible risks associated with the interaction of entities in the event of the purchase and further maintenance of equipment produced in other countries.

Principle of efficiency of technological processes.

In order to reduce the duration and cost of building high-rise buildings in the formation of the concept and design, it is expedient to strive to reduce the number of unique solutions, to unify and combine the technological processes of construction [13].

It is also advisable to consider technological solutions that are most appropriate for the organization of construction on a particular land plot. Within the framework of operation of a multifunctional facility, unification and combination of technological processes related to the functioning of hotel functional units allows achieving synergy effect and, as a result, reducing operational costs and improving project efficiency [9].

Principle of income maximization.

The choice of functional parts within the multifunctional high-altitude facility and the determination of their ratio should be aimed at providing the maximum synergetic effect increasing the performance indicators for the functioning of each individual parts and the object as a whole. In some cases, it is also possible to create separate elements within a multifunctional facility that are not effective in themselves, but increase the efficiency of other elements by increasing the prestige, comfort or inflow of potential consumers [9, 14].

Principle of fund management.

Since insufficient funding is one of the main reasons for the excess of the planned construction dates and the overall efficiency of high-rise construction projects, it is necessary to take into account pessimistic forecasts of the project financial plans and to choose financing strategies in favor of more conservative ones that ensure optimal the ratio of equity and borrowed funds of investors, strategies [15].

Principle of risk management.

For investment projects of commercial real estate of various functional types, various risks that arise under the influence of factors of different economic nature are often characteristic. In this regard, when selecting the functional parts within the multifunctional high-rise facility and determining their ratios, it is expedient to conduct a correlation analysis of risks and on its basis a version of the concept that is characterized by an acceptable level of integrated risk of the project.

Conclusions

The development of high-rise construction in Russia currently has a high dynamics, but there are a number of development directions that are characterized by substantial reserves of improving the efficiency of investment projects, including optimization of town-planning and functional-typological concepts of construction, improving procedures for selecting and justifying technical, technological and organizational design decisions, etc.

The model of formation of an effective concept of investment projects of multifunctional high-rise construction developed by the author can contribute to the development of methodological tools in the field of managing the implementation of high-rise construction projects, taking into account their specificity in the current economic conditions.

References

1. G. Ptichnikova, Procedia Engineering, **165**, 1903-1910 (2016)
2. V. Zhilov, High-rise construction in Russia: present and prospects, Stroyprofile, **2**, 34 (2007)
3. I. Afanasenko, V. Borisova, *Economic logistics* (Peter, Saint Petersburg, 2013)
4. A. Volkov, V. Chulkov, R. Kazaryan, Procedia Engineering, **91**, 373-376 (2014)
5. P. Feng, W. Xingkuan, Procedia Engineering, **21**, 943-947 (2011)
6. T. Ahmad, A. Aibinu, M. Thaheem, Procedia Engineering, **180**, 1695-1704 (2017)
7. J. Tamošaitienė, E. Gaudutis, M. Kračka, Procedia Engineering, **57**, 1151-1155 (2013)
8. A. Murzin, *Real estate: economics, evaluation and development* (Phoenix, Rostov on Don, 2013)
9. E. Kalcheva, A. Taki, Y. Hadi, Procedia - Social and Behavioral Sciences, **216**, 960-973 (2016)
10. G. Savitskaya, *Analysis of economic activity of enterprises* (New knowledge, Minsk, 2004)
11. N. Gohl, A. Loose, S. Stark, H. Drück, Energy Procedia, **48**, 1365-1373 (2014)
12. I. Puķīte, I. Geipele, Procedia Engineering, **172**, 905-912 (2017)
13. G. Sternik, S. Sternik, N. Tulinova, *Real Estate Development* (Moscow, Prospekt, 2016)
14. D. Ben-Shahar, Y. Deng, E. Sulganik, Journal of Housing Economics, **18-1**, 25-33 (2009)
15. A. Yudina, *Technological processes in construction* (Academy, Moscow, 2013)

Improvement of the cost-benefit analysis algorithm for high-rise construction projects

Andrey Gafurov[1], *Oksana* Skotarenko[1], and *Vladimir* Plotnikov[2],*

[1] Murmansk Arctic State University, 15 Kapitana Yegorova str., Murmansk, Russia, 183038
[2] Saint-Petersburg State University of Economics, 21 Sadovaya str., St. Petersburg, Russia, 191023

Abstract. The specific nature of high-rise investment projects entailing long-term construction, high risks, etc. implies a need to improve the standard algorithm of cost-benefit analysis. An improved algorithm is described in the article. For development of the improved algorithm of cost-benefit analysis for high-rise construction projects, the following methods were used: weighted average cost of capital, dynamic cost-benefit analysis of investment projects, risk mapping, scenario analysis, sensitivity analysis of critical ratios, etc. This comprehensive approach helped to adapt the original algorithm to feasibility objectives in high-rise construction. The authors put together the algorithm of cost-benefit analysis for high-rise construction projects on the basis of risk mapping and sensitivity analysis of critical ratios. The suggested project risk management algorithms greatly expand the standard algorithm of cost-benefit analysis in investment projects, namely: the "Project analysis scenario" flowchart, improving quality and reliability of forecasting reports in investment projects; the main stages of cash flow adjustment based on risk mapping for better cost-benefit project analysis provided the broad range of risks in high-rise construction; analysis of dynamic cost-benefit values considering project sensitivity to crucial variables, improving flexibility in implementation of high-rise projects.

1 Introduction

The problems of evaluation of high-rise construction projects in Russia are relevant. This is largely due to the fact that in Russia high-rise construction is still poorly developed. Therefore, project, economic, technological and other solutions are not sufficiently typified. In our country, houses taller than 75 meters are recognized as high-rise. There are almost 3 thousand of them in the country. At the same time, a little more than 100 buildings with a height of more than 100 meters have been built, over 80% of them are located in Moscow.
As the experience in the construction of high-rise buildings shows, their cost compared with conventional facilities is about 25% higher. This increase is due to more expensive engineering systems, complex fire-fighting systems, the use of thicker structures, etc. In addition, the design and construction of high-rise buildings takes longer. This also leads to higher costs for such projects due to higher costs for servicing loans.

* Corresponding author: plotnikov_2000@mail.ru

These factors make it necessary to develop special methods of cost-benefit analysis for high-rise construction projects. In order to evaluate the overwhelming majority of Russia-specific investment projects, a number of methods are used on the basis of dynamic values and the discounted cash flow model (DCF) [1, p. 215]. However, the specific nature of high-rise investment projects entailing long-term construction, high risks, etc. implies a need to improve the standard algorithm of cost-benefit analysis.

2 Methods

The authors used different methods: weighted average cost of capital (WACC), dynamic cost-benefit analysis of investment projects (NPV, IRR, PI, DPP, etc.), risk mapping, scenario analysis, sensitivity analysis of critical ratios, etc. These methods were used to evaluate high-rise construction projects.

3 Results

Putting together an algorithm of cost-benefit analysis for high-rise construction projects on the basis of risk mapping and sensitivity analysis of critical ratios. A typical algorithm of cost-benefit analysis for construction projects comprises the following main stages:

- forecasting reports and projected cash flow;
- background for a discounted cash flow rate;
- calculation of dynamic cost-benefit values.

Nowadays, there exists a certain academic consensus on the range of methods to be used at every stage of the algorithm. The main methods include the following:

- methods of providing a background for a discounted rate (cumulative risk premium assessment, weighted average cost of capital (WACC), etc.) [3, p. 113-120];
- dynamic cost-benefit analysis of investment projects (NPV, IRR, PI, DP, etc) [5, p. 216-223].

In the Russian practice, a similar algorithm is used for high-rise construction projects as well. However, high rise construction is characterised by a number of specific features that may include:

- dependence of project planning on real estate market cycles and market trends;
- significant fluctuations in the scope and structure of project expenses;
- dependence of investment projects on scopes of long-term obligations [9, p. 45];
- lengthy term for a new product (plot selection, topographic and geodesic surveys, design and approval, construction, related infrastructure and landscaping, etc.) [6];
- considerable expenses and highly complex engineering geology survey [2];
- technical complexity of project implementation;
- high variety of risks in such projects.

The combination of the above project-specific features in high-rise construction determines a systemic occurrence of various problems in implementation of the standard algorithm, mostly related to insufficient connection between separate stages of the algorithm.

In order to provide a better reasoning, we need to review the core problems of implementation and probable solutions at every stage of the standard algorithm.

The first stage, "Forecasting reports and projected cash flow", should take into account that high-rise construction projects imply a considerably more varied (in comparison with standard investment projects) range of cash flows that includes [9, p. 52]:

- income from long-term assets;
- income from maintenance;
- income from capital investment;

- income from transactions;
- selling long-term assets.

The considerable variety of cash flow types in combination with a great influence of a broad range of risks requires cash flow adjustment using risk management methods. Modern risk management approaches towards investment projects imply risk management during project implementation and operation [7, p. 864]. However, separating the algorithm of cost-benefit analysis in high-rise construction projects requires a more concise project-specific method.

The risk mapping method (based on preliminary risk identification and assessment) [4, p. 152] may become crucial in this situation, but in implementation of high-rise construction projects there are many various risks that are systematic yet differing in their significance and therefore impact at different project stages (currency rate fluctuation, increasing construction costs, falling demand, etc.). So, to avoid ambiguity, it is reasonable to map risks for each stage of a high-rise construction project, determining risk boundaries for each map separately.

To provide a background for a discounted cash flow rate in an investment project at *the second stage* of cost-benefit analysis in a high-rise construction project, it is reasonable to take into account that, in cost-benefit analysis of any project, there is an issue of coordinated use of methods in risk assessment (to adjust the projected cash flow) and discounted rate assessment (to calculate the project capital accurately). A lack of coordination between these methods leads, first, to discrepancies in assessment of the same risks with different methods, and secondly, to redundant acknowledgement of those risks.

The lengthy time of high-rise construction projects, high dependence of such projects on capital scope and cost, and other risk exacerbate this lack of coordination. To minimise and exclude those errors, it is reasonable to specify the risks to be identified with each of the methods:

- during risk mapping, it is reasonable to assess influence of all risks at each project stage, except the risk of changes in the cost of capital;
- risks of changes in the cost of capital (during determining a discounted rate).

This approach will significantly reduce the number of methods to provide a background for a discounted rate, which, given the complexity of high-rise construction projects, will significantly simplify the calculation. In a situation when most high-rise construction projects require involvement of various stakeholders and a significant share of borrowed capital, the method of weighted average cost of capital (WACC) may be crucial for developing a background for the discounted rate, with bank loan interest rates, inflation levels, etc. being the main parameters for determining it [3, p. 162].

The method of weighted average cost of capital is applied to high-rise construction projects as follows:

- the project is financially isolated but it is being under consideration while a funding source is not yet determined;
- it is a high-risk project affecting a scope and structure of funding.

However, it is typical of most major high-rise construction to be economically isolated (several commercial findings, assets, and financing structure) and led to a detailed evaluation. Stakeholders, their repayment schedule from their cash flow and assets, and their interest and commitment are supposed to be known. It makes a weighted average cost of capital appear inaccurate. For a more accurate way to predict a remaining cash flow and its discount, it is reasonable to calculate a cash flow and its discount by both equity rate (ER method) and expenses from the equity rate and income on equity (Fig. 1).

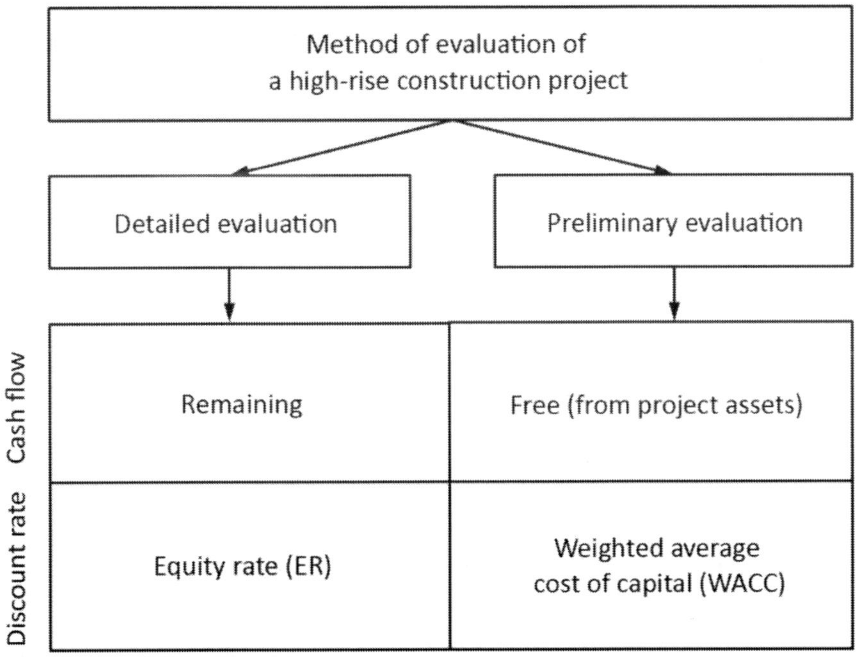

Fig. 1. Methods of evaluation of a high-rise construction project

At *the third stage*, during calculation of dynamic cost-benefit values, it is necessary to take into account that implementation of high-rise construction projects is connected with a significant scope and variety of risks during the project cycle [7, p. 841]. Specifically, the scope and complexity of such projects determine presence of various groups of risks:

- a broad group of external risks (political and economic crises, competition, economic situation, customs fees, etc.);
- prevalence of risks with a large scope and/or high probability of expected loss, etc.;
- a broad group of changes in internal project parameters (fixed and variable expenses, scope of capital expenses, etc.).

The presence of significant external factors (sectoral, administrative, legal, environmental and other risks) requires including project scenario analysis into the standard cost-benefit analysis (including the "with-project" and "without-project" analysis as a specific case of the method) (Fig. 2).

Additionally, it is reasonable to use the method of sensitivity analysis of critical ratios for more precise assessment of internal risks in high-rise construction projects [8, p. 144-147]. The core variables in analysis of high-rise construction projects are annuals sales of high-rise construction facilities, prices per square metre, variable expenses per square metre, fixed costs of building a high-rise facility, corporate tax rates, initial investment costs and implementation time of a high-rise construction project.

Tables 1–3 show results of an analysis of a typical high-rise construction project. The sensitivity was based on rational ranges or dependencies [3, p. 236-237].

The factors of the first zone (I) require a deeper risk investigation, since they are most NPV-sensitive and least predictable. The factors of Zone II require specific attention during a construction project (that is why every crucial factor is supposed to be calculated). Zone III is least risky and not subject to further consideration.

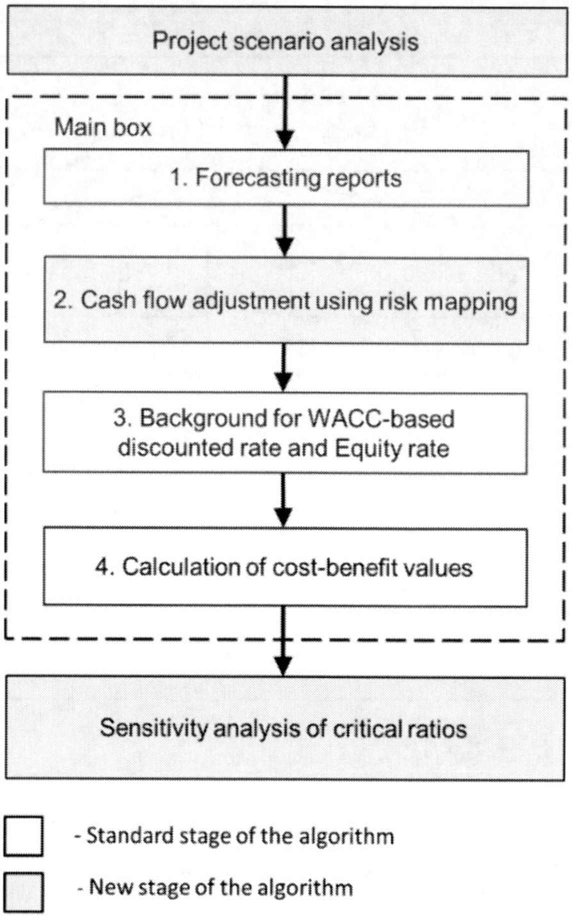

Fig. 2. Cost-benefit analysis algorithm for high-rise construction projects.

Table 1 – Rating key risk-based high-rise construction factors

Variable (X)	Variation X, %	Variation NPV, %	Ratio, variation of NPV to X variation percentage	Rating
Annual amount of high-rise construction, monetary units	10	40	4,0	1
Price for a square metre, monetary units	10	25	2,5	3
Varying costs for a square metre, monetary unit/m^2	10	30	3,0	2
Standing charges for a high-rise building, monetary units	10	15	1,5	5
Corporate tax rate, %	10	5	0,5	7
Initial investment, monetary units	10	10	1,0	6
Project implementation, years	10	20	2,0	4

Table 2 – Sensitivity and predictability in high-rise construction projects

Variable (X)	Sensitivity	Predictability
Annual sales of high-rise buildings	High	Low
Price per square metre	High	Low
Recurring cost per square metre	High	Medium
Permanent expenses for a high-rise building	Low	High
Corporate tax rate	Low	Medium
Initial investment	Medium	Medium
How long it takes	Medium	Low

Table 3 – Matrix of sensibility and predictability of variables

Predictability	Sensitivity		
Low	I	I	I
Medium	I	II	III
High	II	III	III

For a more precise calculation of crucial issues of an investment project, it is reasonable to use the following model depicting the results influenced from within and without:
Critical values are according to (NPV):

$$NPV = (Qp_i - VCQ - FC - tax + D)FM4(r,n) - IC \tag{1}$$

Q is the annual sales;
VCQ is the summary of the variable cost;
FC is the fixed cost
tax is the annual depreciation;
D is the annual corporate tax; and
IC is the initial investment.

The structure means that the cash flow were equal to net profit adjusted to depreciation. Crucial points of an investment project and a sensitive margin (SM) will be determined for each one (Table1):

1) Loss-free sales (QBE):

$$Q_{BE} = \frac{IC - DFM4(r,n)}{(P-VC)(1-tax)FM4(r,n)} + \frac{FC}{P-VC} \tag{2}$$

p is a unit of production

2) Loss-free unit of production (PBE):

$$P_{BE} = \frac{IC - FM4(r,n)[D-(1-tax)FC]}{Q \cdot FM4(r,n)(1-tax)} + VC \tag{3}$$

3) Loss-free variables per unit (VCBE):

$$VC_{BE} = p_i - \frac{OC - FM4(r,n)[D-(1-tax) \cdot FC]}{Q \cdot FM4(r,n)(1-tax)} \tag{4}$$

4) Loss-free annual expenses (FCBE):

$$FC_{BE} = \frac{Q\cdot(p_i - VC)(1-tax)+D}{(1-tax)} - \frac{IC}{(1-tax)FM4} \tag{5}$$

5) Maximum allowed initial investment (ICBE):

$$IC_{BE} = ([Q(p_i - VC) - FC](1 - tax) + D)FM4 \tag{6}$$

6) Maximum corporate tax rate (taxmax):

$$tax_{max} = 1 - \frac{IC - D\cdot FM4(r,n)}{([Q(p_i - VC) - FC]FM4(r,n)} \tag{7}$$

7) Payoff period of an investment project (PP):

$$PP = \frac{IC}{[Q(p_i - VC) - FC](1-tax)+D} \tag{8}$$

A sensitivity calculation for a high-rise construction project is based on the following values:

The highest risk is typical of projects with a lower sensitivity margin (SM) of a certain value. Consequently, a project like this is more negatively affected within and without.

Results of an investment project are most sensitive to the price of one square metre, specific variable construction expenses, and annual sales of high-rise construction facilities. In most high-rise construction projects, the aforementioned variables are characterised by both maximum sensitivity (usually, the value of NPV per cent change ratio to the variable per cent change ratio is above 3.0) and lower foreseeability [7, p. 668]. In a situation like this, development of market-adapted pricing is a priority for marketing and finance services of the real estate developer.

Special attention should be paid to reducing variable expenses per production unit, for instance, by cutting supply prices of materials and accessories, more efficient labour management, etc. A project has a good safety margin as regards changes in fixed expenses. It helps compensate for the higher risk of changes in the price of one square metre with advertising activities aimed at increasing the probability of selling high-rise facilities. Increases in corporate tax rates, considerable changes in equipment costs and other initial investment are not very dangerous for high-rise construction projects.

4 Conclusion

It is therefore reasonable to use the algorithm presented in Fig. 1 in cost-benefit analysis of high-rise construction projects as a whole. Using the algorithm will simplify calculations and improve accuracy of cash flow assessment by using methods that take into account the specific nature of high-rise construction projects. The suggested algorithm considerably improves reliability of cost-benefit project analysis in high-rise construction by using an interlinked list of supplementary risk management instruments.

References

1. Guidelines for cost-benefit analysis of investment projects. *Rossiyskaya gazeta*, 28 June (1999)

2. *Engineering geology and geotechnical survey in high-rise construction*. [Online]. Available: http://www.geoinfo.ru/includes/periodics/electronic_journal/2016/1013/000012240/detail.shtml (04.05.2017)
3. M. A. Limitovsky. *Investment projects and real options in emerging markets* (2017)
4. M. V. Gracheva, A. B. Sekerina (eds). *Risk management in an investment project* (2013)
5. O. V. Skotarenko. *Forming of sustainable development mechanisms of an enterprise on the basis of investment efficiency assessment* (2017)
6. *High-rise construction: nuances and stages of building*. [Online]. Available: http://stroimprosto-msk.ru/stati/stroitelstvo-vysotnyx-zdanij-nyuansy-i-etapy-vozvedeniya (04.05.2017)
7. I. I. Mazur, V. D. Shapiro (eds.). *Project management* (2011)
8. N. A. Kazakova. *Financial environment of entrepreneurship and entrepreneurial risks* (2012)
9. C. Hewlett, G. Kaufmann. *Strategy for Real Estate Companies* (2011)

Project risk management in the construction of high-rise buildings

Boris Titarenko[1*] , *Amir* Hasnaoui[2], *Roman* Titarenko[3], and *Liliya* Buzuk[4]

[1]Moscow State University of Civil Engineering, Yaroslavskoe shosse, 26, Moscow, 129337, Russia
[2]The Finance and Economics Department, La Rochelle Business School, France
[3]Stockholm School of Economics, Russia
[4]Russian State Social University, Moscow

Abstract. This paper shows the project risk management methods, which allow to better identify risks in the construction of high-rise buildings and to manage them throughout the life cycle of the project. One of the project risk management processes is a quantitative analysis of risks. The quantitative analysis usually includes the assessment of the potential impact of project risks and their probabilities. This paper shows the most popular methods of risk probability assessment and tries to indicate the advantages of the robust approach over the traditional methods. Within the framework of the project risk management model a robust approach of P. Huber is applied and expanded for the tasks of regression analysis of project data. The suggested algorithms used to assess the parameters in statistical models allow to obtain reliable estimates. A review of the theoretical problems of the development of robust models built on the methodology of the minimax estimates was done and the algorithm for the situation of asymmetric "contamination" was developed.

1 Introduction

One of the project risk management processes in the construction of high-rise buildings is a quantitative analysis of risks. The quantitative analysis usually includes the assessment of the potential impact of project risks and their probabilities. If to solve the first problem, it is sufficient to attract a qualified specialist in this field and provide him or her with necessary statistical information and tools to estimate the potential damage from the risk occurrence, then to estimate the probability of risks it's necessary to apply more sophisticated and specialized methods.

The main problem at the conceptual stage of project management is the selection of approaches and methods that will be used in project risk management. The traditional meth-

* Corresponding author: boristitarenko@mail.ru

ods are usually based on explicit or implicit assumptions about the normality of the probabilistic laws of indicators distribution (see, e.g., Carvalho and Rabechini [1]; Mbaidheen and Alawneh [2]).

In the case when these distribution laws are far from normal (Gaussian), the application of standard methods can lead to a substantial distortion of the results as was shown by Tukey [3]. He showed that even at small deviations of the model from Gaussian, the traditional estimates quickly lose their optimal features.

The robust methods were developed which allow to find stable estimates in situations of inhomogeneous populations (Huber [3]).

According to Huber "small deviations from the model assumptions should impair the performance only slightly, that is, the latter (in terms of the asymptotic variance of an estimate) should be close to the nominal value calculated at the model". And "for statistics representable as a functional *T* of the empirical distribution, qualitative robustness is essentially equivalent to weak continuity of *T*" (Huber and Ronchetti [3]).

At present, a robust approach is widely used in various branches of mathematical statistics: factor analysis (Yuan and Zhong [4, 5]; Toman [6]), discriminant analysis (Todorov [7]), claster analysis (Nevolainen [8]), regression models and multivariate analysis (Cizak [9]; Agostinelli [10]), general problems (Morgenthaler [11]; Toman, [6]; Mbaidheen and Alawneh [2]).

At the end of the 20th century, it was first suggested to use the robust methods for project risk management. In recent years, studies in this field have been continued (Titarenko [12,13]; Bubnov [14]; Volkov et al. [15]).

Following Huber's approach, these authors have considered not only the estimates with the smallest asymptotic variance, but the estimates with minimum asymptotic bias and suggested to use in the robust algorithms the procedures of both types at the same time, for example, to obtain a robust estimate of location parameter by minimizing the asymptotic variance and a scale parameter by minimizing the asymptotic bias.

The purpose of this paper is to develop a system model of project risk management and to form on its basis such approaches that combine the traditional methods of project risk analysis with the robust methods to ensure stable estimates of risk parameters.

This paper has the following structure. Section 2 applies and extends the robust approach of P. Huber for the tasks of regression analysis of project data. Section 3 shows theoretical problems of the robust models development. Section 4 develops the algorithm for the situation of asymmetric "contamination". Section 5 concludes the paper.

2 Robust risk management approaches

Standard approaches of economic information processing: definition of averages, calculations of regression dependencies, etc. are based on explicit or implicit assumptions that the distribution laws of economic indicators are normal.

In case these distributions are far from normal, the use of standard methods can lead to substantial distortion of the results. Lack of information in most cases does not allow to establish reliably the law of its distribution. Therefore there is a problem to create such methods of information processing that would be less sensitive to the kind of its distribution law and, in particular, would not so much respond to large random deviations. In the 1960-1970s J. Tukey and P. Huber developed an entire branch of statistics dealing with the construction and the study of "robust" procedures.

Probabilistic models used in project management are quite diverse due to a variety of project types. However, the applied classical assessment methods of models parameters, methods of finding links between technical and economical these models. Application of the robust methods allows to obtain sufficiently reliable (robust to possible inconsistencies of the models and the real phenomena) estimates of the models parameters. indexes, methods of experiments planning and etc. are sensitive to possible deviations from the hypothetical models. This dramatically decreases the reliability of the estimates and results received from these models. Application of the robust methods allows to obtain sufficiently reliable (robust to possible inconsistencies of the models and the real phenomena) estimates of the models parameters.

In initiating to establish a robust procedure it's necessary to consider three interrelated aspects of the problem:

1. Description of a supermodel, reflecting any possible deviations from a "true" model.
2. Establishment of procedures among which will be chosen the best.
3. Defining the purpose of the robust procedure, i.e. the definition of "optimal" procedure. The robustness assessment will be in the fact that its characteristics (bias, efficiency, etc.) change very little within the supermodel.

Deviations from the theoretical model caused by the "gross" errors or other reasons lead to the loss of efficiency. Therefore, the most common optimality criterion is connected with a characteristic of the estimates efficiency. The estimate should have a high efficiency for the "theoretical" model to maintain a sufficiently high efficiency within the framework of the supermodel, i.e. within the theoretical model.

In economic research it is often necessary to study the relations between different economic indicators. The dependence between some indicators z_1, ..., z_n and linked with them by another x indicator can be usually submitted in the form of a linear regression equation

$$x = \theta_1 z_1 + \ldots + \theta_n z_n + \xi . \qquad (1)$$

Classical method of solving the problem of finding estimates of the parameters θ_1, ..., θ_n in (1) is a method of least squares that consists in solving the minimization problem

$$\sum_{i=1}^{N} (x_i - \theta_1 z_{1i} - \ldots - \theta_n z_{ni})^2 = \min_{\theta_1, \ldots, \theta_n} \qquad (2)$$

i.e. in choosing such θ_1, ..., θ_n that N observed sets $(x, z_1, \ldots, z_n)$ give the least deviation in terms of (2).

Solution of the problem (2) is equivalent to solving a set n of linear equations

$$\sum_{i=1}^{N} (x_i - \theta_1 z_{1i} - \ldots - \theta_n z_{ni}) z_{ji} = 0 \quad (j = 1, \ldots, n) \qquad (3)$$

Possible way to obtain estimates that are resistant to gross errors is to replace the quadratic function in (2) to another that is less sensitive to large deviations of x_i.

Assume that a random variable ξ in (1) has a distribution $P_\varepsilon(y) = (1-\varepsilon)P(y) + \varepsilon H(y)$ where theoretical and "contaminating" distributions of $P(y)$ and $H(y)$ are symmetric, i.e. $P(y) = 1 - P(-y)$, $H(y) = 1 - H(-y)$. All distribution functions P_ε with such a view let's combine into a single class K_ε. If $P(y)$ is a function of the normal distribution, this model of gross errors shows the situation when the approximately $(1-\varepsilon)N$ deviations ξ_i are the subject to the normal law (as in the classical case), and some εN of them are gross errors. Value ε (intensity of "contamination" or the probability of gross errors occurrence) let consider as a known number.

For such a case P. Huber [3] suggested to use developed by him a common approach for the estimation of the location parameter to obtain sustainable estimates θ_1, ..., θ_n and instead of (2) to solve the problem

$$\sum_{i=1}^{N} F(x_i - \theta_1 z_{1i} - \ldots - \theta_n z_{ni}) = \min_{\theta_1, \ldots, \theta_n} \quad (4)$$

with some properly chosen function F. This problem is reduced to solving a set of equations (as a rule, nonlinear already)

$$\sum_{i=1}^{N} f(x_i - \theta_1 z_{1i} - \ldots - \theta_n z_{ni}) z_{ji} = 0 \quad (j = 1, \ldots, n) \quad (5)$$

where $f(u) = F'(u)$.

3 Theoretical problems of the robust models development

P. Huber managed to prove the consistency, nonbias and asymptotic normality of the estimates resulting from (5), and B. Polyak proved the minimaxity of their asymptotic dispersions.

P. Huber [3] proved the following statement. Let

(a) values ξ_i the realization of the random component ξ are independent and equally distributed with the symmetric distribution function $P_\varepsilon(y) \in K_\varepsilon$;

(b) known vectors of the input data $\mathbf{z_i} = (z_{1i}, \ldots, z_{ni})$ are random, independent from ξ and have the same distribution with a finite non-degenerate covariance matrix **B**, so that $b_{rs} = E(z_{ri} z_{si})$;

(c) $F(y)$ is an arbitrary non-negative symmetric, convex or monotone with y positive function, with some A and any y_1 and y_2 meeting the condition $|F(y_2) - F(y_1)| < A|y_2 - y_1|$;

(d) in class K_ε there is a distribution P_ε^0 with a density $p_\varepsilon^0(y)$ for which the integral of Fisher information $I(p) = \int_{-\infty}^{\infty} \left(\frac{p'(y)}{p(y)} \right)^2 p(y) dy$ takes the lowest value, and thus $F_0(y) = -\ln p_\varepsilon^0(y)$ is a convex function of y.

Then

(a) estimates $\hat{\theta}_i$ resulting from the solution of the problem (4) when $N \to \infty$ are consistent, unbiased and have an asymptotically normal distribution with covariance matrix

$$S(F,p_\varepsilon)=\mathbf{B}^{-1}\frac{\int\limits_{-\infty}^{\infty}(F'(y))^2p_\varepsilon(y)dy}{\left(\int\limits_{-\infty}^{\infty}p_\varepsilon(y)dF'(y)\right)^2}; \tag{6}$$

(b) for all $p_\varepsilon \in K_\varepsilon$

$$\mathbf{S}(F_0,p_\varepsilon)\le \mathbf{S}(F_0,p_\varepsilon^0)=\mathbf{B}^{-1}(I(p_\varepsilon^0))^{-1}\le \mathbf{S}(F,p_\varepsilon^0) \tag{7}$$

where the matrix inequality $\mathbf{A}\ge\mathbf{B}$ means a non-negative certainty of the quadratic form $y^T(\mathbf{A}-\mathbf{B})y$.

Thus, the statement implies that the estimates obtained from (4) with the function $F(y)=F_0(y)$ the largest possible in class K_ε of covariance matrix is the minimum of all possible.

The problem of finding the Huber's robust estimates for the parameters θ_1, ..., θ_n of a linear regression is to solve the problem

$$\sup_{p_\varepsilon}\mathbf{S}(F,p_\varepsilon)=\min_F, \tag{8}$$

which having due regard to (6) is reduced to the problem

$$\sup_{p_\varepsilon}\frac{\int\limits_{-\infty}^{\infty}f^2(y)p_\varepsilon(y)dy}{\left(\int\limits_{-\infty}^{\infty}p_\varepsilon(y)f'(y)dy)\right)^2}=\min_f \tag{9}$$

$$f(x)=\begin{cases}-\dfrac{p'(x)}{p(x)}, & x\in G=\left\{x:\left|\dfrac{p'(x)}{p(x)}\right|<k\right\};\\ k, & x\in G_+=\left\{x:-\dfrac{p'(x)}{p(x)}\ge k\right\};\\ -k, & x\in G_-=\left\{x:-\dfrac{p'(x)}{p(x)}\le -k\right\}.\end{cases}$$

$$\frac{\varepsilon k}{1-\varepsilon}=-\int\limits_{G_+}(kp(x)+p'(x))dx-\int\limits_{G_-}(kp(x)-p'(x))dx. \tag{10}$$

Minimization of the convex function (4) is a simple computationally problem. For its solution well convergent procedures can be applied.

Consider the most common variant of the gross errors model when $P(y)$ is the normal distribution with zero mathematical expectation and σ^2 variation. In this case, the Huber's model is significantly different from the commonly used multivariate regression model.

Firstly, independent variables z_i in the model are considered not as the known constants, but as the random ones. In other words, the obtaining estimates $\hat{\theta}_j$ will be the most accurate not for the given specified z_i, but only for the average for the plurality of different possible combinations of such z_i. Therefore, there is no guarantee that for the specific values z_i the obtaining estimates will be good enough.

Secondly, the Huber's model is adapted only for estimating the regression coefficients $\hat{\theta}_j$ but it is not applicable for the variation estimate σ^2. Moreover, as shown in the above formulas for obtaining estimates $\hat{\theta}_j$ the variation σ^2 should be known in advance.

Let's take the quite simple procedure allowing to estimate θ_j jointly with σ^2. To that end, as a function *f* take

$$f(y) = \max(\min(k, y), -k) \tag{10}$$

where $k = k(\varepsilon)$ is defined as

$$\frac{\varepsilon k}{1-\varepsilon} = \frac{2}{\sqrt{2\pi}}\left(e^{-\frac{-k^2}{2}} - k\int_k^{\infty} e^{-\frac{x^2}{2}} dx \right).$$

The solution of problem (2) by the transition to solve the following set of equations:

$$\sum_{i=1}^{N} f\left(\frac{x_i - \sum_{m=1}^{n} \theta_m z_{mi}}{\sigma} \right) z_{ij} = 0 \;\; (j = 1, \ldots, n); \quad \sum_{i=1}^{N} f^2\left(\frac{x_i - \sum_{m=1}^{n} \theta_m z_{mi}}{\sigma} \right) = \beta \tag{11}$$

where to obtain more accurate and nonbias without "contamination" estimates σ it's possible to choose

$$\beta = (N-n)\int_{-\infty}^{\infty} f^2(y) p(y) dy$$

The procedure is iterative and based on the piecewise linear behavior function $f(y)$.

Let $\theta^{(r)}, \sigma^{(r)}$ be the estimates for $\theta = (\theta_1, \ldots, \theta_n)$ and σ received at *r*-th step of the procedure. Let divide all possible *i* values into three sets:

$$H = \left\{ i : x_i - \sum_m \theta_m z_{mi} < -k\sigma^{(r)} \right\}; \quad C = \left\{ i : \left| x_i - \sum_m \theta_m z_{mi} \right| \le k\sigma^{(r)} \right\};$$

$$B = \left\{ i : x_i - \sum_m \theta_m z_{mi} > k\sigma^{(r)} \right\}.$$

Now the set of equations (11) can be written as

$$\sum_C \left(x_i - \sum_m \theta_m z_{mi} \right) z_{ji} + \left(\sum_B z_{ji} - \sum_H z_{ji} \right) k\sigma = 0, \;\; j = 1, \ldots n; \tag{12}$$

 https://doi.org/10.1051/e3sconf/20183303074

$$\sum_C\left(x_i-\sum_m\theta_m z_{mi}\right)^2+\left(\sum_B 1-\sum_H 1\right)k\sigma^2=\beta\sigma^2 \quad (13)$$

where the letters standing under the index of summation denotes the set of values i which is summed.

4 Computational algorithm

1. Choose the initial values $\theta^{(0)}$ and $\sigma^{(0)}$. Suppose $r=0$.
2. Get sets H, C and B.
3. Calculate matrix **W**, vectors y and d as well as scalar quantities Q and M as follows

$$w_{mi}=\sum_C z_{ji}z_{mi};\ y_m=\sum_C x_i z_{mi};\ d_m=\sum_B z_{mi}-\sum_H z_{mi};$$

$$Q=\sum_C x_i^2;\ M=\sum_B 1+\sum_H 1$$

4. Solve sets of equations

$$\sum_m w_{mj}\eta_m=y_j,\ j=1,\ldots n; \qquad \sum_m w_{mj}\delta_m=d_j,\ j=1,\ldots n$$

relatively to values η_1, ..., η_n and δ_1, ..., δ_n.

5. Values $\theta_m(\sigma)=\eta_m+k\sigma\delta_m$ meet (12) for all σ. If to substitute them into (13), from the obtained equation it is possible to define σ

$$\sigma^{(r+1)}=\sqrt{\frac{Q-\sum_m\eta_m y_m}{\beta-k^2\left[\sum_m\delta_m d_m+M\right]}}.$$

6. Suppose $\theta_m^{(r+1)}=\eta_m+k\sigma^{(r+1)}\delta_m$ and find new sets of H, C and B. If they are the same as the sets found at the previous iteration, then $\theta_m^{(r+1)}, \sigma_m^{(r+1)}$ is the final solution of the problem (11). Otherwise, let's change r to $r+1$ and go to step 3 of the algorithm. Since there is only a finite number of combinations for sets H, C and B, the procedure either ends in a finite number of steps or starts periodically to repeat itself.

To compile the program it is necessary to follow some cautions. So, if C contains very few elements, the denominator in the expression for $\sigma^{(r+1)}$ can become very low or even negative. In this case, it is necessary to increase $\sigma^{(r)}$ and repeat step 2 of the algorithm. It is necessary to limit the total number of iterations to avoid infinite loops.

The procedure gives estimates for θ_1, ..., θ_n and σ. The covariance matrix for θ_1, ..., θ_n is determined approximately by

$$\frac{N\beta\sigma^2}{(N-n)(N-M)}\left(1+\frac{nM}{N(N-M)}\right)\mathbf{W}^{-1} \quad (14)$$

In conclusion, it's necessary to emphasize that this procedure despite the simplicity is heuristic and its optimality is not proved.

We consider that it's more efficient to combine Huber's approach to estimate parameters θ_1, ..., θ_n with the sustainable assessment method σ given above. In this case, the second equation (11) should be replaced by

$$\sigma^2 = Med\left(\frac{\left(x_i - \sum_{m=1}^{n}\theta_m z_{mi}\right)^2}{a^2}\right) \tag{15}$$

where *Med* is a median and $a \approx 0{,}675$ is a 75% quantile of the standard normal distribution.

5 Discussion

Currently the robust methods cover the following range of tasks: regression and correlation analysis, dispersion analysis, factor analysis, planning of experiments, simulation, statistical estimation of models parameters, estimation of systems reliability.

This paper applies and expands the robust approach of P. Huber for the tasks of regression analysis of project data. The suggested algorithms used to estimate the parameters in statistical models allow to obtain more reliable estimates. Theoretical problems of the robust models development built on the methodology of minimax estimates and proposed by P. Huber are shown as well as the algorithm for the situation asymmetric "contamination" is developed. Thus, this paper shows that the optimal solutions largely depend on the methods of information processing on which basis these solutions are taken. Reliability of the solutions can be reached through the use of special robust procedures for the analysis of management information. An adequate model and a robust information technology decision-making represent the necessary and sufficient conditions for effective management.

References

1. M. Carvalho, R.J. Rabechini, *Impact of risk management on project performance: the importance of soft skills*. International Journal of Production Research **53**(2), 321–340 (2015)
2. M. Mbaidheen, A. Alawneh, *Assessing the risk of corporate strategic change*. International Journal of Information, Business and Management **9(**1), 45–56 (2017)
3. P. Huber, E.Ronchetti, *Robust statistics*, second ed. J. Wiley, New Jersey.(2009)
4. K. Yuan, W. Chan, Y. Tian, *Expectation–robust algorithm and estimation equations for means and dispersion matrix with missing data*. Ann. Inst. Math. **68**, 329–351.(2016)
5. K. Yuan,. X. Zhong, *Outliers, leverage observations and influential cases in factor analysis: using robust procedures to minimize their effect*. Sociological Methodology **38**, 329–368(2008)
6. A. Toman, *Robust confirmatory factor analysis based on the forward search algorithm*. Stat. Papers **55**, 233–252 (2014)

7. V. Todorov, *Robust selection of variables in linear discriminant analysis*. Stat. Math. & Appl **15**, 395–407 (2007)

8. J. Nevolainen, S. Datta, H. Ojc, *Inference on the marginal distribution of clustered data with informative cluster size*. Stat. Papers **55**, 71–92 (2014)

9. P. Cizak, *Efficient robust estimation of regression models*. Comput. Stat. Data Anal **55**, 774–788 (2011)

10. C.Agostinelli, A. Leung, V.Yohair, R. Zomer, *Robust estimation of multivariate location and scatter in the presence of cellwise and casewise contamination*. TEST **24**, 441-461 (2015)

11. S. Morgenthaler, *A survey of robust statistics*. Stat. Math. & Appl. **15**, 271–293(2007)

12. B Titarenko, System approach to the project risk management. Vestnik MGSU **5**, 340–343 (2011)

13. B.Titarenko, S.Titov, R. Titarenko, Applied Mechanics and Materials **638–640,** 2338–2341 (2014)

14. G. Bubnov, B. Titarenko, S.Titov. R. Titarenko, *Isorisk curves as a tool for increasing flexibility of risk management in engineering projects*. Contemporary Engineering Sciences **8(21)**, 991–999 (2015)

15. A. Volkov, A Sedov, P Chelyshkov., B.Titarenko,., G.Malyha, , E. Krylov, Research Journal of Pharmaceutical, Biological and Chemical Sciences **7(3),** 2416–2420 (2016)

Formation of integrated structural units using the systematic and integrated method when implementing high-rise construction projects

Ivan Abramov[1,*]
[1] Moscow State University of Civil Engineering, 26 Yaroslavskoe shosse, Moscow, 129337, Russia

Abstract. Development of design documentation for a future construction project gives rise to a number of issues with the main one being selection of manpower for structural units of the project's overall implementation system. Well planned and competently staffed integrated structural construction units will help achieve a high level of reliability and labor productivity and avoid negative (extraordinary) situations during the construction period eventually ensuring improved project performance. Research priorities include the development of theoretical recommendations for enhancing reliability of a structural unit staffed as an integrated construction crew. The author focuses on identification of destabilizing factors affecting formation of an integrated construction crew; assessment of these destabilizing factors; based on the developed mathematical model, highlighting the impact of these factors on the integration criterion with subsequent identification of an efficiency and reliability criterion for the structural unit in general. The purpose of this article is to develop theoretical recommendations and scientific and methodological provisions of an organizational and technological nature in order to identify a reliability criterion for a structural unit based on manpower integration and productivity criteria. With this purpose in mind, complex scientific tasks have been defined requiring special research, development of corresponding provisions and recommendations based on the system analysis findings presented herein.

1 Introduction

Competent HR management ensures higher labor productivity and enhanced reliability of structural units resulting in maximum avoidance of negative developments in the course of implementation of a high-rise construction project (1).

Payroll usually accounts for about 30-40% of total project costs [2], while manpower is deemed to be a decisive resource of any project as a factor contributing to improved productivity and reliability of the business [3, 4].

* Corresponding author: ivan@yandex.ru

Labor productivity (5) is a critical factor of labor cost efficiency. Its level is characterized by the ratio of the volume of completed work to the time spent on its completion. The labor productivity level determines:

Pace of construction;

Revenues of all parties involved in the project;

Cost reduction per square meter.

An important part of construction planning is formation of construction crews staffed depending on particulars of the future construction project and the objectives to be achieved.

These objectives include pursuance of research focused on formation of integrated construction crews at the stage of designing organizational and technological construction processes.

A major contribution to defining and finding solutions to designing organizational and technological construction process-related tasks, including those to be implemented by construction crews, has been made by the following research:

The authors of works [6-9] review efficiency potential models for organizational and technological solutions of a construction project that make it possible to take into account the impact of organizational, technological and managerial solutions in the implementation of a construction project expressed by various factors.

The authors of works [10-13] suggest an approach to tackling the optimization task of resource distribution in the planning of construction and installation works. The resource under review is manpower (crews) with its sufficiency and work load determining the duration of works as a critical planning factor.

Case study [14] offers a quantity assessment methodology for factors affecting labor productivity in masonry construction, including compatibility, conformance to requirements and professionalism.

The authors of work [15] argue that the productivity growth does not depend on fulfilment of as many tasks as possible irrespective of the time schedule, a greater workload or the number of work hours spent.

Productivity improves subject to a more predictable work schedule allowing for a better coordination between work load and the available capacities (man-hours).

2 Findings and Discussion

The author has used his analysis and research findings to develop a model that simulates significant factors affecting the final outcome and identifies a reliability factor of a structural unit based on the integration and production values of the construction crew under review. These significant factors affecting the final outcome include qualification, professional training, experience, productivity, labor condition management, violations of labor discipline, and coordination of production processes performed by various crews.

Unlike the analyzed works and research, this paper reviews a systematic and integrated method to determine the impact of the identified integration and production factors of a construction crew on the reliability of a structural unit.

3 Methods

A great diversity of high-rise construction projects are being undertaken in Russia with unique requirements set to structural units responsible for various project stages (Fig. 1).

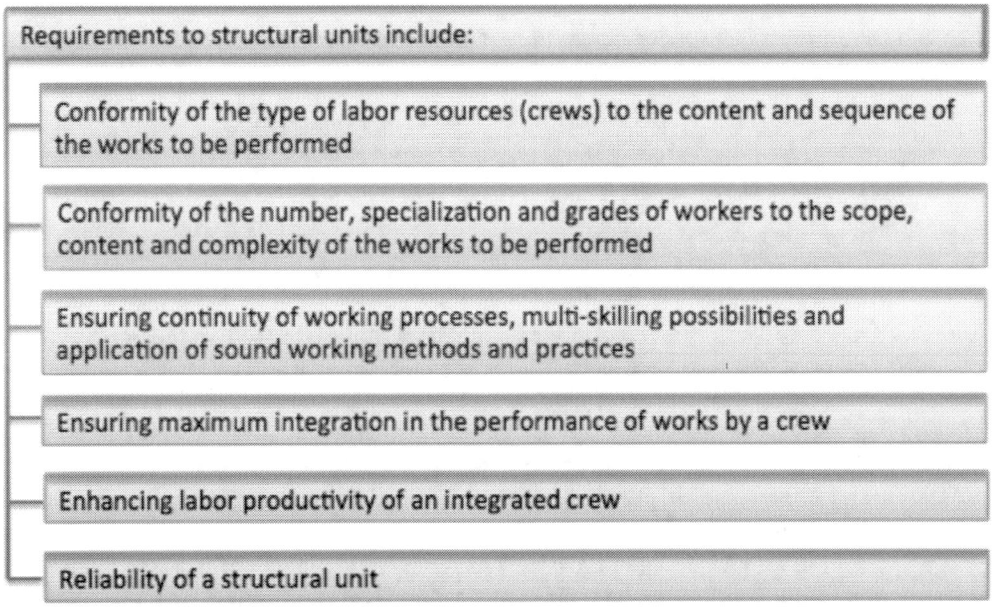

Fig.1. Diversity of high-rise construction projects

To begin with, let us focus on the requirements of ensuring maximum integration in the performance of works by a crew and enhancing labor productivity.
Formation of integrated construction crews supported by corresponding research into factors affecting the productivity of these structural units is a vital task.
The author has developed an algorithm for accomplishing this task that can be schematically represented as follows:

– Identification of factors affecting the integration of a crew as an element of a structural construction unit;
– Assessment of the factors;
– Development of an instrument for defining a measure of integration of an element of a structural construction unit, in particular, a crew;
– Identification of factors affecting the productivity of structural construction units;
– Assessment of the factors;
– Development of an instrument for defining a measure of productivity of an element of a structural construction unit, in particular, a crew;
– Determination of the factors' impact on the final outcome of the works;
– Assessment of the findings of research with the reliability of the structural unit used as the main criteria.

Based on an expert survey, let us identify and describe the main factors affecting integration of a crew:

– Number of interchangeable workers (A);
– Professional levels of workers (B);
– Proficiency of workers in various occupations (C);
– Working time lost in transfer of workloads between workers or units (D);
– Degree of combination of production processes performed by units or workers (F).

Having assessed the factors affecting the outcome (FAR):

Description	A	B	C	D	F
Factor's weight	from -2	from -3	from -2	from -2	from -1

to +2	to +3	to +2	to +2	to +1

We can determine a measure of integration of an integrated construction crew Icc calculated using the following formula:

$$\text{Icc} = k\sum_{i=1}^{n} V_i = k(V_1 + V_2 + \ldots + V_n), \quad (1)$$

where V_i is an aggregate FAR, and k is a coefficient reflecting recommendations of former employers.

Assessment of integration:

-10≤ Icc ≤0 – this crew is excluded from the project;

0< Icc ≤3 – this crew has little chances of participation in the project;

4< Icc ≤6 – this crew has moderate chances of participation in the project;

7< Icc ≤10 – this crew has the highest chances and priority of participation in the project.

Having determined the measure of integration of a construction crew in accordance with the developed algorithm [16, 17], we proceed with the task of determining its productivity.

Research and experience of previous projects have identified the following main factors affecting the productivity of an integrated construction crew:

– Labor distribution within the crew (A): competent placing of workers by the construction manager according to their qualification levels and skills;

– Interchangeability (B): ability to employ workers at any working areas in the event of unforeseen circumstances or upon completion by these workers of their scope of works;

– Human factor – compatibility (C): workers have various individual characteristics, and if they don't find a common language, this may lead to lower productivity of their work;

– Suitability (D): workers have various specializations and can be employed with a higher efficiency in construction of any type of buildings or structures;

– Qualification (F): qualification levels of workers. Workers are able to perform the main types of works at a level conforming to their qualification;

– Self-organization (I): performance of the scope of works irrespective of whether any managerial staff (engineers, technicians or the construction manager) are present on the construction site or not.

Having listed the main requirements to a construction crew and having established a link with the expert assessment, we can rank the factors affecting the final outcome (FAR′) as follows:

Description	A	B	C	D	F	I
Factor's weight	from -3 to +3	from -2 to +2	from -2 to +2	from -2 to +2	from -1 to +1	from -1 to +1

The above analysis should be followed by an assessment of integrated crews using a mathematical model. The productivity indicator of an integrated construction crew Ipcc is calculated according to the following formula:

$$\text{Ipcc} = k\sum_{i=1}^{n} W_i = T(V_1 + V_2 + \ldots + V_n) \quad (2)$$

where W_i is an aggregate FAR′, and T is a coefficient reflecting recommendations of former employers.

Assessment of productivity:

-11 ≤ Ipcc ≤ 0 – this crew is excluded from the project;

0 < Ipcc ≤ 4 – this crew has little chances of participation in the project;

4< Ipcc ≤8 – this crew has moderate chances of participation in the project;

8< Ipcc ≤11 – this crew has the highest chances and priority of participation in the project.

It should be noted that the productivity and integration indicators of a construction crew require development of an instrument to measure the reliability level of a structural construction unit.

Based on the undertaken research and having determined the productivity and integration indicators of a construction crew, let us find a common indicator of reliability of a structural unit depending on the identified productivity and integration indicators.

Let us introduce a reliability indicator R reflecting reliability of a structural unit in general. According to Academician A. A. Gusakov(18), "Organizational and technological reliability (OTR) is the ability of organizational, technological and managerial economic solutions to ensure attainment of the targeted outcome of construction in the conditions of random disturbances inherent to construction as a complex probabilistic system". As we can see, R should depend directly on integration and productivity of a crew. The higher the integration indicator, the more consolidated and harmonious is the crew. This has an obvious positive impact on the reliability indicator. As far as the productivity of a crew is concerned, the more skilled the crew members are, the greater impact it has on the reliability indicator of the structural unit in terms of the quality and duration of works.

Consequently, we can describe reliability as follows:

$$R = f(C_1 * C_{CC} + C_2 * I_{pcc}), \tag{3}$$

where C_1 and C_2 are some positive real number constants that reflect the distinct proportionate impact of both integration and productivity of a crew on the reliability of a structural unit. In our model, we proceed from C_1 = 1/10 and C_2 = 1/11, as the impact of both of these factors on reliability is assumed as being equal.

The f function has a positive first-order derivative over the whole segment, as the higher the linear combination, the greater is the reliability. Also, the f function has a negative second-order derivative over the whole segment, as the growth of productivity and integration leads to a decreasing effect versus scale. In other words, if reliability and integration are high, an increment of the functional argument will produce a lesser effect compared to a situation of low reliability and integration. This is due to the fact that an increase in integration or reliability of an element of a highly reliable structural unit produces a limited effect only. If the reliability factor R of a structural unit is low, an increase of the above-mentioned indicators will boost reliability to a great extent. This function f can be assumed as follows:

$$f(x) = 100 * sgn(x) * \sqrt[3]{|x|}, \tag{4}$$

which meets all conditions described above.

$$f(x) = 100 * sgn(0.25) * \sqrt[3]{|0.25|} = 62.9$$
$$f(x) = 100 * sgn(0.5) * \sqrt[3]{|0.5|} = 79.3$$
$$f(x) = 100 * sgn(1) * \sqrt[3]{|1|} = 100$$
$$f(x) = 100 * sgn(1.5) * \sqrt[3]{|1.5|} = 114.4$$
$$f(x) = 100 * sgn(2) * \sqrt[3]{|2|} = 125.9$$

The graph of this function is shown Fig. 2.

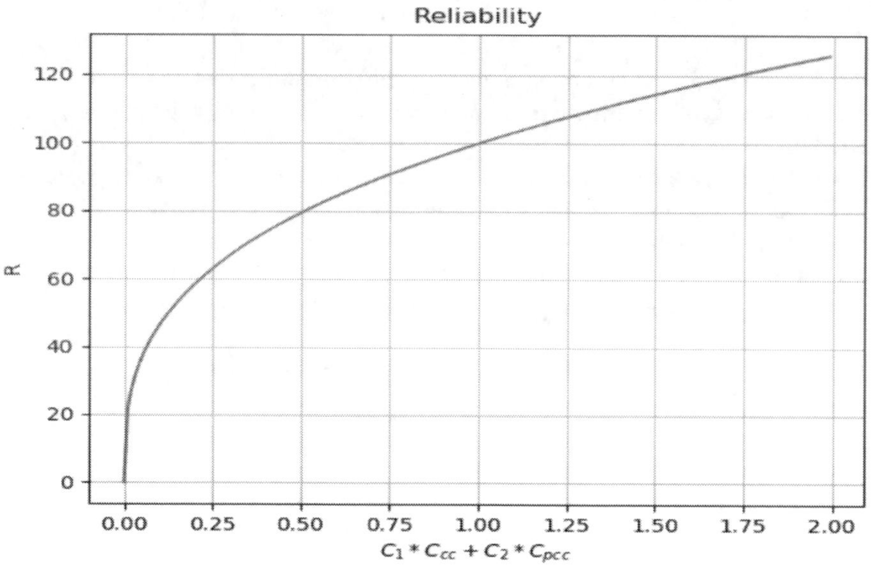

Fig. 2 The graph of function

Diminishing return can be illustrated by the following example:

With Icc = 1 and Ipcc = 2, the argument of this function is equal to 0.282. An increment of each indicator makes this argument equal to 0.473. As a result, the reliability function grows approximately by 12.34. With Icc = 9 and Ipcc = 10, the argument is equal to 1.81. An increment of both indicators makes it equal to 2. In this case, the reliability function grows approximately by 4.12. This is an evidence of decreasing returns of the scale. Also, with Icc = 6 and Ipcc = 7, an increment of each indicator brings us to the "top" group and makes the reliability grow by 5.26. It means that in order to enter the "top" group, we need a greater growth of reliability as compared to that required in a situation within that group.

For the reliability factor to be within the range of 0 to 100, we can additionally divide the function into$f(2)$.

4 Conclusions

The reliability factor of a structural construction unit has an impact on the resultant qualitative characteristics of a high-rise construction project such as duration, quality and other technical and economic indicators.

This article offers theoretical recommendations and scientific and methodological provisions of an organizational and technological nature for competent formation of integrated construction crews with the purpose of enhancing their productivity and identifying a reliability criterion of their parent structural units.

References

1. A. Kazaz, S. Ulubeyli, Building and Environment, **42**, 2132-2140 (2007).
2. K. Harmon, B. Cole, Construction Briefings, **8**, 1–19 (2006).
3. M. Liu, G. Ballard, W. Ibbs, Journal of Management in Engineering, **27**, 236-242 (2011).
4. S. AbouRizk, P. Knowles, U. Hermann, Journal of Construction Engineering and

Management, **127**, 502-511 **(2001).**
5. L. Floreza, J. Cortissoz, Procedia Engineering, **164**, 42-48 (2016).
6. A. Lapidus, M. Yershov, P. Oleynik, et al, Technology and Organization of Construction, **4**, 6-11 (2014).
7. A. Lapidus, I. Abramov, Organizational and process-related design of private low-rise construction projects when developing a calendar plan, Science Bulletin, **4** (**2017**).
8. A. Lapidus, A. Berezhnyi, Bulletin of the Moscow State University of Civil Engineering, **3**, 149-153 (**2012**).
9. A. Lapidus, I. Abramov. Science review, **4**, 6-9 (**2017**).
10. V. Mishchenko, D. Yemelyanov, A. Tikhonenko, Voronezh State University Architecture and Civil Engineering, **6**, 76-78.
11. V. Mishchenko, D. Yemelyanov, A. Tikhonenko, Industrial and Civil Engineering, **10**, 69-71 (**2013**).
12. V. Mishchenko, E. Gorbaneva, Y. Rithy, F. Lin, Application of the flow method in low-rise urban residential development in hot climates, Ho Chi Minh City University of Architecture, 28-38.
13. P. Oleynik, V. Grigoriev, Technology and Organization of Construction, **2,** 42-44 (**2014**).
14. L. Floreza, J. Cortissoz, Procedia Engineering, **164**, 42-48 (**2016**).
15. A. Fini, T. Rashidi, Journal of Construction Engineering and Management, **142** (2015).
16. A. Lapidus, I. Abramov, Science Review, **9** (**2017**).
17. I. Abramov, T. Poznakhirko, A. Sergeev, The analysis of the functionality of modern systems, methods and scheduling tools, MATEC Web of Conferences, **86** (**2016**).
18. A. Gusakov, System techniques in construction, ASV Publishers (**2004**).
19. P. Oleynik, V. Brodskiy, Technology and Organization of Construction, **2**, 40-43 (2013).
20. P. Oleynik, V. Brodskiy, Technology and Organization of Construction, **3**, 35-38 (2013).
21. P. Oleynik, V. Grigoriev, Technology and Organization of Construction, **2**, 42-44 (2014).
22. P. Oleynik, V. Brodskiy, Technology and Organization of Construction, **2**, 40-43 (2013).
23. A. Lapidus, A. Makaraov, Bulletin of the Moscow State University of Civil Engineering, **8**, 150-160 (2015).
24. V. Mischenko, D. Yemelyanov, A. Industrial and civil construction, **10**, 69-71 (2013).
25. A. Lapidus, T. Bidov, International Journal of Applied Engineering Research, **11** (2016).

Peculiarities of solving the problems of modern logistics in high-rise construction and industrial production

Anatoliy E. Rubtsov[1], *Elena V.* Ushakova[1*] , and *Tamara V.* Chirkova[1]

[1]Saint Petersburg University of Management Technologies and Economics, Institute of Management Economics and Information Technology, 190103, Russia

Abstract. Basing on the analysis of the enterprise (construction organization) structure and infrastructure of the entire logistics system in which this enterprise (construction organization) operates, this article proposes an approach to solve the problems of structural optimization and a set of calculation tasks, based on customer orders as well as on the required levels of insurance stocks, transit stocks and other types of stocks in the distribution network, modes of operation of the in-company transport and storage complex and a number of other factors.

1 Introduction

The need to accelerate and expand the implementation of the logistics approach in the practice of economic activities of industrial and construction organizations, and many other industries and business activities is due to a number of objective reasons and, above all, the transition from the seller's market to the consumer (buyer) market, which requires high flexibility of the production systems in responding to the rapidly changing priorities and requirements of consumers.

The key factor in the transition of the world economy from the producer's market to the buyer's market is the formation and implementation of the logistic principles of marketing into economic activities.

This period (early 1990's) was characterized by the rapid development of military logistics, primarily in the United States. By the early 1950s, the United States basically formulated the Theory of Military Logistics. Practical significance of the developed logistics approaches in military construction, production, supply, transportation, etc. was proved by the US Army at the opening of the second front in World War II, when the principles of unified management in the supply, storage, transportation, and construction were successfully applied.

One of the oldest organizations not only in the US but also in the world is the National Association of Purchasing Management Problems, founded in 1915 and transformed in 1967 into the National Association of Procurement Agents. During the 1950s and 1970s, in Western countries, and especially in the United States, it has undergone rapid development

* Corresponding author: ushakovaev@yandex.ru

in business logistics, many of its problems have been resolved both theoretically and in practical applications.

Famous American writer and management consultant Peter Drucker described the activity in the distribution as the most sadly neglected and in the same time most promising sphere of American business.

Along with it, those years and nowadays as well, existed a number of objective economic and technological factors explaining the need for accelerated development of logistics. Among those factors are increase in the share of services and the emergence of a large number of diverse competitive goods. This, in turn, required the search for new ways to improve commodity distribution structures, reducing costs in the processes of physical distribution.

One of the most important factors contributing to the further development of logistics thinking and increasing the efficiency of the functioning of logistics systems and supply chains was the emergence and implementation of modern computer technology along with its practical usage to solve a variety of logistics problems and tasks. On this basis, a huge number of mathematical and other methods, micro-logistics, information systems and technologies for management, planning, production were developed and introduced into logistics, which significantly shortened the time for their implementation, increased the accuracy of the solution, created all necessary conditions for a significant reduction of all types costs.

In the article the essence and possible logistic approaches and methods of solving some logistical problems and tasks are considered. It is worth to mention the insufficient of attention to those approaches so far.

Modern industrial production and high-rise construction are mechanisms of high complexity, including own technological units that produce semi-products, parts, components, assemblies, assembly units from primary goods and raw materials, and subsequently assemble finished products from these elements, both manufactured at the enterprise and purchased from other manufacturing enterprises, as well as a large number of various auxiliary units, usually combined in the term "infrastructure" of the production and, accordingly, the infrastructure of the construction. Thus, the comparison of the structural and logistic characteristics of an industrial enterprise and a construction company alongside with the operating principles and methods of improving logistical processes show the similarity of the processes taking place in the corresponding logistic systems and methods for their improvement. Taking this generalization into account, we will further consider only the peculiarities of solving the problems of production logistics on the industrial enterprise, assuming they are essentially similar to those of solving the problems of construction organization logistics.

At the present time, industrial enterprises, associations, concerns, large construction organizations quite often comprise a significant number of subsidiaries that manufacture individual units, assembly units that assemble finished products, which is clearly seen in the case of automotive companies. Lately it is common for subsidiaries being parts of associations, concerns and such to be situated at significant distances from the main factory, in different countries and even on different continents. This significantly complicates the processes of establishing logistics systems and their optimization due to a range of additional tasks requiring for optimal solution, for example, transportation of goods, creation and management of interstage inventories, optimization of transportation routes and others. Thus, the tasks of inbound logistics in relation to the separate units, assemblies, blocks etc, being parts of the finished products, are in their essence the tasks of outbound logistics in relation to the finished products. These problems and tasks often indicate the importance of an integrated approach to the analysis of the supply objectives: from suppliers of material resources to final consumers of finished products in accordance with their requirements and business objectives.

2 Methods of research

Manufacturing process of any production usually includes a large number of both elementary and complex types of activity, united by the main schedule of the enterprise. From an integrated planning perspective, the production plan is based on the sales plan and the distribution plan. At the same time, the purpose of production should be formed on the basis of corporate and marketing strategies of the enterprise, and in its content it should be outside the production cycle. Consequently, an independent (isolated) consideration of the production logistics is possible only with a business approach with given corporate goals and limitations of the external and internal environments affecting the logistics system. These goals and limitations are the basis for the operational master production scheduling of the release of a certain range of products and the appropriate logistics support for production processes.

Any production process consists of two groups of operations on material flows:

a) operations of transformation of material resources (MR) into finished products (FP), i.e. purely production operations;
b) logistical operations related to the transportation, intermediate storage and handling of MR, work-in-progress (WIP) and FP.

The analysis of these two groups of operations and functions shows that they are so closely intertwined with each other that in production logistics it is more appropriate to talk not about logistical, but about production-logistical operations, functions and systems.

Thus, the following operations, functions and systems related to the information flows can be classified as production-logistical:

- operational calendar planning with a detailed schedule of FP release;
- operational management of production and technological processes;
- quality control, maintenance of product quality standards and standards of logistics services;
- forecasting, planning and rationing of MR and WIP consumption in the production process;
- dispatching of internal technological transport;
- control and management of MR, WIP and FP stocks at all levels of the internal warehouse system and in the technological production process.

This having been said, it should be borne in mind that all operations and functions of this production-logistical complex should be carried out within the framework of the company's logistics strategy to achieve the following goals of its functioning:

- MR, WIP, FP level optimization in the production-technological cycle and warehouse system;
- optimization of the operation of the in-company transport and storage complex;
- reduction of the production-technological cycle duration;
- reduction of the logistical costs in the FP production.

Based on this complex of production-logistical operations, functions and systems and its goals set in the logistics strategy, production logistics can be defined as management of internal materials and information flows that ensures the implementation of the production program with the least total costs, while meeting all the limitations of the logistics system. This formulation is the most understandable to Russian industrial enterprises.

Given the limitations of the external environment, the logistics support of production procedures will be optimized based on the minimization of the total costs associated with the management of MR, WIP, FP, internal transportation, cargo handling, storage, etc. The main factors of the external environment for the manufacturing logistics are the decisions taken to supply MR, the EP distribution system, the final consumer market. The parameters of the external environment determine the numerous limitations (resource, financial, in terms of volumes and terms of delivery of MR and the conditions for their delivery, financing and lending, tax rates, customs duties, etc.) and setting actions that determine the system setting. The main parameter of setting up an internal logistics system is the dynamics of sales and demand for the company's products. In addition, it is necessary to take into account the limitations on production impacts, as well as the fact that volume and time demand indicators are usually formed basing not only on the consumers orders, but also on the required levels of stocks (transit, insurance, etc.) in the distribution network.

Production logistics processes with pre-set tuning parameters and limitations optimize the main indicators, which include:

- frequency of MR and WIP stocks turnover;
- total logistics costs;
- velocity of MR and WIP turnover for individual warehouses;
- costs per unit of FP;
- loading of warehouses and vehicles;
- risk associated with the content of internal stocks, and others.

When looking at the production logistics from the point of view of its interaction with operational management, it is necessary to pay attention to the integrating and coordinating capacity of logistics aimed at supporting the system stability of the enterprise and the entire supply chain by providing quality support for the dynamics of demand for the required volumes of products and services. This is confirmed by numerous concepts and technologies in which, in varying degrees, it is almost always possible to find elements of coordination and integration between production and supply, production and sales or the total integration of all key production-logistical processes [13].

General theory of management argues that the introduction of integrating elements into the control loop contributes to improving the stability of the system as a whole and improving the quality of management. Within that framework the production logistics can be represented as some adaptive management system that tracks the dynamics of demand and optimizes the quality parameters of production processes. Analysis and synthesis of internal logistics systems on the basis of modern management theory make it possible to obtain a lot of constructive solutions for the algorithms of logistics management, decision-making procedures, improving the quality and improving the management of material and related flows.

It is reasonable to analyze interaction of logistics and production from two points of view:

a) from the point of view of internal logistics, i.e. inventory management of MR, WIP and FP in the production process, organization of MR order procedures in structural divisions of the enterprise, management of technological transport and warehouses;
b) from the point of view of external logistics, i.e. management of key logistics functions in the MR supply and FP distribution.

Both main and auxiliary units of any industrial enterprise implement a certain set of functions (elementary and complex) that constitute the subject of production logistics. Obviously, the management of the enterprise shouldn't be artificially divided into the management of the main divisions and the management of the industrial infrastructure of the enterprise, since these components are aimed at the same goal, which is, essentially, to produce finished products in accordance with the specified production schedule and in compliance with the requirements of quality standards, and achieving maximum savings in all types of resources.

Both for integrated logistics system in general and for production logistics, all types of links in the production structure of the enterprise involved in the management of the main and related logistical flows should be maximally coordinated and integrated.

Recently, conversions of MR into the finished products with the use of appropriate equipment, energy, information, labour resources are often called transformation processes, or processes of converting inputs to outputs (FP and/or services).

In the process of transformation, the costs of the transformation processes are added to the input costs, forming the price (prime cost) of the FP (service) production.

Implementation of the operational and logistic control and management of transformation processes requires the use of the reliable information obtained through feedback circuits. In this case the main objects of logistics management are the material and related information and financial flows that circulate between the production units of the enterprise. As there are a lot of such production units at the enterprise, from the point of view of logistics each of them can be a supplier and at the same time a consumer of certain MR, WIP and FP, forming internal logistics chains and channels. The links of these logistic chains and channels include also warehousing, transport, cargo handling and other units related to both main and auxiliary production and ensuring the existence of the material flows and the execution of the production program.

3 Results

The system of interaction between operational and logistics management of an enterprise proves to be optimal only when based on a production schedule that takes into account the requirements and tasks of satisfying the consumer demand and providing answers to questions about who will produce what, where, when and in what quantity. The production schedule developed on the basis of the master production schedule makes it possible to establish the volume and time characteristics of material flows differentiated for each structural production division. A classical interface with operational management for internal logistics can be represented by the duration (length) of the production period (production-technological cycle). One of the main tasks of internal logistics for enterprises using YIT, QR, LP logistic concepts/technologies and the like is the reduction (lead-time) of the production-technological cycle. However, it should be borne in mind that there is no need to accelerate the reduction of the production-technological cycle if there are sufficiently large reserves of FP in the distribution networks of the enterprise. This is how internal and external logistics are linked. Due to the modern market trends the time factor becomes more significant both in production and in sales, but maneuvering opportunities are significantly narrowed in the production compared to those in distribution, as most enterprises use the same advanced technologies of FP production, i.e. flexible automated and robotized equipment. Therefore, in order to reduce the production-technological cycle many enterprises seek to improve existing internal systems and technologies, introducing tools such as APS (advanced planning and scheduling technology).

The interaction of logistics and operational management can also become visible when taking into account the seasonal demand in the production of FP. This demand is not always fully predictable, therefore, in order to insure against the risk of sales loss enterprises seek to maximize the volume of FP release, which at the same time leads to an increase in costs for storage and maintenance of FP stocks both in production and in the distribution network. The task of logistics management for solving this problem is to optimize total costs in production and distribution to create seasonal stocks of FP [4].

What is crucial for the production logistics is the MR procurement management. A modern approach to procurement management should be universally based on YIT concept and minimizing MR stock levels at the manufacturer.

The problems of relations and interaction between logistics and production at enterprises exist in many areas and in many cases remain rather acute up to the present time. For example, for the industrial associations, concerns, etc., enterprises that assemble FP from separate products manufactured at other enterprises of the same association (holding ...), solving the "produce or buy" task is an important and frequently encountered problem of logistics and production.

The influence of the external logistics on the production in many respects also depends on a wide range of factors, which include the availability of warehouses to enterprises, the quality of warehouse equipment, the location and distance of warehouses from manufacturing enterprises and from consumers, the ways of FP transportation and storage, methods of inventory management and others.

4 Conclusions

From the point of view of modern operational management, the acquisition of additional competitive advantages results from the use of advanced production and logistics technologies and integrated information systems, which allow faster, better and more accurate implementation of all the production and logistics operations by means of flexible production integrated information systems. Along with the use of TQM philosophy, advanced technologies integrate production-logistical operations of an enterprise to meet consumer demand with the highest quality of products and services. The synergy of modern opportunities for universal quality management, advanced production technologies and logistics opens the way for the enterprise to achieve competitive advantages in flexibility, which includes rapid re-equipment, shortening of the production cycle, release of a new product range, acceleration of reaction to changes in consumer demand, lowering the production costs and improving the quality of products, related services and productivity.

Thus, such integration can result in an increase in the efficiency and system stability of the enterprise in the market and creation of the opportunities for introducing production innovations and for the successful implementation of the strategic tasks.

References

1. S. Agrawal, R.K. Singh, M. Qasim A literature review and perspectives in reverse logistics, Resources Conservation and Recycling, 97. 76-92 (2015)
2. Anna Sobotka, Joanna Czaja Analysis of the Factors Stimulating and Conditioning Application of Reverse Logistics in Construction Procedia Engineering, **122**, 11-18 (2015)
3. Al., Adbulmohsen and R., Janaka, Proc. of the International Council for research and Innovation in Building and construction (CIB), W78-W102, (2011)
4. B.A. Anikin T.A. Rodkin Logistics and Supply Chain Management Theory and Practice. Prospekt, Moscow, 2017
5. D.J. Bowersox, D.J. Closs Logistics. Integrated supply chain. Trans. with English. – (Moscow: ZAO Olimp-BUSINESS)
6. D. Bowersox, J., Closs David G. *Logistics: an integrated supply chain*. (Moscow, Olimp-Business Publishing House, 2017)

7. VI Berezhnoy et al Economic mechanism for improving the efficiency of transport. Sat. scientific. works. LEEI, 35-40 (1996)
8. D. Dzh. Bauersoks, D.Dzh. Kloss, *Logistika. Integrirovannaya tsep' postavok 2-e izdanie* (ZAO «Olimp-BIZNES», Moskva, 2017)
9. R. Hosseini, N. Chileshe, R. Rameezdeen, S. Lehmann, Reverse Logistics for the Construction Industry: Lessons from the Manufacturing Context international Jurnal of Construction Engineering and Management, 3, 75-90 (2014)
10. V.S. Lukinskiy, E.V. Bereginoy, Logistics of motor transport: a tutorial (Finance and Statistics Moscow 2004)
11. L. M. Juan, C. Martin, R. Puertas Journal of Applied Economics **20**, Issue 1, 169-192 (2017)
12. Ala-Risku, T. Mikko, International Journal of Production Economics 104, 19-29 (2006)
13. V.I. Sergeeva, *Corporate logistics in questions and answers* (INFRA-M, Moscow, 2013)

The methodic of calculation for the need of basic construction machines on construction site when developing organizational and technological documentation

Boris Zhadanovsky[1] and *Sergey* Sinenko[1*]

[1]Moscow State University of Civil Engineering, Yaroslavskoe shosse, 26, Moscow, 129337, Russia

Abstract. Economic indicators of construction work, particularly in high-rise construction, are directly related to the choice of optimal number of machines. The shortage of machinery makes it impossible to complete the construction & installation work on scheduled time. Rates of performance of construction & installation works and labor productivity during high-rise construction largely depend on the degree of provision of construction project with machines (level of work mechanization). During calculation of the need for machines in construction projects, it is necessary to ensure that work is completed on scheduled time, increased level of complex mechanization, increased productivity and reduction of manual work, and improved usage and maintenance of machine fleet. The selection of machines and determination of their numbers should be carried out by using formulas presented in this work.

1 Introduction

The shortage or surplus of machines for construction & installation works significantly affect economic indicators of functioning in construction companies. The rate and quality of construction & installation works and labor productivity largely depend on the technical level of mechanization, as well as the degree of provision of the construction project with machines (level of labor mechanization).

To determine the need for machines for construction it is necessary to consider:

- completion of work on objects in accordance with the scheduled time;
- increasing level of complex mechanization of construction and installation works;
- increasing labor productivity and reducing the proportion of manual labor;
- application of progressive work production methods;
- improving the usage of existing fleet of machines;
- provision of the machine fleet with required rate of renewal.

*Corresponding author: SlavinaAY@mgsu.ru

2 Methods

Calculation of demand and supply of the machines to construction companies needs to account for the production by these organizations of the whole complex of organizational, technological and operational activities aimed at continuous improvement of efficient use of the machine fleet [1, 2, 7, 8, 9, 10].

The need to calculate the demand of construction machines arises at the following levels of planning construction work when determining:

- need for machines during development of work execution design for the construction of a specific object;
- need for machines on the construction & installation works program by a separate construction organization (trust, mechanization management);
- need and justification of supply requests for the supply of machines to construction departments or to the construction industry in whole.

The methodology for determining need of machines for these planning levels is based on the same approach and varies in detail and in the amount of input data. In the first case, the source of the initial data for the calculations is the organizational and technological documentation and in other cases, the average data of construction projects (organizations) – submitted statistical data during the base period (a year).

Formation of the construction machine fleet should be carried out starting with the process of selecting the means of mechanization during preparation of work execution design (PPR) for the construction object and further on the hierarchical structure of management in planning of the development of the machine fleet at the level of trust (department) of mechanization, and the ministry (authorities) [1, 3, 4, 5].

Determination of the need for machines in the construction industry should also begin with the lowest level (object) and reach final form at the highest level of management (construction ministry).

Determining the need for machines at different levels of fleet formation has its own characteristics both in methodical approach and in concrete terms, i.e. the need per unit of work (pieces, m^3, m^2, etc.), or per 1 million rubles of construction and installation work (SMR). This considers specifics of functioning and purpose of the construction machine fleet (fleet of the construction object, fleet of the company or fleet of the ministry).

As noted above, methodological approach in determining the need of machines for the construction object, the trust (department) of mechanization, and the ministry (authorities) has differences in determination of the initial data and in formation of basic information.

When determining the need of construction machines for the construction object (by construction object we mean separately constructed buildings and structures, for example a residential building, a factory) specific operating conditions are considered, which are expressed in variation of ground soil, number of storeys of buildings, cargo size, and real physical volumes of work tied to this object with distribution in their structure (trenches, construction pits, excavations, etc.). In addition, usage time of the machine on the construction object can vary from a few days to several months. After performing a certain amount of work the machine is no longer needed and it can be used on another object, in other operating conditions.

For machines used with one type of work equipment the operating conditions at different sites can differ on many technological factors. For universal construction machinery, which has several types of changeable work equipment, the number of technological factors that influence its utilization which is expressed in the resulting productivity over time at the object may be much longer [2, 8,9,10].

Specifically, this can be shown on the example of single-bucket universal excavator. The following excavation work is required at the site: vertical land leveling of the surface layer, excavation of soil in a construction pit and excavation of several trenches, to loosen several lenses of firm soil, and to fill up hollows. Moreover, these works can be carried out only consequently and it is not economical to involve additional equipment for loosening, for example, a bulldozer-ripper. Hence the need to use the following changeable work equipment: front shovel, backhoe shovel, hydraulic hammer, grappler. The excavator can be used with all this equipment. This should be considered when calculating the work of the excavator on the object. In addition, we should consider not only the operational performance in volume of work of the machine, but also the use of the machine over time.

When determining the need of construction machines for the object, it is necessary to select machines based on the area of their rational use [4,5, 6,7]. One significant difference in the calculation of need of machines or volume is to determine the actual production of machines, in relation to a specific object.

The optimality of choosing a particular type of machine in comparison with other types of machines can be justified by use of these machines in rational areas defined by equations of equal costs.

Calculation of needed construction machines for the mechanization department should be carried out for a year in advance. Only if there is accurate data on the volume and structure of works for a longer period (2-3 years), the calculation can be performed on the same term with subsequent annual updating of demands based on specific planned tasks considering the structure the works such as, for example, earth excavation in table. 1.

Table 1. The earth excavation

N	Technological methods of works production	In % to total volume	
		Actual	Recommended
1	Tractors with mounted excavating equipment	2,2	2
2	Single-bucket excavator	37,6	35
3	Continuous bucket excavator	3,3	7,9
4	Scrapers	9,4	13,6
5	Bulldozers	40,4	32,2
6	Land grader	1,3	2,6
7	Single-bucket loaders	1,1	2,6
8	Other means of mechanization	4,2	5
9	Manually	0,5	0,3

The need for the mechanization department in the means of mechanization is formed based on requests requiring machines for construction sites, as well as to perform unscheduled works (reserve) and works in production base of construction, which can be determined by analogy of the past period. In accordance, the need for machines in mechanization department should be determined by method of optimization of the fleet structure or by using aggregate-based indicators, accounting the analysis of work in the base period (at least 3 years).

The peculiarity of determining the need for construction machinery for a ministry (department) is that it is defined for the construction industry without reference to a certain construction object according to the demand norms with a certain adjustment depending on the level of mechanization of construction for a given period. The need is determined by the annual amount of construction & installation work in monetary terms. It does not specify a specific model of the machine which is replaced by an average machine of the fleet, which is determined by the sum of the main parameters of the machines (m^3-bucket capacity, t - loading capacity, etc.). The norms on which the need is determined have a directive nature and are set for a period of 3-5 years. When calculating norms of the need of construction machines construction indicators (specific physical quantities, the structure of means of mechanization and the annual output of cars) are determined accounting the probabilistic nature of changing during the planning period, based on the analysis of these indicators in the base period. These standards consider the need of ministries (authorities) in the means of mechanization, taking into account the relocation of construction machines from one site to another, being on repair, reserve of machinery to perform unscheduled work, etc.

When planning construction indicators, trends in the development of mechanization work in the planned period are considered, during which standard norms will be taken into account. Concluding from the above, when calculating the need for construction machinery, as the hierarchical level of the organization for which calculation is done, increases, accuracy of the calculation reduces while the complexity of the task increases.

Calculation of the need for construction machines is carried out in the following sequence: the amount of work to be conducted is identified; the structure of mechanization means is determined; the operational hourly productivity of machines is calculated; the required number of machines to perform the specified volumes of work is calculated [1].

Calculating the number (need) of construction machines N using the formula:

$$N = \frac{Q}{b_{ech} * T_t}, \quad (1)$$

where Q- amount of given type of work in kind; b_{ech} - operational productivity of the machine when performing the given type of work; T_t - duration of work of the machine on the given type of work, mach.- h.

The duration of work of the machine on the given type of work is determined by the formula (10):

$$T_t = \frac{(T_{dv} - d_{nB})}{\dfrac{1}{t_{sm} * k_{sm} + DT_p}}, \quad (2)$$

where $T_{дв}$ - specified duration of operation of the machines as determined from the schedule, days;

t_{sm} - average duration of shifts, h,

k_{sm} - average shift factor of the machine;

dnB -average duration to relocate the machine, days;

DTp - duration of stay of machine in maintenance and repair (TM and R), days/mach - hour.

Duration of the machine in maintenance and repair is defined per " Recommendations..."[3]

$$DT_p = ((\frac{T_p}{t_{01}} - \frac{T_p}{t_{02}}) * d_{01} + (\frac{T_p}{t_{02}} - \frac{T_p}{t_{rm}}) * d_{02} + (\frac{T_p}{t_{rm}} - 1) * d_{rm}) * \frac{K_h}{t_{rm}}, \quad (3)$$

where T_p - average life until first major repair, Moto-h;

t01,t02,t_{rm} - period of technical maintenance (TM) and a running maintenance (RM), Moto-h; d01; d02; d_{rm}-duration of stay of machines in and TM and RM, days;

K_h -coefficient of conversion Moto-h in mach.- h.

The specified duration of the machine in accordance with the schedule is determined by the formula:

$$T_{dv} = (T_d - D_v) * \left(1 - \frac{D_1}{T_d}\right), \quad (4)$$

where T_d-planned schedule of construction of the site (calendar), days; D_v-number of days off; D_1 - number of days with unfavorable weather conditions.

When determining the need of universal machines of one size which perform two or more given types of work, the calculation is carried out according to the formula:

$$N_0 = \sum_{i=1} N_i = \sum \frac{Q_i}{b_{echi}} * T_{vi}, \quad (5)$$

where Ni — the need for machines on the i-th type of work; Qi - the amount of work i-th type in kind; T_{vi} - duration of operation of the machine at the i-th type of work; b_{echi} - hourly operational performance of the machine when conducting i-th type of work.

By using formulas (1) and (5) we can calculate the need for leading machines. The need for auxiliary or non-leading machines operating in the technological complex is calculated depending on the performance of the leading machine [3,4].

The number of auxiliary machines relative to leading machines when the performance of machines is expressed in the same measuring units is calculated by the formula:

$$N_{vs} = \frac{N_0 * b_{eo}}{b_{ev}}, \quad (6)$$

N_0 – number (need) of the leading construction machines;

b_{eo}, b_{ev} - hourly operational performance leading and auxiliary machines respectively.

If the performance of the leading and auxiliary machines is expressed in different units, for example, concrete pump - concrete mixer truck, the need of auxiliary machines is determined by the formulas:

$$N_{vs} = N_0 * K_{pr} * \frac{b_{eo}}{b_{ev}}, \quad (7)$$

where K_{pr} - reduction coefficient of the dimension of performance from the auxiliary machine to the leading machine;

$$N_{vs} = N_0 * \frac{b_{eo}}{b_{evp}}, \quad (8)$$

where b_p - hourly operational performance of the auxiliary machine, presented in the measurement unit of performance of the leading machine.

For example, the algorithm for determining the operational performance of a concrete batching complex which works according to the technological scheme should be determined considering the delivery of the concrete mix, the performance of the concrete pump, measured in m^3/h.

To ensure the delivery of concrete, concrete mixers are used. The main feature of concrete mixers is the capacity of the drum (bunker) m^3. The capacity of the mixer truck to deliver material from the place of manufacture to the construction site is determined by the formula:

$$B_{ev} = \frac{V}{t_r + t_n + \frac{S}{V_{tr}} + \frac{S}{V_p} + t_m}, \quad (9)$$

V- volume of the concrete mixer truck m^3-

t_r, t_p -time to unload and load concrete h,

S - the distance from the object to the place of unloading, km;

V_{tr}, V_p - the speed of the concrete mixer truck when loaded and empty km/ h; T_m-time for manoeuvring, h. (according to the guidelines to determine the need for basic construction machines, TSNIOMTP, 1989)

Conclusions

During the development of organizational and technological documentation (Construction method statement, work execution design, method statements), the choice of the main machines and calculation of their number should be derived using the recommendations and formulas given in the article.

References

1. M. N. Yershov [etc.] Organizational technology solutions at reconstruction of the public buildings which are in the operation mode: The monograph (DIA publishing house, 2013)
2. B.V. Zhadanovsky, S. A. Sinenko [and others] development of the projects of organization of construction of industrial buildings and structures: Proc. textbook, (Moscow, Publishing DIA, 2016)
3. B.V. Zhadanovsky, S. A. Sinenko. International Journal of Applied Engineering Research. **11(3)** 1724-1727. (2016)
4. S. A.Mommy, B. V. Zhadanovsky, S. A. Sinenko. Scientific review. **17**, 84-89. (2015)
5. B.V. Zhadanovsky, S. A. Sinenko. Scientific review, **9-2**, 435-438. (2014)
6. B.V. Zhadanovsky, S. A. Sinenko, S.A. Mamochkin. European Conference on Innovations in Technical and Natural Sciences 7th International scientific conference. Austrian Journal of Technical and Natural Sciences. **5-6**, 3-8. (2015)
7. B.V. Zhadanovsky, S. A. Sinenko. Visualization of design, organization of construction and technological solutions. Computing in Civil and Building Engineering Proceedings 2014 International Conference. pp. 137-142. (2014)
8. A. Volkov, V. Chulkov, R. Kazaryan, S. Sinenko. Applied Mechanics and Materials. **584-586**. pp. 2681-2684. (2014)
9. S. A. Sinenko, A. M. Slavin. Scientific review. **1**. S. 98-103. (2016)
10. B.V. Zhadanovsky, S.A. Sinenko, M.F. Cuzin. Technology and organization of building production. **1**. p. 38-40. (2014)

The use of methods of structural optimization at the stage of designing high-rise buildings with steel construction

Andrey Vasilkin [1,*]

[1]Moscow State University of Civil Engineering; 26, Yaroslavskoe Shosse, Moscow, 129337, Russia

Abstract. The more designing solutions at the search stage for design for high-rise buildings can be synthesized by the engineer, the more likely that the final adopted version will be the most efficient and economical. However, in modern market conditions, taking into account the complexity and responsibility of high-rise buildings the designer does not have the necessary time to develop, analyze and compare any significant number of options. To solve this problem, it is expedient to use the high potential of computer-aided designing. To implement automated search for design solutions, it is proposed to develop the computing facilities, the application of which will significantly increase the productivity of the designer and reduce the complexity of designing. Methods of structural and parametric optimization have been adopted as the basis of the computing facilities. Their efficiency in the synthesis of design solutions is shown, also the schemes, that illustrate and explain the introduction of structural optimization in the traditional design of steel frames, are constructed. To solve the problem of synthesis and comparison of design solutions for steel frames, it is proposed to develop the computing facilities that significantly reduces the complexity of search designing and based on the use of methods of structural and parametric optimization.

1 Introduction

One of the orientation of development of computer-aided design systems is the automation of the process of searching for design solutions [1]. Traditionally, it is believed, that the design process cannot be automated because the search and adoption for design solutions is one of the creative tasks of a design engineer, whose decision is largely based on intuition and experience [2]. At the same time, it is possible to identify formally solvable procedures, the implementation of which by means of information technology allows to solve a number of design tasks automated [3].

Such problems include the task of finding a set of variable parameters of project solutions that satisfy specified criteria and constraints [4]. One of the methods of searching parameters in the decision space, which is widely used in the design, is the nonlinear programming [5,6,7]. However, in the practice of building high-rise buildings connected

* Corresponding author: vergiz@mail.ru

with the discrete nature of the variable parameters and the complexity of formalizing the objective function, the use of nonlinear programming is rare, since it requires engineers to have a deep knowledge of optimization theory.

2 Methods

Methodologically, the task of searching for design solutions in the design of high-rise buildings can be considered as a combination of two main tasks: the synthesis of the structure of the design object and the synthesis of the parameters of the design object- the choice of the numerical values of the variable parameters of the structural elements.

To obtain an effective solution, this methodology should be based on a wide use of the capabilities of modern information technologies and optimization methods [8]. The methodology is based on the use of computing methods of search on simulation models of buildings, application of the principles of decomposition of complex models and organization of the system design process. In search designing, the engineer must simultaneously take into account the various design goals and requirements for the design, both possible in quantitative terms and those that cannot be expressed quantitatively, for example the architectural expression that is particularly relevant for high-rise buildings [9]. Many of these goals and requirements may contradict each other, suggesting the need for a decision by the engineer.

To make the frame better in general, the optimization technique should help the engineer to manage increasingly complex building projects, enabling the synthesis of several design options. Taking into account the possibilities of modern software, the use of computing methods in the field of structural and parametric optimization of engineering and architectural designing allows for in-depth analysis and extends the range of the analyzed high-rise buildings in the initial stages of designing, since it increases the speed of search and acceptance of project solutions, taking into account the variability of impacts on the subsequent stages of the life cycle of the structure. These tasks are solved successfully during the synthesis of the topology of structural coverage [10], structural constructions [11], the search for the building layout [4].

Structural and parametric optimization is the methodology that helps to solve the tasks, to meet the needs of engineers and to direct the vector of solutions in alternative design. Structural and parametric optimization is widely used in machine building [12,], aircraft construction [14], algorithms for solving problems and practical tasks in construction have been developed [15, 16]; there are examples of carrying out structural optimization in search conceptual design [11]. Nevertheless, for complex tasks with a large number of variable parameters, which include the construction of high-rise buildings, modern design tools lack functionality to ensure wide use of optimization methods. Also, most optimization algorithms focus more on solving purely engineering problems than on constructive design synthesis and search design [17, 18]. Often, the ability of engineers to synthesize a structure is limited to insufficiently effective feedback, which affects the search for a solution. As a result, such an effective and powerful method as structural and parametric optimization, which allows to significantly reduce the cost and complexity of designing, is used in insufficient capacity in the design process.

3 Results

Thus, there is a need for a study aimed at increasing the availability, ease of use and flexibility of optimization processes to assist engineers in finding effective design solutions in the early stages of designing high-rise structures.

Many domestic and foreign researchers have made a large initial contribution to the methods of parametric and structural optimization [19,20]. Such works as the study in the field of numerical optimization [19], the use of the genetic algorithm [21,22,23] are included in the list. Although these studies point to the enormous potential for structural and parametric optimization being used in the field of design, quite a few studies pay attention to the possibilities of interactive application of optimization methods for solving problems and achieving development goals, which are used by engineers. Also, many current computing devices require complicated preliminary preparation and manipulation, long- time setup, as well as complicated transfer of information between different programs, which reduces the possibility of their application in practice.

The above tasks can be accomplished using information technology supporting search for design solutions, within which the application of evolutionary principles of multi-purpose parametric optimization with complex objective functions is realized [1]. This subsystem includes such functions as finding compromises, providing a variety of solutions, correcting the design goal while searching for solutions, visualizing and exporting results, which allows to synthesize optimal designs in the process of search design. The integration of such systems with the using of CAD (computer-aided)-programs in designing is especially effective when relying on the functionality of existing tools [24, 25].

Structural optimization at the stage of search design. This article describes the application of structural and parametric optimization methods, as well as approaches to their integration into the process of searching for design solutions at the stage of variant design of high-rise buildings. The structure of the projected object is a specific system that describes the totality of the composite elements and the relationships between them. A complete set of elements defining a design object can be represented as an imitation model of an object. Depending on the designing stage, the degree of specification and the working at the designing solution, the simulation model can reflect the various characteristics of the designed object [26]. Description of the simulation model can take different forms, it can be a mathematical dependence, a draft or information (numerical) model.

It is efficiently to select the rational structure of the design object is with the help of structural synthesis. Each structure has a set of parameters that the engineer can accept as variable or constant. The choise of numerical values of the set of variable parameters, for example steel class, section type, section size, etc. is the problem of parametric synthesis. The methodology for finding the structure of the design object is shown in Fig. 1.

When the structure is formed, the designer deals with formally undefined structural relationships and qualitative criteria, therefore at this stage, the formation of the initial structure is carried out taking into account the informal description of the projected object [19].

One of the main principles embodied in most optimization methods is the use of a model that simulates a real design object and organization of a design cycle for it. The process of computer-aided design is based on using the model as a means of analysis and selection procedure of the parameters as a means of automated solution search. In general, the computer-aided design system can have various means for forming a structure and selecting parameters, depending on the design stage, that is, the designer's goals.

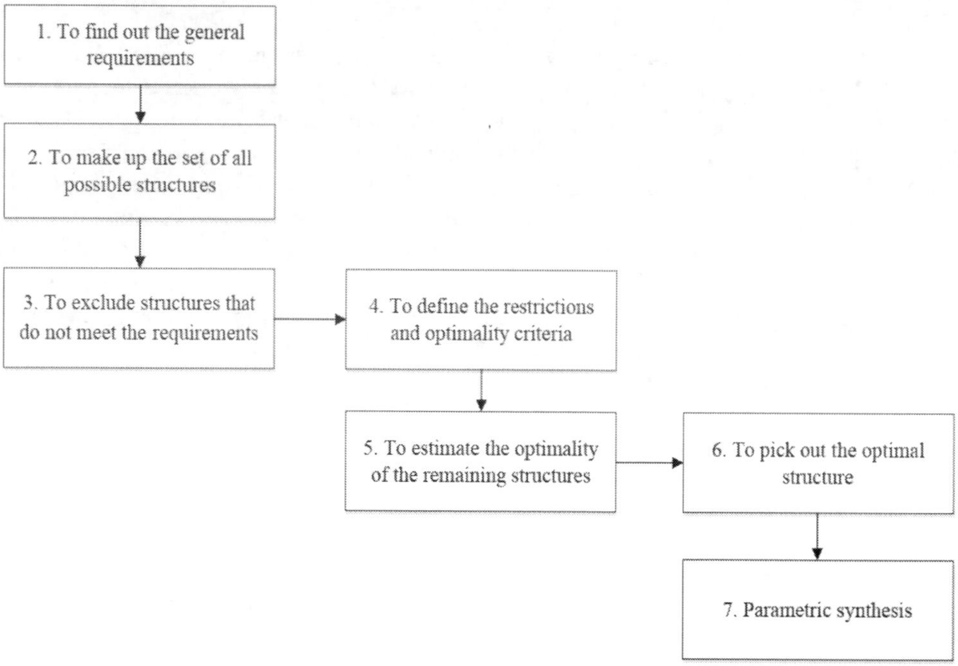

Fig. 1. The methodology of choice of the structure of the design object.

It is necessary to develop a software tool that is automated to search for the optimal structure of the designed object for practical application of the structural optimization method. It is obvious that today an engineer in many ways independently makes decisions, but at the same time it is possible to identify formalized procedures that can be represented in the form of an algorithm. In the future, when solving such problems, it is logical to expect the increasing application of CAD, which is associated with the development of computing technologies and capacities.

The initial data for the design of the object is a technical task that contains an indication of the functional purpose of the object, operating conditions (construction area, technological impacts), process technology for industrial buildings, and the environment description. in which the functioning and designing of the object is carried out in the form of a set of non-variable parameters, which can't be changed by the engineer. Analyzing the technical task the engineer, based of his experience, sets the criteria for the synthesis of the design solution and the imposed restrictions.

The following stages describe the operation of the software model in the synthesis of the designed object structure.

In the developed computing facility, a database is created that contains a group of design objects (beams, industrial buildings, steel-plate constructions, high-rise constructions, etc.) and for each group a list of possible structures and elements is given.

Stage 1. The download manager leaves the list of elements of the projected system from the database to the user's choice.

Stage 2. The user selects a specific element from the list for his designed object.

Stage 3. The user sets the threshold values of the corresponding variables.

Stage 4. The computing means calculates by the embedded algorithm and presents to the user the possible values of the variable parameters of the structure of the object, taking into account the restrictions associated with the shipping clearances, the possibility of manufacturing and assemblage, the peculiarities of operating conditions, etc.

At the last stage, the user specify the optimality criterion, based on which the selects the optimal solution of all the variants proposed by the software. This criterion is a set of output functions of the simulation model and the corresponding restrictions in the form of numerical or qualitative values, which must be met by these figures. One of these functions can be adopted as the target when optimizing. In this case, in the process of synthesis, this function must reach an extremum - a maximum or a minimum.

Stage 5. Then the structure sequentially varies among a visible number of possible variants of its implementation, the procedures of parametric synthesis are realized for each structure. Variants of the structure can be set by the user and thus have a rather small spectrum, or be synthesized using a genetic algorithm.

Consideration of the presented design methodology allows us to present the process of designing steel constructions in the form of a generalized scheme that is invariant to design objects (Fig. 2).

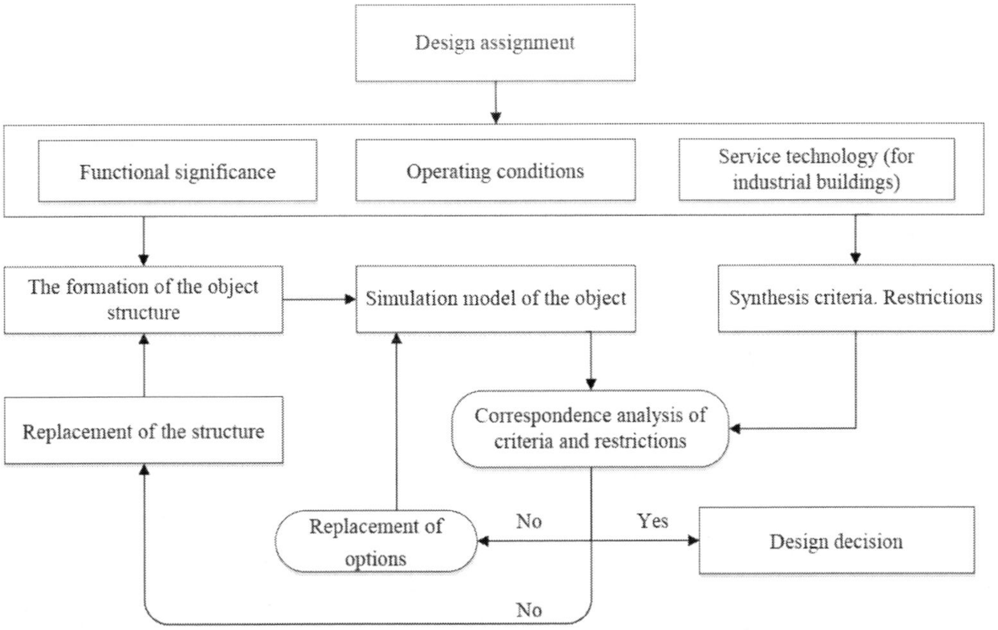

Fig. 2. Generalized design scheme.

Thus, the design process can be viewed as a set of two iterative cycles: the formation and modification of the structure and the exhaustive search of parameters. In parametric synthesis, it is assumed that the structure of the designed system and its simulation model has already been determined, the parameters of the elements have a metric expression.

It is difficult to synthesize the structure, as the engineer has to figure out which parameter and within what limits it will be changed. It is necessary to find out such a parameter in order it will be possible for the synthesized structures differ from each other, the range of possible solutions is wider and then it is hoped that the more the number of variants can be synthesized, the more effective one will be

4 Conclusions

High-rise buildings consist of a complex of constructions, elements and engineering systems, each of which is subject to an individual designing. In a number of cases, the automation of the structure is possible due to the certainty and homogeneity of the original redundant structure. In a number of cases, the automation of the structure is possible due to the certainty and homogeneity of the original redundant structure [21-26]. The design process is of a systemic nature, the main task of which is to check and link the solutions of individual constructions to the setup. In connection with heterogeneity and high dimensionality of search spaces, it is necessary to realize decomposition of a complex task into separate local iterative procedures. At the same time, they are combined into a structure according to certain rules, often these rules are informal, when the knowledge and experience of an engineer are actively used in solving problems of this class.

On the basis of the presented schemes, in the future it is planned to develop a computing means - a software complex, with the help of which the synthesis of project solutions, their visualization and their comparison and analysis is possible. The development of appropriate utilities to the computing means, which was previously developed by the author, can improve the functionality of information technology of supporting search for design solutions. One more important opportunity to increase the productivity of the technique is the use of genetic algorithms to synthesize the structure of the designed object, in this connection the author is currently developing and testing the appropriate tools.

References

1. A.A. Vasilkin, Promyshlennoe i grazhdanskoe stroitel'stvo, 5, 76-80 (2016)
2. D. Dzhons, Mir, 374 (1976)
3. A.V. Ginzburg, Ya.A. Lobyreva, D.A. Semernin, Nauchnoe obozrenie, **16**, 461-464 (2015)
4. A.M. Ibragimov, A.A. Vasilkin, Nauchnoe obozrenie, **13**, 32-36 (2016)
5. Edilson F. Arruda, F. Ourique, A. Almudevar, R. C. Silva, American Journal of Operations Research, **3**, 431-438 (2013)
6. J. McCall, Journal of Computational and Applied Mathematics, **184**, 205-222 (2005)
7. S. Rajeev, C. Krishnamoorthy, J StructEng, **123** (1997)
8. A.A. Volkov, A.A. Vasilkin, Procedia Engineering, **153**, 850-855 (2016)
9. M. Maevskaya, Vysotnye zdaniya, **4**, 20 – 24 (2016)
10. M.N. Kirsanov, Stroitel'naya mekhanika i raschet sooruzhenii, **2**, 60-63 (2010)
11. N.A. Panchukov, Izvestiya vuzov. Stroitel'stvo, **6**, 63-73 (2015)
12. D.V. Dmitrienko, Vestnik gosudarstvennogo universiteta morskogo i rechnogo flota im. admirala S.O. Makarova, **3**, 655-662 (2017)
13. V.E. Kostin, V.N. Tyshkevich, A.V. Sarazov, A.V. Sin'kov, V.F. Belukha, Mezhdunarodnyi zhurnal eksperimental'nogo obrazovaniya, **4-1**, 96-97 (2012)
14. A.P. Erokhin, Novoe slovo v nauke i praktike: gipotezy i aprobatsiya rezul'tatov issledovanii, **6**, 95-99n(2013)
15. I.M. Tumenova, Inzhenernyi vestnik Dona, **2(45)**, 131 (2017)
16. O.A. Shteinbrekher, T.V. Burnysheva, Inzhenernyi zhurnal: nauka i innovatsii, **10(70)** (2017)
17. A.M. Boldyrev, A.A. Svetikov, Metallicheskie konstruktsii, **4**, 263-268 (2008)
18. A.G. Tamrazyan, E.A. Filimonova, Sovremennye problemy rascheta zhelezobetonnykh konstruktsii, zdanii i sooruzhenii na avariinye vozdeistviya, 424-433 (2016)
19. Y.K. Vermishev, *Metody avtomaticheskogo poiska reshenii pri proektirovanii slozhnykh tekhnicheskikh system* (Radio i svyaz', Moscow, 1982)
20. D.J. Wilde, *Globally optimal design* (Wiley, New York, 1978)

21. D.Yu. Zaporozhets, A.M. Ksalov, I.A. Pshenokova, Informatika, vychislitel'naya tekhnika i inzhenernoe obrazovanie, **1(25)**, 14-26.(2016)
22. A.P. Karpenko, S.A. Gadzhiev, Sistemy proektirovaniya, tekhnologicheskoi podgotovki proizvodstva i upravleniya etapami zhiznennogo tsikla promyshlennogo produkta, 421-425 (2016)
23. V. Faghihi, K. F. Reinschmidt, J. H. Kang, Expert Systems with Applications, **16**, 7565-7578 (2014)
24. N. Brown, J. Ochsendorf, C. Mueller, de Oliveira, Proceedings of the 3rd International Conference on Structures and Architecture, 1103–1111 (2016)
25. N. C. Brown, C. T. Mueller, Energyand Buildings, **127**, 748-761 (2016)
26. Yu.V. Sobolev, A.A. Vasil'kin, A.D. Koloskov, Promyshlennoe i grazhdanskoe stroitel'stvo, **12**, 44-45 (2005)

AUTHOR INDEX

AUTHOR INDEX

AUTHOR INDEX

AUTHOR INDEX

AUTHOR INDEX

EDP Sciences – Web of Conferences
17, Avenue du Hoggar
Parc d'Activité de Courtabouf
BP 112
F-91944 Les Ulis Cedex A
France

ISBN 978-1-5108-5945-6

9 781510 859456

CURRAN ASSOCIATES INC.
proceedings.com